Holography and Speckle

Wolfgang Osten • Yuhong Bai
Editors

Holography and Speckle

A Review of 60 Successful Years

Springer

Editors
Wolfgang Osten
Institute for Applied Optics
University of Stuttgart
Stuttgart, Baden-Württemberg, Germany

Yuhong Bai
Chinese Academy of Sciences
Changchun Institute of Optics, Fine Mechanics and Physics
Changchun, China

ISBN 978-3-031-36880-6

Partly previously published open access in the journal "Light – Advanced Manufacturing", Special Edition 2021 and 2022

This Springer imprint is published by the registered company Springer Nature Switzerland AG
The registered company address is: Gewerbestrasse 11, 6330 Cham, Switzerland

Paper in this product is recyclable.

Dedicated to all those who have contributed to the success of holographic technologies.

Preface

"Celebrating Holography after 60 years of successful application"

This collection of papers is an extended edition of a compilation that was originally published in 2021 and 2022 in the Open Access Journal "*Light—Advanced Manufacturing*" on occasion of some remarkable anniversaries of Holography and related techniques such as Holographic Interferometry and Speckle Metrology in the year 2021.

The invention of Holography by Dennis Gabor [1] goes back to the year 1948. Gabor could show that the complete original wavefront of the object both with its amplitude and phase can be reconstructed by adding a coherent background to a wave coming from an object while recording in a storage medium. However, because no adequate source of coherent light was available at this time, the small coherence length of the mercury lamp forced him to arrange everything along one axis. This setup, known as in-line holography, has the disadvantage that the desired reconstruction is in line with disturbing diffraction orders, and can be separated from these disturbances only by focusing. Gabor finally gave up his investigations and mentioned in 1955 to his collaborator Michael W. Haine, "*It was a very ill wind which I let out now almost eight years ago which blew nobody any good, least of all to myself.*"

The holographic principle, on which Gabor had set his high hopes, especially with respect to the enhancement of the resolution of the electronic microscope, seemed to retain only the status of a scientific curiosity. But the breakthrough would be achieved a couple of years later. In the early 1960s, Emmett Leith and Juris Upatnieks, two physicists at the University of Michigan, implemented their off-axis scheme as a smart combination of Gabor's holographic principle with the carrier frequency technique known from side-looking radar. Their famous publication came out 61 years ago in 1961 [2], [3]. In this way the twin image problem could be effectively eliminated. However, it resulted in the requirement of increased temporal coherence which could be satisfied using a powerful coherent light source only. Such a source was soon available. In 1958, Charles Townes at the Columbia University, and a Russian research group led by Nicolay Basov and Aleksandr Prokhorov at the Lebedev Institute for Physics, Moscow, simultaneously and independently analyzed the possibilities of applying the principle of the microwave amplification by stimulated emission of radiation (MASER) for the optical region of the electromagnetic spectrum.

The first operating optical MASER, now known as the laser, was constructed by Maiman and was demonstrated at the Hughes Research Laboratories in Malibu, California, in 1960 [4]. The symbiosis of the holographic and laser principle opened the door for a large variety of new technologies and applications in optical imaging, information processing, and metrology. The 1960s were filled with a variety of successful practical applications and innovations such as display or volume holography [5], computer generated holography [6], holographic interferometry [7], holographic nondestructive testing, holographic pattern recognition, and holographic storage technology. Gabor took an active part in this development. 50 years ago, on the occasion of his Nobel lecture at the Imperial Colleges in 1971, he also tried to take a look into the future of holography [8] Gabor mentioned "However, there are important areas in which we can do much more, and where an improvement is badly needed. This is the area of microholograms, for storing and for display. An even more ambitious scheme, probably even farther in the future, is three-dimensional cinematography, without viewing aids such as

Polaroids." The basics for such a technological leap like digital storage, reconstruction, and synthesis of holograms were already on the way [9], [10] when Gabor made this prediction.

Holography also established the basis for new measurement technologies that play an important role in manufacturing. 1966 Karl A. Stetson and Robert L. Powell published their groundbreaking article "Holographic Interferometry" [7]. A. J. McGovern and J. C. Wyant proposed 1971 a new method for testing optical elements based on computer generated holograms [11]. Later Karl Stetson correctly mentioned „The novelty of Holography is Speckle". This astonishing effect because of the interaction of coherent light with scattering media was observed already many years ago. But to use Speckle as an outstanding tool for non-destructive testing goes back to the early 70th [12] and was also stimulated by holography. The application of video tubes instead of photo plates and electronic recording was a further major step in 1971 to make Speckle metrology called ESPI (Electronic Speckle Pattern Interferometry) to a mature technology [13].

In 2021, 60 years after the publication of the seminal article by Leith and Upatnieks, 55 years after the invention of Holographic Interferometry by Stetson and Powel, and 50 years after Gabor's remarkable Nobel lecture, the editors took the opportunity to celebrate these groundbreaking contributions to modern optics with a special edition of "Light—Advanced Manufacturing". Our call for papers found an impressive echo in the international community. This issue has been extended by four new papers and shapes now the anniversary edition by Springer. Because these contributions address many different topics, we decided to structure them into five chapters:

- 6 papers "Celebrating Holography and its Impact over 60 years"
- **10** papers dealing with the "Digital Transition of Holography"
- 8 papers describing the Potential of Holography for the "Making of new Optical Components"
- **12** papers addressing the "Wide Field of Applications opened by Holography," and
- 7 papers discussing "Perspectives of Holographic Technologies".

Our thanks go to all authors who joined our journey through 60 years of a meanwhile traditional but even young discipline of modern coherent optics. All these contributions express the conviction that this journey has just started and that we can still expect many remarkable innovations in the coming years. Further thanks go to all reviewers who helped us with their comments and proposals to keep the high quality of that edition. We know about the large workload that we transfer to the colleagues with the request for careful assessment of all submissions. Special thanks go to Richard Kowarschik from the Friedrich-Schiller-University Jena who compiled with us the LAM edition in 2022. Finally, we thank all members of the editorial office of LAM and Springer who made it possible to publish that extensive collection of papers in a relatively short time after the acceptance of all submitted papers. Special thanks go to Shuai Ding, Siqiu Guo, both from LAM, Ute Heuser and Vijay Kumar, both from Springer, for their careful assistance and support.

The editors hope that this book will find an interested audience and will be a welcome contribution for another successful 60 years of holography and coherent optics.

Wolfgang Osten
Yuhong Bai

References

1. D. Gabor, "A new microscopic principle", Nature 161, 777–778 (1948)
2. Emmett N. Leith, and Juris Upatnieks, "New techniques in wavefront reconstruction", Journal of the Optical Society of America 51 (1961) 1469
3. Emmeth N. Leith, and J. Upatnieks, "Photography by Laser," Scientific American, 212(1965). 6, 24–35
4. T. H. Maiman, "Stimulated optical radiation in ruby" Nature 187, 493–494 (1960)
5. Yu. N. Denisyuk, "On the Reflection of Optical Properties of an Object in the Wave Field of Light Scattered by It", (in Russian), Doklady Akademii Nauk SSSR 144, No. 6, 1275–1278 (1962)
6. B. R. Brown and A. W. Lohmann, "Complex spatial filtering with binary masks", Applied Optics, vol. 5, no. 6, pp. 967–969, (1966).
7. Karl A. Stetson and Robert L. Powell, "Holographic Interferometry," Journal of the Optical Society of America Vol. 56, Issue 9, pp. 1161–1163 (1966)
8. D. Gabor, "Physics 1971-1980", Nobel Lecture, S. Lundqvist, Ed., World Scientific Publishing Co., Singapore (1992) dedicated to the award of the Nobel Prize in 1971
9. J. W. Goodman and R. W. Lawrence, "Digital image formation from electronically detected holograms", Appl. Phys. Lett. 11(3), 77–79 (1967)
10. T. S. Huang, "Digital holography", Proc. IEEE 59(9), 1335–1346 (1971)
11. A. J. McGovern and J. C. Wyant, "Computer generated holograms for testing optical elements", Applied Optics 16(1971)619
12. J. A. Leendertz, "Interferometric displacement measurement on scattering surfaces utilizing speckle effect", 1970 J. Phys. E: Sci. Instrum. 3, 214
13. . N. Butters, J. A. Leendertz, "Holographic and Video Techniques applied to engineering measurements", In: J Meas Control. Band 4, 1971, S. 349–354

Contents

Part I Celebrating Holography and its Impact over 60 years

The Discovery of Holographic Interferometry, Its Development and Applications 3
Karl A. Stetson

The Language of Holography 13
James D. Trolinger

Contributions of Holography to the Advancement of Interferometric Measurements of Surface Topography 23
Peter J. de Groot, Leslie L. Deck, Rong Su, and Wolfgang Osten

Shearography and Its Applications – A Chronological Review 43
Rajpal Sirohi

55 Years of Holographic Non-destructive Testing and Experimental Stress Analysis: Is There Still Progress to be Expected? 73
Wolfgang Osten and Giancarlo Pedrini

Reflections About the Holographic and Non-holographic Acquisition of Surface Topography: Where are the Limits? 89
Gerd Häusler and Florian Willomitzer

Part II Digital Transition of Holography

Digital Holography in Production: An Overview 101
Markus Fratz, Tobias Seyler, Alexander Bertz, and Daniel Carl

Snap-shot Topography Measurement via Dual-VCSEL and Dual Wavelength Digital Holographic Interferometry 115
Daniel Claus, Igor Alekseenko, Martin Grabherr, Giancarlo Pedrini, and Raimund Hibst

Parallax Limitations in Digital Holography: A Phase Space Approach 127
Ulf Schnars and Claas Falldorf

Resolution Enhancement of Digital Holographic Microscopy via Synthetic Aperture: A Review 135
Peng Gao and Caojin Yuan

Phase Microscopy and Surface Profilometry by Digital Holography 151
Myung K. Kim

Integrated Self-Referencing Single Shot Digital Holographic Microscope and Optical Tweezer 163
Subhash Utadiya, Sachin Joshi, Nimit Patel, Chetna Patel, Mugdha Joglekar, Vani Cahhniwal, Timothy O'Connor, Bahram Javidi, and Arun Anand

Compression Strategies for Digital Holograms in Biomedical and Multimedia Applications 177
Peter Schelkens, Ayyoub Ahar, Antonin Gilles, Raees Kizhakkumkara Muhamad, Thomas J. Naughton, Cristian Perra, Antonio Pinheiro, Piotr Stępień, and Malgorzata Kujawińska

Digital Holography as Metrology Tool at Micronanoscale for Soft Matter 199
Zhe Wang, Lisa Miccio, Sara Coppola, Vittorio Bianco, Pasquale Memmolo, Volodymyr Tkachenko, Vincenzo Ferraro, Ernesto Di Maio, Pier Luca Maffettone, and Pietro Ferraro

A Review of Common-Path Off-Axis Digital Holography: Towards High Stable Optical Instrument Manufacturing 225
Jiwei Zhang, Siqing Dai, Chaojie Ma, Teli Xi, Jianglei Di, and Jianlin Zhao

Machine Learning Techniques Applied to Digital Holography 243
Brad Bazow, Thuc Phan, Thanh Nguyen, Van Lam, Anyu Sun, Lin-Ching Chang, Christopher B. Raub, and George Nehmetallah

Part III Making of New Optical Components

Holographic Optics in Planar Optical Systems for Next Generation Small Form Factor Mixed Reality Headsets 277
Bernard C. Kress and Maria Pace

Hybrid Multimode - Multicore Fibre Based Holographic Endoscope for Deep-Tissue Neurophotonics 309
Yang Du, Sergey Turtaev, Ivo T. Leite, Adrian Lorenz, Jens Kobelke, Katrin Wondraczek, and Tomáš Čižmár

Integrated Binary Hologram to Monitor Cargo Release from a Drug-Eluting Film 319
Arkady S. Abdurashitov, Pavel I. Proshin, Valery V. Tuchin, and Gleb B. Sukhorukov

Holography with High-power CW Coherent Terahertz Source: Optical Components, Imaging, and Applications 329
Yulia Choporova, Boris Knyazev, and Vladimir Pavelyev

Miniaturization of a coherent monocular structured illumination system for future combination with digital holography 347
Andreas Walter Stark, Eugene Wong, Holger Babovsky, Christian Franke, and Richard Kowarschik

Design of broadband terahertz vector and vortex beams: I. Review of materials and components 355
Nikolay V. Petrov, Bogdan Sokolenko, Maksim S. Kulya, Andrei Gorodetsky, and Aleksey V. Chernykh

Design of broadband terahertz vector and vortex beams: II. Holographic assessment 375
Nikolay V. Petrov, Bogdan Sokolenko, Maksim S. Kulya, Andrei Gorodetsky, and Aleksey V. Chernykh

Simulation-based design optimization of the holographic wavefront sensor in closed-loop adaptive optics 389
Andreas Zepp, Szymon Gladysz, Karin Stein, and Wolfgang Osten

Part IV Wide Field of Applications

A Review of the Dual-Wavelength Technique for Phase Imaging and 3D Topography . . . 407
Haowen Zhou, Mallik M. R. Hussain, and Partha P. Banerjee

Multi-positional Image-Based Vibration Measurement by Holographic Image Replication . . . 429
Simon Hartlieb, Michael Ringkowski, Tobias Haist, Oliver Sawodny, and Wolfgang Osten

Melt Pool Monitoring in Laser Beam Melting with Two-Wavelength Holographic Imaging . . . 439
Matthieu Piniard, Beatrice Sorrente, Gilles Hug, and Pascal Picart

A Symmetry Concept and Significance of Fringe Patterns as a Direct Diagnostic Tool in Artwork Conservation . . . 451
Vivi Tornari

Holography in the Invisible. From the Thermal Infrared to the Terahertz Waves: Outstanding Applications and Fundamental Limits . . . 473
Marc Georges, Yuchen Zhao, and Jean-François Vandenrijt

Investigation of human organoid retina with digital holographic transmission matrix measurements . . . 487
Nektarios Koukourakis, Felix Wagner, Stefan Rothe, Mike O. Karl, and Jürgen W. Czarske

Variable shearing holography with applications to phase imaging and metrology . . . 503
Prithiviraj Shanmugam, Alyson Light, Avery Turley, and Konstantinos Falaggis

Ultra-high speed holographic shape and displacement measurements in the hearing sciences . . . 521
Haimi Tang, Pavel Psota, John J. Rosowski, Cosme Furlong, and Jeffrey Tao Cheng

The Application of Digital Holography for Accurate Three-Dimensional Localisation of Mosquito-Bednet Interaction . . . 535
Matthew L. Hall, Katherine Gleave, Angela Hughes, Philip J. McCall, Catherine E. Towers, and David P. Towers

In-process Monitoring in Laser Grooving with Line-Shaped Femtosecond Pulses Using Optical Coherence Tomography . . . 551
Satoshi Hasegawa, Masatoshi Fujimoto, Toshihisa Atsumi, and Yoshio Hayasaki

Iterative Projection Meets Sparsity Regularization: Towards Practical Single-Shot Quantitative Phase Imaging with In-line Holography . . . 561
Yunhui Gao and Liangcai Cao

Shaping Singular Optical Coherence with Coherence Holography . . . 579
Wei Wang and Mitsuo Takeda

Part V Perspectives of Holographic Technologies

Holography, and the Future of 3D Display . . . 593
Pierre-Alexandre Blanche

Holographic Augmented Reality Display with Conical Holographic Optical Element for Wide Viewing Zone 607
Yusuke Sando, Kazuo Satoh, Daisuke Barada, and Toyohiko Yatagai

The State-of-the-Art in Computer Generated Holography for 3D Display 617
David Blinder, Tobias Birnbaum, Tomoyoshi Ito, and Tomoyoshi Shimobaba

Holographic Techniques for Augmented Reality and Virtual Reality Near-Eye Displays 647
Jae-Hyeung Park and Byoungho Lee

Holographic 3D Imaging Through Random Media: Methodologies and Challenges 661
Mitsuo Takeda, Wolfgang Osten, and Eriko Watanabe

Review of Engineering Techniques in Chaotic Coded Aperture Imagers 675
Vijayakumar Anand, Joseph Rosen, and Saulius Juodkazis

Deep Holography 689
Guohai Situ

Index 713

Part 1

Celebrating Holography and its Impact over 60 years

Stetson. *Light: Advanced Manufacturing* (2022)3:2
https://doi.org/10.37188/lam.2022.002

Official journal of the JHL 2689-9620
www.light-am.com

Review Open Access

The discovery of holographic interferometry, its development and applications

Karl A. Stetson*

Abstract

This paper recounts the discovery of holographic interferometry, discusses its development, and itemizes some of its major applications.

Keywords: Holography, Interferometry, Fringes

When I read the book by Sean Johnston, *Holographic Visions*[1], I felt a strong sense of dissatisfaction. The laser hologram was a wonderful discovery, but what did it really accomplish? Johnston spent chapters discussing its role in art etc. and only nine pages on holographic interferometry, which is unquestionably the most profound application of the hologram. Holographic interferometry makes the fullest use of a hologram's ability to reproduce the optical field reflected or transmitted by an object, and it utilizes two unforeseen capabilities of a hologram. First, a hologram can record fields that are incoherent and reconstruct them coherently so that they can interfere in its reconstructed field. Second, a hologram can reconstruct the optical field from an object with such fidelity that it can interfere with the original field from the object when the hologram is relocated in its original position.

In the section of *Holographic Visions* devoted to it[2], Sean Johnston credits the discovery of holographic interferometry to Emmet Leith and Juris Upatnieks at the Willow Run Laboratories of the Institute of Science and Technology of the University of Michigan. He based that on email communications and cites Robert Powell and me as having rediscovered it. From December 24, 1962, when the first laser hologram was made (I was in the laboratory doing other work that day and remember it well) until the publication of 3D holograms in December of 1964[3], this laboratory had the field to itself. On 19 December 1963 they observed a black spot in a hologram reconstruction of a sheet of paper fastened to a plate with a bolt hole in it, and in March 1964 they observed fringes in an unsupported edge of a cardboard sheet cemented to an aluminum block. These are clearly the first observations of phenomena that would be called holographic interferometry once it was discovered and understood,

Johnston's account of the "rediscovery" made by Robert Powell and me is based upon a letter dated 14 Oct. 1976 by Powell to P. Jackson at what was then the Museum of Holography in New York and is now at the MIT museum. Powell's letter describes a hologram with vague fringes in it that was our first observation of an interference effect in a hologram reconstruction. Unfortunately, his memory was in error, and he ascribed the fringes to cooling beeswax, which we used as an adhesive. My memory of this is quite different. I had settled on using bee's wax as an adhesive for our setups because I was concerned that any glue might have an unknown curing time during which it might change an object's position during a recording. The wax formed a good bond with objects, and, once it cooled, it stayed put exactly. There was never any degradation of a hologram due to cooling bee's wax. The hologram to which Powell referred to was made when we were searching for a holder for the holographic plates. I found a U-shaped frame slightly larger than our 4" × 5" plates, and wedged the hologram plate into it against a piece of rubber.

Correspondence: Karl A. Stetson (kastetson@holofringe.com)
Karl Stetson Associates LLC, 2060 South Street, Coventry, CT 06238, USA

This was unstable, and in the hologram reconstruction we saw an ill-defined set of diagonal bands that looked somewhat like interference fringes. I quickly abandoned that holder and we had a proper one made. That hologram was set aside without further consideration at the time. Of the two of us, I was the primary experimentalist and did the bulk of equipment manipulation.

One day, I entered our laboratory and found a colleague, Fred Rotz, using a spectrometer to measure our laser's coherence. He and others were speculating that the laser might not be as coherent as the manufacturers claimed. The Spectra-Physics representative had told me that these lasers emitted 3 longitudinal modes, and thus their longitudinal coherence would be periodic in twice the cavity length, which was 60 cm. I decided we should demonstrate that, and set up an experiment where the unexpanded object beam struck an object set between two parallel mirrors. We positioned our hologram plate so that we could look through it over one mirror into the other and see a sequence of images, each 10 cm further in path length. We recorded several holograms with zero-path length lying at different points in that sequence of dots, and all of these holograms exhibited a periodic variation in image brightness as a function of path length just as we expected. A photograph of one of these reconstructions is shown in Fig. 1.

We noticed in one of these holograms what looked like a horizontal black fringe in the nearest image of the object beam spot. More importantly, this fringe moved as we raised or lowered our view through the hologram, so it was clearly not something on the object itself. We immediately set up an experiment to contain the entire object beam on a white surface, and capture the entire reference beam on the hologram plate. This recording gave us fringes that moved as we raised or lowered the point through which we looked on the hologram plate. A photograph of the reconstruction of this hologram is shown in Fig. 2. I checked the laser and found that it was operating in more than one transverse mode. It was a Spectra-Physics/Perkin-Elmer model 110 laser that had a confocal mirror configuration, a spherical output mirror and a flat high reflectance mirror at the center of curvature of the spherical mirror. The discharge

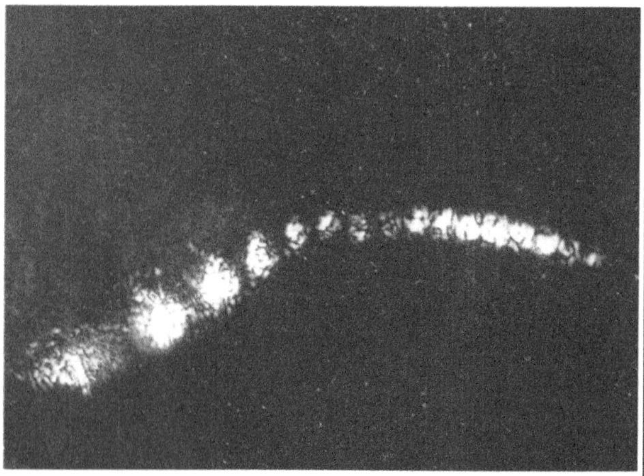

Fig. 1 A reconstruction of a hologram recorded of the object-beam spot seen reflected between two parallel mirrors so that each image has an increased path length of 10 cm[5]. Adapted from Ref. 5 with permission from Optica Publishing Group.

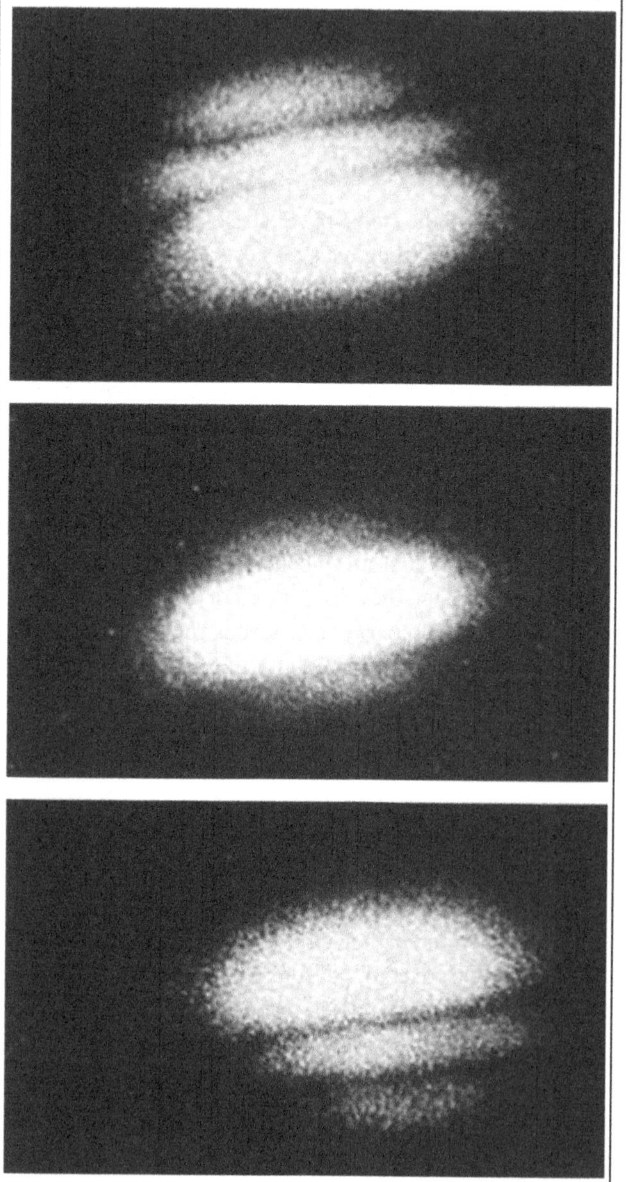

Fig. 2 Reconstructions from a hologram of the object beam illuminating an area on a white background where the laser was operating in a combination of TEM_{00} and TEM_{02} modes[5]. Adapted from Ref. 5 with permission from Optica Publishing Group. The three images are the spots as seen through the top, center, and bottom of the hologram, which contained the entire reference beam.

tube was located between these two mirrors. There was a longitudinal adjustment for the flat mirror, and a single TEM_{00} mode was only obtained when the distance was set correctly. We initially had a combination of a 00 mode and a 20 mode and, after adjusting the laser, we recorded a hologram with a 00 mode and a 10 mode.

The question was: What gave us these fringes? I suggested to Bob that the photographic plate was recording separate holograms for each mode combination of the object and reference beams. When the hologram was reconstructed, however, both recordings reconstructed their respective object beams, but since both fields were now being generated by the same reconstruction beam, the two reconstructed fields could interfere. The shifting of the fringe positions would be due to phase variations between the two transverse mode recordings. Bob found this very disturbing, because it meant that fields that were incoherent during the recording of a hologram were being made coherent in its reconstruction. At this point that I referred back to the hologram we had made with the plate wedged against the rubber pad as proof of what I was suggesting. Bob was still unconvinced, and I said something like, "look, if we record a double-exposure hologram and give the object as small rotation between the exposures, wouldn't we expect to see fringes in the reconstruction?" That question spurred immediate action. It took us some time to learn how much rotation to give the object, hologram plate, or reference beam, but we got this to work and showed double-exposure interference patterns in hologram reconstructions, as shown in Fig. 3.

As with Leith and Upatnieks, our object was a sheet of paper cemented to an aluminum plate, and in one of the double-exposure reconstructions, we noticed a perturbation of the fringes in a region where there was a bolt hole. In retrospect, I'm sure the paper had statically deformed between the two exposures, but at the time we wondered if the effect could be vibration. We set up a 35 mm film can with a solenoid mounted under it, and this generated the remarkable set of J_0 fringe patterns which we presented at the 1965 spring meeting of the Optical Society of America and published in the December 1965 issue of JOSA[4]. Reconstructions of seven holograms showing vibration modes are shown in Fig. 4.

The point is that Robert Powell and I observed a phenomenon, studied it in detail, proposed an explanation, and then tested and confirmed that explanation, in contrast to the mere observations made by Leith and Upatnieks. I feel that observation of a phenomenon does not count as a discovery. Human beings observed for centuries that the planets move relative to the stars, but Copernicus discovered that this was best explained by a heliocentric solar system. People observed lightning for all of human history, but Benjamin Franklin discovered that it was an electrical discharge. So, the question is: Did Leith and Upatnieks perceive the two properties of holograms that I defined in my opening paragraph and thus understand the concept of holographic interferometry? I think not, and the evidence is found in the patent filed under their names for holographic vibration analysis, US patent number 3548643, which I will discuss later.

As we submitted our paper to the OSA for the spring meeting of the Optical Society of America at Dallas, TX, in April 1965, I proposed to Emmett that we should also present a paper on the entire set of experiments we had done defining holographic interferometry. He vetoed the idea saying that one paper was adequate, an unfortunate decision, and the reason for it became clear to me in 1970. We presented this additional material at the OSA annual meeting in Philadelphia in October 1965, and published it in JOSA in September 1966[5].

Bob and I searched for prior work using interferometry for vibration analysis and came up with the paper by Harold Osterberg[6], which identified the fringes we saw in our reconstructions as a zero-order Bessel function of the first kind. It also suggested a method for real-time vibration analysis. Just after the spring meeting of the OSA, it occurred to me that if we could replace a hologram in its holder accurately enough, its reconstruction ought to

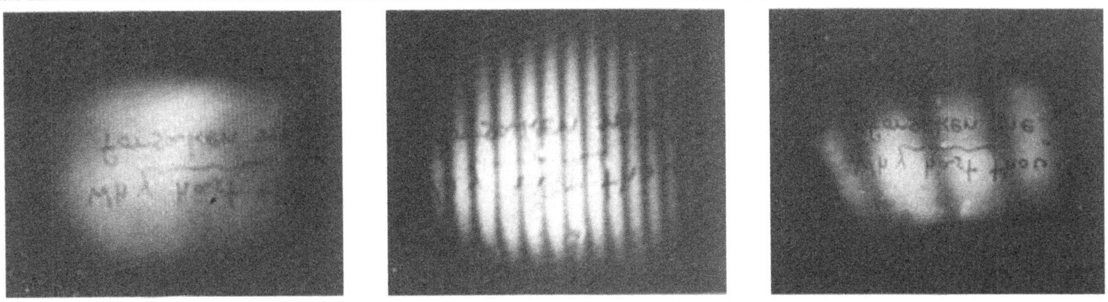

Fig. 3 Reconstructions from double-exposure holograms where the object was given small rotations between exposures. In the first figure, the fringes are too fine to resolve, but were quite visible in the original.

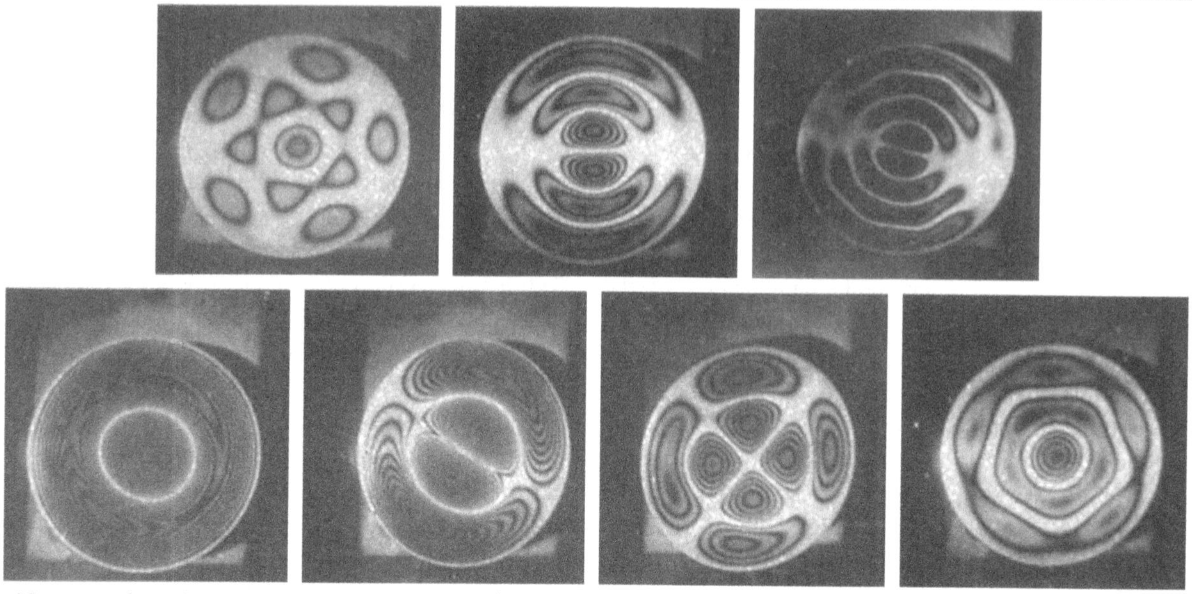

Fig. 4 Reconstructions of a can top vibrating in various modes[4]. Adapted from Ref. 4 with permission from Optica Publishing Group.

interfere with the field from the original object. We tried this and it worked. Interference between the original object and its hologram reconstruction are shown in Fig. 5.

Kenneth Haines and Percy Hildebrand, who also worked in the Willow Run Laboratory, were thinking about holography in April 1965. They borrowed our real-time holography setup to record the holograms they used for real-time interferometry showing deformation of a plate due to bolt tightening, as shown in Fig. 6.

At the time, Bob and I thought that it was good to have more people participating in this work, but it proved otherwise. I have long regretted that we didn't put a dc current through the solenoid with which we had excited vibrations on our 35 mm film can. We could easily have shown its static deformation, both in double-exposure and real-time holographic interferometry.

In April 1966, Haines and Hildebrand published a paper[7] that presented real-time interferometry between an object and its hologram reconstruction as a new method. Our prior work[8] is described by the sentence "A related technique for vibration measurement was used by Powell and Stetson." The primary content of this paper is their analysis of the fringes and fringe localization. Unfortunately, both this paper and a nearly identical paper following it[9] were mathematically dense and never provided insight into fringe localization.

Johnston quotes Hildebrand regarding his communication to Emmett Leith in the spring of 1965 as recounted to patent attorneys in 1970, and this confirms our priority. Percy didn't realize that I had recorded the same ideas in my notebook two weeks earlier and that Leith "did not mention Karl's entry to me when he signed mine because he didn't want a conflict to arise. However, he foresaw the present problem, which of course doesn't help me now." That problem was the division of the initial U of M patent on holographic interferometry.

The initial University of Michigan patent application on holographic interferometry was filed in 1965 with six names: Emmett Leith, Juris Upatnieks, Robert Powell, Karl Stetson, Kenneth Haines, and Percy Hildebrand to reflect the work all of us had done on this discovery. In 1970, I was at the National Physical Laboratory in Teddington, UK

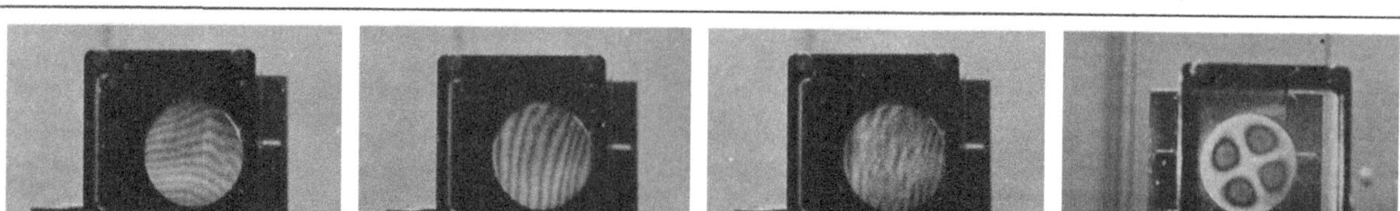

Fig. 5 Photographs of interference between a hologram reconstruction and the original object[8]. The last two images show vibration of the object as seen in real-time and as a hologram reconstruction. Reprinted from Ref. 8 with permission from Optica Publishing Group.

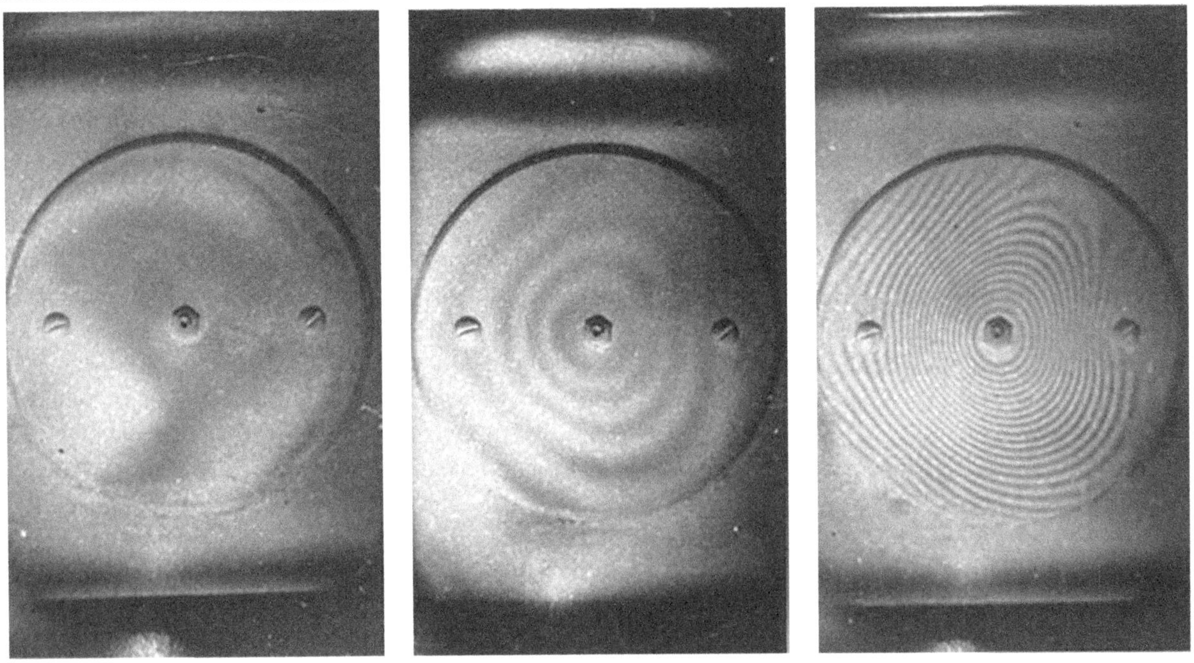

Fig. 6 Real-time fringes due to bold tightening on a circular plate by Haines and Hildebrand[7]. Adapted from Ref. 7 with permission from Optica Publishing Group.

and received a communication from an English solicitor representing the United States lawyers who were handling this patent application. I was told that the initial patent application was being broken up into three applications, each based upon the work initially done by each pair of inventors, Leith & Upatnieks, Stetson & Powell, and Haines & Hildebrand. In particular, the invention of holographic vibration analysis was being filed by Leith & Upatnieks, based on the disclosure they filed in April of 1964. I do not have the papers connected with this, but I am quite sure that Powell and I were credited with the discovery of real-time interferometry between an object and its hologram reconstruction, and Haines and Hildebrand were credited with the concept of holographic contouring by varying laser wavelengths. Leith, Upatnieks, Haines, and Hildebrand had all signed the necessary documents, but Robert Powell was raising objections.

They provided me with a copy of the initial disclosure on vibration analysis by Leith & Upatnieks. After carefully reading this document, I pointed out that it contained no reference to interference within the reconstruction of a hologram, and it did not mention any fringes within the hologram reconstruction. Essentially, the disclosure filed by Leith & Upatnieks described a hologram reconstruction that would be analogous to a Chladni pattern, where small particles on a vibrating surface migrate away from vibration antinodes and collect at vibration nodes. It would be bright at the vibration nodes and dark in the antinodes. Before I would sign their documents, I required that all reference to fringes and all mention of how to measure vibration amplitude within antinode of a vibrating object via its holographic reconstruction be removed from the claims of the patent. This effectively neutered it. There is one mention of fringes in the discussion section, and there are drawings sketching some of the results that Powell and I obtained, but none of this is claimed. In retrospect, I should have insisted on expunging all mention of fringes and any of our work from the entire document, but since the legal teeth of a patent are its claims, I let it go. Powell accepted the modifications I had obtained, and signed the documents also. So far as I know, the other two applications were never awarded patents, and all searches I have performed support that conclusion. I do not know if they were ever filed. Anyone who has worked with lawyers will attest that had there been any way in which the Leith-Upatnieks disclosure or anything in their notebooks could have been used to argue that they understood that the concepts of holographic interferometry, I could never have imposed the restrictions I did. For this reason, I feel justified in saying that Robert Powell and I discovered holographic interferometry. Juris Upatnieks confirmed this for me in an email communication: "We observed the effect and noted it, but did not get into any detail. Our main interest was image quality."

One possible contender for prior discovery was Melvin H. Horman[10]. He presented a paper at the same OSA meeting in Dallas where we presented our vibration work, and published in Applied Optics in March of 1965, with a received date of October 1964. His paper was theoretical with no experimental work, and it proposed that a hologram could replace an object in an otherwise conventional Mach-Zehnder interferometer. To perceive that theoretically was impressive, but there is no mention of the fundamentals of holographic interferometry.

The year 1965 saw a number of independent discoveries of holographic interferometry. Collier, Doherty, and Pennington[11], described interference effects within a hologram reconstruction as a moiré effect. Pulsed laser holographic interference was reported by Brooks, Heflinger, and Wuerker[12], and the interferogram of shock waves around a fired bullet is shown in Fig. 7.

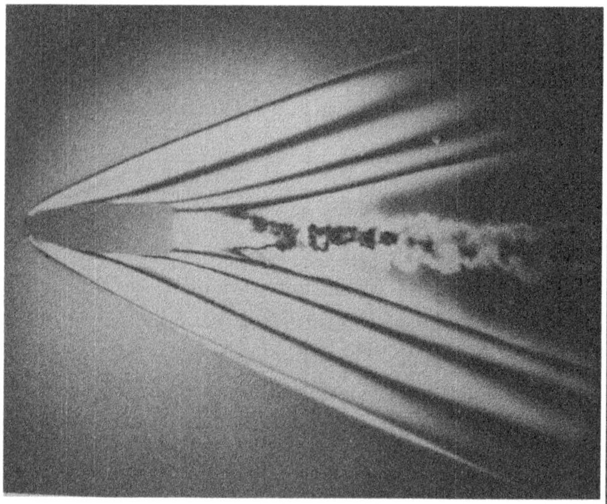

Fig. 7 Fringes formed around a bullet captured in a pulsed laser hologram. By accident, the laser fired a double pulse, the first of which occurred before the bullet was in the field of view.

In England, J. M. Burch published a paper in which he described experiments done at the National Physical Laboratory in Teddington that showed both double-exposure and real-time holographic interferometry[13]. In September of that year, Haines and Hildebrand published a paper proposing that if the wavelength of a laser could be varied, holograms could be used to generate contours of an object surface via interference effects[14]. No experimental demonstration was done because at that time laser wavelengths could not be varied. In the Soviet Union, holographic interferometry was discovered about this same time by Yu I. Ostrovsky, et al[15]. It is clear from these and other publications that year that holographic interferometry was a phenomenon waiting to be discovered by nearly anyone recording holograms.

The search for applications

Robert Powell and I were not mechanical engineers. For example, one of the vibration patterns we recorded of our film can bottom showed a five lobed pattern that was an impossible vibration mode for a circular vibrating structure (the upper left image in Fig. 4). In retrospect, it was a combination of a five-diameter mode and a two-ring circumferential mode and not a mode by itself, but we did not understand that at the time. I communicated information about our discovery quite freely, and one of the people I told was Ralph Grant, a professor of acoustics at the University. I also came in contact with Donald Gillespie, who was very interested in finding a way to exploit holograms financially. Don worked in a laser laboratory and was able to make his own lasers. I helped him set up a holography laboratory in his basement in which he made holograms for demonstration purposes. Eventually he and his brother John Gillespie formed the first company to make equipment specifically designed for holography laboratories, Jodon Engineering.

Ralph Grant teamed up with a lawyer, Joseph Crafton to form Grant-Crafton Optronics, aka GCO. They were fortunate to find that this new interferometry could make bonding flaws visible on laminated structures, such as pneumatic tires, when subjected to changes in pressure, and this became a major business. Fig. 8 shows an example of this[16].

The aircraft industry became a major user of holographic interferometry. Holographic testing required a significant initial investment which added considerable cost to products, and this was practical for items like airplanes

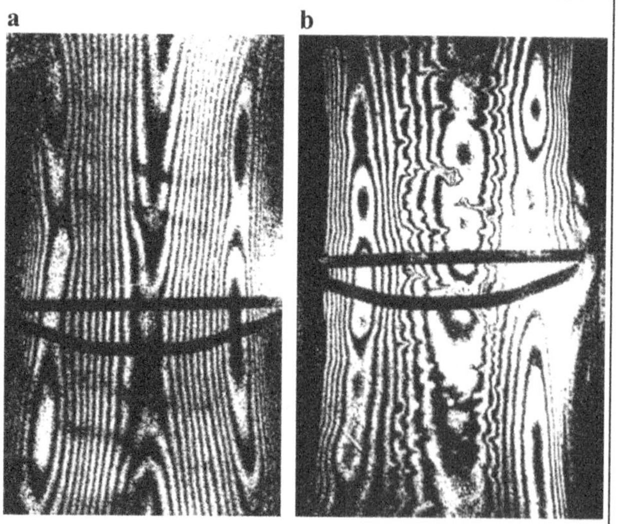

Fig. 8 Holographic Tire Testing. **a** Fringes from pneumatic pressure change for a good tire. **b** Fringes from change in pneumatic pressure showing bonding flaws.

which cost a lot to make and for which failures are very serious. In jet engines, high cycle fatigue due to vibration is a major source of failure for the blades, and they are designed to have resonances that are not excited at the rotation speeds of operation. Holography offered a way to visualize the mode shapes and confirm mathematical analyses. An example is shown in Fig. 9.

This was used very early on by Pratt & Whitney, GE, and Rolls-Royce, and also by Westinghouse for analysis of vibration modes for steam turbine blades. Aircraft structures made of honeycomb cores and metal or composite skins were also inspected by holography for delaminations. Another major area of application has been analysis of air flow and analysis of plasmas, which generally required use of pulsed lasers. A wide survey of techniques can be found in the book, *Holographic Nondestructive Testing*, edited by R. K. Erf[17], and a detailed discussion of holographic interferometry and its applications can be obtained in the books *Holographic Interferometry*, by C. M. Vest[18]. and *Interferometry by Holography*, by Y. I. Ostrovsky, et al[19].

Engineers at Pratt & Whitney also developed a holographic method to inspect abradable seals in jet engine casing for bonding flaws. High efficiency of a jet engine depends upon having a narrow gap between blade tips and the engine casing. An abradable material is bonded to the engine casings so that if the rotating blades come in contact with the casing due to any deformation of the engine as a whole, they will merely scrape away some material and not break off. If there are places where the seal is not well bonded to the casing, these may be knocked out by the scraping blades and result in poorer engine efficiency. To inspect a casing segment, it is excited with ultrasonic noise during a hologram recording. Disbonded areas appear as dark spots in the hologram reconstruction (see Fig. 10). Initially, photographic holograms were recorded of the engine casing, and were developed in place using a monobath developer. These systems were eventually replaced with digital holographic interferometry systems.

Digital Holographic Interferometry

There was a major problem with the fringes of holographic interferometry. They provide excellent graphic displays of object deformation or vibration, but converting

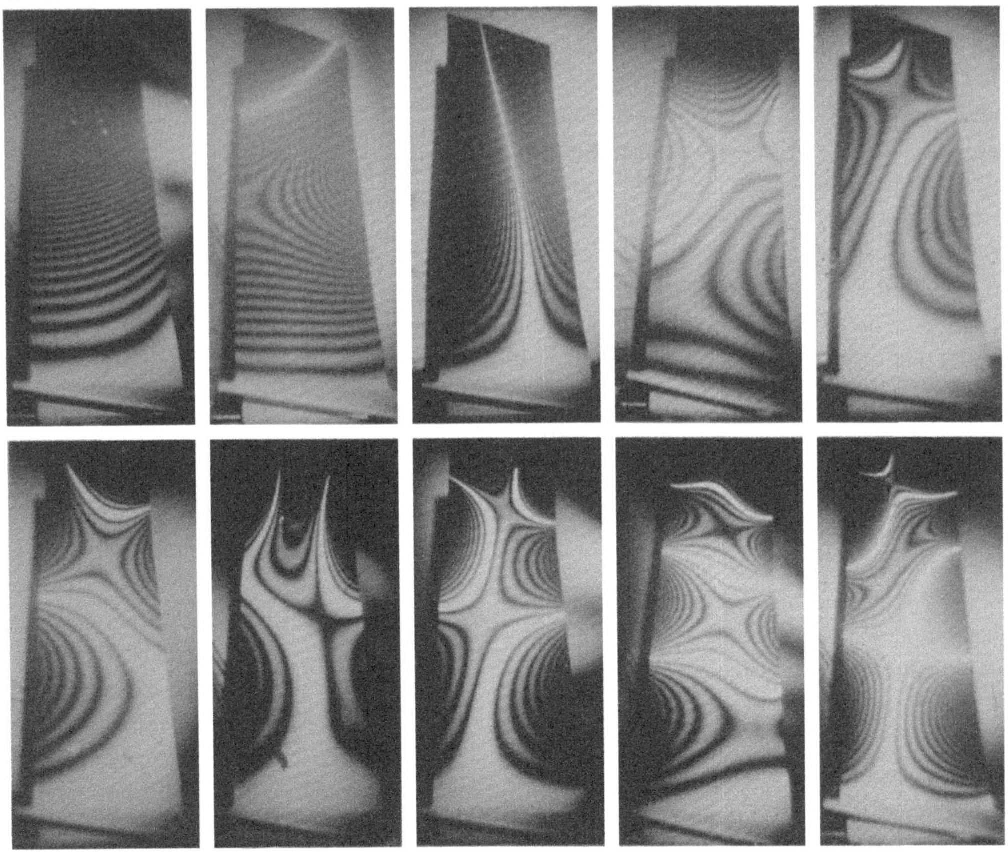

Fig. 9 Ten vibration modes of a jet engine turbine blade (courtesy of Pratt & Whitney). Reproduced with permission from Raytheon Technologies Corporation, Pratt & Whitney division.

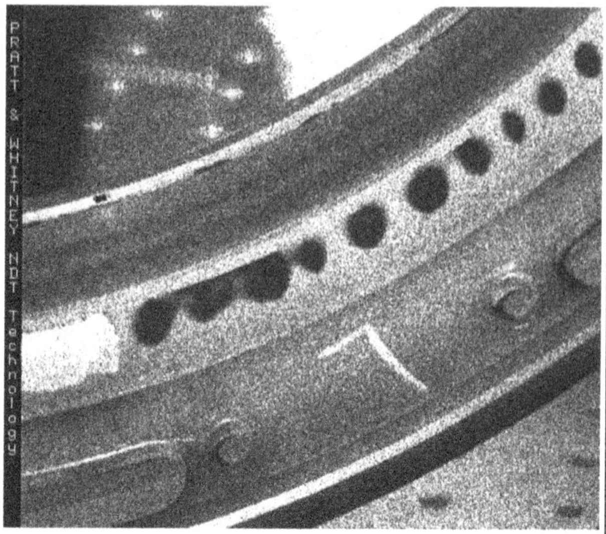

Fig. 10 A hologram reconstruction of an abradable jet engine seal which was excited with ultrasonic noise. The disbonded areas show up as black spots in the reconstruction.

them to numerical data is not easy. Exact deformations can be assigned to the centers of the fringes, but a continuous deformation function can only be obtained by interpolation between fringes. This is a difficult and tedious process at best. Fortunately, a solution eventually presented itself with the development of digital holographic interferometry and ultimately the development of digital holography[20].

Digital holography and holographic interferometry became possible because of three developments. The first was the personal computer, the second was the development of systems for digitizing television camera signals and the eventual development of self-digitizing cameras, and the third was the development of phase-stepping technology. This eventually resulted in systems that could display the reconstruction of a hologram on a monitor in real-time[21]. With the object imaged directly on the TV camera, the resulting display was a hologram reconstruction of the object in real-time. This allowed the real-time display of vibration fringes and the system could also store a hologram of an object, subtract it from the one displayed in real time, and display fringes due to static deformations. These systems could also the display in real-time of the phase of the static deformation fringe function[22]. This phase was obtained by an arctangent calculation, and therefore was cycled, or wrapped every value of 2π. The problem of unwrapping phase maps to obtain continuous phase functions is a general one, fortunately, and numerous methods were developed[23,24].

The problem remained of how to obtain numerical data from the J_0 Bessel function fringes of a time-average hologram. A solution was found[25,26] by recognizing that J_0 fringes were nearly periodic. Applying a bias vibration to the reference beam at the same frequency and phase as the object vibration, while displaying J_0 fringes, allows them to be shifted in the same manner as phase stepping in a conventional interferometer. Acquiring three or four such images with the equivalent of $\pi/3$ or $\pi/4$ phase shifts provided data for a phase calculation. The error that results from using J_0 fringes in place of cosine fringe can be calculated and used to correct the resulting phase calculation. The result is a wrapped phase map for the object vibration mode.

The numerical data is obtained at the pixel locations of the TV camera, and, although this does not provide a perfectly continuous data map on the object, it is fine enough for most practical applications. One application is the comparison of measured mode shapes to finite element calculations. A common method for predicting vibration modes on a structure is to divide it into small segments or elements, for which deformations can be modeled as simple bending or stretching. These segments are then connected in a system of equations that results in an overall matrix that can be solved for the deformations or vibration modes of the structure. Comparison of these theoretical mode shapes with actual data requires normalizing the experimental data to match the theoretical mode shape values, finding the experimental values at the points closest to the node points of the model, and calculating a map of differences.

Another application for vibration mode data comes from the jet aircraft engine industry. Replacement blades in the compressors and turbines of an engine must be certified to have not only the same material and geometry as the blades they replace, they must also match their vibration modes, both in frequency and mode shape. Holographic vibration analysis has provided a way to do this using what are called modal assurance criteria[27,28].

Conclusion

Holographic interferometry has come a very long way from its origin. Before its discovery, interferometry was reserved for the study of polished optical surfaces, but the laser hologram extended it to virtually any object that could reflect or transmit light. It became possible to measure deformation and vibration on engineering structures with a precision defined by the wavelength of light, and it became possible to find flaws in the bonding between structures. What the future holds in the way of new applications only time will tell.

Conflict of interest

The authors declare that they have no conflict of interest.

Received: 22 July 2021 Revised: 15 June 2022 Accepted: 16 June 2022
Published online: 16 June 2022

References

1. Johnston, S. F. Holographic Visions: A History of New Science. (Oxford: Oxford University Press, 2006).
2. Johnston, S. F. Holographic Visions: A History of New Science. (Oxford: Oxford University Press, 2006). pp. 191-200.
3. Leith, E. N. & Upatnieks, J. Wavefront reconstruction with diffused illumination and three-dimensional objects. *Journal of the Optical Society of America* **54**, 1295-1301 (1964).
4. Powell, R. L. & Stetson, K. A. Interferometric vibration analysis by wavefront reconstruction. *Journal of the Optical Society of America* **55**, 1593-1598 (1965).
5. Stetson, K. A. & Powell, R. L. Hologram interferometry. *Journal of the Optical Society of America* **56**, 1161-1163 (1966).
6. Osterberg, H. An interferometer method of studying the vibrations of an oscillating quartz plate. *Journal of the Optical Society of America* **22**, 19-35 (1932).
7. Haines, K. A. & Hildebrand, B. P. Surface-deformation measurement using the wavefront reconstruction technique. *Applied Optics* **5**, 595-602 (1966).
8. Stetson, K. A. & Powell, R. L. Interferometric hologram evaluation and real-time vibration analysis of diffuse object. *Journal of the Optical Society of America* **55**, 1694-1695 (1965).
9. Haines K. A. & Hildebrand, B. P. Interferometric measurements on diffuse surfaces by holographic techniques. *IEEE Transactions on Instrumentation and Measurement* **15**, 149-161 (1966).
10. Horman, M. H. An application of wavefront reconstruction to interferometry. *Applied Optics* **4**, 333-336 (1965).
11. Collier, R. J., Doherty, E. T. & Pennington, K. S. Application of moiré techniques to holography. *Applied Physics Letters* **7**, 223-225 (1965).
12. Brooks, R. E., Heflinger, L. O. & Wuerker, R. F. Interferometry with a holographically reconstructed comparison beam. *Applied Physics Letters* **7**, 248-249 (1965).
13. Burch, J. M. The application of lasers in production engineering (The 1965 Viscount Nullfield Memorial Paper). *Production Engineering* **44**, 431-442 (1965).
14. Haines, K. & Hildebrand, B. P. Contour generation by wavefront reconstruction. *Physics Letters* **19**, 10-11 (1965).
15. Kakos, A. et al. Interferometry holographic investigation of a laser spark. *Physics Letters* **23**, 81-83 (1966).
16. Brown, G. M. Pneumatic tire inspection. in Holographic Nondestructive Testing (ed Erf, R. K.) (New York: Academic Press, 1974), 355-364.
17. Erf, R. K. Holographic Nondestructive Testing. (New York: Academic Press, 1974).
18. Vest, C. M. Holographic Interferometry. (New York: John Wiley & Sons, 1979).
19. Ostrovsky, Y. I., Butusov, M. M. & Ostrovskaya, G. V. Interferometry by Holography. (Berlin: Springer, 1980).
20. Schnars, U. & Jüptner, W. Direct recording of holograms by a CCD target and numerical reconstruction. *Applied Optics* **33**, 179-181 (1994).
21. Stetson, K. A. et al. An electro-optic holography system with real-time arithmetic processing. *Journal of Nondestructive Evaluation* **8**, 69-76 (1989).
22. Stetson, K. A. & Wahid, J. Real-time phase imaging for nondestructive testing. *Experimental Techniques* **22**, 15-17 (1998).
23. Ghiglia, D. C. & Pritt, M. D. Two-Dimensional Phase Unwrapping: Theory, Algorithms, and Software. (New York: John Wiley & Sons, 1998).
24. Stetson, K. A., Wahid, J. & Gauthier, P. Noise-immune phase unwrapping by use of calculated wrap regions. *Applied Optics* **36**, 4830-4838 (1997).
25. Stetson, K. A. & Brohinsky, W. R. Fringe-shifting technique for numerical analysis of time-average holograms of vibrating objects. *Journal of the Optical Society of America A* **5**, 1472-1476 (1988).
26. Pryputniewicz, R. J. & Stetson, K. A. Measurement of vibration patterns using electro-optic holography. Proceedings of SPIE 1162, Laser Interferometry: Quantitative Analysis of Interferograms: Third in a Series; 25 April 1990; San Diego, United States. San Diego, United States: SPIE, 1989, 456-467.
27. Allemang R. J. The modal assurance criterion-twenty years of use and abuse. *Journal of Sound and Vibration* **37**, 14-21 (2003).
28. Stetson, K. A. Calculating modal assurance criteria from electronic holography data. *Sound & Vibration* **6**, 8-11 (2008).

Official journal of the JHL 2689-9620
www.light-am.com

Review | Open Access

The language of holography

James D. Trolinger*

Abstract

Holography has evolved over its 70+ years of history across widespread and diverse communities and locations. This has produced a complex language that is sometimes inconsistent, confusing, and incorrect, resulting in a general public that often appears widely confused and/or ignorant regarding what holograms actually are and are not. Today's holographers employ many types of recording media to record all types of waves, such as light, sound, radar, and simulated waves, and for many different applications and purposes. This study examines the language of holography along with its origins, problems, and possible solutions, while recognizing that certain "errors" in the language are so embedded in society that simple and ideal fixes may be beyond reach. This leaves us with certain questions, as follows. First, can the correct language be restored and should we undertake the task, or are we stuck with every three-dimensional image being called a hologram? Is it our duty to better educate the public with a more useful and consistent language, or should we just go with the flow? In this work, beginning with the insights provided by the pioneers of the field, we attempt to set the stage for a more useful holography language and definitions, specialized so as to be understandable and usable by various audiences, including those interested in non-optical holography. Such knowledge can help the general public to take greater interest and enjoyment in holography; this would also be beneficial to those more involved in holography. Accordingly, this article offers advice for achieving this result.

Introduction

People with relatively little optical background, such the general public and non-optical professionals, are most likely to encounter holograms in museums, magazines, art galleries, holographic displays, and security devices (such as those on credit cards). These are optical holograms, and are produced to create three-dimensional (3D) images of objects. Those who are vaguely familiar with holograms associate them with 3D imaging, whereas others may have no clue what holograms and holography are. After seeing an image in a good hologram for the first time, most viewers are surprised, fascinated, or even stunned, and almost always ask "how was it made"? The discussion immediately gets more difficult, and discretion with the language is needed to keep the further discussion meaningful.

Deepening the discussion on holography requires an understandable language, which, as with any language, becomes more complicated with depth. Teaching the general public interference and diffraction is not practical, necessary, or even useful. The optics community can provide the general public an understandable and accurate language and methods to enhance the enjoyment of holography, while maintaining consistency with the much more comprehensive, technical language required by the technical holography community. However, this requires that the technical community maintain accuracy and consistency in its own language.

Much can be learned from the evolution of holography and its language, which, after a brief emergence in 1947, was followed by a 15-year hibernation in obscurity. Then, breakthroughs sixty years ago (the anniversary we

Correspondence: James D. Trolinger (jtrolinger@metrolaserinc.com)
MetroLaser Incorporated, 22945 Mill Creek Road, Irvine, CA 92653, USA

celebrate in this issue) revived it into an explosive revolution. Owing to those innovations, today's holographers in many different fields employ a wide range of recording and reconstruction processes. They use many types of recording media to study and display all types of waves, including light, sound, radar, x-ray, and simulated waves, and for many different applications and purposes. Since 1947, this evolution across widespread and diverse communities and locations has produced a diverse language which is sometimes inconsistent, confusing, and incorrect, resulting in a general public that appears widely confused and ignorant regarding what holograms actually are and are not. Most novices do not understand the exciting aspects of holograms.

This study examines the language of holography and its origins, problems, and possible solutions, while recognizing that some of the "errors" in the language are so embedded in society that simple and ideal fixes may be beyond reach. This leaves us with certain questions. For example, can the correct language be restored, and should we undertake the task, or are we stuck with every 3D image being called a hologram? Here, beginning with insights provided by the pioneers of the field, we attempt to (a) set the stage for a more useful and enjoyable holography language and definitions specialized so as to be understandable and usable by all audiences, and to (b) establish a foundation for a second comprehensive, general holography language required by the technical community, which does not necessarily need to be understood by the general public.

Historical Background

The concept and initial naming of holography was provided by Dennis Gabor, in 1947[1]. He abandoned holography five years later owing to fundamental technical problems that limited its usefulness, and most of his followers soon left the field as well. In the early 1960's, Emmett Leith, Juris Upatnieks[2,3], and Yuri Denisyuk[4], without awareness of Gabor's earlier work or that of each other, created new forms of holography that removed the problems that had limited Gabor. They created dramatic holograms containing visible, 3D images that gained international public and professional recognition. The resulting explosive revolution cleared the way for many applications, including those in art, display, advertising, security, inspection, diagnostics, computing, data storage, new optical components, spectroscopy, microscopy, and most of the sciences. These advances resulted in a Nobel prize for Gabor in 1971, an honor that many holographers believed the new pioneers deserved to share. Each new application expanded the language of holography, i.e., to describe the different procedures, hardware, software, and products to make them possible.

Gabor's 1948 paper was titled "Microscopy by Reconstructed Wave-fronts", which probably led his early followers and others to appropriately call the process "wavefront reconstruction"; however, Gabor himself rarely referred to the process as wavefront reconstruction. He named the recording "hologram ", as derived from the Greek *holos* (whole) and *gramma* (message), meaning a recording containing the total information required for reconstructing a wave, as this could become a 3D image of an object. The word "holography" was a natural extension of the terminology for the process.

In Gabor's words: "If a diffraction diagram of an object is taken with coherent illumination, and a coherent background is added to the diffracted wave, the photograph will contain the full information on the modifications which the illuminating wave has suffered in traversing the object: the image of the object can be reconstructed from this diagram without calculation. One has only to remove the object, and to illuminate the photograph by the coherent background alone. The wave emerging from the photograph will contain as a component a reconstruction of the original wave, which appears to issue from the object[5]". How to add the background with a separate, off-axis beam, so as to make the process much more useful, remained undiscovered for another 15 years.

Emmett Leith's and Juris Upatnieks' work dramatically improved the process in the early 1960's, and immediately rendering holography as practical. This work grew out of research on "side-looking-radar" and was documented by Leith, beginning in 1956[2,6]. The "Michigan Side-Looking-Radar " document contained the theory of the Leith and Upatnieks holograms. Unfortunately, this document was classified and could not be published for many years. Therefore, their first off-axis hologram appeared to derive from a method of understanding and processing radar waves, as reconstructed with visible light, years before lasers were invented and before Leith knew about Gabor's work[7]. Leith later stated about his work, in a 1986 interview, "It turned out that the theory paralleled Gabor's work which I didn't know then. I first came across a paper on holography, or wavefront reconstruction as it was called then, in the October 1956 Journal of the Optical Society of America. I read the paper and said, 'Oh my gosh, it's already been done'"!

During the same period, Soviet physicist Yu. N. Denisyuk, unaware of the other work in holography, solved Gabor's limitations by combining holography with a photographic recording method developed by Gabriel Lippman, who received the Nobel Prize for the method in

1908. Denisyuk's hologram added the background with a separate beam directed toward the other side of the hologram and makes use of the thickness of the a recording medium to enable wavefront reconstruction with white light reflecting from the hologram. He soon adopted the language originally posed by Gabor and called his process holography, now commonly known as Denisyuk holography. Denisyuk said, "a hologram of an object is the optical equivalent of the object itself". When illuminated by light, the hologram of an object does the same thing to the light that the object itself would do; thus, an onlooker viewing the light emerging from the hologram would see the same thing the onlooker would see when looking at the object, and the onlooker would not be able to distinguish the object from a hologram of it. "A really simple example", he said, "is that a hologram of a convex mirror will act just like a convex mirror when illuminated by a point source[5]". This represents yet another simple way to describe a hologram, assuming that one can relate to the concept of an "optical equivalent".

The inventions and language of these pioneers provide us with three separate categories for optical holograms: the in-line hologram of Gabor, the off-axis hologram of Leith and Upatnieks, and the white light reflection hologram of Denisyuk. These are now commonly known as Gabor holography, Leith Upatnieks holography, and Denisyuk holography.

Adolph Lohman, a highly respected early researcher, described holography in the following way: "The propagation of the light wave (sic. from the object) is interrupted somewhere by the photographic plate on which the hologram is recorded. The light wave is frozen in the hologram. When properly illuminating the hologram later in the second step (sic. the wavefront reconstruction), the light wave 'frozen' on the hologram is revived. It continues its travel to form an image as if the light wave had not been interrupted. Unlike classical methods of image formation, the hologram does not necessarily bear any resemblance to the image which will be formed when it is properly illuminated[8]". The hologram appears as a window through which a viewer can observe a 3D image of an object in the space behind. The image can also extend into the space in front.

Using this concept, a hologram can similarly be described as a window on which information is recorded that enables it to transform light coming from a point source of light into a different light wave, i.e., one that would have come from a 3D object sitting behind the hologram. A viewer looking through the hologram sees a 3D image of the object sitting behind the hologram. Transmission holograms transform light passing through the hologram, and reflection holograms transform reflected light.

Defining Hologram and Holography

Although the words of the early pioneers are useful in the search for useful definitions, they had no way of realizing how broad and complicated the field of holography would become. More useful and understandable definitions and languages are needed by the general public as well as the technical community, and although these do not have to be the same, they must be consistent, compatible, and sufficiently comprehensive to serve the technical community and general public for the experiences they will most likely encounter.

A completely general definition covering all holograms is unlikely to be understandable or even useful to many people who will see and enjoy optical holograms; however, broad definitions and language are needed to cover the newest forms and applications of holograms, especially non-optical holograms, such as acoustical, digital, and real time holograms, as well as the many types and uses of holography. We conclude that an optimal approach is to adopt at least two consistent definitions: one that is comprehensible to the general public, and a second to be understood and used more by the technical community. A general and comprehensive definition can accommodate all types of holograms. Although there is no intent to review the details of all types of holograms here, selected types are named and referenced below. Moreover, we provide additional details on digital holography because of its novelty, emerging importance, and applications in the field.

Many people have never knowingly seen a hologram, or at least a good one. Many people who have not seen (but think they have seen) a hologram, and even those who have had some experience, are not likely to know how widely holography is used. By far, the most commercially successful and ubiquitous holograms are the embossed, cast, or printed holograms appearing on credit cards[9], money, stamps, advertising, decorations, souvenirs, and magazine covers. Billions of such holograms have been produced. Practically everyone owns one or more such holograms, and many owners do not even know their product is a hologram. Embossed, cast, and photopolymer holograms are printed directly into photosensitive materials placed on a surface in various ways. Such holograms, which can be produced for pennies each in mass production, can act effectively as security devices and anti-counterfeiting tools, thereby saving millions of dollars. This is simply because the production process requires expensive equipment, and is far too complicated for most counterfeiters.

Sketches or figures like Figs. 1, 2 can be helpful for answering the inevitable question "How is a hologram made"? A common manifestation of an optical hologram is a photograph of the interference pattern resulting from overlaying two laser light beams originating from the same laser, known as the object and reference beams, as shown in Fig. 1. Lasers are used because proper interference requires coherent light; this can be provided by a laser. The object beam comprises light reflected from an illuminated object, and the reference beam arrives unperturbed from the laser.

When such a transparency of the photograph is later illuminated with a similar reference beam of light (Fig. 2), a visible 3D image of the object can be seen in the space originally occupied by that object. The viewer can move around and continue viewing through the hologram, and the image will continue to be visible from different angles. The hologram acts like a window with the object sitting behind it.

Based on the above discussion, the following is a definition for the general public/layman who will most likely encounter holograms largely for creating 3D images,

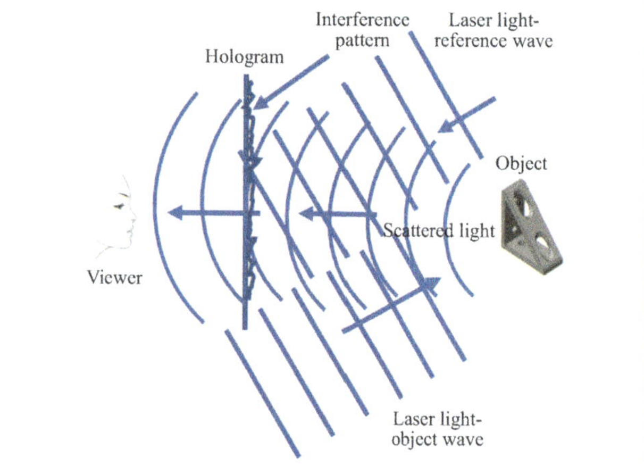

Fig. 1 Recording a hologram of an object.

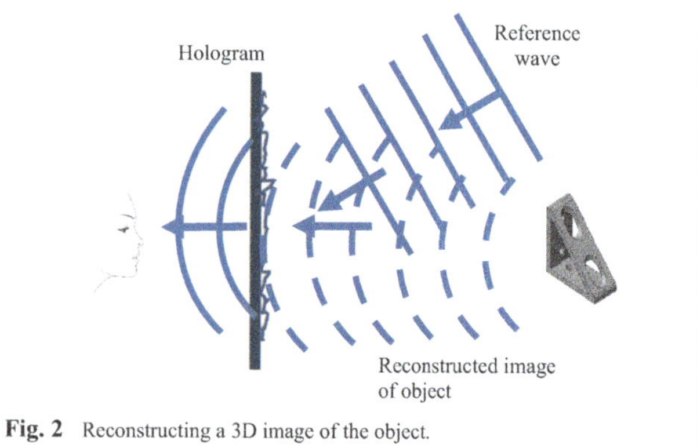

Fig. 2 Reconstructing a 3D image of the object.

e.g., in holographic art galleries, holographic displays, and museums, and on credit cards. For these audiences, an optical hologram can better be understood from a description such as that of Lohman[8].

Optical Hologram (n). A recording of a light wave that behaves like a window in which the light rays and waves coming from an object are frozen (recorded). With appropriate illumination of this window, the recorded waves can be turned on at a later time and can then continue to travel onward beyond the window, such that a viewer sees a 3D image of the original object in its original position. If the original object was a 3D object, the viewer sees it in the hologram as a nearly perfect 3D image of the object; in an excellent hologram, this image cannot be distinguished from the object itself.

3D Imaging Products That Are Not Holograms

Throughout most of the history of humankind, humans have been restricted to recording two dimensional (2D) images in drawings and paintings. One exception was in sculpture, a difficult and time-consuming process. Revolutionary steps occurred when artists discovered and began employing shadows and perspective methods to provide an illusion of 3D images. Not long after photography enabled the 2D recording of live scenery, stereoscopy was invented[10]. It was discovered that by presenting two different views of a scene recorded at different angles to our separate eyes, an illusion of depth was achieved. Stereo viewing created its own language as people began calling this "3D"; this was an incorrect description of what was actually seen, revealing errors in the language of 3D viewing long before holography. Viewing two 2D photographs provides much less of the 3D information seen by two eyes in a real scene or when viewing an image produced by a hologram, and provides only an illusion of a 3D image.

Full 3D imaging provides more than just depth information. With full 3D imaging a viewer perceives depth, focus, and parallax. Parallax (vertical and horizontal) is what is seen when the viewer moves their head up and down or sideways. The nearest objects move relative to the objects further away, allowing the viewer to see around an object in the foreground by moving the eyes up and down or side to side. In true 3D imaging, a viewer can focus the eyes differently at different depths; this also enhances the perspective in an image. This allows a viewer, at any time, to see both close and distant objects with better resolution.

Experienced and knowledgeable hologram viewers spend much more time looking at images in holograms than images in stereograms, because there is much more to see. When experienced viewers observe holograms on display they are often in constant motion to assist with the viewing, a process that is sometimes referred to as the "holography dance". We can teach the general public about the "holography dance" to improve hologram appreciation. By closing one eye, a viewer immediately transforms a stereo image into a 2D image, whereas a single eye provides 3D information when viewing a true 3D image. With one eye closed, moving one's eye up and down can determine if the image seen is a true 3D image, or one photograph from a stereo pair recorded in the hologram.

Because pure hologram images retain all of their 3D properties, distance perceptions, vertical and horizontal parallax, and focus, they are almost always more interesting in appearance than stereoscopic images (and can even be exciting). Nevertheless, owing to our fascination with 3D imagery, many techniques have emerged and continue to emerge for providing 3D illusions. Of these, many, like stereoscopy, are very effective entertainment devices that are not based on holography; nevertheless, they are often mistaken for holography and even passed off deliberately as holograms for publicity. The following are examples of techniques and devices that are not based on holography.

a. Stereo photography and projection imagery
b. Pepper's ghost images
c. Fly's-eye images
d. Lenticular photographs
e. Light field imaging
f. Virtual reality headsets
g. Integral photographs

The first holograms from both Gabor and Leith were holograms of photographs. Since then, holography techniques have often been combined with photography and stereoscopy methods, so as to take advantage of the simplicity, lower cost, and portability of the photography. The illusion of 3D can be retained with holograms of photographs and stereo pairs and with the help of stereography, with tradeoffs in accuracy and the loss of some 3D information. Photography is a simple and inexpensive process, whereas holography is less portable and more expensive.

Taking photos of a person sitting on a stool as a camera is translated in front of the person provides photos with different angles that can be used as stereo pairs to give a 3D illusion. A similar procedure places the individual on a turntable that rotates as a fixed camera takes photos at different angles. Various techniques can transform sets of photos into what appears to be a 3D image of an individual. One non-holographic method, known as integral photography, uses a lens array to produce an effective 3D illusion and is common in souvenir stores. Holograms of such photographic recordings are now recorded with results that sometimes improve upon the non-holographic methods and have become widely used. The trained eye can easily distinguish holograms from stereo photographs presented without holography, because holography can preserve vertical parallax, whereas photography does not.

Holography cannot yet and may never improve all 3D imaging techniques, which are quite effective in providing the desired 3D information/illusion for various applications. Virtual reality (VR) headsets are an example where video stereo photography is being pushed to its very limits in recording and computation, so as to achieve effective illusions of large, dynamic scenes for games and training purposes. Holography may eventually compete with existing VR headsets, but the necessary hardware and software does not yet exist. The same is true for video recordings of rock stars in action, none of which are holograms. Holograms have been made of stationary rock stars, but not as a live video depicting singing and dancing on a stage.

Digital Holography

Digital holography (which became viable only after the more recent advent of high-resolution digital technology) is possibly the fastest growing technical holography topic, with over 100 new publications per year. This may be owing to the possibility of performing advanced holography research without the necessity for expensive lasers, optics, or even laboratories; all that is required is a relatively inexpensive computer. Digital holography also has an important role in (and potential for) hologram production, making it commercially important. Techniques developed by this industry have enabled faster and lower-cost hologram production than anyone had imagined. In science and medicine, digital holography is revolutionizing microscopy.

In digital holography[11,12], information in the hologram is recorded and stored in a digital format as discrete bits in a computer. A hologram of a wavefront can be produced by interfering the wavefront with a reference wave on the sensor of a digital camera or sensor and then storing the information in the camera or in a computer. The fundamental requirement for recording a digital hologram is that the pixel size in the camera must be small enough to enable the recording of the fringes in the hologram.

Some holographers object to referring to a recording stored in a computer as a hologram, rather thinking of a hologram as something that can be seen and used to display a 3D image. Nevertheless, after storing the hologram in a computer, it can be employed, computationally, to reconstruct and propagate the stored wavefront to any place in a virtual space. For example, it can be used to compute and display the intensity, i.e. focus, at some distant plane, and then to display it on the computer monitor. This enables electronic scanning through a 3D image. Thus, a hologram stored in a computer can be used to provide the same advantages as other non-digital holograms.

Fig. 3 illustrates the recording and reconstruction operations for digital holography. In general, the digital hologram can be transferred to a spatial light modulator and displayed like any other hologram, and, therefore, should be called a hologram.

In digital holography, the object is illuminated by coherent light, some of which is scattered and mixed with a reference wave on a digital sensor.

The resulting interference pattern is recorded by a computer. At this stage in digital holography, there are two choices for reconstructing and viewing the wavefront, which are also illustrated in the figure. Most commonly, the recorded hologram is analyzed computationally by applying well-known diffraction and propagation equations, which can propagate the wave to any desired plane in virtual space; then, the intensity at that plane can be displayed on the monitor. In this way, the reconstructed image can be scanned electronically and displayed plane-by-plane.

Alternatively, the stored hologram can be transferred to a spatial light modulator illuminated by a reference wave as shown in the figure, and viewed in the same way analogue holograms are viewed.

With digital holography, holograms can be produced, and holography experiments can be run, numerically, not only without lasers, but also with synthetic wavefronts (without having real wavefronts at all). In this way, holography experiments can be simulated, providing a straightforward way to test an optical setup before it exists as hardware.

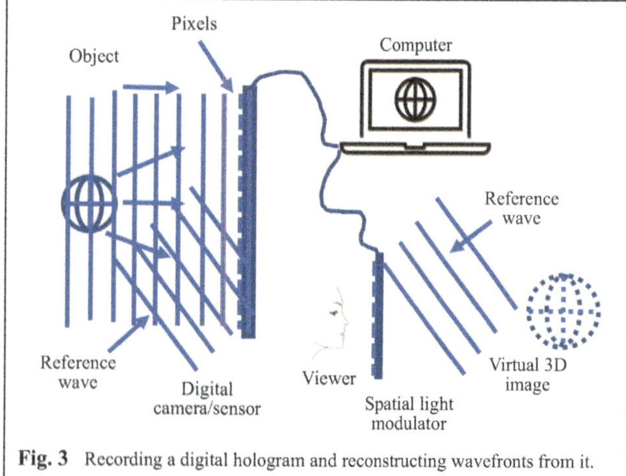

Fig. 3 Recording a digital hologram and reconstructing wavefronts from it.

The data stored in a digital hologram in a computer can be used to produce a variety of different types of holograms. The computer can control an etching machine to reproduce computer-generated holograms. These have a wide range of practical and commercial applications, such as for use in spectroscopy gratings, holographic optical elements, and embossing and casting masters for use in mass production. Holograms can be printed into direct writing materials like photopolymers. These operations result in yet another vocabulary of holography procedures and materials.

Comprehensive and General Definition

In searching for a correct, useful, and optimal definition of holography, the author sought out the opinions and preferences from many friends and associates worldwide. In almost every case, individuals responded with their own different definition. Although some seemed inadequate or not sufficiently general, most provided insights, and were incorporated into the wording and ideas leading to the results arrived at here. Many of the suggestions were too restrictive and left out important features or hologram types, or placed requirements that were not necessary. Accordingly, it is a challenge to obtain universal agreement on the best definition of holography. The author recently participated in a panel discussion with experts, who also found it difficult to arrive at a single definition of holography.

General definitions of holograms and holography should accommodate at least the following properties, features, wavelengths, and applications of holograms: security, mass production[13], computer graphics[14], spaceflight[15], microscopy[16,17], acoustics[18], x-rays[19], microwaves[20], and art.

They should include terms and phenomena understandable by a person with an optics background, and who is also qualified to create a completely general definition of holography. Those who do not understand any of the following points lack knowledge of the important features of holography that must be covered by a general definition.

1. Holograms are recordings that capture and preserve all of the information (i.e., the amplitude and phase) in waves required to reproduce the waves (including light, sound, radar, microwaves, and x-ray waves), in such a way that an accurate replica of the original wave can be reconstructed and propagated.
2. When a visible image of an object is reconstructed from a hologram and viewed through the hologram, it can be seen in 3D; moreover, if the object has depth, it can be focused and "seen" from its sides, and exhibits

parallax.

3. Not all holograms are used for storing and displaying 3D images.
4. A hologram is not an image, and an image is not a hologram.
5. Real-time holograms of time-varying waves can be recorded continuously and used simultaneously to reconstruct a changing wave or its conjugate.
6. Not all holograms are recorded on photographic or photosensitive materials.
7. Holograms can comprise any material, and many different mechanisms can enable the storing of the necessary information for a sufficient time, i.e., to enable reconstructing the original wavefront from which the information was obtained.
8. In digital holography, holograms can be created, stored, processed, and analyzed in a computer without using lasers or optics.
9. In digital holography, wavefronts can be reconstructed, focused, filtered, projected, and interfered numerically in a computer, as if in space.
10. The information stored in a hologram can exist on the surface and/or throughout a volume of recording media, or in a computer.
11. Holograms can exist in air/gas (otherwise known in physics as four-wave mixing).
12. Holograms known as HOEs (holographic optical elements) can replace and/or act as many types of optical elements, such as lenses, mirrors, and beamsplitters.
13. Holograms can be produced without using lasers, coherent light, or any type of real wave. (e.g., embossed, etched, computer-generated holograms, and digital holograms)
14. A phase conjugate mirror is a hologram.
15. Holograms are not required to produce 3D images, as 3D images and 3D effects can be produced by other optical methods.
16. Holograms are used for various applications, as follows.

• 3D imaging
• Security-Mastercard, Visa, currency, passports, stamps, tickets, documents
• Advertising and displays
• Portraiture
• Art
• Interferometry
• Measurement
• Information storage, archival, retrieval, and processing
• Optical elements
• Visualization of radar, sonar, ultrasound, lidar, magnetic resonance imaging, computer-aided tomography scans in 3D
• Stamping masters for producing embossed holograms
• Aberration correction

Holography clearly requires its general definition and language to include much more than the layman's definition given above. More general definitions that are consistent with the "window" definition, but which also cover the entire field of holography, are as follows.

Optical Hologram (n) A recording of a light wave that employs the principles of diffraction and interference phenomena to store information sufficient to enable reconstruction of an almost perfect replica of the original wave.

Hologram (n) A recording of any type of complex wave, such as a light, sound, x-ray, radio, or synthetic wave, that employs the principles of diffraction and interference phenomena to store information sufficient to enable reconstruction of a nearly perfect replica of the original wave.

Holography (n) The process of recording and using holograms; usually a two-step process, in which the first step records or creates the hologram, and the second step employs it to reconstruct and manipulate wavefronts and images.

Notably, the above definition of "hologram" includes the expression "complex wave"; this can be interpreted in two different ways, both of which are correct. A layman may think of a complex wave as another way of saying "complicated wave", which is correct; a scientist is more likely to think of it in mathematical terms, i.e., as a wave described by a complex expression, with a real part providing the amplitude, and the imaginary part representing the phase. This latter definition is also correct, and enables operations like conjugation, a mathematical procedure for changing the sign of an imaginary term to reverse the direction of the wave while maintaining its phase.

The above general definitions are consistent with the layman's definition, and they also rule out other forms of 3D image recording. They include the processes and languages used for the following types of holography.

1. White Light Reflection holograms (Denisyuk holograms)
2. In-line holograms (Gabor holograms)
3. Off-axis holograms (Leith-Upatnieks holograms)
4. Synthetic wavefronts (described below)
5. Embossed holograms
6. Cast holograms
7. Holograms of photographs
8. Holographic optical elements

9. Holographic gratings
10. Image plane holograms
11. Phase conjugate mirrors
12. Real time holograms
13. Digital holograms
14. Computer generated holograms

Existing Incorrect Definitions and Language in Holography

Compounding the language problems today are the many products that are referred to as "holographic", but that certainly do not fall within the ranges of our last two definitions. In legal language, a "holograph" is a handwritten document describing the wishes of the writer, which adds confusion to the definition of a hologram. People sometimes mistakenly use hologram and holograph interchangeably.

An on-line search for the definition of "holography" can partially explain the confusion regarding holography. A Google search for "Hologram Definition" produces incorrect definitions, describing holograms as images. A hologram is emphatically not an image. Others are equally incomplete, or outright wrong. Such definitions appear in legitimate dictionaries. The following are actual definition examples found from a random search.

1. Hologram- a three-dimensional image formed by the interference of light beams from a laser or other coherent light source.
2. Hologram- If you've ever looked through a View-Master, you've seen a hologram — an image or photograph that appears to be three-dimensional.
3. Hologram definition: A hologram is a three-dimensional photographic image created by laser beams.

The incorrect expressions widely accepted in holography, some even by experts, may explain much of the confusion in the public understanding. Incorrect definitions now appear in many classical dictionaries and on-line. Some of this language is so widely used and part of the common language that it may be difficult or impossible to remove. This situation possibly began when many people began referring to the images produced by holograms as "holograms", opening the door for all 3D images, including non-holographic images, to be called "holograms". Some widely used phrases are listed below.

1. "He is a hologram". A person or an image of a person cannot be a hologram; what is seen is a holographic image of a person. 3D images and effects can be produced without using holograms, and existed long before holography.
2. "That image is a hologram". Holograms can be used to produce images, but they, themselves, are not images; rather, holograms are recordings.
3. "Reconstruct the object". Objects are not reconstructed. Images of objects are reconstructed.
4. "I will reconstruct the hologram". With one exception, holograms are not reconstructed; holograms are produced or formed (constructed), and then are used to reconstruct wavefronts and images. The exception is the case where a hologram has been formed and then duplicated (reconstructed) somewhere else, as another hologram with the same information. With this exception, "reconstructing a hologram" literally makes no sense. This error is one that will be especially difficult to correct, because many holography researchers and colleagues use this literally meaningless expression.

Discussion and Conclusions

In the foregoing, we described and analyzed existing problems in the language of holography. Understandably, many people now think all 3D images are holograms. Many companies call their products holograms when they are not, many holographers are unaware of applications of holography besides 3D imagery, and communications in holography can be confusing and incorrect. This is further compounded by incorrect definitions for holograms existing in respectable dictionaries. Is this situation repairable, or is the language so widely accepted that we must live with it? Does the scientific community have an obligation (or even the ability) to better educate the general public regarding holography to help them appreciate and enjoy the richness of holography?

As with the arts and all forms of entertainment, holography can become more enjoyable as one learns more about its structure, what to look for, how to look, what there is to be appreciated, how it can offer more, and what it offers that is different than other forms of 3D imaging. Learning some of the holography language and how to look at holograms can provide depth to improve discussions and entertainment, and to make them more enjoyable. One of the greatest gifts holographers could give to the public is an understandable, useful, consistent, and accurate language that will help others enjoy holography. That would also be valuable to the field of holography.

With hopes of developing definitions that should prove useful for different holography communities, we offer consistent definitions. The first definition for optical holography should be understandable and useful to all, and covers most holograms likely to be encountered by the general public. The second definition is a completely general definition for the scientific community that is consistent with, and more general than, the first; it is especially useful for optical scientists. The general definition accommodates applications in the many diverse industries where holography plays an important role.

Acknowledgements

The author wishes to acknowledge many discussions with my friend, Dr. Kenneth A. Haines, who worked with Leith and Upatnieks at the University of Michigan in the 1960s, and who provided many insights into the historical information included in the paper. Ken, one of the pioneers in commercial holography, founded Simian Corporation and Eidetic Images Incorporated, which was later acquired by American Banknote Corporation, and led the effort to refine and apply digital holography, embossing, and casting techniques that made applications in the security industry practical, and produced the well-known holography cover pages for National Geographic, including both embossed and cast holograms.

Conflict of interest

The author declares that he has no conflict of interest.

Received: 26 August 2021 Revised: 28 November 2021 Accepted: 01 December 2021

Accepted article preview online: 03 December 2021

Published online: 16 December 2021

References

1. Gabor, D. Microscopy by reconstructed wave-fronts. *Proceedings of the Royal Society A:Mathematical, Physical and Engineering Sciences* **197**, 454-487 (1949).
2. Leith, E. N. & Upatnieks, J. Reconstructed wavefronts and communication theory. *Journal of the Optical Society of America* **52**, 1123-1130 (1962).
3. Leith, E. N. & Upatnieks, J. New techniques in wavefront reconstruction. *Journal of the Optical Society of America* **51**, 1469-1473 (1961).
4. Denisyuk, Y. N. Photographic reconstruction of the optical properties of an object in its own scattered radiation field. *Doklady Akademii Nauk SSSR* **144**, 1275-1279 (1962).
5. Denisyuk, Y. N. Fundamentals of Holography, Lecture notes, revised from the 1978 Russian edition, Mir publishers, (1984)
6. Leith, E. N. A data processing system viewed as an optical model of a radar system. Memo to W. A. Blikken of Willow Run Laboratories, University of Michigan, 22 May (1956).
7. Caufield, H. J. The Art and Science of Holography: a Tribute to Emmeth Leith and Yuri Denisyuk. (Bellingham: SPIE Press, 2004).
8. Lohmann, A. W. Optical Information Processing. (Universitätsverlag Illmenau, 2006).
9. Haines, K. A. The credit card holography story (2020). at https://www.phi.xyz/credit-card-holography.
10. Welling. Photography in America. (University of New Mexico Press, 1987).
11. Schnars, U. et al. Digital Holography and Wavefront Sensing. (New York: Springer, 2015).
12. Picart, P. New Techniques in Digital Holography. (London: John Wiley & Sons, 2015).
13. Haines, K. A. Development of embossed holograms. Proceedings of the SPIE 2652, Practical Holography X. San Jose: SPIE, 1996.
14. Haines, K. & Haines, D. Computer graphics for holography. *IEEE Computer Graphics and Applications* **12**, 37-46 (1992).
15. Trolinger, J. D. et al. SHIVA-(Spaceflight Holography in a Virtual Apparatus). Microgravity Materials Science Conference 2000. NASA, 2001.
16. Yu, X. et al. Review of digital holographic microscopy for three-dimensional profiling and tracking. *Optical Engineering* **53**, 112306 (2014).
17. Trolinger, J. D. & Dioumaev, A. K. A new digital holographic microscope for the study of biological systems. Proceedings of the SPIE 11490 Interferometry XX. SPIE, (2020)
18. Williams, E. G. Fourier Acoustics: Sound Radiation and Nearfield Acoustical Holography. (Academic Press, 1999).
19. Faigel, G. & Tegze, M. X-ray holography. *Reports on Progress in Physics* 355-393 (1999).
20. Zhuravlev, A. et al. Shallow depth subsurface imaging with microwave holography. Proceedings of the SPIE 9072, Detection and Sensing of Mines, Explosive Objects, and Obscured Targets XIX. Baltimore: SPIE, 2014. http://dx.doi.org/10.1117/12.2051492.

Official journal of the JHL 2689-9620
www.light-am.com

Review | Open Access

Contributions of holography to the advancement of interferometric measurements of surface topography

Peter J. de Groot[1,*], Leslie L. Deck[1], Rong Su[2] and Wolfgang Osten[3]

Abstract

Two major fields of study in optics—holography and interferometry—have developed at times independently and at other times together. The two methods share the principle of holistically recording as an intensity pattern the magnitude and phase distribution of a light wave, but they can differ significantly in how these recordings are formed and interpreted. Here we review seven specific developments, ranging from data acquisition to fundamental imaging theory in three dimensions, that illustrate the synergistic developments of holography and interferometry. A clear trend emerges, of increasing reliance of these two fields on a common trajectory of enhancements and improvements.

Keywords: holography, topography, interferometry, metrology

Introduction

The historian Sean F. Johnston has observed that "…just as holography allows observers to see an image with multiple perspectives, exploring its history demands a range of viewpoints[1]. " Holography has attracted microscopists, quality control engineers, scientists, and even artists. This paper views holography from the perspective of developers of optical interferometers for measuring surface topography.

Comparing holography with interferometry requires some impressions of how these two technologies differ, and how they are the same. Based on the works of Gabor[2] and Leith[3], and consistent with widely-accepted definitions of the technique[4,5], holography is a two-step process: (a) Recording the amplitude and phase of a propagating wavefront as an intensity image—the hologram—using interference with a reference wavefront, followed by (b) reconstruction of the propagating wavefront from the intensity image using optics or a digital simulation. Fig. 1 summarizes these two steps. The reconstruction of the propagating wavefront enables changing the image focus and perspective, leading to the famous three-dimensional (3D) imaging capabilities of the technique. One of the miracles of holography is that despite the sometimes-indecipherable holographic patterns, it is not only possible to create these 3D images, but also to quantitatively analyze vibrations, deformations and shape using multiple and time-averaged exposures[6].

Interferometry is also a two-step process, but classical interferograms are created and interpreted in ways that are distinguishable from holography. Fig. 2a illustrates interferometry using two nearly flat optical parts illuminated from above with spectrally-filtered light. Fig. 2b shows a traditional conversion to surface topography, where the bright and dark fringes are

Correspondence: Peter J. de Groot (peterd@zygo.com)
[1]Zygo Corporation, Middlefield, CT USA
[2]Shanghai Institute of Optics and Fine Mechanics, Chinese Academy of Sciences, Shanghai, China
Full list of author information is available at the end of the article.

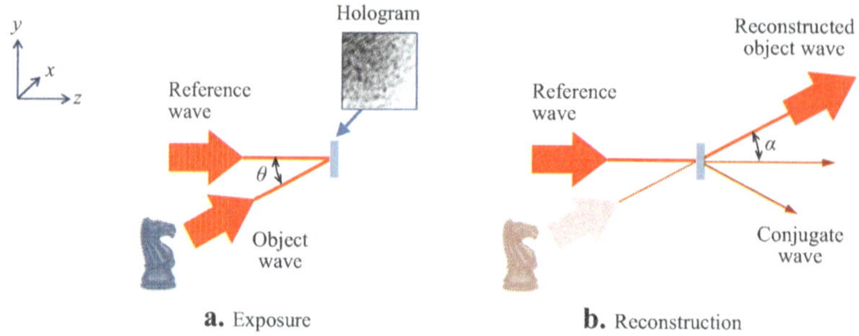

Fig. 1 **a** Creation of a hologram and **b** reconstruction of the propagating object wave by diffraction.

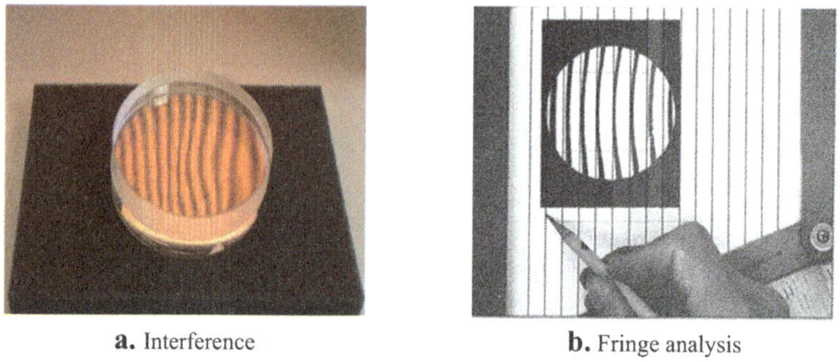

Fig. 2 **a** Formation of interference fringes and **b** manual analysis based on fringe straightness. The fringe analysis image is from a 1977 handbook on interferogram analysis, courtesy of Zygo Corporation.

interpreted as contour lines at increments approximately equal to half the illumination wavelength[7]. Interferometry does not normally include the uniquely "holographic" step of recreating a propagating wavefront from the recording of the original scattered light. For surface topography measurements, the interferogram is most often captured at the position of best focus of an imaging system, and the structure of the interference pattern is mapped directly to surface coordinates[8].

Notwithstanding these distinctive characteristics, interferometry and holography share the common principle of holistically recording the magnitude and phase distribution of propagating light reflected from an object as an intensity pattern, using the interference of reference and object waves[9]. The availability in the 1990s of electronic cameras with sufficient sampling density to record holograms digitally reinforced this commonality, allowing for direct interpretation in magnitude and phase of complex holographic recordings. The physically unique step of holographic reconstruction of a propagating wavefront has become a software model, which can be applied equally well to holograms and to interferograms. This evolution has brought the two techniques even more closely together, and the convergence continues today.

The goal of this paper is to discover concepts and inventions in holography that have contributed to the advancement of classical interferometric methods that are not normally thought of as holography. We have chosen seven illustrative examples of such connections: phase shifting interferometry, carrier fringe interferometry, suppression of coherent noise, digital refocusing, computer generated holograms, measurements of rough surface topography, and 3D imaging theory. We do not attempt here to provide a comprehensive review of any one of these individual topics. Rather, our goal is to highlight some of the significant and perhaps surprising developments in interferometer design, data processing and physical optics theory inspired by holography.

Phase shifting interferometry

The shared fundamental principle of capturing wavefront information in holography and interferometry is also a common difficulty, in that the complex amplitudes of the recorded light field are locked inside an intensity pattern. In modern interferometers for topography measurement, this problem is most often solved by phase shifting interferometry (PSI) or comparable heterodyning techniques[10]. Described most generally, the PSI idea is to record multiple interferograms that are nearly identical apart from a phase shift between them, so as to provide enough information to extract the object light phase and intensity. Although the history of PSI is usually traced to homodyne and heterodyne displacement measurements[11], it can be meaningfully argued that PSI for areal measurements came first to holography and then independently to interferometry.

To view the conceptual development of PSI from the perspective of holographic methods, it will be useful to write down the basic mathematical principles of holography. In the holographic exposure shown in Fig. 1a, we have at the hologram plane a light field reflected from an object defined by the complex representation

$$u_o = \sqrt{I_o}\exp(i\phi_o) \quad (1)$$

and a reference light field represented by

$$u_R = \sqrt{I_R}\exp(i\phi_R) \quad (2)$$

For the realization of Gabor in-line holograms, the recording tilt angle θ shown in Fig. 1a is close to zero. The result after square-law detection is the interference pattern

$$I = |u_o|^2 + |u_R|^2 + u_R u_o^* + u_R^* u_o \quad (3)$$

which reduces to the familiar two-beam interference equation

$$I = (I_o + I_R) + 2\sqrt{I_o I_R}\cos(\phi_o - \phi_R) \quad (4)$$

In both holography and interferometry, what interests us is the phase ϕ_o and intensity I_o of the original object wave, rather than this periodic cosine function. The problem is essentially the same for both technologies; however, the problem manifests itself differently in holography and interferometry.

In film-based holography, the exposed hologram is illuminated with the original reference to create the holographic image. By careful control of the photographic process, the amplitude transmissivity of the hologram can be made proportional to the intensity of the original interference pattern[12]. The synthesis of the holographic image is represented mathematically by multiplying the intensity pattern of Eq. 3 by the complex amplitude of the reference wave:

$$u_R I = \left(u_R|u_o|^2 + u_R|u_R|^2\right) + u_R^2 u_o^* + |u_R|^2 u_o \quad (5)$$

In Eq. 5 it is only the final term containing u_o that is of interest. In the original in-line Gabor holograms, the reference u_R is a diverging wavefront, and the reconstruction includes overlapping, coaxial diffracted beams. The relative divergence is insufficient for separating the original wavefront u_o from the conjugate wavefront u_o^*, resulting in a degradation of the desired image. This is where the phase shifting idea comes into play.

As Gabor and Goss describe in a 1966 retrospective of early developments, a holographic interference microscope constructed during the period 1951-56 isolates the desired wavefront using polarizing optics to introduce phase shifts between two otherwise identical holograms[13]. Gabor's quadrature microscope is shown in the simplified diagrams of Fig. 3. A key component is a special prism with a coating that introduces a quarter-wave phase shift, labeled $\lambda/4$ in Fig. 3a, between reflected and transmitted light. After exposure, we have two holograms with fringe patterns that we can write in simplified form as

$$I_A = \cos(\phi - \phi_R) \quad (6)$$

$$I_B = \sin(\phi - \phi_R) \quad (7)$$

where constant terms have been set aside since these can be determined by other means. If we now illuminate with two copies of the reference wave that are also shifted in phase by a quarter wave as in Fig. 3b, the coherent sum of the reconstructed images is proportional to a net result

$$u_T = \cos(\phi - \phi_R) + i\sin(\phi - \phi_R) \quad (8)$$

where again constant terms have been set aside. The reconstruction is recognized as a complex-valued amplitude, described as a "total wavefront reconstruction." [13] At the same time, the conjugate wavefronts generated by the two holograms cancel, at least in principle. In practice, additional elements are needed to precisely tune the phase shift and to adjust the intensities of the two reconstructed wavefronts to achieve this goal. One of the many remarkable features of this ingenious arrangement is that the calculations leading to Eq. 8 are performed entirely in optics, without electronic detection or digital data processing, in what can be reasonably described as an early all-optical computer of complex mathematical functions.

Although there have been modern implementations of optical quadrature holography[14], the legacy of the apparatus of Fig. 3 is not so much its specialized design, as it is the concept of introducing phase shifts between otherwise identical interference intensity patterns to recover complete wavefront information. This basic idea was soon adopted

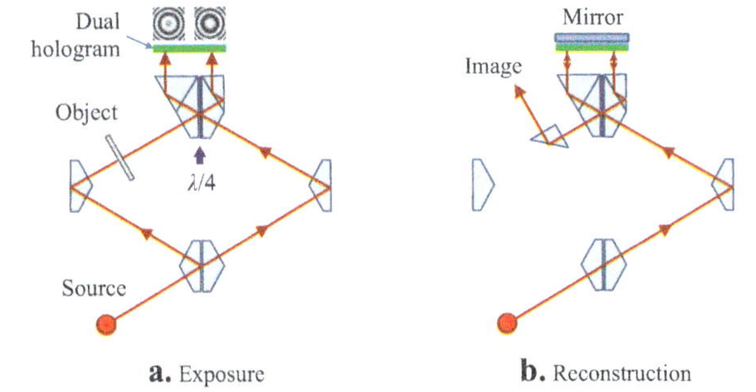

Fig. 3 Quadrature holography proposed by Gabor et al. in 1951. Lenses have been left out for simplicity in the diagram.

for non-holographic interferometry. Interferograms, like holograms, are also created by mixing a reference wave with light that is reflected from the object. Most often it is assumed that there is a linear relationship between the phase ϕ_o and the object surface height h_o as a function of surface coordinates x, y, with a factor of two related to reflection:

$$\phi_o(x,y) = 4\pi h_o(x,y)/\lambda \tag{9}$$

Common interferometer optics are arranged to create an image light field u_I of the an in-focus object light field u_o, resulting in an intensity pattern I similar to Eq. 4:

$$I = (I_I + I_R) + 2\sqrt{I_I I_R}\cos(\phi_I - \phi_R) \tag{10}$$

where the imaged light field is

$$u_I = \sqrt{I_I}\exp(i\phi_I) \tag{11}$$

The challenge is how to extract the phase ϕ_I. The manual pattern analysis of Fig. 2b follows the dark and bright contour lines corresponding to the interference phase. Clearly this has limited resolution and accuracy.

Shack described in 1971 the construction and demonstration of a "direct phase-sensing Interferometer" using polarizing optics such that the orientation of the polarization at the sensor is linearly proportional to the optical-path difference[15]. This approach enables determination of fractional fringe values limited in resolution only by engineering considerations such as electronic signal to noise. By 1983, advances in electronic detection and microprocessors made it possible for Smythe and Moore to use polarization-based phase shifting for measurements over 64 × 64 image points, with a data acquisition time of less than 1 μs. McLaughlin and Horwitz in 1987 introduced pixel-based polarization phase shifting using a 256 × 256 camera mask, eventually leading the design to the diagram in Fig. 4 developed by Millerd, Wyant and colleagues, using a micropolarizer array mask for the camera[16,17].

As is well known, the PSI concept is not limited to the simultaneous imaging of multiple interference patterns using polarization. Gabor and Goss note in their 1966 paper about their quadrature microscope that " … photographs can be of course obtained in succession[13]. " This is the basis of temporal PSI, which has the advantage of nearly identical optical path for each of the multiple phase-shifted images, minimizing error sources. The milestone work in temporal PSI was that of Bruning, who in 1974 measured topography using interferometry and computerized sine-cosine fitting[18,19], following foundational work at Perkin Elmer[20]. After the introduction of commercial mechanical phase shifting systems in the 1980s (Fig. 5), there was a flourishing of developments in PSI algorithm design, that has been extensively reviewed[21,22]. Today temporal PSI methods include adaptive and iterative algorithms to account for environmental disturbances and opto-mechanical imperfections[23].

The use of temporal PSI in holography developed along a parallel path to interferometry. Electronic phase evaluation of holograms in the 1970's included the work of Dändliker et al[24]. However, true digital PSI analysis of holograms had to wait until camera pixel densities were sufficient to resolve the dense fringe patterns of holograms captured far from the in-focus image of the object surface Many advances followed developments of electronic speckle-pattern interferometry, dual-exposure holographic interferometry, and non-destructive testing, all of which are closely related to holography[25-27]. These fields contributed significantly to determining the interference fringe order using phase unwrapping techniques developed for noisy speckle patterns[28].

Today both temporal and polarization-based PSI are employed for optical testing and interference microscopy, as illustrated by the large-aperture, wavelength-tuned PSI system of Fig. 6. While many factors contributed to its development, it can be meaningfully argued that many of the foundational ideas date back to the earliest work in wavefront synthesis using multiple phase-shifted holograms.

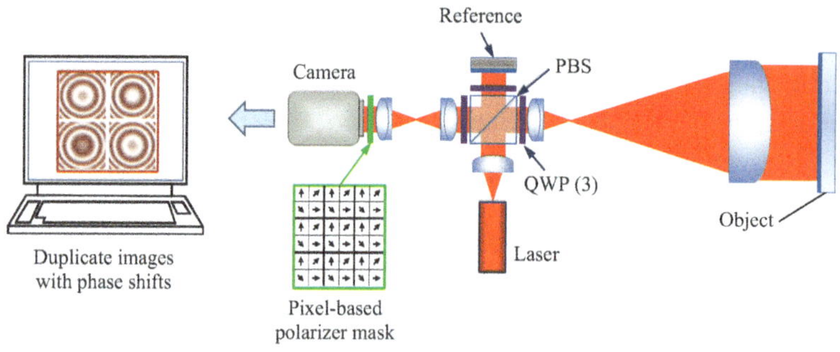

Fig. 4 Modern polarization-based phase shifting interferometer.

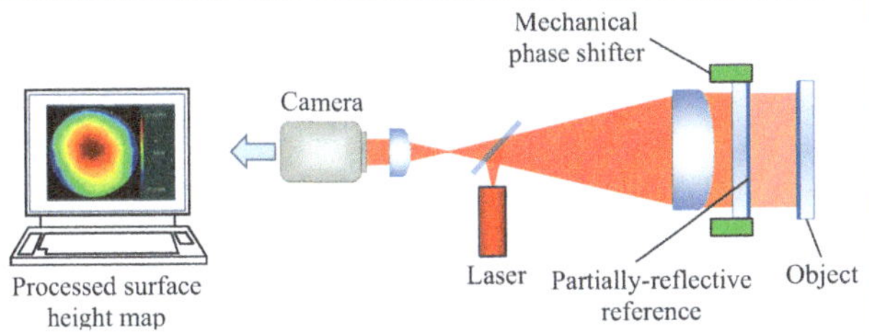

Fig. 5 Laser Fizeau interferometer with a mechanical phase shifting system for the reference flat.

Carrier fringe interferometry

A key breakthrough in holography came when Leith and Upatnieks showed that the overlapping images in Gabor holograms could be separated by using a reference wave at a large enough angle that the reconstructed real and conjugate images formed by the hologram become separable in the far field[29]. This is illustrated in Fig. 1b, where α is this all-important diffraction angle. The result is equivalent to a single-sideband detection by optical methods[30].

By the 1970s, it was realized that the equivalent of far-field separation of the propagating wavefronts could be emulated for interferometry without holographic reconstruction. In the context of improving astronomical images, Roddier and Roddier analyzed wavefronts using direct electronic processing of carrier-fringe interferograms[31]. The concept gained popularity with the seminal work of Takeda in 1982, who described carrier fringe methods for structured light and interferometric measurements of surface topography. In spite of continuous improvements[32,33], for many years the available computers and cameras were not quite up to the task of providing high lateral resolution images quickly using digital Fourier transforms. Mertz[34] and then Freischlad and Küchel[35] developed convolution kernels to speed up the calculations; but the carrier fringe technique was overshadowed in the 1990s by developments in temporal PSI. By the 2000s, enabling technologies had advanced enough for real-time data processing using two-dimensional digital Fourier transforms at high lateral resolution, and the carrier fringe technique was given new life[36].

The principle underlying carrier fringe methods in interferometry follows from communications theory and Lohmann's Fourier analysis of the holographic reconstruction process[30]. To clarify this principle mathematically, we first introduce a nonzero tilt angle θ in the y, z plane defined in Fig. 1 for the reference light u_R, such that Eq. 3 now reads

$$I(y) = |u_o|^2 + |u_R|^2 + u_R u_o^* \exp(2\pi i v y) + u_R^* u_o \exp(-2\pi i v y) \tag{12}$$

where

$$v = \sin(\theta)/\lambda \tag{13}$$

is the spatial frequency of the resulting carrier fringes in the interference pattern. If we simulate a holographic reconstruction with a plane reference wave of unit amplitude and zero phase, the scattered light can be represented by a spectrum $\widetilde{U}$ of plane waves, each of which is the solution to the free-space Helmholtz equation[37,38]. The spectrum for this intensity pattern is

$$\widetilde{U}(f_x, f_y) = \frac{1}{\iint dxdy} \iint I(x,y) \exp[-i2\pi(f_x x + f_y y)]\, dxdy \tag{14}$$

which we write as a Fourier transform

$$\widetilde{U}(f_x, f_y) = \mathcal{F}\{I(x,y)\} \tag{15}$$

In the y, z plane, the Fourier frequency f_y corresponds to the diffraction angle α according to

$$f_y = \sin(\alpha)/\lambda \tag{16}$$

Fig. 6 A 600 mm aperture, wavelength-tuned, temporal phase shifting Fizeau interferometer for testing large optical flats. Photograph courtesy of Zygo Corporation.

From Eq. 12 and 15, we have for the reconstructed light field in frequency space

$$\widetilde{U}(f_x, f_y) = \mathcal{F}\{I_R + I_0\} + \mathcal{F}\{u_o^* u_R\} \otimes \delta(f_y + \nu) + \mathcal{F}\{u_R^* u_0\} \otimes \delta(f_y - \nu) \quad (17)$$

In the far field, whether it is physical or a digital reconstruction in a computer, these three terms are separable by their frequency. It comes as no surprise given Fig. 1 that the last term, which contains the desired information and is centered at a spatial frequency $f_y = \nu$, diffracts at an angle $\alpha = \theta$. After isolating this last term and recentering it about zero frequency, the inverse transform is

$$u_R^* u_0 = \mathcal{F}^{-1}\{\mathcal{F}\{u_R^* u_0\}\} \quad (18)$$

If the reference light u_R has a uniform phase and intensity, or at least if it is known and can be compensated, this process recovers the original complex object field u_0, with separable phase and intensity. Importantly, while the classical holographic reconstruction method involves illuminating the recorded pattern a second time with the reference light u_R, this step is not required if we have a digital recording of the intensity I with sufficient lateral resolution.

In conventional interferometry, we assume that the optical system creates an imaged light field u_I that is as much as possible an in-focus duplicate of the object light field u_o[8]. Because of the suppression of one of the sidebands in frequency space, the digital processing results in complex-valued light field u_I, from which we readily determine the phase

$$\phi_I(x, y) = \arg\{u_I(x, y)\} \quad (19)$$

and corresponding measured surface heights

$$h_I(x, y) = \lambda \phi_I(x, y)/4\pi \quad (20)$$

Fig. 7 summarizes this sequence, from carrier-fringe interferogram to final topography map.

Fig. 8 illustrates a common configuration for creating dense carrier fringes in a Fizeau interferometer. A relative tilt of the reference and object surfaces results in dense interference fringes at the camera, on the order of hundreds of fringes over the viewing aperture. Provided that the instrument has manageable or compensated aberrations when operated off axis, this holographic data acquisition requires few changes in the opto-mechanical hardware of a laser Fizeau system. For this reason, carrier fringe interferometry is often option for systems that also offer mechanical phase shifting as in Fig. 5.

Coherent noise reduction

Major advances in holography date from the invention of visible-wavelength lasers[39,40]. However, there are good reasons for not using lasers in holography and interferometry, because of speckle effects and coherent noise from dust particles and unwanted reflections[41]. Leith notes a benefit of arc lamps to reduce the noise in holograms, describing pre-laser experiments from 1958 to 1962 as "...quite a phenomenal success."[42] Line sources

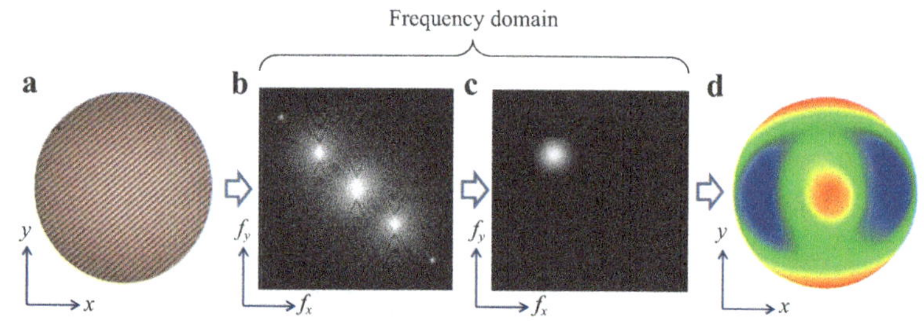

Fig. 7 Data processing for carrier fringe interferometry. **a** Original hologram with high density carrier fringes. **b** The two-dimensional Fourier transform of the hologram. **c** Frequency-domain filtering to isolate one of the Fourier sidebands. **d** The final surface topography map.

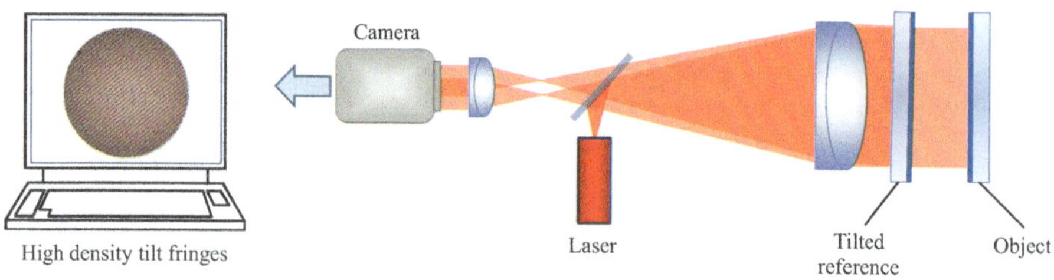

Fig. 8 Laser Fizeau interferometer configured for off-axis, carrier fringe interferometry.

that are incoherent in one direction and limited coherence in the other substantially reduce the noise in carrier fringe holograms of flat objects[29].

With the arrival of the laser, it was possible to produce holograms of three-dimensional objects such as the chess piece in Fig. 1. The results provided astonishing depth and parallax, in a way that would have been impossible with previously-available light sources. Lasers enabled the wide range of holography applications that we know of today; but they came with new problems associated with increased coherent noise.

Even prior to the laser, Kirkpatrick and El-Sum noted in 1956 that with coherent imaging "…careful cleaning of the optical surfaces was found to remove only a minority of the noise sources... The whole nuisance was removed by rotating the entire light-source unit, including the filter, in a continuous manner about the optical axis of the system."[43] If the time of exposure is long, these motions reinforce the desired stationary pattern while averaging away much of the coherent noise[44–47]. Averaging may be achieved in different ways, including rotating an optical element around the axis of the optical system, moving a diffuser across the illuminating beam, changing the incident direction of the illuminating beam with a rotating element, or by moving different masks in the Fourier plane of the imaging system[48,49]. Fig. 9 illustrates a method proposed by Close in 1972 for a portable holographic microscope based on a pulsed ruby laser[46]. The microscope records four holograms, each with an independent speckle pattern corresponding to a rotary position of the prism. The four images formed by the holograms are incoherently superposed to reduce coherent noise and speckle granularity.

Lasers began to appear in unequal-path configurations in the 1960s[50], and averaging methods originally developed for holography to reduce coherent noise were shown to be effective for improving results in interferometry[51]. Fig. 10 illustrates the averaging technique applied to a Fizeau interferometer[52,53]. In this example, each surface height map is the result of a single camera frame data acquisition using the tilt-fringe method[54,55]. Between each acquisition, the direction of illumination by the laser light source is changed by eccentric rotation of a lens near the system focus of the Fizeau geometry. The left-hand surface topography image in Fig. 10a shows high-spatial frequency coherent noise that moves depending on the position of the light source. The average of 16 height maps shown in Fig. 10b for a continuous change in source position is smoother and a better representation of the true surface topography, with much reduced coherent noise.

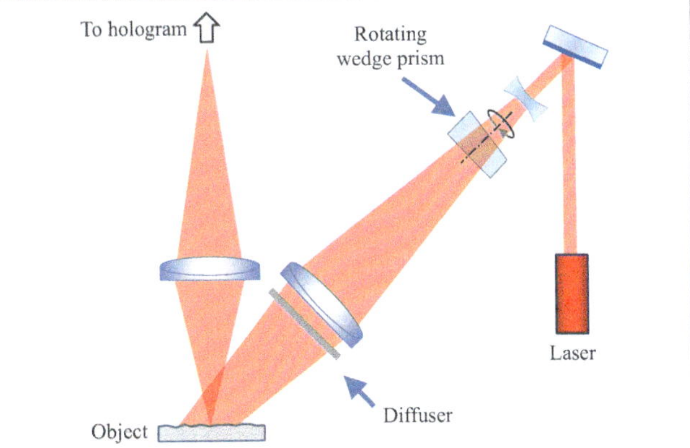

Fig. 9 Coherent noise reduction system using a rotating wedge prism. The diagram shows only the illumination of the object—the reference path is separate.

In place of averaging final topography maps, an alternative method for coherent noise reduction is to synthesize a larger monochromatic source size in real time, using a laser in combination with rapidly-rotating mirrors or diffusing disks. An advantage is that the 'live' camera image is itself improved, and data averaging is not necessary. A basic requirement is that the coherence reduction should not adversely affect the interference fringe contrast, sometimes leading to an adjustable apparent source size[56]. An effective means of preserving fringe contrast with an extended source in a laser Fizeau was invented by Küchel, based on tracing a point source in a circle or ring about the optical axis, which preserves the desired interference fringe pattern while randomizing other coherent effects[57].

In some instruments the reduction in coherence follows a fully coherent portion of the optical system. Commercial Fizeau interferometers often employ coherent imaging to project an interference pattern onto a rapidly-rotating diffuser, resulting in randomized speckle, before a multi-element zoom lens creates the final image on the camera[58]. The benefit is high interference contrast, without coherent artifacts and patterns from the imaging optics and camera cover slips. With the advent of large-format cameras, however, the higher lateral resolution of fully-coherent imaging has made it more attractive to directly image onto the camera[59].

Digital holographic Fizeau interferometer

Gabor's background and research interests led him to think of holography as a new microscopic imaging technique with a large depth of field, allowing the microscopist to examine the different planes of the image at leisure[2]. The ability to refocus the image after it is recorded remains one of the defining characteristics of holography, and frees us from the need to carefully image

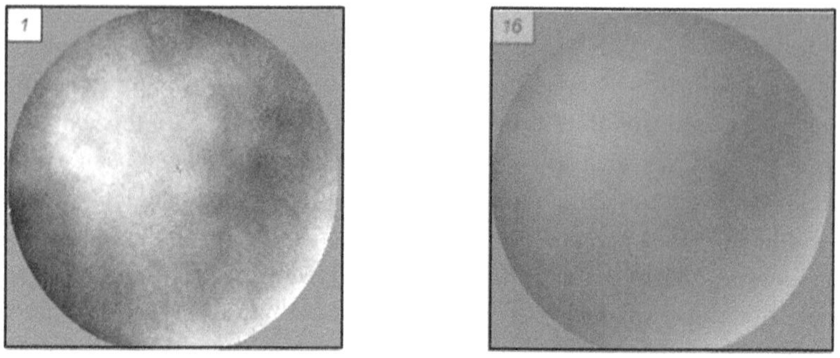

a. Without averaging **b.** With averaging

Fig. 10 a: Grey-scale surface height map by reconstruction of an object surface topography from a single camera frame using carrier-fringe methods. **b**: Improved image quality after averaging 16 topography maps with the light source rotated about the optical axis between data acquisitions. The surface height range for these images is a few nanometers.

objects onto film or a detector. It also enables the recording of measurement volumes, with the ability to sharply image cross sections of three-dimensional data[60]. The capability became even more attractive with digital holography, for which the refocusing is achieved entirely within a computer[61].

The process of digitally refocusing is straightforward, provided of course that we have knowledge of the object light field u_0, gained by holographic reconstruction or by high-resolution digital interferometric recording with PSI or carrier-fringe analysis. To evaluate how this light propagates, as before, we first determine the spectrum of plane waves from the Fourier transform

$$\widetilde{U_o}(f_x, f_y) = \mathcal{F}\{u_o(x,y)\} \quad (21)$$

As the light propagates in the z direction, the spectrum of plane waves evolves according to

$$\widetilde{U_o}(f_x, f_y, z) = \wp(f_x, f_y, z)\,\widetilde{U_o}(f_x, f_y) \quad (22)$$

where the free-space propagator

$$\wp(f_x, f_y, z) = \exp\left[i\frac{2\pi z}{\lambda}\sqrt{1-\lambda^2\left(f_x^2+f_y^2\right)}\right] \quad (23)$$

accounts for how much each plane wave will be shifted in phase in terms of a distance z. The inverse Fourier transform is the focus-shifted object light field:

$$u_o(x,y,z) = \mathcal{F}^{-1}\left\{\widetilde{U_o}(f_x, f_y, z)\right\} \quad (24)$$

These calculations are astonishingly powerful, enabling in some cases the recording of images without any imaging optics—the focusing and real or virtual image formation is performed afterwards in a computer. Hence one of the popular definitions of holography as lensless photography[62].

Although digital refocusing is familiar in digital holographic microscopy[63], it is not usually considered a feature or capability of surface-topography interferometry. Nonetheless, from the preceding mathematical description of the method, it is entirely feasible to refocus conventional interferometry data in the same way after acquisition. As data densities have increased, so has interest in correcting for focus errors to maintain high lateral resolution in interferometry.

Unlike holographic systems, traditional interferometers are arranged such that the object surface is precisely focused onto the camera prior to data acquisition. Fig. 11 illustrates a simplified focus mechanism. Focusing is often a manual process, involving a subjective determination of image sharpness. Because optical surfaces are often featureless by design, a common procedure involves positioning a straightedge as close as possible to the adjusting surface and adjust focus until the straightedge appears sharpest. The combination of cumbersome setup and human error are such that it is reasonable to assert that very few interferometers today are operating at their full potential, simply because of focus errors. Digital refocusing presents an opportunity to solve this problem using software.

To evaluate the benefits of digital holographic refocusing in an interferometer, we first need to define what is meant by lateral resolution. Some familiar examples include the widely-known Rayleigh and Sparrow limits, which in conventional greyscale imaging correspond to the smallest separation between object points for which there are two clearly separable image points[64]. A frequency-domain equivalent is the Abbe limit, which is the highest detectable sinusoidal spatial frequency. These limits have analogous definitions for surface topography measurement[65]. However, it has become common to quantify topographical lateral resolution by means of the instrument transfer function (ITF), which describes an instrument's height response over a range of spatial frequencies[64,66]. The ITF formalism has several limitations,

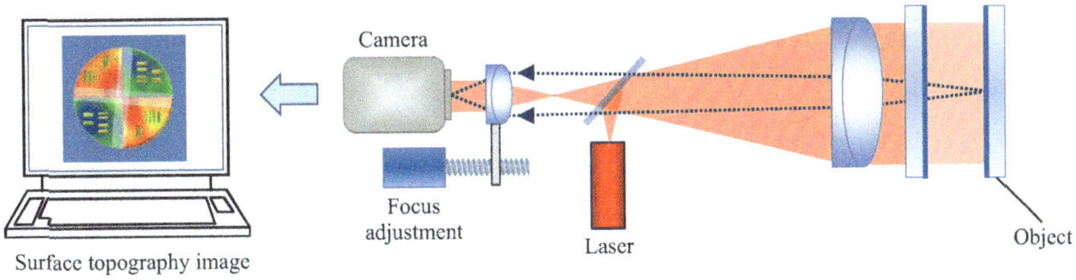

Fig. 11 Focusing mechanism for a laser Fizeau interferometer.

including the requirements of linear height response and shift invariance, but it nonetheless provides a more complete description of lateral resolving power than a single number[67].

The most favorable ITF is achieved at the position of best focus. Fig. 12 shows experimental results for two different focus positions showing the sensitivity of ITF response to focus error. The ITF evaluation consists of a Fourier lateral spatial frequency analysis of the edge spread function, using a specialized ITF artefact with embedded topographical step features. What is most notable about these results is they represent two distinct outcomes for what appears to be nearly identical image sharpness, as shown in Fig. 13. The reason for this is that the higher spatial frequencies—those from 9 to 10 cycles/mm—present the identical impression of sharp edges in a subjective test. A consequence is that a traditional focus adjustment using the eye in place of a complete evaluation of the ITF can easily result in setting the incorrect focus.

The results shown in Fig. 12 motivate the use of digital holographic refocusing to automate the focusing process[68]. Propagating the complex-valued interferogram in object space and integrating the ITF curve for each position results in Fig. 14. In this case, the best focus position is approximately 15 mm away from the initial setup. Using

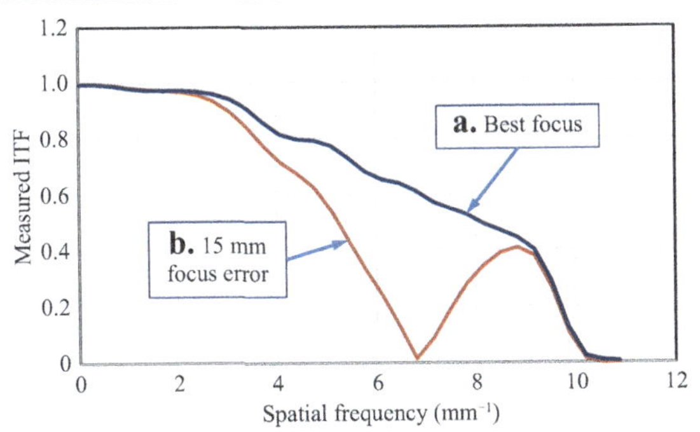

Fig. 12 Height response of a laser Fizeau interferometer as a function of surface spatial frequency for two different focus positions: **a** Best focus. **b** Object surface out of focus by an amount equivalent to moving the object axially 15 mm away from best focus.

this information, an encoded mechanical focus system readjusts the interferometer to the position that provides the best possible lateral resolution, as defined by the ITF and the closely-related Strehl ratio. There remains only to have an automated way of locating the physical location of an object that does not have any surface features—a task that is readily accomplished using a wavelength-tuned laser source to calculate the distance from the reference surface to the object surface[69].

While it is certainly possible to focus the system shown in Fig. 11 without digital holography, this is only because there is a single object with a surface shape that is well within the depth of field. Modern laser Fizeau interferometers are capable of measuring multiple surfaces in an optical system simultaneously using frequency-domain PSI methods[70]. Fig. 15 illustrates this capability. In place of a mechanical or polarization-based phase shift, the laser source is swept in optical frequency by an amount $\Delta\nu$, resulting in a phase shift $\Delta\varphi$ for the interference fringe pattern given by

$$\Delta\varphi = 4\pi L\Delta\nu / c \tag{25}$$

where L is the group-velocity optical path distance to one of the object surfaces, and c is the speed of light. Tuning the laser isolates the signal contributions of each of the object surfaces based on the size of the phase shift as a function of the frequency sweep according to Eq. 25. The phase shifting also enables the reconstruction of the topography for each individual surface[71]. However, for multiple surface testing with a single data acquisition, the best focus position varies with each surface. Thus digital holographic methods are essential to refocusing the extracted interferograms after data acquisition[72]. In this instrument, there is a true fusion of the two measurement principles in what can best be described as a digital holographic Fizeau interferometer.

Computer generated holograms

If he were alive today, Isaac Newton might claim to have been the first person to have captured a hologram experimentally—as a drawing—as well as to have provided a formula for synthesizing a hologram

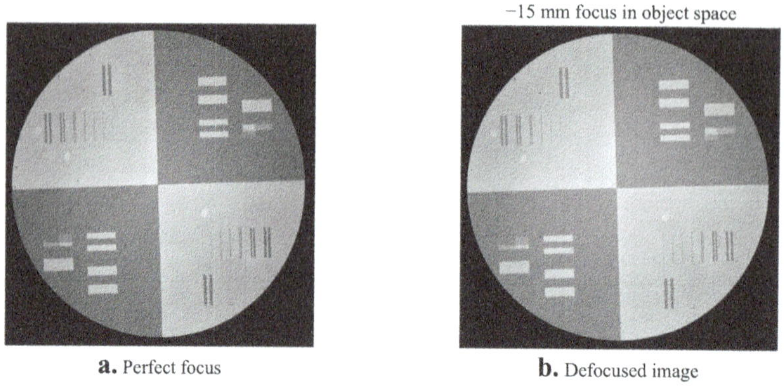

Fig. 13 Greyscale topography images with a surface height range of 50 nm for a 100 mm diameter ITF artefact, for the two difference focus positions **a** and **b**, shown in Fig. 12.

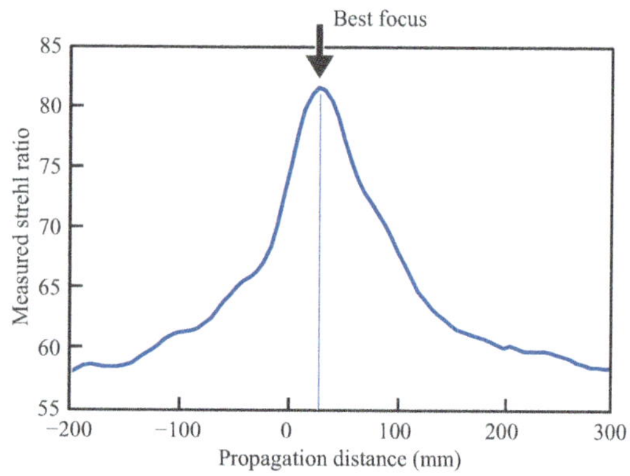

Fig. 14 Integrated ITF values as a function of focus position, obtained with digital holographic propagation in software.

computationally for a variety of wavelengths. The evidence is a drawing in the 2nd book of "Opticks," illustrating the famous experiment that we recognize today as the Newton's rings, formed by the interference of light reflected from a convex lens in contact with a flat glass plate[73]. Hooke might also make a claim of priority if he were alive today, as previous to Newton he had observed a similar effect, and unlike Newton, was open to the idea that the phenomenon could be understood in terms of light waves[74].

To see how the experiments of Hook and Newton relate to the modern technique of computer-generated holograms to assist in interferometry for surface topography measurement, we must advance in time to Gabor's discovery of holography. One of the obstacles to the early acceptance of 'diffraction microscopy' as a promising idea was the difficulty in understanding the theory behind the technique conceptually. In the absence of the compelling demonstrations of 3D images of toy trains and chess pieces that were not technically feasible without a laser, the holographic principle initially had to be appreciated at a theoretical level. The now familiar maths seemed esoteric and inaccessible, even to many physicists[75]. To facilitate understanding of the new method, Rogers in 1950 proposed a conceptual approach in which a Gabor hologram is thought of as a superposition of Fresnel zone patterns, each one of which forms an image point[76]. These individual patterns look very much like Newton's rings, as illustrated in Fig. 16.

Fresnel zone patterns on glass plates have been familiar since the time of Rayleigh in the 19th century as an

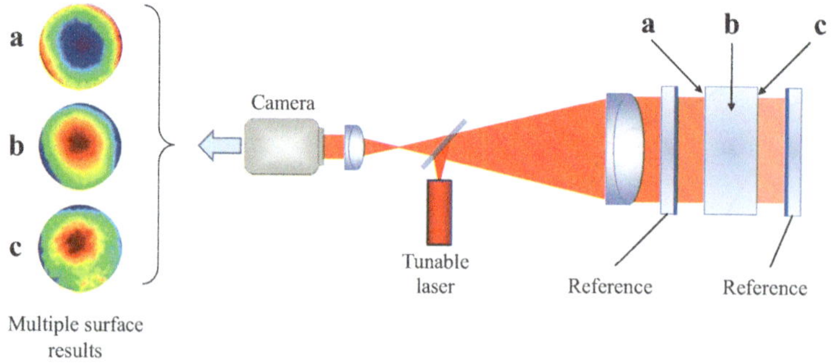

Fig. 15 Multi-surface testing using a wavelength-tunable laser and frequency-domain PSI for measuring **a** the front surface, **b** the material homogeneity and **c** the back surface of an object or optical system.

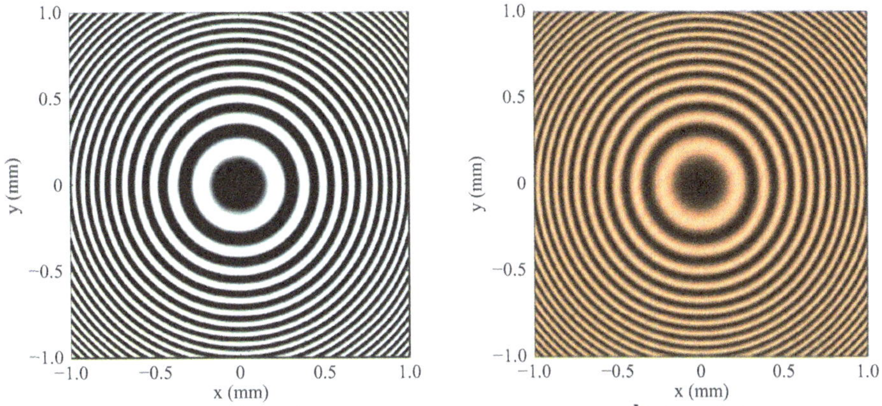

a. Fresnel zone plate **b.** Newton's rings

Fig. 16 a Computed Fresnel zone plate pattern for creating a focus at a distance of 100 mm at a wavelength of 0.5 μm. **b** Newton's rings at the same wavelength for a convex surface with a radius of 200 mm in contact with a flat reference surface, equivalent to a Gabor hologram of a single virtual source point.

elementary kind of diffractive optic[77]. An analysis for the radius ρ of the alternating bright and dark circular zones readily leads to the formula

$$\rho^2 = m\lambda R \tag{26}$$

where R is the distance from the plate to the focal point, and m is the integer index for the alternating zones[78]. This formula, not coincidentally, is nearly identical to the formula for the alternating bright and dark interference fringes in Newton's experiment:

$$\rho^2 = m\lambda R/2 \tag{27}$$

where the distance R is the radius of curvature of a convex spherical surface in contact with a flat glass, and the factor of 2 difference is the result of viewing the spherical glass in reflection. A photograph of Newton's rings is therefore equivalent to a Gabor hologram of a single image point. The humble zone plate therefore shares many of the essential characteristics of holograms, including that any portion of the zone plate can be used to create a focus spot. In Newton's drawing, for example, only half of the ring structure is represented, which has a similar pattern to that of a Leith-Upatnieks off-axis hologram for a single image point[30].

These analogies between zone plates and synthesized or computer-generated holograms (CGHs) were well understood by researchers developing applications for new laser-based, unequal-path interferometers for testing the surface form of optical component in the late 1960s. The driving force in these developments was the need for accurate testing of lenses and mirrors having aspherical surface form. Twyman argued already in 1918 in favor of the increased use of aspheres in optical designs, recognizing that with the aid of an interferometer "…the optician can face without dismay the task of making with precision quite considerable departures from the sphericity … "[79] However, interferometers are best configured as null detectors, providing the highest precision and accuracy when comparing object and reference wavefronts of nearly identical shape. While there are many clever ways to achieve a null test using reflecting and refracting optics for specific species of asphere, such as those represented by a single conic constant, the CGH dramatically increases the solution space by simply altering the formula for the opaque and transparent zones[80].

One of the attractive features of CGH null correctors is that the accuracy of wavefront construction is largely determined by the in-plane location of the diffracting zones, rather than surface heights. Thus, in place of painstakingly polishing an aspherical reference surface to nanometer precision, it is possible to synthesize the reflected wavefront from a precision reference on a more relaxed dimensional scale. In 2007, in a fascinating retrospective of the early days of CGHs, Lohmann describes the use of pen plotters and hand touch-up in the 1960s to create simple grey-scale patterns that were photographically reduced to usable holograms[81]. These experiences with pen plotters and ruling engines for CGHs were duplicated or independently reinvented in several labs worldwide. By the 1970s, pioneering work by Pastor[82], Wyant[83], Schwider[84], and many others rapidly filled in the configuration and application space for CGH-compensated interferometers for optical components[85,86]. The development of advanced photolithography and electron beam writing for semiconductor manufacturing has led to present-day capabilities for testing optics using a configuration similar to that of Fig.17 with far greater departure and higher accuracy than Twyman would have imagined possible[87].

The CGH technique for optical testing has received significant competition from methods that seek greater flexibility than can be provided by a null test. Methods involving axial scanning for rotationally symmetric aspheres, when provided with precision staging systems, achieve uncertainty levels approaching those of custom CGH solutions without a fixed optical configuration for each part type[88]. Reconfigurable geometries based on spatial light modulators assemble zones of low fringe density into final topography maps using a sequence of off-axis illumination angles[89]. Another solution is an optical stylus with precision part motions to trace out full 3D topographies with single-point scanning[90].

To remain competitive, developers have leveraged advances in enabling technologies to make CGH solutions more accessible and practical for fast, production-line testing with simplified part alignment. Fig. 18 illustrates modern use of a CGH for aspheres and freeforms. The instrument is an upward-looking, 633 nm wavelength laser Fizeau, and the test part is a diamond-turned, rotationally symmetric asphere with a radius of 128 mm for the best fit sphere. The aspheric departure is 70.8 μm over a 44 mm diameter. The off-axis, lithographically-produced, binary CGH is positioned just below the part. The CGH and the part are held in precision kinematic fixtures shown in Fig. 18a that allow for part removal and replacement without realigning the instrument to achieve the low fringe density shown in Fig. 18b. The final height map in Fig. 18c shows that the part deviates from design, with a residual 0.223 μm peak-valley (PV) central feature at this stage of machining, after compensating the 70.8 μm departure using the CGH.

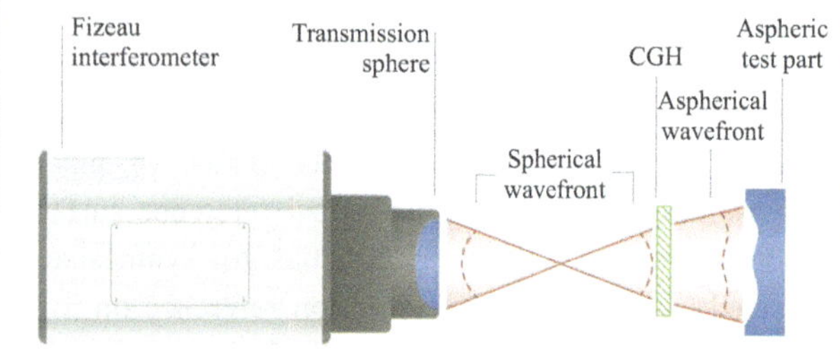

Fig. 17 Configuration for optical testing of an aspheric surface using a laser Fizeau interferometer with a CGH.

Vibrations, deformations, and rough surface topography

The most obvious contribution of holography to interferometry is clear from the name of the technique: 'holographic interferometry.' This discovery by several researchers working in laser-based 3D imaging holography in the mid-1960s ushered in a wide range of applications of high interest to metrologists[91,92]. The term 'discovery' applies perfectly here, as many of the first steps towards this powerful technique were taken experimentally, sometimes by accident, and later explained theoretically[93,94]. There are now many books on the subject, cataloguing methods for quantitative analysis of stress, strain, deformation and overall contour of diffusing objects rendered in 3D by means of holography[95–98].

As we shall see in this section, the discovery of holographic interferometry has had a profound influence on capabilities and interpretations of interferometric techniques[6]. To identify and examine these connections, it is useful to follow the conceptual approach of Heflinger et al. in their 1966 paper on what was at that time the new field of holographic interferometry[99]. We first consider a flat object that is tilted between two holographic exposures of the same hologram. Fig. 19a shows a simulated hologram with the dense carrier fringes characteristic of off-axis holography of plane objects with collimated reference and object illumination. This is sometimes referred to as the "primary image " in holographic interferometry. Fig. 19b illustrates the effect of a double exposure, with the object tilted slightly along a diagonal between exposures. The incoherent superposition of the

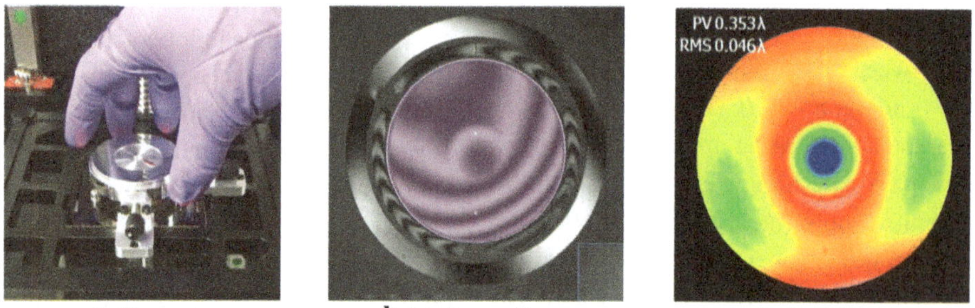

a. Alignment **b.** Interference fringes **c.** Surface height map

Fig. 18 Modern example of a CGH or diffractive-optic null corrector for a testing an asphere. Photographs courtesy of Zygo Corporation and Arizona Optical Metrology (AOM).

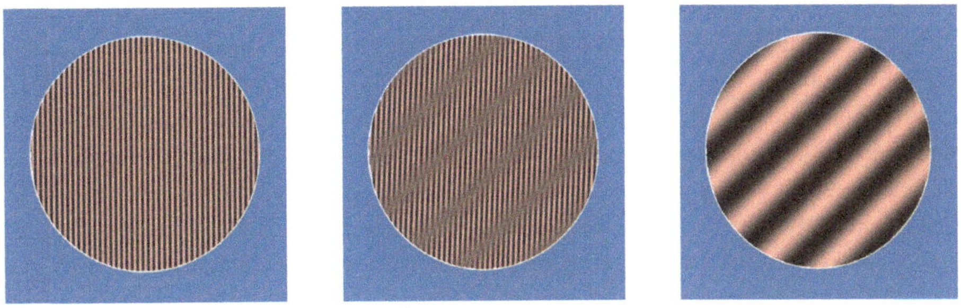

Fig. 19 Simulations illustrating double-exposure holography for a flat part. **a** Carrier fringes resulting from a single off-axis exposure with collimated light; **b** incoherent superposition of two exposures with slightly different tilt angles for the part; **c** the result of creating a part image from the double-exposure hologram by wavefront synthesis.

intensity patterns for the two object orientations results a modulation of the fringe contrast in the hologram, that we readily perceive by eye. When this double-exposure hologram is reilluminated with the reference wave to synthesize the original wavefront from the object, the result is a fringe pattern as shown in Fig. 19c, superimposed on the object image[100]. The fringes in Fig. 19c are the result of variations in diffraction efficiency attributable to the contrast modulation in the hologram[99]. The orientation and density of the fringes informs us how the object has changed between exposures. In this case, the object has been tilted along a diagonal without deforming the object surface, resulting in straight fringes, the density of which is a measure of the amount of tilt.

In the example of Fig. 19, we see that the holographic recreation of a propagating wavefront serves to demodulate the incoherent superposition of interference patterns present in the double-exposure hologram, converting contrast variations to interference fringes representing the difference between the two exposures. Since these superimposed patterns in the hologram are mutually incoherent, they can be generated at different times, with different positions for the component parts of the holographic system, and even with different wavelengths. Hence the astonishingly wide range of applications for the technique.

An intriguing interpretation of holographic interferometry is the concept of temporal wavefront division. When the hologram of Fig. 19b is illuminated with coherent light, the reconstructed wavefront from the first exposure interferes with the wavefront from the second exposure, resulting in the interference pattern of Fig. 19c, even though the exposures were captured at different times[94,99]. A compelling example of this interpretation is the presence of 'live' fringes from a single-exposure hologram with the object illuminated once again and viewed through the hologram. In this case, the fringe formation takes place dynamically, revealing deformations caused by varying loads, vibrations and temperature gradients in real time, by detecting the minute differences between the original exposure and the light reflected from the object[93].

The many extraordinary capabilities of digital holographic methods would seem to be unique to holography. However, there is a link to conventional interferometry in the fringe contrast modulations in image-plane holograms. The effect of a double exposure on a hologram is an interference pattern that can be directly interpreted without the need to recreate propagating wavefronts according to the second step (b) shown in Fig. 1 of a holography experiment[99]. In their 1965 paper on vibration analysis using wavefront reconstruction, Powel and Stetson credit Osterberg for a method developed in 1932 for visualizing the resonant mode analysis of quartz crystals using a Michelson interferometer[101,102]. The contours of the resonant modes in the Osterberg experiments appear as fringe contrast patterns that are interpreted with exactly the same Bessel function analysis used in holographic interferometry.

After the discovery of holographic interferometry, Polhemus[103] developed in 1973 a Twyman Green two-wavelength interferometer using analog electronic demodulation to generate an equivalent-wavelength fringe pattern that follows the surface shape, similar to a double-exposure holographic setup, but without the step of illuminating the hologram and observing the diffraction patterns. In 2003, Bosseboef and Petigrand[104] and Patorski et al[105]. developed methods using a conventional phase shifting interference microscope to demodulate the fringe contrast contours for resonant mode analysis of micro-machined parts. There are many such examples, where demodulation of complicated interference patterns emulates the functionality of holographic interferometry without the physical reconstruction of wavefronts by coherent illumination of holograms.

An interesting and unexpected example of how holographic techniques and signal analyses have informed developments in optical metrology is coherence scanning interferometry (CSI), currently the most widely-used method of interference microscopy for measurements of surface topography[106,107]. The CSI instrument shown in Fig. 20a uses spectrally broadband light and high numerical apertures to generate fringe-contrast variations in the interference pattern shown in Fig. 20b. In a similar way to holography, the interference image is essentially indecipherable, even though the rough object surface is in focus. To make sense of the interference information, the pattern must be demodulated, so as to separate contrast effects from the fringes themselves. In his 1984 PhD thesis, Haneishi[108] adapts the digital holographic carrier-fringe signal processing method described by Takeda[54] for analysis to these CSI fringes. The method relies on scanning the interference objective axially so as to create a signal similar to the one shown in Fig. 21a for a single image point or camera pixel. This signal is Fourier transformed along the z scanning direction to the frequency domain, where the positive-frequency sideband is isolated and shifted to zero frequency prior to inverse transforming to create a new signal corresponding to the fringe contrast envelope alone. Tracing the peak value of the fringe contrast over the field of view creates a topography map similar to the one shown in Fig. 21b. Although software procedures have evolved significantly and independently from holographic interferometry since 1984, a frequency domain analysis of CSI signals is the foundation for many advanced CSI systems today[107,109].

Optical modeling using 3D transfer function

Holographic imaging of 3D objects, many of which are as deep as they are wide, presents a challenge for diffraction theories based on representing a wavefront as a complex amplitude distribution in an x,y plane propagating

a. CSI instrument

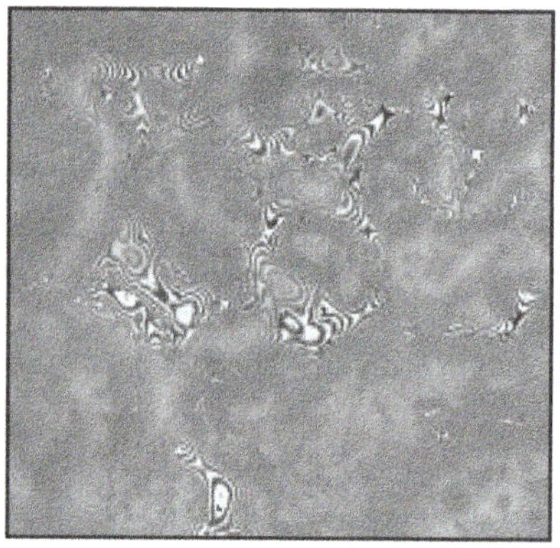

b. CSI image

Fig. 20 a A coherence scanning interferometer based on a microscope platform. **b** A single camera frame image of an interferogram captured with this instrument, for a textured surface.

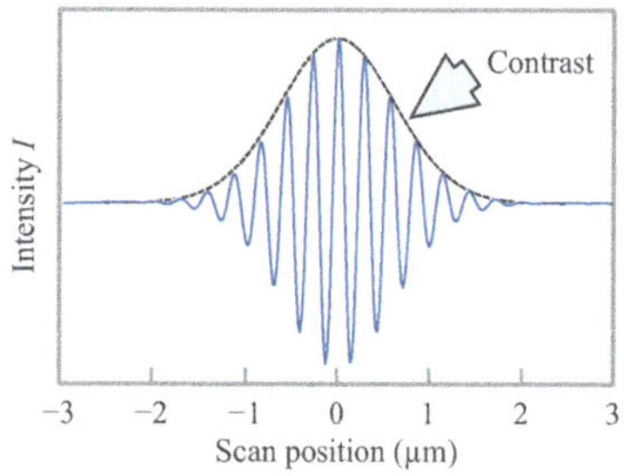

a. CSI signal for one image point

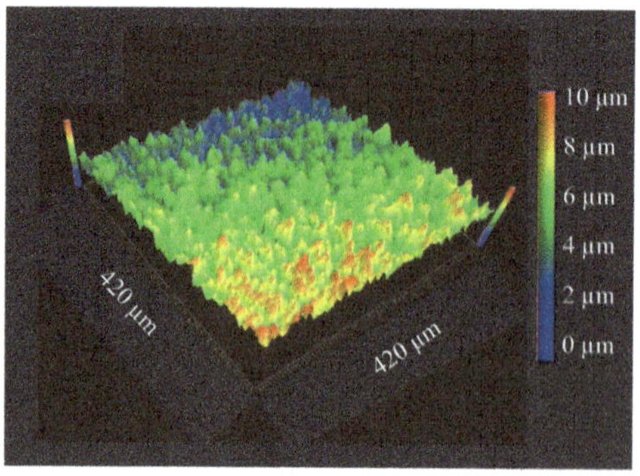

b. CSI image

Fig. 21 a Simulated signal for one image point or camera pixel during an axial scan of the interference objective in a CSI instrument. **b** Experimentally measured topography corresponding to the interference pattern shown in Fig. 20b.

in the z direction. What, exactly, is the structure of a light field scattered from the chess piece in Fig. 1? How can it be represented within a plane, as we have done so far in this paper? When we reilluminate the hologram, how can we calculate the resulting 3D image? These questions about holographic recording and imaging stimulated in the 1960s a reevaluation of the theory of light scattering and imaging in 3D. In this section, we briefly review how holography has influenced the development of 3D diffraction theories for the evaluation and performance improvement of interference microscopes.

Many of the properties of optical instruments can be understood using tradition of Abbe theory and Fourier optics modeling, including the spatial-bandwidth filtering properties of imaging systems.[37,66,110,111] The first step in a Fourier optics model of an interferometer is to simplify the representation of the surface topography to a phase distribution confined to a plane normal to the optical axis[112,113]. The most common approach is to assume a linear relationship between the surface height at the measured phase shift, as in Eq. 9

$$\phi_o(x,y) = 4\pi h_o(x,y)/\lambda \tag{28}$$

from which we calculate the complex-valued object light field in the x, y plane using Eq. 1

$$u_o(x,y) = \sqrt{I_o}\exp[i\phi_o(x,y)] \tag{29}$$

In effect, the surface topography is imprinted upon the reflected wavefront as a variation in phase ϕ_o proportional to the surface height h_o. The corresponding spectrum of plane waves follows from Eq.(21):

$$\widetilde{U}_o(f_x,f_y) = \mathcal{F}\{u_o(x,y)\} \tag{30}$$

To account for the filtering effects of imaging, classical Fourier optics defines a frequency-domain transfer function (TF) $\widetilde{H}$, resulting in an imaged spectrum $\widetilde{U}_I$:

$$\widetilde{U}_I(f_x,f_y) = \widetilde{H}(f_x,f_y)\widetilde{U}_o(f_x,f_y) \tag{31}$$

The simplest transfer function for coherent imaging is a simple frequency bandwidth limit based on the NA A_N of the imaging system:

$$\widetilde{H}(f_x,f_y) = \begin{cases} 1 \ for \ \sqrt{f_x^2+f_y^2} \leqslant f_N \\ 0 \ \ otherwise \end{cases} \tag{32}$$

where in this case the Abbe frequency is

$$f_N = A_N/\lambda \tag{33}$$

The imaged light field is then

$$u_I(x,y) = \mathcal{F}^{-1}\left\{\widetilde{U}_I(f_x,f_y)\right\} \tag{34}$$

In most interferometry applications, we consciously ignore the effects of optical filtering and field-dependent focus effects and write

$$h_I(x,y) = \lambda\phi_I(x,y)/4\pi \tag{35}$$

where the phase

$$\phi_I(x,y) = \arg\{u_I(x,y)\} \tag{36}$$

is determined by PSI or some other means of phase estimation based on interpretations of the interferograms created by mixing the imaged light field u_I with a reference light field u_R.

A more realistic approach to the imaging problem begins by defining the object as a distribution of scattering points in volumetric space, as opposed to height values within an object surface that shift phase values in a complex representation localized to a plane. In 1967, Frieden described a transfer theory in terms of a 3D spectrum of plane waves scattered from such objects, and found that for both incoherent and coherent illumination, the image obeys linear frequency-domain theorems that resemble the familiar results of classical 2D Fourier optics[112]. These theories were expanded in the context of holography by Wolf[114,115] and by Dändliker and Weiss[116]. These 3D Fourier methods are foundational for modern digital holographic microscopy[117,118] and quantitative phase imaging[119].

3D imaging theory for non-holographic interferometry has followed a circuitous path[120]. The traditional modeling of what is fundamentally a 3D problem using 2D complex representation of the object surface as in Eqs. 28,29 implicitly assumes that all surface points can be at the same focus position along the z direction at the same time. The limit to this 2D approximation is that the surface height variations must be small with respect to the depth of field of the imaging system[110,121]. For surface topography measurements using interferometry, this is not so challenging a limitation—common Fizeau interferometers have depths of field on the order of several millimeters and a surface height measurement range of perhaps a few 10s of micrometers. For this reason, adoption of 3D methods has been taken up more quickly in high-magnification microscopy, particularly for confocal microscopy, which works precisely on the principle that surface topography features at high NA cannot all be at the same focus position relative to the depth of field[122,123].

While 2D Fourier optics is adequate for Fizeau interferometers with apertures of 100 mm or more, it is a less satisfying approximation for interference microscopy, for which the sharpness and contrast of interference fringes at high magnifications can vary with just a few micrometers in height variation[124]. Coupland and colleagues have derived the 3D image formation and effective transfer function of CSI based on the Kirchhoff approximation where the surface of a homogeneous medium is represented as a continuous single layer of scattering points[125]. This approach has been demonstrated to be of significant practical value, not only for understanding the origins of measurement errors as a function of slope, curvature, and focus[126–128], but for correcting aberrations as well[129].

Although 3D imaging theory for interferometry shares some common concepts with the 2D Fourier optics model, there are some significant differences. In place of the simple conversion from topography h to phase-shifted wavefront in the x, y plane as in Eqs. 28,29, the object surface is defined by a 3D Dirac delta function Δ_o, which represents an infinitely thin scattering layer under the Kirchhoff approximation. The 3D Fourier transform of this surface model is:

$$\widetilde{\Delta_o}(f_x, f_y, f_z) = \frac{1}{\iiint dxdydz} \iiint \Delta_o(x,y,z) \times \exp[-i2\pi(f_x x + f_y y + f_z z)]\, dxdydz \quad (37)$$

or more compactly,

$$\widetilde{\Delta_o}(f_x, f_y, f_z) = \mathcal{F}_{3D}\{\Delta_o(x,y,z)\} \quad (38)$$

Importantly, this spectrum $\widetilde{\Delta_o}$ does not contain any information at all about the optical system—it is a frequency-domain description of a representation of the object surface itself. All of the information about the optical system is contained in the 3D TF $\widetilde{H_{3D}}$, defined as a linear, shift-invariant frequency-domain function that maps from the object directly to the interference fringes O_I, which for two-beam monochromatic interferometry is the cosine part of Eq. 10[130]. The corresponding 3D frequency representation is

$$\widetilde{O_I}(f_x, f_y, f_z) = \widetilde{\Delta_o}(f_x, f_y, f_z)\, \widetilde{H_{3D}}(f_x, f_y, f_z) \quad (39)$$

from which we calculate

$$O_I(x,y,z) = \mathcal{F}_{3D}^{-1}\left\{\widetilde{O_I}(f_x, f_y, f_z)\right\} \quad (40)$$

It can be shown that this 3D formalism duplicates classical 2D Fourier optics modeling if the object Δ_o is defined as a thin phase object contained entirely within the x, y plane[130].

To gain insight into the geometrical meaning of 3D transfer functions, it is useful to work through the construction of a cross-sectional TF corresponding to the y, z plane of incidence for monochromatic interferometry at high NA[131]. The construction starts in Fig. 22a by defining incident wavevectors $\mathbf{k_i}$ and scattered wavevectors $\mathbf{k_s}$. These wavevectors have a length given by the inverse wavelength $1/\lambda$. In an interferometer, the phase sensitivity to surface topography for a given pair of incident and scattered wavevectors is

$$\mathbf{k} = \mathbf{k_s} - \mathbf{k_i} \quad (41)$$

This is illustrated in Fig. 22b, where the corresponding frequencies are

$$f_y = \mathbf{k} \cdot \hat{y} \quad (42)$$

$$f_z = \mathbf{k} \cdot \hat{z} \quad (43)$$

The range of possible $\mathbf{k_i}$ orientations consistent with the illumination geometry sweeps out an arc. For each incident wavevector, there is a corresponding arc of possible scattered wavevectors $\mathbf{k_s}$, limited by the acceptance cone of the imaging system. The combination of these two arcs defines the range of f_y, f_z values that are detectable using interferometry. Sweeping the $\mathbf{k_i}$ and $\mathbf{k_s}$ vectors through all possible angles results in a magnitude distribution proportional to the number of ways in which these vectors can be oriented to reach the same point in frequency space.

Fig. 23 shows the final 3D TF result for a spatially-incoherent, monochromatic light in an interference microscope. Note that the magnitude of the TF is represented by the color bar. Along the lateral f_y axis, the boundaries are given by the Abbe frequency. Along the vertical f_z axis, we have the range of possible interference fringe frequencies for variations in surface height along the z axis, with the maximum possible frequency for normal

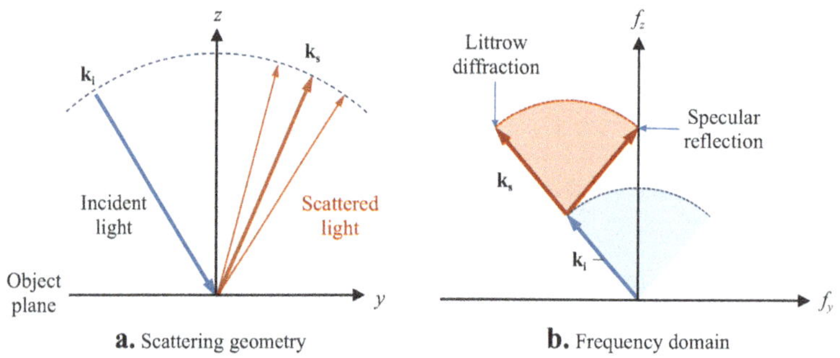

Fig. 22 a Geometry of incident and scattered rays at a surface. **b** The corresponding frequency domain illustrating the process of determining the range of possible spatial frequencies detected by the instrument.

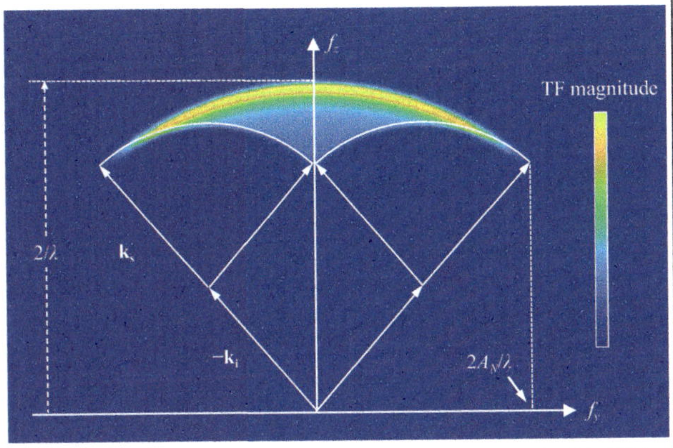

Fig. 23 Geometrical construction of the cross section of a 3D-TF in monochromatic light for an interferometer with a linear pupil filled with incoherent light. The TF magnitude is an arbitrary scale.

incidence and specular reflection of $2/\lambda$.

Conclusions and perspectives

As noted by Reynolds, et al., "A hologram is essentially an interferogram, usually made in a division of amplitude interferometer. "[132] Bryngdahl and Lohmann asserted in 1968 that "…an interferogram is nothing but a special type of a hologram, which we call an 'image hologram'. "[8] Going further, Itoh and Ohtsuka proposed that "…it may even be possible to call all the image reconstruction techniques which are based on interferometric measurements of the spatial coherence function as interferometric imaging techniques."[133] With these observations in hand, it can be tempting to conclude that perhaps 'holography' is just another name for 'interferometry.'[134] Fortunately for both fields of study, this is clearly not the case; even if the common principle is the holistic recording and reconstruction of wavefronts using interference patterns.

The birth 60 years ago of practical, laser-based holography resulted in a rapid series of innovations that found their way from holography into interferometry. These advances include using polarization and temporal phase shifting methods, solutions for coherent noise, digital refocusing, interferometric topography measurements of rough surfaces, dynamic interferometry using the off-axis carrier fringe method, and imaging theory using 3D transfer functions. A particularly interesting example is holographic null testing, using computer-generated or experimental holograms. Arguably, the foundations for holographic null testing were laid well over a century ago, for both fabrication and use of diffractive optics for the testing of aspheres and freeforms. But it was not until the discovery of laser-based holography that the technique became a standard for high-precision testing in research and production.

Of course, the selected examples here are far from being a comprehensive list of contributions of holography. One clear trend is the continuously increasing influence of holography on interferometric methods for surface topography measurements. It is possible that this eventually will lead to a fusion of holography with techniques not normally thought of as holographic. This evolution in applied optical metrology will surely lead to significant new solutions going forward.

Acknowledgements
The authors wish to thank the many contributors to this paper, including Edward LaVilla, who provided the CGH example of Fig. 18, and members of the Zygo Innovations Group, who reviewed the manuscript and participated in excellent discussions regarding the history, meaning and practical use of holographic principles in optics. We offer special thanks to James Wyant, one of the pioneers of holography for interferometric optical testing, for discussions and helpful suggestions.

Author details
[1]Zygo Corporation, Middlefield, CT USA. [2]Shanghai Institute of Optics and Fine Mechanics, Chinese Academy of Sciences, Shanghai, China. [3]University of Stuttgart, Stuttgart, Germany

Conflict of interest
The authors declare no conflict of interest.

Received: 10 September 2021 Revised: 14 January 2022 Accepted: 19 January 2022
Accepted article preview online: 24 January 2022
Published online: 02 April 2022

References

1. Johnston, S. F. A historian's view of holography. in New Directions in Holography and Speckle. (eds Caulfield, H. J. and Vikram, C. S.) (Valencia: American Scientific Publishers, 2008), 3–15.
2. Gabor, D. A new microscopic principle. *Nature* **161**, 777-778 (1948).
3. Leith, E. N. Holography. in The Optics Encyclopedia. (ed Brown, T. G.) (Weinheim: Wiley-VCH, 2007), 773-800.
4. Kreis, T. Handbook of Holographic Interferometry: Optical and Digital Methods. (Weinheim: Wiley-VCH, 2005).
5. Hariharan, P. Optical Holography: Principles, Techniques and Applications. 2nd edn (Cambridge Cambridge University Press, 1996).
6. Wernicke, G., and Osten, W. Holografische Interferometrie - Grundlagen, Methoden und ihre Anwendungen in der Festkörpermechanik. (Leipzig: VEB Fachbuchverlag, 1982).
7. Zygo Corporation. Interferogram interpration and evaluation handbook. 4th edn. Applications note, (Middlefield: Zygo Corporation, 1977)
8. Bryngdahl, O., and Lohmann, A. W. Interferograms are Image Holograms. *Journal of the Optical Society of America* **58**, 141-142 (1968).
9. Goodman, J. W. Analogy between Holography and Interferometric Image Formation. *Journal of the Optical Society of America* **60**, 506-509 (1970).
10. Malacara, D., Servín, M., and Malacara, Z. Phase Shifting

Interferometry. in Interferogram Analysis for Optical Testing. 2nd edn (Boca Raton: CRC Press, 2005)
11. Carré, P. Installation et utilisation du comparateur photoélectrique et interférentiel du Bureau International des Poids et Mesures. *Metrologia* **2**, 13-23 (1966).
12. Waters, J. P. Holography. in Holographic Nondestructive Testing. (ed Erf, R. K.) (New York: Academic Press, 1974), 5-59.
13. Gabor, D., and Goss, W. P. Interference Microscope with Total Wavefront Reconstruction. *Journal of the Optical Society of America* **56**, 849-858 (1966).
14. Liu, J.-P., and Poon, T. -C. Two-step-only quadrature phase-shifting digital holography. *Optics Letters* **34**, 250-252 (2009).
15. Schack, R. V. Direct Phase-Sensing Interferometer. *Journal of the Optical Society of America A* **655A**, MG14 (1971).
16. Millerd, J. E. et al. Pixelated phase-mask dynamic interferometer. Proc. SPIE 5531, Interferometry XII: Techniques and Analysis. Denver, Colorado, United States: SPIE, 2004, 304-314.
17. Wyant, J. C., Advances in Interferometric Metrology. Proc. SPIE 4927, Optical Design and Testing. Shanghai, China: SPIE, 2002, 154-162.
18. Bruning, J. H. et al. Digital Wavefront Measuring Interferometer for Testing Optical Surfaces and Lenses. *Applied Optics* **13**, 2693-2703 (1974).
19. Morgan, C. J. Least-squares estimation in phase-measurement interferometry. *Optics Letters* **7**, 368-370 (1982).
20. Crane, R. New Developments in Interferometry: Interference Phase Measurement. *Applied Optics* **8**, 538-542 (1969).
21. Malacara, D., Servín, M., and Malacara, Z. Phase-Detection Algorithms. in Interferogram Analysis for Optical Testing. 2nd edn (Boca Raton: CRC Press, Taylor & Francis Group, 2005)
22. de Groot, P. Phase Shifting Interferometry. in Optical Measurement of Surface Topography. (ed Leach, R.) (Berlin: Springer Verlag, 2011), 167-186.
23. Deck, L. L. Model-based phase shifting interferometry. *Applied Optics* **53**, 4628-4636 (2014).
24. Dändliker, R., Ineichen, B., and Mottier, F. M. High resolution hologram interferometry by electronic phase measurement. *Optics Communications* **9**, 412-416 (1973).
25. Creath, K. Phase-shifting speckle interferometry. *Applied Optics* **24**, 3053-3058 (1985).
26. Stetson, K. A., 50 years of holographic interferometry. Proc. SPIE 9442, Optics and Measurement Conference 2014. SPIE, 2015, 11.
27. Hariharan, P., Oreb, B. F., and Brown, N. A digital phase-measurement system for real-time holographic interferometry. *Optics Communications* **41**, 393-396 (1982).
28. Ghiglia, D. C., and Pritt, M. D. Two-Dimensional Phase Unwrapping, Theory, Algorithms, and Software. (New York: John Wiley & Sons, 1998).
29. Leith, E. N., and Upatnieks, J. Reconstructed Wavefronts and Communication Theory*. *Journal of the Optical Society of America* **52**, 1123-1130 (1962).
30. Lohmann, A. Optische Einseitenbandübertragung Angewandt auf das Gabor-Mikroskop. *Optica Acta:International Journal of Optics* **3**, 97-99 (1956).
31. Roddier, C., and Roddier, F. Interferogram analysis using Fourier transform techniques. *Applied Optics* **26**, 1668-1673 (1987).
32. Bone, D. J. Fourier fringe analysis: the two-dimensional phase unwrapping problem. *Applied Optics* **30**, 3627-3632 (1991).
33. Kujawinska, M., Spik, A., and Wojciak, J., Fringe Pattern Analysis Using Fourier Transform Techniques. Proc. SPIE 1121, Interferometry '89. Warsaw: 1990, 130-135.
34. Mertz, L. Real-time fringe-pattern analysis. *Applied Optics* **22**, 1535-1539 (1983).
35. Freischlad, K. et al. Real-time wavefront measurement with lambda/10 fringe spacing for the optical shop. Proc. SPIE 1332, Optical Testing and Metrology III: Recent Advances in Industrial Optical Inspection. San Diego, CA, United States: SPIE, 1991,
36. Deck, L. L., Environmentally friendly interferometry. Proc. SPIE 5532, Interferometry XII: Applications. Denver, Colorado, United States: SPIE, 2004, 159-169.
37. Goodman, J. W. Introduction to Fourier Optics. 4th edn (New York: W. H. Freeman Macmillan Learning, 2017).
38. Goodman, J. W. Introduction to Fourier Optics. 2nd edn (New York: McGraw-Hill, 1996).
39. Maiman, T. H. Stimulated Optical Radiation in Ruby. *Nature* **187**, 493-494 (1960).
40. Leith, E. N., and Upatnieks, J. Wavefront Reconstruction with Continuous-Tone Objects. *Journal of the Optical Society of America* **53**, 1377-1381 (1963).
41. Reynolds, G. O. et al. Sources of Coherent Noise and Their Reduction. in The New Physical Optics Notebook: Tutorials in Fourier Optics. (eds Parrent, G. B. and Reynolds, G. O.) (Bellingham, Washington: SPIE Optical engineering press, 1989), 199-219.
42. Leith, E. N. 1 - Introduction. in Handbook of Optical Holography. (ed Caulfield, H. J.) (San Diego: Academic Press, 1979), 1-12.
43. Kirkpatrick, P., and El-Sum, H. M. A. Image Formation by Reconstructed Wave Fronts I Physical Principles and Methods of Refinement. *Journal of the Optical Society of America* **46**, 825-831 (1956).
44. Martienssen, W., and Spiller, S. Holographic reconstruction without granulation. *Physics Letters A* **24**, 126-128 (1967).
45. Thomas, C. E. Coherent Optical Noise Suppression. *Applied Optics* **7**, 517-522 (1968).
46. Close, D. H. High Resolution Portable Holocamera. *Applied Optics* **11**, 376-383 (1972).
47. Cronin, D. J., Pinard, A. I., and Smith, A. E. Coherent optical system with expanded bandwidth and noise suppression. US Patent 3, 770, 340 (1973).
48. Pawluczyk, R. Holographic microinterferometer with noise suppression. *Applied Optics* **28**, 3871-3881 (1989).
49. McKechnie, T. S. Speckle Reduction. in Laser Speckle and Related Phenomena. (ed Dainty, J. C.) (Berlin, Heidelberg: Springer, 1975), 123-170.
50. Houston, J. B., Buccini, C. J., and O'Neill, P. K. A Laser Unequal Path Interferometer for the Optical Shop. *Applied Optics* **6**, 1237-1242 (1967).
51. de Groot, P. J., Applications of optical coherence in interferometric metrology. Proc. SPIE 10834, Speckle 2018: VII International Conference on Speckle Metrology. Janów Podlaski, Poland: SPIE, 2018, 108340.
52. Freischlad, K. R., and Kuechel, M. F., Speckle reduction by virtual spatial coherence. Proc. SPIE 1755, Interferometry: Techniques and Analysis. San Diego, California, United States: SPIE, 1992, 38-43.
53. Zygo Corporation. Coherent artifact reduction system. AN-0065 edn. Application note, (Middlefield: Zygo Corporation, 2014)
54. Takeda, M., Ina, H., and Kobayashi, S. Fourier-transform method of fringe-pattern analysis for computer-based topography and interferometry. *Journal of the Optical Society of America* **72**, 156-160 (1982).
55. Sykora, D. M., and de Groot, P., Instantaneous Interferometry: Another View. OSA Technical Digest, International Optical Design Conference and Optical Fabrication and Testing. Jackson Hole, Wyoming, United States: Optical Society of America, 2010, OMA1.
56. Freischlad, K. Interferometer light source. US Patent 6061133 (2000).
57. Küchel, M., Spatial Coherence in Interferometry: Zygo's new method

to reduce intrinsic noise in interferometers. Proc. Optatec, International trade fair for optical technologies, components and systems. Frankfurt, Germany: P. E. Schall GmbH & Co., 2004,

58. Domeniacali, P. L., and Hunter, G. C. Optical interferometer system with CCTV camera for measuring a wide range of aperture sizes. US Patent 4, 201, 473 (1980).
59. Deck, L. L., and de Groot, P. J., Using the Instrument Transfer Function to Evaluate Fizeau Interferometer Performance. OSA Technical Digest, Optical Design and Fabrication 2017 (Freeform, IODC, OFT). Denver, Colorado, United States: Optical Society of America, 2017, OM2B. 7.
60. Trolinger, J. D., Is holography ready for yet another life? or make holography great again. Proc. SPIE 9960, Interferometry XVIII. San Diego, California, United States: SPIE, 2016, 996008.
61. Goodman, J. W., and Lawrence, R. W. Digital Image Formation from Electronically Detected Holograms. *Applied Physics Letters* **11**, 77-79 (1967).
62. Wu, Y. C., and Ozcan, A. Lensless digital holographic microscopy and its applications in biomedicine and environmental monitoring. *Methods* **136**, 4-16 (2018).
63. Colomb, T., and Kühn, J. Digital Holographic Microscopy. in Optical Measurement of Surface Topography. (ed Leach, R.) (Berlin Heidelberg: Springer, 2011), 209-235.
64. ISO 25178-600: 2019, Geometrical product specifications (GPS) — Surface texture: Areal — Part 600: Metrological characteristics for areal-topography measuring methods. Geneva: International Organization for Standardization (2019).
65. de Groot, P. J., Progress in the specification of optical instruments for the measurement of surface form and texture. Proc. SPIE 9110, Dimensional Optical Metrology and Inspection for Practical Applications III. Baltimore, Maryland, United States: SPIE, 2014, 91100M.
66. de Groot, P. *et al.* , Fourier optics modelling of coherence scanning interferometers. Proc. SPIE 11817, Applied Optical Metrology IV. San Diego, California, United States: SPIE, 2021, 118170M.
67. de Groot, P. J. The instrument transfer function for optical measurements of surface topography. *Journal of Physics:Photonics* **3**, 024004 (2021).
68. Deck, L. L. Method and apparatus for optimizing the optical performance of interferometers. US Patent 10267617 (2019).
69. Deck, L. L., Absolute distance measurements using FTPSI with a widely tunable IR laser. Proc. SPIE 4778, Interferometry XI: Applications. Seattle, WA, United States: SPIE, 2002, 218-226.
70. Deck, L. L. Fourier-transform phase-shifting interferometry. *Applied Optics* **42**, 2354-2365 (2003).
71. Deck, L. L., and de Groot, P. J., The state of the art in swept-wavelength laser Fizeau interferometry. Proc. SPIE 11817, SPIE Optical Engineering + Applications. SPIE, 2021, 118170N.
72. Deck, L. L. Interferometric method and apparatus using calibration information relating a focus setting to a test object position. US Patent 10890428 (2021).
73. Newton, I. Opticks or, a Treatise of the Reflections, Refractions, Inflections, and Colours of Light. 4th edn (London: S. Smith and B. Walford, 1704).
74. Wood, R. W. Physical Optics. (New York: MacMillan, 1911).
75. Johnston, S. F. From white elephant to Nobel Prize: Dennis Gabor's wavefront reconstruction. *Historical Studies in the Physical and Biological Sciences* **36**, 35-70 (2005).
76. Rogers, G. L. Gabor Diffraction Microscopy: the Hologram as a Generalized Zone-Plate. *Nature* **166**, 237-237 (1950).
77. Rayleigh, L. Wave Theory. in Encylopaedia Brittanica vol **24**. 9th edn (New York: Henry G. Allen and Company, 1888), 421-458.
78. Fresnel, A. J. Calcul de l'intensité de la lumière au centre de l'ombre d'un ecran et d'une ouverture circulaires eclairés par un point radieux. in Œuvres Complètes d'Augustin Fresnel. (Paris: Imprimerie Impériale, 1866)
79. Twyman, F. Vl. Interferometers for the experimental study of optical systems from the point of view of the wave theory. *Philosophical Magazine* **35**, 49-58 (1918).
80. Malacara, D. *et al.* Testing of Aspheric Wavefronts and Surfaces. in Optical Shop Testing. 3rd edn (ed Malacara, D.) (Hoboken: John Wiley & Sons, Inc., 2007), 435-497.
81. Lohmann, A. W. A Pre-History of Computer-Generated Holography. *Optics and Photonics News* **19**, 36-47 (2008).
82. Pastor, J. New Developments in Interferometry, Part III: Hologram Interferometry and Optical Technology. *Applied Optics* **8**, 525-531 (1969).
83. Wyant, J. C., and Bennett, V. P. Using Computer Generated Holograms to Test Aspheric Wavefronts. *Applied Optics* **11**, 2833-2839 (1972).
84. Schwider, J., and Burow, R. Testing of asphericsby means of rotational-symmetric synthetic holograms. *Optica Applicata* **6**, 83-88 (1976).
85. Lee, W. -H. III Computer-Generated Holograms: Techniques and Applications. in Progress in Optics vol **16**. (ed Wolf, E.) (Amsterdam: Elsevier, 1978), 119-232.
86. Pruss, C. et al. Computer-generated holograms in interferometric testing. *Optical Engineering* **43**, 2534-2540 (2004).
87. Glatzel, H. et al. Projection optics for EUVL micro-field exposure tools with 0.5 NA. Proc. SPIE 9048, Extreme Ultraviolet (EUV) Lithography V. San Jose, California, United States: 2014, 90481K.
88. Küchel, M. F., Interferometric measurement of rotationally symmetric aspheric surfaces. Proc. SPIE 7389, Optical Measurement Systems for Industrial Inspection VI. Munich, Germany: SPIE, 2009, 738916.
89. Pruss, C. et al. Measuring aspheres quickly: tilted wave interferometry. *Optical Engineering* **56**, 111713 (2017).
90. Berger, G., and Petter, J., Non-contact metrology of aspheric surfaces based on MWLI technology. Proc. SPIE 8884, Optifab 2013. Rochester, New York, United States: 2013, 8884V.
91. Georges, M. Chapter 6 - Holographic Interferometry: From History to Modern Applications. in Optical Holography. (ed Blanche, P. -A.) Elsevier, 2020), 121-163.
92. Stetson, K. A. The discovery of holographic interferometry, its development and applications (accepted for publication). *Light: Advanced Manufacturing*, (2021).
93. Stetson, K. A., and Powell, R. L. Interferometric Hologram Evaluation and Real-Time Vibration Analysis of Diffuse Objects. *Journal of the Optical Society of America* **55**, 1694-1695 (1965).
94. Haines, K. A., and Hildebrand, B. P. Surface-Deformation Measurement Using the Wavefront Reconstruction Technique. *Applied Optics* **5**, 595-602 (1966).
95. Kim, M. K. et al. Digital Holography and 3D Imaging: introduction. *Applied Optics* **53**, DH1 (2014).
96. Nehmetallah, G., and Banerjee, P. P. Applications of digital and analog holography in three-dimensional imaging. *Advances in Optics and Photonics* **4**, 472-553 (2012).
97. Vest, C. M. Holographic Interferometry. (New York: Wiley-Interscience, 1979).
98. Abramson, N. H., Bjelkhagen, H., and Cooke, F. Industrial Holographic Measurements. *Applied Optics* **12**, 2792-2796 (1973).
99. Heflinger, L. O., Wuerker, R. F., and Brooks, R. E. Holographic Interferometry. *Journal of Applied Physics* **37**, 642-649 (1966).
100. Collier, R. J., Doherty, E. T., and Pennington, K. S. Application of moiré techniques to holography. *Applied Physics Letters* **7**, 223-225 (1965).
101. Powell, R. L., and Stetson, K. A. Interferometric Vibration Analysis by

Wavefront Reconstruction. *Journal of the Optical Society of America* **55**, 1593-1598 (1965).

102. Osterberg, H. An Interferometer Method of Studying the Vibrations of an Oscillating Quartz Plate. *Journal of the Optical Society of America* **22**, 19-35 (1932).
103. Polhemus, C. Two-Wavelength Interferometry. *Applied Optics* **12**, 2071-2074 (1973).
104. Bosseboeuf, A., and Petitgrand, S., Application of microscopic interferometry techniques in the MEMS field. Proc. SPIE 5145, Microsystems Engineering: Metrology and Inspection III. Munich, Germany: SPIE, 2003, 1-16.
105. Patorski, K. et al. Simplified time-averaged digital interferometry for vibration studies of microelements. Proc. SPIE 4933, Speckle Metrology. Trondheim, Norway: SPIE, 2003,
106. Dresel, T., Häusler, G., and Venzke, H. Three-dimensional sensing of rough surfaces by coherence radar. *Applied Optics* **31**, 919-925 (1992).
107. de Groot, P. Principles of interference microscopy for the measurement of surface topography. *Advances in Optics and Photonics* **7**, 1-65 (2015).
108. Haneishi, H. Signal processing for film thickness measurements by white light interferometry. Ph. D. Thesis, University of Electro-communications, Chofu, Tokyo, (1984).
109. de Groot, P., and Deck, L. Surface Profiling by Analysis of White-light Interferograms in the Spatial Frequency Domain. *Journal of Modern Optics* **42**, 389-401 (1995).
110. de Groot, P., and Colonna de Lega, X. Fourier optics modeling of interference microscopes. *J. Opt. Soc. Am. A* **37**, B1-B10 (2020).
111. Voelz, D. G. Computational Fourier Optics: A MATLAB Tutorial. (Bellingham: SPIE Press, 2011).
112. Frieden, B. R. Optical Transfer of the Three-Dimensional Object. *Journal of the Optical Society of America* **57**, 56-66 (1967).
113. Mantravadi, M. V., and Malacara, D. Newton, Fizeau, and Haidinger Interferometers. in Optical Shop Testing. 3rd edn (ed Malacara, D.) (Hoboken: Wiley-Interscience, 2007), 1-45.
114. Wolf, E. Determination of the Amplitude and the Phase of Scattered Fields by Holography*. *Journal of the Optical Society of America* **60**, 18-20 (1970).
115. Wolf, E. Three-dimensional structure determination of semi-transparent objects from holographic data. *Optics Communications* **1**, 153-156 (1969).
116. Dändliker, R., and Weiss, K. Reconstruction of the three-dimensional refractive index from scattered waves. *Optics Communications* **1**, 323-328 (1970).
117. Kim, M. K. Digital Holographic Microscopy: Principles, Techniques, and Applications. (New York: Springer, 2011).
118. Kou, S. S., and Sheppard, C. J. R. Imaging in digital holographic microscopy. *Optics Express* **15**, 13640-13648 (2007).
119. Popescu, G. Principles of Biophotonics, Volume 1: Linear systems and the Fourier transform in optics. (Bristol: IOP Publishing, 2018).
120. Chmelík, R. Three-dimensional scalar imaging in high-aperture low-coherence interference and holographic microscopes. *Journal of Modern Optics* **53**, 2673-2689 (2006).
121. Sheppard, C. J. R. Imaging of random surfaces and inverse scattering in the Kirchoff approximation. *Waves in Random Media* **8**, 53-66 (1998).
122. Sheppard, C. J. R., and Matthews, H. J. Imaging in high-aperture optical systems. *Journal of the Optical Society of America A* **4**, 1354-1360 (1987).
123. Sheppard, C. J. R., and Cogswell, C. J. Three-dimensional image formation in confocal microscopy. *Journal of Microscopy* **159**, 179-194 (1990).
124. Chakmakjian, S., Biegen, J. F., and de Groot, P., Simultaneous focus and coherence scanning in interference microscopy. IWI Proceedings, International Workshop on Interferometry. Wako, Saitama, Japan: RIKEN, 1996, 171-172.
125. Coupland, J. et al. Coherence scanning interferometry: linear theory of surface measurement. *Applied Optics* **52**, 3662-3670 (2013).
126. Su, R. et al. On tilt and curvature dependent errors and the calibration of coherence scanning interferometry. *Optics Express* **25**, 3297-3310 (2017).
127. Su, R. et al. Effects of defocus on the transfer function of coherence scanning interferometry. *Optics Letters* **43**, 82-85 (2018).
128. Thomas, M. et al. Surface measuring coherence scanning interferometry beyond the specular reflection limit. *Optics Express* **29**, 36121-36131 (2021).
129. Su, R. et al. Lens aberration compensation in interference microscopy. *Optics and Lasers in Engineering* **128**, 106015 (2020).
130. Su, R. et al. Scattering and three-dimensional imaging in surface-topography measuring interference microscopy. *Journal of the Optical Society of America A* **38**, 27-42 (2020).
131. Zhou, K. C. et al. Unified k-space theory of optical coherence tomography. *Advances in Optics and Photonics* **13**, 462-514 (2021).
132. Reynolds, G. O. et al. Division of Amplitude Interferometry. in The New Physical Optics Notebook: Tutorials in Fourier Optics. (eds Parrent, G. B. and Reynolds, G. O.) (Bellingham: Optical Engineering Press, 1989), 242-263.
133. Itoh, K., and Ohtsuka, Y. Interferometric image reconstruction through the turbulent atmosphere. *Applied Optics* **20**, 4239-4244 (1981).
134. de Groot, P., Holography: Just a fancy word for interferometry? Proc. OSA Digital Holography & 3-D Imaging, Digital Holography and Three0Dimensional Imaging. Heidelberg, Germany: Optical Society of America, 2016, DM3I. 1.

Official journal of the JHL 2689-9620
www.light-am.com

Review Open Access

Shearography and its applications – a chronological review

Rajpal Sirohi*

Abstract

This paper presents the activities in the field of shearography in chronological order and highlights the great potential of this holographic measurement technology. After a brief introduction, the basic theory of shearography is presented. Shear devices, phase-shift arrangements, and multiplexed shearography systems are described. Finally, the application areas where shearography has been accepted and successfully used as a tool are presented.

Keywords: Electronic speckle pattern interferometry, Shearography, Non-destructive testing

Introduction

A grainy pattern formed in space when a laser beam strikes an object is called a speckle pattern. These patterns were considered a bane of holographers because the images reconstructed from the holograms were grainy. Several methods were investigated to reduce this deleterious effect. During the late 1960s, it was discovered that the speckle pattern carries information about the object and hence can be used for measurement[1]. The field of speckle metrology thus came into existence. Both speckle photography and speckle interferometry were investigated to measure the displacement and deformation of the objects. Unlike holographic interferometry, speckle interferometry can be configured to measure the components of the deformation vector. Early recordings were made on photo-emulsions (photographic plates or films) and the technique came to be known as speckle pattern interferometry[2]. Since it was possible to control the size of the speckles in the speckle pattern, electronic recording using a vidicon tube was attempted successfully and the technique was renamed as electronic speckle pattern interferometry (ESPI)[3]. When speckle patterns are recorded on a charge-coupled device (CCD) or a complementary metal-oxide-semiconductor (CMOS) sensor and manipulated digitally, the technique is known as digital speckle pattern interferometry (DSPI)[3]. Using ESPI/DSPI along with phase shifting, the deformation map of an object subjected to an external agency is obtained almost in real time. The technique is whole-field, extremely sensitive, and can be configured to obtain either in-plane components or out-of-plane component, or all the three components of the deformation vector simultaneously[2,3,5–7]. It has also been applied to the study of vibrating objects in both time-averaged and stroboscopic modes.

Material scientists, stress analysts, quality assurance personnel, product developers, and many others are interested not only in deformation but also in strain values. Strain measurements are the basis for predicting how long an object will perform without failure when subjected to certain external influences. There are several techniques for strain measurement, of which some are point-wise, and some are whole-field techniques. Shearography is a whole-field optical technique that has evolved both for qualitative and quantitative strain (difference quotient) measurements. When shear interferometry features are included in speckle pattern interferometry, the technique is known as speckle pattern shear interferometry (SPSI) or speckle shear interferometry (SSI) or simply shearography.

Shearography is a displacement gradient sensitive, full-field optical technique that is resilient to environmental

Correspondence: Rajpal Sirohi (rs_sirohi@yahoo.co.in)
Alabama A&M University Normal, AL 35762, USA

disturbances and vibrations. It does not respond to rigid body translations. Because shearography is an almost common-path technique, the temporal coherence requirement is considerably relaxed. The experimental setup is simple and offers reduced sensitivity to fill the gap between DSPI and moiré techniques[4]. The first paper on shearography was written by Leendertz and Butters[8], who validated the theoretical framework of shearography by comparing twist contours of a plate with those generated as moiré fringes from the holographically obtained displacement pattern and its shifted version. For shearing, a Michelson interferometric arrangement was used. This work was followed by publications on the visualization of strain fields and slopes, the use of different shearing methods, and some theoretical aspects of the technique[9–47]. Recordings were done on photo-emulsions and the information was extracted by Fourier filtering[9,23,55]. In 1980, Nakadate et al.[20] reported the use of a video camera to record the slope contours of static and vibrating objects. In 1988, some publications reported the recording of object images on an electronic detector[48–50]. Thereafter, researchers continued with strain field visualizations, slope contours, and even curvature using both electronic and non-electronic detectors[51–87]. The importance of shearography as a tool for measurement and inspection inspired the organization of several conferences on this topic from 1987 to 2006. These conferences have attracted many researchers to present works demonstrating manifold applications such as nondestructive testing and evaluation (NDT&E), examination of large structures, particularly in aeronautics, and nondestructive testing (NDT) of art and cultural heritage objects. At the same time, there were many publications on theoretical aspects and applications of the technique[88–279]. The first application of phase stepping to obtain quantitative strain was reported by Steinchen et al.[97]. Many researchers have described the measurement of strain in their publications. In reality, shearography measures not the displacement derivative, but the difference quotient – an approximation to a derivative. If the shear is made very small for the difference quotient to approximate a derivative, the sensitivity decreases; for a vanishingly small shear there is no sensitivity. Several mobile and compact shearographic systems have been developed to perform NDT activities in an industrial environment[127,182,202,210,214,215]. After 2005, activity on this topic slowed, perhaps due to the fact that it had evolved as an inspection and NDT technique with several systems on the market. In spite of the maturity, some theoretical aspects continue to be explored, new applications have been developed such as NDT of micro-objects, and research on phase evaluation methods continues[280–379]. Several review articles[226,265,314,360], review-like articles[127,188,211,227,258,288,308] and two books[237,353] have been published on shearography. It is interesting to note that there is renewed interest in developing shearography systems with and without multiplexing capabilities[325,346,349–351,354,361–365,370–372,377–382,384–393].

As the technique evolved, it went through various names. The most common usage has been shearography, with some prefixes to it like digital shearography, TV-shearography, electronic shearography, laser shearography, and digital speckle shearography. In the beginning, shearography was called speckle-shearing interferometry and speckle-shear interferometry. However, some authors have called it electronic speckle pattern shearing interferometry[152–154,157,163,220,255,269], digital speckle pattern shearing interferometry[20,132,155,176], electronic speckle-shearing pattern interferometry[57,140,194,196,251], digital shearing speckle pattern interferometry[177,341,358], and digital speckle shearing interferometry[116,145,149,167,186].

Theoretical Background

A typical experimental setup for shearography is shown in Fig. 1. It consists of a light source, typically a laser, an imaging system, a shear device placed either between the object and the lens or between the lens and the recording medium, and a recording medium such as photo-emulsion or a CCD/CMOS imager. The object is illuminated by a laser beam, and its image is formed on the detector via the shear device. Shearing causes a point on the object to be imaged as two points. Alternatively, two points on the object are imaged as a single point. A wave from one point acts as a reference wave to the wave from the second point; hence, there is no need for a reference wave. Between exposures, the object is subjected to an external loading that deforms the object.

The theoretical framework of shearography is based on the same phase difference equation that is valid for both holographic interferometry and speckle pattern interferometry[2,3]. The optical phase difference equation is given by[7]:

$$\phi = (\boldsymbol{k}_o - \boldsymbol{k}_i) \cdot \boldsymbol{L} = \boldsymbol{k} \cdot \boldsymbol{L} \qquad (1)$$

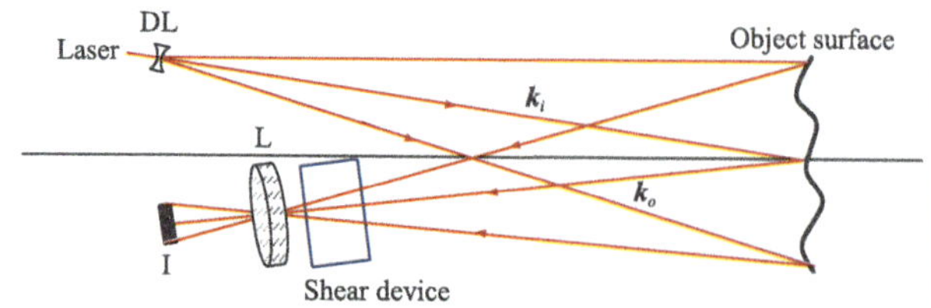

Fig. 1 Schematic of a speckle pattern shear interferometer: DL-diverging lens; L-imaging lens; I-image sensor.

where $\boldsymbol{k}_0$ and $\boldsymbol{k}_i$ are the propagation vectors in the directions of observation and illumination, respectively, and $\boldsymbol{L}$ is the deformation vector at a point on the object. The vector $\boldsymbol{k}$ [$=(\boldsymbol{k}_o-\boldsymbol{k}_i)$] is called the sensitivity vector. The deformation vector is expressed as $\boldsymbol{L}=u\hat{\boldsymbol{i}}+v\hat{\boldsymbol{j}}+w\hat{\boldsymbol{k}}$ with components u, v, and w along the x-, y-, and z-axes.

In shearography, there are two points on the object under consideration: a point P and its sheared counterpart Q. On loading the object, these two points undergo different displacements, as shown in Fig. 2. The point $P(x,y,z)$ is on the undeformed surface, and the point $Q(x+\Delta x,y,z)$ is on the sheared surface. These are very close to each other on the two surfaces with a lateral shear Δx along the x-axis. When the object is deformed, these points are displaced to their new locations $P'(x+u,y+v,z+w)$ and $Q'(x+\Delta x+u+\Delta u,y+v+\Delta v,z+w+\Delta w)$. Point P undergoes a displacement $\boldsymbol{L}=u\hat{\boldsymbol{i}}+v\hat{\boldsymbol{j}}+w\hat{\boldsymbol{k}}$ and point Q undergoes a displacement $\boldsymbol{L}'=(u+\Delta u)\hat{\boldsymbol{i}}+(v+\Delta v)\hat{\boldsymbol{j}}+(w+\Delta w)\hat{\boldsymbol{k}}$. Therefore, $\Delta\boldsymbol{L}$, the change in the displacement vector, is given by $\Delta\boldsymbol{L}=\boldsymbol{L}'-\boldsymbol{L}=\Delta u\hat{\boldsymbol{i}}+\Delta v\hat{\boldsymbol{j}}+\Delta w\hat{\boldsymbol{k}}$.

Following Eq. 1, the phase difference ϕ_P between the waves reaching any point on the detector plane when point P displaces to point P' is given by

$$\phi_P=(\boldsymbol{k}_o-\boldsymbol{k}_i)\cdot\left(u\hat{\boldsymbol{i}}+v\hat{\boldsymbol{j}}+w\hat{\boldsymbol{k}}\right)=k_xu+k_yv+k_zw \qquad (2)$$

where k_x, k_y, and k_z are the components of the vector $\boldsymbol{k}=(\boldsymbol{k}_o-\boldsymbol{k}_i)$, i.e., $\boldsymbol{k}=k_x\hat{\boldsymbol{i}}+k_y\hat{\boldsymbol{j}}+k_z\hat{\boldsymbol{k}}$.

Similarly, the phase difference between the waves reaching the same point on the detector plane when point Q moves to point Q' is given by

$$\begin{aligned}\phi_Q&=\boldsymbol{k}\cdot\left[(u+\Delta u)\hat{\boldsymbol{i}}+(v+\Delta v)\hat{\boldsymbol{j}}+(w+\Delta w)\hat{k}\right]\\&=k_x(u+\Delta u)+k_y(v+\Delta v)+k_z(w+\Delta w)\end{aligned} \qquad (3)$$

Calculations of the phases ϕ_P and ϕ_Q in terms of the coordinates of the source and a point on the detector, and their respective distances, can be found in books[237,353] and in several references[19,265,314].

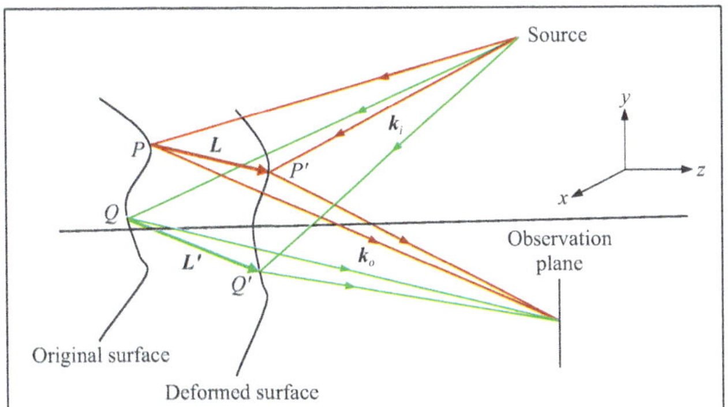

Fig. 2 Original and deformed states of the object showing the location of sheared points on deformation: P is a point on the surface and Q is its sheared counterpart. On loading, these points displace to P' and Q'. Deformation is very small and hence sensitivity vectors remain unchanged.

Because the wave from the sheared point acts as a reference wave to the one from the other point on the surface or vice-versa, there is no need for a reference wave. These two waves produce an interference pattern. Indeed, waves from the scattering points within the resolution element around points P and Q interfere, producing a speckle at the image point. When the entire object is considered, its image is a speckle pattern. A record of the speckle pattern is called a shearogram. Let the amplitude of the waves from point P and point Q at a point (x_i,y_i) at the image plane be

$$a_P(x_i,y_i)\propto a_0(x,y)\,e^{i\theta_P}$$

$$a_Q(x_i,y_i)\propto a_0(x+\Delta x,y)\,e^{i\theta_Q}$$

Both the amplitudes [$a_0(x,y)$ and $a_0(x+\Delta x,y)$] and phases [θ_P and θ_Q] are random variables. The irradiance distribution at the image point can be expressed as

$$\begin{aligned}I(x_i,y_i)&\propto|a_0(x,y)|^2+|a_0(x+\Delta x,y)|^2+\\&\quad 2a_0(x,y)a_0(x+\Delta x,y)\cos\phi;\ \phi=(\theta_Q-\theta_P)\\I(x_i,y_i)&=\mathcal{I}_1(x_i,y_i)+\mathcal{I}_2(x_i,y_i)+2\sqrt{\mathcal{I}_1(x_i,y_i)\mathcal{I}_2(x_i,y_i)}\cos\phi\\&=I_0(1+\gamma\cos\phi)\end{aligned} \qquad (4)$$

where $\mathcal{I}_1(x_i,y_i)$ and $\mathcal{I}_2(x_i,y_i)$ are the irradiances of the waves from points P and Q at the image point (x_i,y_i), I_0 is the total irradiance (sum of irradiances of both waves), and γ is the modulation. Because the two points P and Q are very close to each other, the amplitudes of the waves from these points can be assumed to be equal. In this case, the irradiance distribution can be expressed as

$$I(x_i,y_i)=I_0(1+\cos\phi)$$

Usually, two exposures are made in shearography: the first exposure when the object is in its undeformed state and the second exposure after the application of load, that is, when the object is in its deformed state.

The irradiance distribution in the first exposure can be expressed as

$$I_1(x_i,y_i)=I_0(1+\cos\phi) \qquad (5)$$

On loading, the surface deforms, and points P and Q respectively move to points P' and Q'. The waves from points P' and Q' acquire additional phases ϕ_P and ϕ_Q respectively.

The irradiance distribution recorded now is given by

$$\begin{aligned}I_2(x_i,y_i)&=2a_0^2\left[1+\cos\{(\theta_Q+\phi_Q)-(\theta_P+\phi_P)\}\right]\\&=I_0\left[1+\cos\left(\phi+\Delta\phi_{QP}\right)\right]\end{aligned} \qquad (6)$$

where $\Delta\phi_{QP}=(\phi_Q-\phi_P)$ is the phase difference between the two waves from points Q and P at a point on the recording plane. By rewriting the phase difference $\Delta\phi_{QP}$, we obtain

$$k_x(u+\Delta u)+k_y(v+\Delta v)+k_z(w+\Delta w)-(k_x u+k_y v+k_z w)=k_x\Delta u+k_y\Delta v+k_z\Delta w \quad (7)$$

Assuming that the magnitude of the shear is very small, the phase difference can be expressed as

$$\Delta\phi_{QP}=(\phi_Q-\phi_P)=\left(k_x\frac{\partial u}{\partial x}+k_y\frac{\partial v}{\partial x}+k_z\frac{\partial w}{\partial x}\right)\Delta x$$
$$=(\boldsymbol{k}_o-\boldsymbol{k}_i)\cdot\frac{\partial \boldsymbol{L}}{\partial x}\Delta x=\boldsymbol{k}\cdot\frac{\partial \boldsymbol{L}}{\partial x}\Delta x \quad (8)$$

For a very small shear, $\Delta\boldsymbol{L}/\Delta x$ is expressed as $\partial\boldsymbol{L}/\partial x$. The phase difference depends on the derivatives of the components of the deformation vector if an exceedingly small lateral shear is used. In practice, the difference quotient is measured rather than the derivative. If the magnitude of the shear is made very small such that it approximates a derivative, the deformation phase $\Delta\phi_{QP}$ tends to become very small, and hence the sensitivity of the technique becomes poor. It is possible to employ other shear types, e.g., radial shear and theta shear, and obtain the radial derivative $(\partial w/\partial r)$[28,29] and the theta derivative $(\partial w/\partial\theta)$[28,29].

Although almost all publications discuss the theory of shearography along with other aspects, including applications, several papers are devoted only to the theoretical aspects of the technique[70,71,76,166,168,224].

Photographic Recording

In photographic recording, the irradiance distributions $I_1(x_i, y_i)$ and $I_2(x_i, y_i)$ are recorded sequentially over a time period T. The recorded exposure E is expressed as

$$E=E_1+E_2=2I_0T\left[1+\cos\left(\varnothing+\frac{\Delta\phi_{QP}}{2}\right)\cos\left(\frac{\Delta\phi_{QP}}{2}\right)\right] \quad (9)$$

where T is the time of exposure. Thus, the record contains a speckle pattern modulated by $\cos(\Delta\phi_{QP}/2)$. The term $\cos(\Delta\phi_{QP}/2)$ carries the information of the derivative of the deformation vector that an object has suffered due to loading. This information, which is in the form of a fringe pattern, is extracted by Fourier filtering[9,23,55]. Usually, laser light is used for Fourier filtering, but in some cases, filtering can also be performed with white light[55]. The fringe pattern is speckled, and the fringes have a low contrast with a $\cos^2$-type distribution. The fringes can be sharpened by multiple exposures or by making use of the nonlinearity of the recording process[31,33,36,48]. The contrast of the fringes is improved using apertures in front of the lens, albeit at the expense of loss of light[13,29,34,36].

If the shearogram corresponding to the undeformed state of an object recorded on photo-emulsion is repositioned exactly after development, the deformation derivatives can be observed in real time as the object is deformed[74,93]. The theoretical framework of real-time shearography is similar to that of real-time holographic interferometry: there is a phase change of π in the fringe pattern as compared to double-exposure shearography[168]. Real-time shearography has been used to determine fractional fringe order by the translation of a wedge plate[110]. Instead of recording a shearogram on a photographic plate, the use of photorefractive crystals is suggested for real-time shearography[282]. To extract information pertaining to $\Delta\phi_{QP}$, a doubly exposed shearogram is Fourier filtered[9,13], resulting in a fringe pattern. These fringes can be sharpened by making multiple exposures or by making use of the nonlinearity of the recording process.

Electronic Recording and Digital Processing

In electronic recording, the two exposures corresponding to $I_1(x_i, y_i)$ and $I_2(x_i, y_i)$ are handled independently. In one processing method, the exposure corresponding to $I_2(x_i, y_i)$ is subtracted pixel-by-pixel from $I_1(x_i, y_i)$. The voltage output of the detector is proportional to $[I_1(x_i, y_i)-I_2(x_i, y_i)]$. The brightness on the monitor is proportional to the voltage output of the detector, and therefore it can be expressed as

$$B\propto[I_1(x_i,y_i)-I_2(x_i,y_i)]=I_0\left[\cos\phi-\cos\left(\phi+\Delta\phi_{QP}\right)\right]$$
$$=2I_0\sin\left(\phi+\frac{\Delta\phi_{QP}}{2}\right)\sin\frac{\Delta\phi_{QP}}{2} \quad (10)$$

For negative values of $\sin\Delta\phi_{QP}/2$, the brightness on the monitor will be zero. To avoid this loss of signal, the output signal is squared before being displayed, and therefore the brightness B on the display monitor is expressed as

$$B\propto 4\,I_0^2\sin^2\left(\phi+\frac{\Delta\phi_{QP}}{2}\right)\sin^2\left(\frac{\Delta\phi_{QP}}{2}\right) \quad (11)$$

The term $\sin^2(\phi+\Delta\phi_{QP}/2)$ contains the random phase, and therefore it represents the speckled portion of the brightness distribution on the monitor, which is modulated by the term $\sin^2(\Delta\phi_{QP}/2)$. There would be dark regions, called fringes, wherever $\phi_{QP}/2=m\pi$ for integer values of m. Substituting for $\Delta\phi_{QP}$, we obtain the condition for the formation of dark fringes as

$$\Delta\phi_{QP}=\left(k_x\frac{\partial u}{\partial x}+k_y\frac{\partial v}{\partial x}+k_z\frac{\partial w}{\partial x}\right)\Delta x$$
$$=(\boldsymbol{k}_o-\boldsymbol{k}_i)\cdot\frac{\partial \boldsymbol{L}}{\partial x}\Delta x=\boldsymbol{k}\cdot\frac{\partial \boldsymbol{L}}{\partial x}\Delta x=2m\pi \quad (12)$$

The fringe pattern arises because of the gradient of the

deformation vector. The fringe visibility depends on the correlation between the two interfering waves[20,61,256].

It can be seen from Eq. 10 that the DC term is eliminated by the subtraction process. In another recording scheme, two speckle patterns with a phase difference of $\pi/2$ are obtained simultaneously using polarized beam illumination for the deformed and undeformed states of the object[122]. Because the phase difference is fixed, a fast Fourier transform (FFT) can be used to eliminate the DC terms of the speckle patterns. In the early 1980s, TV cameras were used to record shearograms, but were followed by CCD and CMOS cameras. Ng and Chau examined the noise-related performance of CCD cameras used for shearography[84]. Osten et al.[232] demonstrated that CMOS cameras produce better shearograms than CCD cameras, even for non-cooperative surfaces under examination.

Recording of vibrating objects

Let an object be vibrating sinusoidally with frequency ω and amplitude w. The average irradiance distribution I_{av} recorded over a period much longer than the period of vibration is expressed as[9,63,123,155,226,265,348,366]

$$I_{av}(x_i, y_i) = I_0(1 + \gamma \cos\phi J_0(\Omega)) \tag{13}$$

where $\Omega = (4\pi/\lambda)(\partial w/\partial x)\Delta x$ when the observation and illumination directions are anti-parallel and the object is illuminated normally, Δx is the shear, and $J_0(x)$ is the Bessel function of zero order. The output from the CCD camera will be proportional to the irradiance distribution, and therefore the fringe pattern on the monitor will have a strong speckle background. Furthermore, the zero-order Bessel function causes only a few fringes to be observed. However, stroboscopic illumination circumvents this problem: the fringes are now cosinusoidal and the background can be removed by subtraction, as is done in the static case[117,123,183]. With stroboscopic illumination, the shearogram can be frozen at any instant of illumination by short stroboscopic pulses synchronized with the frequency of the vibrating object, and the phase-shift technique can be applied to obtain the derivative of the vibration amplitude[155,197,251]. Shearography with stroboscopic illumination and a large shear is used to measure small out-of-plane vibration amplitudes over a large range[251]. To measure transient vibrations, Hung et al.[125] used a high-speed camera, and the images were stored in the memory of the high speed acquisition system.

In another method to improve the visibility of time-averaged fringes, time-averaged shearograms are captured sequentially. The processor outputs a signal that produces the brightness on the monitor, which is proportional to[86,120,162,226,314]

$$B(x,y) \propto J_0^2(\Omega) \propto J_0^2\left(\frac{4\pi}{\lambda}\frac{\partial w(x,y)}{\partial x}\Delta x\right) \tag{14}$$

The visibility of the fringes will be maximum where $\partial w(x,y)/\partial x = 0$ and will decrease rapidly as $\partial w(x,y)/\partial x$ increases. Hence, a zero-order Bessel fringe will occur where the amplitude is maximum. The zero-order Bessel fringe can be shifted if one of the mirrors of the shearing Michelson interferometer is excited with the same frequency as that of the object[86] or by using the scheme suggested by Valera and Jones[120], Valera et al.[124], and Chatters et al.[107]. The zero-order Bessel fringe can also be shifted by wavelength modulation[135,162,314]. However, this requires an unbalanced Michelson interferometer for shearing. It has been shown that a single Bessel fringe pattern obtained under subtraction operation could be sufficient to calculate the phase using a phase recovery method, based on genetic algorithms[280]. Methods for improving the visibility of fringes were described by Chen et al.[226]. For the study of transient vibrations such as those arising from impact, the use of a double-pulse laser is suggested[185,220]. Steichen et al.[185] also mention the use of a double-flash CCD camera that may record two shearograms separated in the range of 100 ns to 50 ms. Procedures to use single-pulse shearography and double-pulse shearography for the study of vibrating objects were described by Spooren et al.[80]. Another study reported the use of a double-pulse laser for shearographic recording in which a carrier frequency is generated by changing the curvature of the illumination beam between the impact stressing of the metallic plates[194,196,220]. A procedure to measure damping using shearography was described by Wong and Chan[195]. A method that produces binary phase patterns where the phase changes are related to the zeros of the Bessel function is presented for estimating the vibration amplitudes from the shearographic fringe patterns[366].

Optical Configurations

As pointed out in the introduction, a speckle pattern interferometer can be configured to be sensitive either to an in-plane displacement component or an out-of-plane displacement component. The same is true for a speckle pattern shear interferometer, i.e., it can be configured to yield fringes pertaining either to the in-plane derivative (strain) or out-of-plane derivative (slope).

In-plane gradient sensitive configuration

There are two distinct possibilities, as shown in Fig. 3a, 3b. In one arrangement, the object is illuminated by collimated beams lying either in the y-z plane or x-z plane, which are symmetric to the normal at a point on the object

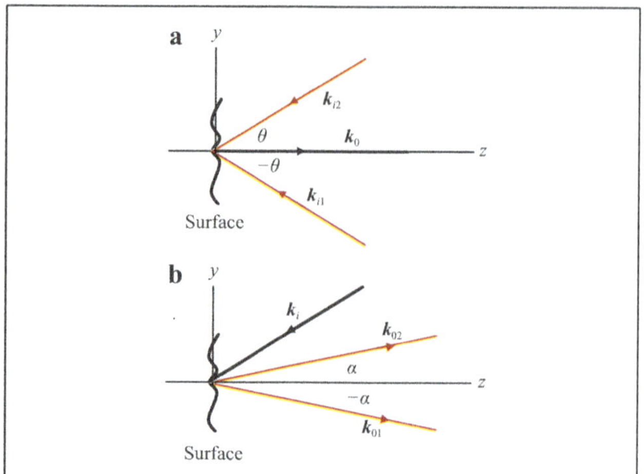

Fig. 3 **a** symmetric illumination directions, **b** symmetric observation directions.

surface, and the observation can be along any direction but preferably along the normal to the surface.

The formation of dark fringes in digital shearography is described by Eq. 12, which has been rewritten as

$$(\boldsymbol{k}_o - \boldsymbol{k}_i) \cdot \frac{\partial \boldsymbol{L}}{\partial x} \Delta x = 2m\pi \tag{15}$$

When two symmetric directions of illumination are used, as shown in Fig. 3a, the fringe formation is governed by

$$(\boldsymbol{k}_o - \boldsymbol{k}_{i2}) \cdot \frac{\partial \boldsymbol{L}}{\partial x} \Delta x - (\boldsymbol{k}_o - \boldsymbol{k}_{i1}) \cdot \frac{\partial \boldsymbol{L}}{\partial x} \Delta x = 2m_1\pi - 2m_2\pi = 2m'\pi$$

$$(\boldsymbol{k}_{i1} - \boldsymbol{k}_{i2}) \cdot \frac{\partial \boldsymbol{L}}{\partial x} \Delta x = 2m'\pi \tag{16}$$

where m' is an integer that takes values 0, ±1, ±2, ±3, $\cdots$. Assuming that the beams lie in the x-z plane and make an angle of $\pm\theta$ with the surface normal, then $\boldsymbol{k}_{i1} = 2\pi/\lambda\left(\sin\theta\hat{\boldsymbol{i}} - \cos\theta\hat{\boldsymbol{k}}\right)$ and $\boldsymbol{k}_{i2} = 2\pi/\lambda\left(-\sin\theta\hat{\boldsymbol{i}} - \cos\theta\hat{\boldsymbol{k}}\right)$. Hence,

$$(\boldsymbol{k}_{i1} - \boldsymbol{k}_{i2}) \cdot \frac{\partial \boldsymbol{L}}{\partial x} \Delta x = \frac{4\pi}{\lambda} \sin\theta \frac{\partial u}{\partial x} \Delta x = 2m'\pi \rightarrow\rightarrow$$

$$\frac{\partial u}{\partial x} = \frac{m'\lambda}{2\sin\theta\Delta x} = \frac{\lambda\Delta\phi_{QP}}{4\pi\sin\theta\Delta x} \tag{17}$$

The sensitivity of the principal strain fringe pattern depends on both the magnitude of the shear and the angle of illumination. This method has been implemented experimentally in several ways, including using a Michelson interferometer[136,152,164,187,206,236,345,351], a birefringent crystal[133,211,265], a birefringent wedge[160], and a Wollaston prism[379] for shearing. When a wedge plate covers half the aperture of an imaging lens or when a two-aperture or multi-aperture mask with wedges is placed in front of an imaging lens, the recording contains information about the in-plane component along with the derivatives of the displacement. This has been studied by Ng and Chau[75], Mohan et al.[91], Mohan and Sirohi[166], and Wang et al.[224].

Similarly, if for a given direction of illumination, two directions symmetric to the surface normal are used for observation, the configuration is in-plane gradient sensitive. Assuming that the directions of observation lie in the x-z plane and make an angle of $\pm\alpha$ with the local normal, the fringe formation is governed by

$$(\boldsymbol{k}_{01} - \boldsymbol{k}_{02}) \cdot \frac{\partial \boldsymbol{L}}{\partial x} \Delta x = \frac{4\pi}{\lambda} \sin\alpha \frac{\partial u}{\partial x} \Delta x = 2m\pi \rightarrow\rightarrow$$

$$\frac{\partial u}{\partial x} = \frac{m\lambda}{2\sin\alpha\Delta x} = \frac{\lambda\Delta\phi_{QP}}{4\pi\sin\alpha\Delta x} \tag{18}$$

In practice, θ can take values approaching 90^0, but the angle α usually takes much smaller values limited by the lens aperture. However, the experimental arrangement can be configured so that angle α can also take large values[169,379,392].

When the shear is along the y-direction and the beams are confined to the y-z plane, the y-gradient of the in-plane component under these configurations is given by

$$\frac{\partial v}{\partial y} = \frac{m\lambda}{2\sin\theta\Delta y} = \frac{\lambda\Delta\phi_{QP}}{4\pi\sin\theta\Delta y} \tag{19}$$

and

$$\frac{\partial v}{\partial y} = \frac{m\lambda}{2\sin\alpha\Delta y} = \frac{\lambda\Delta\phi_{QP}}{4\pi\sin\alpha\Delta y} \tag{20}$$

where Δy is the shear. To realize this experimentally, an opaque plate consisting of several openings is placed before the imaging lens. Each opening carries either a shear plate, or a ground glass plate, or just a plate for compensation[27,32,34,39,169]. This arrangement generates the *carrier frequency* and is well adopted for Fourier filtering whether the recording is photographic or electronic. A three-aperture arrangement with one aperture carrying a wedge plate produces an in-plane fringe pattern and a combination of in-plane and derivative fringes[166]. The influence of in-plane displacement on slope fringes was examined by Mohan et al.[91,166]. It has also been pointed out that a single illumination direction will also give the in-plane derivative when loading is such that there is no out-of-plane deformation or there is a sequential illumination from two directions[163]. It has been mentioned that an in-plane sensitive configuration can be devised that yields all four in-plane strains independent of out-of-plane strains[163].

Out-of-plane gradient sensitive configuration

It is obvious from the phase difference Eq. 12 that the configuration will be sensitive to the gradient of out-of-plane deformation if the directions of illumination and observation are anti-parallel ($\boldsymbol{k}_i = -\boldsymbol{k}_o$)and are along the normal to the surface. In that situation,

$$(\boldsymbol{k}_o - \boldsymbol{k}_i) \cdot \frac{\partial \boldsymbol{L}}{\partial x}\Delta x = \frac{4\pi}{\lambda}\hat{\boldsymbol{k}} \cdot \frac{\partial \boldsymbol{L}}{\partial x}\Delta x = \frac{4\pi}{\lambda}\frac{\partial w}{\partial x}\Delta x \tag{21}$$

Hence

$$\frac{\partial w}{\partial x} = \frac{m\lambda}{2\Delta x} = \frac{\lambda \Delta\phi_{QP}}{4\pi\Delta x} \tag{22}$$

When the shear is along the y-direction, we have

$$\frac{\partial w}{\partial y} = \frac{m\lambda}{2\Delta y} = \frac{\lambda \Delta\phi_{QP}}{4\pi\Delta y} \tag{23}$$

Orthogonal slopes, or a slope with double the sensitivity and curvature, can be obtained using a four-aperture arrangement in which two wedges are appropriately placed at the two opposite apertures and compensating plates on the remaining apertures[34]. Sirohi and Mohan[170] described a two-aperture configuration that is insensitive to in-plane deformation. This configuration is used to obtain the slope fringes. Using a similar configuration, the sensitivity to slope change was increased two-fold by coating the object with a retro-reflective paint[181]. It was shown that the visibility of the correlation fringes increases when the object surface is covered by a microsphere of 10.2μm mean diameter[230]. Using two CCD cameras in a Michelson interferometer for shearing, both $\partial w/\partial x$ and $\partial w/\partial y$ are obtained for the same load on the object[188].

Multi-component Configurations

As shown in Section on optical configurations, we obtain the x-derivative of the displacement when the shear is along the x-direction and the object is illuminated symmetrically by beams lying in the x-z plane, and the y-derivative of the displacement when shear is along the y-direction with illuminating beams in the y-z plane. For a general direction of illumination and shear, e.g., along the x-direction, the phase difference contains information of all three x-derivatives of the displacement. However, if three different directions of illumination are used, it is possible to separate the three derivatives[50,164,204,206,314]. The same is true if three different directions of observation are used. Similarly, using a single direction of illumination and four directions of observation, and a single CCD camera, all three x- or y-gradients of the deformation vector can be obtained[245]. A speckle pattern shearing interferometer has been reported that uses three diode lasers of different wavelengths along with three CCD cameras for the determination of two-dimensional strain distributions[201]. Using three directions of shear, simultaneous measurement of derivatives of displacement in three different directions with phase-shift using Fourier processing has been demonstrated[372]. This method is cumbersome and uses six cube beam splitters. Correto and Perlo[100] suggested the use of multiple laser diodes to illuminate the object surface uniformly, while Kalms and Osten[239] used two mutually incoherent laser sources for illumination. More information about multicomponent configurations is provided in the section on multiplexing.

Thus far, the analysis has assumed well-defined directions for illumination and observation and a flat object. In practice, a real object is far from being a plane surface, and the illumination is usually divergent to cover a larger area for examination. These issues lead to errors in the measured values of deformation derivatives. The effect of divergence was examined by Abdullah and Petzing[264]. There are other factors such as the magnitude of shear, the sensitivity vector, the errors in phase-shifting and the rigid body motion that influence the measurement accuracy of the derivatives of deformation[187]. Many studies have considered these aspects in their theoretical works[269,315]. The creep of the PZT actuator on which the shearing mirror is mounted can change the magnitude of shear during the measurement, leading to an error[332]. A calibration procedure is described that accounts for the local variability of the magnitude of shearing and compensates for variability of the sensitivity vector[344]. Although shearography is said to be immune to rigid body translation, there is a decorrelation resulting in a decrease in the modulation of fringes. The effect of rigid body translation can be compensated with sandwich shearography, which combines shearography and sandwich holography[54], and by flexible mirrors[42]. The in-plane rigid body translation is compensated by using a combination of speckle photography and shearography[256]. The shifting of the sensor by an appropriate amount, and digital shifting of the image, have been suggested to alleviate the effect of rigid body translation[158]. Furthermore, shearography itself has been used to obtain the deformation. To extract deformation from its gradient, the effect of doubling the image must be reduced[137,138]. When the phase distribution is spatially limited, it can be extracted from a single-sheared phase distribution[287]. In this study, out-of-plane deformation and its gradient were captured using ESPI and shearography, respectively. From the shearographic data, the out-of-plane deformation was reconstructed, and matched well with the deformation measured by ESPI. A shearography system was described that measures the orthogonal slope maps from which the deflection profile is generated[343]. A 2D deconvolution technique for images generated by a phase-stepped shearography system was described, which enables the extraction of quantitative displacement data from shearography phase change images[161]. Rosso et al.[299] used holography to record an object on a photorefractive crystal, and the holographically reconstructed image was used as an object in a shearography setup for strain measurement. The technique of simultaneous coherent imaging and strain measurement by coupling photorefractive holography and shearography seems to have potential for depth-resolved strain measurement inside optically diffusing biological tissue.

Some researchers have examined the fringe formation due to static as well as vibratory load on an object with holography vs. shearography[21,67,78,108,240,389], ESPI vs. shearography[68,99,241,250], TV holography vs. shearography[115,131,134,165,360], and digital shearography vs. electronic shearography[281] when these techniques are used for non-destructive testing. Both the holographic interferometry and ESPI respond to deformation and show flaw-induced displacement anomalies in the fringe pattern, whereas shearography shows flaw-induced strain anomalies in the fringe pattern. Because their sensitivities are very different, they can be used sequentially.

Shear Devices

The most commonly used arrangement for shearing is a balanced Michelson interferometer[8,10,78,80,81,97,99,106,152,164,182,206,232,241,251,303,317,339]. The object is imaged via two paths, and the shear is achieved by tilting one of the mirrors, thereby producing two images. The amount of shear is controlled by the angle of the tilt. Parker and Salter[214] described a shearographic system in which a face of the cube beam splitter is mirrored, and a PZT-mounted mirror is used for shearing and phase-shifting. Thus, the system uses an unbalanced Michelson arrangement. They also described the limitations of this system. Instead of a Michelson interferometer, some researchers have used a Mach-Zehnder interferometer for shearing[194,246,301,323,380,386,391]. Many researchers have used a wedge plate[17,19,27,32,34,41,58,60–62,93,97,170,181], a biprism[20,71,75,76,90,111,171,280], parallel plates[9,11], a plane parallel plate in reflection[130,382], a wedge plate in reflection[249], and Dove prisms[24,25,28] in front of the imaging lens. A wedge plate in reflection provides both the shear and the carrier fringes for spatial phase-shifting. A plate with four openings at the corners of a square placed in front of the imaging lens results in sheared images at a defocused plane[13,16,18]. A single grating[15,30,38,100,132], a Ronchi grating[146,147], a cross-grating[43], and a pair of gratings[12,153,318] have also been used for shearing.

A holographically recorded lens has been used for both imaging and shearing[37,331]. A Savart plate[14,143,202,321], a birefringent wedge plate[160], a birefringent crystal[68,133,139], a Wollaston prism[122,148,177,179,262,341,358,369], and a single or double circular polarization grating[329] along with the polarizers have also been used as shear devices. A right-angle prism with its diagonal face coated to separate the TE and TM polarizations along with a glass plate attached to it with an index-matching liquid followed by a liquid crystal phase modulator and a polarizer provides both the shear and the phase-shift[295,299]. A polarization-based shearing device results in an almost common-path speckle pattern shearing interferometer. A split lens obtained by cutting an imaging lens into two halves, one half of which can be shifted laterally or longitudinally, has been used for lateral shear as well as for radial shear[22,23,28,29,33,57,58]. The split lens provided laterally shifted images when one half was translated in its plane, and radially sheared images when translated longitudinally along the optical axis. A multi-segment lens was used to record the out-of-plane derivatives for different loadings[26,33]. However, split lenses or multi-segment lenses have not been used in practice because they are not commercially available. Instead, sheared images were obtained by placing a wedge plate over half the aperture of the lens[19]. In many cases, an opaque screen with two or several openings was placed before the imaging lens, and the shearing elements were placed appropriately over these openings[17,24,27,34,39,181]. By judicious choice, it was possible to obtain the slope and principal strains along the x- and y-directions from the same double-exposure shearogram. Details of various shearing arrangements are given by Sirohi et al.[193]. Radial shear was also achieved in a cyclic shear interferometer[275]. Several authors have used the 4f arrangement in a shearography setup to increase the field of view[317,325,340,345,360,378,359,360]. Falldorf et al.[342] used an air wedge in a 4f configuration that provided variable shear. Zhao and Chung[273] used a grating realized by an electrically addressable liquid crystal SLM.

Ganesan et al.[49] described a shearing arrangement comprising two right-angle prisms instead of two mirrors in a Michelson configuration that can be used to introduce all five shear types. A microprism array placed in front of the camera lens was also used for shearing[225]. The amount of shearing was adjusted by changing the distance between the camera lens and the array. This arrangement was claimed to be light-efficient. Wong[234] described a simpler shearing arrangement consisting of a mirror and a beam splitter. The arrangement suffered from magnification error but had the advantage that it can be turned into an ESPI set up by rotating the mirror. Mohanty et al.[24] used a pair of Dove prisms in front of the lens to introduce rotational shear. Debrus[14] employed a Savart plate between a pair of polarizers to obtain lateral shear. Joenathan and Buerkle[121,153] employed a holographic grating followed by a ground glass for shearing, as did Mihaylova et al.[247,274]. Murty's[130,382] plane parallel plate is a convenient shear element that can be used in reflection. The magnitude of the shear depends on the thickness, refractive index of the plate, and the angle of incidence; therefore, the shear is very difficult to modify. A convenient method in which an air parallel plate was created by two glass plates whose separation can be varied easily was adopted by Mihaylova et al.[255,307] for shearing.

Loading Methods or Excitation Methods

To study the response of an object, it is subjected to an external agency. As mentioned previously, the initial state (undeformed state) of the object is compared with the deformed state caused by loading. The objective of loading is to generate surface displacement gradients in the region where the subsurface flaws are to be detected. The response of an object to an applied load depends on several factors: the material, the size and location of the fault, the stiffness of the construction, and the type of load. However, the only free parameter in the experiment is the load. Consequently, its type and amount should be selected carefully[207]. A number of loading methods have been applied to study object response[105,127,142,139,176,281,387]. These are briefly described below.

Mechanical Loading

This is the most commonly used method to obtain strain data as a function of stress, or to study the failure or fatigue response of a particular system under load. It has been applied to visualize strains, slopes, and twist, as well as for the quantitative measurement of strain, slope, twist, and crack detection[8,11,23,40,145].

Partial Vacuum

The object is placed in a chamber that is slowly evacuated, thereby creating a partial vacuum. It is suggested that the vacuum should be applied first to the sample, and a reference frame should be captured. Vacuum is then slowly bled off to create a pressure difference, and then successive frames/images are captured[81]. Vacuum stressing is the most effective method for detecting unbonds[109] and impact damage on aluminum honeycomb structures[104,182]. Using vacuum stressing, the presence of debonds and delamination in composite and glass fiber reinforced plastic (GRP) laminates is easily visualized[59,67,81,90,108,156,265]. Self-adhesive vacuum stressing is used to test aircraft components and aircraft bodies[210,232,233].

Pressure Loading

This is used to detect wall thickness defects and cracks in pressure vessels, pipes, etc. The object is subjected to internal pressure between the exposures or frames/images. Steinchen et al.[118,176] used this method to detect and characterize microcracks in GRP pipes. It is also used to locate disbonds in GRP pipes[112]. This method has also been used to inspect power plant piping for corrosion defects[90] and to detect programmed flaws in steel pipes[149].

Thermal Loading

This is used to test laminated structures, honeycomb structures, and reinforced structures, and to locate skin-to-core debonds and cracks[176]. Heating is usually provided with infra-red lamps[134,283,329], a hair blower[61,182], or a heat gun[151]. Laser heating is used to test flat-bottomed hole specimens[106] and a thin aluminum plate with a scratch at the back[249]. Impulsive thermal loading is used for crack detection in metal plates and pipes[310], and purely thermal loading with a hair dryer is used to detect cracks on a composite turbine blade[265]. A thermal pad has also been employed for thermal stressing[369].

Microwave Heating

Microwave heating is a special form of thermal loading that can easily detect the presence of water vapor trapped inside the object of study[142]. Microwave excitation causes moisture in the materials to heat up and thus induces highly localized deformation detectable by shearography. This mode of stressing is only applicable to nonmetallic composites.

Induction Current Loading

Induction current or eddy current, which is produced by changing magnetic field, is used to heat materials and is currently being used for defect detection in NDT[387].

Vibratory or Acoustical Loading

Vibratory loading, in addition to studying the modes of an object, has been applied to test welded joints, debonds, and delamination. The object is excited either by a piezo-electric device[107], a shaker[85], or a vibration generator such as a loudspeaker[339]. Generally, the excitation frequencies are much higher than the resonance frequency for visualizing cracks, delamination, or disbands. Directed acoustical waves from a transducer appear to be promising for detecting fatigue cracks in metal plates[368].

Impact or Transient Loading

A transient load is applied with an electro-dynamic hammer (firing pin). The shock waves thus produced excite the defects[246].

Phase-Shifting

Phase shifting, either temporal or spatial, is employed when quantitative data are required about the strain field or defects. In many cases, phase maps are presented even for the visualization of gradients of the deformation vector.

Temporal phase-shifting was first introduced in shearography by Owner-Petersen[61], followed by Kadono et al.[62]. Present-day shearographic instruments are equipped with phase shifting. To obtain the phase values at each pixel a minimum of three irradiance values are required, with different phase values in a 0 to 2π interval. Some algorithms use more than three irradiance values. In temporal phase shifting, these irradiance values are captured sequentially by changing the phase difference between the interfering waves. If three phase steps are used in the interval 0 to 2π such that the phase is stepped by $2\pi/3$, that is, the irradiance distribution in the interference pattern is captured at 0, $2\pi/3$ and $4\pi/3$, then the phase ϕ at each pixel is obtained from[62,97,112,141,248,281,324]

$$\phi = \tan^{-1}\frac{\sqrt{3}(I_3 - I_2)}{(2I_1 - I_2 - I_3)} \tag{24}$$

where I_1, I_2, and I_3 are the irradiance distributions captured at the 0, $2\pi/3$ and $4\pi/3$ phase steps, respectively. When four phase steps are used in the interval 0 to 2π with a phase step of $\pi/2$, the phase ϕ at each pixel is obtained from[61,105,179,197,202,265]

$$\phi = \tan^{-1}\frac{(I_4 - I_2)}{(I_1 - I_3)} \tag{25}$$

where I_1, I_2, I_3, and I_4 are the irradiance distributions captured at the 0, $\pi/2$, π and $3\pi/2$ phase steps, respectively. The phase ϕ is a wrapped phase and must be unwrapped. Unwrapping poses no problem when the shear is small. However, there are situations in which variable shear is required to obtain a derivative map. Brug[198,202] presented a method to obtain derivatives in real time using a Michelson interferometer for shearing and two CCD cameras, where shear varies with time.

The five-step algorithm is also used because it is robust to noise and insensitive to calibration errors[214]. Wu et al.[320] described a method of in situ calibration of phase shift by employing an additional optical element in the Michelson interferometer. Common phase-shift algorithms used in digital shearography were enumerated by Zhao et al.[360].

The phase steps can be obtained by translating a PZT-mounted mirror[97,111,176,338,339], or using a polarization-based phase-shifter[160,202,209,275,284,295,305], or a liquid crystal cell[62,265,273], or a Wollaston prism along with a liquid crystal cell[305], or a Wollaston prism along with HiBi fiber wrapped around a PZT cylinder[96,120,124,162,179], or a combination of a quarter-wave and a rotating half-wave plate[275], or by translation of a diffractive optical element[146,361] or by source wavelength modulation[141,152,223]. A lateral shift of the source with a PZT provides a phase shift that has been incorporated in phase-shift algorithms[149]. It has been shown that rotation of an object introduces an additional phase that varies linearly with the angle of rotation[262]. The rotation of the object has been used for phase shifting.

The temporal phase-shift method is susceptible to external disturbances, such as vibration, temperature fluctuation, or rapid motion of the test object itself. The spatial phase shifting (SPS) technique is a simple method to eliminate external disturbances. In spatial phase shifting (SPS), a carrier frequency is introduced such that there are three or four pixels between the fringes depending on whether a three-step or four-step phase-shift algorithm is used[150,271,283]. In one publication, a four-step error-compensating algorithm was used to generate the phase map[271]. Alternately, the Fourier transform method can be applied to obtain the phase difference between the undeformed and deformed states of an object[325,326,345,351,372,377,380,382]. It is desirable that the zeroth order and the desired first-order spectra should be separated. This determines the spatial frequency of carrier fringes. A shearographic setup in which a Michelson interferometer is embedded for shearing usually employs temporal phase shifting. However, it can be used for spatial phase shifting if a CCD camera with a sufficiently large number of pixels is employed[325].

A study of the mode shapes of a turbine blade excited by a piezoelectric shaker at different frequencies was carried out by DSPI at a wavelength of 10.6 μm and by shearography at a wavelength of 532 nm to assess which of the two techniques could be used in an industrial environment[348]. Some interesting results were presented in this paper[348]. Measurements of the modal rotation fields of an engineering object, e.g., a beam with single and multiple damages, are obtained with shearography, and an optimal sampling technique is used to improve damage localization[330,352,374]. Because the sensitivity of shearography depends on the magnitude of shear, a numerical study of damage localization as a function of shear has also been reported[355]. A new method for structural damage identification using cubic spline interpolation has been described[376]. The method is based on the interpolation of modal rotations measured using speckle shearography. The signal-to-noise ratio (SNR) in shearography as a function of the magnitude of shear on the modal rotation fields has also been investigated[356].

To study time-dependent deformations, it is better to capture information in a single frame. For this reason, the Fourier transform method for processing data is preferred. However, it requires a carrier frequency, so that various Fourier spectra are separated. Several methods have been proposed to provide carrier frequency in the shearogram[47,53,82,126,194,196,222,301,330,370,372,382]. Carrier frequency fringes are generated by changing the curvature of the illuminating beam before the second exposure or the second frame[47,167,221] or by a small rotation of the object[171]. When a Mach-Zehnder interferometer is used as a shearing device, a parallel shift of the mirror introduces shear, and the rotation of the mirror produces the carrier frequency[246,330]. The use of multi-aperture with wedge plates in front of the imaging lens automatically creates a carrier frequency. Bhaduri et al.[291,294] employed three-aperture shearography with two wedge plates to obtain a curvature phase map, and with a single wedge plate both the displacement and the displacement gradient using the Fourier transform method. Carrier frequency fringes have also been used for data reduction[44,53], to locate and size the debonds in GRP plates[69], to obtain surface coordinates and slopes[73,159], to obtain flexural strains[40], and to determine the order of a fringe[47].

Joenathan et al.[190] developed a method in which the object is deformed continuously, and a large number of sheared images of the object deformation are acquired using a high-speed CCD camera. The derivative of the object deformation is then retrieved from this large set of data using Fourier transformation[190,272]. This method is capable of obtaining information for object displacements over 500 μm. In another study, a continuous wavelet transform was applied to extract the phase change from a series of shearograms[276]. In addition to the Fourier transform and the windowed-Fourier transform methods of phase recovery from a shearogram, some researchers have demonstrated other methods such as the curvelet transform for edge detection[328], Hilbert-Huang transform[340], and wavelet transform[358]. A heterodyne shearographic system in which orthogonally polarized beams are frequency shifted by $\pm\omega_s$ is described for obtaining derivative information in real time[341,358]. Frequency-shifted beams illuminate the object, which is imaged through a Wollaston prism sandwiched between two polarizers on the CCD camera. The output of the CCD is a heterodyne signal over which the difference phase rides. The difference phase information is extracted using wavelet transformation and a proper filter. Andhee et al.[260] carried out a comparative study of conventional and phase-stepped shearography. Conventional shearography is good for making qualitative observations in real time, while phase-stepped shearography delivers quantitative information[161]. A procedure to maintain sub-pixel alignment between shearograms obtained with a single camera in a polarization-based phase-stepped two-bucket shearing interferometer has been described[267]. The procedure was based on cross-correlating the two shearograms. A more recent paper discusses an improvement of the two-bucket shearing interferometer when dealing with the unfavorable polarization states that arise when a rough metallic surface is illuminated with a linearly polarized light[284]. Recently, a theory of surface phase-resolved shearography has been presented that considers speckle statistics and delivers less noisy specklegrams[373].

Multiplexing in Shearography

Separation of the in-plane and out-of-plane gradient components of the deformation of an object surface generally requires illumination of the object from three different directions or observation from three different directions. In spatial-division multiplexing (SDM), the object is illuminated from a single direction, and viewed simultaneously from three different directions through three shearing interferometers. The first attempt to obtain all three displacement derivatives was made by Hung and Durelli[18]. Sharma et al.[32,41] used a three-aperture arrangement to obtain slope, curvature, and deformation. A similar arrangement was used by Bhaduri et al.[294] using the Fourier transform method. They used the (1, N) spatial phase-shifting method to obtain derivative phase maps[283]. Filtering from the ±1 order halo from a doubly exposed shearogram, recorded with a holographic grating as a shear element, yields the slope and curvature fringes, and filtering from the ±2 order halo yields the slope fringes with double the sensitivity[38]. Speckle interferometry and speckle shear interferometry have been combined in a simple setup that uses four apertures[39]. Deformation, slope, and curvature fringes can be obtained by filtering from the appropriate halos. Iwahashi et al.[43] used a cross grating to obtain three slope distributions simultaneously. Joenathan et al.[26,33] extracted three out-of-plane derivatives from the same shearogram using a multi-segment lens for imaging. Barrera et al.[365] used an aperture-based spatial phase-shift multi-shearing direction system to acquire information in three different shearing directions simultaneously. They used wedge plates appropriately oriented on the apertures; therefore, both the shear and the carrier frequency were not variable. Dymny et al.[154] demonstrated a configuration that displays *x*- and *y*- derivatives simultaneously using two Michelson interferometers and two CCD cameras. The use of two CCD cameras and two shearing units to simultaneously obtain *x*- and *y*- derivatives is also recommended by Steinchen et al.[176]. Siebert and Schmitz[203] used two CCDs to measure two orthogonal gradients simultaneously in a temporal phase-shift digital shearography system using two orthogonal polarization states. They also suggested that the two orthogonal gradients can be recorded sequentially on a single CCD camera using a liquid crystal cell as a fast polarizer.

Mohan et al.[131] combined TV holography and

shearography into a single setup such that the deformation map and the derivative map are displayed simultaneously, one above the other, on the monitor. This setup can be used to obtain the deformation and derivative maps for both static and vibratory objects. James and Tatam[204] used sequential illumination of the object from three directions and a single CCD camera. The illumination was synchronized with the frame rate of the CCD camera. Each frame contains an image of an object when illuminated from one of the illumination directions. Kästle et al.[201] used wavelength-division multiplexing (WDM) to determine two-dimensional strain distributions. They used three diode lasers to illuminate the object simultaneously with three different wavelengths, and the sheared object images in each wavelength were recorded by three CCD cameras. It was shown that using three directions of illumination and two orthogonal shears, all six derivative components can be measured[164]. A compact system based on this has been reported[206]. Groves et al.[209,223] described a shearography system to measure two orthogonal slopes quasi-simultaneously using polarization multiplexing. They used an unbalanced polarization-sensitive Michelson interferometer to produce sheared images in orthogonal polarizations. They also described a multi-directional viewing configuration with one direction of illumination and a single CCD camera[242,244]. The images were transported by coherent fiber bundles to the CCD via a balanced Michelson interferometer. The quality of correlation fringes was very poor owing to decorrelation of the speckle patterns due to the influence of environmental variables on the fiber bundles. These authors reported another multi-component shearography system that can fully characterize the surface strain using time-division multiplexing of the measurement channels[245]. An application of this system to measure strains near a weld joint in a hydrostatically loaded pipe was presented. Francis et al.[285,300,302] used a similar pulsed laser shearographic system to measure the surface strains on static and rotating objects. Groves et al.[228,279] have also described a multi-directional illumination and single-direction observation system to fully measure surface strain. This system was used to measure crack length and surface strain.

A shearography system was described that employed colinear beams from three different lasers for illumination, a dichroic mirror in the Michelson interferometer, and a single CCD camera[350]. This provided shears along the *x*- and *y*-directions simultaneously and was used to obtain shearograms of the *x*- and *y*-derivatives of out-of-plane displacement simultaneously. Xie et al.[354] described a polarization-based digital shearography system containing two Mach–Zehnder interferometers to record shearograms in two orthogonal directions simultaneously, and applied it for the NDT of honeycomb structures. Gao et al.[380] described a shearography setup that employs a Mach–Zehnder interferometer for shear and spatial frequency generation for the simultaneous measurement of out-of-plane deformation and its first derivative. In a recent paper, Zhao et al.[378] described a combined DSPI and shearography system that yields both the deformation and the slope phase maps simultaneously using Fourier transform processing. In the same paper, a system is described that uses three different lasers (blue, green, and red) to illuminate the object in three different directions, and a CCD with three chips responding to each color that yields 3D deformation and derivative phase maps using the Fourier transform processing. They also described another multiplexed system, similar to that described earlier[380], which uses a Mach-Zehnder interferometer for shear and spatial frequency generation and a single CCD camera to yield the deformation and the derivative phase maps[391]. Instead of a spatial frequency carrier for generating the phase map from a single frame, a pixelated sensor was used[381]. A large number of shearograms were captured when the object underwent deformation. The spatiotemporal filtering algorithm in the complex domain was applied to these shearograms, and phase maps were obtained from the adjacent shearograms. Sun et al.[375] described a system that uses a spatial light modulator (SLM) for both shear and phase shift. Another system used two apertures to generate the spatial frequency, and an SLM for shear[385]. The out-of-plane deformation and its derivatives were obtained using Fourier processing. Zhong et al.[386] used two Mach-Zehnder interferometers in tandem to obtain gradients of deformation in two orthogonal directions along with a twist. Zhang et al.[390] described a shearography system in which one of the mirrors of a Michelson interferometer is replaced by an SLM, which provides both shear and phase shift. It should be noted that SDM, WDM, and polarization multiplexing allow the capture of information of the derivatives and the deformation simultaneously, while TDM accomplishes this sequentially.

An optical arrangement is described in which an ESPI system can be converted to a digital shearography system by rotating one mirror of a Michelson interferometer by 45 degrees[270]. Thus, both the displacement and its derivative can be obtained sequentially. Another arrangement that uses a two-aperture screen in front of the imaging lens has been used to measure the out-of-plane displacement and slope sequentially[271]. Because of the two apertures, a carrier frequency is generated, and therefore Fourier processing is used to obtain the phase map. Several researchers have combined the ESPI/DSPI system and the shearography system into a single unit that can measure the deformation and its derivatives either simultaneously or sequentially[99,134,175,270,326,339,378,391,392]. A single-axis system that comprises a shearography system and a speckle photography system is described that measures two out-of-plane displacement gradients and four in-plane displacement gradients to fully characterize the surface strain[268]. The in-plane displacement component gradients are obtained by numerical differentiation of the in-plane components measured by the speckle photography system. Wang et al.[370] described an optical configuration that uses a Wollaston prism for shear and can yield *x*- and *y*- slopes or *x*-slope and a radial slope simultaneously. The carrier frequency is provided by the tilt of a mirror, and Fourier transform processing is used to generate phase maps. Simultaneous measurement of deformation and its first derivative was demonstrated using the spatial phase shift method followed by Fourier processing[326].

A common-path shearography setup was described for the simultaneous measurement of the in-plane and out-of-plane strain components under dynamic loading[377], using two laser beams with different wavelengths that symmetrically illuminate the test object, a plane parallel plate for the shear, and a single CMOS camera. The phase was recovered from a single frame using Fourier processing. Another method to separate the in-plane and out-of-plane strains is to use a symmetric illumination configuration with a Wollaston prism as a shear element in which sequential recordings are made[379]. A recent paper describes the design and development of a high-speed shearography system and the measurement of surface strain during a high-speed impact on aluminum and composite samples[393]. This study also compares the performance of shearographic systems based on the Michelson and Mach-Zehnder interferometers.

Low Coherence Shearography

The temporal coherence requirement for shearography is not stringent because polarization-based shearographic systems are nearly common-path and the path difference in a balanced Michelson or Mach-Zehnder interferometer is very small. A source of adequate temporal coherence is required only when an unbalanced Michelson or Mach-Zehnder interferometer is employed for shearing, and wavelength modulation is used for phase stepping. In general, a light source with adequate brightness, spatial coherence, and limited temporal coherence is needed for shearography to visualize the deformation gradients[232]. In an earlier study, it was demonstrated that the shearographic fringes were formed on the object even when the coherence length was only 5 mm[80]. Falldorf et al.[238] employed a mercury arc lamp to study the deformation of an alloy strip that was mechanically loaded using a four-step phase stepping method. Wang and Tieu[254] also described the use of a low-coherence source in shearography. A mercury lamp was used to measure the out-of-plane displacement derivative of a centrally loaded flat plate. In another study, two laser diodes without temperature stabilization were used in a shearographic system[186]. It is easier to integrate laser diodes in compact shearographic systems[175]. Further, for the examination of large surface area of an object and to overcome non-uniform illumination due to the Gaussian nature of the laser beam, the use of two or more mutually incoherent laser beams has been suggested[239,357]. The importance of using mutually incoherent sources is obvious from the photographs taken with illumination from two mutually incoherent sources and two mutually coherent sources, as shown in Fig. 4.

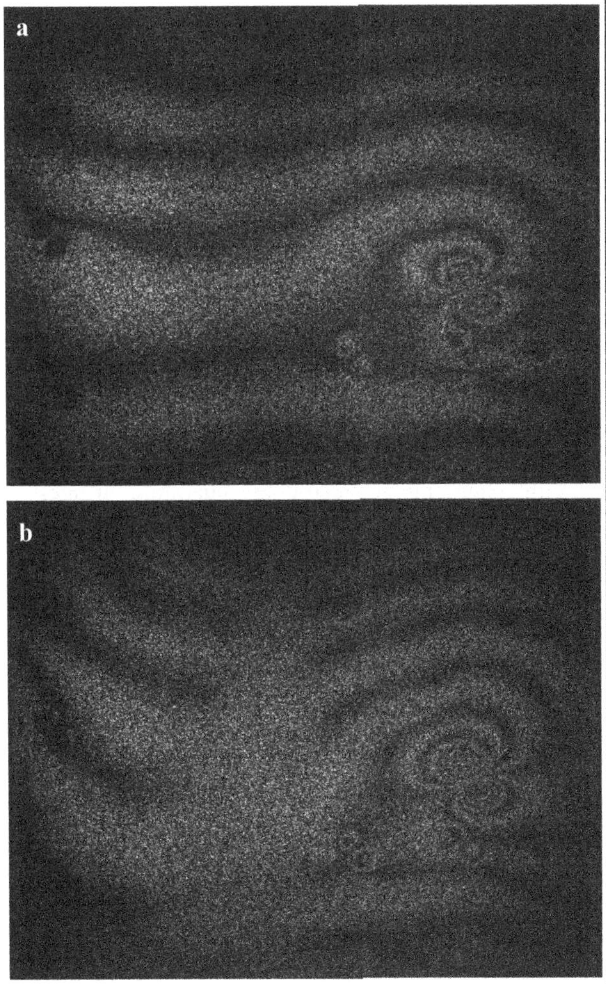

Fig. 4 Shearogram of a part of an Airbus A310 rudder side panel (1 m × 1.5 m), made with **a** two mutually incoherent sources, **b** two mutually coherent sources (Courtesy Prof. Osten).

Applications of Shearography

Shearography is a non-destructive, full-field, non-contact inspection technique that is widely used as a qualitative tool for structural health monitoring, location, and identification of damage and defect sites. It has also been used to obtain quantitative data on strain fields and the location and size of defects. Some well-known applications of shearography are presented in this section.

Visualization and measurement of strain fields, slope, curvature and twist

Early work on shearography was to visualize twist contours[8], derivatives of surface displacement[9,10], in-plane strains[13,164], slope contours[11–17,20,22–29,32,34,36,40,153,160,169,170,181,191,361,365], slope of modal amplitude of vibration[9,16,117], and curvature[15,19,32,35,36,38,79] in plates, beams, and other objects. Slope contours are also obtained by (i) shifting the illumination source[60,94,114,144], (ii) changing the wavelength[113,174], (iii) changing the refractive index of the medium around the object[17], and (iv) rotation of the object[128,156,169,200] between exposures. These methods were reviewed by Shang et al.[218]. It has been pointed out that the generation of slope contours by the shift of the source is quite a complex procedure[248]. Another study showed the dependence of slope and shape determination on the direction of illumination[286]. Owner-Petersen[61] used digital speckle pattern shear interferometry (DSPSI) for both qualitative and quantitative strain evaluation. It was commented that DSPSI is not useful for quantitative evaluation unless the space-bandwidth product is increased. It should be noted that the visualization of strain and slope continued even after photographic recording was replaced by video detection[105].

With photographic recording, curvature fringes are obtained as a moiré of the slope patterns obtained with a three-aperture shearographic setup in which two outer apertures carry wedge plates[32,34,35,41,119,235]. In such a configuration, the displacement and its first derivative influence the curvature fringes[235]. Rastogi[140] obtained the curvature and twist fringes by configuring an experimental setup using four mirrors. Murukeshan et al.[205,229] used two Michelson interferometers in tandem to generate two overlapping slope patterns, thereby producing curvature fringes. In contrast to multiple imaging using a three-aperture shearographic setup, four exposures are sequentially made with the object laterally shifted after two exposures, thereby recording two shifted slope patterns that produce curvature fringes due to the moiré phenomenon[95]. A similar result would be obtained if the photographic plate was shifted in its plane after two exposures. It is shown that the curvature fringes can be sharpened by multiple exposures[31,36,48,219].

Quantitative measurement of strain, slope, curvature, and twist

Quantitative measurement of strain, slope, curvature, and twist requires the measurement of shear in addition to the phase measurement. Methods to obtain the magnitude of shear have been described by Ng[116], Andersson et al.[216], and Khaleghi et al.[338]. Steinchen et al.[97,111,136,188] measured the in-plane strains and slopes of a cranked bar and a disc under compression using a four-step phase-shift algorithm and obtained values that closely matched the theoretical values. They also obtained time-averaged shearograms of a disc vibrating at different frequencies. Winther[50] measured the strain in a pipe clamp. He et al.[200] measured the slopes of various objects by rotating them between frames. Steinchen et al.[97] described a method to obtain pure in-plane strain and pure out-of-plane gradient using a single direction of illumination. They also described another method using two independent directions of illumination[111,212,236]. A method to measure the derivative of the vibration amplitude was described by Valera et al.[124,162] using a fiber-based speckle shear interferometer. The out-of-plane strain in a sandwich plate with full-potted inserts was measured using shearography and compared with analytically obtained results[292,304]. The greatest strains occurred in the vicinity of the boundary edge, but not at the edge.

Bhaduri et al.[291] used a three-aperture arrangement along with Fourier transform processing to obtain curvature contours. Tay et al.[263] obtained the first derivative of displacement by applying a temporal wavelet transform, and the second derivatives, curvature and twist, were extracted by applying a continuous Haar wavelet transform to a series of speckle patterns. Fu et al.[322] captured a large number of frames of a continuously deforming object with a high-speed CCD camera and extracted the curvature and twist contours at any instant using Fourier and windowed-Fourier transforms. Balakrishnan et al.[349] used a derivative-based phase tracker to retrieve the curvature and twist maps directly from a single shearography fringe pattern. The deflection, slope, and curvature contours of bent plates were generated computationally from the speckle patterns recorded in front of the plates before and after deformation[383]. Bhaduri et al.[313] demonstrated a method to determine the curvature and twist of an object quantitatively using two-channel double-aperture digital shearography. The double-aperture mask produces spatial carrier fringes inside the speckle, and two channels of the system produce two sets of slope phase maps that are shifted to obtain the curvature and twist data. To measure the in-plane strains in orthogonal directions, a shearography setup uses two Michelson interferometers to create two shearograms, one in the x shearing direction and the other in the y shearing direction, which are recorded by a single CCD camera[351]. Two lasers with different wavelengths are used for illumination, and the windowed-Fourier transform is used to obtain the phase maps. For measuring the in-plane strains under dynamic loading, the same shearography setup was used, but with two CCD cameras[345]. A speckle strain gauge was developed with a gauge length of approximately a millimeter that was used to measure the biaxial strain and Poisson's ratio of vinyl tape, polyvinylidene fluoride sheet, and rubber[51,66]. It was also used to measure the thermal expansion coefficients of several metals and non-metals.

Non-destructive testing (NDT)

Shearography is accepted as a powerful non-destructive testing technique, owing to its several unique features such as whole-field observation, resiliency to environmental disturbances and vibrations, low coherence requirement, simple optical setup, readiness for shop-floor applications, and ability to examine larger structures, albeit with tricky illumination. However, there are some critical issues to be considered when applying shearography for non-destructive testing[83,134]. An appropriate stressing method must be chosen so that the desired defect or flaw appears. The field of view and resolution are equally important for testing throughput. The sensitivity of shearography depends on the magnitude of shear. However, too large a shear results in speckle decorrelation because the technique does not remain common-path. Conversely, if a large shear is used, the derivatives become the relative displacements between two neighboring points on the object separated by a distance equal to the amount of shearing. The actual relationship between resolution and field of view is application-specific because it depends on the geometry of the test structure, the condition of the test surface, and the available laser power[81,83]. A method to detect flaws by measuring the first derivative and numerically obtaining the second derivative was proposed and demonstrated along with the influence of the magnitude of shear on flaw detection[334]. Theoretical and experimental investigations have been reported regarding the relationship between shear distance and performance issues, such as the immeasurable zone in the target with discontinuity, signal-to-noise ratio, sensitivity, and shear distortion[303]. Furthermore, it has been mentioned that by using a Wollaston prism for shear and 4f optical configuration, the shear can be varied over a large range[384].

The first application of shearography to NDT appears to be credited to Hung[21,240], who applied it to detect cracks in a pressure vessel, and ply separation in a truck tire. There are several publications on this topic originating from him and his research group, dealing with different aspects of non-destructive testing[21,45,52,72,105,142,139,158,165,173,178,180,211,217,240,258,265,288,308,310]. NDT has been applied for defect detection in plates and diaphragms[23,27–29]. It was observed that the theta shear vividly brings out defects when the deflection profile due to loading is circularly symmetric[28]. It has been shown that NDT of difficult-to-reach objects can be conducted by transmitting the sheared images through a coherent fiber bundle to the detector[58,252]. Kim et al.[241] used it to detect internal cracks in pressure pipelines by applying pressure loading. When the magnitude of the shear was smaller than the crack length, shearography predicted its length correctly. When the shear was larger than the crack length, shearography overestimated the length. In contrast to the detection of disbonds, shearography offers less promise for locating and sizing fatigue cracks in riveted aluminum aircraft skins[81]. Shearography is used to estimate the size and depth of disbands in GRP plates[59,67] and helicopter rotor blades[184,243] using vacuum stressing. Time-averaged shearography is used to estimate the size of flaws in 2-ply and 4-ply beams[63]. Shearography offers identical inspection capabilities for aircraft both in production and in the field[103,109,127,202,210,215,239]. It is used to detect cracks and disbands in skins[77,81,104], debonds in lap joints[77], debonds in fan cases[109], disbands in skin panels[85], corrosion in aluminum and composite honeycombs[107] and poor or improper repairs, and impact damage to epoxy graphite or thermoplastic aircraft structures[103,127,189]. Inspection of test panels that simulated Boeing 727 and 737 fuselage construction was carried out using shearography[77,103]. The panels contained defects that represented problems specific to aircraft aging. Principal component analysis (PCA) is incorporated into shearography in which a series of phase maps of a sandwich structure with defects at various depths are obtained as the temperature decreases[336]. PCA improves the defect detection capability of shearography. A good description of the use of shearography to conduct NDT of composites is provided by Newman[363].

Several groups have used compact and mobile shearographic systems for the inspection of aircraft structures[64,103,127,189,202,210,214,215,239,243,250,261,293]. A tripod mounted system is used to inspect (i) honeycomb structures, (ii) monolithic FRP structures, and (iii) metallic structures using 3-step and 4-step phase-shifting, and all known methods of stressing[127]. Another mobile shearography system using two CCD cameras with Savart plates in front of them and phase-shifting by a PZT-activated mirror is described, which captures 25 frames per second[143]. This system and its improved version are used for defect detection in aircraft materials[202]. It records the phase-stepped images simultaneously and performs better in a hostile environment than a system that is sequentially phase-stepped. A shearography system was developed for comprehensive testing of aircraft structures and was used to inspect the rudder of Airbus aircraft for delamination and debonding using vacuum stressing[210,215,239]. The sensing system is tripod mounted, and uses a Nd:YAG laser at 532 nm which is also tripod mounted. Fig. 5a shows a rudder shell from an Airbus A310, constructed from CFRP-Honeycomb, with marked fault regions that were tested by shearography. Fig. 5b is the shearogram obtained with thermal loading, and Fig. 5c is the demodulated phase map.

Surfaces with non-optimal reflectivity are referred to as non-cooperating surfaces. In most cases, they can be treated with paint. However, when any such treatment is forbidden, other optical solutions are employed, such as the use of diffusers and multiple laser diodes[215,333]. The important features of the shearography system are its ability to deal with non-cooperating surfaces and the fringe processing system that outputs excellent final results[215]. The shearography system is used on the shop floor to test aircraft components and whole aircraft at a maintenance facility[215,232,233,239], and for testing automobile components such as CFRP cross bars for internal imperfections[232,239]. For testing large objects, the use of two mutually incoherent lasers or two mutually incoherent beams derived from the same laser is suggested[239]. Comparison with ultrasonics and thermography favors shearography as a versatile inspection technique in industrial settings due to its full-field access, robust mode of operation, and fast delivery of results. The design details of a lightweight mobile shearography system that uses a diode laser for illumination along with its applications for the inspection of aircraft components and systems are given by Kalms and Jüptner[261] and by Kalms[293]. In another comparative study, shearography was found to be the preferred technique for reliable detection of manufacturing or in-service defects such as disbands and impact damage of aircraft structures[278]. Shearography was applied for surface strain measurements of a hydrostatically loaded ABS pipe, and the values obtained were compared with those obtained by Bragg gratings and resistance gauges[289]. In a very specific study to detect defects in thin mirrors, shearography was compared with deflectometry[335]. It was observed that the deflectometry delivered better results. In another study, a specular reflecting surface was illuminated with scattered light illumination, and shearograms with good contrast were recorded[371]. This work advanced the application of shearography for NDT of non-cooperating objects.

Shearography has been used to detect microcracks in poly methyl methacrylate (PMMA)[148], foam-adhesive bonded metal laminates[108], GRP pipes[176,281], internal cracks in pressure pipelines[241], and fatigue cracks in titanium Ti-10-2-3 alloy[279]. Inspection of horizontal stabilizer and aileron for delamination with vacuum stressing, radome for delamination, and rotor blades for debonds with thermal stressing has been reported[182]. It has also been shown that post-processing of the shearogram presents flaw details better against a fringe-free background. A helicopter rotor blade with induced defects at different depths was inspected with shearography using an infrared lamp[250,297,298]. It was observed that the number of fringes in the characteristic bulls-eye pattern decreased with the depth of the defect. Another interesting observation was that the fringe pattern in the shearogram of two partially overlapping identical disbonds was visually similar to that from a single disbond[288]. A comparative study of NDT of composite materials with shearography and thermography has been reported[151,180,288,319]. Shearography is employed to inspect power plant pipes for defects arising from corrosion[90], pipes in the chemical industry[145], automobile dashboards, and some components used in aircraft and helicopters[90]. It has been used to test for debonds on external tanks and rocket boosters of space shuttles at the Kennedy Space Center[98,102].

Defect detection using TV-holography and shearography

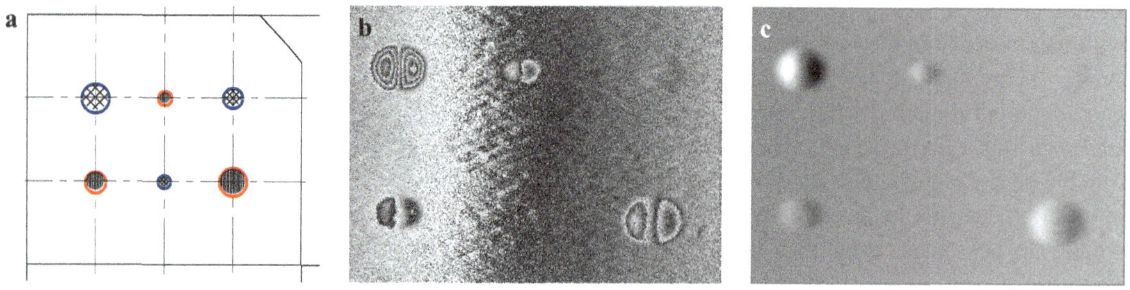

Fig. 5 Shearographic testing of an aircraft panel (500 mm × 500 mm) using thermal load **a** A310 rudder shell, CFRP-Honeycomb, with marked fault regions, **b** shearogram after thermal loading, and **c** demodulated phase map (Courtesy Prof. Osten).

can be very fast, but the depth and size of the defect are difficult to determine accurately. Conversely, ultrasonic techniques can characterize the defects in detail, including both the shape and depth, but they are slow. This observation is based on NDT carried out on a glass-reinforced plastic plate with programmed voids[134]. Assuming thinning of thickness of an aluminum plate as a diaphragm on a rigid boundary, shearography is applied to characterize this[193,199]. Shearography provides a rapid method to check the hermetic seals of microelectronics packages for leaks[178]. It has also been used to measure strains in electronic power devices under normal operating conditions[221]. Indeed, it is a versatile technique for quantitative measurement of vibration amplitudes and for non-destructive testing[257].

Hung[173,211] carried out testing and evaluation of composites using shearography. Hung et al.[217,240] evaluated the soundness of bonding in laminated composite structures using vacuum and vibratory stressing and carried out NDT of an adhesively laminated plate using vacuum stressing. Shearography can be used either in a fringe mode or a phase map mode for non-destructive testing in real time[324]. The phase map mode is more sensitive than the fringe mode for defect detection. Several algorithms for obtaining phase maps have been demonstrated for real-time NDT. An endoscopic shearography system to inspect the inside of cylindrical surfaces for defects was recently reported[362]. This system with radial sensitivity is used for the inner inspection of adhesion faults in composite material pipes. A novel shearographic system in which the image is transmitted through a rigid endoscope is used for NDT of a turbine blade in an automobile turbo-charger, and also of laminates[252]. Recently, shearography has been used to inspect metallic parts reinforced by GRP[387], and to detect the shape and size of defects on metallic substrates under composite repairs[388].

Residual stress

For the measurement of residual stress, stress relief is achieved either by drilling a blind hole or by making a micro-indentation on the object surface. The change in stress will result in a change in the in-plane and out-of-plane displacements, which is detected by shearography[129]. This technique has been used to measure residual stress in plastic and reinforced materials[65,208], concrete[88], and composites[208]. For almost real-time presentation of the results, the recording was performed on a thermoplastic material and the stress was relieved by indentation[56]. A large shear arrangement has been used in the determination of residual stresses using ball indentation[46,142,171,172,211,265] and shallow blind hole-drilling[65,88,172,208]. Viotti et al. described a portable shearographic system to measure residual stresses by means of a hole drilling technique based on a compact in-plane interferometer with only one symmetrical dual-beam illumination[277].

Art Conservation

A shearography system developed by Kalms et al.[210] and Osten et al.[215] was used to explore the structural robustness of XVII century Byzantine icons[231]. In another program, a shearography sensor was developed along with an impact assessment procedure (IAP)[290]. The sensor was used to detect defects in wooden panel paintings and canvas paintings using thermal stressing. A tripod-mounted system was used to examine paintings in the Benaki Museum, Greece[233]. Extended dynamic range (EDR) imaging technology has been implemented in a shearography system[296]. EDR imaging in shearography offers several benefits such as low noise in phase maps, less effort in aligning the illumination beam, and the use of multiple camera shutter speeds. EDR imaging shearography is also used to detect defects in wooden panel paintings and canvas paintings using thermal stressing. In another study, terahertz technology was combined with shearography with thermal loading for defect detection in wooden panel paintings[306,311]. To gain an understanding of the appearance and classification of faults in complex anisotropic objects such as artwork, finite element method (FEM) simulations along with shearographic measurements are carried out[367]. By comparing the simulations (different parameters) with the measurements, the depth and width of the faults can be determined with an accuracy of approximately 200 μm and 1,000 μm respectively. The shearography sensor provides surface and sub-surface information about the artwork, and the terahertz imaging sensor provides information about the bulk of the sample. The work on art conservation is further enhanced by incorporating virtual reality in shearography so that a user can remotely interact with the artwork[312]. Researchers have demonstrated that the false height in virtual reality can be used to display shearography displacement gradient maps. This technique allows the user to view both the displacement gradient measurements and the original colors of the painting.

The effect of environmental variables such as temperature and humidity on oil-on-canvas paintings has been studied continuously over a period of 30 hours in a museum; significant activity was observed when the temperature varied from 73.5 ^{0}F to 75.9 ^{0}F and humidity from 46.6% to 49.8%[338]. Using a combined DSPI and shearography system, the restored art objects were examined with acoustic loading over a range of frequencies to detect delamination and defects[339]. An important factor in detecting delamination, cracks, and holes in artworks, particularly paintings, is the type of loading employed. For example, thermal loading reveals trapped humidity, trapped air, and delamination. Acoustic loading has been used to detect delamination. A detailed description of various types of loading techniques, such as pressure, temperature, humidity, and vibration, and their influence on the sensitivity of flaw detection on wooden and canvas paintings, was provided by Buchta et al.[346]. Fig. 6 shows a painting on canvas of size 290 mm × 250 mm with a narrow wooden frame. It is illuminated with a green laser beam and the double-exposure shearogram is displayed on the monitor, showing a defect in the center. Thermal stressing using infrared lamps was used as a method of excitation.

Fig. 6 **a** A painting on canvas, **b** illuminated painting along with its double e^x^posure shearogram (with permission from Ref. 290).

Special Applications

It has been demonstrated that theta shear can be used to measure the tilt of planar rough surfaces with good accuracy, which depends inversely on the magnitude of the theta shear[24,25]. Stroboscopic shearography has been used to effectively measure ultrasonic waves propagating through aluminum bars[183]. The effect of the direction and magnitude of shear on the propagation direction of ultrasonic waves was also investigated. Shearography has also been used to obtain the flexural strain in a turbine blade[197]. Another application has been to examine large objects such as metal cabinets [337]. It has been demonstrated that stroboscopic shearography not only identifies defects but can also quantify them[388]. A feasibility study to examine bridge response by shearography has been reported[89,101]. Unfortunately, no shearograms were presented in this study.

Quantitative whole-field density measurements have been shown to be possible using shearography in a hostile industrial environment[130]. Essentially, shearography maps the refractive index in the flow, which is then related to the density. Both density and temperature have been mapped in a flow field[177]. Shearography was also applied to measure the temperature distribution in a liquid petroleum gas flame[92]. Shearography is applied to the experimental analysis of high-power ultrasonic machining components using a pulse laser[157].

An interesting application of shearography is the measurement of distance[167]. A fringe pattern is created by the motion of the source, in which the fringe width is related to the distance. Interestingly, this can also be used to measure velocity[192]. A shearographic setup using a Savart plate for shearing generates a fringe pattern in which the fringe width is proportional to the depth of the object, thus turning it into a profilometer that provides absolute distance measurements of rough surfaces in the sub-micrometric range[321].

A conceptual approach to apply shearography to detect abnormal growth in biological cavities was presented by conducting experiments on a half-cylindrical tissue phantom specimen[253]. Features that resemble abnormal growth in body cavities, such as polyps in different parts of the gastrointestinal path, especially in the large intestine, are introduced in different layers of the phantom. It was possible to detect these abnormalities using shearography. Recently, Buchta et al.[364] simulated tumor-like tissue in a silicone phantom and applied shearography to detect regions of different stiffness with pressure loading. Fig. 7a shows a sketch of the silicone sample (with dimensions in mm) in which hard silicone spheres of different sizes are embedded at different depths. Two different types of silicone exhibiting different stiffness properties were chosen: hard silicone representing the cancerous tissue and a softer texture representing normal tissue. Fig. 7b is a photograph of the transilluminated sample. Fig. 7c shows the strain map of a 10 mm sphere obtained with 1.2 mm shear and a 4 mbar pressure change. Shearography was able to discriminate between two types of silicone. However, it was not possible to provide correct information about the size and depth of hard silicone spheres. Nevertheless, this technique may prove to be important for guidance and discrimination during minimally invasive surgery.

Shearography can be configured for the study and testing of small objects if the conventional imaging lens is replaced by a long-focus microscope objective[307,309]. In one study, a pair of parallel plates separated by a distance acted as a shearing device as well as a phase-shifter[307]. A pressure sensor diaphragm (with a size of 1.5 mm × 1.5 mm and thickness of 25 μm) etched in a silicon substrate of size 3 mm × 3 mm and approximately 1 mm thick was studied using a low magnification zoom lens with a longer working

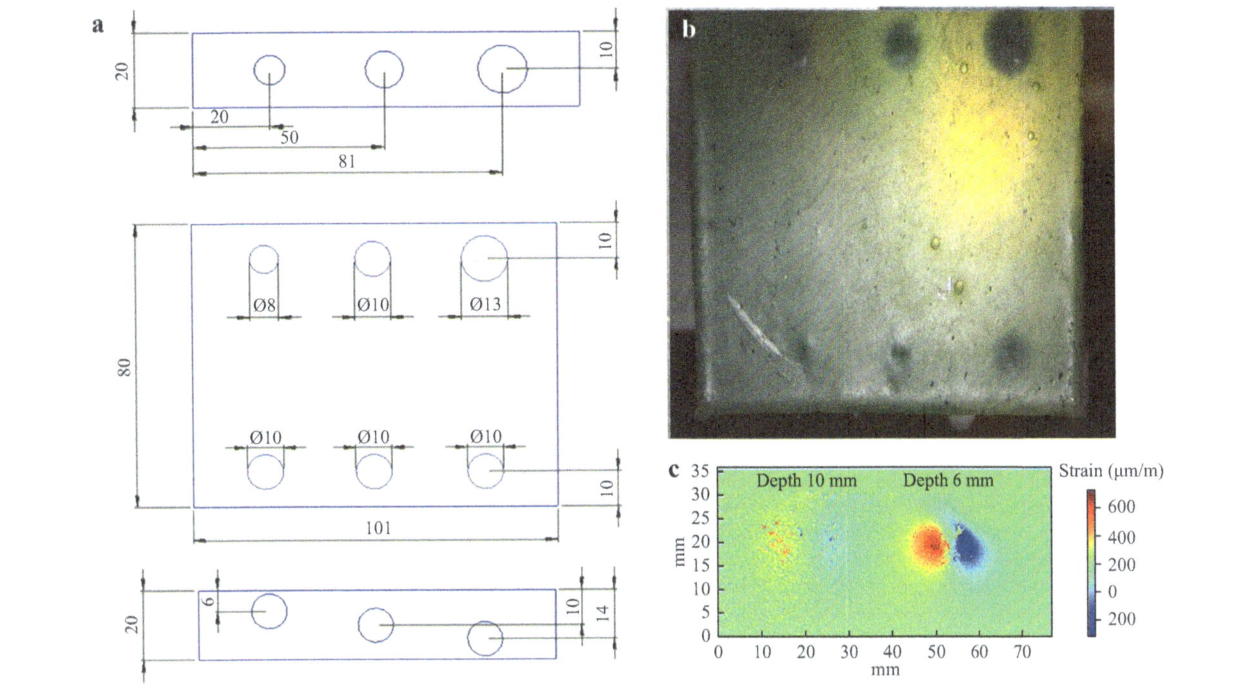

Fig. 7 **a** Sketch of the silicone sample (dimensions in mm), **b** photograph of the transilluminated sample, and **c** processed shearography data for two hard-silicone spheres (with permission from Ref. 364).

distance[309]. It used a conventional Michelson interferometer for shearing and phase-shifting. Udupa et al.[259] investigated a 200 mm Si wafer for defects by thermally loading the wafer with an infrared lamp. Shearography was able to successfully detect defects in the mm range.

Piezo-electrically, laser- and impact-generated Lamb waves in CFRP have been studied using pulse laser shearography employing a Mach-Zehnder interferometer as a shear device; this system was also used to generate carrier fringes for spatial phase shifting[301]. Shearography has been employed to evaluate the thermo-mechanical behavior of various coatings on different substrates, and for the determination of thermal stress and strain as a function of temperature[266,316]. Owing to its low temporal coherence requirement, shearography is able to measure the axial positions of phosphors embedded in a medium[347]. Phosphors embedded in a medium were illuminated by a structured beam from the excitation laser. The phase of the fluorescent wave from the phosphor was measured by shearography using a balanced Michelson interferometer for shearing. The axial position was reconstructed from the measured phase. A combination of fluorescent microscopy and shearography can provide the axial depth of phosphor placed several hundred micrometers in a highly scattering medium with good resolution.

There are several publications that describe the measurement of the shape of an object with shearography[114,227,248,286]. In all of these publications, the slope was integrated to obtain the shape. However, shearography can provide the form of an object. The basic framework of the procedure was provided by Elster and Weingärtner[213]. By applying several shears in the x- and y-directions, several x- and y-derivatives can be measured. The form of the object can be reconstructed by integrating the x- and y-derivatives.

Conclusion

Since its birth in 1973, shearography has evolved as a technique for visualizing and measuring displacement derivatives. Essentially, it measures the difference quotient that approximates a derivative when the shear is small. However, the sensitivity of shearography is linearly dependent on the magnitude of the shear. Because the shearographic configurations are nearly common-path, the requirement for temporal coherence is relaxed, and hence shearography can be performed with low-coherence sources such as a mercury source or an LED. Waves from the object can be transported using imaging fiber optic cables, and therefore inaccessible objects can be examined. Furthermore, the sensitivity also depends on the angle between the local surface normal and the illumination or observation beam, and maximum sensitivity is observed with respect to the derivative of the out-of-plane displacement (slope). Shearography has been accepted as an NDT technique for examining a variety of objects. Several mobile and non-mobile systems have been developed by researchers for their own use and by companies for commercialization. The first commercial 3D-shearographic system was offered by Ettemeyer in the early 1990s through his company. Another system using an air-cooled argon ion laser was developed in 1994. Some of the companies that offer shearography systems are (i) Dantec Dynamics A/S, Denmark; (ii) Optonor, Norway; (iii) isi-sys GmbH, Germany; (iv) Laser Technology, Inc., USA; (v) Laser Optical Engineering Ltd, UK; (vi) Carl Zeiss, Germany.

Acknowledgements

It is my pleasure to thank Prof. Wolfgang Osten for inviting me to prepare this article. Special thanks are to Prof. Jayant Sirohi, who painstakingly went over the manuscript and made critiques that improved the paper considerably. I would like to acknowledge the support provided by the Department of Physics, Alabama A&M University, Huntsville, Alabama, USA. I would also like to thank the reviewers of the manuscript for their constructive suggestions and for bringing out some of the references to my attention.

Conflict of interest

Received: 16 July 2021 Revised: 21 December 2021 Accepted: 25 December 2021

Accepted article preview online: 31 December 2021

Published online: 14 January 2022

References

1. Leendertz, J. A. Interferometric displacement measurement on scattering surfaces utilizing speckle effect. *Journal of Physics E:Scientific Instruments* **3**, 214-218 (1970).
2. Erf, R. K. (Ed.) Speckle Metrology. (New York: Academic Press, 1978).
3. Sirohi, R. S. (Ed.) Speckle Metrology. (New York: CRC Press, 1993).
4. Patorski, K. & Kujawinska, M. Handbook of the Moiré Fringe Technique. (Amsterdam: Elsevier, 1993).
5. Kaufmann, G. H. (Ed.), Advances in Speckle Metrology and Related Techniques-. (Weinheim: Wiley-VCH 2011).
6. Viotti, M. R. & Albertazzi, A. Jr. Robust Speckle Metrology: Techniques for Stress Analysis and NDT. (Bellingham: SPIE Press, 2014).
7. Sollid, J. E. Holographic interferometry applied to measurements of small static displacements of diffusely reflecting surfaces. *Applied Optics* **8**, 1587-1595 (1969).
8. Leendertz, J. A. & Butters, J. N. An image-shearing speckle-pattern interferometer for measuring bending moments. *Journal of Physics E:Scientific Instruments* **6**, 1107-1110 (1973).
9. Hung, Y. Y. & Taylor, C. E. Speckle-shearing interferometric camera -- a tool for measurement of derivatives of surface-displacement. Proceedings of SPIE 0041, Developments in Laser Technology II. San Diego, United States: SPIE, 1973.
10. Hung, Y. Y. A speckle-shearing interferometer: a tool for measuring derivatives of surface displacement. *Optics Communications* **11**, 132-135 (1974).
11. Hung, Y. Y. & Taylor, C. E. Measurement of slopes of structural deflections by speckle-shearing interferometry. *Experimental Mechanics* **14**, 281-285 (1974).
12. Hariharan, P. Speckle-shearing interferometry: a simple optical system. *Applied Optics* **14**, 2563 (1975).
13. Hung, Y. Y., Rowlands, R. E. & Daniel, I. M. Speckle-shearing interferometric technique: a full-field strain gauge. *Applied Optics* **14**, 618-622 (1975).
14. Debrus, S. Speckle shearing interferometer using a Savart plate. *Optics Communications* **20**, 257-261 (1977).
15. Assa, A., Betser, A. A. & Politch, J. Recording slope and curvature contours of flexed plates using a grating shearing interferometer. *Applied Optics* **16**, 2504-2513 (1977).
16. Hovanesian, J. D., Hung, Y. Y. & Durelli, A. J. New optical method to determine vibration induced strains with variable sensitivity after recording. Proceedings of the Conference on Experimental Stress Analysis. Smolenice, Czechoslovakia, 1978.
17. Hung, Y. Y. et al. Optical method for measuring contour slopes of an object. *Applied Optics* **17**, 128-131 (1978).
18. Hung, Y. Y. & Durelli, A. J. Simultaneous measurement of three displacement derivatives using a multiple image-shearing interferometric camera. *The Journal of Strain Analysis for Engineering Design* **14**, 81-88 (1979).
19. Hung, Y. Y. & Liang, C. Y. Image-shearing camera for direct measurement of surface strains. *Applied Optics* **18**, 1046-1051 (1979).
20. Nakadate, S., Yatagai T. & Saito, H. Digital speckle-pattern shearing interferometry. *Applied Optics* **19**, 4241-4246 (1980).
21. Hung, Y. Y. Shearography: a new optical method for strain measurement and nondestructive testing. *Optical Engineering* **21**, 391-395 (1982).
22. Murthy, R. K., Sirohi, R. S. & Kothiyal, M. P. Speckle shearing interferometry: a new method. *Applied Optics* **21**, 2865-2867 (1982).
23. Murthy, R. K., Sirohi, R. S. & Kothiyal, M. P. Detection of defects in plates and diaphragms using a split-lens speckle-shearing interferometer. *NDT International* **15**, 329-333 (1982).
24. Mohanty, R. K., Joenathan, C. & Sirohi, R. S. Speckle-shear interferometry with double dove prisms. *Optics Communications* **47**, 27-30 (1983).
25. Mohanty, R. K., Joenathan, C. & Sirohi, R. S. Measurement of tilt with speckle-shear interferometry. *Journal of Optics* **12**, 118-121 (1983).
26. Joenathan, C., Mohanty, R. K. & Sirohi, R. S. Multiplexing in speckle shear interferometry. *Optica Acta:International Journal of Optics* **31**, 681-692 (1984).
27. Sharma, D. K., Sirohi, R. S. & Kothiyal, M. P. Non destructive testing of a diaphragm by multiaperture speckle shearing technique. *Optik* **66**, 133-136 (1984).
28. Sirohi, R. S. Speckle shear interferometry. *Optics & Laser Technology* **16**, 251-254 (1984).
29. Murthy, R. K., Mohanty, R. K., Sirohi, R. S. & Kothiyal, M. P. Radial speckle shearing interferometer and its engineering applications. *Optik* **67**, 85-94 (1984).
30. Iwahashi, Y., Iwata, K. & Nagata, R. Single-aperture speckle shearing interferometry with a single grating. *Applied Optics* **23**, 247-249 (1984).
31. Mohanty, R. K., Joenathan. C. & Sirohi, R. S. Speckle fringe sharpening in speckle shear interferometry. *Journal of Optics* **13**, 49-51 (1984).
32. Sharma, D. K., Sirohi, R. S. & Kothiyal, M. P. Simultaneous measurement of slope and curvature with a three-aperture speckle shearing interferometer. *Applied Optics* **23**, 1542-1546 (1984).
33. Joenathan, C., Mohanty, R. K. & Sirohi, R. S. On the methods of multiplexing in speckle shear interferometry. *Optik* **69**, 8-12 (1984).
34. Sharma, D. K., Sirohi, R. S. & Kothiyal, M. P. Multiaperture speckle shearing arrangements for stress analysis. *Optics Communications* **49**, 313-317 (1984).
35. Joenathan, C., Mohanty, R. K. & Sirohi, R. S. Curvature by halo

combination in speckle shear interferometry. *Journal of Optics* **13**, 114-119 (1984).
36. Mohanty, R. K., Joenathan. C. & Sirohi, R. S. Fringe sharpening and information coupling in speckle shear interferometry. *Applied Optics* **23**, 4596-4600 (1984).
37. Joenathan, C., Mohanty, R. K. & Sirohi, R. S. Hololens in speckle and speckle shear interferometry. *Applied Optics* **24**, 1294-1298 (1985).
38. Joenathan, C. & Sirohi, R. S. Holographic gratings in speckle shearing interferometry. *Applied Optics* **24**, 2750-2751 (1985).
39. Mohanty, R. K., Joenathan. C. & Sirohi, R. S. Speckle and speckle-shearing interferometers combined for the simultaneous determination of out-of-plane displacement and slope. *Applied Optics* **24**, 3106-3109 (1985).
40. Takezaki J. & Hung, Y. Y. Direct measurement of flexural strains in plates by shearography. *Journal of Applied Mechanics* **53**, 125-129 (1986).
41. Sharma, D. K., Mohan, N. K. & Sirohi, R. S. A holographic speckle shearing technique for the measurement of out-of-plane displacement, slope and curvature. *Optics Communications* **57**, 230-235 (1986).
42. Pyzin, G. P., Artemenko, S. B. & Ignatev, A. G. Compensation for rigid and deformational displacements in speckle-shearing interferometry. *Soviet Physics Technical Physics* **31**, 528-530 (1986).
43. Iwahashi, Y., Iwata, K. & Nagata, R. Simultaneous measurement of three slope distributions with single-aperture speckle shearing interferometer. *Applied Optics* **25**, 328-329 (1986).
44. Templeton, D. W. & Hung, Y. Y. Computerization of data deduction in shearography. Proceedings of SPIE 0814, Photomechanics and Speckle Metrology. San Diego, CA, United States: SPIE, 1987.
45. Hung, Y. Y. Nondestructive evaluation by electronic shearography. Proceedings of the 16th Symposium on Nondestructive Evaluation. San Antonio, Texas, 1987.
46. Hung, M. Y. Y. et al. Fast detection of residual stresses by shearography. Proceedings of SPIE 0955, Industrial Laser Interferometry II. Dearborn, MI, United States: SPIE, 1988.
47. Hung, Y. Y., Hovanesian, J. D. & Takezaki, J. A fringe carrier technique for unambiguous determination of fringe orders in shearography. *Optics and Lasers in Engineering* **8**, 73-81 (1988).
48. Ganesan, A. R., Joenathan. C. & Sirohi, R. S. Sharpening of fringes in digital speckle pattern interferometry. *Applied Optics* **27**, 2099-2100 (1988).
49. Ganesan, A. R., Sharma, D. K. & Kothiyal, M. P. Universal digital speckle shearing interferometer. *Applied Optics* **27**, 4731-4734 (1988).
50. Winther, S. 3D strain measurements using ESPI. *Optics and Lasers in Engineering* **8**, 45-57 (1988).
51. Yamaguchi, I., Takemori, T. & Kobayashi, K. Stabilized and accelerated speckle strain gauge. Proceedings of SPIE 1162, Laser Interferometry: Quantitative Analysis of Interferograms: Third in a Series. San Diego, United States: SPIE, 1989.
52. Hung, Y. Y. Shearography: a novel and practical approach for nondestructive inspection. *Journal of Nondestructive Evaluation* **8**, 55-67 (1989).
53. Templeton, D. W. & Hung, Y. Y. Shearographic fringe carrier method for data reduction computerization. *Optical Engineering* **28**, 30-34 (1989).
54. Lamprecht, H. A. & van Rooyen, E. Sandwich shearography: a technique for the cancellation of unknown rigid body displacements in shearography. Proceedings of SPIE 1162, Laser Interferometry: Quantitative Analysis of Interferograms: Third in a Series. San Diego, United States: SPIE, 1989.
55. Klumpp, P. A. Simple spatial filtering for shearograms. *Optics & Laser Technology* **21**, 105-111 (1989).
56. Hung, Y. Y. & Hovanesian, J. D. Fast detection of residual stresses in an industrial environment by thermoplastic-based shearography. Proceedings of 1990 SEM Spring Conference on Experimental Mechanics, Albuquerque. Bethel: SEM, 1990.
57. Joenathan, C. & Torroba, R. Simple electronic speckle-shearing-pattern interferometer. *Optics Letters* **15**, 1159-1161 (1990).
58. Aiyer, A. A. Shearography: a potential portable on-site nondestructive evaluation inspection tool. Proceedings of SPIE 1212, Practical Holography IV. Los Angeles, CA, United States: SPIE, 1990.
59. Toh, S. L. et al. Application of shearography in nondestructive testing of composite plates. *Journal of Materials Processing Technology* **23**, 267-275 (1990).
60. Tay, C. J. et al. The measurement of slope using shearography. *Optics and Lasers in Engineering* **14**, 13-24 (1991).
61. Owner-Petersen, M. Digital speckle pattern shearing interferometry: limitations and prospects. *Applied Optics* **30**, 2730-2738 (1991).
62. Kadono, H., Toyooka, S. & Iwasaki, Y. Speckle-shearing interferometry using a liquid-crystal cell as a phase modulator. *Journal of the Optical Society of America A* **8**, 2001-2008 (1991).
63. Toh, S. L. et al. Flaw detection in composites using time-average shearography. *Optics & Laser Technology* **23**, 25-30 (1991).
64. Newman, J. W. Shearographic inspection of aircraft structure. *Materials Evaluation* **49**, 1106-1109 (1991).
65. Long, K., Hung, Y. Y. & Der Hovanesian, J. Measurement of residual stresses in plastic materials by electronic shearography. Proceedings of SPIE 1554, Second International Conference on Photomechanics and Speckle Metrology. San Diego, CA, United States: SPIE, 1991.
66. Yamaguchi, I. & Kobayashi, K. Material testing by the laser speckle strain gauge. Proceedings of SPIE 1554, Second International Conference on Photomechanics and Speckle Metrology. San Diego, CA, United States: SPIE, 1991.
67. Shang, H. M. et al. Interpretation of holographic and shearographic fringes for estimating the size and depth of debonds in laminated plates. Proceedings of SPIE 1554, Second International Conference on Photomechanics and Speckle Metrology. San Diego, CA, United States: SPIE, 1991.
68. Hung, Y. Y. Electronic shearography versus ESPI in nondestructive evaluation. Proceedings of SPIE 1554, Second International Conference on Photomechanics and Speckle Metrology. San Diego, CA, United States: SPIE, 1991.
69. Shang, H. M., Soh, C. C. & Chau, F. S. The use of carrier fringes in shearography for locating and sizing debonds in GRP plates. *Composites Engineering* **1**, 157-165 (1991).
70. Qin, Y. W., Wang, J. Q. & Ji, X. H. Study of electronic shearing speckle technique. Proceedings of SPIE 1554, Second International Conference on Photomechanics and Speckle Metrology. San Diego, CA, United States: SPIE, 1991.
71. Xu, B. Q. & Wu, X. P. Fringe formation in speckle shearing interferometry. Proceedings of SPIE 1554, Second International Conference on Photomechanics and Speckle Metrology. San Diego, CA, United States: SPIE, 1991.
72. Hung, Y. Y. Electronic shearography for measurement of time-dependent deformation. Proceedings of the VII International Congress on Experimental Mechanics, vol. 2. Las Vegas, Nevada, 1992.
73. Tay, C. J. et al. Measurements of surface coordinates and slopes by shearography. *Optics & Laser Technology* **24**, 209-213 (1992).
74. Chau, F. S. & Ng, T. W. A real-time digital shearing speckle interferometer. *Measurement Science and Technology* **3**, 381-383 (1992).
75. Ng, T. W. & Chau, F. S. Suppression of fringe-modulating in-plane displacement fields in shearing speckle interferometry. Proceedings

of SPIE 1821, Industrial Applications of Optical Inspection, Metrology, and Sensing. Boston, MA, United States: SPIE, 1992.

76. Chau, F. S. & Ng, T. W. Determining the optimum image recording conditions in shearography based on spatial frequency considerations. Proceedings of SPIE 1821, Industrial Applications of Optical Inspection, Metrology, and Sensing. Boston, MA, United States: SPIE, 1992.
77. Tyson II, J. & Feferman, B. Inspection of Fabricated Fuselage Panels Using Electronic Shearography. Report AD-A257639 (1992).
78. Pryputniewicz, R. J. Electronic shearography and electronic holography working side by side. Proceedings of SPIE 1821, Industrial Applications of Optical Inspection, Metrology, and Sensing. Boston, MA, United States: SPIE, 1992.
79. Tay, C. J. et al. Curvature measurement of hydroformed axisymmetrical shells using shearography. *Experimental Techniques* **17**, 27-30 (1993).
80. Spooren, R., Dyrseth, A. A. & Vaz, M. Electronic shear interferometry: application of a (double-) pulsed laser. *Applied Optics* **32**, 4719-4127 (1993).
81. Deaton, J. B. Jr. & Rogowski, R. S. Applications of electronic shearography for the inspection of airskin structures. Proceedings of SPIE 2001, Nondestructive Inspection of Aging Aircraft. San Diego, CA, United States: SPIE, 1993.
82. Wang, B. S. & Zhang, X. Electronic speckle carrier shearography for automated deformation measurement. Proceedings of SPIE 2003, Interferometry VI: Techniques and Analysis. San Diego, CA, United States: SPIE, 1993.
83. Deaton, J. B. Jr. & Rogowski, R. S. Electronic shearography for nondestructive evaluation: the influence of the field of view and the shearing angle. Proceedings of SPIE 2066, Industrial Optical Sensing and Metrology: Applications and Integration. Boston, MA, United States: SPIE, 1993.
84. Ng, T. W. & Chau, F. S. Performance of charge-coupled devices in digital shearography. Proceedings of SPIE 2066, Industrial Optical Sensing and Metrology: Applications and Integration. Boston, MA, United States: SPIE, 1993.
85. Safai, M. Nondestructive evaluation of aircraft fuselage panels with electronic shearography. Proceedings of SPIE 2066, Industrial Optical Sensing and Metrology: Applications and Integration. Boston, MA, United States: SPIE, 1993.
86. Mohan, N. K., Saldner, H. O. & Molin, N. E. Electronic shearography applied to static and vibrating objects. *Optics Communications* **108**, 197-202 (1994).
87. Ng, T. W. & Chau, F. S. A digital shearing speckle interferometry technique for modal analysis. *Applied Acoustics* **42**, 175-185 (1994).
88. Hung, Y. Y. & Long, K. Evaluation of residual stress in concrete structures by digital shearography. Proceedings of SPIE 2004, Interferometry VI: Applications. San Diego, CA, United States: SPIE, 1993.
89. Maji, A. K., Satpathi, D. & Zawaydeh, S. Electronic shearography for inspecting steel bridges: a feasibility study. Proceedings of SPIE 2004, Interferometry VI: Applications. San Diego, CA, United States: SPIE, 1993.
90. Ettemeyer, A. Nondestructive inspection with shearography. Proceedings of SPIE 2248, Optical Measurements and Sensors for the Process Industries. Frankfurt, Germany: SPIE, 1994.
91. Mohan, N. K. et al. Separation of the influence of in-plane displacement in multiaperture speckle shear interferometry. *Optical Engineering* **33**, 1973-1982 (1994).
92. Shakher, C. & Nirala, A. K. Measurement of temperature using speckle shearing interferometry. *Applied Optics* **33**, 2125-2127 (1994).
93. Hung, Y. Y., Tang, S. & Hovanesian, J. D. Real-time shearography for measuring time-dependent displacement derivatives. *Experimental Mechanics* **34**, 89-92 (1994).
94. Tay, C. J. et al. On the determination of slope by shearography. *Optics and Lasers in Engineering* **20**, 207-217 (1994).
95. Tay, C. J. et al. Direct determination of second-order derivatives in plate bending using multiple-exposure shearography. *Optics & Laser Technology* **26**, 91-98 (1994).
96. Valera, J. D. & Jones, J. D. C. Phase stepping in fiber-based speckle shearing interferometry. *Optics Letters* **19**, 1161-1163 (1994).
97. Steinchen, W. et al. Electronic shearography (ESPSI) for direct measurement of strains. Proceedings of SPIE 2248, Optical Measurements and Sensors for the Process Industries. Frankfurt, Germany: SPIE, 1994.
98. Bird, L. Evaluating laser shearography for nondestructive testing at the Kennedy Space Center. Proceedings of SPIE 2349, Industrial Optical Sensors for Metrology and Inspection. Boston, MA, United States: SPIE, 1994.
99. Schulz, B. Electronic speckle pattern interferometrie through shearography. Proceedings of SPIE 2358, First International Conference on Vibration Measurements by Laser Techniques: Advances and Applications. Ancona, Italy: SPIE, 1994.
100. del Carretto, P. & Perlo, P. P. Compact multibeam shearography for deformation and vibration observation. Proceedings of SPIE 2358, First International Conference on Vibration Measurements by Laser Techniques: Advances and Applications. Ancona, Italy: SPIE, 1994.
101. Satpathi, D. & Maji, A. K. Electronic shearography for bridge inspection. Proceedings of SPIE 2446, Smart Structures and Materials 1995: Smart Systems for Bridges, Structures, and Highways. San Diego, CA, United States: SPIE, 1995.
102. Davis, C. K. et al. Shearographic nondestructive evaluation of Space Shuttle thermal protection systems. Proceedings of SPIE 2455, Nondestructive Evaluation of Aging Aircraft, Airports, Aerospace Hardware, and Materials. Oakland, CA, United States: SPIE, 1995.
103. Newman, J. W. Shearography nondestructive evaluation techniques for aerospace. Proceedings of SPIE 2455, Nondestructive Evaluation of Aging Aircraft, Airports, Aerospace Hardware, and Materials. Oakland, CA, United States: SPIE, 1995.
104. Bohn, C. E. Shearographic nondestructive inspection on the B-2 program. Proceedings of SPIE 2455, Nondestructive Evaluation of Aging Aircraft, Airports, Aerospace Hardware, and Materials. Oakland, CA, United States: SPIE, 1995.
105. Hung, Y. Y. Automated shearography for nondestructive evaluation and strain measurement. Proceedings of SPIE 2455, Nondestructive Evaluation of Aging Aircraft, Airports, Aerospace Hardware, and Materials. Oakland, CA, United States: SPIE, 1995.
106. Spicer, J. W. M. et al. Thermal stressing techniques for flaw characterization with shearography. Proceedings of SPIE 2455, Nondestructive Evaluation of Aging Aircraft, Airports, Aerospace Hardware, and Materials. Oakland, CA, United States: SPIE, 1995.
107. Chatters, T. C., Pouet, B. F. & Krishnaswamy, S. Additive-subtractive phase-modulated shearography with synchronized acoustic stressing. *Experimental Mechanics* **35**, 159-165 (1995).
108. Shang, H. M., Tham, L. M. & Chau, F. S. Shearographic and holographic assessment of defective laminates with bond-lines of different elasticities. *Journal of Engineering Materials and Technology* **117**, 322-329 (1995).
109. Lape, D., Newman, J. W. & Craig, D. Advanced laser shearography inspection of turbo-fan engine composite fan cases. Proceedings of SPIE 2455, Nondestructive Evaluation of Aging Aircraft, Airports, Aerospace Hardware, and Materials. Oakland, CA, United States: SPIE, 1995.
110. Sim, C. W., Chau, F. S. & Toh, S. L. Determination of fractional fringe

order by phase shifting in real-time shearography. *Journal of Modern Optics* **42**, 271-277 (1995).
111. Steinchen, W. et al. Out-of-plane and in-plane strain measured by shearography. Proceedings of SPIE 2509, Smart Structures: Optical Instrumentation and Sensing Systems. Munich, Germany: SPIE, 1995.
112. Yang, L. X. et al. Precision measurement and nondestructive testing by means of digital phase shifting speckle pattern and speckle pattern shearing interferometry. *Measurement* **16**, 149-160 (1995).
113. Huang, J. R. & Tatam, R. P. Optoelectronic shearography: two wavelength slope measurement. Proceedings of SPIE 2544, Interferometry VII: Techniques and Analysis. San Diego, CA, United States: SPIE, 1995.
114. Griffen, C. T., Hung, Y. Y. & Chen, F. Three-dimensional shape measurement using digital shearography. Proceedings of SPIE 2545, Interferometry VII: Applications. San Diego, CA, United States: SPIE, 1995.
115. Hung, Y. Y. & Griffen, C. T. Digital shearography versus TV-holography for vibration measurement. Proceedings of the 15th Biennial Conference on Mechanical Vibration and Noise and 1995 ASME Design Engineering Technical Conferences, Vol. 3. Boston, Massachusetts, 1995.
116. Ng, T. W. Shear measurement in digital speckle shearing interferometry using digital correlation. *Optics Communications* **115**, 241-244 (1995).
117. Chen, F., Griffen, C. T. & Hung, Y. Y. Stroboscopic phase shifting shearography for automated vibration measurement. Proceedings of the 15th Biennial Conference on Mechanical Vibration and Noise and 1995 ASME Design Engineering Technical Conferences, Vol. 3. Boston, Massachusetts, 1995.
118. Steinchen, W. et al. Application of shearography to quality assurance. *Journal of Materials Processing Technology* **52**, 141-150 (1995).
119. Tay, C. J. et al. Multiple-image shearography: a direct method to determine curvatures. *Applied Optics* **34**, 2202-2206 (1995).
120. Valera, J. D. R. & Jones, J. D. C. Vibration analysis by modulated time-averaged speckle shearing interferometry. *Measurement Science and Technology* **6**, 965-970 (1995).
121. Buerkle, L. & Joenathan, C. Electronic speckle pattern shearing interferometer using a holographic grating. Proceedings of SPIE 2622, Optical Engineering Midwest '95. Chicago, IL, United States: SPIE, 1995.
122. Griffen, C. T., Chen, F. & Hung, Y. Y. Measurement of time-dependent displacement using dual speckle pattern phase-shifting TV holography. Proceedings of SPIE 2622, Optical Engineering Midwest '95. Chicago, IL, United States: SPIE, 1995.
123. Steinchen, W., Yang, L. X. & Kupfer, G. Vibration analysis by digital shearography. Proceedings of SPIE 2868, Second International Conference on Vibration Measurements by Laser Techniques: Advances and Applications. Ancona, Italy: SPIE, 1996.
124. Valera, J. D. R., Jones, J. D. C. & Løkberg, O. J. Exact vibration amplitude derivative measurement with TV shearography. *Measurement Science and Technology* **7**, 918-922 (1996).
125. Hung, Y. Y., Griffen, C. T. & Chen, F. High speed shearography for measuring transient deformation and vibration. Proceedings of the VIII International Congress on Experimental Mechanicals. Nashville, Tennessee, 1996.
126. Pedrini, G., Zou, Y. L. & Tiziani, H. J. Quantitative evaluation of digital shearing interferogram using the spatial carrier method. *Pure and Applied Optics:Journal of the European Optical Society Part A* **5**, 313-321 (1996).
127. Bisle, W. J., Scherling, D. & Tober, G. Phase stepping shearography for testing commercial aircraft structures: an application review of advanced image processing techniques for shearography. in Review of Progress in Quantitative Nondestructive Evaluation (eds Thompson, D. O. & Chimenti, D. E.) (Boston: Springer, 1996).
128. Santhanakrishnan, T. et al. Slope change contouring for 3D deeply curved objects by multi-aperture speckle shear interferometry. *Optik* **104**, 27-31 (1996).
129. Long, K. W. 3-Beam phase shift shearography for simultaneous measurement of in-plane and out-of-plane displacements and its applications to residual stress measurements. PhD thesis, Oakland University, 1996.
130. Burnett, M. & Bryanston-Cross, P. J. Measurements of transonic shock structures using shearography. Proceedings of SPIE 2861, Laser Interferometry VIII: Applications. Denver, CO, United States: SPIE, 1996.
131. Mohan, N. K., Saldner, H. O. & Molin, N. E. Recent applications of TV holography and shearography. Proceedings of SPIE 2861, Laser Interferometry VIII: Applications. Denver, CO, United States: SPIE, 1996.
132. Rabal, H., Henao, R. & Torroba, R. Digital speckle pattern shearing interferometry using diffraction gratings. *Optics Communications* **126**, 191-196 (1996).
133. Hung, Y. Y. & Wang, J. Q. Dual-beam phase shift shearography for measurement of in-plane strains. *Optics and Lasers in Engineering* **24**, 403-413 (1996).
134. Chen, L. X. Complementary use of TV-holography/shearography and ultrasonic techniques for NDE of fiber-reinforced plastics. Proceedings of SPIE 2944, Nondestructive Evaluation of Materials and Composites. Scottsdale, AZ, United States: SPIE, 1996.
135. Huang, J. R., Ford, H. D. & Tatam, R. P. Heterodyning of speckle shearing interferometers by laser diode wavelength modulation. *Measurement Science and Technology* **7**, 1721-1727 (1996).
136. Steinchen, W., Yang, L. X. & Schuth, M. TV-shearography for measuring 3D-strains. *Strain* **32**, 49-57 (1996).
137. Waldner, S. Removing the image-doubling in shearography by reconstruction of the displacement field. *Optics Communications* **127**, 117-126 (1996).
138. Waldner, S. Removing the image-doubling in shearography: theory and application. Proceedings of SPIE 2944, Nondestructive Evaluation of Materials and Composites. Scottsdale, AZ, United States: SPIE, 1996.
139. Hung, Y. Y. Shearography for non-destructive evaluation of composite Structures. *Optics and Lasers in Engineering* **24**, 161-182 (1996).
140. Rastogi, P. K. Measurement of curvature and twist of a deformed object by electronic speckle-shearing pattern interferometry. *Optics Letters* **21**, 905-907 (1996).
141. Huang, J. R., Ford, H. D. & Tatam, R. P. Phase-stepped speckle shearing interferometer by source wavelength modulation. *Optics Letters* **21**, 1421-1423 (1996).
142. Hung, Y. Y. Shearography and applications in experimental mechanics. Proceedings of SPIE 2921, International Conference on Experimental Mechanics: Advances and Applications. Singapore, Singapore: SPIE, 1996.
143. Weijers, A. L., van Brug, H. H. & Frankena, H. J. Real-time deformation measurement using a transportable shearography system. Proceedings of SPIE 2921, International Conference on Experimental Mechanics: Advances and Applications. Singapore, Singapore: SPIE, 1996.
144. Tay, C. J., Shang, H. M. & Choong, D. Triple-exposure shearography for the measurement of surface slopes. Proceedings of SPIE 2921, International Conference on Experimental Mechanics: Advances and Applications. Singapore, Singapore: SPIE, 1996.
145. Xie, L., Chau, F. S. & Toh, S. L. Nondestructive testing and evaluation

using phase-shifting electronic shearography. Proceedings of SPIE 2921, International Conference on Experimental Mechanics: Advances and Applications. Singapore, Singapore: SPIE, 1996.
146. Gundlach, A. et al. Speckle shearing interferometry using a diffractive optical beam splitter. *Optical Engineering* **36**, 1488-1493 (1997).
147. Wu, X. P., Fu, S. J. & Pang, L. Y. Special Ronchi grating as the shearing element of shearing ESPI for nondestructive testing. Proceedings of SPIE 2921, International Conference on Experimental Mechanics: Advances and Applications. Singapore, Singapore: SPIE, 1996.
148. Lee, H., Wang, L. S. & Krishnaswamy, S. Fourier processing of shearing interferometric fringe patterns. Proceedings of SPIE 2921, International Conference on Experimental Mechanics: Advances and Applications. Singapore, Singapore: SPIE, 1996.
149. Shu, Y., Chau, F. S. & Toh, S. L. Precision phase measurement in digital speckle shearing interferometry. Proceedings of SPIE 2921, International Conference on Experimental Mechanics: Advances and Applications. Singapore, Singapore: SPIE, 1996.
150. Sirohi, R. S. et al. Spatial phase shifting for pure in-plane displacement and displacement-derivative measurements in electronic speckle pattern interferometry (ESPI). *Applied Optics* **36**, 5787-5791 (1997).
151. Bison, P. G. et al. NDE of composite materials by thermal method and shearography. Proceedings of SPIE 3056, Thermosense XIX: an International Conference on Thermal Sensing and Imaging Diagnostic Applications. Orlando, FL, United States: SPIE, 1997.
152. Patorski, K. & Olszak, A. G. Digital in-plane electronic speckle pattern shearing interferometry. *Optical Engineering* **36**, 2010-2015 (1997).
153. Joenathan, C. & Buerkle, L. Electronic speckle pattern shearing interferometer using holographic gratings. *Optical Engineering* **36**, 2473-2477 (1997).
154. Dymny, G., Kujawinska, M. & Waldner, S. Modified electronic speckle pattern shearing interferometry for simultaneous derivative map measurements. Proceedings of SPIE 3098, Optical Inspection and Micromeasurements II. Munich, Germany: SPIE, 1997.
155. Steinchen, W. et al. Vibration analysis by digital speckle pattern shearing interferometry. Proceedings of SPIE 3098, Optical Inspection and Micromeasurements II. Munich, Germany: SPIE, 1997.
156. Rastogi, P. K. An electronic pattern speckle shearing interferometer for the measurement of surface slope variations of three-dimensional objects. *Optics and Lasers in Engineering* **26**, 93-100 (1997).
157. Petzing, J. N. & Tyrer, J. R. Analysis of Power ultrasonic components using shearing interferometry. *Optics and Lasers in Engineering* **26**, 235-248 (1997).
158. Hung, Y. Y., Wang, J. Q. & Hovanesian, J. D. Technique for compensating excessive rigid body motion in nondestructive testing of large structures using shearography. *Optics and Lasers in Engineering* **26**, 249-258 (1997).
159. Tay, C. J., Shang, H. M. & Choong, D. Cross influence of coordinate and slope related fringes during shearographic profiling. *Optics and Lasers in Engineering* **26**, 259-278 (1997).
160. Nakadate, S. Phase shifting speckle shearing polarization interferometer using a birefringent wedge. *Optics and Lasers in Engineering* **26**, 331-350 (1997).
161. Maas, A. A. M. & Somers, P. A. A. M. Two-dimensional deconvolution applied to phase-stepped shearography. *Optics and Lasers in Engineering* **26**, 351-360 (1997).
162. Valera, J. D. R. et al. Strain and vibration analysis by fibre based speckle shearing interferometry. *Optics and Lasers in Engineering* **26**, 361-376 (1997).
163. Tyrer, J. R. & Petzing, J. N. In-plane electronic speckle pattern shearing interferometry. *Optics and Lasers in Engineering* **26**, 395-406 (1997).
164. Aebischer, H. A. & Waldner, S. Strain distributions made visible with image-shearing speckle pattern interferometry. *Optics and Lasers in Engineering* **26**, 407-420 (1997).
165. Hung, Y. Y. Digital shearography versus TV-holography for non-destructive evaluation. *Optics and Lasers in Engineering* **26**, 421-436 (1997).
166. Mohan, N. K. & Sirohi, R. S. Fringe formation in symmetric three-aperture speckle shear interferometry: an analysis. *Optics and Lasers in Engineering* **26**, 437-447 (1997).
167. Ng, T. W. Optical distance sensing using digital speckle shearing interferometry. *Optics and Lasers in Engineering* **26**, 449-460 (1997).
168. Toh, S. L., Chau, F. S. & Sim, C. W. Theoretical basis for real-time shearography. *Journal of Modern Optics* **44**, 279-286 (1997).
169. Santhanakrishnan, T., Mohan, N. K. & Sirohi, R. S. Oblique observation speckle shear interferometers for slope change contouring. *Journal of Modern Optics* **44**, 831-839 (1997).
170. Sirohi, R. S. & Mohan, N. K. An in-plane insensitive multiaperture speckle shear interferometer for slope measurement. *Optics & Laser Technology* **29**, 415-417 (1997).
171. Hathaway, R. B., Hovanesian, J. D. & Hung, M. Y. Y. Residual stress evaluation using shearography with large-shear displacements. *Optics and Lasers in Engineering* **27**, 43-60 (1997).
172. Hung, M. Y. Y., Long, K. W. & Wang, J. Q. Measurement of residual stress by phase shift shearography. *Optics and Lasers in Engineering* **27**, 61-73 (1997).
173. Hung, Y. Y. Automated nondestructive shearographic inspection of debonds in composites using multi-frequency vibrational stressing. Proceedings of 1997 SEM Spring Conference. Bellevue, Washington, 1997.
174. Huang, J. R., Ford, H. D. & Tatam, R. P. Slope measurement by two-wavelength electronic shearography. *Optics and Lasers in Engineering* **27**, 321-333 (1997).
175. Fomitchov, P. A. & Krishnaswamy, S. A compact dual-purpose camera for shearography and electronic speckle-pattern interferometry. *Measurement Science and Technology* **8**, 581-583 (1997).
176. Steinchen, W. et al. Nondestructive testing of microcracks using digital speckle pattern shearing interferometry. Proceedings of SPIE 3098, Optical Inspection and Micromeasurements II. Munich, Germany: SPIE, 1997.
177. Song, Y. Z. et al. Digital shearing speckle interferometry applied to optical diagnostics in flow. Proceedings of SPIE 3172, Optical Technology in Fluid, Thermal, and Combustion Flow III. San Diego, CA, United States: SPIE, 1997.
178. Hung, Y. Y. & Shi, D. H. Technique for rapid inspection of hermetic seals of microelectronic packages using shearography. *Optical Engineering* **37**, 1406-1409 (1998).
179. Murukeshan, V. M., Seng, O. L. & Asundi, A. Polarization phase shifting shearography for optical metrological applications. *Optics & Laser Technology* **30**, 527-531 (1998).
180. Hung, Y. Y. Computerized shearography and its application for nondestructive evaluation of composites. in Manual on Experimental Methods of Mechanical Testing of Composites (ed. Jenkins, C. H.) (Society for Experimental Mechanics, 1998).
181. Santhanakrishnan, T., Palanisamy, P. K. & Sirohi, R. S. Optical configuration in speckle shear interferometry for slope change contouring with a twofold increase in sensitivity. *Applied Optics* **37**, 3447-3449 (1998).
182. Lu, G. W., Bard, B. A. & Wu, S. D. Real-time portable phase-stepping shearography system for NDE. Proceedings of SPIE 3397, Nondestructive Evaluation of Aging Aircraft, Airports, and Aerospace Hardware II. San Antonio, TX, United States: SPIE, 1998.
183. Bard, B. A., Gordon, G. A. & Wu, S. D. Laser-modulated phase-stepping digital shearography for quantitative full-field imaging of ultrasonic waves. *The Journal of the Acoustical Society of America* **103**, 3327-

3335 (1998).
184. Walz, T. & Ettemeyer, A. Automatic shearography inspection system for helicopter rotor blades. Proceedings of SPIE 3397, Nondestructive Evaluation of Aging Aircraft, Airports, and Aerospace Hardware II. San Antonio, TX, United States: SPIE, 1998.
185. Steinchen, W. et al. Developmental steps for double-pulse shearography. Proceedings of SPIE 3478, Laser Interferometry IX: Techniques and Analysis. San Diego, CA, United States: SPIE, 1998.
186. Steinchen, W. et al. Application of laser diodes in digital speckle pattern shearing interferometry. Proceedings of SPIE 3415, Laser Diodes and Applications III. Quebec, Canada: SPIE, 1998.
187. Steinchen, W. et al. Digital shearography for strain measurement: an analysis of measuring errors. Proceedings of SPIE 3479, Laser Interferometry IX: Applications. San Diego, CA, United States: SPIE, 1998.
188. Steinchen, W. et al. Strain analysis by means of digital shearography: potential, limitations and demonstration. *The Journal of Strain Analysis for Engineering Design* **33**, 171-182 (1998).
189. Steinchen, W. et al. Non-destructive testing of aerospace composite materials using digital shearography. *Proceedings of the Institution of Mechanical Engineers, Part G:Journal of Aerospace Engineering* **212**, 21-30 (1998).
190. Joenathan, C. et al. Novel temporal Fourier transform speckle pattern shearing interferometer. *Optical Engineering* **37**, 1790-1795 (1998).
191. Shang, H. M., He, Y. M. & Tay, C. J. Digital shearography for slope measurement. Proceedings of SPIE 3407, International Conference on Applied Optical Metrology. Balatonfured, Hungary: SPIE, 1998.
192. Sirohi, R. S. et al. An Optical Strain Gauge. Proceedings of SPIE 3407, International Society for Optical Engineering. SPIE, 1998.
193. Sirohi, R. S. et al. Shear ESPI with small objects. Proceedings of SPIE 3407, International Conference on Applied Optical Metrology. Balatonfured, Hungary: SPIE, 1998.
194. Fernandez, J. L. et al. Double-pulsed-carrier speckle-shearing pattern interferometry for transient deformation analysis. Proceedings of SPIE 3478, Laser Interferometry IX: Techniques and Analysis. San Diego, CA, United States: SPIE, 1998.
195. Wong, W. O. & Chan, K. T. Measurement of modal damping by electronic speckle shearing interferometry. *Optics & Laser Technology* **30**, 113-120 (1998).
196. Dávila, A., Kaufmann, G. H., & Pérez-López, C. Transient deformation analysis by a carrier method of pulsed electronic speckle-shearing pattern interferometry. *Applied Optics* **37**, 4116-4122 (1998).
197. Yang, L. X. et al. Vibration analysis by means of digital shearography. *Optics and Lasers in Engineering* **30**, 199-212 (1998).
198. van Brug, H. Temporal phase unwrapping and its application in shearography systems. *Applied Optics* **37**, 6701-6706 (1998).
199. Sirohi, R. S. et al. Nondestructive assessment of thinning of plates using digital shearography. *Optical Engineering* **38**, 1582-1585 (1999).
200. He, Y. M., Tay, C. J. & Shang, H. M. Digital phase-shifting shearography for slope measurement. *Optical Engineering* **38**, 1586-1590 (1999).
201. Kästle, R., Hack, E. & Sennhauser, U. Multiwavelength shearography for quantitative measurements of two-dimensional strain distributions. *Applied Optics* **38**, 96-100 (1999).
202. van Brug, H. H. Real-time speckle shearography system for defect detection in aircraft materials. Proceedings of SPIE 3586, Nondestructive Evaluation of Aging Aircraft, Airports, and Aerospace Hardware III. Newport Beach, CA, United States: SPIE, 1999.
203. Siebert, T. & Schmitz, B. New shearing setup for simultaneous measurement of two shear directions. Proceedings of SPIE 3637, Practical Holography XIII. San Jose, CA, United States: SPIE, 1999.
204. James, S. W. & Tatam, R. P. Time-division-multiplexed 3D shearography. Proceedings of SPIE 3744, Interferometry '99: Techniques and Technologies. Pultusk Castle, Poland: SPIE, 1999.
205. Murukeshan, V. M., Ganesan, A. R. & Sirohi, R. S. Curvature measurement using double shear TV holography. *Optik* **110**, 57-60 (1999).
206. Waldner, S. & Brem, S. Compact shearography system for the measurement of 3D deformation. Proceedings of SPIE 3745, Interferometry '99: Applications. Pultusk Castle, Poland: SPIE, 1999.
207. Osten, W., Kalms, M. K. & Jueptner, W. P. O. Some ways to improve the recognition of imperfections in large-scale components using shearography. Proceedings of SPIE 3745, Interferometry '99: Applications. Pultusk Castle, Poland: SPIE, 1999.
208. Hung, Y. Y. et al. Evaluation of Residual Stresses in Plastics and Composites by Shearography. (SAE International Congress and Exposition, 1999).
209. Groves, R. M., James, S. W. & Tatam, R. P. Polarization-multiplexed and phase-stepped fiber optic shearography using laser wavelength modulation. Proceedings of SPIE 3745, Interferometry '99: Applications. Pultusk Castle, Poland: SPIE, 1999.
210. Kalms, M. K. et al. NDT on wide-scale aircraft structures with digital speckle shearography. Proceedings of SPIE 3824, Optical Measurement Systems for Industrial Inspection. Munich, Germany: SPIE, 1999.
211. Hung, Y. Y. Applications of digital shearography for testing of composite structures. *Composites Part B:Engineering* **30**, 765-773 (1999).
212. Steinchen, W. et al. Determination of strain distribution by means of digital shearography. *Measurement* **26**, 79-90 (1999).
213. Elster, C. & Weingärtner, I. Solution to the shearing problem. *Applied Optics* **38**, 5024-5031 (1999).
214. Parker, S. C. J. & Salter, P. L. A novel shearography system for aerospace non-destructive testing. *Proceedings of the Institution of Mechanical Engineers, Part G:Journal of Aerospace Engineering* **213**, 23-33 (1999).
215. Osten, W. et al. Shearography system for the testing of large-scale aircraft components taking into account noncooperative surfaces. Proceedings of SPIE 4101, Laser Interferometry X: Techniques and Analysis. San Diego, CA, United States: SPIE, 2000.
216. Andersson, A. et al. TV shearography: quantitative measurement of shear-magnitude fields by use of digital speckle photography. *Applied Optics* **39**, 2565-2568 (2000).
217. Hung, Y. Y. et al. Evaluating the soundness of bonding using shearography. *Composite Structures* **50**, 353-362 (2000).
218. Shang, H. M. et al. Surface profiling using shearography. *Optical Engineering* **39**, 23-31 (2000).
219. Mohan, N. K. The influence of multiple-exposure recording on curvature pattern using multi-aperture speckle shear interferometry. *Optics Communications* **186**, 259-263 (2000).
220. Fernandez, A. et al. Measurement of transient out-of-plane displacement gradients in plates using double-pulsed subtraction TV shearography. *Optical Engineering* **39**, 2106-2113 (2000).
221. Dilhaire, S. et al. Measurement of the thermomechanical strain of electronic devices by shearography. *Microelectronics Reliability* **40**, 1509-1514 (2000).
222. Shang, H. M. et al. Generation of carrier fringes in holography and shearography. *Applied Optics* **39**, 2638-2645 (2000).
223. Groves, R. M., James, S. W. & Tatam, R. P. Polarization-multiplexed and phase-stepped fibre optic shearography using laser wavelength modulation. *Measurement Science and Technology* **11**, 1389-1395 (2000).
224. Wang, K. F., Tieu, A. K. & Li, E. B. Influence of in-plane displacement and double-aperture orientation on slope fringe formation in

double-shearing-aperture speckle interferometry. *Opt. Eng. Optical Engineering* **39**, 2124 (2000).

225. Lau, B., Kronthaler, T. & Schilling, R. A microprism array as shearing device for speckle shearing interferometry. *Optics and Lasers in Engineering* **36**, 389-396 (2001).

226. Chen, F. Digital shearography: state of the art and some applications. *Journal of Electronic Imaging* **10**, 240-251 (2001).

227. Rastogi, P. K. Measurement of static surface displacements, derivatives of displacements, and three-dimensional surface shapes—examples of applications to non-destructive testing. in Digital Speckle Pattern Interferometry and Related Techniques (ed Rastogi, P. K.) (New York: Wiley, 2001).

228. Groves, R. M., James, S. W. & Tatam, R. P. Full surface strain measurement using shearography. Proceedings of SPIE 4448, Optical Diagnostics for Fluids, Solids, and Combustion. San Diego, CA, United States: SPIE, 2001.

229. Murukeshan, V. M. et al. Double shearography for engineering metrology: optical and digital approach. *Optics & Laser Technology* **33**, 325-328 (2001).

230. Martínez-Celorio, R. A. et al. Visibility enhancement of carrier fringes in Electronic Speckle Shearing Pattern Interferometry using microspheres for light detection in back reflection. *Optik* **112**, 99-104 (2001).

231. Tornari, V. et al. Laser-based systems for the structural diagnostic of artwork: an application to XVII-century Byzantine icons. Proceedings of SPIE 4402, Laser Techniques and Systems in Art Conservation. Munich, Germany: SPIE, 2001.

232. Osten, W. et al. Progress with the implementation of a shearography system for the testing of technical components. Proceedings of SPIE 4900, Seventh International Symposium on Laser Metrology Applied to Science, Industry, and Everyday Life. Novosibirsk, Russian Federation: SPIE, 2002.

233. Kalms, M. K., Osten, W. & Jueptner, W. P. O. Advanced shearographic system for nondestructive testing of industrial and artwork components. Proceedings of SPIE 4915, Lasers in Material Processing and Manufacturing. Shanghai, China: SPIE, 2002.

234. Wong, W. O. A simple electronic speckle shearing interferometer. *Optics & Laser Technology* **34**, 399-403 (2002).

235. Wang, K. F., Tieu, A. K. & Li, E. B. Influence of displacement and its first- and second-order derivative components on curvature fringe formations in speckle shearography. *Applied Optics* **41**, 4557-4561 (2002).

236. Steinchen, W., Kupfer, G. & Mäckel, P. Full field tensile strain shearography of welded specimens. *Strain* **38**, 17-26 (2002).

237. Steinchen, W. & Yang, L. X. Digital Shearography: Theory and Application of Digital Speckle Pattern Shearing Interferometry. (Bellingham: SPIE Press, 2003).

238. Falldorf, C., Osten, W. & Kolenovic, E. Speckle shearography using a multiband light source. *Optics and Lasers in Engineering* **40**, 543-552 (2003).

239. Kalms, M. K. & Osten, W. Mobile shearography system for the inspection of aircraft and automotive components. *Optical Engineering* **42**, 1188-1196 (2003).

240. Hung, Y. Y., Shang, H. M. & Yang, L. X. Unified approach for holography and shearography in surface deformation measurement and nondestructive testing. *Optical Engineering* **42**(5), 1197-1207 (2003).

241. Kim, K. S. et al. Analysis of an internal crack of pressure pipeline using ESPI and shearography. *Optics & Laser Technology* **35**, 639-643 (2003).

242. Groves, R. M., James, S. W. & Tatam, R. P. Multicomponent shearography using optical fiber imaging-bundles. Proceedings of SPIE 5144, Optical Measurement Systems for Industrial Inspection III. Munich, Germany: SPIE, 2003.

243. Krupka, R., Waltz, T. & Ettemeyer, A. Industrial applications of shearography for inspections of aircraft components. Proceedings of SPIE 5144, Optical Measurement Systems for Industrial Inspection III. Munich, Germany: SPIE, 2003.

244. Groves, R. M., James, S. W. & Tatam, R. P. Multi-component pulsed-laser shearography using optical fiber imaging-bundles. Proceedings of SPIE 5191, Optical Diagnostics for Fluids, Solids, and Combustion II. San Diego, California, United States: SPIE, 2003.

245. Groves, R. M., James, S. W. & Tatam, R. P. Pipe weld investigation using shearography. *Strain* **39**, 101-105 (2003).

246. Santos, F., Vaz, M. & Monteiro, J. A new set-up for pulsed digital shearography applied to defect detection in composite structures. *Optics and Lasers in Engineering* **42**, 131-140 (2004).

247. Mihaylova, E. et al. Electronic speckle pattern shearing interferometer with a photopolymer holographic grating. *Applied Optics* **43**, 2439-2442 (2004).

248. Groves, R. M., James, S. W. & Tatam, R. P. Shape and slope measurement by source displacement in shearography. *Optics and Lasers in Engineering* **41**, 621-634 (2004).

249. Kurtz, R. M. et al. Reflection shearography for nondestructive evaluation. Proceedings of SPIE 5422, Unmanned Ground Vehicle Technology VI. Orlando, Florida, United States: SPIE, 2004.

250. Findeis, D. & Gryzagoridis, J. A comparison of the capabilities of portable shearography and portable electronic speckle pattern interferometry. Proceedings of SPIE 5393, Nondestructive Evaluation and Health Monitoring of Aerospace Materials and Composites III. San Diego, CA, United States: SPIE, 2004.

251. Casillas, F. J. et al. Small amplitude estimation of mechanical vibrations using electronic speckle shearing pattern interferometry. *Optical Engineering* **43**, 880 (2004).

252. Schuth, M., Vössing, F. & Yang, L. X. A shearographic endoscope for nondestructive test. *Journal of Holography and Speckle* **1**, 46-52 (2004).

253. Sujatha, N. U. & Murukeshan, V. M. Nondestructive inspection of tissue/tissue like phantom curved surfaces using digital speckle shearography. *Optical Engineering* **43**, 3055-3060 (2004).

254. Wang, K. F. & Tieu, A. K. Theory and experiment of spatially and temporally partially coherent speckle shearing interferometry. *Optics & Laser Technology* **36**, 43-45 (2004).

255. Mihaylova, E., Whelan, M. & Toal, V. Simple phase-shifting lateral shearing interferometer. *Optics Letters* **29**, 1264-1266 (2004).

256. Barrientos, B. et al. Measurement of out-of-plane deformation by combination of speckle photography and speckle shearing interferometry. *Optik* **115**, 248-252 (2004).

257. Mäckel, P. Die Scherografie-Ein quantitatives Messverfahren zur Schwingungsmessung und zerstörungsfreien Prüfung. *Laser Technik Journal* **1**, 49-54 (2004).

258. Yang, L. X. et al. Digital shearography for nondestructive testing: potentials, limitations, and applications. *Journal of Holography and Speckle* **1**, 69-79 (2004).

259. Udupa, G. et al. Defect detection in unpolished Si wafers by digital shearography. *Measurement Science and Technology* **15**, 35-43 (2004).

260. Andhee, A., Gryzagoridis, J. & Findeis, D. Comparison of normal and phase stepping shearographic NDE. Proceedings of SPIE 5767, Nondestructive Evaluation and Health Monitoring of Aerospace Materials, Composites, and Civil Infrastructure IV. San Diego, CA, United States: SPIE, 2005.

261. Kalms, M. & Jueptner, W. Mobile shearography. Proceedings of SPIE 5852, Third International Conference on Experimental Mechanics and

Third Conference of the Asian Committee on Experimental Mechanics. Singapore: SPIE, 2004.

262. Sun, P. Digital phase-shifting shearography for strain measurement by using a rotating platform system. *Optical Engineering* **44**, 085601 (2005).

263. Tay, C. J. & Fu, Y. Determination of curvature and twist by digital shearography and wavelet transforms. *Optics Letters* **30**, 2873-2875 (2005).

264. Abdullah, W. S. W. & Petzing, J. N. Development of speckle shearing interferometer error analysis as an aperture function of wavefront divergence. *Journal of Modern Optics* **52**, 1495-1510 (2005).

265. Hung, Y. Y. & Ho, H. P. Shearography: an optical measurement technique and applications. *Materials Science and Engineering:R:Reports* **49**, 61-87 (2005).

266. Habib, K. Thermally induced deformations measured by shearography. *Optics & Laser Technology* **37**, 509-512 (2005).

267. Somers, P. A. A. M. & Bhattacharya, N. Maintaining sub-pixel alignment for a single-camera two-bucket shearing speckle interferometer. *Journal of Optics A:Pure and Applied Optics* **7**, S385-S391 (2005).

268. Groves, R. M. et al. Single-axis combined shearography and digital speckle photography instrument for full surface strain characterization. *Optical Engineering* **44**, 025602 (2005).

269. Cordero, R. R. & Labbe, F. Uncertainty evaluation of displacement gradients measured by electronic speckle pattern shearing interferometry (ESPSI). *Measurement Science and Technology* **16**, 1315-1321 (2005).

270. Bhaduri, B., Mohan, N. K. & Kothiyal, M. P. A dual-function ESPI system for the measurement of out-of-plane displacement and slope. *Optics and Lasers in Engineering* **44**, 637-644 (2006).

271. Bhaduri, B. et al. Use of spatial phase shifting technique in digital speckle pattern interferometry (DSPI) and digital shearography (DS). *Optics Express* **14**, 11598-11607 (2006).

272. Groves, R. M. & Osten, W. Temporal phase measurement methods in shearography. Proceedings of SPIE 6341, Speckle06: Speckles, From Grains to Flowers. Nimes, France: SPIE, 2006

273. Zhao, S. A. & Chung, P. S. Digital speckle shearing interferometer using a liquid-crystal spatial light modulator. *Optical Engineering* **45**, 105606 (2006).

274. Mihaylova, E. et al. Photopolymer diffractive optical elements in electronic speckle pattern shearing interferometry. *Optics and Lasers in Engineering* **44**, 965-974 (2006).

275. Bhaduri, B., Mohan, N. K. & Kothiyal, M. P. Cyclic-path digital speckle shear pattern interferometer: use of polarization phase-shifting method. *Optical Engineering* **45**, 105604 (2006).

276. Quan, C., Fu, Y. & Miao, H. Wavelet analysis of digital shearing speckle patterns with a temporal carrier. *Optics Communications* **260**, 97-104 (2006).

277. Viotti, M. R. et al. A portable digital speckle pattern interferometry device to measure residual stresses using the hole drilling technique. *Optics and Lasers in Engineering* **44**, 1052-1066 (2006).

278. Růžek, R., Lohonka, R. & Jironč, J. Ultrasonic C-Scan and shearography NDI techniques evaluation of impact defects identification. *NDT & E International* **39**, 132-142 (2006).

279. Groves, R. M. et al. Full-field Laser Shearography Instrumentation for the Detection and Characterization of Fatigue Cracks in Titanium 10-2-3. *Journal of ASTM International* **3**, 12757 (2006).

280. Ochoa, N. A. & Silva-Moreno, A. A. Fringes demodulation in time-averaged digital shearography using genetic algorithms. *Optics Communications* **260**, 434-437 (2006).

281. Yang, L. X. Recent developments of digital shearography for NDT. *Material Evaluation* **64**, 704-709 (2006).

282. Pauliat, G. & Roosen, G. Continuous monitoring of a surface slope by real-time shearing interferometry with a photorefractive crystal. *Applied Optics* **45**, 993-999 (2006).

283. Bhaduri, B., Mohan, N. K. & Kothiyal, M. P. (1, N) spatial phase-shifting technique in digital speckle pattern interferometry and digital shearography for nondestructive evaluation. *Optical Engineering* **46**, 051009 (2007).

284. Somers, P. A. A. M. & Bhattacharya, N. Handling unfavourable polarization states in a polarization-based shearing speckle interferometer. *Journal of Optics A:Pure and Applied Optics* **9**, S92-S97 (2007).

285. Francis, D., James, S. W. & Tatam, R. P. Surface strain measurement using multi-component shearography with coherent fibre-optic imaging bundles. *Measurement Science and Technology* **18**, 3583-3591 (2007).

286. Anand, A. et al. Fresnel wavefront propagation model for shearography shape measurement. Proceedings of SPIE 6617, Modeling Aspects in Optical Metrology. Munich, Germany: SPIE, 2007.

287. Falldorf, C. et al. Efficient reconstruction of spatially limited phase distributions from their sheared representation. *Applied Optics* **46**, 5038-5043 (2007).

288. Hung, Y. Y. et al. Review and comparison of shearography and pulsed thermography for adhesive bond evaluation. *Optical Engineering* **46**, 051007 (2007).

289. Groves, R. M. et al. Surface strain measurement: a comparison of speckle shearing interferometry and optical fibre Bragg gratings with resistance foil strain gauges. *Measurement Science and Technology* **18**, 1175-1184 (2007).

290. Groves, R. M. et al. Shearography as part of a multi-functional sensor for the detection of signature features in movable cultural heritage. Proceedings of SPIE 6618, O3A: Optics for Arts, Architecture, and Archaeology. Munich, Germany: SPIE, 2007.

291. Bhaduri, B., Kothiyal, M. P. & Mohan, N. K. Curvature measurement using three-aperture digital shearography and fast Fourier transform. *Optics and Lasers in Engineering* **45**, 1001-1004 (2007).

292. Huang, S. J. & Liu Y. F. The out-of-plane strain measurement of composite sandwich plate with fully-potted insert using digital phase-shifting shearography. *ICCES* **4**, 187-193 (2007).

293. Kalms, M. Mobile shearography in applications. Proceedings of SPIE 6762, Two- and Three-Dimensional Methods for Inspection and Metrology V. Boston, MA, United States: SPIE, 2007.

294. Bhaduri, B., Mohan, N. K. & Kothiyal, M. P. Simultaneous measurement of out-of-plane displacement and slope using a multiaperture DSPI system and fast Fourier transform. *Applied Optics* **46**, 5680-5686 (2007).

295. Rosso, V. et al. Almost-common path interferometer using the separation of polarization states for digital phase-shifting shearography. *Optical Engineering* **46**, 105601 (2007).

296. Groves, R. M., Pedrini, G. & Osten, W. Real-time extended dynamic range imaging in shearography. *Applied Optics* **47**, 5550-5556 (2008).

297. Gryzagoridis, J. & Findeis, D. Benchmarking shearographic NDT for composites. *Insight* **50**, 249-252 (2008).

298. Findeis, D., Gryzagoridis, J. & Musonda, V. NDT detection and quantification of induced defects on composite helicopter rotor blade and UAV wing sections. Proceedings of SPIE 7155, Ninth International Symposium on Laser Metrology. Singapore, Singapore: SPIE, 2008.

299. Rosso, V. et al. Simultaneous coherent imaging and strain measurement using coupled photorefractive holography and shearography. *Optics Letters* **33**, 797-799 (2008).

300. Francis, D. Surface strain measurement using pulsed laser shearography with fibre-optic imaging bundles. PhD thesis, Cranfield University, UK, 2008.
301. Focke, O., Hildebrand, A. & von Kopylow, C. Inspection of laser generated lamb waves using shearographic interferometry. Proceedings of the 1st International Symposium on Laser Ultrasonics: Science, Technology and Applications. Montreal, Canada, 2008.
302. Francis, D., James, S. W. & Tatam, R. P. Surface strain measurement of rotating objects using pulsed laser shearography with coherent fibre-optic imaging bundles. *Measurement Science and Technology* **19**, 105301 (2008).
303. Lee, J. R. et al. Investigation of shear distance in Michelson interferometer-based shearography for mechanical characterization. *Measurement Science and Technology* **19**, 115303 (2008).
304. Huang, S. J. & Lin, Y. T. Out-of-plane strain measurement in sandwich plates with single fully potted insert by using digital shearography. *Strain* **44**, 253-258 (2008).
305. Sun, C. R., Chen, J. L. & Lu, H. Improved phase-shifted digital speckle shearography for time-dependent deformation measurement. *Optical Engineering* **47**, 065601 (2008).
306. Groves, R. M. et al. Multi-sensor evaluation of a wooden panel painting using terahertz imaging and shearography. Proceedings of SPIE 7391, O3A: Optics for Arts, Architecture, and Archaeology II. Munich, Germany: SPIE, 2009.
307. Mihaylova, E. & Toal, V. Simple versatile shearing interferometer suitable for measurements on a microscopic scale. *Optics and Lasers in Engineering* **47**, 271-273 (2009).
308. Hung, Y. Y. et al. Review and comparison of shearography and active thermography for nondestructive evaluation. *Materials Science and Engineering:R:Reports* **64**, 73-112 (2009).
309. Kumar, U. P., Kothiyal, M. P. & Mohan, N. K. Microscopic TV shearography for characterization of microsystems. *Optics Letters* **34**, 1612-1614 (2009).
310. Huang, Y. H. et al. NDT&E using shearography with impulsive thermal stressing and clustering phase extraction. *Optics and Lasers in Engineering* **47**, 774-781 (2009).
311. Groves, R. M. et al. 2D and 3D non-destructive evaluation of a wooden panel painting using shearography and terahertz imaging. *NDT & E International* **42**, 543-549 (2009).
312. Groves, R. M. et al. 2.5D virtual reality visualisation of shearography strain data from a canvas painting. Proceedings of SPIE 7391, O3A: Optics for Arts, Architecture, and Archaeology II. Munich, Germany: SPIE, 2009.
313. Bhaduri, B., Tay, C. J. & Quan, C. G. Direct measurement of curvature and twist using two-channel double-aperture digital shearography. *Optical Engineering* **49**, 033604 (2010).
314. Francis, D., Tatam, R. P. & Groves, R. M. Shearography technology and applications: a review. *Measurement Science and Technology* **21**, 102001 (2010).
315. Goto, D. T. & Groves, R. M. Error analysis of 3D shearography using finite-element modelling. Proceedings of SPIE 7718, Optical Micro- and Nanometrology III. Brussels, Belgium: SPIE, 2010.
316. Habib, K. Failure of thin films: optical shearography versus electrochemical impedance spectroscopy. *Optik* **121**, 1317-1323 (2010).
317. Wu, S. J., He, X. Y. & Yang, L. X. Enlarging the angle of view in Michelson-interferometer-based shearography by embedding a 4*f* system. *Applied Optics* **50**, 3789-3794 (2011).
318. Dhanotia, J. et al. Slope measurement of bent plates using double grating shearing interferometry. *Applied Optics* **50**, 2958-2963 (2011).
319. Taillade, F. et al. Shearography and pulsed stimulated infrared thermography applied to a nondestructive evaluation of FRP strengthening systems bonded on concrete structures. *Construction and Building Materials* **25**, 568-574 (2011).
320. Wu, S. J. et al. Digital shearography with in situ phase shift calibration. *Optics and Lasers in Engineering* **50**, 1260-1266 (2012).
321. Frade, M., Enguita, J. M. & Álvarez, I. In situ 3D profilometry of rough objects with a lateral shearing interferometry range finder. *Optics and Lasers in Engineering* **50**, 1559-1567 (2012).
322. Fu, Y., Guo, M. & Liu, H. Determination of instantaneous curvature and twist by digital shearography. *Optical Engineering* **51**, 083602 (2012).
323. Cai, C. Q. & He, L. F. Improved Mach-Zehnder interferometer-based shearography. *Optics and Lasers in Engineering* **50**, 1699-1705 (2012).
324. Zhu, L. Q. et al. Real-time monitoring of phase maps of digital shearography. *Optical Engineering* **52**, 101902 (2013).
325. Xie, X. et al. Michelson interferometer based spatial phase shift shearography. *Applied Optics* **52**, 4063-4071 (2013).
326. Xie, X. et al. Simultaneous measurement of deformation and the first derivative with spatial phase-shift digital shearography. *Optics Communications* **286**, 277-281 (2013).
327. Zastavnik, F. et al. Comparison of shearography to scanning laser vibrometry as methods for local stiffness identification of beams. *Strain* **50**, 82-94 (2014).
328. Zhang, Y., Li, T. & Li, Q. L. Defect detection for tire laser shearography image using curvelet transform based edge detector. *Optics & Laser Technology* **47**, 64-71 (2013).
329. Blain, P. et al. An in-line shearography setup based on circular polarization gratings. *Optics and Lasers in Engineering* **51**, 1053-1059 (2013).
330. Lopes, H. et al. Localization of damage with speckle shearography and higher order spatial derivatives. *Mechanical Systems and Signal Processing* **49**, 24-38 (2014).
331. Sirohi, R. S. Digital shear speckle pattern interferometry with holo-elements. *Asian Journal of Physics* **23**, 509-515 (2014).
332. Zastavnik, F. et al. Errors in shearography measurements due to the creep of the PZT shearing actuator. *Measurement Science and Technology* **25**, 085007 (2014).
333. Xu, N. et al. Shearography for specular object inspection. *Optics and Lasers in Engineering* **61**, 14-18 (2014).
334. Xu, X. et al. Identification of flaws using digital shearography. *Lasers in Engineering* **28**, 201-211 (2014).
335. Huke, P., Burke, J. & Bergmann, R. B. A comparative study between deflectometry and shearography for detection of subsurface defects. Proceedings of SPIE 9203, Interferometry XVII: Techniques and Analysis. San Diego, California, United States: SPIE, 2014.
336. Vandenrijt, J. F., Lièvre, N. & Georges, M. P. Improvement of defect detection in shearography by using principal component analysis. Proceedings of SPIE 9203, Interferometry XVII: Techniques and Analysis. San Diego, California, United States: SPIE, 2014.
337. Zhu, L. Q., Wu, S. J. & Yang, L. X. Stroboscopic digital shearographic system for vibration analysis of large-area object. *Instruments and Experimental Techniques* **57**, 493-498 (2014).
338. Khaleghi, M. et al. Long-term effects of cyclic environmental conditions on paintings in museum exhibition by laser shearography. in Advancement of Optical Methods in Experimental Mechanics, Volume 3 (eds Jin, H. et al.) (Cham: Springer, 2014), 283-288.
339. Krzemień, L. et al. Combining digital speckle pattern interferometry with shearography in a new instrument to characterize surface delamination in museum artefacts. *Journal of Cultural Heritage* **16**, 544-550 (2015).
340. Amar, S. et al. Digital carrier superposition by Hilbert-Huang transform for optical phase recovery in speckle shearing interferometry. *Optical Engineering* **54**, 013101 (2015).

341. Feng, Z. et al. A polarized digital shearing speckle pattern interferometry system based on temporal wavelet transformation. *Review of Scientific Instruments* **86**, 093102 (2015).
342. Falldorf, C., Klattenhoff, R. & Bergmann, R. B. Single shot lateral shear interferometer with variable shear. *Optical Engineering* **54**, 054105 (2015).
343. Bai, P. X., Zhu, F. P. & He, X. Y. Out-of-plane displacement field measurement by shearography. *Optics & Laser Technology* **73**, 29-38 (2015).
344. Zastavnik, F. et al. Calibration and correction procedure for quantitative out-of-plane shearography. *Measurement Science and Technology* **26**, 045201 (2015).
345. Xie, X. et al. Measurement of in-plane strain with dual beam spatial phase-shift digital shearography. *Measurement Science and Technology* **26**, 115202 (2015).
346. Buchta, D. et al. Artwork inspection by shearography with adapted loading. *Experimental Mechanics* **55**, 1691-1704 (2015).
347. Schindler, J. et al. Retrieving the axial position of fluorescent light emitting spots by shearing interferometry. *Journal of Biomedical Optics* **21**, 125009 (2016).
348. Languy, F. et al. Vibration mode shapes visualization in industrial environment by real-time time-averaged phase-stepped electronic speckle pattern interferometry at 10.6 μm and shearography at 532 nm. *Optical Engineering* **55**, 121704 (2016).
349. Deepan, B., Quan, C. G. & Tay, C. J. Determination of slope, curvature, and twist from a single shearography fringe pattern using derivative-based regularized phase tracker. *Optical Engineering* **55**, 121707 (2016).
350. Richoz, G. L. & Schajer, G. S. Simultaneous two-axis shearographic interferometer using multiple wavelengths and a color camera. *Optics and Lasers in Engineering* **77**, 143-153 (2016).
351. Wang, Y. H. et al. Simultaneous dual directional strain measurement using spatial phase-shift digital shearography. *Optics and Lasers in Engineering* **87**, 197-203 (2016).
352. Mininni, M. et al. Damage identification in beams using speckle shearography and an optimal spatial sampling. *Mechanical Systems and Signal Processing* **79**, 47-64 (2016).
353. Yang, L. X. & Xie, X. Digital Shearography: New Developments and Applications. (Bellingham: SPIE Press, 2016).
354. Xie, X. et al. Polarized digital shearography for simultaneous dual shearing directions measurements. *Review of Scientific Instruments* **87**, 083110 (2016).
355. dos Santos, J. V. A., Lopes H. & Moreno-García, P. Numerical study on damage identification using shearography with different shearing amounts. *Procedia Structural Integrity* **5**, 1198-1204 (2017).
356. Lopes, H. et al. Influence of shearing amount and vibration amplitude on noise in shearography. *Procedia Structural Integrity* **5**, 1205-1212 (2017).
357. Falldorf, C. et al. Sparse light fields in coherent optical metrology [Invited]. *Applied Optics* **56**, F14-F19 (2017).
358. Wang, X. et al. Temporal heterodyne shearing speckle pattern interferometry. *Optics and Lasers in Engineering* **93**, 76-82 (2017).
359. Sun, J. F. et al. Dynamic measurement of first-order spatial derivatives of deformations by digital shearography. *Instruments and Experimental Techniques* **60**, 575-583 (2017).
360. Zhao, Q. H. et al. Digital shearography for NDT: phase measurement technique and recent developments. *Applied Sciences* **8**, 2662 (2018).
361. da Silva, F. A. A. et al. Evaluation of a novel compact shearography system with DOE configuration. *Optics and Lasers in Engineering* **104**, 90-99 (2018).
362. Macedo, F. J. et al. Inspection of defects of composite materials in inner cylindrical surfaces using endoscopic shearography. *Optics and Lasers in Engineering* **104**, 100-108 (2018).
363. Newman, J. W. Shearography nondestructive testing of composites. *Comprehensive Composite Materials II* **7**, 270-290 (2018).
364. Buchta, D. et al. Soft tissue elastography via shearing interferometry. *Journal of Medical Imaging* **5**, 046001 (2018).
365. Barrera, E. S. et al. Multiple-aperture one-shot shearography for simultaneous measurements in three shearing directions. *Optics and Lasers in Engineering* **111**, 86-92 (2018).
366. Kirkove, M. et al. Determination of vibration amplitudes from binary phase patterns obtained by phase-shifting time-averaged speckle shearing interferometry. *Applied Optics* **57**, 8065-8077 (2018).
367. Buchta, D. et al. Combination of FEM simulations and shearography for defect detection on artwork. *Strain* **54**, e12269 (2018).
368. Liu, H. J. et al. Directed acoustic shearography for crack detection around fastener holes in aluminum plates. *NDT & E International* **100**, 124-131 (2018).
369. Wang, X. et al. Application of digital shearing speckle pattern interferometry for thermal stress. *Measurement* **125**, 11-18 (2018).
370. Wang, S. J. et al. Dual-directional shearography based on a modified common-path configuration using spatial phase shift. *Applied Optics* **58**, 593-603 (2019).
371. Yan, P. Z. et al. Shearography for non-destructive testing of specular reflecting objects using scattered light illumination. *Optics & Laser Technology* **112**, 452-457 (2019).
372. Yan, P. Z. et al. Spatial phase-shift digital shearography for simultaneous measurements in three shearing directions based on adjustable aperture multiplexing. *Optical Engineering* **58**, 054105 (2019).
373. DeWeert, M. J. et al. Principles of surface-phase-resolved shearography. *Optical Engineering* **58**, 114103 (2019).
374. Katunin, A., Lopes, H. & dos Santos, J. V. A. Identification of multiple damage using modal rotation obtained with shearography and undecimated wavelet transform. *Mechanical Systems and Signal Processing* **116**, 725-740 (2019).
375. Sun, F. Y. et al. The application of SLM in shearography detecting system. *Optics and Lasers in Engineering* **114**, 90-94 (2019).
376. Arco, C. et al. Application of spline interpolation to speckle shearography measurements for damage identification. *Procedia Structural Integrity* **17**, 718-725 (2019).
377. Hooshmand-Ziafi, H., Hassani, K. & Dashtdara, M. Dual-sensitive spatial phase-shifting shearography based on a common-path configuration. *Optical Engineering* **58**, 114104 (2019).
378. Zhao, Q. H. et al. Simultaneous 3D measurement of deformation and its first derivative with speckle pattern interferometry and shearography. *Applied Optics* **58**, 8665-8672 (2019).
379. Dong, J. et al. Real-time dual-sensitive shearography for simultaneous in-plane and out-of-plane strain measurements. *Optics Express* **27**, 3276-3283 (2019).
380. Gao, X. Y. et al. Double imaging Mach-Zehnder spatial carrier digital shearography. *Journal of Modern Optics* **66**, 153-160 (2019).
381. Yan, P. Z. et al. Pixelated carrier phase-shifting shearography using spatiotemporal low-pass filtering algorithm. *Sensors* **19**, 5185 (2019).
382. Hooshmand-Ziafi, H. et al. Common-path spatial phase-shift speckle shearography using a glass plate. *Review of Scientific Instruments* **90**, 105105 (2019).
383. Giordano, A. & Chiang, F. P. Digital laser speckle technique for generating deflection, slope, and curvature contours of bent plates. *Optical Engineering* **59**, 094101 (2020).
384. Dong, J. et al. Shear-unlimited common-path speckle interferometer. *Optics Letters* **45**, 1305-1308 (2020).
385. Sun, F. Y. et al. A spatial-phase-shift-based defect detection

shearography system with independent adjustment of shear amount and spatial carrier frequency. *Optics & Laser Technology* **124**, 105956 (2020).
386. Zhong, S. M. et al. Multi-directional shearography based on multiplexed Mach-Zehnder interference system. *Journal of Modern Optics* **67**, 346-354 (2020).
387. Oliveira, B. C. F. et al. Square-pulse shearography inspections of metallic parts repaired with a glass fiber reinforced polymer using pressure, radiation, vibration, and induction loading methods. *International Journal of Pressure Vessels and Piping* **187**, 104187 (2020).
388. Felipe, D. S. et al. Procedure for detecting the shape and size of defects on metallic substrates under composite repairs using shearography. *Applied Optics* **59**, 8089-8097 (2020).
389. Zhang, B. Y. et al. Overview of digital shearography for NDT. *Materials Evaluation* **78**, 342-351 (2020).
390. Zhang, B. Y. et al. Spatial-light-modulator-based dual shearing direction shearography. *Applied Optics* **59**, 11080-11086 (2020).
391. Zhao, Q. H. et al. A new multiplexed system for the simultaneous measurement of out-of-plane deformation and its first derivative. *Optics Communications* **482**, 126602 (2021).
392. Gu, G. Q. et al. Improved depth characterization of internal defect using the fusion of shearography and speckle interferometry. *Optics & Laser Technology* **135**, 106701 (2021).
393. Anisimov, A. G. & Groves, R. M. Extreme shearography: development of a high-speed shearography instrument for quantitative surface strain measurements during an impact event. *Optics and Lasers in Engineering* **140**, 106502 (2021).

Official journal of the JHL 2689-9620
www.light-am.com

Review Open Access

55 years of holographic non-destructive testing and experimental stress analysis: is there still progress to be expected?

Wolfgang Osten and Giancarlo Pedrini*

Abstract

Holographic methods for non-destructive testing, shape measurement, and experimental stress analysis have shown to be versatile tools for the solution of many inspection problems. Their main advantages are the non-contact nature, the non-destructive and areal working principle, the fast response, high sensitivity, resolution and precision. In contrast to conventional optical techniques such as classical interferometry, the holographic principle of wavefront storage and reconstruction makes it possible to investigate objects with rough surfaces. Consequently, the response of various classes of products on operational or artificial load can be examined very elegantly. The paper looks back to the history of holographic metrology, honors the inventors of the main principles, discusses criteria for the selection of a proper inspection method, and shows exemplary applications. However, the main focus is on modern developments that are inspired by the rapid technological process in sensing technology and digitization, on current applications and future challenges.

Keywords: Non-destructive testing, Holography, Stress analysis, Digital holography, Speckle, interferometry

Introduction

In 1965, Karl Stetson and his colleague Robert Powell discovered the potential of holography for interferometry[1,2]. In contrast to classical interferometry, the holographic principle of wavefront storage and reconstruction made it possible to investigate objects with rough surfaces. The following years were filled with the implementation of various techniques such as the real-time technique, the double-exposure technique, the time-average and stroboscopic technique[3]. An impressive overview about the early progress in implementation and application of principles of holographic nondestructive testing was given in 1974 by Robert K. Erf[4]. Shortly after the discovery of holographic interferometry, the speckle noise, originally considered a blemish for the quality of the holograms, was recognized as a further powerful tool for non-destructive testing[5]. Y.Y. Hung's combination of shearing interferometry with speckle correlation opened the door for coherent metrology to step out of the laboratory and to bring its big potential to the factory floor[6-8]. Another book from Robert K. Erf reports in 1978 also about the great potential of speckle metrology[9]. The holographic tire test[10], which has long served as the most representative industrial application for HNDT, has now been supplemented by numerous other applications with high practical relevance. Early examples are Niels Abramsons holographic study of the deformation of two milling machines during their operation[11], the optimization of the car body with respect to vibration excitation using pulsed laser holography by Armin Felske and Alfons Happe[12], the vibration study of a railway bridge excited by a running train by Hans

Correspondence: Giancarlo Pedrini (pedrini@ito.uni-stuttgart.de)
Institute for Applied Optics ITO University Stuttgart Pfaffenwaldring 9, 70569 Stuttgart, Germany

Steinbichler[13], and the diagnostics of paintings with holographic methods by Salvatore Amadesi et al[14]. Further early applications such as the measurement of material parameters, residual stress measurement, vibration analysis, plasma diagonstics, shape measurement, and fault detection can be found in[4,9, 15-17].

Methods of holographic non-destructive testing, experimental stress analysis and shape measurement have all the advantages provided by optical principles: the non-contact and high speed interaction with the object under test, the scalability of the dimension of the probing tool, the high resolution of the data, the simultaneous availability of numerous information channels and the flexible adaptability of the comparative standard – the wavelength. On the other hand the user is still confronted with a number of serious challenges. Two of the biggest challenges that currently attract high attention in both the technical as well as life sciences are: exceeding the physical limits of resolution and improve the precision of the measurement. Therefore, these methods are subject to constant improvement. The characteristics that give rise to improve the performance of the systems are obviously dependent on the purpose of the measurement and the object under test. However, there are also general features that can be used to assess the performance of a measurement system. Here we refer to its spatial and temporal resolution, its area related resolution, the precision and trueness of the achieved measurement data, the robustness of the system against external influences, its degree of automation, and its ability to work as close as possible to the process to be inspected.

In this contribution, we discuss briefly general and application dependent features for the assessment and selection of optical measurement systems. Afterwards, we show several recent applications where holographic principles are applied with advantage and where the sensors are inspired by digital technologies. Finally, we list some challenges for HNDT that are currently matter of investigation.

Some criteria for selecting a special measurement technology

To write about the significance of metrology in our high-tech world would mean to bring owls to Athens. Metrology and quality are two sides of the same coin and high quality standards are a must for the majority of manufacturers in all industrial branches. The well-known slogan "*If you can't measure it, you can't improve it*", attributed to Lord Kelvin, belongs to the most-important quotes in business and expresses clearly the position that decisions without a reliable data basis are fuzzy. For modern production technologies this phrase can be simply rewritten as following: "*You cannot produce anything better as you can measure it*". Many adaptions of this wisdom are circulating meanwhile, such as "*If you can't measure it, you can't make it*" or "*If you can't measure it, you can't manage it*"①. The most pertinent expression for this causality is in our opinion probably the one attributed to the American statistician W. Edwards Deming who famously said: "*In God we trust, all others must bring data.*" Generating, collecting and interpreting data seems to be more important than ever. In times of all-encompassing digitization, IoT-mania, social networks, and mobile communication, the term Big Data refers mainly to the use of predictive analytics, user behavior analytics, or certain other advanced data analytics methods that extract value from a data system to find new correlations and to spot business trends, prevent diseases, combat crime and so on[18,19]. The analysis of this trend is by no means the subject of our article.

Instead, we focus on data acquisition and evaluation in a very limited environment called optical metrology. The situation here is on the one hand quite different as in the social-economic context of the above mentioned Big Data hype. On the other hand there are some similarities which could be of interest for our topic. In both cases a user or customer, respectively, is involved. While in the first case the client is largely passive and the system examines his behavior in the context of a larger community of acting individuals to come up with suggestions and conclusions, the client in our case is the active player who has a concrete problem (mainly to assess the quality of items or processes) and looks for the best way to solve it. In both cases, the generated data are the primary decision-making basis. These decisions are the subject of the analysis of an inverted causality that is aimed to draw conclusions from the achieved results – the measurement data - to their causes (faults, shapes, deformations, strains, ...), see Fig. 1. The difficulties of the so-called identification problems are: instability, ambiguity, sensitivity to noise, and lack of independent information[20,21].

Metrology is not matter of wishful thinking but only matter of facts. This holds especially for optical metrology where we can usually see what we will get. But the belief in pictures is often seductive and the transformation of optical data (intensities, phases, times of flight, ...) into e.g. dimensional data is not even simple due to the mentioned indirect measuring process. Consequently, the discussion

① Peter Drucker (1909-2005) Austrian-born American management consultant, educator, and author, whose writings contributed to the philosophical and practical foundations of the modern business corporation.

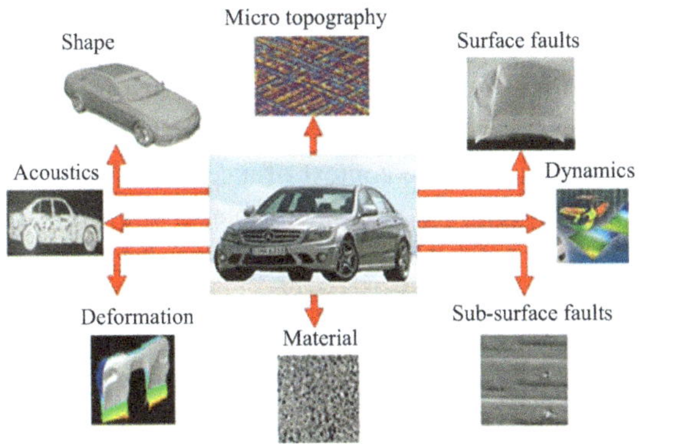

Fig. 1 Critical features and quantities which can be measured effectively by holographic techniques (Reproduced from Ref. 38)

of reliable procedures and measures for the evaluation of the performance of measurement systems and the trueness of the derived data may help to improve the trust in optical measurement systems. However, the subject is far from being simple and an exhaustive treatment of that topic needs more substantial investigations and would fill many pages. Several important players, components and procedures are part of the chain from the given object under test and measurement problem to the selection and task-dependent design of a proper measurement system delivering confiding data. Fig. 2 tries to illustrate the strategy of that selection process.

The user or customer has a specific problem where he needs insights about certain properties of the current sample. Usually he discusses that problem with an expert who transforms the global and often fuzzy task into a specification sheet. This specification sheet contains more or less precise information about the data to be measured directly, the principles and devices used for their measurement, the algorithms executing the identification of the best estimate of the desired data – the identification problem which in the epistemological sense represents a question to the causality[22], and the methods for the estimation of the reliability of the final results. This is in general a straightforward procedure but strictly dependent from the specific measurement problem. There are many recipes and guidelines for the measurement of any data, and for the calculation, presentation and assessment of the final data which quantify the quality of the object under test as well. See for instance the ISO/BIPM Guide for the Expression of Uncertainty in Measurement[23] which is a valuable tool for the estimation of the reliability of the provided data. However, the evaluation of these procedures and guidelines is not the subject of our paper. Our intention is more directed to the discussion of features and measures that can be helpful for the selection and design of a specific measurement system and finally for the estimation and improvement of its performance.

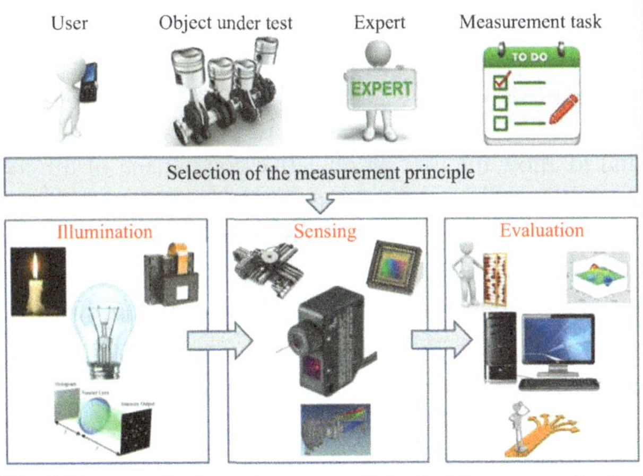

Fig. 2 Players and schematic strategy of the decision process for the task-dependent design of a measurement system: The user has a specific problem concerning the object under test, having a specification sheet he/she consults an expert who is proposing a task dependent system specification and the main components of an optical measurement system: Illumination (light source, beam shaper, light modulator), Sensing (sensor, positioning system, imaging system, detector), Evaluation (analog-digital-conversion, data processing algorithms, software implementation, graphic presentation, decision making) (Reproduced from Ref. 38)

Here we distinguish between general features that can be helpful to characterize the global quality of a measurement system and its 3 main components (see Fig. 2: Illumination, sensing including the selection of the adequate sensors and their path planning[24], and the evaluation of the measurement data including their error analysis), and those features that are characteristic for a particular application. For the latter, the following list of specifications should be considered:

– the object under test, its intended use and use conditions as well,

– the quantities or features to be measured, such as dimensional, mechanical, spectral and kinematic quantities, material parameters, material faults, ...,

– the measuring environment (cooperative or non-cooperative/harsh environment[25]),

– the type of interaction with the object, e.g. the appropriate measuring arrangement, the appropriate load of the object and the choice of an appropriate wavelength or sequence of wavelengths[26,27],

– the proper user interface (human-machine-interface[28], graphical user interface GUI, and programming language[29]),

– the process capability of the measurement system (the process capability is a statistical measure that indicates the probability with which the specified goals can be achieved, the so-called statistical process control SPC[30]),

– the measurement volume and/or the measurement area,

– the surface quality of the object (rough, optically smooth, translucent, …),

– the required measurement speed (e.g. the need to synchronize the measurement with the process timing for in-line measurement processes, or an off-line system decoupled from the process cycle following its own clock rate),

– the required measurement uncertainty[31] (this statistical quantity for the estimation of the precision/repeatability of the results is strongly dependent on the measurement conditions and the properties of the measurement object; a rule of thumb says: the accepted uncertainty has to be smaller as a tenths of the tolerances that can be accepted by the system),

– the way of data presentation and subsequent use of data (quantitative, qualitative, lists, graphics, further processing, …).

There is obviously no all-in-one device suitable for every purpose and consequently there is no general rule for the selection and design of a specific measurement system. The variety of objects, measurement conditions and measurement purposes is far too big. Consequently, it is the responsibility of the expert to select the system that is most appropriate for a specific purpose. However, there are several general rules for the design and features for the assessment of optical measurement systems which can be used to objectively judge its performance and to find ways to improve it. Richard Leach et al have proposed 5 general rules for the design of the measurement pipeline in order to achieve better metrological performance, shorter measurement times, and smaller consumption of resources. They call it *Information-Rich Manufacturing Metrology* (IRM)[32]:

– use information on nominal part geometry,

– use information on manufactured part geometry,

– use information from other measurement,

– use information on measurement process,

– use any other source of additional information.

Additional information is always useful to cope with the ill-posedness of the inverse problem[33]. A successful way for the regularization is the exploitation of the complete information content of a light wave[21]:

– primary information channels: intensity, frequency, phase, angular spectrum, and polarization,

– angular momentum of light: light spin and light orbital angular momentum,

– time of flight (light in flight),

– correlation properties in the light field: temporal and spatial coherence.

These channels can be processed as individual information carriers such as in interferometry where the phase is reconstructed by several intensity measurements using phase-shifting technologies[34], or by a multi-modal data acquisition in the context of a sensor fusion strategy[35]. For the inspection of extended objects, the limited space bandwidth product of existing sensors is often a handicap for the high-resolution detection of existing faults. A smart combination of various techniques allows the skillful fusion of sequential measurements acquired in different scales and consequently a more effective way to find these faults without a time-consuming scan of the entire object with a single high-resolution method[36,37,20].

If one takes into account the big diversity of quantities that can be measured with holographic principles (see Fig. 1) and the always ubiquitous challenge to find the best method suitable for the solution of the concrete problem, then it will be clear that the user needs assistance to make the best possible choice[38]. The understanding of the features of a measurement system, their origin and influence on the quality of the results is the first step on the way to an objective judgement of the data and finally to a system with outstanding performance. Apparently, there is no universally valid and simple recipe for this. This is because the multiplicity of measurement tasks and objects does not allow a generalized rule. Nevertheless, there are meanwhile some serious attempts to tackle this challenge. Here we would particularly like to highlight the initiative "*The Fair Data Sheet*"[39] and the German research project OptAssyst[40]. The Fair Data Sheet aims to contribute to more uniformity in definition and use of features for the characterization of the performance of optical measurement devices for surface metrology, to understand data sheets and to allow the user an objective comparison of different devices and measurement technologies. That is a good basis. However, a number of new questions arise, which are directed to the selection and description of the characteristic features and to the design of future systems.

Alongside the "Fair Data Sheet"[39] some further valuable seminal work should be mentioned: Häusler and Ettl have published an article entitled "About the clever selection and application of optical sensors"[41]. The authors list valuable indicators that should be considered for the selection of a particular sensor. To them belong factors that influence the measurement uncertainty, measures to estimate the information content delivered by the special sensor, and requirements for the surface quality of the measurement object. Peter de Groot et al propose to apply the so-called Instrument Transfer Function (ITF)[42,43] as an object independent measure to determine the limitations of the optical metrology instruments and extend this concept in[44] to the 3D-ITF. They clearly point out that the optical resolution of a surface profiler is in general object-dependent. However, referring to the ITF it is possible to define regimes where its cut-off frequency enables metrics about the capability of a given instrument. In[38] we discussed several measures that can be helpful for the estimation of the performance of an optical measurement system and its components. Furthermore, a system can be qualified by its degree of modularity, flexibility, user-friendliness, robustness against environmental influences, multi-modality performance (property to measure data of various modalities, such as dimensional and kinematic data), consumption of resources, ease of integration, and miniaturization. Established parameters for the assessment of the illumination source are their radiometric and coherence properties (spectral bandwidth, spectral power, coherence length, source dimension, noise level...).

Recent Applications of Holographic and Speckle Techniques for Nondestructive Testing, Experimental Stress Analysis and Shape Measurement

By keeping an eye on the guidelines for the selection of a proper measurement technology discussed in the previous section, we present here recent applications of holographic techniques for the inspection of aircraft and automotive parts, micro-electro-mechanical systems (MEMS), spray-coating processes, erosion monitoring in harsh environment, and artwork inspection. The quantities that are measured directly with holographic and speckle techniques are intensities and phases, while thereby displacements, strains, residual stresses, material faults and shape data are reconstructed.

Inspection of aircraft and automotive parts with digital speckle shearing techniques

Carbon fiber technology (CFRP – Carbon Fiber Reinforced Plastics) and other lightweight materials such as GLARE (Glass Laminate Aluminum Reinforced Epoxy) are used more and more for airplane and automotive parts. For instance modern airplanes are already equipped with such components in the body, the vertical and horizontal stabilizer, the rudder, airbrakes and spoilers. However, the application of new materials is also accompanied by new challenges to an optimal dimensioning. The premature detection of possible damages due to material and construction imperfections is therefore inevitable to guarantee the strength and safety in long-term operation. For this reason, the use of adapted inspection equipment and procedures is a must especially in the phase of construction and testing of prototypes.

Shearography[6-8,13], is an approved and powerful tool for the non-destructive investigation of technical components with respect to material faults and structural imperfections. It is a robust interferometric technique for the detection of damaged regions by applying an appropriate load to the object. Both material and structural imperfections can be recognized as typical fringe patterns in the shearogram. The method is based on the digital correlation of two speckled wave fronts representing two states of the object under test such as the unloaded and the loaded one. A special quality of shearography is the so-called self-reference of each interferometrically sensitive speckle pattern. By shearing of the wavefronts that are scattered from the illuminated object, one of the wavefronts serves as reference for the other. Consequently, the method is very robust against environmental influences and especially rigid body motions of the object have no influence on the speckle correlation. In contrast to holographic interferometry, speckle shearing allows the use of several mutually incoherent light sources for the illumination of larger areas[8].

Considering the system specifications discussed in section 2 and taking into account that large scale technical objects have in general a kind of non-cooperative surface (shiny, curved, varying thickness, changing materials and paints, changing reflectivity, ...), then a careful design of the following system components is needed: the shearography sensor, the illumination unit, the loading system and the evaluation software. When developing a mobile and modular system for aircraft and automotive inspection, these components have to be customized with respect to the boundary conditions in these application fields.

The choice of the appropriate load is of crucial importance for the capability of the inspection system to detect faults under the surface up to a depth of several mm. In our system mainly thermal and mechanical loading are applied[45]. Mechanical loading is performed by changing the pressure within the test environment. To this purpose a self-adhesive vacuum chamber was developed, Fig. 3a. This unit has two pneumatic cycles: one for the load-pressure in the vacuum chamber and the other for a non-destructive grip at the object under test. Furthermore, thermal loading is done with the help of two infrared lamps, Fig. 3b. The connection of that loading devices with the system control of the PC ensures an adaptive variation of the pressure in the vacuum-chamber or the heating of the inspected area by the operator.

The developed system was tested extensively on various aircraft components on the factory floor at Airbus Bremen and in a maintenance facility. A typical test object is the

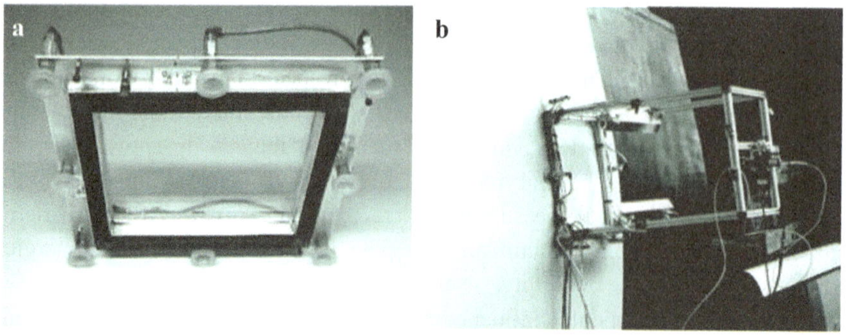

Fig. 3 Loading components of the inspection system for digital speckle shearography. **a** Self-adhesive vacuum chamber for pressure load **b** Self-adhesive thermal loading unit (Reproduced from Ref. 8).

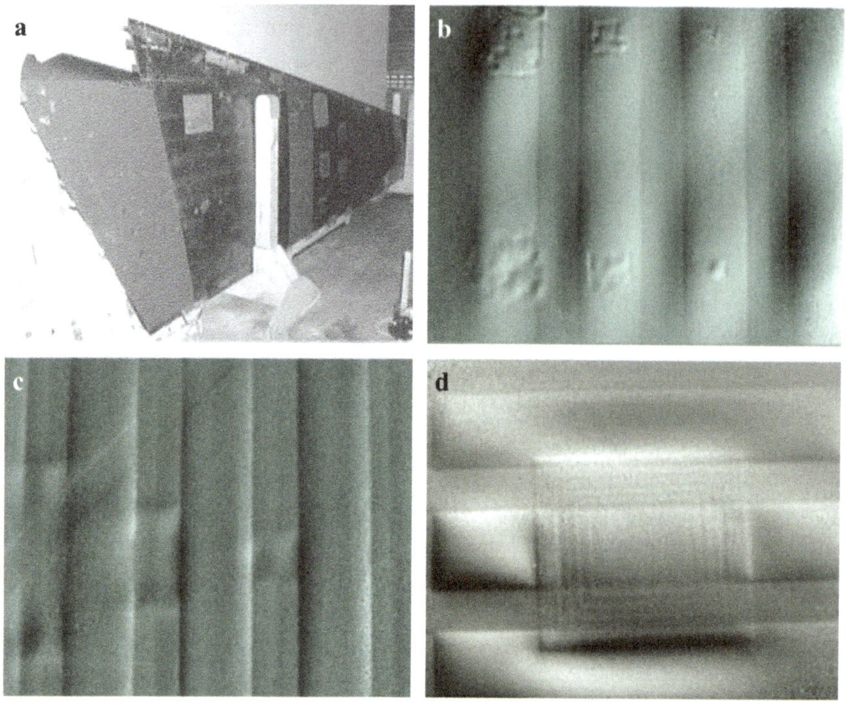

Fig. 4 Part of the inspected Airbus A 320 stabilizer at the Airbus Germany factory in Bremen with some inspection results. (Reproduced from Ref. 8). **a** Inspected CFRP stabilizer **b** Delaminations of the skin **c** Stringer debonding **d** Repaired region.

horizontal stabilizer of the Airbus A320, Fig. 4a. The stabilizer was tested having both the rough CFRP-surface and the painted surface. The inspection goal was the detection of separations of adhesive bonded stringers behind the CFRP-skin with a thickness between 3 mm and 8 mm. Fig. 4a, b show a CFRP-composite structure that was reinforced with stringers and ribs. This shell structure was prepared with skin delaminations and stringer debondings of different sizes. The detected delaminations are in the region of 50 mm, 25 mm and 10 mm edge length, Fig. 4b. The stringer separations have a length of 200 mm, 100 mm und 50 mm, Fig. 4c. Fig. 4d shows the inspection result of a repaired part in the region above two stringers. The different layers are clearly observable.

The very sensitive detection of the faults and their clear appearance in the delivered final shearography images could be demonstrated by a comparison with two other inspection techniques: ultrasonics and thermography. For the verification of the performance of test procedures, aircraft components are prepared with faults of different size and location. Here the delaminated stringer area in the A320 body was investigated using thermal load. Fig. 5a shows the fault image delivered by ultrasonics and Fig. 5b shows the same fault detected with a thermography system. It is obvious to see that the structure of the fault can be resolved more clearly using shearography, Fig. 5c,d.

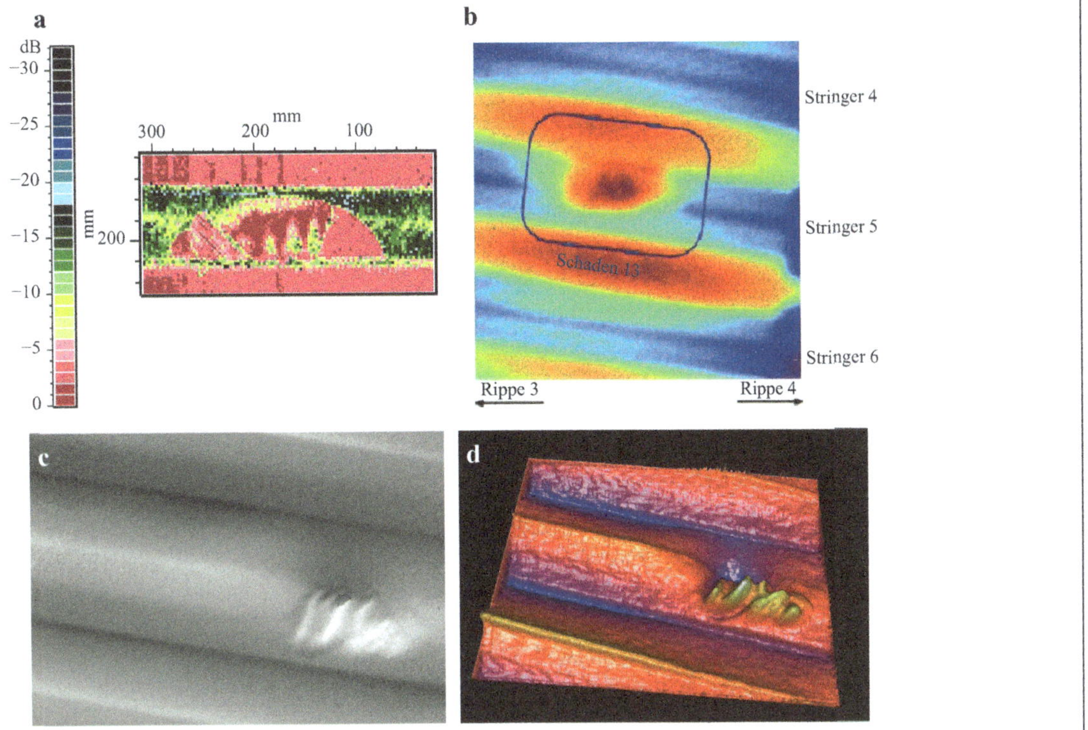

Fig. 5 Inspection of a damage in the test sample of the A320 body by ultrasonics, thermography and shearography. (Reproduced from Ref. 8). **a** Inspection result made with ultrasonics **b** Inspection result made with thermography **c** Inspection result made with the shearography system **d** VRML presentation of the shearographic result.

One important objective for the design of modern cars is to reduce the fuel consumption drastically. Light weight structures made from CFRP are also here an essential part of a bunch of measures. Fig. 6 shows the results of the inspection of a CFRP cross bar member. The cross member was prepared with some internal imperfections to be identified by shearography. For a sensitive investigation the cross member was loaded thermally with infrared radiation and subdivided into 5 inspection areas. The inspection of a single section only takes few seconds. After evaluation, the results corresponding to the 5 inspection areas were stitched together electronically. Fig. 6 gives a clear impression of all internal imperfections of the cross bar.

Taking into account the full-field access, the robust working mode and the fast delivery of the results (the inspection of a 500 mm × 500 mm area takes only a few seconds), shearography recommends itself as powerful industrial inspection technique.

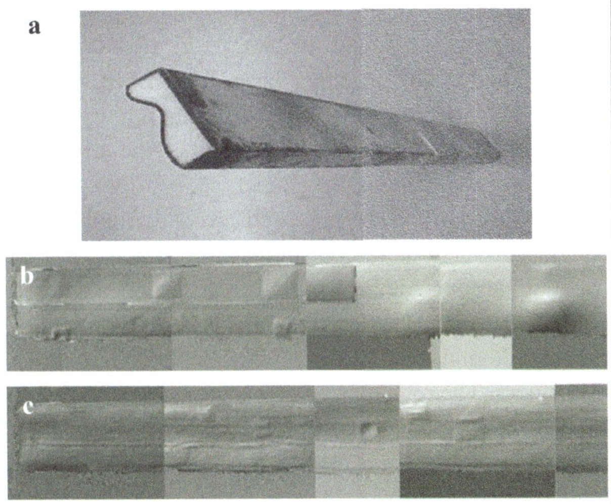

Fig. 6 Inspection of a CFRP automotive cross member (length 100 cm, width 10 cm, depth 4 cm) using shearography (Reproduced from Ref. 8). **a** Cross member **b** Inspection result of the front side **c** Inspection result of the back side.

Inspection of aircraft components using infrared illumination

Aircrafts and automotive components need usually to be investigated in an in-situ industrial environment with strong vibrations, airflows and high temperatures gradients. These conditions make the measurement difficult in particular when interferometric based methods like digital holography (DH) or electronic speckle pattern interferometry (ESPI) are used for NDT of mechanical structures and dynamic analysis. Shearography (see section

3.1) allows to reduce the sensitivity to the perturbation by using a self-referencing arrangement which measure the gradients of the deformations (translations do not influence the measurements). However, for non-destructive testing, aircraft or automotive components are usually exposed to strong loading producing quite large deformations of the structure (e.g. 100 μm). Therefore for measuring such large deformations, the conventional optical systems using visible light are too sensitive. A way to make the system less sensitive to the harsh environmental condition and at the same time measure large deformations, is to use a longer wavelength (e.g. wavelength of 10 μm). In this case ESPI or DH are less sensitive to the perturbations and may be applied. There are many publications describing results obtained by ESPI and DH using visible light but few using long-wavelength infrared. The reason is that its implementation requires a particular equipment. Lasers with long wavelength and the required coherence are available (e.g. CO_2 lasers) but need special care in their use. Infrared detectors with an increasing number of pixels were developed and uncooled micro-bolometer arrays with megapixel format are available. Long wavelengths require optical elements (lenses, beamsplitters) usually made of germanium or zinc selenide. It is expected that the infrared techniques for metrological measurements will develop in the next years since with the growth of the thermal-imaging market, detectors and optical components and detectors can be found in stock at affordable cost. Interferometric NDT systems based on ESPI, DH and shearography were built using the 10.6 μm wavelength emitted by a CO_2 laser[46,47]. Fig. 7 shows the ESPI setup using long wavelength, the geometrical configuration is in practice the same as the one used for visible light but the components are for infrared radiation. It includes a CO_2 laser and an uncooled micro-bolometer array camera with 640 × 480 pixels. The laser beam is split in two parts, the first one illuminates the object, which scatters light towards the camera. The reference beam is coupled into the camera through an injection setup.

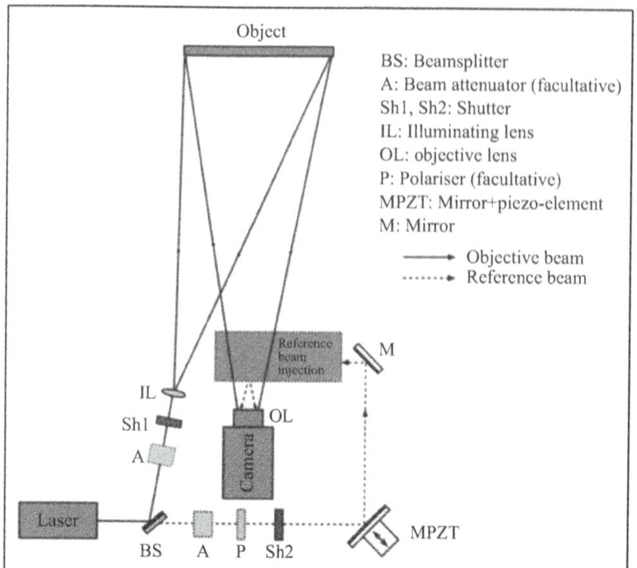

Fig. 7 Scheme of the ESPI setup using long (10.6 μm) wavelength (Reproduced from Ref. 46).

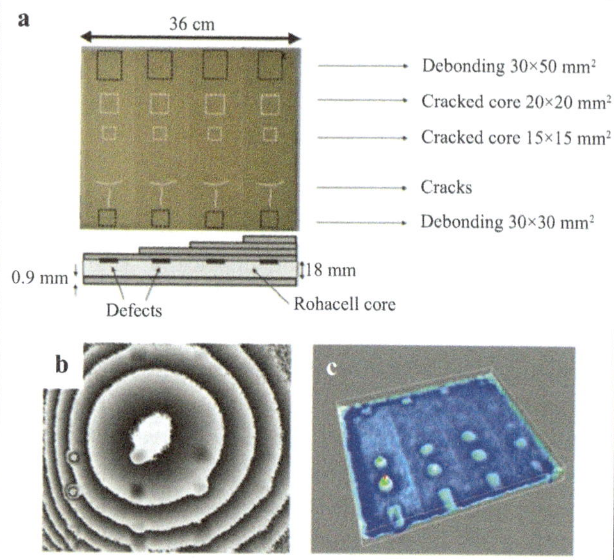

Fig. 8 **a** Definition of the sample used in the experiment, **b** Phase map modulo 2π obtained after heating composite samples. **c** Defect map after postprocessing the phase map (Reproduced from Ref. 46).

Composite samples (glass-fiber reinforced plastic (GFRP), carbon-fiber-reinforced plastic (CFRP), and Kevlar) with various layers and defects of different natures and depths underneath the surface were investigated with the long wavelength ESPI setup. These are the aeronautical components that usually need to be tested on the production site for the assessment of mechanical behavior (displacement measurement) under loads. Fig. 8a shows one of the samples used for the investigations, it has a size of 36 × 30 cm^2 and is composed by a 18 mm thick Rohacell core sandwiched between two Kevlar skins. Defects arranged in four vertical zones with increasing thickness were included between the upper skin and the core. Locations and shapes of the defects are marked on the picture: each zone has debondings (black dots), cracked core (grey dots), and surface cracks (grey lines). The surface roughness of the sample was larger than the wavelength therefore fully developed speckles were produced. Thermal stress using a lamp was applied to the sample and its deformation was measured with the ESPI system shown in Fig. 7. Phase-shifted specklegrams are at first recorded before heating the sample. Then the lamp is switched on for a few tens of seconds. Afterwards new phase-shifted specklegrams were recorded. The phases of the wavefronts recorded before and after loading are calculated from the specklegrams and their differences are computed. Fig. 8b shows the phase map corresponding to the surface deformation. There are circular fringes corresponding to the global deformation of the sample, whereas defects can be seen as local fringe patterns in some places. Fig. 8c shows a result obtained after unwrapping the phase map shown in Fig. 8b, and subtracting the global deformation. The remainder is the contribution of local deformation induced by defects which can be clearly visualized.

Monitoring of spray-coating processes by digital holography

Ceramic coatings are commonly used to improve the wear or heat resistance of many technical components[48], but due to their deposition process, e.g. plasma or high velocity oxygen fuel spraying, rather high residual stresses can build up within the coating and underneath. The reason for that are differences in the coatings and substrates expansion coefficients, inhomogeneous temperature distribution during the process and the quenching of splats. The mechanical hole drilling technique[49] can be used for the detection of residual stresses in coatings, which are locally relieved due to the material removal process. This lead to deformations of the surface around the hole which are measured as relaxed strains through strain gauges rosettes. These deformations in combination with appropriate calibration data (separately determined by simulation for the layer composite), allows the quantitative determination of the residual stress depth profile. The disadvantage of the strain gauges is that they can only be used on flat and relatively smooth surfaces, where the rosette is applied.

In Refs. 50, 51 a method to avoid the mechanical drilling and the application of strain gauges was proposed. A pulsed laser is used for the object machining (ablation process) leading to 3D deformation due to residual stress relaxation which are measured by an optical system based on digital holographic interferometry. The residual stresses at different depth of the coating are calculated from the deformations obtained after incremental loading, the profile (shape, depth) of the machined surface and the material parameters. It was shown in Ref. 51 that the technique can be used for determining residual stresses under industrial conditions. The Atmospheric Plasma Spraying (APS) is placed on a robot arm that can move in the space between the optical measuring system and the object to be coated. The distance between the sample and the measuring system is approximately 0.8 m. The APS and the measuring system are inside a cabin that is hermetically closed during the coating process. The robot for coating and the optical systems are shown in Fig. 9. The purple light is due to the high temperature plasma glow. During the coating process the temperature of the object rises up to 400 ℃. The set-up for laser machining and deformation measurement shown in Fig. 9a includes an infrared pulsed laser (wavelength: 1064 nm, pulse length 10 ns, maximal pulse energy 700 mJ) and a digital holographic system using two cw laser (wavelength 532 nm). The sample is illuminated at the same time from two directions and there are also two reference waves. Multiplexing of holograms allows synchronous measurement of in-plane and out-of-plane deformations. Fig. 10 shows residual stresses inside a coating measured with the holographic method (from 30 μm to 130 μm) and the traditional micro hole drilling method (HDM). In spite of the fact that the DHI measurements seem to

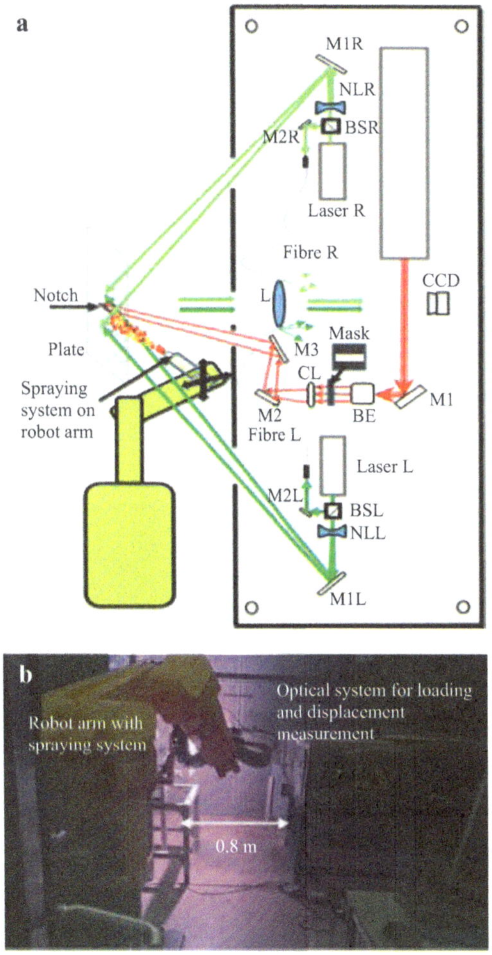

Fig. 9 a Set-up for laser machining and displacement measurement. **b** Coating system mounted on a robot arm and optical system (Reproduced from Ref. 51).

underestimate the residual stresses measured with the HDM, there is a similar behaviour for the common depth interval. In principle it is possible to use laser ablation and digital holography also for depth of 200 μm or more.

Fig. 11a shows a phase map (modulo 2π) corresponding to the in-plane displacements produced by laser loading a notch just after coating. The temperature of the surface monitored by a pyrometer was 360 °C. The plate was then cooled down at 215, 98, 24 °C, and phase maps corresponding to displacements produced by application of the same laser loading were determined (see Figs. 11b–d). From the phase maps, it is possible to calculate the in-plane displacements. In Fig. 11a, there are few fringes and, thus, the displacement and the residual stresses are small. The displacement increases (more fringes in the phase maps) when the temperature decreases, thus the residual stresses are produced when the coating cool down. The phase maps recorded at higher temperature have lower quality (more noise). This can be explained by considering that it was not possible to keep the temperature constant during the measurements. These temperature instabilities produce changes of the coating microstructure degrading the quality of the results.

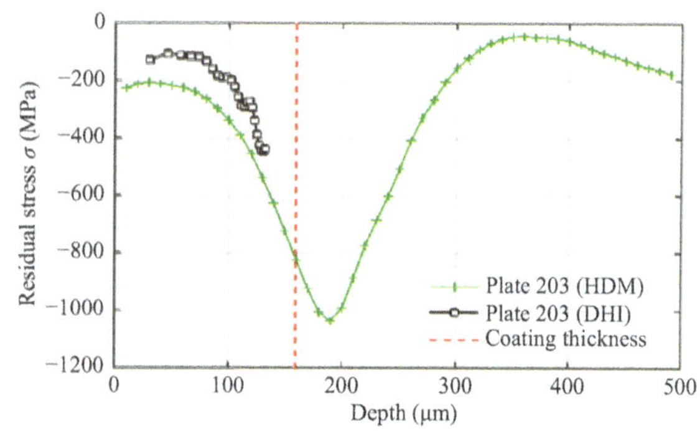

Fig. 10 Residual stresses inside a coating measured by the micro hole drilling method (HDM) and laser ablation combined with digital holographic interferometry (DHI) The coating thickness for this plate is approximately 160 μm (dashed vertical line). Reproduced from Ref. 51.

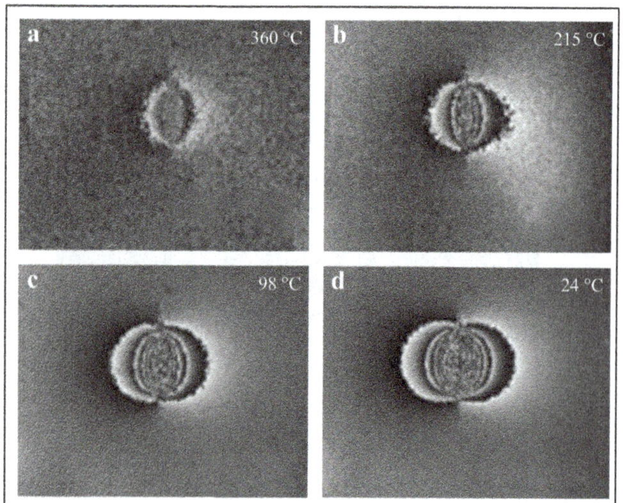

Fig. 11 Phase maps mod (2π) obtained by measurements at different temperatures (Reproduced from Ref. 51).

For obtaining the results shown in Fig. 11, a simple notch was machined in the coating but in Ref. 50 it was shown that more complex structures (cross, circles, rings…) may be machined in the coated surface allowing to locally release and detect residual stresses along different directions. For machining complex structures at different positions of the coating, a spatial light modulator (SLM) together with a pulsed laser were used[50,52].

3D-Measurement of MEMS calibration samples by speckle interferometry

The trend towards miniaturization of micro-electro-mechanical systems (MEMS) and micro-opto-electro-mechanical systems (MOEMS) continues to lead to more and more compact devices. Advantages lie in the integration with control electronics, lower power dissipation, higher sensitivity and better performance. Applications cover a wide range of fields, from optical telecommunications to medicine. Part of the functionality and reliability of the devices is based on the displacement and deformation of micromechanical parts under mechanical, thermal, magnetic or electrostatic loads. The measurement of the deformation of such systems is thus of great importance for confirming analytical and finite element models, accessing material and device properties, detecting potential defects and determining performance. Since typical structures exhibit dimensions in the order of some micrometers, it is necessary to measure their deformation with accuracies down to the nanometer range. Standardized approaches and calibrated setups are therefore essential for the measurement of displacement and deformations in the static and dynamic cases. Standards and guide lines for the use of different full-field optical techniques in the measurement of out-of-plane and in-plane displacements of microsystems were developed[53–55]. This process involved:

- Development of micromechanical reference components designed to deform in a reproducible and precise way when submitted to known loads;
- Calibration of the reference devices by means of optical techniques, performed in laboratory controlled conditions;
- Determination of the measurement uncertainties according to the guide of expression of uncertainty in

measurement (GUM);

Fig. 12a shows a reference test object designed to have a very accurate out-of-plane displacement produced by applying an electrical voltage to the structure. The displacement was measured by using a method based on digital holography. Figs. 12b–d show 3-D representations of the measured out-of-plane deformations when 15, 30, and 50 V are applied to the MEMS. Fig. 13 shows the deformation of the same MEMS at the center of the moving element as a function of the applied voltage. A very good agreement between expected and measured displacement was found. It was shown that the out-of-plane displacements may be performed with an accuracy of ±1.0 nm.

Devices having very accurate in-plane movements were also manufactured and tested[55]. The estimated uncertainty of the in-plane displacement measurements is worse compared with the out-of-plane and depends on the displacement magnitude; it was shown that in-plane displacement measurements in the range of 0–100 nm may be performed with an accuracy of ±5 nm.

Erosion Monitoring in harsh environment by digital holography

In Ref. 56 a feasibility study for erosion measurements under extreme environmental conditions inside the ITER Tokamak is described. The ITER project (Fig. 14a) is the next step in the transition from experimental studies of plasma physics to full-scale electricity-producing fusion power stations. It fuses the hydrogen isotopes deuterium and tritium into helium thereby releasing a high energy neutron. In order to start the fusion reaction the temperature has to be about 150 million Kelvin, creating a plasma. Because there is no material that could withstand such high temperatures, the plasma is guided, contactless, by magnetic fields within the vacuum chamber. However, these fields are not fully closed, resulting in partial plasma contact particularly in the divertor region. This leads to wear effects, affecting the overall performance and reliability of the Tokamak and potentially generating metallic dust. Thus, there is a need for the regular measuring of the erosion and deposition at the wall once the Tokamak starts operating. An erosion and deposition monitor able to measure the changes in the surface shape with a depth resolution of 10 μm is planned. The measurement will be done not on the whole internal surface of the Tokamak but on two surfaces of the divertors that endure high rates of erosion and deposition, each of a size of 10×30 cm^2. Due to the high temperature and radiation it will not be possible to have the measuring system inside the Tokamak, for this reason the

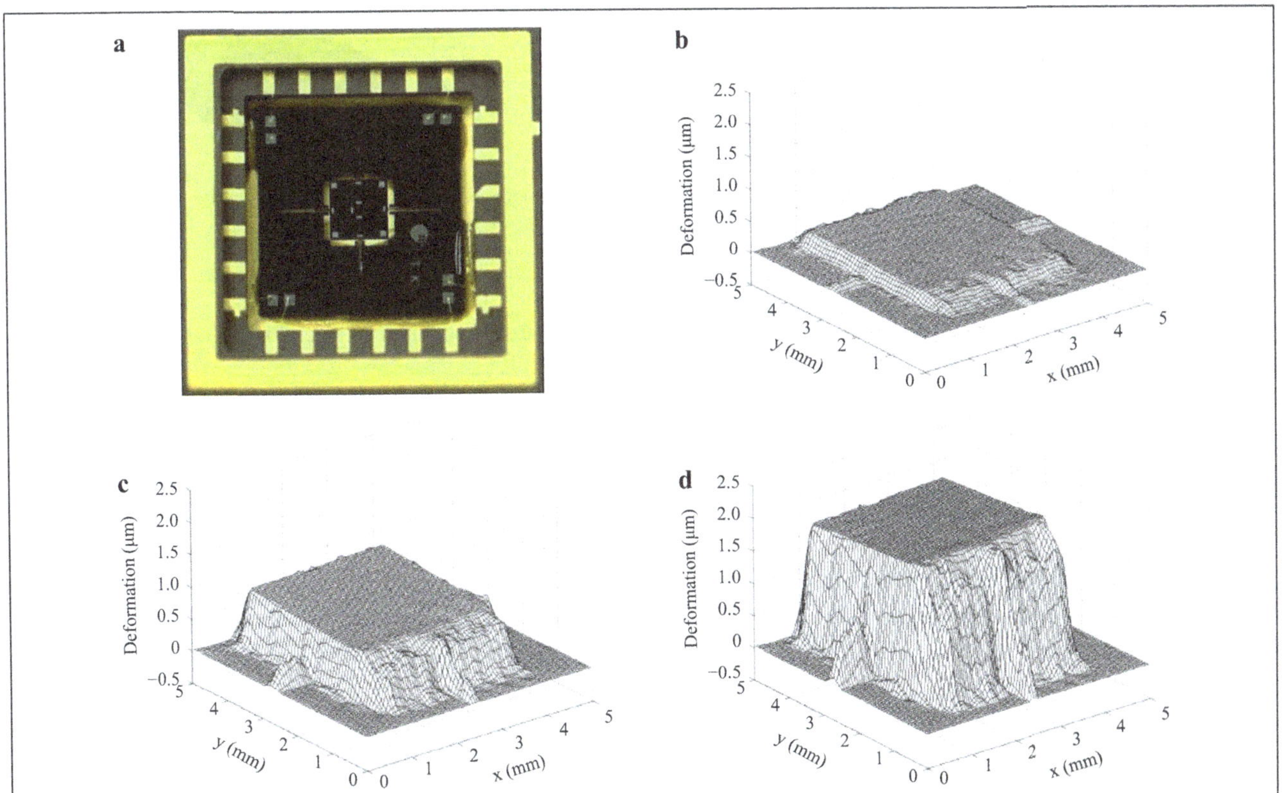

Fig. 12 Photo of the MEMS **a**. Out-of-plane deformations obtained by applying 15 V **b**, 30 V **c**, and 50 V **d**. Reproduced from Ref. 55.

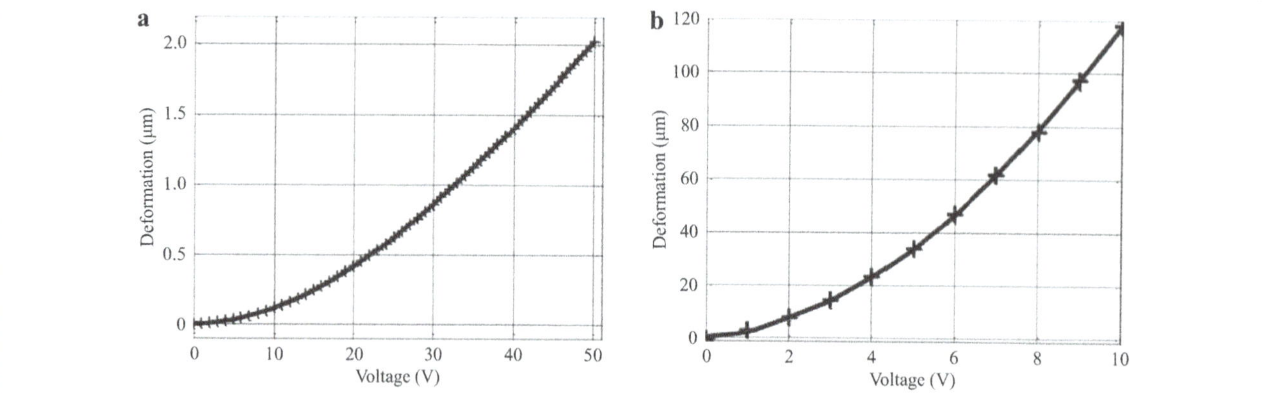

Fig. 13 **a** Measured (+) and expected (solid line) deformation as a function of the applied voltage, **b** detail of **a**. Reproduced from Ref. 55.

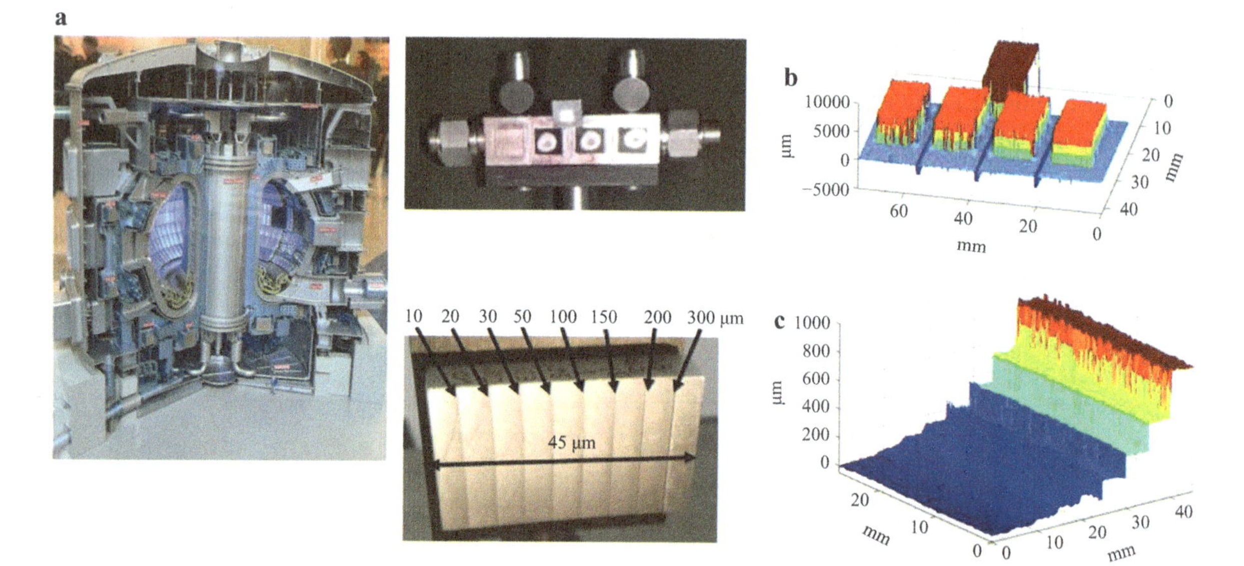

Fig. 14 **a** CAD model of the Tokamak[58]. **b** Shape reconstruction of tungsten sample located at a distance of 23 m from the measuring system. **c** Shape measurement of a rectangular 45 × 27 mm^2 aluminum plate with steps having depths of 10, 20, 30, 50, 100, 150, 200, 300 µm. **b** and **c** were reproduced from Ref. 56.

measurements will be performed remotely. Hence the opto-electronic instruments (detector, laser, controlling electronics) will be located at a distance of about 40 m from the surface to be measured.

In Ref. 56 it was shown that long distance shape measurements in challenging environmental conditions (vibrations and long distance) can be done by two wavelength digital holography[57] and thus this technique could be used for the erosion monitoring inside the Tokamak. Fig. 14b shows the measurement of a tungsten sample located on a non vibration isolated tripod at a distance of 23 m from the measuring system. Fig. 14c shows the measurement of a rectangular 45 × 27 mm^2 aluminum plate with steps having depths of 10, 20, 30, 50, 100, 150, 200, 300 µm. In this last case the sample was investigated by using a 14 m endoscopic system.

Inspection of artwork by digital shearography

Coherent-optical methods have proven that they are also valuable tools for the testing of artwork with respect to many purposes[14,59,60]. Especially, when cultural heritage is moved between various locations the precise monitoring of the current state of the artefact is indispensable[61]. Recently speckle shearography was used for the testing of a famous painting, the Madonna from Stuppach, painted by Matthias Grünwald around 1515[62], Fig. 15a. The painting was borrowed by the township Stuppach for an exhibition in Dresden in 2011. After the return it was evaluated by speckle shearography[63] to help the conservators in the unavoidable process of restoration. Fig. 15 presents some

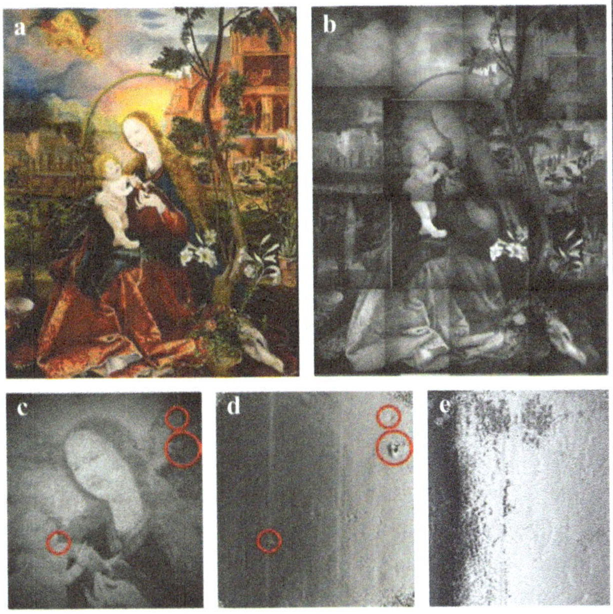

Fig. 15 Investigation of the painting "The Madonna from Stuppach" by digital speckle shearography. **a** the inspected painting before restoration **b** the 16 segments for successive inspection **c** the central segment with marked areas for detected suf-surface defects **d** the demodulated shearogram with the visible defects (bubbles, delaminations and cementations) **e** detected subsurface traces of wood worms.

results where the detected sub-surface defects are clearly visible. The large painting (appr. 1.9 m × 1.5 m) was segmented into 16 subareas. Each of them was investigated separately by making a speckle shearogram. A variety of defects such as bubbles, delaminations, cementations, and impacts of timber worms could be recognized immediately. These results are a valuable guidance for the planing of the restoration work to be done by the conservators. Current activities are directed to the design of an adapted loading procedure[64] and to an improved classification of the faults by a combination of FEM simulation and shearographic measurement[65].

Current challenges and modern approaches for the extension of the application fields of coherent optical metrology

Already in 1981, C.M. Vest published a remarkable study, "Holographic NDE – Status and Future"[66], where he formulated several important tasks with respect to the performance increase of holographic measurement techniques. This kind of wish list was circulated to motivate future research in optical metrology and to improve the unsatisfactorily low use of modern coherent inspection technologies in the industrial practice in the late 1970th. The core wishes are shown in Fig. 16: improve

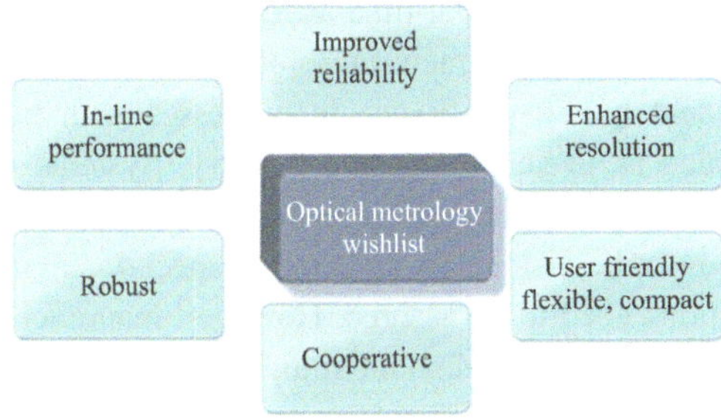

Fig. 16 Optical Metrology: C.M. Vests Wish List from 1981[66] (Reproduced from Ref. 38).

reliability, assure in-line performance, enhance resolution, make the tools more user friendly, flexible and compact, improve robustness against harsh environments and make the technology more cooperative for practical use.

40 years now have passed since Vest has published this study. Visible progress could be achieved mainly by applying new brilliant light sources with tunability of the wavelengths, digital cameras with improved space-bandwidth product and spectral sensitivity; spatial light modulators with both the ability for amplitude as well as phase control, and last but not least by making use of the continuously growing computer power with their amazing GPU performance[21,67,68].

However, the challenges of measurement technology increase continuously. Modern products are getting more miniaturized, more complex and have an increasing number of functionalities. For instance, the critical dimensions of structures written in silicon are meanwhile considerably smaller than the wavelength of the applied light source and this trend is to be sustained with the consistent application of the extreme UV lithography[69]. As the feature sizes are decreasing, so the theoretical and practical constraints of making them and ensuring their quality are increasing[70]. The same holds true for other industrial branches such as optics industry, micro-systems engineering, machine construction and automotive engineering where the surface-quality requirements of critical components and the overall complexity of the products have dramatically grown over the past years. Consequently, modern production and inspection technologies are confronted with a bundle of challenges[21].

Therefore, today's manufacturing industry places huge demands on metrology systems. They must have reasonably high resolution (both spatial and temporal) but yet be robust enough to work in an industrial environment. Systems must be able to be integrated into the production line as close as possible to the place of production and without interference with the production cycle. In-line metrology/defectoscopy is a must for future production systems. Only the real-time feedback of the inspection results into the manufacturing process can contribute to a consistent quality assurance in processes with high cost risk such as in semiconductor industry. Systems must be flexible. One current problem is that in many cases they are unique and custom-made. Often special environment conditions need to be assured. However, manufacturers do not want to have a special metrology system for each single product. They want one system to be able to measure and assess a variety of products. By this, we do not mean that one system has to be able to measure everything from an aspheric lens up to a telescope mirror or a car body. However, a system should be able to measure different versions of one product class such as lenses with various shape that cover spherical, aspherical and free forms. Such systems are already on the way[71]. In contrast to conventional tactile technologies, optical methods do not deliver a direct approach to the quantity of interest such as a tactile stylus that measures a profile height directly. To some extent a certain detour has always to be taken which can make the system tricky and not user-friendly. Thus, only skilled persons are often able to cope with such technologies. Therefore, we discussed in chapter 2 some guidelines that may help the customer to find out the right method and suitable system configuration for solving his or her problem. The mentioned demands are all challenging, but not impossible to achieve. With improved inspection strategies that take advantage of principles of active vision, model-based reconstruction, sensor-fusion with utilizing the full information content of the light field (multi-modal approaches), coherent optical metrology will be able to meet these demands[21,72].

Conclusion

2021 is a remarkable year for coherent optical metrology. 50 years ago Dennis Gabor presented his famous lecture at the London Imperial College on occasion of the awarding of the Nobel prize for the invention of holography. 5 years before, Karl Stetson and Robert Powell published their groundbreaking article "Hologram Interferometry"[1] and opened the door for a whole range of new coherent-optical measurement methods that are widely used meanwhile for non-destructive testing, experimental stress analysis and shape measurement. A first critical review of both the benefits and the problems still to be solved when using holographic principles for metrology was published by Charles Vest 40 years ago. 55 years after Stetsons and Powels article and 40 years after Vests study, we took the opportunity to look back to early days of holographic metrology and honored the inventors of the main principles and pioneering applications. Then we discussed some guidelines for the application dependent selection of a special method and tried to consider these measures for the solution of several exemplary measurement problems. Finally, we listed some challenges for optical measurement methods that are currently matter of investigation.

It still remains to answer our question from the headline. Hopefully we could show, that holographic non-destructive testing is technically mature after 55 years. However, new challenges and more complex boundary conditions require steady improvement and adaptation. Some new approaches where we see great potential, have been discussed. Further progress will certainly be associated with increasing digitization and networking. The Internet of Things (IoT) made it already possible to access remote production and measurement locations as well. The new concepts of remote holographic metrology[73] go far beyond the transmission of digital holograms via the Internet. They also provide an opportunity to communicate with and eventually control the physical setup of a remote metrology system. Furthermore, the metrology system can be modeled in the environment of a three-dimensional (3-D) virtual reality using CAD or similar technology providing a more intuitive interface to the physical setup within the virtual world. An engineer or a scientist wishing to access the remote real-world system can log on to the virtual system, moving and manipulating the setup, and take the desired measurements. The sets of data gained in the measurement process are stored and interpreted automatically for appropriate display within the virtual world providing the necessary feedback to the experimenter. Important metadata, including a description of the setup, the objects under investigation and all relevant system parameters, simultaneously generated by the system, can be recorded automatically to support the processing of the data or the future reproduction of the experiment for verification purposes, even by third parties. Such a system opens up many novel opportunities in industrial inspection such as remote master-sample comparison or the virtual assembly of parts that are fabricated at different locations without the need for transportation and actual physical assembly. Moreover, a multitude of new techniques can be envisaged, and among them advanced ways for the documenting processes, the efficient methods for metadata storage, the possibility for remote reviewing of experimental results, the addition of real experiments to publications by allowing remote access to the metadata and to the experimental setup via the Internet, the presentation of complex experiments in classrooms and lecture halls, the sharing of expensive and complex infrastructure within international collaborations, new ways of remote testing of new devices for their maintenance and service, and many more.

Acknowledgements
The authors gratefully acknowledge the financial support of the German Research Foundation, the Germany Ministry for Research and Education and the European Community. All founded projects are named in the cited publications.

Conflict of interest
The authors declare that they have no conflict of interest.

Received: 19 September 2021 Revised: 16 January 2022 Accepted: 18 January 2022
Accepted article preview online: 08 March 2022
Published online: 10 March 2022

References

1. Stetson, K. A. & Powell, R. L. Hologram interferometry. *Journal of the Optical Society of America* **56**, 1161-1163 (1966).
2. Stetson, K. The discovery of holographic interferometry, its development and applications. *Light – Advanced Manufacturing* (in this issue).
3. Waters, J. P. Interferometric holography. in Holographic Nondestructive Testing (ed Erf, R. K.) (Amsterdam: Elsevier, 1974), 87-103.
4. Erf, R. K. Holographic Nondestructive Testing. (New York: Academic Press, 1974).
5. Leendertz, J. A. Interferometric displacement measurement on scattering surfaces utilizing Speckle effect. *Journal of Physics E:Scientific Instruments* **3**, 214-218 (1970).
6. Hung, Y. Y. A Speckle-shearing interferometer: a tool for measuring derivatives of surface displacements. *Optics Communications* **11**, 132-135 (1974).
7. Hung, Y. Y. & Liang, C. Y. Image-shearing camera for direct measurement of surface strains. *Applied Optics* **18**, 1046-1051 (1979).
8. Kalms, M. K. & Osten, W. Mobile shearography system for the inspection of aircraft and automotive components. *Optical Engineering* **42**, 1188-1196 (2003).
9. Erf, R. K. Speckle Metrology. (New York: Academic Press, 1978).
10. Brown, G. M. Pneumatic Tire Inspection. (Amsterdam: Elsevier, 1974), 355-364.
11. Abramson, N. Sandwich hologram interferometry. 4: holographic studies of two milling machines. *Applied Optics* **16**, 2521-2531 (1977).
12. Felske, A. & Happe, A. Double pulsed laser holography as a diagnostic method in the automative industry. Proceedings of Conference, on Engineering Uses of Coherent Optics, held at the Univ. of Strathclyde 1975. E. R. Robertson (Ed.), Cambridge Univ. Press 1976, pages 595-614. ISBN 0 521 20879 3
13. Steinbichler, H. & Gehring, G. TV-Holography and holographic Interferometry: industrial applications. *Optics and Lasers in Engineering* **24**, 111-127 (1996).
14. Amadesi, S. et al. Holographic methods for painting diagnostics. *Applied Optics* **13**, 2009-2013 (1974).
15. Ostrovsky, Y. I., Butusov, M. M. & Ostrovskaya, G. V. Interferometry by Holography. (Berlin, Heidelberg: Springer, 1980).
16. Vest, C. M. Holographic Interferometry. (New York: Wiley, 1979).
17. Wernicke, G. & Osten, W. Holografische Interferometrie: Grundlagen, Methoden und ihre Anwendung in der Festkorpermechanik. Physik Verlag, Weinheim, 1982. ISBN 10: 3876640660 / ISBN 13: 9783876640662
18. See Wikipedia "Big Data". at https://en.wikipedia.org/wiki/Big_data.
19. Boyd, D. & Crawford, K. Six provocations for big data: social science research network: a decade in internet time: symposium on the dynamics of the internet and society. (2011). at https://papers.ssrn.com/sol3/papers.cfm?abstract_id=1926431.
20. Osten, W. et al. Different approaches to overcome existing limits in optical micro- and nano-metrology. 22nd Congress of the International Commission for Optics: Light for the Development of the World, edited by Ramón Rodríguez-Vera, Rufino Díaz-Uribe, Proc. of SPIE Vol. 8011, 80116O, CCC code: 0277-786X/11/$18. doi: 10.1117/12.905277
21. Osten, W. Optical metrology: the long and unstoppable way to become an outstanding measuring tool. Proceedings of SPIE 10834, Speckle 2018: VII International Conference on Speckle Metrology. Janów Podlaski: SPIE, 2018.
22. Mroczka, J. The cognitive process in metrology. *Measurement* **46**, 2896-2907 (2013).
23. ISO/BIPM Guideline: Uncertainty of Measurement. (1999). at http://www.metrodata.de/papers/resistor_en.pdf.
24. Gronle, M. & Osten, W. View and sensor planning for multi-sensor surface inspection. *Surface Topography:Metrology and Properties* **4**, 024009 (2016).
25. Albertazzi, A. Jr. et al. Speckle interferometry in harsh environments: design considerations and successful examples. Proceedings of SPIE 10678, Optical Micro-and Nanometrology VII. Strasbourg: SPIE, 2018.
26. Osten, W. & Jüptner, W. Measurement of displacement vector fields of extended objects. *Optics and Lasers in Engineering* **24**, 261-285 (1996).
27. Nadeborn, W., Andrä, P. & Osten, W. A robust procedure for absolute phase measurement. *Optics and Lasers in Engineering* **24**, 245-260 (1996).
28. DIN EN ISO 9241. https://docplayer.org/21480943-Deutsche-norm-din-en-iso-9241-110.html
29. Gronle, M. et al. *itom*: an open source metrology, automation, and data evaluation software. *Applied Optics* **53**, 2974-2982 (2014).
30. Mason, B. & Antony, J. Statistical process control: an essential ingredient for improving service and manufacuring quality. *Managing Service Quality:An International Journal* **10**, 233-238 (2000).
31. Pyzdek, T. & Keller, P. A. The Six Sigma Handbook. (New York: McGraw-HillEducation, 2018).
32. Leach, R. et al. Information-rich manufacturing metrology. Proceedings of the 8th IFIP WG 5.5 International Precision Assembly Seminar. Chamonix: Springer, 2019.
33. Hadamard, J. Lectures on Cauchy's Problem in Partial Differential Equations. New Haven: Yale Univ. Press, 1923.
34. Bruning, J. H. et al. Digital wavefront measuring interferometer for testing optical surfaces and lenses. *Applied Optics* **13**, 2693 (1974).
35. Totzeck, M. & Tiziani, H. J. Phase-shifting polarization interferometry for microstructure linewidth measurement. *Optics Letters* **24**, 294-296 (1999).
36. Osten, W. et al. High-resolution measurement of extended technical surfaces with scalable topometry. Proceedings of SPIE 4101, Laser Interferometry X: Techniques and Analysis. San Diego: SPIE, 2000.
37. Kayser, D., Bothe, T. & Osten, W. Scaled topometry in a multisensor approach. *Optical Engineering* **43**, 2469-2477 (2004).
38. Osten, W., Haist, T. & Manske, E. How to drive an optical measurement system to outstanding performance. Proceedings of SPIE 10557, Ultra-High-Definition Imaging Systems. San Francisco: SPIE, 2018.
39. The Fair Data Sheet Initiative. at http://optassyst.de/fairesdatenblatt/.
40. OptAssyst. at http://www.optassyst.de/index.html.
41. Häusler, G. & Ettl, P. Über die kluge Auswahl und Anwendung optischer 3D-Sensoren. *Photonik* **5**, 2-5 (2004).

42. De Lega, X. C. & De Groot, P. J. Lateral resolution and instrument transfer function as criteria for selecting surface metrology instruments. *Imaging and Applied Optics Technical Papers*. http://dx.doi.org/10.1364/OFT.2012.OTu1D.4 (2012).
43. De Groot, P. & De Lega, X. C. Interpreting interferometric height measurements using the instrument transfer function. in Fringe 2005. (ed Osten, W.) (Berlin, Heidelberg: Springer, 2006), 30-37.
44. De Groot, P. et al. Fourier optics modelling of coherence scanning interferometers. Proceedings of SPIE 11817, Applied Optical Metrology IV. San Diego: SPIE, 2021.
45. Osten, W. et al. Shearography system for the testing of large-scale aircraft components taking into account noncooperative surfaces. Proceedings of SPIE 4101, Laser Interferometry X: Techniques and Analysis. San Diego: SPIE, 2000.
46. Alexeenko, I. et al. Nondestructive testing by using long-wave infrared interferometric techniques with CO_2 lasers and microbolometer arrays. *Applied Optics* **52**, A56-A67 (2013).
47. Georges, M. P. et al. Combined holography and thermography in a single sensor through image-plane holography at thermal infrared wavelengths. *Optics Express* **22**, 25517-25529 (2014).
48. Wenzelburger, M., López, D. & Gadow, R. Methods and application of residual stress analysis on thermally sprayed coatings and layer composites. *Surface and Coatings Technology* **201**, 1995-2001 (2006).
49. Schajer, G. S. & Prime, M. B. Use of inverse solutions for residual stress measurements. *Journal of Engineering Materials and Technology* **128**, 375-382 (2006).
50. Pedrini, G. et al. Residual stress analysis of ceramic coating by laser ablation and digital holography. *Experimental Mechanics* **56**, 683-701 (2016).
51. Alekseenko, I. et al. Residual stress evaluation in ceramic coating under industrial conditions by digital holography. *IEEE Transactions on Industrial Informatics* **16**, 1102-1110 (2020).
52. Osten, W. et al. Optical method and arrangement for measuring residual stresses, in particular in coated objects. Patent WO2016184578 (issued 2016). https://patentscope.wipo.int/search/en/detail.jsf?docId=WO2016184578
53. Pedrini, G. et al. Calibration of optical systems for the measurement of microcomponents. *Optics and Lasers in Engineering* **47**, 203-210 (2009).
54. Pedrini, G. et al. Development of reference standards for the calibration of optical systems used in the measurement of microcomponents. *Strain* **46**, 79-88 (2010).
55. Pedrini, G. et al. Measurement of nano/micro out-of-plane and in-plane displacements of micromechanical components by using digital holography and speckle interferometry. *Optical Engineering* **50**, 101504 (2011).
56. Pedrini, G. et al. Feasibility study of digital holography for erosion measurements under extreme environmental conditions inside the international thermonuclear experimental reactor tokamak. *Applied Optics* **58**, A147-A155 (2019).
57. Pedrini, G. et al. Pulsed digital holography for high-speed contouring that uses a two-wavelength method. *Applied Optics* **38**, 3460-3467 (1999).
58. https://en.wikipedia.org/wiki/ITER.
59. Tornari, V. et al. Laser-based systems for the structural diagnostic of artwork: an application to XVII-century Byzantine icons. Proceedings of SPIE 4402, Laser Techniques and Systems in Art Conservation. Munich: SPIE, 2001.
60. Tornari, V. Laser interference-based techniques and applications in structural inspection of works of art. *Analytical and Bioanalytical Chemistry* **387**, 761-780 (2007).
61. EU-Project MultiEncode. Multifunctional Encoding System for Assessment of Movable Cultural Heritage. https://cordis.europa.eu/project/id/6427.
62. Madonna from Stuppach. https://de.wikipedia.org/wiki/Stuppacher_Madonna.
63. Osten, W., Wilke, M. & Pedrini, G. Remote laboratories for optical metrology: from the lab to the cloud. Proceedings of SPIE 8413, Speckle 2012: V International Conference on Speckle Metrology. Vigo: SPIE, 2012.
64. Buchta, D. et al. Artwork inspection by shearography with adapted loading. *Experimental Mechanics* **55**, 1691-1704 (2015).
65. Buchta, D. et al. Combination of FEM simulations and shearography for defect detection on artwork. *Strain* **54**, e12269 (2018).
66. Vest, C. M. Holographic NDE: Status and Future. NGS-GCR-81-318 (Publication of the National Bureau of Standards, Washington DC 1981)..
67. Osten, W. & Jueptner, W. P. O. New light sources and sensors for active optical 3D inspection. Proceedings of SPIE 3897, Advanced Photonic Sensors and Applications. Singapore: SPIE, 1999.
68. Osten, W. et al. Optical metrology: from the laboratory to the real world. Proceedings of SPIE 7387, Speckle 2010: Optical Metrology. Florianapolis: SPIE, 2010.
69. Pirati, A. et al. EUV lithography performance for manufacturing: status and outlook. Proceedings of SPIE 9776, Extreme Ultraviolet (EUV) Lithography VII. San Jose: SPIE, 2016.
70. Den Boef, A. J. Optical wafer metrology sensors for process-robust CD and overlay control in semiconductor device manufacturing. *Surface Topography:Metrology and Properties* **4**, 023001 (2016).
71. Pruss, C. et al. Measuring aspheres quickly: tilted wave interferometry. *Optical Engineering* **56**, 111713 (2017).
72. Osten, W. Different ways to overcome the resolution problem in optical micro and nano metrology, in Optical Imaging and Metrology: Advanced Technologies, eds Osten, W. & Reingand, N., Weinheim: Wiley-VCH, 2012, 327-368.
73. Osten, W., Wilke, M. & Pedrini, G. Remote laboratories for optical metrology: from the lab to the cloud. *Optical Engineering* **52**, 101914 (2013).

Official journal of the JHL 2689-9620
www.light-am.com

Article Open Access

Reflections about the holographic and non-holographic acquisition of surface topography: where are the limits?

Gerd Häusler[1,*] and Florian Willomitzer[2]

Abstract

Recording and (computational) processing of complex wave fields offer a vast realm of new methods for optical 3D metrology. We discuss fundamental similarities and differences between holographic surface topography measurement and non-holographic principles, such as triangulation, classical interferometry, rough surface interferometry and slope measuring methods. Key features are the physical origin of the ultimate uncertainty limit and how the topographic information is encoded and decoded. Besides the theoretical insight, the discussion will help optical metrologists to determine if their measurement results could be improved or have already hit the ultimate limit of what physics allows.

Keywords: Optical metrology, Topography, Physical limits, Holography

Introduction

For the first-ever observers of a holographic recording at the 1964 OSA spring meeting[1], the most intriguing feature was probably the three-dimensional appearance of the reconstructed object. After 60 years of holography, this 'magical' effect still captivates observers who can freely change the viewing perspective and locally focus on the object's surface. The reconstruction seems to be perfect, indistinguishable from the true object, including phase and amplitude. Holography is based on the interferometric superposition of waves, which suggests ultraprecise measurement options, and indeed, holographic interferometry is a paradigm for this option[2], as well as the holographic Null test[3] via computer-generated holograms, invented by Adolf Lohmann[4]. Holographic microscopy flourishes with the availability of high-resolution camera chips[5–7]. A hologram records a complex field, sometimes called a "wavefront", originating from an object under test. This field can be read out optically or computationally. For a proper source encoding, we can modify the recording by placing some optical instrument in front of the holographic plate, e.g., a shearing plate. Furthermore, we can place any optical instrument behind the (analog) hologram to extract specific information about the object. With a camera chip replacing the holographic film, any instruments and much more can be mimicked via computation.

This article discusses the potential and limits of measuring the surface shape or topography, meaning the spatially resolved distance $z(x,y)$. We investigate holography and its major competitors regarding what is similar and different. The topographical information can be deciphered by numerous methods discussed below. We aim to know the physical source of the dominant noise that ultimately limits the achievable uncertainty of measurement, in other words, the lowest possible statistical distance uncertainty δz.

Correspondence: Gerd Häusler (gerd.haeusler@fau.de)
[1]University of Erlangen Nuremberg, Institute of Optics, Information, and Photonics, 91058 Erlangen, Germany
[2]Northwestern University, Department of Electrical and Computer Engineering, Evanston 60208, USA

It will be discussed how holographic methods compare to the established non-holographic methods. Some methods have a close connection to holography; for others, the connection is only indirect. We will look at the problem from a bird's eye view. More details and further references can be found in[8]. The idea of exploiting physical limits to advance optical 3D metrology is described in [9]. From the knowledge of the limits, the user may find out if his measurement results could be improved (for example) by better hardware - or if the results are already hitting the ultimate physical limit. The considerations will also help the manufacturer of 3D metrology find out if the competitor can really and truly satisfy the advertised specifications or if there might be some exaggeration.

We summarize our considerations:

- What is the physical origin of the ultimate measuring uncertainty?

- How is the topography encoded and decoded?

These questions are useful to bring some order into the overabundance of available 3D sensors and help customers understand the limits and the potentials of different sensor principles. Our questions intentionally avoid hardware aspects and solely consider fundamental physical principles.

Optical sensors exploit different kinds of illumination, such as coherent/incoherent, structured/homogeneous, monochromatic/colored, polarized/unpolarized... There are different ways of interaction with the object surface: coherent (Rayleigh scattering), incoherent (thermal, fluorescent), specular/diffuse, surface/volume... The detected modality may be the intensity, the complex amplitude, time-of-flight, polarization, coherence. Permuting all those parameters leads to more than 10,000 possible sensors. Not all of them are physically different, but there are differences significant for the user of 3D sensors. Surprisingly, it turns out that all sensors (so far considered by the authors) belong to only four different measuring principles, which can be classified by the physical origin of the fundamental distance uncertainty. An important parameter for users is the dependence of the distance uncertainty (not the accuracy and lateral resolution) on the aperture (we refer to earlier investigations[8]). The results are condensed in Table 1. The competing measuring principles (see as well Fig. 1) are triangulation (I), classical interferometry (II), rough-surface interferometry (III). There is a category IV that we name "slope-measuring methods". This class comprises methods that intrinsically measure the surface slope or lateral derivative. Table 1 summarizes the fundamental distinction between the categories or classes and displays the recipes to calculate the corresponding limits. Fig. 2 illustrates the systems theoretical simiilarity of hlography with fringe projection triangulation. Fig. 3 displays coherent noise which is the ultimate uncertainty limit of triangulation systems. Figs. 4, 5 illustrate that different measuring principles (here III and IV) display very different limits of the uncertainty of measurement.

In the following, we give a few examples for each class: Amongst others, class I (triangulation) comprises laser triangulation, stereo-photogrammetry, focus search, fringe projection triangulation, or structured illumination microscopy. We note that the wide area of fluorescence methods is not discussed here. Class II (classical interferometry) comprises essentially interferometry at specular surfaces. Whether a surface is 'optically specular' or 'optically rough' depends on the surface *and* on the observer: specular means that the local topography $z(x, y)$ within the optically resolved area (the diffraction disc) varies less (or better much less) in depth than a quarter of the used wavelength to avoid destructive interference and speckles. For a specular surface, the complex signal

Table 1 All known sensors might possibly fit into one of the four categories that differ in terms of the dominant source of noise and its dependence on the observation aperture. The table can be read as well from right to left, to find the correct class. λ is the wavelength, C is the speckle contrast ($C = 1$ for laser illumination), $\sin u_{obs}$ is the observation aperture, θ is the triangulation angle (for focus searching methods, $\theta = u_{obs}$), δx is the lateral resolution, the roughness parameter R_q is the standard deviation of the surface height z, $\delta\alpha$ is the slope uncertainty. SNR is the signal-to-noise ratio.

Class	Physical Principle	Origin of Meas. Uncertainty δz	Lower Bound of Meas. Uncertainty δz	Dependent on Obs. Aperture?
I	Triangulation	Speckle	$\delta z = C\lambda/(2\pi \sin u_{obs} \sin\theta)$ or $\delta z = C\lambda/(2\pi \sin^2(u_{obs}))$ for focus search	Yes
II	Classical Interferometry	Photon Noise	principally, no lower physical bound	No
III	Rough-Surface Interferometry	Surface Roughness	Surface Roughness $< \lvert z - \langle z \rangle \rvert > \approx R_q$	No
IV	Slope-Measuring Methods	Photon Noise	$\delta z \approx \delta x \cdot \delta\alpha \approx \lambda/SNR, \quad \delta x = \lambda/\sin u_{obs}$	Yes

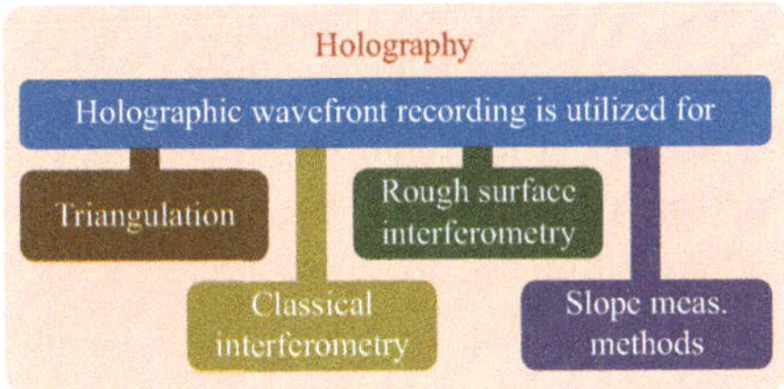

Fig. 1 Holographic acquisition of surface topography: The outstanding feature to record an optical wavefront in a hologram can be utilized in many different ways to measure the surface topography of an object. Respective measurement principles can be categorized into four groups: triangulation, classical interferometry, rough surface interferometry, and slope measuring methods.

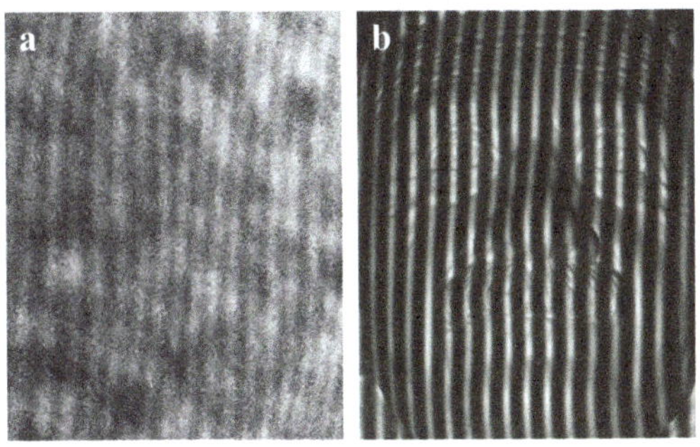

Fig. 2 Systems-theoretical similarity between holographic encoding and decoding in fringe projection triangulation: Both images can be decoded by single-sideband demodulation. **a** Close-up photograph of a transmission hologram recorded on photographic emulsion[12]. **b** Fringe projection triangulation camera image for the measurement of the face of a plaster bust.

degenerates to $exp(ikz) \approx 1 + ikz(x, y)$, with the consequence that the coherent lateral averaging over the diffraction disc leads to a signal corresponding to the mean distance $< z(x, y) >$. For optically rough objects (i.e., for large height variations within the diffraction disc), however, the term $< exp(ikz(x, y)) >$ is a nonlinear and non-monotonic ('chaotic') function of z, leading to the typical random walk of the complex signal and speckle. In class III, we find rough-surface interferometry which includes two-wavelength interferometry and 'coherence scanning interferometry (CSI)'. Unfortunately, the standard term CSI does not distinguish between smooth-surface and rough-surface interferometry, thus obscuring the significant differences of the physical signal generation and the ultimate uncertainty limits depicted in Table 1. To make the difference obvious, we use the term 'rough-surface CSI' or 'coherence radar', as the method was named in the first publication[10]. A modern review of CSI is given in [11].

In class IV, we find incoherent methods such as phase measuring deflectometry, the Hartmann-Shack sensor, photometric stereo, and coherent methods that exploit shearing interferometry with its modifications, such as differential interference contrast. Until today the authors did not find optical 3D sensors that do not fit within one of these categories (the authors encourage the reader to find such sensors).

Now we are ready for the question: where do we find holography? Our fast answer is: it depends on the application.

Holography versus triangulation

Fig. 2 displays a striking similarity between the appearance of a hologram of a rough object and the camera images taken for line- or fringe projection triangulation[13-16]: both display fringes with a carrier frequency f_0 where the fringe phase encodes the local object depth $z(x, y)$.

Even more, the decoding of a hologram is information-theoretically the same as the decoding process in single-shot fringe projection triangulation ("Fourier transform profilometry"[14,17]). Both are decoded by single sideband demodulation, implemented by optical separation of the zero-diffraction order from the $+/-1^{st}$ diffraction orders. Moreover, both methods share the same space-bandwidth

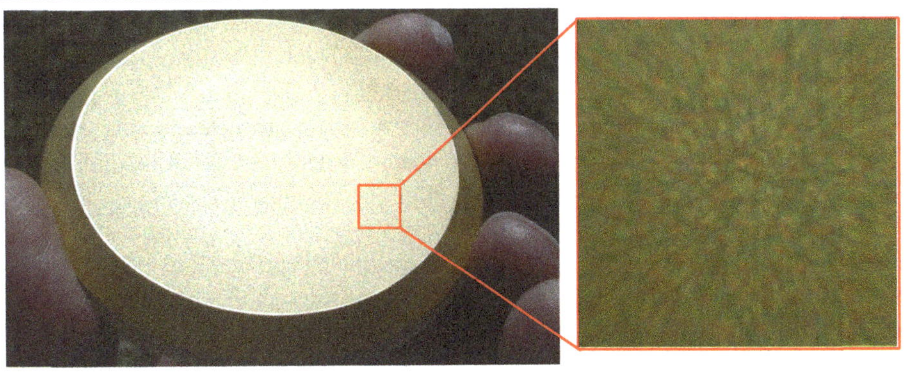

Fig. 3 Groundglass in sunlight: at close distance - with the observation aperture angle close to the illumination aperture angle - speckles can be seen even with a white extended light source.

limitations: only one-third of the available space-bandwidth can be used[15,18]. For the commonly static objects of holography and interferometry, it is standard to replace 2/3 of the expensive space bandwidth by temporal bandwidth via phase shifting[19]. However, temporal phase-shifting requires at least three subsequent exposures, which poses a serious problem for real-time metrology.

Single-shot measurements are possible, but not with a dense 3D point cloud. The simple reason is that one camera pixel cannot deliver the necessary information about the local illumination, the local reflectivity, and the distance $z(x, y)$ of an object point.

Typically, single-shot sensors display point cloud densities much lower than 1/3 of the available pixels[20] as the available space bandwidth also needs to account for the solution of the so-called 'indexing problem'. The "single-shot 3D movie camera"[15,20,21] solves this problem with a trick that (from a systems-theoretical side) shows strong similarities to two-wavelength holography or -interferometry: Instead of a sequence of images taken at different wavelengths (or fringe-frequencies), two cameras simultaneously capture two images of the fringe-encoded surface from two different viewing angles. Interestingly, the decoding algorithms' mathematical structure resembles those used in dual-wavelength interferometry[20,22,23].

So far, about the systems theory of encoding and decoding. Irrespective of the system theoretical similarities, the physics of the encoding is fundamentally different in both principles. In a hologram, the optical path-length and the reference wave encode the (interference) fringe phase, respectively the local distance $z(x, y)$. In the triangulation system, the distance is encoded just by the perspective distortion of the fringe pattern projected onto the object surface, see Fig. 2.

Holograms of rough objects and the reconstructions display speckles which are a fundamental source of the uncertainty, not only for holographic metrology: Watching the seemingly noise-free triangulation image of Fig. 2b more carefully, it turns out that the fringe phase displays a small random error as well. This error is caused by the unavoidable partial spatial coherence of the illumination in combination with the observation aperture. There is always some residual spatial coherence, even with a large white incoherent light source, which results in an ultimate lower limit of the measuring uncertainty δz for the local distance z of an object point[14]. Fig. 3 illustrates the 'ubiquitous spatial coherence': Already partial spatial coherence disturbs measurements of rough surfaces, even if speckles are unnoticed by a distant observer with a small observation aperture. The ultimate uncertainty limit is

$$\delta z = \frac{C}{2\pi} \frac{\lambda}{\sin u_{obs} \sin \theta} , \qquad (1)$$

where θ is the triangulation angle, $\sin u_{obs}$ is the observation aperture and C is the speckle contrast ($C = 1$ for laser illumination, so for most holograms). For focus-searching principles Eq. 1 degenerates to the Rayleigh depth of field (neglecting the $C/2\pi$ factor), as u_{obs} acts as effective triangulation angle:

$$\delta z = \frac{C}{2\pi} \frac{\lambda}{\sin^2 u_{obs}} \qquad (2)$$

For spatially incoherent light sources, the contribution C_s of spatial coherence to the speckle contrast C can easily be estimated by the observation aperture and the illumination aperture[24] via:

$$C_s = \min\left(\frac{\sin u_{obs}}{\sin u_{ill}}, 1 \right) \qquad (3)$$

For the other contributing factors polarization, temporal coherence and averaging via large pixels, see[20,24,25].

An illustrative daily life example: for laser triangulation (with $C = 1$, $\sin u_{obs} = 0.01$, $\sin \theta = 0.2$) we achieve an uncertainty of about 40 µm. With spatially 'incoherent' fringe projection using a large incoherent illumination aperture (see Eq. 3), the same specifications may lead to a uncertainty about four times better.

These beneficial properties made fringe projection triangulation a well-established 'gold standard method' for many macroscopic applications. A well-designed sensor avoiding as much spatial coherence as possible may display a dynamical range of up to 10,000 depth steps. We emphasize that efficient metrology is enabled only by efficient source encoding. In optical metrology, ***illumination*** plays the role of the ***source encoder*** to encode the depth in an information-efficient way (with low redundancy)[26].

Coming back to holography: Can we decipher the surface topography from a hologram? At first glance, this should not be a problem, as we can "see" the 3D reconstruction (this marveling feature can hopefully be exploited for displays in the future). But what about exploitation for metrology? One could suspect that the object surface can be fully reconstructed within the limits of diffraction theory, where the size of the hologram can be seen as the limiting aperture. We approach the answer via an extreme counter-example: a hologram of a diffusely scattering white planar surface. It is impossible to find the surface by focusing through the hologram, as it is impossible to focus a camera onto a white wall in daylight. The deep reason is that the wave field in front of the holographic plate can be restored from the hologram, but (generally) we cannot localize the millions of surface points where the individual spherical waves have been scattered in the direction of the holographic plate. It will be discussed later that holography can access not just a wavefront but, after all, even the coherence function and the surface topography.

In turn, this means that some structure is necessary to acquire the object surface from a simple hologram, either via inherent features (salient points) or structured illumination as used in active triangulation. The structure is necessary for the holographic reconstruction and incoherent methods.

To summarize, we have to admit that although holography - as classical interferometry - encodes distance via the phase of the propagating waves, deciphering of the surface topography is not possible without additional information. Classical interferometry commonly looks at the image plane, so there is a priori information about the object's location. Without such additional information, interferometric measurements can principally measure a wavefront but not the true surface of a remote object.

Even with 'additional information' there is a caveat. Here an illustrative example: from the image of a laser spot projected onto a rough object, the distance z can be found via focus search or shearing interferometry, despite speckle[27,28]. However, it was shown that both methods have to be attributed to class I, which possibly might surprise the reader. The explanation follows in "Holography versus slope measuring methods" section.

Triangulation of a rough surface with laser illumination suffers from a serious measuring uncertainty, determined essentially by the observation aperture: These properties can be looked up in Table 1. Triangulation measurements through a hologram and via indirect measurements by numeric backpropagation must also be attributed to class I. There is one exception: if the object displays fluorescence or is thermally excited, there is no speckle (and no hologram) because of perfect spatial incoherence[29,30]. Microscopy based on fluorescence microscopy can localize molecules with an uncertainty below nanometers.

Consequently, the ultimate uncertainty of holographic (rough surface) topography measurement via focus search (or related methods) is given by Eqs. 1 and 2, and depends strongly on the observation aperture (and not by the much lower photon noise of classical interferometry). Fortunately, by exploiting the large aperture that holography may easily provide, the uncertainty can be comparably high, despite speckle noise.

Holography versus rough-surface interferometry

Phase randomization in speckle patterns forbids the topography measurement of rough surfaces with single-wavelength interferometry. A solution that is well established now is Coherence Scanning Interferometry (CSI). As mentioned in the introduction, the terms 'rough-surface CSI' or 'coherence radar'[10,11,31] are more substantiated.

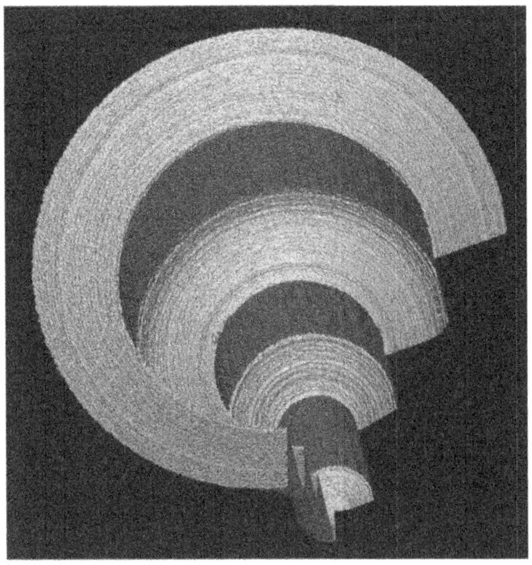

Fig. 4 Gauge for an injection nozzle (about 10 mm long), measured by rough-surface coherence scanning interferometry. Despite the low aperture that allows for measuring within deep holes, the measuring uncertainty is only a few micrometers.

A partial solution that holography can implement is two-wavelength holography, enabling the topography acquisition by contouring: A hologram of the surface is made by illuminating the object and the holographic plate with a wavelength λ_1. In a second step, the holographic plate and the object are illuminated with a wavelength λ_2, slightly different from λ_1. The object recorded at λ_1 is played back with λ_2 and superimposed with the wave from the object illuminated at λ_2. The interference pattern displays contour lines at a distance given by the "synthetic" wavelength[32–36]

$$\Lambda = \frac{\lambda_1 \lambda_2}{|\lambda_1 - \lambda_2|} \tag{4}$$

which is the beat wavelength between λ_1 and λ_2 and hence can be picked orders of magnitudes larger. The light at the optical wavelengths λ_1 and λ_2 acts as a carrier for the synthetic wave. The optical wavelengths are scattered, meaning that the object serves as a "rough mirror" despite the large synthetic wavelength and the object can be seen from all directions. At the same time, the observed interference fringes at the synthetic wavelength do not display phase randomization. It is noteworthy that now, even a surface without features like edges or texture can be reconstructed. Bright contour lines are found where the two waves (with λ_1, λ_2) display the same phase. However, the temporal coherence function is periodic for two-wavelength holography, causing an ambiguous reconstruction. The ambiguity can be avoided by 'multi-wavelength' holography or white-light interferometry[10,35,37].

This remarkable advantage of two-wavelength interferometry and -holography has been utilized recently for the relatively new research field of "Non-Line-of-Sight" imaging, which is concerned with the task of looking around corners and imaging through scattering media. The method is based on the acquisition of a synthetic wavelength hologram which is taken by imaging a remote surface (such as a wall), that can "see" the hidden object and the sensor unit. This remote surface acts as a "virtual holographic plate". The object is reconstructed by holographic backpropagation from the synthetic hologram at the synthetic wavelength[35,38,39]. Following this procedure one can consider the wall as "rough mirror".

After understanding the role of roughness qualitatively, again, the obligatory question: How to categorize two-wavelength holography in terms of noise? The resulting images with more or less 'un-speckled' fringes suggest an attribution to "classical interferometry", characterized by photon noise as the cause of the uncertainty limit.

Combining the advantages of diffuse scattering with visible light illumination, without the drawbacks of speckle, sounds like magic, meaning that carefulness is advisable. And indeed, two-wavelength holography is not attributed to the same class as classical interferometry. Two-wavelength holography and -interferometry belong to class III, where the surface roughness ultimately limits the uncertainty: The reason is that the speckles produced by the two closely spaced wavelengths λ_1, λ_2 display a small phase decorrelation[33,35,40] resulting in a random phase difference $d\varphi_{1,2}$. This phase difference may correspond to a path difference of only $\lambda/100$ (for example). However, in the output signal, the measured path difference is "magnified" by Λ/λ resulting in a random distance error of $\Lambda/100$, instead of $\lambda/100$, in this example. This measuring uncertainty is much larger than the photon-noise limit of classical interferometry.

To estimate the random phase error and its physical cause, we assume that the two speckle patterns at λ_1 and λ_2 are sufficiently spatially correlated, as we select the synthetic wavelength to be much larger than the surface roughness[35,41]. At a certain image point (x', y'), waves are accumulated from a small object area given by the back-projected diffraction point spread function of the observing lens. This area will be approximately the size of a back-projected subjective speckle. The rough surface will have a 'summit' or peak within this area, and there will be the deepest valley point. The peak-valley distance R_t (over a certain length) is one of the common roughness parameters. R_t determines the maximum possible phase difference $d\varphi_{1,2}$ between all waves accumulated within (x', y'). For the two wavelengths λ_1, λ_2, we easily find:

$$d\varphi_{1,2} = 2 \cdot \left(\frac{2\pi R_t}{\lambda_1} - \frac{2\pi R_t}{\lambda_2} \right) = \frac{4\pi R_t}{\Lambda_1} \tag{5}$$

And with $\delta z = \Lambda \cdot d\varphi_{1,2}/2\pi$ we get the measuring uncertainty at the position):

$$\delta z = 2R_t \tag{6}$$

Eq. 6 tells us that the phase decorrelation between the two wavelengths limits the measuring uncertainty to a value given just by the surface roughness. Although an approximation, the physical cause of the measuring uncertainty is found. The same cause, even with a very similar quantitative result, was found in a rigorous analysis for rough-surface CSI[42,43].

A few more clarifying words about rough-surface CSI ('coherence radar')[10,31]: The micro-topography of most "rough" surfaces such as a white wall, a ground glass, or a machined surface cannot laterally be resolved unless we use a high aperture microscope. Nevertheless, even with a small observation aperture, within deep boreholes and from a large distance, we can measure the roughness without laterally resolving the micro-topography. The roughness is given by the statistical measurement error δz. As most daily life surfaces or technical objects have a roughness of only a few micrometers, the uncertainty of rough-surface CSI is much better than that of triangulation. As a further present, this uncertainty is independent of the observation aperture.

With the benefit of hindsight, the similarity of the physical cause of the dominating noise is not surprising, as both methods (scanning white-light interferometry and two-wavelength interferometry/holography), exploit an interplay of two (or more) wavelengths.

We summarize this section by assigning two-wavelength holography and two-wavelength interferometry to class III: These methods exploit just 'time-of-flight' or optical path length variations, so the ultimate limit of height uncertainty is given by the surface roughness and does not depend on the observation aperture[8,43]. Again, rough-surface CSI's ultimate uncertainty is only determined by the object and not by the instrument, which is a remarkable physical and information-theoretical peculiarity.

A short appendix: It is well known that illumination by a wavelength significantly larger than the surface roughness helps measure rough surfaces. One could use far infrared light[44], but this creates a new problem: the surface now acts like a mirror (besides technical issues and the limited availability of infrared detectors). For 3D objects strongly deviating from a plane, the light will scarcely find its way back to the hologram plate/detector. It is important to note that the physical properties of such an experiment must be looked up under category II, "classical interferometry", as the surface is smooth for the long wavelength. We see that changing the illumination may change the physical background dramatically.

Holography versus slope measuring methods

The so far discussed methods intrinsically measure either the lateral position of a local feature (which is translated into distance via triangulation) or they measure the distance via an interferometric measurement of the phase (or time-of-flight) of a coherent (or partially coherent) signal. However, a class of physically completely different methods also exists - where the intrinsic signal is the local slope. 'Intrinsic' means that the slope is not evaluated by a-posteriori differentiation - the slope is already encoded in the optical signal before arriving at the photodetector. This is an invaluable feature, as the encoding is done optically before adding the detector noise. Hence, some of these methods enable sub-nanometer uncertainty for local surface height variation by very simple means. The reason is that intrinsic slope measurement corresponds to a perfect source encoding: thc OTF represents the spatial derivative (or at least an approximation for shearing methods). Among the spatially differentiating methods, we find the so-called phase-measuring deflectometry (PMD)[45–47], which is completely incoherent, and we find classical shearing interferometry[48].

Both methods measure specular surfaces or wavefronts. For rough surfaces, shearing holography comes into play, specifically because of its unlimited options for post-processing of the holographic data. We do not discuss the so-called "photometric stereo" here, as it commonly requires lambertian scatterers and is based on accurate intensity measurements. Hence, "photometric stereo" is sensitive against unavoidable spatial coherence, specifically because small light sources are used to generate shading.

In phase measuring deflectometry[45–47], a large screen with an incoherently radiating sinusoidal fringe pattern is in remote distance from the specular object under test, see Fig. 5. The observer watches the screen that is mirrored by the object. If the object is not planar, the captured fringes in the camera image are distorted.

Local surface deformations beyond the nanometer range can be measured from the distortion which is quantified by phase shifting. There is also a microscopic realization[49], where weak phase objects can be measured in transmission[50]. There is very low spatial coherence with an incoherent (self-luminous) screen, and the method is essentially limited by photon noise. However, the ultimate limit of the height uncertainty depends on the lateral resolution as well, see Table 1. An interesting coupling of the angular uncertainty $\delta\alpha$ and the lateral resolution δx leads to the useful uncertainty product given in Eq. 7. The camera must simultaneously focus on the object surface and the remote screen pattern (sinusoidal fringes). This depth-of-field problem requires a trade-off between angular uncertainty and lateral resolution[51]. More deeply, it is the Heisenberg uncertainty product that does not allow both a small δx and, at the same time, a small $\delta\alpha$, for a single photon. However, the *SNR* and the height resolution are virtually unlimited with many photons. Together with the signal-to-noise ratio *SNR*, we get:

$$\delta z \approx \delta x \delta\alpha \approx \lambda / SNR \tag{7}$$

As an $SNR = 500$ can easily be achieved with standard video cameras, a depth uncertainty of $\delta z = 1$ nm is possible with a piece of few-dollar equipment (λ is 500 nm). Similar considerations prevail to classical shearing interferometry, but this requires more costly equipment and does not deliver a true derivative. The 'scaling' of shearing interferometry is given by the wavelength of light, while some macroscopic fringe generator gives the scaling of

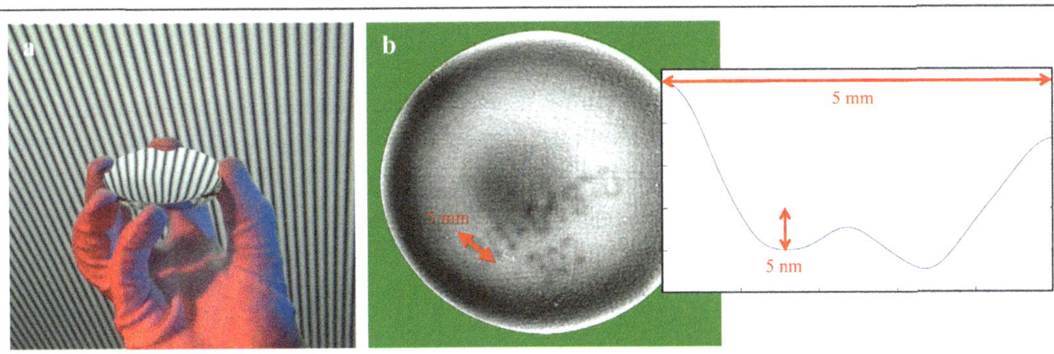

Fig. 5 Phase measuring deflectometry. **a** the basic principle, **b** "ghostwriting" depth map: A lens was marked with a whiteboard marker; later, the numbers were erased. The few nanometer damages of the lens surface can be detected and quantitatively evaluated.

incoherent deflectometry. Nevertheless, incoherent deflectometry can easily compete[51], and both methods are limited by photon noise.

Eq. 7 offers numerous options: for a very low spatial resolution (e.g., for the measurement of a big flat mirror), the angular uncertainty can be in the micro-arcsec range[52].

For the sake of completeness, we also mention the well-established Hartmann-Shack sensor as one more incoherent slope measuring sensor. It allows for direct measurement of wavefronts coming, e.g., from the pupil of a lens under test[53]. A certain limitation is the wavefront sampling by an array of discrete lenslets.

Coherent differentiation is possible as well: Among the approximately "differentiating" methods, there is shearing interferometry and shearing holography[54,55], including its numerous (more or less coherent) implementations such as the Nomarski differential interference contrast[56].

We move on to holographic shearing-interferometry, as it offers the possibility to measure the local slope of a rough surface. A wavefront is compared with a version of itself shifted ('sheared') by a distance s. A precious advantage of shearing methods (via a shearing plate) is the common-path geometry, being insensitive against environmental perturbations. Shearing holography is commonly realized via a shearing plate and captures several sheared interference patterns corresponding to different shears. The different shears correspond to different OTFs in the Fourier space. Hence, a hologram with a desired OTF can be synthesized by combining different interference patterns[57] and evaluated by numerical backpropagation.

To understand the physical limit of rough surface shearing holography, we refer again to the example of a projected laser spot[27,28]: A laser spot with diameter d_{spot} is projected onto the object. In the hologram plane, an objective speckle pattern with speckle diameter $d_{speckle} \approx \lambda \cdot z / d_{spot}$ is generated and superimposed with a laterally shifted copy of itself. As far as the shear s is smaller than the speckle diameter, fringes can be seen, however, with some phase noise, due to the phase decorrelation within each speckle. Earlier investigations reveal that this phase noise leads to a measuring uncertainty $\delta z \approx \lambda / \sin^2 u_{obs}$, which is just the Rayleigh depth of focus (why are we not surprised?). Obviously, shearing interferometry at rough surfaces, as a tool to measure the local distance, displays (only) the same uncertainty limit as focus searching methods. Hence it belongs to class I. This is a result that a naïve observer possibly would not have expected: Shearing holography at rough surfaces is equivalent to triangulation to measure local distance.

The situation is even more difficult for an extended object and a large shear. Principally, each surface point generates a fringe pattern in the hologram, but the phase is random to other surface points, which makes the deciphering of the hologram difficult. Nonetheless, the object can be fully reconstructed from the holograms[57]. The basic idea here is to determine the complex-valued coherence function from the recorded interference patterns, starting from the mixed interference term $E^*(x,y)E(x+s,y)$ of the complex signal E. From there, one can reconstruct finite differences of the wavefield corresponding to positions separated by the shear or by combining several shears, the non-differentiated wavefront.

A real finite difference can be achieved by using "Γ-profilometry", which exploits the temporal coherence function[58]. The method is strongly related to rough-surface CSI, as described in "Holography versus rough-surface interferometry" section. A significant difference to scanning white-light interferometry is that the outcome here is not the surface profile $z(x,y)$ but the difference $\Delta z = z(x,y) - z(x+s,y)$. This property can be an advantage, as for most objects, the depth scanning time will be significantly lower, as the scanning range is limited by the maximum of Δz instead of the full object depth range. Γ-profilometry is an illustrative example of redundancy reduction via source encoding. It follows from these considerations that the roughness of the surface gives the ultimate source of noise for Γ-profilometry with broadband illumination: The method has to be assigned to class III, again with the useful feature that the measurement uncertainty does not depend on the observation aperture. Apparently, it is a fruitful idea to "copy" and possibly improve established methods like rough surface CSI by proper holographic storage and the big toolbox of computational evaluation.

Eventually, the so called 'shearography'[59] has to be mentioned: Speckle shearography takes advantage of the common path geometry of shearing interferometry. The robustness against environmental perturbation is exploited to measure very small (temporal) changes of the surface "slope", the sensitivity depending on the shear. The method is, of course, sensitive against speckle decorrelation. For deformations close to a wavelength, phase decorrelation probably dominates. So only under the assumption of very small surface changes might the uncertainty be limited by photon noise.

Holography versus classical interferometry

We have set aside this consideration until now, hoping that the preceding sections simplify the understanding: As far as specular surfaces are involved, holography belongs to class II, classical interferometry, due to the fact that here the uncertainty is limited by photon noise.

For holographic interferometry at rough objects, the object is compared with a slightly deformed version of itself. As discussed for speckle shearography, the phase in two highly correlated speckles is compared. As far as there is very low decorrelation, the uncertainty is limited by photon noise, as for classical interferometry. However, this is not a binary decision, as phase decorrelation and fringe delocalization occur with increasing deformation.

Conclusion

Holography is a tool to store complex wavefronts and to process the data either optically or by computation. There are numerous implementations - amongst others - to acquire data about the topography of an object's surface, which is the focus of this paper. The comparison of holographic methods with non-holographic methods offers some interesting insights. The first insight is that holographic methods can be assigned to one of four classes defined via the physical cause of the dominant noise and the dependence of the observation aperture. As not every method is investigated, the authors admit that further discussion might get new insights.

Methods based on triangulation are seriously disturbed by speckle, with the consequence that holography is not the first choice to measure surface topography via triangulation, e.g., by focus search or related methods. 3D metrology, based on triangulation, is the turf of incoherent methods, at least for macroscopic objects. Many microscopical objects are weak phase objects, and there is no clear distinction between defocusing noise and speckle. Furthermore, the observation aperture can be extremely high - so with the help of Eq. 2, holographic methods often deliver acceptable results despite some coherent noise.

We think that holographic interferometry aiming to measure sub-λ deformation, is the natural realm of holography, where most incoherent methods fail.

Furthermore, two- or multi-wavelength holography might potentially become a competitor of the corresponding non-holographic methods that we find in the class "rough-surface interferometry". The potential of holography is strongly related to the vast realm of computational options. 'Seeing around the corner' is only one striking example, where several holographic ideas are combined.

Among the methods that intrinsically measure the slope, (incoherent) phase measuring deflectometry has become an invaluable tool to measure virtually any kind of specular surface, even competing with interferometry[54]. Deflectometry displays an extreme sensitivity for local surface defects, with at the same time low hardware requirements. Can holography compete? Maybe for rough surfaces: Among the (approximately) differentiating holographic methods, multi-wavelength shearing-holography seems to be a proper method to measure the slope of rough surfaces. Again, the virtually unlimited options for computational post-processing are an advantage over "purely" optical methods. The uncertainty limit is determined just by the surface roughness, the same limit as rough-surface interferometry.

We conclude by stating that holography offers the presentation of breathtaking 3D images. The underlying storage of a complex wavefield, together with virtually unlimited options for processing by computation, is breathtaking as well. As there will be many not yet invented algorithms, future researchers might profit from knowing and exploiting the fundamental limits to exploit holography better.

Acknowledgements

The authors gratefully acknowledge numerous intriguing discussions with Claas Falldorf (BIAS Bremen) and Christian Faber (Hochschule Landshut), and as well their critical reading of the manuscript.

Conflict of interest

The authors declare no conflict of interest.

Received: 20 December 2021 Revised: 09 March 2022 Accepted: 25 March 2022
Accepted article preview online: 28 March 2022
Published online: 26 April 2022

References

1. Leith, E. N. & Upatnieks, J. Wavefront reconstruction with diffused illumination and threedimensional objects. *Journal of the Optical Society of America* **54**, 1295-1301 (1964).
2. Powell, R. L. & Stetson, K. A. Interferometric vibration analysis by wavefront reconstruction. *Journal of the Optical Society of America* **55**, 1593-1598 (1965).
3. Offner, A. A null corrector for paraboloidal mirrors. *Applied Optics* **2**, 153-155 (1963).
4. Lohmann, A. W. & Paris, D. P. Binary fraunhofer holograms, generated by computer. *Applied Optics* **6**, 1739-1748 (1967).
5. Zhang, T. & Yamaguchi, I. Three-dimensional microscopy with phase-shifting digital holography. *Optics Letters* **23**, 1221-1223 (1998).
6. Kim, M. K. Principles and techniques of digital holographic microscopy. *SPIE Reviews* **1**, 018005 (2010).
7. Kemper, B., Langehanenberg, P. & von Bally, G. Digital holographic microscopy. *Optik & Photonik* **2**, 41-44 (2007).
8. Häusler, G. & Ettl, S. Limitations of optical 3D sensors. in Optical Measurement of Surface Topography (ed Leach, R.) (Berlin Heidelberg: Springer, 2011), 23-48.
9. Häusler, G. Discover better optical sensors - by exploring and exploiting nature' s limits. Computational Optical Sensing and Imaging 2019. Munich, Germany: Optical Society of America, 2019, CTh2A.1.
10. Dresel, T., Häusler, G. & Venzke, H. Three-dimensional sensing of rough surfaces by coherence radar. *Applied Optics* **31**, 919-925 (1992).
11. Su, R. Coherence scanning interferometry. in Advances in Optical

Surface Texture Metrology (ed Leach, R)(Bristol: IOP Publishing, 2020), 2-1-2-27.
12. Epzcaw. https://commons.wikimedia.org/wiki/File:Holographic_recording.jpg (2021).
13. Srinivasan, V., Liu, H. C. & Halioua, M. Automated phase-measuring profilometry of 3-D diffuse objects. *Applied Optics* **23**, 3105-3108 (1984).
14. Takeda, M. & Mutoh, K. Fourier transform profilometry for the automatic measurement of 3-D object shapes. *Applied Optics* **22**, 3977-3982 (1983).
15. Willomitzer, F. & Häusler, G. Single-shot 3D motion picture camera with a dense point cloud. *Optics Express* **25**, 23451-23464 (2017).
16. Huber, F. et al. 3D body scanning with "flying triangulation". Proceedings of 112th DGaO Conference; 30, (2011).
17. Takeda, M., Ina, H. & Kobayashi, S. Fourier-transform method of fringe-pattern analysis for computer-based topography and interferometry. *Journal of the Optical Society of America* **72**, 156-160 (1982).
18. Häusler, G. et al. Why can' t we purchase a perfect single shot 3d-sensor? Proceedings of 113th DGaO Conference; A8, (2012).
19. Bruning, J. H. et al. Digital wavefront measuring interferometer for testing optical surfaces and lenses. *Applied Optics* **13**, 2693-2703 (1974).
20. Willomitzer, F. Single-Shot 3D Sensing Close to Physical Limits and Information Limits. (Berlin Heidelberg: Springer, 2019).
21. Willomitzer, F. et al. Single-shot three-dimensional sensing with improved data density. *Applied Optics* **54**, 408-417 (2015).
22. Cheng, Y. Y. & Wyant, J. C. Two-wavelength phase shifting interferometry. *Applied Optics* **23**, 4539-4543 (1984).
23. Falaggis, K., Towers, D. P. & Towers, C. E. Method of excess fractions with application to absolute distance metrology: theoretical analysis. *Applied Optics* **50**, 5484-5498 (2011).
24. Häusler, G. Speckle and coherence. in Encyclopedia of Modern Optics (eds Guenther, B. D. et al) (Oxford: Academic Press, 2004), 114-123.
25. Dorsch, R. G., Häusler, G. & Herrmann, J. M. Laser triangulation: fundamental uncertainty in distance measurement. *Applied Optics* **33**, 1306-1314 (1994).
26. Wagner, C. & Häusler, G. Information theoretical optimization for optical range sensors. *Applied Optics* **42**, 5418-5426 (2003).
27. Häusler, G. et al. Range sensing based on shearing interferometry. *Applied Optics* **27**, 4638-4644 (1988).
28. Häusler, G. & Herrmann, J. M. Range sensing by shearing interferometry: influence of speckle. *Applied Optics* **27**, 4631-4637 (1988).
29. Thorley, J. A., Pike, J. & Rappoport, J. Z. Super-resolution microscopy: a comparison of commercially available options. in Fluorescence Microscopy: Super-Resolution and Other Novel Techniques (eds Cornea, A. & Conn, P. M.) (Amsterdam: Elsevier, 2014), 199-212.
30. Spellenberg, B., Herrmann, J. M. & Häusler, G. Highly improved range sensingBby reduction of spatial coherence. *Optik* **112**, 299-303 (2001).
31. Caber, P. J. Interferometric profiler for rough surfaces. *Applied Optics* **32**, 3438-3441 (1993).
32. Fercher, A. F., Hu, H. Z. & Vry, U. Rough surface interferometry with a two-wavelength heterodyne speckle interferometer. *Applied Optics* **24**, 2181-2188 (1985).
33. Vry, U. & Fercher, A. F. Higher-order statistical properties of speckle fields and their application to rough-surface interferometry. *Journal of the Optical Society of America A* **3**, 988-1000 (1986).
34. Dändliker, R., Thalmann, R. & Prongué, D. Two-wavelength laser interferometry using superheterodyne detection. *Optics Letters* **13**, 339-341 (1988).
35. Willomitzer, F. et al. Fast non-line-of-sight imaging with high-resolution and wide field of view using synthetic wavelength holography. *Nature Communications* **12**, 6647 (2021).
36. Li, F. Q. et al. SH-ToF: micro resolution time-of-flight imaging with superheterodyne interferometry. Proceedings of 2018 IEEE International Conference on Computational Photography (ICCP). Pittsburgh, PA, USA: IEEE, 2018, 1-10.
37. Wagner, C., Osten, W. & Seebacher, S. Direct shape measurement by digital wavefront reconstruction and multi-wavelength contouring. *Optical Engineering* **39**, 79-85 (2000).
38. Willomitzer, F. et al. High resolution non-line-of-sight imaging with superheterodyne remote digital holography. Computational Optical Sensing and Imaging 2019. Munich, Germany: Optical Society of America, 2019, CM2A.2.
39. Willomitzer, F. et al. Synthetic wavelength holography: an extension of Gabor' s holographic principle to imaging with scattered Wavefronts. Preprint at https: //arxiv.org/abs/1912.11438 (2019).
40. Li, F. Q. et al. Exploiting wavelength diversity for high resolution time-of-flight 3D imaging. *IEEE Transactions on Pattern Analysis and Machine Intelligence* **43**, 2193-2205 (2021).
41. George, N., Jain, A. & Melville, R. D. S. Jr. Experiments on the space and wavelength dependence of speckle. *Applied Physics* **7**, 157-169 (1975).
42. Ettl, P. Studien zur hochgenauen Objektvermessung mit dem Kohärenzradar. MSc Thesis, University of Erlangen, Nuremberg, (1995).
43. Häusler, G. et al. Limits of optical range sensors and how to exploit them. in International Trends in Optics and Photonics (ed Asakura, T.) (Berlin Heidelberg: Springer, 1999), 328-342.
44. Kwon, O., Wyant, J. C. & Hayslett, C. R. Rough surface interferometry at 10.6 µm.. *Applied Optics* **19**, 1862-1869 (1980).
45. Häusler, G. Verfahren und vorrichtung zur ermittlung der form oder der abbildungseigenschaften von spiegelnden oder transparenten objekten. (1999).
46. Knauer, M. C., Kaminski, J. & Hausler, G. Phase measuring deflectometry: a new approach to measure specular free-form surfaces. Proceedings of SPIE 5457, Optical Metrology in Production Engineering. Strasbourg, France: SPIE, 2004, 366-376.
47. Willomitzer, F. et al. Hand-guided qualitative deflectometry with a mobile device. *Optics Express* **28**, 9027-9038 (2020).
48. Bates, W. J. A wavefront shearing interferometer. Proceedings of the Physical Society **59**, 940 (1947)
49. Häusler, G. et al. Microdeflectometryja novel tool to acquire three-dimensional microtopography with nanometer height resolution. *Optics Letters* **33**, 396-398 (2008).
50. Peterhänsel, S. et al. Microdeflectometry in transmission. Proceedings of 110th DGaO Conference; 24, (2009).
51. Faber, C. et al. Deflectometry challenges interferometry: the competition gets tougher! Proceedings of SPIE 8493, Interferometry XVI: Techniques and Analysis. San Diego: SPIE, 2012, 84930R.
52. Ehret, G. et al. Deflectometric systems for absolute flatness measurements at PTB. *Measurement Science and Technology* **23**, 094007 (2012).
53. Hartmann, J. Objektivuntersuchungen. *Zeitschrift für Instrumentenkunde* **24**, 1-21 (1904).
54. Servin, M., Malacara, D. & Marroquin, J. L. Wave-front recovery from two orthogonal sheared interferograms. *Applied Optics* **35**, 4343-4348 (1996).
55. Falldorf, C., von Kopylow, C. & Bergmann, R. B. Wave field sensing by means of computational shear interferometry. *Journal of the Optical Society of America A* **30**, 1905-1912 (2013).
56. Nomarski, G. Microinterferometre differentielle a ondes polarisees. J. Phys. Radium **16**, 9 (1955).
57. Falldorf, C., Agour, M. & Bergmann, R. B. Digital holography and quantitative phase contrast imaging using computational shear interferometry. *Optical Engineering* **54**, 024110 (2015).
58. Falldorf, C. et al. Γ-profilometry: a new paradigm for precise optical metrology. *Optics Express* **29**, 36100-36110 (2021).
59. Francis, D., Tatam, R. P. & Groves, R. M. Shearography technology and applications: a review. *Measurement Science and Technology* **21**, 102001 (2010).

Part II

Digital Transition of Holography

Fratz et al. *Light: Advanced Manufacturing* (2021)2:15
https://doi.org/10.37188/lam.2021.015

Official journal of the JHL 2689-9620
www.light-am.com

Review **Open Access**

Digital holography in production: an overview

Markus Fratz, Tobias Seyler, Alexander Bertz and Daniel Carl*

Abstract

Many challenging measurement tasks in production simultaneously have high requirements for accuracy, measurement field size, lateral sampling, and measurement time. In this paper, we provide an overview of the current state of the art in digital holography for surface topography measurements and present three applications from completely different productions with no alternative to digital holography; we describe the HoloTop sensor family, which has been designed specifically for industrial use, and present the most recent results achieved in real-life industrial applications. All applications address measurement tasks that could not be solved until now, either by optical or tactile means. We start with a description of the first-ever inline integration of a digital holographic measurement system that inspects precision turned parts for the automotive industry. We proceed by presenting measurements performed with a compact sensor that can be placed inside a tooling machine and operated fully wirelessly. In this case, the tool holder was used to position the sensor directly. Integration into a tooling machine places high demands on both robustness and reliability. Finally, the quality control of electronic interconnectors such as microbumps with the highest demand for accuracy and repeatability is demonstrated. All of these applications illustrate the major advantages of digital holographic systems: it is possible to measure a relatively large field of view with interferometric precision and very short acquisition times. Additionally, both reflective and matt surfaces can be measured simultaneously. We end this publication with an assessment of the future potential of this technology and the necessary development steps involved.

Introduction

Quality demands on mechanical and electrical components are increasing steadily, especially for mass-produced parts, for example, in the automotive industry, medical technology, and chip production and assembly in the semiconductor industry. On the one hand, tighter tolerances require very precise measuring methods; on the other hand, the demand for full-field inspection of functional surfaces on 100% of all parts is increasing, and these requirements conflict. Tactile measuring systems are still very common and, in many cases, meet the requirements for accuracy. However, owing to the pointwise tactile recording, a complete surface measurement is, in most cases, not reasonable or possible. Hence, the use of coordinate measurement systems is almost exclusively limited to the inspection of samples in special measurement rooms. Contactless optical measurement methods can overcome these limitations. In optical metrology, a distinction must be made between pointwise and full-field measuring methods. Although optical scanning can be performed orders of magnitude more quickly than tactile measurement, camera-based full-field recording methods have clear advantages in terms of overall acquisition time.

When choosing an optical sensor for a specific application, the question of whether a surface to be measured is suitable for optical inspection is very important. Reflectivity, transmission, scattering, absorption, and slope angle have a strong influence on the signal strength and thus the measurement result. Typically,

Correspondence: Daniel Carl (daniel.carl@ipm.fraunhofer.de)
Fraunhofer Institute for Physical Measurement Techniques IPM, Georges-Köhler-Allee 301, 79110 Freiburg, Germany

reflective and matt surfaces can be distinguished. In the case of reflective surfaces, the phase front of the illumination is significantly influenced only by the shape of the specimen. Matt surfaces, on the other hand, usually destroy the phase relationship through scattering and multiple reflections, a phenomenon which is observable as a speckle effect under coherent lighting.

A large number of previously unresolved measuring tasks have the following similarities:

- Surface reflectivity is often on the edge between reflective and matt or varies rapidly over the surface. Hence, the reflective and matt regions must be measured simultaneously.
- The object's slope has partly steep edges. Owing to the small tolerances, the absolute measurement accuracy must be less than a single micron (three sigma); in some extremely demanding applications, the requisite accuracy is down to the range of 0.1 microns. Fast production cycles and necessary time for part handling often provide no more than 0.1 seconds for measurement, especially in automotive and semiconductor series production.

Nearly every optical 3D sensor based on active illumination can be classified into one of the following five classes: (1) triangulation, (2) coherence scanning interferometry, (3) laser scanning confocal microscopy, (4) reflectometry, and (5) interferometry. To understand the limitations of each class, it is necessary to identify the dominant noise source and the influence of surface properties. Triangulation techniques cannot fulfil the abovementioned accuracy requirements owing to speckle noise and vanishing signals in reflecting regions. Additionally, steep edges are partly shadowed depending on the direction of illumination and imaging, as well as the triangulation angle. Coherence scanning interferometry and laser scanning confocal microscopy are, in principle, capable of measuring reflecting and matt surfaces, but require recording a large number of images, depending on the intended accuracy. Therefore, in both cases, fast measurements of large fields of view are limited by the depth sampling and measurement range. Reflectometry only works on fully mirroring surfaces. In principle, classical interferometry in combination with phase-shifting techniques has the potential to fulfil the requirements for accuracy and measurement time. Depending on the applied phase-shifting technique, only one image (spatial phase shifting) or a handful of images (temporal phase shifting) per field of view need to be captured[1].

Based on three temporal phase-shifted images per wavelength, the number of raw image data points to be taken is at least one order of magnitude smaller than for all methods with comparable accuracy, such as coherence scanning interferometry or laser scanning confocal microscopy[2].

One major limitation of classical interferometry is the restricted unambiguity range of half of the wavelength of the illuminating laser light, which means that only surfaces of optical quality can be measured suitably. Multiple wavelengths can be used to circumvent this limitation. By subtracting the phase maps generated by different wavelengths modulo 2π, a phase map at the beat frequency of the respective wavelengths can be calculated. Several of these artificial (or "synthetic") wavelengths can be generated by recording interferograms at various wavelengths. Subsequently, the reconstructed phase maps can be cascaded such that the smaller synthetic wavelengths compensate for the phase noise of the larger ones. Therefore, the largest synthetic wavelength restricts the unambiguity measurement range, while the smallest one limits the measurement accuracy[3,4]. In the case of mirror-like surfaces, the smallest wavelength can be one of the single laser wavelengths used for recording[5]. In classic interferometry, absolute phase measurements of speckle fields generated by rough surfaces are not possible because the phase values are statistically distributed. By subtracting the phase maps of the correlated speckle fields recorded with an identical optical setup at two slightly shifted wavelengths, this fixed-pattern phase component is eliminated completely as long as the generated synthetic wavelengths are large compared to the roughness of the inspected surface. As a result, the numerically reconstructed phase map shows very little residual speckle noise[4,6].

Extending classical interferometry, digital holography offers another very useful property: the possibility of numerically propagating the reconstructed wavefront. This enables numerical refocusing of the object plane, which has enormous advantages for inline applications. It is well known that precise sample handling in production environments to achieve the best focus position always has its limits (and costs). The application of autofocus techniques on numerically propagated wavefronts, either on amplitude or phase maps, is possible in real time[7].

The content of the following sections provides a short introduction to the theoretical background of multi-wavelength digital holography and a description of the basic design of the HoloTop sensor family. The core of this publication, however, is three state-of-the-art application examples from real-world manufacturing.

We start with the first-ever published inline integration of a sensor based on multi-wavelength digital holography. The results of long-term measurements under rough production conditions on 10 million precision turned parts for the automotive industry are presented.

The demonstration of measurements on machined aluminium surfaces using a milling machine itself as a sensor mount and positioning device pushes the limits even further.

Finally, one of the most challenging applications of digital holography – the quality control of microscopic electronic interconnectors – is shown. The latest camera technology is integrated into the sensor design to fulfil the conflicting requirements of field of view, lateral sampling, and speed. The requirements for electronic device inspection are aggravated by the production cycle times of such devices (within the range of a second), sub-micron repeatability demands, and the high dynamic range of the reflecting surfaces. The most promising measurement results for microbump structures are shown.

To our knowledge, there is no alternative to digital holography for all three use cases that satisfies the simultaneous needs for accuracy, measurement field size, lateral sampling, and measurement time.

Theoretical Background

The term holography is derived from the Greek word *holos*, "whole", and *gráphein*, "writing".

In 1948, Dennis Gábor published his work in the field of minimising aberrations in electron microscopy, for which he was awarded the Nobel Prize in Physics in 1971. Holography was developed as a lensless image method for the reconstruction of complete wavefronts, that is, both amplitude and phase[8].

In digital holography (DH), holographic information is not optically reconstructed but is discretely recorded using semiconductor elements such as cameras[9]. The amplitude and phase of waves reflected by a scattering object contain information about the topography of the illuminated object.

Previously, hologram calculations took several minutes, which initially limited the technology's possibilities to laboratory applications. Only with increasing computing power and the availability of fast, high-resolution cameras, as well as the use of graphics processing units (GPUs), have the requirements for accuracy and production speed of today's production cycles been met; thus, DH has become increasingly relevant for industrial applications[10,11].

During hologram reconstruction, phase-shifting methods are used to convert the nonlinear intensity modulation of the recorded interferograms into relative phase values. In this context, a distinction is made between the two methods of temporal[12–14] and spatial[15,16] phase shifting. Because only systems based on temporal phase shifting are presented in this paper, we refer to the literature for a comparison of both approaches[17].

In holographic measurements, the height information of an object is obtained from the phase of the object wave. When measuring optically rough object surfaces, information about the macroscopic shape of an object is superimposed by an unknown speckle field. However, owing to the spatial coherence of speckle fields of different wavelengths, the macroscopic shape of an object can be extracted by a differential of the phase maps. If holograms of different wavelengths are recorded subsequently, this is referred to as temporal heterodyning[3,4].

In the past, Fraunhofer IPM has developed various digital holographic measurement systems adapted to specific applications. In the following, the architecture and structure of three of these sensors are presented. A detailed description of the workflow used for data acquisition and evaluation can be found in Ref. 4.

Measurement System Setup

Fig. 1 illustrates the general setup of a complete multi-wavelength digital holographic measurement system. It consists of a passive sensor head (A) and a multi-laser control electronics system (B). Variations between the different systems in this contribution are marked with an asterisk.

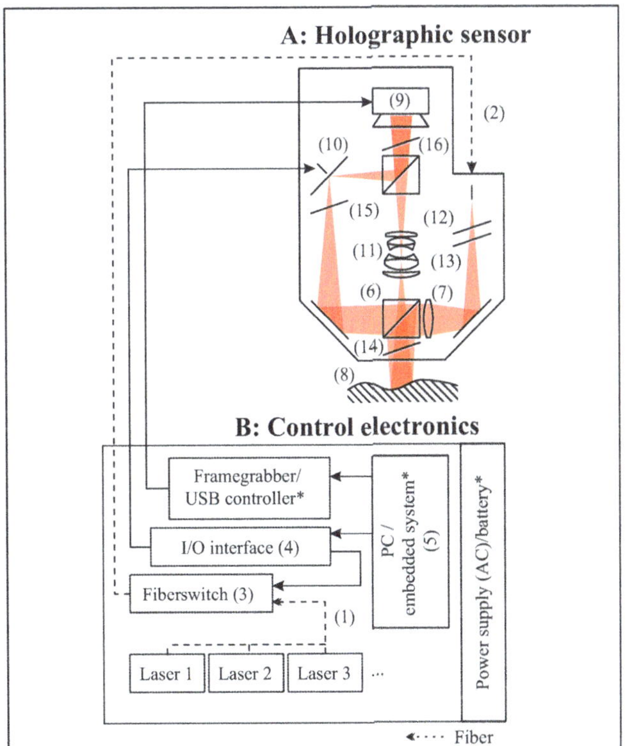

Fig. 1 Sketch of the holographic setup. Multiple laser sources (1) are cycled through by a fiber switch (3), controlled by a computer, inside the control electronics B. Single-frequency laser light is connected to the computer-controlled (5) temporally phase shifting measurement head A by a polarization maintaining fiber (2).

Multiple stabilised lasers (1) (three lasers are shown as examples in Fig. 1) with free-space optical powers of 25–50 mW are used to generate synthetic wavelengths ranging from ~2 μm to 2000 μm. For good interference contrast, the lasers must operate with a single frequency and a single mode. The laser sources (up to eight lasers) are successively coupled into the sensor head using a polarization-maintaining single-mode glass fibre (2) and switched between by a fast micromechanical fibre switch (3). The switch is controlled by an I/O interface (4) which in turn is driven by a PC or embedded system (5) inside the control electronics (B). The switching duration is specified as less than 7 ms.

Inside the sensor head (A), the beam is divided into an object and a reference beam by a polarising beam splitter cube (PBS) (6). One common lens refracts (7) the beams in such a way that they fit the size of the object (8) and the camera chip (9), respectively.

The reference beam is deflected by a mirror mounted on a piezoelectric actuator (10), which is again controlled by the I/O interface of (B). Thus, the phase of the reference beam can be shifted selectively within a single laser wavelength, which is called temporal phase shifting[12].

The light scattered by the sample is imaged onto a camera sensor by an objective lens (11), where it is superimposed with the reference beam. At least three phase-shifted interferograms are recorded for each wavelength. In addition, the following polarising optics are used: a polarising filter (12) is used to minimise the influence of laser polarisation fluctuations. The beam ratio between the object and reference beams is set by a half-wave plate (13). The quarter-wave plate (14) and half-wave plate (15) are required for the beams to pass or be deflected at the following PBS. The final polarising filter (16) enables interference between the object and reference beams.

During sensor development, the measurement field and lateral resolution can be adapted to the measuring task by selecting imaging optics and a camera chip. The following example specifications given in the text refer to the HoloTop 65M sensor unless stated otherwise.

A total of nine images (three lasers used for the measurements presented in this paper, three phase steps) at a resolution of 9344 × 7000 pixels were captured within 200 ms and digitised with 10-bit resolution. The acquisition time for a three-laser configuration is determined by the following: nine images must be recorded by the camera, which is capable of a framerate of 71 frames per second; a fibre switch (Leoni 8 × 2-pm-fiberswitch) must perform two switching operations taking 10 ms each; and between each of the images, the piezo-actuator must be moved for temporal phase shifting, which takes less than 1 ms. The resulting acquisition time is thus 156 ms. A USB controller or framegrabber handles the data being transferred to the computer unit.

Using a parallelised numerical reconstruction on the GPU of the control electronics (a PC or, if available, the internal embedded system) and its Compute Unified Device Architecture (CUDA), the complete reconstruction of the phase maps at the different synthetic wavelengths, along with the evaluation of the height map, takes only 100–300 ms, depending on the filter settings. For very large camera images (65 MP) and filter radii (e.g. 7 × 7 pixels), these processes can take up to 800 ms.

This is done in two steps: first, the complex wavefront in the hologram plane is calculated using a method introduced by Cai et al.[12] that compensates for vibrations and thus changes in the phase between the object and reference beam during data acquisition. Then, the complex wavefront is propagated to the focal plane without reconstruction of the disturbing twin image and zero diffraction order[4,18]. A flowchart of the algorithms used for hologram reconstruction can be found in Ref. 19.

Image acquisition, data evaluation, and feature extraction are done in parallel to meet the requirements of the respective production cycle. For a detailed description and mathematics of the numerical reconstruction procedure of digital multi-wavelength holograms, we refer to[3,4,12,20].

All sensor heads were simulated during development using the finite element method (FEM) to ensure maximum stiffness and thus natural frequencies below externally occurring excitations. For example, a natural frequency of 605 Hz was realised using the HoloTop 20M[21]. This is especially important for the HoloPort system working in the machine tool. With excitation frequencies typically well below 300 Hz, the influence of external stimulations could be minimised by mechanical means at a first natural frequency of 323 Hz[22].

Table 1 shows the specifications of the three holographic measurement systems developed at Fraunhofer IPM, which are presented with an application in this paper. They differ in terms of sampling and optical resolution, depending on the application they were designed for. One remarkable design consideration in the development of holographic sensors is the fact that speckle patterns generated by rough object surfaces must be sampled well, and slight oversampling yields the best results.

In addition to the systems presented here, there are other systems available as well: systems with a postcard-sized field of view, for measurements on moving objects, and with a microscope objective for particularly high lateral resolution[23].

Table 1 Specifications of selected holographic measuring systems developed at Fraunhofer IPM.

	HoloTop 9M	HoloTop 65M	HoloPort
Diffraction limited resolution/μm	13.5	6.4	16.2
Lateral sampling/μm	5.86	1.91	<7
Magnification	−0.94	−1.67	−0.5
Numerical aperture	0.06	0.10	0.04
Camera Sensor	CMOSIS CMV12000	GPixel GMAX3265	Sony IMX253
Number of pixels	3072 × 3072	9344 × 7000	4112 × 3008
Single FOV/mm²	18.0 × 18.0	17.8 × 13.4	20 × 20
Acquisition time per point/s	6.4e-6	3e-9	3.3e-8
Acquisition time per FOV/ms	~ 100	~500	<500

In a slightly modified form, the setups can also be used for deformation measurements using electronic speckle interferometry (ESPI)[24,25].

Accuracy and Reproducibility

For every 3D sensor, the traceability of the measured height values to standards is highly important. For this purpose, we measured the step height of a calibrated height normal. The height normal consisted of two gauge blocks manufactured by Kolb & Baumann GmbH & Co. KG, which nominally differed in height by 20 μm. The real step height was measured by the National Metrology Institute of Germany PTB to be 20.138 μm with an accuracy of 0.013 μm within a temperature range of 19.85 °C to 20.00 °C. We measured the step height using a HoloTop 65M system 25 times.

The laser sources used for these measurements were three fibre coupled frequency doubled solid state lasers with a free space power output of 50 mW each (Cobolt Samba at 532.20 nm and Cobolt Fandango at 517.67 nm and 514.46 nm). The synthetic wavelengths used for measurement here were thus 82.9 μm, 15.5 μm and 15.4 μm.

A height map after the numerical removal of sample tilts is shown in Fig. 2a, and the cross-sections of the height map are shown in Fig. 2b. The height map displays height values from −5 μm (black) to +25 μm (white). Data points with insufficient data quality due to low interference contrast are shown in blue. The cross-sections for all 25 measurements in Fig. 2b are nearly identical despite a low number of single point outliers at the very steep edge of the step resulting in undersampling of fringes in the synthetic phase maps. Calculating the differences between the mean height values on the left and right parts of the height map yielded the measured step height values which are shown in Fig. 2c. The mean value of the measured height was 20.135 μm, and the standard deviation of the measured heights was calculated to be 0.010 μm. Thus, the measured height values were in excellent agreement with the reference measurements by PTB.

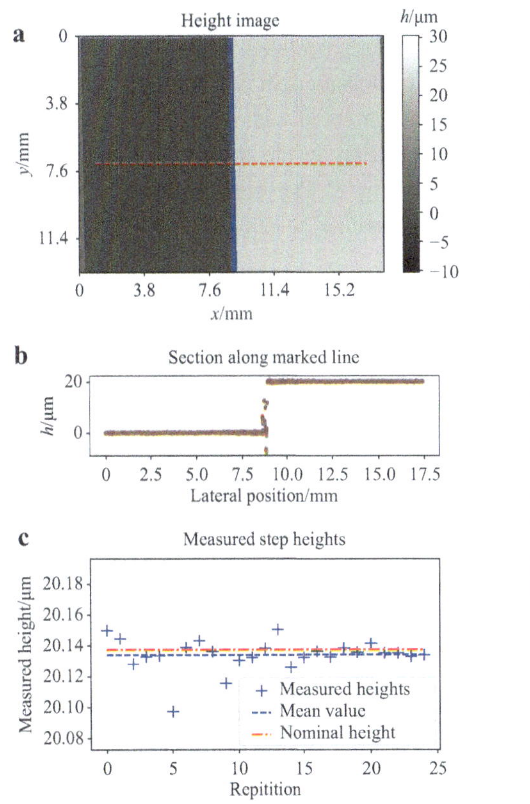

Fig. 2 Measurement results on a calibrated step height standard. **a** Grey value representation of the measured height map consisting of 9344 × 7000 pixels. **b** Cross-section of a along dashed line. **c** Measured values of step height for 25 repetitions.

For high-precision 3D measurements, it is very important to yield equal data quality throughout the field of view. This can be evaluated using a test target with a clearly detectable structure to ensure that the height difference can be calculated between identical regions, even if the sample is moved. The measurement data for one such sample manufactured by VLSI Standards Inc. is shown in Fig. 3. The test target consisted of a quartz substrate with a chromium cover. The grooves of the test target had a depth of 19.320 μm. The uncertainty of the groove depths was given by the manufacturer as +/−0.116 μm. It should be noted that to evaluate the homogeneity of data quality, an extremely tight uncertainty of the reference measurement is not needed because the main focus lies in the reproducibility of the holographic setup; furthermore, its absolute accuracy has been proven with the previous measurement, as shown in Fig. 2. The groove depth of the target was measured 14 times. Between each measurement, the sample was moved arbitrarily within the field of view. Fig. 3a–c show some of these measurements. The depth of the groove was calculated as the difference between the mean value measured in two regions on the top of the groove target, marked with green rectangles, and the mean value of a region inside the groove, marked with a red rectangle in Fig. 3. The dimensions of the regions on the top surface were set to 600 × 400 pixels, and those of the region inside the groove were set to 600 × 450 pixels. The results of this evaluation are shown in Fig. 3d. Each point refers to the groove depth measured at one of the fourteen different target positions. The mean value of the measured groove depths was 19.345 μm, and the standard deviation of all measured groove depths was 0.010 μm.

Thus, the measured groove depths were in excellent agreement with the reference data and showed no recognisable variation across the field of view of the sensor.

Industrial applications

Sealing surface inspection

One of the most prominent industrial applications of multi-wavelength digital holography is the inspection of high-precision sealing surfaces. These components are made from metal and produced with very tight tolerances of 10 μm and below. One manufacturer of such sealing devices for diesel injectors is Werner Gießler GmbH. This company produces approximately 10 million parts per year.

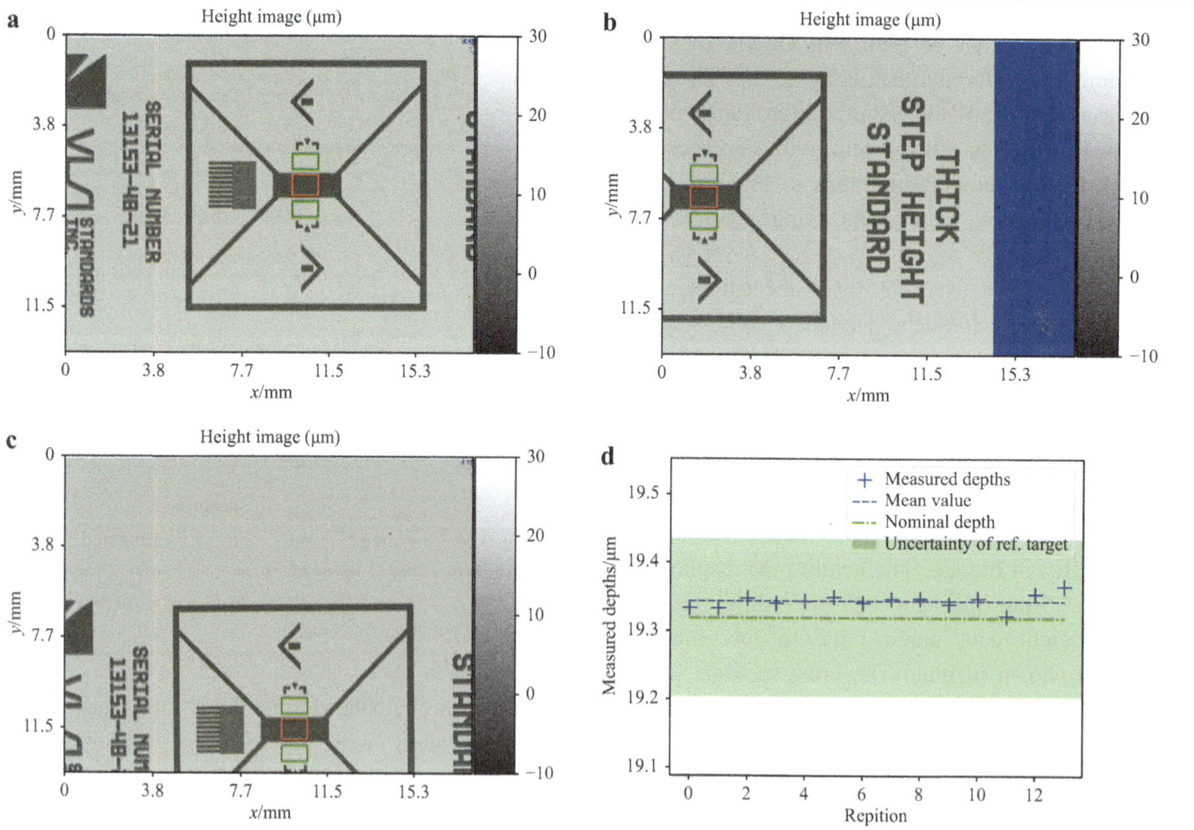

Fig. 3 Measurement results on a groove sample at different positions within the field of view. **a**–**c** height maps for three arbitrary positions of the sample consisting of 9344 × 7000 pixels each. The regions of interest for measurement of the groove depth are marked with red and green frames. **d** Measured height values for 14 arbitrary positions within the field of view of the sensor compared to the reference value from calibration data and uncertainty of reference measurement as given by the manufacturer of the sample.

Since 2015, they have been using a HoloTop 9M sensor for 100% end-of-line inspection, which means that each part produced is inspected with a digital holographic sensor[21,26]. A photograph of the integrated sensor for this inspection task is shown in Fig. 4. The blue box left of the centre of the image contains the interferometer, imaging optics, and camera. The three laser sources used for illumination were installed in a box which was located under the table and is not visible in the photograph. The laser light was guided through polarization-maintaining optical fibres to the sensor. The sensor and the holder on which the samples were placed for measurement were mounted on an actively controlled pneumatically vibration-damped table to decouple measurements from high-frequency floor vibrations (> 20 Hz). The exposure times required for accurate measurements of such samples range from 5 ms to 10 ms and are thus short enough to minimise the influence of lower frequency vibrations.

The laser sources used in this application were three separately stabilised fibre coupled diode lasers with a free space power output of 30 mW each (Toptica Topmode at 637.68 nm, and Toptica DL100 at 633.31 nm and 633.74 nm). Thus, the synthetic wavelengths used for measurement were 922 μm, 103 μm, and 92 μm.

The inspected surfaces were planes and conical surfaces with tilt angles of up to 20°. A representative photograph of the inspected components is shown in the inset of Fig. 4. The region that will be used to evaluate the data quality is marked with a red ellipse. The peak-to-valley deviations from the ideal shape on the most critical surfaces of such a sample must be smaller than ~10 μm, depending on the sample type. The inspection cycle time was 1 s, including data acquisition and handling time.

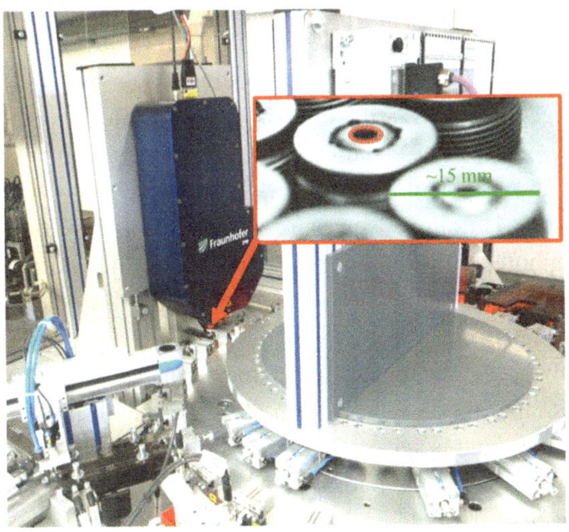

Fig. 4 Photograph of a digital holographic sensor for inline inspection of high-precision sealing surfaces. Inset: representative photograph of samples to be inspected.

The main specifications of the sensor used for this application are listed in Table 1 in the HoloTop 9M column. The field of view of the 18×18 mm^2 sensor was sampled with 3072 × 3072 pixels and was slightly larger than the sample diameter.

The output data were filtered using an amplitude-weighted averaging filter with a filter radius of 3 and a median filter of the same size. As the speckle size in the object plane was in the range of 14 μm, we roughly averaged over 2 × 2 speckles for the final result. Using the open-source image processing toolkits OpenCV and PointCloudLibrary, we separated the individual surfaces and borders of each sample. The functional surfaces were defined as two concentric cones and one plane. From each of the three surfaces, a best-fit geometrical description was calculated for the measurement of slope angles and diameters. The maximum positive and negative distances between each pixel of a surface and the best-fit representation allowed the calculation of the peak-to-valley deviation of each surface, which is closely related to the roughness measure P_T.

A detail of the height map of one of these samples is shown in Fig. 5a. The grey area represents the points that were assigned to the central surface of the sample. The colour map for the grey values ranges from −15 μm to +15 μm. Pixels that were masked out owing to low interference amplitudes are shown in blue. Pixels assigned to a different surface of the sample are shown in yellow.

The sample was measured 25 times. Between each measurement, the sample was picked up by the handling device, arbitrarily slightly rotated, and returned to the measurement position. In Fig. 5b, the cross sections of the 25 measurements along the dashed red line in (a) are depicted. Although there was no numerical realignment of the data, all measured height values lay within a corridor of +/−2 μm which consists of real surface deviation and misalignment due to handling and measurement accuracy.

From each of these 25 measurements, the peak-to-valley deviation of the surface was calculated. The average value of this deviation was 8.1 μm and the standard deviation of the 25 measurements was 0.45 μm.

For evaluation of long-term stability, the measured values of the feature described above were logged throughout the complete production of one year. During this year, more than 9.8 million measurements were carried out. The measured values of the feature are shown in Fig. 6. Each measured value was plotted in a semi-transparent colour to visualise the distribution of the measured value in a representative way. For display and evaluation, outliers have been removed; peak-to-valley deviations larger than

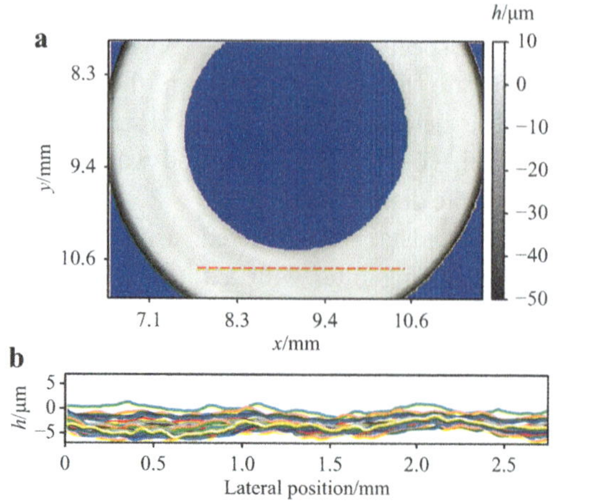

Fig. 5 Measurement results of a sealing sample. a Detail of the height map of the central surface of the sample. **b** Cross-section of 25 measurements along the dashed red line in **a**. For the 25 measurements, the sample was picked up by the handling device and placed back in the measurement position between each of the measurements.

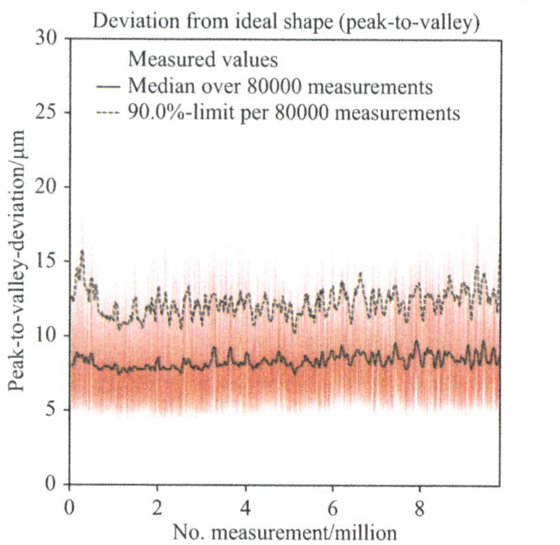

Fig. 6 Measured values for a single feature, peak-to-valley deviations of all data points on one plane surface, over one year. More than 9.8 million parts were measured within that year. All measured values, the median of the measured values over approx. one day (80,000 measurements) and the limit which 90% of all measured values of each day were under are shown.

500 µm cannot be measured realistically because synthetic wavelengths of 1000 µm were used, and these deviations were thus removed (0.02% of all measurements). Such errors appear, for example, if a measurement is performed with a sample placed inaccurately within the field of view or if there is no sample present at all. The black line shows the moving median over 80,000 measurements and the approximate number of measurements per day. The median value of all measured values was 8.3 µm. The median values per day were nearly constant over time and ranged from 7.4 µm to 9.8 µm. Furthermore, for every 80,000 measurements, the measured P_T under which 90% of the respective values lay ranged from 10 µm to 16 µm. These values were plotted as dashed lines in Fig. 6 and were within the range of 10.2 µm to 16.5 µm. The tight boundaries in which those values were found indicate that both the production process and the measurement procedure work with high accuracy as process variations of both contribute to the described variations.

Machine Tool

Machine tools are machines for forming or machining metal or other rigid materials and play a key role in production. Current machinery offers production accuracy in the micrometre range; precision machinery even works in the sub-micrometre range[27].

However, worn tools, incorrectly calibrated sensors, or poorly chosen trajectories can lead to machining results that do not meet the required specifications[28–30].

Therefore, the metrological registration of quality features is an essential part of quality and performance assurance. With a share of almost 90%, geometry inspection (form, dimension, position, and roughness) is highly significant within production metrology. For example, surface topographies, flatness, and microdefects or geometric references of construction elements, such as bores or fittings in the micrometre range, play a major role[31].

Monitoring and, if necessary, controlling the processing quality is a complex iterative process that has not been automated to date. Quality control is usually performed on a random basis outside the machine tool in special measuring rooms[28–30].

For the measurement of high-precision components, there are no measuring devices that allow direct flat three-dimensional measurements in machine tools[31].

Machine tools are complex mechatronic systems that pose a special challenge for high-precision optical measuring systems[19,22].

The HoloPort measuring system developed at Fraunhofer IPM is, to our knowledge, the world's first digital holographic measuring system for direct use in machine tools[19].

Owing to the purely mechanical interface of machine tools, the measuring system was designed for wireless operation and is self-sufficient[22].

Fig. 7 shows a photograph of the measuring system inside a Hermle C32U 5-axis milling machine at Fraunhofer IPM. The system was mounted on the machine's HSK 63 tool taper.

Using the dynamic numerical control (DNC) interface of

Fig. 7 Inspection of a typical milled surface by HoloPort in a Hermle C32U 5-axis milling machine at Fraunhofer IPM[19].

the machine tool, its axis positions can be read and used for computer-aided design (CAD) comparisons of the measured point cloud. Even complex work pieces can potentially be inspected completely and directly within the machine for the first time—the system becomes a full-field coordinate measuring machine (CMM). The accuracy of a complete point cloud of the surface is mainly limited by the lateral sampling of the system's camera and the accuracy of the machine tool axes, which are typically in the single micrometre range[22].

For the following measurements, two lasers (Ondax LMFC 25–50 mW, linewidth 150–300 MHz) with wavelengths of 632 nm and 640 nm were used, resulting in a synthetic wavelength of 58 μm. The reconstructed data were filtered using a 5 × 5 pixel mean filter. All measurements were taken with the machine axis enabled, but with disabled spindle control in order to minimise disturbing excitations[19].

Fig. 8 shows the measurement results of 30 consecutive measurements of the sealing surface of an electric motor housing. A total field of view of 120 × 100 mm with sub-micron precision and lateral sampling of 7 μm was achieved. The cycle was programmed using standard CAD software and converted to an NC program afterwards. The positioning of the measurement fields was chosen so that there was an overlap of approximately 50%, as shown by the black squares. Communication between the machine tool and the measuring system was established via the DNC interface.

In the lower half of the image, a section of the measurement data along the red line is visualised. It can be seen that the general flatness of the sealing surface is only a few micrometres, but a flaw in the material extends by approximately 30 μm.

Fig. 9 shows a single measurement of the blue square, marked in the stitched image in Fig. 8.

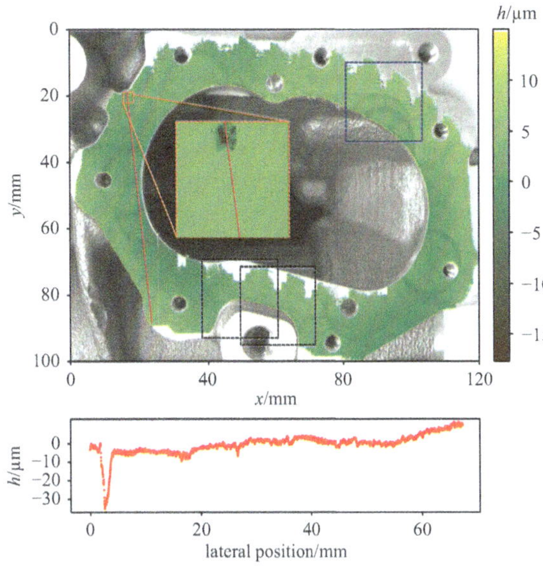

Fig. 8 Stitched measurement, consisting of 30 successively recorded overlapping measurement fields. A total field of view of 120 × 100 mm was recorded with sub-micron precision and lateral sampling of 7 μm. A detailed view (orange) shows the cavity visible in the cross-section.

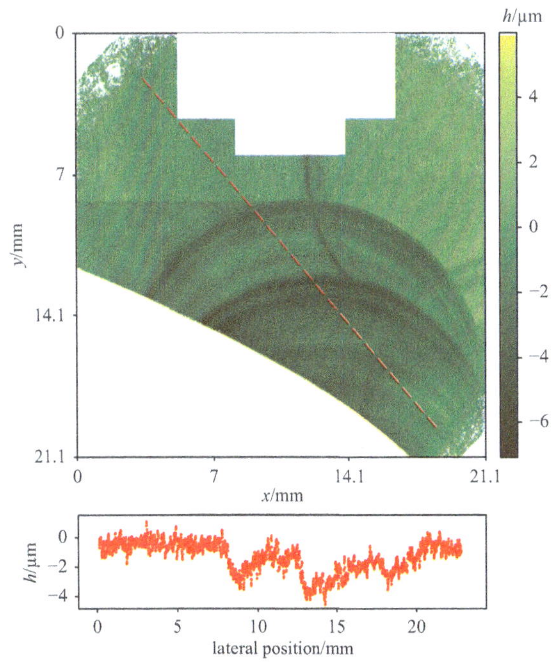

Fig. 9 Single measurement of the stitched measurement data shown in Fig. 8.

In the full-resolution image, one can clearly see several milling traces with a depth in the range of the desired flatness of the sample.

Invalid measurement data such as areas past the measuring object, defocused areas (e.g. the sensor-internal reference target), or weak measuring signals were masked out using white. In the single field of view, this becomes particularly clear in the systematically cut-out areas of the sensor-internal reference target at the upper edge of the image, and the missing data can also be seen for multiple measurements in the stitched data in Fig. 8.

This reference target is used to transform the initial relative measurement data into an absolute scale, if necessary; this reference is not relevant for the flatness within this functional surface but becomes relevant for global references on the part.

Fig. 10 shows one of the two optical planes of the open sensor from below; the reference target and its support stand out through their raw metal appearance.

Digital holography is unique in that the focal plane can be shifted numerically. This means that even samples with an axial extension far beyond the geometric depth of field of the measuring system can be processed to be evaluated in focus across the whole measurement. Fig. 11 shows the numerically refocused measurement data of the reference target at a distance of ~50 mm from the focal plane of the sensor—the actual sample, being far out of focus, has been masked out in this image. The example cross-sectional view shows a continuous transition with an extension with a height of 350 μm.

In the future, an adapted version of the reference target will allow internal wavelength calibration and observation.

Fig. 12 and Fig. 13 show the reproducibility of the complete measuring system, including the machine tool. The area to be measured was targeted by moving the three linear axes of the machine tool ten times successively. With a specified traversing accuracy of 8 μm (without the commercially available additional package for fine calibration), the machine accuracy was in the range of the lateral sampling of the measuring system, and the measured data were very well aligned. Axially, the measured values deviate from each other in the micrometre range. In addition to the system's camera, the machine axes, and external disturbing influences such as vibrations,

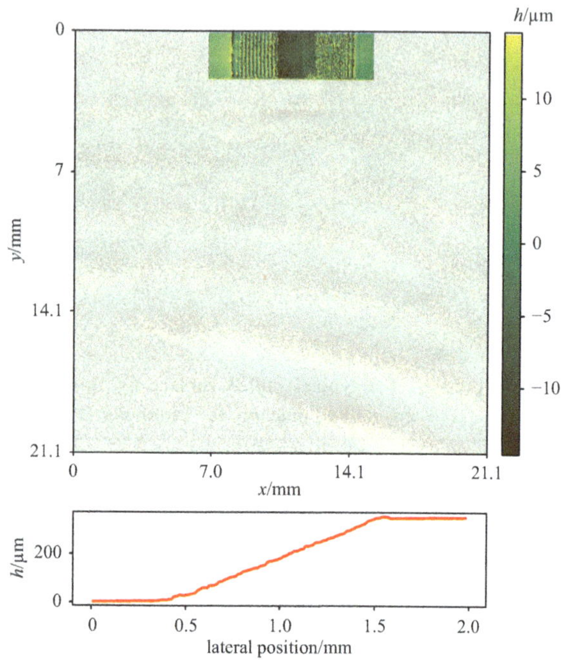

Fig. 11 Numerically refocused reference target with a continuous transition of a height of 350 μm. As the measured part is defocused and shows only noise, it has been masked out using transparency here; only the reference target is fully visible in this view.

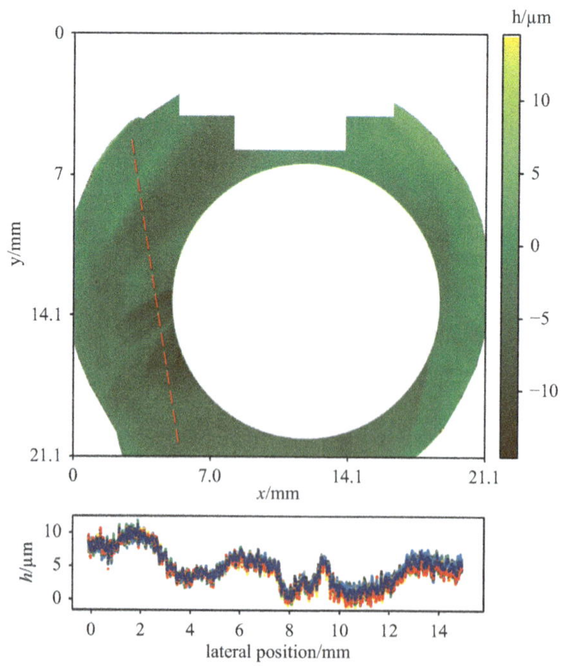

Fig. 12 Illustration of the reproducibility of the measuring system including the machine tool. The area to be measured was targeted by moving the three linear axes of the machine tool ten times successively.

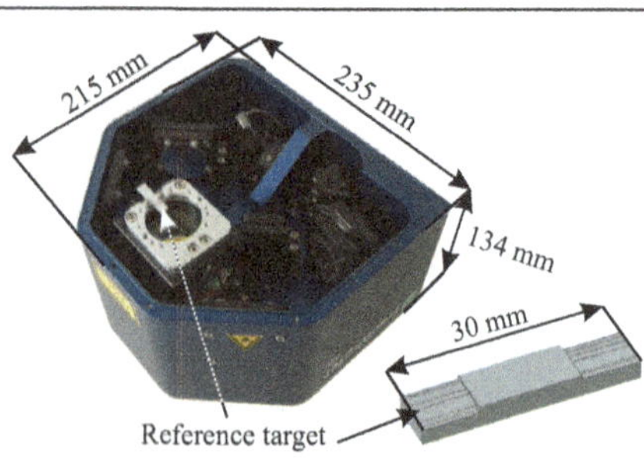

Fig. 10 View into the downside of the sensor head (without electronics) including internal reference body to measure absolute references in the measuring direction.

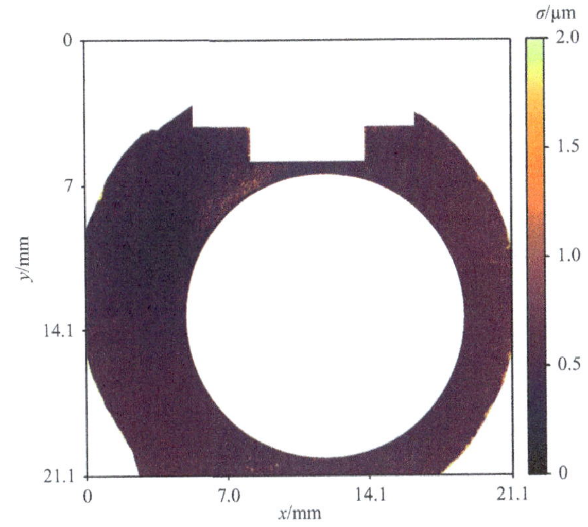

Fig. 13 The standard deviation of the 10 measurements shown in Fig. 12 are mostly below 1 μm. White spaces have been masked, as these areas include either no measuring object or the reference target.

the achievable accuracy depends on the choice of the laser wavelengths and thus the synthetic wavelengths[19,22].

PCB substrates

One of the most interesting applications of digital holography is the quality control of electronic interconnectors, such as ball grid arrays or microbump structures. State-of-the-art microelectronic devices have tens of thousands to hundreds of thousands of electrical contacts that must be connected using reflow soldering or similar technologies. To ensure a proper connection for all contacts of an interconnector, the electrical contacts of the device must fit tolerances in the single micron range. As the production cycle times of such devices are typically in the range of a few seconds or less, a measurement technology that meets the requirements of inline integration must be able to guarantee both sub-micron repeatability and extremely short measurement duration. In addition to the tight accuracy requirements, the high dynamic range between shiny soldering bumps and rough, dark substrates is challenging for optical measurement approaches.

To achieve a high lateral resolution and fast acquisition speed, the sensor system used for this application was equipped with a 65-megapixel camera and a high-resolution objective lens (focal length 88 mm, magnification $\beta' = -1.66$). The sensor system was developed in a joint project with the world's leading semiconductor manufacturer, Intel Corporation.

The laser sources used in this application are three separately stabilised fibre-coupled diode lasers with a free-space power output of 50 mW each (Ondax LMFC, wavelengths: 642.88 nm, 633.82 nm, and 632.915 nm). The synthetic wavelengths used for measurement here are thus 443 μm, 40.9 μm and 40.8 μm. Fig. 14 shows some measurement results of a relevant interconnect sample. Fig. 14a shows the full field of view of the sensor. 10,480 microbumps lie within the field of view and thus were measured in a single measurement. The inset shows a 10x confocal microscope image of a detail of the sample. Fig. 14b shows a detail of (a). Each of the bumps had a diameter of

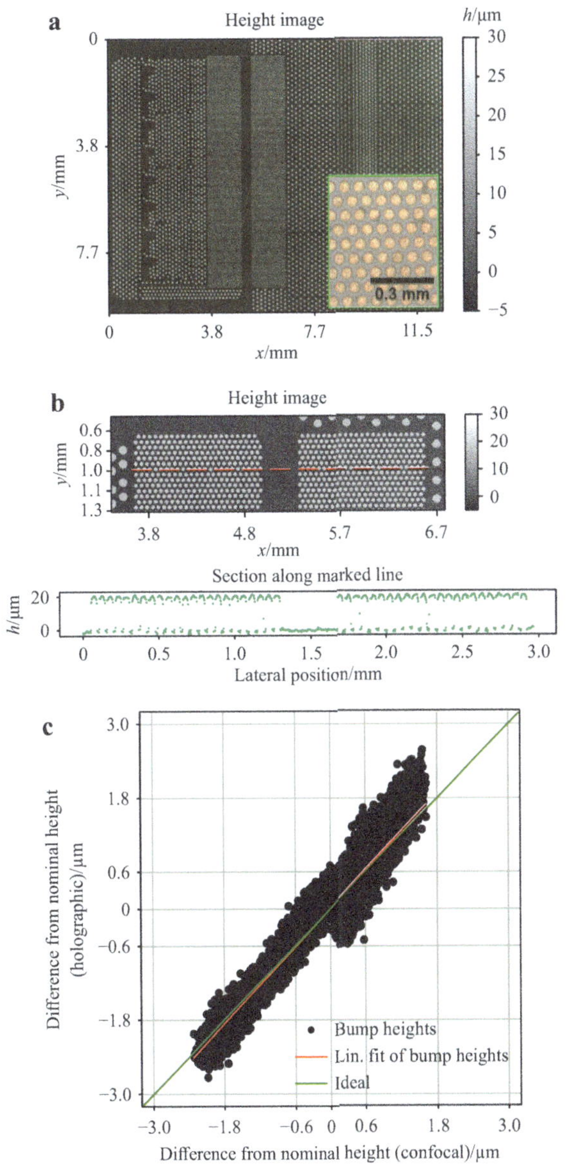

Fig. 14 Measurement results on microbump sample. a Height image of a BGA; inset: microscope image of the sample. **b** Detail of a marked with red. **c** Comparison between holographic data and reference data from a confocal microscope.

50 µm and was approximately 20 µm high. The cross-section through the data given below the image shows excellent data quality. To ensure that the measured heights corresponded to the real heights, reference measurements were performed on the same sample using a confocal microscope. Fig. 14c shows the results of this correlation. Here, for each bump, the height measured with the described holographic setup was plotted against the height measured by the reference tool. The mean value of the measured heights was set to zero. The red line illustrates a linear fit to all measured values. The green line corresponds to an ideal correlation between reference data and measured values (i.e. slope = 1). One can easily see the excellent agreement between the two measurements. The correlation coefficient between both data sets is 96.3% and all holographically measured values lie within a corridor of +/−1.0 µm around the respective reference data. To verify the repeatability of the measurement results, ten successive repeat measurements on the depicted sample were carried out. The heights of 99% of all bumps were measured with standard deviations below 0.108 µm. The heights of all bumps within the field of view were measured with standard deviations equal to or less than 0.155 µm.

Outlook

The above use cases show enormous potential for digital holography with multiple wavelengths. No other 3D measurement technology can capture such a large field of view so quickly with comparable lateral sampling and simultaneously with such high axial accuracy.

When applying these technologies to new fields of application, the major challenge of future work is to further reduce the sensitivity of the interferometric technique to external influences such as relative movements between the sensor head and test specimen. Recently published works on holographic topography measurement of continuously moving objects, such as the measurement of gear flanks during the rotation of a gear, are very promising[32].

Another bottleneck for wide-ranging industrial applications is the extensively frequency-stabilised laser sources. Because the requirements for measurement accuracy and the unambiguity range vary from application to application, it is still necessary to find and characterise an appropriate laser source (e.g. one with long-term stability)for each wavelength needed. Promising lasers with long-time frequency stabilisation are presented, for example, in the following papers[33,34]. However, this is time-consuming and rather uneconomical.

Hence, the current work carried out by Fraunhofer together with the University of Freiburg is concerned with developing a novel light source based on whispering gallery mode resonators. This is intended to enable controlled tuning of the emitted wavelength in the range of 100–300 GHz. For short measurement times, it is necessary to switch between different wavelengths within milliseconds and achieve high reproducibility at the same time.

The goal of all further developments is to open up as many industrial applications as possible on the basis of a common technological platform and to make the technology more reliable and robust. In this context, we want to develop a very compact sensor head for use in multi-axis handling systems such as robots, measuring portals, and coordinate measuring devices.

New manufacturing processes, such as hybrid machining, in which an additively manufactured component is precisely reworked, offer great application potential for process-accompanying measurements using digital holography.

The current unique selling points of the HoloTop family, including robustness, accuracy, and speed, will be further enhanced and extended.

Summary

We showed that the HoloTop sensor family, which is based on multi-wavelength digital holography, is very powerful for tackling demanding industrial topographic measurement tasks with the highest accuracy requirements. The decisive advantage that distinguishes it from other optical measurement methods and commercially available measurement systems is the short measurement time in relation to the amount of measured 3D points per second. Because the raw data quantity per number of 3D points is less than that of comparable methods such as coherence scanning and laser scanning confocal microscopy, we assume that this intrinsic advantage will be retained and that the technology will continue to assert itself.

The presented measurement results also show that the technique works on reflective and matt surfaces simultaneously. This independence from sample roughness and the capability of measuring even steep flank angles shows universal usability. In our opinion, this point in particular limits the practical application of many other optical measuring methods and is particularly responsible for the fact that tactile measuring technology, which is significantly slower, is still widely used.

The outlook describes which future developments can improve the cost-effectiveness of the technology and which fields of application can be addressed with it.

More generally, the described measurement technology can open up those applications in which large measurement fields in the range of several square centimetres must be detected and tested topographically sib-micrometre accuracy. Especially for inline inspection, the comparatively very short measuring times, well below one second, enable a quantitative 100% inspection in cases where currently only sample testing is possible. The latest applications described in this paper prove this point.

Acknowledgements
The authors would like to thank Werner Gießler GmbH, Elzach, Germany for their partnership during the integration of the prototype measuring system HoloTop into series production and the provision of real production data over the period of an entire year. We also thank the Federal Ministry for Economic Affairs and Energy for financial support as part of the Central Innovation Program for small and medium-sized enterprises (ZIM-KA 3398901 LP4). We also thank Intel Corporation for their great confidence in our developments and permission to publish the measurement data. We are also very grateful for the support provided by the precision mechanical workshop of Fraunhofer IPM for sensor integration into a five-axis machine tool.

Conflict of interest
The authors declare that they have no conflict of interest

Received: 23 January 2021 Revised: 25 March 2021 Accepted: 22 May 2021
Accepted article preview online: 27 May 2021
Published online: 18 June 2021

References

1. Leach, R. Optical Measurement of Surface Topography. (Berlin, Heidelberg: Springer, 2011).
2. Malacara, D. Optical Shop Testing. 3rd edn. (Hoboken, NJ: Wiley-Interscience, 2007).
3. Wagner, C., Osten, W. & Seebacher, S. Direct shape measurement by digital wavefront reconstruction and multi-wavelength contouring. *Optical Engineering* **39**, 79-85 (2000).
4. Carl, D. et al. Multiwavelength digital holography with autocalibration of phase shifts and artificial wavelengths. *Applied Optics* **48**, H1-H8 (2009).
5. Lyncée Tec. Digital holography. at https://www.lynceetec.com/digital-holography/ on 2021-04-21.
6. Parry, G. Some effects of surface roughness on the appearance of speckle in polychromatic light. *Optics Communications* **12**, 75-78 (1974).
7. Fonseca, E. S. R. et al. Comparative analysis of autofocus functions in digital in-line phase-shifting holography. *Applied Optics* **55**, 7663-7674 (2016).
8. Gabor, D. A New microscopic principle. *Nature* **161**, 777-778 (1948).
9. Goodman, J. W. & Lawrence, R. W. Digital image formation from electronically detected holograms. *Applied Physics Letters* **11**, 77-79 (1967).
10. Masuda, N. et al. Computer generated holography using a graphics processing unit. *Optics Express* **14**, 603-608 (2006).
11. Shimobaba, T. et al. Real-time digital holographic microscopy using the graphic processing unit. *Optics Express* **16**, 11776-11781 (2008).
12. Cai, L. Z., Liu, Q. & Yang, X. L. Generalized phase-shifting interferometry with arbitrary unknown phase steps for diffraction objects. *Optics Letters* **29**, 183-185 (2004).
13. Greivenkamp, J. E. Generalized data reduction for heterodyne interferometry. *Optical Engineering* **23**, 234350 (1984).
14. Bruning, J. H. et al. Digital wavefront measuring interferometer for testing optical surfaces and lenses. *Applied Optics* **13**, 2693-2703 (1974).
15. Takeda, M., Ina, H. & Kobayashi, S. Fourier-transform method of fringe-pattern analysis for computer-based topography and interferometry. *Journal of the Optical Society of America* **72**, 156-160 (1982).
16. Burke, J. Application and optimisation of the spatial phase shifting technique in digital speckle interferometry. PhD thesis, Carl von Ossietzky University Oldenburg, Oldenburg, 2000.
17. Burke, J. & Helmers, H. Spatial versus temporal phase shifting in electronic speckle-pattern interferometry: noise comparison in phase maps. *Applied Optics* **39**, 4598-4606 (2000).
18. Kreis, T. Holographic Interferometry: Principles and Methods. (Berlin: Akademie Verlag, 1996).
19. Seyler, T. Digitale holographie in der werkzeugmaschine. PhD thesis, BoD, Kaiserslautern, 2020.
20. Kreis, T. Handbook of Holographic Interferometry: Optical and Digital Methods. (Weinheim: Wiley-VCH, 2005).
21. M. Fratz. et al. Inline application of digital holography[Invited]. *Applied Optics* **58**, G120-G126 (2019).
22. Seyler, T. et al. HoloPort – Design and integration of a digital holographic 3-D Sensor in machine tools. *Journal of Sensors and Sensor Systems* **9**, 33-41 (2020).
23. Beckmann, T. Schnelle optische vermessung von mikrostrukturen auf tellergroßen flächen. *Journal für Oberflächentechnik* **60**, 48-51 (2020).
24. Beckmann, T. et al. High-speed deformation measurement using spatially phase-shifted speckle interferometry. Proceedings of SPIE 9006, Practical Holography XXVIII: Materials and Applications. San Francisco, California, United States: SPIE, 2017.
25. Creath, K. Phase-shifting speckle interferometry. *Applied Optics* **24**, 3053-3058 (1985).
26. Fratz, M. et al. Inline application of digital holography[Invited]. *Applied Optics* **58**, G120-G126 (2019).
27. Smith, G. T. Machine Tool Metrology: An Industrial Handbook. (Switzerland: Springer-Verlag, 2016).
28. Brinksmeier, E. et al. Process signatures – an alternative approach to predicting functional workpiece properties. *Procedia Engineering* **19**, 44-52 (2011).
29. Ramesh, R., Jyothirmai, S. & Lavanya, K. Intelligent automation of design and manufacturing in machine tools using an open architecture motion controller. *Journal of Manufacturing Systems* **32**, 248-259 (2013).
30. Möhring, H. C. et al. Intelligent tools for predictive process control. *Procedia CIRP* **57**, 539-544 (2016).
31. Luo, X. C. & Qin, Y. Hybrid Machining: Theory, Methods, and Case Studies. (Amsterdam: Elsevier, 2018).
32. Schiller, A. et al. Motion compensation for interferometric off-center measurements of rotating objects with varying radii. *APL Photonics* **4**, 71301 (2019).
33. Abdelsalam Ibrahim, D. G. & Yasui, T. High-precision 3D surface topography measurement using high-stable multi-wavelength digital holography referenced by an optical frequency comb. *Optics Letters* **43**, 1758-1761 (2018).
34. Abdelsalam Ibrahim, D. G. & Yasui, T. Multi-object investigation using two-wavelength phase-shift interferometry guided by an optical frequency comb. *Applied Physics Letters* **112**, 171101 (2018).

Official journal of the JHL 2689-9620
www.light-am.com

Article Open Access

Snap-shot topography measurement via dual-VCSEL and dual wavelength digital holographic interferometry

Daniel Claus[1,*], Igor Alekseenko[1,2], Martin Grabherr[3], Giancarlo Pedrini[4] and Raimund Hibst[1]

Abstract

In this paper, we propose a dual-wavelength digital holographic interferometry method based on a compact dual vertical-cavity surface-emitting laser (VCSEL) source. The source simultaneously emits light from two highly stabilized coherent light sources with slightly different wavelengths. A highly stabilized and adjustable current source enables the application of digital holographic dual-wavelength techniques to measure the shape of an object with height steps of a few millimeters. The wavelength drift over 12 h over the entire measurement range, which was evaluated using a wavemeter, was smaller than 1 pm. In addition to the low measurement uncertainty at large height jumps, the dual-wavelength digital holographic system distinguishes itself by its robustness to environmental disturbances such as air turbulence, heat load, and/or mechanical vibrations. This is enabled via a fiber-based almost common-path single-shot digital holographic acquisition of the information of the two different wavelengths using angular multiplexing. The experimental setup and data evaluation are discussed, and we present measurements of non-cooperative objects with specular reflective and/or diffuse reflective surfaces having different colors.

Introduction

The motivation of this work stems from an automotive application, where an inline measurement system is required to enable the collection of topographic data of the car body throughout its entire production cycle. Therefore, the system should be capable of delivering data starting from the raw diffuse scattering metal sheets to the colored and polished varnish surfaces of the final car body. Important measures of a car's production quality are the size and depth of the gap between two adjacent car body parts. The same system should preferably be employed during the production cycle to enable the comparability and traceability of any manufacturing errors. The field of view should be at least 10×10 mm^2 with an axial measurement range of 2 mm, a lateral and axial resolution of 20 μm, and an acquisition time of less than 50 ms.

Therefore, a high-speed topographic measurement on non-cooperative surfaces (specular reflective and/or diffuse reflective) needs to be performed, and represents a major challenge in optical metrology. There are surface metrology techniques that can cope with these types of surfaces, such as white light interference microscopy (WLI), confocal microscopy (CM), and conoscopic holography (CH). However, the measurement is very time-consuming, and ranges from a few seconds to minutes, depending on the field of view and resolution. This is because WLI and CM are very detailed, revealing all

Correspondence: Daniel Claus (daniel.claus@ilm-ulm.de)
[1]Institut für Lasertechnologien in der Medizin und Meßtechnik, Helmholzstraße 12, Ulm, D-89081, Germany
[2]Immanuel Kant Baltic Federal University, A. Nevsky str.14, Kaliningrad, 236016, Russia
Full list of author information is available at the end of the article.

surface parameters such as the roughness, waviness, and shape. Furthermore, they can be applied to investigate volume scattering samples because they offer optical sectioning[1]. By contrast, CH represents a polarization optical common path-point interferometer[2], and requires scanning in both lateral dimensions. Therefore, it is based on a more complex setup and cannot satisfy the high demand imposed on the data rate of more than 5 million data points per second. Structured illumination, such as fringe projection, can reveal the shape based on a single imaging frame, but it can only be applied to rough scattering surfaces. Moreover, a projection device, which is often implemented via a spatial light modulator and a certain angle between the projection and observation directions, is necessary to enable triangulation. It is therefore not only limited by the type of surface, but also suffers from shadowing artefacts and cannot be used to investigate deep holes or steep surfaces, which is possible with the aforementioned online systems WLI, CM, and CH.

Therefore, there remains the need for a system that can measure the topography of non-cooperative surfaces at high speed.

One imaging technique that can be employed for such a purpose is dual-wavelength holography (DWH)[3], which can simultaneously be applied to rough scattering[4] and specular reflective surfaces[3]. The measurement system does not rely on triangulation, and can therefore be applied to recover three-dimensional (3D) depth data from deep holes, steep surfaces, or similar geometrical features.

The digital hologram is acquired as an interferometric superposition of the reference wave and object wave on charge-coupled device (CCD) or complementary metal-oxide semiconductor (CMOS) detectors.

Depending on the recording scheme, i.e., on-axis or off-axis, the phase in the hologram plane is recovered via phase stepping/shifting or filtering in the Fourier domain. A detailed discussion of the recording geometries as well as their advantages and disadvantages can be found in Refs.5–7. If a defocused hologram is recorded, reconstruction algorithms[8–10] can be applied to obtain the phase distribution in the object plane. However, in image-plane holography, the digital sensor is positioned in the image plane. Therefore, no refocusing algorithms need to be applied. The phase information in the object plane provides important information about the topography of the object, and is represented as a wrapped phase map with phase values ranging between -π and π. Therefore, the unambiguous evaluation of the phase is limited to a phase difference of less than 2π between neighboring reconstructed object points, which corresponds to a path length change of half a wavelength. Hence, the application of conventional holographic interferometry imposes certain constraints, which hinders the evaluation of strongly curved, inclined surfaces or surface-steps that exceed half the wavelength. This problem can be solved by employing the dual-wavelength method. This method is based on holographic recording and reconstruction at two different wavelengths, λ_1 and λ_2. The phase difference that is calculated from the two wavelength reconstructions can be associated with the synthetic wavelength λ_{syn}:

$$\lambda_{syn} = \frac{\lambda_1 \cdot \lambda_2}{|\lambda_1 - \lambda_2|} = \frac{\lambda_1 \cdot (\lambda_1 + \Delta\lambda)}{\Delta\lambda} \tag{1}$$

which is much larger than that of the individual wavelengths employed. The resulting phase difference map $\Delta\varphi$ consists of contour lines (fringes) with 2π-steps. The theory of fringe formation has been described in more detail in Refs. 11 and 12.

The application of the dual-wavelength method is not only limited to surface metrology but can also be applied for defect detection/evaluation or for comparison of shape differences[13–19].

The topography:

$$T = \frac{\Delta\varphi}{4\pi}\lambda_{syn} \tag{2}$$

can be calculated from the synthetic wavelength and the phase difference $\Delta\varphi$ between the wavelength-corresponding digital holographic reconstructions, in which both the illumination and observation vectors are perpendicularly aligned to the object and sensor surface.

The phase information required to recover a shape that may comprise rapid height jumps within the mm or even cm range can be recovered from a single digital hologram[20]. The imaging speed is limited only by the acquisition speed of the two-dimensional (2D) digital sensor and the power of the light source to supply a sufficiently large amount of scattered and reflected light to enable recordings beyond the noise threshold of the sensor.

Moreover, DWH offers a flexible and lightweight setup, without the need of imaging lenses. This is enabled by the recovery of the complex wavefront in the sensor plane (amplitude and phase) and the application of wave propagation algorithms to refocus from the sensor to the object plane, as discussed in Ref. 21.

However, the measurement uncertainty of the DWH depends on the stability of the previously measured individual wavelengths during the calibration process. To better understand the measurement error introduced by a change in the wavelength difference, we differentiated Eq. 1 with respect to the wavelength difference $\Delta\lambda$, which results in

$$\frac{\partial \lambda_{syn}}{\partial \Delta\lambda} = \frac{\lambda_1^2}{\Delta\lambda^2} \quad (3)$$

For a desired measurement uncertainty of 1/100 of the measurement range T, the standard deviation of the wavelength difference is

$$\partial\Delta\lambda = \frac{\Delta\lambda \cdot (\lambda_1 + \Delta\lambda)}{100 \cdot \lambda_1} \quad (4)$$

Plotting the standard deviation of the wavelength difference against the synthetic wavelength results in the curve shown in Fig. 1, with the first wavelength λ_1= 850 nm. Hence, for a measurement uncertainty of 1/100 of the measurement range of, for instance, 2 mm, the standard deviation of the wavelength difference should not exceed 2 pm. This imposes a high demand on the stability of the light source.

Different methods have been developed to address this issue. In Ref. 22, a white light-emitting diode (LED), which represents a broadband spectral light source, was employed. More than 1 000 holograms were recorded by shifting the piezo in the reference path. By applying the Fourier transform along the z-axis of the recorded stack of holograms, the spectral signal could be recovered. The spectrally resolved holograms were displayed in the k-space, which corresponds to $1/\lambda$, with λ being the corresponding wavelength. The equal spacing in k has an advantage in that phase map pairs with the same difference in k generate the same synthetic wavelength. This effect was used, and 34 equally k-spaced phase difference maps were averaged to obtain a speckle noise-minimized reconstruction. Moreover, when considering the entire complex spectral information, a volume-scattering sample could be investigated, and an optical section that was focused on different axial slices could be applied in the same manner as the previously mentioned WLI. In addition to the spectral self-calibrating and noise minimizing characteristics, the system has a low acquisition speed, which is in the range of several minutes, and which is therefore not applicable for our task.

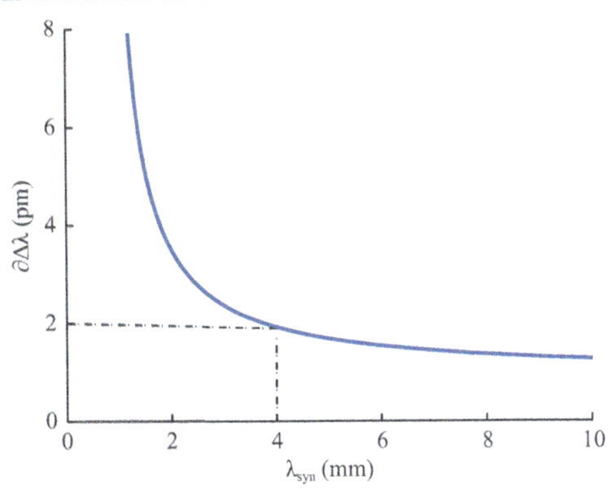

Fig. 1 Synthetic wavelength over the standard deviation of the wavelength difference.

In Ref. 23, a conventional line spectrometer was employed to simultaneously record the individual wavelengths during the experiments. This approach is feasible for short synthetic wavelengths within the μm range, but fails at longer synthetic wavelengths in the millimeter range or beyond, which requires a spectral resolution in the pm or sub-pm regime to ensure a reasonable measurement uncertainty.

In Ref. 24, the detector was employed as a spectrometer using a reflection grating. However, this approach suffers from a limited applicable range of the synthetic wavelength owing to the limited spectral resolution. Moreover, only one point is measured, which requires lateral scanning to obtain 3D information of the entire object.

Another system relies on the application of reference objects of a previously known height or phase delay, which is sequentially moved into the path[25], and hence enables the recovery of the synthetic wavelength from the phase difference map. However, this approach is based on the assumption that the environmental conditions and the wavelength of the employed light source(s) remain unchanged for the subsequent recording of the holograms with two different wavelengths. Furthermore, the reference device needs to be moved in and out of the optical path, which requires additional stages, and hence, increases the complexity of the setup.

Results

The aforementioned issue that impedes the determination of the synthetic wavelength during the measurement is addressed via the application of a highly stabilized coherent light source. We employed a vertical-cavity surface-emitting laser (VCSEL) light source for this task, and although it has a diameter of only 50 μm and a similar height, it has a coherence length of up to 50 cm. Moreover, the radiation profile that is emitted is circularly symmetric at all axial distances from the source, which when compared to the predominantly elliptical radiation profile of edge-emitting devices, makes them perfectly suited for single-mode fiber launching with high coupling efficiency.

Compared to conventional coherent light sources such as He-Ne lasers or edge-emitting laser diodes, VCSELs have a very short laser cavity. Therefore, within a few μs after changing the current, the wavelength is already stabilized, which enables the generation of multiple synthetic wavelengths within a short sequence. Moreover, the temporal coherence covers a few tens of centimeters, which outperforms the typical coherence length of an edge-emitting laser (only a few millimeters) and is comparable to that of He-Ne lasers.

Thermally caused wavelength drifts are very small in comparison to edge-emitting lasers (typically 0.25 $nm \cdot K^{-1}$ (Ref. 26).

In addition, the wavelength can be tuned by a few nm by changing either the current (0.5 $nm \cdot mA^{-1}$) or the temperature (0.06 $nm \cdot K^{-1}$).

Therefore, they cannot be highly wavelength stabilized. Fine adjustment of the wavelength by the application of a precise adjustable current source enables the adaption of the synthetic wavelengths to the object of interest or improvement of the measurement uncertainty using multiple synthetic wavelengths.

The generation of the spectrally highly stabilized dual-wavelength VCSEL source is presented in the Materials and Methods section.

To highlight not only the speed advantage, but also to demonstrate that the results do not depend on the surface roughness, material, and color, the single-shot dual-wavelength method was applied to specular reflective surfaces, diffuse reflective surfaces, and industrial surfaces with different colors. The fiber-based setup is shown in Fig. 2c, and holograms that are based on an off-axis single-shot image plane holographic arrangement are recorded. More details about the setup are discussed in the Methods section. The field of view covered an area of 10 × 10 mm^2. The ILA CMOS image sensor was used, which offers 2000 × 2000 pixels with a pixel size of 6.5 μm and a 16-bit analog-to-digital converter (ADC) with a low readout noise of 2.2 electrons. In this manner, it is possible to tolerate the slightly increased output image noise due to the simultaneous recording of two holograms in a single acquisition, as reported in Ref. 27. After recording the hologram, the wavelength corresponding to the information content must be extracted.

The reconstruction of the complex amplitude was accomplished by filtering in the Fourier plane[9] (see Fig. 3). Owing to the different inclination angles of the two reference beams, which correspond to the two different wavelengths, two differently oriented carrier frequencies having almost the same amplitude were incoherently superimposed in a single acquired hologram. Using a 2D Fourier transformation, the complex object waves for the two different wavelengths can be separated in the Fourier space. Thus, the corresponding information can be filtered and zero-padded to result in the same number of pixels. The complex object wave in the image plane is obtained by applying an inverse Fourier transform.

To compensate for the residual phase, a reference hologram was recorded, as shown in Fig. 3. For the reference hologram, a flat mirror was positioned at the same distance from the sensor at which the object is located. The use of the same reconstruction procedure shown in Fig. 3 results in a residual phase map, which accounts for the phase difference introduced by the imaging lens and the two spherical reference wavefronts that are incident on the sensors at different angles. The reference wavefront is intrinsic to the system for the same choice of synthetic wavelength, and therefore, only has to be recorded once if the synthetic wavelength is not changed. Even if the synthetic wavelength changes, this reference phase map may still be used. A change in the synthetic wavelength results in a different scaling of the residual phase with respect to the synthetic wavelength; with a shortened synthetic wavelength, the residual phase possesses more 2π phase wraps, but the overall shape remains unchanged, as discussed in Ref. 28. After subtraction with the reference phase map, the phase of the object under investigation that corresponds to the synthetic wavelengths becomes clearly visible and free from major disturbances (see Fig. 4).

Specular reflective object: VCSEL calibration, assessment of measurement uncertainty and sensitivity

To evaluate the measurement uncertainty and sensitivity of our proposed device, specular reflective surfaces were chosen, which do not introduce speckle noise and therefore result in a high-quality difference phase map. We used reflective gauge blocks manufactured according to DIN EN ISO 3 650, which offers a thickness manufacturing uncertainty of ± 0.4 μm. The synthetic wavelength that was used was 8.92 mm, and the corresponding 2D height distributions for the two-gauge blocks with heights of 2 mm and 1 mm are shown in Figs. 5a–c. The field of view was covered by both gauge blocks, resulting in a total height of 3 mm. The exposure time used to evaluate an area of 10 × 10 mm^2 was 5 ms. Thus, the minimum required axial measurement range of 2 mm and the measurement time could even be exceeded.

To evaluate the height sensitivity of our proposed device and the measurement uncertainty, blocks with two different gauges and with a height difference of 40 μm were chosen, while maintaining the same synthetic wavelength of 8.92 mm. Fig. 5d shows the topography profile. Because the step-profile is clearly visible, the measurement uncertainty is 9.8 μm, which according to Ref. 29 is obtained from the square root of the sum of the squared standard deviations of the higher (σ = 7.8 μm) and lower surfaces (σ = 6.0 μm); corresponds to 0.22% of the

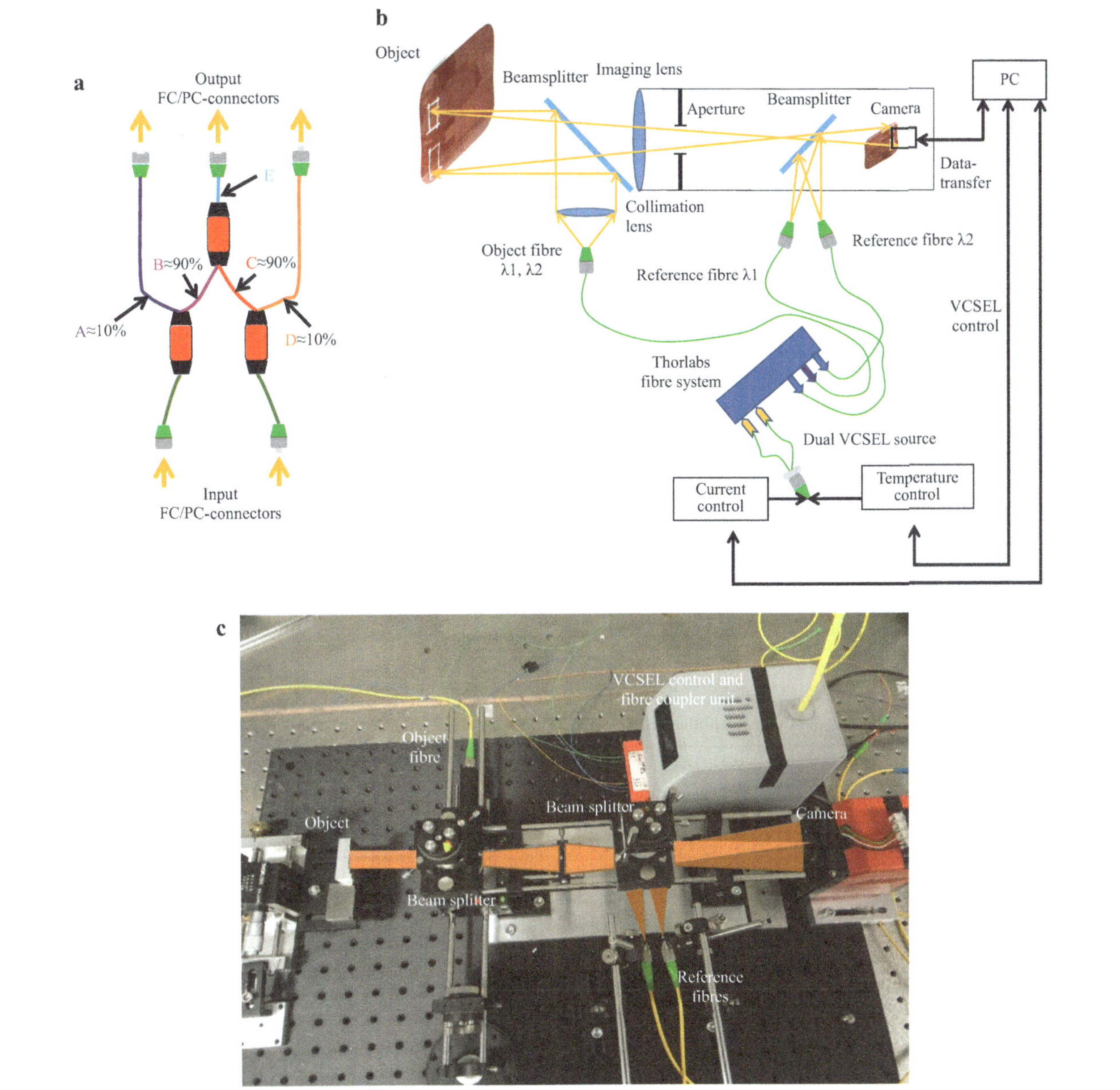

Fig. 2 Experimental setup **a** Schematic presentation of ILM designed fibre system, where fibres A and B correspond to the reference- and object-fibre of the first VCSEL, D and C the reference- and object fibres of the second VCSEL, E the combined object-fibers from object-fibers of the first VCSEL B and second VCSEL C, **b** schematic implemented setup, **c** experimental setup with a step object of sequentially reduced step size.

measurement range of 4.46 mm (half the synthetic wavelength). For smaller step heights such as 20 μm, a clear distinction was no longer possible.

Diffuse reflective step sample

To further prove the versatility of the dual-wavelength method at recovering the topography of any object with respect to the surface quality, an in-house computer numerical control (CNC) milled diffusely scattering object was selected. The sample has eight steps with varying heights of 2 mm, 1 mm, 400 μm, 200 μm, 100 μm, 40 μm, 20 μm, and 10 μm, as shown in Fig. 6. The diffuse reflective step sample was produced with a manufacturing accuracy of ±15 μm provided by the CNC machine.

Because the rough surface scatters light more strongly than the specular surface used in our previous investigation, the exposure time had to be increased to 35 ms. In Fig. 7, the topography map and cross-section plots for the 2-mm and 1-mm steps are represented. The heights measured with our system are 1.96 mm for the manufactured 2-mm step and 0.96 mm for the 1-mm step. The quality difference between the specular reflective and

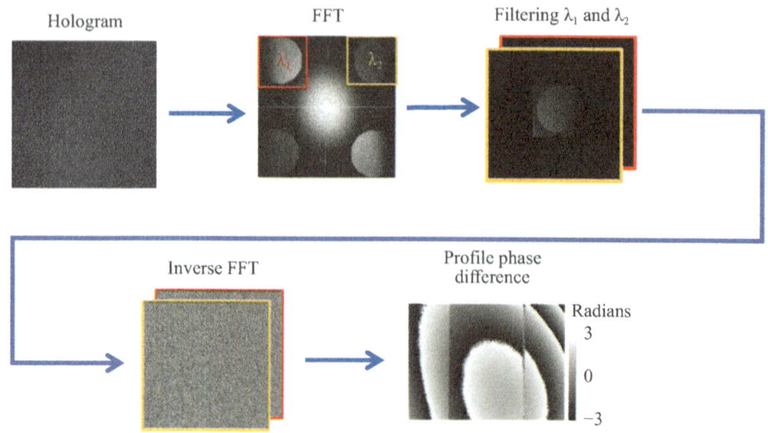

Fig. 3 Phase difference and topography reconstruction algorithm.

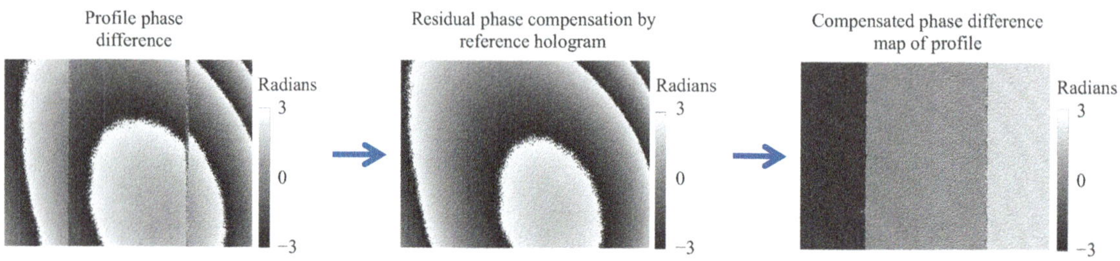

Fig. 4 Residual phase compensation to obtain the undisturbed topography map of the sample under investigation.

diffuse reflective objects is based on a different degree of speckle noise, which is obviously higher for the second one, owing to the increased roughness of the surface.

However, the quality of dual-wavelength digital holography can be increased when employing more than two wavelengths combined with the application of a hierarchical demodulation[30]. However, this comes at the cost of either an increased complexity of the setup to enable single-shot imaging or an increased acquisition time.

Non-cooperative-colored surfaces

To further highlight the advantage of the dual-wavelength method with respect to its independence regarding the color of the object surface, two car body parts, which have the same shape but different colors (red and black), were chosen. These two non-cooperative parts were placed close to each other. The surface of the testing parts consisted of a combination of specular and diffuse reflective surface areas. The upper layer was varnished, whereas the bottom was covered with industrial paint. Because the upper varnished layer is predominantly transparent with a thickness of 30 µm, light passes through it in an almost undisturbed manner. Therefore, a major part of the light interacts only with the paint. In Fig. 8a, non-cooperative surfaces are represented. The yellow rectangle marks the investigated region of 10 × 10 mm^2, where the two parts are joined. Fig. 8b, c represents the topography map and its cross-section, respectively, with the red line representing the average height value. The corresponding average profile height between the touched surfaces was 2.98 mm.

Discussion

A dual-VCSEL-based dual-wavelength digital holographic system was developed. It can deliver a measurement uncertainty of 1/500th of the measurement range for a full-field optical surface topography measurement within a single shot. Hence, it enables the investigation of dynamic events or harsh environmental conditions with more than 100 million recovered topographic data points within 1 s. The challenge of obtaining the required exact knowledge of the two wavelengths during the experiment was addressed via the development of a compact dual-VCSEL light source, which is spectrally highly stabilized using a well-suited precise temperature control (Peltier element) and current control unit. This approach resulted in a wavelength drift of 1 pm over 24 h, and is applicable to the entire tunable wavelength range of approximately 1.5 nm.

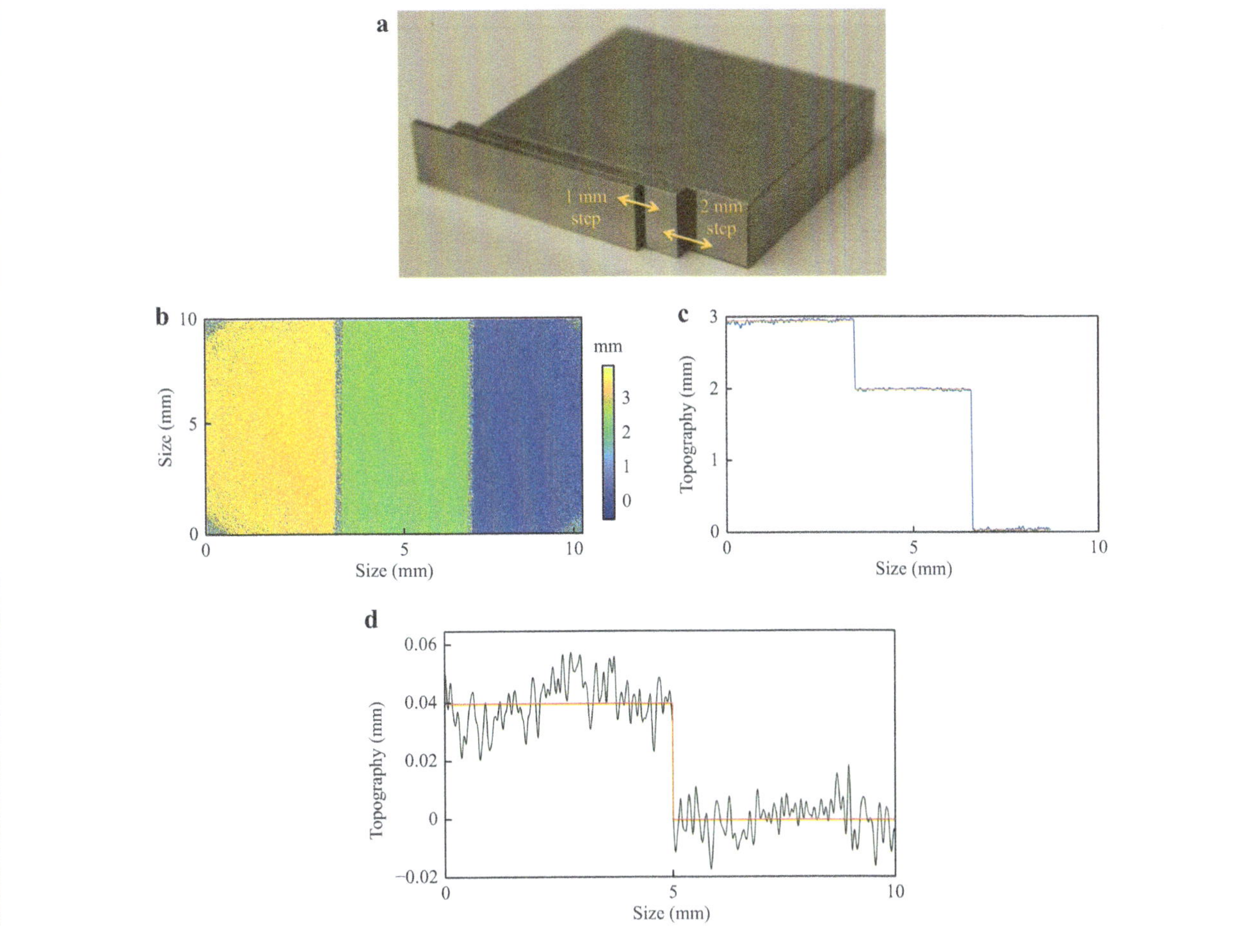

Fig. 5 Measurement results of specular reflective sample a specular reflective height gauges, b topography map, c topography cross-section plot, d topography cross-section for 40-µm step height.

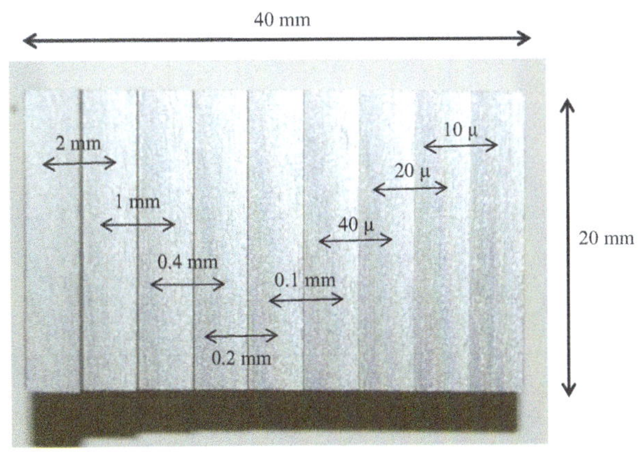

Fig. 6 Diffuse reflective step sample.

In this manner, we could develop a reliable measurement system that delivers a good measurement uncertainty of 0.22% of the measurement range. We demonstrated the versatile applicability of our single-shot VCSEL-based dual-wavelength system using specular reflective, diffuse reflective, and differently colored samples. The dual VCSEL approach that is demonstrated can further be improved, delivering an even better measurement uncertainty. This can be accomplished by either performing multiple recordings while changing the wavelength, resulting in another ideally smaller synthetic wavelengths, or by the application of more than two VCSELs on a common actively temperature-stabilized Peltier element. This element enables the simultaneous recording of information provided by different wavelengths. However, this comes at the cost of increased complexity of the setup and reduced dynamic range for each wavelength in the recording process. The use of additional wavelengths, which are spectrally more distant from each other, may result in speckle decorrelation. In combination with a setup with a high numerical aperture, all micro-topographic parameters (shape, waviness, and roughness profile) can be recovered[31]. To date, in our case, special emphasis has been placed on the recovery of the shape with reduced speckle noise. Therefore, the two speckle-patterns that are recorded

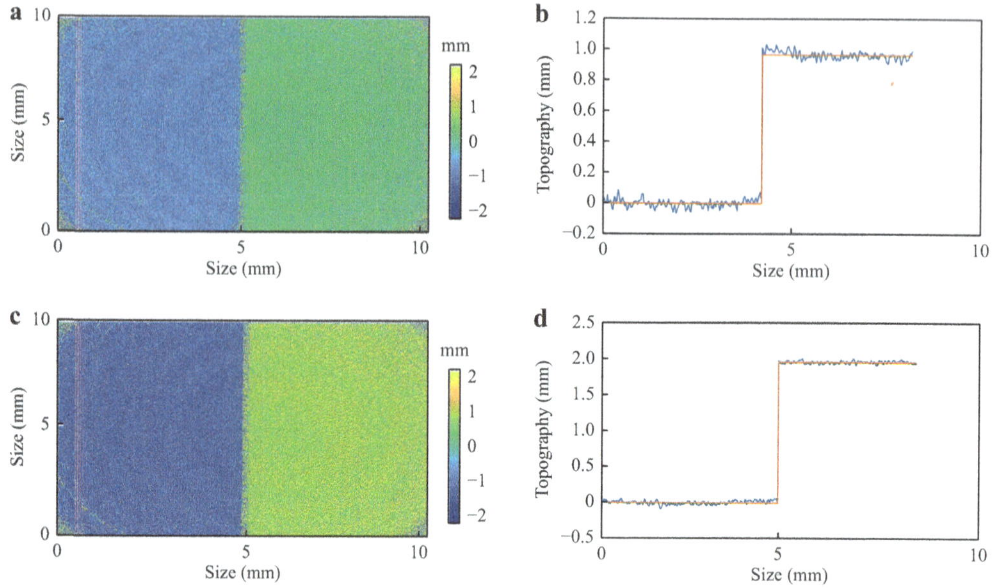

Fig. 7 Measurement results of diffuse reflective sample a topography map representation with a 1-mm step, **b** topography cross-section for a 1-mm step, **c** topography map representation of a 2-mm step, **d** topography cross-section for a 2-mm step.

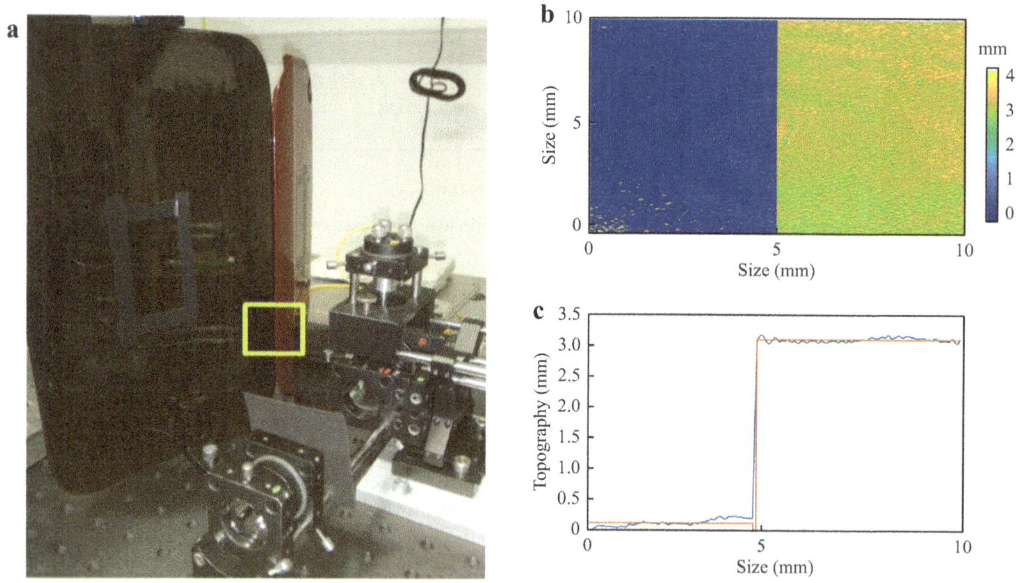

Fig. 8 Non-cooperative-colored surfaces and measurement results a setup with non-cooperative-colored surfaces, **b** topography map, **c** topography cross-section plot.

have a high degree of correlation, which hinders the retrieval of roughness information from the dual-wavelength phase-difference map. The speckle from the individual wavelength reconstruction and its corresponding contrast can only be used to derive roughness-related parameters and not the roughness profile itself. Typical roughness-related parameters that can be extracted from the speckle contrast are the arithmetic mean (Ra) roughness value or correlation length[32]. However, in addition to the recovery of these few mostly integrative roughness parameters, the object under investigation needs to satisfy the weak scattering condition, and pre-calibration becomes necessary. Therefore, there is little industrial benefit when employing speckle-correlated wavelength pairs to recover roughness parameters[33].

Finally, the compact and spectrally highly stabilized VCSEL-based dual-wavelength method can be extended to the multi-wavelength method. This step enables a further reduction in measurement uncertainty, and in the case of wavelength pairs causing speckle-decorrelation, the recovery of the roughness profile, from which all relevant roughness parameters can be derived.

Conclusion

In this study, we demonstrated dual-wavelength digital holographic interferometry based on a highly spectrally stabilized dual-VCSEL light source, which is tunable up to a 2-nm difference in wavelength. Stability is achieved through accurate thermal and current stabilization. Therefore, without any reference or monitoring system, it can be applied for topography measurements with a synthetic wavelength in the millimeter range, while still maintaining a sufficiently high measurement uncertainty. The setup was realized in a compact, lightweight, and robust manner using optical fibers, which were directly bonded to the active emission area of each VCSEL.

The dual-wavelength measurement system was realized in a single-shot manner with an exposure time of only a few milliseconds. This means that both wavelengths are simultaneously exposed to changes in the environment, such as heat drift, vibrations, or air flow. Moreover, owing to the short integration time and the usual low-frequency nature of these changes, the environmental changes are frozen in time. In the phase subtraction process, the influence of the environment is minimized. Hence, the measurement system is well suited for in-line and online monitoring during the production process. Furthermore, it can be applied to non-cooperative surfaces, which may be scattering, specular reflective, or a combination of both. In addition, it is not restricted to any color on the surface. Therefore, it enables the traceability of parts that are subjected to changing surface properties during the production cycle, such as car bodies.

Materials and methods

Dual-VCSEL-based light source

We mounted two VCSELs with an emission wavelength of 850 nm in close proximity to each other, i.e., approximately 1.2 mm, on a common heat sink that was equipped with a thermistor to read the temperature and a thermo-electric cooling element (TEC) to stabilize the temperature, as shown in Fig. 9. To simplify the further handling of the emitted light and to enable separate beam steering for both wavelengths, the active emitter surface of each VCSEL was directly bonded to a single-mode fiber. The fiber bonding was performed using a UV-curing epoxy resin. The final fiber-bonded dual-VCSEL light source is shown in Fig. 10, whereas at each fiber end, light of the corresponding VCSEL with mutually different wavelengths is emitted.

A control circuit based on an Arduino microcontroller and the required electronic components was implemented in the ILM electric engineering stores, by which the temperature was stabilized to 10 mK and the current could be set from 3 mA to 6.2 mA in 20-μA steps with a current stability of 8 μA. Using a USB connection, the current can be set by a computer.

The wavelength current response of both VCSELs was then recorded using a highly precise wavemeter (HighFinesse, Typ:WS7-60) with a measurement uncertainty of 0.13 pm. The wavelength standard deviation for both VCSELS over 12 h over the entire sweep range was less than 1 pm. However, when switching the system again the next day, the starting point of the curves slightly changed, which may be because of a certain amount of mechanical loading as the VCSELs have been touched when connecting them to the control board. Hence, in principle, the system should be very stable once calibrated.

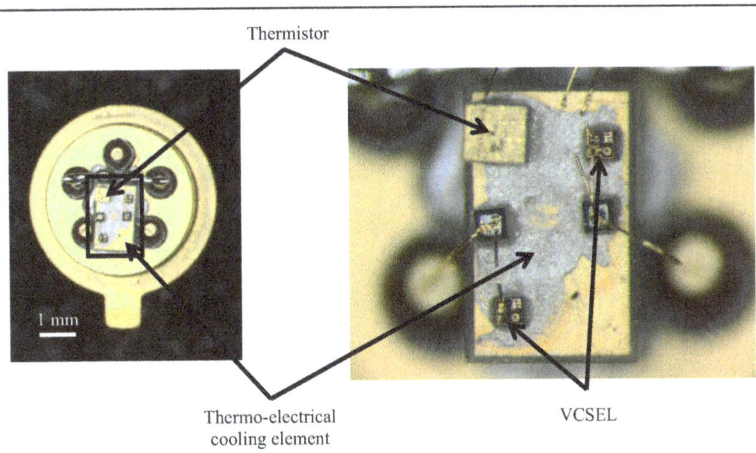

Fig. 9 Two VCSELs mounted on a common TEC.

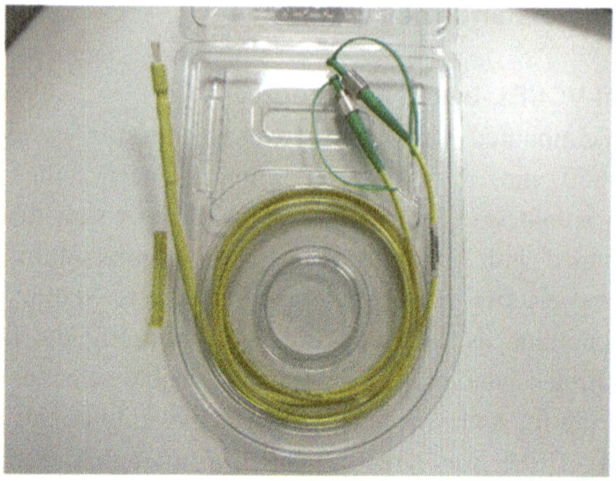

Fig. 10 Dual-VCSEL single-mode fiber light source.

However, to account for any change that may occur daily (mechanical load, changes humidity, changed air pressure), a reference object can be used each day before starting the measurement, as proposed in Ref. 25. The aforementioned reflective gauge blocks (see Fig. 5) were also used for the calibration, which were only applied once a day before the measurement to calibrate the synthetic wavelength; this wavelength can be calculated from the known topography of the reference objects by rearranging Eq. 2:

$$\lambda_{syn} = \frac{4\pi T}{\Delta\varphi} \tag{5}$$

Therefore, we applied a current setting to VCSELs according to the curves shown in Fig. 11, which delivered a synthetic wavelength of more than 8 mm. Using the gauge blocks, a synthetic wavelength of 8.92 mm was confirmed. However, in contrast to Ref. 25, the reference object only needs to be applied once a day before the measurement starts, and not throughout the entire measurement process. In fact, hermetic sealing of the dual-VCSEL source and remote control of the applied current should ensure long-term stability and reproducibility of the emitted wavelength with a spectral drift of below one pm.

VCSEL-based dual-wavelength digital holographic interferometry

To further improve the compactness of the system and to increase the stability toward environmental changes, the VCSEL-based dual-wavelength setup was implemented using an in-house designed fiber system manufactured by Thorlabs, as shown in Fig. 2a. Through a mating sleeve at the end of the two fiber-bonded VCSELs, the light is launched into a fiber system, which directs 10% of the light into the reference fiber and 90% into the object fiber. Then, the two object fibers are merged again into a common fiber output, by which the intensity at the common fiber tip is reduced by 50%, resulting in 45% of the initial light being launched into the fiber of each VCSEL. Hence, the object fiber emits light that is approximately five times more intense than the reference fiber. Thus, it is ensured that even at steep surface gradients or strongly scattering surfaces, there is sufficient scattered light from the object to produce an interference pattern with high-contrast interference fringes. The best interference contrast is obtained if both the reference and object light have the same intensity. The light emitted from the object fiber was collimated and redirected via a beam splitter plate to hit the object perpendicularly, as shown in Fig. 2b, c. Therefore, the sensitivity vector is the same for all object points, and the opposed-directed illumination and observation vectors standing perpendicular to the object plane result in the highest out-of-plane phase sensitivity. Disturbing reflection effects, which usually occur when employing a cube beam splitter, can be minimized by the application of beam splitter plates. Moreover, because of the thin wall of the beam splitter plate, the occurrence of spherical aberrations is reduced compared to that of a beam splitter cube.

Image plane holograms have been recorded, and offer an advantage in that after applying the separation of the corresponding imaging terms in the Fourier-domain, the reconstructed complex object wave is directly obtained. No further numerical refocusing algorithms have to be applied,

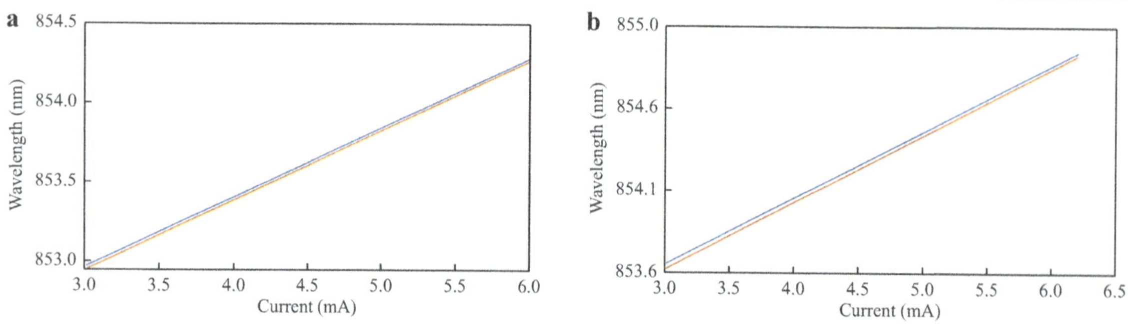

Fig. 11 Current–wavelength diagram over two days, blue first day and orange second day for a VCSEL 1, **b** VCSEL2.

which in the case of the direct Fresnel-transform would result in slightly different pixel sizes for both wavelengths that will have to be compensated prior to further processing. Image plane holography is one of the computationally least expensive approaches to recovering the complex object wave. Moreover, the speckle noise is reduced because the corresponding diffraction orders, at least the zeroth and first order according to Abbe's criterion[34], that are transmitted through the aperture are recombined to generate an imaging point and not a speckle. Hence, the speckle noise can be reduced in image-plane holography, as confirmed in Ref. 35, and can further be reduced using a high numerical aperture, which was reported in Ref. 36 for confocal imaging and was confirmed by Ref. 37 in digital holography.

However, more speckles arise in a defocused recording position because of the interference of diffraction orders that correspond to other object points. Speckles were still clearly recorded in the image plane. However, their origin results from the limited aperture, which for high diffraction angles allows only one of the two necessary diffraction orders to be transmitted; this interferes with the other transmitted diffraction orders with an arbitrary phase offset, and hence in an unrelated manner.

This effect is enhanced at scattering surfaces, which can be considered to be composed of various inclined surfaces and structures, and therefore results in diffraction cones that are directed away from the optical axis. At highly scattering surfaces, a fully developed speckle pattern can be obtained, which means that the speckle is the main carrier of information about the object and its surface. In such cases, it becomes a crucial requirement for the successful application of the dual-wavelength method for the recovery of shape at small NA, in which both wavelengths are close to each other, to ensure that both speckle patterns are not decorrelated. When subtracting the phase of such a correlated speckle pattern, which is encoded via interference with the corresponding reference waves, the object shape can be revealed. The previously randomly appearing phase of the individual wavelength reconstruction becomes homogenized after subtracting the phase reconstruction of the second wavelength; thus, the speckle noise is reduced.

According to Ref. 38, if the surface height fluctuations obey Gaussian statistics, the speckle decorrelation caused by the wavelength change can be expressed by the intensity correlation coefficient as follows:

$$|\mu_T(\Delta\lambda)|^2 = \exp\left[-(4\pi)^2\left(\frac{\sigma_T}{\bar{\lambda}}\right)^2\left(\frac{|\Delta\lambda|}{\bar{\lambda}}\right)^2\right] \qquad (6)$$

The standard deviation of the surface-height fluctuations corresponds to the R_q roughness value. The averaged wavelength and the wavelength difference can both be replaced by the synthetic wavelength, which results in the following simplified expression:

$$|\mu_T(\Delta k)|^2 = \exp\left[-\left(\frac{4\pi R_q}{\lambda_{syn}}\right)^2\right] = \exp\left[-(4\pi R_q \Delta k)^2\right] \qquad (7)$$

where Δk is the difference in wavenumber ($k = 1/\lambda$). An intensity correlation value of zero indicates a complete decorrelation, and "one" indicates complete correlation of the intensity speckle pattern. Therefore, the intensity correlation should be larger than 0.95 to ensure that speckle decorrelation can be neglected. By modifying Eq. 7, the maximum difference in the wavenumber Δk or its reciprocal parameter, λ_{syn} can be calculated to obtain the limit of possible wavelength pairs:

$$\Delta k \leq \frac{\sqrt[2]{-\ln(0.95)}}{4\pi R_q} \qquad (8)$$

The R_q value of our diffusely scattered samples is 1.1 µm, which results in a wavenumber difference of 16.38 mm^{-1}; this corresponds to a synthetic wavelength of 61.03 µm. Our chosen synthetic wavelength is larger than 4 mm, which satisfies the condition set in Eq. 8, resulting in a correlation coefficient of approximately 1.

Because the phase information of the two wavelengths is recorded in a single acquisition, speckle decorrelation or instabilities due to environmental conditions can be neglected.

Acknowledgements

The authors would like to acknowledge the support of BMBF under grant "AuMeRo."

Author details

[1]Institut für Lasertechnologien in der Medizin und Meßtechnik, Helmholzstraße 12, Ulm, D-89081, Germany. [2]Immanuel Kant Baltic Federal University, A. Nevsky str.14, Kaliningrad, 236016, Russia. [3]Priolas GmbH, Kellerhalde 24, Ulm 89081, Germany. [4]Instiut für Technische Optik, Universität Stuttgart, Pfaffenwaldring 9, 70195, Stuttgart, Germany

Author contributions

D. Claus conceptualized and initiated the research and took part in developing the method. I. Alekseenko was responsible for the research and development of the proposed technique. M. Grabherr developed and provided the VCSEL for the study. G. Pedrini and R. Hibst proposed a method to improve the research. The manuscript was discussed and corrected by all authors.

Data availability

The data that support the plots within this paper and other findings of this study are available from the corresponding author upon reasonable request.

Code availability
All codes that support this study are available from the corresponding author upon reasonable request.

Conflict of interest
The authors declare that they have no conflict of interest.

Received: 28 May 2021 Revised: 12 November 2021 Accepted: 18 November 2021
Accepted article preview online: 25 November 2021
Published online: 29 December 2021

References

1. Sheppard, C. J. R. Scanning confocal microscopy. in Encyclopedia of Optical Engineering (ed Driggers, R. G.) (New York: Marcel Dekker, 2003), 2525-2544.
2. Gaburro, N., Marchioro, G. & Daffara, C. A versatile optical profilometer based on conoscopic holography sensors for acquisition of specular and diffusive surfaces in artworks. Proceedings of SPIE 10331, Optics for Arts, Architecture, and Archaeology VI. Munich: SPIE, 2017, 103310A.
3. Wyant, J. C. Testing aspherics using two-wavelength holography. *Applied Optics* **10**, 2113-2118 (1971).
4. Fischer, E. W. et al. Dual-wavelength heterodyne interferometry for rough-surface measurements. Proceedings of SPIE 1319, Optics in Complex Systems. Garmisch: SPIE, 1990.
5. Claus, D., Iliescu, D. & Bryanston-Cross, P. Quantitative space-bandwidth product analysis in digital holography. *Applied Optics* **50**, H116-H127 (2011).
6. Claus, D., Watson, J. & Rodenburg, J. Analysis and interpretation of the Seidel aberration coefficients in digital holography. *Applied Optics* **50**, H220-H229 (2011).
7. Kelly, D. P. & Claus, D. Filtering role of the sensor pixel in Fourier and Fresnel digital holography. *Applied Optics* **52**, A336-A345 (2013).
8. Creath, K. Phase-shifting holographic interferometry. in Holographic Interferometry (ed Rastogi, P. K.) (Berlin: Springer, 1994), 109-150.
9. Takeda, M., Ina, H. & Kobayashi, S. Fourier-transform method of fringe-pattern analysis for computer-based topography and interferometry. *Journal of the Optical Society of America* **72**, 156-160 (1982).
10. Fienup, J. R. Phase retrieval algorithms: a comparison. *Applied Optics* **21**, 2758-2769 (1982).
11. Hildebrand, B. P. & Haines, K. A. Multiple-wavelength and multiple-source holography applied to contour generation. *Journal of the Optical Society of America* **57**, 155-162 (1967).
12. Friesem, A. A. & Levy, U. Fringe formation in two-wavelength contour holography. *Applied Optics* **15**, 3009-3020 (1976).
13. Shin, S. & Yu, Y. H. Three-dimensional information and refractive index measurement using a dual-wavelength digital holographic. *Journal of the Optical Society of Korea* **13**, 173-177 (2009).
14. Di, J. L. et al. Dual-wavelength common-path digital holographic microscopy for quantitative phase imaging of biological cells. *Optical Engineering* **56**, 111712 (2017).
15. Kühn, J. et al. Real-time dual-wavelength digital holographic microscopy for MEMS characterization. Proceedings of SPIE 6716, Optomechatronic Sensors and Instrumentation III. Lausanne: SPIE, 2007, 671608.
16. Gare, S. et al. Dual wavelength digital holography for 3D particle image velocimetry. *Journal of the European Optical Society: Rapid Publications* **10**, 15009 (2015).
17. Rajshekhar, G., Gorthi, S. S. & Rastogi, P. Simultaneous measurement of in-plane and out-of-plane displacement derivatives using dual-wavelength digital holographic interferometry. *Applied Optics* **50**, H16-H21 (2011).
18. Osten, W. et al. Remote shape control by comparative digital holography. Proceedings of 4th International Workshop on Automatic Processing of Fringe Patterns. Berlin: Akademie Verlag, 2001, 373-382.
19. Osten, W., Baumbach, T. & Jüptner, W. Comparative digital holography. *Optics Letters* **27**, 1764-1766 (2002).
20. Kühn, J. et al. Real-time dual-wavelength digital holographic microscopy with a single hologram acquisition. *Optics Express* **15**, 7231-7242 (2007).
21. Schnars, U. & Jueptner, W. Numerical reconstruction. in Digital Holography (Berlin: Springer, 2005), 44-55.
22. Claus, D. et al. Accuracy enhanced and synthetic wavelength adjustable optical metrology via spectrally resolved digital holography. *Journal of the Optical Society of America A* **35**, 546-552 (2018).
23. GmbH, V. Device for detecting a 3D structure of an object. (2013).
24. DeGroot, P. Interferometric laser profilometer. (1992).
25. Carl, D. et al. Multiwavelength digital holography with autocalibration of phase shifts and artificial wavelengths. *Applied Optics* **48**, H1-H8 (2009).
26. Bodendorfer, T. et al. VCSEL in interferometry: a comparison to edge-emitting diode lasers regarding their applicability in speckle-interferometry. *Journal of Optoelectronics and Advanced Materials:Rapid Communication* **6**, 36-39 (2012).
27. Pedrini, G. et al. Feasibility study of digital holography for erosion measurements under extreme environmental conditions inside the International Thermonuclear Experimental Reactor tokamak. *Applied Optics* **58**, A147-A155 (2019).
28. Gross, H. et al. Wave aberrations. in Handbook of Optical Systems Volume 3 Aberration theory and corrections of optical systems (Wiley-VCH, 2007), 80.
29. Taylor, J. R. An Introduction to Error Analysis. (Colorado: University Science Books, 1996), 488.
30. Wagner, C., Osten, W. & Seebacher, S. Direct shape measurement by digital wavefront reconstruction and multi-wavelength contouring. *Optical Engineering* **39**, 79-85 (2000).
31. Kühn, J. et al. Fast noncontact surface roughness measurements up to the micrometer range by dual-wavelength digital holographic microscopy. Proceedings of SPIE 7718, Optical Micro- and Nanometrology III. Brussels: SPIE, 2010, 771805.
32. Fujii, H. & Asakura, T. Roughness measurements of metal surfaces using laser speckle. *Journal of the Optical Society of America* **67**, 1171-1176 (1977).
33. Cheng, C. F. et al. Absolute measurement of roughness and lateral-correlation length of random surfaces by use of the simplified model of image-speckle contrast. *Applied Optics* **41**, 4148-4156 (2002).
34. Abbe, E. Beiträge zur theorie des mikroskops und der mikroskopischen wahrnehmung. *Archiv für mikroskopische Anatomie* **9**, 413-468 (1873).
35. Zeng, Y. N. et al. Characteristics analysis of digital image-plane holographic microscopy. *Scanning* **38**, 288-296 (2016).
36. Claus, D. & Nizami, M. R. Influence of aberrations and roughness on the chromatic confocal signal based on experiments and wave-optical modeling. *Surface Topography:Metrology and Properties* **8**, 025031 (2020).
37. Karray, M., Slangen, P. & Picart, P. Comparison between digital Fresnel holography and digital image-plane holography: the role of the imaging aperture. *Experimental Mechanics* **52**, 1275-1286 (2012).
38. Goodman, J. W. Wavelength and angle diversity. in Speckle phenomena in optics: theory and applications. (Greewood Village: Roberts and Company Publishers, 2007), 159.

Official journal of the JHL 2689-9620
www.light-am.com

Article | Open Access

Parallax limitations in digital holography: a phase space approach

Ulf Schnars[1,*] and Claas Falldorf[2]

Abstract

The viewing direction in Digital Holography can be varied if different parts of a hologram are reconstructed. In this article parallax limitations are discussed using the phase space formalism. An equation for the parallax angle is derived with this formalism from simple geometric quantities. The result is discussed in terms of pixel size and pixel number of the image sensor. Change of perspective is demonstrated experimentally by two numerical hologram reconstructions from different parts of one single digital hologram.

Keywords: Digital holography, Perspective, Parallax, Phase space

Introduction

Holography is a three-dimensional imaging technique. An observer viewing through a hologram can change the depth of focus and the perspective[1].

With the advent of digital cameras, Digital Holography became feasible, which had a substantial impact to optical metrology[2]. Applications range from phase contrast imaging of biological and technical specimen[3], volumetric imaging of particle distributions[4] and precise shape measurement of functional surfaces[5,6]. Furthermore, recorded digital holograms of objects or scenes can be directly used as datasets to be displayed by holographic optical displays[7,8]. The main limiting factor in Digital Holography is the pixel size of the CCD or CMOS device used to record the holograms. The pixel size (distance between the center of neighboring pixels) determines the maximum angle between reference and object wave and thus the recordable object size. This has been widely investigated[9,10].

The focus of this article is on the parallax angle under which an object can be reconstructed from a digitally recorded hologram. First holographic reconstructions showing this effect were already published some decades ago[11]. In this contribution, we will explore the concept of multiple perspective views from the same digital hologram by means of a phase space analysis[12,13], which has been proven useful for describing sampling issues and viewing angle specifics in digital holography before[14,15]. It is well known, that the viewing direction depends on the area of the hologram (i.e. the aperture) through which an observer looks. We will show an easy to comprehend graphical analysis in phase space, which confirms this fundamental property for the case of digital holography as well. Additionally, we will quantitatively investigate how the size and the position of the viewing aperture, and the properties of the camera sensor relate to each other, and which limitations arise from these relations. The result is verified by numerical reconstructions from a digital hologram.

Correspondence: Ulf Schnars (schnars@t-online.de)
[1]private, 27628 Hagen im Bremischen, Germany
[2]BIAS-Bremer Institut für Angewandte Strahltechnik GmbH, Klagenfurter Str. 5, 28359 Bremen, Germany
These authors contributed equally: Ulf Schnars, Claas Falldorf

Definition of Parallax Angle

The parallax angle γ is defined in Fig. 1. An object with square-shaped front surface with side length L is located in a distance d from the hologram plane. S is the side length

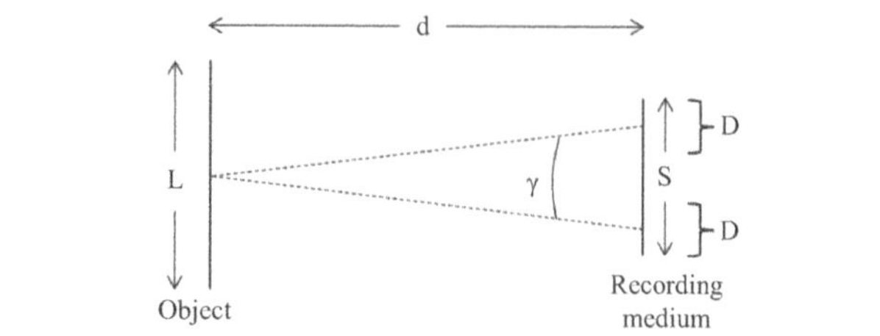

Fig. 1 Definition of the parallax angle γ (case $D=S/3$ is shown).

of the recording device. The reference wave is not shown in this figure.

It is not possible to reconstruct a hologram only from a single ray originating from the edge of the hologram. Therefore apertures with finite side length D are defined, which are located at the edges of the recording medium. In Fig. 1 an example with $D = S/3$ is shown. The arms of the parallax angle γ are originating from the centers of the apertures with side length D, and γ is therefore defined by

$$\gamma \approx \frac{S-D}{d} \tag{1}$$

In practice D has to be chosen small enough to enable a large parallax angle, but high enough to enable high resolution holographic reconstructions. In classical holography using photographic plates with very fine granularity there are no limitations concerning the plate size, so γ can be increased by enlarging the plate dimensions. In Digital Holography S has to be replaced by the side length of the image sensor, which is the product of the pixel number N in one direction and the pixel distance Δp:

$$\gamma \approx \frac{N\Delta p - D}{d} \tag{2}$$

However, the discrete nature of the camera sensor has important consequences regarding the limitation of the parallax angle. In what follows, we will therefore investigate the dependencies between the pixel pitch Δp, the recording distance d, the aperture size D and the number of pixels N. This will provide us the means to understand the intrinsic limitations towards the viewing angle γ in Digital Holography when using Eq. 2. Finally, we will also investigate the relationship between the reconstruction quality and the size of the aperture D and provide experimental validation of the model based on a recorded digital hologram.

Analysis using the Phase Space Formalism

In the following, we will provide a phase space analysis of the concept of viewing directions in Digital Holography. We will not consider the recording process itself, but assume that the complex amplitude $u(x)$ of the object wave is determined across the sensor domain, either using off-axis Digital Holography, or inline phase shifting Digital Holography. For more details on the recording process, the reader might wish to refer to Ref. 2.

Phase space offers the great benefit of describing the spatial frequency structure of the object wave in arbitrary propagation states. We will make use of the joint space-space frequency representation of the Wigner distribution function (WDF)

$$W(x,\nu) = \int_{-\infty}^{\infty} u\left(x+\frac{x'}{2}\right)u^*\left(x-\frac{x'}{2}\right)\exp(-i2\pi\nu x')dx' \tag{3}$$

Amongst other phase space representations, such as the windowed Fourier transform[16], or Gaussian wavelets[17], the support of the WDF allows us to assess the impact of apertures and to derive global parameters of the reconstructed wave field based on a simple and concise graphical approach[18]. We will not revise the mathematical properties of the WDF here. We encourage the interested reader to refer to Bastiaans to study fundamental relations of the WDF in general[19], and Testorf and Lohmann for more details on the mathematical properties of the WDF towards Digital Holography[20], including various recording configurations.

Instead, we will make use of a common heuristic approach which is based on a graphical representation of the support of the WDF. It treats wave field elements as if they were linearly related to their phase space representation[20]. In reality this is not the case, because the WDF contains a product relationship which lets arise cross terms. However, in all cases which are discussed in the following, the presence of the cross terms is negligible, because they always lie inside of the considered support of the WDF and therefore do not alter its shape.

Propagation of the object wave

Phase space depicts space along the x-axis and (spatial) frequency along the ν-axis. Hence, the corresponding diagrams are two-dimensional even for the case of a one-dimensional wave field. To keep the analysis simple, we will therefore restrict the discussion to the one-dimensional case. However, we will take care, that the extension to two-dimensions is straight forward.

Since phase space depicts space and spatial frequencies at the same time, any location x_0 is associated with a corresponding frequency spectrum. In this context, the complex amplitude of a scattered wave field in front of an object can be represented by its support in phase space: Confined along the x-axis by the size of the object L and confined along the ν_x-axis by the maximum propagation frequency $1/\lambda$. This is shown by the grey rectangle in Fig. 2. The area of the rectangle is called the space-

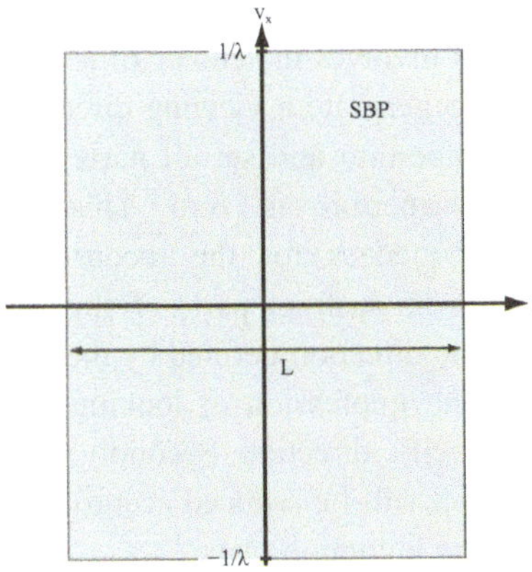

Fig. 2 Support of the complex amplitude of a wave field scattered from an object of size L in phase space. For the sake of simplicity, the diagram refers to a 1-dimensional wave field. Every spatial position is associated with a spectrum. The area covered by the grey rectangle is called the space-bandwidth-product (SBP).

bandwidth-product (SBP)[18].

The frequencies ν_x can be associated with tilted plane waves and therefore with propagation directions. To visualize this relationship, Fig. 3 depicts a plane wave. Looking at the grey triangle, we recognize that

$$\cos(\alpha) = \frac{\lambda}{p_x} = \nu_x \cdot \lambda \tag{4}$$

where α is the x-component of the propagation direction with respect to the object plane (not the optical axis!). In case of a two dimensional object plane, we can accordingly define the y-component of propagation by $\cos(\beta) = \nu_y \cdot \lambda$, since directional cosines are separable.

Fig. 3 is essential to understand how the support of the wave field in phase space changes due to the process of

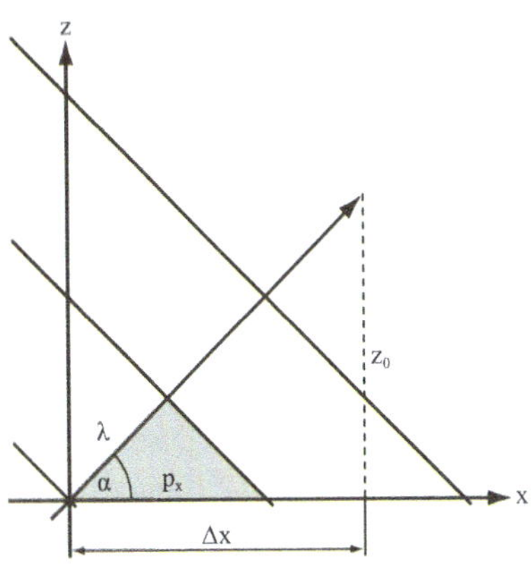

Fig. 3 Propagation of a plane wave: The lines indicate wave fronts. The propagation direction α is associated with the spatial frequency $\nu_x = 1/p_x$ of the complex amplitude in the object plane, as seen from the grey triangle. After propagation of z_0, the wave fronts have also moved laterally by Δx.

free space propagation. From it, we can infer that any wave field component with frequency ν_x will shift in x-direction while propagating along a given distance z_0. Using Fig. 3 and Eq. 4, it follows:

$$\Delta x(\nu_x) = \frac{z_0}{\tan(\alpha)} = \frac{z_0 \cdot \cos(\alpha)}{\sin(\alpha)} \approx \nu_x \cdot \lambda \cdot z_0 \tag{5}$$

where we have used the paraxial approximation in the last step, which is justified by the fact that in Digital Holography we almost always deal with waves traveling close to the optical axis, so that $\alpha \approx \pi/2$ and consequently $\sin(\alpha) \approx 1$.

As we can see from Eq. 5, under the assumption of paraxial approximation, propagation will cause a shear of the support in phase space, as shown in Fig. 4. The shearing angle can be simply derived from the gradient in Fig. 4, which is given by:

$$\tan(\varepsilon) = \frac{\nu_x}{\Delta x} = \frac{\nu_x}{\nu_x \cdot \lambda \cdot z_0} = \frac{1}{\lambda \cdot z_0} \tag{6}$$

This is all we need to describe the spatial frequency distribution of the object wave during propagation and reconstruction in Digital Holography.

As an example, we can use this scheme to calculate the minimum distance d between an object and a camera chip which will ensure the recorded object wave to not contain frequencies above the Nyquist frequency $\nu_N = 1/2\Delta p$. We will assume an object with size L and a camera chip with size S and pixel pitch Δp. The situation in phase space is depicted in Fig. 5. From the geometry and Eq. 6 we directly obtain:

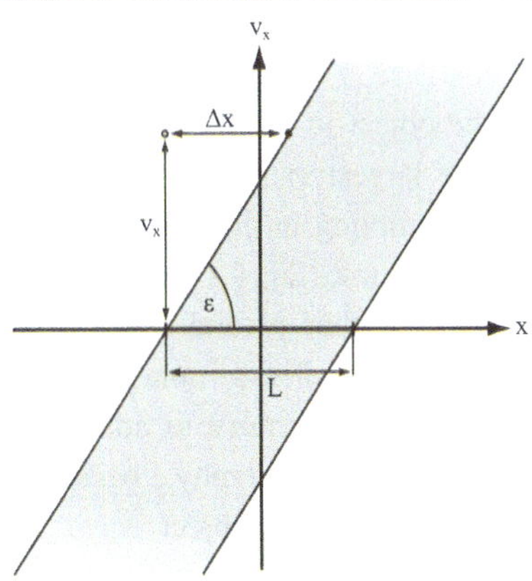

Fig. 4 Propagation in phase space: Since every spectral component also moves laterally during the propagation process, propagation is expressed by a shear of the support. As an example, the spectral component indicated by the disc has been moved from its initial position indicated by the circle by a distance Δx. L denotes the spatial extent of the initial wave field in the object plane. From the geometry we can determine the gradient ε.

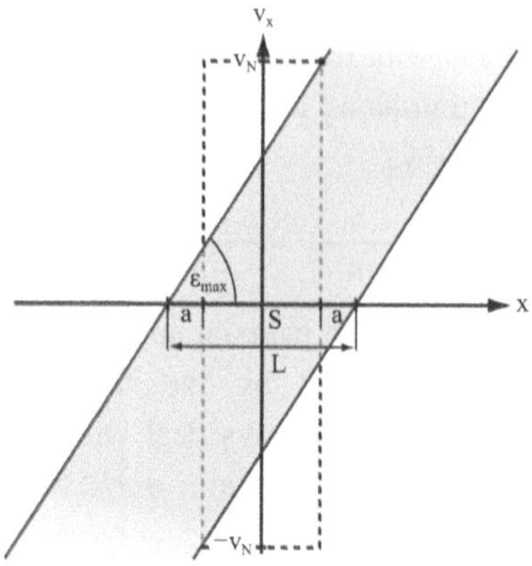

Fig. 5 Minimum propagation distance for Nyquist sampling: The dashed square depicts the SBP of a sensor of size S and Nyquist frequency ν_N. To ensure proper sampling, we have to make sure that the amount of shear is sufficient, so that no frequency incident on the sensor exceeds the Nyquist frequency. Evaluating the corresponding maximum gradient ε_{max} allows to determine the minimum propagation distance. Please, refer to the text for details.

$$\tan(\varepsilon_{max}) = \frac{\nu_N}{a+S} = \frac{1}{\lambda d} \tag{7}$$

Rearranging the equation and using the pixel distance Δp rather than the Nyquist frequency yields

$$d = \frac{2\Delta p}{\lambda}(a+S) \tag{8}$$

From Fig. 5 we see that $a = (L-S)/2$. If we further assume $S = N \cdot \Delta p$, with N the number of pixels in one direction of the sensor, we finally arrive at

$$d = \frac{\Delta p}{\lambda}(L + N\Delta p) \tag{9}$$

Since spatial frequencies are separable in the frequency domain, the extension to two dimensions is again straight forward. In this case, L would rather represent the largest extent L_x of the object in x-direction. Accordingly we can define L_y as the largest extent in y-direction. In order to ensure Nyquist sampling in this case, the larger value has to be inserted as L into Eq. 9 (assuming a square sensor with equal pixel number N and pixel pitch Δp in both directions). However, please note that even if it is convenient, it is not mandatory to adhere to the Nyquist sampling in Digital Holography, because the specific structure of the propagated object wave in phase space allows the application of generalized sampling approaches[21].

Viewing directions in phase space

Up to now, we have not identified any viewing directions. The wave field originating from a digital hologram does not provide any viewing directions per se, since light is simply propagating in all directions at the same time. The concept of a specific viewing direction therefore always involves the choice of a limiting aperture. Consequently, to generate a viewing direction, we have to insert a digital aperture and set all parts of the hologram outside of this aperture to zero. This will have two important consequences for the reconstruction process. Firstly, we only reconstruct parts of the wave field that traveled along the direction defined by the aperture. This is what gives us the impression of looking at the object or scene from a specific direction. Secondly, the resolution of the reconstruction will be reduced according to the size of the aperture in the hologram plane.

As shown in Fig. 6, we can quantify both effects in phase space. Fig. 6a depicts the object wave in phase space across the area given by the SBP of the camera sensor and Fig. 6b after propagation into the object plane. As an example, we have chosen to define two digital apertures of size $D = S/2$, indicated by the letters A and B, which will provide us with view A and view B on the object. The average viewing directions can be simply derived by finding the centers of the apertures. The corresponding frequencies ν_A and ν_B define the viewing directions in accordance with Eq. 4. From Fig. 6a and Eq. 6 we find for example for view A

$$\tan(\varepsilon) = \frac{\nu_A}{x_C} = \frac{1}{\lambda \cdot z_0} \tag{10}$$

where x_C denotes the center of the aperture. Inserting this into Eq. 4 with $\nu_x = \nu_A$ yields

$$\cos(\alpha_A) = \nu_A \cdot \lambda = \frac{x_C}{z_0} \tag{11}$$

The viewing direction α_A so found is defined with respect to the hologram plane. If we want to define it more conveniently with respect to the optical axis, we use $\zeta = \pi/2 - \alpha_A$ and find for small angles ζ

$$\cos(\alpha_A) = \sin(\zeta) \approx \zeta = \frac{x_C}{z_0} \tag{12}$$

View A and view B are only the limiting cases for a hologram with size S. In principle, we can select any digital aperture with size D between the positions $-x_C$ and x_C, which gives us the total viewing range of 2ζ. Indeed, if we insert $x_C = (S-D)/2$ and $z_0 = d$, we can compare Eq. 12 with Eq. 1 to verify $\gamma = 2\zeta$.

It is important to note, that because of the definition of ζ towards the optical axis, Eq. 12 cannot be straight forward extended to 2 dimensions. Instead we will assume, that in the two dimensional case, the aperture is only limiting in x-direction, i.e. the hologram is separated into a left view A and a right view B, so that ζ lies in the x-z-plane, in analogy to the scheme introduced by Fig. 1.

Assuming that no other optical elements, e.g. lenses, are

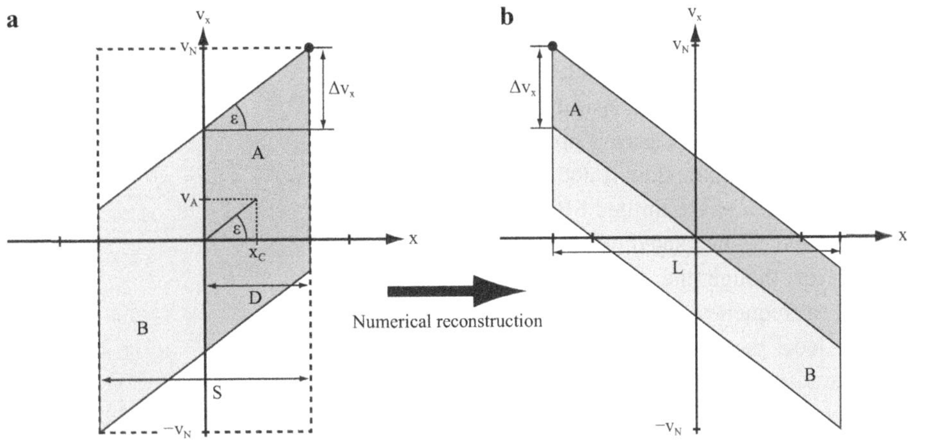

Fig. 6 a The support of the propagated object wave is divided into two views indicated by the different grey values and the letters A and B. Experimentally, this can be realized by a digital aperture, which sets all values of the complex amplitude outside of the aperture to zero. In this example the apertures that provide the views A and B have the size D and are centered around x_C and $-x_C$. S indicates the size of the sensor. **b** The support in phase space after numerical reconstruction of the object wave into the object plane. Δv_x indicates the local bandwidth of the wave field, which can be used to determine the resolution, and L is the size of the object. For more details, please refer to the text.

used, it is clear from Fig. 6a that there are only two possible ways to increase the viewing range: Firstly we could reduce the pixel pitch Δp, which would increase the Nyquist frequency v_N, and secondly we could enlarge $x_C = (S - D)/2$ by reducing the size of the aperture D. The latter, however, will further decrease the resolution of the reconstructed object wave, as we will see below.

Let us now insert Eq. 9 into Eq. 2, in order to quantitatively assess the limitations arising from the discrete recording device:

$$\gamma = \frac{N\lambda}{(L+N\Delta p)}\left(1-\frac{D}{N\Delta p}\right) \tag{13}$$

The side length of the aperture D will always be a fraction κ of the side length of the sensor S, so that $D = \kappa S = \kappa N \Delta p$ with $\kappa \leq 0.5$. This means, that the last term in Eq. 13 is a constant, and we finally yield

$$\gamma = \frac{N\lambda}{(L+N\Delta p)}(1-\kappa) \tag{14}$$

From Eq. 14, we see that the parallax angle γ solely depends on the object size L and on the pixel size of the sensor Δp if N and λ are fixed. This is shown graphically in Fig. 7, where γ is plotted on the y-axis and Δp is plotted on the x-axis. Four curves are shown with $N = 1000$, $N = 2000$, $N = 3000$ and $N = 4000$. For all 4 curves $L =$ 1 cm, $\lambda = 633$ nm and $\kappa = 0.5$ are chosen. It can be seen in all four curves that the parallax angle increases with decreasing pixel size. That means image sensors with small pixel sizes of about 1 μm are advantageous to allow larger changes of perspective in Digital Holography. Today available CMOS sensors have pixel sizes down to 1,6 μm

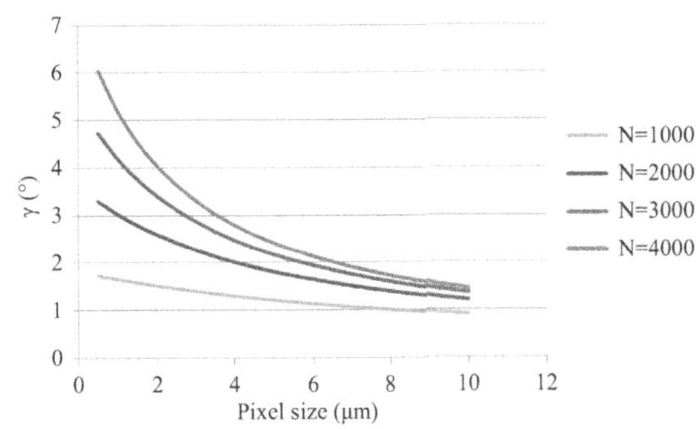

Fig. 7 Parallax angle in degrees (y-axis) versus pixel size in micron (x-axis) for different pixel numbers

and more than 10 million pixels, enabling parallax angles of several degrees.

In addition it can be concluded by comparison of the curves, that increasing the number of pixels N does not lead to a significantly larger parallax angle for large pixel sizes around 10 μm. Significantly larger angles can be achieved by enlarging N only with small pixel sizes of approx. 1 μm. This effect becomes clearer if we focus on small objects ($L << N\Delta p$). In this case, Eq. 14 can be further approximated:

$$\gamma = \frac{\lambda}{\Delta p}(1-\kappa) \tag{15}$$

Now γ is inversely proportional to Δp and independent from the pixel number N. An increase of N has no effect on γ because the necessary larger distance between object and sensor compensates the effect of the higher pixel number. Methods proposed earlier to enlarge the aperture by using several image sensors in parallel[22] are therefore not suitable to increase the parallax angle.

Reduced resolution

We can assess the resolution of the reconstructed object wave, when we perform the numerical reconstruction. In phase space, this means we invert the shearing process of the support as shown in Fig. 6b. Again view A and view B are indicated. As we can see, due to the limited SBP of the aperture, the bandwidth of the reconstructed object wave is significantly reduced. Even though the support of each view covers a large area in frequency space, the resolution is solely defined by the local bandwidth $\Delta\nu_x$. In fact, the inclined structure of the support indicates a wave front chirp, which does not carry any information. Comparing Fig. 6b with Fig. 6a, we see from the geometry, that

$$\Delta\nu_x = D \cdot \tan(\varepsilon) = D \cdot \frac{1}{\lambda \cdot z_0} \tag{16}$$

holds. If we consider the numerical aperture of the aperture $NA_D = D/2z_0$ and the minimum detail to be resolved as the inverse of the bandwidth $d_x = 1/\Delta\nu_x$, Eq. 16 turns into the well known diffraction limited resolution for a rectangular aperture

$$d_x = \frac{\lambda}{2 \cdot NA_D} \tag{17}$$

Since we assumed the digital apertures to be only limiting in x-direction, the point-spread-function of the reconstructed object wave will be elongated in this direction. In y-direction we will need to insert the numerical aperture of the sensor chip $NA_S = S/2z_0$ into Eq. 17 to obtain d_y.

Experimental Results

Phase shifting Digital Holography is used to demonstrate the change of perspective if different parts of a hologram are used for reconstruction. A CMOS sensor with 2054 × 2452 pixel and a pixel pitch of 3.45 μm × 3.45 μm is used. The wavelength is 532 nm. The object is placed in a distance of 0.175 m from the sensor. The effective width of the brick used as object in viewing direction is $L \approx$ 1.7 cm. A plane reference wave and the object wave interfere on the surface of the sensor. The initial phase in the hologram plane is recorded using phase-shifting. The numerical reconstruction into the object area is done with the Fresnel approximation. The resulting image resolution (pixel distance in the reconstructed image) is 11.0 μm × 13.1 μm. These values are not equal due to the different pixel numbers in x- and y-direction. A reconstruction using the entire hologram matrix (2054 × 2452 pixel) is shown in Fig. 8.

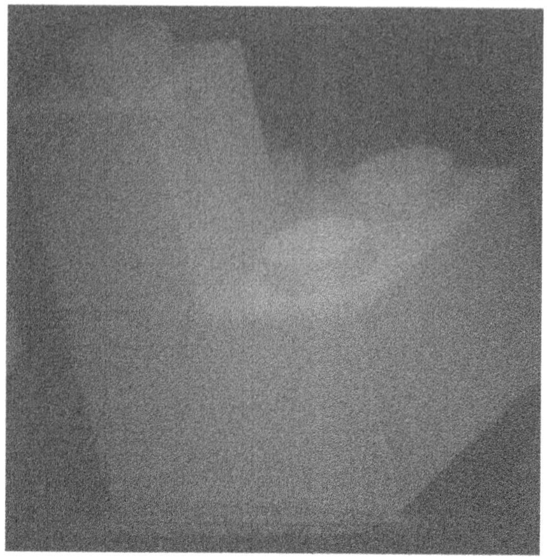

Fig. 8 Reconstruction of a hologram using the entire hologram matrix (2054 × 2452 pixel, image rescaled to quadratic shape).

Views from different angles are generated by two reconstructions with only 2054 × 800 pixel ($\kappa = 0.33$). A view from the left is generated by using only the columns 1 to 800 of the hologram matrix, the grey values of the other columns are set to zero (zero padding). This reconstruction is shown in Fig. 9. A view from the right is generated by using only columns 1652 to 2452; the grey values of the other pixel are set to zero. This view is shown in Fig. 10. The speckle size in both images, Fig. 9 and Fig. 10, is larger than in Fig. 8. This is due to the smaller aperture (lower number of pixel in horizontal direction) used for the reconstructions.

The change of perspective is small but visible by comparing Fig. 9 and Fig. 10. The viewing direction on the knobs on top of the lower brick is slightly different. The real parallax angle according to Eq. 2 is $\gamma = 1.85°$. This is only slightly lower than the theoretical maximum angle calculated with Eq. 14 of $\gamma_{th} = 1.96°$ degrees. The perspective change is visible more clearly in the video provided as supplementary material.

Discussion

Parallax in Digital Holography has been analyzed using a phase space approach. It has been shown that the parallax angle depends on the pixel size of the image sensor. The smaller the pixel size the larger is the parallax angle. Increasing the pixel number leads to a noticeable increase of the parallax angle only for sensors with small pixel sizes. An experimental example shows that different views can be reconstructed from a single hologram. The predicted change of perspective is small, but visible in Digital Holography. The phase space approach is an appropriate tool to calculate parallax angles in Digital Holography.

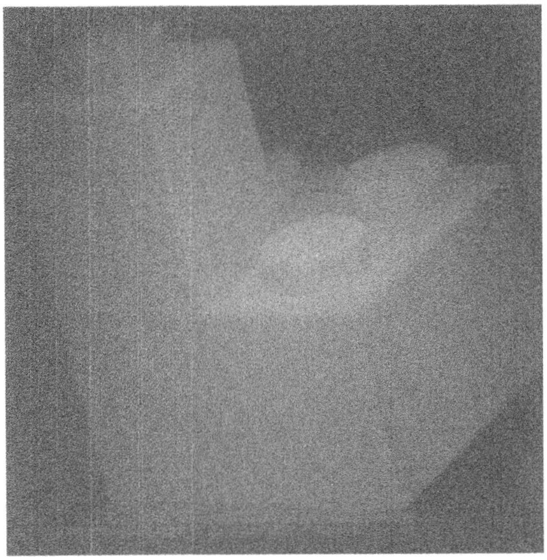

Fig. 9 Reconstruction from the left stripe of the hologram matrix (2054 × 800 pixel).

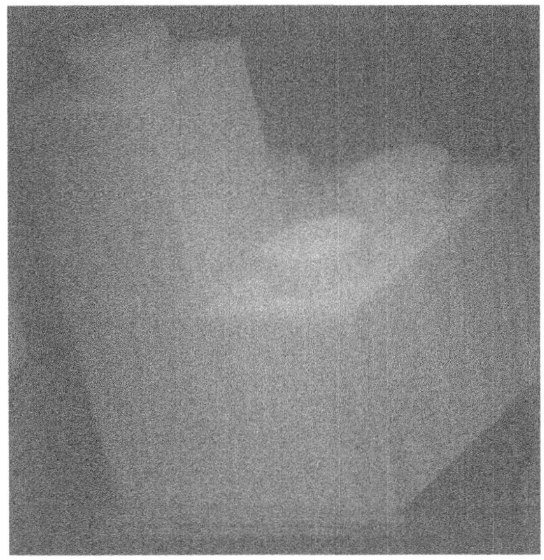

Fig. 10 Reconstruction from the right stripe of the hologram matrix (2054 × 800 pixel).

With state of the art CCD or CMOS devices the viewing angle change is very small. However, future devices might have pixel sizes below one micron. Large parallax changes comparable to those of classical film holography will be possible, enabling new applications in imaging.

Materials and methods

The holographic set-up consists of a Michelson-type interferometer. A plane wave is split at the beam splitter into two partial waves. The reference wave is guided to a mirror in one arm of the interferometer. The other wave illuminates the object and is reflected from there. The plane reference wave and the diffusely reflected object wave are recombined at the beam splitter and interfere at the surface of the CMOS sensor. The initial phase of the wave in the sensor area is generated with the help of a phase shifting unit.

Numerical reconstruction of the wave into the object area is done with the Fresnel approximation. The code is implemented in MATLAB.

More details of the recording process and of the numerical reconstruction are described in the literature, the reader is referred to Ref. 2.

Acknowledgements

We would like to thank Prof. Ralf B. Bergmann for fruitful discussions. The experimental results presented here are based on digital holograms recorded in the laboratories of BIAS - Bremer Institut für angewandte Strahltechnik GmbH. We are grateful to Thomas Meeser for experimental support with the hologram recordings. The recordings were performed within the frame of project 'Real 3D', funded by the EU 7th Framework Program FP7/2007-2013 under agreement 216105.

Author contributions

Both authors contributed to the writing and discussion of the manuscript with 50 percent each.

Conflict of interest

The authors declare no conflicts of interest.

Supplementary information is available for this paper at https://doi.org/10.37188/lam.2022.028.

Received: 26 August 2021 Revised: 05 April 2022 Accepted: 20 April 2022
Accepted article preview online: 28 April 2022
Published online: 20 June 2022

References

1. Hariharan, P. Optical Holography: Principles, Techniques and Applications. (New York: Cambridge University Press, 1984).
2. Schnars, U. et al. Digital Holography and Wavefront Sensing. (Heidelberg: Springer, 2015).
3. Kemper, B. & Von Bally, G. Digital holographic microscopy for live cell applications and technical inspection. *Applied Optics* **47**, A52-A61 (2008).
4. Langehanenberg, P., Von Bally, G. & Kemper, B. Autofocusing in digital holographic microscopy. *3D Research* **2**, 4 (2011).
5. Simic, A. et al. In-line quality control of micro parts using digital holography. Proceedings of SPIE 10233, Holography: Advances and Modern Trends V. Prague: SPIE, 2017, 1023311.
6. Falldorf, C. et al. Reduction of speckle noise in multiwavelength contouring. *Applied Optics* **51**, 8211-8215 (2012).
7. Kujawinska, M. et al. Multiwavefront digital holographic television. *Optics Express* **22**, 2324-2336 (2014).
8. Agour, M., Falldorf, C. & Bergmann, R. B. Holographic display system for dynamic synthesis of 3D light fields with increased space bandwidth product. *Optics Express* **24**, 14393-14405 (2016).
9. Kelly, D. P. et al. Quantifying the 2.5D imaging performance of digital holographic systems. *Journal of the European Optical Society* **6**, 11034 (2011).

10. Kozacki, T. et al. Holographic capture and display systems in circular configurations. *Journal of Display Technology* **8**, 225-232 (2012).
11. Schnars, U. Digitale Aufzeichnung und Mathematische Rekonstruktion von Hologrammen in der Interferometrie. (Duesseldorf: VDI-Verlag, series 8, No 378, 1994).
12. Situ, G. & Sheridan, J. T. Holography: an interpretation from the phase-space point of view. *Optics Letters* **32**, 3492-3494 (2007).
13. Testorf, M. E., Hennelly, B. M. & Ojeda-Castañeda, Phase-Space Optics: Fundamentals and Applications. (New York: McGraw-Hill, 2009).
14. Hennelly, B. M. Digital holography in the light of phase space. *Progress in Electromagnetics Research* **515** (2010).
15. Kozacki, T. et al. Extended viewing angle holographic display system with tilted SLMs in a circular configuration. *Applied Optics* **51**, 1771-1780 (2012).
16. Arnold, J. M. Rays, beams and diffraction in a discrete phase space: Wilson bases. *Optics Express* **10**, 716-727 (2002).
17. Gabor, D. Theory of communication. Part 1: the analysis of information. *Journal of the Institution of Electrical Engineers-Part III:Radio and Communication Engineering* **93**, 429-441 (1946).
18. Lohmann, A. W. et al. Space–bandwidth product of optical signals and systems. *Journal of the Optical Society of America A* **13**, 470-473 (1996).
19. Bastiaans, M. J. Wigner distribution function and its application to first-order optics. *Journal of the Optical Society of America* **69**, 1710-1716 (1979).
20. Testorf, M. & Lohmann, A. W. Holography in phase space. *Applied Optics* **47**, A70-A77 (2008).
21. Stern, A. & Javidi, B. Improved-resolution digital holography using the generalized sampling theorem for locally band-limited fields. *Journal of the Optical Society of America A* **23**, 1227-1235 (2006).
22. Kreis, T. M., Adams, M. & Jueptner, W. P. O. Aperture synthesis in digital holography. Proceedings of SPIE 4777, Interferometry Xi: Techniques and Analysis. Seattle: SPIE, 2002, 69-76.

Official journal of the JHL 2689-9620
www.light-am.com

Review | Open Access

Resolution enhancement of digital holographic microscopy via synthetic aperture: a review

Peng Gao[1,2,*] and Caojin Yuan[3,*]

Abstract

Digital holographic microscopy (DHM), which combines digital holography with optical microscopy, is a wide field, minimally invasive quantitative phase microscopy (QPM) approach for measuring the 3D shape or the inner structure of transparent and translucent samples. However, limited by diffraction, the spatial resolution of conventional DHM is relatively low and incompatible with a wide field of view (FOV) owing to the spatial bandwidth product (SBP) limit of the imaging systems. During the past decades, many efforts have been made to enhance the spatial resolution of DHM while preserving a large FOV by trading with unused degrees of freedom. Illumination modulation techniques, such as oblique illumination, structured illumination, and speckle illumination, can enhance the resolution by adding more high-frequency information to the recording system. Resolution enhancement is also achieved by extrapolation of a hologram or by synthesizing a larger hologram by scanning the sample, the camera, or inserting a diffraction grating between the sample and the camera. For on-chip DHM, spatial resolution is achieved using pixel super-resolution techniques. In this paper, we review various resolution enhancement approaches in DHM and discuss the advantages and disadvantages of these approaches. It is our hope that this review will contribute to advancements in DHM and its practical applications in many fields.

Keywords: Digital holographic microscopy, Resolution enhancement, Computational imaging, Illumination modulation

Introduction

Optical microscopy is a powerful tool for detecting subtle structures of samples, playing an irreplaceable role in many fields. However, traditional optical microscopy can only obtain the amplitude information of samples, and transparent samples, such as living cells cannot be investigated. Although fluorescence microscopy can selectively render and highlight the structures of interest by prior tagging with fluorescent markers, the phototoxicity and photobleaching of fluorescence tagging make it difficult to continuously observe living cells over a long time. Thus, there is an urgent need for label-free microscopic tools able to follow organelles in live cells. Digital holographic microscopy (DHM) [1-6], which combines digital holography and microscopy, is a label-free, quantitative phase microscopy approach. A typical setup for DHM is shown in Fig. 1a, in which a magnified object wave and a reference wave interfere with each other, and the generated off-axis hologram or phase-shifting

Correspondence: Peng Gao (peng.gao@xidian.edu.cn) or Caojin Yuan (yuancj@njnu.edu.cn)
[1]School of Physics and Optoelectronic Engineering, Xidian University, Xi'an 710071, China
[2]Academy of Advanced Interdisciplinary research, Xidian University, Xi'an 710071, China
Full list of author information is available at the end of the article.

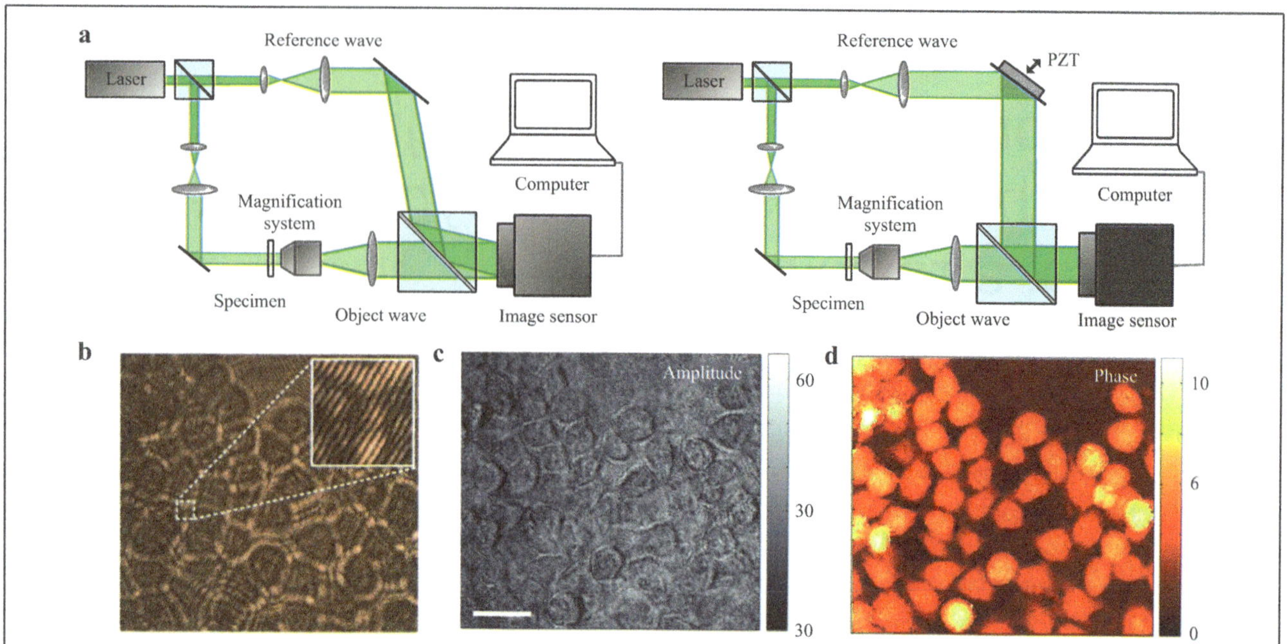

Fig. 1 Principle of digital holographic microscopy (DHM). **a** Schematic setup of off-axis (left) and in-line DHM (right). **b** Exemplary hologram of off-axis DHM. **c, d** Amplitude and phase images of HeLa cells reconstructed from **b**. Scale bar in **c** is adapted with permission from Ref. 76 ©The Optical Society.

holograms are recorded by a CCD camera. From the off-axis hologram (Fig. 1b), both the amplitude and phase images (Fig. 1c, d) of a sample can be obtained simultaneously. As an alternative, inline lensless DHM can help recover the phase information with an iterative algorithm without using an independent reference wave[7,8]. Refocusing of the reconstructed images can further be performed digitally using the obtained complex amplitude of the object wave[9,10]. In general, DHM can not only observe transparent samples with high endogenous contrast, but also quantitatively assess the thickness or refractive index distributions of these samples. Therefore, DHM has been widely applied in industrial inspection[11,12], visualization of liquid/gas flows[13], biomedical imaging[14], etc.

The spatial resolution is of great importance for DHM because it determines the smallest structures that can be resolved. As illustrated in Fig. 2a, when two infinitely small points separated by a Rayleigh distance of $0.61\lambda/NA$ are imaged by a diffraction-limited imaging system, two Airy patterns are generated in the imaging plane, whose complex amplitudes are denoted as O_1 and O_2, respectively. Here, λ and NA indicate the illumination wavelength and the sum numerical aperture (NA) of the imaging system (NA_{imag}) and illumination system (NA_{illum}), respectively. For incoherent illumination, the intensity of the image is linearly related to the intensity emitted from the sample, that is, $|O_1|^2+|O_2|^2$. The two patterns are considered to be resolved if the center of the intensity pattern of O_1 coincides with the first zero of the pattern of O_2 (Fig. 2b). For coherent imaging (e.g., DHM), a linear relationship holds between the input complex amplitude and the output complex amplitude. In this case, the intensity of the image corresponds to $|O_1+O_2|^2$, which is dependent on the phase difference $\Delta\varphi$ between O_1 and O_2. It is evident from Fig. 2c that for $\Delta\varphi = 0$ (in-phase), the Airy patterns of O_1 and O_2 are considered unresolved. However, for $\Delta\varphi = \pi$ (anti-phase), the Airy patterns of O_1 and O_2 are well separated. To conclude, there is no easy criterion for evaluating the resolution of coherent microscopy (DHM), considering that the resolution depends on both the microscope and the phase of the sample. Nevertheless, we recommend the following:

$$\sigma = \frac{\kappa_1 \lambda}{NA} = \frac{\kappa_1 \lambda}{NA_{\text{imag}} + NA_{\text{illum}}} \tag{1}$$

for evaluating the resolving power of optical microscopy. Factor κ_1 is determined by the experimental parameters, such as the coherent noise level and the SNR of the detector, etc[3,15]. Herein, we assign $\kappa_1 = 0.82$[16,17], for which the principal intensity maximum of pattern O_1 coincided with the first minimum of the in-phase O_2. This suggests two options for enhancing the spatial resolution of the DHM. The first is the use of a shorter wavelength, for example, 193-nm[18] light was used to enhance the spatial resolution of the DHM. The other is to enlarge the NAs of both the illumination and the recording systems.

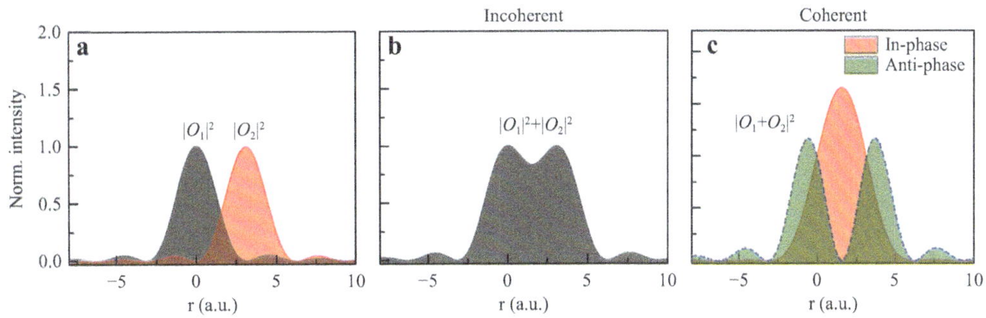

Fig. 2 Resolution criteria in incoherent and coherent systems. **a** Intensity distribution of the Airy patterns for the complex amplitude distributions, O_1 and O_2, obtained by imaging two infinitely small point sources through a diffraction-limited system; the superimposed intensities of O_1 and O_2 in incoherent microscopy **b** and coherent microscopy **c**. In-phase and anti-phase in **c** indicates that the cases O_1 and O_2 have a phase difference of 0 and π, respectively.

In addition to the resolution, the field of view (FOV) is also a vital parameter of an imaging system. The ratio between these two parameters determines the space-bandwidth product (SBP) of the imaging system[19,20]. Often, a DHM system has a limited SBP, and consequently, there is an inherent trade-off between resolution and FOV. This means that a reconstructed image with a higher resolution has a smaller FOV. Just as an example, a standard 20 × /0.4 microscope lens with a resolution limit of 0.8 μm and a circular FOV of 1.1 mm in diameter, provides a SBP of ~7 megapixels[21]. However, microscopists always expect to have a higher SBP in the sense of an increasing resolution while maintaining FOV. There are many degrees of freedom, including spatial/temporal resolution, FOV, polarization, spectrum, etc., in an imaging system. We can convert or trade some unused degrees of freedom (e.g., time) for an increase in SBP. In the past few decades, computational microscopy methods have emerged, generating images with high resolution and a wide FOV by using modulated illumination or applying other physical manipulations during the imaging process. By synthesizing apertures using the object information obtained under different modulation/manipulation operations, the SPB limitation of conventional microscopes (including DHM) can be bypassed[22,23].

This paper aims to review the resolution enhancement approaches of DHM, which are classified into three types: 1. Enlarging the NA of illumination via oblique illumination, structured illumination, and speckle illumination; 2. Enlarging the NA of the recording system via hologram extrapolation, hologram expansion, and pixel super-resolution. 3. Artificial intelligence (AI) assistance resolution enhancement approaches for DHM. This paper will cover the basic principles, implementation schemes, and exemplary results of the different approaches. We conclude by providing a summary and future outlook of resolution-enhancement approaches for DHM.

Results

Enlarging the NA of illumination

A microscopic imaging system is commonly introduced to DHM, which provides a magnified and high-resolution image of the sample. A plane wave is often used for illumination, and only the spatial frequencies diffracted by the sample up to ~ NA_{imag}/λ can be transmitted through the limited aperture of the microscope lens (Fig. 3a, *upper part*), resulting in an upper limit for the spatial resolution in terms of $0.82\lambda/NA_{\text{imag}}$. Therefore, using oblique illumination, structured illumination, or speckle illumination to provide a NA_{illum} is an effective method for further improving the resolution.

DHM with oblique illumination

Since the first implementations of oblique beam illumination in DHM in the 1970s[24–27], many approaches have been extensively reported on for replicating a larger aperture[28–48]. When a sample is illuminated by an oblique wave, the high spatial frequency of the object wave in the opposite direction of the oblique illumination will be downshifted and, thus, will pass through the limited aperture of the imaging system (Fig. 3a, *lower part*). The downshifted frequencies are back-assembled to their original positions in the spectrum of the object, thus, synthesizing a wider spectrum than that of the NA-defined aperture[31]. The final resolution-enhanced image is obtained by an inverse Fourier transform (FT) of the synthesized spectrum. In comparison with conventional lens-based DHM with a resolution limit of 0.82λ/NA, DHM with oblique beam illumination has an enhanced spatial resolution, as described in Eq. 1.

Kuznetsova et al. and Schwarz et al. used oblique beam illumination to downshift the high-frequency components

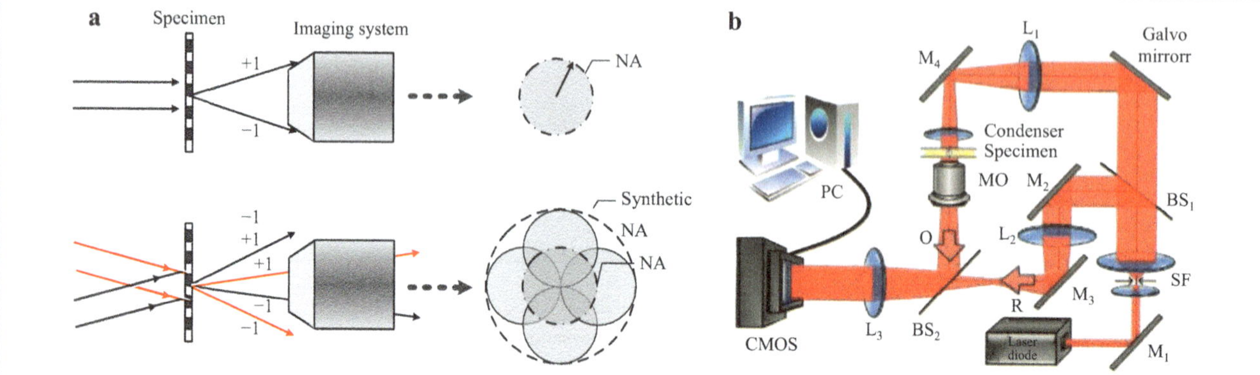

Fig. 3 DHM with oblique illumination. **a** Schematic of synthetic aperture via oblique illumination in different orientations. **b** schematics of setup of off-axis DHM with oblique illumination. Fig. 3b is adopted from Ref. 49.

of the spectrum of the object, and introduced a conjugate carrier phase to upshift the spectrum content back to its proper region[28,50]. This process was repeated sequentially for both the *x*- and *y*-illumination directions, thus allowing for the transmission of extended frequency bandwidths of the 2D object, yielding a resolution gain with a factor of three. Mico et al. utilized a vertical-cavity surface-emitting laser (VCSEL) array to generate multi-directional oblique illumination of DHM and retained a resolution gain with a factor of five[47]. The use of scanning elements has also been reported as a technological improvement for providing more flexible oblique beam illumination[28,33,37,51], which has also been validated[29,31,52] in comparison with the tilting of the illumination direction. For instance, Cheng et al. used a 2D Galvo-scanner to generate oblique illumination for DHM (Fig. 3b), achieving a two-fold isotropic resolution enhancement. [49] Oblique beam illumination for resolution enhancement has also been implemented using a spatial light modulator (SLM) [53-55] or a fiber bunch[56-60]. Meanwhile, single-shot synthetic aperture DHM was realized by using an SLM to generate multiple non-coplanar illumination beams, and using coherence gating to avoid self-interference between these non-coplanar beams[53]. Moreover, wavelength multiplexing techniques[34-36] were utilized to perform multiple oblique illuminations during a single exposure. Oblique beam illumination techniques have recently been implemented in a commercial upright Olympus microscope[39]. Notably, the maximum synthetic numerical aperture is limited by NA<1 in imaging systems with dry objectives, and further resolution improvements to $\lambda/3.7$ were demonstrated using evanescent waves[29]. In addition, axial rather than transversal resolution improvement has also been validated by SA generation[61], and an application of the technique to edge processing was also reported[62]. It is interesting to mention that oblique illumination has also been applied to differential interference contrast (DIC) microscopy[63] and Zernike phase contrast microscopy[64].

Recently, oblique illumination was also applied to obtain 3D refractive index (RI) tomographic images of transparent or translucent samples via optical diffraction tomography (ODT)[49,65]. ODT enables the probing of 3-D RI maps of a sample by recording the complex wavefront diffracted while rotating the sample[66,67] or varying illumination angles[68,69]. Kim[68] adapted DHM to ODT by using a 2D galvanometric mirror to generate different illumination directions and consequently, reconstructed a 3D RI distribution of the samples with high lateral and axial resolutions. Ozcan[70] presented an on-chip ODT scheme that enables the imaging of a large volume of approximately 15 mm^3, with a spatial resolution of < 1 μm ×1 μm × 3 μm. In 2020, Wang et al.[71] reported a mechanical-scanning-free ODT configuration using self-accelerating Airy beams that are tilted along the beam path direction, providing a 3D volumetric view of the cells. Recently, Kus et al.[72] proposed a real-time ODT system for recording a complete set of projections in a single hologram in a multiplexing manner by using a microlens, with which the tomographic images of living cells and flow cytometry could be observed clearly. In addition to advances in ODT hardware, artificial intelligence (AI)-based reconstruction algorithms have been reported to reduce the recording time and improve the quality of the reconstructed images[73].

DHM with structured illumination (SI)

Structured illumination microscopy (SIM) is a super-resolution optical microscopy technique that has the merits of being fast, minimally invasive, and has no requirement for fluorescent labeling[74,75]. Compared with conventional wide-field (WF) microscopy, SIM provides two-fold spatial resolution enhancement by illuminating the sample with a periodic pattern and recording the generated moiré pattern. Because of the moiré effect, the high spatial

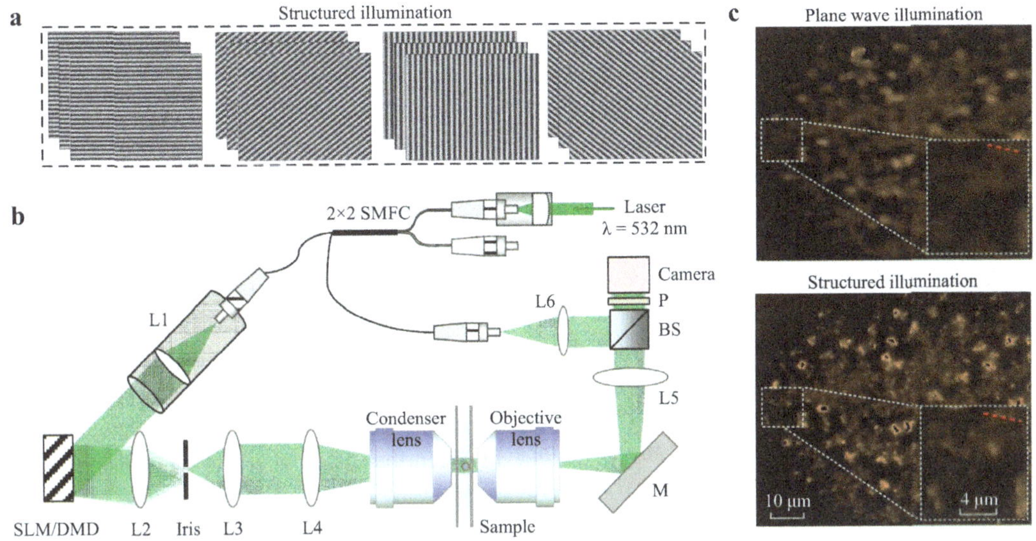

Fig. 4 DHM with structured illumination. **a** Schematic of structured illumination with different orientations and phase shifts. **b** Schematic setup of DHM with structured illumination. **c** Experimental results for resolution enhancement. Reconstructed phase images using plane wave illumination (top) and structured illumination (bottom). **b** is adopted from Ref. 77. and **c** is adopted from Ref. 76.

frequency components of the sample, which are not accessible in conventional microscopy, are shifted into the detectable domain and thus can be observed. However, the application of SIM has been limited to fluorescent samples. In 2010, Mudassar and Hussain[60] applied it to non-fluorescent scattering imaging, for which fringe patterns were generated by the interference between the beams carried by two fibers.

A DHM with structured illumination generated by an electric modulation device has been proposed. The fringe patterns were generated by a spatial light modulator (SLM)[76] in 2013 and by a digital micromirror device (DMD)[77] in 2015, whereby fringes with different orientations and phase shifts were projected without mechanical motion. In the implementation, four binary phase gratings rotated by m×45° (Fig. 4a) and generated by an SLM or a DMD are projected onto the sample, so that the sample is illuminated sequentially by four sinusoidal fringe patterns. After passing through a telescope system comprised of the microscope objective *MO* and lens *L*, the object wave interferes with a tilted reference wave *R*, and the generated holograms are recorded by a CCD camera, as shown in Fig. 4b. We denote with Ψ_{mn} the object wave generated by illuminating the sample with the grating as having the *m*-th orientation and *n*-th phase shifting. Thus, the recorded hologram can be described as $I_{mn} = |R + \Psi_{mn}|^2$. From this intensity, the wave Ψ_{mn} can be reconstructed using standard reconstruction methods, as in off-axis DHM. Generally, Ψ_{mn} can be decomposed into three components, $A_{m,-1}$, $A_{m,0}$, and $A_{m,1}$ corresponding to the -1st, 0th, and $+1^{st}$ diffraction orders of the illumination wave. When we assume that the phase shift increment for each shifting of the grating is α, Ψ_{mn} can be written as:

$$\Psi_{mn} = \gamma_{-1}\exp(-in\alpha)A_{m,-1} + \gamma_0 A_{m,0} + \gamma_1\exp(in\alpha)A_{m,1} \tag{2}$$

where γ_{-1}, γ_0, and γ_1 denote the magnitudes of the diffraction orders. From Eq. 2, we can calculate the reconstructed object waves $A_{m,-1}$, $A_{m,0}$, and $A_{m,1}$ from the different diffraction orders:

$$\begin{bmatrix} A_{m,-1} \\ A_{m,0} \\ A_{m,1} \end{bmatrix} = \begin{bmatrix} \gamma_{-1}\exp(-i\alpha) & \gamma_0 & \gamma_1\exp(i\alpha) \\ \gamma_{-1}\exp(-i2\alpha) & \gamma_0 & \gamma_1\exp(i2\alpha) \\ \gamma_{-1}\exp(-i3\alpha) & \gamma_0 & \gamma_1\exp(i3\alpha) \end{bmatrix}^{-1} \cdot \begin{bmatrix} \Psi_{m1} \\ \Psi_{m2} \\ \Psi_{m3} \end{bmatrix} \tag{3}$$

Then, $A_{m,-1}$ and $A_{m,1}$ under each illumination direction are compensated for by the carrier phase due to the tilted propagation direction along the $\pm 1^{st}$ diffraction orders, and synthesized together with $A_{m,0}$ in the Fourier plane to yield the synthetic spectrum. Finally, a focused image with enhanced resolution was retrieved by the inverse FT of the synthetic spectrum. Assuming that the angular aperture of the imaging system is NA_{imag}, the synthetic NA of the SI DHM is $NA = NA_{imag} + \sin\theta_{illum}$. Here, θ_{illum} denotes the illumination angle of the +1 or −1 diffraction orders, and the highest value of the $\sin\theta_{illum}$ is, in turn, determined by the NA of the condenser lens. Fig. 4c validates the resolution enhancement of the structured illumination in comparison with on-axis illumination. Of note, the phase image at the bottom (obtained with structured illumination) has a better resolution compared with that on the top, and the two particles separated by 1.4 μm became distinguishable.

Recently, structured illumination has been widely applied to different variations of DHM for the purpose of imaging transparent samples with resolution enhancement[76–89]. SI-based DHM was extended for resolution enhancement in the axial direction, aside from the lateral directions[90]. An iterative reconstruction algorithm, as well as a PCA-based algorithm, were employed to obtain resolution-enhanced images using structured illumination with unknown phase shifts or free of phase shifts[82–84, 89]. In 2021, an end-to-end deep-learning-based method, DL-SI-DHM, was proposed for improving the reconstruction efficiency and accuracy of SI-DHM[91]. In addition to synthetic-aperture phase imaging, Shin et al. in 2015 and Chowdhury et al. in 2017 utilized DMD- and SLM-based structured illumination to obtain 3D refractive index tomographic images of samples[77,92].

DHM with speckle illumination

Speckle illumination has been known since the invention of the laser in 1960[93]. Speckle illumination often reduces the signal-to-noise ratio (SNR) in coherent imaging. Speckle fields can be understood as a combination of plane waves with various random illumination directions. These oblique plane waves shift the spectrum of the object in the reciprocal plane, allowing for access to additional spatial frequencies. Thus, speckles can be used to enhance the spatial resolution and reduce the coherent noise in DHM or other coherent imaging[37, 94–97].

Fig. 5 shows the optical setup of DHM with a speckle illumination according to Ref. 95. The speckle field was generated by an SLM and projected using a telescopic system L_1-MO_1 in the sample plane, where $A^i_{Speckle}$ is the ith speckle illumination and A^i_o refers to the complex amplitudes of the object wave when the sample was illuminated with $A^i_{Speckle}$. When A^i_o interferes with a reference wave R, the hologram can be denoted as $I^i_o = |A^i_o + R|^2$. The complex amplitude of the speckle field $A^i_{Speckle}$ can be reconstructed from the hologram recorded in the absence of a sample. Often, the holographic resolution-enhanced image can be obtained by averaging multiple angle-dependent E-field images $O_{ave} = 1/N\sum_{i=1}^{N} A^i_o/A^i_{Speckle}$[96]. It is worth mentioning that the resolution enhancement capability of such an averaging operation will be lower when the speckle field has an unequal weight or power over the entire frequency spectrum.

To overcome this limitation, an iterative reconstruction method[95] was proposed for synthesizing a larger NA. The key idea of the iteration is to propagate the object wave between the image plane and the CCD plane, updating the speckle illumination in the image plane and the object wave intensity (with the recorded one) in the CCD plane. The iterative method enhances not only the resolution but also the SNR when the reconstructed object wavefronts from different speckle-based holograms are averaged. Utilizing SLM for speckle field illumination avoids mechanical movements and allows for a high repeatability in comparison with dynamic devices. Time-changing speckle illumination is similar to structured illumination. The concept in common is the encoding of an object with an a priori known high-resolution pattern that folds the high spatial frequencies into the low bandwidth range. The decoding patterns should be matched to the time-changing encoding patterns.

Similar to speckle illumination, a scattering medium can also be placed before the sample to enhance the spatial resolution and FOV[98,99]. Multiple scattering of the medium is a deterministic process that is described by a transmission matrix (TM) in transmission geometry. TM-based imaging has been introduced in DHM, yielding an improved resolution beyond what an objective-lens-based imaging system can achieve[99]. Similarly, a single

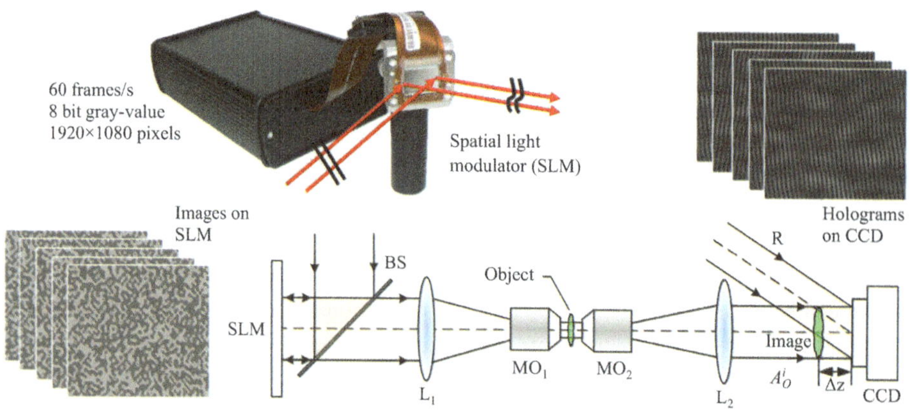

Fig. 5 Schematic of DHM with speckle illumination.

multimode optical fiber was demonstrated for scanner-free and wide-field endoscopic imaging[96]. Beyond resolution enhancement, Baek et al. demonstrated in 2018 that a high-resolution, long working distance, reference-free DHM can be performed by using a scattering layer to replace the conventional imaging lens[100]. By exploiting the randomness of a multiple-scattering layer, it allows for holographic imaging of a microscopic object without introducing a reference beam. The authors experimentally demonstrated a high NA (0.7) with a working distance of 13 mm.

Enlarging the NA of DHM recording systems

Compared to DHM with an imaging system, DHM without using an imaging lens (lens-free DHM) possesses three significant advantages. The first aspect is its compactness and low-cost because it does not employ an objective. The second is a large FOV. The third is the high compatibility and ease of integration with microfluidic and on-chip devices. In the following sections, we review the resolution enhancement of lens-free DHM via hologram extrapolation, hologram synthesis, and pixel super-resolution techniques.

Hologram extrapolation

As a major type of lens-free DHM, lensless DHM can be traced back to an inline architecture proposed by Gabor[8]. In Gabor, in lensless DHM[8,101,102] (Fig. 6a) a sample is illuminated by a spherical wave. The wave diffracted by the sample interferes with the non-diffracted wave, which also acts as a reference wave. Direct reconstruction of the generated hologram suffers from overlapping of the DC term and the twin image. Iterative approaches[7,103,104], and compressive sensing[105] have been invented to retrieve phase information from an oversampled diffraction pattern. The achievable spatial resolution is believed to be limited by the size of the hologram[8,104]:

$$\sigma = \frac{\lambda z}{N\Delta_S} \tag{4}$$

where λ is the wavelength, z is the distance between the object and the screen, N is the number of pixels, and Δ_S is the pixel size of the detector. In 2013, resolution enhancement of lensless DHM was reported by extrapolation of the hologram[104], and later, this method was applied to terahertz inline lensless holography[106,107]. The flowchart of the resolution enhancement routine is shown in Fig. 6b, and includes the following steps:

(1) Initialize the object wave $U(x_S, y_S)$ in the hologram plane by taking the square root of the hologram intensity H_0 of $N_0 \times N_0$ and padding it with zeros to a matrix of $N \times N$ with $N > N_0$ (see Fig. 6a$_1$). In contrast, random noise can be added to the padded area.

(2) Propagate the padded wave $U(x_S, y_S)$ to the object plane by using the propagation integral[7], yielding $o(x_o, y_o)$.

(3) In the object plane, a constraint concerning the finite size and positive absorption of the sample is applied to $o(x_o, y_o)$, resulting in $o'(x_o, y_o)$.

(4) Propagate $o'(x_o, y_o)$ in the hologram plane and replace the amplitude of $o'(x_o, y_o)$ with the square root of the hologram intensity $H0$ within the central $N_0 \times N_0$ area (see Fig. 6a$_2$).

Once the above procedures have been repeated over 500 iterations, the complex amplitude of the object wave can be

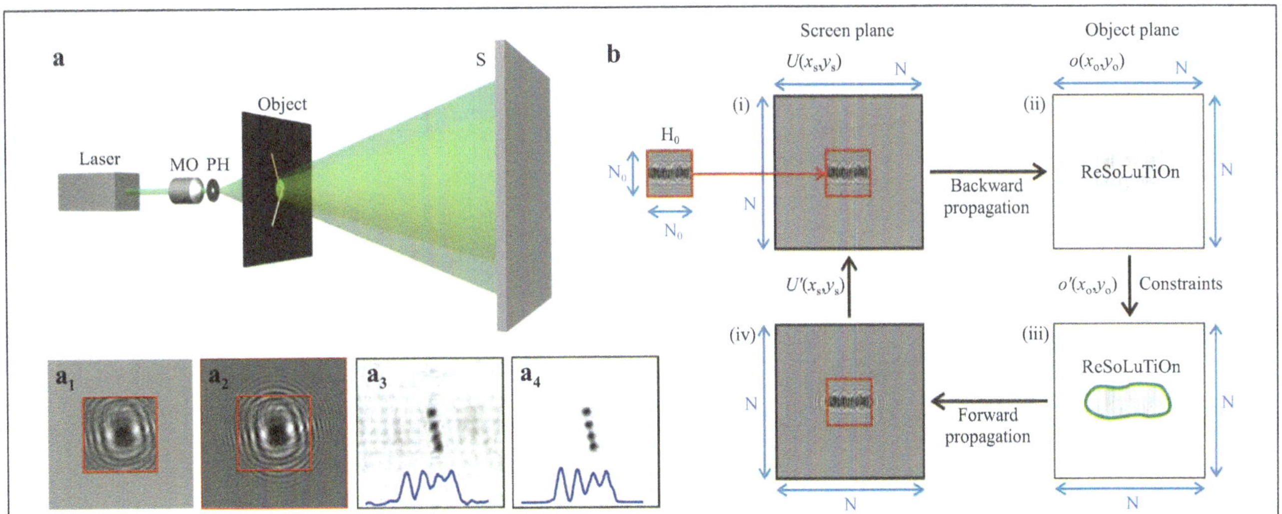

Fig. 6 Resolution enhancement of lensless DHM by self-extrapolation of holograms. **a** Experimental setup and results. a$_1$, hologram with zero-padding. a$_2$, hologram after 100 iterations. a$_3$ and a$_4$ refer to the reconstructed image without and with self-extrapolation, respectively. **b** Illustration of iterative reconstruction of a hologram by self-extrapolation. The image is adopted from Ref. 104.

retrieved, which has a resolution of $\sigma = \lambda z/(N\Delta_S)$ instead of $\lambda z/(N_0\Delta_S)$. It is clear that the image reconstructed with hologram extrapolation (Fig. 6a$_4$) has finer structures than that obtained without using self-extrapolation (Fig. 6a$_3$). Quantitatively, the resolution was enhanced from 5.9 μm to 1.8 μm by padding a hologram with 300 × 300 to 1000 × 1000 pixels[108]. The reason for the resolution enhancement by hologram extrapolation is that the hologram itself contains a series of wavelets constituting the wavefront, which can be fitted better through extrapolation and yield a larger effective hologram[104]. This hypothesis is confirmed by a closer look at the intensity distributions ((iv) in Fig. 6a$_2$) in the hologram plane.

Hologram synthesis

In addition to inline lensless DHM, DHM, which records the interference between an object wave and an independent reference wave can reconstruct the phase of a sample free of twin-image disturbances. Fig. 7a shows a standard lens-free DHM configuration using an independent reference wave[109,110]. Fig. 7b shows a modified lensless DHM configuration by the insertion of an SLM into the classical Gabor architecture[102]. The SLM is used to perform phase shifting of the *DC* term of the object wave. The complex amplitude distribution of the object wave passing through a sample can be recovered using a standard phase-shifting reconstruction algorithm. For the two types of lens-free DHM configurations (with and without an independent reference wave), the spatial resolution is limited by the size of the hologram, as shown in Eq. 4. In 2002, resolution enhancement of lens-free DHM was first demonstrated[111] by generating an expanded hologram with multiple sub-holograms. Each sub-hologram covers a different frequency spectrum of the input object. When the reconstructed image of each hologram is synthesized, an image with an enlarged FOV and an enhanced spatial resolution can be obtained[109,110]. Generally, hologram synthesis of lens-free DHM can be performed shifting the digital camera[109] or sample[110,112], using oblique beam illuminations[113], or inserting a diffraction grating between the object and digital camera[114,115]. The above-mentioned three strategies are equivalent and generate the same output, which is a larger synthetic hologram that contains a larger sample spectrum.

Figs. 7c–e show the experimental results of the lens-free DHM based on hologram synthesis. The synthesized spectrum is shown in Fig. 7c, and the resolution-enhanced image (Fig. 7e) can be obtained from the Fourier transform of the synthesized spectrum. It is clear that the SA image of the blood cells in Fig. 7e has much finer structures than that in the conventional image in Fig. 7d. Furthermore, it was found that the hologram synthesis enhanced the effective NA from 0.13 to 0.75.

Pixel super-resolution for on-chip DHM

The architecture of a standard on-chip DHM[116–119] is shown in Fig. 8a. A partially coherent light source is used for illumination, usually at > 2–3 cm (z_1 distance) above the sample. A semi-transparent sample is placed on top of an image sensor with a typical spacing of < 1 mm (z_2 distance). As a result, the sample casts an inline hologram, which is directly recorded by a CMOS or CCD image

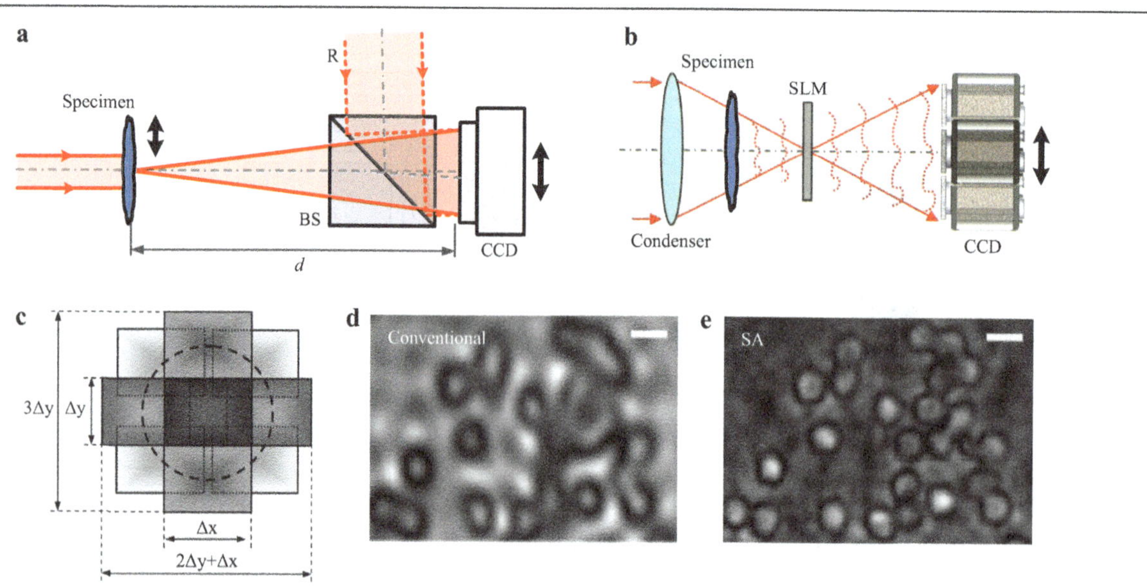

Fig. 7 Resolution enhancement of DHM via synthesizing a larger hologram. **a** Synthetic aperture of lens-free DHM using an independent reference wave. **b** Synthetic aperture of lensless DHM[102], reproduced with permission © IOP Publishing. **c** Synthesized frequency spectrum. **d** and **e**, Images of blood cells without and with applying synthetic aperture, respectively. The images in (**c**–**e**) are adapted from Ref. 113 ©The Optical Society.

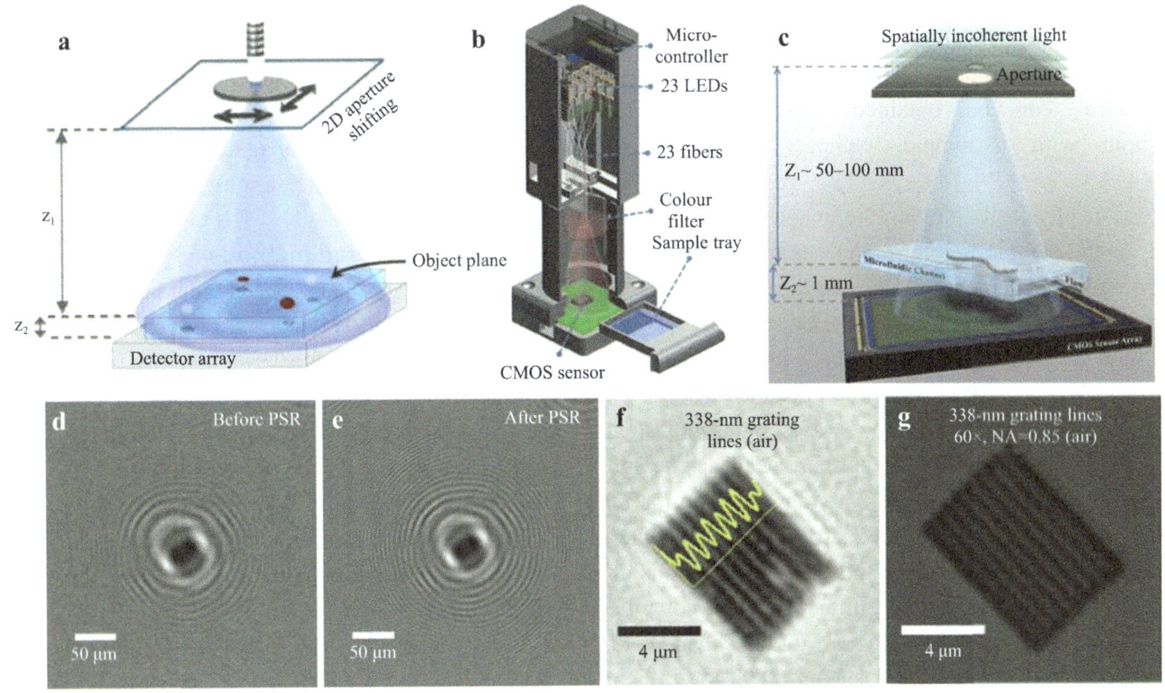

Fig. 8 Pixel super-resolution (PSR) for on-chip lensless DHM. Subpixel shifts in a digital hologram can be achieved through **a** source shifting, **b** an array of static light sources, e.g., fiber-coupled LEDs that are individually controllable, sensor shifting, or **c** sample shifting using e.g., a micro-fluidic flow. **d** An in-line hologram before PSR. **e** Pixel super-resolved hologram of the same object. **f** Reconstructed image of the grating with 338 nm line-width using PSR. **g** Microscope comparison of the same region of interest. The image is adopted from Ref. 118.

sensor, without using any imaging lenses. The magnification time of the on-chip DHM is defined as $M = (z_1 + z_2)/z_1$. For z_1 much larger than z_2, the magnification is approximately 1. The hologram recorded in the on-chip DHM is an in-line hologram (Gabor). The direct reconstruction of object-related holographic information from an inline hologram is often obscured by the twin image and the DC term. Phase retrieval approaches using an iterative approach[7] or additional constraints, such as multiple measurements at different heights[120–123] or illumination angles[124], can be employed to eliminate twin image noise. There are several factors that can limit the resolution of on-chip DHM, including diffraction, pixel size, image chip area, and coherence of the light source. In practice, the primary limitation of lens-free on-chip DHM resolution comes from the pixel size of the sensor (typically 1–2 μm).

To improve the resolution of on-chip DHM, pixel super-resolution (Pixel SR)[125,126] can be adopted by shifting the hologram along the x- and y-directions in subpixel increments. The relative lateral shift of the hologram with respect to the sensor array can also be performed via shifting of the image sensor[127], the illumination source[128] (Fig. 8a, b), or the sample[129] (Fig. 8c), respectively. A low-resolution hologram is captured at each location of the shifted grid. Then, using multiple LR holograms, a higher-resolution hologram can be digitally synthesized by using a "shift-and-add" algorithm, in which the low-resolution (LR) holograms are upsampled, shifted, and digitally added. Alternatively, iterative methods can be used to upsample a high-resolution image from a series of low-resolution images. For this purpose, a cost function can be used for optimization[125]:

$$x^* = \operatorname{argmin} \sum_i \|W_i \cdot x - y_i\|_p^q + \alpha\gamma(x) \tag{5}$$

This equation contains two operations, namely, norm and regulation, where i represents the recording times. y_i is a low-resolution hologram, and x is the desired high-resolution image. W_i represents the digital processes of shifting and down-sampling of an image. This synthesis method is equivalent to recording holograms with smaller pixel sizes and higher spatial density detectors. The second part $\gamma(x)$ is a regularization parameter to maintain/regulate the desired quality in the reconstructed image. By solving this optimization function, an image with a pixel size smaller than that of the digital camera can be obtained by synthesizing a series of sub-pixel-shifted LR images. A comparison of Fig. 8d, e shows that PSR can provide a more refined hologram with high-frequency fringes. PSR yields a high-resolution image (Fig. 8f) of a grating (linewidth ~300 nm), whose resolution is equivalent to an effective NA~0.9, as shown in Fig. 8g[116]. Using the PSR techniques, sub-micrometer resolution over a wide FOV of 20–30 mm^2 can be obtained, providing gigapixel throughput with a simple, compact on-chip DHM design[130,131].

Instead of physically moving the image sensor, the light source, or the sample, virtual pinhole scanning completed by deconvolving a hologram with the measured PSF is an alternative way to improve the resolution of an on-chip DHM with an extended light source[132]. PSR can also be performed by varying the propagation distances (i.e., z_2)[133], or by varying the illumination wavelengths in small increments (e.g., 2–3 nm)[134]. To further speed up the sub-hologram recording process by varying the illumination wavelength, color LR holograms can be recorded, where the holograms of the red, reen, and blue illuminations are multiplexed using a sub-pixel shifted Bayer color filter array[119]. Considering that PSR can only enhance the effective NA of the on-chip DHM to 0.9, a synthetic aperture-based on-chip microscope with which the illumination angle is scanned across the surface of a dome (from −50° to 50°) was proposed, which further increased the effective NA to 1.4, achieving a 250-nm resolution at a wavelength of 700-nm across the very large FOV of 20.5 mm^2[124].

Theoretically, an effective NA of 1+n can be achieved in a lensless on-chip DHM using the PSR/SA approach, assuming that the medium between the source and the sample planes is air. Here, n denotes the refractive index of the medium that fills the space between the sample and sensor planes. However, none of these techniques can be considered to exceed the diffraction limit of light, as they are entirely based on propagating waves that result from coherent light-matter interactions.

Deep-learning based methods

Deep-learning[135,136] has been demonstrated as being a powerful tool for solving various inverse problems by training a network with a large quantity of paired images. Deep learning can map the relationship between the input and target output distributions without any prior knowledge of the imaging model. Deep learning has been widely used for holographic image reconstruction[137–139], auto-focusing[140,141], and resolution enhancement[142,143].

The deep-learning configuration for the resolution enhancement of DHM is shown in Fig. 9. By training a set of matched LR and high-resolution images, the relationships between them can be learned. Thus, a high resolution can be retrieved from one or multiple low-resolution images captured by a similar setup. The establishment of the mapping function between the high-resolution output image and the LR input images is called the objective function and is expressed as follows:

$$R_{\text{learn}} = \underset{R_\theta, \theta \in \Theta}{\operatorname{argmin}} \sum_{n=1}^{N} F\left[x_n, R_\theta(y_n)\right] + g(\theta) \tag{6}$$

where R_{learn} is the optimal solution of the weight parameter

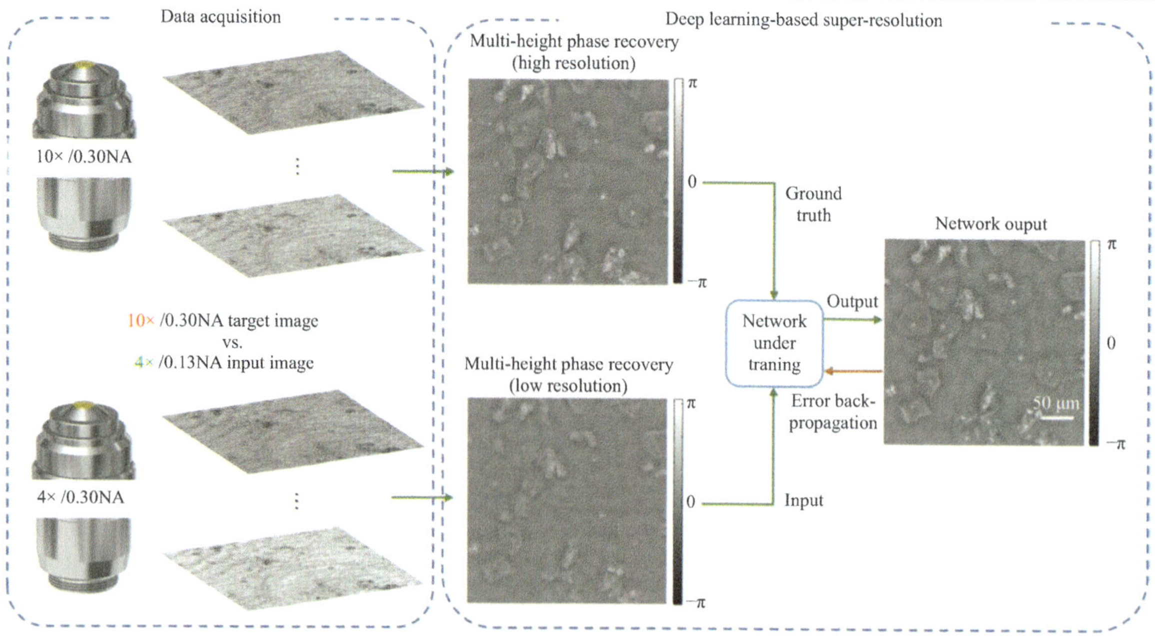

Fig. 9 Schematics of the deep-learning based method for resolution enhancement of DHM. Images were taken from Ref. 144.

of the network, x and y are the training set, F[145] is the loss function for appropriate error metrics, and $g(\theta)$ is a regularization term for the parameter θ to avoid overfitting[146]. By continuously adjusting θ, the error between the output $R_{\theta}(y_n)$ of the neural network and the target image x_n is minimized. Therefore, as long as the establishment of the model is completed through data statistics, it can be applied to solve inverse problems in imaging. Using the network described in Eq. 6, deep learning has also been used to enhance the resolution of data-driven models for the following two systems: (1) lensless DHM and (2) lens-based DHM systems.

In the first case, a deep learning algorithm based on a generative adversarial network (GAN) must be used in an on-chip lensless DHM to achieve sub-pixel resolution[147]. In this method, the LR reconstructed images are used as inputs and the corresponding high-resolution images of different samples obtained by the "shift-and-add" approach are used as the labels for training the network. A well-trained framework can generate high-resolution images from unseen LR data without an iterative engine and prior knowledge of the system. In 2019, Liu et al. demonstrated that a resolution-improved image could be obtained from much fewer (e.g., 1 or 4 instead of 36 in conventional Pixel SR algorithm) low-resolution sub-pixel-shifted images of the same object by using a neural network, which significantly reduced both the number of hologram measurements and the computation time necessary for reconstruction[144].

The network training process for the lens-based microscopic imaging system is shown in Fig. 8. The reconstructed images of the holograms obtained under the low-magnification objective were used as inputs, and the counterparts obtained under the high-magnification objective were used as labels in the network to generate an image with a resolution of the high-magnification objective, but with the same FOV as of the low-magnification objective [144]. In this way, the diffraction limit defined by the NA of the objective and the SBP of the system can be broken, so that the holographic reconstruction image has a higher resolution and the reconstruction efficiency is improved as compared with the iterative process for each image. A deep-learning-based upscaling method was proposed for overcoming the trade-off between a large FOV and a high resolution, yielding an enlarged hologram with a smaller effective pixel size (enhanced sample details), for which inline holograms are used as training material for the convolutional neural network[148]. In this method, the deep-learning method only uses a single image obtained with a low-NA objective for magnification, allowing detailed analysis of a sample with a high resolution and a large FOV. However, considering that LR-input images do not contain high-resolution components physically, the resolution enhancement capability (or the right guessing of unknown details) of the network is limited by the adequacy of the training data[149]. As a remedy, people tend to integrate deep-learning-based strategies and illumination-modulation techniques to enhance the resolution of DHM. For instance, in 2021, Meng et al. demonstrated that the deep-learning algorithm simplifies[91] not only the resolution-enhanced reconstruction process of SI-DHM but also minimizes the recording requirements[91].

In general, for both lensless DHM and lens-based DHM, a deep-learning-based framework can enlarge the SBP of coherent imaging systems using image data and convolutional neural networks, and provides a rapid, non-iterative method for solving inverse image reconstruction or enhancement problems in optics. Furthermore, it has a faster imaging efficiency, even with little prior knowledge[147], and can reconstruct in real time[150]. This is because deep-learning methods use large-scale datasets to impose potential constraints on the resolution enhancement problem. A neural network has learned millions of weight parameters from a large amount of training data. These parameters can establish a high-dimensional nonlinear relationship between the input and output, which is difficult to express through simple formulas. Although classical principled algorithms are often outperformed by deep learning models, they also retain key advantages. First, classical algorithms inherently produce outputs that are consistent with their inputs because they rely exclusively on the underlying physics and principles formulated. Second, classic algorithms can be generalized for any valid measurement because they are not limited by the adequacy of the training data. Third, classic algorithms, in contrast to deep learning models, do not produce widely erratic outputs after minute tweaks to their inputs.[149]. In light of the above analysis, it is our opinion that deep learning results cannot be blindly trusted, something certainly true of any singular piece of evidence. Incorporating physical models into deep neural networks provides a new idea for expanding the generality and reliability of imaging[151], which should be a new trend in deep learning-based resolution enhancement.

Discussion

In this review, we summarize resolution enhancement approaches for digital holographic microscopy (DHM). Generally, by trading with other degrees of freedom, the spatial resolution is enhanced, and a large field of view (FOV) is preserved (Table 1). Resolution enhancement

Table 1 Overview of Different Resolution enhancement techniques for DHM

Technique/Methods		Features/Capabilities	References
DHM using an imaging lens	Oblique illumination Structured illumination Speckle illumination	• Complex configuration due to the usage of an objective, suffers from objective-related aberration • Achievable resolution: $\sigma = \frac{\kappa_1 \lambda}{NA_{imag}+NA_{illum}}$ • Resolution enhancement factor of 2~5, depending on NA_{illum}/NA_{imag}. • FOV is limited by the objective design: 1x1 mm^2 under a 10× objective, etc.	49, 77, 96
DHM without using an imaging lens	Self-extrapolation of holograms	• No extra measurement is needed • Achievable resolution is $\lambda z/(N\Delta_S)$ instead of $\lambda z/(N_0\Delta_S)$, when the hologram is effectively extrapolated from $N_0\Delta_S$ to $N\Delta_S$.	104, 106-108
	Hologram expansion by shifting the CCD or the sample	• Achievable resolution: $\delta=\lambda z/L_{hologram}$ • Effective NA up to 0.3, and FOV: up to 20×16 mm^2	110, 112, 113
	Pixel super-resolution for on-chip DHM	• Subpixel-shifts (e.g.,6×6) of hologram yields an NA of 0.9 • Oblique illumination further enhances NA to 1.4 and improves the SNR. • FOV up to 20 mm^2	116, 118, 125, 128, 129124
AI algorithm based techniques	Deep-learning	• Relies on large-scale data sets, time-consuming. • Reconstruction can be wrong if the training data is inadequate. • Resolution is diffraction limited.	144,148

approaches can be classified into two categories, namely, lens-based DHM and lens-free DHM, based on whether an objective lens is used. In lens-based DHM, oblique illumination, structured illumination, and speckle illumination were used to enhance the spatial resolution of DHM. In sample-distant lens-free DHM, self-extrapolation of holograms can enhance the physical-aperture-limited spatial resolution. Furthermore, the resolution can be further enhanced by synthesizing a larger hologram by scanning the sample, camera, illumination, or inserting a diffraction grating between the sample and the camera. With a lens-free on-chip DHM, resolution enhancement can be obtained by pixel super-resolution and a synthetic aperture by shifting the illumination beam. Recently, deep learning has also been used to enhance the resolution in DHM. Being trained with plenty experimental datasets obtained under well-controlled imaging conditions, deep neural networks enable the enhancement of the spatial resolution or the SBP of DHM in a rapid, non-iterative manner. It is worth pointing out that the resolution enhancement approaches in DHM can surpass the resolution limited by the geometry (NA) of the imaging system in terms of 0.82λ/NA. However, they cannot surpass the physical diffraction limit λ(2NA)[152] in far-field imaging.

In the past decade, intensive efforts have been made to achieve coherent imaging resolution beyond the diffraction limit. There are two coherent imaging approaches that can achieve super-resolution capability. The first is synthetic-aperture DHM using evanescent waves, which can exceed the diffraction limit in the near-field region. The second is super-oscillation illumination or super-oscillatory lens-based optical microscopy[153]. We envisage that there will be an increasing number of investigations concerning the physics of DHM resolution enhancement, and these investigations will widen the applications of DHM in different fields.

Acknowledgements

The authors acknowledge the support from the National Key Research and Development Program of China (2021YFF0700300), the National Natural Science Foundation of China (NSFC 62075177, 62175112), the Natural Science Foundation of Shaanxi Province (2020JM-193 and 2020JQ-324), and the Fundamental Research Funds for the Central Universities (XJS210503, XJS210504, JC2112, and JB210513).

Author details

[1]School of Physics and Optoelectronic Engineering, Xidian University, Xi' an 710071, China. [2]Academy of Advanced Interdisciplinary research, Xidian University, Xi 'an 710071, China. [3]Key Laboratory for Opto-Electronic Technology of Jiangsu Province, Nanjing Normal University, Nanjing 210023, China

Conflict of interest

The authors declare that they have no conflict of interest.

Received: 03 September 2021 Revised: 18 January 2022 Accepted: 19 January 2022

Accepted article preview online: 21 January 2022

Published online: 27 January 2022

References

1. Schnars, U. & Jüptner, W. Direct recording of holograms by a CCD target and numerical reconstruction. *Applied Optics* **33**, 179-181 (1994).
2. Yu, X. et al. Review of digital holographic microscopy for three-dimensional profiling and tracking. *Optical Engineering* **53**, 112306 (2014).
3. Osten, W. et al. Recent advances in digital holography [Invited]. *Applied Optics* **53**, G44-G63 (2014).
4. Micó, V. et al. Resolution enhancement in quantitative phase microscopy. *Advances in Optics and Photonics* **11**, 135-214 (2019).
5. Goodman, J. W. & Lawrence, R. W. Digital image formation from electronically detected holograms. *Applied Physics Letters* **11**, 77-79 (1967).
6. Kim, M. K. Principles and techniques of digital holographic microscopy. *SPIE Reviews* **1**, 018005 (2010).
7. Latychevskaia, T. & Fink, H. W. Solution to the twin image problem in holography. *Physical Review Letters* **98**, 233901 (2007).
8. Gabor, D. A new microscopic principle. *Nature* **161**, 777-778 (1948).
9. Langehanenberg, P., von Bally, G. & Kemper, B. Autofocusing in digital holographic microscopy. *3D Research* **2**, 4 (2011).
10. Gao, P. et al. Autofocusing of digital holographic microscopy based on off-axis illuminations. *Optics Letters* **37**, 3630-3632 (2012).
11. Charrière, F. et al. Characterization of microlenses by digital holographic microscopy. *Applied Optics* **45**, 829-835 (2006).
12. Markus, F. et al. Digital holography in production: an overview. *Light:Advanced Manufacturing* **2**, 15 (2021).
13. Cubreli, G. et al. Digital holographic interferometry for the measurement of symmetrical temperature fields in liquids. *Photonics* **8**, 200 (2021).
14. Kemper, B. & von Bally, G. Digital holographic microscopy for live cell applications and technical inspection. *Applied Optics* **47**, A52-A61 (2008).
15. Born, M. & Wolf, E. Principles of Optics. 7th edn. (Cambridge: Cambridge University, 1999).
16. Cotte, Y. et al. Marker-free phase nanoscopy. *Nature Photonics* **7**, 113-117 (2013).
17. den Dekker, A. J. & van den Bos. A. Resolution: a survey. *Journal of the Optical Society of America A* **14**, 547-557 (1997).
18. Faridian, A. et al. Nanoscale imaging using deep ultraviolet digital holographic microscopy. *Optics Express* **18**, 14159-14164 (2010).
19. Lohmann, A. W. et al. Space-bandwidth product of optical signals and systems. *Journal of the Optical Society of America A* **13**, 470-473 (1996).
20. Cox, I. J. & Sheppard, C. J. R. Information capacity and resolution in an optical system. *Journal of the Optical Society of America A* **3**, 1152-1158 (1986).
21. Zheng, G. A., Horstmeyer, R. & Yang, C. H. Wide-field, high-resolution Fourier ptychographic microscopy. *Nature Photonics* **7**, 739-745 (2013).
22. Micó, V., Zalevsky, Z. & García, J. Optical superresolution: imaging beyond abbe's diffraction limit. *Journal of Holography and Speckle* **5**, 110-123 (2009).
23. Zalevsky, Z., Micó, V. & Garcia, J. Nanophotonics for optical super resolution from an information theoretical perspective: a review. *Journal of Nanophotonics* **3**, 032502 (2009).
24. Ueda, M. & Sato, T. Superresolution by holography. *Journal of the Optical Society of America* **61**, 418-419 (1971).
25. Ueda, M., Sato, T. & Kondo, M. Superresolution by multiple superposition of image holograms having different carrier frequencies. *Optica Acta:International Journal of Optics* **20**, 403-410 (1973).
26. Sato, T., Ueda, M. & Yamagish, G. Superresolution microscope using electrical superposition of holograms. *Applied Optics* **13**, 406-408 (1974).
27. Sato, T., Ueda, M. & Ikeda, T. Real time superresolution by means of an ultrasonic light diffractor and TV system. *Applied Optics* **13**, 1318-1321 (1974).
28. Kuznetsova, Y., Neumann, A. & Brueck, S. R. J. Imaging interferometric microscopy. *Journal of the Optical Society of America A* **25**, 811-822 (2008).
29. Neumann, A., Kuznetsova, Y. & Brueck, S. R. J. Optical resolution below λ/4 using synthetic aperture microscopy and evanescent-wave illumination. *Optics Express* **16**, 20477-20483 (2008).
30. Micó, V. et al. Superresolution digital holographic microscopy for three-dimensional samples. *Optics Express* **16**, 19260-19270 (2008).
31. Hillman, T. R. et al. High-resolution, wide-field object reconstruction with synthetic aperture Fourier holographic optical microscopy. *Optics Express* **17**, 7873-7892 (2009).
32. Bühl, J. et al. Digital synthesis of multiple off-axis holograms with overlapping Fourier spectra. *Optics Communications* **283**, 3631-3638 (2010).
33. Gutzler, T. et al. Coherent aperture-synthesis, wide-field, high-resolution holographic microscopy of biological tissue. *Optics Letters* **35**, 1136-1138 (2010).
34. Granero, L. et al. Single-exposure super-resolved interferometric microscopy by RGB multiplexing in lensless configuration. *Optics and Lasers in Engineering* **82**, 104-112 (2016).
35. Calabuig, A. et al. Single-exposure super-resolved interferometric microscopy by red-green-blue multiplexing. *Optics Letters* **36**, 885-887 (2011).
36. Calabuig, A. et al. Resolution improvement by single-exposure superresolved interferometric microscopy with a monochrome sensor. *Journal of the Optical Society of America A* **28**, 2346-2358 (2011).
37. Choi, Y. et al. Synthetic aperture microscopy for high resolution imaging through a turbid medium. *Optics Letters* **36**, 4263-4265 (2011).
38. Yuan, C. J. et al. Resolution improvement in digital holography by angular and polarization multiplexing. *Applied Optics* **50**, B6-B11 (2011).
39. Picazo-Bueno, J. Á. et al. Superresolved spatially multiplexed interferometric microscopy. *Optics Letters* **42**, 927-930 (2017).
40. Mico, V. et al. Synthetic aperture superresolution with multiple off-axis holograms. *Journal of the Optical Society of America A* **23**, 3162-3170 (2006).
41. Kim, M. et al. High-speed synthetic aperture microscopy for live cell imaging. *Optics Letters* **36**, 148-150 (2011).
42. Micó, V., Zalevsky, Z. & Garcia, J. Superresolved common-path phase-shifting digital inline holographic microscopy using a spatial light modulator. *Optics Letters* **37**, 4988-4990 (2012).
43. Kuznetsova, Y., Neumann, A. & Brueck, S. R. J. Imaging interferometric microscopy - approaching the linear systems limits of optical resolution. *Optics Express* **15**, 6651-6663 (2007).
44. Mico, V., Zalevsky, Z. & García, J. Synthetic aperture microscopy using off-axis illumination and polarization coding. *Optics Communications* **276**, 209-217 (2007).
45. Indebetouw, G. et al. Scanning holographic microscopy with resolution exceeding the Rayleigh limit of the objective by superposition of off-axis holograms. *Applied Optics* **46**, 993-1000 (2007).
46. Mico, V. et al. Superresolved imaging in digital holography by superposition of tilted wavefronts. *Applied Optics* **45**, 822-828 (2006).
47. Mico, V. et al. Single-step superresolution by interferometric imaging.

Optics Express **12**, 2589-2596 (2004).

48. Indebetouw, G., El Maghnouji, A. & Foster, R. Scanning holographic microscopy with transverse resolution exceeding the Rayleigh limit and extended depth of focus. *Journal of the Optical Society of America A* **22**, 892-898 (2005).
49. Cheng, C. J. et al. Superresolution imaging in synthetic aperture digital holographic microscopy. Proceedings of the IEEE 4th International Conference on Photonics. Melaka: IEEE, 2013.
50. Schwarz, C. J., Kuznetsova, Y. & Brueck, S. R. J. Imaging interferometric microscopy. *Optics Letters* **28**, 1424-1426 (2003).
51. Lee, D. J. & Weiner, A. M. Optical phase imaging using a synthetic aperture phase retrieval technique. *Optics Express* **22**, 9380-9394 (2014).
52. Mico, V. et al. Transverse resolution improvement using rotating-grating time-multiplexing approach. *Journal of the Optical Society of America A* **25**, 1115-1129 (2008).
53. Lin, Y. C. et al. One-shot synthetic aperture digital holographic microscopy with non-coplanar angular-multiplexing and coherence gating. *Optics Express* **26**, 12620-12631 (2018).
54. Hussain, A. et al. Simple fringe illumination technique for optical superresolution. *Journal of the Optical Society of America B* **34**, B78-B84 (2017).
55. Hussain, A. et al. Super resolution imaging achieved by using on-axis interferometry based on a Spatial Light Modulator. *Optics Express* **21**, 9615-9623 (2013).
56. Mudassar, A. A. A simplified holography based superresolution system. *Optics and Lasers in Engineering* **75**, 27-38 (2015).
57. Hussain, A. & Mudassar, A. A. Optical super resolution using tilted illumination coupled with object rotation. *Optics Communications* **339**, 34-40 (2015).
58. Hussain, A. & Mudassar, A. A. Holography based super resolution. *Optics Communications* **285**, 2303-2310 (2012).
59. Phan, A. H., Park, J. H. & Kim, N. Super-resolution digital holographic microscopy for three dimensional sample using multipoint light source illumination. *Japanese Journal of Applied Physics* **50**, 092503 (2011).
60. Mudassar, A. A. & Hussain, A. Super-resolution of active spatial frequency heterodyning using holographic approach. *Applied Optics* **49**, 3434-3441 (2010).
61. Micó, V., García, J. & Zalevsky, Z. Axial superresolution by synthetic aperture generation. *Journal of Optics A:Pure and Applied Optics* **10**, 125001 (2008).
62. Micó, V., Zalevsky, Z. & García, J. Edge processing by synthetic aperture superresolution in digital holographic microscopy. *3D Research* **2**, 1 (2011).
63. Kim, M. et al. Three-dimensional differential interference contrast microscopy using synthetic aperture imaging. *Journal of Biomedical Optics* **17**, 026003 (2012).
64. Gao, P. et al. Phase-shifting Zernike phase contrast microscopy for quantitative phase measurement. *Optics Letters* **36**, 4305-4307 (2011).
65. Wolf, E. Three-dimensional structure determination of semi-transparent objects from holographic data. *Optics Communications* **1**, 153-156 (1969).
66. Kim, K., Yoon, J. & Park, Y. Large-scale optical diffraction tomography for inspection of optical plastic lenses. *Optics Letters* **41**, 934-937 (2016).
67. Charrière, F. et al. Living specimen tomography by digital holographic microscopy: morphometry of testate amoeba. *Optics Express* **14**, 7005-7013 (2006).
68. Kim, K., Yoon, J. & Park, Y. Simultaneous 3D visualization and position tracking of optically trapped particles using optical diffraction tomography. *Optica* **2**, 343-346 (2015).
69. Chen, M. et al. Multi-layer Born multiple-scattering model for 3D phase microscopy. *Optica* **7**, 394-403 (2020).
70. Isikman, S. O. et al. Lens-free optical tomographic microscope with a large imaging volume on a chip. *Proceedings of the National Academy of Sciences of the United States of America* **108**, 7296-7301 (2011).
71. Wang, J. et al. Airy-beam tomographic microscopy. *Optica* **7**, 790-793 (2020).
72. Kuś, A. Real-time, multiplexed holographic tomography. *Optics and Lasers in Engineering* **149**, 106783 (2022).
73. Kamilov, U. S. et al. Learning approach to optical tomography. *Optica* **2**, 517-522 (2015).
74. Gustafsson, M. G. L. Surpassing the lateral resolution limit by a factor of two using structured illumination microscopy. *Journal of Microscopy* **198**, 82-87 (2000).
75. Gustafsson, M. G. L. Nonlinear structured-illumination microscopy: wide-field fluorescence imaging with theoretically unlimited resolution. *Proceedings of the National Academy of Sciences of the United States of America* **102**, 13081-13086 (2005).
76. Gao, P., Pedrini, G. & Osten, W. Structured illumination for resolution enhancement and autofocusing in digital holographic microscopy. *Optics Letters* **38**, 1328-1330 (2013).
77. Chowdhury, S. et al. Refractive index tomography with structured illumination. *Optica* **4**, 537-545 (2017).
78. Sánchez-Ortiga, E. et al. Enhancing spatial resolution in digital holographic microscopy by biprism structured illumination. *Optics Letters* **39**, 2086-2089 (2014).
79. Lee, K. et al. Time-multiplexed structured illumination using a DMD for optical diffraction tomography. *Optics Letters* **42**, 999-1002 (2017).
80. Hussain, A. & Fuentes, J. L. M. Resolution enhancement using simultaneous couple illumination. *Journal of Optics* **18**, 105702 (2016).
81. Ganjkhani, Y. et al. Super-resolved Mirau digital holography by structured illumination. *Optics Communications* **404**, 110-117 (2017).
82. Yuan, C. J. et al. Resolution enhancement of the microscopic imaging by unknown sinusoidal structured illumination with iterative algorithm. *Applied Optics* **56**, F78-F83 (2017).
83. Yeh, L. H., Tian, L. & Waller, L. Structured illumination microscopy with unknown patterns and a statistical prior. *Biomedical Optics Express* **8**, 695-711 (2017).
84. Zheng, J. J. et al. Digital holographic microscopy with phase-shift-free structured illumination. *Photonics Research* **2**, 87-91 (2014).
85. Chowdhury, S. & Izatt, J. Structured illumination diffraction phase microscopy for broadband, subdiffraction resolution, quantitative phase imaging. *Optics Letters* **39**, 1015-1018 (2014).
86. Lai, X. J. et al. Coded aperture structured illumination digital holographic microscopy for superresolution imaging. *Optics Letters* **43**, 1143-1146 (2018).
87. Chowdhury, S. & Izatt, J. Structured illumination quantitative phase microscopy for enhanced resolution amplitude and phase imaging. *Biomedical Optics Express* **4**, 1795-1805 (2013).
88. Yuan, C. J., Feng, S. T. & Nie, S. P. Digital holographic microscopy by using structured illumination. *Chinese Journal of Lasers* **43**, 0609003 (2016).
89. Li, S. H. et al. Phase-shifting-free resolution enhancement in digital holographic microscopy under structured illumination. *Opt Express* **26**, 23572-23584 (2018).
90. Chowdhury, S. et al. Structured illumination multimodal 3D-resolved quantitative phase and fluorescence sub-diffraction microscopy. *Biomedical Optics Express* **8**, 2496-2518 (2017).

91. Meng, Z. et al. DL-SI-DHM: a deep network generating the high-resolution phase and amplitude images from wide-field images. *Optics Express* **29**, 19247-19261 (2021).
92. Shin, S. et al. Active illumination using a digital micromirror device for quantitative phase imaging. *Optics Letters* **40**, 5407-5410 (2015).
93. Goodman, J. W. Speckle Phenomena in Optics: Theory and Applications. (Englewood: Roberts & Co, 2006).
94. García, J., Zalevsky, Z. & Fixler, D. Synthetic aperture superresolution by speckle pattern projection. *Optics Express* **13**, 6073-6078 (2005).
95. Zheng, J. J. et al. Autofocusing and resolution enhancement in digital holographic microscopy by using speckle-illumination. *Journal of Optics* **17**, 085301 (2015).
96. Park, Y. et al. Speckle-field digital holographic microscopy. *Optics Express* **17**, 12285-12292 (2009).
97. Liu, Y. et al. Dynamic speckle illumination digital holographic microscopy by doubly scattered system. *Photonics* **8**, 276 (2021).
98. Yilmaz, H. et al. Speckle correlation resolution enhancement of wide-field fluorescence imaging. *Optica* **2**, 424-429 (2015).
99. Choi, Y. et al. Overcoming the diffraction limit using multiple light scattering in a highly disordered medium. *Physical Review Letters* **107**, 023902 (2011).
100. Baek, Y., Lee, K. & Park, Y. High-resolution holographic microscopy exploiting speckle-correlation scattering matrix. *Physical Review Applied* **10**, 024053 (2018).
101. Repetto, L., Piano, E. & Pontiggia, C. Lensless digital holographic microscope with light-emitting diode illumination. *Optics Letters* **29**, 1132-1134 (2004).
102. Micó, V. et al. Superresolved phase-shifting Gabor holography by CCD shift. *Journal of Optics A:Pure and Applied Optics* **11**, 125408 (2009).
103. Miao, J., Sayre, D. & Chapman, H. N. Phase retrieval from the magnitude of the Fourier transforms of nonperiodic objects. *Journal of the Optical Society of America A* **15**, 1662-1669 (1998).
104. Latychevskaia, T. & Fink, H. W. Resolution enhancement in digital holography by self-extrapolation of holograms. *Optics Express* **21**, 7726-7733 (2013).
105. Zhang, W. H. et al. Twin-image-free holography: a compressive sensing approach. *Physical Review Letters* **121**, 093902 (2018).
106. Rong, L. et al. Terahertz in-line digital holography of dragonfly hindwing: amplitude and phase reconstruction at enhanced resolution by extrapolation. *Optics Express* **22**, 17236-17245 (2014).
107. Rong, L. et al. Terahertz in-line digital holography of human hepatocellular carcinoma tissue. *Scientific Reports* **5**, 8445 (2015).
108. Latychevskaia, T. & Fink, H. W. Coherent microscopy at resolution beyond diffraction limit using post-experimental data extrapolation. *Applied Physics Letters* **103**, 204105 (2013).
109. Di, J. L. et al. High resolution digital holographic microscopy with a wide field of view based on a synthetic aperture technique and use of linear CCD scanning. *Applied Optics* **47**, 5654-5659 (2008).
110. Micó, V., Ferreira, C. & García, J. Surpassing digital holography limits by lensless object scanning holography. *Optics Express* **20**, 9382-9395 (2012).
111. Massig, J. H. Digital off-axis holography with a synthetic aperture. *Optics Letters* **27**, 2179-2181 (2002).
112. Bianco, V., Paturzo, M. & Ferraro, P. Spatio-temporal scanning modality for synthesizing interferograms and digital holograms. *Optics Express* **22**, 22328-22339 (2014).
113. Granero, L. et al. Synthetic aperture superresolved microscopy in digital lensless Fourier holography by time and angular multiplexing of the object information. *Applied Optics* **49**, 845-857 (2010).
114. Paturzo, M. et al. Super-resolution in digital holography by a two-dimensional dynamic phase grating. *Optics Express* **16**, 17107-17118 (2008).
115. Liu, C. Super-resolution digital holographic imaging method. *Applied Physics Letters* **81**, 3143-3145 (2002).
116. Greenbaum, A. et al. Imaging without lenses: achievements and remaining challenges of wide-field on-chip microscopy. *Nature Methods* **9**, 889-895 (2012).
117. Mudanyali, O. et al. Compact, light-weight and cost-effective microscope based on lensless incoherent holography for telemedicine applications. *Lab on A Chip* **10**, 1417-1428 (2010).
118. Wu, Y. C. & Ozcan, A. Lensless digital holographic microscopy and its applications in biomedicine and environmental monitoring. *Methods* **136**, 4-16 (2018).
119. Wu, Y. C. et al. Demosaiced pixel super-resolution for multiplexed holographic color imaging. *Scientific Reports* **6**, 28601 (2016).
120. Mitome, M. Transport of intensity equation method and its applications, Microscopy, 70, 69–74 (2021).
121. Pedrini, G., Osten, W. & Zhang, Y. Wave-front reconstruction from a sequence of interferograms recorded at different planes. *Optics Letters* **30**, 833-835 (2005).
122. Allen, L. J. & Oxley, M. P. Phase retrieval from series of images obtained by defocus variation. *Optics Communications* **199**, 65-75 (2001).
123. Fienup, J. R. Reconstruction of an object from the modulus of its Fourier transform. *Optics Letters* **3**, 27-29 (1978).
124. Luo, W. et al. Synthetic aperture-based on-chip microscopy. *Light:Science & Applications* **4**, e261 (2015).
125. Bishara, W. et al. Lensfree on-chip microscopy over a wide field-of-view using pixel super-resolution. *Optics Express* **18**, 11181-11191 (2010).
126. Farsiu, S. et al. Fast and robust multiframe super resolution. *IEEE Transactions on Image Processing* **13**, 1327-1344 (2004).
127. Greenbaum, A. et al. Wide-field computational imaging of pathology slides using lens-free on-chip microscopy. *Science Translational Medicine* **6**, 267ra175 (2014).
128. Bishara, W. et al. Holographic pixel super-resolution in portable lensless on-chip microscopy using a fiber-optic array. *Lab on A Chip* **11**, 1276-1279 (2011).
129. Bishara, W., Zhu, H. Y. & Ozcan, A. Holographic opto-fluidic microscopy. *Optics Express* **18**, 27499-27510 (2010).
130. McLeod, E. et al. *Toward giga-pixel nanoscopy on a chip*: a computational wide-field look at the nano-scale without the use of lenses. *Lab on A Chip* **13**, 2028-2035 (2013).
131. Greenbaum, A. et al. Increased space-bandwidth product in pixel super-resolved lensfree on-chip microscopy. *Scientific Reports* **3**, 1717 (2013).
132. Feng, S. D. & Wu, J. G. Resolution enhancement method for lensless in-line holographic microscope with spatially-extended light source. *Optics Express* **25**, 24735-24744 (2017).
133. Wang, H. D. et al. Computational out-of-focus imaging increases the space–bandwidth product in lens-based coherent microscopy. *Optica* **3**, 1422-1429 (2016).
134. Luo, W. et al. Pixel super-resolution using wavelength scanning. *Light:Science & Applications* **5**, e16060 (2016).
135. LeCun, Y., Bengio, Y. & Hinton, G. Deep learning. *Nature* **521**, 436-444 (2015).
136. Ongie, G. et al. Deep learning techniques for inverse problems in imaging. *IEEE Journal on Selected Areas in Information Theory* **1**, 39-56 (2020).
137. Rivenson, Y., Wu, Y. C. & Ozcan, A. Deep learning in holography and coherent imaging. *Light:Science & Applications* **8**, 85 (2019).
138. Zhang, G. et al. Fast phase retrieval in off-axis digital holographic microscopy through deep learning. *Optics Express* **26**, 19388-19405

(2018).
139. Xiao, W. et al. Adaptive frequency filtering based on convolutional neural networks in off-axis digital holographic microscopy. *Biomedical Optics Express* **10**, 1613-1626 (2019).
140. Rivenson, Y. et al. Phase recovery and holographic image reconstruction using deep learning in neural networks. *Light:Science & Applications* **7**, 17141 (2018).
141. Ding, H. et al. Auto-focusing and quantitative phase imaging using deep learning for the incoherent illumination microscopy system. *Optics Express* **29**, 26385-26403 (2021).
142. Rivenson, Y. et al. Deep learning microscopy. *Optica* **4**, 1437-1443 (2017).
143. Nehme, E. et al. Deep-STORM: super-resolution single-molecule microscopy by deep learning. *Optica* **5**, 458-464 (2018).
144. Liu, T. R. et al. Deep learning-based super-resolution in coherent imaging systems. *Scientific Reports* **9**, 3926 (2019).
145. Ghosh, N. & Bhattacharya, K. Cube beam-splitter interferometer for phase shifting interferometry. *Journal of Optics* **38**, 191-198 (2009).
146. McCann, M. T., Jin, K. H. & Unser, M. Convolutional neural networks for inverse problems in imaging: a review. *IEEE Signal Processing Magazine* **34**, 85-95 (2017).
147. Barbastathis, G., Ozcan, A. & Situ, G. On the use of deep learning for computational imaging. *Optica* **6**, 921-943 (2019).
148. Byeon, H., Go, T. & Lee, S. J. Deep learning-based digital in-line holographic microscopy for high resolution with extended field of view. *Optics & Laser Technology* **113**, 77-86 (2019).
149. Belthangady, C. & Royer, L. A. Applications, promises, and pitfalls of deep learning for fluorescence image reconstruction. *Nature Methods* **16**, 1215-1225 (2019).
150. de Haan, K. et al. Deep-learning-based image reconstruction and enhancement in optical microscopy. *Proceedings of the IEEE* **108**, 30-50 (2020).
151. Wang, F. et al. Phase imaging with an untrained neural network. *Light:Science & Applications* **9**, 77 (2020).
152. Goodman, J. W. Introduction to Fourier Optics. 3rd edn. (Greenwoood Village: Roberts & Company Publishers, 2005).
153. Rogers, E. T. F. et al. A super-oscillatory lens optical microscope for subwavelength imaging. *Nature Materials* **11**, 432-435 (2012).

Kim. *Light: Advanced Manufacturing* (2022)3:19
https://doi.org/10.37188/lam.2022.019

Official journal of the JHL 2689-9620
www.light-am.com

Review Open Access

Phase microscopy and surface profilometry by digital holography

Myung K. Kim*

Abstract

Quantitative phase microscopy by digital holography is a good candidate for high-speed, high precision profilometry. Multi-wavelength optical phase unwrapping avoids difficulties of numerical unwrapping methods, and can generate surface topographic images with large axial range and high axial resolution. But the large axial range is accompanied by proportionately large noise. An iterative process utilizing holograms acquired with a series of wavelengths is shown to be effective in reducing the noise to a few micrometers even over the axial range of several millimeters. An alternate approach with shifting of illumination angle, instead of using multiple laser sources, provides multiple effective wavelengths from a single laser, greatly simplifying the system complexity and providing great flexibility in the wavelength selection. Experiments are performed demonstrating the basic processes of multi-wavelength digital holography (MWDH) and multi-angle digital holography (MADH). Example images are presented for surface profiles of various types of surface structures. The methods have potential for versatile, high performance surface profilometry, with compact optical system and straightforward processing algorithms.

Keywords: Phase, Microscopy, Profilometry, Digital, Holography

Introduction

Surface profilometry is an essential technology in many fields of industry such as micro-electronics fabrication and precision machining, as well as in biomedicine. Traditional mechanical stylus-based scanners are largely replaced by non-contact optical methods, such as using patterned illumination, multi-view triangulation, and white light interference[1]. Digital holography is a new candidate for high speed, high precision profilometry, due to its ability to reconstruct both the amplitude and the phase of the full optical field. The phase profile of the hologram is directly proportional to the optical path length, such as from transmission through the thickness of a material or reflection off a surface[2]. This is the basis of many applications of quantitative phase microscopy by digital holography, such as cellular microscopy[3,4], micro-device diagnostics[5], and microfluidics[6], to name a few.

The complex, i.e. amplitude and phase, field profile is extracted from the interference intensity patterns by off-axis or phase-shifting digital holography[7]. The optical phase is proportional to the optical path length, but only up to modulo 2π, and therefore the measured optical path length is ambiguous modulo one wavelength. The so-called phase unwrapping is a common problem in any phase imaging systems, and many numerical phase unwrapping algorithms have been developed[8], which are typically based on subjective decisions regarding locations of phase discontinuity, and tend to be non-deterministic, unpredictable, and computationally heavy. A technique called optical phase unwrapping has been introduced[9–11], based on combination of several holograms acquired with different wavelengths, that generates phase profiles with

Correspondence: Myung K. Kim (mkkim@usf.edu)
Department of Physics, University of South Florida, Tampa, FL 33620

effective wavelength[12] much larger than the original wavelengths. The method is completely deterministic and computationally straightforward, and works well even when phase topology is complicated.

The effective, or synthetic, wavelength can be made very large by choosing two closely spaced wavelengths, but this entails proportionate amplification of any noise in the resulting optical path profile. It then becomes necessary to acquire and combine additional holograms with a third, fourth, or more wavelengths, and the noise is reduced in an iterative process described below. The larger the height range to be measured or the more severe the noise level, the more number of wavelengths and holograms is required. The multi-wavelength digital holography (MWDH) has been proposed and demonstrated using a dye laser tuned at three different wavelengths[13], a tunable diode laser at seven different wavelengths[14], or three discrete lasers[15], for example.

But the use of more than two or three wavelengths becomes increasingly difficult because of the complexity and cost of the optical system, as well as possible chromatic aberration over large range of wavelengths[16]. On the other hand, it has been noted early on that a shift in object illumination angle can also have a similar effect as a wavelength shift, because of the variation in the optical path length through a given height or thickness of the object[17]. Instead of a set of multiple lasers or scanning of a tunable laser, the angular direction of illumination is scanned to multiple positions to achieve essentially the same effect as with multiple wavelengths.

This article reports a series of studies on optical phase unwrapping based on multi-wavelength digital holography (MWDH) or multi-angle digital holography(MADH) for potential applications in surface profiling with millimeter-scale axial range and micrometer-range axial resolution. In MWDH studies, up to four wavelengths, from as many discrete lasers, are used to generate surface profiles with several mm height range and several μm height resolution. In MADH studies, it is possible to apply as many as eight different effective wavelengths with a much simpler and flexible optical system. The experimental results are summarized in Section 2. Discussions in Section 3 include comparison of the MWDH and MADH systems. These systems are also described in comparison with a few other related but distinct systems that utilize multi-wavelength or multi-angle acquisition of holograms. Section 4 describes the theoretical background and experimental system and processes.

Results

For illustration of basic processes of four-wavelength MWDH in Fig. 1, the object consists of a pair of glass plates with resolution target patterns with the step size between them measured to be 1.54 mm. As detailed in Materials and Methods, a set of four holograms are acquired by phase-shifting digital holography using four lasers with wavelengths $\lambda_1 = 0.639\,616\,\mu\text{m}$, $\lambda_2 = 0.639\,595\,\mu\text{m}$, $\lambda_3 = 0.639\,539\,\mu\text{m}$, and $\lambda_4 = 0.638\,929\,\mu\text{m}$. Fig. 1a, b display the first two phase profiles $\Phi_1(x,y)$ and $\Phi_2(x,y)$. Because of the minute difference of $0.000\,025\,\mu\text{m}$ between the two wavelengths, any difference between the two phase profiles is not easily discernable. Take the difference between the two phase profiles, modulo 2π, and rescale with the synthetic wavelength $\Lambda_{12} = \lambda_1\lambda_2/|\lambda_1 - \lambda_2|$, to obtain the height profile Z_{12}. The three-dimensional shape of the object then becomes apparent, shown in Fig. 1c, as well as the step between the two glass plates in Fig. 2a. The set of parameters used in this experiment is tabulated in Table 1.

The difference of the two phase profiles corresponds to another phase profile with very large synthetic wavelength $\Lambda_{12} = 19\,749\,\mu\text{m}$, amplifying the wavelength by a factor of $\Lambda_{12}/\lambda_1 \approx 3\times10^4$. The difference phase map has the same level of phase noise, other than a factor of order one, as the initial phase maps, meaning that with the amplification of effective wavelength, the noise in the height profile is also amplified by the same factor. In Fig. 2a, the noise in the height profile is measured to be $\delta Z_{12} = 919.1\,\mu\text{m}$, and the 3D profile in Fig. 1c shows significant amount of rough and spiky textures on the surfaces. Significant reduction in noise is achieved by iterative stitching of additional difference phase profiles Z_{13} and Z_{14} with shorter synthetic wavelengths $\Lambda_{13} = 5\,302\,\mu\text{m}$ and $\Lambda_{14} = 595\,\mu\text{m}$, respectively. The stitching process maintains the overall height range to Λ_{12} while reducing the noise to the

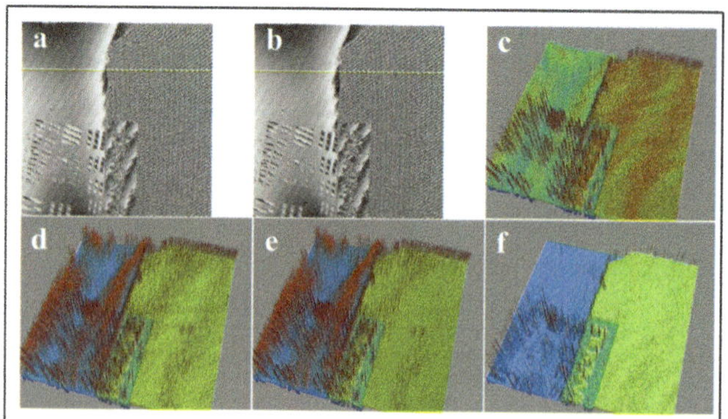

Fig. 1 Basic process of optical phase unwrapping by MWDH using four wavelengths, tabulated in Table 1. The object is a stack of two resolution target surfaces, with field of view $24\times24\,\text{mm}^2$. **a** $\Phi_1(x,y)$; **b** $\Phi_2(x,y)$; **c** $Z_{12}(x,y)$; **d** $Z_{12\cdots3}(x,y)$; **e** $Z_{12\cdots4}(x,y)$; **f** $Z_{proc}(x,y)$. See Fig. 2 for indication of the z-scale. See text for details.

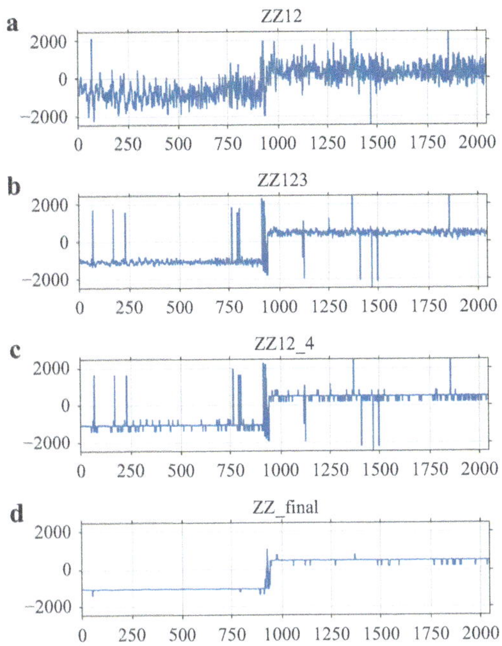

Fig. 2 Graphs of cross-sections of **a** Z_{12}; **b** $Z_{12\cdots3}$; **c** $Z_{12\cdots4}$; **d** Z_{proc}, through the yellow line indicated in Fig. 1a. The vertical scales, in μm, are multiplied by a factor 0.5 to account for the reflection geometry. The horizontal scale is the pixel index. A 3×3 median filter is applied to $Z_{12\cdots4}$ to obtain Z_{proc}. Measured height noise is tabulated in Table 1. The measured step height is 1.55 mm, comparable to the 1.54 mm nominal glass thickness.

Table 1 Parameter used in Fig. 1. λ_n : individual wavelength; Δ_{1n} : difference wavelength; Λ_{1n} : synthetic wavelength; δZ_{1n} : noise in Z_{1n}; 'proc': final processed profile.

n	λ_n, μm	Δ_{1n}, μm	Λ_{1n}, μm	δZ_{1n}, μm
1	0.639 616			
2	0.639 595	0.000 021	19 749.176	919.138
3	0.639 539	0.000 077	5 301.693	564.925
4	0.638 929	0.000 687	594.695	593.199
proc				101.721

proportionately lower levels of Z_{13} and Z_{14}, whose measured noise in Fig. 2d, e are $\delta Z_{13} = 564.9\,\mu$m and $\delta Z_{14} = 593.2\,\mu$m, respectively. The last stitching operation with Z_{14} has no apparent improvement in the measured noise because the reduction in synthetic wavelength is too rapid. The necessary condition is given in the Materials and Methods section, but with the given set of laser wavelengths, there is no flexibility in the choice of wavelengths. On the other hand, as we will see, the generation of effective wavelength can be made very flexible by varying the object illumination angle instead. A common and effective method of reducing spiky noise is using median filters. A mild median filter with 3×3 window is applied to the last profile $Z_{12\cdots4}$ of the iterative series to obtain the final processed profile Z_{proc} in Fig. 1f and Fig. 2d, where the corresponding noise is reduced to $\delta Z_{proc} = 101.7\,\mu$m. The measured step height in Z_{proc} is 1.55 mm, consistent with the nominal thickness of the glass plate.

In Fig. 3, 4, a similar set of parameters is used to image the surface of a US quarter dollar coin, placed on a mirror-

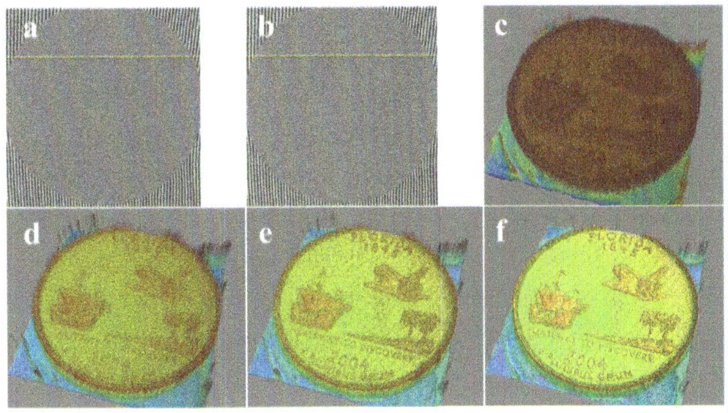

Fig. 3 Example of optical phase unwrapping by MWDH. The object is a US quarter dollar coin, with field of view 24×24 mm^2. A set of four wavelengths are used, such that $\Lambda_{12} = 28\,213\,\mu$m, $\Lambda_{13} = 4\,328\,\mu$m, and $\Lambda_{14} = 485\,\mu$m. **a** $\Phi_1(x,y)$; **b** $\Phi_2(x,y)$; **c** $Z_{12}(x,y)$; **d** $Z_{12\cdots3}(x,y)$; **e** $Z_{12\cdots4}(x,y)$; **f** $Z_{proc}(x,y)$. See Fig. 4 for indication of the z-scale.

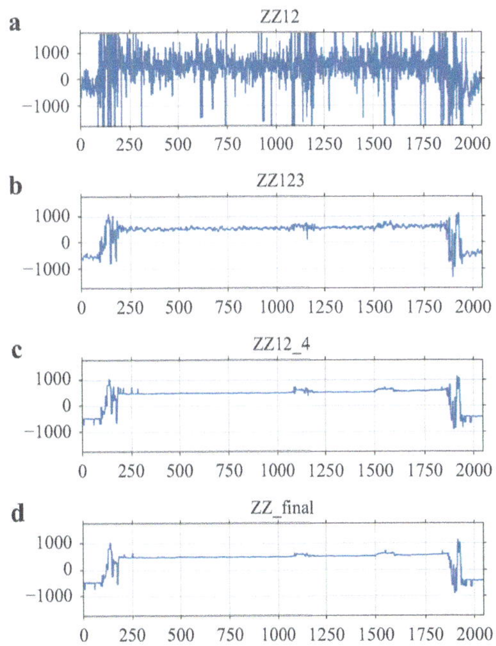

Fig. 4 Graphs of cross-sections of **a** Z_{12}; **b** $Z_{12\cdots3}$; **c** $Z_{12\cdots4}$; **d** Z_{proc}, through the yellow line indicated in Fig. 3a. The vertical scales, in μm, are multiplied by a factor 0.5 to account for the reflection geometry. The horizontal scale is the pixel index. A 3×3 median filter is applied to each of $Z_{12} \sim Z_{12\cdots4}$ and to Z_{proc}. Measured height noise varies as $\delta Z_{12} = 385.0\,\mu$m, $\delta Z_{12\cdots3} = 22.6\,\mu$m, $\delta Z_{12\cdots4} = 3.8\,\mu$m, and $\delta Z_{proc} = 3.2\,\mu$m.

like surface. The synthetic wavelengths used were $\Lambda_{12} = 28\,213\,\mu m$, $\Lambda_{13} = 5\,302\,\mu m$, $\Lambda_{14} = 595\,\mu m$. The single-wavelength phase profiles in Fig. 3a, b have no discernable features, because of the random granular structure of the metallic surface. Yet the surface profile Z_{12} from the difference profile does reveal the overall features of the surface pattern. The iterative processing with 'stitching' is applied as above, and in order to improve the image quality, a 3 × 3 median filter was applied to each step of Z_{12}, Z_{123}, and $Z_{12\cdots4}$. The noise level, measured over a small smooth area of the coin surface, is reduced from $\delta Z_{12} = 385.0\,\mu m$ to $\delta Z_{14} = 3.8\,\mu m$. It is seen that final processed images with just a few micrometer noise is achieved with relatively mild application of median filters.

The process of MADH is illustrated with an example in Fig. 5, 6, where the object consists of a set of three mirror-like surfaces with nominal step heights 1.54 mm and 0.94 mm between them. A set of eight holograms are acquired, using a single HeNe laser of $\lambda_0 = 0.632\,8\,\mu m$, at illumination angles $\theta_n = \theta_0 + \Delta_n$, tabulated in Table 2. In this experiment, an alternate version of the optical apparatus was used, where the reference beam is stationary and only the object illumination is scanned. The starting position $\theta_0 = 18°$ corresponds to the case of the object beam coinciding with the reference beam direction, while Δ_n are precisely controlled by the rotation stage with 0.000 25° resolution. The first two phase profiles Φ_1 and Φ_2 are displayed in Fig. 5a, b. The patterns indicate slight relative tilt between the three surfaces. The angular step $\theta_2 - \theta_1 = 0.0014°$ between the first two holograms determines the synthetic wavelength $\Lambda_{12} = 7\,545\,\mu m$, with the effective wavelengths $\lambda_n = \lambda_0 \cos\theta_n = 0.601\,829\,\mu m$ and $0.601\,781\,\mu m$, for $n = 1$ and 2, repectively. Additional angular steps $\theta_n - \theta_1$, which increase by a factor of two in each step in this example, lead to synthetic wavelengths Λ_{1n} that decrease by a factor of $\alpha \approx 0.50$, each step for $n > 2$. The 'stitching' process is iterably applied with the series of profiles $Z_{13}, \cdots, Z_{18}$. The noise levels δ_{1n} of the 'stitched' maps $Z_{12\cdots n}$, in Fig. 5, 6 are measured as the

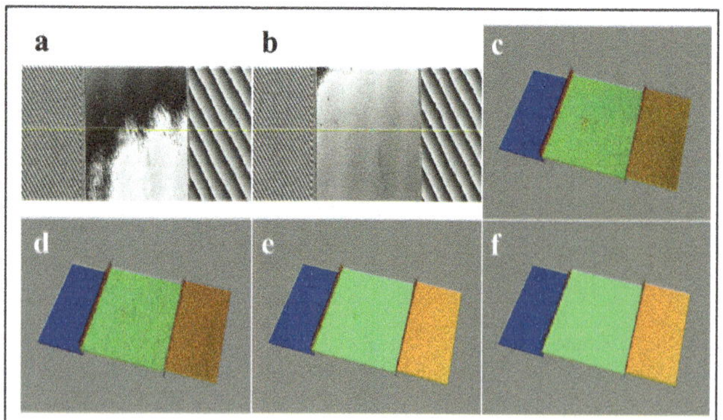

Fig. 5 Basic process of optical phase unwrapping by MADH using eight illumination angles for eight effective wavelengths, tabulated in Table 2. The object is a stack of three mirror-like surfaces, with field of view $9.87 \times 5.55\,mm^2$. **a** $\Phi_1(x,y)$; **b** $\Phi_2(x,y)$; **c** $Z_{12}(x,y)$; **d** $Z_{12\cdots3}(x,y)$; **e** $Z_{12\cdots8}(x,y)$; **f** $Z_{proc}(x,y)$. Plots for $Z_{12\cdots4}$ through $Z_{12\cdots7}$ are not displayed. See text for details. See Fig. 6 for indication of the z-scale.

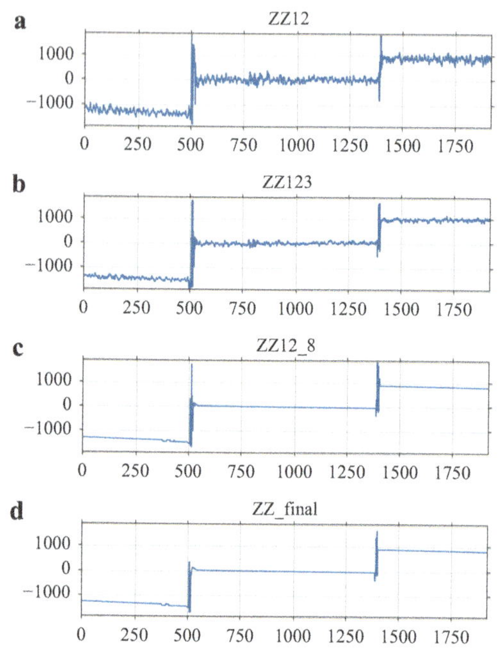

Fig. 6 Graphs of cross-sections of **a** Z_{12}; **b** $Z_{12\cdots3}$; **c** $Z_{12\cdots8}$; **d** Z_{proc}, through the yellow line indicated in Fig. 5a. The vertical scales, in μm, are multiplied by a factor 0.5 to account for the reflection geometry. The horizontal scale is the pixel index. Graphs for $Z_{12\cdots4}$ through $Z_{12\cdots7}$ are omitted. A 3×3 median filter is applied to $Z_{12\cdots8}$ to obtain Z_{proc}. Measured height noise is tabulated in Table 2. The measured step heights are 1.56 mm and 0.97 mm, comparable to the nominal glass thicknesses of 1.54 mm and 0.95 mm.

Table 2 Parameter used in Fig. 5. λ_n: individual wavelength; Δ_{1n}: difference wavelength; Λ_{1n}: synthetic wavelength; δZ_{1n}: noise in Z_{1n}; 'proc': final processed profile.

n	λ_n, μm	Δ_{1n}, μm	Λ_{1n}, μm	δZ_{1n}, μm
1	0.601 829			
2	0.601 781	0.000 048	7 545.193	127.426
3	0.601 733	0.000 096	3 772.296	76.897
4	0.601 637	0.000 192	1 885.847	35.985
5	0.601 445	0.000 384	942.623	20.029
6	0.601 060	0.000 769	470.397	6.961
7	0.600 281	0.001 548	233.376	4.057
8	0.598 697	0.003 132	115.043	3.507
proc				2.018

standard deviation of a small area near the center of the frame. They progressively decrease by a factor close to α each step, while maintaining the overall height range Λ_{12}. The surface profiles rendered as 3D plots in Figs. 5c–f show obvious enhancement of image quality between Z_{12} and $Z_{12\cdots8}$, as the noise decreases from $\delta Z_{12} = 127.4\,\mu m$ to $\delta Z_{18} = 3.5\,\mu m$, with $\delta Z_{18}/\delta Z_{12} = 0.027\,5 = \alpha^{5.18}$, consistent with the expected factor of α^{n-2}. In this case, the noise reduction is quite complete throughout the surface except at the steps because of the beveled edges reflecting insufficient light towards the camera. With application of a mild 3×3 median filter on the last stitched profile $Z_{12\cdots8}$, a final noise level of $\delta Z_{proc} = 2.0\,\mu m$ is achieved over a height range of $\Lambda_{12} = 3\,758\,\mu m$.

Fig. 7, 8 is an example surface profile of a coin (U.S. penny) placed on another (U.S. quarter). Here a set of eight holograms are acquired using the stepping factor $\alpha = 0.50$ for the synthetic wavelengths ranging from $\Lambda_{12} = 2\,911.5\,\mu m$ to $\Lambda_{18} = 44.1\,\mu m$. Because of the microscopically diffuse surface of the coin, as well as the slopes on the boundaries of the letters and relief features, there is significant noise in the raw profile Z_{12}, shown in Fig. 7c, 8a. The relatively small numerical aperture of the system, approx. 0.1, is not able to collect enough light from these sloped boundary surfaces for accurate phase information. The same stitching procedure as above is applied using the series $Z_{13}, \cdots, Z_{18}$, but after each step a 3×3 median filter is applied. This produced reasonably clean surface profile of the coin without noticeable degradation or loss of lateral resolution. The measured noise reduced from $\delta Z_{12} = 147.3\,\mu m$ to $\delta Z_{18} = 10.4\,\mu m$.

As described in Section 4, there is a limit on how fast the iterative noise reduction can progress, i.e. how small the stepping factor α can be. For larger amount of initial noise, the stepping factor cannot be chosen too small, for otherwise the iterative noise reduction can become incomplete. To illustrate, in Fig. 9, the same coin surface data as in Fig. 8 is used, but instead of stitching all of the profiles $Z_{13}, \cdots, Z_{18}$ on to Z_{12}, only the partial set of Z_{12}, Z_{15}, and Z_{18} is used for stitching. Here, $Z(12)$, $Z(12+15)$, and $Z(12+15+18)$, denote Z_{12}, it stitched with Z_{15}, and the last stitched with Z_{18}, respectively. This in effect reduces the stepping factor to $\alpha = 0.125$. Otherwise the same set of processing procedure is applied as with Fig. 8, including the 3×3 median filter at each stitching step. The stepping factor is too small and the iteration is too fast, and significant amount of noise escapes the reduction procedure, the final noise level being 34 μm for $\alpha = 0.125$, compared to 9.0 μm for $\alpha = 0.50$.

A few more examples of surface profiles are given in Fig. 10, by optical phase unwrapping with MWDH or MADH. Imaging parameters of the examples are summarized in Table 3.

Discussion

The median filter is quite effective in removing the spiky noise. For some of the examples presented in Section 2, such as coin surfaces, application of simple 3×3 median

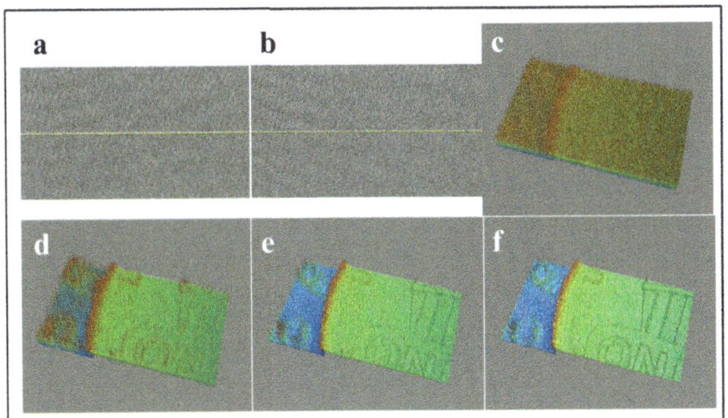

Fig. 7 Example of optical phase unwrapping by MADH. The object is a coin, US penny, placed on top of another, a US quarter, with field of view $9.87\times5.55\,mm^2$. A set of eight wavelengths are used, such that $\Lambda_{1n} = [2\,911.5, 1\,455.5, 727.5, 363.5, 180.9, 89.7, 44.1]\,\mu m$ for $n = 2, 3, \cdots, 8$. **a** $\Phi_1(x,y)$; **b** $\Phi_2(x,y)$; **c** $Z_{12}(x,y)$; **d** $Z_{12\cdots3}(x,y)$; **e** $Z_{12\cdots8}(x,y)$; **f** $Z_{proc}(x,y)$. See Fig. 8 for indication of the z-scale.

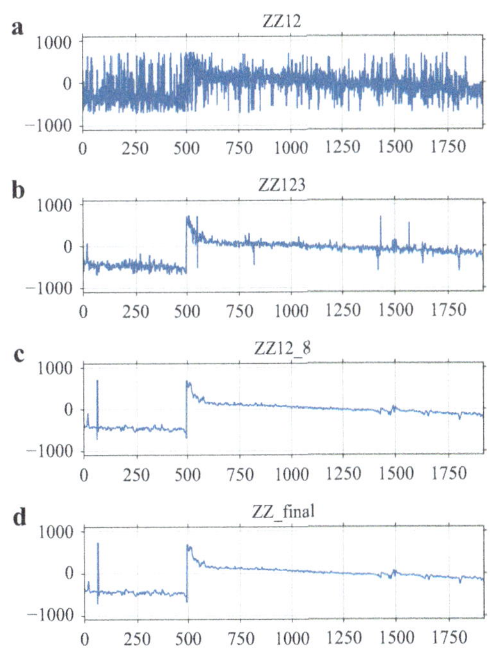

Fig. 8 Graphs of cross-sections of **a** Z_{12}; **b** $Z_{12\cdots3}$; **c** $Z_{12\cdots8}$; **d** Z_{proc}, through the yellow line indicated in Fig. 7a. The vertical scales, in μm, are multiplied by a factor 0.5 to account for the reflection geometry. The horizontal scale is the pixel index. A 3×3 median filter is applied to each of $Z_{12} \sim Z_{12\cdots8}$ and to Z_{proc}. Measured height noise varies as $\delta Z_{12\cdots n} = [147.3, 29.5, 27.2, 23.2, 15.5, 13.9, 10.4]\,\mu m$ for $n = 2, 3, \cdots, 8$, and $\delta Z_{proc} = 9.0\,\mu m$.

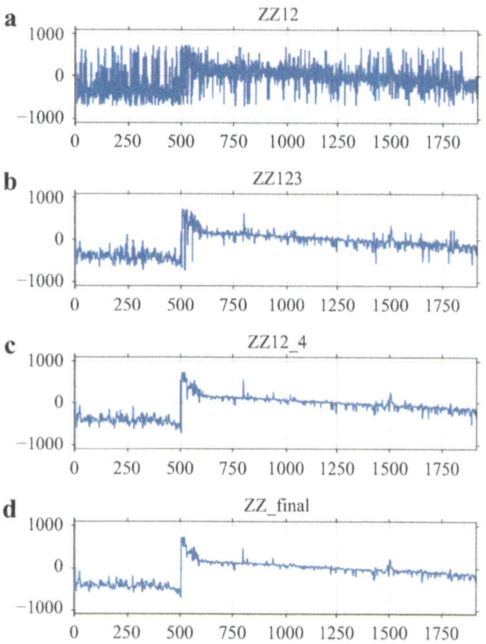

Fig. 9 Effect of iteration being too rapid with $\alpha = 0.125$. Graphs of cross-sections of **a** $Z(12)$; **b** $Z(12+15)$; **c** $Z(12+15+18)$; **d** Z_{proc}. The vertical scales, in μm, are multiplied by a factor 0.5 to account for the reflection geometry. The horizontal scale is the pixel index. A 3×3 median filter is applied to each profile. Measured height noise varies as $\delta Z_{12} = 147.3\,\mu$m, $\delta Z_{12+15} = 58.7\,\mu$m, $\delta Z_{12+18} = 42.2\,\mu$m, and $\delta Z_{proc} = 34.1\,\mu$m. See text for details.

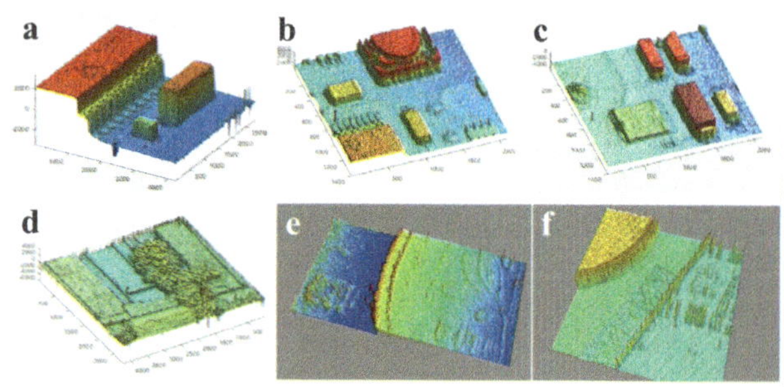

Fig. 10 Examples of MWDH and MADH images. Imaging parameters are tabulated in Table 3.

filter a few times was sufficient to produce good-quality surface profiles. For more rough textured surfaces such as integrated circuit chips and printed circuit boards, heavier dose of filtering was necessary, such as 10 ~ 20 × repeated applications of 3 × 3 median filter. Some loss of surface details is inevitable, but it seems to be justified by the improvement in overall quality of the final image. It is also found useful in some cases to adjust the median filter to larger window size for darker areas of the amplitude image, and vice versa. Such adaptive median filter (AMF) technique was used to obtain the images in Figs. 10a–c, and helped reduce strong noise in corners and metallic electrodes.

Several other denoising methods were attempted from collection of 'conventional' image processing tools, such as windowed Fourier transform (WFT), total variation denoising (TVD), non-local means filtering(NLMF), and block-matching and 3D filtering (BM3D)[18], as well as random mask multi-look (RMML) method for holographic speckle processing[19]. These somewhat casual studies did not find a magic solution for general applications, but each technique tends to be applicable only to specific cases, pose substantial computational load, and cause significant degradation of details. The behavior of spiky noise in this type of stitched phase maps appear very different from typical noise in photographic images or the speckle noise in holographic intensity images. Further systematic study of the behavior of this new type of noise may be of interest[20].

A few points can be made in comparing the methods of MWDH and MADH. First, both methods, when properly set up and carried out, can produce good quality surface profile images. In general, the MADH system has advantage with cost and complexity. The selection of effective wavelengths is very flexible and more controllable. A potential issue is shadowing and obscuration by tall or sharp surface features. The MWDH may be subject to chromatic aberration when the wavelength interval is substantial. The MADH can also be subject to geometric aberration because of the movement of the beam across the optical aperture. If the size of the

Table 3 Parameter used in Fig. 10. N_λ: number of wavelengths; Λ_{12}: first synthetic wavelength; Λ_{1N}: last synthetic wavelength.

Fig	object	method	N_λ	FOV, μm × μm	Λ_{12}, μm	Λ_{1N}, μm
a	PCB	MWDH	4	30 500 × 20 352	13 636.2	242.9
b	PCB	MWDH	4	21 200 × 14 160	18 595.0	243.0
c	PCB	MWDH	4	21 200 × 14 160	18 595.0	243.0
d	brass relief plaque	MWDH	4	30 500 × 20 352	13 636.2	241.1
e	coin, US dime	MADH	8	9 870 × 5 552	1 455.8	22.1
f	three mirror-like surfaces	MADH	4	18 432 × 18 432	19 378.6	2 212.3

aberration is more than a few wavelengths, the reconstructed surface profile may include as much distortion. This type of distortion can be compensated for, if necessary, by acquiring a set of blank holograms with a flat mirror in place of the object plane.

There is an alternate approach to unwrapping of quantitative phase images by angular scanning, where small step phase changes are accumulated over many small step anglular shifts[21]. General type and quality of surface profile images may be similar to the method presented here. In principle, in order to achieve the same level of height range and resolution, both methods of this paper and of[21] start from the same first wavelength λ_1 and scan to the final synthetic wavelength λ_N. The method of[21] takes uniform wavelength steps throughout the scan, while the method of this paper takes progressively larger steps to reach the final wavlength. Direct experimental comparison has not been attempted, but because of the significantly larger number of holographic frames, the acquisition may take longer time and therefore susceptible to instability.

One may also note other related but distinct holographic methods of surface profiling. For example, many holograms are acquired using a range of wavelengths. The stack of holograms is then Fourier transformed along the inverse-of-wavelength axis, which represents the 3D surface topography[22]. Similar result can be obtained by replacing the wavelength scanning with angular scanning[23-25]. These processes in effect synthesize temporal incoherence from diversity of frequencies, i.e. inverse of wavelength, either by wavelength scanning or angular scanning. The axial resolution is limited by the synthesized coherence length and in principle significantly lower than the sub-wavelength axial resolution of phase profile-based methods presented here.

These demonstration experiments validate the basic concept of optical phase unwrapping and iterative noise reduction by multi-wavelength and multi-angle digital holography. With the quantitative phase method presented here, high resolution surface profiles can be obtained with large unambiguous range, while maintaining close to single-wavelength axial resolution. Further work is in progress, in particular with respect to the novel behavior of phase noise in this type of quantitative phase imaging system, as well as development of profilometry applications in device fabrication and biomedical imaging.

Materials and methods

Principle of optical phase unwrapping by MWDH

The basic process of multi-wavelength optical phase unwrapping is illustrated with simulations in Fig. 11. Suppose the object is an inclined plane of height $a_h = 15\,\mu\text{m}$, depicted with the function $h(x)$ in Fig. 11a, and a set of holograms are acquired using a series of wavelengths

$$\{\lambda_n\} = \lambda_1, \lambda_2, \lambda_3, \cdots \quad (1)$$

The complex optical fields

$$E_n(x,y) = a_n(x,y) \cdot \exp[i\Phi_n(x,y)] \quad (2)$$

are computed from the captured holographic interference patterns, where $a_n(x,y)$ and $\Phi_n(x,y)$ are the amplitude and phase profiles of the optical field, respectively. In off-axis holography, the phase is encoded as a modulation of the high-spatial-frequency interference fringe patterns in a single camera frame, whereas in phase-shifting digital holography, three or more camera frames are combined while the phase of reference field is shifted. If the acquired optical field resulted from reflection from or transmission through a structured surface or layer, the phase profile $\Phi_n(x,y)$ is directly proportional to the profile of the optical path length $Z(x,y)$

$$\Phi_n(x,y) = Z(x,y) \cdot \frac{2\pi}{\lambda_n} \quad (3)$$

The wrapped phase problem arises from the fact that the optical phase is defined only up to modulo 2π, and the

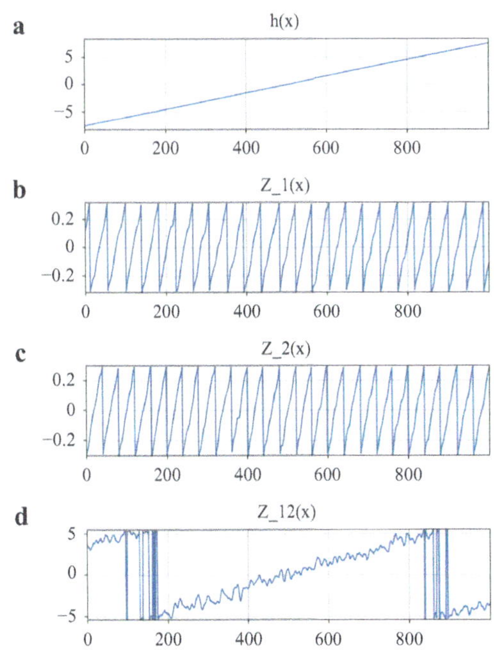

Fig. 11 Simulation of 2-wavelength optical phase unwrapping. **a** $h(x)$; **b** $Z_1(x)$; **c** $Z_2(x)$; **d** $Z_{12}(x)$. The vertical scales are in μm. The horizontal scale is the pixel index. Height of the incline is 15 μm. Assumed wavelengths are $\lambda_1 = 0.635\,\mu\text{m}$, $\lambda_2 = 0.600\,\mu\text{m}$, so that $\Lambda_{12} = 10.886\,\mu\text{m}$. Phase noise of $\varepsilon = 3\%$ results in height noise $\delta Z_{12} = 0.462\,\mu\text{m}$.

derived surface height profile

$$Z_n(x,y) = \Phi_n(x,y) \cdot \frac{\lambda_n}{2\pi} \tag{4}$$

is limited to modulo λ_n. In Fig. 11b the height profile Z_1 acquired using a wavelength $\lambda_1 = 0.635\,\mu m$ has a sawtooth shape of height equal to the wavelength. A phase noise of standard deviation $2\pi \cdot \varepsilon$, with $\varepsilon = 3\%$, is added in the phase profile to simulate random phase noise between hologram acquisitions.

Two-wavelength optical phase unwrapping is based on taking the conjugate product of two optical fields acquired with two slightly different wavelengths, λ_1 and λ_2, as

$$E_{12}(x,y) = E_1 \cdot E_2^* = a_{12} \cdot \exp[i\Phi_{12}(x,y)] \tag{5}$$

where

$$\Phi_{12} = \mathrm{mod}(\Phi_1 - \Phi_2 + \pi, 2\pi) - \pi \tag{6}$$

The particular modulo operation is necessary when the phase angle of a complex number is defined in the range $[-\pi, \pi]$. The difference phase Φ_{12} is equivalent to the phase of an effective, or synthetic, wavelength Λ_{12}, which can be much larger than the original wavelengths:

$$\Phi_{12} = 2\pi\frac{Z}{\lambda_1} - 2\pi\frac{Z}{\lambda_1} = 2\pi\frac{Z}{\Lambda_{12}} \tag{7}$$

where

$$\Lambda_{12} = \frac{\lambda_1\lambda_2}{|\lambda_1 - \lambda_2|} \tag{8}$$

The optical path derived from the difference phase can range up to the synthetic wavelength,

$$Z_{12}(x,y) = \Phi_{12} \cdot \frac{\Lambda_{12}}{2\pi} \tag{9}$$

as illustrated in Fig. 11d with $\lambda_1 = 0.6350\,\mu m$ and $\lambda_2 = 0.6000\,\mu m$, so that $\Lambda_{12} = 10.886\,\mu m$, and the height profile $h(x)$ is properly represented up to the maximum Λ_{12}.

By choosing two closely spaced wavelengths λ_1 and λ_2, the synthetic wavelength Λ_{12} can be made as large as necessary to cover the maximum surface height of the object to be profiled. On the other hand, the extension of the synthetic wavelength comes with amplification of noise in the height profile. For suppose the phase noise in Φ_1 is a fraction of 2π,

$$\delta\Phi_n = 2\pi \cdot \varepsilon \tag{10}$$

where ε is a dimensionless random function that includes both spatial and frame-to-frame variations. The corresponding noise in the derived optical path is

$$\delta Z_n = \delta\Phi_n \cdot \frac{\lambda_n}{2\pi} = \varepsilon \cdot \lambda_n \tag{11}$$

which is approximately $\delta Z_1 = 0.019\,\mu m$ with the assumed values of $\lambda_1 = 0.635\,\mu m$ and $\varepsilon = 3\%$. The noise in the optical height derived from the difference phase and synthetic wavelength is, with $\delta\Phi_{12} \approx \sqrt{2}\delta\Phi_1$,

$$\delta Z_{12} = \delta\Phi_{12} \cdot \frac{\Lambda_{12}}{2\pi} = \sqrt{2}\varepsilon \cdot \Lambda_{12} \tag{12}$$

and the noise in the optical height is amplified in proportion to the amplification in wavelength, to approx. $\delta Z_{12} \approx 0.462\,\mu m$ in Fig. 11d. The amplification in wavelength is accompanied by similar amplification in noise.

The noise can be reduced if a shorter synthetic wavelength, Λ_{13} is used, with a larger difference $\lambda_1 - \lambda_3$. The simulation in Fig. 12 uses $\lambda_3 = 0.5300\,\mu m$, so that $\Lambda_{13} = 3.205\,\mu m$. Start from $Z_{12}(x,y)$ with the range $\Lambda_{12} = 10.886\,\mu m$, Fig. 12a, and make integer steps of Λ_{13}, Fig. 12b. Then, 'stitch' the steps with the new profile $Z_{13}(x,y)$, Fig. 12c. That is,

$$Z_{12\cdots3}(x,y) = \mathrm{round}\left(\frac{Z_{12}(x,y)}{\Lambda_{13}}\right) \cdot \Lambda_{13} + Z_{13}(x,y) \tag{13}$$

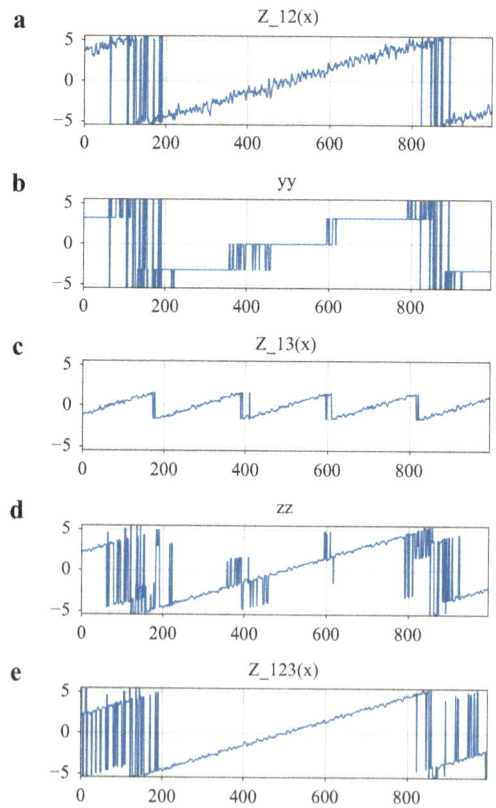

Fig. 12 Simulation of 3-wavelength optical phase unwrapping. **a** Z_{12}; **b** $Y_{12} = \mathrm{round}(Z_{12}/\Lambda_{13}) \cdot \Lambda_{13}$; **c** Z_{13}; **d** $Z'_{12\cdots3} = Y_{12} + Z_{13}$; **e** $Z_{12\cdots3}$. The vertical scales are in μm. The horizontal scale is the pixel index. Assumed third wavelength is $\lambda_3 = 0.530\,\mu m$, so that $\Lambda_{13} = 3.205\,\mu m$. Final height noise is reduced to $\delta Z_{12\cdots3} = 0.136\,\mu m$.

where round() represents integer rounding operation. If the phase noise is included in the above expression,

$$\begin{aligned} Z_{12\cdots3}+\delta Z_{12\cdots3} &= \mathrm{round}\left(\frac{Z_{12}+\delta Z_{12\cdots3}}{\Lambda_{13}}\right)\cdot\Lambda_{13}+(Z_{13}+\delta Z_{13}) \\ &= Z_{12\cdots3}+\Delta\cdot\Lambda_{13}+\delta Z_{13} \\ &= Z_{12\cdots3}+\Delta\cdot\Lambda_{13}+\sqrt{2}\varepsilon\Lambda_{13} \end{aligned} \quad (14)$$

Here $\Delta\cdot\Lambda_{13}$ represents spikes of height Λ_{13} scattered at positions near the step boundaries of round(), as evident in Fig. 12d. These spikes can be suppressed by adding or subtracting Λ_{13} wherever the absolute difference $|Z_{12}-Z_{12\cdots3}|$ is comparable to Λ_{13}. This process is valid with the requirement

$$\frac{\delta Z_{12}}{\Lambda_{13}} \approx \sqrt{2}\varepsilon\cdot\frac{\Lambda_{12}}{\Lambda_{13}} \ll 1 \quad (15)$$

and Λ_{13} should be chosen large enough compared to $\sqrt{2}\varepsilon\cdot\Lambda_{12}$, or

$$\alpha = \Lambda_{13}/\Lambda_{12} \gg \sqrt{2}\varepsilon \quad (16)$$

The overall noise is then reduced from $\sqrt{2}\varepsilon\cdot\Lambda_{12}\approx 0.462\,\mu\mathrm{m}$ to $\sqrt{2}\varepsilon\cdot\Lambda_{13}\approx 0.136\,\mu\mathrm{m}$, as demonstrated in Fig. 12e.

In the above simulation example, the synthetic wavelength is extended to over 10 μm, which can be suitable for profiling microscopically structured surfaces. On the other hand, for applications requiring profiling of macroscopic surfaces with millimeters of height range, the large amplification of wavelength entails similarly large noise. In order to reduce the large noise, it is then necessary to extend the iterative series using a larger number of wavelengths. Thus, another difference phase profile $\Phi_{14}(x,y)$ with synthetic wavelength $\Lambda_{14}<\Lambda_{13}$ is used to 'stitch' the new profile $Z_{14}(x,y)$ on to the previous synthesized profile $Z_{12\cdots3}(x,y)$. If we again let $\Lambda_{14}=\alpha\Lambda_{13}$, the noise level for the new profile $Z_{12\cdots4}$ is $\delta Z_{12\cdots4}=\sqrt{2}\varepsilon\Lambda_{14}=\sqrt{2}\varepsilon\cdot\alpha^2\Lambda_{12}$. The process can continue until $\delta Z_{12\cdots n}=\varepsilon\cdot\Lambda_{1n}=\varepsilon\cdot\alpha^{(n-2)}\Lambda_{12}$ reaches desired height resolution. In the simulation of Fig. 13, the height range is now $\Lambda_{12}=2\,000.0\,\mu\mathrm{m}$ with $\lambda_1=0.635\,000\,\mu\mathrm{m}$ and $\lambda_2=0.634\,798\,\mu\mathrm{m}$, or $\Delta_{12}=\lambda_1-\lambda_2=0.000\,202\,\mu\mathrm{m}$. Three more wavelengths are used with difference wavelengths $\Delta_{13}=0.000\,805\,\mu\mathrm{m}$, $\Delta_{14}=0.003\,209\,\mu\mathrm{m}$, and $\Delta_{15}=0.012\,646\,\mu\mathrm{m}$ for synthetic wavelengths $\Lambda_{13}=500.0\,\mu\mathrm{m}$, $\Lambda_{14}=125.0\,\mu\mathrm{m}$, and $\Lambda_{15}=31.25\,\mu\mathrm{m}$, respectively, with a fixed stepping factor $\alpha=0.25$ for the series. Assuming phase noise with $\varepsilon=3\%$, the initial noise 84.9 μm in Z_{12} is progressively reduced to 21.2 μm, 5.3 μm, and 1.3 μm in Z_{13}, Z_{14}, and Z_{15}, respectively.

If the iteration is made to run too fast with the stepping factor α too small, the noise reduction by stitching as described above can be incomplete and significant amount of noise can remain after the procedure. In Fig. 14a, the noise is increased to $\varepsilon=7\%$ while using the same set of wavelengths $\lambda_1,\cdots,\lambda_5$, with $\alpha=0.25$, as the above example. The condition $\varepsilon\ll\alpha$ is not sufficiently strong that there are many spikes $\Delta\cdot\Lambda_{1n}$ in resulting profiles $Z_{12\cdots n}$. In order to improve the noise figure, we slow the iteration using a larger value of $\alpha=0.50$, and increase the number of wavelengths to eight, so that the final synthetic wavlengths in the two series are the same: $\Lambda_{18}=31.25\,\mu\mathrm{m}$ with $\alpha=0.25$ and $\Lambda_{15}=31.25\,\mu\mathrm{m}$ with $\alpha=0.50$. In Fig. 14b, the stitching process is applied to the series of eight wavelengths to obtain the profiles $Z_{12}, Z_{12\cdots3},\cdots, Z_{12\cdots8}$. The final noise level close to $\delta Z_{12\cdots8}=2.19\,\mu\mathrm{m}$ is achieved without significant number of spike. This simulation result suggests that large height or noisy surfaces may be profiled using slower progression of Λ_{1n}, i.e. larger α, and larger number of wavelengths.

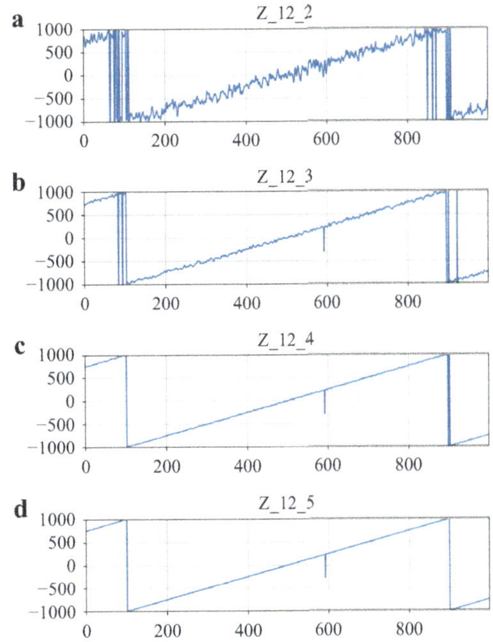

Fig. 13 Simulation of iterative reduction of noise for large height range $\Lambda_{12}=2000.0\,\mu\mathrm{m}$, using a series of five wavelengths. **a** $Z_{12}(x)$; **b** $Z_{12\cdots3}(x)$; **c** $Z_{12\cdots4}(x)$; **d** $Z_{12\cdots5}(x)$. The vertical scales are in μm. The horizontal scale is the pixel index. With a stepping factor $\alpha=0.25$, the synthetic wavlengths are set to $\Lambda_{1n}=[2\,000.0, 500.0, 125.0, 31.25]\,\mu\mathrm{m}$, for $n=2,\cdots,5$, which can be obtained from the wavelengths $\lambda_n=[0.635\,000, 0.634\,798, 0.634\,195, 0.631\,791, 0.62\,354]\,\mu\mathrm{m}$, for $n=1,2,\cdots,5$. With the assumed phase noise of $\varepsilon=3\%$, the height noise progressively dcreases as $\delta Z_{12\cdots n}=[77.7, 20.5, 5.3, 1.2]\,\mu\mathrm{m}$ for $n=2,\cdots,5$.

The 'stitching' process is rather sensative to a few different types of errors and these need to be addressed carefully. The process requires that each of the phase

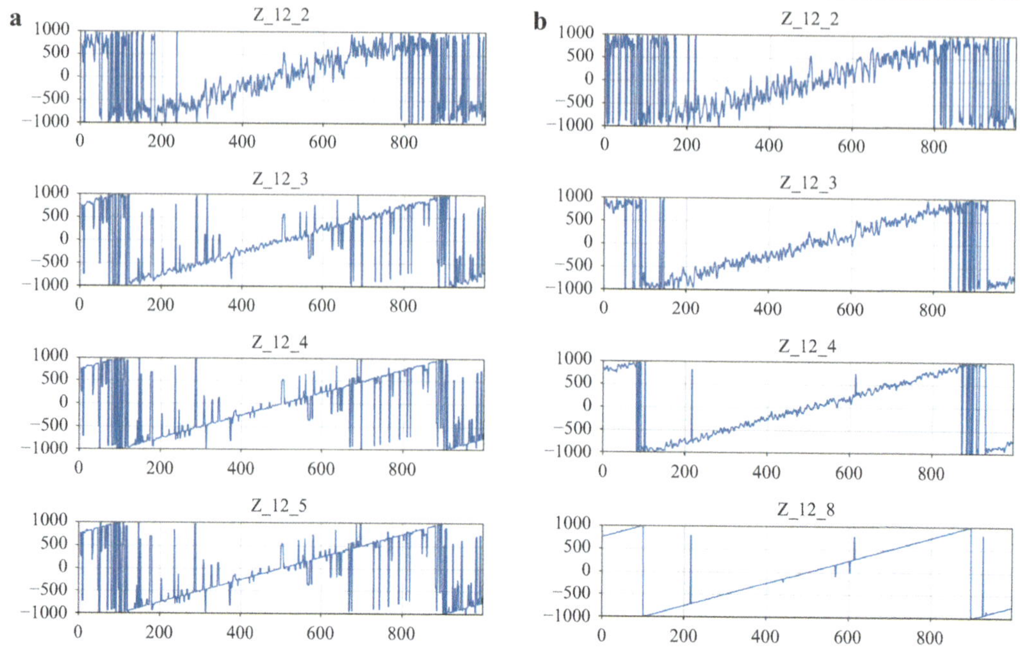

Fig. 14 Effect of stepping factor α for large noise profiles, with $\varepsilon = 7\%$. For the column **a**, the same stepping factor $\alpha = 0.25$ is used as in Fig. 13. The measured height noise is $\delta Z_{12\cdots n} = [110.0, 37.2, 51.6, 50.6]\,\mu m$ for $n = 2, \cdots, 5$. For the column **b**, a slower stepping factor $\alpha = 0.50$ is used with eight wavelengths, so that the final synthetic wavelength $\Lambda_{18} = 31.25\,\mu m$ is the same as Λ_{15} in a). Graphs for $Z_{12\cdots 5} \sim Z_{12\cdots 7}$ are omitted. The vertical scales are in μm. The horizontal scale is the pixel index. Now, the measured height noise is $\delta Z_{12\cdots n} = [99.2, 51.5, 35.5, 13.8, 11.5, 2.9, 2.2]\,\mu m$ for $n = 2, \cdots, 8$. That is, the final noise value of $\delta Z_{12\cdots 8} = 2.2\mu m$ obtains for $\alpha = 0.50$, instead of $\delta Z_{12\cdots 5} = 50.6\mu m$ for $\alpha = 0.25$.

profiles $\Phi_n(x, y)$ is referenced to an identical global phase, which is never the case due to random variations in the laser, optical, and mechanical systems. This is remedied by fixing the difference phase $\Phi_{1n}(x_0, y_0)$ at a suitable, relatively noise-free, reference point to be zero.

Another source of error is that the synthetic wavelength is inversely proportionale to a small difference wavelength. In MWDH, the laser wavelengths drift with temperature and other factors. Although a high precision wave meter is available, the current setup does not incorporate tight monitoring and control of it. In MADH, the angular position is more closely monitored but the precision is limited and any mechanical instability may affect it. In any case, at least in the current setup, it is necessary, during post-processing, to verify and adjust the accuracy of synthetic wavelengths. This is carried out by scanning the synthetic wavelength over a range and looking for optimal values for minimizing the noise spikes.

Principle of optical phase unwrapping by MADH

For example, to achieve a height range of $\Lambda_{12} = 2.0$ mm near $\lambda = 0.635\,\mu m$, the required wavelength difference is $\Delta\lambda \approx 0.2$ nm. And to achieve μm -level height resolution, several wavelengths are needed. For such applications, acquiring and maintaining multiple laser sources with precision-tuned wavelengths quickly becomes prohibitively difficult. The cost and complexity of the optical system can be significantly reduced if, instead of multiple lasers, a single laser is used and the multiple effective wavelengths are generated by changing the illumination angle of the object. The variation of the optical phase through a given height h of the object is $2\pi h / \lambda \cos\theta$, which can be described as illumination from a fixed direction with effective wavelength

$$\lambda_\theta = \lambda \cdot \cos\theta \tag{17}$$

Multiple effective wavelengths λ_n can be generated simply by tuning the illumination angle to various positions θ_n. The synthetic wavelength between the 1st and n-th effective wavelengths is then

$$\Lambda_{1n} = \frac{\lambda_1 \lambda_n}{|\lambda_1 - \lambda_n|} = \lambda \cdot \frac{\cos\theta_1 \cos\theta_n}{|\cos\theta_1 - \cos\theta_n|} \approx \lambda \cdot \frac{\cos\theta_1 \cos\theta_n}{\Delta\theta \cdot \sin\theta} \tag{18}$$

The rest of the theoretical description of MADH is in exact parallel with that of MWDH, but there are significant differences in practical considerations of the two methods. The MWDH requires multiple lasers or tunable lasers as well as a means to combine and switch between them, whereas the MADH requires one good laser and a means for fine control of illumination angle. To achieve 2.0 mm height range as above, the required angle shift is

$\Delta\theta = 0.070°$, when $\theta = 15°$, indicating the level of angular precision needed. The MADH is mostly immune from chromatic aberration but can be affected by spherical or higher order geometric aberrations due to the movement of the illumination across optical apertures. In order to keep Λ_{1n} becoming too large, or the angular step $\Delta\theta$ too small, the angle θ needs to be somewhat away from zero, $\theta \sim 15°$ for example. This can cause shadowing or obscuration of the object surface when the surface has sharp or tall features.

Optical system

A schematic of one version of the optical system is depicted in Fig. 15, based on Michelson interferometer configuration. It can accommodate several operating modes, including multi-wavelength or multi-angle holography, as well as off-axis or phase-shifting holography. The fiber coupler FC introduces laser light into the system. Typical power input at FC is a few mW. Expanded and collimated beam is reflected by the mirror MI. The confocal lens combination L1 (f1 = 100 mm) and L2 (f2 = 200 mm) images the rotating mirror to the object plane, which is in turn imaged on to the camera plane by the lens L4 and a camera lens LC. The reference mirror MR is also conjugate with the camera plane, and is mounted on a piezo actuator for phase shifting. The polarizing beam-splitter PBS and quarter wave plates QO and QR together with the polarizer P controls and optimizes intensity ratio between the object and reference beams. The lenses L2 and L4 and quarter wave plates and polarizer, as well as the polarizing beam splitter are 2 inches in diameter or width. Typical field of view of the object plane is 10 ~ 30 mm.

For multi-wavelength experiments, a set of four lasers (Lasos DPSS) are used with nominal wavelengths $\lambda_1 = 639.617$ nm, $\lambda_2 = 639.592$ nm, $\lambda_3 = 639.416$ nm, and $\lambda_4 = 637.928$ nm. These lasers are fiber-coupled into a four-channel fiber switch (Leoni), before being coupled into the interferometer. All fibers are polarization preserving single mode. The laser wavelengths do drift with changing temperature, and a high finesse wavemeter (Toptica) with 0.1 pm resolution was used to measure the wavelengths in real time before calculating synthetic wavelengths and other parameters.

For multi-angle experiments, a 30 mW HeNe laser fiber-coupled into the interferometer is deflected by the illumination mirror MI mounted on a motorized rotation stage (Thorlabs DDR25), which has encoder resolution of 0.000 25 deg/count. The starting angle $\theta_0 = 18°$ is set with manual estimate using a ruler. Instead of trying to measure this angle with any better precision, the synthetic wavelength can be calibrated according to the known step size of an object. By placing both the object and reference planes conjugate to the camera, it is ensured that the relative angle of incidence between the two beams does not change with the rotation of the mirror.

In another configuration, based on a modified Mach-Zhender interferometer, only the object beam is tilted while the reference beam remains stationary. This produces a slope in the holographic phase profile, but can be easily

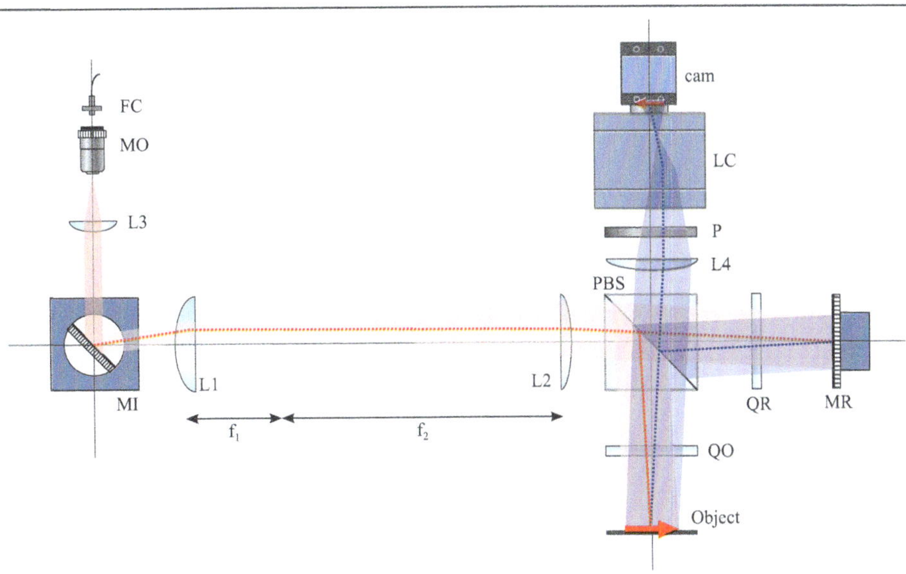

Fig. 15 Schematic of the MWDH/MADH apparatus. FC: fiber coupler; MO: microscope objective; L3: collimating lens; MI: illumination mirror on digital rotation stage; L1 & L2: confocal lens pair to image MI to object and reference planes; PBS: polarizing beam splitter; QO & QR: quarter wave plates; MR: piezo-mounted reference mirror; L4 & LC: camera imaging lens; P: linear polarizer; cam: camera.

compensated for numerically in reconstruction. The reference mirror MR can be tilted by an appropriate amount for off-axis digital holography, or piezo-shifted for phase-shifting digital holography.

Typically, a complex-valued hologram is acquired by combining several camera frames of interference intensity while the reference mirror is piezo-shifted, per standard procedures of phase-shifting digital holography. Then the source wavelength is switched in MWDH or the MI mirror angle is switched in MADH. A series of such holograms are acquired, which are then post-processed to construct the 3D surface profile of the object. Python-based programs are used for most of hardware control, image acquisition, and holographic image processing. Some parts of the software system also use LabVIEW and Matlab.

Acknowledgements

This project has not been externally funded.

Conflict of interest

The author declares no conflicts of interest.

Received: 17 September 2021 Revised: 08 February 2022 Accepted: 18 February 2022

Accepted article preview online: 08 March 2022

Published online: 06 May 2022

References

1. Stout, K. J. & Blunt, L. Three Dimensional Surface Topography. (London: Penton Press, 2000).
2. Cuche, E., Bevilacqua, F. & Depeursinge, C. Digital holography for quantitative phase-contrast imaging. *Optics Letters* **24**, 291-293 (1999).
3. Mann, C. J., et al. High-resolution quantitative phase-contrast microscopy by digital holography. *Optics Express* **13**, 8693-8698 (2005).
4. Marquet, P., Depeursinge, C. & Magistretti, P. J. Exploring neural cell dynamics with digital holographic microscopy. *Annual Review of Biomedical Engineering* **15**, 407-431 (2013).
5. Osten, W. Optical Inspection of Microsystems. (Boca Raton: Taylor & Francis, 2007).
6. Merola, F., et al. Tomographic flow cytometry by digital holography. *Light: Science & Applications* **6**, e16241 (2017).
7. Kim, M. K. Digital Holographic Microscopy. (New York: Springer, 2011).
8. Ghiglia, D. C. & Pritt, M. D. Two-Dimensional Phase Unwrapping: Theory, Algorithms, and Software. (New York: Wiley, 1998).
9. Gass, J., Dakoff, A. & Kim, M. K. Phase imaging without 2*p* ambiguity by multiwavelength digital holograph. *Optics Letters* **28**, 1141-1143 (2003).
10. Parshall, D. & Kim, M. K. Digital holographic microscopy with dual-wavelength phase unwrapping. *Applied Optics* **45**, 451-459 (2006).
11. Khmaladze, A., et al. Simultaneous dual-wavelength reflection digital holography applied to the study of the porous coal samples. *Applied Optics* **47**, 3202-3210 (2008).
12. Cheng, Y. Y. & Wyant, J. C. Two-wavelength phase shifting interferometry. *Applied Optics* **23**, 4539-4543 (1984).
13. Wagner, C., Osten, W. & Seebacher, S. Direct shape measurement by digital wavefront reconstruction and multi-wavelength contouring. *Optical Engineering* **39**, 79-85 (2000).
14. Wada, A., Ka to, M. & Ishii, Y. Large step-height measurements using multiple-wavelength holographic interferometry with tunable laser diodes. *Journal of the Optical Society of America A* **25**, 3013-3020 (2008).
15. Mann, C. J., et al. Quantitative phase imaging by three-wavelength digital holography. *Optics Express* **16**, 9753-9764 (2008).
16. De Nicola, S., et al. Recovering correct phase information in multiwavelength digital holographic microscopy by compensation for chromatic aberrations. *Optics Letters* **30**, 2706-2708 (2005).
17. Yamaguchi, I., Oh ta, S. & Kato, J. I. Surface contouring by phase-shifting digital holography. *Optics and Lasers in Engineering* **36**, 417-428 (2001).
18. Gonzalez, R. C. & Woods, R. E. Digital Image Processing. 3rd edn. (Upper Saddle River: Pearson, 2016)
19. Bianco, V., et al. Quasi noise-free digital holography. *Light: Science & Applications* **5**, e16142 (2016).
20. Xia, H. T., et al. Phase calibration unwrapping algorithm for phase data corrupted by strong decorrelation speckle noise. *Optics Express* **24**, 28713-28730 (2016).
21. Psota, P., et al. Multiple angle digital holography for the shape measurement of the unpainted tympanic membrane. *Optics Express* **28**, 24614-24628 (2020).
22. Kim, M. K. Tomographic three-dimensional imaging of a biological specimen using wavelength-scanning digital interference holography. *Optics Express* **7**, 305-310 (2000).
23. Jeon, Y. & Hong, C. K. Optical section imaging of the tilted planes by illumination-angle-scanning digital interference holography. *Applied Optics* **49**, 5110-5116 (2010).
24. Dong, J., J ia, S. H. & Jiang, C. Surface shape measurement by multi-illumination lensless Fourier transform digital holographic interferometry. *Optics Communications* **402**, 91-96 (2017).
25. Martinez-Carranza, J., et al. Multi-incidence digital holographic profilometry with high axial resolution and enlarged measurement range. *Optics Express* **28**, 8185-8199 (2020).

Utadiya et al. *Light: Advanced Manufacturing* (2022)3:37
https://doi.org/10.37188/lam.2022.037

Official journal of the JHL 2689-9620
www.light-am.com

Article | Open Access

Integrated self-referencing single shot digital holographic microscope and optical tweezer

Subhash Utadiya[1], Sachin Joshi[2], Nimit Patel[1], Chetna Patel[1], Mugdha Joglekar[1], Vani Cahhniwal[1], Timothy O'Connor[2], Bahram Javidi[2] and Arun Anand[1,3,*]

Abstract

Digital holographic microscopy is a single-shot technique for quantitative phase imaging of samples, yielding thickness profiles of phase objects. It provides sample features based on their morphology, leading to their classification and identification. However, observing samples, especially cells, in fluids using holographic microscopes is difficult without immobilizing the object. Optical tweezers can be used for sample immobilization in fluids. The present manuscript provides an overview of our ongoing work on the development of a compact, low-cost microscopy system for digital holographic imaging of optically trapped samples. Integration of digital holographic microscopy system with tweezers is realized by using the optical pickup unit extracted from DVD burners to trap microsamples, which are then holographically imaged using a highly compact self-referencing interferometer along with a low-cost, in-house developed quadrant photodiode, providing morphological and spectral information of trapped particles. The developed integrated module was tested using polystyrene microspheres as well as human erythrocytes. The investigated system offers a multitude of sample features, including physical and mechanical parameters and corner frequency information of the sample. These features were used for sample classification. The proposed technique has vast potential in opening up new avenues for low-cost, digital holographic imaging and analysis of immobilized samples in fluids and their classification.

Keywords: Digital holographic microscopy, Optical tweezers, Red blood cells, Quantitative phase imaging

Introduction

Digital holographic microscopy has emerged as a powerful tool for imaging transparent phase objects, including live cells. It provides single-shot, high contrast images of unstained samples, which are usually transparent to visible electromagnetic radiation[1-3]. The technique also has the added advantage of numerical focusing. Numerically reconstructed digital holograms of the phase objects provide the sample's complex amplitude distribution[4].

The phase extracted from the complex amplitude distribution, can be used for the computation of optical thickness of the investigated object with sub-wavelength axial accuracy[5], refractive index tomography[6], cell migration studies[7], live cell imaging[8-16] including study of human red blood cells[9], cell division[12], study of cardiomyocytes[13], and embryonic stemcells[14]. Digital holographic microscope can also be combined with microfluidic systems for cell identification[15]. Object thickness distribution and its time variation provide many

Correspondence: Arun Anand (aanand-apphy@msubaroda.ac.in)
[1]Optics Laboratory, Applied Physics Department, Faculty of Technology and Engineering, The Maharaja Sayajirao University of Baroda, Vadodara 390001, India
[2]Department of Electrical and Computer Engineering, University of Connecticut, Storrs, CT, 06269-4157, USA
Full list of author information is available at the end of the article.
These authors contributed equally: Subhash Utadiya, Sachin Joshi

sample features that could be used for its characterization, identification, and classification[7-9,12-17].

In transmission mode, digital holographic microscopes are usually constructed using the Mach-Zehnder two-beam interferometer geometry[5], where One of the extracted beams from a laser source trans-illuminates the sample, and carries the object amplitude and phase information, and acts as the object beam. The second beam, which is unmodulated by sample information, serves as the reference beam. The object and the reference beams are made to superpose on a digital array by using beamsplitters, mirrors, and other beam steering optical components leading to the creation of digital holograms, which are numerically evaluated. A simplified version of digital holographic microscopes employing amplitude division self-referencing geometry microscopes have also been demonstrated[18-25] and used in cell classification problems[22-24]. Digital holographic microscopes, when coupled with image processing[9], pattern recognition, and machine learning algorithms[16-17,22-24], provide researchers with a potent noninvasive, label-free, real-time tool to characterize, identify and classify technical and biological phase objects[14,17,22-24].

The realization that a tightly focused laser beam could immobilize microparticles and control their movement opened new research areas[26]. The high intensity gradient at the focus of the laser beam draws the microparticles to the laser beam's center and traps them there[27]. This laser beam can then be used to micro manipulate the trapped particle. Nowadays, optical tweezers find place in many areas of imaging of technical and biological samples. It has been used for the imaging, characterization and isolation of stem cells[28,29]. It has also been demonstrated for trapping and manipulating viruses and bacteria[30] and studying the mechanics of single DNA molecules attached to micro-beads[31]. Optical tweezers were also explored for fertilization of oocytes and development of cell biosensors[32]. The advantage of optical tweezers is that they can trap and manipulate samples with dimensions ranging from nanometers to few hundreds of micrometers. This aspect is especially useful in optical sorting, elasticity, and deformability measurement, and single-cell isolation[33-35]. Optical tweezers are also ideal for exploring, characterizing, and classifying red blood cells (RBCs)[36-42]. It has been used to study the behavior of red blood cells in patients with sickle cell anemia[36]. It was also used to diagnose malaria using the power spectrum of Brownian position fluctuations of the trapped cells[37]. Shape asymmetry (deformation and shape change) in RBCs due to the onset of diseases was also observed using optical traps[38]. Pathological analysis for the determination of biomechanical properties of red blood cells was also reported[40].

The coupling of digital holographic imaging devices to optical traps will further enhance its capabilities in the study and characterization of bio-samples. Such a combination will allow investigating the morphological changes occurring to bio-samples in their natural environment, obtaining sample features otherwise inaccessible using optical tweezers, like sample volume, surface area, shape, and thickness fluctuations. Two-beam interferometer geometry has already been reported for combining optical tweezers and digital holographic microscopy[25,43-45].

Optical tweezers coupled with digital holographic microscopes for quantitative phase imaging of the trapped particles require laser sources for optical trapping and holographic imaging and a light source for bright field imaging. The technique will also need digital arrays for brightfield image acquisition and digital recording of holograms. A quadrant photodiode is also necessary to determine the corner frequency and trap stiffness, which directly depends upon the Brownian fluctuations of the trapped particle. So, integrating the optical trap and digital holographic microscope leads to a bulky and costly device. However, optical traps can be designed using any laser source which has enough power to trap the investigated sample. A DVD Optical Pickup Unit (OPU) has all the components (high power laser, high NA lens mounted on a voice coil) necessary to make a compact optical trap and has already been demonstrated[35].

As mentioned earlier, digital holographic microscopes in transmission mode usaually use a two-beam geometry based on Mach-Zehnder interferometer[5]. Such a microscope is large, bulky, expensive, and difficult to integrate with an optical trap. Also, in Mach-Zehnder geometry the object beam (beam interacting with the sample) and reference beam travels along different paths. The beams will pick up uncorrelated phase variations (due to mechanical noise), which will reflect as phase noise in the cell thickness variation measurements. However, digital holographic microscopes can also be implemented using self-referencing geometry. In such geometry, a portion of the object beam, which is unmodulated by object information, is made to interfere with the portion of the same beam containing object information leading to the formation of holograms[18-20]. A self-referencing digital holographic microscope using can be easily integrated into an optical tweezer. The simplest Self-Referencing Digital Holographic Microscope (SRDHM) utilizes a glass plate to create two laterally sheared versions of the same object wavefront[19,21-24]. When the shear is much larger than the object's size, it leads to the creation of holograms[19]. It has been demonstrated to be robust for label-free cell identification[22,24]. Digital holographic microscope based on self-referencing lateral shearing geometry requires a single glass plate to generate holograms[19,21–24], making it very simple, cost-effective and easy to integrate with optical traps, making the whole device compact. Also in lateral shearing digital holographic microscope, the object and the reference beam travel along identical paths encountering the same set of optical elements (magnifying lens and glass plate), leading to high immunity towards mechanical noise. It should be noted here that single-shot quantitative phase imaging could also be achieved using transport of intensity equations using multifunctional metastructures. However, it might be challenging to integrate this with optical tweezers[46,47].

We have been working on integrating low-cost optical tweezers using DVD optical pickup units and SRDHM modules for quantitative phase imaging of trapped particles for their characterization and identification[48]. This manuscript is an overview of our work on integrating optical traps with digital holographic imaging. We have used self-referencing digital holographic microscope using lateral shearing geometry for quantitative phase imaging of the trapped particles. It is an effective and compact way to integrate digital holographic microscopy and optical traps. In our earlier work[48], we have shown the proof of concept of the integrated device. In the present manuscript, the device is extended to optically trap red blood cells and extract cell features based on its morphology and its time variation, from digital holographic images. These features and the corner frequency information of the trapped particles are supplied to a machine learning model based on Support Vector Machine (SVM)[49] for cell classification.

Results

Fig. 1a shows the schematic of the optical trap integrated with the digital holographic microscope. The trapping mechanism comprises of an OPU containing a diode laser module (λ = 650 nm) and a lens of ~4 mm focal length and ~0.6 NA (Supplementary Fig. S1 shows details of the OPU). One of the advantages of OPU is that the focusing lens is mounted on a steering mechanism allowing axial and lateral displacement of the lens by applying electric current[35,48]. This enables easy focusing and micromanipulation of trapped particles. A blue LED (Luxeon star λ = 470 nm, $\Delta\lambda$ = 30 nm) was integrated to the device for bright field imaging. Output from a green laser diode module (Thorlabs CPS532, Output power 4.5 mW) was coupled to the OPU by a dichroic filter (inbuilt in OPU) for digital holographic imaging. The working of the integrated system is detailed in Supplementary Video V1.

Optical tweezer based on optical pickup unit

To assess the OPU based optical trap, it was tested by trapping 5 μm and 15 μm diameter polystyrene

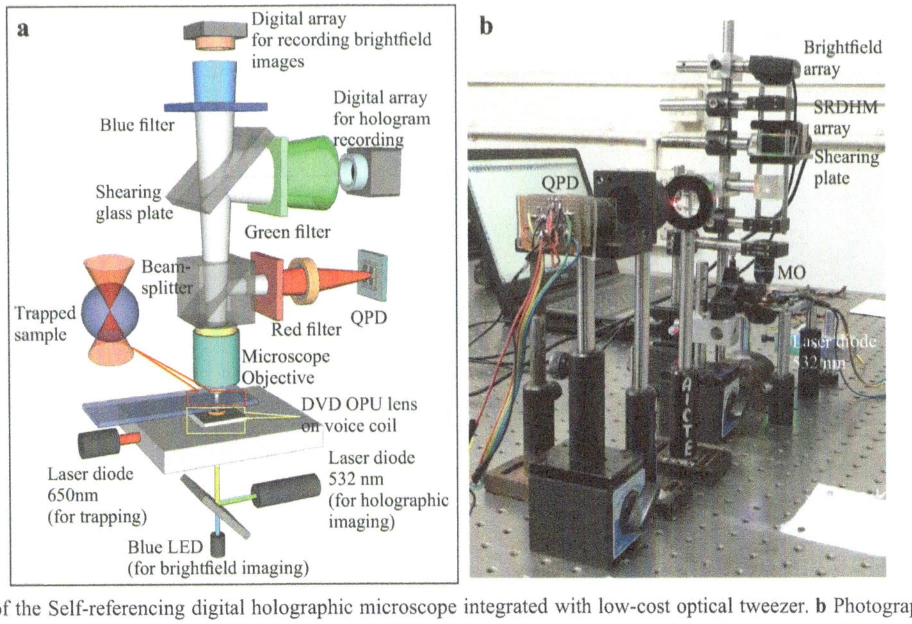

Fig. 1 **a** Schematic of the Self-referencing digital holographic microscope integrated with low-cost optical tweezer. **b** Photograph of the table-top setup. Laser (650 nm) from the DVD optical pickup unit acts as the trapping beam. Laser output from 532 nm laser diode module was used for digital hologram recording. A blue LED (470 nm) was used for brightfield imaging. An in-house developed QPD measures the laser beam fluctuations due to the Brownian motion of the trapped particle.

microspheres (Polysciences Inc.) immersed in deionized water. These trapped microsphers were then manipulated by moving the trapping lens by applying an electric current to the voice coil on which it was mounted. Fig. 2 shows the movement of the 15 μm polystyrene microsphere (refractive index = 1.59) in the trap recorded using the brightfield camera. A blue bandpass filter was used to block the trapping laser beam and hologram laser source reaching the bright field camera, and images was captured at the rate of 15 Hz. A microscope objective lens (40X, NA = 0.65) was used to image the sample on to the bright field sensor.

The measured power of the trap laser at the back focal plane of the OPU lens was less than 40 mW. In Fig. 2, the black arrow depicts the direction of the movement of the trapping beam. Results shown in Fig. 2, confirm that the system can immobilize and manipulate micro-particles[46]. Supplementary Video V2 and video V3 show the micromanipulation of the trapped 15 μm and 5 μm diameter polystyrene microspheres, respectively. Both the videos were recorded at 15 fps for 30 s.

Quantitative Phase imaging with integrated SRDHM

Holographic images of the trapped phase objects were recorded using the lateral shearing self-referencing digital holographic microscope (SRDHM). It comprises of a laser diode module working at 532 nm. The laser output was expanded using a 25 mm focal length lens to ensure that an expanding laser beam illuminates the sample. A 5 mm thick fused silica glass plate kept at an angle of 45° with the incoming laser beam acted as the shearing element. The two laterally sheared versions of the object wavefronts were made to superpose on digital array (Thorlabs DCU224C, 8-bit, 4.65 μm pixel pitch), which recorded the resulting holograms. Fig. 3 shows the portion of the object hologram (hologram recorded with the object in the field of view) of a 15 μm diameter polystyrene microsphere immersed in microscope oil (refractive index = 1.52). A reference hologram (without the object but with the medium surrounding the object in the field of view) was also recorded for phase comparison (subtraction). The recorded holograms were numerically reconstructed using angular spectrum propagation diffraction integral, best suited for short distance propagations[2,3]. The phase difference ($\Delta\phi$) computed by subtracting the reference phase from object phase is related to the object thickness (h) distribution through the constant average refractive index of the sample (n_O) and the surrounding medium (n_R) via the relation $\Delta\phi = (2\pi/\lambda)(n_O - n_R)h$. If the refractive indices of the sample and the surrounding medium are not available, the same relationship yields the sample optical thickness distribution. The phase subtraction also nullifies much of the phase due to aberrations as it remains the same between exposures (as we use same set of optical elements for recording of object and reference holograms), bringing out the sample phase distribution. Fig. 3b shows the sample phase distribution obtained after phase subtraction for polystyrene microsphere of 15 μm diameter. This phase, along with the refractive index values, yield the object thickness variation shown in Fig. 3c. Fig. 3d shows the cross-sectional profile of the thickness distribution along the dashed line shown in Fig. 3b. As can be seen from Fig. 3b, the measured value is close to the manufacturer specified value, showing that SRDHM can be used for

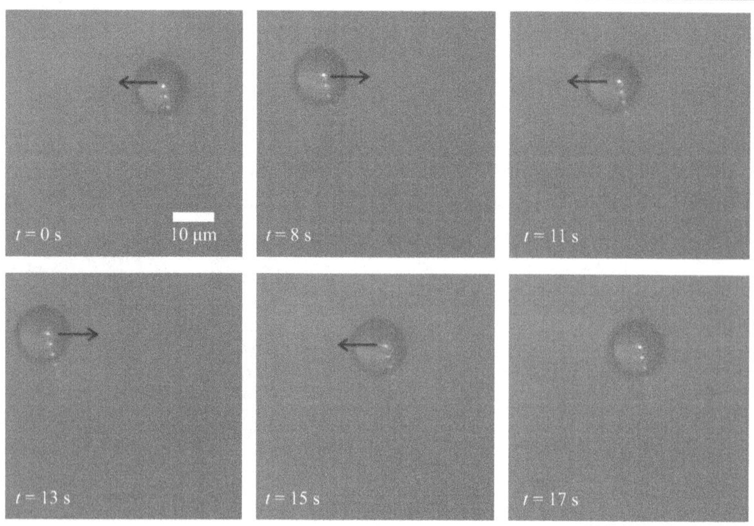

Fig. 2 Brightfield images of micromanipulation of 15 μm diameter polystyrene spheres (Video V2 shows the details). Multiple reflections from the blue filter led to multiple spots in the image.

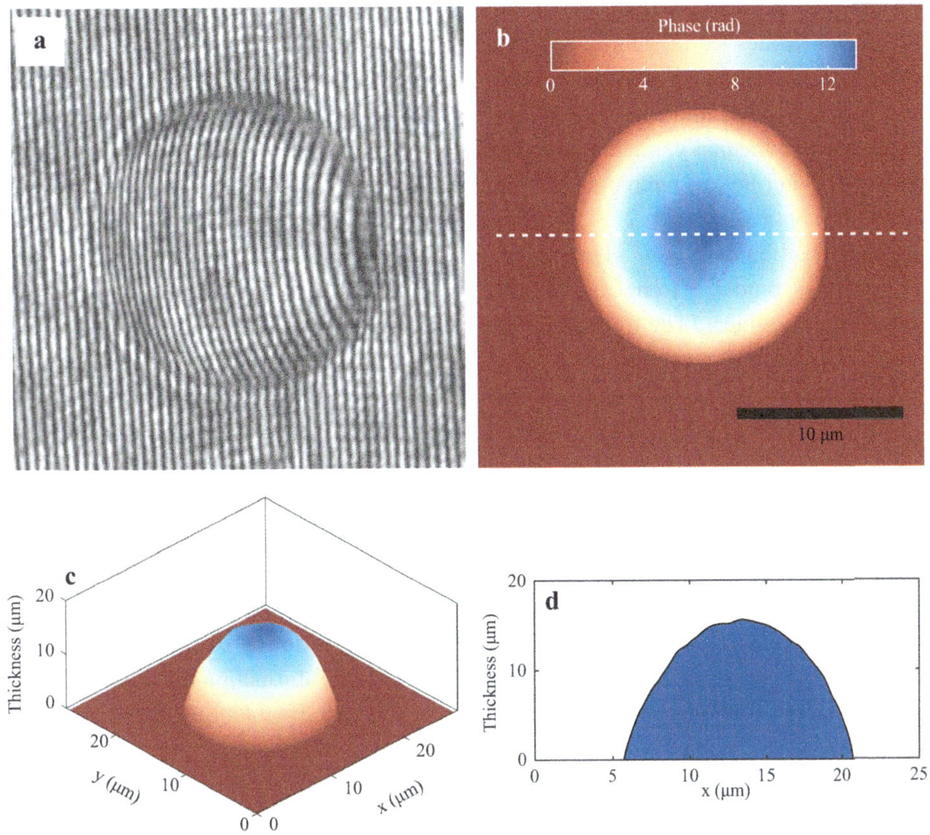

Fig. 3 Quantitative phase imaging using SRDHM module. **a** Portion of the hologram recorded with 15 µm polystyrene microspheres immersed in microscope oil. **b** Reconstructed phase distribution of the test sample obtained after phase subtraction. **c** Three-dimensional rendering of the thickness distribution obtained by using the refractive indices of the sample and the surrounding medium. **d** Cross-sectional thickness profile along the white dashed line in Fig. 3b.

accurate thickness measurement of phase objects.

Motion measurement of trapped particles

The strength of the optical tweezer depends upon the trapping laser power and the characteristics of the trapped particle, including its shape[50,51]. The trapping force is computed by measuring the power spectrum (S) of the Brownian motion of the trapped particle. The Brownian motion of the trapped particle is measured by a quadrant photodiode (QPD), which acts as the position detection device[42]. The strength of the trap is also proportional to the trap stiffness, which in turn is proportional to the corner frequency (f_c) which is computed from the power spectral density (PSD) of the temporal fluctuations of the trapped particle. Therefore, the corner frequency is a reliable measure of trap strength[48,50]. The power spectrum of these Brownian fluctuations in position will have a Lorentzian distribution ($S(f) = A/(f^2 + f_c^2)$, where A is a constant that depends upon the viscous drag coefficient and temperature). By fitting the resulting PSD with a Lorentzian distribution, corner frequencies in x and y directions can be computed. The corner frequency is sensitive to the deformation and shape change of the trapped particle and hence can be used for identification and classification. We have calculated the corner frequencies of the trapped micro-objects from the power spectrum. Corner frequency can be used to determine trap stiffness knowing the viscosity of the medium surrounding the trapped particle and the diameter of the trapped particle. Force can then be calculated using the stiffness value by measuring the displacement of the particle. However, we have used only the corner frequency values in the classification process as it is proportional to trap stiffness.

As mentioned, measurement of the corner frequency (which also measures the force of the trap) requires power spectral analysis of the time-varying position of the trapped sample. QPDs detect the position by measuring the change in intensity of light falling on the detector and are fast compared to arrayed image sensors. The use of commercially available QPDs will increase the cost of the developed device. We used four photodiodes (OPT101) in a 2 × 2 format to develop a cost-effective alternative. However, using these photodiodes led to significant lateral separation (due to the large area of the diodes and circuitry involved) between the individual diodes compared to standard QPDs available commercially, making position determination difficult. This hurdle was overcome by using closely spaced multiomode optical fibers in 2 × 2 format to collect the laser light for position detection[48]. The trapping red laser beam was focused onto the front end of these fibers using a 25 mm focal length lens. The front end of the fibers was positioned such that the beam spot covered all four fibers. The light collected by the optical fibers was then delivered to the photodiodes at the module's backend. This geometry is equivalent to four closely spaced photodiodes. A data acquisition system using Arduino Uno R3 was used for logging the output from the photodiodes at a sampling rate of 10 kHz[48]. Supplementary Fig. S2 gives details of the developed QPD system using multimode optical fibers.

Corner frequency of trapped microspheres

The displacement data was recorded (at the rate of 10 kHz for a period of 20 s) for the polystyrene beads (5 μm and 15 μm diameter) to measure the corner frequency. The obtained power spectrum and the corresponding Lorentzian fit (solid black line) for 5 μm and 15 μm diameter samples are shown in Fig. 4. The corner frequency is determined from the Lorentzian fit to the power spectrum. Fig. 4a, b shows the PSD for the position variation in x and y directions, respectively, for 15μm diameter microsphere. Similarly, Fig. 4c, d show the spectrum for 5 μm diameter sample.

Measured corner frequencies for both classes of microspheres in x and y direction are shown in Fig.5a, b, respectively. The measured corner frequencies for 15 μm diameter microspheres were f_{cx} = 3.87 ± 2.01 Hz and f_{cy} = 4.08 ± 2.12 Hz. The measured corner frequencies for 5 μm diameter microspheres were f_{cx} = 8.63 ± 3.03 Hz and f_{cy} = 8.37 ± 2.97 Hz.

Corner frequency of trapped red blood cells

Position information of human red blood cells (RBC) in phosphate buffer solution (PBS) was also recoded using the QPD and used for determination of corner frequency. The

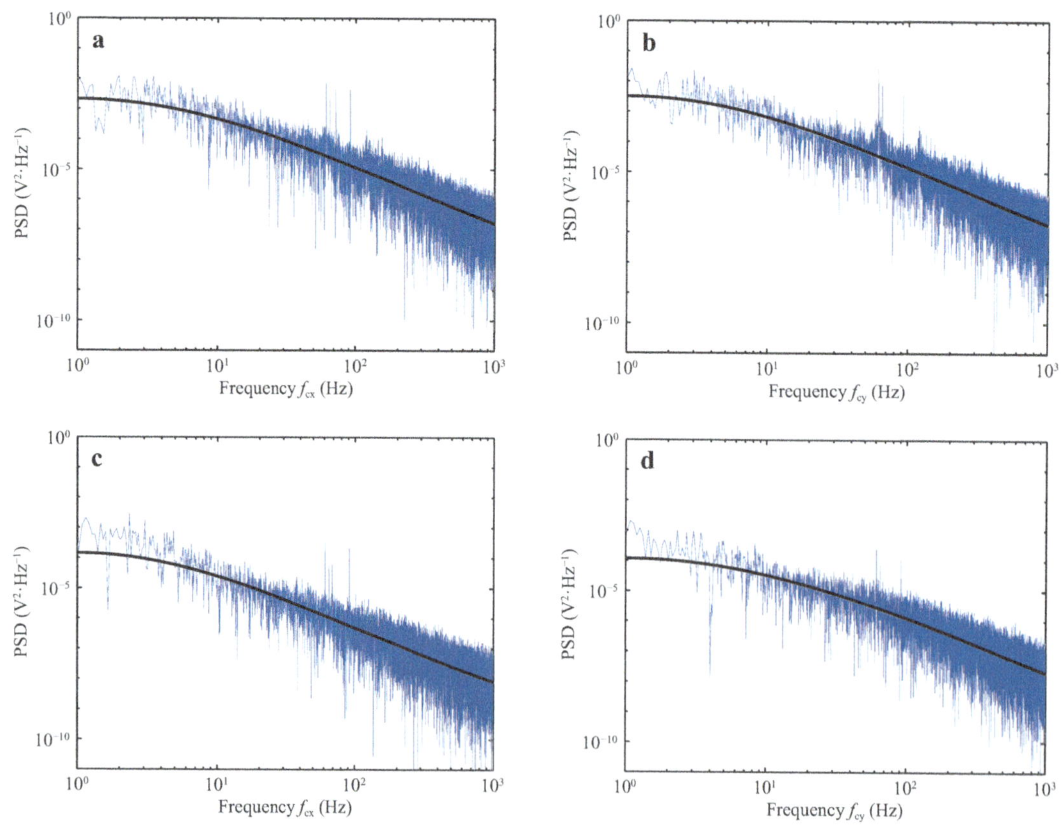

Fig. 4 Power spectrum measured using QPD. **a, b** power spectrum for 15 μm diameter particle obtained for particle Brownian motion in x-direction and y-direction respectively. **c, d** power spectrum for 5 μm diameter particle for motion in x-direction and y-direction respectively. Solid black line represents the Lorentzian fit to the power spectrum.

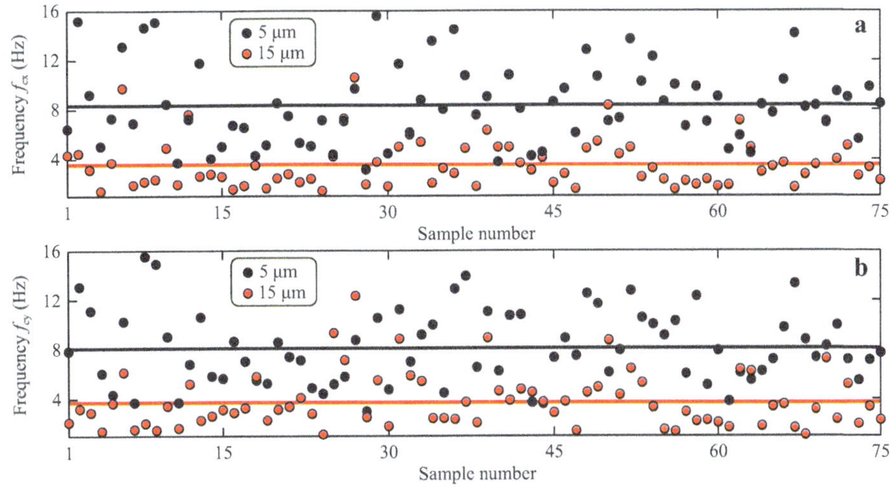

Fig. 5 Measured corner frequencies for 15 μm and 5 μm diameter polystyrene microspheres computed from **a** power spectrum for Brownian motion in *x*-direction and **b** power spectrum for Brownian motion in *y*-direction. Solid black line is the mean corner frequency value for 5 μm particle, and the solid red line represents the mean corner frequency value for 15 μm particle.

studies used red blood cells extracted from healthy blood samples and blood samples tested positive for Sickle Cell Disease (SCD). Fig. 6a, b show the power spectrum for motion in the *x*-direction for healthy and SCD positive red blood cells, respectively. In the case of SCD, red blood cells are deformed compared to heathy red blood cells, which will be reflected as a change in the corner frequency[38].

Fig. 7 shows the measured corner frequency (f_{cx}) for the investigated red blood cell, obtained from healthy sample and sickle cell disease positive sample. The mean values of the corner frequencies are shown as solid lines in the figure. In the case of RBCs, the obtained corner frequency alone may not be enough to determine the class of the cells.

Quantitative phase imaging of trapped microspheres

Quantitative phase imaging of trapped particles was achieved using the integrated SRDHM module shown in Fig. 1a. Using SRDHM a significant number of sample features based on sample thickness and its time variation could be extracted. From the thickness profile 20 biophysical and biomechanical features of the cells were extracted[21–24]. The extracted cell features were diameter, eccentricity, mean thickness, maximum thickness, minimum thickness, median thickness, coefficient of variation of thickness, skewness of thickness distribution, kurtosis of thickness distribution, sphericity, projected area, curved surface area, total surface area, surface area to volume ratio, projected area to volume ratio, amplitude of thickness fluctuation, mean peak frequency of thickness fluctuation, amplitude of volume fluctuation and peak

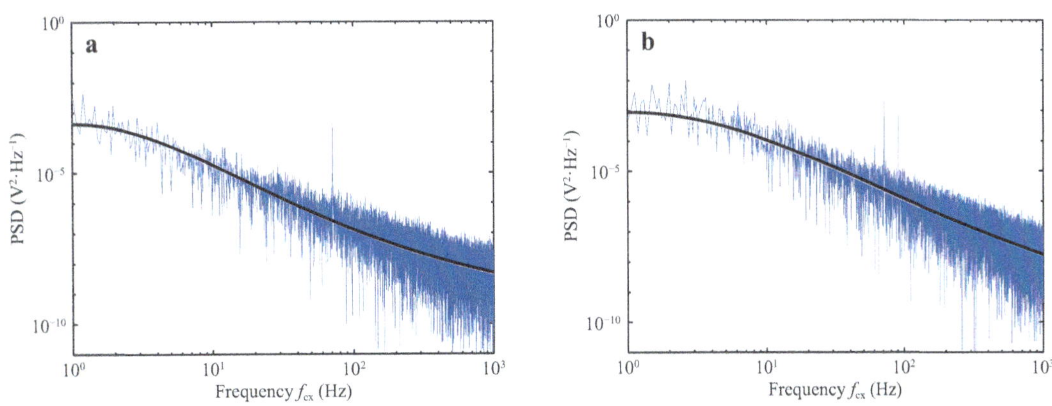

Fig. 6 Power spectrum for red blood cells. **a** Healthy sample. **b** SCD positive sample. The solid black line represents the Lorentzian fit to the power spectrum.

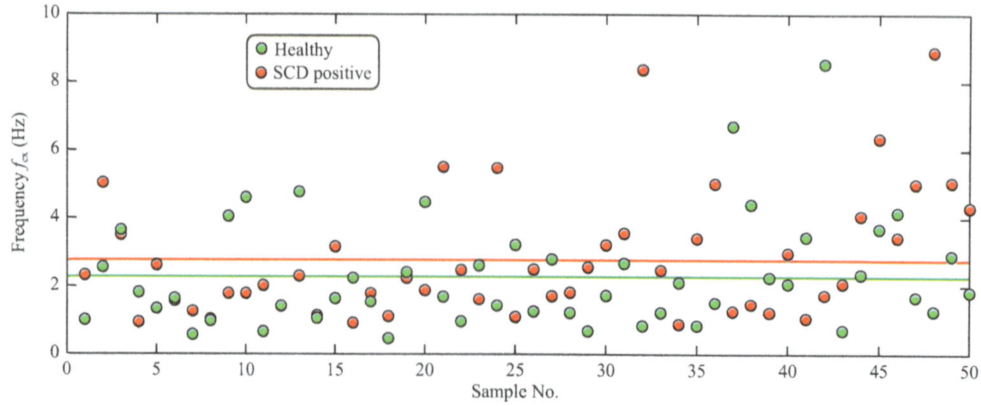

Fig. 7 Measured corner frequencies for healthy and SCD positive red blood cells computed from power spectrum for Brownian motion in x-direction. Solid redline line represents the mean corner frequency value for SCD positive sample, solid green line represents the mean corner value for healthy sample and solid black line represents mean corner frequency value obtained considering both classes of RBCs.

frequency of volume fluctuation. These features contain vital information required for cell classification.

For quantitative phase imaging of microspheres, a video of the resulting holograms was recorded at the rate of 30 Hz for a period of 16 s. Supplementary Video V4 illustrates the time series holograms. Each hologram was extracted from the video and numerically reconstructed individually, and the sample thickness information was computed[19–21]. Fig. 8a shows a hologram of a 15μm diameter polystyrene bead immobilized by the trap at one time instant. Fig. 8b shows the thickness distribution at the same time instant obtained from the reconstructed quantitative phase map. Fig. 8c shows the time varying thickness profile of the trapped 15 μm polystyrene sphere at the points shown in Fig. 8b. The effect of the Brownian motion of the sample is evident from this plot. The variation in the thickness happens since the trapped microspheres shifts its position as can be seen from the Video V4. As trapped microsphere shifts, the optical path length of the light passing through it changes (due to change in angle of incident ray and as well as the length travelled through the sample), leading to change in optical thickness at that spatial point. The physical and mechanical features of the sample extracted from the hologram, along with corner frequency information can lead to better sample classification. The fluctuation profile of the sample is obtained by computing the standard deviation of the thickness variation in every 2 s time window and is shown in Fig. 9.

Quantitative phase imaging of trapped red blood cells

Red blood cells in PBS were immobilized using the developed integrated system and their holograms were recorded. Fig. 10a shows the portion of the recorded hologram of an immobilized healthy RBC. The optical thickness information obtained from the numerically reconstructed hologram is shown in Fig. 10b.

A time series of holograms can also be used for extraction of time varying cell morphology (Supplementary Video V5, shows the time series holograms of trapped RBC). After extracting the cell thickness profile at each time instant, the cell thickness distribution as a function of time can be determined. Fig. 10c shows the cell thickness at different time instances. Supplementary Video V6 provides a detailed view of cell thickness as a function of time.

Quantitative phase maps can also be used to compute the thickness fluctuations of the trapped red blood cell. As in the case of microspheres, in the case of red blood cells also, the time varying thickness at each point can be obtained and analyzed. Fig. 11 shows the optical thickness variation of RBC at the points shown in Fig. 10b. Here the optical thickness variation rather than the real thickness is shown. Optical thickness represents thickness multiplied by refractive index difference between the cell and the surrounding medium. In fact, classification of cells can be achieved using features based on optical thickness negating the need for knowing the exact refractive indices of the cells, which may vary for cells of different classes.

The standard deviation of the thickness variation in a time window provides the cell thickness fluctuation profile as shown in Fig. 12. Supplementary Video V7 shows the fluctuation profile computed from thickness variation occurring in each second.

Discussions

The developed low-cost optical tweezer can trap microparticles and manipulate them by moving the lens of

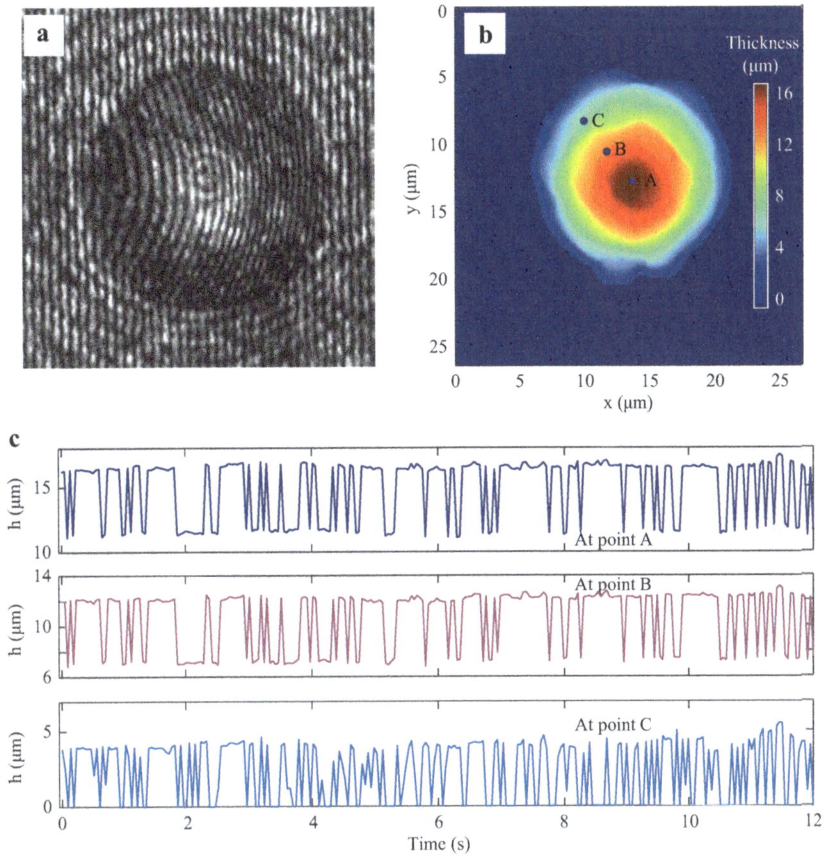

Fig. 8 Digital Holographic imaging of trapped microsphere. **a** Portion of the hologram of the trapped 15 µm diameter polystyrene microsphere. **b** Reconstructed thickness profile. **c** Time varying thickness at the points shown in Fig. 8b, obtained from a series of holograms.

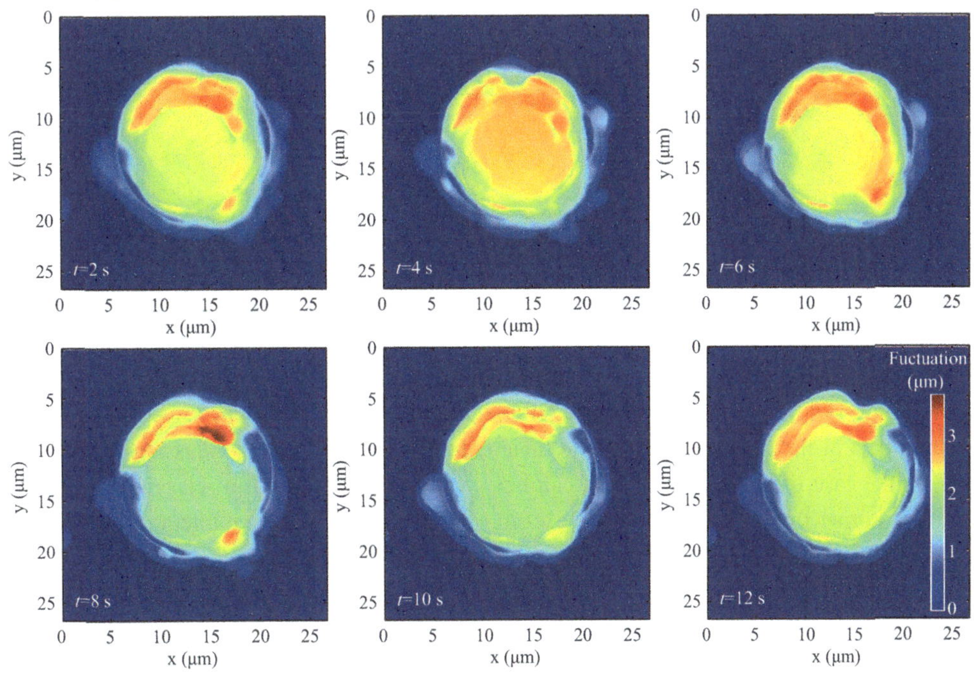

Fig. 9 Thickness fluctuation of the trapped 15 µm polystyrene microsphere obtained from the standard deviation of the time varying thickness profile.

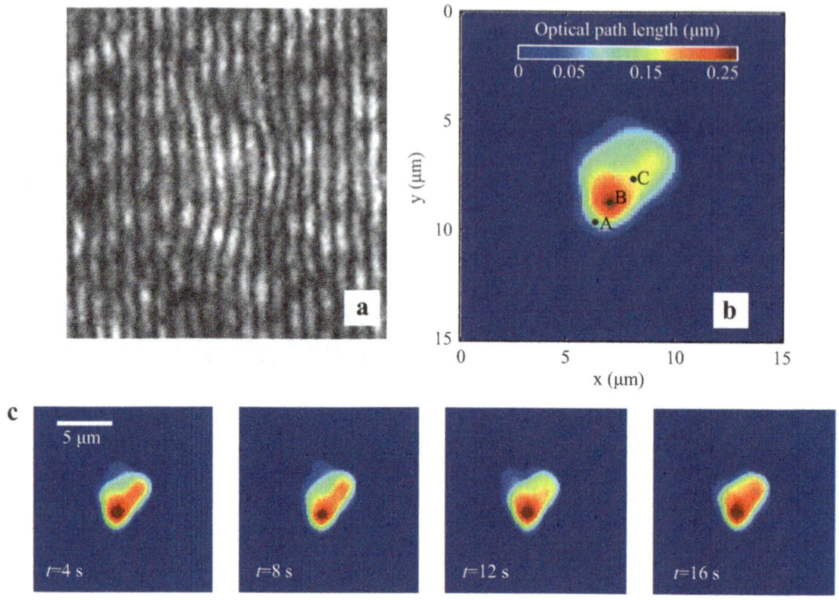

Fig. 10 Digital Holographic imaging of trapped red blood cell. **a** Portion of the hologram of the trapped RBC. **b** Reconstructed thickness profile of RBC. **c** Time varying quantitative phase images of the trapped cell.

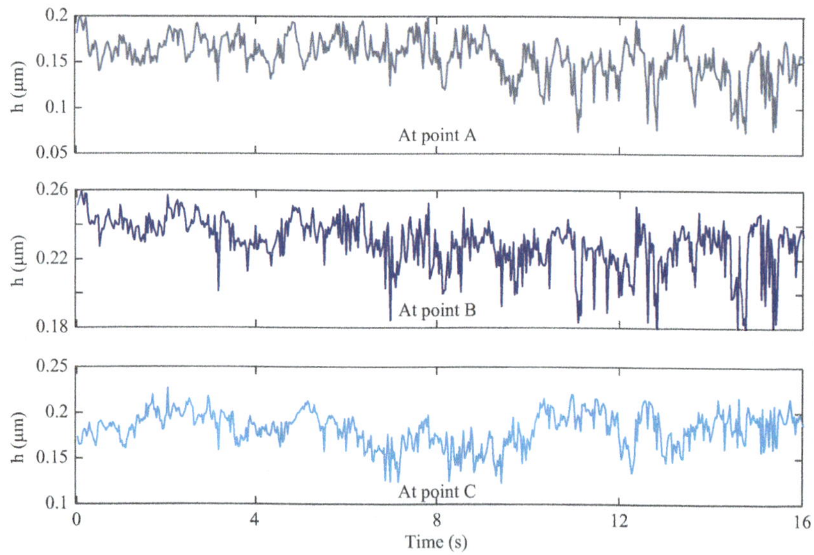

Fig. 11 Time varying optical thickness of the trapped blood cell at the points shown in Fig. 10b. Standard deviation of this thickness variation provides the cell thickness fluctuation.

the optical pickup unit. The in-house developed QPD provides the position data of the trapped particles and their corner frequency. In the case of red blood cells, as can be seen from Fig. 7, only corner frequency many not be useful in cell classification. To check the technique's effectiveness in classifying red blood cells into healthy and SCD, Machine Learning model based on Support Vector Machine (SVM)[49], which is ideal for classification problems, was trained. The training used the extracted feature values of 50 healthy and 50 SCD cells. Initially, only the corner frequency information was used to classify the cells. SVM with a Gaussian Kernel and five-fold cross validation provided a classification accuracy of 62%. However, the proposed integrated optical trap and SRDHM also yield quantitative phase images of trapped micro-particles including living cells and provide a wide range of features based on sample thickness and its time variation, which is normally not available in conventional optical tweezers. These parameters and the Brownian motion spectral information can be used in machine learning

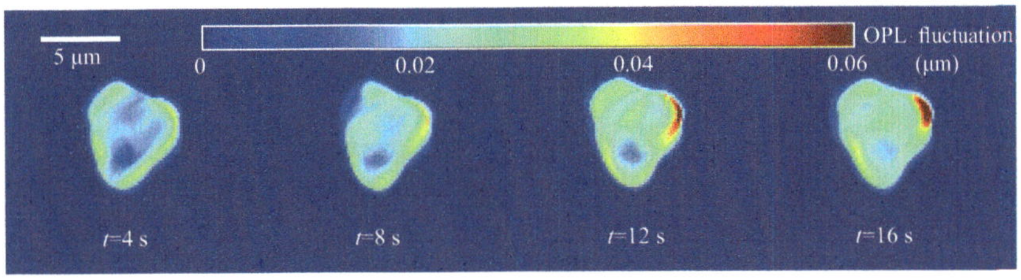

Fig. 12 Cell optical thickness fluctuation of trapped red blood cell at various time instances.

algorithms to classify the cells. The extracted morphological features along with the corner frequency information was then used to train a machine learning model. Fig. 13 provides the classification results in the form of a confusion matrix.

The inclusion of morphological features improved the classification accuracy to 89%. The trap beam (focused laser beam) may also affect the sample morphology and lead to its deformation, which may depend upon the class of cells investigated. This aspect has to be investigated further.

The optical trap, which is derived from the optical pickup unit of the DVD burner provides good enough immobilization of the investigated particles. It is not necessary that a DVD optical pick must be used in the trapping of the microparticles.

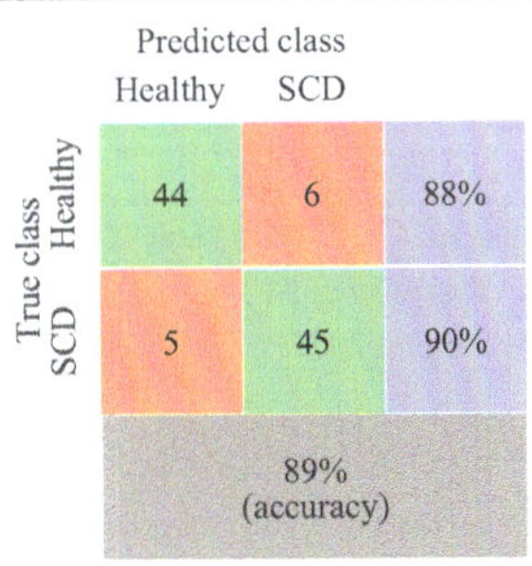

Fig. 13 Confusion matrices for classification of red blood cells.

Low-cost laser diode modules, both in the electromagnetic spectrum's visible and infrared regions, are readily available. These lasers can be coupled with a lens mounted on a voice coil (or a scanning mechanism) to act as the micromanipulation unit. This will be further investigated in the future.

The technique discussed in the manuscript utilized a CCD array to record holograms. However, it is not necessary to use costly CCD arrays to record holograms. Low-cost digital arrays like webcams can also provide high quality quantitative phase images which are comparable to those offered by CCD arrays[48]. Fig. 14a shows the portion of the hologram of a 15 μm diameter polystyrene microsphere recorded using a commercial VGA webcam array. The quantitative phase profile of the sample was reconstructed after subtraction of the phase obtained from reference hologram and is shown in Fig. 14b. The result shows that webcam arrays are suitable candidates for recording holograms and their use further reduces the cost of the device.

Since the described technique is compact, requiring fewer optical and beam steering components, it could be converted into a field-portable device. This will be carried out soon. The method provides a multitude of sample features, which has been demonstrated to provide high accuracy classification of cells. In the next step, in-depth cellular analysis that uses both biophysical and

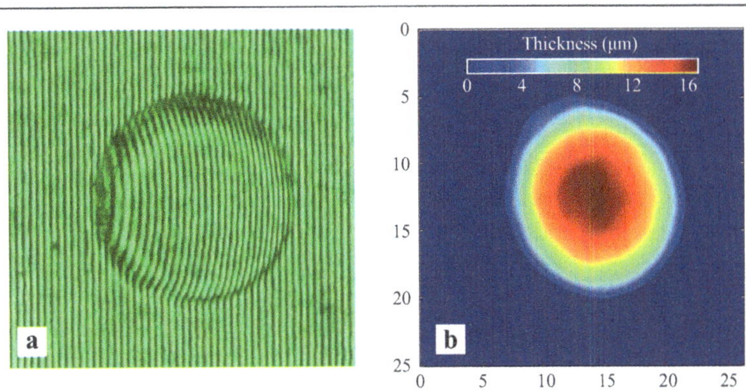

Fig. 14 Use of webcams for recording of holograms **a** Portion of the hologram of a 15 μm polystyrene microsphere recorded by a VGA webcam array. **b** Reconstructed thickness profile of the microsphere.

biomechanical cell properties of immobilized cells for sorting, classification, and identification will be investigated.

Materials and Methods

Samples of Polystyrene beads (refractive index 1.52) of diameters 5 micrometers and 15 micrometers were prepared to calibrate the developed Optical Tweezer setup. The sample was prepared by making the solution of respective poly beads with deionized water. Sodium Lauryl Sulphate (Molecular weight −288.31 g mol^{-1}), which acts as a surfactant, was added to the solution to prevent clustering of the beads and adhesion of the beads to the wall of the container. The solution was thoroughly mixed and used for analysis[48].

The human erythrocytes were also analyzed using the developed optical tweezer setup. The Phosphate Buffer Saline solution (PBS,1X, pH 7.4) used in preparing RBC suspension, was prepared in the laboratory by continuously stirring the mixture of NaCl (8 g), KCl (0.2 g), Na_2HPO_4 (1.15 g), KH_2PO_4 (0.24 g) in 800 ml of deionized distilled water. Further, the solution was made up to 1 liter by adding required deionized distilled water to it[52]. The PBS solution looks after a proper environment for the biological samples by maintaining constant pH and ion concentration values for biological specimens. Further, it maintains the osmolarity of the cells and thus prevents cell rupturing. Then, the RBC suspension was prepared by mixing the 1 ml of the blood sample with the 1 ml of PBS solution.

The solution was centrifuged at a rate of 2000 rpm for 600 s, resulting in separating a layer of white blood cells, plasma, and red blood cells. The red blood cells were isolated from the samples by removing the supernatant and washing the PBS solution again. The protocol was repeated three times to isolate RBCs from other blood constituents[53]. Then a RBC suspension was obtained by mixing a drop centrifuged of blood (containing only RBCs) with the required PBS solution and analyzed.

For cell classification into healthy and SCD positive, Machine Learning model based on Support Vector Machine (SVM)[49], was trained using the extracted cell parameters (biophysical and biomechanical features obtained from SRDHM and the corner frequency values from QPD) as the independent variables and the cell class as the independent variable. The features from 50 healthy and 50 SCD positive samples were used in training of the model. The accuracy of the model was determined from five-fold cross validation.

Acknowledgements
The work was supported by research grants SERB (EMR/2017/002724), DAE-BRNS (2013/34/11/BRNS/504), DST-FIST and DST-PURSE. AA and VC would like to acknowledge Abdus Salam International center for Theoretical Physics (ICTP), Trieste, Italy for Regular Associate fellowship.

Author details
[1]Optics Laboratory, Applied Physics Department, Faculty of Technology and Engineering, The Maharaja Sayajirao University of Baroda, Vadodara 390001, India. [2]Department of Electrical and Computer Engineering, University of Connecticut, Storrs, CT, 06269-4157, USA. [3]Institute of Interdisciplinary Studies, The Maharaja Sayajirao University of Baroda, Vadodara, India

Author contributions
A.A. and B.J. proposed the idea and initiated the project. S.U., S.J. and N.P. designed and implemented table top device. S.U., S.J., C.P. and M.J. perfomed the experiments, collected the data and helped in the analysis. A.A., V.C. and T.O. performed the numerical reconstrutions and analysis of experimental results. SU., A.A., B.J., V.C. and T.O. co-wrote the paper. A.A. and B.J. supervised the project. The manuscript was discussed and corrected by all authors.

Data availability
The data that support the plots within this paper and other findings of this study are available from the corresponding author upon reasonable request.

Code availability
All codes that support this study are available from the corresponding author upon reasonable request.

Conflict of interest
The authors declare that they have no conflict of interest.

Supplementary information is available for this paper at https://doi.org/10.37188/lam.2022.037.

Received: 14 September 2021 Revised: 11 May 2022 Accepted: 23 May 2022
Accepted article preview online: 24 May 2022
Published online: 04 August 2022

References

1. Ferraro, P., Wax, A. & Zalevsky, Z. Coherent Light Microscopy: Imaging and Quantitative Phase Analysis. (Berlin: Springer, 2011).
2. Schnars, U. & Jueptner, W. Digital Holography: Digital Hologram Recording, Numerical Reconstruction, and Related Techniques. (Berlin: Springer, 2005).
3. Goodman, J. W. Introduction to Fourier Optics. (New York: McGraw-Hill, 1996).
4. Cuche, E., Bevilacqua, F. & Depeursinge, C. Digital holography for quantitative phase-contrast imaging. *Optics Letters* **24**, 291-293 (1999).
5. Marquet, P. et al. Digital holographic microscopy: a noninvasive contrast imaging technique allowing quantitative visualization of living cells with subwavelength axial accuracy. *Optics Letters* **30**, 468-470 (2005).
6. Charrière, F. et al. Cell refractive index tomography by digital holographic microscopy. *Optics Letters* **31**, 178-180 (2006).
7. Dubois, F. et al. Digital holographic microscopy for the three-dimensional dynamic analysis of in vitro cancer cell migration. *Journal of Biomedical Optics* **11**, 054032 (2006).
8. Kemper, B. & Von Bally, G. Digital holographic microscopy for live cell

applications and technical inspection. *Applied Optics* **47**, A52-A61 (2008).

9. Rappaz, B. et al. Comparative study of human erythrocytes by digital holographic microscopy, confocal microscopy, and impedance volume analyzer. *Cytometry Part A* **73A**, 895-903 (2008).
10. Faridian, A. et al. Nanoscale imaging using deep ultraviolet digital holographic microscopy. *Optics Express* **18**, 14159-14164 (2010).
11. Anand, A., Chhaniwal, V. K. & Javidi, B. Real-time digital holographic microscopy for phase contrast 3D imaging of dynamic phenomena. *Journal of Display Technology* **6**, 500-505 (2010).
12. Kemper, B. et al. Label-free quantitative cell division monitoring of endothelial cells by digital holographic microscopy. *Journal of Biomedical Optics* **15**, 036009 (2010).
13. Shaked, N. T. et al. Whole-cell-analysis of live cardiomyocytes using wide-field interferometric phase microscopy. *Biomedical Optics Express* **1**, 706-719 (2010).
14. Anand, A., Chhaniwal, V. K. & Javidi, B. Imaging embryonic stem cell dynamics using quantitative 3-D digital holographic microscopy. *IEEE Photonics Journal* **3**, 546-554 (2011).
15. Shin, D. et al. Optofluidic system for three-dimensional sensing and identification of micro-organisms with digital holographic microscopy. *Optics Letters* **35**, 4066-4068 (2010).
16. Anand, A. et al. Automatic identification of malaria-infected RBC with digital holographic microscopy using correlation algorithms. *IEEE Photonics Journal* **4**, 1456-1464 (2012).
17. Yi, F. L., Moon, I. & Javidi, B. Cell morphology-based classification of red blood cells using holographic imaging informatics. *Biomedical Optics Express* **7**, 2385-2399 (2016).
18. Kemper, B. et al. Simplified approach for quantitative digital holographic phase contrast imaging of living cells. *Journal of Biomedical Optics* **16**, 026014 (2011).
19. Singh, A. S. G. et al. Lateral shearing digital holographic imaging of small biological specimens. *Optics Express* **20**, 23617-23622 (2012).
20. Chhaniwal, V. et al. Quantitative phase-contrast imaging with compact digital holographic microscope employing Lloyd's mirror. *Optics Letters* **37**, 5127-5129 (2012).
21. Anand, A., Moon, I. & Javidi, B. Automated disease identification with 3-D optical imaging: a medical diagnostic tool. *Proceedings of the IEEE* **105**, 924-946 (2017).
22. Javidi, B. et al. Sickle cell disease diagnosis based on spatio-temporal cell dynamics analysis using 3D printed shearing digital holographic microscopy. *Optics Express* **26**, 13614-13627 (2018).
23. Anand, A., Chhaniwal, V. & Javidi, B. Tutorial: common path self-referencing digital holographic microscopy. *APL Photonics* **3**, 071101 (2018).
24. O' Connor, T. et al. Deep learning-based cell identification and disease diagnosis using spatio-temporal cellular dynamics in compact digital holographic microscopy. *Biomedical Optics Express* **11**, 4491-4508 (2020).
25. Ebrahimi, S. et al. Digital holographic microscopy with coupled optical fiber trap for cell measurement and manipulation. *Optics Letters* **39**, 2916-2919 (2014).
26. Ashkin, A. Acceleration and trapping of particles by radiation pressure. *Physical Review Letters* **24**, 156-159 (1970).
27. Ashkin, A. et al. Observation of a single-beam gradient force optical trap for dielectric particles. *Optics Letters* **11**, 288-290 (1986).
28. Uchida, N. et al. Direct isolation of human central nervous system stem cells. *Proceedings of the National Academy of Sciences of the United States of America* **97**, 14720-14725 (2000).
29. Kekarainen, T. et al. Optimization of immunomagnetic separation for cord blood-derived hematopoietic stem cells. *BMC Cell Biology* **7**, 30 (2006).
30. Ashkin, A. & Dziedzic, J. M. Optical trapping and manipulation of viruses and bacteria. *Science* **235**, 1517-1520 (1987).
31. Bustamante, C., Bryant, Z. & Smith, S. B. Ten years of tension: single-molecule DNA mechanics. *Nature* **421**, 423-427 (2003).
32. Clement-Sengewald, A. et al. Fertilization of bovine oocytes induced solely with combined laser microbeam and optical tweezers. *Journal of Assisted Reproduction and Genetics* **13**, 259-265 (1996).
33. Applegate, R. W. Jr. et al. Microfluidic sorting system based on optical waveguide integration and diode laser bar trapping. *Lab on a Chip* **6**, 422-426 (2006).
34. Cho, S. H. et al. Human mammalian cell sorting using a highly integrated micro-fabricated fluorescence-activated cell sorter (μFACS). *Lab on a Chip* **10**, 1567-1573 (2010).
35. Kasukurti, A. et al. Single-cell isolation using a DVD optical pickup. *Optics Express* **19**, 10377-10386 (2011).
36. Pellizzaro, A. et al. Direct laser trapping for measuring the behavior of transfused erythrocytes in a sickle cell anemia patient. *Biomedical Optics Express* **3**, 2190-2199 (2012).
37. Paul, A., Padmapriya, P. & Natarajan, V. Diagnosis of malarial infection using change in properties of optically trapped red blood cells. *Biomedical Journal* **40**, 101-105 (2017).
38. Roy, B. et al. Using Brownian motion to measure shape asymmetry in mesoscopic matter using optical tweezers. *Soft Matter* **12**, 5077-5080 (2016).
39. Gautam, R. et al. Optical force-induced nonlinearity and self-guiding of light in human red blood cell suspensions. *Light:Science & Applications* **8**, 31 (2019).
40. Liang, Y. et al. Manipulation and assessment of human red blood cells with tunable "Tug-of-War" optical tweezers. *Physical Review Applied* **12**, 064060 (2019).
41. Avsievich, T. et al. The advancement of blood cell research by optical tweezers. *Reviews in Physics* **5**, 100043 (2020).
42. Zhu, R. X. et al. Optical tweezers in studies of red blood cells. *Cells* **9**, 545 (2020).
43. Memmolo, P. et al. 3D morphometry of red blood cells by digital holography. *Cytometry Part A* **85**, 1030-1036 (2014).
44. Cardenas, N. & Mohanty, S. K. Optical tweezers assisted quantitative phase imaging led to thickness mapping of red blood cells. *Applied Physics Letters* **103**, 013703 (2013).
45. Bernecker, C. et al. Biomechanics of Ex vivo-generated red blood cells investigated by optical tweezers and digital holographic microscopy. *Cells* **10**, 552 (2021).
46. Kwon, H. et al. Single-shot quantitative phase gradient microscopy using a system of multifunctional metasurfaces. *Nature Photonics* **14**, 109-114 (2020).
47. Engay, E. et al. Polarization-dependent all-dielectric metasurface for single-shot quantitative phase imaging. *Nano Letters* **21**, 3820-3826 (2021).
48. Patel, N. et al. Compact and low-cost instrument for digital holographic microscopy of immobilized micro-particles. *Optics and Lasers in Engineering* **137**, 106397 (2021).
49. Abe, S. Support Vector Machines for Pattern Classification. (London: Springer, 2010).
50. Berg-Sørensen, K. & Flyvbjerg, H. Power spectrum analysis for optical tweezers. *Review of Scientific Instruments* **75**, 594-612 (2004).
51. Fischer, M. et al. Active-passive calibration of optical tweezers in viscoelastic media. *Review of Scientific Instruments* **81**, 015103 (2010).
52. Dulbecco, R. & Vogt, M. Plaque formation and isolation of pure lines with poliomyelitis viruses. *Journal of Experimental Medicine* **99**, 167-182 (1954).
53. Dasgupta, R. et al. Optical orientation and rotation of trapped red blood cells with Laguerre-Gaussian mode. *Optics Express* **19**, 7680-7688 (2011).

Official journal of the JHL 2689-9620
www.light-am.com

Review | Open Access

Compression strategies for digital holograms in biomedical and multimedia applications

Peter Schelkens[1, 2*], Ayyoub Ahar[1, 2], Antonin Gilles[3], Raees Kizhakkumkara Muhamad[1, 2], Thomas J. Naughton[4], Cristian Perra[5], Antonio Pinheiro[6, 7], Piotr Stępień[8] and Malgorzata Kujawińska[8]

Abstract

While 60 years of successful application of holography is celebrated in this special issue, efficient representation and compression of holographic data has received relatively little attention in research. Notwithstanding this observation, and particularly due to the digitization that is also penetrating the holographic domain, interest is growing on how to efficiently compress holographic data such that interactive exchange of content, as well as digital storage can be facilitated proficiently. This is a particular challenge, not only because of its interferometric nature and the various representation formats, but also the often extremely large data volumes involved in pathological, tomographic, or high-end visualization applications. In this paper, we provide an overview of the state of the art in compression techniques and corresponding quality metrics for various practical applications in digital holography. We also consider the future by analyzing the emerging trends for addressing the key challenges in this domain.

Keywords: Holography, Compression, Image quality assessment, Standardization

Introduction

In many digital signal applications, compression of the digital data is often required because of the scarce storage available in comparison to the volume of data that needs to be stored, or because the utilized transmission channel has insufficient bandwidth to allow for timely delivery of the desired content. Hence, enormous efforts have been invested by the research and industrial communities in standardization bodies such as the International Standardization Organization (ISO), the International Electrotechnical Committee (IEC) and the International Telecommunication Union (ITU), more recently international industrial consortia such as the Alliance for Open Media (AOM) or individual organization solutions. Leading international committees are the Joint Photographic Experts Group (JPEG) and Moving Picture Experts Group (MPEG) for still image coding and video & audio coding, respectively, both representing working groups of ISO/IEC which are regularly producing joint standards with ITU-T Subgroup 16. The importance of these initiatives, representing huge efforts, cannot be underestimated. The delivered standard frameworks enable interoperability between devices and applications and as such also shape the bases for a whole ecosystem that gives rise to a whole range of (novel) media products & services that are supporting worldwide communication and media distribution platforms.

Important to mention here as well is that the impact of the above standards is not limited to the 'multimedia' application domain, but also impacts biomedical

Correspondence: Peter Schelkens (peter.schelkens@vub.be)
[1]Vrije Universiteit Brussel, ETRO Dept., B-1050 Brussels, Belgium
[2]imec, Digital and User-Centric Solutions (DUCS), B-3001 Leuven, Belgium
Full list of author information is available at the end of the article.

ecosystems. These standards are actually 'horizontal standards', meaning that they are considered as components that can be integrated in several vertical application chains. Digital Imaging and Communications in Medicine (DICOM) has adopted for example several JPEG and MPEG standards that are now being used in medical imaging (CT, US, MRI, etc.), pathological imaging, and so forth.

A bold statement though, but if one desires holography to flourish, one of the key enabling factors — besides naturally the availability of underlying generation, processing, and rendering technologies — will be whether one succeeds in defining efficient ways to exchange holographic content as it is the case for more classical multimedia content. In fact, large data volumes is one of the biggest challenges faced by digital holography, where efficient compression solutions are key to widespread adoption of these techniques.

A single phase measurement in interefometric applications like digital holographic microscopy (DHM) and holographic tomography (HT) requires one (for the Fourier Transform Method of phase retrieval) or a few holograms (for the Phase Shifting Method) captured for example by a popular Sony IMX250 camera (4.8 MB). However, the storage requirements increase significantly in the case of time-lapse, long-term experiments (1 hour, 1 frame per sec – ap. 173 GB), and large fields of view for multi wells cell culture or pathology slides analysis (15 mm × 15 mm – ap. 100 GB/slide). A typical single tomographic measurement requires capturing between 45-180 holograms per measured volume (0.2–0.8 GB). However, nowadays systems utilizing solutions that better allow to understand the microbiological world by taking into account multi-spatial-temporal scales[1]. Such an approach requires capturing tomographic data from large fields of view (stitching multiple volumes) and for thick samples (digital phase histopathology) and/or for long term monitoring of cell cultures or organoids. This brings us to severe storage requirements reaching several TB per measurement sample or time lapse measurements.

The quality of holographic displays powering visualization applications like XR/VR/AR is measured by the space-bandwidth product (SBP)[2,3], which constraints both the size of the display and its field of view (FoV) aperture. Indeed, to achieve their diffractive behaviour and provide a large viewing angle, displays need very small pixels, with pixel pitches in the order of the wavelength of visible light. For instance, a monochromatic green hologram of size 10 cm × 10 cm with a wavelength of 535 nm and a FoV of around 120° would require a pixel pitch of 0.3 μm and a resolution of more than ($300\,K \times 300\,K$), which implies an uncompressed single frame would be larger than 1 TB.

And this represents actually the challenge; few investigations — relatively seen — have been spent on efficient compression and representation of holographic content. Though we can build upon the expertise related to more classical media content, holographic content represents a set of particular challenges, which cannot be easily by-passed; to name a few: its interferometric nature leading to completely different data behaviour and statistics not matching those expected by classical coding engines. Natural imagery is predominantly low pass, where the power spectra tends to be $\propto 1/f^2$, where f is the spatial frequency[4]. In the case of holograms, each component in the frequency domain represents the contribution of plane waves travelling from some corresponding direction incident at the hologram plane. In particular, they contain high-frequency coefficients which play a major role in 3D visualization and cannot be discarded as in typical image coding techniques. The other difference is related to signal representation, as holograms can be described in myriads of forms, e.g. complex-valued or amplitude-phase based representations. Luckily, this demand has been picked-up by the international research community and moreover, the JPEG committee has identified this need and did recently launch a Call for Proposals for Holographic Content Coding[5].

Another related challenge that requires some particular attention is how to assess the quality of holographic content. For regular image content this is rather straightforward; what an observer sees is immediately interpretable by our brain and a photographic representation of the scene. Hence, it is relatively easy for humans to adhere a quality score to the observed content. Measuring whether this appreciation is scientifically correct is more difficult, but sound subjective quality assessment procedures have been designed and validated. Quality scores that can be validated on the 'ground truth' produced by the earlier mentioned subjective quality assessment procedures can also be obtained numerically, by deploying mathematical measurement procedures accounting for the signal's fidelity, typical image features and potentially also aspects of the human visual system. These subjective and objective quality assessment procedures are evidently not perfect and operate within certain probability bounds, but nonetheless provide sufficient bases for codec evaluation procedures and visual quality measurement products.

Unfortunately, holograms are wave-based projections of a 3D scene and hence not immediately interpretable by human observers. Hence, even measuring the quality of a hologram directly on the hologram itself by solely a numerical quality measure is not necessarily telling the full story of the signal's quality. The numerical or optical reconstructions of the hologram might for example depict a completely different quality depending on the viewing angle or viewport at which this content is being consumed. Also here significant steps forward have been made in recent years, at least from a methodological perspective, on which we will elaborate too.

In the next two sections of the paper, we give an overview on two distinct application domains of holography - biomedical imaging and autostereoscopic visualization. Thereafter, the state of the art on compression technology with respect to holographic content is discussed. In the subsequent section, quality assessment procedures are addressed, covering both subjective and objective procedures, after which we dive into a more elaborate discussion on the challenges faced and what to expect or ambition as future endeavours. Finally, this paper is winding up with a conclusion section.

Biomedical imaging

In this section, we discuss a number of application domains that are in need of efficient hologram representation and compression technologies, starting with biomedical imaging that has experienced already a vast deployment of holographic imaging techniques.

Quantitative phase imaging (QPI)[6], is a label-free approach to imaging cells and tissues, that combines qualities found in microscopy, holography and light scattering techniques: nanoscale sensitivity to morphology and dynamics, 2D, 3D and 4D non-destructive imaging of transparent structures, and quantitative signals based on intrinsic (phase) contrast.

DHM and HT are the most popular and powerful techniques applied in QPI[7-9]. It should be noted that microscopy, which deals with small objects, is the best suited application for digital holography, as the space bandwidth product (SBP) of these objects can easily be matched to the SBP of available detectors. All these facts are fully confirmed by the success of many commercial DHM (e.g. Lyncée Tec and Ovizio) and HT systems (Nanolive[10] and Tomocube[11]). Also, it should be noted that the big players on the optical microscopy market (e.g. Carl Zeiss AG, FEI Company, Hitachi High Technologies Corporation, JEOL Ltd, Leica Microsystems, Nikon Corporation, Philips Healthcare, Olympus Corporation) are moving towards digital and quantitative microscopy in which digital holography is a natural future choice.

DHM can efficiently overcome the most severe limitations of microscopy, namely the very limited depth of field as well as enabling the imaging of phase (transmissive) micro-objects. Holographic microscopy records the light wavefront originating from an object instead of the projected image of the object recorded in common microscopy. The viewable image and quantitative (phase) representation of a micro object is created using a numerical reconstruction algorithm[12]. The numerical reconstruction of the digitally recorded holograms makes it possible to obtain the whole complex diffracted wavefront, i.e. both its amplitude and phase, thus offering the possibility of propagating it further. Such numerical propagation provides many advantages including accurate measurements of an integrated phase of an object (dry mass of a cell), performing autofocusing, extending depth of focus and correcting any kind of aberrations. Examples of life science applications of DHM include cell dry mass extraction and monitoring[13], cell culture inspection[14], automated cell counting, recognition, and classification for diagnostic purposes with the use of artificial intelligence[15], digital phase histopathology[16], assessing cellular responses induced by new drugs[17], and performing label-free high content screening[18] to name just a few.

Due to the integrated character of the data delivered by digital holography (each phase value laterally represents the accumulated phase along all points along the optical axis), QPI realized through DHM has limited capabilities for providing full 3D structural information about cells or tissues. To overcome this limitation, tomographic versions of QPI are applied[6]. Among these methods, optical diffraction tomography based on holographic projections[9] demonstrates a wealth of possibilities for 3D imaging in biomedical applications[9,19], not possible with conventional forms of tomography such as computed tomography (CT)[20]. CT scan techniques make use of X-rays to provide an accurate representation of a 3D object. However, since CT is based on ray optics, it assumes non scattering objects where object features are larger than the imaging wavelength. On the other hand, holographic tomography (HT) is based on the diffraction of light and can be used for capturing very fine details and enabling radiation-free biomedical imaging. Furthermore, in HT the reconstruction obtained from the measurements of multiple 2D holograms (projections) is the 3D refractive index (RI) distribution of the weakly scattering sample under study.

The optical setup for holographic tomography consists of a DHM equipped with a unit controlling illumination beams impinging onto a sample by means of one of two alternative schemes: an object rotation (Full Projection Angle HT, FAHT) or illumination beam rotation (Limited Projection Angle HT, LAHT)[9]. In LAHT, the specimen is stationary, while holographic projections are acquired for different object illumination directions. This makes it a perfect tool for measuring single biological cells, cell cultures and tissues directly at Petri dishes or microscopic slides. After all data is captured, specialized tomographic reconstruction algorithms are employed to retrieve the 3D refractive index distribution[21]. Because RI values depend on the number of intracellular biomolecules, LAHT allows label-free, high resolution quantitative 3D morphological mapping of biological specimens including fixed and live cells[19,22], tissues[23] and even organoids and small organisms[24].

The most popular optical configuration of research and commercial digital holographic microscopes applied for transparent objects is a Mach-Zehnder interferometer with an additional microscope imaging setup in the object arm, to image the object on the image sensor with high magnification. The output hologram is captured as the result of interference between reference and object beams.

Depending on the detector location in an object diffraction field, the recorded digital hologram is an image plane (IPH) or Fresnel hologram (FH). An IPH's capture yields several benefits: a) avoiding the need for numerical propagation, therefore simplifying QPI processing and allowing for easy switching between modalities that do not allow for numerical refocusing, such as white light or fluorescent imaging; b) compact design by shortening the object arm[25]; c) concentrating optimally the bandwidth of individual diffraction orders[26]. Also, in the case of IPHs, the amplitude of an object can be captured solely by recording the object beam intensity in the absence of the reference beam. On the other hand, if the object plane cannot be well identified or optical refocusing is required, the registration of a Fresnel hologram is necessary and the complex amplitude of an object is reconstructed by numerical propagation of the captured wavefront to the object plane[12].

IPHs and FHs can be captured in two alternative ways depending on whether or not there is an angular difference between the propagation directions of the object and reference wavefronts: as on-axis or off-axis holograms. The spatial bandwidth product (SBP) of these two types of holograms vary significantly and so impose different requirements on the space-bandwidth of one's detector. In the first case, the detector and object SBPs are comparable, while for off-axis holograms a significant carrier frequency is required, which increases the necessary detector pixel count at least four times. On the other hand, an advantage of off-axis holograms is that the three diffraction orders can be separated in Fourier space thus easily allowing reconstruction of the object wavefront. The off-axis tilt introduces a carrier frequency that allows one to spatially separate +1, −1 and 0 diffraction orders in the Fourier domain. In case of full spectral separation, a fast and robust Fourier filtering method[27,28] is applied to a single hologram.

For on-axis holograms, or holograms with insufficient carrier frequency, the one wanted diffraction order cannot be separated in the Fourier domain. This is acceptable for applications that only require intensity reconstructions, such as for the three-dimensional localization of particles[29]. To obtain quantitative phase or full-complex reconstructions from an on-axis configuration, the concept of phase shifting digital holography (PSDH) can be used[30]. PSDH requires capturing at least three phase shifted holograms (although more provide robustness) of the same object wavefront. This redundancy of information enables numerical elimination of the twin and the zero-order terms and provides an accurate object phase reconstruction. Unfortunately, PSDH cannot be applied (or is difficult to apply) to measure dynamically changing objects or in the case of unstable environments or long time-lapse sessions, as it requires capturing multiple holograms for a single measurement. These are the reasons why off-axis holograms and the Fourier transform method (FTM) of phase retrieval are applied in most of the research and commercial DHM systems[31].

Regardless of the phase retrieval method, the output data and results from DHMs are delivered in two main formats: a) holograms (interference patterns) being real-valued greyscale images representing high spatial frequencies (for off-axis holograms, dense fringes) and relatively low spatial frequencies (for on-axis and quasi on-axis holograms); and b) complex amplitudes (or real-imaginary components) of the object field. The phases are calculated by FTM or PSDH, while the amplitude is additionally captured (image with no reference wave). In FTM, the bandwidth of the first order term that encodes the received wavefield is limited to a well defined region in the Fourier domain[26]. These complex amplitudes represent objects directly (in the case of image plane holograms) or should be numerically propagated to an object plane in the case of Fresnel holograms.

For holographic tomography it is obligatory to deliver a set of complex amplitudes acquired within a range of viewing angles, which are later processed with one of the tomographic solvers[1,21]. HT setups are similar to their DHM counterparts, but utilize multiple consecutive oblique illumination beam incidences instead of the single normal illumination beam incidence used in DHM. The most common optical architecture for HT is a Mach-Zehnder interferometer with an illumination beam scanning module (LATH configuration) and in-plane off-axis holographic projections capture. Such configurations have already been used in commercial solutions such as NanoLive[10] and TomoCube[11] products. The basic output data from HTs are multiple holograms per a single measurement followed by retrieval of complex amplitudes (CA) of holographic projections gathered into CA sinogram. However, the final HT output is a 3D distribution of refractive indices reconstructed in the measurement volume. The most basic tomographic reconstruction algorithm taking diffraction into account, namely direct inversion, is a reconstruction method that utilizes the Fourier Diffraction Theorem. It assumes either the Born or Rytov approximations of single scattering. The latter has no restrictions on sample thickness and is thus more suitable for analysis of biological specimens.

DI allows for the reconstruction of a 3D scattering potential from multiple complex-valued projections. However, when this method is applied to data acquired with a LAHT system, highly distorted results are obtained[21]. Thus, numerous iterative algorithms with constraints applied in each iteration are reported in the literature[1]. The most often presently used approach is the Gerchberg-Papoulis (GP) method with total variation minimization or object support constraints[21]. The tomographic reconstruction algorithms for LAHT are still not fully mature and they are the object of extensive research by many groups, including the application of artificial intelligence methods[1]. Therefore, a compression type that preserves the original complex-valued object wavefield, permitting future changes in the processing pipeline used by reconstruction algorithms, is most suitable (Table 1).

Visualization

Another important use case for Holography is the field of 3D visualization. Indeed, by embedding all the information describing light waves scattered by a scene in 3D space, the holographic signal contains all the data needed to provide the most convincing depth illusion to the human eye. When looking at a hologram properly displayed and illuminated, viewers perceive the recorded 3D scene as if it were physically present, with proper shade, texture and geometry. More specifically, holography reproduces all the depth perception mechanisms, also called Human Visual System (HVS) depth cues, occurring in natural vision, and solves the Vergence-Accommodation Conflict of stereoscopic systems[32].

To display a hologram, the typical approach is to use a beam shaping device called Spatial Light Modulator (SLM). These are electrically or optically addressed panels which reproduce the recorded object wave by modulating the amplitude or the phase of the illuminating light beam. Several SLM technologies can be used for holographic displays, including Liquid Crystal on Silicon (LCoS)[33], Digital Micro-mirror Displays (DMD)[34,35] and Acousto-Optics Modulators (AOM)[36].

Current commercially available SLMs have resolutions typically limited to 1920×1080 or (3840×2160) and pixel pitches of about $4\mu m$; hence commercial holographic displays using SLM technology are now available on small devices such as Head-Mounted Displays[37,38] or automotive windshields. To develop larger holographic TV screens, it is possible to spatially tile multiple SLMs, increasing the size and/or FoV of the displayed hologram[39]. However, building a system of seamless tiled SLMs is very challenging, and the resolution of such systems is still insufficient for very large holograms.

To plot large holograms with a pixel size in the order of

Table 1 Overview of the most common hologram recording setups, their preferred representations, and typical information extracted in biomedical applications

System	Recording set-up	Representation	Information output
DHM	Single DH measurement	Hologram(s) (intensity images)	2D map and its changes over
	PSDH (multiple DHs)	with or without carrier frequency	time at multiple depths
	Multiplexed	Complex amplitude	Phase/dry mass
	Time-lapse	(both real and imaginary terms)	Movement trajectories in three
	Large-sample (FoV)		dimensions
	Time-lapse + large FoV		Visualization
HT	Single HT measurement	Hologram (intensity image)	2D RI (at best focal plane)
	(projections differing through scanning)	with or without carrier	3D RI
	illumination or wavelength	frequency	Changes in time of
	Thick sample (multiple depths)	Complex amplitude sinogram	2D or 3D RI
	Time-lapse + large volume	(or real/imaginary term)	Visualization

the wavelength of visible light, it is possible to print them onto a photosensitive plate using a wavefront printer[40], whose general setup is shown in Fig. 1. These printers typically comprise a laser, controlled by an electronic shutter, an SLM, a demagnifying lens system, and a photosensitive plate mounted on an X-Y moving stage. The large-resolution hologram is segmented into several rectangular tiles, which are displayed one after the other on the SLM. For each displayed tile, the control system opens the shutter to illuminate the SLM by the laser beam. The reconstructed object wave is then demagnified by the lens system and exposed to the photosensitive plate, which is also illuminated by the laser beam from the other side. After exposure, the shutter is closed and the plate is translated by the X-Y stage to the next exposure position, while the next segment of the hologram is displayed on the SLM. This process is repeated until the entire hologram is exposed. In addition to enable the display of large holograms with a wide FoV, wavefront printers can also print Holographic Optical Elements (HOE) combining several custom optical functions to enable extreme miniaturization of complex optical systems[41].

As explained previously, data transmission for holography is a major issue, requiring novel compression algorithms able to significantly reduce the memory and bandwidth consumption of holographic signals. From this observation, one could argue that it would be more efficient to compress and transmit the data representative of the scene – such as Multiview data or 3D meshes – and convert it to a hologram on the display side using Computer-Generated Hologram (CGH) synthesis methods[42-44]. Nevertheless, in the case of biomedical holograms, the 3D information of the scene is not available. Moreover, in this transmission scheme, the client terminal would have to convert the compressed 3D scene data to hologram sequences in real-time, which is still very challenging using state-of-the-art calculation methods, especially for holograms containing billions of pixels. This is particularly challenging in broadcast scenarios, i.e. one-to-many transmission where one attempts to globally optimize the system, and as such having an N times repeat CGH calculation is highly undesirable. This is of course all on the condition that decoders depict a low complexity — which is usually the case — and that the compression bitrate achieved is meeting the channel bandwidth capacity. Holographic data coding is therefore required for both biomedical and visualization scenarios. Table 2 synthesizes the most common hologram visualization methods.

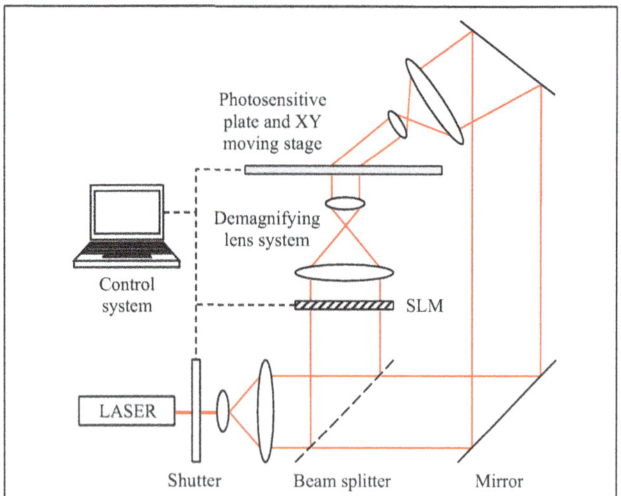

Fig. 1 Wavefront printer setup.

Compressing digital holograms

In this section, we provide an overview of the state-of-the-art in holographic compression solutions. We will start with an evaluation of existing coding standards, which are appealing because of their broad support, but that unfortunately do not cope well with the particular signal properties of holograms. Thereafter, we discuss solutions that have been proposed to handle the signal characteristics of holograms, and we end with a discussion addressing compression solutions in the context of metrological applications, which are important in the context of biomedical imaging.

Deployability of existing coding standards

In this section, we give a brief overview of the various image and video coding standards that are in popular usage for both lossless and lossy compression. Lossless compression bitrates are often significantly larger than for lossy compression and is usually only relevant for critical applications in the medical and metrological domain. For addressing these requirements, ISO launched in 1999 JPEG LS, an image compression standard designed for lossless/near lossless compression of continuous tone

Table 2 Overview of most common hologram visualization methods, their preferred representations and resolutions.

Use case	Visualization device	Representation	Resolutions
Display	Spatial Light Modulator (SLM)	Complex amplitude or real/imaginery term	2×10^6 to 3×10^7
Printing	Photosensitive plate	Complex amplitude or bilevel phase	Over 10^{10}

images captured in biomedical applications. Even when compared to the current state of the art compression standards, JPEG LS usually obtains better compression performance for continuous tone imagery[45] while maintaining a much lower computational burden. Here, the LOCO-1 compression algorithm used is a 'low complexity projection'[46] of the universal context modelling algorithms that deliver the best compression ratios for these use cases. At its core LOCO-1 makes a gradient based prediction, where the prediction residuals are entropy coded using a low-complexity Golomb coder.

For digital holographic microscopy measurements captured with the off-axis setup, it was shown in[31] that JPEG LS obtained the best intra (single frame) lossless compression performance. JPEG 2000, on the other hand, uses a fully reversible wavelet transform with integer coefficients (CDF 5/3) for lossless compression. For lossless binary image compression, the standards JBIG[47] and JBIG2[48] are typically used. Both these standards make use of a context modelling scheme designed for bilevel imagery in conjunction with an adaptive arithmetic coder. More recently, the JPEG XL[49] standard was introduced as a royalty free replacement for legacy JPEG, and is designed to operate on a variety of photographic and synthetic imagery. JPEG XL supports lossless compression of both bi-level and multilevel imagery. Moreover, JPEG XL makes use of the ANS entropy coder that can obtain compression ratios similar to arithmetic coding while having a computational complexity that is comparable to Huffman coding. However, when compared to JPEG XL, JBIG2 delivers a similar coding performance than JPEG XL at an overall lower computation complexity.

Though introduced in 1992, the most popular standard for lossy compression is still JPEG[50]. It was one of the first image compression standards to perform an additional frequency transform operation, before applying quantization and entropy coding. In JPEG, a blockwise DCT (discrete cosine transform) of size 8×8 is applied on the image, where the transform coefficients obtained are quantized using an application specific quantization table. For entropy coding, JPEG supports both Huffman coding and arithmetic coding. Due to the DCT transform used, JPEG was highly effective for compressing natural imagery and evolved to become one of the most widely used image standards for lossy compression. However, for lossless compression, the DCT transform is not performed as it would have lead to more stringent restrictions on the encoder/decoder hardware to ensure lossless transformations. Instead, a simple prediction based lossless compression mode was defined.

Although for lossy compression, JPEG is still a popular standard due to its universal adoption, more recent still image coding standards[51–53] supersede it in terms of compression performance. For example, JPEG 2000[51] uses a discrete wavelet transform (CDF 9/7) with a Mallat style decomposition followed by entropy coding using the EBCOT (Embedded Block Coding with Optimized Truncation) engine, to obtain bitstreams that are not just smaller in size but also comprising of flexible layers supporting different types of progressive decoding modes. Similarly, JPEG XL also supports progressive decoding while having a perceptually optimized reference encoder that gracefully degrades the image quality for low bitrates. Well optimized implementations for JPEG XL are about as fast as legacy JPEG, while offering comparable visual quality to more modern, computationally demanding standards.

JPEG is actually a nice illustration of an 'horizontal' standard that has penetrated deep into so many application ecosystems, that is — while it is being outperformed by many contemporary standards — not being replaced by them, since it would break to many interoperability chains. Its low complexity and intellectual property right uncertainties for newer technologies add in addition to its still high popularity.

However, researchers always face the difficulty of using current lossy image compression standards applied to digital holograms. When the compressing model is applied to the object plane real and imaginary parts, 3D information might be removed, subsequently not allowing the reconstruction of the hologram at different reconstruction distances or different viewing angles with acceptable quality.

Moreover, compressing the hologram plane complex representation faces the problem that the frequencies do not correspond to the frequencies of the reconstructed hologram. As the typical compression technologies reduce the influence of the high frequencies by applying quantization methodologies, the lossy compression in the hologram plane tends to create aliasing. This is particularly evident when JPEG 2000 is applied to the real and imaginary parts of the hologram plane information. The wavelet quantization associated with the subsampling mechanism leads to a very visible aliasing for reconstructed holograms[54].

While we have primarily discussed still image (intra) coding standards till now, the importance of video (inter) coding technologies cannot be understated. At its core, video standards splits the frame to be compressed into macroblocks. Macroblocks can be predicted from other such blocks residing within the same frame or from neighbouring frames, thereby exploiting both spatial redundancy as well as temporal redundancy respectively. The widely used H.264/AVC[55] standard uses blocks up to size 16 × 16, while its successor H.265/HEVC[56] achieves roughly up to 50% better compression performance and supports block sizes of up to 64 × 64. The latest iteration from the JVET group, H.266/VVC, aims to obtain 30–50% better compression rates than HEVC. To achieve this level of compression, the encoding complexity is expected to be several times higher than HEVC. Meanwhile, the video standard VP9[57] from the AOM group is also getting popular due to its royalty free nature.

Unfortunately, all those video codecs suffer from the fact that — as indicated in the previous subsection — the deployed motion estimation and compensation strategies are not compliant with the interferometric nature of holographic content.

Compression solutions targeted to holography

Digital holograms are pixelated 2D patterns; it is thus natural that state-of-the-art holographic data compression methods are largely inspired by conventional image and video coding techniques. Early attempts for holographic data compression only considered the quantization and entropy coding steps. In[58], the authors applied several off-the-shelf lossless and lossy data compression techniques to optically acquired holograms. While lossless coding led to low compression ratios, quantization proved to be a rather efficient technique, achieving compression ratios up to 16. Based on these conclusions, several authors investigated scalar and vector quantization schemes to compress holographic data, such as uniform quantization[59], non-uniform histogram quantization[60], k-means, k-medians and logarithmic quantization[61], and using Lloyd-Max and Linde-Buzo-Gray algorithms[62].

Whereas previously mentioned methods operated on the raw holographic data, it is also possible to use transform coding to improve the efficiency of the quantization and coding steps. Nevertheless, since holograms differ substantially from conventional 2D images, novel transforms are needed to fit holographic data features. In[63], the authors proposed a novel wavelet-bandelet transform to determine the direction of holographic fringes, achieving higher compression ratios than Discrete Cosine Transform (DCT) and Haar wavelets. In[64], the JPEG 2000 codec was modified with two extensions: a fully arbitrary wavelet decomposition and a Directional Adaptive Discrete Wavelet Transform (DA-DWT). Since a significant amount of information is contained in the high frequencies, the fully arbitrary wavelet decomposition enables the high-pass sub-bands to be decomposed with more than three levels, allowing for full packet decomposition. On the other hand, the DA-DWT uses oriented wavelets to better align with holographic fringes. In[65], the authors proposed to extend HEVC with a novel mode-dependent directional transform trained with holographic data to take into account the directionality of the input fringes. In[66] and[67], two separable and non-separable vector lifting schemes were proposed to construct wavelets specifically designed for holographic data. Finally, in[68], the suitability of wave atoms for hologram compression was demonstrated, outperforming HEVC for several test cases thanks to their near-optimal space-frequency localization.

While these methods take into account specificities of holographic signals, they do not exploit their 3D visualization properties. Understanding the correlation between the 3D scene content and holographic data could enable better compression ratios and interesting functionalities such as edition, viewpoint and quality scalability, as well as motion estimation and compensation. Aiming at such functionalities, a few works tried to extract relevant 3D visualization features from hologram pixels. In[69], the author showed that the convolution kernel of Fresnel diffraction forms a family of wavelets whose scale parameter represents the propagation distance. This observation was used in[70] to construct wavelet-like bases called Fresnelets by Fresnel transforming B-spline wavelets. They demonstrated that the Fresnelet transform of a hologram corresponds to its numerical reconstruction at user-defined resolutions. This feature was exploited in[71] to compress digital holograms using a Set Partitioning In Hierarchical Trees (SPIHT) coding algorithm on the Fresnelet coefficients.

More recently, the authors of[72,73] demonstrated that for shallow scenes and using standardized image codecs, encoding the light wave scattered by the hologram in a plane crossing the scene — called object plane coding in Fig. 2 — outperforms the direct encoding of the hologram. Based on this observation, they used a speckle reduction methodology to improve the compression efficiency of hologram coding in the object plane[74]. However, while the above-mentioned techniques are very efficient for semi-flat scenes where all 3D scene content can be brought into reasonable focus in the object plane, their compression performance might be severely decreased when handling deep scenes or holograms with a wide viewing angle. Indeed, in these cases, it is impossible to simultaneously refocus every scene object on a single plane. To overcome these limitations, in[75] the authors proposed a unitary transform to model the propagation of light from non-planar surfaces using a generalization of linear canonical transforms. Using a piecewise approximation of this transform, they propagated the light wave scattered by the

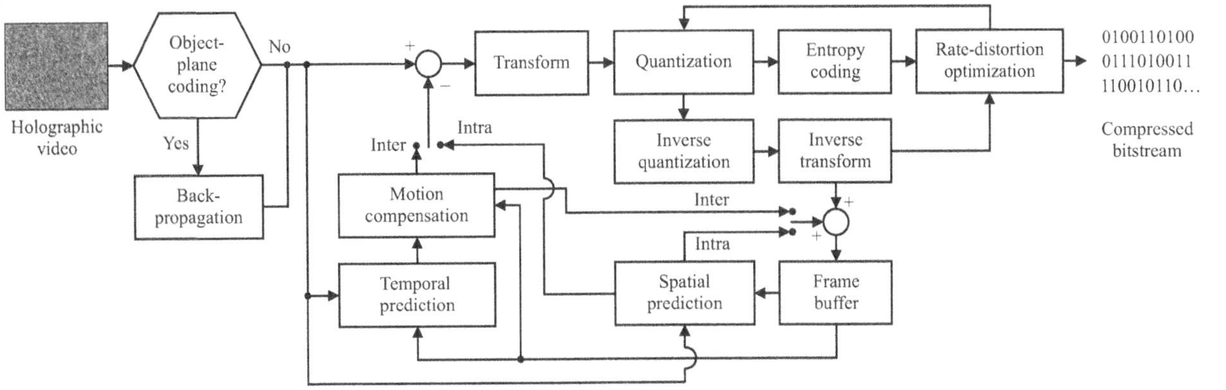

Fig. 2 Generic architecture of a holographic video compression engine.

hologram on a surface fitting the depth profile of the scene to bring every object in focus. The obtained light wave was then encoded using JPEG 2000. This method introduces a new and elegant coding paradigm, but requires the approximated geometry of the scene to be known in advance. Indeed, extracting the depth information from holographic data is still an open issue that remains to be solved.

Another approach to retrieve relevant 3D visualization features from a hologram was proposed in[76]. In this paper, the hologram is divided into rectangular segments which are DCT-transformed to obtain perspective views of the scene. These extracted views are then arranged as a pseudo-video sequence and compressed using the AVC standardized codec. A similar technique was proposed to compress extremely-high resolution holograms in[77]. In this paper, the pixels of the extracted views were rearranged to obtain speckle-reduced orthographic views of the scene, improving the efficiency of the pseudo-video sequence coding with HEVC. While these methods enable interesting functionalities such as viewpoint scalability[78], the extracted views are only valid under the assumption that the viewer is located in the hologram plane, which is never the case in practice. To address these issues and handle arbitrary viewing positions, the authors of[79] exploited the duality between light rays and the local spatial frequencies of a light wave to extract the individual light beams scattered by the hologram. To this end, they used the Matching Pursuit algorithm to decompose the hologram over an over-complete Gabor wavelets dictionary providing an optimal localization in both spatial and frequency domains. The extracted light beams were then compressed using an entropy coder. This method was latter extended to provide viewpoint[80] and quality scalability[81]. This approach presents the advantage of providing a meaningful representation of a hologram's 3D visualization features without any a-priori knowledge on the scene geometry. However, to provide additional functionalities such as edition or motion estimation and compensation, it is necessary to retrieve the scene geometry from the extracted light beams, which is still a very challenging research topic.

As shown in this section, since the early 2000s, several research works have been conducted towards the compression of holographic data. Nevertheless, most of these works were focused on static holograms without taking into account temporal redundancies from one frame to another in holographic sequences. While the static compression methods provide useful tools to understand and extract the most essential features for 3D visualization, they do not yet exploit temporal redundancies from one frame to another, leading to inefficient video compression.

Fig. 2 shows the generic architecture of a holographic video compression engine. As shown in this figure, it mainly comprises of four stages: the transform unit, which aims at concentrating the signal energy in a few number of coefficients, a motion compensation unit to reuse object information from nearby frames, a rate-distortion optimized quantization step to reduce the coefficients' bit depth, and a lossless entropy coding unit to reduce the overall bitstream size. When the hologram is segmented into rectangular blocks, the spatial prediction unit can further removes redundancies from one block to another. Finally, while all these steps can be performed on the raw holographic data, it is possible to use an optional back-propagation step to the object plane discussed previously to improve the overall compression efficiency in some cases.

Developing efficient motion compensation for removing temporal redundancies is crucial to significantly reduce the memory and bandwidth consumption of holographic videos containing billions of pixels per frame. In[82], the authors used MPEG-4 to compress raw holographic data sequences, showing that inter-frame coding performs only slightly better than intra-frame coding. Indeed, because of the non-local nature of holographic signals, classical motion estimation and compensation algorithms based on block matching are inefficient to compress hologram sequences, especially those belonging to visualization use-cases. In[83], a motion compensation solution based on a paraxial approximation was proposed for holograms featuring a singular object. In[84], a motion compensation solution for holography based on the rotational transformation of wavefields[85] was used, where the residuals were further compressed using a STFT transform with an adaptive quantizer, and gains were demonstrated exceeding 20 dB over HEVC. Similarly, a spherical hologram based rotational compensation solution was proposed for an object moving with an arbitrary trajectory described with locally varying curvatures[86]. Unfortunately, all these methods are limited to a single moving object, and require application of object-based hologram segmentation techniques such as[87] to work for scenes with multiple objects. Another open issue yet to be solved is automated motion estimation from holographic videos, which will help to further increase the applicability of these methods. Holographic video coding thus demands a significant research effort.

Compression solutions for metrology

In the context of addressing compression requirements for content typical in holographic metrology, the work in the literature is mostly limited to the use of existing conventional image/video codecs. While custom-made compression solutions may indeed prove to be beneficial, existing codecs are relevant here. Unlike CGH (computer generated holography) generated from deep and complex scenes, holograms captured for metrological purposes typically feature a singular shallow smooth object, where the object wavefield and its reconstruction are image-like after applying the appropriate processing (filtering, demodulation, propagation, and phase unwrapping) steps.

Two approaches to compression are popular here: (1) compressing the final reconstruction and (2) preserving the complex amplitude measurement of the object wavefield. Jaferzadeh et al. in[88] is an example of the former approach where the effectiveness of various JPEG standards in compressing the reconstructed quantitative phase images of red blood cells obtained by off-axis DHM is studied. However, the latter is more applicable for preserving the originally captured data when: a) a need exists to support altering the phase-retrieval pipeline at a later stage; and b) when there is a need for numerical propagation of the optical field[89], which requires the complex amplitude object wavefield.

In this context, a work like[90] is relevant, where the effectiveness of popular wavelet transforms in compressing off-axis hologram measurements, obtained by the camera directly, is studied. Similarly in[91,64] the performance gains obtained by modifying the wavelet transform and decomposition styles used in the JPEG 2000 is investigated. In[31], the fact that the object wavefield was localized to a well-defined region in the Fourier domain for IPH was used to develop efficient compression solutions. Here, the obtained object wavefield after retaining only the required Fourier coefficients is demodulated by the reference carrier wave to obtain a predominantly low-pass signal. This was used to develop spatial bandwidth - optimized lossy and lossless compression solutions that can be implemented using any conventional codec. Similar application of the technique was shown in the context of HT, where it was possible to achieve a bitrate below 1 bit per pixel, while maintaining an RI RMSE less than 10^{-4} for each of the three reconstruction algorithms used[92].

While tailor made solutions for utilizing temporal (inter-frame) redundancy seen in timelapses have yet to be proposed, application of HEVC inter-frame coding after utilizing the spatial bandwidth optimized compression was shown to further reduce bitrate by 15-35%[31]. However for addressing the similarity in different angular projections in HT, application of HEVC inter-frame coding did not bring any improvements[92], underscoring the need for novel algorithms to tackle this problem. In FAHT, the object is rotated with respect to the hologram plane and motion compensation algorithms discussed previously may be utilized, while for LAHT, where the illumination beam is rotated, an appropriate demodulation needs to be developed for compensating for the change happening. The key challenge is developing a complete system that can address all above mentioned sources of redundancies at the same time.

Quality assessment

Visual quality assessment

Accurate visual quality analysis and reliable perceptual quality prediction of holograms is highly demanded. This is due to the rapid advances in developing processing modules of end-to-end holographic visualization pipeline. For example a number of fast numerical methods for CGH have been proposed recently[94-99]. Several approximations and compromises are required in their operations to ease the highly-demanding computations involved. The visual impact of these approximations needs to be analyzed and tested via conducting subjective experiments. Feedback-loops — without the need to reconstruct the object wavefronts — can be created by utilizing high performing objective quality measures to contribute on steering and

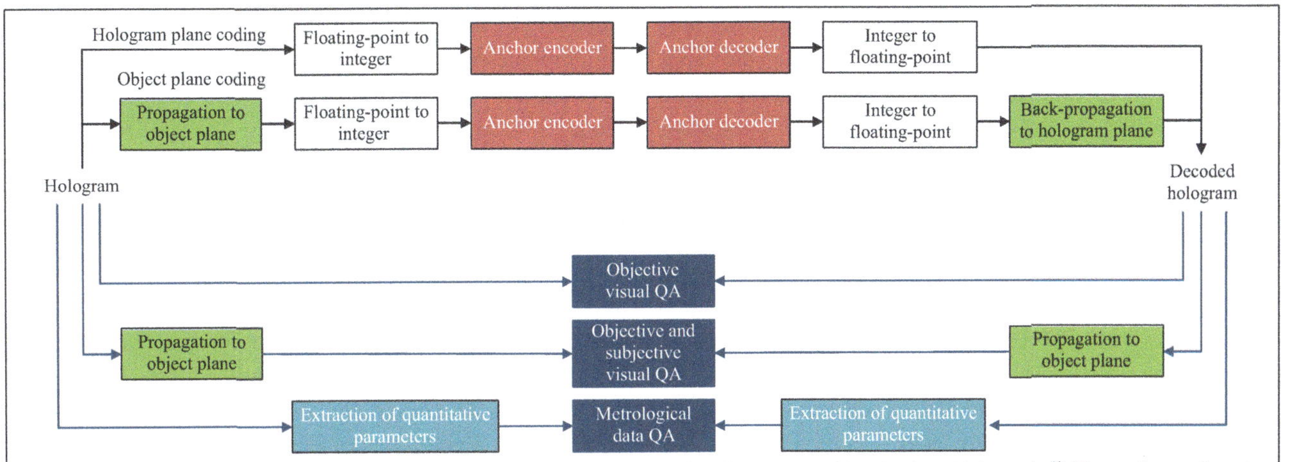

Fig. 3 Pipelines for the assessment of hologram compression methods utilized in the context of JPEG Pleno Holography[93]. The anchor codecs are tested in two pipelines, one performing the encoding in the hologram in the hologram plane, the other in the object plane. Visual quality assessment is performed in both planes, except for the subjective visual quality assessment, which is solely performed in the object plane. Metrological data quality is measured directly on the metrological data extracted from the uncompressed (original) and compressed holograms.

optimizing the performance of these numerical operations. Another example is the need for systematic VQA of compression artifacts on the holograms and having objective measures to predict their visual impact on the reconstructed 3D scenes. Furthermore, high-end solutions for holographic compression can perform more efficiently, by embedding customized evaluation criteria that control their rate-distortion optimization process based on predicting visual quality of the reconstructed 3D scene after decoding. In this section, we will subsequently address subjective quality assessment procedures, objective quality assessment, and, finally, metrological quality assessment. In the latter category, which is particularly useful for the addressed biomedical use case, the impact of compression and hence, the performance of a particular compression technique, is assessed by evaluating the performance of a particular automated task – e.g. phase or refraction index retrieval, or segmentation – after compressing the hologram.

Subjective quality assessment

Conducting systematic subjective experiments for holographic data is the first step in their VQA process. To do so, open-access data sets of holograms recorded or numerically generated from various 3D models and real objects with varied scene-compositions are required. Over the past few years, more than a handful of such public data sets have been created[72,100–104]. The JPEG Pleno database[105], hosts an updated portal with direct links to access these data sets. In[106], a subset of holograms in JPEG Pleno database were chosen and for each one, several 2D reconstructions have been produced.

Next is to identify the testing procedure (illustrated for JPEG Pleno Holography in Fig.3). This mainly depends on the application use case(s) and more importantly the type of the display technology to visualize the holograms for the human observers. Ideally, the reference holograms and their distorted versions have to be visualized via holographic displays. However, holographic displays with acceptable visual characteristics are still rare[107,108]. Moreover, configuring and operating them requires advanced technical skills. One alternative, is to utilize analogue techniques to print them, e.g. on glass plates[109]. This way, very high definition holograms of deep scenes can be printed to illustrate the full potential of the holograms for recreation of a 3D scene[110–115]. However, using hard-copy holograms for subjective experiments is facing serious challenges. Those holograms can not be evaluated by objective quality metrics. From practical point of view, it requires flawless printing of a large number of holograms which normally will be several times bigger than the final hologram set utilized for the subjective test. Because preliminary experiments and mock-up tests are required to shortlist and choose the right scenes and objects, depth of the scenes and distortion levels to span over the full visual quality range. Furthermore, the holograms need to be permuted through and optically aligned multiple times per test subject which will prolong the test far beyond the recommended duration.

Another alternative is to utilize an indirect assessment method, by displaying a single or limited set of reconstructed views and depths on a non-holographic display. This is the most popular option for subjective testing of holograms. Lehtimäki et al.[116] pioneered such efforts by conducting a limited subjective experiment utilizing an auto-stereoscopic display to visualize the holographic reconstructions. In this experiment, they evaluated the depth perception, where front and back reconstruction of each test hologram was generated with multiple aperture resolutions. Consecutively, in[117], the same display technology was utilized to test the perceived quality of holograms produced with different aperture size. In this test, subjects were asked to choose the hologram reconstruction having the lowest distortion in terms of noise, blur (caused from defocus) and depth perception. Next, a 15 inch 2D LCD screen was utilized in[118] to evaluate the relation between numerical error and the perceived quality of the hologram reconstructions. In this test, two simple cases of lossy compression were considered where in one case, a uniform quantization was applied on the hologram and on the other case, the same quantization was applied on the Fourier transform of the holograms. Later, in[119] a methodology to evaluate visual quality of compressed holograms with partial-reconstructions shown on a UHD 2D display was proposed. The holograms were compressed using the real and imaginary parts of the reconstructed holograms on hologram plane. Conventional codecs, namely JPEG, JPEG 2000, H.265/HEVC intra coding were utilized to introduce the compression artifacts in 5 different bitrates. In[120] a subjective experiment using a 2D screen was conducted to investigate the impact of several speckle removal methods. The best performing methods based on the visual appearance of the denoised hologram reconstructions were identified. Nonetheless, in all of these experiments, an important concern regarding the validity of achieved results on non-holographic displays was unresolved; whether or not subjective test results achieved by visualizing holograms on non-holographic displays can replicate at least the overall trend in distribution of the quality scores gathered from a subjective test conducted using an actual holographic display. Ahar et al.[121] conducted the first subjective experiment using a holographic display. They also used the same test stimuli to conduct two separate subjective tests utilizing high-end light field and 2D displays. They also utilized a multi-perspective and multi-depth testing procedure to investigate the compatibility of the quality scores gathered from the 3 displays. In their analysis it was demonstrated that scores gathered from non-holographic displays replicated a very close overall trend compared to the one of holographic display, however observers gave slightly higher scores to the same compressed holograms when displayed on the holographic display than the other two.

All of the experiments using non-holographic displays, only considered reconstructed holograms in the focused position and a single or very limited set of 2D reconstructions were visualized which obviously does not provide information about other 3D aspects of the captured scenes e.g., parallax. A dynamic testing procedure was proposed in[54] to address this issue. In this experiment, pseudo-video sequences were generated from the reconstructed views along a hologram-specific scan path that involved changes in focus and viewing angle. Then, they were displayed using a 4K 2D screen. This method not only allowed for watching the reconstructed scenes from several angles or focal distances, but also provided a better grasp of the parallax and its potential influence from the compression artifacts (see Fig. 4). A similar procedure was also utilized in[122] for evaluating the object plane compression artifacts.

Objective quality assessment

Currently, no objective quality measure is introduced that is specifically designed for evaluating digital holograms. Therefore, hologram quality measurements are carried out by means of currently available measures mainly originated from digital imaging domain. In terms of

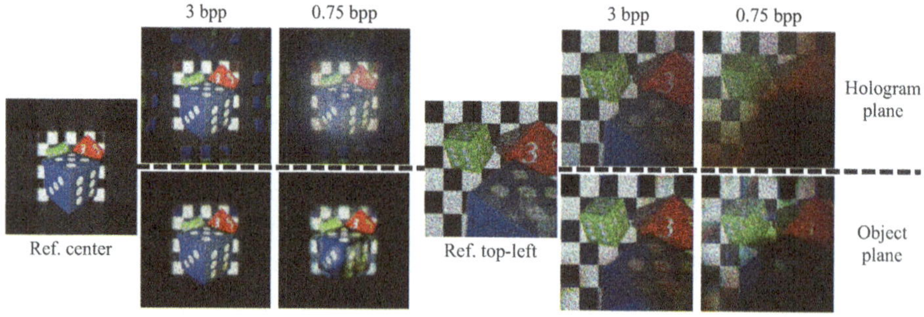

Fig. 4 Representation of JPEG 2000 compression artifacts on Dices16K hologram[102]. Looking at the Center view reconstructions: in hologram plane, the aliasing effect is visible both in high (3 bpp) and low (0.75 bpp) bitrates. Also note combination of the aliasing with heavy distortion in lower bitrate. In object plane, smoothing of the corners of the checker board is visible even in high bitrate (3 bpp). Top-left views for both object plane and hologram plane compression demonstrate significant loss of information and limitation of FoV especially in low bitrate.

functionality, two main groups are typically realized for classification of objective quality measures. It includes methods that essentially measure the mathematical fidelity and those algorithms which are designed to predict the perceptual quality of the visualized content. From the first group, Mean Squared Error (MSE), Signal to Noise Ratio (SNR) and Peak Signal to Noise Ratio (PSNR) are some of the most utilized measures for evaluating the quality of the distorted holograms[118–120,122–124]. In the second group, a quality measure called Structural SIMilarity (SSIM), has been commonly deployed for evaluating digital holograms as well. But since holograms are mostly processed in the form of complex-valued wavefield, quality measures are used separately in their Cartesian components. It is also possible to use the polar components. However, due to non-linear response of the wrapped phase, it was showed that measurements using Cartesian components will be more accurate[72]. The Versatile Similarity Measure (VSM) is a complex-valued fidelity measurement framework that was introduced in[125]. Not only it provides an adaptable bounded measurement for complex-valued data evaluations, but also solves the issue of comparing wrapped-phase data without unwrapping. Also, Sparseness Significance Ranking Measure (SSRM)[126] is a perceptual quality predictor that initially proved its efficiency for digital imaging but due to solely working in the Fourier domain, without any modification could be utilized for complex-valued data as well. Both methods were shown to be more accurate than MSE and PSNR and SSIM, when tested on a limited set of CGHs[127,128]. The prediction accuracy of available objective measures can be analyzed by comparing with the ground-truth visual quality scores obtained from subjective tests. Nonetheless, their performance is influenced by the position in the processing pipeline that prediction of visual quality is required (i.e. quality assessment can be performed in the hologram domain, after numerically reconstructing the captured scene or even after performing post-processing on the reconstructions to remove the speckle noise). A comprehensive performance evaluation of some the most prominent fidelity and image quality measures was performed in[129]. In this experiment, Fourier holograms of DHMulti database[105] were utilized. All holograms were distorted using compression artifacts and later visually annotated from different viewing angles and focal distances. Eventually, quality measures were tested in the hologram domain, right after the compression (QA_1). Additionally, to ensure that the test results are not biased by the characteristics of Fourier holograms, a lossless, numerical conversion to the more conventional Fresnel hologram type was performed and the evaluation was repeated (QA_2). Furthermore, all measures tested on the numerically reconstructed scenes and objects (for the same 2D viewes accompanied with the subjective quality scores) before (QA_3) and after applying a speckle denoising procedure (QA_4). Multiple figures of merit were utilized to evaluate the prediction accuracy, monotonicity and consistency of the tested quality measures. Based on the achieved results for all evaluation criteria and testing points, a usage guideline was provided such that helps interested user to choose the most appropriate method for the holographic VQA tasks until more hologram-oriented quality measures are designed (See Table 3). Experimental results showed that for each test track, a couple of quality measures present a significantly correlated performance compared to the subjectively attained visual scores. Nevertheless, no single measure is robust enough to reliably achieve best results across all 4 testing points. Therefore, none of the available options for objective quality measurement are at the level that confidently alleviate the strong need for design and development of especial VQA algorithms for the holographic data. For more detail on the testing procedure and analysis of the performance of each tested quality measure, interested reader is referred to the original publication in[129].

Metrological quality

Special attention in quality assessment is needed in case of QPI. Techniques such as DHM and HT address strict

Table 3 Usage recommendation of image quality measures for digital holograms, colour coded based on overall performance in each test-track. ✓: advised to be utilized, ⚠: Neutral or results not conclusive within the bounds of this research, ✗: usage is discouraged

Recommendations	Hologram Plane		Object Plane	
Metric	QA_1	QA_2	QA_3	QA_4
MSE	⚠	⚠	✓	✓
NMSE	✓	⚠	✓	✓
PSNR	⚠	⚠	⚠	⚠
SSRM	✗	✗	✓	✓
SSIM	✓	✓	✗	✗
IW-SSIM	⚠	⚠	⚠	⚠
MS-SSIM	⚠	⚠	✗	✗
UQI	✗	⚠	✗	✗
GMSD	✓	⚠	⚠	⚠
FSIM	✗	✗	✗	⚠
NLPD	⚠	⚠	⚠	✗
VIFp	✗	✓	✗	✗

metrological requirements. Popular visual quality assessment metrics such as SSIM are not applicable here, and the impact of compression needs to be measured such that the induced differences are meaningfully expressed in terms of final required values, such as the RMSE calculated on phase/RI values. Presently, in QPI applications of DHM, a user is most often interested in the quantitative phase retrieved from a hologram, which is used for analysis as shown in Fig. 5a. Similarly, in the case of HT, it is more appropriate to measure the error in the refractive index distribution as shown in Fig. 5b.

This requires that the assessment is done after the signal is finally processed after decompression. In DHM, phase unwrapping performed on the complex amplitude signal is a highly non-linear process. It was shown in[31] that even small compression errors in the hologram can lead to large unwrapping errors. Some applications require even more robust phase unwrapping algorithms[130] and the effects of compression should be evaluated in the appropriate framework to ensure relevant results. In the case of HT also, a plethora of reconstruction methods exists, depending on the type of approximation (Born/Rytov) and the regularization being performed. Procedures such as phase unwrapping and tomographic reconstruction algorithms are still being actively developed, posing an additional challenge, since each algorithm requires separate characterization. Therefore, for measuring archival quality, metrics like SNR calculated on the full complex amplitude information may be more relevant.

On the other hand, assessment based on other metrics, such as a performance drop in a specific classification problem, would be suitable for practitioners applying the methods for a particular use case. However, such an approach is not general enough to be considered in research focused only on compression due to vastness of applications of QPI methods and large variety of possible features to be extracted from the images.

General challenges and future trends

In this section, we enumerate important challenges that the holographic community is experiencing when attempting to design efficient codecs that address both the coding performance as functionality requirements of the application domain.

Data representation model

Due to the fact that the holographic modality adheres to a wave-based light propagation model, both amplitude and phase information of the recorded wave front needs to be accounted for. This information can also be represented as real-imaginary (complex-valued) data, and other less popular representations exists. Though during capturing or rendering solely amplitude or phase information might be collected or send to the spatial light modulator (SLM), during processing and also compression of the hologram, the full amplitude-phase or real-imaginary representations will be utilized. Think for example about an amplitude hologram of a 3D scene. In case of non-shallow scenes and while using classical coding engines, it is often preferred to propagate the hologram to an object plane and to apply the

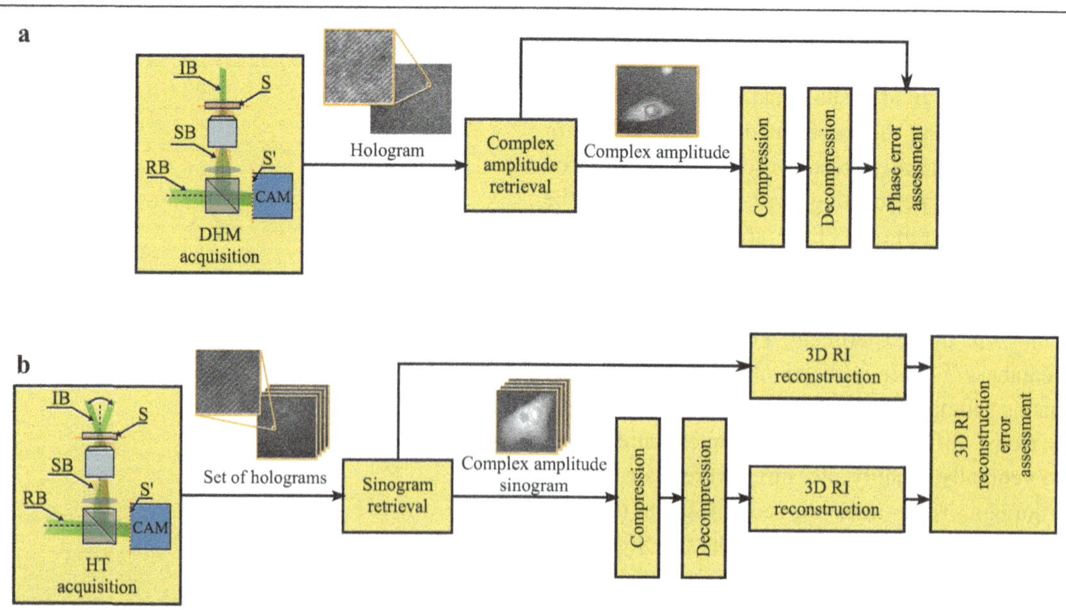

Fig. 5 Pipelines for the assessment of hologram compression in **a** DHM and **b** HT.

compression here to increase coding performance, resulting in the need to compress both real and imaginary information.

A particular case is phase data to be compressed for a phase-modulation SLM. Here the codec should preferably be able to handle phase information. However, this is a particularly hard representation to handle since codecs typically deploy — as discussed earlier — transform- or prediction-based algorithms that attempt to reduce the entropy of the to be encoded information. Often these algorithms result in phase-wrapping issues that in combination with subsequent phase quantization procedures lead to erroneous reconstructed phase information during decoding[44].

The real-imaginary representation is probably the most suitable and generic data representation to be used — at least for the internal representation — in a codec engine, but not always the most effective one since it results in two internal components to represent while the aforementioned phase-modulation case would only require one component. Hence, coding strategies for holograms will have to cope with the variety of representations imposed by the recording, generation, and rendering infrastructure present in the application chain. At this moment, an omnipotent representation that would serve all these use cases in an efficient way from a compression performance perspective is non-existing. Consequently, either compromises at the level of compression efficiency need to be made, either different coding engines have to be deployed, reducing the interoperability.

Multi-component holograms

Colour information is processed very efficiently by classical codecs for natural imagery. Typically, the red, green and blue components are linearly transformed into 1 luminance and 2 chrominance components, the first representing intensity information per pixel and the latter specifying the colour information. For consumer grade applications the colour information is often subsampled by a factor 2 in each image dimension, since the human visual system possesses a significantly lower contrast sensitivity for chrominance information than colour information. This procedure already achieves a compression factor of 2.

Holographic data is quite more cumbersome with this respect since the holograms related to the different colour planes have interference patterns resulting from light interacting at different wavelengths each. Hence, the colour components of a hologram pixel cannot be simply linearly transformed and quantized thereafter without severely distorting the signal. So far no efficient solutions have been proposed to handle this problem.

In addition to this, different hologram components might be related to various polarizations or include also information on fluorescent responses in biomedical imaging. Similar issues might arise here as well with respect to decorrelating information residing in different components (or spectral bands).

Large data volumes

Particularly pathology, tomography and high-end visualization applications result in huge hologram pixel

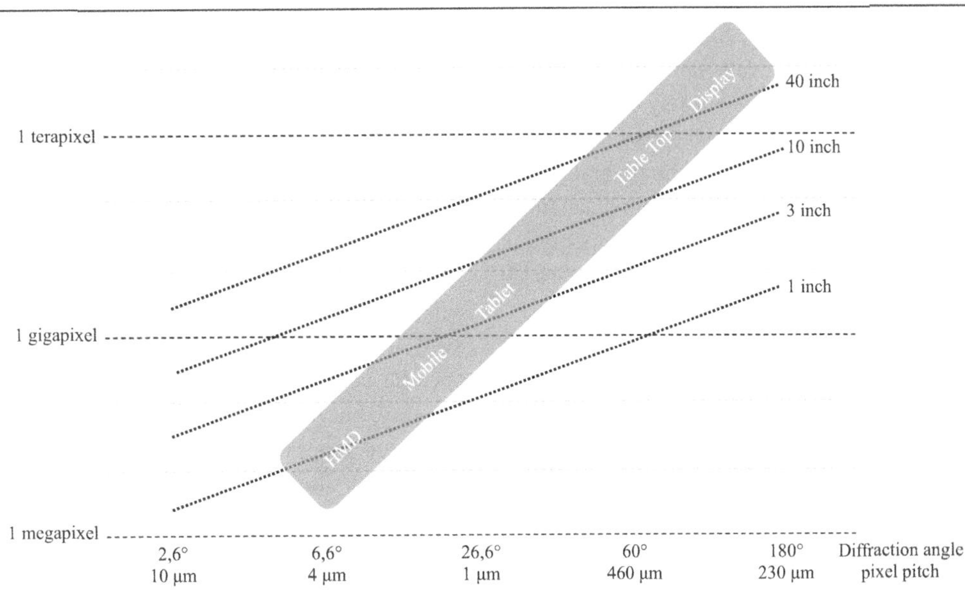

Fig. 6 Simulation of required number of hologram pixels depending on desired display dimensions and pixel pitch (or viewing angle) in case of a wavelength of $\lambda = 460$ (blue light) and a 16:9 screen ratio[44].

counts. In Fig. 6[44] the required numbered of hologram pixels is calculated based on desired display size and viewing angle. Please note that a naive setup is assuming the existance of a patchwork of SLMs. It is clear that the count very quickly explodes to Gigapel and Terapel requirements for larger displays, representing a huge challenge to encode this data within reasonable computational bounds, but also putting huge requirements on storage capacity and transmission bandwidth.

Random access

Random access functionality might in this context reduce the stress in terms of computational complexity, storage capacity and transmission bandwidth by supporting selective consultation of the encoded hologram. This requires the codestream to be compartmented in individually decodable chuncks such that only the required information can be retrieved.

Think for example at large holographic displays where only one or few persons are watching to the screen. In that case it makes a lot of sense to stream only that information to the screen that facilitates the necessary viewports and hence reducing significantly the required communication bandwidth. A logic segmentation to support this kind of functionality would be to deploy partitioning in space-frequency domain, where an immediate connection between the viewports and the segments in this domain exist.

Motion estimation and compensation

Video codec deploy a motion estimation and compensation mechanism to remove the temporal redundancy from a video sequence. Since classical motion estimation and compensation models fail on holographic video because holographic pixels not solely relate to one spatial location in the 3D scene, but are the representing the result of interfering light waves originating from different 3D scene points. Notwithstanding the fact initial attempts have been undertaken to resolve the motion compensation process, more advanced models that allow for object-based motion compensation are not yet available.

Motion estimation is even a more cumbersome problem due to the interferometric nature of holographic content. Therefore current holographic video coding solutions retrieve motion information from the 3D content from which the holograms are generated using CGH techniques. Unfortunately, this is not feasible for content captured with holographic microscopy and tomography systems. Here, the main data cost for these applications arises from the sheer volume of frames being recorded using the camera sensor. As discussed in the use cases section, these frames are typically sequentially correlated as a) in time-lapse, the evolution of an object undergoing small temporal changes is recorded; b) in image stitching, a macroscopic smooth object being captured from overlapping measurements; c) in in tomography, an object being capturing from multiple illumination directions either by introducing small rotations in the object or in the illumination beam, where some applications can be even a combination of these individual modalities. Exploiting this data dependencies would certainly boost the coding performance.

Computational complexity

Given the potentially large data volumes, codec algorithms should depict a relatively low complexity, while offering simultaneously high compression efficiency. Moreover, parallel computation and GPU processing should be intrinsically supported; a single-threaded CPU handling the data decoding for large display would never be able to process timely all data. Hence, multi-threaded, parallel processing and GPU-compliance are conditiones sine qua non for effective coding-decoding mechanisms; a non-evident requirement that conflicts with achieving maximum data decorrelation and effective entropy coding. Typically, the larger the data units the more effective the probability and context modelling and hence, the entropy encoding will perform.

Learning-based solutions

Learning-based solutions, although having been around for a long time to optimize performance of individual codec modules, have recently been penetrating faster in the domain of compression technologies as well, particularly driven by the premise that these techniques potentially improve compression results by deploying 'learned' features to decorrelate a data set rather than more 'rigid' mathematically constructed transform and prediction techniques. In a deep learning context this then finally results in the compression of the produced latent representation.

A first problem is to produce representative training data sets that allow for training the deep neural network (DNN) such that it 'behaves well' for generic or application specific contexts. This is already a challenge for regular imagery, but particularly for holography when deploying it in a more generic context, due to the lack of representative data and its time-consuming production process.

Secondly — and often overlooked for holography — is that DNNs operate very well on classical imagery due to the high correlations typically existing between neighbouring data and the direct physical correlation with the represented 3D scene that can be adhered to the content of neighbouring pixels. Moreover, DNNs mimic surprisingly well part of the visual processing taking place in the human brain. However, this does not hold for holographic content, being interferometric in nature, where the values of the holographic pixels are the result of interference effects between waves originating from all points of the 3D scene represented in the hologram. Naive deployment of DNNs on holographic content will destroy quite some information embedded in the hologram, and the subsequently naive quantization and entropy coding of the latent representation will have effects that will be hard to relate back to the original data.

Hence, significantly more research will be needed in this domain to give rise to DNN-based coding technologies.

Data quality assessment

Subjective quality assessment procedures still suffer from the unavailability of qualitative holographic displays. Hence, alternative solutions such as numerically reconstructing holograms for rendering on 2D or autostereoscopic displays, are currently deployed and can be seen as reasonable alternatives due to their higher intrinsic resolution and image. Since they do not have inherent support for characteristic features of holography, the still remaining vergence-accommodation conflict and lacking depth refocusing functionality negatively influences the potential outcome of the test procedure. Moreover, they suffer from the more severe presence of speckle noise due to the numerical reconstruction, where, as in optical setups, coherence might be slightly reduced due to optical imperfections giving rise to less severe presence of speckle noise. In the absence of a reasonable holographic display, this alternative remains an acceptable choice though the rendering post-processing should be improved to handle e.g. the speckle noise, as well as the subjective testing procedures issued.

The performance of objective quality assessment techniques is still cumbersome as indicated earlier in this article. Few metrics deliver an 'acceptable' behaviour in the hologram domain and/or in the reconstruction domain (object plane). Nonetheless, plenty of space for improvement remains, seen the still poor correlation between perceived visual quality and measured objective quality. The few reasonably performing metrics fail to incorporate support for properties of the human visual system.

Finally, in the context of metrology, JPEG introduced a methodology that is worth further pursuing; measuring the functional quality of (de)compressed content, i.e. to what extend are (1) analysis algorithms — for example reconstructing phase information — affected by the compression task and are (2) the resulting errors still sufficiently bounded. Challenges here remain the dependencies between the nature of the tested content and the feature extractions/analysis algorithms deployed, resulting in more complex dependencies in the evaluations process, and not solely a dependency on the coding algorithm under test.

What next

As should be clear now, the domain of holographic content coding and broader research on holography, needs more cooperation to resolve the above mentioned challenges related to data representation, handling of multiple components, large data volumes, random access functionality, computational complexity, learning-based coding, and data quality assessment. Most of the research actions are too fragmented, and research ecosystems need to be build to resolve the earlier mentioned bottlenecks.

A very fruitful initiative would be the construction of holographic databases shared within the community that are used as reference as well when reporting research results. This exercise should involve the creation of commonly accepted evaluation methodologies as quality assessment procedures.

In that sense, it is enlightening that the JPEG committee is working on a suite of standard specifications, called JPEG Pleno (ISO/IEC 21794) aiming to provide a standard framework for representing new imaging modalities, such as light field, point cloud, and holographic imaging. JPEG Pleno is not just a set of efficient coding tools addressing compression efficiency, it is targeted at creating interoperability in the application chain at the level of the data format. It is a representation framework that will be a fully integrated system for providing advanced functionality support for image manipulation, metadata, random access and interaction, and various file formats. In addition, it should offer privacy protection, ownership rights, and security. Currently, the JPEG Pleno standard defines already the file format for plenoptic modalities, as well as coding technology for light fields (including conformance testing and reference software). In 2021, a call for coding solutions for holographic content has also been issued[5] and the standardization process is currently ongoing. JPEG Pleno Holography will be the first standard aspiring to a versatile solution for efficient compression of holograms for a wide range of applications such as holographic microscopy, tomography, interferometry, printing and display and their associated hologram types. Key functionalities desired include support for both lossy and lossless coding, scalability, and random access and integration within the JPEG Pleno system architecture, with the goal of supporting a royalty free baseline[5].

However, this standard will solely focus on the data exchange format, but not normalization of, for example, the display interfaces. Standardization bodies such as IEC TC 110, specifying components and characteristics of electronic displays, IEC TC 100, focused on audio, video and multimedia systems and equipment, and ISO TC 130, producing specification in the field of printing and graphics technologies, could be instructional in these respects also.

Conclusions

Important steps have been taken to resolve the bottlenecks with respect to the representation and compression of holographic content. This is extremely important to handle efficiently the huge volumes of data produced and consumed by holographic equipment. However, quite a few challenges still need to be tackled more intensively, such as decorrelating the colour information in multi-component holograms, providing random access functionality to the compressed holograms to enable partial decompression supporting desired viewports, motion estimation and compensation models suitable for holography, reducing the computational complexity and increasing the parallelizability, appropriately supporting learning-based applications, and improving the performance of the data quality assessment measures and procedures.

In parallel, interoperability between devised solutions, products, and services should be improved to allow for the creation of application ecosystems handling holographic content. Without this, companies launching holographic products and services will experience additional bottlenecks achieving their market share. Therefore, international standardization activities are of utmost importance to achieve this goal. An initiative to be cherished in this context is the JPEG Pleno Holography standardization effort that aims to provide an exchange and compression framework for holographic content.

Acknowledgements

This work was funded in part by European Research Council (FP7/N.617779–INTERFERE), the French government through the National Research Agency (ANR) Investment referenced ANR-A0-AIRT-07, the Portuguese FCT-Fundação para a Ciência e Tecnologia under the project UIDB/EEA/50008/2020, PLive X-0017-LX-20, and co-financed by the European Regional Development Fund (ERDF) through the Programa Operacional Regional do Centro (Centro 2020) under the project Centro-01-0145-FEDER-000019 - C4 - Centro de Competências em Cloud Computing, the "Convenzione triennale tra la Fondazione di Sardegna e gli Atenei Sardi - Regione Sardegna L.R.7/2007 annualità 2019-DGR DGR 28/21 del 17.05.2015" with project "Formal Methods and Technologies for the Future of Energy Systems", CUP F72F20000350007, call 2019, and by the project Cagliari 2020 (MIUR, PON04a2 00381), Science Foundation Ireland (under grant no. 13/CDA/2224), and co-financed by the European Union under the European Regional Development Fund within the program TEAM TECH/2016-1/4 of Foundation for Polish Science.

Author details

[1]Vrije Universiteit Brussel, ETRO Dept., B-1050 Brussels, Belgium. [2]imec, Digital and User-Centric Solutions (DUCS), B-3001 Leuven, Belgium. [3]Institute of Research & Technology b <>com, Rennes, Franc. [4]Maynooth University, Dept. of Computer Science, Maynooth, Co. Kildare, Ireland. [5]University of Cagliari, DIEE, UdR CNIT, Cagliari, Italy. [6]Universidade da Beira Interior, Covilha, Portugal. [7]Instituto de Telecomunicações, Covilha, Portugal. [8]Warsaw University of Technology, Institute of Micromechanics and Photonics, 02-525 Warsaw, Poland

Conflict of interest

The authors declare that they have no conflict of interest.

Received: 15 September 2021 Revised: 30 June 2022 Accepted: 01 July 2022
Accepted article preview online: 04 July 2022
Published online: 20 July 2022

References

1. Balasubramani, V et al. Roadmap on digital holography-based quantitative phase imaging. *Journal of Imaging* **7**, 252 (2021).
2. Lohmann, A. W. et al. Space-bandwidth product of optical signals and systems. *Journal of the Optical Society of America A* **13**, 470-473 (1996).
3. Huggins, E. Introduction to Fourier optics. *The Physics Teacher* **45**, 364-368 (2007).
4. Burton, G. J. & Moorhead, I. R. Color and spatial structure in natural scenes. *Applied Optics* **26**, 157-170 (1987).
5. ISO/IEC JTC1/SC29/WG1. Final Call for Proposals on JPEG Pleno Holography. 2020 at https://ds.jpeg.org/documents/jpegpleno/wg1n91020-REQ-Final_Call_for_Proposals_on_JPEG_Pleno_Holography.pdf.
6. Park, Y., Depeursinge, C. & Popescu, G. Quantitative phase imaging in biomedicine. *Nature Photonics* **12**, 578-589 (2018).
7. Marquet, P. et al. Digital holographic microscopy: a noninvasive contrast imaging technique allowing quantitative visualization of living cells with subwavelength axial accuracy. *Optics Letters* **30**, 468-470 (2005).
8. Kemper, B. & Von Bally, G. Digital holographic microscopy for live cell applications and technical inspection. *Applied Optics* **47**, A52-A61 (2008).
9. Balasubramani, V. et al. Holographic tomography: techniques and biomedical applications. *Applied Optics* **60**, B65-B80 (2021).
10. Nanolive, S. A. Nanolive. https://nanolive.ch.
11. Tomocube Inc. Tomocube. http://www.tomocube.com/.
12. Kim, M. K. Digital holographic microscopy. in Digital Holographic Microscopy (ed Kim, M. K.). (New York: Springer, 2011), 149–190.
13. Rappaz, B. et al. Noninvasive characterization of the fission yeast cell cycle by monitoring dry mass with digital holographic microscopy. *Journal of Biomedical Optics* **14**, 034049 (2009).
14. Kastl, L. et al. Quantitative phase imaging for cell culture quality control. *Cytometry Part A* **91**, 470-481 (2017).
15. Rubin, M. et al. TOP-GAN: Stain-free cancer cell classification using deep learning with a small training set. *Medical Image Analysis* **57**, 176-185 (2019).
16. Stępień, P., Korbuszewski, D. & Kujawińska, M. Digital holographic microscopy with extended field of view using tool for generic image stitching. *ETRI Journal* **41**, 73-83 (2019).
17. Croft, L. V. et al. Digital holographic imaging as a method for quantitative, live cell imaging of drug response to novel targeted cancer therapies. in Theranostics (eds Batra, J. & Srinivasan. S.). (New York: Springer, 2019), 171–183.

18. Kühn, J. et al. Label-free cytotoxicity screening assay by digital holographic microscopy. *Assay and Drug Development Technologies* **11**, 101-107 (2013).
19. Jin, D. et al. Tomographic phase microscopy: principles and applications in bioimaging. *Journal of the Optical Society of America B* **34**, B64-B77 (2017).
20. Hsieh, J. & Flohr, T. Computed tomography recent history and future perspectives. *Journal of Medical Imaging* **8**, 052109 (2021).
21. Kuś, A. et al. Holographic tomography: hardware and software solutions for 3D quantitative biomedical imaging (Invited paper). *ETRI Journal* **41**, 61-72 (2019).
22. Kim, K. et al. Optical diffraction tomography techniques for the study of cell pathophysiology. *Journal of Biomedical Photonics & Engineering* **2**, 020201 (2016).
23. Hugonnet, H. et al. Multiscale label-free volumetric holographic histopathology of thick-tissue slides with subcellular resolution. *Advanced Photonics*, **3**, 026004 (2021).
24. van Rooij, J. & Kalkman, J. Polarization contrast optical diffraction tomography. *Biomedical Optics Express* **11**, 2109-2121 (2020).
25. Karray, M., Slangen, P. & Picart, P. Comparison between digital Fresnel holography and digital imageplane holography: the role of the imaging aperture. *Experimental Mechanics* **52**, 1275-1286 (2012).
26. Sánchez-Ortiga, E. et al. Off-axis digital holographic microscopy: practical design parameters for operating at diffraction limit. *Applied Optics* **53**, 2058-2066 (2014).
27. Takeda, M., Ina, H. & Kobayashi, S. Fouriertransform method of fringe-pattern analysis for computer-based topography and interferometry. *Journal of the Optical Society of America A* **72**, 156-160 (1982).
28. Cuche, E., Marquet, P. & Depeursinge, C. Spatial filtering for zero-order and twin-image elimination in digital off-axis holography. *Applied Optics* **39**, 4070-4075 (2000).
29. Darakis, E. et al. Microparticle characterization using digital holography. *Chemical Engineering Science* **65**, 1037-1044 (2010).
30. Yamaguchi, I. Phase-shifting digital holography. in Digital Holography and Three-Dimensional Display (ed Poon, T. C.). (Boston: Springer, 2006), 145–171.
31. Stępień, P. et al. Spatial bandwidth-optimized compression of image plane off-axis holograms with image and video codecs. *Optics Express* **28**, 27873-27892 (2020).
32. Hoffman, D. M. et al. Vergence-accommodation conflicts hinder visual performance and cause visual fatigue. *Journal of Vision* **8**, 33 (2008).
33. Kim, S. C. et al. Holographic full-color 3D display system using color-LCoS spatial light modulator. *Proceedings of SPIE* **5742**, Practical Holography XIX: Materials and Applications. San Jose: SPIE (2005), 223–233.
34. Takaki, Y. & Okada, N. Hologram generation by horizontal scanning of a high-speed spatial light modulator. *Applied Optics* **48**, 3255-3260 (2009).
35. Inoue, T. & Takaki, Y. Table screen 360-degree holographic display using circular viewing-zone scanning. *Optics Express* **23**, 6533-6542 (2015).
36. Kollin, J. S., Benton, S. A. & Jepsen, M. L. Real-time display of 3-D computed holograms by scanning the image of an Acousto-optic modulator. *Proceedings of SPIE* **1136**, Holographic Optics II: Principles and Applications. Paris: SPIE (1989), 178–185.
37. Maimone, A., Georgiou, A. & Kollin, J. S. Holographic near-eye displays for virtual and augmented reality. *ACM Transactions on Graphics* **36**, 85 (2017).
38. Yeom, H. J. et al. 3D holographic head mounted display using holographic optical elements with astigmatism aberration compensation. *Optics Express* **23**, 32025-32034 (2015).
39. Zaperty, W., Kozacki, T. & Kujawińska, M. MultiSLM color holographic 3D display based on RGB spatial filter. *Journal of Display Technology* **12**, 1724-1731 (2016).
40. Yoshikawa, H. & Yamaguchi, T. Review of holographic printers for computer-generated holograms. *IEEE Transactions on Industrial Informatics* **12**, 1584-1589 (2016).
41. Yamamoto, K. et al. Hologram printing for nextgeneration holographic display. *Proceedings of SPIE* **10557**, Ultra-High-Definition Imaging Systems. San Francisco: SPIE, 2018.
42. Lucente, M. E. Interactive computation of holograms using a look-up table. *Journal of Electronic Imaging* **2**, 28-34 (1993).
43. Kim, S. C. & Kim E. S. Effective generation of digital holograms of three-dimensional objects using a novel look-up table method. *Applied Optics* **47**, D55-D62 (2008).
44. Blinder, D. et al. Signal processing challenges for digital holographic video display systems. *Signal Processing: Image Communication* **70**, 114-130 (2019).
45. Cai, Q. et al. Lossy and Lossless Intra Coding Performance Evaluation: HEVC, H.264/AVC, JPEG 2000 and JPEG LS. Proceedings of The 2012 Asia Pacific Signal and Information Processing Association Annual Summit and Conference. Hollywood: IEEE, 2012.
46. Weinberger, M. J., Seroussi, G. & Sapiro, G. The LOCO-I lossless image compression algorithm: principles and standardization into JPEG-LS. *IEEE Transactions on Image Processing* **9**, 1309-1324 (2000).
47. Hampel, H. et al. Technical features of the JBIG standard for progressive bi-level image compression. *Signal Processing: Image Communication* **4**, 103-111 (1992).
48. Ono, F. et al. JBIG2-the ultimate bi-level image coding standard.Proceedings 2000 International Conference on Image Processing. *Vancouver: IEEE*, 140-143 (2000).
49. Alakuijala, J. et al. JPEG XL next-generation image compression architecture and coding tools. Proceedings of SPIE 11137, Applications of Digital Image Processing XLII. San Diego: SPIE, 2019, 112-124.
50. Wallace, G. K. The JPEG still picture compression standard. *IEEE Transactions on Consumer Electronics* **38**, xviii-xxxiv (1992).
51. Hannuksela, M. M., Lainema, J. & Vadakital, V. M. The High Efficiency Image File Format Standard[Standards in a Nutshell]. *IEEE Signal Processing Magazine* **32**, 150-156 (2015).
52. Ginesu, G., Pintus, M. & Giusto, D. D. Objective assessment of the WebP image coding algorithm. *Signal Processing: Image Communication* **27**, 867-874 (2012).
53. Skodras, A., Christopoulos, C. & Ebrahimi, T. The JPEG 2000 still image compression standard. *IEEE Signal Processing Magazine* **18**, 36-58 (2001).
54. Ahar, A. et al. Validation of dynamic subjective quality assessment methodology for holographic coding solutions. 2021 13th International Conference on Quality of Multimedia Experience (QoMEX). Montreal: IEEE, 2021.
55. Sullivan, G. J. & Wiegand, T. Video compression - from concepts to the H.264/AVC standard. *Proceedings of the IEEE* **93**, 18-31 (2005).
56. Sullivan, G. J. et al. Overview of the high efficiency video coding (HEVC) standard. *IEEE Transactions on Circuits and Systems for Video Technology* **22**, 1649-1668 (2012).
57. Mukherjee, D. et al. A technical overview of VP9- the latest open-source video codec. *SMPTE Motion Imaging Journal* **124**, 44-54 (2015).
58. Naughton, T. J. et al. Compression of digital holograms for three-dimensional object reconstruction and recognition. *Applied Optics*, **41**, 4124-4132 (2002).
59. Mills, G. A. & Yamaguchi, I. Effects of quantization in phase-shifting digital holography. *Applied Optics* **44**, 1216-1225 (2005).

60. Shortt, A. E., Naughton, T. J. & Javidi, B. Histogram approaches for lossy compression of digital holograms of three-dimensional objects. *IEEE Transactions on Image Processing* **16**, 1548-1556 (2007).
61. Cheremkhin, P. A. & Kurbatova, E. A. Numerical comparison of scalar and vector methods of digital hologram compression. Proceedings of SPIE 10022, Holography, Diffractive Optics, and Applications VII. Beijing: SPIE, 2016, 1002227.
62. Xing, Y. F., Pesquet-Popescu, B. & Dufaux, F. Comparative study of scalar and vector quantization on different phase-shifting digital holographic data representations. 2014 3DTV-Conference: The True Vision-Capture, Transmission and Display of 3D Video (3DTV-CON). Budapest: IEEE (2014).
63. Bang, L. T. et al. Compression of digital hologram for three-dimensional object using Wavelet-Bandelets transform. *Optics Express* **19**, 8019-8031 (2011).
64. Blinder, D. et al. JPEG 2000-based compression of fringe patterns for digital holographic microscopy. *Optical Engineering* **53**, 123102 (2014).
65. Peixeiro, J. P. et al. Holographic Data Coding: Benchmarking and Extending HEVC With Adapted Transforms. *IEEE Transactions on Multimedia* **20**, 282-297 (2018).
66. Xing, Y. F. et al. Vector lifting scheme for phaseshifting holographic data compression. *Optical Engineering* **53**, 112312 (2014).
67. Xing, Y. F. et al. Adaptive nonseparable vector lifting scheme for digital holographic data compression. *Applied Optics* **54**, A98-A109 (2015).
68. Birnbaum, T. et al. Wave atoms for digital hologram compression. *Applied Optics* **58**, 6193-6203 (2019).
69. Onural, L. Diffraction from a wavelet point of view. *Optics letters* **18**, 846-848 (1993).
70. Liebling, M., B lu, T. & Unser, M. Fresnelets: new multiresolution wavelet bases for digital holography. *IEEE Transactions on Image Processing* **12**, 29-43 (2003).
71. Darakis, E. & Soraghan, J. J. Use of fresnelets for phase-shifting digital hologram compression. *IEEE Transactions on Image Processing* **15**, 3804-3811 (2006).
72. Bernardo, M. V. et al. Holographic representation: hologram plane vs. object plane. *Signal Processing: Image Communication* **68**, 193-206 (2018).
73. Bernardo, M. V., Pinheiro, A. M. G. & Pereira, M. Benchmarking coding standards for digital holography represented on the object plane. Proceedings of SPIE 10679, Optics, Photonics, and Digital Technologies for Imaging Applications V. Strasbourg: SPIE 2018.
74. Bernardo, M. V. et al. Efficient coding of experimental holograms using speckle denoising. *Signal Processing: Image Communication* **96**, 116306 (2021).
75. Blinder, D. et al. Unitary Transforms Using Time-Frequency Warping for Digital Holograms of Deep Scenes. *IEEE Transactions on Computational Imaging* **4**, 206-218 (2018).
76. Seo, Y. H., Ch oi, H. J. & Kim, D. W. 3D scanningbased compression technique for digital hologram video. *Signal Processing: Image Communication* **22**, 144-156 (2007).
77. Gilles, A. & Gioia, P. Compression and reconstruction of extremely-high resolution holograms based on hologram-lightfield transforms. Proceedings of SPIE 11510, Applications of Digital Image Processing XLIII. SPIE 2020.
78. Lee, D. H. et al. Viewing angle dependent coding of digital holograms. 2011 19th European Signal Processing Conference. Barcelona: IEEE 2011, 1367–1371.
79. El Rhammad, A. et al. Color digital hologram compression based on matching pursuit. *Applied Optics* **57**, 4930-4942 (2018).
80. El Rhammad, A. et al. View-dependent compression of digital hologram based on matching pursuit. Proceedings of SPIE 10679, Optics, Photonics, and Digital Technologies for Imaging Applications V. Strasbourg: SPIE 2018.
81. El Rhammad, A. et al. Progressive hologram transmission using a view-dependent scalable compression scheme. *Annals of Telecommunications* **75**, 201-214 (2020).
82. Darakis, E. & Naughton, T. J. Compression of digital hologram sequences using MPEG-4. Proceedings of SPIE 7358, Holography: Advances and Modern Trends. Prague: SPIE 2009.
83. Blinder, D., Schretter, C. & Schelkens, P. Global motion compensation for compressing holographic videos. *Optics Express* **26**, 25524-25533 (2018).
84. Muhamad, R. K. et al. Exact global motion compensation for holographic video compression. *Applied Optics* **58**, G204-G217 (2019).
85. Matsushima, K. Formulation of the rotational transformation of wave fields and their application to digital holography. *Applied Optics* **47**, D110-D116 (2008).
86. Cao, H. K. & Kim, E. S. Faster generation of holographic videos of objects moving in space using a spherical hologram-based 3-D rotational motion compensation scheme. *Optics Express* **27**, 29139-29157 (2019).
87. Birnbaum, T. et al. Object-based digital hologram segmentation and motion compensation. *Optics Express* **28**, 11861-11882 (2020).
88. Jaferzadeh, K., Gholami, S. & Moon, I. Lossless and lossy compression of quantitative phase images of red blood cells obtained by digital holographic imaging. *Applied Optics* **55**, 10409-10416 (2016).
89. Langehanenberg, P., von Bally, G. & Kemper, B. Autofocusing in digital holographic microscopy. *3D Research* **2**, (2011).
90. Cheremkhin, P. A. & Kurbatova, E. A. Quality of reconstruction of compressed off-axis digital holograms by frequency filtering and wavelets. *Applied Optics* **57**, A55-A64 (2018).
91. Bruylants, T. et al. Microscopic off-axis holographic image compression with JPEG 2000. Proceedings of SPIE 9138, Optics, Photonics, and Digital Technologies for Multimedia Applications III. Brussels: SPIE, 2014, 128–138.
92. Muhamad, R. K. et al. Off-axis image plane hologram compression in holographic tomography - metrological assessment. *Optics Express* **30**, 4261-4273 (2022).
93. ISO/IEC JTC1/SC29/WG1. Common Test Conditions 6.0 for JPEG Pleno Holography.
94. Symeonidou, A. et al. Computer-generated holograms by multiple wavefront recording plane method with occlusion culling. *Optics Express* **23**, 22149-22161 (2015).
95. Park, J. H. Recent progress in computer-generated holography for three-dimensional scenes. *Journal of Information Display* **18**, 1-12 (2017).
96. Pan, Y. J. et al. A Review of Dynamic Holographic Three-Dimensional Display: Algorithms, Devices, and Systems. *IEEE Transactions on Industrial Informatics* **12**, 1599-1610 (2016).
97. Sugie, T. et al. High-performance parallel computing for next-generation holographic imaging. *Nature Electronics* **1**, 254-259 (2018).
98. Shimobaba, T. & Ito, T. Computer Holography: Acceleration Algorithms and Hardware Implementations. (Boca Raton: CRC Press 2018).
99. Nishitsuji, T. et al. Review of fast calculation techniques for computer-generated holograms with the point light-source-based model. *IEEE Transactions on Industrial Informatics* **13**, 2447-2454 (2017).
100. Blinder, D. et al. Open Access Database for Experimental Validations of Holographic Compression Engines. 2015 Seventh International Workshop on Quality of Multimedia Experience (QoMEX). Pilos: IEEE,

2015.
101. Gilles, A. et al. Hybrid approach for fast occlusion processing in computer-generated hologram calculation. *Applied Optics* **55**, 5459-5470 (2016).
102. Gilles, A. et al. Computer generated hologram from multiview-plus-depth data considering specular reflections. 2016 IEEE International Conference on Multimedia & Expo Workshops (ICMEW). Seattle: IEEE, 2016.
103. Symeonidou, A. et al. Speckle noise reduction for computer generated holograms of objects with diffuse surfaces. Proceedings of SPIE 9896, Optics, Photonics and Digital Technologies for Imaging Applications IV. Brussels: SPIE, 2016.
104. Symeonidou, A. et al. Colour computer-generated holography for point clouds utilizing the phong illumination model. *Optics Express* **26**, 10282-10298 (2018).
105. ISO/IEC JTC1/SC29/WG1. JPEG Pleno Database. https://jpeg.org/jpegpleno/plenodb.html.
106. Corda, R. & Perra, C. A dataset of hologram reconstructions at different distances and viewpoints for quality evaluation. 2019 Eleventh International Conference on Quality of Multimedia Experience (QoMEX). Berlin: IEEE, 2018.
107. Tsang, P. W. M. & Poon, T. C. Review on the stateof-the-art technologies for acquisition and display of digital holograms. *IEEE Transactions on Industrial Informatics* **12**, 886-901 (2016).
108. T. Kreis. 3-D display by referenceless phase holography. *IEEE Transactions on Industrial Informatics* **12**, 685-693 (2016).
109. Lehtimäki, T. M. et al. Using traditional glass plate holograms to study visual perception of future digital holographic displays. Digital Holography and ThreeDimensional Imaging 2016. Heidelberg: Optical Society of America, 2016.
110. Bjelkhagen, H. I. Ultra-realistic 3-D imaging based on colour holography. *Journal of Physics: Conference Series* **415**, 012023 (2013).
111. Su, J. et al. Progress in the synthetic holographic stereogram printing technique. *Applied Sciences* **8**, 851 (2018).
112. Sohn, I. B. et al. Three-dimensional hologram printing by single beam femtosecond laser direct writing. *Applied Surface Science* **427**, 396-400 (2018).
113. Klug, M. A. Display applications of large-scale digital holography. Proceedings of SPIE 4737, Holography: A Tribute to Yuri Denisyuk and Emmett Leith. Orlando: SPIE, 2002.
114. Yang, X. et al. High-resolution Fresnel hologram information simplification and color 3D display. *Optik* **216**, 164919 (2020).
115. Jeon, H. et al. High-resolution binary hologram printing methods. Proceedings of SPIE 11306, Practical Holography XXXIV: Displays, Materials, and Applications. San Francisco: SPIE, 2020.
116. Lehtimäki, T. M. et al. Visual perception of digital holograms on autostereoscopic displays. Proceedings of SPIE 7329, Three-Dimensional Imaging, Visualization, and Display 2009. Orlando: SPIE, 2009.
117. Lehtimäki, T. M. et al. Evaluation of perceived quality attributes of digital holograms viewed with a stereoscopic display. 2010 9th Euro-American Workshop on Information Optics. Helsinki: IEEE, 2010.
118. Lehtimäki, T. M. et al. Comparing numerical error and visual quality in reconstructions from compressed digital holograms. Proceedings Volume 7690, Three-Dimensional Imaging, Visualization, and Display 2010 and Display Technologies and Applications for Defense, Security, and Avionics IV. Orlando: SPIE, 2010.
119. Ahar, A. et al. Subjective quality assessment of numerically reconstructed compressed holograms. Proceedings of SPIE 9599, Applications of Digital Image Processing XXXVIII. San Diego: SPIE, 2015.
120. Fonseca, E. et al. Assessment of speckle denoising filters for digital holography using subjective and objective evaluation models. *Applied Optics* **58**, G282-G292 (2019).
121. Ahar, A. et al. Suitability analysis of holographic vs light field and 2D displays for subjective quality assessment of Fourier holograms. *Optics Express* **28**, 37069-37091 (2020).
122. Amirpourazarian, H. et al. Quality evaluation of holographic images coded with standard codecs. IEEE Transactions on Multimedia, http://dx.doi.org/10.1109/TMM.2021.3096059 (2021).
123. Corda, R. & Perra, C. Hologram domain data compression: Performance of standard codecs and image quality assessment at different distances and perspectives. *IEEE Transactions on Broadcasting* **66**, 292-309 (2019).
124. Corda, R. et al. Investigation of Coding Standards Performances on Optically Acquired and Synthetic Holograms. Proceedings of the 20th International Conference on Advanced Concepts for Intelligent Vision Systems. Cham: Springer, 2020, 396-407.
125. Ahar, A. et al. A new similarity measure for complex amplitude holographic data. Digital Holography and Three-Dimensional Imaging 2017. JeJu Island Republic of Korea: Optical Society of America, 2017.
126. Ahar, A., Barri, A. & Schelkens, P. From sparse coding significance to perceptual quality: A new approach for image quality assessment. *IEEE Transactions on Image Processing* **27**, 879-893 (2018).
127. Ahar, A. et al. Performance evaluation of sparseness significance ranking measure (SSRM) on holographic content. 3D Image Acquisition and Display: Technology, Perception and Applications 2018. Orlando: Optical Society of America, 2018.
128. Ahar, A. et al. A new similarity measure for complex amplitude holographic data. Proceedings of SPIE 10396, Applications of Digital Image Processing XL. San Diego: SPIE, 2017.
129. Ahar, A. et al. Comprehensive performance analysis of objective quality metrics for digital holography. *Signal Processing: Image Communication* **97**, 116361 (2021).
130. Ghiglia, D. C. & Romero, L. A. Robust two-dimensional weighted and unweighted phase unwrapping that uses fast transforms and iterative methods. *Journal of the Optical Society of America A* **11**, 107-117 (1994).

Official journal of the JHL 2689-9620
www.light-am.com

Review | Open Access

Digital holography as metrology tool at micro-nanoscale for soft matter

Zhe Wang[1,2], Lisa Miccio[1,3], Sara Coppola[1,3], Vittorio Bianco[1], Pasquale Memmolo[1], Volodymyr Tkachenko[1], Vincenzo Ferraro[2], Ernesto Di Maio[2], Pier Luca Maffettone[2,3] and Pietro Ferraro[1,3,*]

Abstract

The appearance of the first laser approximately 12 years after the invention of holography by Gabor (1948) revolutionized the field of optical metrology. In fact, the invention of holographic interferometry enabled the exploitation of interferometry on non-mirror surfaces and full-scale objects. The holography-based measurement methods has been implemented to several industrial systems or in support of R&D with the aim of improving new products in many fields (automotive, aerospace, electronics, etc.). To date, holography has been considered an important measurement tool for non-destructive inspection (NDI), strain-stress measurement, and vibration analysis at various engineering sites. Recently, the new paradigm of Industry4.0 has seen the introduction of new technologies and methods of processing materials as well as the development of manufacturing approaches for the realization of innovative products. For example, direct printing, additive, and bottom-up manufacturing processes are expected to involve new ways of making products in future, and most innovative fabrication processes will be based on the manipulation of soft matter (e.g., starting from the liquid phase) that will be shaped at the nanoscale. The inherent characteristics of digital holography (DH) make it a powerful and accurate tool for the visualization and testing of final products, as well as for in situ and real-time monitoring and quantitative characterization of the processes involved during the fabrication cycle. This review aims to report on the most useful applications of soft matter, where the capabilities offered by DH, such as three-dimensional (3D) imaging, extended focus, 3D tracking, full-field analysis, high sensitivity, and a wide range of measurements from nanometers to centimeters, permit completely non-invasive characterizations on a full-scale. Several holographic experimental results of typical samples are reported and discussed where DH plays a primary role as a tool gauge for soft matter.

Keywords: Digital holography, Soft matter, Microscopy, Thin film, Polymers, Liquids, Characterization, Interferometry, Inkjet printing

Correspondence: Pietro Ferraro (p.ferraro@isasi.cnr.it)
[1]CNR-ISASI, Institute of Applied Sciences and Intelligent Systems "E. Caianiello", Via Campi Flegrei 34, 80078 Pozzuoli, Napoli, Italy
[2]Dipartimento di Ingegneria Chimica dei Materiali e della Produzione Industriale, Università degli Studi di Napoli "Federico II", Piazzale Tecchio 80, 80125 Napoli, Italy
Full list of author information is available at the end of the article.
These authors contributed equally: Zhe Wang, Lisa Miccio, Sara Coppola

Introduction

The most important advantage arising from the invention of holography was realized with the development of lasers. The applications of holography has been exploited in many industrial fields as a very sensitive measurement tool, such as aerospace, automotive, energy and cultural heritage, owing to the discovery of holographic interferometry by Powell and Stetson[1,2]. In fact, by employing different holographic modalities, such as holographic interferometry, time-average holography, or the so-called TV and/or digital speckle holography, it was possible to develop accurate optical measuring strategies for non-destructive testing systems and experimental mechanics[3,4]. The above holography-based measurement systems were applied successfully to guarantee the quality of industrial products (e.g., turbine blades, tires, composite materials, photovoltaic solar panels, automobile components, and electronic circuits). Later, with the microelectronic revolution in the 80s and 90s as well as the surge of capabilities of micro-nanofabrication techniques, holography-based methods were successfully applied by scaling their probing-gauge down to microsystems (i.e., MEMS, MOEMS, and NEMS)[5-10]. In the past 20 years, the introduction of powerful PCs and high-performance solid-state light sensors has resulted in the achievement of further developments of holographic technology in digital-modality toward biological, biomedical, and industrial applications, which has led to the widespread use of such systems[11-16].

Recent progress in bottom-up self-assembly fabrication approaches and direct printing for production processes, especially at the micro and nanoscales, has motivated the development of holographic optical tools for achieving accurate and full-field characterization of systems based on soft matter materials, such as liquids and polymers. These are expected to be the basis of the next revolution in fields of biotechnology and flexible electronics systems, among others. Indeed, the most common imaging modality for such microfabrication techniques is direct optical imaging, which is a video-based method. It enables the monitoring of the dynamic evolution of the processes and measuring features related to the geometrical dimensions of the microstructure[17,18]. High-speed and high-resolution cameras were used for the detection of small particles during inkjet printing. Although such imaging systems can be complex, significant information is completely lost as in the case of the detection and measurement of out-of-focus structures and particles. Furthermore, not all of the information related to the phase shift introduced by the samples can be evaluated. Many microstructures are transparent or semi-transparent, and it is critical to measure the phase delay introduced in the light path, as in the case of the evaluation of the optical quality of optical components, such as microlenses and thin-film thickness mapping at high spatial resolution. For these reasons, the inspection of soft matter and related fabrication processes takes advantage of the noteworthy capabilities of DH as a metrology tool. As summarized below, the main attractive features of DH enable the achievement of the necessary capability and flexibility in the measurements:

1) The possibility of numerically managing the complex wavefront scattered or transmitted by the sample under investigation allows the extraction of all of the information (intensity and phase) through a full-digital modality.

2) Flexibility in achieving object imaging well-in-focus (or to obtain good focus of any portion of it) that enables correct DH microscopic measurements to be obtained, even though the object could have a three-dimensional (3D) shape behind the depth of focus of the optical imaging system.

3) Potential to retrieve phase-contrast maps that enable quantitative measurements of the sample in full-field mode and 3D.

4) The possibility to manage and remove aberrations in the optical system using simple and flexible methods, thus simplifying the optical apparatus and measurement operations.

Owing to the above unique features of DH, this paper reviews what are considered the most significant examples to demonstrate the capability of DH to measure soft matter structures and the related fabrication techniques. In particular, we focused on different types of challenging cases.

• thin liquid films and membranes;

• ink-jet printing process for directly printed structures;

• analysis and quantitative measurements of bottom-up self-assembling processes of polymer and/or liquids;

• microfluidic rheology and study of solid-liquid inte-rfaces.

The four different above-mentioned typical cases enable us to demonstrate the aptitude of DH systems in combining metrology and 3D imaging capabilities and by furnishing quantitative measurements in challenging circumstances. We show that 3D dynamic monitoring in real-time or in situ using DH enables quantitative measurements and accurate characterization.

It should be noted that biological matter, both cell populations and tissues, benefits from the above-mentioned characteristics, so DH in recent years has been extensively developed, enabling its incorporation in life sciences[19-26]. However, we focus on and limit the present review to fabrication topics, thus supplying a complete overview of DH in this area without including cells of biological matter that have different peculiarities.

This review is organized into three sections. The first focuses on the measurement of the thin-film thickness over a wide field of view (FoV). Various experimental configurations are described, and their applicability to different materials is provided. For such applications, the thickness measurement ranges from ~50 nm to ~30 μm, with a resolution at the nanometer scale on a large FoV of a circular area with a radius of ~20 mm. The second section is centered on a specific microfabrication technique based on the exploitation of the pyroelectric effect of ferroelectric crystals, where liquids and polymers can be manipulated by electric fields. This enables us to provide many examples where DH has proven to be a suitable metrological tool not only for transparent solid microstructures but also for the full characterization of dynamic processes such as the self-assembly of liquids for the realization of microstructures. The main results show that high-resolution two-dimensional (2D) and 3D measurements can be achieved by exploiting the quantitative phase content encoded in the digital holograms. The third section presents DH to investigate the rheological properties, monitor stress, and track particles embedded in solution or liquid film. All three sections contain a detailed description of the related state-of-the-art. Different setups and target materials, in addition to diverse observables, are described and discussed in this manuscript.

Digital holography for studying thickness evolution of thin liquid films

From soap bubbles for children to semiconductors for electronics, thin films have accompanied humankind through countless centuries, and are an important category of soft matter. Over the past century, with the improvement in industrial production technologies, a large number of products related to thin-film materials have been produced, especially in the field of daily chemistry. With the informatization and intelligence of human society, wearable and in-body devices are becoming the general direction of future technology applications. Because of its special physical and chemical properties, film material has become the best carrier to support future devices, and a detailed measurement of its characteristics is highly desirable.

The mapping of the thin-film thickness is an open and long-term challenge[27]. One of the well-known methods for mapping film thickness is the interferometric approach[28], which uses the optical path difference between the beam's reflection of the upper and lower surfaces of the film to create an interferometric pattern from which the thickness information of the thin film can be determined[29,30]. The interferometry method has been used for a long time. However, conventional interferometric measurements have some shortcomings, such as a small measurement FoV, limited measuring range, and non-real-time measurement. To overcome these drawbacks, a series of interferometric techniques has been proposed in recent decades, namely colorimetric interferometry[31,32], multiwavelength interferometry[33], phase-shift interferometry[34], and scanning interferometry[35], etc. However, for the dynamic thickness measurement of a thin liquid film under a large FoV, interferometric methods hardly meet all recording requirements, especially for films with high viscosity coefficients. In 2019, for the first time, Ferraro et al. implemented DH to map the thin-film thickness during the evolution of liquid bubbles[36]. In 2021, Wong et al. applied DH to measure beer bubble morphology[37]. The height of the beer bubbles and water surface can be revealed simultaneously by the back-propagation process of DH. In fact, most film materials are good samples for holographic measurements because of their transparent or semi-transparent properties[38]. However, the existence of micron-level deformation in the film evolution generates scattering and/or refraction for object beam, which significantly affects holographic recording. This problem was solved by introducing a telescope structure before the camera, which could collect the beams and adjust the magnification. As a result, real-time holographic measurements of thin-film thickness were successfully realized. Many thin-film measurement experiments have been conducted using holography in different scenarios. In the following subsections, we systematically introduce the applications of DH in thin-film measurements and explore the feasibility of the technique in the measurement of soft matter.

Basic principle and experimental setup of digital holographic film imaging

The primary problem faced by DH in thin-film measurement is the limited thickness measurement range in a single shot[39]. There is a specific measurement range for the holographic recording of laser sources in the visible spectral range. Moreover, owing to the influence of surface tension, the thickness distribution of thin-film materials is usually continuous, enabling numerical holographic phase unwrapping. Considering the real-time requirements of film thickness measurement, the off-axis DH system could be a good candidate for testing[14,40]. In preliminary optical geometry, an off-axis Mach-Zehnder recording system was used to measure the dynamic film thickness evolution process[36,41]. A telescope structure was set before the camera to adjust the FoV and suppress the influence of film scattering (the setup and experimental results are shown in Fig. 1.

Because thin films represent imaging objects with rich and varied high-frequency information, it is difficult for a

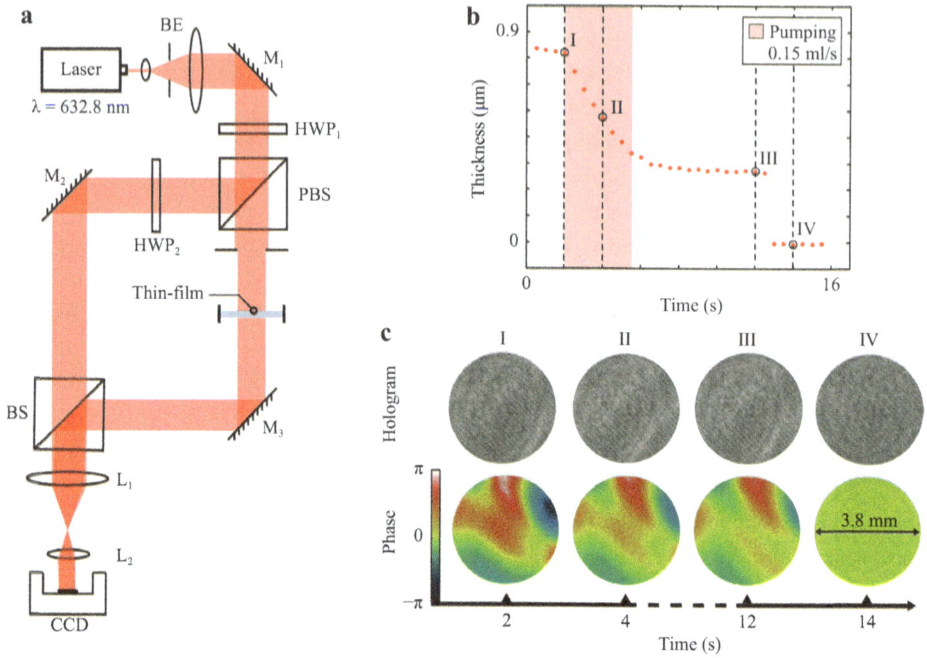

Fig. 1 Conventional holographic thin-film thickness mapping, setup, and preliminary results. **a** Digital holographic setup based on Mach-Zehnder structure. **b** Thickness change before and after rupture in the central area of a polymer film. **c** Full FoV thickness mapping of selected timepoints in **b**. Reproduced from Ref. 36 © 2021 Springer Nature.

lens-less system to collect all information during the recording process. Therefore, a telescope structure (composed of two lenses with different focal lengths) represents a strategy for collecting object information. As shown in Fig. 1a, lenses L_1 and L_2 (focal length of L_1 is longer than that of L_2) were set to construct a telescope lens group for zooming out. The light-collecting ability of the 4-Focus lens group allowed the collection of scattered object beams. In other words, this setup allowed the enlargement of the FoV and the simultaneous recording of high-frequency information. In the initial testing, a film close to rupture was selected as the sample. The thickness change in the central area of the film was measured[41], as shown in Fig. 1b. It can be seen that for thin-film materials with a known refractive index (RI), DH is a good tool for measuring the thickness of the film, and it does not require other reference images. Generally, in interferometry, it is necessary to clearly observe the appearance of a common black film[42] to obtain absolute thickness measurements. However, for holographic measurements, when the thickness of the film is less than twice the recording wavelength, the following phase-thickness conversion formula can be used directly to extract the thickness information from the phase map:

$$h = \frac{\varphi\lambda}{2\pi n} \tag{1}$$

where h is the thickness, φ is the phase, λ is the recording wavelength, and n is the RI of the film. For the range of thickness measurement in a holographic recording system with a specific laser wavelength, the lower limit depends on the wavelength, whereas the upper limit can be continuously increased owing to the holographic phase unwrapping algorithm. Here, PUMA[43] phase unwrapping is a good candidate for extracting film thickness information because of its good phase-unwrapping capabilities for a continuously changing phase. However, in the case of films where the maximum thickness reaches a few micrometric levels or even higher, conventional holographic reconstruction will not be able to implement absolute thickness measurements owing to the limitation of the wrapping phase. To solve this problem, a full life cycle measurement method was proposed[36]. To do this, the entire thickness evolution process of the film, from the initial formation to the final rupture, needs to be recorded. Once the film is close to rupture, the starting area of the rupture tends to be a common black film, indicating that it has the lowest phase value in the full FoV. Taking advantage of this feature, a time variable can be added to the phase-unwrapping process to deal with the phase jump in all recorded frames, as shown in Fig. 2.

As a film thickness extraction scheme, the full life cycle recording method has certain limitations; in fact, the measured film must have a clear natural rupture process. For most free-standing thin liquid films, the natural rupture

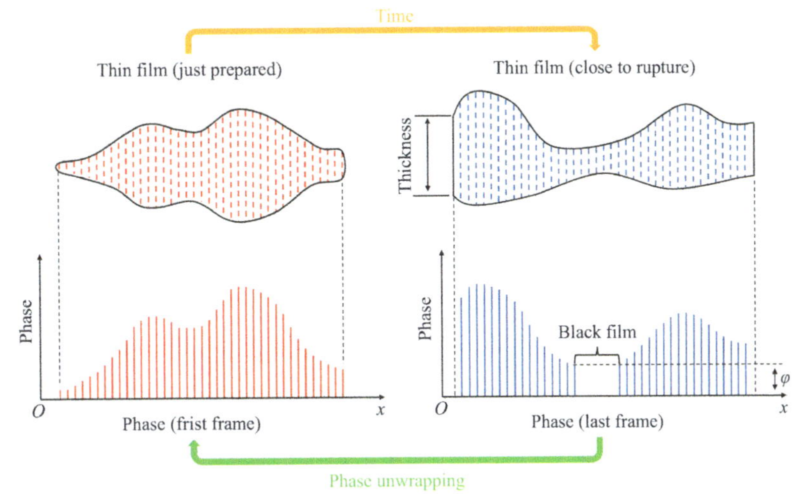

Fig. 2 Schematic diagram of thin-film full life cycle recording phase unwrapping method.

process will occur owing to gravitational drainage[44], which enables us to use the proposed strategy for extracting film thickness information. Therefore, compared to other existing optical measurement methods, DH offers high precision, a large measurement range, and real-time characteristics, representing a good tool for thin liquid film thickness mapping, as discussed in detail in the following sections.

Joint-measurement between white light interferometry and digital holography

DH is based on the numerical diffraction reconstruction of digitally recorded interference fringes to obtain the spatial information of the recorded object[45,46], whereas white-light interferometry (WLI) uses the direct identification of the number and color of interference fringes to obtain thickness information[28,29]. These two methods have certain limitations in real-time thin-film thickness mapping, but for some specific materials, both can be used, for example, surfactant film. Thus, the complementary characteristics of WLI and DH enable comparison and fusion. To implement simultaneous recording for one liquid film using the two methods, joint imaging was designed by combining the conventional off-axis DH geometry and WLI geometry. To ensure that the WLI recording did not affect the intensity contrast of the hologram, oblique illumination was implemented[47]. Owing to the combination of WLI and DH, it is possible to achieve high-precision real-time thickness measurements while expanding the imaging FoV and range[48]. The experimental setup, which combines the two interferometric geometries, is shown in Fig. 3a.

In the experiment reported in Fig. 3b, a surfactant film was selected as the experimental sample, and its RI was determined in a previous study[49]. Meanwhile, for the WLI measurement, the thickness-color correspondence of the surfactant film is known[50]. The surfactant film was formed in a fixed metal ring and simultaneously illuminated by white light and a red laser beam. It can be seen from the experimental results in Fig. 3b that when the film was close to rupture, both DH and WLI could map the thickness of the full FoV. Based on the comparison, it can also be found that the measurement results of DH and WLI maintain good consistency in the range below 1 μm. Moreover, for a thickness greater than 1 μm, WLI hardly provides the right thickness value, while DH still works. For the results at the early stage after film formation, the WLI interferogram showed complicated patterns; thus, it was difficult to interpret and retrieve the effective thickness. The colored interference fringes in such a range of film thicknesses were spatially and periodically modulated by the complex film structure, which led to an impossible reading. On the other hand, the thickness distribution information can be clearly identified in the results of DH. This is the greatest advantage of holographic measurement: for films with complex structures, holographic measurement can still provide an accurate thickness distribution. In conclusion, for surfactant films with a maximum thickness of less than 1 μm and a smooth surface structure, both DH and WLI can be used. However, for films with a higher thickness or complex surface structure, DH would be a better choice. Therefore, the reliability of the DH measurements can be verified through WLI experiments.

However, in the case of film with thickness less than half the wavelength, transmission holography has difficulties in providing accurate thickness measurements. Therefore, one strategy that has been effectively adopted is to use the complementary characteristics of WLI and DH

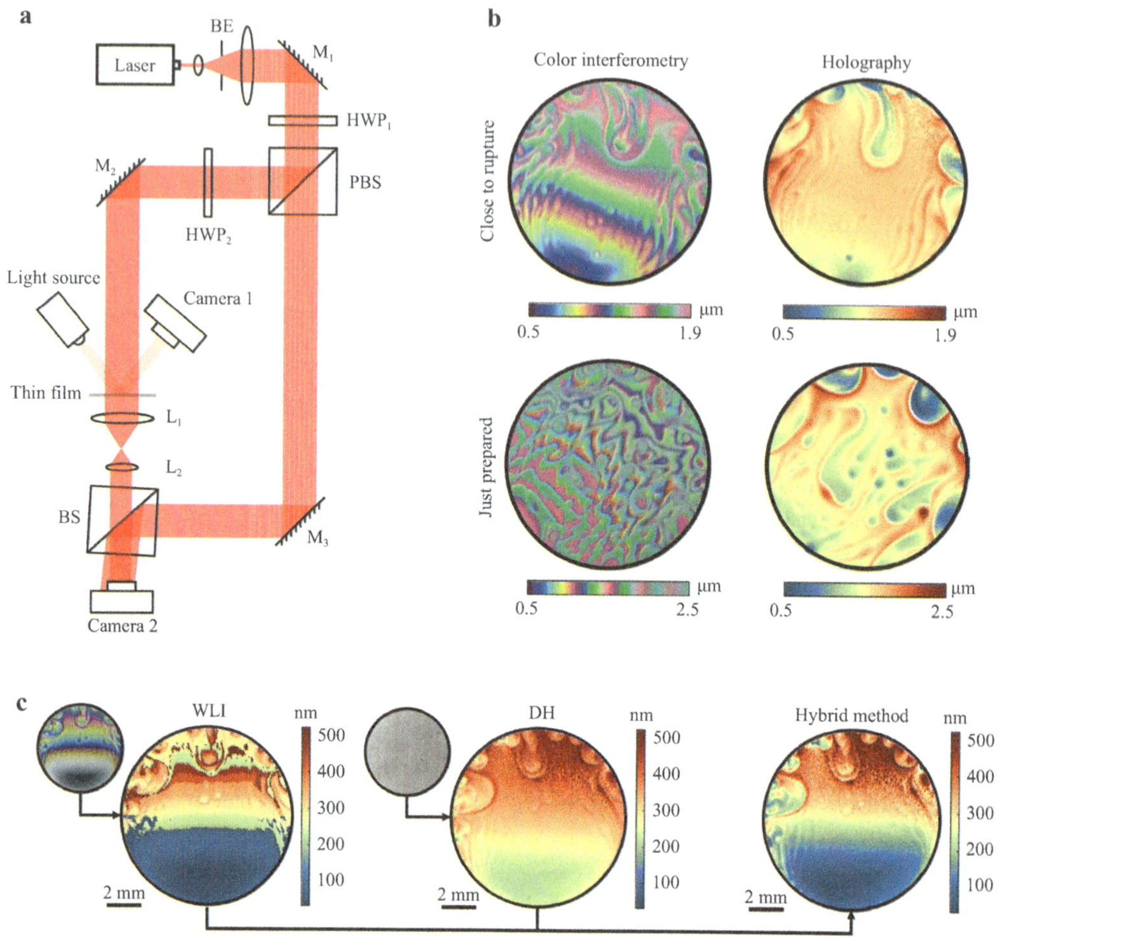

Fig. 3 Optical geometry and results of DH and WLI hybrid imaging. **a** Experimental setup for simultaneous recording. **b** Comparison of measurement results at two different time points. **c** The complementary characteristics of white light interferometry and digital holography will provide high-precision and wide-range thin-film thickness measurement. Adapted with permission from Ref. 48 Copyright 2021 American Chemical Society.

for a new hybrid measurement modality[48]. Thus, phase correction was achieved by comparing the two different measurement results. From the results in Fig. 3c, the deviation in the low-thickness area was significantly improved; essentially, the hybrid recording and reconstruction completed the phase calibration of the film and achieved an accurate thickness distribution. As a tool for improving microscopic resolution, holographic imaging can be easily integrated with other technologies, creating a wider range of possibilities.

Suitability of digital holography in different thin-film experimental arrangements

In the above two sections, conventional imaging setups and methods of DH in thin-film imaging were discussed for the measurement of the thickness level. Furthermore, the difference between the results of WLI and DH for the same film was compared, proving that DH is a reliable measurement method for thin films. In conventional interferometric experiments, the measurement methods and systems are limited because of the need to keep the sample as smooth as possible. In addition, because the distance between the camera and the sample has certain requirements for creating interference fringes, it is difficult to measure the thickness distribution when the film is moving. However, these limitations can be overcome for DH recording. The DH geometry has good flexibility in structural settings; therefore, it can deal with different experimental scenarios[51]. Meanwhile, thanks to the numerical diffraction propagation process, DH can present real-time thickness mapping for the film in motion. Fig. 4 shows the time-lapse recording results obtained by DH in different film motion modes, including pumping bubble, thin film lifting, and thin film opening. Thickness mapping for these processes appeared to be difficult during interferometric recording, but could be completed in DH recording.

In an experiment on pumping liquid bubbles, a flat thin film was prepared at one end of a customized tube, and air was pumped into the tube to promote bubble growth. Finally, a semicircular bubble was formed[36]. The transmissive DH can effectively record the thickness of the bubble during the growth process, and then the thickness information extraction can be completed through numerical refocusing and phase distortion calibration. Fig. 4a shows the measurement results of related experiments. In this experiment, a polyacrylamide (PA) solution was used to create a thin liquid film. For this kind of film, the reconstruction of thickness mapping could be provided using the full life cycle recording method. In addition, owing to the phase distortion calibration, the hemispherical film thickness distribution can be projected as a horizontal mode, which is convenient for researchers to analyze the drainage process. For the lifting thin-film experiment, the film was prepared in the middle of a customized tube and then lifted by pumping air into the tube from the bottom. Because of the centimeter-level movement of the film inside the tube, conventional interferometric methods were unable to provide the thickness distribution during the movement; however, DH could solve this problem. Fig. 4b shows the results of related experiments. The drainage phenomenon in the process of film uplift is related to the velocity of movement, which introduces the potential for controllable drainage. This will help to better study the physical model of the drainage process. The results showed the distribution of the film structure at different moving speeds, and the difference in drainage patterns was revealed. For the film-opening experiments, an optical iris was conceived and manufactured to open the film[52]. Then, an adjusted telescope structure was used to reach a large FoV (~40 mm diameter), which ensured the recording of the entire film area during the film-opening process. This is also a significant advantage of holographic measurement; the FoV can be changed by adjusting the lens group, thereby realizing continuous holographic recording with a large FoV. The opening process of the film included a potential change in the overall film thickness from high to low. This experimental system ensured that the total amount of solution used to prepare the film remained constant, thereby facilitating the analysis of the hydrodynamic process. Fig. 4c shows the results of related experiments.

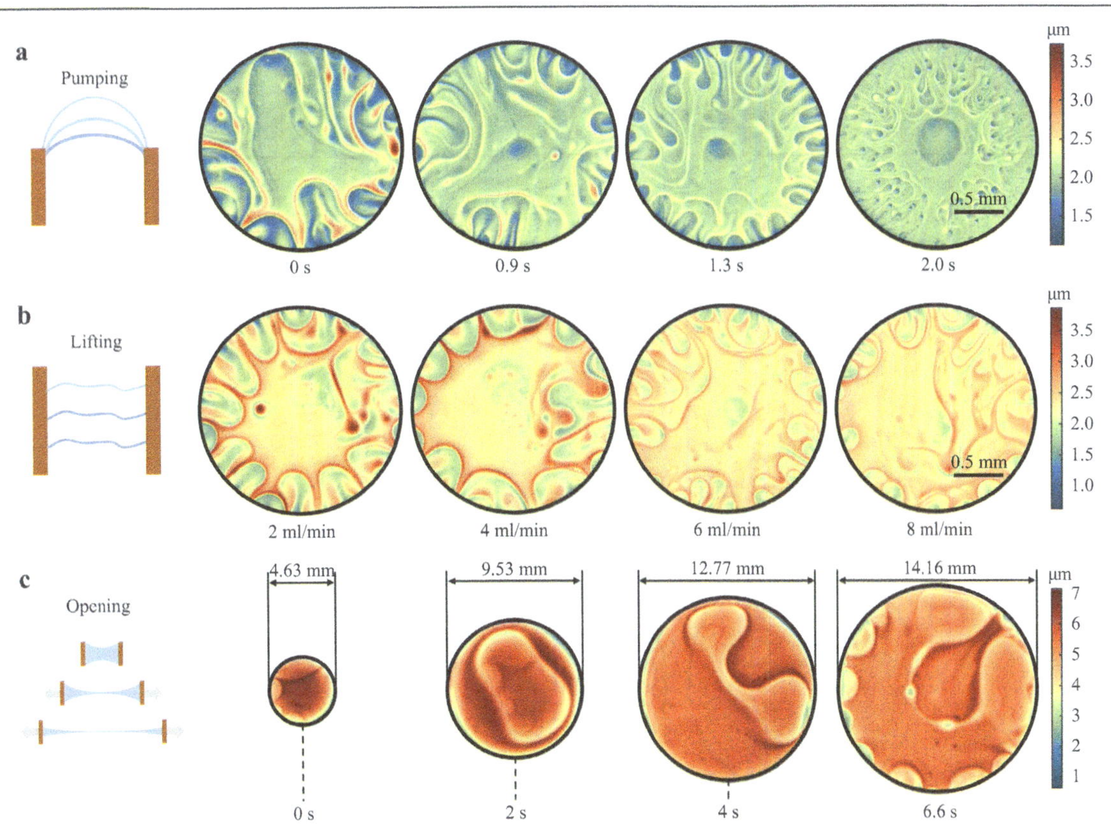

Fig. 4 Digital holography allows thin-film thickness measurements to be produced under different scenarios, breakthrough the limitation for film forming requirement in interferometry. **a** Measurements for pumping bubble. **b** Measurements for thin liquid film lifting in tube. **c** Measurement for film opening.

In the above three experimental scenarios, holographic measurements showed good thickness measurement capabilities. For most experimental environments, the holographic setup can be easily changed to meet the requirements of recording, enabling its application in the field of fluid mechanics testing.

Digital holography performances on different materials

The term "thin film" is ubiquitous in our daily lives, and includes a broad range of materials, from solid to liquid films. In general, under the conditions of interferometric measurements, different configurations are used to complete experiments on different matters. For example, tilted WLI is mainly used for liquid films[20], whereas a vertical interferometric system is required for solid films, such as semiconductors of the indium tin oxide (ITO) layer[53]. In this section, we show the measurement results for films made of different matter to demonstrate the good adaptability of the DH technology in film measurements.

The main problem with conventional interferometric measurements is that it is difficult to realize real-time imaging with a large FoV. In recent years, some methods have been proposed to solve this problem[54], but certain restrictions on the RI and film thickness have been added. However, DH recordings are not affected by these limitations. Fig. 5 shows liquid films prepared using three different solutions: surfactants[48], polyacrylamide (PA)[36], and polydimethylsiloxane (PDMS)[52].

Although we can observe differences in the thickness and morphology of the different films, as shown in Fig. 5, the DH system successfully recorded thickness information. For transmissive holographic recording, light transmittance and polarization of the recording object must be considered. For thin-film materials, most liquid films have good light transmittance, and their polarization effects are so small that they can be ignored. Therefore, holographic measurements can be used for most transparent thin liquid films. Compared with liquid films, solid films are more difficult to image; a main reason is the molecular spacing of the solid films is closer than that of the liquid film, which means that the RI of solid films is generally greater than that of liquid films. Currently, the measurement technology for solid nonuniform thin films remains an unresolved challenge. Many factors limit conventional measurement methods[55], although some initial results have been obtained[30]. From the above experiments, holography has shown excellent measurement capabilities for different materials, is not limited to physical properties, and can successfully provide measurements for most of thin-film materials. This is a very interesting feature that challenges the stringent requirements of interferometry for samples.

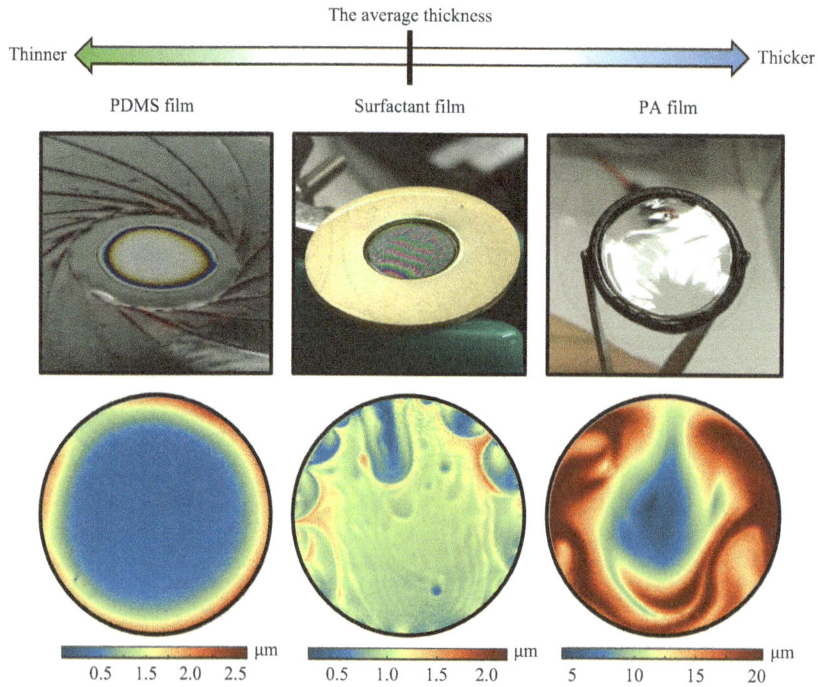

Fig. 5 Digital holography has excellent adaptability to the measurement of transparent thin liquid films with different thickness and surface morphology.

High-speed holographic recording for thin-film rupture

In the previous sections, the film imaging process at a standard recording speed was discussed. However, for holographic recording systems, owing to the excellent real-time imaging capabilities in off-axis geometry, it is possible to use a high-speed CMOS camera to dynamically follow the thickness evolution. Among these, the rupture process of films is a widely studied issue. For a long time, there has been no reliable measurement technology to capture the edge morphology during film rupture, and DH technology can therefore fill this gap.

The film rupture process can be easily imaged using a conventional off-axis holographic setup with a high-speed camera (980 fps). The related process was modelled to describe the rupture mechanism[33]. From the time–space model in Fig. 6c, it can be observed that after using the needle to break the film, it resists the rupture process. These results were revealed using holographic technology for the first time. We believe that relevant experimental data will provide important data support for researchers in fluid mechanics. In fact, many microfluidic phenomena occur very quickly and can be analyzed with the help of DH.

Digital holography for inkjet printing, additive manufacturing and advanced fabrication techniques

The electrohydrodynamic (EHD) manipulation of liquids is becoming an increasingly popular topic owing to its application in physical, chemical, biological, and engineering research areas, such as microdetectors, the micro-dispensing of small sub-nanoliter volumes of fluids for sensors, flat panel displays, and biochips[56]. EHD inkjet printing relies on the generation of droplets at or near a nozzle aperture, followed by noncontact deposition on a substrate with high spatial control. Recently, an interesting approach based on the use of a virtual nozzle on the pyroelectric forces has been introduced and applied for the manipulation and dispensing of liquid inks ranging from low to high viscosity. The pyroelectric effect involves the delivery of materials to surfaces at micrometer and nanometer length scales without nozzles or electrodes. The local heating induced on pyroelectric substrates, such as lithium niobate (LN), generates an electric field through the well-known pyroelectric effect[57–60]. Beyond a threshold electric potential, this field pulls ink from the free liquid surface of the reservoir to form a conical jet similar to the Taylor cone. Controlling the translation of an additional substrate inserted between the LN and the opposing

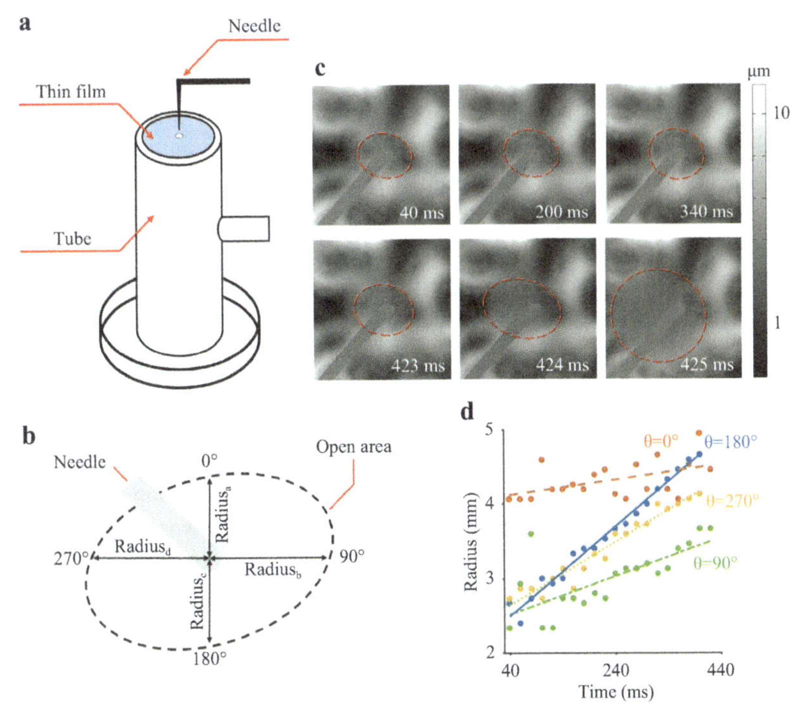

Fig. 6 Digital holographic high-speed imaging for recording thin-film rupture process, experimental scenarios, results, and modeling. **a** Implementation of the film rupture process. **b** Description of parameters of rupture area. **c** Holographic thickness reconstruction of the fracture process. **d** Modeling the process of film resistance to rupture. Reproduced from Ref. 36 © 2021 Springer Nature.

substrate with a liquid reservoir allows additive patterning of the ink materials. The use of pyroelectric effects has provided new capabilities that have been demonstrated in different application contexts, from high-resolution printing and patterning to the fabrication of customized polymeric microstructures[61–64]. Pyro-inkjet and EHD-printing in general have an interesting physical mechanism leading to diverse jetting behaviors according to the operating conditions and physicochemical properties of the liquid, such as surface tension, viscosity, electric conductivity, and relative permittivity[65]. Considering that EHD may be applied to different fields according to various ejection conditions, the visualization and understanding of the printing mode would result in better control of the entire fabrication process. Commonly, most of these modes are observed using only one camera, which is only appropriate for stable symmetric modes such as dripping, microdripping, and cone-jet modes, and not for other unstable asymmetric modes. In the case of unstable trajectories, such as the multispindle mode, oscillating-jet mode, precession, and multijet mode[66,67], 3D visualization is necessary to follow the trajectory and exact locations of the ejected droplets. To address this limitation, a 3D tomographic technique was proposed to study the unstable ejection phenomena of droplets[68]. In addition, the shadowgraphic tomography technique was implemented to reconstruct jetting three-dimensionally[69–71]. Recently, a numerical method for reconstructing the 3D phenomena was proposed to solve the problem of the invisible area by multiplicative line-of-sight (MLOS) and ellipse estimation; however, the experimental procedures are still long-lasting and complicated[69]. Even if good results have been obtained by non-intrusive measurement tomography, the setup required for the measurement is still very expensive, requiring more than two cameras, optical components, and dedicated software. In fact, to reconstruct acceptable results from limited projection data, different tomographic algebraic algorithms have been combined, which require time and computational resources[70,71]. Recently, quantitative phase microscopy (QPM) has been demonstrated to be a valid tool for the evaluation of liquid and polymer droplet characteristics. In Ref.[72], the authors reported on real-time quantitative phase imaging for measuring the evaporation dynamics of sessile microdroplets. The surface wetting properties were retrieved by 3D topographic profiles with nanometer accuracy. QPM has also been used to demonstrate the ability to observe nanosized, semi-transparent, structured coatings involving two materials by combining high-resolution maps of optical path differences and electrical impedance variations[73]. In this framework, DH microscopy was used to investigate the wetting properties of ink dots obtained by direct writing on a glass plate[74]. Even though the proposed method could track the temporal evolution of the height, volume, and contact angle of the printed ink dots of approximately 200 mm from gelation to solidification, no report exists that explains the quantitative visualization of 3D viscous polymers, which usually have a high aspect ratio and a trajectory that could span long distances in the FoV[75]. The quantification of polymeric objects, which rely heavily on manual operations, is labor-intensive and lacks accuracy. In the literature, we found some experiments involving simple cases such as needle and fiber entities, in which the fiber is straight and the tilted angle of the fiber is known[76]. A point cloud method, where a low gray-scale threshold was used to retrieve the focused part of the objects, was demonstrated; however, this method is applicable only to straight objects[77]. Recently, an alternative method based on DH was used to automatically characterize an irregular 3D polymeric object. Holograms of a stringy object were recorded using a high-speed DH system. A depth-of-field extended image (DEI) was first obtained by applying an image-fusion algorithm to reconstructed images at different depths[78]. The edge and skeleton of an object are extracted in 2D space[79]. The object was divided into small sections along the skeleton, and a hybrid method[80] was applied to measure the size and 3D location of each section. All sections were stitched together to retrieve the entire object. DH can simplify the 3D visualization of the manipulation process and simultaneously improve the understanding of unstable and asymmetric physical phenomena. Controlling the jetting of liquids and the manipulation of polymers by DH microscopy would extend the potential of direct writing technologies in different fields where high accuracy is required for fabrication procedures.

Application of digital holography to monitoring of fibers drawn by electrospinning

The manipulation of polymeric fibers is at the base of an innovative 3D printing fabrication technique, in which the pyroelectric effect is used for direct writing and the fabrication of overprinted scaffolds at the microscale. The printing process was activated directly from a reservoir polymeric drop, avoiding the use of nozzles and external voltages. Under the activation of the tethered pyro-electrospinning (TPES) process, the droplet started to deform, creating an elongated tip from which the fibers were drawn[81,82]. It is very important to follow the 3D movement of the polymeric fiber during the elongation process to achieve high precision in deposition and final resolution of the structure. The DH tool can be improved for 3D tracking, even in the case of polymer elongation. The results reported below refer to the activation of a polymeric drop of PDMS under the action of the pyroelectric effect. Drop activation can be described by three main steps: drop deformation, cone formation, and cone elongation. During the final step, the polymer fiber could be unstable, and the movement of the final part with the tip could be synchronized with the upper activation zone. Using a conventional visualization line made of a light source, camera, and objective, it would be difficult to follow the evolution in time and space of the polymeric cone. In fact, the cone can move along the vertical axis from the focus of the visualization. Eventually, a tracking technique could enable the reconstruction of more complex and dynamic structures. In particular, under the instability of the pyroelectric effect used for the manipulation of the material, it is possible to observe the activation of Rayleigh instability. Under this condition, beads-on-a-string (BOAS) are generated and can flow over the polymeric fiber. As investigated in our previous work, the elongation, deformation, and eventually the final breakup of viscoelastic filaments pass through the formation of temporary pearls interconnected by a thin thread, the so-called BOAS structures[83,84]. The evolution in time and space, together with the continuous movement of the fiber and beads flowing on it makes very difficult to observe this process and its characterization.

In this study, a transmissive off-axis holographic recording setup was used to track the movement of polymeric fibers. Owing to the refocusing process of DH, the nonlinear motion process of polymeric fibers can be effectively tracked. Fig. 7 shows the fiber tracking results for 25 frames. Under the action of the pyroelectric effect, PDMS forms fibers that can be recorded, which are so thin that the laser light can pass through them to achieve an effective holographic recording. This significantly improves the accuracy of 3D positioning. However, the numerical diffraction propagation for the entire FoV was difficult to obtain because the fiber showed an increasing trend in the z-axis direction, its top was far away from the recording plane, and its bottom was close to the recording plane. Therefore, we divided the fiber into 17 sub-strips

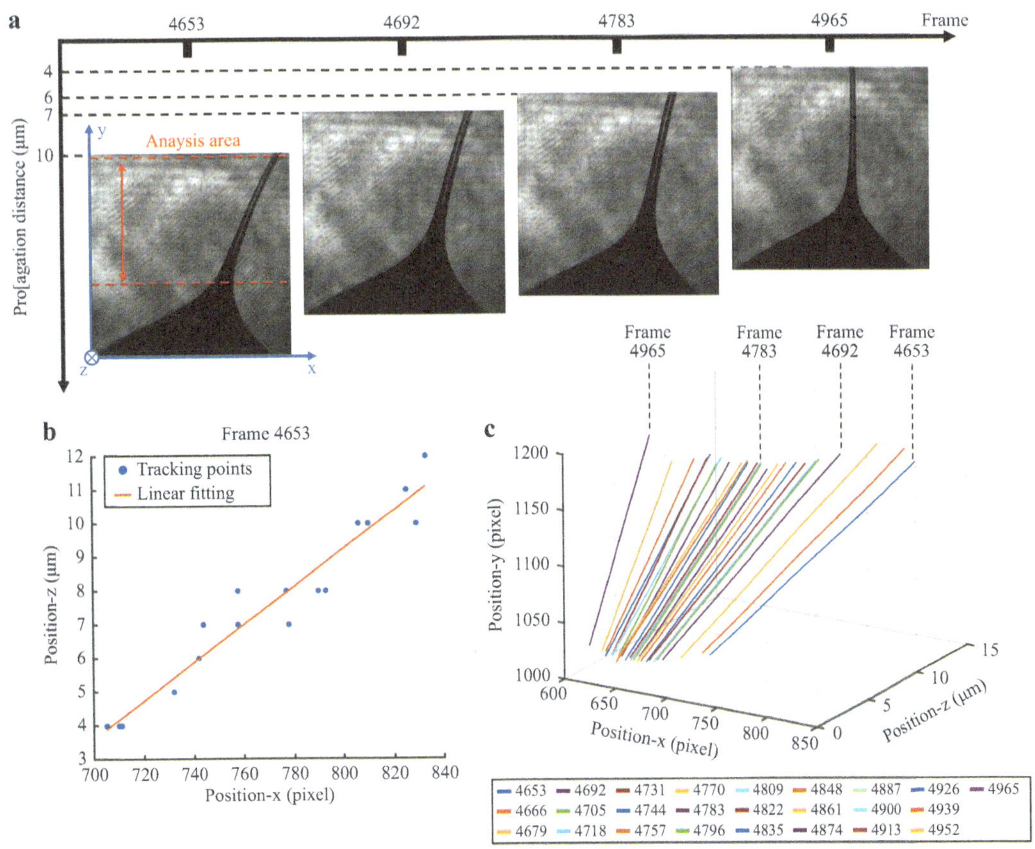

Fig. 7 a Experimental frames of polymer jetting, showing the movement of the fiber in time. **b** Tracking of the fiber position in frame 4653, the blue dots represent the focus position. **c** Three-dimensional fitting of discrete fiber position with a straight line for the experimental frames within the range 4653–4952.

and performed separate holographic reconstruction and Tamura auto-focusing on each strip to determine the best z-axis position. In Fig. 7b, we report the result of frame 4653 as an example, where the blue dots indicate the focal position of each band. The feature correlation recognition method was used to track the x-axis position of the fiber. Here, the y-axis position of the fiber is the absolute position, and the x- and z-axes are the measurement positions. Therefore, the spatial coordinate distribution has a certain discrete type under an overall uniform trend. To facilitate the identification, we fitted the 3D discrete fiber position identification points with a straight line for all steps of fiber movement, as shown in Fig. 7c.

Application of digital holography to monitoring of beads-on-a-string over fiber drawn by EHD

DH also enables us to accurately locate BOAS structures over thin polymeric fibers. As shown in Fig. 8, we selected three frames from a series of BOAS generation processes to reconstruct the position and size of each bead. The spatial location and size of each bead are shown in Fig. 8b. Here, multi-steps diffraction propagation and stripes reconstruction process were used to extract the bead information. Furthermore, in the 2D phase as shown in Fig. 8a, which is reconstructed by a uniform propagation distance, we could see that the beads were arranged vertically on the fiber, but after holographic 3D positioning, the size and spatial position of each bead, including z position, could be retrieved, providing important data for further quantitative analysis.

Digital holography and characterization of polymeric lenses printed by EDH ink jet printing

The process of cone formation at the base of the polymer elongation described in the previous section can also be adopted for the jetting of the polymeric drop. In particular, once the electric charges accumulated on the tip of the polymeric cone overcome the surface charges of the material, small volumes can be dispensed and collected onto a commercial target. In this configuration, the starting reservoir polymeric drop was deposited on a pillar base, while the pyroelectric crystal was mounted over the coverslip upon which the drop was supported, and was heated locally together with the base droplet, thus allowing the generation of pyroelectric surface charges on the crystal. The liquid droplet started to deform under the action of a sufficiently strong electric field, and was released from the conical tip microdroplets onto a moving substrate. By controlling the starting volume, contact angle of the substrate, and by selecting the polymer of interest, droplets of different geometries and volumes could be dispensed by direct writing (see Fig. 9).

Once cured, the polymeric structures could be used as optical microlenses, and interferometric characterization based on DH microscopy was carried out to evaluate the optical aberrations of the microstructures and their

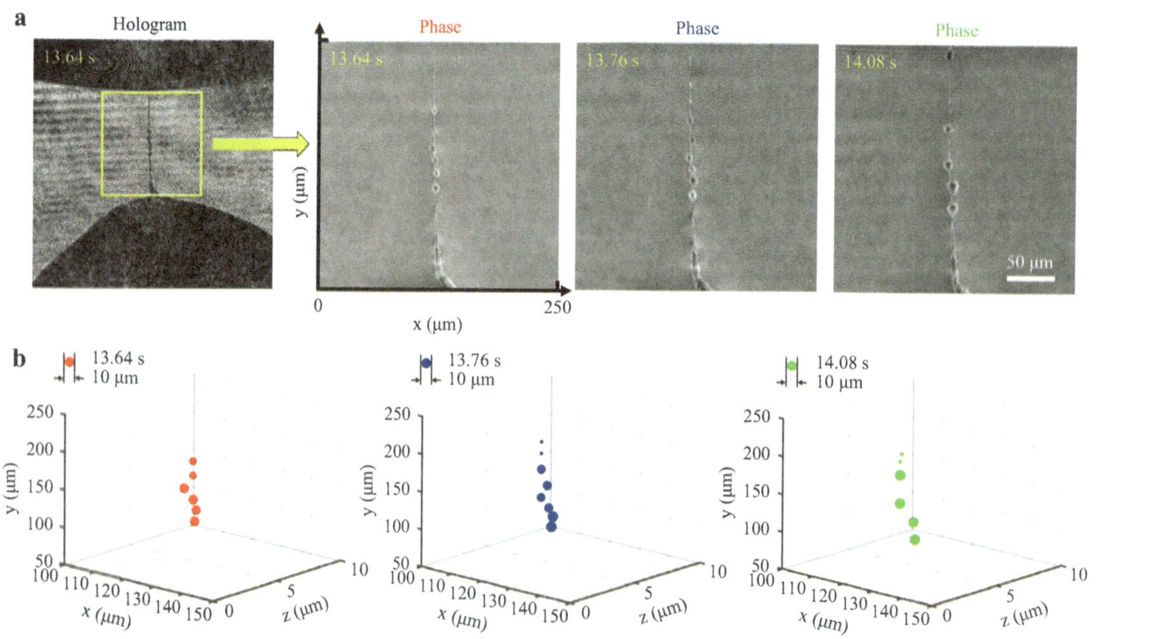

Fig. 8 **a** Hologram and phase reconstruction of experimental frames, the yellow region indicates the region of interest with the polymeric beads. **b** Evolution in time of tracking position of the beads flowing on the fiber.

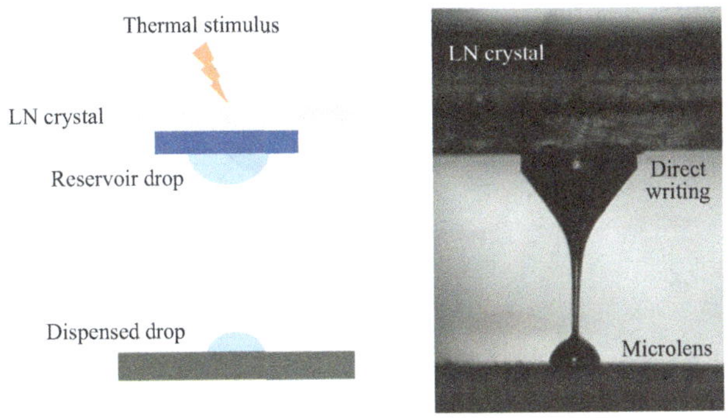

Fig. 9 a Outline of the direct writing set-up: a thermal stimulus is applied to the LN crystal for the activation of the pyro-EHD effect. **b** Direct writing of a polymeric microlens onto a commercial PMMA target.

focusing properties[85,86]. The optical setup consisted of a classical Mach-Zehnder interferometer in the transmission configuration, as shown in Fig. 10.

The beam emitted by a continuous-wave (CW) laser with wavelength $\lambda = 632.8$ nm is first split into two parts, constituting the reference and the object arm of the interferometer. Both encounter a 5 × microscope objective on their path toward the sensor (a 1024 × 1024 CCD camera with pixel pitch $\Delta x = \Delta y = 4.4$ μm). The object beam is directed toward the microfluidic chip where a sample microlens is deposited and it then recombines to the reference in the acquisition plane to produce an interference pattern. Once the hologram is recorded, numerical propagation provides the object complex field in any plane along the optical axis z, from which the intensity and phase distributions of the optical wavefield transmitted

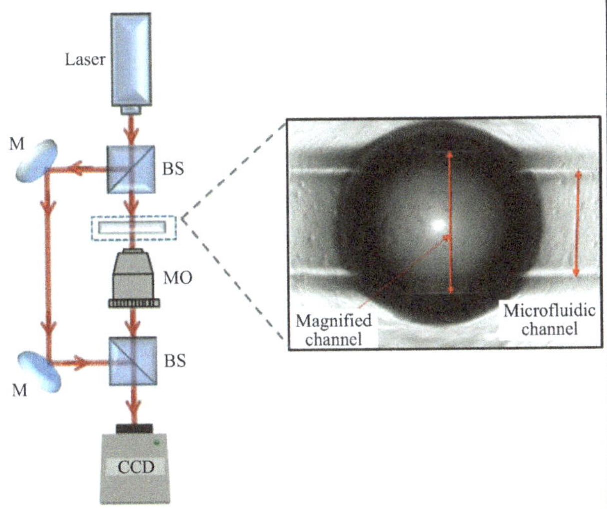

Fig. 10 Mach-Zehnder interferometer in transmission configuration used for the characterization of the polymeric lens fabricated on the microfluidic channel by direct writing. Fig. 10 was reproduced from Ref. 85 with permission from the Royal Society of Chemistry.

by the sample can be extracted[87]. A double-exposure method was adopted to compensate for the aberrations of the optical components in the setup, as well as the phase delay introduced owing to the passage of the object beam through the chip. Hence, the recovered phase-contrast map directly yields information on the optical behavior of the lens. The analysis was performed by exploiting the Zernike fitting to evaluate the contributions of the defocus and other aberration terms. The results of the Zernike analysis are shown in Fig. 11.

Digital Holography for characterizing self-assembled liquid microlenses

The pyroelectric effect has also been investigated to change the wettability of micro-engineered surfaces. In particular, the wettability of LN crystals is patterned by inducing a pyroelectric effect on the domain-engineered samples. As previously mentioned, this technique enables wettability patterning to be achieved by an electrode-less configuration, in contrast with conventional electro-wetting (EW) experiments. In fact, the surface charges inducing this effect are pyroelectrically generated and spatially modulated by the domain grating. A micro-array of thousands of sessile droplets is obtained on a single chip, and using a DH apparatus, it is possible to characterize such microstructures, demonstrating their optical quality as microlenses.

Wettability patterning at solid surfaces is an important aspect for different applications, such as the self-assembly and alignment of particles, as well as for all integrated microfluidic and optofluidic systems where liquid actuation is fundamental[88–91]. EW allows the alteration of the wetting properties of a substrate by means of free electrical charges[92,93]. Among all the application fields, EW has been demonstrated to be a key technology for realizing liquid lenses[94,95]. Usually, EW-based experiments require complex electrode geometries to actuate and control the surface wettability. Therefore, the possibility to pattern surface wettability by activating specific properties of the material without using external electrodes is desirable.

In this section, it is shown that by activating the pyroelectric effect in periodically poled LN (PPLN) substrates, it is possible to pattern and control the surface wettability without the need for external electrodes and circuits. This technique can be used to generate an array of sessile droplets according to the reversed domain structure of the PPLN substrate. LN is a rhombohedral crystal that exhibits pyroelectricity at certain condition. The spontaneous polarization, P_S, changes according to $P_i = \xi_i \Delta T$, where ξ_i is the pyroelectric coefficient and ΔT is the temperature variation. At equilibrium, all P_S in the

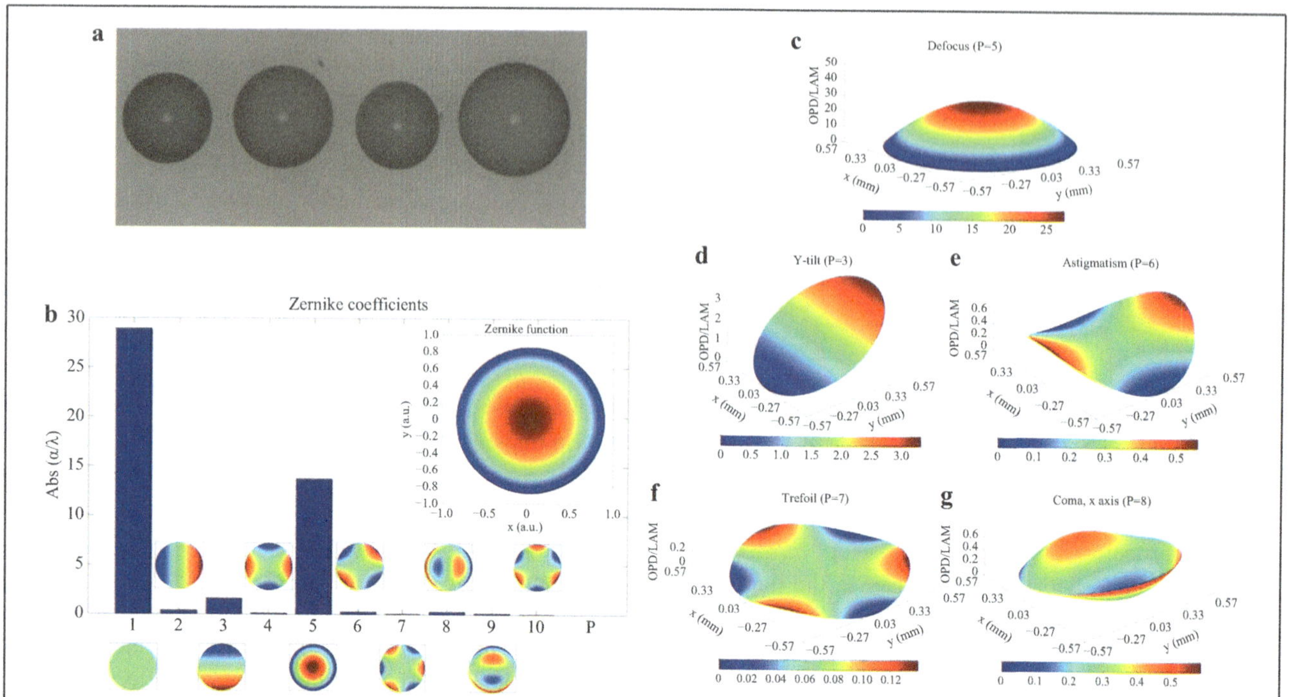

Fig. 11 **a** Microscope image of polymeric lenses. **b** Normalized Zernike coefficients derived from the lens phase-contrast map; for each order P, the corresponding Zernike polynomial is shown. Surface distribution of the main Zernike polynomials: **c** Spherical factor. **d** Tilt along the y-axis. **e** Astigmatism. **f** Trefoil aberration. **g** Coma aberration, x-axis. Fig. 11 was reproduced from Ref. 85 with permission from the Royal Society of Chemistry.

crystal are fully screened by the external screening charge, and no electric field exists[96]. A temperature change modifies P_S, and consequently, such equilibrium is perturbed, causing a lack or excess of surface charge, thus generating a high electric field. The spontaneous polarization of LN was inverted by photolithography and electric field poling to obtain a periodic square array of hexagonal reversed domains. Fig. 12a shows an image of the PPLN crystal. Fig. 12b shows a schematic of the patterning effect induced by the temperature change. A thin layer of an oily substance was spread onto the crystal surface. The sample was made of a crystal substrate, and the oil film was heated and then cooled down, the whole explored range of temperatures is from 25 to 100 °C[97]. All the reported results were obtained using carboxylic acid (pentanoic acid - C5H10O2) as the oily substance spread onto the z-cut LiNbO3 substrates. The liquid film experienced only a smooth shape modification during heating, with the liquid-air interface exhibiting a wave-like (WLR) profile following the geometry of the domain array (top of Fig. 12b). Conversely, pronounced wettability patterning is observed during cooling, which is called the separated lens regime (SLR), as shown in the bottom part of Fig. 12b, when the liquid film starts to break in different regions, forming isolated sessile droplets. The final result consists of an array of sessile droplets that follow the geometry of the PPLN domain structure (see the bottom part of Fig. 12d).

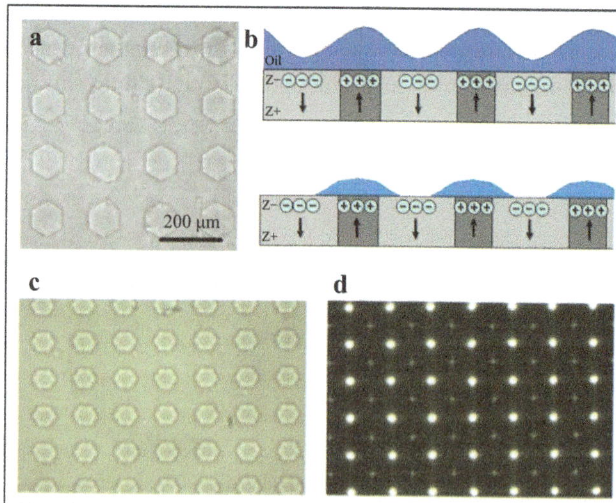

Fig. 12 **a** Optical microscope image of a typical PPLN with a square array of hexagonal reversed domains; **b** schematic views of the sample cross-section corresponding to the wave-like lens regime and to the separated lens regime of the micro-lens array, respectively. The black arrows indicate the orientation of the spontaneous polarization. **c** and **d** two corresponding images of a PPLN sample before and after the liquid reshaping on the surface, respectively. Sessile drops are present corresponding to each hexagon. Reprinted with permission from Refs. 98, 99 © The Optical Society.

Both regimes, i.e., WLR and SLR, were investigated using a DH microscope set-up to characterize the focusing behavior and optical aberrations of the microlenses. Fig. 13a shows a schematic view of the interferometric apparatus used for the measurements based on the Mach-Zehnder interferometer. The calculated wrapped phase map of the lens microarray is shown in Fig. 13b, and after applying an unwrapping algorithm, the continuous-phase image for the microlens array is retrieved, as shown in Fig. 13c.

The temperature of the thermo-controlled plate under the sample was varied at a rate of 10 °C/min, and it was controlled with an accuracy of 0.1 °C. Interferometric characterization was performed by registering the temperature value for each digital hologram captured during thermal treatment. Two different types of analyses were carried out to evaluate the temperature dependence of the focal length and optical aberrations of the lenses, thus characterizing both the tunability and optical performance of the device. The temperature dependence of the focal length was independently investigated during the cooling and heating treatments. The aberrations of the lenses were evaluated through a deeper analysis of the wavefronts transmitted by the lenses themselves, and this was done by exploiting the Zernike fitting[98–101].

The wrapped phase maps exhibit different behaviors in the case of heating and cooling. When the temperature increases, oil shaping occurs only up to a temperature value of approximately 65 °C, and consequently, the number of fringes increases. After this temperature, the lens shape stabilized. Conversely, during the cooling process, the number of fringes increased during the whole process. An unwrapping numerical procedure was applied to phase mode 2π to measure the real phase distribution of the wavefront. Subsequently, a one-dimensional (1D) fitting procedure was performed to evaluate the focal length of each microlens in the array. Fig. 14a, b show the experimental and fitted profiles of the unwrapped phase corresponding to the two different processes. The focal length trend vs. temperature variation is reported in Fig. 14c, d for both the heating and cooling experiments.

The optical behaviors of the microlenses were also characterized in terms of optical aberrations by applying a 2D fitting algorithm to digital holograms of a 4 × 4 lens array acquired during stationary conditions, that is, when the number of fringes is stable in the wrapped phase distribution, while the fitting procedure was performed for each lens of the array and for lenses of several arrays at different time instants. The function used for the fitting process is a linear combination of Zernike polynomials by considering the coefficients for the tilt, astigmatism, defocus, and third-order spherical aberration terms. The coma and higher-order coefficients were neglected because some orders of magnitude were minor.

Fig. 15 shows the surface distribution of each polynomial employed for Zernike fitting, and it enables us to visually compare the focus term with the other terms that describe the wavefront curvature. The main term is the defocus term, which allows the state of the quality of such a microstructure, and the spherical aberration term has a quite high value because the curvature changes from convex to concave between two adjacent microlenses.

Application of digital holography to the characterization of other printed micro-structures

DH can be used to characterize microfluidic structures. In fact, the direct printing previously described could be implemented for the fabrication of high-resolution microchannels with transversal sizes ranging from several

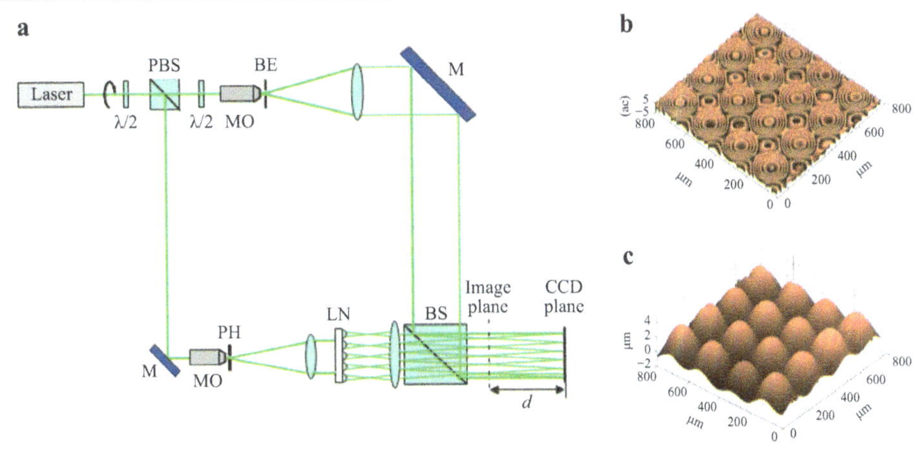

Fig. 13 Holographic liquid microlenses characterizing. Optical geometry is shown in **a**. Two-dimensional representation of **b** the wrapped and **c** the unwrapped phase map corresponding to the 4 × 4 liquid microlens array. Reprinted with permission from Ref. 99 © The Optical Society.

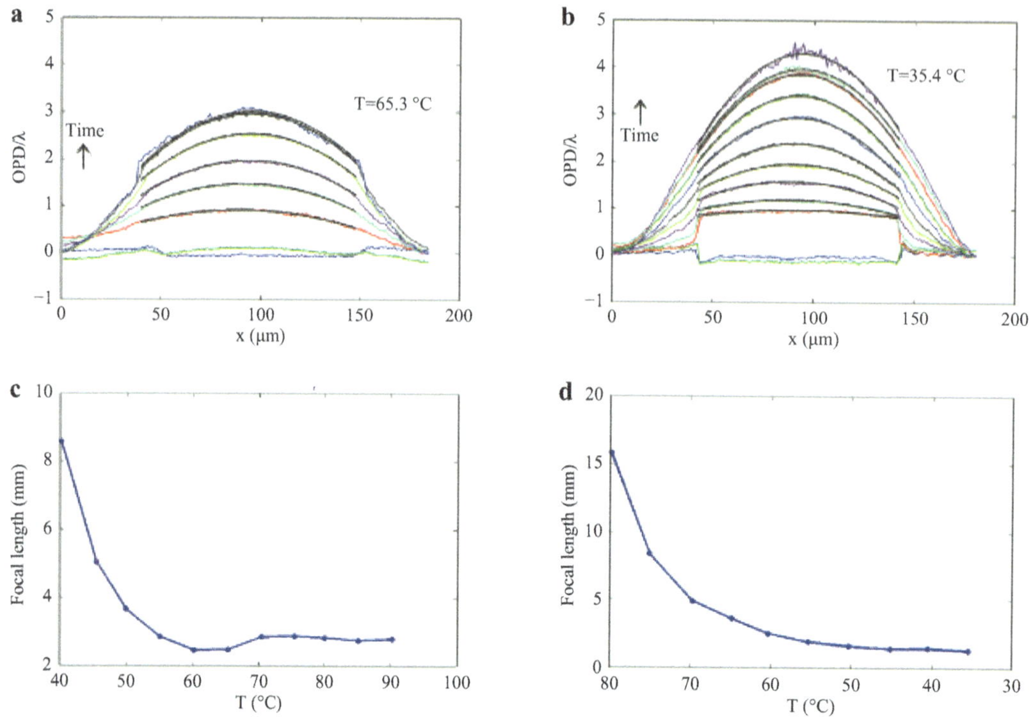

Fig. 14 Experimental and fitted profiles of the unwrapped phase distribution corresponding to **a** the heating and **b** cooling process; **c**, **d** focal length variation as a function of temperature in case of the heating and the cooling process. Reprinted with permission from Ref. 99 © The Optical Society.

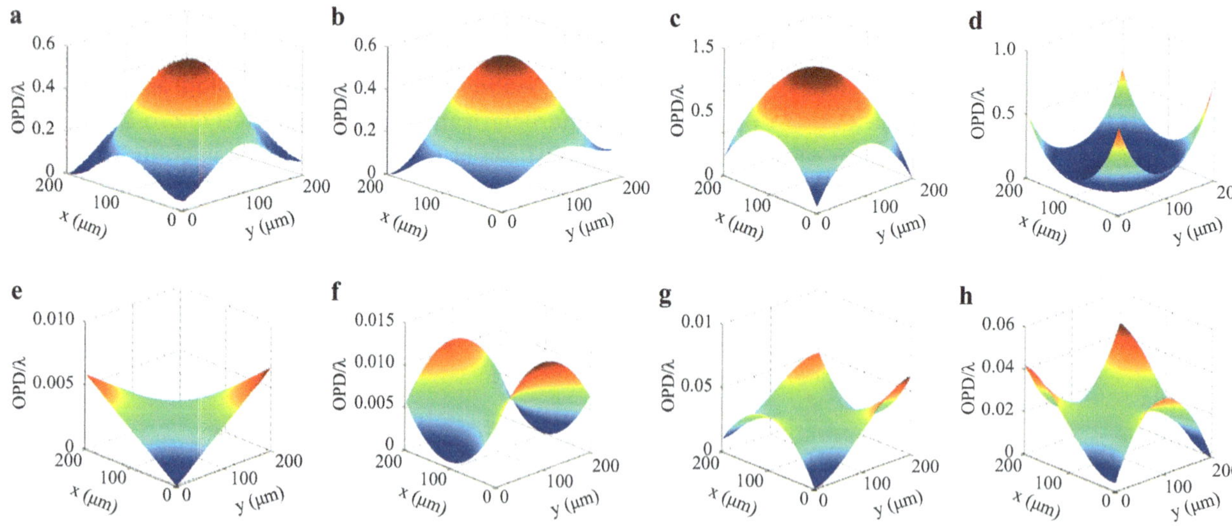

Fig. 15 **a** Measured phase distribution and **b** corresponding Zernike fitted surface; **c** surface distribution of the focus term, **d** the third order spherical aberration, **e** the astigmatism at 45° term, **f** the astigmatism at 90°, **g** the triangular astigmatism on the x base, **h** the triangular astigmatism on the y base. Reprinted with permission from Ref. 99 © The Optical Society.

hundreds of nanometers to hundreds of microns[76]. The printing resolution that has been demonstrated is well below the resolution limit of standard methods, even in the case of processing biocompatible and biodegradable materials, so the final devices are useful candidates for biological analysis because of their intrinsic absence of toxicity and immunogenicity (Fig. 16a, b). In the case of a straight channel (length 1.5 cm and diameter limited to approximately 2 μm), interferometric characterization based on DH microscopy was carried out to evaluate the

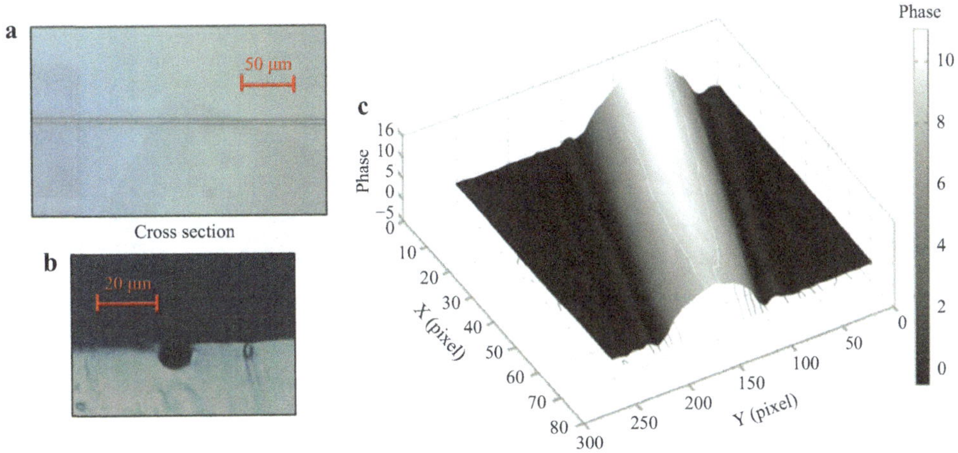

Fig. 16 a Top view of the PDMS device demonstrating the good control of direct printing of a single 3D printed footpath. **b** The channel cross-section is displayed by side view imaging and **c** characterized by digital holography. Adapted with permission from Ref. 102 Copyright 2021 American Chemical Society.

morphology and quality of the footpath finish, as shown in Fig. 16c. The cross-section of the channel is cylindrical, showing the 3D core of the pyro-electrospun microfiber.

Closed-loop fabrication of micro-scaffold combination of digital light processing and digital holography

Currently, the field of artificial biological tissues represents an area of much research interest, and applications range from artificial skin to liver lobules. Similar to the way that a complete LEGO set is made up of bricks, artificial biological tissues can be assembled on the cellular micro-scaffolds. These micro/nanoscale scaffolds are the navigators of cell growth; they can guide the cells' behavior by their spatially heterogeneous architecture, and further influence the functional expression of the engineered tissues directly[103]. Researchers have developed different micro/nano scaffold fabrication methods to advance techniques used when manufacturing different artificial tissues[104–106]. The possibility of monitoring in real-time and in situ high-resolution fabrication processes through accurate characterization remains a significant challenge. Recently, a closed-loop control system for the fabrication of high-fidelity micro-scaffolds was proposed[107]. A combination of digital light processing (DLP) and digital holographic microscopy (DHM) has been used to precisely manufacture cell-encapsulated hydrogel micro-scaffolds under real-time monitoring. Poly(ethylene glycol) diacrylate (PEGDA) and gelatin methacrylate (GelMA) hydrogels were used as experimental materials to fabricate micro-scaffolds using the DLP system. Meanwhile, the 3D imaging capability of DH provided support for the quantitative monitoring of liquid-solid conversion by ultraviolet (UV) light. Fig. 17a shows the proposed set-up used for the holographic feedback-based photocuring system. Herein, DLP guided the fabrication of hydrogels under UV light, while DHM furnished quantitative observation of the forming material. The UV beam, which is modulated by a digital micro-mirror device (DMD), illuminated the hydrogel in a designed manner. Meanwhile, the hydrogel was processed by the UV beam, and it was also irradiated by the laser beam from the DH setup. The object beam carrying its information interferes with the reference beam on the camera screen to form a digitally recorded hologram.

The experimental results are shown in Fig. 17b. To verify the calibration capability of the DH for the fabrication process, comparative experiments were conducted. Herein, the first micro-scaffold (test 1) was fabricated using the DLP printing system without DH monitoring, whereas the second micro-scaffold (test 2) was fabricated under the feedback control of the DHM. Each micro-scaffold was divided into seven blocks by dashed lines to fit the average thickness of each block. The expected thickness of the fabrication template was 280 μm, as indicated by the green line in the plot. Comparing the results, micro-scaffold 2 exhibited smoother surface topography. In general, the fluctuation range of the fitted curve for micro-scaffold 1 was 65.4 μm, which was larger than the fluctuation range obtained for micro-scaffold 2, which was 12.7 μm. From the cross-sectional view, it can be more clearly observed that DHM played a key role in improving the manufacturing accuracy and stability of the DLP system. Using the same approach, the relationship between the Young's modulus and exposure duration was investigated. The above experimental results effectively

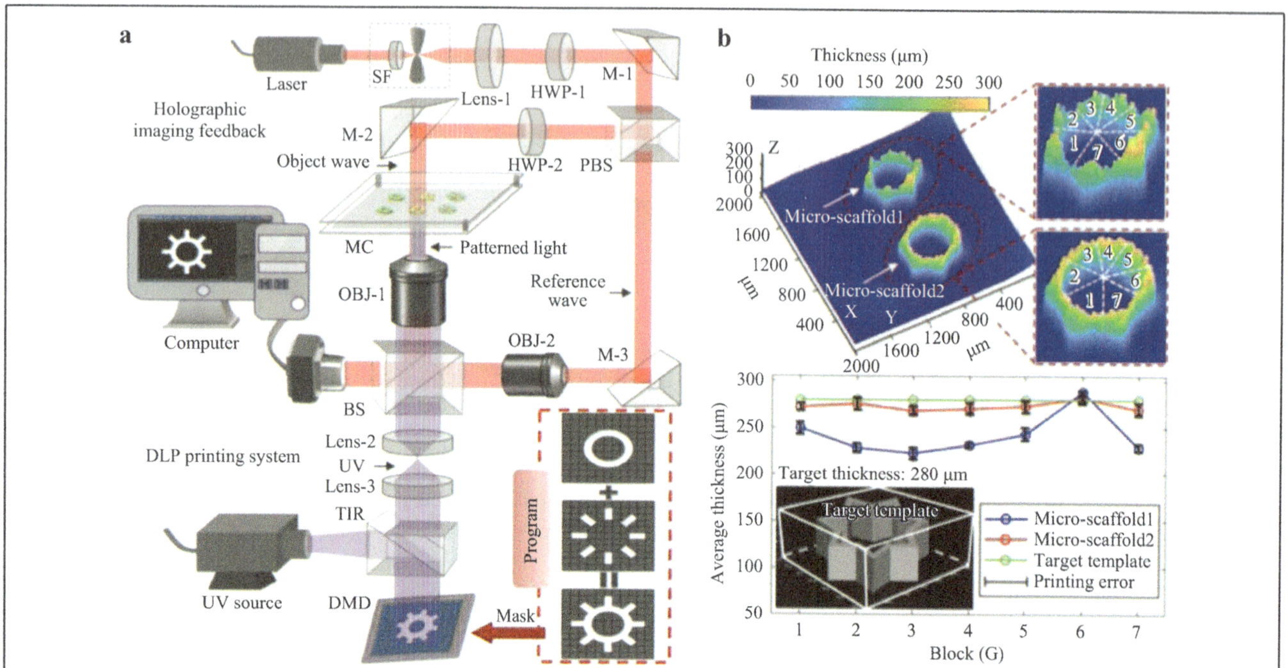

Fig. 17 Closed-loop control fabrication system and experimental results for hydrogel micro-scaffolds. **a** Optical setup. **b** Digital holographic reconstruction results of 3D polygonal micro-scaffold. © [2021] IEEE. Reprinted, with permission, from Ref. 107.

prove the positive effect of DHM on hydrogel micro-scaffold fabrication for constructing a heterogeneous architecture inside a single integrated micro-scaffold. The use of DHM as a closed-loop control method to improve the fabrication accuracy will be an important topic in the near future, especially for the processing of biological materials. The organic combination of transmissive, micron-level, real-time, and 3D imaging in DHM is expected to lead to the development of a good tool for the fast and high-precision fabrication of artificial biological tissues.

Assisting two-photon polymerization fabrication process by digital holography

The increase in miniaturization from medical engineering to consumer electronics requires the development of advanced fabrication methods. Recently, 3D printing was proposed for the fabrication of micro-optics and miniaturization of spectroscopic compact and integrated measurement systems[108]. However, among the existing manufacturing methods, 3D lithography by two-photon polymerization shows intriguing functionalities for printing complex 3D structures of different geometries, including curvilinear features, with a submicrometer resolution[109]. Recently, there has been increased attention on the characterization of samples fabricated by such technology. In fact, it is challenging to address specific goals for two-photon lithography and even more useful manufacturing approaches, such as the non-destructive online monitoring of photoinduced polymerization. Holographic monitoring of photopolymerization dynamics has recently been demonstrated by characterizing the evolution of the photoresist RI during the photopolymerization process[110,111], please see Fig. 18.

Monitoring creation of surface relief gratings using digital holography

Complex surface relief patterns can be manufactured by the light-induced surface structuring of azobenzene-containing films. Such relief patterns, which have different heights and topography patterns, are difficult to achieve using conventional lithography systems, and understanding their formation dynamics and response to different types of light irradiation is still an open task. Recently, DH has been explored to investigate the manufacturing processes. In particular, significant results have been reported on the potential for the development of holographic tools that can write the desired surface relief grating (SRG) structures, and furnish the quasi real-time and in-situ monitoring of the process. In one configuration, a Lloyds interferometer was used to create the SRG pattern, while a DH microscope, working with a He-Ne laser emitting at 632.8 nm, was used to monitor its formation and dynamic evolution[112]. In a more recent paper, the interference pattern for SRG writing was monitored using a DHM operating with a 488-nm light input and a polarization-maintaining

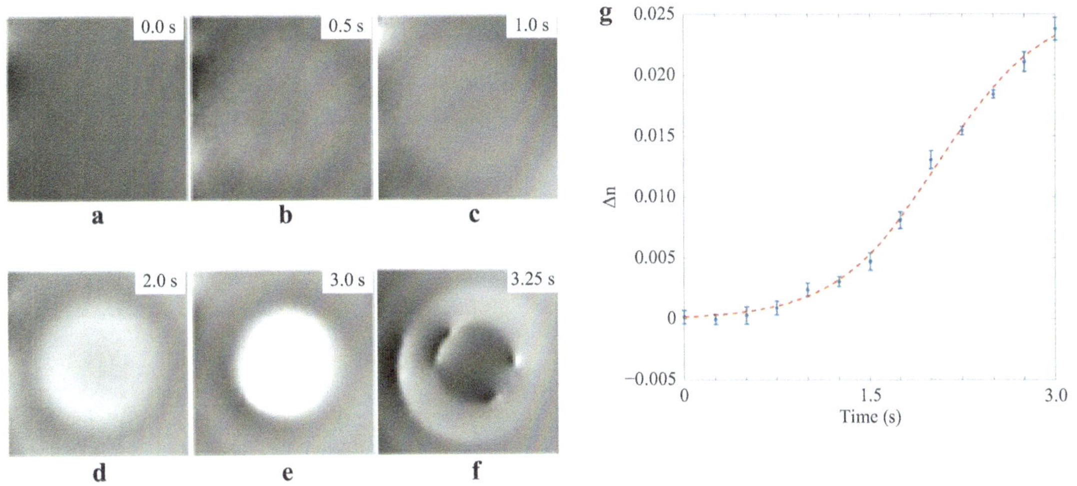

Fig. 18 Online holographic monitoring of the refractive index evolution: phase maps corresponding to **a** the initial state of the sample before the exposure started, and exposure times of **b** 0.5 s, **c** 1.0 s, **d** 2.0 s, **e** 3.0 s. **f** Explosion of the polymer after illumination for 3.0 s. All figures represent the same area of 400 × 400 pixels. **g** Change in the material 's refractive index Δn during the photoinduced polymerization. Red dashed line represents the fitting curve for the experimental data with error bars. Adapted with permission from Ref. 110 © The Optical Society.

fiber[113]. The results reported in this paper prove that DH can provide an optimal measuring instrument for monitoring the formation of SRG patterns. The experimental setup and results are shown in Fig. 19.

Toward the investigation of rheology, stress monitoring, and particle tracking by digital holography

Stress monitoring in soft materials using digital holography

In the pioneering era of DH, many applications have been explored for quantitative analysis and to obtain accurate measurements in transparent materials. The analysis of liquids, gases, plasma, and photoelastic materials has been demonstrated. In fact, owing to the intrinsic features of holographic methods, it was possible to study stress formation in photopolymers during the live recording of holograms. A recent paper[114] demonstrated that it is possible to detect and retrieve stress formation during the recording and bleaching of photopolymer holograms as a result of the distortion of the internal diffractive structure. This is particularly important in the case of holographic solar concentrators or other holographic optical elements that must meet stringent specifications. DH measured the stress formation in photopolymers during hologram acquisition by recording many holograms at successive exposure time intervals.

Another example of the capability of DH to measure deformation was presented in Ref. 115. In this research, it was shown that the deformation of cylindrical nanostructures made of silica aerogels can be analyzed using DH. In particular, two-beam phase-shifting digital holographic interferometry was applied to investigate in-plane displacements. While the structures under investigation were mechanically loaded with various loads from the top surface, phase-shifted holograms of the silica aerogels with and without deformation were acquired using a CCD camera. Reconstruction of the complex amplitude of aerogel samples using the Fresnel diffraction integral was adopted to analyze the reordered holograms. In-plane displacements were obtained from the phase distribution of the deformed aerogels by measuring the change in the two-phase distributions, enabling the observation of their elastic behavior. The deformation of silica aerogels was observed within the range of 0.02058 cm to 0.0017 cm, corresponding to an increase in bulk density from 0.01742 gm cc^{-1} to 0.1952 gm cc^{-1}.

A further example of the application of DH in measuring mechanical properties is the investigation of the Young's modulus of polyacrylamide gels[116]. DH is shown to be a convenient technique for measuring the Young's modulus of soft materials for cellular adhesion. The interaction between an elastic substrate and adherent cells plays a key role in regulating functions and behaviors in the concentration, migration, and proliferation of the cells. Polyacrylamide (PAA) is a soft polymeric gel and a flexible substrate with specific mechanical properties commonly used for testing the mechanical properties of cells. However, to quantitatively measure the stress induced by cells, it is necessary to determine their mechanical response with high accuracy. The accurate

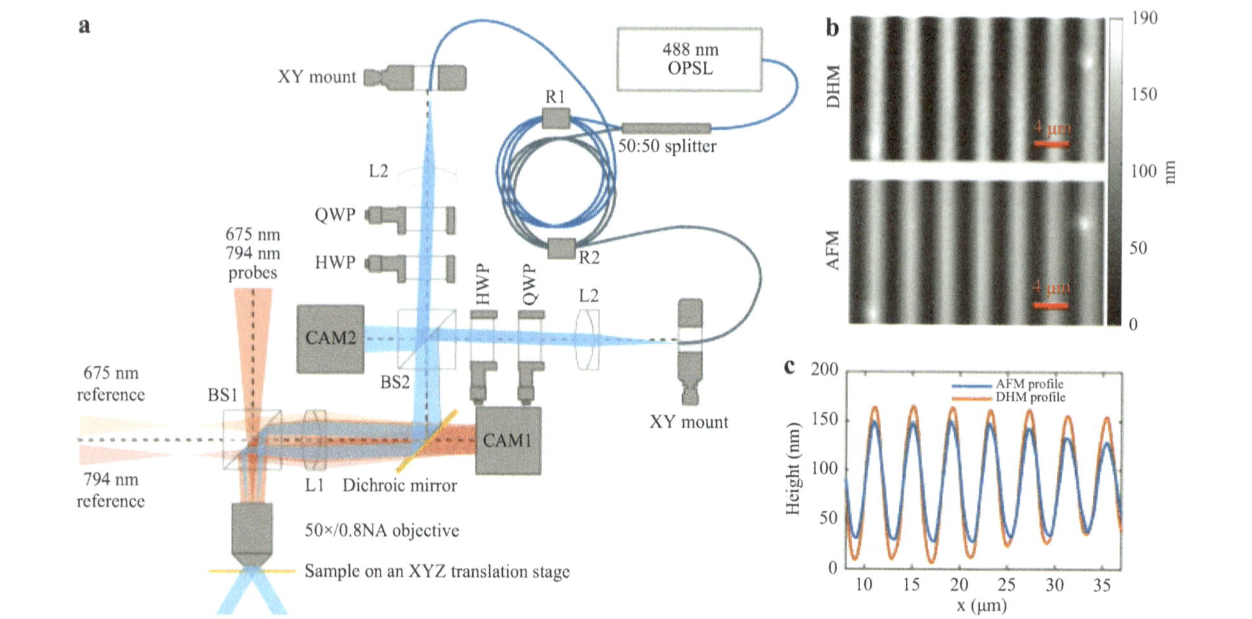

Fig. 19 Optical measurement setup for simultaneous surface height measurement and SRG patterning. Reproduced from Ref. 113 © 2021 Springer Nature.

measurement of the Young's modulus can be achieved through DH by retrieving the optical phase of the light transmitted through the transparent gel.

Investigating microfluidic rheology and liquid-solid interfaces using digital holography

To the best of our knowledge, Lorenz-Mie analysis has been widely used in in-line DHM to track and measure colloidal particles[117], which is also known as holographic particle characterization (HPC). Owing to the Lorenz-Mie theory of light scattering, the 3D tracking of particles with nanometer precision while simultaneously measuring each sphere's radius and refractive index could be achieved. This technique can be used not only for a single particle but also for measuring multiple colloidal particles, as reported by Fung et al.[118]; in this study, up to six colloidal spheres were accurately modeled based on superposition solutions. In the past decade, HPC with Lorenz-Mie analysis has been applied to different areas of colloidal particles. In 2014, Krishnatreya et al. successfully measured the Boltzmann constant of a single particle[119], which means that individual colloidal particles can be tracked over large distances with high precision. In 2017, Cheong et al. applied HPC to colloidal particles in turbid media[120], for the first time verifying its applicability in translucent media. In addition to the above applications, HPC can be used to guide the development of synthesis protocols for colloidal particles[121]. The monodisperse TPM sphere fabrication process was examined by HPC to study the influence of protocol choices on the size and refractive index.

In a recent study, DH microscopy in conjunction with particle tracking velocimetry (PTV) was implemented to measure microfluidic shear rheology and the wall slip of viscoelastic fluids[122,123]. The technique involves introducing non-deformable microparticles into the fluid, recording the holograms, and identifying the particle centroid locations in 3D with a digital reconstruction of the particle scattering field for an imposed driving pressure. Starting from the PTV, it is possible to retrieve the velocity and characterize the shear rheology of the viscoelastic fluids. Then, using a small sample volume (approximately 20 μl), it was possible to characterize the shear rheology of viscoelastic fluids across a wide range of shear rates. The results were compared with those of simulations and classical macrorheometry. Moreover, a recent study used DH to monitor surface slippage to investigate microparticle sedimentation[124]. Understanding the behavior of microparticle sedimentation in fluids is of fundamental importance for many industrial processes and natural phenomena. Digital holographic microscopy was used to measure the effect of the proximity of a flat microfluidic chamber wall on the sedimentation of colloidal microspheres. The numerical refocusing feature of DH was employed to track multiple particles instantaneously, obtaining quantitative images of the samples. By analyzing the 3D paths and speeds of multiple sedimentation microparticles, it was shown that the proximity to flat walls of higher slip lengths causes faster sedimentation. This effect was driven by the ratio of the particle size to the slip length and its distance from the walls. Moreover, the dependence of the 3D trajectories on the internal coatings provided to the chamber walls was also retrieved. In Fig. 20, the paths followed by different microparticles are shown, as they are retrieved by DHM.

Holographic technology has a wide range of applications in microfluidic imaging. A typical example is the application of holographic particle image velocimetry (HPIV) to track particles in fluids[125,126], especially when the particle density is high and the 3D spatial distribution is complex. Here, HPIV technology based on an off-axis DH setup was used to track the particles inside the thin films and realize the tracking of large-scale 3D liquid flow[36]. As the twin technology of in-line DH with Lorenz-Mie analysis, off-axis HPIV focuses on particle tracking for undefined spaces with ultralong-distance movement.

In this experiment, particles with a diameter of 10 μm were injected into the PA solution, which moved with the film during the pumping of the bubble. Fig. 21 shows the trajectories of the three particles, from which the gravitational drainage process of the thin film can be clearly observed. The most notable among them is particle A, which first raised gradually with the film and then quickly moved to the edge with a decrease in height. These datasets can be used to better understand the liquid flow movement process on the surface of the film and convert complicated liquid flow tracking into simple particle tracking. In fact, this technology can be applied in many experiments on film motion processes in the study of membrane gravitational drainage and rupture processes.

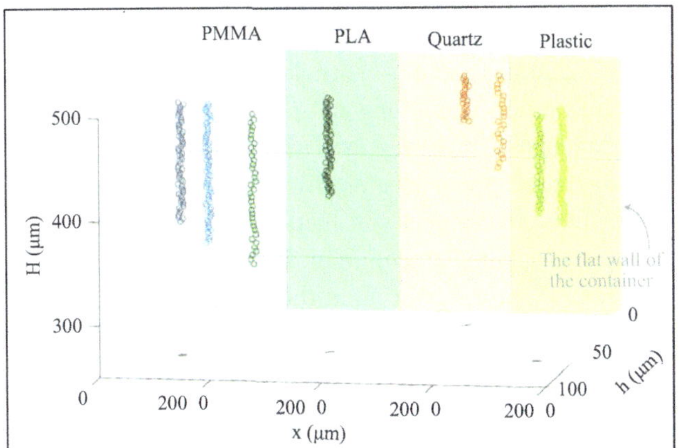

Fig. 20 3D trajectories of several polystyrene micro-particles at different distances from the flat wall of the container within 5 s and in different levels of hydrophobicity wall effects (PMMA, PLA, Quartz, Plastic). The different materials and the sedimenting micro-particles near them are shown in different colors. The different colored circles show the sedimenting micro-particle distances to the flat wall; lighter colors indicate farther ones. Reproduced from Ref. 124 © 2021 Springer Nature.

Application of deep learning to soft matter holographic measurement

The integration of deep learning with DH imaging is an emerging and widely studied topic. In the digital holographic morphology of soft matter, deep learning can also be used to enhance recording and numerical reconstruction processes, especially for colloid aggregation[127]. In 2020, Altman et al. proposed a method for characterizing and tracking colloids holographically (CATCH) in colloidal dispersions[128]. In this study, in-line holograms of colloids were studied using a neural network to acquire the properties of the particles. Although the CATCH output did not achieve the full precision of the conventional Lorenz-Mie analysis, it successfully provided approximate results in a shorter time, providing a new solution for the analysis of the colloidal dynamics process. In 2021, Midtvedt et al. used deep learning to enhance the optical characterization of subwavelength particles[129]. A neural network was applied to link the reconstructed results with the real radius and refractive index of individual subwavelength particles. In this case, the holographic scattering patterns of single particles contain sufficient information to reveal the subwavelength size and the precise refractive index. As a result, the aggregation kinetics of a sample consisting of polystyrene nanoparticles were monitored by combining DH and deep learning without prior knowledge of the physical properties of the surrounding medium. Furthermore, this method could be extended to other application scenarios; for example, tracking the movement of aggregates in special microfluidic channels, and it will open new possibilities for characterizing particle interactions and properties[130]. Neural networks can also be used to improve the accuracy of phase reconstruction for the morphology of time-varying objects, such as fluids and soft matter. Lu et al. proposed an Ynet-ConvLSTM network for extracting phase information during dynamic holographic imaging[131]. The time-related information in the hologram sequence was learned by the neural network, and it was then used to improve the reconstruction accuracy and suppress the background noise. The author implemented the proposed method on quantitative holographic droplet alcohol evaporation imaging, and a comparison between different reconstruction methods was presented. From the results obtained, the phase information obtained by the Ynet-ConvLSTM network reveals higher precise object information compared with the results obtained from the conventional Unet method and two-step phase-shifting

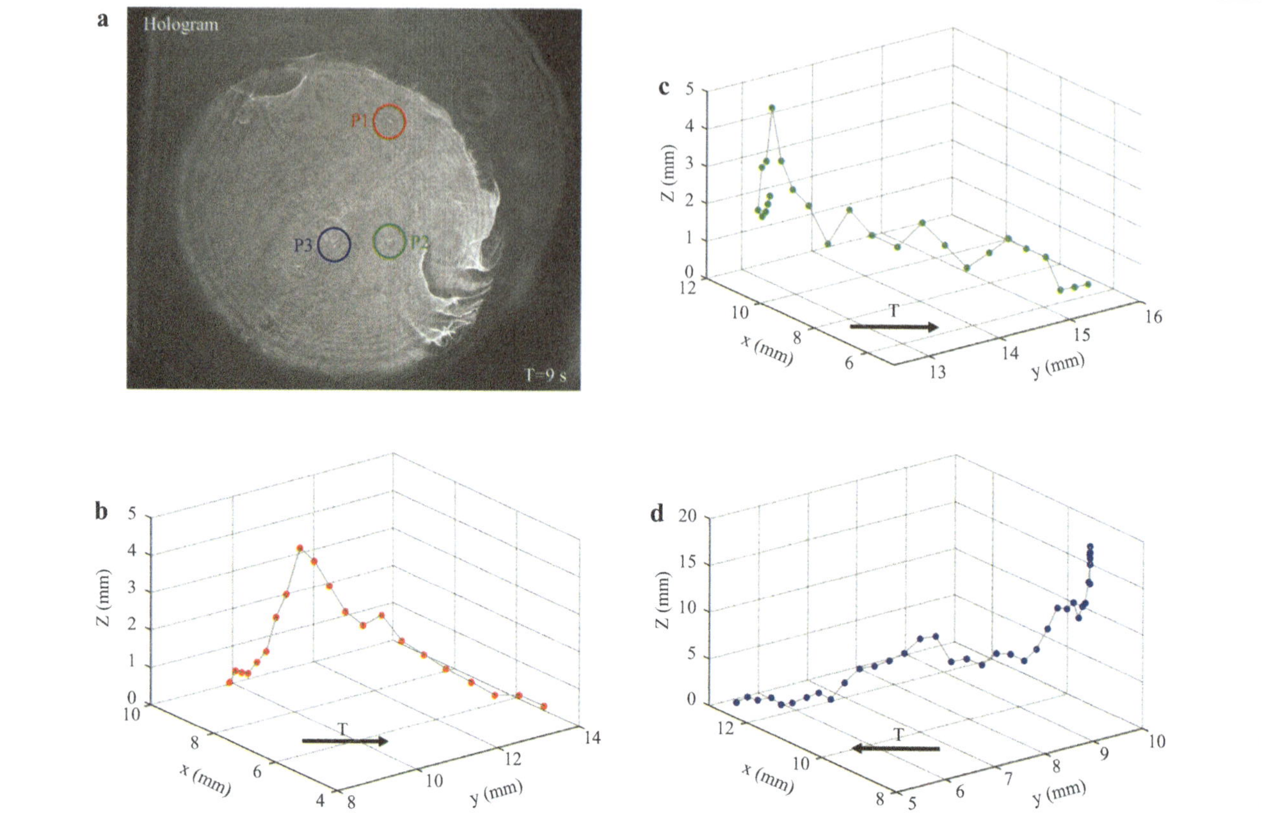

Fig. 21 Holographic particle image velocimetry of particles in pumping bubble. **a** the hologram particles could be clearly observed. **b**, **c**, **d** Three-dimensional trajectory of particle movement. Reproduced from Ref. 36 © 2021 Springer Nature.

algorithm. Through the combination of conventional holographic recording and deep learning reconstruction, phase extraction under a neural network will challenge the conventional numerical diffraction reconstruction process to achieve dynamic high-precision low-noise imaging. In the future, the combination of the DH method based on the deep learning reconstruction process and wavefront shaping[132] may create more possibilities and expand the application scenarios of holographic soft-matter measurement.

Conclusions

We have presented, reviewed, and extensively discussed how DH can be a suitable metrology tool for use in a variety of challenging cases dealing with soft matter, that is, with liquid/polymer matter and fast dynamically evolving processes such as inkjet printing. DH can successfully provide engineering solutions for many applications and basic scientific questions. Even though this review is not exhaustive, we believe that we have reported the most significant cases in which such measurement techniques can be effectively applied to soft matter. We have intentionally excluded "biological matter", as it is outside the scope of this review. In particular, we reported on the applications of DH measurement for different types of soft matter, such as thin films, EHD drawn fibers, polymeric lenses, micro-channels, hydrogel micro-scaffolds, photoresists, surface relief gratings, and particles. Table 1 lists the implementation conditions and parameters of the relevant measurements, where we identified the optimal measurement approaches for different materials/processes.

For most transparent soft matter, DH is one of the best candidates for implementing measurements. Among them, the special case is for ultra-thin liquid films, meaning liquid films with an average thickness of less than 1 μm; WLI will be the best method in theory because it has excellent resolution for smooth sub-micron thickness distribution. Meanwhile, we indicate the possible measurement ranges for different materials, but these ranges are only reasonable references. In soft matter holographic measurements, the precise measurement range is related to the choice of recording wavelength and sample morphology. For an interferometric imaging system, the horizontal resolution depends on the selected magnification lens, while the axial resolution depends on the recording

Table 1 Cross-overview over the different materials and DH presented approaches

Soft matter	Method	Light Source	Measurement range	Sample feature
Thin-film	WLI	White light	~50 nm to ~1.5 μm	Smooth liquid film with thickness less than 1.5 microns.
	Off-axis DH	Visible laser	~300 nm to ~100 μm	Transparent film with continuous morphology.
	Hybrid approach	White light Visible laser	~50 nm to ~100 μm	All types of liquid film.
EHD drawn fiber Polymeric lens Micro-channel Hydrogels micro-scaffold Photoresist Surface Relief Grating	Off-axis DH	Visible laser	According to morphology and phase unwrapping.	Transparent or translucent soft- matter micro-structures with continuous thickness distribution and no-hollow structure.
Particles	In-line DH	Visible laser	Nanoscale particles moving in microscale.	Solution embedded particles.
	Off-axis DH	Visible laser	Microscale particles moving in milliscale.	Solution or films embedded particles.

wavelength and the minimum pixel size of the camera. Therefore, one of the advantages of holographic measurement is its adjustable measurement range and optimized measurement accuracy. Although holographic phase errors can be analyzed and calibrated through specific methods, the main measurement uncertainty still comes from the light source itself[133], which is different from interferometric measurement. The measurement error of interferometry arises from the number of observed interference fringes and the position of the sample[134-136]. This also implies that one of the reasons for choosing DH as the main measurement method for soft matter is that the placement of the sample does not affect the measurement accuracy.

The results reviewed here show that DH is robust in many challenging situations as it does not require frequent calibration while maintaining the attractive features of DH, that is, non-invasive, no-contrast agent, full-field imaging, high spatial resolution, and temporal high-frequency, up to the MHz range. We believe that DH has a bright future in the field of soft matter and advanced fabrication processes owing to its intrinsic capability to perform 3D imaging combined with quantitative analysis[137]. This idea is also supported by the increase in technological applications of thin films, from electronic to cosmic protective clothing and packaged water in gravity-free environments. Several other examples and published papers can be found discussing other challenging applications, we apologize if it was not possible to include all of them in this review.

Acknowledgements
This work is partially supported by the MIUR project, "Piattaforma Modulare Multi Missione" (PM3), ARS01_01181.

Author details
[1]CNR-ISASI, Institute of Applied Sciences and Intelligent Systems "E. Caianiello", Via Campi Flegrei 34, 80078 Pozzuoli, Napoli, Italy. [2]Dipartimento di Ingegneria Chimica dei Materiali e della Produzione Industriale, Università degli Studi di Napoli "Federico II", Piazzale Tecchio 80, 80125 Napoli, Italy. [3]NEAPoLIS, Numerical and Experimental Advanced Program on Liquids and Interface Systems, Joint Research Center CNR Università degli Studi di Napoli "Federico II", Napoli, Italy

Conflict of interest
The authors declare that they have no conflict of interest.

Received: 22 September 2021 Revised: 20 January 2022 Accepted: 20 January 2022
Accepted article preview online: 28 January 2022
Published online: 18 February 2022

References

1. Stetson, K. A. & Powell, R. L. Interferometric hologram evaluation and real-time vibration analysis of diffuse objects. *Journal of the Optical Society of America* **55**, 1694-1695 (1965).
2. Rastogi, P. K. Holographic Interferometry: Principles and Methods. (Berlin: Springer, 2013).
3. Goodman, J. W. & Lawrence, R. W. Digital image formation from electronically detected holograms. *Applied Physics Letters* **11**, 77-79 (1967).
4. Schnars, U. & Jüptner, W. Direct recording of holograms by a CCD target and numerical reconstruction. *Applied Optics* **33**, 179-181 (1994).
5. Singh, V. R. & Asundi, A. In-line digital holography for dynamic metrology of MEMS. *Chinese Optics Letters* **7**, 1117-1122 (2009).

6. Lai, Y. W. et al. Integrity of micro-hotplates during high-temperature operation monitored by digital holographic microscopy. *Journal of Microelectromechanical Systems* **19**, 1175-1179 (2010).
7. Osten, W. Optical microsystems metrology. *Optics and Lasers in Engineering* **36**, 75-76 (2001).
8. Pedrini, G. et al. Calibration of optical systems for the measurement of microcomponents. *Optics and Lasers in Engineering* **47**, 203-210 (2009).
9. Matrecano, M. et al. Improving holographic reconstruction by automatic Butterworth filtering for microelectromechanical systems characterization. *Applied Optics* **54**, 3428-3432 (2015).
10. Pagliarulo, V. et al. Numerical tools for the characterization of microelectromechanical systems by digital holographic microscopy. *Journal of Micro/Nanolithography* **14**, 041314 (2015).
11. Grilli, S. et al. Whole optical wavefields reconstruction by digital holography. *Optics Express* **9**, 294-302 (2001).
12. Ferraro, P. et al. Compensation of the inherent wave front curvature in digital holographic coherent microscopy for quantitative phase-contrast imaging. *Applied Optics* **42**, 1938-1946 (2003).
13. Cuche, E., Bevilacqua, F. & Depeursinge, C. Digital holography for quantitative phase-contrast imaging. *Optics Letters* **24**, 291-293 (1999).
14. Kim, M. K. Principles and techniques of digital holographic microscopy. *SPIE Reviews* **1**, 018005 (2010).
15. Colomb, T. et al. Extended depth-of-focus by digital holographic microscopy. *Optics Letters* **35**, 1840-1842 (2010).
16. Fratz, M. et al. Digital holography in production: an overview. *Light:Advanced Manufacturing* **2**, 134-146 (2021).
17. Yang, L. et al. Multi-material multi-photon 3D laser micro-and nanoprinting. *Light:Advanced Manufacturing* **2**, 1-17 (2021).
18. Lee, A. et al. Optimization of experimental parameters to determine the jetting regimes in electrohydrodynamic printing. *Langmuir* **29**, 13630-13639 (2013).
19. Kim, M. K. Applications of digital holography in biomedical microscopy. *Journal of the Optical Society of Korea* **14**, 77-89 (2010).
20. Memmolo, P. et al. Breakthroughs in photonics 2013: holographic imaging. *IEEE Photonics Journal* **6**, 0701106 (2014).
21. Miccio, L. et al. Perspectives on liquid biopsy for label-free detection of "circulating tumor cells" through intelligent lab-on-chips. *View* **1**, 20200034 (2020).
22. Shaked, N. T. Quantitative phase microscopy of biological samples using a portable interferometer. *Optics Letters* **37**, 2016-2018 (2012).
23. Bishitz, Y. et al. Optical-mechanical signatures of cancer cells based on fluctuation profiles measured by interferometry. *Journal of Biophotonics* **7**, 624-630 (2014).
24. Park, Y. K., Depeursinge, C. & Popescu, G. Quantitative phase imaging in biomedicine. *Nature Photonics* **12**, 578-589 (2018).
25. Blum, O. & Shaked, N. T. Prediction of photothermal phase signatures from arbitrary plasmonic nanoparticles and experimental verification. *Light:Science & Applications* **4**, e322 (2015).
26. Kuschmierz, R. et al. Ultra-thin 3D lensless fiber endoscopy using diffractive optical elements and deep neural networks. *Light:Advanced Manufacturing* **2**, 1-10 (2021).
27. Mysels, K. J. Soap films and some problems in surface and colloid chemistry. *The Journal of Physical Chemistry* **68**, 3441-3448 (1964).
28. Hartl, M. et al. Thin film colorimetric interferometry. *Tribology Transactions* **44**, 270-276 (2001).
29. Israelachvili, J. N. Thin film studies using multiple-beam interferometry. *Journal of Colloid and Interface Science* **44**, 259-272 (1973).
30. Kim, S. W. & Kim, G. H. Thickness-profile measurement of transparent thin-film layers by white-light scanning interferometry. *Applied Optics* **38**, 5968-5973 (1999).
31. Chen, H. et al. Structural coloration by internal reflection and interference in hydrogel microbubbles and their precursors. *Advanced Optical Materials* **9**, 2100259 (2021).
32. Afanasyev, Y. D., Andrews, G. T. & Deacon, C. G. Measuring soap bubble thickness with color matching. *American Journal of Physics* **79**, 1079-1082 (2011).
33. Kitagawa, K. Thin-film thickness profile measurement by three-wavelength interference color analysis. *Applied Optics* **52**, 1998-2007 (2013).
34. Vannoni, M. et al. Measuring the thickness of soap bubbles with phase-shift interferometry. *Optics Express* **21**, 19657-19667 (2013).
35. Gao, F., Muhamedsalih, H. & Jiang, X. Q. Surface and thickness measurement of a transparent film using wavelength scanning interferometry. *Optics Express* **20**, 21450-21456 (2012).
36. Mandracchia, B. et al. Quantitative imaging of the complexity in liquid bubbles' evolution reveals the dynamics of film retraction. *Light:Science & Applications* **8**, 20 (2019).
37. Wong, W. S. Y. et al. Super liquid repellent surfaces for anti-foaming and froth management. *Nature Communications* **12**, 5358 (2021).
38. Osten, W. Optical Inspection of Microsystems. (Boca Raton: CRC Press, 2018).
39. Ferraro, P. et al. Extended focused image in microscopy by digital holography. *Optics Express* **13**, 6738-6749 (2005).
40. Zhang, J. et al. A review of common-path off-axis digital holography: towards high stable optical instrument manufacturing. *Light:Advanced Manufacturing* **2**, 1-17 (2021).
41. Wang, Z. et al. Interferometric measurement of film thickness during bubble blowing. Proceedings of SPIE 10333, Optical Methods for Inspection, Characterization, and Imaging of Biomaterials III. Munich, Germany: SPIE, 2017.
42. Joye, J. L., Hirasaki. G. J. & Miller, C. A. Dimple formation and behavior during axisymmetrical foam film drainage. *Langmuir* **8**, 3083-3092 (1992).
43. Bioucas-Dias, J. M. & Valadao, G. Phase unwrapping via graph cuts. *IEEE Transactions on Image Processing* **16**, 698-709 (2007).
44. Sett, S., Sinha-Ray, S. & Yarin, A. L. Gravitational drainage of foam films. *Langmuir* **29**, 4934-4947 (2013).
45. Schnars, U. et al. Digital holography. in Digital Holography and Wavefront Sensing: Principles, Techniques and Applications (eds Schnars, U. et al.) (Berlin, Heidelberg: Springer, 2014), 39-68.
46. Huang, T. S. Digital holography. *Proceedings of the IEEE* **59**, 1335-1346 (1971).
47. González-Cano, A. & Bernabéu, E. Automatic interference method for measuring transparent film thickness. *Applied Optics* **32**, 2292-2294 (1993).
48. Ferraro, V. et al. Full-field and quantitative analysis of a thin liquid film at the nanoscale by combining digital holography and white light interferometry. *The Journal of Physical Chemistry C* **125**, 1075-1086 (2021).
49. Hoyt, L. F. & Verwiebe, A. Determination of the concentration of liquid soaps by the immersion refractometer. *Industrial & Engineering Chemistry* **18**, 581-582 (1926).
50. Glassner, A. Soap bubbles: part 2. *IEEE Computer Graphics and Applications* **2**, 99-109 (2000).
51. Osten, W. et al. Recent advances in digital holography. *Applied Optics* **53**, G44-G63 (2014).
52. Ferraro, V. et al. Axisymmetric bare freestanding films of highly viscous liquids: preparation and real-time investigation of capillary leveling. *Journal of Colloid and Interface Science* **596**, 493-499 (2021).
53. Kim, Y. et al. Surface measurement of indium tin oxide thin film by wavelength-tuning Fizeau interferometry. *Applied Optics* **54**, 7135-

7141 (2015).
54. Zhang, Y. R. et al. Nanoscopic terraces, mesas, and ridges in freely standing thin films sculpted by supramolecular oscillatory surface forces. *ACS Nano* **10**, 4678-4683 (2016).
55. Piegari, A. & Masetti, E. Thin film thickness measurement: a comparison of various techniques. *Thin Solid Films* **124**, 249-257 (1985).
56. Onses, M. S. et al. Mechanisms, capabilities, and applications of high-resolution electrohydrodynamic jet printing. *Small* **11**, 4237-4266 (2015).
57. Ferraro, P. et al. Dispensing nano–pico droplets and liquid patterning by pyroelectrodynamic shooting. *Nature Nanotechnology* **5**, 429-435 (2010).
58. Coppola, S. et al. Self-assembling of multi-jets by pyro-electrohydrodynamic effect for high throughput liquid nanodrops transfer. *Lab on a Chip* **11**, 3294-3298 (2011).
59. Gennari, O. et al. Investigation on cone jetting regimes of liquid droplets subjected to pyroelectric fields induced by laser blasts. *Applied Physics Letters* **106**, 054103 (2015).
60. Coppola, S. et al. Nanocomposite polymer carbon-black coating for triggering pyro-electrohydrodynamic inkjet printing. *Applied Physics Letters* **106**, 261603 (2015).
61. Ruggiero, F. et al. Electro-drawn polymer microneedle arrays with controlled shape and dimension. *Sensors and Actuators B:Chemical* **255**, 1553-1560 (2018).
62. Coppola, S. et al. On the spraying modality of liquids by pyroelectrohydrodynamics. *ACS Omega* **3**, 17707-17716 (2018).
63. Madugani, R. et al. Terahertz tuning of whispering gallery modes in a PDMS stand-alone, stretchable microsphere. *Optics Letters* **37**, 4762-4764 (2012).
64. Russo, P. et al. Single fibres of pyro-electrospinned PVDF-HFP/MWCNT unveal high electrical conductivity. *Polymer* **159**, 157-161 (2018).
65. Hayati, I., Bailey, A. & Tadros, T. F. Investigations into the mechanism of electrohydrodynamic spraying of liquids: II. Mechanism of stable jet formation and electrical forces acting on a liquid cone. *Journal of Colloid and Interface Science* **117**, 222-230 (1987).
66. Jaworek, A. & Krupa, A. Jet and drops formation in electrohydrodynamic spraying of liquids. A systematic approach. *Experiments in Fluids* **27**, 43-52 (1999).
67. Jaworek, A. & Krupa, A. Generation and characteristics of the precession mode of EHD spraying. *Journal of Aerosol Science* **27**, 75-82 (1996).
68. Gim, Y. et al. Development of limited-view and three-dimensional reconstruction method for analysis of electrohydrodynamic jetting behavior. *Optics Express* **25**, 9244-9251 (2017).
69. Atkinson, C. & Soria, J. An efficient simultaneous reconstruction technique for tomographic particle image velocimetry. *Experiments in Fluids* **47**, 553-568 (2009).
70. Lu, W. K. & Yin, F. F. Adaptive algebraic reconstruction technique: adaptive algebraic reconstruction technique. *Medical Physics* **31**, 3222-3230 (2004).
71. Verhoeven, D. Limited-data computed tomography algorithms for the physical sciences. *Applied Optics* **32**, 3736-3754 (1993).
72. Edwards, C. et al. Measuring the nonuniform evaporation dynamics of sprayed sessile microdroplets with quantitative phase imaging. *Langmuir* **31**, 11020-11032 (2015).
73. Polonschii, C. et al. High-resolution impedance mapping using electrically activated quantitative phase imaging. *Light:Science & Applications* **10**, 20 (2021).
74. Yokota, M. & Aoyama, F. Drying process of an ink-dot analyzed using both digital holographic microscopy and tackiness measurement. *Microelectronic Engineering* **241**, 111543 (2021).
75. Yao, L. C. et al. Three-dimensional dynamic measurement of irregular stringy objects via digital holography. *Optics Letters* **43**, 1283-1286 (2018).
76. Lebrun, D. et al. Particle field digital holographic reconstruction in arbitrary tilted planes. *Optics Express* **11**, 224-229 (2003).
77. Kempkes, M. et al. Three dimensional digital holographic profiling of micro-fibers. *Optics Express* **17**, 2938-2943 (2009).
78. Wu, Y. C. et al. Wavelet-based depth-of-field extension, accurate autofocusing, and particle pairing for digital inline particle holography. *Applied Optics* **53**, 556-564 (2014).
79. Busse, F. et al. Digital interference microscopy and density reconstruction of picosecond infrared laser desorption at the water-air interface. *Journal of Applied Physics* **124**, 094701 (2018).
80. Guildenbecher, D R. et al. Digital holography simulations and experiments to quantify the accuracy of 3D particle location and 2D sizing using a proposed hybrid method. *Applied Optics* **52**, 3790-3801 (2013).
81. Coppola, S. et al. Tethered pyro-electrohydrodynamic spinning for patterning well-ordered structures at micro-and nanoscale. *Chemistry of Materials* **26**, 3357-3360 (2014).
82. Coppola, S. et al. Layered 3D printing by tethered pyro-electrospinning. *Advances in Polymer Technology* **2020**, 1252960 (2020).
83. Grilli, S. et al. 3D lithography by rapid curing of the liquid instabilities at nanoscale. *Proceedings of the National Academy of Sciences of the United States of America* **108**, 15106-15111 (2011).
84. Mecozzi, L. et al. Easy printing of high viscous microdots by spontaneous breakup of thin fibers. *ACS applied materials & interfaces* **10**, 2122-2129 (2018).
85. Vespini, V. et al. Forward electrohydrodynamic inkjet printing of optical microlenses on microfluidic devices. *Lab on a Chip* **16**, 326-333 (2016).
86. Bianco, V. et al. Endowing a plain fluidic chip with micro-optics: a holographic microscope slide. *Light:Science & Applications* **6**, e17055 (2017).
87. Camou, S., Fujita, H. & Fujii, T. PDMS 2D optical lens integrated with microfluidic channels: principle and characterization. *Lab on a Chip* **3**, 40-45 (2003).
88. Liu, S. H. et al. Assembly and alignment of metallic nanorods on surfaces with patterned wettability. *Small* **2**, 1448-1453 (2006).
89. Mugele, F. & Herminghaus, S. Electrostatic stabilization of fluid microstructures. *Applied Physics Letters* **81**, 2303-2305 (2002).
90. Psaltis, D., Quake, S. R. & Yang, C. Developing optofluidic technology through the fusion of microfluidics and optics. *Nature* **442**, 381-386 (2006).
91. Jones, T. B. et al. Dielectrophoretic liquid actuation and nanodroplet formation. *Journal of Applied Physics* **89**, 1441-1448 (2001).
92. Mugele, F. & Baret, J. C. Electrowetting: from basics to applications. *Journal of Physics:Condensed Matter* **17**, R705-R774 (2005).
93. Wang, D. B. et al. Local wettability modification by thermochemical nanolithography with write-read-overwrite capability. *Applied Physics Letters* **91**, 243104 (2007).
94. Hayes, R A. & Feenstra, B. J. Video-speed electronic paper based on electrowetting. *Nature* **425**, 383-385 (2003).
95. Moran, P. M. et al. Fluidic lenses with variable focal length. *Applied Physics Letters* **88**, 041120 (2006).
96. Bourim, E. M. et al. Investigation of pyroelectric electron emission from monodomain lithium niobate single crystals. *Physica B:Condensed Matter* **383**, 171-182 (2006).
97. Ferraro, P. et al. Wettability patterning of lithium niobate substrate by modulating pyroelectric effect to form microarray of sessile droplets.

Applied Physics Letters **92**, 213107 (2008).
98. Grilli, S. et al. Liquid micro-lens array activated by selective electrowetting on lithium niobate substrates. *Optics Express* **16**, 8084-8093 (2008).
99. Miccio, L. et al. Tunable liquid microlens arrays in electrode-less configuration and their accurate characterization by interference microscopy. *Optics Express* **17**, 2487-2499 (2009).
100. Miccio, L. et al. Hemicylindrical and toroidal liquid microlens formed by pyro-electro-wetting. *Optics Letters* **34**, 1075-1077 (2009).
101. Grimaldi, I. A. et al. Graded-size microlens array by the pyro-electrohydrodynamic continuous printing method. *Applied Optics* **52**, 7699-7705 (2013).
102. Coppola, S. et al. Direct writing of microfluidic footpaths by pyro-EHD printing. *ACS Applied Materials & Interfaces* **9**, 16488-16494 (2017).
103. Ricotti, L. et al. Quantification of growth and differentiation of C2C12 skeletal muscle cells on PSS–PAH-based polyelectrolyte layer-by-layer nanofilms. *Biomedical Materials* **6**, 031001 (2011).
104. Aubin, H. et al. Directed 3D cell alignment and elongation in microengineered hydrogels. *Biomaterials* **31**, 6941-6951 (2010).
105. Ahadian, S. et al. Moldable elastomeric polyester-carbon nanotube scaffolds for cardiac tissue engineering. *Acta Biomaterialia* **52**, 81-91 (2017).
106. Kim, S. et al. Fabrication and characterization of magnetic microrobots for three-dimensional cell culture and targeted transportation. *Advanced Materials* **25**, 5863-5868 (2013).
107. Li, X. *et al.* Holographic display-based control for high-accuracy photolithography of cellular micro-scaffold with heterogeneous architecture. *IEEE/ASME Transactions on Mechatronics*. http://dx.doi.org/10.1109/TMECH.2021.3081769 (2021).
108. Toulouse, A. et al. 3D-printed miniature spectrometer for the visible range with a 100× 100 μm^2 footprint. *Light:Advanced Manufacturing* **2**, 2 (2021).
109. Saha, S. K. et al. Scalable submicrometer additive manufacturing. *Science* **366**, 105-109 (2019).
110. Besaga, V. R. et al. Monitoring of photochemically induced changes in phase-modulating samples with digital holographic microscopy. *Applied Optics* **58**, G41-G47 (2019).
111. Zhao, X. & Rosen, D. W. Real-time interferometric monitoring and measuring of photopolymerization based stereolithographic additive manufacturing process: sensor model and algorithm. *Measurement Science and Technology* **28**, 015001 (2016).
112. Pagliarulo, V. et al. Direct quantitative imaging of the writing stage in a photosensitive azopolymer by digital holography. *Soft Matter* **15**, 7809-7813 (2019).
113. Rekola, H. et al. Digital holographic microscopy for real-time observation of surface-relief grating formation on azobenzene-containing films. *Scientific Reports* **10**, 19642 (2020).
114. Jith, A., Kumar, P. T. A. & Kumaran, R. K. Digital holographic method to study stress formation in photopolymer during live recording of holograms. *Optical Engineering* **60**, 035105 (2021).
115. Chikode, P. P. et al. Deformation studies of cylindrical nanostructured silica aerogels by using phase shifting digital holographic interferometry. *Materials Today:Proceedings* **46**, 2298-2306 (2021).
116. Yu, X. et al. Measurement of Young' s modulus of polyacrylamide gel by digital holography. Digital Holography and Three-Dimensional Imaging 2011. Tokyo Japan: Optical Society of America, 2011.
117. Wang, C. et al. Celebrating *Soft Matter's* 10th Anniversary: monitoring colloidal growth with holographic microscopy. *Soft Matter* **11**, 1062-1066 (2015).
118. Fung, J. et al. Imaging multiple colloidal particles by fitting electromagnetic scattering solutions to digital holograms. *Journal of Quantitative Spectroscopy and Radiative Transfer* **113**, 2482-2489 (2012).
119. Krishnatreya, B. J. et al. Measuring Boltzmann's constant through holographic video microscopy of a single colloidal sphere. *American Journal of Physics* **82**, 23-31 (2014).
120. Cheong, F. C. et al. Holographic characterization of colloidal particles in turbid media. *Applied Physics Letters* **111**, 153702 (2017).
121. Middleton, C. et al. Optimizing the synthesis of monodisperse colloidal spheres using holographic particle characterization. *Langmuir* **35**, 6602-6609 (2019).
122. Gupta, S. & Vanapalli, S. A. Microfluidic shear rheology and wall-slip of viscoelastic fluids using holography-based flow kinematics. *Physics of Fluids* **32**, 012006 (2020).
123. Koukourakis, N. et al. Wavefront shaping for imaging-based flow velocity measurements through distortions using a Fresnel guide star. *Optics Express* **24**, 22074-22087 (2016).
124. Panahi, M. et al. 3D monitoring of the surface slippage effect on micro-particle sedimentation by digital holographic microscopy. *Scientific Reports* **11**, 12916 (2021).
125. Wu, Y. C. et al. Quantifying bubble size and 3D velocity in a vortex with digital holographic particle tracking velocimetry (DHPTV). *Flow Measurement and Instrumentation* **76**, 101826 (2020).
126. Meng, H. et al. Holographic particle image velocimetry: from film to digital recording. *Measurement Science and Technology* **15**, 673-685 (2004).
127. Gao, Z. Y. et al. Distortion correction for particle image velocimetry using multiple-input deep convolutional neural network and Hartmann-Shack sensing. *Optics Express* **29**, 18669-18687 (2021).
128. Altman, L. E. & Grier, D. G. CATCH: characterizing and tracking colloids holographically using deep neural networks. *The Journal of Physical Chemistry B* **124**, 1602-1610 (2020).
129. Midtvedt, D. Deep learning enhanced digital holography for characterization of nanoparticles and soft matter. Proceedings of SPIE 11804, Emerging Topics in Artificial Intelligence (ETAI) 2021. San Diego, California, United States: SPIE, 2021.
130. Midtvedt, B. et al. Fast and accurate nanoparticle characterization using deep-learning-enhanced off-axis holography. *ACS Nano* **15**, 2240-2250 (2021).
131. Lu, S. Y. et al. Dynamic quantitative phase imaging based on Ynet-ConvLSTM neural network. *Optics and Lasers in Engineering* **150**, 106833 (2022).
132. Forouhesh Tehrani, K. et al. In situ measurement of the isoplanatic patch for imaging through intact bone. *Journal of Biophotonics* **14**, e202000160 (2021).
133. Thurman, S. T. & Fienup, J. R. Phase-error correction in digital holography. *Journal of the Optical Society of America A* **25**, 983-994 (2008).
134. Lowman, A. E. & Greivenkamp, J. E. Interferometer errors due to the presence of fringes. *Applied Optics* **35**, 6826-6828 (1996).
135. Berger, R., Sure, T. & Osten, W. Measurement errors of mirrorlike, tilted objects in white light interferometry. Proceedings of SPIE 6616, Optical Measurement Systems for Industrial Inspection V. Munich, Germany: SPIE, 2007.
136. Pruss, C. et al. Computer-generated holograms in interferometric testing. *Optical Engineering* **43**, 2534-2540 (2004).
137. Sheridan, J. T. et al. Roadmap on holography. *Journal of Optics* **22**, 123002 (2020).

Zhang et al. *Light: Advanced Manufacturing* (2021)2:23
https://doi.org/10.37188/lam.2021.023

Official journal of the JHL 2689-9620
www.light-am.com

Review | Open Access

A review of common-path off-axis digital holography: towards high stable optical instrument manufacturing

Jiwei Zhang, Siqing Dai, Chaojie Ma, Teli Xi, Jianglei Di* and Jianlin Zhao*

Abstract

Digital holography possesses the advantages of wide-field, non-contact, precise, and dynamic measurements for the complex amplitude of object waves. Today, digital holography and its derivatives have been widely applied in interferometric measurements, three-dimensional imaging, and quantitative phase imaging, demonstrating significant potential in the material science, industry, and biomedical fields, among others. However, in conventional off-axis holographic experimental setups, the object and reference beams propagate in separated paths, resulting in low temporal stability and measurement sensitivity. By designing common-path configurations where the two interference beams share the same or similar paths, environmental disturbance to the two beams can be effectively compensated. Therefore, the temporal stability of the experimental setups for hologram recording can be significantly improved for time-lapsing measurements. In this review, we categorise the common-path models as lateral shearing, point diffraction, and other types based on the different approaches to generate the reference beam. Benefiting from compact features, common-path digital holography is extremely promising for the manufacture of highly stable optical measurement and imaging instruments in the future.

Introduction

Gabor invented holography in 1948, attempting to correct the spherical aberrations of the electron microscope[1]. Based on Gabor's idea, the light field information including the amplitude and phase distributions could be retrieved by recording the interferogram of the object and reference waves. However, owing to the lack of highly coherent light sources and bottleneck of separating the twin images in the on-axis holography, holography did not receive serious attention at that time. With the invention of the high coherence laser and proposal of off-axis holography[2], holography has since developed rapidly. However, limited by the complex physical and chemical processing of photographic plates or films, conventional holography has faced difficulties in real-time measurements. In 1967, Goodman introduced electronic-based imaging methods into hologram-recording and computer technologies into the image reconstruction[3]. Subsequently, digital holography was developed to realise the dynamic and quantitative measurement of different samples with high efficiency[4]. Based on diffraction theory, it was determined that the object light field could be numerically reconstructed using the digitally diffracted light wave from the digital hologram "illuminated" with a simulated reference light wave.

A potential problem of digital holography is that the

Correspondence: Jianglei Di (jiangleidi@nwpu.edu.cn) or Jianlin Zhao (jlzhao@nwpu.edu.cn)
Key Laboratory of Light Field Manipulation and Information Acquisition, Ministry of Industry and Information Technology, and Shaanxi Key Laboratory of Optical Information Technology, School of Physical Science and Technology, Northwestern Polytechnical University, Xi'an 710129, China

mismatch between the numerically simulated reference wave and experimentally used one could induce reconstruction errors such as phase aberration. Hence, different numerical methods were developed to address this problem[5]. Based on the double-exposure principle, digital holographic interferometry (DHI) can eliminate phase aberration through the phase subtraction of the object waves reconstructed by the holograms with and without the sample. Exploiting the capability of measuring physical parameters associated with the object wave phase, DHI has been applied to visualising different complex flow fields[6] such as Karman vortex street in water flow[7], temperature distribution of heat conduction process[8], solution concentration variation[9], liquid diffusion process[10], sound pressure distribution[11], pulse laser ablation process[12], and sound waves[13]. In Fig. 1a–d, four examples are displayed.

In principle, the reconstructed object image can be numerically focused at any position[16] for flexible microscopic imaging. Therefore, by combining digital holography with optical microscopy, digital holographic microscopy (DHM) has become a popular phase imaging method for semi-transparent and transparent (weak scattering) or reflective phase objects. Upon the reflection by the sample surface or transmission through the sample, the phase distribution of the probing beam carries the information of the sample's profile or refractive index (or thickness), respectively. Consequently, DHM enables high-contrast imaging of small specimens as well as quantitative characterisation of the three-dimensional (3D) morphology and refractive index. Owing to the advantages of label-free, non-contact, and wide-field measurement, DHM and its derivatives have found wide application in industrial inspection, biomedical research, and other fields in the past decades[17]. However, the phase information obtained by DHM is an integral along the light propagation direction. To acquire the information inside the test specimen, digital holographic tomography (DHT) has achieved rapid progress and different approaches have been proposed to realise tomographic measurement[14,15,18]. We display two application examples of DHT in Fig. 1e–k. In addition, 3D phase and fluorescence images have been simultaneously obtained using a coherent and incoherent integrated common-path DHM system[19].

Another fundamental problem of digital holography is resolution improvement. Similar to other wide-field optical imaging modalities, the spatial resolution of digital holography is restricted by the spatial resolution of the imaging device and also obeys Abbe's diffraction limit[20]. To improve the imaging resolution, numerous approaches such as oblique illumination[21,22], synthetic aperture[23,24], structured illumination[25], and speckle illumination[26] have been introduced[27]. To further promote the application of digital holography, advanced computation techniques, for example, deep learning, have been introduced in digital holography for numerical reconstruction[28–31], fast image autofocusing[32,33], phase unwrapping[34], phase aberration

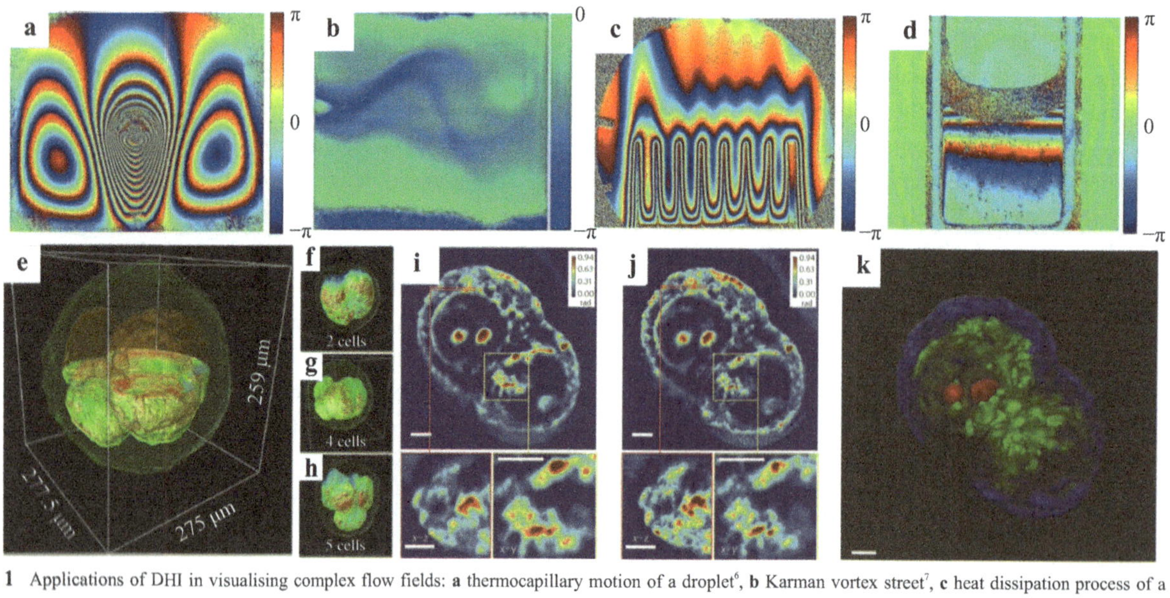

Fig. 1 Applications of DHI in visualising complex flow fields: **a** thermocapillary motion of a droplet[6], **b** Karman vortex street[7], **c** heat dissipation process of a heat sink[8], and **d** protein-lysozyme solution crystallisation process[9]. Applications of DHT in 3D biomedical imaging: **e** an embryo cut through the centre and **f–h** different development stages of bovine embryos[14]. Measurement results of HT29 cells[15]: **i** *Z*-slice (top), cross-section of the region marked by the red box (bottom left) and enlargement of the region marked bythe yellow box (bottom right), **j** deconvolved results of **i**, and **k** 3D rendering of the deconvolution result. Scale bars: 5 μm. Images reprinted with the following permissions: a[6], b[7], d[9] from The Optical Society; c[8] from AIP; c–k[14] from Springer Nature.

compensation[35], and other uses. Deep learning has also been combined with digital holography to achieve additional functionality such as screening of anthrax spores[36], classification of cell morphology[37], and viral particle detection and classification[38].

Regarding the experimental configurations of digital holography and its derivatives, two basic types, namely in-line (on-axis, coaxial) and off-axis geometries are typically used. In-line geometry[39–41] can fully use the camera bandwidth and phase shifting techniques can be used to remove the zero-order frequency component of the hologram[42]. To address the time-consuming problem in the phase-shifting operation, different parallel phase-shifting approaches have been developed[43–46]. However, the corresponding experimental configurations are typically complex and not easily aligned. Therefore, off-axis geometry is more frequently considered, especially for practical instrument constructions. For traditional holographic interferometry, off-axis configurations based on Mach–Zehnder and Michelson interferometers are commonly established for (semi-) transparent and reflective test objects, respectively. Based on these two interference models, a Swiss company, Lyncée Tec SA (https://www.lynceetec.com/) pioneered the commercialisation of digital holographic microscopes and is leading this instrument industry. They have invented different types of microscopes that can be applied in dynamic topography, living cell imaging, MEMS testing, automated measurements, and other areas. Another company from the Republic of Korea, Tomocube (http://www.tomocube.com/), invented holotomographic microscopes based on the principle of DHT to measure the 3D refractive index distributions of biological cells and thin tissues. Our group has also been working on the instrumentation of off-axis holographic experimental setups, including instruments for measuring both large-size and microscopic objects. Fig. 2a–c are the transmission, reflection, and transmission and reflection integrated digital holographic microscopes invented by our group, respectively. Among these, the one in Fig. 2c was built based on the mechanical structure of an inexpensive commercial microscope, making it easy to build and cost-effective.

In general, the object and reference beams in an off-axis holographic configuration interfere via separated paths to form the hologram. Consequently, any mechanical vibration or air disturbance, even marginal, can cause different influences on the two separated interference beams. This induces phase noise and results in low temporal stability. If the object and reference beams propagate along the same path in a common-path configuration[47], the environmental disturbance on the two interference beams can be effectively compensated with significantly enhanced temporal stability. In addition, the compact and robust common-path configuration is advantageous in optical instrument manufacturing. In this review, we summarise the design principle of the common-path off-axis digital holographic setups and present examples of their applications, providing a prospect in high stable optical instrument manufacturing.

Configuration design of common-path off-axis digital holography

First, we outline the basic principle of off-axis digital holography[48]. As indicated in Fig. 3a, the object beam carrying the object information and reference beam, which is typically a uniform plane wave, interfere on the CCD camera with an appropriate angle. For simplicity, we assume that the object beam propagates along the y-axis and the optical axis of the reference beam has an angle of θ relative to the y-axis. The complex amplitudes of the two beams on the hologram-recording plane can be expressed as

$$e_O(x, y) = a_O(x, y)\exp[\mathrm{j}\phi_O(x, y)] \tag{1}$$

$$e_R(x, y) = a_R(x, y)\exp\left(\mathrm{j}2\pi\frac{\sin\theta}{\lambda}y\right) \tag{2}$$

where, $a_O(x, y)$ and $a_R(x, y)$ are the amplitudes and $\phi_O(x, y)$ is the phase distribution of the object wavefront. The

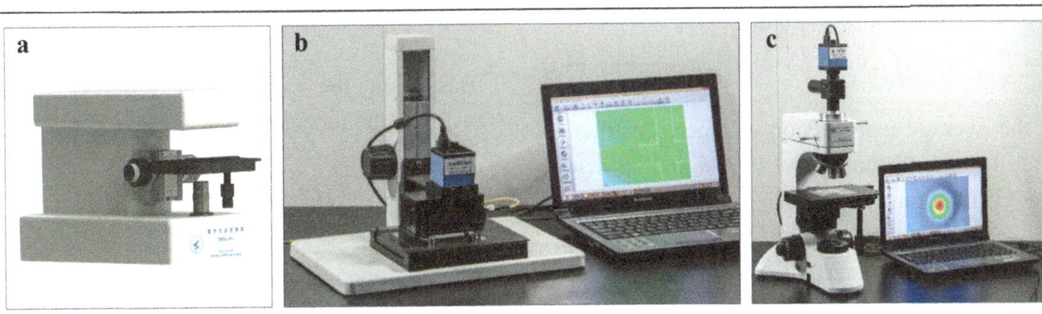

Fig. 2 **a** Transmission, **b** reflection, and **c** transmission and reflection integrated digital holographic microscopes invented by our group.

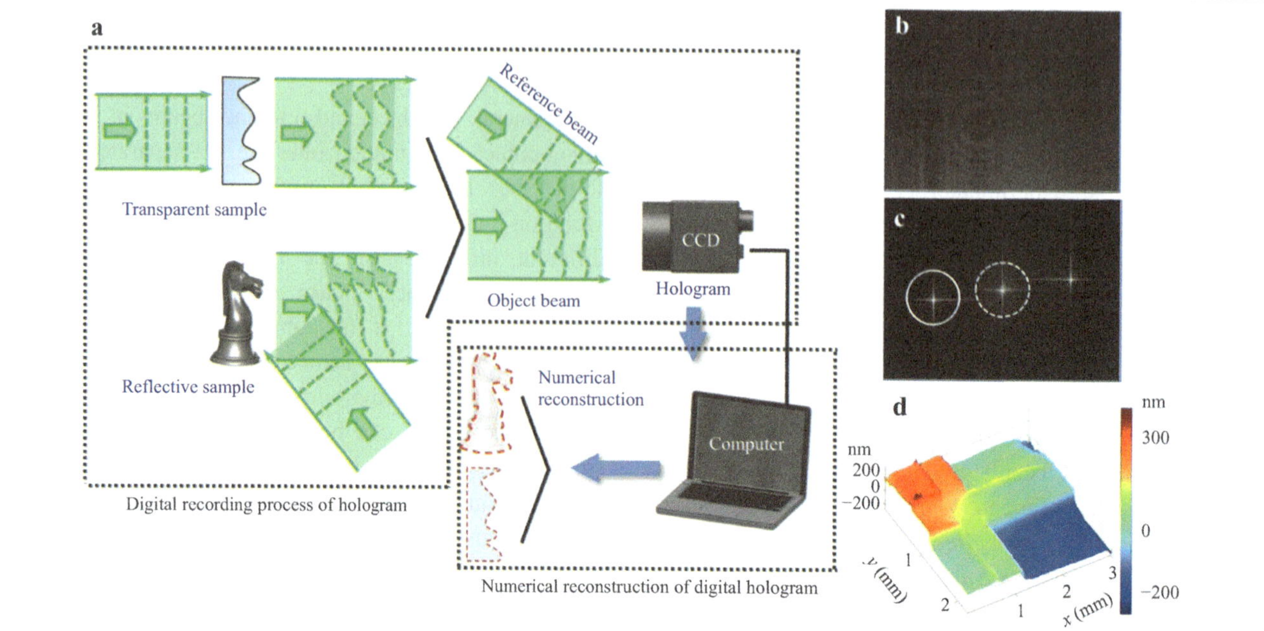

Fig. 3 **a** Operating procedure of off-axis digital holography. Coherent light beam passing through transparent samples or reflected by reflective samples acts as the object beam. The object and reference beams interfere with each other on the CCD with a small angle to form the off-axis digital hologram. **b** Hologram recorded by off-axis digital holography. **c** Fourier transform of the hologram in **b**. **d** Reconstructed 3D height profile of a stair-structured sample.

intensity distribution of the hologram, i.e., the interferogram between the two beams (see example in Fig. 3b) is

$$I(x,y) = |e_O(x,y)+e_R(x,y)|^2 = |e_O(x,y)|^2 + |e_R(x,y)|^2 + e_O(x,y)a_R^*(x,y)\exp\left(-j2\pi\frac{\sin\theta}{\lambda}y\right) + e_O^*(x,y)a_R(x,y)\exp\left(j2\pi\frac{\sin\theta}{\lambda}y\right) \quad (3)$$

For reconstruction, the conjugate of the original reference beam $e_R^*(x, y)$ is used to illuminate the hologram. The diffraction includes four parts:

$$U_1(x,y) = Te_R^*(x,y)|e_O(x,y)|^2 \quad (4a)$$

$$U_2(x,y) = Te_R^*(x,y)|e_R(x,y)|^2 \quad (4b)$$

$$U_3(x,y) = Te_R^*(x,y)e_O(x,y)a_R^*(x,y)\exp\left(-j2\pi\frac{\sin\theta}{\lambda}y\right) \quad (4c)$$

$$U_4(x,y) = Te_R^*(x,y)e_O^*(x,y)a_R(x,y)\exp\left(j2\pi\frac{\sin\theta}{\lambda}y\right) \quad (4d)$$

where T is the average transmission of the reconstruction wave passing through the hologram. Eq. 4a and 4b denote the 0th order diffraction components and Eq. 4c and 4d denote the separated twin images due to the carrier frequency of the reference beam. For numerical reconstruction of the object wave, the Fourier transform-based convolution method is typically used. The $\pm1^{st}$ order Fourier spectra of the hologram corresponding to Eq. 4c and 4d are separated in the Fourier space. The circles in Fig. 3c mark the 0th and 1st order Fourier spectra of the hologram in Fig. 3b. Because only the term in Eq. 4c contains the original object information $e_O(x, y)$, it can be reconstructed through properly selecting the 1st order spectral component in the Fourier space.

The complex amplitude of the diffraction wave carrying the object information at position $z = d$ from the hologram plane can be expressed as

$$u(\xi',\eta') = \int_{-\infty}^{\infty}\int_{-\infty}^{\infty} U_3(x,y)g_{PSF}(x-\xi',y-\eta')\,dxdy = U_3(x,y) * g_{PSF}(\xi',\eta') \quad (5)$$

where the symbol "*" denotes the convolution calculation and $g_{PSF}(\xi', \eta')$ is the point spread function (PSF), which can be expressed as

$$g_{PSF}(\xi',\eta') = \frac{j}{\lambda}\frac{\exp\left(-jk_0\sqrt{d^2+\xi'^2+\eta'^2}\right)}{\sqrt{d^2+\xi'^2+\eta'^2}} \quad (6)$$

Based on the convolution property of the Fourier transform, the reconstructed object wave can be expressed as

$$u(\xi',\eta') = F_{x,y}^{-1}[F_{x,y}(U_3)F_{x,y}(g_{PSF})] \equiv F_{x,y}^{-1}[F_{x,y}(U_3)G_{PSF}] \quad (7)$$

where the operations "$F_{x,y}$" and "$F_{x,y}^{-1}$" denote the two-dimensional Fourier transform and inverse Fourier transform, respectively, and G_{PSF} is the Fourier transform of g_{PSF} and acts as the transfer function of the imaging system. Taking the discrete forms, the reconstruction can be performed digitally. Fig. 3d displays the calculated 3D height profile of a stair-structured sample from the hologram in Fig. 3b.

To record the off-axis hologram, the object beam must interfere with a uniform reference beam at a certain angle. The key issue to design a common-path configuration is to generate a uniform reference beam allowing the two interference beams to pass along similar paths. The greater the level of similarity, the greater the stability of the optical setup. From the perspective of the approach to generate the reference beam, we categorise the common-path models as lateral shearing, point diffraction, and other types.

Common-path configuration based on lateral shearing

The lateral shearing interferometer was invented to measure the asphericity of an optical wavefront[49]. The wavefront under test is first doubled using specific optical elements and the two wavefronts create a lateral shear or displacement. The overlapping area of the two wavefronts forms the interferogram for further analysis. The two wavefronts propagate along similar paths, which inspires the design of lateral shearing-based common-path configurations for off-axis digital holography. The concept of lateral shearing-based off-axis digital holography is illustrated in Fig. 4. The illumination beam passes through a (semi-) transparent sample (or is reflected by an opaque sample) and an imaging system. A beam duplicator reproduces the beam into two beams that are projected on the CCD target at a proper angle. Note here that both beams carry the same sample information. However, as discussed in the previous section, a uniform reference beam containing no sample information is required for the hologram formation. Therefore, the test sample in the lateral shearing-based common-path configuration must be sparse in the field of view (FOV), or only a portion of the beam illuminates the sample. In this manner, the effective interference area, i.e., the overlapping area of the two beams with and without the sample information, can be realised. This feature is different from that of a traditional lateral shearing interferometer.

Using a glass plate

Multiple optical components such as glass plate[50-53], grating[54-58], and beam splitter (BS)[59-62] are typically employed as the beam duplicator in the lateral shearing-based common-path configurations. Fig. 5 displays an example using a glass plate[50]. A plane wave passing through the sample is expanded by a microscope objective (MO) to become a divergent beam. The beam is then reflected at the front and back surfaces of a glass plate oriented at approximately 45° relative to the optical axis of the beam. Consequently, the incident beam is doubled to create a lateral shearing, as indicated in Fig. 5(a). The two beams are divergent satisfying the recording condition of the off-axis geometry. This configuration can be easily integrated with a commercial microscope for practical applications. Glass microspheres were used as the test samples. The hologram was recorded when only one sphere appeared in the FOV. The height of the sphere surface $d = \phi\lambda / (2\pi\Delta n)$ was retrieved from the phase image, where ϕ is the phase value, λ is the light wavelength, and Δn is the index difference between the sample and its surrounding media. Fig. 5b displays the retrieved height distribution of the sphere surface. Note here that the seemingly "two" spheres with positive and negative heights are displayed in the image. In fact, it represents the information of the same sphere. Two spheres appear in the FOV because the beam is reflected twice by the glass plate, and the bending direction of the interference fringes of the left sphere is opposite to that of the right sphere. The duplicated image can limit the effective FOV of this experimental setup. To address this problem, the illumination beam size must be sufficiently large such that the object size is smaller than the interference area and only one copy of the object information is imaged onto the CCD target. Fig. 5c displays the 3D height profile of the

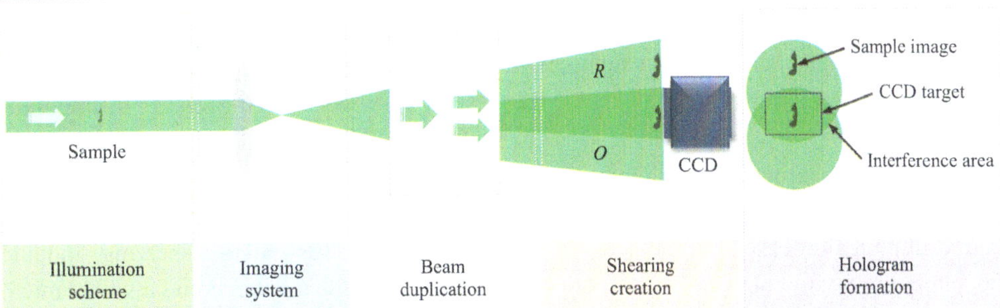

Fig. 4 Concept of common-path configuration based on lateral shearing interference.

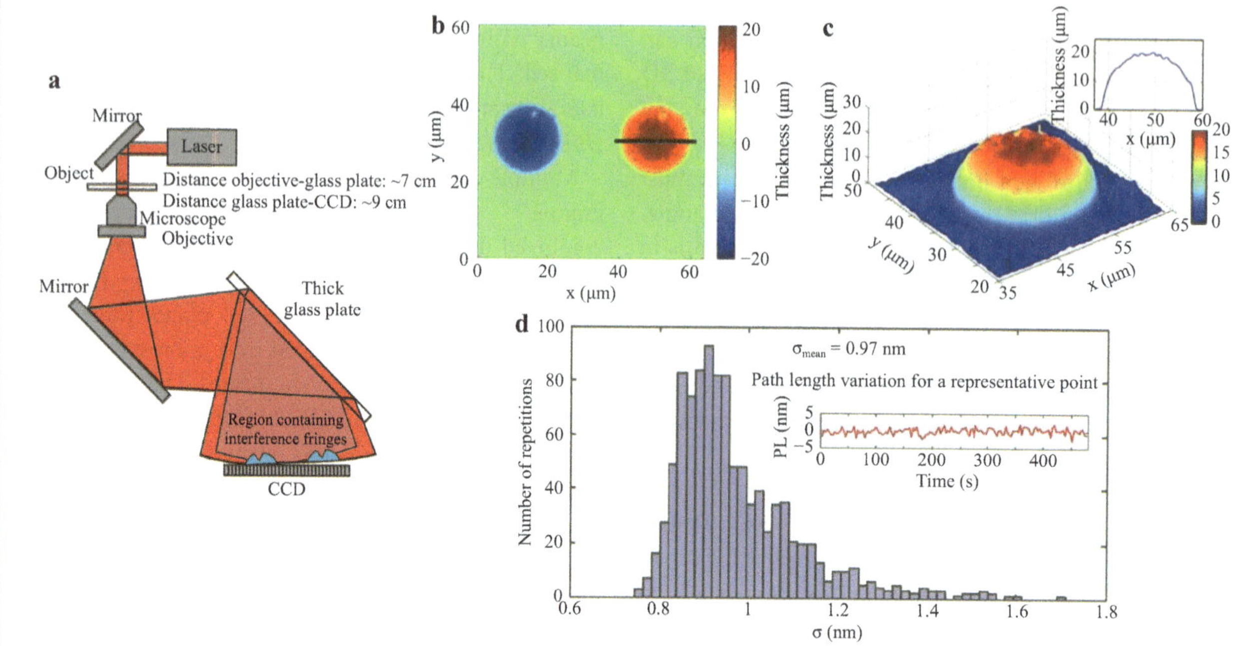

Fig. 5 **a** Experimental setup of common-path DHM based on lateral shearing using a glass plate. **b** Retrieved height distribution of the sphere surface. **c** 3D and cross-sectional thickness profile of the sphere on the right side of **b**. **d** Histogram for the standard deviations of fluctuations[50]. Images reprinted with the permission from The Optical Society.

sphere surface on the right side of Fig. 5b, and the inset indicates the corresponding cross-sectional profile.

The common-path holographic configuration can improve the temporal stability of the setup, especially the phase measurement stability of the object wave. The stability characterisation should be performed without vibration compensation. Using the setup in Fig. 5a as an example, 1200 holograms without objects were recorded over an area of 37.5 μm × 37.5 μm at a speed of 2.5 frames per second for 8 minutes. Phase images of all the holograms were reconstructed and the optical path difference was obtained by subtracting the phase values with that of the previously recorded hologram. Then, 1024 pixel positions were randomly selected in the FOV and the standard deviations of the fluctuations of these positions were calculated. Fig. 5d displays the histogram of the fluctuations with an average value less than 1 nm at a wavelength of 632.8 nm. This indicates an optical path fluctuation of only 0.01 rad. This small fluctuation demonstrates the high temporal stability of the experimental setup.

Based on the lateral shearing-based common-path configuration using a glass plate, researchers developed a compact structured-illumination-based DHM setup[53]. By illuminating the sample with structured intensity patterns, the imaging resolution of the amplitude- and phase-contrast images can be effectively enhanced. This design reduces the complexity of the setup and provides high temporal stability.

Using a grating

In the previous glass plate-based configuration, the divergent beam from the MO reaches the CCD target without an imaging lens. The object cannot be directly imaged on the hologram-recording plane and thus it needs to set an appropriate reconstruction distance. To record the image plane hologram, alternative optical components can be employed. Fig. 6 displays an example of using a grating as the beam duplicator[58]. A plane wave illuminates the sample. The transmitted beam is expanded by an MO and then passes through an imaging tube lens TL, by which the sample is imaged on the image plane IP. A grating G is placed at a distance z after the IP to reproduce multiple replicas of the sample image. Then, the intermediate sample images are projected onto the camera target *via* a 4f unit with lenses L_1 and L_2. During this process, a mask B for spatial filtering is placed at the Fourier plane of lens L_1, such that only the 0th and $+1^{st}$ diffraction orders can pass through it. These two orders are Fourier transformed again by lens L_2 to produce lateral shearing interference (shearing distance l) and form an off-axis image plane hologram. This lateral shearing module can also be integrated with a commercial inverted microscope.

In general, there is a tradeoff between the FOV and off-axis angle that should be balanced in practical applications.

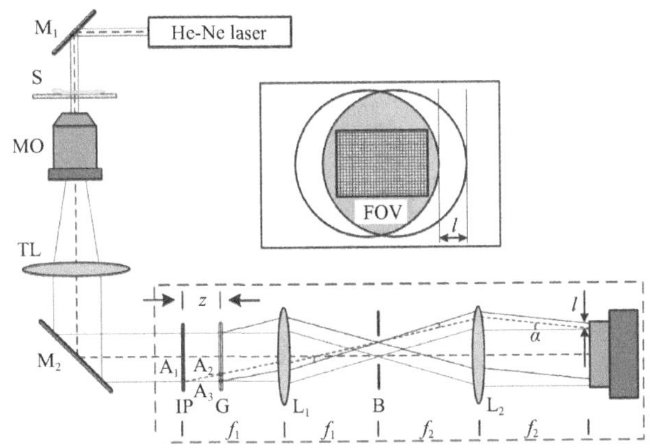

Fig. 6 Experimental setup of common-path DHM based on lateral shearing using a 4*f* imaging unit and diffraction grating[58]**.** M_1, M_2: mirrors, S: sample, MO: microscope objective, TL: tube lens, IP: image plane, G: diffraction grating, L_1, L_2: lenses, B: mask for spatial filtering. Image reprinted with the permission from The Optical Society.

Conversely, in the common-path configuration of Fig. 6, the off-axis angle is solely determined by the grating period d and focal lengths of lenses L_1 and L_2, i.e., $\alpha = f_1\lambda / f_2 d$. The shearing distance $l = f_2\lambda z / f_1 d$ is dependent on the distance z, which can be adjusted by moving the grating along the optical axis freely. Consequently, the off-axis angle related to the interference fringe frequency is not required to match the shearing distance. The flexible selection of the shearing distance does not lose the spatial frequency of the sample.

In addition to the one-dimensional grating, a specific two-dimensional grating has been applied to create four replicas of the object image. Combining the setup in Fig. 6 with multiplexing techniques, researchers have recorded a polarisation hologram and reconstructed the wavefronts of two orthogonally polarised components containing the sample information using a single camera shot[55,51].

Using a BS

A single BS can also be employed to construct lateral shearing-based common-path DHM configurations[59–62]. The output beam from a microscopic imaging system contains the half-object and half-reference parts. Then, the entire beam is further divided by a BS to create the sheared beams. Finally, the half-object and half-reference beams interfere with each other at a certain angle. Using the setup in Fig. 7a, which uses a slight trapezoid Sagnac interferometer and was developed by our group[60] as an example, a polarised BS (PBS) splits the object beam into two counter-propagating components with orthogonal polarisation states. The two components pass through the identical optical elements and then reach the CCD target. The off-axis recording condition can be achieved by marginally rotating the PBS with the transmitted *p*-polarised component maintaining the propagation direction perpendicular to the CCD target. Consequently, the *p*-polarised component acts as the object beam, whereas the *s*-polarised component contains no sample information and serves as the reference beam. A polariser P_2 is used before the CCD to modulate the fringe contrast of the hologram. Note that the functionality of the BS used here is different from that in the Mach–Zehnder and Michelson interferometer-based setups. The two interference beams in Fig. 7a pass along the same paths, ensuring the high temporal stability in the configuration. To verify its high temporal stability compared with that of a Mach-Zehnder interferometer-based configuration, a similar characterisation method with that of Fig. 5d was used. The results demonstrated that the mean phase fluctuation of the common-path setup and Mach-Zehnder interferometer was 0.011 rad (Fig. 7b) and 0.106 rad, respectively. That is, the temporal stability of this common-path configuration was approximately one order of magnitude greater than that of the uncommon-path interferometer configuration. Owing to the high stability, we monitored a water evaporation process that has wide application in microfluidics, surface self-cleaning, and other areas[63]. Moreover, two static measurements were also experimentally demonstrated. The first was a pit on a silica glass damaged by high-power laser irradiation. The retrieved pit depth is displayed in Fig. 7c. The other was a living Hela cell, which is in the human cervical cancer cell lines. The reconstructed quantitative phase image of the cell is displayed in Fig. 7d.

Other special optical components such as beam displacer[64], Wollaston prism (WP)[65], Rochon polariser[66], gradient-index lens[67], and biprism[68] can also be utilised in the lateral shearing-based common-path configurations to construct compact and stable holographic experimental setups.

Common-path configuration based on point diffraction

In the lateral shearing-based common-path off-axis holographic experimental setups, sparse samples in the FOV or undisturbed portion of the illumination beam are required. Moreover, it has a potential problem of a duplicated image appearing on the camera target if the shearing distance is not sufficiently large, which could limit its practical applications. Alternatively, another common-path model referred as diffraction phase microscopy (DPM) was developed by Gabriel[69].

As discussed in the previous section, the key issue to design a common-path configuration is to generate a uniform reference beam allowing the two interference beams to propagate along the same or similar paths. Here,

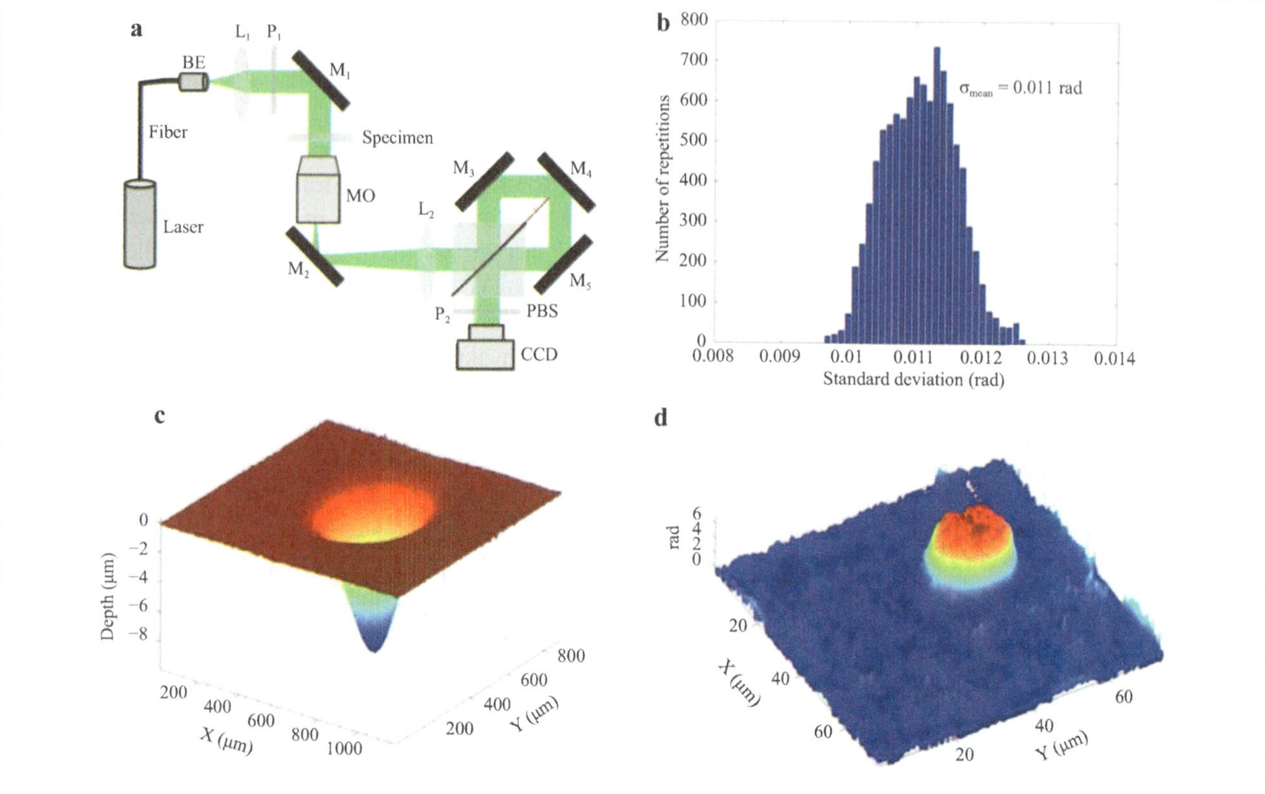

Fig. 7 **a** Typical experimental setup of lateral shearing-based common-path DHM using a BS[60]. **b** Histogram of standard deviation of the common-path setup in **a**. **c** 3D topography of a pit ablated by high-power laser irradiation. **d** Quantitative phase image of a Hela cell. Images reprinted with the permission from The Optical Society.

we introduce the principle of generating the reference beam using the point diffraction model. As depicted in Fig. 8[69], a magnified sample image is obtained in an inverted microscope. To illustrate the imaging process, we start from a virtual point source (VPS) representing the object plane. At the image plane (IP), a grating G replicates the microscopic image and generates several diffraction orders. Then, the diffractions enter into a 4*f* unit with lenses L_1 and L_2. The 0th order beam is low-pass filtered by a pinhole at the focal plane of L_1, resulting in a uniform reference beam. Simultaneously, one of the 1st order beams containing complete sample information passes through the unit and becomes the object beam. Finally, the two beams interfere with each other on the CCD target at an appropriate off-axis angle. As this common-path configuration generates the uniform reference beam by diffracting the object beam through a "point" pinhole, we call this model the "point diffraction model". Unlike the typical Mach–Zender interferometer, here the interference beams pass through the identical optical elements, ensuring high temporal stability. The average standard deviation of the optical path length in the full FOV was measured as 0.7 nm at a wavelength of 532 nm and the temporal

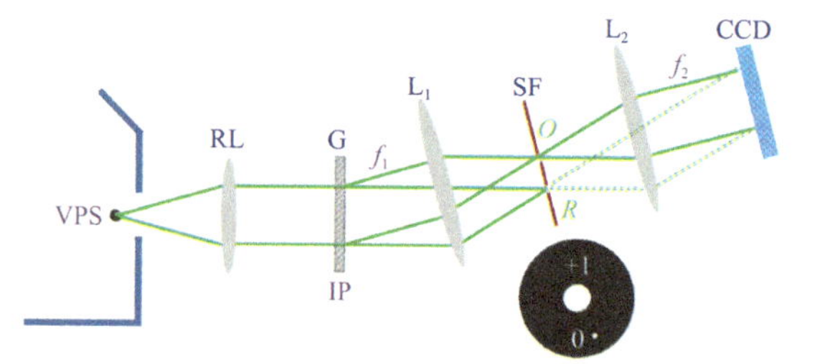

Fig. 8 Basic configuration of point diffraction-based common-path off-axis digital holography[69]. VPS: virtual point source, RL: relay lens, G: grating, IP: image plane, $L_{1,2}$: lenses, SF: spatial filter.

standard deviation was 0.04 nm[69]. Similar to the lateral shearing-based common-path configurations, the common-path module in Fig. 8 can also be integrated with a commercial microscope.

Laser-based illumination

The module in Fig. 8 using a grating for beam duplication, however, has a low utilisation efficiency of the light power and is also somewhat bulky as a 4*f* unit is included. Hence, alternative designs have been proposed to realise the point diffraction[70–72]. For example, a so-called τ interferometer is mounted at the output port of a microscope (Fig. 9a)[73]. This interferometer reproduces the object beam using a cubic BS and two mirrors reflect the beams. The pinhole is pasted on one of the mirrors to eliminate the sample information and create a uniform reference beam. The triangular interferometer-based setup in Fig. 9(b) also uses a cubic BS for the beam duplication, however, the pinhole is implemented in the transmission mode[74]. A biprism can effectively double the object beam, not only compatible in the lateral shearing-based but also in the point diffraction-based common-path configurations (Fig. 9c)[75]. Unlike the former configurations, a digital micromirror device (DMD)-based setup reproduces the uniform beam before passing through the sample, as depicted in Fig. 9d[76]. A DMD positioned at the conjugate plane of the MO's back-aperture plane generates two plane waves for the illumination. Moreover, a glass plate, typically used in a lateral shearing-based configuration with a limited FOV, can also reproduce the object beam in the point diffraction-based configuration with high utilisation efficiency of the light power[77].

Owing to the high stability and compact features, the point diffraction-based common-path DHM has been widely used in biological studies, material science applications, and other applications[78]. For example, the thickness distribution of a 2 × 2 array of microlenses was measured, as displayed in Fig. 10a. In addition, researchers have constructed a DHM setup that operates in reflection mode for opaque samples and dynamically monitored nanoscale topographic changes during semiconductor etching. Fig. 10b displays the reconstructed logo of the University of Illinois when a GaAs wafer was being etched with a solution[79]. Fig. 10c and 10d present the measured topography images of a multi-step stair structure and Archimedean spiral, respectively[80].

White light illumination

Speckle noise caused by the high coherence of lasers can typically contaminate reconstructed object images, resulting in a decrease of the phase measurement sensitivity and limiting the capabilities of investigating samples with subcellular features. Hence, white light illumination with low coherence is applied in point diffraction-based common-path DHM to decrease the spatiotemporal noise effectively[15,81,82]. The corresponding setup is developed from that in Fig. 8, as indicated in Fig. 11a[82]. One of the differences is that a low-coherent halogen lamp with a condenser is used for the illumination. By closing down the condenser aperture to the minimum, the entire illumination field is spatially coherent. Another difference is that digital masks on a spatial light modulator (SLM) perform the spatial filtering to generate the object and reference beams. This common-path configuration makes the optical path difference between the object and

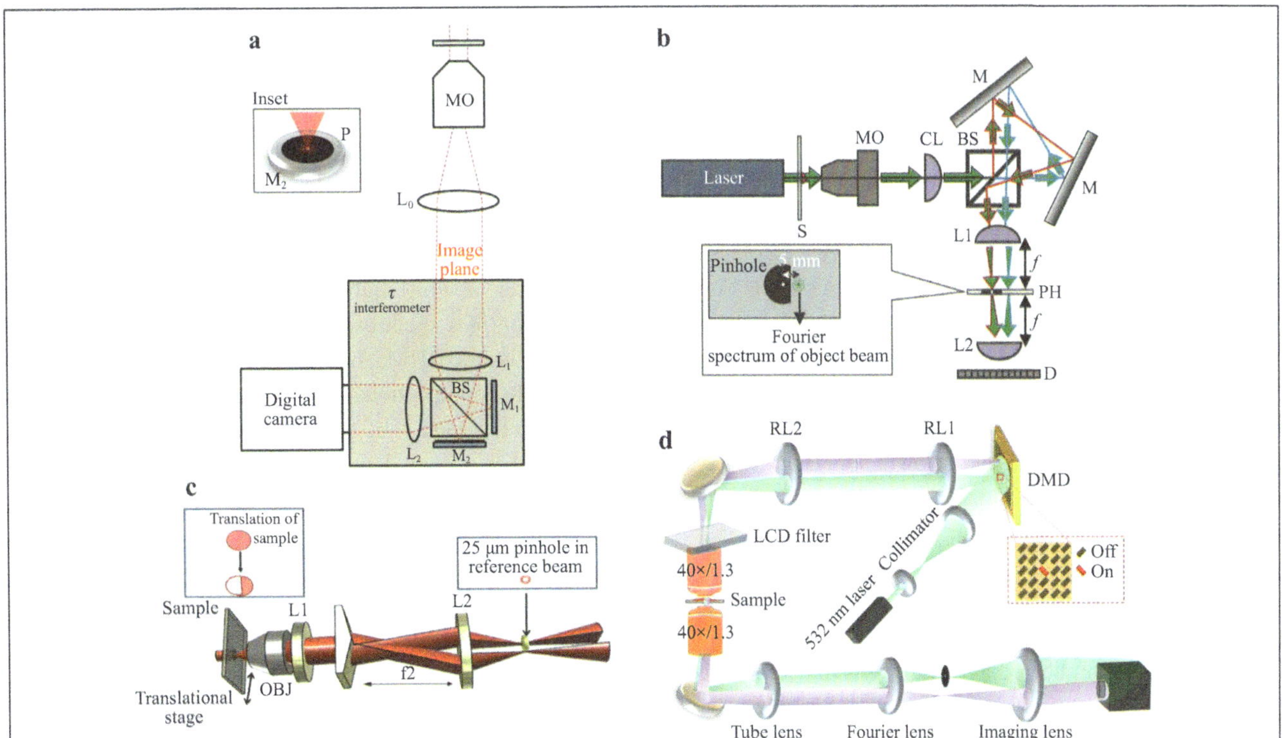

Fig. 9 Experimental setups of point diffraction-based common-path DHM using a τ interferometer[73], b triangular interferometer[74], c biprism[75], and d DMD[76]. Images reprinted with the permission from The Optical Society.

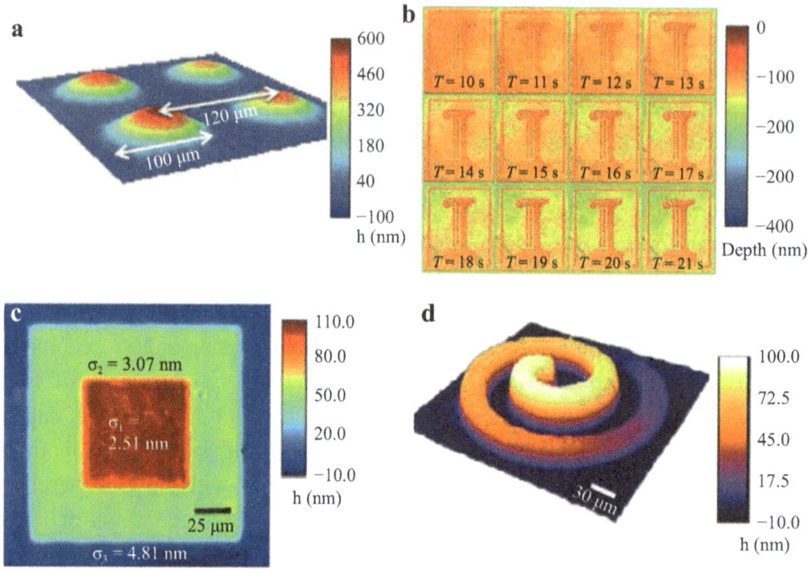

Fig. 10 Applications of point diffraction-based common-path DHM in material science[79,80]: **a** thickness distribution of a microlenses array, **b** real-time imaging of a wet etching process, and topography images of **c** a stair structure and **d** an Archimedean spiral. Images reprinted with the following permissions: a, b[79] from Springer Nature; c, d[80] from The Optical Society.

reference beams extremely small, such that the two beams are coherent and can interfere with each other regardless of the wavelength used. In addition to the SLM, a liquid crystal display device has also been applied to provide the digital filter masks[83].

The white light illumination-based setup exhibits high temporal stability owing to the common-path configuration and high phase measurement sensitivity due to the low-coherent illumination. It has found wide applications in the biomedical field including red blood cell (RBC) membrane fluctuation measurement[81], cardiomyocyte cell beating measurement, and cell growth study[84]. In the following, we provide application examples. Fig. 11b displays the reconstructed phase image of a polystyrene bead. Fig. 11c displays the phase image of three live RBCs where the typical discocyte shapes are clearly present. It can be observed that the phase background of the image is considerably uniform owing to the low coherent illumination. Moreover, the growth of Hela cells was monitored. The dots in Fig. 11d indicate the variations of the cell's dry mass with time. It can be observed that the test cell grows rapidly in the first six hours. Then, its growth slows and finally saturates. Fig. 11e–f are the phase images of the test cell at a time of 2 hours and 16 hours, respectively. As indicated by the arrows, the cell nucleolus in Fig. 11f is becoming larger than that in Fig. 11e.

Regardless of whether the illumination is laser-based or white light-based, each type has its advantages and disadvantages. It was found that laser-based illumination is superior for material science applications where the samples have fine structural features or sharp edges. Conversely, white light-based illumination is more advantageous in biological applications where the sample features are smooth. Moreover, researchers have recently developed an endoscopic phase microscope[85] that combines a gradient-index-lens-based endoscope probe with a point diffraction-based common-path module, and allows single-cell-level resolution phase imaging. This instrument has broadened the application fields of quantitative phase imaging and has demonstrated significant potential in on-site studies.

The common-path configurations discussed above use a single laser or white light source, which cannot attain the distribution of a spectroscopic object light field. However, in transmission DHM, the measured phase information of the sample is proportional to the optical thickness, which is the integral product of the sample thickness and index. Owing to this dual dependence, the thickness and index information of the sample is ambiguous. Hence, more than one measurement is required to decouple these two physical parameters. By performing two measurements with different surrounding media, the thickness and index of the sample could be retrieved through a decoupling procedure[86]. However, the sample state could be disturbed if the surrounding medium is changed. Alternatively, measuring phases of the object wave at different wavelengths can address this problem based on the dispersion property of the surrounding medium[87]. For

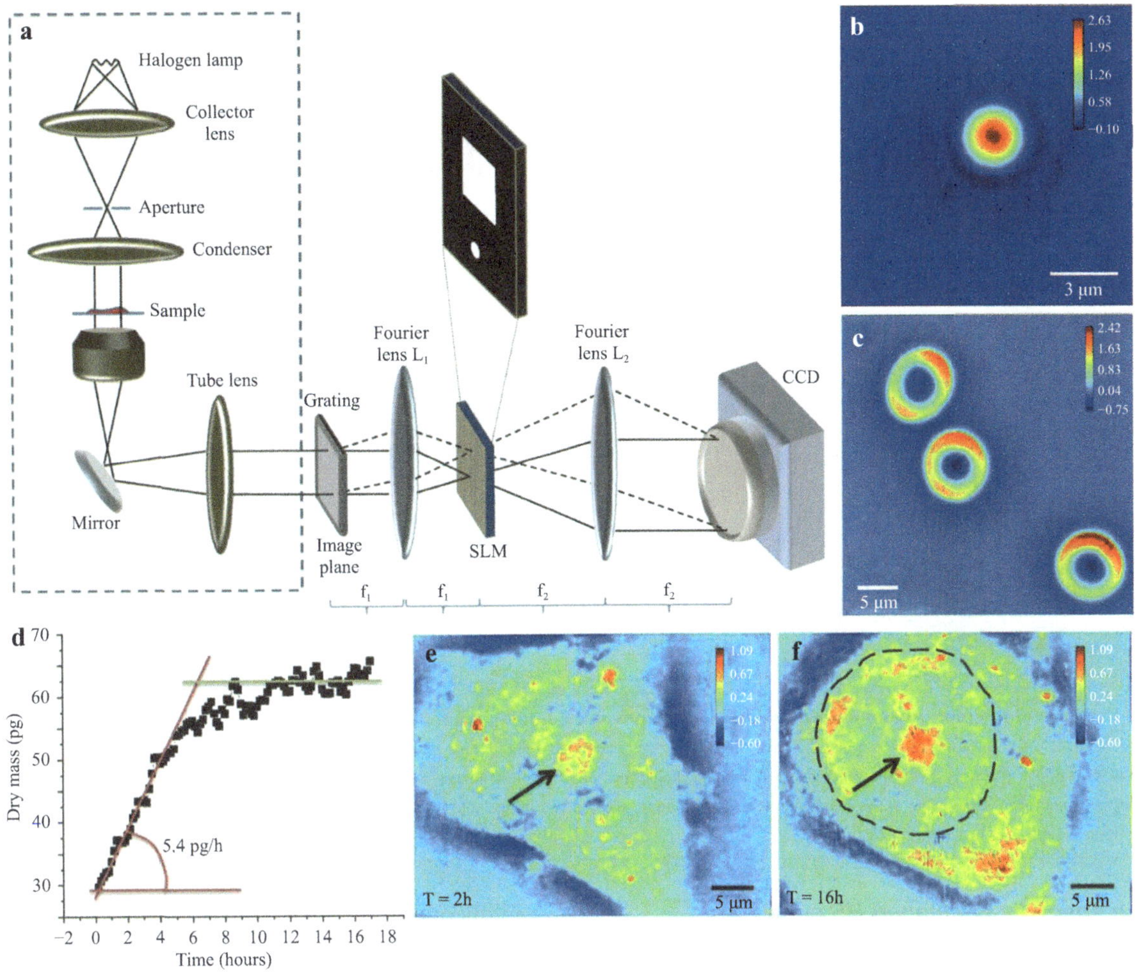

Fig. 11 **a** Experimental setup of point diffraction-based common-path DHM using white light illumination[82]. Reconstructed phase images of **b** a polystyrene bead and **c** live RBCs. **d** Variations of the dry mass for a HeLa cell during its growth. Quantitative phase images of the test HeLa cell at a time of **e** 2 hours and **f** 16 hours. All of the colour bars represent the phase in radians[81]. Images reprinted with the permission from The Optical Society.

example, spectroscopic phase microscopy has been introduced in point diffraction-based common-path DHM configurations[88–90].

Other common-path configurations

Apart from the two types discussed above, other advanced designs have also been proposed to construct common-path off-axis holographic experimental setups[91].

Using a folding mirror

One example uses a self-referencing Lloyd's mirror configuration. As indicated in Fig. 12a, the laser beam passing through the sample is first expanded via an MO. Then, a mirror folds a part of the divergent beam onto the CMOS target. The straight propagating beam carrying the object information and unperturbed reflection beam interfere with each other to form the off-axis hologram[92]. This DHM setup is compact, portable, stable, and easy to implement. Its temporal stability was measured to be approximately 0.9 nm without any vibration compensation, making it suitable for monitoring the changes of a cell profile. Fig. 12b displays an experiment result of the thickness profile of RBCs.

Fig. 12c displays a similar design applied in continuous-wave terahertz (CW-THz) holography[93]. The inset indicates that the sample is illuminated by a half part of the terahertz wave. A pair of mirrors (M_1 and M_2) with an angle of α reflect the entire wavefront into two beams with an angle of 2α. Because the two beams are created from the same wavefront, their power and path length are virtually the same, forming interference patterns with high contrast. Unlike digital holography with visible light illumination, terahertz digital holography is a viable candidate for measuring optically opaque samples. Using the common-path setup in Fig. 12c, a polystyrene cup lid and biological samples such as insect wings are measured. The experiment results are presented in Fig. 12d–g and 12h–k, respectively.

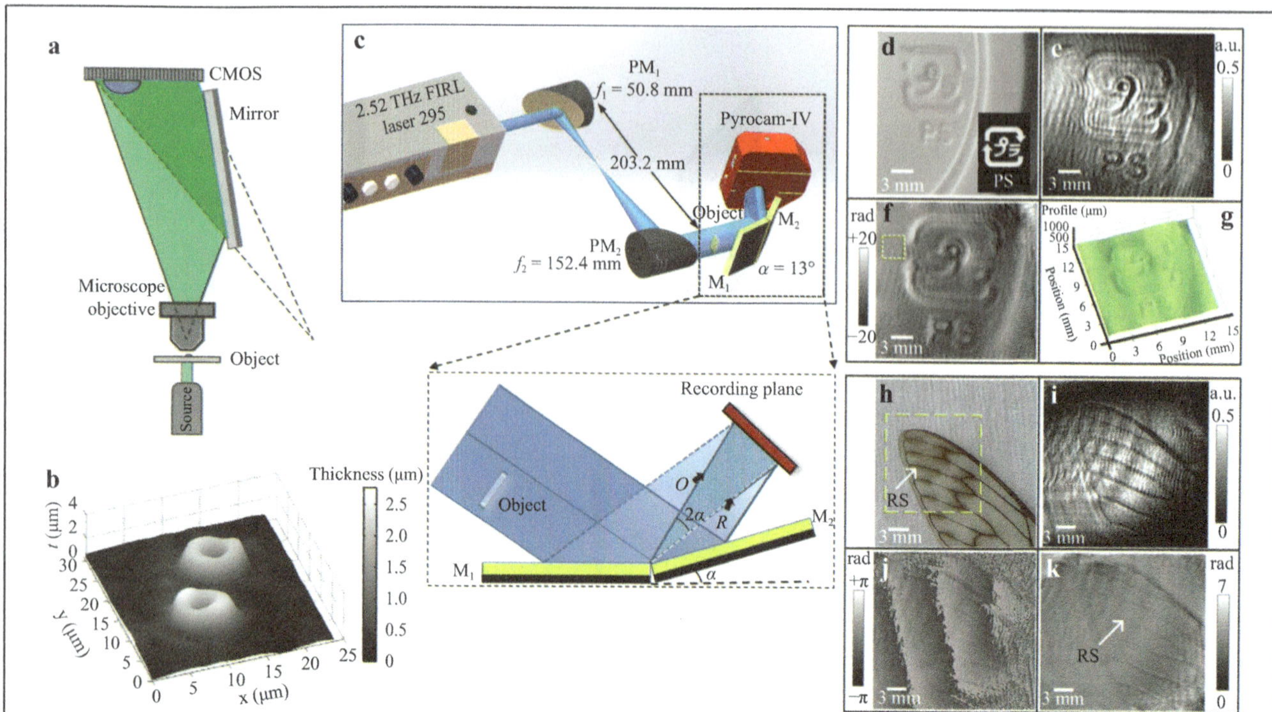

Fig. 12 Experimental setups of common-path digital holography using folding mirrors and their applications[92,93]. **a** Setup using a Lloyd's mirror. **b** Thickness profile of RBCs using the setup in **a**. **c** Setup for CW-THz self-referencing digital holography. **d–g** Experiment results of measuring a polystyrene cup lid: **d** photo of the sample, reconstructed **e** amplitude-contrast image, **f** unwrapped phase-contrast image, and **g** calculated 3D thickness profile. **h–k** Experiment results of measuring a cicada forewing: **h** photo of the sample, reconstructed **i** amplitude-contrast image, and **j** wrapped and **k** unwrapped phase-contrast image. Images reprinted with the permission from The Optical Society.

Using a specially positioned BS cube

Another example uses a single BS cube positioned in a special manner[94–96]. Fig. 13a presents an experimental setup of DHI based on total internal reflection (TIR)[97] to dynamically measure the refractive index distribution of liquid samples. The sample is located at a half part of the prism surface. Only one portion of the reflected beam carries the sample information, the other acts as the reference beam. The entire beam is incident on a BS cube with the semi-reflecting layer marginally inclined relative to the optical axis. The object beam transmits through the layer and the reference beam is reflected. The off-axis angle of the two interference beams is controlled by rotating the BS. The usage of a single BS cube creates a symmetrical common-path configuration. The spherical phase curvature of the illumination beam can be compensated physically by the reference beam during the interference. Owing to the advantages of this common-path holographic configuration, we built a prototype instrument of this setup using the 3D printing technology, as displayed in Fig. 13b–c. This instrument is compact with a total physical length of only 25 mm (Fig. 13c). More importantly, it possesses high measurement stability for on-site applications.

Furthermore, we applied an MO with a long working distance (LWDMO) in the BS cube-based DHM for surface plasmon resonance (SPR) imaging[94,95]. As indicated in Fig. 13d [94], a *p*-polarised beam is reflected at the interface of an SPR configuration consisting of a prism, thin metal film, and dielectric layer. When the incident angle is greater than the critical angle of the TIR, the evanescent wave penetrates from the prism into the metal film. Once the wave vectors of the evanescent wave and surface plasmon wave match with each other at a certain angle, SPR occurs. The test sample acting as the dielectric layer is located at a half part of the metal film, which is the same with that in the setup displayed in Fig. 13a. Then, the entire beam is magnified by the LWDMO and passes through an imaging lens. Subsequently, the beam enters into the BS cube-based common-path configuration. The proposed common-path holographic experimental setup based on SPR (SPRHM) has advanced configuration simplicity as it contains fewer optical components. Moreover, high temporal stability can be realised to achieve highly sensitive SPR measurements. Using the setup in Fig. 13d, we performed SPR imaging of an onion tissue. The captured hologram in Fig. 13e indicates that the cell structures are clearly presented owing to the high

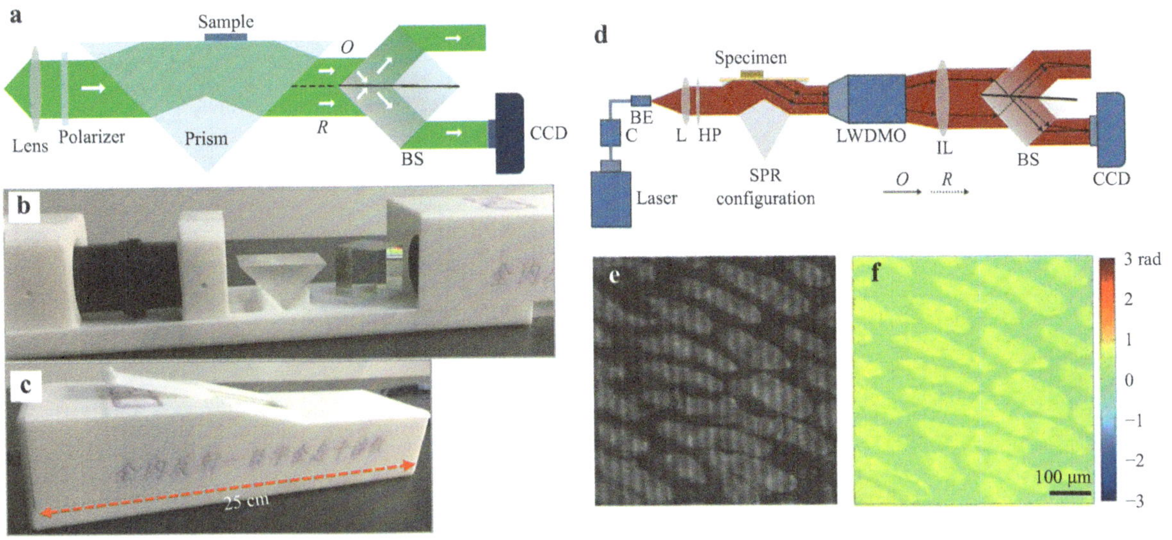

Fig. 13 **a** Common-path DHI setup based on TIR. **b**, **c** 3D-printed prototype instrument of the experimental setup in **a**. **d** Experimental setup of common-path DHM based on prism-coupling SPR[94]. C: fiber coupler, BE: beam expander, L: lens, HP: half-wave plate, LWDMO: long working distance microscope objective, IL: imaging lens, and BS: beam splitter. **e** Captured hologram of an onion tissue when SPR occurs using the setup in **d**. **f** Reconstructed SPR phase image of the tissue. Images reprinted with the following permissions: d–f[94] from The Optical Society.

sensitivity of SPR. The reconstructed phase image in Fig. 13f has a high contrast. Owing to the high sensitivity of SPR and high stability of the common-path configuration, the SPRHM setup in Fig. 13d can image samples with low-contrast index distribution.

Using a WP

The previous common-path SPRHM setup uses a prism-coupling configuration for SPR excitation. Typically, to realise high-resolution SPR imaging, MOs with high magnification and numerical aperture (NA) are required after the prism. However, the working distance of this kind of MO is typically short and incompatible with the prism-coupling SPR configuration because of the geometry limitation of the prism. Furthermore, the direction of the reflected light changes as the SPR excitation angle varies, which requires the realignment of the experimental setup. For this reason, another common-path SPRHM setup based on the MO-coupling configuration was developed[98] using an immersion MO with NA greater than the one for SPR excitation.

As exhibited in Fig. 14a, a focusing lens focuses a linearly polarised plane light beam at the back focal plane (BFP) of an MO (MO_2) with high NA. The output beam collimated by MO_2 illuminates the SPR configuration created with a coverslip deposited with gold film. The *p*-polarised component excites SPR at a certain angle θ. The *s*-polarised component cannot excite SPR. By moving the BS horizontally, the focal position d of the incident beam on the BFP is changed and the incident angle θ can be adjusted accordingly. The incident angle is determined by the sine relationship of an aplanatic imaging system, $d = f\sin\theta$ (f is the focal length of MO_2). The reflected light carrying the object information is recorded by the CCD camera after passing through the imaging lens (IL).

The bottom inset of Fig. 14a displays the off-axis hologram-recording process. The reflected beam from MO_2 is split into two orthogonally polarised components by a WP. The *p*-polarised component acts as the object beam carrying the SPR information of the sample. The *s*-polarised component acts as the reference beam carrying no sample information. The two interference beams possess the same polarisation state after passing through a polariser P and then interfere with each other. The approach of generating the uniform reference beam is different from the previously discussed common-path configurations. Compared with the lateral shearing-based type, the effective FOV of this setup is maintained without loss. Moreover, the level of common-path is greater than that of the point diffraction-based type.

Using the MO-coupling SPRHM setup in Fig. 14a, amplitude- and phase-contrast SPR images with high spatial resolution can be simultaneously obtained, by which the thin film thickness and refractive index distributions can be mapped unambiguously. Moreover, owing to the high temporal stability, this common-path SPRHM setup can be widely applied for dynamically measuring the small variations of refractive index or thickness. For example, we have measured the thickness of multilayer graphene film located in the near field of a metal surface[98]. The measurement resolution can achieve sub-nanometre levels.

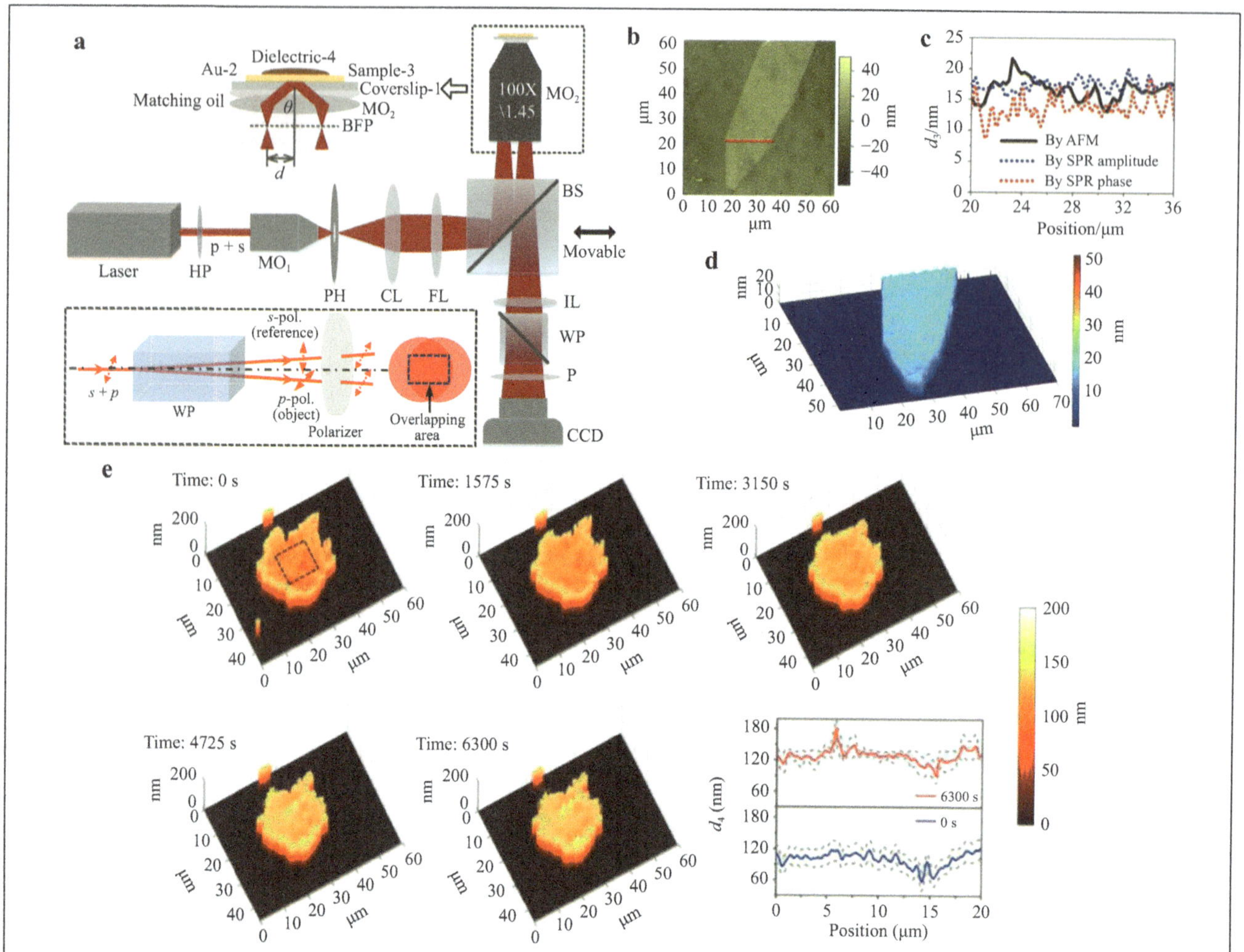

Fig. 14 **a** Experimental setup of common-path SPRHM based on MO-coupling configuration[98]. HP: half-wave plate, $MO_{1;2}$: microscope objectives, PH: pinhole, CL: collimating lens, FL: focusing lens, BS: beam splitter, IL: imaging lens, WP: Wollaston prism, P: polarizer. Inset displays the principle of beam splitting using WP and the common-path recording process of the hologram. **b** AFM image of the test multilayer graphene film. **c** Thickness profile measured by AFM and SPRHM. **d** Teconstructed 3D thickness distribution of the film by SPRHM. **e** Measured adhesion gap distributions of a human breast cancer cell when $t = 0$, 1575, 3150, 4725, and 6300 s, during the apoptosis process by SPRHM[99]. Images reprinted with the following permissions: a–d[98] from The Optical Society; e[99] from Elsevier.

Fig. 14b–d display the experiment results. Furthermore, we also measured the adhesion gap distributions of human breast cancer cells during the apoptosis process in a dynamic and label-free manner (Fig. 14e)[99].

Discussion and outlook

We reviewed common-path configuration designs that aim to improve the temporal stability of off-axis holographic experimental setups. The common-path models were categorised into lateral shearing, point diffraction and other types. For each type, the design principle was given and the related techniques were summarised. For the lateral shearing-based type, the beam carrying the object information is first doubled using a glass plate, grating, or BS. Then, the portions of the two beams with and without the sample information create the shearing interference. This type has a simpler and more compact design compared with that of other types. However, it requires an undisturbed portion of the illumination beam to generate the reference beam, which could reduce the FOV. Moreover, it has the potential problem of a duplicated image appearing on the camera target. This type is typically applied to spatially sparse samples. Conversely, the point diffraction-based type creates a uniform reference beam from the object beam by low-pass filtering in the Fourier domain. This type does not have the problems of duplicated image and limited FOV. However, it typically has a complex configuration and the two beams are separated at the spatial filtering position, resulting in a low level of common-path. This type is suitable for measuring small samples with high spatial frequency, as it is easy to generate the uniform reference beam by spatial filtering. Finally, other advanced designs such as using folding mirrors, specially positioned BS

Table 1 Comparison of three types of common-path configurations

Type	Advantages	Disadvantages	Applications
Lateral shearing	■ **Simple & compact** configuration	■ **Limited** FOV ■ **Duplicated** image (potential)	■ Spatially **sparse** samples
Point diffraction	■ **Large** FOV	■ **Complex** configuration ■ **Low** level of common-path	■ Small samples with **high** spatial frequency
Others	Specific design dependent		

cubes, and WPs were introduced. These designs could avoid the disadvantages existing in the former two types. For clarification, we summarise the advantages and disadvantages of the three types of common-path configurations in Table 1.

The critical point for each type is to allow the object and reference beams to propagate along the same or similar optical paths. The greater the level of similarity, the better the temporal stability the setup can possess. It is worth noting that each design has its own characteristics based on the properties of the measured sample and application environment. The majority of the current common-path designs are limited to a small number of application scenarios. Therefore, boosting the applications of common-path off-axis holographic configurations is an important research direction for the future. Moreover, combining this with other imaging techniques and achieving more functionality by common-path digital holography will attract tremendous interest from different fields.

Furthermore, the reviewed common-path configurations typically include fewer optical components than traditional interferometers. Therefore, the common-path configuration is the best choice in manufacturing compact and highly stable off-axis holographic instruments. From the viewpoint of instrument investigation, it is better to design common-path interference modules that can be easily integrated with commercial setups. In this manner, significant cost and time could be saved in developing commercial instruments. The instrumentation of common-path digital holography will lead to new interdisciplinary research directions in the fields of optical, electronic, and mechanical engineering. Owing to the high stability of the common-path holographic instruments, they can be used out of the laboratories and should find wide applications in industry, chemical, biomedicine, and other fields.

Acknowledgements

Support from the National Natural Science Foundation of China (NSFC) (61927810, 62075183, 62005219) and Fundamental Research Funds for the Central Universities (310202011qd004) is acknowledged.

Author contributions

J. Zhao supervised the project. J. Zhang, C. Ma, and T. Xi performed the literature research. J. Zhang wrote the initial manuscript. S. Dai, C. Ma, J. Di, and J. Zhao revised and commented on the manuscript.

Conflict of interest

The authors declare no conflicts of interest.

Received: 08 March 2021 Revised: 10 August 2021 Accepted: 12 August 2021
Accepted article preview online: 13 August 2021
Published online: 15 September 2021

References

1. Gabor, D. A new microscopic principle. *Nature* **161**, 777-778 (1948).
2. Leith, E. N. & Upatnieks, J. Wavefront reconstruction with diffused illumination and three-dimensional objects. *Journal of the Optical Society of America* **54**, 1295-1301 (1964).
3. Goodman, J. W. & Lawrence, R. W. Digital image formation from electronically detected holograms. *Applied Physics Letters* **11**, 77-79 (1967).
4. Schnars, U. & Jüptner, W. Direct recording of holograms by a CCD target and numerical reconstruction. *Applied Optics* **33**, 179-181 (1994).
5. Di, J. L. et al. Phase aberration compensation of digital holographic microscopy based on least squares surface fitting. *Optics Communications* **282**, 3873-3877 (2009).
6. Di, J. L. et al. Reconstruction of axisymmetric liquid flow field in digital holographic interferometry. *Digital Holography 3D Imag. Mtg.*, 2015, doi: 10.1364/DH.2015.DTh2A.3.
7. Sun, W. W. et al. Real-time visualization of Karman vortex street in water flow field by using digital holography. *Optics Express* **17**, 20342-20348 (2009).
8. Wu, B. J. et al. Visual investigation on the heat dissipation process of a heat sink by using digital holographic interferometry. *Journal of Applied Physics* **114**, 193103 (2013).
9. Zhang, Y. Y. et al. Real-time monitoring of the solution concentration variation during the crystallization process of protein-lysozyme by using digital holographic interferometry. *Optics Express* **20**, 18415-18421 (2012).
10. Wang, J. et al. Dynamically measuring unstable reaction–diffusion process by using digital holographic interferometry. *Optics and Lasers in Engineering* **57**, 1-5 (2014).
11. Chen, X. et al. Measurement and reconstruction of three-dimensional configurations of specimen with tiny scattering based on digital holographic tomography. *Applied Optics* **53**, 4044-4048 (2014).
12. Wang, J. et al. Visual measurement of the pulse laser ablation process on liquid surface by using digital holography. *Journal of Applied Physics* **115**, 173106 (2014).
13. Rajput, S. K. et al. Sound wave detection by common-path digital holography. *Optics and Lasers in Engineering* **137**, 106331 (2021).
14. Nguyen, T. H. et al. Gradient light interference microscopy for 3D imaging of unlabeled specimens. *Nature Communications* **8**, 210

(2017).
15. Kim, T. et al. White-light diffraction tomography of unlabelled live cells. *Nature Photonics* **8**, 256-263 (2014).
16. Grilli, S. et al. Whole optical wavefields reconstruction by Digital Holography. *Optics Express* **9**, 294-302 (2001).
17. Park, Y., Depeursinge, C. & Popescu, G. Quantitative phase imaging in biomedicine. *Nature Photonics* **12**, 578-589 (2018).
18. Merola, F. et al. Tomographic flow cytometry by digital holography. *Light: Science & Applications* **6**, e16241 (2017).
19. Kumar, M. et al. Common-path multimodal three-dimensional fluorescence and phase imaging system. *Journal of Biomedical Optics* **25**, 032010 (2020).
20. Abbe, E. Beiträge zur Theorie des Mikroskops und der mikroskopischen Wahrnehmung. *Archiv für Mikroskopische Anatomie* **9**, 413-468 (1873).
21. Cotte, Y. et al. Marker-free phase nanoscopy. *Nature Photonics* **7**, 113-117 (2013).
22. Simon, B. et al. Tomographic diffractive microscopy with isotropic resolution. *Optica* **4**, 460-463 (2017).
23. Alexandrov, S. A. et al. Synthetic aperture Fourier holographic optical microscopy. *Physical Review Letters* **97**, 168102 (2006).
24. Maire, G. et al. Experimental demonstration of quantitative imaging beyond abbe' s limit with optical diffraction tomography. *Physical Review Letters* **102**, 213905 (2009).
25. Lai, X. J. et al. Coded aperture structured illumination digital holographic microscopy for superresolution imaging. *Optics Letters* **43**, 1143-1146 (2018).
26. Choi, Y. et al. Synthetic aperture microscopy for high resolution imaging through a turbid medium. *Optics Letters* **36**, 4263-4265 (2011).
27. Micó, V. et al. Resolution enhancement in quantitative phase microscopy. *Advances in Optics and Photonics* **11**, 135-214 (2019).
28. Wu, Y. C. et al. Extended depth-of-field in holographic imaging using deep-learning-based autofocusing and phase recovery. *Optica* **5**, 704-710 (2018).
29. Rivenson, Y. et al. Phase recovery and holographic image reconstruction using deep learning in neural networks. *Light: Science & Applications* **7**, 17141 (2018).
30. Ren, Z. B, Xu, Z. M. & Lam, E. Y. M. End-to-end deep learning framework for digital holographic reconstruction. *Advanced Photonics* **1**, 016004 (2019).
31. Wu, Y. C. et al. Deep learning enables high-throughput analysis of particle-aggregation-based biosensors imaged using holography. *ACS Photonics* **6**, 294-301 (2019).
32. Ren, Z. B., Xu, Z. M. & Lam, E. Y. Learning-based nonparametric autofocusing for digital holography. *Optica* **5**, 337-344 (2018).
33. Pinkard, H. et al. Deep learning for single-shot autofocus microscopy. *Optica* **6**, 794-797 (2019).
34. Dardikman-Yoffe, G. et al. PhUn-Net: ready-to-use neural network for unwrapping quantitative phase images of biological cells. *Biomedical Optics Express* **11**, 1107-1121 (2020).
35. Nguyen, T. et al. Automatic phase aberration compensation for digital holographic microscopy based on deep learning background detection. *Optics Express* **25**, 15043-15057 (2017).
36. Jo, Y. et al. Holographic deep learning for rapid optical screening of anthrax spores. *Science Advances* **3**, e1700606 (2017).
37. Li, Y. et al. Classification of cell morphology with quantitative phase microscopy and machine learning. *Optics Express* **28**, 23916-23927 (2020).
38. Goswami, N. et al. Rapid SARS-CoV-2 detection and classification using phase imaging with computational specificity. *bioRxiv* (in the press), doi: 10.1101/2020.12.14.422601.
39. Rosen, J. & Brooker, G. Digital spatially incoherent Fresnel holography. *Optics Letters* **32**, 912-914 (2007).
40. Rosen, J. & Brooker, G. Non-scanning motionless fluorescence three-dimensional holographic microscopy. *Nature Photonics* **2**, 190-195 (2008).
41. Kelner, R., Katz, B. & Rosen, J. Common path in-line holography using enhanced joint object reference digital interferometers. *Optics Express* **22**, 4995-5009 (2014).
42. Yamaguchi, I. & Zhang, T. Phase-shifting digital holography. *Optics Letters* **22**, 1268-1270 (1997).
43. Rodriguez-Zurita, G. et al. One-shot phase-shifting phase-grating interferometry with modulation of polarization: case of four interferograms. *Optics Express* **16**, 7806-7817 (2008).
44. Gao, P. et al. Parallel two-step phase-shifting digital holograph microscopy based on a grating pair. *Journal of the Optical Society of America A* **28**, 434-440 (2011).
45. Lin, M. et al. Parallel phase-shifting digital holography with adaptive function using phase-mode spatial light modulator. *Applied Optics* **51**, 2633-2637 (2012).
46. Shan, M. G. et al. Parallel two-step spatial carrier phase-shifting common-path interferometer with a Ronchi grating outside the Fourier plane. *Optics Express* **21**, 2126-2132 (2013).
47. Anand, A., Chhaniwal, V. & Javidi, B. Tutorial: common path self-referencing digital holographic microscopy. *APL Photonics* **3**, 071101 (2018).
48. Nehmetallah, G. & Banerjee, P. P. Applications of digital and analog holography in three-dimensional imaging. *Advances in Optics and Photonics* **4**, 472-553 (2012).
49. Bates, W. J. A wavefront shearing interferometer. *Proceedings of the Physical Society* **59**, 940-950 (1947).
50. Singh, A. S. G. et al. Lateral shearing digital holographic imaging of small biological specimens. *Optics Express* **20**, 23617-23622 (2012).
51. Di, J. L. et al. Dual-wavelength common-path digital holographic microscopy for quantitative phase imaging based on lateral shearing interferometry. *Applied Optics* **55**, 7287-7293 (2016).
52. Kim, B. M., Park, S. J. & Kim, E. S. Single-shot digital holographic microscopy with a modified lateral-shearing interferometer based on computational telecentricity. *Optics Express* **25**, 6151-6168 (2017).
53. O' Connor, T., Doblas, A. & Javidi, B. Structured illumination in compact and field-portable 3D-printed shearing digital holographic microscopy for resolution enhancement. *Optics Letters* **44**, 2326-2329 (2019).
54. Patorski, K., Zdańkowski, P. & Trusiak, M. Grating deployed total-shear 3-beam interference microscopy with reduced temporal coherence. *Optics Express* **28**, 6893-6908 (2020).
55. Han, L. et al. Double-channel angular-multiplexing polarization holography with common-path and off-axis configuration. *Optics Express* **25**, 21877-21886 (2017).
56. Primot, J. & Guérineau, N. Extended Hartmann test based on the pseudoguiding property of a Hartmann mask completed by a phase chessboard. *Applied Optics* **39**, 5715-5720 (2000).
57. Hillman, T. R. et al. Near-common-path self-reference quantitative phase microscopy. *IEEE Photonics Technology Letters* **24**, 1812-1814 (2012).
58. Guo, R. L. et al. Quantitative phase imaging by wide-field interferometry with variable shearing distance uncoupled from the off-axis angle. *Optics Express* **28**, 5617-5628 (2020).
59. Schubert, R. et al. Enhanced quantitative phase imaging in self-interference digital holographic microscopy using an electrically focus tunable lens. *Biomedical Optics Express* **5**, 4213-4222 (2014).
60. Ma, C. J. et al. Lateral shearing common-path digital holographic microscopy based on a slightly trapezoid Sagnac interferometer. *Optics Express* **25**, 13659-13667 (2017).
61. Varghese, A., Das, B. & Singh, R. K. Highly stable lens-less digital

holography using cyclic lateral shearing interferometer and residual decollimated beam. *Optics Communications* **422**, 3-7 (2018).
62. Sun, T. F. et al. Single-shot two-channel Fresnel bimirror interferometric microscopy for quantitative phase imaging of biological cell. *Optics Communications.* **426**, 77-83 (2018).
63. Edwards, C. et al. Measuring the nonuniform evaporation dynamics of sprayed sessile microdroplets with quantitative phase imaging. *Langmuir* **31**, 11020-11032 (2015).
64. Di, J. L. et al. Quantitative and dynamic phase imaging of biological cells by the use of the digital holographic microscopy based on a beam displacer unit. *IEEE Photonics Journal* **10**, 6900510 (2018).
65. Lee, K. & Park, Y. Quantitative phase imaging unit. *Optics Letters* **39**, 3630-3633 (2014).
66. Kim, K. et al. Diffraction optical tomography using a quantitative phase imaging unit. *Optics Letters* **39**, 6935-6938 (2014).
67. Serabyn, E. et al. Compact, lensless digital holographic microscope for remote microbiology. *Optics Express* **24**, 28540-28548 (2016).
68. Ebrahimi, S. et al. Stable and simple quantitative phase-contrast imaging by Fresnel biprism. *Applied Physics Letters* **112**, 113701 (2018).
69. Popescu, G. et al. Diffraction phase microscopy for quantifying cell structure and dynamics. *Optics Letters* **31**, 775-777 (2006).
70. Girshovitz, P. & Shaked, N. T. Doubling the field of view in off-axis low-coherence interferometric imaging. *Light: Science & Applications* **3**, e151 (2014).
71. Blum, O. & Shaked, N. T. Prediction of photothermal phase signatures from arbitrary plasmonic nanoparticles and experimental verification. *Light: Science & Applications* **4**, e322 (2015).
72. Kumar, M. et al. Single-shot common-path off-axis dual-wavelength digital holographic microscopy. *Applied Optics* **59**, 7144-7152 (2020).
73. Shaked, N. T. Quantitative phase microscopy of biological samples using a portable interferometer. *Optics Letters* **37**, 2016-2018 (2012).
74. Mahajan, S. et al. Highly stable digital holographic microscope using Sagnac interferometer. *Optics Letters* **40**, 3743-3746 (2015).
75. Singh, V., Tayal, S. & Mehta, D. S. Highly stable wide-field common path digital holographic microscope based on a Fresnel biprism interferometer. *OSA Continuum* **1**, 48-55 (2018).
76. Zheng, C. et al. Digital micromirror device-based common-path quantitative phase imaging. *Optics Letters* **42**, 1448-1451 (2017).
77. Vora, P. et al. Wide field of view common-path lateral-shearing digital holographic interference microscope. *Journal of Biomedical Optics* **22**, 126001 (2017).
78. Edwards, C. et al. Diffraction phase microscopy: monitoring nanoscale dynamics in materials science [Invited]. *Applied Optics* **53**, G33-G43 (2014).
79. Edwards, C. et al. Optically monitoring and controlling nanoscale topography during semiconductor etching. *Light: Science & Applications* **1**, e30 (2012).
80. Edwards, C. et al. Digital projection photochemical etching defines gray-scale features. *Optics Express* **21**, 13547-13554 (2013).
81. Bhaduri, B. et al. Diffraction phase microscopy with white light. *Optics Letters* **37**, 1094-1096 (2012).
82. Pham, H. V. et al. Fast phase reconstruction in white light diffraction phase microscopy. *Applied Optics* **52**, A97-A101 (2013).
83. Zheng, C. et al. Diffraction phase microscopy realized with an automatic digital pinhole. *Optics Communications* **404**, 5-10 (2017).
84. Bhaduri, B. et al. Diffraction phase microscopy: principles and applications in materials and life sciences. *Advances in Optics and Photonics* **6**, 57-119 (2014).
85. Hu, C. F. et al. Endoscopic diffraction phase microscopy. *Optics Letters* **43**, 3373-3376 (2018).
86. Cardenas, N. & Mohanty, S. Decoupling of geometric thickness and refractive index in quantitative phase microscopy. *Optics Letters* **38**, 1007-1009 (2013).
87. Rappaz, B. et al. Simultaneous cell morphometry and refractive index measurement with dual-wavelength digital holographic microscopy and dye-enhanced dispersion of perfusion medium. *Optics Letters* **33**, 744-746 (2008).
88. Park, Y. et al. Spectroscopic phase microscopy for quantifying hemoglobin concentrations in intact red blood cells. *Optics Letters* **34**, 3668-3670 (2009).
89. Fu, D. et al. Quantitative dispersion microscopy. *Biomedical Optics Express* **1**, 347-353 (2010).
90. Pham, H. et al. Spectroscopic diffraction phase microscopy. *Optics Letters* **37**, 3438-3440 (2012).
91. Lee, B. et al. Single grating reflective digital holography with double field of view. *IEEE Transactions on Industrial Informatics* **15**, 6155-6161 (2019).
92. Chhaniwal, V. et al. Quantitative phase-contrast imaging with compact digital holographic microscope employing Lloyd' s mirror. *Optics Letters* **37**, 5127-5129 (2012).
93. Wang, D. Y. et al. Continuous-wave terahertz self-referencing digital holography based on Fresnel' s mirrors. *Optics Letters* **45**, 913-916 (2020).
94. Zhang, J. W. et al. Common-path digital holographic microscopy for near-field phase imaging based on surface plasmon resonance. *Applied Optics* **56**, 3223-3228 (2017).
95. Zhang, J. W. et al. Wavelength-multiplexing surface plasmon holographic microscopy. *Optics Express* **26**, 13549-13560 (2018).
96. Qu, W. J. et al. Transmission digital holographic microscopy based on a beam-splitter cube interferometer. *Applied Optics* **48**, 2778-2783 (2009).
97. Zhang, J. W. et al. Dynamical measurement of refractive index distribution using digital holographic interferometry based on total internal reflection. *Optics Express* **23**, 27328-27334 (2015).
98. Zhang, J. W. et al. Compact surface plasmon holographic microscopy for near-field film mapping. *Optics Letters* **42**, 3462-3465 (2017).
99. Dai, S. Q. et al. Real-time and wide-field mapping of cell-substrate adhesion gap and its evolution via surface plasmon resonance holographic microscopy. *Biosensors and Bioelectronics* **174**, 112826 (2021).

Review Open Access

Machine Learning Techniques Applied to Digital Holography

Brad Bazow[1], Thuc Phan[1], Thanh Nguyen[1], Van Lam[2], Anyu Sun[1], Lin-Ching Chang[1], Christopher B. Raub[2] and George Nehmetallah[1,*]

Due to the recent advances of machine learning (ML) techniques in many domains, this Chapter focuses on three topics illustrating how these techniques are used in digital holography. In Section I, we discuss how automatic phase aberration compensation in digital holographic microscopy (DHM) occurs through deep learning (DL) techniques. Then in Section II we discuss how to perform computational multi-wavelength (MW) phase synthesis using Convolutional Neural Networks (CNNs). Section III describes multi-class cell line classification using DHM and ML. We conclude with a discussion of the current utility, challenges and prospects of applying ML to digital holography.

I. Automatic phase aberration compensation for DHM based on DL background detection

Three-dimensional image retrieving techniques are important for many applications[1-5]. These techniques can roughly be divided into interferometric and non-interferometric techniques. Digital Holographic Microscopy (DHM) is an interferometric non-invasive technique for acquiring real-time quantitative phase images which has an enormous impact in many fields such as biology of living cells[6-9], neural science[10], nanoparticle tracking[11], biophotonics, bioengineering and biological processes[12], microfluidics[13], and metrology[14,15]. A traditional DHM system records a digital hologram optically using a microscope objective (MO) and the image reconstruction is performed digitally using optical propagation techniques[16,17]. However, the use of an MO introduces phase aberrations which can be superposed over the biological sample (object). A successful image reconstruction requires very tedious alignment and precise measurement of the system parameters such as reference beam angle, reconstruction distance, and MO's focal length which are often difficult to achieve in a laboratory environment. To overcome these difficulties, in a previous analysis, the use of multiple-wavelength DHM and telecentric DHM configurations were employed which allowed canceling the bulk of optical phase aberration due to the MO and the reference beam[18]. Residual aberrations could be compensated digitally by using Principal Components Analysis (PCA)[19,20] or Zernike polynomial fitting (ZPF)[18,21,22]. However, the use of a multi-wavelength source makes the system setup more complicated and expensive. Also, the existing digital compensation techniques still have other drawbacks. The ZPF requires background information to find the phase residual which is detected semi-manually by cropping background area to perform the fitting[18,21,22]. PCA, on the other hand, automatically predicts phase residual by creating a self-conjugated phase to compensate for the aberrations but assuming that the phase aberrations have only linear and spherical components and leaving higher order phase aberrations unaccounted for. Therefore, an automatic detection of the background areas in DHM would be highly desired. Many segmentation techniques have been proposed which can be divided into semi-automatic techniques such as active contour[23], region growing[24], graph

Correspondence: George Nehmetallah (nehmetallah@cua.edu)
[1]Department of Electrical Engineering and Computer Science, The Catholic University of America, Washington DC 20064
[2]Department of Biomedical Engineering, The Catholic University of America, Washington DC 20064

cut[25], and random walker[26], which require predefined seeds, and fully automatic segmentation techniques such as edge-based[27], region-based[27], split and merge[28], and watershed techniques[29], However, in the case of DHM, these existing methods are not reliable because of the overwhelming phase aberrations and speckle noise in the images.

In this section, we discuss a fully automatic technique to obtain aberration free quantitative phase imaging in Digital Holographic Microscopy (DHM) based on DL[30]. Hence, a combination of ZPF with a fully automatic background detection using deep learning convolutional neural network (CNN) is used to compensate for all the phase aberrations without human intervention such as manual cropping. This technique works with single and multiple wavelength telecentric configurations. Theoretical and experimental results will be obtained to quantitatively assess growth and migratory behavior of invasive cancer cells.

I.1 Bi-telecentric DHM optical setup

Figure 1 shows the bi-telecentric digital holographic microscopy (BT-DHM) system in vertical transmission mode suited for biological sample analyses. A similar setup can also be used in reflection mode as well. This setup is used in an afocal configuration, where the back focal plane of the MO coincides with the front focal plane of the Tube lens($f_o \equiv f_{TL}$), with the object placed at the front focal plane of the MO, resulting in the cancellation of the bulk of the spherical phase curvature normally present in traditional DHM systems. The optical beam from a HeNe laser travels through a spatial filter and periscope system and is collimated with a collimating lens to produce a plane wave beam. The collimated beam is split into a reference beam and an object beam which is focused on the biological sample using an afocal configuration. The two beams which are tilted by a small angle (<1°) from each other are recombined using a beam splitter and interfere with each other on a CCD to generate an off-axis hologram. The magnification of the BT-DHM system is$M = -f_{TL}/f_o$.

The most common numerical reconstruction algorithms used in constructing digital holograms are the discrete Fresnel transform, the convolution approach, and the reconstruction by angular spectrum which is defined as[15,17]:

$$H(f_x, f_y) = \mathcal{F}\{h(x,y)\} = \int_{-\infty}^{\infty}\int_{-\infty}^{\infty} h(x,y)\exp\{-2\pi j(xf_x + yf_y)\}dxdy \quad (1)$$

$$U(f_x, f_y) = H(f_x, f_y)\exp\{2\pi j f_z d\} \quad (2)$$

$$u(\xi,\eta) = \mathcal{F}^{-1}\{U(f_x, f_y)\} = \int_{-\infty}^{\infty}\int_{-\infty}^{\infty} U(f_x, f_y)\exp\{2\pi j(\xi f_x + \eta f_y)\}df_x df_y \quad (3)$$

where d is the distance between image plane and CCD,$h(x,y)$ is the hologram, $u(\xi,\eta)$ is the reconstructed image, $\mathcal{F}$ is the Fourier transform operator, λ is the wavelength, and $f_x, f_y, f_z = \sqrt{\frac{1}{\lambda^2} - f_x^2 - f_y^2}$ are the spatial frequencies. In DHM we introduce a MO to increase the spatial resolution which is computed according to Eq. (4). Due to the magnification 'M' introduced by the MO the pixel size in the image plane,$\Delta\xi_{mag}$, and ,$\Delta\eta_{mag}$ scale according to:

$$\Delta\xi_{mag} = \frac{\lambda d}{N\Delta x M}, \quad \Delta\eta_{mag} = \frac{\lambda d}{N\Delta y M} \quad (4)$$

where N is the number of pixel in one dimension, and Δx, Δy denote the sampling intervals or pixel size $\Delta x = \Delta y = L/N$, $L \times L$ are the dimensions of the CCD

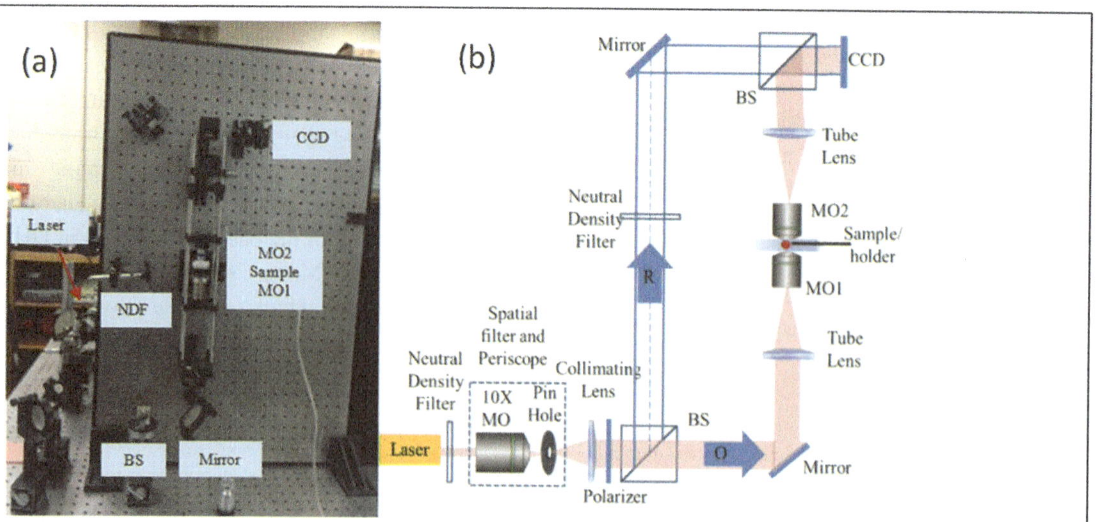

Fig. 1 The BT-DHM system in transmission mode: (a) vertical tabletop setup and (b) optical schematic[30].

sensor. This is intuitively understood by realizing that the holographic recording is now simply a recording of the geometrically magnified virtual image located at distance *d*. Thus, the pixel resolution is automatically scaled accordingly. For a transmissive phase object on/between transmissive surface/s, the phase change (optical thickness *T*) due to the change in index Δn can be calculated as:

$$T(\xi,\eta) = \frac{\lambda}{2\pi}\frac{\phi_{ob}(\xi,\eta)}{\Delta n} \tag{5}$$

where the phase due to the biological sample is expressed as: $\phi_{ob}(\xi,\eta) = \phi(\xi,\eta) - \frac{jk}{2R}(\xi^2+\eta^2)$, where *R* is the radius of curvature of the spherical curvature of the MO and $\phi(\xi,\eta)$ is the total phase of the object beam without using the bi-telecentric configuration.

Conventional image reconstruction using Eq. (3) contains phase aberrations which can be mitigated with the image reconstruction pipeline shown in Fig. 2. First, the hologram is converted into Fourier domain, and the +1 order component is extracted. Then the wrapped phase image is obtained by extracting the phase of the inverse Fourier transform of the cropped spectrum, to be followed by phase unwrapping. The unwrapped phase can be fed into a trained Convolutional Neural Network (CNN) model to determine the background areas. The background information is then fed into the ZPF to calculate the conjugated phase aberration. Phase compensation is done in spatial domain by multiplying the Inverse Fourier Transform of the cropped +1 spectrum order with the complex exponential term which contains the conjugated phase aberration. Then, the Fourier Transform of the compensated hologram is centered and zero padded to the original image size. Finally, the angular spectrum reconstruction technique is performed to obtain the phase height distribution of the full-sized, aberration-free reconstructed hologram, as shown in Fig. 2. One of the crucial steps in the image reconstruction pipeline stated above is to train the proposed deep learning CNN which requires a training dataset of sub-sampled phase aberration images and their corresponding ground truth (label) images. Section I.2 describes in details the data preparation steps for training the CNN model and Section I.3 describes how the CNN model is implemented.

I.2 Biological sample and data preparation

The cancer cells from the highly invasive MDA-MB-231 breast cancer cell line were seeded on type I collagen hydrogels, polymerized at 4 mg/ml and a temperature of 37°C in 35 mm glass-bottomed petri dishes. The cells on collagen were incubated for 24 hours in DMEM medium containing 10% fetal bovine serum, in standard tissue culture conditions of 37°C and 5% CO_2, and 100% humidity. Then, cells were taken from the incubator and imaged with the bi-telecentric DHM setup described above to produce phase reconstruction maps.

Figure 3 shows the steps of data collection/preparation using a PCA method[19]. This process contained three separate parts: (a) collecting background phase aberration

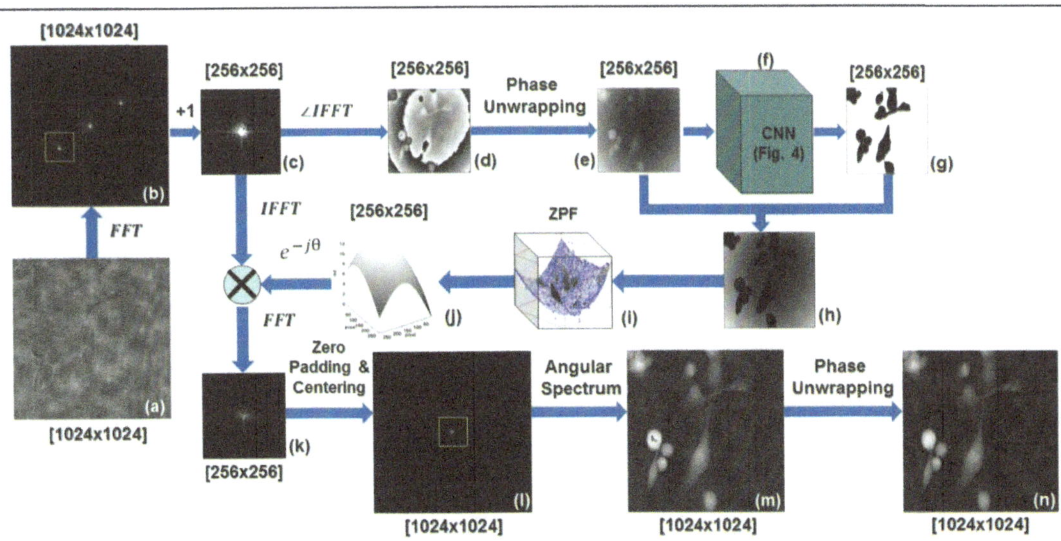

Fig. 2 Image reconstruction pipeline with phase aberration compensation based on CNN + ZPF: (a, b, c) Hologram, Fourier Spectrum of the hologram, and cropped +1 order spectrum, respectively. (d, e) Wrapped phase and unwrapped phase of (c) respectively. (f) CNN trained model, (g) output binary segmentation, (h) visualization of background detection, (i) Zernike polynomial fitting, (j) phase aberration calculated from ZPF, (k) Fourier Spectrum after phase aberration compensation, (l) Fourier Spectrum with zero padding and centering, (m) reconstructed phase map using angular spectrum, and (n) final unwrapped phase map[30].

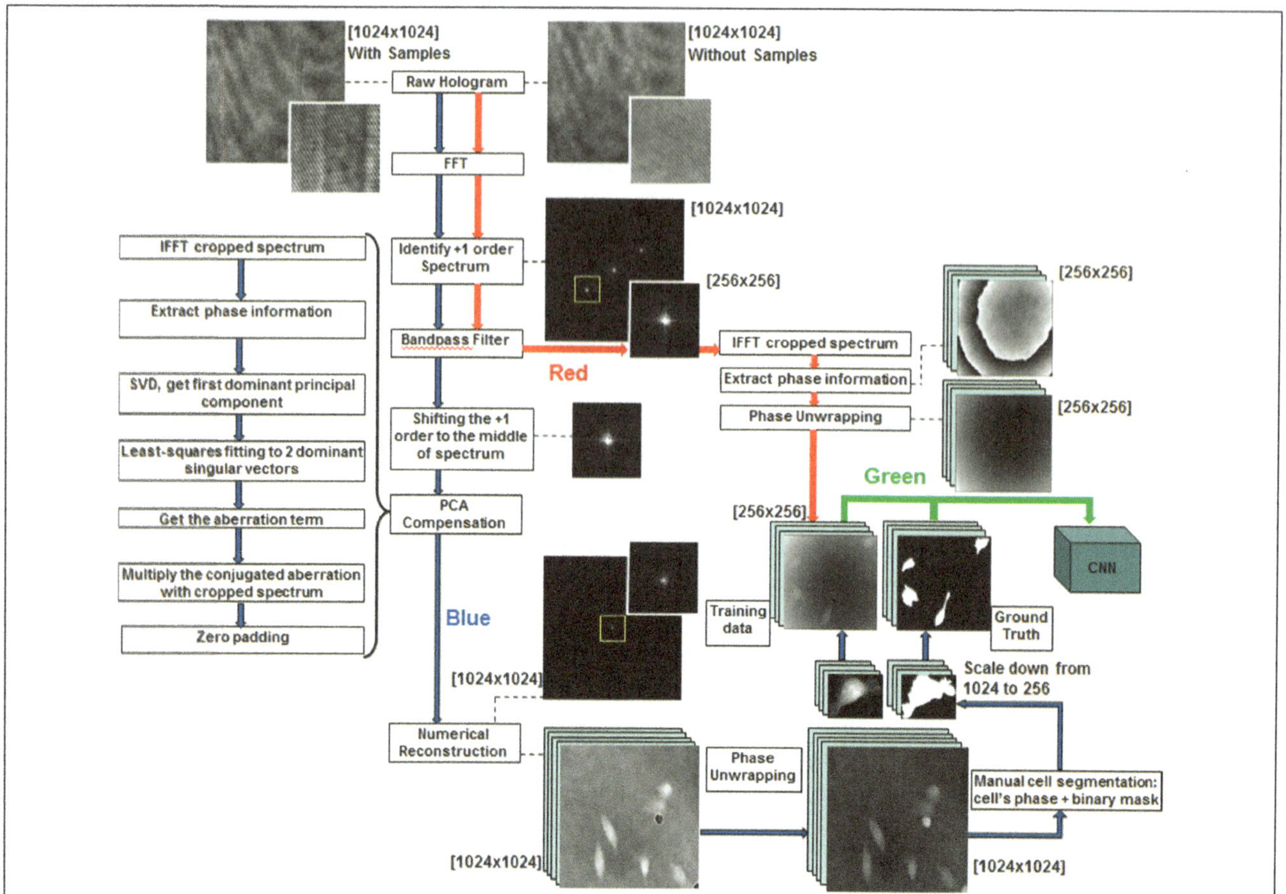

Fig. 3 The pipeline of the data preparation process to be used to train the CNN model. The blue path collects data for single cell segmentations and binary masks. The red path collects sub-sampled phase aberration. The green path shows how the data is fed to the CNN model[30].

(red path), (b) collecting a single cell dataset which includes phase height distribution and binary mask (blue path), and (c) what data is input to the proposed CNN model (the green path). Along the red path, random background phase aberrations were obtained while there was no sample present in the BT-DHM system. As shown in Fig. 1, MO1 and MO2 were both shifted up, down, and rotated to create different phase aberrations. Two-hundred and ten holograms without a sample present were captured and reconstructed, using angular spectrum method. The background sub-sampled (256×256) phase aberration was reconstructed after using a band-pass filter around the +1 order (virtual image location) by using an inverse Fourier transform and phase unwrapping.

Along the blue path, forty holograms containing cancer cells were also reconstructed using the PCA method. For the training stage of the deep-learning CNN, 306 single cells were manually segmented from those forty reconstructed holograms to obtain real phase distribution images and corresponding ground truth binary images (0 for background, 1 for cells). Then, each of cell's phase distribution images, binary masks, and subsampled phase aberration images were augmented using flipping (horizontally and vertically) and rotating (90°, 180°, and 270°). Therefore, 1836 single cell phase distribution images, corresponding to 1836 single cell binary masks and 1 260 sub-sampled background phase aberration were obtained. In order to create the training data set, 4 – 10 real phase maps of cells were randomly added into each of the 1 260 phase aberration images that contain no samples at random positions. It should be noted that the total phase is the integral of the optical path length (OPL). These phase maps were preprocessed with a moving average filter [5×5] to smooth out the edges due to the manual segmentation. Similarly, and corresponding to the same 4 – 10 random positions of the real phase maps, the ground truth binary masks were added to a zero background phase map to create the labeled dataset. Notice that, different types of cells can produce different shapes. In our case, a future objective would be to quantitatively assess the growth and migratory behavior of invasive cancer cells, and hence cells from the invasive MDA-MB-231 breast cancer line were used for this purpose.

Note that, for each type of cells, manual segmentation is

only performed once. Hence, the manual segmentation is only performed in the data preparation stage. Usually, deep learning CNN techniques require a certain amount of training data to produce good results. This additional overhead to collect and prepare the training data can be expensive. However, by augmenting 210 phase images (without sample present) and 306 cell images through flipping and rotation, we are able to create a training dataset of 1260 phase aberration images and their corresponding ground truths images. Eighty percent of these images were randomly selected for training, and the rest of images were used for validation.

I.3 Deep learning convolutional neural network training

In this section, we describe the implementation of deep learning CNN for automatic background detection for digital holographic microscopic images. The deep learning architecture contains multiple convolutional neural network layers[31,32], including max pooling layers, unpooling layers with rectified linear unit (ReLU) activation function, and batch normalization (BN) function, similar to the architecture used in[33]. Let us denote by $x^{(i)}$, $x'^{(i)}$, and $y^{(i)}$ to be the input data volume corresponding to the initial group of phase aberration images, the currently observed volume data at a certain stage of the CNN model, and the output data volume of the CNN model, respectively. The input and output data volume along with the ground truth images have a size of (*batchSize* × *imageWidth* × *imageHeight* × *channel*), where *batchSize* is the number of images in each training session. In our model, the input volume has a size (8×128×128×1) (1 channel indicates a grayscale image), whereas the output volume has a size (8×128×128×2) (2 channels for 2 classes obtained from the one-hot-encoding of the ground truth images[34]). The model shown in Fig. 4 is a simpler version of the U-Net model[33]. An output neuron in the U-net model is computed through convolution operations (which we define as a convolution layer) with the preceding neurons connected to it such that these input neurons are situated in a local spatial region of the input. Specifically, each output neuron in a neuron layer is computed by the dot product between their weights and a connected small region of the input volume, with an addition of the neuron bias:

$$x'^{(i)}_{l} = \sum_{j=0}^{M} W^{(j)}_{l} x'^{(j)}_{l-1} + B^{(j)}_{l}, i = 1, 2, \ldots, N \tag{6}$$

where W is the weight, B is the bias, j is the index in the local spatial region M which is the total number of elements in that region, N is the total number of neurons in each layer which can be changed depending on the architecture, and l is the layer number.

The U-net model contains two parts: Down-sampling (left half of Fig. 4) and up-sampling (right half of Fig. 4). After each convolutional layer, ReLU activation function[35] and BN function[36] were applied to effectively capture non-linearities in data and speedup the training. In the down-sampling path, feature extraction is done through

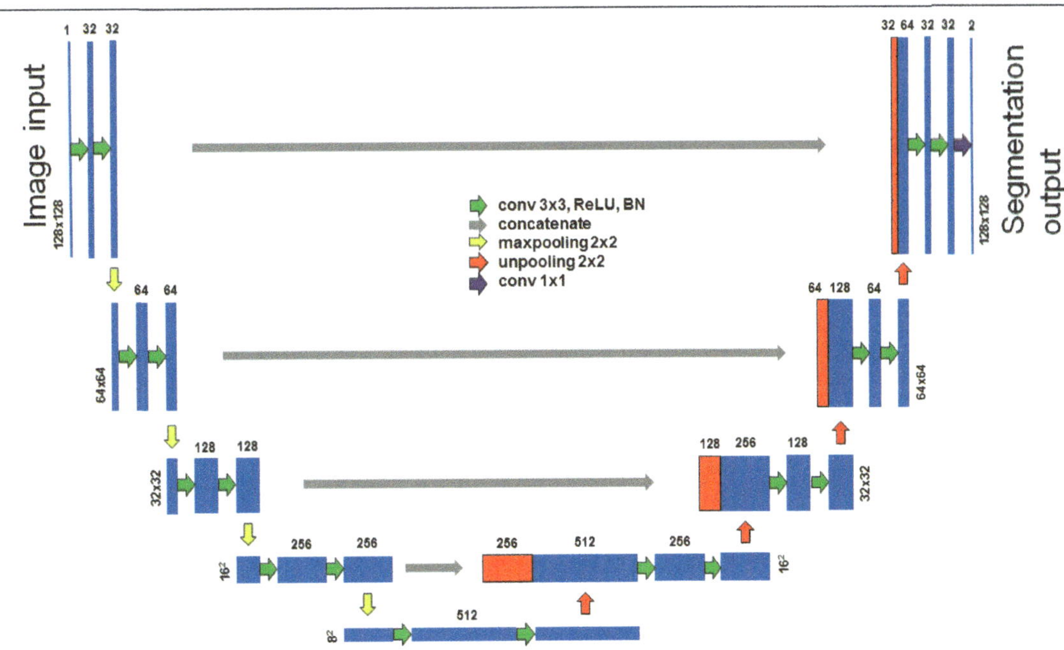

Fig. 4 U-net Convolutional Neural Network model[30].

convolution which transforms the input image to a multi-dimensional feature representation[32,37]. On the other hand, the up-sampling path is a shape generator that produces object segmentation from the extracted features obtained from the convolution path. ReLU activation improves the computational speed of the training stage of the neural networks and prevents the issue of "vanishing gradient" while employing the sigmoidal function traditionally used for this purpose. The ReLU activation function we used is defined as in[35]:

$$f\left(x'^{(i)}\right) = \begin{cases} x'^{(i)} & if\, x'^{(i)} > 0 \\ 0 & otherwise \end{cases}, i = 1,2,\ldots,N \tag{7}$$

where $x'^{(i)}$ is the ith pixel in the volume data under training, N is the total number of pixels in the volume data: $N = batchSize \times layerwidth \times layerheight \times channel$, where *layerwidth* and *layerheight* is the width and height of the image at the lth layer, and the *channel* is the number of weights W in the lth layer.

On the other hand, batch normalization allows the system to: (a) have much higher learning rates, (b) be less sensitive to the initialization conditions, and (c) reduce the internal covariate shift[36]. BN can be implemented by normalizing the data volume to make it zero mean and unit variance as defined in Eq. (8):

$$\widehat{x}'^{(i)} = \gamma \frac{x'^{(i)} - \mu[x'^{(i)}]}{\sqrt{\sigma^2 + \epsilon}} + \beta \tag{8}$$

where $\mu\left[x'^{(i)}\right] = \frac{1}{N}\sum_{i=1}^{N} x'^{(i)}, \sigma^2 = \frac{1}{N}\sum_{i=1}^{N}(x'^{(i)} - \mu[x'^{(i)}])^2$, ϵ is a regularization parameter used to avoid the case of uniform images), γ is a scaling factor, β is the shifting factor ($\gamma = 1, \beta = 0$), and $\widehat{x}'^{(i)}$ is the output of the BN stage.

The down-sampling and up-sampling were done using max pooling and unpooling, respectively. Max pooling is a form of non-linear down-sampling that eliminates non-maximal values, and helps in reducing the computational complexity of upper layers by reducing the dimensionality of the intermediate layers. Also, max pooling was done in part to avoid over fitting. The unpooling operation is a non-linear form of up sampling a previous layer by using nearest neighbor interpolation of the features obtained by max pooling, and resulting gradually in the shape of the samples. Our deep learning CNN model has a symmetrical architecture with max pooling and unpooling filters both with a 2×2 kernel size.

In our experiment, the *Softmax* function, a popular linear classifier defined in Eq. (9), was used in the last layer to calculate the prediction probability of background/cell potential as:

$$S(y^{(i)}) = \frac{e^{y^{(i)}}}{\sum_{i=1}^{N} e^{y^{(i)}}}, i = 1,2,\ldots,N \tag{9}$$

where N (8×128×128×2) is the number of pixels (neurons) needed to be classified in the segmentation process.

An error is a discrepancy measure between the output produced by the system and the correct output for an input pattern. A loss value is the average of errors between the predicted probability $S(y^{(i)})$ and the corresponding ground truth pixel $L^{(i)}$. In our study, the loss function was measured by using the cross entropy function which is defined as:

$$\varepsilon = -\frac{1}{N}\sum_{i=1}^{N} L^{(i)} log\left(S(y^{(i)})\right) \tag{10}$$

The training algorithm was performed by iterating the process of feeding the phase aberration images in batches through the model and calculating the error using an optimizer to minimize the error. The Stochastic Gradient Descent (SGD) optimizer was employed in the back propagation algorithm. Instead of evaluating the cost and the gradients over the full training set, SGD evaluates the values of these parameters using less training samples. The learning rate was initially set to 1e-2, the decay to 1e-6, and the momentum to 0.96[38]. Other parameters used in this section: *batchsize* of 8, image size of 128×128 instead of 256×256 to avoid memory overflow (images will be resized at the end), depth channel of 32 at the first layer, the deepest channel is 512, and training with 360 epochs. The proposed model was implemented in Python using TensorFlow/Keras framework[39] and the implementation was GPU-accelerated with NVIDIA GeForce 970M. Figure 5 shows the training and the validation loss obtained from 360 epochs. Each epoch contains 126 batches of training data. The parameters were updated after each training batch. The training loss and the validation loss started at 0.48 and 0.291 6, respectively. The results suggest that the loss value decreases quickly (i.e., learned quickly) during the first 50 epochs of the training, and the validation loss value decreases with random oscillations (i.e., transitory period) in the first 50 epochs. Note that the validation loss value was slightly less than the value of the

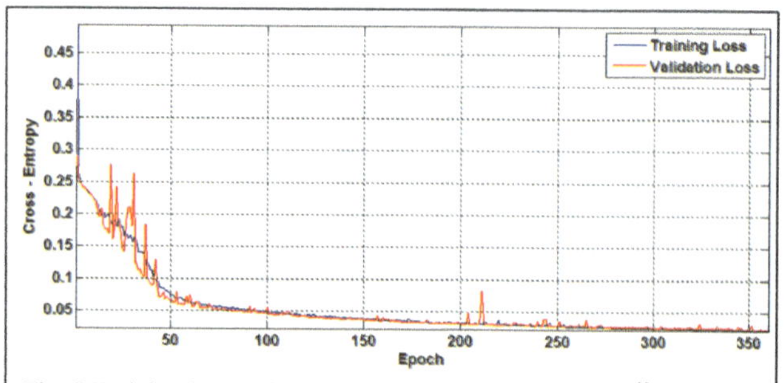

Fig. 5 Training loss and validation loss for the 360 epochs[30].

training loss value during epoch 50 to 220 which implies that the model learned slowly in this period. Between epochs 220 and 360 the validation loss value was slightly higher than the training loss value. Both values decreased slowly to 0.0256 and 0.0237, respectively.

I.4 Testing data and phase aberration compensation

To evaluate the performance of the proposed deep neural network and ZPF technique, 30 holograms recorded by the BT-DHM system (see Fig. 1) and reconstructed using the pipeline in Fig. 2, were tested. The background of a phase aberration image was first located, so the background pixel representation can be used in the ZPF model[18]. As shown in Fig. 2, the unwrapped phase[40] is passed through the trained CNN model discussed in Section I.3 (Fig. 2(g)) to produce the mask prediction $y^{(i)}$ in Eq. (9). The output of the model is normalized in the [0, 1] range and the threshold is set to 0.5 to classify the background and cell area as described by the following equation:

$$B^{(i)} = \begin{cases} 1 & if\, y^{(i)} \le 0.5 \\ 0 & otherwise \end{cases}, i = 1, 2, \ldots, N \qquad (11)$$

Figure 6 shows exemplary channels of selected layers in the down-sampled (top row) and up-sampled (bottom row) paths (see Fig. 4), respectively, for visualization effect. The image on the top left is the raw phase aberration. The next five images on the top row are the outputs of consecutive down-sampling layers. The first five images on the lower row are the outputs of the up-sampled layers, and the lower right image is the binary mask using the threshold function defined in Eq. (11). The down-sampled layers contain the strong features of the image such as the parabolic intensities, and edges, etc., while the up-sampled layers contain the shape of the cells.

In order to measure the conjugated background phase aberration, the pixels from the raw phase image are chosen corresponding to the background pixels' locations obtained from the binary image where ($B^{(i)} = 1$), then converted to a 1D vector to perform the polynomial fitting[18]. Then, the polynomial fitting was implemented using a 5th order polynomial with 21 coefficients as:

$$S(x, y) = \sum_{i=0}^{5} \sum_{j=0}^{5} p_{ij} x^i y^j, i + j \le 5 \qquad (12)$$

where p_{ij} are the coefficients, i and j are polynomial orders, x and y present pixel coordinates. Let the arrays$P = [p_{00} p_{10} \ldots p_{ij} \ldots p_{05}]$ and $A = [a_0 a_1 \ldots a_{10} \ldots a_{20}]$, hold the polynomial model's coefficients and the Zernike model's coefficients. The 21 coefficients of the P polynomial are used to calculate the coefficients of the Zernike polynomial as shown in the following equation:

$$A = z_{i,j,p}^{-1} P \qquad (13)$$

The $z_{i,j,p}$ matrix consists of coefficients corresponding to each order of the Zernike polynomials:

$$\begin{bmatrix} a_0 \\ a_1 \\ \vdots \\ a_{10} \\ \vdots \\ a_{20} \end{bmatrix} = \begin{bmatrix} z_{0,0,0} & z_{0,0,1} & \cdots & z_{0,0,10} & \cdots & z_{0,0,p-1} & z_{0,0,p} \\ z_{1,0,0} & z_{1,0,1} & \cdots & z_{1,0,10} & \cdots & z_{1,0,p-1} & z_{1,0,p} \\ \vdots & \vdots & \ddots & \vdots & \ddots & \vdots & \vdots \\ z_{4,0,0} & z_{4,0,1} & \cdots & z_{4,0,10} & \cdots & z_{4,0,p-1} & z_{4,0,p} \\ \vdots & \vdots & \ddots & \vdots & \ddots & \vdots & \vdots \\ z_{0,5,0} & z_{0,5,1} & \cdots & z_{0,5,10} & \cdots & z_{0,5,p-1} & z_{0,5,p} \end{bmatrix}_{p=20}^{-1} \times \begin{bmatrix} p_{00} \\ p_{10} \\ \vdots \\ p_{40} \\ \vdots \\ p_{05} \end{bmatrix} \qquad (14)$$

The Zernike polynomial model is used to construct the conjugated phase, as:

$$P_{conjugated} = exp\left(-j \sum_{k=0}^{20} a_k Z_k\right), k = 1, 2, \ldots, 21 \qquad (15)$$

where Z_k coefficients are expressed according to Zemax® classification. After obtaining the background area from

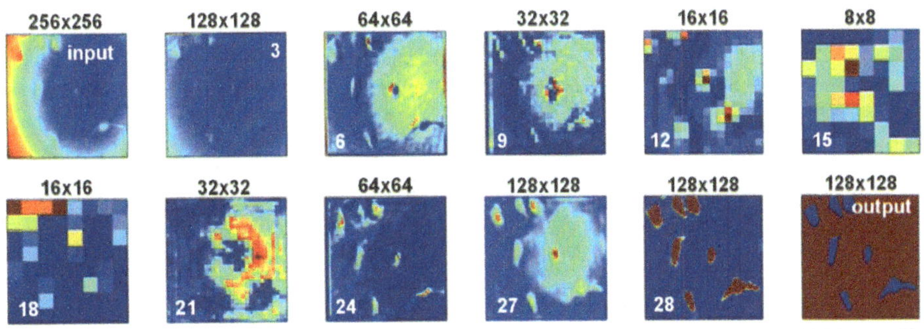

Fig. 6 Visualization of outputs of a selected channel from the following layers: 3, 6, 9, 12, 15, 18, 21, 24, 27, and 28 in CNN[30].

CNN, the conjugated phase aberration was calculated using ZPF, and then multiplied with the initial phase. To obtain the full size aberration compensated reconstructed image, zero padding and spectrum centering was performed on the Fourier transform of the aberration compensated hologram. Then, the angular spectrum reconstruction technique was performed to obtain the phase height distribution of the full-sized, aberration-free reconstructed hologram, as shown in Fig. 2. Figure 7(a) and 7(b) show a typical manual and CNN model's segmentation on the test image in Fig. 2. Figure 7(c) shows the Dice's Coefficient (DC) or F1 score of background area and cell area of 9 typical cases in test data. DC is computed according to the following equation:

$$DC = \frac{2|A \cap A'|}{|A| + |A'|} \tag{16}$$

where |.| denotes the area, A and A' are the segmented areas of a test data based on deep learning CNN and manual segmentation, respectively. Background's DC (0.9582 – 0.9898) is much higher than cell's DC (0.7491 – 0.8764) because of the larger common area in the background. This will lessen the effect of true negative and false positive scenarios in ZPF.

Fig. 8(a) shows a typical reconstruction of real data wrapped phase with aberrations. It is worth noting that the cells in this image do not appear in the training dataset. This means that these holograms were not segmented in the data preparation process. Fig. 8(b) shows the result of background detection using the deep learning CNN classification process. In this example, considerable differences between the training data and the real data were observed. Cells obtained from real data have smoother edges than the ones obtained in the training data. The CNN produces an intentional over segmentation of the cell area which is actually beneficial for background detection. Figure 8(c) is the result of applying (through multiplication) the binary mask on the unwrapped reconstructed phase containing aberrations. Then the phase aberration in the background region was fitted using ZPF to compute the residual phase as shown in Fig. 8(d). Figure 8(e) shows the phase distribution after compensating in the spatial domain according to Fig. 2. Fig. 8(f) shows the final result after phase unwrapping. Figure 9 shows the comparison between PCA and CNN+ZPF techniques. The CNN+ZPF technique produces better results than the PCA technique in approximating the conjugated residual phase. Figs. 9(a) and 9(c) show the phase compensation using PCA and CNN + ZPF, respectively. Fig. 9(b) and Fig. 9(d) are the wrapped conjugated residual phases computed using PCA and CNN+ZPF, respectively. When the PCA's technique is used, the residual phase which contains elliptical concentric pattern was fitted using the least-square method for the two dominant singular vectors corresponding to the first two dominant principal components. This will not compensate for all the distorted regions of the phase distribution.

However, the CNN + ZPF technique takes advantage of the background area; the majority of background information was fitted with higher order (up to 5th order). Hence, the conjugated phase aberration looks more distorted because of those higher orders. Fig. 9(e) shows the Zernike coefficients of the phase difference between CNN+ZPF method and PCA method indicating the error in phase compensation while using the PCA method. Fig. 9(f) shows the profiles of a diagonal dashed line (from bottom left to top right) of PCA's result of Fig. 9(a) and CNN + ZPF's result of Fig. 9(c). The two profiles have different bias phases; the background phase of CNN + ZPF has better flatness (1.35 rad and 0.65 rad) than PCA's background (corresponding to 2.4 rad and 0.95 rad) which can be seen inside the blue dashed rectangle.

II. Computational Multi-wavelength (MW) Phase Synthesis Using Convolutional Neural Networks (CNNs)

Multi-wavelength digital holographic microscopy (MWDHM) has many applications including phase unwrapping for large objects with sharp edges[18,41–43] and refractive index (RI) decoupling[44–47]. With regards to the latter, it has been shown that the refractive index can be estimated from series of independent phase measurements,

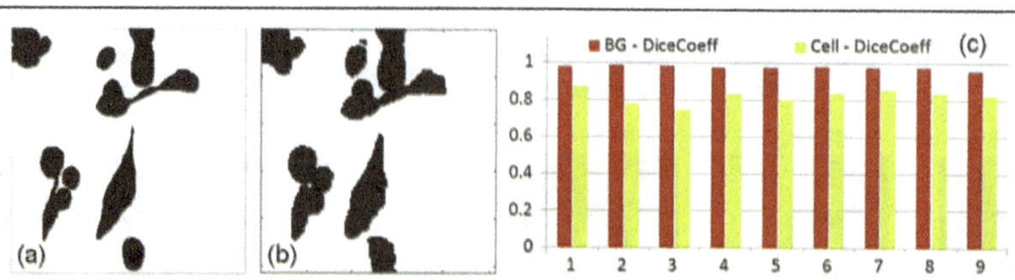

Fig. 7 (a) Typical manual segmentation on the test image of Fig. 6, (b) CNN model's segmentation, and (c) background (BG) dice coefficient and cell dice coefficient on 9 cases of the test data[30].

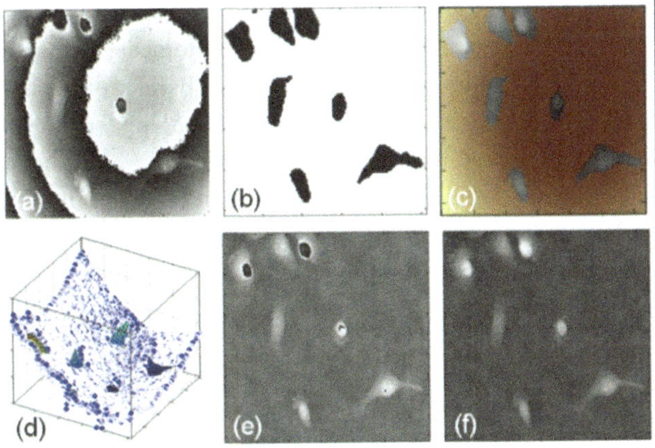

Fig. 8 (a) Wrapped phase with aberration [256 × 256], (b) background detection after CNN [256×256], (c) CNN 's binary mask where background (colored portion) is fed into ZPF [256×256], (d) residual phase [256×256], (e) phase map after phase compensation [1024×1024], and (f) phase unwrapping [1024×1024][30].

e.g. MWDHM can provide an indirect measurement of the integral refractive index given independent phase measurements at two different wavelengths[44,45]. In recent years, deep learning convolutional neural networks (DL/CNN) have proven to be powerful tools in biological imaging applications including holographic phase reconstruction, phase unwrapping, etc.[33,44–58]. Phase retrieval in multi-wavelength unwrapping[41] and multi-wavelength phase-shifting interferometry[47] have been demonstrated using DL/CNN. In this section we demonstrate direct phase retrieval using DL/CNN in MWDHM to provide high-speed single shot measurements of the integral refractive index given a single holographic image by inferring the holographic reconstruction at another wavelength and thus alleviating increased hardware and experimental complexity; and eliminating the need for a wavelength tunable laser source.

Single shot techniques are desirable for cell imaging due to the rapid dynamics of cultured cells samples. For this reason, amongst others, we consider MWDHM in favor of other techniques that have been demonstrated to provide indirect estimates of refractive index. For instance, holographic optical diffraction tomography can provide an estimate of the 3D refractive index given adequate snapshots[59–67], however this requires a more complicated optical setup and multiple unique images acquired successively. Single shot estimates of the 2D refractive index can be obtained using simultaneous multi-wavelength emission sources. The DL/CNN based method for multi-wavelength phase retrieval only requires multi-wavelength holographic imaging for network training, a network is trained once per cell-line and subsequent single wavelength recordings provide single-shot estimates of the refractive index.

To test the feasibility of this concept we consider holographic recordings and reconstructions of simulated

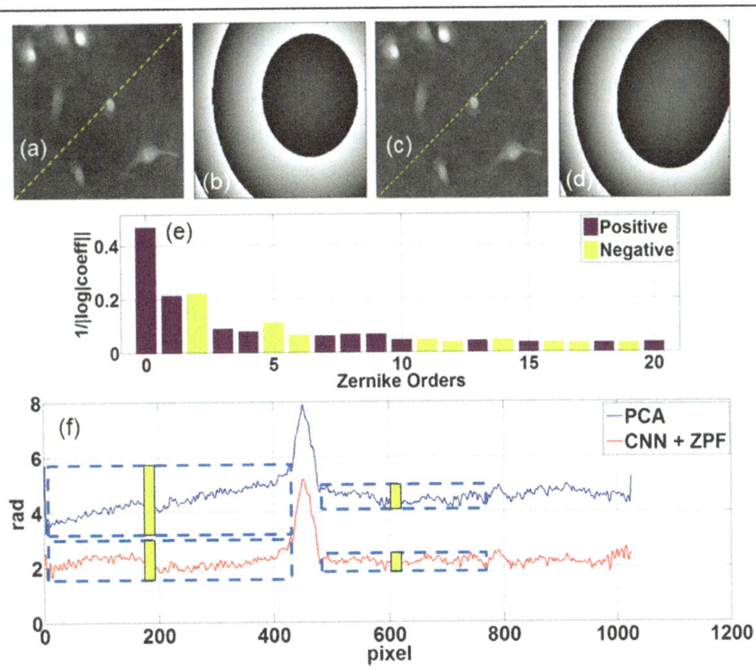

Fig. 9 (a) Phase compensation with PCA, (b) conjugated residual phase of (a), (c) phase compensation using CNN + ZPF, (d) conjugated residual phase of (c), (e) Zernike coefficients of phase difference between CNN+ZPF technique and PCA technique using $1/\|log|ak\|$ scale, and (f) profiles of yellow dash lines in (a) and (c) corresponding to blue and red line, respectively. Yellow bars denote the flatness of region of interest[30].

cellular phase images. We demonstrate dual-wavelength hologram and phase synthesis via DL/CNN, as well as a direct prediction of RI from a single phase, can provide accurate estimates of the integral refractive index assuming an adequately trained network. Specifically, in this section we employed data-driven computational techniques to perform accurate dual-wavelength hologram synthesis (hologram-to-hologram prediction), dual-wavelength phase synthesis (unwrapped phase-to-phase prediction), direct phase-to-index prediction using a single wavelength, hologram-to-phase prediction, and 2D phase unwrapping with sharp discontinuities (wrapped-to-unwrapped phase prediction). Hence, we demonstrate that the full phase reconstruction process can be implemented using a CNN; 2D phase unwrapping can be learned independently or a portion of the full reconstruction. We adopt the *pix2pix* conditional generative adversarial network (cGAN) implementation for image to image prediction[50].

The section is divided as follows. In Section II.1, the fundamentals of RI estimation based on dual-wavelength decoupling are discussed. Sections II.2-3 detail the simulation of cell phantoms and the holographic recording and reconstruction processes. Section II.4 reviews fundamentals of generative adversarial networks (GAN) and provides a detailed description of the specific conditional GAN implementation and parameters considered. Section II.5 presents phase accuracy results, and RI accuracy when relevant, for the prediction schemes discussed above.

II.1 RI estimation using MW decoupling

The optical path length difference of light passing through a transparent object, measured as optical phase, is comprised of the sample topology in addition to local internal refractive index variations. The optical path length is given by

$$L(x,y) = \frac{\lambda}{2\pi}\varphi(x,y) = h(x,y)\left[\bar{n}(x,y) - n_0\right] \tag{17}$$

where φ is the optical phase, h is the transverse dimension of the sample, n_0 is the refractive index of the medium, and $\bar{n}$ is the integral refractive index of the sample, i.e. the path integral of the instantaneous RI along the optical axis, given by

$$\bar{n}(x,y) = \frac{1}{h(x,y)}\int_0^{h(x,y)} n(x,y,z)dz \tag{18}$$

For a single phase measurement these terms are coupled ambiguously unless one of the terms is known *a priori*. Several techniques for interpreting these terms have been proposed and demonstrated[68], all of which require multiple independent phase measurements. Multi-wavelength imaging is an attractive option for biological samples as co-registered independent data can be obtained in a single shot without disturbing the sample or changing the angle of illumination. Given phase estimates at distinct wavelengths, the coupled height and sample refractive index can be solved for in terms of the optical path length for each wavelength as

$$h(x,y) = \frac{L(x,y,\lambda_1)}{\bar{n}(x,y,\lambda_1) - n_0(\lambda_1)} \tag{19}$$

$$\bar{n}(x,y,\lambda_2) = n_0(\lambda_2) + \frac{L(x,y,\lambda_2)}{h(x,y)} \tag{20}$$

The unambiguous analytic solution is only obtainable by making the approximation that the refractive index of the sample itself is equivalent for each wavelength, corresponding to the assumption that the medium is more dispersive than the sample and sample dispersion can be ignored[44]. Assuming $\bar{n}(x,y,\lambda_1) = \bar{n}(x,y,\lambda_2) = \bar{n}(x,y)$, substituting Eq. (20) into Eq. (19), and rearranging terms we obtain the decoupled height profile given by

$$h(x,y) = \frac{L(x,y,\lambda_1) - L(x,y,\lambda_2)}{\Delta n_0} \tag{21}$$

where $\Delta n_0 = n_0(\lambda_2) - n_0(\lambda_1)$ is refractive index mismatch of the medium. Substituting Eq. (21) into Eq. (17) for λ_1 and rearranging terms yields the decoupled refractive index given by

$$\bar{n}(x,y) = n_0(\lambda_1) + \frac{L(x,y,\lambda_1)\Delta n_0}{L(x,y,\lambda_1) - L(x,y,\lambda_2)} \tag{22}$$

Adding highly dispersive agents to the extracellular medium has been demonstrated to improve the accuracy of the analytic solution for closely spaced wavelengths, e.g. Boss et al. demonstrated using a 20 mM fast green FCF dying agent to accurately derive important biophysical parameters for a variety of biological specimens[45] and Rappaz et al. demonstrated using a 144 mOsm/kg H_2O hypotonic solution to measure the dynamics of neurons during hypotonic stress[46].

II.2 Simulation of holographic cell phantoms

Transmissive cell phase imaging is an excellent application for MW refractive index decoupling as cells are highly optically transparent with valuable information contained in the optical path length induced via propagation through the cell; we will consider off-axis digital holographic microscopy for measurements of optical phase with submicron spatial resolution and cm^2 FOV[6,9,69-80]. Cell phantoms, consisting of 3D distributions of refractive index and absorption, are generated stochastically guided by measured optical properties of cells[42,46,67,81–83]. In this section we utilized quantitative phase maps of the MDA-MB-231 breast cancer cell line imaged through a telecentric DHM configuration. Cell segmentations were extracted from these phase maps and used to generate distributions of relevant phase derived features, e.g. maximum thickness, area, volume, and principal axes lengths, etc.[83]. The spherical cells, shown as a 3D distribution of refractive index in Fig. 10, contain random distributions of sub-cellular organelles modeled as 3D ellipsoids. The instantaneous refractive index of the cytoplasm (1.4) and the concentric sub-cellular organelles (1.38-1.52) are drawn from standard normal distributions centered about expected values shown in Fig. 10. We acknowledge that the optical properties of the cells are a first-order approximation; however the cell phantoms provide excellent data for concept validation with the availability of ground truth to bench mark prediction performance.

The intensity and phase in the transverse dimensions are computed using a numerical path integral along the optical axis of the object beam. The complex field is then input to holographic simulation described in Section II.3. The holographic cell phantoms, tapped at various points in the reconstruction process, are used to train and test CNN prediction networks. The dataset consists of 512 synthetic holograms and phase reconstructions at λ=488 nm and 505 nm; stored as 256×256 32 bit images. We consider closely spaced wavelengths in order to minimize error associated with dispersive samples. For each dataset, the phase is generated using a single lens DHM configuration subject to the parabolic lens aberration, the equivalent telecentric configuration, and a quasi-impulse invariant system where the point spread function (PSF) of a limited aperture detector is not enforced. The recording process for an arbitrary single lens DHM configuration is depicted in Fig. 10. Each system is modeled to provide 5x magnification and holographic simulations are recorded on a 256 × 256 focal plane array (FPA) with 7 μm pixel pitch in each dimension. A detailed description of the holographic imaging systems and parameters was described previously[42].

II.3 Simulation of holographic imaging

In this section we provide a brief overview of the simulation of holographic recording and reconstruction used to generate the raw holograms and phase reconstructions for the cell phantoms described above. The optical transformation applied to the object beam of any holographic imaging system was generalized using Collins *ABCD* formulation of scalar diffraction with ray matrix optics[42,85–86], given by

$$\psi_p(x_2,y_2;z) = \frac{jk_0}{2\pi B} e^{-jk_0 z} e^{-j\frac{k_0}{2B}D(x_2^2+y_2^2)} \times \int_{-\infty}^{\infty}\int_{-\infty}^{\infty} \psi_{p0}(x_1,y_1)\, e^{-j\frac{k_0}{2B}A(x_1^2+y_1^2)} e^{j\frac{k_0}{B}(x_1x_2+y_1y_2)} dx_1 dy_1 \tag{23}$$

where the input wavefield $\psi_{p0}(x_1,y_1)$ is propagated through a series of optical transformations to produce the output wavefield $\psi_p(x_2,y_2;z)$ at a distance z. To obtain a numerical solution to Eq. (23), we apply the convolution theorem and the inverse Fourier transform to represent the output wavefield in the spatial domain. After considerable algebra, we arrive at the following cases[42]:

$$\psi_p(x_2,y_2;z) = \begin{cases} \frac{jk_0}{2\pi B} e^{-jk_0 z} e^{-j\frac{BD}{2k_0}(k_x^2+k_y^2)} \\ \mathcal{F}\left[\psi_{p0}(x_1,y_1)\right]\Big|_{k_x=\frac{k_0x_2}{B},k_y=\frac{k_0y_2}{B}}, & A=0, B\neq 0 \\ e^{-jk_0 z} e^{-j\frac{k_0C}{2A}(x_2^2+y_2^2)} \mathcal{F}^{-1} \\ \left[\mathcal{F}\left[\frac{1}{A}\psi_{p0}\left(\frac{x_2}{A},\frac{y_2}{A}\right)\right] e^{j\frac{BA}{2k_0}(k_x^2+k_y^2)}\right], & A\neq 0 \end{cases} \tag{24}$$

where $\mathcal{F}$, $\mathcal{F}^{-1}$ are the Fourier and inverse Fourier transform operators respectively, x_1, y_1 are the transverse coordinates at the input plane with corresponding coordinates at the output plane given by $x_2 = k_x B/k_0$,

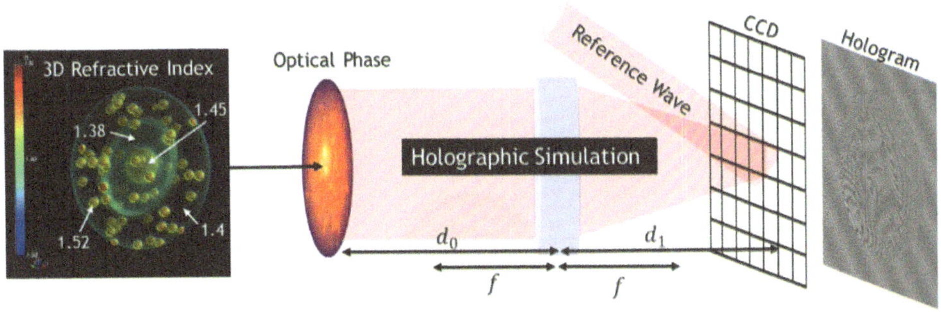

Fig. 10 3D rendering of the internal refractive index of a representative synthetic cell phantom (left). Cell phantoms are integrated to represent the 2D complex optical field at the object plane of an arbitrary holographic imaging system (right)[84].

$y_2 = k_y B/k_0$, which correspond to spatial frequencies k_x, k_y. The object field at the recording plane of any holographic system described by a matrix optics transformation can be computed using Eq. (24). For an off-axis holographic setup, the object wave is interfered with a reference wave at the recording plane. The reference wave is identical to the incident wave, and travels the same optical path length to the interference plane, excluding path length differences introduced by the sample. Denoting the complex phasor amplitude as $\psi = E$, the interference of the object and reference waves measured as intensity, i.e. the hologram, is given by

$$I_H(x_2, y_2) = |E_R + E_O|^2 = |E_R|^2 + |E_O|^2 + E_R^* E_O + E_O^* E_R \quad (25)$$

The hologram is composed of 4 terms, the direct terms $|E_R|^2 + |E_O|^2$ correspond to the zeroth order diffraction or zero-order beam. The terms $E_R^* E_O + E_O^* E_R$ define the interferogram, which contain higher order diffraction as well as the phase information. The term $E_O^* E_R$ forms the real image, and $E_R^* E_O$ forms the virtual or twin image. The hologram is then subsampled on a detector array, which we assume to be a CCD or CMOS array with no loss of generality. The sub-sampled intensity distribution is given by

$$I_H(m,n) = I_H(x_2, y_2)\, rect\left(\frac{x}{L}, \frac{y}{L}\right) \sum_m \sum_n \delta(x - m\Delta x, y - n\Delta y) \quad (26)$$

Fig. 11 shows the general procedure for digital holographic recording and reconstruction for a representative cell phantom. The sample is represented as the complex field derived from 3D distributions of refractive index and absorption. The field at the recording plane is computed using the generalized optical transformation and the hologram is sub-sampled on a CCD. Holographic reconstruction is performed using standard phase retrieval methods[18,42] including digital propagation, spatial/spatial frequency filtering, etc. Aberration compensation is performed using automated procedures based on principal component analysis (PCA) and Zernike polynomials[18,19,42]. 2D digital phase unwrapping is performed using a standard approach based on graph cuts[87].

II.4 cGAN network architecture

To acquire image-to-image predictions we adopted the *pix2pix* cGAN architecture[50], a well-known model for high resolution image to image translation. GANs consist of two adversarial models; a generative model *G* learns the data distribution and is trained to produce outputs that cannot be distinguished from "real" images by an adversarially trained discriminator model *D*. The discriminator is trained to identify "fake" images produced by the generator and estimates the likelihood that a sample belongs to the training data; i.e. GANs learn a mapping from random a noise vector *z* to an output image y, i.e. $G : z \rightarrow y$[48]. Conditional GANs use auxiliary information to condition the generator and discriminator to continuously optimize network parameters and learning; cGANs learn a mapping from an observed image *x* and random noise vector *z*, to an output image *y*, i.e. $G : \{x, z\} \rightarrow y$[49]. Fig. 12 depicts the training procedure for the cGAN architecture used for this section. The generator output *G*(*x*) and input image *x* are concatenated and input to the discriminator network to generate a mapping function *D1*. The input image and the conditional *y*, in this case consisting of the label images or desired generator outputs, are concatenated and input to the discriminator network to generate a mapping *D2*. The feature mappings *D1* and *D2* are used to construct the loss function of the discriminator. The generator output,

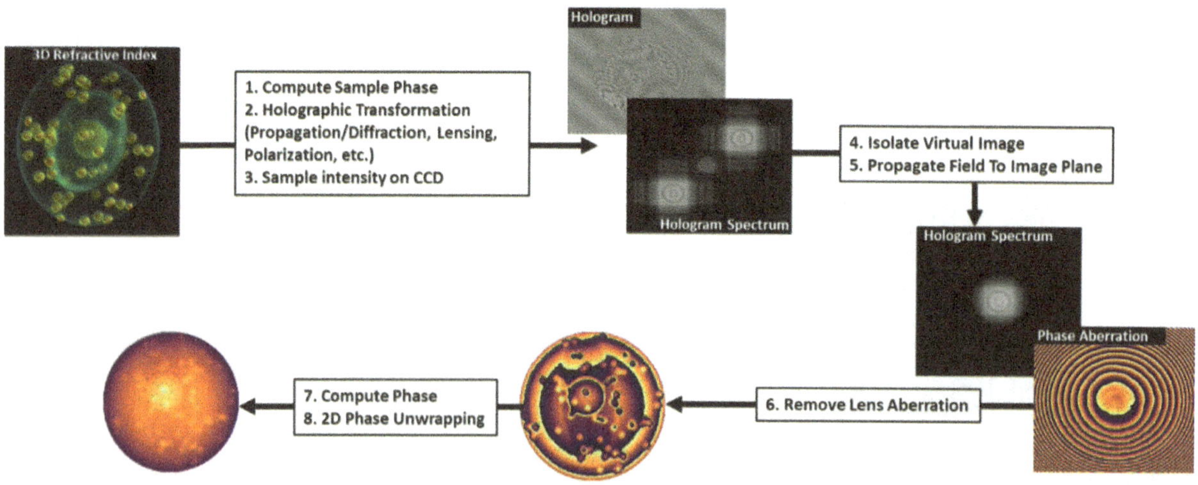

Fig. 11 General procedure for digital holographic recording and reconstruction for a simulated cell phantom[84].

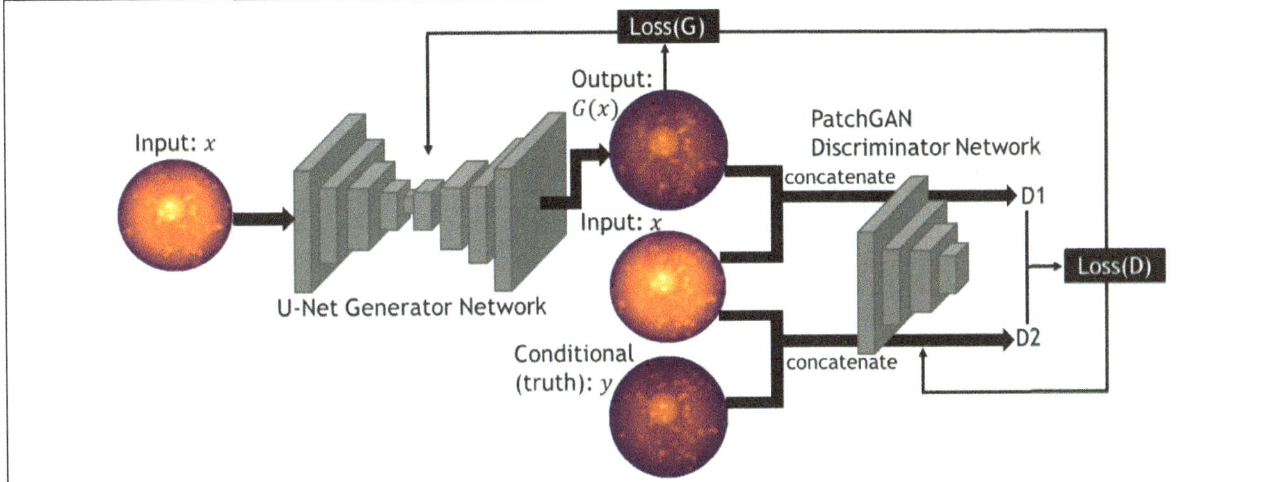

Fig. 12 Schematic illustration of the training procedure for the *pix2pix* cGAN used for image to image prediction. The generator network (U-Net) provides image to image translation, where the training is optimized using a discriminator network (PatchGAN) and conditioned using image labels[84].

conditional, and feature mapping *D1* are used to construct the loss function of the generator.

The composite objective of a conditional GAN can be expressed as[49-50]

$$\mathcal{L}_{cGAN}(G,D)=\mathbb{E}_{x,y}[\log D(x,y)]+\mathbb{E}_{x,z}[\log(1-D(x,G(x,z)))] \quad (27)$$

where $\mathbb{E}$ is the expectation operator. The generator attempts to minimize the objective against an adversarial discriminator that tries to maximize it, i.e.

$$G^{*}=\arg\min_{G}\max_{D}\mathcal{L}_{cGAN}(G,D) \quad (28)$$

The use of the discriminator in the *min-max* objective function provides an excellent model for image to image prediction as the conditioned network can represent sharp features, high resolution/high frequency content, even degenerate distributions whereas simpler models with commonly used pixel-wise loss functions require blurry distributions. It has been proven beneficial to regularize the objective function using the mean absolute error (MAE) as the *L1* norm will further discourage blurring[50]. The final objective is given by

$$G^{*}=\arg\min_{G}\max_{D}\mathcal{L}_{cGAN}(G,D)+\lambda\mathcal{L}_{MAE}(G) \quad (29)$$

where the *L1* loss function is given by

$$\mathcal{L}_{MAE}(G)=\mathbb{E}_{x,y,z}[y-G(x,z)_{1}] \quad (30)$$

We use λ=100 as a regularization parameter to penalize mismatch with the conditional labeled data. The cGAN network trained using *L1* minimization can learn to create a solution that resembles realistic high-resolution images with high-frequency details. Network prediction is achieved by applying the trained generator to testing images. Several CNN architectures were considered including non-adversarial residual neural networks (resNet) and U-Net frameworks for connected layers. While we found that the *pix2pix* framework for conditional GANs using a U-Net generator network yielded the highest accuracy across the cases considered, particularly accurate learning of high resolution detail, this section was not intended to provide an exhaustive comparison of network structures.

The networks are trained using 412 samples, and tested using 100 hold-out samples. Each sample consists of an input/output image pair. For each training epoch, i.e. when each training sample has propagated through the network at least once, the model parameters are updated for each batch of training images using gradient descent optimization and back propagation. The *Adam* algorithm is used for stochastic optimization[88] with a learning rate of 0.0002 and momentum parameters $\beta_1=0.5$, $\beta_2=0.999$.

A. U-Net generator network

The generator network implementation is the well-known U-Net architecture based on the fully convolutional network[33,50], shown in Fig. 13. It consists of a contracting path or encoder and an expansive path or decoder. The encoder/decoder follow typical CNN architectures. The contracting path consists of repeated applications of 4x4 spatial convolutions with stride 2 for downsampling, each followed by a leaky rectified linear unit (LReLU) activation function. At each downsampling step the number of feature channels is doubled from 64 to 512. Every step in the expansive path consists of an up-sampling of the feature map followed by a 4×4 convolution ("up-convolution") with single pixel stride that halves the number of feature channels from 1 024 to 128. At layer

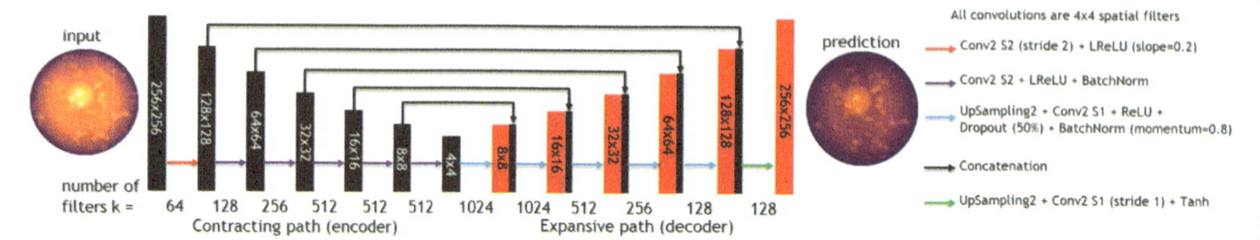

Fig. 13 U-Net generator network consisting of convolutional encoder layers and de-convolutional decoder layers; skip connections between mirrored layers, depicted as black arrows, provide access to additional low level information[84].

interfaces in the expansive path ReLU activation functions are enforced with 50% dropout rate. After the last decoder layer, a convolution is applied to map to the number of output channels and a *Tanh* activation function is applied to acquire the image prediction. The network consists of 12 total layers; batch normalization is performed at each layer interface in the encoder and decoder paths to standardize the data, stabilize and improve network training[36]. The operations performed at each layer interface are detailed in the legend of Fig. 13. While classical CNN solutions require information to flow through the entire encoder/decoder structure; the "U-Net" implementation provides skip connections between mirror symmetric layers in the encoder/decoder. Skip connections are a way to protect the information from loss during transport in neural networks and provide access to low level information which is crucial for image representation[33,50].

B. PatchGAN discriminator network

To further enforce high spatial frequency representation pix2pix adopts the well-known PatchGAN discriminator framework[50], which only penalizes structure at the scale of local image patches. The discriminator is run for a series of convolution patches across the image, averaging all responses to provide an aggregate likelihood that each $N \times N$ patch is a real image produced by the generator. By assuming independence between pixel separations which are large relative to the patch size, the network will learn higher order image texture[50]. The discriminator network, shown in Fig. 14, mimics the first four convolutional layers in the contracting path of the generator network; the number of filters doubles from 64 to 512. After the last layer, a convolution is applied to map to a 1-dimensional output, followed by a *Sigmoid* activation function. The discriminator network is applied independently to 16×16 image patches.

C. cGAN implementation and timing

The cGAN architecture is implemented using the TensorFlow framework version 2.6.0 based on Python 3.8.8. The network training and testing is performed on a PC with Core i9-9940X CPU (3.30 GHz) and 11 GB of RAM and NVIDIA GeForce RTX 2080Ti GPU. The code takes ~2 min per epoch to train using the GPU and ~10 min per epoch using the CPU, where each epoch consists of 412 images of size 256×256. After training, image-to-image prediction takes 135 ms per image using the GPU and 270 ms per image using the CPU.

II.5 Typical results of using DL to acquire phase estimates

In this section we consider several DL/CNN based

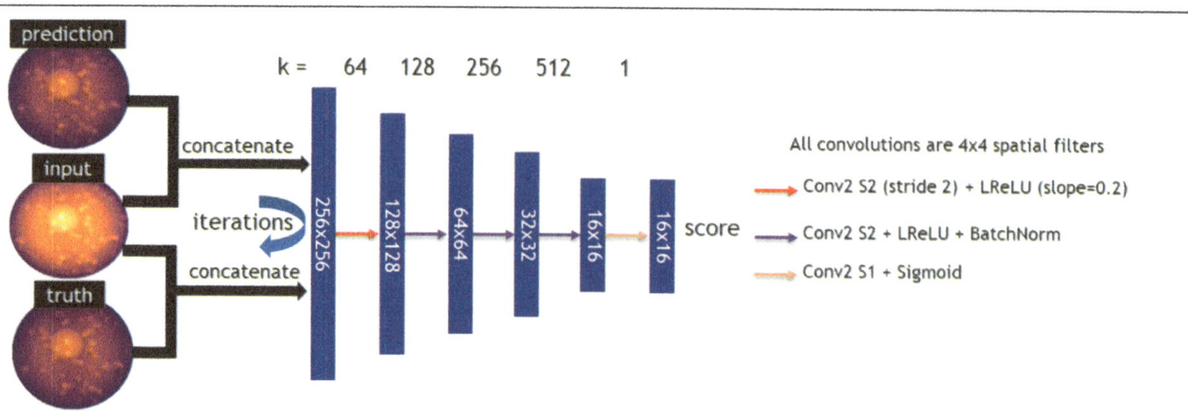

Fig. 14 PatchGAN discriminator network, which mimics the first four convolutional layers of the generator encoding path, is run for a series of convolution patches across the image and only penalizes structure at the scale of local image patches[84].

prediction modalities that can be applied both independently and in conjunction during holographic reconstruction to acquire dual-wavelength phase estimates. We consider dual-wavelength phase estimates acquired using a direct phase-to-phase prediction, which we will refer to as dual-wavelength phase synthesis. Analogously we consider dual-wavelength hologram synthesis, i.e. using raw holographic recordings at one wavelength to predict the raw hologram at another wavelength. These approaches require dual-wavelength refractive index decoupling to obtain estimates of the refractive index, as discussed in Section II.1. We also consider using DL/CNN to acquire the refractive index from a single phase measurement. Independent of multi-wavelength prediction and decoupling we consider holographic phase reconstruction and 2D phase unwrapping based on DL/CNN, which can be components of the proposed RI estimation framework and significant steps towards a fully network based holographic processing system.

The network structure and parameters discussed in Section II.4 were designed to provide high resolution image-to-image prediction and are robust across the variety of applications considered for this work. Networks are trained for the variety of prediction modalities discussed above by altering the input/output of network described in Section II.4; by recording dual-wavelength training data a pre-trained network for each of the prediction modalities considered. Given a single holographic recording, wrapped phase, or unwrapped phase, the phase at two distinct wavelengths can be predicted using a fully learning based approach where pre-trained networks described above are applied in series. Once the phase at two wavelengths are acquired, the refractive index and sample thickness are decoupled analytically using Eqs. (21-22). Fig. 15 shows three distinct procedures for acquiring the phase at multiple wavelengths. The solid arrows represent steps that can be performed using either DL/CNN or traditional techniques, e.g. phase reconstruction and unwrapping. The transparent arrows represent steps that require predictive wavelength synthesis. The transparent blue and red arrows represent the direct unwrapped phase-to-phase prediction discussed in Section II.5A. The absolute unwrapped phase, required as input to network considered in Section II.5A, is estimated from the wrapped phase using either conventional or network based 2D digital unwrapping. Phase unwrapping is represented as a solid blue arrow in Fig. 15; a network based approach to 2D phase unwrapping is discussed in Section II.5E. For holographic imaging applications, phase reconstruction is a prerequisite to multi-wavelength phase synthesis. The solid red arrow represents hologram-to-phase reconstruction, implemented using neural networks in Section II.5D. The transparent green arrow depicts the direct hologram-to-hologram prediction discussed in Section II.5B; followed by an independent hologram-to-phase reconstruction at each wavelength (solid green). Finally, estimates of the refractive index are

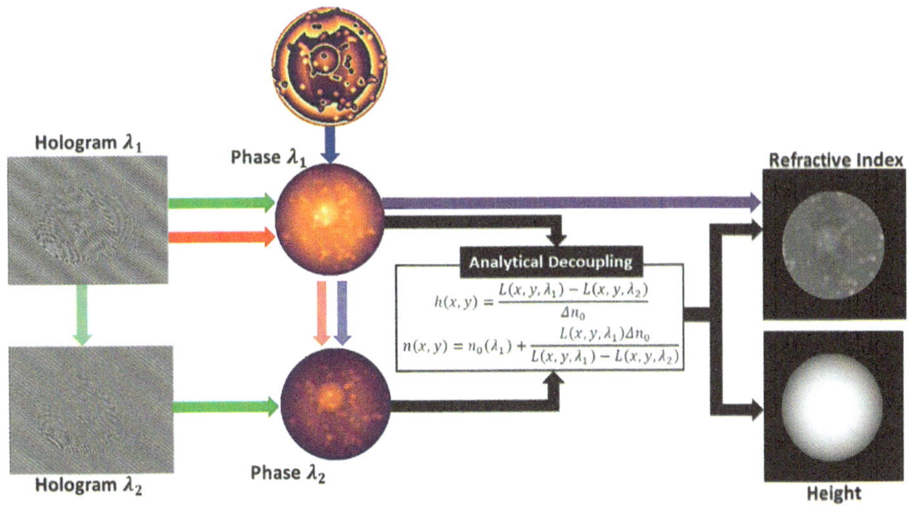

Fig. 15 Dual-wavelength decoupling procedures; each color represents a unique path to dual phase retrieval. The transparent arrows represent processing steps that require predictive wavelength synthesis; the solid arrows can be realized using DL or traditional techniques. Phase synthesis (blue) uses a direct unwrapped phase-to-phase prediction and analytical decoupling to provide an estimate of the RI; absolute unwrapped phase is estimated from wrapped phase using either conventional or network based 2D digital unwrapping. Specific to holographic phase imaging, the hologram-to-phase reconstruction or prediction (red) provides the holographic phase estimates required for phase synthesis. Hologram synthesis (green) consists of a direct hologram-to-hologram prediction; followed by an independent hologram-to-phase reconstruction or prediction at each wavelength. Independent of decoupling, networks are trained to predict RI directly from a single phase input (magenta)[84].

acquired independent of multi-wavelength decoupling using a direct phase-to-index prediction (magenta), discussed in Section II.5C.

We do expect cascading error when applying multiple networks in series; the error from each network will create more mismatch with the data which was used to train the subsequent network. Therefore for each prediction modality we will only consider subsequent conventional processing, e.g. when we demonstrate hologram-to-hologram prediction we will use non-learning based phase reconstruction and unwrapping. We present the prediction accuracy in terms of phase, index, and height for hologram and phase synthesis however we do not consider decoupling for phase reconstruction and unwrapping.

A. Dual-wavelength phase synthesis

In this section we consider obtaining an indirect estimate of the integral refractive index and sample thickness via phase-to-phase prediction. Given phase measurements at one wavelength, we seek to demonstrate accurate prediction of the phase at an alternate wavelength that can be used in conjunction with Eqs. (21-22) to provide accurate predictions of the integral refractive index and thickness. The direct network input/output are used to estimate the refractive index; therefore an unwrapped phase estimate must be available as input *a priori*. For this example we consider a telecentric holographic imaging system; the quadratic lens aberration associated with traditional DHM single lens imaging systems is removed in the optics of the telecentric lensing thus leaving the phase prediction unbiased by errors associated with digital aberration cancellation. The optical field at the reconstruction plane is the magnified image convolved with the PSF of the coherent optical system. We will not impose the PSF of the finite sized detector; this will provide sharp features for the network to train on and will test the networks ability to predict high spatial frequency/high resolution detail. Fig. 16 shows multi-wavelength unwrapped phase synthesis based on convolutional neural networks for a single realization. The simulated phase at 505 nm (a) is input to the trained cGAN which yields the predicted phase at 488 nm (c). The true phase at 488 nm is shown in Fig 16(b), after 100 training epochs the RMSE phase error is <2% for this realization and the images are near identical upon inspection.

Fig. 17 shows the true (a) and predicted (b) height profiles and refractive index (c-d) approximated using Eqs. (21-22). Refractive index and sample thickness are not directly measurable optically; therefore the Ground truth is obtained from true phase and the approximation given in Eqs. (21-22). Fig. 18 shows horizontal slices through the true and predicted phase (a), height profile (b), and refractive index (c). The root mean square error (RMSE) averaged over 100 testing realizations for the phase, height, and refractive index prediction are shown in Fig. 19, the dashed lines represent the 95% confidence intervals. After 100 training epochs the RMS phase, height, and index prediction errors were 0.16 radians, 0.61 microns, and 0.006; corresponding to 0.14σ, 0.34σ, and 1.27σ where σ is the standard deviation of the true phase, height, and index images respectively.

We have demonstrated phase synthesis prediction accuracy that is clearly sufficient from a quantitative phase perspective, as shown in Fig. 16 and Fig. 19 (a). The corresponding decoupled refractive index accuracy is degraded as phase prediction errors are amplified in RI estimation. More specifically Eq. (22) is highly sensitive to phase mismatch due to the dual wavelength path length difference in the denominator of the dominant term. The dashed red line in Fig. 19 (c) represents the refractive index error due solely to the decoupling approximation; i.e. the error that incurs from decoupling the true phase at each wavelength assuming a non-dispersive sample. In simulation we are able to determine the RI estimation accuracy floor to be 0.004, for the simulated optical phantoms considered, by accessing the true phase and refractive index at each wavelength, thus appropriately

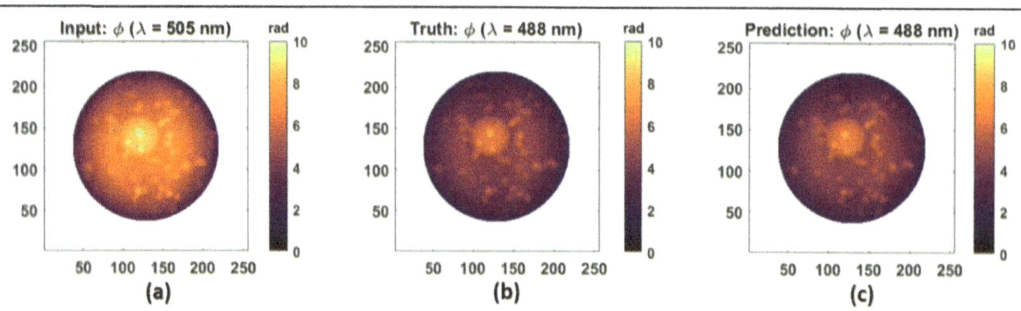

Fig. 16 (a) Unwrapped phase at λ=505 nm used as the input to the network. (b) Ground truthGround truth for the unwrapped phase at λ=488 nm. (c) Unwrapped phase at λ=488 nm predicted by the network[84].

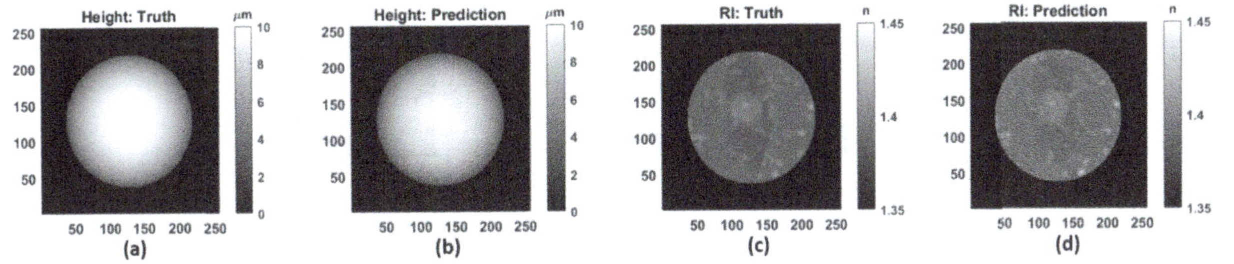

Fig. 17 True (a) and predicted (b) height profiles computed using the predicted phase in Fig. 16 and Eq. (21). True (c) and predicted (d) integral refractive index computed using the predicted phase in Fig. 16 and Eq. (22)[84].

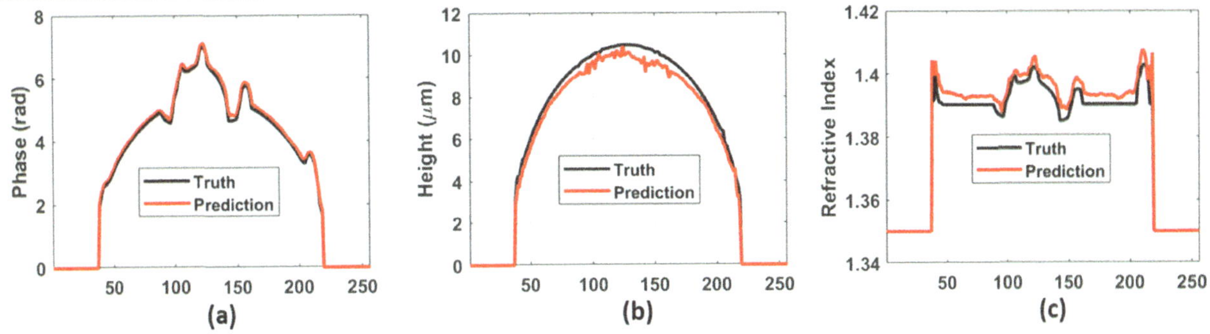

Fig. 18 Comparison of slices of the true and predicted phase (a), sample thickness (b), and refractive index (c)[84].

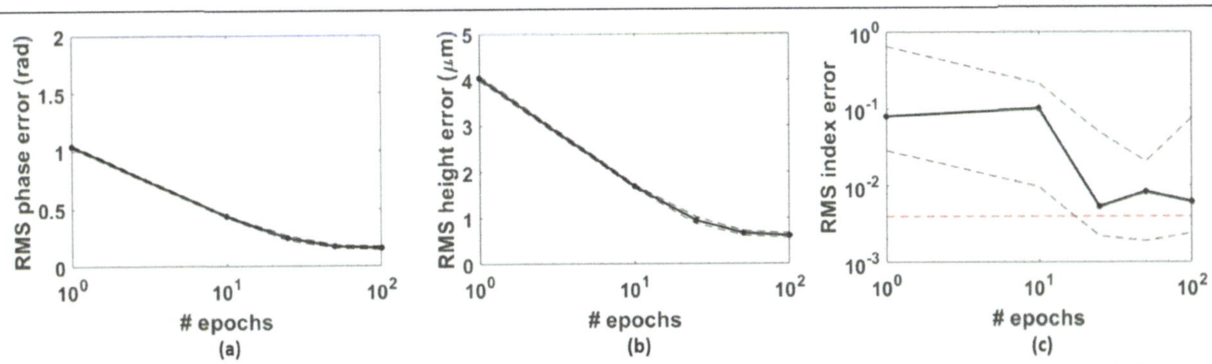

Fig. 19 RMSE averaged over 100 testing samples as a function of training epochs for the phase (a), sample height (b), and refractive index (c). The dashed black lines represent the 95% confidence intervals and the dashed red line in (c) represents the component of the RMS error due solely to the decoupling approximation[84].

benchmarking the performance of RI decoupling based on DL/CNN. While we acknowledge that analytical decoupling could benefit from regularization of the decoupled terms, we consider the achieved RI prediction accuracy to be acceptable as it approaches the limitation set by Eq. (22).

Similarly, it is import to benchmark the sensitivity of Eq. (22) to phase noise fluctuations arising from optical speckle or other sources. Fig. 20 shows the decoupled refractive index for increasing levels of random phase noise, uniformly distributed over the interval $(-\varphi_{s,max}, \varphi_{s,max})$ to represent optical speckle from an object which is rough relative to the optical wavelength[42]. The results are quantified in terms of the ratio of maximum speckle phase to mean sample phase (φ_{mean}). Fig. 20 (e) shows that the RI estimation error due to random phase fluctuations is less than the error induced by the decoupling approximation (red) if the maximum phase noise fluctuations are < 1% of the average sample phase. The phase noise tolerance of refractive index decoupling is less than that of cGAN network prediction, however denoising should be considered as a pre-processing step[89–92].

B. Dual-wavelength hologram synthesis

In this section we consider obtaining indirect estimates of the integral refractive index and sample thickness via hologram-to-hologram prediction. Given raw holographic

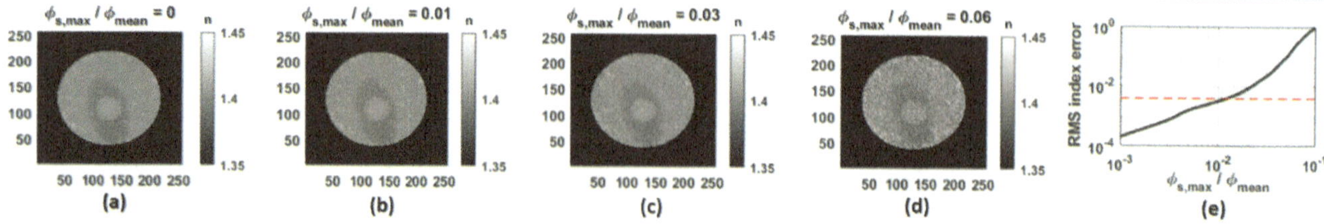

Fig. 20 Decoupled refractive index using the true phase at each wavelength with increasing noise (a-d). The noise is defined to be relative to the mean noise of the simulated phase maps (φ_{mean}). Refractive index decoupling error for increasing noise (e), the red line represents the noise-free error produced by the decoupling approximation[84].

recordings at one wavelength, we seek to demonstrate accurate predictions of raw holographic recordings at an alternate wavelength. Phase reconstructions from each hologram can be used in conjunction with Eqs. (21-22) to provide predictions of the integral refractive index and thickness. The quality of learning increases with the SNR of the interference pattern encoded in the hologram, i.e. the fringe contrast. For decreasing fringe contrast, increasing bit depth will be required in the holographic images as well as increased hidden layers in the neural network. In fact, we found that the number of layers required to learn interference pattern to interference pattern transformations decreased in the presence of a simulated optical lens aberration that increased the fringe contrast in the hologram. Therefore, we consider a telecentric holographic imaging system as an appropriate baseline. Fringe contrast is also increased within limitation by increasing the ratio of object to reference beam power; we assumed that the power in the reference beam was approximately 5 times as large as the power in the object beam. To further decrease fringe contrast we impose the PSF of the limited aperture detector, which acts as a lowpass spatial filter and decreases fringe contrast in the hologram plane.

Fig. 21 shows the mean subtracted raw hologram at 505 nm (a) input to the network, the true holographic recording at 488 nm (b), and the predicted holographic recording at 488 nm (c) for a single testing realization. The RMS hologram error after 100 training epochs is <6% for the realization shown in Fig. 21. Estimates of the optical phase at each wavelength are acquired using non-learning based phase reconstruction approaches discussed in Section II.2. Fig. 22 shows the phase at 488 nm reconstructed using the

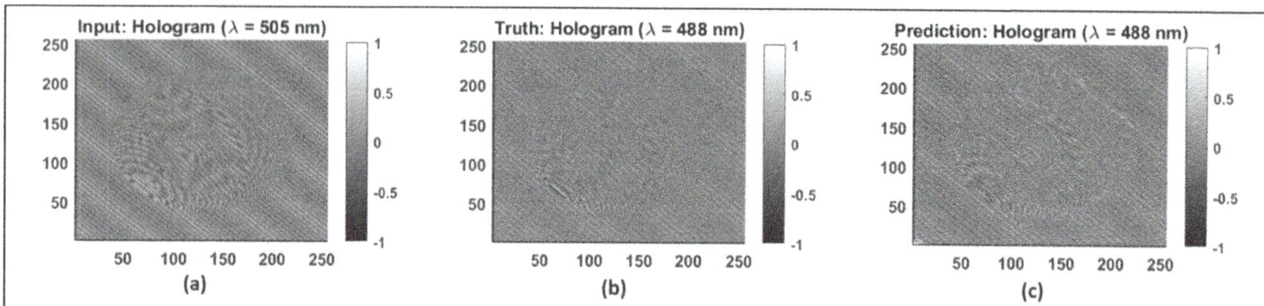

Fig. 21 (a) Raw hologram at λ=505 nm used as the input to the network. (b) Ground truthGround truth for the hologram at λ=488 nm. (c) hologram at λ=488 nm predicted by the network[84].

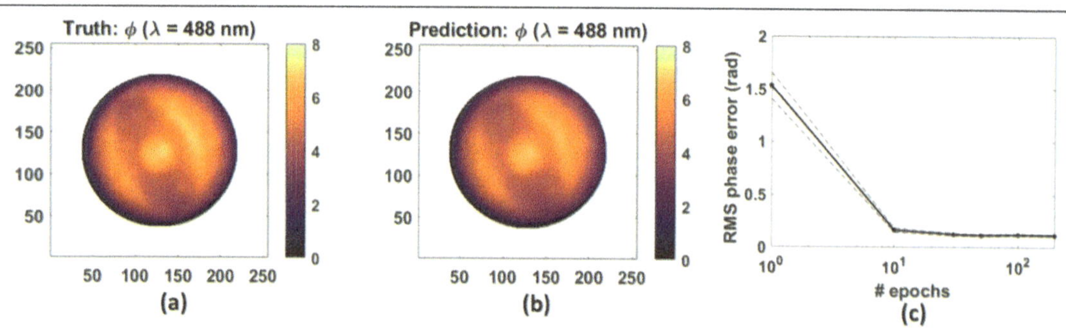

Fig. 22 Reconstructed phase derived from the true (a) and predicted (b) holograms. (c) RMSE averaged over 100 testing realizations as a function of training epochs[84].

true (a) and predicted (b) holograms. After only 10 epochs the RMS phase error was less than 0.1 radians (0.14σ), indicating that the high resolution fringe detail of the synthesized holograms accurately capture the interference encoding required for phase reconstruction. Estimates of the refractive index, shown in Fig. 23, are acquired using the decoupling approximation and the phase reconstructions shown in Fig. 22. The refractive index and height error improved slightly after 10 epochs ultimately reaching 0.003 (0.57σ) and 0.45 microns (0.36σ) respectively. The refractive index estimation error approaches the floor imposed by analytical decoupling, represented by the red line on Fig. 23 (c); therefore demonstrating accurate refractive index estimation based on MW synthetization of raw holographic recordings using DL/CNN. This may be a preferred alternative to MW phase synthesis as acquiring a second holographic recording allows for an independent and controllable phase retrieval processes at each wavelength.

C. Direct network prediction

In this section we consider acquiring an estimate of the refractive index directly from the phase at a single wavelength. This approach does not require the dispersion assumption discussed in Section II.1, and therefore the refractive index can be wavelength dependent. We will use the same optical phantoms and imaging system considered in Section II.5A, as to produce a direct comparison. Fig. 24 shows the direct refractive index prediction (phase-to-index) based on convolutional neural networks for a single realization. The simulated phase at 505 nm (a) is input to the trained cGAN which yields the refractive index at the same wavelength (c).

Fig. 25 (a) shows slices of the true and predicted refractive index at 505 nm, the RMSE as function of training epochs is shown in Fig. 25 (b). The dashed red line again represents the error due solely to the decoupling approximation, although this approach is not subject to such a limitation. After 200 training epochs the refractive index prediction error is 0.009, corresponding to 1.9σ where σ is the standard deviation of the true refractive index images. Comparison with Fig 19 (c) shows that direct refractive index predictions are more accurate than RI decoupling applied to multi-wavelength phase synthesis when the number of training epochs is small, however after ~10 training epochs the phase predictions are sufficiently accurate to produce decoupled RI images which are more accurate than those produced by the direct prediction. We believe that direct refractive index prediction from a single

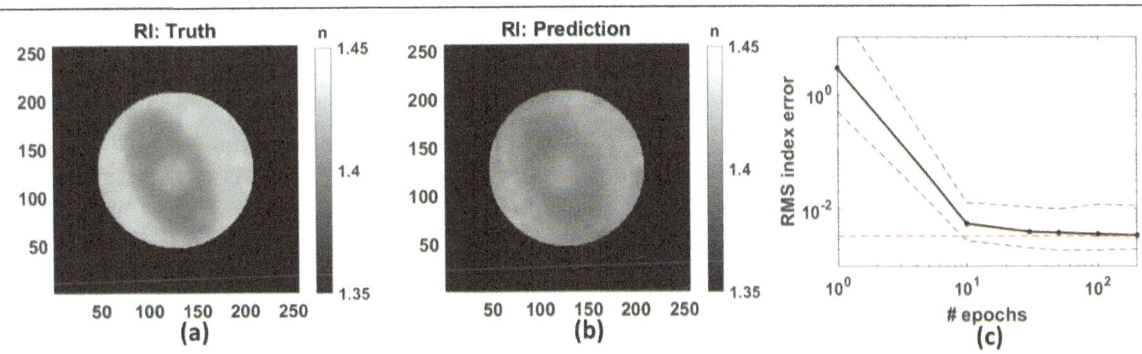

Fig. 23 Refractive index derived from the true (a) and predicted (b) holograms. (c) RMSE averaged over 100 testing realizations as a function of training epochs. The dashed red line in (c) represents the component of the RMS error due solely to the decoupling approximation[84].

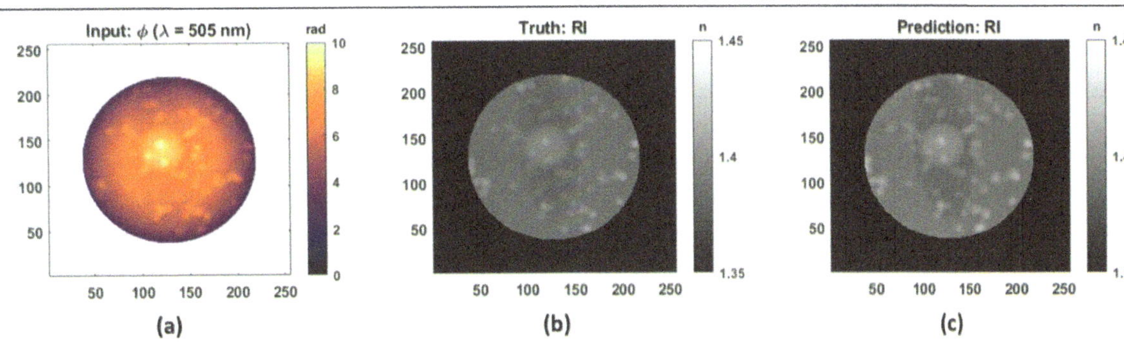

Fig. 24 (a) Unwrapped phase at λ=505 nm used as the input to the network. Ground truth (b) and prediction (c) for the refractive index at λ=505 nm[84].

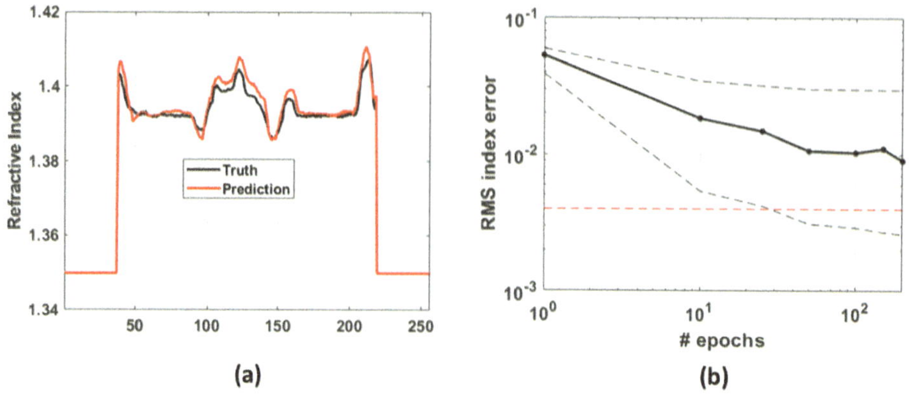

Fig. 25 (a) Comparison of slices of the true and predicted refractive index. (b) Refractive index RMSE averaged over 100 testing samples as a function of training epochs. The dashed red line in (b) represents the component of the RMS error due solely to the decoupling approximation, shown for comparison only[84].

wavelength is challenged because the mapping from a single phase to RI is very ambiguous, and refractive index estimation requires at least 2 degrees of freedom to represent phase as the composition of RI and sample thickness. We found the estimation error to be most severe near edges where the sample thickness is small. We did however demonstrate accurate predictions of the refractive index from a single phase and we expect this approach to outperform MW decoupling for samples that severely violate the dispersion assumption. Another benefit of this approach is that the refractive index images used to train the network are not restricted to multi-wavelength imaging and decoupling, but rather can be derived from any RI estimation technique.

D. Network based holographic phase reconstruction

In this section we consider holographic phase reconstruction based on DL/CNN, where the network layers predict unwrapped phase from raw holographic recordings. The trained network layers encompass the phase reconstruction steps shown in Fig. 11; i.e. the spatial frequency shift associated with the re-introduction of a digital replica of the reference wave, spatial/spatial frequency filtering, conversion to complex angle, aberration compensation, 2D phase unwrapping, etc. We consider a single lens DHM imaging system with 5x magnification such that the native phase reconstruction is obscured by the parabolic phase aberration introduced by the lens. Predicting aberration free absolute phase from a raw hologram recorded in the presence of optical aberrations offers significant advantages as the phase reconstruction process is often most burdened by errors associated with digital unwrapping and aberration compensation. Fig. 26 shows the mean-subtracted hologram (a), the reconstructed phase prior to aberration compensation (b), the true phase predicted by an equivalent telecentric lens which cancels the lens aberration exactly (c), the reconstructed phase as predicted by cGAN (d), and the phase estimate obtained using the standard phase retrieval methods shown in Fig. 11 (e). After 100 training epochs the RMS phase error for phase estimates acquired using cGAN and conventional digital reconstruction are 0.18 radian (0.13σ) and 0.68 radian (0.49σ) respectively, as shown in Fig. 27.

While we have demonstrated improved phase retrieval accuracy using DL/CNN, it is important to note the obvious and important shortcoming of this technique; the network will perform only as well as the approach used to reconstruct the phases used to train the network. Nonetheless, network based reconstruction is a viable alternative as it provides faster and fully automated aberration compensation, whereas conventional approaches are slow and may require user input for each image, e.g. aberration compensation based on Zernike polynomials requires user selected areas of flat phase[18,42].

E. Network based 2D phase unwrapping

In this section we consider 2D phase unwrapping based on deep learning as an alternative to traditional 2D digital phase unwrapping. The problem is to infer absolute phase from modulo-2π phase. The fundamental assumption for phase unwrapping is the maximum path length discontinuity between adjacent pixels is $<\lambda/2$, equivalently the phase is $<\pi$. To increase the effective optical path length, we increased the differential refractive index at the interfaces of the cytoplasm and various sub-cellular concentric layers, as to create sharp $\lambda/2$ discontinuities and a sample that has wrapped several times ambiguously about 2π. The refractive index of the smallest sub-cellular organelles is increased from 1.52 to 2.2, producing a random distribution of sharp path length discontinuities each on a limited pixel scale. These pixel limited, high

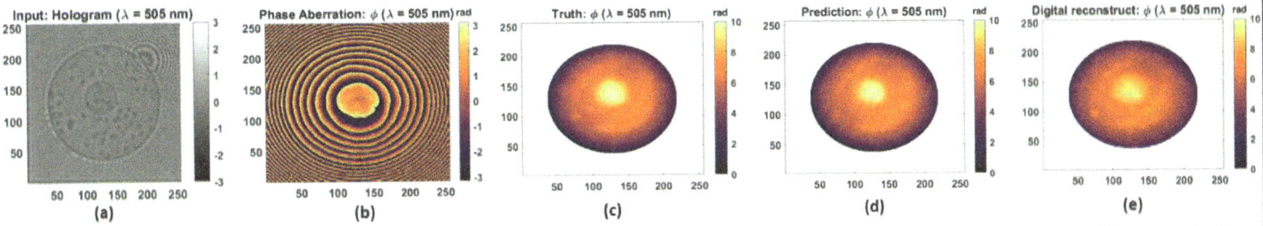

Fig. 26 (a) Mean-subtracted hologram, (b) the reconstructed phase prior to aberration compensation, (c) the true phase predicted by an equivalent telecentric, (d) the unwrapped phase as predicted by cGAN, (e) the phase estimate provided by the phase reconstruction shown in Fig. 11[84].

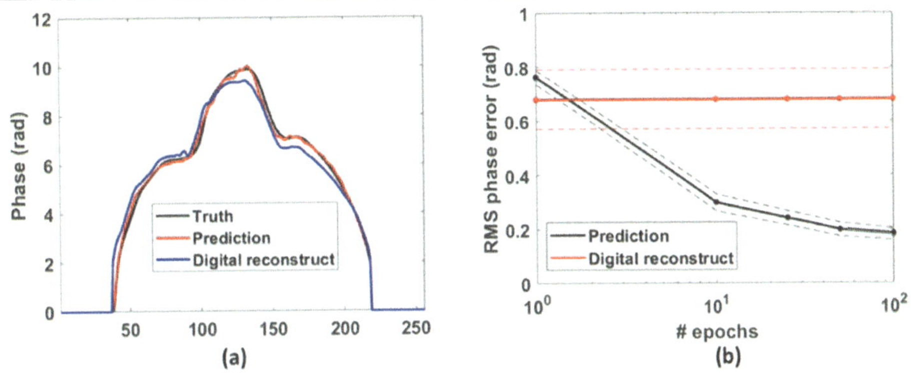

Fig. 27 (a) Comparison of slices of the reconstructed phase used in training (black), the phase reconstruction from cGAN (red), and the phase reconstruction using standard phase retrieval methods (blue). (b) RMSE error as a function of training epochs for cGAN (black) and the phase reconstruction procedure shown in Fig. 11 (red)[84].

refractive index sub-cellular phantoms can be representative of silica colloidal nanoparticles, which could be used to test this approach experimentally. Fig. 28 shows the full 3D refractive index of a representative cell phantom realization.

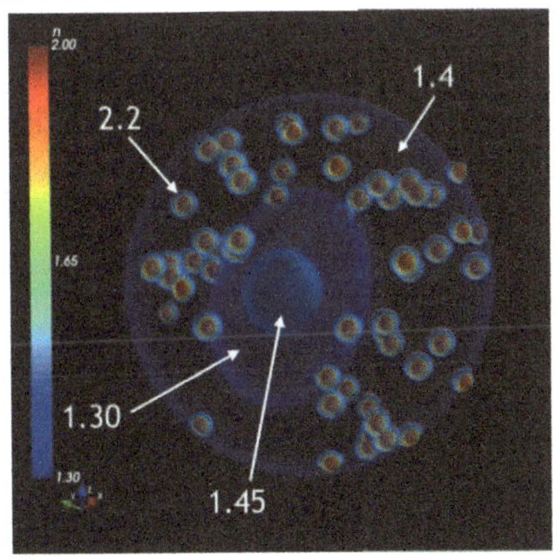

Fig. 28 3D rendering of the internal refractive index of a representative synthetic cell phantom used to test 2D phase unwrapping[84].

We compared several 2D digital phase unwrapping approaches[40,87,93,94] and found the phase unwrapping max-flow/min-cut (PUMA) algorithm based on graph cuts[87] provided the best performance for this dataset. A cGAN was trained using true absolute phase and corresponding ambiguous 2π wrapped phase shown in Fig. 29(a-b). Unwrapped estimates of the absolute phase using the trained cGAN and the PUMA digital unwrapping are shown in Fig. 29(c-d). While both approaches can accurately unwrap smoothly wrapped phase; the learning based approach successfully unwrapped locally and discontinuously wrapped phase in which traditional approaches failed. This is clearly seen in Fig. 30(a) which shows slices through the phase estimates in Fig. 29. The RMSE using PUMA is 2.57 radians (0.88σ), averaged over 100 testing realizations. After 25 training epochs the phase is more effectively unwrapped using a trained cGAN, after 200 training epochs the RMSE is 0.8 radian (0.2σ).

We have demonstrated accurate 2D phase unwrapping using DL/CNN to perform a direct wrapped-to-unwrapped phase prediction; and showed that the network based approach is not limited by $\lambda/2$ discontinuities that often produce errors when using traditional 2D unwrapping approaches. While phase unwrapping has previously been demonstrated using CNN[95], accurate 2D unwrapping of severely discontinuous phase using cGAN has not, to our knowledge, previously been demonstrated. Similar to phase reconstruction, it is important to note that the network will perform only as well as the approach used to unwrap the

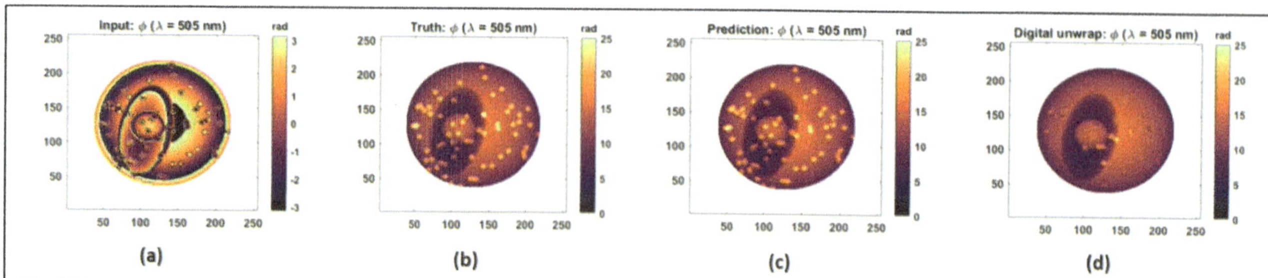

Fig. 29 Wrapped phase input to the network (a), true unwrapped phase (b), unwrapped phase as predicted by (c), and PUMA unwrapped phase estimate (d)[84].

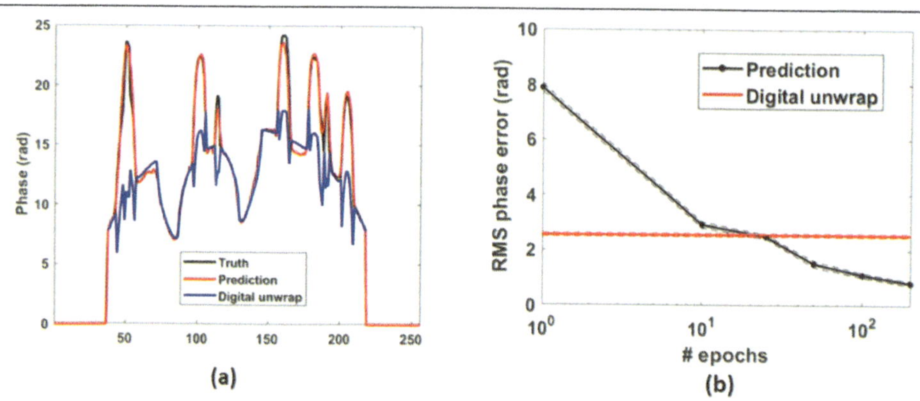

Fig. 30 (a) True absolute phase (black), unwrapped phase prediction using cGAN (red), and unwrapped phase estimate using PUMA (blue). (b) Unwrapped phase RMSE averaged over 100 testing realizations as a function of training epochs[84].

data used to train the network. The accuracy of prediction is expected to decrease when tested against data more severely wrapped than the training set, data that a traditional approach is more likely to fail on. Network prediction can be orders of magnitude more efficient than conventional 2D unwrapping approaches; despite taking considerable time and computing resources to train. The unwrapping time with PUMA ranged from 500-1 500 ms, for the wrapped samples considered in this section. Phase unwrapping using cGAN was approximately 2-5 times faster, making the network based solution particularly advantageous for high frame rate real time processing.

Dual-wavelength phase unwrapping is another viable alternative to digital unwrapping techniques when imaging samples with large height profiles and many λ/2 phase discontinuities. Unwrapping is realized using a dual-wavelength phase subtraction equivalent to imaging at a synthetic beat wavelength, typically much longer than either fundamental wavelengths[18,41-43,47,96]. Wrapped to wrapped phase predictions using cGAN have previously been demonstrated with sufficient accuracy for unwrapping based on dual-wavelength subtraction[41,47]. We have demonstrated accurate unwrapping of discontinuous samples using a direct network prediction, thus avoiding the phase subtraction used in dual-wavelength unwrapping which is extremely sensitive to small prediction errors.

III. Multi-class Cell Line Classification using Digital Holographic Microscopy and Machine Learning

Quantitative phase imaging (QPI) is an invaluable method to investigate cells. It is a label-free technique that provides contrast by quantifying phase shift when light propagates through a thin transparent specimen. Quantitative parameters collected from optical phase maps characterize not only cell shape, but also additional information pertaining to the intracellular distribution of refractive objects, and each cell's geometric thickness profile[97]. Digital holographic microscopy (DHM) is one form of QPI in which a hologram projected from a microscope objective is digitally recorded for subsequent reconstruction. Applications of DHM in biology include monitoring cell cycle status[98], investigating cellular response to novel drug targets[99] or environmental stress[100]. For more generalized applications, QPI assisted with machine learning is an automated, high-throughput potentially robust tool for the classification of imaged objects.

Non-invasive classification of cultured cells is an important task for *in vitro* drug panel screening, phenotypic profiling for manufacturing therapeutic cell lines/proteins, and the evaluation of culture health and cross-contamination. Holographic microscopy and machine learning (ML) are attractive tools for these applications but must address two challenges: high performance in multi-class classification problems (> 2 classes), and identification of important image-derived features for a given classification problem.

Previous studies have shown that a linear support vector machine (SVM) can be used to build a binary epithelial–mesenchymal classifier using two non-cancer cell lines as a modeling dataset (i.e., human gingival fibroblast, HGF, and human gingival keratinocytes, GIE no3B11)[101]. The trained model was then generalized and used to classify two unseen cancer cell lines (MDA-MB-231 and MCF-7). This binary classifier model did not classify multiple cell types, but instead ranked individual cells on an epithelial-mesenchymal morphological axis. Therefore, there was no power to distinguish cancer from noncancer cells.

Specifically, in this section we aim to achieve high performance classification of four cell lines, two breast cancer (MDA-MB-231 and MCF-7) and two noncancer (human gingival fibroblast, HGF; and human gingival keratinocytes, GIE no3B11), with varied epithelial and mesenchymal morphologies; to determine the best machine learning model for this classification; and to identify features most strongly influencing model performance. We trained and evaluated three ML algorithms: Support Vector Machines (SVMs) with various kernels, Random Forest, and AdaBoost, using a previously defined set of 17 features derived from holographic microscopy images; selected components after Principal Component Analysis (PCA); and a subset of original features after feature selection. Grid searching was conducted to determine the optimal set of hyperparameters for each machine learning algorithm before training. The multi-class ML model created using the Random Forest algorithm was reliable and had an average F1 score of 0.89, 0.86, 0.84, and 0.84 for GIE, HGF, MCF7, and MDA cell lines respectively. Moreover, with the feature selection technique, the model performance for each cell line can be further improved. Top important features were geometric (area, perimeter, eccentricity), histogram-based (skew) and textural (contrast, correlation, homogeneity).

Hence, in this section, we aim to develop a 4-way classifier to solve this problem by investigating selected machine learning algorithms fed with image-derived features to achieve high performance in multi-class classification problems. Furthermore, we explore several feature engineering techniques to check feature influence on model performance.

III.1 Methods

Section I describes the setup behind the bi-telecentric DHM used in capturing the holograms and the theory of obtaining the quantitative phase images[18,97]. The general approach of cell culture, imaging, phase height map reconstruction, parameter extraction, and machine learning is outlined in Figure 31. Each model was trained using 5-

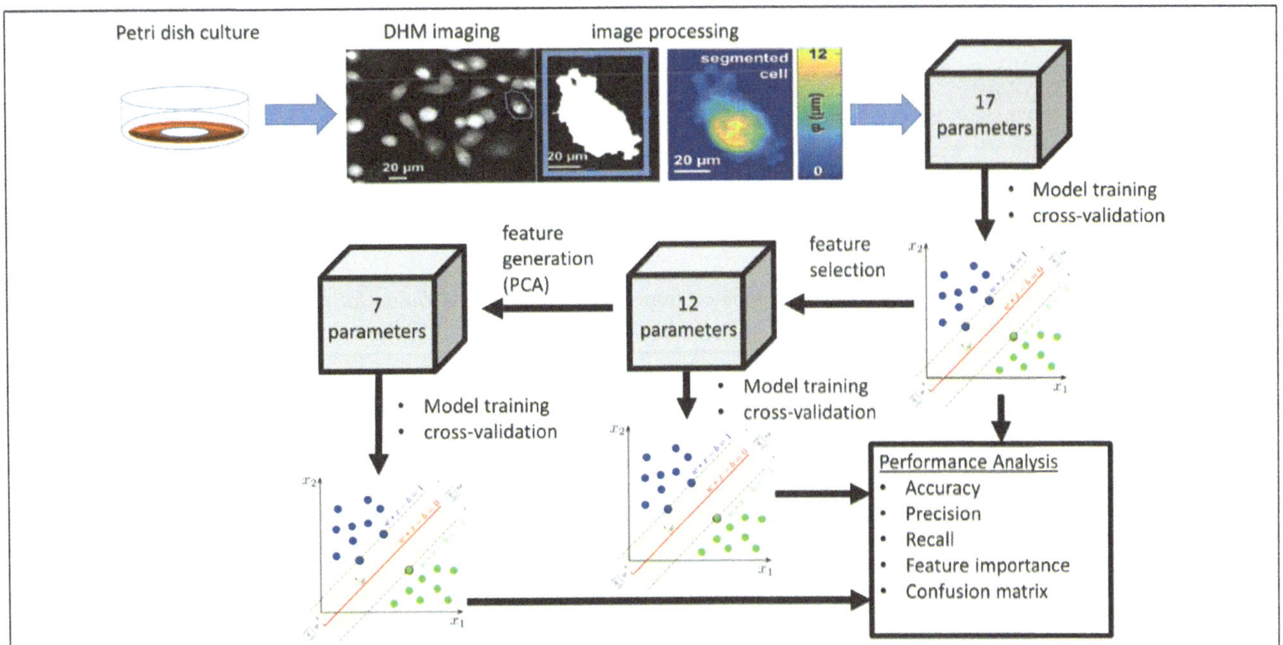

Fig. 31 Flowchart of image processing and machine learning steps[102].

fold cross validation and tested using 100 experimental cells from each class. Performance analysis metrics included accuracy, precision, and recall. Select folds were analyzed through a confusion matrix. After feature selection (see details below), feature importance was evaluated and the five most important features were depicted graphically for pairwise comparisons between the cell lines.

III.2 Data and data pre-processing

All cell lines (MDA-MB-231, MCF-7, HGF, and GIE-no3B11) were cultured in standard tissue culture incubation conditions (37 °C, 5% CO_2, 100% humidity) using a HERAcell 150i (ThermoFisher Scientific, Waltham, MA), and nutrient media as previously described. Before imaging, cells were trypsinized, counted and replated in tissue culture treated, 35 mm diameter Petri dishes. Cells were replated at relatively low density as previously described and allowed to adhere to the tissue culture treated coverglass surface of the Petri dish for 24-48 hours before imaging. Images were collected in 15-20 minute intervals between tissue culture incubation of several Petri dishes to eliminate effects on the cells potentially caused by a drop in temperature and CO_2 levels of the nutrient media exposed to ambient conditions. The DHM was operated at least 30 minutes after turning the laser on, to achieve the most stable laser output. Periodic calibration of the DHM was performed using an optical phase target based on the 1951 USAF resolution target sputter-coated with Molybdenum to a phase height of 350 nm. The accuracy of phase height reconstruction was typically within 1-2%. Fig. 32 shows typical phase height maps of the 4 cell lines.

Reconstructed phase maps from each of the four cell lines previously collected in[5] were segmented using a custom-written code in MATLAB (version R2015a, updated for version R2021a), which was described previously[97]. Seventeen phase parameters/were extracted per cell, to be used as features of the segmented cells for machine learning (see Table 3). A total of 1 295 cells from four different cell lines were used in this study, including 332 cells of GIE, 309 cells of HGF, 307 cells of MCF-7, and 347 cells of MDA-MB-231. The computed 17 features were derived from holographic microscopy images (phase height maps) and have different ranges in values and units. Therefore, feature scaling was performed using the min-max normalization prior to machine learning. Without a proper feature scaling, the feature with a higher value range dominates the training process when calculating distances and the model created would be biased.

Several feature selection techniques were also examined to determine the optimal subset of features leading to no worse or improved model performance, compared to a model built from the full feature set. In this study, we explored 4 different feature selection techniques including Recursive Feature Elimination, Sequential Forward

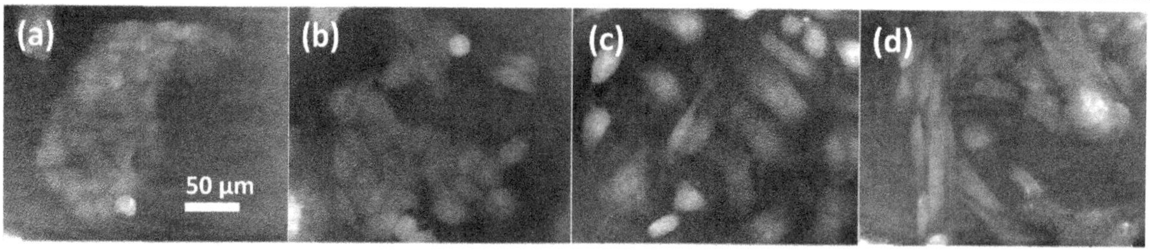

Fig. 32 Phase height maps of the 4 cell lines: (a) GIE, (b) MCF-7, (c) MDA-MB-231, and (d) HGF cells. Scale bar is indicated[102].

Table 1 Average accuracy, precision and recall computed from the 5-fold cross validation for multi-class cell lines classification using different machine learning algorithms.

Algorithm	Accuracy (%, µ ± SD)	Precision (%, µ ± SD)				Recall (%, µ ± SD)			
		GIE	HGF	MCF-7	MDA	GIE	HGF	MCF-7	MDA
SVM–Linear	87.7±1.5	89.2±2.3	87.4±1.2	85.2±4.1	86.8±3.5	87.4±4.2	88.6±2.2	86.6±2.3	86.8±2.5
SVM- Polynomial	87.3±1.2	89.8±2.4	87.8±1.8	86.0±3.5	87.2±3.7	88.6±4.2	88.6±3.8	86.6±2.5	86.8±2.7
SVM– RBF	87.8±0.6	88.2±2.2	88.2±2.2	86.0±2.0	87.4±4.5	88.8±2.4	89.4±1.9	85.0±6.2	86.8±4.1
SVM- Sigmoid	87.2±1.2	89.0±2.5	88.4±2.3	84.4±3.3	86.4±3.2	88.2±2.2	87.6±2.5	86.0±2.8	86.8±2.3
Random Forest	86.1±1.8	89.6±2.6	87.0±2.9	84.0±2.4	84.4±4.8	88.4±2.4	86.0±4.4	85.8±3.8	85.0±3.7
AdaBoost	83.2±1.2	88.0±6.8	81.0±6.4	81.8±6.4	80.6±5.0	88.2±4.5	82.8±7.0	81.2±6.4	78.2±7.6

Selection, Sequential Backward Selection, and SelectKbest. Furthermore, the original set of 17 features are not independent. Therefore, Principal Component Analysis (PCA) was also used to generate a new set of independent variables. Notably, PCA uses orthogonal transformation to create new independent variables called principal components. Each new independent variable is a linear combination of the old features. Therefore, PCA reduced the dimensionality of the feature space while maintaining a certain level of variation in the data. Features directly influence predictive model performance. Therefore, in addition to determining the best machine learning model, feature engineering techniques were evaluated to identify features most strongly influencing model performance.

III.3 Machine learning algorithms implemented

The main purpose of this study was to develop a machine learning model to accurately classify 4 cell lines simultaneously. The implementation was carried out using Python and related libraries including Scikit-learn, Pandas, Numpy, and Matplotlib. Three machine learning (ML) algorithms were investigated: Support Vector Machine (SVM), Random Forest (RF), and AdaBoost (AB) classifiers. To evaluate the performance of the models 5-fold cross-validation was used. Data was shuffled and divided into 5 disjoint subsets, and 80% of the data was used in training and 20% of the data was used for testing in each fold. In addition, grid search was performed on the training data only in each fold for tuning the hyper-parameters for each ML algorithm. Averaged classification accuracy, precision and recall were computed for each class (cell line) in each ML model as performance metrics. Precision measures the ratio of how many cells in a given cell line were correctly classified (i.e., true positive) to how many cells were actually classified as that cell line. Recall measures the ratio of how many cells in a given cell line were correctly classified to how many cells were actually in that cell line.

III.4 Results and discussion

Table 2 summarizes the performance of the 4-way classification models using the selected three machine learning algorithms and the original 17 image-derived features. As mentioned earlier, hyper-parameters for each machine learning algorithm were optimized by using grid search on the training data, and 5-fold cross validation was performed to avoid under-fitting or over-fitting. As shown in Table 2, all the models have compatible performance with average classification accuracies ranged from 83.2% to 87.8%. The accuracy from SVM with RBF kernels is

Table 2 Confusion matrix from SVM classifier with a linear kernel: result of one fold.

Pred	SVM – Linear			
	GIE	HGF	MCF-7	MDA
True				
GIE	298	7	25	2
HGF	13	270	1	25
MCF-7	18	4	265	20
MDA	3	25	15	304

slightly higher compared to SVM with other kernels, Random Forest and AdaBoost. The standard deviation of accuracy is small in all of the models indicating the trained models are stable and reliable without overfitting.

The precision and recall for each cell line were also reported in Table 2. The precision of GIE cell assignment was 88% to 89.8%; of HGF cell assignment, 81% to 88.4%; of MCF-7 cell assignment, 81.8% to 86%; and of MDA cell assignment, 80.6% to 87.4%. The recall of GIE cell assignment was 87.4% to 88.8%; of HGF cell assignment, 82.8% to 89.4%; of MCF-7 cell assignment, 81.2% to 86.6%; and of MDA cell assignment, 78.2% to 86.8%. The best performance with 86.6% recall of MCF-7 cells is from linear and polynomial SVMs, and with 86.8% recall of MDA-MB-231 cells is from any tested SVM classifiers. Table 3 presents multi-class classification performance by absolute numbers of cells in the confusion matrix from the SVM model with the linear kernel, for one training/testing fold. Note that MCF-7 cells have the highest chance (6.5%) to be misclassified as MDA-MB-231 cells, and MDA-MB-231 cells have the highest chance to be misclassified as HGF cells (7.2%).

To further improve model performance, feature selection techniques were evaluated, to identify a possible optimal subset of features for machine learning. Table 4 shows the feature importance scores in percentage using the Random Forest and selectKbest algorithm from the Scikit learn. The result (Table 4) shows that using selectKbest, 12 features have nontrivial importance scores (>1%). Using the Random Forest algorithm, all features had nontrivial importance scores. These nontrivial features were selected for use based on this analysis. The number of PCs used as predictors to our algorithms was also important. Principal Component Analysis (PCA) was also conducted to generate a new set of features. PCA not only reduced the number of dimensions but also reduced the noise in the training dataset. Seven PCs that contained 95% of data variation were selected for use in the modeling.

In the Random Forest algorithm, the feature importance

Table 3 The original 17 image-derived features, their symbols and importance in percentage obtained from the selectKBest algorithm and the Random Forest algorithm.

	Extracted Feature	Symbol	Feature Importance from the selectKBest algorithm	Feature Importance from the Random Forest algorithm
1	Cell Phase Mean	μ	1.6%	2.4%
2	Cell Phase Standard Deviation	Sd	1.9%	2.3%
3	Cell Skewness	Sk	7.5%	6.5%
4	Cell Kurtosis	K	0.2%	2.1%
5	Cell Area	A	15%	16.5%
6	Cell Perimeter	P	20.9%	17.2%
7	Cell Eccentricity	E	7.9%	16.5%
8	Cell Correlation	Co	5.4%	4.8%
9	Cell Contrast	Cn	4.3%	4.3%
10	Cell Engergy	En	1%	2%
11	Cell Homogeneity	Hm	4.7%	3.4%
12	Nuclear Phase Mean	μ_n	0.8%	1.8%
13	Nuclear Phase Standard Deviation	sd_n	0.9%	1.9%
14	Nuclear Phase Maximum	μ_{nmax}	1%	1.8%
15	Nuclear Kurtosis	k_n	7.1%	2.6%
16	Nuclear Skewness	sk_n	3.9%	3.7%
17	Nuclear Area	A	15.9%	10.2%

can be measured by computing the average impurity decrease from all decision trees in the forest. In the SelectKBest algorithm, univariate feature selection divides the features into independent and dependent variables and calculates the importance score of each feature. In order to better compare importance of the features across classification pairs and methods, we transformed the scores into percentages. The top five features ranked by importance score (both Random Forest and separately, SelectKBest) for each pairwise classification were mapped in Fig. 33.

Using the Random Forest importance metric for the GIE vs. MCF-7 classification, Cell Area (A) has the highest importance at >20% contribution to the predictions of the model, followed by Cell Eccentricity (E) and Nuclear Area (a) with 10-20% importance, and with Cell Contrast (Cn) and Cell Perimeter (P) both having less than 10% contribution. In contrast, using the SelectKBest importance metric for the GIE vs. MCF-7 classification, Cell Area (A) and Nuclear Area (a) both have more than 20% importance score, followed by Nuclear kurtosis (Kn) and Perimeter (P) both having importance scores between 10-20%, followed by Cell Eccentricity(E) at less than 10% of the total score. Both evaluations agree in 4 of the top 5 most important features, albeit with a different ranking.

Using the Random Forest importance metric for the GIE vs. HGF classification, Cell Perimeter (P), Cell Area (A) and Nuclear Phase Mean (μn) had highest importance at more than 20% contribution to the predictions of the model, followed by Cell Eccentricity (E) at 10-20% contribution and Cell skew (sk) at less than 10% contribution. Using the SelectKBest importance metric for the GIE vs. HGF classification, Cell Perimeter (P), Cell Area (A) and Nuclear Area (a) have more than 20% of the total score, followed by Cell Eccentricity (E) and Nuclear kurtosis (Kn) at 10-20% of total score. Both evaluations agree in 3 of the top 5 most important features.

Using the Random Forest importance metric for the HGF vs. MDA-MB-231 classification, Cell Eccentricity (E) had highest importance at more than 20% contribution to the predictions of the model, followed by Cell Contrast (Cn), Cell skew (sk), and Cell Correlation (Co) at 10-20% contribution and Cell Homogeneity (Hm) at less than 10% contribution. Using SelectKBest, Cell Correlation (Co) had more than 20% contribution to the predictions of the model, followed by Cell Contrast (Cn), Cell Homogeneity (Hm) and Cell Eccentricity (E) at 10-20% contribution to the model performance, and Cell skew (sk) at less than 10% contribution. Both evaluations agree in 5 of the top 5 most important features, albeit with a different ranking.

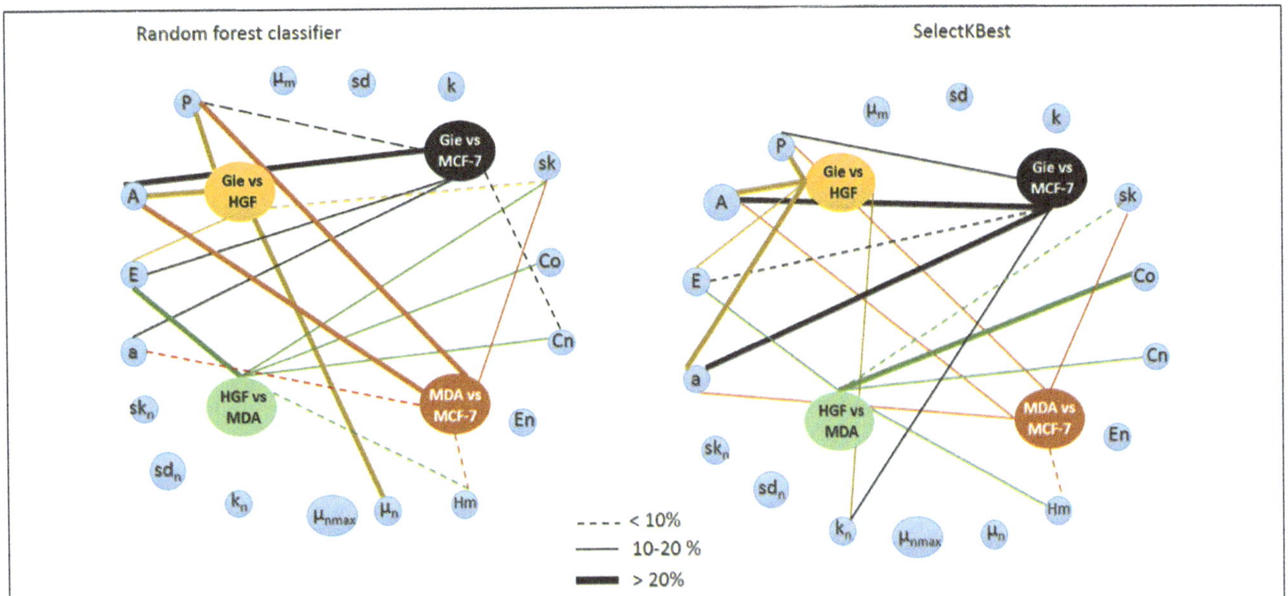

Fig. 33 Feature importance based on calculations from the random forest classifier (left), and selectKBest (right). The top five features important to each pairwise classification were indicated by lines, dashed indicating <10% contribution, thin solid indicating 10-20% contribution, and bold solid indicating >20% contribution to the model performance[102].

Using the Random Forest importance metric for the MDA-MB-231 vs. MCF-7 classification, Cell Area (A) and Cell Perimeter (P) both had more than 20% contribution to the predictions of the model, followed by Cell skew (sk) at 10-20% contribution and Cell Homogeneity (Hm) and Nuclear Area (a) at less than 10% contribution. Using the selectKBest, Cell Area (A), Cell Perimeter (P), Cell skew (sk), and Nuclear Area (a) had 10-20% importance to the total prediction performance, and Cell Homogeneity (Hm) had less than 10% contribution. Both evaluations agree in 5 of the top 5 most important features, albeit with a different ranking.

IV. Conclusion and Future Perspectives

In Section I, we have described a combination of Deep Learning Convolutional Neural Network with Zernike polynomial fitting as a technique to automatically compensate for the phase aberration in a DHM system. The technique benefits from the ability of PCA to obtain the training data for the deep learning CNN model. The trained CNN model can be used as an automatic and *in situ* process of background detection and full phase aberration compensation. From the data testing stage, we noticed that even with images having new features that didn't appear in the training process, CNN managed to detect the background with a high precision. While, many image segmentation techniques are not robust when applied to DHM images due to the overwhelming phase aberration, CNN segments the background spatially based on features regardless to the number of cells and their unknown positions. Thus, the trained CNN technique in conjunction with the ZPF technique is a very effective tool that can be employed in real time for autonomous phase aberration compensation in a DHM system.

In Section II, we have described accurate refractive index estimation using dual-wavelength synthesis based on DL/CNN. The approach was validated using high fidelity simulations of holographic cell recordings. We showed accurate refractive index decoupling by synthesizing the phase as well as the raw hologram. It should be noted that multi-wavelength phase and hologram synthesis are not limited to predicting a single alternate wavelength; N distinct networks can be trained to predict N distinct wavelengths assuming the availability of training data at each wavelength. We demonstrated accurate refractive index predictions using single phase input; although this approach was less accurate than dual-wavelength hologram and phase synthesis we believe the direct RI prediction will outperform approaches that rely on decoupling for dispersive samples. We also demonstrated accurate phase reconstruction and phase unwrapping based on DL/CNN; accurate aberration cancellation and unwrapping of sharp discontinuous phases are shown. We found the *pix2pix* framework for conditional GANs was able to provide accurate predictions of higher order image texture; however, this work was not intended to provide an exhaustive comparison of network structures. In addition, we acknowledge that the variance of the data used to train and test the networks is significantly lower than experimental trials across various conditions, optical hardware, cell-lines, etc.; we expect cross-experiment degradation in prediction accuracy as the training data becomes less representative of subsequent testing images. In addition, we acknowledge that image-to-image prediction and refractive index decoupling will suffer from degrading factors such as speckle and phase noise, and may require low noise conditions and/or speckle mitigation techniques. For instance, 2D phase unwrapping may not alleviate unwrapping errors arising from speckle or other spatially localized phase noise fluctuations. The various approaches should be tested versus the degree of spatially uncorrelated phase noise. This does however serve as a proof of concept for refractive index estimation using wavelength synthesis based on DL/CNN.

In Section III, we achieved high performance classification of four cell lines, two breast cancer and two noncancer, with varied epithelial and mesenchymal morphologies. We determined the best machine learning model (SVMs) and identified the features that most strongly influenced the model performance, which varied by pairwise classification but frequently included cell area, cell perimeter, and cell phase histogram skew, but also several texture and other features. We trained and evaluated three ML algorithms using a previously defined set of features derived from DHM images. We concluded that the multi-class ML models created using the SVM and Random Forest algorithms were most reliable and having accuracy in the 83.2–87.8% range. Feature selection identified a broad range of parameters contributing to the classification performance, including cell geometric features (area, perimeter), phase height histogram features (skew) and textural features (correlation, contrast, homogeneity). Additional textural features might further improve multiclass classification performance, or alternatively, the use of neural network methods such as multi-layered, multilayer perceptron networks. The approach of machine learning to classify multiple groups in quantitative image data is of general interest to the biomedical, defense and commercial sensing communities.

Machine learning and deep neural networks are computational methods that have started to play an important role in digital holographic microscopy and in microscopy research over the past decade. The ultimate aim of machine learning applied to 2D and 3D microscopy image data, is to obtain new scientific knowledge by improving image quality, automating detection, 3D phase unwrapping, performing segmentation and cell classification, auto-refocusing, aberration correction, super-resolution, vibration mitigation, 3D tracking of objects, merging information from different imaging modalities, and creating new images of certain modalities based on other modalities. We envision that machine learning will have a great potential to push new scientific discoveries by allowing us to observe and track smaller and faster objects in higher dimensions and this will have an important impact on a wide cross-disciplinary audience in the physical sciences and life sciences.

In general, machine learning approaches often require many samples for robust and accurate classification; this becomes problematic for cell imaging due to the time intensive process of cell culture preparation and imaging. Researchers have demonstrated high classification accuracy using a fully synthetic training database tested against experimental cells. In addition, it has been demonstrated that accuracy can be improved by augmenting training data with a small number of samples with synthetic samples. In conjunction with a high fidelity simulation of holographic recording and reconstruction, the augmentation can used for cell classification for a variety of holographic configurations.

With all this excitement, some challenges remain to be solved namely, robustness, quantifying uncertainty, and the need for high quality and representative datasets. Merging known physical quantities and constraints into neural networks is promising in addressing some of these challenges. Finally, there should be an interdisciplinary collaborative research effort between machine learning scientists, optical engineers and physicists, and biomedical engineers and biologists to solve some of these challenges and bring in a new generation of more capable, machine-learning-based computational microscopes[103].

Conflict of interest

References

1. H. Nguyen, D. *Nguyen, Z. Wang, H. Kieu, and M. Le", Real-time, high-accuracy 3D imaging and shape measurement", Appl. Opt* **54**, A9-A17 (2015).
2. K. Hien, T. *Pan, Z. Wang, M. Le, H. Nguyen, and M. Vo", Accurate 3D shape measurement of multiple separate objects with stereo vision", Measurement Science and Technology* **25**(3), 035401-7 (2014).
3. T. T. A. Nguyen, H. *N. D. Le, M. Vo, Z. Wang, L. Luu, and J. C. R. Roman", Three-dimensional phantoms for curvature correction in spatial frequency domain imaging", Biomed. Opt. Express* **3**, 1200-1214 (2012).
4. T. Nguyen, G. Nehmetallah, D. Tran, A. Darudi, and P. Soltani, "Fully automated, high speed, tomographic phase object reconstruction using the transport of intensity equation in transmission and reflection configurations, " Appl. Opt. **54**, 10443-10453 (2015).
5. T. Nguyen, G. Nehmetallah, "Non-interferometric tomography of phase objects using spatial light modulators, " *Journal of Imaging* 2(4), **30** (2016).
6. N. Pavillon, J. Kühn, C. Moratal, P. Jourdain, C. Depeursinge, P. J. Magistretti, and P. Marquet, "Early cell death detection with digital holographic microscopy, " PLoS ONE 7(1), e30912 (2012).

7. J. Kuhn, F. Montfort, T. Colomb, B. Rappaz, C. Moratal, N. Pavillon, P. Marquet, and C. Depeursinge, Submicrometer tomography of cells by multiple-wavelength digital holographic microscopy in reflection, Opt. Lett. **34**(5), 653–655 (2009).
8. J. Kühn, E. Shaffer, J. Mena, B. Breton, J. Parent, B. Rappaz, M. Chambon, Y. Emery, P. Magistretti, C. Depeursinge, P. Marquet, and G. Turcatti, "Label-free cytotoxicity screening assay by digital holographic microscopy," Assay Drug Dev. Technol. **11**(2), 101–107 (2013).
9. B. Rappaz, E. Cano, T. Colomb, J. Kühn, C. Depeursinge, V. Simanis, P. J. Magistretti, and P. Marquet, "Noninvasive characterization of the fission yeast cell cycle by monitoring dry mass with digital holographic microscopy," J. Biomed. Opt. **14**(3), 034049-1-5 (2009).
10. N. Pavillon, A. Benke, D. Boss, C. Moratal, J. Kühn, P. Jourdain, C. Depeursinge, P. J. Magistretti, and P. Marquet, "Cell morphology and intracellular ionic homeostasis explored with a multimodal approach combining epifluorescence and digital holographic microscopy," J. Biophoto. **3**(7), 432–436 (2010).
11. N. Warnasooriya, F. Joud, P. Bun, G. Tessier, M. Coppey-Moisan, P. Desbiolles, M. Atlan, M. Abboud, and M. Gross, "Imaging gold particles in living cell environments using heterodyne digital holographic microscopy," Opt. Exp., **18**(4), 3264–3273 (2010).
12. Y. Fang, C. Y. Y. Iu, C. N. P. Lui, Y. Zou, C. K. M. Fung, H. W. Li, N. Xi, K. K. L. Yung, and K. W. C. Lai, "Investigating dynamic structural and mechanical changes of neuroblastoma cells associated with glutamate-mediated neurodegeneration," Scientific Reports **4**: 7074, 1-7 (2014).
13. R. John, and V. *P. Pandiyan, "An optofluidic bio-imaging platform for quantitative phase imaging of lab on a chip devices", Appl. Opt* **55**(3), A54-A59 (2016).
14. L. Williams, P. *Banerjee, G. Nehmetallah, and S. Praharaj, "Holographic volume displacement calculations via multiwavelength digital holography", Appl. Opt* **53**(8), 1597-1603 (2014).
15. G. Nehmetallah, and P. *P. Banerjee, "Applications of digital and analog holography in 3D imaging", Adv. Opt. and Photon* **4**(4), 472-553 (2012).
16. E. Cuche., *P. Marquet, and C. Depeursinge, "Simultaneous amplitude contrast and quantitative phase-contrast microscopy by numerical reconstruction of Fresnel off-axis holograms", Appl. Optics* **38**, 6994-7001 (1999).
17. G. Nehmetallah, R. Aylo and L. Williams, *Analog and Digital Holography with MATLAB®*, SPIE Press, Bellingham, Washington, 2015.
18. T. Nguyen, G. Nehmetallah, C. Raub, S. t Mathews, and R. Aylo, "Accurate quantitative phase digital holographic microscopy with single- and multiple-wavelength telecentric and nontelecentric configurations," Appl. Opt. **55**, 5666-5683 (2016).
19. C. Zuo, Q. *Chen, W. Qu, and A. Asundi, "Phase aberration compensation in digital holographic microscopy based on principal component analysis", Opt. Lett* **38**(10), 1724-1726 (2013).
20. H. ND Le, M. S. Kim, and D-H Kim, "Comparison of singular value decomposition and principal component analysis applied to hyperspectral imaging of biofilm," IEEE Photonics Conference (IPC), 23-27 Sept. (2012).
21. T. Colomb, F. Montfort, J. Kühn, N. Aspert, E. Cuche, A. Marian, F. Charrière, S. Bourquin, P. Marquet, and C. Depeursinge, "Numerical parametric lens for shifting, magnification, and complete aberration compensation in digital holographic microscopy," J. Opt. Soc. Am. A23, 3177-3190 (2006).
22. T. Colomb, E. Cuche, F. Charrière, J. Kühn, N. Aspert, F. Montfort, P. Marquet, and C. Depeursinge, "Automatic procedure for aberration compensation in digital holographic microscopy and applications to specimen shape compensation," Appl. Opt. 45, 851-863 (2006).
23. X. Bresson, S. Esedoḡlu, P. Vandergheynst, J. P. Thiran, and S. Osher. Fast global minimization of the active contour/snake model. *Journal of Mathematical Imaging and Vision* **28**(2), 151-167 (2007).
24. S. A. Hojjatoleslami and J. *Kittler", Region growing:a new approach", IEEE Transactions on Image processing* **7**(7), 1079-1084 (1998).
25. Y. Boykov and F. *L. Gareth", Graph cuts and efficient ND image segmentation", International Journal of Computer Vision* **70**(2), 109-131 (2006).
26. L. Grady, "Random walks for image segmentation," IEEE transactions on Pattern Analysis and Machine Intelligence **28**(11), 1768-1783 (2006).
27. W. Khan, "Image segmentation techniques: A survey," Journal of Image and Graphics **1**(4), 166-170 (2013).
28. R. H. Laprade", Split-and-merge segmentation of aerial photographs. " *Computer Vision, Graphics, and Image Processing* **44**(1), 77-86 (1988).
29. JBMT Roerdink and A Meijster, "The watershed transform: Definitions, algorithms and parallelization strategies." *Fundamenta informaticae* **41**(1-2) 187-228 (2000).
30. T. Nguyen, V. Bui, V. Lam, C. B. Raub, L-C Chang, and G. Nehmetallah, "Automatic phase aberration compensation for digital holographic microscopy based on deep learning background detection," Opt. Exp., 25(13) 15043-15057 (2017).
31. V. Bui and L-C Chang, "Deep learning architectures for hard character classification," Proc. Int. Conf. Art if. Int. p. 108, WorldComp (2016).
32. J. Long, E. Shelhamer, and T. Darrell, "Fully convolutional networks for semantic segmentation." Proceedings of the IEEE Conference on Computer Vision and Pattern Recognition, 2015.
33. O. Ronneberger, P. Fischer, and T. Brox, "U-net: Convolutional networks for biomedical image segmentation," *International Conference on Medical Image Computing and Computer-Assisted Intervention*. Springer International Publishing, 18 May (2015). [arXiv: 1505.04597].
34. David Harris and Sarah Harris, *Digital design and computer architecture* (2nd ed.) (San Francisco, Calif.: Morgan Kaufmann. p. 129).
35. L. Maas, A. Y. Hannun, and A. Y. Ng, "Rectifier nonlinearities improve neural network acoustic models," Proc. of the 30th Intern. Conf. on Machine Learning (ICML), JMLR: W&CP volume 28, Atlanta, Georgia (2013).
36. S. Ioffe, and C. Szegedy. "Batch normalization: Accelerating deep network training by reducing internal covariate shift," [arXiv preprint arXiv: 1502.03167], 11 Feb (2015).
37. H. Noh, S. Hong, and B. Han, "Learning deconvolution network for semantic segmentation," Proceedings of the IEEE International Conference on Computer Vision, 17 May (2015). [arXiv: 1505.04366]
38. Sutskever, J Martens, GE Dahl, GE Hinton, "On the importance of initialization and momentum in deep learning." ICML (3)28, 1139-1147 (2013).
39. M. Abadi, P. Barham, J. Chen, Z. Chen, A. Davis, J. Dean, M. Devin, S. Ghemawat, G. Irving, M. Isard, M. Kudlur, J. Levenberg, R. Monga, S. Moore, D. G. Murray, B. Steiner, P. Tucker, V. Vasudevan, P. Warden, M. Wicke, Y. Yu, X. Zheng, "TensorFlow: A system for large-scale machine learning," Proc. of the 12th USENIX Symposium on Operating Systems Design and Implementation (OSDI), Savannah, Georgia, 31 May (2016).
40. M. A. Herráez, D. *R. Burton, M. J. Lalor, and M. A. Gdeisat", Fast two-dimensional phase-unwrapping algorithm based on sorting by reliability following a noncontinuous path", Appl. Opt* **41**, 7437-7444 (2002).
41. J. Li, Q. Zhang, L. Zhong, J. Tian, G. Pedrini, and X. Lu, "Quantitative phase imaging in dual-wavelength interferometry using a single wavelength illumination and deep learning," Opt. Express 28, 28140-28153 (2020).

42. B. Bazow, T. Phan, T. Nguyen, C. Raub, and G. Nehmetallah, "Simulation of digital holographic recording and reconstruction using a generalized matrix method, " Appl. Opt. 60, A21-A37 (2021).
43. L. Williams, P. Banerjee, G. Nehmetallah, and S. Praharaj, "Holographic volume displacement calculations via multiwavelength digital holography, " Appl. Opt. 53, 1597-1603 (2014).
44. M. R. Jafarfard, S. Moon, B. Tayebi, and D. Y. Kim, "Dual-wavelength diffraction phase microscopy for simultaneous measurement of refractive index and thickness, " Opt. Lett., 39(10) 2908-2911 (2014).
45. D. Boss, J. Kühn, P. Jourdain, C. D. Depeursinge, P. J. Magistretti, and P. P. Marquet, "Measurement of absolute cell volume, osmotic membrane water permeability, and refractive index of transmembrane water and solute flux by digital holographic microscopy, " J. Biomed. Opt. 18(3) 036007 (13 March 2013)
46. B. Rappaz, P. Marquet, E. Cuche, Y. Emery, C. Depeursinge, and P. J. Magistretti, "Measurement of the integral refractive index and dynamic cell morphometry of living cells with digital holographic microscopy, " Opt. Exp., 13(23), 9361-9373 (2005).
47. X. Xu, M. Xie, Y. Ji, and Y. Wang, "Dual-wavelength interferogram decoupling method for three-frame generalized dual-wavelength phase-shifting interferometry based on deep learning, " J. Opt. Soc. Am. A 38, 321-327 (2021).
48. I. Goodfellow, J. Pouget-Abadie, M. Mirza, B. , Xu, D. Warde-Farley, S. Ozair, A. C. Courville, and Y. Bengio, Generative adversarial networks, https://arxiv.org/abs/1406.2661 (2014).
49. M. Mirza and S. Osindero. "Conditional generative adversarial nets, " *ArXiv* abs/1411.1784 (2014): n. pag.
50. P. Isola, J-Y. Zhu, T. Zhou, and A. A. Efros, "Image-to-Image Translation with conditional adversarial networks, " *2017 IEEE Conference on* Computer Vision and Pattern Recognition (CVPR) 5967-5976 (2017); arXiv: 1611.07004 [cs. CV]
51. D. Yin, Z. Gu, Y. Zhang, F. Gu, S. Nie, J. Ma, and C. Yuan, "Digital Holographic Reconstruction Based on Deep Learning Framework With Unpaired Data, " in IEEE Photonics Journal 12(2), 1-12, April (2020), Art no. 3900312,doi: 10.1109/JPHOT.2019.2961137.
52. M. Rubin, O. Stein, R. Giryes, D. Roitshtain, and N. T. Shaked, "Quantitative Phase Maps of Live Cells Classified By Transfer Learning and Generative Adversarial Network (GAN), " *Computational Optical Sensing and Imaging* (COSI), CW2E (CW2E. 7), Orlando, Florida United States, 25–28 June 2018.https://doi.org/10.1364/COSI.2018.CW2E.7
53. K. He, X. Zhang, S. Ren and J. Sun, "Deep Residual Learning for Image Recognition, " *2016 IEEE Conference on Computer Vision and Pattern Recognition (CVPR)*, 2016, pp. 770-778,doi: 10.1109/CVPR.2016.90.
54. D-A. Clevert, T. Unterthiner, and S. Hochreiter, "Fast and accurate deep network learning by exponential linear units (ELUs), " ICLR (2016);https://arxiv.org/abs/1511.07289
55. I. Moon, K. Jaferzadeh, "Automated digital holographic image reconstruction with deep convolutional neural networks, " Proc. SPIE 11402, Three-Dimensional Imaging, Visualization, and Display 2020, 114020A (22 April 2020);doi: 10.1117/12.2554533
56. Y. Rivenson, Y. *Zhang, H. Günaydın, D. Teng and A. Ozcan, "Phase recovery and holographic image reconstruction using deep learning in neural networks", Light Sci. Appl* **7**, 17141 (2018).
57. R. Zhenbo, X. Zhimin, E. Y. M. Lam, "End-to-end deep learning framework for digital holographic reconstruction, " Adv. Photon. 1(1) 016004 (28 January 2019)https://doi.org/10.1117/1.AP.1.1.016004
58. G. Zhang, T. Guan, Z. Shen, X. Wang, T. Hu, D. Wang, Y. He, and N. Xie, "Fast phase retrieval in off-axis digital holographic microscopy through deep learning, " Opt. Express 26, 19388-19405 (2018).
59. D. Jin, R. Zhou, Z. Yaqoob, and P. So, "Tomographic phase microscopy: principles and applications in bioimaging [Invited], " J. Opt. Soc. Am. B 34, B64-B77 (2017).
60. V. Balasubramani, A. Kuś, H. Tu, C. Cheng, M. Baczewska, W. Krauze, and M. Kujawińska, "Holographic tomography: techniques and biomedical applications [Invited], " Appl. Opt. 60, B65-B80 (2021).
61. R. Guo, I. Barnea, and N. Shaked, "Limited-angle tomographic phase microscopy utilizing confocal scanning fluorescence microscopy, " Biomed. Opt. Express 12, 1869-1881 (2021).
62. S. Chowdhury, M. Chen, R. Eckert, D. Ren, F. Wu, N. Repina, and L. Waller, "High-resolution 3D refractive index microscopy of multiple-scattering samples from intensity images, " Optica 6, 1211-1219 (2019).
63. Y. Lin, H. Chen, H. Tu, C. Liu, and C. Cheng, "Optically driven full-angle sample rotation for tomographic imaging in digital holographic microscopy, " Opt. Lett. 42, 1321-1324 (2017).
64. J. Jung, K. Kim, J. Yoon, and Y. Park, "Hyperspectral optical diffraction tomography, " Opt. Express 24, 2006-2012 (2016).
65. Y. Sung, W. Choi, C. Fang-Yen, K. Badizadegan, R. Dasari, and M. Feld, "Optical diffraction tomography for high resolution live cell imaging, " Opt. Express 17, 266-277 (2009).
66. M. Baczewska, K. Eder, S. Ketelhut, B. Kemper, and M. Kujawinska, "Holotomographic investigation of an influence of PFA cell fixation process on refractive index of cellular organelles in epithelial cells, " *SPIE-BIOS Quantitative Phase Imaging VI;* 11249(112491L), San Francisco, California, United States (2020).https://doi.org/10.1117/12.2544660.
67. P. Y. Liu, L. K. Chin, W. Ser, H. F. Chen, C. -M. Hsieh, C. -H. Lee, K. -B. Sung, T. C. Ayi, P. H. Yap, B. Liedberg, K. Wang, T. Bourouina, and Y. Leprince-Wang, "Cell refractive index for cell biology and disease diagnosis: Past, present and future, " Lab Chip, 16(4), 634–644 (2016).
68. G. Dardikman, N. T. Shaked, "Review on methods of solving the refractive index–thickness coupling problem in digital holographic microscopy of biological cells, " *Opt. Commun.*, vol. 422, pp. 8–16, 2018.
69. T. Nguyen, V. Bui, D. Tran, G. Nehmetallah, "Deep learning bi-telecentric digital holographic microscopy for aberration compensation applied to cancer cells, " Digital Holography and Three-Dimensional Imaging. Heidelberg, Germany: The Optical Society (OSA); 2017. p Tu2A. 5.
70. B. Rappaz, B. Breton, E. Shaffer, G. Turcatti, "Digital holographic microscopy: A quantitative label-free microscopy technique for phenotypic screening, " Comb Chem High Throughput Screen 2014;17: 80–88.
71. B. Kemper, A. Bauwens, A. Vollmer, S. Ketelhut, P. Langehanenberg, J. Müthing, H. Karch, and G. Von Bally, "Label-free quantitative cell division monitoring of endothelial cells by digital holographic microscopy, " J. Biomed. Opt. 15(3), 036009 (2010);doi:10.1117/ 1.3431712.
72. B. Kemper, D. Carl, J. Schnekenburger, I. Bredebusch, M. Schäfer, W. Domschke, and G. Von Bally, "Investigation of living pancreas tumor cells by digital holographic microscopy, " J. Biomed. Opt. 11(3), 34005 (2006);doi:10.1117/1.2204609.
73. K. J. Chalut, A. E. Ekpenyong, W. L. Clegg, I. C. Melhuish, and J. Guck, "Quantifying cellular differentiation by physical phenotype using digital holographic microscopy, " Integr Biol (Camb) 4(3), 280–284 (2012);doi:10.1039/c2ib00129b.
74. Z. El-Schich, A. Molder, H. Tassidis, P. Harkonen, M. F. Miniotis, and A. G. Wingren, "Induction of morphological changes in death-induced cancer cells monitored by holographic microscopy, " J. Struct. Biol. 189(3), 207–212 (2015);doi:10.1016/j.jsb.2015.01.010.
75. P. Langehanenberg, L. Ivanova, I. Bernhardt, S. Ketelhut, A. Vollmer, D. Dirksen, G. Georgiev, G. Von Bally, and B. Kemper, Automated three-dimensional tracking of living cells by digital holographic microscopy, J. Biomed. Opt. 14(1), 014018 (2009);doi:10.1117/

1.3080133.

76. F. Dubois, C. Yourassowsky, O. Monnom, J. C. Legros, O. Debeir, P. Van Ham, R. Kiss, and C. Decaestecker, Digital holographic microscopy for the three-dimensional dynamic analysis of in vitro cancer cell migration, J. Biomed. Opt. 11(5), 054032 (2006);doi:10.1117/1.2357174.
77. A. V. Belashov, A. A. Zhikhoreva, T. N. Belyaeva, E. S. Kornilova, N. V. Petrov, A. V. Salova, I. V. Semenova, and O. S. Vasyutinskii. "Digital holographic microscopy in label-free analysis of cultured cells' response to photodynamic treatment, " Opt. Lett. 41(21), 5035–5038(2016);doi:10.1364/OL.41.005035.
78. N. Benzerdjeb, C. Garbar, P. Camparo, and H. Sevestre, "Digital holographic microscopy as screening tool for cervical cancer preliminary study, " Cancer Cytopathol 124(8), 573–580 (2016);doi:10.1002/cncy.21727.
79. D. Roitshtain, L. Wolbromsky, E. Bal, H. Greenspan, L. Satterwhite, N. Shaked, "Quantitative phase microscopy spatial signatures of cancer cells, " Cytometry A 2017;91A: 482–493. doi:10.1002/cyto.a.23100.
80. Y. Wu, A. Ozcan, "Lensless digital holographic microscopy and its applications in biomedicine and environmental monitoring, " Methods 2017; Available at:https://doi. org/10.1016/j.ymeth.2017.08.013
81. V. Lam, T. Nguyen, B. -M. Chung, L. -C. Chang, G. Nehmetallah, C. B. Raub, "Quantitative determination of epithelial and mesenchymal qualities of cancer cells using machine learning and quantitative phase imaging, " J. Biomed. Opt. , 25(2), 026002-1-17 (2020).
82. J. D. Wilson, W. J. Cottrell, and T. H. Foster, "Index-of-refraction-dependent subcellular light scattering observed with organelle-specific dyes, " J. Biomed. Opt. , 12(1), 014010-1-10 (2007).
83. B. Bazow, T. Phan, V. Lam, C. Raub, G. Nehmetallah, "Synthetic training of machine learning algorithms for holographic cell imaging, " Proc. SPIE 11731, Computational Imaging VI, 1173102 (12 April 2021);
84. Brad Bazow, Thuc Phan, Christopher B. Raub, and George Nehmetallah, "Computational Multi-wavelength (MW) Phase Synthesis Using Convolutional Neural Networks (CNNs), " [Invited], Applied Optics, 61(5), B132-B146 (2022).
85. S. A. Collins, "Lens-System Diffraction Integral Written in Terms of Matrix Optics, " J. Opt. Soc. Am. 60(9), 1168-1177 (1970).
86. R. Castañeda and J. Garcia-Sucerquia, "Single-shot 3D topography of reflective samples with digital holographic microscopy, " Appl. Opt. , 57(1) A12-A18 (2018).
87. J. M. Bioucas-Dias and G. Valadao, "Phase Unwrapping via Graph Cuts, " in *IEEE Transactions on Image Processing*, vol. 16, no. 3, pp. 698-709, March 2007,doi: 10.1109/TIP.2006.888351.
88. D. Kingma, J. Ba, "Adam: A Method for Stochastic Optimization, " *CoRR* abs/1412.6980 (2015): n. pag.
89. V. Bianco, P. *Memmolo, M. Leo, et al. , "Strategies for reducing speckle noise in digital holography. Light Sci Appl* 7, 48 (2018).
90. W. Zhou, S. Liu, H. Zhang, Y. Yu, and T. Poon, "A Deep Learning Approach for Digital Hologram Speckle Noise Reduction, " in Imaging and Applied Optics Congress, The Optical Society (Optical Society of America, 2020), paper HTu5B. 5.
91. M. Haouat, J. Garcia-Sucerquia, A. Kellou, et al. , "Reduction of speckle noise in holographic images using spatial jittering in numerical reconstructions, " Optics Letters 42, 1047 -1050 (2017).
92. K. Zhang, W. Zuo, L. Zhang, "FFDNet: Toward a fast and flexible solution for CNN -based image denoising, " IEEE Transactions on Image Processing 27, 4608 -4622(2018).
93. J. Sánchez-Paredes, J. Castro-Ramos, S. Vázquez-Montiel, "A comparison between 2D phase unwrapping techniques, " Proc. SPIE 6292, Interferometry XIII: Techniques and Analysis, 62921K (14 August 2006);doi: 10.1117/12.681385
94. J. C. Estrada, M. Servin, and J. Vargas, "2D simultaneous phase unwrapping and filtering: A review and comparison, " Opt. Lasers Eng. 50(8), 1026-1029, (2012);
95. K. Q. Wang, Y. Li, Q. Kemao, J. L. Di and J. L. Zhao, "One-step robust deep learning phase unwrapping, " Opt. Express 27(10), 15100–15115 (2019).
96. G. Nehmetallah, "Multi-wavelength digital holographic microscopy using a telecentric reflection configuration, " Topical Meeting in Digital Holography and Three-Dimensional Imaging (DH), DM3A. 7, Shanghai, China, 24 - 28 May (2015).
97. Van K Lam, Thanh C Nguyen, Byung M Chung, George Nehmetallah, Christopher B Raub, "Quantitative assessment of cancer cell morphology and motility using telecentric digital holographic microscopy and machine learning. " Cytometry Part A 93.3 (2018): 334-345.
98. Maria F. Miniotis, et al. "Digital Holographic Microscopy for Non-invasive Monitoring of Cell Cycle Arrest in L929 Cells. " Plos One 9.9 (2014): e106546.
99. Lu Xin, et al. "Label-Free Assessment of the Drug Resistance of Epithelial Ovarian Cancer Cells in a Microfluidic Holographic Flow Cytometer Boosted through Machine Learning. " ACS Omega 6.46 (2021): 31046-31057.
100. Alexandre Joushomme, et al. "Label-Free Study of the Global Cell Behavior during Exposure to Environmental Radiofrequency Fields in the Presence or Absence of Pro-Apoptotic or Pro-Autophagic Treatments. " Int J Mol Sci 8.23 (2022): 658.
101. Van Lam, Thanh Nguyen, Vy Bui, Byung Min Chung, Lin-Ching Chang, George Nehmetallah, Christopher Raub "Quantitative scoring of epithelial and mesenchymal qualities of cancer cells using machine learning and quantitative phase imaging. " Journal of biomedical optics 25.2 (2020): 026002.
102. Anyu Sun, Van Lam, Thuc Phan, Lin-Ching Chang, George Nehmetallah, Christopher Raub, "Multi-Classification of Cell Lines using DHM and Machine Learning, " DH and 3D Imaging, W5A. 23, Cambridge UK Aug1-4 (2022).
103. Volpe G, Wählby C, Tian L, Hecht M, Yakimovich A, Monakhova K, Waller L, Sbalzarini IF, Metzler CA, Xie M, Zhang K, Lenton ICD, Rubinsztein-Dunlop H, Brunner D, Bai B, Ozcan A, Midtvedt D, Wang H, Sladoje N, Lindblad J, Smith JT, Ochoa M, Barroso M, Intes X, Qiu T, Yu LY, You S, Liu Y, Ziatdinov MA, Kalinin SV, Sheridan A, Manor U, Nehme E, Goldenberg O, Shechtman Y, Moberg HK, Langhammer C, Špačková B, Helgadottir S, Midtvedt B, Argun A, Thalheim T, Cichos F, Bo S, Hubatsch L, Pineda J, Manzo C, Bachimanchi H, Selander E, Homs-Corbera A, Fränzl M, de Haan K, Rivenson Y, Korczak Z, Adiels CB, Mijalkov M, Veréb D, Chang YW, Pereira JB, Matuszewski D, Kylberg G, Sintorn IM, Caicedo JC, Cimini BA, Bell MAL, Saraiva BM, Jacquemet G, Henriques R, Ouyang W, Le T, Gómez-de-Mariscal E, Sage D, Muñoz-Barrutia A, Lindqvist EJ, Bergman J. Roadmap on Deep Learning for Microscopy. ArXiv [Preprint]. 2023 Mar 7: arXiv: 2303.03793v1. PMID: 36945686; PMCID: PMC10029045. "Roadmap on Deep Learning for Microscopy, " arXiv: 2303.03793 [physics. optics]

Part III

Making of New Optical Components

Official journal of the JHL 2689-9620
www.light-am.com

Review | Open Access

Holographic optics in planar optical systems for next generation small form factor mixed reality headsets

Bernard C. Kress* and Maria Pace

Abstract

Helmet Mounted Displays (HMDs), such as in Virtual Reality (VR), Augmented Reality (AR), Mixed reality (MR), and Smart Glasses have the potential to revolutionize the way we live our private and professional lives, as in communicating, working, teaching and learning, shopping and getting entertained. Such HMD devices have to satisfy draconian requirements in weight, size, form factor, power, compute, wireless communication and of course display, imaging and sensing performances.

We review in this paper the various optical technologies and architectures that have been developed in the past 10 years to provide adequate solutions for the drastic requirements of consumer HMDs, a market that has yet to become mature in the next years, unlike the existing enterprise and defense markets that have already adopted VR and AR headsets as practical tools to improve greatly effectiveness and productivity.

We focus specifically our attention on the optical combiner element, a crucial element in Optical See-Through (OST) HMDs that combines the see-through scene with a world locked digital image. As for the technological platform, we chose optical waveguide combiners, although there is also a considerable effort today dedicated to free-space combiners.

Flat and thin optics as in micro-optics, holographics, diffractives, metasurfaces and other nanostructured optical elements are key building blocks to achieve the target form factor.

Keywords: Holography, Helmet mounted displays, Virtual reality, Augmented reality, Mixed reality

Introduction

To implement the display generation and optical combining functionality of AR, VR and MR HMDs, traditional optical elements and optical architectures have been developed throughout the past decades, especially for the defense market[1] as the enterprise and consumer markets were just burgeoning. Early attempt at providing consumer grade smart glass display solution have had mixed results as in Google Glass V1, North Focus glasses, and other HMD systems from now defunct companies such as ODG, DAQRI, META, etc…

Traditional optics in HMDs include conventional refractive and reflective optics (circularly symmetric or freeform) as in Google Glass, ODG, META, but also conventional volume holographic optics as in North Focus, Intel Vaunt or Sony Smart Glasses. Although impressive optical performances have been demonstrated with such traditional optics and optical architectures for defense HMDs, the form factor, weight, size and wall plug efficiency of such systems were not at a level to provide adequate solutions for prosumer and consumer HMDs.

Correspondence: Bernard C. Kress (bernard.kress@gmail.com)
Microsoft Corp. HoloLens Team, 1 Microsoft Way, Redmond, 98052 WA. USA

There are two main pillars[2] defining the MR experience for the consumer market today: comfort and immersion. **Comfort** comes in various forms:

- **wearable comfort** –weight, size, center of gravity, thermal issues, …
- **visual comfort** –natural 3D cues over large FOV and high angular resolution
- **vestibular comfort** –stable and realistic world locked digital displays
- **social comfort** –socially acceptable form factor and operations

Immersion is defined as a multisensory perceptual experience[3] (from audio to display, from gesture sensing to haptics). Immersion conveys to the user a high sense of realism[4] which is a key feature to implement one of tomorrow's key use cases for MR: the sense of presence[5] of a virtual avatar interacting with the wearer of the HMD. Fig. 1 shows how immersion has increased dramatically when migrating from traditional flat screen displays to Near-To-Eye (NTE) displays.

In OST-HMS, the optical combiner is a key feature which acts as a transparent mirror or a transparent periscope shifting and expanding the entrance pupil generated by the display engine as depicted in Fig. 2.

Periscope operation mode is best suited for goggle or helmet type HMDs where the display engine is located on the world side (temple or forehead) and mirror operation mode is best suited for smart glasses HMDs where the display engine is located in the glasses frames.

Both modes can provide a highly transparent operation and a generous eyebox by replicating the small pupil generated by the display engine. Key features of a waveguide combiner in either mode is optical efficiency, Modulation Transfer Function (MTF) decrease, transparency, ghosts and other parasitic optical artifacts that may be triggered with specific waveguide coupler technologies.

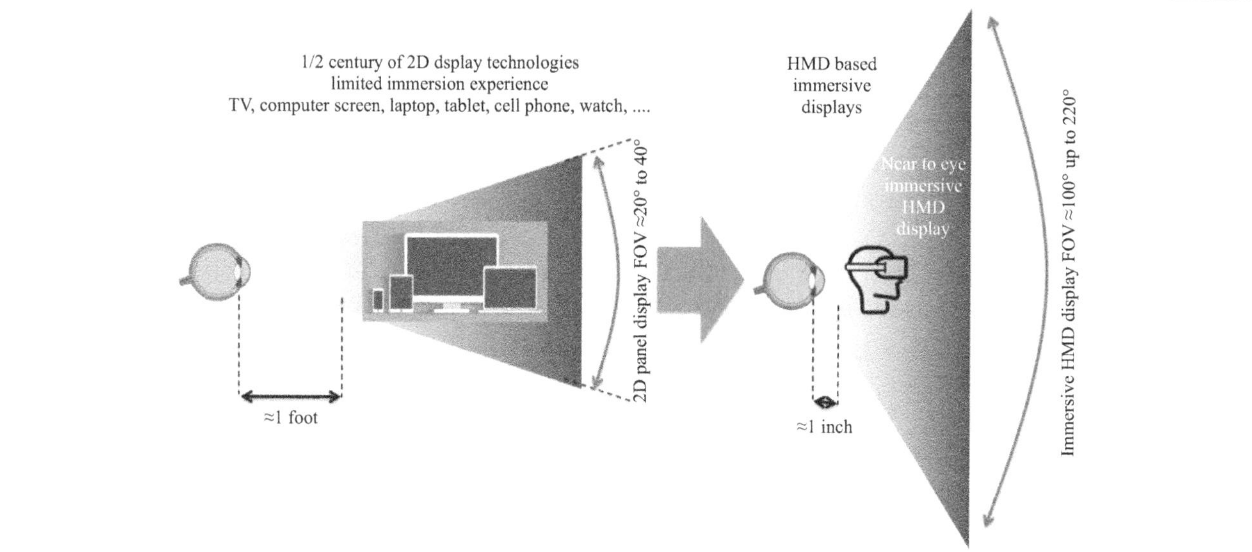

Fig. 1 Immersive NTE displays: a paradigm shift in personal information display.

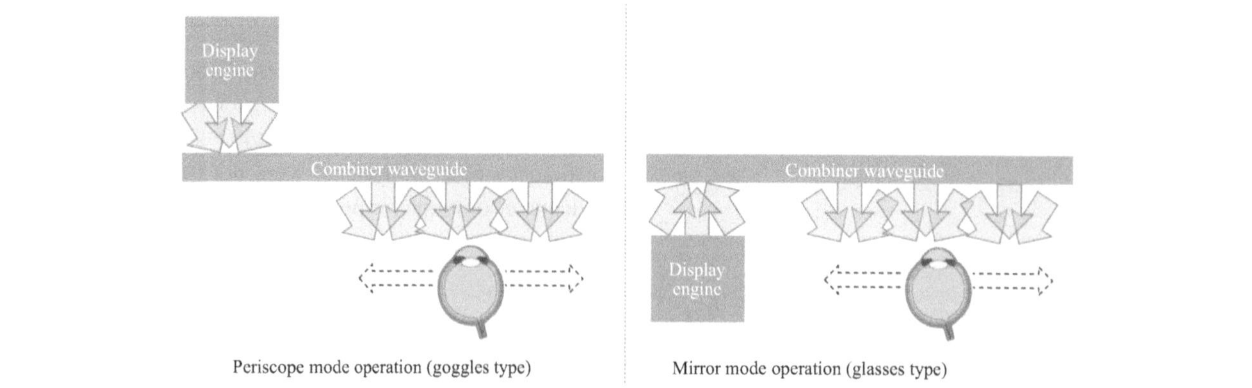

Fig. 2 The optical combiner: periscope mode or mirror mode with pupil shifting/expansion in transparent lightguide.

Flat optics and flat optical systems for display and imaging

Considering the stringent form-factor, weight and size requirements to implement the display generation and optical combiner functionality, flat optics seems to be a de facto choice to implement the combiner task shown in Fig. 2.

However, simply considering flat optics might not be the best way to start the problem as flat optics do no necessarily mean flat optical systems, as described in Fig. 3

Although flat optics can replace advantageously bulky refractive/reflective optics to reduce its overall weight, the size of the system might only be mildly improved. Flat optical system are our goal, and there are many ways to achieve this, by folding the space or redefining the imaging/display functionality.

Flat optics might seem also to be the cure to all refractive/reflective troubles when it comes to weight and size, and sometimes also to mass production via today's optical lithographic or nanoimprint lithographic (NIL) systems. However, flat optical elements bare often much worse aberrations than their traditional curved optical elements counterparts as shown in Fig. 4.

The other downsize of novel flat optics are their limited efficiency or throughput which, sometimes, reduces dramatically the wall plug efficiency of an HMD system, especially in OST mode. Below is a simple comparison of the maximum diffraction efficiency of a surface relief diffractive element when compared to its refractive counterpart (see Fig. 5), for many different lithographic fabrication techniques.

We review in the next two sections how flat optics have been implemented in both VR and AR HMDs to reduce the weight/size, improve the form factor, but also keep similar optical performances and efficiency, two key features of any wearable display.

Flat optics and flat optical systems in VR HMDs

Traditional direct view VR systems as they have been implemented since the early 70s, and still used today in many devices such as the latest Oculus Quest 2, comprise a

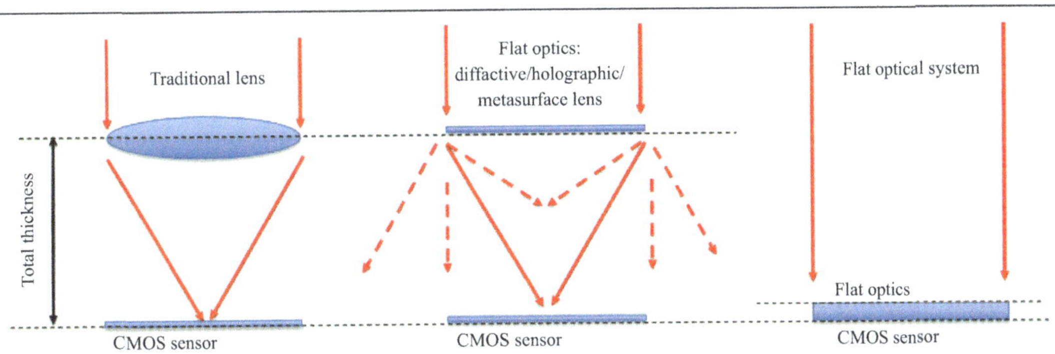

Fig. 3 Flat optics and flat optical systems.

	Curved refractive lens	Flat diffractive lens
Spherical	$S_{\mathrm{I}}=\frac{y^4\theta^4}{4}\left(\left(\frac{n}{n-1}\right)^2+\frac{n+2}{n(n-1)^2}E^2+4\frac{(n+1)}{n(n-1)}ET+\frac{3n+2}{n}T^2\right)$	$S_{\mathrm{I}}=\frac{y^4\theta^4}{4}\left(1+B^2+4BT+3T^2\right)-8\lambda By^4$
Coma	$S_{\mathrm{II}}=\frac{-y^2\theta^2H}{2}\left(\frac{n+1}{n(n-1)}E+\frac{2n+1}{n}T\right)$	$S_{\mathrm{II}}=\frac{-y^2\theta^2H}{2}\left(B+2T\right)$
Astigmatism	$S_{\mathrm{III}}=H^2\theta$	$S_{\mathrm{III}}=H^2\theta$
Petzval curvature	$S_{\mathrm{IV}}=\frac{H^2\theta}{n}$	$S_{\mathrm{IV}}=0$
Distortion	$S_{\mathrm{V}}=0$	$S_{\mathrm{V}}=0$
Bending parameter	$E=\frac{c_1+c_2}{c_1-c_2}$	$B=\frac{2c_{\mathrm{sub}}}{\theta}$
Conjugate parameter	$T=\frac{u+u'}{u-u'}=\frac{m+1}{m-1}$	$T=\frac{u+u'}{u-u'}=\frac{m+1}{m-1}$

Fig. 4 Aberrations of curved refractive and flat diffractive lenses.

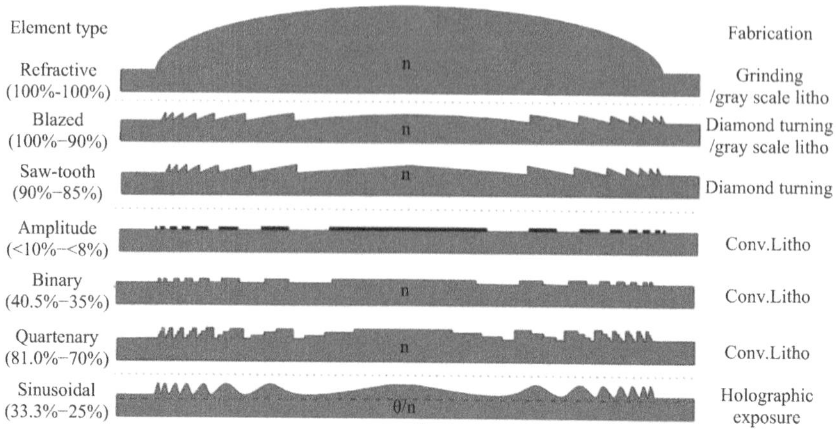

Fig. 5 Efficiency of curved refractive and flat surface relief diffractive lenses.

magnifier lens and a display panel. However, that lens has been optimized in various ways to reduce its weight and size for ever increasing Field of Views (FOVs) as shown in Fig. 6. As one can see the weight and size is improved, and the lens gets flatter. However, the space between the display panel and the lens is unchanged, thus acting very little on the size of the overall system.

In order to reduce the size of the overall system, the display architecture has to be altered by either increasing the Numerical Aperture (NA) of the system and reducing the size of the display or by introducing more lenses, either stacked or spatially multiplexed as described in Fig. 7.

Interestingly, Fig. 7 shows how a thicker lens (polarization pancake lens) can actually reduce the system thickness, which is totally counterintuitive and does not lean towards "thin optics" as a solution. However, thin optics as in compound lenses and Micro-Lens Arrays (MLAs) are indeed showing thinner optics providing thinner optical systems.

Other recent attempts have hybridized some of these concepts in very flat optical systems, by using micro-lenses and holographic polarization lenses with high NAs (see Fig. 8).

Pancake lenses are folding the space with strong polarization optics (holograms) or layers (dichroics). However, a path further would be to fold the space independently from polarization, as it has been demonstrated by Prof Joseph Ford from UCSD with "origami lenses". One of this architecture is depicted in Fig. 9.

Although this is a space folding which reduces the aperture (all lens apertures are centered and toroidal), when going off-axis one can increase further the compactness of the system as it has been proposed very early on by Prof Juergen Jahns from Aachen University (1995). This space folding uses either metal reflective coatings or Total

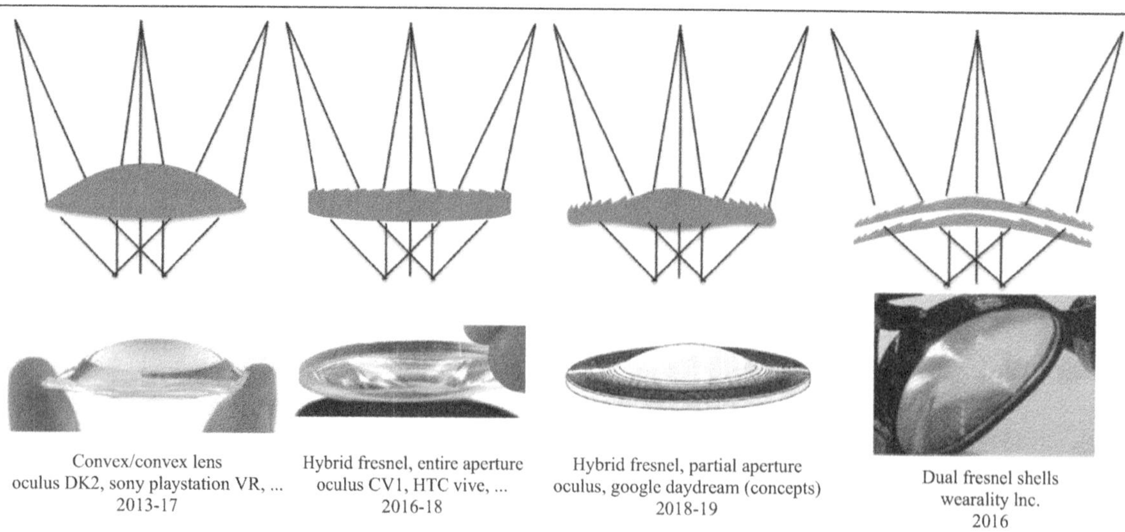

Fig. 6 Reducing the thickness of a VR lens.

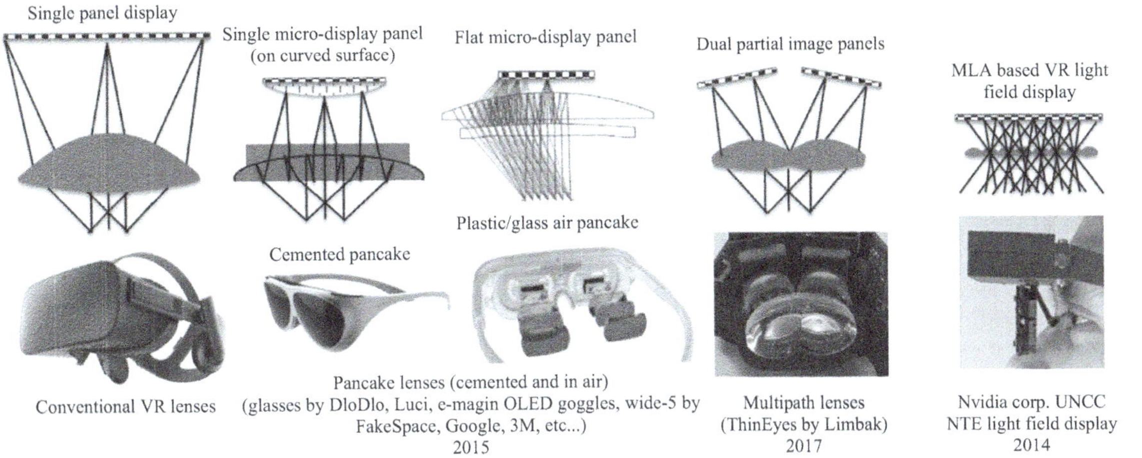

Fig. 7 Reducing the thickness of a VR display system.

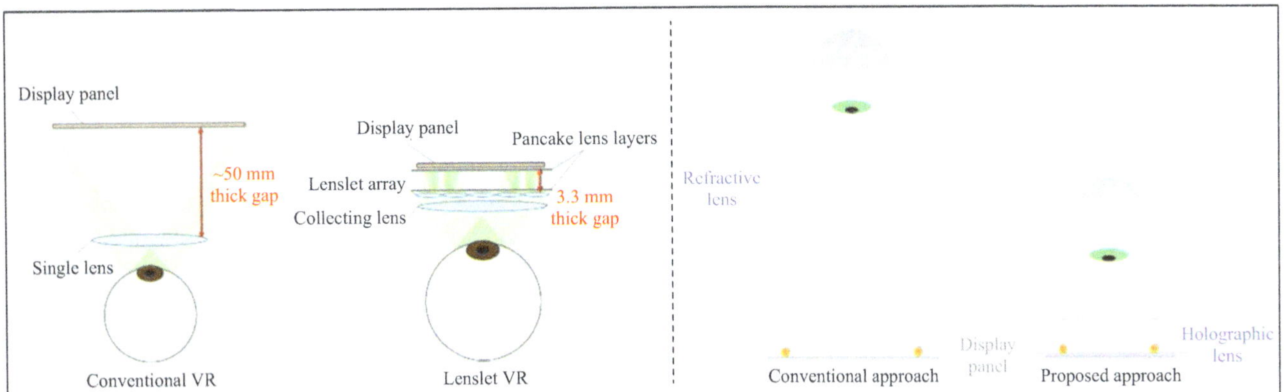

Fig. 8 Flat VR system using micro-lens arrays (left, Byoungho Lee, Seoul National University-2021), or holographic pancake optics (right, Andrew Maimone, Facebook reality Labs-2020)

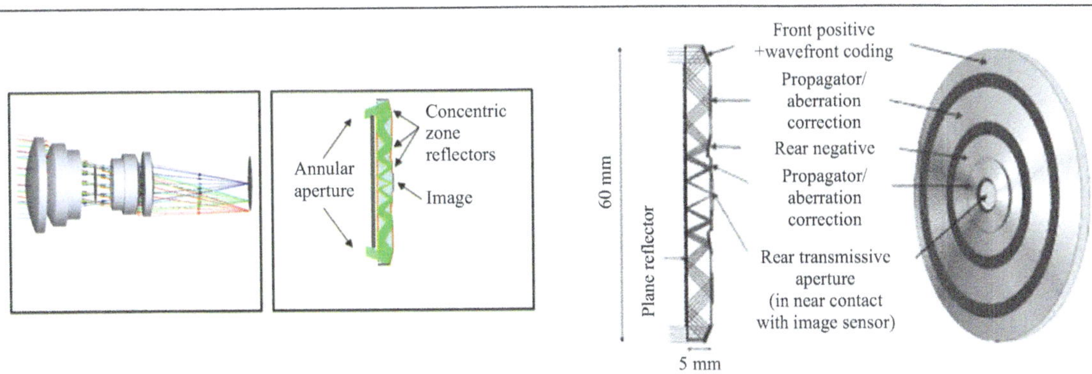

Fig. 9 Origami lenses (Prof. Joseph Ford, UCSD, 2008)

Internal Reflection (TIR) trapping via grating in-couplers as shown in Fig. 10.

One other key feature of flat optics is that the efficiency can be modulated as per the number of surface reliefs in the fabrication process or the duty cycle of the fringes (in both surface relief and holographic), the depth of the surface relief elements, or the thickness of the holographic layer/emulsion. In traditional optical elements (such as in implementing the conventional Bokeh feature in photography), one may use an apodiser in conjunction with a refractive lens. If that lens is to be fabricated as a hologram or a surface relief diffractive (or even a metalens), the efficiency can thus be carefully modulated in order to produce a similar apodization effect but in a single

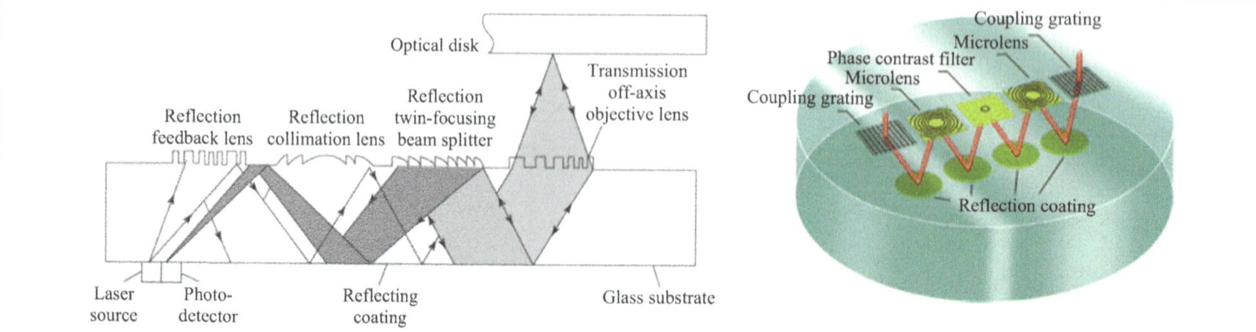

Fig. 10 Planar optical systems (Prof. Juergen Jahns, 1995)

element this time. Efficiency modulation provides a complex lens aperture function which in turn can then yield a specific point spread function, to enhance certain aspects of the image. Gaussian apodizations are used widely, but other more complex apodization can be used in order to combine high resolution and extended depth of focus (as depicted in Fig. 11).

The drawback of that technique is that the undiffracted field might reduce the signal to noise ratio and increase scatter from an increased zero order or unwanted propagating orders.

Fig. 12 shows an example of a space multiplexed planar optical system implementing both structured illumination and extended depth of focus imaging functionality for 3D spatial scanning in a minimal form factor. We fabricated this element via a contact copy mask aligner with optical front/back alignment potential (Suss MA6-BA aligner) followed by ion beam etching of both surfaces.

This planar optical system is aligned lithographically during its fabrication at sub-micron level and, thus, it is virtually impossible to misalign in the field since it is etched on the same fused silica substrate. Such accurate alignment and alignment resilience to environmental disturbances (temperature, shocks, vibrations) makes it perfect candidate for systems to be used in harsh environments.

In order to come up with the optimal waveguide combiner architecture for OST-HMDs, especially targeted to the consumer market, we will take advantage of such space folding architectures in planar waveguide systems. The next sections will be dedicated to such architectures.

Planar Optical Waveguide Combiners

The optical combiner is often the most complex and most costly optical element in the entire MR display architecture: it is the one component seen directly by the user and the one seen directly by the world. It often defines the size and aspect ratio of the entire headset. It is the

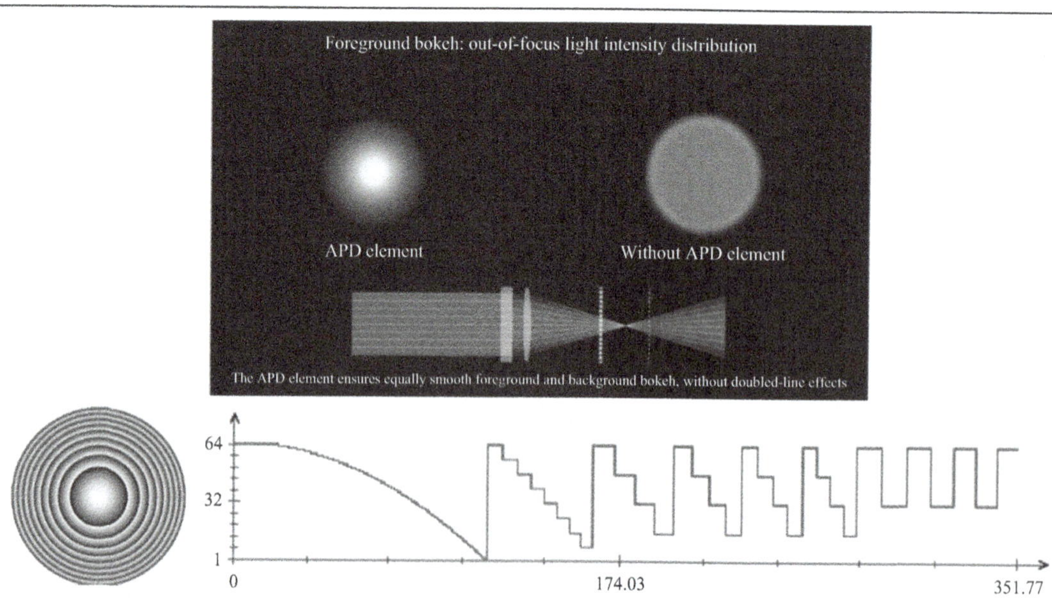

Fig. 11 Apodization of a diffractive lens by modulating the efficiency radially as a function of number of phase levels.

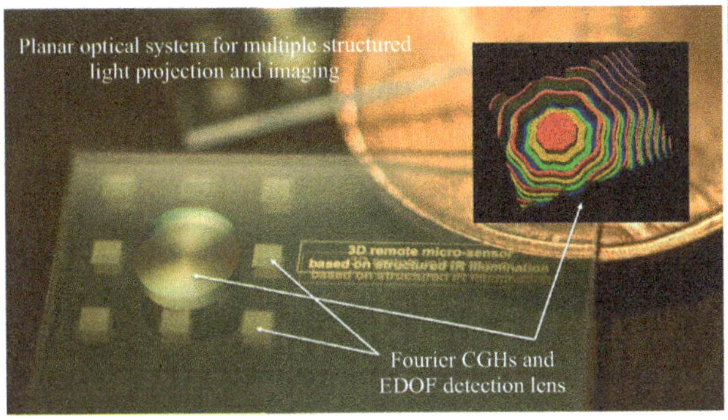

Fig. 12 Etched planar optical system including various CGHs and dual front-back extended depth of focus (EFOD) lens with anodization as a modulation of surface reliefs.

critical optical element that reduces the quality of the see-through and the one that defines the eyebox size (and in many cases, also the FOV).

There are three main types of optical combiners used in most MR/AR/smart glasses today:

- Free-space optical combiners[6],
- TIR prism optical combiners[7] (and compensators), and
- Waveguide based optical combiners[8–10].

When optimizing an HMD display system, the optical engine must be optimized in concert with the combiner engine[11]. Usually, a team that designs an optical engine without fully understanding the limitations and specifics of a combiner engine designed by another team, and vice versa, can result in a suboptimal system or even a failed optical architecture, no matter how well the individual optical building blocks might be designed.

Freeform TIR prism combiners are at the interface between free space and waveguide combiners. When the number of TIR bounces increases, one might refer to them as waveguide combiners. Waveguide combiner architectures are the topic of this review paper.

Waveguide combiners are based on TIR propagation of the entire field in an optical guide, essentially acting as a transparent periscope with a single entrance pupil and often many exit pupils[12].

The primary functional components of a waveguide combiner consists of the input and output couplers. These can be either simple prisms, micro-prism arrays, embedded mirror arrays, surface relief gratings, thin or thick analog holographic gratings, metasurfaces, or resonant waveguide gratings. All of these have their specific advantages and limitations, which will be discussed here. Waveguide combiners have been used historically or tasks very different from AR combiner, such as planar optical interconnections[13] and LCD backlights[14,15].

Waveguide combiners are an old concept, some of the earliest IP dates back to 1976 and applied to HUDs. Fig. 13a shows a patent by Juris Upatnieks dating back 1987, a Latvian/American scientist and one of the pioneers of modern holography[16], implemented in a di-chromated gelatin (DCG) holographic media. A few years later, one-dimensional eyebox expansion (1D EPE) architectures were proposed as well as a variety of alternatives for in- and out- couplers technologies, such as surface relief grating couplers by Thomson CSF (Fig. 13b). Fig. 13c shows the original 1991 patent for a waveguide-embedded partial mirror combiner and exit pupil replication (All of these original waveguide combiner patents have been in the public domain for nearly a decade[17–19]).

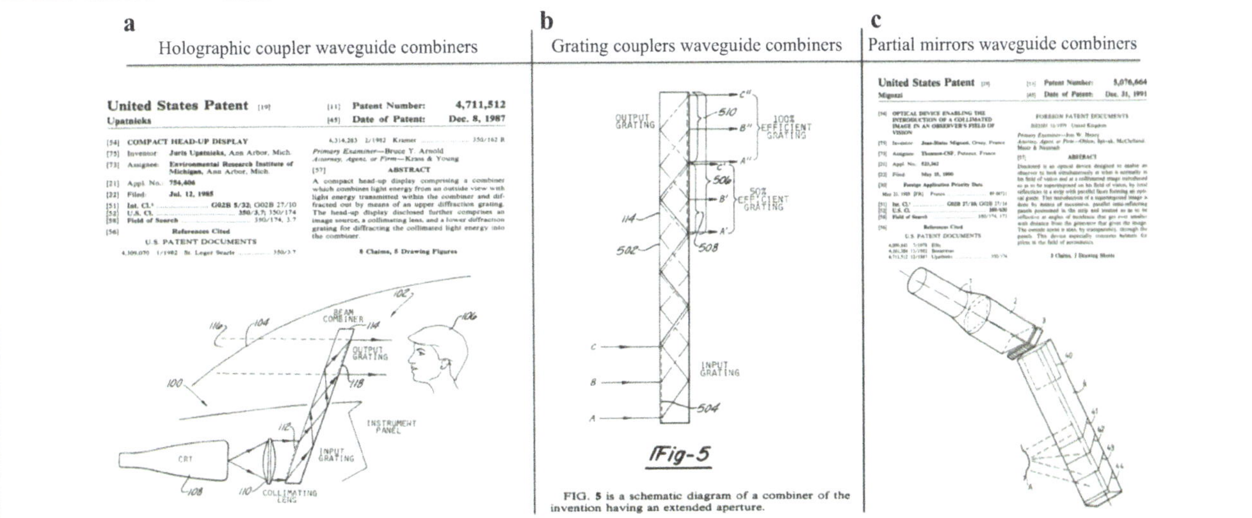

Fig. 13 a Original waveguide combiner patents including holographic (1987), **b** surface relief grating (1989), and **c** partial mirrors (1991) for HUD and HMD applications.

Reflective waveguide architectures

One can take the concept of a flat waveguide with a single curved extractor mirror (Epson Moverio BT300) or freeform prism combiner, or a curved waveguide with curved mirror extractor, to the next level by multiplying the mirrors to increase the eyebox (see the Lumus LOE waveguide combiner) or fracturing metal mirrors into individual pieces (see the Optinvent ORA waveguide combiner or the LetinAR waveguide combiner using fractured mirrors instead of cascaded partial mirrors[20–23]).

While fracturing the same mirror into individual pieces can increase see-through and depth of focus, the use of more mirrors to replicate the pupil is a bit more complicated, especially in a curved waveguide where the two exit pupils need to be spatially de-multiplexed to provide a specific mirror curvature to each pupil to correct for image position: this limits the FOV in one direction so that such overlap does not happen.

Fig. 14 summarizes some of the possible design configurations with such waveguide mirror architectures. Note that the grating- or holographic-based waveguide combiners are not listed here; they are the subject of the next sections.

Fig. 14 shows that many of the waveguide combiner architectures mentioned in this section can be listed in this table[24,25]. Mirrors can be half-tone (Google Glass, Epson Moverio), dielectric (Lumus LOE), have volume holographic reflectors (Luminit or Konica Minolta), or the lens can be fractured into a Fresnel element (Zeiss Tooz Smart Glass). In the Optinvent case, we have a hybrid between fractured metal mirrors and cascaded half-tone mirrors. In one implementation, each micro-prism on the waveguide has one side fully reflective and the other side transparent to allow see-through.

In the LetinAR case, all fractured mirrors are reflective, can be flat or curved, and can be inverted to work with a birdbath reflective lens embedded in the guide.

Even though the waveguide might be flat, when using multiple lensed mirrors, the various lens powers will be different since the display is positioned at different distances from these lensed extractors. When the waveguide is curved, everything becomes more complex, and the extractor mirror lenses need to compensate for the power imprinted on the TIR field at each TIR bounce in the guide. In the case of curved mirrors (either in flat or curved waveguides), the exit pupils over the entire field cannot overlap since the power to be imprinted on each exit pupil (each field position) is different (Moverio BT300 and Zeiss Tooz Smart Glass). This is not the case when the extractors are flat and the field is collimated in the guide (Lumus LOE).

Two-Dimensional Eyebox Expansion

Two-dimensional eyebox expansion is desired (or required) when the input pupil cannot be generated by the optical engine over an aspect ratio tall enough to form the 2D eyebox, due to the FOV (etendue limitations) and related size/weight considerations. A 2D exit pupil expansion (2D EPE) is therefore required (see Fig. 15). 2D expansion can now be performed with either diffractive, holographic or even reflective couplers as in the Lumus Maximus waveguide (2020).

While holographic recording or holographic volume gratings are usually limited to linear gratings, or gratings

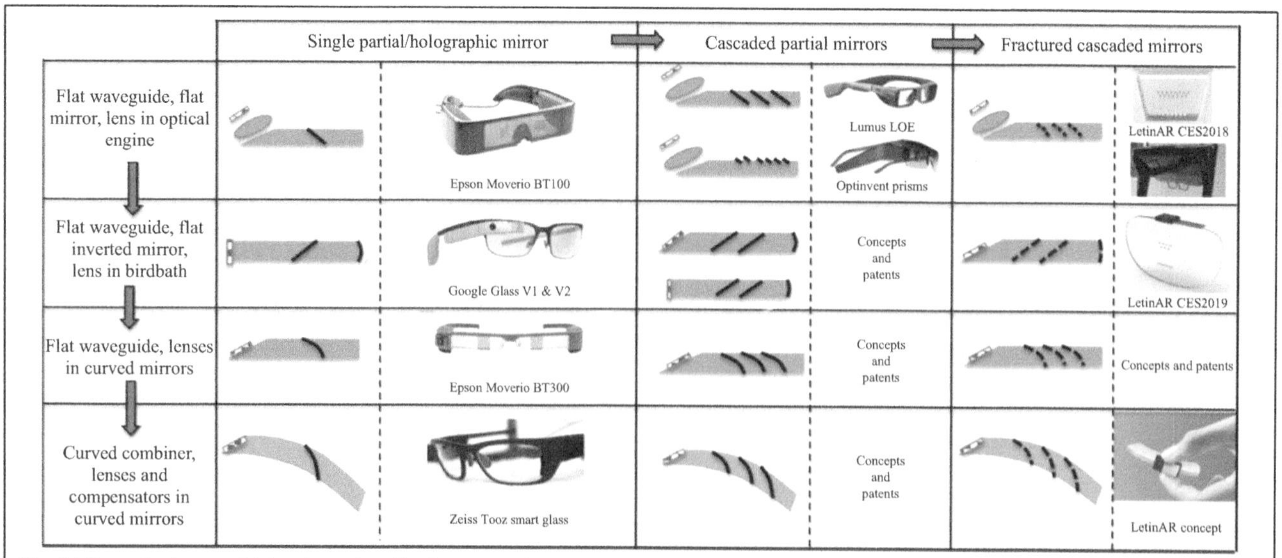

Fig. 14 From single to multiple mirrors in reflective waveguides.

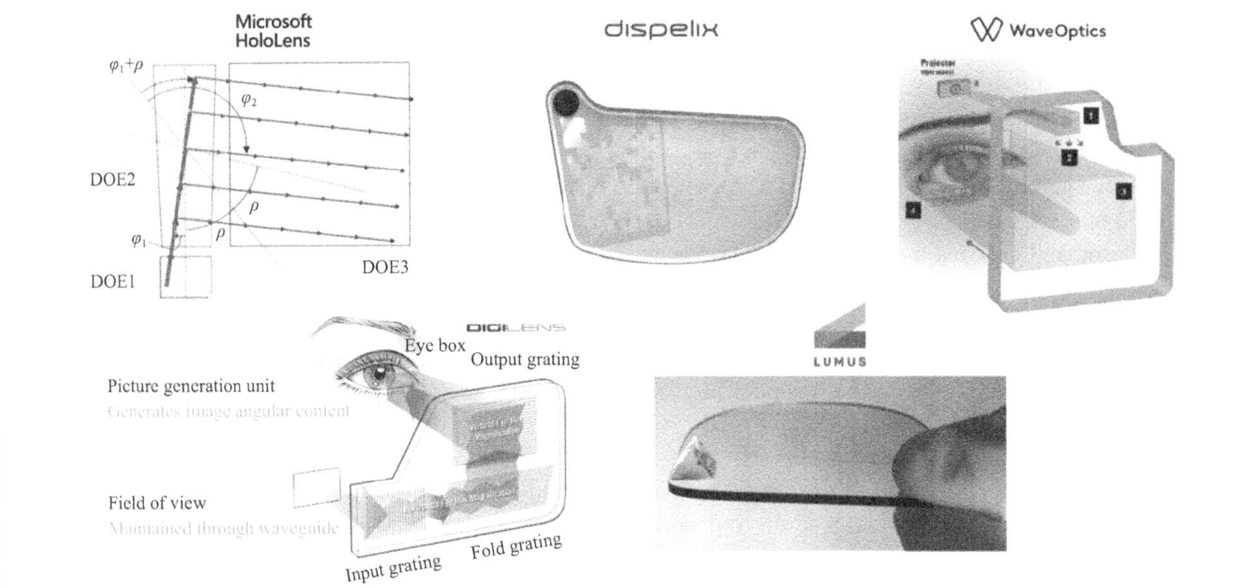

Fig. 15 2D pupil replication architectures in planar optical waveguide combiners.

with slow power (such as off-axis diffractive lenses), surface relief gratings can be either 1D or 2D, linear or quasi arbitrary in shape. Such structures or structure groups can be optimized by iterative algorithms (topological optimization) rather than designed analytically.

Holographic recording, sometimes also called interference lithography when used to create patterns in a resin to be etched in a substrate (such as a nanoimprint master for waveguides) are not limited to linear fringes, but can also produce much more complex structures, such as 2D gratings or even computer generated holograms (CGH) when a digital CGH is used as a reference beam.

Choosing the Right Waveguide Coupler Technology

The coupler element is the key feature of a waveguide combiner. The TIR angle is dictated by the refractive index of the waveguide, not the refractive index of the coupler nanostructures. Very often, the index of the coupler structure (grating or hologram) prescribes the angular and spectral bandwidth over which this coupler can act, thus impacting the color uniformity over the FOV and eyebox.

Numerous coupler technologies have been used in industry and academia to implement the in-coupler and out-couplers, and they can be defined either as refractive/reflective or diffractive/holographic coupler elements.

Refractive/reflective couplers elements

Macroscopic prism

A prism is the simplest TIR in-coupler, and can be very efficient. A prism can be bounded on top of the waveguide, or the waveguide itself can be cut at an angle, to allow normal incident light to enter the waveguide and be guided by TIR (depending on the incoming pupil size). Another way uses a reflective prism on the bottom of the waveguide (metal coated). Using a macroscopic prism as an out-coupler is not impossible, and it requires a compensating prism for see-through, with either a reflective coating or a low-index glue line, as done in the Oorym (Israel) lightguide combiner concept.

Embedded cascaded mirrors

Cascaded embedded mirrors with partially reflective coatings are used as out-couplers in the Lumus (Israel) Lightguide Optical Element (LOE) waveguide combiner. The input coupler remains a prism. As the LOE is composed of reflective surfaces, it yields good color uniformity over the entire FOV. As with other coupler technologies, intrinsic constraints in the cascaded mirror design of the LOE might limit the FOV[26]. See-through is very important in AR systems: the Louver effects produced by the cascaded mirrors in earlier versions of LOEs have been reduced recently thanks to better cutting/polishing, coating, and design. Two dimensional pupil expansion in LOEs have been recently succcesfully demonstrated by Lumus in their latest Maximus X-Lens and Z-lens waveguide designs. High MTF, great color uniformity and high efficiency are key attributes of all reflective waveguide combiners.

Embedded microprism array

Micro-prism arrays are used in the Optinvent (France) waveguide as out-couplers[20]. The in-coupler here is again a prism. Such microprism arrays can be surface relief or index matched to produce an unaltered see-through experience. The micro-prisms can all be coated uniformly with a half-tone mirror layer or can have an alternance of totally reflective and transmissive prism facets, provide a resulting 50% transmission see-through experience. The Optinvent waveguide is the only flat waveguide available today as a plastic guide, thus allowing for a consumer-level cost for the optics. The micro-prism arrays are injection molded in plastic and bounded on top of the guide. Focus lens can also be integrated in a flat waveguide by adding a slight curvature to the extraction prisms. This is possible as such elements are pressure molded in a mold using free-form diamond turned metal inserts rather than cut in glass as embedded cascaded mirrors.

Diffractive/holographic couplers elements

Thin reflective holographic coupler

Transparent volume holograms working in reflection mode—as in di-chromated gelatin (DCG), bleached silver halides (Slavic or Ultimate Holography by Yves Gentet), or more recently photopolymers such as Bayfol® photopolymer by Covestro/Bayer, (Germany)[27], and photopolymers by DuPont (US), Polygrama (Brazil), or Dai Nippon (Japan)—have been used to implement in- and out-couplers in waveguide combiners. Such photopolymers can be sensitized to work over a specific wavelength or over the entire visible spectrum (panchromatic holograms).

Photopolymer holograms do not need to be developed as DCG, nor do they need to be bleached like silver halides. A full-color hologram based on three phase-multiplexed single-color holograms allows for a single plate waveguide architecture, which can simplify the combiner and reduce weight, size, and costs while increasing yield (no plate alignment required). However, the efficiency of such full-RGB phase-multiplexed holograms are still quite low when compared to single-color photopolymer holograms.

Also, the limited index swing of photopolymer holograms allows them to work more efficiently in reflection mode than in transmission mode (allowing for better confinement of both the wavelength and angular spectrum bandwidths).

Examples of photopolymer couplers include Sony LMX-001 Waveguides for smart glasses and the TrueLife Optics (UK) process of mastering the hologram in silver halide and replicating it in photopolymer.

Replication of the holographic function in photopolymer through a fixed master has proven to be possible in a roll-to-roll operation by Bayer/Covestro (Germany). Typical photopolymer holographic media thicknesses range from 16–70 microns, depending on the required angular and spectral bandwidths.

Covestro photopolymer replication technologies such as roll to plate or roll to roll can be used for volume production of such holographic optics, and used in coupler applications, both for AR, but also for larger areas such as in automotive/avionics HUDs and smart windows.

Thin transmission holographic coupler

When the index swing of the volume hologram can be increased, the efficiency gets higher, and the operation in transmission mode becomes possible. This is the case with Digilens' proprietary holographic polymer dispersed liquid crystal (H-PDLC) hologram material[28]. Transmission mode requires the hologram to be sandwiched between two plates rather than laminating a layer on top or bottom of the waveguide as with photopolymers, DCG, or silver halides. Digilens' H-PDLC has the largest index swing today and can therefore produce strong coupling efficiency over a thin layer (typically four microns or less). H-PDLC material can be engineered and recorded to work over a wide range of wavelengths to allow full-color operation.

Thick holographic coupler

Increasing the index swing can optimize the efficiency and/or angular and spectral bandwidths of the hologram. However, this is difficult to achieve with most available materials and might also produce parasitic effects such as haze. Increasing the thickness of the hologram is another option, especially when sharp angular or spectral bandwidths are desired, such as in telecom spectral and angular filters. This is not the case for an AR combiner, where both spectral and bandwidths need to be wide (to process a wide FOV over a wide spectral band such as LEDs). However, a thicker hologram layer also allows for phase multiplexing over many different holograms, one on top of another, allowing for multiple Bragg conditions to operate in concert to build a wide synthetic spectral and/or angular bandwidth, as modeled by the Kogelnik theory[29]. This is the technique used by Akonia, Inc. (a US start-up in Colorado, formerly InPhase Inc., which was originally funded and focused to produce high-density holographic page data-storage media, ruled by the same basic holographic phase-multiplexing principles[30]).

Thick holographic layers, as thick as 500 microns, work well in transmission and/or reflection modes, but they need to be sandwiched between two glass plates. In some specific operation modes, the light can be guided inside the thick hologram medium, where it is not limited by the TIR angle dictated by the index of the glass plates. As the various hologram bandwidths build the final FOV, one needs to be cautious in developing such phase-multiplexed holograms when using narrow illumination sources such as lasers.

Replication of such thick volume holograms are difficult in roll-to-roll operation, as done with thinner single holograms (Covestro Photopolymers, H-PDLC), and require multiple successive exposures to build the hundreds of phase-multiplexed holograms that compose the final holographic structure. However, this can be relatively easy with highly automated recording set-ups as the ones developed by the now-defunct holographic page data-storage industry (In-Phase Corp., General Electric, etc.).

Note that although the individual holograms acting in slivers of angular and spectral bandwidth spread the incoming spectrum like any other hologram (especially when using LED illumination), the spectral spread over the limited spectral range of the hologram is not wide enough to alter the MTF of the immersive image and thus does not need to be compensated by a symmetric in- and out-coupler as with all other grating or holographic structures[31]. This feature allows this waveguide architecture to be asymmetric, such as having a strong in-coupler as a simple prism: a strong in-coupler is always a challenge for any grating or holographic waveguide combiner architecture, and a macroscopic prism is the best coupler imaginable.

Fig. 16 shows both thin and thick volume holograms operating in reflection and/or transmission modes. The top part of the figure shows a typical 1D EPE expander with a single transmission volume hologram sandwiched between two plates. When the field traverses the hologram downwards, it is in off/Bragg condition, and when it traverses the volume hologram upwards after a TIR reflection, it is in an on/Bragg condition (or close to it), thereby creating a weak (or strong) diffracted beam that breaks the TIR condition.

A hologram sandwiched between plates might look more complex to produce than a reflective or transmission laminated version, but it has the advantage that it can operate in both transmission and reflection modes at the same time (for example, to increase the pupil replication diversity).

Surface-relief grating couplers (SRGs)

Fig. 17 reviews the various surface-relief gratings used in industry today (blazed, slanted, binary, multilevel, and analog), and how they can be integrated in waveguide combiners as in-coupling and out-coupling elements[31–35].

Covering a surface-relief grating with a reflective metallic surface (see Fig. 17) will increase dramatically its efficiency in reflection mode. A transparent grating (no coating) can also work both in transmission and reflection modes, especially as an out-coupler, in which the field has a strong incident angle.

Increasing the number of phase levels from binary to quarternary or even eight or sixteen levels increases its efficiency as predicted by the scalar diffraction theory, for normal incidence. However, for a strong incidence angle and for small periods, this is no longer true. A strong out-coupling can thus be produced in either reflection or transmission mode.

Slanted gratings[36,37] are very versatile elements and their spectral and angular bandwidths can be tuned by the slant angles. Front and back slant angles in a same period (or from period to period) can be carefully tuned to achieve the desired angular and spectral operation.

Surface relief gratings have been used as a commodity technology since mastering and mass replication techniques technologies were established and made available in the early 1990s. Typical periods for TIR grating couplers in the visible spectrum are below 500 nm, yielding nanostructures of just a few tens of nanometers if multilevel structures are required. This can be achieved by either direct e-beam write, i-line (or DUV) lithography, or even interference lithography (holographic resist exposure). Surface relief grating structures can be

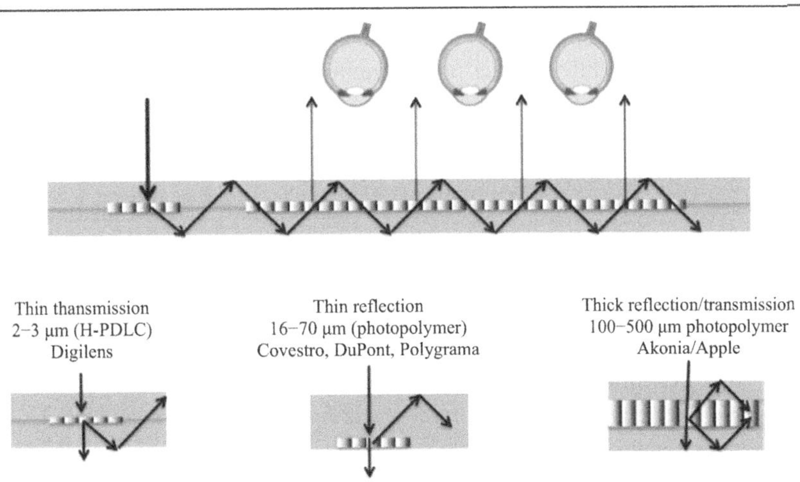

Fig. 16 Different types of volume holograms acting as in- and out-couplers in waveguide combiners.

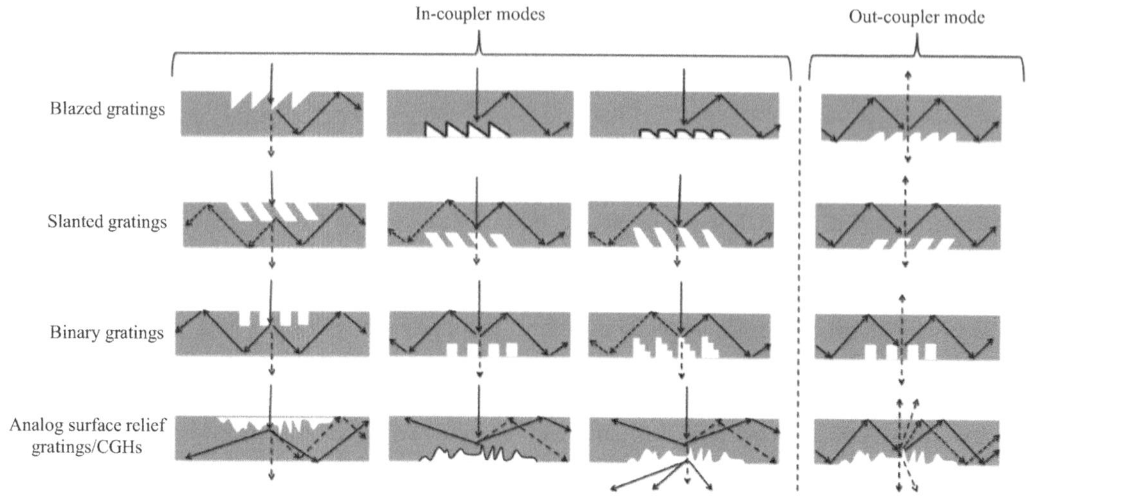

Fig. 17 Surface-relief grating types used as waveguide combiner in-couplers and out-coupler. Solid lines indicate reflective coatings on the grating surface, and dashed lines indicate diffracted orders.

replicated in volumes by nano-imprint, a micro-lithography wafer fabrication technology developed originally for the IC industry[38,39]. Going from wafer-scale fabrication to panel-scale fabrication will reduce costs, allowing for consumer-grade AR and MR products.

Fig. 18 and Fig. 19 illustrate how some of the surface relief gratings shown in Fig. 17 have been applied to the latest waveguide combiners such as the Microsoft HoloLens 1 and Magic Leap One. Multilevel surface relief gratings have been used by companies such as Dispelix Oy, and quasi-analog surface relief CGHs have been used by others, such as WaveOptics Ltd (now part of Snap Inc).

Fig. 18 shows the waveguide combiner architecture used in the Microsoft HoloLens 1 MR headset (2015). The display engine is located on the opposite side of the eyebox. The single input pupil carries the entire image over the various colors at infinity (here, only two colors and the central field are depicted for clarity), as in a conventional digital projector architecture. The in-couplers have been chosen to be slanted gratings for their ability to act on a specific spectral range while letting the remaining spectrum unaffected in the zero order, to be processed by the next in-coupler area located on the guide below, and to do this for all three colors. Such uncoated slanted gratings work both in transmission and reflection modes but can be optimized to work more efficiently in a specific mode. The out-couplers here are also slanted gratings, which can be tuned to effectively work over a specific incoming angular range (TIR range) and leave the see-through field quasi-unaffected. The part of the see-through field that is indeed diffracted by the out-couplers is trapped by TIR and does not make it to the eyebox. These gratings are modulated in depth to provide a uniform eyebox to the user. Note the symmetric in- and out-coupler configuration compensating the spectral spread over the three LEDs bands.

The redirection gratings are not shown here. Input and output grating slants are set close to 45 deg and the redirection grating slants at half this angle. The periods of the gratings are tuned in each guide to produce the right TIR angle for the entire FOV for that specific color (thus the same central diffraction angle in each guide for each RGB LED color band).

Fig. 19 depicts the waveguide combiner architecture used in the Magic Leap One MR headset (2018). The display engine is located on the same side as the eyebox. The input pupils are spatially color-demultiplexed, carrying the entire FOV at infinity (here again, only two colors and the central field are depicted for clarity).

Today, typical yields for slanted gratings produced by Nano-Imprint Lithography (NIL) are typically in the 90% for indices lower than 1.8, for which the process is well defined. For refractive indices of 1.9 and 2.0, the yields are quite lower today. Various contract manufacturers are today using NIL equipment (EVG, Suss, Canon, etc…) to produce volume quantities of waveguides on either plates or wafers.

Spatial color de-multiplexing can be done conveniently with a color sequential LCoS display mode for which the illumination LEDs are also spatially de-multiplexed. In this configuration, the input grating couplers are strong blazed gratings, coated with a reflective metal (such as Al). They do not need to work over a specific single-color spectral width since the colors are already de-multiplexed. The out-couplers are simple top-down binary gratings, which are also depth modulated to produce a uniform eyebox for the user. These binary gratings are shallow, acting therefore

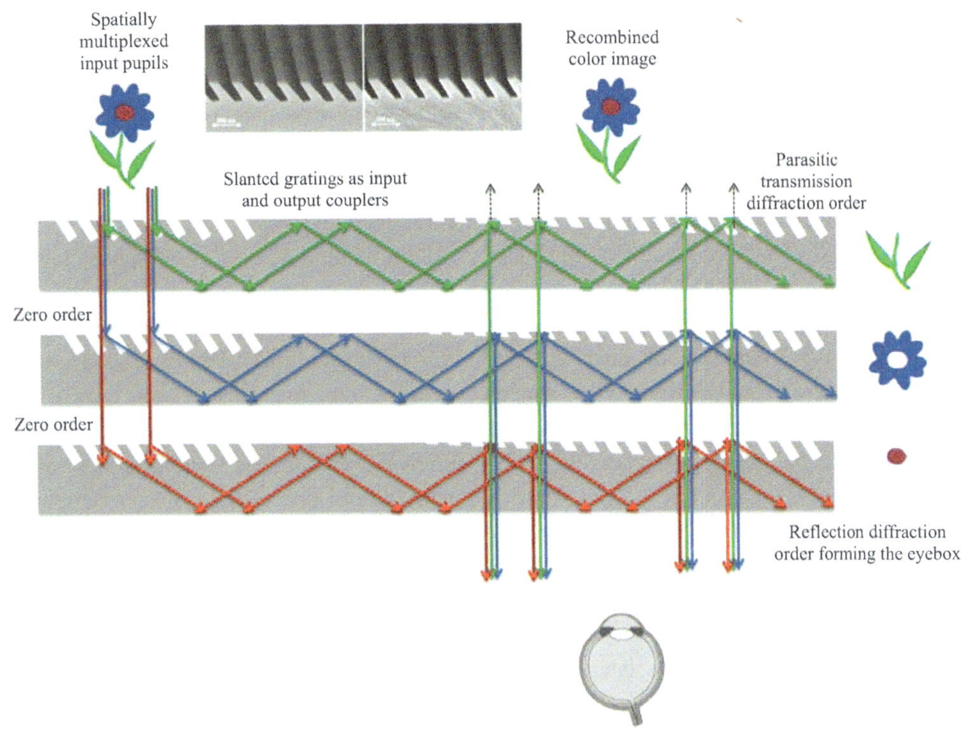

Fig. 18 Spatially color-multiplexed input pupils with slanted gratings as in- and out-couplers working in transmission and reflection mode (HoloLens 1 MR headset).

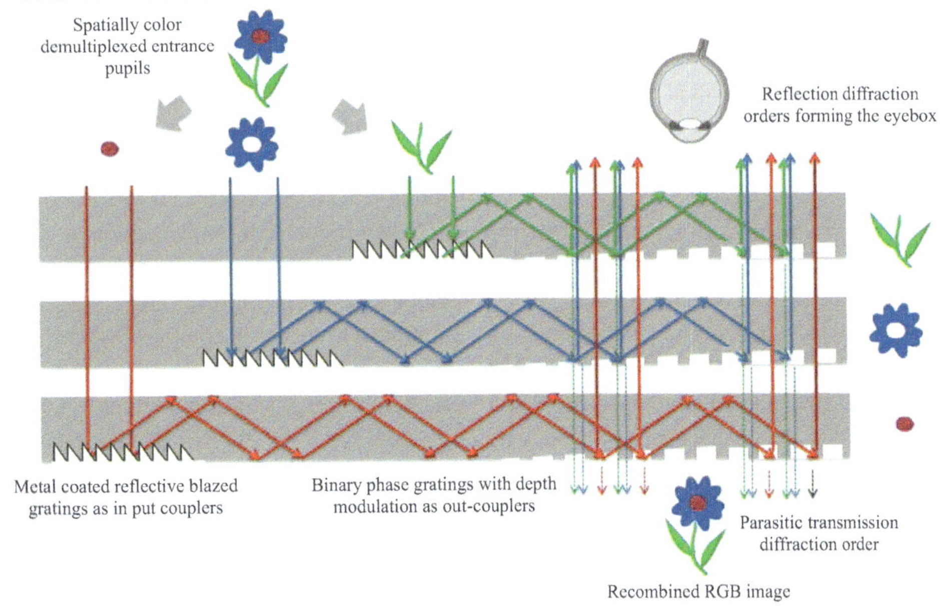

Fig. 19 Spatially color-de-multiplexed input pupils with 100% reflective blazed gratings as in-couplers and binary phase gratings as out-couplers (Magic Leap One MR headset).

very little on the see-through, but they have much stronger efficiency when working in internal reflection diffraction mode, since the optical path length in this case is longer by a factor of $2n\cos(\alpha)$ than in transmission mode, (where n is the index of the guide, and α is the angle if there is incidence in the guide). As in the HoloLens 1, most of the see-through field diffracted by the out-couplers is trapped by TIR.

The redirection gratings (not shown here) are also composed of binary top-down structures. The periods of the gratings are tuned in each guide to produce the right TIR angle for the entire FOV for that specific color (same central diffraction angles for each RGB LED color band).

Other companies use multilevel and/or quasi-analog surface relief diffractive structures to implement in- and out-couplers (see Fig. 17, Fig. 15). This choice is mainly driven by the complexity of the extraction gratings, acting both as redirection gratings and out-coupler gratings, making them therefore more complex than linear or slightly curved (powered) gratings, similar to iteratively optimized CGHs[39]. Allowing multilevel or quasi-analog surface relief diffractive structures increases the space bandwidth product of the element to allow more complex optical functionalities to be encoded with relatively high efficiency.

Resonant waveguide grating couplers (RWGs).

Resonant waveguide gratings (RWGs), also known as guided mode resonant (GMR) gratings or waveguide-mode resonant gratings[40,41], are dielectric structures where these resonant diffractive elements benefit from lateral leaky guided modes. A broad range of optical effects are obtained using RWGs such as waveguide coupling, filtering, focusing, field enhancement and nonlinear effects, magneto-optical Kerr effect, or electromagnetically induced transparency. Thanks to their high degree of optical tuning (wavelength, phase, polarization, intensity) and the variety of fabrication processes and materials available, RWGs have been implemented in a broad scope of applications in research and industry. RWGs can therefore also be applied as in- and out-couplers for waveguide gratings[42].

Fig. 20 shows an RWG on top of a lightguide (referred often incorrectly through the popular AR lingo as a "waveguide"), acting as the in- and out-couplers.

Roll-to-roll replication of such grating structures can help bring down overall waveguide combiner costs. The CSEM research center in Switzerland developed the RWG concept back in the 1980s, companies are now actively developing such technologies[43].

Metasurface couplers

Metasurfaces are a hot topic in research[44]: they can implement various optical element functionality in an ultra-flat form factor by imprinting a specific phase function over the incoming wavefront in reflection or transmission (or both) so that the resulting effect is refractive, reflective, or diffractive, or a combination of them. This phase imprint can be done through a traditional optical-path-difference phase jump or through Pancharatnam-Berry (PB) phase gratings/holograms.

Due to their large design space, low track length, and ability to render unconventional optical functions, metalenses could grow out of the lab to become a unique item in the engineer's bag of tools. If one can implement in a fabricable metasurface an optical functionality that cannot be implemented by any other known optical element (diffractives, holographics, or Fresnels), it is particularly interesting. For example, having a true achromatic optical element is very desirable not only in imaging but also in many other tasks such as waveguide coupling. Another example is ultra-low track length focal stack for IR cameras from Metalenz Corp. Additionally, if one can simplify the fabrication and replication process by using metasurfaces, the design for manufacturing (DFM) can be compelling. However, optical efficiencies, design tools, and large scale fabrication will need to continue to improve and find their way into product.

Achromatic coupler technologies

Waveguide combiners could benefit greatly from a true achromatic coupler functionality: in- and/or out-coupling RGB FOVs, and matching each color FOV to the maximum angular range (FOV) dictated by the waveguide TIR condition. This would reduce the complexity of multiple waveguide stacks for RGB operation.

When it comes to implementing a waveguide coupler as a true achromatic grating coupler, one can either use embedded partial mirror arrays (as in the Lumus LOE combiner), design a complex hybrid refractive/diffractive prism array, or even record phase-multiplexed volume holograms in a single holographic material. However, in the first case, the 2D exit pupil expansion implementation remains complex, in the second case, the microstructures can get very complex and thick, and in the third case the diffraction efficiency can drop dramatically (as in the Konica Minolta or Sony RGB photopolymers combiners,

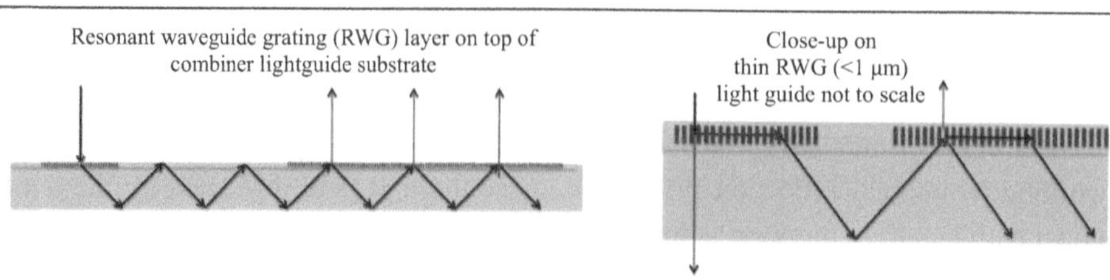

Fig. 20 Resonant Waveguide Gratings as in- and out-couplers on a waveguide combiner.

or in the thick Akonia holographic dual photopolymer combiner, now part of Apple, Inc.).

It has been recently demonstrated in literature that metasurfaces can be engineered to provide a true achromatic behavior in a very thin surface with only binary nanostructures. It is easier to fabricate binary nanostructures than complex analog surface relief diffractives, and it is also easier to replicate them by nanoimprint lithography (NIL) or soft lithography and still implement a true analog diffraction function as a lens or a grating. The high index contrast required for such nanostructures can be generated by either direct imprint in high index inorganic spin-on glass or by NIL resist lift-off after an atomic layer deposition (ALD) process. Direct dry etching of nanostructured remains a costly option for a product.

It is important to remember that metasurfaces or thick volume holograms are not inherently achromatic elements, and never will be. However, when many narrow band diffraction effects are spatially or phase multiplexed in a metasurface or a thick volume hologram, their overall behavior over a much larger spectral bandwidth can effectively lead the viewer to think they are indeed achromatic: although each single hologram or metasurface operation are strongly dispersive, their cascaded contributions may result in a broadband operation which looks achromatic to the human eye (e.g. the remaining dispersion of each individual hologram or metasurface effect affecting a spectral spread that is below human visual acuity-one arcmin or smaller). It is also possible to phase multiplex surface relief holograms to produce achromatic effects, but more difficult than with thick volume holograms or thin metasurfaces.

Mirrors are of course perfect achromatic elements and will therefore produce the best polychromatic MTF (such as with Lumus LOE combiners or LetinAR pin mirror waveguides).

World side light leakage and rainbow effects with waveguide combiners

As we have reviewed in this section, world side leakage (also commonly referred to as "eye glow") is a parasitic display feature that in some instances can block the eye contact when the display is on, and furthermore provide a very unnatural glow replacing the headset wearer's eyes. Not all waveguides have same world side leakage. We classify in order the various waveguide combiner technologies below.

Waveguide technologies classification for world side leakage (and world side display rainbow effect):

- Surface relief gratings with binary profiles (about 50% of the eye side leakage).
- Surface relief gratings with slanted profiled (about 40% of the eye side leakage)
- Holographic grating couplers (about 8-10% of the eye side leakage).
- Reflective waveguides (5% without AR coating and <1% with AR coating)

Summary of waveguide coupler technologies

Table 1 summarizes the various waveguide coupler technologies reviewed here, along with their specifics and limitations.

Although this table shows a wide variety of optical couplers, most of today's AR/MR/smart glass products are based on only a handful of traditional coupler technologies such as thin volume holograms, slanted surface-relief gratings, and embedded half-tone mirrors. The task of the optical designer (or rather the product program manager) is to choose the right balance and the best compromise between coupling efficiency, color uniformity over the eyebox and FOV, mass production costs, and size/weight.

Fig. 21 shows the various coupler elements and waveguide architectures grouped in a single table, including surface relief grating couplers, thin holographic couplers, and thick holographic couplers in three, two, and single flat guides. For geometric waveguide combiners that use embedded mirrors or other reflective/refractive couplers (such as micro-prisms).

Design and Modeling of Optical Waveguide Combiners

Designing and modeling a waveguide combiner is very different from designing and modeling a freespace optical combiner. As conventional ray trace in standard optical CAD tools such as Zemax™, CodeV™, Fred™ or TracePro™ are sufficient to design effective free space and even TIR prism combiners, to design waveguide combiners, especially when using diffractive or holographic couplers, a hybrid ray-trace / rigorous Electro-Magnetic diffraction mode is usually necessary.

The modeling efforts is shared between two different tasks:

- Local, rigorous EM light interaction with micro- and anno-optics couplers (gratings, holograms, metasurfaces, RWGs, …)
- Global architecture design of the waveguide combiner, building up FOV, resolution, color and eyebox, by the use of more traditional ray trace algorithms.

Waveguide Coupler Design, Optimization and Modeling

Coupler/light interaction model

Modeling of the angular and spectral Bragg selectivity of

Table 1 Benchmark of various waveguide coupler technologies.

Waveguide coupler tech	Operation	Reflective coupling	Transmission coupling	Efficiency modulation	Lensed out-coupler	Spectral dispersion.	Color uniformity	Dynamically tunable	Polarization maintaining	Mass production	Company/ Product
Embedded mirrors	Reflective	Yes	No	Dielectric coatings	No	Minimal	Good	No	Yes	Slicing, coating, polishing,	Lumus Ltd.DK 50, Maximus 2D EPE
Micro-prisms	Reflective	Yes	No	Dielectric coatings	Yes	Minimal	Good	No	Yes	Injection molding	Optinvent, ORA
Surface relief slanted grating	Diffractive	Yes	Yes	Depth, Duty cycle, slant	Yes	Strong	Needs comp.	Possible with LC	No	NIL (wafer, plate)	Microsoft HoloLens, Vuzix Inc, Nokia,...
Surface relief blazed grating	Diffractive	Yes	No	Blaze/antiblaze angles	No	Strong	Needs comp.	Possible with LC	No	NIL (wafer, plate)	Magic Leap One,
Surface relief binary grating	Diffractive	Yes	Yes	Depth, Duty cycle	Yes	Strong	Needs comp.	Possible with LC	No	NIL (wafer, plate)	Magic Leap One
Multilevel surface relief grating	Diffractive	Yes	Yes	Depth, Duty cycle	Yes	Strong	Needs comp.	Possible with LC	Possible, but difficult	NIL (wafer,plate)	WaveOptics Ltd, BAE, Dispelix,
Thin photopolymer hologram	Diffractive	Yes	Yes	Index swing	Yes, but difficult	Strong	Needs comp.	Possible with shear	No	Contact print roll to roll	Sony Ltd, TruelifeOptics Ltd,
H-PDLC volume holographic	Diffractive	No	Yes	Index swing	Yes, but difficult	Strong	Need comp.	Yes (electrical)	No	Exposure	Digilens Corp. (MonoHUD)
Thick photopolymer hologram	Diffractive	Yes	Yes	Index swing	Yes, but difficult	Minimal	Need com;.	No	No	Multiple exposure	Akonia Corp (now Apple Inc.)
Resonant Waveguide Grating	Diffractive	Yes	Yes	Depth, Duty cycle	Yes	Can be mitigated	NA	Possible with LC	Possible	Roll to roll NIL	CSEM / Resonannt Screens
Metasurface coupler	Mostly diffractive	Yes	Yes	Various	Yes	Can be mitigated	Needs comp.	Possible with LC	Possible	NIL (wafer, plate)	Metalenz Corp.

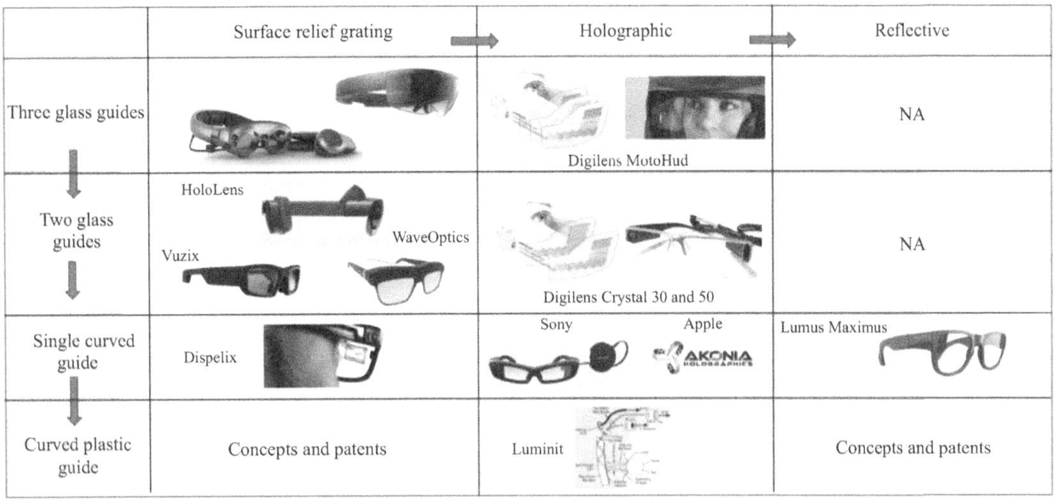

Fig. 21 Summary of waveguide combiner architectures with 1D or 2D EPE schemes.

volume holograms, thin or thick, in reflection and transmission modes, can be performed with the couple wave theory developed by Kogelnik in 1969[29].

Similarly, modeling of the efficiency of surface relief gratings can be performed accurately with rigorous coupled-wave analysis (RCWA)[33,34], especially the Fourier modal method (FMM). The finite difference time domain (FDTD) method—also a rigorous EM nanostructure modeling method—can in many cases be a more accurate modeling technique but also much heavier and more CPU time consuming. However, the FDTD will show all the diffracted fields, the polarization conversions, and the entire complex field, whereas the Kogelnik model and the RCWA will only give efficiency values for particular diffraction orders.

The FDTD can model non-periodic nanostructures, while RCWA can accurately model quasi-periodic structures. Thus, the FDTD might help with modeling *k*-vector variations (rolled *k*-vector) along the grating, slant, depths, and duty cycle variations, as well as random and systematic fabrication errors in the mastering and replication steps. The Kogelnik theory is best suited for slowly varying index modulations with moderate index swings (i.e., photopolymer volume holograms).

Free versions of the RCWA-FMM[35] and FDTD[32-35] codes can be found on the internet. Kogelnik theory can be easily implemented as a straightforward equation set for transmission and reflection modes. Commercial software suites implementing FDTD and RCWA are R-Soft from Synopsys and Lumerical.

These models predict the efficiency in each order for a single interaction of the light with the coupler element. In order to model the entire waveguide combiner, especially when a pupil replication scheme is used, conventional ray tracing optical design software can be used, such as Zemax, or more specific light-propagation software modules, such as the ones by LightTrans, Germany[37] (see Fig. 22 for ray tracing through 2D EPE grating waveguides).

The interaction of the EM field with the coupler regions (surface relief structures or index modulations) modeled through the RCWA or Kogelnik can be implemented via a dynamically linked library (DLL) in conventional optical design software based on ray tracing (e.g., C or Matlab code). As the FDTD numerical algorithm propagates the entire complex field rather than predicting only efficiency values (as in the RCWA or Kogelnik model), it is therefore more difficult to implement as a DLL.

Raytrace optimization of the high-level waveguide combiner architecture with accurate EM light/coupler interactions modeling are both required to design a combiner with good color uniformity over the FOV, a uniform eyebox over a target area at a desired eye relief, and high efficiency (in one or both polarizations). Inverse propagation from the eyebox to the optical engine exit pupil is a good way to simplify the optimization process. The design process can also make use of an iterative algorithm to optimize color over the FOV/eyebox and/or efficiency, or even reduce the space of the grating areas by making sure that no light gets lost outside the effective eyebox.

Waveguide couplers have specific angular and spectral bandwidths that affect both the FOV and the eyebox uniformity. A typical breakdown of the effects of a 2D EPE waveguide architecture on both spectral and angular bandwidths on the resulting immersive display is shown in Fig. 23.

Fig. 24 shows that the coupler's spectral and angular

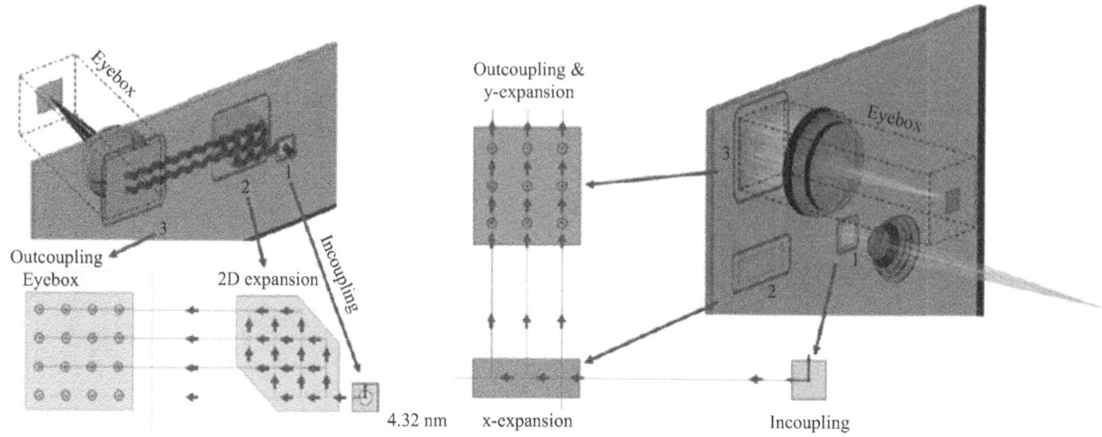

Fig. 22 Waveguide grating combiner modeling by LightTrans (Germany) in 2D EPE version.

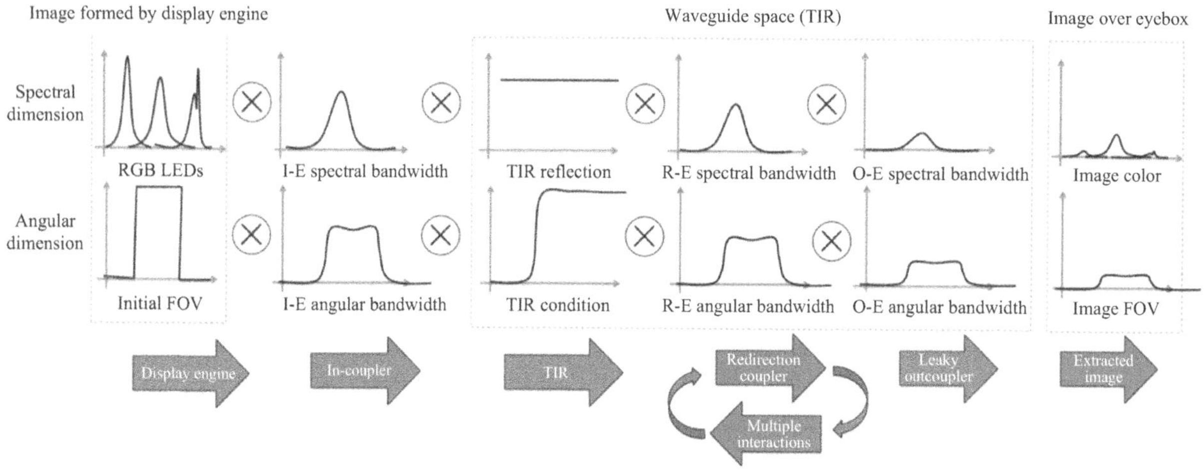

Fig. 23 Cascaded effects of the field/coupler interactions on the FOV uniformity.

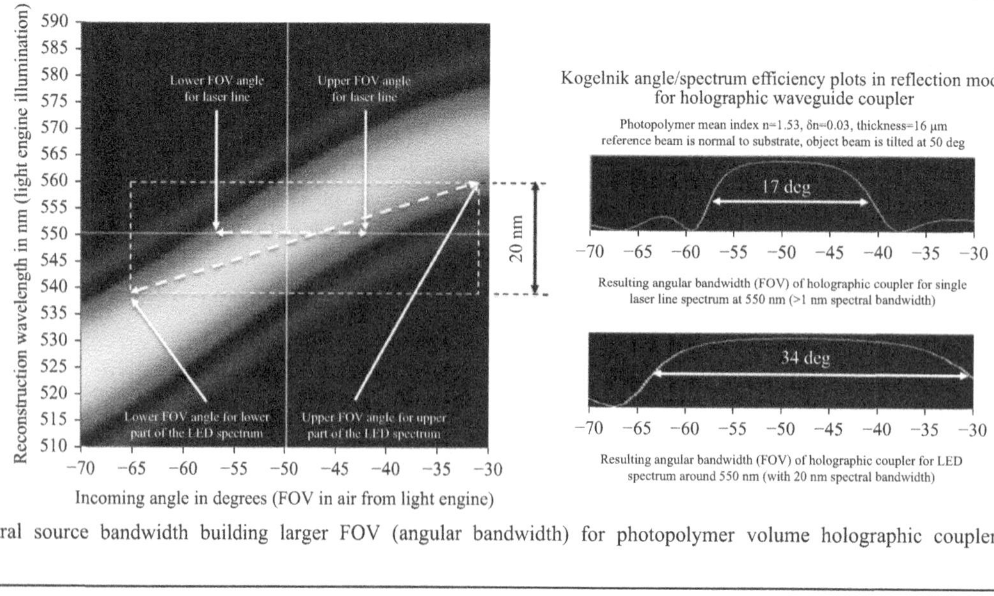

Fig. 24 Spectral source bandwidth building larger FOV (angular bandwidth) for photopolymer volume holographic coupler in waveguide combiners.

bandwidths are critical to the FOV uniformity, especially color uniformity. While embedded mirrors and micro-prisms have a quasi-uniform effect on color and FOV, others do not, such as gratings and holograms. Therefore, it is interesting to have the flattest and widest spectral and angular bandwidths possible. For volume holograms, this means operating in reflection mode and having a strong index swing (Kogelnik), and for surface gratings, this means a high index (as predicted b the RCWA-FMM or FDTD). The angular bandwidth location can be tuned by the slant angle in both holograms and surface gratings. Multiplexing bandwidths can help to build a larger overall bandwidth, bot spectral and angular, and is used in various implementations today. Such multiplexing can be done in phase, in space or in time, or a combination of the above. Finally, as spectral and angular bandwidths are closely linked, altering the spectral input over the field can have a strong impact on FOV and vice versa.

Polarization and degree of coherence are two other dimensions one should need to investigate especially when lasers or VCSELS are used in the optical engine or if polarization maintaining (or rather polarization conversion) is required. The multiple interactions in the R-E regions can produce multiple miniature Mach–Zehnder interferometers, which might modulate the intensity of the particular fields.

Increasing FOV by using illumination spectrum

The ultimate task for a holographic or grating coupler is to provide the widest FOV coupling possible, matching the FOV limit dictated by the TIR condition in the underlying waveguide (linked to the refractive index of the waveguide material).

We have seen that volume holographic combiners have been used extensively to provide a decent angular in and out-coupling into the guide. However, most of the available holographic materials today have a low index swing and thus yield a relatively small angular bandwidth in the propagation direction. In this case the FOV bottleneck is the coupler, not the TIR condition in the waveguide.

A typical Kogelnik efficiency plot in the angular/spectral space for a reflection photopolymer volume holographic coupler is show in Fig. 24 (spectral dimension vertical and angular dimension horizontal).

The hologram specifications and exposure set-up in Fig. 24 are listed below:

- Mean holographic material index: 1.53,
- Holographic index swing: 0.03,
- Photopolymer thickness: 16 microns,
- Operation mode: reflective,
- Polarization: ('s' but very little change when moving to 'p' polarization),
- Design wavelength: 550 nm,
- Reconstruction wavelength: LED light from 540–560 nm (20-nm bandwidth),
- Normal incidence coupling angle: 50 deg in air.

When using a laser (<1-nm line) as a display source (such as in a laser MEMS display engine), the max FOV is the horizontal cross-section of the Kogelnik curved above (17-deg FWHM). However, when using the same color as an LED source (20 nm wide, such as in an LED-lit LCoS micro-display light engine, the resulting FOV is a slanted cross-section (in this case increased to 34-deg FWHM), and a 2× FOV gain is achieved without changing the waveguide index or the holographic coupler, only the illumination's spectral characteristics.

However, this comes at the cost of color uniformity: the lower angles (left side of the FOV) will have more contributions from the shorter wavelengths (540 nm), and the higher angles (right side of the FOV) will have more contributions from the longer wavelengths (560 nm). This slight color non-uniformity over the FOV is typical for volume holographic couplers.

Increasing FOV by optimizing grating coupler parameters

Unlike holographic couplers, which are originated and replicated by holographic interference in a phase change media (see previous section), surface-relief gratings (SRGs) are rather originated by traditional IC lithographic techniques and replicated by nano-imprint lithography (NIL) or soft lithography. The topological structure of the gratings can therefore be optimized digitally to achieve the best functionality in both spectral and angular dimensions. Topological optimization needs to account for DFM (design for manufacturing) and typical lithographic fabrication limitations. The angular bandwidth of an SRG coupler (i.e., the FOV that can be processed by this SRG) can be tuned by optimizing the various parameters of such a grating structure, such as the front and back slant angles, the grating fill factor, the potential coating(s), the grating depth, and of course the period of the grating (Fig. 25). Additional material variables are the refractive indices of the grating structure, grating base, grating coating, grating top layer, and underlying waveguide.

The SRG grating parameters can be optimized to provide a larger FOV, albeit with a lower overall efficiency, matching better the available angular bandwidth provided by the TIR condition in the guide. Lower efficiency is okay over the out-couplers since they are tuned in the low-efficiency range to produce a uniform eyebox (the in-coupler, however, needs to be highly efficient since there is only one grating interaction to couple the entire field into TIR mode).

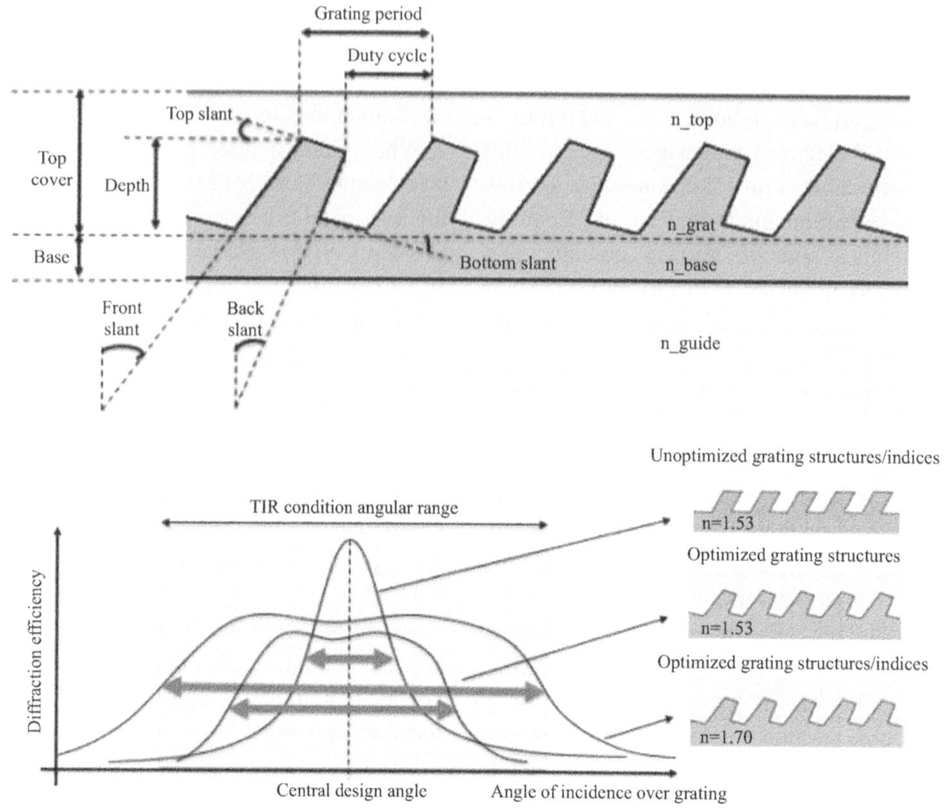

Fig. 25 Optimizing the grating parameters to optimize color uniformity over the FOV.

Calculations of coupling efficiency have been carried out with an RCWA FMM algorithm and topological optimization by a steepest descent algorithm. Note that both unoptimized and optimized gratings have the same grating periods as well as the same central slant angle to position respectively the spectral and the angular bandwidths on identical system design points (with the FOV generated by the display engine and wavelength of the illumination source).

The bottleneck in FOV with the unoptimized grating structure is not the TIR condition (i.e., the index of the waveguide) but rather the grating geometry and the index of the grating. The angular bandwidth of the optimized grating should overlap the angular bandwidth of the waveguide TIR condition for best results over the largest possible FOV. Also, a "top hat" bandwidth makes the color uniformity over the FOV less sensitive to systematic and random fabrication errors in the mastering and the NIL replication of the gratings. Increasing the index of the grating and reducing the back slant while increasing the front slant angle can provide such an improvement.

Additional optimizations over a longer stretch of the grating can include depth modulations, slant modulations (rolling *k*-vector), or duty cycle modulations to produce an even wider bandwidth over a large, uniform eyebox.

Using dynamic couplers to increase waveguide combiner functionality

Switchable or tunable TIR couplers can be used to optimize any waveguide combiner architecture, as in

- increasing the FOV by temporal sub-FOV stitching at double the refresh rate,
- increasing the brightness at the eye by steering a reduced size eyebox to the pupil position (thus increasing the perceived eyebox size), and
- increasing the compactness of the waveguide combiner by switching multiple single-color couplers in color sequence in a single guide.

Dynamic couplers can be integrated in various ways: polarization diversity with polarization-dependent couplers (the polarization switching occurring in the optical engine), reconfigurable surface acoustic wave (SAW) or acousto-optical modulator (AOM) couplers, electro-optical (EO) modulation of buried gratings, switchable surface-relief gratings in an LC layer, switchable metasurfaces in an multilayer LC layer, tunable volume holograms (by shearing, pressure, pulling), or switchable holographic-polymer dispersed liquid crystals (H-PDLC), as in Digilens' volume holographic couplers.

High-Level Waveguide-Combiner Design

The previous section discussed ways to model and optimize the performance of individual couplers, in either grating or holographic form. We now go a step further and look at how to design and optimize the overall waveguide combiner architectures.

Choosing the waveguide coupler layout architecture

We have seen that couplers can work in either transmission or reflection mode to create a more diverse exit-pupil replication scheme (producing a more uniform eyebox) or to improve the compactness of the waveguide by using both surfaces, front and back. The various couplers might direct the field in a single direction or in two or more directions, potentially increasing the FOV that can propagate in the waveguide without necessarily increasing its index.

Fig. 26 shows how the optical designer can expand the functionality of in- or out-couplers, with architectures ranging from bi-dimensional coupling to dual reflective/transmission operation in the same guide with sandwiched volume holograms or top/bottom grating couplers.

More complex and more functional coupler architectures have specific effects on MTF, efficiency, color uniformity, and FOV. For example, while the index of the guide allows for a larger FOV to propagate, the index of the grating structures in air would increase the spectral and angular bandwidths to process a larger FOV without compromising color uniformity or efficiency. The waviness of the waveguide itself will impact the MTF as random cylindrical powers added to the field. Multiple stacked waveguides might be efficient at processing single colors, but their misalignment will impact the MTF as misaligned color frames. Similarly, hybrid top/bottom couplers will affect the MTF if they are not perfectly aligned (angular alignment within a few arc seconds).

Building a uniform eyebox

As the TIR field gets depleted when the image gets extracted along the out-coupler region, the extraction efficiency of the out-coupler needs to gradually increase in the propagation direction to produce a uniform eyebox. This complicates the fabrication process of the couplers, especially when the gradual increase in efficiency needs to happen in both pupil replication directions.

For volume holograms, the efficiency can be increased by a stronger index swing in the photopolymer or PDLC (through a longer exposure or a thickness modulation). For surface relief gratings, there are a few options, as shown in Fig. 27. This is true for the redirection grating (R-E) as well as the out-coupler (O-E).

Groove depth and duty cycle modulation can be performed on all type of gratings, binary, multilevel, blazed, and slanted. Duty cycle modulation has the advantage of modulating only the lateral structures, not the

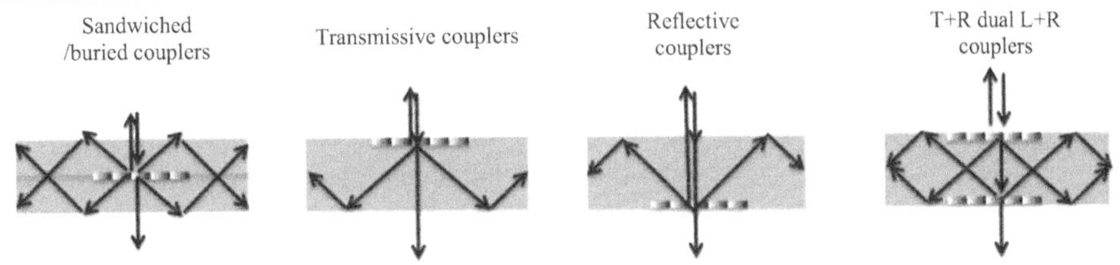

Fig. 26 More functional coupler architectures that yield compact and efficient waveguide combiners.

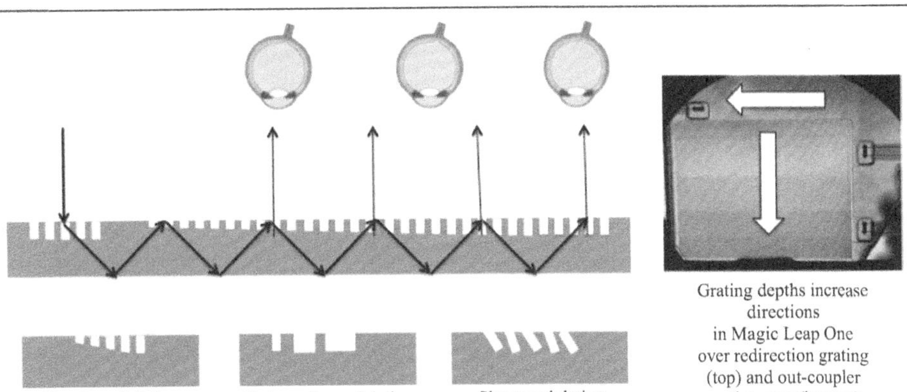

Fig. 27 Modulation of the outcoupling efficiency to build up a uniform eyebox.

depth, which makes it an easier mastering process. Modulating the depth of the gratings can be done in binary steps (as in the Magic Leap One, Fig. 28– right) or in a continuous way (Digilens waveguide combiners).

Grating front- and back-slant angle modulation (in a single grating period or over a larger grating length) can change the angular and spectral bandwidths to modulate efficiency and other aspects of the coupling (angular, spectral, polarization). Periodic modulation of the slant angles is sometimes also called the "rolling *k*-vector" technique and can allow for larger FOV processing due to specific angular bandwidth management over the grating area. Once the master has been fabricated with the correct nanostructure modulation, the NIL replication process of the gratings is the same no matter the complexity of the nanostructures (caution is warranted for slanted gratings where the NIL process must resolve the undercut structures; however, the slanted grating NIL process (with slants up to 50 deg) has been mastered by many foundries around the world)[37].

Spectral spread compensation in diffractive waveguide combiners

Spectral spread comes to mind as soon as one speaks about gratings or holographic elements. It was the first and still it is the main application pool for gratings and holograms: spectroscopy. Spectral spread is especially critical when the display illumination is broadband, such as with LEDs (as in most of the waveguide grating combiner devices today, such as the HoloLens 1, Vuzix, Magic Leap, Digilens, Nokia, etc.), with a notable difference in the HoloLens 2 (laser MEMS display engine). The straightforward technique to compensate for the inevitable spectral spread is to use a symmetric in-coupler/out-coupler configuration, in which the gratings or holograms work in opposite direction and thus compensate in the out-coupler any spectral spread impacted in the in-coupler (Fig. 28).

Although the spectral spread might be compensated, one can notice on Fig. 28 that the individual spectral bands are spatially de-multiplexed at the exit ports while multiplexed at the entry port. Strong exit-pupil replication diversity is thus required to smooth out any color non-uniformities generated over the eyebox.

This symmetric technique might not be used to compensate for spectral spread across different colors (RGB LEDs) but rather for the spread around a single LED color. The spread across colors might stretch the RGB exit pupils too far apart and reduce the FOV over which all RGB colors can propagate by TIR.

The pupil replication diversity can also be increased by introducing a partial reflective layer in the waveguide (by combining two plates with a reflective surface), thus producing a more uniform eyebox in color and field.

Field spread in waveguide combiners

The different fields propagating by TIR down the guide are also spread out, no matter the coupler technology (mirrors, prisms, gratings, holograms, etc.), see Fig. 29.

A uniform FOV (i.e., all fields appearing) can be formed over the eyebox with a strong exit pupil diversity scheme. This is a concept often misunderstood as in many cases only one field is represented when schematizing a waveguide combiner. Fig. 29 shows the field spread occurring in a diffractive waveguide combiner. The number of replicated fields is also contingent on the size of the human eye pupil. If the ambient light gets bright, i.e., the human eye pupil gets smaller, then only part of the FOV might appear to the user, missing a few fields.

Focus spread in waveguide combiners

When a pupil replication scheme is used in a waveguide combiner, no matter the coupler, the input pupil needs to be formed over a collimated field (image at infinity/far field). If the focus is set to the near field instead of the far field in the display engine, each waveguide exit pupil will produce an image at a slightly different distance, thereby producing a mixed visual experience, overlapping the same image with different focal depths. It is quasi-impossible to compensate for such focus shift over the exit pupils because of both spectral spread and field spread over the

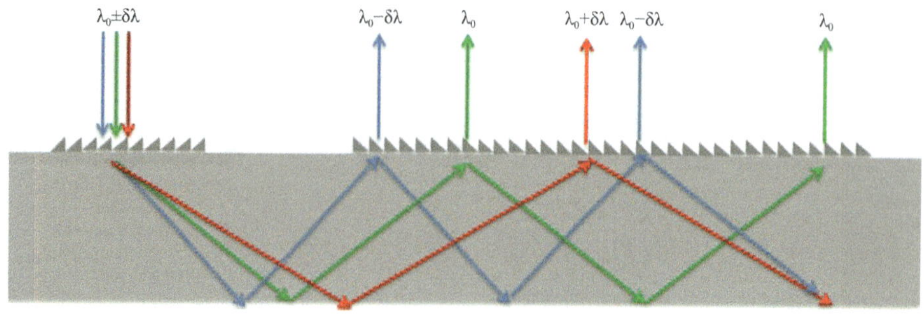

Fig. 28 Spectral spread compensation in a symmetric in-coupler/out-coupler waveguide combiner.

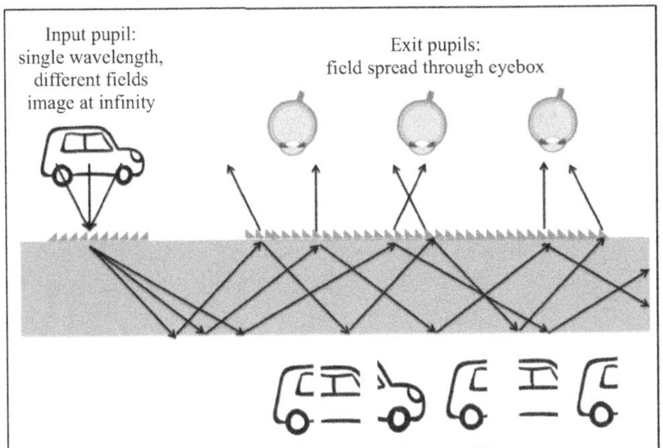

Fig. 29 Fractional field spread in a waveguide combiner.

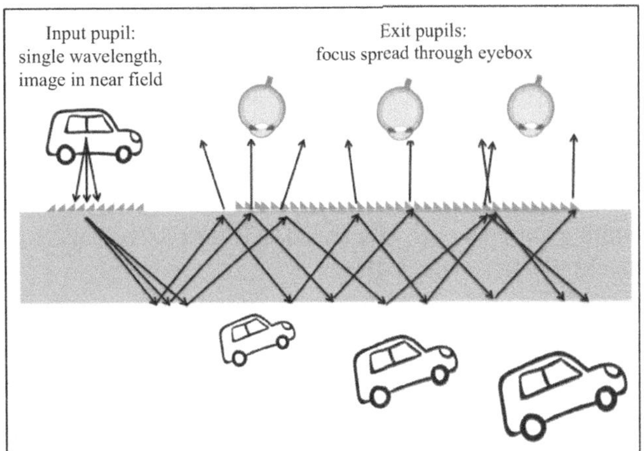

Fig. 30 Focus spread in a waveguide combiner with a non-collimated input field.

exit pupils, as discussed previously. Fig. 30 shows such a focus spread over the eyebox from an input pupil over which the image is formed in the near field.

The image over the input pupil can, however, be located in the near field when no pupil replication scheme is performed in the guide, such as in the Epson Moverio BT300 or in the Zeiss "Tooz" Smart Glass (yielding a small FOV and small eyebox).

When pupil replication is used in the guide, the virtual image can be set at a closer distance for better visual comfort by using a static (or even tunable) negative lens acting over the entire eyebox. For an unperturbed see-through experience, such a lens needs to be compensated by its conjugate placed on the world side of the combiner waveguide. This is the architecture used in the Microsoft HoloLens 1 (2015)[37].

Another, more compact way would introduce a slight optical power in the O-E, so that this coupler takes the functionality of an off-axis lens (or an off-axis diffractive lens) rather than a simple linear grating extractor or linear mirror/prism array. Although this is difficult to implement with a mirror array (as in an LOE), it is fairly easy to implement with a grating or holographic coupler. The grating lens power does not affect the zeroth diffraction order that travels by TIR down the guide but affects only the out-coupled (or diffracted) field. The see-through field is also not affected by such a lensed out-coupler since the see-through field diffracted by such an element would be trapped by TIR and thus not enter the eye pupil of the user.

All three configurations (no lens for image at infinity, static lens with its compensator, and powered O-E grating) are shown in Fig. 31. The left part of the eyebox shows an extracted field with image at infinity, the center part shows

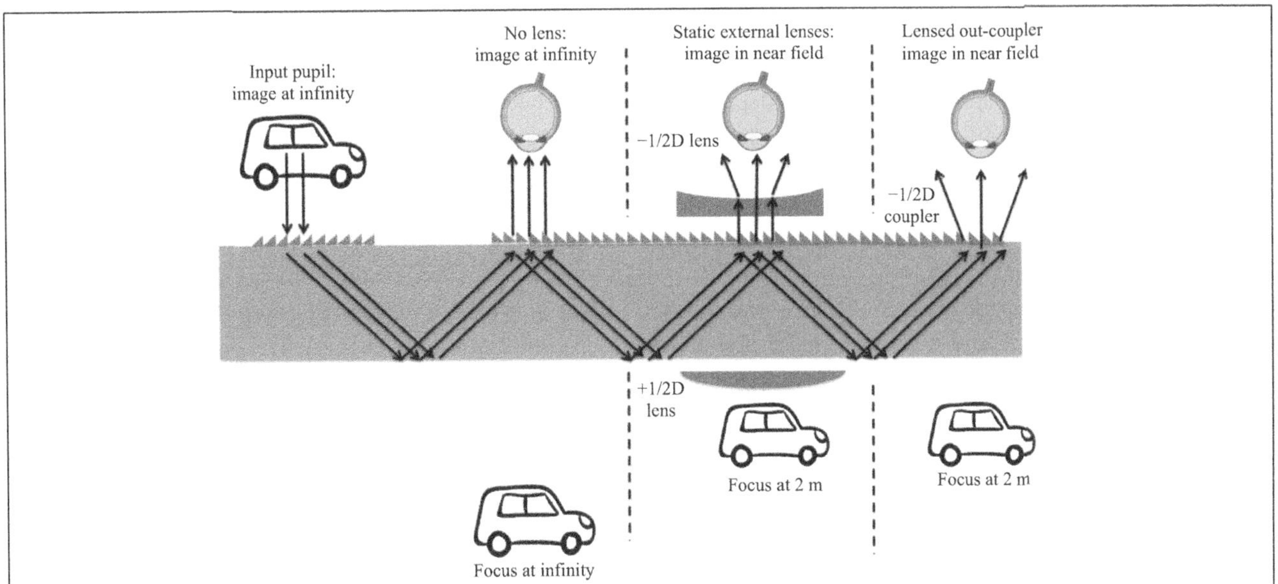

Fig. 31 Two out-coupler architectures positioning the virtual image in the near field over all exit pupils.

an extracted field with image at infinity that passes through a negative lens to form a virtual image closer to the user and its counterpart positive lens to compensate for see-through (as in the Microsoft HoloLens 1, 2015), and the right part of the eyebox shows an extracted field with the image directly located in the near field through a powered grating extractor (as with an off-axis diffractive lens, e.g., the Magic Leap One, 2018).

For example, a half-diopter negative lens power would position the original extracted far field image to a more comfortable 2-m distance, uniformly over the entire eyebox.

A powered out-coupler grating might reduce the MTF of the image, especially in the direction of the lens offset (direction of TIR propagation), since the input (I-E) and output (O-E) couplers are no more perfectly symmetric (the input coupler being a linear grating in both cases, and the out-coupler an off-axis diffractive lens). Thus, the spectral spread of the image in each color band cannot be compensated perfectly and will produce LCA in the direction of the lens offset. This can be critical when using an LED as an illumination source, but it would affect the MTF much less when using narrower spectral sources, such as lasers or VCSELs.

One of the main problems with such a lensed out-coupler grating configuration when attempting to propagate two colors in the same guide (for example, a two-guide RGB waveguide architecture is the generation of longitudinal chromatic aberrations (due to the focus changing with color since the lens is diffractive). Using a single color per guide and a laser source can greatly simplify the design task.

Propagating full color images in the waveguide combiner over a maximum FOV

We have seen in the previous paragraphs that the spectral spread of grating and holographic couplers can be perfectly compensated with a symmetric in- and out-coupler configuration. This is possible over a single color band but will considerably reduce the FOV if used over the various color bands (assuming that the couplers will work over these various spectral bands).

In order to maximize the RGB FOV in a waveguide combiner, one solution is to use stacked guides optimized each for a single color band (see Fig. 32), each coupling a maximum FOV by tuning the diffraction angle of the in- and out-couplers accordingly. This is the architecture used in both HoloLens 1 and Magic Leap One although the position of the input pupil (light engine) is opposite in both devices.

Air gaps between all plates are required to produce the TIR condition. Such gaps also allow for additional potential filtering in between plates for enhanced performance (such as spectral and polarization filtering).

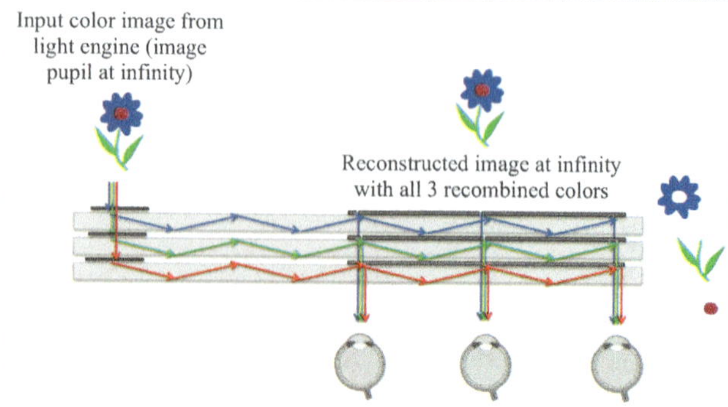

Fig. 32 Stacked waveguides combiners that provide the largest FOV TIR propagation over three colors.

Fig. 33 shows the functional diagram of such a single-color plate as a top view as well as its *k*-vector space depiction. Here again, I-E refers to the in-coupler, R-E refers to the leaky 90-deg redirection element, and O-E refers to the leaky out-coupler that forms the final eyebox (for 2D pupil replications).

Note that the entire FOV is shown on the *k*-vector diagram, but only a single field (central pixel in the FOV, with entry normal to the guide) is shown in the eyebox expansion schematic..

The FOV in the direction of the in-coupling can be increased by a factor of two when using a symmetric in-coupling configuration[44] in which the input grating or hologram (or even prism(s)) would attempt to couple the entire FOV to both sides.

As the TIR angular range does not support such an enlarged FOV, part of the FOV is coupled to the right and part of the FOV is coupled to the left. Due to the opposite directions, opposite sides of the FOV travel in each direction. If such TIR fields are then joined back with a single out-coupler, the original FOV can be reconstructed by overlapping both partial FOVs, as in Fig. 34.

In the orthogonal direction, the FOV that can be coupled by TIR remains unchanged. This concept can be taken to more than one dimension, but the coupler space on the waveguide can become prohibitive.

Waveguide-coupler lateral geometries

We have reviewed the various coupler technologies that can be used in waveguide combiners, as well as the 2D exit pupil expansion that can be performed in waveguide combiners. Waveguide combiners are desirable since their thickness is not impacted by the FOV, unlike other combiner architectures such as free-space or TIR prisms. However, the lateral dimensions of the waveguide (especially the redirection coupler and out-coupler areas over the waveguide) are closely linked to size of the in-

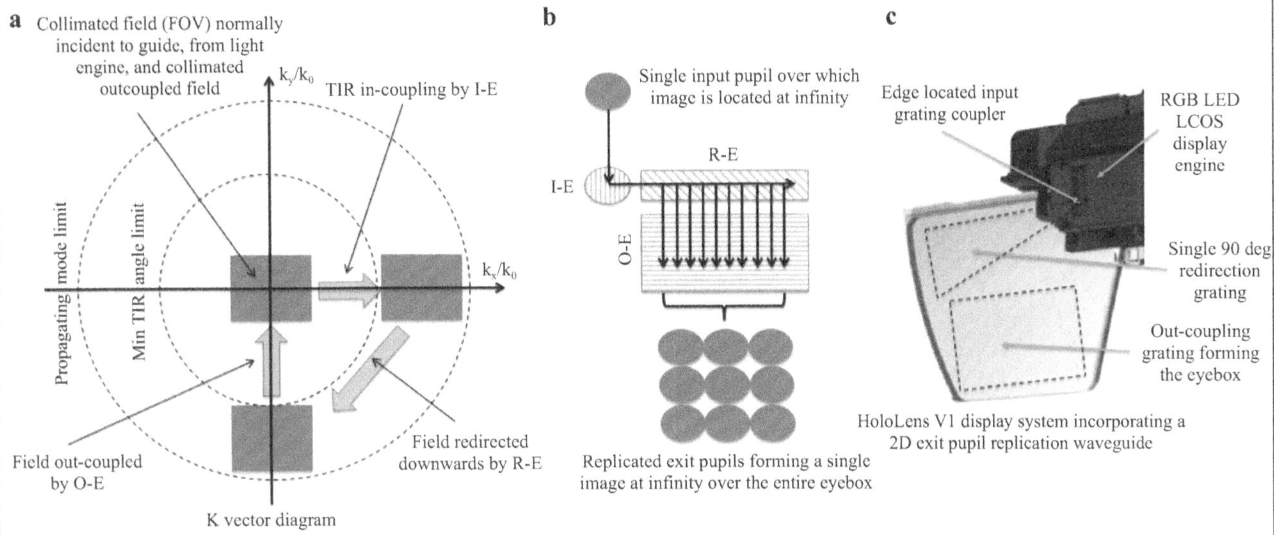

Fig. 33 k-vector diagram and lateral pupil replication layout for a single guide and single color.

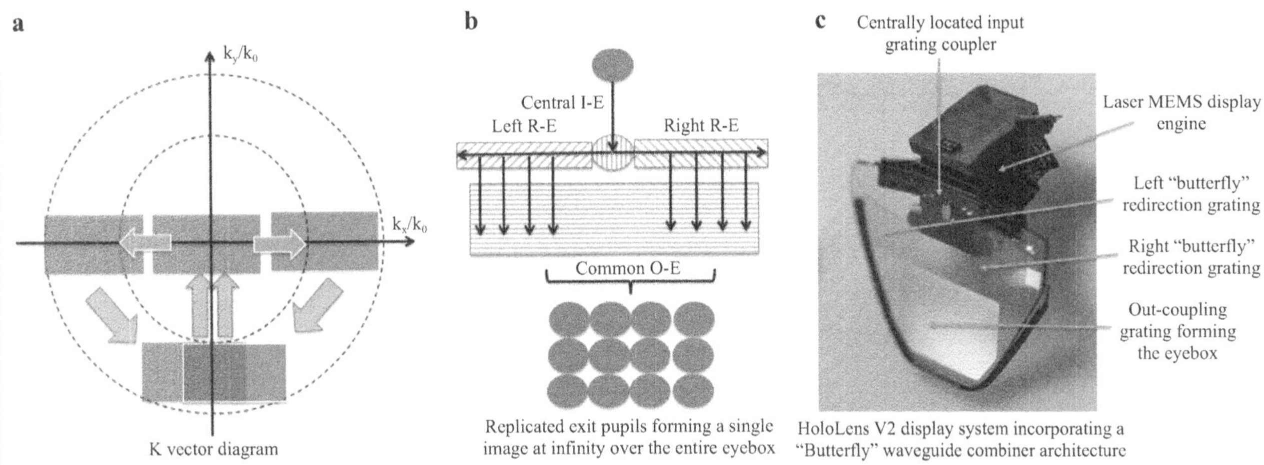

Fig. 34 Symmetric in-coupling for FOV increase in the direction of in-coupling.

coupled FOV, as shown in Fig. 35. For example, the R-E region geometry is dictated by the FOV in the waveguide medium: it expands in the direction orthogonal to the TIR propagation, forming a conical shape.

The largest coupler area requirement is usually the out-coupler element (center), aiming at processing all FOVs and building up the entire eyebox. Eye relief also strongly impacts this factor. However, its size can be reduced in a "human-centric optical design" approach: the right part of the FOV at the left edge of the eyebox as well as the left part of the FOV at the right edge of the eyebox can be discarded, thus considerably reducing the size of the O-E without compromising the image over the eyebox. Note that in Fig. 34 the *k*-vector diagram (a) shows the FOV, whereas the lateral schematics of the waveguide in (b) and (c) show the actual size of the coupler regions.

Reducing the input pupil can help to reduce the overall size and thickness of the combiner. However, the thickness of the guide must be large enough not to allow for a second I-E interaction with the incoming pupil after the first TIR bounce. If there is a second interaction, then by the principle of time reversal, part of the light will be out-coupled and form a partial pupil (partial moon if the input pupil is circular) propagating down the guide instead of the full one. This is more pronounced for the smallest field angle, as depicted in Fig. 36.

However, if the polarization of the field is altered after the first TIR reflection at the bottom of the guide, the parasitic outcoupling can be reduced if the I-E is made to be highly polarization sensitive.

Reducing the waveguide thickness can also produce stronger pupil diversity over the eyebox and thus better

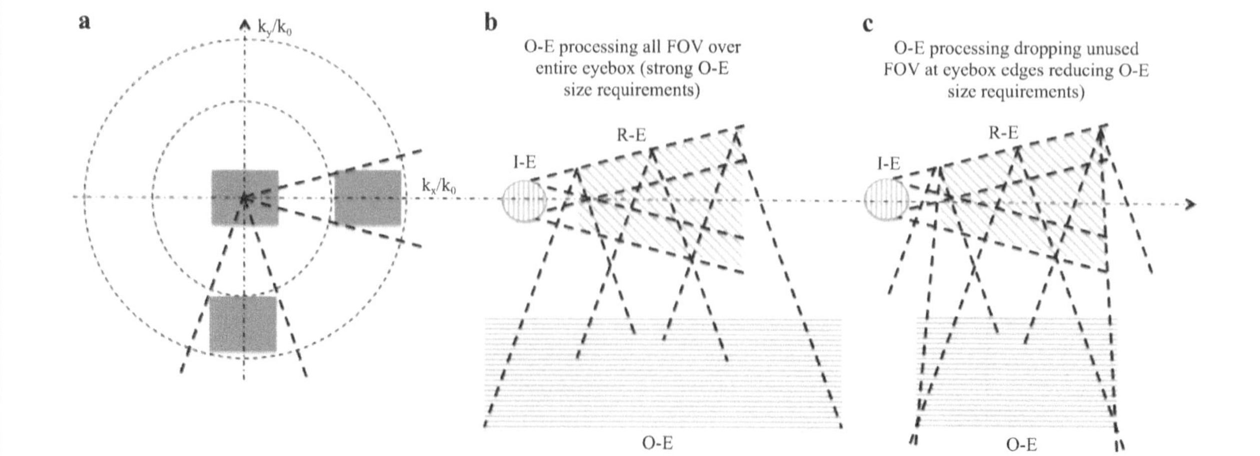

Fig. 35 Redirection and out-coupler areas as dictated by the in-coupled FOV.

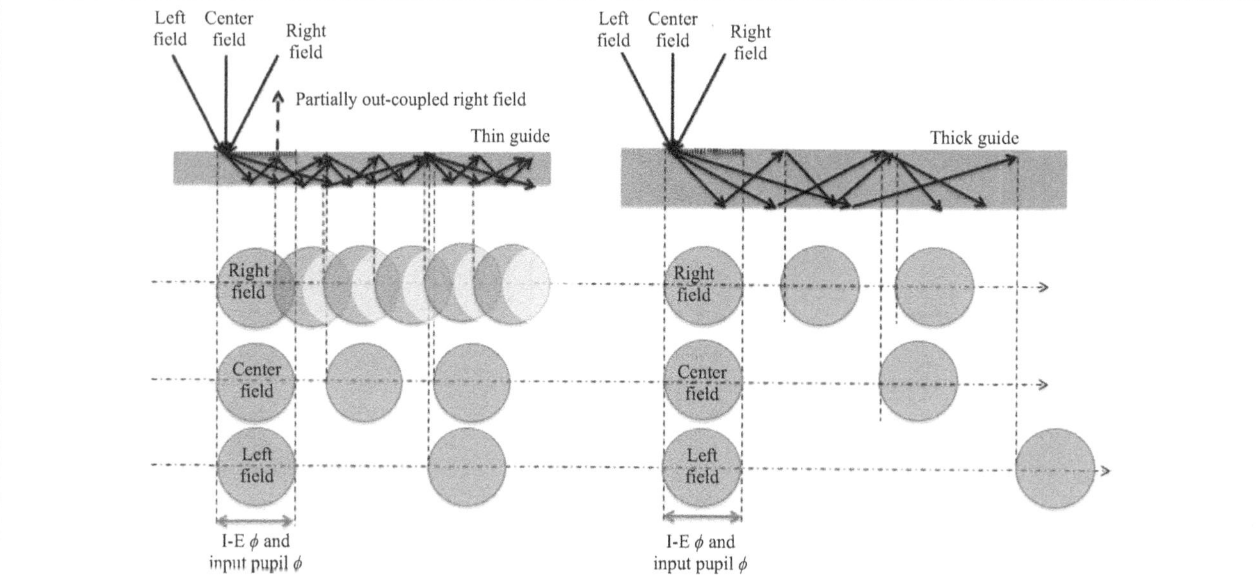

Fig. 36 Effects of the input pupil size (and size of the I-E) and thickness of guide on a single field TIR pupil bouncing down the guide.

eyebox uniformity. If reducing the guide is not an option (for parasitic out-coupling of the input pupil and also for etendue limitations in the display engine), a semi-transparent Fresnel surface can be used inside the guide (as in two guides bounded together), which would reflect only part of the field and leave the other part unperturbed, effectively increasing the exit pupil diversity.

Fig. 37 shows how the space of the out-coupler grating is dictated solely by the FOV and the eyebox. Note that many fields can be cancelled at the edges and towards the edges of the eyebox, as they will not enter the eye pupil (right fields on the left eyebox edge and left fields on the right eyebox edge). This can also reduce the size of the redirection grating considerably. This holds true for both eyebox dimensions.

Reducing the number of plates for RGB display and maintain FOV reach

Reducing the number of plates without altering the color of the image while propagating the maximum FOV allowed by the index of the guide is a desirable feature since it reduces the weight, size and complexity of the combiner, and make it also less prone to MTF reductions due to guide misalignments. Both lateral and longitudinal angular waveguide misalignments will contribute to a reduction of the MTF built by the display engine. Waveguide surface flatness issues are yet more cause for MTF reduction.

Due to the strong spectral spread of the in-coupler elements (gratings, holograms, RWGs, or metasurfaces), the individual color fields are coupled at higher angles as

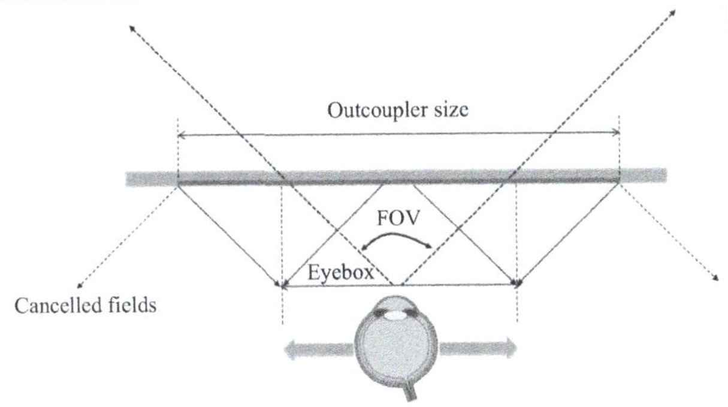

Fig. 37 Eyebox and FOV dictate the size of the out-coupler area.

the wavelength increases, which reduces the overall RGB FOV overlap that can propagate in the guide within the TIR conditions (smallest angle dictated by the TIR condition and largest angle dictated by pupil replication requirements for a uniform eyebox). This issue is best depicted in the *k*-vector diagram on the right side of Fig. 38.

A lower spectral spread, such as through a prism in-coupler, would increase the RGB FOV overlap in a single guide, such as in an LOE (embedded partial mirrors out-couplers) from Lumus or in the micro-prism array couplers from Optinvent.

The left configuration in Fig. 38 acts as a hybrid spatial/spectral filter, filtering the left part of the blue FOV, allowing the entire green FOV to be propagated (if the grating coupler periods have been tuned to match the green wavelength), and filtering the right part of the red FOV. The configuration in Fig. 38 propagates the entire RGB FOV (assuming the couplers can diffract uniformly over the entire spectrum) at the cost of the FOV extending in the direction of the propagation. However, when considering binocular vision, this limitation could be mitigated by engineering a symmetric color vigneting in each eye (particularly on blue and red), providing a uniform stereo color vision in a single RGB guide with high FOV (e.g., Dispelix Oy).

Recently, two plate RGB waveguide combiner architectures have been investigated, reducing by one-third the weight and size of traditional three-guide combiners, where the green FOV is shared between the top and bottom layer (see Fig. 39). Various companies are using this two-plate RGB waveguide combiner architecture today, including Vuzix, WaveOptics and Digilens.

However, this requires the grating (or holograms, RWGs, or metasurfaces) to be efficient over a larger spectral band, which implies that surface relief gratings are to be replicated in a higher refractive index, widening their spectral (and angular) bandwidths. High-index grating replication by NIL stretches the traditional wafer-scale NIL resin material science (inclusion of TiO_2 or ZrO_2 nanofiller particles). Nano-imprint at a Gen2 panel size of higher-index inorganic spin-on glass material might be the best fit, which also solves the resin or photopolymer reliability issues over various environmental conditions (temperature, pressure, shear, UV exposure, and humidity).

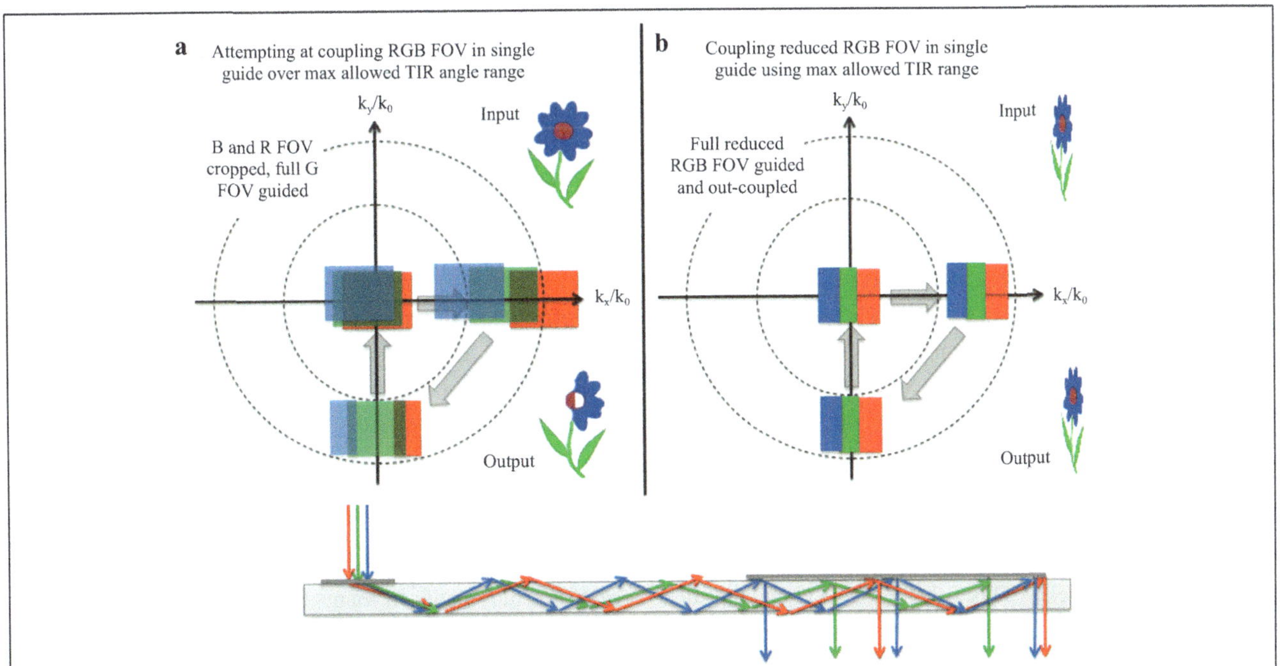

Fig. 38 k-vector diagram of a single-plate waveguide combiner using **a** RGB FOV coupling over a single-color TIR angular range condition and **b** RGB reduced FOV sharing the same TIR range.

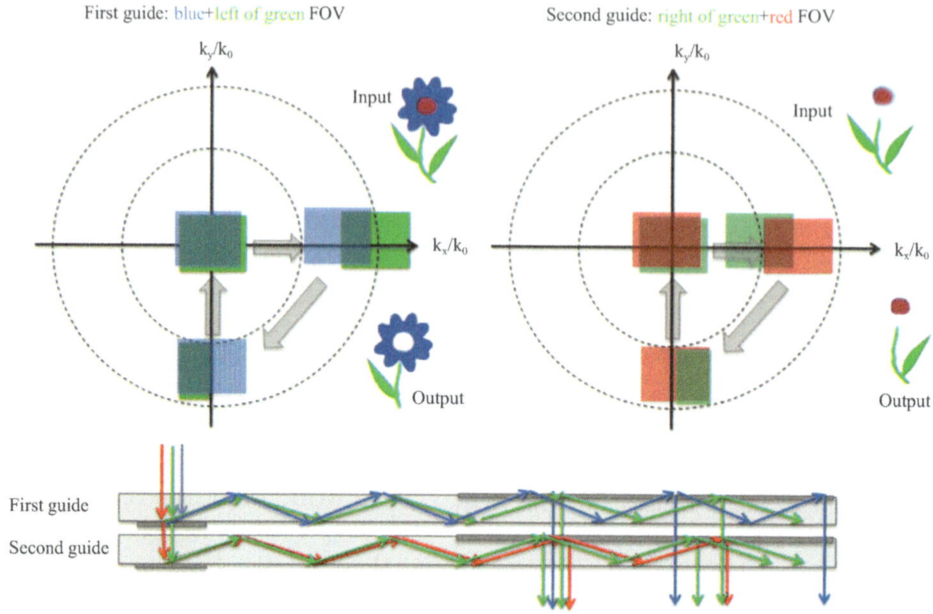

Fig. 39 Two-guide RGB waveguide combiner configuration.

This two-guide RGB configuration splits the green FOV in two at the in-coupler region and merges them again over the out-coupler region. For good color uniformity over the FOV and the eyebox, especially in the green field, this technique requires perfect control of the two-guide efficiency balance. Pre-emphasis compensation of the guide mismatch is possible using the display dynamic range, but this requires precise calibration, reduces the final color depth, and does not solve the stitching region issue where the two fields overlap.

An alternative to the architecture uses the first guide to propagate green and blue FOVs and the second guide to propagate only the red FOV, as green and blue are closer spectrally to each other than red. This change, however, slightly reduces the allowed FOV traveling without vignetting but solves the green FOV stitching problem.

Although going from three plates to two plates brings a small benefit in size, weight, and cost, the added complexity of the color split geometry and the resulting color non-uniformities over the eyebox might overshadow the initial small benefits.

A single-plate RGB waveguide combiner would provide a much stronger benefit, as there is no need to align multiple guides anymore, because everything is aligned lithographically by NIL inside the single plate (potentially also front and back). This would also yield the best possible MTF and the lowest costs.

One single-plate solution is to phase multiplex three different color couplers with three different periods into a single layer, and then tune it so that there is no spectral overlap (no color ghost images over the eyebox). Such phase multiplexing is theoretically possible in volume holograms. This might be achieved in the Akonia (now Apple) thick holographic material (500 microns). If a thinner photopolymer (less than 20 microns) is desired for better reliability and easier mass production, a large holographic index swing is required. Standard photopolymers can be panchromatic and can also be phase multiplexed, but the resulting efficiency remains low, and color cross-contamination between holograms is an additional issue. This is also theoretically possible with surface relief gratings, but it is difficult to simultaneously achieve high efficiency and a high extinction ratio over the three color bands. Metasurfaces and RWGs can theoretically produce such phase-multiplexed layers but with the same limitations.

Multifunctional waveguide combiners

Smart glasses are pushing for smaller form factors and thus asking to integrate more functionality of the waveguide than simple optical combining. This is particularly true in smart glasses as the waveguide is the largest area available to the optical designer on the entire system.

Such added functionality on the same waveguide platform may come in many flavors, and a non exhaustive list is shown below:

- Optical combiner (original functionality, either in mirror or periscope modes
- Focus lens to bring the far field image at a more

comfortable position to reduce Vergence Accommodation Conflict (VAC), such as in Magic Leap 1.

- Integration of prescription (Rx) lens as it is done in the Zeiss Tooz curved waveguide (low index glue separating the TIR waveguide and the Rx lens).
- Eye tracking illumination (LEDs) and eye tracking sensors (CMOS sensors)
- Head tracking sensors
- Gesture sensing
- Face tracking sensors (both optical and non optical)
- Wireless antennas (wifi, BT, 5G)
- Resolution enhancement features (optical wobulation).

Optical wobulation is an old concept which is again highly desirable especially since novel microdisplay panel technologies are being developed (microLED panels) with high brightness but still low resolution for the coming years. Optical wobulation can thus take advantage of such high brightness microdisplays are producing from a low Pixel Per Inch (PPI) a higher Pixel Per Degree (PPD) in the angular space through time domain multiplexing (see Fig. 40).

In a smart glass form factor, optical wobulation can be done in various ways, in the illumination system of the display engine, on the display panel itself, in the pupil forming system of the display engine or directly in the waveguide combiner, as shown in Fig. 41.

Today, various implementations have been investigated, from illumination system wobulation to display panel wobulation (Optotune XPR) to waveguide wobulation (Digilens Trex). They are described in Fig. 42.

Although pupil switching, as done in Magic Leap 1, was not intended for optical wobulation but rather for waveguide focus switching, it could be nevertheless a good candidate assuming that various spatially demultiplexed input pupil couplers have slightly different grating periods providing a fraction of a pixel shift in the TIR field.

Optotune XPR is a moving part wobulator producing a display shift at the display panel plane.

Digilens Trex waveguide optical wobulator is providing a true optical wobulation without the need for spatially demultiplexing input pupils as with Magic Leap 1 illumination switch. Also, it does not alter the dimension of the display engine since all is done in the waveguide. It is therefore a good candidate for small form factor wobulation in smart glasses.

The cost saving by using an off the shelve smaller display panel (LCOS, DLP or microLED) is desirable especially if the wobulator does not affect the overall size of the display engine. The power savings by using a smaller panel are however compensated in part by the additional power required to drive the wobulation system (expect for illumination switch wobulation).

One step further would be to increase the angular shift to an integer number of pixels on top of a sub-pixel shift so that an optical foveation feature can be implemented, producing a foveated high PPD experience in the

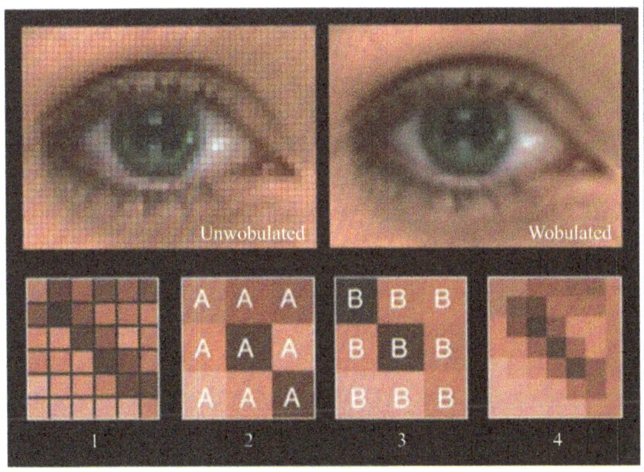

Fig. 40 Optical wobulation creating a high PPD experience from a low PPI panel.

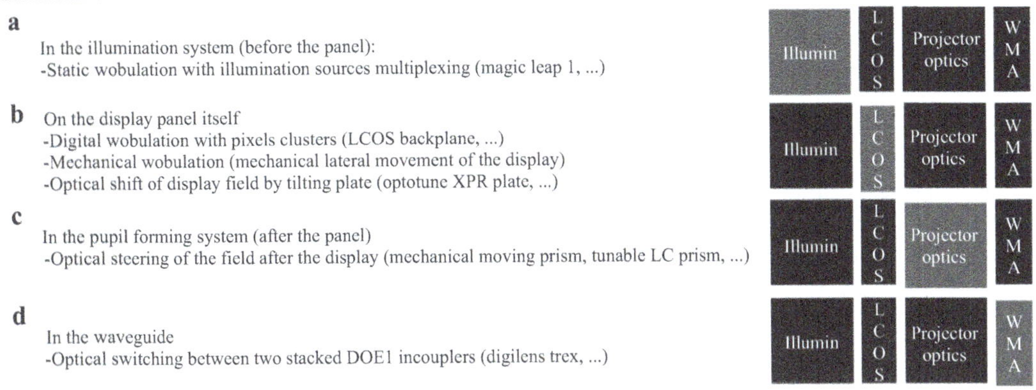

Fig. 41 Optical wobulation architectures for AR HMDs.

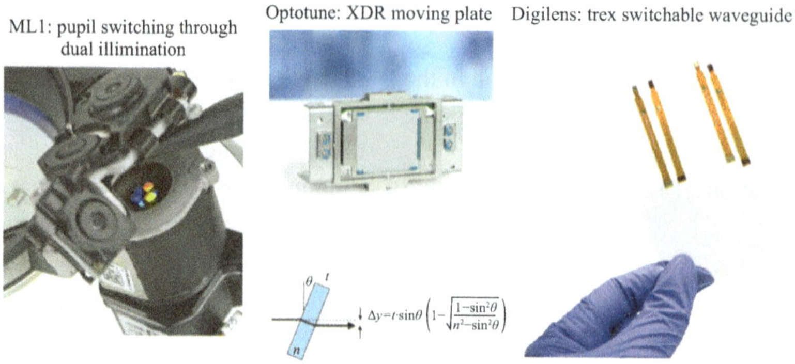

Fig. 42 Optical wobulation architectures for AR HMDs.

angular/time space with a uniform low resolution PPI display panel.

Gaze contingent optical foveation could also be implemented in such waveguide by the use of additional switchable input gratings multiplexed in various ways on the waveguide.

We will certainly see in the coming years more ways to increase the functionality of the waveguide combiner, especially for smart glasses, as it is the largest area available to the designer on the glasses to implement both display, field combining, sensing and communication functions all in a transparent medium.

Conclusion

The aim of this paper is to capture the state of the art in waveguide combiner systems for AR and MR headsets, especially as diffractive waveguide combiners (surface relief diffractives, volume holographic and others such as metasurfaces, resonant gratings etc…). We also reviewed the various reflective (or geometric) waveguides combiner architectures which can provide greater efficiency, great color uniformity and high MTF. Lately, such reflective waveguide architectures have been integrating two-dimensional pupil replication, as did previously their diffractive counterparts.

For optimal system efficiency, the lightguide, display engine and sensors need to be co-designed as a global system to match the human visual system through human-centric optical design.

The coming years will be an exciting time for AR VR MR hardware development. A full ecosystem to allow for mass production and lower costs of waveguide grating combiners is growing worldwide, including high index ultra-flat glass blanks, high index resin material developers, process equipment developers, nanoimprint lithography equipment developers, novel non-contact surface relief grating replication process, and also dedicated software design tools developers allow finally this technology to emerge as a viable option for the upcoming consumer Mixed Reality and smart glass market.

One key aspect of next waveguide developments is to include more functionality in a single waveguide, not only for field combining, but also for focus tuning, prescription lens integration, sensing and even optical wobulation and gaze contingent optical foveation.

However, one has to remember that delivering on the promises of the ultimate wearable display hardware is only one aspect of the trial and opportunity ahead for Mixed Reality, delivering on strong use cases, especially for consumer markets, will be the other critical item to consider.

Conflict of interest

The authors declare that they have no conflict of interest.

Received: 06 December 2021 Revised: 27 June 2022 Accepted: 28 June 2022
Accepted article preview online: 05 July 2022
Published online: 02 August 2022

References

1. Withrington R. J. Optical Design Of A Holographic Visor Helmet-Mounted Display. Proceedings Volume 0147, Computer-Aided Optical Design; (1978)
2. Jerald, J. The VR Book: Human-Centered Design for Virtual Reality. (New York: Association for Computing Machinery, M & C, Morgan & Claypool, 2016).
3. Barfield, B. Fundamentals of Wearable Computers and Augmented Reality. 2nd edn. (Boca Raton: CRC Press, 2016).
4. Inzerillo, L. Augmented reality: past, present, future. Proceedings of SPIE 8649, the Engineering Reality of Virtual Reality 2013. Burlingame, California, United States: SPIE, 2013.
5. Azuma, R. T. A survey of augmented reality. *Presence: Teleoperators and Virtual Environments* **6**, 355-385 (1997).
6. Cakmakci, O. & Rolland, J. Head-worn displays: a review. *Journal of Display Technology* **2**, 199-216 (2006).
7. Rolland, J. & Cakmakci, O. Head-worn displays: the future through new eyes. *Optics and Photonics News* **20**, 20-27 (2009).
8. van Krevelen, D. W. F. & Poelman, R. A survey of augmented reality technologies, applications and limitations. *International Journal of*

Virtual Reality **9**, 1-20 (2010).
9. Low, K. L. et al. Combining head-mounted and projector-based displays for surgical training. Proceedings of the IEEE Virtual Reality, 2003. Los Angeles, CA, USA: IEEE, 2003.
10. Amitai, Y., Friesem, A. A. & Weiss, V. Holographic elements with high efficiency and low aberrations for helmet displays. *Applied Optics* **28**, 3405-3416 (1989).
11. Kress B. et al. Optical Architectures for Augmented-, Virtual-, and Mixed-Reality Headsets, SPIE Press, SBN: 9781510634336, Volume: PM316 (2020)
12. Kress, B. C. & Cummins, W. J. 11-1: Invited *p*aper: towards the ultimate mixed reality experience: hololens display architecture choices. SID Symposium Digest of Technical Papers **48**, 127-131 (2017).
13. Miller, J. M. et al. Design and fabrication of binary slanted surface-relief gratings for a planar optical interconnection. *Applied Optics* **36**, 5717-5727 (1997).
14. Kimmel, J. et al. A novel diffractive backlight concept for mobile displays. *Journal of the Society for Information Display* **16**, 351-357 (2008).
15. Kimmel, J. et al. 55.3: Diffractive backlight light guide plates in mobile electrowetting display applications. *SID Symposium Digest of Technical Papers* **40**, 826-829 (2009).
16. Zhang, N. N. et al. Improved holographic waveguide display system. *Applied Optics* **54**, 3645-3649 (2015).
17. Yoshida, T. et al. A plastic holographic waveguide combiner for light-weight and highly-transparent augmented reality glasses. *Journal of the Society for Information Display* **26**, 280-286 (2018).
18. Mukawa, H. et al. 8.4: Distinguished *p*aper: a full color eyewear display using holographic planar waveguides. SID Symposium Digest of Technical Papers **39**, 89-92 (2008).
19. Oku, T. et al. 15.2: High-luminance see-through eyewear display with novel volume hologram waveguide technology. SID Symposium Digest of Technical Papers **46**, 192-195 (2015).
20. Sarayeddine K. et al. Key challenges to affordable see-through wearable displays: the missing link for mobile AR mass deployment. Proceedings Volume 8720, Photonic Applications for Aerospace, Commercial, and Harsh Environments IV; 87200D (2013)
21. Äyräs, P. , Saarikko, P. & Levola, T. Exit pupil expander with a large field of view based on diffractive optics. *Journal of the Society of Information Display* **17**, 659-664 (2009).
22. Levola, T. Diffractive optics for virtual reality displays. *Journal of the Society for Information Display* **14**, 467-475 (2006).
23. Kress, B. & Shin, M. Diffractive and holographic optics as optical combiners in head mounted displays. Proceedings of the 2013 ACM Conference on Pervasive and Ubiquitous Computing Adjunct Publication. Zurich, Switzerland: ACM, 2013.
24. Cameron, A. A. Optical waveguide technology and its application in head-mounted displays. Proceedings of SPIE 8383, Head- and Helmet-Mounted Displays XVII; and Display Technologies and Applications for Defense, Security, and Avionics VI. Baltimore, Maryland, United States: SPIE, 2012.
25. Homan, M. The use of optical waveguides in head up display (HUD) applications. Proceedings of SPIE 8736, Display Technologies and Applications for Defense, Security, and Avionics VII. Baltimore, Maryland, United States: SPIE, 2013.
26. Cheng, D. W. et al. Design of an ultra-thin near-eye display with geometrical waveguide and freeform optics. *Optics Express* **22**, 20705-20719 (2014).
27. Jurbergs, D. et al. New recording materials for the holographic industry. Proceedings of SPIE 7233, Practical Holography XXIII: Materials and Applications. San Jose, California, United States: SPIE, 2009.
28. Alastair, G. DigiLens: Design and Fabrication Considerations for Holographic Waveguide AR Displays. Proceedings of SPIE 11764, SPIE AVR21 Industry Talks II, SPIE, 2021.
29. Kogelnik, H. Coupled wave theory for thick hologram gratings. *Bell Labs Technical Journal* **48**, 2909-2947 (1969).
30. Curtis, K. & Psaltis, D. Cross talk in phase-coded holographic memories. *Journal of the Optical Society of America A* **10**, 2547-2550 (1993).
31. Golub, M. A., Friesem, A. A. & Eisen, L. Bragg properties of efficient surface relief gratings in the resonance domain. *Optics Communications* **235**, 261-267 (2004).
32. Moharam, M. G. et al. Stable implementation of the rigorous coupled-wave analysis for surface-relief gratings: enhanced transmittance matrix approach. *Journal of the Optical Society of America A* **12**, 1077-1086 (1995).
33. Estepa, L. A. et al. Corrected coupled-wave theory for non-slanted reflection gratings. Proceedings of SPIE 8171, Physical Optics. Marseille, France: SPIE, 2011.
34. Liu, V et al. S4: A free electromagnetic solver for layered periodic structures, Computer Physics Communications 183 (2012) 2233–2244, Elsevier 2021.
35. da Silva Ferreira A. et al. Development of a computational environment for MIT electromagnetic equation propagation simulator. *2017 IEEE XXIV International Conference on Electronics, Electrical Engineering and Computing (INTERCON)*, 2017, pp. 1-4, doi: 10.1109/INTERCON.2017.8079691.
36. Levola, T. et al. Replicated slanted gratings with a high refractive index material for in and outcoupling of light. *Opt. Express* **15**, 2067-2074 (2007).
37. Hellmann C. et al. Physical-optics analysis of lightguides for augmented and mixed reality glasses. Proceedings Volume 11062, Digital Optical Technologies SPIE 2019.
38. Farn, M. W. Binary gratings with increased efficiency. *Applied Optics* **31**, 4453-4458 (1992).
39. Kress, B. C. & Meyrueis, P. Applied Digital Optics: From Micro-Optics to Nanophotonics. (Chichester: John Wiley and Sons Publisher, 2009).
40. Quaranta, G. et al. Steering and filtering white light with resonant waveguide gratings. Proceedings of SPIE 10354, Nanoengineering: Fabrication, Properties, Optics, and Devices XIV. San Diego, California, United States: SPIE, 2017.
41. Basset, G. Resonant screens focus on the optics of AR (Conference Presentation). Proceedings of SPIE 10676, Digital Optics for Immersive Displays. Strasbourg, France: SPIE, 2018.
42. Genevet, P. et al. Recent advances in planar optics: from plasmonic to dielectric metasurfaces. *Optica* **4**, 139-152 (2017).
43. Capasso, F. The future and promise of flat optics: a personal perspective. *Nanophotonics* **7**, 953-957 (2018).
44. Chen, W. T. et al. Broadband achromatic metasurface-refractive optics. *Nano Letters* **18**, 7801-7808 (2018).

Official journal of the JHL 2689-9620
www.light-am.com

Article   Open Access

Hybrid multimode - multicore fibre based holographic endoscope for deep-tissue neurophotonics

Yang Du[1], Sergey Turtaev[1], Ivo T Leite[1], Adrian Lorenz[1], Jens Kobelke[1], Katrin Wondraczek[1] and Tomáš Čižmár[1,2,3,*]

Abstract

In-vivo microendoscopy in animal models became a groundbreaking technique in neuroscience that rapidly expands our understanding of the brain. Emerging hair-thin endoscopes based on multimode fibres are now opening up the prospect of ultra-minimally invasive neuroimaging of deeply located brain structures. Complementing these advancements with methods of functional imaging and optogenetics, as well as extending its applicability to awake and motile animals constitute the most pressing challenges for this technology. Here we demonstrate a novel fibre design capable of both, high-resolution imaging in immobilised animals and bending-resilient optical addressing of neurons in motile animals. The optimised refractive index profile and the probe structure allowed reaching a spatial resolution of 2 μm across a 230 μm field of view for the initial layout of the fibre. Simultaneously, the fibre exhibits negligible cross-talk between individual inner-cores during fibre deformation. This work provides a technological solution for imaging-assisted spatially selective photo-activation and activity monitoring in awake and freely moving animal models.

Keywords: Multimode fibre, Multicore fibre, Fibre-based imaging, Microendoscopy, Endomicroscopy, Holographic endoscopy

Introduction

Neuroscience nowadays routinely utilises optogenetic toolsets to stimulate and silence functional activity within neuronal networks and even individual neurons *in-vivo*[1–5]. The efficient application of modern optogenetics relies on precise spatial-temporal control of light. Achieving this non-invasively is, however, challenging when targeting deeply-located brain regions, since tissue scattering prevents sufficient localisation (focusing) of optical power. Separately, invasive solutions often cause high risks of tissue damage, growing with tissue depth. Optical fibres are commonly seen as a solution for minimally-invasive light delivery, enabling stimulation and activity recording in deep brain regions due to their miniature footprint[6–8]. Advanced fibre probes, such as multi-channel probes[9], fibre bundle[10] or tapered fibre[11,12] have been developed for targeting multiple regions.

Recently, holographic endoscopes, relying on imaging through hair-thin multimode fibres, were successfully demonstrated in *in-vivo* deep-brain imaging achieving cellular and even subcellular resolution[13–15]. An ongoing

Correspondence: Tomáš Čižmár (Tomas.Cizmar@leibniz-ipht.de)
[1]Leibniz Institute of Photonic Technology, Albert-Einstein-Straße 9, Jena, 07745, Germany
[2]Institute of Scientific Instruments of CAS, Královopolská 147, Brno, 612 64, Czechia
Full list of author information is available at the end of the article.

challenge for holographic endoscopy in freely moving animals, which has attracted significant attention of the relevant research community[16–19], remains to maintain the highest imaging performance while the fibre undergoes bending deformations. Amongst the multitude of competing strategies, possible solutions were proposed by exploitation of complex numerical and machine-learning algorithms[20,21], exploitation of guide-stars[22,23] and bending resilient fibres[24,25].

In this work, we designed and manufactured a hybrid, multimode – multicore fibre (M^3CF), which features 61 corelets spaced with $\approx 4\times$ the density of neurons in typical brain tissue (see Fig. 1a for the description of its anatomy). Thanks to the cascaded refractive index in its structure, the fibre allows two regimes of light transport resembling either a multimode fibre or a fibre bundle, with the possibility to switch from one regime to another by controlling the numerical aperture (NA) of the light coupled into the structure. The multimode fibre regime is to be exploited in the initial phase of an *in-vivo* experiment in order to navigate the distal end of the endoscope towards the desired brain structure and map out the layout and the connectivity of the neurons in the circuits under investigation. Following the implantation, anchoring the distal end with respect to the desired neuronal circuit and mapping out the connectivity context (all to be performed in immobilised models under anaesthesia), the fibre-bundle regime is to ensure light transport to and from each individual neuron in the field of view despite the motion of the awaken and behaving animal model. The proposed application scenario, therefore, consists of two parts - acute high-resolution imaging session (multimode regime) for guided implantation in sedated animal and later chronic studies in the motile model utilising low-resolution (bundle regime) stimulation and readout of preselected areas of interest.

To verify the desired qualities of the manufactured M^3CF waveguide, it was utilised in a testing holographic endoscopy system, which has been optimised to provide control over the light transport in both regimes. While kept in stationary configuration, the best achievable resolution of imaging, based on raster-scanning a diffraction-limited focus, has been estimated. Further, the M^3CF has been subjected to a series of tests under which the impact of fibre bending on the imaging (multimode fibre regime) and targeted light delivery (fibre bundle regime) has been assessed.

Results

The M^3CF, presented in Fig. 1a, was developed using the stack and draw technique, as described in the Methods section. The cladding and the pulp, with diameters of 278 µm and 232 µm respectively, form a step-index multimode waveguide with a refractive index difference corresponding to an NA of 0.1. The corelets with a diameter of 11.3 µm are equidistantly distributed across the core with a spacing of 9.8 µm. Size-wise, the design is balanced around two aspects - respecting morphological features of the neuron cell structures to be imaged and minimising cross-talk between the corelets and the pulp. Particularly, the core size and inter-core separation are comparable to the typical rodents neuron cell body, making it rather unlikely to have no overlap between cores and underlaying cell bodies during the experiment. An increased refractive index of the corelets, with respect to the surrounding material of the core, denoted in this paper as ‘the pulp’, is tailored to form an additional step in the index profile, resulting in an NA of 0.1. The cascaded index profile plays a key role in the presented design, enabling the core and the corelets to be addressed independently by changing the coupling NA.

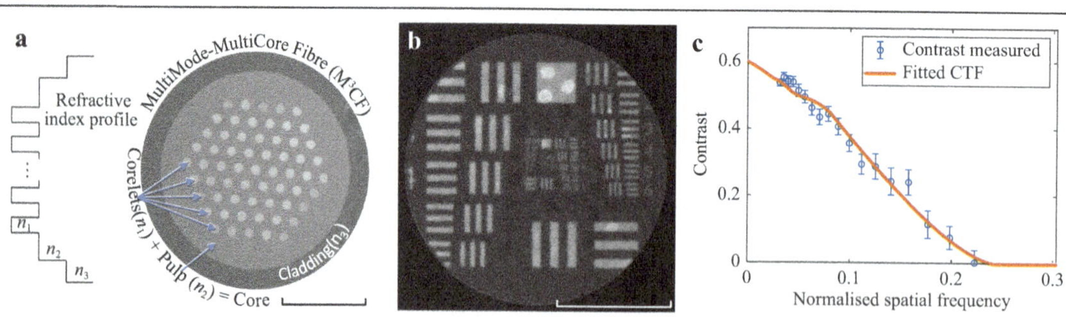

Fig. 1 Probe design and optical resolution assessment. **a** Microscope image of the cross-section of the manufactured M^3CF. **b** Image of a negative USAF 1951 target acquired through the M^3CF by the means of holographic endoscopy. **c** Image contrast as a function of normalised spatial frequency and the fitted transfer function (CTF) corresponding to an aberration-free imaging system with a circular aperture. Scale bars indicate a length of 100 µm.

Multimode fibre regime: Imaging quality assessment

Light coupled into the area of the corelets under NA below 0.1 will remain confined in the corelets. In contrast, signals falling into the NA interval of 0.1 and 0.2 would cross the index barriers between corelets and the pulp, which on its own only supports light transport below the 0.1 NA threshold. Although individually indexes of the corelets and pulp correspond to 0.1 NA only, light propagating in the pulp causes an imaging resolution enhancement in the corelets region as shown in previous studies[26,27].

Therefore, when implemented in the wavefront shaping system, light propagation is supported across the whole core, and despite leading to different resolution across the pulp and the corelets' areas, it enables high-contrast imaging across the entire core area. The wavefront shaping technique relying on an acquisition of the fibre transmission matrix[28–31] is utilised in this work, allowing the formation of focal points at desired positions across the chosen focal plane behind the distal fibre facet. A sequential illumination of the negative USAF 1951 test target with accordingly generated foci while collecting light passing through, allowed the assessment of the resolution of the system as presented in Fig. 1b. Visibly non-uniform intensity across the fibre facet is likely due to higher wavefront shaping fidelity achieved at corelets regions. Featuring remarkable circular symmetry, the latter can preserve a circular polarisation state of the propagating light[20], allowing for generating high-contrast focal points, while controlling only a single polarisation state at the input side of the fibre as performed in our study. Note that in our assessment, we subject the complete recorded image to the analysis, deliberately making ourselves blind to whether the local response originates from the core or the pulp. The contrast transfer function (CTF) analysis based on the target element groups 6-8 resulted in the effective NA of 0.121 ± 0.006, and corresponding resolution limit (Abbe criterion) of 2.0 ± 0.1 μm, as descried in[32].

Multimode fibre regime: Performance under bending

The developed hybrid probe, calibrated as part of the holographic imaging system, was exposed to a series of bending tests to evaluate the imaging performance degradation caused by fibre deformation. Once the calibration procedure is done, allowing for raster-scanning at proximity to the fibre facet, the USAF 1951 resolution target was placed at the focal plane of the probe. In the first experimental settings, a 33 cm-long segment of M^{3}CF was gradually bent, forming a smooth layout close to a right angle arc, as presented in Fig. 2a. Fig. 2b–d present images of a target element (group 4, no. 2), obtained for initially calibrated, straight shape and significantly bent layouts. The high-contrast image of the bars, obtained through the unperturbed fibre, is turned into a uniform background across the pulp area and only parts of the object overlapping with corelets resemble the original image. Being a standard step-index multimode fibre corelets are still capable of preserving imaging capability over small

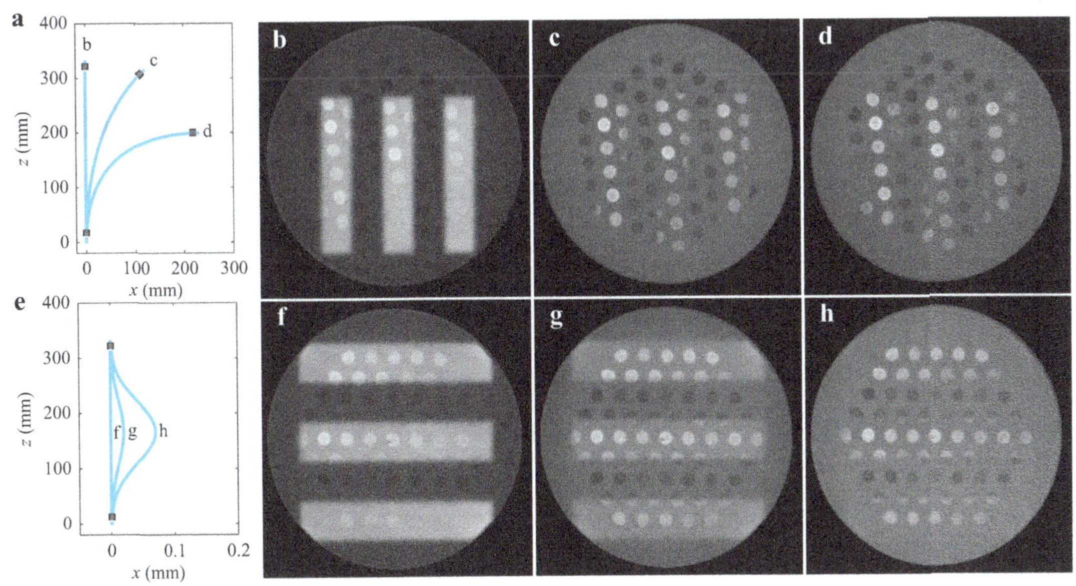

Fig. 2 a–d Imaging performance of the M^{3}CF undergoing gradual fibre deformations. **a** Layouts of experimental fibre contortions. **b–d** Images obtained at different layouts as denoted in **a**. **e–h** Imaging performance of the M^{3}CF under minor fibre deformations. **e** Layouts of experimental fibre contortions. **f–h** Images obtained at different layouts as denoted in **e**.

bending deformations[33].

An additional bending test was designed to observe practical boundaries for applying holographic endoscopy through the developed probe. This new setting involves the displacement of the distal end of the fibre probe towards its proximal end using a linear translation stage, as depicted in Fig. 2e. Precision control over fibre deformation in this scheme allowed observing the degradation process of imaging quality, presented in Fig. 2f–h (performed at element 6 of group 3), from partially diminished contrast to complete loss of recognition similar to the previous experiment.

As expected, the performance within the pulp area is highly susceptible to fibre deformation, which limits utilising high-resolution imaging modality to a nearly rigid configuration suitable for neuroimaging applications only in immobilised animal models. In contrast, the corelets provide a low resolution but stable imaging performance even under significant bending deformations, acting like a fibre bundle. Another aspect worth mentioning in the scope of the proposed application of imaging-assisted implantation is sensitivity to local temperature changes. During a feasibility check, about 20 mm of the fibre tip was gradually heated up to 18°C above room temperature. No significant loss of imaging performance has been observed throughout the whole interval.

Fibre bundle regime: Cross-talk evaluation

The experimental setup geometry, introduced in the methods section and utilised for holographic endoscopy imaging, allows for generating diffraction-limited focal spots at any desired position across the proximal fibre facet. With a digital micromirror device (DMD) placed in a far-field plane to the input fibre facet, the effective coupling NA can be controlled by constraining the area of the displayed DMD patterns within an appropriately sized circular aperture. Therefore, without modifying the experimental setup (shown in Materials and methods section), all 61 corelets can be addressed at will by projecting corresponding holograms on the DMD, as conceptually illustrated in Fig. 3a. Limiting the NA to 0.1 ensures light signals are spatially confined within the targeted corelets, thereby maintaining the light transport through the M[3]CF in the fibre bundle regime. To evaluate the cross-talk between corelets, we recorded high dynamic range images of the output fibre facet while coupling light to individual corelets, one at a time. The confusion matrices (cross-talk maps) were compiled from images of

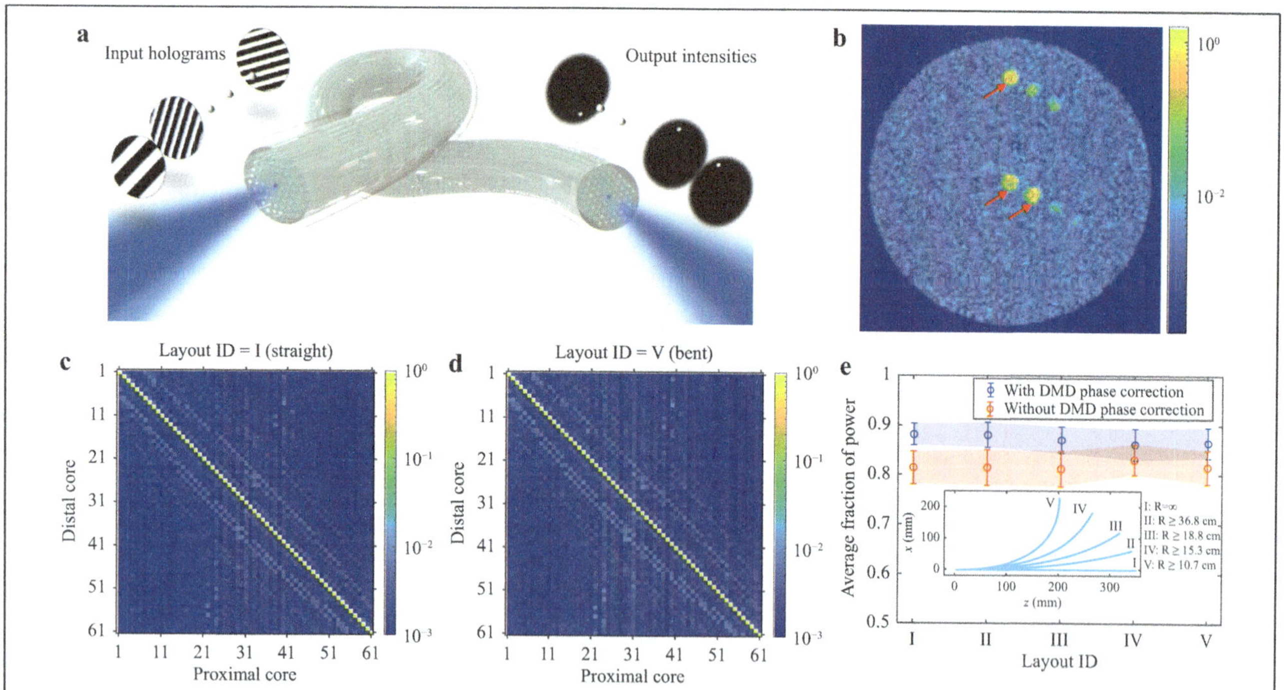

Fig. 3 Cross-talk assessment of the M[3]CF. **a** Illustration of the experimental concept. Addressing individual corelets is performed via displaying corresponding grating patterns on the DMD, placed in a far-field plane of the proximal fibre facet. **b** Image of the distal facet of the fibre, while three corelets (marked with red arrow) are simultaneously addressed with a single DMD pattern. **c, d** Confusion matrices, representing inter-core cross-talk, for both straight **c** and significantly bent **d** fibre layouts. **e** Averaged fraction of power preserved in the addressed corelets at various fibre bending positions. An experiment was performed with and without active correction of aberrations inherent to DMD. The inset depicts experimental fibre layouts for different bending positions and corresponding bending radii.

the distal M[3]CF facet. Two examples of such matrices are shown in Fig. 3c and d for the fibre under different bending conditions. As an integrated measure for spatial confinement of light, we use the average fraction of the power confined in the desired corelet with respect to the overall power delivered through the whole waveguide. This metric is plotted in Fig. 3e for different bending layouts. It is noteworthy that a portion of stray signal can appear in the neighbouring cores already at the coupling stage due to significant aberration in the system, in our case, caused mainly by the inherent curvature of the DMD, which has been utilised as a spatial modulator. A wavefront correction algorithm[34] has been applied to counteract the aberration, thus facilitating an effective light coupling to the corelets and improved light confinement to the desired corelets at the fibre output, as shown in Fig. 3e. Overall, the measurements indicate a negligible influence of the corelets decoupling with the fibre at different layouts. Finally, we utilised a single combined hologram for coupling of signal to three corelets simultaneously. The resulting output, presented in Fig. 3b, confirms the ability to reach multiple regions of interest at once.

To summarise the independent validations of the multimode and fibre-bundle operating regimes, we performed a demonstration to exemplify the proposed applicational scenario involving both the modes mentioned above. A coverslip with a complex arrangement of 6 μm in diameter green fluorescent particles was placed at the focal plane of the calibrated, ready-to-image holographic endoscope equipped with an M[3]CF probe. At the first stage, while the fibre probe remains static in its initial precalibrated position, a high-resolution imaging session allowed navigation across the coverslip to select a region of interest with complex and dense distribution of the beads presented in Fig. 4a. Based on this high-resolution recording and known corelets layouts, a predicted low-resolution image, shown in Fig. 4b, was acquired as a reference for the following tests. Once navigation is done and the object of interest is in the field of view, the system can be switched to the fibre-bundle regime. Consequent display of 61 DMD patterns, pre-selected to efficiently couple light in the corelets, allowed to probe every core individually (Fig. 4e for scanning order) and record the fluorescent response via a photomultiplier tube. For clarity of comparison, the recorded intensities were mapped to the corelet layout of the probe. We demonstrated the fibre-bundle regime while bending and twisting the M[3]CF probe under test simply by hand, roughly mimicking

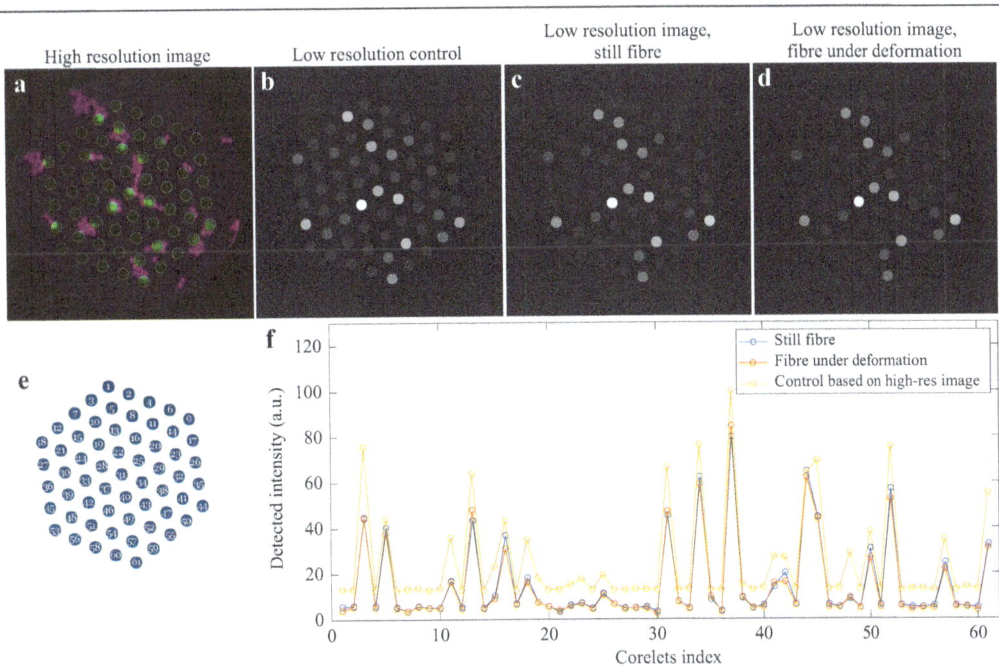

Fig. 4 Demonstration of application scenario. **a** High-resolution image of flusorescent beads obtained in the multimode fibre regime. Cyan colour corresponds to regions of the pulp, green highlights the regions of the corelets. **b** Predicted fluorescence response in fibre-bundle regime (probing corelets), obtained from the high-resolution image in **a**. **c, d** Fluorescent response obtained by probing individual corelets in both initial, pre-calibrated and pronouncedly deformed (bend+twist) positions correspondingly. **e** Distal fibre corelets layout with numbers representing scanning order. **f** Recordings obtained via sequential scan across the corelets for initial (corresponding to **c**), deformed **d** positions, as well as for the predicted result based on the high-resolution recording.

deformations expected in field experiments. Records of the detected intensities (Fig. 4f) for static and dynamically deformed fibres carry the same set of features and show a close match in magnitude. The normalised intensities mapped in Fig. 4c, d for both instances are visually indistinguishable.

Discussion

This work demonstrates a fibre probe designed specifically for imaging-assisted deep-brain activity monitoring and optogenetics. The proposed design relies on a structure with a cascaded refractive index profile. Together with the cladding, the core forms a highly multimode waveguide. At the same time, it maintains independent light guidance within 61 equidistantly spaced corelets. This arrangement allowed performing raster-scanning imaging at the distal fibre facet with an optical resolution of about 2 μm across the 230 μm field of view when employed in a holographic endoscope system. The embedded multicore structure demonstrated robust optical access, which remains stable even under severe bending deformation when holographic imaging performance is significantly compromised. The optical scheme with a DMD placed in a far-field plane to the proximal fibre facet was arranged for the laser power to be efficiently redistributed between the corelets, enabling high-speed sequential or even simultaneous channelling of light via multiple corelets at will.

These operating modes delivered through a single minimally invasive probe can offer synergetic benefits in animal studies, allowing high-resolution visual guidance for precise probe implantation and initial acute observations, followed by chronic monitoring and stimulation of the pre-selected regions of interest in freely moving animals.

Although integrated into a single probe, it is important to highlight that large multimode core and multicore imaging structures can be utilised independently. The infused corelets do not obstruct the field of view or compromise the imaging quality when performing holographic endoscopy. At the same time, light coupled to the small cores remains spatially confined when propagating through the fibre.

Such direct access to the sample plane could allow spatially resolved photostimulation without the necessity of wavefront shaping and highly coherent laser sources. Moreover, operational wavelengths of optogenetic tools for light-activated control and reporting neuronal activity are often detuned from each other, enabling their simultaneous application. As a result, a budget-friendly set of an LED array, relay optics, and dichroic mirrors could extend a holographic endoscopy system for deep-brain imaging with optogenetics modality. Moreover, once implantation is done, the comprehensive wavefront shaping system can be replaced with such LED-based module, not only avoiding permanent need in an expensive piece of equipment, but also enabling a switch to any desired excitation wavelength for following chronic experiments.

We envision that the design of highly versatile yet practical M^{3}CF probes presented in this work will become an enabling attribute giving the relevant neuroscience community access to the deepest brain regions while being equipped with the most powerful portfolio of neurophotonics tools for studying neuronal activity.

Materials and methods

Fibre design and manufacture

The M^{3}CF fibre was fabricated by using the stack-and-draw technology. The preform design and stack are shown in Fig. 5a, b. The primary preform was assembled in a four ring hexagonal stack of Germanium doped rods. The corelet rods having an outer diameter of 1 mm are prepared by using modified chemical vapor deposition (MCVD) preforms with 24 doped core layers of 2.5 mol% GeO_2 and a core-clad-ratio of 0.5. The designed NA of the corelet rods is 0.1. The MCVD substrate tube material and the container tube material (outer diameter 11.2 mm; inner diameter 9.5 mm) was high purity silica Heraeus F300. For the minimization of geometrical deformations, the volume between the 61 hexagonally arranged corelet rods and the container tube was filled with Heraeus F300 silica rods of diameter 0.7 mm and 0.47 mm respectively. The resulting preform was overcladded with a depressed index silica tube (outer diameter 14 mm; inner diameter 12 mm). The material of the overcladding tube was Heraeus F520 (Δn -5×10^{-3}, 1 mol% SiF_4). The designed NA for the core / cladding interface was 0.1.

The M^{3}CF fibre was fabricated from the above described preform inside the in-house fibre drawing tower. The preform was evacuated during the drawing process to sinter any remaining-interspace volume. The drawing process was conducted at the speed of 6.5 m/min, leading to M^{3}CF of 278 μm in diameter. The germanium doped corelets were manufactured with average size of 11.3 μm in diameter. For protecting the fibre from environmental impacts, a single layer coating of UV curing acrylate was applied to the M^{3}CF, which has been removed during the light-transport experiments.

Light-transport geometry

The experimental setup is illustrated in Fig. 6. A single

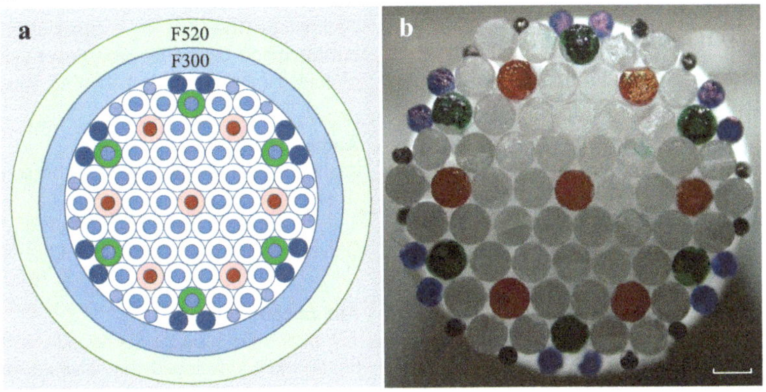

Fig. 5 Schematic diagram of M[3]CF preform design. The primary preform was assembled in a four ring hexagonal stack of Germanium doped rods with a total of 61 elements. The primary jacket tube was high purity silica Heraeus F300. Silica rods (Heraeus F300) with two different diameters were filled between hexagonal arranged rods and the primary overcladding tube. A secondary tube Heraeus F520 was applied as outer cladding. **b** A photograph of preform stacking as described in **a**. The colours labelled on the glass rods are used to ensure we have the correct structure arrangement during the stacking procedure. Scale bar indicates length of 1 mm.

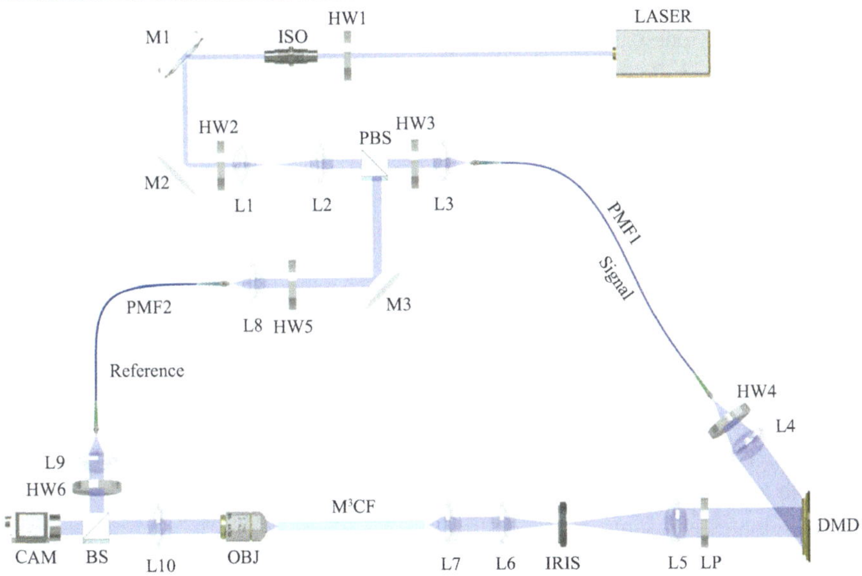

Fig. 6 Scheme of multicore fibre-based imaging setup. LASER: Coherent Sapphire SF 488; HW1-HW6: Thorlabs WPH05M-488; ISO: Thorlabs IO-3-488-HP; M1-M3: Thorlabs BB1-E02; L1: Thorlabs AC254-050-A-ML; L2, L4: Thorlabs AC254-100-A-ML; L3, L7, L8, L9: Thorlabs A240TM-A; L5: Thorlabs AC254-200-A-ML; L6: Thorlabs AC254-200-A-ML; L10: Thorlabs AC254-125-A-ML;PMF1: Thorlabs P5-488PM-FC-1; PMF2: Thorlabs P5-488PM-FC-2; DMD: ViALUX V7001; LP: Thorlabs LPVISE100-A; IRIS: Thorlabs SM2D25D; OBJ: Olympus Plan Achromat 10X NA 0.25; PBS: Thorlabs PBS101; BS: Thorlabs BS010; CAM: Basler acA640-750um. Beam-blocks and Neutral-density filters are not shown in the scheme.

frequency continuous wave laser at wavelength of 488 nm was used in all experiments. Linearly polarised light beam from the laser is expanded by a telescope formed from two achromatic lenses L1 and L2. The light beam is separated by a polarizing beam splitter PBS to a signal and a reference beams. A half wave plate HW1 and an optical isolator ISO are added to control the overall power of the two beams. A half wave plate HW2 is inserted to alter the power ratio between the two beams. These two beams are further coupled into single mode polarisation maintaining fibres PMF1 and PMF2 via achromatic lenses L3 and L8, respectively. Half wave plates HW3 and HW5 align the two beams with either a fast or slow axis of the polarization-maintaining fibres. A half wave plate HW4, together with a linear polarizer LP are applied to ensure a linearly polarized light to be coupled into the M[3]CF fibre. In the signal path, the beam is collimated by an achromatic lens L4 to overfill the active area of a digital micromirror device DMD. The DMD is installed in an off-axis regime at an angle of 24° with respect to the incident light beam.

The farfield of the first-order diffracted signal formed by holograms applied on the DMD is relayed by lenses L5, L6 and L7 to the proximal end of M^3CF. The focal lengths of L5, L6 and L7 are carefully selected to adapt to the NA of M^3CF. An iris is set at the Fourier plane of L5 for separating the first diffraction order from other contributions. Microscope objective OBJ and tube lens L10 image the distal end of fibre facet onto a CMOS camera. In the reference path, a lens L9 collimates the beam exiting from PMF2. A half wave plate HW6 is chosen to adjust the linear polarization angle of the reference beam for maximizing the contrast of the interference signal. A non-polarizing beam splitter BS is placed to combine the signal and the reference beams.

We adopt the transmission matrix (TM) algorithm for controlling the light transport through the M^3CF[20,35] and image acquisition. The TM represents a linear relationship between input and output optical modes of a given medium, and with its availability, one can synthesize any desired output field despite the medium's complexity. The DMD (chip size: 1024 × 768 pixels) is a programmable component employed in the system for light beam shaping through the fibre. Although the DMD is a binary amplitude light modulator, it enables providing phase modulation when it is applied in the off-axis regime[15]. The M^3CF supports approximately 12490 modes at the wavelength of 488 nm. The input modes are chosen as a square grid of 191 × 191 diffraction-limited foci. Prior to the TM acquisition, a primary alignment is performed by integrating the output field received from the camera for each individual input mode. This step is to ensure that the fibre is correctly positioned and oriented with respect to the imaging system and it eliminates any input modes falling outside the spatial constraints of the fibre, reducing their number to ~26 000. At the distal end of the fibre, the output modes are analogously occupying a square grid of 352 × 352 focal points with an interval of ~0.7 μm to fully cover and significantly oversample the entire M^3CF core during acquisition. Further, during image acquisition, the number of output modes is analogously reduced to those ~87000 falling into the circular area of the core.

Acknowledgements

The authors acknowledge the support from the European Research Council (724530), Ministry of Education, Youth and Sports (CZ.02.1.01/0.0/15_003/0000476), Thüringer Ministerium für Wirtschaft, the European Regional Development Fund (CZ.02.1.01/0.0/15_003/0000476), Wissenschaft und Digitale Gesellschaft, the Federal Ministry of Education and Research, Germany (BMBF), and the Thüringer Aufbaubank. The authors would like to acknowledge Dr. Jörg Bierlich, Claudia Aichele, Anne Matthes for preform preparation and fibre drawing; Anka Schwuchow for taking microscope photographs for Fig. 1a; Dr. Dirk E. Boonzajer Flaes for fruitful discussions. The multicore fibre used in this work was drawn at fibre drawing facility at Leibniz Institute of Photonic Technology. The USAF 1951 target used in all imaging experiments was custom-made by the Electron Beam Lithography group of the Institute of Scientific Instruments of the CAS, v.v.i.

Author details

[1]Leibniz Institute of Photonic Technology, Albert-Einstein-Straβe 9, Jena, 07745, Germany. [2]Institute of Scientific Instruments of CAS, Královopolská 147, Brno, 612 64, Czechia. [3]Institute of Applied Optics, Friedrich Schiller University Jena, Fröbelstieg 1, 07743 Jena, Germany

Author contributions

Y.D. performed all experiments. Y.D. and I.T.L. analysed the results. Y.D., S.T. and I.T.L. developed the software for experiments. A.L., J.K. and K.W. designed and fabricated the multicore fibre. T. Č. conceived the idea and led the project. Y.D., S.T., J.K. and T. Č. wrote the manuscript with assistance from all authors.

Conflict of interest

The authors declare no conflicts of interest.

Received: 31 August 2021 Revised: 06 April 2022 Accepted: 21 April 2022
Accepted article preview online: 28 April 2022
Published online: 21 June 2022

References

1. Deisseroth, K. Optogenetics: 10 years of microbial opsins in neuroscience. *Nature neuroscience* **18**, 1213-1225 (2015).
2. Häusser, M. Optogenetics: the age of light. *Nature methods* **11**, 1012-1014 (2014).
3. Shemesh, O. A., et al. Temporally precise single-cellresolution optogenetics. *Nature neuroscience* **20**, 1796-1806 (2017).
4. Frank, J. A., Antonini, M. -J. & Anikeeva, P. Nextgeneration interfaces for studying neural function. *Nature biotechnology* **37**, 1013-1023 (2019).
5. Robinson, N. T. M., et al. T Targeted activation of hippocampal place cells drives memory-guided spatial behavior. *Cell* **183**, 1586-1599 (2020).
6. Aravanis, A. M., et al. An optical neural interface: in vivo control of rodent motor cortex with integrated fiberoptic and optogenetic technology. *Journal of neural engineering* **4**, S143-S156 (2007).
7. Sparta, D. R., et al. Construction of implantable optical fibers for long-term optogenetic manipulation of neural circuits. *Nature protocols* **7**, 12-23 (2012).
8. Miyamoto, D. & Murayama, M. The fiber-optic imaging and manipulation of neural activity during animal behavior. *Neuroscience research* **103**, 1-9 (2016).
9. Sych, Y., et al. High-density multi-fiber photometry for studying large-scale brain circuit dynamics. *Nature methods* **16**, 553-560 (2019).
10. Kim, C. K., et al. Simultaneous fast measurement of circuit dynamics at multiple sites across the mammalian brain. *Nature methods* **13**, 325-328 (2016).
11. Pisanello, F., et al. Dynamic illumination of spatially restricted or large brain volumes via a single tapered optical fiber. *Nature neuroscience* **20**, 1180-1188 (2017).
12. Pisano, F., et al. Depth-resolved fiber photometry with a single tapered optical fiber implant. *Nature methods* **16**, 1185-1192 (2019).
13. Ohayon, S., et al. Minimally invasive multimode optical fiber microendoscope for deep brain fluorescence imaging. *Biomedical optics express* **9**, 1492-1509 (2018).
14. Vasquez-Lopez, S. A., et al. Subcellular spatial resolution achieved for deep-brain imaging in vivo using a minimally invasive multimode

fiber. *Light: science & applications* **7**, 110 (2018).
15. Turtaev, S., et al. High-fidelity multimode fibre-based endoscopy for deep brain in vivo imaging. *Light: science & applications* **7**, 92 (2018).
16. Orth, A., et al. Optical fiber bundles: Ultra-slim light field imaging probes. *Science advances* **5**, eaav1555 (2019).
17. Shin, J., et al. A minimally invasive lens-free computational microendoscope. *Science advances* **5**, eaaw5595 (2019).
18. Choi, W. et al. Fourier holographic endoscopy for label-free imaging through a narrow and curved passage. Preprint at https://arxiv.org/abs/2010.11776 (2020).
19. Badt, N. & Katz, O. Label-free video-rate microendoscopy through flexible fibers via fiber bundle distal holography (FiDHo). Proceedings of Digital Holography and Three-Dimensional Imaging 2021. Washington, DC, USA: Optica Publishing Group, 2021.
20. Plöschner, M., Tyc, T. & Čižmár, T. Seeing through chaos in multimode fibres. *Nature photonics* **9**, 529-535 (2015).
21. Borhani, N., et al. Learning to see through multimode fibers. *Optica* **5**, 960-966 (2018).
22. Li, S. H., et al. Compressively sampling the optical transmission matrix of a multimode fibre. *Light: science & applications* **10**, 1-15 (2021).
23. Gordon, G. S. D., et al. Characterizing optical fiber transmission matrices using metasurface reflector stacks for lensless imaging without distal access. *Physical review X* **9**, 041050 (2019).
24. Flaes, D. E. B., et al. Robustness of light-transport processes to bending deformations in graded-index multimode waveguides. *Physical review letters* **120**, 233901 (2018).
25. Tsvirkun, V., et al. Flexible lensless endoscope with a conformationally invariant multi-core fiber. *Optica* **6**, 1185-1189 (2019).
26. Kim, Y., et al. Adaptive multiphoton endomicroscope incorporating a polarization-maintaining multicore optical fibre. *IEEE Journal of Selected Topics in Quantum Electronics* **22**, 6800708 (2015).
27. Gordon, G. D., et al. Coherent imaging through multicore fibres with applications in endoscopy. *Journal of Lightwave Technology* **37**, 5733-5745 (2019).
28. Vellekoop, I. M. & Mosk, A. P. M Focusing coherent light through opaque strongly scattering media. *Optics letters* **32**, 2309-2311 (2007).
29. Popoff, S., et al. Image transmission through an opaque material. *Nature communications* **1**, 81 (2010).
30. Mosk, A. P., et al. Controlling waves in space and time for imaging and focusing in complexmedia. *Nature photonics* **6**, 283-292 (2012).
31. Rotter, S. & Gigan, S. Light fields in complex media: Mesoscopic scattering meets wave control. *Reviews of modern physics* **89**, 015005 (2017).
32. Leite, I. T., et al. Observing distant objects with a multimode fiber-based holographic endoscope. *APL Photonics* **6**, 036112 (2021).
33. Leite, I. T., et al. Three-dimensional holographic optical manipulation through a high-numerical-aperture softglass multimode fibre. *Nature Photonics* **12**, 33-39 (2018).
34. Čižmár, T., Mazilu, M. & Dholakia, K. In situ wavefront correction and its application to micromanipulation. *Nature photonics* **4**, 388-394 (2010).
35. Čižmár, T. & Dholakia, K. Shaping the light transmission through a multimode optical fibre: complex transformation analysis and applications in biophotonics. *Optics express* **19**, 18871-18884 (2011).

Abdurashitov et al. *Light: Advanced Manufacturing* (2022)3:30
https://doi.org/10.37188/lam.2022.030

Official journal of the JHL 2689-9620
www.light-am.com

Article | Open Access

Integrated binary hologram to monitor cargo release from a drug-eluting film

Arkady S. Abdurashitov[1,*], Pavel I. Proshin[1], Valery V. Tuchin[2,3] and Gleb B. Sukhorukov[1,4,5]

Abstract

Nowadays, holography translates from a pure technical tool for recording the phase and amplitude of the light wave to a widely applicable research-based method. Holographic devices are used for security enhancement, entertainment, 3D display technologies and augmented reality. Binary computer-generated holograms (amplitude or phase-based) are of specific interest. They are easy to compute and their manufacturing methods are fast and robust. In this work, a method of manufacturing amplitude-only binary holograms out of biopolymers films is proposed. Opaque cargo bits made out of different bioactive substances (antibiotics, dyes, etc.) absorb or scatter specific parts of the incoming light wave. Cargo release was conducted by submerging the produced holograms into the aqueous solution at specific pH and temperature levels. Drug elution from the film was traced out by the numerical analysis of the far-field diffraction pattern, which otherwise could only be measured post-factum using sophisticated spectroscopic or chromatography devices. This approach broadens the application of general holography over to the field of biomedical research relevant to quantitative monitoring of the drug elution.

Keywords: Computer generated holograms, Biopolymers, Release monitoring, Payload holograms

Introduction

Holography per se is a way to record both the intensity and phase of a light wave. Unlike traditional photography, which maps 3D wave-front onto the 2D plane through intensity-only recording, holographic image capturing allows for depth-perception and different viewing angels[1,2]. more importantly, a mathematical framework which was derived to describe the physical principle of the hologram[3-5] could be used to simulate light wave propagation and diffraction pattern from specific diffractive optics elements (DOEs). Given desired diffraction pattern, corresponding DOE could be computed and processed for manufacturing. The rapid growth in computing power of modern personal computers allows computer-generated holograms (CGH) to be digitally produced in a fraction of a second[6,7].

The Iterative Fast Fourier Transform Algorithm (IFFTA) is one of the best known mathematical platforms for evaluating the DOE configuration to obtain the desired far-field diffraction pattern. IFFTA is suitable for getting binary amplitude-only or phase-only CGHs as well as for multi-level phase-only holograms[8,9]. The algorithm's core is repeatedly altering between spatial and frequency domains, substituting the phase factor from the previous iteration to the next one. Usually, a root mean squared error (RMSE) is used to compare the calculated far-filed diffraction pattern to the desired image and stop the iteration process. The crucial step of the IFFTA is the phase quantization process.

Correspondence: Arkady S. Abdurashitov (a.abdurashitov@skoltech.ru)
[1]Center for Neurobiology and Brain Restoration, Skolkovo Institute of Science and Technology, Moscow 121205, Russia
[2]Science Medical Center, Saratov State University, Saratov 410012, Russia
Full list of author information is available at the end of the article.
These authors contributed equally: Arkady S. Abdurashitov, Pavel I. Proshin, Valery V. Tuchin, Gleb B. Sukhorukov

All numerical calculations are usually performed using 32 or 64-bits float numbers, but manufacturing and displaying technologies operate with a much lower bit number (1-8 bits). The quantization step is required to reduce the number of bits. In this work, we will consider and produce a binary hologram (1 bit). In this case, the phase and amplitude holograms are mathematically equivalent. The straightforward approach is to divide phase interval from 0 to 2π into two equal segments and use π value as a threshold for phase quantization. However, more complicated algorithms were proposed[10–12] to find the optimal threshold value and obtain the best possible far-field diffraction outcome. After the quantization step, the resulted hologram is ready to be manufactured.

Polymers are promising materials for the production of DOEs[13–15]. Their optical properties are well known and could be tuned to match optical needs[16–18]. Some polymers are bio-compatible/bio-degradable such as polylactic-co-glycolic acid (PLGA), poly(ε-caprolactone) (PCL), etc.[19] or even has an Food and Drug Administration (FDA) approval for in vivo usage like polylactic acid (PLA)[20]. PLA is transparent within the visible spectral range, and the performance of the amplitude holograms is mostly independent of the illumination wavelength (only image scale is affected). These properties make it possible to use a large variety of laser sources to reconstruct diffraction patterns. Such attributes makes PLA one of the best materials for manufacturing DOEs, specifically for biomedical applications[19].

There are three leading technologies for polymer-based DOEs production:

1. *Micro-hot embossing.* In this approach, a master mold usually made of nickel or rigid thermo-resistive polymer is pressed into the polymer film. Processed film is heated up to its glass transition temperature[21–23]. Under pressure and heat, the master mold pattern is transferred onto the surface of the processed polymer.

2. *Laser ablation and engraving.* The ultra-short laser pulses (nanoseconds and less) perform local polymer removing to form the desired surface pattern of DOEs[24–26]. Typically, UV lasers were used as most polymers have moderate to significant absorption coefficient in this spectral range.

3. *Casting.* This method implies using solvents. At first, the polymer is dissolved in the suitable solvent. Then it is casted into a special mold, or the mold itself is dip-coated into the dissolved polymer[27,28]. Usually, molds are made out of inert silicone-based polymers like polydimethylsiloxane (PDMS). Master molds should be wettable by the solvent; otherwise, fine surface patterns will not be replicated.

All of the above-mentioned manufacturing methods produce phase-only DOEs patterns either by redistribution polymer volume (micro-hot embossing and casting) or permanently removing local parts of the polymer film (laser ablation or engraving). In order to produce amplitude-only hologram, usually a spot of opaque ink is deposited onto the substrate's surface. In all of the described methods, the produced DOE only carries information in terms of their diffraction pattern and not physically. In principle, useful bioactive substances could be doped into the ink, but in this case, the cargo mass will be much smaller compared to the proposed approach, and compatibility between light-blocking ink and the drug should be addressed.

We introduce an additive approach of producing biocompatible DOEs using direct drug printing (DDP). Bio -active substances are hot-printed onto the surface of flat polymer film as cargo bits. Spatial distribution of cargo bits forms the precomputed DOE. In our approach, the bioactive substance does not interact with any organic solvents and is placed "as is" onto the inert biopolymer, ensuring no changes in the pharmacological effect of the drug. The resulted film acts as a transmissive amplitude-only hologram. This cargo packaging system, that creates a clear far-field diffraction pattern when illuminated by a coherent light source, can find its place in various biomedical problems. One direct applications of payload holograms is tracking the lifespan of colorless content and measurement of the characteristic release time of active substances under various environmental conditions. The most likely application of our technology is to complement the standard antibiotic testing procedure. Adding a visual channel for tracking the drug release, which requires only the presence of a coherent light source, will significantly improve the assessment of the effect of the antibiotic and allow to control visually the release time and amount of eluted cargo. The central novelty of the paper is the fact that proposed holograms not only carry information in terms of precomputed diffraction patterns but also have a physical useful load in the form of light-blocking cargo-bits made of different bioactive substances. As far as we know, our paper is the first to demonstrate the possibility of using holograms as a payload carrier.

Results

Computer simulation of the far-field diffraction pattern in cases of different cargo release types

The release of the cargo can be divided into two main types. The first type is an accidental release from individual sealed cargo bits. In this case, release occurs consequently at random spots of produced hologram due to sealing defects or natural degradation of the polymer film. The second type is a continuous dissolving of unsealed cargo. In this case, the release coincides across the entire area of the hologram at a specific rate, governed by the cargo amount and its solubility in a given solvent. Fig. 1 schematically illustrates these release types and the main idea of payload DOE.

The first release type was simulated by randomly removing 10% of the initial opaque cargo bits from the hologram unless the desired release mark was reached. The second type was implemented as iterative reduction of overall hologram contrast by 10% unless hitting the same release mark. Root mean square contrast (RMSC) was used to measure the visibility of the far-field diffraction pattern for both types of the release. Results of numerical modeling presented in Fig. 2 clearly shows the difference between described cargo releases schemes.It is a well-known fact that by its nature, each elementary part of the hologram contains the information about the whole scene that is being recorded[4]. It means that a small uncorrupted part of the initial hologram is sufficient to reconstruct the entire recorded scene with limited contrast. But if the hologram loses its opaqueness uniformly across the whole area at a given rate, it will lead to a faster decay in terms of output image contrast with respect to the remaining payload. Our simulations show that, in principle, up to 70% of the cargo is traceable in the case of local accidental release and up to 40% in the case of simultaneous release from the entire area of the hologram.

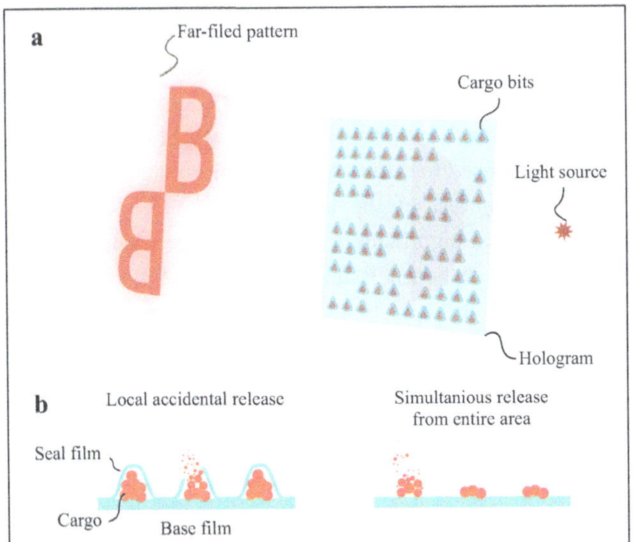

Fig. 1 a Main concept of payload holograms. Opaque cargo bits are placed in a specific order onto the base film to form pre-computed pattern. By illuminating this film with a coherent light source, a desired far-field diffraction pattern could be obtained. **b** Graphical representation of abovementioned cargo release types.

Manufacturing of amplitude-only holograms loaded with bioactive substances

Binary amplitude-only holograms are manufactured using the direct drug printing method (DDP). This method is designed to work primarily with water-soluble cargos as the most bioactive substances have this property. PDMS mold with wells located according to IFFTA outcome is first loaded with a fine powder of active substance. Most of the dry excess is then removed using a silicone putty knife (SPK). The loaded mold is placed into the freezer at −20°C for 2 minutes. After the mold is taken out of the freezer, a certain amount of condensation would immediately form on its surface, wetting the loaded powder in the well and fully dissolves a small amount of excessive substance between them. Dissolved excess material could be easily removed using the same SPK as before. Wetting of powder that sits in the wells allows it to clunk and form a single bit that can be extracted from the mold. If we neglect the wetting process, the crystals inside the well will be in an unbound state and only a small part of them will be removed from the wells during printing. Highly water-soluble loads like NaCl could form a monocrystalline at this stage.

To extract cargo bits, the mold is rolled onto a 15 μm thick PLA film heated up to 100°C. During the rolling process, viscous and sticky polymer picks up the cargo bits and extracts them from the mold's wells. The result of such rolling is an amplitude-only hologram consisted of a flat polymer film on which opaque cargo bits are located in a certain arrangement. In this study, polylactide acid (PLA, Sigma Aldrich) was used as a base polymer for hologram manufacturing. PLA is biocompatible and biodegradable as well as FDA approved for in vivo application. Still, other polymers with low melting points and/or glass transition temperature like polycaprolactone (PCL) or poly lactic-co-glycolic acid (PLGA) could also be used as base material. The final step is to seal the cargo bits to prevent them from fast dissolving. A flat and soft PDMS sheet is first dipped into the 1% (w/w) solution PLA in chloroform to form a thin (1 μm) film at the PDMS's surface. After that step, flat-coated PDMS is rolled onto the preheated hologram (70°C) to seal exposed cargo bits. The total area of the produced hologram is 1 cm^2. The minimum size of the hologram is 2 × 2 mm (limited by the reconstruction beam size), and maximum size 5 × 5 cm (limited by the manufacturing equipment). But the 1 × 1 cm proved to be optimal from the perspective of production simplicity and speed, ease of use, and sufficiency of the loaded substance. Graphical representation of DDP method and manufactured holograms with different cargo is illustrated in Fig. 3. For details of hologram computing and mold manufacturing,

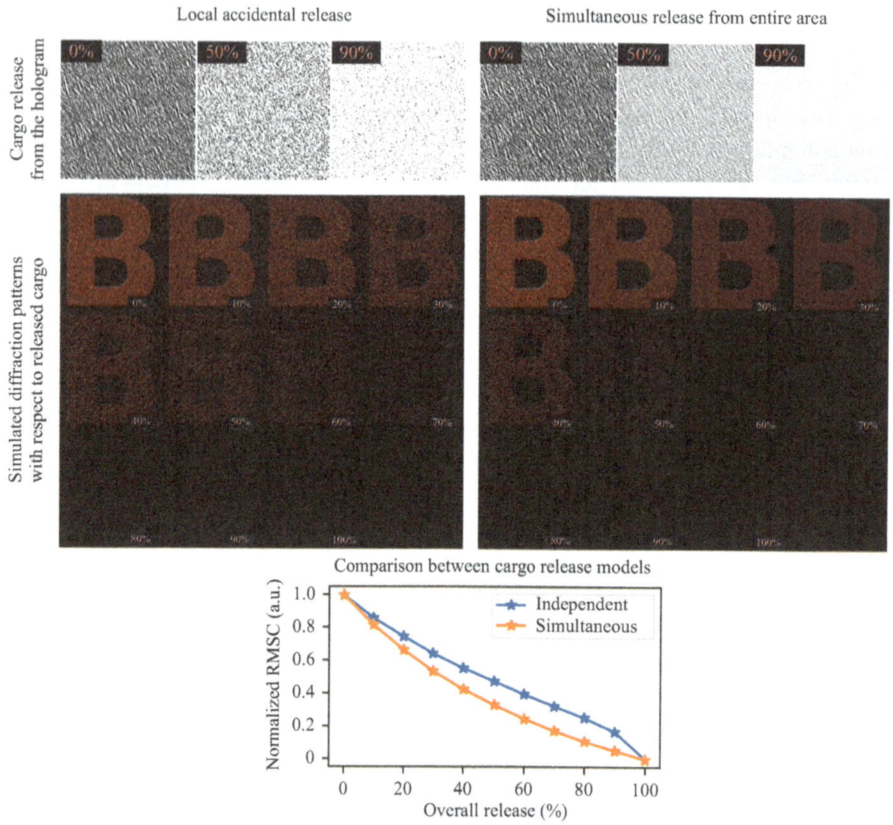

Fig. 2 Computer simulation of different cargo release modes and their effect on the resulted far-field diffraction pattern. The first row shows the appearance of holograms as they release load in various fashions. The local accidental release mode preserves the light-blocking ability of individual cargo bits and their placement, but significantly reduces their overall quantity. Simultaneous release across the whole area of the hologram keeps the total number of cargo bits and their placement unchanged. Instead, the total transparency of the hologram increases. Root mean square contrast of a "B" letter produced in a far-filed zone demonstrates that the local accidental release mode has greater dynamic range when assessing the release of a bioactive substance. Up to 70% of initial cargo could be tracked accurately compared to 40% in the case of simultaneous release from the entire hologram.

please refer to the materials and methods section.

Cargo release from the loaded hologram and its effect on the far-field diffraction pattern

To verify the effect of cargo release on the far-field diffraction pattern, a set of holograms were manufactured using the abovementioned DDP method. As for cargo, we chose vancomycin hydrochloride as it is a well water-soluble (100 mg/ml) powder medication generally used to treat a number of bacterial infections[29,30].

Cargo release was conducted by submerging manufactured holograms into 100 ml of water solution of KOH (pH 8.5) at 37°C. It is a known fact that PLA is resistant to acidic environments (pH 3-5) and progressively increases its degradation rate with higher pH values (pH > 7)[31,32]. After 24 hours of slow stirring, holograms were extracted from the solution, washed out with deionized water, and dried with compressed air (Fig. 4a).

To check the diffraction patterns before and after cargo release, a simple test setup was arranged. Red diode laser (650 nm, 5 mW) was used as a light source. Black screen was placed at 4 meters from the hologram plane. A black screen was used to dim the diffracted light and increase the apparent contrast of the hologram. The use of a white screen gave a significantly reduced image contrast due to diffuse light scattering within the screen. CCD camera with a 25 mm photo lens was focused onto the screen's surface (Fig. 4b). As expected, the cargo release process leads to an absence of a far-field diffraction pattern. SEM images of a processed hologram show no cargo bits and presence of large number of pores and craters. This is a consequence of the actively proceeding process of polymer degradation and release of a bioactive substance (Fig. 4c). Experimental results clearly show that the far-field diffraction pattern has the ability to reflect the current state of cargo release process. A clearly distinguishable letter "B" was obtained in the far-field diffraction zone from the initial hologram. At 6 and 12 hours release marks, a gradual decrease in the

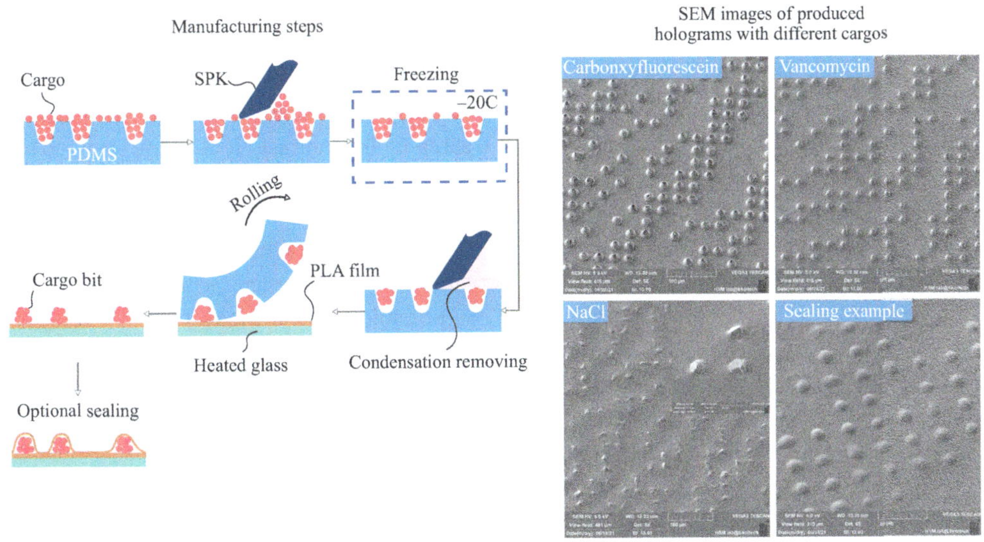

Fig. 3 Schematic representation of the manufacturing process and scanning electron microscopy (SEM) of the produced holograms. The critical steps in the manufacturing process are excess powder removing and freezing. Freezing allows for formation of condensation onto the surface of the PDMS mold. This condensate dissolves the small amount of excess powder left after the initial scurrying process. This thin layer of water solution could be easily removed. The result is almost no redundant powder left in-between the cargo bits that could corrupt the far-field diffraction pattern. The size of a cargo bit is governed by the physical dimensions of the well in the mold and the physical and chemical properties of the loaded powder. Some highly water-soluble cargo like NaCl could produce monocrystalline structures within the mold indentations.

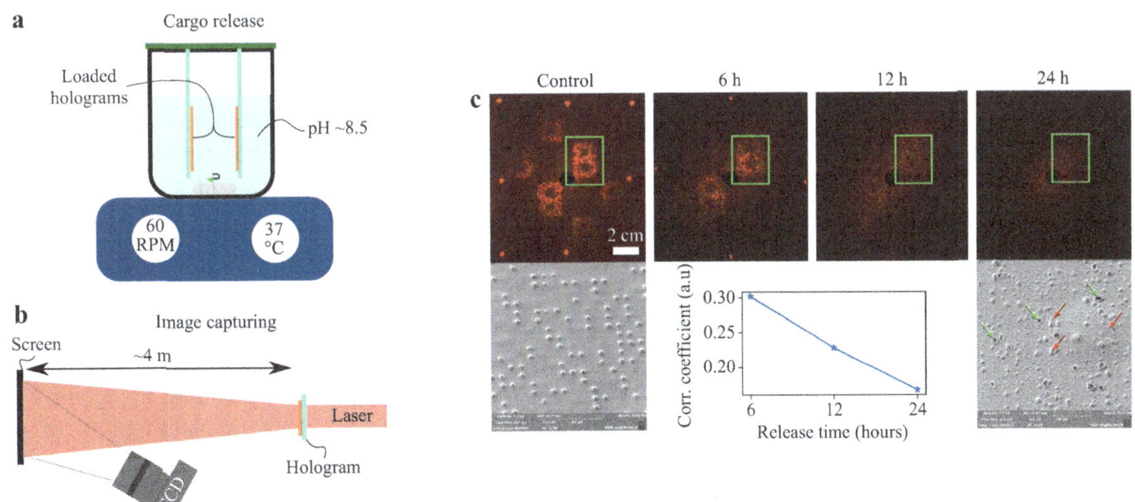

Fig. 4 Experimental setups for **a** cargo release inducing and **b** far-field diffraction pattern capturing. Cargo release was performed in a chemical beaker covered with a plastic cap to prevent vapor escaping and keep the total volume of the solution unchanged. In order to uniformly distribute releasing cargo and avoid concentration spikes at the surface of the hologram, slow stirring (60 RPM) was applied. **c** Far-field diffraction patterns obtained at different timestamps of a cargo release process and corresponding SEM images of hologram's surfaces at the initial state-'control' and 24 hours mark since the start point of the release. The correlation coefficient was computed between the diffraction images at different release times within the ROI indicated by the green rectangle.

quality of the diffraction pattern is observed. After 24 hours of the cargo release procedure, no clear diffraction pattern was observed. A slight halo around the zero-order diffraction spot was produced due to the formation of many micro-pores and craters correspondingly marked by the red and green arrows. These micro-pores and craters are formed because of the polymer degradation process. Correlation analysis of acquired diffraction images presented at the graph (Fig. 4c) shows the linear dependence between the correlation coefficient and the release time which is consistent with simulation data, presented in Fig. 2 for the local accidental release mode. In the real case, the release mechanism will tend to lean into the random regime, as the polymer will degrade unpredictably along its surface while being exposed to the inhomogeneous aqueous environment of the living system[33]. Computer-generated holograms as micro-packaging system opens up a potential opportunity for such devices to be used as a quantitative screening tool for cheap and fast cargo lifespan monitoring.

Discussion

An extension of polymer-based binary holograms to biomedical applications is proposed. We note that there have been recent reports on the DOEs manufacturing out of biopolymers such as PLA[34]. The noticeable pool of articles dedicated to developing biopolymer drug packaging and delivery microsystems in the form of microchambers placed on a thin film[35–39]. These works focused on the problems of encapsulation of different types of cargo, especially low molecular-weight ones, while aiming to propose a way for controlling cargo release dynamic in various environments. Making these packaging systems sensitive to external stimuli such as ultrasound waves and high-power laser radiation could be done to accelerate the cargo release process[40]. However, the effects of the interaction of low-energy laser radiation with array of micro-containers and their ability to form complex diffraction patterns were practically not considered in these works. To our knowledge there is only one article dedicated to that problem, but the authors confined themselves to considering only simple 2D diffraction grating[41]. We think that there are potential benefits of manufacturing these packaging and delivery polymer microsystems in such a way that they can produce a complex diffraction pattern. This will open a new way of low-cost and efficient monitoring of cargo lifespan. In that case, diffraction efficiency has to be taken into account. Diffraction efficiency is the ratio between the amount of light incident on the hologram and the amount of light which forms diffraction pattern. The main goal of any packaging system is to achieve a high mass of an active substance per area unit while maintaining needed mechanical and chemical properties. In a case of amplitude-only holograms, some parts of the film must be cargo-free to allow light beam passage. The area of these non-payload parts could be minimize by choosing a proper far-field diffraction pattern.

Cargo-bits are small, separated micro-containers. The size of such micro-containers does not exceed several tens of micrometers. They are practically indistinguishable to the naked eye. Of course, it is possible to observe the release of drugs from micro-containers by their optical image. However, for this it is necessary to use an optical system with at least a tenfold magnification. In this case, the field of view of the optical system will not exceed a few square millimeters, which will require scanning of the sample. In contrast, it is quite easy to expand the laser beam to the required size (several square centimeters), which will cover the entire area of the hologram. To observe the diffraction pattern, it will be sufficient to use a simple converging lens with the required focal length, or it is possible to do without it by observing diffraction in the far zone at a distance of several meters from the hologram. These considerations applies to simple cases, such as in-vitro testing of the characteristic release times of antibiotics and other drugs.

Yet, we have showed only analog way to reconstruct the far-field diffraction pattern using the laser. In principle, the same pattern could be obtained numerically from images of the hologram's surface. For example, a direct reconstruction method via laser will not work if a payload hologram is placed in a highly scattering environment, for example, subcutaneously. However, the recent outstanding progress in the field of optical clearing of biological tissues[42–48] makes us confident that payload holograms embedded in the human body will provide the necessary information through skin. Optical clearing techniques significantly reduces the scattering coefficient of a tissue, making it much more transparent within a reasonable amount of time. This virtual transparency window could be used to obtain optical images of the holographic surface. Numerical reconstruction could then performed to compute the far-field diffraction pattern.

In this work, we focused on the amplitude-only binary holograms as they are most suited to perform as cargo-carriers. Amplitude holograms, in general, possess the lowest diffraction efficiency among hologram types. The binary nature forces them to block $\approx$50% of incoming optical power. Another half is directed to form zero-order and precomputed diffraction patterns correspondingly. It is considered that for amplitude hologram, maximum efficiency is 0.125 (1 is perfectly efficient)[49]. In our case, the main factor influencing the image quality is the process of infilling stamp's wells with the cargo. In order for the amplitude hologram to operate near its limit, it is necessary to ensure complete opaqueness of the specific parts of the hologram. Due to the different physical and chemical properties of loaded powders as well as the manual loading process, there is a drift in the opaqueness of cargo-bits along the surface of the hologram. Such drift could be minimized by the development of the automatic cargo-loading machine that ensures uniform infill of the wells as well as utilizing different additives that improve the cohesive properties of the drug to help form a dense light-blocking bit.

We have to mention that in the scope of our paper, we dominantly described the fact that conventional holograms could be used as a payload carrier. From this perspective, it is important to achieve a high cargo mass per unit area of the hologram. The geometry of the container (individual diffractive bit) highly affects the quality of produced image[50] and it is a trade-off between high useful volume, ability to be loaded with active substances, and acceptable diffraction image quality. As we mentioned above, in the framework of the proposed case, the resulting contrast of the diffraction pattern can be reduced for the sake of increasing the overall drug loading and ease of manufacturing. To achieve the best diffraction output, the cargo bit must be flattened out, and space between them should be eliminated to form light-blocking islands instead of individual dots. However, this will greatly reduce the overall mass of carried load and complicate the manufacturing process as well as disturb the release kinetics.

The impact of various parameters of the system, such as the pitch of the cargo-bits, their size, and film thickness on the holograms, is predictable. For instance, the pitch will affect the scale of the produced diffraction pattern, pitch and size of the cargo-bits will produce a certain surface density, which is directly related to the cargo-mass per unit area of the hologram. The increase in thickness of the sealing film will delay the drug-eluting process in time accordingly. These factors could be used to tune the hologram for the best performance for a specific drug and characteristic release times, but they are not decisive in demonstrating the effect itself.

Theoretically, there is a direct relationship between the cargo-bit distribution in the plane of the hologram and the far-field diffraction pattern. In practice, degradation of the polymer film occurs simultaneously with the drug-eluting process from the cargo bits. Degradation of the polymer film manifests itself in the form of the appearance of a large number of micro-pores of different sizes. These release-associated micro-pores make a significant contribution to the reduction of the contrast of the diffraction pattern and make quantitative cargo release analysis hard to implement. one way around is to first calibrate the degradation process of the base film and its effect on the far-field diffraction pattern and then apply this data to correct the observed precomputed far-field image. A few bold assumptions have to be addressed at this point, namely: polymer degradation is isotropic, and micro-pores morphology is the same at different parts of the films and should not vary much from film to film.

In summary, we have described the method for producing payload holograms. The proposed direct drug printing technique works with water-soluble compounds, but can be adapted to other types of substances if necessary. Numerically and experimentally, we have shown the ability of the diffraction pattern to reflect the amount of cargo remaining in the hologram after a certain amount of time. This unique feature of holographic packaging systems can potentially be used in a wide range of tasks requiring monitoring of the amount of emitted substance. One such application could be microbiotic testing. A visual channel for tracking the amount of released substance can significantly increase the amount of information received during testing and more accurately characterize the effect of an antibiotic.

Materials and methods

Computer generated hologram calculating using iterative Fast Fourier Transform Algorithm

The self-written algorithm in Python code was used to compute computer-generated holograms. To compare the produced diffraction pattern and desired image at each iteration root mean square error was calculated. 256 iterations were done to obtain the final hologram. At the quantization stage π value was chosen as a single threshold. There is a direct relation between the distribution of binary transmission and the distribution of a binary phase redundancy[51]. During the computation of the hologram we can consider the binary amplitude hologram as a binary phase transmittance which can only have the complex values $exp(-i0) = 1$ and $exp(-i\pi) = -1$. Offsetting these values with a DC-term transforms the resulting binary phase distribution into a binary amplitude distribution between 0 and 1[52]. 128×128 pixels images were processed using this Python script.

Computer simulation of the drug-elution effect on the visibility of the diffraction pattern and experimental data analysis

A custom Python script was used to simulate the spontaneous local release and continuous dissolving of the cargo bits and their effect on the resulting far-field diffraction pattern. For a quantitative assessment of the visibility of the diffraction pattern, the value of RMSC was calculated by the formula:

$$RMSC = \sqrt{\frac{1}{MN}\sum_{n=1}^{N}\sum_{m=1}^{M}(I_{n,m} - \bar{I})^2} \qquad (1)$$

there M, N are the number of pixels along the width and

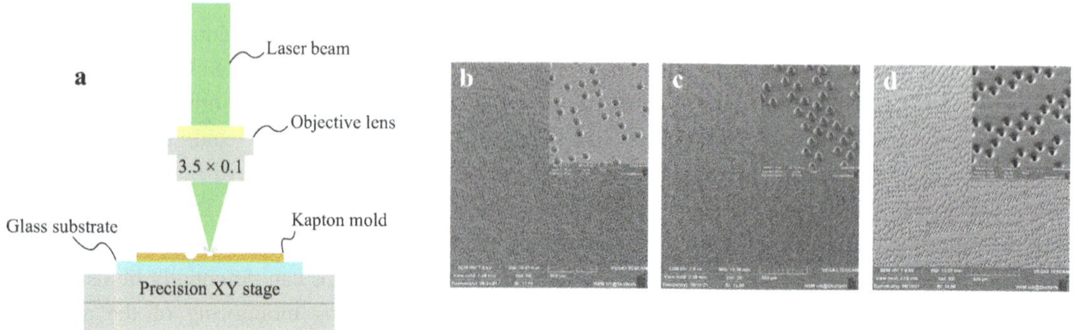

Fig. 5 **a** The most significant aspects of the experimental laser ablation setup. A collimated laser beam fully illuminates the back entrance pupil of the objective lens to ensure diffraction-limited spot size at the front focal plane. **b** SEM image of manufactured Kapton master mold. **c** SEM image of the negative PDMS mold featuring surface quality of produced indentations via laser ablation process. **d** SEM image of the positive PDMS mold used for direct drug printing method of hologram manufacturing.

height in the region of interest (ROI) which fully enclose the "B" letter, $I_{n,m}$ and $\overline{I}$ are the intensity of a given pixel and average intensity within the ROI correspondingly. In order to evaluate the effect of cargo release in a real experiment, a set of images were taken at different time-points of a drug-eluting process, namely 6, 12 and 24 hours. Correlation coefficient between the initial diffraction pattern (letter "B") and various stages of release was computed using predefined function *corrcoef* from the *Numpy* module to quantify the effect of a drug release process.

Mold manufacturing process

In the first stage, the master mold was manufactured out of Kapton polymer using laser ablation method. Cobolt Tor XS (532 nm, 50 µJ, 1.9 ns) was used as a light source. The microscopic objective lens with a numerical aperture of 0.1 focuses laser light onto the Kapton's surface. 25 laser pulses were used to form one indentation. The pulse repetition rate was set to 1 kHz. To control the spatial position of the master mold, the precise XY stage driven by the custom software was utilized. Fig. 5a highlights the most critical aspects of a custom-made laser ablation system. Resulted indentations are in a cone form with diameter 22 ± 2 µm, height 20 ± 2 µm and angle of 35°. Step between indentations was set to 40 µm. The second step is to replicate Kapton mold features into the PDMS mold, suitable for direct drug printing protocol. Sylgard 184 (DOWSIL) was mixed in the manufacturer prescribed ratio (10:1 w/w), then degassed in the vacuum chamber for 30 minutes and poured onto the surface of the patterned Kapton film. The whole sample was placed into the oven at 100°C for 35 minutes. After the curing process negative PDMS mold (with pillars) was detached from the Kapton film. The third step is PDMS to PDMS casting. This casting cannot be done directly because during the curing process, two PDMS molds will stick together and form a monolithic unit unless a separating layer is used. Negative PDMS mold was dip-coated by the 0.5% (w/w) solution of PLA in chloroform to form a separating film. A new batch of PDMS was mixed using the same protocol as before and poured onto PLA-covered negative PDMS mold. PDMS was cured in the oven at 60°C for 4 hours. Such low curing temperature was chosen to prevent the thermal softening of polylactide film. After curing, two PDMS molds could be easily separated from each over. Resulted positive PDMS mold was used for payload hologram manufacturing.

Acknowledgements

VVT was supported by the grant No. 13.2251.21.0009 of the Ministry of Science and Higher Education of the Russian Federation (agreement No. 075-15-2021-942); the work was also partly supported by the Siberian State Medical University's development program Priority 2030.

Author details

[1]Center for Neurobiology and Brain Restoration, Skolkovo Institute of Science and Technology, Moscow 121205, Russia. [2]Science Medical Center, Saratov State University, Saratov 410012, Russia. [3]A.N. Bach Institute of Biochemistry, Research Center of Biotechnology of the Russian Academy of Sciences, Moscow, 119071, Russia. [4]School of Engineering and Materials Science, Queen Mary University of London, London E1 4NS, United Kingdom. [5]Siberian State Medical University, Moskovskiy Trakt, 2, Tomsk 634050, Russia

Author contributions

Authors contributes equally to this work.

Conflict of interest

Authors declares no conflict of interests.

Received: 01 September 2021 Revised: 12 April 2022 Accepted: 20 April 2022
Accepted article preview online: 26 April 2022
Published online: 19 May 2022

References

1. Collier, R. Optical Holography. (Amsterdam: Elsevier, 2013).
2. Goodman, J. W. Introduction to Fourier Optics. 3rd edn. (Englewood: Roberts & Co., 2005).
3. Hariharan, P. Optical Holography: Principles, Techniques, and Applications. 2nd edn. (Cambridge: Cambridge University Press, 1996).
4. Schnars, U. et al. Digital holography. in Digital Holography and Wavefront Sensing (eds Schnars, U. et al.) (Berlin, Heidelberg: Springer, 2015), 39-68.
5. Poon, T. C. & Liu, J. P. Introduction to Modern Digital Holography: with MATLAB. (Cambridge: Cambridge University Press, 2014).
6. Shimobaba, T. et al. Fast calculation of computer-generated-hologram on AMD HD5000 series GPU and OpenCL. *Optics Express* **18**, 9955-9960 (2010).
7. Takada, N. et al. Fast high-resolution computer-generated hologram computation using multiple graphics processing unit cluster system. *Applied Optics* **51**, 7303-7307 (2012).
8. Ripoll, O., Kettunen, V. & Herzig, H. P. Review of iterative Fourier-transform algorithms for beam shaping applications. *Optical Engineering* **43**, 2549-2548 (2004).
9. Soïfer, V. A. Methods for Computer Design of Diffractive Optical Elements. (New York: John Willey & Sons, Inc., 2002).
10. He, Z. H. et al. Optimal quantization for amplitude and phase in computer-generated holography. *Optics Express* **29**, 119-133 (2021).
11. Ruffato, G. et al. Design, fabrication and characterization of computer generated holograms for anti-counterfeiting applications using OAM beams as light decoders. *Scientific Reports* **7**, 18011 (2017).
12. Long, P. & Hsu, D. Quantization and sampling considerations of computer-generated hologram for optical interconnection. Proceedings of SPIE 1461, Practical Holography V. San Jose, CA, United States: SPIE, 1991.
13. Khan, M. S. et al. Polymer-based diffractive optical elements for rear end automotive applications: design and fabrication process. *Applied Optics* **57**, 9106-9113 (2018).
14. Schauer, S. et al. Tunable diffractive optical elements based on shape-memory polymers fabricated via hot embossing. *ACS Applied Materials & Interfaces* **8**, 9423-9430 (2016).
15. Winfield, R. J. , et al. Production of polymer diffractive optics by contact printing. Proceedings of SPIE 4876. SPIE, (2003).
16. Aziz, S. B. et al. A comprehensive review on optical properties of polymer electrolytes and composites. *Materials* **13**, 3675 (2020).
17. Hutchinson, M. H. et al. Optical properties of polylactides. *Journal of Polymers and the Environment* **14**, 119-124 (2006).
18. Hossain, S. Optical properties of polymers and their applications. Theses 1685. (2019). at https://digitalcommons.njit.edu/theses/1685/.
19. Arif, U. et al. Biocompatible polymers and their potential biomedical applications: a review. *Current Pharmaceutical Design* **25**, 3608-3619 (2019).
20. PLA is an FDA-approved generally recognized as safe (GRAS) polymer. at https://www.accessdata.fda.gov/cdrh_docs/pdf8/K082276.pdf.
21. Peng, L. F. et al. Micro hot embossing of thermoplastic polymers: a review. *Journal of Micromechanics and Microengineering* **24**, 013001 (2014).
22. Deshmukh, S. S. & Goswami, A. Hot Embossing of polymers-a review. *Materials Today:Proceedings* **26**, 405-414 (2020).
23. Sun, J. Y. et al. Development and application of hot embossing in polymer processing: a review. *ES Materials & Manufacturing* **6**, 3-17 (2019).
24. Ravi-Kumar, S. et al. Laser ablation of polymers: a review. *Polymer International* **68**, 1391-1401 (2019).
25. Treviño-Palacios, C. G., Zapata-Nava, O. J. & Olivares-Pérez, A. Optical damage as a computer generated hologram recording mechanism. *Journal of Applied Research and Technology* **13**, 591-595 (2015).
26. Wang, Z. P. et al. High-quality micropattern printing by interlacing-pattern holographic femtosecond pulses. *Nanophotonics* **9**, 2895-2904 (2020).
27. Rajput, D. et al. Solution-cast high-aspect-ratio polymer structures from direct-write templates. *ACS Applied Materials & Interfaces* **5**, 1-5 (2013).
28. Brinker, C. J. Dip coating. in Chemical Solution Deposition of Functional Oxide Thin Films (eds Schneller, T. et al.) (Vienna: Springer, 2013), 233-261.
29. Levine, D. P. Vancomycin: a history. *Clinical Infectious Diseases* **42 Suppl 1**, S5-S12 (2006).
30. Álvarez, R. et al. Optimizing the clinical use of vancomycin. *Antimicrobial Agents and Chemotherapy* **60**, 2601-2609 (2016).
31. Xu, L. B., Crawford, K. & Gorman, C. B. Effects of temperature and pH on the degradation of poly (lactic acid) brushes. *Macromolecules* **44**, 4777-4782 (2011).
32. Lazzari, S. et al. Modeling the pH-dependent PLA oligomer degradation kinetics. *Polymer Degradation and Stability* **110**, 80-90 (2014).
33. Sharma, S. K. & Mudhoo, A. A Handbook of Applied Biopolymer Technology: Synthesis, Degradation and Applications. (Cambridge: RSC Publishing, 2011).
34. Cambiasso, J., Goyanes, S. & Ledesma, S. Holographic gratings recorded in poly (lactic acid)/azo-dye films. *Optical Materials* **47**, 72-77 (2015).
35. Gai, M. Y. et al. Polylactic acid sealed polyelectrolyte multilayer microchambers for entrapment of salts and small hydrophilic molecules precipitates. *ACS Applied Materials & Interfaces* **9**, 16536-16545 (2017).
36. Gai, M. Y. et al. Polylactic acid nano-and microchamber arrays for encapsulation of small hydrophilic molecules featuring drug release *via* high intensity focused ultrasound. *Nanoscale* **9**, 7063-7070 (2017).
37. Sindeeva, O. A. et al. Polylactic acid-based patterned matrixes for site-specific delivery of neuropeptides on-demand: functional NGF effects on human neuronal cells. *Frontiers in Bioengineering and Biotechnology* **8**, 497 (2020).
38. Zhang, J. X. et al. Microchamber arrays made of biodegradable polymers for enzymatic release of small hydrophilic cargos. *Soft Matter* **16**, 2266-2275 (2020).
39. Sindeeva, O. A. et al. Effect of a controlled release of epinephrine hydrochloride from PLGA microchamber array: in vivo studies. *ACS Applied Materials & Interfaces* **10**, 37855-37864 (2018).
40. Kurochkin, M. A. et al. Laser-triggered drug release from polymeric 3-D micro-structured films via optical fibers. *Materials Science and Engineering: C* **110**, 110664 (2020).
41. Zhang, J. X. et al. Stimuli-responsive microarray films for real-time sensing of surrounding media, temperature, and solution properties via diffraction patterns. *ACS Applied Materials & Interfaces* **12**, 19080-19091 (2020).
42. Tuchin, V. V. Optical clearing of tissues and blood using the immersion method. *Journal of Physics D: Applied Physics* **38**, 2497-2518 (2005).
43. Zhu, D. et al. Recent progress in tissue optical clearing. *Laser & Photonics Reviews* **7**, 732-757 (2013).
44. Bashkatov, A. N. et al. Measurement of tissue optical properties in the context of tissue optical clearing. *Journal of Biomedical Optics* **23**, 091416 (2018).
45. Oliveira, L. M. C. & Tuchin, V. V. The Optical Clearing Method: a New Tool for Clinical Practice and Biomedical Engineering. (Cham: Springer, 2019).
46. Yu, T. T. et al. Physical and chemical mechanisms of tissue optical clearing. *iScience* **24**, 102178 (2021).

47. Sdobnov, A. Y. et al. Recent progress in tissue optical clearing for spectroscopic application. *Spectrochimica Acta Part A:Molecular and Biomolecular Spectroscopy* **197**, 216-229 (2018).
48. Tuchin, V. V., Zhu, D. & Genina, E. A. Handbook of Tissue Optical Clearing: New Prospects in Optical Imaging. (Boca Raton: CRC Press, 2022).
49. Wyrowski, F. Diffraction efficiency of analog and quantized digital amplitude holograms: analysis and manipulation. *Journal of the Optical Society of America A* **7**, 383-393 (1990).
50. Wlodarczyk, K. L. et al. Laser microsculpting for the generation of robust diffractive security markings on the surface of metals. *Journal of Materials Processing Technology* **222**, 206-218 (2015).
51. Wyrowski, F. Characteristics of diffractive optical elements/digital holograms. Proceedings of SPIE 1211, Computer and Optically Formed Holographic Optics. Los Angeles, CA, United States: SPIE, 1990.
52. Cirino, G. A. et al. Digital holography: computer-generated holograms and diffractive optics in scalar diffraction domain. in Holography-Different Fields of Application (ed Monroy, F.) (Rijeka: IntechOpen, 2011).

Official journal of the JHL 2689-9620
www.light-am.com

Review Open Access

Holography with high-power CW coherent terahertz source: optical components, imaging, and applications

Yulia Choporova[1,2*], Boris Knyazev[1,2] and Vladimir Pavelyev[3,4]

Abstract

This paper presents the results of 15 years of studies in the field of terahertz holography at the Novosibirsk free electron laser. They cover two areas: research on obtaining holographic images in the terahertz range and the use of diffractive optical elements to form high-power terahertz radiation fields with specified characteristics (intensity, phase, and polarization), using well-studied and widely applied in the optical range methods of optical (analog), digital, and computer-generated holography. All experiments were performed with the application of high-power coherent monochromatic frequency-tunable radiation from the Novosibirsk free electron laser. The features of hologram registration in the terahertz range are described. Methods, technologies, and optical materials for terahertz holographic elements are discussed. A wide range of promising applications of high-power terahertz fields with a given spatial structure is considered. The results of the study of terahertz holograms recorded as digital holograms, as well as radiation-resistive optical elements realized as computer-synthesized holograms, are presented.

Keywords: Terahertz range, Holography, Diffractive optical elements, Imaging, Free electron laser

Introduction

In recent decades, special attention in modern optics has been paid to studies of the terahertz region of the spectrum. After a long period, during which research in this spectral region was limited by the lack of suitable radiation sources and detectors, the emergence of broadband terahertz radiation sources based on the use of femtosecond lasers[1,2] has led to a burst of research and development of applications in this field. Most of the works used the time-domain spectroscopy technique. An advantage of this method is the possibility of simultaneous measurement of the amplitude and phase of the probe radiation, which makes it possible to implement a unique measurement method. On the other hand, this technique does not allow investigating fast (transient) processes and implementing some of the optical schemes well-known in the visible range.

The recent appearance of monochromatic sources of terahertz radiation (quantum cascade lasers[3], diode lasers[4], gyrotrons[5] and free electron lasers[6,7]), some of which have high average power, makes it possible to expand the variety of implemented optical systems in this range. Terahertz monochromatic radiation is applied to examination of the structure of materials and spectroscopy[8,9], construction of scanning and imaging systems[10], biomedicine[11], process materials[12], creation of

Correspondence: Yulia Choporova (yu.yu.choporova@gmail.com)
[1]Novosibirsk State University, 630090 Novosibirsk, Russia
[2]Budker Institute of Nuclear Physics, 630090 Novosibirsk, Russia
Full list of author information is available at the end of the article.
These authors contributed equally: Yulia Choporova, Boris Knyazev, Vladimir Pavelyev

remote sensing systems (lidars)[13] and telecommunication systems[14,15], as well as in fundamental research, for example, of a gas discharge[16]. In a significant part of applications[8,11–13,16], the use of high-power terahertz radiation is required.

In contrast to the visible range, holography is one of the least developed areas of terahertz optics. A search of the Webofscience Core Collection reveals that only 0.5% of the publications related to the terahertz range are devoted to holography, most performed using time-domain spectroscopy (see, for example in Ref. 17). Only 4% of them were carried out with the use of monochromatic radiation sources (see, e. g. in Refs. 18–20).

In this review, we summarize the results of fifteen years of research in the field of holography[21], carried out at the Novosibirsk free electron laser (NovoFEL)[13,22]. These studies covered both digital holography (analog-to-digital conversion), that is, the holographic recording and reproduction of images of objects with subsequent registration of images with a digital camera, and computer holography (digital-to-analog conversion), that is, the conversion of wave fields using diffraction optical elements (DOEs), which played the role of computer-synthesized holograms in this case.

NovoFEL is one of the facilities of the Siberian Center for Synchrotron and Terahertz Radiation. At present, it is a source of monochromatic terahertz radiation with the highest average power, and in a routine mode, depending on the wavelength, it generates radiation with a power of 10 to 200 W. If necessary, the power can be reduced both by an electronic shutter, which blocks the generation of a given fraction of pulses in the electron accelerator, and by ordinary attenuators. The laser operates in a quasi-continuous mode with a frequency of 5.6 MHz, emitting pulses with a duration of 100 ps, which limits the coherence length in holographic measurements. The Gaussian laser beam is linearly polarized. The divergence of a beam with a profile close to Gaussian is equal to $3 \cdot 10^{-3}$ rad. The laser has three separate resonators, which can generate wavelength-tunable radiation in the ranges of 8–10, 40–80, and 90–340 μm. Most of the experiments were carried out at wavelengths of 130–150 μm; some experiments were carried out at other wavelengths. The lasing spectrum width is 0.3–1% of the wavelength.

The important direction was the development, manufacture, research, and use of diffractive optical elements for the transformation of FEL radiation into specified light fields. DOEs were created to convert a Gaussian beam into beams of a different mode composition, as well as DOEs with the help of which the beam was focused into specified 2D and 3D regions. In the visible range, this problem is easily solved with spatial light modulators, but there are no commercially available modulators in the terahertz range. In addition, the DOEs must be resistant to the high radiation power in strong light fields. In addition to holographic applications, the development of diffractive elements of terahertz and submillimeter ranges resistant to high-power radiation is necessary, for example, for optical systems of gyrotrons[5], used, in particular, for heating thermonuclear plasma[23]. The long wavelength enables production of DOEs using technologies that are not widely used in the optical range - laser ablation[24] and 3D printing[25] for the manufacture of terahertz photonic elements. Below, we present the characteristics of some DOEs investigated in experiments at NovoFEL, fabricated using original techniques.

Another important application of NovoFEL is carrying out full-scale experiments to simulate DOEs designed for operation in the short-wavelength range. The possibilities of experimentally studying the interaction of a light wave with a subwavelength 2D or 3D structure in the optical range are limited by the resolving power of micro- and nanostructuring technologies in the manufacture of DOEs. A terahertz laser with a tunable wavelength makes it possible to study the diffraction of a light wave by objects of complex topology and with a large ratio of the wavelength to the characteristic size of inhomogeneity.

Imaging devices

In the terahertz range, radiation detectors are key elements in optical systems requiring image recording. In this part we present holography as an imaging technique for amplitude and phase objects. An image appears after numerical reconstruction of a digital hologram described in "Classical holography imaging" section below. Here we leave out of our scope single-channel detectors, hot-electron detectors[26], and semiconductor detectors[27], which can be applied to recording of holograms by scanning techniques[20]. All matrix detectors we employed to record terahertz images in different experiments are based on the detection of the thermal effect of radiation, but with different techniques of registration of the thermal imprint.

In more than 15 years of our research in that area[28, 29], which can be called holography in the broad sense of the word, five imaging systems have been created or adapted to the terahertz range by our effort[30]. None of them is universal, combining high sensitivity, high spatial and temporal resolutions, as well as a large working area, which is important for achieving a high Nyquist frequency, particularly in the case of long-wavelength radiation. For this reason, the choice of recorder is determined by requirements of each particular experiment.

The recorders are shown in Fig. 1, and their characteristics are summarized in Table 1. The first three devices are not directly sensitive to terahertz radiation, and they could only register its heat trace. The very first primitive Fresnel holograms, diffraction patterns of amplitude masks with a reference hole for the reference beam[28], were captured with the use of a thermal imager (TI) sensitive to near-infrared radiation. Due to the low spatial and temporal resolution of the holograms recorded in this way, as well as the strong thermal noise at the laboratory at a wavelength of 3 μm, later we used other methods.

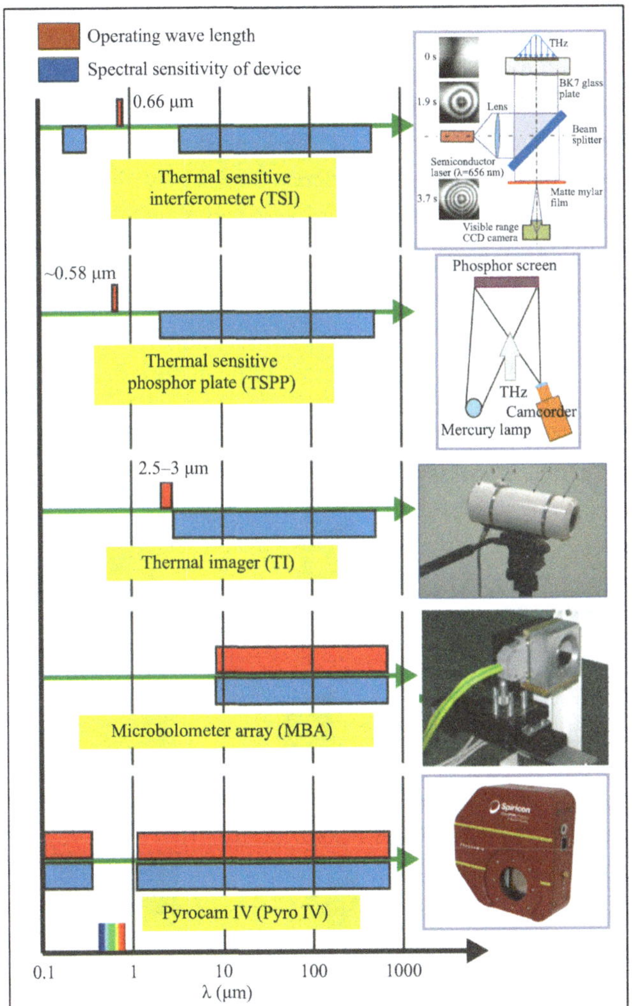

Fig. 1 Imaging devices used in terahertz range. Blue bars: ranges of spectral sensitivity of devices. Red bars: operating wavelength of final recorders.

The other two devices, in which radiation of the visible range was used for registration of terahertz images, had a higher spatial resolution. The large area of holograms obtained with their help improves the quality of their reconstruction. In a thermal sensitive interferometer (TSI), a video camera records the dynamics of an interference pattern in a plane-parallel glass plate illuminated by a collimated visible diode laser beam. The investigated terahertz radiation illuminates the plate, which is not transparent in the terahertz range, from the reverse side. The change in the refractive index in the thin glass layer due to the heating increases locally the optical path length. Since it is known a priori that when the plate is exposed to terahertz radiation after opening of the shutter, the phase incursion at each point can only be positive, the interferogram can be easily converted numerically into the intensity distribution of the terahertz beam. The thermo-optical properties of BK7 glass are standard and well known. In Ref. 31, we showed that such an interferometer is an absolute power meter. It does not require any calibration, because a shift by one fringe depends only on thermal-optical characteristics of the plate, and for K7 glass it corresponds to a locally absorbed power of 5.1 J/cm^2. Unlike all other visualizers, this device is used in the single measurement mode. Re-measurement is possible only after the plate has cooled down. The possible size of the image is determined by the size of the illuminated area, which in our case was equal to 60 mm.

Another large-area recorders, eight Macken Instruments thermosensitive phosphor plates (TSPPs), were capable of recording holograms with an area of 75 by 75 mm. The principle of their operation is based on thermal quenching of luminescence excited by a mercury lamp. We have shown[32] that the response to heating is linear up to a quenching value of about 50%. Table 1 shows the

Table 1 Characteristics of imaging devices at wavelength of 140 μm (asterisk denotes pixel size)

Device	TSI	TSPP	TI	MBA	Pyrocam IV
Mode	active	active	passive	direct	direct
Δx (mm)	0.3	0.25/1.5	2	0.051*	0.08*
Δt (s)	0.3	0.2/1.3		0.02	0.02
Repetition rate (fps)	–	4/1	25	20	25
Area (mm^2)	50 × 50	75 × 75	–	16.32 × 12.24	25.6 × 25.6
Sensitivity	medium	low	medium	very high	high

characteristics of plates No. 7 and No. 8.

For the first time, matrix receivers capable of directly detecting terahertz radiation in real time were used in the mid-2000s[33,34]. Those were microbolometer arrays (MBAs) with vanadium oxide sensors designed for the mid-infrared range. Their sensitivity threshold in the THz range was 15 times less than that in the IR range. Nevertheless, it was rather high (10^{-3} W/cm^2 or 33 nW/pixel). It was soon discovered that they were polarization-sensitive, and the antenna effect in the lead wires turned out to be a mechanism for absorbing terahertz radiation[35]. The antenna effect is now employed in newly designed microbolometer arrays[36]. The resolution of array receivers is restricted by the wavelength limit since the size of their sensitive elements is smaller than the wavelength. A relative disadvantage of these cameras, for example, when shooting holograms, is their small physical size. Recently, commercially available Pyrocam IV pyroelectric camera with a 320×320 matrix, but with sensitivity lower than the MBA sensitivity, have appeared. Thus, at present, there is some possibility of choosing imaging devices in the terahertz range. Examples of holograms recorded by different imaging devices and reconstructed images are shown in Fig. 2.

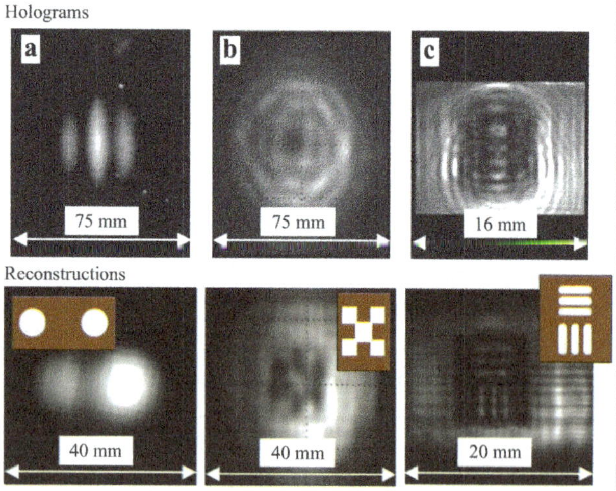

Fig. 2 Holograms (top row) recorded **a** by thermal imager (two circular apertures with diameter of 6 mm and spacing of 14 mm), **b** thermal sensitive phosphor plate, **c** microbolometer matrix, and respective reconstructed images (bottom row); objects (metal masks). Figure **a** is adapted with permission from Ref. 28; Figures **b** and **c** are adapted with permission from Ref. 29.

Classical holography imaging

Because of the lack of media that store the intensity distribution in the THz range, holograms can be recorded only digitally, and thus they can be reconstructed either digitally or by means of a second light source (e.g. visible range laser source). There are several approaches to calculation of back propagation, overviewed in Ref. 29. We consider back propagation as the most adequate technique for image reconstruction. In the terahertz range, the pixel size to the wavelength ratio limits the applicability of reconstruction techniques. Here, given our experience with sampling limitations, we will show reconstruction using only the Rayleigh-Sommerfeld convolution (RSC) approach since it works the best with our geometries. Let us briefly recall that to perform the reconstruction numerically, one needs to multiply the hologram transmission function $H(\xi, \eta)$ by the complex conjugate reference wave $E_R(\xi, \eta)$ in the hologram plane $\{\xi, \eta\}$ (Fig. 3a). We used for the reconstruction a plane wave with uniform intensity,

$$E_R(\xi, \eta) = E_R^*(\xi, \eta) = 1. \quad (1)$$

The output wave propagates through free space in the backward direction and forms a wave field in the object

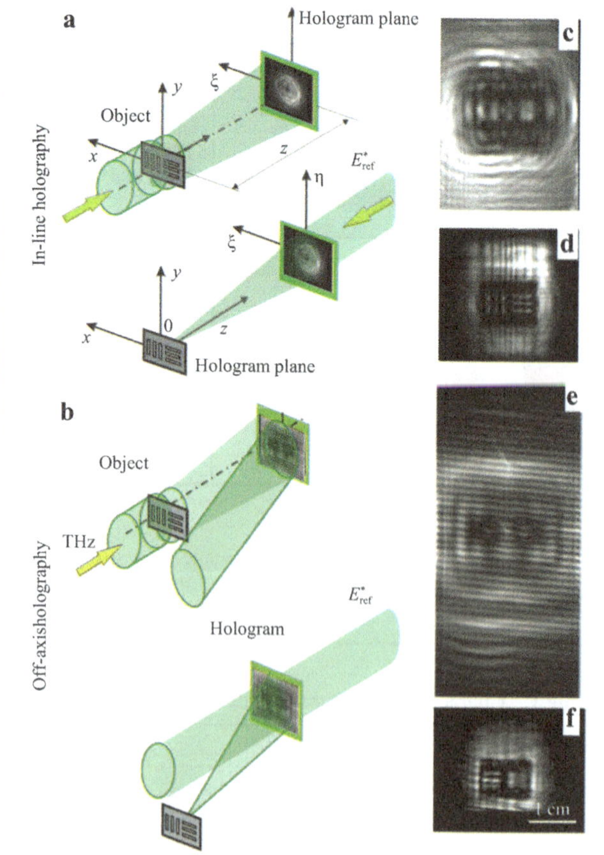

Fig. 3 Images of example object showing finest achieved resolution for inline and off-axis geometries. Recording and reconstruction schemes **a**, **b**. Holograms of metal mask **c**, **e** and reconstructed images **d**, **f** with RSC approach, for in-line and off-axis scheme respectively.

plane $\{x,y\}$, which is described by the Rayleigh-Sommerfeld diffraction integral

$$E(x,y)=\frac{i}{\lambda}\iint H(\xi,\eta)\frac{\exp(-ikr)}{r}\cos(\mathbf{n},\mathbf{r})d\xi d\eta, \quad (2)$$

where $H(\xi,\eta)$ is the hologram transmission function, and λ is the wavelength. The distance $r(\xi,\eta,x,y)$ is determined as follows:

$$r=z^2+(\xi-x)^2+(\eta-y)^2. \quad (3)$$

$$\cos(\mathbf{n},\mathbf{r})=z/r, \quad (4)$$

is the reconstructed wave. Solution of Eq. 2 enables derivation of the intensity and phase distributions in the object plane using the following expressions:

$$I(x,y)=|E(x,y)|^2, \quad (5)$$

$$\varphi(x,y)=\arg\frac{Im[E(x,y)]}{Re[E(x,y)]}. \quad (6)$$

Eq. 2 can be solved directly without any approximations, or, if the experimental configuration allows, it can be simplified and solved by other methods, which include one or several (Fast) Fourier transforms. For the Fraunhofer diffraction zone ($z>k[(\Delta x)^2+(\Delta y)^2]_{max}/2$), the hologram can be reconstructed with a single Fourier transform. Since in a real holographic experiment, the terahertz hologram is almost always located in the near zone, hereinafter we will consider only the Fresnel diffraction.

By virtue of the convolution theorem and Eq. 2, $E(x,y)$ can be written down as

$$E(x,y)=\mathcal{F}^{-1}\{\mathcal{F}[H(\xi,\eta)]\cdot\mathcal{F}[h(\xi,\eta))]\}, \quad (7)$$

which is referred below to as the RSC, where

$$h(\xi,\eta)=\frac{iz\exp(-ik\sqrt{\xi^2+\eta^2+z^2})}{\lambda(\xi^2+\eta^2+z^2)} \quad (8)$$

is the impulse response function.

Reconstruction of holographic images without aliasing requires fulfillment of the conditions of the sampling theorem. In the case of the RSC, such conditions, depending on four dimensionless parameters, were derived in Ref. 29.

The RSC approach also works for off-axis holograms recorded in the scheme of the Mach-Zehnder interferometer. To compare the off-axis and inline holography resolutions, we created an image of a target with horizontal and vertical slits of 0.6 mm (Fig. 3). For higher spatial resolution, the off-axis hologram was made of five frames by scanning in a plane perpendicular to the direction of beam propagation. It can be seen that this resolution is limited and unattainable in the absence of an expansion of the hologram recording field.

To conclude, in-line holography systems have a larger effective field of view and a higher imaging resolution as compared with off-axis ones[37] for the terahertz range. However, there are systems with a strong zero order diffraction, like imaging in attenuated total reflection systems, which require using an off-axis recording technique[38,39]. Besides, it should be mentioned that information about phase in an inline scheme in general can be obtained with additional phase retrieval algorithms, while off-axis holography easily provides it. At the same time, specifically to the THz frequency range, where, as it was mentioned earlier (Table 1), the pixel size can be less than the wavelength, the angles between the reference and object waves can be larger. Thus, it is possible to form subwavelength interference fringes, which will not degrade the resolution during the reconstruction, similarly, as it was done in numerical postprocessing in work Ref. 10.

Diffractive optical elements

In practical holography of the terahertz range, two main directions can be distinguished. In the first[38-40], as in the optical range[41], there is a real object, the amplitude-phase image of which is to be reproduced. In this case, it is necessary to use the methods of interference recording of the THz hologram of the object, described in the previous sections. On the other hand, one can pose the problem of forming a terahertz field given only by its calculated characteristics (intensity or amplitude-phase distribution). In this case, it is necessary to numerically solve the inverse problem of the theory of diffraction[42,43] with the subsequent fixation of the result on an optical material transparent in the terahertz range. At high radiation power, an additional requirement is a sufficiently high radiation resistance of the material. The approach that consists in computer synthesis of a hologram of an object, given only its calculated light distribution, is known as computer holography.

The methods of computer holography have much in common with the methods of diffractive computer optics, which is widely used for control of characteristics of coherent beams in the optical range[42-44].

One of the most important characteristics of diffractive elements is their efficiency. Further, when describing a DOE, we will use two terms characterizing this quantity. By the diffraction efficiency (DE) of element we mean the passed radiation energy fraction that is involved in the formation of the given (desired) light distribution. The energy efficiency (EE) of element is the incident beam energy fraction that produces the given distribution. Accordingly, the energy efficiency takes into account the Fresnel reflection, absorption, and scattering.

Materials and technologies for DOE manufacturing

In the terahertz range, there are several approaches to diffractive optics manufacturing: polymer lenses, which demonstrate low absorption and are fabricated by mechanical milling[45], 3D printing[46], imprinting with a silicon master[47] or by hot pressing[48]. However, an intense radiation field can damage these elements[30]. High-resistivity silicon or diamond is more resistant[49], which is especially important for high power radiation. We will show by the example of diffractive lens (DL) how silicon and diamond diffractive optical elements can be manufactured. DLs (spherical and cylindrical ones) are a promising alternative to commercially available refractive lenses and reflective metal mirrors[50], since they are typically much thinner and have lower weight.

Binary and multilevel[51] silicon DLs have been fabricated via the technology of lithography. A binary DOE micro-relief fabricated on the surface of high-resistive silicon substrate via single reactive ion etching technology (Bosch process) is considered in Ref. 52. The Bosch process in an inductively coupled plasma source configuration (ICP–RIE)[52] in an atmosphere of C_4F_8/Ar (passivation) and SF_6/Ar (etching) was used. To decrease the Fresnel losses after fabrication of diffractive microrelief, the double-sided antireflection coating of Parylene C[52] has been applied. Binary elements are made for a certain wavelength, for example, for $\lambda = 141$ μm; the micro-relief height is $h = 29.1$ μm, and the silicon refractive index is $n = 3.42$. (Fig. 4a). Binary lenses are easier to make, but they have limited diffraction efficiency (a theoretical estimation of binary Fresnel lens diffractive efficiency is 41%[42]). Losses due to the Fresnel reflection at the "silicon-air" interfaces can be reduced with application of double-sided antireflection coating of Parylene C[52].

The laser ablation technology developed at GPI RAS has been first applied to creation of a silicon Fresnel lens with four-level relief[53]. The laser ablation process in Ref. 53 was based on the use of femtosecond Yb:YAG laser. Later the laser ablation technology has been further improved for fabrication of lenses continuous relief (Fig. 4b)[24]. A multimode laser system (VaryDisk50 from Dausinger +

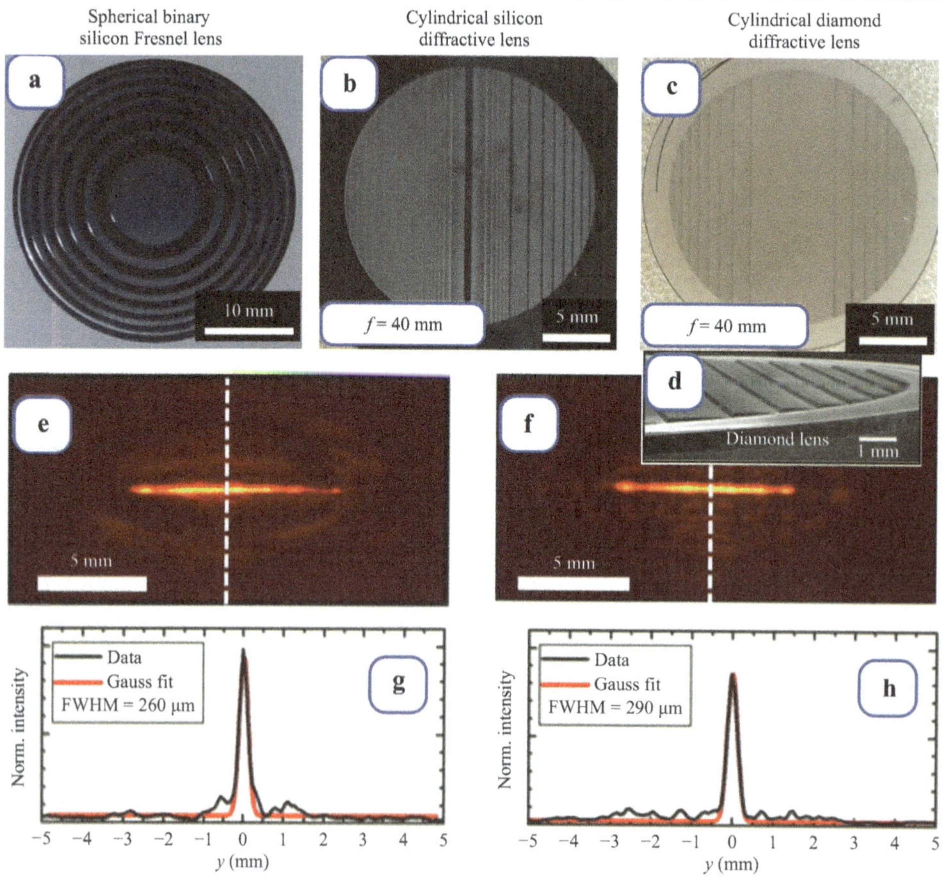

Fig. 4 Diffractive lenses designed for $\lambda = 141$ μm: silicon binary spherical Fresnel lens **a**, silicon cylindrical diffractive lens with continuous profile **b**, diamond cylindrical diffractive lens with continuous profile **c**, **d**. Intensity distributions in focal planes of silicon **e**, **g** and diamond **f**, **h** lenses, respectively.

Giesen GmbH, Germany) generating 1 ps pulses at a wavelength of 1030 nm has been used for fabrication of continuous relief by laser ablation in Ref. 24. The measured diffractive efficiency of silicon diffractive cylindrical lens with continuous relief was 93%[24].

Diamond films (DFs) is another promising optical material for high-power terahertz application[54]. The thermal conductivity of DF is larger than the thermal conductivity of silicon; a DF has a smaller refractive index ($n = 2.4$), and therefore diamond optical elements have lower Fresnel losses. Fabrication of infrared diamond lenses by laser ablation are described in Ref. 55. The technology of terahertz diffractive microrelief fabrication by laser structuring of silicon surface[56] with subsequent diamond film chemical vapour deposition (CVD) on the structured silicon surface was proposed first in Ref. 57. The silicon substrate was removed after diamond film deposition. A THz diamond cylindrical diffractive lens with continuous relief was fabricated (Fig. 4c). The intensity distribution measured in the focal plane of the fabricated cylindrical lenses is presented in Figs. 4e–h. The measured diffractive efficiency of the diamond diffractive cylindrical lens with continuous relief was 95%[57].

Binary and multilevel DOEs for focusing into given 2D and 3D areas

A number of applications (THz range imaging and radioscopy[58], soft-tissue ablation[11], material processing[12], etc.) require focusing of THz range radiation into pre-given 2D areas. The formation of uniform intensity distribution in the THz range will allow us to use THz ablation more effectively, give up point-by-point scanning, and develop a new generation of laser beam scanners.

Results of focusing a THz laser Gaussian beam into a square area with uniform intensity distribution are presented in Fig. 5[59]. The binary two-level microrelief of the element shown in Fig. 5a was calculated by a stochastic procedure[42] with the following parameters: the aperture size $D = 30$ mm; the wavelength $\lambda = 141$ μm; the focal length $f = 368$ mm; the radius of the incident Gaussian beam $w = 9$ mm; the square focal area size $a = 8.6$ mm. The calculated diffractive efficiency of the element was $e = 55.6\%$.

A binary micro-relief (Fig. 5c) was fabricated via single reactive ion etching of silicon with subsequent application of a double-sided antireflection coating of Parylene C. The experiment has shown strong dependence of the element performance on the deviation of the illuminating beam mode radius value from the estimated one. The intensity distribution in the focal plane of the element is shown in Fig. 5e.

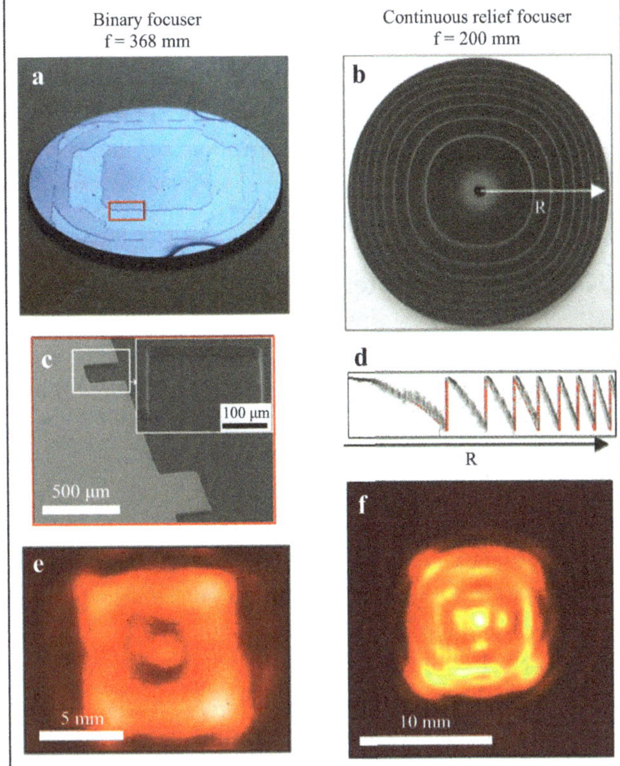

Fig. 5 Silicon DOEs with binary **a** and continuous **b**[61] microrelief focusing Gaussian beam into square. **c** Binary microrelief recorded by electron microscopy. **d** Profile of continuous microrelief measured with white light interferometry (black plot) and simulated profile (red line). Intensity distributions in focal plane of silicon focusing DOEs with binary microrelief **e** and with continuous microrelief **f**.

A DOE with a continuous profile (Fig. 5b, d) for focusing a Gaussian beam into a square-shaped region was designed with the ray optics approach[60]. This DOE was fabricated by means of the technology of laser ablation. The element was calculated for the wavelength $\lambda = 130$ μm and had the focal length $f = 200$ mm. The intensity distribution in the DOE focal plane is presented in Fig. 5f. The measured diffraction efficiency of the element[61] was equal to $(97 \pm 2)\%$.

Practical applications, such as terahertz 3D scanning and imaging (including elongated objects), ablation, optical discharge generation, etc., require focusing of terahertz radiation, often with a large focal depth. In Ref. 62, the results of studies of a silicon binary DOE intended for focusing a Gaussian beam into a paraxial segment are presented. The binary phase function of the element was calculated by a stochastic procedure[63] modification based on a genetic algorithm. The calculated paraxial segment length was $z = 30$ mm, and the DOE efficiency $e = 19.3\%$ in average varied in narrow boundaries for different paraxial segment cross-sections. With antireflection coating, the energy efficiency values measured in different planes during the experiment were in the range of 17.4–18.6%. The "self-reproduction" of the focused Gaussian beam during its propagation along the paraxial segment is worth noting[62]. The application of the fabricated optical element enabled the formation of the paraxial intensity distribution in the given space boundaries of 110 mm to 140 mm.

DOE for mode content control

Hermite-Gauss beams

The appearance of powerful terahertz gyrotrons and their modifications opens up the possibility of creating terahertz lidar systems and communication systems in free space[64,65]. Investigation of the propagation of high-power terahertz beams with different transverse structures in inhomogeneous media at the Novosibirsk variable-wavelength free electron laser makes it possible to carry out full-scale modeling and optimization of such systems. For sources like free electron laser it, is nearly impossible to tune to higher modes inside an optical resonator.

Computer-synthesized holograms were proposed to form single-mode Gauss-Laguerre and Gauss-Hermite beams from an illuminating laser beam in the visible or infrared range[66,67]. For reduction of complex transmission function to a pure phase or purely amplitude form, coding methods based on the introduction of a modulated carrier into the phase (amplitude) of a digital hologram were applied. However, this approach led to a low diffraction efficiency of the hologram, which was associated with the emergence of a large number of higher diffraction orders. As shown in Ref. 42, in the case of low-order Gauss-Hermite modes, a transparent element the phase function of which is the portrait of the mode to form enables creation of a beam with a required mode content of more than 80% when the element is illuminated with a Gaussian beam.

This approach was used in Ref. 68 for production of elements intended for the formation of single-mode Gauss-Hermite beams from an illuminating beam of a high-power terahertz laser. Since the phase function of the Gauss-Hermite mode over the entire cross-section area of the beam takes one of two values, that differ by π (Fig. 6), the corresponding element will have a binary microrelief, created in Ref. 68 by the technology of single reactive ion etching (Bosch process) of the surface of a silicon substrate. This method was previously used in Ref. 52.

Fig. 6a shows the phase of the element manufactured in Ref. 68, intended for the formation of the Gauss-Hermite mode (1,0) (white color: phase value of π, black color : of 0). The intensity of the beam formed by the manufactured element in the focal plane of a TPX lens installed directly behind the element is shown in Fig. 6d for simulation and in Fig. 6g for the experiment. Figs. 6b, e, and h show similar results of a study of a silicon binary element[68], which forms a single-mode Gauss-Hermite beam (1,1). Fig. 6b shows the phase function of the element, and Fig. 6h depicts the intensity of the beam formed by the manufactured element[68], at the focal distance.

It is known that the use of diffractive optical elements (computer-synthesized holograms) makes it possible to obtain beams of gas, semiconductor and solid-state lasers with a given composition of transverse modes at a wavelength corresponding to one of the longitudinal laser modes[42]. DOEs, the manufacturing methods of which are described in Ref. 68, enable formation of beams of coherent radiation with a given transverse-mode composition in the terahertz range. An important property of NovoFEL for various experiments is the ability to tune its wavelength over a wide range[69], whereas DOEs with continuous relief and binary DOEs are calculated to operate at a given wavelength. In our studies, we have successfully used various phase elements calculated for the wavelength $\lambda = 141$ μm in experiments performed at wavelengths $\lambda/(2m-1)$, in particular, at $\lambda/3 = 47$ μm. We also experimentally demonstrated that diffractive elements, for example, helical binary axicons, are capable of forming a given transverse mode composition even when the wavelength deviates from the calculated one by 25–30%. In this case, the main unwanted mode is the zero-order diffraction, which can be easily filtered out.

Bessel beams

Light beams with a topological charge are considered in reviews[70,71]. In Ref. 42, computer-synthesized holograms were applied to form beams with a topological charge ("vortex beams") of the visible and infrared ranges. For calculation of binary diffractive optical elements that form a terahertz Bessel beam with a topological charge, the same approach has been applied as in Refs. 72–74. The phase function of the element is described by the formula

$$\Phi(r,\phi) = (\pi/2)\mathrm{sign}(\sin(l\phi - \kappa r)), \tag{9}$$

where $p = 2\pi/\kappa$ is the period.

The element forms a beam, which amplitude is described by the Bessel function of the first kind:

$$E(r,\phi,z,t) = E_0 J_{|l|}(k_r r)\exp[i(l\phi + k_z z - \omega t)], \tag{10}$$

where E_0 is the maximum amplitude of oscillations, ϕ is the azimuthal angle, ω is the cyclic frequency of oscillations, k is the wavenumber, z is the distance from the source to the object, and l is a positive or negative integer, called the topological charge, which determines the degree of "twisting" of the beam.

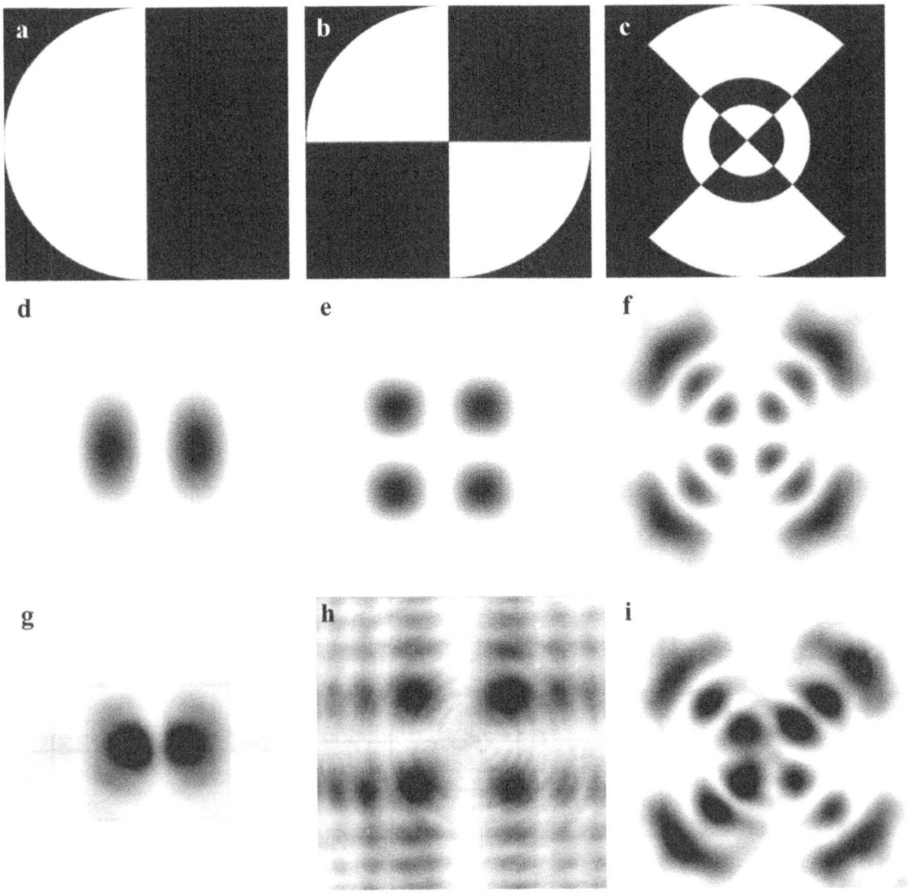

Fig. 6 Phase functions of DOEs generating Gauss-Hermite (1,0) beam **a**, Gauss-Hermite (1,1) beam **b**, beam consisting of the sum of Gauss-Laguerre modes (2,2) and (2,-2) with equal weights **c** (black colour corresponds to phase of 0 and white colour corresponds to π); intensity distributions **d–f** in cross-sections of beams generated by DOEs with phase functions **a–c** respectively, as result of computer simulation; intensity distributions **g–i** in cross-sections of beams generated by DOEs with phase functions **a–c** respectively, as result of natural experiment.

Longitudinal and cross sections of beams with orbital angular momentum formed from a NovoFEL Gaussian beam by binary spiral axicons are shown in Fig. 7. The phase relief of the axicons is shown in the insets (black colour corresponds to a phase of 0, and white colour corresponds to π). Cross-sections of the beams can be described well by the square of the Bessel function of the first kind $(J_{|l|}(k_r r))^2$ with the topological charges $l = 1$ and $l = 2$. The intensity distribution in the cross-section maintains within a distance of about 180 mm, after which, due to the limited aperture of the axicons, it begins to diverge. Therefore, the beams can be considered "diffraction-free". The formed beam had an annular structure with the diameters of the rings increasing with the topological charge. The value and sign of the beam topological charge can be detected with the Mach-Zehnder interferometer shown in the bottom of the figure. This scheme is, in fact, a holographic one.

Examples of applications in the terahertz range

In the optical range, the use of software-controlled spatial optical modulators makes it possible to assemble sophisticated optical systems. In the terahertz range, especially in the case of high power radiation, it is necessary to use computer-synthesized holograms (DOEs). The lateral feature sizes of the structures are large enough, and it is not necessary to use high-resolution lithographical masks. In this section, we will give several examples of using DOEs to form beams in solving several applied problems.

Multichannel free space communication and lidar

The previous sections of this article present the results of manufacture and study of binary silicon elements intended for the formation of single-mode Gauss-Hermite beams[68], as well as single-mode Bessel beams with orbital angular momentum (or "vortex beams")[75,76] of high-power coherent terahertz radiation. The creation of such elements made it

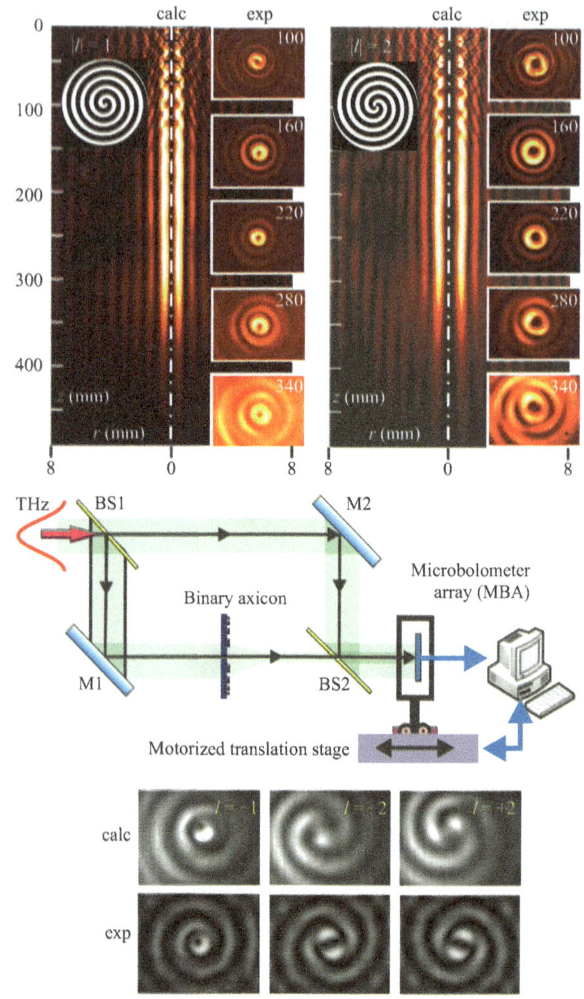

Fig. 7 Bessel beams with orbital angular momentum $l = 1$ and $l = 2$ formed from NovoFEL Gaussian beam by binary spiral axicons (see insets) and system for detection of beam topological charge; figures are adapted with permission from Refs. 74, 75.

possible to obtain new experimental results on the propagation of rotating (diffraction-free) beams in homogeneous[75] and inhomogeneous[77] media (which is important for telecommunication and lidar applications[78]), as well as in the study of the possibility of constructing multichannel systems for information transmission in free space of the terahertz range with transverse-mode channel multiplexing[79].

Typical results of experiments to study self-recovery of beams with orbital angular momentum in free space after their passing through an inhomogeneous medium[77] are shown in Fig. 8. Figs. 8a–c show the intensity distributions of the Bessel vortex beam[80] formed by this element, the Bessel beam after passing through a slab of foamed polyethylene, as well as after propagation of this beam in free space at a distance of 100 mm from the film, respectively. The radiation wavelength was 141 μm. Comparing Fig. 8b, c, one can conclude that terahertz Bessel beams with orbital angular momenta are self-healing after passing through an inhomogeneous medium.

In Refs. 66, 81 an approach to multiplexing channels in the optical range, based on selective power control of individual transverse waveguide modes, Mode Division Multiplexing (MDM), is considered. In Ref. 66, it was proposed to use DOEs for the formation of individual mode beams and analysis of the transverse-mode composition of waveguide radiation in the problem of constructing an MDM in the optical range. Paper[79] presents the results of a study of a full-scale model of a two-channel communication system in the free space of the terahertz range (wavelength $\lambda = 141$ μm). Single-mode Gauss-Hermite (1, 0) and Gauss-Hermite (0, 1) beams were used as carriers of individual channels formed by the elements[68]. For determination of the presence of one of these modes in the output beam, a silicon binary element matched to a given mode (correlation filter) was used[79]. The element was designed via the approach previously applied in Ref. 66 for optical range. Fig. 8d shows the intensity distribution in the correlation field of this element in the case of presence of a given mode in the beam. The correlation peak is clearly visible in the center of the image. Fig. 8e shows the intensity distribution in the correlation field of this element in the absence of a given mode in the beam. Thus, the creation of diffractive optical elements that make it possible to control the transverse-mode composition of the terahertz laser beam enables creation of multichannel systems in the terahertz range for information transmission in free space with transverse-mode multiplexing.

Papers[75,76] were devoted to control of the transverse-mode composition of beam without change in its polarization state. However, a significant part of actual lidar and telecommunication applications[78] requires formation of beams with a given transverse-mode composition and polarization state. Work[82] describes the manufacture and study of a silicon element (subwavelength axicon) to form from a linearly polarized Gaussian beam of a terahertz laser (operating wavelength $\lambda = 129$ μm) a beam with an intensity distribution close to that of Bessel beam, with radial polarization order 2. The element (Fig. 8f) was manufactured by the technology of single reactive-ion etching (RIE) of the silicon surface (Bosch process). Electron microscopy images of the central fragment of the manufactured element are presented in Fig. 8g, h. The results of an experimental study of the manufactured element are in good agreement with theoretical expectations and the results of computer simulation[82]. Thus, the creation of subwavelength terahertz elements makes it possible to form beams with a given

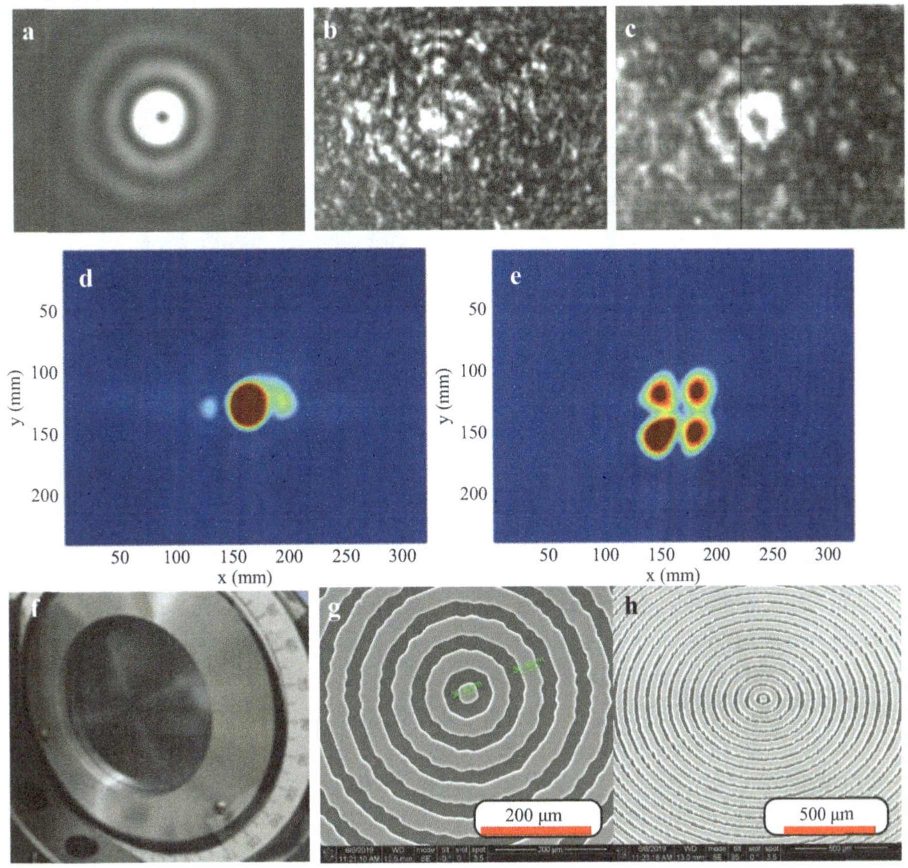

Fig. 8 Bessel vortex beam formed by DOE **a**, disturbed Bessel beam after passing through slab of foamed polyethylene **b**, disturbed Bessel beam after propagating in free space at distance of 100 mm from slab **c**, measured intensity distribution in output plane of correlator corresponding to presence of pre-given mode in illuminating beam **d**, measured intensity distribution in output plane of correlator corresponding to absence of pre-given mode illuminating beam **e**, fabricated subwavelength axicon **f**, and central fragments of axicon microrelief measured by a scanning electron microscope **g**, **h**.

transverse-mode composition and polarization state.

Vortex microbeam arrays with a given topological charge in the Talbot configuration

It has long been known[83] that periodic amplitude gratings are the simplest diffractive optical elements behind which when the grating is illuminated by a plane wave, images of the grating and its higher orders are formed in the Talbot planes. The position of the Talbot planes is described by the formula $z(j,v,u)=Z_T(j+u/v)$, where j is the number of the main plane, v is an integer, u is a natural number ($u<v$), and $Z_T=2p^2/\lambda$ is the Talbot distance. The diameter of the images of openings is equal in both the principal and fractional Talbot planes, but the period of the grating image in the fractional planes is a fraction of the grating period. The Talbot effect is used in numerous applications including advanced lighting systems, metrology, and more. The modern theory of the effect was developed in a series of works[84–87]. An overview of the current state of research can be found in Ref. 88.

In work Ref. 80, we first studied experimentally the diffraction of vortex Bessel beams[89] by two-dimensional periodic gratings. It was found that behind the grating, in the same Talbot planes and with the same spatial frequency, periodic distributions of the wavefield were formed, but instead of images of holes, rings were observed with the same topological charge as that of the illuminating beam.

The theory of the Talbot effect under the illumination of gratings with Bessel beams of arbitrary order was developed in our paper[90]. It was shown, in particular, that zero-order beams also created ring structures. The radii of the formed rings depend, in addition to the parameters mentioned above, on the transverse wave number of the Bessel beam κ: $\rho=(\kappa p^2/\pi)(j+u/v)$. Experiments were carried out on linearly polarized Bessel beams of orders from 0 to 2. Fig. 9 shows the images observed in several Talbot planes behind the gratings illuminated by Gaussian

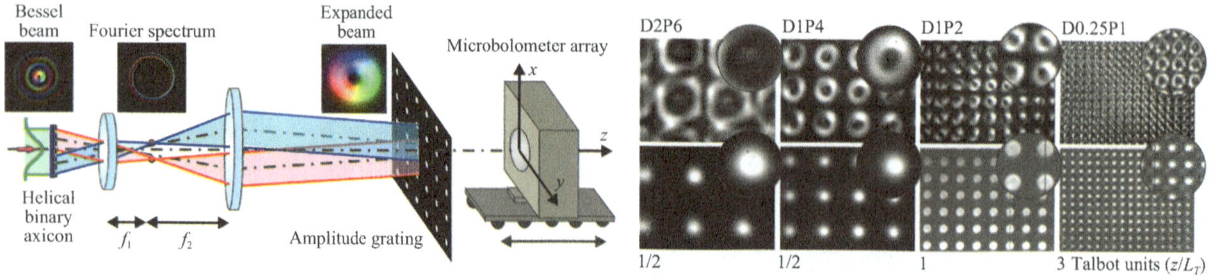

Fig. 9 Talbot effect. Principal optical scheme and images recorded with the microbolometer array in main and fractional Talbot planes[80]. Gratings were illuminated with Gaussian beam (low row) and Bessel beam of the first order (upper row) at a wavelength of 141 μm; $f_2/f_1 = 4.5$. Phase distribution is shown with artificial colors from $\phi = 0$ (blue) to $\phi = 2\pi$ (red). DxPy is diameter of openings and grating period in mm. Frame size of MBA is 16.32 × 12.24 mm^2.

beam and Bessel beams of the first order. It is interesting that the calculated pattern is observed even for gratings with a hole diameter of only two wavelengths (see the last column of the images, for which the opening diameters were equal to 250 μm). The amplitude and phase in the experimentally observed arrays of the vortex beamlets well agreed with the theory. Recently, similar experiments in the visible range and corresponding theoretical calculations were published in papers[91,92]. The results obtained in these papers are close to our results.

Thus, the use of spiral axicons to modify the classical optical experiment yielded a new physical effect.

Vector vortex beams

Vortex beams, such as those described in "Bessel beams" section, have numerous applications, but in some cases, it becomes necessary to specify a special spatially-inhomogeneous polarization distribution over the beam cross-section. Such beams are called vector beams[93,94]. They usually have at least one singularity. Beams with spatially inhomogeneous phase and polarization distributions are called vector vortex beams. In this section, we will describe a technique for the generation of vector vortex beams in the terahertz range.

In the experiments, we used binary spiral axicons of 50 mm in diameter, which created Bessel beams with topological charges 3 and 9. The phase relief of the axicon producing a beam with topological charge $l = -9$ is shown in Fig. 10a. The sign of the topological charge of the beams can be changed by rotation of the axicon about th vertical axis by 180°. At illumination with a linearly polarized Gaussian beam of the NovoFEL (Fig. 10d), the Bessel beam was formed at a distance of 120 mm behind the axicon. The simulated intensity distribution of the beam is shown in Fig. 10b. The beam phase distribution is shown with artificial colors on the intensity distribution. The intensity distribution over the beam cross-section, obtained in the experiment (Fig. 10f), was in good agreement with the theoretical one.

To obtain a Bessel vector beam, a segmented half-wave plate[95], consisting of eight sectors of a birefringent crystal, was installed in 50 mm in front of the axicon (Fig. 10e). The optical axis of each next sector is rotated by an angle of 22.5°. This plate is essentially a discrete version of the half-wave q-plate[96–98] with the distribution of local optical axes of $\alpha(r,\varphi) = q\varphi + \alpha_0$, which is commonly used to convert a circularly polarized beam into a beam with an orbital angular momentum. In our case, the plate, which was an analog of a q-plate with $q = 1/2$ and $\alpha_0 = 0$, was illuminated by horizontally or vertically linearly polarized radiation, which made it possible to obtain, respectively, a radially or azimuthally polarized beam, while its orbital angular momentum was conserved. The intensity distribution in the Bessel beam (Fig. 10h) shows eight discontinuities, which apparently appeared on the lines of gluing of the plate segments. However, the cross-section of the beam did not change significantly.

The fact that the Bessel mode is largely conserved becomes clear when the Fourier image of the beam is obtained via its focusing with a high-resistance silicon lens. Experimentally obtained images in the focal plane of a lens with a focal length of 150 mm are shown in Fig. 10g for linearly polarized beams and in Fig. 10i for radially polarized beams. The simulated beam is shown in Fig. 10c. Beams created in this way are called "perfect" beams[99]. Their practical significance is determined by the fact that for Bessel beams with the same value of the transverse wave number, the diameters of the rings formed by the lenses are the same for any values of the topological charge. In the case of an ideal Bessel beam, non-existent in the real world, an annular intensity distribution will be observed in the focal plane. A Bessel beam created by binary axicon and its Fourier transform differ from ideal ones. In work Ref. 89, this is shown for three types of

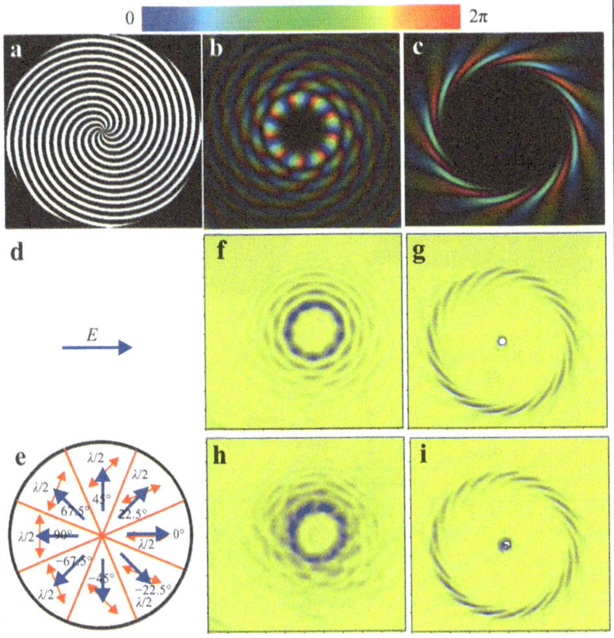

Fig. 10 Formation of vector vortex beams with topological charge $l = -9$ using binary spiral axicon[89]. **a** Spiral axicon. **b, c** Intensity and phase distribution of Bessel beam formed by axicon and its Fourier transform (the latter is also called "perfect beam"). **d** Polarization of NovoFEL beam, $\lambda = 141$ μm. **e** Polarization of laser beam passed through segmented half-wave plate; red arrows: optical axis directions. **f, g** Bessel and perfect beams obtained using linearly polarized radiation; **h, i** the same with radially polarized radiation. Frame size in **f–i** is 25.6 × 25.6 mm^2.

diffractive axicons: binary and with continuous phase distributions. In particular, in the case of a binary axicon, a perfect beam is a collection of $2|l|$ short spirals located along the ring[100]. In our case, it can be seen that even the "damage" to a Bessel beam by a segmented waveplate practically does not change the perfect beam.

Generation of vortex surface plasmon polaritons using perfect beams

An example of the use of perfect beams is the demonstration of the generation of terahertz surface plasmon polaritons (SPPs) with orbital angular momentum on an axisymmetric, singly connected transmission line[101, 102]. This approach is a first pace to the development of multiplex plasmonic transmission lines in the terahertz range[103]. The experimental setup is shown in Fig. 11. The input optical system enables formation of vortex perfect beams with $|l| = 3$ or 9 with radial and azimuthal polarizations. The annual beam radii were equal to the input diameter of the cylindrical line. Plasmons on the cylinder were excited by diffraction of a perfect beam on the front edge (end-fire coupling technique).

The plasmons excited propagated, rotating, along the cylindrical line and at the exit again turned into a free-space wave. The fact of the generation of plasmons and their rotation during propagation along the cylinder has been reliably confirmed by investigation of the free-space wave outcoupled of the rear end of the cylinder. The output wave was observed when the wave illuminating the front end had radial polarization (Fig. 11c), and its intensity dramatically decreased in a case of azimuthal polarization (Fig. 11d). Residual radiation is observed due to a small fraction of radially polarized radiation generated by the segment plate. Since plasmons can only exist as TE waves, the observed effect unambiguously confirms their existence.

The experiments showed as well that in a sequence of investigating transformations "vortex wave → rotating plasmon → outcoupled free-space wave", the transverse component of the Poynting vector is conserved for all carriers in this chain. Since, as mentioned above, perfect beams have the same diameters, simultaneous generation of plasmons with different orbital angular momenta is possible. It can be expected that the wave emitted due to plasmon diffraction at the rear end of the line will retain the combination of topological charges and the information can then be demultiplexed. Demultiplexing the signal will also require the use of diffractive elements, such as a Dammann grating[104], or elements for correlation analysis[105].

Conclusion

Summarizing the results, let us list the possibilities offered by the use of a powerful monochromatic frequency-tunable terahertz radiation source for holographic research.

1. We have demonstrated the efficiency of digital and computer holography methods in the terahertz range, using the Novosibirsk free electron laser as a radiation source. The techniques for recording holograms we created or adapted to the terahertz range made it possible to record digital holograms and reproduce images, in certain cases with a resolution close to the wave limit. The high power of NovoFEL and the possibility of recording terahertz images with an MBA with a frequency of up to 20 fps enable recording series of holograms of dynamic processes. Diffractive optical elements synthesized on the computer and made of high-resistance silicon or artificial diamond make it possible to form 2D and 3D light fields with specified intensity or amplitude-phase distributions and with a specified polarization state, which is often necessary in development of optical systems in the terahertz range.

2. For the calculation and modeling of diffractive optical elements (computer-generated holograms) of the terahertz range, methods for solving direct and inverse problems of

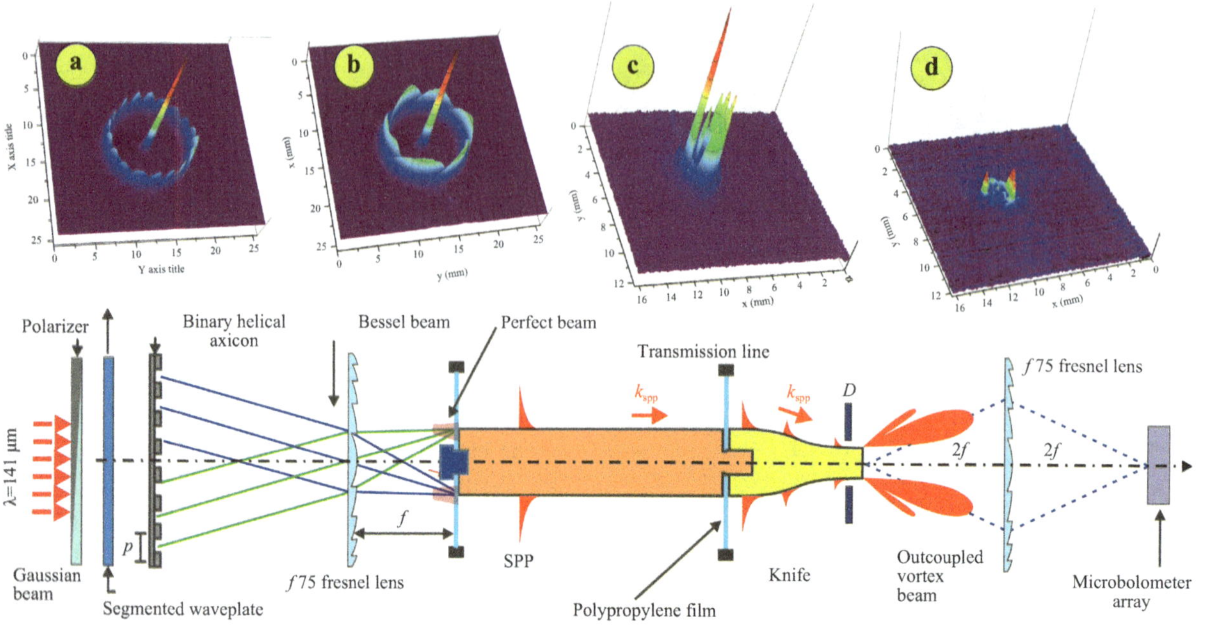

Fig. 11 Experimental setup for the generation of surface plasmon polaritons with orbital angular momentum[101]. The front end of the transmission line was illuminated by one of the perfect beams with topological charge $|\ell| = 3$ or 9. Their intensity distributions recorded by a Pyrocam IV are shown in insets **a** and **b**. Surface plasmon polaritons where excited by the end-fire coupling technique and propagated along a brass line covered with a 1 mm thick ZnSe layer. At the rear end of the line, plasmons transformed into vortex-free space waves and were recored with a microbolometer array. **c**, **d** Images of the outcoupled waves in-plane of the cylinder rear end, taken for radial **c** and azimutal **d** polarization of the input perfect beam with $|\ell| = 3$.

the theory of diffraction, previously developed and studied in the optical wavelength range, were successfully used. At the same time, for the manufacture of calculated diffractive optical elements, new technologies were developed for the formation of a phase diffractive microrelief on substrates of terahertz optical material with high radiation resistance.

3. A free electron laser in the terahertz range with a wavelength tunable in a wide range makes it possible to use the fabricated diffractive optical elements with a continuous profile for experiments in various regions of the terahertz range with a wavelength step corresponding to a change of $(2m)\pi$ $(m = 0, 1.2...)$ in the phase shift on the element microrelief. Binary elements, in addition, are able to work correctly when the radiation wavelength is detuned from the calculated one by 10 - 15%.

4. Due to the large (relative to the optical range) wavelength, a terahertz laser with a tunable wavelength and high radiation power, such as NovoFEL, allows full-scale simulation of the interaction of coherent radiation with substantially subwavelength 2D and 3D structures. This approach is free from the disadvantages of computer modeling[105]. Note that analogs of some experiments, the results of which are presented in this article, are practically impossible in the optical range.

Acknowledgements

The studies described in "Imaging devices", "Vortex microbeam arrays with a given topological charge in the Talbot configuration", "Vector vortex beams", "Generation of vortex surface plasmon polaritons using perfect beams" sections were supported by the Russian Science Foundation grant No. 19-12-00103, "Classical holography imaging" section by RFBR No. 18-32-00407 and "Diffractive optical elements", "Material and technologies for DOE manufacturing", "Binary and multilevel DOEs for focusing into given 2D and 3D areas", "DOE for mode content control" and "Multichannel free space communication and lidar" sections by RSF No. 19-72-20202. The authors are grateful to V. I. Konov, G. N. Kulipanov, V. A. Soifer, and N. A. Vinokurov for their support of the research underlying this review, as well as V. S.Cherkassky, V. V. Gerasimov, O. E. Kameshkov, M. S. Komlenok, T. V. Kononenko, M. S. Mitkov, N. D. Osintseva, A. K. Nikitin, K. N. Tukmakov, A. S. Reshetnikov, coauthors of our articles on holography and diffractive optics. We also thank the NovoFEL team, Ya. V. Getmanov, Ya. I. Gorbachev, M. A. Scheglov, D. A. Skorokhod, and O. A. Shevchenko for technical support. The experiments were carried out at the Novosibirsk Free Electron Laser Facility, which is part of "the Siberian Synchrotron and Terahertz Radiation Center". This manuscript has been prepared with the help of I. G. Sokolova and N. Dessmann.

Author details

[1]Novosibirsk State University, 630090 Novosibirsk, Russia. [2]Budker Institute of Nuclear Physics, 630090 Novosibirsk, Russia. [3]Samara National Research University, 443086 Samara, Russia. [4]Image Processing Systems Institute of the RAS – Branch of the FSRC "Crystallography and Photonics" RAS, 443001 Samara, Russia

Author contributions

Boris Knyazev, Yulia Choporova, Vladimir Pavelyev contributed equally.

The manuscript was written, discussed, and corrected by all authors.

Data availability

The data that support the findings of this study are available from the authors upon reasonable request.

Conflict of interest

The authors declare that they have no conflict of interest.

Received: 09 September 2021 Revised: 25 April 2022 Accepted: 25 April 2022
Accepted article preview online: 29 April 2022
Published online: 09 June 2022

References

1. Xu, L., Zhang, X. C. & Auston, D. H. Terahertz beam generation by femtosecond optical pulses in electro-optic materials. *Applied Physics Letters* **61**, 1784-1786 (1992).
2. Grischkowsky, D. et al. Far-infrared time-domain spectroscopy with terahertz beams of dielectrics and semiconductors. *Journal of the Optical Society of America B* **7**, 2006 (1990).
3. Williams, B. S. Terahertz quantum-cascade lasers. *Nature Photonics* **1**, 517-525 (2007).
4. Köhler, R. et al. Terahertz semiconductor-heterostructure laser. *Nature* **417**, 156-159 (2002).
5. Glyavin, M. Y., Luchinin, A. G. & Golubiatnikov, G. Y. Generation of 1.5-kW, 1-THz Coherent Radiation from a Gyrotron with a Pulsed Magnetic Field. *Physical Review Letters* **100**, 015101 (2008).
6. Carr, G. L. et al. High-power terahertz radiation from relativistic electrons. *Nature* **420**, 153-156 (2002).
7. Kulipanov, G. N. et al. Novosibirsk Free Electron Laser—Facility Description and Recent Experiments. *IEEE Transactions on Terahertz Science and Technology* **5**, 798-809 (2015).
8. Cherkassky, V. S. et al. Introscopy of solids at novosibirsk terahertz free electron laser. In *2006 Joint 31st International Conference on Infrared Millimeter Waves and 14th International Conference on Teraherz Electronics*, pages 320–320. IEEE, 2006.
9. Hübers, H. W., Richter, H. & Wienold, M. High-resolution terahertz spectroscopy with quantum-cascade lasers. *Journal of Applied Physics* **125**, 151401 (2019).
10. Behnken, B. N. et al. Real-time imaging using a 28 THz quantum cascade laser and uncooled infrared microbolometer camera. *Optics Letters* **33**, 440 (2008).
11. Weightman, P. Prospects for the study of biological systems with high power sources of terahertz radiation. *Physical Biology* **9**, 053001 (2012).
12. Agranat, M. B. et al. Damage in a Thin Metal Film by High-Power Terahertz Radiation. *Physical Review Letters* **120**, 085704 (2018).
13. Kulipanov, G. N. et al. Experimental study of the interaction between terahertz radiation from the Novosibirsk free-electron laser and water aerosol. *Atmospheric and Oceanic Optics* **28**, 165-168 (2015).
14. Pavelyev, V. S. et al. Towards multichannel terahertz telecommunication based on mode division multiplexing. *AIP Conference Proceedings* **2299**, 030002 (2020).
15. Grant, P.D. et al. Terahertz free space communications demonstration with quantum cascade laser and quantum well photodetector. *Electronics Letters* **45**, 952 (2009).
16. Kubarev, V. V. et al. Threshold Conditions for Terahertz Laser Discharge in Atmospheric Gases. *Journal of Infrared, Millimeter, and Terahertz Waves* **38**, 787-798 (2017).
17. Turov, A. T. et al. Resolution and contrast in terahertz pulse time-domain holographic reconstruction. *Applied optics* **58**, G231-G240 (2019).
18. Choporova, Y. Y., Cherkassky, V. S. & Knyazev, B. A. In-line and reference-beam holography experiments on Novosibirsk free electron. Proceedings of 2011 International Conference on Infrared,Millimeter, and Terahertz Waves, IEEE Microwave Theory & Tech Soc; NASA, California Inst Technol, Jet Prop Lab; Univ Wollongong, IEEE, 2011.
19. Heimbeck, M. S. et al. Terahertz digital holographic imaging of voids within visibly opaque dielectrics. *IEEE Transactions on Terahertz Science and Technology* **5**, 110-116 (2014).
20. Heimbeck, M. S. & Everitt, H. O. Terahertz digital holographic imaging. *Advances in Optics and Photonics* **12**, 1 (2020).
21. Choporova, Y. Y., Knyazev, B. A. & Mitkov, M. S. Holography as imaging technique for the THz range. Proceedings of 41st International Conference on Infrared, Millimeter, and Terahertz waves (IRMMW-THz), IEEE, 2016.
22. Shevchenko, O. A. et al. The novosibirsk free electron laser facility. *AIP Conference Proceedings* **2299**, 020001 (2020).
23. Denisov, G. G. et al. The concept of a gyrotron with megawatt output at both first and second cyclotron harmonics for plasma heating in spherical tokamaks. *Radiophysics and Quantum Electronics* **63**, 345-353 (2020).
24. Kononenko, T. V. et al. Silicon kinoform cylindrical lens with low surface roughness for high-power terahertz radiation. *Optics & Laser Technology* **123**, 105953 (2020).
25. Siemion, A. The magic of optics—an overview of recent advanced terahertz diffractive optical elements. *Sensors* **21**, 100 (2021).
26. Semenov, A. D., Gol'tsman, G. N. & Sobolewski, R. Hot-electron effect in superconductors and its applications for radiation sensors. *Superconductor Science and Technology* **15**, R1-R16 (2002).
27. Komiyama, S. Single-Photon Detectors in the Terahertz Range. *IEEE Journal of Selected Topics in Quantum Electronics* **17**, 54-66 (2011).
28. Cherkassky, V. S. et al. Imaging techniques for a high-power THz free electron laser. *Nuclear Instruments and Methods in Physics Research Section A: Accelerators, Spectrometers, Detectors and Associated Equipment* **543**, 102-109 (2005).
29. Choporova, Y. Y., Knyazev, B. A. & Mitkov, M. S. Classical Holography in the Terahertz Range: Recording and Reconstruction Techniques. *IEEE Transactions on Terahertz Science and Technology* **5**, 836-844 (2015).
30. Knyazev, B. A. et al. Real-Time Imaging Using a High-Power Monochromatic Terahertz Source: Comparative Description of Imaging Techniques with Examples of Application. *Journal of Infrared, Millimeter, and Terahertz Waves* **32**, 1207-1222 (2011).
31. Vinokurov, N. A. et al. Visualization of radiation from a high-power terahertz free electron laser with a thermosensitive interferometer. *Technical Physics* **52**, 911-919 (2007).
32. Knyazev, B. A. & Kubarev, V. V. Wide-field imaging using a tunable terahertz free electron laser and a thermal image plate. *Infrared Physics & Technology* **52**, 14-18 (2009).
33. Lee, A. W. M. et al. Real-time terahertz imaging over a standoff distance (>25meters). *Applied Physics Letters* **89**, 141125 (2006).
34. Dem' yanenko, M. A. et al. Imaging with a 90 frames/s microbolometer focal plane array and high-power terahertz free electron laser. *Applied Physics Letters* **92**, 131116 (2008).
35. Dem' yanenko, M. A. et al. Microbolometer detector arrays for the infrared and terahertz ranges. *Journal of Optical Technology* **76**, 739 (2009).
36. Gou, J. et al. Spiral Antenna-Coupled Microbridge Structures for THz Application. *Nanoscale Research Letters* **12**, 91 (2017).
37. Xu, L. et al. Imaging analysis of digital holography. *Optics Express* **13**, 2444 (2005).

38. Choporova, Y. Y. & Knyazev, B. A. Holography as an ATR THz imaging technique. In *2018 43rd International Conference on Infrared, Millimeter, and Terahertz Waves (IRMMW-THz)*, pages 1–1. IEEE, sep 2018.
39. Wang, D. Y. et al. Dynamic full-field refractive index distribution measurements using total internal reflection terahertz digital holography. *Photonics Research* **10**, 289 (2022).
40. Petrov, N. V. et al. Terahertz phase retrieval imaging in reflection. *Optics Letters* **45**, 4168-4171 (2020).
41. Hariharan, P. *Optical Holography*. Cambridge University Press, jul 1996.
42. Soifer, V. A. *Computer Design of Diffractive Optics*. Woodhead Publishing, 2012.
43. Wyrowski, F. Digital holography: diffractive optics on the basis of scalar diffraction theory. Proceedings of Workshop on Digital Holography, Vol. 1718 Prague, Czech Republic: Society of Photo-Optical Instrumentation Engineers, 1993.
44. Schnars, U. & Jüptner, W. P. Digital recording and numerical reconstruction of holograms. *Measurement science and technology* **13**, R85 (2002).
45. Sypek, M. et al. Highly efficient broadband double-sided Fresnel lens for THz range. *Optics Letters* **37**, 2214 (2012).
46. Furlan, W. D. et al. 3D printed diffractive terahertz lenses. *Optics Letters* **41**, 1748 (2016).
47. Walsby, E. D. et al. Imprinted diffractive optics for terahertz radiation. *Optics Letters* **32**, 1141 (2007).
48. Vedernikov, V. M. et al. Diffractive elements for a free electron laser. *Optoelectronics, Instrumentation and Data Processing* **46**, 365-375 (2010).
49. Rogalin, V. E., Kaplunov, I. A. & Kropotov, G. I. Optical Materials for the THz Range. *Optics and Spectroscopy* **125**, 1053-1064 (2018).
50. Agafonov, A. N. et al. Elements of the Terahertz Power Reflective Optics with Free-Form Surfaces. *Optoelectronics, Instrumentation and Data Processing* **55**, 148-153 (2019).
51. Walsby, E. D. et al. Multilevel silicon diffractive optics for terahertz waves. *Journal of Vacuum Science & Technology B: Microelectronics and Nanometer Structures* **20**, 2780 (2002).
52. Agafonov, A. N. et al. Silicon diffractive optical elements for high-power monochromatic terahertz radiation. *Optoelectronics, Instrumentation and Data Processing* **49**, 189-195 (2013).
53. Komlenok, M. S. et al. Fabrication of a multilevel THz Fresnel lens by femtosecond laser ablation. *Quantum Electronics* **45**, 933-936 (2015).
54. Sukhadolau, A. V. et al. Thermal conductivity of CVD diamond at elevated temperatures. *Diamond and Related Materials* **14**, 589-593 (2005).
55. Kononenko, V. V. et al. Diamond diffraction optics for CO 2 lasers. *Quantum Electronics* **29**, 9-10 (1999).
56. Komlenok, M. S. et al. Silicon diffractive optical element with piecewise continuous profile to focus high-power terahertz radiation into a square area. *Journal of the Optical Society of America B* **38**, B9 (2021).
57. Komlenok, M. et al. Diamond diffractive lens with a continuous profile for powerful terahertz radiation. *Optics Letters* **46**, 340-343 (2021).
58. Cherkassky, V. S. et al. Techniques for introscopy of condense matter in terahertz spectral region. *Nuclear Instruments and Methods in Physics Research Section A: Accelerators, Spectrometers, Detectors and Associated Equipment* **575**, 63-67 (2007).
59. Agafonov, A. N. et al. Optical elements for focusing of terahertz laser radiation in a given two-dimensional domain. *Optical Memory and Neural Networks* **23**, 185-190 (2014).
60. Duparré, M. et al. Investigation of computer-generated diffractive beam shapers for flattening of single-modal CO_2 laser beams. *Applied Optics* **34**, 2489 (1995).
61. Tukmakov, K. N. et al. A continuous-profile diffractive focuser for terahertz radiation fabricated by laser ablation of silicon. *Computer Optics* **42**, 941-946 (2018).
62. Agafonov, A. N. et al. Focusing of Novosibirsk Free Electron Laser (NovoFEL) radiation into paraxial segment. *Journal of Modern Optics* **63**, 1051-1054 (2016).
63. Kachalov, D. G. et al. Application of the direct search in solving a problem of forming longitudinal distribution of intensity. *Journal of Modern Optics* **58**, 69-76 (2011).
64. Idehara, T. et al. The gyrotrons as promising radiation sources for thz sensing and imaging. *Applied Sciences* **10**, 980 (2020).
65. Kumar, N. et al. A review on the sub-thz/thz gyrotrons. *Infrared Physics & Technology* **76**, 38-51 (2016).
66. Soifer, V. A. & Golub, M. A. *Laser beam mode selection by computer generated holograms*. CRC Press, 1994.
67. Golub, M. A. et al. Spatial phase filters matched to transverse modes. *Soviet Journal of Quantum Electronics* **18**, 392-393 (1988).
68. Agafonov, A. N. et al. Control of transverse modal spectrum of terahertz laser irradiation by binary silicon optical elements. *Computer Optics* **38**, 763-769 (2014).
69. Knyazev, B. A., Kulipanov, G. N. & Vinokurov, N. A. Novosibirsk terahertz free electron laser: instrumentation development and experimental achievements. *Measurement Science and Technology* **21**, 054017 (2010).
70. Yao, A. M. & Padgett, M. J. Orbital angular momentum: origins, behavior and applications. *Advances in Optics and Photonics* **3**, 161 (2011).
71. Molina-Terriza, G., Torres, J. & Torner, L. Twisted photons. *Nature Physics* **3**, 305-310 (2007).
72. Alekseev, A. N. et al. Conversion of Hermite-Gaussian and Laguerre-Gaussian beams in an astigmatic optical system. 1. Experiment. *Technical Physics Letters* **24**, 694-696 (1998).
73. Khonina, S. N. et al. Trochoson. *Optics Communications* **91**, 158-162 (1992).
74. Knyazev, B. A. et al. Generation of terahertz surface plasmon polaritons using nondiffractive bessel beams with orbital angular momentum. *Physical review letters* **115**, 163901 (2015).
75. Choporova, Y. Y. et al. High-power Bessel beams with orbital angular momentum in the terahertz range. *Physical Review A* **96**, 023846 (2017).
76. Volodkin, B. et al. Fabrication and characterization of diffractive phase plates for forming high-power terahertz vortex beams using free electron laser radiation. *Optical and Quantum Electronics* **48**, 223 (2016).
77. Knyazev, B. A. et al. Transmission of high-power terahertz beams with orbital angular momentum through atmosphere. *Proceedings of 41st International Conference on Infrared, Millimeter, and Terahertz waves (IRMMW-THz)*; IEEE, 2016.
78. Rubinsztein-Dunlop, H. et al. Roadmap on structured light. *Journal of Optics* **19**, 013001 (2017).
79. Choporova, Y. Y. et al. Two channel terahertz communication based on spatial mode multiplexing. *Proceedings of 44th International Conference on Infrared, Millimeter, and Terahertz Waves (IRMMW-THz)*, IEEE, 2019.
80. Knyazev, B. et al. Quasi-Talbot effect with vortex beams and formation of vortex beamlet arrays. *Optics Express* **26**, 14174 (2018).
81. Shinmura, Y., Ezoe, H. & Yoshikawa, M. Observation of mode in graded-index optical fibers with bending and cross talk in MDM. *IEICE Trans. Electron.* **E80-C**, 828-830 (1997).
82. Pavelyev, V. S. et al. Silicon subwavelength axicons for terahertz beam polarization transformation. *Journal of Physics: Conference Series*

1745, 012022 (2021).

83. Talbot, H. F. Facts relating to optical science. No. IV. *The London, Edinburgh, and Dublin Philosophical Magazine and Journal of Science* **9**, 401-407 (1836).
84. Cowley, J. M. & Moodie, A. F. Fourier Images: II -The Out-of-focus Patterns. *Proceedings of the Physical Society. Section B* **70**, 497-504 (1957).
85. Winthrop, J. T. & Worthington, C. R. Theory of Fresnel Images I Plane Periodic Objects in Monochromatic Light. *Journal of the Optical Society of America* **55**, 373 (1965).
86. Montgomery, W. D. Self-Imaging Objects of Infinite Aperture. *Journal of the Optical Society of America* **57**, 772 (1967).
87. Lohmann, A. W., Knuppertz, H. & Jahns, J. Fractional Montgomery effect: a self-imaging phenomenon. *Journal of the Optical Society of America A* **22**, 1500 (2005).
88. Wen, J. M., Zhang, Y. & Xiao, M. The Talbot effect: recent advances in classical optics, nonlinear optics, and quantum optics. *Advances in Optics and Photonics* **5**, 83 (2013).
89. Knyazev, B., Cherkassky, V. & Kameshkov, O. "Perfect" Terahertz Vortex Beams Formed Using Diffractive Axicons and Prospects for Excitation of Vortex Surface Plasmon Polaritons. *Applied Sciences* **11**, 717 (2021).
90. Kotelnikov, I. A., Kameshkov, O. E. & Knyazev, B. A. Diffraction of bessel beams on 2D amplitude gratings—a new branch in the talbot effect study. *Journal of Optics* **22**, 065603 (2020).
91. Ikonnikov, D. A. et al. Two-dimensional talbot effect of the optical vortices and their spatial evolution. *Scientific Reports* **10**, 1-10 (2020).
92. Ikonnikov, D. A. et al. 3d optical vortex lattices. *Annalen der Physik* **533**, 2100114 (2021).
93. Zhan, Q. W. Cylindrical vector beams: from mathematical concepts to applications. *Advances in Optics and Photonics* **1**, 1 (2009).
94. Suzuki, M. et al. Comprehensive quantitative analysis of vector beam states based on vector field reconstruction. *Scientific Reports* **9**, 9979 (2019).
95. Saito, Y. et al. z-polarization sensitive detection in micro-raman spectroscopy by radially polarized incident light. *Journal of Raman Spectroscopy: An International Journal for Original Work in all Aspects of Raman Spectroscopy, Including Higher Order Processes, and also Brillouin and Rayleigh Scattering* **39**, 1643-1648 (2008).
96. Marrucci, L., Manzo, C. & Paparo, D. Optical spin-to-orbital angular momentum conversion in inhomogeneous anisotropic media. *Physical review letters* **96**, 163905 (2006).
97. Marrucci, L. et al. Spin-to-orbital conversion of the angular momentum of light and its classical and quantum applications. *Journal of Optics* **13**, 064001 (2011).
98. Rubano, A. et al. Q-plate technology: a progress review. *JOSA B* **36**, D70-D87 (2019).
99. Ostrovsky, A. S., Rickenstorff-Parrao, C. & Arrizón, V. Generation of the "perfect" optical vortex using a liquid-crystal spatial light modulator. *Optics Letters* **38**, 534 (2013).
100. Kotlyar, V. V., Kovalev, A. A. & Porfirev, A. P. Optimal phase element for generating a perfect optical vortex. *Journal of the Optical Society of America A* **33**, 2376 (2016).
101. Gerasimov, V. V. et al. Vortex surface plasmon polaritons on a cylindrical waveguide: generation, propagation, and diffraction. *Journal of Optics* **23**, 10LT01 (2021).
102. Knyazev,B. A.et al. Experiments on Generation of Vortex Surface Plasmon Polaritons on Plane and Cylindrical Conductors in Mid-Infrared and THz Ranges. Proceedings of 45th International Conference on Infrared, Millimeter, and Terahertz Waves (IRMMW-THz), IEEE, 2020.
103. Knyazev, B. A. et al. Feasibility of generating surface plasmon polaritons with a given orbital momentum on cylindrical waveguides using diffractive optical elements. *Computer Optics* **43**, 992-1000 (2019).
104. Wang, Z. X., Zhang, N. & Yuan, X. C. High-volume optical vortex multiplexing and demultiplexing for free-space optical communication. *Optics Express* **19**, 482 (2011).
105. Soifer, V. A. *Methods for Computer Design of Diffractive Optical Elements*. New York: John Wiley & Sons, Inc., 2001.

Stark et al. *Light: Advanced Manufacturing* (2022)3:34
https://doi.org/10.37188/lam.2022.034

Official journal of the JHL 2689-9620
www.light-am.com

Article **Open Access**

Miniaturization of a coherent monocular structured illumination system for future combination with digital holography

Andreas Walter Stark[1,*], Eugene Wong[1], Holger Babovsky[2], Christian Franke[1] and Richard Kowarschik[1]

Abstract

Holographic and 3D-measurement processes are an often-used tool in industry, medicine, and scientific applications. While small deviations of objects can be visualized by holographic means with high accuracy, optical systems with active structured illumination are a reliable source of absolute 3D-information in these fields. The combination of digital holography with structured illumination allows to simultaneously measure deformations and absolute 3D coordinates but also requires coherent light and has already been demonstrated in principle with a stereo camera setup. Multi-camera systems are limited to certain setup sizes given by the volume and distance of the detectors. Reducing the system to a one-camera (monocular) setup reduces space and acquisition costs. By using a multi-aperture illumination source an extremely high projection rate could be realized and reduced to a monocular approach with a novel voxel-calibration technique, while the projection system itself still requires a large amount of space. In this paper we present a miniaturized, monocular 3D-measurement system that works with repeatable, coherent speckles, generated by a fiber-coupled laser whose light was distributed by a fiber-switch to a diffuser plate connected with a measurement-head, also including a camera. By addressing different fibers through the switch, varying but repeatable patterns are generated. The size of the device (diameter < 3 cm) is now mainly limited by the volume of the camera. A first 3D-reconstruction of an object and an outlook for a combination of this system with digital holography is given, allowing absolute 3D-coordinates and relative deviations of object points to be measured simultaneously.

Keywords: 3D-measurement, Structured illumination, Speckle projection, Monocular 3D-measurement, Combination 3D-measurment and holography

Introduction

3D-measurement systems based on structured illumination as well as the methods of digital holography have been successfully used in many areas of science and technology for many years[1-5]. The advantages of the structured illumination lie in the fast and accurate determination of absolute 3D-object coordinates[6-9], but photogrammetric 3D-measurement techniques are limited in their axial resolution, not surpassing 1 μm in deviation. Digital holography is capable of tracking deviations (displacement vectors) in the range of the used wavelength and below but requires coherent light as illumination source[10-14]. A combination of the measurement of absolute

Correspondence: Andreas Walter Stark (andreas.stark@uni-jena.de)
[1]Institute of Applied Optics and Biophysics, Friedrich Schiller University Jena, 07743 Jena, Germany
[2]University Hospital, 07743 Jena, Germany

position and relative (height) variation seems therefore quite useful. In structured illumination, there have been methods for some years that work with coherent light (e. g. speckles) that can therefore be well linked to digital holography, which in principle has already been demonstrated for stereo camera systems[15,16]. However, these combined systems of stereophotogrammetry and digital holography are relatively large and therefore not suitable for many interesting applications in science and technology that require geometrically small and compact measuring heads. These kinds of measurements are usually carried out with miniaturized or endoscopic systems[17-20]. For instance, most medical applications must work in a very close range to the test object. On the other hand, similar industrial applications require a certain working distance from measurement instrument to object, such as pick-and-place tasks or product control applications. With this premise in mind a technique has been miniaturized, that was first developed for incoherent Multi-Aperture-Array-Projectors (MAAP)[21-23] and then used with coherent illumination as well[24].

In the following text a short description of the basic concepts of 3D-measurement with MAAPs, a recap of the use of repeatable speckles for 3D-measurement and the combination of 3D-measurement and holography is given. We then present our novel findings regarding details about the miniaturization of a 3D-measurement system utilizing repeatable speckles as well as results and a discussion about potentials and drawbacks of the technique.

Basics

3D-Measurement with Multi-Aperture-Array-Projectors (MAAP)

Specialized cameras are often the priciest components of a stereophotogrammetric setup. This holds true for high-speed measurements[6,25] as well as underwater[26] or infrared[27] applications. Avoiding some of the costs and at the same time reducing the required space by using only a one-camera or monocular setup can therefore be highly advantageous. A monocular system is from our point of view therefore each system making use of only one imaging path. The possibility to reduce a setup in such a way is dependent on the quality of the projector unit which includes the repeatability of the projected patterns. For single aperture projectors such as digital light processing (DLP) projectors, the calibration technique is a well-known practice[28-30]. However, these techniques should not be confused with "depth from single image"[31].

Since single aperture off the shelf projectors are subject to certain speed limitations, Multi-Aperture-Array-Projectors (MAAPs) have been developed, utilizing multiple projection centers and thus allowing a higher projection rate of up to 3 kHz[6]. In the stereophotogrammetric approach to gather 3D-information a correlation search is performed, leading to a correlation coefficient ρ with a typical threshold of 0.9. The correlation coefficient can be calculated by

$$\rho(I,J) = \frac{\sum_{n=1}^{N}\left(I_n - \bar{I}\right)\left(J_n - \bar{J}\right)}{\sqrt{\sum_{n=1}^{N}\left(I_n - \bar{I}\right)^2}\sqrt{\sum_{n=1}^{N}\left(J_n - \bar{J}\right)^2}} \tag{1}$$

Here I_n, J_n represent the gray values in pixels I and J in each of one of the two cameras, in the n-th image of the measurement sequence. $\bar{I}, \bar{J}$ represent the average gray values over the sequence at the pixels I and J. These MAAPs have also been used to calibrate a monocular 3D-measurement system with the help of a voxel calibration technique[21].

Here a linear translation stage has been used to move a reference plane through the measurement volume as depicted in Fig. 1. The patterns projected by the MAAP have been captured separately at every stage position, so that the volume was calibrated voxel-wise. The reconstruction process is based on the comparison of the stack of reference gray values gained from the calibration process and measurement images of the object under test. The same formula (1) could be used to distinguish correlating positions, even though the cross-correlation threshold had to be set to 0.99, as empirically found by Wong et al.[21] to avoid false reconstructions. The threshold differs from typical values known from stereo photogrammetry (for example 0.9[32]) especially due to the different type of correlation search, that is performed in axial direction here. The number of images depends on the

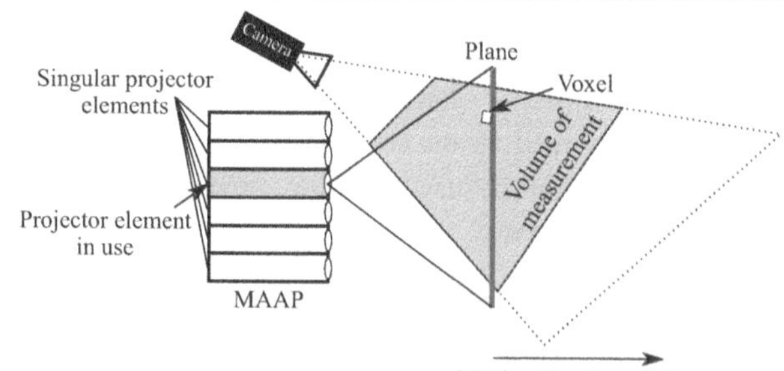

Fig. 1 Sketch of the calibration of a monocular camera system with a Multi-Aperture-Array-Projector (MAAP) for 3D-measurement. The plane can be moved to 100 – 10,000 positions separated by a distance between 1 μm - 100 μm, in dependence on the desired depth and quality of the calibration. At each position the gray values of all the patterns are captured and stored, creating a stack of reference gray values per voxel.

number of possible patterns to be generated by the MAAP (e.g., six by Wong et al.[21]). The calibrated system was then used to measure the shape of 3D-objects in high speed and with precise and accurate results. The MAAPs themselves however are as large or even larger than the cameras used and not miniaturizable without extensive effort.

Repeatable Speckles

When a coherent beam is statistically scattered areas of constructive and destructive interference are generated, which are called speckles. These patterns can be used as illumination for active stereophotogrammetry[32]. Apart from the coherent beam, speckle projection itself only requires a scattering element or surface of the size of the beam diameter. By illuminating a diffuser plate with coherent light led through fibers fixed on the diffuser plate the same speckle distribution could be reproduced, as often as the respective fiber was addressed by the coherent beam.

With a fiber switch and multiple fibers, each fixed to a diffuser plate, a number or sequence of such patterns could be repeatedly generated. We successfully applied this technique to generate 3D-reconstructions, using galvanometric mirrors to switch between fibers[24]. Here the above-described calibration approach for MAAPs was used, resulting in an effective measurement field of 10 cm × 10 cm × 5 cm. An image of a test object (relief of the German physicist Ernst Abbe) and of its 3D-reconstruction is depicted in Fig. 2.

From these experiments it became clear that the effectively required projector size, derived from the illuminated spots on the diffuser, was much smaller than the size of the used projector (5 cm × 5 cm). A miniaturization was therefore easier to realize compared with an incoherent projector that would rely on optical components for the projection process.

Combining 3D-measurement and holography

As previously mentioned, 3D-measurement techniques such as active stereophotogrammetry can deliver accurate and precise values of the absolute position of an objects surface, while holography, such as off-axis holography, can detect small relative changes of object surfaces, even below the wavelength in use. The requirement for holographic measurement is the coherence of the light used to illuminate the objects. The scattered light and a reference wave from the same laser source have to overlap, so that interference patterns are generated. The interference patterns can then be used to describe height variations of the surface compared to a master hologram or to the interference fringes induced by strain, pressure, or temperature changes. In earlier works these two techniques, 3D-measurement and off-axis holography, have been combined so that absolute and relative object information could be gathered simultaneously[15,16] (see Fig. 3).

Here, a stereophotogrammetric technique has been adapted, that involved coherent structured illumination. A sequence of 30 stereo images was taken, each with a different coherent pattern projected on the object. 3D-points were found by a correlation search between the gray value stacks of the images taken from both measurement cameras. By calculating the correlation coefficient (see Eq. 1) it was possible to find best corresponding pixels in both camera views and hence to reconstruct a 3D-point cloud (Fig. 4a).

The coherent light scattered by the object has been overlapped with a reference wave from the same laser source on the camera chips, leading to interference patterns in the images, so called off-axis holograms. These patterns changed when the surface under test was deformed even slightly, for example, by temperature variations. The test object was a table tennis ball with a radius of about 18.9 mm, whose reconstructed 3D-point cloud can be seen in Fig. 4a. Depicted in Fig. 4b are interference patterns mapped on the point cloud. Remarkably the system could find deformations with absolute values of up to 1.3 µm.

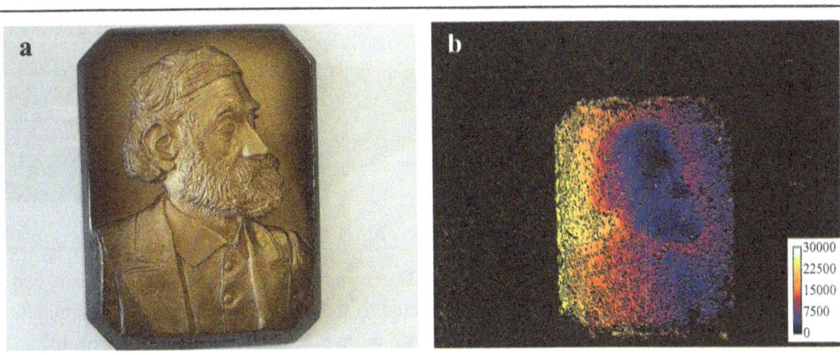

Fig. 2 a Test object, relief of German physicist Ernst Abbe, and **b** 3D-reconstruction using a macroscopic but monocular speckle projector as used in[24]. The false coloring represents the z-value of each reconstructed point in µm. Please note that unreconstructed areas have been set to 0.

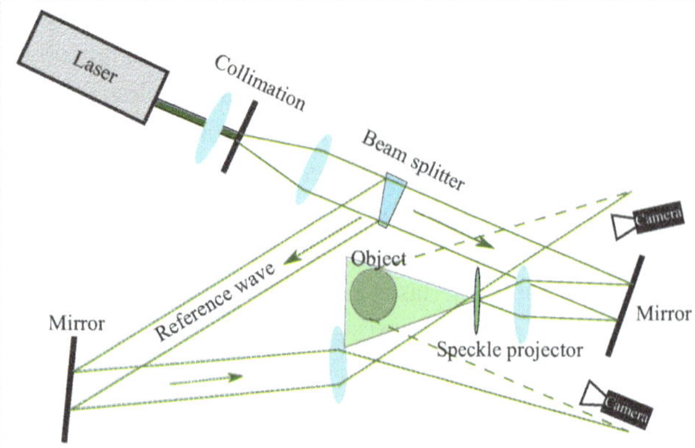

Fig. 3 Schematic drawing of the experimental setup used to combine stereophotogrammetric 3D measurement and holographic deformation measurement in a previous paper[16].

With the addition of more cameras, it was also possible to calculate the complete deformation vectors in space. The next step to effectively combine the two techniques and enable scenarios in hard accessible volumes where temperature changes and pressure/strain deformations are measurable simultaneously, requires a miniaturization of the 3D-measurement technique. In order to achieve that goal, we next combined optical fibers with a diffuser plate not much larger than the illumination spots to generate repeatable laser speckles and a small camera, calibrated after using the technique from Wong et al.[21].

Miniaturization of the speckle projection technique

The described miniaturization for the speckle generation process[24] introduces new proportions to the design of such monocular measurement systems: The limiting factor of the setup size is now, in principle, not the projector itself, but the camera (see Fig. 5).

To miniaturize the setup as much as possible while keeping a stable setup, that allows not only strict reproducibility of speckles but also fixed distances from camera to speckle generation spots and a good beam quality, a fiber switch (Leonie, type: "FO Switch eol 1 × 12 VIS HP") and a fiber coupled laser (Integrated Optics, type: 0532-13B/532 nm DPSS, MatchBox series 2) have been connected to a specifically designed measurement head (Fig. 6a) with a diameter of 2.6 cm, without fibers. The fibers themselves have been chosen to be single mode fibers (Thorlabs; type: "P3-460B-FC-1"), so that only the intensity of the output and not the profile of the illumination spot would be changed due to twists of the

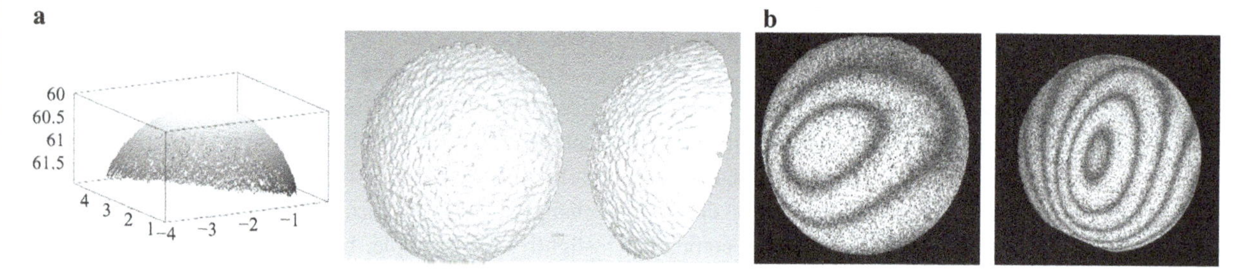

Fig. 4 **a** Image of the reconstructed 3D point cloud of a table tennis ball (values in mm) taken from a previous publication[13]. Reprinted/Adapted with permission from [Optics Letters 35(8) (2010)] © The Optical Society. **b** The interference patterns have been mapped on the point cloud, taken from a previous publication[16]. [Reprinted/Adapted] with permission from [Optics Letters 36(23) (2011)] © The Optical Society.

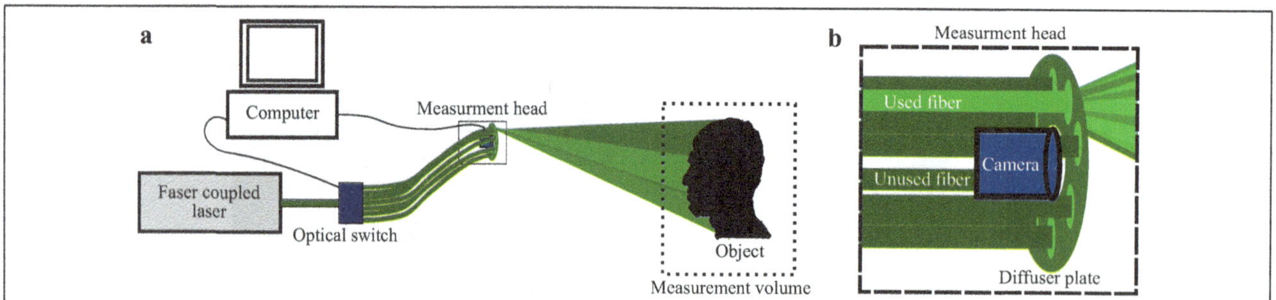

Fig. 5 Sketch of the miniaturized setup for speckle-based monocular 3D measurement. A fiber coupled laser generates coherent light that is distributed over a fiber switch to different single mode fibers, all connected to the measurement head. To illuminate the object only one fiber is addressed at a time and for a measurement a sequence of images, each with a separate pattern had to be taken. **b** The measurement head connects the fibers, the diffuser plate and the camera in a stable fashion, so that the same patterns are repeatedly filling the same volumes as seen from the camera.

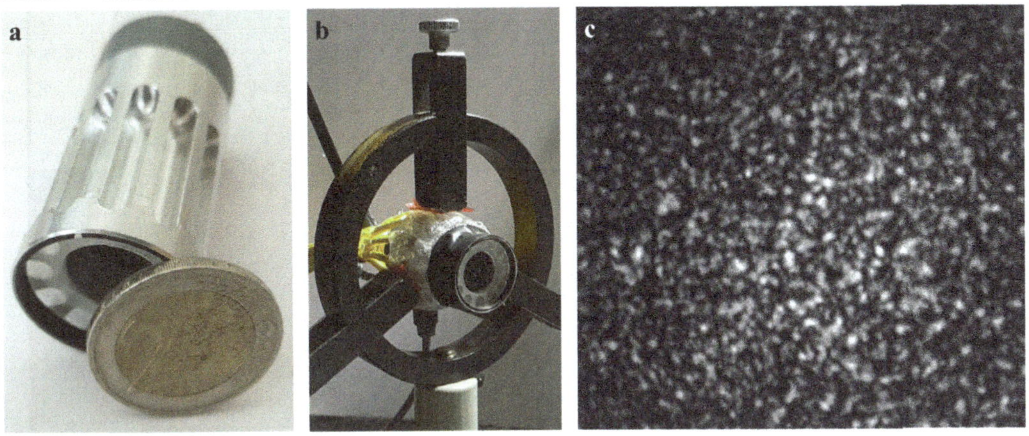

Fig. 6 **a** Photography of the measurement head compared to a 2 € - coin without fibers but with diffuser plate and camera. The openings to lead the single mode fibers can be seen on the sides of the head, leading to the diffuser plate. **b** The completed head with connected fibers in a three-point holder is depicted. **c** One of the speckle patterns projected on the reference plane for the calibration.

fibers. Additionally, a circular diffuser plate with a diameter of 2.5 cm, containing a centered, circular opening with a diameter of 1.6 cm, was fitted to the measurement head. This diffuser plate was used to scatter the light coming out of the fibers, while the centered camera maintained a clear view of the measurement field. The miniature camera and objective (XIMEA MU9PM-MH, CMOS) was chosen due to its small size (Edmund Optics 7,2 mm FL f/1.8 Red M12) and the relatively small distortion caused in relation to its size and working distance. Ten fibers were arranged and fixed on the measurement head around the camera, using a UV-hardened plastic glue, so that a strict reproducibility of the speckle patterns was possible. This arrangement led to the problem that a change of position of the fibers relative to the diffuser plate required each fiber to be removed and newly connected. The complete measurement head, also including the fibers, with a diameter of now 2.8 cm and a length of about 5 cm is shown in Fig 6b.

The working distance was set to be 10 cm, comparable to the working distance used with the 3D-measurement setups utilizing the MAAP[21-23]. One of the speckle patterns captured by the camera when projected on the reference plane is shown in Fig. 6c. The calibration was carried out as explained in the basics: A reference plane was mounted on a linear stage and moved through the measurement volume in 500 steps of 100 µm each, resulting in a total measurement volume depth of 5 cm. With the number of calibration steps the amount of stored data and the calculation effort increase. The best trade-off for in the regard of the authors was therefore the here presented measurement volume depth. At each position of the plain an image was captured for every projected set of objective speckles. The stack of reference images was used to find the objects z-position by comparing the measurement images to the gray values stored in the calibration stack as previously shown by Wong et al.[22] It has to be remarked that the calibration process itself was not altered or adapted to fit the coherent illumination by objective speckles in difference to the incoherent projection process, that was originally addressed with this technique. The axial size of the speckles had therefore no influence on the calibration process. Only the resulting gray value distribution have been stored and compared. The coherent noise that develops when coherently illuminated surfaces are imaged, so called subjective speckles, were also present in the measurements as well as during the calibration process. A technique to reduce this noise by spatially filtering the images[33] did not lead to an improvement of the reconstruction results.

Results

Most importantly, the main task of the above-described miniaturized projection system was achieved, i.e. the strictly repeatable generation of speckle patterns in order to robustly apply the calibration method mentioned in the basics according to Wong et al. As a proof of concept for possibility of a 3D-measurement with the here proposed miniaturized system, a 3D object (miniature gypsum bust of the German poet Johann Wolfgang von Goethe, height 5 cm, Fig. 7a, a measurement image can be seen in Fig. 7b) could be reconstructed (Fig. 7c). Notably, this result constitutes the first successful implementation of this novel speckle-based 3D measurement with a miniaturized, monocular system. However, the actual reconstruction is relatively imprecise (a ratio between standard deviation and measurement field diagonal of 10^{-3} compared to existing methods, which allow a ratio of 10^{-5} [6, 25-27, 32,33]), but

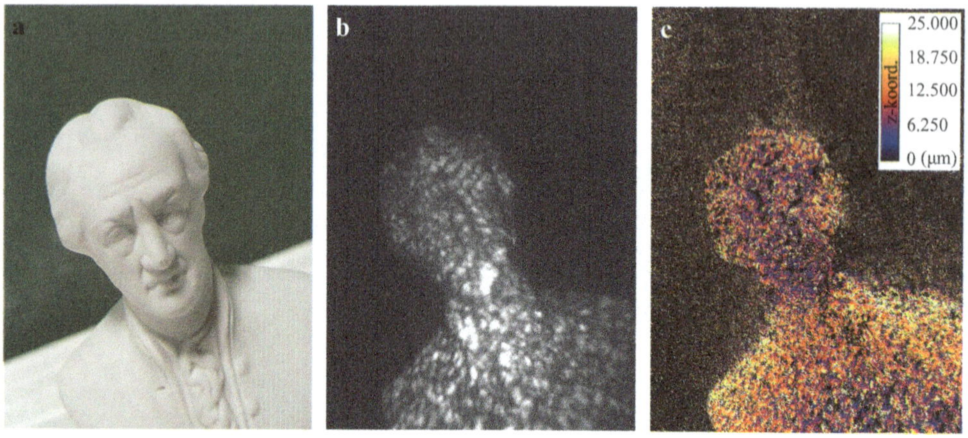

Fig. 7 **a** Photography of a miniature gypsum bust (height 5 cm) of German poet Johann Wolfgang von Goethe. **b** Single measurement image with the miniaturized speckle based monocular setup. **c** 3D-reconstruction from 6 patterns. The z-values are depicted in false colors. Please note that unreconstructed points have been depicted with value 0 as well.

remarkably complete. From the authors' point of view, the reconstruction quality of this first prototype can and will be enhanced to a considerable extent. For instance, there is significant potential for improvement in the structure size of the speckle patterns, which were probably too large in their lateral extension, relative to the intended resolution. A typical value is here a lateral spread of 10 pixel per objective speckles, as seen by the measurement camera[32]. Due to the comparatively large working distance for an endoscopic approach of 10 cm, only a small angle between the illumination and the observation direction was given, which may have had a negative effect on the point assignment and thus on the 3D-point reconstruction. An additional drawback is coherent noise, so called subjective speckles, that occurs when coherently illuminated surfaces are imaged. The here presented images have also been filtered with different kinds of spatial low pass filters, such as Gaussian and sinusoidal functions. A technique that allowed considerable improvements in speckle-based stereophotogrammetry[33]. Unfortunately, this method did only worsen the results to the point, where the bust couldn' t be recognized at all. In the future other reduction methods will have to be found to decrease the influence of this noise. The authors believe that the angle between the illumination and observation and the structure size of the speckles in their present form, are causing the majority of the errors and should be dealt with in a first step of improvement.

In our upcoming work, these parameters shall be investigated and will be varied by designing the basic body of the structure in such a way that adjustments of this kind will be possible without the mechanical disassembly of the device itself. The first step will be an increased number of projected patterns. Previous publication showed that a larger number of patterns reduces the standard deviation[32]. This will increase the measurement time as well as the effort of reconstruction but should be worth the effort. A decrease in distance from measurement head to object should allow for a larger angle between illumination and observation. Furthermore, the size of the projected speckles has to be optimized be it the number given by[32] or away from it.

Outlook

To the knowledge of the authors the here presented method of 3D-reconstruction has not yet been realized in this way and offers considerable potential for improvement. In addition to these obvious adjustments, however, the use of coherent light also allows the connection of 3D-measurement with holography. If one of the illumination channels or one of the fibers were used to illuminate the camera chip, while the object was illuminated by another fiber at the same time, holograms could also be recorded with this setup. A possible implementation, in this case by off-axis holography, is shown in Fig. 8. Here, a mirror element is used to guid the

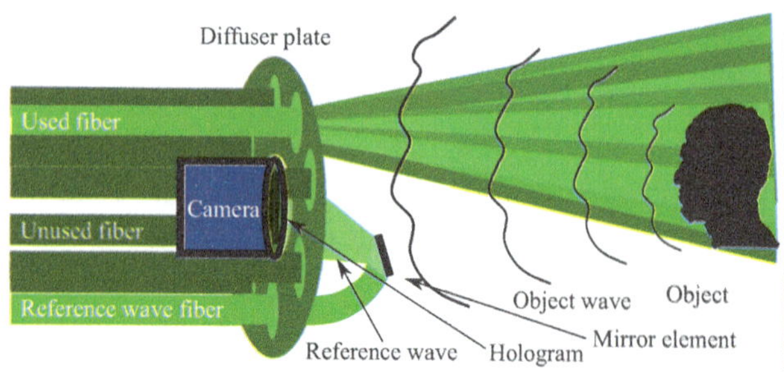

Fig. 8 Sketch of a combination of the presented 3D-measurement technique with an off-axis holographic approach.

light of the reference wave to the detector chip. Obviously, holography requires coherent interference between the reference wave and the object wave. To fulfil this condition the fiber generating the reference wave should be longer than the fibers creating speckles to illuminate the object. The path length difference induced in the reference wave fiber, needed to allow reference and object waves to interfere, would depend on the working distance of the device and should at least as long as the light path to the object and back to the detector. In case the coherence length of the laser in use is long enough, path length matching is not mandatory. Also, the intensity of the reference wave has to be adapted to the intensity of the object wave.

Such a measuring device could, as already shown in previous publications for a stereophotogrammetric setup[15,16], record height changes in the sub-micrometer range in addition to the 3D-shape. In this way, complex and fast processes, such as the cooling of components or the change in shape of an object under pressure or stress, could be observed and evaluated in real time. And with this device these measurements could be carried out in hardly accessible volumes of interest. Future research projects will therefore be oriented in this direction.

Discussion

In this publication, a novel, miniaturized, monocular 3D-measurement technology was presented and, in addition to the necessary basics, the voxel-wise calibration of a measurement volume and the repeatability of speckle patterns. First results were presented that prove the feasibility of the principle and its miniaturization. Compared to other 3D-measurement methods, the reconstruction quality can still be significantly improved, especially in terms of precision, which can only be classified in the millimeter range, for example by better adapting the structure size of the speckles in use to the measurement volume and reducing the working distance. Hence the presented system offers the possibility of including holography using coherent light for illumination. This would potentially enable the visualization of smallest height variations, such as caused by changes in heat or load on an object, and their precise assignment to absolute 3D-coordinates even in hard to access measurement volumes. Future work of the authors will focus on improving the method presented here in this direction and on the adaptation of the 3D-measurement method for high-resolution fluorescence microscopy.

Acknowledgements

We acknowledge the support by the machine shop (M2) of the faculty of physics and astronomy of the university of Jena helping to design and constructing the measurement head. We also gladly acknowledge the funding by the German federal ministry of education and research (BMBF) under the grant number 03ZZ0473.

Author contributions

A.W.S. and R. K. had the idea for the manuscript. The manufacturing of the setup was done by A.W.S. and the team from M2 (see above) and the experiments were done by A.W.S. and E.W. The reconstruction process was done by E.W. The project was supervised by R.K. A.W.S., E.W., H.B., C.F. and R.K. all contributed to writing the paper. The manuscript was discussed and corrected by all authors.

Data availability

The data that support the plots within these paper and other findings of this study are available from the corresponding author upon reasonable request.

Code availability

The code is available from the corresponding author upon reasonable request.

Conflict of interest

The authors declare that they have no conflict of interest.

Received: 15 September 2021 Revised: 15 April 2022 Accepted: 28 April 2022
Accepted article preview online: 13 May 2022
Published online: 15 July 2022

References

1. Fratz, M. et al. Digital holography in production: an overview. *Light: Advanced Manufacturing* **2**, 283-295 (2021).
2. Nolte, D. D. Cancer holography for personalized medicine. *Optics and Photonics News* **32**, 42-49 (2021).
3. Harte, J. M., et al. Chest wall motion analysis in healthy volunteers and adults with cystic fibrosis using a novel Kinect-based motion tracking system. *Medical & Biological Engineering & Computing* **54**, 1631-1640 (2016).
4. Wang, Q. & Kim, M. K. Applications of 3D point cloud data in the construction industry: a fifteen-year review from 2004 to 2018. *Advanced Engineering Informatics* **39**, 306-319 (2019).
5. Haleem, A. & Javaid, M. 3D scanning applications in medical field: a literature-based review. *Clinical Epidemiology and Global Health* **7**, 199-210 (2019).
6. Heist, S. et al. GOBO projection for 3D measurements at highest frame rates: a performance analysis. *Light: Science & Applications* **7**, 71 (2018).
7. Furukawa, R. et al. Single-shot dense active stereo with pixel-wise phase estimation based on grid-structure using CNN and correspondence estimation using GCN. Proceedings of the IEEE/CVF Winter Conference on Applications of Computer Vision. Waikoloa, USA: IEEE, 2022.
8. Willomitzer, F. & Häusler, G. Single-shot 3D motion picture camera with a dense point cloud. *Optics Express* **25**, 23451-23464 (2017).
9. Zuo, C. et al. Micro Fourier transform profilometry (μFTP): 3D shape measurement at 10, 000 frames per second.. *Optics and Lasers in Engineering* **102**, 70-91 (2018).
10. Zhong, F. Q., Kumar, R. & Quan, C. G. RGB laser speckles based 3D profilometry. *Applied Physics Letters* **114**, 201104 (2019).
11. Montrésor, S. et al. Error analysis for noise reduction in 3D deformation measurement with digital color holography. *Journal of the Optical Society of America B* **34**, B9-B15 (2017).

12. Marar, A. &. Kner, P. Three-dimensional nanoscale localization of point-like objects using self-interference digital holography. *Optics Letters* **45**, 591-594 (2020).
13. Etchepareborda, P., Moulet, M. H. & Melon, M. Random laser speckle pattern projection for non-contact vibration measurements using a single high-speed camera. *Mechanical Systems and Signal Processing* **158**, 107719 (2021).
14. Nobukawa, T. et al. Bimodal incoherent digital holography for both three-dimensional imaging and quasi-infinite–depth-of-field imaging. *Scientific Reports* **9**, 3363 (2019).
15. Grosse, M. et al. 3D shape measurement of macroscopic objects in digital off-axis holography using structured illumination. *Optics Letters* **35**, 1233-1235 (2010).
16. Babovsky, H. et al. Stereophotogrammetric 3D shape measurement by holographic methods using structured speckle illumination combined with interferometry. *Optics Letters* **36**, 4512-4514 (2011).
17. Furukawa, R. et al. Fully auto-calibrated active-stereo-based 3D endoscopic system using correspondence estimation with graph convolutional network. Proceedings of the 2020 42nd Annual International Conference of the IEEE Engineering in Medicine & Biology Society (EMBC). Montreal: IEEE, 2020.
18. Guo, C. L., Urner, T. & Jia, S. 3D light-field endoscopic imaging using a GRIN lens array. *Applied Physics Letters* **116**, 101105 (2020).
19. Lukic, A. et al. Endoscopic fiber probe for nonlinear spectroscopic imaging. *Optica* **4**, 496-501 (2017).
20. Matz, G. et al. Chip-on-the-tip compact flexible endoscopic epifluorescence video-microscope for in-vivo imaging in medicine and biomedical research. *Biomedical Optics Express* **8**, 3329-3342 (2017).
21. Wong, E. et al. Calibration of an array projector used for high-speed three-dimensional shape measurements using a single camera. *Applied Optics* **57**, 7570-7578 (2018).
22. Wong, E. et al. View-synthesized 're-calibration' of an array projector for 3D measurement from an arbitrary monocular view. *Optics and Lasers in Engineering* **141**, 106559 (2021).
23. Wong, E. et al. Optimization-based extrinsic calibration of a three-dimensional sensor composed of an array projector and a single camera. *Optical Engineering* **58**, 104109 (2019).
24. Stark, A. et al. Repeatable speckle projector for single-camera three-dimensional measurement. *Optical Engineering* **57**, 120501 (2018).
25. Heist, S. et al. 5D hyperspectral imaging: fast and accurate measurement of surface shape and spectral characteristics using structured light. *Optics Express* **26**, 23366-23379 (2018).
26. Landmann, M. et al. High-speed 3D thermography. *Optics and Lasers in Engineering* **121**, 448-455 (2019).
27. Kühmstedt, P. et al. GOBO projection for underwater 3D measurement technique. Proceedings of SPIE 10329, Optical Measurement Systems for Industrial Inspection X. Munich: SPIE, 2017.
28. Zhang, S. Recent progresses on real-time 3D shape measurement using digital fringe projection techniques. *Optics and Lasers in Engineering* **48**, 149-158 (2010).
29. Zhang, S. & Huang, P. S. Novel method for structured light system calibration. *Optical Engineering* **45**, 083601 (2006).
30. Legarda-Sáenz, R., Bothe, T. & Jüptner, W. P. O. Accurate procedure for the calibration of a structured light system. *Optical Engineering* **43**, 464-471 (2004).
31. Hu, J. J. et al. Revisiting single image depth estimation: toward higher resolution maps with accurate object boundaries. Proceedings of the 2019 IEEE Winter Conference on Applications of Computer Vision (WACV). Waikoloa: IEEE, 2019.
32. Schaffer, M., Grosse, M. & Kowarschik, R. High-speed pattern projection for three-dimensional shape measurement using laser speckles. *Applied Optics* **49**, 3622-3629 (2010).
33. Stark, A. W., et al. Subjective speckle suppression for 3D measurement using one-dimensional numerical filtering. *Applied Optics* **34**, 9473-9483 (2019).

Official journal of the JHL 2689-9620
www.light-am.com

Review Open Access

Design of broadband terahertz vector and vortex beams: I. Review of materials and components

Nikolay V. Petrov[1,*], Bogdan Sokolenko[2], Maksim S. Kulya[1], Andrei Gorodetsky[3] and Aleksey V. Chernykh[1]

Abstract

In this paper, we review the existing approaches for vortex and vector beam shaping and generation in the terahertz frequency range. The particular focus of this review is on the possibility of homogeneous topological charge formation in the ultra-wide spectral interval inherent to ultrashort terahertz pulses. We review the available materials and components, analyse proposed and potentially possible solutions for broadband terahertz vortex and vector beam shaping, compare all developed approaches, and put forward a unified concept for constructing passive shapers of such beams from the existing component base.

Keywords: Terahertz radiation, Vortex beams, Vector beams, Achromatic wave plates, Broadband terahertz radiation, Digital holography, Terahertz holography, Total internal reflection, Metasurface, THz liquid crystals, Geometric phase

Introduction

Terahertz (THz) beam shaping[1] attracts more and more attention in the last decade. Among the structures obtained by the optical beam shaping, a large niche is occupied by cylindrical vector[2] and vortex[3] beams with inhomogeneous polarisation states and optical vortices with phase singularities[4]. The formation of such beams in the THz range becomes a hot topic in current research, as they are gaining much attention due to their amazing properties, such as high efficiency coupling ability with bare metal wires[5]. THz beams are involved in numerous applications, such as imaging[6], telecommunications[7], and material characterization[8]. There are also other tasks, such as toroidal moment generation[9] and waveguide particle acceleration[10].

The ability of THz radiation to excite rotational and vibrational degrees of freedom of atoms and substances[11,12] makes it possible to utilise THz beams to precisely drive certain molecular states. For example, longitudinally polarised THz beams[13] can be used as efficient particle accelerators[14,10]. Due to the possibility of achieving orbital angular momentum (OAM) with a large number of eigenstates[15–17], the involvement of THz vortex beams are considered as a perspective approach to increase the bandwidth of communication channels[7,18,19]. Recently, super-resolution pump-probe imaging with THz vortex beam was used to observe saturable absorption behaviour in bilayer graphene[20]. THz beam shaping is also of high interest for imaging applications: for example, the use of THz Bessel beams shows better contrast and resolution of the image[6].

Correspondence: Nikolay V. Petrov (n.petrov@niuitmo.ru)
[1]Digital and Display Holography Laboratory, ITMO University, St. Petersburg 197101, Russia
[2]V.I. Vernadsky Crimean Federal University, Simferopol 295007, Russia
Full list of author information is available at the end of the article.

Quite wide variety of transverse beam modes has been experimentally demonstrated in the THz range: Laguerre-Gaussian THz beams[21], Bessel-Gaussian THz beams[22,23], Airy THz beams[24], circular polarisation Airy THz beams[25], and even more complex transverse beam modes. However, a significantly smaller proportion of publications on vector and vortex beams are associated with ultrafast pulsed broadband radiation. Among the latter, the various specific polarisation states and the spiral wavefront structure are presented at a limited range of spectral components that comprise the wave packet.

Ultrashort pulses possessing a homogeneous vortex structure across broad spectral range are known as broadband uniformly topologically charged (BUTCH)[26] beams and belong to the class of topological charge dispersion-free beams[27–32]. Such beams in THz frequency range are already finding unique applications, e.g., as sensors of magnetic properties[33], or as a carrier in the information encoding[34–37]. The formation of these beams is a complicated task due to the necessity to provide a homogeneous cyclic phase swirl for a large number of dispersive spectral components. These gradually increasing applications of such beams have prompted a demand for further development of techniques for their formation and assessment, which led to the appearance of this work.

We decided to divide the material into two interconnected manuscripts with numbered titles containing a common part "Design of broadband terahertz vector and vortex beams". This paper is the first part of this set of two interrelated papers. It contains an exhaustive review of the state-of-the-art approaches for broadband THz vortex and vector beam shaping, and comparative analysis of their pros and cons. It includes the results available to date from various research groups. In contrast, Part II[38], reports on the development and application of our proposed holographic approach for the analysis of the ultrabroadband THz wavefront evolution during it propagation through the partially-achromatic elements of designed beam shapers. Ultrafast spatially inhomogeneous wave trains are known to demonstrate strong spatio-temporal couplings[39,40] and their comprehensive assessment is usually a nontrivial task[41,42] due to the necessity of their characterisation in multiple modalities. At the same time, such diversity opens up a large scope for the formation of new beam types[43–46] and expands the potential of their applications. In the second part of this set of papers[38], we will focus on the capabilities of THz pulse time-domain holography (THz PTDH)[47–50] to describe and visualise the formation and propagation of such beams. For a better understanding of the possibilities offered by a detailed analysis of the beam properties, in the second paper of this paired set we consider few examples of the broadband performance of the THz vortex beams.

As far as this article is concerned, it presents not only a detailed review, systematisation, and comparison of existing and potential approaches to THz broadband vector and vortex beam shaping. We begin with an analysis of existing and promising components applicable as modulators for the formation of such beams. Finally, we aggregate the available approaches to the manufacturing of advanced beam shapers and other possible mechanisms for the tailoring of vector or/and vortex broadband radiation in THz frequency range.

Current state: developed approaches for BUTCH beams formation

While reviewing the current methods for vortex beams formation in the THz range, we can not help mentioning the existing review papers[3,51,52], describing most of the approaches proposed and developed to date. Unfortunately, despite the vast range of methods for vortex beam formation considered therein, none of them are focused on the broadband achromatic vortex beams, thus making this study useful and unique.

Several research groups[3,53] suggest dividing all the existing methods of vortex beams formation into two main categories: passive and active THz vortex beam shaping. Under the passive approach, the vortex beams are formed from pre-generated smooth wavefront THz pulses with Gaussian transverse profile of amplitude, while under the active one the authors assume direct THz vortex generation upon nonlinear downconversion from infrared radiation. We fully support this division, and it remains valid for the subset of broadband beams. However, we would suggest addressing these categories as *THz beam shaping* and *direct THz beam generation* (Fig. 1) to avoid ambiguity with terminology accepted in metasurface design, where the term "active" implies the possibility of tuning, as a rule, by an external field. Turning to the previously available scientific literature on THz vector and vortex beams formation, we should mention that, since most holographic methods of vortex beam formation (e.g. diffractive optical elements, computer-synthesised holograms) target to monochromatic radiation only they are outside of the scope of our consideration.

Among the categories of broadband optical vortex formation methods considered in Fig. 1, historically, passive approaches for the visible frequency range were the first to be developed. Subsequently, they were adapted for the THz frequency range. We note that most of the techniques for obtaining broadband THz vortex and vector beams demonstrated to date implement a geometric method

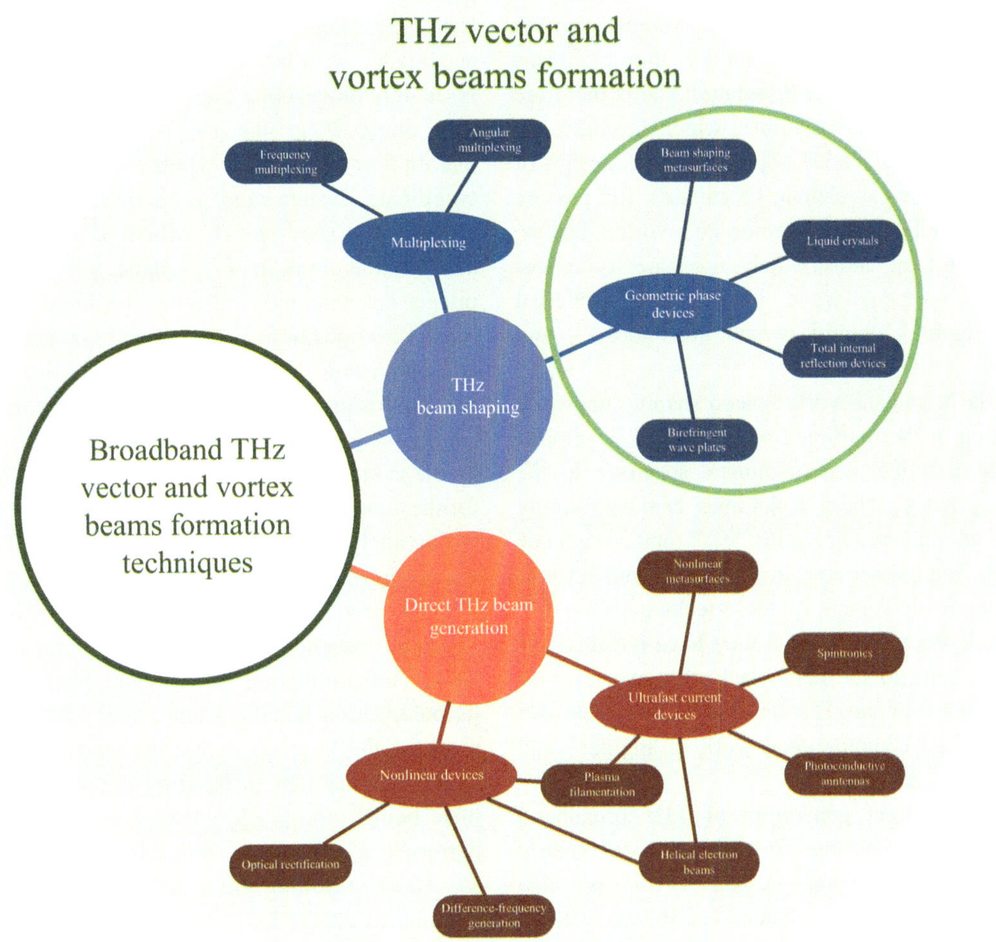

Fig. 1 Schematic art view of different techniques of THz broadband vector and vortex beam formation.

of phase transformation by rotating the polarisation vector. Let us look at the underlying principles of these techniques through a historical background.

In the early 1930s, it was shown that a circularly polarised paraxial beam transfers a spin angular momentum (SAM) S[54] associated with the polarisation of a wave capable of imparting rotation to anisotropic optical elements. Later, in the 1990s, Allen et al.[55,56] demonstrated a paraxial light beam carrying orbital angular momentum (OAM) L which is directly associated with optical vortices. These momenta are conserved for the paraxial beam spreading in any direction in the free space or in a homogeneous and isotropic transparent medium. In their seminal research, Ciattoni et al.[57] proved that in free space, the sum of the projections of these momenta in the direction of propagation z is an invariant of the wave motion, and hence the balance of the total angular momentum flux is conserved: $S_z + L_z = 1$[58]. However, optically anisotropic and inhomogeneous media intercouple both beam polarisation and wavefront shape, thus, consequently, can change both SAM and OAM, independently[59]. The physics behind the effect is based on a quantum phenomenon known as the geometric phase of Pancharatnam-Berry[60,61]. This principle is fundamental for optical devices that use full SAM-OAM conversion[59,62]. The origin of this term is associated with the geometric transformation of the polarisation state defining the corresponding cyclic adiabatic process, which appears to be an evolution path mapped on the surface of the polarisation Poincaré sphere[63]. Therefore, an arbitrary wavefront and phase reshaping will correspond to half of the solid angle enclosed by the path of a continuous sequence of polarisation transformations, which is determined only by the geometry of the polarisation path[64]. Polarisation elements, based on this principle, are commonly named Pancharatnam-Berry optical elements

(PBOE) and are not limited to helical waves shaping but can be extended to unbound wavefront control. Another feature of PBOE is subject to phase modification by means of only a space-variant retarder orientation and is therefore insensitive to wavelength variation[65], which provides the basis for a practically valuable approach to the design of achromatic phase optical elements[66] suitable for passive formation of broadband THz vector and vortex beams. Among all PBOE, the most widely used are: q-devices made with birefringent wave plates[67], total internal reflection principle[68,69], liquid crystal analogues[70,71], and metasurfaces[72–75].

Another THz broadband vortex beam shaping approach is the adaptation of the multiplexing techniques developed for other ranges of the electromagnetic spectrum to the THz frequency range. These techniques can be roughly divided into angular multiplexing[28,45,76–80] and frequency multiplexing[81–89] and, as a rule, use off-axis and on-axis schemes, respectively. Generally, these techniques are less energy-efficient; therefore, to date, they have not received widespread use, due to the acute lack of powerful THz sources. And today in the THz frequency range we can single out only one computational work in angular[80] and few in frequency[87,88] multiplexing.

In the case of direct generation of THz broadband vortices and vector beams, the principle of the geometric phase is also fundamental. However, it is also supplemented by the specific features of the conversion mechanisms from ultrashort femtosecond pump pulses into THz radiation. Direct generation approaches can be roughly divided into nonlinear and ultrafast current-based devices. The first category includes optical rectification[5,90–94], difference-frequency generation[95–98], helical electron beams[99,100], and plasma filamentation[101–108]. Helical electron beams and plasma filamentation on their inherent mechanisms of downconversion are also related to the second group where they supplement the photoconductive antennas[98,109–111], spintronic emitters[112] and nonlinear metasurfaces[113], as presented in Fig. 1. It should be noted that the approach to plasma generation of high-intensity THz radiation in liquid[114,115] is currently showing great potential. Such THz source may open a new direction in the design of vortex and vector beam generators.

In this work, our attention will be focused on "passive" THz broadband beam shaping techniques, among which the geometric phase approach highlighted with a green circle in Fig. 1 is the most promising to controlling the beam structure in a broad spectrum.

Q-devices: principle of operation

The principle of the geometric phase discussed in this paper is based on the use of the so-called q-plates, so we will first consider the principle of their operation. In accordance with the generally accepted definition, q-plate is an inhomogeneous anisotropic medium, which optical axis angle θ in the (x,y) plane exhibits a uniform distribution over the whole surface defined in this coordinate system and is described by the following relation[64]: $\theta(\psi) = q\psi + \theta_0$, where $\psi = \arctan(y/x)$ is the azimuthal angle in the (x,y) plane, q is an integer or semi-integer constant and θ_0 denotes the angle between "zeroth" optical axis position at $(\psi = 0)$ and x-axis and may take an arbitrary value. The q-plate got its title due to the fact that the polarisation of a passing wave is experiencing a continuous sequence of spatial turns which is determined by the value q[116,59]. Various approaches have been proposed for the construction of such devices[117–121].

The analysis of the balance of SAM and OAM for each photon in a circularly polarised wave ($S_z = \pm\hbar$, where $\pm$ denotes circularity handedness and $\hbar$ is a reduced Planck constant) passing through this q-plate demonstrates a mutual transformation of momenta when the light inverts its polarisation helicity from $+\hbar$ to $-\hbar$ and therefore the resulting OAM arises, while the overall variation of total angular momentum suffered by the photon passing the q-plate being[62]: $\Delta J_z = (2q-1)\hbar - \hbar = 2\hbar(q-1)$. The q-plate performs a wavefront tailoring with phase given by $\varphi = \pm 2q\psi$, resulting in a helical vortex structure with topological charge $\mathsf{L} = \pm 2q$. The handedness of the vortex phase helicity and a sign of the topological charge are controlled by the polarisation state of the input field.

The manufacture of the q-plate does not always reproduce the smooth rotation of the optical axis. The design of the discrete q-plate can be composed of a set of sectors with individual inclinations of the optical axis. Thus, it is possible to implement the constant phase delay in each sector to obtain the resulting phase shift throughout the plate $\varphi = 2\pi\mathsf{L}$.

For the completeness of the description, it should be added that q-plates are a special case of the so-called j-plates[122,65], devices, which can convert an arbitrary (which means is not limited by two circular polarisation states) input SAM into two arbitrary output total angular momentum states. However, since the concept of j-plates is currently more developed only for the visible range and is not widespread in the THz frequency range, we will use the q-devices notation for this paper.

In the general case, different combinations of the polarisation state of the THz beam and the charge of the q-plate are possible, leading to various vector and vortex beams. One approach for cylindrical vector beam shaping and three main approaches to the formation of the vortex beams are shown in Fig. 2, while the total number of these states is greater[123]. In the classification under consideration, two columns define two possible approaches that differ in the order of the main elements in the shaper. The type I approach (Figs. 2a–d) assumes the initial conversion of the input linearly polarised THz field into a circularly polarised one using a quarter-wave plate (QWP), followed by passing through the q-device. In contrast, the type II scheme (Figs. 2e–h) implies a straightforward action of the q-device on linearly polarised radiation. The following notation is also used here: 'c' means 'Circularly polarised' 'x' denotes the 'vorteX', 'v' denotes 'Vector'. In most canonical cases (Icx and Ix), a circularly polarised pulsed THz beam incident normally to the surface of the q-device is converted by the half-wave retarder into a beam with axial phase singularity and inverse circularity of polarisation. According to scenario IIv which is illustrated in Fig. 2e–f the vector beams with radial and azimuthal polarisations can be formed from the initial linearly polarised wave with the single q-device[2]. Another type of conversion from the linearly polarised non-singular beam to the vector beam carrying spiral wavefront (scenario IIx) is illustrated in Fig. 2g–h. In this case, polarisation of the incoming beam is first converted, for example, to radial state by q-plate with $q = 1/2$, and then phase delay required by vortex beam is introduced with a QWP. To turn the resulting polarisation to a homogeneous linear state, a polariser is used. It should be mentioned that this sophisticated configuration of vortex beam shaping is also viable. The closest implementations of scenario IIx for THz frequency range were introduced by Imai et al.[93] and Lin et al.[96], but the proposed approaches belong to the category of active methods. All four approaches are summarised in Table 1.

Discrete q-plates are well established tools for vector and vortex beam shaping in the optical domain. Nevertheless, the most vital limitation of q-devices is the requirements for the incident beam[52]. In most cases, the design of the q-plates is done for the collimated beam with flat wavefront. In the non-paraxial case, the incoming beam wavefront has a complex angular spectrum, resulting in the inhomogeneities of the phase delay distribution. Moreover, such devices usually operate in the limited spectral band. In the second paper[38] of this paired set, we discuss that these

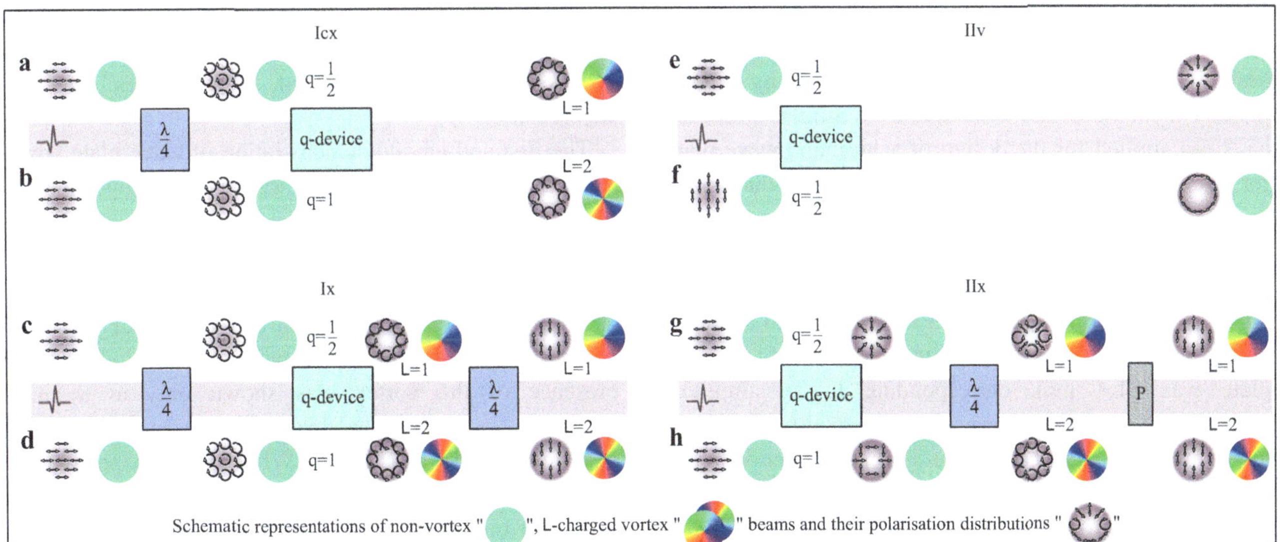

Fig. 2 Two main scheme types (I and II) for THz vector (v) and vortex (cx, x) beam shaping, respectively. $L = \pm 1$ **a**, **c**, **g** and $L = \pm 2$ **b**, **d**, **h** charged beam with circular cx **a**, **b**, and linear x **c**, **d**, **g**, **h** polarisation, respectively; cylindrical v beams with radial **e** and azimuthal **f** polarisations respectively. The notations of polarisation elements are used: $\lambda/4$ is a quarter-wave plate; P denotes a linear polariser.

Table 1 Types of approaches for THz vortex and/or vector beams shaping

Type	Input polarisation	First element	Second element	Third element	Output polarisation
Icx	Linear	QWP	Q-plate	—	Circular
Ix	Linear	QWP	Q-plate	QWP	Linear
IIv	Linear	Q-plate	—	—	Azimuthal / Radial
IIx	Linear	Q-plate	QWP	Polariser	Linear

features in the spatial distribution of the phase delay can be tracked using THz PTDH.

To overcome the wave plate and q-device chromatic limitation to be applicable for broadband THz radiation, materials, and devices with dispersion properties less sensitive to the incident radiation frequency are required. This demand causes the task of developing of achromatic quarter- and half-wave plates design for a broad frequency range. Several approaches to the design of achromatic wave plates and devices have been proposed: multilayer stacks of birefringent material layers[124–128], total internal reflection[68], liquid crystals[129], and metasurfaces[74,130–132] (Fig. 3). In subsequent section we will briefly overview all of them.

Achromatic birefringent wave plates

Natural birefringent media, highly transparent within a broad THz spectral range, such as crystalline quartz and sapphire, are well suited for the manufacture of wave plates. However, bare substrates will introduce linear chromatic dispersion.

The method of achromatic wave plates engineering for a certain visible wavelength range has been known for over 50 years[137]. To even out phase retardation at all wavelengths of interest, a stack of birefringent plates is designing, where the achromaticity is achieved by tailoring plates' thicknesses and orientations[138]. Similar procedure has been applied for the design of achromatic wave plates for the THz range, using both crystalline quartz and sapphire as birefringent media[124–127].

Masson and Gallot were the first to present an achromatic quarter-wave plate, consisting of six layers and operating in the range from 0.25 to 1.75 THz to achieve a phase delay of $\pi/2 \pm 3\%$[124]. The total thickness of the wave plate was 31.4 mm, corresponding to an amplitude transmission at 1 THz measured to be equal to 74%, most losses being introduced by the Fresnel reflection at the wave plate interfaces. The following study[125] provided design of an achromatic quarter-wave plate with retardance stability of $\pm 3°$ around $\pi/2$ in the frequency range of 1.3 – 1.8 THz. This wave plate was composed of nine layers of crystalline quartz (Fig. 3a$_1$) and had a total thickness of 51.72 mm. It should be noted that the use of crystals with strong anisotropy, for example, sapphire (Al_2O_3) with a large birefringent contrast (n_o = 3.39, n_e = 3.07), allows thinner layers and, hence, results in thinner wave plates. Taking this fact into account, Wu et al.[128] designed and fabricated a thinner achromatic quarter-wave plate based on stack birefringent sapphire discs (approximately 3 mm total thickness) with an overall phase error of a few percent in the 0.1 – 0.8 THz range (Fig. 3a$_2$). The transmittance of this plate is between 60% and 40% in the range of 0.1 to 0.8 THz, and again most losses are present due to the Fresnel losses what is cause by the higher values of the sapphire refractive index. However, according to our simulation which is reported in the second paper[38] from this paired set, these sapphire wave plates may be inefficient for vortex beams formation, since they do not yield the desired phase difference $\Delta\varphi = \varphi_x - \varphi_y = \pi/2$ and a perfect circularity.

Despite the high transparency of crystalline quartz for THz radiation, which is greater than that of many plastics, it multilayer slab can significantly (at times) attenuate THz radiation. Therefore, alternative possibilities for creating achromatic wave plates are also being explored. Tangible progress here was provided by the work of C.-J. Yang et al.[139], in which it was shown that (110) and (001) cuts of $DyScO_3$ crystals have a fairly wide frequency range of minimal dichroism, which makes it possible to use 50 and 370 μm-thick crystals as QWP in the frequency ranges 0.5 – 0.7 and 0.5 – 0.61 THz, respectively, for each of the cuts.

Multilayer achromatic wave plates (Fig. 3a$_3$) can be employed to manufacture achromatic q-plates for the formation of broadband THz vortex fields. Stacks of several q-plates[140] will allow introduction of an arbitrary shape of the geometric phase into the passing THz beam and reproduce a large set of OAM values and their superposition[141].

The first and unique demonstration of wave plate vortex beam shapers in the THz range was published by Minasyan et al.[67], where using a single-layer eight-sector quartz q-plate with a charge of $q = 0.5$ installed in scheme accordingly to Fig. 2c, they demonstrated the possibility of obtaining a vortex vector beam at a frequency of 1 THz. As the wave plates they used were not achromatic, the presence of the vortex was shown only at a single frequency, a narrow bandpass filter was used in the experimental demonstration. While the perspectives of using achromatic wave plates in the same setup and achieving broadband THz vortex beams were discussed in the aforementioned paper, experimentally such generation has not been demonstrated to date. As was shown by Gecevicius et al.[142], for the broadband beam passing through the narrowband chromatic q-plate introducing half-wave retardation at a specific frequency ν_0 only, a very limited part of circularly polarised radiation outside a design frequency $\nu \neq \nu_0$ changes its helicity and transforms into a vortex beam, while most radiation is converted into some unstructured vector field.

To clean all non-vortex beam components, polarisation filtering[142] should be performed. Provided that the input THz field has circular polarisation uniform over the entire

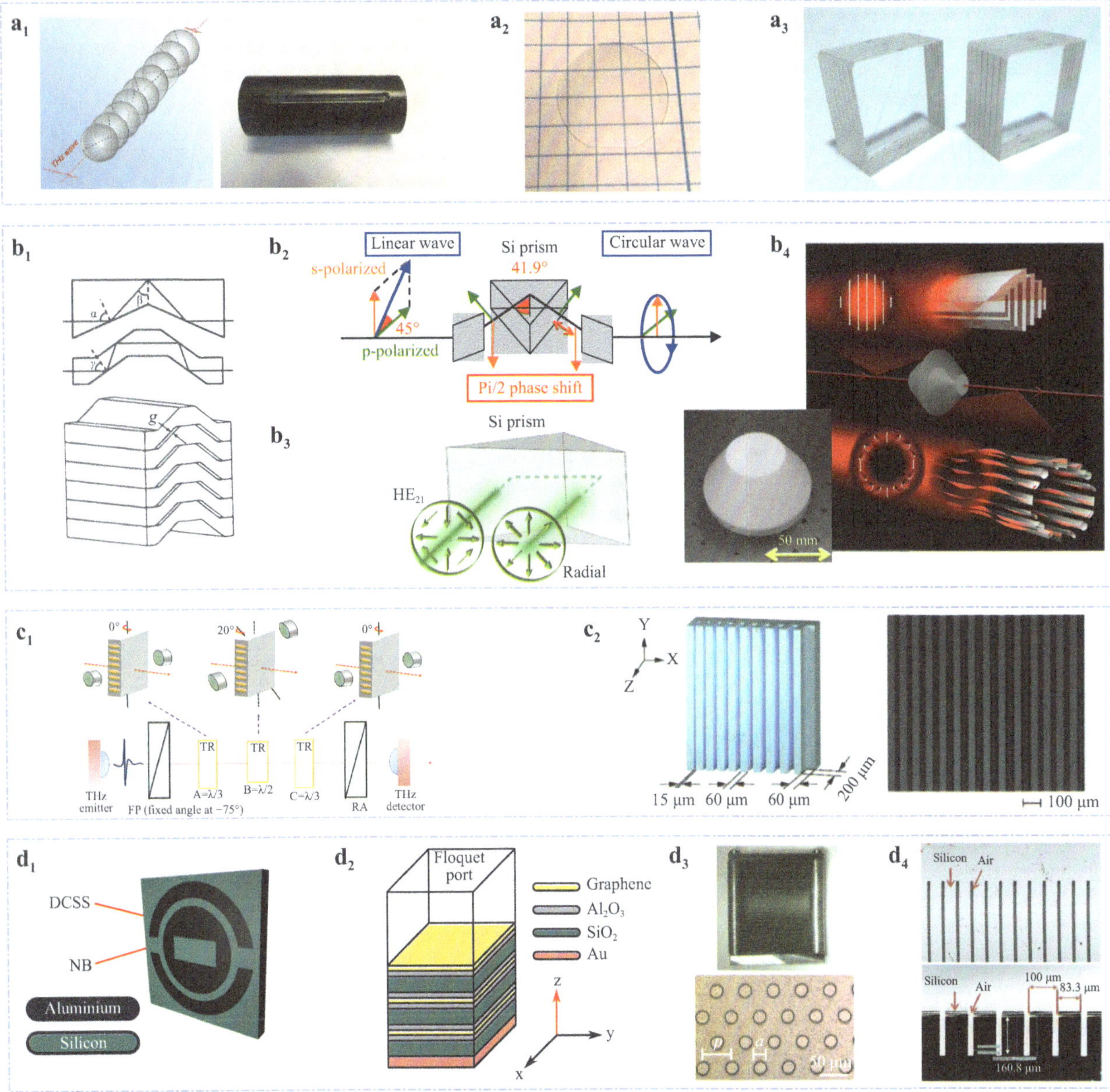

Fig. 3 Examples of THz achromatic polarising components: **a** multilayer birefringent media, **b** total internal reflection devices, **c** liquid crystals, **d** multilayer metasurfaces. $\mathbf{a}_1$ The achromatic QWP made from crystalline quartz plates[125]. $\mathbf{a}_2$ The birefringent sapphire disc[128]. $\mathbf{a}_3$ The achromatic HWPs made from crystalline quartz plates (our work). $\mathbf{b}_1$ The three-, four-reflection type prisms and the multistacked prism-type wave plate[68]. $\mathbf{b}_2$ The scheme for the linear-to-circularly polarised THz radiation conversion with the total reflection in the Si prism[133]. $\mathbf{b}_3$ The triangular Si prism as HWP[112]; $\mathbf{b}_4$ (main) the illustration of THz vectorial vortex control[134] (CC BY 4.0) and $\mathbf{b}_4$ (inset) the polytetrafluoroethylene axially symmetric wave plate[53]. $\mathbf{c}_1$ The scheme of the LC-based achromatic wave plate with three magnetically tuned elements[135]. $\mathbf{c}_2$ The sub-wavelength dielectric gradient grating[136]. $\mathbf{d}_1$ The geometric phase building block[74] and $\mathbf{d}_2$ the graphene-based unit cell[75] for the metasurface vortex plate. $\mathbf{d}_3$ The parallel metal plate waveguides with a pillar array[131]. $\mathbf{d}_4$ The achromatic silicon-grating-based QWP[132]. Images adapted with following permissions: $\mathbf{a}_1$[125] from Elsevier; $\mathbf{a}_2$[128], $\mathbf{b}_4$ (inset)[53] from Springer Link; $\mathbf{b}_1$[68], $\mathbf{b}_2$[133], $\mathbf{b}_3$[112], $\mathbf{c}_1$[135], $\mathbf{c}_2$[136], $\mathbf{d}_3$[131], $\mathbf{d}_4$[132] from The Optical Society.

spectrum, spectral density after such conversion is modulated by the following frequency dependence: $\left(1-\exp\left(i\pi(\nu/\nu_0)\right)\right)$. Spectral efficiency of such conversion by monochromatic q-plate, optimized for the frequency of $\nu_0 = 0.75$ THz, we calculated, as shown on Fig. 4a$_1$. The narrow bandwidth of the output vortex suggests that an achromatic q-plate should be employed for the formation of a THz vector vortex beam with higher purity and better conversion efficiency. We have designed such a q-plate. Its concept and the transformation of the structure of the THz field after its passage are discussed in detail in the second paper[38] from this paired set.

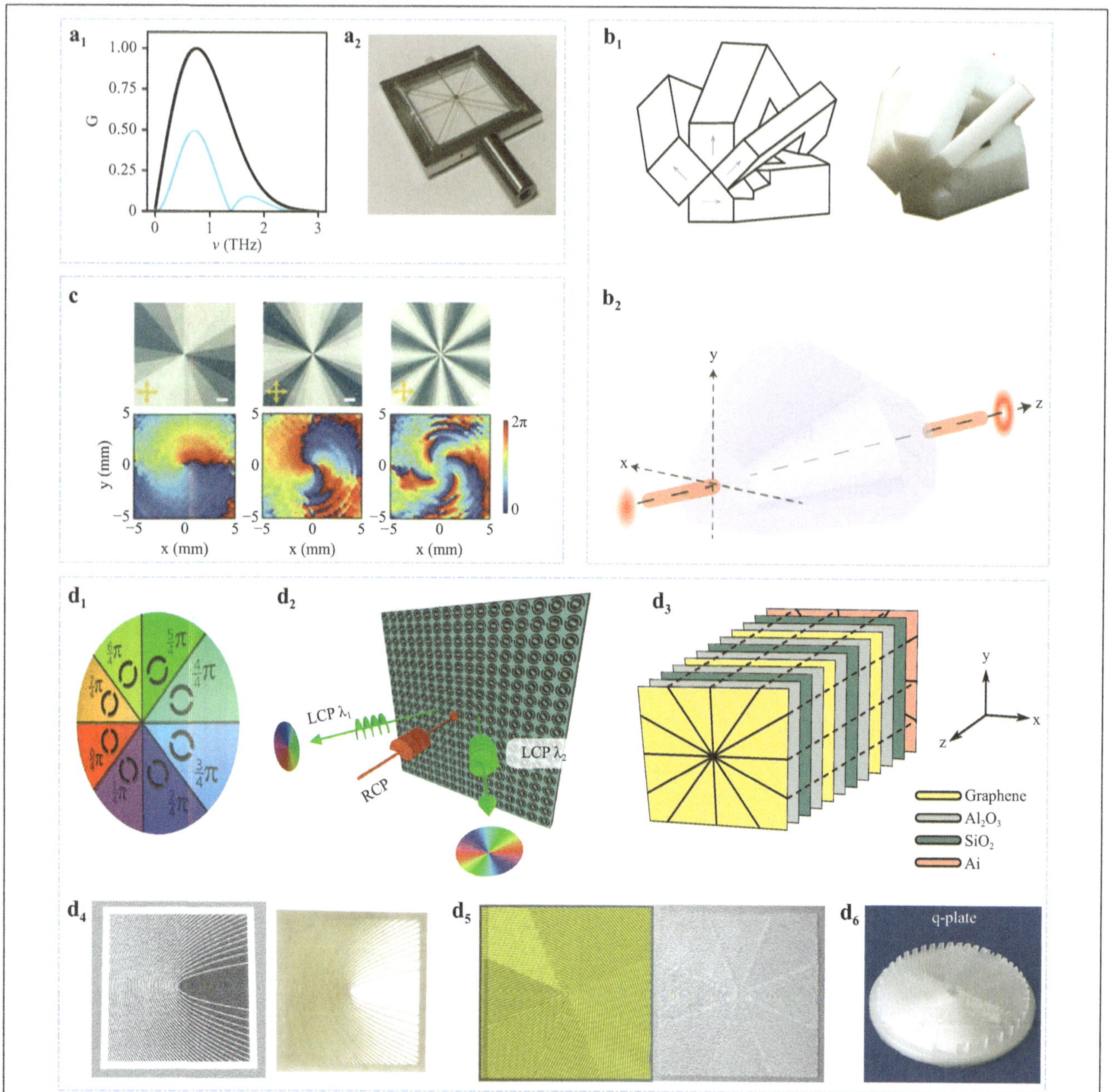

Fig. 4 Examples of geometric phase devices for THz frequency range based on: **a** achromatic birefringent wave plates, **b** total internal reflection, **c** liquid crystals and **d** metasurfaces. $\mathbf{a}_1$ The black and cyan curves show the original and modified after passing through the chromatic q-plate the modulus of the temporal spectrum (as described in Section "Achromatic birefringent wave plates"[67, 70, 71]) and $\mathbf{a}_2$ the image of sectorial chromatic q-plate (the photo was courtesy provided by Emmanuel Abraham). $\mathbf{b}_1$ Schematic representation of segmented achromatic q-device for broad THz range and its image[69], $\mathbf{b}_2$ the achromatic non-axially-symmetric continuous body q-device[69] (adapted, CC BY 4.0). **c** Photos and measured phase distribution at $\nu = 1$ THz from a liquid crystal polymer q-plates with charges L = 1, 2, and 4[71] (adapted with permissions from John Wiley and Sons). $\mathbf{d}_1$ The phase distribution in the THz beam reflected from the corresponding sectors of metasurface[145]. $\mathbf{d}_2$ Dual-band reflective[74] and $\mathbf{d}_3$ multilayer transparent[75] metasurface vortex plates. $\mathbf{d}_4$ The continuous, $\mathbf{d}_5$ polylactide[146] and $\mathbf{d}_6$ cyclic olefin copolymer[147] (adapted with permissions from The Optical Society) sectorial 3d-printed sub-wavelength q-plates with $q = 1/2$.

Total internal reflection devices

One of the most promising candidates for achromatic devices in the broad frequency range are devices based on total internal reflection (TIR), as there are plenty of materials with flat refractive index dispersion in the THz range[133]. Among the achromatic TIR components developed to date for the THz range, one can find Fresnel prisms[33,133,112] (Fig. 3b$_4$) and non-axially symmetric wave plates introducing angular variation of the beam phase shift[69] (Fig. 4b$_1$, b$_2$). We note that polarisation tailoring by

TIR inside the cone[143] can also be used for such purpose, though it has not been realised in the THz range yet.

Arbitrary TIR local change polarisation state in broadband THz beam can be performed by several complete internal reflections within a prism when, for instance, linearly polarised THz radiation is converted to circularly polarised radiation (Fig. 3b$_2$). Importantly, a total reflection phase-shift is independent of the wavelength due to the flat dispersion of prism material (Si) in the THz frequency range. Increasing the incidence angle up to 45° gives approximately π of phase shift between orthogonal polarisation components, thus a half-wave retardation has been performed (see Fig. 3b$_3$). Implementation of this approach for the visible range was experimentally demonstrated in TIR device by Wakayama et al. in 2012[144] and then reproduced with metasurfaces by F. Bouchard et al. in 2014[63]. The same year, Kawada et al.[68] designed and manufactured TIR based element for the broadband THz wavelength range, which included half- and quarter-wave plates (Fig. 3b$_1$). A stack of cyclo-olefin polymer prismatic elements of three- and four-reflection type prisms were assembled to wave plates, which demonstrate achromatism over a wide region of the THz spectrum from 0.1 to 2.5 THz. Later, this group proposed a cylindrical symmetric TIR element in the form of two glued hollow truncated axicons[134,53] (Fig. 3b$_4$), which provides rotation of the polarisation due to Fresnel reflection, causing a change in the phase of the vortex due to the SAM-OAM coupling. It modulates the polarisation states of the beam depending on the azimuthal angle, which leads to the transformation of a uniformly polarised Gaussian beam into a vortex one with a uniform state of polarisation along its cross-section.

Later the same group of researchers[69] proposed two variants of non-axially-symmetric half-wave plate, equivalent in their principle of operation to a q-plate with $q = 0.5$. The first of the configurations designed and manufactured for the THz range, consists of four segments oriented at azimuthal angles of $\psi = 0, \pi/4, \pi/2$ and $3\pi/4$ (Fig. 4b$_1$). Each segment of this device is formed by a pair of Fresnel rhombuses providing a certain number of TIR from their faces. In contrast to the work of Kawada et al.[68], relative shift between orthogonal polarisations, giving a phase delay equal to π, was provided after four TIRs. Since the paper also considers the prospects of using non-axially-symmetric elements for shorter wavelength spectral ranges, where higher precision requirements for processing and assembly complexity make it difficult to implement the proposed solution, non-axially-symmetric half–wave plate realised in a continuous body (Fig. 4b$_2$) was proposed. Such an element is an assembly of two components obtained by rotating the Fresnel rhombus at an angle of 90° along the beam propagation axis.

Liquid crystals

Liquid crystals (LC) are another promising class of anisotropic material suitable for manipulating the THz wavefront by achieving a geometric phase shaping and thus suitable for application in vortex and vector beam modulators. They are attractive due to the controllability of the director distribution (e.g. fast optical axis) and sensitivity to the external electric[148–152], and electromagnetic[153] fields, as well as temperature[52,154]), allowing to consider them as one of the promising approaches for OAM dynamic control, needed, for example, in wireless and optical telecommunications[52]. To date, various LC based THz polarisation devices have been proposed: phase-shifters[148,150,155,156], wave plates[152,157,135], gratings[158,159], q-plates[70] and other beam shaping plates[160,71].

Among commonly used LC materials, the most interesting for the THz beam shaping applications are nematic LC E7 (by Merck) with a birefringence of 0.130 – 0.148 in the frequency range between 0.2 THz to 1.2 THz[150,135,136]; MDA-00-3 461 (by Merck) with birefringence is about 0.2 in the range between 0.3 – 1.4 THz[129] and LC NJU-LDn-4 with an average birefringence of 0.306 in the range between 0.5 to 2.5 THz[152,159,157,70,160]. High birefringence allows for quite subtle polarisation control. This feature is important because it is quite difficult to perform homogeneous or desired prealignment in thick cells and high operating voltage. Using remarkably narrow LC cell gap of 125 μm thickness with nematic LC NJU-LDn-4 in the 0.5 – 2.5 THz band, Wang et al. developed electrically controlled both transmissive[152] and reflective[157] phase retarders. Proposed structure with sub-wavelength metal grating electrodes allowed a controlled phase delay in two operational modes: tuneable quarter-wave plate, covering the range between 1.1 – 2.5 THz, and half-wave plate for 2.2 – 2.5 THz range. In addition to bias control, phase delay after the reflecting LC cell can be adjusted by the beam angle of incidence, thereby realising a more flexible control of the output polarisation state[157]. Each of the materials has a response time of the order of several milliseconds, making it possible to create fast controlled LC converters with a given phase delay on its basis. For vectorial THz broadband beam shaping, cells providing $\pi/2$ or π phase retardation in a broad frequency range are required.

Regarding the achromaticity requirement, we note that all of the above devices provide required phase delay only at a certain frequency. For the phase and polarisation conversion purposes of broadband THz beams, it is necessary to tune the LC cells continuously within the operating range. Thus, the main disadvantage of such devices is the dispersion of phase retardation. At the same time, there are practically no works aimed at achieving an achromatic phase shift in the THz spectral range at the moment, with very few exceptions[129,135,136].

On the basis of an approach similar to that used for achromatic crystalline birefringent wave plates, a broader tuneable range can be achieved by stacking LC cells, thus cancelling dispersion and improving conversion efficiency over the entire THz range. In two recent works[129,135] retarders consisting of three discrete LC cells with different thicknesses demonstrated achromatic tuneable THz quarter-wave plate in a relatively wide band from 0.20 to 0.80 THz[129] and from 0.20 to 0.70 THz[135]. The remarkable advantage proposed by Hsieh et al. retarder[135] is in applying magnetic fields to tune LC without the use of transparent electrodes (Fig: 3c_1), usually not too transparent for THz radiation. The overall loss of the proposed retarder is about 8 dB. However, other researchers note that devices controlled by a magnetic field are space-consuming, expensive, and high energy demanding[152]. Thus, it seems quite unfeasible that LC achromatic devices with spatially inhomogeneous polarising properties will be implemented in the near future. In addition to the achromaticity problem, the resulting cells demonstrate a complex dependence of the orthogonal polarisation components transmittance on the frequency, leading to an uneven radial distribution of the output beam at individual spectral components. Yang et al.[150] compensated this disadvantage by using a nanowhiskers-structured substrate obtained from obliquely evaporated indium titanium oxide (ITO) by electron beam glancing angle deposition, which limits the electrically tuneable LC cell. This allowed greater transparency (77.8% and 75.4%) in the range of 0.2 to 1.2 THz. However, 517 μm thick cell provided a phase shift of $\pi/2$ for frequencies of 1.05 THz and above only. Furthermore, the combination of subwavelength gratings with variable spacing between grooves (Fig. 3c_2) filled with mixture of polymer solvent epoxy resin and E7 LC allowed expanding the cell working range and equalising the transmittance of the two orthogonal components to 50%[136]. Phase shift of the proposed tuneable 500 μm thick THz wave plate cell can vary from π to 0 rad near 0.8 THz, when the bias is increased from 0 to 80 V, thus realising both half-wave and quarter-wave plate properties. However, the authors succeeded in experimentally demonstrating the transformation of linear polarisation into elliptical polarisation at a given frequency only. It is also worth noting that in this work, the authors experimentally found that it is possible to introduce an achromatic phase delay $\Delta\varphi = \pi \pm 20\%$ in the range of 0.55 to 1.05 THz with a Si grating whose anisotropy property originates from the gradients in the fill rate of Si between the gratings.

Summarizing the state-of-the-art of LC based broadband THz vortex shaping, it should be noted that due to the technology immaturity for providing achromatic anisotropy, it is still problematic to form wave packets containing a BUTCH beam. However, taking into account the possibility of magnitude and introduced phase shift dynamic adjustment, it is possible to consistently form a vortex wave front at various spectral components over a wide THz spectrum.

LC-formed vortex structures were characterised by a holographic imaging system[70], which is very similar by design to our detection module for THz PTDH manufactured on the basis of 3D-models with open-source codes[161,162]. Despite the fact that LC cells were irradiated by broadband THz radiation, vortex phase distributions are demonstrated for a separate spectral component of 1 THz (or 1.2 THz) only. In the case of LC q-plate, with a 250 μm cell gap needed to obtain $\pi/2$ phase retardation at 1.0 THz (Fig. 4c), the recorded conversion efficiency reaches 50%. Due to the phase delay dispersion, the proposed LC elements could work effectively without full photopatterned realignment at a single frequency only, which puts them on a par with birefringent q-devices[67]. Another particular solution for shaping the THz vortex beam with LC cells was proposed by Shen et al.[160]. Digital micro-mirror device was adopted for dynamic micro-lithography to control the azimuthal angle inhomogeneity of LC directors to obtain the Bessel-like beam phase diagram with axial singularity. A cell gap was established at 400 μm to meet maximum performance at 1.2 THz with half-wave retardation, when the output field exhibits characteristics of both vortex and Bessel beams. The works published to date on this topic use the scheme shown in Fig. 2c, d where micro-lithography photopatterned LC cells[70,71] are used as q-plates. The other solution for vortex beam formation with LC cell is the use of polarisation controllable geometric phase forked grating[159]. Further development towards achromatic tuneable LC based retardation devices will involve combination of LCs and metasurfaces[163].

Beam shaping metasurfaces

Extremely elevated interest towards metasurfaces research in the recent years can be deducted from the number of review papers published in this topic over the past 5 years[65,164–174]. Among the listed reviews, some are specifically focussed onto THz metasurfaces[164,165,169,172], while others are devoted exclusively to the IR range[173]. Such interest is driven by the great opportunities offered by the metasurfaces: extreme tunability through variation of their material and geometry, and possibility to fabricate dynamically controlled metasurfaces, extremely required for the development of compact, integrated photonic and optoelectronic devices.

Metasurfaces can be utilized in almost any optical component – either for enhancing its properties, or reducing the fabrication cost, size, and the amount of the material used. Based on the physical principle, metasurfaces can be divided into two main categories[3,172]: the first category includes metasurfaces comprised of resonant subwavelength antennas which predominantly interact with linearly polarised radiation, while metasurfaces of the second category operate with circularly polarised radiation and based on geometric phase principle. The presence of resonance imposes an additional limitation on the frequency response of the first type metasurfaces. By contrast, the second type metasurfaces operate under the spin-orbit conversion principle, as discussed earlier, and such mechanism is well suited to provide achromatic properties[65]. In terms of vector and vortex beam shaping tasks, THz beam shaping metasurfaces can be divided into three classes: q-plate type devices, isotropic phase shifter[175,75], and radial polarisers. Only the first of these relate to the formation of vortex beams with geometric phase stitching will be discussed here. While the latter two types relate to multiplexing technologies and are beyond the scope of this paper. However, one cannot ignore the existence of the prominent phase shifter: the broadband tuneable reflector of a graphene-based metamaterial[75]. A schematic drawing of the composite metasurface is shown in Fig. $4d_3$. It is divided into 12 sectors, each filled with the same multilayer units (Fig. $3d_2$). In order to adjust the three chemical potentials and the corresponding surface impedances of the graphene layers, an external DC voltage is applied between the graphene layer and the SiO_2 layer in each sandwich structure. The insulating layer is Al_2O_3; the THz beam is reflected from the Au base. The advantage of the proposed vortex beam modulator is the wide operating range 1.8 – 2.8 THz and the dynamic switching between different modes with OAM $L = \pm1, \pm2, \pm3$.

Metasurface-based q-devices, similarly to natural birefringent material-based components, employ geometric phase conversion for vector and vortex beam shaping in the THz range. In work[145] 3-bit active Pancharatnam–Berry coding metasurfaces are developed to modulate the amplitude of reflected THz beams. The reflection phase distribution of the coding particles which is shown in Fig. $4d_1$. Each meta-atom has a specific local axis rotation by a certain angle within $[0, \pi]$ range. Nevertheless, to date, numerous solutions[72,146,147,176–181], operating either in narrow frequency band or having flat enough birefringence to be applicable for wider spectral range devices, still introducing a frequency dependent phase shift though[149]. Among these plethora of solutions, easy-to-manufacture 3d-printed plates shine out (see Fig. $4d_5$, d_6)[177,178,146,180,147]. The grating structure has a period of 0.4 – 0.7 mm and wall thickness of 0.22 ± 0.02 mm made of COC-filaments. In dependence of the grating height, a quarter- or half-wave retardance can be achieved. Some solutions, similarly to liquid crystals, can dynamically change their properties thus allowing for tuneable beam shaping devices[181–183].

Due to the large amount of research aimed at metasurface design, the state-of-the-art in achromatic meta-based wave plates and q-devices design is significantly more diverse and developed than all other previously mentioned approaches. In[184] the design of a flat achromatic reflective quarter-wave retarder operating in the frequency range between 1.8 – 2.8 THz is proposed. The achromatic quarter-wave plate based on a stack of overlapped parallel metal plates with sub-wavelength holes is demonstrated in[130]. This waveguide with a controlled gap between the plates is an effective medium without significant losses, allowing for easy dispersion control of the entire birefringent structure via chemically etched array of through holes. Operating range of the proposed achromatic wave plate occupies about one octave (0.67 – 1.21 THz). Subsequently, to shift the achromatic operating range to higher frequencies (2.0 – 3.0 THz), the authors replaced the holes with periodic rough structures (pillars) of $3.7 \pm 0.5\ \mu m$ height and diameter around 20 μm[131]. The stack of plates that functions as an achromatic quarter-wave retarder (Fig. $3d_3$) with a 7% deviation from the ideal $\pi/2$ was experimentally demonstrated. In addition, other metallic multi-functional broadband metasurfaces were proposed[185,186], that can be employed as achromatic polarisation elements for the formation and dynamic modulation of THz optical vortices and vector beams, were proposed.

In work[187], the metamaterial design of two waveguide arrays is discussed. In the numerical experiment the phase difference of $\pi/2$ was achieved in the range between 0.89 – 1.22 THz.

All-dielectric subwavelength gratings offer better transmittance characteristics than metal plates[188]. In work[189], two achromatic wave plate designs are proposed, with operating ranges of 0.67 – 1.35 THz, and 0.70 – 0.85 THz, respectively. Birefringence effect[190] can also be used to design achromatic wave plates[132]. Three achromatic plate designs are proposed in the referenced work, operating in the ranges 0.7 ± 0.3 THz, 1.0 ± 0.45 THz and 5.0 ± 2.16 THz (Fig. $3d_4$).

Some broadband metasurface solutions proposed for the visible range are not yet adapted for THz wavelengths[191,192], though potentially can be. To make the picture complete, we would like to note the work by Lin et al.[193], containing an overview of methods for the formation of broadband vortex beams in the SHF (microwave) radio wave range[194,195].

By locally tailoring the metasurface structure, Zhang et al.[89] designed and fabricated polarisation-independent transmission-type spiral phase plate with 8, 16, and 32 sections. Proposed metasurface consists of multi-sized silicon pillars which perform formation of $L = 1$, $L = 2$, and $L = 4$ optical vortices at 1.0 THz with efficiencies up to 75.2%. In the work by He et al.[21], vortex beam shaping by a metasurface with V-shaped resonators at 0.75 THz is considered, but its operational spectral width is not discussed. Furthermore, vortex beam shaper for the THz wave range with spatially arranged aluminium building blocks of reflective metasurfaces is proposed by Li et al.[196]. By rotating a double-C-shaped slot (DCSS) and nano-bar (NB) resonators in the unit cell (Fig. 3d_1), the geometrical phase shift is imprinted onto the incident circularly polarised wave and thus convert it into the wave with opposite helicity[74]. From the units, the metalens (Fig. 4d_2) with a helicoidal phase delay is constructed. These reflective metasurface devices operate in 0.3 – 0.7 THz range and generate optical vortices with topological charges of $L = \pm 1$ and $L = \pm 2$. The sign of the output charge depends on the incident circular polarisation handedness. Due to the reflective design, the conversion efficiency of the metasurface may reach 80% and 92% at 0.45 THz and 0.7 THz, respectively[74]

By introducing vanadium dioxide (VO_2)[197,198] into the metasurface, Wang et al.[199] designed temperature tuneable metasurface for Pancharatnam-Berry phase modulation of circularly polarised THz waves as well as linearly polarised waves retardance. Proposed device operates at 0.62 – 1.3 THz band and forms vortex wavefront with the unit topological charge. Promising solutions for switchable metasurfaces by changing electrical parameters[200] or chemical potential of graphene layer were proposed[175,75]. From the tailoring approach point of view, such metasurfaces do not behave as PBOEs, as, for example, ones in[89], but are employing scalar wavefront conversion method, as SPPs.

Comparison of the characteristics of the mentioned materials suitable for the manufacturing of the achromatic THz wave plates and other THz optical elements is presented in Table 2, while the Table 3 summarise the relevant information about beam converters, demonstrated to the date.

Comparison of the approaches

After looking carefully at the collection of all reviewed approaches to achromatic device design for the THz frequency range, and noticing a great number of solutions, put together for easier assessment and comparison, we can try drawing a whole picture of the state-of-the-art and perspectives in this area of research.

The first thing to notice, all the proposed approaches can be classified into either Type I or Type II shaping, without offering any novel solutions, and search for such new shaping schemes may be the next direction of research in this area.

Multilayer achromatic wave plates from birefringent materials, being historically the first proposed solution, still offer the best operational spectral range, and have significantly progressed, becoming thinner and easier in production, due to the designs including layers of the identical thickness. Further development of metasurface based materials with exceptional birefringence, may give this approach a new kickstart, due to the controllability and compactness of the resulting devices.

Total internal reflection devices are quite accessible, as they can be directly 3d-printed, allow for relatively simple implementation of smooth gradient introduction, and offer exceptional spectral bandwidth, however, are rather bulky and thus are hard to integrate into the setup. With further development of higher refractive index polymers for 3d-printing, they potentially will become very competitive approach to achromatic vector and vortex beam shaping in the THz frequency band.

Liquid crystal-based devices, expectedly, allow for extra fine tunability, and potentially could be used in a multilayer device, however, the losses, introduced by the liquid crystals are often a limiting factor for their use as effective beam shaping approach for the THz frequency range.

Metasurfaces, allowing for extreme engineering of their properties and being still optically thin, are the main candidates for the future broadband THz beam shapers. Active metasurfaces will allow on-line adjustment of the wavepltes to the needs of the beamshaper, and possibly even tune the output beam topological charge and/or polarisation. Combination of metasurfaces of a different kind with other approaches, such as multilayering and liquid crystals, can possibly become an ultimate solution for compact, efficient, and tuneable BUTCH beam shapers in the THz frequency range.

Conclusion

In this part of our paired set of papers, we reviewed materials, methods, and solutions for geometric phase and

Table 2 Achromatic (A) and tuneable (T) THz wave plates

Ref.	Material	Thickness, mm	Range, THz	Retardance	Tolerance	Type	Notes
			Multilayer wave plates				
124	Quartz	31.4	0.25–1.75	$\pi/2$	3%	A	6 layers
125	Quartz	51.72	1.3–1.8	$\pi/2$	±3 deg.	A	9 layers
126	Quartz	—	1.0–5.0	$\pi/2$, π	10%	A T	3 layers
127	Quartz	—	1.11–1.87	$\pi/2$, π	±5 deg.	A T	3 layers
128	Al_2O_3	2.453–3.419	0.2–2.0 0.1–0.8	$\pi/2$	0.5% (simul.) 4.7% (exper.)	A	6–8 layers
Our work	Quartz	17.66 22.96	0.4–1.4 0.4–1.4	$\pi/2$, π	0.5% (simul.)	A	7 layers
139	$DyScO_3$	0.05 0.37	0.5–0.7 0.5–0.61	$\pi/2$, $\pi/2$	±3% (phase) ±10% (ampl.)	A	1 layer
			Total internal reflection wave plates				
68	Si and plastic	122	0.1–2.5	$\pi/2$; π	—	A	Transmissive
			Liquid crystal-based wave plates				
150	E7	~0.52	1.05–1.2	$\pi/2$	—	T	Transmissive
152	NJU-LDn-4/1	0.25	1.1–2.5 2.1–2.5	$\pi/2$ π	—	T	Transmissive
152	NJU-LDn-4/2	0.5	0.5–2.5 0.9–2.5	$\pi/2$ π	—	T	Transmissive
157	NJU-LDn-4	~0.13	1.1–2.5 2.2–2.5	$\pi/2$ π	—	T	Reflective
70	NJU-LDn-4	0.25	>2.0	π	—	T	Transmissive
129	MDA-00-3 461/4	0.8–1.0	0.35–0.5 0.2–0.9	$\pi/2$	—	A T	3 LC plates
135	E7	2.56–3.86	0.2–0.5 0.3–0.7 0.8–1.0	$\pi/2$ $\pi/2$ π	±9 deg.	A T	3 LC plates in magnetic field
136	Si sweep grating with E7 & epoxy	0.5 + 0.5	0.375–1.1 0.8–1.1	$\pi/2$, π	—	T	Transmissive
			Metasurface wave plates				
136	Si grating with gradient period	0.015–0.06	0.55–1.05	π	~20%	A	Transmissive
184	Polyimide on Au	0.02–0.04	1.8–2.8	$\pi/2$	±2 deg.	A	Reflective
130	Metal	0.06–0.1	0.67–1.21	$\pi/2$	5%	A	Transmissive
131	Metal	0.02–0.04	2.0–3.1	$\pi/2$	7%	A	Transmissive
185, 186	Metal on polyimide	0.066	1–1.4; 1.4–1.8	$\pi/2$	±5 deg.	A T	Transmissive 2 + 3 layers
187	Al on polymer	0.02–0.07	0.89–1.24	$\pi/2$	±10 deg.	A	Transmissive
189	Al	0.02–0.2	0.7–0.85 0.67–1.35	$\pi/2$, π	±5 deg.	A T	Transmissive
190	PP loaded TiO_2	0.15–0.25	0.15–0.375	$\pi/2$	—	A	Transmissive
132	Si grating	0.14–0.15	0.47–0.8	$\pi/2$	±3%	A	Transmissive
147	Olefin copolymer	0.41-0.66	0.23-0.4	$\pi/2$	±10%	A	Transmissive 2 layers

Table 3 THz beam converters

Ref.	Material	Thickness, mm	Range, THz	Type	Retardance	Beam type	Notes
			Segmented birefringent beam converters				
67	SiO_2 single layer	—	1.0	Ix, IIa	π	Vectorial, Vortex $L=\pm1$	Chromatic
38	SiO_2 multi layer	—	0.1–1.4, 1.9–2.6	Icx, Ix	π	Vortex $L=\pm1$	Achromatic
			Total internal reflection beam converters				
134, 53	$(C_2F_4)_n$	—	0.1–1.6	IIv	$\pi/2$ 163 deg	Vectorial	Achromatic
69	$(C_2F_4)_n$	—	0.1–1.6	IIv	π	Vectorial, Vortex	Achromatic
			Liquid crystal beam converters				
159	NJU-LDn-4	0.25	1.0	Icx	$\pi/2$	Vortex $L=\pm1$	Chromatic
70	NJU-LDn-4	0.25	1.0	Ix	$\pi/2$	Vortex $L=\pm1$ $L=\pm4$	Tuneable
160	NJU-LDn-4	0.4	1.2–1.4	Icx	π	Bessel-Vortex $L=\pm2$	Broadband
			Metasurface beam converters				
196	Al on polymide	0.05	0.3–0.45	Icx	$0–2\pi$	Vortex $L=\pm1$ $L=\pm2$	Achromatic
74	Al on Si	0.06	0.45 and 0.7	Icx	$0–2\pi$	Vortex $L=\pm1$ $L=\pm2$	Tuneable
75	Graphene on Al_2O_3/SiO_2	0.02	1.8–2.8	–	$0–2\pi$	Vortex $L=\pm1$ $L=\pm2$ $L=\pm3$	Tuneable (not a q-device)
197	Au on polyimide with VO_2	0.03–0.05	0.62–1.3	Icx	$0–2\pi$	Vectorial, Vortex $L=\pm1$	Achromatic Tuneable
145	Ge on polyimide	0.01–0.05	1.0–1.2	Icx	$0–\pi$	Vortex $L=\pm1$	Tuneable
146	PLA	2.3	0.14–0.16	IIv	π	Vectorial	Chromatic
147	Olefin copolymer	0.22–1.5	0.325	IIv	π	Vectorial	Chromatic

beam shaping and polarisation control in the THz frequency range. We have outlined the basic schemes for achromatic vortex and vector beam shaping and brought together the available approaches for the development of birefringent components. We classified, described, and characterised most existing different beam-shaping techniques and discussed the possibility of creating the new setups and schemes for these purposes. We also overviewed and compared materials and methods, with special focus on their broadband operation, to be applicable for pulsed THz beam shaping.

The effective way to expand the range of achromaticity that has already been implemented in three of the four types of geometric phase devices discussed in this review is the concept of multilayer composition. The use of a layered structure makes it possible to compose materials with the wide operating spectral range, in which the differences between the boundary frequencies can be as much as tenfold differential. However, adding more layers leads to the degradation of the integral transmittance and increases the cost of the product. Furthermore, other experimental factors, such as thickness neglect, the impact of the Fabry-Perot effect between the layers, or large wavelength differences at the limits of the designed spectral range), which are weakly expressed with the small number of layers, become appreciable.

Within the considered framework, layered crystalline wave plates are promising for the manufacture of high-efficiency retarders. The process of designing a multilayer achromatic structure for the case of using natural birefringent media is the most developed. The need to use precision polishing and assembly equipment, accompanied by proper control, acts as a factor that increases the cost of production.

Liquid crystals and metamaterials are the most perspective way components from the point of view of the possibility to achieve switchability for different THz ranges. Solutions were proposed for both reflection or transmission modes of operation, $0-2\pi$ phase conversion, and high fidelity. To date, they are still limited in continuous THz range coverage and require reconfiguration or re-writing of the cells to maintain a specified phase delay at the desired working frequency. Considering the concept of multilayer elements for chromatism cancelation, we should expect the development of broadband LC and metamaterials toward multilayer structures, which is already outlined in some research today.

Devices based on total internal reflection stand apart from the materials discussed, because a completely different physical principle is involved. A distinctive feature of plastic materials is their sub-THz transparency; combined with cheap 3D printing, it will allow one to create wave plates and elements based on them in a certain range of frequencies. Along with this, total internal reflection elements made of Si and plastic materials extend the operating range of retarders. Already existing work shows that this is a promising solution for vector field transformation. A fundamental limitation in the selection of materials for achromatic beam converters based on total internal reflection is also their limited transmission and absorption characteristics.

Acknowledgements
The Authors thank Dr. Yaroslav Grachev for the photo of achromatic quartz wave plates, shown in Fig. $3a_3$. Russian Science Foundation, Project No 19-72-10147. The work of A.G. is supported by EPSRC Grant No. EP/S018395/1

Author details
[1]Digital and Display Holography Laboratory, ITMO University, St. Petersburg 197101, Russia. [2]V.I. Vernadsky Crimean Federal University, Simferopol 295007, Russia. [3]School of Physics and Astronomy, University of Birmingham, Birmingham B15 2TT, UK

Conflict of interest
The authors declare no conflicts of interest.

Received: 18 September 2021 Revised: 30 June 2022 Accepted: 08 July 2022
Accepted article preview online: 11 July 2022
Published online: 02 August 2022

References

1. Siemion, A. Terahertz diffractive optics—smart control over radiation. *Journal of Infrared, Millimeter, and Terahertz Waves* **40**, 477-499 (2019).
2. Zhan, Q. Cylindrical vector beams: from mathematical concepts to applications. *Advances in Optics and Photonics* **1**, 1-57 (2009).
3. Wang, H. et al. Recent advances in generation of terahertz vortex beams and their applications. *Chinese Physics B* **29**, 097404 (2020).
4. Desyatnikov, A. S., Kivshar, Y. S. & Torner, L. Optical vortices and vortex solitons. *Progress in Optics* **47**, 291-391 (2005).
5. Zheng, Z. et al. Efficient coupling of propagating broadband terahertz radial beams to metal wires. *Optics Express* **21**, 10642-10650 (2013).
6. Minkevičius, L. et al. Bessel terahertz imaging with enhanced contrast realized by silicon multi-phase diffractive optics. *Optics Express* **27**, 36358-36367 (2019).
7. Zhu, L. et al. Experimental demonstration of basic functionalities for 0.1-THz orbital angular momentum (OAM) communications. *Proceedings of the OFC 2014 M3K.7* (2014).
8. Nobahar, D. & Khorram, S. Terahertz vortex beam propagation through a magnetized plasma-ferrite structure. *Optics & Laser Technology* **146**, 107522 (2022).
9. Wätzel, J. & Berakdar, J. Open-circuit ultrafast generation of nanoscopic toroidal moments: the swift phase generator. *Advanced Quantum Technologies* **2**, 1970011 (2019).
10. Hibberd, M. T. et al. Acceleration of relativistic beams using laser-generated terahertz pulses. *Nature Photonics* **14**, 755-759 (2019).
11. Mittleman, D. M. Twenty years of terahertz imaging [Invited]. *Optics Express* **26**, 9417-9431 (2018).
12. Dhillon, S. S. et al. The 2017 terahertz science and technology roadmap. *Journal of Physics D: Applied Physics* **50**, 043001 (2017).
13. Wang, M. et al. Longitudinal component properties of circularly polarized terahertz vortex beams. *Frontiers in Physics* **9**, 1-8 (2021).
14. Nanni, E. A. et al. Terahertz-driven linear electron acceleration. *Nature Communications* **6**, 8486 (2015).
15. Zhang, C. et al. Control of the Spin Angular Momentum and Orbital Angular Momentum of a Reflected Wave by Multifunctional Graphene Metasurfaces. *Materials* **11**, 1054 (2018).
16. Kampfrath, T., Tanaka, K. & Nelson, K. A. Resonant and nonresonant control over matter and light by intense terahertz transients. *Nature Photonics* **7**, 680-690 (2013).
17. Koenig, S. et al. Wireless sub-THz communication system with high data rate. *Nature Photonics* **7**, 977-981 (2013).
18. Wei, X. et al. Orbit angular momentum encoding at 0.3 THz via 3D printed spiral phase plates. Proceedings of SPIE, 9275. Infrared, MillimeterWave, and Terahertz Technologies III. Beijing, China: SPIE, 2014.
19. Su, X. et al. Modal purity and LG coupling of an OAM beam reflected by a rough surface for NLoS THz links. 2021 IEEE International Conference on Communications Workshops (ICC Workshops). Montreal, Canada: IEEE, 2021.
20. Miyamoto, K. et al. Highly intense monocycle terahertz vortex generation by utilizing a Tsurupica spiral phase plate. *Scientific Reports* **6**, 38880 (2016).
21. He, J. et al. Generation and evolution of the terahertz vortex beam. *Optics Express* **21**, 20230-20239 (2013).
22. Wu, Z. et al. Vector characterization of zeroorder terahertz Bessel beams with linear and circular polarizations. *Scientific Reports* **7**, 13929 (2017).
23. Wu, Z. et al. Vectorial diffraction properties of THz vortex Bessel beams. *Optics Express* **26**, 1506-1520 (2018).
24. Liu, C. et al. 3D-printed diffractive elements induced accelerating terahertz Airy beam. *Optics Express* **24**, 29342-29348 (2016).
25. Liu, C. et al. Terahertz circular Airy vortex beams. *Scientific Reports* **7**, 3891 (2017).
26. Kulya, M. et al. Spatio-temporal and spatiospectral metrology of terahertz broadband uniformly topologically charged vortex beams. *Applied Optics* **58**, A90-A100 (2019).
27. Bezuhanov, K. et al. Vortices in femtosecond laser fields. *Optics Letters* **29**, 1942-1944 (2004).
28. Wright, A. J. et al. Transfer of orbital angular momentum from a super-continuum, white-light beam. *Optics Express* **16**, 9495-9500 (2008).
29. Tokizane, Y., Oka, K. & Morita, R. Supercontinuum optical vortex pulse generation without spatial or topological-charge dispersion. *Opt. Express* **17**, 14517-14525 (2009).
30. Shvedov, V. G. et al. Efficient beam converter for the generation of high-power femtosecond vortices. *Optics Letters* **35**, 2660-2662 (2010).
31. Yamane, K., Toda, Y. & Morita, R. Ultrashort opticalvortex pulse generation in few-cycle regime. *Optics Express* **20**, 18986-18993 (2012).
32. Huang, L. et al. Dispersionless phase discontinuities for controlling light propagation. *Nano Letters* **12**, 5750-5755 (2012).
33. Sirenko, A. A. et al. Terahertz vortex beam as a spectroscopic probe of

magnetic excitations. *Physical Review Letters* **122**, 237401 (2019).

34. Kulya, M. S. et al. Propagation dynamics of ultrabroadband terahertz beams with orbital angular momentum for wireless data transfer. Proceedings of SPIE, 11307. Broadband Access Communication Technologies XIV. San Francisco, California, United States: SPIE, 2020.
35. Petrov, N. V. et al. On the features of the interference of a set of broadband uniformly topologically charged beams. Proceedings of SPIE, 11499. Terahertz Emitters, Receivers, and Applications XI. Online Only: SPIE, 2020.
36. Tsiplakova, E. G. et al. Interference enabled binary data encoding within broadband uniformly topologically charged terahertz beams. Proceedings of OSA Imaging and Applied Optics Congress 2021. Washington, DC: OSA, 2021.
37. Tsiplakova, E. G. et al. Investigation of spectral encoding with terahertz broadband uniformly topologically charged beams. Proceedings of the 46th International Conference on Infrared, Millimeter and Terahertz Waves (IRMMW-THz). Chengdu, China: IEEE, 2021.
38. Petrov, N. V. et al. Design of broadband terahertz vector and vortex beams: II. Holographic assessment. *Light: Advanced Manufacturing Manuscript* **3**, 44 (2022).
39. Akturk, S. et al. Spatio-temporal couplings in ultrashort laser pulses. *Journal of Optics* **12**, 093001 (2010).
40. Belashov, A. V. et al. Effect of object thickness on ultrashort pulse diffraction. *Applied Optics* **58**, 9434-9442 (2019).
41. Borot, A. & Quéré, F. Spatio-spectral metrology at focus of ultrashort lasers: a phase-retrieval approach. *Optics Express* **26**, 26444-26461 (2018).
42. Sheridan, J. T. et al. Roadmap on holography. *Journal of Optics* **22**, 123002 (2020).
43. Pariente, G. & Quéré, F. Spatio-temporal light springs: extended encoding of orbital angular momentum in ultrashort pulses. *Optics Letters* **40**, 2037-2040 (2015).
44. Kondakci, H. E. & Abouraddy, A. F. Diffractionfree space–time light sheets. *Nature Photonics* **11**, 733-740 (2017).
45. Chong, A. et al. Generation of spatiotemporal optical vortices with controllable transverse orbital angular momentum. *Nature Photonics* **14**, 350-354 (2020).
46. So, I. A., Plachenov, A. B. & Kiselev, A. P. Simple unidirectional finite-energy pulses. *Physical Review A* **102**, 063529 (2020).
47. Bespalov, V. G. & Gorodetskiĭ, A. A. Modeling of referenceless holographic recording and reconstruction of images by means of pulsed terahertz radiation. *Journal of Optical Technology* **74**, 745-749 (2007).
48. Petrov, N. V., Gorodetsky, A. A. & Bespalov, V. G. Holography and phase retrieval in terahertz imaging. Proceedings of SPIE, 8846. Terahertz Emitters, Receivers, and Applications IV. San-Diego, California, United States: SPIE, 2013.
49. Petrov, N. V. et al. Application of Terahertz Pulse Time-Domain Holography for Phase Imaging. *IEEE Transactions on Terahertz Science and Technology* **6**, 464-472 (2016).
50. Kulya, M. S. et al. Vectorial terahertz pulse timedomain holography for broadband optical wavefront sensing. Proceedings of SPIE, 11279. Millimeter, and Submillimeter-Wave Technology and Applications XIII. San Francisco, California, United States: SPIE, 2020.
51. Headland, D. et al. Tutorial: Terahertz beamforming, from concepts to realizations. *APL Photonics* **3**, 051101 (2018).
52. Yang, H. et al. Terahertz orbital angular momentum: Generation, detection and communication. *China Communications* **18**, 131-152 (2021).
53. Wakayama, T., Higashiguchi, T. & Otani, Y. Generation of the determined vectorial vortex beams by use of an achromatic axially symmetric waveplate. *Optical Review* **24**, 449-461 (2017).
54. Beth, R. A. Mechanical detection and measurement of the angular momentum of light. *Physical Review* **50**, 115-125 (1936).
55. Allen, L. et al. Orbital angular momentum of light and the transformation of Laguerre-Gaussian laser modes. *Physical Review A* **45**, 8185-8189 (1992).
56. Dholakia, K. et al. Second-harmonic generation and the orbital angular momentum of light. *Physical Review A* **54**, R3742-R3745 (1996).
57. Ciattoni, A., Cincotti, G. & Palma, C. Angular momentum dynamics of a paraxial beam in a uniaxial crystal. *Physical Review E* **67**, 036618 (2003).
58. Desyatnikov, A. et al. Spatially engineered polarization states and optical vortices in uniaxial crystals. *Optics Express* **18**, 10848-10863 (2010).
59. Marrucci, L., Manzo, C. & Paparo, D. Optical spin-to-orbital angular momentum conversion in inhomogeneous anisotropic media. *Physical Review Letters* **96**, 163905 (2006).
60. Pancharatnam, S. Generalized theory of interference, and its applications - Part I. Coherent pencils. *Proceedings of the Indian Academy of Sciences - Section A* **44**, 247-262 (1956).
61. Berry, M. The adiabatic phase and Pancharatnam' s phase for polarized light. *Journal of Modern Optics* **34**, 1401-1407 (1987).
62. Marrucci, L. Generation of helical modes of light by spin-to-orbital angular momentum conversion in inhomogeneous liquid crystals. *Molecular Crystals and Liquid Crystals* **488**, 148-162 (2008).
63. Bouchard, F. et al. Achromatic orbital angular momentum generator. *New Journal of Physics* **16**, 123006 (2014).
64. Marrucci, L., Manzo, C. & Paparo, D. PancharatnamBerry phase optical elements for wave front shaping in the visible domain: Switchable helical mode generation. *Applied Physics Letters* **88**, 221102 (2006).
65. Puentes, G. Spin-orbit angular momentum conversion in metamaterials and metasurfaces. *Quantum Reports* **1**, 91-106 (2019).
66. Niv, A. et al. Polychromatic vectorial vortex formed by geometric phase elements. *Optics Letters* **32**, 847-849 (2007).
67. Minasyan, A. et al. Geometric phase shaping of terahertz vortex beams. *Optics Letters* **42**, 41-44 (2017).
68. Kawada, Y. et al. Achromatic prism-type wave plate for broadband terahertz pulses. *Optics Letters* **39**, 2794-2797 (2014).
69. Wakayama, T. et al. Demonstration of a terahertz pure vector beam by tailoring geometric phase. *Scientific Reports* **8**, 8690 (2018).
70. Ge, S. et al. Terahertz vortex beam generator based on a photopatterned large birefringence liquid crystal. *Optics Express* **25**, 12349-12356 (2017).
71. Shen, Z. et al. Planar terahertz photonics mediated by liquid crystal polymers. *Advanced Optical Materials* **8**, 1902124 (2020).
72. Zhou, H. et al. Generation of terahertz vortices using metasurface with circular slits. *IEEE Photonics Journal* **6**, 5900107 (2014).
73. Mun, S.-E. et al. Broadband circular polarizer for randomly polarized light in few-layer metasurface. *Scientific Reports* **9**, 2543 (2019).
74. Wang, T. et al. Dual-band high efficiency terahertz meta-devices based on reflective geometric metasurfaces. *IEEE Access* **7**, 58131-58138 (2019).
75. Shi, Y. & Zhang, Y. Generation of wideband tunable orbital angular momentum vortex waves using graphene metamaterial reflectarray. *IEEE Access* **6**, 5341-5347 (2017).
76. Mariyenko, I. G., Strohaber, J. & Uiterwaal, C. J. G. J. Creation of optical vortices in femtosecond pulses. *Optics Express* **13**, 7599-7608 (2005).
77. Zeylikovich, I. et al. Ultrashort Laguerre – Gaussian pulses with angular. *Optics Letters* **32**, 2025-2027 (2007).
78. Schwarz, A. & Rudolph, W. Dispersioncompensating beam shaper for femtosecond optical vortex beams. *Optics Letters* **33**, 2970-2972

(2008).
79. Hancock, S. W. et al. Free-space propagation of spatiotemporal optical vortices. *Optica* **6**, 1547-1553 (2019).
80. Wang, L. et al. Terahertz angle-multiplexed metasurface for multi-dimensional multiplexing of spatial and frequency domains. *Advanced Theory and Simulations* **3**, 2000115 (2020).
81. Zhan, Q. & Leger, J. R. Interferometric measurement of the geometric phase in space-variant polarization manipulations. *Optics Communications* **213**, 241-245 (2002).
82. Zhou, Z., Tan, Q. & Jin, G. Cylindrically polarized vortex beams generated by subwavelength concentric Al metallic gratings. *Journal of Optics* **13**, 075004 (2011).
83. Chen, W. et al. Generating cylindrical vector beams with subwavelength concentric metallic gratings fabricated on optical fibers. *Journal of Optics* **13**, 015003 (2011).
84. Kang, M. et al. Twisted vector field from an inhomogeneous and anisotropic metamaterial. *Journal of the Optical Society of America B* **29**, 572-576 (2012).
85. Bock, M., Jahns, J. & Grunwald, R. Few-cycle highcontrast vortex pulses. *Optics Letters* **37**, 3804-3806 (2012).
86. Lin, J. et al. Nanostructured holograms for broadband manipulation of vector beams. *Nano Letters* **13**, 4269-4274 (2013).
87. Wang, X. et al. Full vector measurements of converging terahertz beams with linear, circular, and cylindrical vortex polarization. *Optics Express* **22**, 24622-24634 (2014).
88. Xie, Z. et al. Generation of terahertz vector beams with a concentric ring metal grating and photogenerated carriers. *Optics Letters* **40**, 359-362 (2015).
89. Zhang, H. et al. Polarization-independent all-silicon dielectric metasurfaces in the terahertz regime. *Photonics Research* **6**, 24-29 (2018).
90. Amer, N. et al. Generation of terahertz pulses with arbitrary elliptical polarization. *Applied Physics Letters* **87**, 221111 (2005).
91. Chang, G. et al. Generation of radially polarized terahertz pulses via velocity-mismatched optical rectification. *Optics Letters* **32**, 433-435 (2007).
92. Sato, M. et al. Terahertz polarization pulse shaping with arbitrary field control. *Nature Photonics* **7**, 724-731 (2013).
93. Imai, R. et al. Generation of broadband terahertz vortex beams. *Optics Letters* **39**, 3714-3717 (2014).
94. Dhaybi, A. A. et al. Terahertz vortex beam generation by infrared vector beam rectification. *Journal of the Optical Society of America B* **36**, 12-18 (2019).
95. Saito, K., Tanabe, T. & Oyama, Y. Concept of annular vector beam generation at terahertz wavelengths via a nonlinear parametric process. *Applied Optics* **54**, 2769-2775 (2015).
96. Lin, Q. et al. Generation of terahertz vortex pulses without any need of manipulation in the terahertz region. *Optics Letters* **44**, 887-890 (2019).
97. Miyamoto, K. et al. Generation of highquality terahertz OAM mode based on soft-aperture difference frequency generation. *Optics Express* **27**, 31840-31849 (2019).
98. Pi, H. et al. Integrated vortex beam emitter in the THz frequency range: Design and simulation. *APL Photonics* **5**, 076102 (2020).
99. Zhang, H. et al. Coherent terahertz radiation with orbital angular momentum by helically microbunched electron beam. *AIP Advances* **11**, 055115 (2021).
100. Zhang, H. et al. Narrowband terahertz emission with tunable orbital angular momentum by vortex laserbeam interaction. *IEEE Photonics Journal* **14**, 1-8 (2022).
101. Sobhani, H. et al. Terahertz twisted beams generation in plasma. *The European Physical Journal D* **70**, 168 (2016).
102. Wang, H. et al. Terahertz necklace beams generated from two-color vortex-laser-induced air plasma. *Physical Review A* **98**, 013857 (2018).
103. Ivanov, M. et al. Intensity modulated terahertz vortex wave generation in air plasma by two-color femtosecond laser pulses. *Optics Letters* **44**, 3889-3892 (2019).
104. Ivanov, M. et al. Terahertz wave generation in air by femtosecond optical vortex pulses. Proceedings of SPIE, 11124. Terahertz Emitters, Receivers, and Applications X. San Diego, California, United States: SPIE, 2019.
105. Sobhani, H. & Dadar, E. Terahertz vortex generation methods in rippled and vortex plasmas. *Journal of the Optical Society of America A* **36**, 1187-1196 (2019).
106. Wang, H. et al. Generation and evolution of different terahertz singular beams from long gasplasma filaments. *Optics Express* **29**, 996-1010 (2021).
107. Wang, H. et al. Spatiotemporal instabilities of terahertz OAM beams from air plasma via chirping a few-cycle vortex pump field. *Journal of Optics* (2022).
108. Wang, H. et al. Local OAM manipulation of a terahertz wave from the air filament by chirping the few-cycle vortex pump laser. *Optics Express* **30**, 9727-9744 (2022).
109. Winnerl, S. et al. Terahertz Bessel-Gauss beams of radial and azimuthal polarization from microstructured photoconductive antennas. *Optics Express* **17**, 1571 (2009).
110. Kan, K. et al. Radially polarized terahertz waves from a photoconductive antenna with microstructures. *Applied Physics Letters* **102**, 221118 (2013).
111. Kaltenecker, K. J. et al. Gouy phase shift of a tightly focused, radially polarized beam. *Optica* **3**, 35-41 (2016).
112. Niwa, H. et al. Switchable generation of azimuthallyand radially-polarized terahertz beams from a spintronic terahertz emitter. *Optics Express* **29**, 13331-13343 (2021).
113. Chen, Y., Yang, X. & Gao, J. Spin-selective secondharmonic vortex beam generation with babinetinverted plasmonic metasurfaces. *Advanced Optical Materials* **6**, 1800646 (2018).
114. Ponomareva, E. A. et al. Double-pump technique – one step closer towards efficient liquid-based THz sources. *Optics Express* **27**, 32855-32862 (2019).
115. Ponomareva, E. A. et al. Varying pre-plasma properties to boost terahertz wave generation in liquids. *Communications Physics* **4**, 4 (2021).
116. Biener, G. et al. Formation of helical beams by use of Pancharatnam–Berry phase optical elements. *Optics Letters* **27**, 1875-1877 (2002).
117. Slussarenko, S. et al. Tunable liquid crystal q-plates with arbitrary topological charge. *Optics Express* **19**, 4085-4090 (2011).
118. Naidoo, D. et al. Controlled generation of higherorder Poincaré sphere beams from a laser. *Nature Photonics* **10**, 327-332 (2016).
119. Ji, W. et al. Meta-q-plate for complex beam shaping. *Scientific Reports* **6**, 25528 (2016).
120. Larocque, H. et al. Arbitrary optical wavefront shaping via spin-to-orbit coupling. *Journal of Optics* **18**, 124002 (2016).
121. Shu, W. et al. Polarization evolution of vector beams generated by q-plates. *Photonics Research* **5**, 64-72 (2017).
122. Devlin, R. C. et al. Arbitrary spin-to-orbital angular momentum conversion of light. *Science* **358**, 896-901 (2017).
123. Cardano, F. et al. Polarization pattern of vector vortex beams generated by q-plates with different topological charges. *Applied Optics* **51**, C1-C6 (2021).
124. Masson, J.-B. & Gallot, G. Terahertz achromatic quarter-wave plate. *Optics Letters* **31**, 265-267 (2006).
125. Chen, Z. et al. Terahertz achromatic quarter wave plate: Design,

fabrication, and characterization. *Optics Communications* **311**, 1-5 (2013).
126. Kaveev, A. K. et al. Terahertz polarization conversion with quartz waveplate sets. *Applied Optics* **52**, B60-B69 (2013).
127. Kaveev, A. K. et al. Tunable wavelength terahertz polarization converter based on quartz waveplates. *Applied Optics* **53**, 5410-5415 (2014).
128. Wu, L. et al. A Compact Broadband Terahertz Range Quarter-Wave Plate. *Journal of Infrared, Millimeter, and Terahertz Waves* **41**, 642-654 (2020).
129. Yang, C.-S. et al. Liquid-crystal-enabled electrically tunable terahertz achromatic-wave plate. Proceedings of 2016 41st International Conference on Infrared, Millimeter, and Terahertz waves (IRMMW-THz). Copenhagen, Denmark: IEEE, 2016.
130. Nagai, M. et al. Achromatic THz wave plate composed of stacked parallel metal plates. *Optics Letters* **39**, 146-149 (2014).
131. Nagai, M. et al. Achromatic wave plate in THz frequency region based on parallel metal plate waveguides with a pillar array. *Optics Express* **23**, 4641-4649 (2015).
132. Zhang, B. & Gong, Y. Achromatic terahertz quarter waveplate based on silicon grating. *Optics Express* **23**, 14897-14902 (2015).
133. Hirota, Y. et al. Polarization modulation of terahertz electromagnetic radiation by four-contact photoconductive antenna. *Optics Express* **14**, 4486-4493 (2006).
134. Wakayama, T. et al. Determination of the polarization states of an arbitrary polarized terahertz beam: Vectorial vortex analysis. *Scientific Reports* **5**, 9416 (2015).
135. Hsieh, C.-F. et al. Liquid-crystal-based magnetically tunable terahertz achromatic quarter-wave plate. *Optics Express* **27**, 9933-9940 (2019).
136. Zhang, X. et al. Tunable terahertz phase shifter based on dielectric artificial birefringence grating filled with polymer dispersed liquid crystal. *Optical Materials Express* **10**, 282-292 (2020).
137. Pancharatnam, S. Achromatic combinations of birefringent plates. *Proceedings of the Indian Academy of Sciences - Section A* **41**, 137-144 (1955).
138. Beckers, J. M. Achromatic linear retarders. *Applied Optics* **10**, 973-975 (1971).
139. Yang, C.-J. et al. Birefringence of orthorhombic DyScO 3: Toward a terahertz quarter-wave plate. *Applied Physics Letters* **118**, 223506 (2021).
140. Peet, V. Laser beam shaping by conical refraction in biaxial crystals. Proceedings of SPIE, 8429. Optical Modelling and Design II. Brussels, Belgium: SPIE, 2012.
141. Rubano, A. et al. Q-plate technology: a progress review. *J. Opt. Soc. Am. B* **36**, D70-D87 (2019).
142. Gecevicius, M. et al. Toward the generation of broadband optical vortices: extending the spectral range of a q-plate by polarization-selective filtering. *Journal of the Optical Society of America B* **35**, 190-196 (2018).
143. Radwell, N. et al. Achromatic vector vortex beams from a glass cone. *Nature Communications* **7**, 10564 (2016).
144. Wakayama, T. et al. Achromatic axially symmetric wave plate. *Optics Express* **20**, 29260-29265 (2012).
145. Li, J. et al. Amplitude modulation of anomalously reflected terahertz beams using all-optical active Pancharatnam–Berry coding metasurfaces. *Nanoscale* **11**, 5746-5753 (2019).
146. Dong, X. et al. Sub-terahertz wideband vector beam generator based on superwavelength lattice dielectric grating. *Optik* **193**, 162991 (2019).
147. Rohrbach, D., Kang, B. J. & Feurer, T. 3D-printed THz wave- and phaseplates. *Optics Express* **29**, 27160-27170 (2021).
148. Hsieh, C.-F. et al. Voltage-controlled liquid-crystal terahertz phase shifter and quarter-wave plate. *Optics Letters* **31**, 1112-1114 (2006).
149. Piccirillo, B. et al. Photon spin-to-orbital angular momentum conversion via an electrically tunable q -plate. *Applied Physics Letters* **97**, 241104 (2010).
150. Yang, C.-S. et al. Liquid crystal terahertz phase shifters with functional indium-tin-oxide nanostructures for biasing and alignment. *Applied Physics Letters* **104**, 141106 (2014).
151. Chen, P. et al. Arbitrary and reconfigurable optical vortex generation: a high-efficiency technique using director-varying liquid crystal fork gratings. *Photonics Research* **3**, 133-139 (2015).
152. Wang, L. et al. Broadband tunable liquid crystal terahertz waveplates driven with porous graphene electrodes. *Light: Science & Applications* **4**, e253 (2015).
153. Sasaki, T. et al. Optical control of polarized terahertz waves using dye-doped nematic liquid crystals. *AIP Advances* **8**, 115326 (2018).
154. Vieweg, N. et al. Molecular properties of liquid crystals in the terahertz frequency range. *Optics Express* **18**, 6097-6107 (2010).
155. Yang, C.-S. et al. Voltage-controlled liquidcrystal terahertz phase shifter with indium–tin–oxide nanowhiskers as transparent electrodes. *Optics Letters* **39**, 2511-2513 (2014).
156. Yang, C.-S. et al. High-transmittance 2π tlectrically tunable terahertz phase shifter with CMOS-compatible driving voltage enabled by liquid crystals. *Applied Sciences* **9**, 271 (2019).
157. Wang, L. et al. Tunable reflective liquid crystal terahertz waveplates. *Optical Materials Express* **7**, 2023-2029 (2017).
158. Lin, C.-J. et al. Manipulating terahertz wave by a magnetically tunable liquid crystal phase grating. *Optics Express* **16**, 2995-3001 (2008).
159. Ge, S.-J. et al. Generating, separating and polarizing terahertz vortex beams via liquid crystals with gradient rotation directors. *Crystals* **7**, 314 (2017).
160. Shen, Y. et al. Photopatterned liquid crystal mediated terahertz Bessel vortex beam generator[Invited]. *Chinese Optics Letters* **18**, 080003 (2020).
161. Grachev, Y. V., Kokliushkin, V. A. & Petrov, N. V. An open-source 3D-printed terahertz pulse time-domain holographic detection module for broadband beam inspection. Proceedings of 2020 45th International Conference on Infrared, Millimeter, and Terahertz Waves (IRMMW-THz). IEEE, 2020.
162. Grachev, Y. V., Kokliushkin, V. A. & Petrov, N. V. Open-source 3D-printed terahertz pulse time-domain holographic detection module. *Applied Optics* **61**, B307-B313 (2022).
163. Shen, Y. et al. Electrically tunable terahertz focusing modulator enabled by liquid crystal integrated dielectric metasurface. *Crystals* **11**, 514 (2021).
164. Al-Naib, I. & Withayachumnankul, W. Recent progress in terahertz metasurfaces. *Journal of Infrared, Millimeter, and Terahertz Waves* **38**, 1067-1084 (2017).
165. Hashemi, M. R., Cakmakyapan, S. & Jarrahi, M. Reconfigurable metamaterials for terahertz wave manipulation. *Reports on Progress in Physics* **80**, 094501 (2017).
166. Nemati, A. et al. Tunable and reconfigurable metasurfaces and metadevices. *Opto-Electronic Advances* **1**, 180009 (2018).
167. Pachava, S. et al. Generation and decomposition of scalar and vector modes carrying orbital angular momentum: a review. *Optical Engineering* **59**, 041205 (2019).
168. Hajian, H. et al. Active metamaterial nearly perfect light absorbers: a review[Invited]. *Journal of the Optical Society of America B* **36**, F131-F143 (2019).
169. Yaxin, Z. et al. Terahertz smart dynamic and active functional electromagnetic metasurfaces and their applications. *Philosophical Transactions of the Royal Society A: Mathematical, Physical and Engineering Sciences* **378**, 20190609 (2020).

170. Wei, Q. et al. Optical wavefront shaping based on functional metasurfaces. *Nanophotonics* **9**, 987-1002 (2020).
171. Scheuer, J. Optical metasurfaces are coming of age: short- and long-term opportunities for commercial applications. *ACS Photonics* **7**, 1323-1354 (2020).
172. Zang, X. et al. Metasurfaces for manipulating terahertz waves. *Light: Advanced Manufacturing* **2**, 10 (2021).
173. Cao, G. et al. Infrared metasurface-enabled compact polarization nanodevices. *Materials Today* **50**, 499-515 (2021).
174. Ding, F., Tang, S. & Bozhevolnyi, S. I. Recent advances in polarization-encoded optical metasurfaces. *Advanced Photonics Research* **2**, 2000173 (2021).
175. Chang, Z. et al. A reconfigurable graphene reflectarray for generation of vortex THz waves. *IEEE Antennas and Wireless Propagation Letters* **15**, 1537-1540 (2016).
176. Chiang, Y.-J. & Yen, T.-J. A compositemetamaterial-based terahertz-wave polarization rotator with an ultrathin thickness, an excellent conversion ratio, and enhanced transmission. *Applied Physics Letters* **102**, 011129 (2013).
177. Hernandez-Serrano, A. I., Castro-Camus, E. & Lopez-Mago, D. q-plate for the generation of terahertz cylindrical vector beams fabricated by 3D printing. *Journal of Infrared, Millimeter, and Terahertz Waves* **38**, 938-944 (2017).
178. Hernandez-Serrano, A. I., Castro-Camus, E. & Lopez-Mago, D. Pancharatnam-Berry phase optical elements fabricated by 3D printing for shaping terahertz beams. Proceedings of SPIE, 10347. Optical Trapping and Optical Micromanipulation XIV. San Diego, California, United States: SPIE, 2017.
179. Xu, S.-T. et al. Terahertz polarization mode conversion in compound metasurface. *Applied Physics Letters* **111**, 031107 (2017).
180. Dong, X.-P. et al. Wideband sub-THz half-wave plate using 3D-printed low-index metagratings with superwavelength lattice. *Optics Express* **27**, 202-211 (2019).
181. Zeng, H. et al. Broadband terahertz reconfigurable metasurface based on 1-bit asymmetric coding metamaterial. *Optics Communications* **458**, 124770 (2020).
182. Wang, D. et al. Multiband switchable terahertz quarter-wave plates via phase-change metasurfaces. *IEEE Photonics Journal* **8**, 5500308 (2016).
183. Zhang, R. et al. Broadband and switchable terahertz polarization converter based on graphene metasurfaces. *Optics Express* **29**, 24804-24815 (2021).
184. Sun, L. et al. Achromatic terahertz quarter-wave retarder in reflection mode. *Applied Physics B* **106**, 393-398 (2012).
185. Cong, L. et al. Highly flexible broadband terahertz metamaterial quarter-wave plate. *Laser & Photonics Reviews* **8**, 626-632 (2014).
186. Cong, L. et al. A tunable dispersion-free terahertz metadevice with Pancharatnam-Berry-phase-eabled modulation and polarization control. *Advanced Materials* **27**, 6630-6636 (2015).
187. Ke, X. et al. Double-stacked hyperbolic metamaterial waveguide arrays for efficient and broadband terahertz quarter-wave plates. *Scientific Reports* **7**, 574 (2017).
188. Yuan, Y. et al. Terahertz dual-band polarization control and wavefront shaping over freestanding dielectric binary gratings with high efficiency. *Optics and Lasers in Engineering* **143**, 106636 (2021).
189. Mu, Q. et al. Broadband phase shift engineering for terahertz waves based on dielectric metasurface. *Optics Communications* **434**, 12-18 (2019).
190. Scheller, M., Jördens, C. & Koch, M. Terahertz form birefringence. *Optics Express* **18**, 10137-10142 (2010).
191. Kruk, S. et al. Invited Article: Broadband highly efficient dielectric metadevices for polarization control. *APL Photonics* **1**, 030801 (2016).
192. Song, Q. et al. Broadband decoupling of intensity and polarization with vectorial Fourier metasurfaces. *Nature Communications* **12**, 3631 (2021).
193. Lin, Z., Ba, Z. & Wang, X. Broadband high-efficiency electromagnetic orbital angular momentum beam generation based on a dielectric metasurface. *IEEE Photonics Journal* **12**, 4600611 (2020).
194. Tang, S. et al. High-efficiency broadband vortex beam generator based on transmissive metasurface. *Optics Express* **27**, 4281-4291 (2019).
195. Dharmavarapu, R. et al. Dielectric cross-shapedresonator-based metasurface for vortex beam generation at mid-IR and THz wavelengths. *Nanophotonics* **8**, 1263-1270 (2019).
196. Li, J.-S. & Zhang, L.-N. Simple terahertz vortex beam generator based on reflective metasurfaces. *Optics Express* **28**, 36403 (2020).
197. Makarevich, A. et al. Hydrothermal epitaxy growth of self-organized vanadium dioxide 3D structures with metal–insulator transition and THz transmission switch properties. *CrystEngComm* **22**, 2612-2620 (2020).
198. Kaydashev, V. E. et al. Electrically reduced optical switching threshold of VO2-based THz metasurface. *arXiv* 2204.13427 (2022).
199. Wang, H., Ling, F. & Zhang, B. Tunable metasurfaces for independent control of linearly and circularly polarized terahertz waves. *Optics Express* **28**, 36316-36326 (2020).
200. Soleimani, H. & Oraizi, H. Programmable terahertz vortex beam reflectarray antenna based on a graphene phoenix unit cell. *Journal of Physics D: Applied Physics* **54**, 165302 (2021).

Official journal of the JHL 2689-9620
www.light-am.com

Review Open Access

Design of broadband terahertz vector and vortex beams: II. Holographic assessment

Nikolay V. Petrov[1,*], Bogdan Sokolenko[2], Maksim S. Kulya[1], Andrei Gorodetsky[3] and Aleksey V. Chernykh[1]

Abstract

In this paper, we demonstrate the capabilities of the terahertz pulse time-domain holography in visualisation, simulation, and assessment of broadband THz vortex beam formation dynamics upon its shaping by elements of beam converter, and further propagation and manipulation. By adding Jones matrix formalism to describe broadband optical elements, we highlight the differences in the spatio-spectral and spatio-temporal structure of the formed vortex and vector beams dependence on the modulator used and visualise their modal features. The influence of diffraction field structure from each element in the broadband vortex modulator is revealed in numerical simulation and the formed beams are analysed against the simplified Laguerre-Gaussian beam model.

Keywords: Terahertz radiation, Vortex beams, Vector beams, Achromatic wave plates, Broadband terahertz radiation, Digital holography, Terahertz holography

Introduction

Vortex and vector terahertz (THz) beams[1] became quite a hot topic in the last decade due to their importance for wireless data transfer multiplexing[2-4], imaging[5], including in highly scattering and absorbing media[6], ultrafast current and (de)magnetisation processes probing[7] and even waveguide electron acceleration by a longitudinal polarisation mode[8,9]. Despite the growing application-driven demand for shaped vector and vortex beams[10], especially for ultrafast THz radiation, the amount of approaches to formation and generation, as well as the number of techniques to conveniently assess and characterise such beams is rather limited since they were mainly adopted from the visible range[11].

In the first paper[12] of this paired set we provided a comprehensive review of the perspectives for vector and vortex THz beam shaping with geometric phase effect[13]. We have outlined two main approaches characterised by the different order of the main elements in the shaper. Possible variants for the arrangement of polarising components in the modulator for the formation of a circularly, linearly polarised vortex and cylindrical vector beams are shown in Fig. 1. In general, an overall number of degrees of freedom and achievable states are much greater[14]. Type I shaping approach shown in the first column of Fig. 1 consists in an action of a q-plate on a circularly polarised THz beam after it has passed a quarter-wave plate (QWP), while the type II scheme features the direct action of the q-plate on the linearly polarised incoming beam. As a result, the top row presents the concept of the singular beams formation, namely the circularly polarised vortex for type I[15], and radially (or azimuthally, depending on the incoming beam polarisation) polarised vector beam with a flat phase surface for type II of the shaping approach. If all the components demonstrate

Correspondence: Nikolay V. Petrov (n.petrov@niuitmo.ru)
[1]Digital and Display Holography Laboratory, ITMO University, St. Petersburg 197101, Russia
[2]V.I. Vernadsky Crimean Federal University, Simferopol 295007, Russia
Full list of author information is available at the end of the article.

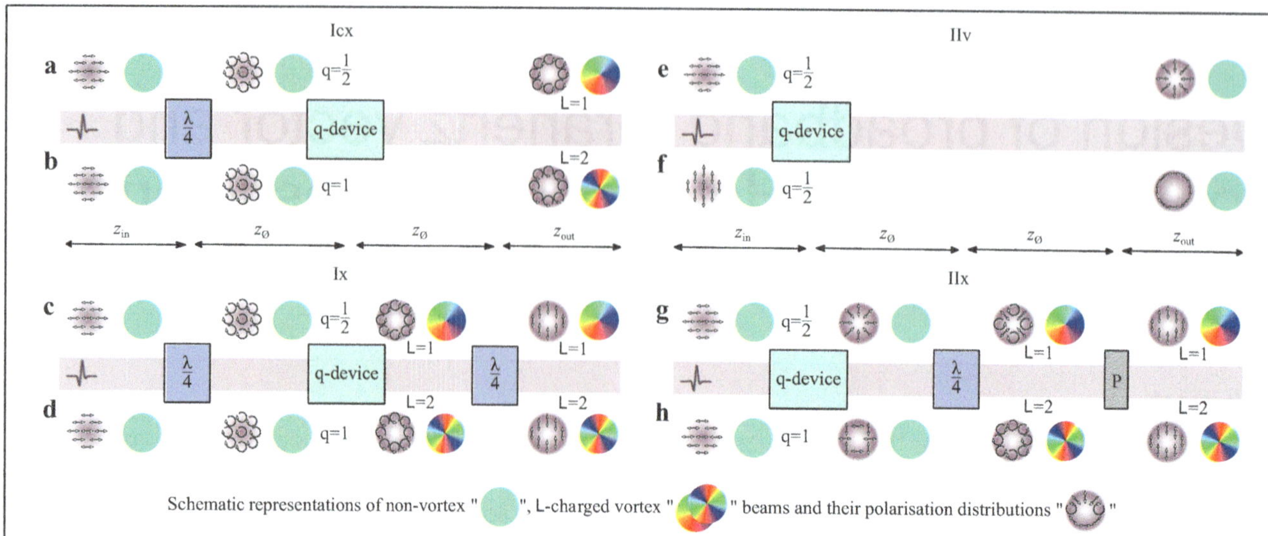

Fig. 1 Two main scheme types (I and II) for THz vector (v) and vortex (cx, x) beam shaping, respectively. $L = \pm 1$ **a**, **c**, **g** and $L = \pm 2$ **b**, **d**, **h** charged beam with circular cx **a**, **b**, and linear x **c**, **d**, **g**, **h** polarisation, respectively; cylindrical v beams with radial **e** and azimuthal **f** polarisations respectively. z_{in}, $z_{\varnothing}$ and z_{out} are the distances from the Gaussian beam waist to the modulator, between the devices, and from the modulator to the receiver, respectively. The notations of polarisation elements are used: $\lambda/4$ is a QWP; P denotes a linear polariser.

achromaticity in the overlapping spectral regions, THz broadband uniformly topologically charged (BUTCH) beams[6] within the overlap of achromaticity of all shaper elements, will be eventually formed.

Let us look more carefully into the process of vector and vortex beam formation, illustrated on Fig. 1. The incident linearly polarised pulsed THz field passes through various combinations of optical components shown in the corresponding schemes. Beam structure transformation schematically displayed with round insets. Due to the presence of various polarisation components in vector beams and the complexity of unambiguous representation of the corresponding them phase distribution we designate the absence and presence of transverse energy circulation by cyan and rainbow circles, respectively. According to the type I (Fig. 1a, b), a circularly polarised pulsed beam from a QWP, incident normally to the surface of a q-plate, is converted into a 'cx'–type beam (notation 'cx' is used as a reference to 'Circularly polarised vorteX') with opposite circularity and projections of orbital angular momenta (OAM) in the direction of propagation z: $L_z = \pm 1$ and $L_z = \pm 2$, depending on a q-plate value, and axial phase singularity. To make the beam linearly polarised again, another QWP is used after the q-plate on the scheme Ix (Fig. 1c, d). Note that scheme Ix supports another way of circularly-to-linearly polarised vortex conversion which may be done using a linear polariser instead of the second QWP. But in this case, a part of the energy corresponding to the orthogonal polarisation component will be cut off, which will lead to greater losses. The second type of shaping, designated as IIv ('v' denotes 'Vector') allows one to convert a linearly polarised non-singular beam to a vector beam with radial or azimuthal polarisation (Fig. 1e, f). To perform vector-to-vortex beam conversion carrying helical wavefront, formed vector beams must be further directed on the QWP to obtain the topological charge (Fig. 1g, h, the appearance of transverse energy circulation in the linear polarised component of the vector beams after its passage through the QWP is shown by a rainbow circular insets, the conversion to linear polarisation by a polariser P is also illustrated) in accordance with IIx scenario. Such configuration of elements still was not directly validated in the THz frequency range, but was used to prepare vortex pulses involved in the subsequent active generation of vortex THz beams[16,17].

A vast variety of polarisation components can be used as constituents of the beam shapers, and we reviewed a state-of-the-art of these components in the first paper[12] of this paired set. Detailed analysis of achromatic components revealed numerous limitations of their applicability, mostly due to non-perfect achromaticity and geometric limitations. Therefore, when designing a beam shaper, before embedding any component, it is necessary to take into account all the specifics of its operation. The optimal tool for this task is THz pulse time-domain holography (PTDH)[18–20], since this technique provides the ultimate opportunity to track the dynamics of all characteristics of the broadband THz field in the spatial, temporal, spectral and angular coordinates during its propagation, both in free space and in optical systems with known characteristics[6,21–23].

THz PTDH is a modern technique that includes extensive experimental capabilities for the physical measurement of broadband THz wavefronts with their subsequent time-reversal and aimed at solutions of a wide spectrum of practical tasks. In addition, it also can be exploited as an effective 'virtual instrument' to predict specific properties of ultrafast pulses after their propagation through or reflectance from the specific elements/components/samples. In this paper, we will pay attention to opportunities for numerical wavefront shapers design, while the description of experimental works[20,23,24], and manufacturing of the measurement unit based on 3D-models in open-source codes[25,26], remains beyond the scope of our current consideration.

Numerical research using virtual instruments is currently gaining popularity for solving various technical problems. Virtual instruments are already being effectively used in metrology for measuring the uncertainty estimation in coordinate measuring machines[27-29], microscopic 3D surface measurements[30,31]. Numerical methods are also in demand in such tasks as beam shaping[32] and the design and prediction of optical functionalities of various optical components, such as metasurfaces[33]. The main condition that guarantees success when using numerical tools is the presence of an accurate validated physical model of the simulated process. In the tasks of developing elements to form an electromagnetic field with desired structure, as a rule, methods of classical electrodynamics are used to estimate an impact of variable element parameters. However, such approaches as finite-difference time-domain (FDTD) are quite resource-intensive, and their application is not always justified. In this paper, we combined the Jones matrix formalism with the mathematical model of the THz PTDH method based on the scalar theory of diffraction, describing the propagation of the orthogonal polarisation components of the ultrafast THz field. The resulting technique allows tuning the beam shaping components to ensure maximally possible matching between the characteristics of formed and desired beam.

In this part of our paired set of papers, we consider some examples of THz vortex beam formation in details, and demonstrate the capabilities of a THz PTDH in assessment of the impact of the designed achromatic modulator's components on the incoming broadband THz field. Using THz PTDH, we visualise the spatio-temporal and spatio-spectral evolution of vectorial ultra-broadband THz wavefields upon their process of diffraction in type Ix and Icx modulators and in free space at their output. This technique allows for taking into account the transformation of characteristics of these wavefields by the optical elements and further optimisation of the mentioned wavefront modulators in sequential calculations. We reveal the capabilities of the proposed technique by providing the design of beam shapers that introduce not only a positive but also a negative uniform topological charge in a certain limited range of the THz spectrum and show the spectral and temporal characteristics of the formed beams.

Theoretical description and methods

This section describes the applicability of THz PTDH for numerical and experimental studies of achromatic vortex beam modulators. We will demonstrate the capabilities of the PTDH on the example of multilayer quartz wave plates with a certain achromaticity in wavefront retardation in the broad THz range, which have been mentioned in the first paper[12] of this paired set. However, we note that we propose our technique not only as a method for the quartz plate-based beam shapers design, but for any component-based beam shapers, with any known spectral amplitude and phase characteristics, either full or partially achromatic, such as those reviewed in the first paper of this paired set[12]. The eligible components can exploit the total internal reflection principle, liquid crystals, metasurfaces and on others principles (such as twisted fibers[34], if they are adapted for the THz range). The only necessary condition for the element to be taken into account is the spectral dependence of its action on the incoming beam presented either in the form of spatial amplitude and phase mask, or, for polarising elements, in the form of Jones or Müller matrices.

Jones matrix formalism for the achromatic wave plates design

In the framework of the Jones formalism, linearly polarised along the x-axis broadband THz field propagating orthogonal to the director axis of a birefringent medium can be represented by the following Jones vector:

$$\mathbf{G}_{\text{in}}(x,y,\nu)=\begin{pmatrix} G_{\text{in}_x}(x,y,\nu) \\ 0 \end{pmatrix} \tag{1}$$

and the action of an optical element onto the electric field vector is given by the Jones matrix.

$$J(\nu)=\begin{pmatrix} A(\nu) & B(\nu) \\ C(\nu) & D(\nu) \end{pmatrix} \tag{2}$$

resulting in a certain frequency-dependent change of the broadband THz field:

$$\begin{pmatrix} G_{\text{out}_x}(x,y,\nu) \\ G_{\text{out}_y}(x,y,\nu) \end{pmatrix}=\begin{pmatrix} A(\nu) & B(\nu) \\ C(\nu) & D(\nu) \end{pmatrix}\cdot\begin{pmatrix} G_{\text{in}_x}(x,y,\nu) \\ 0 \end{pmatrix} \tag{3}$$

One of the simplest and most developed approaches to

design of achromatic wave plates essential for broadband THz beam manipulation is to stack N plates of birefringent material and adjust their thicknesses and orientation to even out the introduced phase retardation of the ordinary wave relative to the extraordinary wave for all spectral components of interest (Fig. 2a). The phase delay introduced by a single j–th birefringent layer, coorientated with the incoming wave polarisation, can be written as:

$$\delta_j(\nu, h_j) = \varphi_x - \varphi_y = 2\pi h_j (n_e - n_o)\frac{\nu}{c} \tag{4}$$

where n_o and n_e are refractive indices for ordinary and extraordinary waves, respectively; h_j is the thickness of the j-th plate.

The influence of each birefringent layer acting as a wave plate in the multilayer achromatic plate on transmitted broadband radiation can be described by the frequency-dependent Jones matrix[35]:

$$J_j(\nu) = \begin{pmatrix} A_j(\nu) & B_j(\nu) \\ -B_j^*(\nu) & A_j^*(\nu) \end{pmatrix} \tag{5}$$

where

$$\begin{cases} A_j(\nu) = \cos\left(\dfrac{\delta_j(\nu)}{2}\right) + i\cos(2\theta_j)\cdot\sin\left(\dfrac{\delta_j(\nu)}{2}\right) \\ B_j(\nu) = i\sin(2\theta_j)\cdot\sin\left(\dfrac{\delta_j(\nu)}{2}\right) \end{cases} \tag{6}$$

Here θ is the angle of the j-th layer optical axis orientation with respect to the x axis. The resulting matrix of the composite wave plate can be obtained by multiplying the matrices J_j of each layer in reverse order:

$$J_a = \prod_{j=N}^{1} J_j \tag{7}$$

Adjusting the orientation θ_j and thickness h_j, it is possible to provide any relatively uniform achromatic phase delay $\delta_a(\nu) \approx$ constant in a multilayer structure for a relatively broad frequency range[35], thus allowing for relatively achromatic wave plate operation of the resulting stack. If the target wave plate is a QWP, the necessary retardation is $\delta_a = \pi/2$ in the designed achromaticity interval. Similarly, for the half-wave plate (HWP) the target delay will be $\delta_a = \pi$.

For the designed assembly to work as a q-plate, the director of an HWP must be sequentially changed in a circular roundabout of the optical axis. In practise, this is usually achieved by splitting the wave plate into a finite number of sectors. Often, upon manufacture, it is not always possible to reproduce a smooth rotation of the optical axis of a q-plate. In such cases, q-plate is divided into a discrete number of sectors, each having its own discrete angle ψ of the optical axis following the above-mentioned rule. This approach allows the introduction of a constant phase delay in each segment $\Delta\varphi = \pm(2q\psi/M)$, where M is a number of segments selected in such a way that the value of the resulting phase shift throughout the entire plate is equal to $\varphi = 2\pi L$, where L is a vortex topological charge. In this case, the dependence $\theta(\psi)$ on each m-th segment will take the form of a piecewise smooth function, representable by a set of functions in the following form:

$$\Phi(\psi) = \begin{cases} qm\Delta, & \psi \in (m\Delta, (m+1)\Delta) \\ 0, & \psi \notin (m\Delta, (m+1)\Delta) \end{cases} \tag{8}$$

where $\Delta = 2\pi/M$ is the angular measure of each sector, $m = 0, 1, 2, \ldots M-1$. Thus, the resulting director's rotation with respect to the plate can be written as the following sum:

$$\theta(\psi) = \sum_{m=0}^{M-1} \Phi(\psi) \tag{9}$$

Further extrapolating the idea of designing multilayered wave plates introducing spectrally-isotropic response for THz pulses, we propose the design of an achromatic q-plate created from eight sectors of a quartz birefringent achromatic HWP ($\delta = \pi$) as shown in Fig. 3c in the first paper[12] of this paired set. Being located in the type I beam shaping scheme (Figs. 1a–d) this plate will introduce uniform topological charge across its whole operational

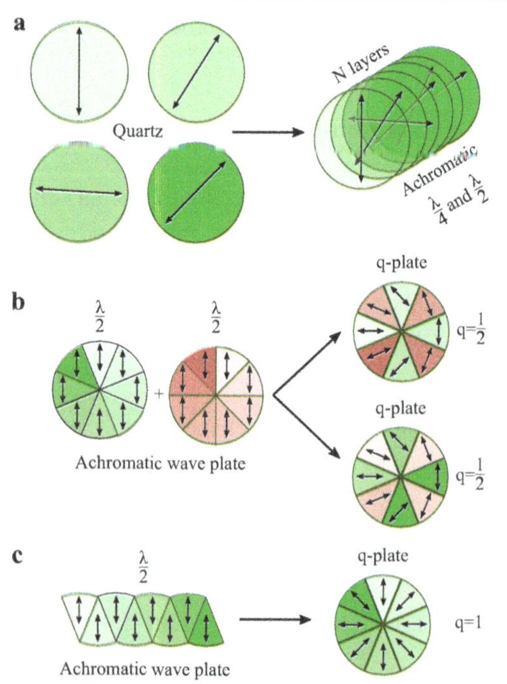

Fig. 2 Layouts for achromatic wave plate **a**, and q-plates with $q = 1/2$ **b** and $q = 1$ **c**.

spectrum, and the beam after passing such wave plate becomes a BUTCH beam. Schematic examples of such 8-sector q-plates for L = 1 and L = 2 topological charges and the corresponding cuts of the parent HWPs are shown in Fig. 2b, c.

Achromatic modulator components design

Our group designed 7-layer achromatic wave plates (Fig. 2) providing a reasonably flat response in the frequency range 0.4 – 1.4 THz. To calculate design, we took the data for refractive indices used in the Ref.[35], and minimised the following criteria for QWP and HWP correspondingly:

$$\sum_{\nu=0.4\text{THz}}^{1.4\text{THz}} \|\delta(\nu)-\frac{\pi}{2}\|+\|\Delta\varphi(\nu)-\frac{\pi}{2}\|+\|\varepsilon(\nu)-1\| \qquad (10)$$
$$\sum_{\nu=0.4\text{THz}}^{1.4\text{THz}} \|\delta(\nu)-\pi\|+\|\Delta\varphi(\nu)-\pi\|+\|\varepsilon(\nu)\|$$

Here

$$\delta(\nu)=2\tan^{-1}\sqrt{\frac{|\text{Im}\,A|^2+|\text{Im}\,B|^2}{|\text{Re}\,A|^2+|\text{Re}\,B|^2}},\ \Delta\varphi=\varphi_y-\varphi_x,\text{and}$$
$$\varepsilon=\frac{|G_x|^2+|G_y|^2-\sqrt{|G_x|^4+|G_y|^4+2|G_x|^2|G_y|^2\cos 2\Delta\varphi}}{|G_x|^2+|G_y|^2+\sqrt{|G_x|^4+|G_y|^4+2|G_x|^2|G_y|^2\cos 2\Delta\varphi}}.$$

are the chromatic retardation, direct phase delay and the ellipticity, correspondingly.

Wavefield evolution during the propagation through the modulator

To track in detail the changes that occur in the spatial frequency structure of the broadband THz vector and/or vortex field, generated by the combination of wave plates and q-plates, introduced in Figs. 1a–d, we will use the THz PTDH for a stepwise numerical simulation of the wavefront propagation process[36], and supplement it with the effect of the wave plates.

Ideally, all beam shaping components must be stacked together to minimise diffraction and Fresnel losses. However, in practise, it becomes a difficult task. Thus, to avoid superimposition of the Fresnel reflections of the time-domain signal from the wave plate-air interfaces, all components of the beam shaper must be separated by a distance $z_\varnothing > 0.5\cdot c\cdot T_w$, where T_w is the time interval of observation.

In PTDH, the equations of scalar diffraction theory are solved for each spectral and polarisation component (see Eqs. (1–3), (6–11) in work[6], Eqs. (1–10) in paper[37]). While detailed description of the method can be found elsewhere[22,37–39], for the sake of simplicity, the change of the broadband scalar field during propagation to the optical path z can be denoted with propagation operator $\mathcal{F}_z$. It should be noted that in the propagation process from the source to the first modulator element, the broadband THz field is noticeably transformed[23], and therefore the evolution of the field at the distance z_{in} must also be taken into account by applying the operator $\mathcal{F}_{z_{\text{in}}}$ to the field at the source to obtain the vector field in the input plane of the first modulator element:

$$\mathbf{G}_{\text{in}}(x,y,z_{\text{in}},\nu)=\mathcal{F}_{z_{\text{in}}}(\mathbf{G}_0(x,y,0,\nu)) \qquad (11)$$

Then, the process of propagation of the THz incoming vector field $\mathbf{G}_{\text{in}}(x,y,z_{\text{in}},\nu)$ through the Icx-type beam shaper (shown in Fig. 1a) can be described as:

$$\mathbf{G}_{\text{out}}(x,y,z_\Sigma,\nu)= =\mathcal{F}_{z_{\text{out}}+z_{\text{Icx}}+z_\varnothing}\left(J_q\mathcal{F}_{z_\varnothing}\left(J_{\frac{\lambda}{4}}\mathbf{G}_{\text{in}}(x,y,z_{\text{in}},\nu)\right)\right) \qquad (12)$$

Here, J_q and $J_{\lambda/4}$ are the Jones matrices of q-plate and QWP, correspondingly; $z_\Sigma=z_{\text{out}}+z_{\text{Icx}}+2z_\varnothing+z_{\text{in}}$ is the aggregate optical beam path from THz source to the observation plane; $z_\varnothing$ is the distance between wave plates in the modulator; $z_{\text{Icx}}=h_{\lambda/4}/n_e+h_q/n_e$ is the optical beam path inside achromatic wave plates. For simplicity, we take into account the optical beam path in the wave plates after the matrix transformation of the field, assuming that the change in the structure of the beam when propagating inside these wave plates is insignificant in comparison with the aggregate optical beam path ($z_{\text{Icx}}\ll z_\Sigma$).

Similarly, for the type IIx beam shaper, the field at the output can be written as:

$$\mathbf{G}_{\text{out}}(x,y,z_\Sigma,\nu)= =\mathcal{F}_{z_{\text{out}}+z_{\text{IIx}}}\left(J_p\mathcal{F}_{z_\varnothing}\left(J_{\frac{\lambda}{4}}\mathcal{F}_{z_\varnothing}(J_q\mathbf{G}_{\text{in}}(x,y,z_{\text{in}},\nu))\right)\right) \qquad (13)$$

with $z_{\text{IIx}}=h_q/n_e+h_{\lambda/4}/n_e+z_p$ the optical thickness of type IIx modulator components. Here $z_p\to 0$ is the polariser optical thickness and J_p is the Jones matrix of the linear polariser.

Thus, by varying the distance z_{out}, one can numerically track the evolution of the formed complexly structured beams during their further propagation in free space. These possibilities of numerical research, after the beam shaper manufacturing, can be supplemented by the potential of physical experimental research. Indeed, having an electro-optical detection module[40] available or fabricating it using additive manufacturing methods based on open source 3D models[25,26], it is possible to record the spatio-temporal dependence of the THz field strength in the diffraction zone. Then, by excluding a possible measurement system response (e.g. excluding noise[37,41]), one can perform time-reversal of broadband THz wavefront and track the reverse dynamics of the pulse propagation, as was repeatedly implemented[20,23,42].

Important factors to take into account

When designing beam modulators, a number of important features should be taken into account. There are 4 factors that should be considered: (i) the limited spectral bandwidth of anisotropic phase delay in broadband achromatic wave plates imposes a restriction on the spectral interval at which the homogeneous topological charge is formed; (ii) the spectrally-resolved ellipticity of broadband polarised radiation, which characterises achromatic QWPs in the composition of the modulator affects the azimuthal homogeneity in the spectrally-resolved amplitude distribution of the formed beam; (iii) q-plate introduces the spiral phase shift in the modal distribution of input radiation which leads to the appearance of a singular point, which, in turn, is the cause of an additional diffraction perturbation; (iv) the segmented q-plate with eight sectors forms the non-smooth phase distribution with eight phase steps, the traces of which, during the wavefront propagation, also appear in amplitude.

Holographic assessment results

Parameters used in calculations

This section presents the results of numerical simulations of the BUTCH beam-shaping process using the THz PTDH method. For the numerical experiment, the following parameters were set: refractive indices of crystalline quartz $n_o = 2.116$, $n_e = 2.165$, the thicknesses of multilayer wave plates are $h_{\lambda/4} = 17.7$ mm and $h_q = 23.0$ mm. The first QWP was placed $z_{in} = 180$ mm after the incoming plane, optical distance between the plates in the beam shaper $z_{\varnothing} = 15$ mm, $z_{\Sigma} = 350$ mm. The optical beam paths z_{type} for various scheme types (i.e. z_{Icx} and z_{Ix}) were taken into account: $z_{out} + z_{type} = 140$ mm. The other parameters are: transverse spatial grid size $N_x = N_y = 128$ pixels, physical size of transverse spatial grid grid $D_x = D_y = 40, 80,$ and 120 mm, temporal grid size $N_t = 2048$ points, temporal window $T_w = 100$ ps, THz pulse duration $\tau = 0.3$ ps (the corresponding temporal THz spectrum was in the $0.01 - 3$ THz range with a central frequency of 0.75 THz). The initial THz field is modelled as the Gaussian beam with the waist parameter $\rho_0 = 4.7$ mm[43], which is appropriate for the beam size of the THz source (TERA-AX module by Avesta Project Ltd.).

Features of BUTCH beams formation

Various designs of achromatic modulator components were prepared and numerically examined in a virtual assembly to gain insight into the properties of the formed vortex beam. Fig. 3 represents the characteristics of calculated linear and circular polarised vortex beams formed by the best set of achromatic components.

Let us consider the obtained results in more detail. Meticulous handling of all the details that are usually ignored reveals new features for accurate beam shapers design. The first 3 rows of Fig. 3 show the phase difference of the transverse spatial distributions $\Delta\varphi = \varphi_y(x,y) - \varphi_x(x,y)$ for the orthogonal field components after the first QWP (row 1), the q-plate (row 2) (Supplementary S1.mp4). The phase maps shown in the first row are in accordance with the QWP phase delay plot presented immediately above in this figure. Then, we can observe the formation and evolution of the vortex structure as the THz field passes through the next modulator element, namely, q-plate. In row 2 it can be seen that in the frequency range 0.4 – 1.4 THz, the spatial distribution of the phase difference between the y- and x-polarisation components is quite uniform, and their values themselves correspond to $-\pi/2$. These homogeneous relationships in the phase delay induced by the first two modulator elements are also maintained in the detection zone (row 3).

Rows 4–7 of Fig. 3 demonstrate the spatial transverse distributions of the electric field modulus $|G_x(x,y)|$ and $|G_y(x,y)|$ for both x and y polarisations of the denoted spectral components and their corresponding phase distributions $\varphi_x(x,y)$ and $\varphi_y(x,y)$ (Supplementary S2.mp4). On the spectral components that fall within the indicated range of achromaticity of the beam shaper elements, a donut-shaped distribution of the field amplitude is expectedly observed (rows 4 and 6), and spatial phase distributions of the field at these frequencies, shown in rows 5 and 7, clearly reveal vortex structure rotated by $-\pi/2$ with respect to each other. Thus, we can confirm that the designed achromatic QWP and the segmented q-plate, assembled from engineered achromatic HWPs, make a perfect circularly polarised THz BUTCH beam shaper. All these results witness the operability of the scheme of Icx-type.

To obtain a linearly polarised THz BUTCH beam, the circularly polarised BUTCH beam described above is sent through another achromatic QWP, for Ix-type shaping, where the circular polarisation is converted to linear one in the operational range of QWP, depending on its orientation, as outlined in Fig. 1c and discussed above. Row 8 shows the modulus $|G_y(x,y)|$, while row 9 depicts the phase $\varphi_y(x,y)$ of the y-component of the electric field for the corresponding frequencies at the virtual detection plane beyond the second QWP when a linearly polarised

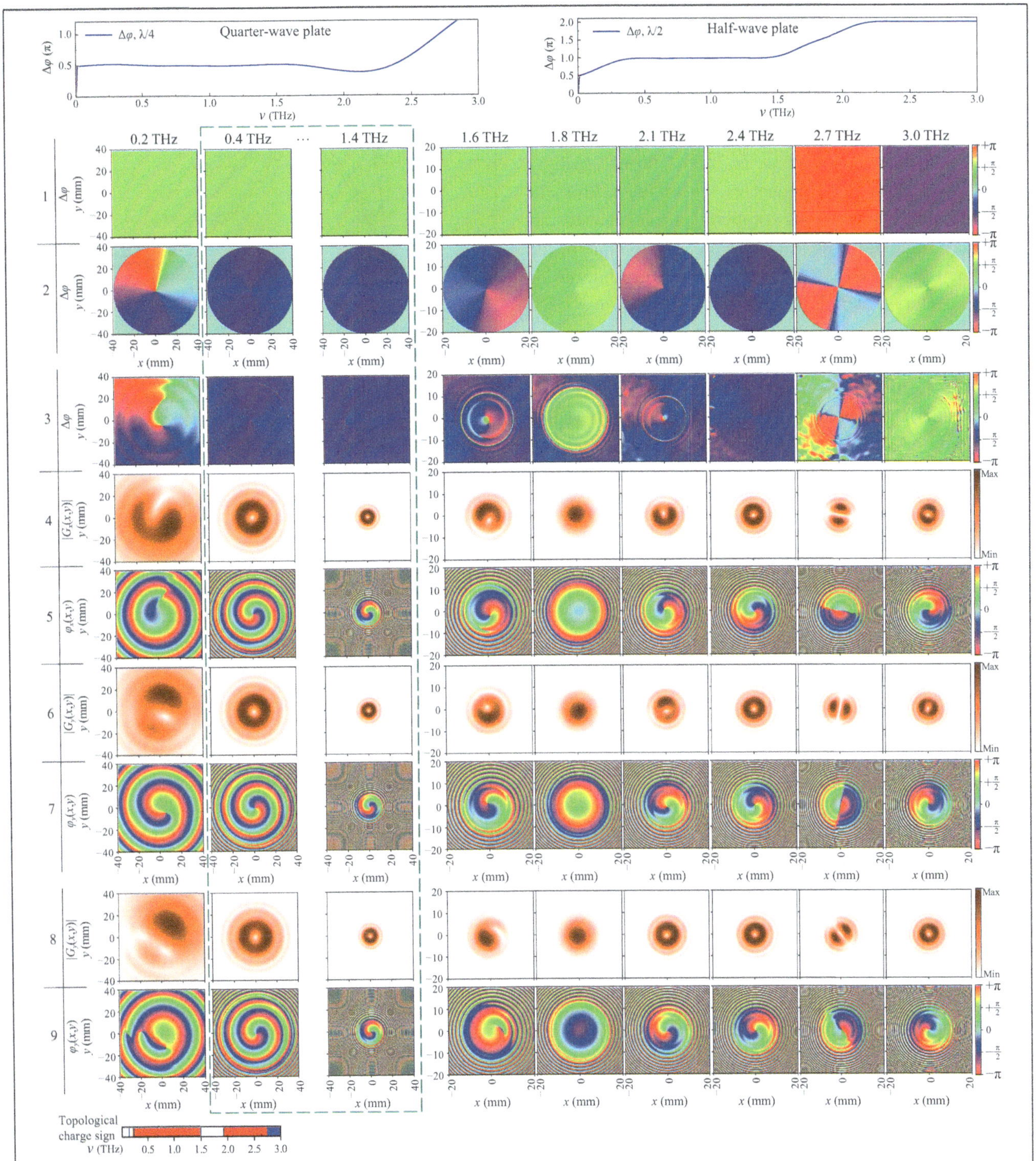

Fig. 3 Frequency-resolved characteristics of the THz field formed by sets of achromatic components arranged in Icx (rows 1-7) and Ix (rows 1-2, 8-9). Rows 1-3 are the phase differences $\Delta\varphi$ between transverse THz field polarisation components, directly after the propagation QWP (row 1), q-plate (row 2) and in the detection plane z_{Σ} for Icx-type (row 3); rows 4-9 are spectral density modulus (amplitude) (rows 4, 6 and 8) and the phase (rows 5, 7 and 9) images in the detection plane z_{Σ} for Icx- (rows 4, 5 for x- and 6, 7 for y- polarisation) and Ix- (rows 8-9) types. Top insets: the dependencies of the phase shift $\Delta\varphi$ on the THz radiation frequency introduced by the QWP (left graph) and HWP (right graph).

vortex in the range of achromaticity of the modulator components is completely formed (see Supplementary S3.mp4).

THz PTDH allows to visualise and analyse the behaviour of the characteristics of a broadband THz field even in more complex cases when the formation of a spectrally homogeneous vortex beam structure does not occur. For our designed modulator's achromatic components, such a regime takes place outside their working frequency. Here the vortex structure is partially broken. Instead of fractional charges typical of monochromatic vortex beams, or beams formed using a spiral phase plate, at frequencies where the achromatic components stop working properly, local changes in curvature slope of the helical phase profile curvature occur. Near the frequency $\nu = 2.8$ THz (see phase distributions on Supplementary materials S2.mp4 and S3.mp4, or rows 5, 7 and 9 of Fig. 3 for 3.0 THz), the beam acquires a topological charge of opposite sign to the charge sign of the other spectral components. It should be noted here that we consider the topological charge in a bounded region around a singularity. The presence of singular points outside the beam limits is not taken into account. The presence and sign of the topological charge for the linearly polarised vortex beam are plotted of Fig. 3, where the red colour denotes the positive and blue – negative charge, respectively. The frequency range, where the charge is positive (0.25 – 1.45 THz) covers the calculated operational range of the wave plates (0.4 – 1.4 THz). We note that the beam has zero charge in the range between 1.5 – 1.9 THz, getting a positive charge again at the higher frequencies (2.0 – 2.8 THz). Above 2.8 THz, the topological charge flips its sign. Previously discussed orthogonal components of circularly polarised beam, has the similar spectral topological charge distribution, but the still present at $\nu = 1.6$ THz.

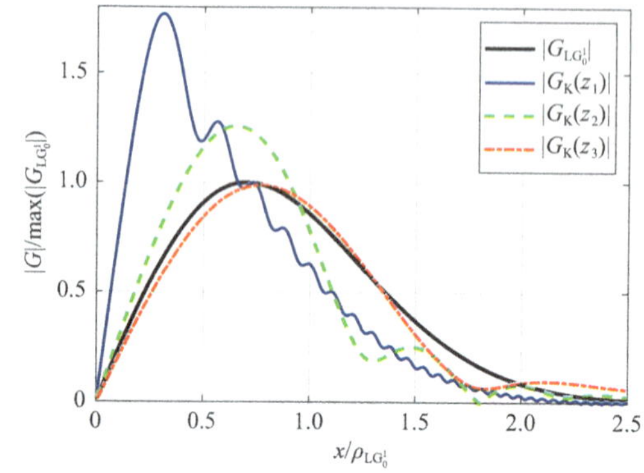

Fig. 4 Cross-section distribution of the Kummer beam amplitude modulus $|G_K|$ normalised to the $G_{LG_0^1}$ mode at the distances $z_1 = 210$, $z_2 = 350$, and $z_3 = 1000$ mm for $\nu_0 = 0.7$ THz.

Structural Features of Kummer Beams

In a common case, the formation of vortex beams is reduced to the fact that a helical phase shift is added to the initial field of the fundamental Gaussian mode. In most works devoted to the studies of BUTCH beams, it is often assumed that the complex amplitude distributions of individual spectral components, which are physically formed in the experimental setup, correspond to Laguerre-Gauss (LG) modes However, this is not completely correct. The formed beams are more accurately represented as Kummer beams[44–47]. They show a specific self-diffraction behaviour upon free propagation, different from that of the LG modes. The radial distribution of the Kummer beams amplitude (spectral density modulus) in the cross section $y = 0$ for central frequency $\nu_0 = 0.7$ THz of broadband wave train $|G_K(x, y = 0, z = z_\Sigma, \nu = \nu_0)|$ is shown in Fig. 4. We give three distributions, each of which normalised to the LG mode $G_{LG_0^1}(x, y, z = z_\Sigma, \nu = \nu_0)$ at the corresponding distance by two parameters: the maximum value of the amplitude $\max(G_{LG_0^1}(x, y, z = z_\Sigma, \nu = \nu_0))$ and the transverse beam dimension $\rho_{LG_0^1}(z) = \sqrt{1 + z^2/z_{R^2}}$ (where z_R is the associated Rayleigh range)[43,48]. Since the LG beam retains self-similarity during propagation, with this normalisation the amplitude distribution LG function at any distance is unchanged with the unit maximum at the radial coordinate $\sqrt{1/2}$. In such consideration, the diffraction spreading effect is compensated and the comparison of the two beams becomes clearer. In the vicinity of the q-plate the position of the Kummer beam amplitude maximum is closer to the axis than that of the LG beam. Concentric field diffraction perturbations are well visible in the near and middle zones of the Kummer beams. Further, as the field propagates away from the modulator, the spectrally-resolved diffraction structure of Kummer's BUTCH beams loses high spatial frequencies and becomes increasingly similar to the structure described by the LG functions. A detailed review of such and LG beams structure for different topological charges is provided in Ref. 45,46. From the results of research on similar diffraction vortex structures in monochrome visible radiation, it is known that they have a slower stepwise decrease in amplitude at the periphery of the beam and a higher divergence, increasing with the value of the topological charge of the optical vortex being formed[45].

The main reasons to consider Kummer beams are their distinct divergence and non-trivial field distribution. Since the ability to conserve energy flow plays an important role in information transmission, further we draw attention to the advisability of taking into account the features of diffraction propagation and analysing the evolution of the field structure. Also, we will study the effect of the finite number of sectors in the q-plate that usually was disregarded, although it significantly affects the entire beam configuration.

BUTCH beams propagation detailed characterisation

It should be noted that the results presented up to this point illustrate the features of the formation of a vortex beam in accordance with only the first three factors indicated earlier. Indeed, the frequency-resolved characteristics in Fig. 3 clearly show the influence of the achromaticity range of the elements (factor (i)), violation of the concentric structure in the spatial amplitude distribution within the spectral interval where the transmission of orthogonal transverse components of the vector THz field

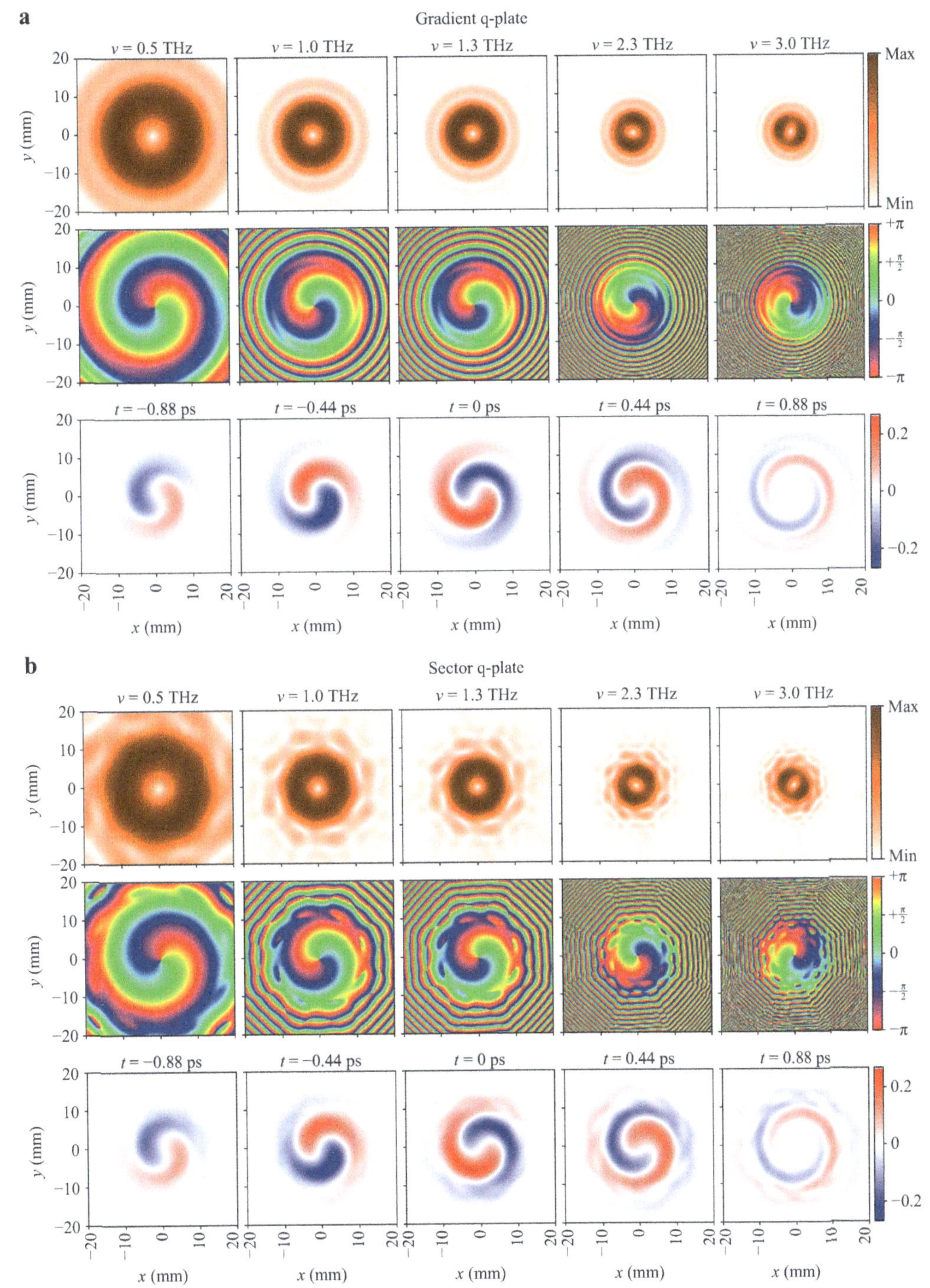

Fig. 5 Transverse spatial structure of the output diffracted field in spectral (upper two rows in **a** and **b**) and temporal (lower row in **a** and **b**) resolutions. Simulated vortex beam characteristics according to the two mathematical models: the gradient q-plate **a** and the sector q-plate **b**. Frequency (in THz) and time (in ps) are shown on top of the images.

is inhomogeneous (e.g. on frequency $\nu = 2.7$ THz, factor (ii)). The diffraction impact of a singularity introduced into the beam (factor iii), which leads to the formation of an amplitude distribution well approximated by the Kummer functions, was also seen in the frequency range 0.4 – 1.4 THz, and analysed in details in Fig. 4. Let us now turn to the influence of the factor (iv). In the case of the gradient q-plate, as shown earlier, the vortex beam has only concentric perturbations in the amplitude distribution (Fig. 5a). However, when the q-plate is composed of sectors, additional field disturbances inevitably occur from phase jumps at their radial junctions. In that case, the formed vortex beams have the 8-fold rotational symmetric field structure with specific lobes from each of the sectors (Fig. 5b).

The increased energy density in the Kummer beams for the near diffraction zone is promising for the tasks of wireless data transmission at short distances, as it allows to gain the level of the transmitted signal. Fig. 6 shows the evolution of the spatial-temporal distribution (first row) and the cross-sectional amplitude and phase distributions for the central frequency (second and third row, respectively). It is conditionally possible to distinguish three characteristic forms in the structure of the beam, as its diffraction evolution proceeds. The first form exists in the vicinityoftheq-plateandispresentedatthedistance$z = 210$mm, Fig. 6, when the observation plane is 15 mm away from the q-plate. In the near diffraction zone, the amplitude distribution is modulated by small speckles that have some characteristic radial direction. The phase distribution reveals an eight-sector structure. As the beam propagates further ($z = 250$ and 300 mm in Fig. 6), it is possible to observe the THz field transformation into the second form, where the smoothing of the granular field structure begins: (i) speckles become larger and rotate according to the azimuth, in the direction of the energy circulation vector; (ii) simultaneously, an octagonal low-intensity zone is formed around the central beam point; (iii) the phase distribution reveals the births of a large number of singular point dipoles (pairs of helical phase dislocations with opposite topological charges)[49,50]. Finally, the speckled beam structure disappears, and the amplitude-phase distribution of the field becomes smoother. At $z = 600$ mm it can be seen that the central vortex remains surrounded by only eight singular point dipoles. In the far zone (here $z =$ 1000 mm), the singularity dipoles annihilate each other and the low-intensity octagonal contour becomes blurred and ceases to be visible in the background of the whole beam.

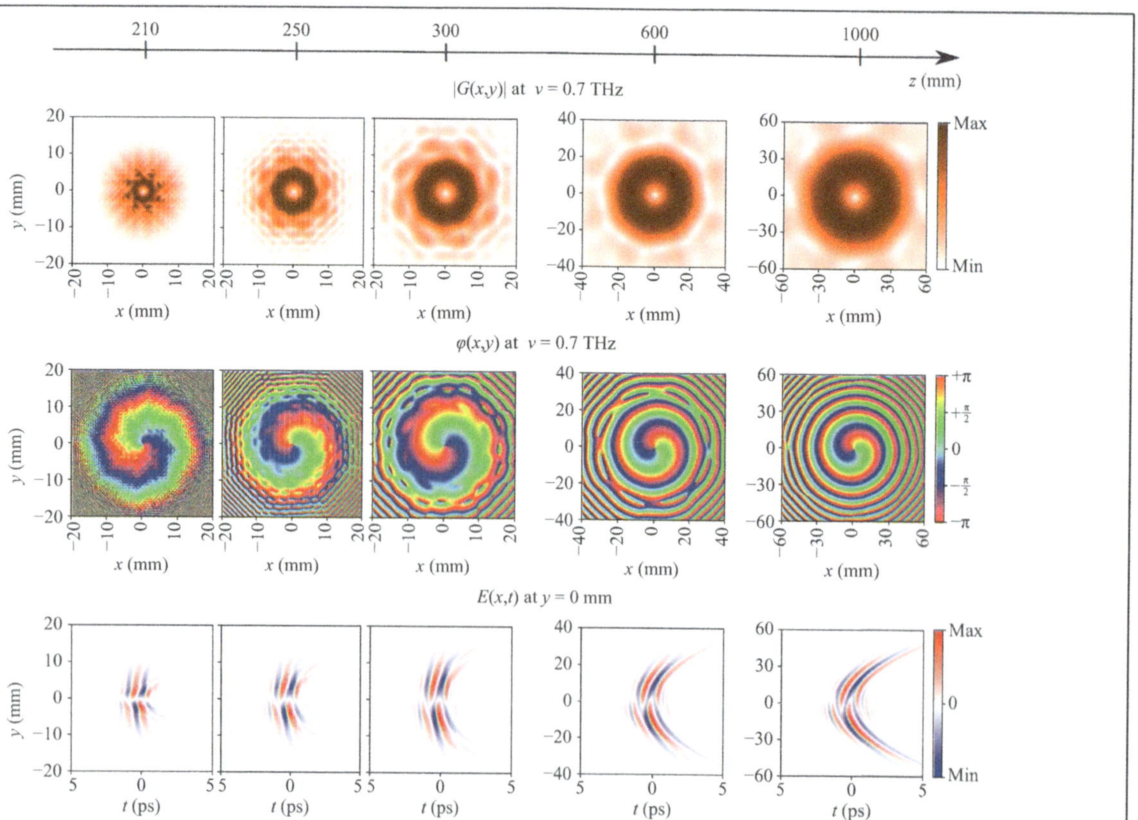

Fig. 6 Evolution of the structure of the Kummer's BUTCH-beam field formed by the developed modulator.

The described free space diffraction processes correlate well with the Kummer beam diffraction nature (Fig. 4). However, an interesting feature is imposed by a discrete structure of the q-plate: a large number of vortex dipoles are formed, which in the process of beam propagation annihilate each other, and finally only one central vortex remains.

We can also see that the vortex THz field formed by the beam shaper with 8-sectorial q-plate transforms significantly during propagation. However, near the axis, it retains quite a regular structure.

Discussion and conclusion

In the second paper from this paired set, we have demonstrated the possibilities of theoretical analysis of the ultrabroadband THz wavefront behaviour as it passing through the partially-achromatic elements of beam converters. Based on the example of using multilayer quartz wave plates as key functional elements, we have shown how to relate the earlier developed mathematical model of THz PTDH, responsible for the simulation of wavefront propagation between polarisation elements with the Jones matrix formalism, describing the specific mechanism of polarisation transformation carried out by each layer of such wave plates. Thus, it becomes possible to study the integral transformation of the characteristics of the THz field during its interaction with all components of the modulator, as well as its further evolution during propagation in free space. It should also not be forgotten that the THz PTDH allows us to solve the additional problem of validation of beam characteristics after physical manufacturing of the designed modulator. It can be achieved through the experimental measurements of THz field in diffraction zone with its optionally subsequent numerical time-reversal to the modulator output plane, and even further backward propagation through the modulator components. Consequently, the proposed holographic approach allows carrying out numerical and physical experiments aimed at refining the design of the beam shapers in order to ensure the maximally possible matching between the characteristics of formed and desired beams.

To clearly illustrate the indicated possibilities, in this paper we have provided several examples of visualisation of the structure of scalar and vector beams formed by modulators based on a certain combination of quarter-wave and q-plate combinations with achromatic properties in the range of 0.4 – 1.4 THz. In particular, the features of the stage-by-stage process of formation of the vortex structure were revealed. It was also established that for a certain anisotropic phase delay formed by the wave plates outside the designed achromatic spectral interval, in a certain limited range of the high-frequency THz spectrum, it is possible to form a vortex structure with a topological charge sign opposite to the charge sign of vortex spectral components, which are formed within the working range of polarising elements. This difference in the sign of the topological charge revealed in the numerical study may not seem significant. Indeed, taking into account that the operating range of the achromaticity of these elements corresponds to the spectral interval of the maximum efficiency of the hypothesised THz radiation source, the opposite charged spectral components will have very low intensity, less than the noise level, and will be indistinguishable in a real experiment. However, it should be taken into account that there are possibilities for controlling the spectral characteristics of the elements making up the modulator, including the fact that reducing or shifting the operating range towards lower frequencies is allowable. Thus, the integral working characteristic of the modulator can also be modified to cover the spectral region where the formation of a charge of the opposite sign at moderate power will take place. That means, this observation and the use of the approach proposed in this paper open up great opportunities for the engineering of a new class of broadband beam shapers with a precisely controlled vector and vortex THz field structure within the operating interval.

The relationship between the characteristics of the achromatic components of the modulator and the formed vortex beams has also been discussed and demonstrated, such as the impact of the ellipticity induced by the quarter-wave plates on the azimuthal homogeneity of the amplitude distribution. The influence of axial singular point introduction into the beam by q-plate, which leads to the mode formation describing by the hypergeometric Kummer function together with the influence of the sector q-plate structure that triggers the appearance of radial jumps on phase distributions and reveals births of a large number of singular point dipoles in the middle diffraction zone have also been shown.

The spectrally and temporally resolved visualisation of the distributions of shaped THz field during its further free-space propagation have also been carried out.

We believe that the proposed approach will find applications in the design and advanced manufacturing of achromatic wave plates and beam modulators with custom characteristics.

Acknowledgements

Russian Science Foundation, Project No 19-72-10147. The work of A.G. is supported by EPSRC Grant No. EP/S018395/1

Author details
[1]Digital and Display Holography Laboratory, ITMO University, St. Petersburg 197101, Russia. [2]V.I. Vernadsky Crimean Federal University, Simferopol 295007, Russia. [3]School of Physics and Astronomy, University of Birmingham, Birmingham B15 2TT, UK

Conflict of interest
The authors declare no conflicts of interest.

Supplementary information is available for this paper at https://doi.org/10.37188/lam.2022.044.

Received: 18 September 2021 Revised: 13 July 2022 Accepted: 13 July 2022
Accepted article preview online: 13 July 2022
Published online: 02 August 2022

References

1. Siemion, A. Terahertz diffractive optics—smart control over radiation. *Journal of Infrared, Millimeter, and Terahertz Waves* **40**, 477-499 (2019).
2. Zhu, L. et al. Experimental demonstration of basic functionalities for 0.1-THz orbital angular momentum (OAM) communications. Proceedings of Optical Fiber Communication Conference. San Francisco, California, United States: Optica Publishing Group, 2014, M3K.7.
3. Kulya, M. S. et al. Propagation dynamics of ultrabroadband terahertz beams with orbital angular momentum for wireless data transfer. Proceedings of SPIE, 11307. Broadband Access Communication Technologies XIV. San Francisco, California, United States: SPIE, 2020, 113070J.
4. Petrov, N. V. et al. On the features of the interference of a set of broadband uniformly topologically charged beams. Proceedings of SPIE, 11499. Terahertz Emitters, Receivers, and Applications XI. Online Only: SPIE, 2020, 114990I.
5. Minkevičius, L. et al. Bessel terahertz imaging with enhanced contrast realized by silicon multi-phase diffractive optics. *Optics Express* **27**, 36358-36367 (2019).
6. Kulya, M. et al. Spatio-temporal and spatiospectral metrology of terahertz broadband uniformly topologically charged vortex beams. *Applied Optics* **58**, A90-A100 (2019).
7. Sirenko, A. A. et al. Terahertz vortex beam as a spectroscopic probe of magnetic excitations. *Physical Review Letters* **122**, 237401 (2019).
8. Nanni, E. A. et al. Terahertz-driven linear electron acceleration. *Nature Communications* **6**, 8486 (2015).
9. Hibberd, M. T. et al. Acceleration of relativistic beams using laser-generated terahertz pulses. *Nature Photonics* **14**, 755-759 (2020).
10. Miyamoto, K. et al. Direct observation of the topological charge of a terahertz vortex beam generated by a Tsurupica spiral phase plate. *Applied Physics Letters* **104**, 261104 (2014).
11. Beijersbergen, M. et al. Helical-wavefront laser beams produced with a spiral phaseplate. *Optics Communications* **112**, 321-327 (1994).
12. Petrov, N. V. et al. Design of broadband terahertz vector and vortex beams: I. Review of materials and components. *Light: Advanced Manufacturing Manuscript* **3**, 43 (2022).
13. Suzuki, M. et al. Analysis of the Pancharatnam-Berry phase of vector vortex states using the Hamiltonian based on the Maxwell-Schrödinger equation. *Physical Review A* **94**, 043851 (2016).
14. Cardano, F. et al. Polarization pattern of vector vortex beams generated by q-plates with different topological charges. *Applied Optics* **51**, C1-C6 (2012).
15. Bekshaev, A., Bliokh, K. Y. & Soskin, M. Internal flows and energy circulation in light beams. *Journal of Optics* **13**, 053001 (2011).
16. Im ai, R. et al. Generation of broadband terahertz vortex beams. *Optics Letters* **39**, 3714-3717 (2014).
17. Lin, Q. et al. Generation of terahertz vortex pulses without any need of manipulation in the terahertz region. *Optics Letters* **44**, 887-890 (2019).
18. Bespalov, V. G. & Gorodetskĭ, A. A. Modeling of referenceless holographic recording and reconstruction of images by means of pulsed terahertz radiation. *Journal of Optical Technology* **74**, 745-749 (2007).
19. Petrov, N. V., Gorodetsky, A. A. & Bespalov, V. G. Holography and phase retrieval in terahertz imaging. Proceedings of SPIE, 8846. Terahertz Emitters, Receivers, and Applications IV. San-Diego, California, United States: SPIE, 2013, 88460S.
20. Petrov, N. V. et al. Application of Terahertz Pulse Time-Domain Holography for Phase Imaging. *IEEE Transactions on Terahertz Science and Technology* **6**, 464-472 (2016).
21. Wang, X. et al. Coaxial waveguide mode reconstruction and analysis with THz digital holography. *Optics Express* **20**, 7706-15 (2012).
22. Kulya, M. S. et al. On terahertz pulsed broadband Gauss-Bessel beam free-space propagation. *Scientific Reports* **8**, 1390 (2018).
23. Liu, X. et al. Spectral Fresnel filter for pulsed broadband terahertz radiation. *AIP Advances* **10**, 125104 (2020).
24. Balbekin, N. S. et al. Nondestructive monitoring of aircraft composites using terahertz radiation. Proceedings of SPIE, 9448. Saratov Fall Meeting 2014: Optical Technologies in Biophysics and Medicine XVI; Laser Physics and Photonics XVI; and Computational Biophysics. Saratov, Russia: SPIE, 2015, 94482D.
25. Grachev, Y. V., Kokliushkin, V. A. & Petrov, N. V. An open-source 3D-printed terahertz pulse time-domain holographic detection module for broadband beam inspection. Proceedings of 2020 45th International Conference on Infrared, Millimeter, and Terahertz Waves (IRMMW-THz). Buffalo, United States: IEEE, 2020.
26. Grachev, Y. V., Kokliushkin, V. A. & Petrov, N. V. Open-source 3D-printed terahertz pulse time-domain holographic detection module. *Applied Optics* **61**, B307-B313 (2022).
27. Trenk, M., Franke, M. & Schwenke, H. The "Virtual CMM" a software tool for uncertainty evaluation – practical application in an accredited calibration lab. Proceedings of ASPE. Uncertainty Analysis in Measurement and Design. State College, Pennsylvania, United States: ASPE, 2004, 1–6.
28. Küng, A. et al. Application of a virtual coordinate measuring machine for measurement uncertainty estimation of aspherical lens parameters. *Measurement Science and Technology* **25**, 094011 (2014).
29. Giusca, C. L., Leach, R. K. & Forbes, A. B. A virtual machine-based uncertainty evaluation for a traceable areal surface texture measuring instrument. *Measurement* **44**, 988-993 (2011).
30. Moroni, G., Sy am, W. P. & Petrò, S. A simulation method to estimate task-specific uncertainty in 3D microscopy. *Measurement* **122**, 402-416 (2018).
31. Su, R. & Leach, R. Physics-based virtual coherence scanning interferometer for surface measurement. *Light: Advanced Manufacturing* **2**, 120 (2021).
32. Sharma, A. K., Tsang, T. & Rao, T. Theoretical and experimental study of passive spatiotemporal shaping of picosecond laser pulses. *Physical Review Special Topics - Accelerators and Beams* **12**, 033501 (2009).
33. Wu, K. et al. Modelling of free-form conformal metasurfaces. *Nature Communications* **9**, 3494 (2018).
34. Alexeyev, C. N. et al. Topological charge conversion and localization in defected heterogeneous multihelicoidal optical fibers. *Journal of Optics* **21**, 085601 (2019).
35. Masson, J.-B. & Gallot, G. Terahertz achromatic quarter-wave plate.

Optics Letters **31**, 265-267 (2006).
36. Belashov, A. V. et al. Effect of object thickness on ultrashort pulse diffraction. *Applied Optics* **58**, 9434-9442 (2019).
37. Kulya, M. et al. Hyperspectral data denoising for terahertz pulse time-domain holography. *Optics Express* **27**, 18456-18476 (2019).
38. Balbekin, N. et al. The modeling peculiarities of diffractive propagation of the broadband terahertz two-dimensional Field. Physics Procedia, 73. 4th International Conference of Photonics and Information Optics, PhIO. Moscow, Russian Federation: Elsevier, 2015, 49–53.
39. Kulya, M. S. et al. Fast Terahertz Spectroscopic Holographic Assessment of Optical Properties of Diabetic Blood Plasma. *Journal of Infrared, Millimeter, and Terahertz Waves* **41**, 1041-1056 (2020).
40. Jiang, Z. & Zhang, X.-C. Terahertz imaging via electrooptic effect. *IEEE Transactions on Microwave Theory and Techniques* **47**, 2644-2650 (1999).
41. Kulya, M. et al. Terahertz pulse time-domain holography with balance detection: complex-domain sparse imaging. *Applied Optics* **58**, G61-G70 (2019).
42. Kulya, M. S. et al. Vectorial terahertz pulse timedomain holography for broadband optical wavefront sensing. Proceedings of SPIE, 11279. Millimeter, and Submillimeter-Wave Technology and Applications XIII. San Francisco, California, United States: SPIE, 2020, 112790D.
43. Soskin, M. S. & Vasnetsov, M. V. Singular optics. Progress in Optics, 42, 219–276. 2001.
44. Bekshaev, A. et al. Optical vortex generation by volume holographic elements with embedded phase singularity: Effects of misalignments. *Ukrainian Journal of Physical Optics* **14**, 171-186 (2013).
45. Bekshaev, A. & Karamoch, A. Spatial characteristics of vortex light beams produced by diffraction gratings with embedded phase singularity. *Optics Communications* **281**, 1366-1374 (2008).
46. Bekshaev, A. Y. et al. Structure of optical vortices produced by holographic gratings with "fork" geometry: Kummer beams. arXiv 0906.2619 (2009).
47. Pas' ko, V. A. et al. Analysis of optical vortex beams with integer and fractional topological charge. Proceedings of SPIE, 5477. Sixth International Conference on Correlation Optics, Chernivtsi, Ukraine: SPIE, 2004, 83–88.
48. Chernykh, A. V. & Petrov, N. V. Optical vortex trajectory of the edge-diffracted single-charged Laguerre-Gaussian beam. *Optics and Lasers in Engineering* **139**, 106504 (2021).
49. Khoroshun, A. et al. Singular skeleton of a Laguerre–Gaussian beam transformed by the doublephase-ramp converter. *Applied Optics* **56**, 3428-3434 (2017).
50. Bekshaev, A. et al. Controllable singular skeleton formation by means of the Kummer optical-vortex diffraction at a rectilinear phase step. *Journal of Optics* **23**, 034002 (2021).

Official journal of the JHL 2689-9620
www.light-am.com

Article Open Access

Simulation-based design optimization of the holographic wavefront sensor in closed-loop adaptive optics

Andreas Zepp[1,*], Szymon Gladysz[1], Karin Stein[1] and Wolfgang Osten[2]

Abstract

Adaptive optics systems are used to compensate for wavefront distortions introduced by atmospheric turbulence. The distortions are corrected by an adaptable device, normally a deformable mirror. The control signal of the mirror is based on the measurement delivered by a wavefront sensor. Relevant characteristics of the wavefront sensor are the measurement accuracy, the achievable measurement speed and the robustness against scintillation. The modal holographic wavefront sensor can theoretically provide the highest bandwidth compared to other state of the art wavefront sensors and it is robust against scintillation effects. However, the measurement accuracy suffers from crosstalk effects between different aberration modes that are present in the wavefront. In this paper we evaluate whether the sensor can be used effectively in a closed-loop AO system under realistic turbulence conditions. We simulate realistic optical turbulence represented by more than 2500 aberration modes and take different signal-to-noise ratios into account. We determine the performance of a closed-loop AO system based on the holographic sensor. To counter the crosstalk effects, careful choice of the key design parameters of the sensor is necessary. Therefore, we apply an optimization method to find the best sensor design for maximizing the measurement accuracy. By modifying this method to take the changing effective turbulence conditions during closed-loop operation into account, we can improve the performance of the system, especially for demanding signal-to-noise-ratios, even more. Finally, we propose to implement multiple holographic wavefront sensors without the use of additional hardware, to perform multiple measurement at the same time. We show that the measurement accuracy of the sensor and with this the wavefront flatness can be increased significantly without reducing the bandwidth of the adaptive optics system.

Keywords: Adaptive optics, Holography, Holographic wavefront sensor, Karhunen-Loève modes

Introduction

Applications such as free-space laser communications and directed energy suffer from the influence of the atmosphere. In addition to transmission losses due to absorption and scattering, these systems are particularly limited by optical turbulence. Adaptive optics (AO) systems are designed to improve the turbulence-affected pointing accuracy, data rate, resolution or focusability. This is usually done with three core components: A wavefront sensor measures the wavefront distortion. A control computer converts this measurement to a control signal.

Correspondence: Andreas Zepp (andreas.zepp@iosb.fraunhofer.de)
[1]Fraunhofer IOSB, Ettlingen, Fraunhofer Institute of Optronics, System Technologies and Image Exploitation, Gutleuthausstraße 1, 76275 Ettlingen, Germany
[2]University of Stuttgart, Institute for Applied Optics ITO, Pfaffenwaldring 9, 70569 Stuttgart, Germany

The control signals are sent to an adaptable device such as a deformable mirror (DM) or a deformable lens to compensate for the measured wavefront deformations[1]. Since the atmosphere is in motion, measurement and correction have to be performed in an open or closed loop. Depending on the turbulence strength, distance and application, the required loop rate can be a few hundred or even several thousand Hertz[2].

Many different technologies and types of wavefront sensors are available or under development to meet these diverse scenarios and requirements. The Shack-Hartmann wavefront sensor[3], pyramid wavefront sensor[4], curvature sensor[5] as well as interferometric methods[6] are widely used. Closed-loop turbulence correction was also demonstrated using a plenoptic camera[7] as wavefront sensor and the stochastic parallel gradient descent (SPGD) algorithm for wavefront sensorless AO[8,9]. Work is also being carried out on wavefront sensors based on the holographic principle[10]. The optical reconstruction of a hologram's real image depends heavily on the wavefront of the reconstruction laser[11]. It is therefore a very promising approach to use the intensity distribution of the real image to obtain information about the wavefront of the reconstructing laser.

Several methods for modal as well as zonal holographic wavefront sensing have been published. The most prominent is the modal holographic wavefront sensor (HWFS). It enables direct measurement of individual aberration modes. Core of modal HWFS is formed by two holographic gratings for each aberration of interest, which are usually multiplexed in a single hologram plate[12] or a single computer-generated hologram (CGH)[13] to be displayed on a spatial light modulator (SLM). The sensor principle is described in detail in the next section. The advantages of this measuring method are the high achievable measuring speed (MHz bandwidths are possible) since no time-consuming calculations are needed, and the easy scalability. The same is true for the zonal HWFS[14]. Instead of decomposing the wavefront into individually measured aberration modes, it is sampled with a number of measuring points[15]. An alternative approach for holographic wavefront sensing is the use of multiplexed Fourier holograms[16]. With a classical Van-der-Lugt correlator[17], the cross-correlation between the incident wavefront and a number of reference waves can be generated optically and the strength of aberration modes determined[18].

1The HWFS enables the fastest wavefront estimation, in principle up to three orders of magnitude faster than the Shack-Hartmann wavefront sensor. However, the measurement accuracy of the sensor suffers greatly from so-called inter-modal crosstalk: different aberration modes present in the laser beam influence each other's measurements[19]. Several improvements have been proposed, particularly for the modal sensor, to minimize the impact of crosstalk. Neil et al[20]. as well as Yao et al[21]. proposed a sensitivity matrix to take the crosstalk into account while Kong et al[22]. as well as Konwar and Boruah[23] implemented novel modifications to the wavefront sensing scheme. Dong et al. proposed avoiding the crosstalk by first closing the AO loop with the help of a second sensor of a different type[24,25]. All of these approaches were able to reduce the crosstalk effect and improve the measurement accuracy to a certain degree for a limited number of modes. The HWFS has been successfully implemented in closed-loop AO systems and the high measurement speed as well as the correction capability have been proven[26–28]. However, the measurements have been performed for static aberrations only and realistic optical turbulence with a high number of modes has not been addressed, yet.

In this paper, we evaluate the performance of a closed-loop AO system based on HWFS under realistic conditions: We simulate dynamic turbulence including more than 2500 modes with time-evolving amplitudes and we consider shot noise in our measurements. To reduce crosstalk effects for these challenging scenarios, we investigate a sensor optimization, which has been developed in order to increase the measurement accuracy in open-loop applications[29]. We show that minor adaptions of the optimization procedure can improve the closed-loop AO performance even more, especially in high noise scenarios.

Theory

The modal HWFS is based on the phase biasing method[20,30]: The phase of the incident beam with the unknown wavefront deformation is manipulated by adding a phase bias, as shown in Fig. 1. The bias corresponds to the aberration mode for which the HWFS is trained. The total strength of the aberration mode is the sum of the strength contained in the unknown wavefront and the strength of the bias. When the biased beam is focused, the intensity distribution in the detector plane depends on the unknown wavefront as well as on the phase bias. If the residual strength of the biased mode is zero, that is, when the bias cancels out the aberration present in the unknown wavefront, the focus of the beam is narrowest (top path in Fig. 1). In all other cases the focus is widened. The greater the residual strength of the mode, the more distributed is the light and the lower the focus intensity is in the center of the focus. Accordingly, a simple intensity measurement in a small central area around the focus (red circles on detectors in Fig. 1) gives information about the strength of

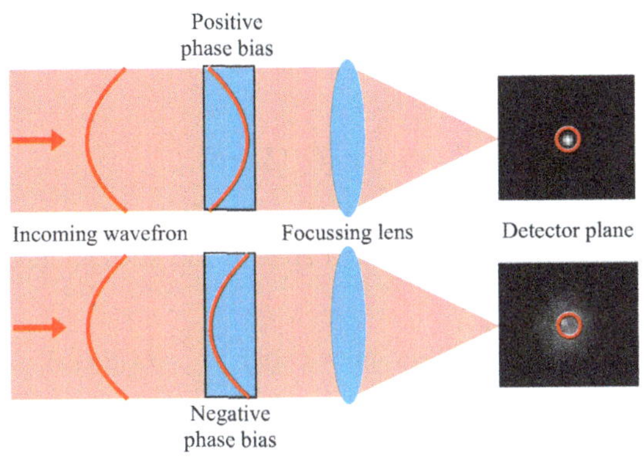

Fig. 1 Schematic representation of the phase biasing principle: A phase bias corresponding to the aberration of interest is added to the incoming wavefront. When the light is focused onto a detector, the intensity distribution depends on the residual wavefront: If the bias reduces the amplitude of the present aberration, more light is concentrated in a small central region (red circles).

the mode within the unknown wavefront[31].

However, to remove the dependency on the total intensity of the laser beam and to increase the linearity of the sensor response, a second measurement with the same bias but opposite sign is carried out (lower path in Fig. 1): Assuming a proper calibration, the contrast of the two intensity measurements gives the strength of the aberration mode sought. To measure multiple aberrations present in the wavefront, the two measurements have to be performed for each mode. By splitting the beam into two partial beams for each aberration mode and positioning a corresponding phase plate in each partial beam, all measurements can be carried out simultaneously. However, it has been proven to be very practical to generate the biased partial beams with diffraction gratings.

For the modal HWFS such diffraction gratings are produced holographically: A reference wave and an object wave are superimposed in the plane of the hologram plate. The interference pattern is recorded and acts as diffraction grating after chemical post-processing. To enable the sensor to measure aberration mode $G_n(x,y)$, two holograms are needed: For the first one the reference wave R_{n+} is chosen to contain only the aberration of interest with the positive bias b_n, while the object wave O_{n+} is a spherical wave that converges at a point $A_{n+}(x_{n+}, y_{n+}, z_A)$ behind the hologram plane (see Fig. 2a). For the second hologram, the reference wave R_{n-} contains the same aberration of bias b_n, but with negative sign. The spherical object wave O_{n-} converges at point $A_{n-}(x_{n-}, y_{n-}, z_A)$ (see Fig. 2b). Accordingly, the time-independent complex amplitudes of these waves in the hologram plane, that is the x/y-plane with $z = 0$, can be written as

$$O_{n+}(x,y) = \exp\{-jk\sqrt{(x-x_{n+})^2 + (y-y_{n+})^2 + z_A{}^2}\} \quad (1)$$

$$O_{n-}(x,y) = \exp\{-jk\sqrt{(x-x_{n-})^2 + (y-y_{n-})^2 + z_A{}^2}\} \quad (2)$$

$$R_{n+}(x,y) = \exp\{+jb_nG_n(x,y)\} \quad (3)$$

$$R_{n-}(x,y) = \exp\{-jb_nG_n(x,y)\} \quad (4)$$

where $k = 2\pi/\lambda$ is the wave number and j is the imaginary unit.

The interference patterns of R_{n+} and O_{n+} as well as of R_{n-} and O_{n-} are given by

$$\begin{aligned} H_{n\pm}(x,y) &= |O_{n\pm}(x,y) + R_{n\pm}(x,y)|^2 \\ &= O_{n\pm}O^*_{n\pm} + O^*_{n\pm}R_{n\pm} + O_{n\pm}R^*_{n\pm} + R_{n\pm}R^*_{n\pm} \end{aligned} \quad (5)$$

where * denotes the complex conjugate. For the sake of readability, the spatial coordinates are omitted. The two holograms $H_{n+}(x,y)$ and $H_{n-}(x,y)$ are recorded with a photosensitive hologram plate. After chemical post-processing both holograms are stored as diffraction gratings. The transmission $\tau_{n\pm}$ of the holographic gratings

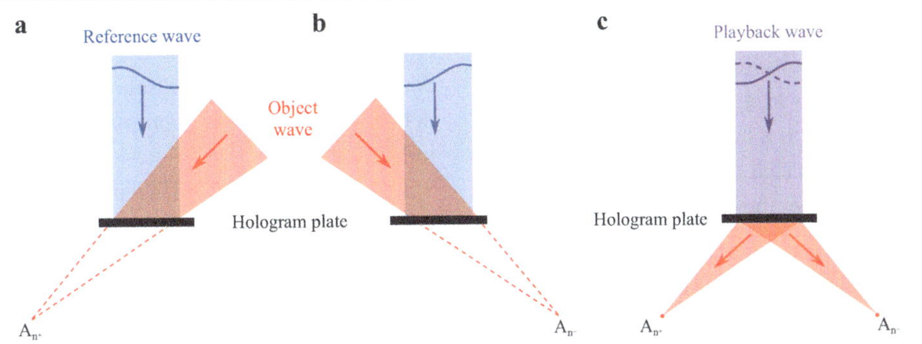

Fig. 2 Recording and reconstruction of a multiplexed hologram for holographic wavefront sensing; **a** first hologram with object beam converging to point A_{n+} and reference beam containing the aberration of interest with positive amplitude, **b** second hologram with object beam converging to point A_{n-} and reference beam containing the aberration of interest with negative amplitude, **c** reconstruction of the object beams when illuminated with a playback beam.

is proportional to these recorded intensity distributions.

By illuminating the multiplexed hologram with a playback, or reconstruction, wave $P(x,y)$, four waves are created for each hologram:

$$P(x,y)\cdot\tau_{n\pm}(x,y)\sim P\cdot(O_{n\pm}O_{n\pm}^{*}+O_{n\pm}^{*}R_{n\pm}+O_{n\pm}R_{n\pm}^{*}+R_{n\pm}R_{n\pm}^{*}) \tag{6}$$

The first and the last terms are the zeroth diffraction orders of the playback wave, only affected in its real amplitude by $|R_{n\pm}|^2$ or $|O_{n\pm}|^2$. The second terms reconstruct the phase-conjugated object waves. Only the third terms are of interest for the HWFS, since they reconstruct the two object waves and with this, create two spots at the positions A_{n+} and A_{n-} (see Fig. 2c). The intensity distributions of the two reconstructed spots in the detector plane ($z=z_A$) are given by

$$I_{n\pm}(x,y)\sim\left|P\cdot O_{n\pm}\cdot R_{n\pm}^{*}\right|^2 \tag{7}$$

The playback wave with the wavefront $W(x,y)$ can be represented by the weighted sum of the aberration modes as

$$P(x,y)=\exp\{jW(x,y)\}=\exp\{j\sum\nolimits_{i=1}^{\infty}a_iG_i(x,y)\} \tag{8}$$

where a_i is the strength or amplitude of mode $G_i(x,y)$. With this, the spot intensity integrated over a central detector area F can be written as

$$\begin{aligned}I_{n+}&\sim\iint_F\left|\exp\left\{j\sum\nolimits_{i=1}^{\infty}a_iG_i(x,y)\right\}\cdot\exp\{-jb_nG_n(x,y)\}\cdot\exp\left\{-jk\sqrt{(x-x_{n+})^2+(y-y_{n+})^2}\right\}\right|^2dF\\&=\iint_F\left|\exp\left\{j\left[\sum\nolimits_{i=1,\,i\neq n}^{\infty}a_iG_i(x,y)+(a_n-b_n)G_n(x,y)-k\sqrt{(x-x_{n+})^2+(y-y_{n+})^2}\right]\right\}\right|^2dF\end{aligned} \tag{9}$$

$$\begin{aligned}I_{n-}&\sim\iint_F\left|\exp\left\{j\sum\nolimits_{i=1}^{\infty}a_iG_i(x,y)\right\}\cdot\exp\{+jb_nG_n(x,y)\}\cdot\exp\left\{-jk\sqrt{(x-x_{n-})^2+(y-y_{n-})^2}\right\}\right|^2dF\\&=\iint_F\left|\exp\left\{j\left[\sum\nolimits_{i=1,\,i\neq n}^{\infty}a_iG_i(x,y)+(a_n+b_n)G_n(x,y)-k\sqrt{(x-x_{n-})^2+(y-y_{n-})^2}\right]\right\}\right|^2dF\end{aligned} \tag{10}$$

The last term in Eqs. 9,10, $-k\sqrt{(x-x_{n\pm})^2+(y-y_{n\pm})^2}$, is responsible for the focusing of the beam in the points $A_{n\pm}$. The second term, $(a_n\pm b_n)G_n(x,y)$, contains the information about the unknown amplitude a_n of the mode of interest. The first term, $\sum_{i=1,\,i\neq n}^{\infty}a_iG_i(x,y)$, describes the wavefront apart from the aberration mode of interest. If this term equals zero that means no other aberrations are present in the wavefront but the one of interest. Then the unknown amplitude a_n can be determined from the contrast $S_{\text{mes},n}$ of the two intensity measurements I_{n+} and I_{n-} as

$$a_n\cong S_{\text{mes},n}=c_n\frac{I_{n+}-I_{n-}}{I_{n+}+I_{n-}} \tag{11}$$

for $-b_n\leqslant a_n\leqslant b_n$. The proportionality factor c_n depends on the bias b_n and the detector size F. It can be obtained by measurements as described in a later section. If $a_n=b_n$, the second term in Eq. 9 is canceled out and the intensity I_{n+} is maximum. At the same time, the intensity I_{n-} is minimum. If $a_n=-b_n$, the reverse happens: The second term in Eq. 10 is canceled out and the intensity I_{n-} is maximum while I_{n+} is minimum. The relationship between a_n and $S_{\text{mes},n}$ within these limits is approximately linear.

However, if the unknown wavefront $W(x,y)$ is distorted by atmospheric turbulence, the term $\sum_{i=1,\,i\neq n}^{\infty}a_iG_i(x,y)$ is non-zero and affects the intensity measurements. Therefore, the measurement of a particular mode is not independent of the presence of other modes. This inter-modal crosstalk causes Eq. 11 to be inaccurate. The stronger the turbulence, the less precise the measurements. This effect is particularly noticeable when measuring higher order modes. In Kolmogorov turbulence, they have on average significantly lower amplitudes than low-order modes. Accordingly, their small a_n is totally swamped in the phase background $\sum_{i=1,\,i\neq n}^{\infty}a_iG_i(x,y)$. The measurement accuracy suffers significantly. To reduce crosstalk, careful design of the modal HWFS is required[29]. Three key design parameters can be identified in Eqs. 9,10: the modal basis $G(x,y)$, the bias b and the detector size F. If the HWFS is designed to measure multiple aberration modes, bias and detector size have to be optimized for each aberration mode individually.

For practical and efficiency reasons, a multi-mode HWFS uses hologram multiplexing to do the beam splitting and biasing with a single diffractive optical element (DOE). All holograms, two for each mode, are recorded one after another in the same hologram plate. By illuminating this plate afterwards with the unknown wavefront, two spots for each mode are reconstructed in the detector plane. The spot pairs corresponding to a particular mode can be detected independently of the other spots. For spot detection a photodiode array or an area detector with a dedicated region of interest (ROI) for each spot can be used. The different detector sizes can be implemented by varying the sizes of the photodiodes, the chosen ROI or corresponding masks (pinholes) in front of the detector.

Methods

In the following we describe how we simulate the modal HWFS. First, we show the sensor implementation. Afterwards, we describe the simulation of HWFS-based, closed-loop AO and discuss how we generate dynamic turbulence as a sequence of correlated phase maps. Finally, we show the optimization procedure of the sensor and present adaption methods to account for its use in closed-loop configurations.

HWFS implementation

The simulation of the modal HWFS includes the holograms and the detectors. As described before, in an experimental setup, it is advantageous to implement a multi-mode sensor in a single DOE by multiplexing all sub-holograms in one multiplex-hologram. However, the multiplexing method, the number of multiplexed modes as well as the used geometry of the spots in the detector plane can influence the sensor performance[32]. To keep the results independent of these design-specific sources of error we treat and illuminate each sub-hologram individually and do not perform multiplexing. Accordingly, we simulate an optical setup where the incoming beam of interest is split into two partial beams per mode and each of these equal partial beams illuminates a different sub-hologram. This corresponds to the original phase biasing method.

We simulate each sub-hologram as thin phase hologram on a 256×256 grid. We use a circular aperture. The HWFS is coded to detect 42 modes (modes number 4 to 45 according to Noll's indexing[33]). Accordingly, the sensor consists of 2×42 sub-holograms. The optimal number of modes depends on the turbulence scenario, the performance of the AO system, in which the sensor is used, and the improvement goal[34]. In order to obtain the best possible comparability of the results, we do not optimize the number of modes, but keep it constant for all simulations. The sub-holograms create 84 biased spots in the detector plane. We use Eqs. 9,10 to calculate the intensities of the spots. We optimize the bias for each mode individually as we will discuss in a later section. We simulate each spot in the detector plane with 256×256 pixels. The pixel scale corresponds to $\lambda/2D$, D is the aperture of the optical system and $\lambda = 1550$ nm the wavelength of the laser beam. We normalize the spots to have a total power of one by dividing each pixel value by the sum of all pixel values. Afterwards, we multiply each pixel by the number of signal photons focused into the spot to get the number of photons per detector pixel. To determine I_{n+} and I_{n-}, all photons inside the particular ROI are summed for each spot. We optimize the ROI for each mode individually as described in a later section. We choose the unit of the ROI to be λ/D. Hence, the results are independent of specific detector designs and dimensions.

In our simulations, we consider nine signal-to-noise ratio (SNR) scenarios. We implement the different SNRs by limiting the average number of signal photons per spot and introducing shot noise[35]. Shot noise can be modeled as a Poisson process; its variance equals the average number of photons N_{Ph} and the SNR is given by

$$SNR = \frac{N_{Ph}}{\sqrt{N_{Ph}}} = \sqrt{N_{Ph}} \tag{12}$$

For each scenario, the average number of photons and the corresponding SNR is listed in Table 1. The first scenario does not include noise. We assume that the diffraction efficiency is identical for all sub-holograms and that the average number of signal photons is equal for all spots. However, the number of photons inside the specific ROIs differs depending on the particular bias and the aberrations present in the beam.

To generate the noisy signal, we consider shot noise only. Other sources of noise, such as readout noise, depend on the detector. Especially the number of pixels is crucial to determine the total amount of noise in the presence of readout noise. Accordingly, it would be necessary to perform the simulations for a particular detector. In contrast, shot noise can be treated independently of most of the specifications of the detector (we assume a quantum efficiency of 100% and do not consider the dynamic range of the detector). The noise is applied by replacing the photon sums I_{n+} and I_{n-} with a Poisson random draw based

Table 1 Overview of SNR scenarios

Scenario	1	2	3	4	5	6	7	8	9
SNR	∞	34.64	31.62	28.28	24.50	20.00	14.14	10.00	7.07
Average number of photons per spot	1200	1200	1000	800	600	400	200	100	50

on mean value given by these photon sums. Afterwards we compute their weighted contrast to determine the amplitude of the particular mode (Eq. 11). The needed calibration factor c_n is obtained during the optimization process, as described in a later section.

We implement the HWFS based on two different modal bases: Zernike modes denoted with Z and Karhunen-Loève modes denoted with K. The advantage of Zernike modes is their simple analytic expression. They are widely used in AO and are the basis of choice in nearly all publications about holographic wavefront sensing. However, Zernike modes are not the best choice to describe Kolmogorov turbulence statistics[34]. Some Zernike modes are correlated which involves redundancy when representing turbulence. Karhunen-Loève modes are designed to remove this redundancy. They can be expressed as linear combination of Zernike modes

$$K_k(x,y) = \sum_j u_{jk} Z_j(x,y) \tag{13}$$

with the coefficients u_{jk}[36,37]. All Karhunen-Loève modes are statistically independent. Hence, if a finite number of modes is used to represent a wavefront, Karhunen-Loève modes contain the maximum amount of information. If the same number of Zernike and Karhunen-Loève modes are perfectly corrected by an ideal AO system, the residual phase variance of the wavefront is lower for the Karhunen-Loève-based system[38]. We will compare HWFS based on Karhunen-Loève modes and Zernike modes. The interested reader can find a detailed description of how we calculate the Karhunen-Lòeve modes in Ref.[9] and a comparison of the first 50 Zernike and Karhunen-Karhunen-Loève modes in Fig. 3 of Ref. 39.

HWFS-based closed-loop AO

We simulate a basic closed-loop AO system based on modal HWFS. We assume post-compensation AO with wavefront measurement and correction at the receiver side, as it is widely used in astronomical and laser communications applications. Nevertheless, the results can also be transferred to a pre-compensation scheme, where wavefront measurement and correction are performed at the transmitter. Pre-compensation is of special relevance for directed energy applications with non-cooperative targets[40]. The principle of our post-compensation AO system is illustrated in Fig. 3: After passing the atmospheric turbulence, the laser light is reflected by the DM to be focused onto a detector. In a real AO system, this detector is the receiver for scientific measurements. We use this detector to monitor the AO performance. In the following, we refer to it as the science detector. Part of the light is directed to the HWFS by a beam splitter (BS). The

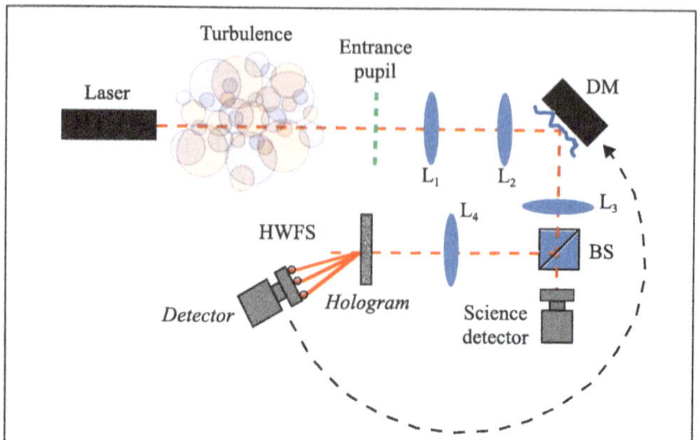

Fig. 3 Principle design of a basic post-compensation closed-loop AO system. The entrance pupil is imaged with lenses L_1 and L_2 in the DM plane and with lenses L_3 and L_4 in the hologram plane. The beamsplitter BS directs some light to the wavefront sensor and transmits most of it to be focused on the science detector. Wavefront sensor signal is used as control signal for the DM.

entrance pupil of the optical system, the DM and the hologram plate are conjugated planes. The light is diffracted at the multiplexed holographic grating into the spots. The spots are sensed by the HWFS detector and the amplitudes of the aberration modes are determined. The wavefront, approximated by the HWFS measurement, is composed by summing the modes multiplied by the measured amplitudes. To correct for this distorted wavefront, it is conjugated, multiplied by the AO gain and passed to the DM as control signal to reshape its reflective surface accordingly. Measurement and correction are performed in a loop with the AO bandwidth to adapt to the temporally-evolving turbulence. In a closed-loop system, no absolute measurement of the wavefront at the entrance pupil is performed, instead the wavefront sensor measures the residual wavefront distortion after the light has passed the DM. As control scheme we assume a traditional pure integrator, where the measured residual wavefront W^{res} is multiplied by the gain g and subtracted from the prior surface shape of the DM W^{DM}_{m-1} to obtain the desired new shape of the DM W^{DM}_m[41].

$$W^{DM}_m = W^{DM}_{m-1} - gW^{res} \tag{14}$$

where the AO correction step m.

To simulate the effect of atmospheric turbulence, we use pupil-plane phase maps. To represent dynamic turbulence, the phase maps are generated as a correlated sequence.

Wavefronts propagating through the atmosphere are usually affected by several turbulence layers moving at different speeds. This leads to wavefront deforming over time, so-called "wavefront boiling"[42]. Hence, for a realistic simulation of the wavefront dynamics, a turbulence profile, represented by the refractive index structure constant $C_N{}^2$, as well as a wind speed profile along the beam path is needed. We created a C_N^2 profile based on models and measurements by Montoya et al[43]. and Zhang and Wang[44] and the wind speed profile from García-Lorenzo et al[45]. Both profiles apply to Canary Islands during daytime. The profiles are discretized in nine layers. The heights and the median wind speeds of each layer are listed in Table 2.

We represent each of the nine layers by a single phase strip. To fill the strips with phase values according to the turbulence profile of the individual layer, we use the FFT-based method of McGlamery[46]. The layers are virtually aligned above the entrance pupil of our AO system. We define the turbulence sampling frequency to be 4 kHz. We translate the phase strips according to the average wind speed of the corresponding layer (Table 2) from right to left. Fig. 4 shows the procedure for four phase strips as an example. The lengths of the yellow arrows represent the median wind speed of each layer. After the translation, we crop the 256 × 256 pixel region above the pupil of each phase strip (red boxes in Fig. 4) and sum their phase values to obtain the phase map in the pupil-plane. After 4000 translations we obtain a sequence of 4000 correlated phase maps representing in total one second of dynamic turbulence. We generate phase map sequences for the normalized turbulence strengths $D/r_0 = 4$ to $D/r_0 = 20$, where D is the aperture and r_0 is the Fried parameter for plane wave.

In our AO simulations we assume perfect tip and tilt correction. Thus, we remove global tip and tilt in all phase maps.

We simulate the closed-loop correction of this dynamic turbulence as sketched in Fig. 5. The top row shows the phase maps applied to the system at its sample bandwidth

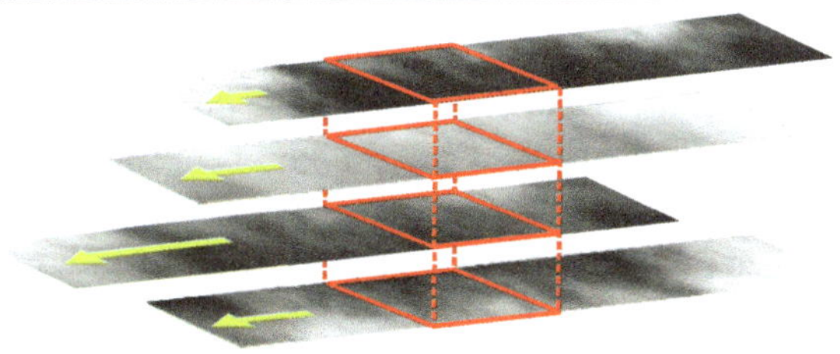

Fig. 4 To simulate a phase map sequence to represent dynamic turbulence, each turbulence layer is represented by a phase strip. Phase strips are translated according to the median wind speed of their layer (represented by yellow arrow) and the phase values inside the red boxes are added to obtain the pupil-plane phase map.

of 4 kHz. The bottom row shows the measurements of the HWFS as weighted sum of the 42 detected modes. The second row represents the shape of the DM. Before the first measurement, the surface is flat. Afterwards it is shaped according to the HWFS measurement. We assume the DM to be ideal and to assume the prescribed shape perfectly. The shape of the DM is updated only after a HWFS measurement. Since the AO bandwidth is lower than the bandwidth with which we sample the dynamic atmosphere (in the sketched example three times smaller), measurement and correction are not performed for each phase map. That means that the turbulence evolves between the AO iterations while the DM surface remains unchanged. To monitor the performance of the AO system, we simulate the point spread function (PSF) of the focused beam in the plane of the science detector with 256x256 pixels of pixel size $\lambda/2D$. This corresponds to Nyquist sampling of a diffraction limited PSF and is obtained by zero padding of the wavefront[47]. A cross-section of the PSFs is displayed in the third row of Fig. 5 in blue. The red line represents the Strehl ratio over time. We use the Strehl ratio as quality metric and calculate it as peak of the PSF divided by the peak of the diffraction-limited PSF[48]. Both PSFs are normalized to have the same total power.

HWFS optimization

Phase bias and ROI are key design parameters of the modal HWFS. The combination of the two has a major impact on the occurrence of inter-modal crosstalk. Accordingly, a proper choice of bias and ROI can improve the measurement accuracy significantly. The best combination, however, differs for different modal bases, individual modes and turbulence strengths. To determine the optimal combination of bias and ROI, we use an optimization method that aims to find the parameters with the weakest crosstalk[29]. The workflow of the optimization procedure is shown in Fig. 6. The optimization has to be performed for each turbulence strength (D/r_0) individually and on a mode-by-mode basis.

It makes sense that the measuring range is based on the expected range of variability of the mode of interest $G_n(x,y)$ for the given turbulence strength. We choose the range to be three times the modes' standard deviation σ_n in Kolmogorov turbulence[33]. For all relevant combinations p of bias b_n and ROI F_n, a crosstalk-free sensor response

Table 2 Heights above the telescope and the corresponding median wind speed of the 9-layer model. Heights are given for the location of the top of each layer.

Height above the telescope [m]	500	2600	4000	5600	7600	8600	10000	12600	16000
Median wind speed [m/s]	5.11	5.97	7.08	8.64	11.26	12.94	15.04	13.21	8.62

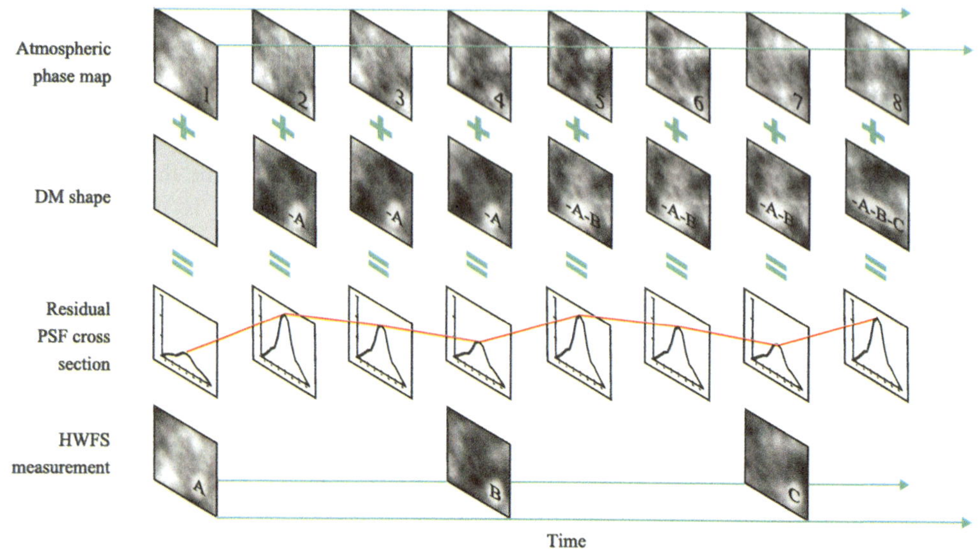

Fig. 5 Depiction of HWFS-based AO simulation with dynamic turbulence. Top row: atmospheric phase map sequence is applied to the system at its sample rate (4 kHz). Second row: DM shape is updated after each HWFS measurement. The measured residual wavefront is subtracted from the current DM shape. Third row: the corrected wavefront is focused onto the science detector. The cross-section of the PSF is shown in blue. Red line represents the Strehl ratio over time. Bottom row: HWFS measurement of the residual wavefront, performed at the AO bandwidth (AO bandwidth < turbulence sample rate).

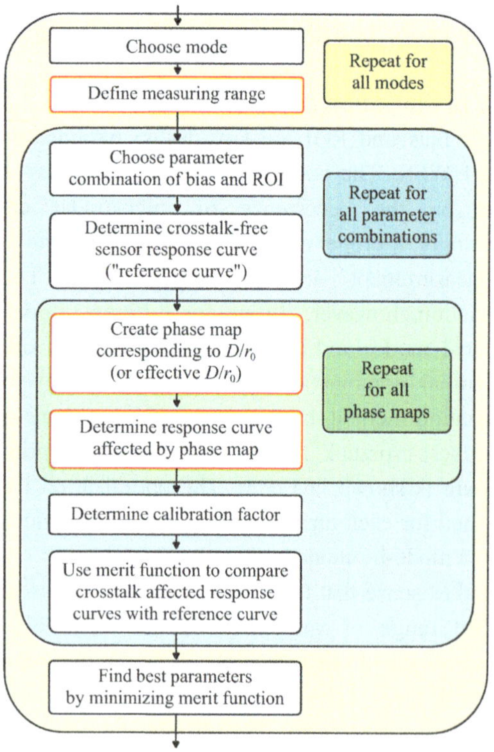

Fig. 6 Workflow of the optimization procedure for finding the best combination of bias and ROI to reduce crosstalk effects.

curve $S_{\mathrm{ref},n,p}$ is determined as a reference. The sensor response curve is the result of M measurements (Eq. 11), whereby the wavefront $W(x,y)$ of the playback beam contains the mode of interest $G_n(x,y)$ only. The amplitude a_n of this mode is successively increased for each of the M measurements to cover the entire measuring range, while $a_{i\neq n}=0$ (see Eq. 8). Accordingly, the term $\sum_{i\neq n}^{\infty} a_i G_i(x,y)$ vanishes in Eqs. 9,10 and the reference response curve cannot be affected by crosstalk. The dashed black curve in Fig. 7a shows an example of a reference response curve.

In the next step, a response curve $S_{\mathrm{mes},n,p}$ is measured for a playback beam that is distorted by realistic atmospheric turbulence. We simulate the distorted wavefront $W(x,y)$ as the weighted sum of 2555 Zernike modes. We exclude the first three modes, piston, tip and tilt, from this phase map. The weighting factors are normally distributed random variables. Their standard deviation σ_n is obtained with the turbulence strength and the diagonal elements c_{nn} of the covariance matrix of Zernike modes in Kolmogorov turbulence[49]:

$$\sigma_n = \sqrt{c_{nn}(D/r_0)^{\frac{5}{3}}} \tag{15}$$

The weighting factor of the mode of interest is not set as random number like for the other modes. Instead, we increase a_n successively for each of the M measurements within the measurement range to determine the response curve. The obtained response curve differs from the reference curve, since the high number of modes present in the phase map creates crosstalk. For stronger turbulence as well as for modes of higher order, the crosstalk increases

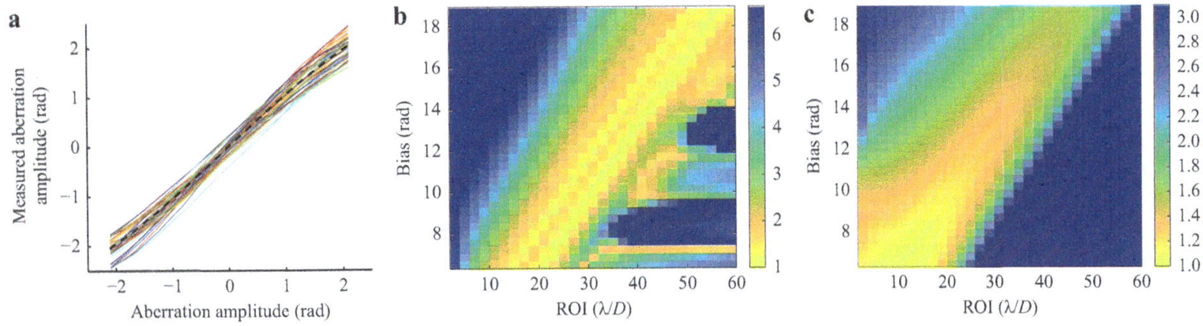

Fig. 7 a Optimization results of HWFS to measure Karhunen-Loève mode 4 at turbulence strength of D/r_0=10; dashed black curve: crosstalk-free reference sensor response; solid colored curves: crosstalk-affected response curves, obtained from different independent phase maps representing D/r_0=10; **b** J-map: merit function for bias and ROI combinations, values greater than the median are set to the median; **c** c-map: calibration factors for bias and ROI combinations, values greater than the median are set to the median.

and the curve deviates more from the reference. However, the measurement of a single crosstalk-affected response curve alone is not sufficient, since the amplitudes of the aberrations are random and the crosstalk can therefore vary greatly. We repeat the measurements k times to obtain k crosstalk-affected response curves. For each response curve a different phase map is created and applied. The phase maps are independent and uncorrelated. For sensor calibration, we calculate the mean of all k response curves to obtain the average response curve. We approximate the proportionality or calibration factor c_n (see Eq. 11) of the HWFS as the reciprocal slope of a linear fit to the average response curve. The smaller the calibration factor, the higher the sensitivity of the sensor. The colored curves in Fig. 7a display 40 calibrated crosstalk-affected response curves for mode 4 (defocus).

To compare the crosstalk-affected response curves with the reference curve, we calculate the merit function $J_{n,p}$ for the current mode and parameter combination as

$$J_{n,p} \cong \sqrt{\frac{1}{M}\sum_{m=1}^{M}[\langle S_{\text{mes},n,p}(a_{n,m})\rangle - S_{\text{ref},n,p}(a_{n,m})]^2} + \frac{1}{M}\sum_{m=1}^{M}\sqrt{\langle [S_{\text{mes},n,p}(a_{n,m}) - S_{\text{ref},n,p}(a_{n,m})]^2\rangle} \quad (16)$$

where the symbol $\langle\ \rangle$ denotes sample averaging over the k measurements. For the simulations presented in this paper we set $k = 40$ and $M = 11$. The first term of the merit function J calculates the deviation of the average response curve from the reference curve. The second term determines the deviation of each of the k crosstalk-affected response curves from the reference. Smaller J means that the crosstalk-affected curves are more similar to the reference and with this that the crosstalk is smaller.

After repeating this procedure for all relevant combinations of bias and ROI, we can summarize the results in two maps: The J-map contains the results of the merit function for all parameter combinations and the c-map contains the determined calibration factors for the same combinations. A normalized J-map is shown in Fig. 7b, the corresponding normalized c-map in Fig. 7c. Normalization is performed by dividing J and c by their smallest value. Bias is plotted on the y-axis, ROI on the x-axis. Outliers are suppressed by replacing all values that are greater than the median with the median (blue color). A small value (yellow color) in the J-map means that the crosstalk is comparatively low for this combination and a small value in the c-map means that the sensitivity of the sensor is comparatively high for this combination. We obtain individual J- and c-maps for different modes and turbulence strengths.

To find the optimal HWFS design, i.e. the best combination of bias and ROI, we minimize the merit function. Specifications of the used detector like dynamic range or scenarios in the starved-photon regime might limit the acceptable sensitivity of the HWFS. In that case we can take the c-map into account to exclude combinations that do not fulfill the sensitivity requirements.

It has been proven that choosing a sensor design based on this optimization procedure increases the measurement accuracy of the modal HWFS significantly[29]. The sensor is designed for a certain turbulence strength (D/r_0) and takes into account the turbulence statistics and the expected amplitudes of the modes. Using the sensor in turbulence scenarios that it was not optimized for will most likely still give good results, albeit not as good as when the sensor is used in conditions it was designed for. The same can be expected when implementing the optimized HWFS in a closed-loop AO system. After closing the loop and starting the correction, the residual amplitudes of the measured modes, and with this the sizes of the spots, are reduced. As a consequence, bias and ROI are no longer optimal for the new, residual conditions. Additionally, the crosstalk that emanates from the corrected modes is reduced. To account for changing conditions during closed-loop operation, we propose to modify the optimization procedure. The modified steps are outlined in red in the workflow chart (Fig. 6).

1. The sensor is not optimized for the turbulence strength D/r_0 of the expected turbulence scenario, but for an effective D/r_0 that corresponds to the conditions during closed-loop operations. Thus, phase maps to represent the turbulence strength have to be created for the effective D/r_0.

2. With smaller residual amplitudes of the modes, the measuring range can be reduced to increase the accuracy in that smaller region.

3. The expected SNR and photon numbers are considered by applying both to the response curve measurements.

Results

First, we present the performance analysis of the modal HWFS in closed-loop operation in realistic turbulence. We compare results from Zernike- and Karhunen-Loève-based AO systems with and without optimization. We then show that adapting the sensor optimization to the expected SNR scenario and to closed-loop operation increases its performance. Finally, we propose the use of a multi-ROI HWFS to improve the measurement accuracy of the sensor in closed-loop operation.

Performance of HWFS in closed loop operation

To evaluate the performance of an AO-system based on HWFS, we simulate closed loop operation as described before. We use the Strehl ratio of the PSF on the science detector as performance metric. The results for the turbulence strength $D/r_0 = 10$ are shown in Fig. 8. The HWFS is coded to measure the modes 4 to 50. We set the AO bandwidth to 2 kHz and the AO gain to 0.3. Fig. 8a presents the results of the Zernike-based system, Fig. 8b of the Karhunen-Loève-based system. The x-axes show the time, sampled with 4000 correlated phase maps of dynamic turbulence. The y-axes show the Strehl ratio. The solid lines represent the instantaneous Strehl ratios, while the dashed lines of the corresponding color visualize their mean value. The red curves in both panels show the Strehl ratio if no AO correction is performed. Since the same turbulence sequence is used, the two Strehl ratios without AO are identical. The black curves represent the AO performance with perfect HWFS measurement, i.e. without measurement errors. This is the maximum the system can achieve for the given AO bandwidth, gain and number of modes. Due to the higher variances of low-order Karhunen-Loève modes than their Zernike counterparts and their statistic independency, the mean Strehl ratio of the perfect Karhunen-Loève-based system (black dashed line in panel b) is higher than the one of the perfect Zernike-based system (black dashed line in panel a). Karhunen-Loève-based AO is capable of achieving higher Strehl ratios. The blue curves correspond to the use of HWFS without optimization. Instead of performing the optimization method to find the best combination of bias and ROI, we set the two parameters according to simple, common-sense guidelines: We chose the bias to define the measurement range and set it to 3σ of the modes. We keep the ROI relatively small to maximize the sensor sensitivity and set it to five times the size of the diffraction limited PSF. Compared to the case of no correction (red curves), a clear improvement can be seen for both bases. While the

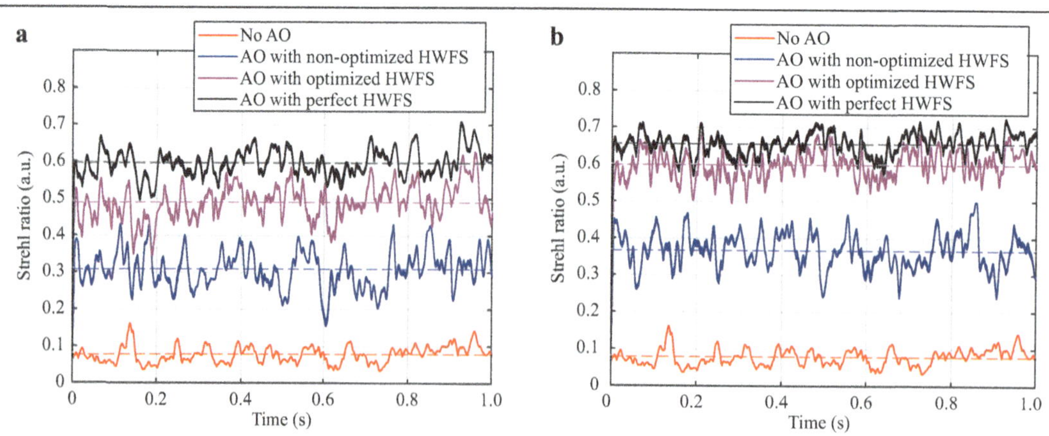

Fig. 8 Strehl ratio over time for closed-loop AO with Zernike-based HWFS **a** and Karhunen-Lòeve-based HWFS **b**. The turbulence strength is $D/r_0 = 10$, the AO bandwidth is 2 kHz and the AO gain is 0.3. 45 modes are measured and corrected. Dashed lines: mean values of the instantaneous Strehl ratios.

results in open-loop operation suffer from the limited measurement accuracy due to crosstalk[29], the results in closed-loop operation are less affected. The reason is that by decreasing the amplitudes of the modes during operation, their crosstalk to other modes decreases as well. Nonetheless, there is still potential for improvement compared to the case of perfect correction. The magenta curves show the performance of the optimized HWFSs. Here we chose bias and ROI according to the optimization results for the specific D/r_0. This optimization leads to significant improvement of the performance. For the Zernike sensor the mean Strehl ratio is increased from 0.31 without optimization to 0.49 with optimization. For the Karhunen-Loève sensor the mean Strehl ratio is increased even more from 0.37 to 0.6. When comparing the results of both modal bases with optimization (magenta curves in panel a and b), it can be seen that the Karhunen-Loève-based AO achieves a higher Strehl ratio. Additionally, the difference between optimized HWFS AO and perfect HWFS AO is smaller. This shows that the optimization is more efficient for Karhunen-Loève modes. Accordingly, this modal basis is the best choice for HWFS in closed-loop operation.

Sensor adjustment to closed-loop conditions

Although the optimized HWFS achieves very good results in closed-loop operation, these are likely not the best possible results. The measurement environment for the HWFS changes during closed-loop operation: the variances of the modes decrease as well as the spot sizes. Hence, the conditions differ from those during the optimization process. Especially in challenging scenarios like high D/r_0 or poor SNR, we expect that there is potential for further improvement. As described before, we adapt the optimization method to account for the closed-loop operation. In order to simplify the distinction, we refer to the original optimization as "classic optimization" and to the adapted optimization as "modified optimization".

Fig. 9 shows the measurement error of Karhunen-Loève-based HWFSs as a function of SNR. We use the HWFSs to measure the amplitudes of 42 modes in 100 independent, uncorrelated phase maps. The phase maps represent the turbulence strength of interest. To define the accuracy, we do not look at the absolute measurement error but the measurement error relative to the standard deviation σ_n of the measured mode. Therefore, we divide the root mean square (RMS) error of these 100 amplitude measurements by σ_n and average over all modes. Multiplied by 100 the cost function E is given in percent by

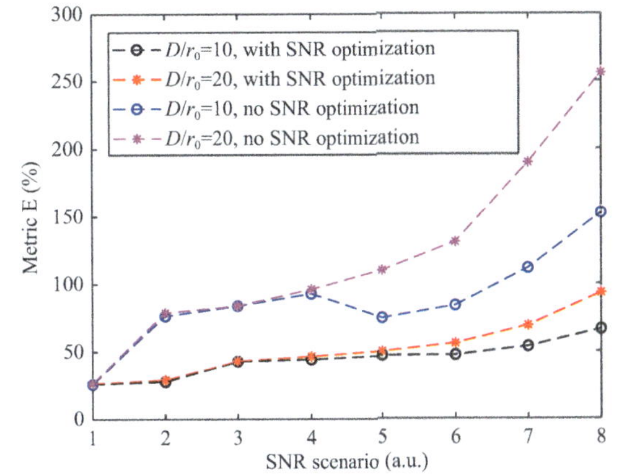

Fig. 9 Cost function E as function of SNR scenario for classic optimization (blue and magenta curve) and modified optimization (black and red curve). HWFS is coded to measure 42 Karhunen-Loève modes. Results are based on 100 measurements of independent, static phase maps corresponding to D/r_0=10 (black and blue curve) and D/r_0=20 (red and magenta curve).

$$E = 100\frac{1}{N}\sum_{n=4}^{N}\left[\frac{1}{\sigma_n}\sqrt{\frac{1}{M}\sum_{m=1}^{M}(U_{\mathrm{mes},n,m} - U_{\mathrm{truth},n,m})^2}\right] \quad (17)$$

where N is the number of the highest mode, M the number of independent phase screens, $U_{\mathrm{mes},n,m}$ the measured amplitude and $U_{\mathrm{truth},n,m}$ the ground truth of the nth mode at the mth phase map.

The higher the cost function, the less accurate the measurements are. The blue curve in Fig. 9 shows the cost function (metric E) of a classically optimized HWFS designed for $D/r_0 = 10$ and measuring at $D/r_0 = 10$. The optimization did not take SNR into account. Accordingly, the optimal bias and ROI for SNR scenario 1 (no noise present) is used for all SNR scenarios. The magenta curve shows the error of a HWFS measuring at $D/r_0 = 20$, designed for this turbulence strength but without SNR optimization. The black and the red curves correspond to the metric E of HWFSs where we took the SNR during the optimization into account: Bias and ROI are optimized for each SNR independently. For SNR scenario 1 the displayed metric is about 25% for all four sensors, i.e. the average measurement error is about 25% of the modes' standard deviation. For the classically optimized sensors the measurement error increases sharply as the signal-to-noise ratio decreases. A reliable, reproducible measurement hardly seems possible in demanding scenarios. The optimal parameters for noiseless measurements are not the best choice for noisy measurements. Modifying the process to optimize each scenario independently increases the accuracy significantly.

To account for the changing conditions during closed-loop operation, we do not optimize for the initial turbulence strength, which is given by the turbulence scenario, but for an effective turbulence strength given by the performance of the locked-in AO system. To define the effective turbulence strength, we assume an AO system based on a perfect HWFS and analyze the residual amplitudes of the modes during the closed-loop operation. Since with ideal HWFS no measurement errors occur and the perfectly measured amplitudes are corrected perfectly, these residual amplitudes are the lowest that can be achieved. However, because of the limited AO bandwidth and an AO gain smaller than 1, the residual amplitudes are nonzero. Fig. 10a shows results for closed-loop measurements with perfect HWFS with different gain and bandwidth combinations for $D/r_0 = 10$. The presented relative mean residual amplitudes $RMRA_n$ are defined for each mode number n as

$$RMRA_n = 100\frac{1}{\sigma_n}\frac{1}{M}\sum_{m=1}^{M}\left|a_{\mathrm{res},n,m}\right| \tag{18}$$

where M is the number of measured correlated phase maps of the sequence and $a_{\mathrm{res},n,m}$ is the measured residual amplitude of the nth mode at the mth phase map. In Eq. 18 the absolute values of all residual amplitudes of the sequence are averaged and divided by the standard deviation of the corresponding mode. Because higher order modes suffer more from crosstalk, the relative mean residual amplitudes increase with increasing mode number. We see that a perfect AO system with a gain of 0.3 and a bandwidth of 1 kHz at $D/r_0 = 10$ can reduce the amplitudes to roughly 10% to 40% of σ_n (solid black curve). We use this information to determine our expected effective turbulence strength after closing the loop. We define a correction factor f_n ascending from 10% to 40% for each mode (shown as black dots in Fig. 10a). We use these factors to create phase maps for the effective turbulence strength: we generate phase maps for the initial turbulence strength D/r_0 but multiply the amplitudes of the modes that we want to measure and correct by the corresponding factor f_n. Additionally, we reduce the measurement range we want to optimize for from $3\sigma_n$ to $2\sigma_n$. This modified optimization is referred to as "adapting the initial D/r_0".

An alternative approach to generate an effective D/r_0 is to find a D/r_0 (lower than the initial one) by matching the phase variance. The pupil-averaged phase variance $\langle\varepsilon^2\rangle$ can be calculated as the sum of the variances of all modes. With Eq. 15 and the assumption that piston is ignored and tip and tilt are perfectly corrected, this can be written as[50]

$$\begin{aligned}\left\langle\varepsilon^2\right\rangle &= \sum_{n=4}^{N} f_n^{\,2} c_{nn}(D/r_0)^{\frac{5}{3}} + \sum_{n=N+1}^{\infty} c_{nn}(D/r_0)^{\frac{5}{3}} \\ &\approx \sum_{n=4}^{N} f_n^{\,2} c_{nn}(D/r_0)^{\frac{5}{3}} + 0.2944N^{-\sqrt{3}/2}(D/r_0)^{\frac{5}{3}}\end{aligned} \tag{19}$$

where N is the index of the highest mode, that is measured and corrected. The phase variance is split in two terms: the first contains the modes that can be corrected by the AO system and the second includes all other modes. The second term is approximated by an asymptotic formula found by Noll[33]. In Fig. 10b the residual phase variance is shown as function of D/r_0. The red curve shows the residual phase variance when no AO correction is performed, i.e. when we set the correction factor $f_n = 1$ for all modes. This corresponds to the phase variance of the initial, uncorrected turbulence strength.

The black curve represents a perfect correction of the first N modes, i.e. when we set the correction factor $f_n = 0$

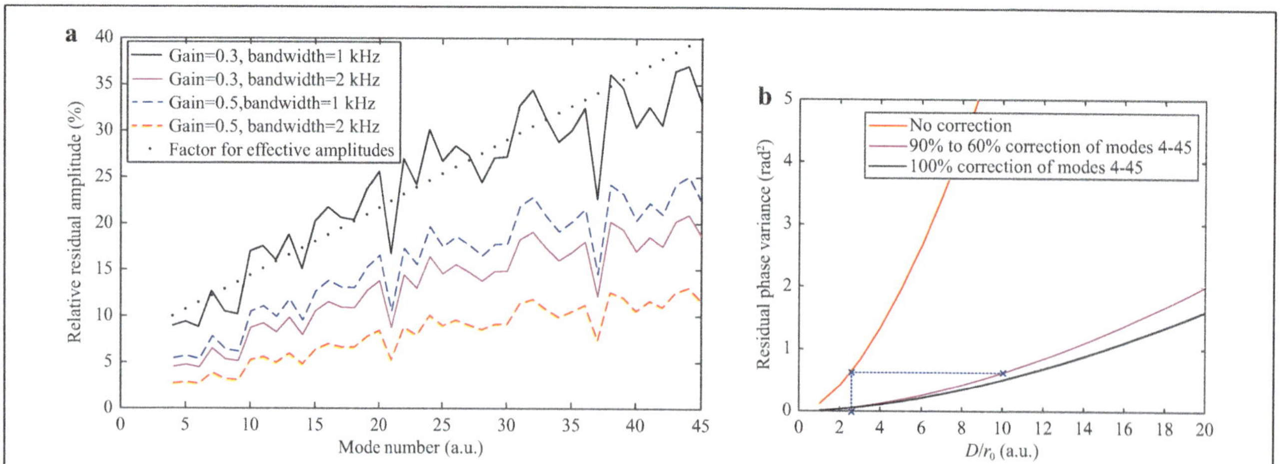

Fig. 10 **a** Relative residual mean amplitude of closed-loop AO with perfect HWFS at D/r_0=10 for different gain and bandwidth combinations. The black dots correspond to a linear fit applied to the black graph. **b** Pupil-averaged residual phase variance as function of D/r_0.

for all modes. To create the magenta curve, we chose an ascending correction factor, increasing from 0.1 to 0.4 with increasing mode number. Thus, the curve represents a partial correction of the modes, corresponding to the residual amplitudes we determined for AO with perfect Karhunen-Loève-based HWFS, gain of 0.3 and bandwidth of 1 kHz at $D/r_0 = 10$. As illustrated with the blue dotted lines, an uncorrected turbulence strength of $D/r_0 = 3$ has roughly the same phase variance as the partially corrected turbulence strength of $D/r_0 = 10$. Hence, we can define $D/r_0 = 3$ as the effective turbulence strength for closed-loop operation at $D/r_0 = 10$. Effective turbulence strengths can be found for other turbulence scenarios accordingly. The optimization is performed for the effective D/r_0 instead of the initial one. This modified optimization is referred to as "changing to different D/r_0".

In Fig. 11 the classic optimization and the two modified optimizations, i.e. adapting and changing D/r_0 while considering SNR conditions, are compared. The optimized HWFSs were used to measure 42 Karhunen-Loève modes in a closed-loop AO system with a gain of 0.3 and a bandwidth of 1 kHz. The mean Strehl ratio of the PSF on the science detector is shown for the SNR scenarios 1, 4 and 7. Results are given for the turbulence strengths $D/r_0 = 10$ (dashed lines) and $D/r_0 = 20$ (dotted lines). HWFSs with classic optimization are plotted in red, with modified optimization based on adapting the D/r_0 in black and with modified optimization based on changing the D/r_0 (changing to $D/r_0 = 3$ for $D/r_0 = 10$ scenario and to $D/r_0 = 5$ for $D/r_0 = 20$ scenario) in magenta. The mean Strehl ratios for no AO and perfect AO cases are presented as horizontal lines in blue color for $D/r_0 = 10$ scenario and in green color for $D/r_0 = 20$ scenario. Using modified optimization always yields an increased Strehl ratio. For SNR scenario 1, optimization based on adapting D/r_0 performs best and can raise the Strehl ratio by 0.03 for both turbulence conditions. For SNR scenarios 4 and 11, the absolute Strehl ratios decrease, but the improvement by using modified optimization is significant. In conclusion, in nearly all cases the Strehl ratio can be increased to above 10%, which is the reported threshold of usefulness of AO correction for laser communications[51].

Although the modified optimization method is ideal for HWFS in closed-loop operation, it is obviously not ideal when closing the loop. Fig. 12 shows the instantaneous Strehl ratios over time for AO systems correcting 45 Zernike modes with a gain of 0.5 and a bandwidth of 2 kHz. The turbulence strength is $D/r_0 = 10$. The black curve represents AO with an HWFS that is optimized for this initial turbulence strength. The Strehl ratio increases almost directly to the level of the mean Strehl ratio. The blue curve represents AO with an HWFS that is optimized for the effective D/r_0. The measurement accuracy is lower initially. Hence, it takes more iteration steps and time to increase the Strehl ratio and reach the plateau. After this longer start-up phase, however, the Strehl ratio is higher. For very challenging scenarios with low signal and SNR and high D/r_0, the system that is optimized for the effective D/r_0 might not be able to close the loop at all. To avoid this in these special situations or to minimize the time to reach the maximum Strehl ratio, sensor design can be adjusted during operation: When starting AO, the optimal parameters for the initial D/r_0 are used and after closing the loop the sensor design is switched to the optimal parameters for the effective D/r_0. The dashed magenta curve in Fig. 12 shows the Strehl ratios when adjusting the parameters after 5 ms (marked by vertical dotted grey line). The Strehl ratio is raised rapidly and

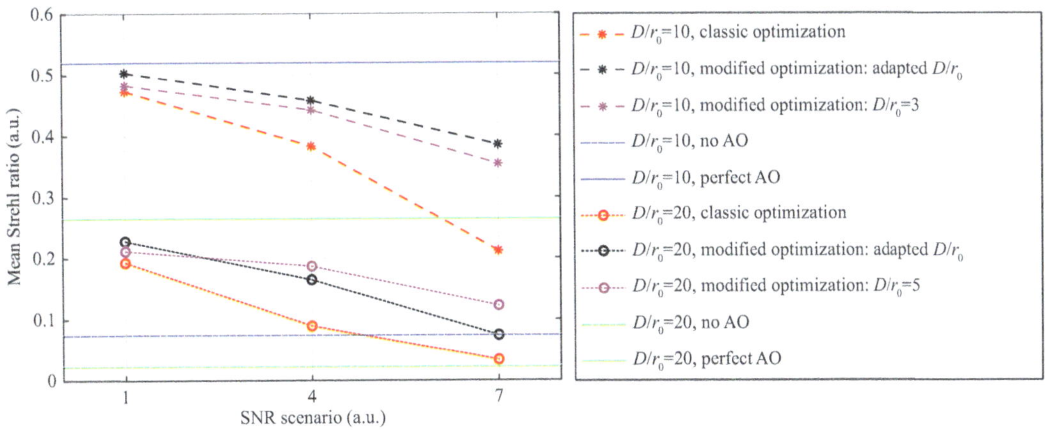

Fig. 11 Mean Strehl ratio for closed-loop AO with Karhunen-Lòeve-based HWFS, AO bandwidth of 1 kHz and AO gain of 0.3 when measuring and correcting 42 modes, shown for different SNR scenarios.

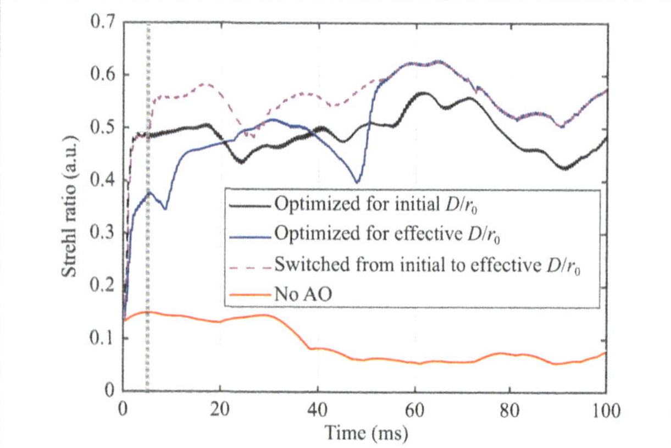

Fig. 12 Strehl ratio over time for closed-loop AO with Zernike-based HWFS, AO bandwidth of 2 kHz and AO gain of 0.5, when measuring and correcting 45 modes. The vertical dotted grey line marks the time when the sensor design is switched from optimized for initial to effective D/r_0 (dashed magenta curve).

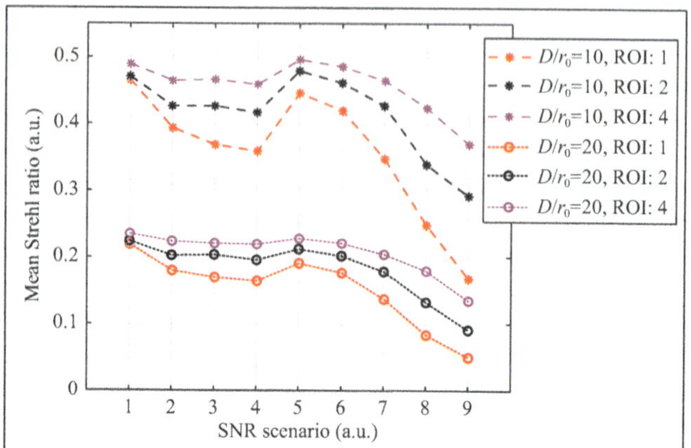

Fig. 13 Mean Strehl ratio for closed-loop AO with optimized Karhunen-Lòeve-based HWFS, AO bandwidth of 1 kHz and AO gain of 0.3 when measuring and correcting 42 modes, shown for different SNR scenarios. Dashed lines represent measurements at D/r_0=10 and dotted lines at D/r_0=20. The colors correspond to different number (1, 2 or 4) of used digital ROI.

further increased to the higher level. The ability to adjust the sensor design during operation improves the stability of the AO system.

Evaluation of Multi-ROI HWFS

Measurement accuracy, especially in noisy scenarios, can be increased by averaging over multiple successively performed measurements or by increasing the exposure time to improve the SNR. It has to be noted though that the increased measurement accuracy is accompanied by a reduction in the possible AO bandwidth. However, this problem can be avoided by designing the HWFS to carry out multiple measurements at the same time rather than one after another. When using an area detector for spot detection, multiple different HWFSs can be implemented without changing the number of spots or the biases for the modes. The spots are detected with different digital ROIs, i.e., for each HWFS the spot intensity is summed over a different number of detector pixels. The detector has to be read-out only once. Afterwards the sensor output can be calculated for several sensors with different digital ROIs and accordingly different calibration factors (Eq. 11). The total number of signal photons and the SNR differ for all measurements so that averaging over all measurements can reduce the influence of noise. The J-map, obtained during the optimization process, can be used to find the best combination of a number r ROIs with the same bias. For each row of the J-map the r smallest values are summed and the row with the smallest sum provides the bias (see Fig. 7b). With the corresponding calibration factors gained from the c-map, the r HWFSs can be used nearly simultaneously.

Fig. 13 shows results of closed-loop AO with a Karhunen-Loève HWFS using 1 (red), 2 (black) and 4 (magenta) digital ROIs. The mean Strehl ratio of the closed-loop operation is given for the 9 SNR scenarios (see Table 1). The AO gain is 0.3, the bandwidth 1 kHz. Results are presented for $D/r_0 = 10$ (dashed lines) and $D/r_0 = 20$ (dotted lines). Multiple-ROI measurements always lead to a higher Strehl ratio. The improvement of the AO performance is significant. For SNR scenario 9, the Strehl ratio is increased by a factor of 1.7 for both turbulence strengths when using only two different ROIs. Even for scenario 1 without noise, the AO performance benefits noticeably when using four ROIs.

Discussion

The simulations show that a closed-loop AO system based on a HWFS is capable of correcting atmospheric turbulence. This is true even when bias and ROI are chosen based on common sense instead of optimization results. However, when using the optimization method that is intended for open-loop operation, the improvement of the Strehl ratio can be increased significantly. As expected, the optimized parameters are a good choice, but not necessarily the best choice for the changing conditions of closed-loop operation, especially for demanding SNR scenarios. Optimizing for the effective turbulence strength, i.e. the residual phase variance after closing the AO loop, has been found to be very effective. However, closing the loop can be problematic if a high number of modes is corrected and the difference between initial turbulence strength and the expected effective turbulence strength is too big. As shown in the previous section, adaption of the parameters to the current conditions increases the stability visibly. Changing bias and ROI during operation is not feasible for all sensor implementations. The holograms multiplexed in a hologram plate are fixed and cannot be changed afterwards. When using an SLM as DOE, the displayed holograms can be exchanged easily during the operation. The number of modes and with this the number of signal photons per spot can be adjusted. If an area detector like a CCD camera is used, the ROI can be changed for each mode at any time. With this combination the HWFS design is extremely flexible and can be adjusted to the turbulence conditions. On the other hand, the measurement speed of such a HWFS is limited by the readout speed of the detector. If the high bandwidth of the sensor is required, a fast photodiode array is usually used for spot detection. In this case, the detector size can be adjusted by means of a variable aperture in front of the photodiodes. Of course, this would result in an increase in complexity of HWFS. Instead of adapting the sensor design from initial to effective turbulence strength, we can choose a design that matches both conditions. By overlaying the J-maps of both optimizations, we can find a combination of bias and ROI that provides satisfactory results for both cases. Comparing J-maps of increasing D/r_0, we see that the combinations with the best results overlap. It is therefore possible to find a good compromise that ensures good stability even for challenging scenarios.

In addition to the modified optimization, the use of multiple ROIs increases the performance of the system noticeable. Nevertheless, this approach is also limited to the use of an area detector. Implementations with photodiodes are conceivable but significantly less practical. Using multiple ROI is particularly effective for low SNR scenarios. However, in these scenarios it is beneficial to avoid the use of many pixels in order to keep the readout noise of the detector as low as possible. When using multiple ROI, pixel binning is only possible to a limited extent. Therefore, the gain from the additional apertures must be weighed against additional noise. The limitations of the detector, such as dynamic range, have to be considered as well when designing a real system. If the detector has enough light available so that noise is not critical, it is always useful to implement multiple ROI detection at almost no additional cost. Especially high-order modes profit from the increased measurement accuracy.

Conclusion

The modal holographic wavefront sensor enables high-speed measurement of aberration modes in parallel. After detecting raw intensities, post-processing is limited to a fast contrast calculation for each mode. However, the measurement of one mode is influenced by the presence of additional modes. Especially when dealing with atmospheric turbulence, where the number of modes is infinite, this crosstalk effect reduces the measurement accuracy of the sensor. To investigate the usability of the sensor as part of a closed-loop AO system, we simulated realistic atmospheric turbulence. Choosing the design parameters according to simple, common-sense guidelines enables a respectable correction of the wavefront distortions. When applying an optimizing procedure that is developed for open-loop operation, we achieve a significantly better correction. However, especially for challenging SNR scenarios, we identify the potential for improvement. We modified the optimization procedure to take the new, residual conditions, i.e., the effective turbulence strengths, after closing the loop into account. This leads to a significant increase in performance and makes the HWFS usable even under difficult conditions. By using multiple HWFSs with the same bias but different digital ROIs simultaneously, we achieve an additional improvement and reduce the influence of noise. In the next step we will verify these results experimentally. We will create the HWFS as multiplexed computer-generated hologram displayed on an SLM and implement it in a closed-loop AO system. To validate the efficiency of the sensor optimization, we will use an 800 m free-space path through the atmosphere and monitor the improvement of the Strehl ratio during closed-loop operation.

Acknowledgements

This work was sponsored by WTD 91 (Technical Center of Weapons and Ammunition) of the Federal Defence Forces of Germany – Bundeswehr in the project ABU-SLS and by the Office of Naval Research Global under award no. N62909-17-1-2037.

Conflict of interest

The authors declare that they have no conflict of interest.

Received: 23 September 2021 Revised: 28 March 2022 Accepted: 01 April 2022
Accepted article preview online: 02 April 2022
Published online: 21 June 2022

References

1. Tyson, R. K. Adaptive Optics Engineering Handbook (New York: Dekker, 1999).
2. Conan, J. M., Rousset, G. & Madec, P. Y. Wave-front temporal spectra in high-resolution imaging through turbulence. *Journal of the Optical Society of America* **12**, 1559-1570 (1995).
3. Platt, B. C. & Shack, R. History and principles of Shack-Hartmann wavefront sensing. *Journal of Refractive Surgery* **17**, S573-S577 (2001).
4. Ragazzoni, R. Pupil plane wavefront sensing with an oscillating prism. *Journal of Modern Optics* **43**, 289-293 (1996).
5. Roddier, F. Curvature sensing and compensation: a new concept in adaptive optics. *Applied Optics* **27**, 1223-1225 (1988).

6. Murty, M. V. R. K. The Use of a Single Plane Parallel Plate as a Lateral Shearing Interferometer with a Visible Gas Laser Source. *Applied Optics* **3**, 531-534 (1964).
7. Wu, C. S., Ko, J. & Davis, C. C. Lossy wavefront sensing and correction of distorted laser beams. *Applied Optics* **59**, 817-824 (2020).
8. Vorontsov, M. A. & Sivokon, V. P. Stochastic parallel-gradient-descent technique for high-resolution wave-front phase-distortion correction. *Journal of the Optical Society of America* **15**, 2745-2758 (1998).
9. Segel, M. & Gladysz, S. Optimal, blind-search modal wavefront correction in atmospheric turbulence part I: simulations. *Optics Express* **29**, 805-820 (2021).
10. Solovev, M. A. & Venediktov, V. Y. Holographic wavefront sensors: state of the art and prospects. Proceedings of SPIE 9508, Holography: Advances and Modern Trends IV. Prague, Czech Republic: SPIE 2015.
11. Gabor, D. Holography - the reconstruction of wavefronts. *Electronics & Power* **12**, 230-234 (1966).
12. Zepp, A., Gładysz, S. & Stein, K. Holographic wavefront sensor for fast defocus measurement. *Advanced Optical Technologies* **2**, 433-437 (2013).
13. Palomo, P. M., Zepp, A. & Gladysz, S. Characterization of the digital holographic wavefront sensor. Proceedings of SPIE 9242, Remote Sensing of Clouds and the Atmosphere XIX; and Optics in Atmospheric Propagation and Adaptive Systems XVII. Amsterdam, Netherlands: SPIE, 2014.
14. Andersen, G., Ghebremichael, F., Gaddipati, R. & Gaddipati, P. Fast computer-free holographic adaptive optics. Proceedings of SPIE 8447, Adaptive Optics Systems III. Amsterdam, Netherlands: SPIE, 2012.
15. Ghebremichael, F., Andersen, G. P. & Gurley, K. S. Holography-based wavefront sensing. *Applied Optics* **47**, A62-A69 (2008).
16. Kovalev, M. S. et al. Measurement of wavefront curvature using computer-generated holograms. *Optics Express* **27**, 1563-1568 (2019).
17. Lugt, A. V. Signal Detection By Complex Spatial Filtering. *IEEE Transactions on Information Theory* **10**, 139-145 (1964).
18. Gorelaya, A., Orlov, V. & Venediktov, V. Holographic wavefront sensor, based on diffuse Fourier holography. Proceedings of SPIE 10425, Optics in Atmospheric Propagation and Adaptive Systems XX. Warsaw, Poland: SPIE, 2017.
19. Dong, S. et al. Response analysis and experimental results of holography-based modal Zernike wavefront sensor. Proceedings of SPIE 8165, Unconventional Imaging, Wavefront Sensing, and Adaptive Coded Aperture Imaging and Non-Imaging Sensor Systems. San Diego, California, United States: SPIE, 2011.
20. Neil, M. A. A., Booth, M. J. & Wilson, T. New modal wave-front sensor: a theoretical analysis. *Journal of the Optical Society of America* **17**, 1098-1107 (2000).
21. Yao, K. N. et al. Closed-loop adaptive optics system with a single liquid crystal spatial light modulator. *Optics Express* **22**, 17216-17226 (2014).
22. Kong, F. P. & Lambert, A. Improvements to the modal holographic wavefront sensor. *Applied Optics* **55**, 3615-3625 (2016).
23. Konwar, S. & Boruah, B. R. Improved linear response in a modal wavefront sensor. *Journal of the Optical Society of America* **36**, 741-750 (2019).
24. Dong, S. H., Haist, T. & Osten, W. Hybrid wavefront sensor for the fast detection of wavefront disturbances. *Applied Optics* **51**, 6268-6274 (2012).
25. Dong, S. H. et al. Hybrid curvature and modal wavefront sensor. Proceedings of SPIE 9227, Unconventional Imaging and Wavefront Sensing 2014. San Diego, California, United States: SPIE, 2014.
26. Neil, M. A. A., Booth, M. J. & Wilson, T. Closed-loop aberration correction by use of a modal Zernike wave-front sensor. *Optics Letters* **25**, 1083-1085 (2000).
27. Andersen, G. et al. Fast, compact, autonomous holographic adaptive optics. *Optics Express* **22**, 9432-9441 (2014).
28. Yao, K. N. et al. Analysis of a holographic laser adaptive optics system using a deformable mirror. *Applied Optics* **56**, 6639-6648 (2017).
29. Zepp, A. et al. Optimization of the holographic wavefront sensor for open-loop adaptive optics under realistic turbulence. Part I: simulations. *Applied Optics* **60**, F88-F98 (2021).
30. Ribak, E. & Ebstein, S. A fast modal wave-front sensor. *Optics Express* **9**, 152-157 (2001).
31. Booth, M. J. Direct measurement of Zernike aberration modes with a modal wavefront sensor. Proceedings of SPIE 5162, Advanced Wavefront Control: Methods, Devices, and Applications. San Diego, California, United States: SPIE, 2003.
32. Liu, C. H. et al. Performance analysis of multiplexed phase computer-generated hologram for modal wavefront sensing. *Applied Optics* **50**, 1631-1639 (2011).
33. Noll, R. J. Zernike polynomials and atmospheric turbulence. *Journal of the Optical Society of America* **66**, 207-211 (1976).
34. Roddier, F. Adaptive Optics in Astronomy. (Cambridge: Cambridge University Press, 1999).
35. Schottky, W. Small-shot effect and flicker effect. *Physical Review* **28**, 74-103 (1926).
36. Dai, G. M. Modal wave-front reconstruction with Zernike polynomials and Karhunen–Loève functions. *Journal of the Optical Society of America* **13**, 1218-1225 (1996).
37. Dai, G. M. Modal compensation of atmospheric turbulence with the use of Zernike polynomials and Karhunen–Loève functions. *Journal of the Optical Society of America* **12**, 2182-2193 (1995).
38. Anzuola, E., Zepp, A., Gladysz, S. & Stein, K. Holographic wavefront sensor based on Karhunen-Loève decomposition. Proceedings of SPIE 9979, Laser Communication and Propagation through the Atmosphere and Ocean V. San Diego, California, United States: SPIE, 2016.
39. Scharmer, G. B. et al. Adaptive optics system for the new Swedish solar telescope. Proceedings of SPIE 4853, Innovative Telescopes and Instrumentation for Solar Astrophysics. Waikoloa, Hawai' i, United States: SPIE, 2003.
40. Banet, M. T. & Spencer, M. F. Compensated-beacon adaptive optics using least-squares phase reconstruction. *Optics Express* **28**, 36902-36914 (2020).
41. Kulcsár, C. et al. Optimal control, observers and integrators in adaptive optics. *Optics Express* **14**, 7464-7476 (2006).
42. Roddier, F., Gilli, J. M. & Lund, G. On the origin of speckle boiling and its effects in stellar speckle interferometry. *Journal of Optics* **13**, 263-271 (1982).
43. Montoya, L. et al. Modeling day time turbulence profiles: Application to teide observatory. Proceeding of the Adaptive Optics for Extremely Large Telescopes 5. 2017.
44. Zhang, X. F. & Wang, L. Q. Improvement in the performance of solar adaptive optics. *Research in Astronomy and Astrophysics* **14**, 471-484 (2014).
45. García-Lorenzo, B. et al. Astronomical site ranking based on tropospheric wind statistics. *Monthly Notices of the Royal Astronomical Society* **356**, 849-858 (2005).
46. McGlamery, B. L. Computer simulation studies of compensation of turbulence degraded images. Proceedings of SPIE 0074, Image Processing. Pacific Grove, United States: SPIE, 1976.
47. Iftekharuddin, K. M. & Awwal, A. Field Guide to Image Processing (Bellingham: SPIE Press, 2012).
48. Roberts, Jr., L. C. et al. Is that really your Strehl ratio?. Proceedings of SPIE 5490, Advancements in Adaptive Optics. Glasgow, United Kingdom: SPIE, 2004.
49. Roddier, N. A. Atmospheric wavefront simulation using Zernike polynomials. *Optical Engineering* **29**, 1174-1180 (1990).
50. Tyson, R. K. Principles of adaptive optics. 4th edn. (Boca Raton: CRC Press, 2015).
51. Levine, B. M. et al. Horizontal line-of-sight turbulence over near-ground paths and implications for adaptive optics corrections in laser communications. *Applied Optics* **37**, 4553-4560 (1998).

Part IV

Wide Field of Applications

Official journal of the JHL 2689-9620
www.light-am.com

Review Open Access

A review of the dual-wavelength technique for phase imaging and 3D topography

Haowen Zhou[1,2], Mallik M. R. Hussain[3] and Partha P. Banerjee[1,*]

Abstract

Optically transmissive and reflective objects may have varying surface profiles, which translate to arbitrary phase profiles for light either transmitted through or reflected from the object. For high-throughput applications, resolving arbitrary phases and absolute heights is a key problem. To extend the ability of measuring absolute phase jumps in existing 3D imaging techniques, the dual-wavelength concept, proposed in late 1800s, has been developed in the last few decades. By adopting an extra wavelength in measurements, a synthetic wavelength, usually larger than each of the single wavelengths, can be simulated to extract large phases or height variations from micron-level to tens of centimeters scale. We review a brief history of the developments in the dual-wavelength technique and present the methodology of this technique for using the phase difference and/or the phase sum. Various applications of the dual-wavelength technique are discussed, including height feature extraction from micron scale to centimeter scale in holography and interferometry, single-shot dual-wavelength digital holography for high-speed imaging, nanometer height resolution with fringe subdivision method, and applications in other novel phase imaging techniques and optical modalities. The noise sources for dual-wavelength techniques for phase imaging and 3D topography are discussed, and potential ways to reduce or remove the noise are mentioned.

Keywords: Topography, 3D imaging, Phase unwrapping, Dual wavelength

Introduction

An essential problem in interferometry and holography is that when the distances to be measured are larger than the source or illumination wavelength, interferometry and holography may fail to unwrap the correct phase information. This problem arises from the failure in unambiguously determining the interferometric orders with single-wavelength measurement. A key to solve this problem is to employ an extra wavelength to generate a synthetic wavelength in the measurement. This is often referred to as the dual-wavelength technique (DWT) or two-wavelength technique. The concept of utilizing different wavelengths to extend the absolute distance/phase measurement range of using a single wavelength was introduced by Michelson and Benoit in the 1890s[1-3]. They measured a 1 cm etalon with a Michelson interferometer and four principal colors in cadmium emissions (red, green, blue and violet, with wavelengths of 643.85, 508.58, 479.99 and 467.82 nm, respectively). The excess fractions of interferometric fringes for each wavelength were then calculated based on the known wavelength. This method is also named as the method of excess fractions. The same method, after four decades, found its application in standard measurement for gauge blocks using cadmium

Correspondence: Partha P. Banerjee (pbanerjee1@udayton.edu)
[1]Department of Electro-Optics and Photonics, University of Dayton, Dayton, OH 45469, USA
[2]Department of Electrical Engineering, California Institute of Technology, Pasadena, CA, 91125, USA
Full list of author information is available at the end of the article.

light source in the interferometer[4,5].

In 1947, Forrester et al. discussed the possibility of experimentally using two nearly the same visible wavelengths to create a beat wavelength that is larger than original ones[6]. The concept of using two wavelengths in holography was first introduced by Wyant[7] as a convenient alternative to test aspheric optical elements. Interferometry is generally the preferred method, because the complete surface is covered in a single measurement, unlike a scanning probe, which measures the contour a cross-section at a time. The major problem in using interferometry for testing a largely deformed aspheric is the large number of fringes in the resulting interferogram. A method of reducing the number of fringes in the interferogram would be to use a longer wavelength light source in the interferometer. However, there were three main disadvantages in using a longer wavelength, nonvisible light source: (a) ordinary refractive elements cannot be tested this way, (b) standard photographic films could not be used to record the interferogram directly, and (c) not being able to see the radiation could cause added experimental difficulty. These problems could be solved by using DWT in holography. DWT together with holography provides a means of using only visible light to obtain an interferogram identical to the one that would be obtained if a longer wavelength were used. Due to the large number of different wavelengths obtainable from commercially available lasers, a wide range of equivalent wavelengths could be obtained using DWT. Around the 1970s and 1980s, this dual-wavelength concept has also been successfully applied to interferometry[8-12] and holography[13,14].

In recent three decades, DWT has been developed in various directions. Dispersion management can be an annoying issue in dual-wavelength or white-light interferometry. Ideally, the geometric paths in two arms of the interferometer should be the same, however, the imperfection of beam splitters or dispersion nature of the material may introduce unwanted phase accumulations in the interferometer. In other words, the wavelength in the medium which may differ from the laser wavelength can lead to errors in the topography results. Pfortner and Schwider developed a numerical algorithm to compensate the dispersion errors in white-light interferometry[15]. Another common phenomenon when dealing with different wavelengths is chromatic aberrations. The methods of eliminating chromatic aberrations in DWT have been reported in Refs.[16-18]. One way to eliminate the chromatic aberration is to convert the aberrations into scaling effect in lateral plane. Nicola et al.[16] managed the chromatic aberrations elimination by rescaling the pixel size of digital holograms at reconstruction plane. Another way is by using digital refocusing in dual-wavelength digital holography (DWDH). Since digital holography (DH) can reconstruct the complex wavefront of the object or sample, the wavefront at in-focus or out-of-focus planes can be obtained easily by numerically propagating the wavefront to the target plane. Ferraro et al.[18] employed this idea in chromatic aberration compensation. By digital refocusing the out-of-focus optical field (the defocus is introduced by chromatic aberration in using different wavelengths), the optical field can be brought back to the right focal plane. Retrace errors, coming from the mismatch between test ray and reference ray, is an issue in aspherical testing[19]. Iterative algorithms can be adopted to remove the retrace errors in dual-wavelength aspherical testing[20,21]. Researchers also tried to go beyond the phase ambiguity limit of the synthetic wavelength in DWT. De Groot reported a method to extend the unambiguous range measurement using DWT in interferometers[22]. The method first generates a beat fringe pattern and then compares the beat pattern with one of the single wavelength fringe patterns. When the two patterns share a larger period, an extended synthetic wavelength can be obtained, which can be numerical estimated and give rise to a longer absolute range measurement. Further improvements have been done by Falaggis et al.[23,24]. Also, they developed methods in more general techniques in multi-wavelength modalities[23-25]. Other developments and improvements in DWT was benefited by the advancement of laser technology. Tunable laser sources together with nonlinear optics[26-31] give DWT more degrees of freedom in wavelength choice and free the researchers from aligning an extra laser beam in the optical system. The stability of wavelength is also a nontrivial issue in high precision optical metrology. Monitoring and controlling the wavelength drifts can enhance the performance of DWT[32-35]. Super-sensitive DWT with the sum of the two phase maps has been proposed[36-38], however, unfortunately, ignored by researchers over a few decades. Recently, super-sensitive DWT has drawn researchers' attention and developments have been reported in Refs.[39-43]. The super-sensitive DWT will be further discussed in later sections.

A brief review on the historical development of DWT described above shows the technique's diverse advancements, especially in interferometric techniques. In fact, the DWT is not limited to interferometry and holography. It can also be applied to incoherent techniques, such as fringe projection profilometry (FPP)[41, 43-48] or deflectometry[49,50]. Abundant DWT technique development and implementation has been studied in FPP and a few good review papers have been published[41,44-47]. Our paper will put less efforts on reviewing the history and early applications of DWT in FPP, but we will still review a recent work on combining super-sensitive DWT with FPP, which has not been included in the review papers mentioned above. Need to mention that FPP can be considered as a category of methods in structured light field[51–54]. There are also techniques sharing similar concepts with DWT. For instance, the Moiré pattern, which is a beat pattern formed by the superposition of two fringe patterns with different spatial frequencies, can extract phase information of the object with optical projection and numerical reconstructions[55–57]. The other category is non-interferometric coherent imaging techniques, e.g. coherent diffractive imaging (CDI)[58], ptychography[59], Fourier ptychography[60]. DWT has potential in implementing in these imaging methods to boost their performance in 3D topography and phase imaging. The related applications of DWT in coherent imaging techniques and novel optical imaging modalities will also be covered in the selected applications of DWT section of this paper.

Throughout the literature search, we notice that DWT appears as different expressions in research article titles, though the mathematical backgrounds are essentially the same. This sets an obstacle for students or new researchers in this field to read through previous research work or search related papers with keywords. Here we list an alternative keywords involving the DWT to help new researchers and students in this field or intended to work on this technique. The straightforward technical equivalent terms for DWT are two-wavelength holography/interferometry/FPP[7,8,10,12,61,62], variable wavelength interferometry[9], multiple-wavelength holography/interferometry[11,13,29,31], two-color interferometry[32], dual-wavelength holography[34,40,63–70], two-wavelength phase-metrology[43], dual/two-wavelength phase unwrapping[71,72], multiwavelength holography[17,18,73–76], synthetic wavelength holography/interferometry[35,77–84], two/dual-frequency FPP[85–87]. An important statement should be made here that actually the concept of multi-wavelength techniques is different from DWT. Multi-wavelength technique (MWT) employs a set of wavelengths to form a very large synthetic wavelength, which can achieve phase unwrapping free feature[23–25]. Some papers titled with "multi-wavelength" related techniques, however, their number of wavelengths in use is two, which still fall in the scope of DWT. MWT and DWT are closely related with similar mathematical framework, but different implementations. Though the review of MWT is not in the scope of this paper, MWT developments deserve researchers' attention. A few references are cited here for some research advances and reviews on MWT[23–25, 41, 88–100]. The other category of keywords for DWT is related to the feature or applications of DWT: unambiguous range measurement, absolute distance/height measurement, extended range metrology, large step-height/displacement measurement, extended range contouring[9,22,29,34,35,62,101]; nonnull aspherical testing, gauge block measurement, the method of excesses fractions[4,5,7,19–21,96]. All the keywords mentioned above are in the mathematical framework using the phase difference from the phase maps of two wavelengths. This has been extensively studied over the last five decades. The other mathematical framework is to use the sum and the difference of the two phase maps. This methodology can also be considered as DWT and has been reported in 1970s[36–38]. Unfortunately, it did not draw much attention over a few decades. Recently, this method has found its way in super-sensitive measurement applications[39–43].

In this review, we present the principle of DWT with an example in DH and introduce the usage of phase sum in DWT to achieve high sensitivity. Examples of different scales of synthetic wavelengths are discussed with various experimental realizations and applications. DWT in combination of novel coherent imaging techniques is summarized. Then, the performance of DWT and noise sources are analyzed. In the end, we briefly conclude this article and point out the potential future directions of DWT.

Principle of DWT

Here, we introduce the principle of DWT with an example in holographic imaging. Consider an interference between a plane wave and a point source, which is the simplest in-line hologram. The intensity pattern is

$$I_0(x,y:z) = I_a + I_b cos\left(k_0 \frac{x^2+y^2}{2z}\right) = I_a + I_b cos\left(\frac{\pi(x^2+y^2)}{\lambda_0 z}\right) \quad (1)$$

where z is the distance from the point source to the measurement plane, and $k_0 = 2\pi/\lambda_0$ is the propagation constant with λ_0 being the wavelength. Assume now, that there are two wavelengths $\lambda_{1,2}$ making this interference pattern, then the respective intensity patterns are

$$I_{1,2}(x,y:z) = I_a + I_b cos\left(\frac{k_{1,2}(x^2+y^2)}{2z}\right) \quad (2)$$

where $k_{1,2} = 2\pi/\lambda_{1,2}$ and the constants I_a, I_b have been left unchanged for the two wavelengths for simplicity. If these two intensity patterns are added, the sum of the cosines yields

$$cos\left(\frac{k_1(x^2+y^2)}{2z}\right) + cos\left(\frac{k_2(x^2+y^2)}{2z}\right) = 2cos\left(\frac{(k_1+k_2)(x^2+y^2)}{2z}\right) cos\left(\frac{(k_1-k_2)(x^2+y^2)}{2z}\right) \quad (3)$$

The addition and subtraction of the propagation constants can be recast in the form $k_1 \pm k_2 = 2\pi(\lambda_2 \pm \lambda_1)/(\lambda_1\lambda_2)$. Of relevance here is the second term, expressible as $cos(\pi(x^2+y^2)/(\Lambda z))$, where

$$\Lambda = \frac{\lambda_1\lambda_2}{\lambda_2 - \lambda_1} \quad (4)$$

In other words, it is as if the interference pattern is now being recorded by a wavelength which is neither $\lambda_1$nor λ_2, but a synthetic or artificially created wavelength Λ, which, typically, is expected to be much larger than $\lambda_{1,2}$. The resulting advantage is that there are far less number of Newton's rings.

In the first demonstration of dual-wavelength holography (DWH), the fringe pattern obtained from light passing through an optical element such as a lens (i.e., a *phase object*) under test, using a wavelength λ_1 in an interferometer such as the Mach-Zehnder type as shown in Fig. 1, was photographed. This photographic recording of the fringe pattern (hologram) was then developed and replaced in the interferometer in the exact position it occupied during exposure, and it was illuminated with the fringe pattern obtained by testing the optical element using a different wavelength λ_2. The moiré pattern obtained is identical to the interferogram that would have been obtained if the optical element were tested using a synthetic wavelength Λ as defined in Eq. 4. A simple way to see this is to note that if the first hologram is "read" using the interfering waves from the second wavelength, note that $cos(k_1(x^2+y^2)/(2z))$ multiplied by the diverging wave from the point source $exp(-jk_2(x^2+y^2)/(2z))$ contains a term $exp(j(k_1-k_2)(x^2+y^2)/(2z))$, where j denotes the square root of -1 in engineering convention. The intensity pattern comprising the interferogram contains $cos((k_1-k_2)(x^2+y^2)/(2z))$, which is similar to the second term on the RHS of Eq. 3, and therefore can be expressed in terms of the synthetic wavelength given in Eq. 4.

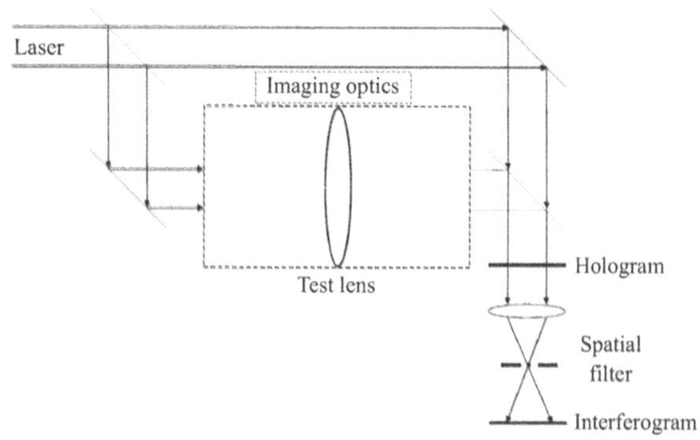

Fig. 1 Mach-Zehnder configuration for hologram recording using wavelength λ_1, followed by reconstruction using wavelength λ_2, yielding the interferogram or moire pattern generated using the synthetic wavelength Λ. This schematic is adapted from the original figure in Wyant[7].

DWH involves finding the difference between a fringe pattern recorded at one instant of time and a fringe pattern existing at some later instant of time. If the two fringe patterns are different for reasons other than wavelength change, e.g., air turbulence, incorrect results result. The effect of air turbulence can be reduced by recording the two interferograms resulting from the two wavelengths simultaneously. When this interferogram (hologram) is illuminated with a plane wave, spatially filtered, and reimaged, one obtains an interferogram identical to that obtained using the first method of DWH described above. Since both fringe patterns are recorded simultaneously, and dispersion is small, air turbulence is essentially the same as if a long wavelength light source were used in the interferometer.

With the advent of digital recording, DWH could be modified so that two holograms are either simultaneously or sequentially recorded using wavelengths $\lambda_{1,2}$, followed by numerically processing the phase data after reconstruction. The methodology of DH is slightly different from conventional holography. While in analog holography the interference pattern is recorded on a film and developed, in DH the intensity is recorded in a digital camera sensor. The hologram is numerically reconstructed using the laws of Fresnel propagation (or back-propagation), depending on whether the virtual or real image is being reconstructed.

Consider the hologram of a point source (see Eq. 1), which is slightly modified to take into account any angle that may be introduced between the reference and the object:

$$h(x,y) \propto I_0(x,y:z) = I_a + I_b cos\left(k_0\frac{x^2+y^2}{2z}\right)cos(Kx) \quad (5)$$

where $K = k_0 sin\theta$, with θ representing the angle in the $x-z$ plane between the reference and the object waves. If this is "illuminated" by a plane wave and back propagated, the optical field after a distance z of back-propagation can be expressed as

$$\Gamma(x,y) = h(x,y) * g(x,y;-z) \propto h(x,y) * \exp\left(+jk_0\frac{x^2+y^2}{2z}\right) \quad (6)$$

where $g(x,y;z)$ is the impulse response for propagation, * denotes the convolution operator. This convolution can be computed with a single Fourier transform method:

$$\Gamma(x,y) = \iint_{-\infty}^{\infty} h(x',y') exp\left(+jk_0 \frac{(x-x')^2+(y-y')^2}{2z}\right) dx'dy'$$
$$\propto \iint_{-\infty}^{\infty} h(x',y') exp\left(jk_0 \frac{x'^2+y'^2}{2z}\right)$$
$$exp\left(j\left(\frac{k_0}{z}\right)(xx'+yy')\right) dx'dy'$$
$$= F_{x,y}\left[h(x',y') exp\left(jk_0 \frac{x'^2+y'^2}{2z}\right)\right]\Bigg|_{k_{x,y}=k_0(x,y)/z} \quad (7)$$

In Eq. 7, $F_{x,y}$ is the Fourier transform operator. For the simple hologram function stated above, it can be shown that after propagating a distance z, one of the terms in $h(x,y)$ will yield the (virtual) image of the point source, with the other terms corresponding to the dc (zeroth order) and the defocused (real or twin) image of the object. If there is an angle between the reference and the object while recording the hologram, the different components will be spatially separated. Alternatively, $h(x,y)$ can be Fourier transformed, filtered around the spatial frequency K, and centered before back-propagation; this removes the dc (zeroth-order) and the twin-image components.

From the reconstructed complex field $\Gamma(x,y)$ in Eq. 7, the intensity and phase of the object can be obtained using

$$I(x,y) = |\Gamma(x,y)|^2 \quad (8)$$

$$\phi(x,y) = arctan\left(\frac{Im[\Gamma(x,y)]}{Re[\Gamma(x,y)]}\right) \quad (9)$$

The phase $\phi(x,y)$ is proportional to the object surface feature height and the illumination angle θ, theoretically, $0 \le \theta < 90^{\circ}$[102,103]. The reconstructed phase in Eq. 9 is a 2D map of the wrapped phase, exhibiting modulo 2π fringe spacing of the wavelength λ, which must be unwrapped to yield the absolute phase. The resulting unwrapped phase is the phase profile for a phase object, and plotted in either 2D or 3D. A commonly used unwrapping technique is the phase-unwrapping max-flow/min-cut (PUMA) algorithm, based on the graph-cut technique developed by Bioucas-Dias and Valadao[104]. The PUMA algorithm is commonly chosen for its ready availability and good performance in the presence of significant phase noise.

Just as light passing through a phase object acquires additional phase, light reflected off a contoured object also acquires phase, and can be similarly studied, as shown schematically in Fig. 2. Holographic topography is the measurement of surface shape by means of holographic recording and reconstruction. When recorded digitally, the phase contours may be post-processed to yield the object height up to 1/100th of the synthetic wavelength[105]. Digital holographic topography is a more general 2D case of holographic profilometry, where only a cross-section of a given topography can be extracted, and avoids the need for scanning.

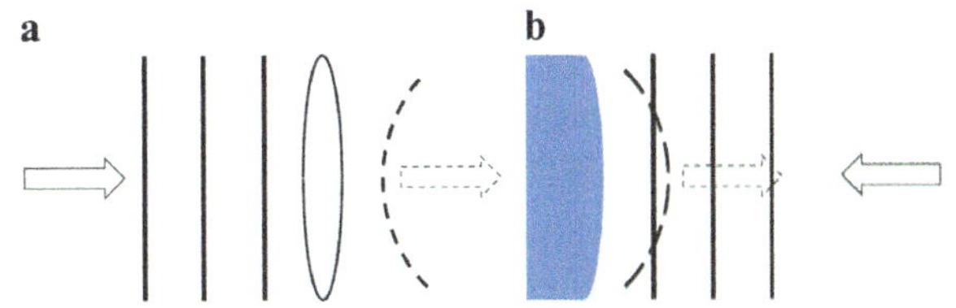

Fig. 2 Schematic showing that light transmitted through a phase object as in **a** or reflected from a surface as in **b** acquire phase profiles, characteristic of the refractive index profile or shape of the phase object as in **a** or the topography of the surface as in **b**, respectively.

DWDH has been explored extensively to quantify surface topography and displacement measurements for both fixed and time-varying objects[30,63,74,102,103,105]. By choosing different wavelengths, unambiguous phase unwrapping can be achieved to resolve different scales of depth jumps. Typically, the difference between the two wavelengths, $\Delta\lambda = |\lambda_2 - \lambda_1|$, is of the order of $10-50\,nm$, which yields synthetic wavelengths Λ of approximately $30-10\,\mu m$, respectively. Under the assumption that topographic resolution is of the order of $1/100$ of Λ, this allows feature heights of the order of $1-3\,\mu m$ to be resolved reasonably well[63]. However, considerable work has also been performed using much longer synthetic wavelengths to measure *mm*- or even *cm*-scale features[106]. Proper choice of λ_1, λ_2, such that λ_1 λ_2, yields very long synthetic wavelengths. For instance, using acousto-optics and RF frequency of $\sim 1\,GHz$, synthetic wavelengths in the order of tens of cm can be achieved using a 514 nm laser source[78].

For large height variations, typically tens to hundreds of wavelengths, the height variation corresponds to several multiples of 2π of phase. Due to the *arctan* function used in the phase reconstruction process (Eq. 9), the phase information is wrapped in the modulo of 2π, which creates a 2π-ambiguity when unwrapping the phase. This limits the maximum absolute distance measurement along axial direction within $\lambda/2n$, where λ is the wavelength and n is the refractive index. Large height variation situations can be improved by using the DWT by carefully selecting two different wavelengths, using them to record and reconstruct holograms separately, and generating a synthetic wavelength to reconstruct the combined hologram. The synthetic wavelength is usually much larger than the optical wavelength, thus, it can resolve larger height variations. The single wavelength technique (SWT) is not an ideal solution to very small height variations[105,107]. For small variations, use of the DWT together with fringe subdivision, nanometer scale axial resolution is possible[64].

The essence of DWDH is to have two wavelengths accumulating different phases in a given distance and generate the beat pattern from phase difference. The principles governing DWDH and synthetic wavelength generation[73,105,107] are briefly summarized here. DWDH requires two holograms to be recorded at two separate wavelengths, $\lambda_{1,2}$, and the phase difference to be subsequently computed. For λ_1, the phase difference of two adjacent fringes can be expressed as $\phi_1 = (2\pi/\lambda_1)\Delta d$, where Δd is axial height separation. Need to mention that the illumination angle and geometry of interferometer have been taken into consideration in Δd. Similarly, for λ_2, $\phi_2 = (2\pi/\lambda_2)\Delta d$. The phase difference can be expressed as the height variation multiplied by a synthetic wavenumber K corresponding to a synthetic wavelength Λ:

$$\phi_1 - \phi_2 = \frac{2\pi}{\frac{\lambda_1\lambda_2}{\lambda_1 - \lambda_2}}\Delta d = \frac{2\pi}{\Lambda}\Delta d \qquad (10)$$

Thus, the synthetic wavelength originates from the phase accumulation of coherent light wave, which is dependent on the wavelength or the longitudinal spatial frequency of the light source. With a given height jump, the phase accumulated by different wavelengths are different. The subtraction of the two phases yields a beat frequency or equivalently a beat (or synthetic) wavelength that can extend the axial measurement range.

Typically, two holograms for two different wavelengths are required to be captured sequentially or by two cameras. The two holograms are reconstructed separately to get two complex fields of the object at the reconstruction plane. Note that due to different wavelengths used in the holograms, a pixel mismatch problem exists. To calculate the phase difference map for synthetic requires a pixel-to-pixel subtraction. The mismatch comes from the wavelength dependence of the pixel size in the reconstruction plane[103],

$$\Delta x_{r_i} = \frac{\lambda_i \Delta d}{N\Delta x} i = 1,2 \qquad (11)$$

where Δx_{r_i} is the reconstruction pixel size, N is the number of the pixels along x-axis, Δx is the pixel pitch of the camera, i is the index of the wavelengths. The reconstruction pixel size is linearly related to the wavelengths. In Eq. 11, the height Δd and the pixel pitch cannot be changed for a fixed system. To achieve pixel mismatch correction, in other words, $\Delta x_{r_1} = \Delta x_{r_2}$; the only way is to increase or decrease the number of pixels. A typical way to match the pixels is zero-padding the longer wavelength hologram before reconstruction; and after reconstruction, zero-padding the shorter wavelength hologram or cropping the longer wavelength hologram[16,78]. The padsize is the number of zero elements to be added symmetrically to each edge of the hologram, rounded to the nearest integer value, and is given by

$$\text{padsize} = round\left[\frac{N}{2}\left(\frac{\lambda_1}{\lambda_2} - 1\right)\right] \qquad (12)$$

assuming $\lambda_1 > \lambda_2$. After pixel matching, the two phase maps are pixel-wise subtracted according to the recipe:

$$\Delta\Phi = \begin{cases} \phi_1 - \phi_2 & if \quad \phi_1 > \phi_2 \\ \phi_1 - \phi_2 + 2\pi & if \quad \phi_1 \le \phi_2 \end{cases} \qquad (13)$$

The phase difference map $\Delta\Phi$ is equivalent to the phase reconstructed from a hologram recorded by the synthetic wavelength. Then upon phase unwrapping and phase-to-height conversion ($\Delta d = \Delta\Phi/2\pi\Lambda$), the 3D topography of the surface can be deduced.

The principle shown above is simple and straightforward. This is to give readers, especially for those who beginning research in this field, a general idea of how to implement DWT in an application. Note that the zero padding is needed to match the scaling effect from different wavelengths or for chromatic aberration compensation. There are also many other techniques or methods in DWDH reconstruction, such as transfer function approach[102], Shen-Wang convolution method[108], chirp Fourier transform[109], etc. These methods still need "zero-padding" due to sampling requirement of fast Fourier transform algorithm. However, this is different from the zero-padding mentioned in Eq. 12. Also, the holographic imaging is only one of the methods which can implement DWT. These methods including interferometry, CDI, ptychography, Fourier ptychography, FPP, will be introduced in the next section of this paper.

It is worth mentioning that phase-shifting techniques are also commonly adopted in DWDH[110,111]. By modulating the reference path with several constant phase shifts, a phase-shifting algorithm can be applied to retrieve the phases of static objects and remove the twin image, zeroth-order components, with trade-off in extra captures of holograms or interferograms.

The other direction of DWT is to use the sum of two phase maps. The sum of the phase map, similarly can generate a wavelength $\Lambda_+ = \lambda_1\lambda_2/(\lambda_1 + \lambda_2)$ smaller than each of the wavelengths. Noted that when $\lambda_1 \approx \lambda_2$, the $\Lambda_+ \approx \lambda_1/2$. Together with the synthetic wavelengths generated from the phase difference (Eq. 4), Λ_+ can be used to retrieve the phase with high accuracy or low noise performance, in the meantime, still can achieve the extended height/phase measurement[43]. To retrieve the phase, a two-sensitivity temporal phase-unwrapping technique has been developed[112]. The essence of the algorithm lies in the formula:

$$\tilde{\phi}_S = G\tilde{\phi}_D + W\left[\phi_S - G\tilde{\phi}_D\right] \tag{14}$$

where $\left\{\phi_D, \phi_S - G\tilde{\phi}_D\right\} \in (-\pi, \pi]$; $\tilde{\phi}_S$ and $\tilde{\phi}_D$ are unwrapped phases of phase sum and phase difference wavelengths (Λ_+, Λ), respectively; $G = \Lambda/\Lambda_+$; ϕ_S is the wrapped phase of the phase sum wavelength; $W[x] = \arg\{e^{jx}\}$ is a wrapping function. From Eq. 14, the unwrapped phase can be obtained. Need to mention that even though the technique is named as a phase unwrapping technique, the phase map resulting from Eq. 14 may still need extra phase unwrapping step, when the phase/height jump is larger than the synthetic wavelength Λ. Despite this, the two-sensitivity temporal phase-unwrapping technique can possibly achieve a better signal-to-noise ratio (SNR) than the traditional phase difference mathematical framework. There is a SNR gain of G^2 between phase sum and phase difference[43].

Selected applications in DWT

The abundant choice of wavelengths in the visible range from laser sources has facilitated a wide range of synthetic wavelengths. In 1984, Cheng and Wyant used Argon-ion and He-Ne lasers to create synthetic wavelengths from 6.45 to 28.5 μm in phase shifting interferometry[12]. Abeywickrema et al. applied dual-wavelength technique to the 3D topography for fingerprints[75]. Fingerprints were preserved by depositing 1 000-nm-thick columnar thin films (CTFs) of chalcogenide glass of nominal composition $Ge_{28}Sb_{12}Se_{60}$. Argon ion laser with wavelengths 514.5 *nm* and 457.9 *nm* are used to generate a synthetic wavelength of 4.16 μm. With this synthetic wavelength the level-3 feature (pores on the ridges of the fingerprints) can be resolved with a depth around 1-2 μm, as shown in Fig. 3. This work reported both transmission and reflection mode setups. The transmission mode is achieved by a Mach-Zehnder interferometer; while the reflection mode is designed with a Michelson-type interferometer[75]. Gass et al. used a He-Ne laser and a ND:YAG laser with wavelengths of 632.8 *nm* and 532 *nm*, respectively[113]. The 3.3 μm synthetic wavelength is used to reconstruct the 3D topography of reflective resolution targets and spherical mirrors. The same wavelengths from a He-Ne laser and a diode-pumped solid state laser are adopted in common-path digital holographic microscopy[65]. The wrapped phase map corresponding to the synthetic wavelength has sparser fringes than single wavelength case for ablated pits and grooves specimens (Fig. 4). This can help reduce the computational efforts in the phase unwrapping process. The above examples discussed are using dual wavelengths to generate a micron scale synthetic wavelength.

For micron scale synthetic wavelengths, the two wavelengths can be easily obtained from different lasers. For a millimeter scale or even centimeter scale synthetic wavelength, a narrow spacing between different wavelengths is required. A common way is to use the dual-longitudinal-mode He-Ne laser[64,83,114] around 632.8 *nm* with a 1 *GHz* spectrum separation, which yields a synthetic wavelength around 30 *cm*. By using tunable lasers or acoustic-optic methods[29,74,77,79], similar synthetic wavelengths can be achieved, suitable for 3D imaging of

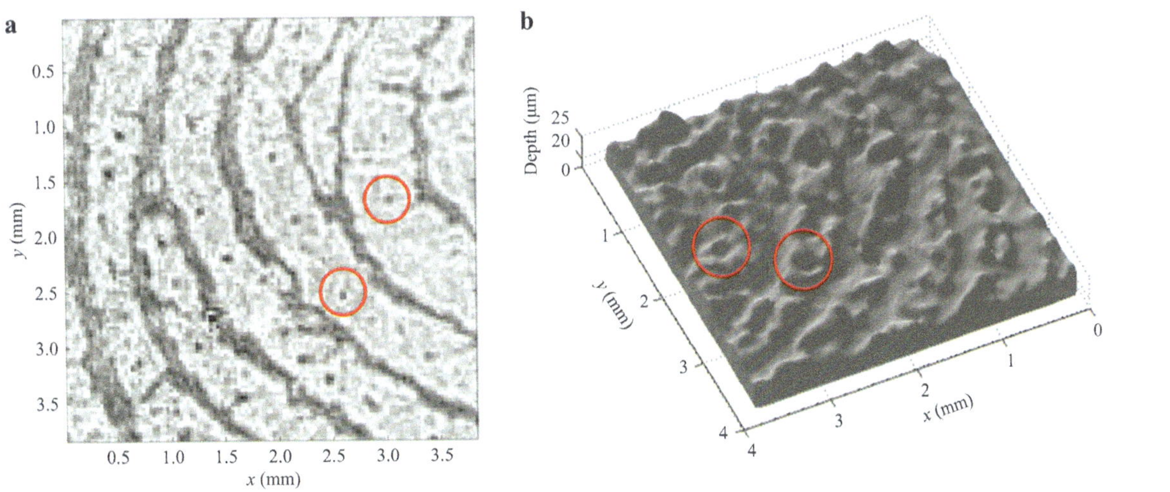

Fig. 3 a Reconstructed amplitude $|E_{\lambda 1}(x,y)|$, at wavelength λ_1 =514.5 nm of a CTF developed fingerprint on a blackened glass slide and **b** topogram of the CTF-developed fingerprint. The red circles identify level-3 details (pores). The hologram was recorded at a recording distance d = 40 cm using the Michelson geometry in the reflection mode (Fig. 7 in ref. 75).

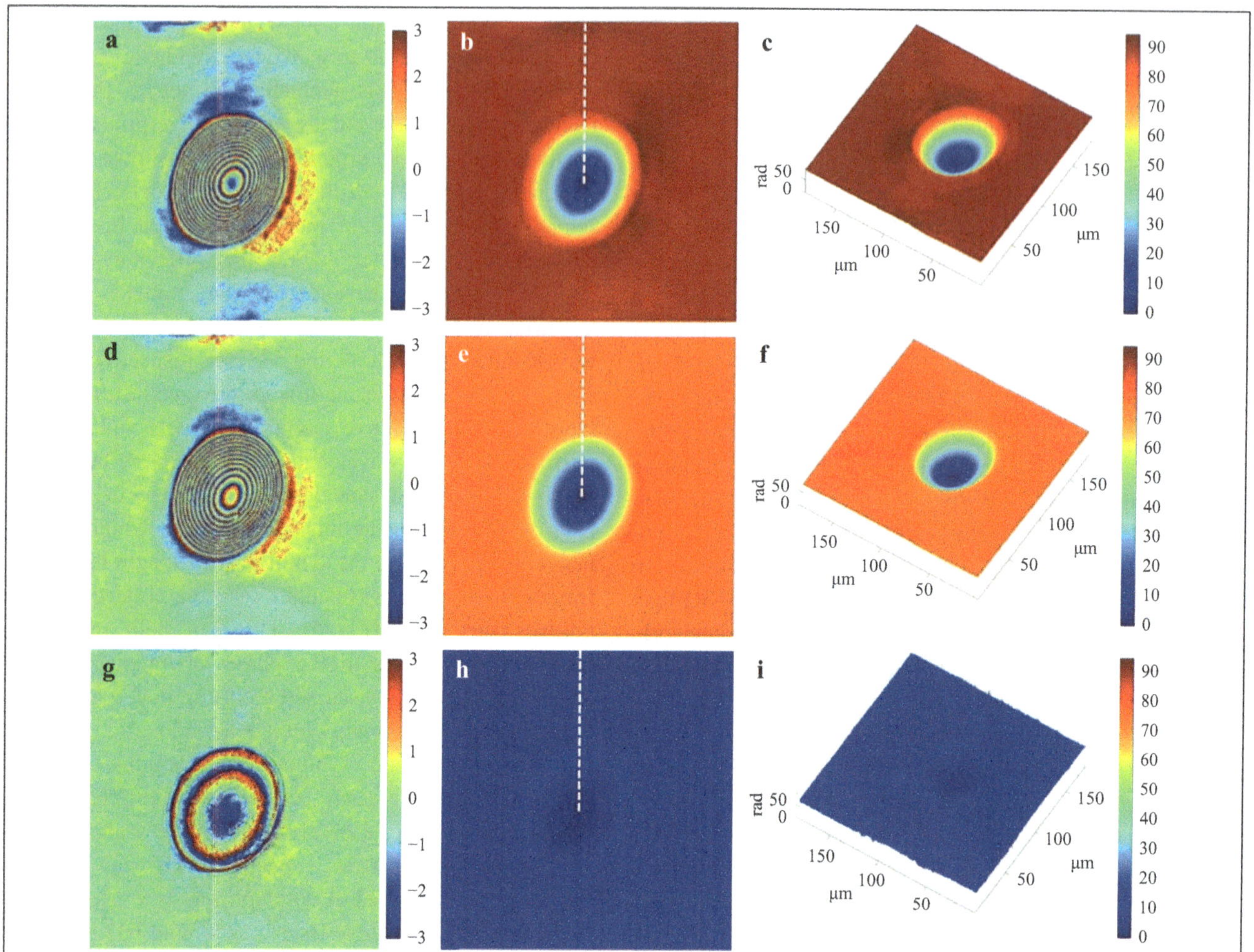

Fig. 4 Measurement results for the ablated pit specimen. **a**, **d**, **g** Reconstructed wrapped phase maps for λ_1 =532 nm, λ_2 =632.8 nm, and Λ = 3339.8 nm; **b**, **e**, **h** unwrapped phase maps for λ_1, λ_2 and Λ; and (c, f, i) 3D surface topography of the specimen for λ_1, λ_2 and Λ (Fig. 4 in ref. 65).

large object features. However, the stability and linewidth of the laser modes can be an impediment to achieving very high synthetic wavelengths. In addition, structured light imaging methods has a significant advantage in imaging macro objects, by projecting wide fringe patterns to the targets[51,52,56].

In the works described above, dual-wavelength acquisitions were performed sequentially. To further apply DWT in high-throughput applications, single capture dual-wavelength 3D topography has been studied. In 1998, Onodera et al. proposed a single acquisition method in Fourier interferometry with a single interferogram recorded in two different wavelengths[115]. In 2007, Kuhn et al. applied the single acquisition in real-time DWDH applications[66]. The idea is based on spatial multiplexing of hologram by incoherent addition of two single-wavelength interferograms on the same image sensor. The optical setup has two Mach-Zehnder interferometers built around the sample and creating two separate reference paths for two wavelengths (see Fig. 5). The camera obtains two off-axis interferograms of the sample simultaneously with two fringe patterns in a 90-degree rotation relation. This multiplexing method has also been developed for dynamic measurement[67,76] and a dual-wavelength interferometry external module for simultaneous phase unwrapping[71]. Single capture of the dual-wavelength hologram can also be achieved by using color channels of a Bayer mosaic camera[116]. From different color channels, the interferograms can be recorded simultaneously. Polarization multiplexing is the other way to perform real time DWT[69]. A combination of Bayer filter and linear micro-polarizers has also been proposed recently to obtain a single snapshot[91].

The other limit of using DWT is to resolve nanometer/sub-nanometer features. These nanometer scale features are hard to reconstructed by single wavelength, due to the small variations in height displays tiny displacement in lateral fringes. Therefore, noise can be

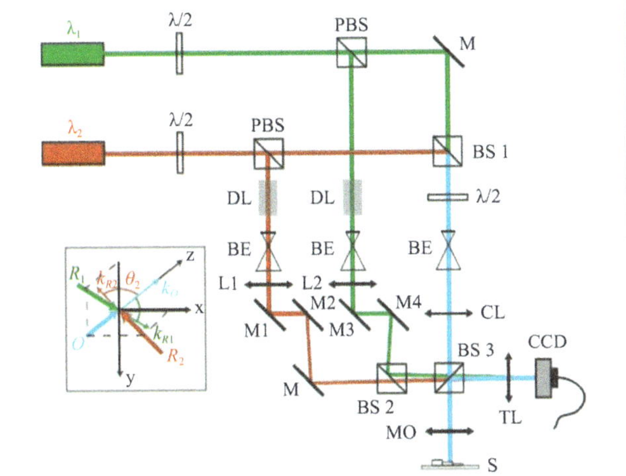

Fig. 5 Two-wavelength DH microscopy setup. λ/2: half-waveplate, M: mirrors, PBS: polarizing beamsplitter, BS: non-polarizing beamsplitter, DL: delay line, BE: beam-expander, L: lens, CL: condenser lens, MO: x3 achromatic microscope objective, 0.1 NA, S: specimen, TL: tube lens; Inset: 3D distribution of the incident waves propagation directions upon the CCD plane, $\boldsymbol{k}_{R1}$ and $\boldsymbol{k}_{R2}$ are the propagation vector directions of the reference waves $\boldsymbol{R}_1$ for wavelength λ_1 and $\boldsymbol{R}_2$ for λ_2 (Fig. 1 from ref. 66).

dominant in topography maps. Theoretically, Eq. 4 implies that the synthetic wavelength cannot be smaller than either of the single wavelength. Using a fringe subdivision method[117] together with synthetic wavelength method, the nanometer level accuracy reconstruction can be achieved. In 2002, Chen et al. proposed to employ a large synthetic wavelength to subdivide fringes formed by a small single wavelength[64]. Two Michelson interferometers for two different wavelengths from two orthogonal polarized He-Ne laser modes are used to obtain the phase difference, $\Delta\Phi = 2\pi((L/\Lambda) - (L_2/\lambda_2))$. A piezoelectric transducer (PZT) is used to introduce nanometer scale optical path difference for one of the wavelengths; an optical path difference is also introduced to the other interferometer to keep the phase difference $\Delta\Phi$ as a constant. Then, from setting $2\pi(L/\Lambda - L_2/\lambda_2) = 2\pi((L+\Delta L)/\Lambda - (L_2+\Delta l_2)/\lambda_2)$, $\Delta l_2 = \lambda_2/\Lambda\Delta L$. This yields to a fringe subdivision factor λ_2/Λ. From the fringe subdivision factor the small fringe displacement from single wavelength can be amplified in the large synthetic wavelength. A similar method has also been applied together with self-mixing interference or signal processing to achieve nanometer accuracy[83].

DWT can also be adapted to CDI techniques. CDI requires no lenses or reference beams and only collects diffracted intensity patterns[118]. CDI was first developed for electron beams for which aberration-free focusing tools did not exist. For this reason, CDI technique can also be very useful in the wavelength range where no good optics exist, e.g., X-ray, deep UV. Since CDI only collects the intensity patterns off an object due to diffraction, the technique uses the physics of beam propagation to iteratively calculate the phase pattern. CDI-related techniques can be categorized as single-shot CDI and Ptychography/Fourier Ptychography.

The single-shot CDI method to retrieve phase is well-known in optics, X-rays, electron imaging and crystallography[118–121]. In single-shot CDI, only one diffraction pattern is collected at the far-field. The phase is retrieved by iteratively performing Fourier/inverse-Fourier transforms. In each iteration the intensity pattern is substituted with the collected intensity together with the constrained values. The algorithm finally converges to an optimal result. However, single-shot CDI heavily depends on the efficacy of the chosen algorithm[122]. Recently, dual-wavelength CDI has been realized by exploiting the wavelength-dependent scaling of the support domain to recover separate reconstructions for each wavelength[123]. Other phase retrieval algorithms[124–131] are based on the Gerchberg–Saxton (GS) algorithm[124], for example, error reduction (ER)[125], hybrid input output (HIO)[124], etc.

Ptychography records multiple diffraction patterns from the same object by shifting the aperture placed before the object (Fig. 6a)[132]. Collimated light hits the aperture and the field distribution after the aperture is called the 'probe field', $P_n(x, y)$, where n represents the nth position of the aperture. For an object transparency $T(x, y)$, the 'exit field' is $E(x, y) = P_n(x, y)T(x, y)$. After propagation to the far-field, this transforms into the nth detected field $D_n(x, y)$. Fig. 6b shows a modified setup for ptychography that can be used for reflective phase objects. Wigner distribution deconvolution method[133] can be used to extract the phase of $T(x, y)$. However, this method requires a large n with smaller step-size from aperture to aperture. A widely used algorithm, the extended ptychographic iterative engine, iteratively determines the phase of $T(x, y)$. In Fourier ptychography, the aperture scans the Fourier space instead of the object space[134]. It also has other modality to achieve synthetic aperture by angular illumination[60]. DWTs have also been adopted in ptychography[135] and aperture scanning Fourier ptychography[136]. By applying DWT, the absolute height or phase measurement can be greatly enhanced. DWT can also be a good combination with other modern imaging modalities to extend the height and phase measurement range, e.g. non-line-of-sight imaging[77]. Other possible directions of applications will be further discussed in the conclusion and future direction section.

As mentioned in the introduction and principle sections, the other mathematical framework of DWT is to use the sum of the two phase maps. The two-sensitivity temporal phase-unwrapping technique has been applied to phase

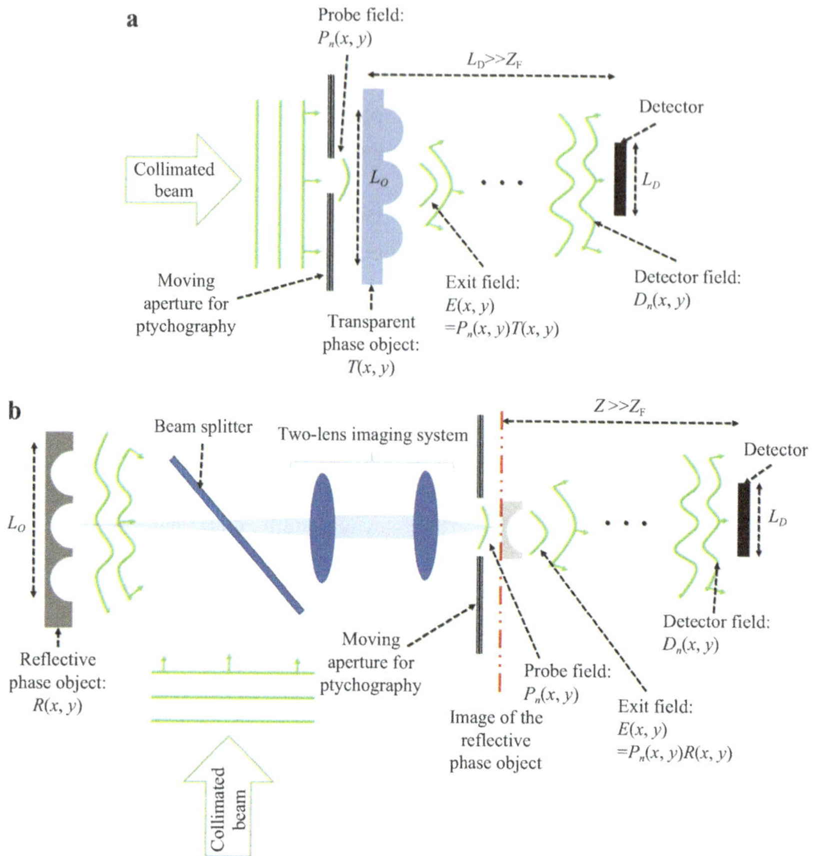

Fig. 6 a An experimental setup for ptychographic CDI for a transparent (phase) object. **b** A modified experimental setup for ptychographic CDI for a reflective object (with no absorption).

metrology with FPP. The algorithm is shown in Eq. 14. The SNR for the measurement of a fluorescent spiral lamp using phase sum is higher than the phase difference by a factor of $G^2 = (\Lambda/\Lambda_+)^2 = 289$. Fig. 7 (adopted from Figs. 9 and 11 in Ref. 43) shows the difference and the sum of the phase maps, and the super-sensitive phase-sum method shows a clear depression in noise (blue solid line). The flexibility and the ease of implementation of this method is promising in the future applications.

DWT performance

The DWT shows great advantage in avoiding phase unwrapping ambiguity, achieving sub-wavelength longitudinal resolution imaging with fringe subdivision, single-shot imaging, and adapting to various coherent/incoherent imaging techniques. However, it still has limitations. Here, we analyze:

(a) the limit of maximum resolvable phase or height based on selection of system parameters (*Itoh condition*),

(b) choice of optimum detector size based on truncation ratio, which is the ratio of the object size to the detector size (*truncation condition*), and

(c) possible noise sources degrading the reconstruction results.

To systematically investigate the first two conditions mentioned above, the following assumptions are made:

(1) to quantify the performance improvement due to dual-wavelength measurements, the phase retrieval algorithms are assumed to be accurate. For this analysis, the PUMA algorithm developed by Bioucas-Dias and Valadao[104] is used; however, any phase unwrapping algorithm could have been used. In actual experiments, the final performance has dependency on both phase retrieval and phase unwrapping algorithms;

(2) for any wavelength λ_1, an object of size L_O and a detector placed at a distance of Z away from the object (Fig. 8). The object can be transparent (Fig. 8a) so that it can be placed directly in the path of a collimated beam to collect the diffracted light (represented by the deformed wavefronts), or reflective (Fig. 8b) so that a beam splitter can be used to illuminate it and then collect the diffracted light. The assumption is: $Z \gg Z_F$, where the Fraunhofer distance, $Z_F = 2{L_O}^2/\lambda_1$. For a detector of size L_D placed in the far-field region ($Z \gg Z_F$), the minimum detectable

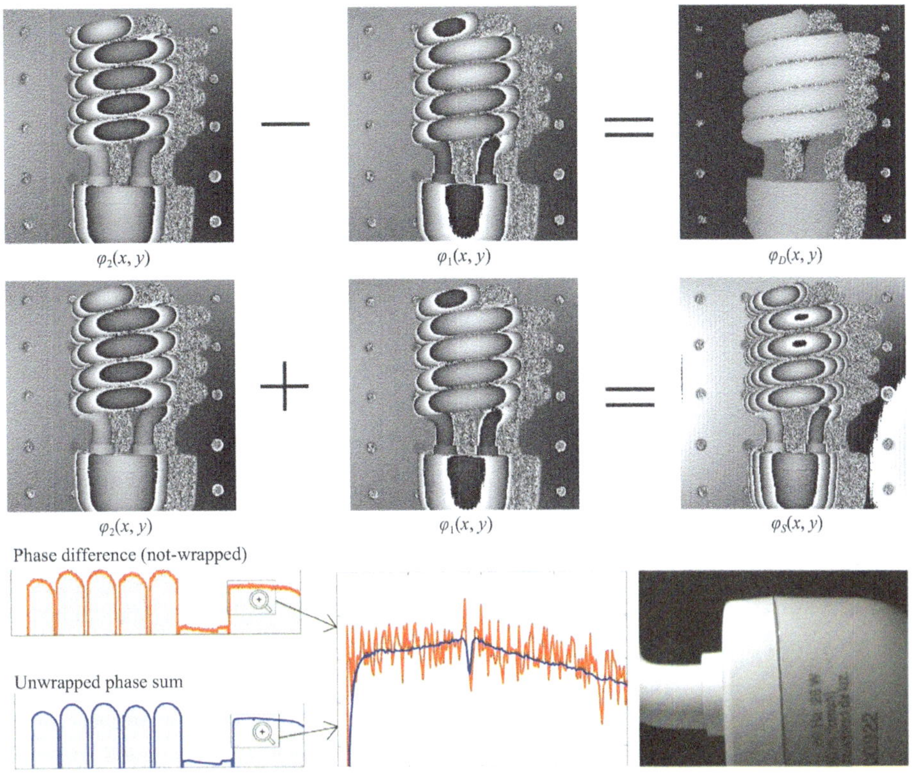

Fig. 7 The upper row shows the non-wrapped phase-difference while the middle row shows the highly wrapped phase-sum. The phase-sum has much higher sensitivity, so it is highly wrapped. The lower row: comparsion of the SNR between the phase-difference (in red), and the super-sensitive phase-sum (in blue). We have zoomed-in to clearly see the depression at the joining of the two plastic pieces shown in the photograph-detail (Figs. 9 and 11 in ref. 43).

feature size of the object is, $\Delta x_O = (\lambda_1 Z)/(N\Delta x_D)$ (from Eq. 11). Here, $L_D = \Delta x_D N$ is the detector size, $\Delta x_D$is the pixel pitch of the detector and N is the number of pixel along x-axis. Note that this can also be valid in near field using Eq. 7. For simplicity, the simulations below are performed assuming far-field. Since features smaller than a pixel also cannot be detected (due to aliasing), Δx_O is equivalent to pixel size in object plane.

As long as the above assumptions are satisfied, the following two conditions must be met for successful unwrapping using synthetic wavelength.

Itoh condition (system condition)

The Nyquist limit[137], also known in literature as the *Itoh condition*[104,138], states that the maximum phase slope (i.e. change of phase per pixel) that can be successfully unwrapped is

$$\tilde{S}_{Itoh} = \pi \left[\frac{rad}{pixel}\right] \tag{15}$$

A π radian phase can be achieved by propagation of a distance $\lambda_1/2$ along the z-axis. Meanwhile, Eq. 11 provides the effective pixel pitch on the object plane. For single-wavelength measurement, this becomes

$$S_{Itoh,\lambda_1} = \frac{\lambda_1/_2}{Z*\lambda_1/_{L_D}} = \frac{L_D}{2Z}\left[\frac{\mu m}{\mu m}\right] \tag{16}$$

The maximum phase slope anywhere in the object plane must be smaller than the Itoh phase slope, i.e. $S_{max} \leq S_{Itoh}$. For a given optical system design, the Itoh slope, S_{Itoh} (in units of $\mu m/\mu m$), serves as a system parameter since it can be determined in terms of wavelengths, detector size, object-to-detector distance etc. A higher S_{Itoh} corresponds to a more capable optical system that can handle surfaces with a relatively sharper height jump.

From Eq. 16, S_{Itoh,λ_1} has no dependency on the wavelength used for measurement. The benefit of using a larger λ_1 could be a better phase resolution in the z-axis. However, this benefit is counteracted by the larger pixel size in the transverse plane. In the far-field, the diffraction pattern is the Fourier transform of the object phase distribution. Experimentally, only a portion of the diffraction pattern can be collected due to the finite size of the detector. The process of omitting the higher spatial frequencies (at the outskirts of the diffraction pattern) is called "apodization". From Eq. 16, it is apparent that both a

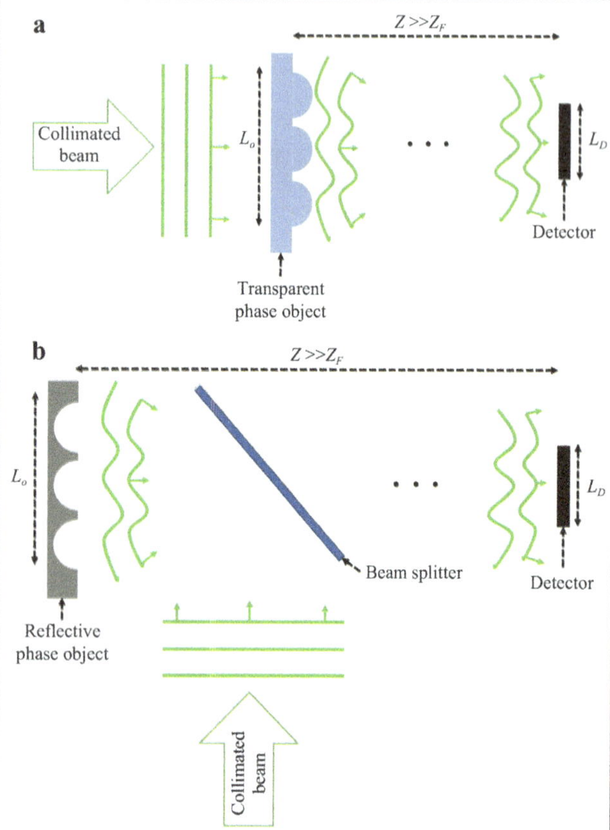

Fig. 8 A simplified setup that records the diffraction pattern off of a **a** transparent (phase) object and **b** reflective object (with no absorption) illuminated with a collimated beam. For both cases the object-to-detector distance Z is such that $Z \gg Z_F$. The reflective setup uses a beam splitter to illuminate the object. The features on the objects are arbitrarily chosen.

larger detector size L_D and a smaller object-to-detector distance Z (assuming $Z \gg Z_F$ still holds) reduce apodization and cause a higher value of S_{Itoh,λ_1}.

DWT uses a synthetic wavelength Λ (Eq. 4). For this case, the new Itoh condition becomes

$$S_{Itoh,\Lambda} = \frac{\Lambda/2}{Z*\lambda_2/L_D} = \frac{L_D}{2Z}\frac{\lambda_1}{\lambda_1 - \lambda_2}\left[\frac{\mu m}{\mu m}\right] \tag{17}$$

For a given detector size and distance, DWT improves the maximum detectable slope limit by a factor of $\lambda_1/(\lambda_1 - \lambda_2)$. In other words, a larger λ_1 and, correspondingly, an increased z-axis resolution, give a higher value of $S_{Itoh,\Lambda}$. Also, closer separation of λ_1 and λ_2 leads to a larger $S_{Itoh,\Lambda}$. This improvement is illustrated in the following example, typical of a realistic experiment.

Assume that a detector is placed $Z = 100\ mm$ away from a phase object of size L_O (see Fig. 8). The detector has 512×512 pixels and the pixel pitch is $5.2\ \mu$m. An object size of $L_{ox}, L_{oy} = 2\ \mu$m is used and the two wavelengths used for measurement are $0.501\ \mu$m and $0.496\ \mu$m. For this scenario, the Fraunhofer distance $Z_F = 15.96\ \mu$m is much smaller compared to the total propagation distance Z. Thus, for SWT, $S_{Itoh,\lambda_1} = 0.13312$ and for DWT, $S_{Itoh,\Lambda} = 1.33386$. The object comprises of a Gaussian height profile $h_{in}(x, y) = h_0 e^{(x^2+y^2)/\sigma_0{}^2}$ with a peak height, $h_0 = 10\ \mu m$, and a width $\sigma_0 = 400\ \mu m$ as in Fig. 9a. The truncation ratio T is defined as

$$T = \frac{\sigma_x}{L_{ox}} \tag{18}$$

For these parameters of the optical system, the performance of PUMA is simulated. In a physical experiment, the diffraction pattern would be collected by a detector. Since the detector has finite size, the object plane is effectively pixelated, according to Eq. 11. The pixel size depends on the illumination wavelength. Analogous to the experiment, the object plane is sampled to match the effective pixelation. A resampling algorithm is used to achieve slightly different pixelations at two wavelengths. Experimentally, using some phase retrieval algorithms, a wrapped phase profile on the object plane can be recovered. For simulation purpose, a modulo-2π function is used to wrap the resampled input phase profile. Then, PUMA is used to unwrap the phase. Using appropriate wavelength, the phase profile is converted to a height profile and compared against the input height profile. The SNR is defined using the relation

$$SNR = 10\log_{10}\left(\frac{h_{in,rms}}{h_{err,rms}}\right) \tag{19}$$

where, $h_{in,rms}$ is the root-mean-squared (*RMS*) height of the object and $h_{err,rms}$ is the *RMS* error between actual height $h_{in}(x, y)$ and measured height $h_{out}(x, y)$:

$$h_{in,RMS} = \left(\frac{1}{L_{ox}L_{oy}}\iint_{x=0,y=0}^{x=L_x,y=L_y}(h_{in}(x,y))^2dxdy\right)^{1/2} \tag{20}$$

$$h_{err,RMS} = \left(\frac{1}{L_{ox}L_{oy}}\iint_{x=0,y=0}^{x=L_x,y=L_y}(h_{out}(x,y) - h_{in}(x,y))^2dxdy\right)^{1/2} \tag{21}$$

For a given truncation ratio ($T = 0.2$ for this case), the peak of the height profile is varied and the maximum slope of that height profile S_{max} (i.e. the slope at the inflection point which is $1\sigma_0$away from the peak location) is determined. For each peak height, the *SNR* is also calculated using Eqs. (19-21). As predicted by the Itoh condition, the *SNR* drops when $S_{max} > S_{Itoh}$. Fig. 9b demonstrates these phenomena for single wavelength measurement. For one input height profile (Fig. 9a), the measured height profile is shown in Fig. 9c. The measurement fails (as indicated by a very low *SNR* in Fig. 9b) due to $S_{max} > S_{Itoh,\lambda_1}$. Then, the measurement is repeated using dual wavelengths. This

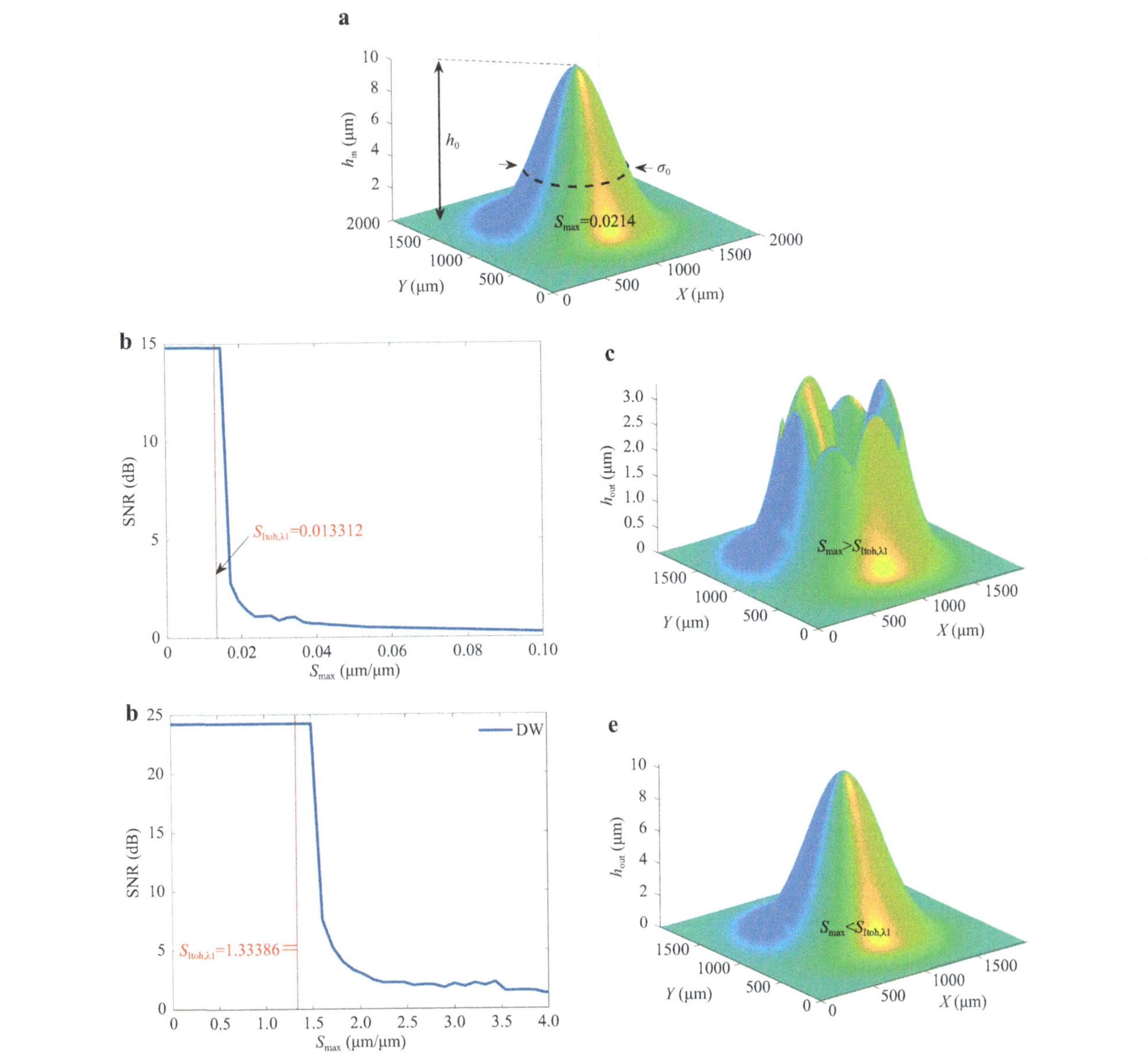

Fig. 9 **a** True height profile for a reflective object at the object plane. It is a Gaussian height profile with peak height $h_0 = 10\,\mu m$, and width $\sigma_0 = 400\,\mu m$. The black dotted line denotes the waist of the Gaussian where the maximum slope S_{max} is located. **b** For single wavelength measurement, SNR vs S_{max}. For $S_{max} > S_{Itoh,\lambda_1}$, the SNR drops. **c** When $S_{max} > S_{Itoh}$, the phase unwrapping fails. **d** The SNR vs S_{max} with dual wavelength measurement. Under the condition $S_{max} > S_{Itoh,\Lambda}$, SNR drops. **e** When $S_{max} < S_{Itoh}$, the phase unwrapping was successful.

increases the S_{Itoh} limit to $S_{Itoh,\Lambda} = 1.33386$. Fig. 9d indicates that the *SNR* keeps at a higher value until $S_{Itoh,\Lambda}$. Fig. 9e shows the measured height profile with high *SNR*. This example illustrates how the dual wavelength measurement pushes the S_{Itoh} limit higher and makes a high *SNR* measurement possible for a height profile that clearly has low *SNR* for a single wavelength measurement.

Truncation condition (input condition)

For a successful phase unwrapping ($SNR > 10\,dB$), the Itoh condition (i.e.$S_{max} < S_{Itoh}$) is a necessary but not sufficient condition. The truncation ratio T also needs to be within certain range. From Fig. 10, the SNR peaks within the range $0.05 < T < 0.5$. For an optical system (i.e. given S_{Itoh} limit) and certain input object size, a smaller T produces less resolved object plane causing lower SNR. On the other hand, for the same system and input object size, a larger T introduces apodization artifacts such as ripples at the edge of the measured object height. The range of T will be further investigated in the future.

Fig. 11 shows an actual and computed height profile when the two conditions (i.e. the Itoh condition and truncation condition) are met. Four Gaussian height profiles are placed in an array to mimic a physical object

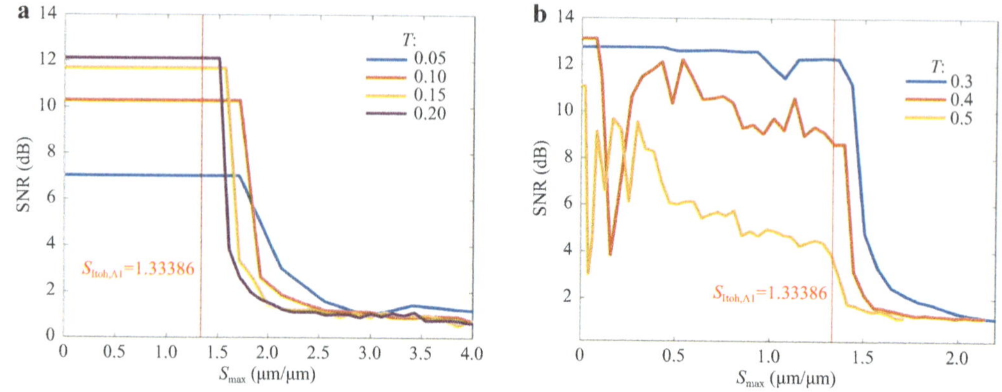

Fig. 10 The SNR vs S_{max} with dual wavelength measurement for different truncation ratios **a** $T = 0.05, 0.10, 0.15$ and 0.20 **b** $T = 0.3, 0.4$ and 0.5.

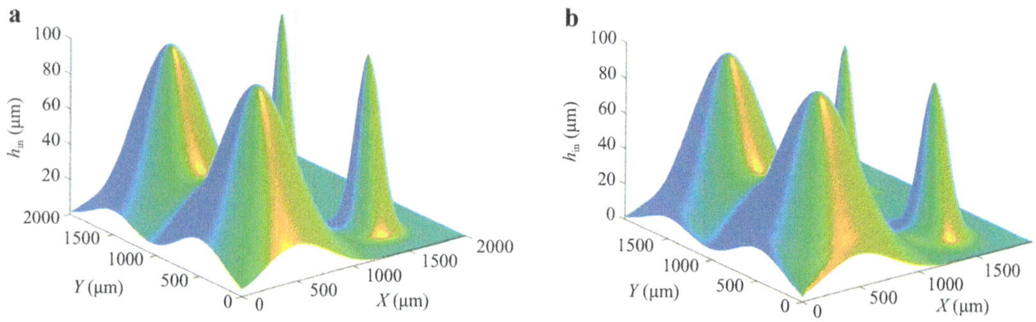

Fig. 11 a The actual input height profile and **b** the computed height profile using DWT. Both Itoh condition ($S_{max} < S_{Itoh}$) and truncation condition ($0.05 < T < 0.5$) are met. Measured SNR is 12.1 dB.

with surface variations. The output SNR is ~12 dB for this simulation which represents successful unwrapping.

Noise performance

Compared to SWT, DWT usually has a noisier performance[30,68], since using the synthetic wavelength imposes an amplification to the noise. The error sources are briefly summarized here. Possible reasons for noises are wavelength drifts (linewidth of each of the illumination wavelength), chromatic aberrations, dispersion, speckle noise, speckle decorrelation noises, retrace errors, and air turbulence. The noise sources also become limiting factors to achieve very large unambiguous distance measurements experimentally. Currently, the reported achievable synthetic wavelengths are in the order of tens of centimeters[78,114].

The stability of laser central wavelengths may not directly affect the noise performance, however, this effects the accuracy of the synthetic wavelength's calculation[139]. The accuracy of the synthetic wavelength becomes essential especially for large synthetic wavelengths. Fabry–Perot etalons have been used to perform mode selection/stabilization at the laser output[89]. The expected error σ_Λ in Λ due to wavelength selection can be determined using[73]

$$\sigma_\Lambda{}^2 = \left(\frac{\partial \Lambda}{\partial \lambda_1}\right)^2 \sigma_{\lambda_1}{}^2 + \left(\frac{\partial \Lambda}{\partial \lambda_2}\right)^2 \sigma_{\lambda_2}{}^2 \tag{22}$$

where $\sigma_{\lambda_1,\lambda_2}$ denote the errors in λ_1, λ_2. This yields

$$\left(\sigma_\Lambda / \Lambda^2\right)^2 = \left(\sigma_{\lambda_1} / \lambda_1{}^2\right)^2 + \left(\sigma_{\lambda_2} / \lambda_2{}^2\right)^2 \tag{23}$$

indicating that the error increase quadratically with Λ, This wavelength selection error can lead to significant uncertainty in the synthetic wavelength as $\lambda_1 \to \lambda_2$. Thus, for short synthetic wavelengths, any single phase (height) measurement can be assumed to have reasonably high accuracy, limited primarily by shot and coherence noise[66]. However, wavelength drift errors quickly begin to dominate at long synthetic wavelengths. The fractional error, which is proportional to σ_Λ/Λ, increases linearly with Λ.

For contribution to errors from the laser linewidth, since DWT usually has a small separation between different wavelengths, each of the illuminating lasers is supposed to have a narrow bandwidth. For SWT, the phase corresponding to a height h is given by $\phi_\lambda = 2\pi h/\lambda$. Hence, the magnitude of error in phase is $|\delta\phi_\lambda| = 2\pi h/\lambda^2 |\delta\lambda|$. Therefore, fractional error in measurement of phase is $|\delta\phi_\lambda|/\phi_\lambda = (2\pi h/\lambda^2)/(2\pi h/\lambda)|\delta\lambda| = |\delta\lambda|/\lambda$, where $|\delta\lambda|$ may arise from the linewidth of the laser. In DWT, the phase corresponding to a height h is given by $\phi_\Lambda = 2\pi h/\Lambda$. Hence, $|d\phi_\Lambda| = 2\pi h/\Lambda^2 |\delta\Lambda|$. The fractional error in measurement of phase is $|\delta\phi_\Lambda|/\phi_\Lambda = |\delta\Lambda|/\Lambda$. We need to re-express this in terms of $|\delta\lambda|$ for a fair comparison with the result for SWT. Now $\Lambda = \lambda_1\lambda_2/(|\lambda_1 - \lambda_2|) \approx \lambda^2/\Delta\lambda$, as two similar wavelengths are commonly used. Then, $d\Lambda = \left((2\lambda\Delta\lambda - \lambda^2(d\Delta\lambda/d\lambda))/(\Delta\lambda)^2\right)d\lambda$, and the magnitude of the maximum error is $|\delta\Lambda| = ((2\lambda\Delta\lambda + \lambda^2|d\Delta\lambda/d\lambda|)/(\Delta\lambda)^2)|\delta\lambda|$. Therefore, $|\delta\phi_\Lambda|/\phi_\Lambda = |\delta\Lambda|/\Lambda = 2|\delta\lambda|/\lambda + |d\Delta\lambda/d\lambda|(|\delta\lambda|/\Delta\lambda)$. Note that $|d\Delta\lambda/d\lambda| \sim 1$, since the change in $\Delta\lambda = |\lambda_1 - \lambda_2|$ is equal to the change in either wavelength $\lambda_{1,2}$. The ratio of the fractional error of DWT to that of SWT is

$$\frac{Error_{DWT}}{Error_{SWT}} = \frac{\dfrac{|\delta\phi_\Lambda|}{\phi_\Lambda}}{\dfrac{|\delta\phi_\lambda|}{\phi_\lambda}} = \frac{2\dfrac{|\delta\lambda|}{\lambda} + \dfrac{|\delta\lambda|}{\Delta\lambda}}{\dfrac{|\delta\lambda|}{\lambda}} = 2 + \frac{\lambda}{\Delta\lambda} \tag{24}$$

Using $\lambda = 500\,nm$, $\Delta\lambda = 20\,nm$, Eq. 24 shows that the error in DWT due to laser linewidth is at least 25 times more than the error in SWT.

The usage of more than one wavelength is usually accompanied with chromatic aberrations and dispersions in the optical system. These factors have also become the noise sources of DWT. Strategies have been proposed to correct chromatic aberrations. One way is to convert the wavelength-dependent phase term into the scaling effect in the lateral plane. As discussed in the "Principle of DWT" section, the pixel size at the reconstruction plane is rescaled by Eq. 11 and then matched for pixel-to-pixel subtraction with zero-padding (Eq. 12). The advantage of this method is it is convenient and easy to determine, but may suffer from the imperfection of pixel match. As shown in Eq. 12, due to the discrete nature of pixels, the calculation results of padsize need to be rounded to the nearest integer. This rounding process also contribute to the noise. This can be reduced by carefully selecting the number of pixels of the phase maps and also the wavelengths to make $N/2(\lambda_1/\lambda_2 - 1)$ close to an integer in sacrifice of imposing another constrain to the wavelength selection. Thanks to the advancement of signal processing techniques, the matching issue can be resolved by rescaling the pixel size with chirped scaling algorithms and fast Fourier transform[140,141]. The other way to deal with the chromatic aberration is to digitally refocus the optical field of the corresponding wavelength to the focused plane[18]. The nature of chromatic aberration is that the focal plane is shifted along optical axis for different wavelengths. This can be interpreted as a spherical aberration is added to each of the wavelength. Since the intensity and phase of the complex field can be obtained by all the quantitative phase imaging methods, the digital propagation can be performed to refocus the complex optical field to the focal plane. This technique is easy to implement and free of pixel matching, but determining when the object is in focus is tricky and may vary among human visual identifications or different in-focus evaluation metrics[142,143]. The dispersion errors, as the other wavelength-dependent error source caused by the dispersion of optical components (e.g. beam splitter), can be monitored and managed by numerical algorithms mentioned above or by modeling the static dispersion of the optical components and compensating with numerical calculations[15].

Speckle noise exists in all coherent imaging techniques. Due to the noise amplification shortcoming of DWT, the speckle noise has been significantly amplified in DWT, especially for large synthetic wavelengths. Thus, to mitigate the speckle noise effect is of great importance. A common way to reduce the speckle noise in SWT is to take multiple shots and average them, or equivalently, using a rotating diffuser whose spinning period is much smaller than the camera's exposure time. For DWT, averaging two phase difference maps is possible[84]. Other methods to reduce speckle noise in SWT[144] can also be applied to DWT. As we discussed the speckle noise here, it is worth mentioning that the DWT can be applied to both diffuse or specular reflecting surfaces. We have used DWT with reference beams using diffuse surfaces such as screwheads, dents, fingerprints, etc. using synthetic wavelengths in the order of tens, hundreds and thousands of microns[73,75,145], partly diffuse and partly specular surfaces such as silicon wafer samples with photoresist features coated with aluminum using synthetic wavelengths in the order of microns[145], of as well as for (specular) reflecting surfaces, such as mirrors separated by distances in the order of centimeters using synthetic wavelengths in the order of tens of centimeters (achieved using acousto-optic modulators at 1 Ghz for frequency shifting)[78].

Speaking of diffuse surfaces, the speckle decorrelation noise is common during diffuse surface measurement. There are a few possible origins of the speckle decorrelation noise: the modifications at the object surface between two measurements (e.g. mechanical shifts, heating), laser wavelength changes, reduced number of camera pixels, defocusing in reconstruction (mostly from inaccurate propagation distance), saturation of camera pixels in hologram recording, insufficient bit-depth of the camera (shot noise induced) and so on[146-150]. For DWT, the major cause of speckle decorrelation noise in diffuse surface measurement is from recording two holograms with two different wavelengths. Possible ways of reducing the speckle decorrelation noises are: averaging over multiple captures[148], or digital filtering. Numerous digital filtering techniques have been proposed to reduce the speckle decorrelation noise: spatial filtering (viz., moving average filter, medium filter, Gaussian filter), Wiener filter, adaptive filter[151], wavelet approaches[152,153], and so on. An exhaustive performance analysis and comparison among these digital filter techniques (a number of 34 algorithms) in reducing speckle decorrelation noise have been made quantitatively by Montresor and Picart[150]. Retrace errors, existing in specular surface testing, are due to presence of nonnull aspherical surface during interferometric testing, and specially in systems like the traditional interferometers, where the test surface is at a focal distance past the focus of a perfect focusing lens. This type of noise can be calibrated by ray tracing the testing system model[19,154]. Air turbulence is another source of noise. The turbulence in air can lead to the inhomogeneous distribution of refractive index in space. This non-uniform distribution will add different phase maps to the object and reference beam phases. Currently, no perfect solution has been proposed to solve this issue. One potential way to compensate for air turbulence is to store a hologram of the turbulence and read it out with the optical field from the object under investigation that passes through the same turbulence[155]. Other correction methods using sharpness metrics have also been employed in DH and in holographic aperture ladar[156,157]. The other way can possibly be using adaptive optics methods to compensate the turbulence induced phase in real-time[158-161]. Attempts have been made to compensate for errors introduced due to non-uniform and random refractive index changes of air with post processing of data from holographic interferometry measurements. One approach, which uses a purely temporal interference filter, reduced the noise from the refractive index errors from air disturbance over a rather limited spatial region[162]. A more complete spatio-temporal filter requires that the spatio-temporal statistics of the noise from the air can be measured[163]. The method has been applied to the case of locally homogenous/isotropic refractive index fluctuations in a wind tunnel.

Conclusions and Discussions

In this paper, we have reviewed the early and recent developments of DWT and its applications in different scenarios. The essence of DWT is to employ an extra wavelength in the imaging process and form a large synthetic wavelength. The significant advantage of the synthetic wavelength is to avoid the phase ambiguity in the phase retrieval and phase unwrapping. This feature can extend the measurement ability from optical wavelength scale to tens of centimeters scale. In the meantime, with the help of fringe subdivision method, DWT can also achieve nanometer-level accuracy in 3D topography. The large measurement dynamic range is appealing to numerous industry applications. For high-throughput applications, single-shot DWT has been achieved by using spectra-multiplexing, polarization-multiplexing, or different color channels of a color camera. For early developments and applications of DWT, the technique mainly focuses on the implementation in holography and interferometry. With fast advances in 3D/phase coherent imaging techniques, DWT can also be adapted to various novel optical modalities, viz. CDI, ptychography, aperture-scanning Fourier ptychography, non-line-of-sight imaging, etc. In this review, a comprehensive summary of DWT performance and error sources is made. Error sources from such as system design (Itoh condition), wavelength drift, chromatic aberration, dispersion, speckle noises, retrace errors, and air turbulence, have been discussed. Also, the existing or potential correction and compensation methods have been pointed out.

The unambiguity range measurements with DWT, as mentioned before, can be varied from nanometer scale to tens of centimeters scale with synthetic wavelengths from micro level to tens of centimeters level. These measurement ranges can satisfy most of the industry applications. For even larger height jumps (meter scale), structured light techniques, time-of-flight, or FPP techniques can be adopted. The future direction of DWT, instead of increasing the synthetic wavelength, should aim to achieve large unambiguous height measurement range and high accuracy at the same time, i.e. a kind of low-noise DWTs. This is critical to phase imaging and 3D topography, and its biomedical or industry applications. One potential way to fulfill the demand for large unambiguous height measurement and high accuracy simultaneously is to employ the idea of phase sum, as discussed in the "Principle of DWT" section. This idea was implemented five decades ago in holography and interferometry[36-38], and recently only used in FPP[43]. No recent work has reported its application in holography, interferometry or any other coherent imaging techniques. Further investigations on phase sum and difference usage can be done to achieve high sensitivity and large unambiguous measurement range. Another direction to establish low-noise DWT is to minimize the noises from different noises sources. As discussed in the "Noise performance" section, there are some error sources that can be solved with reported method, but air turbulence remains a challenge. Adaptive optics methods can potentially help with the turbulence correction in real time. Instead of real-time correction, the other way to compensate turbulence can be to fix the fluctuation first, and then correct it. Speckle interferometry can freeze the turbulence variations in short exposure time[164]. DWT can also be applied to speckle interferometry. Two-wavelength microscopic speckle interferometry using a single-chip color camera for characterization of micro level samples has been demonstrated by Upputuri et al.[165]. The system allows quantification of large deformations and shapes of specimens with a rough surface. Further research, such as dealing with air turbulence with the help of speckle imaging, should be considered.

From an application perspective, DWT can be adapted to different novel optical modalities. Aperture scanning type Fourier ptychography together with DWT, as an example, using a LED source and appropriate band-pass filters has been proposed for topographic measurement[136]. The use of DWT extends the axial measurement range, provides a speckle-free phase image, and reduces the possibility of a systematic error, which yields a high-quality topographic image. The proposed system can measure the surface topography of nano- to micro-structures. Because the LED has a relatively wide spectral bandwidth, it is easy to obtain the two wavelengths required for the proposed method by applying appropriate bandpass filters. Need to mention that other aperture scanning type Fourier ptychography modalities[166] can also be used to implement DWT; however, angular illuminated Fourier ptychography may not be suitable for thick samples. Another example is non-iterative coherent complex imaging based on Kramers-Kronig relations. The methodology has been recently proposed and developed[167]. It can achieve large space-bandwidth-product with non-iterative reconstruction and a minimum of two captures[168]. This newly proposed method can be a good candidate for topography applications with DWT.

Some other non-traditional applications and future directions are now discussed. DWT can potentially be used for aberration corrections. This recent work by Willomitzer et al.[77] is inspired by Gabor's original idea of "analysis, synthesis, and correction" for improving the resolution of electron microscopes. Gabor envisioned recording electron waves with aberrated electron microscope optics (analysis at electron wavelength λ_e), then reconstructing this electron image by playing the hologram back with an optical wave (*synthesis* at optical wavelength $\lambda \gg \lambda_e$) while exploiting optical wavefront correction (*correction*, at wavelength λ). Likewise, DWT can be used to correct unknown wavefront aberrations introduced when visible light is transported through scenes with strong scattering. By capturing two holograms at two closely spaced wavelengths (*analysis*), each showing random aberrations and computationally beating the two signals together, a low frequency synthetic wavelength hologram (correction) can be achieved. The synthetic wavelength hologram, which contains information of the order of the synthetic wavelength Λ, is virtually impervious to the effects of scattering at the optical wavelengths. Similar to Gabor's idea, during *synthesis*, the object is reconstructed by playing back the computationally corrected hologram with the much larger synthetic wavelength Λ. As shown by Willomitzer et al.[77], DWT also potentially offers a "novel solution to image hidden objects through scatters or non-line-of-sight geometries at the physical space-bandwidth limit ". By applying full-field lock-in detector technology, it is possible to capture a synthetic hologram of the hidden object within milliseconds. Since the synthetic wavelength can be readily changed, the captured hologram provides a window to fully reconstruct the 3D light field of the hidden scene at high space-bandwidth-product. The idea of DWT may also be implemented in holography with applications in dynamic interferometry, beam-steering, holographic display etc. Here are some additional thoughts listed:

(a) one can write a hologram at one wavelength and reconstruct at certain shifted wavelength, for dynamic interferometry. Optically addressed liquid crystal spatial light modulators can be used to record thin dynamic holograms for dynamic two-wavelength holography[159]. One can thus realize a dynamic interferometer, providing arbitrary scaling of the wavefront distortions. Such interferometers can be useful in adaptive optics, viz., for simplification in measurement of wavefront distortions, for recording of dynamic holographic correctors, working in spectral ranges where direct holographic recording is impossible, in particular, mid-IR, etc.;

(b) another idea is writing a transmission grating at one wavelength in a volume holographic recording medium (e.g., photorefractive material), and reading out using a longer wavelength in a reflection grating geometry, with possible applications in beam-steering[169], image processing, etc.

Author details

[1]Department of Electro-Optics and Photonics, University of Dayton, Dayton, OH 45469, USA. [2]Department of Electrical Engineering, California Institute of Technology, Pasadena, CA, 91125, USA. [3]Halo Industries Inc., Palo Alto, CA, 95303, USA

Conflict of interest

The authors declare that they have no conflict of interest.

Received: 01 September 2021 Revised: 28 January 2022 Accepted: 15 February 2022
Accepted article preview online: 08 March 2022
Published online: 02 April 2022

References

1. Michelson, A. A. & Benoit, J. R. Détermination expérimentale de la valeur du mètre en longueurs d' ondes lumineuses. (Paris: Gauthier-Villars et fils, 1894).
2. Benoît, R. Application des phénomènes d' interférence à des déterminations métrologiques. *Journal de Physique Théorique et Appliquée* **7**, 57-68 (1898).
3. Born, M. & Wolf, E. Principles of optics: electromagnetic theory of propagation, interference and diffraction of light. (Cambridge: Cambridge University Press, 1999).
4. Poole, S. P. & Dowell, J. H. Application of interferometry to the routine measurement of block gauges. in Optics and Metrology. (ed Mollet, P.) (New York: Pergamon, 1960).
5. Engelhard, E. Precise interferometric measurement of gauge blocks. Proceedings of the Symposium on Gauge Blocks. 1957.
6. Forrester, A. T., Parkins, W. E. & Gerjuoy, E. On the possibility of observing beat frequencies between lines in the visible spectrum. *Physical Review Journals Archive* **72**, 728 (1947).
7. Wyant, J. C. Testing aspherics using two-wavelength holography. *Applied Optics* **10**, 2113-2118 (1971).
8. Polhemus, C. Two-wavelength interferometry. *Applied Optics* **12**, 2071-2074 (1973).
9. Bien, F. et al. Absolute distance measurements by variable wavelength interferometry. *Applied Optics* **20**, 400-403 (1981).
10. Dändliker, R., Thalmann, R. & Prongué, D. Two-wavelength laser interferometry using superheterodyne detection. *Optics Letters* **13**, 339-341 (1988).
11. Cheng, Y. Y. & Wyant, J. C. Multiple-wavelength phase-shifting interferometry. *Applied Optics* **24**, 804-807 (1985).
12. Cheng, Y. Y. & Wyant, J. C. Two-wavelength phase shifting interferometry. *Applied Optics* **23**, 4539-4543 (1984).
13. Hildebrand, B. P. & Haines, K. A. Multiple-wavelength and multiple-source holography applied to contour generation. *Journal of the Optical Society of America* **57**, 155-162 (1967).
14. Zelenka, J. S. & Varner, J. R. A new method for generating depth contours holographically. *Applied Optics* **7**, 2107-2110 (1968).
15. Pförtner, A. & Schwider, J. Dispersion error in white-light Linnik interferometers and its implications for evaluation procedures. *Applied Optics* **40**, 6223-6228 (2001).
16. De Nicola, S. et al. Recovering correct phase information in multiwavelength digital holographic microscopy by compensation for chromatic aberrations. *Optics Letters* **30**, 2706-2708 (2005).
17. Javidi, B. et al. Three-dimensional image fusion by use of multiwavelength digital holography. *Optics Letters* **30**, 144-146 (2005).
18. Ferraro, P. et al. Quantitative phase microscopy of microstructures with extended measurement range and correction of chromatic aberrations by multiwavelength digital holography. *Optics Express* **15**, 14591-14600 (2007).
19. Liu, D. et al. Practical methods for retrace error correction in nonnull aspheric testing. *Optics Express* **17**, 7025-7035 (2009).
20. Zhang, L. Q. et al. Measurement of steep aspheric surfaces using improved two-wavelength phase-shifting interferometer. Proceedings of SPIE 10458 AOPC 2017: 3D Measurement Technology for Intelligent Manufacturing. Beijing: SPIE, 2017, 10458.
21. Zhang, L. et al. Non-null annular subaperture stitching interferometry for steep aspheric measurement. *Applied Optics* **53**, 5755-5762 (2014).
22. De Groot, P. J. Extending the unambiguous range of two-color interferometers. *Applied Optics* **33**, 5948-5953 (1994).
23. Falaggis, K., Towers, D. P. & Towers, C. E. Multiwavelength interferometry: extended range metrology. *Optics Letters* **34**, 950-952 (2009).
24. Falaggis, K. et al. Multi-wavelength phase unwrapping: a versatile tool for extending the measurement range, breaking the Nyquist limit, and encrypting optical communications. Proceedings of SPIE 10749 Interferometry XIX. San Diego: SPIE, 2018, 1074913.
25. Falaggis, K., Towers, D. P. & Towers, C. E. Algebraic solution for phase unwrapping problems in multiwavelength interferometry. *Applied Optics* **53**, 3737-3747 (2014).
26. Friesem, A. A. & Levy, U. Fringe formation in two-wavelength contour holography. *Applied Optics* **15**, 3009-3020 (1976).
27. Sodnik, Z. et al. Two-wavelength double heterodyne interferometry using a matched grating technique. *Applied Optics* **30**, 3139-3144 (1991).
28. Fercher, A. F., Vry, U. & Werner, W. Two-wavelength speckle interferometry on rough surfaces using a mode hopping diode laser. *Optics and Lasers in Engineering* **11**, 271-279 (1989).
29. Wada, A., Kato, M. & Ishii, Y. Large step-height measurements using multiple-wavelength holographic interferometry with tunable laser diodes. *Journal of the Optical Society of America A* **25**, 3013-3020 (2008).
30. Mann, C. J. et al. Quantitative phase imaging by three-wavelength digital holography. *Optics Express* **16**, 9753-9764 (2008).
31. Wada, A. Multiple-wavelength holographic interferometry with tunable laser diodes. in Advanced Holography – Metrology and Imaging (ed Naydenova, I.) (Rijeka: InTech, 2011).
32. De Groot, P. & Kishner, S. Synthetic wavelength stabilization for two-color laser-diode interferometry. *Applied Optics* **30**, 4026-4033 (1991).
33. Wang, C. L., Chuang, Y. H. & Pan, C. L. Two-wavelength interferometer based on a two-color laser-diode array and the second-order correlation technique. *Optics Letters* **20**, 1071-1073 (1995).
34. Chen, K. H. et al. Alternative method of wavelength drift free dual-wavelength heterodyne interferometry for the absolute distance measurement. *Optical Review* **16**, 492-494 (2009).
35. Wu, G. H. et al. Synthetic wavelength interferometry of an optical frequency comb for absolute distance measurement. *Scientific Reports* **8**, 4362 (2018).
36. Mustafin, K. S. & Seleznev, V. A. Methods of increasing the sensitivity of holographic interferometry. *Soviet Physics Uspekhi* **13**, 416 (1970).
37. Weigi, F. A generalized technique of two-wavelength, nondiffuse holographic interferometry. *Applied Optics* **10**, 187-192 (1971).
38. Weigl, F. Two-wavelength holographic interferometry for transparent media using a diffraction grating. *Applied Optics* **10**, 1083-1086 (1971).
39. Di, J. L. et al. Dual wavelength digital holography for improving the measurement accuracy. Proceedings of SPIE 8769 International Conference on Optics in Precision Engineering and Nanotechnology (icOPEN2013). Singapore: SPIE, 2013, 87690G.
40. Di, J. L. et al. Improvement of measurement accuracy in digital holographic microscopy by using dual-wavelength technique. *Journal of Micro/Nanolithography, MEMS, and MOEMS* **14**, 041313 (2015).
41. Zuo, C. et al. Temporal phase unwrapping algorithms for fringe projection profilometry: a comparative review. *Optics and Lasers in Engineering* **85**, 84-103 (2016).
42. Xiong, J. X. et al. Improved phase retrieval method of dual-

wavelength interferometry based on a shorter synthetic-wavelength. *Optics Express* **25**, 7181-7191 (2017).
43. Servin, M., Padilla, M. & Garnica, G. Super-sensitive two-wavelength fringe projection profilometry with 2-sensitivities temporal unwrapping. *Optics and Lasers in Engineering* **106**, 68-74 (2018).
44. Zhang, S. Recent progresses on real-time 3D shape measurement using digital fringe projection techniques. *Optics and Lasers in Engineering* **48**, 149-158 (2010).
45. Gorthi, S. S. & Rastogi, P. Fringe projection techniques: whither we are?. *Optics and Lasers in Engineering* **48**, 133-140 (2010).
46. Su, X. Y. & Zhang, Q. C. Dynamic 3-D shape measurement method: a review. *Optics and Lasers in Engineering* **48**, 191-204 (2010).
47. Su, X. Y. & Chen, W. J. Fourier transform profilometry: a review. *Optics and Lasers in Engineering* **35**, 263-284 (2001).
48. Burke, J. et al. Reverse engineering by fringe projection. Proceedings of SPIE 4778 Interferometry XI: Applications. Seattle: SPIE, 2002.
49. Hausler, G. et al. Deflectometry vs. interferometry. Proceedings of SPIE 8788 Optical Measurement Systems for Industrial Inspection VIII. Munich: SPIE, 2013, 87881C.
50. Shin, S. & Yu, Y. Measuring the both surfaces profiles of optical element using transmission deflectometry with liquids. 2015 11th Conference on Lasers and Electro-Optics Pacific Rim (CLEO-PR). Busan: IEEE, 2015, 1-2.
51. Geng, J. Structured-light 3D surface imaging: a tutorial. *Advances in Optics and Photonics* **3**, 128-160 (2011).
52. Forbes, A., De Oliveira, M. & Dennis, M. R. Structured light. *Nature Photonics* **15**, 253-262 (2021).
53. Zhou, H. W. et al. Digital correlation of computer-generated holograms for 3D face recognition. *Applied Optics* **58**, G177-G186 (2019).
54. Salvi, J., Pagès, J. & Batlle, J. Pattern codification strategies in structured light systems. *Pattern Recognition* **37**, 827-849 (2004).
55. Sciammarella, C. A. The moiré method **B** a review. *Experimental Mechanics* **22**, 418-433 (1982).
56. Guo, H. Y., Zhou, H. W. & Banerjee, P. P. Single-shot digital phase-shifting Moiré patterns for 3D topography. *Applied Optics* **60**, A84-A92 (2021).
57. He, F. et al. Moiré patterns in 2D materials: a review. *ACS Nano* **15**, 5944-5958 (2021).
58. Miao, J. W. et al. Extending the methodology of X-ray crystallography to allow imaging of micrometer-sized non-crystalline specimens. *Nature* **400**, 342-344 (1999).
59. Hoppe, W. Beugung im inhomogenen primärstrahlwellenfeld. I.Prinzip einer phasenmessung von elektronenbeungungsinterferenzen. *Acta Crystallographica Section A* **25**, 495-501 (1969).
60. Zheng, G. A., Horstmeyer, R. & Yang, C. Wide-field, high-resolution Fourier ptychographic microscopy. *Nature Photonics* **7**, 739-745 (2013).
61. Lin, C. et al. Spatial pattern-shifting method for complete two-wavelength fringe projection profilometry. *Optics Letters* **45**, 3115-3118 (2020).
62. Deng, W. B. Absolute phase recovery based on two-wavelength fringes using a defocusing technique in fringe projection profilometry. *Optical Engineering* **59**, 064109 (2020).
63. Abdelsalam, D. G., Magnusson, R. & Kim, D. Single-shot, dual-wavelength digital holography based on polarizing separation. *Applied Optics* **50**, 3360-3368 (2011).
64. Chen, B. Y., Cheng, X. H. & Li, D. C. Dual-wavelength interferometric technique with subnanometric resolution. *Applied Optics* **41**, 5933-5937 (2002).
65. Di, J. L. et al. Dual-wavelength common-path digital holographic microscopy for quantitative phase imaging based on lateral shearing interferometry. *Applied Optics* **55**, 7287-7293 (2016).
66. Kühn, J. et al. Real-time dual-wavelength digital holographic microscopy with a single hologram acquisition. *Optics Express* **15**, 7231-7242 (2007).
67. Fu, Y. et al. Dual-wavelength image-plane digital holography for dynamic measurement. *Optics and Lasers in Engineering* **47**, 552-557 (2009).
68. Zhou, H. W. et al. Performance analysis of phase retrieval using transport of intensity with digital holography. *Applied Optics* **60**, A73-A83 (2021).
69. Abdelsalam, D. G. & Kim, D. Real-time dual-wavelength digital holographic microscopy based on polarizing separation. *Optics Communications* **285**, 233-237 (2012).
70. Guo, R. L. et al. Phase unwrapping in dual-wavelength digital holographic microscopy with total variation regularization. *Optics Letters* **43**, 3449-3452 (2018).
71. Turko, N. A. & Shaked, N. T. Simultaneous two-wavelength phase unwrapping using an external module for multiplexing off-axis holography. *Optics Letters* **42**, 73-76 (2017).
72. Karako, L. et al. Flipping interferometric module for simultaneous dual-wavelength unwrapping of quantitative phase maps of biological cells. *Frontiers in Physics* **9**, 667023 (2021).
73. Williams, L. et al. Holographic volume displacement calculations via multiwavelength digital holography. *Applied Optics* **53**, 1597-1603 (2014).
74. Wagner, C., Osten, W. & Seebacher, S. Direct shape measurement by digital wavefront reconstruction and multi-wavelength contouring. *Optical Engineering* **39**, 79 (2000).
75. Abeywickrema, U. et al. High-resolution topograms of fingerprints using multiwavelength digital holography. *Optical Engineering* **56**, 034117 (2017).
76. Haus, J. W. et al. Instantaneously captured images using multiwavelength digital holography. Proceedings of SPIE 8493 Interferometry XVI: Techniques and Analysis. San Diego: SPIE, 2012, 84930W.
77. Willomitzer, F. et al. Fast non-line-of-sight imaging with high-resolution and wide field of view using synthetic wavelength holography. *Nature Communications* **12**, 6647 (2021).
78. Abeywickrema, U. et al. Holographic topography using acousto-optically generated large synthetic wavelengths. Proceedings of SPIE 9771 Practical Holography XXX: Materials and Applications. San Francisco: SPIE, 2016, 97710C.
79. Lu, S. H. & Lee, C. C. Measuring large step heights by variable synthetic wavelength interferometry. *Measurement Science and Technology* **13**, 1382-1387 (2002).
80. Hase, E. et al. Multicascade-linked synthetic-wavelength digital holography using a line-by-line spectral-shaped optical frequency comb. *Optics Express* **29**, 15772-15785 (2021).
81. Weimann, C. et al. Synthetic-wavelength interferometry improved with frequency calibration and unambiguity range extension. *Applied Optics* **54**, 6334-6343 (2015).
82. De Groot, P. J. & McGarvey, J. A. Laser gage using chirped synthetic wavelength interferometry. Proceedings of SPIE 1821 Industrial Applications of Optical Inspection, Metrology, and Sensing. Boston: SPIE, 1993.
83. Chen, J. B. et al. Synthetic-wavelength self-mixing interferometry for displacement measurement. *Optics Communications* **368**, 73-80 (2016).
84. Claus, D. et al. Accuracy enhanced and synthetic wavelength adjustable optical metrology via spectrally resolved digital holography. *Journal of the Optical Society of America A* **35**, 546-552

(2018).
85. Su, W. H. & Liu, H. Y. Calibration-based two-frequency projected fringe profilometry: a robust, accurate, and single-shot measurement for objects with large depth discontinuities. *Optics Express* **14**, 9178-9187 (2006).
86. Qiao, N. & Quan, C. Dual-frequency fringe projection for 3D shape measurement based on correction of gamma nonlinearity. *Optics & Laser Technology* **106**, 378-384 (2018).
87. Li, J. L. et al. Optimized two-frequency phase-measuring-profilometry light-sensor temporal-noise sensitivity. *Journal of the Optical Society of America A* **20**, 106-115 (2003).
88. Tahara, T. et al. Three-wavelength digital holography using spatial frequency-division multiplexing and dual reference arms. Proceedings of SPIE 9720 High-Speed Biomedical Imaging and Spectroscopy: Toward Big Data Instrumentation and Management. San Francisco: SPIE, 2016, 972009.
89. Barbosa, E. A. et al. Enhanced multiwavelength holographic profilometry by laser mode selection. *Optical Engineering* **46**, 075601 (2007).
90. Nadeborn, W., Andrä, P. & Osten, W. A robust procedure for absolute phase measurement. *Optics and Lasers in Engineering* **24**, 245-260 (1996).
91. Tian, X. B. et al. Snapshot multi-wavelength interference microscope. *Optics Express* **26**, 18279-18291 (2018).
92. Zhong, J. G. & Wang. M. Phase unwrapping by lookup table method: application to phase map with singular points. *Optical Engineering* **38**, 2075-2080 (1999).
93. Towers, C. E., Towers, D. P. & Jones, J. D. C. Optimum frequency selection in multifrequency interferometry. *Optics Letters* **28**, 887-889 (2003).
94. Towers, C. E., Towers, D. P. & Jones, J. D. C. Absolute fringe order calculation using optimised multi-frequency selection in full-field profilometry. *Optics and Lasers in Engineering* **43**, 788-800 (2005).
95. Falaggis, K., Towers, D. P. & Towers, C. E. Unified theory of phase unwrapping approaches in multiwavelength interferometry. Proceedings of SPIE 8011 22nd Congress of the International Commission for Optics: Light for the Development of the World. Puebla: SPIE, 2011, 80117H.
96. Falaggis, K., Towers, D. P. & Towers, C. E. Method of excess fractions with application to absolute distance metrology: wavelength selection and the effects of common error sources. *Applied Optics* **51**, 6471-6479 (2012).
97. Zuo, C. et al. High-speed three-dimensional shape measurement for dynamic scenes using bi-frequency tripolar pulse-width-modulation fringe projection. *Optics and Lasers in Engineering* **51**, 953-960 (2013).
98. Porras-Aguilar, R. & Falaggis, K. Absolute phase recovery in structured light illumination systems: sinusoidal vs. intensity discrete patterns. *Optics and Lasers in Engineering* **84**, 111-119 (2016).
99. Xu, Y. et al. Multi-frequency projected fringe profilometry for measuring objects with large depth discontinuities. *Optics Communications* **288**, 27-30 (2013).
100. Fei, L. H. et al. Single-wavelength phase retrieval method from simultaneous multi-wavelength in-line phase-shifting interferograms. *Optics Express* **22**, 30910-30923 (2014).
101. Saucedo, T. et al. Simultaneous two-dimensional endoscopic pulsed digital holography for evaluation of dynamic displacements. *Applied Optics* **45**, 4534-4539 (2006).
102. Goodman, J. W. Introduction to Fourier Optics. 3rd edn. (Greenwoood Village: Roberts & Company Publishers, 2005).
103. Schnars, U. & Jueptner, W. Digital Holography: Digital Hologram Recording, Numerical Reconstruction, and Related Techniques. (Berlin: Springer, 2005).
104. Bioucas-Dias, J. M. & Valadao, G. Phase unwrapping via graph cuts. *IEEE Transactions on Image Processing* **16**, 698-709 (2007).
105. Kreis, T. Handbook of Holographic Interferometry: Optical and Digital Methods. (Weinheim: Wiley, 2005).
106. Morimoto, Y. et al. Subnanometer displacement measurement by averaging of phase difference in windowed digital holographic interferometry. *Optical Engineering* **46**, 025603 (2007).
107. Nehmetallah, G. T., Aylo, R. & Williams, L. Analog and Digital Holography with MATLAB. (Bellingham: SPIE Press, 2015).
108. Shen, F. B. & Wang, A. B. Fast-Fourier-transform based numerical integration method for the Rayleigh-Sommerfeld diffraction formula. *Applied Optics* **45**, 1102-1110 (2006).
109. Xia, X. G. Discrete chirp-Fourier transform and its application to chirp rate estimation. *IEEE Transactions on Signal Processing* **48**, 3122-3133 (2000).
110. Yamaguchi, I. & Zhang, T. Phase-shifting digital holography. *Optics Letters* **22**, 1268-1270 (1997).
111. Tahara, T. et al. Dual-wavelength phase-shifting digital holography selectively extracting wavelength information from wavelength-multiplexed holograms. *Optics Letters* **40**, 2810-2813 (2015).
112. Servin, M. et al. Temporal phase-unwrapping of static surfaces with 2-sensitivity fringe-patterns. *Optics Express* **23**, 15806-15815 (2015).
113. Gass, J., Dakoff, A. & Kim, M. K. Phase imaging without 2π ambiguity by multiwavelength digital holography. *Optics Letters* **28**, 1141-1143 (2003).
114. Chen, B. Y. et al. Development of a laser synthetic wavelength interferometer for large displacement measurement with nanometer accuracy. *Optics Express* **18**, 3000-3010 (2010).
115. Onodera, R. & Ishii, Y. Two-wavelength interferometry that uses a Fourier-transform method. *Applied Optics* **37**, 7988-7994 (1998).
116. Rinehart, M. T. et al. Simultaneous two-wavelength transmission quantitative phase microscopy with a color camera. *Optics Letters* **35**, 2612-2614 (2010).
117. Birch, K. P. Optical fringe subdivision with nanometric accuracy. *Precision Engineering* **12**, 195-198 (1990).
118. Chapman, H. N. et al. High-resolution *ab initio* three-dimensional X-ray diffraction microscopy. *Journal of the Optical Society of America A* **23**, 1179-1200 (2006).
119. Chapman, H. N. et al. Femtosecond diffractive imaging with a soft-X-ray free-electron laser. *Nature Physics* **2**, 839-843 (2006).
120. Williams, G. J. et al. Three-dimensional imaging of microstructure in Au nanocrystals. *Physical Review Letters* **90**, 175501 (2003).
121. Marchesini, S. et al. X-ray image reconstruction from a diffraction pattern alone. *Physical Review B* **68**, 140101 (2003).
122. Latychevskaia, T. Iterative phase retrieval in coherent diffractive imaging: practical issues. *Applied Optics* **57**, 7187-7197 (2018).
123. Malm, E., Fohtung, E. & Mikkelsen, A. Multi-wavelength phase retrieval for coherent diffractive imaging. *Optics Letters* **46**, 13-16 (2021).
124. Gerchberg, R. W. & Saxton, W. O. A practical algorithm for the determination of phase from image and diffraction plane pictures. *Optik* **35**, 237-250 (1971).
125. Fienup, J. R. Phase retrieval algorithms: a comparison. *Applied Optics* **21**, 2758-2769 (1982).
126. Oszlányi, G. & Süto, A. *Ab initio* structure solution by charge flipping. *Acta Crystallographica Section A* **60**, 134-141 (2004).
127. Luke, D. R. Relaxed averaged alternating reflections for diffraction imaging. *Inverse Problems* **21**, 37-50 (2004).
128. Martin, A. V. et al. Noise-robust coherent diffractive imaging with a single diffraction pattern. *Optics Express* **20**, 16650-16661 (2012).
129. Shechtman, Y. et al. Phase retrieval with application to optical

imaging: A contemporary overview. *IEEE Signal Processing Magazine* **32**, 87-109 (2015).
130. Ou, X. Z., Zheng, G. A. & Yang, C. Embedded pupil function recovery for Fourier ptychographic microscopy. *Optics Express* **22**, 4960-4972 (2014).
131. Tian, L. et al. Multiplexed coded illumination for Fourier ptychography with an LED array microscope. *Biomedical Optics Express* **5**, 2376-2389 (2014).
132. Faulkner, H. M. L. & Rodenburg, J. M. Movable aperture lensless transmission microscopy: a novel phase retrieval algorithm. *Physical Review Letters* **93**, 023903 (2004).
133. Bates, R. H. T. & Rodenburg, J. M. Sub-ångström transmission microscopy: a Fourier transform algorithm for microdiffraction plane intensity information. *Ultramicroscopy* **31**, 303-307 (1989).
134. Ou, X. Z. et al. Aperture scanning Fourier ptychographic microscopy. *Biomedical Optics Express* **7**, 3140-3150 (2016).
135. Claus, D. et al. Dual wavelength optical metrology using ptychography. *Journal of Optics* **15**, 035702 (2013).
136. Choi, G. J. et al. Dual-wavelength Fourier ptychography using a single LED. *Optics Letters* **43**, 3526-3529 (2018).
137. Greivenkamp, J. E. Sub-Nyquist interferometry. *Applied Optics* **26**, 5245-5258 (1987).
138. Itoh, K. Analysis of the phase unwrapping algorithm. *Applied Optics* **21**, 2470 (1982).
139. Schuhler, N. et al. Frequency-comb-referenced two-wavelength source for absolute distance measurement. *Optics Letters* **31**, 3101-3103 (2006).
140. Amein, A. S. & Soraghan, J. J. A new chirp scaling algorithm based on the fractional Fourier transform. *IEEE Signal Processing Letters* **12**, 705-708 (2005).
141. Sukhoy, V. & Stoytchev, A. Numerical error analysis of the ICZT algorithm for chirp contours on the unit circle. *Scientific Reports* **10**, 4852 (2020).
142. Attota, R. et al. Application of through-focus focus-metric analysis in high resolution optical metrology. Proceedings of SPIE 5752 Metrology, Inspection, and Process Control for Microlithography XIX. San Jose: SPIE, 2005.
143. Tamamitsu, M. et al. A robust holographic autofocusing criterion based on edge sparsity: comparison of Gini index and Tamura coefficient for holographic autofocusing based on the edge sparsity of the complex optical wavefront. Proceedings of SPIE 10503 Quantitative Phase Imaging IV. San Francisco: SPIE, 2018, 105030J.
144. Bianco, V. et al. Strategies for reducing speckle noise in digital holography. *Light: Science & Applications* **7**, 48 (2018).
145. Williams, L. et al. Volume displacement measurement via multi-wavelength digital holographic surface topography at the microscopic level. Proceedings of SPIE 9006 Practical Holography XXVIII: Materials and Applications. San Francisco: SPIE, 2014, 90060K.
146. Yamaguchi, I., Yamamoto, A. & Kuwamura, S. Speckle decorrelation in surface profilometry by wavelength scanning interferometry. *Applied Optics* **37**, 6721-6728 (1998).
147. Lehmann, M. Decorrelation-induced phase errors in phase-shifting speckle interferometry. *Applied Optics* **36**, 3657-3667 (1997).
148. Schnars, U. et al. Digital Holography and Wavefront Sensing: Principles, Techniques and Applications. 2nd ed. (Berlin: Springer-Verlag, 2015).
149. Poittevin, J. et al. Quality assessment of combined quantization-shot-noise-induced decorrelation noise in high-speed digital holographic metrology. *Optics Express* **23**, 30917-30932 (2015).
150. Montresor, S. & Picart, P. Quantitative appraisal for noise reduction in digital holographic phase imaging. *Optics Express* **24**, 14322-14343 (2016).
151. Lee, J. S. Digital image enhancement and noise filtering by use of local statistics. *IEEE Transactions on Pattern Analysis and Machine Intelligence* **PAMI-2**, 165-168 (1980).
152. Mallat, S. A Wavelet Tour of Signal Processing. (New York: Academic Press, 1999).
153. Donoho, D. L. De-noising by soft-thresholding. *IEEE Transactions on Information Theory* **41**, 613-627 (1995).
154. Wang, D. D. et al. Compact snapshot multiwavelength interferometer. *Optics Letters* **44**, 4463-4466 (2019).
155. Banerjee, P. P., Beresnev, L. A. & Vorontsov, M. A. Cancellation of effects of large phase distortions on images by dynamic holography using ferroelectric liquid crystal spatial light modulators. Proceedings of SPIE 3760 High-Resolution Wavefront Control: Methods, Devices, and Applications. Denver: SPIE, 1999, 83-87.
156. Farriss, W. E. et al. Sharpness-based correction methods in holographic aperture ladar (HAL). Proceedings of SPIE 10772 Unconventional and Indirect Imaging, Image Reconstruction, and Wavefront Sensing 2018. San Diego: SPIE, 2018, 107720K.
157. Tippie, A. E. & Fienup, J. R. Multiple-plane anisoplanatic phase correction in a laboratory digital holography experiment. *Optics Letters* **35**, 3291-3293 (2010).
158. Rodríguez, C. et al. An adaptive optics module for deep tissue multiphoton imaging in vivo. *Nature Methods* **18**, 1259-1264 (2021).
159. Venediktov, V. Y. et al. Two-wavelength dynamic holography and its application in adaptive optics. in Adaptive Optics for Industry and Medicine (ed Love, G. D.) (World Scientific, 1999), 317-322.
160. Hampson, K. M. et al. Adaptive optics for high-resolution imaging. *Nature Reviews Methods Primers* **1**, 68 (2021).
161. Saha, S. K. Modern optical astronomy: technology and impact of interferometry. *Reviews of Modern Physics* **74**, 551 (2002).
162. Lycksam, H. et al. Wiener filtering of interferometry measurements through turbulent air using an exponential forgetting factor. *Applied Optics* **47**, 2971-2978 (2008).
163. Lycksam, H., Sjödahl, M. & Gren, P. Measurement of spatiotemporal phase statistics in turbulent air flow using high-speed digital holographic interferometry. *Applied Optics* **49**, 1314-1322 (2010).
164. Labeyrie, A. Attainment of diffraction limited resolution in large telescopes by Fourier analyzing speckle patterns in star images. *Astron. Astrophys* **6**, 85-87 (1970).
165. Upputuri, P. K. et al. Two-wavelength microscopic speckle interferometry using colour CCD camera. Proceedings of SPIE 9302 International Conference on Experimental Mechanics 2014. Singapore: SPIE, 2014, 93023K.
166. Zheng, G. A. et al. Concept, implementations and applications of Fourier ptychography. *Nature Reviews Physics* **3**, 207-223 (2021).
167. Baek, Y. et al. Kramers-Kronig holographic imaging for high-space-bandwidth product. *Optica* **6**, 45-51 (2019).
168. Shen, C. et al. Non-iterative complex wave-field reconstruction based on Kramers-Kronig relations. *Photonics Research* **9**, 1003-1012 (2021).
169. Scott, A., Banerjee, P. & Slagle, J. Non-mechanical beam steering using reflection-mode readout of volume gratings in a photorefractive material. Proceedings of SPIE 11498 Photonic Fiber and Crystal Devices: Advances in Materials and Innovations in Device Applications XIV. SPIE, 2020, 114980G.

Hartlieb et al. *Light: Advanced Manufacturing* (2021)2:32
https://doi.org/10.37188/lam.2021.032

Official journal of the JHL 2689-9620
www.light-am.com

Article | Open Access

Multi-positional image-based vibration measurement by holographic image replication

Simon Hartlieb[1,*], Michael Ringkowski[2], Tobias Haist[1], Oliver Sawodny[2] and Wolfgang Osten[1]

Abstract

In this study we present a novel and flexibly applicable method to measure absolute and relative vibrations accurately in a field of 148 mm × 110 mm at multiple positions simultaneously. The method is based on imaging in combination with holographic image replication of single light sources onto an image sensor, and requires no calibration for small amplitudes. We experimentally show that oscillation amplitudes of 100 nm and oscillation frequencies up to 1000 Hz can be detected clearly using standard image sensors. The presented experiments include oscillations of variable amplitude and a chirp signal generated with an inertial shaker. All experiments were verified using state-of-the-art vibrometers. In contrast to conventional vibration measurement approaches, the proposed method offers the possibility of measuring relative movements between several light sources simultaneously. We show that classical band-pass filtering can be omitted, and the relative oscillations between several object points can be monitored.

Introduction and state of the art

Precise measurement of deformations and vibrations is required in a wide range of industrial applications. Classical approaches such as laser Doppler vibrometers (LDVs) offer the possibility of measuring object vibrations at high temporal and spatial resolution. However, as soon as multiple simultaneous measurements are required, the use of single LDVs quickly becomes impractical and costly. The market-driven need for simultaneous vibration measurements at multiple positions manifested itself in the development of LDVs with multi-beam[1–4] and multi-sensor[5–7] applications. In addition, commercial solutions consisting of multiple sensor heads are available[8], but these systems are either limited by the flexibility of beam orientation or costly in terms of adjustment of sensor heads, signal synchronization, and especially in terms of price. Scanning laser interferometers[9,10] can be used to measure full-field vibrations, but the obtained signals cannot be acquired simultaneously, making it impossible to measure transient signals.

Image-based vibration measurement techniques inherently offer the possibility of measuring the movement of an object remotely at multiple positions as well as measuring the displacement of several objects simultaneously.

In many cases, planar, passive targets are used, reaching accuracies in the range of 0.6 mm at a distance of 100 m and 0.1 mm for a distance of 15 m and only low-frequency vibrations are detected[11,12]. As an alternative to the planar target, active elements (LEDs) and edge detection can be used[13,14]. Especially in the field of edge detection, a lot of research regarding motion magnification, a technique that

Correspondence: Simon Hartlieb (Hartlieb@ito.uni-stuttgart.de)
[1]Institute for Applied Optics, University of Stuttgart, Pfaffenwaldring 9, Stuttgart 70569, Germany
[2]Institute for System Dynamics, University of Stuttgart, Waldburgstraße 17/19, Stuttgart 70563, Germany

allows magnifying small displacements, has been conducted recently. A promising approach was presented by Wu et al., where small movements were magnified in video sequences by applying a bandwidth filter in the spatial frequency domain[15]. This method was used in the vibration analysis of structures and achieved impressive accuracies of 0.006 pixel[16,17]. Nevertheless, the method is limited to stationary vibrations and can produce misleading artifacts[18]. The time-varying motion filtering proposed by Liu et al. can be a solution for the problem of limited bandwidth[19].

Higher frequency vibrations of up to 1000 Hz were analyzed by several authors[20–22], using techniques such as phase-based optical flow and pattern-matching. The reported standard uncertainties are 0.032 mm for vibration frequencies of 100–300 Hz and 0.013 mm for 400–600 Hz using passive targets and a pattern-matching technique[22]. This corresponds to a dynamic range of 12000:1 (measurement range/uncertainty).

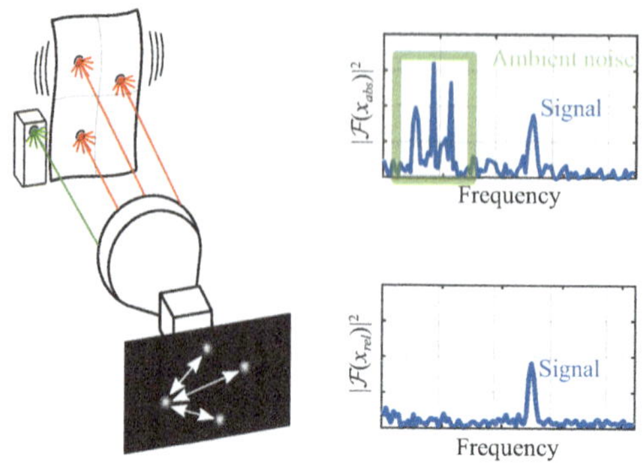

Fig. 1 Scheme of the vibration measurement method, including the spectra of the absolute and relative spot position signals. Point light sources on a vibrating object (red) and a static object (green) are imaged to a camera. The center of the imaged spot is used as the position signal. Environmental vibrations (marked with a green square) are present in the absolute position spectrum x_{abs} of the red light sources. By calculating the relative movement x_{rel} between the red and green light sources, ambient vibrations can be removed.

Principle and methods

Our approach for measuring the vibration of an object is shown in Fig. 1. Multiple light sources (shown in red) were attached to a vibrating object and imaged to a camera. By monitoring the position of each object light source with a fixed acquisition frame rate, movements between each frame can be ascribed to a superposition of the object vibration and ambient vibrations of the laboratory room, neglecting other error sources such as air turbulence and sensor noise. One static light source (green) is not attached to the vibrating object; therefore, it is only exposed to ambient vibrations. By calculating the relative displacement between an object and a static light source, these unwanted vibrations can be compensated for. This is schematically visualized in the two spectra shown in Fig. 1. The upper spectrum is given for the absolute movement of an object light source x_{abs} in the presence of unwanted ambient vibrations. The lower shows the spectrum of the relative displacement x_{rel} between a vibrating and a static light source without ambient vibrations.

In the following, this method is improved by using holographical image replication to reach very high resolution, and we present experimental results to support the ability to compensate for ambient vibration in relative position measurement.

In our measurement system, the position of each light source is defined by the location of the intensity distribution (spot) on the camera sensor. Therefore, it is crucial for our measurement system to detect the spot position as accurately as possible. To this end, the holographic multipoint method is applied[23]. In the following, we provide a brief introduction to the underlying spot detection.

Holographic multipoint method

The accuracy of subpixel spot position detection is limited by several factors such as photon noise, electronic noise[24], discretization and quantization of the camera sensor[25], and the choice of centroiding algorithm[26]. The number of collected photons that contribute to the intensity distribution of a spot is directly correlated with the uncertainty of its position determination. Starting from a single object point, if M photons are imaged to the camera and individual position measurements are made for each photon, the mean over all individual measurements leads to a positional standard deviation that is reduced by a factor of $\sqrt{M}$[27]. Temporal averaging is one way to increase the number of collected photons and therefore improve accuracy. However, temporal averaging reduces neither the fixed pattern noise of the image sensor nor the discretization error. In terms of image-based vibration measurements, temporal averaging would also reduce temporal resolution.

The principle of our vibration measurement method is based on spatial averaging. The spot of a single light source is holographically replicated into a cluster of N = 21 spots. The spot cluster is generated using a lithographically fabricated diffractive optical element (DOE)[28], which is placed in the Fourier plane of a bi-telecentric lens, as shown in Fig. 2. If the light source is moved, all replicated spots move by the same amount (Fourier shift theorem). As described above, by making the light source brighter, the number of photons and

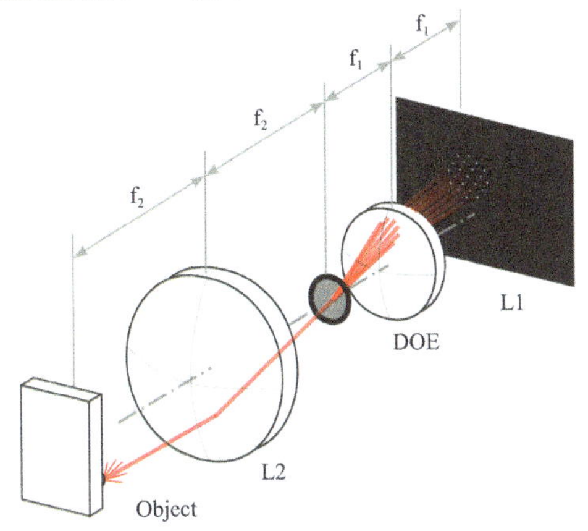

Fig. 2 Scheme of a bi-telecentric lens with f_1, f_2 as the focal lengths of the lenses L1 and L2. A diffractive optical element (DOE) is placed in the aperture (Fourier) plane. The DOE replicates a single light source into a predefined cluster of $N = 21$ spots.

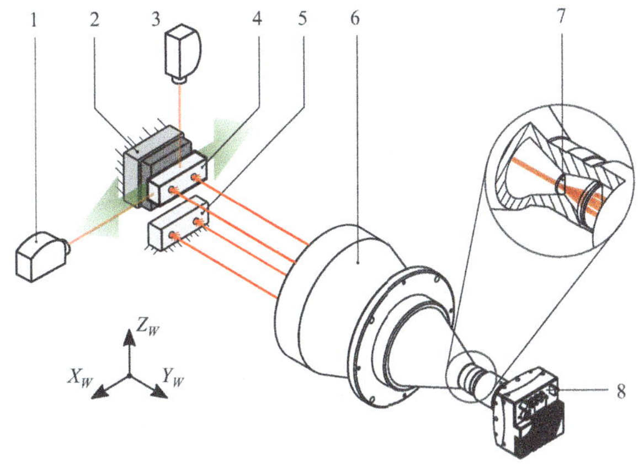

Fig. 3 Experimental setup: A telecentric lens (6) images the light sources of a moving (4) and a static (5) object using a high-speed camera (8). Each spot of a light source is replicated using a DOE (7). A piezo stage (2) is used to actuate an object (4), and its displacement is measured by two vibrometers (1,3).

simultaneously the number of pixels that sample the intensity distribution are increased. By calculating the mean of all spot centers per cluster, the position accuracy of the light source can be improved ideally by a factor of $\sqrt{N}$. This has been shown by Haist et al.[23], where a single light source was replicated into a cluster of $N = 16$ spots. The detection accuracy was improved in experiments from 0.01 pixels to 0.0028 pixels, which corresponds to an improvement factor of 3.6 (theoretical factor of 4).

Measurement setup

A schematic of the measurement setup is depicted in Fig. 2. We use a camera system consisting of a bi-telecentric lens (6) (Vico DTCM110-150-AL, $NA = 0.0085$, $|\beta'| = 0.11$, distortion < 0.1%) and a high-speed camera (8) (Mikrotron EoSens 4CXP, 4 Mpix, pixel size $d_p = 7$ µm, 563 fps full frame). In the aperture plane of the bi-telecentric lens (6), a DOE (7) is mounted to perform the multipoint replication. Light sources (Roithner LaserTechnik, SMB1N-D630, $\lambda = 630$ nm, FWHM = 16 nm, radiant intensity = 120 mW/sr) are attached to a static object (5) and to an object (5) that is mechanically actuated by a piezo stage (2) (Physik Instrumente P-611.2S, X/Y-travel range = 100 µm, resolution = 0.2 nm). In front of each light source, a pinhole ($D = 200$ µm) is mounted to obtain comparable radiation profiles and small spots on the camera. The displacement of the object (4) is measured in the X_W- and Z_W-directions using two vibrometers (1) and (3) (Polytec NLV 2 500). In this study we use only the X_W-direction.

The piezo stage is operated in a closed-loop with a commercial controller (PI E-727.3SDA, 20 kHz). The position references for generating the vibrations are supplied from a dSpace DS-1005 1 GHz PowerPC (2 kHz/3 kHz sampling rate), which is also used for signal acquisition. The camera is hardware-triggered using a frequency generator (Rigol DG1022Z). The maximum acquisition frame rate of the camera is defined by the size of the captured image region of interest (ROI). For an ROI size of 900 pixels × 576 pixels, the maximum frame rate is 2000 frames per second (fps). The size of one cluster in the X- and $Y-$ directions is 170 pixels × 170 pixels, so a total of 15 object points (clusters) can be monitored simultaneously at 2000 fps. For an image size of 700 pixels × 300 pixels, the maximum frame rate can be increased to 3000 fps.

The two-dimensional measurement field is defined by the camera sensor size divided by the magnification of the imaging lens. For the measurement setup presented in this study, the field size is 148 mm × 110 mm. This enables the simultaneous monitoring of a total number of 130 equally spaced object points with a frame rate of 563 Hz.

Image processing

To measure the vibration of each light source, it is necessary to calculate the position of each corresponding cluster in the image. The procedure for determining the subpixel cluster position is summarized in Fig. 4.

Step 1: In the first step, an image region containing one cluster is selected, saved, and used for cross correlation with the captured images. In this template image, the coarse position of each spot is localized by convolution with a blur kernel and by the application of a local maxima algorithm. The coarse spot positions (marked as blue crosses) are given in the local template coordinate system and are stored in a vector (spot map). As a consequence of

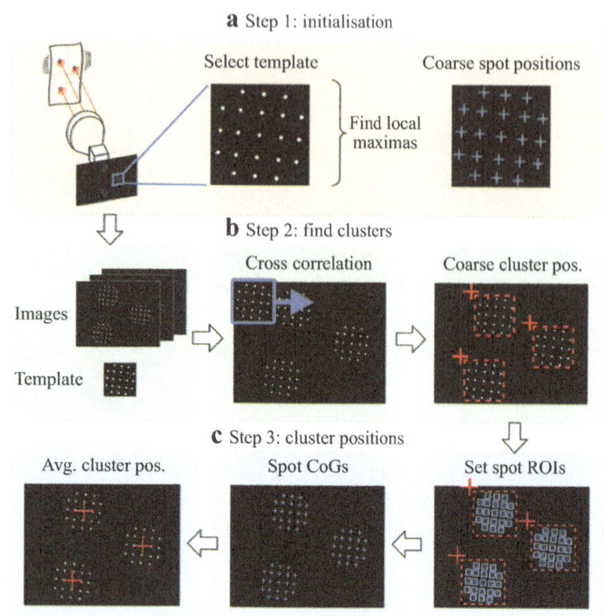

Fig. 4 Three-step image processing procedure: **a** Initialization with spot map; **b** Find clusters; **c** Calculate averaged subpixel cluster positions.

the Fourier-Shift-Theorem, all replicated spots move by the same amount and, therefore, the relative position between all spots per cluster is in good approximation constant over the whole image. Therefore, once the spot map is generated, it can be applied to all clusters in all images of the image stack.

Step 2: In the second step each cluster position is localized coarsely. The upper left corners of each cluster are found using cross-correlation with the template from step 1. The correlation result is blurred, and local maxima are obtained by applying a local maxima algorithm. For vibration measurements, it is sufficient to detect the cluster positions only for the first image and to keep those positions for all subsequent images of the stack, because the cluster movement does not exceed a few pixels.

Step 3: The coordinates of the spot map (step 1) are combined with the coarse cluster positions from step 2. An ROI can be applied around each single spot per cluster. The ROIs are marked in Fig. 4c using blue squares. The spot diameter is approximately 8 to 10 pixels, so the square ROI around each spot is selected to be 30 pixels wide, which corresponds to 1.9 mm in the object space.

The subpixel position (x_{sp}, y_{sp}) of one spot inside the ROI is calculated as

$$(x_{sp}, y_{sp}) = \left(\frac{\sum_x \sum_y xI(x,y)}{\sum_x \sum_y I(x,y)}, \frac{\sum_x \sum_y yI(x,y)}{\sum_x \sum_y I(x,y)} \right) \tag{1}$$

using the gray value weighted center of gravity (CoG), where $I(x, y)$ is the intensity at position (x, y). All intensity values below a threshold are set to zero to remove background noise. The position of one cluster containing N spots is given by

$$(x_{cl}, y_{cl}) = \left(\frac{1}{N} \sum_{i=1}^{N} x_{sp,i}, \frac{1}{N} \sum_{i=1}^{N} y_{sp,i} \right) \tag{2}$$

The position of each cluster is given in pixel coordinates. A conversion to a metric length unit can be achieved using the magnification β' of the bi-telecentric lens and the pixel size d_p of 7 µm. The conversion factor k is given by

$$k = \frac{d_p}{\beta'} = \frac{7\mu\text{m}}{0.11} = 63.64 \frac{\mu\text{m}}{\text{Pixel}} \tag{3}$$

This linear mapping can be applied only for small movements on the sensor, because the lens distortion introduces an error that is position dependent. However, for vibrations, linear mapping is a valid assumption.

Experiments and results

The goal of the presented experiments is to obtain an overview of the capabilities of the proposed vibration measurement method in terms of accuracy improvement and resolution limit. To investigate the sensitivity of the proposed measurement method over a wide frequency range, an inertial shaker is used to create a chirp signal up to high frequencies.

Accuracy improvement

The measurement setup shown in Fig. 3 is used to demonstrate the accuracy improvement using the multipoint method. The piezo stage is actuated in the X_W-direction with a sinusoidal oscillation of amplitude 1 µm and frequency of 50 Hz for a time period of 0.4 s. The camera image has a size of 900 pixels × 576 pixels, and the sampling rate of the camera is set to 1 000 fps.

In Fig. 5, the positional signals of the vibrometer and the camera are given in micrometer and pixel units. The reference signal of the X-vibrometer is plotted in the middle (red). For clarity, the camera signals have an additional offset. The signal is evaluated using a conventional single spot (upper, blue signal) as described in Eq. 1, shifted by +3 µm. The averaged cluster position using the multipoint method (lower, green signal), calculated according to Eq. 2, is shifted by -3 µm. Both signals are high-pass filtered (30 Hz cutoff frequency). In the lower plot, the difference between the two camera signals and the reference signal is shown. It is clearly visible that the multipoint position signal (x_{cl}) matches the vibrometer signal (x_{vib}) better than the conventional single-spot evaluation (x_{sp}) by a factor of 4. The standard

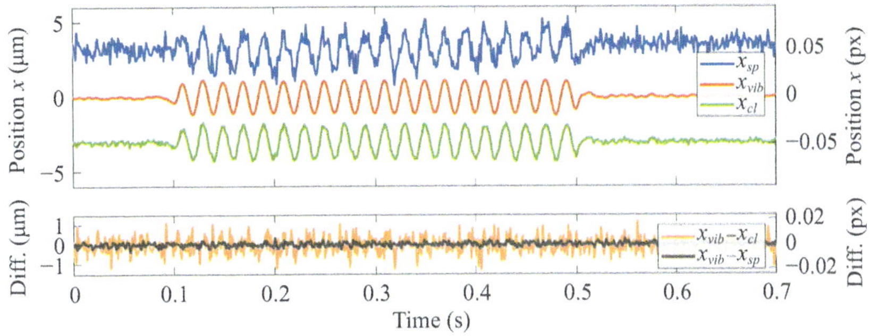

Fig. 5 Accuracy improvement of multipoint method: X–oscillation measurement (frequency 50 Hz, 1 µm amplitude) with the proposed camera system and a vibrometer as a reference. The upper signal x_{sp} (blue) is evaluated using the position signal of a single spot (Eq. 1) and the lower signal x_{cl} (green) using the averaged cluster position (Eq. 2) (shown with ±3 µm offsets). The difference with respect to the reference signal x_{vib} is plotted in the lower chart.

deviation of the difference signals is σ_{sp} = 0.402 µm for a single spot, and σ_{cl} = 0.104 µm for the averaged cluster position. Applying the conversion factor k (Eq. (3)), this corresponds to σ_{sp} = 0.0069 pixels and σ_{cl} = 0.0017 pixels.

Resolution limit

To analyze the lower bound on the measurable amplitudes, we use the same experimental setup as in the previous section.

Our experiments show that an amplitude of 100 nm can still be measured using this method. Together with the measurement range of 148 mm this corresponds to a dynamic range of 1480000:1. In Fig. 6a the camera (red) and the vibrometer (blue) signals are depicted for an oscillation amplitude of 100 nm. Both signals are high-pass filtered (30 Hz cutoff frequency) to obtain the pure motion signals detected by the vibrometer and the camera. Despite the noise of the camera signal, the oscillation frequency is clearly visible and matches the vibrometer signal. An amplitude of 100 nm in the object space can be converted to the imaging sensor using the magnification β' or the conversion factor k. This corresponds to 11 nm (0.0016 pixels) on the sensor. The difference between the vibrometer and the camera is plotted in the lower chart. The standard deviation of the difference signal in the image plane is σ_d = 0.095 µm or σ_d = 0.0015 pixels.

Fig. 6b shows the amplitude spectrum of the camera (upper plot) and the vibrometer (lower plot). The filtered-out ambient vibration spectrum (below 30 Hz) is shown in light blue. It can be observed that both measured frequency spectra show good correspondence. The detected peak frequency of the proposed measurement system as well as that of the reference system is 49.62 Hz. Small stimulations at a frequency of approximately 100 Hz are also detected with both systems.

Ambient vibration

Ambient vibrations are present in almost every vibration measurement scenario. Conventionally, they are suppressed using a high-pass filter, as done in the previous

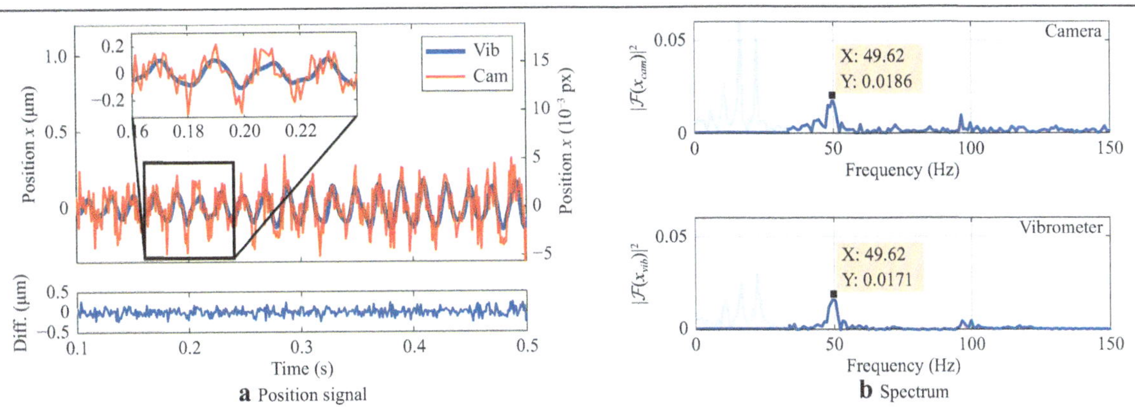

Fig. 6 Piezo stage oscillation (X_W-direction) of 50 Hz and 100 nm amplitude measured by the proposed camera setup and a vibrometer as reference. a Position signals over time of camera (red) and vibrometer (blue); the difference is plotted beneath. Both signals are high-pass filtered (30 Hz cutoff frequency). **b** Amplitude spectrum of the camera signal (upper) and the vibrometer signal (lower). Both have a peak frequency at 49.62 Hz. The filtered-out ambient vibration spectrum is shown in light blue.

sections. The frequency peaks of the ambient vibrations in our laboratory room are 10.4 Hz, 16.2 Hz, and 22.1 Hz. In the case of the vibrometer, those disturbing vibrations are not easy to avoid, whereas with the camera system, they can be suppressed almost entirely. Therefore, as described in the introduction, we use the relative movement between the moving and static light sources in the image.

A typical image containing four clusters, C_0 to C_3, is depicted in Fig. 7a. The blue clusters C_0 and C_1 belong to the light sources that are attached to the piezo stage, and the yellow clusters C_2 and C_3 belong to the static light sources (see Fig. 3a).

The relative position is calculated as

$$x_{rel} = \frac{x_{C_0} + x_{C_1}}{2} - \frac{x_{C_2} + x_{C_3}}{2} \tag{4}$$

with $x_{c,i}$ being the X-coordinates of clusters C_i, i = 0,1,2,3.

A spectral analysis was performed for the vibrometer, the absolute position signal of cluster C_0, and the relative position signal according to Eq. 4. The results are shown in Fig. 7b. Ambient vibrations introduced by the laboratory environment are marked in the spectrum of the absolute cluster position C_0 (middle plot) and in the vibrometer spectrum (lower plot) with green squares. The upper plot shows the spectrum of the relative position signal obtained using Eq. 4. It can be seen that all three peaks of the ambient vibrations are compensated almost entirely. We would like to emphasize that with this method, the signal to be measured does not have to be spectrally separated from ambient vibrations. The reason we have chosen a signal frequency higher than the ambient vibration frequencies was to be able to compare our results with the LDV.

High frequency measurement

We also use a chirp signal to verify the sensitivity of the proposed measurement method over a wide frequency range. The chirp signal consisted of frequencies that linearly increased from 200 Hz to 1000 Hz at a constant amplitude. An inertial shaker is used to convert the chirp signal to a movement in the X–direction, and a single light source is attached to the shaker tip. The camera frame rate is set to 3000 fps, requiring a reduced image resolution of 700 pixels × 300 pixels. The vibrometer signal is also acquired at 3000 Hz.

Fig. 8 shows the measurement results of the vibrometer and the camera, including the position signals in the time domain and their difference (a), as well as the short-time Fourier transform (STFT) of each signal in (b) and (c). The vibrometer and the camera signals are high-pass filtered with 30 Hz and both are compensated for positional and time offset. The time-domain plot demonstrates that the change in amplitude and frequency can be accurately measured by the proposed system. The standard deviation of the difference signal is σ_d = 0.242 µm or σ_d = 0.0038 pixels.

Discussion

All presented measurement results were achieved by applying a (linear) conversion factor k from the image domain to real-world units. However, as the vibrational amplitude increases, the influence of lens distortions on the position signal increases. To evaluate the introduced error due to the linear mapping, an accurate calibration of the camera system would be necessary. However, if larger amplitudes are to be measured, a resolution of 100 nm often is not necessary. In [29, 30], it was reported that the standard deviation of the reprojection error for a linear

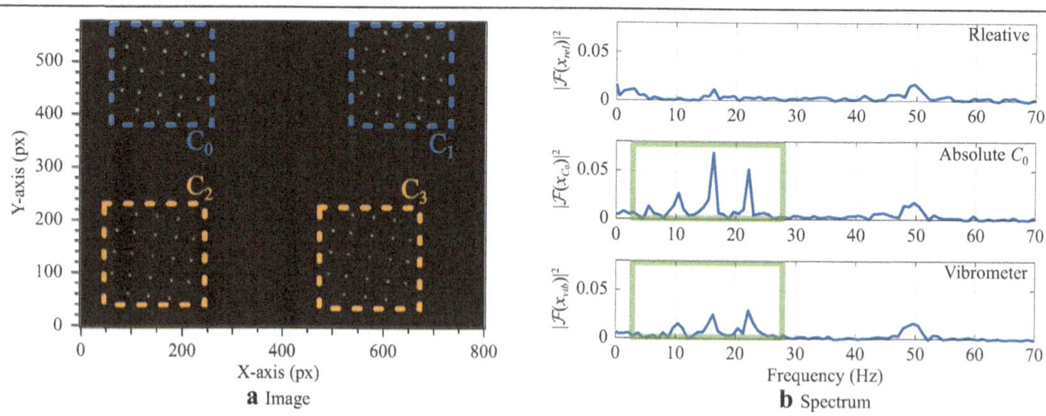

Fig. 7 **a** The camera image of the experimental setup of Fig. 3 containing four clusters. The blue clusters C_0 and C_1 are actively moved by the piezo stage, and the yellow clusters C_2 and C_3 are static. **b** The spectrum of an X-oscillation (50 Hz, 100 nm amplitude) for the mean relative position $(x_{c_0} + x_{c_1})/2 - (x_{c_2} + x_{c_3})/2$ (upper), the absolute position x_{c_0} (middle), and the vibrometer position signal (lower). The absolute position signal of the camera and vibrometer show environmental vibrations (green squares) that necessitate the use of a high-pass filter. In the relative signal (upper chart), these vibrations are intrinsically compensated and require no filtering.

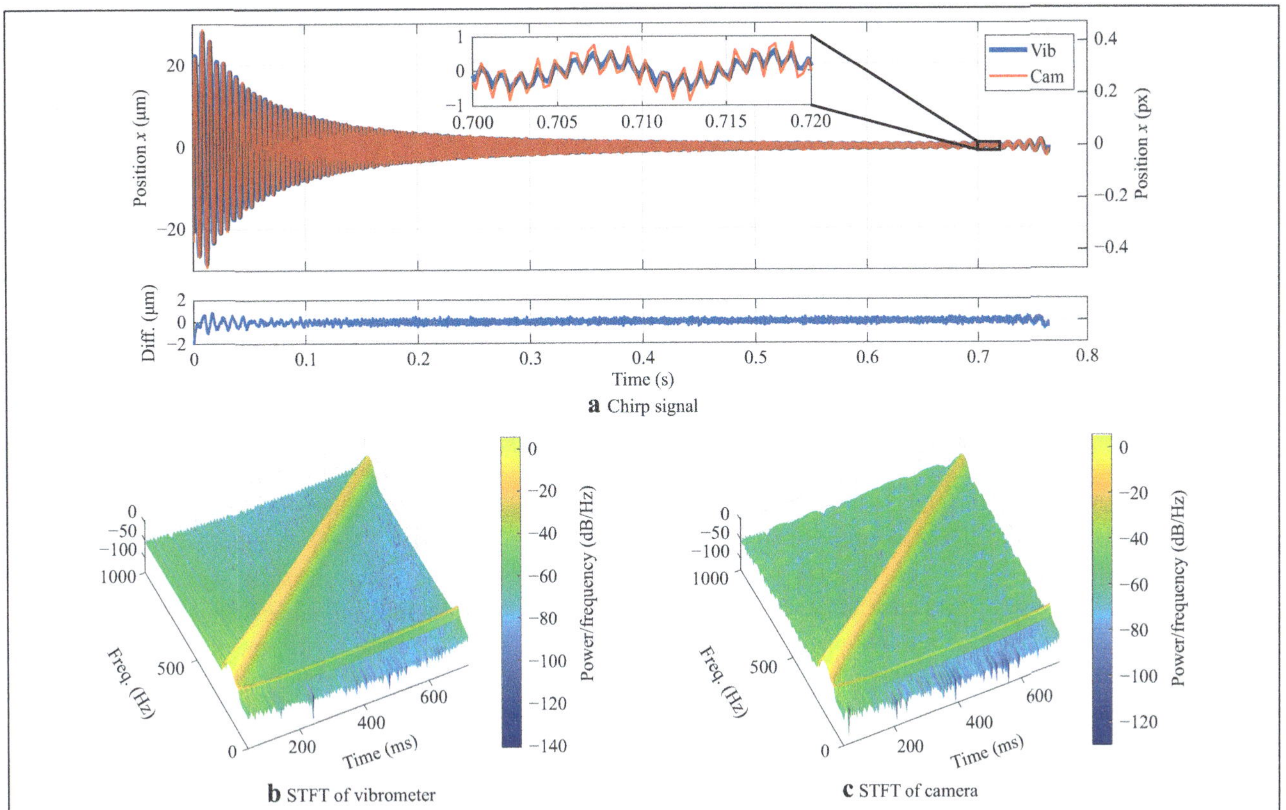

Fig. 8 High frequency measurement. **a** Chirp signal of an inertial shaker with linearly increasing frequency from 200 Hz to 1000 Hz measured with a camera and vibrometer. **b** and **c** are the short-time Fourier transform (STFT) of each signal.

conversion factor is 16 μm (standard deviation) for a measurement field of 100 mm × 74 mm. Simulations using 1000 different measurements with amplitudes ranging from 0.1 mm to 50 mm show that the standard deviation is in good approximation linearly dependent on the amplitude. Therefore, depending on the application (vibration amplitude size, desired accuracy), it has to be decided whether calibration is necessary. For our experiments, this error was negligible because the measured amplitudes were very small.

Similar to any other vibration measurement method, the proposed system has certain advantages and disadvantages. The precision of the proposed method depends on the amount of light available at the object points. To achieve a high signal-to-noise ratio, we applied active light sources with commercial pinholes (Thorlabs, P200K) mounted in front to obtain small spots and a similar intensity distribution for all clusters. A commercial pinhole is rather inconvenient for use in industrial applications because it makes the light sources bulky and difficult to attach. Here, a blackened aluminum foil with small holes can be used as the aperture, placed in front of the LEDs. Together with a small coin cell, the LED modules can become very small, easy to attach, and can operate independently. Another disadvantage compared to an LDV is the need for active light modules that must be attached to the object. One solution for this issue is the use of fluorescent particles or micro reflectors (spheres) that are attached to the object, in combination with external illumination. The resulting effect on measurement resolution, in particular for small amplitudes and high frequencies, will be discussed in a following publication.

On the other hand, the advantages of the system include its high resolution and ease of use regarding application and signal processing. Furthermore, with the proposed camera-based method, it is possible to measure the vibration amplitudes directly in two dimensions without the need for time-based integration of the acceleration signal, as would be necessary for LDVs. This also offers the opportunity to measure the relative movements between multiple target points.

The concept of averaging over a certain lateral domain to increase the measurement uncertainty is used not only in our method, but also in digital image correlation (DIC), which is a common technique for measuring full-field vibrations. In DIC, the resolution as well as the accuracy of lateral (in-plane) and axial (out-of-plane) position measurements depend on the size of the correlation patch.

Using large correlation patches, the in-plane and out-of-plane resolution of DIC can be very good. In [31], for example, a sensitivity of 3.3 μm was reported for a field of view of 100 mm × 80 mm. Large patches, however, also lead to reduced lateral resolution and sampling of this lateral and axial position measurement. This is a problem for high lateral resolution measurements, which are necessary, especially for objects that deform. In DIC, even slight deformations over the extent of the correlation patch lead to inaccuracies in the correlation. Our method avoids this problem by averaging while maintaining the highest lateral resolution of the object by using localized small object points. Averaging is realized in the image space instead of the object space.

To date, the multipoint method has been used for the measurement of building deformation and 2D/3D coordinate measurements[29,30,32,33]. In the case of building deformation measurement, accelerometers are often employed because of their high resolution and ease of use. However, accelerometers measure only changes in velocity and, therefore, tend to drift, especially for slow movements. In addition, they often need wiring for power supply and data transfer, which makes them more inconvenient to apply, compared to an independent light module.

The application scope of our method is mainly seen in the industrial sector, for example, in measuring vibrations of car engines, car body parts (e.g., brakes), or other objects where it is helpful to measure at multiple positions simultaneously. Applications that are not practicable for the proposed method are those in which the mass or stiffness of the measured object is changed significantly by the applied light modules, for example in mini- or microsystems or perhaps in ultra-lightweight constructions. For larger applications such as bridges or buildings, the multipoint method can also be used, but error influences resulting from air turbulence and thermally induced refraction index changes must be considered. As long as these restrictions are considered, this measurement system offers a cheap and simple way to measure a large range of amplitudes with very high resolution.

Summary and conclusion

The proposed vibration measurement system is based on the detection of active light markers attached to a vibrating object. A diffractive optical element is used to replicate each light source holographically into a cluster of $N = 21$ spots in the image plane. Spatial averaging of all spot centers per cluster improves the detection accuracy of the light marker position ideally by a factor of $\sqrt{N}$.

The proposed vibration measurement setup is able to detect vibration amplitudes of 100 nm clearly in both the spatial and frequency domains with a standard deviation of $\sigma_d = 0.095$ μm, which corresponds to $\sigma_d = 0.0015$ pixels. To the best of our knowledge, this is the highest reported accuracy for imaging-based spot position measurements. The dynamic range of the proposed system is 1480000:1. High frequencies were analyzed by attaching a light source to an inertial shaker. A chirp signal from 200 Hz to 1000 Hz with amplitudes from 28 μm to 0.3 μm was measured. Both spectrograms (STFT) of the camera and the vibrometer show good correspondence.

We also showed that ambient vibrations can be compensated almost entirely by calculating the relative movement between the light markers attached to the environment and to the object, making it possible to avoid the use of band-pass filters. In addition, the concept of band-pass filtering may also be combined with the aforementioned subtraction method. In this way, the signal quality in terms of noise of the specimen, which vibrates with a frequency in the range of the ambient, can be improved even further. The main advantages of the proposed method are its simplicity and cost regarding components and signal processing, and in the high resolution that can be achieved even under external disturbances.

Acknowledgements

We gratefully thank the German Research Foundation (DFG - Deutsche Forschungsgemeinschaft) for funding the projects "Dynamische Referenzierung von Koordinatenmess- und Bearbeitungsmaschinen" (OS 111/42-2 and SA 847/16-2).

Author contributions

Simon Hartlieb performed the mechanical and optical design as well as the measurements, evaluations, and article writing. Michael Ringkowski implemented the measurement software and supported the measurements and evaluations. Tobias Haist, Oliver Sawodny, and Wolfgang Osten provided advice throughout the work, including reviewing and editing of the article.

Conflict of interest

The authors declare no conflicts of interest.

Received: 13 April 2021 Revised: 25 November 2021 Accepted: 28 November 2021
Accepted article preview online: 06 December 2021
Published online: 21 December 2021

References

1. Y. Fu et al. Spatially encoded multibeam laser doppler vibrometry using a single photodetector. *Optics Letters* **35**, 1356-1358 (2010).
2. P. B. Phua et al. Multi-beam laser doppler vibrometer with fiber sensing head. *AIP Conference Proceedings* **1457**, 219-226, 06 (2012).
3. T. Haist et al. Characterization and demonstration of a 12-channel laser-doppler vibrometer. volume **8788**. Proceedings of SPIE, 2013.

4. C. Yang et al. A multi-point laser doppler vibrometer with fiber-based configuration. *Review of scientific instruments* **84**, 121702 (2013).
5. R. D. Burgett et al. Mobile mounted laser Doppler vibrometer array for acoustic landmine detection. In *Detection and Remediation Technologies for Mines and Minelike Targets VIII*, volume **5089**, pages 665–672. Proceedings of SPIE, 2003.
6. W. N. MacPherson et al. Multipoint laser vibrometer for modal analysis. *Applied Optics* **46**, 3126-3132 (2007).
7. J. M. Kilpatrick & Markov V. Matrix laser vibrometer for transient modal imaging and rapid nondestructive testing. In *Eighth International Conference on Vibration Measurements by Laser Techniques: Advances and Applications*, volume **7098**. Proceedings of SPIE, 2008.
8. Polytec GmbH. https://www.polytec.com/de/vibrometrie/produkte/full-field-vibrometer/mpv-800-multipoint-vibrometer. (2021-11-04).
9. D. Kim et al. 3-d vibration measurement using a single laser scanning vibrometer by moving to three different locations. *IEEE Transactions on Instrumentation and Measurement* **63**, 2028-2033 (2014).
10. K. Kokkonen & M. Kaivola. Scanning heterodyne laser interferometer for phase-sensitive absoluteamplitude measurements of surface vibrations. *Applied Physics Letters* **92**, (2008).
11. S. Kim et al. A vision system for identifying structural vibration in civil engineering constructions. In *Proceedings of 2006 SICE-ICASE International Joint Conference*. IEEE, 2006.
12. D. Ribeiro et al. Non-contact measurement of the dynamic displacement of railway bridges using an advanced video-based system. *Engineering Structures* **75**, 164-180 (2014).
13. A. M. Wahbeh et al. A vision-based approach for the direct measurement of displacements in vibrating systems. *Smart Materials and Structures* **12**, 785-794 (2003).
14. S. Patsias & W. J. Staszewskiy. Damage detection using optical measurements and wavelets. *Structural Health Monitoring* **1**, 5-22 (2002).
15. H. Y. Wu et al. Eulerian video magnification for revealing subtle changes in the world. *ACM Transactions on Graphics* **31**, (2012).
16. J. G. Chen et al. Modal identification of simple structures with high-speed video using motion magnification. *JournalofSoundandVibration* **345**, 58-71 (2015).
17. J. G. Chen et al. Video camera-based vibration measurement for civil infrastructure applications. *Journal of Infrastructure Systems* **23**, B4016013 (2017).
18. N. Wadhwa et al. Phase-based video motion processing. *ACM Transactions on Graphics* **32**, (2013).
19. Z. Liu et al. Time-varying motion filtering for visionbased nonstationary vibration measurement. *IEEE Transactions on Instrumentation and Measurement* **69**, 3907-3916 (2020).
20. Z. Liu et al. Vision-based vibration measurement by sensing motion of spider silk. *Procedia Manufacturing* **49**, 126-131 (2020).
21. L. P. Yu & B. Pan. Single-camera high-speed stereo-digital image correlation for full-field vibration measurement. *Mechanical Systems and Signal Processing* **94**, 374-383 (2017).
22. G. D' Emilia, L. Razzè & E. Zappa. Uncertainty analysis of high frequency image-based vibration measurements. *Measurement* **46**(8), 2630-2637 (2013).
23. T. Haist et al. Multi-image position detection. *Optics express* **22**, 14450-14463 (2014).
24. R. D. Gow et al. A comprehensive tool for modeling cmos image-sensor-noise performance. *IEEE Transactions on Electron Devices* **54**, 1321-1329 (2007).
25. B. F. Alexander & K. C. Ng. Elimination of systematic error in subpixel accuracy centroid estimation [also Letter 34(11)3347-3348(Nov1995)]. *Optical Engineering* **30**, 1320-1331 (1991).
26. M. R. Shortis, T. A. Clarke & T. Short. Comparison of some techniques for the subpixel location of discrete target images. In *Videometrics III*, pages 239–250. Proceedings of SPIE 2350, 1994.
27. S. Thomas. Optimized centroid computing in a shackhartmann sensor. In *Advancements in Adaptive Optics*, volume **5490**. Proceedings of SPIE, 2004.
28. F. Schaal et al. Applications of diffractive optical elements for optical measurement techniques. In *Holography, Diffractive Optics, and Applications VI*, volume **9271**, pages 1–7. SPIE, 2014.
29. S. Hartlieb et al. Hochgenaue kalibrierung eines holografischen multi-punkt positionsmesssystems. *tm - Technisches Messen* **87**, 504-513 (2020).
30. S. Hartlieb et al. Highly accurate imaging based position measurement using holographic point replication. *Measurement* **172**, 108852 (2021).
31. T. Schmidt, J. Tyson & K. Galanulis. Full-field dynamic displacement and strain measurement using advanced 3d image correlation photogrammetry: part 1. *Experimental Techniques* **27**, 47-50 (2003).
32. F. Guerra et al. Precise building deformation measurement using holographic multipoint replication. *Applied Optics* **59**(9), 2746-2753 (2020).
33. S. Hartlieb et al. Accurate 3D coordinate measurement using holographic multipoint technique. In *Optics and Photonics for Advanced Dimensional Metrology*, volume **11352**, pages 1–12. Proceedings of SPIE, 2020.

Official journal of the JHL 2689-9620
www.light-am.com

Article Open Access

Melt pool monitoring in laser beam melting with two-wavelength holographic imaging

Matthieu Piniard[1], Beatrice Sorrente[1], Gilles Hug[2] and Pascal Picart[3,*]

Abstract

Over the past two decades, laser beam melting has emerged as the leading metal additive manufacturing process for producing small- and medium-size structures. However, a key obstacle for the application of this technique in industry is the lack of reliability and qualification mainly because of melt pool instabilities during the laser-powder interaction, which degrade the quality of the manufactured components. In this paper, we propose multi-wavelength digital holography as a proof of concept for *in situ* real-time investigation of the melt pool morphology. A two-wavelength digital holographic setup was co-axially implemented in a laser beam melting facility. The solidified aluminum tracks and melt pools during the manufacturing of 316L were obtained with full-field one-shot acquisitions at short exposure times and various scanning velocities. The evaluation of the complex coherence factor of digital holograms allowed the quality assessment of the phase reconstruction. The motion blur was analyzed by scanning the dynamic melt pool.

Keywords: Laser beam melting, Optical diagnostic, Digital holography, Multi-wavelength holography

Introduction

Additive manufacturing (AM) is a key technology that allows building structures by the progressive addition of materials with complex geometries that cannot be achieved with conventional machining techniques. For example, complex honeycomb structures, called "lattices", can be realized through topological optimization. AM represents a unique opportunity to optimize and lighten structures and thus reduce energy consumption. Metal additive manufacturing processes, mainly used in the aerospace, energy, automotive, and medical sectors, are categorized into two main families: powder bed processes (LBM: laser beam melting; SLM: selective laser melting; EBM: electron beam melting; and MBJ: metal binder jetting) and directed energy deposition processes (WAAM: wire-arc additive manufacturing; EBAM: electron beam additive manufacturing; and LENS: 3D metal printing by laser deposition and fusion). Mass production of customized structures can be realized with AM. This property is particularly interesting for art and design, but also for medicine and dentistry: prostheses or implants can be adapted to each patient.

In the LBM process, metal or ceramic powder layers are successively melted to manufacture 3D mechanical parts[1]. A CAD model of the object to be built is created from a dedicated 3D modeling software and then sliced. In the case of metal parts, LBM is mainly carried out by melting a bed of metal powder using a high-power laser. First, a thin layer of powders is deposited on a substrate plate, and a very powerful laser (few hundred watts) melts the fine metal powders (≃ 40-60 microns in diameter) following the pattern determined by the computer. A roller (the re-coater)

Correspondence: Pascal Picart (pascal.picart@univ-lemans.fr)
[1]DOTA, ONERA, Université Paris Saclay, 91123 Palaiseau, France
[2]Laboratoire d'Étude des Microstructures, ONERA-CNRS, UMR 104, 29 Avenue de la Division Leclerc, 92320 Châtillon, France
Full list of author information is available at the end of the article.
These authors contributed equally: Matthieu Piniard, Beatrice Sorrente, Gilles Hug, Pascal Picart

then applies a new layer of powder. The laser draws the next layer. These steps are repeatedly applied until the object is completely realized. The structure is thus built by the successive melting of powder layers.

Owing to the complex physical phenomena involved in the laser-matter interaction, melt-pool morphology instabilities can affect the final quality of the structure and remain challenging to foresee by simulation. Everton et al[2]. listed defects that may occur during the process: pores (gas), *balling*, inflated powder, and cracks. Sames et al[3]. identified the same defects and delved deeper, compared to Everton et al., into the relationships between process inputs and defects. Grasso and Colosimo[4] proposed the classification of defects into six categories: porosity, *balling*, geometric defects, surface defects, residual stresses/cracks/debonding, and microstructure inhomogeneity.

The lack of repeatability and stability of the process slows the democratization of AM in industry. An AM cycle may take several hours or days. For example, an unacceptable defect that appears in the first few minutes of the cycle means wasting the investment in raw material, energy, and time to produce a non-standard part. Therefore, *in situ* monitoring and quality control are key issues in the development of AM. X-ray tomography is undoubtedly the most successful of the existing monitoring techniques for metal parts[5]. However, this is an *ex situ* technique that requires specific facilities. Therefore, to achieve the performance and safety requirements of the aerospace industry, significant efforts have been made over the last ten years to develop monitoring techniques that can be implemented in AM machines and take advantage of the ability to monitor and control manufacturing layer by layer. Several monitoring approaches based on the melt-pool process radiation or with a secondary illumination source have been developed to measure diverse dimensions: length, width, height, and area[6–16]. However, these methods cannot measure the 3D shape (called topography) of the melt pool. Nonetheless, the final track morphology is influenced by volumetric and capillary forces applied to the melt pool[13,17]. Therefore, the shape of the melt-pool surface is of primary interest for monitoring its stability. Because of its intrinsic heterogeneity, its movement over the powder bed, and its own dynamics, the measurement of the melt-pool surface shape requires a single-shot full-field acquisition with a short exposure time of a few microseconds. Indeed, the traveling of the processing laser can reach 1 $m \cdot s^{-1}$ in industrial processes. In this paper, the term full-field means that the image of the area under interest is obtained with no requirement concerning surface scanning. This reveals the ability of the optical setup to yield a collection of data points organized as a fine mesh of the inspected area.

Among the optical approaches for surface shape measurement, two-wavelength digital holography proved to be a relevant tool for desensitized testing. Examples include steep optical surfaces (aspheric mirrors and lenses)[18], large deformation of structures[19], and surface shape profiling[20]. Such an approach can also be used for surface roughness measurement when the roughness is large compared to the wavelength[21]. A wide range of applications of multi-wavelength holography have also been demonstrated, such as endoscopic imaging[22,23], investigation of mechanical structures[24], erosion measurements[25], and in-line industrial inspection[26]. Regarding off-axis digital holography and spatial multiplexing of two-wavelength digital holograms, the operation becomes real-time, in the sense that the surface shape can be measured at each instant at which the holograms are recorded. In this paper, the term real-time refers to the ability of the measurement system to record useful data of the phenomenon of interest at its own temporal scale. It follows that the camera frame rate must be in the range of several thousand Hertz.

In this paper, we propose multi-wavelength digital holography as a proof of concept for *in situ* investigation of the melt pool during metal additive manufacturing. The paper is organized as follows. First, the principles of digital image-plane holography are discussed. Then, the experimental setup employed in this study is detailed. Finally, we present experimental results. Several experiments were conducted: static tracks, moving tracks, and melt pool without and with powder. A discussion on the experimental results concludes the paper.

Basics of spatially multiplexed two-wavelength digital holographic interferometry

Principles of digital image-plane holography

Digital holography was experimentally established in the 90's[27–29]. Recently, many fascinating possibilities have been demonstrated: the focus can be chosen freely[30], a single hologram can provide amplitude-contrast and phase-contrast microscopic imaging[31], image aberrations can be compensated[32], properties of materials can be investigated[33], and digital color holography[28,34] and time averaging are also possible[35]. Digital holography can be realized through various architectures such as Fresnel holography, Fourier holography, lens-less Fourier holography, and image-plane holography[36,37]. Although these architectures present differences in their implementation, they exhibit the same spatial frequency bandwidth when arranged according to an off-axis approach. In this study, we analyzed a case in which off-axis digital image-plane holograms were recorded. This choice is the most suitable for LBM diagnostics. Indeed, recording a single hologram per instant constitutes a powerful tool for studying dynamic events and carrying out high-speed acquisition. In case of image-plane holography, an imaging system is associated with a variable aperture to produce an image of the scene of interest as close as possible to the sensor area (object projected onto the sensor plane). Note that a slight defocus could be also of interest. At the sensor plane, the reference wave $\mathcal{R}$ is mixed with the wave from the imaging system $\mathcal{A}'$ to produce the digital hologram, expressed as

$$\mathcal{H} = |\mathcal{R}|^2 + |\mathcal{A}'|^2 + \mathcal{R}^*\mathcal{A}' + \mathcal{R}\mathcal{A}'^* \quad (1)$$

The term $\mathcal{R}^*\mathcal{A}'$ usually refers to the $+1$ order of the hologram, whereas $\mathcal{R}\mathcal{A}'^*$ refers to the -1 order. In the off-axis configuration, the reference wave is inclined and impacts the sensor at a certain incidence angle such that its complex-valued amplitude can be written as follows (a_R is considered a constant and (u_0, v_0) denotes spatial frequencies):

$$\mathcal{R}(x,y) = a_R \exp(2i\pi(u_0 x + v_0 y)) \quad (2)$$

The recovery of the complex-valued amplitude of the image of the object is obtained by filtering the $+1$ order in the Fourier spectrum of the hologram. Filtering can thus be written as follows (FT means Fourier transform):

$$A'_r \simeq \mathcal{R}^*\mathcal{A}' = \mathrm{FT}^{-1}[\mathrm{FT}[\mathcal{H}] \times G] \quad (3)$$

Eq. 3 is a convolution formula, and the transfer function G is given by the bandwidth-limited angular-spectrum transfer function in the Fresnel approximation:

$$G(u,v) = \begin{cases} \exp(-i\pi d_r \lambda((u-u_0)^2 + (v-v_0)^2)) \\ \quad \text{if } ((u-u_0)^2 + (v-v_0)^2) \le R_u^2 \\ \quad 0 \text{ if not} \end{cases} \quad (4)$$

In Eq. 4, λ is the wavelength of light, R_u is the cut-off spatial frequency of the imaging system, and d_r is the refocus distance when the projected image is not perfectly focused at the sensor plane. Note that if the image is perfectly in-focus, then $d_r = 0$ and the transfer function G is simply a binary filter centered at spatial frequency (u_0, v_0) with bandwidth related to R_u. The main parameter of interest in the extracted $+1$ order is the optical phase, ψ, which can be used for metrology purposes, such as surface height measurements.

Multiplexed two-wavelength digital holography

Two-wavelength digital holography has many advantages over single-wavelength holography. With a unique wavelength, the measured surface height is ambiguous when it is larger than the wavelength. Given that the surface of the object is also rough, the surface shape cannot be reconstructed because the phase is extracted from the filtering results from a speckle pattern. Thus, the ambiguity and randomness of the phase can be mitigated by using a second wavelength, leading to the so-called synthetic wavelength. Consequently, the ambiguity is bypassed and the measurement range is increased from microns to millimeters (or larger). With two wavelengths, namely λ_1 and λ_2, the synthetic wavelength is given by $\Lambda = \lambda_1\lambda_2/|\lambda_1 - \lambda_2|$[38]. The surface height of the object, $h(x,y)$, is calculated using the following equation (here, the illumination and observation of the surface are at normal incidence):

$$h(x,y) = \frac{\Lambda}{4\pi}[\psi_2(x,y) - \psi_1(x,y)] = \frac{\Lambda}{4\pi}\Delta\psi \quad (5)$$

where (ψ_1, ψ_2) denotes the optical phases extracted from the two holograms at the two wavelengths.

To obtain real-time measurements of the surface height at the melt-pool area, the simultaneous recording of the two phases (ψ_1, ψ_2) is required (f_{ex} and f_{ey} refer to the spatial sampling frequencies of the sensor, i.e., $f_{ex} = 1/p_x$ and $f_{ey} = 1/p_y$, where p_x and p_y are the sampling pitches of the sensor). This can be achieved by spatial multiplexing of the two holograms at the two wavelengths, so that a single hologram is recorded for both wavelengths. At the sensor plane, the hologram at λ_2 is recorded simultaneously with that at λ_1, and the spatial frequencies of its reference wave are adjusted so that its $+1$ order is localized in the Fourier spectrum, separately from the $+1$ order at λ_1. Fig. 1 illustrates the spectral distribution in the Fourier spectrum of spatially multiplexed digital holograms at the two wavelengths. The basic arrangement for producing spatially multiplexed digital holograms is shown in Fig. 2. The two reference beams are inclined with different

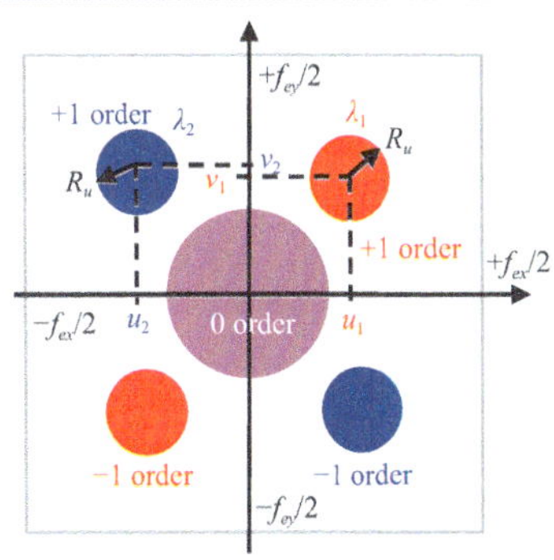

Fig. 1 Spectral distribution of the multiplexed off-axis digital holograms in the Fourier domain at two wavelengths.

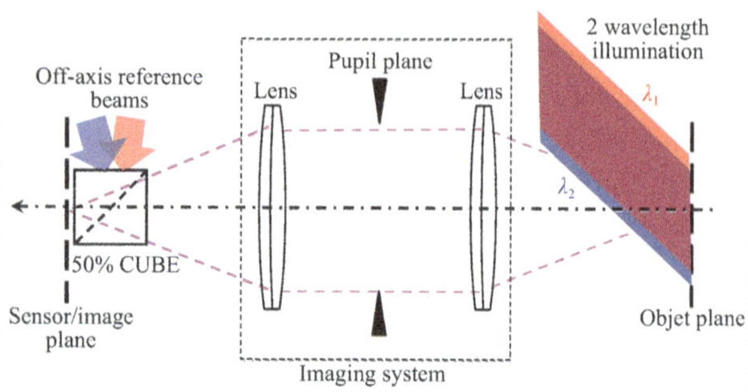

Fig. 2 Imaging of the object area through the image-plane holographic system with spatial multiplexing of the two holograms at two wavelengths.

incidence angles to produce the spatial frequencies (u_1, v_1) for λ_1 and (u_2, v_2) for λ_2.

From a computational point of view, the phase difference expressed in Eq. 5 is obtained by calculating $\Delta\psi = \arg[C_{12}] = \arg[A'_{r2}A'^*_{r1}]$, where A'_{r1} and A'_{r2} are the two image fields from the two wavelengths. According to Eq. (3), each of the recovered complex-valued A'_{r1-2} image includes a bias due to each spatial carrier frequency. More specifically, this bias is due to the presence of the term $\mathcal{R}^*$ in the output of the Fourier filtering, which must be compensated. This point is discussed in the next subsection.

Compensation of spatial carrier frequencies

The compensation of the bias in the phase difference requires knowing the spatial frequencies of the two wavelengths, i.e., (u_1, v_1) and (u_2, v_2). These couples can be estimated with spectral analysis when evaluating the centroids of the two +1 orders in the hologram spectrum (refer to Fig. 1). This yields approximate values that can be used to start the compensation procedure. The method proposed in this paper is to partially compensate for the two carrier frequencies and then refine their residue by using a power spectrum density (PSD) analysis. Thus, partial compensation retains a portion of the frequencies in the complex data C_{12}. The ratio of the spatial frequencies is denoted by α (typ. $\alpha = 0.5$). The partially compensated complex-valued data C_p can be calculated using the following expression:

$$\begin{aligned} C_p =& A'_{r2}\exp(2i\pi\alpha(u_2x + v_2y))\times \\ & A'^*_{r1}\exp(-2i\pi\alpha(u_1x + v_1y)) \\ & \propto \mathcal{A}'_2\mathcal{A}'^*_1\exp(2i\pi(\Delta ux + \Delta vy)) \end{aligned} \quad (6)$$

PSD analysis is carried out with data C_p by independently calculating the average PSD along each line and column. Then, the peak in each PSD is detected, and its associated spatial frequency is estimated. For accurate peak estimation, zero-padding of the lines and columns is applied up to 2^{16} data points. These two estimations over lines and columns yield $(\Delta u, \Delta v)$. Finally, the residue is compensated according to the expression $C = C_p\exp(-2i\pi(\Delta ux + \Delta vy))$, and the bias-free phase difference is obtained from $\Delta\psi = \arg[C] = \arg[\mathcal{A}'_2\mathcal{A}'^*_1]$. This compensation is systematically applied to digital holograms. The next section elaborates on the experimental setup.

Experimental setup

Diagnostic and melt modules

The optical setup used to perform topographic measurements on a melt pool is shown in Fig. 3. This includes two modules. In the first module, the holographic module (Fig. 3a), is equipped with two linearly polarized lasers: a 20 mW He-Ne laser and a 50 mW CrystaLaser CRL-DL635-050-SO-Optical-Isolator laser diode characterized by a coherence length of 10 m. These sources respectively emit at 632.8 nm and 634.4 nm, thus yielding a synthetic wavelength equal to 286.74 µm. Both laser beams are expanded before crossing polarizing beam splitters (PBS1 and PBS2). The transmitted beams generate two reference beams, while the reflected beams are superimposed to form a unique probe beam with cube C1 and mirror M3. The probe beam is shaped through beam expander BE3, PBS3, and lens L1, whose focal length is equal to 300 mm, to produce an illumination spot between 0.5 mm and 1mm in diameter. This large illumination allows the exploration of the melt pool and the surrounding area, which is still hot. The quarter wave plate (QWP) between PBS3 and M7 modifies the linear s-polarization into a circular one. The probe beam impacts the surface of interest at normal incidence. After retro-diffusion at the object plane, the QWP modifies the circular polarization of the probe beam into a p-polarization and crosses PBS3 to be directed towards the imaging system composed of the $2f_1 - 2f_2$ telecentric system and a magnification adjustable as $G_t = 2$ and $G_t = 4.29$. This system includes a circular pupil with a 20-mm diameter to set the spatial frequency bandwidth R_u of the holograms. Thus, the imaging system operates with $R_u = 27$ mm^{-1}. Such aperture is compatible with those commonly used in galvanometer scan heads employed in AM machines (typ. f#15).

The probe beam is mixed with the reference beams at the sensor plane using cube C3. The two reference beams pass through C3, whereas the two probe beams are reflected by C3 to the sensor area. Along the reference paths[29], the light impacts the sensor with a slight tilt angle to provide off-axis recording of the holograms. The two reference beams

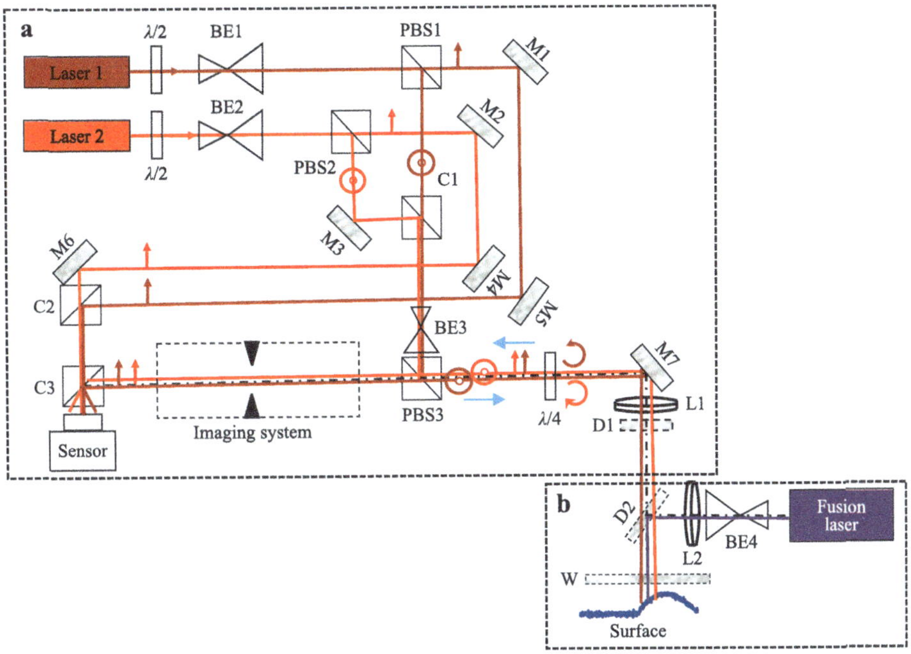

Fig. 3 **a** Two-wavelength holographic setup for LBM monitoring, **b** laser fabrication machine (BE1-3: beam expanders, PBS1-3: polarizing beam splitters, M1-7: mirrors, C1-3: beam splitter cubes, D1: chromatic filter, D2: dichroic plates, W: protection window, L1-2: lenses).

are combined using a splitter cube C2. At the image plane, an 8-bit Photron Fastcam MC2 (pixel pitch of 10 μm) is used to record sequences of two-wavelength digital holograms. The digital holographic system is coupled to a simplified LBM machine, as depicted in Fig. 3b. The two modules are coupled through the dichroic plate D2 characterized by a 980-nm cut-off wavelength so that the light from the processing laser cannot be transmitted to the imaging system. The processing laser is a continuous monomode laser with power adjustable from 50 W to 500 W (SPI laser SM-S00472-8) emitting at 1080 nm and equipped at the end of the fiber with a QBH connector associated with a collimator. The laser power can be adjusted from 50 W to 500 W. The processing laser is equipped with a low-power coaxial laser emitting at 632.8 nm (not shown in Fig. 3b). This alignment laser allows checking that the probe beam and the processing laser are properly superimposed on the powder bed plane. The beam expander BE4 and focusing lens L2 produce, after reflection on the dichroic plate D2 with a cut-off wavelength of 980 nm in the melting zone, a spot whose diameter ranges from 70 μm to 200 μm. The object to be analyzed is installed on a motorized XY stage that permits the alignment and scanning of the processing laser. Note that the dichroic plate D2 operates the first filtering of the thermal radiation from the melting area, while a narrow filter D1 (10 nm) centered around 633 nm drastically reduces this noise source. As a result, a very weak flux of thermal photons may illuminate the sensor. In addition, the light flux from the heated powder bed due to laser fusion is incoherent and cannot interfere with the probe beams. Thus, it is naturally filtered in off-axis digital holography because it is localized in the zero-order lobe (Fig. 1). Nevertheless, the minimization of its contribution at an acceptable level is strategic to avoid image saturation or loss of fringe contrast, which would destroy the useful information contained in the +1 order. The light back-scattered towards the sensor by the processing laser is also filtered by the same optical devices and procedure. Information contained in the zero-order lobe could be exploited to monitor other process signatures, but this is not the purpose of this study. Note that the analysis of the standard deviation of noise in surface-shape data from two-wavelength spatially-multiplexed digital holograms was discussed in[39]. The influence of noise in the measurements of the surface shape was described using an analytical approach. For further details, please refer to[39].

Spatial resolution

The distance between the powder bed and the last lens in an additive manufacturing machine (DFM system or F-θ lens) is generally large because the optics must be protected from spatters or vapor plumes resulting from laser processing. An optical window is inserted to ensure protection. The typical working distances range from 100 mm to 300 mm. Considering that the melt-pool dimensions in LBM processes typically span between 250 μm and 500 μm, the spatial resolution of the holographic system is a key element for defect diagnosis. The circular aperture inserted in the imaging system (refer to Fig. 3a) plays the role of the aperture diaphragm. Thus, this diaphragm is located in the Fourier plane of the imaging system. Its diameter is adjusted to ensure the non-overlap of the diffraction orders in the hologram spectrum (refer to Fig. 2). In addition, it controls the size of the speckle grain in the image plane. As mentioned previously, the aperture is equivalent to a binary spatial frequency filter with a cut-off frequency of R_u. It was demonstrated that the lateral resolution in the object plane ρ_x of a coherent imaging system is given by $\rho_x \simeq 0.82/R_u/G_t$[40], where G_t is the magnification between the object and the image plane. Consequently, the average size of the speckle gain at the object plane is equal to ρ_x. With $R_u \simeq 27\ mm^{-1}$ and $G_t = 2$ for a wavelength of 632.8 nm in the holographic system, the spatial resolution of the probe module is $\rho_x \simeq 15.2$ μm. This is equivalent to $f\# \simeq 15$. Given that the shape of a typical melt pool is characterized by low spatial frequencies[41], approximately ten sampling points should be sufficient to detect defects and classify them.

Results

Topography of manufactured aluminum tracks

To qualitatively appraise the quality of the topographic measurement in real conditions, two parallel aluminum alloy tracks realized with the LBM machine were considered. These are remarkably steady experimental conditions because the manufactured object was static, and its structure was stable after cooling. The hologram was acquired with an exposure time τ = 145 μs, and the optical magnification was $G_t = 2$. From the spatially multiplexed digital hologram, the complex-valued amplitudes for both wavelengths were extracted, and the phase difference was obtained. With the moduli of the complex amplitudes, the autocorrelation of the speckle field was performed to estimate the spatial resolution in the image plane. Fig. 4a shows the autocorrelation function of the speckle field for wavelength λ_1, and Fig. 4b shows its profile along the horizontal line. Note that the autocorrelation function has a width of 0.017 mm, thus confirming that the spatial resolution in the object plane is approximately 15-17μm, which is in good agreement with the theoretical estimation.

Owing to the surface roughness at the object plane, the phases and the Doppler phase (difference between the two phases ψ_1 and ψ_2) extracted from the holograms became corrupted by speckle decorrelation noise. This phase noise can be estimated using a noise estimator[42]. Fig. 4c shows

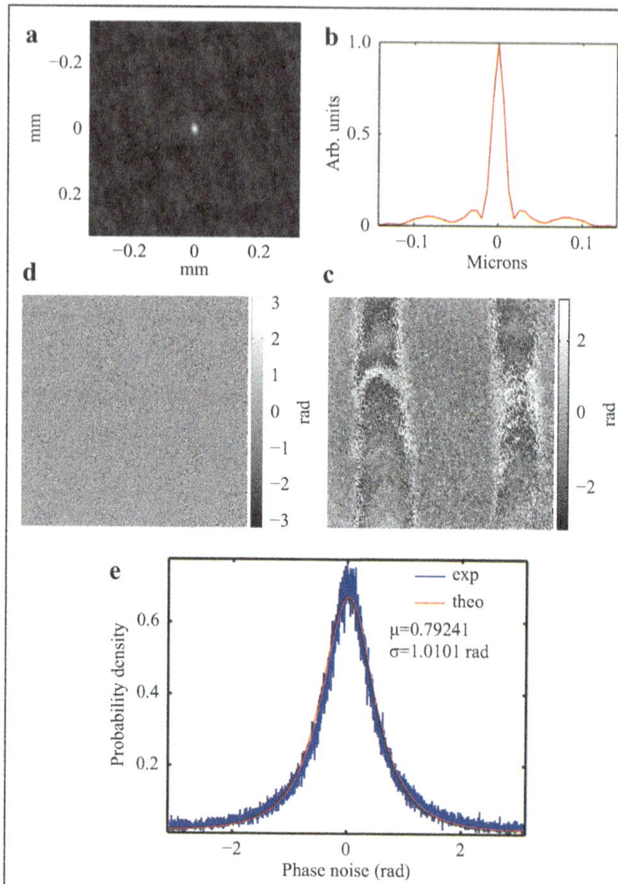

Fig. 4 **a** Autocorrelation function of the speckle field at wavelength λ_1; **b** horizontal profile of **a**; **c** raw Doppler phase map extracted from the phases at λ_1 and λ_2; **d** noise map from **c**; **e** probability density functions of the phase noise from **d** (blue line: experiments, red line: theory).

the raw phase map for the aluminum tracks, and Fig. 4d depicts the noise map estimated from the raw Doppler phase. The phase noise is in the range $[-\pi, +\pi]$ and exhibits non-Gaussian statistics. Fig. 4e shows the probability density functions (PDFs) calculated from the noise map in Fig. 4d. By fitting the experimental PDF with L2-norm minimization, the value of $|\mu|$ can be estimated from the theoretical equation by using the following relation[43–48] (with ϵ denoting the phase noise induced by the decorrelation noise and $\beta = |\mu|\cos(\epsilon)$):

$$p(\epsilon) = \frac{1-|\mu|^2}{2\pi}\left(1-\beta^2\right)^{-3/2}\left(\beta\sin^{-1}\beta + \frac{\pi\beta}{2} + \sqrt{1-\beta^2}\right) \tag{7}$$

The coherence factor is a quality metric that indicates the reliability of the phase estimation. Vry et al. [49] considered that the standard deviation $\sigma_\epsilon \leq \pi/4$ is the criterion required for high-quality measurement of the Doppler phase, which leads to the condition $|\mu| \geq 0.85$. For other authors[50], $|\mu|$ should be higher than 0.37, which is more tolerant to noise but requires a powerful filtering algorithm. In Fig. 4d, the noise standard deviation was found to be $\sigma_\epsilon \simeq 1.01$ rad $=\pi/3.11$, whereas the modulus of the coherence factor was estimated to be $|\mu| \simeq 0.79$. Thus, with $|\mu| \simeq 0.79$, the measurement can be considered to be of correct quality for the experimental parameters.

Fig. 5 shows the experimental results concerning the topography of the two aluminum tracks obtained with the holographic setup. Fig. 5a displays a photograph of the top-view of the aluminum tracks, and Fig. 5b, c show the topography of the tracks. In Fig. 5d, the region of the tracks represented by the dashed black line in Fig. 5c corresponding to the dashed white line in Fig. 5a, b, is zoomed.

Topography of dynamic aluminum tracks

When carrying out measurements during the additive manufacturing process, the effect of motion blur on the image quality has to be investigated. This effect is observed along the length of the tracks, which is parallel to the displacement of the motorized stage. If V_0 is the velocity of the scan at the melt pool and B is the blur effect expressed as a fraction of the spatial resolution, then we have the following expression:

$$B = \frac{V_0 \tau}{\rho_x} \tag{8}$$

Motion blur is problematic if B is close to or larger than one during exposure time τ. Considering an exposure time τ = 25 µs and a spatial resolution of 15 µm, then for $B = 0.2$, the maximum allowed scan velocity is 150 mm·s^{-1}. For higher velocities, the exposure time must be reduced according to Eq. 8. For V_0 = 1 m·s^{-1}, τ should not exceed 3 µs. In this case, the monitoring laser power must be sufficient to reach an adequate level of the signal-to-noise ratio of the measurement. To qualitatively check the quality of the measurement for various velocities of the scan, different velocities were set: 100 mm·s^{-1}, 250 mm·s^{-1}, and 380 mm·s^{-1}. For V_0 = 380 mm·s^{-1} and τ = 25 µs, we obtain $B = 0.63$; and for V_0 = 250 mm·s^{-1} and τ = 25 µs, we obtain $B = 0.41$. These values of B were slightly beyond the limit estimated using Eq. 8. However, for V_0 = 100 mm·s^{-1} and τ = 25 µs, we have $B = 0.16$, which is reasonable. When processing the digital holograms for the three velocities, the values of $|\mu|$ were systematically estimated to check the overall quality of the holographic measurements in terms of phase noise. For each experiment, the values of $|\mu|$ remained higher than 0.85, which means that for this set of velocities, the noise quality criterion was respected. However, this criterion did not prevent the measurement from presenting other artifacts such as motion blur. Fig. 6 shows the manufactured aluminum tracks at the beginning, middle, and end of the process for the three selected velocities. Figs. 6a–c show, for V_0 = 100 mm·s^{-1}, the topography of the tracks at 4.25 ms

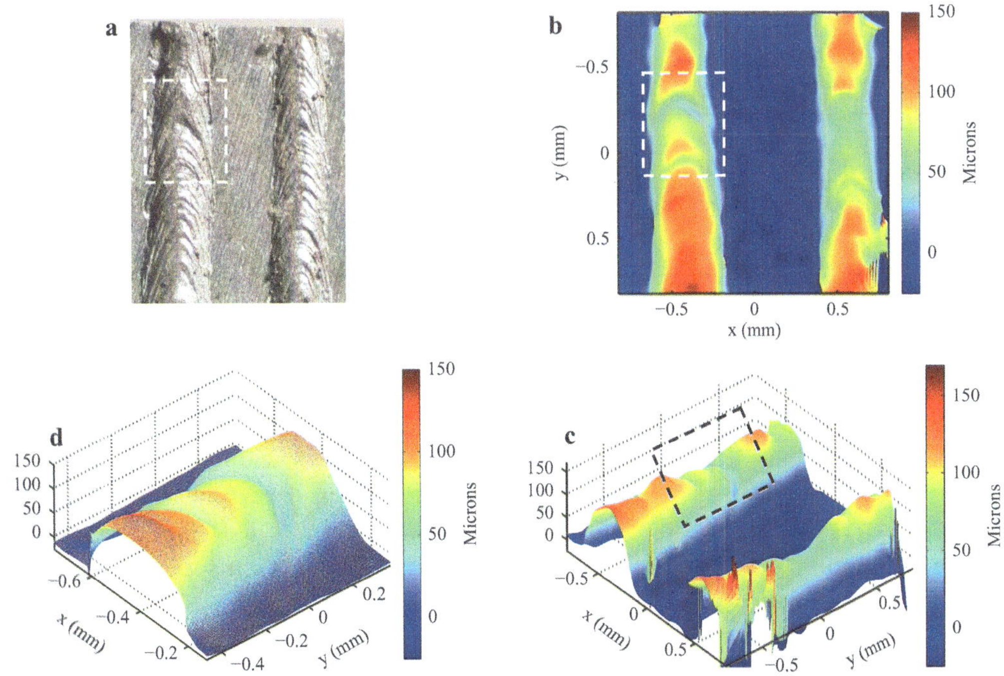

Fig. 5 a Photograph of the top-view of the aluminum tracks for testing the topography measurement; **b** topography of tracks measured with the setup; **c** 3D view of **b**; **d** zoom of the topography within the dashed black line in **c** corresponding to the dashed white line in **a** and **b**.

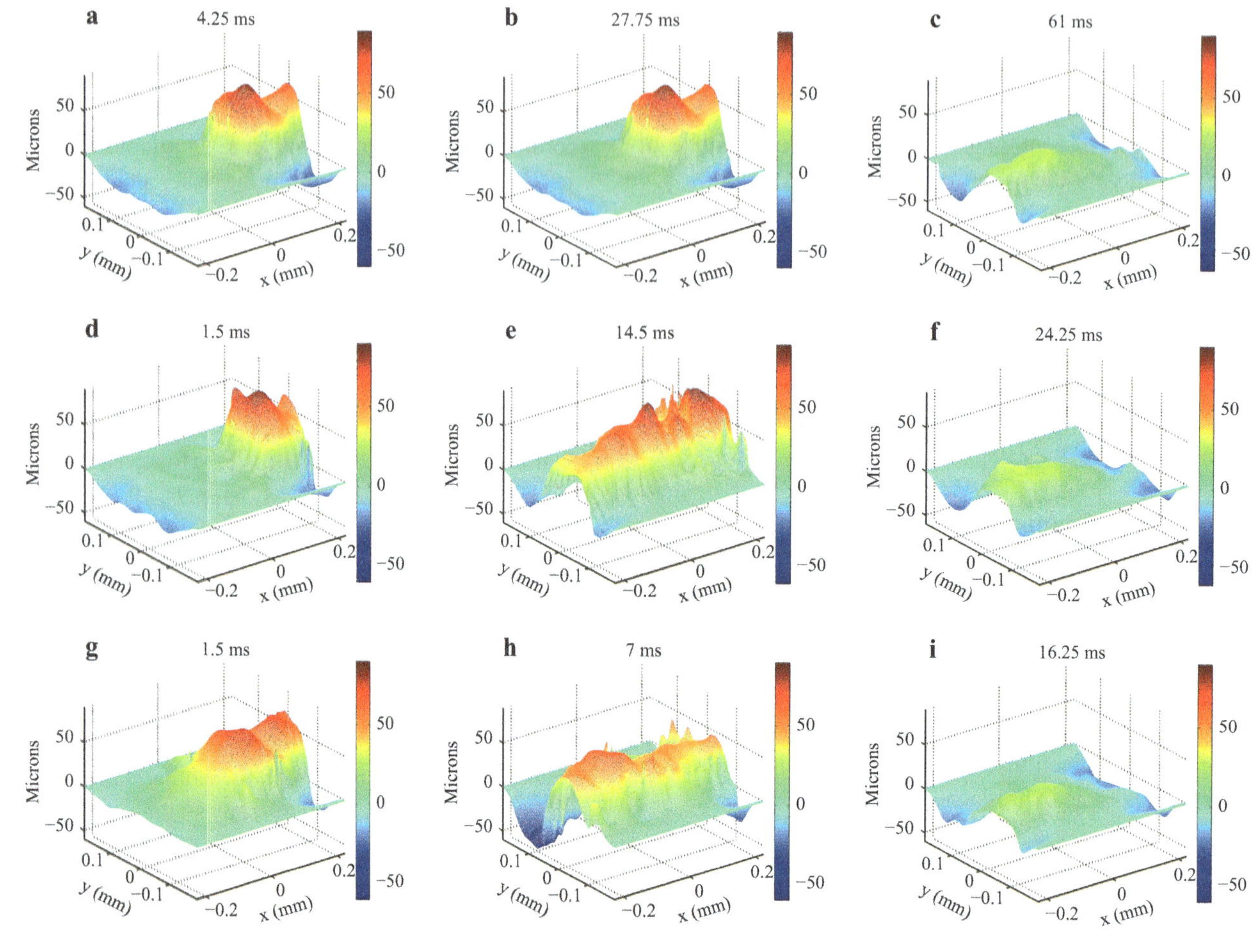

Fig. 6 Topography of dynamic aluminum tracks. **a** Topography for V_0 = 100 mm·s^{-1} at 4.25 ms after the beginning of the recording, **b** at 27.75 ms, **c** at 61 ms; **d** topography for V_0 = 250 mm·s^{-1} at 1.5 ms after the beginning of the recording, **e** at 14.5 ms, **f** at 24.25 ms; **g** topography for V_0 = 380 mm·s^{-1} at 1.5 ms after the beginning of the recording, **h** at 7 ms, **i** at 16.25 ms. Movies: visualization of dynamic aluminum tracks for V_0 = 100 mm·s^{-1}, V_0 = 250 mm·s^{-1}, and V_0 = 380 mm·s^{-1}. Movie 1: visualization of the dynamic tracks for V_0 = 100 mm·s^{-1}; Movie 2: dynamic tracks for V_0 = 250 mm·s^{-1}; Movie 3: dynamic tracks for V_0 = 380 mm·s^{-1}.

(just after the beginning of the recording), at 27.75 ms (middle of the time sequence), and at 61 ms (end of the sequence). Figs. 6d–f depict the topography of the tracks for $V_0 = 250$ mm·s^{-1} at 1.5 ms (just after the beginning of the recording), at 14.5 ms (middle of the time sequence), and at the end at 24.25 ms (end of the sequence). Figs. 6g–i show the topography of the tracks for $V_0 = 380$ mm·s^{-1} at 1.5 ms (just after the beginning of the recording), at 7 ms (middle of the time sequence), and at 16.25 ms (end of the sequence).

Monitoring of the melt pool

Experiments were carried out in open air, and an argon flow from a nozzle was sprayed above the fusion area to reduce the oxidation of the substrate and powder. This means that the melt pool generated during the experiments tended to be unstable. Given that the goal of this study was to realize a proof of concept, it was interesting to evaluate the ability of the holographic system to detect such instabilities. To ensure uniform velocity during fusion, the laser processing was initialized after the start of the motorized stage to avoid any acquisitions during the acceleration phase.

Melt pool on 316L substrate

A sheet of 316L plate was used. The exposure time was set to $\tau = 6.25$ µs, and the frame rate was 4000 fps. This allowed freezing the dynamic of the melt pool and avoided motion blur in the holograms. The optical magnification was set at $G_t = 4.29$, the spatial resolution at the melt pool area was approximately 7.5 µm, obtaining $B = 0.08$. Therefore, blur motion was irrelevant. The processing laser had a power of 75 W for a spot diameter between 100 and 200 µm and a velocity of $V_0 = 100$mm·s^{-1}. Under such conditions, a keyhole regime may be observed.

Fig. 7 shows two 3D plots of the topography of the melt pool obtained at two different instants during the laser fire. Fig. 7a displays the melt pool at 87.25 ms, whereas Fig. 7b depicts the melt pool at 88 ms. Fig. 8 shows bottom views corresponding to 3D plots in Fig. 7a, b.

Melt pool on 316L substrate with powder

In this case, a 316L sheet was used with powder 316L. This powder had a thickness of $\simeq 100\ \mu m$ and absorbed the laser energy. The melt pool was composed of the fusion of the powder and the fusion of the superficial layer of the substrate. The exposure time was set to $\tau = 25$ µs, and the frame rate was 4000 fps. The increase in the exposure time was due to the non-cooperative surface of the 316L powder, which weakly reflects photons. It follows that the hottest area where the keyhole might appear could not be observed owing to the dynamics of the melt pool. The processing laser had a power of 150 W for a spot diameter between 100 and 200 µm and a velocity of $V_0 = 100\ mm{\cdot}s^{-1}$. The optical magnification was set to $G_t = 4.29$. Thus, the observed area was approximately $400 \times 400\ \mu m^2$. These experimental parameters led to $B = 0.33$, which means that the recording was not fully immune to motion blur; this was imposed because of the limited power of the two lasers of the holographic setup.

Fig. 9 shows six pictures of the topography of the melt pool obtained at different instants during the laser melting. The tail and body formation of the melt pool was defined as shown in Fig. 9g.

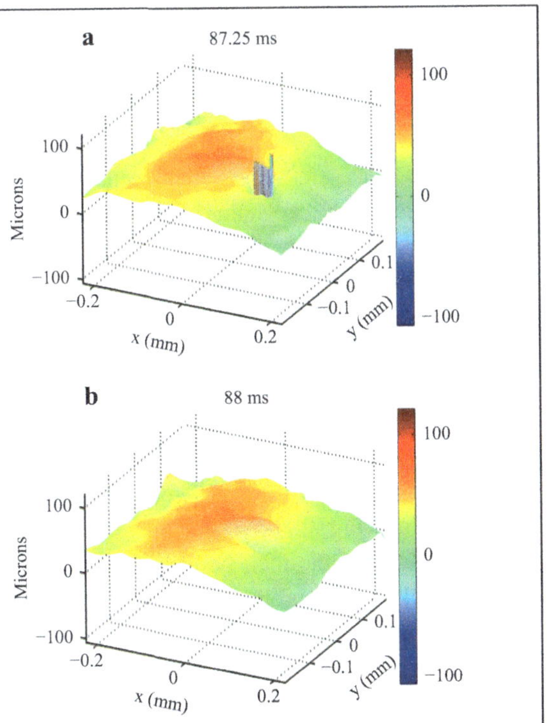

Fig. 7 Monitoring of melt pool on 316L substrate. **a** 3D plot of melt pool with keyhole at 87.25 ms after the beginning of the recording; **b** 3D plot of melt pool with keyhole at 88 ms after the beginning of the recording. Laser scanning is along the x direction.

Discussion

The comparison between the photograph and the tracks in Fig. 5 shows that the topography measured by two-wavelength holography is faithful. The herringbone structure is clearly observable in both the photograph and the topography measurement.

In the following comments on experiments, keep in mind that the processing laser scanned along the x

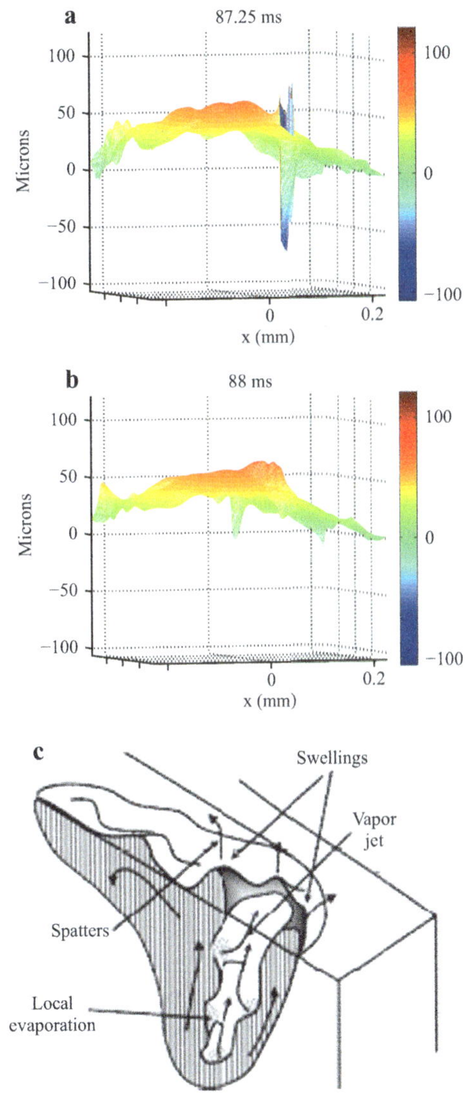

Fig. 8 Keyhole in melt pool on 316L substrate. **a** Bottom view of melt pool with keyhole at 88 ms after the beginning of the recording; **b** bottom view of the melt pool with a keyhole at 88 ms after the beginning of the recording; **c** illustration of the Rosenthal keyhole regime (adapted from Ref. 51). Movie 4: visualization of the keyhole regime.

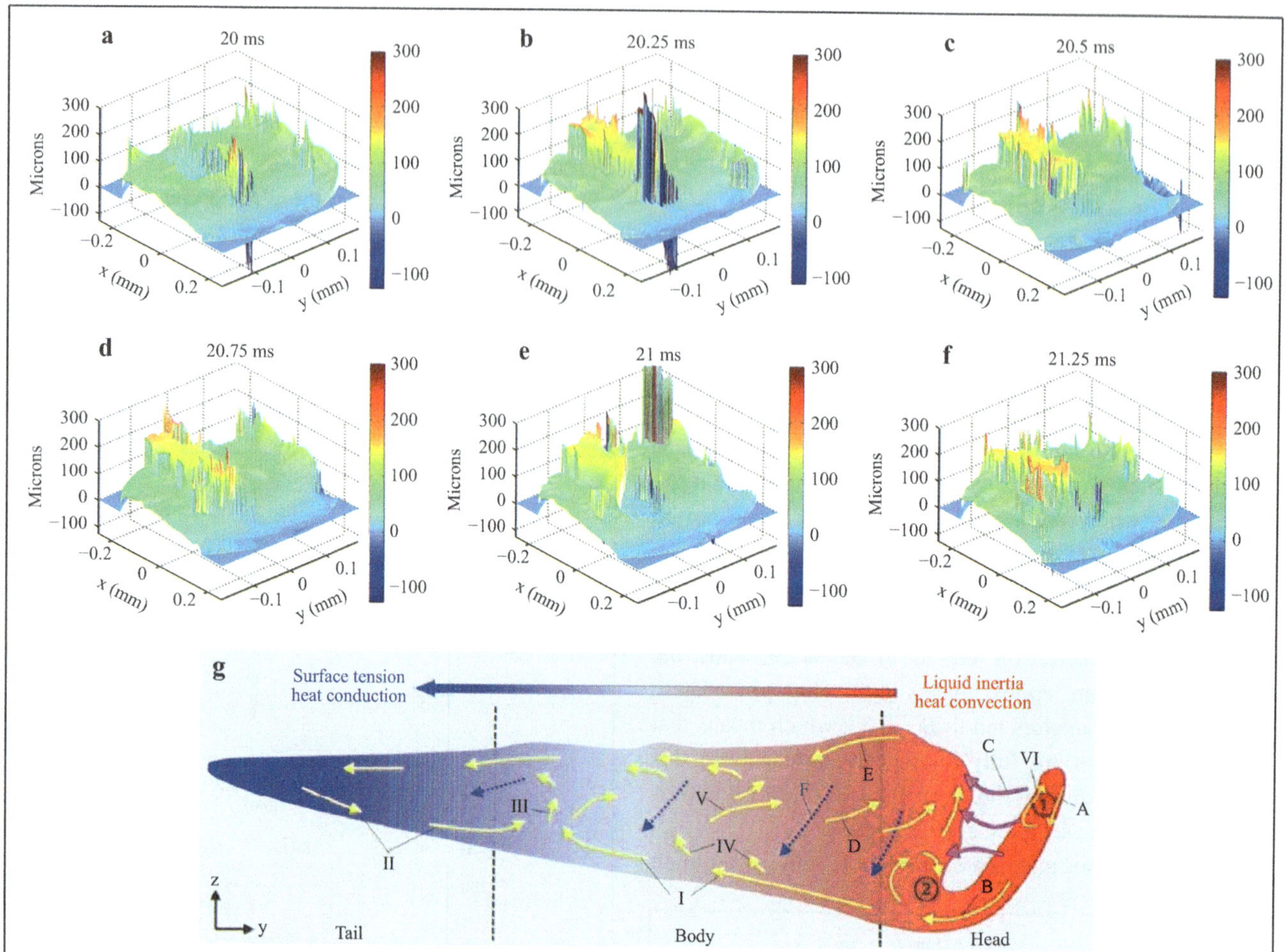

Fig. 9 Topography of melt pool on 316L substrate with 316L powder, **a**–**f** several 3D plots at different instants during laser melting, **g** illustration of melt pool in keyhole mode (adapted from Ref. 41). Movie 5: visualization of the melt pool on 316L substrate with powder

direction from left to right.

The topographies obtained for the dynamic tracks are shown in Fig. 6, and the respective supplementary movies demonstrate that the holographic setup can yield measurements when the track is scanned at a moderate velocity. Indeed, when the velocity is higher than 100 $mm \cdot s^{-1}$, motion blur contributes to the degradation of the quality of the measurements. This could be avoided by using more laser power to reduce the exposure time and increase the velocity up to that of industrial processes ($\simeq 1\ m \cdot s^{-1}$).

Figs. 7–8 permit to visually appraise the depression zone that indicates that the melt pool included the keyhole regime. This regime represents a crucial defect that cannot be identified from thermal radiation monitoring techniques. This regime may be unstable, as can be seen in Fig. 8b, where, at 88 ms, the keyhole regime has disappeared. These results are in good agreement with the description of the keyhole regime proposed by Ge et al[52]. for similar processing conditions. In Fig. 7a, a large variety of surface tension gradients are shown. A deep depression zone characteristic of the keyhole regime was formed at the bottom of the melt pool. The general shape of the melt pool is in agreement with the Rosenthal model represented in Fig. 8c. In Fig. 7b, the melt pool presents a dome shape close to that of the previous view but then enters a transition mode without depression. During the sequence, the melt pool remained unstable and oscillating between these two states[14,53].

When considering melting of the 316L powder on the 316L substrate, Fig. 9 and the supplementary movie show the fluctuations of the melt pool and the establishment of the hot track. These preliminary results demonstrate that monitoring of the melt pool during melting is possible with digital holography, although the results in Fig. 9 are possibly influenced by the motion blur ($B = 0.33$) owing to the increase in the exposure time to 25 μs.

Conclusions

This paper proposes two-wavelength digital holography

as a proof of concept for real-time *in-situ* full-field monitoring of melt pools in metal additive manufacturing processes. The holographic system was coupled to a simplified LBM machine, and a high-speed camera was employed to record spatially multiplexed two-color holograms at both temporal and spatial scales of the laser melting process. The surface topography at the melt-pool plane depends on the synthetic wavelength generated by the two wavelengths. The effect of motion blur on the dynamic measurements is discussed. Experimental results are provided: double aluminum tracks on an aluminum substrate, monitoring of dynamic tracks at several velocities, visualization of the keyhole regime on 316L substrate, and visualization of the dynamic melt pool with 316L powder on 316L substrate. To the best of our knowledge, these are the first results of *in situ* full-field melt pool and track topography in LBM. Although the proof of concept was successful, several improvements are required to increase the performance of the setup. The main point is related to the increase in the laser power of the two probe beams. Indeed, more power for the holographic measurement will achieve better efficiency because more photons could be collected from the melt-pool area, especially if the target has a non-cooperative surface (e.g., the 316L powder used in this study). This is also in good agreement with the possibility of reducing the exposure time of the sensor to a few hundred nanoseconds in order to be fully immune to motion blur. This would be of great interest to reduce the motion blur from the displacement as well as that from the intrinsic dynamics of the melt pool along the z axis. From a computational point of view, processing of digital holograms, including phase extraction, noise reduction, and unwrapping, requires optimization through the use of a GPU to reduce the calculation time and to rapidly deliver useful data after recording. This study demonstrates that digital holography provides a new opportunity to optimize *in situ* additive manufacturing processes and develop reliable numerical models to understand the dynamics of the melting process.

Acknowledgements
This study was carried out in the framework of a research program at ONERA, the French Aerospace Lab, and a PhD program co-funded by Direction Générale de l'Armement (DGA). It was also carried out in collaboration with the Laboratory Procédés et Ingénierie en Mécanique et Matériaux (CNRS 8006) from Arts et Metiers Institute of Technology (Paris, France). The authors are grateful to Frédéric Coste, Pierre Lapouge, and Patrice Peyre for helpful discussions. The authors would like to thank Joseph Montri for his help during the experiments. Special acknowledgement to Alain Appriou for his valuable support.

Author details
[1]DOTA, ONERA, Université Paris Saclay, 91123 Palaiseau, France. [2]Laboratoire d'Étude des Microstructures, ONERA-CNRS, UMR 104, 29 Avenue de la Division Leclerc, 92320 Châtillon, France. [3]Laboratoire d'Acoustique de l'Université du Mans, CNRS 6613, Institut d'Acoustique - Graduate School (IAGS), Le Mans Université, Avenue Olivier Messiaen, 72085 Le Mans, France

Author contributions
B.S., P.P. and G.H. directed the project; M.P., B.S., and P.P. performed the experiments. M.P., P.P., G.H. and B.S. analyzed the experimental results. All authors contributed to the discussions and the preparation of the paper.

Conflict of interest
The authors declare no conflicts of interest.

Supplementary information is available for this paper at https://doi.org/10.37188/lam.2022.011.

Received: 03 September 2021 Revised: 19 January 2022 Accepted: 20 January 2022
Accepted article preview online: 26 January 2022
Published online: 02 March 2022

References

1. Cao, S. et al. Review of laser powder bed fusion (LPBF) fabricated Ti-6Al-4V: process, post-process treatment, microstructure, and property. *Light: Advanced Manufacturing* **2**, 20 (2021).
2. Everton, S. K. et al. Review of in-situ process monitoring and in-situ metrology for metal additive manufacturing. *Materials & Design* **95**, 431-445 (2016).
3. Sames, W. J. et al. The metallurgy and processing science of metal additive manufacturing. *International Materials Reviews* **61**, 315-360 (2016).
4. Grasso, M. & Colosimo, B. M. Process defects and in situ monitoring methods in metal powder bed fusion: a review. *Measurement Science and Technology* **28**, 044005 (2017).
5. Zhao, C. et al. Real-time monitoring of laser powder bed fusion process using high-speed X-ray imaging and diffraction. *Scientific Reports* **7**, 3602 (2017).
6. Doubenskaia, M. et al. Comprehensive optical monitoring of selective laser melting. *JLMN-Journal of Laser Micro/Nanoengineering* **7**, 236-243 (2012).
7. van Gestel, C. *Study of physical phenomena of selective laser melting towards increased productivity*. PhD thesis, Swiss Federal Institute of Technology in Lausanne, Lausanne, Switzerland, 2015.
8. Calta, N. P. et al. High speed hyperspectral thermal imaging of the melt pool dynamics during metal additive manufacturing. Conference on Lasers and Electro-Optics. San Jose, CA USA: Optical Society of America, 2017, ATh4B.2.
9. Craeghs, T. et al. Determination of geometrical factors in Layerwise Laser Melting using optical process monitoring. *Optics and Lasers in Engineering* **49**, 1440-1446 (2011).
10. Thombansen, U. & Abels, P. Observation of melting conditions in selective laser melting of metals (SLM). Proceedings of SPIE 9741, High-Power Laser Materials Processing: Lasers, Beam Delivery, Diagnostics, and Applications V. San Francisco, CA USA: SPIE, 2016, 97410S.
11. Rombouts, M. et al. Fundamentals of Selective Laser Melting of alloyed steel powders. *CIRP Annals* **55**, 187-192 (2006).
12. Clijsters, S. et al. In situ quality control of the selective laser melting process using a high-speed, real-time melt pool monitoring system. *The International Journal of Advanced Manufacturing Technology* **75**,

1089-1101 (2014).
13. Kanko, J. A., Sibley, A. P. & Fraser, J. M. In situ morphology-based defect detection of selective laser melting through inline coherent imaging. *Journal of Materials Processing Technology* **231**, 488-500 (2016).
14. Gunenthiram, V. et al. Analysis of laser–melt pool–powder bed interaction during the selective laser melting of a stainless steel. *Journal of Laser Applications* **29**, 022303 (2017).
15. DePond, P. J. et al. In situ measurements of layer roughness during laser powder bed fusion additive manufacturing using low coherence scanning interferometry. *Materials & Design* **154**, 347-359 (2018).
16. Fleming, T. G. et al. Tracking and controlling the morphology evolution of 3D powder-bed fusion in situ using inline coherent imaging. *Additive Manufacturing* **32**, 100978 (2020).
17. Kruth, J. P. et al. Consolidation phenomena in laser and powder-bed based layered manufacturing. *CIRP Annals* **56**, 730-759 (2007).
18. Wyant, J. C., Or eb, B. F. & Hariharan, P. Testing aspherics using two-wavelength holography: use of digital electronic techniques. *Applied Optics* **23**, 4020-4023 (1984).
19. Ninane, N. & Georges, M. P. Holographic interferometry using two-wavelength holography for the measurement of large deformations. *Applied Optics* **34**, 1923-1928 (1995).
20. Pedrini, G. et al. Shape measurement of microscopic structures using digital holograms. *Optics Communications* **164**, 257-268 (1999).
21. Lehmann, P. Surface-roughness measurement based on the intensity correlation function of scattered light under speckle-pattern illumination. *Applied Optics* **38**, 1144-1152 (1999).
22. Kolenovic, E. et al. Miniaturized digital holography sensor for distal three-dimensional endoscopy. *Applied Optics* **42**, 5167-5172 (2003).
23. Kandulla, J. et al. Two-wavelength method for endoscopic shape measurement by spatial phase-shifting speckle-interferometry. *Applied Optics* **43**, 5429-5437 (2004).
24. Khodadad, D. et al. Fast and robust automatic calibration for single-shot dual-wavelength digital holography based on speckle displacements. *Applied Optics* **54**, 5003-5010 (2015).
25. Pedrini, G. et al. Feasibility study of digital holography for erosion measurements under extreme environmental conditions inside the International Thermonuclear Experimental Reactor tokamak[invited]. *Applied Optics* **58**, A147-A155 (2019).
26. Fratz, M. et al. Inline application of digital holography[invited]. *Applied Optics* **58**, G120-G126 (2019).
27. Schnars, U. & Jüptner, W. Direct recording of holograms by a CCD target and numerical reconstruction. *Applied Optics* **33**, 179-181 (1994).
28. Yamaguchi, I., Matsumura, T. & Kato, J. I. Phase-shifting color digital holography. *Optics Letters* **27**, 1108-1110 (2002).
29. Picart, P. New Techniques in Digital Holography. (Hoboken: John Wiley & Sons, Inc., 2015).
30. Yamaguchi, I. et al. Image formation in phase-shifting digital holography and applications to microscopy. *Applied Optics* **40**, 6177-6186 (2001).
31. Cuche, E., Bevilacqua, F. & Depeursinge, C. Digital holography for quantitative phase-contrast imaging. *Optics Letters* **24**, 291-293 (1999).
32. De Nicola, S. et al. Recovering correct phase information in multiwavelength digital holographic microscopy by compensation for chromatic aberrations. *Optics Letters* **30**, 2706-2708 (2005).
33. Picart, P. et al. Investigation of fracture mechanisms in resin concrete using spatially multiplexed digital Fresnel holograms. *Optical Engineering* **43**, 1169-1176 (2004).
34. Tankam, P. et al. Real-time three-sensitivity measurements based on three-color digital Fresnel holographic interferometry. *Optics Letters* **35**, 2055-2057 (2010).
35. Picart, P. et al. Time-averaged digital holography. *Optics Letters* **28**, 1900-1902 (2003).
36. Kreis, T. M., Adams, M. & Jüptner, W. P. O. Methods of digital holography: a comparison. Proceedings of SPIE 3098, Optical Inspection and Micromeasurements II. Munich, Germany: SPIE, 1997, 224-233.
37. Karray, M., Slangen, P. & Picart, P. Comparison between digital Fresnel holography and digital image-plane holography: the role of the imaging aperture. *Experimental Mechanics* **52**, 1275-1286 (2012).
38. Kreis, T. Handbook of Holographic Interferometry: Optical and Digital Methods. (Weinheim: Wiley-VCH, 2005).
39. Piniard, M. et al. Modelling of the photometric balance for two-wavelength spatially multiplexed digital holography. Proceedings of SPIE 11783, Modeling Aspects in Optical Metrology VIII. Munich, Germany: SPIE, 2021, 1178308.
40. Latychevskaia, T. Lateral and axial resolution criteria in incoherent and coherent optics and holography, near- and far-field regimes. *Applied Optics* **58**, 3597-3603 (2019).
41. Guo, Q. L. et al. In-situ full-field mapping of melt flow dynamics in laser metal additive manufacturing. *Additive Manufacturing* **31**, 100939 (2020).
42. Montresor, S. & Picart, P. Quantitative appraisal for noise reduction in digital holographic phase imaging. *Optics Express* **24**, 14322-14343 (2016).
43. Lehmann, M. Decorrelation-induced phase errors in phase-shifting speckle interferometry. *Applied Optics* **36**, 3657-3667 (1997).
44. Lehmann, M. Phase-shifting speckle interferometry with unresolved speckles: a theoretical investigation. *Optics Communications* **128**, 325-340 (1996).
45. Lehmann, M. Optimization of wavefield intensities in phase-shifting speckle interferometry. *Optics Communications* **118**, 199-206 (1995).
46. Middleton, D. An Introduction to Statistical Communication Theory. (New York: McGraw-Hill, 1960).
47. Davenport, W. B. & Root, W. L. An Introduction to the Theory of Random Signals and Noise. (New York: McGraw-Hill, 1958).
48. Dainty, J. C. Laser Speckle and Related Phenomena. (Berlin, Heidelberg: Springer, 1975).
49. V ry, U. & Fercher, A. F. Higher-order statistical properties of speckle fields and their application to rough-surface interferometry. *Journal of the Optical Society of America A* **3**, 988-1000 (1986).
50. Chang, N. A. & George, N. Speckle in the 4f optical system. *Applied Optics* **47**, A13-A20 (2008).
51. Fabbro, R. Developments in Nd: YAG laser welding. in Handbook of Laser Welding Technologies: A Volume in Woodhead Publishing Series in Electronic and Optical Materials (ed Katayama, S.) (Philadelphia: Woodhead Publishing, 2013), 47-72.
52. Ge, W. J., Fuh, J. Y. H. & Na, S. J. Numerical modelling of keyhole formation in selective laser melting of Ti6Al4V. *Journal of Manufacturing Processes* **62**, 646-654 (2021).
53. Patel, S. & Vlasea, M. L. Melting mode thresholds in laser powder bed fusion and their application towards process parameter development. Solid Freeform Fabrication Symposium 2019: Proceedings of the 30th Annual International. Austin, 2019, 1190-1199.

Tornari. *Light: Advanced Manufacturing* (2022)3:18
https://doi.org/10.37188/lam.2022.018

Official journal of the JHL 2689-9620
www.light-am.com

Article   Open Access

A symmetry concept and significance of fringe patterns as a direct diagnostic tool in artwork conservation

Vivi Tornari*

Abstract

Previous collaborative studies have shown the main fringe patterns and their typical classification with regard to defects. Nevertheless, the complexity of the results prevents defect detection automation based on a fringe pattern classification table. The use of fringe patterns for the structural diagnosis of artwork is important for conveying crucial detailed information and dense data sources that are unmatched compared to those obtained using other conventional or modern techniques. Hologram interferometry fringe patterns uniquely reveal existing and potential structural conditions independent of object shape, surface complexity, material inhomogeneity, multilayered and mixed media structures, without requiring contact and interaction with the precious surface.

Thus, introducing a concept that from one hand allows fringe patterns to be considered as a powerful standalone physical tool for direct structural condition evaluation with a focus on artwork conservators' need for structural diagnosis while sets a conceptual basis for defect detection automation is crucial. The aim intensifies when the particularities of ethics and safety in the field of art conservation are considered.

There are ways to obtain the advantages of fringe patterns even when specialized software and advanced analysis algorithms fail to convey usable information. Interactively treating the features of fringe patterns through step-wise reasoning provides direct diagnosis while formulates the knowledge basis to automate defect isolation and identification procedures for machine learning and artificial intelligence (AI) development. The transfer of understanding of the significance of fringe patterns through logical steps to an AI system is this work's ultimate technical aim. Research on topic is ongoing.

Keywords: Holography, Holographic interferometry, Speckle interferometry, Fringe patterns, Artwork, Cultural heritage, Heritage science, Symmetry

Introduction

Preliminary update

In optical and digital phase-shifting holographic interferometry, a highly coherent TEM00 diverged laser beam illuminates a surface undergoing static or dynamic displacement. The rest and new positions of the displaced surface are recorded and archived, allowing double—or due to recent developments in PC-driven algorithms—multiple, continuous, or interval selective monitoring of surface reactions. Recording depends on the problem being explored and the related methodology, with the goal of revealing substantial information. PC-driven recording is

Correspondence: Vivi Tornari (vivitor@iesl.forth.gr)
Institute of Electronic Structure and Laser / Foundation for Research and Technology-Hellas, Nikolaou Plastira 100, Voutes, Heraklion, Crete, Greece

related to monitoring procedures based directly on a large number of surface datasets in the form of interferograms, with interference patterns characterizing the specific displacement of the surface in space and time under specific operating parameters and naturally or artificially induced excitation.

The displacement of a surface from an initial rest position indicates the rearrangement of the points at new positions. In the structural diagnosis of artwork, in order for displacement to be harmless in terms of the object's physicomechanical properties and reversible elasticity, the excitation force needs to be very small (i.e., induce very little displacement) yet sufficient enough to induce displacement between all particles in the body relative to their reference position at rest in order to furnish high spatial information content. Ethically, minimal displacement in resolvable conditions in terms of the recording system and identifiable information in terms of diagnostic potential are prerequisites for artwork investigation.

Regarding artwork monitored under vibration-free strict stability requirements, displacement (strain) is not caused by changes due to rigid body motions but only due to changes in a) surface morphology due to local defects and b) surface size due to whole body dimensional movements, for example, shrinkage/expansion. The resulting displacement on the negative or positive z-axis is recorded in the direction normal to the axis, forming interference fringe patterns representing deformation with scalar optical path units of lengths (μm).

Fringe patterns are visualized as combined reactions of 1. "partial" or spatially limited displacements ("local" defects) and 2. "whole" extended areas of the surface ("whole" body spatial displacements, for example, shrinkage/swelling), and if referring strictly to dimensional (normal) deformation, the other components are negligible. The areas indicated by local fringe patterns represent "limited" or "extended locations/areas" with common deformation characteristics within "whole-body" displacement, a partial to whole body combination of information to become knowledge which is capable to describe the reactions and condition of the materials or layered constructions composing the artwork.

In the above context, the methodology presented here exploits how each interference fringe pattern reveals a physical tool to assess unique dynamic conditions and an approach that entails exploring them as representative information for the direct diagnosis of artwork, independent of other actions such as interactive measurements, unwrapping phase processes, FT or engineering stress and strain analysis. The objective of the methodology is to provide recognition steps for achieving direct diagnosis and automation reasoning.

In general, interference fringes are formed as a result of processes involving critical physical phenomena concerning wave physics and field superposition realized under the strict boundary conditions of the relevant physical laws. These laws are mathematically described, providing fringe pattern analysis with a direct quantitative information source indicative of the magnitude of the effects that the phenomena involved can reveal. The method allows for a highly precise quantitative study of the investigated surface.

The concept of incorporating holographic interferometry geometry for artwork structural analysis is justified by its advantages, starting from the scale of measurement, which, due to the incorporation of lasers, is of the order of a few micrometers, allowing small loads to be intentionally applied to successfully provoke small displacements without endangering the precious surface. The wavelength of light is added as a highly sensitive measurement for studying locally expressed invisible defects.

Implementation of hologram interferometry in structural diagnosis is based on the inherent property of field superposition in creating hologram interferometry, which allows a light wave diffusely scattered by an object to be holographically recorded and reconstructed with precision such that it can be interferometrically compared with light scattered by the same object at another instance in time[1–6]. Interferometric precision denotes the accuracy and non-destructivity of the measurement, which are both critical properties in cultural heritage (CH) conservation research. Indeed, holographic interferometry, which provides an interferometric comparison (a holographic interferogram) of two or more waves, at least one of which is holographically reconstructed, is a technique that introduced laser applications into the CH field. Furthermore, it has been the basis of and provides a standard reference for the comparative development of techniques based on the principles of optical coherent metrology[7–15].

Historical update: Holographic interferometry in cultural heritage

With regard to CH research, holography was introduced to art conservation and restoration applications not long after the invention of the laser.

Although there was intense interest in the application of lasers to CH soon after the laser's invention, especially for cleaning and spectroscopy, research on the structural diagnostic capabilities of the holographic interferometry method to detect hidden anomalies was not progressing at the same pace, and there were only a few laboratories researching challenging topics related to holography for structural diagnosis of artworks.

In general, in the art conservation field, conventional conservation practices for structural examination require either well-equipped heavy installation facilities or time consuming and limited resolution techniques, for example, the commonly used X-ray imaging and finger knocking, respectively. The most commonly employed method to deliver a structural condition evaluation report entails transportation of portable artwork to an X-ray facility for examination OR by stereomicroscope and manual inspection. Even though these approaches have long proven to be effective, it is certain that traditional practices in the restoration of CH treasures, monuments, and museum collections are limited, and professionals are increasingly demanding new tools and practices in response to the impact of the deteriorating environment and the highly increased cultural demand for artwork exhibitions. Demand for new methods that are faster than the existing time-consuming procedures, integrated and non-fragmented instrumentation, sensitivity, accuracy, resolution and repeatability are all factors driving a new era in conservation tools that will allow for the transfer of technology know-how from laser research into the field of art conservation.

CH research that uses holography originated from certain laboratories that introduced holography and holographic interferometry techniques mainly to structural diagnostic applications.

Beginning in the 70s and 80s, Profs. Franco Gori and Domenica Paoletti published literature related to optical holographic interferometry and sandwich holography for painting and artwork structural diagnosis and continued later with electronic speckle pattern interferometry (ESPI) experimentation. Affiliated with the University of Aquila in Italy, the professors were very productive at applying holographic interferometry to panel paintings for defect detection and developing techniques for phase control, such as sandwich holography in which a carrier frequency input modulates the formation of interference fringes in the interferometric reconstruction process by using single optical holograms in temporally different acquisitions. It could be said that after Prof. Asmus' first attempt in Venice, Prof. Paoletti and colleagues were the first to realize the power of holographic interferometry non-destructive testing (HINDT) in CH structural diagnosis and recognize holographic interference fringes as a unique tool to trace invisible hidden defects at determined locations and possibly correlate these with the defect type present in the artwork[16-21].

In the 80s, Prof. Pierre Boone at Ghent University in Belgium extended the use of holography and HINDT from application to mechanical problems to the study of defective ceramic and museum objects with video holography[22]. In several works, Prof. Boone explored three-dimensional (3D) representation applications of holography involving pulse lasers for producing high-quality portrait and naturalistic transmission and reflection holograms. In these 3D representation applications, important contributions advanced progress one step further. The 3D display can be attributed to Prof. Yuri Denisyuk and colleagues of the Russian Academy of Science, for their revelation of the unique 3D display properties of single-beam holographic representation in replicating museum objects for transport exhibitions via Lippman holography, which produced a large number of unique, quality single-beam holograms of museum objects, facilitating the founding of a holography museum in Saint Petersburg[23-25]. Prof. Markov has also shown special interest in the 3D properties of holography and has contributed significantly to the idea of holography display applications for museums, while Prof. Boone worked diligently in the field during the 90s[26,22]. In another approach involving potential color holography, Prof. Von Bally and colleagues from Carl von Ossietzky University of Oldenburg, H. Bjelkhagen, and other scientists and PhD students contributed to color holography, aspects of display, and technical holography applications in museum practice, with particular attention to the quality of the reconstructed image, holographic non-destructive methods, memory, and image synthesis[27,28]. Prof. Von of Bally University can be credited for cuneiforms in artwork documentation. Oldenburg made further advances in the holographic remote study of archaeological objects[29]. In the same period, Profs. Klaus Hinsch and G. Gulker, along with their colleagues from Carl von Ossietzky Oldenburg University, studied the use of speckle interferometry for the weathering deterioration assessment of monuments, an application critical for immovable CH as well as for the transfer of know-how from structural engineering applications to CH[29-31].

Regarding the contribution of coherent metrology and digital image processing to structural diagnosis, in general, significant advancements can be attributed to Profs. Juptner and Wolfgang Osten as well as their colleagues at BIAS, who elaborated on coherent metrology, surface inspection, and digital holography. Prof. Osten at BIAS and Institute Technische Optic University of Stuttgart, a visionist and organizer of international FRINGE conferences where the most internationally recognized holography experts gathered, made a unique contribution that has shaped coherent metrology applications, holographic interferometry, shearography, fringe projection applications, and phase-shifting interferometry. The development of BIAS's FRINGE PRO software is specially mentioned here, as well as the mathematical analysis presented by Dr. Ulrike Mieth from BIAS, which, under the supervision of Profs. Juptner and Osten, elaborated the mathematical formulation of holographically-generated fringe patterns[32-36].

During the 90s, collaboration between Prof. Osten of BIAS and the author of this paper initiated systematic research on system investigation for CH, especially with regard to correlating the theoretical and mathematical models of interference fringe pattern formation with experimental data findings from actual defective and non-defective samples simulating structural defects in artwork[37-39]. The joint effort produced an artwork defect classification table listing fringe pattern appearance and morphological symptoms indicating the presence of identifiable hidden defects. Hence, detection and definition through the interpretation and validation of interference fringe patterns facilitated a thorough classification of structural symptoms indicating types of defects based on holographic interferometry fringe patterns[40].

The author's personal research since the early 90s started at Holography Laboratories at the RCA in conjunction with the V&A conversation department and the optical laboratories of ICSTM. After the elaboration of advanced holography display techniques, research was focused on the development of double pulse and double to multiple exposure holographic interferometry sequential recording methodologies to monitor reactions and study the environmental effects of dimensional changes in artwork. This research continued at IESL/FORTH with the development of defect detection and identification protocols, multi-exposure comparative holographic interferometry, and digital holographic speckle pattern interferometry in a lightweight portable remote access system termed the Digital Holographic Speckle Pattern Interferometry or DHSPI that monitors the environmental impact and traces defect conditions in order to study defect types and classification, defect growth and deterioration mechanisms, physical aging and interventive laser processes, preventive deterioration methodologies in direct monitoring in climate chambers, the role of defects in equilibrium processes, software development, and hybrid DHSPI instrumentation. Special interest was focused on interference fringe patterns as a unique tool in direct structural diagnostics for complex formations as a result of complex artwork constructions. In this paper, accumulated experience with the use of fringe patterns for defect and material reaction interpretation and implementation in direct structural diagnosis is presented and explained to guide future developments in AI[41-62].

Complementary relevant research on exploring the properties of speckle and shear interferometry has increased over the years, especially in the last decade, at a number of different laboratories that have produced a wealth of experimental and computer-simulated investigations to benefit the development of the optical coherence field in CH applications[49-51, 63-87].

The aim of this article is to highlight the unique properties of fundamental fringe patterns as physical tools and demonstrate their significance in revealing the structural condition and mechanical reactions of artwork, possibly allowing for automated evaluation.

Conceptual methodology – Significance of fringe patterns

Measurement – Localization concept of interference fringe formation

Despite the widely applied displacement analysis approach that entails the calculation of engineering strain along with unwrapping interferograms, in artwork application, the direct approach requires in short measuring the fringe number from the zero-order fringe to calculate the relative displacement through fringe density per area and using differences in the interferograms' fringe density to extract the "whole field" vs. tracing and classifying "partial fields", thereby indicating "local" defect positions and types. A study on wrapped interferograms followed this procedure.

Interference fringe patterns are expressions of optical path changes for each fringe pair equal to half the wavelength with primary spatial frequency z = 0, depending on the angle of observation and wavelength.

$fy = sin\theta/\lambda$, in mμ and/or lines/mm

The fringes wrap the phase of the wavefront carrying the displacement at the recording point. Fringes are directly quantitative, and their density is a number expressing the total displacement of the examined area of the surface in numbers.

Visual interference fringe patterns are regarded as a "physical tool" to measure the structural integrity of an investigated surface. This concept is straightforward. The displacement patterns follow symmetry laws. Symmetrical patterns are easily recognizable. Any homogeneous and isotropic surface under a normal small force generates "whole body" displacement recognized in interference by the "symmetric" intensity distribution. These are the symmetry *architectures* that can be used to define symmetry vs. the symmetry deviations of non-homogeneous and isotropic bodies, as is the case for any artwork that generates symmetric deviations closely related to specific archetypes. Local disruptions breaking the symmetry in the "whole body" symmetric intensity distribution indicate a "partial" loss of surface continuity generated by locally defined discontinuities, resulting in visible in fringe patterns local asymmetries. These locally defined asymmetries are also close to an easily recognizable symmetry archetype. Each indicates a specific type of discontinuity.

Thus, fringe patterns can be considered partial or whole body intensity distributions characterized by spatially limited defined symmetry and spatially extended defined symmetry, respectively. In an interferogram, the denser the localized fringe pattern formation, the larger the areas of spatially limited defined symmetry. These indicate the distinct subsurface and bulk characteristics that interrupt the spatially extended but defined symmetry indicating whole body displacement.

These local interruptions represent invisible defects hidden under the examined surface that are interferometrically traced and revealed due to their impact on the examined surface. Expert conservators familiar with the artwork's construction and materials are usually needed to identify their cause, and once the local anomalies due to hidden defects are traced, the conservator studies and evaluates the artwork's condition. To inform the conservator about the artwork's structural condition, a defect map is generated based on the fringe pattern distribution covering the full surface, and the map initially serves as an on-site visual qualitative evaluation. The defect map could be sufficient to evaluate the conservation conditions and restoration needs through visualization of the corresponding defective surface areas. When distinct fringe patterns are visualised, as shown in Fig. 1, where it is seen that the whole field distribution, here of concentric fringes, is interrupted by locally distinct fringe patterns (features that interrupt the whole body fringes locally), this visual performance allows tracing of the defective areas while facilitates on-site evaluation of the risk posed to the precious surface.

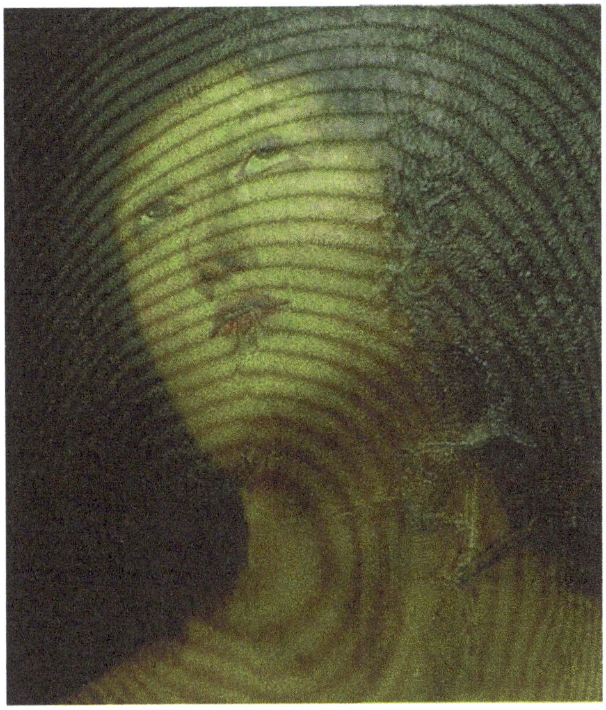

Fig. 1 Examination of a painting for structural diagnosis evaluation, where defect detection is clearly visualized in the surface abnormal distribution of fringe patterns. (Painting attributed to Rafael, in cooperation with the National Gallery of Athens).

Significance of fringe patterns

According to the laws of interference physics, if there is measurable displacement within the scale of the measuring device, interference fringes are always formed, and the task is to record them. If locally distinct fringe patterns are not formed, then there is always a visible whole body fringe pattern covering the surface representing the reaction of the total surface area as a whole. The whole body fringe pattern signifies and measures the surface's condition, its cohesion properties, and the overall integrity of the structure to a specific load.

Here, clarification of terms is essential to avoid confusion in the rest of the text. A "whole body" fringe pattern differs from a "rigid body" fringe pattern because of intentionally provoked displacement. The whole body fringe pattern is formed due to the surface's overall reaction to deformation, whereas the rigid body fringe pattern is formed as a result of background noise affecting the x-y system's stability, represented by linear distribution of the fringe pattern. The rigid body is linearly symmetric but distinguishable, as it clearly differs from the symmetry characterizing the formation of the whole body fringe pattern. Rigid-body fringes with stable equidistant fringes represent noise frequency. Experimentally, rigid-body fringes should be avoided by securing stability of recording system; however, there are applications where they can be useful as control for stability tests, for example, in paper interferometry where fringes on paper are expected to form within ns exposures using laser pulses the stability prerequisites tend to be underestimated since in such short exposures the rigid body fringes are rarely formed, unless recording system is fully unstable.

It is thus evident that holographic interferometry fringes are a reliable witness offering evidence of a successful interferometric experiment, as they demonstrate the efficiency of various excitation sources in terms of either naturally or artificially induced displacement. If excitation is capable of interacting with the investigated structure's hidden defects, fringes are revealed on the surface, while the same fringes simultaneously provide information about the overall condition of the structure and about the setup interferometric efficiency. As shown in Table 1, the fringes are assessed:

Table 1

a) the influence of the displacement cause -the load
b1) the condition of the investigated structure – the whole surface impact
b2) the condition of the investigated defects – the local impact
c) the condition of measurement – the stability of measuring system

Using fringe patterns for diagnosis

It has been discussed how fringe patterns are assigned as 1.whole-body and 2. defects from the fringe distribution caused by displacement. In the presented diagnostic approach, the focus is on the direct diagnosis offered by secondary interference fringe patterns. The significance and manipulation of fringe patterns for direct structural diagnosis is based in understanding fringe pattern morphology and its use as a physical tool. Hence, the fundamental aim is to understand the fringe patterns as a picture or an image that, through special characteristics, provides information relevant to the structural diagnosis of artwork such that any conservator can directly proceed with diagnostic evaluation. Operation Diagram 1 shows visible fringe patterns for art diagnosis. These are termed secondary fringe patterns and are the fringes formed by at least two exposures and seen covering the surface. All the fringes, whole or partial body, which are extended all over the surface or limited to small areas, are secondaryinterference fringes that upon formation are visible to the naked eye.

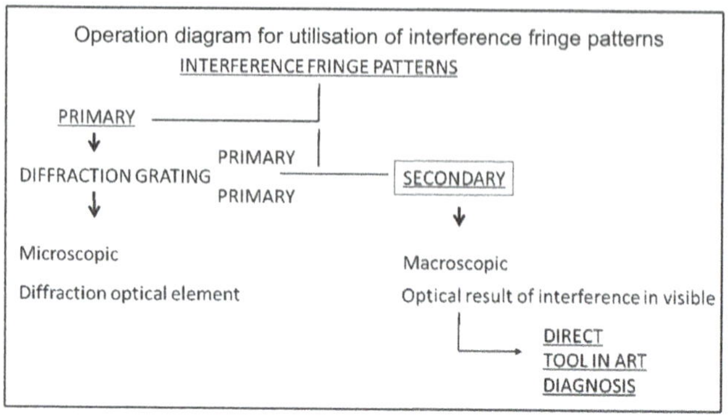

The method is very versatile and applicable to many different types of artwork with completely different characteristics. It can therefore become a universal tool in conservation, which means that understanding it as a direct physical tool is important. Interference fringe patterns' unique properties and advantages as a non-destructive direct source of information suited for diverse applications and a variety of objects for which they function as a direct visual qualitative examination tool and a direct quantitative assessment tool are all inherently instituted in the interference phenomenon. Interested readers should refer to mathematical and experimental presentations of the classification table of fringe patterns for common defect types and an explanation of its validation in artwork[36,37,41].

Fringe density

Artwork and decorative art usually refer to complex constructions of multilayered superpositions. Each layer represents a different material or mixture of materials, potentially both organic and inorganic and commonly with different compositions and thicknesses. The schematic is shown in Fig. 2.

When such a construction, considered to be inhomogeneous and anisotropic, and due to heterogeneity, lacking a representative volume element (RVE), is forced to move slightly by a few micrometers from its initial position predominantly in the +z or −z direction, that is, when it is forced to shrink or swell due to an artificial thermal increase or microclimate variation in humidity or temperature, the different elements forming separate density matrices move with different displacement

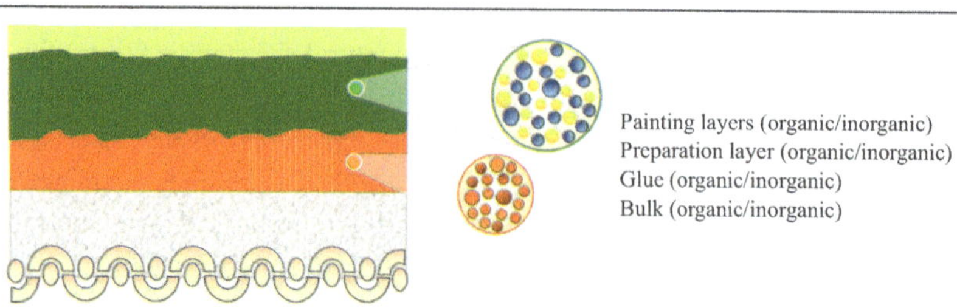

Fig. 2 Indicative cross-section schematic of a painting-like artwork.

amplitudes.

In our methodology, we are interested in the artwork's exhibited displacement amplitude as a whole body and not each constituent layer's ideal theoretical displacement amplitude. It is important that this is crystal clear, since the challenge in the art conservation field is to satisfy longstanding needs by adapting or creating engineering methods and practices for real, everyday CH applications.

The above description of artwork refers to flawless artwork, that is, artwork without defects, providing an idealized approach regarding materials and an even more idealized one with respect to the construction of artwork, which, by definition, suffers extensive flaws and defects. The fringe pattern approach, which is presented next, was used for defect detection via interferometry.

Artwork is composed of heterogeneous material opposed to the engineering of composite materials not typical to its mixture, as can be seen in the insert in Fig. 2, and since each artwork is a unique piece made of a pallet of ingredients, there is an insufficient number of common properties to support methodologies that embrace and evaluate the artwork as whole, unless one chooses to either destroy the artwork in order to study each layer separately (although this does not represent the artwork in its actual multilayered condition) or study samples and computer simulations[88–91].

In contrast, the holographic interference direct diagnosis approach investigates the artwork as a complete object. The fringe density (Δd/interferogram or area) allows us to obtain relative surface values of displacement that account for macroscopically uniform characteristics with respect to the applied boundary conditions of spatial heterogeneity and the RVE without destructive interaction with the constituent layers and materials[92–95]. The fringe density/interferogram per area is calculated based on fringe number measurements using the concept of the zero-order fringe. The zero-order fringe is self-evident in interferograms because it is the RVE of the continuum of the material without any movement, so that in vibration isolation conditions or using relatively stable areas under in situ conditions with large or immovable artwork, the zero-order fringe provides a means to evaluate deformation within the mechanical description of the behavior of solid objects. This definition of motionless points as the RVE of zero-order fringes is related to art materials, as they are assumed to be a continuous mass, despite the fact that their matter consists of mixtures of discrete particles. The artwork is considered to be a continuum solid object allowing full-field measurements of small scale to be assigned to length scales far exceeding the interatomic distances, which, in turn, allows the study of physical phenomena such as the displacement reaction of macroscopic objects where displacement at different length scales cannot affect each other. This approach is theoretically steady because it is based on the interference field's statistical ability to interfere with close and distant neighbor waves, producing a statistical effect representing surface displacement in the form of a fringe pattern.

In HINDT, the examined surface is considered to be a continuum surface acting as a diffuser when illuminated by an expanded coherent laser beam. The twice diffusively reflected light before and after a homogeneous force is applied interferes with a twin unmodulated reference. The optical path of the laser beam changes according to the whole body (excluding rigid body) displacement of the artwork's surface in which each point of the surface is considered as a collection of points reflecting photons obeying coherent wave propagation, diffraction, and interference principles. The reflected photons statistically interfere with neighboring photons involved in the same path and change to produce displacement of the same phase change value, which gives rise to interference fringes extended in surface for as far as the phase value remains same. See an example of the principle of fringe superposition from interferogram reconstruction in Fig. 3a for a primary pattern (diffraction grating hologram) and 3b for secondary fringe patterns (interferogram).

In Fig. 3a, a coherent plane wave field is deflected by a double-slit screen to emanate two mutually coherent spherical waves that spontaneously interfere, generating a primary modulated light intensity pattern forming a diffraction grating that is captured by a photosensitive screen at distance z. This formation gives rise to a fringe pattern (primary), and if illuminated, represents the interference field between the interfering waves. The secondary interference pattern that gives rise to visible holographic interferometry fringes reveals whole body displacement and partial fields; thus, potential defects are the superposition of two fields, where one of which has been provoked by a slight displacement force (not shown in this schematic). The change in dynamic requires continuous frame capturing and foresees a series of images from secondary patterns captured in sequential time events during relaxation of applied force or during continuous displacement toward reaching equilibrium with the environment, as shown in example Fig. 3c from a cooling down monitoring.

This is the sequential capturing utilized in the developed monitoring methodology, where hundreds of interferograms of each area as time windows of the area's reactions are recorded and processed. The first wavefront U_0 represents the wave at time $t = 0$, where φ_0 represents

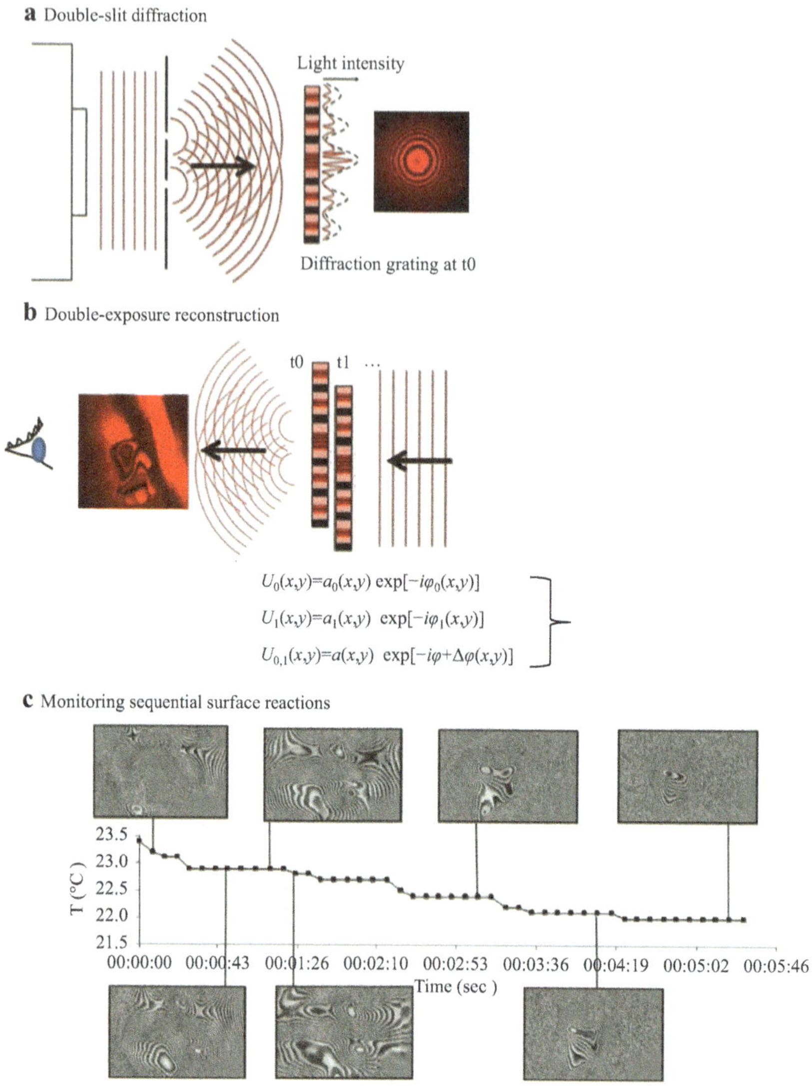

Fig. 3 Schematic representation of the developed real-time monitoring methodology of full field surface sequential displacement: **a** generation of a double-slit diffraction grating, **b** generation of interference between two diffraction gratings, and **c** monitoring dynamic displacements through the sequential capturing of interference fringes.

the phase modulation from the surface of the object, while the interference between the consequent wavefronts is represented by the phase difference Δφ. In the first wave, the wavefront's phase change is due only to the surface characteristics, whereas when the initial wavefront interferes with the wavefronts generated from the same surface after displacement, the phase change is formed according to the path differences that occur temporally in the surface wavefront.

As the surface wavefront changes over time, the visible secondarily formed interference fringes change the density following the change occurring on the surface of the object $\Delta d = d_2 - d_1$. If, as in the case of thermal load, the object's surface has undergone an initial transient impact, reaching its maximum displacement at the start time followed by lower displacement positions during relaxation, the same profile would be expected from the exhibited interference fringe density. The fringe density is shown to be the highest at the beginning of the measurement, with increasingly lower values until full surface relaxation is exhibited. The end of the process is indicated by a "no-fringe" surface, which signifies the end of the motion, ending the measurement. The displacement curve resulting from this process produces a fringe measurement, which follows the Newton law of cooling. Fig. 4 is the known graph of the Newton law of cooling, where the slope of the curve characterizes the rate of the cooling phenomenon dT/dt that takes place on the surface after thermal excitation. In the thermally-induced diagnostic methodology, velocity $v(t)= \Delta x/\Delta t \rightarrow dx/dt$ dominates the

a

Time (sec)	Surface T (°C)	Lab T (°C)	Lab RH
Reference	23,5	22,2	62,2
00:00:00	24,0	22,0	62,1
00:00:08	23,8	22,2	62,2
00:00:15	23,7	22,0	62,1
00:00:23	23,7	22,1	62,2
00:00:30	23,6	21,9	62,2
00:00:38	23,6	22,0	62,2
00:00:45	23,7	22,0	62,2
00:00:53	23,6	21,8	62,2
00:01:00	23,6	21,9	62,3
00:01:08	23,5	21,8	62,3

b

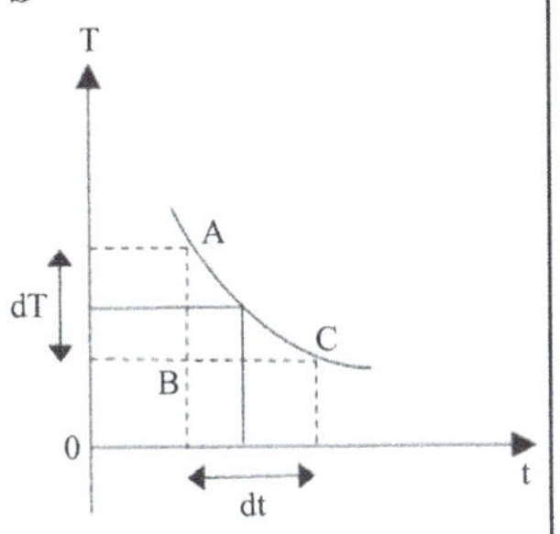

Fig. 4 Characteristic real experimental value table and theoretical mean curve slop during cooling.

fringe density in sequential interferograms and reveals acceleration of the surface and changing conditions toward equilibrium.

Fig. 4 provides an example of the laboratory experiments. The relevant thermal load values of the time and temperature of the surface are shown in the table in Fig. 4 in the first two columns, while the lab's relative humidity and temperature are shown for the control. If the thermal load is applied as an excitation force, the methodology for the measurement procedure starts with a $\Delta T = 0.5$ °C, and in this example, it takes 68 s to return the surface to the initial value, with the mean acceleration of surface-temperature decay providing the rate of change of velocity

$$\alpha(t) = \Delta v/\Delta t \rightarrow dv/dt$$

affecting fringe formation, so that it takes 8 s to 15 s for a 0.1°C change. In addition, as shown in the table, the very characteristic initial fast heat diffusion of surface loss in the first 8 s at 0.2 °C justifies the higher fringe density appearing in the first interferogram(s). These values, although dependent a) on the material and its condition as well as b) the parameterization of induced heat, specifically i) the type of induced heat, ii) the distance from the surface, and iii) the angle of the heat loader, represent a characteristic surface reaction after thermal loading. This method is implemented to generate the required fringe density for the defect trace, which, in turn, characterizes a) an artwork's sensitivity reaction to the applied thermal load and b) the ability of thermal excitation to trace hidden defects.

The excitation duration measured in seconds is experimentally more suitable than the temperature difference (ΔT) and is used in unknown cases not tested before since the ΔT is not *a priori* known for a variety of artwork surfaces and mixed materials. To reveal the most possible information about defects and structural integrity, the excitation methodology requires extending the excitation duration that is applied step-wise from the minimum to the maximum in the interest of the artwork's safety. Fringe density is again the criterion that will decide when the excitation increase will stop; the criterion is that the interferogram reaches irresolvable fringe density values, signifying unnecessary load excess. An example is shown in Fig. 5a–d, where a sample of a Byzantine icon is subjected to thermal excitation to perform a structural diagnostic procedure for defect detection, and the fringe density is increased at each step. In Fig. 5, only 5b is suitable for fringe counting, either manually or using automated software, because c and d have a higher irresolvable fringe density.

The fringe contrast in Fig. 5b–d seems to decrease because the fringe density increases, making the fringes

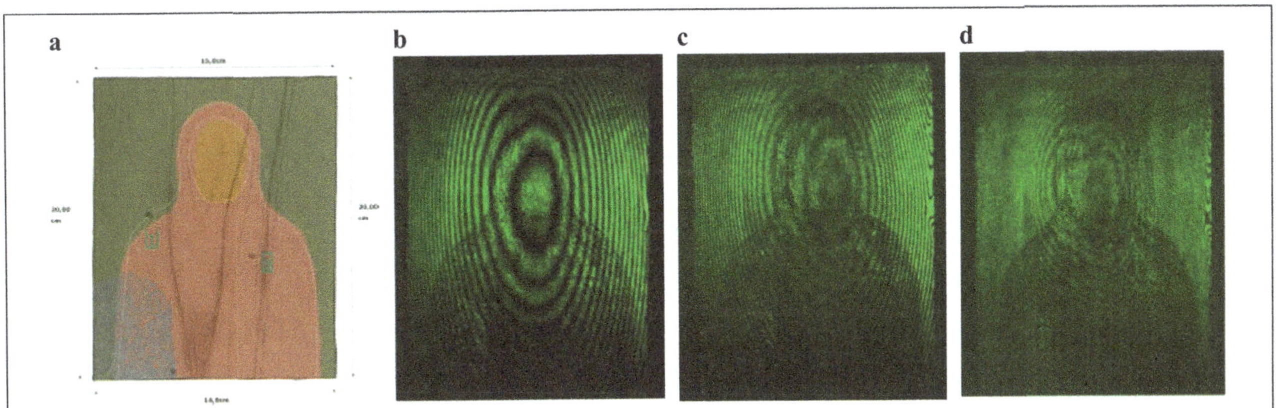

Fig. 5 Exemplary fringe density change as reaction to different parameters for applied surface acceleration using homogenised thermal load in **a** the byzantine icon sample; **b** 1 s, **c** 2 s, and **c** 5 s of applied excitation duration.

irresolvable with regard to being measured, photographed and studied. The interferograms show the effect of fringe density as a product of optical path changes, such as increasing displacement provocation.

Procedure: The fringe density is measured depending on the complexity of the interferogram on one to three axes on the raw interferogram either by a) visually counting the fringes from zero-order or b) using an automated software to measure after unwrapping the number of fringes from the fringe skeleton, with the operator interactively setting the zero-order point from which measurement starts. Quantitative evaluation of interferogram fringe counting via either method is a prerequisite. To start measuring a) manually, after the zero-order point is established, it is essential to recognize the characteristic symmetry of the fringe patterns, which entails the controller directly ascertaining the fringe number and density measurements without duplicating the values. If the symmetry of the displacement (whole field) is not recognized, the error of doubly adding fringes is possible. In cases of extreme aging and complicated fringe patterns, which are common in artwork diagnostic examinations, requiring a deep understanding of fringe pattern formation, the rule of whole field symmetry may collapse, and symmetry recognition should be applied locally on each pattern, indicating an isolated defect. When b) measuring using software, if unwrapping is possible, *a priori* knowledge is required to set the zero-order point, and fringe counting is performed automatically, but in complex cases involving aging, unwrapping collapses, and symmetry recognition in the whole or partial field is necessary to evaluate the interferograms.

The fringe density exhibited in the visible interferogram is the secondary spatial frequency of the two interferograms that witnesses the body's reaction to the induced load, here thermal load. Thermal load is applied in the present diagnostic procedure, to generate controlled surface displacement for the structural diagnostic purposes. As the surface's temperature increase is internally absorbed, it influences heterogeneous construction by exerting different impacts on its constituent elements, which have diverse thermal coefficients. These, in turn, are differentially displaced, provoking diverse displacement impact on the monitored surface. Hence, if there are hidden defects in the subsurface, even when they are attributable to different/ materials, if affected by the load, they will exert a visible impact on the surface.

With the recording methodology of sequential monitoring applied throughout the thermal cooling of the artwork's excited body, differential displacement is captured in time as it influences the body, starting from its surface, and diffuses into the depths of its construction. As time passes after excitation and while the body cools down, thermal excitation diffuses into the structure from the surface to the lowermost layer, visualizing the impact of the defects on different layers at different depths. The longer the duration for which a defect impact appears on the interferogram, the further it extends in terms of depth, revealing extended disintegration of the subsurface layers and bulk; conversely, the shorter the duration of its appearance, the less pervasive it is through the object's volume, and the more restricted its influence to a certain depth is. The critical defect diagnostic parameters are when and for how long the defect appears. Hence, the defects' sequential recording positions the defects in the depth of the construction and defect depth is indirectly evaluated.

Risk index: To gain preliminary information about each artwork, we perform zero-order fringe identification (defined as the part of the object with no displacement) either manually or via interactive software on one or more axes. In direct (manual) fringe number measurement, a different fringe density for each direction axis in the interferogram, either as a fringe number or as a product of wavelength $N\lambda/2$ (N number of fringes in micrometers), is assigned, and the mean value is extracted. The mean value changes in the presence of a defect, allowing numerical differentiation of the defect and the whole body.

The whole body fringe density measurements, in a direct defect diagnosis investigation with homogenized transient thermal load over the surface, fit as expected in the theoretical graph, as shown in the graphical representation given in Fig. 4, based on experimental values. A theoretical graphical representation based on experimental values taken from a sample with a known defect was verified. The data were obtained from measurements taken both from the whole body fringe pattern and from partial localized defects. The result visualizes the differences between whole and partial field reactions due to existing defects, as presented in Table 2 as values and graphically in Fig. 6. The partial body representing the areas with defects clearly exhibits a higher number of fringes, producing an intense local fringe density either in excitation duration parameterization Fig. 6a and during the cooling down after excitation Fig. 6b.

Table 2 shows that excitation duration is a more controllable and repeatable parameter than ΔT. Fringe density changes can differentiate displacement along the same symmetry axis, distinguishing whole from partial body and indicating the defect area. In the particular example of defect detection based on fringe density, the detected defect is located 5 mm under the surface and has a strong impact on the surface in almost all experimental

Table 2 Fringe Density spatial differentiation

Samples number	Excitation parameters				Whole-body	Partial-body
	Excitation, sec	Initial T °C	Final T °C	$\Delta T = T_1 - T_0$	Fringe density ($N_n = \Delta d/dx$ μ/cm)	Fringe density ($N_n = \Delta d/dx$ μ/cm)
1	9	26,5	30,8	4,3	0,26	2,10
	12	26,4	31,8	5,4	0,31	2,93
	15	26,7	32,8	6,1	0,36	3,33
	20	26,8	34,8	8	0,41	4,41

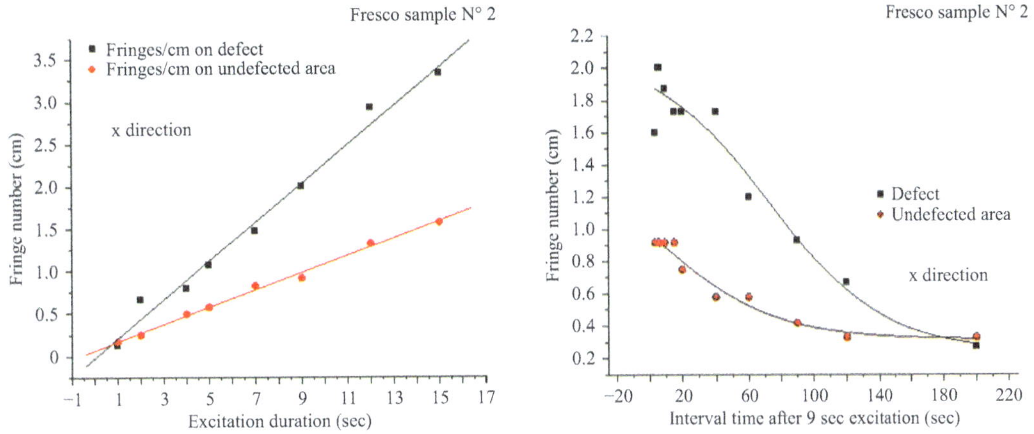

Fig. 6 Graphical representation of fringe density for partial vs. whole body differentiation.

thermal ranges. The defect isolation and measuring procedure is possible for as many defects as are isolated in the whole field interference fringe distribution, resulting in a surface risk map with an itemised defect risk index, as shown in the example given in Fig. 7a, b (painting attributed to Rafael; cooperation with National Gallery of Athens).

Fringe number measurement, either from zero-order fringes in lab vibration isolation conditions or relatively in situ, large surfaces and immovable conditions, that is, the defect isolation method, facilitates risk prioritization to serve as setting priorities in the restoration strategy. Following this map, a conservator can decide which defect exerts the most significant impact on the precious surface at the outset, considerably reducing restoration time. Fringe measurement can be performed in situ directly from the visible raw data after establishing the non-excited area (zero order fringe) and can be further post-processed in the lab using other software tools to study other defect characteristics. The fringe density, if needed, can be analyzed further by applying locally developed post-processing software routines, which provide additional information about the defect's impact on the precious surface, detailed coordinates, shape, micromorphology, defect interconnection, etc. Thereafter, the itemized defect risk index can be generated and utilized as an early warning prioritization tool in everyday conservation practices, assisting conservators in their decision-making interventions.

Fringe pattern

In addition to the significant parameter of fringe density, that is, the method of realizing measurements of the displacement of the artwork's surface based on fringe density, which has been shown to provide evidence of differential reactions within the surface, allowing for the location of intensity distribution abnormalities termed whole or rigid body vs local defects and assigning risk index of the impact on the surface, there is another important feature of interference termed fringe patterns that produce visual data of intensity distribution formation and provide significant evidence of local anomaly or the type of defect. In this process, symmetry deviation is the significant parameter.

Fringe patterns, especially dense fringe patterns, may appear to be random abstract formations, but they are not. Indeed, Dr. Ulrike Mieth (PhD Univ. Bremen – BIAS collaboration) has proven that they can be described through linear and non-linear partial differential equations, and she provided a thorough mathematical explanation of each of the most commonly appearing fringe patterns[88]. Based on this study we later confirmed experimentally the

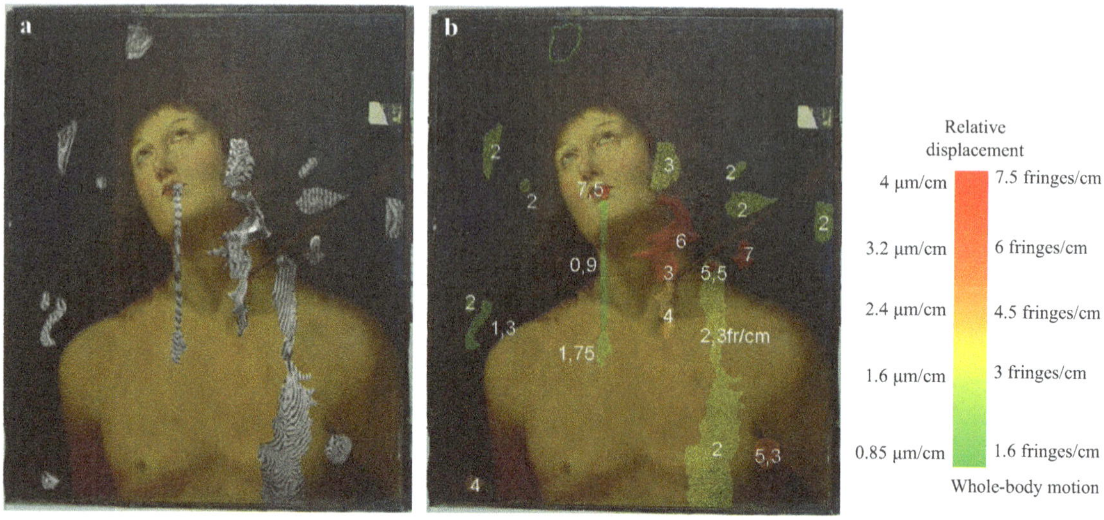

Fig. 7 Investigation for assessment of risk damage in painting surface. In **a** defect isolation and **b** defect risk index resulting from measurement of whole body vs. defect fringe density. (*NGA courtesy)

mathematical and simulation approach, with greatest possible fidelity of the resulting correlation. All different approaches resulted in the same pattern classes that were used to define the type of defect with high accuracy[35,41].

Fringe patterns are described as visual mathematical entities exhibiting symmetry, order, and limitation, the qualities of which are found in symmetry property exploration[88–101].

Some fringe patterns exhibit reflection symmetry, where two halves of the pattern are mirror-like or perfectly coinciding if folded, as shown in Fig. 8, which provides a recognizable characteristic example from a typically symmetrical shearography pattern representing the deformation gradient. It should be clarified here that the "mirror-symmetry" of shearography wrapped patterns due to self-reference is characterized by dual mirror-type symmetry, independent of the existence of defects.

Scale symmetry is exhibited in fringe pattern formation without constrained object dimensions, which are observed to appear as the same, independent of scale, meaning that the same fringe pattern is exhibited whether from a tiny object made of any material or a larger object made of any material (material effect fringe pattern). The scale symmetry pattern is a remarkable effect, as it ignores the object's dimensions and scale. Examples of this symmetry exist in interference fringe pattern formation during the investigation of artwork such as wall paintings, panel paintings, and sculptures; large and small objects exhibit

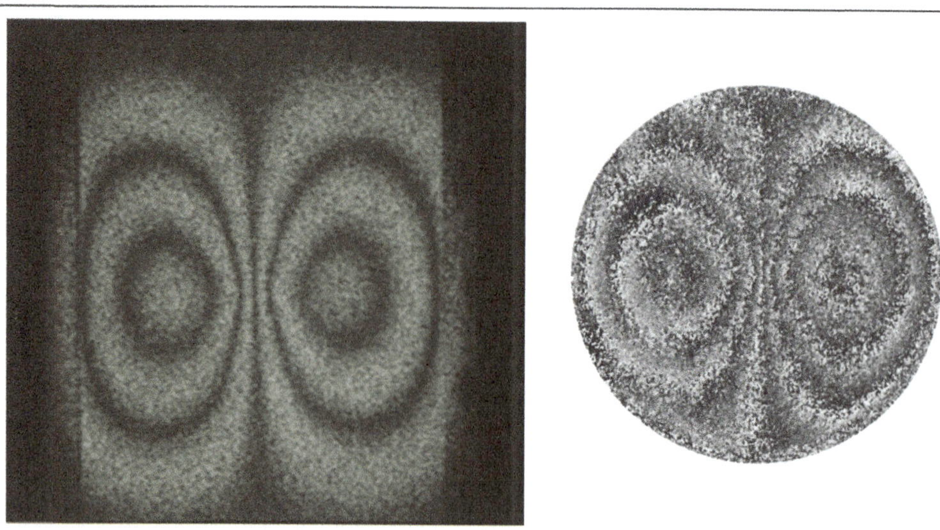

Fig. 8 Mirror symmetry exhibited from interfering two twin coherent beams, with one wavefront sheared with respect to the other, to assess deformation gradient from shearography pattern.

the same symmetry under displacement[35–37, 41]. One would think that fractal mathematical laws play a role in the generation of these fringe patterns.

Complex fringe patterns prompt fractal geometry, exhibiting a form of scale symmetry and exact or approximate self-similarity; that is, the entire fringe pattern has the same shape as one or many or all of the smaller parts constituting the fringe pattern, as shown in Fig. 9.

Smaller parts that are similar in shape to larger parts, regardless of dimensions and scale, which is common in mathematical fractal generation, are typically found in hologram interferometry fringe patterns. This symmetry is observed in complex deteriorated artwork where holographic interferometry patterns are produced during the displacement of aged artwork carrying multiple internal defects. These interferograms are the most complicated, exhibiting similar fringe patterns in multiple dimensions within the same but larger-scale fringe pattern; these are mostly exhibited in artwork-induced displacement for diagnostic purposes, as shown in Fig. 9. Disorganized matter generates similar displacement patterns, and the greater the degradation of materials, the more complex the fringe patterns that need to be unwrapped and evaluated. Hence, automation algorithms fail in such cases, and the whole to partial body and symmetry approach helps to locate and isolate the defective areas.

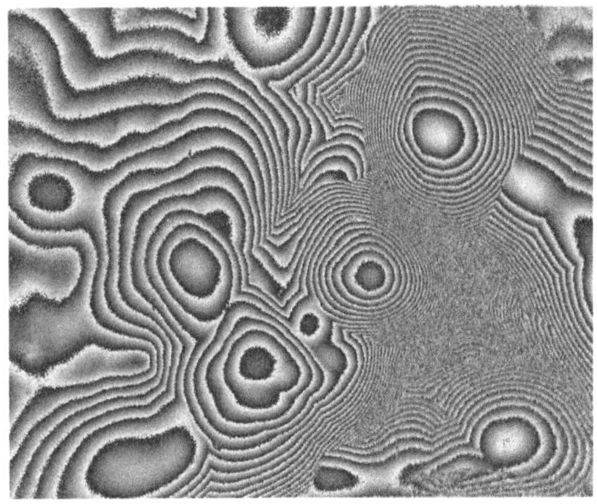

Fig. 9 Wrapped interferogram from a wall painting exhibiting multiple similar-symmetry circular-type fringe patterns of different fringe densities.

In Fig. 10a–c, characteristic cases of symmetrical whole body interference fringe pattern formation are shown for three different types of coherent interferometry techniques. Each interferometry technique generates an interference phenomenon that affects the formation of the fringe pattern; however, symmetry remains a common property in all techniques.

Revealing the "whole from a part", as in optical holography, describes an idealized symmetry property. If

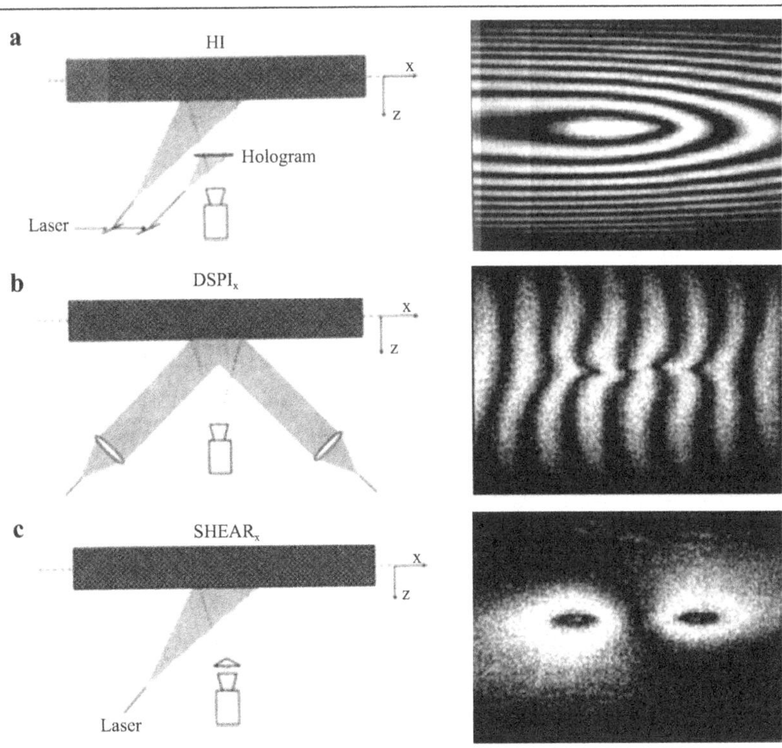

Fig. 10 Examples of the exhibited symmetry investigation of the same sample in **a** real-time holographic interferometry, **b** speckle interferometry, and **c** shearography.

just a portion of a developed optical hologram is illuminated the whole of the recorded image is reconstructed. By changing the illumination direction, the whole image is reconstructed following the different viewing angle.

In natural objects, as in the impressive phenomenon of the structure of snowflakes, a symmetrical portion of the snowflake's tip is repeated throughout the whole snowflake. This description reminds us of the unique feature of holography recording, where one viewer can see the entire recorded scene from just a small part of the optical hologram. This is the "interconnection" property of the primary fringes of holograms forming a diffraction grating from any random object if the object has been coherently illuminated and the recording medium receives the reflection from the surface. In an optical hologram, interconnection and reconstruction of information about the entire scene lead to visualization of one of the exotic properties of optical holography. This property can be demonstrated via a holographic film: if a hologram can be torn apart in small pieces, and each single piece, no matter how small, is illuminated with coherent light, each small piece can then be reconstructed to reveal the whole field, each from the viewpoint of the film's light-receiving angle at the time of recording, with parallax limited by the angle of and information in the piece. It may be that the tiny piece of the hologram can be regarded as holding all the information about the whole, as with the tip of a snowflake.

It is suggested here that the holographic interconnection property be considered a type of *paraxial symmetry* with a *conserved* quantity, where "symmetry" is defined by *the axis of propagation for angle of observation φ in the case of horizontal parallax and θ in the case of vertical parallax*. In holograms, regarding symmetrical paraxial quality preservation in holographic paraxial viewing, the entire scene is conserved despite the change in the viewing angle, suggesting that perhaps even seemingly random and complex objects and scenes are folded symmetries, or, according to Noether's theorem, "Any symmetry in nature has a corresponding conserved quantity"[102].

The property of fractal organization defined as "self-similarity independent of scale, in all scales" is a highly extended property in hologram interferometry, in interference fringe patterns where self-similarity is observed as the same symmetry patterns formed interferometrically from different objects made of diverse materials with unrelated constructions and in different scales. Whether a tiny object or large scale object is investigated same variance of symmetry patterns arise making impossible to extract the scale of the object without prior knowledge.

Careful experimental or computer simulated investigation of various materials under homogeneous thermal load reveals that inorganic materials predominantly exhibit in-plane displacement with parallel isopach isoline fringe patterns, while organic materials predominantly exhibit out-of-plane fringe patterns in concentric circular interference fringe pattern arrangements.

Wood is a particularly interesting organic material because of wood cutting. The wood cut determines fringe pattern formation, regardless of the type of wood. The wood cut makes distinct fringe patterns that are distinguishable from other wood cuts. When different wood cuts are interferometrically tested, they exhibit characteristic fringe patterns that indicate the type of wood cut. However, characterizing wood material through the fringe pattern is more complicated, and a database generated with strict boundary measuring conditions is required.

Despite the physical interest that the exhibited symmetry holds for revealing order in complex systems or even providing clues to the deeper order of the universe, the idea of our methodology is to use the expected symmetry of whole field interference fringe patterns to trace, through asymmetries or local collapses of symmetry, the hidden impact of defects on the surface, independent of the pattern's complex organization and use the fringe pattern's appearance, regardless of how small or distorted it may be, to locate and identify, through fringe pattern classifications, the type of hidden defect. In this context, the most common computer simulated and experimentally derived fringe patterns have been investigated using known defect samples. Experimental results have delivered fringe patterns that have been shown to be fully correlated to load and defect.[36,38,41].

Given that symmetry is a way to order nature, and fringe patterns exhibit symmetry, they can help to order highly complex interference responses. In physical phenomena studied under symmetry theorems, when symmetry breaks, there is a cause, and the same can be expected in examined cases of hidden defects' impact on an artwork's surface because any localized impact on the surface breaks the whole body symmetry of the surface displacement.

Topology analogue of symmetry

For use with artwork specifically, direct structural evaluation, which is actually the practical aim of the development of the concepts and methodologies presented here, via the analogous geometric topology approach is implemented in order to exploit the idea of the "symmetry of whole body fringe patterns" and use it as the basis to distinguish "whole field displacement" from "partial field displacement" to identify localized irregularities. Using this approach, the interferometry fringe pattern of a displaced surface described geometrically falls within one of three possible geometries, as shown in Table 3.

Table 3 Topology geometries to surface geometries

negative curvature	positive curvature	zero curvature
hyperbolic	spherical	flat

The above classification fits the intensity distribution in the formation of fringe patterns and can be implemented safely for use in everyday evaluation to perform direct, fast structural diagnosis.

When a plane surface (d_x, d_y) undergoes a homogenized induced or natural displacement (d_z), as previously described, the recorded whole body secondary interference fringe patterns correspond to the description of possible geometry topology surfaces in Table 3. *Nematic* field topology geometry has been observed in high-complexity fringe patterns, as in aged and highly deteriorated primarily organic multilayered artwork[101–102]. However, this case has not been classified here because it cannot be experimentally proven since this complexity cannot be repeated to create a classifiable example. Given that age or material cannot be considered representative of a sample, and a database does not exist, the case is not included in the experiments. The main difficulty in including the case is that a plethora of real samples or known defect samples are needed, and these are not easily found or constructed with precision. Thus, the nematic field was excluded from the classification.

Whole body to partial body procedure

It is clarified here that in most experimental cases, the displacement direction is known, and it can be determined via classic displacement vector analysis relating the motion direction to one-, two-, or three-dimensional surface displacement[10–13]. The displacement vectors describing the surface displacement are mostly suited to engineering analysis for position determination. However, by applying the symmetry approach to the distinction of whole and partial fields for direct diagnosis and defect detection, the process becomes visual. In this approach, topology and continuous symmetry breaking in soft matter are more suitable and relevant mathematical theories that better satisfy the presented methodology[101,102].

The delivered fringe patterns of surfaces exhibiting displacement, without exhibiting the impact of any internal defect, are covered by intensities that have a normalized distribution interchanging between the maxima and minima of the light intensities forming the fringes (recorded either analogically or digitally). This is shown in the identity function *id(x)* taken from the optical interferogram given in Fig. 11. Interferogram symmetry is evident, with minor deviations from symmetry, and can be idealized in surface geometry (Figure 11ciii, diii bottom). For interferometrically visualized surface displacement, as in the case depicted in Fig. 11ciii, if direction is important, vector analysis is needed because there are two possible negative and positive displacements of the geometric topology, as shown in Fig. 11dii, diii.

The other case for vector analysis if direction is important is on-line displacement, that is, x or y displacement, corresponding to flat surface geometry (Fig 11bi, ii–di). If a load is mechanically induced, it simplifies the evaluation because the direction is known. However, even with a 3D load, as in the case of thermally induced or environmental provocation of a natural impact, the displacement is deduced if the measurements of temperature (T) and relative humidity (RH) are taken during the experiment, and an association can be established, as shown in Fig. 12a–c. There, an increase or decrease in RH is monitored, and the surface geometry is deduced from concentric fringes that correspond to an increase in positive curvature, while a decrease is negative. Due to this additional information, the direction is known (+z increase: swelling, −z decrease: shrinking).

Hence, one-to-one correspondence between points is used to determine the type of displacement by identifying the surface geometry, as shown in Fig. 11a–d. This fully applies to the linear in-plane displacements seen in Fig. 11bi, ii–ci, and ii–di, where the dominant whole body displacements are parallel fringes arranged in the x and y directions.

The correspondence between mirror points of symmetry in the concentric fringe pattern related to dimensional displacement generates a spherical or hyperbolic curvature in the fringe pattern, as shown in Fig. 11biii–ciii–diii. It is implied that this determines the type of displacement in the two cases given in Fig. 11diii, given the known dimensional load and monitoring of the values, which can correspond to fringe pattern formation.

The whole field fringe pattern is then used to extract partial fields to locate defects. If the data is too big for manual exploration, the developed software next dscribed and shown in Fig. 13 is used to automatically scan the wrapped raw data.

Automation – Scanning Software

For interferometry fringes to find a special place in art diagnosis, the automatic detection of defects is very relevant to any effort. In this context, an effort to a scanning algorithm developed to automatically locate

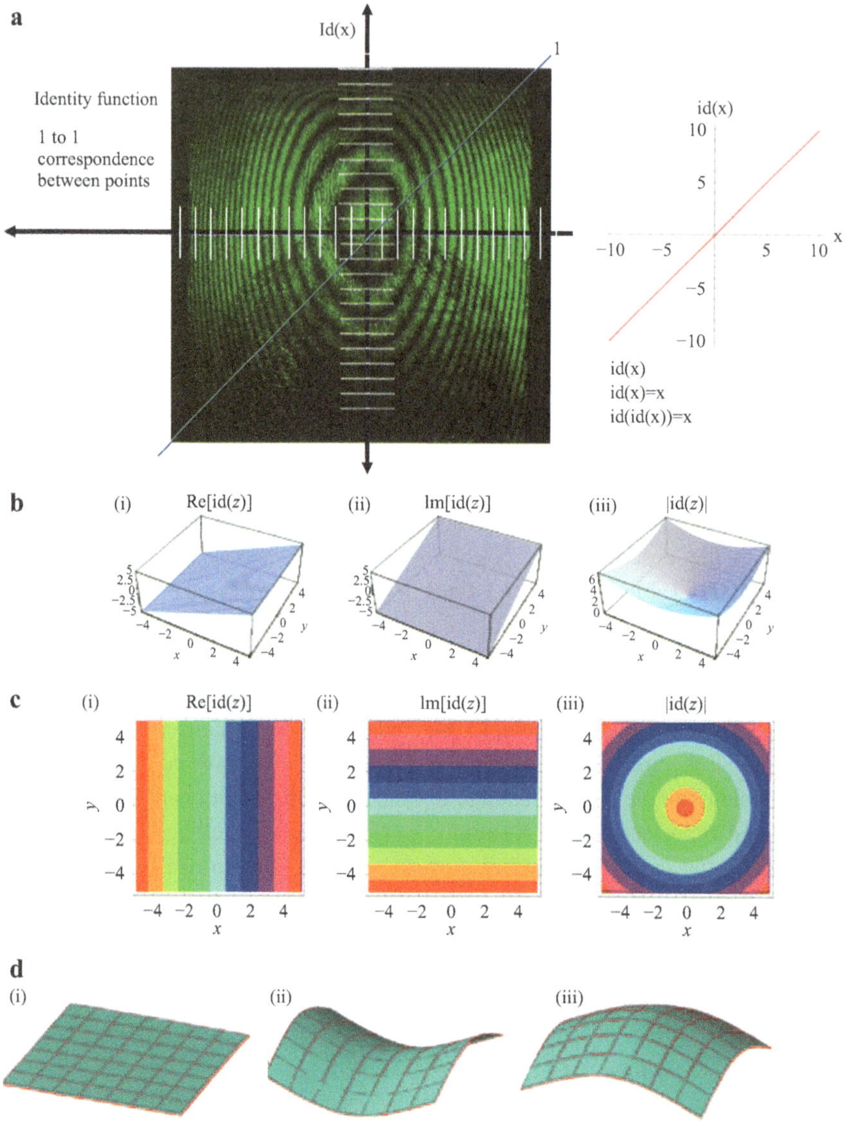

Fig. 11 Fringe pattern symmetry to surface topology: a surface geometry analogy of whole body displacement.

defects is shown in Fig. 13a, used directly on a real example of a defective symmetric fringe pattern; the whole/partial field fringe density is directly evaluated, as explained and defect is extracted as shown in Fig. 13b, and extracted. The procedure is repeated for as many axes as required to minimize error. The procedure starts with a chosen axis such as x or y and the diagonal. Each chosen axis is then scanned as many times as required to deliver the data, as in the manual case shown in the last column of Table 2. Scanning can also be performed manually on interferograms that directly exhibit whole vs. partial body displacement, but using processing tools in post-processing software, interferograms can be uploaded automatically and evaluated in sequence.

A scanning example of a characteristic symmetry break due to defects is shown in Fig. 13a, b.

The determination of whole- vs. partial-body fringes and the quantification of effects due to load or defect are performed by assigning spatial fringe density, as shown in Table 2. The aim here is to use automatic evaluation and scanning to accurately localize the defect's appearance. ω_0 is the overall induced frequency that generates the whole body, and ω_{def} is the deviation of the overall displacement visualizing the partial field. The step-wise process is described below.

The first step is to count the fringes of the interferogram with line profiles on both the x and y axes (Fig. 13). The step in each axis is user-defined and depends on the interferogram's estimated fringe density/area in cm or mm. This is the method of extracting the sum of fringes in each

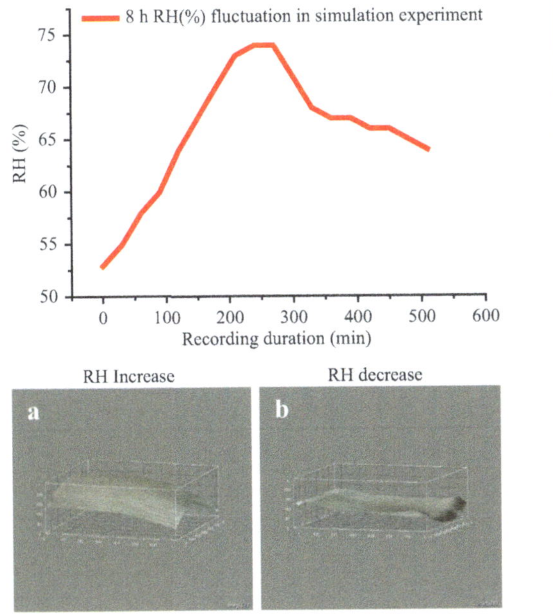

Fig. 12 Increase or decrease in RH is monitored, and the surface geometry is deducted from the fringe pattern.

line profile along the x and y axes and then calculating the average localization value ω_0. Here,

$$\omega_{0x} = sum\ of\ all\ fringes\ along\ the\ X\ axis / number\ of\ line\ profiles\ along\ X$$

$$\omega_{0Y} = sum\ of\ all\ fringes\ along\ the\ Y\ axis / number\ of\ line\ profiles\ along\ Y$$

The next step is to compute the sum of each line profile exceeding (or not) ω_{0x} (for X scanning) and ω_{0Y} (for Y scanning); their sums are ωX^+ and ωX^- (X axis) and ωY^+ and ωY^- (Y axis). The equations describing the procedure are as follows.

For the X axis,

$$\sum_{\omega x}^{+} = \int^{\omega x,y} d\omega_{x,y}/d\omega_0 = \omega X^+$$

$$\sum_{\omega x}^{-} = \int^{\omega x,y} d\omega_{x,y}/d\omega_0 = \omega X^-$$

For the Y axis,

$$\sum_{\omega Y}^{+} = \int^{\omega x,y} d\omega_{x,y}/d\omega_0 = \omega Y^+$$

$$\sum_{\omega Y}^{-} = \int^{\omega x,y} d\omega_{x,y}/d\omega_0 = \omega Y^-$$

where $\omega_{x,y}$ are the edge coordinates along each line profile across X and Y scanning, respectively.

Once we have found the numerical values for ωX^+, ωX^-, ωY^+, and ωX^-, we are ready to calculate the basic equation that indicates the localized deformation deviation. The result of this equation is a number that shows whether the structure has a local deformation. This number must be greater than 1 in order to assume a defective image. We use the following equations:

$$\text{For the X axis}: \quad \omega X^+ - \omega X^- / \omega X^- = \omega_{defY}\ \%\ \omega_{0X}$$

$$\text{For the Y axis}: \quad \omega Y^+ - \omega Y^- / \omega Y^- = \omega_{defY}\ \%\ \omega_{0Y}$$

The development follows the order shown in Table 4:

The values stored in programs' memories are resynthesized at the end of the interferogram procedure to emphasize the deviation from the norm in a separate result file, as shown in the example given in Fig. 14.

The example shows that part of the defect escaped detection, and it is likely that the algorithm will not identify all or may identify points that do not belong to the defect region; in this case, the operator can interactively correct the coordinates to fully assign the region. To optimize the software, further development is needed, with the aim of automatic detection in a number of sequential complex interferograms through additional algorithm

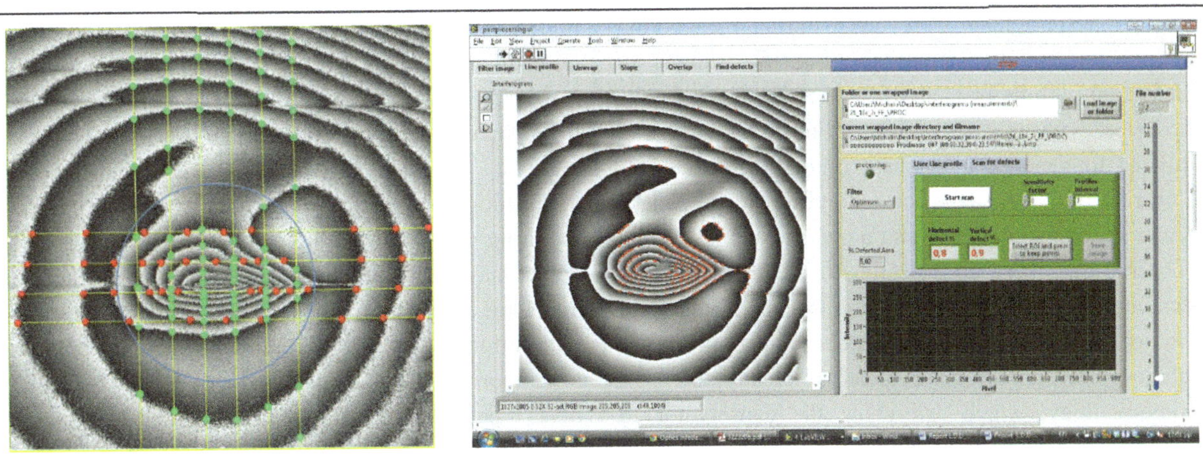

Fig. 13 Example of the developed automatic tool to extract defects from whole-body fringes.

Table 4 Stepwise scanning for defect localisation

1	The application renders the image size and whole -body geometry.
2	The wrapped filtered interferogram is scanned with line profiles with user defined intervals depending on the whole-body geometry (e.g. every 10 pixels) posing as a criterion the expected minimum distance between fringes derived by fringe density. This method, scan the intensity values of every line and if detects reversed values recognizes a fringe change.
3	The scan crosses in both axis X and Y. Each line profile records fringe number (density) and reversed positions (color). Each line profile obtains a total number. At end of each line measurement an average between all line profiles is calculated (spatial average). A spatial average of all lines is the basis for local comparison.
4	Then line profiles are compared, any line profile is greater or lower than the spatial average value, summing the greater and the lower separately and it keeps them in its memory to continue to next interferogram
5	Every change of the X,Y fringe coordinate in every line across X and Y scanning is recording and comparing each other

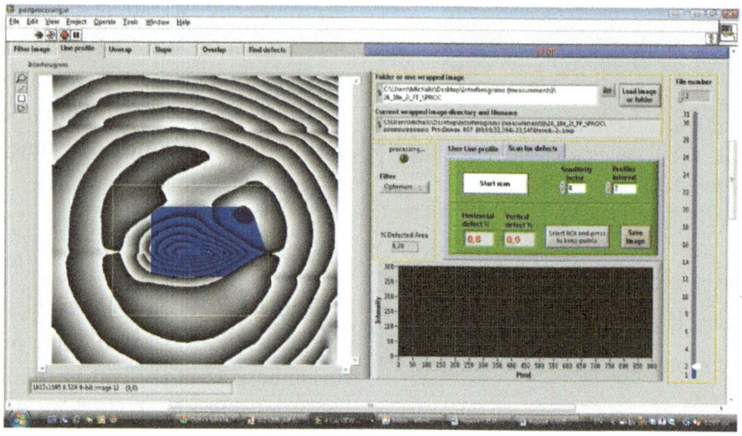

Fig. 14 Example of an automatic detection tool to extract partial field defects from whole-body fringes.

clustering and K-means to better elaborate the defined defect areas. However, defect detection via automated software has been continuously disappointing in artwork structural diagnosis.

The following example shows an unknown stone sample investigation. The sample was homogeneously excited in a microwave oven for 15 min, and after exit from the oven, sequential interferograms were taken during the cooling down process at 10 s intervals. Fig. 15a shows an interferogram with its FFT, and Fig. 15b shows the details of the defect with its FFT. Although Fourier analysis can be utilized to generate a database of different materials with characteristic Fourier spectra, if sufficient samples of the same structural condition and aging are available, one can observe that these do not offer the required detailed information for the structural condition of the sample and the buried defect. Defect size, shape, location, defect morphology, defect coordinates, defect microstructure, and micromorphology, which can guide the conservator to apply protective measures and consolidation, are lost. Hence, the most essential information is not applicable for use in light of the conservation demand for the precise and accurate documentation of defect topography and the evaluation of structural conditions.

Automatic defect detection through wrapped interferograms has features that hold very important information to artwork conservation, but it poses a long-term challenge since it has been attempted in its fully traditional form in other applications via software tools without success or with very disappointing results with respect to artwork. The reason for this requires reference to the particular requirements of the structural diagnosis of artwork, especially the very nature of preventive application to aged artwork with a plethora of defects requiring analysis. In artwork conservation, any small defect may be important and crucial for structural integrity and long-term preservation. Defects should be described with information such as exact coordinates, shape and micromorphology, impact on surface, depth inside the bulk, type of defect, and relationship with other defects in order to provide a helpful diagnostic tool to ensure that protective interventions are properly applied.

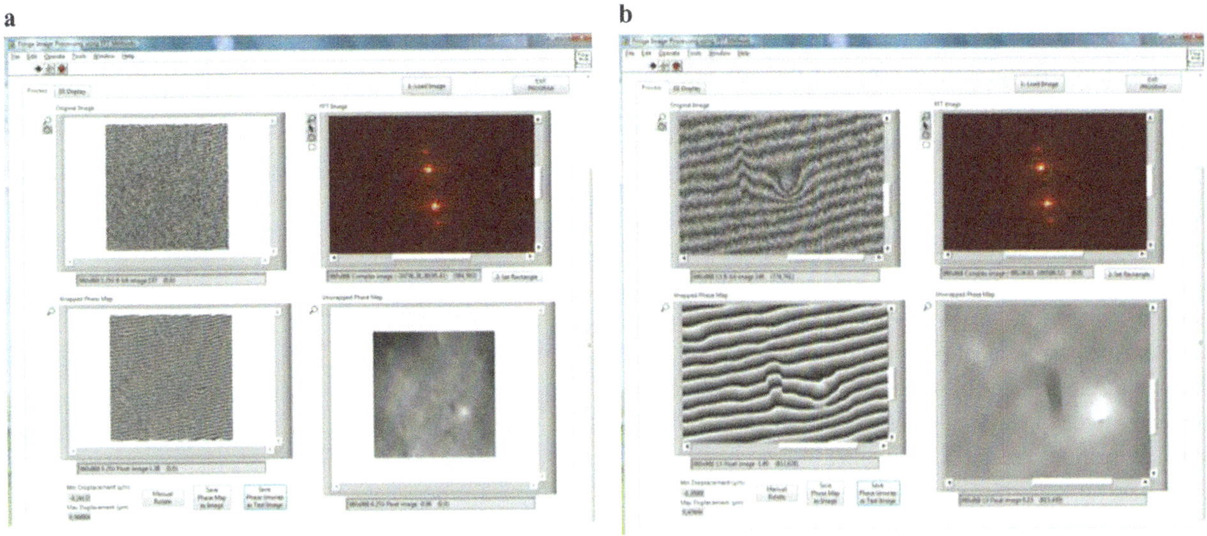

Fig. 15 Example of stone sample in **a** an interferogram of the whole surface of the sample and its FFT; **b** details of the defect and its FFT.

Work is ongoing regarding this aspect.

Conclusions

In this paper it is described the conceptual path taken to support the evidential conclusion that fringe patterns exhibit scale-independent self-similarity and finally symmetry. They are stepwise described the experimental observations that paved the way to show the fringe pattern formation obeys symmetry laws, order and repeatability. Interference fringe patterns are manifestations of symmetry.

In this effort the aim is simple and straightforward based on the striking experimental data. The challenge and the future aim of the concept it is to develop algorithms capable of tracing automatically defects within complex fringe patterns. An advancement that could facilitate conservation and prevent deterioration of Cultural Heritage. Final aim is detection instruments based on high resolution measurements resolved by artificial intelligent supported algorithms for tracing and processing the information of interference fringe patterns which hold tremendous information worth exploring for conservation strategies and maintenance.

Holographic interferometry fringes provide reliable and accurate evidence of the surface displacement characterizing the condition of artwork. They are a valuable physical tool, even in cases of complex fringe formation, given their ability to facilitate direct structural diagnosis and evaluation. In this paper, we briefly present the methodology followed to first understand the significance of the main characteristics of fringe patterns, such as fringe density and fringe appearance, to detect physical reactions and defects in artwork based on raw interference fringe patterns before the unwrapping procedure. Unwrapping complex interferometric patterns, the most common result in artwork investigation using hologram interferometry-related techniques, can be very complex and impossible to accomplish via automatic software processing, as most methods fail to render important interferometric data useful to conservators, unless the data are unscrambled through external interaction and a priori knowledge of operator and then post-processed. In this context, instead of following the unwrapping and skeletonization method, we introduce and follow another path to accomplish direct visual evaluation through the study of the interference fringe characteristics of density and symmetry, as well as the study of secondary spatial frequency and the appearance of fringe patterns.

Research efforts are ongoing to accumulate the required knowledge of the reactions and effects that visualize and form interference fringes in artwork investigation to potentially facilitate know-how transfer and the development of algorithms for automated defect detection, localization, and isolation toward the delivery of an automatic defect detection process and automatic risk-index classification.

The unique holographic interferometry properties and image qualities that are only successfully formed and provided under strict boundary measuring conditions are the keys to a physically broader approach and understanding of the significance and utility of interferometry fringe patterns in art conservation, as opposed to strict technical approaches, as in most engineering disciplines, which result in less complex fringe patterns. Several physical analogies are employed to serve the specific objective of evaluating the problems encountered in structural diagnostics in the artwork field. The exhibited symmetry and its corresponding topology approach combined with the fringe density measurements can be implemented to deliver algorithms capable to deliver a direct defect topography map without losing information delivered by the interferogram. In this study, we proved the suitability of whole vs. partial field, surface topology geometries, and symmetry exhibited significance and evaluation. Implementing some or all of these physical analogies in the context of novel software development as approaches to deep learning and AI education may hold the key to automation. Research is ongoing.

Acknowledgements
I am grateful to my lab team and especially to my lab assistant, Michalis Andrianakis, as well as to all my colleagues throughout the years, co-authors and project partners, and especially the EC-funded projects LASERART, LASERACT, and MULTIENCODE, which allowed in-depth study of defect detection problems and ambiguities in the context of artwork. I would also like to acknowledge CLIMATE for CULTURE, which allowed further development of the monitoring of natural environmental effects.

Conflict of interest
The author declares no conflicts of interest.

Received: 23 August 2021 Revised: 11 February 2022 Accepted: 15 February 2022
Accepted article preview online: 06 April 2022
Published online: 16 June 2022

References

1. Gabor, D. A new microscopic principle. *Nature* **161**, 777-778 (1948).
2. Gabor, D. Microscopy by reconstructed wave-fronts. *Proceedings of the Physical Society A* **197**, 454-487 (1949).
3. Stroke, G. W. An Introduction to Coherent Optics and Holography. 2nd edn. (New York: Academic Press, 1969).
4. Hariharan, P. Optical Holography. (Cambridge: Cambridge University Press, 1984).
5. Abramson, N. The Making and Evaluation of Holograms. (London: Academic Press, 1981).
6. Leith, E. N. & Upatnieks, J. Wavefront reconstruction with diffused illumination and three-dimensional objects. *Journal of the Optical Society of America* **54**, 1295-1301 (1964).
7. Powell, R. L. & Stetson, K. A. Interferometric vibration analysis by wavefront reconstruction. *Journal of the Optical Society of America* **55**, 1593-1598 (1965).
8. Kock, W. E. Engineering Applications of Lasers and Holography. (New York: Plenum Press, 1975).
9. Okoshi, T. Three-Dimensional Imaging Techniques. (New York: Academic Press, 1976).
10. Vest, C. M. Holographic Interferometry. (New York: Wiley, 1979).
11. Jones, R. & Wykes, C. Holographic and Speckle Interferometry: A Discussion of the Theory, Practice, and Application of the Techniques. 2nd edn. (Cambridge: Cambridge University Press, 1989).
12. Haines, K. A. & Hildebrand, B. P. Surface-deformation measurement using the wavefront reconstruction technique. *Applied Optics* **5**, 595-602 (1966).
13. Stetson, K. A. Fringe interpretation for hologram interferometry of rigid-body motions and homogeneous deformations. *Journal of the Optical Society of America* **64**, 1-10 (1974).
14. Hecht, E. & Zajac, A. Optics. 2nd edn. (Boston: Addison-Wesley, 1987).
15. Leith, E. N. & Upatnieks, J. Reconstructed wavefronts and communication theory. *Journal of the Optical Society of America* **52**, 1123-1130 (1962).
16. Asmus, J. F. et al. Holography in the conservation of statuary. *Studies in Conservation* **18**, 49-63 (1973).
17. Amadesi, S. et al. Holographic methods for painting diagnostics. *Applied Optics* **13**, 2009-2013 (1974).
18. Paoletti, D., Spagnolo, G. S. & D'Altorio, E. A. The state of art of holographic non destructive testing in work of art diagnostics. *Revue de Physique Appliquée* **24**, 389-399 (1989).
19. Amadesi, S., D'Altorio, A. & Paoletti, D. Single-two hologram interferometry: a combined method for dynamic test on painted wooden statues. *Journal of Optics* **14**, 243-146 (1983).
20. Amadesi, S., DoAltorio, A. & Paoletti, D. Sandwich holography for painting diagnostics. *Applied Optics* **21**, 1889-1890 (1982).
21. Amadesi, S., D' Altorio, A. & Paoletti, D. Real and nonreal time holographic nondestructive test (HNDT) for painting diagnostics. Proceedings of SPIE 0369, Max Born Centenary Conference. Edinburgh, UK: SPIE, 1983, 497-501.
22. Boone, P. M. & Markov, V. B. Examination of museum objects by means of video holography. *Studies in Conservation* **40**, 103-109 (1995).
23. YN Denisyuk, On the reflection of optical properties of an object in a wave field of light scattered by it, - Doklady Akademii Nauk SSSR, en-academic.com (1962).
24. Denisyuk, Y. N. Art holography in thick materials based on Lippmann photographic plates. *Journal of Technical Physics* (1978).
25. Denisyuk, Y. N. The manifestation of the optical properties of an object in the wave field of the radiation it scatters. *Doklady Akademii Nauk SSSR* **144**, 1275-1278 mathnet.ru (1962).
26. Markov, V. B. Holography in museums. in Optical Technologies in the Humanities (eds Dirksen, D. & von Bally, G.) (Berlin, Heidelberg: Springer, 1997), 31-41. http://dx.doi.org/10.1007/978-3-642-60872-8_3.
27. Von Bally, G. et al. Recording of color holograms on PFG-03Ts. *Technical Physics Letters* **21**, 667-668 (1995).
28. Bjelkhagen, H. I. & Vukicevic, D. Lippmann color holography in a single-layer silver-halide emulsion. Proceedings of SPIE 2333, Fifth International Symposium on Display Holography. Lake Forest, IL, United States: SPIE, 1995. http://dx.doi.org/10.1117/12.201881.
29. Gülker, G. et al. Deformation mapping and surface inspection of historical monuments. *Optics and Lasers in Engineering* **24**, 183-213 (1996).
30. Gülker, G., Hinsch, K. D. & Kraft, A. Deformation monitoring on ancient terracotta warriors by microscopic TV-Holography. *Optics and Lasers in Engineering* **36**, 501-513 (2001).
31. Trolinger, J. D. Civil and structural engineering measurements with interferometry, shearography, and holography.. *Optics and Lasers in Engineering* **24**, 87-88 (1996).
32. Schnars, U., Kreis, T. M. & Jüptner, W. P. O. Digital recording and numerical reconstruction of holograms: reduction of the spatial frequency spectrum. *Optical Engineering* **35**, 977-982 (1996).
33. Wagner, C. et al. Digital recording and numerical reconstruction of lensless Fourier holograms in optical metrology. *Applied Optics* **38**, 4812-4820 (1999).

34. Osten, W. Optical microsystems metrology. *Optics and Lasers in Engineering* **36**, 75-76 (2001).
35. Osten, W., Kalms, M. K. & Jueptner, W. P. O. Some ways to improve the recognition of imperfections in large-scale components using shearography. Proceedings of SPIE 3745, Interferometry'99: Applications. Pultusk Castle, Poland: SPIE, 1999, 244-256.
36. Osten, W., Jueptner, W. P. O. & Mieth, U. Knowledge-assisted evaluation of fringe patterns for automatic fault detection. Proceedings of SPIE 2004, Interferometry VI. San Diego, USA: SPIE, 1994, 256-268.
37. Osten, W., Elandaloussi, F. & Mieth, U. The Bias Fringe Processor- A useful tool for the automatic processing of fringe patterns in optical metrology. Proceedings of the 3rd International Workshop in Optical Metrology, Series in Optical Metrology. Bremen: Akademie Verlag, 1997, 98-107.
38. Tornari, V. et al. Laser-based systems for the structural diagnostic of artwork: an application to XVII- century Byzantine icons. Proceedings of SPIE 4402, Laser Techniques and Systems in Art Conservation. Munich, Germany: SPIE, 2001.
39. Tornari, V., Pedrini, G. & Osten, W. Remote photonic metrology in the conservation of cultural heritage. Proceedings of SPIE 8790, Optics for Arts, Architecture, and Archaeology IV. Munich, Germany: SPIE, 2013. http://dx.doi.org/10.1117/12.2021338.
40. Tornari, V., Osten, W. Shearography as part of a multi-functional sensor for the detection of signature features in movable cultural heritaged. Proceedings of SPIE 6618, O3A: Optics for Arts, Architecture, and Archaeology. Munich, Germany: SPIE, 2007. http://dx.doi.org/10.1117/12.727497.
41. Tornari, V., Tsiranidou, E. & Bernikola, E. Interference fringe-patterns association to defect-types in artwork conservation: an experiment and research validation review. *Applied Physics A* **106**, 397-410 (2012).
42. Bernikola, E., Nevin, A. & Tornari, V. Rapid initial dimensional changes in wooden panel paintings due to simulated climate-induced alterations monitored by digital coherent out-of-plane interferometry. *Applied Physics A* **95**, 387-399 (2009).
43. Tornari, V. et al. Fully-non-contact masking-based holography inspection on dimensionally responsive artwork materials. *Sensors* **8**, 8401-8422 (2008).
44. Tornari, V. Laser interference-based techniques and applications in structural inspection of works of art. *Analytical and Bioanalytical Chemistry* **387**, 761-780 (2007).
45. Thizy, C. et al. Photorefractive holographic interferometry for movable artwork assessment. Proceedings of Photorefractive Effects, Photosensitivity, Fiber Gratings, Photonic Materials and More 2007. Squaw Creek, USA: Optical Society of America, 2007, MB49.
46. Chaban, A., Deiana, R. & Tornari, V. Wall mosaics: a review of on-site non-invasive methods, application challenges and new frontiers for their study and preservation.. *Journal of Imaging* **6**, 108 (2020).
47. Yan, K. T. et al. Deep learning-based wrapped phase denoising method for application in digital holographic speckle pattern interferometry. *Applied Sciences* **10**, 4044 (2020).
48. Zhou, W. J. et al. Tip crack imaging on transparent materials by digital holographic microscopy. *Journal of Imaging* **5**, 80 (2019).
49. Tornari, V. et al. Impact of relative humidity on wood sample: a climate chamber experimental simulation monitored by digital holographic speckle pattern interferometry. *Journal of Imaging* **5**, 65 (2019).
50. Tornari, V. et al. Heat transfer effects on defect boundaries captured by digital holographic interferometry and infrared thermography workstation: an overview on experimental results. *Experimental Techniques* **44**, 59-74 (2020).
51. Chaban, A. et al. A combined non-invasive approach to the study of a mosaic model: first laboratory experimental results. *Journal of Imaging* **5**, 58 (2019).
52. Tornari, V. On development of portable Digital Holographic Speckle Pattern Interferometry system for remote-access monitoring and documentation in art conservation. *Strain* **55**, e12288 (2019).
53. Kosma, K. et al. Digital holographic interferometry for cultural heritage structural diagnostics: a coherent and a lowнcoherence optical setнup for the study of a marquetry sample. *Strain* **54**, e12263 (2018).
54. Tornari, V. et al. Interference fringe patterns in documentation on works of art: application on structural diagnosis of a fresco painting. *American Journal of Art and Design* **2**, 1-15 (2017).
55. Tornari, V. et al. Complimentarity of digital holographic speckle pattern interferometry and simulated infrared thermography for Cultural Heritage structural diagnostic research. *International Journal of Engineering Research & Science* **2**, 129-141 (2016).
56. Zitek, P. et al. Diffusion-model-Based risk assessment of moisture originated wood deterioration in historic buildings. *Building and Environment* **94**, 218-230 (2015).
57. Tornari, V. et al. Preventive deformation measurements on cultural heritage materials based on non-contact surface response of model samples. *Studies in Conservation* **60**, S143-S158 (2015).
58. Tornari, V., Tsiranidou, E. & Bernikola, E. Crack-growth on canvas paintings during transport simulation monitored with digital holographic speckle interferometry. *Advances in Research* **2**, 967-986 (2014).
59. Tornari, V. Delocalized photomechanical effects of UV ns laser ablation on polymer substrates captured by optical holography workstation: an overview on experimental result. *Advances in Optics* **2014**, 105482 (2014).
60. Márton, Z. et al. Holographic testing of possible mechanical effects of laser cleaning on the structure of model fresco samples. *NDT & E International* **63**, 53-59 (2014).
61. Tornari, V. et al. Synchronized deformation monitoring in laser cleaning: an application for cultural heritage conservation. *Universal Journal of Physics and Application* **7**, 149-159 (2013).
62. Tornari, V. et al. Micro-mapping of defect structural micro-morphology in the documentation of fresco wall paintings. *International Journal of Heritage in the Digital Era* **2**, 1-23 (2013).
63. Tornari, V. Spatial coordinates in interferometry fringes: a timeless artwork multipurpose documentation. *Journal of Basic and Applied Physics* **1**, 39-48 (2012).
64. Krzemień, L. et al. Combining digital speckle pattern interferometry with shearography in a new instrument to characterize surface delamination in museum artefacts. *Journal of Cultural Heritage* **16**, 544-550 (2014).
65. Accardo, G. et al. The use of speckle interferometry in the study of large works of art. *The Journal of Photographic Science* **33**, 174-176 (1985).
66. Young, C. R. T. Quantitative measurement of in-plane strain of canvas paintings using ESPI. Proceeding of Institute of Physics Applied Optics and Optoelectronics Conference. Brighton: IOP, 1998, 79-84.
67. Pfeifer, T. et al. Strain/stress measurements using electronic speckle pattern interferometry. Proceedings of SPIE 3520, Three-Dimensional Imaging, Optical Metrology, and Inspection IV. Boston, MA, United States: SPIE, 1998, 262-271.
68. Wegner, R. & Ettemeyer, A. Miniaturization of speckle interferometry for rapid strain analysis. Proceedings of SPIE 3824, Optical Measurement Systems for Industrial Inspection. Munich, Germany: SPIE, 1999, 30-37.
69. Young, C. Measurement of the biaxial properties of nineteenth

century canvas primings using electronic speckle pattern interferometry. *Optics and Lasers in Engineering* **31**, 163-170 (1999).

70. Del Sette, F. et al. Automated displacement measurements on historical canvases. *Heritage Science* **5**, 21 (2017).
71. Buchta, D. et al. Lock-in-shearography for the detection of transport-induced damages on artwork. Proceedings of SPIE 10331, Optics for Arts, Architecture, and Archaeology VI. Munich, Germany: SPIE, 2017, 103310G.
72. Buchta, D. et al. Combination of FEM simulations and shearography for defect detection on artwork. *Strain* **54**, e12269 (2018).
73. Zhao, Q. H. et al. Digital shearography for NDT: phase measurement technique and recent developments. *Applied Sciences* **8**, 2662 (2018).
74. Bylund Melin, C. Wooden Objects in Historic Buildings: Effects of Dynamic Relative Humidity and Temperature. (Gothoburgensis: Göteborgs Universitet, 2017).
75. Sirohi, R. S. Optical Methods of Measurement: Wholefield Techniques. 2nd edn. (Boca Raton: CRC Press, 2018).
76. Anisimov, A. G., Serikova, M. G. & Groves, R. M. 3D shape shearography technique for surface strain measurement of free-form objects. *Applied Optics* **58**, 498-508 (2019).
77. Anaf, W. et al. Real-time wood behaviour: the use of strain gauges for preventive conservation applications. *Sensors* **20**, 305 (2020).
78. Pagliarulo, V. et al. Full-field NDT methods for investigation of paintings on poplar. Proceedings of SPIE 11785, Multimodal Sensing and Artificial Intelligence: Technologies and Applications II. SPIE, 2021, 1178511.
79. Tornari, V. & Andrianakis, M. Exemplary case studies on movable and immovable cultural heritage investigation using portable system based on digital holographic speckle pattern interferometry. Proceedings of SPIE 10827, Sixth International Conference on Optical and Photonic Engineering. Shanghai, China: SPIE, 2018, 108271X.
80. Madruga, F. J. et al. Measuring the water content in wood using step-heating thermography and speckle patterns-preliminary results. *Sensors* **20**, 316 (2020).
81. Chaban, A. et al. Comparison of induced thermal change to climate chamber simulated environmental change in mosaic model by digital holographic speckle pattern interferometry (DHSPI). Proceedings of the SPIE 11058, Optics for Arts, Architecture, and Archaeology VII. Munich, Germany, SPIE, 2019, 110580I.
82. Kupczak, A. et al. Processing relative humidity data using discrete Fourier transform to control strain in art objects. *Strain* **55**, e12311 (2019).
83. dePolo, G. et al. After the paint has dried: a review of testing techniques for studying the mechanical properties of artistso paint. *Heritage Science* **9**, 68 (2021).
84. Daffara, C. et al. A simple method for artworks monitoring by simultaneous speckle interferometry (ESPI) and speckle photography. Proceedings of SPIE 11784, Optics for Arts, Architecture, and Archaeology VIII. SPIE, 2021, 117840I.
85. Huet, C. Application of variational concepts to size effects in elastic heterogeneous bodies. *Journal of the Mechanics and Physics of Solids* **38**, 813-841 (1990).
86. Sab, K. On the homogenization and the simulation of random materials. *European Journal of Mechanics. A. Solids* **11**, 585-607 (1992).
87. Omairey, S. L., Dunning, P. D. & Sriramula, S. Development of an ABAQUS plugin tool for periodic RVE homogenisation. *Engineering with Computers* **35**, 567-577 (2019).
88. Mieth, U. Untersuchungen zum erscheinungsbild von materialfehlern in holografischen interferogrammen. PhD thesis, University of Bremen, Bremen, 1998.
89. Sanei, S. H. R. & Fertig III, R. S. Uncorrelated volume element for stochastic modeling of microstructures based on local fiber volume fraction variation. *Composites Science and Technology* **117**, 191-198 (2015).
90. Spencer, A. J. M. Continuum Mechanics. (London: Longman, 1980), 83.
91. Roberts, A. J. A One-Dimensional Introduction to Continuum Mechanics. (World Scientific, 1994).
92. Tiennot, M., Iannuzzi, D. & Hermens, E. Evolution of the viscoelastic properties of painting stratigraphies: a moisture weathering and nanoindentation approach. *Heritage Science* **9**, 77 (2021).
93. Wigner, E. P. Symmetries and Reflections. (Bloomington: Indiana University Press, 1967).
94. Anderson, P. W. More is different: broken symmetry and the nature of the hierarchical structure of science. *Science* **177**, 393-396 (1972).
95. Cucker, F. Manifold Mirrors: The Crossing Paths of the Arts and Mathematics. (Cambridge: Cambridge University Press, 2013), 77-78, 83, 89, 103.
96. Gouyet, J. F. Physics and Fractal Structures. (Paris: Masson Springer, 1996).
97. Wilczek, F. A Beautiful Question: Finding Nature' s Deep Design. (New York: Penguin Press, 2015).
98. Noether, E. Invariante Variationsprobleme, Nachrichten der Königlichen Gesellschaft der Wissenschaften zu Göttingen. (Mathematisch-Physikalische Klasse, 1918), 235-257.
99. Seifi H. Topologu optimization and additive manufacturing of structural nodes of gridshell structures, PhD thesis, RMIT University, 2019
100. Chaikin, P. M. & Lubensky, T. C. Principles of Condensed Matter Physics. (Cambridge: Cambridge University Press, 1995).
101. Kim, Y. K., Shiyanovskii, S. V. & Lavrentovich, O. D. Morphogenesis of defects and tactoids during isotropicBnematic phase transition in self-assembled lyotropic chromonic liquid crystals. *Journal of Physics: Condensed Matter* **25**, 404202 (2013).
102. Warner, M. & Mostajeran, C. Nematic director fields and topographies of solid shells of revolution. *Proceedings of the Royal Society A: Mathematical, Physical and Engineering Sciences* **474**, 20170566 (2018).

Official journal of the JHL 2689-9620
www.light-am.com

Review | Open Access

Holography in the invisible. From the thermal infrared to the terahertz waves: outstanding applications and fundamental limits

Marc Georges*, Yuchen Zhao and Jean-François Vandenrijt

Abstract

Since its invention, holography has been mostly applied at visible wavelengths in a variety of applications. Specifically, non-destructive testing of manufactured objects was a driver for developing holographic methods and all related ones based on the speckle pattern recording. One substantial limitation of holographic non-destructive testing is the setup stability requirements directly related to the laser wavelength. This observation has driven some works for 15 years: developing holography at wavelengths much longer than visible ones. In this paper, we will first review researches carried out in the infrared, mostly digital holography at thermal infrared wavelengths around 10 micrometers. We will discuss the advantages of using such wavelengths and show different examples of applications. In nondestructive testing, large wavelengths allow using digital holography in perturbed environments on large objects and measure large deformations, typical of the aerospace domain. Other astonishing applications such as reconstructing scenes through smoke and flames were proposed. When moving further in the spectrum, digital holography with so-called Terahertz waves (up to 3 millimeters wavelength) has also been studied. The main advantage here is that these waves easily penetrate some materials. Therefore, one can envisage Terahertz digital holography to reconstruct the amplitude and phase of visually opaque objects. We review some cases in which Terahertz digital holography has shown potential in biomedical and industrial applications. We will also address some fundamental bottlenecks that prevent fully benefiting from the advantages of digital holography when increasing the wavelength.

Keywords: Long wave infrared, Terahertz waves, Digital holography, Speckle interferometry

Introduction

The hologram of an object or a scene consists of the recording of the interference between two mutually coherent waves, arising from a laser beam split into two parts, the first travelling via the object (the object beam) and the second one not (reference beam). At the early stage of holography, holograms were recorded on photochemical recording materials[1] (nowadays called "analogue holography"), which are still of great interest for the 3D rendering of portraits or scenes[2]. Soon in the mid-'60s, holographic interferometry (HI) surfaced. It uses the interference of two or more wavefronts read out through holograms recorded at different states of a moving object, yielding observation of interference patterns (interferograms) after the holographic plate[3]. These patterns can be related to phase change in the object or scene. HI was applied since then in an ever-increasing number of applications, among which fluid mechanics, solid mechanics, nondestructive testing (NDT, e.g.

Correspondence: Marc Georges (mgeorges@uliege.be)
Centre Spatial de Liège, STAR Research Institute, Liège Université, Liege Science Park, 4031 Angleur (Liège), Belgium

damage/crack detection), among others[4,5]. The rapid need for industrial applicability of HI devices led to the introduction of simplified hologram recording modalities[6]. Mainly, electronic recording devices were used to store holograms directly on the matrix sensor, owing to the speckle effect (in which case holograms are called specklegrams). The speckle pattern interferometry can be used to retrieve an interference pattern similar to HI for monitoring object changes[7,8]. Later digital holography (DH) was proposed also based on electronic recording[9]. Contrarily to speckle interferometry, which requires a lens to image the subjective speckle on a matrix sensor, DH can be used under a wide variety of optical schemes[10]. In particular, lensless schemes are possible. The object wavefront is then calculated using free space propagation integral on the digitally recorded hologram pattern up to the object location by numerical refocusing. One popular configuration is the off-axis DH using the Fresnel approach under paraxial approximation, applying simply fast Fourier transforms (FFT)[11]. The other one is inline DH with phase-shifting (PS), which allows eliminating unwanted diffraction orders superposed to the object image[12].

Whether one considers analogue holography, speckle interferometry or DH, the laser wavelength λ plays a central role. Soon at the beginning, its impact on setup stability requirements, HI measurement range and the type of object's reflectivity were experienced.

Most of the holography applications are performed at visible wavelengths. This is due to the fact that many components are readily available in visible (lasers, cameras, recording materials, optical components), but also simply because many applications are intended to display images for the human eye. Despite this, applying holography at non-visible wavelengths can provide fascinating outcomes.

In this paper, we will present developments and results obtained by several groups with DH and speckle interferometry at long wavelengths. Mainly, we will concentrate on long-wave infrared (LWIR) and beyond, the Terahertz (THz) waves. Shorter IR wavelengths (SWIR and MWIR) are not considered here, mainly because long coherence length lasers are still difficult to find with sufficient power and affordable prices. To clarify the ideas, we consider LWIR extending from 7 μm to 15 μm. Then comes the far-infrared (FIR), starting at 15 μm (20 THz frequency) and generally considered as the start of the THz spectrum. The sub-THz range extends from 300 μm to 3 mm (1 to 0.1 THz). Also, although some analogue recording materials were demonstrated in the past (reviewed elsewhere[13,14]), they proved to be unpractical. Therefore, we will focus on digital hologram recording.

After briefly reviewing the main components necessary in these ranges, we will see that long wavelengths can solve some issues experienced in the visible. Meantime, other issues will appear as well as interesting features that can be used advantageously. The paper is not intended to be an exhaustive review. Instead, we will emphasize the limits of using long wavelengths while showing their benefits through some representative applications in both the LWIR and the THz.

Main components for long wavelengths

In holography, the main components needed are lasers, camera matrix sensors, beamsplitters and combiners, and a series of more classical components for beamforming. Exhaustive reviews of existing technologies were given elsewhere: for the LWIR, see Georges[13], and for THz, see, for instance, Valzania et al. [15]. Here we give only a short introduction for understanding the rest of the paper. Regarding sources, since we need good coherence lengths, we limit the discussion to continuous ones.

In LWIR, the main source of interest is the CO_2 laser, emitting lines around 10μm, powerful (a few watts), relatively stable and with hundred meters coherence lengths[13]. This source is the one used the most in LWIR, although quantum cascade lasers (QCL) are also studied for holography[16].

Concerning cameras, two leading technologies exist. The cooled cameras are developed with either the quantum-well infrared photodetectors (QWIP) or the Mercury Cadmium Telluride (MCT), the latter being the most available commercially. They can be found in large formats and show very low noise, but their cooling system generates vibrations, which we found incompatible with a proper holographic recording. Therefore, uncooled technologies will be preferred. In these, we find the pyroelectric and the microbolometer technologies. The former was used for the very first demonstration of LWIR DH[17] but had limited pixel numbers. Microbolometer cameras are now well established in the thermography market. While having lower signal quality than cooled cameras, they have ever-increasing pixel numbers (up to megapixel format) and were used in most applications discussed in the paper.

In THz-FIR, the widespread coherent sources are the gas lasers, with up to 500 mW power at 118.8 μm (295-FIR laser by Edinburgh Instruments). It is composed of a 200W CO_2 laser, pumping a cell filled with various gases in order to obtain different wavelengths, between 90 μm to 500 μm. QCL lasers make rapid progress towards similar powers. In the sub-THz range, impact ionization avalanche transit-time (IMPATT) diodes and Gunn diodes are compact sources in this category[15]. These sources are highly coherent. Therefore, the coherence length is not an issue for interferometry with the object at meter distances.

THz cameras are primarily based on microbolometer technologies adapted to the THz range. Interestingly, microbolometers can be used at FIR wavelengths because of some residual response, although only under the lensless scheme (without LWIR imaging lens)[18].

Other optical components (lenses, mirrors, beamsplitters, etc.) can be found easily on catalogues by various manufacturers. The reader should refer to previous reviews papers for more details[13,15].

Specific features of long wavelengths on holography

The impact of camera parameters versus wavelength

The interference fringes in front of the detector plane are continuous and infinite. However, during the digital hologram recording, they are sampled by the camera with discrete pixels and a finite extent of the imaging area. Visible, LWIR and THz cameras typically have µm-scale pixel pitch p and tens of mm-scale imaging area $Np \times Np$ (N number of pixels), whereas the working wavelengths differ by 2-3 orders of magnitude. Naturally, the optimum recording conditions to be respected for faithful hologram recording vary. We now discuss these concerns and their impacts based on the three typical cases summarized in Table 1.

The first aspect is that the spatial sampling condition should be fulfilled: the finest fringe should be correctly sampled by the pixels, which limits a maximum off-axis angle[14]:

$$\theta_{max} = 2\sin^{-1}\left(\frac{\lambda}{4p}\right) \tag{1}$$

This implies that, when observing an object with lateral dimension D, a minimum recording distance should be respected, following

$$z_{min} \approx \frac{2p(D+Np)}{\lambda} \tag{2}$$

For a given recording distance, larger objects can be observed when working with a longer wavelength. The maximum off-axis angle at visible (λ = 532 nm) with a typical pitch p = 4 µm is 3.8 degrees. A 2m-sized object should be set at least 15 m away from the camera. The maximum off-axis angle at LWIR (λ = 10.6 µm) with pitch p = 17 µm is relaxed to 18 degrees. The minimum recording distance is reduced to 3.2 m. At THz range, especially when working with typical bolometer cameras, the off-axis angle is no longer limited since $\lambda > 4p$ hold in many cases. The relaxation of maximum off-axis angle is advantageous for situations where a large field of view is expected. Coincidentally, at the LWIR range, the high output power of CO_2 lasers guarantees sufficient flux for illuminating large areas. On the other hand, although the sampling condition is no longer an issue for THz DH, macroscopic objects imaging is unsuitable due to the lack of powerful THz sources. Thus, LWIR DH is the best choice for large object or scenes imaging, as is demonstrated by Ferraro's group in a series of papers[19-23]. They noticed that, at such wavelengths and with setup parameters discussed above, the speckle grains become very large and affect the reconstructed image quality. Nevertheless, special techniques described elsewhere can mitigate this[24].

The second concern is the size of the imager area. The intrinsic resolution is diffraction-limited, i.e., the lateral resolution depends on the wavelength and the numerical aperture (NA) of the imaging system. Under lensless DH configuration, the NA is determined by the size of imager Np and the recording distance between the object plane and the detector plane:

$$d_r = \frac{\lambda}{2NA} = \frac{\lambda}{2\sin\alpha} = \frac{\lambda}{2}\sqrt{\left(\frac{2z}{Np}\right)^2 + 1} \approx \frac{\lambda z}{Np} \tag{3}$$

The high-frequency components of object wave are diffracted outside of the imager zone. Therefore, they are lost during DH recording. As a result, the attainable intrinsic resolution decreases proportionally with working wavelength. LWIR DH with a slightly lower lateral resolution is acceptable for macroscopic DH applications. However, the diffraction is severe when using THz wavelength. Achieving a sub-mm level resolution regarded as a minimum requirement in most THz imaging cases

Table 1 Typical parameters of cameras and impact of wavelength on distance object-detector and resolution

	λ (µm)	p (µm)	N	θ_{max} (deg)	z_{min}* (m)	Intrinsic resolution d_r
Visible	0.532	4	1000 × 1000	3.8	15	0.0001 z
LWIR	10.6	17	640 × 480	18	3.2	0.001 z
THz	118.8	17	640 × 480	–	–	0.01 z

*For a 2 m object

remains challenging due to the large wavelength. Thus minimizing the object-detector distance is crucial for THz imaging. The recording distance is often squeezed to less than 5 cm when working with the 1cm^2-sized imagers at THz wavelength. The geometrical configuration often sets the lower limit of object-detector distance.

For THz off-axis DH setup, one should leave enough space for reference wave injection: the object-detector distance can be reduced to 1 cm. For inline DH or other reference-free phase retrieval setup, the recording distance can be reduced further. An extensive detection area is needed for Sub-THz DH due to the mm-level wavelength. Suitable 2D array detectors are not yet available. Therefore Sub-THz DH is realized by a raster-scanned detector. For instance, in Heimbeck et al.[25], when working at 0.495 THz (λ = 606 μm), for a sample located 16.3 cm away from the detector, a detection area of 20 × 20 cm^2 is needed to achieve a lateral resolution comparable to the wavelength. The short recording distance in THz DH translates to a high Fresnel number. Therefore, the angular spectrum method (ASM)[10] and Rayleigh-Sommerfeld convolution[10] are more suitable than the paraxial Fresnel approximation for THz DH reconstruction. Consequently, the discrete Fresnel transform method, which is widely used in visible and LWIR DH for its simplicity (only one FFT is required) and the flexibility of reconstructing large FOV, is not appropriate for THz DH.

Phase measurement range and resolution

HI is used for measuring variations of the phase $\varphi = 2\pi nl/\lambda$ of the objects by subtracting the phase from holograms recorded at different object states. Phase variations $\Delta\varphi$ can occur through any change related to the object wave, i.e. in the optical path length (the product of the refractive index n and the geometrical path length l) or the wavelength λ. Regardless of the nature of phase change, an interference pattern (interferogram) is observed in the image plane (x, y). It is expressed as

$$I(x,y) = I_0(x,y)\,[1 + m(x,y)\cos(\Delta\varphi(x,y))] \tag{4}$$

with I_0 the average irradiance and m the contrast. The goal is to retrieve the phase change and to interpret the optical path difference (OPD) which induces the phase change. In DH, the phase is easily obtained in the reconstruction process, while in some other versions of holography (analogue holography, speckle interferometry), phase quantification techniques must be considered[5,6,26]. In any case, once the phase difference $\Delta\varphi$ is computed, it can be related to the OPD and the wavelength through $\Delta\varphi = (2\pi/\lambda)\,OPD$.

In transmission mode, the light passes through the transparent object. The OPD between two points relates to the variation of thickness Δt or refractive index Δn. For a homogeneous transparent object with refractive index n, the thickness difference between two points is written as $\Delta t = \Delta\varphi\lambda/[2\pi(n-1)]$. For an object with uniform thickness t, the refractive index variation is $\Delta n = \Delta\varphi\lambda/[2\pi(t-1)]$.

Under reflection configuration, the light is reflected by the surface of the object under investigation. The phase difference reveals the object's 3D profile. The phase difference is proportional to the height variation Δd of the surface profile.

The measurable range and resolution are two aspects that should be noticed when relating the phase distribution and the measurand. First off, the obtained phase distribution is wrapped in modulo 2π. A more extensive measurement range of variation corresponds to 2π phase variation thanks to the large working wavelength. As illustrated in Fig. 1, a deformation ranging from 0 to 50 μm is observed using reflection configuration under normal incidence illumination. In this case, the relationship between the deformation Δd and the phase map $\Delta\varphi$ obeys the following

$$\Delta\varphi = (4\pi/\lambda)\,\Delta d \tag{5}$$

Since the phase retrieved is always obtained modulo 2π, there the deformation is too large to be observed in visible wavelength (Fig.1b) because the resolution criteria of Nyquist-Shannon is not fulfilled to distinguish two neighbouring fringes. The number of fringes reduces with the increase of working wavelength. However, the measurement precision decreases when working with a large wavelength since the measurand is proportional to the wavelength. It is considered as a shortcoming when working with THz wavelength. In the same example, if the phase resolution, i.e., the minimum distinguishable phase variation, is 0.2 rad for three wavelengths, the minimum measurable displacement will be $\lambda/63$, which corresponds to 0.17 μm in Fig. 1c whereas 1.9 μm in Fig. 1d. If prior knowledge of the deformation is known, unwrapping the phase obtained with LWIR in Fig. 1c will give the optimum result among the three wavelengths.

Reflectivity type (scattering vs specular)

The fact that an object appears scattering or specular depends on the ratio between its roughness R and λ[27]. If $R \ll \lambda$, the object appears purely specular, while for $R \gg \lambda$, it appears purely scattering, favoring the speckle effect. In practice, things are not so simple; some surfaces may appear partly scattering and specular when the roughness is close to λ.

In the case of solid reflecting objects, holography is

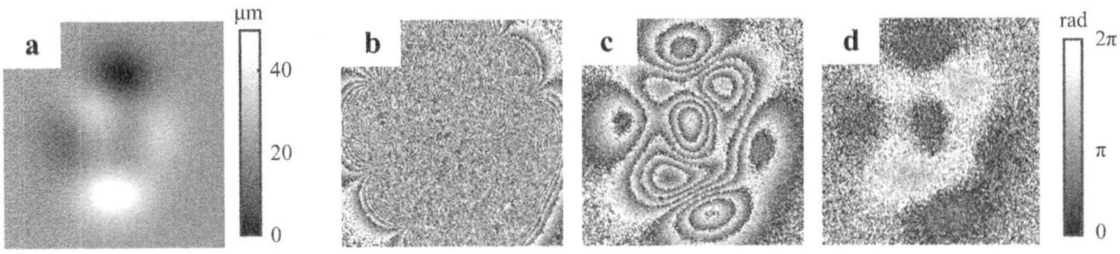

Fig. 1 Simulation of phase variation under different working wavelengths. **a** shows the simulated deformation map. **b–d** shows the wrapped phase obtained at resp. 532 nm, 10.6 µm, 118.8 µm based on Eq. 5.

easily applicable when the surface appears scattering. The object reflects part of the light to the camera, where it interferes with the reference beam. As long as a good balance of power between reference and object beams is guaranteed, there is always a way to maximize the hologram contrast at recording. This allows the development of lensless DH configurations, where imaging lenses are not necessary for image focusing. When the object is specular, this is more complicated to achieve because the light reflected by the object can fall well out of the hologram recording location. The light collection is necessary through schemes adapted explicitly to the object shape. This is feasible in a very limited number of cases (both spherical and aspheric optics cases), but complex specular surface shapes could be too challenging. Artificially introducing a scattering property is feasible: removable powder is often applied on objects. However, this is not always feasible or recommended in actual industrial applications.

Working at long λ further complicates the problem since surfaces scattering in visible become specular in LWIR. Even some objects present neighbouring surfaces with either specular or scattering behaviour. Therefore this reflectivity issue has to be considered in developing a LWIR interferometer and strongly depends on the targeted objects.

One application discussed later concerns the NDT of carbon fibre-based composite structures. We found that many composites have a roughness of the order or more significant than the wavelength of CO_2 laser[28]. Therefore, the object is sufficiently scattering to allow DH or speckle interferometry. On that basis, we were able to use a mobile LWIR speckle interferometer on aeronautical parts[29].

In another application, the target objects are large aspheric reflectors for space that are purely specular. The removable powder cannot be applied. In some specific cases (concave or flat surfaces), such mirrors can be illuminated by an extended diffuser which generates artificial scattering reflected by the mirror towards the camera sensor, enabling hologram recording. This was studied in detail elsewhere[30] and will be presented in section applications.

Sometimes objects have neighbouring areas with either specular or scattering behaviour. This requires specific illumination arrangements for dealing with both at the same time[31].

Holographic setup stability

Recording holograms with larger wavelengths makes the setup more tolerant to external perturbations. Setup stability is a non-trivial requirement for interferometric technologies, including digital holography. Any perturbation in the environment can lead to the variation of the phase of beams reaching the detector (mainly the one from the object), blurring the interference fringes. As a rule of thumb, the optical path in the interferometric setup should vary no more than a fraction of the laser wavelength during hologram recording. For that reason, holography at visible wavelengths requires setup to be mounted on vibration-isolated tables and preferably with tabletops to avoid air turbulence or alternatively making use of short laser pulses.

In some cases, working out-of-the lab is necessary and using long-wavelength alleviates the stability requirement for DH setup. This feature makes LWIR DH more resilient to the inevitable perturbations. This was demonstrated by our group in industrial plant conditions[29] and even outdoors by others[32]. For the same reason, the THz DH is further immune against external perturbations. For instance, one can envisage hologram recording via raster-scanning a single-point detector where the acquisition can take tens of minutes.

Thermal self-emission

The MWIR and LWIR ranges are characterized by the thermal radiation emitted by bodies at temperatures around the ambient or corresponding to temperature elevation of a few tens of degrees, which are typical to those used in NDT. In particular, Planck's Law shows that the maximum emission of ideal black bodies around 20°C corresponds to the main emission lines of CO_2 laser (around 10 µm). The direct consequence is that thermal emission will be present as incoherent background into the hologram.

Like any other noise, this can be filtered out if needed, what we showed in Ref. 14. For some DH configurations, like the inline with phase-shifting, the computation naturally eliminates the thermal background (provided it does not vary during the phase-shifting acquisition)[33,34].

However, keeping the thermal background can be of interest for some NDT applications. Indeed, the thermal infrared camera used in LWIR DH can also be used for thermography. On the one hand, the latter is a well-known NDT method that allows observing structures' thermal behaviour under stress or revealing subsurface anomalies by tiny surface temperature differences. On the other hand, HI measures the mechanical behaviour or allows observing subsurface defects by local fringe variations. Being complementary, it can be of high interest to combine both[35].

In order to observe the thermal image, image-plane holography configurations are required due to the incoherent nature of thermal radiation (it cannot be propagated by usual DH numerical algorithms). For that reason, LWIR speckle interferometry was developed[34]. We will describe this application later.

Penetration of materials

Performing holography through materials can also be of significant interest, not only for measuring refractive index variations as consistently demonstrated on many occasions, but also for reconstructing images (amplitude and phase) of objects occluded by visually opaque materials. Because of the wide variety of materials, their state (solid, liquid, gas) and the broadness of the spectrum, it is impossible to have an exhaustive list of potentialities.

However, let us recall that the atmosphere transmits well the LWIR wavelengths. Although absorptions lines exist for H_2O or CO_2, the lines of the CO_2 laser around 10 μm can pass through the smoke or even flames. This property was used advantageously in an outstanding application[36] that we will discuss in a later section.

Going further at long wavelengths appears the THz domain where many materials become transparent[37,38] and where a tremendous activity in new technologies for imaging is ongoing[39,40]. THz radiation's non-ionization penetration capability is the highlight of THz imaging technologies, making it a safe alternative to X-ray imaging. Meanwhile, the penetration ability of THz radiation is more selective than X-ray. It depends mainly on the materials under investigation and the working wavelength. For various dielectrics materials, the absorption coefficient increases significantly when the frequency exceeds 1 THz, including standard three-dimensional (3D) printable materials such as Acrylonitrile butadiene styrene (ABS), Polylactic acid (PLA), and Nylon, most adhesives and certain industrial plastics such as Polyvinyl Chloride (PVC) and Polycarbonates (PC). Therefore, THz DH should be carried out with sub-THz wavelength when investigating these materials. Polymers including polypropylene (PP), polyethylene (PE), polytetrafluoroethylene (PTFE, Teflon), cyclic olefin polymers (Topas and ZEONEX), dehydrated tissues, and thin textile remain highly transparent at FIR THz range. Therefore, it is preferable to image these samples at FIR frequency with a smaller wavelength to achieve a better resolution for imaging concerns.

Review of LWIR holography applications

This chapter will present some typical applications that can be achieved with LWIR holography, taking advantage of one or several of the specific features discussed in the previous section.

Seeing through smoke and flames by LWIR DH[36]

A thermographic camera performs imaging in a classical way, like any other imaging system with an objective lens. The situation is depicted in Fig. 2a. If a hot source (a flame, a hot plume) is present in front of the subject to the image (here a statuette), the image will be blinded by it (Fig. 2b). Considering LWIR DH in the lensless configuration allows solving this problem. Part of the CO_2 laser beam illuminates the whole scene, and some rays travel from the object to the sensor, although disturbed by the flame (substantial variations of refractive index), Fig. 2c. Despite this, by the holographic reconstruction process, it is possible to numerically refocus the object behind the disturbance, as is shown in Fig. 2d.

Here the authors use the off-axis DH configuration. They can isolate the object in a part of the reconstruction plane and reconstruct the image thanks to a single hologram. This configuration is well suited for dynamic events, where the thermo-optical disturbance is rapidly changing. By its incoherent nature, the thermal emission is present in the zero-order, thus not reconstructed in the image after the spectral filtering.

The authors also used the same principle to image an object (the statuette) surrounded by smoke, as is shown in the bottom left medaillon. The object is enclosed in a plexiglass box (transparent at 10 μm), Fig. 2e. It is possible to record a thermographic image in these conditions, Fig. 2f. When the box is filled with smoke, Fig. 2g, the object is not visible anymore, although it is possible to reconstruct the holographic image, Fig. 2h.

Finally, the most impressive application concerns the

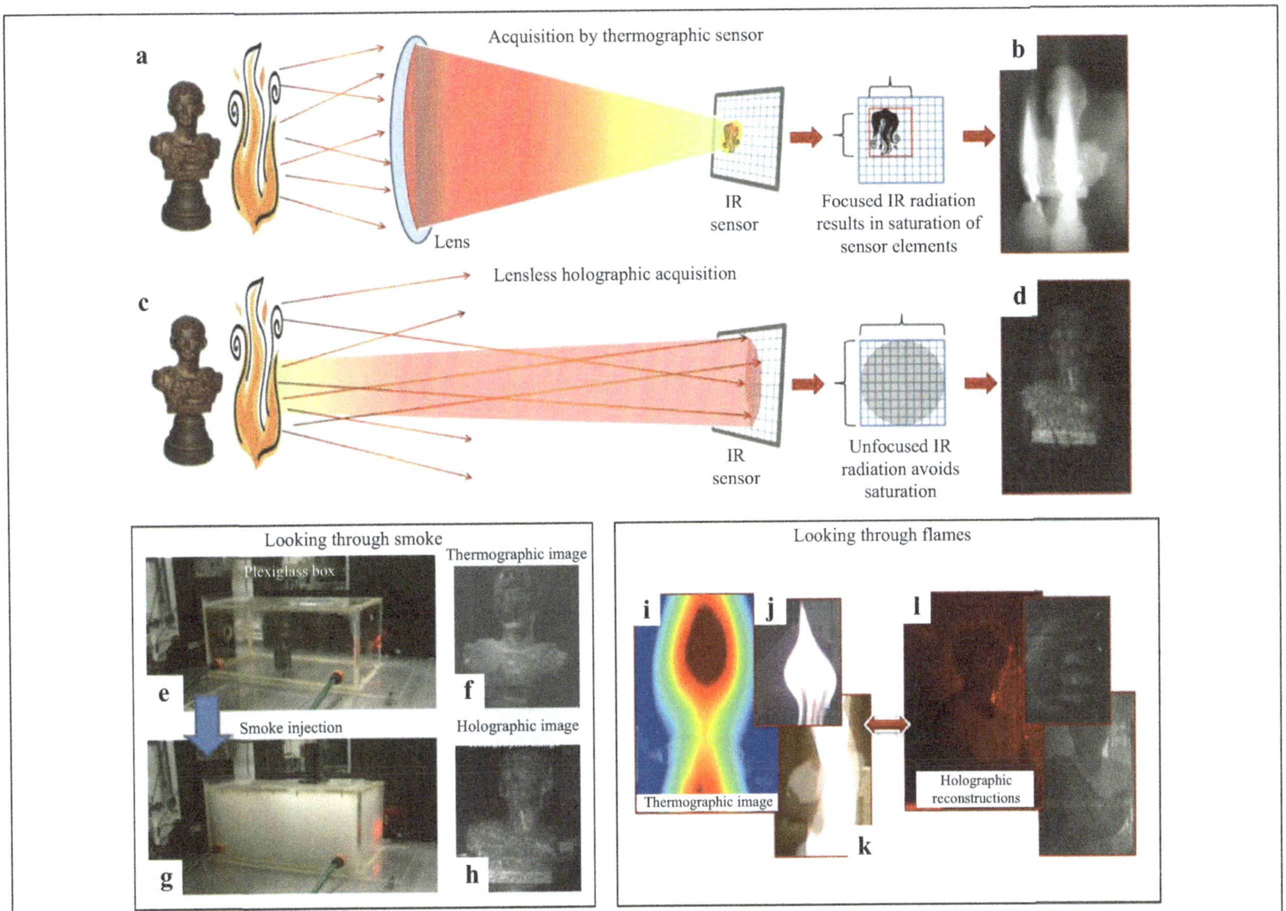

Fig. 2 Looking through smoke and flames. **a** Classical thermographic imaging scheme; **b** LWIR lensless DH acquisition scheme; **c** A thermographic image obtained with the statuette behind a flame; **d** Holographic reconstruction after numerical refocusing at the distance of the object. (The figure is adapted from figures of Ref. 36, by permission of OSA)

observation of a living human behind a flame, as is shown in the bottom right medaillon. Fig. 2i shows the thermographic image: the subject is not observable because the image is blinded by the flame. This is also the case with images taken in visible, Fig. 2j, k where the subject is barely visible. As is shown in Fig. 2l, applying LWIR DH allows the reconstruction of the image of the subject behind the flame and to recognize it is a human being.

The large size of reconstructed objects, as already stressed earlier, is put in evidence in this application. Recently the group demonstrated a self-referenced configuration easing setup portability[41].

Measurement of large deformations of space components by LWIR DHI

HI is a method that can be used for monitoring the thermo-mechanical behaviour of space structures undergoing temperature changes, similar to those experienced during space flight. One project funded by the European Space Agency (ESA) aimed at measuring up to hundred-micrometre level deformations that occur on the meter-sized space aspheric reflectors. The idea was to use DH to perform phase measurement at different temperatures during thermal cycling in a vacuum chamber simulating space environments[14].

We already stressed that, in LWIR, objects appear more specular than in the visible range. In the ESA application, reflectors intended for far-infrared/sub-mm wavelengths were to be tested. They mostly appeared specular in the visible. Therefore, we needed to consider the design of a DH interferometer for specular objects. A constraint was that the laser and the camera are not vacuum-compatible and must be kept out of the vacuum chamber.

Fig. 3a shows the design of the DH interferometer. Details of its development are given elsewhere[14]. In brief, the configuration is inline DH, which maximizes the resolution in the reconstructed image. The expanded and collimated laser beam enters the vacuum chamber through a window W. The beam is split by BS1 into two parts. The reference beam travels to sensor S through the beam

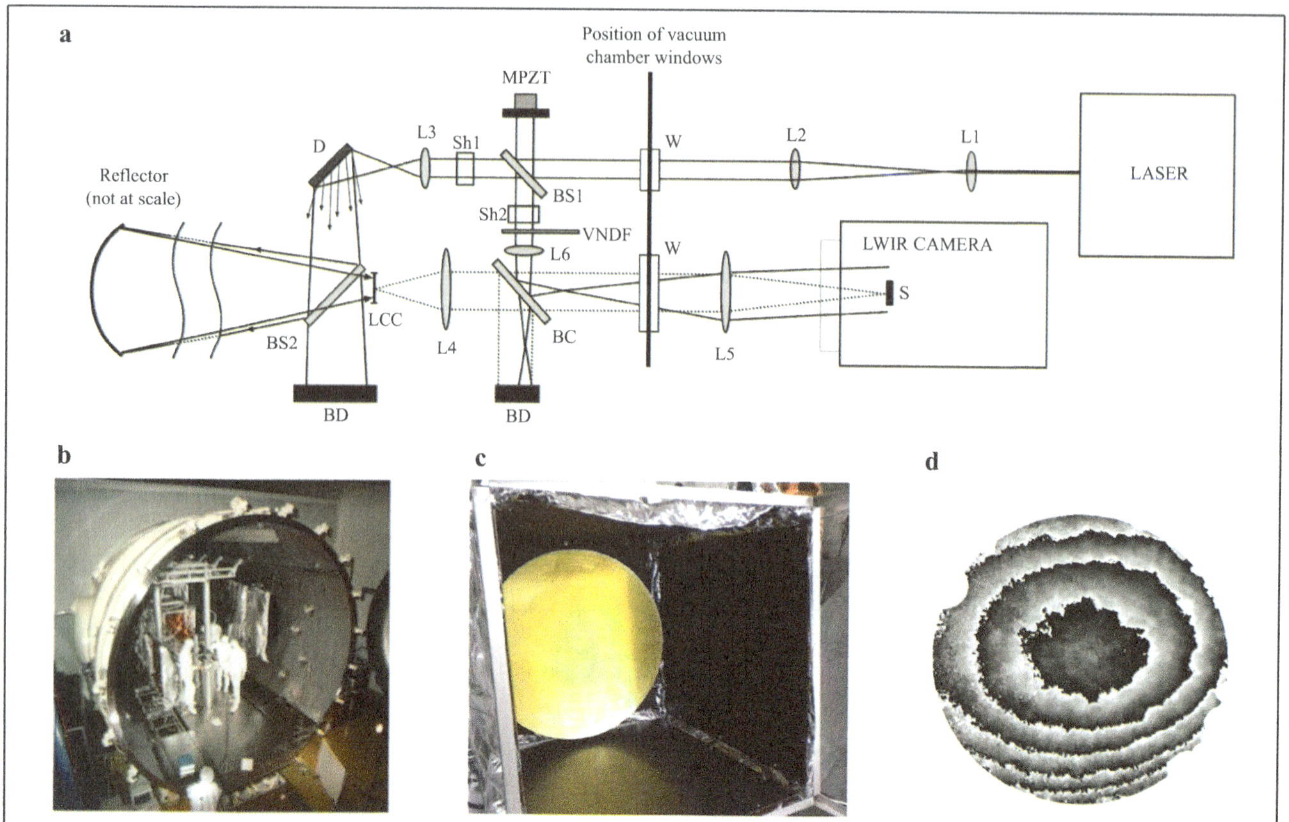

Fig. 3 Measuring large deformation of large space reflectors by inline LWIR digital holographic interferometry. **a** Sketch of the interferometer adapted to vacuum-chamber operation; **b** View of the vacuum chamber; **c** The 1.1m diameter gold-coated reflector in a thermal shroud for radiative transfer; **d** Phase map mod 2π showing the deformation of the reflector from 224 K to 107.5 K.

combiner BC and passes by a folding mirror mounted on a piezo translator (MPZT) for applying phase-shifting. After BS1, the object beam illuminates a reflective diffuser D. The scattered light illuminates the reflector, a parabolic mirror with a 1.1m diameter and 1.58m focal length. The mirror reflects specularly the speckled wavefront towards the camera sensor through the beam combiner and a second window W. A set of lenses L1 to L6 are in the various paths for adapting the dimensions of the beams while relaying the images from inside the chamber to the camera.

Fig. 3b shows a picture of the 5 m diameter vacuum chamber where the experiment was implemented. Fig. 3c shows the gold-coated aspheric mirror in a thermal shroud allowing temperature variations by radiative transfer. The rest of the interferometer (except the camera, the laser and three lenses) was placed inside the chamber and kept at ambient temperature. Fig. 3d shows the phase difference modulo 2π due to the deformation of the mirror between 224 K and 107.5 K. Excellent agreement was found between this measurement and another taken with a classical infrared interferometer adapted explicitly to the reflector.

A second application of the same technique concerned the cryogenic test of a 4×4 mosaic of detectors used of the EUCLID space mission, Fig. 4a. It was required to measure the global deformation of the ensemble, the deformation of each detector as well as the displacement of each of them with respect to their neighbors[42]. The range of deformation was 20 μm which is compatible with LWIR DHI. Also, since relative piston movements between sub-assemblies are to be sought, we incorporated the temporal phase unwrapping to follow the phase of each pixel independently to their neighbours. In this project, the setup was redesigned for coping with the new object. The latter is roughly flat, and the diffuser D to be used must be at least as large as the object. The setup is shown in Fig. 4b, with again the laser and the camera outside the vacuum chamber. Details about the experiment and the results can be found in Ref. 42. Fig. 4c shows a 3D plot of the residual deformation of the detectors mosaic.

Mobile LWIR speckle interferometer for simultaneous deformation and temperature variation measurement

As discussed in the thermal self-emission subsection, it is possible to record both the thermal background image of

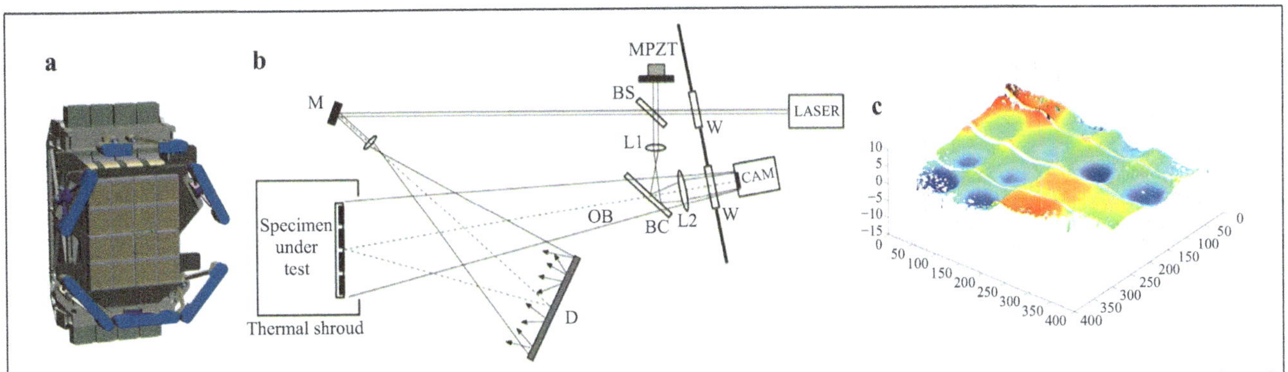

Fig. 4 Measuring deformations and rigid body motions of detectors arranged in a 4 × 4 mosaic for the EUCLID space observatory. **a** Drawing of the mosaic of detectors in its mechanical holder; **b** Sketch of the interferometer adapted to vacuum-chamber operation; **c** Phase map mod 2π showing the residual deformation from 293 K to 90 K.

an object together with its hologram. The only condition is to use an image holography configuration, i.e. with a lens focusing the object image on a sensor. In that sense, speckle interferometry responds to this condition. It is related to digital holography, except that the processing does not make use of numerical refocusing.

Using phase-shifting allows retrieving phase in the usual way (for the holographic part), and the temperature background is obtained through the addition of the phase-shifted holograms[34]. Doing so for different object states, we can compute the respective optical phases and the thermal backgrounds and calculate their differences. Once proven in the lab, this concept was implemented in a mobile instrument for in-field NDT[29,43]. Fig. 5a–c shows the principle of the instrument. Details are given elsewhere[29]. In brief, it is composed of a lower bench Fig. 5a, with the CO_2 laser and a polarization beamsplitter (PBS), detailed in Fig. 5b. A part of the beam is used to illuminate the object. The remaining reference beam is directed to the upper bench, Fig. 5c, where it is directed to the camera through a beam combiner BC and a retroreflecting mirror CM. The camera receives the light through BC as well as the reference beam. The instrument is shown in blue in Fig. 5d.

The bottom right medallion shows the first laboratory demonstration of simultaneous deformation and temperature variation measurement. Fig. 5e shows the object: a composite panel with a repaired part (dark area). The structure is heated thanks to infrared lamps on both sides of the object. Fig. 5f, g show respectively the phase map related to the deformation and the temperature change. We have both information at every pixel simultaneously, Fig. 5h, which is an advantage compared to separate systems measuring deformation and temperature independently. This was used in a variety of applications for determining the thermo-mechanical behaviour of materials.

The most impressive application was the use of the instrument at an aircraft testing plant on a full-scale composite fuselage, as seen in Fig. 5d. The zone observed is seen coated with scattering removable powder. The convex shaped surface could not allow a homogeneous scattering profile.

The fuselage incorporates some artefacts simulating internal damages. Fig. 5i shows the deformation (phase map $\Delta\varphi$) and the temperature change obtained at different instants during a step heating of a few tens of seconds. We can see defects appearing in the deformation signal (defects circled in red) while they are not visible in the thermal signal, and conversely (defects circled in yellow).

The instrument was later adapted to vibration measurement[44], but temperature variations were not found sufficiently high to be put in evidence.

Roadmap of THz digital holography towards industrial applications

Developing THz DH offers another way of performing THz imaging, a flourishing realm over the last two decades[40,45,46]. The first development of DH with THz radiation originated from the low-frequency (sub-THz) range. Dated back to 2004, an off-axis THz digital hologram was recorded with a 100 GHz Gunn diode oscillator and a single-point Schottky Barrier Diodes (SBD) detector[47]. In 2008, Tamminen et al.[48] proposed a THz reflective off-axis DH setup using a vector network analyzer as transmitter and receiver at 310 GHz to image a metallic target placed 1.5 m away from the detector. In 2011, Heimbeck et al.[49] realized THz off-axis DH with a tunable frequency-multiplied microwave synthesizer and an SBD single-point detector in the range 0.66–0.76 THz. The frequency tunability of the THz source allowed them to apply the well-known dual-wavelength approach to

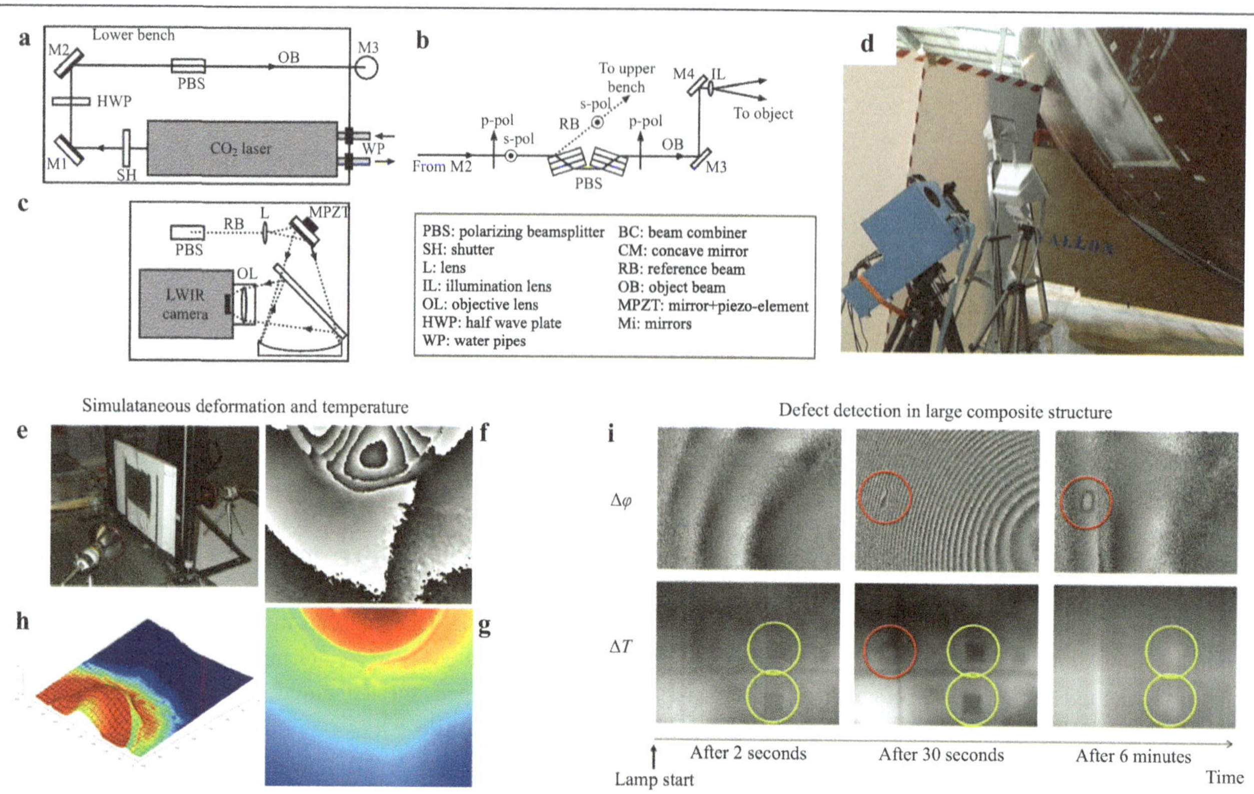

Fig. 5 Mobile LWIR speckle interferometer: **a** lower bench with laser, **b** upper bench with camera, **c** beam separation and object illumination. **d** The system when used for NDT experiment on large composite aircraft structure. **e** Helicopter composite structure with central repair, **f** interferogram due to deformation after thermal heating, **g** simultaneous temperature variation, **h** hybrid view of deformation and temperature variation. **i** Defect detection on large aircraft composite structure by simultaneous deformation and temperature variation measurement.

eliminate the 2π phase ambiguity problem.

Full-field was the inevitable trend for THz DH and other possible coherent lensless imaging techniques. However, at the sub-THz range, the long wavelength requires very large 2D arrays with adequate sensitivity for sub-THz radiation[50].

On the other hand, despite the selective penetration ability in the FIR range, which limits the further application, the band is more suitable for investigating emerging coherent lensless imaging techniques with commercially available 2D thermal FPA arrays in single-shot without tedious raster-scanning using a single-point detector. In 2011, Ding et al. reported the first full-field off-axis THz DH setup with a FIRL laser and a pyroelectric camera at 2.52 THz with a lateral resolution of 3.4λ[51]. The first THz inline DH was performed in 2012 by the same research group[52].

Since this first successful demonstration, the THz coherent lensless imaging became an active research area. Thermal FPAs paired with FIR CW sources became the mainstream equipment for investigating full-field DH imaging techniques.

After the above proof-of-concept demonstrations, the development of THz DH devoted itself to further improving the imaging performance, notably in resolution enhancement. In terms of DH recording, besides minimizing recording distance for resolution enhancement, the synthetic aperture (SA) technique was successfully implemented with both off-axis[53,54] and in-line[55,56] configurations, making up for the insufficient size of THz detectors. For inline THz DH, efforts have been mainly made on reconstruction algorithms towards better reconstruction quality[57,58,59].

As the imaging ability is maturing, THz DH has been gradually carried out towards application. Similar to visible wavelengths, due to the specificity of the setup, inline THz DH was solely implemented in digital holographic microscopy (DHM) fashion for biomedical applications, seeing through sparse samples. The testing samples have transitioned from dehydrated insect specimens to cancerous tissue. In Ref. 60, the reconstruction of a human hepatocellular carcinoma tissue slice (Fig. 6a) proved that early signs of liver cancer and diseases could be traced using THz inline DH.

The reconstructed phase after resolution enhancement revealed a sign of tissue fibrosis (the blue arrow in the phase map in Fig. 6c), making THz inline DH a promising

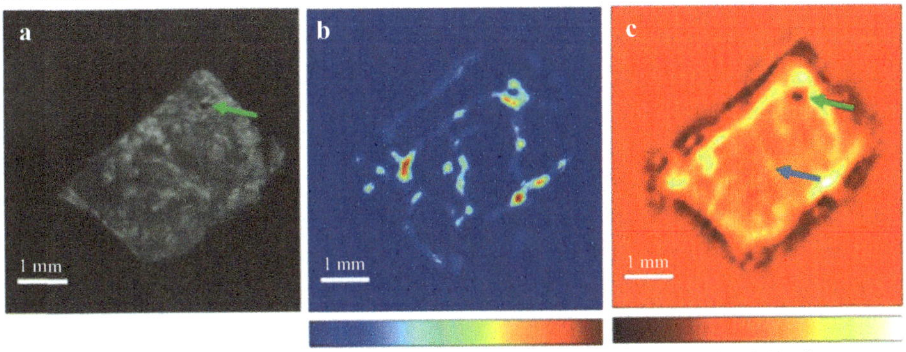

Fig. 6 THz inline digital holographic reconstructions. **a** Photograph of the object, a human hepatocellular carcinoma tissue; **b** Reconstructed absorption and **c** phase shift after numerically extrapolating the holograms recorded via sub-pixel registration method. The green arrow indicates a cut across a vessel or a region damaged after freezing the object. The blue arrow indicates a vertical line which is a sign of tissue fibrosis (Figure adapted from Fig 3 and Fig 4 of Ref. 60, by permission of Sci. Rep.)

tool for cancer diagnosis. Off-axis THz DH offers a wider choice of samples, thus more application scenarios, thanks to the independent reference wave. The samples can be measured in both transmission and reflection mode. In transmission mode, thickness measurement of optically opaque materials at FIR range was carried out in Refs. 61, 62. At the Sub-THz range, visualizing the void in a 3D-printed sample has been demonstrated at 0.495 THz[25,63]. A depth of 45 μm was clearly distinguished from the phase image. On the other hand, imaging concealed objects using off-axis DH under reflection mode paves the way to applications in non-destructive inspections. In Refs. 64, 65, real-time imaging of moving metallic objects hidden by a thin yet rigid plate made of optically opaque yet highly transparent (PE or PP) has been successfully demonstrated, as shown in Fig. 7c, d adapted. However, despite the encouraging results, the perturbation caused by the interaction between complex cover materials and the samples still needs to be tackled before meeting the challenge of real-world industrial inspection applications[66]. The FOV of the reconstruction scene remained on a cm-scale. Moreover, the reflectivity and the transmittance of both the imaging target as well as the cover materials is also an important concern in terms of applications[67].

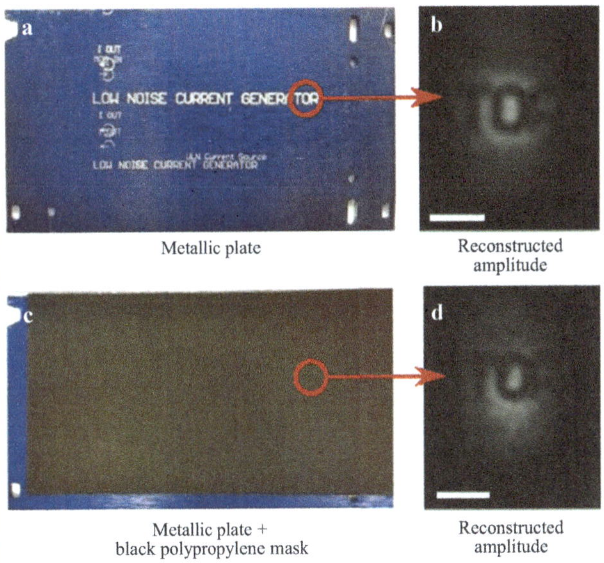

Fig. 7 Real-time imaging concealed object using reflective off-axis THz DH. **a** Photograph of the object, a metallic plate with inscriptions; **b** Reconstructed amplitude of the object at one position during its movement; **c** Photograph of the plate cover; **d** Reconstructed amplitude of the concealed inscription. (Figure adapted from Fig. 1 Ref. 64, by permission of Sci. Rep.)

Discussion-Conclusion

This overview shows that using wavelengths much longer than the visible spectrum in holography can bring many advantages, tackling some apparent issues. In particular, all applications in LWIR show that holography can work efficiently under industrial conditions, even in a harsh outdoor environment. Furthermore, wavelength also plays a central role in metrology where it sets the sensitivity of the holographic interferometry method: large displacements can be measured more easily with a decent resolution. This was put in evidence in real-life projects for testing large space structures undergoing significant temperature variations.

In addition, we have seen that particular behaviour appears with long wavelengths. First, objects tend to be specular, which is not an advantage. However, this can be mitigated up to some point by using removable scattering powder, when possible, or illuminating objects through diffusers. Second, specifically in LWIR, thermal emission appears, corresponding to ambient temperature. This incoherent background can be naturally filtered out, but it can also be used appropriately to perform simultaneous deformation and temperature measurements. It was used advantageously in aeronautical NDT on the sizeable composite fuselage.

We have seen that long wavelengths penetrate the matter better. In LWIR, human-size objects illuminated through smoke or flames can be refocused under a lensless DH configuration. In THz, light can penetrate a wide range of solid materials. Thus, DH with THz radiation allows imaging objects concealed by these visually opaque materials.

Despite all these advantages and outstanding applications, there is a price to pay. We discussed a fundamental limit that is related to the ratio of the pixel pitch of state-of-art cameras and the wavelength. These aspects hold true from LWIR ranges to THz ranges. In LWIR, larger objects can be observed with acceptable resolution when working at wavelengths longer than visible wavelengths for a given recording distance, assuming lensless Fresnel DH. However, due to the limited size of array detector in THz, keeping acceptable resolution requires placing the objects very close to the detector, limiting the applicability of Fresnel DH. Other DH reconstruction methods, such as ASM or convolution, can be used on objects of centimetric sizes, such as those found in the biomedical field, as was demonstrated in cancer detection, which is by far the most appropriate application of THz DH in the FIR regime. On the other hand, the Sub-THz regime is the band where we can make the utmost use of the penetration ability to develop seeing-through applications. However, the lack of sufficiently large and sensitive array detectors limits the development of Sub-THz DH towards application. More industrial applications can be envisaged with the availability of large and sensitive array detectors for Sub-THz.

In the visible range, wavefront shaping components such as spatial light modulators (SLMs) will offer more possibilities in DH, such as self-interference incoherent DH. For instance, a proof-of-concept multispectral incoherent DH experiment has been demonstrated in IR range[68]. The maturing of IR and THz SLMs technologies will push these methods towards applications.

Acknowledgements

M.G. and Y.Z. acknowledge European Regional Development Fund/Wallonia region project (TERA4ALL); MG. and JF.V. acknowledge the support of ESA (GSTP project Contract No. 22540/09/NL/SFe) and EU (FP7 European project FANTOM ACP7-GA-2008-213457).

Author contributions

M.G. initiated the researches on LWIR and THz DH. JF.V. made his PhD thesis on LWIR DH, and results presented here were obtained during several projects led by M.G. or JF.V. Y.Z. made her PhD thesis on THz H and results presented here were obtained during the PhD. Discussion about fundamental limits was mainly studied by Y.Z. under the supervision of JF.V. and M.G. The manuscript was discussed and corrected by all authors.

Conflict of interest

The authors declare that they have no conflict of interest.

Received: 15 September 2021 Revised: 07 March 2022 Accepted: 08 March 2022
Accepted article preview online: 16 March 2022
Published online: 11 April 2022

References

1. Bjelkhagen, H. & Brotherton-Ratcliffe, D. Ultra-Realistic Imaging: Advanced Techniques in Analogue and Digital Colour Holography. (Boca Raton: CRC Press, 2016).
2. Gentet, Y. & Gentet, P. CHIMERA, a new holoprinter technology combining low-power continuous lasers and fast printing. *Applied Optics* **58**, G226-G230 (2019).
3. Powell, R. L. & Stetson, K. A. Interferometric vibration analysis by wavefront reconstruction. *Journal of the Optical Society of America* **55**, 1593-1598 (1965).
4. Vest, C. M. Holographic Interferometry. (New York: John Wiley & Sons, 1979).
5. Kreis, T. Handbook of Holographic Interferometry: Optical and Digital Methods. (Weinheim: Wiley, 2005).
6. Georges, M. Holographic interferometry: from history to modern applications. in Optical Holography: Materials, Theory and Applications (ed Blanche, P. A.) (Amsterdam: Elsevier, 2020), 121-163.
7. Leendertz, J. A. Interferometric displacement measurement on scattering surfaces utilizing speckle effect. *Journal of Physics E:Scientific Instruments* **3**, 214-218 (1970).
8. Jones, R. & Wykes, C. Holographic and Speckle Interferometry. 2nd edn. (Cambridge: Cambridge University Press, 1989).
9. Schnars, U. & Jüptner, W. Direct recording of holograms by a CCD target and numerical reconstruction. *Applied Optics* **33**, 179-181 (1994).
10. Picart, P., Gross, M. & Marquet, P. Basic fundamentals of digital holography. in New Techniques in Digital Holography (ed Picart, P.) (Hoboken: Wiley, 2015), 1-66.
11. Picart, P. & Leval, J. General theoretical formulation of image formation in digital Fresnel holography. *Journal of the Optical Society of America A* **25**, 1744-1761 (2008).
12. Zhang, T. & Yamaguchi, I. Three-dimensional microscopy with phase-shifting digital holography. *Optics Letters* **23**, 1221-1223 (1998).
13. Georges, M. Long-wave infrared digital holography. in New Techniques in Digital Holography (ed Picart, P.) (Hoboken: Wiley, 2015), 219-254.
14. Georges, M. P. et al. Digital holographic interferometry with CO_2 lasers and diffuse illumination applied to large space reflector metrology [invited]. *Applied Optics* **52**, A102-A116 (2013).
15. Valzania, L. et al. THz coherent lensless imaging. *Applied Optics* **58**, G256-G275 (2019).
16. Ravaro, M. et al. Mid-infrared digital holography and holographic interferometry with a tunable quantum cascade laser. *Optics Letters* **39**, 4843-4846 (2014).
17. Allaria, E. et al. Digital holography at 10.6 µm. *Optics Communications* **215**, 257-262 (2003).

18. Hack, E. et al. Comparison of thermal detector arrays for off-axis THz holography and real-time THz imaging. *Sensors* **16**, 221 (2016).
19. Paturzo, M. et al. Optical reconstruction of digital holograms recorded at 10.6 μm: route for 3D imaging at long infrared wavelengths. *Optics Letters* **35**, 2112-2114 (2010).
20. Pelagotti, A. et al. Digital holography for 3D imaging and display in the IR range: challenges and opportunities. *3D Research* **1**, 6 (2010).
21. Paturzo, M. et al. Infrared digital holography applications for virtual museums and diagnostics of cultural heritage. Proceedings of SPIE 8084, O3A: Optics for Arts, Architecture, and Archaeology III. Munich, Germany: SPIE, 2011, 80840K.
22. Geltrude, A. et al. Infrared digital holography for large objects investigation. Proceedings of SPIE 8082, Optical Measurement Systems for Industrial Inspection VII. Munich, Germany: SPIE, 2011, 80820C.
23. Pelagotti, A. et al. An automatic method for assembling a large synthetic aperture digital hologram. *Optics Express* **20**, 4830-4839 (2012).
24. Bianco, V. et al. On-speckle suppression in IR digital holography. *Optics Letters* **41**, 5226-5229 (2016).
25. Heimbeck, M. S. et al. Terahertz digital holographic imaging of voids within visibly opaque dielectrics. *IEEE Transactions on Terahertz Science and Technology* **5**, 110-116 (2015).
26. Robinson, D. W. & Reid, G. T. Interferogram Analysis: Digital Fringe Pattern Measurement Techniques. (Philadelphia: Institute of Physics, 1993).
27. Yamaguchi, I. Fundamentals and applications of speckle. Proceedings of SPIE 4933, Speckle Metrology 2003. Trondheim, Norway: SPIE, 2003, 1-8.
28. Vandenrijt, J. F. et al. Electronic speckle pattern interferometry at long infrared wavelengths. Scattering requirements. in Fringe 2009-6th International Workshop on Advanced Optical Metrology (eds Osten, W. & Kujawinska, M.) (Berlin, Heidelberg: Springer, 2009), 596-599.
29. Vandenrijt, J. F. et al. Mobile speckle interferometer in the long-wave infrared for aeronautical nondestructive testing in field conditions. *Optical Engineering* **52**, 101903 (2013).
30. Vandenrijt, J. F. et al. Long-wave infrared digital holographic interferometry with diffuser or point source illuminations for measuring deformations of aspheric mirrors. *Optical Engineering* **53**, 112309 (2014).
31. Georges, M. P. et al. Digital holographic interferometry and speckle interferometry applied on objects with heterogeneous reflecting properties. *Applied Optics* **58**, G318-G325 (2019).
32. Poggi, P. et al. Remote monitoring of building oscillation modes by means of real-time Mid Infrared Digital Holography. *Scientific Reports* **6**, 23688 (2016).
33. De Nicola, S. et al. Wave front reconstruction of Fresnel off-axis holograms with compensation of aberrations by means of phase-shifting digital holography. *Optics and Lasers in Engineering* 37, 331-340 (2002).
34. Georges, M. P. et al. Combined holography and thermography in a single sensor through image-plane holography at thermal infrared wavelengths. *Optics Express* **22**, 25517-25529 (2014).
35. Georges, M. P. Comparison between thermographic and holographic techniques for nondestructive testing of composites: similarities, differences and potential cross-fertilization. Proceedings of SPIE 9660, SPECKLE 2015: VI International Conference on Speckle Metrology. Guanajuato, Mexico: SPIE, 2015, 966002.
36. Locatelli, M. et al. Imaging live humans through smoke and flames using far-infrared digital holography. *Optics Express* **21**, 5379-5390 (2013).
37. Naftaly, M. & Miles, R. E. Terahertz time-domain spectroscopy for material characterization. *Proceedings of the IEEE* **95**, 1658-1665 (2007).
38. Piesiewicz, R. et al. Properties of building and plastic materials in the THz range. *International Journal of Infrared and Millimeter Waves* **28**, 363-371 (2007).
39. Dhillon, S. S. et al. The 2017 terahertz science and technology roadmap. *Journal of Physics D:Applied Physics* **50**, 043001 (2017).
40. Guerboukha, H., Nallappan, K. & Skorobogatiy, M. Toward real-time terahertz imaging. *Advances in Optics and Photonics* **10**, 843-938 (2018).
41. Bianco, V. et al. Off-axis self-reference digital holography in the visible and far-infrared region. *ETRI Journal* **41**, 84-92 (2019).
42. Vandenrijt, J. F. et al. Digital holographic interferometry in the long-wave infrared and temporal phase unwrapping for measuring large deformations and rigid body motions of segmented space detector in cryogenic test. *Optical Engineering* **55**, 121723 (2016).
43. Georges, M. P. Speckle interferometry in the long-wave infrared for combining holography and thermography in a single sensor: applications to nondestructive testing: the FANTOM project. Proceedings of SPIE 9525, Optical Measurement Systems for Industrial Inspection IX. Munich, Germany: SPIE, 2015, 952557.
44. Languy, F. et al. Vibration mode shapes visualization in industrial environment by real-time time-averaged phase-stepped electronic speckle pattern interferometry at 10.6 μm and shearography at 532 nm. *Optical Engineering* **55**, 121704 (2016).
45. Mittleman, D. M. Twenty years of terahertz imaging [Invited]. *Optics Express* **26**, 9417-9431 (2018).
46. Valušis, G. et al. Roadmap of terahertz imaging 2021. *Sensors* **21**, 4092 (2021).
47. Mahon, R., Murphy, A. & Lanigan, W. Terahertz holographic image reconstruction and analysis. in Infrared and Millimeter Waves, Conference Digest of the 2004 Joint 29th International Conference on 2004 and 12th International Conference on Terahertz Electronics. Karlsruhe, Germany: IEEE, 2004, 749-750.
48. Tamminen, A., Ala-Laurinaho, J. & Raisanen, A. V. Indirect holographic imaging at 310 GHz. Proceedings of 2008 European Radar Conference. Amsterdam, Netherlands: IEEE, 2008, 168-171.
49. Heimbeck, M. S. et al. Terahertz digital holography using angular spectrum and dual wavelength reconstruction methods. *Optics Express* **19**, 9192-9200 (2011).
50. Gorodetsky, A., Freer, S. & Navarro-Cía, M. Assessment of cameras for continuous wave sub-terahertz imaging. Proceedings of SPIE 11499, Terahertz Emitters, Receivers, and Applications XI. SPIE, 2020, 114990Y. (Checked all online materials, but the publication location was not found, please contact the author for confirmation – confirmed, here is the doihttps://doi.org/10.1117/12.2568516)
51. Ding, S. H. et al. Continuous-wave terahertz digital holography by use of a pyroelectric array camera. *Optics Letters* **36**, 1993-1995 (2011).
52. Xue, K. et al. Continuous-wave terahertz in-line digital holography. *Optics Letters* **37**, 3228-3230 (2012).
53. Zolliker, P. & Hack, E. THz holography in reflection using a high resolution microbolometer array. *Optics Express* **23**, 10957-10967 (2015).
54. Valzania, L. et al. Resolution limits of terahertz ptychography. Proceedings of SPIE 10677, Unconventional Optical Imaging. Strasbourg, France: SPIE, 2018, 1067720.
55. Huang, H. C. et al. Synthetic aperture in terahertz in-line digital holography for resolution enhancement. *Applied Optics* **55**, A43-A48 (2016).
56. Deng, Q. H. et al. High-resolution terahertz inline digital holography based on quantum cascade laser. *Optical Engineering* **56**, 113102 (2017).
57. Rong, L. et al. Terahertz in-line digital holography of dragonfly

hindwing: amplitude and phase reconstruction at enhanced resolution by extrapolation. *Optics Express* **22**, 17236-17245 (2014).
58. Huang, H. C. et al. Continuous-wave terahertz multi-plane in-line digital holography. *Optics and Lasers in Engineering* **94**, 76-81 (2017).
59. Chen, G. H. & Li, Q. Markov chain Monte Carlo sampling based terahertz holography image denoising. *Applied Optics* **54**, 4345-4351 (2015).
60. Rong, L. et al. Terahertz in-line digital holography of human hepatocellular carcinoma tissue. *Scientific Reports* **5**, 8445 (2015).
61. Yamagiwa, M. et al. Real-time amplitude and phase imaging of optically opaque objects by combining full-field off-axis terahertz digital holography with angular spectrum reconstruction. *Journal of Infrared, Millimeter, and Terahertz Waves* **39**, 561-572 (2018).
62. Zhao, Y. C. et al. Iterative phase-retrieval-assisted off-axis terahertz digital holography. *Applied Optics* **58**, 9208-9216 (2019).
63. Heimbeck, M. S. & Everitt, H. O. Terahertz digital holographic imaging. *Advances in Optics and Photonics* **12**, 1-59 (2020).
64. Locatelli, M. et al. Real-time terahertz digital holography with a quantum cascade laser. *Scientific Reports* **5**, 13566 (2015).
65. Humphreys, M. et al. Video-rate terahertz digital holographic imaging system. *Optics Express* **26**, 25805-25813 (2018).
66. Valzania, L., Zolliker, P. & Hack, E. Topography of hidden objects using THz digital holography with multi-beam interferences. *Optics Express* **25**, 11038-11047 (2017).
67. Appleby, R. & Wallace, H. B. Standoff detection of weapons and contraband in the 100 GHz to 1 THz region. *IEEE Transactions on Antennas and Propagation* **55**, 2944-2956 (2007).
68. Anand, V. et al. Exploiting spatio-spectral aberrations for rapid synchrotron infrared imaging. *Journal of Synchrotron Radiation* **28**, 1616-1619 (2021).

Koukourakis et al. *Light: Advanced Manufacturing* (2022)3:23
https://doi.org/10.37188/lam.2022.023

Official journal of the JHL 2689-9620
www.light-am.com

Article Open Access

Investigation of human organoid retina with digital holographic transmission matrix measurements

Nektarios Koukourakis[1,2*], Felix Wagner[3,4], Stefan Rothe[1], Mike O. Karl[3,4*] and Jürgen W. Czarske[1,2,5,6,*]

Abstract

Advanced manufacturing of retinal organoid samples from human induced pluripotent stem cells represents a promising way to study the development of retinal diseases. The retina is an epithelium composed of different cell layers with unique optical properties and detects light by photoreceptor neurons for visual function. There are still many challenges in detecting early and distinct cellular changes in retinal disease. In this paper, we study the capability of the optical transmission matrix, which fully describes the transition of a light field propagating through a scattering sample. Despite its rich information content, the transmission matrix is commonly just used for light delivery through scattering media. Digital holography is employed to measure the complex light-field information of the transmitted light. We demonstrate that singular value decomposition of the transmission matrix allows to discriminate phantom tissues with varying scattering coefficient. We apply these findings to retinal organoid tissues. Application of the protonophore carbonyl cyanide m-chloro-phenylhydrazone (CCCP), a known inducer of retinal damage in animals, caused cell death and structural changes in human retinal organoids, which resulted in distinct changes in the transmission matrix. Our data indicate that the analysis of the transmission matrix can distinguish pathologic changes of the retina towards the development of imaging-based biomarkers.

Keywords: Transmission matrix, Digital holography, Retina, Organoids, Neurodegeneration, Cell death

Introduction

The transport of light through strongly scattering media is complex. Local refractive index distributions and structural inhomogenities inside the medium induce multiple scattering events that prohibit the imaging of objects or the targeted light transport through the sample. Consequently, a focused wave of monochromatic light sent into the medium will appear as a speckled output. However, the seemingly random speckle pattern is deterministic, which opens the possibility for addressing the scattering effect by adequately shaping the wavefront by employing a spatial light modulator (SLM).

The huge technological and methodological progress of wavefront shaping (WS) approaches[1,2] allows today to deliver light or transport information into or through thick biological tissues in real-time and with high spatial resolution. However, the precondition is that the mask displayed on the SLM fits to the desired task.

Most WS techniques have in common that digital holography is the fundamental key for both extracting the

Correspondence: Nektarios Koukourakis (nektarios.koukourakis@tu-dresden.de) or Mike O. Karl (Mike.Karl@dzne.de) or Jürgen W. Czarske (juergen.czarske@tu-dresden.de)
[1]Laboratory for Measurement and Sensorsystem Technique, TU Dresden, Dresden, Germany
[2]Competence Center for Biomedical Laser Systems (BIOLAS), TU Dresden, Dresden, Germany
Full list of author information is available at the end of the article.
These authors contributed equally: Nektarios Koukourakis, Felix Wagner

amplitude and phase information of the light field that interacted with the sample and for tailoring the holographic amplitude or phase mask displayed on the SLM. Although in general there are various techniques to measure the phase information of light[3–5], most of them cannot deal with the phase of speckled light fields or suffer of too low spatial or temporal resolution, which means in practice that digital holography is the working horse in this field.

Knowledge on the desired phase mask can be gained in various ways. The fastest method which requires only one measurement is digital optical phase conjugation (DOPC), which is a digital implementation of optical phase conjugation known from analogue holography[6]. It was proven that by using DOPC, light can be focused through scattering tissue of about 100 transport mean free path, which corresponds to 9.6 cm tissue thickness[7]. However, DOPC requires that a guide star is introduced into the scattering sample, which emits a known wavefront. In contrast to iterative WS techniques using, for example, photoacoustic guide star signals[8], DOPC enables to create a focus at the guide star position in single-shot. In this way, focusing through polystyrene[9], mouse skull[10], chicken breast[7,11] or multimode optical fibres (MMF)[12–14] was demonstrated. WS especially revolutionised the applicability of MMF, as they enabled remotely controlling the light propagation through the waveguide. MMFs in principle behave comparable as biological samples, which is why they can be regarded as the waveguide equivalent to bulk random media[13]. MMFs have the advantage that the number of free modes is known from scratch[15] and that the medium is designed for transporting light, which makes it a light efficient scattering sample. Full information on the scattering medium can be obtained by determining the input-output relation, the so-called transmission matrix (T_M), which links an orthonormal set of inputs to those at the output. Knowledge of the T_M enables to control light propagation through the scattering medium and to form a desired light field at the output[16,17]. Besides applications in light delivery, the T_M allows to use the scattering medium as imaging optics[18]. Different approaches to measuring a T_M have been presented and tested predominantly at MMFs in the past decade[19–34]. There are only few examples where the T_M was exploited to study scattering samples. One first example was reported by Popoff et al. at ZnO deposit[35]. In other work, it was shown that by employing the T_M, for example, transmission eigenchannels can be identified to optimise the intensity transport through the medium under test[36–39]. Furthermore, the T_M bares information of scattering properties, like the optical memory effect[40].

In this paper, we present first T_M measurements on human retinal organoids. The retina is a tissue at the back of our eye providing vision by sensing light, and it is easily monitored live in patients. The retina has several unique optical properties with reflecting and scattering structures[41–44], and its physiological functions and changes upon pathology are incompletely understood. Various inherited and acquired retinal degenerative diseases cause changes and loss of tissue structure and cells which may result in vision loss. Degeneration of the light detecting cells, photoreceptors, is commonly mediated by cell death. To study this, several animal models and experimental approaches have been developed. Recently, 3D retina organ-like systems, retinal organoids, have been developed that could serve as a human system for fundamental and preclinical translational research[45–47]. Multimodal imaging technologies are routinely applied to clinically assess the structure and function of the retina in human patients[48,49]. However, there are still many challenges in detecting early and distinct cellular changes in retinal diseases by imaging. For example, high resolution imaging, and analysis in 3D across all five major retinal tissue layers is currently still limited, and it is still difficult to obtain contrast on various specific tissue structures. Organoids provide a preclinical human ex vivo system that might help advancing imaging technologies[45,50,51], i.e. to experimentally study the underlying physiologic and pathologic tissue properties that correlate with distinct imaging signals and its changes thereof in patients by ex vivo and in vivo correlations[52].

In this paper, we analyze the capabilities of T_M to distinguish different tissue samples. The approach is established at phantom tissues, and applied for the analysis of human tissue. Specifically, we investigate whether differences in tissue properties can be distinguished in human retinal organoids with and without experimentally induced pathology.

Transmission Matrix

Measurement of the Transmission Matrix

The transmission matrix T_M fully describes the propagation of an input light field E_{in} (in amplitude, phase and polarisation) through a complex scattering medium for a given wavelength. The entries of the matrix are occupied by coefficients t_{mn} characterizing the light transmission between the mth output free mode and the nth free mode at the input. Therefore, the outgoing light field E^m_{out} on the mth mode can be described by $E^m_{\mathrm{out}} = \sum_n (t_{mn} E^n_{\mathrm{in}})$, which is a linear superposition of n modes represented by incoming light fields E^n_{in}.

In this work, the objective is to investigate the light transmission properties of retinal organoids in a defined sample region. It is reasonable to choose scanning focal points as input modes inspired by Refs. 53, 54. By appropriate shifting of the focal points, an orthogonal basis can be guaranteed. Within our investigations, pixels of holographically captured complex light fields inside the selected observation window are chosen as output modes. In this configuration, it is relatively easy to control light transport into each output mode when using scanning foci at the input. To ensure a pixel-to-pixel projection, and especially to create a simple model for T_M inversion, the scanning foci are treated as shifting Fourier pixels of the spatial domain. Foci, i.e. Fourier pixels are generated with the help of digital tilts, which are displayed on the SLM. The required phase masks are calculated using simply a Fourier transform. Experimentally, focal scanning is performed by propagation through a microscope objective. The resulting scan width depends on the optical system and the displayed tilt angle, i.e. location of the Fourier pixel. After hologram reconstruction, complex values inside the selected observation window are reshaped to column vectors, which are normalised by their Euclidean norm, filling the transmission matrix. In Fig. 1, the T_M measurement procedure is illustrated.

Experimental setup

The experimental realization of the holographic setup is displayed in Fig. 2. A single-mode laser emitting at 532 nm wavelength (Laser Quantum Torus) is used as a light source. Polarization optics are employed to control the intensity in both object and reference beams. One beam is directed towards an SLM (HoloEye Gaea, 10 mega pixel), which is used as a simple scanner during measurement of the T_M. A 4f-geometry is used to relay the image of the SLM, which is Fourier transformed by a further lens. This Fourier plane coincides with the image plane of the exit port of a commercial inverted microscope, with a custom-built transmission detection path. Displaying a plain phase on the SLM results in a simple on-axis focusing with a microscope objective onto the sample. The light which propagates through the scattering sample is collected by a second microscope objective and is imaged onto a digital camera. On the camera sensor, the object beam interferes with the reference beam recording an off-axis hologram.The hologram is Fourier transformed to get the spatial frequency spectrum. Due to the off-axis configuration the DC and the twin-image components are well-separated. Filtering of the sample spectrum and performing an inverse Fourier transform yields the light-field information. Hence, the amplitude and phase can be measured with single shot[55,56]. Adding a tilt on the phase displayed on the SLM results in a lateral shift of the focus. For each shifted focus-position the output light-field is measured and stored as described in the previous section.

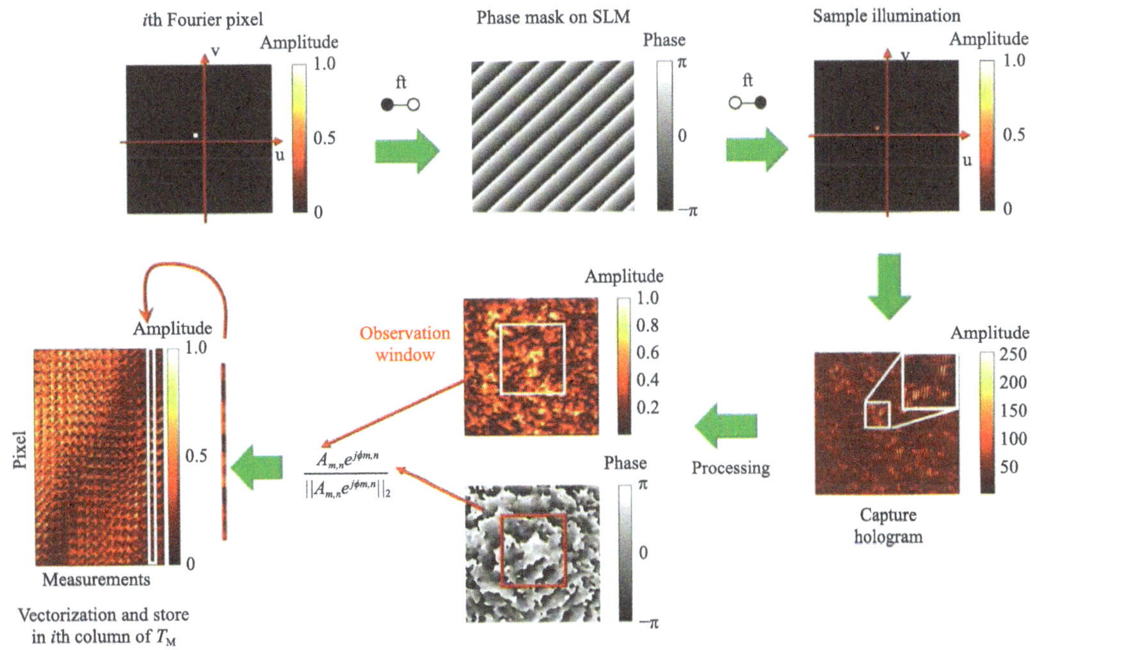

Fig. 1 Procedure of the transmission matrix measurement. Scanning foci are used to measure the T_M. A focus is created by a microscope objective. The scan is accomplished by displaying tilts on the SLM. As the SLM is located in the Fourier plane, each tilt corresponds to a certain shift in the image plane. The focus is scattered by the sample and for each shift-position the output speckle pattern is captured holographically. After reconstruction, the 2D amplitude and phase values are shaped to a normalised vector and positioned as a column in the T_M.

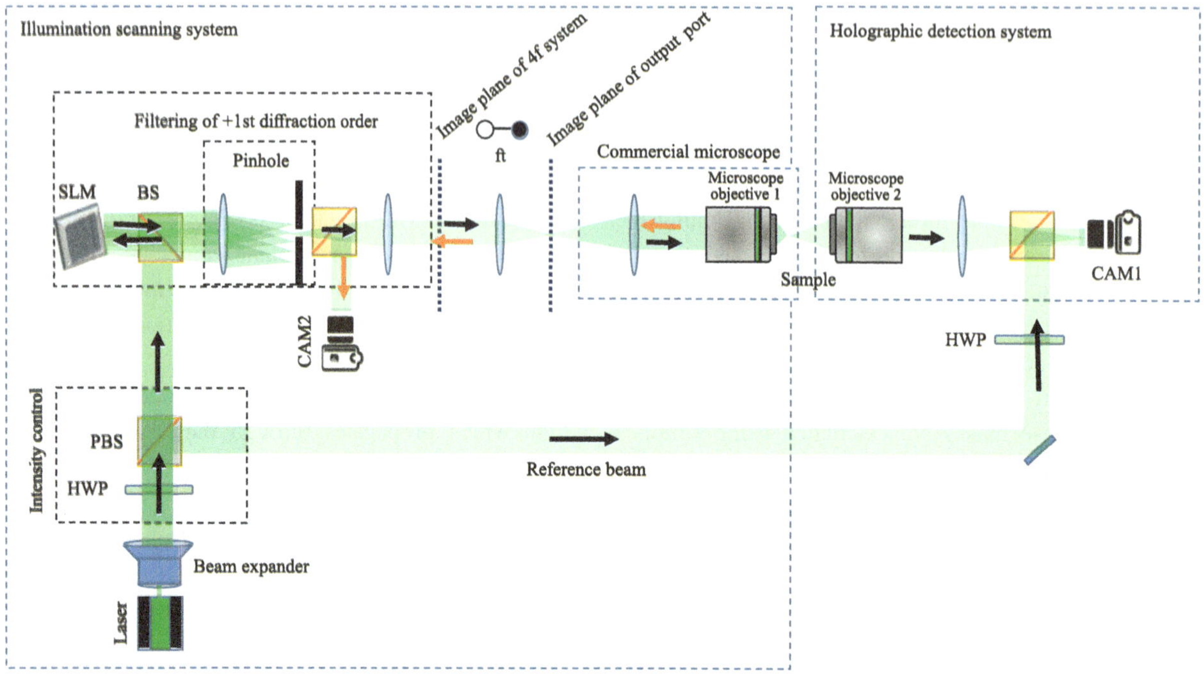

Fig. 2 Sketch of the digital holographic setup for the measurement of the T_M. Laser: Torus from Laser Quantum (532 nm), SLM: spatial light modulator HoloEye Gaea, HWP: half-wave plate, PBS: polarizing beam splitter, BS: beam splitter, FT: Fourier transform, Microscope Objective 1: Nikon 40x, 0.6 NA, Microscope Objective 2: Leica Long working distance 40x, NA: 0.6.

Inversion of the Transmission Matrix

The T_M has a large information content, which is underlined by the fact, that knowledge of the T_M enables to control the output light field through a scattering sample. Commonly, in such a case the aim is to control a large observation window at the output by using a minimal number of measurements. Therefore, the matrices measured are usually not square, but rather rectangular. In this work, the number of camera pixels inside the observation window is much larger than the number of focal points used. Thus, the measurement system is underconstrained. Due to mathematical limitations, it is not possible to perform a straightforward matrix inversion, even with perfect measurements. This is why an inversion approximation is used. In the field of holographic measurement technology, the Tikhonov inversion[57] has become established for solving inverse problems for any type of system[24,58,59]. Tikhonov inversion compensates for singularities by singular value management. In straightforward inversion procedures, singularities occur due to noise amplification caused by inverting measurement noise. With the help of the singular value decomposition (SVD), the complex T_M of dimension $M \times N$ and rank r can be described as

$$T_M = U \cdot \Sigma \cdot V^H \tag{1}$$

U is a unitary $M \times M$ matrix and V is a unitary $N \times N$ matrix, whereas H denotes the Hermitian transpose. The column vectors of U $u_i, i = 1, \ldots M$ are called left-singular vectors, the column vectors of V $v_i, i = 1, \ldots N$ are called right-singular vectors of T_M. Σ is a real $M \times N$ matrix occupied by positive diagonal entries $\sigma_i, i = 1, \ldots, r$, which are called singular values of T_M. If T_M is both square and regular, it has an inverse, such that $T_M \cdot T_M^{-1} = I$, where I is the identity matrix. Under these assumptions, the inverse can be expressed by:

$$T_M^{-1} = V \cdot \Sigma^+ \cdot U^H \tag{2}$$

Σ^+ contains the singular values σ_i from Σ with the same index, but inverted on the diagonal entries $\sigma_i^+ = 1/\sigma_i$, $i = 1, \ldots, r$. Within the investigations we consider, the T_M measured is usually rectangular and does not have an inverse, accordingly. By adjusting the correct number of rows and columns of Σ^+, however, a pseudo inverse $T_M^\dagger$ can be found via SVD providing $T_M^\dagger \cdot T_M \approx I$. Additionally, following the Tikhonov approach, noise amplification during the inversion process can be attenuated. Due to measurement noise, the difference between the largest and smallest singular value can be significantly large and entries in Σ^+ can reach extremely high values. Tikhonov inversion, however, performs singular value management according to a regularization parameter λ, such that:

$$\sigma_i^+ = \frac{\sigma_i}{\sigma_i^2 + \lambda^2}, i = 1, \ldots, r \tag{3}$$

For proper inversion, λ is set to zero and $T_M^{\dagger} = T_M^{-1}$ under square and regular conditions of T_M. If $\lambda > 0$, singular values are compressed for noise reduction, but also the relationship $T_M^{\dagger} \cdot T_M$ deviates more and more from I. The goal is to keep λ as small as possible, but as large as necessary. Within our investigations, λ was empirically set to 10%. We have carried out inversions according to Tikhonov varying the extent of different numbers of rows and columns.

Results

The classical realization of T_M measurements aims to control light transmission through the sample at the output. However, the T_M contains rich sample information, which we want to extract, in order to categorise various biological scattering samples. To study various possibilities, we will take also non-classical realization of the T_M measurement into account. The samples used are phantom tissues of different thickness and scattering properties and various retinal tissues (see Materials & Methods section).

Matrix inversion on tissue mimicking phantoms

In a first step, we perform T_M inversion proving that the T_M we measured is really the transmission matrix of the scattering sample. 1,600 input tilts are employed to measure the T_M of a scattering phantom of 1 TMFP (see Fig. 3).

The number of pixels to be controlled are 160×160, which means that the measured T_M has $M = 25{,}600$ columns. With the effective magnification of the system, we calculate the size of the observation window to be $\approx 12\ \mu m \times 12\ \mu m$.

Far fewer tilts than camera pixels are used to avoid temporal instabilities or drifts. Thus, the T_M measured is rectangular with N columns and $N < M$. After Tikhonov inversion, it is possible to create any desired light field at the output, albeit with a reduced peak to background ratio (PBR) [10] compared to the optimum, due to the low number of tilts used (see Fig. 4a). Note that the T_M enables establishing a focus scan at the output, which is not dependent on the optical memory effect.

How to extract sample information from the transmission matrix

As described above, we use a scanning focus at the sample input and measure the light field after interacting with the sample at the output using a camera. Thus, the recorded data contains information on:

1. **attenuation** of the light level transmitted through the sample, which can be attributed to backscattering, absorption and shadowing.

2. **scattering**, which leads to redistribution of the intensity distribution and speckles.

3. **local differences** of the attenuation and scattering.

Extraction of sample information from the T_M requires the characteristics of the illumination to be known. Interaction of the scanning focus with the sample will lead to a change of the recorded output light field, which bears the desired information. Thus, it is required to measure a reference, which calibrates the systematic aberrations and the system behavior. However, the appropriate choice of a calibration strategy can be challenging. The straightforward approach is to determine the T_M of the

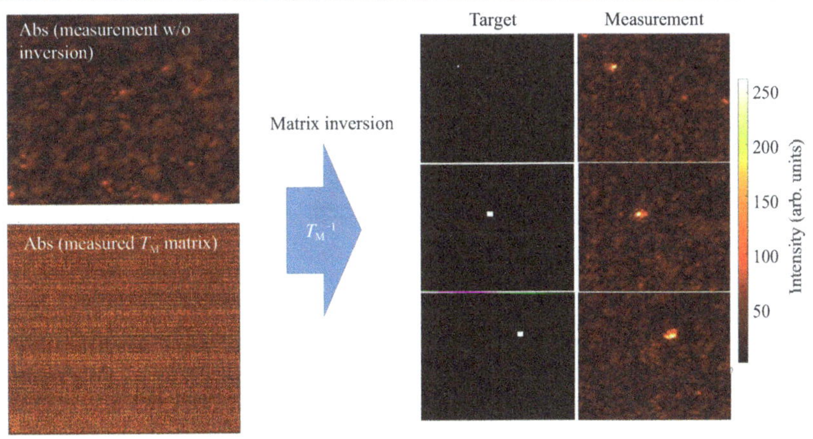

Fig. 3 Example for a T_M inversion. The output light field distribution without wavefront pre-shaping is a simple speckle-field (left, top). Knowing the T_M of the scattering medium (left,bottom) allows for targeted light field generation at the output using Tikhonov inversion. The right part of the Figure shows three exemplary target positions and the corresponding experimentally created foci.

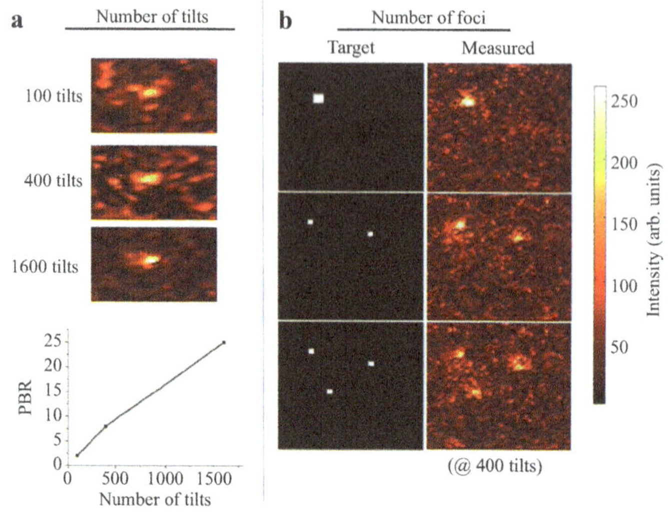

Fig. 4 Details on matrix inversion. The resulting quality of the inversion both scales with decreasing the number of tilts **a** and increasing the number of desired foci generated in the observation window **b**. However, matrix inversion and the creation of a focus also works with only 100 tilts compared to 25,600 pixels controlled in the observation window.

scanning focus without any sample in the beam path, which means that the T_M of a cover glass is measured as depicted in Fig. 5a. However, this reference is not suitable for the case which is commonly present for T_M measurements, where the light field is measured at the output of the scattering sample, as shown in Fig. 5b. The microscope objective MO2 needs to be refocused to the output, which will consequently result in a defocusing compared to the scanning focus reference. Hence, if a reference is required, either a defocused calibration needs to be recorded, or an alternative approach for the measurement procedure needs to be investigated.

The defocused reference is sample-dependent and needs to be recorded for each measurement separately. A disadvantage is, that a comparison of samples with different thicknesses strongly depends on the repositioning precision. Additionally, for thick samples (i.e. large defocus), the whole camera is illuminated and aberrations play a bigger role (see Fig. 5c). In order to use the scanning focus as a reference properly, as shown in Fig. 5a, the scanning focus needs to be recorded through the scattering tissue, as shown in Fig. 5d. This results in another kind of transmission matrix which we refer to as $T_{M,\text{focus}}$. Usually, $T_{M,\text{focus}}$ is not suitable for inversion creating a tailored output light field. However, the information contained is suitable for aberration correction on the detection camera. The spatial rearrangement of the Gaussian input focus by sample illumination can be compensated by applying $T_{M,\text{focus}}$. The applicability of this approach is limited by the working distance of MO2 and the optical path length of the

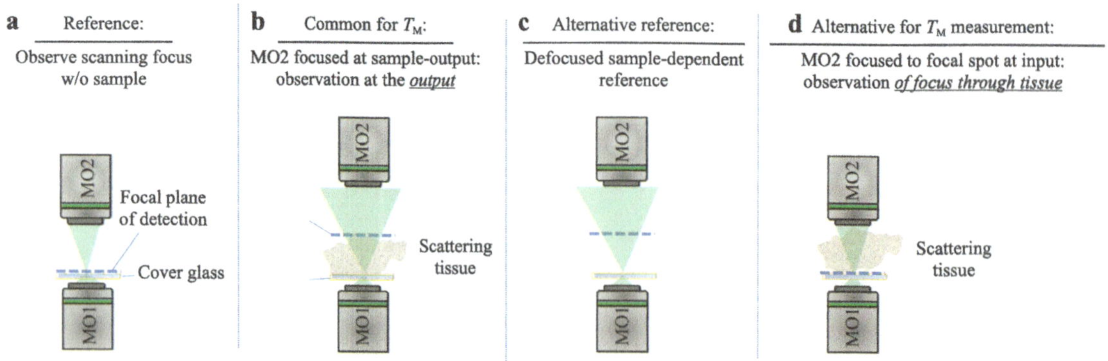

Fig. 5 Various experimental approaches for providing a suitable reference for T_M measurement. **a** The calibration measurement is performed on a cover glass slide, on which the scattering sample is located during measurement. The microscope objective for the detection focuses on the cover glass. **b** For the common approach of T_M measurement, the microscope objective is focused on the output of the scattering sample. **c** Either an alternative reference or **d** an alternative procedure for the matrix measurement observing the focus through the sample is required.

scattering sample, as it is required that the focus is observable through the tissue. Depending on the actual application, the sample can be characterised by using a reference tissue. In the following, we performed measurements on phantom tissues, which we fabricated according to the recipe described in [60]. These tissues are based on polydimethylsiloxane (PDMS) with various glycerol concentration that change the scattering coefficient of the samples (compare Table 1). All samples have the same thickness of 500 μm ± 20 μm, which enables to take the sample with pure PDMS as a reference.

Table 1 Overview of the specifications of the phantom tissues used, for 532 nm wavelength.

Sample	TMFP	MFP	thickness
PDMS	0	0	500 μm
s1	0.1	4	500 μm
s2	0.3	12	500 μm
s3	1	40	500 μm

Matrix processing

Knowledge about the transmission matrix enables to control the light propagation through scattering tissue. This shows that the matrix contains rich sample-specific information. Yet, implementation of matrix inversion does not rely on sample characteristics, which is why they are commonly not of interest. With suitable mathematical operations, however, various figures of merit can be obtained. A powerful tool to determine sample characteristics is SVD (Eq. 1). The course of the singular values (SV) on the diagonal of Σ is based on specific laws and depends on properties of the sample observed. An evaluation of the distribution of singular values enables analysing scattering properties of the sample under test. Popoff et al. have shown that the random-like scattering behavior of an ideal ZnO deposit is mathematically well described[35]. According to the random matrix theory[61], the SV distribution of mean-free matrices occupied by uncorrelated entries of variance 1 is following the so-called "quarter-circle law". However, it should be mentioned that the course of the SV significantly depends on the setup and therefore, for example, additional averaging over the camera pixels or removal of correlated entries is necessary in order to observe an SV distribution according to the quarter-circle-law.

Mathematically, if the SVD is applied to T_M (or $T_{M,\text{focus}}$), it tries to minimise its rank. In this way, the SVD can be used to identify essential information channels that provide information content, but with minimum number. For example, performing SVD to an exemplary T_M with rank $r = 800$ using Eq. 1, 800 SVs can be determined, as shown in Fig. 6a. The original T_M can be reconstructed by multiplication of the three extracted matrices U, Σ, V^H. Here, all SV are used. However, if the entries on the diagonal of Σ are considered as independent information channels, certain properties can be assigned to them. In most cases, SV are sorted by magnitude in ascend (or descend) order, whereby several orders of magnitude can lie between the strongest and weakest SV. Strong SV correspond to a strong flow of information through these channels. However, the information content of the T_M can be reproduced by a number of information channels which is smaller than its rank. There is a point between the strongest and the weakest SV where a distinction between essential information channels and non-essential ones is possible. Channels with substantial information content can

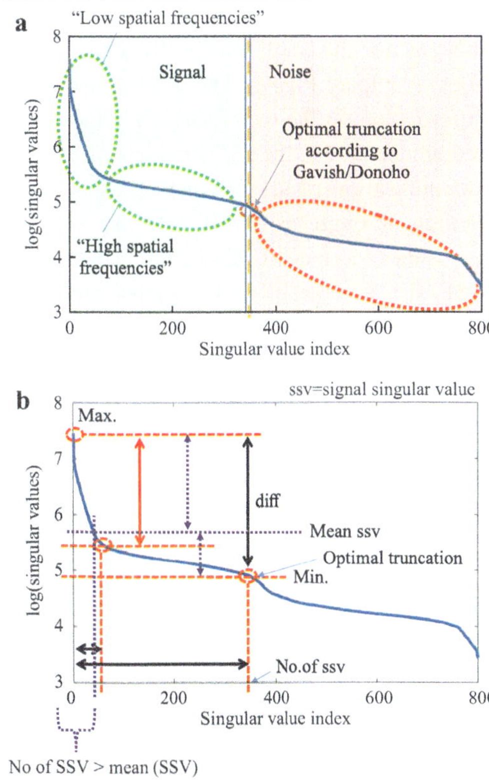

Fig. 6 Explanation of typical SVD-curve. We measure the T_M and perform the SVD. **a** The resulting sorted singular values (blue curve) are further evaluated according to the method described in Ref. 62, which allows to determine the optimal truncation point τ. This point separates the singular values corresponding to the signal from the noise. The remaining singular values of the truncated SVD correspond to the spatial frequencies occuring in the T_M. **b** Besides τ, the course of the SVD allows to determine various characteristic values. In the following, we will just concentrate on the dynamic range (difference max (SSV)-min (SSV)), the ratio r max (SSV)/mean (SSV), the position of the "knee" and the number of SSVs.

be considered as signal and the others as noise (see Fig. 6a). Gavish and Donoho have introduced a strategy where the optimal point of truncation between signal and noise can be identified[62]. Considering any rectangular $M \times N$ data matrix Y with unknown noise level and $M \neq N$, the optimal threshold $\omega(\beta)$ can be determined by

$$\omega(\beta) \approx 0.56\beta^3 - 0.95\beta^2 + 1.82\beta + 1.43 \tag{4}$$

with $\beta = M/N$. Hence, the truncation point τ can be derived:

$$\tau = \omega(\beta) \cdot \sigma_{\text{med}} \tag{5}$$

with σ_{med} being the median SV of Y. At this transition point, the SV vector is truncated, keeping the signal singular values (SSV). The characteristic levels of magnitude in relation to the truncation point τ are illustrated in Fig. 6b. The SSV are an analogon to a Fourier transform. While the high value SSV correspond to low spatial-frequencies (and high intensities), the lower value SSV belong to the high spatial-frequencies (and low intensities) (compare Fig. 7). The general level of the SSV bares information on the overall transmittance. In the field of telecommunications, the T_M provides information about real communication channels when using space division multiplexing through an MMF. In the course of the singular values, a cut-off frequency can be identified, above which a number of usable spatial channels can be determined[21]. As mentioned above, the SVD strongly depends on setup parameters. This finding relates to the aforementioned reference problem. If we measure the T_M for 5 axial positions of the detection objective MO2 *without* any scattering tissue in the beam-path, we obtain strongly changed SVD courses (see Fig. 8). However, in order to obtain relevant sample information, a reference is required to observe the sample induced changes in the T_M.

Phantom tissues with constant thickness and varying scattering coefficient

We keep the experimental conditions constant, such as laser power, exposure time and only change the phantom tissue under test increasing the scattering coefficient. As all tissues have the same thickness, we take the pure PDMS sample as the reference. First, we measure the corresponding $T_{M,\text{focus}}$ of all samples, as shown in Fig. 9. After SVD evaluation (see Fig. 10), we observe that with increasing scattering τ shifts to a lower SV index and thus, the signal is spread over a smaller SSV range. Hence, as a first relevant magnitude, the shift of τ can be used to compare the relative scattering behaviour of the sample under test. Additionally, the course of the SSV depends on sample characteristics. These are among others: the ratio r between the maximum SSV and the mean $r = \max(SSV)/\text{mean}(SSV)$, the dynamic range $d = \max(SSV) - \min(SSV)$, the number of SSV with $SV > \text{mean}(SSV)$, the number of remaining SSV after truncation and the position of the 'knee', which divides high SSV and low SSV (compare explanation shown in Fig. 6b).

The measurement on the phantom tissues was repeated at N=9 different lateral positions. As shown in Fig. 10,

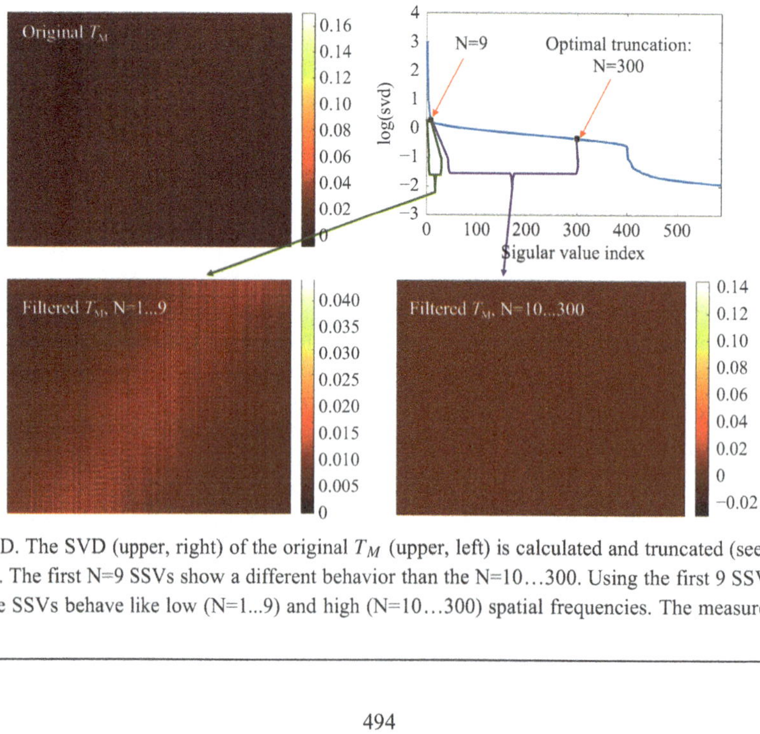

Fig. 7 Truncation of the SVD. The SVD (upper, right) of the original T_M (upper, left) is calculated and truncated (see Eq. 5). The signal is spread over N=300 singular values. The first N=9 SSVs show a different behavior than the N=10…300. Using the first 9 SSVs for reconstruction and the N=10 to 300, shows that the SSVs behave like low (N=1...9) and high (N=10…300) spatial frequencies. The measurement was performed on the phantom tissue 's3'.

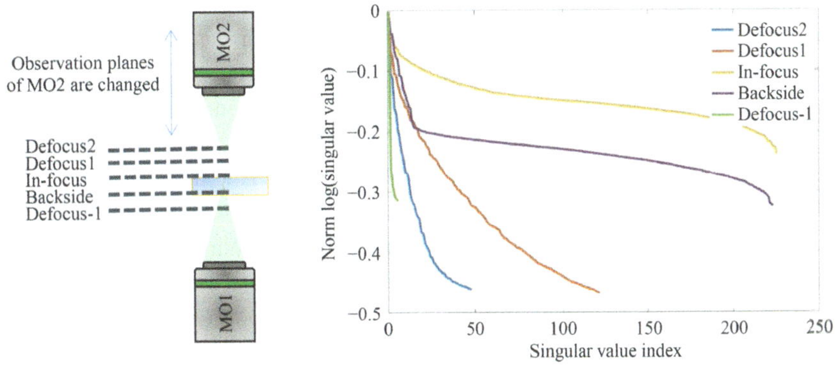

Fig. 8 Influence of defocus on the SVD. Defocusing the detection objective yields strong changes of the SVD course. Additionally, the number of SSV decreases and the dynamic range increases. However, for too strong defocusing the SSVs nearly vanish.

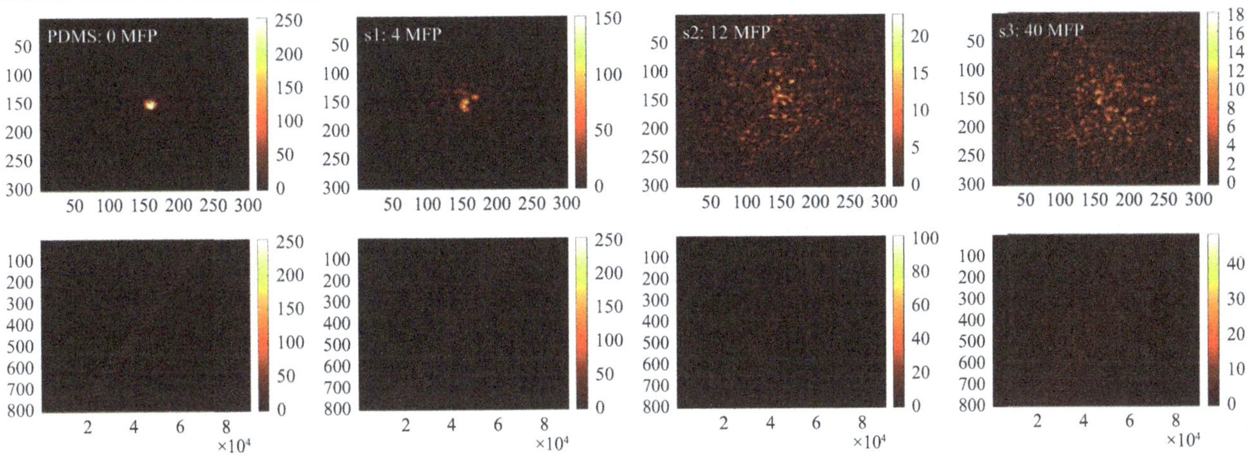

Fig. 9 Measurements on different phantom tissues. Top row (from left to right): exemplary focus visualizations through PDMS (0 MFP), s1 (4 MFP), s2 (12 MFP), and s3 (40 MFP). Bottom row: Corresponding $T_{M,\text{focus}}$.

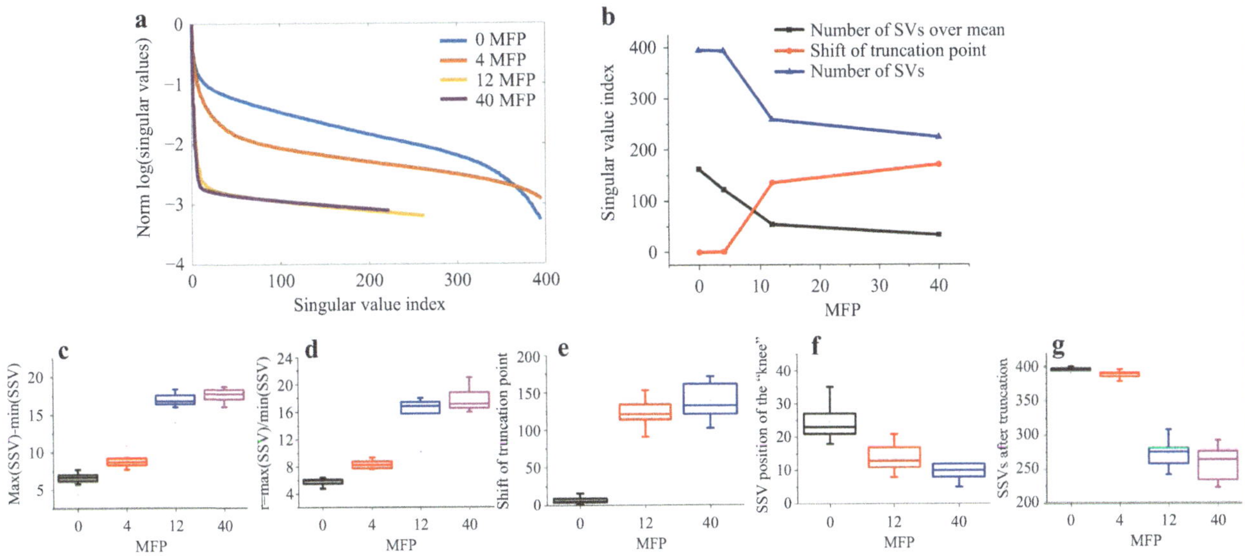

Fig. 10 SVD evaluation of the measurements on different phantom tissues. **a** SSVs of the four phantom tissues. **b** The number of SVs after truncation decreases and the truncation point τ shifts to lower index with increased scattering. **c** The dynamic range $d = \max(SSV) - \min(SSV)$ and **d** the ratio $r = \max(SSV)/\text{mean}(SSV)$ increase with scattering. Both **e** the truncation point and **f** the position of the "knee", that divides low and high spatial frequencies, shift to lower index numbers. **g** Finally, the remaining signal is spread over a reduced number of SSVs.

clear trends with increase of scattering are observable. As the experimental conditions are kept constant, the signal-to-noise-ratio (SNR) decreases with increasing scattering, resulting in a reduction of the overall transmittance, which reduces the mean SSV level. As there are still high SSV values present and the minimum SSV decreases, the dynamic range of the SSV is increased.

These results show that it is possible to distinguish samples with different scattering properties by SVD evaluation. However, a reference is required and the experimental conditions have to be kept constant. τ is strongly dependent on the SNR of the measurement. If, for example, a measurement is repeated for different exposure times, a shift of τ to lower index can be observed. The same holds, if the conditions are kept constant, but areas of the sample block (or strongly attenuate) the light propagation. Repeating the measurements on the phantom tissues and adjusting the laser power, to keep the signal levels nearly constant, τ does not shift significantly (compare Fig. 11). However, along with sample changes, i.e. increasing scattering and blocking, the characteristic behavior does not change, i.e. d and r still increase and the position of the knee shifts, while the pure change of the exposure time leads to decreasing d and r.

Changing the measurement geometry from focus observation to the classical T_M geometry comes along with changes of the SSV as shown in Fig. 12. The dynamic range increases, while the number of SSV decrease. The measurement of T_M in the classical geometry, strongly depends on the SNR. Hence, it was not possible to keep the laser power and exposure time constant for a comparable measurement series, as the signal strength is strongly decreasing by the transmission through the samples with increasing scattering. However, the same trends are present for observation of the focus through the tissue and the classical T_M measurement geometry. There is a strong crosstalk between the SVD method and changing experimental conditions. Hence, it is difficult to compare the SVD of different measurement series quantitatively. Surely, the phantom tissue is a special case, where samples are compared which have a highly different dynamic range of scattering. While the reference PDMS sample and the low-scattering s1 show a transition behavior, the samples s2 and s3 with larger scattering coefficient show a different behavior due to the strongly reduced SNR. If just a comparison of s2 and s3 would be targeted, choosing adjusted experimental conditions and a different reference would have a strong impact on the SVDs and would increase the differences. Nevertheless, the results show that indeed it is possible to find differences in the course of SSV.

Reproduction of human retinal pathology in an organoid system

Human retinal organoids were generated from hiPSC derived from healthy individuals, and show complete retinal cell generation at about 150 days[46]. Organoids could be maintained in culture for more than 220 days (Fig. 13a). Histological analysis of immunostained retinal organoid cryosections showed a layered retinal structure based on a cell nuclei distribution (Fig. 13a). The majority of cells are photoreceptor neurons in the most outer retinal layer, indicated by staining for RCVRN (Fig. 13b), whereas other neurons and Müller glia are located in the inner retina as previously shown[46]. Analysis of retinal sections by TUNEL assay showed rarely any positive cells indicating absence of any major cell death in this organoid system under control conditions (Fig. 13c). To establish a human retinal

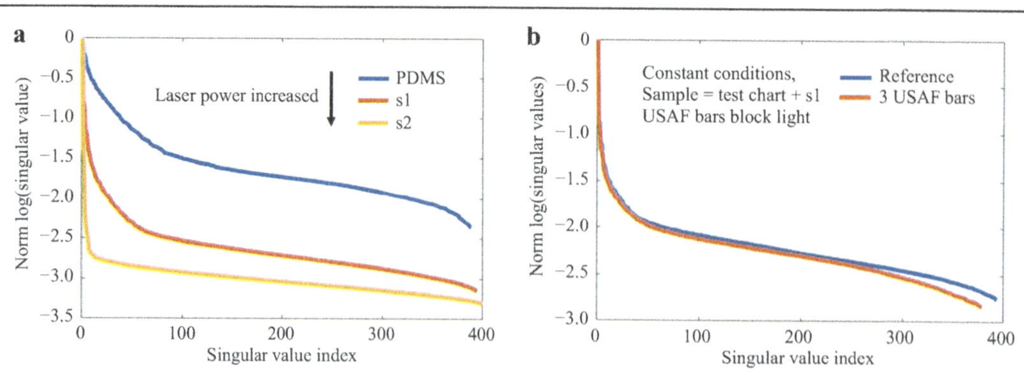

Fig. 11 Crosstalk between varying experimental conditions and SVD evaluation. **a** The measurement of $T_{M,\mathrm{focus}}$ on the phantom tissues is repeated. The used laser power is adjusted to keep a constant signal level despite an increase of scattering. This still results in increasing d and r and a shift of the "knee", however τ is not shifted. **b** The phantom tissue s1 is placed on the bars of an USAF test chart. If the focus is scanned across the bars, the light is blocked, mimicking strong absorption or e.g. bone. As a reference, we perform a scan with a laterally shifted test chart. Other conditions are kept constant. Again, d and r increase, while τ shifts to smaller index.

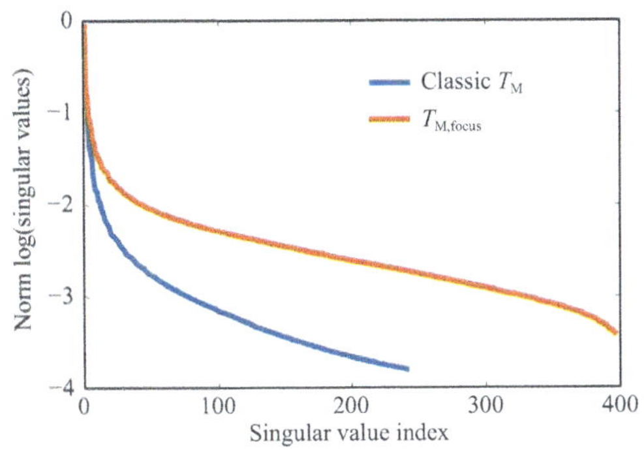

Fig. 12 Influence of changing the measurement geometry from focus observation to classical. Changing the geometry at otherwise constant conditions leads to a reduction of the SNR for each camera acquisition. In the presented example (measurement on s1), this leads to a shift of the truncation point and an increase of the dynamic range. The general behavior remains.

organoid model reproducing pathologic changes in tissue structure and cell death, we tested the application of the protonophore carbonyl cyanide m-chloro-phenyl-hydrazone (CCCP, 5 μM), which is used to uncouple oxidative phosphorylation, an essential component of cell energy metabolism. CCCP was previously shown to induce apoptosis of photoreceptors from salamanders[63], and mitochondrial damage in human retinal ganglion cells derived from hiPSC[64]. Organoids were treated with CCCP for 11 days in comparison to solvent controls (N = 5 per variable). Microscopic observation of organoids live in culture showed that all CCCP-treated but not control organoids underwent major changes in color and shape (Fig. 13a, d). Histological analysis confirmed this, CCCP caused a loss in layered retinal structure based on nuclei staining, and photoreceptor cells became displaced from the outer layer to the inner retina. Further, some cell nuclei showed a condensed morphology, indicative of cell death, and analysis by TUNEL assay confirmed this. Thus, application of CCCP not only damages isolated retinal cells in cell culture and animals in vivo[63,64], but is also sufficient to induce a severe pathology in a 3D complex human retina organoid system providing an efficient experimental pathology model.

Transmission matrix analysis of healthy and pathologic human retina organoids

We prepared tissue slices derived from human retinal organoids (Fig. 13a) in order to investigate whether different anatomical locations in healthy control organoids can be discriminated by T_M measurements, and whether there are any differences in comparison to CCCP-induced pathologic organoids. Measurements were performed at four evenly spaced radial positions across the retinal epithelium, spanned from the outer (apical) border of the epithelium up to the presumptive inner (basolateral) epithelial border (Fig. 14a), and applied the classical T_M geometry for analysis. Exemplary results are shown in Fig. 14b, c. With increasing distance from the outer

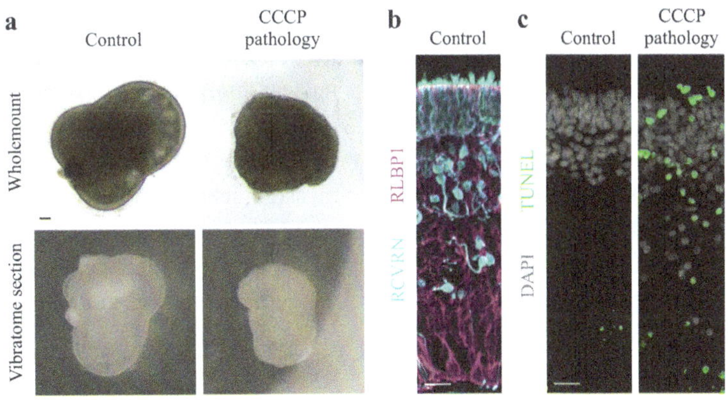

Fig. 13 CCCP induces a severe pathology in human retinal organoids: **a** Brightfield microscopic images of human retinal organoid shown as wholemount in cell culture either 11 days after challenge with CCCP (pathologic) or without (solvent healthy control) and as cross-sectional vibratome section. In control samples, the outer layer of the retinal epithelium is visible as a bright band around the entire organoid circumference. Upon CCCP-treatment, this epithelium becomes reduced in thickness, and the otherwise smooth epithelial (organoid) surface becomes pathologically uneven, indicating outer retinal degeneration. Scale bar 100 μm. **b** Healthy human retinal organoids show layered structure: organoid cross-section fluorescently labeled for photoreceptor neurons (RCVRN) located in the most outer retinal layer and Müller glia (RLBP1) spanning the entire retinal epithelium. Scale bar 20 μm. **c** Degenerating human retinal organoids show structural loss and increased cell death: Human retinal organoid cross-section under control conditions and after CCCP treatment (day 10 of treatment) were green fluorescently labeled by TUNEL assay, which indicates cell death. All cell nuclei are counterstained with DAPI, which shows that control retinas are organized into an outer nuclear layer with densely packed cell nuclei and slightly separated to an inner nuclear layer below with less densely packed cell nuclei. In CCCP treated retina, DAPI also shows that some cell nuclei become small roundish, most of which also are TUNEL positive. Scale bar 20 μm.

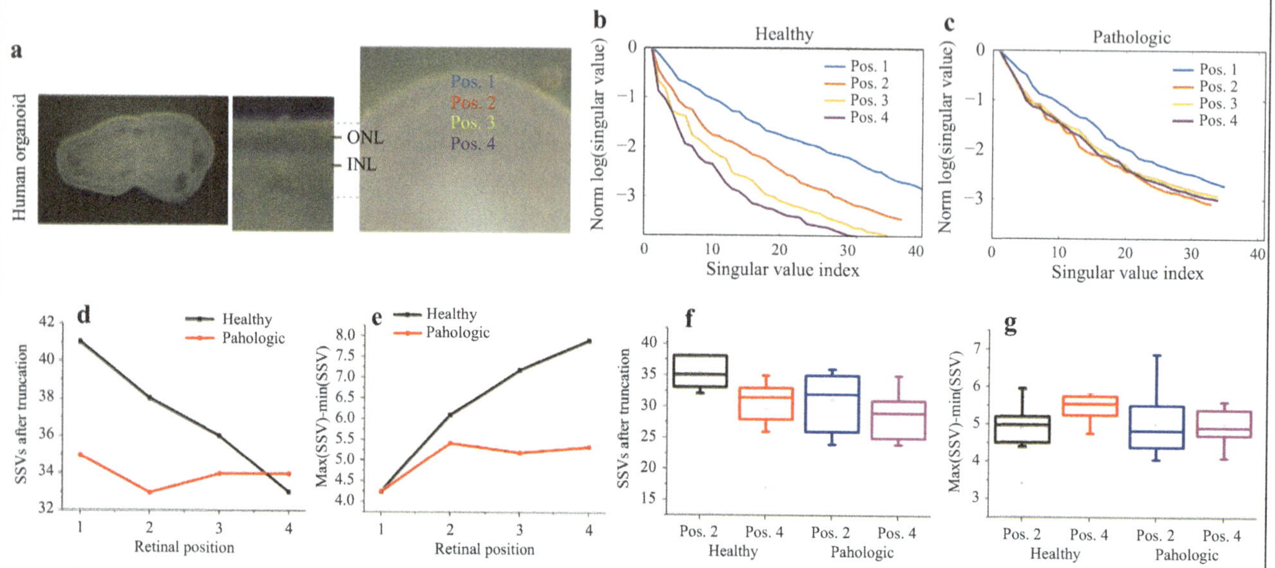

Fig. 14 SVD analysis of T_M from human retinal organoids with and without CCCP-induced neurodegeneration. **a** Locations of the measurement positions. **b, c** The SSV courses for the healthy and degenerated retinal tissue show clear different behaviour with changing distance from the sample surface. **d** The truncation point shifts and the remaining signal is spread over a reduced number of SSV, while the **e** dynamic range increases, which indicates an increase of scattering in healthy tissue with increasing distance from the surface. The pathologic tissue behaves differently, and shows higher scattering in the data, which could have its origin in structural loss. **f** and **g** The difference of maximum and minimum SSV (left) and the number of SSVs after truncation (right) show differences between positions 2 and 4.

(apical) retinal surface, the SSV course changed, indicating an increase of scattering, when the assumptions on SSV as described above are considered. For the pathologic organoid, the measurements looked different compared to controls: the SSV course remained at a comparably constant level. In sum, the observation that light-scattering tissue properties increase with greater distance from the healthy retinal surface was consistent for all measurements (Fig. 14f, g). Further, overall the measurements on the pathologic tissues were more variable and showed an increase in scattering compared to controls (Fig. 14f, g).

A statistical ANOVA analysis of the differences between the positions Pos2 and Pos4 for healthy and pathologic samples supports these observations (ANOVA at the 0.05 level). The differences between the two positions for the healthy group reach significance levels of $p_{\mathrm{SSV,healthy}} = 0.015$ and $p_{\mathrm{diff,healthy}} = 0.049$ for the two shown metrics. For the pathologic samples, no statistically relevant differences between the two positions are observed, with the significance levels of $p_{\mathrm{SSV,patho}} = 0.12$ and $p_{\mathrm{diff,patho}} = 0.97$. Together, our data indicate that changes in T_M might reflect specific tissue properties, like cell composition, and changes thereof upon pathology.

Discussion

For the first time worldwide, the human organoid retina was examined with digital holographic measurement of the transfer matrix, which is promising for the distinction between healthy and diseased tissue. These studies are crucial for in vitro organoid manufacturing and have potential also for in vivo biomarker monitoring. In our experiments, we measured and observed differences in T_M at several different anatomical positions. These differences were lost in organoids treated with CCCP, which induced histopathological changes in retinal layer structure, cell nuclear morphology, and cell death, indicative of retinal degeneration. Since structural landmarks visible by brightfield microscopy were lost upon pathology, the measurement positions might not correlate any more to the same positions like in control tissues. However, since all positions in the pathologic tissue showed rather comparable values, this might not matter. Whether latter changes in T_M are caused by the redistribution of cells within the retinal epithelium, differences in cell density, or other pathologic changes is still unclear. Taken together, our data indicate that T_M measurements might discriminate tissue intrinsic properties and pathologic changes thereof. However, evaluation via SVD exhibits significant crosstalk with changing environmental and experimental conditions. Future studies with living retina organoid tissues, retina from healthy and diseased animal models and human donor tissues, as well as healthy and pathologic tissues of other organs, will be useful to confirm and extend our observations.

Wavefront shaping using an SLM is a key to acquiring the T_M in the basis of scanning focal spots. By using $160 \times 160 = 25,600$ camera pixels, a relatively large number of rows were used for T_M measurement. Since temporal stability has to be ensured and the SLM is quite slow in practice with < 10 Hz repetition rate, we decide to use underconstrained conditions. This means that less foci are scanned than camera pixels inside the observation window are used. Performing T_M inversion with underconstrained conditions leads to focus generation through scattering media, albeit with reduced quality compared to a fully defined system. This could also be a possible source of uncertainty in the SVD studies carried out.

This can be overcome using real-time capable modulators with higher repetition rate, such as a digital micromirror device[65] or binary phase modulators[66]. Note, that SVD evaluation does not allow identification of local differences in the samples, but can be understood as the extractor of the mean scattering of the scanned area.

The results presented lead to a promising direction towards the development of a technology that could be used to develop new clinical imaging biomarkers, as well as quality controls for organoid production and tissue maintenance in cell culture. However, in the current configuration, optical access to both the input and output of the sample is required when measuring the T_M. For *in vivo* applications or situations where the target region is located inside the scattering medium, the determination of the T_M is challenging as the output is commonly not accessible. A more relevant practical solution could be to determine the reflection matrix of the scattering medium[67–70]. Here, recent approaches employ time-resolved reflection matrix determination using coherence gating to address a target-depth inside the scattering medium[70].

Materials and methods

Fabrication of phantom tissues

The phantoms were prepared freshly. The fabrication of tissue-mimicking phantoms has been explained in detail in[60]. Briefly, first, the two components of silicone elastomer polydimethylsiloxane (PDMS) (Dow Corning GmbH) were mixed together in quantities of 10:1 parts by volume, a ration of the base PDMS material to the curing agent volumes. Subsequently, glycerol (Sigma-Aldrich GmbH) in 0.5 parts per volume was included carefully by preventing the excessive amounts of air bubbles. The mixture was thoroughly mixed and poured into petri dishes (Henke Sass, Wolf GmbH) with controlled thicknesses. Air bubbles were evacuated from the mixture in a vacuum chamber. The mixture was left at room temperature to solidify for 24 hours.

Advanced manufacturing of human retinal organoids: hiPSC maintenance and human retinal organoidogenesis

Previously established human induced pluripotent stem cells (hiPSC) lines iPS-5A and CRX-H2B-mCherry[46,71,72] were used in this study. hiPSCs were maintained in mTeSR1 medium (Stem Cell Technologies) on Matrigel-coated culture dishes and passaged using ReLeSR (Stem Cell Technologies). Human retinal organoids were differentiated using a protocol adapted from previous publications[46]. Briefly, hiPSCs were dissociated into small clusters and suspended in Matrigel (growth-factor reduced, BD Bioscience). The almost solid gel was resuspended into small clumps for floating culture in N2B27 medium (1:1 Dulbecco's Modified Eagle Medium (DMEM)/F12: Neurobasal A medium, 1% B27+vitamin A, 0.5% N2, 1% penicillin /streptomycin, 1% GlutaMAX, and 1 mM 2-mercaptoethanol). The developing neuroepithelial cysts were plated on Matrigel-coated dishes on day 5. Intact cell sheets were detached from the cell culture plates at day 14 using dispase (Stem Cell Technologies) and transferred to floating culture in a B27 supplemented medium (DMEM/F12, 1% B27 without vitamin A, 1% penicillin/streptomycin, 1% GlutaMAX, 1% NEAA, and 0.1% amphotericin B). Developing retinal domains were manually isolated at D25-30 using surgical tweezers (Fine Science Tools, Dumont No. 5). From D25 on, 10% FBS was added to the B27 medium. On D100, medium was changed to N2+FBS. EC23 supplement (0.3 μM) was added from D25 to D120. Half medium was replaced every 2–3 days. All procedures involving hiPSC were performed in accordance with the ethical standards of the institutional and/or national research committee, as well as with the 1964 Helsinki declaration and its later amendments; and approved by the ethical committee at the TU Dresden.

Retinal organoid sample preparation for transmission matrix measurements

For comparison of healthy and degenerating retina, 220 day old human retinal organoids were treated daily with 5 μM CCCP (Tocris, 0452) for 11 days or received solvent control. Half medium was replaced every other day. 5 organoids (N) per treatment were analyzed. All tissues were fixed in 4% PFA for 30 min at room temperature and stored in PBS at 4 °C. Transmission matrix measurements were performed across a radial line with evenly chosen distances from the apical organoid boundary to the inner retina as depicted in Fig. 14.

Fabrication of retinal organoids: immunohistochemistry and microscope imaging analysis

For immunohistochemistry of tissue cryosections, samples were fixed in 4% PFA for 30 min, cryoprotected using a graded series of sucrose solutions (10% for 1 h, 30% for 3 h, 50% overnight), embedded in O.C.T. (Sakura Finetech) and stored at −20 °C. Tissues were sectioned at 12 μm thickness using a cryostar NX70 cryostat (Thermo Scientific), mounted on Superfrost Plus slides (Thermo Scientific) and stored at −80 °C. Sections were washed 10 min in PBS and incubated for 60 min at room temperature in blocking solution (0.5% BSA and 0.3% Triton-X-100 in PBS). Primary antibodies against RCVRN (Recoverin, Millipore AB5585, 1:2500) and RLBP1 (CRALBP, Invitrogen MA1-813, 1:200) diluted in blocking solution were added for 48 h at 4 °C. Tissue was washed in PBS (3x 10 min) before species-specific secondary antibodies conjugated to fluorophores (488, Cy3, 647; Dianova, diluted 1:1000 in blocking solution) were added for 1h at room temperature. Cell nuclei were counterstained with DAPI. For cell death visualization, a TUNEL assay was performed using In Situ Cell Death Detection Kit Fluorescein (Sigma-Aldrich, Roche products) according to the manufacturer's instructions. Samples were imaged on a widefield microscope equipped with an ApoTome2 (Zeiss). Images were analyzed and optimised by making minor changes to contrast, and cropped in Adobe Photoshop CC and arranged using Adobe Illustrator CC.

Acknowledgements

This work was financially supported by the Funding Programs for DZNE Helmholtz (M.K.); TU Dresden CRTD (M.K.); Bundesministerium für Bildung und Forschung (BMBF) ReSight (01EK1613A), and DFG KA2794/5-1 SPP2127 (M.K.). Furthermore, it was partially supported by DFG Cz55/40-1 and DFG Cz55/42-1. The hiPSC line was kindly provided by Andreas Hermann (DZNE Dresden, Uni Rostock). We thank all MOKALAB and BIOLAS members for their excellent support.

Author details

[1]Laboratory for Measurement and Sensorsystem Technique, TU Dresden, Dresden, Germany. [2]Competence Center for Biomedical Laser Systems (BIOLAS), TU Dresden, Dresden, Germany. [3]Center for Regenerative Therapies Dresden (CRTD), TU Dresden, Dresden, Germany. [4]German Center for Neurodegenerative Diseases (DZNE) Dresden, Dresden, Germany. [5]Faculty of Physics, TU Dresden, Dresden, Germany. [6]Excellence Cluster Physics of Life (PoL), TU Dresden, Dresden, Germany

Conflict of interest

The authors declare that they have no conflict of interest.

Received: 16 September 2021 Revised: 22 February 2022 Accepted: 08 March 2022

Accepted article preview online: 10 March 2022

References

1. Vellekoop, I. M. & Mosk, A. P. Focusing coherent light through opaque strongly scattering media. *Optics Letters* **32**, 2309-2311 (2007).
2. Vellekoop, I. M. Feedback-based wavefront shaping. *Optics Express* **23**, 12189-12206 (2015).
3. Parthasarathy, A. B. et al. Quantitative phase imaging using a partitioned detection aperture. *Optics Letters* **37**, 4062-4064 (2012).
4. Wang, W. J. et al. Real-time monitoring of adaptive lenses with high tuning range and multiple degrees of freedom. *Optics Letters* **45**, 272-275 (2020).
5. Platt, B. C. & Shack R. History and principles of shack-Hartmann Wavefront sensing. *Journal of Refractive Surgery* **17**, S573-S577 (2001).
6. Leith, E. N. & Upatnieks, J. Holographic imagery through diffusing media. *Journal of the Optical Society of America* **56**, 523 (1966).
7. Shen, Y. C. et al. Focusing light through biological tissue and tissue-mimicking phantoms up to 9.6 cm in thickness with digital optical phase conjugation. *Journal of Biomedical Optics* **21**, 085001 (2016).
8. Lai, P. X. et al. Photoacoustically guided wavefront shaping for enhanced optical focusing in scattering media. *Nature Photonics* **9**, 126-132 (2015).
9. Cui, M. and Yang, C. Implementation of a digital optical phase conjugation system and its application to study the robustness of turbidity suppression by phase conjugation. *Optics Express* **18**, 3444-3455 (2010).
10. Tehrani, K. F. et al. In situ measurement of the isoplanatic patch for imaging through intact bone. *Journal of Biophotonics* **14**, e202000160 (2021).
11. Shen, Y. C. et al. Focusing light through scattering media by full-polarization digital optical phase conjugation. *Optics Letters* **41**, 1130-1133 (2016).
12. Papadopoulos, I. N. et al. Focusing and scanning light through a multimode optical fiber using digital phase conjugation. *Optics Express* **20**, 10583-10590 (2012).
13. Czarske, J. W. et al. Transmission of independent signals through a multimode fiber using digital optical phase conjugation. *Optics Express* **24**, 15128-15136 (2016).
14. Scharf, E. et al. Video-rate lensless endoscope with self-calibration using wavefront shaping. *Optics Letters* **45**, 3629-3632 (2020).
15. Gloge, D. Weakly guiding fibers. *Applied Optics* **10**, 2252-2258 (1971).
16. Popoff, S. et al. Image transmission through an opaque material. *Nature Communications* **1**, 81 (2010).
17. Flaes, D. B., Štolzová, H. & Čižmár, T. Time-averaged image projection through a multimode fiber. *Optics Express* **29**, 28005-28020 (2021).
18. Kim, M. et al. Transmission matrix of a scattering medium and its applications in biophotonics. *Optics Express* **23**, 12648-12668 (2015).
19. Di Leonardo, R. & Bianchi, S. Hologram transmission through multi-mode optical fibers. *Optics Express* **19**, 247-254 (2011).
20. Choi, Y. et al. Scanner-free and wide-field endoscopic imaging by using a single multimode optical fiber. *Physical Review Letters* **109**, 203901 (2012).
21. Conkey, D. B., Caravaca-Aguirre, A. M. & Piestun, R. High-speed scattering medium characterization with application to focusing light through turbid media. *Optics Express* **20**, 1733-1740 (2012).
22. Carpenter, J., Eggleton, B. J. & Schröder, J. 110x110 optical mode transfer matrix inversion. *Optics Express* **22**, 96-101 (2014).
23. Loterie, D. et al. Digital confocal microscopy through a multimode fiber. *Optics Express* **23**, 23845-23858 (2015).
24. Plöschner, M., Tyc, T. & Čižmár, T. Seeing through chaos in multimode fibres. *Nature Photonics* **9**, 529-535 (2015).
25. Caravaca-Aguirre, A. M. & Piestun, R. Single multimode fiber endoscope. *Optics Express* **25**, 1656-1665 (2017).
26. N' Gom, M. et al. Mode control in a multimode fiber through acquiring its transmission matrix from a reference-less optical system. *Optics Letters* **43**, 419-422 (2018).
27. Turtaev, S. et al. High-fidelity multimode fibre-based endoscopy for

deep brain in vivo imaging. *Light:Science & Applications* **7**, 92 (2018).

28. Caramazza, P. et al. Transmission of natural scene images through a multimode fibre. *Nature Communications* **10**, 2029 (2019).
29. Lee, S. Y., Bouma, B. & Villiger, M. Confocal imaging through a multimode fiber without active wave-control. 2019 IEEE Photonics Conference (IPC). San Antonio, USA: IEEE, 2019.
30. Rothe, S. et al. Transmission matrix measurement of multimode optical fibers by mode-selective excitation using one spatial light modulator. *Applied Sciences* **9**, 195 (2019).
31. Chen, H. S. et al. Remote spatio-temporal focusing over multimode fiber enabled by single-ended channel estimation. *IEEE Journal of Selected Topics in Quantum Electronics* **26**, 7701809 (2020).
32. Lee, S. Y. et al. Measuring the multimode fiber transmission matrix from only the proximal side. 2020 IEEE Photonics Conference (IPC). Vancouver, Canada: IEEE, 2020.
33. Resisi, S. et al. Wavefront shaping in multimode fibers by transmission matrix engineering. *APL Photonics* **5**, 036103 (2020).
34. Singh, S., Labouesse, S. & Piestun R. Tunable mode control through myriad-mode fibers. *Journal of Lightwave Technology* **39**, 2961-2970 (2021).
35. Popoff, S. M. et al. Measuring the transmission matrix in optics: an approach to the study and control of light propagation in disordered media. *Physical Review Letters* **104**, 100601 (2010).
36. Vellekoop, I. M. & Mosk, A. P. Universal optimal transmission of light through disordered materials. *Physical Review Letters* **101**, 120601 (2008).
37. Choi, W. et al. Transmission eigenchannels in a disordered medium. *Physical Review B* **83**, 134207 (2011).
38. Kim, M. et al. Maximal energy transport through disordered media with the implementation of transmission eigenchannels. *Nature Photonics* **6**, 581-585 (2012).
39. Yılmaz, H. et al. Transverse localization of transmission eigenchannels. *Nature Photonics* **13**, 352-358 (2019).
40. Judkewitz, B. et al. Translation correlations in anisotropically scattering media. *Nature Physics* **11**, 684-689 (2015).
41. Solovei, I. et al. Nuclear architecture of rod photoreceptor cells adapts to vision in mammalian evolution. *Cell* **137**, 356-368 (2009).
42. Labin, A. M. et al. Müller cells separate between wavelengths to improve day vision with minimal effect upon night vision. *Nature Communications* **5**, 4319 (2014).
43. Franze, K. et al. Müller cells are living optical fibers in the vertebrate retina. *Proceedings of the National Academy of Sciences of the United States of America* **104**, 8287-8292 (2007).
44. Reichenbach, A. & Bringmann, A. New functions of müller cells. *Glia* **61**, 651-678 (2013).
45. Capowski, E. E. et al. Reproducibility and staging of 3D human retinal organoids across multiple pluripotent stem cell lines. *Development* **146**, dev171686 (2019).
46. Völkner, M. et al. Optimized adeno-associated virus vectors for efficient transduction of human retinal organoids. *Human Gene Therapy* **32**, 694-706 (2021).
47. Völkner, M. et al. Retinal organoids from pluripotent stem cells efficiently recapitulate retinogenesis. *Stem Cell Reports* **6**, 525-538 (2016).
48. Guymer, R. & Wu, Z. C. Age-related macular degeneration (AMD): More than meets the eye. The role of multimodal imaging in today' s management of AMD—a review. *Clinical & Experimental Ophthalmology* **48**, 983-995 (2020).
49. Rahman, N. et al. Macular dystrophies: clinical and imaging features, molecular genetics and therapeutic options. *British Journal of Ophthalmology* **104**, 451-460 (2020).
50. Scholler, J. et al. Dynamic full-field optical coherence tomography: 3D live-imaging of retinal organoids. *Light:Science & Applications* **9**, 140 (2020).
51. Browne, A. W. et al. Structural and functional characterization of human stem-cell-derived retinal organoids by live imaging. *Investigative Ophthalmology & Visual Science* **58**, 3311-3318 (2017).
52. Stockinger, P. et al. Correlation of in vivo/ex vivo imaging of the posterior eye segment. *Ophthalmologe* **118**, 153-159 (2021).
53. Bianchi, S. & Di Leonardo, R. A multi-mode fiber probe for holographic micromanipulation and microscopy. *Lab on a Chip* **12**, 635-639 (2012).
54. Plöschner, M. et al. GPU accelerated toolbox for real-time beam-shaping in multimode fibres. *Optics Express* **22**, 2933-2947 (2014).
55. Cuche, E., Marquet, P. & Depeursinge, C. Spatial filtering for zero-order and twin-image elimination in digital off-axis holography. *Applied Optics* **39**, 4070-4075 (2000).
56. Koukourakis, N. et al. Photorefractive two-wave mixing for image amplification in digital holography. *Optics Express* **19**, 22004-22023 (2011).
57. Tikhonov, A. N. On the solution of ill-posed problems and the method of regularization. *Doklady Akademii Nauk* **151**, 501-504 (1963).
58. Popoff, S. M. et al. Controlling light through optical disordered media: transmission matrix approach. *New Journal of Physics* **13**, 123021 (2011).
59. Rothe, S. et al. Physical layer security in multimode fiber optical networks. *Scientific Reports* **10**, 2740 (2020).
60. Wróbel, M. S. et al. Nanoparticle-free tissue-mimicking phantoms with intrinsic scattering. *Biomedical Optics Express* **7**, 2088-2094 (2016).
61. Marčenko, V. A. & Pastur, L. A. Distribution of eigenvalues for some sets of random matrices. *Mathematics of the USSR-Sbornik* **1**, 457-483 (1967).
62. Gavish, M. & Donoho, D. L. The optimal hard threshold for singular values is $4/\sqrt{3}$. *IEEE Transactions on Information Theory* **60**, 5040-5053 (2014).
63. Yang, J. H. et al. Apoptotic cell death of cultured salamander photoreceptors induced by cccp: CsA-insensitive mitochondrial permeability transition. *Journal of Cell Science* **114**, 1655-1664 (2001).
64. Das, A. et al. Programmed switch in the mitochondrial degradation pathways during human retinal ganglion cell differentiation from stem cells is critical for RGC survival. *Redox Biology* **34**, 101465 (2020).
65. Turtaev, S. et al. Comparison of nematic liquid-crystal and DMD based spatial light modulation in complex photonics. *Optics Express* **25**, 29874-29884 (2017).
66. Schmieder, F. et al. Optogenetic stimulation of human neural networks using fast ferroelectric spatial light modulator—based holographic illumination. *Applied Sciences* **8**, 1180 (2018).
67. Gérardin, B. et al. Full transmission and reflection of waves propagating through a maze of disorder. *Physical Review Letters* **113**, 173901 (2014).
68. Popoff, S. M. et al. Exploiting the time-reversal operator for adaptive optics, selective focusing, and scattering pattern analysis. *Physical Review Letters* **107**, 263901 (2011).
69. Kim, M. et al. Exploring anti-reflection modes in disordered media. *Optics Express* **23**, 12740-12749 (2015).
70. Choi, Y. et al. Measurement of the time-resolved reflection matrix for enhancing light energy delivery into a scattering medium. *Physical Review Letters* **111**, 243901 (2013).
71. Kim, S. et al. Generation, transcriptome profiling, and functional validation of cone-rich human retinal organoids. *Proceedings of the National Academy of Sciences of the United States of America* **116**, 10824-10833 (2019).
72. Gasparini, S. J. et al. Extensive incorporation, polarisation and improved maturation of transplanted human cones in a murine cone degeneration model. *bioRxiv*. http://dx.doi.org/10.1101/2021.08.26.457641 (2021).

Shanmugam et al. *Light: Advanced Manufacturing* (2022)3:16
https://doi.org/10.37188/lam.2022.016

Official journal of the JHL 2689-9620
www.light-am.com

Article Open Access

Variable shearing holography with applications to phase imaging and metrology

Prithiviraj Shanmugam, Alyson Light, Avery Turley and Konstantinos Falaggis*

Abstract

We report variable shear interferometers employing liquid-crystal-based geometric-phase (GP) gratings. Conventional grating-based shear interferometers require two lateral shifts of the gratings to enable phase-shifting capabilities in x- and y- direction and two axial shifts of the gratings to adjust the shear amount in x- and y-direction, i.e., these systems need control of four axes mechanically. Here we show that the GP gratings combined with a pixelated polarization camera give instantaneous-phase shifting so that no mechanical movement is necessary to obtain phase shifts. Furthermore, we show that a single fixed shear with a rotational shear axis provides a more robust selection of shears while requiring the control of only one mechanical axis. We verify this statement using spatial domain and frequency domain criteria. We further show that the resolution of the reconstructed wavefield depends not only on the numerical aperture of the imaging system, the pixel size of the detector, or the spatial coherence of the source but also on the ability to determine the shear amount accurately. To improve this, we report a methodology to accurately detect the shear amounts using the second derivative of the autocorrelation function of the measured holograms, which further relaxes the requirements on mechanical stability.

Keywords: Metrology, Shearing interferometry, Phase retrieval, Wavefront reconstruction, Shearing holography

Introduction

Single-beam wavefront reconstruction techniques are exciting alternatives to interferometry. These techniques are inherently robust to vibrations and often can be designed in a compact optical setup. Recent research expands those techniques to partial coherent sources, where LEDs are employed in place of lasers. Another interesting field is the area of bio-imaging, where existing microscopes are converted into phase measuring devices using a single-beam phase wavefront reconstruction add-on. A further appealing aspect is that many single-beam systems can be designed to work with no beamsplitter, allowing the sensor to be closer to the sample and enabling simpler optical measurements with higher-numerical aperture. In phase retrieval, there are two representative families of wavefront reconstruction techniques. Firstly, techniques based on deterministic models[1–5], and secondly, techniques that employ an iterative solver that utilizes wave-propagation techniques[6,7] or Fourier relationships[8,9].

With the advent of faster computer systems, new cameras are often developed with a higher pixel count, diminishing many advances of the higher computational power. Optical measurement technologies need to adapt to survive this trend through more innovative algorithms, new measurement concepts, or both.

For instance, in the field of phase retrieval, the

Correspondence: Konstantinos Falaggis (kfalaggi@uncc.edu)
University of North Carolina at Charlotte, Charlotte, North Carolina, USA

convergence may be improved using multi-resolution and relaxation techniques[10]. Another possibility is to alter the illumination[11] or combine deterministic and iterative algorithms[12].

However, the elephant in the room of all methods mentioned above is the inverse problem resulting from the data produced by the unmodulated optical beam. In other words, often, the data is too similar, and significant efforts need to be made to reduce the similarity (e.g., repeating the measurement at a large defocus distance[3]).

Recent advances in computational shear interferometry[13,14] allow actively modulating an incident beam as it interferes with itself. Modulation of the beam increases instantly the diversity in the data. In particular, when employed with phase-shifting techniques[15], it is possible to obtain the complex crossterm of two interfering wavefields directly. Despite this sophistication, the system by Falldorf[13] requires a spatial light modular (SLM), which significantly contributes to the system costs. The wavefront reconstruction technique reported by Falldorf et al.[13] is based on a gradient descent approach for non-symmetric shearing systems. However, compared to the non-symmetric case, deriving the grading of a that non-holomorphic function for the case of a "symmetric " shearing system is not straightforward. Hence other approaches like the alternating projections by Konijnenberg[16] are preferred reconstruction algorithms.

Recent advances in geometric phase (GP) elements lead to the development of GP gratings and GP lenses that can be employed in lateral and radial shearing interferometry[17-19], with polarization phase-shifting capabilities[20,21]. The remaining quest is to recover the correct wavefront from the captured phase-shifted interferograms.

This paper proposes a variable shearing interferometer that consists of a series of GP gratings and polarization optics. This setup enables measuring a series of phase-shifted interferograms for a given shear amount, as shown in Fig. 1. In a subsequent step, we reconstruct the wavefield using a modified version of the alternating projections algorithm, developed by Konijnenberg[16], and a series of interferograms with different amounts of shear. The discussed systems are the first experimentally demonstrated GP grating-based variable shear interferometers for complex wavefield reconstructions. We discuss various possible configurations of these instruments and demonstrate the success of two distinct shear selection strategies (employing one or two grating pairs). We analyze these findings using the "frequency information density" and the "spatial information density" analyses based on transfer function and support function concepts introduced by Servín in Ref. 22. We further show

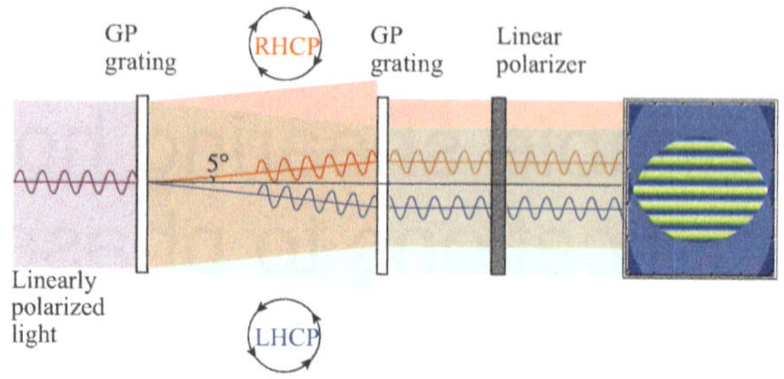

Fig. 1 Principle of GP grating based shearing interferometry for the case of linearly polarized light.

that a GP grating configuration with a fixed shear amount, but an adjustable rotational shear axis (one mechanical rotational axis) is sufficient to obtain robust measurements while maintaining phase-shifting capabilities using a pixelated polarization camera. We also investigate the effect of other error sources (e.g., imperfect shear estimation), which affects the resolution of the reconstructed wavefront. This effect is not commonly known (for coherent systems, the resolution is often a result of numerical aperture, pixel size, or spatial coherence of the source). To mitigate this source of error, we report a shear detection technique using the peaks of the second derivative of the autocorrelation function of the holograms. Measuring the shear amount directly from the hologram is different from relying on a well-calibrated system and increases the system's accuracy.

Results

Shear Selection

Polarization gratings, also known as geometric phase (GP) gratings, are liquid crystal polymer-based optical gratings with a diffraction angle that depends on the polarization and wavelength of light. GP gratings are designed to diffract right-handed circularly polarized light (RHCP) and left-handed circularly polarized light (LHCP) at a negative angle and positive diffraction angle, respectively. The value of this angle depends on the prescription; commercially available gratings with various groove densities result in diffraction angles of the first-order between 3 and 10 degrees. The diffraction angle can be related directly to the wavelength and groove density using the well-known equation $m\lambda = d\sin(\theta)$.

A lateral shear in the spatial domain is achieved when two GP gratings are placed consecutively[17], as shown in Fig. 1. In that configuration, both identical GP gratings are aligned along the same axis and orientation. In this way, the first GP grating deflects the incoming beams at the angles +/− α, and the second GP grating compensates this deflection by diffraction the beam into −/+ α. The properties of these GP gratings in this configuration are outlined in Ref. 17. Adjusting the distance between the first and the second GP grating allows adjusting the shear amount of the two interfering waves, as shown in Fig. 2, enabling unique properties that can be used for adjustable shearing interferometry.

As shown in Fig. 2, shearing along the y-axis is possible by flipping the gratings by 90 degrees. An instrument employing a series of four GP gratings enables shearing in both x- and y-direction. This configuration shears along the x- and y-axis independently. This system uses 4 gratings in total: one pair for x-shear and the other pair for y-shear.

A different possible configuration is to maintain a fixed shear between two beams and modulate the fringes by rotating the shear axis about the optical axis, as shown in Fig. 3. This configuration requires only one GP gratings pair and is more compact than the other configuration of Fig. 2c.

Furthermore, as the linearly polarized light at the instrument's input is converted into left and right circular polarized components, there is a need to employ a linear polarizer to make these two waves interfere. Notably, this configuration provides geometric phase-shifting capability[23], i.e., by rotating the axis of the polarizer, it is possible to create a series of phase-shifted interferograms, as shown in Fig. 4.

In the following sections, we propose using GP gratings for multi-shear variable shearing interferometry with phase-shifting capability. We further evaluate the performance of the instrument with mentioned above configurations using multi-shear wavefront reconstruction techniques.

Wavefront reconstruction algorithm for multi-shear variable shear interferometry

The previous section showed how GP gratings could obtain a series of phase-shifted interferograms for various shear amounts. These interferograms allow calculating the complex cross-terms of the two-interfering waves at each shear. The challenge is to reconstruct the complex wavefield from this data. In this section, we present an iterative algorithm based on the approach of Konijnenberg[16] that was inspired by the alternating projection algorithms[9,24].

Consider a general wavefield u with the amplitude A and the phase ϕ as

$$u = A \cdot \exp(i\phi) \tag{1}$$

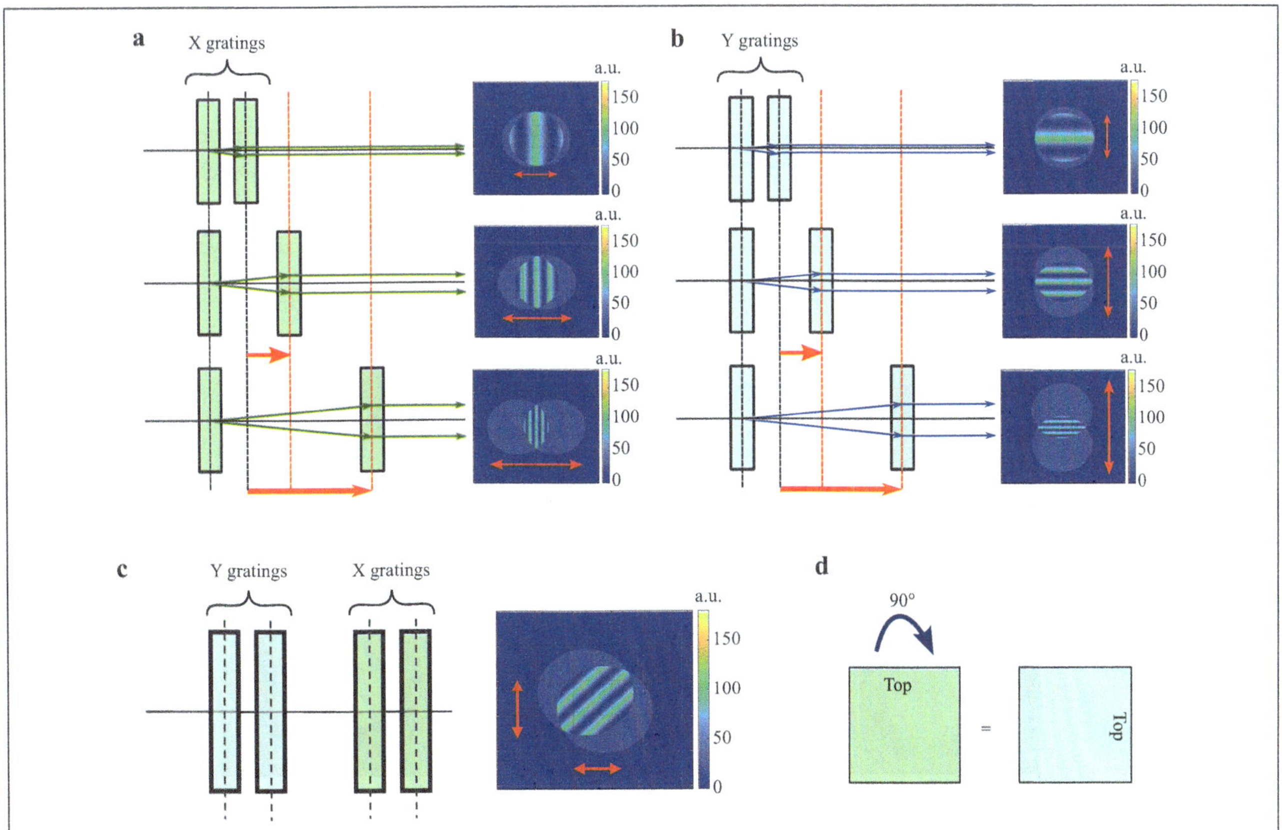

Fig. 2 Examples of interferograms for the case of a spherical wave at the distance z = 350 mm with different shear amounts in x- and y- direction.

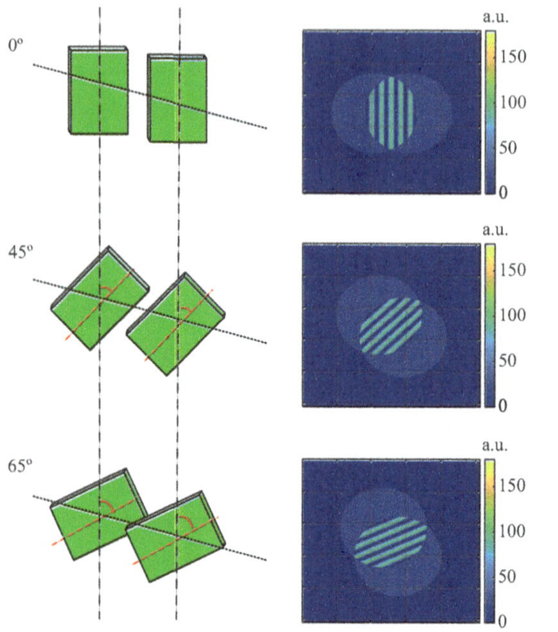

Fig. 3 Examples of interferograms for the case of a spherical wave at the distance z = 250 mm with constant shear and different grating rotational angles.

which produces the following interferogram $I_{n,\delta}(\vec{x})$ in shearing, as

$$I_{n,\delta}(\vec{x}) = \left|u\left(\vec{x} + \vec{ds_n}\right)\right|^2 + \left|u\left(\vec{x} - \vec{ds_n}\right)\right|^2 + 2\mathcal{R}\left\{C_n(\vec{x})\exp(-i\delta)\right\} \quad (2)$$

Where $\vec{ds} = [ds_x, ds_y]$ and, ds_x and ds_y is the shear in x- and y- direction, δ is the phase shift induced by the rotating polarizer, and $C_n(\vec{x})$ is the complex cross-term

$$C_n(\vec{x}) = u^*\left(\vec{x} - \vec{ds_n}\right) \cdot u\left(\vec{x} + \vec{ds_n}\right) \quad (3)$$

The term $C_n(\vec{x})$ is estimated using phase-shifting techniques, which for the four-bucket algorithm results in

$$C_n(\vec{x}) = \left(I_{n,0}(\vec{x}) - I_{n,\pi}(\vec{x})\right) + i\left(I_{n,\frac{\pi}{2}}(\vec{x}) - I_{n,\frac{3\pi}{2}}(\vec{x})\right) \quad (4)$$

The reconstruction algorithm uses the term $C_n(\vec{x})$ that is obtained at different shears '$\vec{s_n}$' to reconstruct the object wavefield through an iterative process. In this work, we employ an alternating projections (AP) approach developed by Konijnenberg[16] for the x-ray community. The algorithm applies two distinct constraint sets, where each constraint allows only a limited family of solutions. The objective of the wavefield reconstruction algorithm is to identify the unique solution that satisfies all constraints (and for all shears). These constraints are satisfied by employing alternating projections between the two solution domains, in which we apply both, so-called "object constraints" and so-called "measurement constraints".

Switching between "object constraints" and "measurement constraints " is applied continuously until the algorithm converges to a specific wavefield. The convergence can be conveniently estimated using the following loss function

$$L = \sum_{n=1}^{S}\left(\left|f^M_{n,+}(\vec{x}) - f^O_{n,+}(\vec{x})\right|^2 + \left|f^M_{n,-}(\vec{x}) - f^O_{n,-}(\vec{x})\right|^2\right) \quad (5)$$

where subscript n denotes that the wavefield uses the shear $\vec{ds_n}$, S is the total number of shears, and for convenience, we define the estimated wavefield after N iterations as $f(\vec{x})$ to distinguish it from ideal wavefield $u(\vec{x})$. The

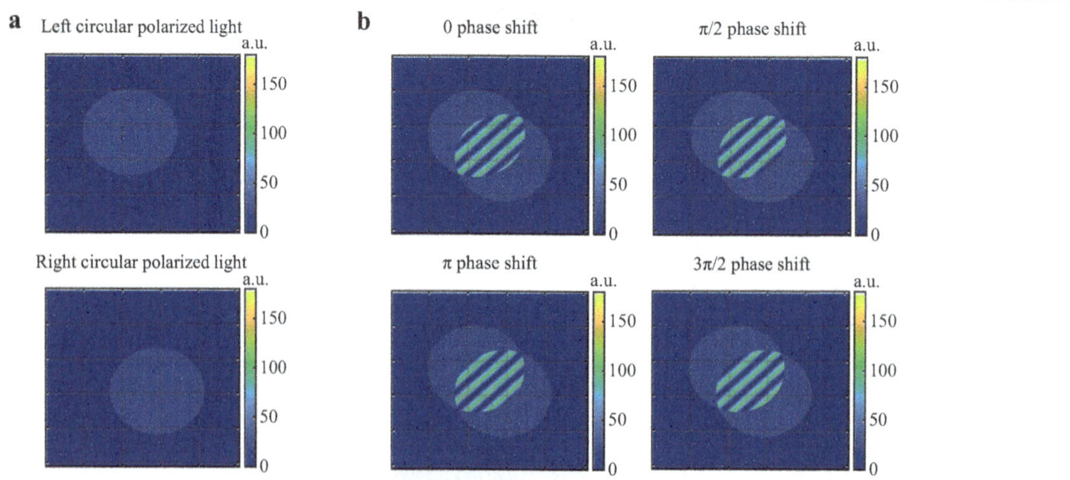

Fig. 4 **a** Intensity ditribution of a spherical wavefront with constant amplitude before reaching the linear polarizer. **b** Intensity after the linear polarizer for four different angles of the polarizer axis to produce a phase shift between interfering waves of 0, π/2, π, and 3π/2.

notation $f(\vec{x})$ is used as a general indication to represent estimated wavefields from both constraints, where the notation $f^M(\vec{x})$ denotes the estimate of the wavefield directly after the measurement constraint is applied, and $f^O(\vec{x})$denotes the estimate of the wavefield directly after the object constraint is applied. For further simplification, the symbols '$n,+$' and '$n,-$' in the subscript are used to represent positive and negative shears, i.e., $f_{n,\pm}(\vec{x}) = f(\vec{x} \pm \vec{ds_n})$. Consequently, the estimated complex cross amplitude term can be presented as

$$C_{\mathrm{est,n}}(\vec{x}) = f^*(\vec{x} - \vec{ds_n}) \cdot f(\vec{x} + \vec{ds_n}) = f_{n,-}(\vec{x})^* \cdot f_{n,+}(\vec{x}) \tag{6}$$

and can be generated using the estimated wavefield $f^M{}_n(\vec{x})$ at any shear $\vec{S}_n$.

Notably, in the presence of noise, there is no exact solution. However, the wavefield reconstruction algorithm estimates the best fit solution that matches the experiment data. The loss function serves as a metric to be observed for convergence with an increasing number of iterations.

Fig. 5 shows the steps of the wavefield reconstruction algorithm. After an initial guess, the algorithm continuously applies the so-called object and measurement constraints alternatingly until the loss function reaches the desired level. The application of those constraints is discussed in more detail below.

Estimation of $f^M_{n,\pm}(\vec{x})$ using the initial guess

This step of the algorithm uses only the initial guess $f(\vec{x})$ in the first iteration to estimate $f^M_{n,\pm}(\vec{x})$. An initial guess is assumed for $f(\vec{x})$ (e.g., a spherical wave, a plane wave, or a random wavefield) and is laterally sheared by the same shears used in the experiment. For a wavefield $f(\vec{x})$, the shearing is done in the Fourier domain using the following equations.

$$F(\vec{k}) = \mathcal{F}\{f(\vec{x})\} \tag{7}$$

$$f^M_{n,\pm}(\vec{x}) = \mathcal{F}^{-1}\left\{F(\vec{k}).\exp\left(\mp 2\pi \cdot i \cdot \vec{k} \cdot \vec{ds_n}\right)\right\} \tag{8}$$

Assuming that the initial guess wavefield is from the measurement constraint domain, the initial guess and its sheared components are indicated as $f^M_{n,\pm}(\vec{x})$. A wavefield with constant phase and unit amplitude is used as an initial guess for the reconstructions done in this paper.

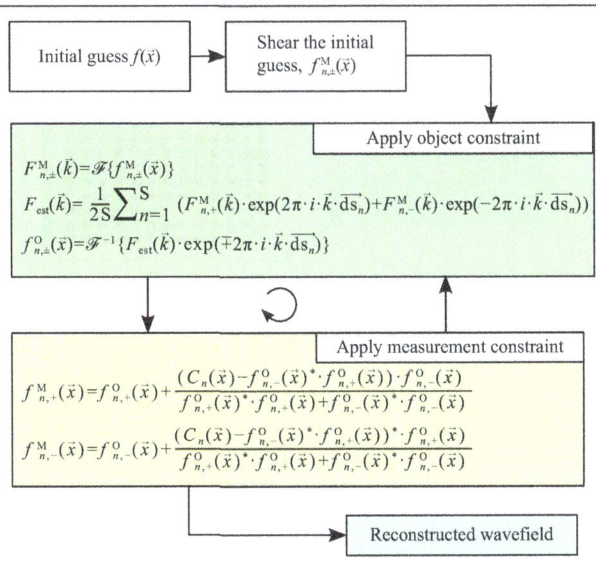

Fig. 5 Principle of multi-shear wavefront reconstruction algorithm.

Application of the Object constraint

The object constraints are applied by transforming the sheared components of the estimated wavefield to the Fourier domain and compensating the shear for each $f^M_{n,\pm}(\vec{x})$ to obtain $f(\vec{x})$ using the Fourier Shift theorem. All calculated values for $F(\vec{k})$ are averaged to obtain a more accurate result.

$$F^M_{n,\pm}(\vec{k}) = \mathcal{F}\{f^M_{n,\pm}(\vec{x})\} \tag{9}$$

$$F_{est}(\vec{k}) = \frac{1}{2S}\sum_{n=1}^{S}\left(F^M_{n,+}(\vec{k})\cdot\exp\left(2\pi\cdot i\cdot\vec{k}\cdot\vec{ds_n}\right) + F^M_{n,-}(\vec{k})\cdot\exp\left(-2\pi\cdot i\cdot\vec{k}\cdot\vec{ds_n}\right)\right) \tag{10}$$

Finally, $F_{est}(\vec{k})$ is transformed back into the spatial domain using the inverse Fourier transform. However, because the measurement constraints (applied in the second part of the iteration) require the wavefield for different shears, we apply again the Fourier shift theorem to obtain $f^O_{n,\pm}(\vec{x})$ as

$$f^O_{n,\pm}(\vec{x}) = \mathcal{F}^{-1}\left\{F_{est}(\vec{k})\cdot\exp\left(\mp 2\pi\cdot i\cdot\vec{k}\cdot\vec{ds_n}\right)\right\} \tag{11}$$

Application of the Measurement constraint

The measurement constraint equations use the estimated wavefield from the object constraint $f^O_{n,\pm}(\vec{x})$ and the measured complex cross-term $C_n(\vec{x})$(obtained from phase shifting) to estimate the wavefield solution in the measurement constraint domain. The following equations are a modified version of the equations used by Konijnenberg in[16] to estimate the wavefield.

$$f^{M}_{n,+}\left(\vec{x}\right) = f^{O}_{n,+}\left(\vec{x}\right) + \frac{\left(C_n\left(\vec{x}\right) - f^{O}_{n,-}\left(\vec{x}\right)^{*} \cdot f^{O}_{n,+}\left(\vec{x}\right)\right) \cdot f^{O}_{n,-}\left(\vec{x}\right)}{\left|f^{O}_{n,+}\left(\vec{x}\right)\right|^{2} + \left|f^{O}_{n,-}\left(\vec{x}\right)\right|^{2}} \tag{12}$$

$$f^{M}_{n,-}\left(\vec{x}\right) = f^{O}_{n,-}\left(\vec{x}\right) + \frac{\left(C_n\left(\vec{x}\right) - f^{O}_{n,-}\left(\vec{x}\right)^{*} \cdot f^{O}_{n,+}\left(\vec{x}\right)\right)^{*} \cdot f^{O}_{n,+}\left(\vec{x}\right)}{\left|f^{O}_{n,+}\left(\vec{x}\right)\right|^{2} + \left|f^{O}_{n,-}\left(\vec{x}\right)\right|^{2}} \tag{13}$$

Behavior of the algorithm

The algorithm uses two sets of constraint equations, object constraint Eq. 10, and measurement constraint Eqs. 12, 13. Each set of constraint equations has a solution set. The algorithm starts with an initial guess projected onto the object constraint solution set by using the respective equations. This produces an estimated wavefield which is a projection on the object constraint solution set. This wavefield is then projected onto the measurement constraint solution set using the respective equations, and the process repeats until convergence. Ideally, the solution we are looking for is the one that is at the intersection of these two solution sets. However, in practice, due to noise and other external factors, such intersection does not exist, and the algorithm converges when the projections oscillate between the two solution sets with solutions at the closest proximity to each other.

Simulation using synthetic data

The performance of the algorithm was evaluated by simulations using MATLAB. Preliminary simulations were made for a resolution of 256 × 256 pixels. For this simulation, a point source was generated with a wavelength of 600 nm at a z-distance of 87.5 mm and was sampled with a pixel size of 3.45 μm. The reconstruction algorithm uses FFTs to shear the estimated wavefields at different steps in the reconstruction. The use of FFTs, in turn, implicitly imposes a periodic boundary condition. Therefore, we use an aperture for both simulations and experiments to mitigate problems during reconstruction that might arise from periodic properties of FFTs. For this reason, a circular aperture mask with a diameter of 180 pixels was applied. Future reconstruction algorithms may exclude samples that exceed the computational window, especially at larger shear. Those samples need to be reconstructed with measurements taken at smaller shear amounts.

A series of four phase-shifted intensities were generated for each shear position. The shears generated for this simulation are shown in Fig. 6a. Additive white Gaussian noise (AWGN) with a standard deviation of 20% of the peak intensity has been added to the generated intensities to model the effect of noise in these simulations. Afterward, to realistically model the effect of cameras, an 8-bit discretization has been applied to the intensities.

For this simulation, a total of 48 interferograms have been generated corresponding to 12 different shears with 4 phase-shifted interferograms. The ideal wavefront and reconstructed wavefront are shown in Fig. 6b, c, respectively. The corresponding phase error of the reconstructed wavefront is shown in Fig. 6d. Further investigation of the error map showed the presence of high-frequency artifacts, as highlighted in Fig. 7.

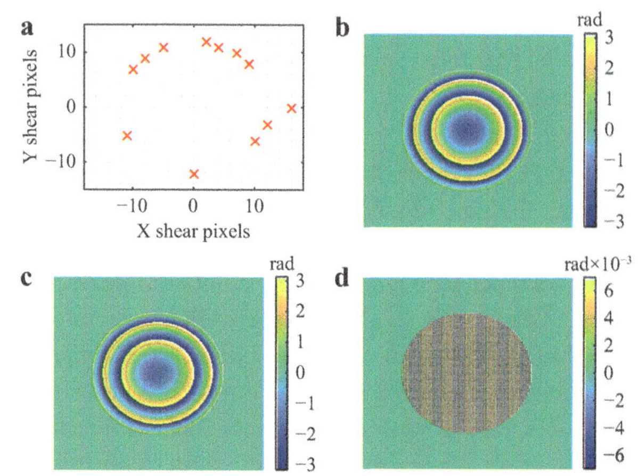

Fig. 6 (Simulation) Results of reconstruction with simulation parameters: FoV 256 × 256 pixels, Pixel size 3.45 μm, Wavelength 0.6 μm, Aperture size 90 pixels, Z distance 87.5 mm. **a** Shear selection. **b** Ideal phase map. **c** Reconstructed phase map. **d** Error map between ideal and reconstructed data.

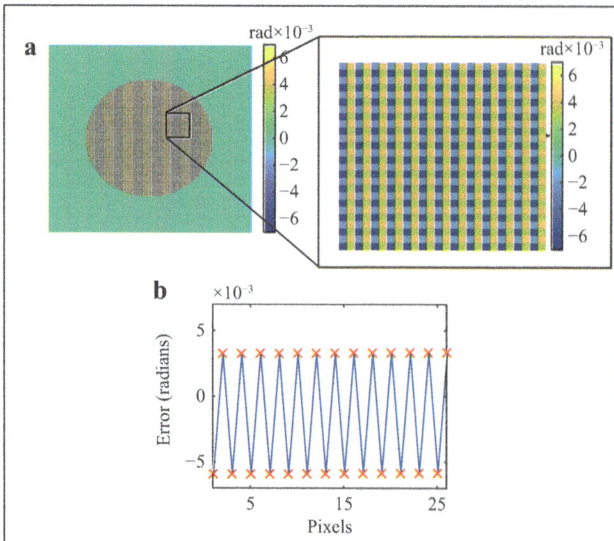

Fig. 7 High frequency artifact from reconstruction algorithm. **a** Error map zoomed. **b** Error map profile.

These high-frequency artifacts occur at Nyquist frequency and appear to be a product of the reconstruction algorithm. Further analysis is required to investigate the source of these artifacts. For the reconstructions discussed in this paper, we use a low-pass Fourier filter to mitigate the formation of these artifacts.

Fig. 8 shows error maps generated by mapping the difference between the ideal and reconstructed wavefield before and after applying the filter. A cross-section of the error profile before and after filtering is shown in Fig. 8c, d, respectively. This simple procedure reduces the magnitude of the RMS error by one order of magnitude. Because a Fourier low-pass filter has been employed with a corner frequency of $k_c = (4/5) \cdot (1/\text{pixel size})$, an artifact is produced along the boundary of the aperture. However, other filtering techniques may be applicable depending on the application, and even Zernike polynomials may be fitted to the data. The use of different filtering techniques is beyond the scope of this work.

Evolution of loss function

As mentioned earlier, the evolution of the loss function indicates how close the algorithm is to achieving the best solution or achieving convergence. The loss function is evaluated in these experiments by calculating the change between the solutions generated by the two constraints on positive and negative sheared estimated wavefields. The loss function was already defined in Eq. 5 and is reiterated here in Eq. 14 for convenience. Fig. 9a shows the evolution of loss function during reconstructions done in Fig. 8 before the filter was applied.

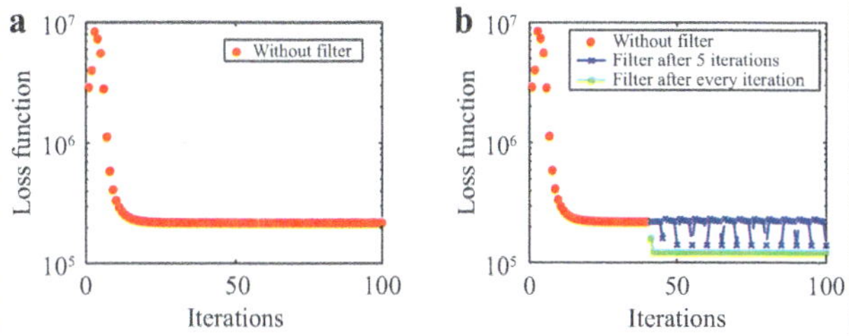

Fig. 9 Evolution of loss function. **a** without filter. **b** with filter.

$$L = \sum_{n=1}^{S} \left(\left| f^{M}_{n,+}(\vec{x}) - f^{O}_{n,+}(\vec{x}) \right|^2 + \left| f^{M}_{n,-}(\vec{x}) - f^{O}_{n,-}(\vec{x}) \right|^2 \right) \quad (14)$$

Fig. 9a shows that the loss function converges after 30 iterations of the reconstruction without a Fourier filter. Given the filter's effectiveness in Fig. 8, we have investigated the use of the filter within the reconstruction algorithm. The sudden surge at the start of the loss function in Fig. 9 is a characteristic feature of phase retrieval and reconstruction algorithms. The temporary increase of the value of the Loss function is expected because the solver climbs out of the local minimum. Other phase retrieval algorithms such as the "hybrid input-output algorithm"[8] show a similar behavior.

Two cases were considered for this study: (i) using the filter after every 5 iterations and (ii) applying the filter in every iteration. Fig. 9b shows that applying a filter enables a drop in the convergence values in both cases, thus indicating an improvement in the reconstruction process. Interestingly, the reconstruction algorithm for case (i) consistently converges back to the solution containing the high-frequency artifact. Therefore, we recommend employing a Fourier filter within the wavefront reconstruction algorithm to suppress all undesired artifacts continuously.

Please note, the algorithm described is a modified version of the alternating projections algorithm developed by Konijnenberg et al. in Ref. 16. For better clarity we summarize the differences to the original algorithm from Ref. 16.

- The data recorded from experiments in Ref. 16 are in the Fourier domain. In this paper, all recorded data from experiments are in the spatial domain.
- Eq. 10 is the modified version of the object constraint equation. The original equation (Eq. 10 in Ref. 16) was defined in the spatial domain. In this paper, this equation is defined in the Fourier domain.
- Eqs. 12,13 are the modified versions of the measurement constraint equations. The original equations (Eq. 15 in Ref. 16) were defined in the Fourier domain. In this paper, these equations are defined in the spatial domain.

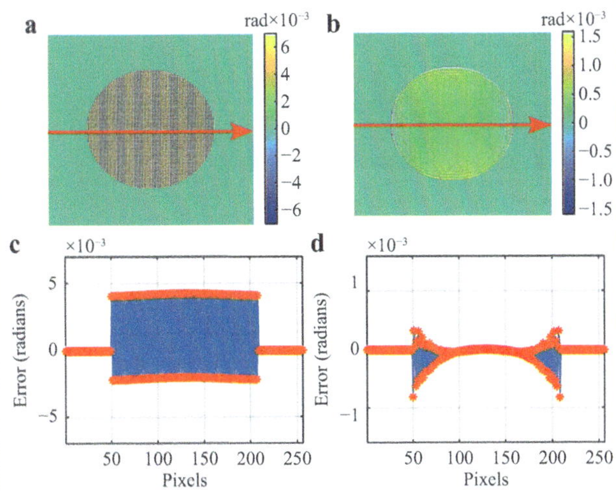

Fig. 8 a Error map before filter (RMS error: 4.1E-3 rad). **b** Error map after filter (RMS error: 0.42E-3 rad). **c** Profile of error map before filter. **d** Profile of error map after filter.

• The algorithm in Ref. 16 uses two sets of equations in measurement constraint (Eqs. 15,16 in Ref. 16), rough-reconstruction equations for initial convergence (Eq. 16 in Ref. 16) and fine-reconstruction equations for final refinement (Eq. 15 in Ref. 16). This paper only uses the modified versions of the fine reconstruction equations.

• After initial convergence, we include a Fourier filter in the measurement constraint equations to mitigate the effects of high-frequency artifacts.

• Multi-resolution technique[10] is incorporated into this algorithm to mitigate the effects of initial guess and improve convergence.

Comparison of the two configurations: (Simulation)

In the cartesian coordinate shear system, there is the freedom to choose the shear in the X and Y direction independently. However, the optical configurations shown in Fig. 2 are restricted to one quadrant due to geometrical constrains. In principle, this limitation may be overcome using a 4f- optical setup (obtain positive and negative shears for a grating pair). However, this configuration increases the size of the system. There is freedom in the polar coordinate shear system to choose shears across all four quadrants, but these shears lie on a circle with a constant radius. When these shear systems are implemented experimentally, the shear amount cannot be controlled directly and estimated.

The choice of shear selection has an impact on the reconstruction of the wavefield. In general, increasing the number of shears provides more information on the object wavefield for the algorithm to reconstruct the wavefield. However, improper selection of shears could also lead to the algorithm not having sufficient frequencies to reconstruct the object wavefield, thus resulting in artifacts. We want to highlight this typical behavior of this system via simulation using a few exemplary sets of shears, as shown in Fig. 10.

Each simulation was performed with 24 shears, where the selections are shown in Figs. 10a–c. The selections analyze two possible xy-shearing configurations shown in Fig. 10a, b (see Fig. 2 for configuration) and the fixed shear with variable shear axis shown in Fig. 10c (see Fig. 3 for configuration).

In test 1 (Fig. 10a), the shears have been selected along

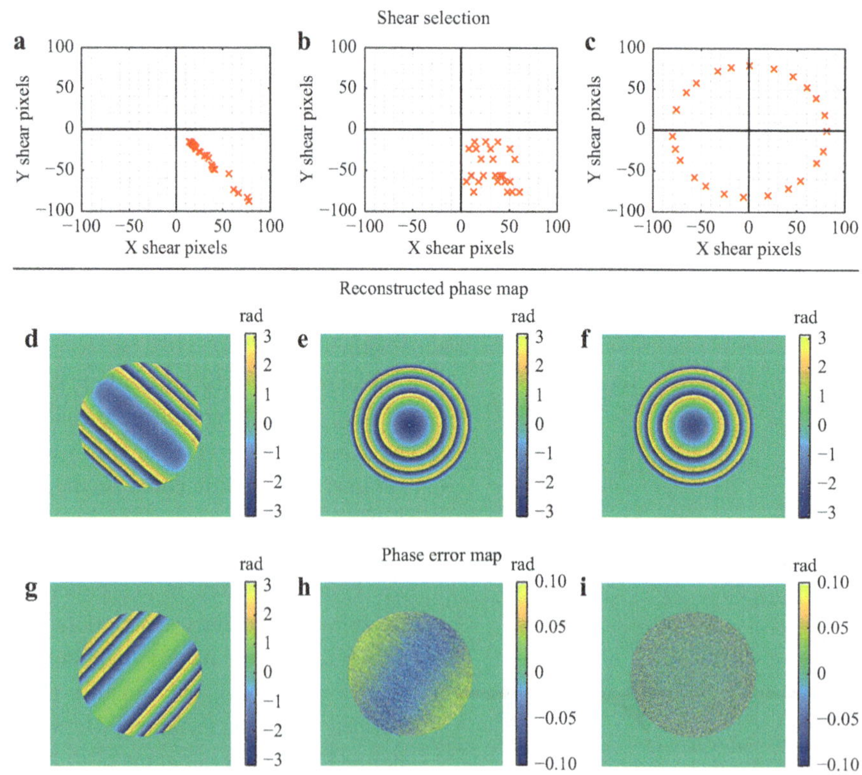

Fig. 10 Comparison of reconstruction results between different configurations. **a** Shear selection for test 1 in cartesian coordinate configaration. **b** Shear selection for test 2 in cartesian coordinate configuration. **c** Shear selection for polar coordinate configuration. **d** Reconstructed phase map from test 1. **e** Reconstructed phase map from test 2. **f** Reconstructed phase map from polar coordinate configuration. **g** Phase error map from reconstruction using test 1 parameters. **h** Phase error map from reconstruction using test 2 parameters. **i** Phase error map from reconstruction using polar coordinate configuration.

an approximately linear trend. The reconstructed phase map from this shear selection showed that some of the frequencies were not being reconstructed correctly, which is a consequence of the ambiguity that arises from the measured data. The reconstruction results are shown in Fig. 10d, g. When the selection of shears was scattered across one quadrant (Fig. 10b) as in test 2, the reconstruction algorithm had sufficient data to reconstruct most of the point-source wavefield, as shown in Fig. 10e. However, the error map in Fig. 10h shows a curvature, indicating the possibility of some missing frequencies. This slowly varying error is critical as it produces a so-called form error that is difficult to filter out when observing aperture size measurement objects (e.g., mirrors or lenses under test). This error may be less critical when the sample is small (e.g., in the case of a cell under a microscope). The shear selection shown in Fig. 10c is based on a variable shearing axis but with a fixed shear amount. The latter system allows accessing all four quadrants, and the shears can be selected at all possible rotation angles of the shear axis. The reconstruction is shown in Fig. 10f. The error map shown in Fig. 10i shows that the error map is dominated only by high-frequency noise patterns, which can be filtered more easily. These results demonstrate that selecting shears across all quadrants results in better reconstruction because it provides more information for a successful reconstruction.

Transfer function analysis (see also Ref. 22), to estimate frequency domain information densities

To better understand the results and characteristics features in the error maps of Fig. 10, further study was done using transfer function analysis to investigate the effectiveness of the selected shear sets to transmit information for wavefield reconstruction. Extensive studies have been done by Falldorf[25] for non-symmetric shears and Servin[22] for symmetric shears in understanding the frequency response of spatially shearing systems. Inspired by Servin's description[22] for the transfer function, we define the frequency information density as

$$T(k_x,k_y)=\sum_{n=1}^{S}\sin^2\left[2\pi\vec{k}.\vec{ds_n}\right] \qquad (15)$$

where $\vec{k}=[k_x,k_y]$ is the frequency coordinate and $\vec{ds}=[ds_x, ds_y]$ is the shear in x- and y- directions. This transfer function analysis estimates the "frequency information density" at each spatial frequency coordinate for a given set of shears. These functions were evaluated and mapped for all three configurations as shown in Fig. 11.

Figs. 11a–c are the frequency domain information

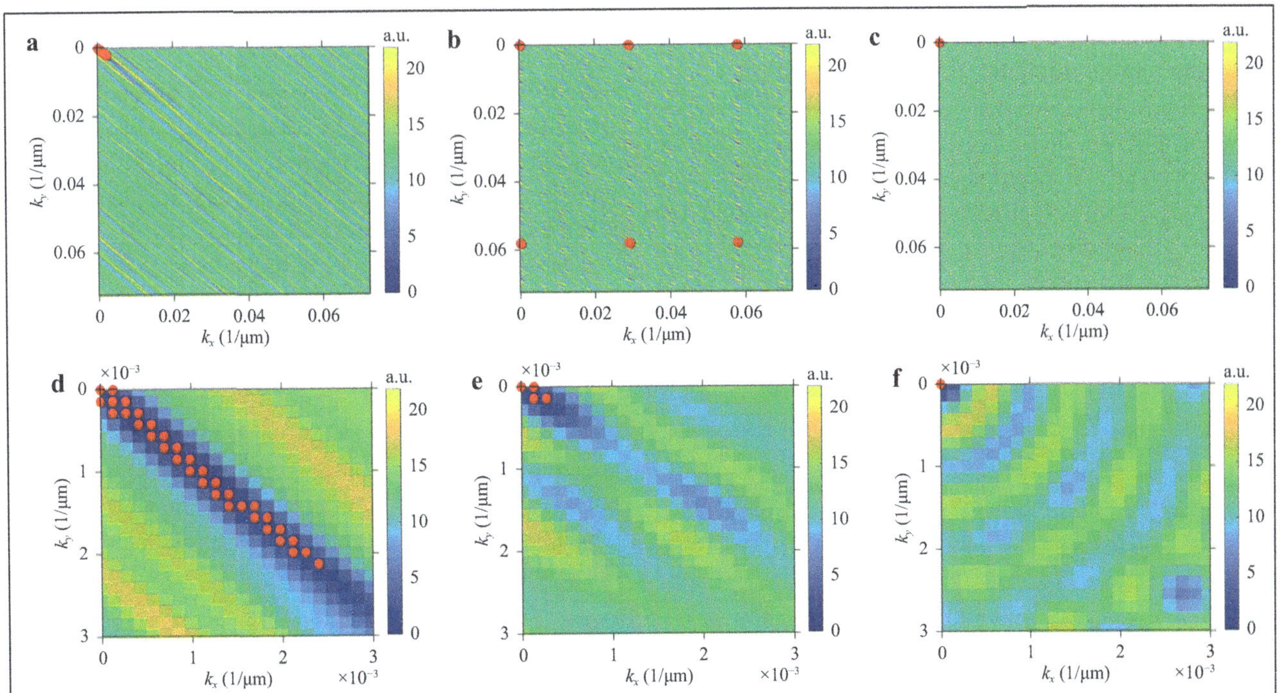

Fig. 11 Transfer function analysis (Frequency information density plots with points below threshold 2): **a** Test 1in cartesian coordinate configuration (34 datapoints below threshold). **b** Test 2 in cartesian coordinate configuration (21 datapoints below threshold). **c** Polar coordinate configuration (1 datapoint below threshold). **d** Test 1 plot in cartesian coordinate configuration zoomed near small frequency regions. **e** Test 2 plot in cartesian coordinate configuration zoomed near small frequency regions. **f** Polar coordinate configuration plot zoomed near small frequency regions.

density maps for the three shear sets described in Fig. 10. Figs. 11d–f are the same maps but zoomed in near low spatial frequencies to visualize better the behavior at these regions, where regions below the threshold of 2 are highlighted in red. Fig. 11d shows a significantly high amount of frequency sets that have transfer function values below the threshold. These cluster appear to dominate in regions where kx ≈ ky. This indicates that a large number of frequencies along the 45-degree angle will have poor reconstruction in the estimated wavefield at longer spatial wavelengths. This conclusion is in agreement with Fig. 10g. A similar behavior is observed in Fig. 11b, e, however, the number of frequency sets with transfer function values below the threshold of 2 are located at both low and high frequencies, with a minimal presence near low spatial frequencies. The locations of these cluster indicate that very low spatial frequencies will appear in the error map but will have better reconstruction than the previous shear set. This is observed in Fig. 10h, but there are also high frequency errors present due to low values of the transfer function at higher spatial frequencies. The polar configuration in Fig. 11f has no transfer function values below the threshold of 2, except for the (trivial) frequency at (0,0). This indicates good reconstruction of the estimated wavefield at all spatial frequencies.

Spatial support analysis (see also Ref. 22), to estimate spatial domain information densities

The following section demonstrates how spatial overlaps between different shears impact the reconstruction of the final wavefield. When an object wavefield passes through the gratings, it splits in two, represented as beam 1 and 2 in Fig. 12a. Depending on the shear amount, there is an overlap region between the two beams (overlap region is indicated in yellow). Since the overlap region produces an interference pattern, the phase information from the overlap region feeds into the algorithm. Fig. 12b shows which areas of the input beam provide phase information to the reconstruction algorithm (regions indicated in yellow).

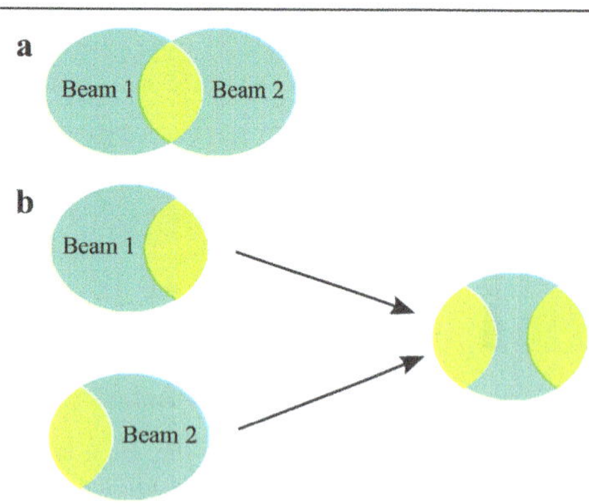

Fig. 12 Demonstration of data recorded by each shear: **a** Beam 1 and 2 overlapping. **b** Summation of overlapped regions indicating regions that have data for reconstruction (yellow regions).

This idea is further extended to multi-shear interferometry in Fig. 13. Fig. 13a shows the overlap regions from an object beam with a diameter of 600 pixels sheared by 175 pixels along the x-direction. Fig. 13b shows the overlap regions if a second shear is added to the measurement in the orthogonal direction by the same amount. In Fig. 13b the yellow areas indicate regions with two overlaps, green areas indicate regions with one overlap, and the dark blue areas indicate regions with no overlap. This idea is extended over 24 shears in Fig. 13c, showing a significantly higher count of overlaps (or number of effective measurements). This case also shows an example of a situation where the algorithm will not completely reconstruct all spatial frequencies near the wavefront's center because the chosen fixed shear is too large (175 pixels). Fig. 13d shows an alternative set of 24 shears (i.e., with a small shear amount compared to (c) with 80.5pixels). There, all of the spatial locations have been measured at least 20 times. The case presented in Fig. 13d has the highest potential to successfully reconstruct the wavefield due to maximum overlap across the whole area.

It should be noted that (as discussed by Servin[22]) the wavefront may be reconstructed even at regions where there is no overlap. This phenomenon is especially true for lower spatial frequencies because they do not need to be sampled everywhere. However, the reconstruction is increasingly difficult at higher spatial frequencies at a given location if no interference exists.

The same concept is applied to the three configurations / shear sets discussed in Fig. 10 and the overlap distribution is mapped in Fig. 14. Fig. 14a, b show fewer overlaps along the boundary at 45-degree angle. The number of effective measurements drops from 24 to 13. On the contrary, Fig. 14c shows uniform distribution of overlaps covering the majority area of the object beam. At the border, there are still 20 effective measurements recorded, which indicates better reconstruction across the whole area of the object.

Effect of pixel shear errors on the resolution of the reconstruction (Monte Carlo simulation)

Any error present in the estimated shear value propagate through the reconstruction process and result in an error in the final reconstructed wavefield. This mismatch between actual and assumed shear amounts can result in error artifacts in the reconstructed wavefield or missing

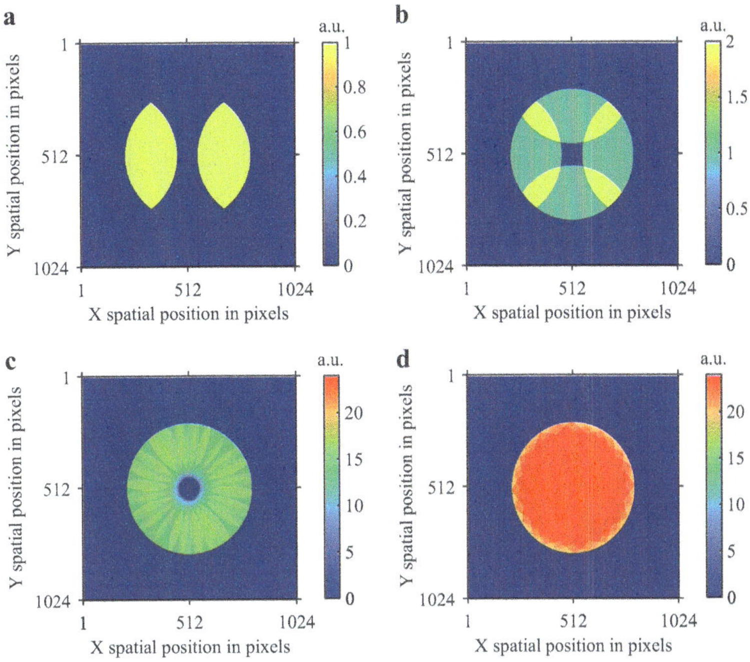

Fig. 13 Spatial information density distribution for different shear patterns on 1024 × 1024 pixel maps with aperture diameter of 600 pixels. **a** Single shear of 175 pixels. **b** Two orthogonal and equal shears of 175 pixels. **c** 24 shear in polar configuration with a fixed shear of 175 pixels. **d** 24 shear in polar configuration with a fixed shear of 80.5 pixels.

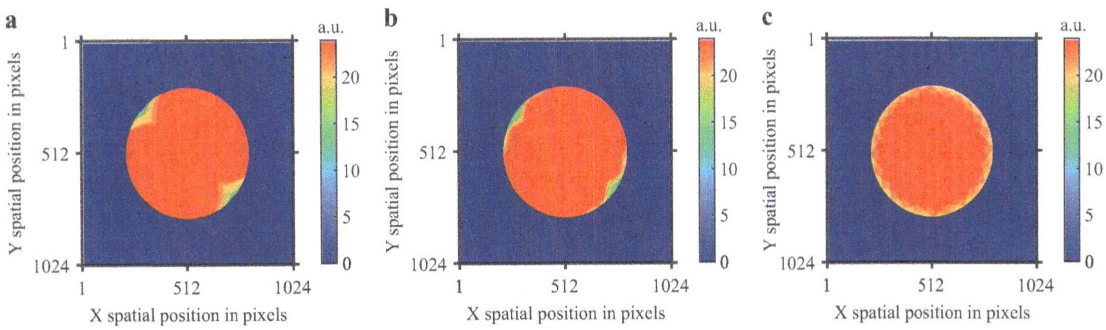

Fig. 14 Spatial information density distribution across all three configurations with 24 shears. **a** Test 1 in cartesian coordinate system (linear pattern). **b** Test 2 in cartesian coordinate (shears in scattered pattern). **c** Polar coordinate system with shears at unequal angles.

frequencies from the original wavefield. Monte-Carlo simulations were carried out to evaluate the effect of errors in the estimated shear pixels in reconstructing the wavefield by adding normal distributed random errors to the shear values in X and Y direction for the shear configuration in Fig. 10c (100 simulations for each case). Furthermore, a total of 20% intensity noise (of the peak value) has been added to all interferograms to include the effects of noise in realistic measurement conditions.

This series of simulations provides an overview of how the phase errors in reconstructed wavefields are affected by the imperfect shears. The results of the series of simulations have been summarized in a box plot, as shown in Fig. 15. As expected, when observing the RMS error in the reconstructed phase, there is a clear correlation with the increasing amount of shear error. However, it is interesting to notice that the different spatial frequencies are affected in different ways; as the standard deviation of the shear error increases, the ability of the algorithm to reconstruct higher spatial frequencies decreases. Figs. 15b–d show the measurement of a USAF target where the Monte Carlo simulation used a shear error of 1-, 3-, and 5-pixels standard deviation, respectively.

Additional processing of measurement data needed for experimental data

Shear detection

The reconstruction algorithm requires the shear amount

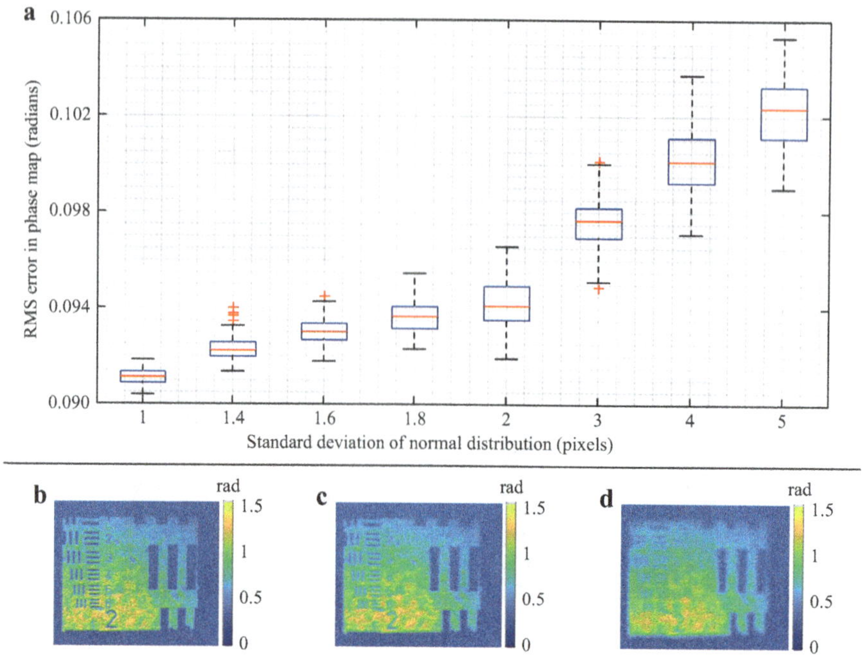

Fig. 15 **a** Box-plot showing a summary of the simulations. **b** Reconstructed phase map at standard deviation 1 pixel. **c** Reconstructed phase map at standard deviation 3 pixels. **d** Reconstructed phase map at standard deviation 5 pixels.

to be known. As concluded from the Monte Carlo simulations, the better the estimation of the shear amount, the lower the reconstruction error and the higher the spatial resolution. In this work, we have estimated the shear amount by calculating the auto-correlation of the recorded intensity maps. To increase the robustness of this estimation, we have removed the sinusoidal fringe modulation by averaging all four measured phase-shifted intensity patterns. One examplary measurement is shown in Fig. 16a. The second-order derivative of the auto-correlation map has been computed along one direction to further enhance the peak in the auto-correlation map. The resulting spikes for the case in Fig. 16a are shown in Fig. 16b, and they are prominently identifiable. The second-order derivative of the auto-correlation map comprises three peaks: a center peak and a peak on either side of the center peak, which is located at twice the value of the actual positive and negative shear amounts. The shear amount is estimated by tracking one of the side peak positions to the center peak. The difference in X and Y coordinates of the side and center peak gives twice the shear value for each averaged intensity map. The actual shear can be detected down to an accuracy of 0.5 pixels.

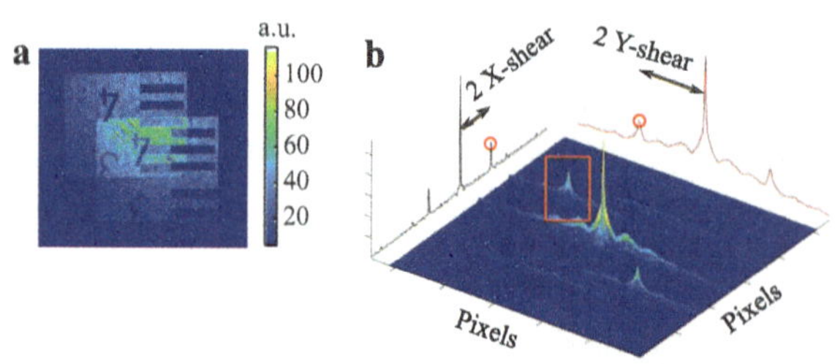

Fig. 16 **a** Intensity map with sheared images of object. **b** Second derivative of auto-correlation map showing peaks corresponding to shear in X and Y direction.

Additional compensation techniques for the polarization camera

When working with left and right circular polarized waves, it is convenient to employ pixelated polarization cameras to record four phase-shifted intensities simultaneously. The polarization camera enables single-shot phase-shifting during the measurement process for each shear. The single-shot phase-shifting capability offers an advantage in reducing the number of translation axes, enhanced stability, and mitigation of alignment errors. The reduction of the number of translation axes further ensures repeatability.

Commonly, each super-pixel on the detector of the polarization camera has four pixels that record the image at four different polarizer angles. As the two incident waves have a left-circular and right-circular polarization, it is possible to obtain simultaneously four interferograms at four different phase shifts. The orientation of this pixelated array is shown in Fig. 17, which also shows an example of a defocused point source measured at a single fixed shear

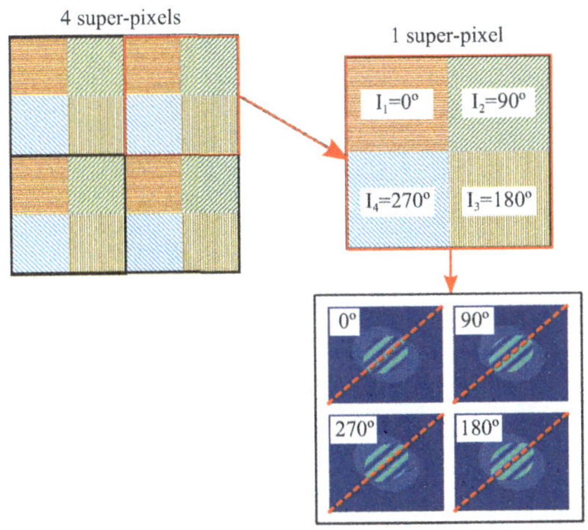

Fig. 17 Schematic of the polarization camera sensor and intensity maps demonstrating the phase shifts.

with 0°, 90°, 180°, and 270° phase shifts. The recorded intensities have a resolution of 1024 × 1024 pixels.

One critical correction is needed before the phase can be calculated. The individual frames are often clustered in a super-pixel and can be separated without difficulty. However, all four frames are not located at the same spatial position. An additional interpolation is needed (e.g., including the Fourier shift theorem) to obtain all four interferograms at the same spatial location. In this work, we have compensated this spatial mismatch for the interferograms I2, I3, and I4 (but not I1), to obtain an interferogram that is shifted by 1/2 super-pixels.

Experimental setup

Cartesian Coordinate Shear configuration

Fig. 18 shows the experimental setup for the Cartesian coordinate shear setup, along with a schematic showing the arrangement of optical components for the setup. The light source illuminating the object is a class 3B, 520 nm wavelength, 30 mW green laser. The object used for the experiments is a USAF target. Two sets of lenses with 4f configuration were used to keep the aperture 1 and the object in focus. Aperture 2 was used to limit the effects of stray light, back reflections, and the higher-order diffraction orders. A polarization camera was used to capture the four phase-shifted interferograms simultaneously (FLIR Blackfly-S model BFS-U3-51S5P-C). The grating arrangement is shown in detail in Fig. 19.

The cartesian coordinate shear system in Fig. 19a has four polarization gratings, two on the left-hand side (blue) to shear the wavefield in the vertical direction (y-direction), and the two on the right-hand side (green) to shear the wavefield in the horizontal direction (x-direction). Two leadscrew stages are connected to the two outermost X and Y polarization gratings which translate along the Z-axis to create the shears. The leadscrew stages utilize stepper motors and encoders to accurately determine the stages' position and limit switches for homing the system. Additive manufacturing (3D printing) was used to construct the stage, and accompanying electrical circuitry and software were developed for the closed-loop control system. The two inner gratings are fixed. Since only one

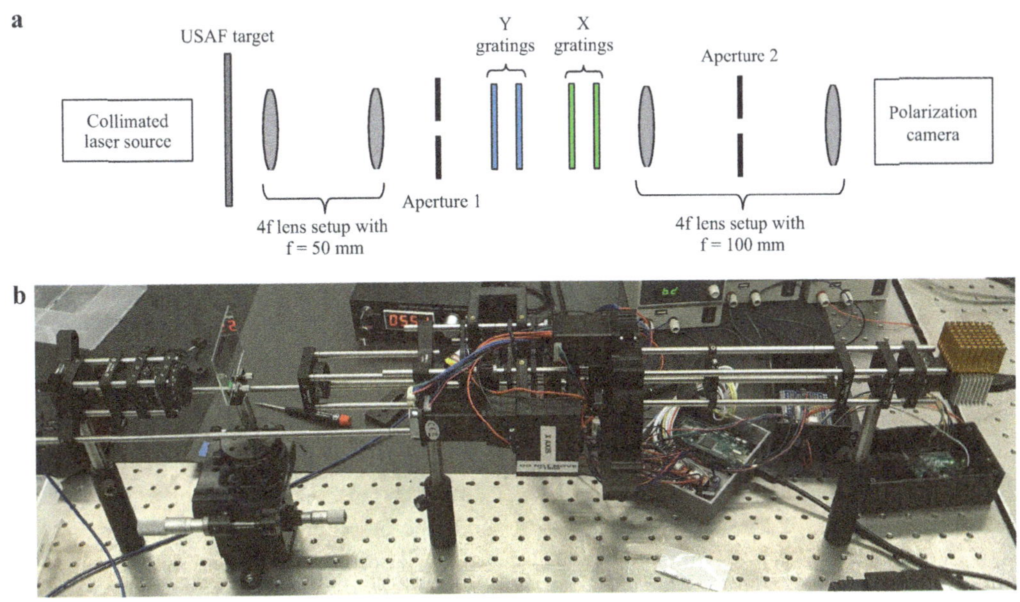

Fig. 18 Experimental setup for Cartesian coordinate shear setup **a** schematic **b** experiment.

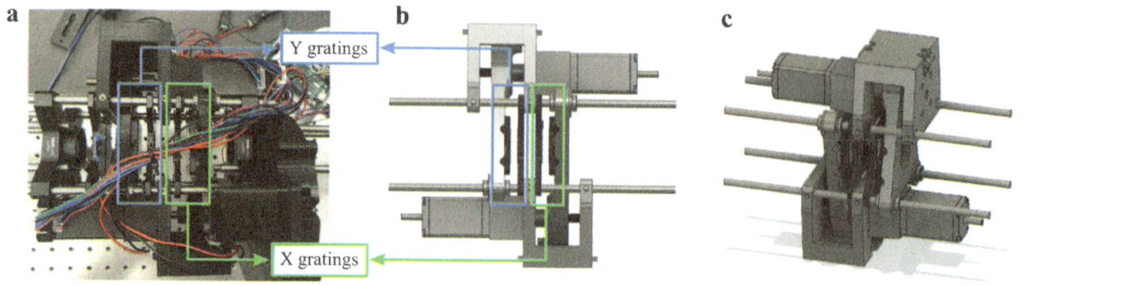

Fig. 19 Grating arrangement in Cartesian coordinate shear setup **a** experimental setup **b** schematic of the setup **c** 3D schematic of the setup.

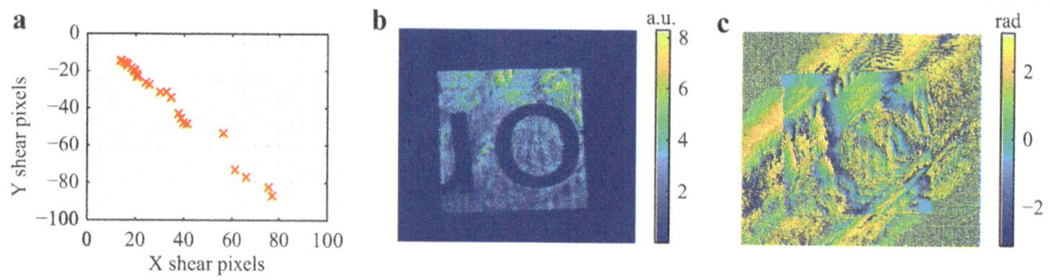

Fig. 20 Reconstructed results from cartesian coordinate shear experiment **a** shear selection **b** amplitude map **c** phase map.

grating can move in each grating pair, the shears are limited to only one quadrant (see Fig. 10a, b).

An experiment was carried out to validate the previous findings regarding the shear configuration in Fig. 10a. The results for the first configuration are shown in Fig. 20, where a total of 24 different shear amounts have been applied (see Fig. 20a) using an instrument that can independently adjust the x- and y- shears. The object used in this experiment is a simple section of an amplitude USAF target, where the remaining sections are blacked out due to aperture 1 (see Fig. 18a).

The shears were estimated from experimental data using the auto-correlation method, as discussed in the previous section. Once the shears were estimated, they were given as input to the algorithm to reconstruct the wavefield. A multi-resolution technique[10] was implemented within the reconstruction algorithm to mitigate the effects of initial guesses and improve the convergence. For our dataset with 1024 × 1024 pixels, a 5-level multi-resolution reconstruction starting from the coarsest grid with 64 × 64 pixels was implemented. The estimated wavefield is then up-sampled by a factor of 2, and the reconstruction process happens at a higher resolution (i.e., 128 × 128 pixels) using the newly up-sampled wavefield as the initial guess. This process repeats until the estimated wavefield reaches the original resolution, i.e., 1024 × 1024 pixels. The results for this 5-layer multi-resolution approach are shown in Fig. 20.

The reconstructed amplitude map (Fig. 20b) shows the object features that would be expected. However, the phase map in Fig. 20c shows the presence of parasitic signals. This reconstruction is expected given the outcome of previous simulations, see Fig. 10. This is a direct consequence of the shear selection: most of the shears selected for this experiment are dominant along the -45° angle. This selection has generated parasitic signals along the 45° angle, as seen in Fig. 20c.

Constant shearing Polar coordinate shear configuration

The other shear configuration that produced consistently better results was the configuration of a fixed shear with a variable shear axis, see Fig. 10c. An experimental implementation is shown in Fig. 21, where the xy-shear system in Fig. 18 is replaced with a simple two grating set up on a rotational mount. Fig. 22 shows a more detailed view of the shearing mechanism. Notably, the 2-grating setup can be realized in a more compact optical setup; however, we chose to keep the 4f configuration to compare the two setups better. The results for the configuration in Fig. 21, using 24 different shear amounts, are shown in Fig. 23. Similar to the case of Fig. 20, a portion of the USAF target has been selected for the measurement, and the field of view is limited by aperture 1 (see Fig. 21).

Fig. 23a, b show that the solver successfully reconstructs both the amplitude and the phase map, free of parasitic fringes. For better visualization, a mask has been generated (see Fig. 23c) using the amplitude information and applied to the phase data (see Fig. 23d).

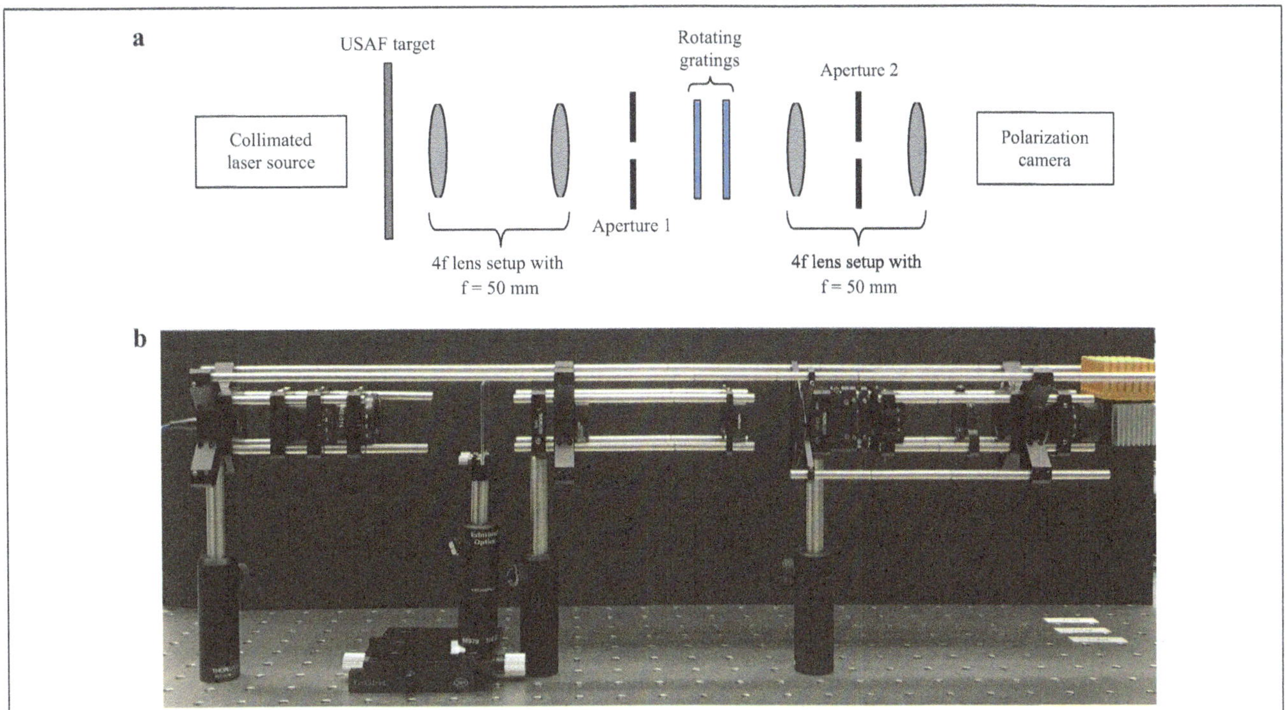

Fig. 21 Experimental setup for polar coordinate shear system **a** schematic, **b** experiment.

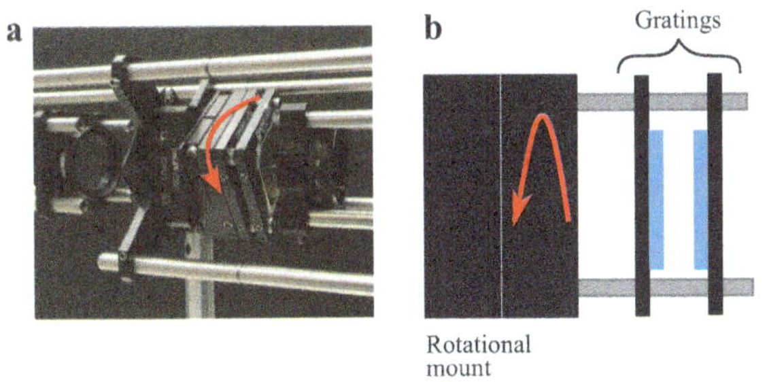

Fig. 22 Gratings arrangement in polar coordinate shear configuration **a** Experimental setup showing the rotation of gratings. **b** Schematic of the grating arrangement.

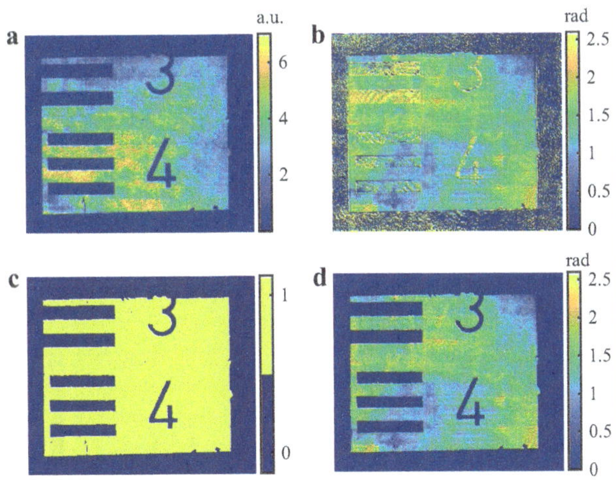

Fig. 23 Reconstructed results **a** amplitude map **b** phase map **c** mask from amplitude map **d** phase map with mask applied.

A further experiment has been conducted to verify that the instrument in Fig. 21 can capture and reconstruct the wavefield at a defocused distance. For this purpose, we shift the camera's position by a distance of ~170 mm, so that both the object and the aperture get defocused by the same amount.

The reconstructed wavefield is shown in Fig. 24a, c. After reconstruction, the wavefield was numerically refocused to the visually best reconstruction distance (z=162.94 mm) using the Angular Spectrum method[26,27]. The results of the refocused wavefield are shown in Fig. 24b, d. For better visualization, various regions of the reconstructions have been zoomed in and are presented in Fig. 25.

Materials and methods

The light source used for the experiment is an FC-520-040-PM laser system (520 nm, 30 mW) from CivilLaser. The object sample is a combined USAF 1951 and Dot Grid Target (Edmund Optics #62-465). The GP gratings have been purchased from Edmund Optics (#12-677, 96% diffraction efficiency at 550nm, 159grooves/mm, 25 × 25 × 0.45 mm). The light illuminating the object is linearly polarized to produce 50% RHCP and 50% LHCP at the GP grating, maximizing the fringe visibility. The camera used for imaging is a FLIR Blackfly-S model BFS-U3-51S5P-C polarization camera. The lenses used for the 4f-configuration have an effective focal length of 50 mm.

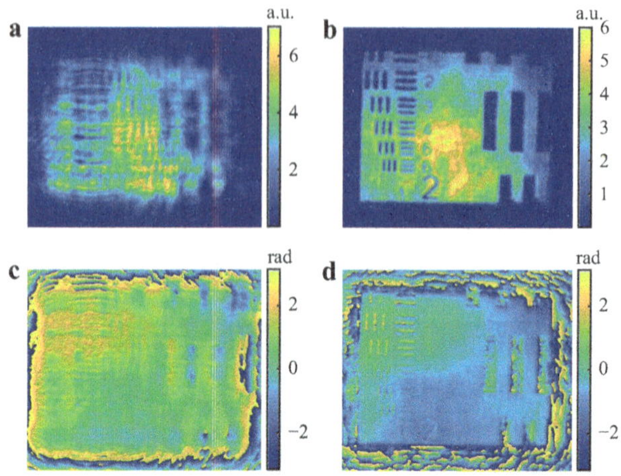

Fig. 24 Reconstructed results from defocused object **a** amplitude map before numerical refocusing. **b** Amplitude map after numerical refocusing. **c** Phase map before numerical refocusing. **d** Phase map after numerical refocusing.

Conclusion

In this work, a GP grating-based shearing holography system has been presented. The GP gratings used in this work have the unique ability to have a polarization-sensitive diffraction angle, where the RHCP and LHCP light beams are diffracted in two different directions. Figs. 1–3 show how these properties can be exploited with a linear polarizer to create an adjustable shearing interferometer. This paper presented two distinct configurations, (i) one that allows for adjustable xy-shearing, and (ii) one that keeps the shear amount constant but rotates the shear axis. Another aspect of the GP grating-based shearing interferometer is that a series of phase-shifted interferograms can be obtained by either rotating the polarizer (see Fig. 4) or using a pixelated polarization camera (see Fig. 17).

The ability to produce a series of phase-shifted interferograms obtained under various shear amounts opens the possibility to obtain high fidelity data that can be processed in a wavefront reconstruction algorithm. In this work, we have implemented the approach of Konijnenberg[16] outlined in Fig. 5 but added a filtering procedure to remove algorithm-specific artifacts (see Fig. 6d, Figs. 7–9). We further investigated the effect of the shear selection. Fig. 10 showed that choosing the shears to be on a circle in the sx-sy plane provides consistently good results, where the error is of high-frequency nature. These findings are supported using Servin's[22] transfer function analysis and spatial domain support that were referred to as "frequency domain information density" (see Fig. 11) and "spatial domain information density" (see Fig. 14). One critical step is the estimation of the exact shear amount. Any discrepancy between the actual shear amount and the values used within the wavefront reconstruction algorithms will increase, to some degree, the RMS in the reconstructed phase, see Fig. 15. However, the results in Fig. 15 also show that an imperfect shear estimate value has a strong impact on the resolution of the reconstructed wavefield (even if the RMS increases only marginally by 10%, for the case of typical levels of intensity noise). To ensure accurate results, we have reported a shear estimation technique using the second derivative of the autocorrelation function, where the spikes of that function estimate shear amount down to an accuracy of 0.5 pixels, as shown in Fig. 16. The reconstruction

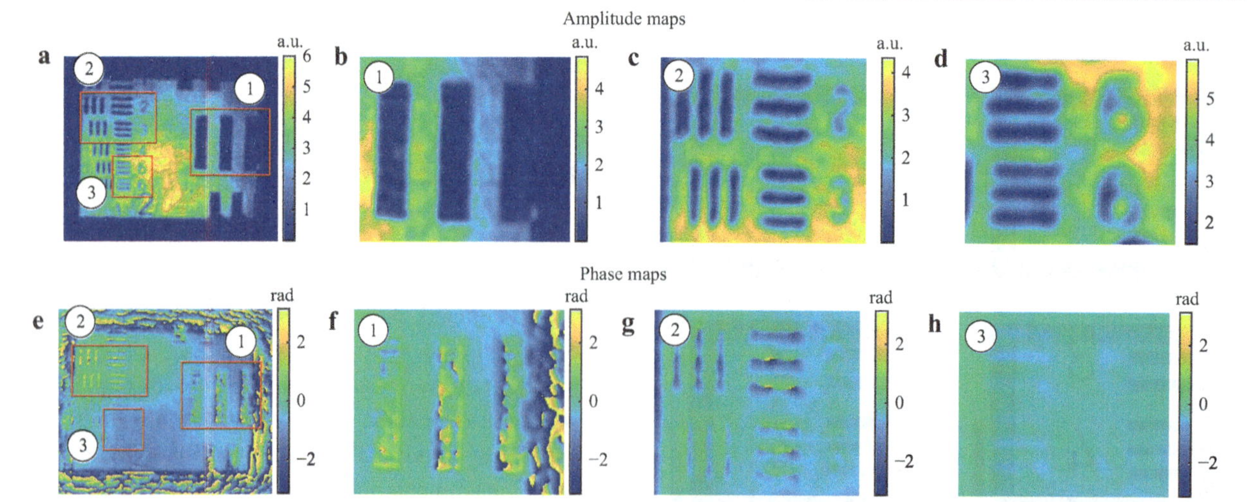

Fig. 25 Reconstructed wavefield of defocused object (zoomed) after post processing: **a** Amplitude map of the full object. **b** Region 1 amplitude map. **c** Regions 2 amplitude map. **d** Regions 3 amplitude map. **e** Phase map of the full object. **f** Region 1 phase map. **g** Region 2 phase map. **h** Region 3 phase map.

algorithm presented in this paper does not limit the possible applied shears. The algorithm can process sub-pixel shearing as the algorithm primarily operates on FFTs. However, the shear detection technique presented in this paper is limited in identifying shears. The procedure detects twice the values of shears for a sheared interferogram as integer, and hence the shears can be estimated down to integer multiples of 0.5 pixels. More sophisticated shear detection algorithms (using weighted mass algorithms or techniques based on ideas from Guizar-Sicairos M. et al[28].) might enable finer shearing detection down to 0.01 pixels and are subject to future research.

The previous aspects have been verified experimentally for both shearing configurations, as shown in Fig. 18,19 for the xy-shearing component and Fig. 21,22 for the configuration that rotates the shear axis. As expected, the system configuration that keeps both gratings fixed while the grating pair rotates about the optical axis provides the visually best reconstruction result with no parasitic fringes, as shown in Fig. 20,23. A further advantage of the setup with the variable shear axis is that it requires only two GP gratings and can be, in principle, incorporated into a more compact setup. There are also several motorized rotating mounts available that could automate this process.

Finally, we demonstrate the feasibility of this approach by reconstructing the wavefield of a defocused object at the distance z = 162.94 mm. Refocusing using the Angular Spectrum method[26,27] produces Fig. 24,25.

These findings show the advantage of GP grating-based shear interferometers compared to conventional dual grating (or dual double-grating) configurations. A conventional dual-axis grating-based variable shearing interferometer with phase shift capabilities requires two independent axial shifts of the x- and y- gratings (to adjust the shear) and lateral shifts of the gratings in x- and y-direction (for phase-shifting). Having to control four mechanical axes independently is a significant system complexity, where the freedom to adjust all parameters appropriately is often compromised due to space constraints. In contrast, the results of this work show that a GP grating configuration with a fixed shear amount, but an adjustable rotational shear axis can provide the shear selections for robust measurements while maintaining phase-shifting capabilities using a pixelated polarization camera. In other words, robust variable shearing interferometry can be made possible by controlling only one mechanical axis (instead of four). This finding has implications for future compact interferometer designs.

Acknowledgements

The authors acknowledge the support and funding from the Center for Precision Metrology (CPM).

Author contributions

K Falaggis is the Principal Investigator and proposed the idea and initiated the project. P Shanmugam made the codes, conducted numerical simulations, performed experimental setup assembly, conducted experiments, and processed the data. A Turley fabricated components for the experiments, performed electric circuit design and connections. A Light assisted in the experiments and generated visualizations for the manuscript. A Turley and A Light contributed equally to this manuscript. The manuscript was discussed by all authors and written by P Shanmugam and K Falaggis.

Conflict of interest

The authors declare that they have no conflict of interest.

Received: 16 September 2021 Revised: 08 February 2022 Accepted: 15 February 2022
Accepted article preview online: 21 February 2022
Published online: 25 March 2022

References

1. Nugent, K., Paganin, D. & Gureyev, T. A phase odyssey. *Physics Today* **54**, 27-32 (2001).
2. Guigay, J. P., Langer, M., Boistel, R. & Cloetens, P. Mixed transfer function and transport of intensity approach for phase retrieval in the Fresnel region. *Optics Letters* **32**, 1617-1619 (2007).
3. Falaggis, K., Kozacki, T. & Kujawinska, M. Optimum plane selection criteria for single-beam phase retrieval techniques based on the contrast transfer function. *Optics Letters* **39**, 30-33 (2014).
4. Martinez-Carranza, J., Falaggis, K. & Kozacki, T. Multi-filter transport of intensity equation solver with equalized noise sensitivity. *Optics Express* **23**, 23092-23107 (2015).
5. Reed Teague, M. Deterministic phase retrieval: a Green' s function solution. *Journal of the Optical Society of America* **73**, 1434-1441 (1983).
6. Almoro, P., Pedrini, G. & Osten, W. Complete wavefront reconstruction using sequential intensity measurements of a volume speckle field. *Applied Optics* **45**, 8596-8605 (2006).
7. Zalevsky, Z., Mendlovic, D. & Dorsch, R. G. Gerchberg-Saxton algorithm applied in the fractional Fourier or the Fresnel domain. *Optics Letters* **21**, 842-844 (1996).
8. Fienup, J. R. Phase retrieval algorithms: a comparison. *Applied Optics* **21**, 2758-2769 (1982).
9. Fienup, J. R. Phase retrieval algorithms: a personal tour [Invited]. *Applied Optics* **52**, 45-56 (2013).
10. Falaggis, K., Kozacki, T. & Kujawinska, M. Accelerated single-beam wavefront reconstruction techniques based on relaxation and multiresolution strategies. *Optics Letters* **38**, 1660-1662 (2013).
11. Almoro, P. F., Maallo, A. M. S. & Hanson, S. G. Fast-convergent algorithm for speckle-based phase retrieval and a design for dynamic wavefront sensing. *Applied Optics* **48**, 1485-1493 (2009).
12. Falaggis, K., Kozacki, T. & Kujawinska, M. Hybrid single-beam reconstruction technique for slow and fast varying wave fields. *Optics Letters* **40**, 2509-2512 (2015).
13. Falldorf, C., von Kopylow C. & Bergmann, R. B. Wave field sensing by means of computational shear interferometry. *Journal of the Optical Society of America A* **30**, 1905-1912 (2013).
14. Falldorf, C., Agour, M. & Bergmann, R. B. Advanced wave field sensing using computational shear interferometry. Proceedings of SPIE 9204, Interferometry XVII. San Diego: SPIE, 2014.

15. Servín, M., Quiroga, J. A. & Padilla, J. M. Fringe Pattern Analysis for Optical Metrology: Theory, Algorithms, and Applications. (Weinheim: Wiley-VCH Verlag GmbH & Co. KGaA, 2014).
16. Konijnenberg, A. P. et al. Phase retrieval algorithms for lensless imaging using diffractive shearing interferometry. *Journal of the Optical Society of America A* **37**, 914-924 (2020).
17. Blain, P. et al. An in-line shearography setup based on circular polarization gratings. *Optics and Lasers in Engineering* **51**, 1053-1059 (2013).
18. Ch oi, K., Y im, J. & Min, S. W. Achromatic phase shifting self-interference incoherent digital holography using linear polarizer and geometric phase lens. *Optics Express* **26**, 16212-16225 (2018).
19. Alemán-Castaneda, L. A. et al. Shearing interferometry via geometric phase. *Optica* **6**, 396-399 (2019).
20. Millerd, J. E. & Wyant, J. C. Simultaneous phase-shifting Fizeau interferometer. U.S. patent US7057738B2 (2003).
21. Ne al, R. M. & Wyant, J. C. Polarization phase-shifting point-diffraction interferometer. *Applied Optics* **45**, 3463-3476 (2006).
22. Servín, M., Cywiak M. & Dávila A. Lateral shearing interferometry: theoretical limits with practical consequences. *Optics Express* **15**, 17805-17818 (2007).
23. Hariharan, P. Optical Interferometry. 2nd edn. (Amsterdam: Elsevier, 2003).
24. Youla, D. Generalized image restoration by the method of alternating orthogonal projections. *IEEE Transactions on Circuits and Systems* **25**, 694-702 (1978).
25. Falldorf, C. Measuring the complex amplitude of wave fields by means of shear interferometry. *Journal of the Optical Society of America A* **28**, 1636-1647 (2011).
26. Kozacki, T., Falaggis, K. & Kujawinska, M. Computation of diffracted fields for the case of high numerical aperture using the angular spectrum method. *Applied Optics* **51**, 7080-7088 (2012).
27. Falaggis, K., Kozacki, T. & Kujawinska, M. Computation of highly off-axis diffracted fields using the band-limited angular spectrum method with suppressed Gibbs related artifacts. *Applied Optics* **52**, 3288-3297 (2013).
28. Guizar-Sicairos, M., Thurman, S. T. & Fienup, J. R. Efficient subpixel image registration algorithms. *Optics Letters* **33**, 156-158 (2008).

Official journal of the JHL 2689-9620
www.light-am.com

Article Open Access

Ultra-high speed holographic shape and displacement measurements in the hearing sciences

Haimi Tang[1,2,4,*], Pavel Psota[3], John J. Rosowski[4,5], Cosme Furlong[1,2,4,5] and Jeffrey Tao Cheng[4,5]

Abstract

The auditory system of mammals enables the perception of sound from our surrounding world. Containing some of the smallest bones in the body, the ear transduces complex acoustic signals with high-temporal sensitivity to complex mechanical vibrations with magnitudes as small as tens of picometers. Measurements of the shape and acoustically induced motions of different components of the ear are essential if we are to expand our understanding of hearing mechanisms, and also provide quantitative information for the development of numerical ear models that can be used to improve hearing protection, clinical diagnosis, and repair of damaged or diseased ears.

We are developing digital holographic methods and instrumentation using an ultra-high speed camera to measure shape and acoustically-induced motions in the middle ear. Specifically we study the eardrum, the first structure of the middle ear which initializes the acoustic-mechanical transduction of sound for hearing. Our measurement system is capable of performing holographic measurement at rates up to 2.1 M frames per second. Two shape measurement modalities had previously been implemented into our holographic systems: (1) a multi-wavelength method with a wavelength tunable laser; and (2) a multi-angle illumination method with a single wavelength laser. In this paper, we present a third method using a miniaturized fringe projection system with a microelectromechanical system (MEMS) mirror. Further, we optimize the processing of large data sets of holographic displacement measurements using a vectorized Pearson's correlation algorithm. We validate and compare the shape and displacement measurements of our methodologies using a National Institute of Standards and Technology (NIST) traceable gauge and sound-activated latex membranes and human eardrums.

Keywords: Ultra-high speed digital holography, Microelectromechanical systems (MEMS), Miniaturized Fringe Projection, Middle-ear mechanics, Shape and displacement measurements.

Introduction

The human ear is normally sectioned into three parts: the outer, middle, and inner ear, Fig. 1. The outer ear contains the auricle and external auditory canal. The middle ear is an air-filled space separated from the outer ear by the tympanic membrane (TM), or eardrum, with three ossicular bones (malleus, incus, and stapes) connecting the TM to the inner ear (cochlea). The cochlea is a fluid-filled structure that transduces acousto-mechanical motions into

Correspondence: Haimi Tang (htang3@wpi.edu)
[1]Center for Holographic Studies and Laser micro-mechaTronics, Worcester Polytechnic Institute, Worcester, MA USA, 01609
[2]Mechanical Engineering Department, Worcester Polytechnic Institute, Worcester, MA USA, 01609
Full list of author information is available at the end of the article.

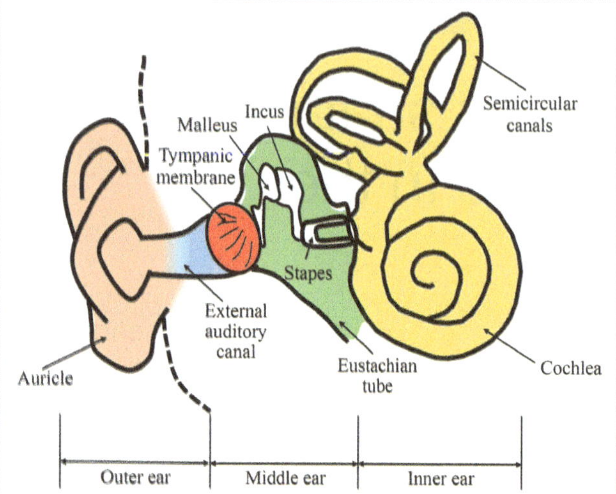

Fig. 1 Schematic of the outer ear, middle ear, and inner ear.

neural signals to the brain[1]. During normal hearing, sound travels in the form of acoustic wave in the air and enters the external auditory ear canal to vibrate the TM[2]. The vibration is then transmitted to the cochlea through the ossicles. Inside the cochlea, the sensory hair cells are set in motion to trigger nerve signals that are then received by the brain[3].

The thin semitransparent TM has unique anatomical and physical features that are critical to the hearing process[4,5]. In particular, the adult human TM has an elliptical shape, measuring 9–10 mm horizontally and 8–9 mm vertically, with inhomogeneous, multi-layered, and anisotropic material characteristics[6]. It also has a flattened conical shape 2–3 mm in depth with apex directed inward towards the middle ear cavity. Characterization of the TM's morphology and vibration induced by acoustic excitation is essential to understand middle-ear structure and function in normal ears and in ears with pathologies such as otitis media with effusion, traumatic TM perforation, formation of retraction pocket and tympanosclerosis[7–18].

We have been developing holographic methodologies utilizing ultra-high speed cameras up to 2.1 million frames per second[19,20] to quantitatively study the structure and function of the TM under different excitation conditions[21–27]. As the frame rate of the camera is inversely related to the pixel density, in this study, we choose to run the camera at 67,200 frames per second to provide a field of view of the TM with a resolution of 512 pixels by 512 pixels. The 67,200 frames per second meets the Nyquist criterion for measuring the typical hearing frequency range from 20 Hz to 20 kHz[28,29]. We have developed two high-speed techniques for TM shape measurements: the multi-wavelength technique[30] and the multi-angle technique[31]. The multi-wavelength technique uses a tunable laser source (to resolve optical phase differences to identify the shape) from a fixed angle of illumination[30]; whereas the multi-angle technique uses a single wavelength laser source but with a dynamically changing angle of illumination (to introduce optical phase changes)[31].

In this paper, we report our development of a new high-speed shape measurement approach that is based on fringe projection techniques to improve the optical efficiency and image quality during high-speed image acquisition for TM measurements. Also, because of the large amount of data that are collected during measurements, we report improved approaches to evaluate the optical phase. Our multi-wavelength and multi-angle techniques calculate the optical phase change during high-speed image acquisition using Pearson's correlation coefficient. Our previous coefficient calculations[30–33] included two levels of nested loops, which took a significant amount of the processing time. In this paper we report an optimized vectorized algorithm that removes the nested loops, greatly speeding up the Pearson's coefficient calculation.

This paper consists of introduction, results, discussion, methods and conclusion and future works sections. The physiology of the middle ear and the past developments of holography techniques on shape and vibration measurements for the middle ear research are discussed in the introduction section. The results section shows preliminary results of shape and vibration measurements of human TM, as well as National Institute of Standards and Technology (NIST) shape measurement comparison of the three shape measurement techniques. The methods section presents the principles of holographic vibration measurement utilizing Person phase correlation and shape measurement based on multi-angle and multi-wavelength configurations. Moreover, a miniaturized fringe projection is also discussed in the methods section and its schematics of the optical setups. The merits of different shape measurement techniques in applications for high-speed TM shape measurement are discussed in the discussion section. At last, the conclusion and future work are discussed.

Results

Representative shape measurements

The surface of a NIST traceable gauge was analyzed using high-speed fringe projection (HSFP) to demonstrate its working principle. A single fringe was projected and scanned across the sample's surface by the scanning MEMS (Micro-ElectroMechanical System) mirror during image acquisition. Fig. 2 shows the experiment setup combining the HSFP with the high-speed digital holography (HDH) system. More details will be presented

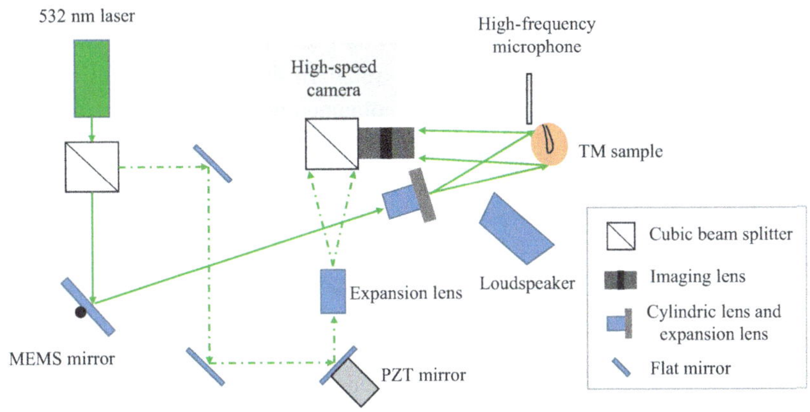

Fig. 2 Schematic of high-speed fringe projection and holographic displacement configuration. The MEMS mirror scans the sample's surface with a single fringe generated by the cylindrical and expansion lens for shape measurement. The PZT (piezoelectric ceramic material) mounted mirror is used to change the optical path for phase-shifting during displacement measurement. The middle ear sample is excited by transient click-like sounds produced by the loudspeaker. The sound wave near the surface of the TM is measured simultaneously by a calibrated broad range microphone.

in the method section. Fig. 3a shows a single projected fringe. The single fringe images were grouped to form the four-step phase-shifted images based on the phase-shifting orders, as shown in Fig. 3b. Fig. 3b also shows multiple raw fringe images $(I_1^1, I_1^2, I_1^3, I_1^4, \ldots I_q^m)$ at different MEMS mirror positions, where m is the phase-shifting order and q is the fringe order. Finally, the optical phase of the sample's shape, as shown in Fig. 3c, was obtained from the four reconstructed phase-shifted images.

The dimensions of a NIST gauge are shown in Fig. 4a. The optical phase and 3D reconstructed shape measured by the Fringe Projection method is shown in Fig. 4b; Fig. 4c is the optical phase and 3d reconstructed shape measured by the multi-angle method. Fig. 4c depicts the optical phase and 3d reconstructed shape measured by the multi-wavelength method. In all 3D reconstructed shapes, the red line traces the height (in z-axis) of the middle section of the sample, where the lowest middle segment of NIST gauge is aligned to zero (z = 0). The average measured height profiles of the two end segments of the NIST gauge measured by three methods are 0.24 mm, 0.25 mm, 0.25 mm and 1.52 mm, 1.51 mm, 1.49 mm respectively. The standard deviations (σ) for the three segments are 0.042 mm, 0.037 mm, 0.014 mm; 0.053 mm, 0.023 mm, 0.014 mm; and 0.113 mm, 0.020 mm, 0.014 mm. The mean results agree with the dimensions provided by the manufacture of 0.25 mm and 1.50 mm, respectively, but for each position the standard deviations are smaller for the multi-wavelength method. The fringe-projection technique produced the largest σs.

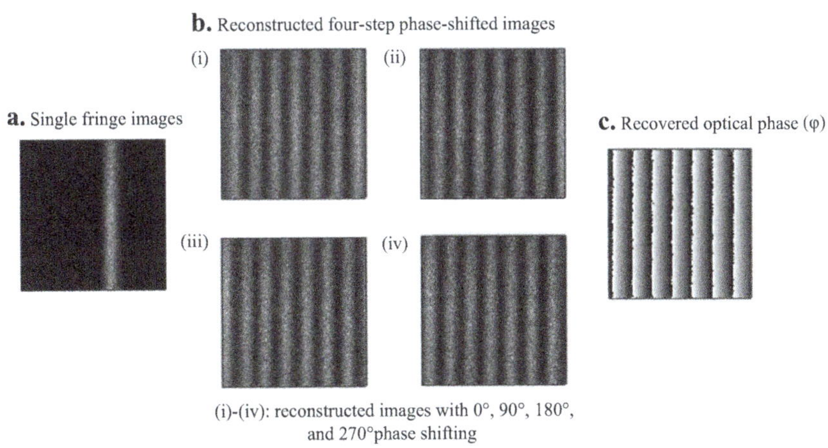

Fig. 3 Fringe projection measurement of a flat surface. **a** Raw single fringe images (I_q^m) captured by a high-speed camera, where m indicates the phase-shifting order and q indicates the fringe order. **b** Reconstruction of four-step phase-shifted images; each multi-fringe image is the summation of I_q^m of the same phase order. **c** Optical phase of the flat surface obtained by the four-step phase-shifting method.

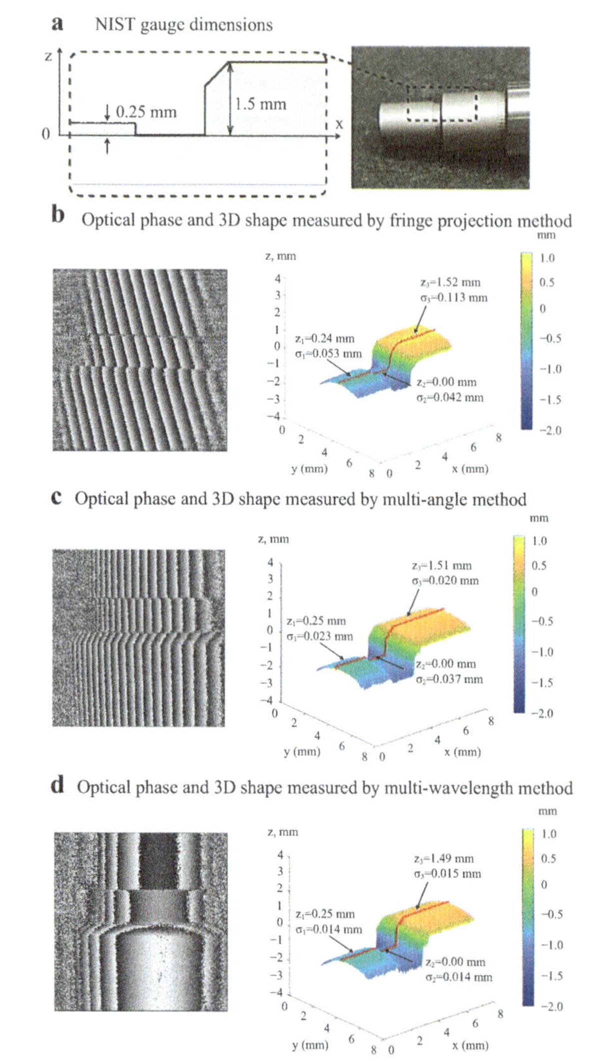

Fig. 4 Measurements of a NIST traceable gauge by three methods. **a** NIST gauge dimensions. **b** The reconstructed optical phase image and the 3D shape measurement result of NIST gauge by fringe projection method. **c** The optical phase and the 3D shape measurement of NIST gauge by the multi-angle method. **d** The optical phase and the 3D shape measurement result of NIST gauge by the multi-wavelength method. Z is the average height, and σ is the standard deviation of the given segment of each red line. All the Z_2 are aligned to the zero position of the Z direction.

The three abovementioned shape measurement methods were used to measure the shape of one of three tympanic membranes (TM 1-3) from three cadaveric human temporal bones. Fig. 5 shows the three TM shapes measured using the HSFP, multi-angle, and multi-wavelength techniques, respectively. To assess variability within these measurement methods, a smoothing filter based on a moving kernel of 5x5 pixels size was applied and the standard deviation between the raw and the smoothed shape results were computed. Fig. 5a(i) shows the wrapped optical phase of TM1 obtained by the HSFP, and Fig. 5a(ii) depicts the smoothed 3D shape profile. The standard deviation between the raw and smoothed shapes obtained with HSFP is 0.11 mm, and the measurement was done by capturing 72 images in 25 ms. Fig. 5b(i, ii) shows the shape of TM2 measured by the multi-angle shape measurement method. The standard deviation is 0.15 mm, and the measurement was performed by capturing 1200 images in about 18 ms. Two pairs of four-phase-shifted images (total of 8 images) out of the 1200 images were chosen to reconstruct the shape at the optimal fringe density for phase unwrapping. Finally, Fig. 5c(i, ii) shows the shape measurement results of TM3 obtained with the multi-wavelength method, with a standard deviation of 0.13 mm. The measurement took about 150 ms due to the limitation of the wavelength tuning speed.

Representative displacement measurements

To demonstrate the accuracy of the HDH displacement measurement, the surface displacements of a latex drum head of diameter 5 mm were measured with the HDH and compared with the displacement at the center of the drum (P1 in Fig.6) simultaneously obtained by an LDV (Polytec VFX-1-140). The latex drum head was excited using an impulsive sound generated from a loudspeaker. Fig. 6a shows a representative optical phase at 4.16 ms after the excitation. Fig. 6b compares the time-waveform of the displacement at *P1* measured using the LDV and the HDH, showing a large overlap. The correlation between the two methods was found to be greater than 99%. Fig. 6c shows the Root–Mean–Square (RMS) pattern of the vibration from 3.5 ms to 7.5 ms after the excitation, with three markers: *P1* is at the center of the drum where the LDV measures the displacement; *P2* and *P3* were chosen arbitrarily to show the lead and lag of the individual displacement on the sample's surface. Fig. 6d depicts the temporal evolution of the wavefronts of the three markers (*P1* to *P3*) obtained using the HDH.

Once validated, the HDH measured the displacements of a human cadaveric TM (TM1) induced by an acoustic click. The optical phase of the displacement was calculated using the vectorized Pearson's correlation coefficient algorithm to speed up the data processing. The vectorized algorithm's performance was compared with the previous nested-loop algorithm. Both algorithms generated the same optical phase shown in Fig. 7a. However, the vectorized algorithm took 0.17 s, whereas the nested loop algorithm took 3.54 s, using the same kernel size of 3 by 3 pixels (Fig. 7b). Clearly the vectorized algorithm outperformed the nested loop algorithm.

A previous study demonstrated that the dominant

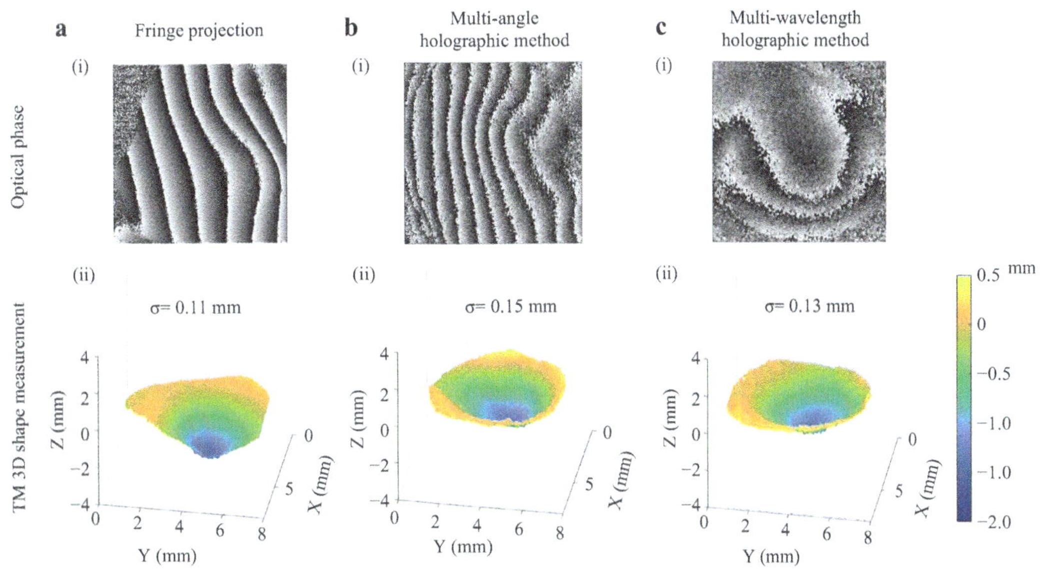

Fig. 5 Representative shape measurement results: Panel **a** is the shape result of TM1 measured by the high-speed fringe projection method, **b** is the TM2 shape measurement result by the multi-angle method, and panel **c** shows the shape measurement result of TM3 using the multi-wavelength method. The top panels (i) show optical phase maps and the bottom panels (ii) show 3D surface plots of the shape. The resolution in terms of standard deviation (σ) for the three shape measurement results is 0.11, 0.15, and 0.13 mm, respectively.

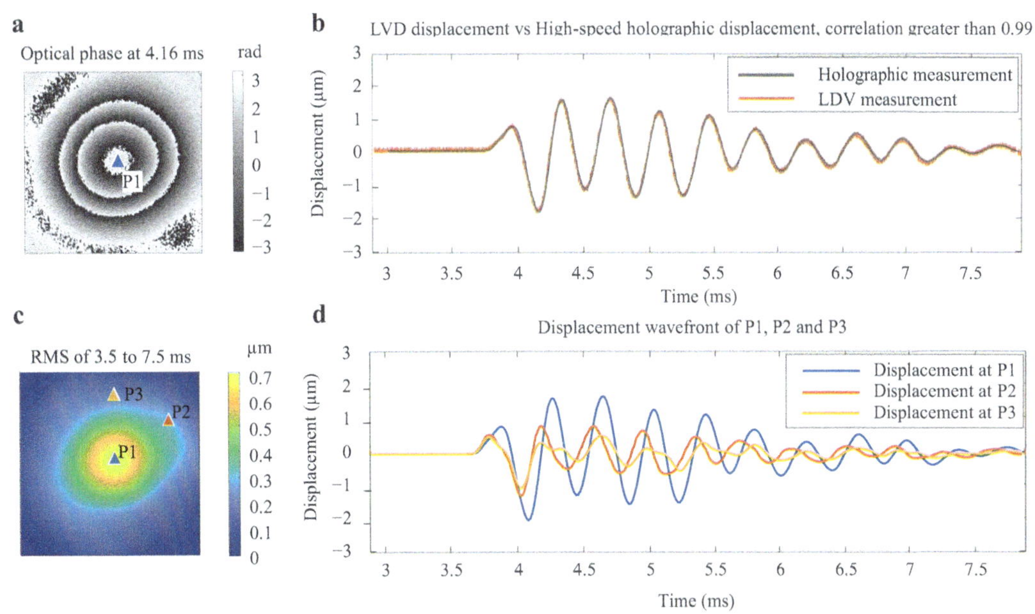

Fig. 6 High-speed displacement measurement results of a latex drum membrane and its comparison with the LDV result. The drum membrane is excited by a short acoustic click. The surface displacement of the membrane is captured by the high-speed holographic method, while the displacement of the center of the membrane is measured by an LDV. **a** The wrapped optical phase of the surface displacement at 4.16 ms after excitation. **b** Comparison of the displacement at the membrane center (P1 in **c**) measured by the LDV and the high-speed holographic method. **c** Root–Mean–Square (RMS) of the displacement 3.5 to 7.5 ms after excitation. The RMS is calculated based on the holographic displacement measurements along the temporal axis **d** Three individual displacement waveforms are marked in **c** from the holographic measurement results. The first 3 ms of the experiment is for the PZT phase shifter to scan the optical path to provide a 90-degree phase-shifted image, while the excitation starts at the 3.72 ms[30]

displacement of the TM is perpendicular to its surface[32]. Therefore, the TM shape information was used to calculate the TM's surface normal vectors, which were combined with the holographic displacement measurements to determine the surface normal displacement. As a representative example, Fig. 8 shows the normal

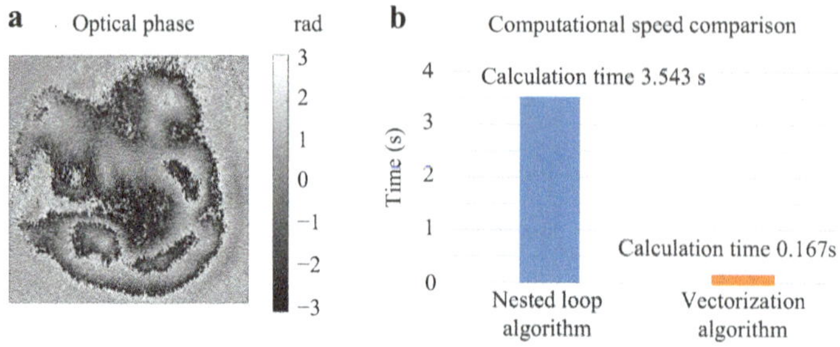

Fig. 7 Comparison of Pearson's correlation coefficient calculation of Optical Phase(OP) from a TM's High-Speed Displacement (HSD) measurement. **a** The OP of HSD measurement was obtained by the Nested Loop Algorithm (NLA) and the Vectorization Algorithm(VA). **b** the comparison of the computational speed shows the NLA takes 3.54 s and the VA takes 0.17 s to calculate the same OP.

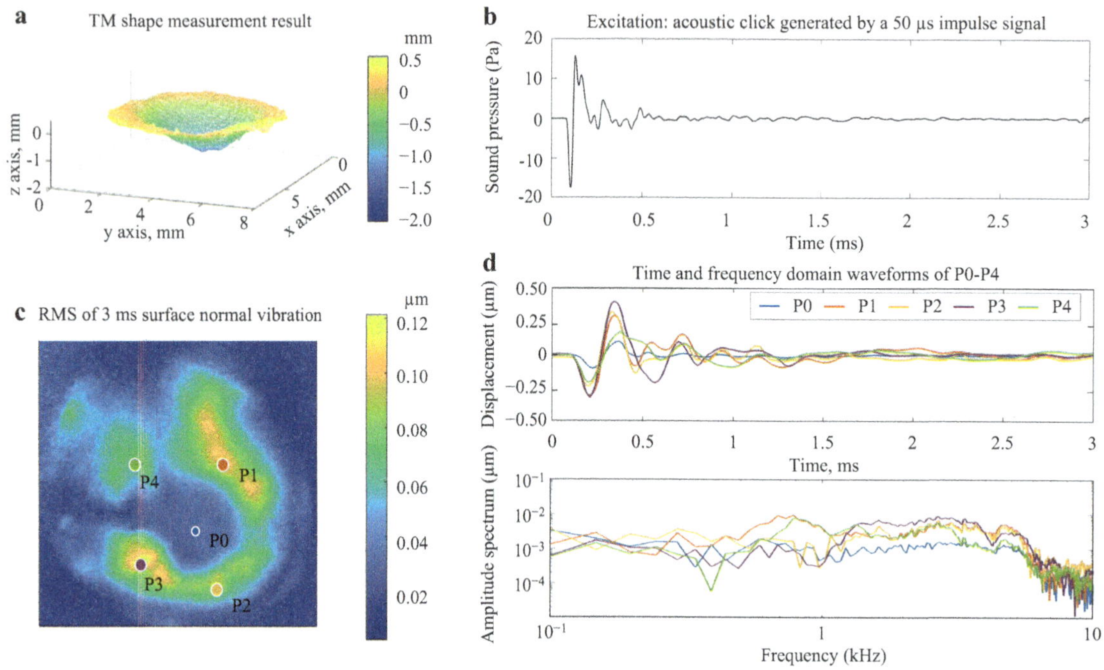

Fig. 8 High-speed holographic surface normal displacement results of a cadaveric TM. **a** The 3D shape measurement used in the calculation of surface normal vector. **b** The excitation signal to deform the TM is an acoustic click generated by a 50 μs impulse. **c** The RMS map of the first 3 ms of vibration normal to the TM surface after the excitation. The RMS is calculated based on the holographic displacement measurements along the temporal axis from 0 to 3 ms after the TM is excited by the acoustic wave. **d** Time-domain and frequency-domain wavefronts of the five individual points. The locations of these selected points are marked in **c**.

displacement component of a human TM sample when it was under an acoustic click excitation. Fig. 8a shows the 3D TM shape measurement results used to calculate the TM's surface normal vectors. Fig. 8b depicts the excitation signal, the click-like sound pressure near the TM produced by a loudspeaker. The RMS map of the surface normal displacement of the TM is shown in Fig. 8c. *P0* is marked at the center of the ossicular attachment to the TM, and *P1* to *P4* mark the maximum RMS value in each quadrant. The individual wavefronts of *P0* to *P4* are shown at the top of Fig. 8d, while the bottom of the figure illustrates the magnitude of the computed frequency response at each location. These figures demonstrate that the amplitudes, cycles of vibrations, the leading and the lagging of the initial vibrations, and the time for each point to settle down are all different at different locations of the TM. The frequency and impulse responses computed from the stimulus and local displacements characterize location dependent TM mechanical properties and help differentiate various middle ear conditions corresponding to middle ear pathologies[34–36].

Discussion

The discussion section compares the three-shape

measurements based on their performance on a NIST gauge and applications on real TM samples, followed by discussion of the practicability of the three methods in TM measurements.

Table 1 compares the three shape measurement methods based on the results reported in this paper. The optical setup for the NIST gauge measurements was configured to mimic that used in the TM measurements, which met constraints like magnification requirements, the limited volume for triangulation, and available spatial resolution for the high-speed camera. The computed resolution is better (of lower value) in the NIST gauge measurements compared to the TM measurements because the NIST gauge is more reflective and does not need the higher illumination power required by the TM. Within the NIST gauge measurements, the resolution of the multi-wavelength method is better than the multi-angle, which is better than HSFP.

Performing high-speed imaging on semi-transparent samples such as TM requires a large illumination density, which significantly affects the optical phase quality of the results. The HSFP and multi-wavelength methods use the same 210 mW power laser light source, whereas the tunable laser output is 10 mW. Furthermore, the HSFP focuses all the available optical power on a single fringe, and can achieve the highest illumination power density of the three methods. In contrast, the multi-wavelength method has the lowest optical phase quality due to the tunable laser's lower illumination power. The image acquisition times for the fringe projection and multi-angle method are limited by the scanning speed of the MEMS mirror, and they have a similar measurement time of about 20–30 ms. The measurement time for the multi-wavelength method is limited by the wavelength tuning time of 150 ms. The processing time of the fringe projection method is much less than the multi-angle and multi-wavelength methods because the latter two need more computational power to track the four-step phase-shifted frames at each illumination position or wavelength.

The resolution of the shape measurement method is related to the optical phase quality as well as the sensitivity of the method's configuration. A comparison of factors that affect the sensitivity of the shape measurements is presented in Table 2. The sensitivity of the fringe projection method depends upon the projected fringe density and the triangulation of the illumination and observation direction. The sensitivity of the multi-angle method depends upon the illumination power as well as the illumination and observation direction and the angle difference of the illumination tuning. The multi-wavelength method's sensitivity depends upon the illumination and observation direction and the wavelength change. The fringe projection method has the highest resolution in terms of standard deviation among the TM measurement results and the lowest resolution among the NIST measurements. This may be because the fringe projection method can use the higher available illumination power when measuring the semi-transparent TM, whereas the additional power is not needed when measuring the NIST.

Table 1 Comparison of the three high-speed shape measurement methods based on TM and NIST measurement

	Fringe projection		Multi-angle		Multi-wavelength	
Illumination density	High		Moderate		Low to moderate	
Acquisition time for the entire measurement (multiple images were taken)	18 ms		26 ms		150 ms	
Processing time	Fast		Moderate		Moderate	
Temporal phase unwrapping	Not compatible		Compatible		Compatible	
Triangulation volume	Needed		Needed		Not needed	
Resolution	TM	NIST	TM	NIST	TM	NIST
	0.11 mm	0.11 mm	0.15 mm	0.04 mm	0.13 mm	0.01 mm

Table 2 Comparison of factors that affect the sensitivity of the three high-speed shape measurement methods

Method	Factors that affect the sensitivity	
Fringe projection method	Illumination and observation direction	Fringe density
Multi-angle method	Illumination and observation direction	Angle difference
Multi-wavelength method	Illumination and observation direction	Wavelength difference

The HSFP method used in this paper is not compatible with temporal phase unwrapping, which requires a series of fringe images, where the fringe width gradually decreases to increase the fringe density. In HSFP the diameter of the laser beam determines the width of the fringe, and it cannot be changed automatically during the measurement in the current configuration. The multi-angle and multi-wavelength methods are suitable for temporal phase unwrapping since the spatial frequency of the optical phase of both methods changes gradually as the tuning of the illumination or the wavelength changes. By introducing temporal phase unwrapping, the resolution of the measurement results can be further improved[36].

The HSFP and multi-angle method require triangulation volume. Fringes projected perpendicular to the sample's surface are not sensitive to the shape of the sample along the projection direction. Thus, for both methods, there must be a triangulation volume to allow the scanning of the illumination. The human eardrum is located at the end of an approximately 2.5 centimeter long and 0.7 centimeter diameter ear canal, and the limited space favors the multi-wavelength method. However, the use of a higher power multi-wavelength laser would be advisable. For measurement of the cadaveric human eardrum, the ear canal can be removed to expose widely the TM, so we propose the multi-angle and fringe projection methods using the high-power fixed wavelength laser to measure the deformation and shape of the cadaveric human eardrum.

The HSFP applies a phase-stepping algorithm that utilizes four phase-stepped images. However, the maximum MEMS mirror's scanning speed is 3 kHz which is not fast enough to scan more than three fringes onto the sample during image acquisition by the high-speed camera since the longest exposure time of the camera is 0.001 s (1 kHz). Therefore, the system cannot capture the completed multi-fringe image in one frame. To overcome this limitation, the completed multi-fringe images are reconstructed after the image acquisition by summing up a series of single fringe images. The number of single fringe images needed is related to the width of the projected fringe and the desired fringe density. In this paper, a total of 72 images are captured to measure the shape of the human eardrum. The processing time of these images for a modern computer is less than 1 second allowing rapid examination of measurement quality.

The multi-angle method measures the shape of a sample by scanning the illumination across the sample. A PZT actuator is used as the phase shifter for the four-step phase-shifting algorithm. Locating the correct frames for the four-step phase-shifting method at each scanning position is done by scanning the Pearson correlation computed for each image to determine the shifted phase. This process for a total of 1200 images is more computationally expensive than the process of HSFP. Furthermore, with the multi-angle method, the illumination beam needs to be collimated and expanded to cover an area larger than the sample surface to maintain the illumination during the scanning and lowering the illumination density for high-speed imaging.

The multi-wavelength method utilizes a tunable laser for shape measurement. The wavelength is tuned during the high-speed image acquisition with a PZT actuator for phase shifting. A typical multi-wavelength shape measurement captures 50 high-speed images at 12 wavelengths. Within these 50 frames, four frames with 90-degree phase-shifted are selected via the calculation of the person correlation[37-40] to calculate the optical phase at a given wavelength. Recording continuously while the phase shifter is moving linearly minimizes unwanted ringing within the phase shifter improving the accuracy of the phase shifting. It is possible to implement the method with 100 high-speed images at 12 wavelengths to improve the phase-shifting accuracy. The processing time of the multi-wavelength is similar to that of the multi-angle shape measurement. The output power and the tuning speed of the tunable laser limit the illumination power density and the measurement time. In our setup, the tunable laser only provides a minimal illumination power for the shape measurement, with a tuning time of 150 ms. A moving average filter has to be used to extract the optical phase due to the minimal illumination power reducing the spatial resolution of the shape measurement.

In summary, the multi-wavelength method does not require triangulation volume to measure the shape, but the optical power of these lasers are limited and the method generally requires painting of the TM. The Multi-angle method enables the shape measurement on the unpainted TM with more optical power, and the HSFP allows rapid shape measurement of the unpainted TM with the least post-processing time. However, both the Multi-angle and HSFP methods have constraints with the small triangulation volume. Therefore, the ideal approach to measure the TM through the intact ear canal in live ears will be to apply the Multi-wavelength method with a tunable laser with the desired optical power and tuning speed for high-speed imaging.

Materials and methods

Digital holography (DH) can record and reconstruct an optical wavefield (amplitude and phase) by comparing the optical path difference of the reference and object beams captured by an optical sensor (camera). For a general holographic interference setup, the measured optical phase ϕ is given by the fringe locus function[36]:

$$\phi = \boldsymbol{d} \cdot \boldsymbol{e} \quad (1)$$

where $\boldsymbol{d}$ is the optical path difference vector and $\boldsymbol{e}$ is the sensitivity vector defined as[38]

$$\boldsymbol{e} = \frac{2\pi}{\lambda}[\boldsymbol{b} - \boldsymbol{s}] \quad (2)$$

where λ is the wavelength of the laser, **b** is the observation unit vector, and **s** is the illumination unit vector. Generally, the object beam is reflected by a diffusive scattering object, and the measured optical phase contains high spatial varying speckle noise. Therefore, the optical path difference cannot be retrieved from a single optical wavefield. However, the change in the measured optical phase ($\Delta\phi$) from the two camera exposures is expressed as

$$\Delta\phi = \boldsymbol{d} \cdot \boldsymbol{e}_2 - \boldsymbol{d} \cdot \boldsymbol{e}_1 \quad (3)$$

where $\boldsymbol{e}_1$ and $\boldsymbol{e}_2$ are the sensitivity vectors at two camera exposures. The optical phase change ($\Delta\phi$) can be retrieved if the speckle distributions at the two exposures correlate with each other. Eq. 3 is the foundation of digital holographic shape and deformation measurement[36,37].

For shape measurement, assuming there is no object deformation during the measurement, the height information (shape) of the measured subject can be obtained by modulating either the magnitude or the direction of the sensitivity vector. The magnitude of the sensitivity vector can be changed by tuning the wavelength of the laser source, which is basically the multi-wavelength method. On the other hand, it is possible to modulate the direction of the sensitivity vector by altering the angle of the object beam, which is the multi-angle shape measurement method. The details about the experimental setups for high-speed multi-wavelength shape and multi-angle shape measurement methods can be found in our previous publications[29,30,34–36] and are briefly described below.

Gauges traceable to the National Institute of Standards and Technology (NIST) were used to validate the shape measurement capabilities of the methods presented in this paper. We then applied these methods to measure the shape of human TMs and compare the measurement results obtained by the three methods, including illumination density, image acquisition speed, optical phase quality, resolution, processing time, temporal phase unwrapping compatibility, and the need for triangulation volume.

To verify the accuracy of the displacement measurement, a latex drum head subjected to an impulsive sound stimulus was simultaneously measured by the High-speed Digital Holographic system (HDH) and a Laser Doppler Vibrometer (LDV). The acoustically induced displacements of a human TM measured with HDH are presented in the results section to demonstrate the capability of the HDH system.

Multi-wavelength shape measurement method

A tunable laser can be used to change the magnitude of the sensitivity vector by changing the wavelength of the laser source, as described in[29]. The optical phase difference ($\Delta\phi$) of a holographic measurement of an object with two different wavelengths results in

$$\Delta\phi = \boldsymbol{d} \cdot \boldsymbol{e}_2 - \boldsymbol{d} \cdot \boldsymbol{e}_1 = \boldsymbol{d} \cdot \left[\frac{2\pi}{\lambda_2}(\boldsymbol{b} - \boldsymbol{s}) - \frac{2\pi}{\lambda_1}(\boldsymbol{b} - \boldsymbol{s})\right] \quad (4)$$

In Eq. 4, the optical phase ($\Delta\phi$) is the difference measured from the interference pattern captured by the camera. The wavelengths λ_1 and λ_2 are known experimental parameters for the tunable laser, and $\boldsymbol{b}$ and $\boldsymbol{s}$ are the observation and illumination vectors, respectively, which can be measured directly or obtained by calibration. The optical path difference vector $\boldsymbol{d}$ can be solved to extract shape information. In the simplified 2D case shown in Fig. 9a, **P** is a random point on the subject, **B** is the observation point, and **S** is the illumination point. In this case, the amplitude of the optical path difference vector $\boldsymbol{d}$ can be expressed as:

$$|\boldsymbol{d}| = |\boldsymbol{PB} - \boldsymbol{SP}| = \sqrt{(\boldsymbol{P}_x - \boldsymbol{S}_x)^2 + (\boldsymbol{P}_y - \boldsymbol{S}_y)^2} + \sqrt{(\boldsymbol{B}_x - \boldsymbol{P}_x)^2 + (\boldsymbol{B}_y - \boldsymbol{P}_y)^2} \quad (5)$$

where P_y is the user-defined pixel coordinates in the y-direction. P_x, which contains the shape information in the x-direction, is the only unknown component of Equation (5) but can be solved for. Eqs. 4,5 show that as the wavelength change ($\lambda_1 - \lambda_2$) increases, the synthetic wavelength ($\lambda_1\lambda_2/(\lambda_1 - \lambda_2)$) increases, and thus uncertainty of the shape measurement decreases.

Multi-angle shape measurement method

The shape of the TM can also be measured by mechanically changing the direction of the illumination vector $\boldsymbol{s}$. From Eq. 3,

$$\Delta\phi = \boldsymbol{d} \cdot \boldsymbol{e}_2 - \boldsymbol{d} \cdot \boldsymbol{e}_1 = \boldsymbol{d} \cdot \left[\frac{2\pi}{\lambda}(\boldsymbol{b} - \boldsymbol{s}_2) - \frac{2\pi}{\lambda}(\boldsymbol{b} - \boldsymbol{s}_1)\right] \quad (6)$$

where $\boldsymbol{s}_1$ and $\boldsymbol{s}_2$ are the illumination vectors at two camera exposures that can be measured or calibrated[30]. The shape information is obtained by solving the optical phase difference using Eq. 6.

For the experimental arrangement shown in Fig. 9b, by introducing a known angle change (dθ) to the illumination

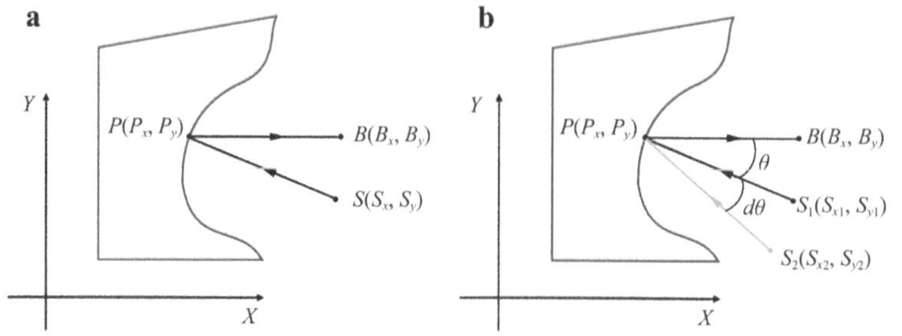

Fig. 9 2D Holographic schematic for shape measurements: **a** multi-wavelength experimental setup; **b** multi-angle wavelength experimental setup. The observation unit vectors are $\boldsymbol{PB}$ and the illumination vectors are $\boldsymbol{SP}$ for the two methods. S_1 and S_2 are the illumination points before and after changing the illumination angle in the multi-angle method.

vector, the height in the x direction can be determined by the following relation[36]:

$$x = \Delta\phi\left(\frac{2\pi}{\lambda}\sin\theta d\theta\right)^{-1} \tag{7}$$

The sensitivity of the method increases as the angle change increases; moreover, it also depends on the non-zero illumination angle.

Digital fringe projection shape measurement method

In this paper a third shape measurement method called fringe projection method was also implemented in our setup, which has several advantages comparing with the other two methods as being discussed in the Results and Discussion sections. A MEMS mirror (Brand: Mirrorcle) with a cylindrical lens was used to project a single fringe onto the sample, which was set up at its maximum scan rate of 3 kHz, and the frame rate and exposure time of the high-speed camera (Brand: Photron Fastcam SAZ, with a resolution of 512 by 512 pixels) were set at 67, 200 frames per second and 1/67200 second, respectively. The camera was configured to be triggered by the position of the MEMS mirror to capture one image of a single fringe at each mirror position. All the single fringe images were summed up numerically to generate a synthesized, completed multi-fringe image after the high-speed acquisition.

The experimental setup for the high-speed fringe projection is previously shown in Fig. 2. The light source used is a 532-nm fixed wavelength laser (Oxxius). The laser beam is divided into a reference beam and an illumination beam. Two mirrors direct the reference beam to reach the expansion lens and are combined with the object beam at the beam splitter in front of the High-speed camera. The illumination beam first reaches the MEMS mirror, which has two rotation axes. The vertical axis rotation is used to switch the illumination to the cylindrical lens for fringe projection, and it is also used to switch the expansion lens for displacement measurements. The cylindrical lens is used to generate a single vertical fringe onto the sample. The horizontal rotation of the MEMS mirror is used to scan the single fringe across the surface of the sample during the shape measurement. The mechanical shutter in the reference beam is activated to block the reference beam for fringe projection during the shape measurement. The mechanical shutter is deactivated during the high-speed displacement measurement, and the MEMS mirror fixes the illumination beam to the expansion lens. The reflected light from the sample interferes with the reference beam on the beam splitter in front of the camera. Finally, the resultant speckle pattern is recorded.

The position of the fringes was based on the size of the sample and the configuration of the optical setup. The first single fringe image was taken at the left-most side of the field of view. The second MEMS mirror position was programmed to be at 90° phase-shifted position of the first fringe, the third at the 180° phase-shifted position, and the fourth at the 270° phase-shifted position. (Four-phase shifting was chosen for its balance of shifting speed and overall accuracy; however, any phase shifting strategy can be implemented.) This process was repeated until the entire sample was scanned. The repeatability of the MEMS mirror is examined by investigating the signal fringe image. The MEMS is programmed to project a signal fringe to a fixed position (Fig. 10a) for high-speed imaging and move the fringe out of the field of view back to its zero position. This procedure is repeated five times. A Gaussian curve is horizontally fit to the intensity image row by row, and the peak of the Gaussian profile is used to determine the position of the projected fringe. The difference in fringe position of the five consecutive measurements is 0.63 pixels, and the averaged fringe width is 100 pixels. The error introduced by the MEMS mirror in repositioning the fringe is less than 1% of the fringe width, and is assumed to be negligible.

Once all the single fringe images were captured, they

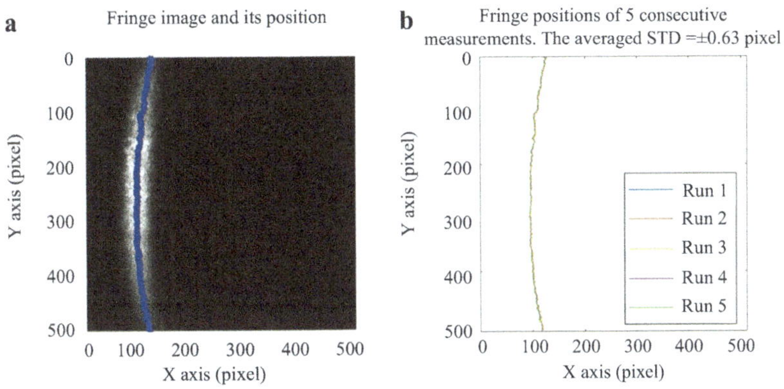

Fig. 10 The repeatability of the MEMS mirror is examined by checking the repeatability of the fringe position projected by the MEMS mirror during 5 consecutive TM measurements. **a** One fringe image with the Fringe position determined by averaged pixel position of the intensity image of each row. **b** the fringe position of 5 consecutive measurements and its averaged standard deviation is ± 0.63 pixels with averaged fringe width of 100 pixels. The reposition error introduced by the MEMS mirror is less the 1% of the fringe width and is assumed to be negligible.

were divided into four groups based on the angles of phase-shifting (0°, 90°, 180°, and 270°). The phase-shifting is calibrated based on the optical setup by making the distance between each projected fringe to be $W/4$, where W is the width of the single fringe. The four-step phase-shifted multi-fringe images represented the numerical summation of each image group. The optical phase of the shape of the sample was calculated using Eq. 8.

The fringe projection shape measurement method is based on the projection of well-defined fringe patterns onto the sample. In the high-speed fringe configuration, the high-speed camera captures the *4n* single fringe images sequentially, $I_1^1, I_1^2, I_1^3, I_1^4, I_2^1, I_2^2, I_2^3, I_2^4 \ldots I_n^1, I_n^2, I_n^3, I_n^4$, where the superscript indicates the phase-shifting order and the subscript indicates the fringe order. Numerical summations $(I_1^1 + I_2^1 \ldots I_n^1)$, $(I_1^2 + I_2^2 \ldots I_n^2)$, $(I_1^3 + I_2^3 \ldots I_n^3)$, and $(I_1^4 + I_2^4 \ldots I_n^4)$ are calculated to reconstruct the completed multi-fringe images corresponding to the four-step phase-shifting frames. The optical phase φ of the shape of the sample is[37]

$$\phi = arctan\frac{(I_1^4 + I_2^4 \ldots I_n^4) - (I_1^2 + I_2^2 \ldots I_n^2)}{(I_1^1 + I_2^1 \ldots I_n^1) - (I_1^3 + I_2^3 \ldots I_n^3)} \quad (8)$$

High-speed digital holographic displacement measurement method

The displacement measurement method measures the full-field-of-view (> 200,000 points at 67,200 camera frame rate) transient displacements[29,31,32–35]. The Pearson's correlation coefficients between the deformed frame and two undeformed reference frames with 0 and 90° phase-shifting are calculated to quantify the optical phase of the sample's displacement. Assuming I_r is the intensity of the reference beam and I_o is the intensity of the object beam, and φ is the optical phase difference between the object and the reference beams. The interference patterns of the reference (undeformed) I_{ref} and deformed I_{def} states can be expressed as shown in Eqs. 9,10[29,38], respectively:

$$I_{ref} = I_r + I_o + 2\sqrt{I_r I_o}\cos(\phi) \quad (9)$$

$$I_{def} = I_r + I_o + 2\sqrt{I_r I_o}\cos(\phi + \Delta\phi) \quad (10)$$

where $\Delta\phi$ is the optical phase change due to the sample's deformation. To extract $\Delta\phi$, we calculate the Pearson's correlation coefficient of a small window (e.g., 3 by 3 pixels) between the deformed and reference frames by using Eq. 11[40]

$$\rho(I_{ref}, I_{def}) = \frac{\langle I_{ref} I_{def}\rangle - \langle I_{ref}\rangle\langle I_{def}\rangle}{\left[\langle I_{ref}^2\rangle - \langle I_{ref}\rangle^2\right]^{\frac{1}{2}}\left[\langle I_{def}^2\rangle - \langle I_{def}\rangle^2\right]^{\frac{1}{2}}} \quad (11)$$

where operator <> presents the expected value. Since the speckle pattern captured by the camera sensor is randomly distributed and the two same-sized matrices can be correlated, the expected value of the speckle pattern is equivalent to its numerical average[39]. Assume φ is uniformly distributed and $\Delta\phi$ is constant within the small window, then the coefficient can be simplified in the following format by inserting (9) and (10) into (11)[32]:

$$\rho = \frac{(r+1)^2 + 4r\cos(\Delta\phi)}{(r+1)^2 + 4r} \quad (12)$$

where $r = I_o/I_r$ is the intensity ratio of the object and reference beams. Therefore, the optical phase difference due to the deformation of the sample can be calculated by[29]

$$\Delta\phi = arctan\frac{\rho(I_{ref}, I_{def})}{\rho\left(I_{ref+\pi/2}, I_{def}\right)} \quad (13)$$

where $\left(I_{ref+\frac{\pi}{2}}\right)$ is the 90° phase-shifted reference frame of the undeformed frame. Khaleghi et al. have demonstrated that the dominant deformation of the TM is along the surface normal direction[40]. Razavi[40] demonstrated that the sensitivity vector of the holographic system varies when the sample is located close to the imaging lens; however, the magnitude of displacement (*S*) along the surface normal direction, can be compensated by the sensitivity vector variation ***K***(x,y) and the surface normal vector of the sample ***n***(x,y), can be measured as $\Delta\phi(x,y,t)$[34]

$$S\,(x,y,t) = \frac{\Delta\phi(x,y,t)}{\boldsymbol{K}(x,y)\cdot\boldsymbol{n}(x,y)} \tag{14}$$

where *x* and *y* are the pixel coordinates of the camera and *t* is the time.

Pearson's correlation vectorization

Pearson's correlation coefficient was used to calculate the optical phase between two camera exposures. The previous algorithm used a sliding window technique on each recorded image. The process consists simply of moving a sliding window (e.g., 3 by 3 pixels) from point to point in the horizontal and vertical directions. Calculating the Pearson's coefficient of two of 512 by 512-pixel images requires more than 250,000 iterations, which takes about 3.5 seconds on a modern computer. For a typical displacement measurement, more than 1000 images are captured, which takes about one hour to finish the process. The long processing time makes accessing the measurement quality during the measurement impractical. The algorithm is optimized by a vectorization procedure to remove the nested loops to improve the computational performance. Vectorization is the process of converting an algorithm from operating on a single value at a time to operate on a set of vectors or matrices at one time utilizing the vector operation capabilities of modern CPUs.

To generalize the Pearson's correlation coefficient calculation for n by n kernel size (n=3,5,7 ...), defining $I^{'}(i,j)$ are n^2 translated matrices of the original image matrix I, where i= 1,2,3...n and j = 1,2,3...n. and $I_{x,y}$ is the matrix element at (x,y) coordinate, $I^{'}(i,j)$ is obtained by translating $I_{x,y}$ by the following Equation:

$$I^{'}(i,j) = I_{(x-\frac{n-1}{2}+i-1).(y-\frac{n-1}{2}+j-1)},$$
$$i = 1,2,3\ldots n \text{ and } j = 1,2,3\ldots n \tag{15}$$

Both the undeformed frame I_{ref} and deformed frame I_{def} are translated to $I_{ref}^{'}(i,j)$ and $I_{def}^{'}(i,j)$ by Eq. 15. The Pearson's correlation coefficient of I_{ref} and I_{def} can be calculated by Eq.11, and simplified expression is:

$$\rho(I_{ref},I_{def}) = \frac{\sum_i^n\sum_j^n X\cdot Y}{[\sum_i^n\sum_j^n X^2\cdot\sum_i^n\sum_j^n Y^2]^{\frac{1}{2}}} \tag{16}$$

where $X = I_{ref}^{'}(i,j) - \frac{1}{n^2}\sum_i^n\sum_j^n[I_{ref}^{'}(i,j)]$ and $Y = I_{def}^{'}(i,j) - \frac{1}{n^2}\sum_i^n\sum_j^n[I_{def}^{'}(i,j)]$. The fastest calculation is the 3 by 3 pixels kernel version because it requires the minimum number of translated matrices, which takes about 0.15 s to finish on a modern computer. This 3 by 3 pixel kernel of the Pearson's coefficient is suitable for displaying the results immediately after the measurement to assess the optical phase quality to determine if the measurement needs to be repeated.

The high-speed holographic displacement measurement method described in this paper utilizes the optimized Pearson's correlation algorithm to calculate the optical phase change introduced by the surface deformation of the sample. With the optimized algorithm, the calculation becomes 21 times faster using the same computation power. The improved processing speed makes it possible to display the optical phase of the high-speed image immediately after the acquisition for us to examine the measurement quality between the measurements, which streamlines the experiment process and makes measurements on live ears more practicable.

Conclusion and future work

This paper presents a newly developed shape measurement method using a miniaturized fringe projection system with a Microelectromechanical systems (MEMS) mirror. We compared this method to measure the shapes of a NIST target and human TMs with two previously developed methods: (1) multi-wavelength method and (2) multi-angle of illumination method. We also present vibration measurements of human TMs made with an upgraded holographic displacement system, which is validated by comparing the holographic results of a vibrating drum head to simultaneously measured LDV results. For the shape measurement, the multi-wavelength method performs best on measuring the NIST target, while the measurement quality decreases on measuring the human TM due to the lower power of the tunable laser. The fringe projection and multi-angle methods can measure the unpainted TM due to the adaptation of a higher power laser. The fringe projection method requires fewer captured images than the multi-angle methods, making it a more practical shape measurement for live animal study. However, the fringe projection method and multi-angle method require triangulation volume, posing challenges to measurements made through an intact ear canal.

In the future, we will continue developing the displacement and shape measurement in the direction of utilizing the multi-wavelength method by identifying a suitable high power tunable laser for this application. The future implementation will be combined with the HSH system with an endoscopic configuration to allow measurements of the eardrum shape and responses through the intact ear canal in live ears. We have experimented with the Dynamic Time wrap algorithms to integrate the time series data for displacement pattern extraction from the holographic measurements. We will apply more artificial intelligence and data mining technologies to automate and streamline the data process.

Acknowledgements
Grant support from the US National Institute on Deafness and Other Communication Disorders (NIDCD R01 DC016079) is gratefully acknowledged. We would also like to acknowledge partial support by the Center for Holographic Studies and Laser micro-mechaTronics (CHSLT) at WPI. We also thank all of the reviewers of this manuscript for their insightful suggestions and input.

Author details
[1]Center for Holographic Studies and Laser micro-mechaTronics, Worcester Polytechnic Institute, Worcester, MA USA, 01609. [2]Mechanical Engineering Department, Worcester Polytechnic Institute, Worcester, MA USA, 01609. [3]Faculty of Mechatronics, Informatics and Interdisciplinary Studies, Technical University of Liberec, Liberec, Czech Republic, 461 17. [4]Eaton-Peabody Laboratory, Massachusetts Eye and Ear, Boston, MA USA,02114. [5]Department of Otolaryngology-Head and Neck Surgery, Harvard Medical School, Boston, MA, USA,02115

Author contributions
Conceptualization, JJR, CF, JTC; methodology, HT, PP; software, HT, PP; validation, HT; formal analysis, HT; investigation, HT; resources, JJR, CF, JTC; data curation, HT, writing —original draft preparation, HT; writing —review and editing, HT, CF, JJR, JTC; visualization, HT; supervision, JJR, CF, JTC; project administration, JJR, CF, JTC; funding acquisition, JJR, CF, JTC.

Conflict of interest
The authors declare that they have no conflict of interest.

Received: 30 September 2021 Revised: 28 January 2022 Accepted: 15 February 2022
Published online: 14 March 2022

References

1. Yost, W. A. Fundamentals of Hearing: An Introduction. 5th edn. (New York: Academic Press, 2006).
2. Fastl, H. & Zwicker, E. Information processing in the auditory system. in Psychoacoustics (eds Fastl, H. & Zwicker, E.) (Berlin, Heidelberg: Springer, 2007). https://doi.org/10.1007/978-3-540-68888-4_3.
3. Volandri, G. et al. Biomechanics of the tympanic membrane. *Journal of Biomechanics* **44**, 1219-1236 (2011).
4. Rosowski, J. J. Outer and middle ears. in Comparative Hearing: Mammals (eds Fay, R. R. & Popper, A. N.) (New York: Springer, 1994), 172-247. https://doi.org/10.1007/978-1-4612-2700-7_6.
5. Geisler, C. D. From Sound to Synapse: Physiology of the Mammalian Ear. (New York: Oxford University Press, 1998).
6. Lim, D. J. Structure and function of the tympanic membrane: a review. *Acta Oto-Rhino-Laryngologica Belgica* **49**, 101-115 (1995).
7. De Greef, D. et al. Details of human middle ear morphology based on micro-CT imaging of phosphotungstic acid stained samples. *Journal of Morphology* **276**, 1025-1046 (2015).
8. van der Jeught, S. et al. Full-field thickness distribution of human tympanic membrane obtained with optical coherence tomography. *Journal of the Association for Research in Otolaryngology* **14**, 483-494 (2013).
9. Aernouts, J., Aerts, J. R. M. & Dirckx, J. J. J. Mechanical properties of human tympanic membrane in the quasi-static regime from *in situ*, point indentation measurements. *Hearing Research* **290**, 45-54 (2012).
10. Rosowski, J. J. et al. Computer-assisted time-averaged holograms of the motion of the surface of the mammalian tympanic membrane with sound stimuli of 0.4-25 kHz. *Hearing Research* **253**, 83-96 (2009).
11. Rosowski, J. J. Models of external- and middle-ear function. in Auditory Computation (eds Hawkins H.L. et al.) (New York: Springer, 1996), 15-61.
12. Lim, D. J. Human tympanic membrane: an ultrastructural observation. *Acta Oto-Laryngologica* **70**, 176-186 (1970).
13. Wang, X. L. et al. Motion of tympanic membrane in guinea pig otitis media model measured by scanning laser Doppler vibrometry. *Hearing Research* **339**, 184-194 (2016).
14. Fay, J. P., Puria, S. & Steele, C. R. The discordant eardrum. *Proceedings of the National Academy of Sciences of the United States of America* **103**, 19743-19748 (2006).
15. Aernouts, J. et al. Elastic characterization of membranes with a complex shape using point indentation measurements and inverse modelling. International Journal of Engineering Science 48.6 (2010).
16. Cheng, T. et al. Viscoelastic properties of human tympanic membrane. *Annals of Biomedical Engineering* **35**, 305-314 (2007).
17. Fay, J. et al. Three approaches for estimating the elastic modulus of the tympanic membrane. *Journal of Biomechanics* **38**, 1807-1815 (2005).
18. Milazzo, M. et al. The path of a click stimulus from ear canal to umbo. *Hearing Research* **346**, 1-13 (2017).
19. Kakue, T. et al. High-speed phase imaging by parallel phase-shifting digital holography. *Optics Letters* **36**, 4131-4133 (2011).
20. Fuller, P. W. W. An introduction to high speed photography and photonics. *The Imaging Science Journal* **57**, 293-302 (2009).
21. Goode, R. L. et al. Laser Doppler Vibrometer (LDV)Ba new clinical tool for the otologist. *The American Journal of Otology* **17**, 813-822 (1996).
22. Goode, R. L. et al. New knowledge about the function of the human middle ear: development of an improved analog model. *The American Journal of Otology* **15**, 145-154 (1994).
23. Gan, R. Z., Wood, M. W. & Dormer, K. J. Human middle ear transfer function measured by double laser interferometry system. *Otology & Neurotology* **25**, 423-435 (2004).
24. Rosowski, J. J., Nakajima, H. H. & Merchant, S. N. Clinical utility of laser-Doppler vibrometer measurements in live normal and pathologic human ears. *Ear and Hearing* **29**, 3-19 (2008).
25. Decraemer, W. F., Khanna, S. M. & Funnell, W. R. J. Vibrations at a fine grid of points on the cat tympanic membrane measured with a heterodyne interferometer. Proceedings of EOS/SPIE International Symposia on Industrial Lasers and Inspection, Conference on Biomedical Laser and Metrology and Applications. Munchen: 1999, 1-4.
26. de La Rochefoucauld, O. & Olson, E. S. A sum of simple and complex motions on the eardrum and manubrium in gerbil. *Hearing Research* **263**, 9-15 (2010).
27. Cheng, J.T. et al. Motion of the surface of the human tympanic membrane measured with stroboscopic holography. *Hearing Research* **263**, 66-77 (2010).
28. Kim, S. et al. Effect of age on binaural speech intelligibility in normal hearing adults. *Speech Communication* **48**, 591-597 (2006).
29. Desoer, C. & Wang, Y.T. On the generalized Nyquist stability criterion. *IEEE Transactions on Automatic Control* **25**, 187-196 (1980).
30. Razavi, P. et al. Combined high-speed holographic shape and full-field displacement measurements of tympanic membrane. *Journal of Biomedical Optics* **24**, 031008 (2018).
31. Psota, P. et al. Multiple angle digital holography for the shape

measurement of the unpainted tympanic membrane. *Optics Express* **28**, 24614-24628 (2020).
32. Dobrev, I. et al. High-speed digital holography for transient response of the human tympanic membrane. in Advancement of Optical Methods in Experimental Mechanics, Volume 3 (eds Jin, H. et al.) (Cham: Springer, 2015). https://doi.org/10.1007/978-3-319-06986-9_39.
33. Khaleghi, M. et al. In-plane and out-of-plane motions of the human tympanic membrane. *The Journal of the Acoustical Society of America* **139**, 104-117 (2016).
34. Tang, H. M. et al. High-speed holographic shape and full-field displacement measurements of the tympanic membrane in normal and experimentally simulated pathological ears. *Applied Sciences* **9**, 2809 (2019).
35. Tang, H. et al. Analyses of the tympanic membrane impulse response measured with high-speed holography. *Hearing Research* **410**, 108335 (2021).
36. Tang, H. et al. High speed Holographic Shape and Vibration Measurement of the Semi-transparent Tympanic Membrane. *Mechanics of Biological Systems and Materials & Micro-and Nanomechanics & Research Applications*, Conference Proceedings of the Society for Experimental Mechanics Series. Springer, Cham. https://doi.org/10.1007/978-3-030-59765-8_12(2021).
37. Jones, R. & Wykes, C. Holographic and Speckle Interferometry. 2nd edn. (Cambridge: Cambridge University Press, 1989).
38. Kreis, T. Handbook of Holographic Interferometry: Optical and Digital Methods. (Weinheim: Wiley-VCH, 2005).
39. Dobrev, I. T. Full-field vibrometry by high-speed digital holography for middle-ear mechanics. PhD thesis, Worcester Polytechnic Institute, Worcester, 2014.
40. Razavi, P. Development of high-speed digital holographic shape and displacement measurement methods for middle-ear mechanics in-vivo. PhD thesis, Worcester Polytechnic Institute, Worcester, 2018.

Official journal of the JHL 2689-9620
www.light-am.com

Article Open Access

The application of digital holography for accurate three-dimensional localisation of mosquito-bednet interaction

Matthew L Hall[1,*], Katherine Gleave[2], Angela Hughes[2], Philip J McCall[2], Catherine E Towers[1] and David P Towers[1]

Abstract

Understanding mosquito interaction with long-lasting insecticidal bednets is crucial in the development of more effective intervention methods to protect humans from malaria transmission. As such, a 240 × 240 × 1000 mm laboratory setup for the in-line recording of digital holograms and subsequent in-focus reconstruction and 3D localisation of mosquitoes is presented. Simple bednet background removal methods are used to accurately localise a mosquito obscured by a bednet in 3D coordinates. Simulations and physical data demonstrate that this method is suitable for mosquitoes positioned 3–1000 mm behind a bednet. A novel post-processing technique, involving a cascade-correlation of a Tamura of Intensity focus metric extracted from digitally reconstructed scenes, accurately localises mosquitoes positioned 35–100 mm behind a bednet from a single digital hologram. The result of this study is a scalable digital holographic methodology to examine mosquito-bednet interaction in 3D at a level of accuracy previously only seen in 2D imaging of mosquitoes in a much smaller volume.

Keywords: Digital Holography, Metrology, Entomology

Introduction

Digital holography has previously been used in the 3D metrology of particles, oil droplets, plankton, and other sea creatures[1-5]. Expanding the recording volume to obtain information on 3D mosquito flight behaviour behind a bednet is of interest to entomology researchers investigating malaria vector control[6]. Long-lasting insecticidal bednets (LLINs) are the most effective method of control against disease carrying mosquitoes[7] and understanding the precise modes of action of LLINs in terms of mosquito-bednet interaction is crucial in their ongoing development, as rates of insecticide resistance within mosquito populations continue to increase[8]. Until now, a solution to accurately examine 3D flight behaviour of a mosquito obscured by a bednet has not been achieved.

Previous similar metrology applications of digital holography have focused on smaller fields of view[9], often in a lensless configuration[10], although Malkeil et al. achieved a 55 mm diameter recording volume[2]. Larger field-of-view digital holography to reconstruct bigger objects[11,12], including reconstruction of an object through a diffuse surface[13], has generally been reserved for improving the quality of a reconstructed image rather than for exact localisation of an object. Singh et al examined the concept of viewing objects obscured by a diffuse surface using time-averaged digital holography to improve deep tissue imaging[13,14]. Other novel methods to reconstruct and

Correspondence: Matthew L Hall (m.l.hall@warwick.ac.uk)
[1]Optical Engineering Group, School of Engineering, University of Warwick, Coventry, CV4 7AL, UK
[2]Vector Biology Department, Liverpool School of Tropical Medicine, Liverpool, L3 5QA, UK

localise an object obscured by diffuse surfaces, solid objects, and multiple-object scattering include a technique to reconstruct images of shrapnel during an explosion[15], and a femtosecond pulsed laser to reconstruct objects deep inside a dense spray[16]. Further techniques have been developed for imaging an object through optically thick diffusive media[17,18] using the Fourier domain shower curtain effect or multiple spatially shifted realisations between two objects, e.g. the bednet and mosquito. Spatially continuous diffusive material is more challenging than the bednet of relevance here, however, the techniques to reconstruct 'hidden' objects require hundreds to thousands of images as input for the reconstruction algorithm and exemplar data is given for time stationary objects. In the case of the mosquito, its orientation and morphology (legs, wings, body) are time variant during flight and hence such methods cannot be applied directly to this case.

Non-holographic methods for extracting the 3D position of small objects often involve stereo-pair imaging, as used when Butail et al. examined 3D swarming behaviour of mosquitoes in the field[19]. However, the inherent problem with matching stereo image pairs led to short tracks (5–7 seconds), necessitating time consuming human supervision to combine into long time-series data. Kroner et al. modified a 2D retro-reflective screen imaging arrangement[20] by adding a small separation between the light source and camera to give a quasi-stereo setup where a mosquito's secondary shadow on the retro-reflective screen and its primary shadow on the camera formed a stereo-image pair[21]. Missing data points occur due to the poor shadow-background contrast and are amplified by the presence of a bednet. In all but the last approach, non-occluded access to the measurement space from multiple views is required with angularly well separated cameras. In their review of *Photonic Monitoring of Atmospheric and Aquatic Fauna*, Brydegaard and Svanberg suggest that, when examining flying insects, a focus strategy is necessary to achieve high sensitivity while maintaining a high depth-of-field[22]. Digital holography offers the ability to refocus a scene during post-processing of data, so is a valid approach which could be applied to other insects, such as biting midges, moths, and flower pollinators[23].

Digital holography offers unique solutions to 3D imaging problems, and a telecentric in-line configuration is an attractive proposition when examining mosquito interaction with insecticide soaked walls and nets. The *XY* position of an object is easily resolvable using conventional image processing techniques, and the *z*-axis position is calculated via reconstruction of the object scene and evaluation of suitable focus metrics[24–26]. Digital re-focusing of a mosquito hologram gives the potential for 3D tracking of mosquito flight in a wide range of experimental configurations involving bednets.

In this paper, a scalable digital holographic methodology to accurately localise a mosquito in 3D coordinates in a large field of view is presented. The paper first determines the *z*-axis resolution limits at larger lateral imaging resolutions (effective pixel size, EPS, Δx_e, mm/pixel) compatible with an increased Field of View (FoV), expanding the scope of previous reports[27,28] and for the case of multiple object localisation. The *particle signature function* focus metric[25,28] proposed by this previous work performs well at resolving multiple objects, but is less effective in the case of a mosquito obscured by a bednet due to the disturbed reference wave diffracting through the opaque lattice-structure, resulting in unacceptable positional uncertainties ($z_{err} > 10$ mm). A suitable recording setup for forward scattering in-line digital holography is then introduced, along with a methodology for the reconstruction and localisation of mosquitoes. A cross-correlation method for the identification and *XY* localisation of free-flying mosquitoes and mosquitoes obscured by a bednet is then presented along with a novel *z*-axis localisation technique to improve separation and localisation of the bednet and mosquito *z*-axis positions for the obscured mosquito case. The method involves a two-step cascade correlation method of a Tamura-of-Intensity (ToI) focus metric, previously described by Tamamitsu et al[29]. The result is a method described as the Cascade-Correlation of Focus Metric (CCoFM) technique. A simple bednet-background removal technique is also introduced. Large field-of-view physical data is provided to demonstrate that the system is capable of refocusing fast-moving mosquitoes with a high frequency wing-beat (~ 700 Hz) using a continuous-wave laser source. Data involving mosquitoes obscured by a bednet are then investigated. Simulations are used to determine the limits of the CCoFM and bednet-background removal techniques with regards to the mosquito position relative to the bednet, and physical data is used to validate the *z*-axis resolution limits close to the bednet.

In terms of digital holography applied to similar underwater organism tracking experiments[2,3], the field of view in this paper is approximately 5 × larger, and is therefore defined in the following sections as 'Large Field-of-View', although it is smaller than 2D mosquito flight imaging experiments previously reported[30]. However, other 100 × 100 × 100 mm flight-chamber laboratory bio-assays[31], involving a single layer of bednet in front of a mosquito, indicate that the techniques presented in the following paper can be applied to entomology research.

Effect of Imaging Resolution (Effective Pixel Size) on Localisation Accuracy

The effective pixel size (EPS, Δx_e), also known as image lateral spatial resolution or sampling resolution, is defined as the recorded volume transverse distance (x_r, y_r) divided by the number of pixels on the recording device (M, N) so that $\Delta x_e = \min(x_r/M, y_r/N)$. The EPS is increased by using fewer pixels to record a given volume, or by increasing the recorded volume for a given number of pixels. The demagnification factor of a two-lens system between the object and camera is used to alter the EPS.

Work previously conducted concluded that a trade-off between EPS and localisation accuracy exists, and that the minimum resolvable distance between two objects increases as EPS is increased[28]. As a progression on this previous work[28], the minimum resolvable z-axis distance between two physical objects (2.4 mm diameter spheres) was evaluated for discrete EPS values of $\Delta x_e = 58.60$ μm, $\Delta x_e = 78.13$ μm, and $\Delta x_e = 97.67$ μm, presented in Fig. 1. The higher EPS values yield a larger minimum resolvable distance for three focus metrics assessed in the previous paper (edge-gradient based, Tamura of intensity, particle signature function), although the minimum resolvable distance is different for each of the three focus metrics. Since the minimum resolvable distance for two objects in close proximity appears to increase for larger EPS values, it was anticipated that the same trend would be observed for resolving a mosquito obscured by a bednet. However, the particle signature function focus metric previously reported works well for resolving multiple objects, but is less effective in the case of a mosquito obscured by a bednet due to the disturbed reference wave diffracting through the lattice-structure. As such, the Tamura of Intensity focus metric provides the best balance between multiple-object resolvability, and working with a disturbed reference wave through a bednet. An EPS of ~ 80 μm was chosen for the physical data reported in this manuscript due to the larger change in minimum resolvable distance for the Tamura of intensity focus metric between EPS values of 78.13 μm and 96.67 μm in Fig. 1.

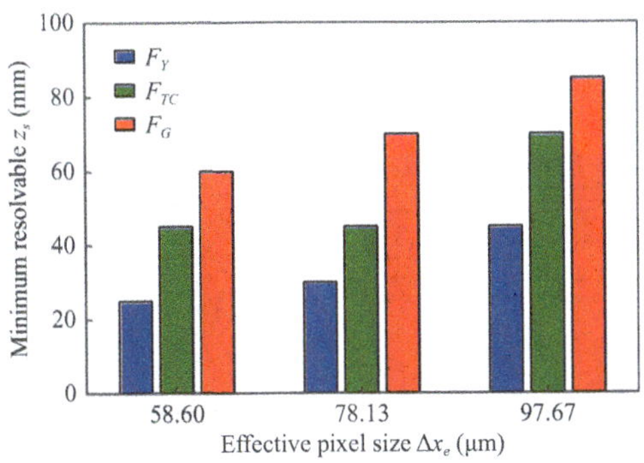

Fig. 1 Effect of EPS, Δx_e, on minimum resolvable distance of two mosquitoes in close proximity for EPS values of $\Delta x_e = 58.60$ μm, 78.13 μm, and 96.67 μm. Extension of data in previous paper[28].

Results

Large Field of View Mosquito Reconstructions

A flight chamber was constructed and the front was placed at a perceived distance of 0.9 m away from the back-focal point of a focusing off-axis parabolic (OAP) mirror and plano-convex lens telecentric demagnification system of focal lengths $f_o = 2540$ mm, $f_i = 175$ mm, demagnification factor $M_D = 14.53\times$, and an EPS of $\Delta x_e = 79.89$ μm. Twin-image suppression in digital holography has been a key topic for many researchers when absolute phase values or high-accuracy metrology and profilometry is necessary[32,33]. For positional metrology of a small object, a simple method to alleviate the effect of the twin-image on reconstruction is to use a large propagation distance, as the primary and twin-image are equidistant either side of the hologram plane. Therefore, a larger propagation distance from the object to hologram plane produces a more defocused twin-image at the plane of best focus of the reconstructed object, meaning the twin-image will have a smaller contribution to the focus metric calculations. During experimentation, this was physically achieved by placing the front of the flight chamber at a known z-axis distance from the back focal plane of the mirror-lens demagnification system ($z = 100$ mm) and moving the CCD back from the focal plane of the imaging lens by 4.26 mm (in this case). At the CCD-side of the mirror-lens demagnification system, the growth of the forward scattered first order diffraction ring increases at a rate proportional to the square of the lateral demagnification of the two lens system (with respect to z-axis distance), but the plane reference wave remains collimated. This allows for a more compact optical setup, as a 1 mm movement of the CCD amounts to a 211 mm movement of the apparent object position for the case of a lateral demagnification of 14.53×.

An exposure time of 150 μs was chosen for recording, based on the calculated expected movement of a mosquito wing during exposure. Using approximate values of a mosquito wing: length of 2.4 mm, beat frequency of 700 Hz, and amplitude of 45 degrees, it was calculated that the tip of the wing would move through an arc length of 0.19 mm, corresponding to approximately 2.5 pixels when using an EPS value of 79.89 μm. As a result, motion blur was kept to a minimum. The forward-scattering nature of the experimental setup reduces issues associated with phase-shift through the exposure time due to the movement of the object, as would be the case in a reflective holographic setup.

Fig. 2 shows a 3D flight track of a mosquito along with four best-focus intensity reconstructions of the flying mosquito in the flight chamber. The mosquito can be seen to fly through the flight chamber in the *z*-direction (from the back at ~ −1000 mm to the front at ~ −0 mm) before probing at the glass surface of the front of the flight chamber (4–7.5 seconds). Several features of the mosquito in these specific intensity reconstructions, such as the legs, body, and wings, can be resolved and identified, and the orientation of the mosquito could be calculated if required. However, the relatively low frame-rate of the camera (30 Hz) compared to the wing beat (700 Hz) means that the captured positional-phase in the wing-beat cycle is random. Despite this, in a time-series of images, some statistical moments could be extracted from each image and linked to mosquito orientation against time. This could provide potential scope for future experiments to assess mosquito orientation with respect to plumes of CO_2, attractants, and repellents.

As previously mentioned, there is a keen interest in mosquito-bednet interaction. As such, an investigation into obscured mosquito localisation of a mosquito released behind a bednet was performed. Fig. 3 shows reconstructions of two mosquitoes obscured by a bednet: one in full flight (Fig. 3a, b), and the other crawling on the net (Fig. 3c, d). For each of the reconstructions, a moving background subtraction and a stationary background subtraction was performed (both described later in Background Removal of Bednet). A moving subtraction works well for free-flying mosquitoes and eliminates the requirement for a perfectly still bednet during the recording. However, Fig. 3c shows that the moving background subtraction is not suitable for use with a mosquito resting or crawling at a low speed on the net. The stationary background subtraction reconstructions feature a small amount of speckle noise compared to the moving background subtraction reconstructions due to the small movement of the bednet between the background net recording and the frame of interest.

Mosquitoes Obscured by a Bednet

Examining laboratory-based mosquito-host interaction has previously provided useful insights into quantifying mosquito flight behaviour[31], and indicates that useful data can be extracted from experiments which are not full-room-scale in size. The following techniques can be applied to similar experimental setups comprising a mosquito situated behind a single layer of bednet in a laboratory setting. Further work on methods to localise a mosquito obscured by multiple layers of bednet will be the subject of future research.

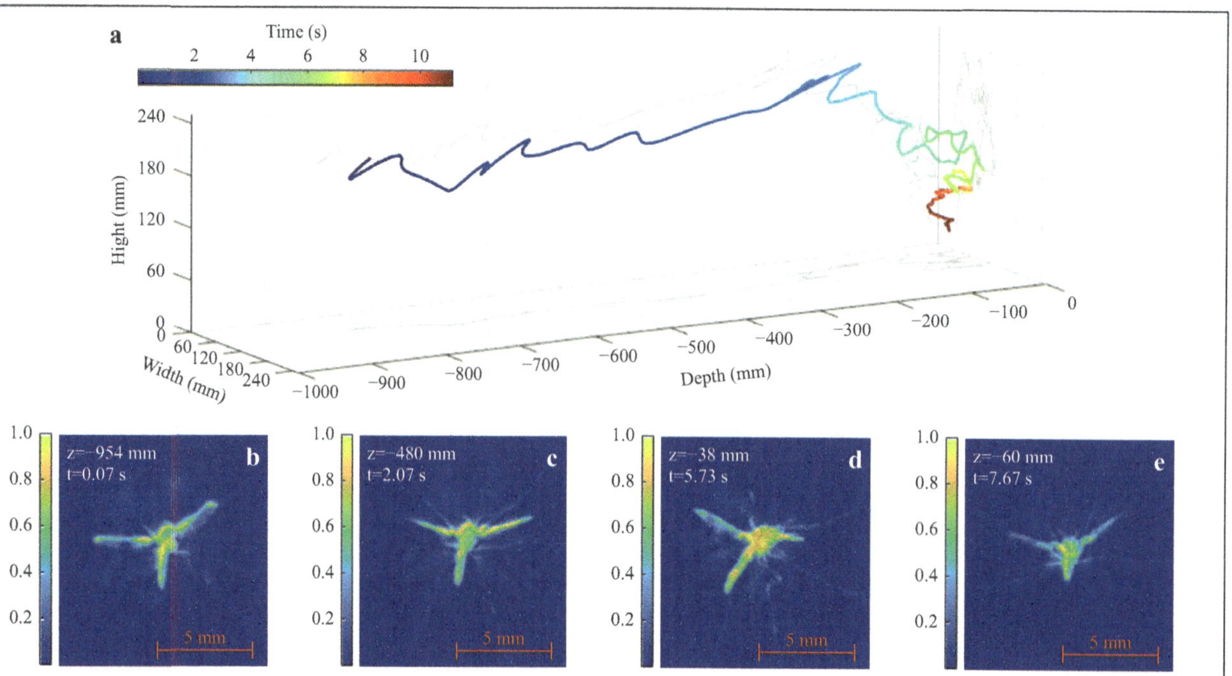

Fig. 2 **a** Experimental 3D flight track of a mosquito flying through the chamber and probing at the glass (grey denotes 2D projections). **b–e** Best-focus intensity reconstructions of mosquito at specified times (colour denotes normalised intensity in range [0 1]). *Note: Animations of mosquito flight tracks are provided as Supplementary Information.*

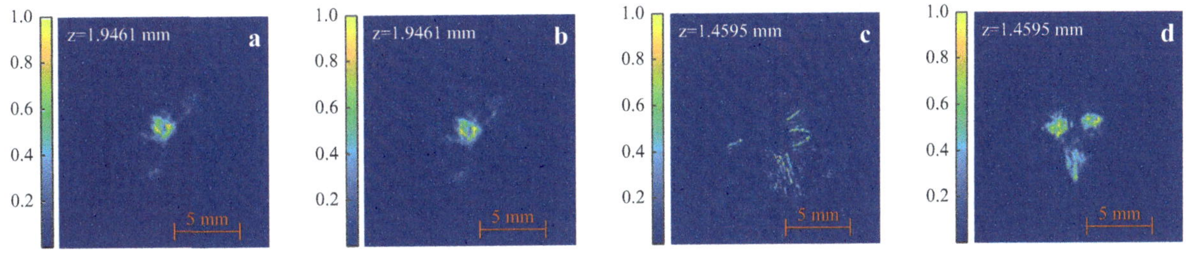

Fig. 3 Intensity reconstructions of a mosquito obscured by a bednet (color denotes normalised intensity). Flying mosquito with **a** moving frame and **b** stationary background subtraction. Mosquito crawling on net with **c** moving frame and **d** stationary background subtraction.

For efficiency in data collection, simulations were performed to assess the concepts introduced, and then physical validation data was collected on a smaller field-of-view to examine the *z*-axis resolution limits, but at the same EPS as the large FoV data described in Large Field of View Mosquito Reconstructions.

Background Removal of Bednet

For a more detailed examination of mosquito-bednet interaction close to the net, the obvious solution is to orientate the optical recording setup such that the bednet does not obscure the mosquitoes with respect to the camera. However, while examining interaction with a barrier-style bednet where there are multiple net interfaces orthogonal to one another[6], or in the case of examining mosquito-bednet-hole interaction[34], this may not be feasible. In this case, background subtraction of the bednet is necessary - either from a pre-recorded stationary bednet-only background, or using a moving-frame subtraction technique. The background removal technique exploits the well-known phenomena where partial holograms can be used to reconstruct an entire scene[35,36]. The partial holograms in this case are from the sections of the bednet that do not block diffracted light from the mosquito (i.e. the holes). Using a stationary background for subtraction offers the advantage of localising mosquitoes that have landed, although a significant constraint of this method is that the net must remain stationary between recording the bednet-only background hologram and the recording of mosquito interaction with the bednet. A moving background-subtraction technique (i.e. $I_F - I_{(F-n)}$ where I is the intensity hologram, F is the frame of interest and n is the number of frames prior) allows for some bednet movement during recording, but the diffraction pattern of a stationary mosquito is lost by the subtraction as can be seen previously in Fig. 3c. Further techniques to join tracks that end and start in the same position could be implemented, but reconstruction of mosquito orientation would be lost during this time.

To implement this background removal, a number of assumptions were made. Firstly, it is assumed that the detector is linear and that the optical fields sit within the linear range of the detector. Background removal (or net removal) can either be computed by subtraction ($I_O - I_{BG}$) or division (I_O / I_{BG}). Removal by division yields a more mathematically correct technique, as the wavefronts of the object and background add coherently at the hologram plane. However, the subtraction approach has been found to work well for obscured object localisation, resulting in a smoother and less-noisy focus metric curve with respect to the *z*-axis.The negatively valued pixels (using subtraction) in the mosquito position on the hologram produce a positive intensity on back propagation to the object space. The reconstructed mosquito is positively valued, while the background is close to zero (except for the diffracted waves). This makes the positively-valued reconstructed object contribute more to the focus metric calculations – while the close-to-zero background contributes less to the focus metric calculation – resulting in a more robust focus metric curve. For the background division method, the resultant background is close to 1, so any variations in the background due to noise and/or the twin-image will have a higher contribution to the focus metric calculation.

To demonstrate this, Fig. 4 shows the Tamura of intensity focus metric curves of a mosquito flying behind a bednet using both background removal methods, for two frames in the time domain. For both frames, the subtraction-based focus metric (solid line) retains a single maximum close to the plane of best-focus. As such, the background subtraction technique was preferred for obscured object localisation. However, for the second frame, the dashed blue line indicates that the division-based method has failed to identify the plane of best-focus of the mosquito. Evidence of a local variation in the curve at the plane of best focus implies that further image processing of holograms using the divide method may yield successful results, and investigating the interplay between focus metric calculations and background removal methods is the subject of future study (i.e. testing different combinations).

From the non-obscured mosquito data in Supplementary Video 2 (see Supplementary Information), the mean *z*-axis

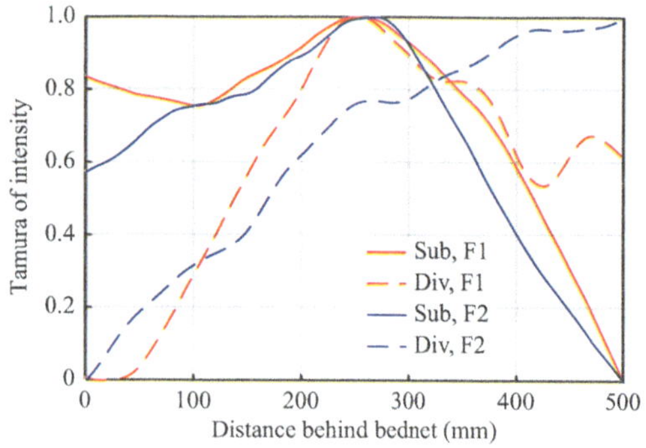

Fig. 4 Tamura of Intensity focus metric curves *vs* distance behind a bednet for an obscured mosquito using the subtraction (Sub) and division (Div) background removal techniques for two video frames (F1, F2).

position and standard deviation for frame-to-frame position difference for 180 frames (6 s) of mosquito resting at the end of the video was −566.622 ± 1.729 mm and −567.465 ± 1.222 mm for the division and subtraction methods, respectively. The z-axis reconstruction plane increments were 2 mm, so the standard deviation is less than the distance between two reconstructed planes for both methods. However, the perceived detail and quality of intensity reconstruction for the subtraction method was deemed to be better than background removal using division, and so the subtraction method was preferred in both the case of a mosquito not obscured by a bednet and of a mosquito obscured by a bednet.

Fig. 5a shows the simulated results of localisation error (in z) *vs* z-axis position of a mosquito positioned 0–1000 mm behind a bednet, while Fig. 5b shows the focus metric curve (after bednet subtraction) of a mosquito positioned with a mosquito-bednet separation of $z =$ 1000 mm. The mean localisation errors were 1.16 mm, 5.17 mm, and 7.11 mm for EPS values of $\Delta x_e = 58.60$ μm, $\Delta x_e = 78.13$ μm, and $\Delta x_e = 97.67$ μm, respectively.

Fig. 6a shows the difference hologram obtained when an intensity recorded bednet-only hologram is subtracted from a mosquito-bednet hologram at the hologram plane (both recorded at an EPS of $\Delta x_e = 78.13$ μm). The subtraction at the hologram plane is more computationally efficient than subtracting at reconstructed planes, as it is a direct intensity-only subtraction. The difference hologram is then back-propagated, rather than computing two sets of back-propagation operations from the bednet-only hologram and combined mosquito-bednet hologram before subtraction. There is no mathematical difference in the complex reconstruction between a difference hologram being back propagated to a specific plane, as compared to both holograms being back propagated to the same plane before subtraction.

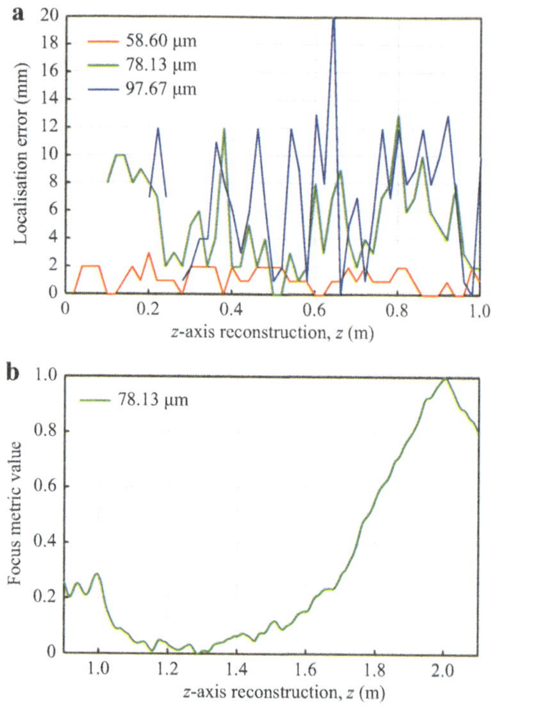

Fig. 5 a Axial placement error, z_{err} *vs* mosquito-bednet z-axis separation, z_s, for three effective pixel sizes: $\Delta x_e = 58.60, 78.13$, 97.67 μm using bednet subtraction - simulated data. **b** Focus metric curve (ToI) for a mosquito-bednet z-axis separation of 1000 mm at $\Delta x_e = 78.13$ μm.

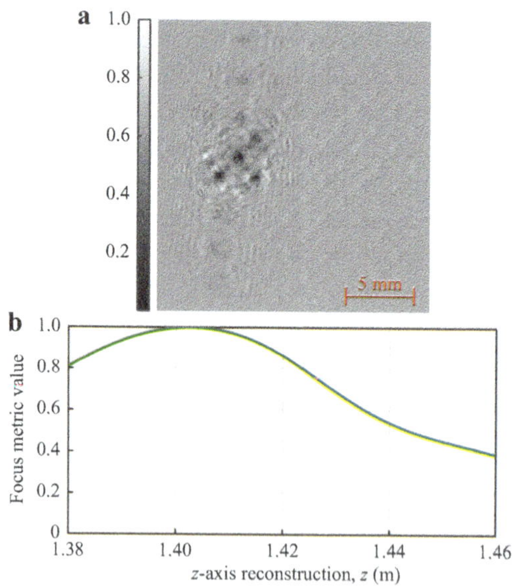

Fig. 6 a Difference hologram between mosquito-bednet hologram and bednet-only hologram (greyscale denotes normalised intensity in range [0 1]). Mosquito $z_s = 3$ mm behind a bednet positioned at $z = 1.40$ m and **b** resulting ToI focus metric curve.

In this case, the mosquito was positioned 3 mm behind a bednet positioned at $z = 1.40$ m. The ToI focus metric curve in Fig. 6b shows the calculated mosquito z-axis position. The maxima of the focus metric curve localised the mosquito at $z = 1.405$ m, yielding an error of $z_{err} = 2$ mm.

Fig. 7a shows physical results for the z-axis localisation error *vs* mosquito-bednet separation for each EPS value. The graph shows no definite trend as z_s increases, although the method failed to resolve mosquitoes for $z_s < 45$ mm for an EPS of $\Delta x_e = 97.67$ μm. The mean localisation error for $z_s \geq 45$ mm was calculated for fair comparison between EPS values. Errors of 2.75 mm, 4.75 mm, and 9.04 mm were obtained for EPS values of 58.60 μm, 78.13 μm, and 97.67 μm, respectively. Fig. 7b shows the intensity reconstruction at the plane of best focus of a mosquito positioned $z_s = 3$ mm behind the bednet. Qualitatively, the reconstruction using an intensity-subtraction appears much sharper than the reconstruction using the CCoFM method, and some features of the mosquito can be identified, such as the wings (top right and left), body (centre), and legs (bottom left).

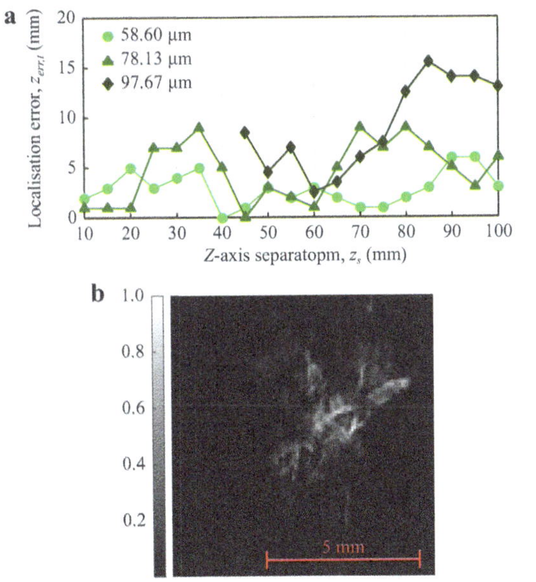

Fig. 7 a Axial placement error, z_{err} *vs* mosquito-bednet z-axis separation, z_s, for three effective pixel sizes: $\Delta x_e = 58.60, 78.13, 97.67$ μm using the background removal z-axis localisation. **b** Intensity reconstruction of a mosquito $z_s = 3$ mm behind the bednet (greyscale denotes normalised intensity in range [0 1]).

Alternative Method: Cascade-Correlation of Focus Metric (CCoFM)

The Cascade-Correlation of Focus Metric (CCoFM) is suitable for applications where the bednet may move during recording and accurate localisation below a minimum resolvable z-axis distance is not required as long as the 'time-in-zone' can be attained by continuing to track the mosquito in 2D XY coordinates. This method could also be used to resolve multiple occluded mosquitoes in the axial direction with potential future applications in recording mosquito swarming behaviours.

Previous work on multi-object localisation[28] indicates that the ToI and edge gradient focus metric curves appear to merge into an elongated curve with a single peak as the z-axis separation of multiple objects decreases. With a mosquito at 35 mm behind the bednet, the ToI focus metric curve becomes elongated (for the case where the bednet is not subtracted) and, at an EPS of 58.60 μm, no well separated maxima are observed. More detail on this feature is given in Materials and Methods.

Fig. 8a shows the z-axis localisation error *vs* mosquito-bednet z-axis separation for each of the EPS values obtained from physical data. As previously discussed, a different mosquito-only mask was used for each EPS value during the cross-correlation process, as the Q-value of the focus metric curve decreases as EPS increases. The CCoFM method failed to resolve the mosquito and bednet as separate objects in the z-direction for z-axis separations of 35 mm, 50 mm, and 60 mm for EPS values of 58.60 μm, 78.13 μm, and 97.67 μm, respectively. For fair comparison across all EPS values, the mean localisation error for $z_s \geq 60$ mm was calculated, yielding results of 9.22 mm, 9.23 mm, and 10.78 mm. Fig. 8b shows the reconstructed image at the plane of best focus ($z = 1.45$ m) of a mosquito

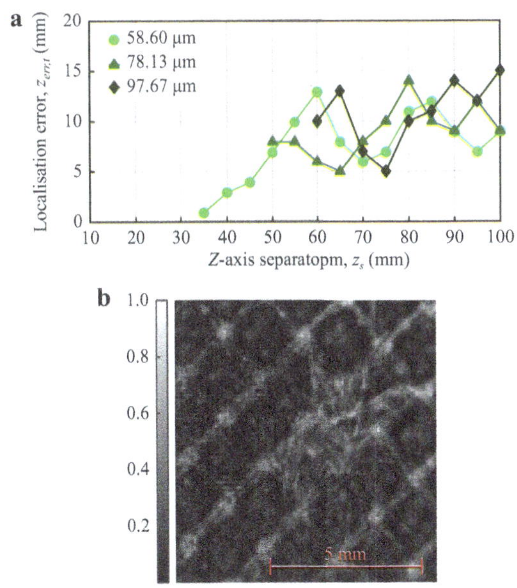

Fig. 8 a Axial placement error, z_{err} *vs* mosquito-bednet z-axis separation, z_s, for three effective pixel sizes: $\Delta x_e = 58.60, 78.13,$ 97.67 μm using the CCoFM z-axis localisation - physical data. **b** intensity reconstruction of a mosquito $z_s = 50$ mm behind the bednet (greyscale denotes normalised intensity in range [0 1]).

50 mm behind the bednet positioned at $z = 1.40$ m, recorded with an EPS of 78.13 μm as an example of the z-axis resolution limit for this EPS using the CCoFM method. Qualitatively, it is hard to define the shape of the mosquito and the resultant sharpness of the mosquito reconstruction. However, the *XY* localisation method demonstrated in XY Localisation combined with the CCoFM z-axis method successfully identifies and determines the mosquito 3D positional coordinates.

Discussion and Conclusions

A combination of physical and simulated results have been used to demonstrate that digital holography provides a suitable method for the 3D localisation of mosquitoes in a large field of view compared to similar previously reported digital holographic applications. The expansion on previous work on the assessment of effective pixel size *vs* localisation error and minimum resolvable distance between two objects provided the parameters to design a physical experimental large field of view setup (240 × 240 × 1000 mm). Intensity reconstructions of a free-flying mosquito recorded using this setup indicate that finer features of a moving mosquito can be defined, and the mosquito can be localised in 3D coordinates in a large field of view. Simulations allowed for a large number of data points to be collected on object localisation of a mosquito obscured by a bednet, and the upper z-axis resolution limit of the background removal technique was defined. The use of physical data to validate the method and assess the lower z-axis resolution limits indicate that the methods presented in this paper perform well with real data.

The use of an effective pixel size (imaging resolution) throughout this manuscript allows the recording volume to be scalable in terms of optical hardware components. When scaling the optical hardware, the effective pixel size, wavelength, and mosquito size are retained. However, by using a camera with a larger number of pixels, high-quality Fresnel lenses or holographic optical elements may provide a method to scale the recording volume to a larger size. However, it is suspected that a Fresnel lens would produce a large amount of forward scattering onto the hologram that would likely impact on the reconstruction and localisation accuracy of a mosquito. A feasibility study on the use of Fresnel lenses and/or holographic optical elements as a hardware option for increasing the recording volume will be conducted in future work.

Table 1 summarises the physical results presented earlier in the paper in terms of minimum resolvable distance between a bednet and a mosquito and mean localisation error for the two proposed z-axis localisation methods and each of the EPS values.

Table 1 Largest unresolvable z-axis separation and mean localisation error for the two proposed z-axis localisation methods and three effective pixel sizes - physical results.

Method	CCoFM		Background Removal	
	Resolvable $\geqslant z_s$, mm	Mean z_{err}, mm	Resolvable $\geqslant z_s$, mm	Mean z_{err}, mm
$\Delta x_e = 58.60$ μm	35	9.22	3	2.75
$\Delta x_e = 78.13$ μm	50	9.23	3	4.75
$\Delta x_e = 97.67$ μm	60	10.78	45	9.04

The background removal of bednet can be used when precise mosquito-bednet 3D tracking very close to the bednet is required, but is strictly limited by the condition that the bednet cannot move between the bednet-only hologram being recorded and the end of the main recording of mosquito-bednet interaction. This limitation could be relaxed using a moving frame subtraction technique, although this method would not be suitable if the mosquito is resting, crawling, or moving very slowly in the lateral direction, as the diffraction patterns will subtract from one another. Examining the interaction between mosquitoes and holes in the surface of a bednet (a typical fault after regular bednet usage) is one potential application. A previous study used a method of having a very short depth of field that only covered the thickness of the net in the z-direction[34]. The mosquito would therefore go in and out of sharp focus as it crosses the *XY* plane of the hole. This paper examines the time between a mosquito entering a dwelling to successfully passing through the bednet hole but does not explore the flight kinematics in detail. The lack of true 3D positional data makes it difficult to determine if the mosquito is randomly bouncing or searching the bednet through subsequent contacts, or not making contact at all. The background removal method could improve a study such as this. The background removal has other novel biological applications: the process enables the 3D kinematics of a mosquito approaching a bednet to be accurately described up to the surface of a net (2D imaging has an error in the unknown third axis), as well as accurately recording insecticide dosages to a mosquito freely interacting with an insecticide-soaked net (rather than lab-based *in-vitro* insecticide dosage tests).

Fig. 9a, b show reconstructed images of the same mosquito from a mosquito-bednet hologram after bednet subtraction and without subtraction, respectively. In both cases, the mosquito was positioned $z_s = 20$ mm behind the bednet and recorded at an EPS of $\Delta x_e = 78.13$ μm. Accurate reconstruction of the finer features of the mosquito, such as its wings, body, and legs, may make

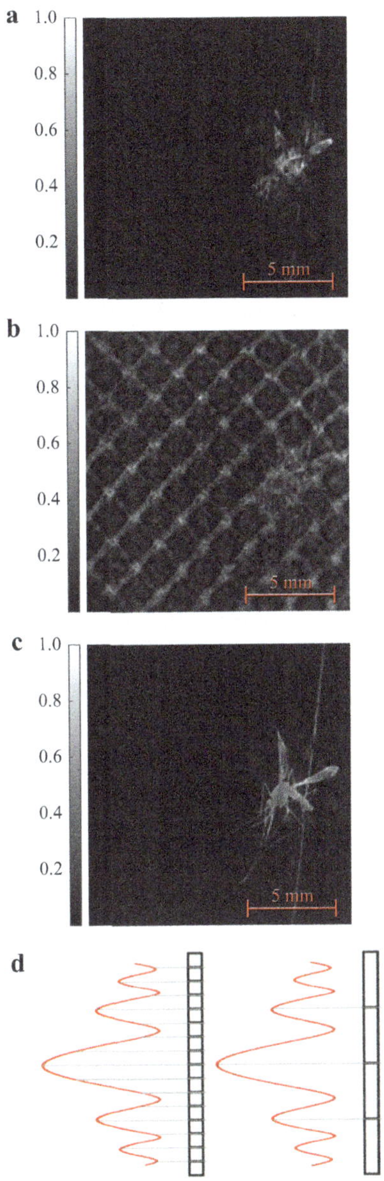

Fig. 9 Intensity reconstruction of a mosquito located $z_s = 20$ mm behind a bednet positioned at $z = 1.40$ m **a** with and **b** without background net subtraction. **c** Reconstruction of the same mosquito recorded with no bednet (mosquito-only hologram). **d** Diagram to show sampling of diffracted wavefront with smaller and larger EPS.

orientation calculations possible in future work, which may give further insights into mosquito-bednet-hole interaction. Fig. 9c, of the same mosquito not obscured by a bednet, is shown for qualitative comparison.

It is interesting to note the large change in minimum resolvable distance for the background removal method between EPS values of $\Delta x_e = 78.13$ μm and $\Delta x_e = 97.67$ μm. As previously discussed, the EPS determines the sampling rate for which an interference pattern of a hologram is recorded and stored - Fig. 9d shows this qualitatively. The recorded diffraction pattern on the hologram plane is the superposition of the diffracted waves from: the mosquito, the bednet, plus the undisturbed reference wave. It is well known that the spatial extent in the *XY* plane of a diffraction pattern for a given object size increases as propagation distance increases. Therefore, increasing the object-to-hologram distance or, in this case, the mosquito-to-bednet distance may provide better distinction at the hologram between bednet and mosquito. Recording at a lower sampling rate (higher EPS) averages the combined signal, and if the Nyquist limit lies at an EPS value between $\Delta x_e = 78.13$ μm and $\Delta x_e = 97.67$ μm, the finer multi-object interference fringes between the mosquito and the bednet at lower *z*-axis separations may be lost. Further investigation into this, to determine if a critical Δx_e exists, may be conducted in future work.

The minimum resolvable distance for the CCoFM method increases between each of the EPS values of 58.60 μm, 78.13 μm, and 97.67 μm, which is in-line with the result shown in Fig. 1 and was expected due to the decreasing *Q*-value in the single object focus metric curves as EPS increases. With a decreased *Q*-value, the superposition of two distinct curves into a single curve occurs at a larger *z*-axis separation and so a larger minimum resolvable distance is expected.

The CCoFM provides a methodology that is inherently temporally stable, as bednet background removal is not conducted and therefore movement of the bednet throughout the duration of a hologram time-series video recording will not influence the reconstruction and calculation of the focus metric curve. The downside of this method is that there are minimum resolvable distances between the bednet and mosquito, so accurate determination of close contact mosquito-bednet interaction is lost as the CCoFM second-pass focus metric becomes a curve with a single maxima. Fig. 10a shows the *XY* position of a dead mosquito manually moved in the *XY* plane over 30 frames (1 second) of a recorded video, where the *z*-axis position of the mosquito relative to the bednet was approximately $z_s = 20$ mm throughout the recording. Fig. 10b shows the CCoFM second-pass curves corresponding to three distinct frames, indicating a single peak for the mosquito-bednet reconstruction, meaning that resolving the two separate objects in the *z*-axis was unsuccessful using this method at this *z*-axis separation.

Although the *z*-axis separation was not resolved in the case above, it can be concluded that a mosquito localised in *XY* featuring a single maxima in the second-pass CCoFM curve is located at $z_{mos} < z_{res}$, where z_{res} is the minimum resolvable distance of that particular setup as shown in Table 1. A time-series mosquito position dataset with this

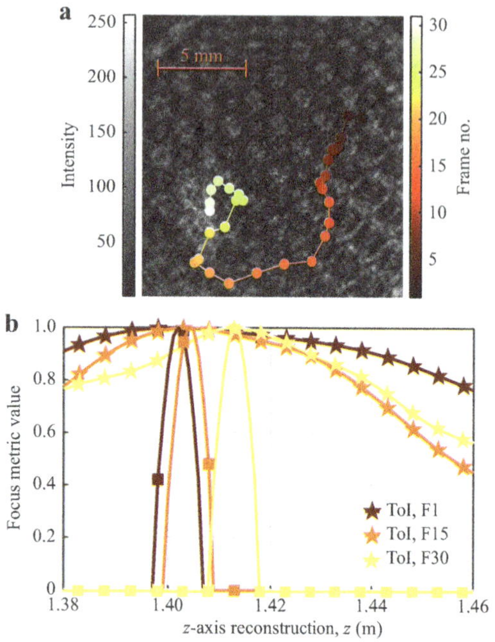

Fig. 10 a Time-series *XY* positions of a manually moved mosquito located $z_s \approx 20$ mm behind a bednet positioned at $z = 1.40$ m. **b** Tamura of Intensity focus metric (stars) and second-pass cross-correlation (squares) displaying a single peak (Frames 1, 15, 30).

condition included would allow a single mosquito to be tracked pre-, during, and post- bednet interaction even though the precise 3D interaction for $z_{mos} < z_{res}$ would not be determined. In previous studies, mosquito behaviour has been quantified into distinct categories, with movement close to the bednet below a certain velocity threshold classified as 'resting'[37]. The CCoFM method combined with an appropriate tracking algorithm could therefore be useful in accurately determining resting time(s) and estimating a pseudo-dosage of insecticide that a mosquito receives.

In this paper, we have demonstrated that digital holography offers the potential to examine 3D mosquito flight behaviour in a large volume where conventional imaging systems may be less effective. A physical experimental setup for the large field of view (240 × 240 × 1000 mm) recording of in-line digital holograms was presented. Intensity reconstructions of mosquitoes in full-flight were also presented, demonstrating that 3D localisation of a flying mosquito and reconstruction of smaller fast-moving features, such as the legs, body, and wings, was possible using digital holographic techniques. Applying this methodology to a time-series of holograms allows the accurate determination of mosquito tracks in 3D coordinates. Simulations and physical data were used to examine the concept of localisation of a mosquito obscured by a bednet, as accurately determining mosquito-bednet interaction is of particular biological importance in the development of intervention methods to reduce malaria transmission in humans. A novel Cascade-Correlation of Focus Metric was introduced for the accurate reconstruction and localisation of a mosquito occluded by a bednet and compared to a simple bednet subtraction technique. The benefits and limits of each approach were considered with regards to, experimental design, *z*-axis resolution limits, and upper limits for mosquito-bednet distance. Potential biological applications were also discussed, and it is anticipated that biologically significant experiments will be performed in future work. The result of this study is a scalable digital holographic methodology to examine mosquito flight in 3D at a level of accuracy previously only seen in 2D imaging of a smaller volume, and the ability to resolve 3D mosquito-bednet interaction of mosquitoes occluded by a bednet for the first time. For accurate and reliable kinematic analysis of mosquito flight close to a bednet, detailed 3D imaging is crucial to correct errors in the unknown third axis that are present when using 2D imaging, and digital holography offers a potential solution.

Materials and Methods

Recording Setup

A continuous wave 130 mW CrystaLaser DL785-150-SO with a wavelength of 785 nm provides illumination of the scene by expanding and spatially filtering the beam through an objective lens and aperture (expansion optics, EO). Two neutral density filters are placed between the laser head and spatial filter, consisting of OD = 1.5 and OD = 0.3, which reduces transmission of light to 1.7%, equating to approximately 2 mW of power. This would allow a theoretical increase in area of 64 × (8 × radius) if utilising the full 130 mW of laser power available for the same power density interacting with the object.

The collimating optic (CO) provides plane wave illumination through the object volume. The laser light is subjected to forward scattering at the mosquito, and the combined mosquito-object wave and plane reference wave then propagates through free space. This combined wave is focused through a focusing optic (FO) and imaging optic (IO) in a two-component telecentric system onto a sensor (CCD), where the interference pattern is recorded as a digital hologram. A schematic of the collimated cylindrical probe volume and telecentric demagnification system is shown in Fig. 11.

In the large-scale experiment (see Large Field of View Mosquito Reconstructions), the collimating optic (CO) and focusing optic (FO) were 355 mm diameter off-axis

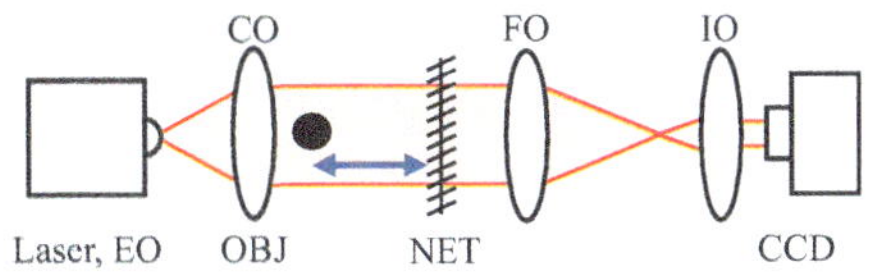

Fig. 11 Experimental setup for optical recording of a digital in-line hologram with a demagnifying two-lens telecentric system. CO: collimating optic, FO: focusing optic, IO: imaging optic, CCD: sensor array.

parabolic mirrors with a focal length of 2540 mm, set up in a Z-type interferometer configuration. The imaging optic was a 25.4 mm diameter plano-convex lens with a focal length of 175 mm, providing a demagnification factor (M_D) of 14.53 × and an effective pixel size of 79.93 μm in the recording volume. The sensor was a Ximea CB120RG-CM 4096 × 3072 pixel camera with pixel size of 5.5 μm.

For the bednet validation experiments (see Mosquitoes Obscured by a Bednet), the collimating optic was a 25.4 mm plano-convex lens with a focal length of 500 mm, and the two-component telecentric system consisted of 25.4 mm plano-convex lenses with 500 mm and 50 mm, 400 mm and 30 mm, and 500 mm and 30 mm focal lengths to yield effective pixel sizes of 58.60 μm ($M_D = 10.00$ ×), 78.13 μm ($M_D = 13.33$ ×), and 97.67 μm ($M_D = 16.67$ ×), respectively. These EPS values were chosen based on the results shown in Fig. 1 and previously conducted work[28]. The sensor was a Dalsa Genie Nano M1920 1936 × 1216 camera with a pixel size of 5.86 μm. These validation experiments were conducted before access to the OAP mirrors was possible, and are included to provide proof-of-concept physical data of the obscured-mosquito localisation techniques at a comparable EPS to the larger field-of-view experiments in Large Field of View Mosquito Reconstructions. This smaller system can be scaled to a larger recording volume by increasing the number of pixels on the sensor and using larger collimating and focusing optics, while maintaining the effective pixel size, wavelength, and object size.

Noise Analysis of Images

To assess the noise levels in the background images, the first 100 frames of data were assessed from the 12-bit Ximea recordings with a time stationary scene. The mean intensity across the frames was 1865 with a spatial standard deviation of 255 gray levels. However, with background subtraction, the standard deviation decreased to 80 grey levels. Methods to reduce this speckle such as using a shorter coherence length light source to reduce multiple reflections and scattering between the two panes of glass in the flight chamber will be the subject of further study. The number of pixels above 80% of the saturation (i.e. 80% of 4095) and below 20% are 1 and 1182, respectively (out of 9 million), meaning that the assumption that the recording is within the linear range of the CCD/CMOS array is valid. When a mosquito is introduced, these numbers are 1 and 3782, respectively, due to the mosquito blocking out some of the light, and hence the larger number of dark pixels. Along the time-axis of the first 100 images, the pixel-to-pixel standard deviation, for a random sample of 5 pixels, is 17.51, 6.46, 14.20, 14.22 and 21.16 grey levels, yielding a noise level of 1.2% for the standard deviation of 21.16 compared to the same pixel's mean intensity.

Fig. 12 shows a typical cross-section of a mosquito signal after a background subtraction. The grey level minimum of the mosquito signal is −1621 with diffraction patterns giving local minima at −913 and −596 visible on the left hand side, significantly above the temporal noise level of approx 20 grey levels and spatial noise level of approx 80 grey levels.

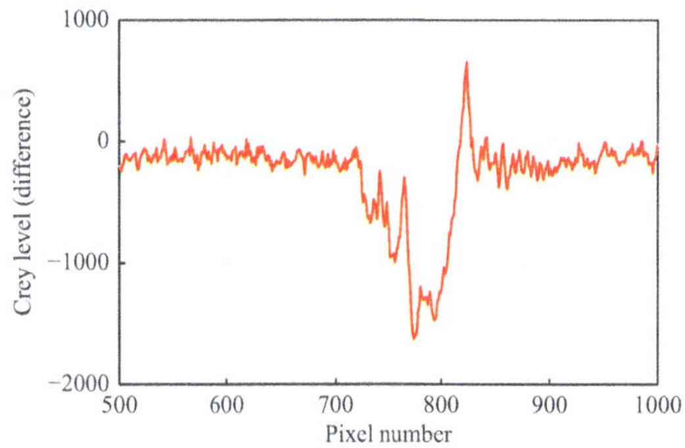

Fig. 12 Typical cross section of a mosquito hologram subject to background subtraction, indicating the grey level and pixel number.

Optical Properties of a Bednet and Mosquito

Bednets are composed of nylon fibres, which have an index of refraction of n = 1.5 and an approximate diameter of 100 μm (or approximately 125λ at 785 nm), which are twisted into threads of approximately 4-5 fibres each. The threads are woven into a regular net with an aperture size of approximately 2.5 mm (approx. 3000λ). The aperture size is set to block passage of mosquitoes whilst allowing flow of air to the occupants. The nylon fibres are optically opaque and hence the optical field after transmission is governed by the components passing through the apertures in the net structure. Hence the bednet can be defined as a transmission function where $T_B(x, y) = 0$ for all locations B corresponding to the nylon fibres and $T(x, y) = 1 \ \forall \ (x, y) \notin B$. Similarly, the mosquito can be considered as a transmission function $T_M(x, y) = 0$ for all points M corresponding to the mosquito and $T(x, y) = 1 \ \forall \ (x, y) \notin M$. The validity of these assumptions is supported by the comparisons between experimental and synthetic data used to evaluate the performance of the focus metric algorithms and in which the optical fields are manipulated by the aforementioned transmission functions.

The smallest feature of interest in the physical experiments are the legs of a mosquito, which have a diameter of 80 μm (or 100λ at 785 nm)[38]. The forward scattering numerical aperture is given by $\theta = 1.22\lambda/d$, where λ is the wavelength, and d is the diameter of the object. This yields a forward scattering numerical aperture of NA ~ 0.012. The much larger numerical aperture of the off-axis parabolic mirrors (NA ~ 0.07) means that the full scattering angle is captured, and the most relevant limiter of the optical system performance in this paper is the effective pixel size, which is a direct result of the discretised sampling of a diffraction pattern on a digital CCD/CMOS sensor.

Reconstruction of a Scene

The reconstruction of a scene from a digitally recorded hologram requires the calculation of the resultant reconstruction images by applying the discrete propagation formulae between multiple planes in small increments (z_{inc}) away from the hologram. The degree of focus of a reconstructed scene is highest at the z-axis position of the axial object location at the time of recording. The 3D position of the object can therefore be determined from a single hologram.

Digital Reconstruction

Reconstruction using the recorded digital holograms was achieved by performing a two-step thin-lens fourier transform operation to simulate propagation through the two-component telecentric system in reverse, with specific focal lengths according to the optics used to record the hologram. Back-propagation into the object space was performed using a Fresnel transfer function method[39]. The scene was reconstructed on multiple planes with a z-axis increment of $z_{inc} = 2$ mm and focus metrics were calculated for each mosquito found in a reconstructed plane.

XY Localisation

For a non-obscured free-flying mosquito, a simple thresholding technique can be used to identify the XY position of mosquitoes from a hologram. The hologram is thresholded to reveal distinctly bounded areas from the diffraction patterns of forward scattered objects, where the centre of mass of each area is the XY position of a mosquito.

However, for a scene involving a mosquito behind a bednet, additional image processing techniques are used to extract the lateral object position(s). In this case, a two-pass cascade-correlation technique is used. To explain the cross-correlation process, two intensity images need to be defined: a cropped mosquito-only mask, I_M, (shown by the dashed line in Fig. 13a); and a combined mosquito and bednet recorded hologram, $I_{M,B}$ (Fig. 13b). The first step is that I_M is cross-correlated against $I_{M,B}$, shown by:

$$C_1 = I_M * I_{M,B} \tag{1}$$

Where C_1 is the first-pass cross-correlation map shown in Fig. 14a. The primary peak in this cross-correlation map indicates the XY position of the mosquito, although several smaller peaks exist due to the holes in the bednet. Thresholding of the cross correlation map removes these smaller peaks, resulting in a thresholded map $C_{1,t}$. A threshold value of 0.18 was adequate to remove these smaller peaks without removing the primary peak. This threshold was found by cross-correlating the mask, I_M, with a bednet-only hologram, I_B, and the peak of this correlation map was found. The mask selection, in terms of mosquito orientation and distance from the hologram plane, is not as important as maybe expected. For example, for the case of a mosquito placed 100 mm behind a bednet positioned at $z = 900$ mm, a mask from a mosquito 1400 mm away from the hologram plane is more similar to the mosquito behind the bednet ($z = 1000$ mm) than the bednet itself ($z = 900$ mm). The absolute similarity between the mask and target is not as important as the difference in correlation between the bednet and the mosquito - the cross-correlation peak indicating the mosquito position can

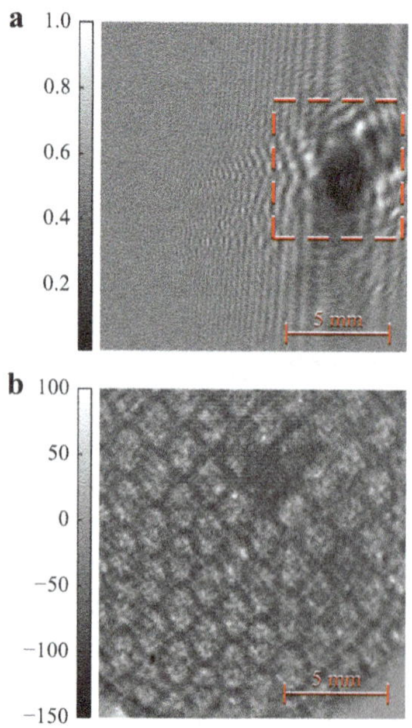

Fig. 13 a Mosquito-only hologram (mask outlined). **b** Mosquito-bednet hologram of mosquitoes 40 mm (left) and 55 mm (right) behind a bednet at $z = 1.40$ m.

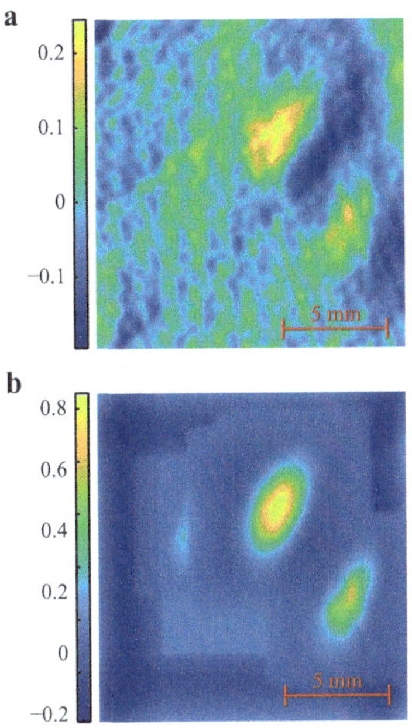

Fig. 14 **a** First-pass cross-correlation between mask and mosquito-bednet hologram. **b** Second-pass cross-correlation map, indicating the mosquito *XY* positions.

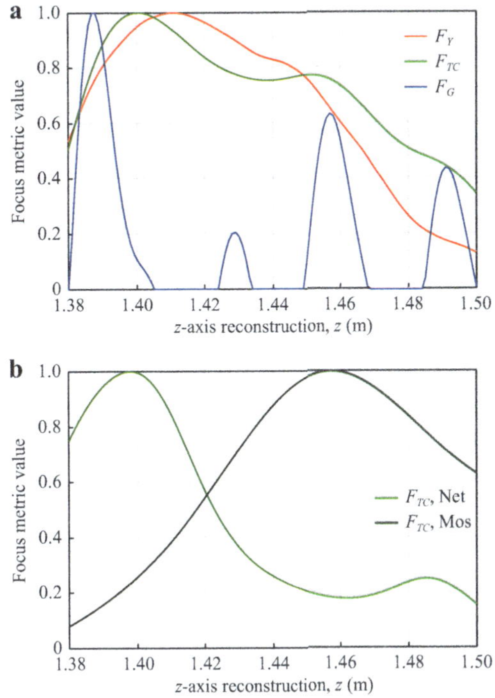

Fig. 15 **a** Focus metric curves from reconstruction of **a** mosquito-bednet hologram (net at $z = 1.40$ m, mosquito at 1.455 m), and **b** bednet-only and mosquito-only holograms.

be isolated from the background correlation map, especially after thresholding. Therefore, this mask can also be created using a simulated mosquito-only hologram or a pre-recorded reconstruction of an object of a similar size to a mosquito recorded at the same EPS and with the same incremental distance between reconstructed *z*-planes. To improve contrast and further refine the lateral position, a secondary mask is formed by an auto-correlation of I_M, such that:

$$I_{M2} = I_M * I_M \quad (2)$$

This step calculates a 2D gaussian function using the original mask, and this secondary mask is cross-correlated against the thresholded first-pass cross correlation by:

$$C_2 = I_{M2} * C_{1,t} \quad (3)$$

where C2 is the second-pass cross correlation map as shown in Fig. 14b. This correlation map indicates the position of two mosquitoes behind the bednet. Intuitively, the easier to spot mosquito in Fig. 13b corresponds to a stronger cross-correlation in Fig. 14b.

Focus Metrics for Z-axis localisation

The peak of a focus metric determines the *Z*-axis position of a reconstructed object. Previous work indicates that the proposed *particle signature function* focus metric[25,28] works well for resolving multiple objects, as shown in Effect of Imaging Resolution on Localisation Accuracy, but is less effective in the case of a mosquito obscured by a bednet due to the disturbed reference wave diffracting through the lattice-structure. Fig. 15a, below, shows the focus metric curves of a dead *anopheles gambiae* mosquito placed $z = 55$ mm behind a bednet positioned at $z = 1.40$ m (mosquito on the right in Fig. 13b). Examining the edge-gradient and Tamura-of-Intensity focus metric curves reveals that these methods are less susceptible to the phase profile created by the disturbed reference wave through the lattice-structured bednet and that the ToI focus metric has two maxima in the approximate *z*-axis positions of the bednet and mosquito. The phase-based *particle signature function* focus metric curve contains additional peaks which give false-positive mosquito positions.

Fig. 15b shows the focus metric curves obtained via reconstruction of a bednet-only hologram (Net) and mosquito-only hologram (Mos), where all three focus metrics localise the bednet and mosquito *z*-axis positions, but only the data for the ToI metric is displayed. Comparing this to Fig. 15a, it is apparent that the mosquito-bednet graph appears to resemble a superposition of the two separate curves in the ToI and edge-gradient based focus metrics. The later method proposed in Alternative Method: Cascade-Correlation of Focus Metric (CCoFM) aims to resolve these combined curves as they merge into a single elongated peak at lower *z*-axis separations between the bednet and mosquito.

Simulation

Simulations were used in Mosquitoes Obscured by a Bednet to determine the limits of the proposed methods. Simulated holograms were created by taking in-focus images of a mosquito and bednet separately. The images were both binarised, and the mosquito image was forward propagated using a Fresnel Transfer Function[39] up to the position of the bednet (for any desired mosquito-bednet separation). The resultant complex hologram was multiplied by the binarised bednet image to simulate the strands of the bednet (as 0 intensity) blocking some parts of the forward propagating light. This combined image was then propagated the desired distance to the back focal plane of a two-component telecentric system. A pair of fourier transform operations simulated propagation through the system and onto a CCD. The image was then resized onto the correct number of pixels as required by the demagnification factor, and the absolute value of each pixel was determined to simulate intensity data (rather than complex-valued amplitude-phase data). The intensity was remapped in integer values [0 4095] to simulate a 12-bit recording device. For back-propagation and object reconstruction, the same process was used as for the physical data.

Cross-Correlation of Focus Metric

The CCoFM follows a similar methodology as the *XY* localisation technique described in XY Localisation. First, a mosquito-only focus metric mask, F_M, is defined by cropping a Tamura of Intensity focus curve from the reconstruction of a mosquito-only hologram at the dashed lines shown in Fig. 16a. This mask can also be obtained from simulated data of an object of similar size, or from a mosquito positioned in a different *z*-axis position. An important point to note on mask selection is that the *z*-axis reconstruction increment (z_{inc}) and effective pixel size (Δx_e) must be the same for both the mask and reconstruction of interest, and the mask should be appropriately symmetrical. A combined mosquito and bednet focus metric, $F_{M,B}$, as shown in Fig. 16b, appears to be the superposition of the bednet-only focus metric and mosquito-only focus metric shown in Fig. 16a. Cross correlating F_M and $F_{M,B}$ yields the cross correlation map F_{C1}:

$$F_{C1} = F_M * F_{M,B} \quad (4)$$

This cross-correlation map is then thresholded to retain values > 0.4 as shown by CCM1 in Fig. 16c and assigned as $F_{C1,t}$. An autocorrelation of the mask F_M yields a secondary mask, F_{M2}:

$$F_{M2} = F_M * F_M \quad (5)$$

The secondary mask, F_{M2}, is cross-correlated against the thresholded first-pass cross correlation map $F_{C1,t}$ to yield a final cross-correlation map, F_{C2}, with two well-separated maxima indicating the bednet and mosquito *z*-axis positions as shown by CCM2 in Fig. 16c.

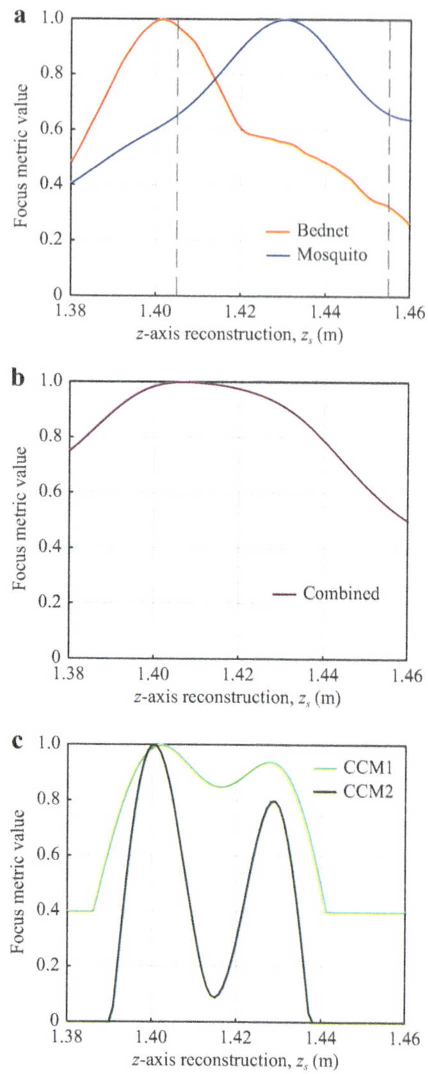

Fig. 16 Tamura-of-Intensity focus metric *vs* *z*-axis reconstruction distance: **a** separate single object curves of mosquito and bednet; **b** obtained curve of mosquito behind bednet; **c** cross-correlation curves demonstrating improved *z*-axis localisation and separation of the two objects.

Acknowledgements

This work was supported by the Engineering and Physical Sciences Research Council under Grant Number EP/N509796/1. A UK Medical Research Council award (MRC: MR/M011941/1) also contributed to this work.

Conflict of interest

The authors declare no conflicts of interest.

Supplementary information is available for this paper at https://doi.org/10.37188/lam.2022.020.

There are three supplementary documents attached to this manuscript. An animated version of Fig. 2 is provided, which shows the mosquito reconstruction at each frame. Another animation is also provided of a different mosquito flight track, where the mosquito eventually comes to rest on the ceiling of the flight chamber. A supplementary document on the simulation steps for modelling forward propagation onto a detector plane is also included.

Received: 28 October 2021 Revised: 23 February 2022 Accepted: 24 February 2022
Published online: 09 June 2022

References

1. Owen, R. B. & Zozulya, A. A. In-line digital holographic sensor for monitoring and characterizing marine particulates. *Optical Engineering* **39**, 2187-2197 (2000).
2. Malkiel, E., Abras, J. N. & Katz, J. Automated scanning and measurements of particle distributions within a holographic reconstructed volume. *Measurement Science and Technology* **15**, 601-612 (2004).
3. Gopalan, B., Malkiel, E. & Katz, J. Experimental investigation of turbulent diffusion of slightly buoyant droplets in locally isotropic turbulence. *Physics of Fluids* **20**, 095102 (2008).
4. Merola, F. et al. Tomographic flow cytometry by digital holography. *Light: Science & Applications* **6**, e16241 (2017).
5. Dyomin, V. et al. Holography of particles for diagnostics tasks [Invited]. *Applied Optics* **58**, G300-G310 (2019).
6. Murray, G. P. D. et al. Barrier bednets target malaria vectors and expand the range of usable insecticides. *Nature Microbiology* **5**, 40-47 (2020).
7. Bhatt, S. et al. The effect of malaria control on Plasmodium falciparum in Africa between 2000 and 2015. *Nature* **526**, 207-211 (2015).
8. World Health Organization. World Malaria Report 2020: 20 Years of Global Progress and Challenges. (Geneva: World Health Organization, 2020).
9. Sheng, J. et al. Digital holographic microscopy reveals prey-induced changes in swimming behavior of predatory dinoflagellates. *Proceedings of the National Academy of Sciences of the United States of America* **104**, 17512-17517 (2007).
10. Lu, J. et al. Lagrangian particle tracking in three dimensions via single-camera in-line digital holography. *New Journal of Physics* **10**, 125013 (2008).
11. Mundt, J. & Kreis, T. M. Digital holographic recording and reconstruction of large scale objects for metrology and display. *Optical Engineering* **49**, 125801 (2010).
12. Memmolo, P. et al. Refocusing criterion via sparsity measurements in digital holography. *Optics Letters* **39**, 4719-4722 (2014).
13. Singh, A. K. et al. Looking through a diffuser and around an opaque surface: a holographic approach. *Optics Express* **22**, 7694-7701 (2014).
14. Singh, A. K. et al. Exploiting scattering media for exploring 3D objects. *Light: Science & Applications* **6**, e16219 (2017).
15. Mazumdar, Y. C. et al. Megahertz-rate shock-wave distortion cancellation via phase conjugate digital in-line holography. *Nature Communications* **11**, 1129 (2020).
16. Trolinger, J. D. et al. Spatially and temporally resolved diagnostics of dense sprays using gated, femtosecond, digital holography. Proceedings of SPIE 10373, Applied Optical Metrology II. San Diego: SPIE, 2017, 25.
17. Edrei, E. & Scarcelli, G. Optical imaging through dynamic turbid media using the Fourier-domain shower-curtain effect. *Optica* **3**, 71-74 (2016).
18. Valzania, L., Zolliker, P. & Hack, E. Coherent reconstruction of a textile and a hidden object with terahertz radiation. *Optica* **6**, 518-523 (2019).
19. Butail, S. et al. Reconstructing the flight kinematics of swarming and mating in wild mosquitoes. *Journal of the Royal Society Interface* **9**, 2624-2638 (2012).
20. Voloshin, V. et al. Diffuse retro-reflective imaging for improved video tracking of mosquitoes at human baited bednets. *Royal Society Open Science* **7**, 191951 (2020).
21. Kröner, C. et al. 3D tracking of mosquitoes: a field compatible technique to understand malaria vector behaviour. 3D Image Acquisition and Display: Technology, Perception and Applications 2016. Heidelberg Germany: OSA, 2016, TW5A. 4.
22. Brydegaard, M. & Svanberg, S. Photonic monitoring of atmospheric and aquatic fauna. *Laser and Photonics Reviews* **12**, 1800135 (2018).
23. Kirkeby, C., Wellenreuther, M. & Brydegaard, M. Observations of movement dynamics of flying insects using high resolution lidar. *Scientific Reports* **6**, 29083 (2016).
24. Dubois, F. et al. Focus plane detection criteria in digital holography microscopy by amplitude analysis. *Optics Express* **14**, 5895-5908 (2006).
25. DeJong, J. & Meng, H. Digital holographic particle validation via complex wave. *Applied Optics* **46**, 7652-7661 (2007).
26. Gao, J. et al. Uncertainty characterization of particle depth measurement using digital in-line holography and the hybrid method. *Optics Express* **21**, 26432-26449 (2013).
27. Hall, M. L., Towers, C. E. & Towers, D. P. Correction of 3D localisation error of multiple objects in close-proximity in digital holography. Adaptive Optics: Analysis, Methods & Systems 2020. Washington, DC: OSA, 2020, JW2A. 14.
28. Hall, M. L. et al. Improved three-dimensional localization of multiple small objects in close proximity in digital holography. *Applied Optics* **60**, A285-A295 (2021).
29. Tamamitsu, M. et al. A robust holographic autofocusing criterion based on edge sparsity: comparison of Gini index and Tamura coefficient for holographic autofocusing based on the edge sparsity of the complex optical wavefront. Proceedings of SPIE 10503, Quantitative Phase Imaging IV. San Francisco, California, United States: SPIE, 2017.
30. Angarita-Jaimes, N. C. et al. A novel video-tracking system to quantify the behaviour of nocturnal mosquitoes attacking human hosts in the field. *Journal of the Royal Society Interface* **13**, 20150974 (2016).
31. Hughes, A. et al. Quantifying late-stage host-seeking behaviour of Anopheles gambiae at the insecticidal net interface using a baited-box bioassay. *Malaria Journal* **19**, 140 (2020).
32. Ling, H. J. & Katz, J. Separating twin images and locating the center of a microparticle in dense suspensions using correlations among reconstructed fields of two parallel holograms. *Applied Optics* **53**, G1-G11 (2014).
33. Oe, K. & Nomura, T. Twin-image reduction method using a diffuser for phase imaging in-line digital holography. *Applied Optics* **57**, 5652-5656 (2018).
34. Dickerson, A. K., Olvera, A. & Luc, Y. Void entry by Aedes aegypti (Diptera: Culicidae) mosquitoes is lower than would be expected by a randomized search. *Journal of Insect Science* **18**, 9 (2018).
35. Vest, C. M. Holographic Interferometry. (Berkeley: Wiley, 1979).
36. Hariharan, P. Basics of Holography. (Cambridge: Cambridge University Press, 2002).
37. Parker, J. E. A. et al. Infrared video tracking of Anopheles gambiae at insecticide-treated bed nets reveals rapid decisive impact after brief localised net contact. *Scientific Reports* **5**, 13392 (2015).
38. Wu, C. W., Kong, X. Q. & Wu, D. Micronanostructures of the scales on a mosquitoos legs and their role in weight support. *Physical Review E* **76**, 017301 (2007).
39. Poon, T. C. & Liu, J. P. Introduction to Modern Digital Holography with MATLAB. (Cambridge: Cambridge University Press, 2014).

Official journal of the JHL 2689-9620
www.light-am.com

Article Open Access

In-process monitoring in laser grooving with line-shaped femtosecond pulses using optical coherence tomography

Satoshi Hasegawa[1,*], Masatoshi Fujimoto[2], Toshihisa Atsumi[2] and Yoshio Hayasaki[1]

Abstract

A line-shaped beam is useful for increasing the processing speed in laser grooving and scribing. In laser grooving, depth control of the processed structure is important for performing precise processing. In this paper, in-process monitoring of the depth of a structure formed by femtosecond laser processing with a line-shaped beam using swept-source optical coherence tomography (SS-OCT) was demonstrated. In the evaluation of the SS-OCT system, the depth resolution, measurement accuracy, and axial measurable range were 15.8 μm, ±2.5 μm and 5.3 mm, respectively. In laser grooving, the structural shape and the distribution of deposited debris were successfully monitored. The measured depth agreed well with the depth obtained using a laser confocal microscope. The proposed method will be effective for precise laser processing with feedback control of the laser parameters based on in-process monitoring of the processed structure.

Keywords: Optical coherence tomography, Computer-generated hologram, Femtosecond laser processing, In-process monitoring

Introduction

A line-shaped beam formed by a cylindrical lens and diffractive optical elements (DOEs) has been employed for increasing the processing speed in laser grooving of silicon wafers[1], laser scribing of indium tin oxide (ITO) films[2] and thin-film solar cells[3], and laser cleaving of glass substrates[4]. In laser grooving and scribing using a conventional Gaussian beam, undesired melting of the material due to thermal conduction, called the heat-affected zone (HAZ), occurs around the processed area if the intensity at both ends of the Gaussian profile is below the ablation threshold of the substrate material. This becomes more remarkable when nanosecond and picosecond pulses are employed. Also, the center peak intensity of the Gaussian beam may cause serious damage to the substrate. A line-shaped beam is useful for reducing the HAZ and suppressing damage to the substrate because of its uniform intensity profile.

In laser grooving and scribing, control of the structural depth is important for performing precise processing. To obtain the desired depth of the processed structure, optimal processing parameters (e.g., wavelength, pulse energy, pulse duration, pulse repetition rate, beam spot size and its focal position, and scanning speed) should be explored according to the target material. Due to the lack of a general theoretical framework for the physical phenomena involved in laser processing, a complex set of parameters is usually explored experimentally. Verification in post-processing using scanning electron microscopy (SEM) and atomic force microscopy (AFM) provides a wealth of

Correspondence: Satoshi Hasegawa (hasegawa_s@cc.utsunomiya-u.ac.jp)
[1]Center for Optical Research and Education, Utsunomiya University, 7-1-2 Yoto, Utsunomiya 321-8585, Japan
[2]Hamamatsu Photonics K.K., 5000 Hirakuchi, Hamakita-ku, Hamamatsu 434-8601, Japan

information to optimize the set of parameters based on a high-resolution image of the processed structure. The post-processing is iteratively performed until the desired structure is obtained. Accordingly, significant time and production costs are required. Also, the post-processing yields only a static image of the processed structure and does not provide access to dynamic processes during laser processing.

To overcome these issues, in-process monitoring of the depth of the processed structure in laser drilling of silicon with picosecond pulses has been demonstrated[5]. The laser processing system was combined with a transmission optical microscope to obtain a side view image of the processed structure. There are some potential advantages to in-process monitoring. The first advantage is that monitoring the advance of the ablation front enables appropriate real-time feedback control of the processing parameters. Therefore, the processing of unknown materials could be done without a priori knowledge related to optimal processing parameters. Another advantage is the ability to directly observe the processed structure without complex sample preparation such as cut-and-polish post-processing. However, with in-process monitoring using an optical microscope, the target samples are limited to materials that are transparent to the illumination light of the microscope. Also, in observing using a microscope, processing is limited to drilling close to the side surface of the sample to prevent diffraction from the edge of the sample[6].

Other methods for measuring the depth of the processed structure are laser-scanning confocal microscopy (LCM), white-light interferometry (WLI)[7], optical coherence tomography (OCT)[8], ultrasound and X-ray tomography, and magnetic resonance imaging (MRI). The spatial resolution of ultrasound and X-ray tomography and MRI as medical imaging techniques is from sub-millimeter to a few millimeters. On the other hand, LCM, WLI and OCT, with a resolution from sub-micron to a few microns, are suitable for laser microfabrication. An advantage of OCT compared with LCM and WLI is the ability to perform non-mechanical depth scanning (A-scan) and single-shot measurement. Accordingly, it is possible to perform laser processing while simultaneously measuring the depth of the processed structure. Another advantage is the measurement speed and the measurable range of a few millimeters. Recently, OCT with an A-scan rate of a few megahertz has been demonstrated for clinical applications[9]. Therefore, OCT is superior for in-process monitoring of laser processing. Previously, in-process monitoring of the depth of the processed structure in laser processing of stainless[10,11], glass[12], acrylic[13] and ABS resins[14], and biomaterials[15–17] has been demonstrated using OCT. In particular, the improvement of hole depth accuracy in laser processing using OCT has been reported[18].

In this paper, we demonstrated, for the first time, in-process monitoring of structures processed with line-shaped femtosecond pulses generated by a computer-generated hologram (CGH) displayed on a spatial light modulator (SLM)[19–21]. The in-process monitoring was performed using swept-source OCT (SS-OCT)[22]. The proposed method is suitable for in-process OCT measurement during femtosecond laser processing because the shaped beam processes a large area at high speed as shown in the processing results using the line-shaped beam. In addition, the variability of the CGH by the SLM is useful for feedback control of the intensity distribution of the shaped beam based on the in-process OCT measurement. The spatial resolution of laser processing using ordinary focused femtosecond pulses is generally higher than the lateral measurement resolution of the OCT. Therefore, the OCT may not provide valid measurement result depending on the relationship between the resolution of laser processing and OCT measurement.

In the Principle section, the advantage and basis of SS-OCT and the calibration process of the signal obtained by OCT are described. In the Experimental Setup section, the optical system used for the experiments is shown. In the Experimental Results section, to evaluate the performance of the SS-OCT system, the depth resolution, measurement accuracy, and measurable range were estimated. Furthermore, to verify the feasibility of SS-OCT for in-process monitoring of the depth of the processed structure in femtosecond laser processing, multi-ring drilling was performed using a conventional Gaussian beam. Finally, in-process monitoring of the depth of the processed structure in laser grooving with line-shaped femtosecond pulses was performed. To our knowledge, this is the first demonstration of its kind. From the results, not only the depth of the processed structure but also the shape of the structure and the distribution of deposited debris in the processed groove were simultaneously monitored. The measured depth agreed well with the results obtained using LCM. In the Conclusion section, we summarize our paper.

Principle

Advantage of SS-OCT

OCT is classified into time-domain OCT (TD-OCT)[8] and frequency-domain OCT (FD-OCT)[23]. FD-OCT is characterized by relatively faster and higher signal-to-noise ratio (SNR) measurement compared to TD-OCT. FD-OCT is further classified into spectral-domain OCT (SD-OCT) and SS-OCT. In SD-OCT, a broadband source with a center wavelength of approximately 840 nm is often used for the measurement, where the light reflected from the sample is detected by a spectrometer. The spectrometer usually employs a silicon-based line sensor because the other compound semiconductor sensors have practical difficulties in terms of cost and sensitivity. Therefore, the wavelength range, which is easily applicable to SD-OCT, is less than about 1 μm. In SS-OCT, a broadband wavelength swept source with a center wavelength of approximately 1.06 or 1.30 μm is often employed, where the light is detected by a point photodetector. Currently, photodetectors with sensitivity to a variety of wavelengths are available, so there is a high degree of freedom in selecting the wavelength of the light source.

In this experiment, SS-OCT with a center wavelength of 1.30 μm was employed because SS-OCT has several advantages over SD-OCT. The first advantage is a simple optical system with a photodetector. In particular, the use of a balanced photodetector allows high-SNR measurement. The second advantage is that since a 1.3 μm light has been widely used in optical communications, many fiber components are available at a relatively low cost. In addition, since the group velocity dispersion of the fiber components is zero at a wavelength of 1.3 μm, the interferometer can be constructed simply. The third advantage is that the wavelength for the measurement can be easily separated from the wavelength for ultrashort pulsed laser processing using broadband light sources (e.g., Ti:sapphire, 800 nm; Nd:YAG, 1064 nm; and Yb:silica fiber, 1030 nm). Therefore, since the two beams for processing and the measurement are coaxially radiated onto the sample, in-process monitoring of the depth of the processed structure during laser processing is possible. Another advantage is less impact on the measurement by debris generated by laser processing because the wavelength of SS-OCT is longer than that of SD-OCT.

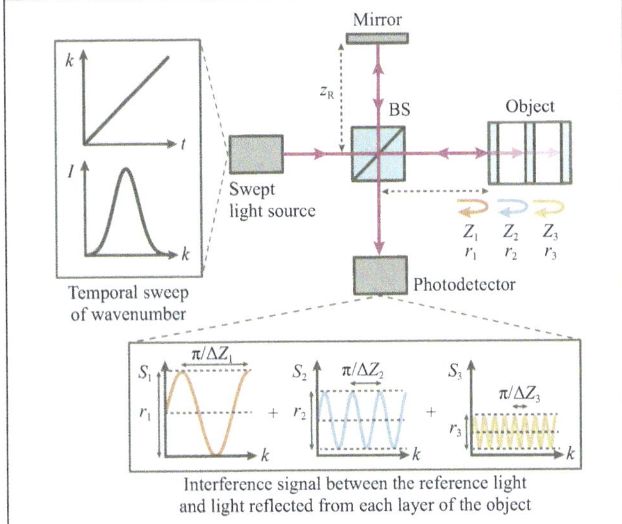

Fig. 1 Optical setup of SS-OCT.

Basis of SS-OCT

The beam emitted from a swept-light source in SS-OCT is split by a beam splitter (BS) into object and reference light beams, as shown in Fig. 1. The signal detected by the photodetector, $s(k)$, is based on interference between the object and reference beams and is described as

$$\begin{aligned} s(k) &= |r_R a(k)\exp(ik\cdot 2Z_R) + \sum_{n=1} r_n a(k)\exp(ik\cdot 2Z_n)|^2 \\ &= |r_R a(k)|^2 + \left|\sum_{n=1} r_n a(k)\right|^2 \\ &\quad + r_R a^2(k)\sum_{n=1} r_n \exp\{ik\cdot 2(Z_R - Z_n)\} + c.c. \\ &= (r_R^2 + \sum_{n=1} r_n^2)I(k) + 2r_R I(k)\sum_{n=1} r_n \cos(k\cdot 2\Delta Z_n) \qquad (1) \end{aligned}$$

where the indexes R and n indicate light reflected from the reference mirror and each reflection layer into the object, respectively; $k = 2\pi/\lambda$ is the wavenumber; a is the amplitude; r is the amplitude reflectivity, which is assumed to be real for simplicity; Z is the optical path length from the beam splitter to the reference mirror and each reflection layer into the object; $c.c.$ indicates complex conjugation; and $I(k) = a^2(k)$ is the power spectrum. The contribution due to interference between object beams is ignored when r_n is sufficiently small. In the second term of the equation, the frequency of the cosine wave is proportional to the optical path difference $\Delta Z_n = Z_R - Z_n$, which is related to the depth of the structure. To determine ΔZ_n, the detected signal $s(k)$ is analyzed by inverse Fourier transformation $\mathcal{F}^{-1}$ with the following equation

$$\begin{aligned} S(Z) &= \mathcal{F}^{-1}\{s(k)\} = (r_R^2 + \sum_{n=1} r_n^2)C(Z) \\ &\quad + \sum_{n=1} r_R r_n\{C(Z) \circledast \delta(Z \pm 2\Delta Z_n)\} \qquad (2) \end{aligned}$$

where $C(Z) = \mathcal{F}^{-1}\{I(k)\}$ is an autocorrelation function; δ is the delta function; and $\circledast$ indicates a convolution operation. In the equation, the first term means the DC component (0th order signal). The second term (n-th order signal) includes information regarding ΔZ_n. According to the equation, the width of the peak spectrum in the n-th order signal is a function of $C(Z)$ appearing as a point spread function (PSF). Therefore, the measurement resolution of the depth of the structure is limited by the spectral width of the light source.

Calibration of interference signal

The wavenumber k should ideally be proportional to the sweep time t of the light source, as shown in Fig. 1. However, in an actual light source, the relationship between k and t is nonlinear. This nonlinearity causes a decrease in measurement resolution. Therefore, calibration of the relationship is essential to perform accurate measurement. Fig. 2 shows the calibration process based on the Fourier transform method[24,25]. In the process, a gold mirror was used as an object, and $\Delta Z_1 = Z_R - Z_1$ was set to 481 μm. Fig. 2a, b show the optical interference signal $s(k)$ and its Fourier spectrum $S(Z)$ before the calibration, respectively. In Fig. 2b, a blue line indicates the 1st-order spectrum of the interference signal, corresponding to the optical path difference between object and reference arms. The spectrum was broadband because $s(k)$ had distortion due to the nonlinear relationship between k and t. Fig. 2c, d indicate the intensity and phase of the interference signal obtained by Fourier transformation of the 1st-order spectrum, respectively. In Fig. 2d, the relationship between the phase, namely k, and time t is nonlinear. The sampling time interval for obtaining $s(k)$ is normally set to be constant for t. In the experiment, the time interval was calibrated to be an equal interval for the phase, as shown in the inset. Then, $s(k)$ for the calibrated time interval is newly obtained by linear interpolation. Fig. 2e, f show $s(k)$ and $S(Z)$ after the calibration, respectively. With this calibration process, the peak of the Fourier spectrum was sharper than that before the calibration (Fig. 2b).

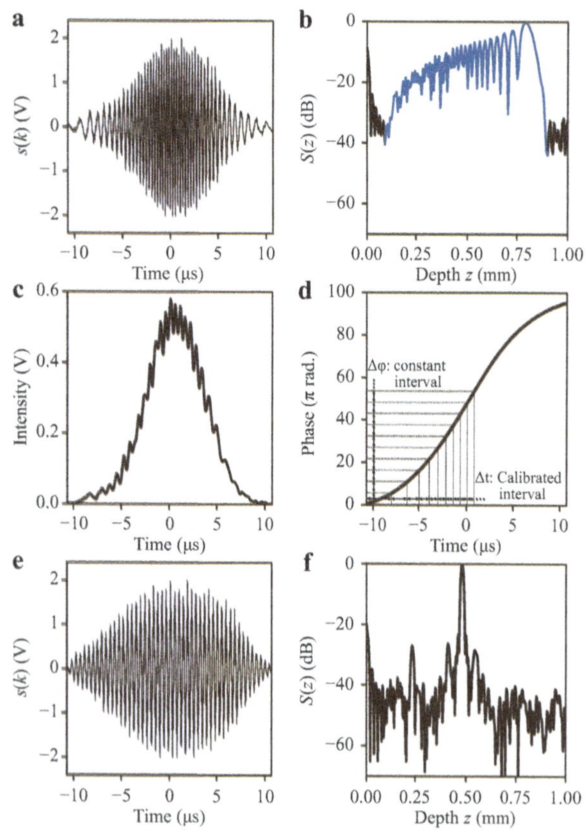

Fig. 2 a Optical interference signal $s(k)$ and **b** its Fourier spectrum $S(Z)$ before calibration. A blue line corresponds to the 1st-order spectrum of $s(k)$. **c** The intensity and **d** phase obtained by Fourier transformation of the 1st-order spectrum. **e** $s(k)$ and **f** $S(Z)$ after calibration.

Experimental setup

Fig. 3 shows the experimental setup. The setup was mainly composed of a femtosecond (fs) laser processing system, an OCT system, and an optical microscope. Femtosecond pulses emitted from an amplified Ti:sapphire femtosecond laser system (COHERENT, Micra and Legend Elite) had a center wavelength of 800 nm, a spectral width of 30 nm (full width at half-maximum (FWHM)), a pulse duration of 50 fs, and a repetition frequency of 1 kHz, and were radiated onto the LCOS-SLM (Hamamatsu Photonics, X10468-02) through a pulse energy controller composed of a half-wave plate (HWP) and a polarizing beam splitter (PBS), and 2× beam expanding optics. The pulse was diffracted by a CGH displayed on the SLM and was transformed to a line-shaped beam which was introduced to laser processing optics. The plane of the SLM was imaged onto a galvanometer mirror (GM) (Thorlabs, GVS212) and the pupil plane of an objective lens (OL, ×5, NA = 0.13) (Sigma, EPL-5) using a 4f imaging system. A white-light-emitting diode (LED), an infrared (IR) cut filter, and a

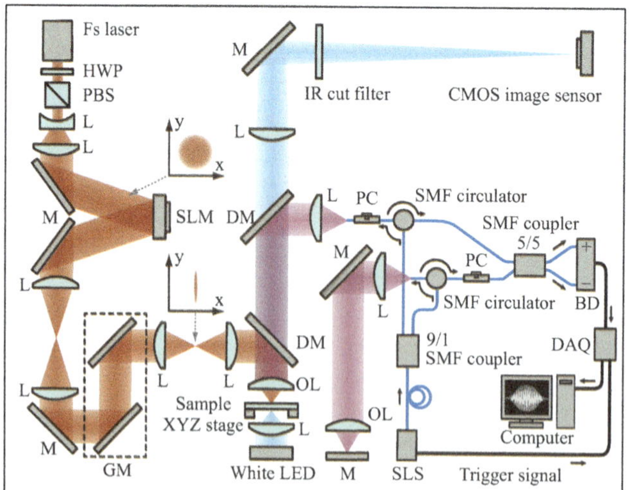

Fig. 3 Experimental setup: L, lens; M, mirror; HWP, half-wave plate; PBS, polarizing beam splitter; GM, galvanometer mirror; DM, dichroic mirror; OL, objective lens; SLS, swept-light source; SMF, single-mode fiber; PC, polarization converter; BD, balanced detector; DAQ, data acquisition device.

complementary metal-oxide-semiconductor (CMOS) image sensor (The Imaging Source, DMK33UP2000) were used to obtain the image of the processed structure. The OCT system, composed of fiber optics, was used to measure the depth of the processed structure in laser processing. The beam from a swept-light source (SLS)[25] with a center wavelength of 1.3 μm, a spectral width of 80 nm, and a repetition frequency of 20 kHz in frequency scanning was split into the object and reference arms by a single-mode fiber (SMF) coupler (Thorlabs, TW1300R2A1) with a coupling ratio of 9:1. Since the light source employs an electro-optic crystal of potassium tantalate niobate ($KTa_{1-x}Nb_xO_3$, KTN) as a driver for sweeping the output wavelength, it has no mechanically moving components and can stably be used in industrial applications that require quantitative analyses. The object beam was radiated onto the sample coaxially with the fs laser beam through a dichroic mirror (DM) via an SMF circulator (Thorlabs, CIR1310-APC) with a core diameter of 8.2 μm. The reference beam also illuminated the reference mirror. A polarization converter (PC) (Thorlabs, CPC900) was used to regulate the polarization of the beam. An optical interference signal between the reflected object and reference beams was detected by a balanced detector (BD) (Hamamatsu, C12668-02) with an SMF coupler (Thorlabs, TW1300R5A1) with a coupling ratio of 5:5. The interference signal was transferred to a computer (Intel Core i7-10700, 2.9 GHz, 8 core, Main memory: 32 GB, GPU: GeForce GTX1660Ti 6 GB) via a data acquisition device (DAQ) (Pico Technology, picoscope3205A) with an electric high-pass filter (Thorlabs, EF517) synchronized by a trigger signal from the SLS. The depth of the processed structure was quantitatively evaluated by performing an inverse Fourier transform of the interference signal. The times required to capture the optical interference signal and to calculate the Fourier spectrum (A-scan profile) were 3 and 10 ms, respectively. A fused silica substrate (thickness of 2 mm) was mounted on a three-axis translation motor stage (Thorlabs, ZFS13B) as a sample. A laser confocal microscope (Olympus, OLS4000) was used to observe the processed structure in detail.

Experimental results

Performance evaluation of OCT system

The depth resolution, measurement accuracy, and measurable range in the OCT system were evaluated. Unprocessed glass was used as the sample. Fig. 4a, b show a typical $s(k)$ with a Hanning window and $S(Z)$, respectively. The window function was used to remove unnecessary high-frequency components generated by the

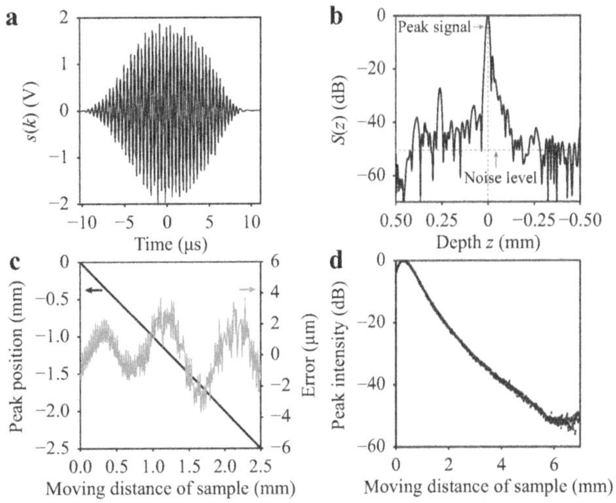

Fig. 4 **a** $s(k)$ with Hanning window and **b** $S(Z)$. **c** Change in peak position of the Fourier spectrum and measurement error versus moving distance of the sample in the axial direction. **d** Change in peak intensity of the Fourier spectrum versus moving distance of the sample.

Fourier transform. In Fig. 4b, the peak spectrum corresponded to the light reflected from the glass surface. The depth Z of the peak spectrum was defined as position zero. The full width at half maximum (FWHM) of the peak spectrum, namely, the depth resolution Z_r, was estimated to be 15.8 μm. If the spectrum of the light source is an ideal Gaussian distribution, a theoretical Z_r is defined with the following equation:

$$Z_r = \frac{2\ln 2}{\pi} \frac{\lambda_c^2}{n\Delta\lambda} \quad (3)$$

where λ_c and $\Delta\lambda$ are the center wavelength and the spectral width of the light at FWHM, respectively; and n is the refractive index of the sample. By using Eq.3, the theoretical Z_r was calculated to be 10 μm. The lower actual depth resolution compared with the theoretical one is mainly due to applying the window function to $s(k)$. The SNR of the peak spectrum was 50 dB. In Fig. 4c, the black line indicates the change of the peak position of the spectrum versus the moving distance of the glass sample. The sample was moved along the optical axial direction with a velocity of 1 mm/s and a step interval of 10 μm while measuring the peak position of the spectrum. The gray line indicates the difference between the moving distance and the measured peak position, that is, the measurement error. Two types of frequency components were observed. One was a high-frequency component, which might be derived from the positioning accuracy of the motor stage. The other was a low-frequency component, which might be derived from the vibration of the interference optical setup. The peak-to-valley (PV) heights in the high- and low-frequency components were about ±1 μm and ±2.5 μm, respectively. These values were relatively smaller compared with the depth resolution Z_r of 15.8 μm. Fig. 4d shows the change of the peak intensity of the spectrum versus the moving distance. The distance from the peak to the intensity of −50 dB was 5.3 mm. From the result, the axial measurable range was estimated to be 5.3 mm. The measurable range is related to the coherence length of the light source and a confocal effect depending on the relationship between the NA of the objective lens and the core diameter of the circulator. The coherence length was 3 mm at FWHM, corresponding to −6 dB in the specifications of the light source. Therefore, the measurable range was mainly limited to the confocal effect.

In-process monitoring in multi-ring drilling with Gaussian beam

At first, to verify the feasibility of the proposed method, in-process monitoring of the depth of the processed structure in multi-ring drilling was performed using a Gaussian beam. Figs. 5a–c show the depth of the structure in multi-ring drilling when the pulse energy E was set to 4.3, 10 and 43 μJ, respectively. The focused fs beam was circularly scanned by the galvanometer mirror, as shown in the inset. In the beam scanning, eight rings with a radius interval of 6.25 μm were sequentially drawn from the inside to the outside. One beam scan was defined as drawing eight rings. The time required for each beam scan S_n was 900 ms. The measurement position for OCT was set to the center of the processed structure (single-point measurement). The graphs corresponding to each optical microscope image indicate $S(Z)$ (A-scan profile) obtained by OCT. The dashed line indicates the depth position of the glass surface. The arrow indicates the depth position of the peak spectrum. The depth position of the peak spectrum was defined as the maximum value of $S(Z)$ in the depth range from the glass surface ($Z = 0$ mm) to $Z = -0.5$ mm. From the results, the depth position of the peak spectrum was shifted to the deeper side with increasing S_n. In Fig. 5a with near threshold pulse energy, the shift of the peak spectrum was small. In a comparison between Fig. 5b, c, a difference in the intensity of the peak spectrum was observed. The difference originated from the decrease of the backscattering light due to debris deposited on the bottom of the structure in multi-ring drilling. In Fig. 5c with larger pulse energy, the intensity of the peak spectrum was one order of magnitude larger than that in Fig. 5b because the deposited debris was blown away by the shockwave generated by the fs laser irradiation[26,27].

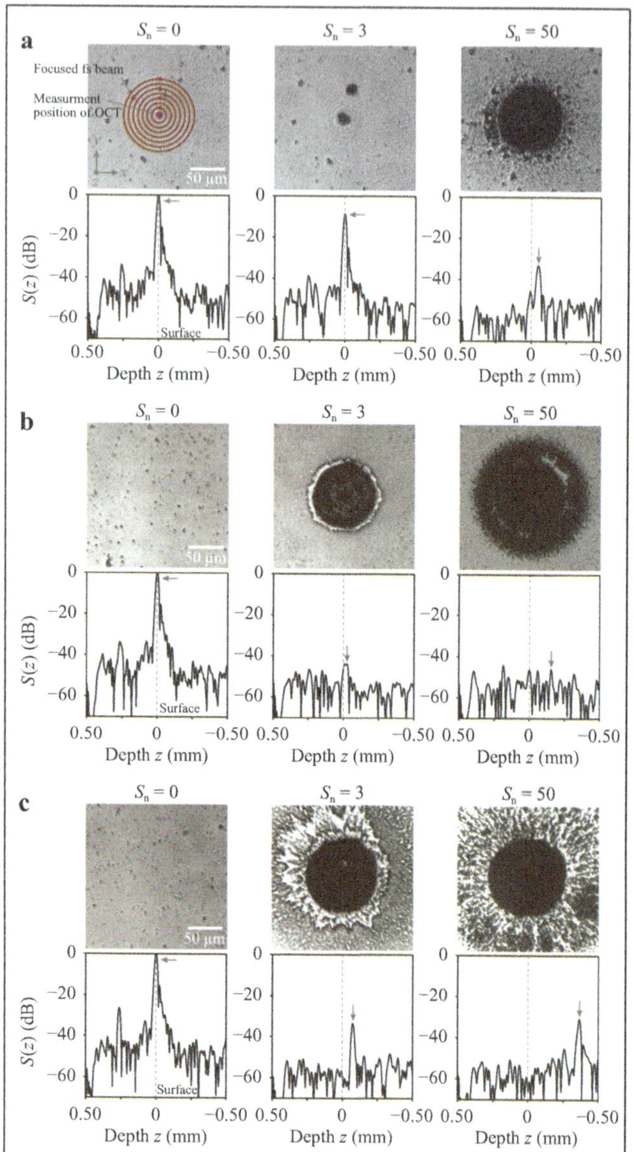

Fig. 5 In-process monitoring of depth of processed structure in multi-ring drilling. Optical microscope image of the processed structure and $S(Z)$ obtained by OCT versus number of scans S_n when E was set to **a** 4.3, **b** 10, and **c** 43 μJ, respectively. Arrows indicate the peak spectrum.

Figs. 6a–c show the change of the A-scan profile of the processed structure in multi-ring drilling versus S_n and the cross-sectional depth profile of the final structure when E was set to 4.3, 10, and 43 μJ, respectively. In the change of the A-scan profile, the white dashed line indicates the surface position of the glass sample. In the cross-sectional profile, the black dot and the gray dashed line indicate the depth measured by LCM and OCT, respectively. From the result, the depth obtained by OCT agreed well with the maximum depth obtained by LCM. In Fig. 6b, the intensity of the peak spectrum was gradually decreased with the increase of S_n because accumulated debris continuously pooled at the bottom of the structure. Therefore, the SNR was decreased due to the decrease of the backscattering

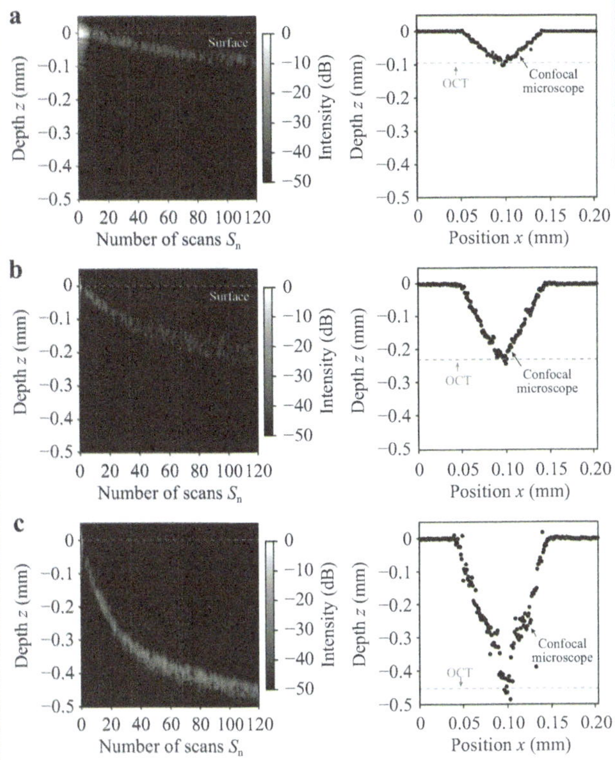

Fig. 6 Change of A-scan profile of the processed structure in multi-ring drilling versus S_n and cross-sectional depth profile of the final structure when E was set to **a** 4.3, **b** 10 and **c** 43 μJ. White dashed lines indicate the surface position of the sample. Black dots indicate the depth profiles measured by LCM. Gray dashed lines indicate the depth measured by OCT.

light; however, in-process monitoring of the depth of the processed structure was successfully performed. On the other hand, in Fig. 6c, the intensity of the peak spectrum rapidly increased when S_n was larger than 10. The increase in SNR was related to the effect of the debris removal by the shockwave.

Fig. 7 shows the change of the depth of the structure processed by multi-ring drilling versus E as S_n was changed. The dashed line indicates the depth of focus of the fs laser beam, which is determined by the NA of the objective lens. When E was less than 8 μJ, there was no significant change in the difference between the depths of the structures as a function of S_n because E was near the threshold. When E was around 10 μJ, the depth of the structure approximately linearly increased with the increase of S_n. On the other hand, when E was larger than 30 μJ, the increase in the depth of the structure showed nonlinearity as S_n was increased. In particular, in the early stages of multi-ring drilling (S_n < 10), the depth of the structure rapidly increased because the depth of focus of the beam was close to the depth of the structure. However, after the

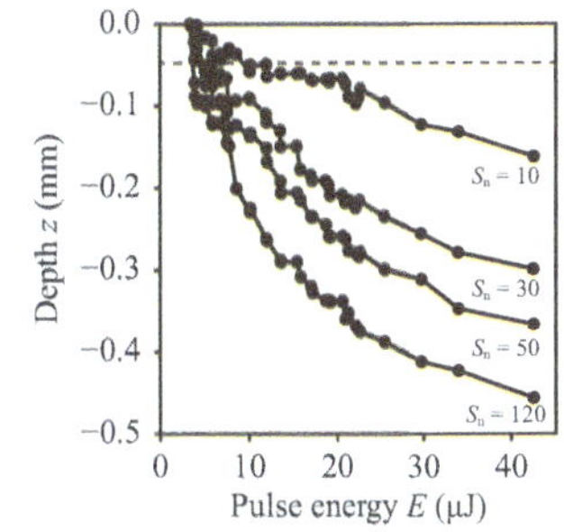

Fig. 7 Change of the depth of processed structure in multi-ring drilling versus E as a function of S_n. Dashed line indicates the depth of focus of the fs laser beam.

intermediate stage of multi-ring drilling (50 $\leqslant S_n$), the depth of the structure approximately saturated. Therefore, a focus shift of the fs laser beam according to the depth of the structure will be necessary from the viewpoint of improving the efficiency of multi-ring drilling.

In-process monitoring in laser grooving with line-shaped beam

In-process monitoring of the depth of the processed structure in laser grooving with a line-shaped beam was demonstrated. Fig. 8a shows the CGH used to generate the line-shaped beam and its computer reconstruction. Fig. 8b shows an optical microscope image of the structure processed by the line-shaped beam with E = 206 μJ. The glass sample was repeatedly scanned along the positive and negative directions on the x-axis with a constant velocity of 50 μm/s while being irradiated with the line-shaped beam. The number of scans S_n was set to 10. Fig. 8b corresponds to the optical microscope image of the structure when S_n = 2. When each scan S_n was finished, the glass sample was scanned again along the x-direction with a velocity of 1 mm/s without irradiation with the line-shaped beam to measure the depth of the structure by OCT. The measurement position for OCT was set to the center of the structure shown as a dashed line in the figure. The measurement interval of the x-axis in the OCT was 5 μm.

Figs. 9a–c show the cross-sectional depth profile (B-

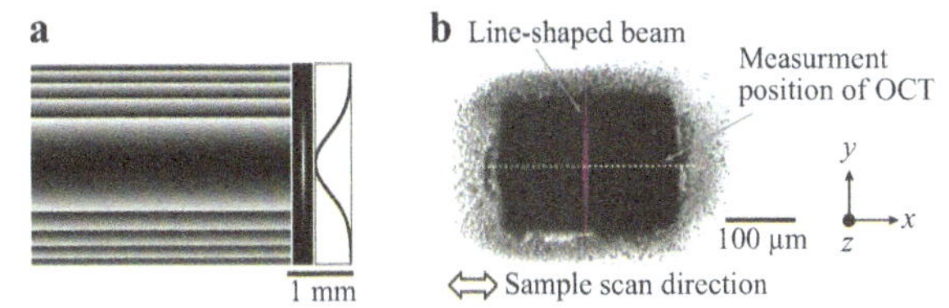

Fig. 8 **a** CGH used to generate the line-shaped beam and its computer reconstruction. **b** Optical microscope image of the structure processed by the line-shaped beam.

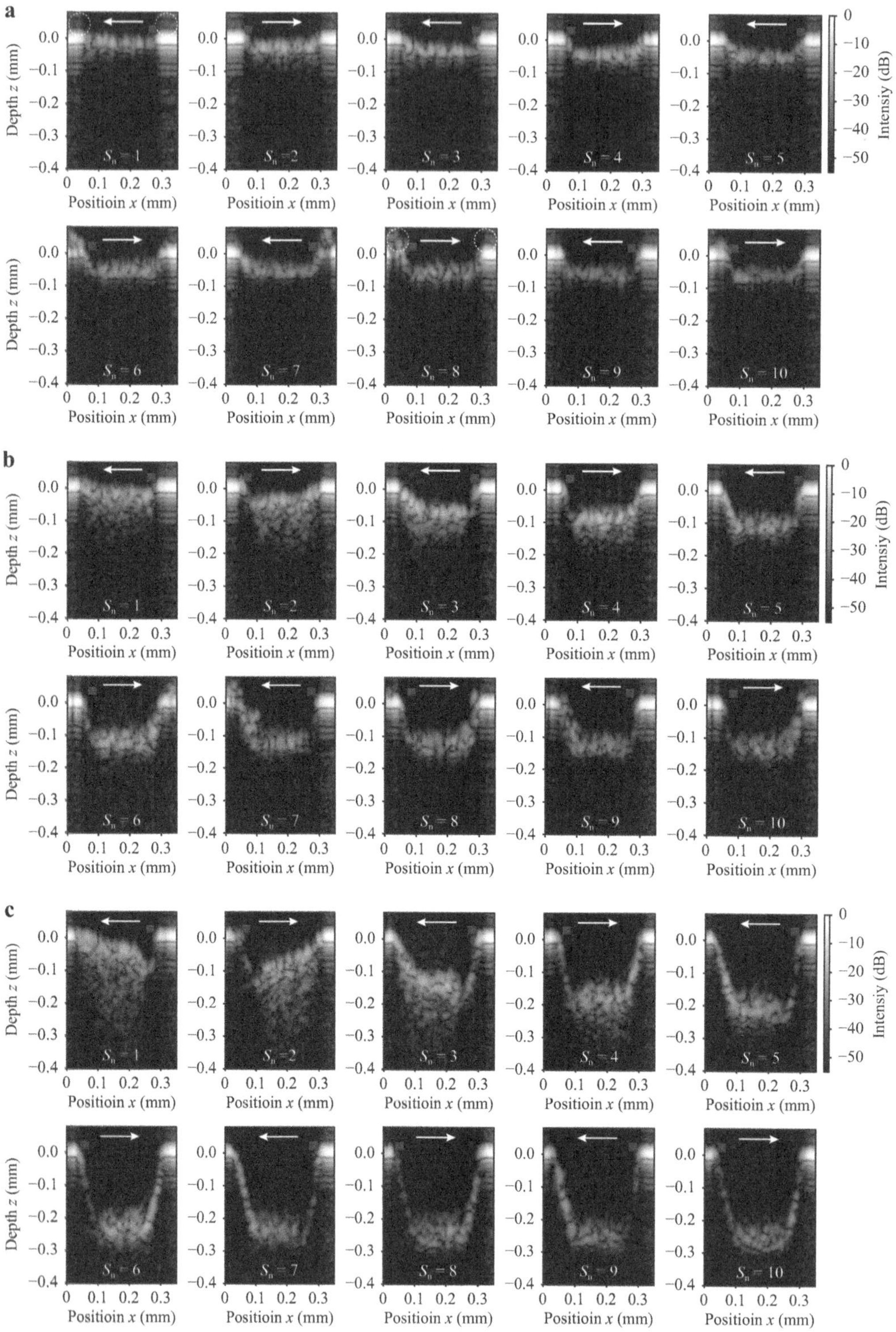

Fig. 9 Crossectional depth profile (B-scan image) of the structure processed by the line-shaped beam versus S_n when E was set to **a** 32, **b** 74 and **c** 206 μJ. Arrows indicate the scanning direction of the glass sample. Red points indicate the position of the fs laser beam at the end of the scanning.

scan image) of the structure processed by the line-shaped beam versus S_n when E was set to 32, 74 and 206 μJ, respectively. The B-scan image was generated from a series of the A-scan profiles measured at different x positions. Arrows indicate the scanning direction of the sample. Red points indicate the position of the fs laser beam at the end of the scanning. In the case of low E (Fig. 9a), debris was deposited at both ends of the structure (the area shown in the dashed circle) as S_n was increased. After $S_n = 8$, deposited debris was not observed. The absence of debris is expected to be related to the shockwave caused by the laser irradiation and the decrease in debris production by the saturation of the depth of the processed structure. In the case of middle E (Fig. 9b), the depth became deeper with increasing S_n. In the case of high E (Fig. 9c), debris deposited in the processed structure was obviously moved according to the scanning direction due to the shockwave caused by the laser irradiation. In particular, the processed structure was deeper at the end position of the scanning because the debris was blown away by the shockwave. On the inside of the processed structure, the A-scan profile was observed to become elongated along the axial direction. The elongation is due to the multiple reflections of backscattered light by the debris because the multiple reflections resulted in the optical path difference longer compared with the reflection from the surface structure. In comparison to the elongation in Fig. 9a, the elongation in Fig. 9c was clearly observed. From the results, in-process monitoring of the depth and shape of the processed structure and the distribution of deposited debris was successfully performed.

Fig. 10a shows the change in the depth of the groove processed by the line-shaped beam versus S_n when E was set to 32, 74, and 206 μJ. The depth was defined as the deepest position of the peak spectrum in the B-scan image. The dashed line indicates the depth of focus of the fs laser beam. When $4 \leqslant S_n$, the structural depth was almost saturated with the increase of S_n. Fig. 10b shows the cross-sectional depth profile of the processed structure in $S_n = 10$ when E was set to 32, 74 and 206 μJ. Black dots and gray lines indicate the depth profiles measured by OCT and LCM, respectively. In OCT, the position of the peak

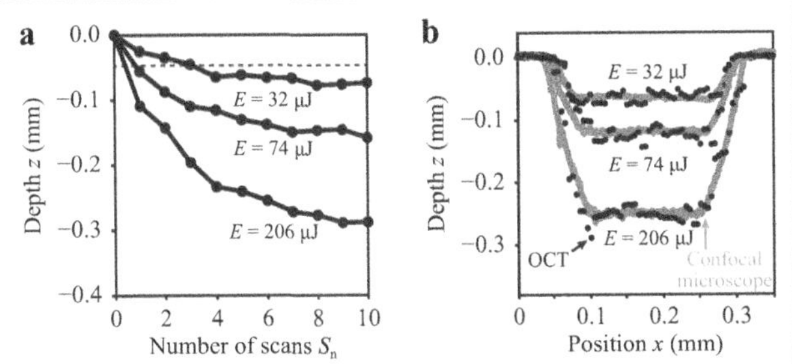

Fig. 10 a Change of the structural depth in laser grooving with the line-shaped beam versus S_n. Dashed line indicates the depth of focus of the fs laser beam. **b** Cross-sectional depth profile of the processed structure when $S_n = 10$. Black dots and gray lines indicate the depth profiles measured by OCT and LCM, respectively.

spectrum of each A-scan profile was plotted. From the result, both profiles agreed well. On the bottom of the structure, a relatively flat surface was successfully obtained.

Fig. 11 shows the change of the depth of the structure processed by the line-shaped beam versus E as S_n was varied. The dashed line indicates the depth of focus of the fs laser beam. When E was less than 30 μJ, there was no significant change in the difference between the depths of the structures as a function of S_n because E was near the threshold. When E was larger than 30 μJ, the increase in the depth of the structure showed nonlinearity with the increase of S_n. In particular, in the early stages of laser grooving ($S_n < 3$), the depth of the structure rapidly increased. On the other hand, after the intermediate stage of laser grooving ($5 \leqslant S_n$), the depth of the structure approximately saturated. As in the case of laser grooving with line-shaped, a focus shift of the beam according to the structural depth will also be effective for improving the efficiency of laser grooving.

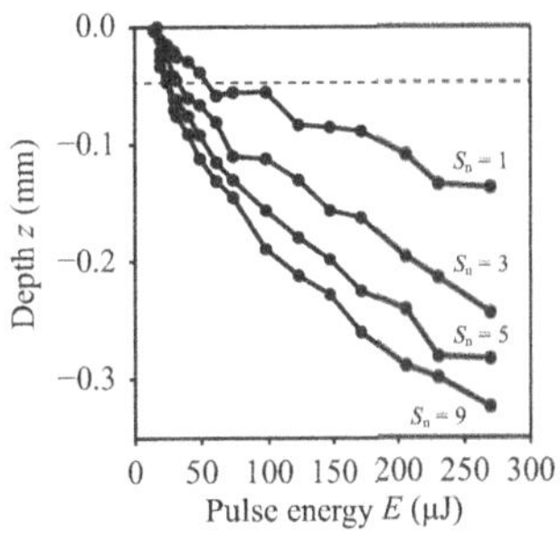

Fig. 11 Change of the depth of the processed structure in laser grooving with the line-shaped beam versus E as a function of S_n. Dashed line indicates the depth of focus of the fs laser beam.

Conclusion

In-process monitoring of the depth of a processed structure in laser grooving with a line-shaped fs laser beam was demonstrated by using SS-OCT. To evaluate the performance of the OCT system, the depth resolution, measurement accuracy, and measurable range were estimated to be 15.8 μm, ±2.5 μm, and 5.3 mm, respectively. In addition, to verify the feasibility of OCT for in-process monitoring of the depth of processed structures in laser processing, multi-ring drilling was performed. In the case of drilling with low pulse energy, the SNR of the A-scan profile was significantly decreased by the debris deposited on the bottom of the structure; however, in-process monitoring of the depth of the processed structure was successfully performed. On the other hand, in the case of drilling with high pulse energy, the SNR was rapidly increased with the increase of S_n. The reason for this originated from the effect of the debris removal due to the shockwave generated by the fs laser irradiation. Finally, in-process monitoring of the depth of a processed structure in laser grooving with a line-shaped beam was performed. From the results, not only the depth of the structure but also the shape of the structure and the distribution of deposited debris in the processed groove were successfully monitored. The measured depth agreed well with the result obtained using LCM. The proposed method will be useful for precise laser processing with feedback control of the laser parameters based on in-process monitoring of the structure.

Acknowledgements
This work was supported by JSPS KAKENHI Grant Number JP18K14142 and was performed for the Council for Science, Technology and Innovation (CSTI), Cross-ministerial Strategic Innovation Promotion Program (SIP), "Photonics and Quantum Technology for Society 5.0" (Funding agency: QST).

Conflict of interest
The authors declare no conflicts of interest.

Received: 31 August 2021 Revised: 27 April 2022 Accepted: 28 April 2022
Accepted article preview online: 04 May 2022
Published online: 17 May 2022

References

1. Bovatsek, J. M. & Patel, R. S. Highest-speed dicing of thin silicon wafers with nanosecond-pulse 355nm q-switched laser source using line-focus fluence optimization technique. Proceedings of SPIE 7585, Laser-based Micro- and Nanopackaging and Assembly IV. San Francisco: SPIE, 2010, 75850K.
2. Rung, S. et al. Laserscribing of thin films using top-hat laser beam profiles. *JLMN-Journal of Laser Micro/Nanoengineering* **8**, 309-314 (2013).
3. Račiukaitis, G. et al. Laser processing by using diffractive optical laser beam shaping technique. *JLMN-Journal of Laser Micro/Nanoengineering* **6**, 37-43 (2013).
4. Wa ng, Y. Z. & Lin, J. Characterization of the laser cleaving on glass sheets with a line-shape laser beam. *Optics & Laser Technology* **39**, 892-899 (2007).
5. Döring, S. et al. In situ imaging of hole shape evolution in ultrashort pulse laser drilling. *Optics Express* **18**, 20395-20400 (2010).
6. Dietrich, J. et al. Investigation of increased drilling speed by online high-speed photography. *Optics and Lasers in Engineering* **46**, 705-710 (2008).
7. Leith, E. N. & Swanson, G. J. Achromatic interferometers for white light optical processing and holography. *Applied Optics* **19**, 638-644 (1980).
8. Huang, D. et al. Optical coherence tomography. *Science* **254**, 1178-1181 (1991).
9. Matten, P. et al. MHz swept-source OCT for clinical applications. *Investigative Ophthalmology & Visual Science* **61**, 2539 (2020).
10. Webster, P. J. L. et al. Automatic real-time guidance of laser machining with inline coherent imaging. *Journal of Laser Applications* **23**, 022001 (2011).
11. Hayashi, N. et al. In-process measurement of a keyhole using a low-coherence interferometer with a high repetition rate. *Optics Express* **29**, 32169-32178 (2021).
12. Tomlins, P. H. et al. Femtosecond laser micro-inscription of optical coherence tomography resolution test artifacts. *Biomedical Optics Express* **2**, 1319-1327 (2011).
13. Kamiya, M. & Aoshima, S. I. Real-time monitoring of processed hole depth under femtosecond laser processing. *The Review of Laser Engineering* **33**, 685-689 (2005).
14. Fujimoto, M. et al. Depth monitoring system during laser processing using KTN-based wavelength-swept light source of 1.3-µm wavelength band. Proceedings of SPIE 1169, High-Power Laser Materials Processing: Applications, Diagnostics, and Systems X. SPIE, 2021, 1167908.
15. Kuranov, R. V. et al. Optical coherence tomography for in situ monitoring of laser corneal ablation. *Journal of Biomedical Optics* **7**, 633-642 (2002).
16. O h, W. Y. et al. Ultrahigh-speed optical frequency domain imaging and application to laser ablation monitoring. *Applied Physics Letters* **88**, 103902 (2006).
17. Massow, O., Will, F. & Lubatschowski, H. Optical coherence tomography controlled femtosecond laser microsurgery system. Proceedings of SPIE 6627, Optical Coherence Tomography and Coherence Techniques III. Munich: SPIE, 2007, 228-233.
18. Webster, P. J. L. et al. *In situ* 24kHz coherent imaging of morphology change in laser percussion drilling. *Optics Letters* **35**, 646-648 (2010).
19. Hayasaki, Y. et al. Variable holographic femtosecond laser processing by use of a spatial light modulator. *Applied Physics Letters* **87**, 031101 (2005).
20. Hasegawa, S., Hayasaki, Y. & Nishida, N. Holographic femtosecond laser processing with multiplexed phase fresnel lenses. *Optics Letters* **31**, 1705-1707 (2006).
21. Hasegawa, S., Shiono, K. & Hayasaki, Y. Femtosecond laser processing with a holographic line-shaped beam. *Optics Express* **23**, 23185-23194 (2015).
22. Chinn, S. R., Swanson, E. A. & Fujimoto, J. G. Optical coherence tomography using a frequency-tunable optical source. *Optics Letters* **22**, 340-342 (1997).
23. Fercher, A. F. et al. Measurement of intraocular distances by backscattering spectral interferometry. *Optics Communications* **117**, 43-48 (1995).
24. Yasuno, Y. et al. Three-dimensional and high-speed swept-source optical coherence tomography for in vivo investigation of human anterior eye segments. *Optics Express* **13**, 10652-10664 (2005).
25. Fujimoto, M. et al. Stable wavelength-swept light source designed for industrial applications using KTN beam-scanning technology. Proceedings of SPIE 10110, Photonic Instrumentation Engineering IV. San Francisco: SPIE, 2017, 101100Q.
26. Sakuma, K. et al. Holographic laser sweeper for in-process debris removal. *Applied Physics B* **119**, 533-538 (2015).
27. Abe, T. et al. In-process debris removal in femtosecond laser processing. *Applied Physics A* **123**, 700 (2017).

Official journal of the JHL 2689-9620
www.light-am.com

Article **Open Access**

Iterative projection meets sparsity regularization: towards practical single-shot quantitative phase imaging with in-line holography

Yunhui Gao[1] and Liangcai Cao[1,*]

Abstract

Holography provides access to the optical phase. The emerging compressive phase retrieval approach can achieve in-line holographic imaging beyond the information-theoretic limit or even from a single shot by exploring the signal priors. However, iterative projection methods based on physical knowledge of the wavefield suffer from poor imaging quality, whereas the regularization techniques sacrifice robustness for fidelity. In this work, we present a unified compressive phase retrieval framework for in-line holography that encapsulates the unique advantages of both physical constraints and sparsity priors. In particular, a constrained complex total variation (CCTV) regularizer is introduced that explores the well-known absorption and support constraints together with sparsity in the gradient domain, enabling practical high-quality in-line holographic imaging from a single intensity image. We developed efficient solvers based on the proximal gradient method for the non-smooth regularized inverse problem and the corresponding denoising subproblem. Theoretical analyses further guarantee the convergence of the algorithms with prespecified parameters, obviating the need for manual parameter tuning. As both simulated and optical experiments demonstrate, the proposed CCTV model can characterize complex natural scenes while utilizing physically tractable constraints for quality enhancement. This new compressive phase retrieval approach can be extended, with minor adjustments, to various imaging configurations, sparsifying operators, and physical knowledge. It may cast new light on both theoretical and empirical studies.

Keywords: Phase retrieval, Quantitative phase imaging, Compressive sensing, Digital holography

Introduction

The phase of an optical field contains various sources of information. Holography, invented by Gabor in 1947[1], provides access to phase information and has been used as a powerful imaging technique[2,3]. At the optical frequency, however, electromagnetic waves oscillate so fast that the detection devices can directly record only the average incident power. The loss of phase information poses unique challenges for the numerical inversion of the imaging process, leading to the so-called "twin-image" effect that severely degrades the imaging quality[4,5]. As a result, phase retrieval, or the recovery of the phase from intensity-only measurements, remains a longstanding issue and lies at the heart of digital holography and many related optical imaging techniques[6]. In recent years, phase retrieval has

Correspondence: Liangcai Cao (clc@tsinghua.edu.cn)
[1]State Key Laboratory of Precision Measurement Technology and Instruments, Department of Precision Instruments, Tsinghua University, Beijing 100084, China

regained increasing research interests following the development of computational imaging, which leverages advancements in signal processing theory, numerical algorithms, hardware design, and computation power[7].

Existing phase retrieval methods for in-line holography can be broadly classified into two categories, namely diversity phase retrieval and compressive phase retrieval. The former approach requires taking multiple images with some data diversity that translate phase information into intensity distribution. This can be achieved, for example, by varying the imaging distance[8–10], illumination wavelength[11–14], probe position[15–17], illumination angle[18–20], and modulation pattern[21–24]. While this approach can achieve quantitative phase imaging with increased data diversity and redundancy, it sacrifices the temporal resolution, system cost, and complexity. As an alternative, compressive phase retrieval takes a different path by exploring additional prior knowledge of the wavefield, which could potentially achieve phase recovery beyond the information-theoretic limit or even from a single shot.

Historically, early works tackled the phase ambiguity problem by utilizing simple yet effective physical constraints such as non-negativity[25,26], support[27–29], absorption[30], histogram[31], and atomicity[32]. These constraints are typically enforced via projections in the complex Euclidean space, and the corresponding phase retrieval algorithms proceed by projecting the estimated wavefield back in an iterative manner[33,34]. Since the seminal Gerchberg-Saxton[35] and Fienup's error-reduction algorithms[36], iterative projection algorithms have played a pivotal role in phase retrieval for decades[37–39]. However, this classical approach relies on carefully handcrafted physical constraints, which may fail to fully extract the inherent features of complex scenes under limited measurements, particularly in the case of single-shot phase retrieval[40]. Consequently, it has mostly been demonstrated for simple objects. The quantitative phase analysis of complex samples, such as dense biological specimens, remains a challenging task.

The concept of compressive phase retrieval was formally established by an explosion of theoretical and empirical studies following the development of compressed sensing[41–43]. Sparsity priors in various domains such as the spatial[44–49], gradient[50–61], wavelet[62] and other domains[63,64] or with a dictionary-learned transform[65–67] have been demonstrated as effective regularizers for phase recovery. More recently, implicit image priors from advanced denoisers such as BM3D[68–73] or represented by deep neural networks[74–93] have also been studied in the context of phase retrieval, resulting in improved reconstruction quality. Although the regularization techniques generally yield superior imaging performance, they are highly dependent on the scene and may require manual fine-tuning to achieve satisfactory results.

In pursuit of a compressive phase retrieval method for in-line holography that features both robustness and high fidelity, we propose a novel computational framework that encapsulates physical constraints and sparsity priors as a unified optimization problem. Specifically, we consider the well-known absorption and support constraints and gradient-domain sparsity, leading to a constrained complex total variation (CCTV) model. This is because TV-regularized problems can be solved using computationally efficient algorithms and are thus highly scalable to large-scale problems[94,95]. The piece-wise smoothness underlying the TV model can also characterize multidimensional data such as the three-dimensional refractive index distribution, offering good generalizability to various applications[96–99]. Furthermore, TV-regularized inversion algorithms offer mathematically tractable convergence behaviors, which is important for both theoretical soundness and practical implementation.

Based on the CCTV model, compressive phase retrieval is formulated as a regularized inverse problem and is solved via an accelerated proximal gradient algorithm together with an efficient denoiser for the proximal update. Unlike conventional methods that rely on a heuristic basis, we provide a rigorous mathematical analysis that guarantees the convergence of the proposed algorithms using prespecified parameters. We experimentally demonstrated quantitative phase imaging of various samples from a single in-line hologram, illustrating the effectiveness of the CCTV model in characterizing complex real-world scenes while leveraging physically tractable constraints for quality enhancement. We further elaborate on algorithmic behaviors based on extensive numerical studies, corroborating the theoretical results while providing general guidance for practitioners.

Results

Phase retrieval via CCTV

For electromagnetic waves at optical frequencies, sensors can only respond to the average power of the incident wavefield. Thus, phase information, i.e., the complex argument of the signal, is lost during the measurement process. The forward model can be abstractly expressed, regardless of the imaging configuration, in a vectorized form as

$$\boldsymbol{y} = |\boldsymbol{A}\boldsymbol{x}| \tag{1}$$

where $\boldsymbol{y} \in \mathbb{R}^m$ is the observed measurement, $\boldsymbol{A} \in \mathbb{C}^{m \times n}$ is the sampling matrix, and $x \in \mathbb{C}^n$ is the underlying complex signal that represents the transmission of the object. In the vectorized formulation, column vectors x and y are raster-scanned from the corresponding two-dimensional complex object transmission function and hologram amplitude, respectively. It is important to note that, the purpose of vectorization is to simplify the algorithm derivation and notations. Vectorization is not required when implementing the algorithm in practice.

Generally, the sampling matrix may involve various linear physical processes such as probe illumination, free-space propagation and mask modulation, all of which are determined by the optical configuration of the imaging system. In this work, we consider the lensless in-line holographic imaging model for illustration, which is shown in Fig. 1a. The object is illuminated by a coherent plane wave, such that the retrieved complex field directly reveals the absorption and phase shift introduced by the object. An

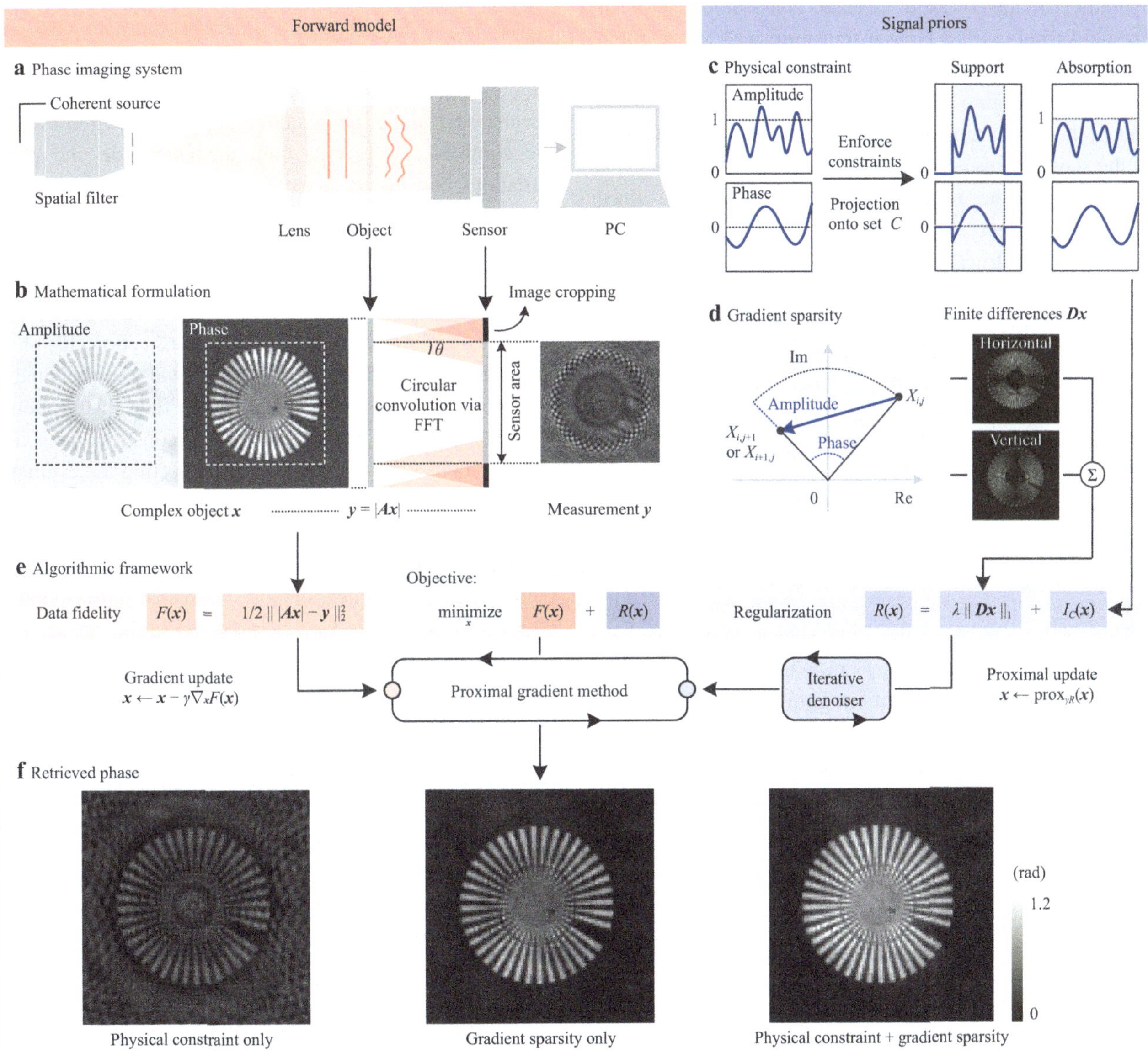

Fig. 1 Schematic of the proposed compressive phase retrieval framework. **a** The optical configuration of an in-line holographic imaging system. **b** The angular spectrum diffraction model adopted for numerical calculation. The diffraction angle θ and the corresponding convolution kernel size are determined by the sampling frequency (or, equivalently, the sensor pixel size). **c** Physical knowledge, such as object support or absorption, is often enforced as hard constraints via projection operations. **d** The complex TV regularizer promotes sparsity in both the gradients of amplitude and phase. **e** We combine the forward model and signal priors into an optimization problem, which is solved via the proximal gradient method. An iterative denoiser is introduced for the proximal update step. **f** Retrieved phase from experimental data. Combining physical constraints and gradient sparsity yields the best performance.

imaging sensor was placed close to the object, and the in-line hologram was recorded. Such a simple optical configuration enables low-cost and portable system design for various imaging applications[100,101]. The physical process involves only free-space propagation of the wavefield, which is calculated using the angular spectrum method[102]. Therefore, the sampling matrix of our imaging model can be expressed as

$$A = CQ \tag{2}$$

where $Q \in \mathbb{C}^{n \times n}$ denotes the free-space propagation operator calculated as a circular convolution with the Fresnel kernel via fast Fourier transforms (FFTs), $C \in \mathbb{R}^{m \times n}$ denotes an image cropping operation to model the finite size of the imaging sensor[103]. Notice that we have $m < n$, that is, the object dimension is larger than the pixel number of the sensor. This is because out-of-field objects may also contribute to the recorded image owing to the linear convolution effect[54]. This process is illustrated in Fig. 1b.

Phase reconstruction from a single in-line hologram is inherently challenging due to the existence of ambiguous solutions. Therefore, compressive phase retrieval takes an inverse problem approach that utilizes regularization techniques to ensure well-posedness. In this work, we introduce the integration of a complex TV and a hard constraint as the regularizer, leading to an optimization problem in the following form:

$$\min_{x} \underbrace{\frac{1}{2}\||Ax| - y\|_2^2}_{F(x)} + \underbrace{\lambda\|Dx\|_1 + I_C(x)}_{R(x)} \tag{3}$$

where $\|\cdot\|_p$ denotes the ℓ_p vector norm or the corresponding matrix norm. $F(x)$ is the data-fidelity term and $R(x)$ is the regularization term. The fidelity term ensures that the solution does not deviate significantly from the forward model in Eq. 1. Meanwhile, the regularization term introduces additional prior knowledge of the wavefield to improve the reconstruction quality further.

There are various choices for the fidelity function $F(x)$ in the phase retrieval literature, such as the intensity-based formulation $F(x) = (1/2)\||Ax|^2 - y^2\|_2^2$[104–106] and the amplitude-based formulation $F(x) = (1/2)\||Ax| - y\|_2^2$[107–109]. The main reasons for choosing the amplitude-based residual here are threefold. First, from a physical perspective, it is closely related to the maximum likelihood estimation under Poisson noise, which captures the noise statistics more accurately than the Gaussian noise model[110]. Second, on the practitioner's side, using the amplitude-based formulation ensures the convergence of the iterative algorithms with a prespecified algorithm step size, obviating the need for complicated line searches and simplifying the implementation. Third, many empirical studies have observed faster convergence when minimizing amplitude-based fidelity functions[111,112].

The proposed regularization function $R(x)$ comprises two separate terms. The first term is the sparsity regularizer with the parameter $\lambda > 0$. Specifically, we considered the anisotropic complex TV seminorm:

$$\begin{aligned}\|Dx\|_1 &= \mathrm{TV}(X) \\ &= \sum_{i=1}^{n_\xi - 1}\sum_{j=1}^{n_\upsilon} |X_{i+1,j} - X_{i,j}| + \sum_{i=1}^{n_\xi}\sum_{j=1}^{n_\upsilon - 1} |X_{i,j+1} - X_{i,j}|\end{aligned} \tag{4}$$

where $X \in \mathbb{C}^{n_\xi \times n_\upsilon}$ with $n_\xi n_\upsilon = n$ denotes the non-vectorized two-dimensional image corresponding to x. $D \in \mathbb{C}^{2n \times n}$ denotes the spatial finite-difference operator for the vectorized x along the two directions. As shown in Fig. 1d, the complex TV reveals both the amplitude and phase variations between adjacent sampling points of the object transmission function and thus serves as a good sparsifying transform for complex-valued images[113]. Apart from the complex TV, we introduce a second regularizer $I_C(x)$, which is an indicator function of the constraint set C. This is motivated by conventional iterative projection (IP) phase retrieval algorithms, where phase retrieval is formulated as a feasibility problem, and the unknown field distribution is assumed to belong to certain physical constraint sets. The IP algorithms typically proceed by projecting the estimated wavefield onto constraint sets in an iterative manner[37,38]. Here, we consider the support and absorption constraints in particular because they are commonly adopted for reconstructing complex objects, and their corresponding constraint sets are both convex. Fig. 1c presents a conceptual illustration of these physical constraints, and the mathematical definitions are presented in Table 1. The proposed regularizer $R(x)$ comprises a complex TV and a hard physical constraint and is thus termed the constrained complex total variation regularizer. As a special case, when $C = \mathbb{C}^n$, CCTV reduces to the complex TV function without physical constraints, and when $\lambda = 0$, it reduces to

Table 1 Physical constraints for phase retrieval

Constraint	Definition of C	Projection $x' = \mathcal{P}_C(x)$
Support	$x \odot \mathbf{1}_P = \mathbf{0}$	$x_i' = \begin{cases} 0, & i \in P \\ x_i, & i \notin P \end{cases}$
Absorption	$\|x\| \leq 1$	$x_i' = \begin{cases} x_i/\|x_i\|, & \|x_i\| > 1 \\ x_i, & \|x_i\| \leq 1 \end{cases}$

[1] The symbol $\odot$ denotes the element-wise (Hadamard) multiplication operator. $\mathbf{0}$ and $\mathbf{1}$ denote the zero vector and all-ones vector, respectively. P denotes the set of pixels outside the support region, and 1_P denotes the indicator vector of P.

the classical problem of phase retrieval from a modulus measurement and a physical constraint.

Quantitative phase imaging applications

We built a proof-of-principle in-line holographic imaging system for optical experiments, as shown in Fig. 2a. A 660 nm diode-pumped solid-state laser (Cobolt Flamenco 300) was used as the source for coherent illumination. The optical beam generated by the laser passes through a spatial filter that comprises an objective lens and a pinhole at the back focal plane. The filtered wave was then collimated using a convex lens to illuminate the object. The diffraction pattern was recorded using a CMOS sensor (QHY174GPS, pixel size 5.86 μm, resolution 1920×1200) which was placed a few millimeters away from the object. The captured raw intensity image was then digitally processed using a personal computer for holographic reconstruction. In all optical experiments, we consider only the absorption constraint because the objects have extended field distributions and thus the support constraint is inapplicable. Before running the iterative phase retrieval algorithm, a preprocessing step is required to determine the imaging distance, which can be achieved by either manual tuning or using some autofocusing

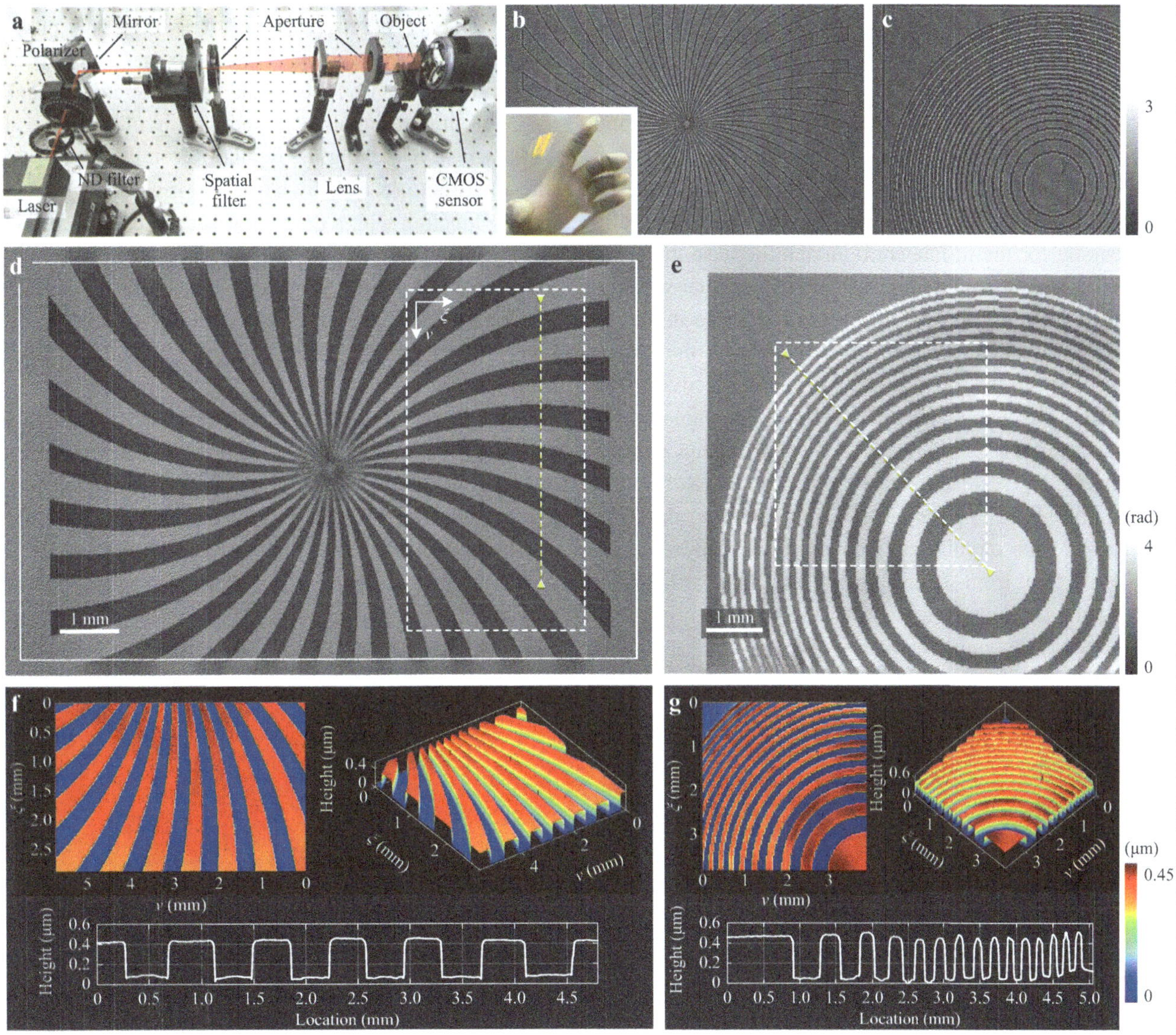

Fig. 2 Experimental characterization of a fabricated phase plate. **a** Optical setup of the in-line holographic imaging system. **b** and **c** show the captured in-line holograms of two transparent patterns. The inset shows an image of the phase plate. **d** and **e** show the reconstructed phase distribution from **b** and **c**, respectively. The white solid rectangle in **d** indicates the sensor's field of view. **f** and **g** visualize the rendered surface profiles and the cross-sectional height profiles of **d** and **e**, respectively. The white dashed rectangles indicate the enlarged areas, and the yellow dashed lines indicate the cross-sections. The same applies to other figures.

algorithms[114–116]. Each captured intensity image is normalized by the background image which is obtained using the same experimental conditions, only with the object removed. This calibration process ensures the validity of the absorption constraint[39].

We demonstrated the quantitative phase imaging capability of the proposed compressive phase retrieval method for real-world objects. Fig. 2 shows the imaging results of a phase plate, which was fabricated by etching multiple self-designed binary patterns onto a JGS1 quartz glass substrate. The retrieved phases of the spiral pattern and Fresnel zone plate are shown in Fig. 2d, e, respectively. Smooth staircase profiles with sharp edges can be accurately retrieved. Artifacts on the surface were preserved, indicating a high spatial resolution. Given that the object has a uniform refractive index distribution and that the fabricated features lie only on the surface of the substrate, the surface profiles can be directly calculated and characterized with knowledge of the medium refractive index, as shown in Fig. 2f, g. It is important to note that recovering the low-frequency phase is inherently challenging for the in-line configuration because the weak phase transfer function diminishes to zero together with the spatial frequency[117]. Hence, slowly varying phase structures hardly contribute to the intensity variations at the sensor plane, rendering them particularly difficult to recover. This has been the main drawback for in-line holography compared to off-axis holography[118,119]. Our findings suggest that the CCTV regularizer may be a competitive solution for this problem. In addition, because we have addressed the issue of the circular convolution model, the retrieved image is free from boundary artifacts, and the object distribution outside the sensor area can be partially recovered, as shown in Fig. 2d.

We imaged a muscle tissue section to validate the proposed method on complex objects, as shown in Fig. 3. The twin image was effectively eliminated from the retrieved amplitude and phase, which visually demonstrated the high-fidelity imaging performance of the method. The retrieved phase reveals thickness variations inside the tissue slice, facilitating label-free and low-cost clinical studies or diagnoses[120,121].

Ablation experiments

We conducted ablation studies based on experimental data to quantify the performance improvements obtained by the physical constraints and the complex TV regularizer respectively. For comparison, we implemented compressive phase retrieval algorithms using only physical constraints or gradient sparsity. The former, referred to as the IP method, can be viewed as an accelerated variant of the conventional iterative projection algorithm. This was implemented by setting the regularization parameter $\lambda = 0$. The algorithm proceeds by iteratively minimizing the data-fidelity function with a gradient descent step and enforcing physical constraints with a projection step. The latter, referred to as the CTV method, uses only the complex TV regularizer without any physical constraints. It is implemented by defining the constraint set $C = \mathbb{C}^n$ and thus the projection onto set C is reduced to identity mapping.

Both the IP and CCTV methods can utilize various physical constraints. To better differentiate among different methods, we use the letter 'A' to represent the method that uses the absorption constraint, the letter 'S' for the support constraint, and 'AS' for both constraints. For example, 'IP-A' represents an iterative projection algorithm that uses only the absorption constraint.

We first performed phase imaging of a quantitative

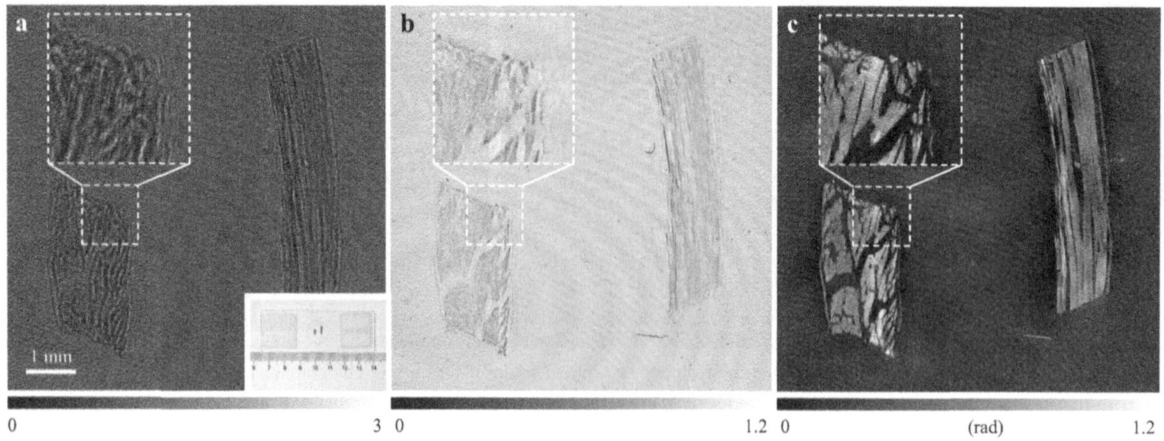

Fig. 3 Holographic imaging results of a muscle tissue section. **a** Experimentally recorded intensity image. The inset is an image of the sample slide. **b** and **c** are the retrieved amplitude and phase, respectively.

phase microscopy target (Benchmark Technologies) using three methods, namely IP-A, CTV, and CCTV-A, as shown in Fig. 4. Note that the object has an extended distribution and thus the support constraint is inapplicable. The IP algorithm fails to eliminate the twin-image artifact completely, and the phase structures are not clearly revealed, which implies that the physical constraint alone is insufficient to suppress the phase ambiguities. The CCTV method achieved better imaging quality than CTV, as illustrated by visual and quantitative comparisons. The ground-truth phase values were calculated using the structures' height and medium refractive index, which are both provided by the manufacturer. The root-mean-square errors (RMSEs) of the cross-sectional phase profiles

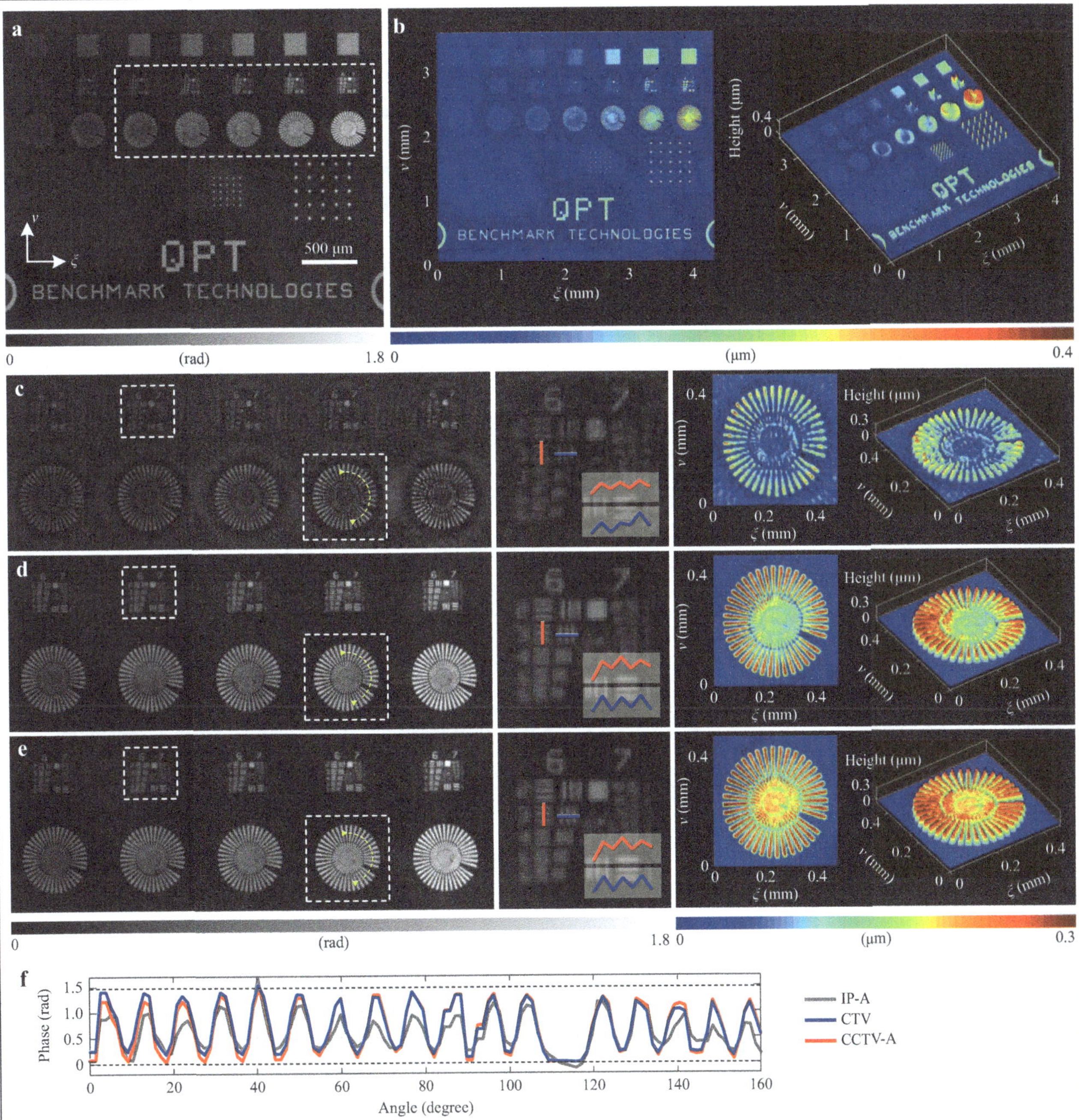

Fig. 4 Experimental comparison of different phase retrieval algorithms with a quantitative phase resolution target. **a** Full field-of-view phase reconstruction by the proposed CCTV-A method. **b** The calculated surface profile corresponding to **a**. **c**, **d** and **e** show the retrieved phase and height profiles by the IP-A, CTV, and CCTV-A methods, respectively. **f** The cross-sectional phase profiles, where the dashed line indicates the calculated theoretical phase value.

retrieved using the IP-A, CTV, and CCTV-A methods were 0.880, 0.463, and 0.418 rad, respectively. In addition, the finest resolved features of the USAF 1951 resolution target are highlighted. Group 6, Element 3 is resolved by all three methods, indicating a spatial resolution of 80.6 lines/mm, which agrees with the maximum resolution of 85.3 lines/mm of the sensor. All three methods achieve the same sampling-limited resolution, which implies that the TV-based regularizer does not sacrifice the spatial resolution for quality enhancement. The resolution can be further improved beyond the Nyquist sampling limit, however, if one takes into account the down-sampling effect of the sensor pixels[112,122–126].

Similar improvements in the CCTV method over the IP and CTV methods were also observed in the imaging results of an amplitude object. Fig. 5 compares the amplitudes of a positive 1951 USAF test target (Thorlabs, Inc.) retrieved by the IP-A, CTV, and CCTV-A methods. Quantitatively, the RMSEs of the retrieved amplitude profiles 1/2 in Fig. 5d are 0.331/0.338, 0.240/0.239, 0.238/0.232 for the IP-A, CTV and CCTV-A methods, respectively. Furthermore, as indicated by a qualitative visual comparison, the IP method suffers from high reconstruction noise, and the CCTV model improves the performance of CTV by suppressing artifacts with no loss of spatial resolution.

Materials and methods

The proximal gradient framework

In the regularized inverse problem of Eq. 3, the fidelity function $F(\boldsymbol{x})$ is a (almost everywhere) differentiable function whereas the regularizer $R(\boldsymbol{x})$ is a non-differentiable function. Because we are dealing with real-valued functions over complex-valued variables, CR-calculus is adopted[127]. The gradient of $F(\boldsymbol{x})$ is given by[112]

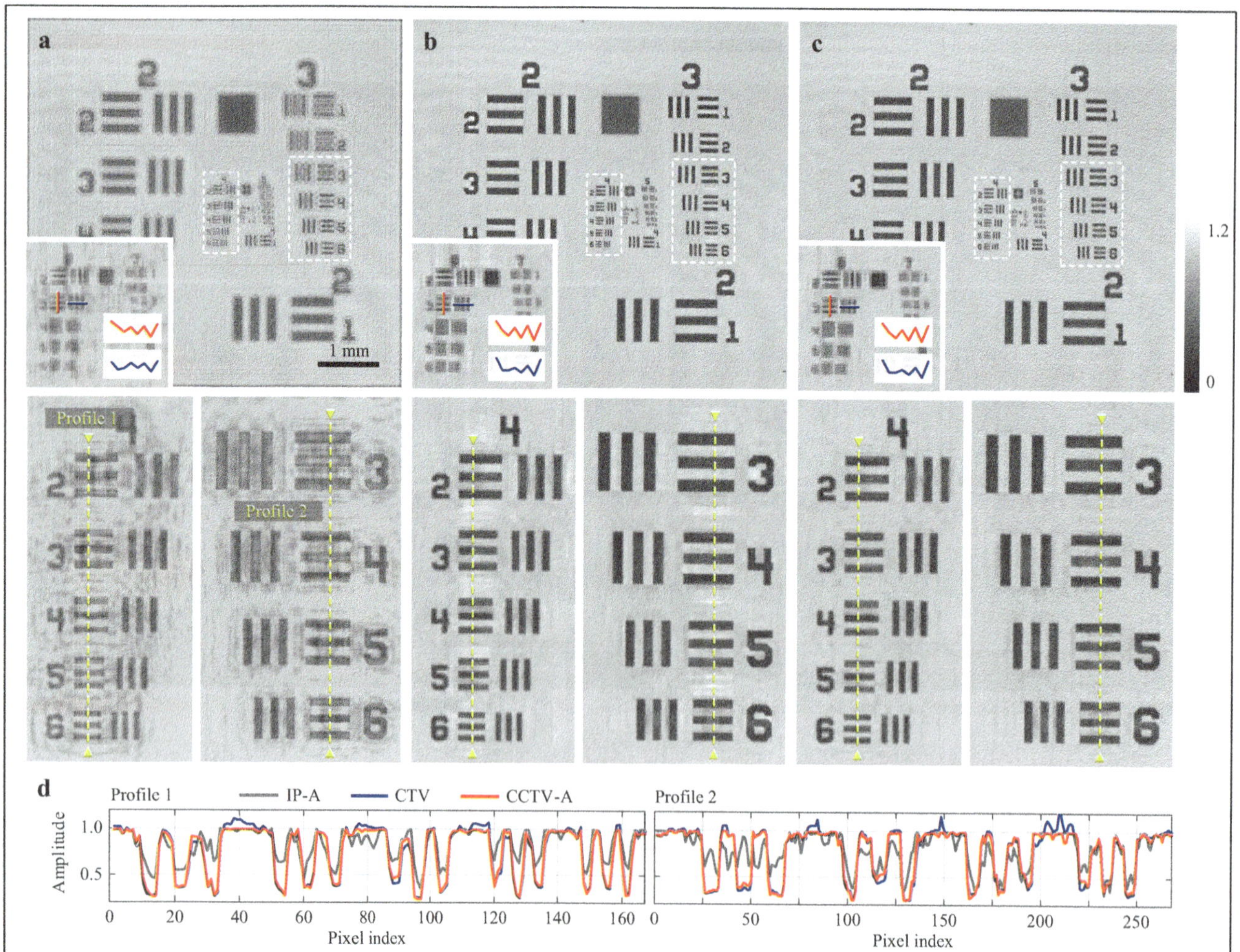

Fig. 5 Experimental comparison of different phase retrieval algorithms with an amplitude resolution target. **a**, **b** and **c** show the retrieved amplitude using the IP-A, CTV, and CCTV-A algorithms, respectively. The inset indicates the finest structures resolved by the corresponding method. Below are the corresponding enlarged areas. **d** Cross-sectional profiles.

$$\nabla_x F(x) = \frac{1}{2} A^{\mathsf{H}} \mathrm{diag}\left(\frac{Ax}{|Ax|}\right)(|Ax| - y) \tag{5}$$

where $(\cdot)^{\mathsf{H}}$ denote the conjugate transpose (Hermitian) operator. Non-smoothness only occurs when Ax has zero entries, which, as we will show below, does not affect the algorithm's convergence behavior.

The difficulty in minimizing the objective function mainly arises from the non-differentiability of $R(x)$. The proximal gradient method provides a general framework for solving composite non-smooth optimization problems in the form of Eq. 3 by minimizing the non-differentiable $R(x)$ via its proximal operator $\mathrm{prox}_{\gamma R}$, which is defined as[128]

$$\mathrm{prox}_{\gamma R}(v) = \underset{x}{\mathrm{argmin}}\, R(x) + \frac{1}{2\gamma}\|x - v\|_2^2 \tag{6}$$

where $\gamma > 0$ is the step size. Notice that Eq. 6 coincides with an image denoising problem, with $R(x)$ serving as a regularization function. For the CCTV regularizer, the denoising subproblem can be solved efficiently using the gradient projection algorithm introduced in the next subsection. The proximal gradient algorithm proceeds by iteratively applying a gradient update and proximal update. A variant using the Nesterov's acceleration technique was adopted[129], as shown in Algorithm 1. The accelerated proximal gradient (APG) algorithm is also known as the fast iterative shrinkage thresholding algorithm (FISTA)[130]. APG is well known for its superior convergence speed over the basic proximal gradient algorithm in the convex setting. Through numerical studies, we empirically demonstrate that it also helps to speed up convergence for the nonconvex phase retrieval problem.

Algorithm 1 Accelerated proximal gradient algorithm

Input: Initial guess $x^{(0)}$, step size γ, $\beta_t = t/(t+3)$, and iteration number T.
1: $u^{(0)} = x^{(0)}$
2: **for** $t = 1, 2, \ldots, T$ **do**
3: $\quad v^{(t)} = u^{(t-1)} - \gamma \nabla_u F(u^{(t-1)})$ ▹ Gradient update
4: $\quad x^{(t)} = \mathrm{prox}_{\gamma R}(v^{(t)})$ ▹ Proximal update via Alg. 2
5: $\quad u^{(t)} = x^{(t)} + \beta_t (x^{(t)} - x^{(t-1)})$
6: **end for**

CCTV denoising algorithm

The proximal update step in Eq. 6 involves solving a CCTV-regularized image denoising subproblem. It is a convex optimization problem when C is closed and convex. No exact closed-form solution to this problem is available. Thus, an iterative solver should be invoked at every iteration. We followed the approach proposed in Refs. 95,131 and consider solving the primal problem via its dual, which is given by

$$\min_{w \in S} \left\{ G(w) \equiv -\|\mathcal{H}_C(v - \gamma D^{\mathsf{H}} w)\|_2^2 + \|v - \gamma D^{\mathsf{H}} w\|_2^2 \right\} \tag{7}$$

where $w \in \mathbb{C}^{2n}$ denotes the dual variable, $\mathcal{H}_C(x) \stackrel{\text{def}}{=} x - \mathcal{P}_C(x)$ with $\mathcal{P}_C$ representing the Euclidean projection operator on C, and S denotes the convex set which contains all $w \in \mathbb{C}^{2n}$ that satisfy

$$\|w\|_\infty \le \lambda \tag{8}$$

where $\|w\|_\infty = \max_i |w_i|$. As shown in the Supplementary Document, the desired primal optimal solution $x^\star$ is related to the dual optimal solution $w^\star$ through the following equation:

$$x^\star = \mathcal{P}_C(v - \gamma D^{\mathsf{H}} w^\star) \tag{9}$$

The problem of Eq. 7 can be effectively solved using the gradient projection algorithm, where the dual variable is iteratively updated by a gradient step followed by a projection operation $\mathcal{P}_S(\cdot)$ onto the set S. An accelerated variant of the algorithm is presented in Algorithm 2. The Wirtinger gradient of the dual objective function $G(w)$ is given by[131]

$$\nabla_w G(w) = -\gamma D \mathcal{P}_C(v - \gamma D^{\mathsf{H}} w) \tag{10}$$

Algorithm 2 Accelerated gradient projection algorithm

Input: Observation v, parameter γ, step size η, $\beta_t = t/(t+3)$, and subiteration number T_{sub}.
1: $w^{(0)} = \mathbf{0}$, $z^{(0)} = w^{(0)}$
2: **for** $t = 1, 2, \ldots, T_{\mathrm{sub}}$ **do**
3: $\quad w^{(t)} = \mathcal{P}_S\left(z^{(t-1)} - \eta \nabla_z G(z^{(t-1)})\right)$
4: $\quad z^{(t)} = w^{(t)} + \beta_t (w^{(t)} - w^{(t-1)})$
5: **end for**
6: $x = \mathcal{P}_C(v - \gamma D^{\mathsf{H}} w^{(T_{\mathrm{sub}})})$

Convergence analysis

In the above subsections, we introduce the proximal gradient algorithm and the corresponding denoising algorithm for the proximal update. We now show that the algorithm parameters, namely the step sizes γ and η, can be predetermined based on the measurement scheme and the sparsifying transform.

Considering the nonconvexity of the problem of Eq. 3, we present a weaker result by showing the convergence of the basic (non-accelerated) proximal gradient algorithm, which is summarized by the following theorem. The detailed proof is provided in the Supplementary Document. Furthermore, extensive numerical studies have demonstrated that the accelerated algorithm converges following the same step size selection rule.

Theorem 1 *The basic proximal gradient algorithm (with* $\beta_t \equiv 0$ *in Algorithm 1) for the problem of Eq. 3 converges to* *a stationary point using a fixed step size γ that satisfies*

$$\gamma \leq \frac{2}{\rho(\mathbf{A}^{\mathsf{H}}\mathbf{A})} \tag{11}$$

where $\rho(\cdot)$ denotes the spectral radius.

The above theorem is an extension of the gradient descent algorithm[132] to the more general class of the proximal gradient algorithms.

The sampling matrix $\mathbf{A}$ depends only on the physical model. Thus, the step size can be known a priori. Recall that the sampling matrix comprises a circular convolution operator $\mathbf{Q}$ and an image cropping operator $\mathbf{C}$. Circular convolution can be efficiently calculated via FFTs as $\mathbf{Q} = \mathbf{F}^{-1}\text{diag}(\mathbf{h})\mathbf{F}$, where $\mathbf{F}$ is the Fourier matrix and $\mathbf{h}$ is the transfer function. Note that $|\mathbf{h}| = 1$. Thus, $\mathbf{Q}$ is a unitary matrix that satisfies $\mathbf{Q}^{\mathsf{H}}\mathbf{Q} = \mathbf{I}$, where $\mathbf{I}$ is the identity matrix. It can be easily verified that $\rho(\mathbf{A}^{\mathsf{H}}\mathbf{A}) \leq 1$, which, according to Theorem 1, implies that we should use step size $\gamma = 2$ for the proximal gradient algorithm. This choice of step size is in fact nontrivial and generally applicable to any passive measurement scheme.

The next theorem establishes the convergence of the accelerated gradient projection algorithm to solve the convex denoising subproblem via its dual (Eq. 7). The proof can be found in the Supplementary Document.

Theorem 2 *Assuming that the constraint set C is closed and convex, the accelerated gradient projection algorithm for the problem of Eq. 7 converges to the global optimum using a fixed step size η that satisfies*

$$\eta \leq \frac{1}{\gamma^2\rho(\mathbf{D}^{\mathsf{H}}\mathbf{D})} \tag{12}$$

Theorem 2 suggests that the step size for the denoising algorithm can also be determined beforehand based on γ given by Theorem 1 and the sparsifying transform $\mathbf{D}$. Specifically, for the finite-difference operator, we show in the Supplementary Document that $\rho(\mathbf{D}^{\mathsf{H}}\mathbf{D}) \leq 8$, which, together with $\gamma = 2$, suggests $\eta = 1/32$ for this particular case. The convergence established by Theorem 1 assumes that an exact solution to the denoising subproblem is obtained at each iteration. The algorithm may converge to a suboptimal solution when the proximal update is inexact, as with the CCTV regularizer. Readers can refer to Ref. 133 and references therein for convergence results regarding inexact proximal gradient algorithms. Therefore, the iteration number used for the gradient projection algorithm should be chosen as a balance between reconstruction quality and computation efficiency.

Discussion

We performed numerical studies based on the same in-line holographic imaging model as described in the experiment section to quantitatively evaluate the algorithmic behaviors and further shed light on our empirical observations and theoretical results. The wavelength was set to 500 nm, the pixel size was 5 μm, and the imaging distance was 5 mm. Eight standard test images of size 256×256 were used to generate the virtual phase objects for the simulation. The phase of the object was assumed to have a range of $[0, \pi]$. The objects have a uniform amplitude distribution and were zero-padded to avoid boundary artifacts. Additive white Gaussian noise was added to each hologram such that the signal-to-noise ratio (SNR) was 20 dB by default. We calculated the phase RMSE as a metric to quantitatively compare the reconstruction quality of different methods.

In all numerical studies, the algorithms were initialized with an initial estimate of $\mathbf{x}^{(0)} = \mathbf{A}^{\mathsf{H}}\mathbf{y}$. The algorithms were implemented using MATLAB on a personal computer with Core i7-11700F CPU @ 2.50GHz (Intel) and 16 GB of RAM. The MATLAB code is publicly available in Ref. 134.

Hyperparameter selection

We first investigated the influence of the subiteration number T_{sub} of the gradient projection denoising algorithm on the convergence of the proximal gradient algorithm. As previously pointed out, T_{sub} is a key hyperparameter that should be determined beforehand. We ran the APG algorithm for 1000 iterations using different T_{sub} values ranging from one to ten. Meanwhile, four regularization parameters, $\lambda = 1 \times 10^{-3}$, 3×10^{-3}, 1×10^{-2}, and 3×10^{-2} were considered for comparison. To quantify the convergence behavior of the inexact APG algorithms, we calculated the relative error of the final objective function value for each T_{sub} relative to that with $T_{\text{sub}} = 10$. Fig. 6a indicates that the relative error decreases monotonically as T_{sub} increases, and thus, a more accurate proximal update can be obtained by increasing the number of subiterations. We empirically selected the smallest T_{sub} that yielded an error below 1% as the optimal choice. Another observation made from Fig. 6a is the dependence of T_{sub} on regularization parameter λ. As λ increases, one needs more subiterations for the denoising algorithm to obtain an adequate solution. This seems quite natural, considering the nonlinear nature of the denoising operation. A larger λ imposes a stronger regularization force on the image, thus requiring more linear iterative steps to approximate the solution.

Another hyperparameter of great importance is the regularization parameter λ, which offers a tradeoff between model consistency and signal priors. We compared different selections of λ in terms of both the reconstruction quality and computational efficiency. Fig. 6b plots the

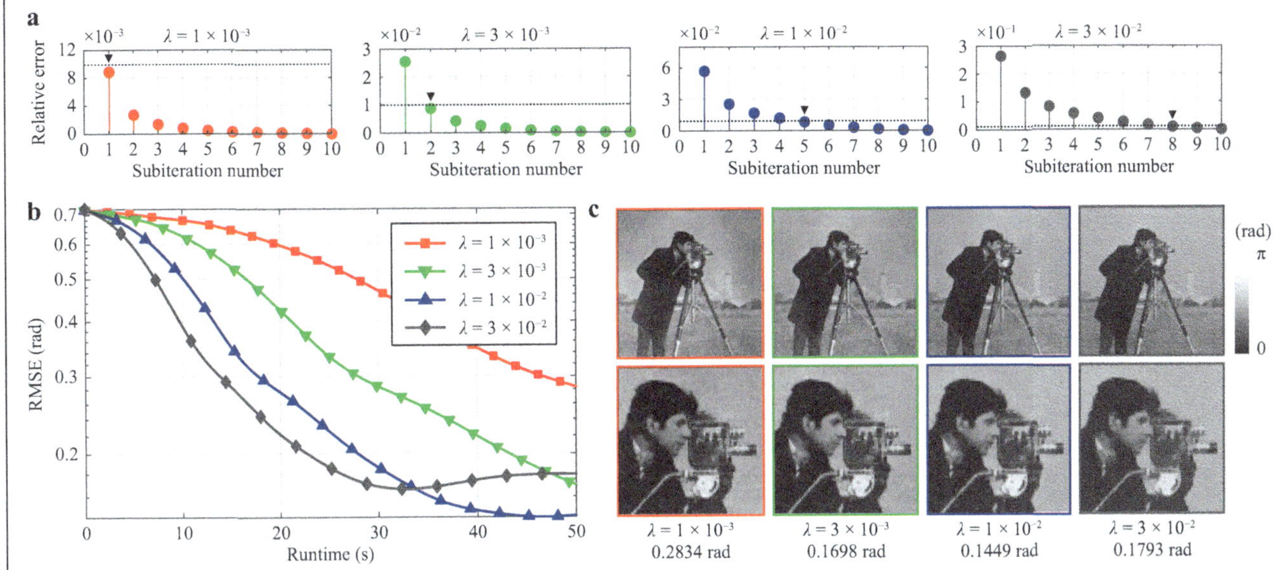

Fig. 6 Comparison of different hyperparameter selections based on simulations. **a** Relative error against subiteration number T_{sub} for different regularization parameters λ. The dashed lines indicate the value of 1%, and the downward-pointing triangles indicate the optimal choice of subiteration number that satisfies the error criterion. **b** Root-mean-square error against runtime for different λ when using the corresponding optimal T_{sub} in **a**. **c** Phase obtained by running the algorithms for 50 s. The second row shows the enlarged areas. The regularization parameters and RMSEs are shown below. No physical constraints are considered in this simulation.

convergence curves against runtime, where the optimal T_{sub} was used for each λ, that is, $T_{\mathrm{sub}} = 1$ for $\lambda = 1\times10^{-3}$, $T_{\mathrm{sub}} = 2$ for $\lambda = 3\times10^{-3}$, $T_{\mathrm{sub}} = 5$ for $\lambda = 1\times10^{-2}$, and $T_{\mathrm{sub}} = 8$ for $\lambda = 3\times10^{-2}$. A larger regularization parameter can help accelerate the convergence at the beginning stage. However, the final result tends to be over-smoothed, as shown in Fig. 6c. Therefore, we set $\lambda = 1\times10^{-2}$ for balanced performance in terms of speed and accuracy.

Comparison of physical constraints

Having determined the optimal hyperparameters for the complex TV, we evaluated the effect of incorporating additional physical constraints. Fig. 7 shows the results obtained using no constraints, support constraint, absorption constraint, or both. For the support constraint, the support region was set to the central 256×256 pixels of the image. The introduction of the absorption constraint can significantly improve the overall quality, while the support constraint can help suppress artifacts at the boundaries. Incorporating physical constraints also helps accelerate convergence.

Influence of measurement noise

The inherent ill-posedness of single-shot phase retrieval leads to high sensitivity to measurement noise. To quantitatively compare the phase reconstruction quality under noisy conditions, we added different levels of additive white Gaussian noise with an SNR ranging from 5 to 30 dB to the simulated hologram, as shown in Fig. 8a. The corresponding phases retrieved by the CCTV-AS method are presented in Fig. 8c. The CCTV method can achieve high-quality phase retrieval under moderate-noise conditions. However, when the hologram is severely degraded by measurement noise, a larger regularization parameter λ is required to suppress reconstruction artifacts. The retrieved phase may then exhibit a staircase artifact when the regularization parameter is too large, which is an inherent problem for TV-based regularizers. Therefore, more advanced image priors are required to achieve better reconstruction quality under noisy conditions.

Generalization to different scenes

To further examine the generalizability of the CCTV model, we tested the algorithm on various test images and compared it with IP and CTV methods. The results are shown in Fig. 9. The conventional IP algorithm generally suffers from a high reconstruction error because the physical constraints alone are insufficient to suppress the twin-image artifact. On the other hand, the CTV model often yields visually appealing results, yet its performance is highly dependent on the sparsity of the scene. For example, the CTV model tends to perform well for scenes with very sparse gradients, such as the *House* image, but it may be unsuitable for retrieving complex scenes, such as the *Goldhill* and *Mandrill* images. In all cases, the CCTV model offers the best reconstruction quality owing to the

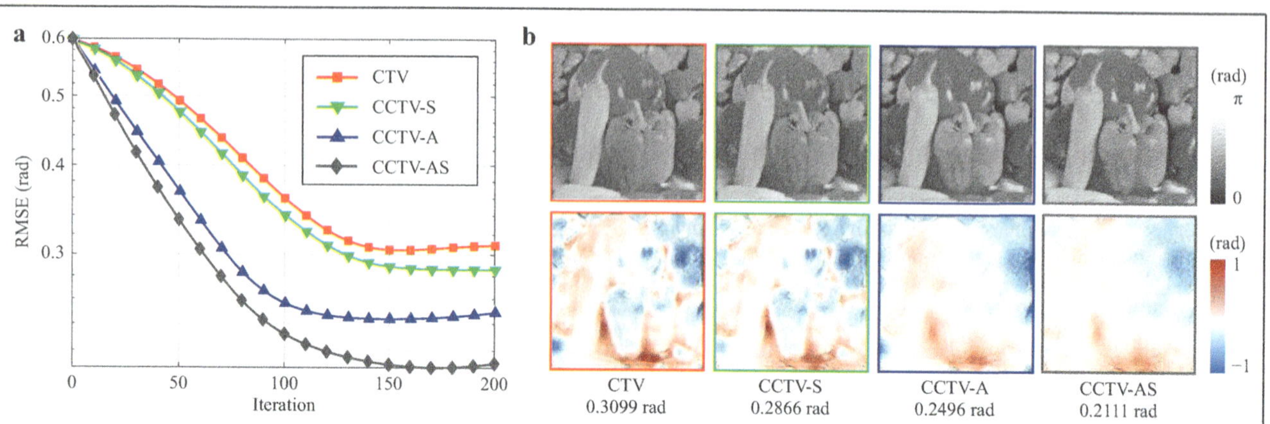

Fig. 7 Comparison of convergence speed and reconstruction quality with different physical constraints. **a** The RMSE is plotted against the iteration number. **b** The upper and lower rows show the reconstructed phase after 200 iterations and the corresponding phase residuals, respectively. The calculated phase RMSEs are shown below.

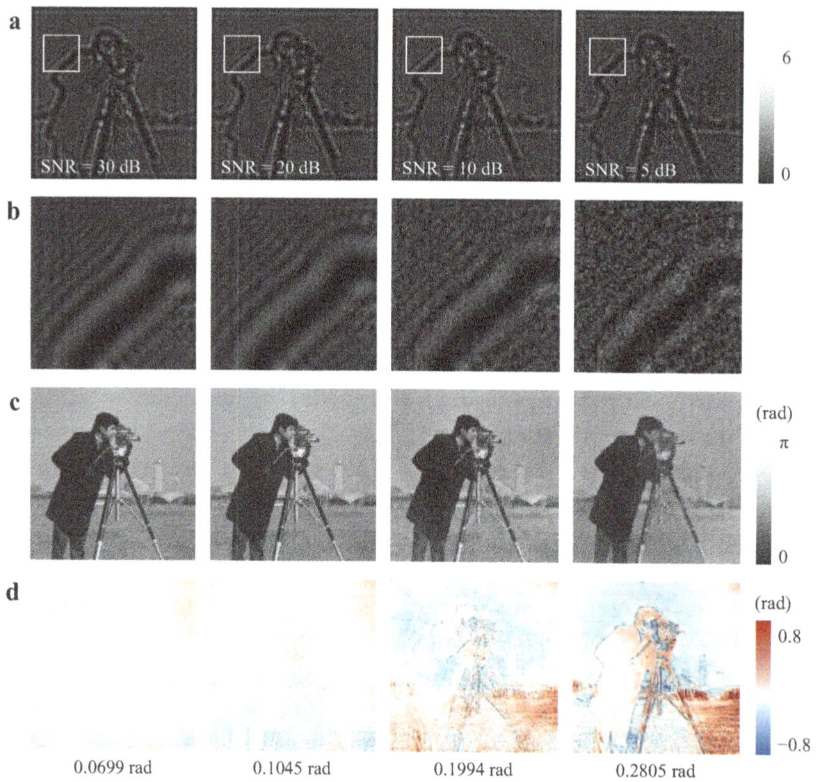

Fig. 8 Effect of measurement noise on reconstruction quality. **a** Simulated noisy holograms with different SNRs. **b** Corresponding enlarged areas of **a**. **c** Retrieved phase by the CCTV-AS method. **d** Phase residuals of **c**. The calculated phase RMSEs are shown below.

combination of sparsity priors and physical constraints. We also set $\lambda = 1 \times 10^{-2}$ as fixed throughout the simulation experiments. This implies that the selection of λ is not very sensitive to the scene content.

Nesterov acceleration for nonconvex optimization

We also empirically validated the effectiveness of Nesterov's method for the nonconvex phase retrieval problem by comparing the APG algorithm with the non-accelerated version in both unconstrained and constrained cases, as shown in Fig. 10a, b, respectively. Nesterov's method significantly improved the convergence speed in both cases. The APG algorithm involves only an additional extrapolation step per iteration, whose time consumption is negligible compared with the gradient and proximal update steps. The empirical success of Nesterov's method can be

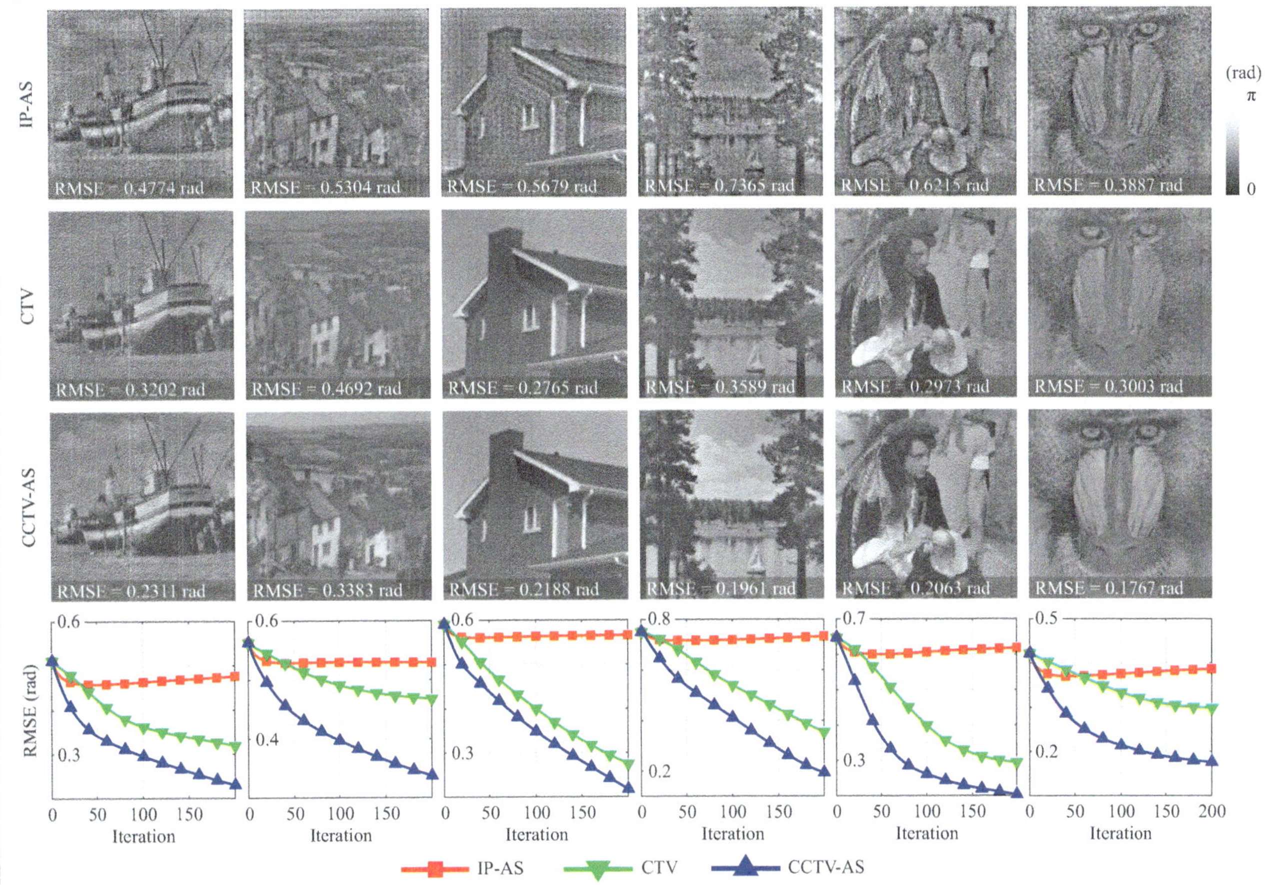

Fig. 9 Comparison of the IP-AS, CTV, and CCTV-AS methods for various test images (from left to right: *Ship*, *Goldhill*, *House*, *Lake*, *Male*, *Mandrill*). The upper three rows show the reconstructed phase by the three methods. The bottom row plots the RMSE against the iteration number.

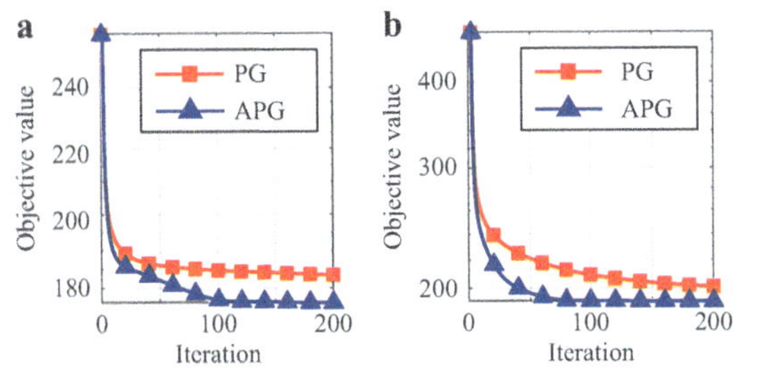

Fig. 10 Convergence of the proximal gradient (PG) algorithm and its accelerated variant (APG) in **a** unconstrained case and **b** constrained case.

explained partly by the similarity between the amplitude-based fidelity function and the convex least-squares formulation[107].

Conclusion

We have proposed a unified compressive phase retrieval framework for in-line holography that combines the complex total variation and physical knowledge of the wavefield. The resulting CCTV model can characterize the sparse nature of complex real-world scenes while leveraging physically tractable constraints to further improve imaging quality, as confirmed by both simulated and optical experiments. In addition, we conducted a detailed theoretical analysis from a mathematical viewpoint, which provides general guidance for algorithm parameter selection to ensure convergence, obviating the need for laborious manual tuning. We emphasize that the proposed framework is generally applicable to various optical configurations and can easily adapt to other physical constraints and sparsifying operators.

Although numerically extending the proposed algorithmic framework to coherent diffraction imaging is straightforward, retrieving complex-valued images from a single far-field diffraction pattern remains challenging. Future work will explore more advanced image priors that enable snapshot coherent diffraction imaging of complex samples[135]. Another direction for future research is to achieve quantitative phase imaging of objects with a large phase shift, as only weak phase objects have been tested in this work. Comparison of the proposed method with other existing quantitative phase imaging methods also remains to be studied in the future[136,137].

Acknowledgements
The authors acknowledge the National Natural Science Foundation of China (Grant No. 61827825) for financial support. The authors would also like to thank the anonymous reviewers for their constructive comments and suggestions, which improved the presentation of the paper. Part of this paper was posted as a preprint[138] before submission.

Conflict of interest
The authors declare no conflicts of interest.

Supplementary information is available for this paper at https://doi.org/10.37188/lam.2023.006.

Received: 07 June 2022 Revised: 01 February 2023 Accepted: 03 February 2023
Accepted article preview online: 04 February 2023
Published online: 11 March 2023

References

1. Gabor, D. A new microscopic principle. *Nature* **161**, 777-778 (1948).
2. Javidi, B. et al. Roadmap on digital holography [Invited]. *Optics Express* **29**, 35078-35118 (2021).
3. Sheridan, J. T. et al. Roadmap on holography. *Journal of Optics* **22**, 123002 (2020).
4. Liu, G. & Scott, P. D. Phase retrieval and twin-image elimination for in-line Fresnel holograms. *Journal of the Optical Society of America A* **4**, 159-165 (1987).
5. Barton, J. J. Removing multiple scattering and twin images from holographic images. *Physical Review Letters* **67**, 3106-3109 (1991).
6. Shechtman, Y. et al. Phase retrieval with application to optical imaging: a contemporary overview. *IEEE Signal Processing Magazine* **32**, 87-109 (2015).
7. Dong, J. et al. Phase retrieval: from computational imaging to machine learning: a tutorial. *IEEE Signal Processing Magazine* **40**, 45-57 (2023).
8. Almoro, P., Pedrini, G. & Osten, W. Complete wavefront reconstruction using sequential intensity measurements of a volume speckle field. *Applied Optics* **45**, 8596-8605 (2006).
9. Waller, L., Ti an, L. & Barbastathis, G. Transport of intensity phase-amplitude imaging with higher order intensity derivatives. *Optics Express* **18**, 12552-12561 (2010).
10. Zuo, C. et al. Transport of intensity equation: a tutorial. *Optics and Lasers in Engineering* **135**, 106187 (2020).
11. Bao, P. et al. Phase retrieval using multiple illumination wavelengths. *Optics Letters* **33**, 309-311 (2008).
12. Bai, Y. R. et al. Ptychographic microscopy via wavelength scanning. *APL Photonics* **2**, 056101 (2017).
13. Fan, Y. et al. Single-shot isotropic quantitative phase microscopy based on color-multiplexed differential phase contrast. *APL Photonics* **4**, 121301 (2019).
14. Gao, Y. H. & Cao, L. C. Projected refractive index framework for multi-wavelength phase retrieval. *Optics Letters* **47**, 5965-5968 (2022).
15. Faulkner, H. M. L. & Rodenburg, J. M. Movable aperture lensless transmission microscopy: a novel phase retrieval algorithm. *Physical Review Letters* **93**, 023903 (2004).
16. Zhang, Z. B. et al. Invited article: mask-modulated lensless imaging with multi-angle illuminations. *APL Photonics* **3**, 060803 (2018).
17. Jiang, S. W. et al. Wide-field, high-resolution lensless on-chip microscopy via near-field blind ptychographic modulation. *Lab on a Chip* **20**, 1058-1065 (2020).
18. Zheng, G. A., Horstmeyer, R. & Yang, C. Wide-field, high-resolution Fourier ptychographic microscopy. *Nature Photonics* **7**, 739-745 (2013).
19. Tian, L. et al. Computational illumination for high-speed *in vitro* Fourier ptychographic microscopy. *Optica* **2**, 904-911 (2015).
20. Zheng, G. A. et al. Concept, implementations and applications of Fourier ptychography. *Nature Reviews Physics* **3**, 207-223 (2021).
21. Zhang, F. C., Pedrini, G. & Osten, W. Phase retrieval of arbitrary complex-valued fields through aperture-plane modulation. *Physical Review A* **75**, 043805 (2007).
22. Katkovnik, V. et al. Computational super-resolution phase retrieval from multiple phase-coded diffraction patterns: simulation study and experiments. *Optica* **4**, 786-794 (2017).
23. Wu, Y. C., Sharma, M. K. & Veeraraghavan, A. WISH: wavefront imaging sensor with high resolution. *Light: Science & Applications* **8**, 44 (2019).
24. Li, R. J. & Cao, L. C. Complex wavefront sensing based on alternative structured phase modulation. *Applied Optics* **60**, A48-A53 (2021).
25. Fienup, J. R. Reconstruction of an object from the modulus of its Fourier transform. *Optics Letters* **3**, 27-29 (1978).
26. Tian, Y. C. & Fienup, J. R. Phase retrieval with only a nonnegativity constraint. *Optics Letters* **48**, 135-138 (2023).
27. Fienup, J. R. Reconstruction of a complex-valued object from the modulus of its Fourier transform using a support constraint. *Journal of the Optical Society of America A* **4**, 118-123 (1987).
28. Marchesini, S. et al. X-ray image reconstruction from a diffraction pattern alone. *Physical Review B* **68**, 140101 (2003).
29. Abbey, B. et al. Keyhole coherent diffractive imaging. *Nature Physics* **4**, 394-398 (2008).
30. Latychevskaia, T. & Fink, H. W. Solution to the twin image problem in holography. *Physical Review Letters* **98**, 233901 (2007).
31. Zhang, K. Y. J. & Main, P. Histogram matching as a new density modification technique for phase refinement and extension of protein molecules. *Acta Crystallographica Section A: Foundations of Crystallography* **46**, 41-46 (1990).
32. Elser, V. Solution of the crystallographic phase problem by iterated projections. *Acta Crystallographica Section A: Foundations of Crystallography* **59**, 201-209 (2003).
33. Levi, A. & Stark, H. Image restoration by the method of generalized projections with application to restoration from magnitude. *Journal of the Optical Society of America A* **1**, 932-943 (1984).
34. Elser, V. Phase retrieval by iterated projections. *Journal of the Optical Society of America A* **20**, 40-55 (2003).
35. Gerchberg, R. W. & Saxton, W. O. A practical algorithm for the determination of phase from image and diffraction plane pictures. *Optik* **35**, 237-246 (1972).
36. Fienup, J. R. Phase retrieval algorithms: a comparison. *Applied Optics* **21**, 2758-2769 (1982).
37. Bauschke, H. H., Combettes, P. L. & Luke, D. R. Phase retrieval, error reduction algorithm, and Fienup variants: a view from convex optimization. *Journal of the Optical Society of America A* **19**, 1334-1345 (2002).
38. Marchesini, S. Invited article: a unified evaluation of iterative projection algorithms for phase retrieval. *Review of Scientific Instruments* **78**, 011301 (2007).
39. Latychevskaia, T. Iterative phase retrieval for digital holography: tutorial. *Journal of the Optical Society of America A* **36**, D31-D40 (2019).
40. Rong, L. et al. Twin image elimination from two in-line holograms via

phase retrieval. *Chinese Optics Letters* **10**, 060902 (2012).
41. Brady, D. J. et al. Compressive holography. *Optics Express* **17**, 13040-13049 (2009).
42. Rivenson, Y., Stern, A. & Javidi, B. Compressive Fresnel holography. *Journal of Display Technology* **6**, 506-509 (2010).
43. Rivenson, Y., Stern, A. & Javidi, B. Overview of compressive sensing techniques applied in holography. *Applied Optics* **52**, A423-A432 (2013).
44. Moravec, M. L., Romberg, J. K. & Baraniuk, R. G. Compressive phase retrieval. Proceedings of SPIE 6701, Wavelets XII. San Diego, CA, USA: SPIE, 2007, 712–722.
45. Shechtman, Y., Be ck, A. & Eldar, Y. C. GESPAR: Efficient phase retrieval of sparse signals. *IEEE Transactions on Signal Processing* **62**, 928-938 (2014).
46. Jaganathan, K., Oymak, S. & Hassibi, B. Sparse phase retrieval: uniqueness guarantees and recovery algorithms. *IEEE Transactions on Signal Processing* **65**, 2402-2410 (2017).
47. Denis, L. et al. Inline hologram reconstruction with sparsity constraints. *Optics Letters* **34**, 3475-3477 (2009).
48. Wang, G. et al. Sparse phase retrieval via truncated amplitude flow. *IEEE Transactions on Signal Processing* **66**, 479-491 (2018).
49. Pacheco, C. et al. Adaptive sparse reconstruction for lensless digital holography via PSF estimation and phase retrieval. *Optics Express* **30**, 33433-33448 (2022).
50. Kostenko, A. et al. Phase retrieval in in-line x-ray phase contrast imaging based on total variation minimization. *Optics Express* **21**, 710-723 (2013).
51. Horisaki, R. et al. Single-shot phase imaging with a coded aperture. *Optics Letters* **39**, 6466-6469 (2014).
52. Horisaki, R., Egami, R. & Tanida, J. Single-shot phase imaging with randomized light (SPIRaL). *Optics Express* **24**, 3765-3773 (2016).
53. Zhang, W. H. et al. Twin-image-free holography: a compressive sensing approach. *Physical Review Letters* **121**, 093902 (2018).
54. Momey, F. et al. From Fienupos phase retrieval techniques to regularized inversion for in-line holography: tutorial. *Journal of the Optical Society of America A* **36**, D62-D80 (2019).
55. Jolivet, F. et al. Regularized reconstruction of absorbing and phase objects from a single in-line hologram, application to fluid mechanics and micro-biology. *Optics Express* **26**, 8923-8940 (2018).
56. Chang, H. B. et al. Total variationBbased phase retrieval for Poisson noise removal. *SIAM Journal on Imaging Sciences* **11**, 24-55 (2018).
57. Guo, C. et al. Lensfree on-chip microscopy based on dual-plane phase retrieval. *Optics Express* **27**, 35216-35229 (2019).
58. Wu, J. C., Ya ng, F. & Cao, L. C. Resolution enhancement of long-range imaging with sparse apertures. *Optics and Lasers in Engineering* **155**, 107068 (2022).
59. Gaur, C., Mohan, B. & Khare, K. Sparsity-assisted solution to the twin image problem in phase retrieval. *Journal of the Optical Society of America A* **32**, 1922-1927 (2015).
60. Villanueva-Perez, P. et al. Contrast-transfer-function phase retrieval based on compressed sensing. *Optics Letters* **42**, 1133-1136 (2017).
61. Galande, A. S. et al. Quantitative phase imaging of biological cells using lensless inline holographic microscopy through sparsity-assisted iterative phase retrieval algorithm. *Journal of Applied Physics* **132**, 243102 (2022).
62. Rivenson, Y. et al. Sparsity-based multi-height phase recovery in holographic microscopy. *Scientific Reports* **6**, 37862 (2016).
63. Loock, S. & Plonka, G. Phase retrieval for Fresnel measurements using a shearlet sparsity constraint. *Inverse Problems* **30**, 055005 (2014).
64. Pein, A. et al. Using sparsity information for iterative phase retrieval in x-ray propagation imaging. *Optics Express* **24**, 8332-8343 (2016).
65. Tillmann, A. M., Eldar, Y. C. & Mairal, J. DOLPHInjdictionary learning for phase retrieval. *IEEE Transactions on Signal Processing* **64**, 6485-6500 (2016).
66. Krishnan, J. P., Bioucas-Dias, J. M. & Katkovnik, V. Dictionary learning phase retrieval from noisy diffraction patterns. *Sensors* **18**, 4006 (2018).
67. Li, X. Y. et al. Dictionary-based compressive Fourier ptychography. *Optics Letters* **47**, 2314-2317 (2022).
68. Heide, F. et al. ProxImaL: Efficient image optimization using proximal algorithms. *ACM Transactions on Graphics* **35**, 84 (2016).
69. Katkovnik, V. & Astola, J. Phase retrieval via spatial light modulator phase modulation in 4f optical setup: numerical inverse imaging with sparse regularization for phase and amplitude. *Journal of the Optical Society of America A* **29**, 105-116 (2012).
70. Metzler, C. A., Maleki, A. & Baraniuk, R. G. BM3D-PRGAMP: compressive phase retrieval based on BM3D denoising. Proceedings of 2016 IEEE International Conference on Multimedia & Expo Workshops. Seattle, WA, USA: IEEE, 2016, 2504–2508.
71. Shi, B. S. et al. Constrained phase retrieval: when alternating projection meets regularization. *Journal of the Optical Society of America B* **35**, 1271-1281 (2018).
72. Jin, X. Y. et al. Iterative denoising phase retrieval method for twin-image elimination in continuous-wave terahertz in-line digital holography. *Optics and Lasers in Engineering* **152**, 106986 (2022).
73. Guo, C. et al. Lensfree on-chip microscopy based on single-plane phase retrieval. *Optics Express* **30**, 19855-19870 (2022).
74. Sinha, A. et al. Lensless computational imaging through deep learning. *Optica* **4**, 1117-1125 (2017).
75. Nguyen, T. et al. Deep learning approach for Fourier ptychography microscopy. *Optics Express* **26**, 26470-26484 (2018).
76. Rivenson, Y., Wu, Y. C. & Ozcan, A. Deep learning in holography and coherent imaging. *Light: Science & Applications* **8**, 85 (2019).
77. Hand, P., Leong, O. & Voroninski, V. Phase retrieval under a generative prior. Proceedings of the 32nd International Conference on Neural Information Processing Systems. Montréal, Canada: Curran Associates Inc., 2018, 9154–9164.
78. Metzler, C. et al. prDeep: robust phase retrieval with a flexible deep network. Proceedings of the 35th International Conference on Machine Learning. Stockholm, Sweden: PMLR, 2018, 3501–3510.
79. Işıl, Ç., Oktem, F. S. & Koç, A. Deep iterative reconstruction for phase retrieval. *Applied Optics* **58**, 5422-5431 (2019).
80. Wang, Y. T., Sun, X. H. & Fleischer, J. When deep denoising meets iterative phase retrieval. Proceedings of the 37th International Conference on Machine Learning. Online, PMLR, 2020, 10007–10017.
81. Shi, B. S., Li an, Q. S. & Chang, H. B. Deep prior-based sparse representation model for diffraction imaging: A plug-and-play method. *Signal Processing* **168**, 107350 (2020).
82. Wang, F. et al. Phase imaging with an untrained neural network. *Light: Science & Applications* **9**, 77 (2020).
83. Bostan, E. et al. Deep phase decoder: self-calibrating phase microscopy with an untrained deep neural network. *Optica* **7**, 559-562 (2020).
84. Chang, X. Y., Bi an, L. H. & Zhang, J. Large-scale phase retrieval. *eLight* **1**, 4 (2021).
85. Zhang, F. L. et al. Physics-based iterative projection complex neural network for phase retrieval in lensless microscopy imaging. Proceedings of 2021 IEEE/CVF Conference on Computer Vision and Pattern Recognition. Nashville, TN, USA: IEEE, 2021, 10518–10526.
86. Zhang, X. Y., Wa ng, F. & Situ, G. H. BlindNet: an untrained learning approach toward computational imaging with model uncertainty. *Journal of Physics D: Applied Physics* **55**, 034001 (2021).

87. Yang, D. Y. et al. Coherent modulation imaging using a physics-driven neural network. *Optics Express* **30**, 35647-35662 (2022).
88. Chen, Z. J. et al. Phase recovery with deep complex-domain priors. *IEEE Signal Processing Letters* **29**, 887-891 (2022).
89. Kang, I. et al. Simultaneous spectral recovery and CMOS micro-LED holography with an untrained deep neural network. *Optica* **9**, 1149-1155 (2022).
90. Zhu, H. et al. DNF: diffractive neural field for lensless microscopic imaging. *Optics Express* **30**, 18168-18178 (2022).
91. Tian, Z. M. et al. Lensless computational imaging with a hybrid framework of holographic propagation and deep learning. *Optics Letters* **47**, 4283-4286 (2022).
92. Luo, H. X. et al. Diffraction-Net: a robust single-shot holography for multi-distance lensless imaging. *Optics Express* **30**, 41724-41740 (2022).
93. Situ, G. H. Deep holography. *Light: Advanced Manufacturing* **3**, 278-300 (2022).
94. Rudin, L. I., Osher, S. & Fatemi, E. Nonlinear total variation based noise removal algorithms. *Physica D: Nonlinear Phenomena* **60**, 259-268 (1992).
95. Chambolle, A. An algorithm for total variation minimization and applications. *Journal of Mathematical Imaging and Vision* **20**, 89-97 (2004).
96. Fabian, Z. et al. 3D phase retrieval at nano-scale via accelerated Wirtinger flow. Proceedings of the 2020 28th European Signal Processing Conference. Amsterdam, Netherlands: IEEE, 2021, 2080–2084.
97. Berdeu, A. et al. Comparative study of fully three-dimensional reconstruction algorithms for lens-free microscopy. *Applied Optics* **56**, 3939-3951 (2017).
98. Chen, M. et al. Multi-layer Born multiple-scattering model for 3D phase microscopy. *Optica* **7**, 394-403 (2020).
99. Sun, D. et al. Influence of sparse constraint functions on compressive holographic tomography. *Applied Optics* **60**, A111-A119 (2021).
100. Daloglu, M. U. et al. Low-cost and portable UV holographic microscope for high-contrast protein crystal imaging. *APL Photonics* **4**, 030804 (2019).
101. Valzania, L. et al. THz coherent lensless imaging. *Applied Optics* **58**, G256-G275 (2019).
102. Goodman, J. W. Introduction to Fourier Optics. 3rd edn. (Greenwoood Village: Roberts and Company Publishers, 2005).
103. Matakos, A., Ramani, S. & Fessler, J. A. Accelerated edge-preserving image restoration without boundary artifacts. *IEEE Transactions on Image Processing* **22**, 2019-2029 (2013).
104. Candès, E. J., Li, X. D. & Soltanolkotabi, M. Phase retrieval via Wirtinger flow: theory and algorithms. *IEEE Transactions on Information Theory* **61**, 1985-2007 (2015).
105. Sun, J., Qu, Q. & Wright, J. A geometric analysis of phase retrieval. *Foundations of Computational Mathematics* **18**, 1131-1198 (2018).
106. Isernia, T., Leone, G. & Pierri, R. Radiation pattern evaluation from near-field intensities on planes. *IEEE Transactions on Antennas and Propagation* **44**, 701 (1996).
107. Zhang, H. S. & Liang, Y. B. Reshaped Wirtinger flow for solving quadratic system of equations. Proceedings of the 30th International Conference on Neural Information Processing Systems. Barcelona, Spain: Curran Associates Inc., 2016, 2630–2638.
108. Wang, G., Giannakis, G. B. & Eldar, Y. C. Solving systems of random quadratic equations via truncated amplitude flow. *IEEE Transactions on Information Theory* **64**, 773-794 (2017).
109. Yang, C. et al. Iterative algorithms for ptychographic phase retrieval. (Berkeley, CA, USA: Lawrence Berkeley National Lab., 2011).
110. Fannjiang, A. & Strohmer, T. The numerics of phase retrieval. *Acta Numerica* **29**, 125-228 (2020).
111. Yeh, L. H. et al. Experimental robustness of Fourier ptychography phase retrieval algorithms. *Optics Express* **23**, 33214-33240 (2015).
112. Gao, Y. H. & Cao, L. C. Generalized optimization framework for pixel super-resolution imaging in digital holography. *Optics Express* **29**, 28805-28823 (2021).
113. Aghamiry, H. S., Gholami, A. & Operto, S. Complex-valued imaging with total variation regularization: an application to full-waveform inversion in visco-acoustic media. *SIAM Journal on Imaging Sciences* **14**, 58-91 (2021).
114. Zhang, Y. B. et al. Edge sparsity criterion for robust holographic autofocusing. *Optics Letters* **42**, 3824-3827 (2017).
115. Ren, Z. B., Xu, Z. M. & Lam, E. Y. Learning-based nonparametric autofocusing for digital holography. *Optica* **5**, 337-344 (2018).
116. Trusiak, M. et al. Darkfocus: numerical autofocusing in digital in-line holographic microscopy using variance of computational dark-field gradient. *Optics and Lasers in Engineering* **134**, 106195 (2020).
117. Zhang, J. L. et al. Resolution analysis in a lens-free on-chip digital holographic microscope. *IEEE Transactions on Computational Imaging* **6**, 697-710 (2020).
118. Huang, Z. Z. et al. Dual-plane coupled phase retrieval for non-prior holographic imaging. *PhotoniX* **3**, 3 (2022).
119. Tang, J. et al. Comparison of common-path off-axis digital holography and transport of intensity equation in quantitative phase measurement. *Optics and Lasers in Engineering* **157**, 107126 (2022).
120. Rivenson, Y. et al. PhaseStain: the digital staining of label-free quantitative phase microscopy images using deep learning. *Light: Science & Applications* **8**, 23 (2019).
121. Jiang, S. W. et al. Resolution-enhanced parallel coded ptychography for high-throughput optical imaging. *ACS Photonics* **8**, 3261-3271 (2021).
122. Onural, L. Sampling of the diffraction field. *Applied Optics* **39**, 5929-5935 (2000).
123. Kelly, D. P. et al. Resolution limits in practical digital holographic systems. *Optical Engineering* **48**, 095801 (2009).
124. Kelly, D. P. et al. Quantifying the 2.5D imaging performance of digital holographic systems. *Journal of the European Optical Society* **6**, 11034 (2011).
125. Luo, W. et al. Propagation phasor approach for holographic image reconstruction. *Scientific Reports* **6**, 22738 (2016).
126. Gao, Y. H., Ya ng, F. & Cao, L. C. Pixel super-resolution phase retrieval for lensless on-chip microscopy via accelerated Wirtinger flow. *Cells* **11**, 1999 (2022).
127. Kreutz-Delgado, K. The complex gradient operator and the CR-calculus. arXiv: 0906.4835 (2009).
128. Parikh, N. & Boyd, S. Proximal algorithms. *Foundations and Trends' in Optimization* **1**, 127-239 (2014).
129. Nesterov, Y. E. A method for solving the convex programming problem with convergence rate \$O(1/k.2)\$. *Soviet Mathematics Doklady* **269**, 543-547 (1983).
130. Beck, A. & Teboulle, M. A fast iterative shrinkage-thresholding algorithm for linear inverse problems. *SIAM Journal on Imaging Sciences* **2**, 183-202 (2009).
131. Beck, A. & Teboulle, M. Fast gradient-based algorithms for constrained total variation image denoising and deblurring problems. *IEEE Transactions on Image Processing* **18**, 2419-2434 (2009).
132. Xu, R. et al. Accelerated Wirtinger flow: a fast algorithm for ptychography. arXiv: 1806.05546 (2018).
133. Schmidt, M., Le Roux, N. & Bach, F. Convergence rates of inexact

proximal-gradient methods for convex optimization. Proceedings of the 24th International Conference on Neural Information Processing Systems. Granada, Spain: Curran Associates Inc., 2011, 1458–1466.

134. Gao, Y. H. & Cao, L. C. Compressive phase retrieval via constrained complex total variation regularization (CCTV). at https://github.com/THUHoloLab/CCTV-phase-retrieval.

135. Ye, Q. L., Wang, L. W. & Lun, D. P. K. Towards practical single-shot phase retrieval with physics-driven deep neural network. arXiv: 2208.08604 (2022).

136. Baek, Y. S. & Park, Y. K. Intensity-based holographic imaging via space-domain KramersBKronig relations. *Nature Photonics* **15**, 354-360 (2021).

137. Zhang, J. W. et al. A review of common-path off-axis digital holography: towards high stable optical instrument manufacturing. *Light: Advanced Manufacturing* **2**, 333-349 (2021).

138. Gao, Y. H. & Cao, L. C. A complex constrained total variation image denoising algorithm with application to phase retrieval. arXiv: 2109.05496 (2021).

Official journal of the JHL 2689-9620
www.light-am.com

Open Access

Shaping singular optical coherence with coherence holography

Wei Wang[1] and Mitsuo Takeda[2]

Abstract

As a new branch of modern statsitcal optics, singular coherence deals with a wide class of optical coherence effects associated with phase singularities as well as with the topological phenomena in optical coherence function of stochastic optical fields. With the aid of an emerging technology referred to as coherence holography with its unique capabilities for coherence shaping, stochastic optical fields with desired phase singularities in coherence function have been synethsized. The local structures and some propagation properties of coherence vortices associated with the topology of coherence function are also reviewed in this chapter.

Keywords: Coherence Holography, Statistical Optics, Phase Singularities, Coherence Vortex, Coherence Dynamics

Introduction

Optical fields are inherently of a statistical nature. It has long been recognized that the cross-correlation of the fluctuating fields, known as the mutual coherence function, has played a fundamental role in describing the stochastical properties of light[1-3]. The concept of optical coherence has laid a foundation where predictions of the outcomes of optical experiments can be treated in a unified way, and considerable progress has been made due to its theoretical importance and practical interest.

On the other hand, singular optics is a branch of modern physical optics to deal with a class of effects associated with phase singularities as well as the topology phenomena in light-related fields of at least two variables[4-6]. When light waves travelling in different directions intersect, destructive interference can lead to complete cancellation with zero intensity and ill-defined phase at points in two dimensions. To date, most work related to beams with phase singularities has been associated with monochromatic light that is both temporally and spatially coherent, however this needs not necessarily be the case. Phase singularities of the coherence function occur for the generalization of Young's double-slit experiment to partially coherent light[7,8] and a new term of *coherence vortex* has been introduced to indicate phase singularity of the complex coherence function[9]. Subsequently, phase singularities of the correlation functions of optical fields have been examined theoretically[10,11], and some early experiments have been conducted to verify the existence of phase singularities of the temporal and spatial coherence functions when a spiral phase plate is illuminated with partially coherent beams [12-14].

In this review, we will shed light on creation and detection of singular coherence effects with the aid of unconventional holography referred to as coherence holography[15-17]. The arrangement of this review will be as follows. We will first introduce some concepts associated with coherence vortex in spatial coherence function and review the principle of coherence holography for synthesizing statistical optical fields with desired spatial

[1]Institute of Photonics and Quantum Sciences, School of Engineering and Physical Sciences Heriot-Watt University, Edinburgh, EH14 4AS, UK
[2]Center for Optical Research and Education (CORE), Utsunomiya University, 7-1-2, Yoto, Utsunomiya, Tochigi, 321-8585, Japan

coherence function. Next, we will discuss in more detail the creation and detection of phase singularity in mutual intensity. Afterwards, we will discuss the local structure of coherence vortex and morphology of phase singularity points and lines. Finally we briefly review some topological reactions between coherence vortices during the propagation of mutual coherence.

Admittedly it is beyond the ability of the authors to present a comprehensive review that covers all the work relevant to the subject of singular coherence phenomena without bias. We therefore abandon such efforts from the outset and restrict ourselves to a review that focuses on some of our recent work to develop coherence holography for shaping singular coherence effects of statistical optical fields.

Foundations of singular coherence theory

Before presenting some experimental studies on coherence vortices, we first introduce some elementary concepts and theories related to the singular coherence phenomena in stochastic optical field. The optical coherence function $\Gamma(\mathbf{r}_1,t_1;\mathbf{r}_2,t_2)$ is a quantity of great importance, and plays a fundamental role in holography and interferometry. In analogy to the energy conservation law of scalar wave field[18], the *conservation law of mutual coherence* has been derived, which takes the form[17]

$$\frac{\partial W_\Gamma(\mathbf{r}_1,t_1;\mathbf{r}_2,t_2)}{\partial t_1}+\nabla_1\cdot\mathbf{T}_\Gamma(\mathbf{r}_1,t_1;\mathbf{r}_2,t_2)=0 \qquad (1)$$

where

$$W_\Gamma(\mathbf{r}_1,t_1;\mathbf{r}_2,t_2)=\alpha\left[\frac{1}{c^2}\frac{\partial\Gamma^*}{\partial t_1}\frac{\partial\Gamma}{\partial t_1}+\nabla_1\Gamma^*\cdot\nabla_1\Gamma\right]$$

$$\mathbf{T}_\Gamma(\mathbf{r}_1,t_1;\mathbf{r}_2,t_2)=-\alpha\left[\frac{\partial\Gamma^*}{\partial t_1}\nabla_1\Gamma+\frac{\partial\Gamma}{\partial t_1}\nabla_1\Gamma^*\right] \qquad (2)$$

c is the speed of light in vacuum, and α is a positive constant. Eq. 1 is a continuity equation in which the scalar quantity W_Γ represents an energy-density-like quantity of mutual coherence (which we term the *coherenergy density*), and the vector quantity term $\mathbf{T}_\Gamma$ represents a current-density-like quantity of mutual coherence (which we term the *coherence current density*). Note the fact that $\Gamma(\mathbf{r}_1,t_1;\mathbf{r}_2,t_2)$ is a function of two sets of spatio-temporal points. Without loss of generality, we will restrict our discussion to the case for the variation of point $\mathbf{r}_1$ and time t_1 by keeping $\mathbf{r}_2$ and t_2 fixed as our reference point. Another set of equation that involves the variation of r_2 and t_2 can also be derived in a similar way.

Borrowing from the general theory for vector analysis[19], it is possible to rewrite the vector of coherence current density as the sum of two parts based on the Helmholtz decomposition theorem, that is $\mathbf{T}_\Gamma=-\nabla_1\varphi_\Gamma+\nabla_1\times\mathbf{A}_\Gamma$, where φ_Γ is the *coherence scalar potential* introducing the irrotational part of the coherence current and $\mathbf{A}_\Gamma$ is the *coherence vector potential* providing the solenoidal vector component. The significance of this decomposition can be explained as follows. Just as the circulating energy flow of complex scalar wave field creates a conventional optical vortex[4-6,20-21], the rotational components of the coherence current density $\mathbf{T}_\Gamma$ derived from $\mathbf{A}_\Gamma$ are the sole source for the existence of phase singularity in mutual coherence function. Therefore, we are able to define the associated *angular coherence momentum density* as $\mathbf{M}_\Gamma=\mathbf{r}_1\times\mathbf{T}_\Gamma/c$. Another central role will be played by the *vorticity of coherence current density* associated with $\mathbf{T}_\Gamma$, namely $\mathbf{\Omega}_\Gamma=(\nabla_1\times\mathbf{T}_\Gamma)/2$. The vector $\mathbf{\Omega}_\Gamma$ is important because its norm is proportional to the rate of the rotation for coherence current and it points along the coherence vortex line, around which the phase of mutual coherence function increases by 2π in a positive sense with respect to $\mathbf{\Omega}_\Gamma$.

In practice, one is often interested in interference effects under quasi-monochromatic conditions, where the coherence function can be rewritten as $\Gamma(\mathbf{r}_1,t_1;\mathbf{r}_2,t_2)\approx J(\mathbf{r}_1,\mathbf{r}_2)\exp(-j2\pi\bar{\nu}\tau)$ with $J(\mathbf{r}_1,\mathbf{r}_2)$ being the mutual intensity, $\bar{\nu}$ being the mean frequency of the narrowband light and $\tau=t_2-t_1$ being the time difference. The coherence current density $\mathbf{T}_\Gamma$ in Eq. 2 becomes $\mathbf{T}_J(\mathbf{r}_1,\mathbf{r}_2)=2\alpha\bar{k}c|J|^2\nabla_1\chi$ with $\bar{k}=2\pi\bar{\nu}/c$. Here, we have replaced $\mathbf{T}_\Gamma(\mathbf{r}_1,t_1;\mathbf{r}_2,t_2)$ by $\mathbf{T}_J(\mathbf{r}_1,\mathbf{r}_2)$ to stress that the quantity of $\mathbf{T}_J$ is independent of time. The differential form of the coherence conservation law in Eq. 1 now becomes $\nabla_1\cdot\mathbf{T}_J=0$, and the vector potential of mutual intensity exists with $\mathbf{T}_J=\nabla_1\times\mathbf{A}_J$, since $\mathbf{T}_J$ is solenoidal. Meanwhile, the angular momentum density for the mutual intensity may be found through $\mathbf{M}_J=\mathbf{r}_1\times\mathbf{T}_J/c$. Similarly, the vorticity associated with $\mathbf{T}_J$ can be defined as $\mathbf{\Omega}_J=(\nabla_1\times\mathbf{T}_J)/2$, which is also a solenoidal field due to the fact $\nabla\cdot\mathbf{\Omega}_J=0$. Another quantity of most interest is the line integration of $\mathbf{T}_J$ around a closed curve ℓ, termed as *coherence circulation*, giving the flux of vorticity for the mutual intensity current embraced by a surface S, i.e. $\Phi_J=\oint_\ell\mathbf{T}_J\cdot d\ell=2\iint_S\mathbf{\Omega}_J\cdot\mathbf{n}dS$. As we will see later, the coherence circulation plays an important role in describing the interaction between coherence vortices.

On the other hand, phase singularities are phenomena in a physical field of at least two variables[20], where the physical quantity represented by the field can naturally be presented in a plane, such as the plane of complex number. In the coherence theory, the mutual coherence function $\Gamma(\mathbf{r}_1,t_1;\mathbf{r}_2,t_2)$ of stochastic optical field can be considered naturally as the complex field with modulus and phase information. For this complex field of great physical interest, we can rewrite

$$\begin{aligned}\Gamma(\mathbf{r}_1,t_1;\mathbf{r}_2,t_2) &= \mathrm{Re}\{\Gamma(\mathbf{r}_1,t_1;\mathbf{r}_2,t_2)\} + j\mathrm{Im}\{\Gamma(\mathbf{r}_1,t_1;\mathbf{r}_2,t_2)\} \\ &= |\Gamma(\mathbf{r}_1,t_1;\mathbf{r}_2,t_2)|\exp[j\chi(\mathbf{r}_1,t_1;\mathbf{r}_2,t_2)]\end{aligned} \quad (3)$$

where $\mathrm{Re}\{\cdots\}$ and $\mathrm{Im}\{\cdots\}$ are the real and imaginary parts and χ is the phase. Just as the angle of polar coordinates is not defined at the origin, the phase of optical coherence function is not defined when $\Gamma = 0$, and near such a point, the whole 2π range of phases occurs. This means that there is a net charge of phase in mutual coherence function in a circuit C enclosing the zero point, quantized in units of 2π:

$$q = \frac{1}{2\pi}\oint_C \nabla_1\chi(\mathbf{r}_1,t_1;\mathbf{r}_2,t_2)\cdot dr_1 \quad (4)$$

where the integer q is called the topological charge of the phase singularity in mutual coherence function and the sign of q is called the sign of the phase singularity, positive if the phase of mutual coherence function increases in a right-handed sense. From Eq. 3, the phase singularity in mutual coherence function can also be understood as the point of intersection of zero crossing lines for the real and imaginary parts of Γ (i.e., $\mathrm{Re}\{\Gamma\} = 0$ and $\mathrm{Im}\{\Gamma\} = 0$). In three dimensions, the complex zeros are curved lines occuring along the intersection of the real and imaginary zero surfaces.

Principle of coherence holography

The principle of coherence holography was inspired by the formal analogy between the coherence function $\Gamma(\mathbf{r}_1,t_1;\mathbf{r}_2,t_2)$ and the optical field $u(r,t)$ both of which obey the same wave equations[1-3]:

$$\begin{pmatrix}\text{Coherence}\\ \text{Function}\end{pmatrix} \quad \nabla_1^2\Gamma(\mathbf{r}_1,t_1;\mathbf{r}_2,t_2) = \frac{1}{c^2}\frac{\partial^2\Gamma(r_1,t_1;r_2,t_2)}{\partial t_1^2}$$

$$\begin{pmatrix}\text{Optical}\\ \text{Field}\end{pmatrix} \quad \nabla^2 u(r,t) = \frac{1}{c^2}\frac{\partial^2 u(r,t)}{\partial t^2}. \quad (5)$$

As a consequence of this analogy, a formal similarity can also be found between the diffraction integral and the formula of van Cittert-Zernike theorem. As an illustrative example, let us consider the characteristic of the field created by Fraunhofer diffraction from a circular aperture. If the aperture is illuminated by a spatially coherent plane wave, the familiar Airy pattern of the form $u(r) \propto 2J_1(ar)/\alpha r$ will be observed in the far field as the Fourier transform of the circular aperture, with α and J_1 being a scale factor and the first-order Bessel function of the first kind, respectively. Alternatively, if the aperture is illuminated by spatially incoherent light, the Airy pattern will be replaced by a nearly uniform intensity distribution without spatial structure. Does this mean that the information about the circular aperture structure has been lost? Of course, it is not. As we know, the structural information about the circular aperture is encoded into the coherence function of the diffraction field. By using an appropriate interferometer for field correlation, we can detect the coherence function that has exactly the same form $\Gamma(\Delta r) \propto 2J_1(a\Delta r)/(a\Delta r)$, with Δr being the separation between two points on the observation plane. Historically this has served as the principle for determining the angular size of a star with a Michelson stellar interferometer. Let us now change the circular aperture with a hologram created with coherent light and appropriate geometry. If we illuminate the hologram with a spatially coherent plane wave, we will observe twin (original and conjugate) images reconstructed from the hologram, having a bright spot of the DC component in the center. This conventional holographic reconstruction is familiar to all. What happens if we illuminate the hologram with spatially incoherent quasi-monochromatic light? In similarity to the case of the circular aperture, we cannot see any reconstructed image in the far field. However, with the aid of an appropriate interferometer, we can reconstruct the recorded image as the mutual coherence function of the optical field in the observation plane. This is the basic principle of coherence holography. Just as a conventional computer-generated hologram (CGH) can create an arbitrary 3-D optical field, a computer-generated coherence hologram (CGCH) can synthesize a stochastic optical field with a desired 3-D coherence function.

A careful reader may wonder how the spatially extended source of the incoherently illuminated hologram can create an optical field that has high spatial coherence between desired points. The principle of coherence holography for a general case can be explained on the basis of reciprocity in spatial coherence and holographic recording [15].

Referring to Fig. 1a, let us first consider recording of a conventional hologram with coherent illumination. We consider a point at O on a 3-D object recorded with a reference beam from a point source at R. Suppose a point at A on the hologram is located on one of the bright fringes resulting from constructive interference between the object and reference waves from the points O and R. Then the optical path difference between the two waves becomes integer multiples of the wavelength λ, such that $|\mathbf{r}_{OA} - \mathbf{r}_{RA}| = n\lambda$. Similarly, for point B on another bright fringe, we also have $|\mathbf{r}_{OB} - \mathbf{r}_{RB}| = m\lambda$ with m being an integer. Reciprocally, we illuminate the hologram from behind with spatially incoherent quasi-monochromatic light as shown in Fig. 1b. This generates an extended incoherent source with its irradiance distribution

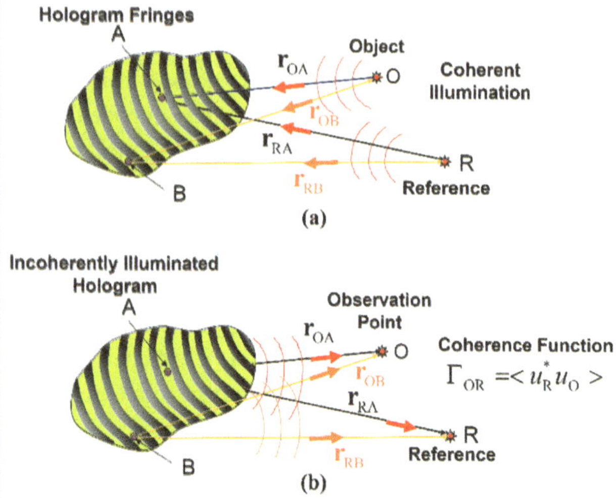

Fig. 1 **a** Recording of a coherence hologram with coherent illumination. **b** Reconstruction of the coherence hologram with incoherent illumination. Object is reconstructed as a coherence function. Reproduced with permission from SPIE. © SPIE.

proportional to the intensity transmittance of the hologram, which is assumed to be proportional to the recorded fringe pattern. Let us now consider a reciprocal process in which light emitted from point A on the source reaches points O and R to create fields u_O and u_R, respectively. From reciprocity of wave propagation and because of $|\mathbf{r}_{OA} - \mathbf{r}_{RA}| = n\lambda$, u_O and u_R are always in phase with the phase difference $2n\pi$, irrespective of the initial phase and the position of the point source A, as far as A is located on one of the bright fringes; this condition is automatically satisfied for all points on the hologram, such as point B, because no lights are emitted from the location of the dark fringes which would make u_O and u_R out of phase and thereby reduce the coherence of the fields between points O and R. Thus, the incoherently illuminated hologram creates an optical field that is in phase and has high coherence between this particular pair of points at O and R. If we fix one point as a reference at R and move the other point O over the 3-D space, we can reconstruct the object as a 3-D distribution of the coherence function $\Gamma_{OR} = < u_R^* u_O >$, which can be detected using a suitable interferometer. It should be noted that, if the distances $|\mathbf{r}_{OA}|$ and $|\mathbf{r}_{RA}|$ are much larger than the distance between the object and the reference point, the field can be regarded as statistically stationary in wide sense so that the coherence function depends only on the difference of the coordinates between points O and R.

To have a specific idea about how to synthesize stochastic optical field with a desired coherence distribution, let us consider a simple example of coherence holography that can generate a high coherence peak at a desired axial position. First, we record a Gabor-type (on-axis) hologram with mutually coherent diverging spherical waves from an object and a reference source at points O and R on axis, so that we have a Fresnel zone-plate hologram as shown in Fig. 2a. If we illuminate the Fresnel zone plate coherently from behind with a spherical wave converging into R, we can reconstruct a spherical wave that focuses into the object point O, where it creates a high amplitude optical field $u(\mathrm{O})$. Now we illuminate the Fresnel zone plate hologram with spatially incoherent quasi-monochromatic light. Even though the focus spot disappears and the field becomes a nearly uniform random field, we can observe (with an appropriate interferometer) the coherence function $\Gamma(\mathrm{O}, \mathrm{R})$ that takes a high peak value at position O with reference to the field $u(R)$ at point R. This is because the light arriving at the points O and R from any one of the point sources on the bright-fringe rings are in phase with phase difference $2n\pi$ despite the random initial phase on the incoherent ring source. Remember that a Fresnel zone plate serves as a lens for a coherent read-out wave and the position of the focus spot O can be moved along the optical axis by changing the power of the Fresnel zone-plate lens. Similarly, the Fresnel zone plate illuminated by spatially incoherent light serves as a lens for the coherence wave, and the position of the focus spot O with a high coherence peak can be moved along the optical axis by changing the power of the Fresnel zone-plate lens; this enables spatial coherence tomography and profilometry with a quasi-monochromatic light.

As discussed above, the optical field with a desired spatial coherence function can be generated with a

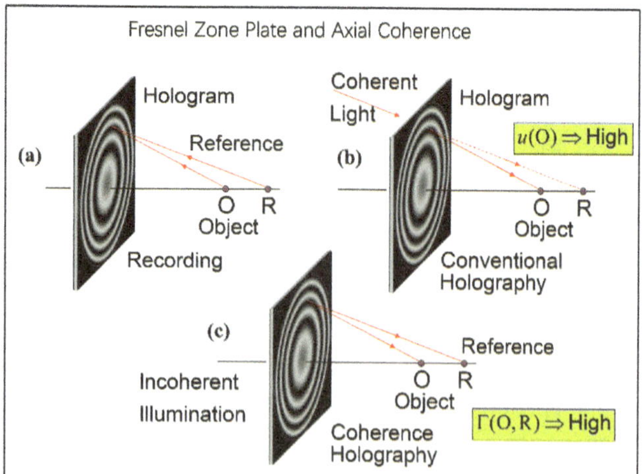

Fig. 2 **a** Recording of a zone plate hologram with coherent diverging spherical waves from an object and a reference source on axis. **b** Reconstruction of the hologram with a coherent spherical wave converging to the reference source point. **c** Illumination of the hologram with spatially incoherent light from behind. Reproduced with permission from JSAP. ©JSAP

computer-generated coherence hologram (CGCH). Therefore, we will apply the principle of coherence holography to explore the singular coherence function for stochastic field.

Generation and detection of coherence vortex

Generating an optical vortex with phase singularity by use of a computer-generated hologram (CGH) is a well-known technique[24–25]. The formal analogy between the optical field and the coherence function (or coherence wave) seen in Eq. 5 permits one to use CGCH for the generation of a coherence vortex with singularity in the phase of the coherence function[16].

Here we introduce an example of coherence vortex generation using a Gabor-type coherence hologram[16]. Instead of the Fresnel zone plate hologram for a spherical wave front (shown in Fig. 2), we generate a spiral Fresnel zone plate hologram by adding a spiral phase to the spherical wave front, as shown in Fig. 3a. The CGCH is illuminated from behind with spatially incoherent light generated by passing laser light through a rotating ground glass. The light from the CGCH is introduced into the interferometer through lens L1 in such a manner that the light from each point source S on the CGCH becomes a weakly divergent spherical wave with its radius of curvature controllable with the relative shift Δf of the hologram from the front focal plane of the lens L1. The axial separation Δz between the two mirrors M1 and M2 introduces a longitudinal optical path difference corresponding to the axial distance $OR = 2\Delta z$ in Fig. 2c, which can be used to detect the axial coherence. The axial separation Δz introduces not only an axial shear $\delta z = 2\Delta z$ but also a radial shear $\delta r \propto \Delta z r$ caused by the beam divergence. Seen from the side of L2, the radial expansion of the divergent beam reflected from M1 at z is smaller than that of the beam reflected from M2 at $z + \Delta z$, because the former beam travels shorter distance than the latter when they are virtually superposed on M1 for interference and imaged onto CCD by L2. It is this radial shear that permits the visualization of the lateral coherence distribution, in addition to the axial coherence distribution obtained from the axial shear.

Shown in Fig. 4a-g are synthesized coherence functions visualized as fringe visibility at different axial locations; a small mirror tilt is introduced to help visual interpretation and for the quantitative Fourier fringe analysis[26]. Referring to Fig. 2c with $OR/2 = 1.0$ mm, we can interpret the fringes in Fig. 4 as follows. When the observation plane is off from both the object point O and the reference point R in Fig. 2c, we observe the low-contrast fringe patterns as shown in Fig. 4a, c, e, g. When the axial position of the observation plane coincides with the object point O (at $\Delta z = -1.0$ mm) or its conjugate image point O' (at $\Delta z = 1.0$ mm, not shown in Fig. 2c) located in symmetry to the reference point R, we observe a high contrast fringe pattern with a fringe dislocation representing an optical vortex as shown in Fig. 4b, f with the opposite sign. When the observation plane coincides with the reference point R (at $\Delta z = 0.0$ mm), a high contrast fringe pattern also appears but we cannot observe the coherence vortex for the reason that we have no sensitivity for the lateral phase distribution because the redial shear $\delta r \propto \Delta z r$ reduces to zero at $\Delta z = 0$. The coherence vortices have the degree of coherence equal to zero without fringe contrast, while the intensities of the field do not vanish; this is quite different from the traditional optical vortices in the coherence field. The helical phase distribution of the coherence vortex in Fig. 4h was obtained by applying Fourier fringe analysis[26] to the high contrast fringe pattern with a fringe dislocation. Generation and detection of a coherence vortex by

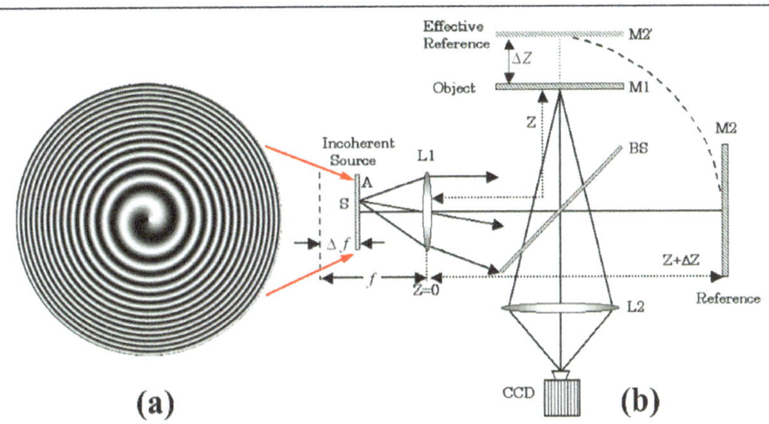

Fig. 3 **a** Computer-generated coherence hologram (CGCH) for coherence vortex generation. **b** Interferometer for visualization of a coherence vortex. Reproduced with permission from APS. ©APS

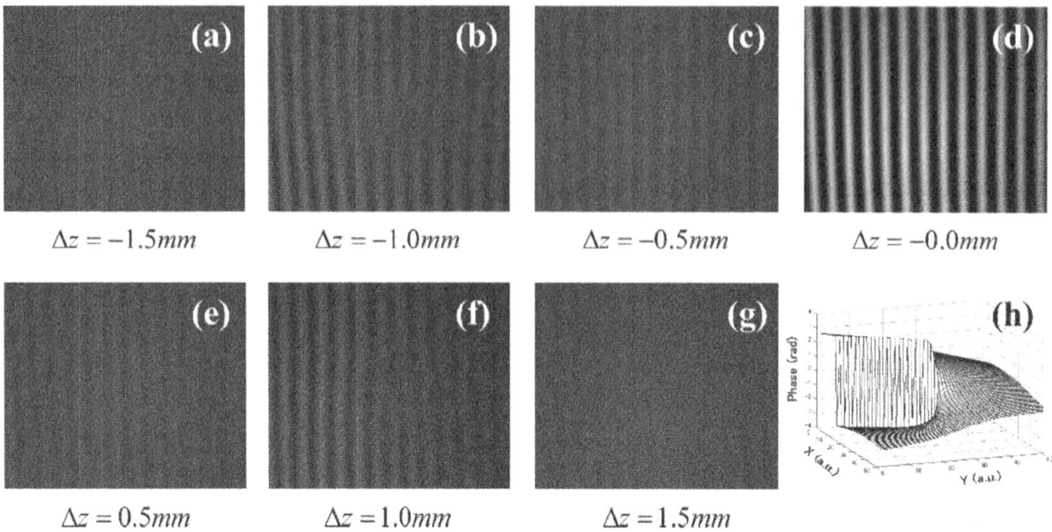

Fig. 4 a-g Synthesized coherence function visualized as fringe visibility at different axial location. **h** Spiral phase distribution obtained from the fringe pattern with a dislocation. Reproduced with permission from APS. ©APS

coherence holography has led us to the new realm of fluid dynamics for coherence current, and the discovery of the conservation law of coherence linear momentum and coherence orbital angular momentum as shown in the next section.

Core structure of coherence vortex

Because the mathematical behaviour of the coherence vortices is essentially identical to the behaviour of traditional optical vortices, we can experimentally investigate the local properties of the coherence vortex by directly referring to the results already derived for an optical vortex. Note that, in contrast to the discontinuous and complicated phase structure around a phase singularity, the real and imaginary parts have an extremely simple structure consisting of a smooth monotonic surface [27]. Thus, we can obtain the finer details of the phase profile and explore the core structure around a coherence vortex from the interpolated real and imaginary parts of the mutual coherence function as shown in Fig. 5.

With the aid of a two-dimensional interpolation of the real and imaginary parts of a coherence function, we investigate the local anisotropy of coherence vortex points. Fig. 6a shows an example of the distribution of magnitude of coherence function $|\Gamma|$ in the neighbourhood of a coherence vortex, and Fig. 6b depicts the corresponding local contours of the coherence current T_Γ with the zero-crossing lines $\mathrm{Re}\{\Gamma\} = 0$ and $\mathrm{Im}\{\Gamma\} = 0$ inserted. As expected, the coherence vortex, occurring at the crossing of the zero-contour lines, is a point in the plane of the coherence function. Similar to the case of an optical vortex[28-30], elliptical local contours for the modulus of coherence function and circular flow lines of the coherence current are observed around the coherence vortex, which is located at the center of the contour ellipses and contour circles, respectively. In analogy to the conventional current for the optical field that represents the average energy flow (Poynting vector), we may also characterize the flow of contrast through the coherence current (fringe visibility flow) defined by Eq. 2. Fig. 6b provides a direct experimental verification of the contrast flow enclosing the phase singularity of the complex coherence function.

Around the singular point, the phase of coherence function doesn't change uniformly. The local anisotropy of a coherence vortex can be also represented in a Keplerian sense[30]. Taking the coherence vortex as the origin, and introducing a cylindrical coordinate system $(\hat{r}, \hat{\varphi}, \hat{z})$, the new version of Keppler's law takes the form $r^2 \partial\varphi / \partial\chi = |\Gamma|^2 / |\boldsymbol{\Omega}_\Gamma(0)|$ with $\Omega_\Gamma(0)$ being the vorticity on the vortex. This relation can be interpreted as follows. Along the contour line of $|\Gamma|$, equal area sectors of the ellipse are swept in equal intervals of phase χ of coherence function as shown in Fig. 6a. The interpolation method introduced above allow us to demonstrate the conservation of the angular momentum-like quantity in the coherence domain defined as $r^2 \partial\varphi / \partial\chi$.

Coherence current, angular coherence momentum and coherence circulation

Vortices are inherent to many dynamic phenomena as a direct consequence of the spiral motion of a physical quantity[4-6]. In fluid dynamics, they are fluid vortices originating from a rotary flow velocity field [31]. In optics, they manifest themselves as optical vortices with a

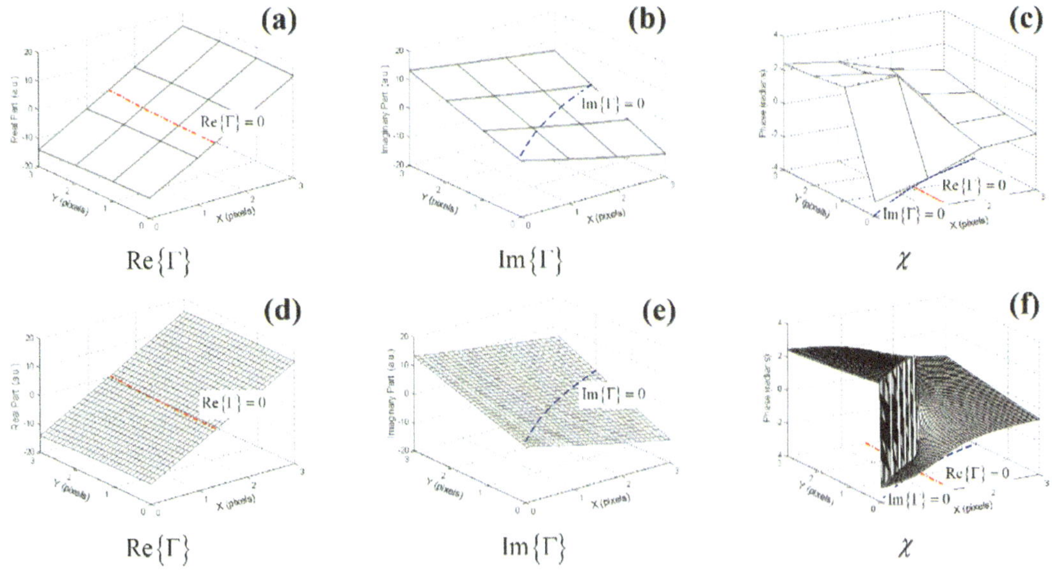

Fig. 5 a-f Real and imaginary parts of coherence function, and the corresponding phase structure around a phase singularity. Upper row: before interpolation; Lower row: after interpolation.

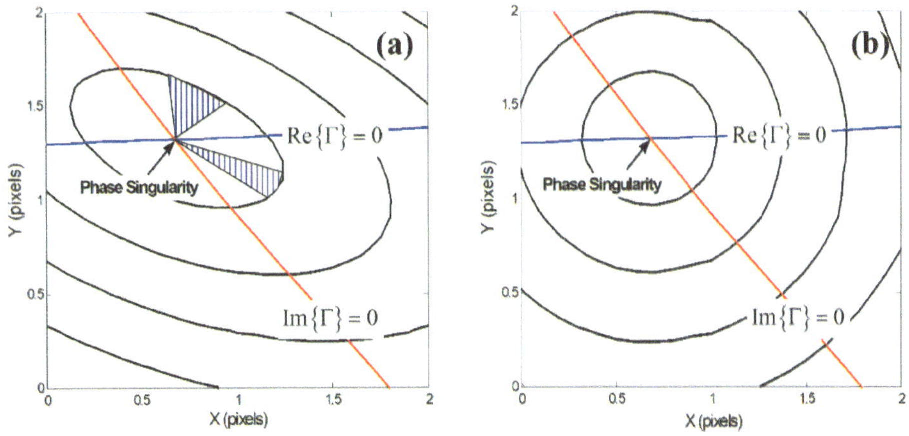

Fig. 6 Core structure around a coherence vortex with zero crossings of real and imaginary parts of coherence function. **a** Contours of modulus of coherence function; **b** Coherence current contours. Reproduced with permission from APS. ©APS

spiralling Poyning vector (energy flow) [32,21]. In the previous section, we have provided a direct experimental verification of the coherence current (fringe visibility flow) enclosing a phase singularity in coherence function. Similar to optical vortex and angular momentum of light, the coherent current has an azimuthal component in the immediate vicinity of a vortex and hence there is a net flow of both coherenergy and coherence current around the vortex line, which in turn gives an angular coherence momentum directed long the vortex line. To investigate the evolution of angular coherence momentum associated with a coherence vortex, experiments have been conducted based on the same optical geometry as shown in Fig. 3b. By moving the position of reference mirror, we change the optical path difference in 0.05-mm step between the two arms of the Michelson interferometer and measure the visibility of the fringes by using the Fourier transform method[26] to investigate the spatial evolution of the coherence current enclosing the phase singularity of coherence function.

Fig. 7a shows an example of the recorded fringes at the position $\Delta z = 0.90$ mm, where the complex coherence function is represented by the visibility and the shift of the fringes. Fig. 1c shows the streamlines of coherence current with its labels identifying flow strength, which was calculated from the definition of $\mathbf{T}_J$ using the values of the complex degree of coherence reconstructed from the interferogram within the area of the dashed line frame of

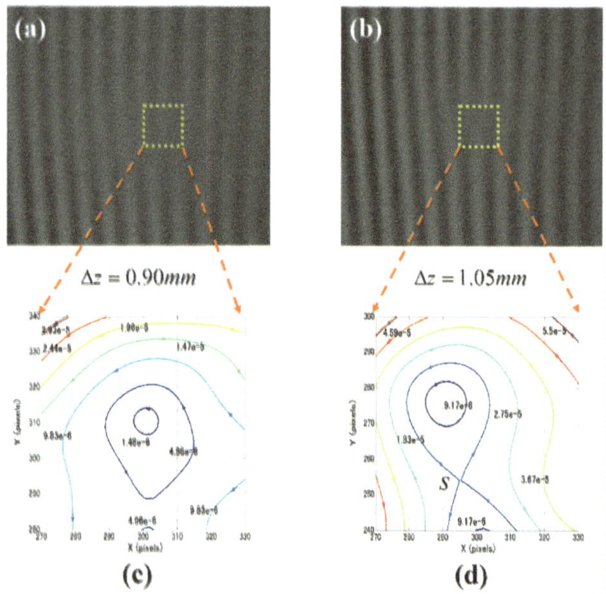

Fig. 7 The recorded interferograms and the corresponding coherence currents within the area of the dashed line frames for different optical path difference. **a**, **c**: $\Delta z = 0.90$ mm; **b**, **d** $\Delta z = 1.05$ mm. Reproduced with permission from APS. ©APS

Fig. 1a. The fringes recorded for $\Delta z = 1.05$ mm is shown in Fig. 7b, and the Fig. 7d is the corresponding coherence current distribution. In all of these figures, we can find that the regions of interferograms with high fringe contrast are always labelled with large value of coherence current, whose magnitude is a measure of the coherenergy. In analogy to the critical points with zeros of Poynting vector for the optical field[30], the phase of coherence function not only has a phase singularity, which is located at the center of the circular coherence current, but also has another critical point (saddle point) indicated by a letter s in Fig. 7d, where $\Delta\chi$ vanishes with zeros of coherence current.

Following the convention of fluid dynamics[31], we have introduced some geometrical concepts to characterize the spatial evolution of the coherence vortex[17]. If a closed curve ℓ was chosen as one contour for the modulus of the mutual intensity $|J|$, this contour line is tangential to the vorticity vector Ω_J serving as the boundary of a small planar area and a collection of these lines can form a coherence vortex tube. Making use of the definition for the topological charge of a phase singularity, we have the coherence circulation around a coherence vortex $\Phi_J \propto |J|^2$, which is analogous to the Helmholtz's first theorem in fluid dynamics[31]. This expression indicates that the coherence circulation Φ_J, or the flux of the vorticity, is constant at all cross sections of the coherence vortex tube if the closed curve ℓ is chosen as one contour of $|J|$.

Fig. 8a-f shows the recorded interferograms with synthesized coherence functions as fringe visibilities for different optical path difference. Fig. 8g gives a cross-section view of a coherence vortex tube embracing the coherence vortex line indicated in red crosses, which is determined by the intersection of the zero crossing of real and imaginary parts of coherence function. As expected, a tube-like structure is observed, whose wall is defined by a contour line of degree of coherence equal to 0.1, i.e., $|\mu| = 0.01$. By applying Fourier fringe analysis[26] to the

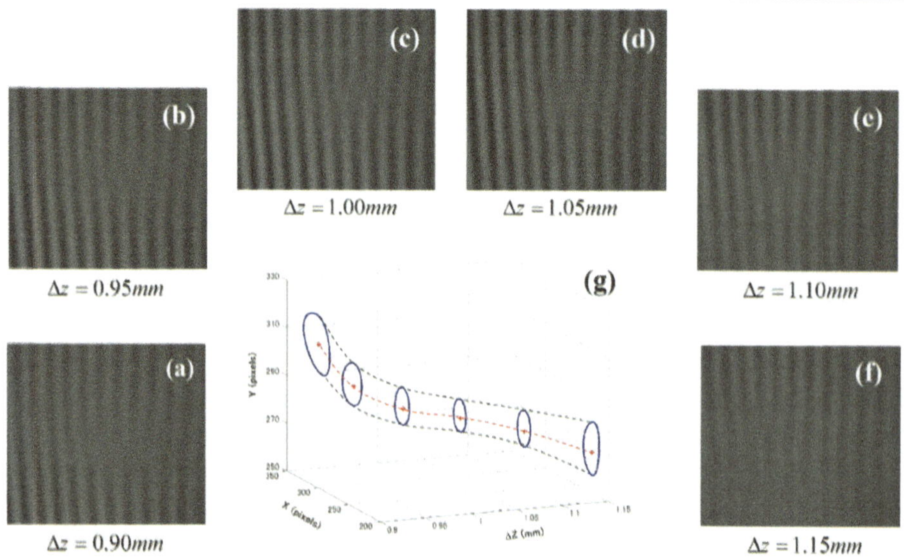

Fig. 8 **a-f** Synthesized coherence function visualized as fringe visibility for different optical path difference. **g** A cross-sectional view of a coherence vortex tube whose wall is defined by an equi-fringe visibility surface embracing the coherence vortex line. Reproduced with permission from APS. ©APS

recorded interferograms and the interpolation method for the reconstructed real and imaginary parts of the coherence function, we calculate the cross-section area S for coherence vortex tube and the corresponding vorticity $\mathbf{\Omega}_J$ based on its definition. With an increase in the value S^{-1}, the magnitude of the coherence vorticity increases linearly. This demonstrates that the coherence circulation around a coherence vortex tube is constant. When the coherence vortex tube comes to a waist, where the tube cross-sectional area is minimal, the average vorticity over that cross-section must be maximal, and conversely for the broadening of the tube. Meanwhile, a related observation is that coherence vortex tubes cannot terminate in coherence domain, because, otherwise, the constancy of coherence circulation would not be achieved. Therefore, coherence vortex tubes are constrained to form closed loops within the finite spatial volume of the coherence function, or to terminate at infinity.

Fujiwhara effect in optical coherence function

The Fujiwhara effect[33] is a meteorological phenomenon that occurs when two nearby cyclonic vortices circle each other around their common center. Binary interaction of smaller circulations can cause the development of a larger cyclone, or cause two cyclones to merge into one. In the previous section, we have studied the evolution of coherence current and the corresponding coherence circulation for a single coherence vortex. Here, we report our observation of the fluid-like rotation of a pair of coherence vortices with the same topological charge, which shares a formal analogy to the Fujiwhara effect of vortex filaments in a fluid.

Referring to Fig. 9a, let us first understand the principle of Fujiwhara effect in terms of coherence function. Let Φ^k_J be the coherence circulation near the coherence vortex k (for $k = A, B$). From Biot-Savart law, the induced coherence current at a point for a vortex filament with infinite length can be written as $\mathbf{T}^k_J = \Phi^k_J/(2\pi\rho_k)$ with ρ_k being the perpendicular distance between the point of interest and the vortex line. Following the theorem developed for fluid dynamics[34], we can show that the centers of two coherence vortex filaments remain at a constant distance $L = \rho_A + \rho_B$ and rotate with an angular velocity $\omega = (\Phi^A_J + \Phi^B_J)/(2\pi L^2)$ around the center of gravity for the two-body system marked as the red spot.

To study the interaction of the coherence currents associated with two coherence vortices of the same topological charges, an experiment has been conducted to synthesize stochastic field with a pair of phase singularities in optical coherence function. Fig. 9b shows the experimental geometry for coherence synthesis and visualization based on the principle of coherence holography[35]. A computer-generated coherence hologram encoded with vortex related information is projected onto a rotating ground glass to serve as a spatially incoherent light source. For coherence visualization, two telescope systems with different magnifications have been used for two arms of the Michelson interferometer to introduce the radial shearing[36]. By moving the position of the reference mirror, we change the optical path difference and investigate the spatial evolution and interaction of coherence currents enclosing two coherence vortices with experimental results shown in Fig. 10.

The upper row of Fig.10 shows the interferograms recorded at different optical path differences from$\Delta z = 0$ mm to $\Delta z = -4$ mm in 1-mm step, respectively. As expected, there are two coherence vortices located at the centers of fork-like fringe patterns with zero fringe contrast. The middle row of Fig.10 is the moduli for the reconstructed optical coherence functions, where two dark holes in each frame indicate the locations of coherence vortices. The contour lines for the corresponding coherence current magnitudes are shown in the lower row of Fig. 10, where two concentric cycles can be observed

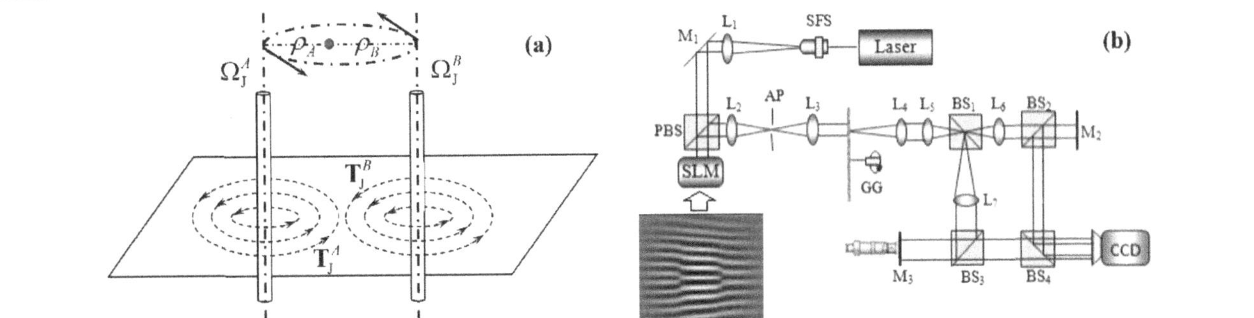

Fig. 9 **a** Schematic diagram for the Fujiwhara effect in optical coherence function. **b** Experimental setup for visualization of the Fujiwhara effect in coherence function. Reproduced with permission from Optica. ©Optica

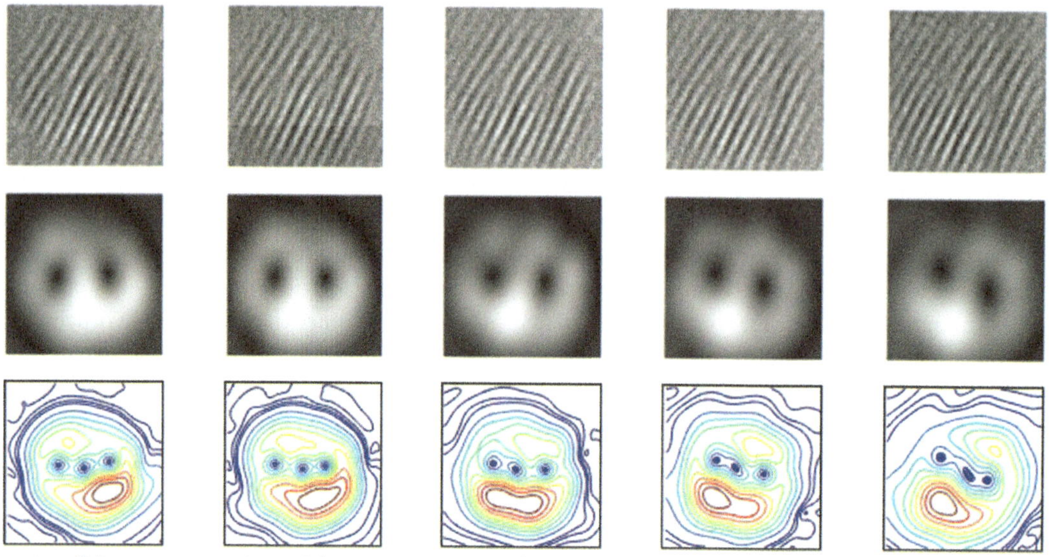

Fig. 10 Upper row: Recorded interferograms with synthesized coherence functions at different optical path differences; Middle row: Distributions of the moduli of the coherence functions |Γ| reconstructed from the corresponding interferograms for different optical path difference; Lower row: Contour lines of the coherence current magnitudes $|\mathbf{T}_J|$ for the synthesized coherence functions at different locations Reproduced with permission from Optica. ©Optica

around two coherence vortices as expected. It's interesting to note that a point with zero coherence current can also be observed in the middle area of these two coherence vortices due to the fact of cancellation of the coherence currents induced by two coherence vortex filaments. When these two coherence vortices are in proximity of one another, their centers begin orbiting cyclonically about a point between the two vortices due to the interaction of coherence currents induced by coherence circulations from two coherence vortices. Therefore, Fig. 10 provides a direct experimental evidence of the Fujiwhara effect in optical coherence function.

Conclusions

In summary, we have reviewed some theory developed to understand the coherence vortex and basic principle and some of the unique functionalities of coherence holography with examples of our own work related to the singular coherence phenomena. However, this review covers only part of the singular optical coherence and the various extensions and potentials of coherence holography, with many other relevant subjects being left over. The focus of the review has been placed on the application of the coherence holography for synthesizing stochastic fields with singular spatial coherence, which leads a new concept of coherence dynamics for statistical optics.

Conflict of interest

References

1. Mandel, L. & Wolf, E. Optical Coherence and Quantum Optics. (Cambridge: Cambridge Univ. Press, 1995).
2. Goodman, J. W. Statistical Optics. 2nd edn. (Hoboken: John Wiley & Sons, Inc., 2015).
3. Wolf, E. Introduction to the Theory of Coherence and Polarization of Light. (Cambridge: Cambridge Univ. Press, 2007).
4. Soskin, M. S. and Vasnetsov, M. V. Singular Optics. *Prog. Optics* **42**, 219-276 (2001).
5. Dennis, M. R., O' Holleran K. & Padgett M. J. Singular optics: optical vortices and polarization singularities. *Prog. Optics* **53**, 293-363 (2009).
6. Gbur, G. J. Singular Optics. (Boca Raton: CRC Press, 2016).
7. Schouten, H. F., Gbur, G. Visser, T. *D. & Wolf, E. Phase singularities of the coherence functions in Young' s interference pattern, Opt. Lett* **28**, 968-970 (2003).
8. Bogatyryova, G. V. et al. Partially coherent vortex beams with a separable phase. *Opt. Lett* **28**, 878-880 (2003).
9. Gbur, G. & Visser, T. D. Coherence vortices in partially coherent beams. *Opt. Commun* **222**, 117-125 (2003).
10. Maleev, I. D., Palacios, D. M., Marathay, A. S. & Swartzlander, G. A. Spatial correlation vortices in partially coherent light: theory. *J. Opt. Soc. Am. B* **21**, 1895-1900 (2004).
11. Fischer, D. G. & Visser, T. D. Spatial correlation properties of focused partially coherent light. *J. Opt. Soc. Am. A* **21**, 2097-2102 (2004).
12. Swartzlander, G. A. & Schmit, J. Temporal correlation vortices and topological dispersion. *Phys. Rev. Lett* **93**, 093901 (2004).
13. Palacios, D. M., Maleev, I. D., Marathay, A. S. & Swartzlander, G. A. Spatial correlation singularity of a vortex field. *Phys. Rev. Lett* **92**, 143905 (2004).
14. Motzek, K., Kivshar, Y. S., Shih, M. F. & Swartzlander, G. A. Spatial coherence singularities and incoherent vortex solitons. *J. Opt. Soc. Am. B* **22**, 1437-1442 (2005).
15. Takeda, M, Wang, W. , Duan, Z. H. & Miyamoto, Y. Coherence holography, **13**, 9629-9635 (2005).
16. Wang, W., Duan, Z. H., Hanson, S. G., Miyamoto, Y. & Takeda, M.

Experimental study of coherence vortices: local properties of phase singularities in a spatial coherence function. *Phys. Rev. Lett* **96**, 073902 (2006).
17. Wang, W. & Takeda, M. Coherence current, coherence vortex, and conservation law of coherence. *Phys. Rev. Lett* **96**, 223904 (2006).
18. Born, M. & Wolf, E. Principle of Optics, 7th edn. (Cambridge : Cambridge Univ. Press, 2005).
19. Arfken, G. B. & Weber, H. J. Mathematical methods for physicists, 4th edn. (Waltham: Academic Press, 1995).
20. Nye, J. F. & Berry, M. V. Dislocations in wave trains. *Proc. R. Soc. Lond. A* **336**, 165-190 (1974).
21. Allen, L. Barnett, S. M. & Padgett, M. J. Optical Angular Momentum (Bristol: Inst. of Phys. , 2004).
22. Takeda, M, Wang, W. & Naik, D. N. Coherence holography and photon-correlation holography: Marriage between holography and statistical optics, Proc SPIE 8122: 812207 (2011).
23. Takeda, M. Synthesis and control of three-dimensional spatial coherence fields. *Oyo Butsuri* (*Applied Physics in Japanese*) **82**, 46-50 (2013).
24. Dholakia, A. J., Allen, L. & Padgett, M. J. The production of multi-ringed Laguerre-Gaussian modes by computer-generated holograms. *J. Mod. Opt.* **45**, 1231-1237 (1998).
25. Li, S. & Wang, Z. Generation of optical vortex based on computer-generated holographic gratings by photolithography. *Appl. Phys. Lett* **103**, 141110 (2013).
26. Takeda, M., Ina. H. & Kobayashi, S. Fourier-transform method of fringe-pattern analysis for computer-based topography and interferometry. *J. Opt. Soc. Am* **72**, 156-160 (1982).
27. Wang, W., Hanson, S. G., Miyamoto, Y. & Takeda, M. Experimental invesigation of local properties and statistics of optical vortices in random wave fields. *Phys. Rev. Lett* **94**, 103902 (2005).
28. Schouten, H. F. & Shamir, J. Parameterization and orbital angular momentum of anisotropic dislocations. *J. Opt. Soc. Am. A* **13**, 967-973 (1996).
29. Berry, M. V. and Dennis, M. R. Phase singularities in isotropic random waves. *Proc. R. Soc. Lond. A* **456**, 2059-2079 (2000).
30. Dennis, M. R. Local properties and statistics of phase singularities in generic wavefields, Proc. of SPIE 4403 *Singular Optics: Fundamentals and Applications*, SPIE Proc. 13-23 (2001).
31. Batchelor, G. K. An Introduction to Fluid Dynamics (Cambridge: Cambridge Univ. Press, 1967).
32. Allen, L., Beijersbergen, M. W. Spreeuw, R. *J. C. & Woerdman, J. P. Orbital angular-momentum of light and the transformation of Laguerre-Gaussian laser modes, Phys. Rev. A* **45**, 8185-8189 (1992).
33. Fujiwhara, S. The natural tendency towards symmetry of motion and its application as a principle in meteorology. *Q. J. R. Met. Soc* **47**, 287-293 (1921).
34. Lamb, H. Hydrodynamics, 6th Edn. (Cambridge: Cambridge Univ. Press, 1945).
35. Zhao, J. Chen, B. C. , Takeda, M. & Wang, W. Observation of the Fujiwhara effect in optical coherence function by using coherence holography, Image and Applied Optics Congress 2019 (Munich, German). Paper No. ITh2B. 6. (2019).
36. Zhao, J. & Wang, W Experimental study of spatial coherence diffraction based on full-field coherence visualization, J. Opt. *Soc. Am. A* **31**, 2217-2222 (2014).

Part V

Perspectives of Holographic Technologies

Blanche. *Light: Advanced Manufacturing* (2021)2:28
https://doi.org/10.37188/lam.2021.028

Official journal of the JHL 2689-9620
www.light-am.com

Review | Open Access

Holography, and the future of 3D display

Pierre-Alexandre Blanche*

Abstract

The pioneers of holography, Gabor, Leith, Upatnieks, and Denisyuk, predicted very early that the ultimate 3D display will be based on this technique. This conviction was rooted on the fact that holography is the only approach that can render all optical cues interpreted by the human visual system. Holographic 3D displays have been a dream chased after for many years, facing challenges on all fronts: computation, transmission, and rendering. With numbers such as 6.6×10^{15} flops required for calculations, 3×10^{15} b/s data rates, and 1.6×10^{12} phase pixels, the task has been daunting. This article is reviewing the recent accomplishments made in the field of holographic 3D display. Specifically, the new developments in machine learning and neural network algorithms demonstrating that computer-generated holograms approach real-time processing. A section also discuss the problem of data transmission that can arguably be solved using clever compression algorithms and optical fiber transmission lines. Finally, we introduce the last obstacle to holographic 3D display, which is is the rendering hardware. However, there is no further mystery. With larger and faster spatial light modulators (SLMs), holographic projection systems are constantly improving. The pixel count on liquid crystal on silicon (LCoS) as well as microelectromechanical systems (MEMS) phase displays is increasing by the millions, and new photonic integrated circuit phased arrays are achieving real progress. It is only a matter of time for these systems to leave the laboratory and enter the consumer world. The future of 3D displays is holographic, and it is happening now.

Introduction

Not long after the first demonstration of holographic images by Leith and Upatnieks[1], as well as by Denisyuk[2], a similar technique was used by De Bitetto and others to display motion picture holograms[3,4]. As the name indicates, a motion picture hologram uses a rapid succession of static holographic images to reproduce the effect of movement in 3D. Because of these successes, it was hoped at the time that a holographic television would be developed soon after. Unfortunately, more than 50 years later, there is still no holographic television in sight.

Compared to motion picture, a holographic display system needs to capture, transmit, and render 3D information. In the case of an interactive display system, such as a computer screen, there is the additional constraint that real-time manipulation of 3D data is required. This makes a universal holographic 3D display much more challenging to develop than the simple successive projection of pre-recorded holograms.

To provide a sense of how difficult a holographic display is compared to other forms of telecommunication, it is useful to consider the commonly used data rate metric. Even though focusing on the data rate does not take into account some other technological aspects, such as rendering complexity, it does allow comparison over a large range of techniques. For a holographic display to have an acceptable field of view of perhaps ±45° (diffraction angle θ), the law of diffraction indicates that the diffractive element size d for a central wavelength of $\lambda = 500$ nm should be on the order of 350 nm.

Correspondence: Pierre-Alexandre Blanche (pablanche@optics.arizona.edu)
Wyant College of Optical Sciences, University of Arizona, Tucson, AZ, USA

$$d = \frac{\lambda}{2sin\theta} \quad (1)$$

It should be noted that elements should be approximately 10 times smaller than predicted by equation 1 to achieve high-efficiency blazed diffraction gratings.

Given a 70 cm diagonal-screen television set (i.e., 50 cm × 50 cm), the number of active elements must be on the order of 2×10^{12}. To accommodate human vision perception, the display requires a minimum refresh rate of 60 Hz (corresponding to the flicker vision threshold), and at least three colors to fill the eye chromaticity gamut. The gray-level resolution for a conventional display is generally at least 8 bits (256 levels). For the sake of comparison, we will use the same number of phase levels for a hologram, although phase levels are related to efficiency and not shades of gray in holograms. It is even possible to reconstruct a 3D image with a binary hologram by sacrificing some spatial bandwidth and efficiency[5]. The resulting data rate for such a display, excluding any type of compression algorithm, would be

$$\text{pxl nbr} \times \text{rep. rate} \times \text{res.} \times \text{colors} = 2 \times 10^{12} \times 60 \times 8 \times 3 \approx 3 \times 10^{15}\ \text{b/s}. \quad (2)$$

Fig. 1 plots the data rate in bits per second for different telecommunication systems according to the time of their introduction. Starting with the optical telegraph (or Chappe's semaphore) presented to Napoleon Bonaparte in 1798, the optical telegraph had a typical rate of transmission of approximately 2 to 3 symbols (196 different types) per minute, or 0.4 b/s. Consequently, the electrical telegraph, popularized in the early 1840s using Samuel Morse's code, achieved a rate of approximately 100 b/s. Graham Bell's telephone was introduced in 1876 and supported voice frequency transmission up to 64 kb/s[6]. The early NTSC black and white electronic television, available in the 1940s, had 525 interlaced lines and displayed images at a rate of 29.97 frames per second at a bit rate of 26 Mb/s[7]. The color NTSC format was introduced 10 years later and tripled the black and white bandwidth to accommodate red, green, and blue channels[8]. More recently, the digital video format makes it easier to establish the bit rate based on pixel count (excluding compression) with HDTV 720p@1.33 Gb/s in 1990, ultra-HDTV 2160p(4K)@12.7 Gb/s in 2010, and currently 4320p(8K)@47.8 Gb/s. Note that these values are for uncompressed data feeds, and for the sake of comparison, do not include any type of compression algorithm.

The evolution of the bit rate for telecommunication devices plotted in Fig. 1 shows a trend that can be extrapolated to predict the emergence of holographic displays with a data rate of 3×10^{15} b/s. This extrapolation estimates the emergence of a commercial holographic display by 2100. Although this extrapolation is indicative of the difficulties ahead, it is also very encouraging. The date of 2100 is by no means an inescapable natural law. Similar to the doubling of the transistor count on chips every year, this prediction can be affected in one way or another by the amount of effort invested in the research and development of holographic display technologies.

In this manuscript, we investigate the reasons why holography is still perceived to be the ultimate technique to develop a commercial 3D display, review the progress that has been accomplished toward this goal, and discuss the missing technologies that are still needed to promote the emergence of such a 3D display.

3D Human Vision and 3D displays

Understanding the human visual system and how it

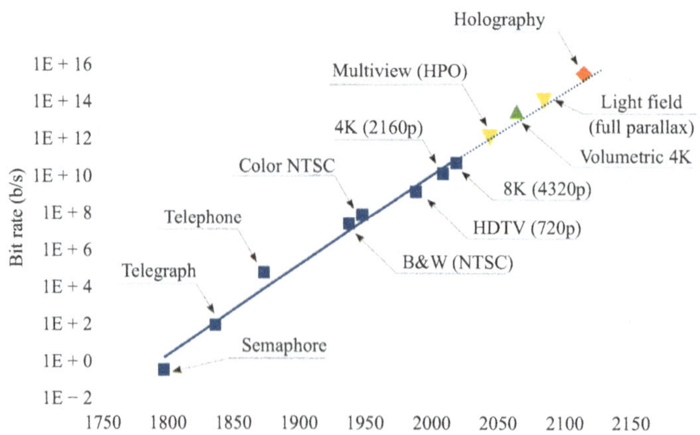

Fig. 1 Stairway to holography: approximate bit rate magnitude of various telecommunication devices according to their year of introduction.

perceives the third dimension is key to developing a 3D display[9–11]. The human visual system takes input from many different cues to determine depth perception. It should be noted that most of these cues originate from 2D phenomena. Among these are shading, shadowing, perspective, relative size, occlusion, blurriness, and haze. The example presented in Fig. 2 shows how three simple discs, presented on whatever 2D display you are reading this article on, are interpreted as 3D balls owing to these cues.

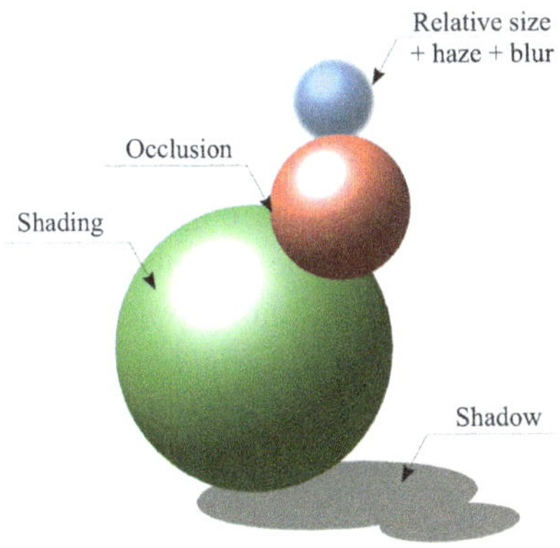

Fig. 2 Examples of some of the 2D visual cues that affect depth perception.

Because these 2D cues are processed by the human visual system to determine the depth of a scene, then a painting, a photograph, or a movie is intelligible as long as these cues are correctly reproduced. When they are not, this leads to optical illusions such as infinite staircases and other impossible shapes.

The same applies to any 3D display system, which must, first and foremost, represent these 2D cues before introducing any additional cues. Additional 3D cues are stereo disparity, motion parallax, and accommodation. We briefly review these cues in the following sections.

Stereo Disparity

Stereo disparity is the change in parallax of the scene observed between the left and right eyes. It only requires that two images be reproduced, and as such, is the most technologically manageable 3D cue. It is so manageable in fact that the introduction of stereoscopic displays pre-dates the invention of photography. The first system was invented by Sir Charles Wheatstone in the early 1830s using drawn images[12]. This was then followed by taking pictures from two positions, or with a camera having two objectives.

When a stereo projection is meant for a single individual, such as a head-worn display, it is relatively easy to keep the left and right views separated. Images are separated by simply introducing a physical partition between both eyes[13]. For a larger audience, the separation between left and right views is often achieved by having the viewers use eyewear with different left and right lenses. The left and right image coding can be achieved using color (anaglyphs), orthogonal polarization, or alternating shutters[14,15].

From a user perspective, the eyewear requirement for stereo display has been accepted in special venues such as theaters, where large productions continue to be released in stereoscopic 3D. However, the commercial failure of stereoscopic 3D television seems to indicate that for everyday experience, the public is not enthusiastic about wearing special glasses in their own living rooms[16].

Autostereoscopy

Autostereoscopic displays achieve stereoscopy without the need for special glasses. The left and right views are directly projected toward the viewer's intended eyes using parallax barriers or a microlens array[17–19]. To ensure that the correct eye intersects the correct projection, autostereoscopic systems require that the viewer be located at a particular position. This inconvenience has proven sufficient to limit the adoption of autostereoscopic 3D television by the consumer market[20]. It should also be noted that autostereoscopic systems with an eye tracking mechanism that mitigates the fixed viewer zones have been developed, but have not achieved wide popularity[21–23].

Motion Parallax

Motion parallax requires many views to be projected, allowing the viewer to see the correct parallax even when moving in front of the display. The density of the different views that are projected needs to be such that the autostereoscopic information is correctly reproduced. Therefore, at least two views per inter-pupillary distance are required. However, to achieve a smooth transition from one perspective to the next, a much larger density of views is required[24]. The optimum view density depends on the exact configuration of the display and the expected viewer distance, but numbers are on the order of one view per degree[25–27].

In most of the literature, a display that reproduces motion parallax is called a "multiview" or "multi-view" display while a "light-field" display reconstructs 3D images based on the concept of ray-optics and integral imaging[28–32]. In a multiview display, the display is designed such that the motion parallax can be reproduced smoothly when a viewer's position changes. This is considered a multiview-type autostereoscopic display. However, when the display is also capable of reconstructing virtual or real images, it is usually called a light-field display.

We can apply the same data rate computation introduced earlier for the different types of telecommunication devices (see Fig. 1) to a multiview (or light-field) display that reproduces motion parallax. In this case, we find that for a display with 2160p(4K) lateral resolution to reproduce motion parallax with a ±45° field of view, the bit rate is on the order of $12.7 \times 90^2 = 10^5$ Gb/s. The square factor arises from the fact that both horizontal and vertical parallaxes are considered in this case.

Because the human visual system involves a mostly horizontal inter-pupillary distance, and lateral movement is favored over vertical movement, horizontal parallax is more important than vertical parallax. The latter is often discarded in multiview displays to allow a lower data rate of $12.7 \times 90 = 10^3$ Gb/s.

When the viewer remains motionless in front of a multiview display, the observed parallax provides an experience similar to that of autostereoscopic displays[33]. However, because of the much larger number of views, a light-field display is not subject to the same limited number of view zones as autostereoscopic systems[34]. Therefore, user experience is much better, and acceptance is more likely.

Considering their somewhat achievable data rate and advantages over auto-stereoscopy, multi-view and light-field displays are currently the subject of intense research[35-40]. This technology certainly represents the next 3D display platform that will appear in the marketplace, and some specialized applications have already started to emerge[41].

The Vergence-Accommodation Conflict

The vergence-accommodation conflict is the Achilles heel of all the display systems that we have introduced thus far: stereoscopic, autostereoscopic, multiview, and light-field (with some exception for the latter) and occurs when mismatched visual 3D cues are presented to an observer. The vergence-accommodation conflict occurs because the images projected by these displays are located at a fixed distance, thus producing a constant accommodation cue that cannot be adjusted, whereas vergence is provided by the parallax, which can be reproduced, and thus may vary within a scene. Disparity between accommodation and vergence cues can create a conflict in perception. This conflict leads to some visual discomfort, which is well documented in the literature[42-45].

Light-field displays can reproduce some amount of accommodation when the ray density is sufficiently large. This condition is often referred to as a super-multiview[46,47]. Accommodation occurs in a light-field display because the image plane can be moved in and out of the display plane. This is achieved by directing the light rays from different sections of the panel toward one voxel region, as shown in Fig. 3a.

However, there is some belief that if the view density keeps increasing in a light-field display, the accommodation distance can be extended at will. This belief arises from the extrapolation that light-field displays approximate a wavefront curvature by using line segments. If these segments are sufficiently small, they may become indistinguishable from the true wavefront curvature. Unfortunately, this ray-tracing simplification does not occur because diffraction along the pixel edges takes place, limiting the voxel resolution. Even with a pixel density in the 100s per degree, when an object is projected too far from the plane of the light-field display, it becomes blurry because of the diffraction among pixels. This diffraction effect cannot be avoided and intrinsically reduces the depth resolution and accommodation of light-field displays[48,49].

To eliminate the diffraction phenomena experienced with smaller pixel sizes, strong coherence among pixels is required so that the light-field display becomes indistinguishable from holography.

The difficulty of reproducing accommodation induces visual discomfort by having to limit the display depth of the field. To reproduce a voxel out of the plane of the display, the light should be focused at that point by the optical system. Without the capability to refocus subpixels at will, the light-field display can only produce a flat wavefront from the emission plane. As presented in Fig. 3a, when a light-field display attempts to reproduce a voxel that is too far away from the emission plane, the voxel invariably becomes blurry.

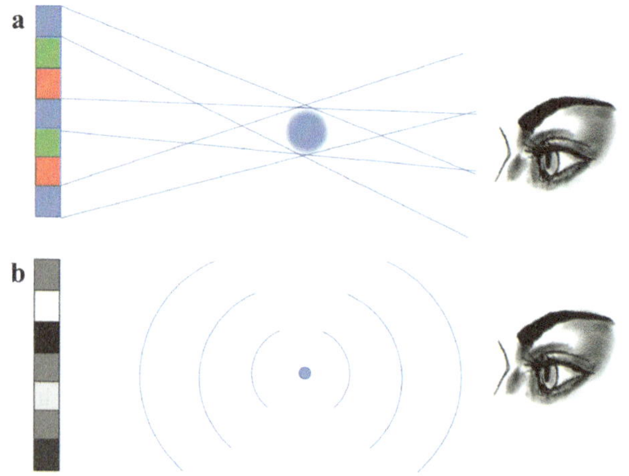

Fig. 3 Illustration of the projection of a voxel out of the emission plane by a a light-field display, **and b** a holographic display.

To address this problem, researchers have developed multiplane light-field displays[50–52]. This is possible because the plane of emission can be refocused by optical elements and moved along the view depth. However, this requires some multiplexing to generate different planes in time or space, which increases the bandwidth required by the system. Another aspect that should not be overlooked is that occlusions between different planes are difficult to control when there are many view zones[53].

Volumetric Displays

Volumetric displays have voxels located in 3D space and are affected by the same occlusion problem as a multi-plane light-field display. For both systems, the occlusions can only be correctly reproduced for one point of view[54,55]. Some systems (both volumetric and light-field) use an eye tracking mechanism to re-calculate the occlusions and present the correct image wherever the viewer is located[56]. However, only one correct perspective can be achieved, precluding its application for multiple observers.

In a volumetric display, the occlusion problem occurs because the emission of the voxel is omnidirectional, and there is no absorptive voxel. Nevertheless, volumetric displays have the advantage of being able to reproduce the field depth without resolution loss. They can be somewhat more natural to view when they do not use a screen to display an image. In this case, the image appears floating in thin air, which has a dramatic effect on the viewer's perception[55,57,58].

Volumetric displays also have the disadvantage of not being capable of projecting images outside a limited volume. The image depth is bounded by that volume, and a deep landscape or object that seemingly reaches out of the display cannot be reproduced[54].

The mathematical computation of the bit rate for a volumetric display is as simple as multiplying the resolution of a 2D screen by the third dimension, refresh rate, and dynamic range. In Fig. 1, the data rate for a 4K volumetric display is

$$\begin{aligned} &x \times y \times z \times \text{rep.rate} \times \text{res.} \times \text{colors} \\ &= 4096 \times 2160 \times 1000 \times 60 \times 8 \times 3 = 1.3 \times 10^{13}\ \text{b/s}. \quad (3) \end{aligned}$$

However, because volumetric display setups are easily scalable, lower-resolution systems can be readily used to showcase the potential of the technology[59–61].

Holographic 3D Display

Since the studies by Leith, Upatnieks, and Denisyuk, static holograms have demonstrated that the technique is capable of reproducing all the cues that the human visual system uses to comprehend the three dimensions[1,2]. By using high quality photosensitive materials, it is now possible to copy existing artifacts and display convincing holographic reproductions in full color[62,63]. The question that remains is how to do the same with a refreshable display.

There are three fundamental problems to be solved to create a holographic television: the *computation* of the holographic pattern from the 3D information, the *transmission* of the data from where it is captured to where it needs to be displayed, and the *reproduction* of the holographic pattern on the screen to display the 3D image.

Computed Generated Holograms

The calculation of the diffraction pattern from a 3D image is based on the physical model of light propagation. The Kirchhoff and Fresnel diffraction integrals provide the value of the field at a distance z from an aperture of arbitrary shape[5]. Early studies on computer generated holograms (CGHs) were undertaken in the late 1960s by Lohmann, Brown, and Lesem[64–66]. Although the mathematical equations have been well-known for a very long time, their actual computation is far from trivial and has often been referred to as the "computational bottleneck"[67]. Even today, the number of operations per second (flops) needed to compute the size and location of the diffraction elements in real time is beyond the capabilities of any computer.

To provide a sense of scale, and following the approach by Slinger et al.[68], a brute force approach computation for a 720p (1280×720) holographic display requires 100×100 diffractive elements per pixel for full parallax, and 4000 multiplication and accumulation calculations per element at a rate of 60 Hz, and for three colors, this computes to $1280 \times 720 \times 100 \times 100 \times 4000 \times 3 \times 60 =$ 6.6 petaflops. Simplifications in the way that holograms are computed had to be made.

The strongest simplification is to move away from the Fresnel integral and work in the far field with the Fraunhofer integral, which is a Fourier transform. By eliminating the z coordinate, the Fraunhofer integral yields the solution of the diffraction pattern for a given intensity field at infinity. Owing to the fast Fourier transform algorithm, a solution can be efficiently computed[69]. However, Fourier holograms have only a single image plane, so this simplification sacrifices image depth, which is not desirable for a 3D display.

It should also be noted that the Fourier transform yields a real and an imaginary solution. These two components correspond to the amplitude and phase values of the hologram. Most of the time, the element used to display the diffractive pattern (such as a spatial light modulator) can only reproduce one or the other, but not both. This means that the result from a single Fourier transform will have a significant amount of noise when reproducing an image. Other sources of noise in holograms originate from the quantization error of the phase levels, diffraction in the pixel structure, and speckle caused by the random phase[70].

To reduce the noise and boost the signal, Gerchberg and Saxton developed an iterative algorithm (GS) in 1972[71]. However, the GS algorithm only works for 2D input images and does not accept 3D information. Nevertheless, to obtain some image depth, it is possible to compute individual holograms for different discrete planes[72–74]. This solution renders both vergence and accommodation. However, because the holograms for the different image planes have been computed separately from one another, and not as a whole 3D scene, occlusions can only be reproduced for a single perspective. This is the same problem previously mentioned regarding volumetric displays. More recently, some algorithms have been developed to address the occlusion problem in multi-plane hologram computations. Such algorithms have demonstrated the capability to render correct occlusions in a limited view zone[75–77].

For 3D displays, hologram computations directly based on 3D models can be accomplished[78]. Although the detailed description of these algorithms is beyond the scope of this article, it should be noted that these algorithms can be separated into two broad categories: wavefront-based methods, and ray-based methods.

In the ray-based technique, the hologram is calculated from incoherently captured 2D image planes of a 3D scene, and relies on the geometric optics formalism of light propagation. The ray-based methods include two distinct categories: holographic stereogram (HS) and multiple viewpoint projection (MVP)[79–81]. Because they do not compute the wavefront propagation, HS and MVP techniques are much faster than wavefront-based methods and can render photorealistic images[82]. However, because the full wavefront of the object is not taken into account, ray-based methods can have difficulties rendering some of the 3D optical cues. Moreover, HS techniques experience some limitations in depth of field because the different views are combined incoherently[48]. Another drawback of the MVP approach is the need to capture or render a large number of images involving small increments in the location of the camera. Otherwise, the motion parallax can be jumpy, and occlusions are not represented well. In a sense, HS and MVP holograms are hybrids falling between light-field displays and holographic displays.

In the wavefront-based methods, the propagation of the light wave is computed starting from a point light source that is illuminating either a point cloud or a polygon representation of the object. The CGH is calculated by simulating the interference between the wavefront coming from the object and another point light source or reference beam. The advantage of these methods is that they natively consider occlusion and parallax cues, so their rendering is accurate. However, this accuracy comes at the cost of high processing demand, as described previously[48,83,76]. This processing demand is driven by the fact that each point of the CGH must take into account the interference between each and every beam projected from the portion of the object that is visible from that location in the hologram plane. When the point on the hologram moves, the perspective of the object also moves correspondingly.

Some of the computations for generating the CGH can be stored upfront in a look-up table[84]. This information can be retrieved much faster than if it is computed multiple times. There is a fine balance to achieve between large memory storage and computational complexity. To accomplish this balance, different types of algorithms have taken advantage of look-up tables at different stages of CGH computations[85–87].

To handle the massive amount of data required to compute 3D holograms in a reasonable amount of time, computations are sometimes performed on specially built hardware accelerators[88]. This hardware is dedicated to the calculation of the Fresnel phase pattern[86,88–91].

Despite recent advances in the field of computer holography, it seems that from the published images available, the realism of the projected 3D images computed with wavefront-based algorithms requires significant improvement to become convincing (see Fig. 4)[75,92]. This is by no means a criticism of the notable achievements accomplished in that discipline, it is rather a testament to how difficult it is to reproduce a fully detailed holographic image.

In many cases, the holographic image computed using a wavefront-based method lacks texture (see Fig. 4(2)). This should not be surprising, considering that texture rendering considers the finest details of the material surface. This level of detail cannot yet be achieved by computer. A parallel can be drawn with the more familiar world of 2D animation, where early movies depicted blocky characters lacking luster but in modern times, the production team can use any level of photorealism that suits the artistic needs of the story (see for example,[93]).

Techniques such as machine learning, neural networks, and artificial intelligence have recently been applied with great success to the computation of holograms[94–98]. In the most general sense, the algorithms associated with these techniques work by training a computing network with

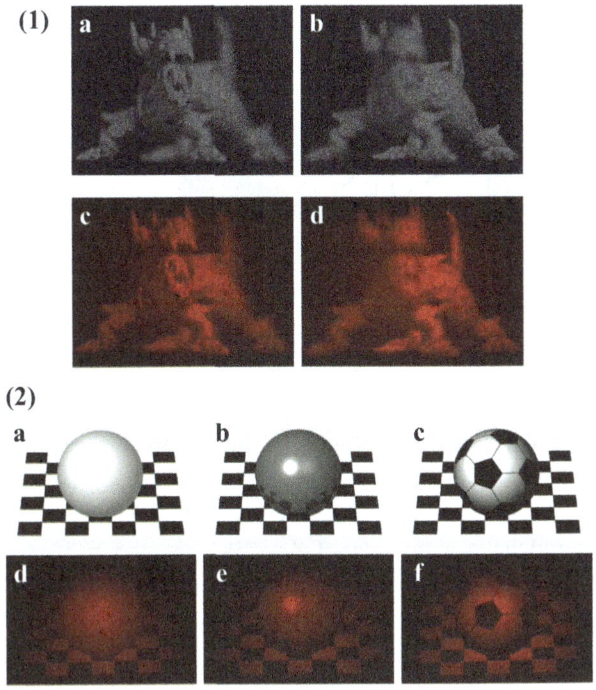

Fig. 4 Examples of optical reconstructions of wavefront-based computer generated holograms from recently published articles. **(1)** Reproduced from [75]. Numerical and optical reconstruction results when focusing on the a, c head and b, d tail of the dragon. **(2)** From [93] presenting rendering images and optical reconstruction images of different surfaces a, d rough surface b, e smooth surface c, f rough surface with texture.

some input images (ground truth) and use a camera to observe holograms generated by an optical system. The parameters of the code are optimized by the algorithm to minimize a loss function that forces the holograms to converge toward the original images. Once training is complete, the parameters are frozen, and the algorithm is eventually capable of calculating the hologram of any input image. These techniques are particularly effective at performing fast computations once the training period is complete and at solving the problems associated with texture rendering[99,96]. However, for the most part, these algorithms work with 2D images, but are expected to soon be extended to 3D images.

Transmission of Holograms

The image captured for a holographic display can satisfy the minimum requirements of the human eye and does not have to use coherent illumination and resolve nanometric interference fringes, as is the case with static holograms.

To suit the human eye accommodation, the 3D information to be reproduced can have a depth resolution of only a few centimeters, instead of the nanometers achieved with holography[10]. Such an image can even be compressed into a 3D mesh model overlaid by a texture pattern, as that used in modern video games. Video games handle this information together with the location of a virtual camera to display a 2D image. In the same way, the game engine could display a 3D image if the display required it. As a case in point, video games can be adjusted to and played using stereoscopic virtual reality headsets.

We know that the amount of data to be transmitted to the display system is not overwhelming and can easily be accommodated by today's technology. However, from a 3D image, the computation of its hologram dramatically increases the amount of information because the diffraction pattern cannot be scaled to accommodate the lateral resolution of the human eye ($\approx$ 1 arcminute or 0.3 mrad), but must instead be dimensioned according to the wavelength of visible light ($\approx$ 500 nm). This scaling can be appreciated in Fig. 1 by comparing the data rate of a volumetric display ($\approx 10^{13}$ b/s) to a holographic display ($\approx 10^{15}$ b/s)).

Because of this increase in data size, it may be much more efficient to transmit the 3D image/model rather than the holographic pattern[100,101]. In this case, the computation of the hologram should be performed at the client (receiver) location. This model is named the "thick client" because the computation is performed locally to avoid overwhelming the long-distance transmission medium. This means that the local site requires significant computational power to support this decoding (see the above section regarding Computed Generated Holograms).

Alternatively, a remote bank of servers can be used to compute the hologram quickly and efficiently. In this configuration, only a "lean client" with limited processing power is needed at the receiver site, whose function is to process the local transmission for display purposes. Under this scenario, the near full amount of data (10^{14} b/s) would need to be transmitted by the server to the client through a "fat pipe" network.

A model of the transmission and reproduction of holographic images is presented in Fig. 5, along with the different orders of magnitude of the computation and data rate needed at each stage.

We are already familiar with this type of lean client/fat pipe architecture as that used for cell phones and cloud computing. To support holography, the entire transmission network would have to be upgraded to support a 10^6 increase in data flow (from HDTV to holography in Fig. 1). This increase is not unreasonable considering that we have already experienced five cellphone network generations over the past 20 years[102,103].

There is no clear advantage between the thick and lean client models. The main reason is that the need for the transmission of 3D holographic images and movies does

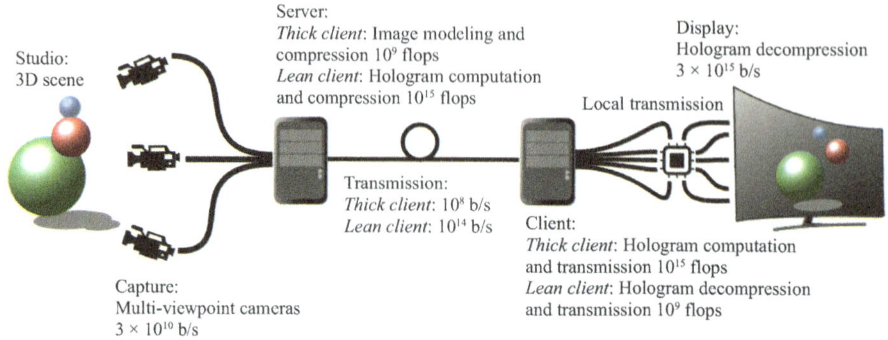

Fig. 5 Schematic of a holographic television transmission process. Comparison between thick client and lean client architectures.

not yet exist. However, it should be noted that the compression algorithm for hologram storage and transmission is not as effective as algorithms for natural images, such as JPEG and MPEG. This arises from the fact that the resolution of a diffraction pattern cannot be decreased without destroying the light interference it is suppose to generate and therefore the holographic image. Diffraction patterns need to be compressed using a near-lossless algorithm[104–107].

Another important point regarding the transmission of holograms is that the computation of the interference pattern is specific to the display architecture. For the proper reproduction of a hologram, the computation of the interference pattern must consider whether the display is operating in full parallax versus horizontal parallax only, what are the exact illumination wavelengths, and what is the pixel density (among other parameters). Likewise, legacy displays that will be operating at the same time, such as 2D televisions, stereoscopic, auto-stereoscopic, and eventually volumetric displays, must be considered. To ensure compatibility among all devices, a lean client configuration would have to send the various display parameters to the server and receive the pre-calculated data in return. In the case of a thick client architecture, the server can invariably send the same model to the client, which is then further transformed locally. From this perspective, the thick client is simply another type of display that can be integrated into a lean client network, making these two concepts complementary rather than antagonistic.

Holographic Display Setups

Spatio-Temporal Product

In the context of an imaging device, the spatio-temporal product (STP) is the product of the number of pixels and the refresh rate. The introduction of the refresh rate in the equation is important because temporal multiplexing schemes can be applied to improve the display resolution (holographic or not). Multiplied by the pixel dynamic, the STP can be directly compared to the data rates used in Fig. 1. Given that we established that a holographic display should have a bit rate of 3×10^{15} b/s, the STP of the display device should have the same order of magnitude.

It is technically feasible to create a large holographic display by tiling multiple spatial light modulators (SLMs). Using 4K SLMs that have a bit rate adapted for 2D imaging, i.e. 12.7 Gb/s, 230,000 SLMs would be required to reach 3×10^{15} b/s, and 15,000 personal computers would be required to operate the number of screens involved.

These ludicrous numbers again point to the difficulty of the task at hand. Nevertheless, researchers have demonstrated that such a tiling approach actually works, although on a smaller scale[108–111].

Reduction of the STP can be achieved by reproducing the horizontal parallax only (HPO) and scanning the image vertically. HPO offers a reduction of the STP by a factor of approximately 10^3 compared to full parallax, and HPO displays do not need to maintain coherence between the different horizontal lines composing the 3D image. Because the human eye disparity is mostly horizontal, the loss of the vertical parallax in HPO holograms does not significantly impede 3D perception[112]. However, some other artifacts could be introduced if an HPO holographic display is used, such as astigmatism or the need for the viewer to stand at a given distance. Despite these issues, many researchers have taken advantage of the STP reduction offered by HPO to demonstrate holographic projection using a variety of systems[113–115].

Another possible approach to reduce the STP of a full holographic system is to limit the eye box in which the hologram is projected. Using this technique, the light is directed toward the viewer using an eye-tracking system or a head-mounted display (AR/VR headset)[116–120]. Knowing the location of the viewer dramatically reduces the calculation of the hologram because only a limited number of viewpoints need to be taken into account. Likewise, if the viewer is within a predefined region, the angular extent of the hologram (its diffraction angle) can be narrowed, and the number of diffractive pixels decreases. The advantage of this technique is that it does not sacrifice image quality or 3D cues.

Spatial Light Modulators and Phase Array Devices

There is no question that the introduction of the liquid crystal on silicon (LCoS) SLM has helped considerably in advancing holographic displays. LCoS SLMs have many convenient features for displaying diffraction patterns. With a pixel pitch down to a few microns, the diffraction angle can be as large as 10° (see Eq. 1). SLMs can process large pixel counts, which helps to achieve high-resolution holograms. LCoS also has the advantage of being able to modulate the phase, which, along with an 8 bit phase-level resolution, achieves high diffraction efficiency. Unfortunately, the viscoelasticity of liquid crystals limits the refresh speed of LCoS to a few milliseconds. This speed is adequate for imaging purposes, even in color, but ultimately limits the STP of the LCoS SLM with respect to holographic applications[121].

To increase the STP of SLMs, it is possible to move away from the LCoS technology and use microelectromechanical systems (MEMS) instead. MEMS are composed of micro-mirrors that can be tilted or moved to interact with light. They can have the same number of pixels and approximately the same pixel pitch as LCoS. However, their refresh rate can be orders of magnitude higher[122,123]. This increases their STP by the same factor, reducing the number of units needed to create a holographic display[124–126].

Early MEMS examples include micro-ribbons developed by Sony that were used to construct a diffractive light modulator (or grating light valve). This technology boasted an impressive switching speed of 20 ns[127,128]. However, micro-ribbons are one-dimensional, which requires another scanning mechanism to form a 2D image.

At about the same time, Texas Instruments experimented with a phase modulator that moved pixels up and down to modulate the phase[129]. Unfortunately, this MEMS modulator was not commercialized. Instead, Texas Instruments invested in one of the most popular MEMS, the digital light processor (DLP)[130].

The DLP works by flipping its mirrors in one direction or the other to redirect light. This MEMS was developed for imaging purposes, such as projectors and televisions, and is very efficient in these configurations[131]. However, for holography, the DLP can only display amplitude holograms with maximum efficiency of 10%. Nevertheless, the DLP STP can reach an impressive 47.7 Gpixels/s with a resolution of 1920 × 1080 pixels and a refresh rate of 23 kHz. Some chipsets have an even larger number of pixels supporting the 4K UHD resolution of 3840 × 2160 pixels, but their refresh rate is slower (60 Hz), reducing the STP to 0.5 Gpixels/s.

Most recently, Texas Instruments has rejuvenated its earlier attempt at a phase modulator and is developing a piston MEMS capable of achieving a much higher efficiency[132–134]. This phase light modulator (PLM) should be extremely useful in the development of holographic 3D display systems. If the PLM is capable of operating at 20 kHz as some DLPs can, it will increase the STP of this MEMS by a factor of 100 compared to typical LCoS SLMs.

Another approach that can increase the intensity of a hologram using low-efficiency devices is to use a refreshable holographic material. Refreshable materials, such as photorefractive polymers, can record the wavefront generated by the SLM and, owing to their high diffraction efficiency, amplify the intensity of the hologram. Video-rate holographic projection, as well as large holographic displays, have been demonstrated using this type of material[34,135,136]. However, it should be noted that these materials currently rely on an electronically addressable device (SLM, DLP, or other) to display the initial holographic pattern.

Considering that the STP is the key to unlocking a practical holographic 3D display, the approach taken by a group of researchers at MIT and BYU (initiated before the DLP was available) was to start with the device that had the largest STP at the time, acousto-optic material (AOM)[137,138]. Regarding acousto-optic materials, the propagation of a sound wave creates a density modulation that diffracts light. If the sound wave is correctly programmed, the diffracted light can form a holographic image. In its waveguide format, the acousto-optic modulator allows for a longer interaction length between light and the acoustic wave generated, which further increases the STP[139–141]. A single leaky acousto-optic waveguide can have a 50 MHz usable bandwidth per color, which corresponds to 1.67 Mpixels at 30 Hz. By fabricating multiple waveguide channels in a single crystal, these numbers can easily be increased by several thousands to reach an STP of 50 Gpixels/s. Although the initial demonstrations using AOM provided horizontal parallax only, it is theoretically feasible to feed the different waveguides using a single laser source and control the phase such that horizontal and vertical coherent beam steering can be achieved[142,143].

Another high-STP device is the phased array photonic integrated circuit (PIC)[144,145]. In this approach, nanophotonic phased arrays are built by recording branching waveguides on a photonic wafer (see Fig. 6). The waveguides are organized such that they distribute light projected from a single source over a 2-dimensional grid. The phase at the end of each waveguide can be adjusted by an electro-optic or thermo-optic phase regulator. The light is extracted orthogonally from the wafer by a grating outcoupler terminating each waveguide. Analogous to phased array radar, the grating outcoupler is also called an optical antenna.

The advantage of PIC phased array technology is the very high frequency at which the phase can be modulated. Electro-optic modulators can reach one hundred GHz[146,147]. This inherently elevates the data rate to the 10^{10} b/s order. Using an array with 300 × 300 antennae would achieve the 10^{15} b/s required for a holographic display. As we will discuss, the present difficulties of photonic phased arrays are the wafer material, the gaps between antennae, and the phase accuracy between antennae.

The preferred material for PIC is silicon, which does not transmit visible light. Other materials with better transmission in the visible wavelengths should be used for display purposes. Silicon nitride or silica platforms have already been explored for optical phased arrays in the literature, but remain in their experimental phase[145,146,148,149].

Compared to MEMS and LCoS that have a fill factor above 90%, the fill factor of the phased array is fairly low, at approximately 25%. The fill factor affects the diffraction efficiency owing to the presence of side-lobe emissions that cannot be canceled if the antennae are too far apart. This separation is due to the limited turn radius of the waveguide and the required separation between waveguide elements to avoid cross-coupling[150]. Both factors, turn radius and waveguide separation, are dictated by the difference in the index of refraction between the inside and outside of the waveguide. A larger index difference would allow for a larger fill factor.

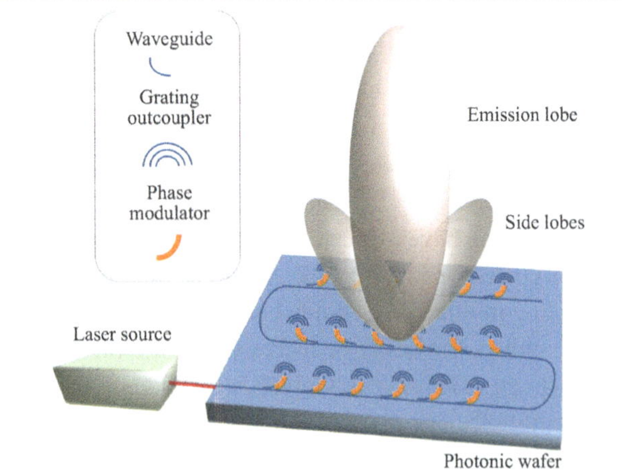

Fig. 6 Schematic of a photonic integrated circuit optical phased array. A single coherent laser source is directed inside a waveguide, from which light is extracted by multiple grating couplers (acting as light antennae). The phase at each antenna can be tuned using a phase modulator to create a hologram.

The phase control of the pixels is better in LCoS than in both MEMS and phased arrays. The LCoS phase is analog and proportional to the applied voltage and is therefore uniform across pixels. In contrast, current MEMS micro-mirror levels are discrete and limited to 4 bits, and exhibit some nonlinearity[134]. For phased arrays, the phase control is analog and accurate, but has to be characterized for each element individually owing to manufacturing inconsistencies[145].

In summary, none of the current SLM technologies is sufficiently mature to meet all the criteria required for a large-size, high-definition holographic 3D display. This should not overshadow the considerable progress that has been made over the past years, bringing the end goal ever closer.

Conclusion

Holography is still considered as the ultimate technology that will enable rendering of all the optical cues needed for the human visual system to see projected images in 3D. All other technologies, such as (auto) stereoscopy, light-field, or volumetric displays suffer from trade-offs that limit 3D rendering. Nonetheless, these technologies will likely prove to be stepping stones leading to better visual comfort until holographic displays are achieved.

Some of the doors that were preventing holographic television from being made possible only a few years ago have already been unlocked. The fast computation of 3D holograms to properly control occlusions and parallax is now within reach as is a solution to the problem of data transmission. The exact architecture of the network (thick or lean client) is unclear, but higher compression rates and ever faster telecommunication infrastructures supporting the Internet mobile communications make streaming the data for a holographic television feasible, if not yet accessible.

However, some challenges remain to be solved. The two main obstacles at the time this manuscript was written are the computation of photorealistic 3D holograms in a reasonable amount of time, and a suitable electronic device for the reproduction of large holographic 3D images with high resolution.

For the pioneers of holography, Gabor, Leith, Upatnieks, and Denisyuk, the challenge of controlling diffractive pixels could only have been a physicist's dream. This dream has now been transformed into an engineering nightmare, as devices that can project holographic images exist, but scaling the format to project large images can be surprisingly difficult. Computational time becomes prohibitive, and controlling trillions of pixels currently requires an extremely large number of graphic cards.

Nonetheless, the difficulty of projecting large holograms will soon no longer be in the engineering realm, but will rather transform into an economic problem. Initially, the hardware needed to build a holographic television will be too expensive to be successfully commercialized as a television set. Once price becomes affordable, the holographic television will face the same hurdle that each new media is bumping into, which is the availability of correctly formatted content. There is no advantage in owning a holographic television if creators are still producing 2D movies exclusively. For these reasons, it is likely that the market will move incrementally, starting with HPO multiview, expanding to light-field and full parallax, and finally reaching holography.

To achieve this dream of the perfect display, we should remember that the bit rate progression reported in Fig. 1 is not an immovable fact of nature. It is a testament to human ingenuity and hard work. This exponential growth can be influenced in one way or another by our own actions. Ultimately, where there is a will, there is a way, and the desire for a truly immersive visual experience is ingrained in human nature. It is this desire that will make the holographic television a reality sooner than later.

Conflict of interest

There are no conflicts to declare.

Received: 06 May 2021 Revised: 05 November 2021 Accepted: 19 November 2021
Accepted article preview online: 25 November 2021
Published online: 20 December 2021

References

1. Leith, E. N. & Upatnieks, J. New techniques in wavefront reconstruction. *Journal of the Optical Society of America* **51**, 1469-1473 (1961).
2. Denisyuk, Y. N. On the reflection of optical properties of an object in a wave field of light scattered by it. *Doklady Akademii Nauk SSSR* **144**, 1275-1278 (1962).
3. De Bitetto, D. A holographic motion picture film with constant velocity transport. *Applied Physics Letters* **12**, 295-297 (1968).
4. Jacobson, A. D., Evtuhov, V. & Neeland, J. K. Motion picture holography. *Applied Physics Letters* **14**, 120-122 (1969).
5. Blanche, P. A. Introduction to holographic principles. in Optical Holography: Materials, Theory and Applications (ed Blanche, P. A.) Ch. 1 (Amsterdam: Elsevier, 2020), 1-39.
6. Ciciora, W. et al. Cable telephony. in Modern Cable Television Technology: Video, Voice, and Data Communications 2nd edn (eds Ciciora, W. et al) Ch. 6 (Amsterdam: Morgan Kaufmann, 2004), 229-284.
7. Plonus, M. Digital systems. in Electronics and Communications for Scientists and Engineers 2nd edn (ed Plonus, M.) Ch. 9 (Oxford: Butterworth-Heinemann, 2020), 355-480.
8. Pritchard, D. H. U.S. color television fundamentals: a review. *SMPTE Journal* **86**, 819-828 (1977).
9. Dunn, D. et al. Stimulating the human visual system beyond real world performance in future augmented reality displays. Proceedings of 2020 IEEE International Symposium on Mixed and Augmented Reality (ISMAR). Porto de Galinhas, Brazil: IEEE, 2020, 90-100.
10. Howard, I. P. Perceiving in Depth, Volume 1: Basic Mechanisms. (Oxford: Oxford University Press, 2012).
11. Park, M. C. & Mun, S. Overview of measurement methods for factors affecting the human visual system in 3D displays. *Journal of Display Technology* **11**, 877-888 (2015).
12. Mayo, H. Outlines of Human Physiology. 4th edn. (London: Renshaw, 1837).
13. Cakmakci, O. & Rolland, J. Head-worn displays: a review. *Journal of Display Technology* **2**, 199-216 (2006).
14. Urey, H. et al. State of the art in stereoscopic and autostereoscopic displays. *Proceedings of the IEEE* **99**, 540-555 (2011).
15. Woods, A. J. Crosstalk in stereoscopic displays: a review. *Journal of Electronic Imaging* **21**, 040902 (2012).
16. Read, J. C. A. & Bohr, I. User experience while viewing stereoscopic 3D television. *Ergonomics* **57**, 1140-1153 (2014).
17. Sexton, I. & Surman, P. Stereoscopic and autostereoscopic display systems. *IEEE Signal Processing Magazine* **16**, 85-99 (1999).
18. Dodgson, N. A. Autostereoscopic 3D displays. *Computer* **38**, 31-36 (2005).
19. Iizuka, K. Engineering Optics. 3rd edn. (New York: Springer, 2008).
20. Rotter, P. Why did the 3D revolution fail?: the present and future of stereoscopy [commentary]. *IEEE Technology and Society Magazine* **36**, 81-85 (2017).
21. Chen, Y. S. et al. Video-based eye tracking autostereoscopic displays. *Optical Engineering* **40**, 2726-2734 (2001).
22. Yi, S. Y., Chae, H. B. & Lee, S. H. Moving parallax barrier design for eye-tracking autostereoscopic displays. Proceedings of 2008 3DTV Conference: the True Vision-Capture, Transmission and Display of 3D Video. Istanbul, Turkey: IEEE, 2008, 165-168.
23. Yoon, K. H. et al. Autostereoscopic 3D display system with dynamic fusion of the viewing zone under eye tracking: principles, setup, and evaluation [Invited]. *Applied Optics* **57**, A101-A117 (2018).
24. Yu, X. B. et al. Large viewing angle three-dimensional display with smooth motion parallax and accurate depth cues. *Optics Express* **23**, 25950-25958 (2015).
25. Kara, P. A. et al. The key performance indicators of projection-based light field visualization. *Journal of Information Display* **20**, 81-93 (2019).
26. Kara, P. A. et al. The interdependence of spatial and angular resolution in the quality of experience of light field visualization. Proceedings of 2017 International Conference on 3D Immersion (IC3D). Brussels, Belgium: IEEE, 2017, 1-8.
27. Stern, A., Yitzhaky, Y. & Javidi, B. Perceivable light fields: matching the requirements between the human visual system and autostereoscopic 3-D displays. *Proceedings of the IEEE* **102**, 1571-1587 (2014).
28. Watanabe, Y. & Kakeya, H. A full-HD super-multiview display based on adaptive time-division multiplexing parallax barrier. *ITE Transactions on Media Technology and Applications* **8**, 230-237 (2020).
29. Wan, W. Q. et al. Super multi-view display based on pixelated nanogratings under an illumination of a point light source. *Optics*

and Lasers in Engineering **134**, 106258 (2020).
30. Wu, G. C. et al. Light field image processing: an overview. *IEEE Journal of Selected Topics in Signal Processing* **11**, 926-954 (2017).
31. Martínez-Corral, M. & Javidi, B. Fundamentals of 3D imaging and displays: a tutorial on integral imaging, light-field, and plenoptic systems. *Advances in Optics and Photonics* **10**, 512-566 (2018).
32. Huang, H. K. & Hua, H. High-performance integral-imaging- based light field augmented reality display using freeform optics. *Optics Express* **26**, 17578-17590 (2018).
33. Rogers, B. & Graham, M. Similarities between motion parallax and stereopsis in human depth perception. *Vision Research* **22**, 261-270 (1982).
34. Blanche, P. A. et al. Holographic three-dimensional telepresence using large-area photorefractive polymer. *Nature* **468**, 80-83 (2010).
35. Yang, L. et al. Demonstration of a large-size horizontal light-field display based on the LED panel and the micro-pinhole unit array. *Optics Communications* **414**, 140-145 (2018).
36. Yu, X. B. et al. Dynamic three-dimensional light-field display with large viewing angle based on compound lenticular lens array and multi-projectors. *Optics Express* **27**, 16024-16031 (2019).
37. Liu, B. Y. et al. Time-multiplexed light field display with 120-degree wide viewing angle. *Optics Express* **27**, 35728-35739 (2019).
38. Gao, X. et al. Full-parallax 3D light field display with uniform view density along the horizontal and vertical direction. *Optics Communications* **467**, 125765 (2020).
39. Kara, P. A. et al. Cinema as large as life: large-scale light field cinema system. Proceedings of 2017 International Conference on 3D Immersion (IC3D). Brussels, Belgium: IEEE, 2017, 1-8.
40. Wang, P. R. et al. A full-parallax tabletop three dimensional light-field display with high viewpoint density and large viewing angle based on space-multiplexed voxel screen. *Optics Communications* **488**, 126757 (2021).
41. Looking Glass Factory. (2021). at https://lookingglassfactory.com/.
42. Lambooij, M. T., IJsselsteijn, W. A. & Heynderickx, I. Visual discomfort in stereoscopic displays: a review. Proceedings of SPIE 6490, Stereoscopic Displays and Virtual Reality Systems XIV. San Jose, CA, United States: SPIE, 2007, 64900I.
43. Hoffman, D. M. et al. Vergence–accommodation conflicts hinder visual performance and cause visual fatigue. *Journal of Vision* **8**, 33 (2008).
44. Kim, J., Kane, D. & Banks, M. S. The rate of change of vergence–accommodation conflict affects visual discomfort. *Vision Research* **105**, 159-165 (2014).
45. Daniel, F. & Kapoula, Z. Induced vergence-accommodation conflict reduces cognitive performance in the Stroop test. *Scientific Reports* **9**, 1247 (2019).
46. Watanabe, Y. & Kakeya, H. Time-division and color multiplexing light-field display using liquid-crystal display panels to induce focal accommodation. *Applied Optics* **60**, 1966-1972 (2021).
47. Li, T. T. et al. View-dependent light-field display that supports accommodation using a commercially-available high pixel density LCD panel. *SID Symposium Digest of Technical Papers* **51**, 1013-1016 (2020).
48. St Hilaire, P. et al. Are stereograms holograms? A human perception analysis of sampled perspective holography. *Journal of Physics: Conference Series* **415**, 012035 (2013).
49. Huang, H. K. & Hua, H. Effects of ray position sampling on the visual responses of 3D light field displays. *Optics Express* **27**, 9343-9360 (2019).
50. Zhan, T. et al. Multifocal displays: review and prospect. *PhotoniX* **1**, 10 (2020).
51. Balram, N. & Tošic, I. Light-field imaging and display systems. *Information Display* **32**, 6-13 (2016).
52. Jo, Y. et al. Tomographic projector: large scale volumetric display with uniform viewing experiences. *ACM Transactions on Graphics* **38**, 21 (2019).
53. Chang, J. H. R., Kumar, B. V. K. V. & Sankaranarayanan, A. C. Towards multifocal displays with dense focal stacks. *ACM Transactions on Graphics* **37**, 198 (2018).
54. Smalley, D. et al. Volumetric displays: turning 3-D inside-out. *Optics and Photonics News* **29**, 26-33 (2018).
55. Smalley, D. E. et al. A photophoretic-trap volumetric display. *Nature* **553**, 486-490 (2018).
56. Jones, A. et al. Rendering for an interactive 360° light field display. *ACM Transactions on Graphics* **26**, 40-es (2007).
57. Hirayama, R. et al. A volumetric display for visual, tactile and audio presentation using acoustic trapping. *Nature* **575**, 320-323 (2019).
58. Kumagai, K., Hasegawa, S. & Hayasaki, Y. Volumetric bubble display. *Optica* **4**, 298-302 (2017).
59. Rogers, W. & Smalley, D. Simulating virtual images in optical trap displays. *Scientific Reports* **11**, 7522 (2021).
60. Suzuki, K., Fukano, Y. & Oku, H. 1000-volume/s high-speed volumetric display for high-speed HMD. *Optics Express* **28**, 29455-29468 (2020).
61. Baek, H. et al. Wheel screen type lamina 3D display system with enhanced resolution. *Current Optics and Photonics* **5**, 23-31 (2021).
62. Sarakinos, A. & Lembessis, A. Color holography for the documentation and dissemination of cultural heritage: optoclones™ from four museums in two countries. *Journal of Imaging* **5**, 59 (2019).
63. Gentet, P. et al. Evaluation of the realism of a full-color reflection H2 analog hologram recorded on ultra-fine-grain silver-halide material. *Open Physics* **17**, 449-457 (2019).
64. Lohmann, A. W. & Paris, D. P. Binary fraunhofer holograms, generated by computer. *Applied Optics* **6**, 1739-1748 (1967).
65. Lesem, L. B, Hirsch, P. M. & Jordan, J. A. The kinoform: a new wavefront reconstruction device. *IBM Journal of Research and Development* **13**, 150-155 (1969).
66. Brown, B. R. & Lohmann, A. W. Computer-generated binary holograms. *IBM Journal of research and Development* **13**, 160-168 (1969).
67. Cameron, C. D. et al. Computational challenges of emerging novel true 3D holographic displays. Proceedings of SPIE 4109, Critical Technologies for the Future of Computing. San Diego, CA, United States: SPIE, 2000, 129-140.
68. Slinger, C., Cameron, C. & Stanley, M. Computer-generated holography as a generic display technology. *Computer* **38**, 46-53 (2005).
69. Cooley, J. W. & Tukey, J. W. An algorithm for the machine calculation of complex Fourier series. *Mathematics of Computation* **19**, 297-301 (1965).
70. Bianco, V. et al. Quasi noise-free digital holography. *Light: Science & Applications* **5**, e16142 (2016).
71. Gerchberg, R. W. & Saxton, W. O. A practical algorithm for the determination of phase from image and diffraction plane pictures. *Optik* **35**, 237-250 (1972).
72. Makowski, M. et al. Iterative design of multiplane holograms: experiments and applications. *Optical Engineering* **46**, 045802 (2007).
73. Makowski, M., Sypek, M. & Kolodziejczyk, A. Colorful reconstructions from a thin multi-plane phase hologram. *Optics Express* **16**, 11618-11623 (2008).
74. Velez-Zea, A. Iterative multiplane hologram generation with mixed constraint. *Applied Optics* **60**, 224-231 (2021).
75. Zhang, H., Cao, L. C. & Jin, G. F. Computer-generated hologram with occlusion effect using layer-based processing. *Applied Optics* **56**, F138-F143 (2017).

76. Liu, J. P. & Liao, H. K. Fast occlusion processing for a polygon-based computer-generated hologram using the slice-by-slice silhouette method. *Applied Optics* **57**, A215-A221 (2018).
77. Liu, S. T. et al. Occlusion calculation algorithm for computer generated hologram based on ray tracing. *Optics Communications* **443**, 76-85 (2019).
78. Sahin, E. et al. Computer-generated holograms for 3D imaging: a survey. *ACM Computing Surveys* **53**, 32 (2020).
79. Su, Y. F. et al. Projection-type multiview holographic three-dimensional display using a single spatial light modulator and a directional diffractive device. *IEEE Photonics Journal* **10**, 7000512 (2018).
80. Wang, Z. et al. Resolution-enhanced holographic stereogram based on integral imaging using moving array lenslet technique. *Applied Physics Letters* **113**, 221109 (2018).
81. Zhang, X. et al. Resolution-enhanced holographic stereogram based on integral imaging using an intermediate-view synthesis technique. *Optics Communications* **457**, 124656 (2020).
82. Shi, L. et al. Towards real-time photorealistic 3d holography with deep neural networks. *Nature* **591**, 234-239 (2021).
83. Nishitsuji, T. et al. Review of fast calculation techniques for computer-generated holograms with the point-light-source-based model. *IEEE Transactions on Industrial Informatics* **13**, 2447-2454 (2017).
84. Tsang, P. W. M., Poon, T. C. & Wu, Y. M. Review of fast methods for point-based computer-generated holography [Invited]. *Photonics Research* **6**, 837-846 (2018).
85. Lucente, M. E. Interactive computation of holograms using a look-up table. *Journal of Electronic Imaging* **2**, 28-34 (1993).
86. Sato, H. et al. Real-time colour hologram generation based on ray-sampling plane with multi-GPU acceleration. *Scientific Reports* **8**, 1500 (2018).
87. Pi, D. P. et al. Acceleration of computer-generated hologram using wavefront-recording plane and look-up table in three-dimensional holographic display. *Optics Express* **28**, 9833-9841 (2020).
88. Wang, Z. et al. Simple and fast calculation algorithm for computer-generated hologram based on integral imaging using look-up table. *Optics Express* **26**, 13322-13330 (2018).
89. Wang, Y. C. et al. Hardware implementations of computer-generated holography: a review. *Optical Engineering* **59**, 102413 (2020).
90. Sugie, T. et al. High-performance parallel computing for next-generation holographic imaging. *Nature Electronics* **1**, 254-259 (2018).
91. Shimobaba, T. & Ito, T. Computer Holography: Acceleration Algorithms and Hardware Implementations. (Boca Raton: CRC Press, 2019).
92. Yu, C. Y. et al. Four-image encryption scheme based on quaternion Fresnel transform, chaos and computer generated hologram. *Multimedia Tools and Applications* **77**, 4585-4608 (2018).
93. Sun, M. Y. et al. Acceleration and expansion of a photorealistic computer-generated hologram using backward ray tracing and multiple off-axis wavefront recording plane methods. *Optics Express* **28**, 34994-35005 (2020).
94. Desiderio, K. & Phillips, I. How Pixar' s animation has evolved over 24 years, from 'toy story' to 'toy story 4' . (2019). at https://blog.adafruit.com/2019/06/30/how-pixars-animation-has-evolved-over-24-years-from-toy-story-to-toy-story-4/.
95. Horisaki, R., Takagi, R. & Tanida, J. Deep-learning-generated holography. *Applied Optics* **57**, 3859-3863 (2018).
96. Lindsay, M. et al. Machine learning assisted holography. Proceedings of SPIE 11731, Computational Imaging VI. SPIE, 2021, 1173103.
97. Peng, Y. F. et al. Neural holography with camera-in-the-loop training. *ACM Transactions on Graphics* **39**, 185 (2020).
98. Chakravarthula, P. et al. Learned hardware-in-the-loop phase retrieval for holographic near-eye displays. *ACM Transactions on Graphics* **39**, 186 (2020).
99. Lee, J. et al. Deep neural network for multi-depth hologram generation and its training strategy. *Optics Express* **28**, 27137-27154 (2020).
100. Wu, J. C. et al. High-speed computer-generated holography using an autoencoder-based deep neural network. *Optics Letters* **46**, 2908-2911 (2021).
101. Yanagihara, H. et al. Real-time three-dimensional video reconstruction of real scenes with deep depth using electro-holographic display system. *Optics Express* **27**, 15662-15678 (2019).
102. Zhao, Y. et al. Fast calculation method of computer-generated hologram using a depth camera with point cloud gridding. *Optics Communications* **411**, 166-169 (2018).
103. Sofana, R. S. et al. Future generation 5G wireless networks for smart grid: a comprehensive review. *Energies* **12**, 2140 (2019).
104. Bohli, A. & Bouallegue, R. How to meet increased capacities by future green 5G networks: a survey. *IEEE Access* **7**, 42220-42237 (2019).
105. El Rhammad, A. et al. Color digital hologram compression based on matching pursuit. *Applied Optics* **57**, 4930-4942 (2018).
106. Jiao, S. M. et al. Compression of phase-only holograms with JPEG standard and deep learning. *Applied Sciences* **8**, 1258 (2018).
107. Stepień, P. et al. Hologram compression in quantitative phase imaging. Proceedings of SPIE 11249, Quantitative Phase Imaging VI. San Francisco, California, United States: SPIE, 2020, 112491Q.
108. Liu, M. J., Yang, G. L. & Xie, H. Y. Method of computer-generated hologram compression and transmission using quantum back-propagation neural network. *Optical Engineering* **56**, 023104 (2017).
109. Zeng, Z. X. et al. Full-color holographic display with increased-viewing-angle [Invited]. *Applied Optics* **56**, F112-F120 (2017).
110. Lum, Z. M. A. et al. Increasing pixel count of holograms for three-dimensional holographic display by optical scan-tiling. *Optical Engineering* **52**, 015802 (2013).
111. Onural, L., Yaraş, F. & Kang, H. Digital holographic three-dimensional video displays. *Proceedings of the IEEE* **99**, 576-589 (2011).
112. An, J. et al. Slim-panel holographic video display. *Nature Communications* **11**, 5568 (2020).
113. Takaki, Y. & Yokouchi, M. Accommodation measurements of horizontally scanning holographic display. *Optics Express* **20**, 3918-3931 (2012).
114. Kozacki, T. et al. Fourier horizontal parallax only computer and digital holography of large size. *Optics Express* **29**, 18173-18191 (2021).
115. Henrie, A. et al. Hardware and software improvements to a low-cost horizontal parallax holographic video monitor. *Applied Optics* **57**, A122-A133 (2018).
116. Sando, Y. et al. Real-time interactive holographic 3D display with a 360° horizontal viewing zone. *Applied Optics* **58**, G1-G5 (2019).
117. Maimone, A., Georgiou, A. & Kollin, J. S. Holographic near-eye displays for virtual and augmented reality. *ACM Transactions on Graphics* **36**, 85 (2017).
118. Duan, X. H. et al. Full-color see-through near-eye holographic display with 80° field of view and an expanded eye-box. *Optics Express* **28**, 31316-31329 (2020).
119. Zaperty, W., Makowski, P. L. & Kozacki, T. Multi-SLM color holographic 3D display of real-world holographic content with numerical data adaptation. Frontiers in Optics 2017. Washington, D.C. United States: Optical Society of America, 2017, FTu4C.3.
120. Leister, N. et al. Full-color interactive holographic projection system for large 3D scene reconstruction. Proceedings of SPIE 6911, Emerging Liquid Crystal Technologies III. San Jose, California, United States: SPIE, 2008, 202-211.

121. Häussler, R., Leister, N. & Stolle, H. Large holographic 3D display for real-time computer-generated holography. Proceedings of SPIE 10335, Digital Optical Technologies 2017. Munich, Germany: SPIE, 2017, 177-187.
122. Lazarev, G. et al. LCOS spatial light modulators: trends and applications. in Optical Imaging and Metrology: Advanced Technologies (eds Osten, W. & Reingand, N.) Ch. 1 (Weinheim: Wiley, 2012), 1-29.
123. Wang, Y. M. et al. 2D broadband beamsteering with large-scale MEMS optical phased array. *Optica* **6**, 557-562 (2019).
124. Dudley, D., Duncan, W. M. & Slaughter, J. Emerging digital micromirror device (DMD) applications. Proceedings of SPIE 4985, MOEMS Display and Imaging Systems. San Jose, CA, United States: SPIE, 2003, 14-25.
125. Chlipala, M. & Kozacki, T. Color LED DMD holographic display with high resolution across large depth. *Optics Letters* **44**, 4255-4258 (2019).
126. Son, J. Y. et al. Holographic display based on a spatial DMD array. *Optics Letters* **38**, 3173-3176 (2013).
127. Jiao, S. M. et al. Complex-amplitude holographic projection with a digital micromirror device (DMD) and error diffusion algorithm. *IEEE Journal of Selected Topics in Quantum Electronics* **26**, 2800108 (2020).
128. Bloom, D. M. Grating light valve: revolutionizing display technology. Proceedings of SPIE 3013, Projection Displays III. San Jose, CA, United States: SPIE, 1997, 165-171.
129. Saruta, K. et al. Nanometer-order control of MEMS ribbons for blazed grating light valves. Proceedings of the 19th IEEE International Conference on Micro Electro Mechanical Systems. Istanbul, Turkey: IEEE, 2006, 842-845.
130. Lin, T. H. Implementation and characterization of a flexure-beam micromechanical spatial light modulator. *Optical Engineering* **33**, 3643-3648 (1994).
131. Douglass, M. DMD reliability: a MEMS success story. Proceedings of SPIE 4980, Reliability, Testing, and Characterization of MEMS/MOEMS II. San Jose, CA, United States: SPIE, 2003, 1-11.
132. Monk, D. W. Digital light processing: a new image technology for the television of the future. 1997 International Broadcasting Convention IBS 97. Amsterdam, Netherlands: IET, 1997, 581-586.
133. Bartlett, T. A., McDonald, W. C. & Hall, J. N. Adapting Texas instruments DLP technology to demonstrate a phase spatial light modulator. Proceedings of SPIE 10932, Emerging Digital Micromirror Device Based Systems and Applications XI. San Francisco, California, United States: SPIE, 2019, 109320S.
134. Bartlett, T. A. et al. Recent advances in the development of the Texas instruments phase-only microelectromechanical systems (MEMS) spatial light modulator. Proceedings of SPIE 11698, Emerging Digital Micromirror Device Based Systems and Applications XIII. SPIE, 2021, 116980O.
135. Ketchum, R. S. & Blanche, P. A. Diffraction efficiency characteristics for MEMS-based phase-only spatial light modulator with nonlinear phase distribution. *Photonics* **8**, 62 (2021).
136. Blanche, P. A. Photorefractive Organic Materials and Applications. (Cham: Springer, 2016).
137. Li, X. et al. Highly photorefractive hybrid liquid crystal device for a video-rate holographic display. *Optics Express* **24**, 8824-8831 (2016).
138. Benton, S. A. Experiments in holographic video imaging. Proceedings of SPIE 10308, Holography. Tatabánya, Hungary: SPIE, 1990, 103080C.
139. St-Hilaire, P. et al. Color images with the MIT holographic video display. Proceedings of SPIE 1667, Practical Holography VI. San Jose, CA, United States: SPIE, 1992, 73-84.
140. Matteo, A. M., Tsai, C. S. & Do, N. Collinear guided wave to leaky wave acoustooptic interactions in proton-exchanged $LiNbO_3$ waveguides. *IEEE Transactions on Ultrasonics, Ferroelectrics, and Frequency Control* **47**, 16-28 (2000).
141. Smalley, D. E. et al. Anisotropic leaky-mode modulator for holographic video displays. *Nature* **498**, 313-317 (2013).
142. Qaderi, K. & Smalley, D. E. Leaky-mode waveguide modulators with high deflection angle for use in holographic video displays. *Optics Express* **24**, 20831-20841 (2016).
143. Jolly, S. et al. Progress in transparent flat-panel holographic displays enabled by guided-wave acousto-optics. Proceedings of SPIE 10558, Practical Holography XXXII: Displays, Materials, and Applications. San Francisco, California, United States: SPIE, 2018, 105580L.
144. Bove, V. M. Jr. Holographic television. in Optical Holography: Materials, Theory and Applications (ed Blanche, P. A.) Ch. 4 (Amsterdam: Elsevier, 2020), 73-82.
145. Sun, J. et al. Large-scale nanophotonic phased array. *Nature* **493**, 195-199 (2013).
146. Sun, J. et al. Large-scale silicon photonic circuits for optical phased arrays. *IEEE Journal of Selected Topics in Quantum Electronics* **20**, 8201115 (2014).
147. Mahrous, H. et al. A compact 120 GHz monolithic silicon-on-silica electro-optic modulator. *Optical and Quantum Electronics* **52**, 111 (2020).
148. Jarrahi, M. et al. Optical switching based on high-speed phased array optical beam steering. *Applied Physics Letters* **92**, 014106 (2008).
149. Porcel, M. A. G. et al. [INVITED] Silicon nitride photonic integration for visible light applications. *Optics & Laser Technology* **112**, 299-306 (2019).
150. Su, T. H. et al. Experimental demonstration of interferometric imaging using photonic integrated circuits. *Optics Express* **25**, 12653-12665 (2017).
151. Wang, H. J. et al. Broadband silicon nitride nanophotonic phased arrays for wide-angle beam steering. *Optics Letters* **46**, 286-289 (2021).
152. Larocque, H. et al. Beam steering with ultracompact and low-power silicon resonator phase shifters. *Optics Express* **27**, 34639-34654 (2019).

Sando et al. *Light: Advanced Manufacturing* (2022)3:12
https://doi.org/10.37188/lam.2022.012

Official journal of the JHL 2689-9620
www.light-am.com

Article | Open Access

Holographic augmented reality display with conical holographic optical element for wide viewing zone

Yusuke Sando[1,*], Kazuo Satoh[1], Daisuke Barada[2,3] and Toyohiko Yatagai[2]

Abstract

In this study, we propose a holographic augmented reality (AR) display with a wide viewing zone realized by using a special-designed reflective optical element. A conical holographic optical element (HOE) is used as such a reflective optical element. This conical HOE was implemented to reconstruct a diverging spherical wave with a wide spread angle. It has a sharp wavelength selectivity by recording it as a volume hologram, enabling augmented reality (AR) representation of real and virtual 3D objects. The quality of the generated spherical wave and the spectral reflectivity of the fabricated conical HOE were investigated. An optical superimposition between real and virtual 3D objects was demonstrated, thereby enhancing the validity of our proposed method. A horizontal viewing zone of 140° and a vertical viewing zone of 30° were experimentally confirmed. The fabrication procedure for the conical HOE is presented, and the calculation method of the computer-generated hologram (CGH) based on Fermat's principle is explained in detail.

Keywords: Holographic 3D display, Holographic optical element, Conical hologram, Wide viewing zone

Introduction

In recent times, the product development process using augmented reality (AR) and virtual reality (VR) has been actively advanced[1]. However, to enhance the realistic and immersive sensation in displays, a natural three-dimensional (3D) representation is inevitable[2-4]. Holographic 3D displays are suitable for this purpose because they can artificially reproduce wavefronts that are physically compatible with real 3D objects[5,6]. Thus, no conflict arises between accommodation and convergence responses, which means that they can display natural 3D objects with less eye fatigue than conventional 3D displays. Moreover, both binocular parallax and continuous motion parallax can be obtained in principle. Recently, some research groups have successfully performed high-quality holographic 3D reconstructions[7-9]. A major factor of these methods is the extremely large number of pixels of computer-generated holograms (CGHs). However, because these CGHs were fabricated via laser lithography, animated 3D movies were not feasible. In this regard, different approach using a high-speed spatial light modulator (SLM) was proposed[10]. This method is based on the time-division method, where the patterns displayed on the SLM are synchronised with a scanning mirror, such as a galvano mirror. Despite the limited number of pixels of the SLM, the time-division synchronisation enabled the display performance to be equivalent to that of the CGHs with an extremely large number of pixels. Thus, this method has

Correspondence: Yusuke Sando (sando@tri-osaka.jp)
[1]Osaka Research Institute of Industrial Science and Technology, 2-7-1 Ayumino, Izumi, Osaka 594-1157, Japan
[2]Center for Optical Research & Education, Utsunomiya University, 7-1-2 Yoto, Utsunomiya, Tochigi 321-8585, Japan
Full list of author information is available at the end of the article.

the potential to realise animated 3D movies if the CGHs can be calculated and transferred in real-time.

However, in the aforementioned methods, the shape of the holograms is planar, which results in a fundamental limitation in the enlargement of the viewing zone. As shown in Fig. 1a, a whole 3D object is observable only by the viewers in front of the hologram. For viewers looking at the hologram from an angle, the field of view (FOV) is limited depending on the angle from the normal to the hologram. Furthermore, no view is possible when observing from the side, top, bottom, or backward directions. This limitation stems from the shape of the hologram and is unavoidable in conventional planar holograms.

A straightforward approach to overcome this limitation is by producing curved holograms. For example, cylindrical holograms have been proposed to enlarge the horizontal viewing zone to 360°[11,12], as shown in Fig. 1b. Computer-generated cylindrical holograms were also demonstrated[13,14]. Although these methods could produce sufficient motion parallax in the horizontal direction, the vertical viewing zone was never improved and remained the same as that of the conventional planar holograms. To realise an omnidirectional observation, only the concept and calculation method for spherical holograms, as shown in Fig. 1c, were proposed[15–17]. However, non-planar SLMs are not currently available; however, SLMs that can dynamically change the pattern of the CGHs are necessary to display animated 3D movies. Thus, neither the cylindrical nor spherical holograms are practical solutions to overcome the limitations of planar holograms.

Our research group addressed this problem and proposed the use of a convex parabolic mirror[18]. As shown in Fig. 2, the convex parabolic mirror is placed between the CGH and viewers. If the CGH is designed appropriately[19], the 3D object is reconstructed inside the convex parabolic mirror as a virtual image. Because the convex parabolic mirror reflects the incident wavefront radially with a wide-spread angle, the viewing zone can be enlarged in both the horizontal (azimuthal) and vertical (elevation) directions. In our previous study, azimuthal and elevation viewing zones of 150° and 60°, respectively, were experimentally demonstrated. This viewing zone is much larger than that of other preceding viewing zone enlargement methods[10,20,21]. Moreover, our previously reported method has a high potential to realise animated 3D movies while keeping the viewing zone large because only planar holograms were used and can equally be replaced by SLMs.

However, the use of a metallic convex parabolic mirror narrows its scope of applications. In particular, it is impossible to place real 3D objects inside a parabolic mirror and observe them from the outside. Thus, the superimposition of virtual 3D objects with real 3D objects, that is, AR representation, is not feasible. Moreover, the effect of interactive-physical activities, such as manipulating a virtual object with the human hand, is lost. In this study, a method to realise an AR representation is proposed using a conical holographic optical element (HOE) instead of a metallic parabolic mirror. By recording the HOE as a volume hologram, narrow-band wavelength selectivity appears. This is the see-through optical property and is indispensable for realising a holographic AR display[22,23]. In this study, a conical HOE is fabricated to convert an incident plane wave into a diverging spherical wave. A 3D object is reconstructed inside the conical HOE as a virtual image. Owing to the see-through optical

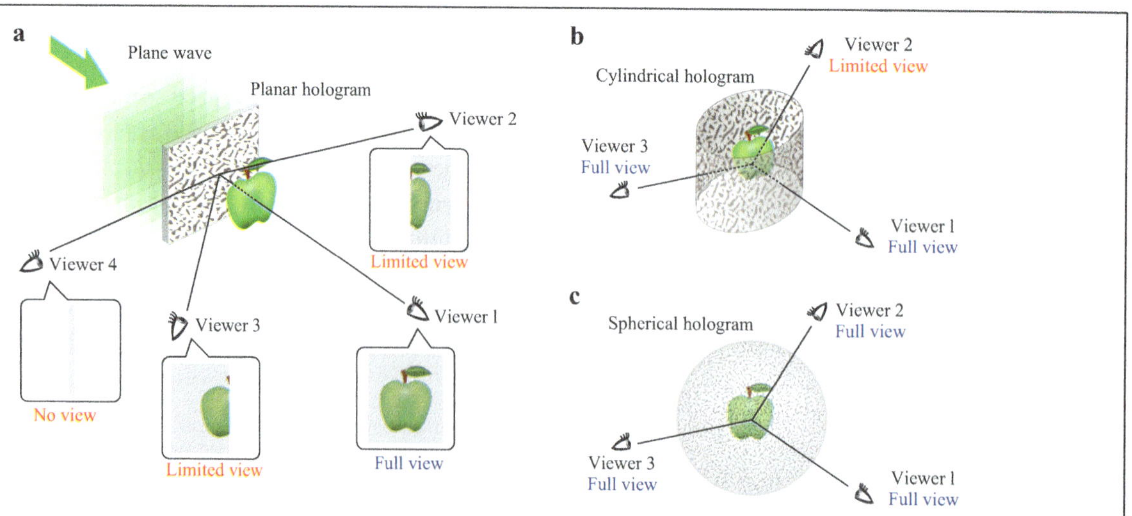

Fig. 1 Illustration of viewing zone in **a** a planar hologram, **b** cylindrical hologram and **c** spherical hologram.

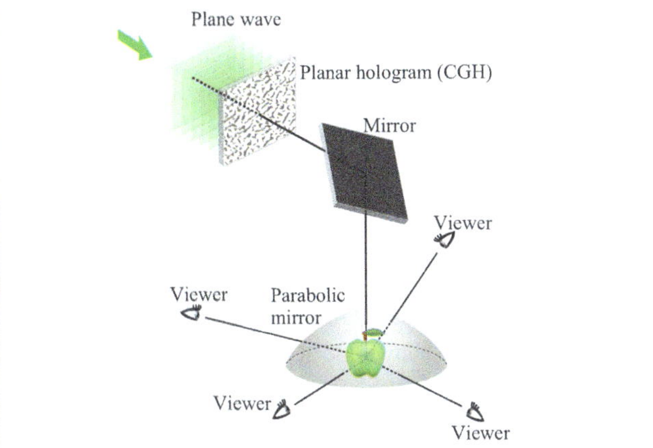

Fig. 2 Concept of our previous optical system with a convex parabolic mirror.

property, a holographic AR display with a real 3D object is demonstrated, and the viewing zone can also be enlarged in the same manner as the convex parabolic mirror.

Results

Optical property of conical HOE

Our proposed method is based on the reflection on a conical HOE. Thus, its optical property is very important. As shown in Fig. 3, our conical HOE converts the incident plane wave entering vertically downward into a spherical wave diverging outside. The maximum diverging angle ω is determined by an objective lens used in recording the HOE. In this study, an objective lens with a numerical aperture (NA) of 0.85 was used, and the maximum diverging angle was estimated to be approximately 58°. The wavelength used for the HOE recording was 532 nm. The apex angle was 90°, and the distance d between the apex and the divergence point O of a spherical wave, which was the origin of the conical HOE, was set to 20 mm.

First, the generation of spherical waves by the conical HOE was investigated. As shown in Fig. 4a, a picture of the conical HOE in generating a spherical wave was captured by a digital camera. In this figure, a plane wave enters the conical HOE vertically downward, although it is invisible, and a bright spot appears inside as a virtual image. Fig. 4b shows a cross-section profile of the bright spot. This profile was obtained by capturing a bright spot using a camera with a telecentric lens and extracting a lateral line passing through the bright spot. This bright spot was a divergence point of the spherical waves, and the focus of the camera was set to this point. Thus, the full width at half maximum (FWHM) of the bright spot can be used as an evaluation index of the quality of the generated spherical waves. Because the FWHM can depend on the elevation angle θ, which is the angle from the horizon to the optical axis of the digital camera, the relationship between the FWHM and θ was examined, as shown in Fig. 4c. The vicinity of the apex was distorted in forming a conical shape, although a small hole with a diameter of 5 mm was created to relieve this shape distortion. Thus, in θ = 70° and 80°, the FWHM is slightly larger than that at other elevation angles, and the maximum elevation angle is limited to approximately θ = 80°. If θ is less than 40°, a bright spot does not occur, that is, no spherical wave is generated, which is valid considering the maximum diverging angle ω.

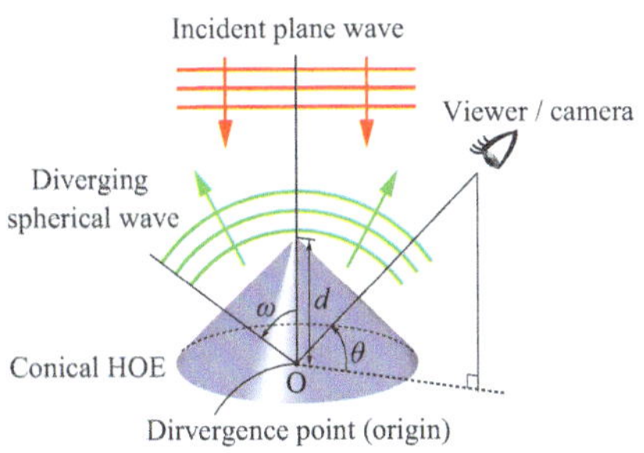

Fig. 3 Optical behaviour of a conical HOE. The incident wavefront and outgoing wavefront are depicted in red and green, respectively.

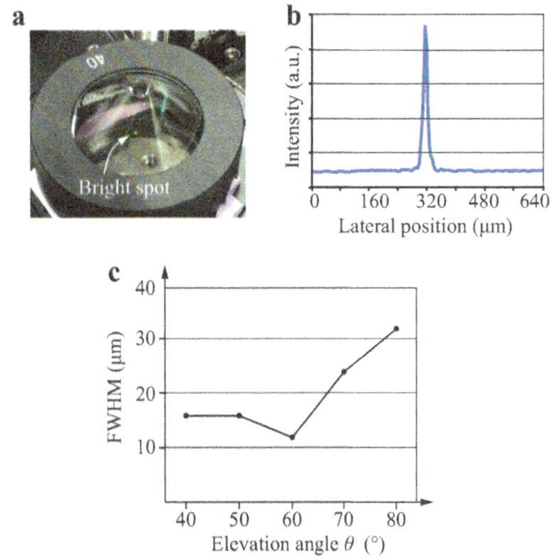

Fig. 4 Generation of diverging spherical waves. **a** Photograph of a conical HOE, **b** cross-section profile and **c** dependency of full width at half maximum (FWHM) on the elevation angle θ.

Next, the diffraction efficiency of the recorded HOE was measured. In this measurement, a light wave from a laser light source with a wavelength of 532 nm entered the conical HOE directly without passing through a collimator. Thus, the illuminated area on the conical HOE was as small as the diameter of the laser light source and it was approximately 1.2 mm. An optical power meter (3664, manufacture by HIOKI E.E. Corporation) was used to measure light intensity diffracted from the conical HOE. Fig. 5 shows the diffraction efficiency versus the elevation angle θ. The diffraction efficiency was defined as the ratio of the diffracted light intensity to the input light intensity. The diffraction efficiency had the strong dependency on the elevation angle θ, and its peak appeared at $\theta = 60°$. On the other hand, the diffraction efficiency in $\theta = 80°$ was significantly low. This needs to be improved by optimizing experimental conditions.

Then, the wavelength selectivity of the conical HOE was also examined. White light emitted from an LED light source was shaped into a quasi-plane wave when passing through a lens. This quasi-plane wave entered the conical HOE parallel to its rotationally symmetric axis. A spectrometer (QEPro, manufactured by Ocean Optics) picked up the light reflected on the conical HOE. For example, in the case of $\theta \cong 60°$, the spectral reflectivity is shown in Fig. 6. This spectral reflectivity was calculated by dividing the spectral intensity obtained using the spectrometer by the spectrum of the incident light and normalizing it based on a maximum value. Fig. 6 shows the sharp peak near the wavelength of 532 nm, which corresponds to the wavelength used in recording the conical HOE. The FWHM of this sharp peak is approximately 11 nm. The spectral bandwidth of 11 nm is considered to be proper in this study because a light wave with this bandwidth can be treated substantially as a monochromatic light. In addition, moderate bandwidth is necessary to relax incident light conditions in reconstruction. Thus, this wavelength selectivity is indispensable for realising our holographic AR display.

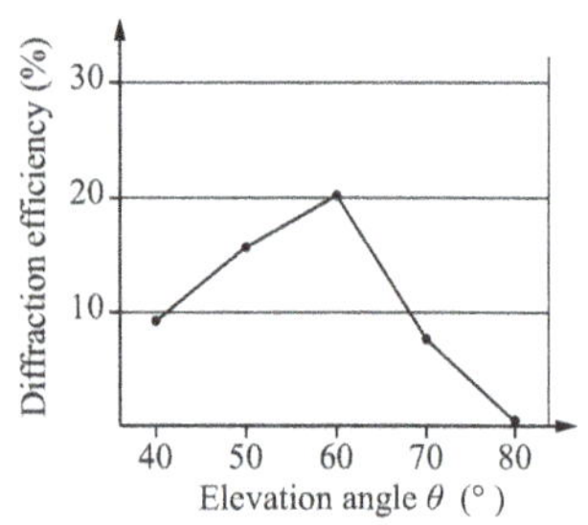

Fig. 5 Diffraction efficiency of a conical HOE.

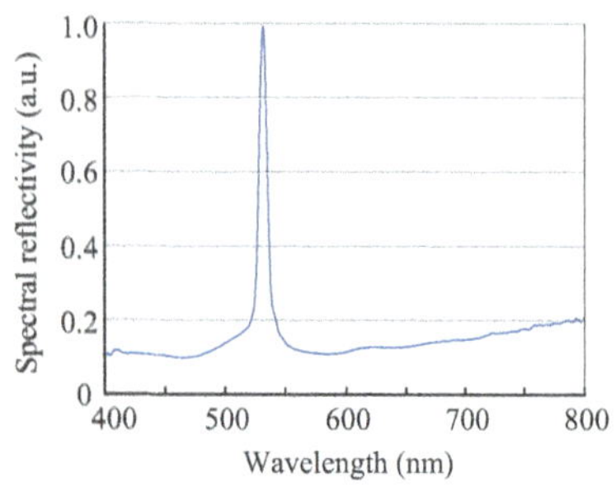

Fig. 6 Wavelength selectivity of a conical HOE.

Demonstration of our holographic AR display

A holographic AR display system with a very large viewing zone was constructed. A schematic of the optical system with a conical HOE is shown in Fig. 7. A fiber-output laser with a wavelength of 532 nm was used as the light source. After passing through lens L_1, the collimated wave enters the CGH as a plane wave. Then, the wavefront modulated by the CGH was optically Fourier-transformed by lens L_2 with a focal length f of 150 mm. On the Fourier plane of lens L_2, a spatial filter was inserted and half of the wavefront was filtered to remove unwanted waves such as the 0th-order and conjugated waves because the CGH in this study is a binary-amplitude type. The filtered wavefront entered the conical HOE vertically downward, was reflected in very wide directions, and then arrived at the viewers. If the CGH pattern is designed properly by considering the reflection on the conical HOE, then the viewers can recognise a reconstructed 3D object inside the conical HOE as a virtual image.

In this study, a wire-frame cube was used as a virtual 3D object, as shown in Fig. 8a. Four line segments were added at the four corners of the frontal face to distinguish it from

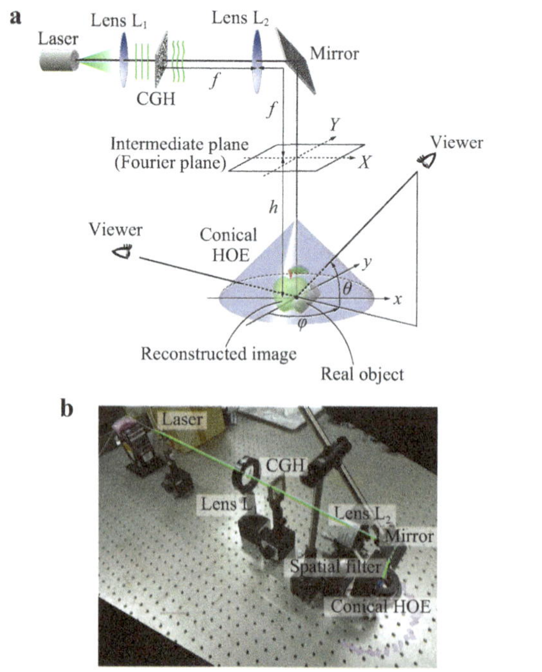

Fig. 7 **a** Schematic and **b** a picture of our holographic AR display system constructed.

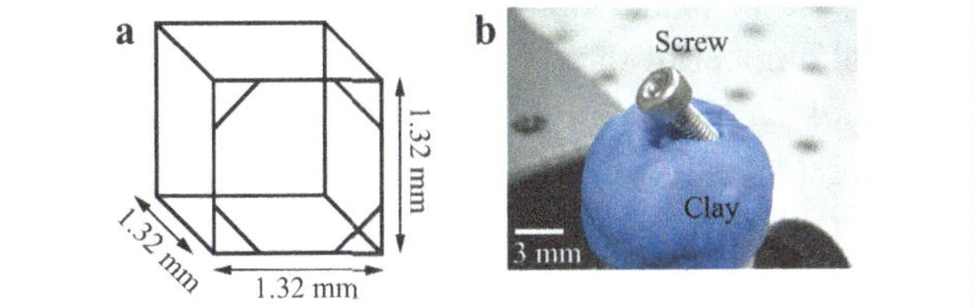

Fig. 8 A 3D object used for the demonstration. **a** Schematic of the virtual 3D object and **b** picture of the real object.

the back face. The size of the wire-frame cube was 1.32 mm × 1.32 mm × 1.32 mm. The number of pixels and pixel pitch of the CGH were 8,192 × 8,192 and 3.5 μm × 3.5 μm, respectively. The number of the point light sources constituting the 3D object was 420. The total calculation time was approximately 12.5 hours. The binary-amplitude pattern of the CGH was formed on a glass substrate by direct laser writing and wet etching. The distance h between the back Fourier plane of lens L_2 and the origin of the conical HOE was set to 25 mm. The real object, which was superimposed with the virtual 3D object in the optical reconstruction, was a metallic screw, as shown in Fig. 8b, and it was located by the virtual 3D object.

Under these conditions, the optical reconstruction of the virtual 3D object with the real object was demonstrated. Some images captured from various directions are shown in Fig. 9. In the experiment, the rotation angle of the conical HOE was adjusted per photograph to avoid using the unavailable areas of the conical HOE because some flaked parts between the photopolymer film and the conical plastic mould were found. Moreover, the shape of the mould, especially in the vicinity of the apex, was never geometrically correct. However, this treatment is not as significant to verify the validity of our proposed method because the conical HOE is rotationally symmetric. In Fig. 9, the horizontal and vertical positions of each image indicate the azimuthal φ and elevation θ angles, respectively, as is shown in Fig. 7a. Because the wire-frame cube is reconstructed as a virtual image, its image is properly reconstructed without disturbance by the real object; therefore, an AR reconstruction is realised. As shown in this figure, the reconstructed images change appropriately as the viewing direction change. Thus, the realisation of motion parallax and 3D reconstruction are successfully demonstrated. Concerning the viewing zone, our proposed method can present a very wide viewing zone, as well as our previous method using a convex parabolic mirror[18,19]. The azimuthal range of 150° and elevation range of 30° are experimentally confirmed. In Fig. 9, there is no image at the upper right or upper left.

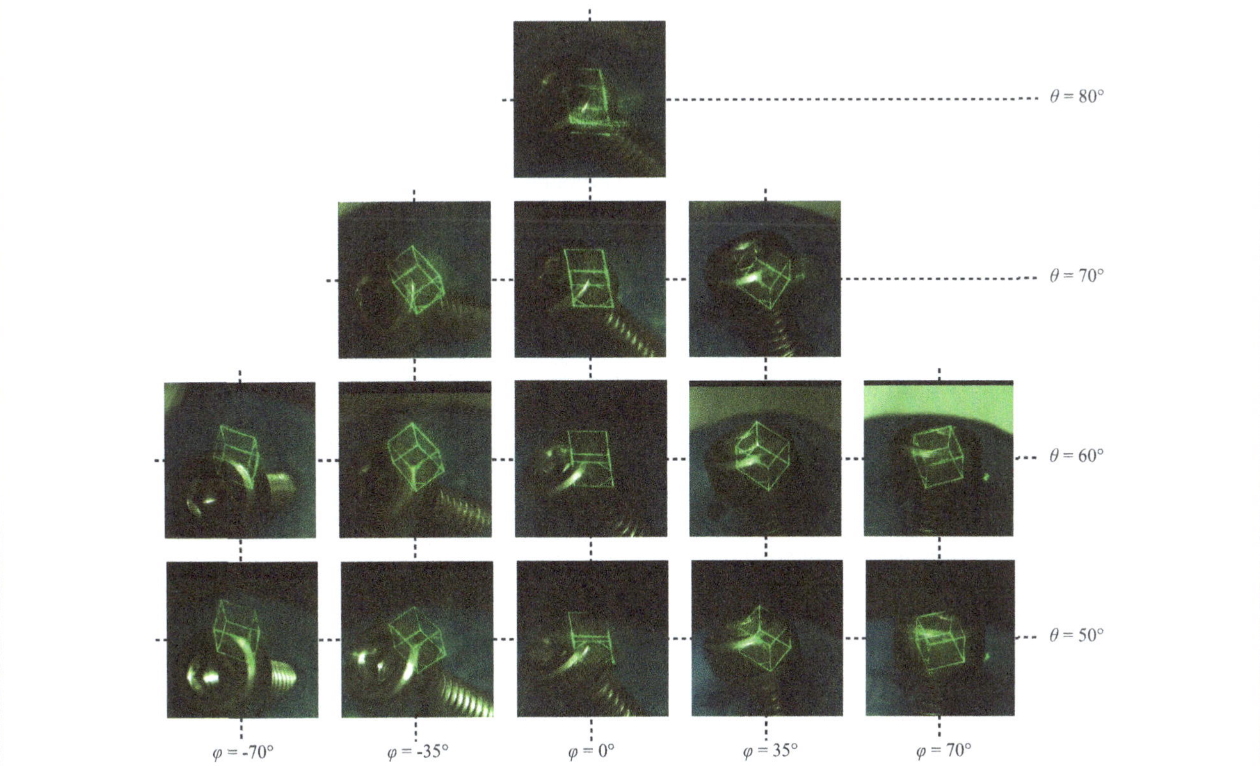

Fig. 9 Photographs of AR representation of holographically reconstructed images with a real object of a metallic screw. Each image was captured from the direction indicated by θ and φ.

Even in these directions, the 3D object could be reconstructed in the same way as the other viewing directions, although these reconstructed images could not be photographed because of the physical contact of the digital camera with a mirror, as shown in Fig. 7.

As can be seen from Fig. 9, the reconstructed 3D objects are distorted. The degree of the distortion varies with reconstruction positions and observation directions. This distortion stems from the misalignment of the optical system, especially the optical property of the conical HOE. The deviation from ideal conical shape results in the distortion of the reconstructed images. This distortion should be evaluated and modified for further development. Moreover, the reconstructible object space is important in holographic 3D displays. The reconstructible object space is not a cuboid or a cone, and it is very difficult to theoretically derive and determine that space because conventional diffraction theory is based on paraxial approximation and can not be applied to a conical HOE with a very wide spread-angle used in this study. Thus, by analysing a spatial-frequency bandwidth of the generated CGH, the reconstructible object size was numerically estimated to be roughly 2 mm. Further analysis will be required to evaluate our proposed method correctly.

Discussion

Notably, the main advantage of our proposed method using a conical HOE is the wide viewing zone. However, the maximum angle of the viewing zone was limited by some factors. One of these was the maximum diverging angle ω of the objective lens, as shown in Fig. 3. Because the conical HOE was recorded using a spherical wave as an object wave, the maximum angle of the vertical viewing zone was limited by ω. The other factor was the illumination area of the incident wavefronts to the conical HOE. To widen the viewing zone, the outer area of the conical HOE needs to be illuminated. Thus, the lateral size of the incident wavefront also determines the viewing zone because the size of it is smaller than that of the conical HOE in this study. In our optical system shown in Fig. 7a, to enlarge the lateral size, it is necessary to reduce the pixel pitch of the CGH or to use a lens with a large focal length. However, in the same way as the conventional holographic 3D display, there is a trade-off relation between the object size and the viewing zone. For example, if a lens with a longer focal length enlarges the viewing zone, the reconstructible object size reduces, and vice versa. Thus, increasing the number of pixels of CGHs is necessary to realise large object reconstructions and wide viewing zones simultaneously. On this point, our proposed method is advantageous because the shape of the hologram is planar. In the case of planar holograms, it is possible to increase the total number of the CGH or SLM pixels via the spatial tiling of multiple SLMs[24,25]. If these methods are incorporated into our holographic 3D display, both large object reconstructions and wide viewing zones can be realised.

In contrast, the azimuthal viewing zone φ is, in principle, 360° because the conical HOE is rotationally symmetric. However, it was less than 180° in our optical demonstration. This is because a spatial filter is inserted, as shown in Fig. 7b, and only the half area of the conical HOE was illuminated. Thus, if a complex-modulation encoding method is used instead of a simple binary-amplitude hologram[26–28], inserting a spatial filter is no longer necessary, and the azimuthal viewing zone becomes 360°.

Materials and methods

Fabrication of a conical HOE

A conical HOE is a holographic optical element. A transparent conical-shaped mould was formed by bending a fan-shaped acrylic plastic under an atmosphere temperature of 120 °C and sticking it. A photopolymer film (Bayhol HX200, manufactured by Covestro AG) was used as a photosensitive material, and it adhered closely to the top side of the conical-shaped mould. The thickness of the conical-shaped mould was 1 mm. This thickness was selected by considering both the effect on the wavefront passing through it and the deformation of the conical-shaped mould in recording. The optical properties of an HOE are determined by an optical system recording holographic interference fringes. Fig. 10 shows the schematic of the optical system to record a conical HOE in this study. A wavefront emitted from a laser light source was collimated by lens L_1, and then split into two plane waves by a beam splitter. One was used as the reference wave. The other became a diverging spherical wave by passing through an objective lens. These two waves interfered each other, and resultant interference fringes

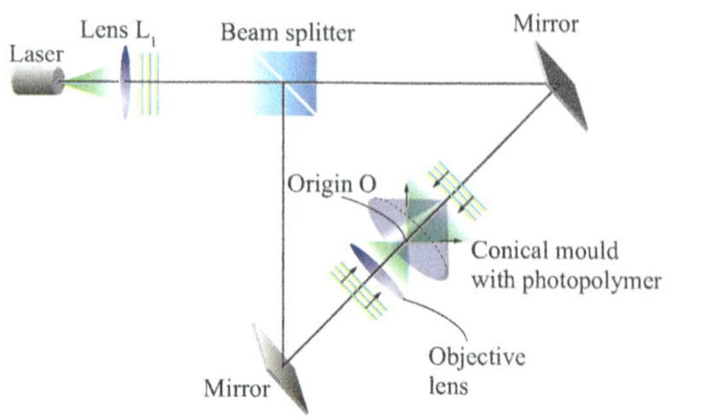

Fig. 10 Schematic of the optical system to record a conical HOE.

were recorded in the photopolymer film. The exposure time was set to 50 s, and the irradiance of the reference wave was 1.0 mW/cm^2. Because the propagation directions of these two waves were opposite, their interference fringes form 3D structure in the photopolymer film on the conical-shaped mould. The hologram recorded in this setup was called a volume-type hologram and had a sharp wavelength selectively. After bleaching the recorded photopolymer film with white light, the conical HOE was completed. In the reconstruction process, the conical HOE behaved as a reflective curved mirror whose focal length was d and NA was 0.85, as shown in Fig. 3. When the conical HOE was illuminated with an incident wave propagating parallel to its rotationally symmetric axis, the diverging spherical wave was reconstructed after the diffractive reflection on the conical HOE.

Calculation method of CGH

In our proposed method, a 3D object is reconstructed inside the conical HOE after reflection. Thus, the pattern of the CGH must be properly designed by considering the reflection on the conical HOE. Although various diffraction calculation methods including Fresnel diffraction have been proposed, they can be applied only to planar surfaces. In this study, the calculation method was originally developed based on geometrical optics by modifying the calculation method reported in our previous study[18]. Our calculation method presents the diffraction from a virtual 3D object to the intermediate plane located over the conical HOE, as shown in Fig. 7a. This intermediate plane is h = 25 mm away from the origin of the conical HOE and is placed such that it becomes the Fourier plane of lens L_2.

First, the virtual 3D object is assumed to be an aggregation of point light sources emitting spherical waves isotropically. The optical path from a single point light source S of the 3D object to a sampling point M on the intermediate plane through the reflection on the conical HOE is considered herein. A schematic of the configuration of the calculation method is shown in Fig. 11, where the origin O of the conical HOE is set at the divergence point of the spherical wave. The optical path is divided into two parts. The first is from S to the reflection point R on the conical surface. The second is from the reflection point R to M. Because the 3D object is reconstructed inside the conical HOE as a virtual image, the first part diffraction is in the forward direction, whereas the second part diffraction is in the backward direction. Thus, the optical path length from S to M is obtained as the difference between these two path lengths. Moreover, to obtain the optical path length, it is necessary to precisely

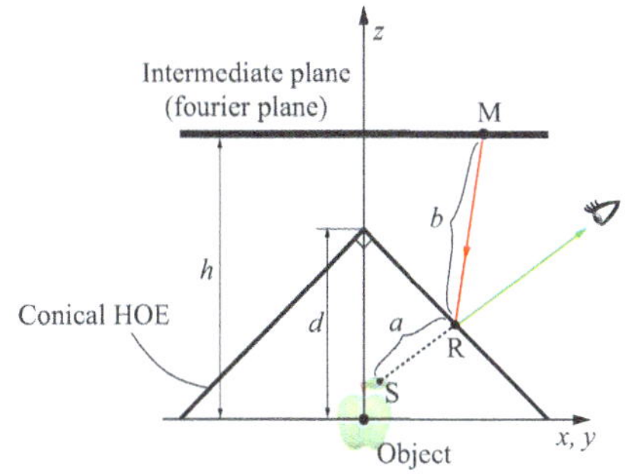

Fig. 11 Schematic of the conical HOE reflection for the CGH calculation.

determine the spatial position of the reflection point R on the conical HOE. When the spatial coordinates of the three points S, R, and M are introduced as (x_S, y_S, z_S), (x_R, y_R, z_R), and (x_M, y_M, z_M), respectively, the optical path length l from S to M is expressed by

$$l(x_R, y_R, z_R) = a(x_R, y_R, z_R) - b(x_R, y_R, z_R) + \phi(x_R, y_R, z_R) \quad (1)$$

where a and b represent the optical path lengths of the first and second part diffractions, respectively, and they are given by

$$a = \sqrt{(x_R - x_S)^2 + (y_R - y_S)^2 + (z_R - z_S)^2} \quad (2)$$

$$b = \sqrt{(x_M - x_R)^2 + (y_M - y_R)^2 + (z_M - z_R)^2} \quad (3)$$

ϕ corresponds to the phase retardation as a result of reflecting on the conical HOE. Because this phase retardation stems from the optical path difference between a reference wave and an object wave in recording the conical HOE, ϕ is formulated as

$$\phi(x_R, y_R, z_R) = (h - z_R) - \sqrt{x_R^2 + y_R^2 + z_R^2} \quad (4)$$

The first and second terms represent the optical path lengths of the reference and object waves, respectively. According to Fermat's principle, the light ray traces the shortest optical path[29], which means that the spatial position (x_R, y_R, z_R) of point R is obtained as the solution which minimises the optical path length l. Note that the reflection point R is constrained on the conical surface as follows:

$$z_R = d - \sqrt{x_R^2 + y_R^2} \quad (5)$$

where d is the distance from the origin O to the apex of the conical HOE, as shown in Fig. 11. The spatial position (x_R, y_R, z_R) is obtained by solving the minimisation problem with this constraint. In this study, this minimisation problem was solved using the Lagrange multiplier method[30]. The Lagrange function $\mathcal{L}(x_R, y_R, z_R, \lambda)$ is defined by

$$\mathcal{L}(x_R, y_R, z_R, \lambda) = l(x_R, y_R, z_R) - \lambda g(x_R, y_R, z_R) \quad (6)$$

where λ is a Lagrange multiplier and $g(x_R, y_R, z_R)$ is

$$g(x_R, y_R, z_R) = z_R - d + \sqrt{x_R^2 + y_R^2} \tag{7}$$

Because the solution of (x_R, y_R, z_R) minimises $l(x_R, y_R, z_R)$, the following equations should be satisfied:

$$\frac{\partial \mathcal{L}}{\partial x_R} = \frac{x_R - x_S}{a} + \frac{x_M - x_R}{b} - \frac{x_R}{s} - \frac{\lambda x_R}{r} = 0 \tag{8}$$

$$\frac{\partial \mathcal{L}}{\partial y_R} = \frac{y_R - y_S}{a} + \frac{y_M - y_R}{b} - \frac{y_R}{s} - \frac{\lambda y_R}{r} = 0 \tag{9}$$

$$\frac{\partial \mathcal{L}}{\partial z_R} = \frac{z_R - z_S}{a} + \frac{z_M - z_R}{b} - \frac{z_R}{s} - 1 - \lambda = 0 \tag{10}$$

$$\frac{\partial \mathcal{L}}{\partial \lambda} = -g(x_R, y_R, z_R) = -z_R + d - r = 0 \tag{11}$$

where

$$s = \sqrt{x_R^2 + y_R^2 + z_R^2} \tag{12}$$

$$r = \sqrt{x_R^2 + y_R^2} \tag{13}$$

As can be seen from Eqs. 10, 11, both z_R and λ can be rewritten as explicit functions of the variables x_R and y_R. Therefore, by substituting the explicit functions z_R and λ in Eqs. 8, 9, the above minimisation problem with the constraint is reduced to a nonlinear simultaneous equation with respect to two variables x_R and y_R. However, this simultaneous equation cannot be solved analytically. In this study, Newton's method with two variables was used to numerically solve the simultaneous equation[31].

When the position of the reflection point R is determined, the optical path length from S to M via R can easily be calculated using Eq. 1, and the complex amplitude at point M contributed by point S is obtained. By repeating this procedure for all the point light sources constituting the 3D object and for all the sampling points on the intermediate plane, it becomes possible to calculate the complex wavefront on the intermediate plane diffracted from the 3D object to be reconstructed. Therefore, by the inverse Fourier transformation of this complex wavefront, the wavefront on the CGH plane is obtained because the intermediate plane is located at the Fourier plane of lens L_2, as shown in Fig. 7a. Note that the lateral half of the complex wavefront on the intermediate plane must be set to zero before the inverse Fourier transformation. Finally, the real part of the resultant wavefront on the CGH is extracted and binarized because the CGH fabricated in this study is of the binary-amplitude type.

Acknowledgements
This research was partially supported by JSPS KAKENHI Grant Number JP20K11914.

Author details
[1]Osaka Research Institute of Industrial Science and Technology, 2-7-1 Ayumino, Izumi, Osaka 594-1157, Japan. [2]Center for Optical Research & Education, Utsunomiya University, 7-1-2 Yoto, Utsunomiya, Tochigi 321-8585, Japan. [3]Graduate School of Engineering, Utsunomiya University, 7-1-2 Yoto, Utsunomiya, Tochigi 321-8585, Japan

Author contributions
Y. Sando: Optical experiment, program development, and writing of the original draft.K. Satoh: Fabrication of computer-generated holograms by photolithography and review of the draft.D. Barada: Fast calculation algorithm for diffraction and review of the draft.T. Yatagai: Diffraction calculation via a nonplanar HOE and review of the draft.

Conflict of interest
The authors declare that they have no conflict of interest.

Received: 16 August 2021 Revised: 10 February 2022 Accepted: 15 February 2022
Accepted article preview online: 22 February 2022
Published online: 02 March 2022

References

1. Li, X. et al. A critical review of virtual and augmented reality (VR/AR) applications in construction safety. *Automation in Construction* **86**, 150-162 (2018).
2. Hong, J. et al. Three-dimensional display technologies of recent interest: principles, status, and issues. *Applied Optics* **50**, H87-H115 (2011).
3. Geng, J. Three-dimensional display technologies. *Advances in Optics and Photonics* **5**, 456-535 (2013).
4. Zhang, et al. Three-dimensional display technologies in wave and ray optics: a review. *Chinese Optics Letters* **12**, 060002 (2014).
5. Pan, Y. et al. A Review of Dynamic Holographic Three-Dimensional Display: Algorithms, Devices, and Systems. *IEEE Transactions on Industrial Informatics* **12**, 1599-1610 (2016).
6. Nozaki, A. et al. Dynamic visual responses of accommodation and vergence to electro-holographic images. *Optics Express* **25**, 4542-4551 (2017).
7. Matsushima, K. & Nakahara, S. Extremely high-definition full-parallax computer-generated hologram created by the polygon-based method. *Applied Optics* **48**, H54-H63 (2009).
8. Igarashi, S. et al. Efficient tiled calculation of over-10-gigapixel holograms using ray-wavefront conversion. *Optics Express* **26**, 10773-10786 (2018).
9. Blinder, D. & Shimobaba, T. Efficient algorithms for the accurate propagation of extreme-resolution holograms. *Optics Express* **27**, 29905-29915 (2019).
10. Takaki, Y. & Nakaoka, M. Scalable screen-size enlargement by multi-channel viewing-zone scanning holography. *Optics Express* **24**, 18772-18781 (2016).
11. Jeong, T. H. Cylindrical Holography and Some Proposed Applications. *Journal of the Optical Society of America* **57**, 1396-1398 (1967).
12. Soares, O. D. D. & Fernandes, J. C. A. Cylindrical hologram of 360° field of view. *Applied Optics* **21**, 3194-3196 (1982).
13. Sakamoto, Y. & Tobise, M. Computer-generated cylindrical hologram. Proceedings of the SPIE 5742 Practical Holography XIX: Materials and Applications. San Jose: SPIE, 2005, 260-273.
14. Yamaguchi, T., Fujii, T. & Yoshikawa, H. Fast calculation method for computer-generated cylindrical holograms. *Applied Optics* **47**, D63-D70 (2008).

15. Sando, Y. et al. Spherical-harmonic-transform-based fast calculation algorithm for spherical computer-generated hologram considering occlusion culling. *Applied Optics* **57**, 6781-6787 (2018).
16. Li, G. et al. Acceleration method for computer generated spherical hologram calculation of real objects using graphics processing unit. *Chinese Optics Letters* **12**, 060016 (2014).
17. Li, G. et al. Synthesis of computer-generated spherical hologram of real object with 360° field of view using a depth camera. *Applied Optics* **52**, 3567-3575 (2013).
18. Sando, Y. et al. Super-wide viewing-zone holographic 3D display using a convex parabolic mirror. *Scientific Reports* **8**, 11333 (2018).
19. Sando, Y. et al. Fast calculation method for parabolic-mirror-reflection holographic 3D display using wavefront segmentation. *Applied Optics* **59**, 8211-8216 (2020).
20. Mishina, T., Okui, M. & Okano, F. Viewing-zone enlargement method for sampled hologram that uses high-order diffraction. *Applied Optics* **41**, 1489-1499 (2002).
21. Sasaki, H. et al. Image size scalable full-parallax coloured three-dimensional video by electronic holography. *Scientific Reports* **4**, 4000 (2014).
22. Duan, X. et al. Full-color see-through near-eye holographic display with 80° field of view and an expanded eye-box. *Optics Express* **28**, 31316-31329 (2020).
23. Su, Y. F. et al. Binocular dynamic holographic three-dimensional display for optical see-through augmented reality using two spatial light modulators. *Optik* **217**, 164918 (2020).
24. Stanley, M. et al. 100 mega-pixel computer generated holographic images from active tiling–a dynamic and scalable electro-optic modulator system. Proceedings of the SPIE 5005 Practical Holography XVII and Holographic Materials IX. Santa Clara: SPIE, 2003, 247-258.
25. Lim, Y. et al. 360-degree tabletop electronic holographic display. *Optics Express* **24**, 24999-25009 (2016).
26. Ulusoy, E., Onural, L. & Ozaktas, H. M. Full-complex amplitude modulation with binary spatial light modulators. *Journal of the Optical Society of America A* **28**, 2310-2321 (2011).
27. Zhu, L. & Wang, J. Arbitrary manipulation of spatial amplitude and phase using phase-only spatial light modulators. *Scientific Reports* **4**, 7441 (2014).
28. Park, S. et al. Characteristics of complex light modulation through an amplitude-phase double-layer spatial light modulator. *Optics Express* **25**, 3469-3480 (2017).
29. Hecht, E. Optics. (Boston: Addison Wesley, 2001).
30. Bertsekas, D. P. Constrained Optimization And Lagrange Multiplier Methods. (New York: Academic Press, 1982).
31. Butt, R. Introduction to Numerical Analysis Using MATLAB. (Hingham: Infinity Science Press 2007).

Review Open Access

The state-of-the-art in computer generated holography for 3D display

David Blinder[1,2,*], Tobias Birnbaum[1,2], Tomoyoshi Ito[3] and Tomoyoshi Shimobaba[3,*]

Abstract

Holographic displays have the promise to be the ultimate 3D display technology, able to account for all visual cues. Recent advances in photonics and electronics gave rise to high-resolution holographic display prototypes, indicating that they may become widely available in the near future. One major challenge in driving those display systems is computational: computer generated holography (CGH) consists of numerically simulating diffraction, which is very computationally intensive. Our goal in this paper is to give a broad overview of the state-of-the-art in CGH. We make a classification of modern CGH algorithms, we describe different algorithmic CGH acceleration techniques, discuss the latest dedicated hardware solutions and indicate how to evaluate the perceptual quality of CGH. We summarize our findings, discuss remaining challenges and make projections on the future of CGH.

Keywords: Digital holography, Holographic display, Computer generated holography, High-performance computing, FPGA, Deep learning, Visual quality assessment

Introduction

Holography is a methodology based on coherent light that can fully describe the optical field. Although, the concept was invented in the late 40's, optical holography did only advance with the development of the laser in the 60's. Holography is a two step method, consisting of an interferometric recording and a reconstruction step. Its ability to capture, measure and reproduce any given wavefield, has made it useful in a broad range of applications. Examples are extended depth-of-field microscopy via digital holographic microscopy[1]; state-of-the-art surface characterization of complex objects (e.g. large mirrors, aspheric optics, biological specimens) via holographic interferometry[2], and the visualization of 3D content[3].

Correspondence: David Blinder (david.blinder@vub.be) or Tomoyoshi Shimobaba (shimobaba@faculty.chiba-u.jp)
[1]Department of Electronics and Informatics (ETRO), Vrije Universiteit Brussel (VUB), Pleinlaan 2, B-1050 Brussel, Belgium
[2]IMEC, Kapeldreef 75, B-3001 Leuven, Belgium
Full list of author information is available at the end of the article.
These authors contributed equally: David Blinder, Tomoyoshi Shimobaba

Holographic visualization

Most relevant for the latter case, holography has the remarkable property of being able to reproduce all known natural visual depth cues[4], such as:

occlusion provides depth information as front objects occlude more distant objects.

eye accommodation derives depth from the ring muscle tension around the lens of the human eye.

(con-)vergence derives depth from the gaze angle of the eyeballs.

stereopsis derives depth from the differences in between the two monocular images of our eyes. The differences are due to the inter-pupillary distance.

All depth cues can be observed in holographic displays with the naked eye and can therefore, in theory, reproduce a faithful 3D representation of any given scene without restrictions on the scene contents. This sets holography apart from other plenoptic imaging methods. Namely, stereoscopic or lightfield imaging fundamentally suffer from the accommodation-vergence conflict, caused by a mismatch between the fixed focal distance of the optical/virtual image (accommodation cues) produced by the display, and the perceived rendered scene-dependent 3D object distances (vergence cues). Volumetric imaging does not have this conflict, but on the other hand can only offer constrained finite render volumes and requires workarounds for occlusions in the rendered volume.

With the advent of the digital revolution, digital holograms (DH)s became possible. DHs consist of matrices of typically scalar complex numbers, which are mapped on some surface upon creation and reconstruction. They are created either through capture in optical setups, for example by charged-coupled-devices based camera sensors, or through numerical calculations on a computer, i.e computer generated holography (CGH). Subsequently, limited viewing or partial rendering is possible on conventional 2D or lightfield displays through numerical reconstruction of specific view-points. For a 3D scene reproduction, the modulation of light dictated by a DH needs to be discretized and printed[5,6] or programmed into a refreshable spatial light modulator devices (SLM). An important requirement for the consumption of 3D scenes, is a sufficiently high space-bandwidth product (SBP) of the modulation medium. The SBP is defined as the product of maximal viewing angle and maximal scene size. Thus with a small SBP either the viewing angle or scene size remains limited. For a modulation pattern of $N \times N$ features with inter-feature distance p, the SBP is equal to N^2, as the spatial extent of the observed scene is limited by the size of the modulated area Np, whilst the maximal spatial frequency given as p^{-1}.

Holographic display technologies

The most relevant technology is refreshable SLM-based holographic displays, as they allow for holographic user interfaces and convenient consumption of holographic content. Since holography leverages diffraction, p is required to be of the order of the wavelength λ of the illuminating laser (i.e typically ranging from $\sim 0.5\lambda$ to $\sim 10\lambda$) — while ensuring a sufficiently large SBP. This poses significant technical challenges, for the signal processing backbone, the display support electronics as well as for the SLM itself. Therefore, three display usage scenarios[3] have emerged.

Head-mounted displays ease the SBP requirements by bringing the SLM close to the eye(s) of a single observer. In that case, only a small area is required per eye – covering the entire eye-box. On top of that, the SLMs in this configuration only need to support a limited field-of-view due to the limited motion range of the eye. Examples are detailed in[4,7,8], particularly Microsoft's HoloLens 1.

Eye-tracking supported displays combine an SLM of limited SBP with beam steering facilitated by eye-tracking and active adaptive optics in order to emulate a large SLM for a single to few observer(s). A famous exemplary implementation is by SeeReal[9].

Full-parallax displays do not reduce the SBP and are providing the maximal supported field-of-view to a theoretically unlimited number of observers. As the required SBP for this scenario is in the order of giga to tera-pixels, they became approachable through first prototypes of metasurface DHs[10] only recently.

Aside from these principal scenarios various variations and combinations exist. The geometry can be planar, curved, or cylindrical; this can be realized both by the display device[11] or by the coherent illumination source wavefront shape[12]. A full-parallax displays may take the shape of a flat "window", not unlike a television[13], or of a 360° table-top display[14]. Important to mention is that many unrelated display technologies are incorrectly referred to as "holographic", such as the "Pepper's ghost illusion" or "holographic pyramid". These consist of conventional projection systems and do not show 3D images nor use any modulation with coherent light.

Optically recording digital holograms

Digital holography is a non-contact measurement technique enabling highly accurate 3D object measurements. It can capture the complete complex-valued optical field, allowing for e.g. post-capture numerical processing such as refocusing or aberration correction. This has many applications besides recording 3D objects for display, such as 3D measurements for microscopy[15], 3D particle and flow measurements[16], biological specimens[17], and 3D object recognition[18].

The basic setup diagram is shown in Fig. 1, where an expanded laser beam goes through a beamsplitter and gets divided into a reference beam and an object beam. The latter will illuminate the object of interest to be recorded. Both beams are then optically recombined and form an interference pattern that can be registered nowadays with a CMOS or CCD camera sensor. This encoded information can be numerically processed to obtain the complex-valued hologram.

These setups come in many forms[2], with various capabilities and trade-offs. Among many others, there are setups using holographic tomography[19], combining recorded holograms with CGH[20], single pixel imaging[21], Fourier horizontal parallax for high resolutions[22], and single-shot multiplexing[23].

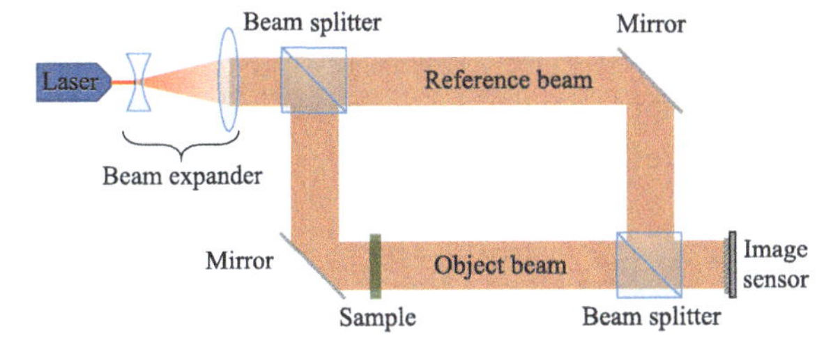

Fig. 1 Simplified diagram of a holographic on-axis recording setup in transmission mode[3].

Unfortunately, recording 3D objects that way for subsequent display on a holographic displays is not very flexible: (1) limitations on scene types prevent the capture of large objects, especially outdoors or dynamic scenes with moving objects, which would have to be illuminated with a stable coherent laser source without intrusion of other light; (2) no virtual content is supported; (3) the resolution capabilities of CCD/CMOS chips are modest, requiring complex multiplexing solutions to reach the requisite sample counts for detailed 3D scenes; and (4) the holographic display setup parameters vary a lot between systems (pixel pitch, resolution, wavelength, optics, etc.). CGH can address all these shortcomings, which will be the main focus of this paper.

Challenges of CGH

Irrespective of the used display, vast numbers of suitable DHs can only be created through CGH which models the propagation of light emitted by a given scene onto a virtual hologram (recording) plane. In this paper, we review the challenges that CGH faces and the state-of-the-art approaches that have been proposed to counter various of those challenges. The principal trade-off is between rendering the most realistic scene and achieving the objective in an acceptable time. Realistic rendering comprises textures, structural surface properties, non-homogeneous lighting, and occlusions required to improve realism and the immersion of the observer. Various CGH techniques are limited in the effects they can faithfully model; in general, graphical effects have to be re-designed for CGH due to substantially different intrinsic representation of numerical diffraction compared to their conventional computer graphics (CG) counterparts. The most realistic CGH is currently achieved by leveraging ray-tracing, but effects such as refractive optics or true mirrors, may require new conceptual approaches altogether.

Highly realistic CG for photographic imagery is already computationally challenging, but CGH is unfortunately still several orders of magnitude more complex to compute. This is because of three main reasons. Firstly, the diffraction integral, which describes said propagation of light, is a highly non-local many-to-many mapping leading to lots of random memory accesses and updates. Secondly, state-of-the-art DHs of acceptable quality require large resolutions of at least 10^8 pixels. For example, a single monochrome high quality hologram of size 10 × 10 cm and a 60° field of view would require the modulation, computation, and signaling of 326000 × 326000 pixels. Thirdly, CGH operates most of the time internally with a high-bitdepth complex-valued representation for accurately modeling light propagation.

In face of these resolutions even current CPU-based hardware is computationally-, memory- and/or, bandwidth-bound. Therefore, specialized, alternative hardware architectures such as graphic processor units (GPU)s and field programmable gate arrays (FPGA)s are increasingly studied for CGH.

Outline

In this paper, we review the state-of-the-art in CGH without limiting ourselves to any particular class of CGH algorithms. That said, there is no unique way of classifying the algorithms in existence. We propose to discern between the basic element representation (e.g. wavelets, points, polygons), algorithmic acceleration (e.g. look-up tables, sparsity based approaches, deep learning based accelerations) and hardware accelerations (e.g. low precision calculations, GPU or FPGA based architectures). Using this, we arrive at a simple classification of CGH methods based on representations in the following section. Thereafter, we study concepts of algorithmic accelerations, which may be found or combined with several of the aforementioned CGH methods. In the next section, we discuss the joint optimization of hardware and CGH approaches before discussing the closely related challenges of preparing content for SLM-based holographic displays. This includes the compensation of SLM specific limitations, complex-amplitude encoding (on typically only real-valued modulation SLMs), image quality enhancement, and visual quality assessment. The paper will end with a conclusion of our findings.

Classification of CGH algorithms

The behavior of light in free space (or any linear, isotropic, homogeneous, non-dispersive medium) is governed by the vectorial Helmholtz equation. Moreover in holography, we are primarily working in a regime, where the diffracting aperture and propagation distances are large compared to the wavelength. In that case *scalar diffraction theory* is sufficiently accurate. For a monochromatic light source (e.g. a laser), the electromagnetic wavefield can be characterized by a scalar field, assigning a single complex-valued number to every point in space. The amplitude corresponds to the magnitude of the electric field, while the phase described the relative phase delay w.r.t. the reference wave. This assumption can be made because the light source is highly monochromatic, so the relative phase delay can be considered constant during the recording of a static hologram or a single holographic video frame.

Scalar diffraction can be modelled accurately by the Rayleigh-Sommerfeld model[24], where the object wavefield in the hologram plane H can be calculated for every $\mathbf{x} \in H$:

$$H(\mathbf{x}) = \frac{1}{i\lambda}\iint_S U(\mathbf{p})\frac{e^{ik\|\mathbf{p}-\mathbf{x}\|}}{\|\mathbf{p}-\mathbf{x}\|^2}\mathbf{n}\cdot(\mathbf{p}-\mathbf{x})d\mathbf{p} \qquad (1)$$

where $\mathbf{p} \in S$ are emitting points on a surface S with amplitude function U over which is integrated. $\mathbf{n}$ is the surface normal of S in $\mathbf{p}$, k is the wave number and i is the imaginary unit. The object wavefield is up to a global phase constant identical to the full hologram, if the reference wave is a plane wave with normal incidence angle. For simplicity and convenience, we consider in this paper the frequently used assumption of the reference wave to be a normally incident plane wave identical to 1.

To account for occlusions, one can utilize the Huygens–Fresnel principle, where secondary point source wavelets get stimulated at occlusion boundaries. Unfortunately, this generic formulation is not useful in practice, as the many-to-many mapping between all emitters $\mathbf{x}$ and hologram points $\mathbf{p}$ grows exponentially. Compounded with the large resolutions and video frame rates needed for digital holograms for display systems, this direct approach is too computationally demanding or even intractable for all but the simplest holograms.

To achieve efficient CGH suitable for (near) real-time computation, one must introduce various approximations. Multiple solutions to this problem were proposed and refined over the past decades, trading off visual fidelity, supported visual effects and calculation time. In the remainder of this section, we classify these different approaches based on their primary decomposition into simpler constituent elements.

Point-cloud methods

One of the most widely used CGH techniques is the *point cloud method* [25]. This is achieved by discretizing Eq. 1 by means of sampling the surfaces S and amplitudes U. Given points $\{\mathbf{p}_1, ..., \mathbf{p}_N\} \in S$ with corresponding amplitudes $\{a_1, ..., a_N\} \in \mathbb{C}$, we get

$$H(\mathbf{x}) = H(x_h, y_h, z_h) = \sum_{j=1}^{N}\frac{a_j \exp(ik\|\mathbf{p_j}-\mathbf{x}\|)}{\|\mathbf{p_j}-\mathbf{x}\|} \qquad (2)$$

where we assume without loss of generality that the hologram plane coincides with the xy plane, so that $\mathbf{x} = (x_h, y_h, z_h = 0)$. These spherical wavefronts can often be accurately approximated by parabolic wavefronts. This is known as Fresnel approximation

$$H(x_h, y_h) = \sum_{j=1}^{N} a_j \exp\left(\frac{\pi i}{\lambda z_j}[(x_h - x_j)^2 + (y_h - y_j)^2]\right) \qquad (3)$$

where $\mathbf{p}_j = (x_j, y_j, z_j)$, ignoring the attenuation factor $\exp(2\pi i/(\lambda z_j))/i\lambda z_j$ in most point-cloud methods, as this will generally not noticeably affect the perception of the hologram; it can optionally be included for improved accuracy. The wavelength is denoted λ, so that $k = 2\pi/\lambda$. The Fresnel approximation has the important property of being separable, facilitating its use to analytically solve more complex diffraction problems, besides being considerably easier to compute. This approximation can be obtained using a truncated Taylor expansion, valid whenever $z \gg \lambda$ and when considering relatively small diffraction angles θ where $\theta \approx \sin\theta$ holds. This will be satisfied in practice whenever the pixel pitch p is at least a few times larger than the wavelength (roughly when $p \geq 4\lambda$), or when the subtended angle of objects on the hologram is sufficiently small, i.e under paraxial viewing conditions. Otherwise, the paraxial approximation introduces noticeable distortions.

In principle, any 3D scene can be approximated to arbitrary accuracy using enough points. Thanks to their generality and simplicity, point-cloud methods are one of the most widely used methods. An example diagram is shown on Fig. 2. Nevertheless, naively executing Eq. 2 is still very costly, as every sampled point $\mathbf{p_j}$ will result in updates to each hologram pixel. This process can be sped up by using different *acceleration techniques*[26-28] as described in a latter section. However, infinitesimal points are not ideal for describing surfaces to model occlusions

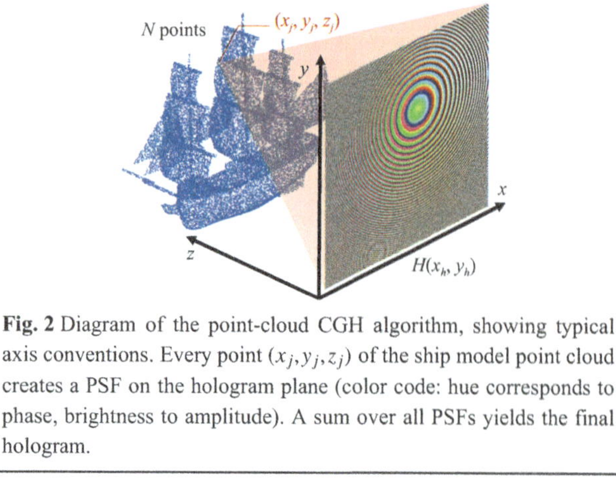

Fig. 2 Diagram of the point-cloud CGH algorithm, showing typical axis conventions. Every point (x_j, y_j, z_j) of the ship model point cloud creates a PSF on the hologram plane (color code: hue corresponds to phase, brightness to amplitude). A sum over all PSFs yields the final hologram.

and light reflection properties. Moreover, individually calculating points can be sub-optimal in cases where objects can be well-described by a few more complex primitives.

Geometric primitives & basis functions

Holograms can be generated with elements more complex than point spread functions (PSFs). In many cases, 2D and 3D objects can be decomposed into a relatively small number of basis elements which can be efficiently computed. This is achieved by formulating solutions for some restricted set of shapes S and complex amplitude distributions U in Eq. 1. We consider two approaches: *analytical methods* and *precomputed basis function* methods.

Analytical methods attempt to solve Eq. 1 analytically (sometimes with approximations) for certain shapes or surfaces, such as lines[29,30], curves[31], and polygons[32,33]. The latter have been augmented with partial support for textures as well[34,35]. This allows for directly calculating the corresponding hologram wavefield in each pixel, which would otherwise require summing thousands to millions of points for achieving the same effect.

One drawback is that analytical models will necessarily limit the system's degrees of freedom, precluding the use of highly complex textures or light emission patterns. We posit that these methods are best used as scene building blocks in conjunction with other methods rather than in isolation.

Another similar technique precomputes diffraction patterns, e.g. as emitted textured surface patches with different light distributions, and stores them in look-up tables (LUT)s. This is often combined with layer-based methods or wavefront recording planes (see next sections) to reduce the LUT memory footprints. In order to increase realism and supported effects, instances of elements can be precomputed at many orientations, axial positions[36], light distributions[37], and in multiple phase noise instantiations thereby reducing unwanted visual artifacts[38]. They can even be used for precomputing occlusion patterns, although these are more limited and less accurate due to being a non-linear and non-stationary filter effect. We will discuss more details in a subsequent section. For the right scene compositions and hologram display properties, this can achieve highly realistic effects. One main limitation is the inherent discretization of the phase space and the exponential growth of look-up table sizes with the number of variables under consideration.

Layer-based methods

When S is a plane, we can rewrite Eq. 1 as a convolution because of the translational symmetry along the xy-plane. This is desirable, since convolutions can be computed efficiently using the Fast Fourier Transform (FFT). For a wavefield parallel to the xy-plane displaced at a distance $d \neq 0$ with the complex-valued amplitude function U, we have that

$$H(x,y,0) = U(x,y,d) * \frac{\exp\left(ik\sqrt{x^2+y^2+d^2}\right)}{\sqrt{x^2+y^2+d^2}} \tag{4}$$

where * is the convolution operator. The transfer function can be computed directly in Fourier space; this method is called the *angular spectrum method* (ASM)[24]: $H(x,y,0) =$

$$\mathcal{F}^{-1}\left\{\mathcal{F}\{U(x,y,d)\}\exp\left(i2\pi d\sqrt{\lambda^{-2}-\omega^2-\eta^2}\right)\right\} \tag{5}$$

where $\mathcal{F}\{\cdot\}$ and $\mathcal{F}^{-1}\{\cdot\}$ respectively denote the forward and reverse Fourier operators, transforming between spatial (x,y) and frequency (ω,η) coordinates.

Multiple variants and extensions to convolutional propagation exist besides the ASM. The Fresnel approximation can also be used here to obtain a separable quadratic phase kernel[41]. Shifted diffraction algorithms[42] allow for off-axis diffraction of plane segments, variants allowing for different pixel pitches at source and destination planes, or even tilted diffraction algorithms[43] where source and destination planes do not need to be parallel.

Layer-based algorithms subdivide the scene into multiple layers, where each scene element is allocated to a specific plane, typically the closest one (cf. Fig. 3). Then, the plane wavefield can be numerically diffracted efficiently to the hologram plane by means of

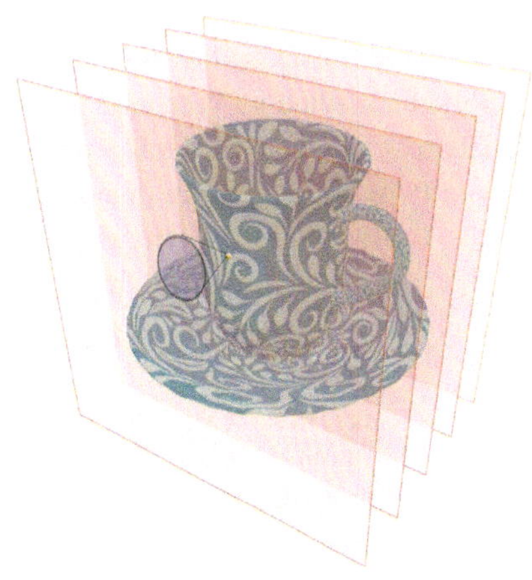

Fig. 3 Diagram of layer-based CGH, showing how a 3D object is segmented into layers, where every scene element is assigned to its closest intermediate wavefront. Here, an exemplary point with its PSF influence region on the matching layer is shown. (3D model and texture sources[39,40])

convolutional propagation. Because of the elements' proximity, only a fraction of the emitted light angles will reach the hologram plane, thereby requiring to only update a small subset of all hologram pixels, benefiting calculation speed (this is closely tied to the usage of wavefront recording planes, see later sections). Moreover, since one has access to the wavefield volume at multiple depth planes, one can easily simulate occlusion and transparency operators by suppressing the amplitude at the sections where the planes intersect with the virtual objects.

Examples of these approaches are RGB+depth methods[44], utilizing a quantized depth map as a guide to assign points to their respective layers or multi-layer propagation methods. The latter precomputes basis functions and occluders[36] thereby optimizing both the number of layers and their mutual distances and reusing the precomputed propagation kernels. Another related way of achieving variable depth is to non-uniformly resample a layer for a more continuous change in depth[45].

Layer based methods are efficient CGH techniques, often requiring local updates and providing more straightforward ways of applying occlusion. In their base form, they are less suitable for extreme-resolution holograms due to their linearithmic complexity, as well as having difficulties to efficiently model effects at steep viewing angles.

General Polygon-based methods

Polygon-based CGH methods are inspired by conventional computer graphics, where objects are represented by 3D meshes of triangles with associated textures maps and material properties. Because polygons are by definition confined to a plane, one can use convolutional diffraction methods to propagate the polygonal wavefield towards the hologram plane. We designate a polygon method as "general" to differentiate it from the analytical polygon methods[32] described in the geometric primitives section. Two major challenges in polygon-based CGH are the efficient and accurate calculation of (1) the source wavefield in the polygon plane and (2) the polygon interactions, e.g. due to mutual occlusions.

In[46], polygons are first expressed in their local coordinates. Their light emission depends on the combination of surface material and the relative position of virtual light sources. It will result in diffuse and specular light components, which have a particular spectral envelope (cf. Fig. 4). This can be realized by using e.g. random phase to induce a uniform spectral response, with subsequent filtering for the specular component[47]. Once the polygon wavefield is calculated, it needs to be aligned to

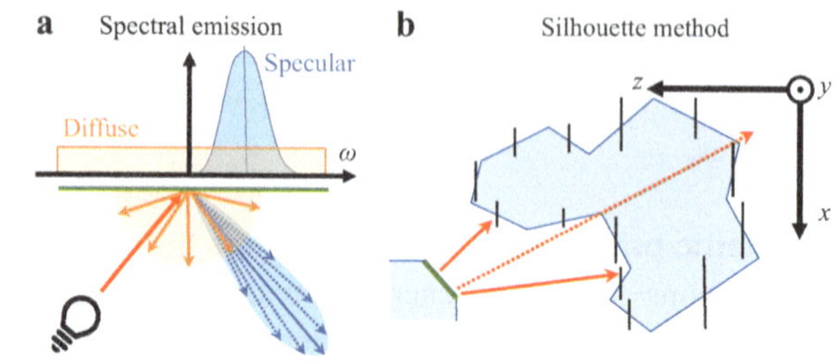

Fig. 4 Concepts from polygonal CGH. **a** diffuse (orange) and specular (blue) emission functions with corresponding spectral envelopes; **b** an exemplary polygon (green) may emit light rays that will erroneously evade occlusion (dotted red ray).

be parallel with the hologram plane before convolutional propagation is possible. This requires a rotational transform of the wavefield, which can typically be expressed by resampling the Fourier domain of the wavefield[43].

To model occlusion, the wavefield can be propagated from back to front, progressively occluding pieces of the wavefield like layer-based methods. The silhouette method[48] does this by creating a mask at the center of every polygon, being the orthogonal projection of each (rotated) polygon's shape along the optical axis parallel to the xy-plane. This creates small "billboards" shielding the light of the advancing wavefield. The approach works well for holograms with small viewing angles and mostly convex shapes, but can cause in more general holograms with complex 3D models light to leak into the final DH. The issue can be addressed e.g. with the "switch-back" method[49] which, in turn, requires more calculations.

Polygon-based CGH can create realistic graphics with accurate occlusion. Its computational efficiency is highly dependent on the polygon count. Because it accurately models diffraction, it can be combined with other CGH methods. A drawback is the requirement of many FFTs, which is hard to scale and optimize when computing high-resolution holograms. It may be somewhat mitigated by only propagating over small distances, to limit the spreading of light across the hologram. Another drawback is that complex lighting effects such as global illumination, soft shadows, ambient occlusion, refraction, etc. are hard to model, as they would require many costly propagations back-and-forth between polygons.

Ray-tracing

Ray tracing is a computer graphics technique for modeling light transport. The main idea is to track individual rays of light bouncing through the virtual scene and interacting with materials, accurately computing the amount of light reaching every pixel of the virtual camera. With the right models, one can achieve photo-realistic rendered images indistinguishable from actual photographs nowadays.

These techniques can be leveraged in CGH to model light transport as well. However, it cannot be utilized straightforwardly, since holography is fundamentally wave-based, differing substantially from ray-based models.

One major problem is that unlike in computer graphics where a single view is rendered, in holography we aim to obtain a continuum of viewpoints, requiring to model incoming light from many angles rather than only from the camera pinhole. A second major problem is the lack of phase coherence: the phase is sensitive to distance changes which are only a fraction of the wavelength, i.e. several nanometers. This means that cast rays from neighboring pixels at similar angles will almost never reach the exact same point, causing mismatching phase which will significantly degrade visual quality. This is exacerbated by the limited precision of 32-bit floating-point numbers, whose machine epsilon is $\approx 10^{-7}$. That means that those ray-tracing methods have to be adapted or combined with other CGH algorithms to be utilized effectively.

In[52], the hologram was computed with backward ray-tracing from *hogels* (sub-holograms containing angular ray information) outward, or in combination with multiple WRPs[53]. That way, advanced shading effects can be leveraged with a similar pipeline to conventional CG, creating realistic holograms. However, this approach does not address the second problem of phase coherence.

More recently, a solution was proposed combining point-based CGH with path tracing to address the phase coherence problem[50]. The scene was adaptively sampled into points, after which the light distribution was measured in every point using the ray-tracing simulation. That way, complex effects such as soft shadows, reflections and ambient occlusion could be accounted for, all with sharp focal cues. Then, modulated PSFs with occlusion were propagated to the hologram plane. (see some examples in Fig. 5)

The main advantage of ray-tracing CGH is the ability to reach high realism, and its compatibility with modern graphics hardware such as the latest NVIDIA RTX series with ray-tracing units. These DHs can serve as reference data or ground-truth. One important drawback is the high computational cost, making it unsuitable for real-time calculation of large holograms or complicated scenes without significant approximations in the near future.

Fig. 5 Examples of virtual scenes (a,c) and views taken from the corresponding CGH (b,d). (a,b) is a Cornell box, showing the use of are light, with soft shadows and global illumination (Source:[50,51]). (c,d) is a grass and forest scene created from detailed 3D meshes, showcasing intricate occlusions and shadows.

Challenges

Aside from computational challenges, which will be considered in the next section, CGH methods suffer from a structural problem. That is, visual effects need to be frequently re-implemented separately for each CGH method and rarely any single method can offer all relevant effects. Depending on the CGH method, occlusion, shadows, material properties such as textures/surface roughness, and non-uniform lighting such as via the Phong model[37] are more or less challenging to implement — see Table 1. For example, occlusions and shadows are a problem when working with points of infinitesimal extent. In contrast, it's easy to model point-wise non-uniform lighting as well as material properties – described in each point typically by a bi-directional reflection distribution function (BRDF).

Advanced effects such as true reflections, that is reflections coming into focus at an out-of-mirror plane upon reconstruction of the DH, and refraction by objects, for example focusing lenses are currently not well studied with any method. The reason is that they require modeling of reflecting and refracting wavefields in complicated ways. A naive approach would require the computation of an interim hologram of the entire visible scene at each surface.

Table 1 CGH algorithm classification summary

Category	Method				
	Point-clouds	Geometric primitives & basis functions	Layer-based	Polygon-based	Ray-tracing
Texture support	Yes	Limited	Yes	Yes	Yes
Non-uniform scene lighting	Medium	Hard	Limited	Hard	Easy
Shadows	Hard	Hard	Limited	Medium	Easy
Occlusion	Hard	Hard	Easy	Medium	Easy
Primitive super-sampling	No	Yes	Yes	Yes	Yes
Supported scene complexity (object count)	Low	High	High	Low-Medium	High
Support object detail	High	Medium	High	Low	High
Support for wide & viewing angles	Yes	Limited	No	Yes	Yes
Computation speed	Slow-Medium	Fast-Very Fast	Fast	Medium	Slow
Implementable on specialized hardware	Easy	Easy	Hard	Hard	Limited

Another challenge is the typically large number of degrees of freedom in each CGH method. Because of insufficiently correlated objective quality metrics in holography (see last section of the paper) and unclear parameter bounds, it still remains unclear which parameter ranges correspond to which visual quality and where trade-offs for the sake of computational complexity can be done. Especially so, when porting methods to different hologram and scene parametrizations.

Finally, as was pointed out before, hybrid CGH methods also hold big promise in terms of visual quality versus computational complexity, since they can combine the best aspects of several methods[54].

Algorithmic CGH acceleration techniques

As the previous section already indicated, one major consideration in using any CGH algorithm is computational complexity. Therefore, several acceleration techniques of the CGH methods have been developed. Any of the acceleration techniques below has been combined in literature already with one or more of the CGH algorithm classes defined in the previous section. In this section we discuss: several ways to reduce the signal complexity in specific, so-called sparse, representations; the re-use of precomputed results; dynamic CGH acceleration; and deep-learning supported CGH.

Sparsity

Sparsity is a concept in signal processing, whereby a signal can be modelled by a relatively small number of significant coefficients when expressed in the right transform space.

This can be used for sparse approximations, where a few percent of the total coefficient count account for the vast majority of the signal's energy. The sparsity concept is used in remote and compressed sensing, in the compression of signals, in deep learning, signal filtering (most common: low-pass filtering), and finally also efficient/sparse computation.

In CGH, the notion of sparsity can be leveraged to significantly reduce computational requirements because of the nature of diffraction as expressed in Eq. 1. Generally all pixels in the hologram plane will be affected by elements in the scene, requiring many updates. But by computing the hologram in a transform domain where the signal is sparse, only a few coefficients need to be updated to compute CGH elements. For a transform $\mathcal{T}$, a linear combination of elements $E_j(x,y), \forall j \in \{1,..,N\}$ (e.g. PSFs, basis functions, polygons, etc.) can be rewritten as

$$H(x,y) = \mathcal{T}^{-1}\{\mathcal{T}\{H(x,y)\}\} = \mathcal{T}^{-1}\left\{\sum_{j=1}^{N}\mathcal{T}\{E_j(x,y)\}\right\} \quad (6)$$

For effective sparse CGH several requirements must be met.

High sparsity, i.e. the ratio of needed transform coefficients to total hologram pixel count must be small, so to accurately approximate the target signal with few updates.

Efficient computation of the coefficient values of the different elements $\mathcal{T}\{E_j(x,y)\}$ directly in transform space. They should not be much more costly than evaluating $E_j(x,y)$ values in the hologram plane for acceleration to be possible.

Efficient (final) inverse transform $\mathcal{T}^{-1}$, with a computational complexity that is low compared to conventional CGH algorithms.

An important distinction to make is that it is not necessarily the hologram itself that needs to be sparse in transform space, but rather the *individual elements* E_j themselves. E.g. in point-cloud CGH, the individual PSFs must be sparse in $\mathcal{T}$ for acceleration to be possible, but their superposition does not have to be. Note, that one can use multiple different transforms, in parallel or cascaded, as well as overcomplete transforms (i.e. with more transform coefficients than hologram pixels) so long as the above requirements are met.

Although not originally characterized a such, many CGH acceleration techniques can be retroactively considered sparse techniques. We make a further classification depending on the chosen transform $\mathcal{T}$ and how the coefficients are computed.

Wavefront recording planes

Interim wavefront recording planes (WRP)s are one of the most widely used CGH acceleration techniques, which is a sparse CGH method where $\mathcal{T}$ is a convolutional diffraction operator. When the distance of virtual scene elements to the WRP are small, the diffraction pattern will be energetically highly spatially localized, meaning only few coefficients need to be updated. For point cloud algorithms, the relevant pixels to be updated can be found using the relationship $\sin(\theta) = \lambda\nu$ between a diffraction angle θ w.r.t. the hologram plane normal and the signal's spatial frequency ν. This means the maximum diffraction angle $\theta_{\max}$ is determined by the Nyquist rate,

$$\sin(\theta_{\max}) = \lambda\nu_{\max} = \frac{\lambda}{2p} \tag{7}$$

governed by the pixel pitch p and the illuminating wavelength λ. WRP coefficients at steep angles need not to be updated because either (1) they induce frequencies in surpassing $\nu_{\max}$ which would cause aliasing, or (2) they produce signals which will never reach the final hologram. In the former case, one can use the geometric relationships between the emitting point and its distance d to the WRP (cf. Fig. 6a) to determine the corresponding PSF width on the WRP to be equal to

$$2|d|\tan(\theta_{\max}) = 2|d|\tan\left(\sin^{-1}\frac{\lambda}{2p}\right) = \frac{2|d|\lambda}{\sqrt{4p^2-\lambda^2}} \tag{8}$$

In the second case, the limiting angle can be found analogously.

Typically, WRP methods[26,56] are point-based but that is not a necessary limitation as shown for example in[36].

Stereogram approximations

Holographic stereograms CGH are comprised of techniques which approximate generated holograms by mapping discretized light field coefficients into specific wavelet-like elements. Each element typically consists of small plane wave segments. The incidence of the ray position on the wavefield plane will determine the position of the centroid of the wavelet's energy, while the incidence angle of the ray will determine the modulation frequency, according to the $\sin(\theta) = \lambda\nu$ relationship.

These plane wave segments can be modeled in general by a *short-time Fourier transform* (STFT) for $\mathcal{T}$. The STFT operator is defined as a family of apodized functions S with different combinations of translations (τ, υ) and

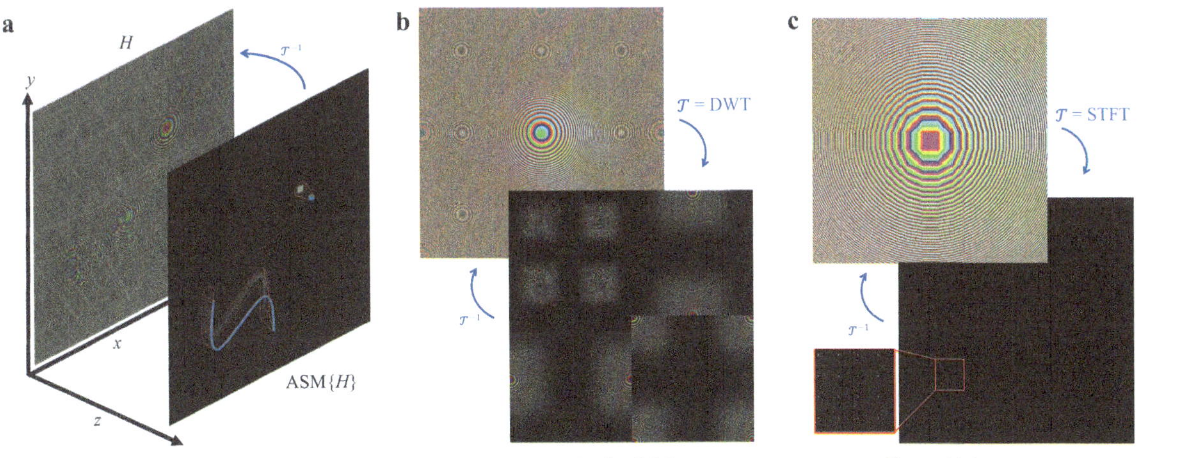

Fig. 6 Examples of sparse CGH methods. **a** The blue 3D curve and point are close to the WRP, only affecting nearby coefficients delineated in the red regions. This contrasts with the hologram plane H, where all pixels are affected. **b** A PSF and its corresponding sparse 2-level Daubechies-4 wavelet transform[55]. **c** Accurate PAS on 16 × 16 coefficient blocks with redundancy 2. Although the sparsity of 1/256 is high, it introduces noticeable distortions.

frequency modulations (ω,η). Formally, we have

$$\begin{aligned}\mathrm{STFT}\{H(x,y)\}(\tau,\upsilon,\omega,\eta) &\equiv S(\tau,\upsilon,\omega,\eta)\\ &= \iint_{-\infty}^{\infty} H(x,y)w(x-\tau,y-\upsilon)e^{-i(\omega x+\eta y)}dxdy \end{aligned} \quad (9)$$

where w is a window function. Typically a rectangular function is used in stereogram CGH, but it can also be e.g. a Hamming or Gaussian window. For a given hologram, the family of functions S is sampled rectilinearly along all 4 dimensions depending on the hologram and scene properties.

The problem with the faithful conversion of rays to wave signals is the Heisenberg uncertainty principle: though a ray has an exact incidence position and angle, it is mathematically impossible for a signal to be simultaneously perfectly localized in space and frequency (i.e. diffraction angle). Thus there always will be an inherent approximation. By widening the window w, one can improve the frequency resolution but lower the spatial resolution, and vice versa.

The source of the emitted rays can be any number of scene elements covered in the previous sections, but they are most commonly point sources and *discrete light fields*. We can define a light field as being a 4D function $L(x,y,\theta_x,\theta_y)$ describing the light intensity of incoming rays at the hologram plane at coordinates (x,y) forming angles (θ_x,θ_y) w.r.t. the hologram normal (here the z-axis) along the x and y axes, respectively. This light field function is discretized, also sampled rectilinearly along its 4 dimensions. This yields a one-to-one correspondence between light field rays and STFT functions. The light field values can be obtained by rendering a scene with conventional computer graphics software, sampling the scene from multiple different virtual camera angles.

The basic stereogram method is primarily concerned with amplitude values, either assigning random phase values[57], or using structured phase patterns[58]. A major cause for quality loss is the lack of phase coherence across neighboring blocks: these will lead to frequent sudden jumps of the signal which degrades the overall quality of the reconstructions. This was first addressed in[59] with the use of *phase-added stereograms* (PAS). By taking the distance a ray has traveled into account, one can modulate the phase as well to ensure coherence.

Originally, these stereograms were computed by subdividing the holograms into blocks and computing plane wave coefficients spatially. This was later accelerated and optimized by applying a block-wise FFT with compensated phase[60]. This is the special case where w is rectangular and the amount of STFT elements in S is exactly equal to the pixel count. This will cause the frequency space to be coarsely quantized. By increasing the sampling density along (ω,η) in S, we get the *accurate phase added stereogram*[27], cf. Fig. 6c leading to more frequency coefficients per block, whose finer quantization substantially improves visual quality.

There are also examples of non-rectangular window use in stereograms. In[61], Hann windows are used for w, sacrificing some calculation performance due to the overlapping windows to get significantly better visual quality than most typical stereogram methods. Thanks to the overlapping windows, the space/frequency resolution trade-off poses less of a problem, making it nearly invariant to hogel size and scene geometry.

The advantage of stereograms is their high compatibility with conventional computer graphics software, allowing for the use of advanced rendering software to calculate all the ray intensities (and depths). Moreover, thanks to the high sparsity of needing to update only one coefficient per ray, and the efficient use of small FFTs, orders of magnitude speedups over conventional CGH can be obtained. The main drawback is the loss of quality. The stereogram approximation generally leads to apparent view discretization and loss of focal sharpness. Scene elements can also often only be placed in a constrained virtual 3D volume, leading to unacceptable visual quality losses otherwise.

Coefficient shrinking methods

Coefficient shrinking is more closely related to the traditional notion of sparsity in e.g. compression and denoising. A signal of interest is transformed with some candidate $\mathcal{T}$, after which all coefficients below some threshold are set to zero. This threshold can be set to be some minimum energy, or to limit the maximum number of non-zero coefficients. For CGH, this principle can be applied on e.g. a point spread function, after which some predetermined fraction of the most significant coefficients are either precalculated and stored in a look-up table or computed on-the-fly.

One of the first works calculating sparse CGH this way was done in[62], leveraging Fourier domain sparsity. A point cloud is divided in clusters, where for each one high-magnitude PSFs coefficients are computed on a WRP. Then, the convolutional Fresnel diffraction in computed with the sparse FFT transform, which is faster than the regular FFT when the fraction of non-zero coefficients is small. A variant of this method was proposed by the same authors using precomputed coefficients as well[63]. Given the large contiguous zero regions in typical layer-based CGH, another sparse "SS-2DFFT" was also proposed[64] to accelerate those CGH calculations.

Wavelets can also be used for $\mathcal{T}$. In[55], the WAvelet ShrinkAge-Based superpositIon (WASABI) method computes the significant Daubechies-4 wavelet coefficients for PSFs, with up to 30-fold speedups over the standard point cloud method (see Fig. 6b). By combining the wavelet transform with a convolutional diffraction operator, the algorithm is sped up further[65]. In later work, a similar shrinkage method was proposed using the the STFT instead, achieving a 2dB PSNR gain and better view preservation over wavelet-based methods[66]. This method was further improved and sped up by analytically computing coefficents rather than storing the precomputed values[67].

A major advantage of coefficient shrinking is the controllable sparsity, allowing for fine-grained control trading-off quality and calculation speed. But because wavefield signals have to be expressed directly in the transform domain, computing coefficients and the memory access patterns for making updates tend to be more complex.

Look-up tables

Many diffraction patterns generated by 3D scene elements recur throughout the hologram. Instead of recomputing them each time, they can be computed in advance and stored in look-up tables (LUT)s which are optimized for fast memory access.

Most LUT CGH methods are designed for encoding PSFs. Precomputed wavefronts were already proposed in the first digital holographic display systems in the early 90's[68], binning 3D points together into cells, each with a matching 1D wavefront for creating horizontal parallax CGH. Unfortunately, straightforwardly extending this principle to full-parallax CGH does not work, since storing 2D diffraction patterns for every possible 3D position would require far too much memory to be practical. Most of the research has thus focused on reducing LUT sizes by exploiting various symmetries.

The novel LUT method (N-LUT)[28] only stores different axial realizations (along z) of PSFs, since we have translational symmetry in holography: moving a point laterally will accordingly shift the induced pattern by the same amount. This was later extended with the "run-length coding method", allowing for grouping points together in a single run when applying LUTs, leveraging scene symmetries since points tend to cluster together. The split LUT (S-LUT) and compressed LUT (C-LUT) techniques[69,70] on the other hand utilize a separable representation of PSFs along the x and y axes, while the other methods[71,72] leverage the PSF's radial symmetry by only storing the radial PSF function, and drawing out a whole PSF quadrant; the latter can then be duplicated and rotated by 90° repeatedly to obtain the complete PSF.

But LUTs are not only used to store PSFs. Precomputed patterns have been used for calculating line patterns[73], or even surface elements with complex lighting patterns and occlusion masks[37].

LUTs are very flexible, since in principle any values can be precomputed and stored for facilitating real-time CGH, representing objects of arbitrary complexity. A disadvantage in many cases is the limited number of encodable parameters, which otherwise leads to a combinatorial explosion of the LUT size in memory. LUTs are also not conducive to caching beyond a certain size, which can cause severe performance degradation; larger LUT may also not be suitable for FPGA or application-specific integrated circuit (ASIC) implementations.

Dynamic CGH acceleration techniques

When generating successive hologram frames in a holographic video display, one can not only leverage redundancies within each frame, but also leverage the temporal correlation between successive frames. The goal of dynamic CGH acceleration methods is to repurpose the data generated in prior frames to generate the current frame to lower the computational cost. This can be accomplished with *motion compensation*: given prior frame(s) and the known ground-truth motion of objects in the virtual scene, the frames can be transformed accordingly to match the scene composition in the current frame, which can be significantly less calculation-intensive than computing each frame from scratch. However, this compensation is never entirely accurate: even if the motion model is perfect, there is still some inherently missing information due to previously invisible object segments either entering the frame or becoming unoccluded by other objects.

An important complication with holographic motion compensation is the nature of diffraction itself. Unlike with conventional video, small object motion will alter all pixels of the hologram. That is why e.g. block-based motion compensation methods as found in modern video codecs fail for temporal decorrelation in holographic video compression[3].

So far, only few attempts have been made to tackle this problem. In[74-76], translational motion parallel to the hologram plane was compensated with the use of look-up tables storing wavefield segments. In[77], 3D rotational motion was compensated using a formulation based on spherical holograms. General rigid body motion compensation was proposed for CGH in[78], where new CGH frames were only recomputed when the compensation error surpassed some threshold. In the recent[79], motion parallax was achieved in dynamic CGH compensation.

These approaches are unfortunately still limited: they can only compensate lateral translational motion parallel to the hologram plane, or only model global scene motion or rigid body motion of single objects. To be generally applicable, two main problems need to be addressed: (1) compensating arbitrary motion of multiple independently moving and mutually occluding objects; and (2) selective differential CGH rendering.

For (1), a first solution was proposed in[80], by means of segmenting and masking holograms in Gabor space, together with global motion compensation operators. Mutually occluding objects with complex shapes and holes could successfully be compensated. For (2), general differential CGH methods are needed that only update the missing parts of the CGH signal to minimize computation costs. Because these differential signals are not directly expressible in the hologram plane efficiently, potential solutions may require transform-domain coefficient updates, similar to sparse CGH.

Deep-learning based acceleration

Deep learning is a class of machine-learning algorithms based on artificial neural networks. The "deep" adjective refers to the many concatenated neural network layers, allowing for learning and encoding highly sophisticated patterns. Deep learning has had a significant impact on many problems in the field of signal processing[91].

In CGH, deep learning has been used as an accelerator or even substitute for different algorithmic components, cf. Fig. 7. Deep learning is particularly useful for complex non-linear optimization processes. It has been successfully used for optimizing the computation of phase-only holograms to minimize distortion w.r.t unapproximated complex-valued counterparts. This was first achieved for 2D scenes[92], and soon after for more complex 3D scenes[86,93,94], reporting significantly higher calculation performance over Gerchberg–Saxton[95] and even Wirtinger holography[96].

Deep learning has proven useful for extracting speckle-free views from generated holograms[88], or for the creation of binary hologram patterns for DMD-based displays optimized to minimize the error with the ground-truth wavefield[97]. It has recently been used to optimize the output for phase-only holographic displays[90], and even a full CGH pipeline based on deep learning was devised[85] that can render realistic color holograms from RGB+depth input, especially suited for head-mounted holographic display systems with low power requirements.

Using deep learning has many benefits, flexibility of updating and tuning the neural network architecture and training algorithms, unparalleled optimization efficiencies, as well as optimized hardware systems (e.g. GPU) for both training and inference. A disadvantage of deep learning is its black box nature, i.e. it's hard to garner insights on how they work. It is also hard to guarantee whether a system can generalize to different or more complex 3D objects and non-homogeneous scene lighting, different display capabilities with e.g. larger viewing angles. Another open question is often also the robustness to outlier input data.

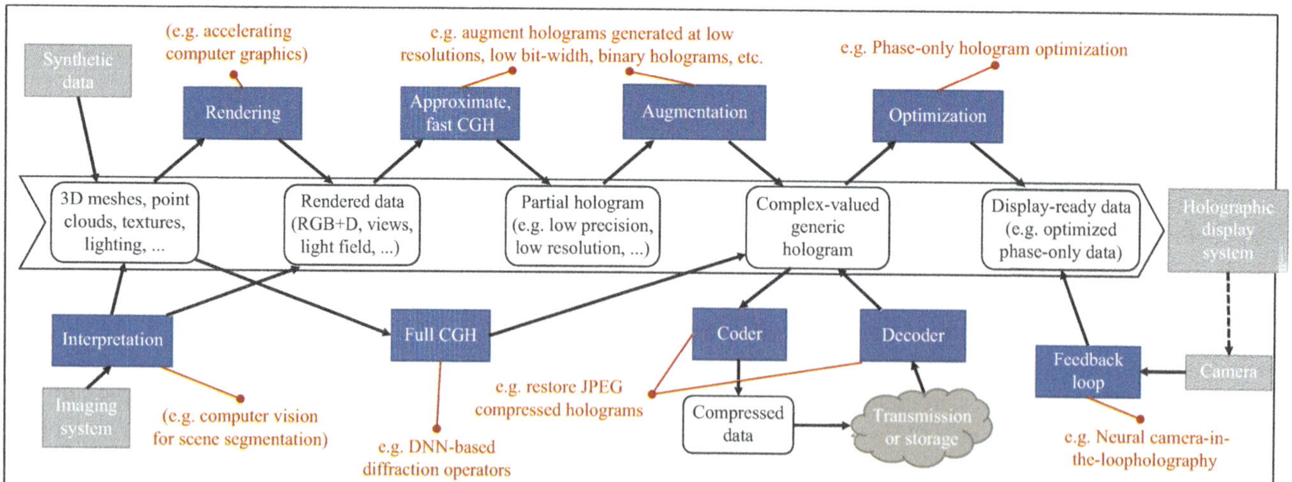

Fig. 7 Diagram of a general CGH pipeline and the utility of neural holography with several examples. Starting from captured or synthetic input data, the data is transformed in multiple phases, where DNN can improve, accelerate or even substitute algorithms. At the start, DNN can be leveraged for accelerating computer graphics rendering[81] or computer vision[82]. DNN can augment data generated at low resolution[83] or low bit-width[84]; replace diffraction operators[85,86]; compress holograms[84,87]; do speckle denoising[88] and account for SLM limitations by optimizing phase-only hologram patterns[89] or using DNN camera-in-loop holography[90].

Challenges

The main challenge for algorithmic accelerations is the streamlining of the data flow, such that neither a computational, memory, or bandwidth bottlenecks limits the rendering speed. Therefore, increasingly many methods are found that combine multiple acceleration methods such as LUTs and WRP sparsity[35,36]; WRPs and wavelets[65]; or stereograms, LUTs and WRPs[98]. In the end, the decision on which accelerations to choose for a given CGH algorithm always depends on the selected computing architecture (capabilities) and the desired visual effects. Nonetheless, the CGH algorithms in combination with the various accelerations techniques result in ranges of computational speed and it is unlikely that these will change in the foreseeable future, cf. Table 1.

Jointly optimizing hardware and CGH

Given the high requirements of high-end holographic video displays, purely algorithmic optimization with solely abstract complexity analysis are not enough. CGH algorithms need to be aware of the instructions and memory size and transport capabilities to reach optimality; furthermore, hardware can be tailored to the specific needs of CGH to increase power and computational efficiency. We will elaborate on many of these aspects in this section.

Leveraging CPUs and GPUs

Since CGH is currently an active research field, most developments target consumer CPU and GPU hardware for system prototyping. A thorough overview of used hardware platform for CGH can be found in[99]. Since CGH requires the computation of millions up to multiple billions of pixels, and because the diffraction operator is linear, they lend themselves very well to massively parallel implementations and architectures. Examples of such implementations are the C++ "Computational wave optics library"[41], the real-time color holographic display systems using a 12-GPU system[100] and the open-source "HoloGen" CGH library[101].

Because of the increasing complexity of modern hardware, proof-of-concept prototyping in e.g. MATLAB or Python may not suffice to ensure deployability for high-performance software implementations or competitiveness with existing solutions after optimization. One such aspect which has received little attention in CGH is *caching*: small data structures with predictable memory access patterns can be read an updated several orders of magnitude faster than large precomputed wavefront data. E.g. this can lead to the counter-intuitive result that on-the-fly calculation of PSFs can be faster than writing PSF LUT to the hologram plane.

An example of the importance of caching can be seen in[102], where a straightforwardly optimized implementation on GPU of the "accurate PAS" CGH algorithm on GPU with a sparsity of about 0.1% only gives a speedup of 2.6 × over the reference point-cloud CGH, which has a far lower expected efficiency than the ≈ 1000-fold reduction in coefficient calculations would indicate. By grouping coefficients together through the use of "lozenge cells"[103], the memory access patterns are optimized leading to a speedup of 86.5 × over the reference with identical results to the base PAS version.

Low precision and bit-width calculations

In conventional displays high bit depths per pixel are important to faithfully represent colors, with modern displays having at least 8 bits per color channel. This is different for holography; the same property of diffraction, spreading the information content all over the hologram plane making CGH computation challenging, also makes holograms more robust to noise with limited visual quality loss. Since signal quantization is roughly equivalent to inserting uniform noise into a signal, holographic displays can still provide high visual quality with lower bit depths.

This principle was demonstrated in[12,104], showing that under the right circumstances, 2-bit phase SLM displays will show only minor signal degradation compared to the 8-bit reference phase signal. That is why even binary holographic displays using DMDs can still be effectively used in high-quality holographic display systems[105].

However, this property of holograms can therefore also be exploited for CGH: calculations can be made using lower precision representations, such as fixed-point integers or the half-precision floating-point format. The accumulated errors will mostly be concentrated in the least significant bits of the wavefield solution, which will have little to no visual impact.

The FFT is an instrumental algorithm in CGH, whose precision optimization would bring important benefits. To this end, an analysis of the fixed-Point accuracy of a low-precision FFT was made[106], modeling the error propagation in for the purpose of CGH. A low-precision point-cloud CGH algorithm was proposed[107] that computes Eq. 2 mostly using 1 byte per coefficients for intermediate computations. It relies i.a. on the fact that phase values encoded as fixed-point integers can safely overflow when represented as a multiple of 2π. It utilizes GPU vector instructions and other intrinsics on packed integers, allowing for high throughput, gaining a 3-fold speed up over an optimized floating-point GPU implementation and a 15dB PSNR accuracy increase over a reference FPGA-based fixed-point integer solution. Low-precision optimizations are inherently used in field-programmable gate array (FPGA) due to the extensive use of fixed-point arithmetic, which is covered in the next section.

Research on low-precision algorithms in the context of CGH is still relatively new, with much potential for further improvements as it is largely orthogonal to the other aforementioned algorithmic optimizations. For example, deep learning CGH solutions could in principle be optimized as well by reducing the representation precision and outputs of (some of) their layers.

Field-programmable gate arrays and application-specific integrated circuit for CGH

A FPGA is a large-scale integration (LSI) that allows designers to freely rewrite logic circuits. An FPGA is composed of logic blocks (a look-up table that can configure any logic circuit), flip-flops, internal memories, switches between logic blocks, and input/output pins for external circuits. The designers can use a hardware description language to implement the desired circuit in the FPGA.

While recent CPUs and GPUs have operating frequencies on the order of GHz, FPGAs have operating frequencies on the order of several hundred MHz, thereby being an order of magnitude slower. However, if a large number of arithmetic circuits can be implemented in FPGAs and run in parallel, FPGAs can surpass CPUs and GPUs in terms of performance for specialized tasks. To maximize FPGA resource utilization, arithmetic circuits with small hardware footprints are required. CPUs and GPUs typically use floating-point operations, which require a lot of hardware resources. Contrarily, FPGAs can use fixed-point operations with small resource requirements. In addition, new hardware-oriented algorithms are required.

MIT Media Lab developed a media processor specialized for image processing called Cheops[108,109] using FPGAs and digital signal processing LSIs. Cheops was capable of performing common image processing operations (matrix-vector multiplication, filtering, color space conversion, and so on) at a high speed. In[109], Cheops was used as a CGH computational processor for holographic displays using acousto-optic modulators (AOMs). It was about 50 times faster than a workstation and about three to five times faster than a supercomputer (Connection Machine-2) during those days.

HORN (HOlographic ReconstructioN)[110] is an FPGA-based computer series dedicated to CGH calculations. It is currently under development, up to its 8th model, and implements the point cloud algorithm on an FPGA, as defined in Eq. 3.

HORN-1[110] and HORN-2[111], which directly implement Eq. 3, were fabricated using discrete ICs and hand wiring, not using an FPGA. HORN-3[112–119] and later versions were designed with FPGAs.

Fig. 8a presents a circuit designed to calculate Eq. 3 using adders, multipliers, and trigonometric look-up tables (LUTs). This circuit is designed by a pipeline circuit. The input data is (x_j, y_j, z_j, a_j), and the data are sequentially fed into the pipeline circuit. The data is processed and advanced to the next stage at every clock cycle. If the number of stages is N_s, the CGH calculation of the first input data is completed after N_s clock cycles. Thereafter, the object data after $j \geq 2$ is accumulated every clock cycle.

In general, since $N \gg N_S$, the computation of a single CGH pixel takes only about N clock cycles, which is a big advantage. Although there are techniques, such as software pipelining for CPUs and GPUs, the internal circuitry of CPUs and GPUs is difficult to fully pipeline. Let us assume that the computation of $a_j \exp(\cdot)$ in Eq. 3 requires 20 floating-point operations. In this case, the CPU and GPU require $20N$ clocks to compute Eq. 3. Contrarily, although the FPGA clock is an order of magnitude slower than the CPU and GPU, the FPGA is fully pipelined; thus, the computation takes only N cycles. This is sufficient to compensate for the difference in the operating frequency. In addition, since Eq. 3 can be computed independently for each pixel, the pipeline circuits can be prepared for each pixel to be computed in parallel.

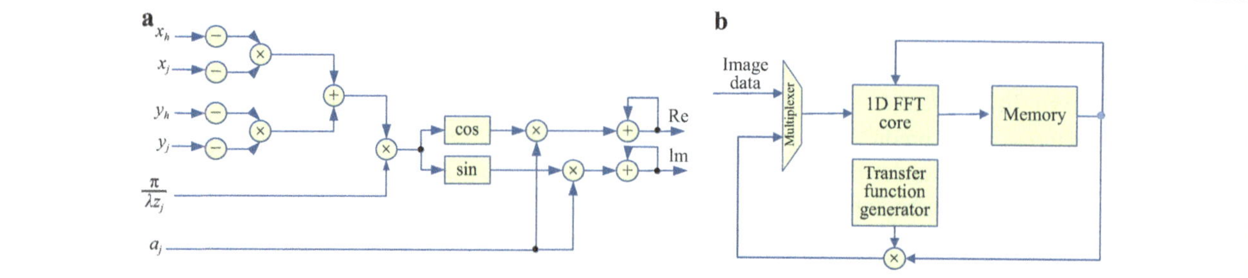

Fig. 8 CGH calculation circuits: **a** calculation pipeline for Eq. 3, **b** FFT-based diffraction calculation circuit.

However, Eq. 3 requires a large number of multipliers. Since multipliers require more circuit resources than adders, it is desirable to minimize their usage. The recurrence formula method can calculate the phase of Eq. 3 using only additions[113,114]. The method developed by[114] requires fewer additions than the method developed by[113]. By cascade-connecting the pipeline circuits that compute this recurrence formula, a large number of CGH computation circuits can be implemented. The latest HORN-8 has 4,500 CGH computation circuits mounted on an FPGA board, and by running them in parallel, the computation speed was six times faster than the latest GPU at the time.

An FPGA implementation with a modified data flow for the recurrence formula method has also been proposed[120]. Reference[121] implemented the point cloud method in an application-specific integrated chipset (ASIC). The ASIC implementation did not use the recurrence formula method to simplify the design but instead used a pipeline circuit with Eq. 3. By arranging this pipeline circuit in two dimensions, 8×8 CGH pixels could be calculated in parallel. The ASIC was fabricated in 0.18 μm CMOS technology and operated at 200 MHz. Although the performance of the ASIC, which was several generations old, was not as high as that of FPGAs which use state-of-the-art semiconductor processes, it demonstrated the potential of ASICs for CGH computation.

Other than Eq. 3, sparsity-based CGH calculation[67] has been implemented in an FPGA[122]. The light wave distribution of an object point on a CGH becomes sparse in the short-time Fourier transform (STFT) domain. Since the STFT requires multiple FFT operations, reference[67] obtained an analytical solution in the STFT domain without FFT operations. A pipeline circuit to calculate this analytical solution was implemented in an FPGA[122]. The error and trigonometric functions included in this analytical solution were implemented in LUTs in the FPGA due to the large circuit size.

The polygon method has a good affinity with three-dimensional (3D) graphics, and CGHs can be calculated by calculating light propagation from polygons. Light waves from polygons are generally calculated using fast Fourier transforms (FFTs), but if the amplitude and phase of the polygons are constant, we can deduce an analytical solution. In[123], the analytical solution of Fraunhofer diffraction from polygons was implemented in an FPGA. Since light waves from the polygons can be computed independently, parallelization of this computational circuit is expected to speed up the polygon method. However, this analytical solution has many calculations that are complicated and costly to compute (e.g., exponentiations, trigonometric functions, square root calculations, and divisions), which increase the required circuit area. It is difficult to install a large number of these calculation circuits.

Diffraction calculations are important in CGH calculations and are performed using FFTs. In[124–126], the FFT-based diffraction calculation was implemented in an FPGA. Fig. 8b presents a schematic of the circuit. Since the two-dimensional (2D) FFT is separable, it can be computed by performing a row-wise 1D FFT followed by a column-wise one-dimensional (1D) FFT. In addition, since it is difficult to design a high-performance FFT circuit, FPGA vendors provide an intellectual property core (IP core) to calculate the FFT, and it can be used to design the circuit for diffraction calculation. FFT calculation results have a large dynamic range. CPUs and GPUs can use floating-point operations, so overflow is unlikely to occur, but FPGAs should use fixed-point operations to perform FFT where overflow must be considered. FFT IP cores usually have a scaling parameter to prevent such a problem. Novel solutions, including scaling information propagation and saturation, to further optimize efficiency and accuracy have been developed[127]. After computing the 2D FFT, the transfer function of Fresnel diffraction or angular spectral method is multiplied. The inverse FFT can then be computed using the same FFT circuit to obtain the diffraction results. In reference[126], researchers were able to perform Fresnel diffraction of 512×512 pixels at 949 frames per second.

The layered method is a relatively simple way to generate high-quality CGH using RGB and depth images. In[128], the layered method was implemented on FPGA. The layered method requires an iterative FFT-based diffraction calculation for the number of shades in the depth image. References[124–126] were designed for digital holography, whereas reference[128] was designed for the layered method. This circuit is also designed with fixed-point operations to reduce the circuit size. The units that perform the diffraction calculations are processed in parallel using 16 1D-FFT circuits. After the 1D FFTs are executed in the horizontal direction, the 1D FFTs are repeated in the vertical direction to calculate the 2D FFT. However, intermediate data created after the FFTs are executed in one direction cannot be stored in the FPGA memory; thus, it is stored in an external DRAM. This circuit was capable of computing full-HD CGH at 16 frames per second. Reference[129] improved on this circuit by incorporating 32 FFT cores; it was capable of computing 4K-resolution CGHs at 30 frames per second.

To reduce speckle noise in holographic projections, there are two primary solutions: the error diffusion method and time averaging of CGHs computed at different random phases. Both methods will be explained later. References[130] and[131] implemented the computationally expensive error diffusion method and a speckle reducing method by time-averaging CGHs computed at different random phases on FPGAs, respectively. As in[124–126,128,129], a 2D FFT is computed by repeatedly using 1D FFTs to generate a Fourier CGH. Six of these computation circuits are installed in one FPGA. The input images to this circuit are given different random phases, and six different CGHs can be obtained simultaneously. Although small FPGAs were used, the speedup was 2.5 times faster than a representative CPU of the time.

Driving holographic displays

As mentioned in the introduction, a main key component of every holographic display is a spatial light modulator (SLM). In this section we are first going to discuss the few principal types of SLMs in existence, including their advantages and limitations before the remaining subsections explain in which ways DHs can be processed in order to achieve a better overall quality, before supplying them to a given SLM. In the last subsection, we are going to briefly summarize how visual quality of holographic displays can be assessed.

Spatial light modulators and their limitations

To compute CGHs, we propagate light virtually from a 3D scene to a hologram plane on which we obtain the complex-valued amplitude of the CGH; i.e. typically called complex(-valued) CGH. However, commonly available SLMs are either amplitude- or phase-modulated. These limitations will cause reductions in visual quality compared to a fully complex-valued modulation, as illustrated in Fig. 9. Furthermore, each SLM is classified into two types: those that can display grayscale CGHs and those that can only display binary CGHs. Liquid crystal displays (LCDs) are often used as SLMs that can display CGHs with a grayscale, whereas digital micromirror devices (DMDs) often can display only binary CGHs. The resolution of available SLM types is similar, but the switching speed of a DMD is about two orders of magnitude faster than that of LCDs, although it can only display binary CGHs.

We have to convert a complex CGH into an amplitude CGH or phase-only CGH (also called: "kinoform"). However, this causes loss of some of the original information of the complex-valued CGH, resulting in a degradation of the reconstructed image.

Complex CGHs can, in principle, produce almost perfectly reconstructed images. It is relatively easy to obtain a grayscale reconstructed image with amplitude CGHs, but zero-order and conjugate wavefields are visible in the reconstructed image, see Fig. 9c. Phase-only CGHs have the advantage that the zero-order wavefield can be avoided, see Fig. 9b. However, obtaining a grayscale reconstructed image is difficult with these CGHs due to the loss of amplitude information. In general, edges of a reconstructed image are emphasized in a phase-only CGH. This is because light diffuses widely and is widely distributed on the CGH when light hits the high-frequency components of the original image. On the other hand, the light hitting the low-frequency components of the original image does not diffuse light as much and is distributed in a part of the CGH. In phase-only CGH, the amplitude of the object light at the CGH is set to a constant. Because many CGH have signal properties resembling the frequency domain, this unit amplitude has a relative high-pass filtering effect, resulting in an edge-enhanced reconstructed image. Binary CGH also tends to produce an edge-enhanced reconstructed image for the same reason as phase-only CGH.

Random phase can be employed to better spread the object light information. This is often used in holographic displays with a wide viewing angle. The use of random

Complex modulation | Phase-only | Amplitude-only | Binary modulation

Fig. 9 "Sphere1A" hologram from the Interfere-III dataset[51] in off-axis configuration with different modulation modes. Depending on the limitations of the SLM capabilities zeroth and conjugate orders or quantization noise may lead to visual quality degradation, for a fixed pixel pitch and pixel count.

phase can also reduce edge enhancement problems in the phase-only and binary CGHs described above. However, this will generated random interference on the reconstructed image, resulting in strong speckle noise. In the following sections, we discuss how to address these problems.

Complex-amplitude encoding

Although SLMs that can directly modulate complex amplitudes are not common, research has been conducted on achieving complex CGH modulation on SLMs. The main methods can be categorized into complex-amplitude encoding methods using one or two SLMs.

Reference[132,133] proposed a method for displaying complex CGH using two SLMs. In reference[133], two SLMs (an amplitude- and a phase-modulated SLM) are used to display a complex CGH. There are two formats for displaying complex CGHs, the polar format $u = A\exp(i\phi)$, where A is the magnitude and ϕ is the argument of the complex amplitude, and the Cartesian format $u = R + iI$, where R and I denote the real and imaginary parts, respectively. This method uses the polar format. Namely, A is displayed on the amplitude-modulated SLM, and ϕ is displayed on the phase-modulated SLM. By optically projecting the phase distribution of the phase-modulated SLM onto the amplitude-modulated SLM using a relay optical system, the complex CGH can be displayed. The problem with the use of multiple SLMs is the alignment of the two SLMs at the pixel order (micrometer order) and the aberration caused by the lens and other factors[134]. On the other hand, the method using multiple SLMs can display higher-resolution complex CGHs compared with the method using a single SLM (described below).

Many complex-amplitude encodings have been proposed that reconstruct complex amplitudes with only one SLM. In[135], the display area of a single amplitude-modulated SLM is divided into two parts to display the complex CGH; this method uses the Cartesian form $u = R + iI$. The real part, R, and the imaginary part, I, of the complex CGH are used as two amplitude CGHs, which are displayed in the divided SLM area. When a sinusoidal grating is placed on the Fourier plane of a 4f optical system, the two CGHs are optically transformed into a complex-amplitude CGH by spatially combining them on the output plane of the 4f optical system. Although the resolution of the complex CGH is halved, alignment is easier compared with the method developed by[134], which uses two SLMs.

The single sideband method with the half-zone plate method to obtain complex amplitudes using a single amplitude-modulated SLM has been proposed[136,137]. The amplitude CGH in this proposed method has, in principle, direct and conjugate light. The amplitude CGH is generated by superimposing real-valued point spread functions (PSFs), where the upper half of the PSF spectrum corresponds to the object light component and the lower half to the conjugate light component (or vice versa). If the lower half of the spectrum and the direct light component are removed in the Fourier plane of the 4f optical system, the complex amplitude of the object light can be obtained at the output of the 4f optical system. However, if half of the spectrum in the vertical direction is discarded, the vertical viewing area and the resolution of the reconstructed image are halved.

In recent years, the double-phase hologram[138] has become a major method for reconstructing complex amplitudes in holographic displays using a phase-modulated SLM. This method is easy to understand by considering complex numbers as vectors. The complex CGH is represented as $u = A\exp(i\phi)$, and its one pixel is represented as a vector. The double-phase hologram considers this vector as the sum of two vectors on the unit circle (the amplitude is always on the unit circle since the phase-modulated SLM cannot modulate the amplitude). Each vector is given by

$$\begin{aligned}&\exp(i(\phi + \cos^{-1}(A/2)) \\ &\exp(i(\phi - \cos^{-1}(A/2)) \qquad (10)\end{aligned}$$

Note that A should be normalized. Although two phase-modulated SLMs are required Eq. (10), a multiplexing method can be used to synthesize complex CGHs using only one SLM[139]. The double-phase hologram can produce a reconstructed image with good quality, but the diffraction efficiency is low, resulting in a darker reconstructed image.

Binary amplitude encoding[140] encodes the complex amplitude into a phase-modulated SLM by approximating the amplitude, A, of the object light in the CGH plane as a binary value using the error diffusion method. Here, $A \approx A_W + A_B$, where A_W and A_B denote the on-pixel and off-pixel, respectively. In this way, the complex amplitude, $A\exp(i\phi)$, of the object light can be approximated as

$$A\exp(i\phi) \approx A_W\exp(i\phi) + A_B\exp(i\phi) \qquad (11)$$

$$= \exp(i\phi A_W) + A_B\exp(i\phi). \qquad (12)$$

The second term is for the off-pixel ($A_B = 0$). Since phase-modulated SLMs cannot modulate the amplitude to zero, we generate a cancelling wave by alternating the phase of 0 rad and π rad to achieve $A_B = 0$. Binary amplitude encoding produces a brighter reconstructed image with almost the same quality as a double-phase hologram.

We describe how to realize complex CGH using a binary SLM (DMD). Since a DMD can only control on/off pixels, it is more difficult to control both amplitude and phase compared with amplitude- and phase-modulated SLMs. The reference[141] assigns a different phase to each pixel of the DMD. Each pixel in the DMD is grouped in a block (this block is called a superpixel, also known as a macro pixel). The complex amplitude is represented by the on/off sub-pixels in the superpixel. For example, if we consider a superpixel with 4×4 sub-pixels, we can represent 2^{16} complex amplitudes. However, if the size of the superpixel is large, the number of pixels in the complex CGH becomes small. There is a tradeoff between the number of pixels and complex-amplitude accuracy. The reference[142] improves over[141] by augmenting it with an error diffusion method.

Image restoration and image estimation methods

In CGH calculations, a random phase is added to widely diffuse the object light, but random interference occurs in the reconstructed image plane, resulting in speckle noise. In addition, when a complex-valued CGH is simply converted to phase-only or binary CGH, the amplitude information of the object light is lost, which may result in an edge-enhanced reconstructed image. This leads to image quality degradation. Noise reduction techniques used in digital holography and CGH are well summarized in[143]. In digital holography that capture holograms of real existing objects, sophisticated noise reduction algorithms, such as[144], can be used on reconstructed images in a computer. In CGH, however, such algorithms cannot be applied to optically reconstructed images. In this section, we describe the following speckle noise reduction and image quality enhancement techniques for CGH: (1) Time-averaging methods, (2) Non-iterative methods and (3) Iterative methods. The complex-amplitude encoding introduced in the previous section is also effective for image quality enhancement.

Time-averaging methods

Time-averaging methods reduce speckle noise by preparing multiple CGHs and reconstructing them in a time-averaged manner. In[145], the speckle noise in the reconstructed image is time-averaged using an SLM that can rapidly switch CGHs generated with different random phases. The intensity of the reconstructed image of a CGH at time t is denoted by I_t. By rapidly switching between CGHs with different random phase instantiations, the human eye can observe the time-averaged reconstruction $I_{\text{avg}} = \sum_t^N I_t$ where N is the number of CGHs. When N CGHs are time-averaged, the signal-to-noise ratio increases in proportion to $\sqrt{N}$. This is also known as one-step phase retrieval (OSPR)[146]. These methods are time-consuming because they require calculating many CGHs. In[147], two SLMs can be prepared and one SLM can be switched in time to display N random phase patterns, and is illuminated with light to produce low-spatial coherence light. By irradiating the low-spatial coherence light to the other SLM (which displays the hologram), we can reduce the speckle noise in the reconstructed image. The SNR is improved proportionally to $\sqrt{N}$, as in[145].

In[148], speckle noise was reduced by generating one random phase CGH, calculating N new CGHs by only changing the phase of this CGH by $\exp(2\pi i(n-1)/N)$, $n \in \{1, \ldots, N\}$ and then performing a time-averaging reconstruction. Since this method requires computing only one CGH, it reduces the generation time of CGH compared with the method developed by[145,146].

A reconstructed pixel from a CGH is not infinitesimal but a spot of finite size. Pixel separation methods[149–151] prepare object data thinned to a certain pixel spacing so that the spots with a finite size do not interfere with each other, generating a partition of multiple CGHs from this thinned object data (whereby the sum of this CGH data partition becomes the original object data). By the switching fast between these CGHs, an averaged reconstructed image with reduced speckle can be obtained since the adjacent pixels do not interfere. The averaged reconstructed image is expressed as $I_{pix} = \sum_t^N I_t$ where I_t denotes the reconstructed image of a CGH at time t. In recent work[152], a light sculpting technique named "DCGH" was proposed modulating light both temporally and spatially at the same time. Because the optimization happens jointly in space and time, better speckle reduction for 3D holograms can be achieved rather than treating them separately.

Non-iterative methods

Speckle reduction techniques without iterative optimization have been proposed. The complex-amplitude encoding described earlier is one of the non-iterative methods. In this section, we discuss error diffusion methods, sampling grid methods, and random phase-free methods.

The error diffusion method reduces the quantization noise of an image by spreading it to the surrounding pixels. This method is often used in the image processing field (e.g., halftoning), but it has also been widely used in CGH for a long time[130,153]. When applying the error diffusion method to CGH, we first calculate the error $e(x,y) = u(x,y) - u'(x,y)$ between pixel $u(x,y)$ in the original complex CGH and pixel $u'(x,y)$ in the transformed CGH (amplitude, phase-only, or binary CGH). We multiply this error by certain weight coefficients and spread it to the surrounding pixels. For example, if we use the commonly implemented Floyd–Steinberg weight coefficient, we can perform the error diffusion method as follows:

$$u(x+1,y) \leftarrow u(x+1,y) + \omega_1 e(x,y) \quad (13)$$

$$u(x-1,y+1) \leftarrow u(x-1,y+1) + \omega_2 e(x,y) \quad (14)$$

$$u(x,y+1) \leftarrow u(x,y+1) + \omega_3 e(x,y) \quad (15)$$

$$u(x+1,y+1) \leftarrow u(x+1,y+1) + \omega_4 e(x,y), \quad (16)$$

where $\omega_1 = 7/16$, $\omega_2 = 3/16$, $\omega_3 = 5/16$, and $\omega_4 = 1/16$. By performing this operation on all the pixels, the error-diffused CGH (amplitude, phase-only, or binary CGH) is obtained. Raster scans are common, but serpentine scans have also been proposed[154]. Holograms that implement serpentine scans are also called bidirectional error diffusion (BERD) holograms, and they have been shown to improve image quality compared with raster scans. The weight coefficients can also be optimized using a genetic algorithm, but these weight coefficients are image-dependent[155]. In addition to the Floyd–Steinberg weight coefficients, many other weight coefficients have been proposed. In[156], each weight coefficient and threshold method were comprehensively surveyed. The error diffusion method has also been used to improve the gray-level display characteristics of reconstructed images[157,158]. In CGH, the error diffusion method is widely used as a basic technique to obtain good-quality phase and binary CGH with low computational cost, but it darkens the reconstructed image.

The sampling grid method multiplies the original object $o(x,y)$ by a sampling grid $d(x,y)$ to reduce the edge enhancement problem of binary[159] and phase-only[160] CGHs. In[160], the sampled object $o_d(x,y)$ is expressed by $o_d(x,y) = o(x,y)d(x,y)$ and we can obtain the final binary or phase-only CGH simply by propagating it to the CGH plane. This method does not require any prior information about the original object and CGH and is also computationally inexpensive. Binary and phase CGHs require that the amplitude of the object light in the CGH plane is uniform; however, the amplitude of the object light is usually not uniform. By adding random phase, the object light distribution can be made uniform, but strong speckle noise is generated. In the sampling grid method, sampling grids act as diffraction gratings and diffuse the object light, which helps make the amplitude of the object light on the CGH plane more uniform. This method uses a periodic sampling grid with 1s and 0s, resulting in a large number of voids. A method to obtain the optimal sampling grid with +1s and −1s using a stochastic binary search has been proposed[161]. Once the optimal sampling grid is obtained, it can be applied to other object images. The problem of many voids[159,160] in the reconstructed image was solved by this method.

In reference[162], a tiled phase mask is obtained using an iterative optimization method introduced in the next section. In this method, a small phase mask with random phase is first optimized by iterative optimization. Then, the optimized small mask is tiled to the same size as the original image to obtain the final mask. The final mask is then multiplied by the original object, and a diffraction calculation is performed to obtain the CGH with low speckle.

A method to improve the image quality of CGH by applying a virtual spherical wave (quadratic phase) to the original object has also been proposed. The random phase-free method[163,164] multiplies the original image by a virtual spherical wave $w(x,y) = \exp(i(\pi/\lambda f)(x^2+y^2))$. The focal length f of the spherical wave is determined from the original object size S_o and the size of the CGH S_h as $f = z/(1 - S_h/S_o)$ where z is the distance between the CGH and the original object. The random phase-free method can produce a reconstructed image where the original image is larger than the size of the CGH without the use of random phases. Therefore, the random phase-free method can be used in lensless holographic projection. As another option, we can consider a random phase-free method that uses plane waves instead of spherical waves. Planar waves also have a smooth phase distribution, so speckle noise can be suppressed. However, when plane waves are used, only a reconstructed image of the same size as the hologram can be obtained, which cannot be used for lens-less holographic projection. Since the phase distribution of the reconstructed image is smooth, there is no unwanted interference between the reconstructed pixels, and the reconstructed image has a low speckle noise. When the random phase-free method is applied to the amplitude CGH, we can obtain a high-quality reconstructed image. However, when this method is applied to phase-only CGH, the complex amplitude of the CGH plane should be converted to the phase-only CGH using the error diffusion method to obtain a high-quality image[164]. As aforementioned, the error diffusion method results in a particularly dark reconstructed image; therefore, references[165,166] use iterative optimization, described in the next section, to calculate the optimal spherical wave. This optimal spherical wave can be applied to an arbitrary original object, and the image quality of the reconstructed image can be improved.

Iterative methods

The Gerchberg–Saxton (GS) algorithm[95,167] was

developed for phase recovery problems in metrology. The GS algorithm uses known information in the spatial and Fourier domains as constraints and recovers the desired information via iteration using diffraction calculations. The GS algorithm is an iterative optimization method that is most commonly used in CGH. In optics, the lens-based GS algorithm is also called the iterative Fourier transform algorithm (IFTA) as lens-based light propagation can be simulated by FFT. Free-space propagation (Fresnel diffraction or the angular spectrum method) can also be used in CGH. It is intuitive that CGH can be optimized using the GS algorithm, but it was not well mathematically understood for some time. Later, it was found that the GS algorithm is a kind of nonconvex alternating projection algorithm. The history and theoretical background are detailed in[168].

The optimization flow using the basic GS algorithm consists of the following steps.

1. Calculate the diffraction calculation from the object plane to the CGH plane.

2. Apply a constraint condition to the CGH plane (in the case of phase-only CGH, only the argument of the complex amplitude is taken. In the case of binary CGH, the complex amplitude is binarized).

3. Calculate diffraction from the CGH plane to the object plane.

4. Apply a constraint condition to the object plane (replace the amplitude component of the complex amplitude of the reconstructed image with the original object).

The starting point for optimization can be either the object plane or the CGH plane. Here, the object plane is used as the starting point. First, we set the original object with an appropriate phase (random phase is often used). We calculate the complex amplitude in the CGH plane by performing diffraction calculations. If the final desired CGH is an amplitude, phase, or binary CGH, we convert the complex amplitude by taking the real part, calculating the argument, or converting to binary, respectively. This transformation is the constraint on the CGH plane. Then, the inverse diffraction calculation is performed to obtain the reconstructed image. In the object plane, the phase of the reconstructed image is used as a free parameter, and the amplitude of the reconstructed image is replaced by the original object. By iterating this process, the desired CGH can be gradually obtained. An extension of the GS algorithm to 3D objects has also been performed[169,170]. For 3D objects, the GS algorithm can be applied to each layer with different depths, similar to the layer method[169]. Alternatively, we can record the complex amplitude of the 3D object in any plane and use that complex amplitude as the object plane constraint[170] (similar to the double-constraint GS method described below). Increasing the free parameters makes the optimization easier. The Fidoc algorithm[171] provides a free parameter area outside the original object (this area is also called the noise area).

Although the GS algorithm is easy to implement, it is prone to local minimum solutions. An approach to solve this problem is introducing weighted feedback. The weighted GS algorithm in[172] uses weights calculated by $w^{(k)}(x,y) = \exp(A_o(x,y) - A_r^{(k)}(x,y))$, where A_o is the amplitude of the original object and A_r is the amplitude of the reconstructed image. This method prepares a noise area[171] and imposes the following amplitude constraints on the object area (S) and noise area ($(x,y) \notin S$) as object plane constraints.

$$A^{(k+1)}(x,y) = \begin{cases} w^{(k)}(x,y)A_o(x,y) & (x,y) \in S \\ A_r(x,y) & (x,y) \notin S \end{cases} \tag{17}$$

where k denotes the index of iterations. In this method, the quadratic phase[163–166] is introduced to the initial phase of the object plane to facilitate optimization. In the other weighted GS algorithm of[173], a noise area was prepared[171], and the following amplitude constraints on the object region (S) and noise area as object plane constraints were imposed.

$$A^{(k+1)} = \begin{cases} A^{(k)}(x,y)(A_o(x,y)/A'^{(k)}(x,y))^{\beta_k} & (x,y) \in S \\ A'^{(k)}(x,y) & (x,y) \notin S \end{cases} \tag{18}$$

where $A^{(k)}(x,y)$ denotes the amplitude of the previous optimized object plane, $A_o(x,y)$ denotes the amplitude of the original object, $A'^{(k)}(x,y)$ denotes the reconstructed amplitude from the CGH, and $\beta_k = \sqrt{\beta_{k-1}}$ denotes the adaptive parameter.

The GS algorithms described above use only the amplitude information of the original object as the object plane constraint. The double-constraint GS algorithm[174] uses both the amplitude and phase of the original object as constraints in the object plane. Since this eliminates the free parameter region, a noise area[171] is introduced around the original object. This method can be thought of as complex-amplitude encoding using the GS algorithm.

The GS method can be used for binary CGH optimization, but a direct binary search[175] is also often used. In this method, each pixel of a binary hologram is inverted. If the quality of the reconstructed image is improved by the inversion, the inverted pixel is adopted; otherwise, the original pixel is used. This is done for all the binary pixels and is one iteration. By repeating this iteration, we can obtain a binary CGH that improves the quality of the reconstructed image. Direct binary search requires a large number of diffraction calculations and thus takes a long time to optimize. The stochastic direct binary search[176] speeds up convergence by probabilistically inverting each pixel. The original binary direct search[175] speeds up the computation by adding and subtracting the analytical solution for a single pixel change, without diffraction calculations. A similar idea was proposed to speed up the computation by storing the result of the diffraction calculation for a single pixel in a LUT and then adding and subtracting the data in the LUT according to the inverted pixel from the reconstructed image[177]. The direct binary search is also prone to local minimum solutions. To get rid of the local minimum, simulated annealing is employed, which is a method that probabilistically changes the adoption of inverted pixels. Simulated annealing is a heuristic optimization method that has been used to optimize phase-only CGH[178].

The gradient descent method iteratively optimizes the CGH using the gradient of a differentiable loss function $\mathcal{L}$ (also called the objective function). A basic gradient descent method is expressed as

$$u \leftarrow u - \alpha \frac{\partial \mathcal{L}}{\partial u} \tag{19}$$

where u represents a CGH or object to be optimized; α denotes a learning parameter; $\leftarrow$ means an operator that updates the calculated value on the right side to the variable on the left side. The choice of loss function is arbitrary, but those most commonly used are mean squared error (MSE) and total variation.

Wirtinger holography[96], which applies the gradient descent method with complex derivatives to CGH, has succeeded in obtaining better image quality than the GS algorithm and double-phase hologram. Wirtinger holography minimize the error $\mathcal{L} = \| |\mathcal{P}(\phi)|^2 - I \|$ using the gradient descent method of Eq. (19), where $\mathcal{P}\{\cdot\}$ denotes a diffraction calculation, e.g. we can use Eq. 5 as the diffraction calculation; ϕ is the phase-only hologram; I is the ground truth target image. The loss function involves complex value operations, but fortunately, current deep learning frameworks[179,180] can easily handle such complex derivatives.

In[181], the optimization of a phase CGH generated by the layered method is performed using the gradient descent method. Since it is difficult to perform optimization for each layer of a 3D scene, the complex amplitude of the 3D scene is recorded on a virtual plane, and this is the ground truth, O. The loss function is defined as

$$\mathcal{L}_M(\Re\{O\}, \Re\{\mathcal{P}(H)\}) + \mathcal{L}_M(\Im\{O\}, \Im\{\mathcal{P}(H)\}) + \beta \mathcal{L}_M(|O|_b, |\mathcal{P}(H)|_b), \tag{20}$$

where O is the complex amplitude of the 3D scene; H is the CGH to be optimized; $\mathcal{L}_M(a,b)$ is the MSE of a and b; $\Re$ and $\Im$ are operators that take real and imaginary parts, respectively; $|\cdot|_b$ is the background region of the argument; and β is a parameter that represents the importance of the background region and is an empirically determined hyperparameter. All operations in this loss function are differentiable. By differentiating the loss function with respect to H, the minimum loss with respect to O (i.e., the optimal H) is obtained using the gradient descent method. The gradient descent method is time-consuming as it requires iterative optimization, but it produces reconstructed images with better quality than the double-phase hologram.

Camera-in-the-loop holography

The GS algorithm and Wirtinger holography described above work well for ideal optical systems. However, in real optical systems there are aberrations and SLM nonlinearities. Without correcting for these, it is not possible to obtain the desired accurate reconstructed image. Some techniques have been proposed for manually correcting these aberrations, but they require a lot of work for every new setup configuration. Recently, camera-in-the-loop-holography has been proposed to solve this problem[85,182,183]. In camera-in-the-loop-holography, optical aberrations and aberration-corrected CGHs can be obtained by minimizing the error $\mathcal{L}$ between the actual reconstructed image captured by a camera and the ground-truth image using the gradient descent method of Eq. 19. The study[182] constructed a DNN that can infer the aberrations of a real optical system. This DNN is trained using a dataset of numerically reconstructed images from CGHs and optically reconstructed images captured by a camera (including aberrations of the optical system).

The camera-in-the-loop-holography of[85] optimizes the CGH ϕ so that the aberrations are corrected using gradient descent Eq. 19. The gradient descent method performs the following iterations: $\phi \leftarrow \phi - \alpha(\partial\mathcal{L}/\partial\phi)$, where $\mathcal{L}$ is the error between the actual reconstructed image captured by a camera and the ground-truth image. The gradient $\partial\mathcal{L}/\partial\phi$ can be expressed as

$$\frac{\partial \mathcal{L}}{\partial \phi} = \frac{\partial \mathcal{L}}{\partial \mathcal{P}'} \frac{\partial \mathcal{P}'}{\partial \phi} \tag{21}$$

where $\mathcal{P}'$ represents the actual light propagation in the optical system including aberrations. Since this light propagation operation is unknown, $\partial\mathcal{P}'/\partial\phi$ cannot be calculated directly. By approximating $\mathcal{P}'$ with $\mathcal{P}$, the light propagation of the ideal optical system in Eq. 5, $\partial\mathcal{L}/\partial\phi$ can be calculated as

$$\frac{\partial \mathcal{L}}{\partial \phi} \approx \frac{\partial \mathcal{L}}{\partial \mathcal{P}'} \frac{\partial \mathcal{P}}{\partial \phi} \tag{22}$$

Using this gradient, we can optimize the CGH. This method is unique in that it does not require prior calibration of aberrations in the optical system, nonlinearities of the SLM, alignment errors, etc.; and these are automatically corrected while optimizing the CGH. Since this method requires re-running this optimization calculation for each image, the study[85] showed that by training a DNN with camera-in-the-loop-holography, an optimized CGH can be inferred from the DNN quickly.

In[90], a neural network is trained by camera-in-the-loop holography, and after training, it can output CGHs for any 3D scene. While this training takes a long time, the inference is very quick. Reference[183] incorporates two SLMs into a Michelson interferometer to reduce undiffracted light generated by the SLMs by interference. This method is capable of obtaining a high-quality reconstructed image. The CGH that is displayed on each SLM is H_1, H_2. If the image reconstructed (optically captured) by the Michelson interferometer is R, the loss function can be expressed as $\mathcal{L} = \mathcal{L}_M(R, O)$. From the gradient descent method, we iteratively optimize H_1, H_2 using the following equation:

$$H_1 \leftarrow H_1 - \alpha \partial \mathcal{L} / \partial H_1 \tag{23}$$

$$H_2 \leftarrow H_2 - \alpha \partial \mathcal{L} / \partial H_2. \tag{24}$$

This method produced reconstructed images with better quality than double-phase holograms, Wirtinger holography, and camera-in-the-loop holography using a single SLM.

Visual quality assessment

Visual quality assessment (VQA) plays a key role for any signal processing pipeline that drives a display system. This is true for any display setups, but especially for comparing developing display prototypes and parametrizing algorithms. The perceived quality of a visualized CGH-DH is determined by a combination of the scenes render quality, as well as the preservation of its various 3D focal cues. All of which can be impaired for many reasons in a holographic display.

A reduced quality can be due to the chosen CGH method, namely its capabilities, approximations, or its parametrization. Examples are: a reduced range of supported visual effects (e.g. no shadows, no specular lighting), the use of computational accelerations (e.g. LUTs, coarse occlusion masks, approximations to the diffraction integral) and e.g. a limited render-primitive count or erroneous/low resolution 3D input data.

Visual quality may also be impaired due to further steps of the holographic signal processing pipeline[3], such as lossy compression, hologram tiling, or modulation-narrowing transforms to match the supported screen capabilities (e.g. phase-only SLMs, binary SLMs, 8 bit dynamic range 2D displays).

Finally, optical components of the display setup may further impair quality — such as through: low contrast and/or limited spatial-resolution SLMs, reference wavefront errors, optical aberrations, alignment errors.

Whenever all components should be taken into consideration, only *subjective assessment* by human observers is a viable strategy. However, upon the development of each individual component, typically non-visual, targeted *objective quality assessment* is predominantly used due to its fast and automated evaluation.

Subjective quality assessment

For subjective VQA of DHs various distorted and ground-truth reconstructions are presented to several dozens of human observers. Each of them is asked to grade the quality for which various schemes exist, see e.g.[184]. The subsequently formed mean opinion scores serve as the ultimate ground-truth of visual quality. The assembly of a test dataset that results in meaningful scores requires a lot of experience.It can even be impossible as in the case where visual defects are not very comparable. For example, information missing in a reconstruction versus a blurred version of the entire scene. Another hindrance for subjective tests, esp. on DH display setups, can be hardware limitations[184] that do not allow for visualizing the complete content. These limitations can be for example a too narrow angular field-of-view, limited reconstruction size, limited depth-refocusability, or low contrast. Subjective tests always involve a lot of administrative and organizational overhead and easily take several weeks to months to complete. To ensure for some degree of reproducibility, they require well-defined testing procedures and monitoring the visual conditions of the test subjects (e.g. test for visual acuity) and the test environment (e.g. ambient light intensity, ambient light color temperature, observer distance). In short, subjective tests are a heavy-weight, accurate tool and most importantly they cannot be utilized for real-time monitoring purposes.

Objective quality assessment

Objective quality assessment of DHs entails the evaluation of mathematical functions on the numerical reconstructions or wavefields. These functions are called *objective quality metrics*, should have a known relation to the perceived visual quality of the reconstructed DH. They may be reference-based or reference-free. No-reference objective quality metrics require an advanced knowledge of the assessed data statistics and properties and therefore much harder to develop. For VQA of DHs, almost exclusively full reference-based metrics are used, which require some sort of ground-truth to compare to. Examples of some of widely used image quality metrics, borrowed from 2D image VQA, are the peak signal-to-noise ratio (PSNR) and structural similarity index measure (SSIM)[185]. The great advantages of objective measures are reproducibility across different setups and the small effort of evaluation allowing for fast quality monitoring. The disadvantages are that they: necessarily require a ground-truth, can be internally limited in dynamic range, and are only an approximation to the subjectively perceived quality. Nonetheless, developing efficient objective measures to specifically cope with complexities of evaluating DH and to provide exclusive assessment of DH 3D aspects is greatly sought after and currently an open topic in the field.

Quality assessment problems for DHs

Some properties of DHs cause additional complexities and prevent simple re-use of QA strategies as used for videos and images. One problem that both VQA strategies share for any plenoptic content, including DHs, is that some form of score pooling is required for the different viewpoints (gaze angles and focal distances). One and the same hologram may appear heavily distorted only at certain viewpoints and the importance of all supported viewpoints has to be weighted. The pooling allows for example to prioritize good quality center view reconstructions over reconstructions from large viewing angles or vice versa. For DHs the pooling strategy is currently often chosen as a simple arithmetic mean. Over time best practices will be established likely for all plenoptic imaging techniques, as this problem is also studied for point cloud imaging/volumetric displays[186–189] and lightfields[190,191]).

Aside from practical limitations such as limited compute capabilities or difficulties arising from multi-terabyte-sized test datasets, there are several problems specific to DH. Publicly available data sets including both optically recorded and CGH holograms from various 3D models/real objects with diverse scene-composition and object characteristics are currently scarce. Only few of such public data sets have been created in recent years[37,38,54,192–195].

True optical holographic display setups are currently limited in their maximal resolutions and screen sizes supported by modern spatial light modulators. In addition, only few labs combine the required expertise of optical engineering and VQA. Therefore, frequently numerical reconstructions are displayed during subjective VQA on alternative display types, such as conventional 2D displays and light field displays; see for example Fig. 10. Fortunately, the test sensitivity of 2D displays seems to be the highest at least for compression artefacts as found in a first study[184].

Because holography is a coherent imaging technique it suffers from a multiplicative, uncorrelated speckle noise, similar to radar[196] and ultrasound imaging[197]. Speckle noise is a problem for subjective VQA and even more for element-wise applied objective VQA[198,199]. The noise is less visible and comparatively easy to suppress in optical setups by trading off some resolution, e.g. through the use of partially coherent illumination in the display setup. In numerical reconstructions however, speckle denoising has to be performed per viewpoint and incurs a large computational cost. An excellent review of optical and algorithmic speckle denoising techniques is provided in[143]. Speckle denoising methods were evaluated in the context of metrological applications in e.g.[200], and for visualization of macroscopic scenes in[199,201,202].

Fig. 10 Comparison of display setups used for subjective VQA of DHs. Approximate screen diagonals are 30 cm, 200 cm, and 5 cm, respectively. Reprinted with permission from[184] © The Optical Society.

Since currently numerical reconstructions are a major tool for both VQA tracks, their calculation in large numbers becomes a problem, as it is extremely costly for high-resolution color DHs. It can take CPU-weeks for large databases when evaluating multiple test combinations.

To avoid the need for pooling and the need for interim reconstructions, objective metrics have also been studied in direct application to wavefields. As most metrics expect a given dynamic range, non-negative content, or real-valued inputs their direct application to DHs is however not evident either. Nonetheless, correlations with subjective quality scores were investigated in[198,203,204] but no general metric has emerged as a reliable choice yet and more research is needed.

The most notable progress in the past years on the VQA of non-metrological applications can be mainly attributed to the JPEG Pleno Holography[205] standardization effort on the development of novel DH compression schemes. It has already lead to several publications and a first consensus of several leading research groups on a first subjective[184,199,206] and objective[198,199,207] evaluation pipeline.

Summary

Although VQA has a high importance for macroscopic display applications, there exists only early research results which are limited to the preliminary assessment of compression artifacts and denoising applications. More complex problems associated to VQA of binary and colour holograms as well as dynamic holography, are almost completely untouched till now[208]. Therefore, significant additional research is needed for the advancement of holographic signal processing pipelines[3], including CGH methods.

Conclusion

Tremendous advances in CGH algorithms in terms of both visual quality/photorealism and computational complexity have been made, especially in recent years. Multiple classes of CGH techniques have been developed with different trade-offs, which should be chosen depending on the needs of the holographic display system's application. Given the high computational requirements of holography, care was taken to optimize many aspects of CGH, starting from approximating the laws of diffraction, compact mathematical representation of graphical basis elements, algorithm design with acceleration structures tailored to modern computing hardware, optimizing for the display capability limitations of SLM devices and even hardware software co-design with dedicated ASIC for CGH.

Despite all these optimizations, the computation of high-resolution, wide viewing angle photo-realistic DHs in real-time remains an important challenge to be tackled. Nevertheless, existing real-time CGH implementations show that driving high quality holographic displays is already feasible and ahead of current displaying hardware resolutions.

In the near term, we expect head-mounted-display type fully diffractive displays to be the first commercially available holographic display systems given their relatively lower technical requirements, likely before the end of the decade. Current CGH algorithms can already drive these display types in real-time, but doing so at high visual quality and with low energy consumption is still an active domain of research. Large full-parallax holographic display systems with high SBP will probably take longer to be widely available due to their much higher computational and display requirements.

In the future, we believe that the further close integration of CGH algorithms with dedicated hardware systems will likely yield the biggest computational gains, enabling immersive holographic television.

Acknowledgements

This research was funded by the Research Foundation - Flanders (FWO), Junior postdoctoral fellowship (12ZQ220N), the joint JSPS-FWO scientific cooperation program (VS07820N) and the Japan Society for the Promotion of Science (19H04132 and JPJSBP120202302). We would like to thank Ayyoub Ahar for advising us on visual quality assessment.

Author details

[1]Department of Electronics and Informatics (ETRO), Vrije Universiteit Brussel (VUB), Pleinlaan 2, B-1050 Brussel, Belgium. [2]IMEC, Kapeldreef 75, B-3001 Leuven, Belgium. [3]Graduate School of Engineering, Chiba University, 1-33 Yayoi-cho, Inage-ku, Chiba, Chiba, Japan.

Conflict of interest

The authors declare that they have no conflict of interest.

Received: 16 September 2021 Revised: 28 April 2022 Accepted: 07 May 2022
Accepted article preview online: 23 May 2022
Published online: 10 June 2022

References

1. Matrecano, M., Paturzo, M. & Ferraro, P. Extended focus imaging in digital holographic microscopy: a review. *Optical Engineering* **53**, 112317 (2014).
2. Tahara, T. et al. Digital holography and its multidimensional imaging applications: a review. *Microscopy* **67**, 55-67 (2018).
3. Blinder, D. et al. Signal processing challenges for digital holographic video display systems. *Signal Process. Image* **70**, 114-130 (2019).
4. Kress, B. C. Optical Architectures for Augmented-, Virtual-, and Mixed-Reality Headsets. 270 pp. (Bellingham: SPIE, 2020).
5. Yoshikawa, H. & Takei, K. Development of a compact direct fringe printer for computer-generated holograms. in Practical Holography XVIII: Materials and Applications. **5290** (San Jose: SPIE, 2004), 114-121.

6. Jeon, H. et al. High-resolution binary hologram printing methods. in Practical Holography XXXIV: Displays, Materials, and Applications. **11306** (San Francisco: SPIE, 2020), 122-127.
7. Duan, X. et al. Full-color see-through near-eye holographic display with 80£ field of view and an expanded eye-box. *Optics Express* **28**, 31316-31329 (2020).
8. Kress, B. C. & Cummings, W. J. Optical architecture of HoloLens mixed reality headset. in Digital Optical Technologies 2017. **10335** (SPIE, 2017), 124-133.
9. Häussler, R. et al. Large real-time holographic displays: from prototypes to a consumer product. 2009.
10. Jiang, Q., Jin, G. & Cao, L. When metasurface meets hologram: principle and advances. *Advances in Optics and Photonics* **11**, 518-576 (2019).
11. Finke, G., Kujawińska, M. & Kozacki, T. Visual perception in multi SLM holographic displays. *Applied Optics* **54**, 3560-3568 (2015).
12. Shi, L. et al. Near-eye Light Field Holographic Rendering with Spherical Waves for Wide Field of View Interactive 3D Computer Graphics. *ACM Trans. Graph.* **36**, 236ú:1-236:17 (2017).
13. Chlipala, M. & Kozacki, T. Color LED DMD holographic display with high resolution across large depth. *Optics Letters* **44**, 4255-4258 (2019).
14. Choo, H.-G. et al. Fourier digital holography of real scenes for 360£ tabletop holographic displays. *Applied Optics* **58**, G96-G103 (2019).
15. Cuche, E., Marquet, P. & Depeursinge, C. Simultaneous amplitude-contrast and quantitative phasecontrast microscopy by numerical reconstruction of Fresnel off-axis holograms. *Applied Optics* **38**, 6994-7001 (1999).
16. Murata, S. & Yasuda, N. Potential of digital holography in particle measurement. *Optics & Laser Technology* **32**, 567-574 (2000).
17. Xu, W. et al. Digital in-line holography for biological applications. *Proceedings of the National Academy of Sciences of the United States of America* **98**, 11301-11305 (2001).
18. Javidi, B. & Tajahuerce, E. Three-dimensional object recognition by use of digital holography. *Optics Letters* **25**, 610-612 (2000).
19. Noda, T., Kawata, S. & Minami, S. Three-dimensional phase-contrast imaging by a computed-tomography microscope. *Applied optics* **31**, 670-674 (1992).
20. Paturzo, M. et al. Synthesis and display of dynamic holographic 3D scenes with real-world objects. *Optics Express* **18**, 8806-8815 (2010).
21. Clemente, P. et al. Single-pixel digital ghost holography. *Physical Review A* **86**, 041803 (2012).
22. Kozacki, T. et al. Fourier horizontal parallax only computer and digital holography of large size. *Optics Express* **29**, 18173-18191 (2021).
23. Kozacki, T. et al. Single-shot digital multiplexed holography for the measurement of deep shapes. *Optics Express* **29**, 21965-21977 (2021).
24. Goodman, J. W. Introduction to Fourier Optics (ed Goodman, J. W.) (New York: W. H. Freeman and Company, Aug. 17, 2017).
25. Tsang, P., Poon, T.-C. & Wu, Y. Review of fast methods for point-based computer-generated holography. *Photonics Research* **6**, 837-846 (2018).
26. Shimobaba, T., Masuda, N. & Ito, T. Simple and fast calculation algorithm for computer-generated hologram with wavefront recording plane. *Optics Letters* **34**, 3133-3135 (2009).
27. Kang, H., Yamaguchi, T. & Yoshikawa, H. Accurate phase-added stereogram to improve the coherent stereogram. *Applied Optics* **47**, D44-D54 (2008).
28. Kim, S.-C. & Kim, E. -S. Effective generation of digital holograms of three-dimensional objects using a novel look-up table method. *Applied Optics* **47**, D55-D62 (2008).
29. Blinder, D. et al. Analytic computation of line-drawn objects in computer generated holography. *Optics Express* **28**, 31226-31240 (2020).
30. Blinder, D., Nishitsuji, T. & Schelkens, P. RealTime Computation of 3D Wireframes in ComputerGenerated Holography. *IEEE Transactions on Image Processing* **30**, 9418-9428 (2021).
31. Nishitsuji, T. et al. GPU-accelerated calculation of computer-generated holograms for line-drawn objects. *Optics Express* **29**, 12849-12866 (2021).
32. Pan, Y. et al. Fast polygon-based method for calculating computer-generated holograms in threedimensional display. *Applied Optics* **52**, A290-A299 (2013).
33. Zhang, Y.-P. et al. Fast generation of full analytical polygon-based computer-generated holograms. *Optics Express* **26**, 19206-19224 (2018).
34. Lee, W. et al. Semi-analytic texturing algorithm for polygon computer-generated holograms. *Optics Express* **22**, 31180-31191 (2014).
35. Ji, Y.-M., Yeom, H. & Park, J. -H. Efficient texture mapping by adaptive mesh division in mesh-based computer generated hologram. *Optics Express* **24**, 28154-28169 (2016).
36. Symeonidou, A. et al. Computer-generated holograms by multiple wavefront recording plane method with occlusion culling. *Optics Express* **23**, 22149-22161 (2015).
37. Symeonidou, A., Blinder, D. & Schelkens, P. Colour computer-generated holography for point clouds utilizing the Phong illumination model. *Optics Express* **26**, 10282-10298 (2018).
38. Symeonidou, A. et al. Speckle noise reduction for computer generated holograms of objects with diffuse surfaces. in Optics, Photonics and Digital Technologies for Imaging Applications IV. **9896** (Brussels: SPIE, 2016), 67-76.
39. User "artmaxwell" . Porcelain Teacup. (2021). at https://www.turbosquid.com/3d-models/free-max-mode-porcelain-teacup/486730.
40. User "macrovector" . Seamless floral pattern on uniform background. (2021). at https://www.freepik.com/vectors/background.
41. Shimobaba, T. et al. Computational wave optics library for C++: CWO++ library. *Computer Physics Communications* **183**, 1124-1138 (2012).
42. Muffoletto, R. P., Tyler, J. M. & Tohline, J. E. Shifted Fresnel diffraction for computational holography. *Optics Express* **15**, 5631-5640 (2007).
43. Matsushima, K., Schimmel, H. & Wyrowski, F. Fast calculation method for optical diffraction on tilted planes by use of the angular spectrum of plane waves. *J. Opt. Soc. Am. A* **20**, 1755-1762 (2003).
44. Okada, N. et al. Band-limited double-step Fresnel diffraction and its application to computer-generated holograms. *Optics Express* **21**, 9192-9197 (2013).
45. Chang, C. et al. Simple calculation of a computergenerated hologram for lensless holographic 3D projection using a nonuniform sampled wavefront recording plane. *Applied Optics* **55**, 7988-7996 (2016).
46. Matsushima, K. & Nakahara, S. Extremely highdefinition full-parallax computer-generated hologram created by the polygon-based method. *Applied Optics* **48**, H54-H63 (2009).
47. Nishi, H. & Matsushima, K. Rendering of specular curved objects in polygon-based computer holography. *Applied Optics* **56**, F37-F44 (2017).
48. Matsushima, K. Introduction to Computer Holography: Creating Computer-Generated Holograms as the Ultimate 3D Image (Springer Nature, 2020).
49. Matsushima, K., Nakamura, M. & Nakahara, S. Silhouette method for hidden surface removal in computer holography and its acceleration using the switch-back technique. *Optics Express* **22**, 24450-24465

(2014).
50. Blinder, D. et al. Photorealistic computer generated holography with global illumination and path tracing. *Optics Letters* **46**, 2188-2191 (2021).
51. Website of the INTERFERE project ERC-CG-2013-PE7, European Research Council consolidator grant http://erc-interfere.eu.
52. Ichikawa, T., Yamaguchi, K. & Sakamoto, Y. Realistic expression for full-parallax computer-generated holograms with the ray-tracing method. *Applied Optics* **52**, A201-A209 (2013).
53. Sun, M. et al. Acceleration and expansion of a photorealistic computer-generated hologram using backward ray tracing and multiple off-axis wavefront recording plane methods. *Optics Express* **28**, 34994-35005 (2020).
54. Gilles, A. et al. Hybrid approach for fast occlusion processing in computer-generated hologram calculation. *Applied Optics* **55**, 5459-5470 (2016).
55. Shimobaba, T. & Ito, T. Fast generation of computergenerated holograms using wavelet shrinkage. *Optics Express* **25**, 77-87 (2017).
56. Pi, D. et al. Acceleration of computer-generated hologram using wavefront-recording plane and lookup table in three-dimensional holographic display. *Optics Express* **28**, 9833-9841 (2020).
57. Wang, Z. et al. Resolution priority holographic stereogram based on integral imaging with enhanced depth range. *Optics Express* **27**, 2689-2702 (2019).
58. Takaki, Y. & Ikeda, K. Simplified calculation method for computer-generated holographic stereograms from multi-view images. *Optics Express* **21**, 9652-9663 (2013).
59. Yamaguchi, M. et al. Phase-added stereogram: calculation of hologram using computer graphics technique. *Proceedings of SPIE* **1914**, 25-31 (1993).
60. Kang, H. et al. Compensated phase-added stereogram for real-time holographic display. *Optical Engineering* **46**, 11-11-11 (2007).
61. Padmanaban, N., Peng, Y. & Wetzstein, G. Holographic Near-Eye Displays Based on Overlap-Add Stereograms. *ACM Trans. Graph.* **38**, (2019).
62. Kim, H. G., Jeong, H. & Ro, Y. M. Acceleration of the calculation speed of computer-generated holograms using the sparsity of the holographic fringe pattern for a 3D object. *Optics Express* **24**, 25317-25328 (2016).
63. Kim, H. G. & Ro, Y. M. Ultrafast layer based computer-generated hologram calculation with sparse template holographic fringe pattern for 3-D object. *Optics Express* **25**, 30418-30427 (2017).
64. Jia, J., Si, J. & Chu, D. Fast two-step layer-based method for computer generated hologram using subsparse 2D fast Fourier transform. *Optics Express* **26**, 17487-17497 (2018).
65. Arai, D. et al. An accelerated hologram calculation using the wavefront recording plane method and wavelet transform. *Optics Communications* **393**, 107-112 (2017).
66. Blinder, D. & Schelkens, P. Accelerated computer generated holography using sparse bases in the STFT domain. *Optics Express* **26**, 1461-1473 (2018).
67. Blinder, D. Direct calculation of computer-generated holograms in sparse bases. *Optics Express* **27**, 23124-23137 (2019).
68. Lucente, M. E. Interactive computation of holograms using a look-up table. *Journal of Electronic Imaging* **2**, 28-34 (1993).
69. Pan, Y. et al. Fast CGH computation using S-LUT on GPU. *Optics Express* **17**, 18543-18555 (2009).
70. Jia, J. et al. Reducing the memory usage for effectivecomputer-generated hologram calculation using compressed look-up table in full-color holographic display. *Applied Optics* **52**, 1404-1412 (2013).
71. Yang, Z. et al. A new method for producing computer generated holograms. *Journal of Optics* **14**, 095702 (2012).
72. Jiao, S., Zhuang, Z. & Zou, W. Fast computer generated hologram calculation with a mini look-up table incorporated with radial symmetric interpolation. *Optics Express* **25**, 112-123 (2017).
73. Nishitsuji, T. et al. Fast calculation of computergenerated hologram of line-drawn objects without FFT. *Optics Express* **28**, 15907-15924 (2020).
74. Kim, S.-C. et al. Fast generation of video holograms of three-dimensional moving objects using a motion compensation-based novel look-up table. *Optics Express* **21**, 11568-11584 (2013).
75. Dong, X.-B., Kim, S.-C. & Kim, E. -S. MPEG-based novel look-up table for rapid generation of video holograms of fast-moving three-dimensional objects. *Optics Express* **22**, 8047-8067 (2014).
76. Kwon, M.-W., Kim, S.-C. & Kim, E. -S. Threedirectional motion-compensation mask-based novel look-up table on graphics processing units for videorate generation of digital holographic videos of threedimensional scenes. *Applied Optics* **55**, A22-A31 (2016).
77. Cao, H.-K. & Kim, E. -S. Faster generation of holographic videos of objects moving in space using a spherical hologram-based 3-D rotational motion compensation scheme. *Optics Express* **27**, 29139-29157 (2019).
78. Symeonidou, A. et al. Efficient holographic video generation based on rotational transformation of wavefields. *Optics Express* **27**, 37383-37399 (2019).
79. Shimobaba, T. et al. Motion Parallax Holograms Generated from an Existing Hologram. *Applied Sciences* **11**, (2021).
80. Birnbaum, T. et al. Object-based digital hologram segmentation and motion compensation. *Optics Express* **28**, 11861-11882 (2020).
81. Sitzmann, V. et al. Light field networks: Neural scene representations with single-evaluation rendering. *Advances in Neural Information Processing Systems* **34**, (2021).
82. Voulodimos, A. et al. Deep learning for computer vision: A brief review. *Computational intelligence and neuroscience* **2018** (2018).
83. Shimobaba, T. et al. Deep-learning-assisted hologram calculation via low-sampling holograms. in 2019 8th International Congress on Advanced Applied Informatics (IIAI-AAI) (2019), 936-941.
84. Shimobaba, T. et al. Dynamic-range compression scheme for digital hologram using a deep neural network. *Optics letters* **44**, 3038-3041 (2019).
85. Shi, L. et al. Towards real-time photorealistic 3D holography with deep neural networks. *Nature* **591**, 234-239 (2021).
86. Horisaki, R. et al. Three-dimensional deeply generated holography [Invited]. *Applied Optics* **60**, A323-A328 (2021).
87. Jiao, S. et al. Compression of phase-only holograms with JPEG standard and deep learning. *Applied Sciences* **8**, 1258 (2018).
88. Park, D.-Y. & Park, J. -H. Hologram conversion for speckle free reconstruction using light field extraction and deep learning. *Optics Express* **28**, 5393-5409 (2020).
89. Ishii, Y. et al. Optimization of phase-only holograms calculated with scaled diffraction calculation through deep neural networks. *Applied Physics B* **128**, 1-11 (2022).
90. Peng, Y. et al. Neural holography with camera-in-theloop training. *ACM Transactions on Graphics (TOG)* **39**, 1-14 (2020).
91. Goodfellow, I., Bengio, Y. & Courville, A. Deep Learning (Cambridge: MIT Press, 2016).
92. Horisaki, R., Takagi, R. & Tanida, J. Deep-learninggenerated holography. *Applied Optics* **57**, 3859-3863 (2018).
93. Eybposh, M. H. et al. DeepCGH: 3D computergenerated holography using deep learning. *Optics Express* **28**, 26636-26650 (2020).
94. Wu, J. et al. High-speed computer-generated holography using an autoencoder-based deep neural network. *Optics Letters* **46**, 2908-2911 (2021).

95. Gerchberg, R. W. & Saxton, W. O. A practical algorithm for the determination of phase from image and diffraction plane pictures. *Optik* **35**, 237-246 (1972).
96. Chakravarthula, P. et al. Wirtinger holography for near-eye displays. *ACM Transactions on Graphics (TOG)* **38**, 1-13 (2019).
97. Goi, H., Komuro, K. & Nomura, T. Deep-learningbased binary hologram. *Applied Optics* **59**, 7103-7108 (2020).
98. Dai, P. et al. Acceleration of fully computed hologram stereogram using lookup table and wavefront recording plane methods. *Applied Optics* **60**, 1814-1820 (2021).
99. Wang, Y. et al. Hardware implementations of computergenerated holography: a review. *Optical Engineering* **59**, 1-30 (2020).
100. Ikawa, S. et al. Real-time color holographic video reconstruction using multiple-graphics processing unit cluster acceleration and three spatial light modulators. *Chinese Optics Letters* **18**, 010901 (2020).
101. Christopher, P. J. et al. HoloGen: An open-source toolbox for high-speed hologram generation. *Computer Physics Communications* **270**, 108139 (2022).
102. Blinder, D. & Schelkens, P. Accelerating phase-added stereogram calculations by coefficient grouping for digital holography. in Proceedings of SPIE 11353, Optics, Photonics and Digital Technologies for Imaging Applications VI (Online Only, France: SPIE, 2020), 1–10.
103. Blinder, D. & Schelkens, P. Phase added substereograms for accelerating computer generated holography. *Optics Express* **28**, 16924-16934 (2020).
104. Zhao, Y. et al. Accurate calculation of computergenerated holograms using angular-spectrum layeroriented method. *Optics Express* **23**, 25440-25449 (2015).
105. Lee, B. et al. Wide-angle speckleless DMD holographic display using structured illumination with temporal multiplexing. *Optics Letters* **45**, 2148-2151 (2020).
106. Dong, D. et al. Fixed-Point Accuracy Analysis of 2D FFT for the Creation of Computer Generated Holograms. in 2019 IEEE Global Conference on Signal and Information Processing (GlobalSIP) (2019), 1-5.
107. Blinder, D. & Schelkens, P. Fast Low-Precision Computer-Generated Holography on GPU. *Applied Sciences* **11**, (2021).
108. Bove, V. M. & Watlington, J. A. Cheops: A reconfigurable data-flow system for video processing. *IEEE Transactions on Circuits and Systems for video Technology* **5**, 140-149 (1995).
109. Watlington, J. A. et al. Hardware architecture for rapid generation of electro-holographic fringe patterns. in Practical Holography IX. **2406** (1995), 172-183.
110. Ito, T. et al. Special-purpose computer HORN-1 for reconstruction of virtual image in three dimensions. *Computer physics communications* **82**, 104-110 (1994).
111. Ito, T. et al. Special-purpose computer for holography HORN-2. *Computer physics communications* **93**, 13-20 (1996).
112. Shimobaba, T. et al. Special-purpose computer for holography HORN-3 with PLD technology. *Computer physics communications* **130**, 75-82 (2000).
113. Shimobaba, T. & Ito, T. An efficient computational method suitable for hardware of computer-generated hologram with phase computation by addition. *Computer Physics Communications* **138**, 44-52 (2001).
114. Shimobaba, T., Hishinuma, S. & Ito, T. Specialpurpose computer for holography HORN-4 with recurrence algorithm. *Computer physics communications* **148**, 160-170 (2002).
115. Ito, T. et al. Special-purpose computer HORN-5 for a real-time electroholography. *Optics Express* **13**, 1923-1932 (2005).
116. Ichihashi, Y. et al. HORN-6 special-purpose clustered computing system for electroholography. *Optics Express* **17**, 13895-13903 (2009).
117. Okada, N. et al. Special-purpose computer HORN-7 with FPGA technology for phase modulation type electro-holography. in 19th International Display Workshops in Conjunction with Asia Display 2012, IDW/AD 2012 (2012), 1284-1287.
118. Nishitsuji, T. et al. Special-purpose computer HORN-8 for phase-type electro-holography. *Optics express* **26**, 26722-26733 (2018).
119. Sugie, T. et al. High-performance parallel computing for next-generation holographic imaging. *Nature Electronics* **1**, 254-259 (2018).
120. Seo, Y.-H. et al. An architecture of a high-speed digital hologram generator based on FPGA. *Journal of Systems Architecture* **56**, 27-37 (2010).
121. Seo, Y.-H., Lee, Y.-H. & Kim, D. -W. ASIC chipset design to generate block-based complex holographic video. *Applied optics* **56**, D52-D59 (2017).
122. Yasuki, D. et al. Dedicated processor for hologram calculation using sparse Fourier bases. *Applied Optics* **59**, 8029-8037 (2020).
123. Pang, Z.-Y. et al. Hardware architecture for full analytical Fraunhofer computer-generated holograms. *Optical Engineering* **54**, 095101 (2015).
124. Masuda, N. et al. Special purpose computer for digital holographic particle tracking velocimetry. *Optics Express* **14**, 587-592 (2006).
125. Abe, Y. et al. Special purpose computer system for flow visualization using holography technology. *Optics express* **16**, 7686-7692 (2008).
126. Cheng, C.-J. et al. Efficient FPGA-based Fresnel transform architecture for digital holography. *Journal of Display Technology* **10**, 272-281 (2013).
127. Liu, X. et al. Bitwidth-Optimized Energy-Efficient FFT Design via Scaling Information Propagation. in Proceedings of the 58th ACM/EDAC/IEEE Design Automation Conference (IEEE Press, San Francisco, CA, USA, 2021).
128. Kim, H. et al. A single-chip FPGA holographic video processor. *IEEE Transactions on Industrial Electronics* **66**, 2066-2073 (2018).
129. An, J. et al. Slim-panel holographic video display. *Nature communications* **11**, 1-7 (2020).
130. Buckley, E. Real-time error diffusion for signal-tonoise ratio improvement in a holographic projection system. *Journal of Display Technology* **7**, 70-76 (2011).
131. Dong, D. et al. Cost-optimized heterogeneous FPGA architecture for non-iterative hologram generation. *Applied Optics* **59**, 7540-7546 (2020).
132. Neto, L. G., Roberge, D. & Sheng, Y. Full-range, continuous, complex modulation by the use of two coupled-mode liquid-crystal televisions. *Applied Optics* **35**, 4567-4576 (1996).
133. Makowski, M. et al. Complex light modulation for lensless image projection. *Chinese Optics Letters* **9**, 120008 (2011).
134. Makowski, M. in Holographic Materials and Applications (ed Kumar, M.) chap. 5 (Rijeka: IntechOpen, 2019).
135. Liu, J.-P. et al. Complex Fresnel hologram display using a single SLM. *Applied optics* **50**, H128-H135 (2011).
136. Bryngdahl, O. & Lohmann, A. Single-sideband holography. *JOSA* **58**, 620-624 (1968).
137. Takaki, Y. & Tanemoto, Y. Band-limited zone plates for single-sideband holography. *Applied optics* **48**, H64-H70 (2009).
138. Hsueh, C.-K. & Sawchuk, A. A. Computer-generated double-phase holograms. *Applied optics* **17**, 3874-3883 (1978).
139. Mendoza-Yero, O., Mınguez-Vega, G. & Lancis, J. Encoding complex fields by using a phase-only optical element. *Optics letters* **39**, 1740-1743 (2014).
140. Shimobaba, T. et al. Simple complex amplitude encoding of a phase-

only hologram using binarized amplitude. *Journal of Optics* **22**, 045703 (2020).
141. Goorden, S. A., Bertolotti, J. & Mosk, A. P. Superpixel-based spatial amplitude and phase modulation using a digital micromirror device. *Optics express* **22**, 17999-18009 (2014).
142. Jiao, S. et al. Complex-amplitude holographic projection with a digital micromirror device (DMD) and error diffusion algorithm. *IEEE Journal of Selected Topics in Quantum Electronics* **26**, 1-8 (2020).
143. Bianco, V. et al. Strategies for reducing speckle noise in digital holography. *Light: Science & Applications* **7**, 1-16 (2018).
144. Bianco, V. et al. Quasi noise-free digital holography. *Light: Science & Applications* **5**, e16142-e16142 (2016).
145. Amako, J., Miura, H. & Sonehara, T. Speckle-noise reduction on kinoform reconstruction using a phaseonly spatial light modulator. *Applied optics* **34**, 3165-3171 (1995).
146. Buckley, E. 70.2: Invited paper: holographic laser projection technology. in SID Symposium Digest of Technical Papers. **39** (2008), 1074–1079.
147. Wang, D. et al. Holographic Display System to Suppress Speckle Noise Based on Beam Shaping. in Photonics. **8** (2021), 204.
148. Christopher, P. J. et al. Improving performance of single-pass real-time holographic projection. *Optics Communications* **457**, 124666 (2020).
149. Takaki, Y. & Yokouchi, M. Speckle-free and grayscale hologram reconstruction using time-multiplexing technique. *Optics express* **19**, 7567-7579 (2011).
150. Makowski, M. Minimized speckle noise in lens-less holographic projection by pixel separation. *Optics express* **21**, 29205-29216 (2013).
151. Mori, Y., Fukuoka, T. & Nomura, T. Speckle reduction in holographic projection by random pixel separation with time multiplexing. *Applied optics* **53**, 8182-8188 (2014).
152. Curtis, V. R. et al. DCGH: Dynamic Computer Generated Holography for Speckle-Free, High Fidelity 3D Displays. in 2021 IEEE Virtual Reality and 3D User Interfaces (VR) (2021), 1-9.
153. Hauck, R. & Bryngdahl, O. Computer-generated holograms with pulse-density modulation. *JOSA A* **1**, 5-10 (1984).
154. Tsang, P. W. M. & Poon, T. -C. Novel method for converting digital Fresnel hologram to phase-only hologram based on bidirectional error diffusion. *Optics express* **21**, 23680-23686 (2013).
155. Yang, G. et al. Error diffusion method with optimized weighting coefficients for binary hologram generation. *Applied optics* **58**, 5547-5555 (2019).
156. Cheremkhin, P. A. et al. Comparative analysis of offaxis digital hologram binarization by error diffusion. *Journal of Optics* (2021).
157. Matsumoto, Y. & Takaki, Y. Improvement of grayscale representation of horizontally scanning holographic display using error diffusion. *Optics letters* **39**, 3433-3436 (2014).
158. Takaki, Y., Matsumoto, Y. & Nakajima, T. Color image generation for screen-scanning holographic display. *Optics express* **23**, 26986-26998 (2015).
159. Tsang, P. et al. Computer generation of binary Fresnel holography. *Applied optics* **50**, B88-B95 (2011).
160. Tsang, P., Chow, Y.-T. & Poon, T. -C. Generation of phase-only Fresnel hologram based on downsampling. *Optics express* **22**, 25208-25214 (2014).
161. Tsang, P. et al. Optimal sampled phase-only hologram (OSPOH). *Optics Express* **29**, 25488-25498 (2021).
162. Velez-Zea, A. & Torroba, R. Optimized random phase tiles for non-iterative hologram generation. *Applied optics* **58**, 9013-9019 (2019).
163. Shimobaba, T. & Ito, T. Random phase-free computergenerated hologram. *Optics express* **23**, 9549-9554 (2015).
164. Shimobaba, T. et al. Random phase-free kinoform for large objects. *Optics express* **23**, 17269-17274 (2015).
165. Nagahama, Y. et al. Speeding up image quality improvement in random phase-free holograms using ringing artifact characteristics. *Applied optics* **56**, F61-F66 (2017).
166. Chen, L. et al. Non-iterative phase hologram generation with optimized phase modulation. *Optics express* **28**, 11380-11392 (2020).
167. Fienup, J. R. Phase retrieval algorithms: a comparison. *Applied optics* **21**, 2758-2769 (1982).
168. Bauschke, H. H., Combettes, P. L. & Luke, D. R. Phase retrieval, error reduction algorithm, and Fienup variants: a view from convex optimization. *JOSA A* **19**, 1334-1345 (2002).
169. Makowski, M. et al. Three-plane phase-only computer hologram generated with iterative Fresnel algorithm. *Optical Engineering* **44**, 125805 (2005).
170. Pang, H. et al. Accurate hologram generation using layer-based method and iterative Fourier transform algorithm. *IEEE Photonics Journal* **9**, 1-8 (2016).
171. Georgiou, A. et al. Aspects of hologram calculation for video frames. *Journal of Optics A: Pure and Applied Optics* **10**, 035302 (2008).
172. Wu, Y. et al. Adaptive weighted Gerchberg-Saxton algorithm for generation of phase-only hologram with artifacts suppression. *Optics Express* **29**, 1412-1427 (2021).
173. Chen, L. et al. Weighted constraint iterative algorithm for phase hologram generation. *Applied Sciences* **10**, 3652 (2020).
174. Chang, C. et al. Speckle-suppressed phase-only holographic three-dimensional display based on doubleconstraint Gerchberg-Saxton algorithm. *Applied optics* **54**, 6994-7001 (2015).
175. Seldowitz, M. A., Allebach, J. P. & Sweeney, D. W. Synthesis of digital holograms by direct binary search. *Applied optics* **26**, 2788-2798 (1987).
176. Chhetri, B. B., Yang, S. & Shimomura, T. Stochastic approach in the efficient design of the direct-binarysearch algorithm for hologram synthesis. *Applied optics* **39**, 5956-5964 (2000).
177. Liu, J.-P., Yu, C.-Q. & Tsang, P. W. Enhanced direct binary search algorithm for binary computergenerated Fresnel holograms. *Applied optics* **58**, 3735-3741 (2019).
178. Yoshikawa, N. & Yatagai, T. Phase optimization of a kinoform by simulated annealing. *Applied optics* **33**, 863-868 (1994).
179. Abadi, M. et al. TensorFlow: A System for LargeScale Machine Learning. in Proceedings of the 12th USENIX Conference on Operating Systems Design and Implementation (USENIX Association, Savannah, GA, USA, 2016), 265-283.
180. Paszke, A. et al. in Advances in Neural Information Processing Systems 32 (eds Wallach, H. *et al.*) 8024-8035 (Curran Associates, Inc., 2019).
181. Chen, C. et al. Multi-depth hologram generation using stochastic gradient descent algorithm with complex loss function. *Optics Express* **29**, 15089-15103 (2021).
182. Chakravarthula, P. et al. Learned Hardware-in-theloop Phase Retrieval for Holographic Near-Eye Displays. *ACM Transactions on Graphics (TOG)* **39**, 186 (2020).
183. Choi, S. et al. Optimizing image quality for holographic near-eye displays withmichelson holography. *Optica* **8**, 143-146 (2021).
184. Ahar, A. et al. Suitability analysis of holographic vs light field and 2D displays for subjective quality assessment of Fourier holograms. *Optics Express* **28**, 37069-37091 (2020).
185. Wang, Z. et al. Image quality assessment: from error visibility to structural similarity. *IEEE Transactions on Image Processing* **13**, 600-612 (2004).
186. Dumic, E., Duarte, C. R. & da Silva Cruz, L. A. Subjective evaluation and objective measures for point clouds — State of the art. in 2018 First

International Colloquium on Smart Grid Metrology (SmaGriMet) (2018), 1-5.

187. Alexiou, E. & Ebrahimi, T. Exploiting user interactivity in quality assessment of point cloud imaging. in 2019 Eleventh International Conference on Quality of Multimedia Experience (QoMEX) (2019), 1–6.

188. Meynet, G. et al. PCQM: A Full-Reference Quality Metric for Colored 3D Point Clouds. in 2020 Twelfth International Conference on Quality of Multimedia Experience (QoMEX) (2020), 1-6.

189. Liu, Q. et al. PQA-Net: Deep No Reference Point Cloud Quality Assessment via Multi-View Projection. *IEEE Transactions on Circuits and Systems for Video Technology* **31**, 4645-4660 (2021).

190. Viola, I., Řeřábek, M. & Ebrahimi, T. A new approach to subjectively assess quality of plenoptic content. in Applications of Digital Image Processing XXXIX. **9971** (San Diego: SPIE, 2016), 287-299.

191. Mahmoudpour, S. & Schelkens, P. On the performance of objective quality metrics for lightfields. *Signal Processing: Image Communication* **93**, 116179 (2021).

192. Blinder, D. et al. Open Access Database for Experimental Validations of Holographic Compression Engines. in Proceedings of the 2015 Seventh International Workshop on Quality of Multimedia Experience. (Pilos: IEEE, 2015).

193. Gilles, A. et al. Computer generated hologram from multiview-plus-depth data considering specular reflections. in 2016 IEEE International Conference on Multimedia & Expo Workshops (ICMEW) (2016), 1-6.

194. Bernardo, M. V. et al. Holographic representation: Hologram plane vs. object plane. *Signal Processing: Image Communication* **68**, 193-206. (2018).

195. ISO/IEC JTC1/SC29/WG1. JPEG pleno database. (2021). at https://jpeg.org/jpegpleno/plenodb.html.

196. Singh, P. et al. A Review on SAR Image and its Despeckling. *Archives of Computational Methods in Engineering* **28**, 4633-4653 (2021).

197. Kumar Pal, S., Bhardwaj, A. & Shukla, A. A Review on Despeckling Filters in Ultrasound Images for Speckle Noise Reduction. in 2021 International Conference on Advance Computing and Innovative Technologies in Engineering (ICACITE) (2021), 973-978.

198. Ahar, A. et al. Comprehensive performance analysis of objective quality metrics for digital holography. *Signal Processing: Image Communication* **97**, 116361 (2021).

199. Fonseca, E. et al. Assessment of speckle denoising filters for digital holography using subjective and objective evaluation models. *Applied Optics* **58**, (2019).

200. Montresor, S. & Picart, P. Quantitative appraisal for noise reduction in digital holographic phase imaging. *Optics Express* **24**, 14322-14343 (2016).

201. Birnbaum, T. et al. Speckle Denoising of ComputerGenerated Macroscopic Holograms. in Digital Holography and Three-Dimensional Imaging 2019 (Optical Society of America, 2019), W3A.1.

202. Kumar, M. et al. Speckle denoising techniques in imaging systems. *Journal of Optics* **22**, 063001 (2020).

203. Corda, R. et al. An exploratory study towards objective quality evaluation of digital hologram coding tools. in Proceedings of SPIE 11137, Applications of Digital Image Processing XLII (San Diego: SPIE, 2019), 358-367.

204. Amirpourazarian, H. et al. Quality evaluation of holographic images coded with standard codecs. *IEEE Transactions on Multimedia*, 1-1 (2021).

205. Muhamad, R. K. et al. JPEG Pleno holography: scope and technology validation procedures. *Applied Optics* **60**, 641 (2021).

206. Ahar, A. et al. Validation of dynamic subjective quality assessment methodology for holographic coding solutions. in 2021 13th International Conference on Quality of Multimedia Experience (QoMEX) (2021), 7-12.

207. ISO/IEC JTC1/SC29/WG1. Common Test Conditions 6.0 for JPEG Pleno Holography. WG1N100035, 93rd JPEG Meeting, (2021).

208. Ahar, A. Perceptual quality prediction and analysis for digital holography. PhD thesis (Vrije Universiteit Brussels, 2021).

Park et al. *Light: Advanced Manufacturing* (2022)3:9
https://doi.org/10.37188/lam.2022.009
Official journal of the JHL 2689-9620
www.light-am.com

Review Open Access

Holographic techniques for augmented reality and virtual reality near-eye displays

Jae-Hyeung Park[1,*] and Byoungho Lee[2,*]

Abstract

Near-eye displays are the main platform devices for many augmented reality (AR) and virtual reality (VR) applications. As a wearable device, a near-eye display should have a compact form factor and be lightweight. Furthermore, a large field of view and sufficient eyebox are crucial for immersive viewing conditions. Natural three-dimensional (3D) image presentation with proper focus cues is another requirement that enables a comfortable viewing experience and natural user interaction. Finally, in the case of AR, the device should allow for an optical see-through view of the real world. Conventional bulk optics and two-dimensional display panels exhibit clear limitations when implementing these requirements. Holographic techniques have been applied to near-eye displays in various aspects to overcome the limitations of conventional optics. The wavefront reconstruction capability of holographic techniques has been extensively exploited to develop optical see-through 3D holographic near-eye displays of glass-like form factors. In this article, the application of holographic techniques to AR and VR near-eye displays is reviewed. Various applications are introduced, such as static holographic optical components and dynamic holographic display devices. Current issues and recent progress are also reviewed, providing a comprehensive overview of holographic techniques that are applied to AR and VR near-eye displays.

Keywords: Holography, Augmented display, Virtual reality

Introduction

Augmented reality (AR) and virtual reality (VR) are emerging fields in both academia and industry. Virtual images are presented in an immersive and interactive manner, thereby providing a new user experience for various applications. The main platform device for AR and VR is the near-eye display (NED)[1,2]. NEDs are glass- or goggle-type wearable display devices. They present virtual digital images that fill the entire visual field of a user or exist in combination with the see-through view of a real scene, enabling VR and AR experiences.

NEDs are composed of a micro-display panel and imaging optics. Light from the micro-display panel, which is placed close to the eye, is collimated by the imaging optics, thereby forming a virtual image at a far distance where the eye can focus comfortably. In AR NEDs, an additional component; that is, an optical combiner, is used to superimpose the virtual images of the micro-display panel onto the see-through light from a real scene in front of the user. As display devices, NEDs have traditional performance metrics, such as resolution and luminance. However, as wearable devices that are close to the eye, NEDs also need to have a compact form factor, light weight, wide field of view (FoV), and large eyebox. Moreover, optical matching of the displayed virtual images to real objects with natural focus cues is important for

Correspondence: Jae-Hyeung Park (jh.park@inha.ac.kr) or Byoungho Lee (byoungho@snu.ac.kr)
[1]Department of Information and Communication Engineering, Inha University, Incheon, 22212, Korea.
[2]School of Electrical and Computer Engineering, Seoul National University, Seoul, 08826, Korea

natural interaction between the virtual images and user. Various optics have been developed to implement AR/VR NEDs[2] to satisfy these requirements. However, bulk lenses and mirrors, which are typically used in NED optics, impose limitations on the achievable weight and form factor, making it difficult to realize wearable and aesthetically plausible devices comfortably. The simple two-dimensional (2D) image presentation of each micro-display panel also fixes the virtual image plane to a single distance, thereby hindering the exact optical matching of the virtual images onto the real scene and causing an uncomfortable viewing experience owing to limited monocular focus cues[3-6].

In recent years, the application of holographic techniques to AR/VR NEDs has attracted significant attention. Holography records and reconstructs the wavefront of a desired light. As static optical components, holographic devices can replace the bulk optics of conventional NEDs[7,8]. Thin and light film-shaped holographic devices contribute significantly to reducing the overall system form factor of NEDs. The high angular and wavelength selectivity of the holographic optical components enables the selective application of different optical powers to spatially overlaid light, thereby realizing optical see-through capability for AR and a single component that is multiplexed with multiple optical functions.

As a dynamic device, the holographic display module can replace the 2D micro-display panel of conventional NEDs[9]. The holographic display module presents holographic three-dimensional (3D) images to each eye of the user. Full monocular focus cues that are provided by holographic 3D images realize the exact optical matching of the displayed images onto real objects and a comfortable viewing experience, without vergence-accommodation conflict (VAC)[3-6, 9]. The wavefront reconstruction capability of holographic displays can be used to compensate for the aberrations of NED optics, thereby reducing the required number of optical components in the NED[10-12]. This feature can also compensate for eye aberrations, and thus, it can enable vision-correcting displays that present clear images without additional vision-correcting lenses[13].

In this article, we provide a brief overview of the holographic techniques that are applied to AR and VR NEDs. We explain the basic principles and features of various holographic applications, such as static holographic optical components and dynamic holographic display devices. Current issues and recent progress are also discussed, thereby providing a comprehensive review of the research prospects of holographic techniques in AR and VR applications.

Static hologram optical device

Holographic optical element

A representative static optical component that uses holographic techniques is the holographic optical element (HOE)[7,8,14-16]. HOEs are diffraction gratings that are fabricated using holographic techniques. During the recording process, a reference beam and signal beam are illuminated on the recording medium, and the interference pattern is recorded. As a recording medium, a photopolymer that records the interference pattern as a spatial refractive index modulation is frequently used. In recent years, liquid crystals (LCs) have also been used to record the interference pattern as a form of the spatial distribution of LC director orientations[8]. Once the pattern has been recorded, the HOE diffracts the incident reference beam, which is usually known as the probe beam in the reconstruction process, to reconstruct the original signal beam. Depending on the signal beam that is used in the recording process, HOEs can offer various optical functions, including diffusers, lenses, and mirrors. The thickness of HOEs is typically less than several tens of micrometers, which enables thin and light form factors of the overall system. However, HOEs are sufficiently thick to exhibit volume hologram characteristics, resulting in high diffraction efficiency and strong angular/wavelength selectivity. These features make HOEs versatile for various AR/VR NEDs.

Fig. 1 depicts typical examples of HOE applications in AR NEDs. In Fig. 1a, the HOE acts as a forward-scattering screen for the projected images[17]. Owing to the angular and wavelength selectivity, the HOE only scatters the light from the projector, while allowing the light originating from the real scene to pass through unaffected. This selective screening function of the HOE enables on-axis AR NED optics with a simple configuration. Fig. 1b presents an example of an HOE working as an optical combiner[18] for a projection-type AR NED. As illustrated in Fig. 1b, the light from the laser projector is collimated and projected onto the HOE that is placed in front of the eye. The HOE acts as a parabolic mirror and focuses the projected light onto a spot on the eye pupil plane so that the image is directly written onto the eye retina plane. Again, the wavelength and angular selectivity of the HOE allow the see-through view of the real scene to be transmitted to the eye unaltered, thereby enabling an optical see-through AR configuration. Fig. 1c depicts another typical application of the HOE; that is, the in- and out-couplers of waveguide-type AR NEDs[19]. The light from a micro-display panel is coupled to the waveguide by the HOE, which diffracts the incident light at an angle for the total

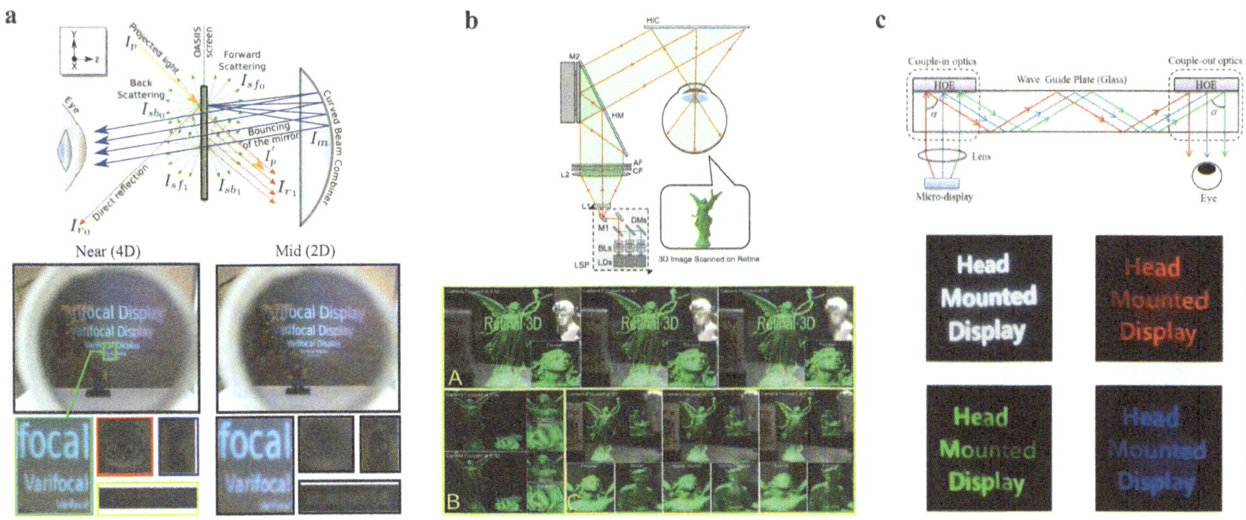

Fig. 1 Representative applications of HOEs to AR NEDs: **a** transparent screen[17], **b** reflecting combiner[18], and **c** in-/out-coupler of waveguide[19]. Images reprinted with the following permissions: a[17], b[18] from ACM (Association for Computing Machinery); c[19] from The Optical Society of Korea.

internal reflections inside the waveguide. Subsequently, the out-coupler HOE in front of the eye diffracts the guided light directly towards the eye, thereby presenting virtual images.

In all of the examples in Fig. 1, the angular and wavelength selectivity of the HOEs make them work only for the projected images, remaining transparent to the real scene light. However, the selectivity can be further used to provide different optical functions for the projected light. Fig. 2a, b show examples of the angular selectivity. In Fig. 2a, two HOEs work as scattering screens, but only for light that is projected from the designed directions[20]. The images from the two projectors are scattered only from the corresponding HOE screens, thereby forming double-layer images. In Fig. 2b, a single HOE is recorded by multiplexing two optical functions; that is, a scattering screen and a micro-mirror array, using two angularly separated reference beams[21]. The recorded HOE works as a screen for the light from a projector, while it works as a micro-mirror array for the light from another projector, thereby enabling 2D and 3D screens, respectively. Fig. 2c presents another example of HOE selectivity. Unlike photopolymer-based HOEs with spatial refractive index modulation grating, LC-based HOEs have polarization selectivity, exhibiting different optical functions to orthogonal polarizations. The two HOEs in Fig. 2c work as concave mirrors only for left circularly polarized (LCP) light and are transparent for right circularly polarized (RCP) light. The example configuration in Fig. 2c uses this feature to switch the exit pupil of the system by focusing the projected beam with different concave mirror HOEs selectively[22].

The thin and light form factors, high diffraction efficiency, and strong selectivity of HOEs, as introduced above, have attracted significant attention and have resulted in the application of HOEs to various AR and VR NED research topics. Fig. 3 depicts several examples in which HOEs are used to expand the eyebox[23–26], enlarge the FoV[27,28], increase the number of image planes[29], and implement foveated displays[30,31]. Figs. 3a–c show the eyebox expansion of Maxwellian AR NEDs. The Maxwellian AR NED focuses the image beam onto a single spot in the eye pupil plane. Owing to the reduced exit pupil of the system, the displayed images have a large depth of field (DoF), thereby providing always-focused images to the user, regardless of the focal power of the eye lens. A drawback of the Maxwellian display is that the images are visible only when the focal spot falls inside the eye pupil, which limits the effective eyebox of the system. In the technique shown in Fig. 3a, the eyebox limitation is mitigated by using a multiplexed HOE[24]. The HOE in Fig. 3a functions as multiple concave mirrors, thereby creating multiple focal spots with separation of the average eye pupil size. One of the focal spots falls inside the eye, so that the images are presented in the extended eyebox area. Fig. 3b extends the eyebox using a similar principle, but with LC-based HOEs[25]. Collimated and linearly polarized light is diffracted along the x-axis by the first HOE and it is separated into three beams; that is, two oppositely diffracted beams with LCP and RCP polarizations, and one non-diffracted DC beam. These beams are converted back into linear polarization by a quarter-wave plate and

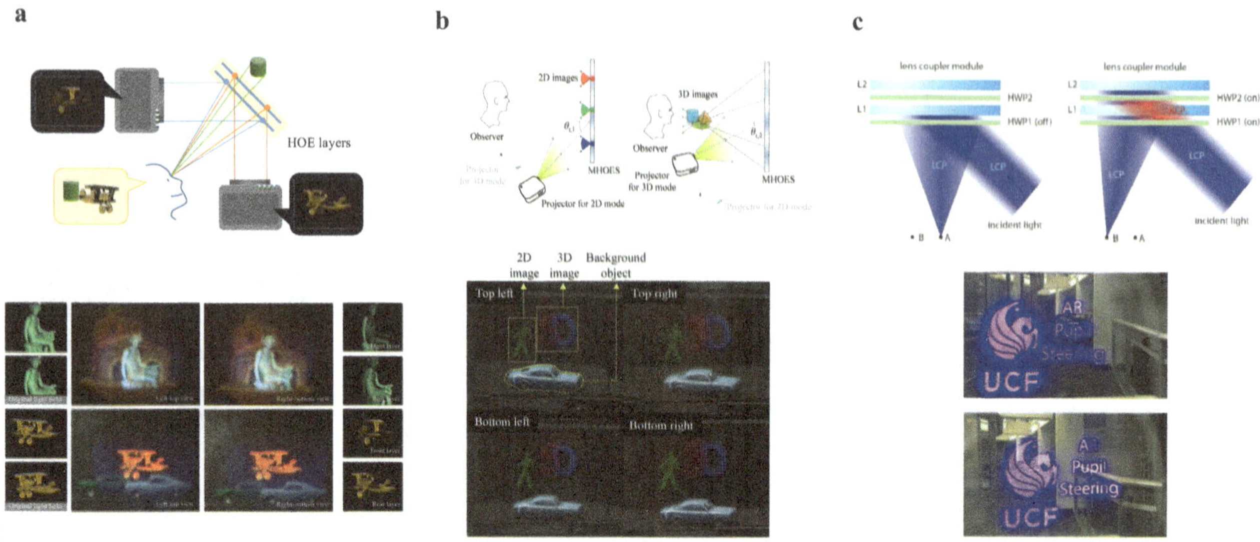

Fig. 2 Selectivity of HOEs: **a** two HOEs working as screens only for the corresponding projectors using angular selectivity[20], **b** an HOE with dual functions that are multiplexed using angular selectivity[21], and **c** LC-based HOEs with polarization sensitivity[22]. Images reprinted with the following permissions: a[20] from ACM (Association for Computing Machinery); b[21], c[22] from Optica Publishing Group.

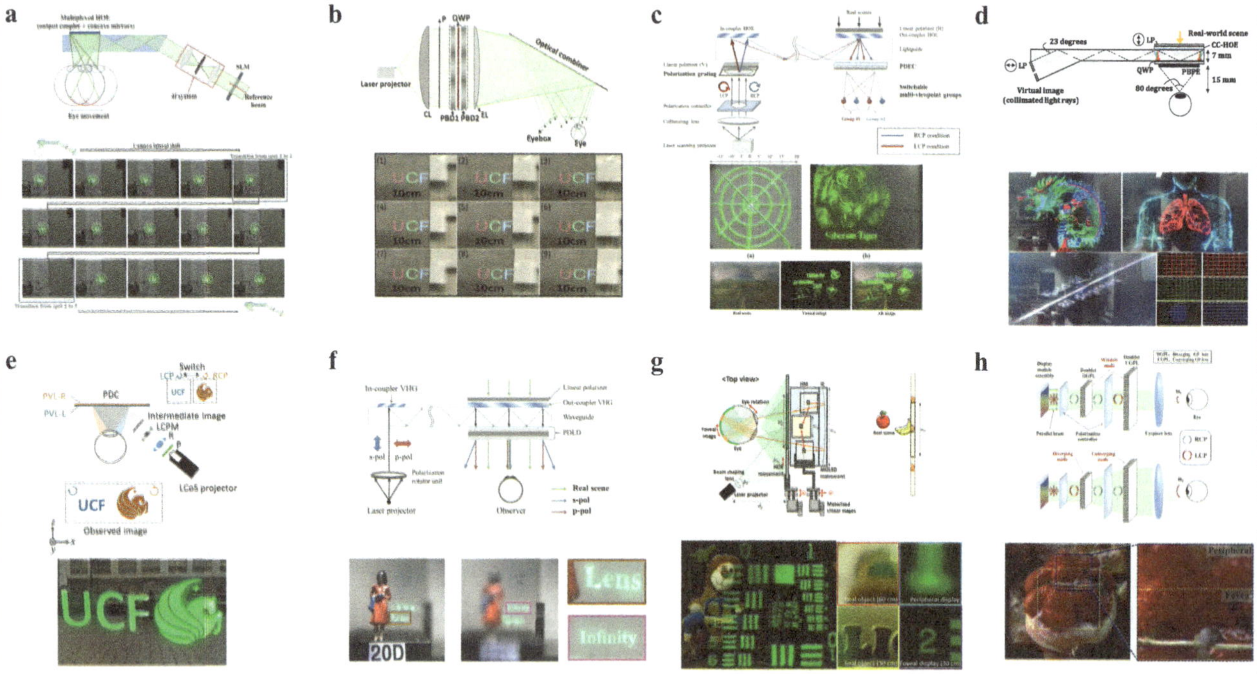

Fig. 3 Recent examples of HOE applications for AR NEDs: **a** eyebox replication using a multiplexed HOE[24], **b** eyebox replication using polarization-dependent HOEs[25], **c** viewpoint switching using polarization-dependent HOEs[26], **d** FoV enhancement using polarization-dependent transmission-type HOEs[27], **e** FoV enhancement using polarization-dependent reflection-type HOEs[28], **f** dual focal plane formation using polarization-dependent HOEs[29], **g** foveated display using an HOE[30], and **h** foveated display using a polarization-dependent HOE[31]. Images reprinted with the following permissions: a[24],b[25], c[26], e[28], f[29], h[31] from Optica Publishing Group; d[27] (CC BY 4.0); g[30] from ACM (Association for Computing Machinery).

diffracted again into three directions on the *y*-axis by the second HOE. Finally, 3 × 3 focal spots are created in the eye pupil plane, resulting in 2D expansion of the eyebox. In Fig. 3c, a polarization grating, multiplexed HOE, and geometric phase lens, which is a type of LC-based HOE, are combined to switch between two sets of interleaved multiple focal spots by controlling the input beam polarization[26].

Fig. 3d, e depict the FoV expansion of AR NEDs using HOEs. In the technique[27] shown in Fig. 3d, an eye-piece element, which is an RCP converter that is sandwiched between a pair of geometric phase lenses, is used. The eye-piece element focuses on the RCP projected images while allowing the LCD real-scene light to pass through unchanged, which enables an AR experience. The chromatic aberration of the geometric phase lenses is pre-compensated by a wavelength-multiplexed out-coupler HOE, which has different focal lengths for the red, green, and blue wavelengths. The dual geometric phase lenses and optical power of the HOE enable a short effective focal length, achieving a large FoV of up to 80°. In the technique[28] shown in Fig. 3e, two LC-based HOEs are stacked to diffract the incident RCP and LCP light to different FoV areas selectively, thereby doubling the overall FoV of the system.

Fig. 3f presents an example of an HOE application with dual image planes[29]. In conventional NEDs, virtual images are formed on an image plane that is fixed at a single distance. The fixed 2D image plane is contrasted by the varying-depth stereoscopic 3D image content, which creates a VAC problem. The VAC problem can be partially mitigated by increasing the number of image planes. In Fig. 3f, the image planes are formed at two distinct distances using a polarization-dependent eye-piece element that is similar to that shown in Fig. 3d. The eye-piece element in Fig. 3f has zero and negative optical powers for the P- and S-polarized light, respectively, thereby forming dual image planes at optical infinity and at a close distance, depending on the input light polarizations.

Finally, the examples in Fig. 3g, h show the implementation of foveated displays using HOEs. In Fig. 3g, the HOE works as an optical combiner, and the peripheral images are presented in the Maxwellian display configuration with a wide FoV[30]. High-resolution foveal images are presented using a micro-OLED panel. The high selectivity and thin shape of the HOE are effectively used in implementing the wearable foveated AR NED prototype with a compact form factor. In Fig. 3h, the polarization-selective focal length of the LC-based HOE is used to provide different magnifications of the virtual images, thereby realizing a time-multiplexed foveated display using a single display panel[31].

Despite many advantages and possible applications to AR/VR NEDs, HOEs still exhibit issues including material reliability, reduced diffraction efficiency for full color, and aberrations. Aberrations are of particular importance because the incident light is usually highly slanted in NED optics. In holographic NEDs, which will be discussed in a later section, these aberrations can be pre-compensated in the hologram content; that is, by means of a computer-generated hologram (CGH). However, in other types of NEDs, the HOE must be recorded with properly adjusted reference and signal beams to minimize aberrations. In recent years, HOE recording techniques using a holographic printer[32,33] or freeform optics[33] have been reported, as illustrated in Fig. 4, which are expected to contribute to the image quality enhancement of AR and VR NEDs using HOEs.

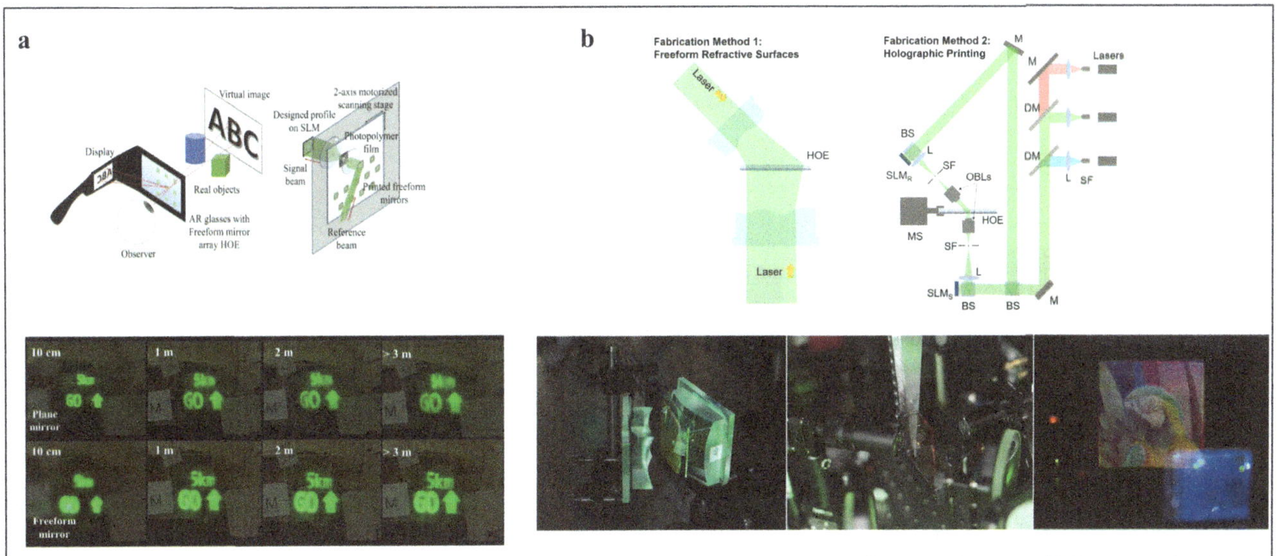

Fig. 4 HOE quality enhancement by recording freeform HOEs using: **a** and **b** holographic printers[32] or **b** freeform optics[33]. Images reprinted with the following permissions: a[32] (CC BY 4.0); b[33] from ACM (Association for Computing Machinery).

Metasurface lens

A metasurface is a photonic structure with artificial subwavelength scatterers; that is, meta-atoms that are distributed in a 2D plane. The geometry of an individual meta-atom can be designed to control the phase, amplitude, and polarization of the transmitted or reflected light, and thus, it exhibits various optical functions. Full 2π range phase control with a subwavelength thickness structure enables ultra-thin and light optics with a high numerical aperture (NA), which is ideal for AR and VR NEDs. Recent progress in metasurface optics has made the application of metasurface lenses to AR and VR NEDs a viable option, based on which research has been initiated[34–36].

Fig. 5 presents examples of the metasurface lens applied to AR and VR optics. In the work[37] depicted in Fig. 5a, a transmission-type metasurface lens made of a 2D array of dielectric nano-rods is used. The width and height of the nano-rods are designed to provide appropriate anisotropy, which determines the transmission efficiency of the co-polarized and cross-polarized light. The orientation angle of the nano-rods is designed to provide a proper phase profile to the cross-polarized light, following the geometric phase or Pancharatnam–Berry (PB) phase principle. Unlike conventional metasurface lenses, the direct transmission of co-polarized light is deliberately maintained to enable a see-through view of the real scene. The fabrication is conducted using a nano-imprinting process with the implementation of a 20 mm diameter metasurface lens. Owing to the high 0.61 NA and transmission-type geometry of the device, the implemented system provides 90° FoV full-color AR images, which demonstrates the potential of metasurface lens application to AR and VR NEDs. In this work, the metasurface lens works in broadband but the chromatic aberration remains, which requires additional optics for compensation. Fig. 5b depicts a work that reported an achromatic metasurface lens for AR and VR applications[38]. The lens aperture is divided into multiple radial zones as in the conventional Fresnel lens, and meta-atoms are designed for each zone. Unlike conventional Fresnel lenses, in which the dispersion is determined solely by the intrinsic material dispersion, the meta-atoms in each zone are designed to provide an appropriate phase profile, dispersion, and phase discontinuity at the zone boundary, which enables a common focal length at three wavelengths; that is, red, green, and blue. The fabricated device has a 2 mm aperture and 0.7 NA, and was verified with full-color AR and VR NED configurations. Finally, the metasurface lens design and fabrication on a freeform surface[39] has recently been reported, as illustrated in Fig. 5c. The overall optical power can be divided and assigned to the freeform surface geometry and metasurface lens with a balance considering the design, fabrication, and tolerance issues. The fabrication of a 2 mm × 1.5 mm metasurface mirror on a toroidal surface has been demonstrated experimentally. Although issues[35] remain, including the fabrication of large lenses and high-order aberrations such as coma, the ultra-thin form factor, high NA, and high design flexibility make the metasurface lens attractive in AR and VR applications, and progress in this area is expected to be fast.

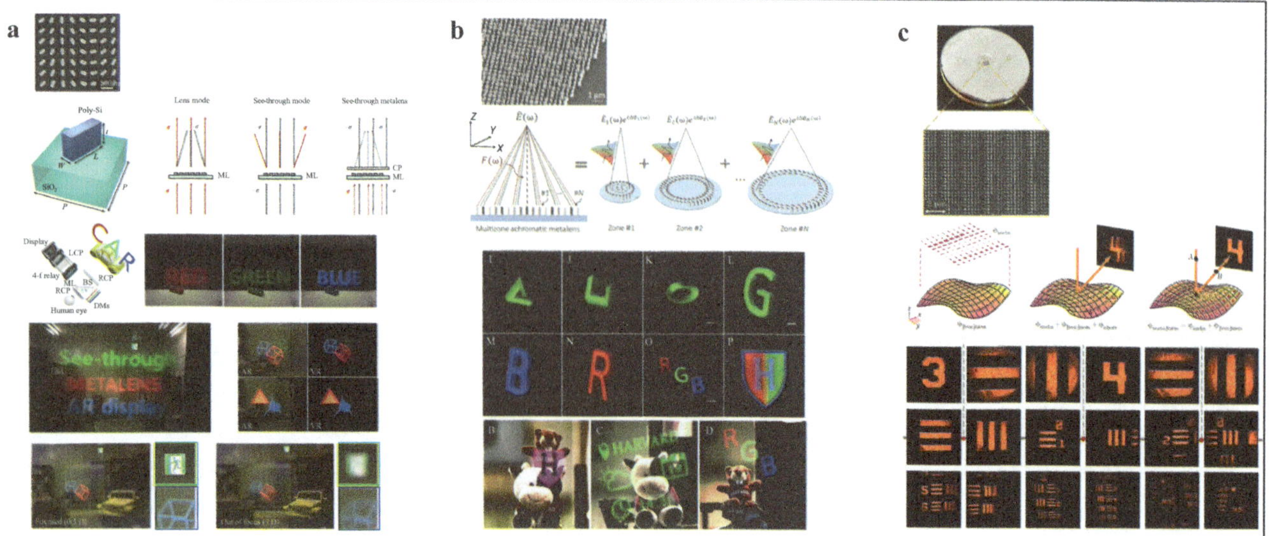

Fig. 5 Metasurface lenses applied to AR/VR NEDs: **a** PB phase metasurface lens[37], **b** achromatic metasufrace lens with radial zones[38], and **c** metasurface fabricated on a freeform surface[39]. Images reprinted with the following permissions: a[37] (CC BY 4.0); b[38], c[39] (CC BY NC 4.0)

Dynamic holographic display

Advantages of holographic display for AR/VR NEDs

The micro-display panels of AR and VR NEDs reproduce the spatial amplitude distribution on the panel surface, thereby presenting 2D images. Holographic displays reconstruct the wavefront of the desired 3D images by modulating the amplitude or phase of the light using a spatial light modulator (SLM). Holographic NEDs refer to those that are configured with a holographic display module instead of the usual micro-display panel[9].

Fig. 6 depicts the differences between the typical AR/VR NEDs and holographic NEDs. In the typical AR/VR NEDs shown in Fig. 6a, 2D images are displayed on a micro-display panel. Diverging light from each pixel of the micro-display panel is refracted by the optics, forming virtual images on a plane that is conjugate to the display panel. CGHs corresponding to the desired 3D images are loaded onto the SLM in the holographic AR/VR NEDs, as illustrated in Fig. 6b. Coherent light illuminating the SLM is modulated in amplitude or phase by the SLM, which reconstructs the wavefront of the 3D images.

An apparent advantage of holographic NEDs is the presentation of 3D images. In conventional NEDs, a stereoscopic image pair is presented to the left and right eyes, and the user perceives 3D depth from the binocular parallax in the stereoscopic images, as illustrated in Fig. 7a. However, the optical virtual image that is seen by each eye is still in the 2D plane that is conjugated to the micro-display panel, and the focal power of the eye lens is accommodated to that distance. As the accommodation distance is fixed at the optical conjugate plane, although the perceived distance or convergence distance varies according to the presented stereoscopic image content, a mismatch occurs between the two distances, which is known as VAC. VAC is one of the major causes of dizziness and discomfort in AR and VR experiences[3-6]. Although VAC is common in many 3D display applications, it is of particular importance in AR and VR applications, as it is severe in the distance range of AR and VR. Fig. 7b depicts the degree of VAC measured in the size of the circle of confusion. When the stereoscopic 3D image distance is larger than the optical virtual image distance of 2 m, the VAC is not significant. However, when the 3D image is closer to the eye, the VAC becomes severe. In AR and VR applications, images that are close to the eye, usually within the arm's length, are important, because they allow direct interaction with the user. As illustrated in Fig. 7b, the VAC in this distance range is significant and requires appropriate mitigation. Various techniques, including Maxwellian displays[40-44], varifocal optics[45], multiple image planes[46-48], and light field displays[49-54], have been developed in this regard. However, holographic displays[9,55-58] provide the most complete solution by reconstructing optical 3D images to each eye with a continuous wavefront. Fig. 8a depicts the 3D image presentation of holographic NEDs with an example of an experimental demonstration[11]. Owing to the optical 3D image formation that is provided by the holographic display, each part of the 3D images is separately focused and blurred according to the camera focus.

Although 3D image presentation without the VAC problem is a prominent advantage of holographic NEDs, the wavefront reconstruction capability of holographic NEDs offers additional benefits, an example of which is the DoF control shown in Fig. 8b[59]. In conventional AR/VR NEDs, the DoF of the displayed virtual images is determined by the NA of the optics and dynamic control is not usually possible. In contrast, in holographic AR/VR NEDs, the DoF is determined by the CGH, which enables

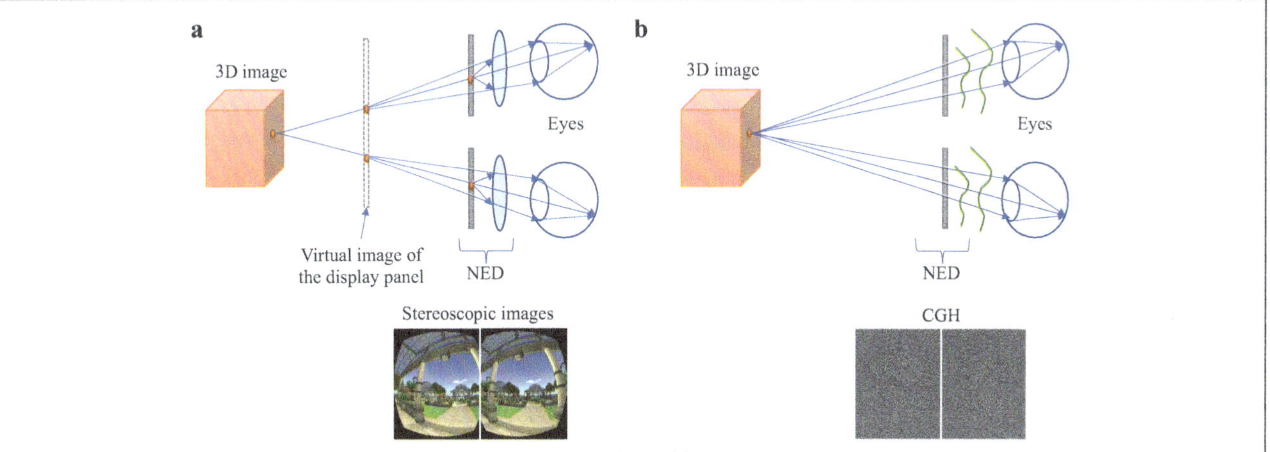

Fig. 6 Comparison between: **a** conventional stereoscopic NED and **b** holographic NED.

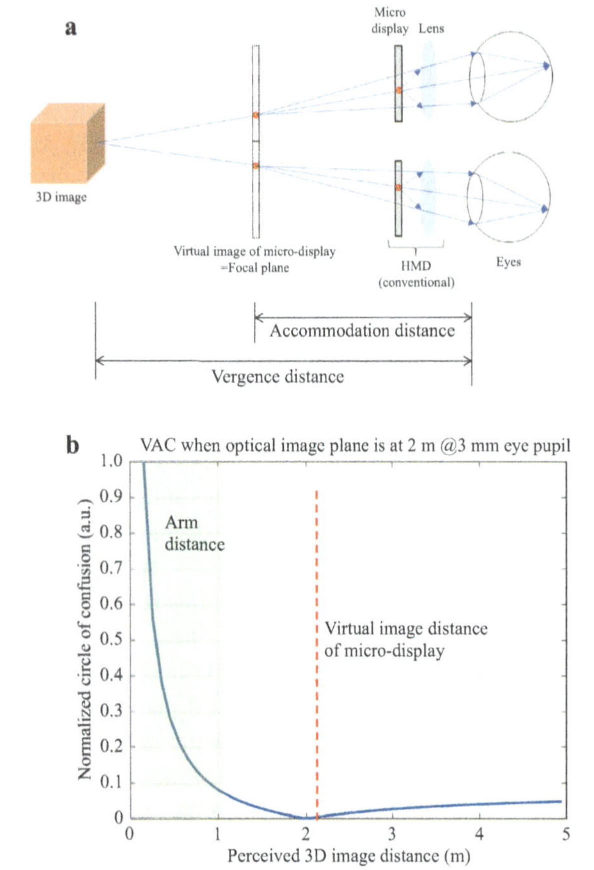

Fig. 7 VAC: **a** conceptual illustration, **b** degree of VAC represented in circle of confusion.

for realistic scene representation, while providing a large DoF to alarming text information for always-focused presentation.

Aberration compensation by the CGH is another important benefit of holographic NEDs. In conventional NEDs, aberrations of the optics need to be corrected using additional optical components, which significantly increases the system weight and size. In contrast, in holographic NEDs, CGH can be synthesized to reconstruct the pre-distorted wavefront[9–13], which compensates for aberrations in the optics, as illustrated in Fig. 8c. Therefore, holographic NEDs can provide clear images without using additional optics, even though optical aberrations exist. This feature is a significant benefit in AR and VR NED designs, in which the compact form factor is critical. Note that the aberration that can be compensated by the CGH is not only the aberration of the NED optics, but also that of the eyes[13]. Once the eye aberration of the user is known, holographic NEDs can present sharp images to a low-vision user who is not wearing vision-correcting glasses. This vision-correcting feature of holographic NEDs has recently been successfully demonstrated, as shown in Fig. 8c.

The benefits of holographic NEDs make them attractive for AR/VR applications, on which research has been actively conducted. In the following section, the most recent progress in holographic NED techniques is introduced.

pixel-wise dynamic control, as shown in Fig. 8b. This feature can be used in AR and VR applications; for example, by providing a shallow DoF to the 3D image part

Eyebox and FoV expansion

The eyebox and FoV are important performance parameters of AR/VR NEDs. They are not usually

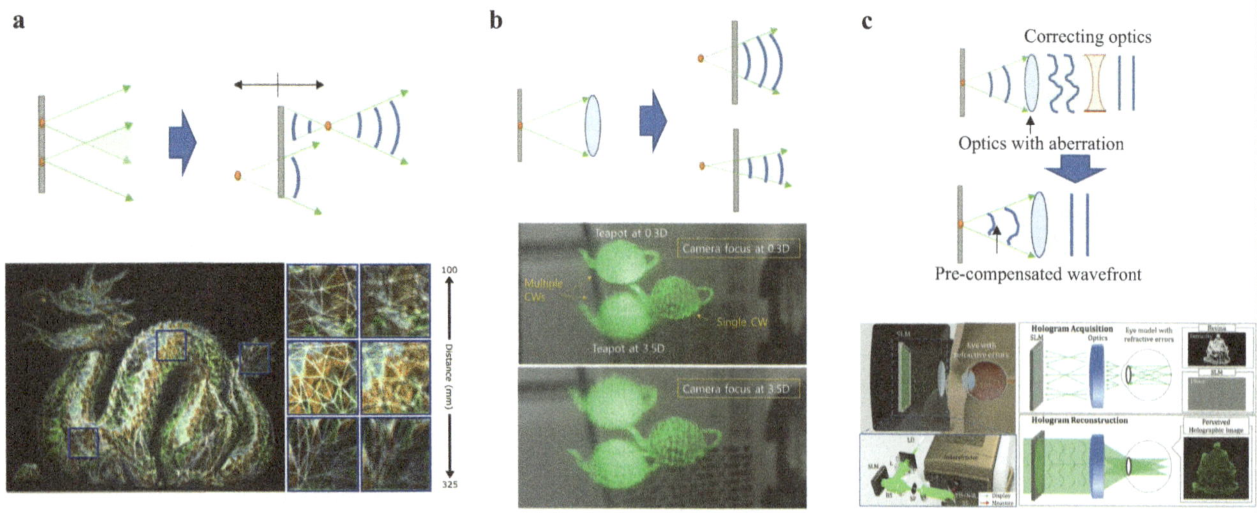

Fig. 8 Features of holographic NEDs: **a** 3D image presentation[11], **b** DoF control[59], and **c** aberration compensation[13]. Images reprinted with the following permissions: a[11] from ACM (Association for Computing Machinery); b[59], c[13] from Optica Publishing Group.

independent, but are correlated to have a tradeoff relationship following the etendue conservation law[1,2]. Despite their many benefits, holographic NEDs have a highly limited etendue that is defined by the space bandwidth product (SBP) of the SLM, which poses limitations on the eyebox and FoV[60–63]. Various techniques have been proposed to overcome these limitations. Fig. 9 presents several examples of recent reports.

Typical approaches for expanding the eyebox include eyebox steering and eyebox replication, as illustrated in Fig. 9a, b, respectively. In the eyebox steering method[64] depicted in Fig. 9a, the laser illumination direction on the SLM is dynamically controlled to steer the eyebox position that is formed around the focal spot of the combiner HOE in front of the eye. A laser beam is reflected by a micro-electromechanical system (MEMS) mirror and diffracted by an HOE (PSHOE), where an array of point sources is recorded. The diffracted light diverges from the selected point source position in the PSHOE and illuminates the SLM. In eyebox steering, the MEMS mirror angle is controlled such that different point sources are selected in the PSHOE, thereby illuminating the SLM in the corresponding directions.

Fig. 9b depicts the eyebox replication approach[65]. The light from the SLM is first filtered using 4-f optics to pass only a horizontal single sideband, so as to remove the DC, conjugate, and horizontal high-order diffraction terms. Unlike typical holographic displays, vertical high-order terms are maintained deliberately. The resultant vertically long eyebox is horizontally replicated and tiled together using an HOE combiner. The HOE combiner works as a horizontally multiplexed concave mirror, which forms three replicas of the eyebox. The spacing between the focal spots of the HOE is designed such that the eyebox is tiled without overlapping or gaps. The resultant 2D array of elementary eyeboxes offers a wide range of eye positions. The distortion that is observed when the eye is placed at the boundary with the elementary eyebox is remedied by wrapping the angular spectrum of the CGH accordingly.

Although the eyebox steering and replication approaches illustrated in Fig. 9a, b expand the effective eyebox, they require eye position information, which necessitates an additional eye pupil tracking system. Fig. 9c presents another approach that increases the eyebox without requiring the eye position information. In holographic NEDs, the eyebox is primarily determined by the maximum diffraction angle of the SLM, which is provided by the pixel pitch. As the pixel pitch cannot be reduced arbitrarily, the SLM diffraction angle and eyebox of the NED are limited. In the work[66] depicted in Fig. 9c, a photon sieve is attached to the SLM to reduce the effective pixel pitch. The tiny randomly distributed apertures in the photo sieve increase the diffraction angle, thereby expanding the eyebox of the system. Note that because the eyebox is expanded over the SBP limit of the SLM, the reconstruction is inevitably distorted and degraded if the usual CGH techniques that assume an ideal reconstruction system are used. This quality degradation is mitigated by solving the CGH optimization problem, which minimizes the difference between the ideal target 3D images and reconstructed images. Unlike eyebox steering and replication, this approach does not require the eye position,

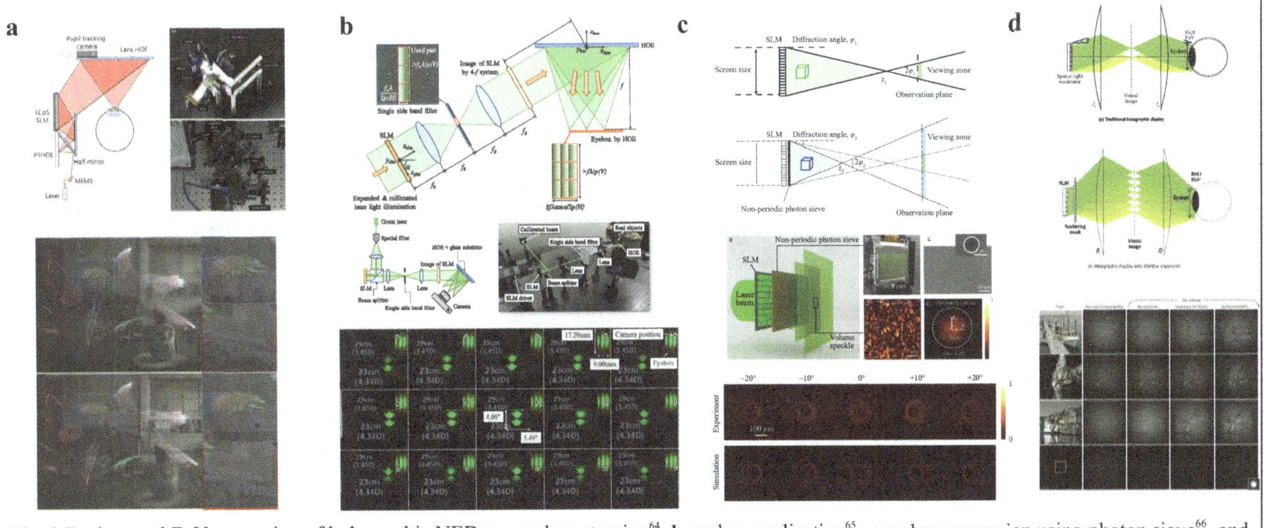

Fig. 9 Eyebox and FoV expansion of holographic NEDs: **a** eyebox steering[64], **b** eyebox replication[65], **c** eyebox expansion using photon sieve[66], and **d** FoV expansion using random phase mask[67]. Images reprinted with the following permissions: a[64], d[67] from ACM (Association for Computing Machinery); b[65] from Optica Publishing Group; c[66] (CC BY 4.0).

which is advantageous.

A similar approach can also be applied to FoV expansion[67], an example of which is presented in Fig. 9d. A random phase mask is placed on the conjugate plane of the SLM, thereby increasing the diffraction angle above the SBP limit. The increased diffraction angle results in an enhancement of the FoV, so that larger 3D images are presented to the user. CGH synthesis is again formulated as an optimization problem, providing 3D images in a larger FOV with minimized quality degradation. An enhancement of the FoV of up to 16× has been demonstrated using this approach.

Foveated holographic displays

The human visual system has a non-uniform resolution distribution across the FoV. Most cone photoreceptors are concentrated in the fovea, thereby providing high resolution only in that area. Based on this observation, foveated displays present high-resolution images only around the eye gaze point, while displaying the peripheral areas with low resolution[68,69]. The required number of pixels to cover the entire FoV is reduced, which enables more efficient and faster operation.

Foveated display techniques have recently been applied in holographic AR/VR NEDs. The studies applied the foveated display concept either to CGH synthesis[70–73] or to the system configuration[74,75]. Fig. 10a, b provide examples of foveated CGH synthesis for holographic AR/VR NEDs. CGH synthesis is a computationally intensive task, for which various acceleration techniques have been proposed. Foveated CGH synthesis reduces the computational load by reducing the target image resolution in the peripheral areas. As illustrated in Fig. 10a, this is achieved by synthesizing the full FoV CGH of the 3D scene that is represented in the low-resolution triangular mesh model, and subsequently updating it progressively to high-resolution meshes only within the foveated area[72]. In the work[73] depicted in Fig. 10b, a similar concept is applied to the layered image model. The foveal and peripheral images are prepared at different sampling rates, and the corresponding holograms are synthesized and combined, thereby reducing the overall computation time.

The techniques depicted in Fig. 10c, d apply a foveated display to the system configuration. In addition to the holographic display module, which accounts for the foveal display, a display panel for the peripheral area is prepared. During operation, the holographic display module presents high-resolution 3D images with a small FoV around the foveated area, whereas the 2D display panel presents peripheral images in a wide FoV with low resolution. In implementation, the dynamic optical steering of the foveated holographic display area following the gaze direction is a challenging issue. The work[74] in Fig. 10c uses an optical configuration that locates the SLM image at the rotation center of the eyeball. Because the SLM image remains in the eyeball regardless of the eyeball rotation, the holographic 3D images follow the gaze within the full FoV. In another method[75], as depicted in Fig. 10d, a MEMS mirror is controlled to steer the holographic foveal area following the gaze. Moreover, an LC deflector is used to expand the steering range further. Although efficient and compact implementation remains an issue, this type of foveated holographic display is attractive and exhibits high potential in that it presents wide FoV images in an efficient manner, without sacrificing the perceived 3D image quality and focus cues of the holographic displays.

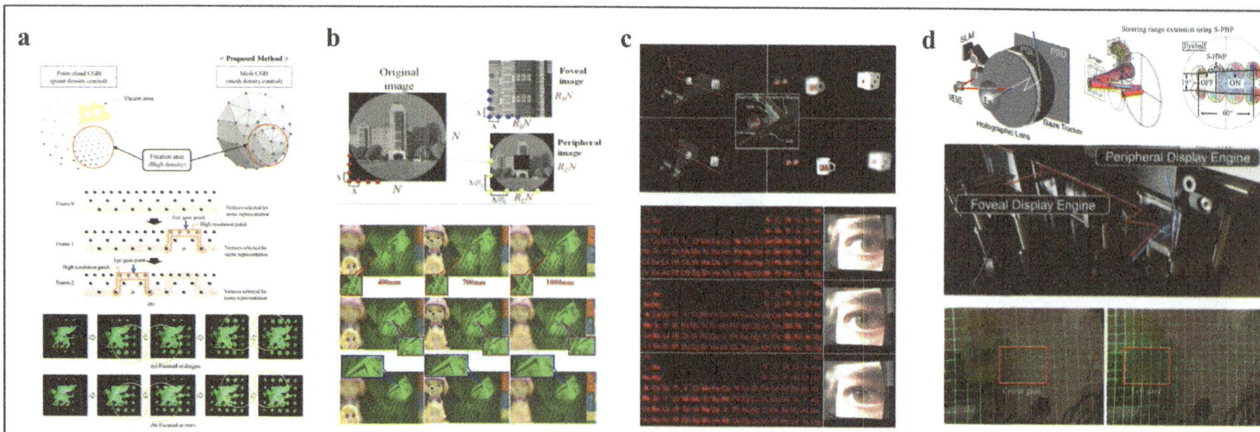

Fig. 10 Foveated holographic displays: **a** foveated CGH for 3D mesh models[72], **b** foveated CGH for layered images[73], **c** foveated holographic display with an SLM image at the eyeball rotation center[74], and **d** foveated holographic display using a MEMS mirror and LC device[75]. Images reprinted with the following permissions: a[72], b[73] from Optica Publishing Group; c[74], d[75] (CC BY 4.0).

Speckle reduction

Speckle noise is a long-standing problem in holographic display. A speckle is a grain-shaped random spatial fluctuation in intensity. In holographic displays, the complex field within the resolution spot of the eye interferes, resulting in speckle noise, which degrades the image quality. Because speckle noise is inherent to a coherent imaging system, its suppression at the perceivable level is of practical importance.

Fig. 11 depicts recent methods for suppressing speckle noise. In Fig. 11a, a temporal averaging technique of the random speckle pattern is used[76]. The same 3D images are generated with different random phases in each sub-frame, creating uncorrelated speckle patterns. These speckle patterns are averaged by temporal multiplexing, thereby reducing the perceived speckle contrast. The speckle contrast reduction is proportional to the square root of the number of sub-frames in the temporal multiplexing. In the work shown in Fig. 11a, a fast SLM, namely a digital micro-mirror device, is used to accommodate fast temporal multiplexing along with structured illumination to expand the viewing area. Speckless holographic video presentation at 60 Hz has been successfully demonstrated. In the method[77] illustrated in Fig. 11b, the object points of the 3D image are divided into spatially interleaved groups and displayed sequentially. The spacing between the object points in the same group is adjusted to be larger than the resolution spot of the eye, which prevents interference. Interleaving can also be applied to the angular spectrum domain[78], as shown in Fig. 11c. In each frame, the holographic 3D image is reconstructed with a set of angularly interleaved plane carrier waves. The angular separation of the plane carrier waves in each frame is set to be larger than the eye pupil, which again prevents speckle noise. With temporal multiplexing, the plane carrier waves scan the entire angular range, providing a natural defocus blur. The spatial and angular interleaving techniques depicted in Fig. 11b, c require temporal multiplexing, but the required number of sub-frames is much less than that of the random speckle pattern averaging technique. Finally, instead of a coherent laser, a partially coherent light source or an LED can be used to suppress speckle noise. The extended linewidth and emitting area of the LED reduce the temporal and spatial coherence of the light, thereby suppressing the speckle noise. However, the reduced temporal and spatial coherence also blurs the reconstruction, resulting in a tradeoff relationship between the speckle noise and image sharpness. In the work[79] shown in Fig. 11d, this tradeoff relationship was thoroughly analyzed, with the proposal of an optimal partially coherent light source.

Real-time photorealistic holographic displays using neural network

Real-time CGH synthesis and photorealistic image quality are crucial for the practical and widespread application of holographic AR/VR NEDs. In recent years, significant progress has been made in this regard with the aid of emerging neural network techniques.

The application of a neural network to CGH synthesis was pioneered by the work[80] shown in Fig. 12a. A basic U-Net architecture is trained to output the CGHs of the target images without a physical wave propagation calculation, which significantly reduces the computation time. The neural network-based CGH synthesis is subsequently further developed to generate CGHs of multi-depth 3D objects with a reconstruction quality that is comparable to the physical wave calculation, as shown in Fig. 12b[81]. The image quality and processing speed were significantly enhanced in the work[82] depicted in Fig. 12c. In this method, a neural network is trained with a random scene generator, which creates a large training dataset, thereby contributing

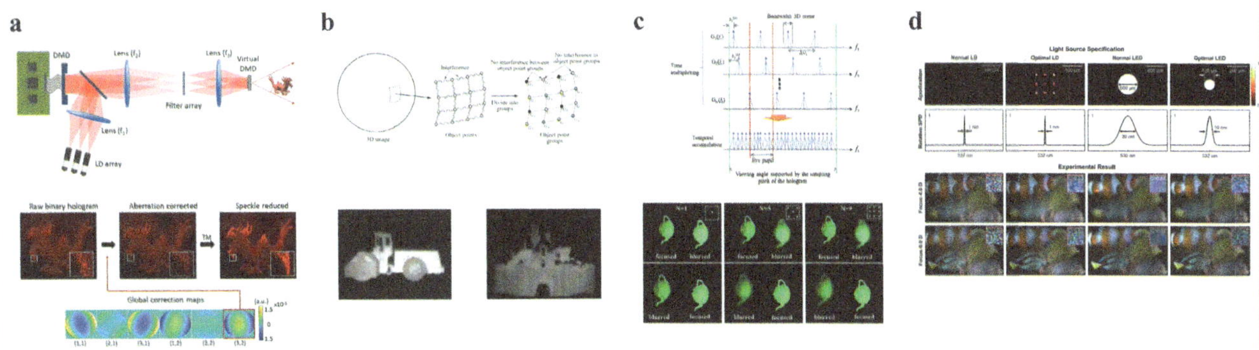

Fig. 11 Speckle reduction techniques: **a** temporal averaging of uncorrelated speckle patterns with structured illumination[76], **b** spatial interleaving of object points[77], **c** angular spectrum interleaving[78], **d** partial coherence of light source optimization[79]. Images reprinted with the following permissions: a[76], b[77], c[78] from Optica Publishing Group; d[79] (CC BY 4.0).

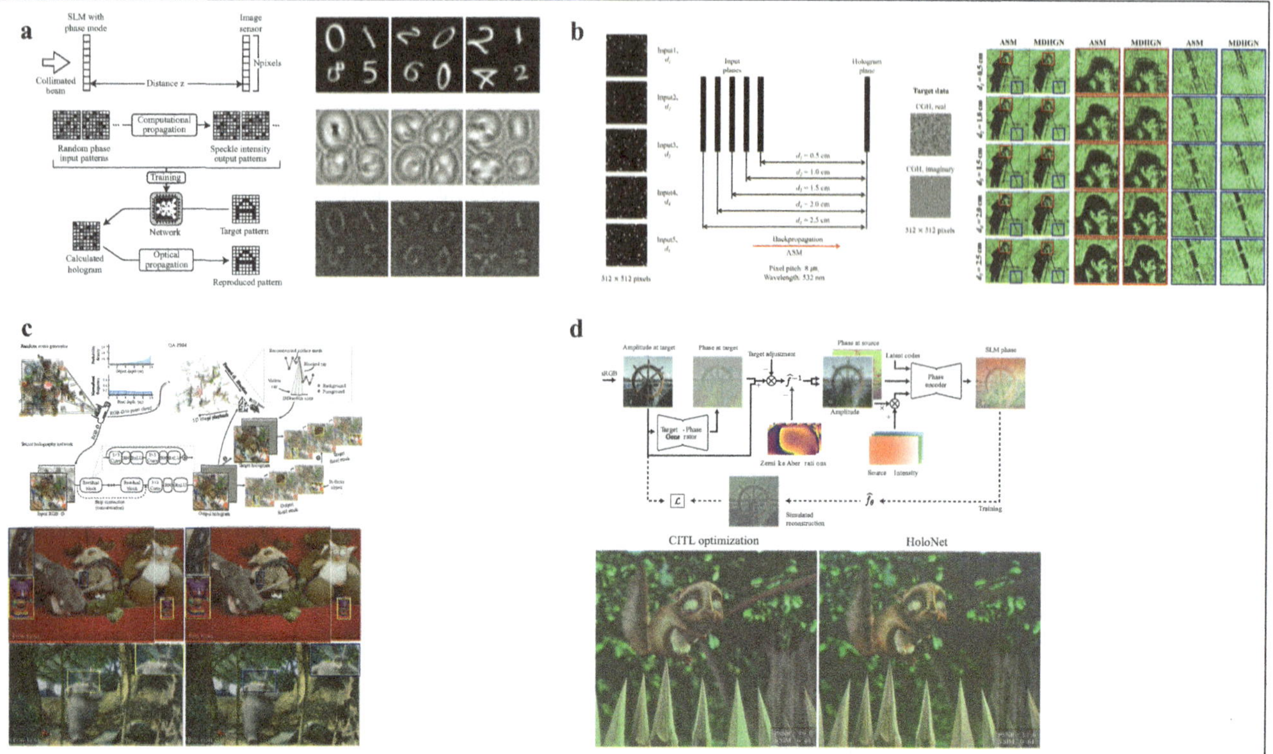

Fig. 12 Image quality and CGH synthesis speed enhancement using neural network: **a** CGH synthesis using a U-Net[80], **b** multi-depth CGH synthesis[81], **c** real-time high-quality CGH synthesis for RGBD input[82], **d** propagation model calibration using camera-in-the-loop feedback and neural holography[83]. Images reprinted with the following permissions: a[80], b[81] from Optica Publishing Group; c[82] from Springer Nature; d[83] from ACM (Association for Computing Machinery).

to the enhancement of the image quality. A sophisticated training strategy, including the designed depth distribution in the synthesized random scene, an occlusion-capable hologram synthesis algorithm, and a loss function that measures the quality in both the reconstructed 3D image volume and the intermediate hologram plane, enables high image quality reconstruction, which has rarely been observed in previous holographic displays. Real-time operation at 60 Hz and with 1080 p resolution is another benefit of this study.

Most conventional CGH techniques, including those shown in Figs. 12a–c, assume an ideal display system. However, a specific implementation of a physical display system deviates from the ideal one in many aspects and causes significant image quality degradation. The work[83] presented in Fig. 12d tackled this problem by incorporating an active feedback configuration. As observed in the figure, a camera is placed in the eyebox to capture the reconstruction. In each iteration, the CGH is updated to reduce the error between the target and reconstructed images. This iterative update of the CGH does not rely on the ideal wave propagation model, but directly considers the physical system and exhibits very high-quality holographic reconstructions. This technique has been further developed to construct an interpretable propagation model considering non-uniform laser illumination, per-pixel phase nonlinearity of the SLM, and aberrations of the system optics. This model was subsequently used to build and train a neural network that outputs the optimized CGH without iterative optimization. Using this technique, 40 Hz CGH synthesis at 1080 p resolution with high image quality was demonstrated. Although the work in Fig. 12d is limited to the target image in a single plane, its extension to multi-depth 3D images[84] has also been reported in recent work. The neural network-based approaches introduced in this section provide a novel breakthrough of the limitations posed by the physics-based calculation of the ideal wave propagation model, and significant progress is expected in this field.

Conclusions

In this article, various holographic techniques that have been applied to AR and VR NEDs have been introduced. Holographic techniques can be used for either static or dynamic optical components. As a static optical component, the holographically recorded volume grating; that is, the HOE, offers a wide range of applications. HOEs are thin and light, which is crucial in AR and VR NEDs;

moreover, the angular, wavelength, and polarization selectivity enable various optical functions, making them an ideal replacement for conventional bulk optics. The recently emerging metasurface lenses have provided an even thinner form factor with a higher NA, demonstrating their potential as AR and VR optics. Holographic displays have been applied to AR/VR NEDs as dynamic displays. The unique wavefront reconstruction capability of holographic displays is used to present natural 3D images with full monocular focus cues, while correcting aberrations in the NED optics and eye lenses. Recent progress has enhanced holographic NEDs in various aspects, including eyebox and FoV expansion, speckle suppression, and foveated presentation. Finally, the combination of holographic techniques with neural networks has exhibited significant performance enhancement in terms of the CGH synthesis speed and reconstruction image quality, making real-time and photorealistic holographic NEDs feasible in the near future.

Acknowledgements
National Research Foundation of Korea (NRF-2017R1A2B2011084).

Author contributions
B. Lee and J.-H. Park supervised this project. J.-H. Park conducted the literature review and wrote the manuscript. The manuscript has been revised and corrected by all authors.

Conflict of interest
The authors declare that they have no conflict of interest.

Received: 25 September 2021 Revised: 19 January 2022 Accepted: 20 January 2022
Accepted article preview online: 28 January 2022
Published online: 22 February 2022

References

1. Cakmakci, O. & Rolland, J. Head-worn displays: a review. *Journal of Display Technology* **2**, 199-216 (2006).
2. Kress, B. C. & Chatterjee, I. Waveguide combiners for mixed reality headsets: a nanophotonics design perspective. *Nanophotonics* **10**, 41-74 (2021).
3. Yano, S. et al. A study of visual fatigue and visual comfort for 3D HDTV/HDTV images. *Displays* **23**, 191-201 (2002).
4. Hoffman, D. M. et al. Vergence-accommodation conflicts hinder visual performance and cause visual fatigue. *Journal of Vision* **8**, 33 (2008).
5. Shibata, T. et al. The zone of comfort: predicting visual discomfort with stereo displays. *Journal of Vision* **11**, 11 (2011).
6. Padmanaban, N. et al. Optimizing virtual reality for all users through gaze-contingent and adaptive focus displays. *Proceedings of the National Academy of Sciences of the United States of America* **114**, 2183-2188 (2017).
7. Lee, B., Yoo, C. & Jeong, J. Holographic optical elements for augmented reality systems. Proceedings of SPIE 11551, Holography, Diffractive Optics, and Applications X. Online: SPIE, 2020.
8. Xiong, J. H. et al. Holographic optical elements for augmented reality: principles, present status, and future perspectives. *Advanced Photonics Research* **2**, 2000049 (2021).
9. Chang, C. L. et al. Toward the next-generation VR/AR optics: a review of holographic near-eye displays from a human-centric perspective. *Optica* **7**, 1563-1578 (2020).
10. Yeom, H. J. et al. 3D holographic head mounted display using holographic optical elements with astigmatism aberration compensation. *Optics Express* **23**, 32025-32034 (2015).
11. Maimone, A., Georgiou, A. & Kollin, J. S. Holographic near-eye displays for virtual and augmented reality. *ACM Transactions on Graphics* **36**, 85 (2017).
12. Nam, S. W. et al. Aberration-corrected full-color holographic augmented reality near-eye display using a Pancharatnam-Berry phase lens. *Optics Express* **28**, 30836-30850 (2020).
13. Kim, D. et al. Vision-correcting holographic display: evaluation of aberration correcting hologram. *Biomedical Optics Express* **12**, 5179-5195 (2021).
14. Lee, S. et al. Analysis and implementation of hologram lenses for see-through head-mounted display. *IEEE Photonics Technology Letters* **29**, 82-85 (2017).
15. Bang, K., Jang, C. & Lee, B. Curved holographic optical elements and applications for curved see-through displays. *Journal of Information Display* **20**, 9-23 (2019).
16. Bigler, C. M., Mann, M. S. & Blanche, P. A. Holographic waveguide HUD with in-line pupil expansion and 2D FOV expansion. *Applied Optics* **58**, G326-G331 (2019).
17. Akşit, K. et al. Near-eye varifocal augmented reality display using see-through screens. *ACM Transactions on Graphics* **36**, 189 (2017).
18. Jang, C. et al. Retinal 3D: augmented reality near-eye display via pupil-tracked light field projection on retina. *ACM Transactions on Graphics* **36**, 190 (2017).
19. Piao, J. A. et al. Full color holographic optical element fabrication for waveguide-type head mounted display using photopolymer. *Journal of the Optical Society of Korea* **17**, 242-248 (2013).
20. Lee, S. et al. Additive light field displays: realization of augmented reality with holographic optical elements. *ACM Transactions on Graphics* **35**, 60 (2016).
21. Yeom, J. et al. Three-dimensional/two-dimensional convertible projection screen using see-through integral imaging based on holographic optical element. *Applied Optics* **54**, 8856-8862 (2015).
22. Xiong, J. H. et al. Aberration-free pupil steerable Maxwellian display for augmented reality with cholesteric liquid crystal holographic lenses. *Optics Letters* **46**, 1760-1763 (2021).
23. Jo, Y. et al. Eye-box extended retinal projection type near-eye display with multiple independent viewpoints. *Applied Optics* **60**, A268-A276 (2021).
24. Kim, S. B. & Park, J. H. Optical see-through Maxwellian near-to-eye display with an enlarged eyebox. *Optics Letters* **43**, 767-770 (2018).
25. Lin, T. G. et al. Maxwellian near-eye display with an expanded eyebox. *Optics Express* **28**, 38616-38625 (2020).
26. Yoo, C. et al. Retinal projection type lightguide-based near-eye display with switchable viewpoints. *Optics Express* **28**, 3116-3135 (2020).
27. Moon, S. et al. Compact augmented reality combiner using Pancharatnam-Berry phase lens. *IEEE Photonics Technology Letters* **32**, 235-238 (2020).
28. Yin, K. et al. Doubling the FOV of AR displays with a liquid crystal polarization-dependent combiner. *Optics Express* **29**, 11512-11519 (2021).
29. Yoo, C. et al. Dual-focal waveguide see-through near-eye display with polarization-dependent lenses. *Optics Letters* **44**, 1920-1923 (2019).
30. Kim, J. et al. Foveated AR: dynamically-foveated augmented reality display. *ACM Transactions on Graphics* **38**, 99 (2019).
31. Yoo, C. et al. Foveated display system based on a doublet geometric phase lens. *Optics Express* **28**, 23690-23702 (2020).
32. Jeong, J. et al. Holographically printed freeform mirror array for augmented reality near-eye display. *IEEE Photonics Technology Letters* **32**, 991-994 (2020).

33. Jang, C. et al. Design and fabrication of freeform holographic optical elements. *ACM Transactions on Graphics* **39**, 184 (2020).
34. Sung, J., Lee, G. Y. & Lee, B. Progresses in the practical metasurface for holography and lens. *Nanophotonics* **8**, 1701-1718 (2019).
35. Lee, G. Y., Sung, J. & Lee, B. Metasurface optics for imaging applications. *MRS Bulletin* **45**, 202-209 (2020).
36. Jiang, Q., Jin, G. F. & Cao, L. C. When metasurface meets hologram: principle and advances. *Advances in Optics and Photonics* **11**, 518-576 (2019).
37. Lee, G. Y. et al. Metasurface eyepiece for augmented reality. *Nature Communications* **9**, 4562 (2018).
38. Li, Z. Y. et al. Meta-optics achieves RGB-achromatic focusing for virtual reality. *Science Advances* **7**, eabe4458 (2021).
39. Nikolov, D. K. et al. Metaform optics: bridging nanophotonics and freeform optics. *Science Advances* **7**, eabe5112 (2021).
40. Westheimer, G. The Maxwellian view. *Vision Research* **6**, 669-682 (1966).
41. Takaki, Y. & Fujimoto, N. Flexible retinal image formation by holographic Maxwellian-view display. *Optics Express* **26**, 22985-22999 (2018).
42. Von Waldkirch, M., Lukowicz, P. & Tröster, G. Defocusing simulations on a retinal scanning display for quasi accommodation-free viewing. *Optics Express* **11**, 3220-3233 (2003).
43. Wu, Y. H. et al. Design of retinal-projection-based near-eye display with contact lens. *Optics Express* **26**, 11553-11567 (2018).
44. Konrad, R. et al. Accommodation-invariant computational near-eye displays. *ACM Transactions on Graphics* **36**, 88 (2017).
45. Dunn, D. et al. Membrane AR: varifocal, wide field of view augmented reality display from deformable membranes. Proceedings of ACM SIGGRAPH 2017 Emerging Technologies. New York: ACM, 2017.
46. Liu, S. X. et al. Reverse-mode PSLC multi-plane optical see-through display for AR applications. *Optics Express* **26**, 3394-3403 (2018).
47. Cui, Wei. & Gao, L. Optical mapping near-eye three-dimensional display with correct focus cues. *Optics Letters* **42**, 2475-2478 (2017).
48. Klug, M. A., Cahall, S. C. & Chung, H. Separated pupil optical systems for virtual and augmented reality and methods for displaying images using same. (2016).
49. Huang, H. K. & Hua, H. High-performance integral-imaging-based light field augmented reality display using freeform optics. *Optics Express* **26**, 17578-17590 (2018).
50. Lanman, D. & Luebke, D. Near-eye light field displays. *ACM Transactions on Graphics* **32**, 220 (2013).
51. Chou, P. Y. et al. Hybrid light field head-mounted display using time-multiplexed liquid crystal lens array for resolution enhancement. *Optics Express* **27**, 1164-1177 (2019).
52. Huang, F. C., Chen, K. & Wetzstein, G. The light field stereoscope: immersive computer graphics via factored near-eye light field displays with focus cues. *ACM Transactions on Graphics* **34**, 60 (2015).
53. Ueno, T. & Takaki, Y. Super multi-view near-eye display to solve vergence-accommodation conflict. *Optics Express* **26**, 30703-30715 (2018).
54. Yeom, J., Son, Y. & Choi, K. Crosstalk reduction in voxels for a see-through holographic waveguide by using integral imaging with compensated elemental images. *Photonics* **8**, 217 (2021).
55. Wakunami, K. et al. Projection-type see-through holographic three-dimensional display. *Nature Communications* **7**, 12954 (2016).
56. Li, X. et al. 3D dynamic holographic display by modulating complex amplitude experimentally. *Optics Express* **21**, 20577-20587 (2013).
57. Moon, E. et al. Holographic head-mounted display with RGB light emitting diode light source. *Optics Express* **22**, 6526-6534 (2014).
58. Zhang, Z. Q. et al. A full-color compact 3D see-through near-eye display system based on complex amplitude modulation. *Optics Express* **27**, 7023-7035 (2019).
59. Park, J. H. & Kim, S. B. Optical see-through holographic near-eye-display with eyebox steering and depth of field control. *Optics Express* **26**, 27076-27088 (2018).
60. Kozacki, T. et al. Extended viewing angle holographic display system with tilted SLMs in a circular configuration. *Applied Optics* **51**, 1771-1780 (2012).
61. Hahn, J. et al. Wide viewing angle dynamic holographic stereogram with a curved array of spatial light modulators. *Optics Express* **16**, 12372-12386 (2008).
62. Li, J., Smithwick, Q. & Chu, D. P. Scalable coarse integral holographic video display with integrated spatial image tiling. *Optics Express* **28**, 9899-9912 (2020).
63. An, J. et al. Slim-panel holographic video display. *Nature Communications* **11**, 5568 (2020).
64. Jang, C. et al. Holographic near-eye display with expanded eye-box. *ACM Transactions on Graphics* **37**, 195 (2018).
65. Choi, M. H., Ju, Y. G. & Park, J. H. Holographic near-eye display with continuously expanded eyebox using two-dimensional replication and angular spectrum wrapping. *Optics Express* **28**, 533-547 (2020).
66. Park, J., Lee, K. & Park, Y. Ultrathin wide-angle large-area digital 3D holographic display using a non-periodic photon sieve. *Nature Communications* **10**, 1304 (2019).
67. Kuo, G. et al. High resolution étendue expansion for holographic displays. *ACM Transactions on Graphics* **39**, 66 (2020).
68. Guenter, B. et al. Foveated 3D graphics. *ACM Transactions on Graphics* **31**, 164 (2012).
69. Patney, A. et al. Towards foveated rendering for gaze-tracked virtual reality. *ACM Transactions on Graphics* **35**, 179 (2016).
70. Hong, J. et al. Gaze contingent hologram synthesis for holographic head-mounted display. Proceedings of SPIE 9771, Practical Holography XXX: Materials and Applications. San Francisco: SPIE, 2016.
71. Wei, L. J. & Sakamoto, Y. J. Fast calculation method with foveated rendering for computer-generated holograms using an angle-changeable ray-tracing method. *Applied Optics* **58**, A258-A266 (2019).
72. Ju, Y. G. & Park, J. H. Foveated computer-generated hologram and its progressive update using triangular mesh scene model for near-eye displays. *Optics Express* **27**, 23725-23738 (2019).
73. Chang, C. L., Cui, W. & Gao, L. Foveated holographic near-eye 3D display. *Optics Express* **28**, 1345-1356 (2020).
74. Cem, A. et al. Foveated near-eye display using computational holography. *Scientific Reports* **10**, 14905 (2020).
75. Lee, S. et al. Foveated near-eye display for mixed reality using liquid crystal photonics. *Scientific Reports* **10**, 16127 (2020).
76. Lee, B. et al. Wide-angle speckleless DMD holographic display using structured illumination with temporal multiplexing. *Optics Letters* **45**, 2148-2151 (2020).
77. Takaki, Y. & Yokouchi, M. Speckle-free and grayscale hologram reconstruction using time-multiplexing technique. *Optics Express* **19**, 7567-7579 (2011).
78. Ko, S. B. & Park, J. H. Speckle reduction using angular spectrum interleaving for triangular mesh based computer generated hologram. *Optics Express* **25**, 29788-29797 (2017).
79. Lee, S. et al. Light source optimization for partially coherent holographic displays with consideration of speckle contrast, resolution, and depth of field. *Scientific Reports* **10**, 18832 (2020).
80. Horisaki, R., Takagi, R. & Tanida, J. Deep-learning-generated holography. *Applied Optics* **57**, 3859-3863 (2018).
81. Lee, J. et al. Deep neural network for multi-depth hologram generation and its training strategy. *Optics Express* **28**, 27137-27154 (2020).
82. Shi, L. et al. Towards real-time photorealistic 3D holography with deep neural networks. *Nature* **591**, 234-239 (2021).
83. Peng, Y. F. et al. Neural holography with camera-in-the-loop training. *ACM Transactions on Graphics* **39**, 185 (2020).
84. Choi, S. et al. Neural 3D holography: Learning accurate wave propagation models for 3D holographic virtual and augmented reality displays. *ACM Transactions on Graphics* **40**, 240 (2021).

Review Open Access

Holographic 3D imaging through random media: methodologies and challenges

Mitsuo Takeda[1,*], Wolfgang Osten[2] and Eriko Watanabe[3,*]

Abstract

Imaging through random media continues to be a challenging problem of crucial importance in a wide range of fields of science and technology, ranging from telescopic imaging through atmospheric turbulence in astronomy to microscopic imaging through scattering tissues in biology. To meet the scope of this anniversary issue in holography, this review places a special focus on holographic techniques and their unique functionality, which play a pivotal role in imaging through random media. This review comprises two parts. The first part is intended to be a mini tutorial in which we first identify the true nature of the problems encountered in imaging through random media. We then explain through a methodological analysis how unique functions of holography can be exploited to provide practical solutions to problems. The second part introduces specific examples of experimental implementations for different principles of holographic techniques, along with their performance results, which were taken from some of our recent work.

Keywords: Holography, Imaging, Scattering, Random Media, Statistical Optics, Coherence, Speckles, Correlation, Digital Holography

Introduction

Over many years, imaging objects through random media has been a major technological challenge for many researchers[1,2]. This is because the techniques for imaging through random media constitute an essential common technological basis in a broad range of fields of science and technology, ranging from telescopic imaging through atmospheric turbulence in astronomy[3] to microscopic imaging through scattering tissues in biology[4]. Reflecting the diversity and complexity of random media in the real world, a variety of techniques have been proposed to deal with the various characteristics of random media, as found in review articles[5–9]. To meet the scope of the special *Anniversary Issue on Holography*, this review places special emphasis on holographic techniques and their unique functionality, which play a pivotal role in imaging through random media. This review comprises two parts. The first part is intended to be a mini tutorial in which we first identify the nature of the problems encountered in imaging through random media. Through a methodological analysis, we then explain how the special functions that are unique to holography can be exploited to provide simple practical solutions to problems. In the second part, we introduce specific examples of experimental implementations and results based on different principles of holographic techniques, where the examples are taken from some of our recent work.

How random media disturb imaging

To understand the nature of the problem, let us first consider a simplified physical model to explain the way in

Correspondence: Mitsuo Takeda (takeda@opt.utsunomiya-u.ac.jp) or Eriko Watanabe (eriko.watanabe@uec.ac.jp)
[1]Center for Optical Research and Education, Utsunomiya University, 7-1-2, Yoto, Utsunomiya, Tochigi, 321-8585, Japan
[2]Institut für Technische Optik, Universität Stuttgart, Pfaffenwaldring 9, 70569 Stuttgart, Germany
Full list of author information is available at the end of the article.

which random media disturb imaging. For simplicity, we divide random media into two types based on their physical characteristics: *inhomogeneous* and *scattering* media for convenience of explanation. It should be noted that objects are not always outside the random medium, such as stars in astronomical imaging; they can be inside the random medium, such as cells in biomedical imaging, which makes the problem more complex. As illustrated in Fig. 1, the inhomogeneous medium is made of randomly distributed relatively mild refractive index variations on a macroscopic scale with a large spatial correlation length so that it remains transparent, such as heat shimmers caused by atmospheric refraction. Although the inhomogeneous medium introduces aberrations by distorting the wavefront of the light from an object, it preserves the validity of the concept of the wavefront defined for a bundle of rays of geometrical optics. On the other hand, the scattering medium is made of randomly distributed strong refractive-index variations at the microscopic scale with a short spatial correlation length. Incident light is scattered by the microstructure in the medium into many secondary wave components whose direction is determined by the phase function, creating a scrambled optical field that no longer has a well-defined wavefront in terms of geometrical optics. Thus, a typical scattering medium appears opaque, like ground glass.

In addition to the intrinsic physical properties of the media, its thickness has a significant influence on imaging performance. For example, with an increase in the thickness of a random medium, the shift invariance (i.e., isoplanatism or memory effect) of the imaging property tends to be lost. It is common to categorize scattering media into an optically thin scattering layer (e.g., ground glass) and a thick scattering volume (e.g., a tank of milk) according to the optical thickness (i.e., the number of scattering events occurring during light propagation). The concept of random media can be extended to include reflective optical surfaces, such as a randomly deformed mirror and a diffusely reflecting surface, which serve as an inhomogeneous medium and a scattering medium with extremely small thickness, respectively.

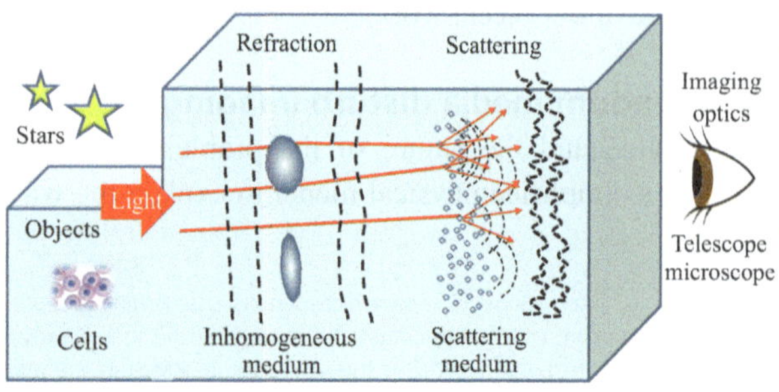

Fig. 1 Simplified model of random media with two typical physical characteristics.

How random media disturb imaging depends not only on the characteristics of the random media, but also on their location between the object and the imaging optics. For example, ground glass severely disrupts the image when it is placed on the pupil plane of imaging optics, but it has a negligible influence when placed on the image plane to serve as a focus screen, such as a camera view finder.

It should also be noted that how the scale of the refractive index change in the media influences imaging performance is not determined by the geometric scale *per se*, but it is instead affected by its relative scale with respect to the imaging optics. For instance, the resolution of a ground telescope is determined by the relative scale of the correlation length of the atmospheric index changes (i.e., the Fried parameter) to the diameter of the telescope.

Methodologies: Problems and Solutions

According to the simplified model of random media, the object light exiting from a random medium becomes a mixture of the three components, namely, unperturbed original light, non-scattered light with distorted wavefronts, and scattered light without well-defined wavefronts. The mixture ratio of the three components varies depending on the specific characteristics of the medium of interest. Because the problem arises from multiple sources of different natures, one must take a multifaceted approach to the problem. Here, we introduce the basic principles of the representative techniques and review them from a methodological viewpoint, with a particular focus on how the unique functionality of holography is effectively used in these techniques.

In the first step of searching for a solution to the problem, a choice can be made from the two options of the basic strategy: *rejection or correction of bad light*. The *rejection* strategy discriminates good light (unperturbed original light) from bad light (scattered light and aberrated light with distorted wavefronts) and rejects bad light in order only for good light to take part in imaging. Good light, however, is reduced rapidly as photons travel through the scattering medium, obeying Beer's exponential law. The *correction* strategy does not reject bad light, but turns it into good light, so that nearly all light takes part in imaging. Holography plays a key role in both strategies and offers the additional merit of 3D imaging with full information of both amplitude and phase.

Holographic rejection of scattered light: imaging only with good light

The best-known example of the rejection-based

technique is optical coherence tomography[10] (OCT), which rejects bad light and makes use of ballistic photons or unperturbed light that passes through the diffusive medium without being scattered. Ballistic photons take straight paths shorter than zigzags, followed by scattered photons. This difference in the light arrival time permits the discrimination between the good and bad light either by coherence gating with short temporal coherence light, as in OCT, with the low spatial coherence of light[11], or by time gating using a short optical pulse, as with femto-photography[12], in which the time-of-flight is detected with a streak camera. Since the early work of time gating by Duguay and Mattick[13], various gating techniques[14–16] have been proposed for rejecting scattered light.

Holography has an intrinsic gating function in its principle, *per se*. Specifically, the interference between the object and reference beams permits the reference beam to serve as a gating signal for time gating[17], temporal coherence gating[18], spatial coherence gating[19], spatial frequency gating[20], and Doppler-frequency-shift gating[21,22]. As illustrated in Fig. 2, an object hidden behind a scattering medium is recorded in a hologram using an ultrashort optical pulse or low-coherence continuous-wave (CW) light. Holographic recording occurs only when both the object and reference lights arrive at the recording point on the hologram precisely in synchronization within the time window set by the width of the light pulse or the coherence length of the CW light. Thus, by controlling the time delay of the reference light, one can set the desired time gate on the object light to be recorded on the hologram. For example, a series of time gates, spatially arranged in chronological order on the hologram, can be created by using a reference beam that illuminates the hologram at a grazing angle with its arrival time delay linearly mapped onto the spatial location on the hologram. This technique of time gating, known as light-in-flight recording[17], enables the recording of high-speed holographic motion pictures of ultrafast phenomena. The time delay is introduced to the object beam not only by multiple scattering in the random medium, but also by the distance traveled by the object beam between the object and the hologram, which provides information about the object depth. Therefore, holographic time/coherence gating can be used for two purposes, namely, rejection of scattered lights, and tomographic sectioning of 3D objects, as in OCT.

Although the technique of time/coherence gating is useful for rejecting multiple scattering light (and tomographic sectioning) with a large time delay, it cannot reject non-scattered aberrated lights (or snake photons) with distorted wavefronts. This is because the time delays corresponding to the wavefront aberrations of non-scattered light are too small to be discriminated by time/coherence gating, so that the aberrated light is accepted as good light. For this reason, wavefront aberration correction with adaptive optics is now one of the hot topics[23] in OCT.

In addition to the limited rejection of aberrated light, the strategy of bad light rejection has a fundamental problem in that all useful information carried by the scattered and/or aberrated light is discarded, even though these bad light have more optical power than the good light in most random media. Therefore, an alternative strategy is necessary for this solution.

Holographic correction of perturbed light: turning bad light into good light

Before discussing holographic techniques, we begin with a quick overview of non-holographic techniques. A well-known example of the non-holographic technique for turning bad light into good light is adaptive optics[24], in which an aberrated wavefront is detected with a wavefront sensor and corrected adaptively with a deformable mirror. With the advent of high-resolution spatial light modulators (SLMs) for phase modulation, various techniques for correcting distorted wavefronts with more complex phase structures have been demonstrated[25,26]. It should be noted, however, that adaptive optics works effectively only within

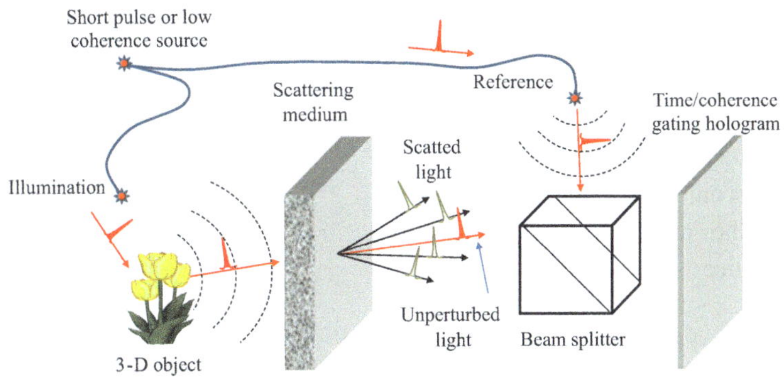

Fig. 2 Schematic illustration of the basic concept and principle of holographic time/coherence gating.

a restricted field of view (i.e., the isoplanatic patch) over which the aberration does not change; in other words, the point spread function of the overall imaging system, including the random medium, retains shift invariance or isoplanatism. Astronomers are now making efforts to solve this problem using multi-conjugate adaptive optics[27]. A random medium is said to have a *memory effect*[28] when its scattered field has a shift invariance with respect to the angle shift of the incident light. The memory effect is known to be closely related to the thickness of the random medium; a strongly scattering medium can have a memory effect as long as its scattering layer is thin (like ground glass), but it loses the memory effect as its thickness increases. The concept of the memory effect also plays an important role in other techniques, which is explained later.

Another type of non-holographic technique takes a mathematical approach to the problem, where the object information is retrieved numerically from the bad light by solving a sort of inverse problem in a wide sense. Specific techniques include model-based parametrization of the physical process for the solution search using optimization algorithms[29–30], single-pixel imaging based on orthogonal pattern projection[31,32], machine learning[33], and neural networks[34,35]. These techniques are now enjoying rapid progress, and their success depends on how well the physical process and prior knowledge are integrated into the mathematical model.

Wavefront correction by holographic phase conjugation

The basic functions of adaptive optics (i.e., wavefront detection and correction) are included in the principle of holography. Wavefront detection is performed in the natural course of hologram recording because the hologram is nothing but an interferogram in which the wavefront, $\phi_O(\boldsymbol{r})$, of the object beam is recorded as

$$\begin{aligned} t(\boldsymbol{r}) \propto |u_O(\boldsymbol{r}) + u_R(\boldsymbol{r})|^2 &= |u_O(\boldsymbol{r})|^2 + 1 \\ &+ |u_O(\boldsymbol{r})| \exp[i\phi_O(\boldsymbol{r})] \exp(-i\boldsymbol{k}\cdot\boldsymbol{r}) \\ &+ |u_O(\boldsymbol{r})| \exp[-i\phi_O(\boldsymbol{r})] \exp(i\boldsymbol{k}\cdot\boldsymbol{r}) \end{aligned} \tag{1}$$

where $t(\boldsymbol{r})$ is the amplitude transmittance of the hologram, $u_O(\boldsymbol{r})$ is the object beam, and $u_R(\boldsymbol{r})$ is the reference plane wave with a unit amplitude, $|u_R(\boldsymbol{r})| = 1$, and phase, $\phi_R = \boldsymbol{k}\cdot\boldsymbol{r}$, for off-axis holography. Also recorded in the hologram is the phase-conjugated object beam, $u_O* = |u_O| \exp(-i\phi_O)$, with the reversed sign of phase, which plays a key role in the wavefront correction, as explained below.

As shown schematically in Fig. 3, the light from a coherently illuminated object is perturbed by a random medium and then recorded with a reference plane wave, $\exp(i\boldsymbol{k}\cdot\boldsymbol{r})$, in a hologram with the amplitude transmittance given by Eq. 1. When the hologram is read out with a counter-propagating plane wave, $\exp(-i\boldsymbol{k}\cdot\boldsymbol{r})$, a phase-conjugated object beam, $u_O* = |u_O| \exp(-i\phi_O)$, is generated from the last term in Eq. 1. Physically, it is a time-reversed object beam that propagates back into the random medium to retrace its path backward through it, like watching a time-reversed movie. As long as the time reversal symmetry holds with the physical law of wave propagation in the random medium, the light exiting from the random medium becomes the original unperturbed object light, although with the sign of phase reversed, which converges to form a real image of the object. Thus, bad light turned into good light.

The unique feature of holographic phase conjugation compared with adaptive optics is as follows. Referring to Fig. 3, we consider two points, A and B, at different locations on the object. The lights from these points as point sources (indicated by rays in black and green) pass through different parts of the random medium and receive different perturbations or wavefront distortions. In conventional adaptive optics, wavefront sensing is performed for a single point source (e.g., a guide star in astronomy), and wavefront correction is made for this single point in the field of view. This means that conventional adaptive optics cannot perform wavefront corrections simultaneously for points A and B from two isoplanatic patches with different perturbations. On the other hand, in holographic phase conjugation, every object point in the field of view serves as its own guide star for imaging itself with light backtracking through the random medium by virtue of time reversal. Therefore, holographic phase conjugation can have a wider field of view for thick and complex random media than conventional adaptive optics. Note, however, that because this advantage is brought about by the coherent holographic recording of objects, the technique of holographic phase conjugation cannot be applied directly to the imaging of incoherently

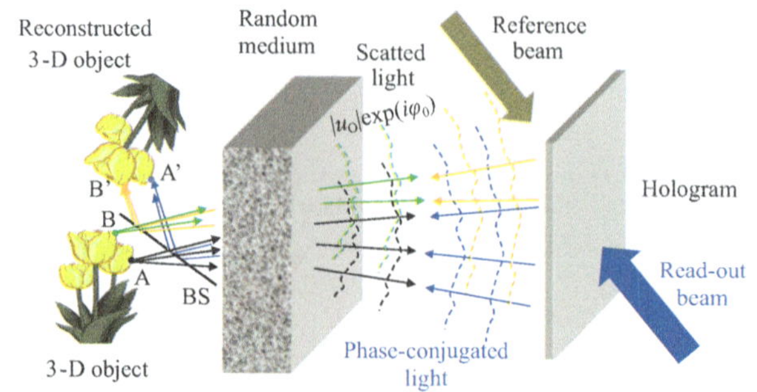

Fig. 3 Schematic illustration of the concept and principle of wavefront correction by holographic phase-conjugation.

illuminated objects, such as stars in astronomy and fluorescence-labeled cells in biology.

The technique of wavefront correction by holographic phase conjugation has a long history going back to early seminal work in the 1960s by Kogelnik[36] and Leith and Upatnieks[37]. With the availability of novel holographic recording materials (e.g., photorefractive crystals) and high-resolution SLMs, new implementations of holographic phase conjugation have been proposed and demonstrated for imaging through random media, including optical phase conjugation using an Fe-doped LiNbO3 photorefractive crystals[38], and a digital optical phase conjugation system[39,40] based on the combination of a high-resolution image sensor and an SLM for digital wavefront sensing and correction.

Wavefront correction by common-path holographic recording

The holographic phase conjugation described above requires duplication of the randomly perturbed object beam, with its phase sign reversed in the reconstruction process. It further requires precise alignment between the hologram and the random medium for the reconstructed phase-conjugated beam to precisely retrace the same path backwards through the random medium. Also, the corrected final image is not available downstream of the hologram, but is formed upstream of the original object itself; a solution to this third problem was given in Kogelnik and Pennington[41].

In their seminal paper in 1966, Goodman et al. proposed yet another holographic technique[42] (i.e., common-path holographic recording), which is free from the problems mentioned above. Fig. 4 shows the concept and principle of wavefront correction via quasi-common path holographic recording. For simplicity, the object is assumed to be a single point source at O on the object, although the principle applies to other points on the object. A reference point source is placed at point R, such that the distance,

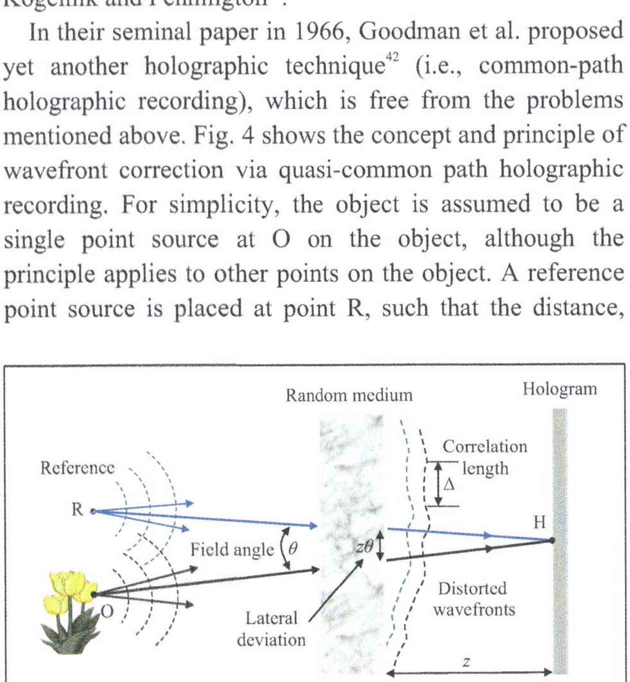

Fig. 4 Schematic illustration of the concept and principle of wavefront correction by common-path holographic recording.

OR, between the object and reference points subtends a small field angle, θ, seen from a recording point, H, on the hologram. Suppose that the random medium is placed at distance, z, from the hologram, then the object and reference rays received at the recording point, H, have emerged from the random medium with relative lateral shift $z\theta$ at the exit. Let Δ represent the lateral correlation length of the perturbed optical field exiting from the random medium, such that Δ is the average linear dimension over which the wavefront, emerging from the random medium, is nearly constant. If the angular distance between the object and the reference satisfies $\theta < \Delta/z$, then the object and reference rays have passed through identical (i.e., highly correlated) portions of the random medium. In other words, it realizes a geometry for common-path holographic recording in which the common distortions of the reference and object waves cancel each other through interference. The condition, $\theta < \Delta/z$, sets the maximum available field angle. Given a correlation length, Δ, (determined by the physical characteristic of the random medium), the field of view can be maximized for $z \approx 0$ by recording the hologram immediately behind the random medium. From the hologram recoded in this technique, the image can be reconstructed in the same manner as with a conventional hologram without special attention to alignment, and it can be observed downstream of the hologram.

After the early experimental demonstration of long-distance holographic imagery through air turbulence by Goodman et al.[43], the concept and principle of common-path holographic recording were combined with digital holography to make new advancements[44-46].

Holographic correloscopy for imaging through a scattering medium

As preliminaries for the explanation of holographic correloscopy[7], let us start with the principle of non-holographic imaging through a thin scattering medium based on speckle intensity correlation[47,48]. As shown in Fig. 5, the light from an object is scattered by a thin but strong scattering medium placed at a distance, $\hat{z}$, from the object, and the scattered field is observed on an observation plane at a distance, z, from the scattering medium. If the object is illuminated with spatially incoherent quasi-monochromatic light, and the scattering medium has a memory effect, the intensity distribution on the observation plane is given by

$$I(\boldsymbol{r}) = S(\boldsymbol{r},\hat{\boldsymbol{r}}) * O(\hat{\boldsymbol{r}}) = \int S(\boldsymbol{r}-\hat{\boldsymbol{r}})\,O(\hat{\boldsymbol{r}})\,d\hat{\boldsymbol{r}} \qquad (2)$$

where $S(\boldsymbol{r},\hat{\boldsymbol{r}})$ is the shift-invariant point spread function or the intensity speckle pattern created by a point source at point $\hat{\boldsymbol{r}}$ on the object, $O(\hat{\boldsymbol{r}})$ is the object intensity

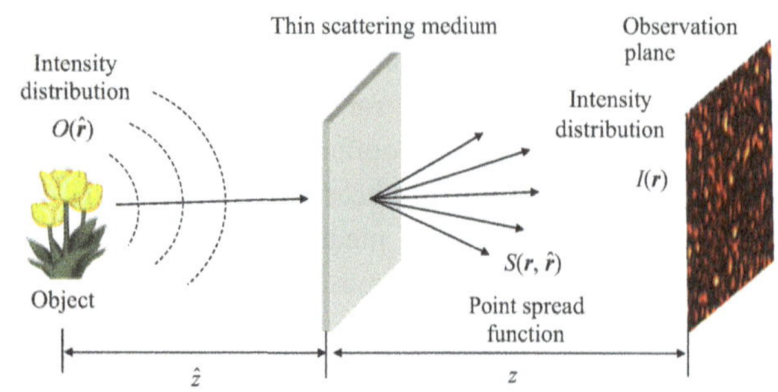

Fig. 5 Non-holographic imaging through a scattering medium based on speckle intensity correlation.

distribution, and $*$ is the convolution operation with the angularly scaled coordinates, $\boldsymbol{r} = z\boldsymbol{\theta}$, and $\hat{\boldsymbol{r}} = -\hat{z}\boldsymbol{\theta}$; the convolution represents the angular memory effect. We can now compute the autocorrelation of the observed intensity distribution.

$$\begin{aligned} I(\boldsymbol{r}) \otimes I(\boldsymbol{r}) &= [S(\boldsymbol{r},\hat{\boldsymbol{r}}) * O(\hat{\boldsymbol{r}})] \otimes [S(\boldsymbol{r},\hat{\boldsymbol{r}}) * O(\hat{\boldsymbol{r}})] \\ &= [S(\boldsymbol{r},\hat{\boldsymbol{r}}) \otimes S(\boldsymbol{r},\hat{\boldsymbol{r}})] * [O(\hat{\boldsymbol{r}}) \otimes O(\hat{\boldsymbol{r}})] \end{aligned} \quad (3)$$

where $\otimes$ denotes autocorrelation; the passage to the second line of Eq. 3 can be easily confirmed from their relationship in the spectral domain. Remembering that $S(\boldsymbol{r},\hat{\boldsymbol{r}})$ is the speckle pattern created by a point source at point $\hat{\boldsymbol{r}}$ on the object, we note that $S(\boldsymbol{r},\hat{\boldsymbol{r}}) \otimes S(\boldsymbol{r},\hat{\boldsymbol{r}})$ has a sharp delta-correlation peak with its width corresponding to the average speckle size, which stands on a constant background resulting from the nonnegativity of $S(\boldsymbol{r},\hat{\boldsymbol{r}})$. By writing

$$S(\boldsymbol{r},\hat{\boldsymbol{r}}) \otimes S(\boldsymbol{r},\hat{\boldsymbol{r}}) \approx \delta(\hat{\boldsymbol{r}}) + \text{const.} \quad (4)$$

we obtain form Eq. 3,

$$I(\boldsymbol{r}) \otimes I(\boldsymbol{r}) \approx O(\hat{\boldsymbol{r}}) \otimes O(\hat{\boldsymbol{r}}) + \text{const.} \quad (5)$$

which states that, apart from the contrast reduction caused by the constant background, the intensity correlation of the scattered field gives the autocorrelation function of the intensity distribution of the hidden object. Hence, the remaining problem is determining the object intensity from its autocorrelation function. This is mathematically equivalent to the phase recovery problem of the power spectrum studied in the field of X-ray diffraction. Bertolotti et al.[47] and Katz et al.[48] used iterative phase-retrieval algorithms, such as the Fienup algorithm[49], which make use of *a priori* knowledge about the object, as constraints in the solution search.

We are now ready to explain holographic correloscopy[7,50] as another solution for phase recovery, remembering that holography was invented as a technique to preserve phase information that was lost by intensity recording in traditional photography. We add a reference point source at point $\boldsymbol{r}_0$, as shown in Fig. 6, so that the intensity distribution in the object space becomes

$$I_O(\hat{\boldsymbol{r}}) = O(\hat{\boldsymbol{r}}) + \delta(\hat{\boldsymbol{r}} - \hat{\boldsymbol{r}}_0) \quad (6)$$

Note that this is the intensity distribution under spatially incoherent illumination, and it differs from the familiar complex amplitude distribution of coherent illumination, such as that shown in Fig. 4. Replacing $O(\hat{\boldsymbol{r}})$ in Eq. 5 with $I_O(\hat{\boldsymbol{r}})$, we have

$$\begin{aligned} I(\boldsymbol{r}) \otimes I(\boldsymbol{r}) &\approx [O(\hat{\boldsymbol{r}}) + \delta(\hat{\boldsymbol{r}} - \hat{\boldsymbol{r}}_0)] \otimes [O(\hat{\boldsymbol{r}}) + \delta(\hat{\boldsymbol{r}} - \hat{\boldsymbol{r}}_0)] + \text{const.} \\ &\approx O(\hat{\boldsymbol{r}} + \hat{\boldsymbol{r}}_0) + O(-(\hat{\boldsymbol{r}} - \hat{\boldsymbol{r}}_0)) \\ &\quad + O(\hat{\boldsymbol{r}}) \otimes O(\hat{\boldsymbol{r}}) + \delta(\hat{\boldsymbol{r}}) + \text{const.} \end{aligned} \quad (7)$$

As in conventional holography, the first term, $O(\hat{\boldsymbol{r}} + \hat{\boldsymbol{r}}_0)$, gives a laterally shifted image of the object, whereas the second term, $O(-(\hat{\boldsymbol{r}} - \hat{\boldsymbol{r}}_0))$, gives a symmetrically shifted image. We regard the remaining terms as noise appearing in the center and background. Thus, we can reconstruct the image using speckle intensity correlation. Singh et al.[50] demonstrated full 3D holographic correloscopy by making use of the memory effect in the axial direction.

Eq. 7 suggests that the real essence of the principle lies in the cross-correlation term between the object and reference, $\delta(\hat{\boldsymbol{r}} - \hat{\boldsymbol{r}}_0) \otimes O(\hat{\boldsymbol{r}})$, which generates the original image corresponding to the interference between the reference and object beams in conventional holography. Then, an idea naturally arises that we record the speckle pattern, $S(\boldsymbol{r},\hat{\boldsymbol{r}})$, or the point spread function, in advance with a reference point source only. We would then compute its cross-correlation with the speckle pattern, $I(\boldsymbol{r})$, of the object:

$$\begin{aligned} S(\boldsymbol{r},\hat{\boldsymbol{r}}) \otimes I(\boldsymbol{r}) &= S(\boldsymbol{r},\hat{\boldsymbol{r}}) \otimes [S(\boldsymbol{r},\hat{\boldsymbol{r}}) * O(\hat{\boldsymbol{r}})] \\ &= [S(\boldsymbol{r},\hat{\boldsymbol{r}}) \otimes S(\boldsymbol{r},\hat{\boldsymbol{r}})] * O(\hat{\boldsymbol{r}}) \\ &\approx O(\hat{\boldsymbol{r}}) + \text{const.} \end{aligned} \quad (8)$$

where use is made of the relation in Eq. 4. In fact, this idea was devised by Freund[51] in 1990 in his seminal paper on *wall lens*, which largely influenced recent research in scatter-based lensless microscopy[52]. This cross-correlation technique requires two steps of speckle recording; therefore, it is not suitable for imaging through dynamic

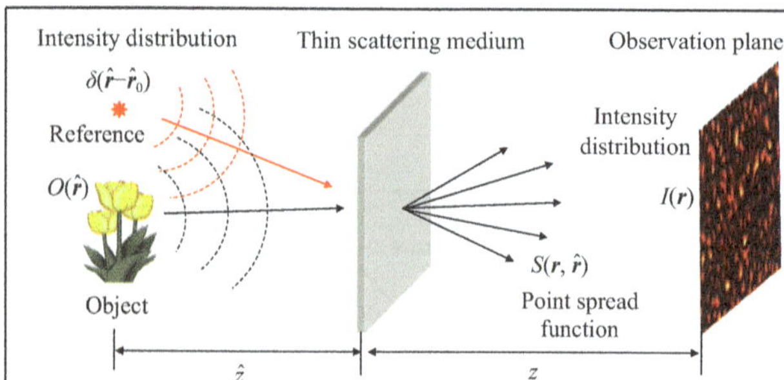

Fig. 6 Holographic correloscopy for imaging through a scattering medium based on speckle intensity correlation.

random media, but it has the advantage of removing noise terms. The principles explained so far are based on intensity correlation under spatially incoherent illumination. Next, we consider the case of a field correlation under coherent illumination.

Referring to Fig. 6, a similar discussion can be made by replacing the intensity variables (denoted by uppercase letters) with the complex-amplitude variables (denoted by the corresponding lowercase letters, apart from the field on the observation plane denoted by $u(\boldsymbol{r})$). Assuming the memory effect, the complex field on the observation plane is given by

$$u(\boldsymbol{r}) = s(\boldsymbol{r},\hat{\boldsymbol{r}}) * u_O(\hat{\boldsymbol{r}}) \tag{9}$$

where $s(\boldsymbol{r},\hat{\boldsymbol{r}})$ is the amplitude point spread function, and $u_O(\hat{\boldsymbol{r}})$ is the field resulting from the superposition of the object and the reference point source.

$$u_O(\hat{\boldsymbol{r}}) = o(\hat{\boldsymbol{r}}) + \delta(\hat{\boldsymbol{r}} - \hat{\boldsymbol{r}}_0) \tag{10}$$

The autocorrelation function (i.e., the spatial coherence function) of the field on the observation plane is given by

$$\begin{aligned}\Gamma(\boldsymbol{r}) &= u(\boldsymbol{r}) \otimes u(\boldsymbol{r}) \\ &= [s(\boldsymbol{r},\hat{\boldsymbol{r}}) * u_O(\hat{\boldsymbol{r}})] \otimes [s(\boldsymbol{r},\hat{\boldsymbol{r}}) * u_O(\hat{\boldsymbol{r}})] \\ &= [s(\boldsymbol{r},\hat{\boldsymbol{r}}) \otimes s(\boldsymbol{r},\hat{\boldsymbol{r}})] * [u_O(\hat{\boldsymbol{r}}) \otimes u_O(\hat{\boldsymbol{r}})]\end{aligned} \tag{11}$$

An important difference from the previous intensity case is that the autocorrelation of the complex-valued speckle field does not have an unwanted constant background.

$$s(\boldsymbol{r},\hat{\boldsymbol{r}}) \otimes s(\boldsymbol{r},\hat{\boldsymbol{r}}) \approx \delta(\hat{\boldsymbol{r}}) \tag{12}$$

so that we have from Eqs. 10,11,

$$\begin{aligned}\Gamma(\boldsymbol{r}) &\approx [o(\hat{\boldsymbol{r}}) + \delta(\hat{\boldsymbol{r}} - \hat{\boldsymbol{r}}_0)] \otimes [o(\hat{\boldsymbol{r}}) + \delta(\hat{\boldsymbol{r}} - \hat{\boldsymbol{r}}_0)] \\ &\approx o(\hat{\boldsymbol{r}} + \hat{\boldsymbol{r}}_0) + o(-(\hat{\boldsymbol{r}} - \hat{\boldsymbol{r}}_0)) + o(\hat{\boldsymbol{r}}) \otimes o(\hat{\boldsymbol{r}}) + \delta(\hat{\boldsymbol{r}})\end{aligned} \tag{13}$$

Thus, the object hidden behind the scattering medium can be reconstructed from the spatial coherence function of the scattered optical field. The most important difference from the intensity correlation is that we can reconstruct the complex amplitude of the object, including the phase information, which permits the holographic interferometry of an object hidden behind the scattering medium[44].

So far, the principle of correloscopy has been explained without providing a clear explanation of the definition of the correlation operation denoted by ⊗. Mathematically, the correlation between statistical signals is defined using the ensemble average as $g(\boldsymbol{r}) \otimes g(\boldsymbol{r}) = < g(\boldsymbol{r}' + \boldsymbol{r}) g^*(\boldsymbol{r}') >$. In practice, however, the ensemble average, <...>, needs to be replaced either by the temporal average or by the spatial average, assuming the stationarity of the statistical process[53]. Although the temporal average is suitable for imaging with time-varying speckle fields through a dynamic scattering medium, such as rotating ground glass, the spatial average is used in imaging through a static scattering medium, as is often the case with speckle intensity correlation.

Here, we show two different ways to implement the principle of field-correlation correloscopy. Fig. 7 shows an example of a fully optical implementation based on a shearing interferometer. The spatial coherence function of the scattered field is detected and visualized in real time as the contrast and shift of interference fringes using a radial shearing interferometer[54]. This scheme, which originated from coherence holography[55], is suitable for imaging through a dynamic scattering medium that permits spatial coherence detection through the time averaging of the speckle fields with the image sensor as an integrator.

For the case of a dynamic scattering medium, such as rotating ground glass, we can find a much simpler solution than the interferometric implementation (shown in Fig. 7) in which the coherence function, $\Gamma(\boldsymbol{r})$, of the scattered field was detected on the observation plane, so as to follow the result in Eq. 13. Let us first note that the thin but strong scattering medium in fast motion destroys the coherence of both the object and the reference beams and serves as a spatially incoherent extended source with the light intensity distribution, $I_S(\tilde{\boldsymbol{r}})$. Next, let us note further that, from the van Cittert-Zernike theorem[1], the coherence function of the scattered field, $\Gamma(\boldsymbol{r})$, can be obtained from the Fourier transform of the incoherent source intensity distribution, $I_S(\tilde{\boldsymbol{r}})$. Now, we have reached a simpler solution[44]. Instead of detecting the coherence function, we detect the intensity distribution of the field exiting the scattering medium by using imaging optics focused on the scatter surface, as illustrated in Fig. 8. Because the scatter medium is thin, the intensity distribution, $I_S(\tilde{\boldsymbol{r}})$, represents the lensless Fourier transform hologram created by the object and point source. Hence, we can understand, without knowledge of the van Cittert-Zernike theorem, that the object can be reconstructed by computing the inverse Fourier transform of the intensity distribution, $I_S(\tilde{\boldsymbol{r}})$, detected with an image sensor; the time-averaged operation takes place over one frame period of the image sensor, which averages out speckles.

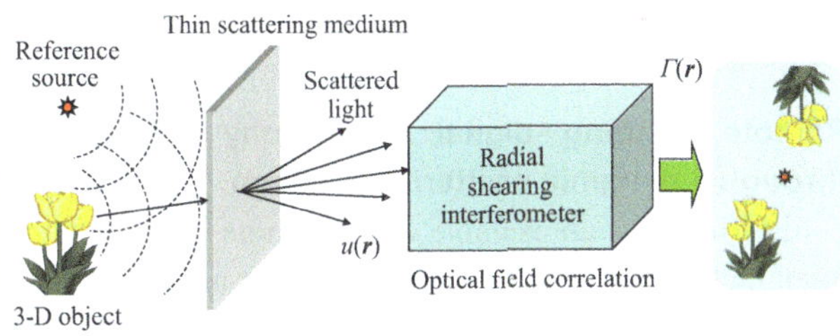

Fig. 7 Interferometric implementation of holographic correloscopy for imaging through a scattering medium.

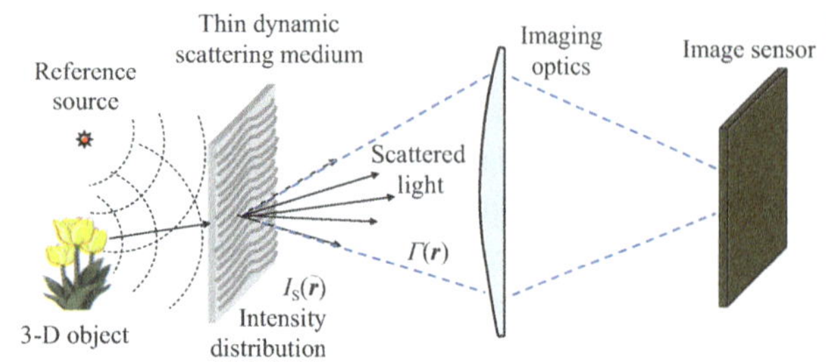

Fig. 8 Simpler implementation of holographic correloscopy for imaging through a scattering medium.

It is of interest to note that this simple implementation (remote imaging digital holography[7]) of holographic correloscopy can be interpreted as an extreme case of common-path holography (shown in Fig. 4), in which the locations of the random medium and hologram are identical with $z = 0$, so that the perfect common-path geometry is realized.

Examples of implementations and experimental results

To provide an idea of how the principles explained in the previous sections are implemented in actual physical systems, we next introduce some specific examples from our past work alongside experimental results.

Holographic correloscopy based on spatial coherence detection with a radial shearing interferometer

Fig. 9 shows a specific example of a fully optical implementation of holographic correloscopy by Naik et al.[54], which corresponds to the principle illustrated in Fig. 7. The optical setup at the top of Fig. 9a creates a lensless Fourier transform hologram of an object pattern, UEC, on a rotating ground glass, which hinders direct observation of the object through it. The light scattered from the ground glass was introduced into a Sagnac radial shearing interferometer, which functions as a real-time optical field correlator that visualizes the complex-valued spatial coherence function directly in the form of the fringe contrast and the fringe shift, as shown in Fig. 9b. The reconstructed image of the hidden object was quantified by fringe analysis, as shown in Fig. 9c.

Remote imaging digital holography for imaging through a dynamic scattering medium

Fig. 10 shows an example of the remote imaging digital holography proposed by Singh et al.[44] for imaging through a dynamic scattering medium. The photo in Fig. 10a shows the experimental setup for implementing the scheme illustrated in Fig. 8. Fig. 10b shows a magnified view of an object and a reference point source, which creates a lensless Fourier transform hologram on a rotating diffuser. Although the rotating diffuser hides the object from the camera, it permits the intensity distribution of the hologram on the diffuser to be imaged by the camera onto a high-resolution CCD image sensor, which also functions as a time integrator to average out speckles during one frame period to capture the hologram. In Fig. 10, (c) shows the hidden object reconstructed from the hologram, and (d) demonstrates the holographic interferometry of the hidden object, in which the left and right photographs show, respectively, the amplitude and phase of the object under thermal deformation for holographic interferometry.

Holographic correloscopy for imaging through a scattering medium based on speckle intensity correlation

Fig. 11 shows an example of the experimental results obtained by Singh et al.[50] using the principle of holographic correloscopy based on the speckle intensity correlation shown in Fig. 6. Fig. 11a shows a microscope image of an original object (a USAF test target), which was placed together with a reference point source at a location hidden behind a diffuser and illuminated with spatially incoherent quasi-monochromatic light. Fig. 11b shows the auto-covariance (i.e., the autocorrelation with the average subtracted) of the speckle intensity of the light scattered by the diffuser. Note the small twin images reconstructed in the top-left and bottom-right corners in symmetry to the central block, stopping the strong central correlation peak. Fig. 11c shows a magnified view of the reconstructed image of the object hidden behind the diffuser. The images in Fig. 11, from (d) through (g), demonstrate imaging through a biological sample (a 1.5-mm-thick chicken breast tissue), where (d) is an original object pattern "H" hidden behind the biological sample, (e) the speckle intensity distribution of the light scattered from the tissue, (f) the auto-covariance of the speckle pattern with the twin images of "H" being reconstructed in the corners, and its magnified image is shown in (g).

Scatter-plate microscope for lensless diffraction-limited microscopy based on intensity cross-correlation.

The concept of Freund's wall lens[51], based on the cross-correlation between the speckle point spread function and the speckle intensity distribution from an object, as in Eq. 8, was extended by Singh et al.[52] to give rise to a scatterplate microscope that makes full use of the magnification property of the memory effect[28,51].

As shown in Fig. 12, a traditional microscope has an

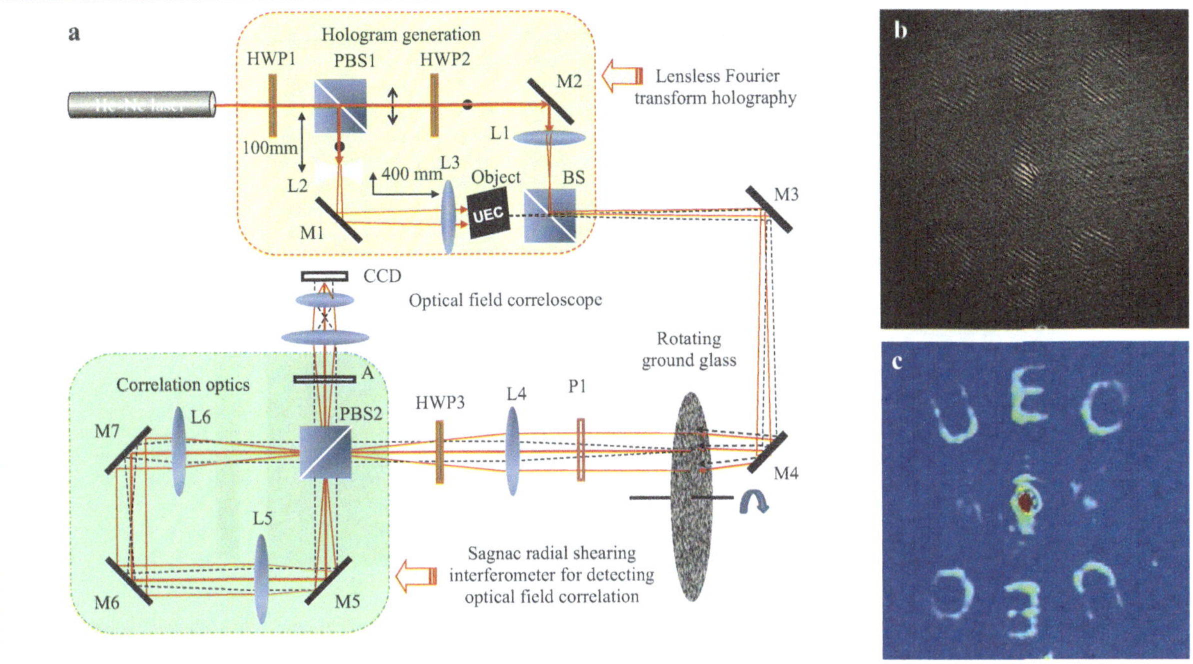

Fig. 9 Holographic correloscopy based on field correlation with a radial shearing interferometer[54]. **a** Experimental setup. **b** Raw fringe pattern. **c** Fringe contrast (modulus of coherence function). Reproduced with permission from OSA. ©OSA

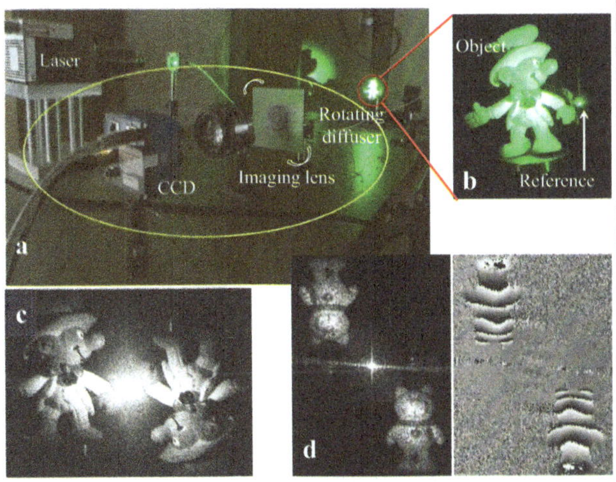

Fig. 10 Remote imaging digital holography for imaging through a dynamic scattering medium[44]. **a** Experimental setup. **b** Magnified view of an object and a reference source. **c** Reconstruction of the hidden object. **d** Holographic interferometry of the hidden object. Reproduced with permission from OSA. ©OSA

objective lens that forms sharp images of a point source and an object, as shown in (a) and (b). If we replace the objective lens with a thin scatter plate, we have a point spread speckle pattern, as shown in (c) and an object speckle pattern, as shown in (d). Because of the angular shift invariance of the memory effect, a shift of the point source on the object space results in a corresponding shift in the speckle pattern magnified by $m = -z/\hat{z}$ times; thus, the scatter plate can serve as a scattering microscope if the original object image is retrieved from the intensity cross-correlation between the two scattered fields shown in (c) and (d). In Fig. 12, (e) is an original object, the numeral 6, (f) the point spread speckle pattern of the scatter plate, (g) the object intensity speckle pattern, and (h) is the image reconstructed from the cross-correlation between the speckle patterns in (f) and (g). Figs. 12i–k demonstrate the application of the scatter-plate microscope to a biological sample (a section of a pine wood stem). In Fig. 12, (i) and (j) are images observed with a conventional microscope (NA = 0.16) illuminated with monochromatic and white light, respectively, and (k) is the image obtained with the scatter-plate microscope (NA = 0.16); the scale bar in the bottom right is 20 μm.

3D microscopic imaging through scattering media based on in-line phase-shift digital holography

Although remote imaging digital holography[44] in Fig. 10 has the advantage of reconstructing the complex field (including phase information) of an object hidden behind a random medium, its application is limited to a dynamic scattering medium. Kodama et al.[45] proposed a scheme for in-line phase-shift digital holography that can image through a static scattering medium. Illustrated in Fig. 13a is an experimental setup in which the light from an object, O, and the phase-shifted reference light are superposed in line to create broad holographic fringes on the static scattering

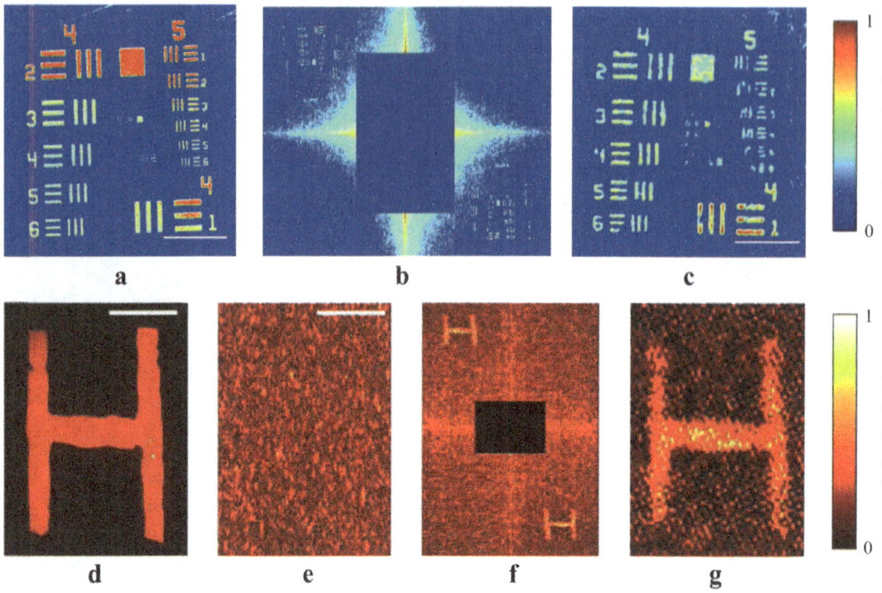

Fig. 11 Experimental results of holographic correloscopy based on speckle intensity correlation[50]. **a** Original object. **b** Auto-covariance of speckle intensity pattern with its central correlation peak blocked. **c** Reconstructed object. **d** Original object. **e** Speckle intensity of light scattered by biological tissue. **f** Auto-covariance of speckle intensity pattern with its central correlation peak blocked. **g** Reconstructed object. Reproduced with permission from Springer Nature. © Springer Nature

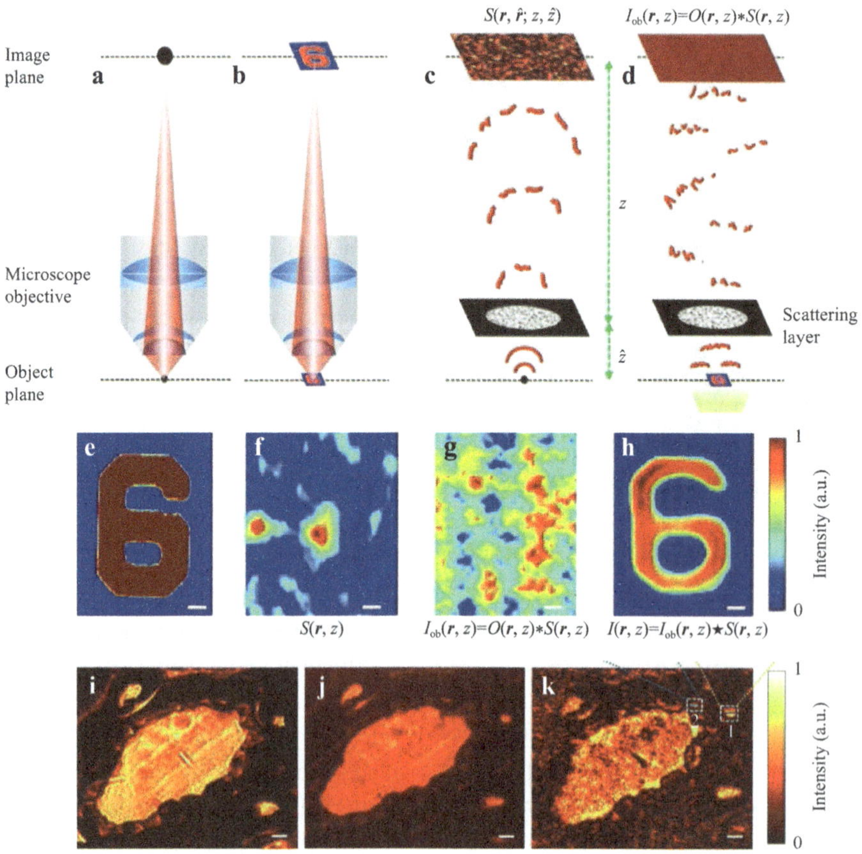

Fig. 12 Scatter-plate microscope based on speckle intensity cross-correlation[52]. **a** Point spread function of microscope. **b** Image of microscope. **c** Point spread function of scatter plate. **d** Scattered object light. **e** Object. **f** Point spread function of scatter plate. **g** Scattered object light. **h** Reconstructed object. **i** Monochromatic microscope image. **j** White-light microscope image. **k** Reconstructed scatter-plate image. Reproduced with permission from Springer Nature. © Springer Nature

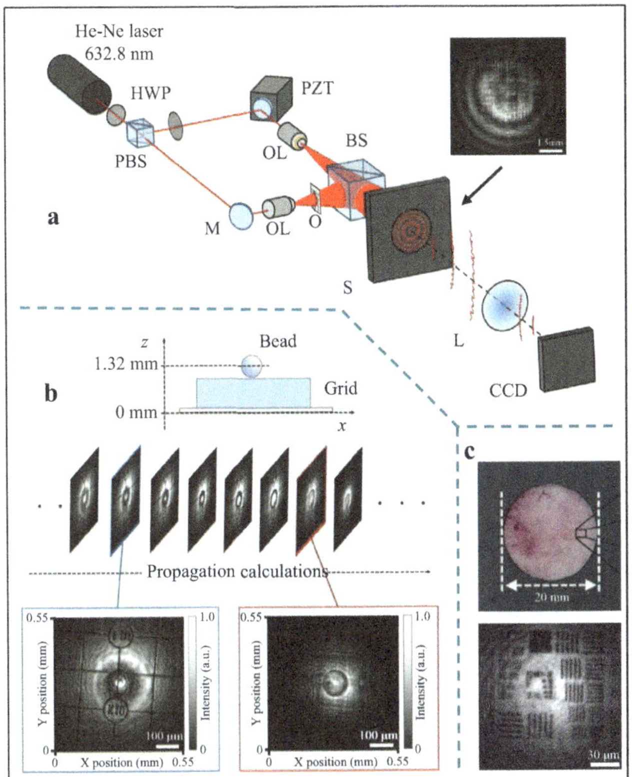

Fig. 13 Imaging through scattering media based on in-line phase-shift digital holography[45]. **a** Experimental setup. **b** Imaging 3-D object through a diffuser. **c** Imaging a test chart though a rat skin. Reproduced with permission from OSA. © OSA.

medium, S, as shown in the inset. Because the in-line holographic fringes with no spatial carrier frequency are much broader than the scale of subjective speckles caused by the finite aperture of the imaging lens L on the scatter medium, the hologram is robust to subjective speckle noise. Therefore, in-line phase-shift digital holography permits reconstruction of the image without time averaging the speckles introduced by the scattering medium. Fig. 13b shows a demonstration of the 3D imaging through a static scattering medium. A 3D object is made of a bead (for phase object) and a grid pattern (for amplitude object) placed 1.32-mm apart from each other, and its 3D image was reconstructed numerically by moving the focal plane, as shown in the middle. The figures at the bottom of Fig. 13b show the grid pattern (left) and the bead (right) in focus. Quantitative phase imaging was also realized by submerging the beads in an index-matched liquid. Fig. 13c demonstrates imaging a test chart through a shaved rat skin with a thickness ~0.5 mm (shown in top). The bottom picture shows the reconstructed image of the test chart hidden behind the rat skin.

Common-path holographic recording for imaging through a thick inhomogeneous medium

Attempts are now being made to explore the possibility of imaging through an inhomogeneous medium with increased thickness using common-path holographic recording (shown in Fig. 4). Fig. 14 shows an example of numerical experiments conducted by Kodama et al.[46] A thick 3D inhomogeneous medium was modeled as a sequence of multiple thin layers with inhomogeneous

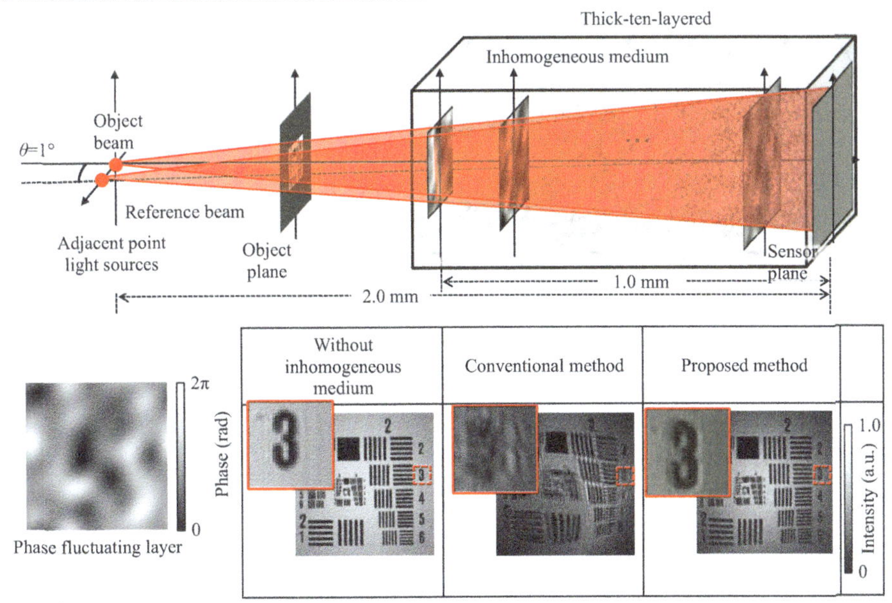

Fig. 14 Imaging through thick inhomogeneous medium by common-path holographic recording; a numerical experiment[46]. Reproduced with permission from OSA. © OSA.

refractive index distributions. As shown at the top of Fig. 14, 10 inhomogeneous layers were placed at depths z = 0.1~1.0 mm in the optical path with a separation of 0.1 mm between layers. Each layer introduces phase fluctuations over the range [0, 2π] with a spatial correlation generated by low-pass filtering (see bottom left in Fig. 14). The object and reference beams passed through the 10 layers with a relative angle of θ=1°; thus, the common-path condition, $\theta \leqslant \Delta/z$, (see Fig. 4) is satisfied for all layers in the inhomogeneous medium. Referring to the three framed pictures at the bottom of Fig. 14, the left shows the image reconstructed without the inhomogeneous medium, and the center and the right show, respectively, the image reconstructed by the conventional (non-common path) digital holography technique and the image reconstructed by the proposed common-path digital holography technique. This shows that common-path holographic recording is promising. At the time of this writing, the efficacy of common-path holographic recording for imaging through a thick and complicated heterogeneous medium has been confirmed experimentally[57].

Conclusion

We have reviewed the principles and techniques for imaging through random media, with a special focus on holographic techniques to meet the scope of the special Anniversary Issue on Holography. Although we have seen how the potential of holography has been exploited from past to present, there still remain many technological challenges for future topics in this research field, such as polarization imaging through birefringent random media, including the vectorial nature of optical fields, and multi-spectral color imaging through dispersive random media. One of the potential solutions may be found in the interdisciplinary integration of physics and informatics, such as in the marriage between optics/photonics and artificial intelligence[56], which is already in progress[33–35].

This paper is not intended to provide a comprehensive, balanced review. Rather, we presented our own view of the similarities and differences of the basic principles of various imaging techniques through random media. This is particularly applicable to the mini-tutorial in the first part of the review, in which we did not necessarily explain the principles in the same way as they were explained in the original papers. This choice was made just for our interest in identifying hidden links between different techniques, but we hope that it will also be of interest to the reader. The second part was to introduce part of the joint efforts made in this field by three universities: Universität Stuttgart (ITO), the University of Electro-Communications (UEC, Tokyo), and Utsunomiya University (CORE). The authors are grateful to all the members who have joined our joint endeavors.

Acknowledgements

E. Watanabe acknowledges support from a Grant-in-Aid for Transformative Research Areas(A) Grant Number A20H05888.

Author details

[1]Center for Optical Research and Education, Utsunomiya University, 7-1-2, Yoto, Utsunomiya, Tochigi, 321-8585, Japan. [2]Institut für Technische Optik, Universität Stuttgart, Pfaffenwaldring 9, 70569 Stuttgart, Germany. [3]The University of Electro-Communications, 1-5-1, Chofugaoka, Chofu, Tokyo 182-8585, Japan

Conflict of interest

The authors declare that they have no conflict of interest.

Received: 09 September 2021 Revised: 26 January 2022 Accepted: 08 February 2022
Accepted article preview online: 07 March 2022
Published online: 01 May 2022

References

1. Goodman, J. W. Statistical Optics. 2nd edn. (Hoboken: John Wiley & Sons, Inc., 2015), 333-414.
2. Goodman, J. W. Speckle Phenomena in Optics: Theory and Applications. 2nd edn. (Bellingham: SPIE Press, 2020).
3. McKechnie, T. S. Light propagation through the atmosphere and the properties of images formed by large ground-based telescopes. *Journal of the Optical Society of America A* **8**, 346-365 (1991).
4. Tuchin, V. V. Tissue Optics: Light Scattering Methods and Instruments for Medical Diagnostics. 3rd edn. (Bellingham: SPIE Press, 2015).
5. Mosk, A. P. et al. Controlling waves in space and time for imaging and focusing in complex media. *Nature Photonics* **6**, 283-292 (2012).
6. Yu, H. et al. Recent advances in wavefront shaping techniques for biomedical applications. *Current Applied Physics* **15**, 632-641 (2015).
7. Takeda, M. et al. Holographic correloscopy – Unconventional holographic techniques for imaging a three-dimensional object through an opaque diffuser or via a scattering wall: A review. *IEEE Transactions on Industrial Informatics* **12**, 1631-1640 (2016).
8. Rosen, J. et al. Roadmap on Chaos Inspired Imaging Technologies. *Applied Physical B* **128**, 1-26 (2022).
9. Gibson, A. P., Hebden, J. C. & Arridge, S. R. Recent advances in diffuse optical imaging. *Physics in Medicine & Biology* **50**, R1-R43 (2005).
10. Huang, D. et al. Optical coherence tomography. *Science* **254**, 1178-1181 (1991).
11. Leith, E. N. et al. Imaging through scattering media using spatial incoherence techniques. *Optics Letters* **16**, 1820-1822 (1991).
12. Velten, A. et al. Recovering three-dimensional shape around a corner using ultrafast time-of-flight imaging. *Nature Communications* **3**, 745 (2012).
13. Duguay, M. A. & Mattick, A. T. Ultrahigh speed photography of picosecond light pulses and echoes. *Applied Optics* **10**, 2162-2170 (1971).
14. Yoo, K. M., Xing, Q. R. & Alfano, R. R. Imaging objects hidden in highly scattering media using femtosecond second-harmonic-generation cross-correlation time gating. *Optics Letters* **16**, 1019-1021 (1991).
15. Hebden, J. C., Kruger, R. A. & Wong, K. S. Time resolved imaging through a highly scattering medium. *Applied Optics* **30**, 788-794 (1991).

16. Toida, M. et al. Two-dimensional coherent detection imaging in multiple scattering media based on the directional resolution capability of the optical heterodyne method. *Applied Physics B* **52**, 391-394 (1991).
17. Abramson, N. H. & Spears, K. G. Single pulse light-in-flight recording by holography. *Applied Optics* **28**, 1834-1841 (1989).
18. Leith, E. et al. Imaging through scattering media with holography. *Journal of the Optical Society of America A* **9**, 1148-1153 (1992).
19. Indebetouw, G. & Klysubun, P. Imaging through scattering media with depth resolution by use of low-coherence gating in spatiotemporal digital holography. *Optics Letters* **25**, 212-214 (2000).
20. Li, S. P. & Zhong, J. G. Dynamic imaging through turbid media based on digital holography. *Journal of the Optical Society of America A* **31**, 480-486 (2014).
21. Stetson, K. A. Holographic fog penetration. *Journal of the Optical Society of America* **57**, 1060-1061 (1967).
22. Paturzo, M. et al. Microscopy imaging and quantitative phase contrast mapping in turbid microfluidic channels by digital holography. *Lab on a Chip* **12**, 3073-3076 (2012).
23. Pircher, M. & Zawadzki, R. J. Review of adaptive optics OCT (AO-OCT): principles and applications for retinal imaging [Invited]. *Biomedical Optics Express* **8**, 2536-2562 (2017).
24. Roddier, F. Adaptive Optics in Astronomy. (London: Cambridge University Press, 1999).
25. Vellekoop, I. M. & Mosk, A. P. Focusing coherent light through opaque strongly scattering media. *Optics Letters* **32**, 2309-2311 (2007).
26. Katz, O., Small, E. & Silberberg, Y. Looking around corners and through thin turbid layers in real time with scattered incoherent light. *Nature Photonics* **6**, 549-553 (2012).
27. Béchet, C. Overview of multi-conjugate adaptive optics reconstructors. Proceedings of SPIE 10703, Adaptive Optics Systems VI. Austin: SPIE, 2018, 107031B.
28. Freund, I., Rosenbluh, M. & Feng, S. C. Memory effects in propagation of optical waves through disordered media. *Physical Review Letters* **61**, 2328-2332 (1988).
29. Antipa, N. et al. DiffuserCam: lensless single-exposure 3D imaging. *Optica* **5**, 1-9 (2018).
30. Liu, H. Y. et al. 3D imaging in volumetric scattering media using phase-space measurements. *Optics Express* **23**, 14461-14471 (2015).
31. Tajahuerce, E. et al. Image transmission through dynamic scattering media by single-pixel photodetection. *Optics Express* **22**, 16945-16955 (2014).
32. Xiao, Y., Zhou, L. N. & Chen, W. Direct single-step measurement of Hadamard spectrum using single-pixel optical detection. *IEEE Photonics Technology Letters* **31**, 845-848 (2019).
33. Horisaki, R., Takagi, R. & Tanida, J. Learning-based imaging through scattering media. *Optics Express* **24**, 13738-13743 (2016).
34. Lyu, M. et al. Learning-based lensless imaging through optically thick scattering media. *Advanced Photonics* **1**, 036002 (2019).
35. Li, S. et al. Imaging through glass diffusers using densely connected convolutional networks. *Optica* **5**, 803-813 (2018).
36. Kogelnik, H. B.S.T.J. briefs: Holographic image projection through inhomogeneous media. *The Bell System Technical Journal* **44**, 2451-2455 (1965).
37. Leith, E. N. & Upatnieks, J. Holographic imagery through diffusing media. *Journal of the Optical Society of America* **56**, 523 (1966).
38. Yaqoob, Z. et al. Optical phase conjugation for turbidity suppression in biological samples. *Nature Photonics* **2**, 110-115 (2008).
39. Cui, M. & Yang, C. Implementation of a digital optical phase conjugation system and its application to study the robustness of turbidity suppression by phase conjugation. *Optics Express* **18**, 3444-3455 (2010).
40. Hillman, T. R. et al. Digital optical phase conjugation for delivering two-dimensional images through turbid media. *Scientific Reports* **3**, 1909 (2013).
41. Kogelnik, H. & Pennington, K. S. Holographic imaging through a random medium. *Journal of the Optical Society of America* **58**, 273-274 (1968).
42. Goodman, J. W. et al. Wavefront-reconstruction imaging through random media. *Applied Physics Letters* **8**, 311-313 (1966).
43. Goodman, J. W. et al. Experiments in long-distance holographic imagery. *Applied Optics* **8**, 1581-1586 (1969).
44. Singh, A. K. et al. Looking through a diffuser and around an opaque surface: a holographic approach. *Optics Express* **22**, 7694-7701 (2014).
45. Kodama, S. et al. Three-dimensional microscopic imaging through scattering media based on in-line phase-shift digital holography. *Applied Optics* **58**, G345-G350 (2019).
46. Kodama, S. et al. Three-dimensional imaging through thick phase-fluctuating medium based on phase-shift digital holography with two adjacent light sources for common-path geometry. OSA Imaging and Applied Optics Congress 2021. Washington: Optical Society of America, 2021.
47. Bertolotti, J. et al. Non-invasive imaging through opaque scattering layers. *Nature* **491**, 232-234 (2012).
48. Katz, O. et al. Non-invasive single-shot imaging through scattering layers and around corners via speckle correlations. *Nature Photonics* **8**, 784-790 (2014).
49. Fienup, J. R. Phase retrieval algorithms: a comparison. *Applied Optics* **21**, 2758-2769 (1982).
50. Singh, A. K. et al. Exploiting scattering media for exploring 3D objects. *Light: Science & Applications* **6**, e16219 (2017).
51. Freund, I. Looking through walls and around corners. *Physica A: Statistical Mechanics and its Applications* **168**, 49-65 (1990).
52. Singh, A. K. et al. Scatter-plate microscope for lensless microscopy with diffraction limited resolution. *Scientific Reports* **7**, 10687 (2017).
53. Takeda, M. et al. Spatial statistical optics and spatial correlation holography: a review. *Optical Review* **21**, 849-864 (2014).
54. Naik, D. N. et al. Real-time coherence holography. *Optics Express* **18**, 13782-13787 (2010).
55. Takeda, M. et al. Coherence holography. *Optics Express* **13**, 9629-9635 (2005).
56. Takeda, M. Towards happy marriage between optics/photonics and artificial intelligence: A review with a historical perspective. Proceedings of SPIE 11142 Optical Technology and Measurement for Industrial Applications Conference Yokohama: SPIE 1114201 (2019).
57. Ohta, M. et al. 3D imaging through a highly heterogeneous double-composite random medium by common-path phase-shift digital holography. *Optics Letters* **47**, 1170-1173 (2022).

Official journal of the JHL 2689-9620
www.light-am.com

Review Open Access

Review of engineering techniques in chaotic coded aperture imagers

Vijayakumar Anand[1,2,*], Joseph Rosen[3,*] and Saulius Juodkazis[1,4,*]

Abstract

Coded aperture imaging (CAI) is a technique to image three-dimensional scenes with special controlled abilities. In this review, we survey several recently proposed techniques to control the parameters of CAI by engineering the aperture of the system. The prime architectures of these indirect methods of imaging are reviewed. For each design, we mention the relevant application of the CAI recorders and summarize this overview with a general perspective on this research topic.

Keywords: Coded aperture imaging, Holography, Chaos engineering, Incoherent optics, Diffractive optics

Introduction

Coded aperture imaging (CAI) technologies offer imaging capabilities and dimensionalities beyond the limits of conventional lens-based imagers[1]. In a lens-based imager, an object point is mapped to a distinct point on the sensor if the object point lies within the depth of focus. Since this is a point-to-point mapping, resulting in the formation of an image of the object directly on the sensor, the lens-based imagers are called direct two-dimensional (2D) imagers. For this reason, it is crucial to produce high-quality lenses with as minimum aberrations as possible. In CAI methods, every point of an object is mapped by an optical modulator to a unique 2D intensity distribution. Any change in the location of the point changes the 2D intensity distribution. Only linear, shift-invariant systems are considered here, and so any change in the lateral direction only shifts the 2D intensity distribution, and a change in the axial direction changes the 2D intensity distribution. This linearity exists for the intensity of light in the case of incoherent imagers. This study is focused only on imaging systems with spatially incoherent and high temporally coherent illumination that includes sources such as light- emitting diodes, fluorescent sources, and also natural light with chromatic filters. In all the experiments discussed next, only critical illumination is considered. When a thick object is made up of multiple points in three-dimensional (3D) space with P observed planes along the Z-axis, the resulting object intensity distribution is formed by the summation of shifted and scaled point spread intensity distributions given as $I_O = \sum_{p=1}^{P} I_{PSF,p} \otimes O_p$, where O_p is the object intensity at plane p, '$\otimes$' is a 2D convolution operator and $I_{PSF,p}$ is the point spread intensity distribution at the plane p. Consequently, the object information at a plane p can be reconstructed as $O_p = I_O * \hat{I}_{PSF,p}$, where '*' is the 2D correlation operator, and $\hat{I}_{PSF,p}$ is a processed version of $I_{PSF,p}$. The above relation can be modified as $O_p = \sum_{k=1}^{P} I_{PSF,k} \otimes O_k * \hat{I}_{PSF,p}$. So, the reconstructed image of the object consists of two correlation terms, namely $I_{PSF,p} * \hat{I}_{PSF,p}$ and $\sum_{m \neq p}^{P} I_{PSF,m} * \hat{I}_{PSF,p}$, which are the autocorrelation and cross-correlation terms respectively, where m is non-zero integer sampling the out-of-focus

Correspondence: Vijayakumar Anand (vijayakumar.anand@ut.ee) or Joseph Rosen (rosenj@bgu.ac.il) or Saulius Juodkazis (sjuodkazis@swin.edu.au)
[1]Optical Sciences Centre and ARC Training Centre in Surface Engineering for Advanced Materials (SEAM), School of Science, Swinburne University of Technology, Hawthorn, VIC 3122, Australia
[2]Institute of Physics, University of Tartu, 50411 Tartu, Estonia
Full list of author information is available at the end of the article.

planes. For 3D imaging, the point response should yield a sharp autocorrelation delta-like peak and low cross-correlation values, enhancing the strength of the object information at the plane of interest and suppressing the object information from other planes. Such a property is present with scattered light, where the speckles formed by scattering in the far-field have an average size of the diffraction-limited point spread function of a direct imaging system with the same numerical aperture. This last feature makes CAI attractive for 3D imaging[2]. Based on the interesting properties of scattered light, most of the recent CAI methods converted the light diffracted from an object into a random wave to encode 3D information of the object in a 2D matrix[3-5]. The introduction of chaos in the form of a random phase (or amplitude) mask converts the light from an object into a speckle pattern and transforms a 2D imager into a 3D one. In most of the previous studies[2-4], the engineering of chaotic apertures is not discussed in detail. While some of the methods[3,5] did not engineer the apertures, it is evident from the previous studies[1,4], that engineering of chaotic apertures is crucial to control the imaging characteristics of CAI systems. Some of the review articles[1,6] mentioned aperture engineering. However, a detailed review of chaotic aperture engineering methods and their role in controlling imaging characteristics have not been discussed. In this article, we review some of the engineering approaches of chaotic apertures and describe how the characteristics of chaos manipulate the imaging characteristics of the system.

Chaos dependent imaging characteristics

The optical configuration of the CAI method is shown in Fig. 1a. Light from an object is incident on an optical modulator that scatters the object wave, and the scattered intensity distribution is recorded by an image sensor. The object and image distances are z_s and z_h, respectively. Let us consider a single point in the object plane at $\overline{r_o} = (x_o, y_o)$ with an intensity I_o. The complex amplitude reaching the optical modulator is given as $C_1 \sqrt{I_o} L(\overline{r_o}/z_s) Q(1/z_s)$, where (x, y) are the modulator's coordinates, $L(\overline{r_o}/z_s) = \exp[j2\pi(x_o x + y_o y)/(\lambda z_s)]$ and $Q(1/z_s) = \exp[j\pi(x^2 + y^2)/(\lambda z_s)]$ are the linear and quadratic phase functions, respectively, and C_1 is a complex constant. The complex amplitude of the optical modulator is $M\exp(j\Phi)$, where M is a constant for a phase-only modulator and for a pinhole array Φ is a constant and $M = \sum_{n=1}^{N} \delta(x - x_n, y - y_n)$, where (x_n, y_n) are set of random locations, and N is the number of pinholes. The complex amplitude after the optical modulator is $C_1 \sqrt{I_o} L(\bar{r}_o/z_s) Q(1/z_s) M\exp(j\Phi)$. The scattered intensity distribution, which is the PSF for a point at z_h from the optical modulator over the plane $\bar{r}_h$, is,

$$I_{\mathrm{PSF}}(\overline{r_h}; \overline{r_o}, z_s) = \left| C_1 \sqrt{I_o} L\left(\frac{\overline{r_o}}{z_s}\right) Q\left(\frac{1}{z_s}\right) M\exp(j\Phi) \otimes Q\left(\frac{1}{z_h}\right) \right|^2 \tag{1}$$

Lateral and axial resolution

The lateral resolution limit of a direct imaging system is the width of the diffraction-limited spot obtained in response to a point in the input, which is $1.22\lambda z_s/D$. In the case of CAI methods, the lateral resolution is given by lateral correlation length L_l given as twice that of the average speckle size[2]. Note that the resolution is defined in the object plane rather than the observation plane. Therefore, the scaling factor z_s/z_h between the (virtual) speckle pattern on the object plane and the (real) speckle pattern on the observation plane is applied. The correlator measures similarities between two patterns: the higher the similarity, the higher the correlation value, and vice versa. The cross-correlation operation of the intensity distribution obtained for a point object over the replications of the same pattern is illustrated in Fig. 1b. The width of the correlation peak is twice as much as the average speckle size, i.e., $L_l = 2.44\lambda z_s/D$. So, the lateral resolution of CAI methods is only half of that of a direct imager. The regular correlation operation, which is equivalent to spatial filtering with the matched filter, has a lateral correlation length equal to

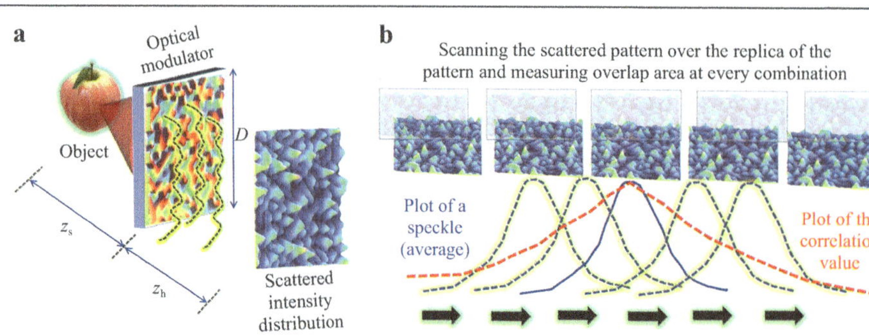

Fig. 1 **a** Optical configuration of simplified coded aperture imaging method. **b** Autocorrelation operation.

twice that of the diffraction-limited spot in the far-field. The matched filter can be mathematically expressed as $I_R = \left|\mathcal{F}^{-1}\left\{\left|\tilde{I}_{\text{PSF}}\right|^{\alpha} \exp\left[iarg\left(\tilde{I}_{\text{PSF}}\right)\right]\left|\tilde{I}_o\right|^{\beta} \exp\left[-iarg\left(\tilde{I}_o\right)\right]\right\}\right|$, where $\tilde{I}_{\text{PSF}}$ and $\tilde{I}_o$ are Fourier transforms of I_{PSF} and I_o, respectively, $\alpha = 1$ and $\beta = 1$. It was shown in 1984 that a phase-only filter when $\alpha = 0$ and $\beta = 1$ generated a sharper correlation peak than the matched filter[7]. Most recently, a non-linear correlation method was developed where the values of α and β were tuned until a minimum entropy was obtained[8]. The entropy is given as $-\sum\sum\phi(m,n)\, log[\phi(m,n)]$, where $\phi(m,n) = |C(m,n)| / \sum_M \sum_N |C(m,n)|$, (m,n) are the indexes of the correlation matrix, and $C(m,n)$ is the correlation distribution. This approach was able to reach the diffraction-limited spot size during reconstruction with a computational resolving power of twice that of a matched filter. The non-linear correlation is manifested by choosing $\beta \neq 1$ since it is easy to see that the operation $|\tilde{I}_o|^{\beta}$ for $\beta \neq 1$ does not satisfy the superposition principle.

The above analysis can be directly extended to the axial resolution. In CAI methods, the axial resolution is given by the axial correlation length L_a. As the lateral correlation length changes in the same fashion as a diffraction-limited spot, the axial correlation curve follows the same profile as that of a direct imager. Hence, the axial correlation length is the same as the axial resolution of a direct imager. The procedure for obtaining the axial correlation curve is shown in Fig. 2. However, one outcome is that neither the lateral nor axial resolutions of the CAI method are dependent upon the scattering degree of the optical modulator defined as the ratio between the length of the scattered pattern b to the maximum length B in the Fourier plane. This phenomenon is not unexpected because the influence of the scattering degree of the modulator in the far-field is expressed by the overall size of the light intensity pattern. On the other hand, the average speckle size at the far-field is dictated by the aperture size of the modulator, and consequently, the lateral and axial resolutions are determined by this aperture size. Let us demonstrate this concept by simulation of the optical fields using Eq. 1 for different levels of chaos. In this study, only thin scatterers are considered, and the scattering degree is characterized by the ratio b/B shown in the Gerchberg-Saxton algorithm (GSA) in Fig. 3a[9]. In this simulation, a positive lens function was integrated into the optical modulator so that the far-field is obtained at the focal plane of the lens and the near-field (Fresnel region) is some range between the focal plane and the optical modulator. The I_{PSF} was simulated for σ variation from 0 to 1 in steps of 0.01, and the L_l was measured from the limits of the Fresnel diffraction region to the far-field regime. The map of variation of L_l for different σ and distances measured from the optical modulator to the sensor is shown in Fig. 3b. The plot of axial correlation values obtained at the origin ($x = 0$, $y = 0$) of $I_{\text{PSF}}(z{=}0) * I_{\text{PSF}}(z)$ for different σ is shown in Fig. 3c. From Fig. 3b, c, it is seen that in the far-field, there is no influence of σ on the imaging resolutions. It must be noted that the above conditions of lateral and axial resolutions are independent of the modulator type: amplitude or phase. In the far-field, the lateral and axial resolutions depend only on the aperture diameter of the modulator.

Super-resolution by sub-diffraction speckles

In this article, the terms super-resolution and resolution enhancement have been used interchangeably to indicate an increase in resolving power obtained by modifying the

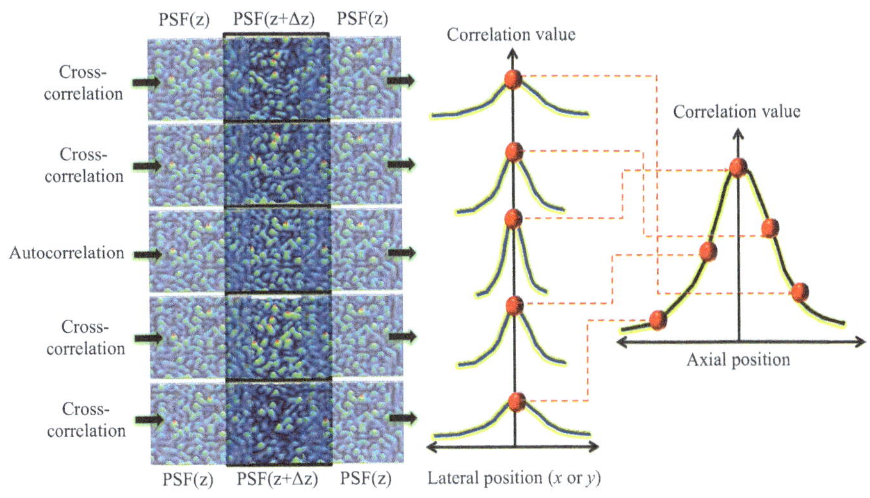

Fig. 2 Procedure for obtaining the axial correlation curve for CAI methods.

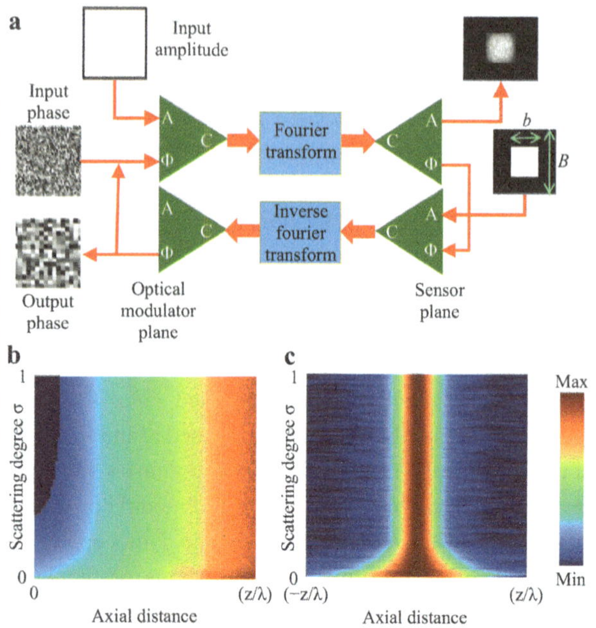

Fig. 3 a Schematic of the phase retrieval algorithm for synthesizing phase masks with a certain σ. **b** Map of L_l versus the scattering degree and axial distance from the optical modulator to the sensor. **c** Axial correlation curve obtained by a cross-correlation of the speckle pattern obtained for different locations of a point object with a reference location indicated as $z = 0$ at the origin ($x = 0$, $y = 0$) for different values of σ.

MTF in two ways: Extending the bandwidth by increasing the cut-off frequency of the modulation transfer function (MTF) and improving the MTF response for higher spatial frequencies but without changing the spatial bandwidth. This section deals with the improvement of the resolution by magnifying the response for higher spatial frequencies. The next section deals with extending the bandwidth. One of the straightforward methods to increase the imaging resolutions of CAI systems is to decrease the average speckle size. A phase retrieval algorithm was designed to generate sub-diffraction spots by introducing a π phase shift between consecutive pixels in the sensor plane and redistributing part of the light away. This approach enabled the generation of sub-diffraction spots. A phase mask was designed such that the diffraction from the mask generated randomly distributed sub-diffraction spots. When the CAI method was applied using this mask, a resolution enhancement of 1.26 times compared to direct imaging was obtained[10,11]. This method was called resolution enhanced coded aperture correlation holography (RE-COACH). The optical configuration of RE-COACH is similar to the one shown in Fig. 1a. The images of the phase mask and intensity distribution on the sensor plane are shown in Fig. 4a–d, respectively. The imaging results of RE-COACH, interferenceless COACH (I-COACH), and direct imaging of a United States Air Force (USAF) resolution

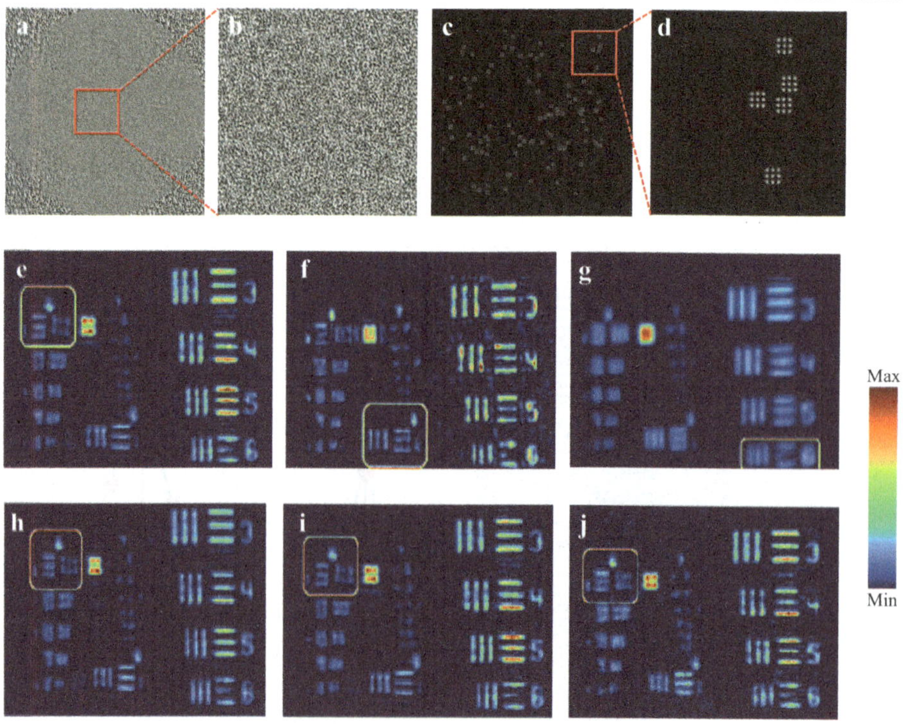

Fig. 4 a, **b** Images of the phase mask that generates randomly distributed sub-diffraction spots as in **c** and **d**. Imaging results from **e** RE-COACH, **f** I-COACH, and **g** direct imaging. Imaging results of RE-COACH for spot density of **h** 0.009, **i** 0.036 and **j** 0.09.

target (Group 5 element 6, Group 6 element 1 and 2) are shown in Figs. 4e–g respectively. The lateral resolution of RE-COACH is seen higher (6.96 μm, Group 6 element 2) than both I-COACH (7.82 μm, Group 6 element 1) and direct imaging (8.77 *μm*, Group 5 element 6). The spot density mentioned in Fig. 4 is defined as the ratio between a number of arranged spots to a number of potential spot positions. It was noted here that if the spot density increased, the lateral resolution decreased. So, it was crucial to control chaos such that the spots do not overlap and increase the lateral correlation lengths. The imaging results of RE-COACH for different values of spot density (0.009, 0.036, and 0.09) are shown in Fig. 4h–j respectively. Beyond a spot density of 0.09, the resolution began converging towards that of I-COACH. In the above study, a non-linear correlation method was applied, resulting in a resolution of the same as the width of the sub-diffraction spot. In summary, in CAI methods, one direct method to improve the lateral and axial resolution is to reduce the size of the fundamental building block – speckle of the intensity distribution. The design of the phase mask can be achieved by implementing a sub-diffraction generation approach as in Ref 10 followed by the random point generator by the GSA. By combining the sub-diffraction spots and the random point generator, random sub-diffraction spots can be generated.

Resolution enhancement by manipulating scattering

The previous subsection shows that introducing a chaotic phase mask into an imaging system can improve the lateral and axial resolutions without altering the spatial bandwidth. In this section, the resolution enhancement based on extending the bandwidth has been discussed. A similar concept was first demonstrated by Choi et al. by introducing turbid media into a lens-based imaging system[12]. The reconstruction mechanism was cumbersome, involving the recording of intensity patterns at various angles. A different approach was attempted by Kashter et al[13]., where a chaotic phase mask was introduced into a well-established incoherent holography technique called Fresnel incoherent correlation holography (FINCH)[14]. This technique is called coded FINCH (C-FINCH). In this study, a thin scatterer was introduced in between the object and a FINCH system, and the resolution of the system was studied when the scattering degree was increased. Later the concept was generalized to regular lens-based imaging systems[15]. The optical configuration of the imaging system is shown in Fig. 5a. The light diffracted from an object consists of information of different spatial frequencies of the object. The smaller features of the object diffract light at larger angles while the larger features diffract at smaller angles. The finite diameter *D* of the entrance pupil can collect only until a certain level of spatial frequencies, which sets the resolution limit of the system. By introducing a thin scatterer in between the object and the aperture of the entrance pupil, it is possible to scatter the higher spatial frequencies, which are beyond the numerical aperture, back into the system and by that to increase the lateral resolution. In the current configuration, it is possible to decrease L_l from $1.22\lambda(z_1+z_2)/D_1$ to $1.22\lambda z_1/D_2$, and by tuning the scattering degree of the modulator σ between 0 to 1, it is possible to reach new resolution limits. Five objects (8.98 *lp/mm* to 14.25 *lp/mm*) of USAF resolution target were imaged. The direct imaging results and reconstruction results for σ = 0.1042 and 0.1458 are shown in Figs. 5b–d, respectively. A notable resolution enhancement is seen by introducing the chaos and increasing its rank. FINCH has the capability to reach a resolution enhancement of 1.5 times that of an equivalent incoherent imaging system. The application of the above method to FINCH increased the resolution by seven times that of an equivalent incoherent imager and five times that of FINCH without chaos. FINCH has lower axial resolution compared to direct imagers, and introducing chaos significantly improved the axial resolving power[13]. Note that this resolution enhancement method does not change the imaging system aperture. Instead, the technique

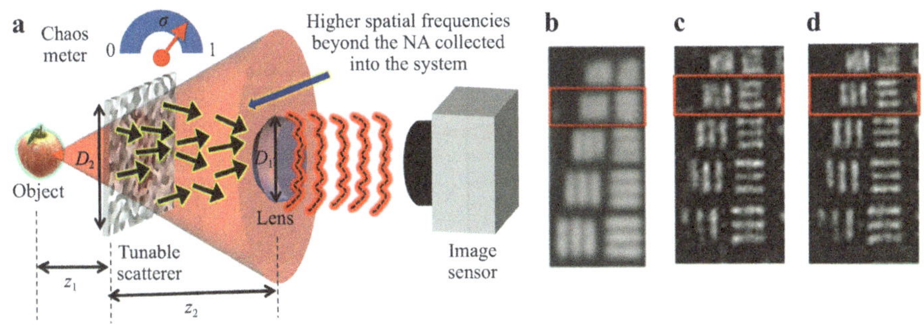

Fig. 5 a Optical configuration of the imaging system with a scatterer placed between the object and entrance pupil of the imaging system. **b** Direct imaging. Reconstruction results for **c** σ = 0.1042 and **d** σ = 0.1458. **b**–**d** are adapted from Ref. 15.

manipulates the light in the free space between the object and the entrance aperture of a given unchangeable system. In that sense, the method is equivalent to the technique of structured illumination[16], in which the light between the source and the object is manipulated.

Super-resolution in partial aperture imaging systems

Even though the above two methods dealt with improvement in lateral resolution, the concepts are completely different. The first method described previously tailored the system's PSF, while the latter brought in higher spatial frequencies beyond the limits of the numerical aperture. In some astronomical imaging systems, the loss of resolution is due to the use of an annular aperture instead of the full aperture[17]. Imaging using annular optical elements suppresses the spatial frequencies higher than zero frequency except for mild enhancement near the cut-off frequency[17]. The PSF and MTF for two widths of annular aperture $w = r$ and $0.04r$ are shown in Fig. 6b, where r is the radius of the full aperture. It is seen that the decrease in the width of the annular aperture suppresses the higher spatial frequencies resulting in a loss of the lateral resolution. The intensity distribution of the PSF is similar to that of a magnitude square a Bessel function of the first order of the first kind. It has been found[18] that if a chaotic phase mask was attached to the annular lens, then the higher spatial frequencies were not suppressed as much as in the case of only an annular spherical (or parabolic for high radius value) mirror. The PSF and MTF of the partial aperture imaging system (PAIS) for w=0.04r are shown in the third line of Fig. 6b. By comparing the MTF of the annular spherical mirror and PAIS, it is seen that PAIS has a better response to higher spatial frequencies. An experiment was carried out[18] to compare direct imaging and PAIS using a spatial light modulator (SLM) consisting of 1920 × 1080 pixels and a pixel pitch of 8 μm. The imaging results for National Bureau of Standards (NBS) objects (14 *lp/mm* and 16 *lp/mm*) obtained for full aperture and a ring of 40 μm width for direct imaging and PAIS are shown in Fig. 7. It is seen that the introduction of chaos has significantly improved the resolving power of the system. It must be noted that this resolution enhancement occurs without altering the bandwidth of the imaging system. The method of PAIS was extended to synthetic aperture imaging where instead of using the full aperture, only the periphery of the aperture was scanned, but the images were reconstructed with the resolution of the full aperture[19].

Enhanced contrast by PSF engineering

The above three subsections showed how introducing and controlling the chaos could improve the lateral resolution in three different ways. In addition to improving the lateral and axial resolutions, in some cases, such as microscopy, it is useful to improve the contrast[20]. In this approach, two recordings were made and subtracted one from the other to improve the contrast of the reconstructed images. One recording was with a vortex filter and the other without it. Edge enhancement approaches in many incoherent holography techniques such as FINCH involved recording multiple holograms with vortex filters[21]. In all the above methods, the PSF of the system was recorded and utilized for reconstructing the image. In contrast, a synthetic PSF generation was proposed recently to improve the contrast of the output images[22]. The synthetic synthesis is a post-engineering approach, where a synthetic PSF is generated using a phase retrieval algorithm as shown in Fig. 8a. The cross-correlation with the recorded PSF produces a donut-shaped intensity distribution. Therefore, when the synthetic PSF was cross-correlated with the object intensity distribution, edge enhanced image of the object was obtained. The edge enhanced image is subtracted from the original reconstruction to generate the contrast-enhanced image of the object. If this approach is implemented to direct imaging methods, additional experiments are required, whereas here[22], the contrast can be improved offline without any additional experiments.

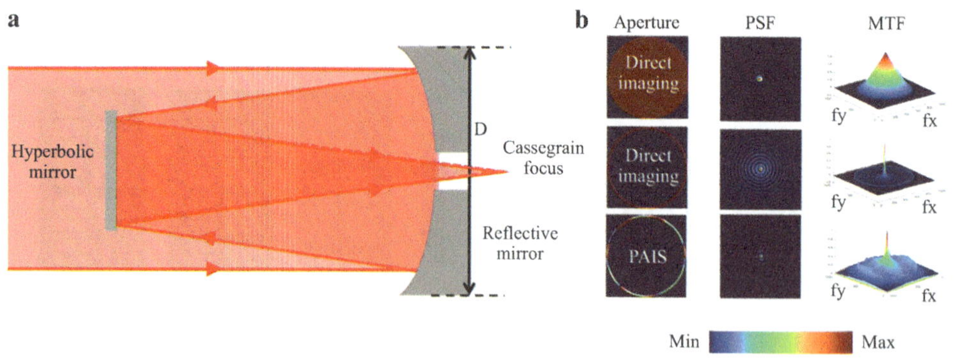

Fig. 6 **a** Optical configuration of Cassegrain telescope. **b** Images of aperture, PSF, MTF for direct imaging ($w = r$ and $w = 0.04r$), and partial aperture imaging system (PAIS) ($w = 0.04r$).

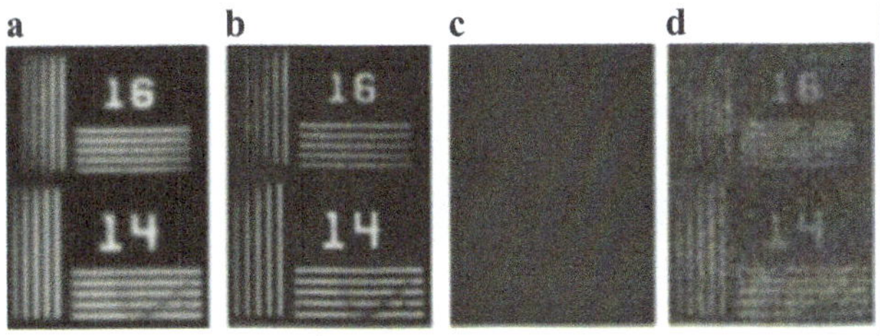

Fig. 7 a Direct imaging and **b** PAIS for full aperture. **c** Direct imaging and **d** PAIS for an annular aperture with a width of 40 μm. Adapted from Ref. 18.

The phase retrieval algorithm starts with a random pure phase function $e^{j\theta}$ which is multiplied with the complex conjugate of the Fourier transform of the recorded PSF and Fourier transformed. The resulting complex amplitude's magnitude (I_D) was replaced by a donut intensity obtained by a Fourier transform of a spiral phase function while the phase was retained. The resulting complex amplitude was inverse Fourier transformed and multiplied with the inverse of the complex conjugate of Fourier transform of PSF. The resulting complex amplitude was converted into a phase-only function, and the loop was iterated until the optimal synthetic PSF was obtained. The images of the PSF and object intensity distributions, phase of Fourier transform of synthetic PSF, original reconstruction, direct imaging result, edge and contrast-enhanced images of a section of a wing of an insect are shown in Fig. 8b–h, respectively.

Resolutions of imaging with a random pinhole array

All the above studies were carried out using phase-only scatterers. The oldest and least expensive scatterer is the pinhole array[23]. A random pinhole array (RPA) maps a point in the object space to a random array of points in the far-field, but in the near field scattering (Fresnel region) of light is usually observed. The lateral resolution in the far-field should be the same for different degrees of chaos. The earlier studies[1] used a phase-only scatterer generated in an active device such as SLM, and so adding a lens function was simpler. In an RPA, without the lens, in the Fresnel region, the lateral resolution is highly dependent upon the scattering degree, which is controlled by the diameter of the pinholes in the RPA. A simulation was carried out with the following parameters: z_s = 10 cm, z_h = 10 cm, diameter of RPA mask D = 2 mm, λ = 600 nm, and n = 100, where n is the number of pinholes, for pinhole radius varying from 10 μm to 50 μm (Fig. 1a). The simulated images of the RPA for radius r_1 = 50 μm to 10 μm, and the corresponding images of the I_{PSF}s are shown in Fig. 9a–j, respectively. It is seen from the figures that the speckle size decreases with the diameter of the pinholes. The plot of the autocorrelation function with a phase-only filter is shown in Fig. 9k. The full width at half maximum (FWHM) of the autocorrelation function is plotted in Fig. 9l. The axial resolution of the system is studied next, in a similar way shown in Fig. 2. The cross-correlation values at the origin (x=0, y=0) were calculated for the cases r_1 = 50 μm to 10 μm and are shown in Fig. 10a. The FWHM of the axial correlation versus the pinhole radius is shown in Fig. 10b. Once again, a linear relationship is obtained between the radius of the pinholes and the axial resolution, and when the pinhole diameter decreases, it approaches the case of direct imaging. Unlike the far-field case where the speckle sizes were the same for all degrees of chaos, with RPA, the

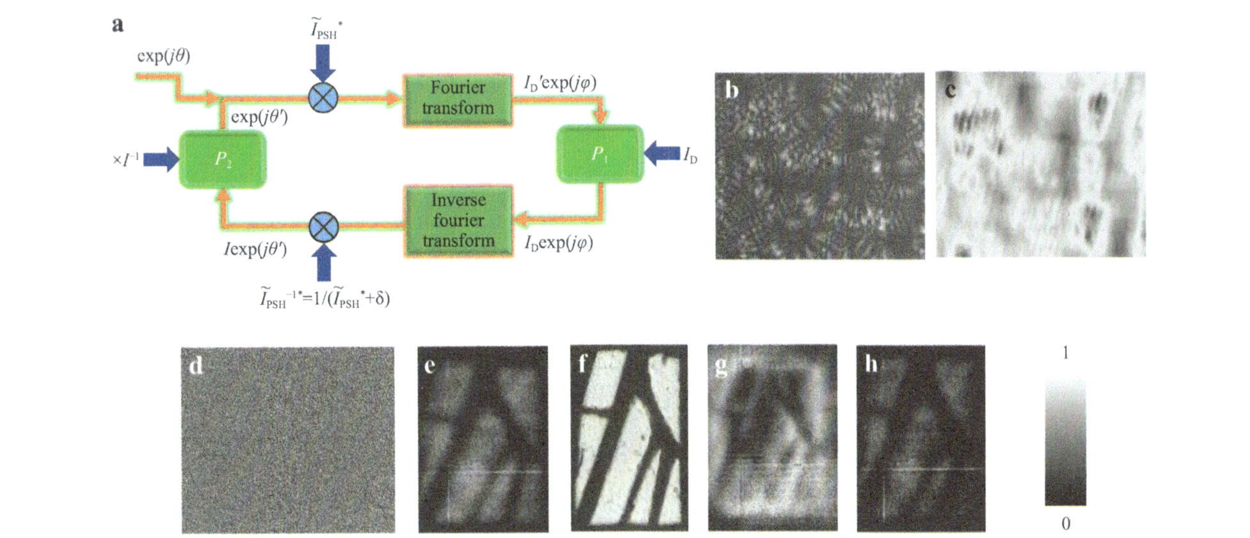

Fig. 8 a Phase-retrieval algorithm for generating the synthetic PSF. Image of the **b** PSF, **c** object intensity distribution, **d** synthetic PSF, **e** original reconstruction, **f** directly recorded image, **g** edge enhanced image and **h** contrast-enhanced image. Adapted from 22.

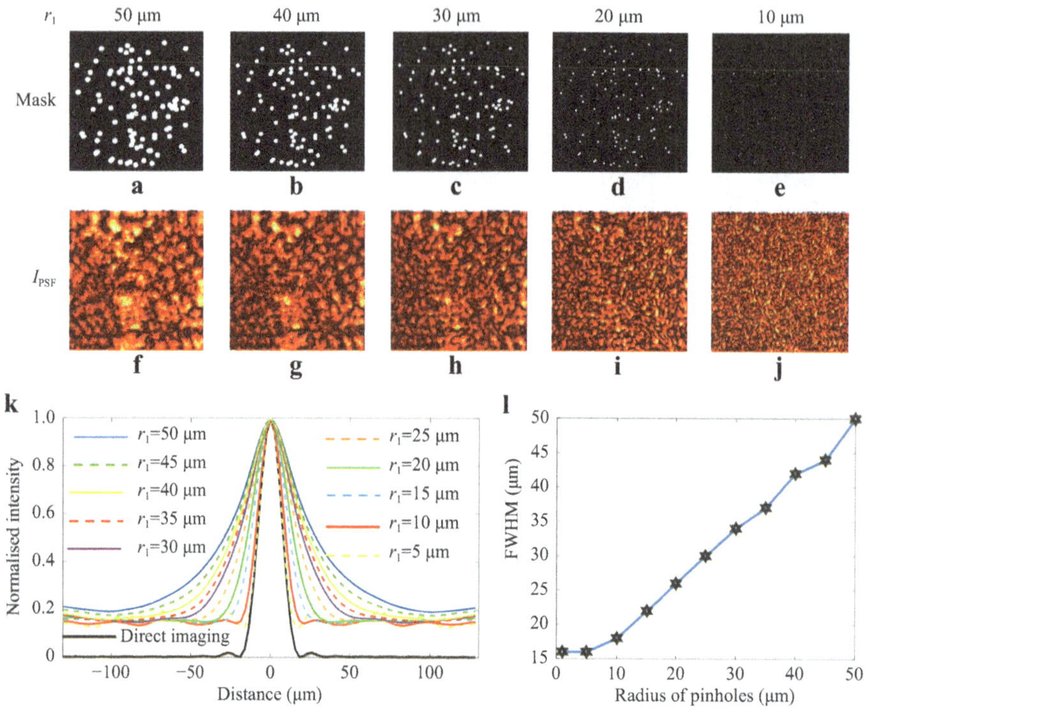

Fig. 9 Image of the RPA mask with r_1 = **a** 50 μm, **b** 40 μm, **c** 30 μm, **d** 20 μm and **e** 10 μm and the corresponding speckle patterns **f–j**. **k** The plot of the autocorrelation function for r_1 = 50 to 5 μm in steps of 5 μm. **l** The plot of the FWHM of the autocorrelation function for r_1 = 50 to 5 μm in steps of 5 μm.

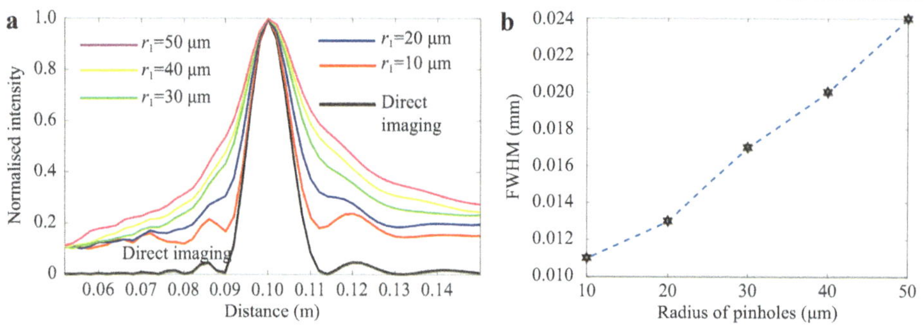

Fig. 10 **a** Plot of the axial correlation curve for r_1 = 50 to 10 μm in steps of 10 μm. **b** Plot of the FWHM of the axial correlation peaks versus the diameter of the pinholes in RPA.

lateral resolution has a secondary bottleneck in addition to NA, which is the diameter of the pinhole that must be taken into consideration while designing the optical system and RPA mask. The dependence of the speckle size in the radius of the pinholes, and thus in the scattering rank, is different from the behavior of the systems of the previous sections. The difference is because the sensor plane is located in the near-field zone of the pinhole mask, where the speckles are obtained by light interference from neighbor pinholes. As much as the pinholes are small, the angle between interfering rays is large, and speckles become small.

Spectral resolving power

One of the main applications of CAI systems is to resolve spectral information. There were many spectral imagers based on CAI[24,25]. In general, the spectral resolution of a direct imaging system depends upon the spectral scanning system installed either in the source or the detector, as dispersion-based spectral sensitivity of refractive lenses is low. Three cases are considered for the same optical configuration: refractive lens, diffractive lens, and RPA. The following design parameters are considered for the comparison. z_s = 10 cm, z_h = 10 cm, D = 2 mm, λ = 0.4–0.8 μm (0.6 μm as design wavelength), f = 5 cm and n = 100. A refractive lens made up of silica is considered.

The Sellmeier equation based on dispersion for silica is given as $\mu^2 = 1 + B_1\lambda^2/(\lambda^2 - C_1) + B_2\lambda^2/(\lambda^2 - C_2) + B_3\lambda^2/(\lambda^2 - C_3)$, where μ is the refractive index and the Sellmeier coefficients[26] for silica are B_1 = 0.6961663, B_2 = 0.4079426, B_3 = 0.8974794, C_1 = 0.0046791 μm^2, C_2 = 0.0135120 μm^2, C_3 = 97.934002 μm^2. The variation of the refractive index with wavelengths is shown in Fig. 11a. The refractive index of silica for λ = 0.6 μm is n = 1.458. The radius of curvature R using Lens maker's formula for a biconvex lens is given as $R = 2f(n - 1)$ = 4.58 cm. The focal length is plotted with respect to the wavelength λ = 0.4 − 0.8 μm in Fig. 11a. The radius of zones of a Fresnel lens is given as $\rho_n \cong \sqrt{2nf\lambda}$ and so $f \cong \rho_n^2/2n\lambda$. For λ = 0.6 μm and f = 5 cm, the radius of the first zone is given as ρ_1 = 245 μm. The focal length $\left(6\times10^{-8}/2\lambda\right)$ is plotted as a function of wavelength in Fig. 11a. As expected, the chromatic aberration is higher for the diffractive lens than for a refractive lens. A point object is imaged using a refractive lens, diffractive lens, and an RPA and for λ = 0.4–0.8 μm. The variation of intensity at ($x = 0$, $y = 0$, z_h = 10 cm) as a function of wavelength is plotted in Fig. 11b for direct imaging with refractive and diffractive lenses. The spectral resolution dependency on the diameter of the pinholes is studied. $I_{PSF}(\lambda$ = 600 nm) is calculated and cross-correlated with $I_{PSF}(\lambda)$ for λ = 401 nm to 800 nm in steps of 1 nm. The plot of the cross-correlated intensity matrix $I_R(x = y = 0)$ for different wavelengths and for different radii of the pinholes r_1 = 10 μm to 50 μm in steps of 10 μm are also shown in Fig. 11b.

From the plots, it is seen that the spectral resolutions are improved with scattering angles and match with that of a diffractive lens when $r_1 \leqslant$ 10 μm. The spectral resolution of a refractive lens made of silica was the least among all three cases. Therefore, it is impossible to resolve spectral information in lens-based direct imaging systems without additional spectrum resolving components. On the positive side, the direct imaging system suffers less than other cases from chromatic aberrations. Chromatic aberration is essential for wavelength discrimination without a colour camera. In some of the recent studies, it was found that the image reconstruction quality was significantly affected when the number of spatio-spectral channels increased in CAI[27]. Therefore, by selecting the scattering degree, it is possible to select the number of resolvable spatio-spectral channels and control the reconstruction quality. The bias levels in the lateral, axial, and spectral correlation curves in comparison to the direct imaging are due to the correlation effects of two positive functions. The above study can also be applied to phase-only scatterers[28]. One of the previous studies showed that the spectral resolution of the CAI method is similar to that of a diffractive lens with the same NA[28].

Field of view

One of the important characteristics of imaging next to lateral and axial resolutions is the field-of-view (FOV) of an imaging system. The FOV of an imaging system is limited by the size of the image sensor and the magnification of the system given as $\sim(sz_s/z_h)$, where s is the size of the sensor. In a direct imaging system, the points of the object that lies outside the FOV are not recorded by the system, as shown in Fig. 12a. The FOV in a CAI system is different owing to the indirect mode of recording and reconstruction using a PSF. In the CAI system, even when an object point lies outside the FOV, there is always some partial scattered pattern that is collected within the sensor area, as shown in Fig. 12b. Considering a typical case where the scattered intensity distribution fills at least the sensor area, the FOV of CAI systems is twice as much as a direct imager with the same magnification and identical image sensor. Therefore, by controlling the scattering degree, the FOV of the imaging

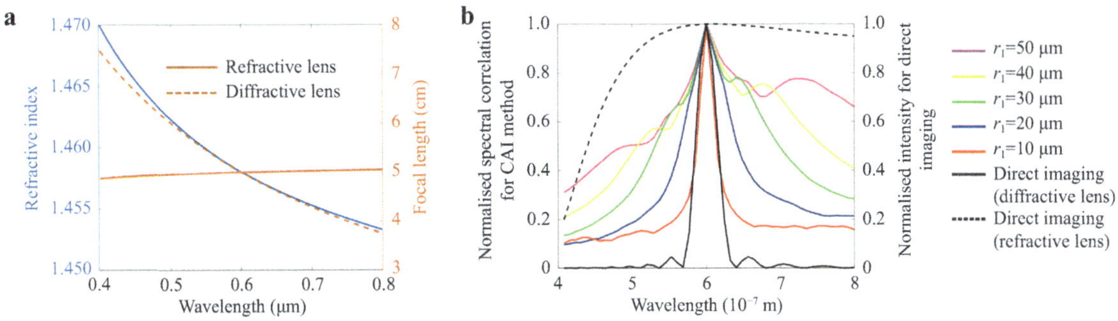

Fig. 11 **a** Plot of the refractive index (blue) and focal length of the refractive lens (brown-solid line) and diffractive lens (brown-dotted line) as a function of wavelength λ = 0.4–0.8 μm. **b** Plot of the $I_R(x = y = 0)$ for λ = 401 nm to 800 nm for r_1 = 10 μm to 50 μm in steps of 10 μm for RPA and direct imaging using a refractive and diffractive lens.

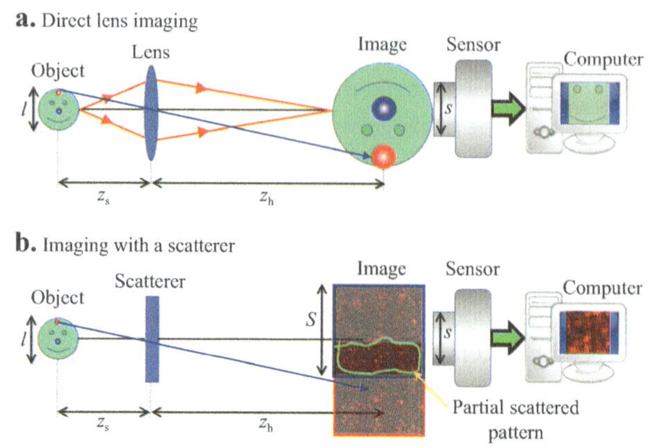

Fig. 12 Schematics of **a** direct imaging and **b** CAI for understanding FOV limitations.

system can be controlled without any limit. However, it is necessary to record the PSF beyond the sensor area by shifting the location of the sensor one time in the calibration phase.

The proposed technique can be applied for extending the FOV of imaging systems[29,30] and the dynamic range of optical lever[31–33] based on coded apertures. In both cases, the scattering degree was increased such that the area of the PSF is larger than the area of the sensor. This can be achieved by changing the window size in the GSA shown in Fig. 3a. The schematic of scanning the sensor across the PSF and recording multiple intensity distribution is shown in Fig. 13a. The resulting multiple patterns are stitched together into a synthetic PSF. The image of the recorded central part of the PSF and synthetic PSF formed by computational stitching of multiple areas of PSF recorded by scanning the sensor over the large area PSF are shown in Fig. 13b, c, respectively. The recorded object intensity distribution is shown in Fig. 13d. The reconstruction results using the regular PSF and synthetic PSF are shown in Fig. 13e, f, respectively. In conclusion, the limit of the field of view can be extended from ~(sz_s/z_h) in direct imaging to ~$(S + s)z_s/z_h$ in indirect imagers based on chaos when ($S > s/2$), where S is the length of the side of the scattered intensity distribution in the sensor plane.

Signal-to-noise ratio

One of the main challenges in CAI approaches is the low signal-to-noise ratio (SNR). The point-to-point mapping of direct imaging has optimal SNR. In CAI, every object point is mapped to a scattered intensity distribution. Consequently, the signal is sometimes not significantly higher than the noise level of the detector. The engineering aspects related to improving SNR are discussed next.

Engineering of phase-only modulators

The engineering of a phase-only modulator is carried out using GSA, as shown in Fig. 3a. To control the scattering degree, it is sufficient to set a constraint in the sensor plane that the signal outside the window is zero. However, in most of the previous studies[1], in addition to defining the boundary in the sensor plane, the signal within the window was set to be constant. This is a crucial factor as the background noise generated due to the correlation of two positive functions tends to be more uniform, which is often easier to deal with during computational processing than random noise. Depending upon the requirement of the imaging system, a trade-off exists between FOV and SNR. A larger FOV disperses the light to a larger area while the SNR decreases. On the other hand, a high SNR at the detector demands the light to be concentrated in a smaller area. To balance between the above contradictions, a modified configuration was selected where the light in the sensor plane was collected in the form of sparsely randomly distributed points[34–36]. In these studies, it was noted that as less as 5–10 dots were sufficient for imaging

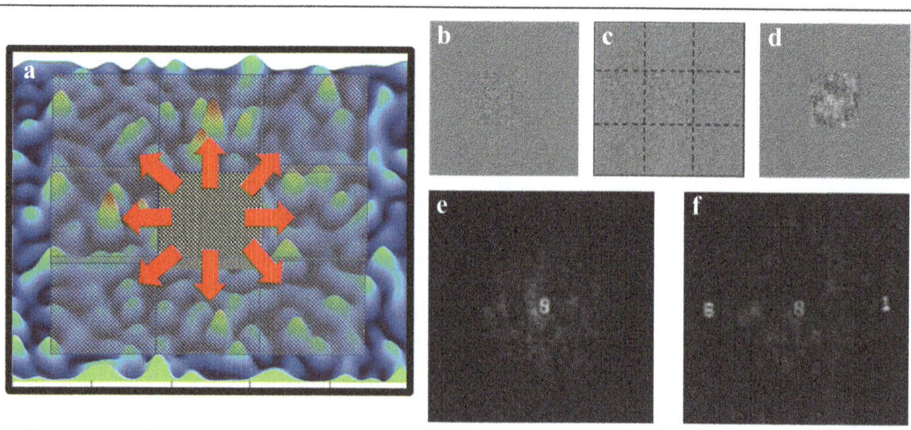

Fig. 13 a Schematic of sensor scanning approach for recording multiple areas of the large PSF. Image of **b** central part of PSF, **c** synthetic PSF, **d** central part of object intensity. Reconstruction results with **e** PSF captured within the sensor area and **f** synthetic PSF. **b–f** are adapted from 29.

with significantly higher efficiency and SNR than the previous studies. However, in this case, at least two camera shots were needed to generate bipolar matrices, which during autocorrelation does not have an offset due to the presence of both positive and negative values.

Engineering of RPA

A random pinhole array was synthesized using two uncorrelated random variables x_i, y_i uniformly distributed on $[-C_2, C_2]$. To reduce the background noise, quasi-random variables u_i, v_i were synthesized such that the mask consisting of the pinholes at the new locations yields images with improved SNR. The SNR is defined as Signal/(Average background noise), which becomes 1/(Average background noise) upon normalizing the reconstructed intensity pattern. The two-step optimization procedure is shown in Fig. 14a[33]. The random variables x_i, y_i are iterated over N times, and the random variables u_i, v_i corresponding to maximum SNR are selected and given as input to the second stage optimization procedure. In the next stage, the location of the pinholes is shifted along the X and Y directions with the limits $-L \leq \Delta u_i \leq L$ and $-L \leq \Delta v_i \leq L$ both in steps of Δ and the SNR is calculated at every step, and the quasi-random variables u_i', v_i' corresponding to the maximum SNR are determined. The second optimization, therefore, runs over $2L/\Delta$ iterations for every location along X or Y direction. Therefore, the total number of iterations in the second optimization is $4nL/\Delta$.

A specific design condition of object distance (u = 10 cm) and image distance (v = 10 cm), wavelength (λ = 617 nm), pinhole diameter (d = 80 μm), mask diameter (D = 8 mm) and number of pinholes (n = 2000) is considered. During every iteration, the light from a point object propagates by 10 cm and is modulated by the RPA mask, and the modulated light propagates by another distance of 10 cm. The intensity distribution at 10 cm from the quasi RPA (QRPA) mask is correlated with a phase-only filtered version of I_{PSF} such that the correlation is

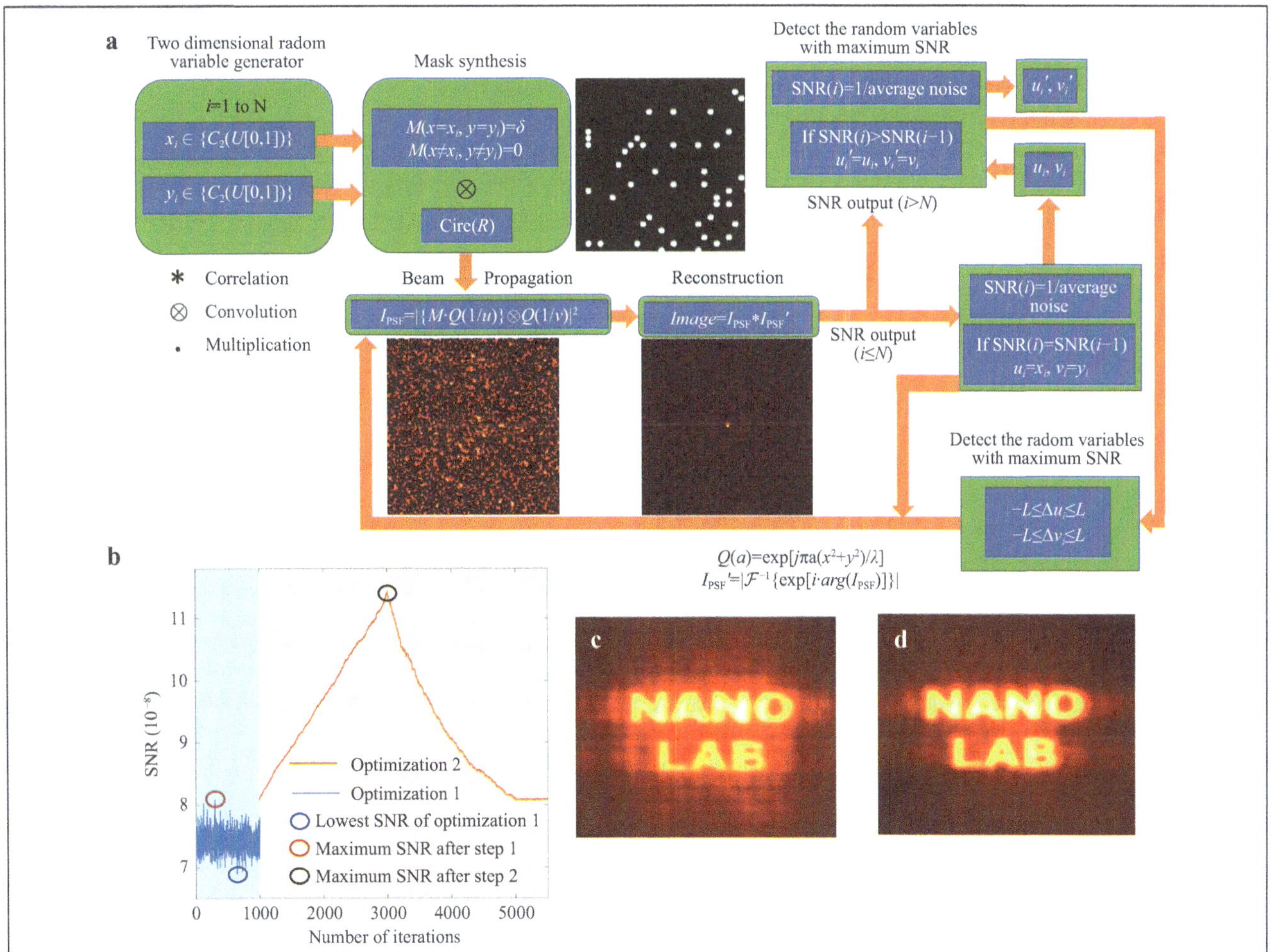

Fig. 14 a Optimization algorithm for the RPA. **b** Plot of SNR with a number of iterations for the two optimization steps. Reconstructed images of a test object 'NANO LAB' **c** lowest SNR after optimization 1 and **d** highest SNR after optimization 2.

$I_{PSF} * I'_{PSF}$, where $I'_{PSF} = \left|\mathcal{F}^{-1}\left\{exp\left[iarg\left(\tilde{I}_{PSF}\right)\right]\right\}\right|$, and $\tilde{I}_{PSF}$ is Fourier transform of I_{PSF}. The SNR was calculated as 1/(Average background noise) after normalizing the maximum intensity value of the correlation. The plot of the SNR with the number of iterations is shown in Fig. 14b. A test object, 'NANO LAB' was tested using the two masks with lowest and highest SNR, and the results are shown in Fig. 14c, d, respectively. It must be noted that there are numerous approaches available for optimizing the RPA. Some approaches involve designing coded apertures based on blue and green noise-based statistical models.[24,37,38] The GSA can also be adapted where the generated optimal phase profile can be binarized and used as an amplitude mask[39]. The RPA can be either sparse or dense depending upon the application. While a high light throughput is desirable in most imaging systems, for monitoring high-intensity events such as industrial processes and sparks, a sparse RPA is desirable[40].

Conclusion and future perspectives

In this review, some of the engineering methods of coded aperture imaging systems that influence the different facets of the imaging, such as lateral, axial, and spectral resolutions, the field-of-view, and signal-to-noise ratio, are discussed. The coded aperture imaging is an umbrella term including both deterministic apertures such as uniformly redundant arrays (URA)[23], modified URA[41], Cassegrain objectives[42], and randomly coded ones. This review focused mainly on the design and engineering approaches of randomly coded apertures. We have also excluded randomly coded apertures used for algorithms of compressed imaging[43]. Imaging through opaque scattering layers[44,45] also does not belong to the present topic since these scattering layers are given apertures without the ability to code or engineer them. The multi-spectral, multi-dimensional, extended field of view and resolution enhancement imaging results demonstrating the above characteristics can be found in the original articles in the reference list. The CAI methods have many advantages over direct imaging systems, and the engineering requirements must be considered carefully to achieve optimal performance. The previously reported coded aperture imagers for spectral imaging consisted of multiple optical components such as lenses, coded apertures, and dispersive elements. In the study, the focus has been on single element coded apertures where the coded aperture acts as the dispersive element. While the single element-based coded apertures are compact, lightweight, and flexible, complicated coding abilities present in multiple component systems are difficult to achieve[46]. Aperture engineering approaches can also be used for improving the performances of noninvasive imaging through scattering layers[44,45]. As discussed in the previous sections, the lateral resolution of an imaging system is given by the width of the autocorrelation function, which is dependent upon the scattering degree of the modulator in the near field. Noninvasive imaging through scattering layers employs a phase-retrieval algorithm whose performance relies on many factors, including the scattering degree of the diffuser. In some of the studies, the scattering degree of the diffuser was increased by projecting the diffuser plane onto an SLM with a strong scattering mask using a 4f system[47]. In some recent studies, the need for scattering to see through obscuration based on phase-retrieval algorithms[48]. In some cases, the poor design of the optical modulators and design conditions will demand additional camera shots to perform statistical averaging or more complicated computational reconstruction algorithms. In any imaging system, the phase mask and optical architecture are crucial and may significantly affect the concept. A notable example is the FINCH which violates the Lagrange invariant condition at special architectures and with engineered phase masks. Another notable finding in one of the recent studies is the realization of a coherent CAI system with a simpler optical architecture as that of an incoherent CAI system by the engineering of chaotic apertures[36]. We believe that in the future, in addition to amplitude and phase modulations, polarization modulations in aperture engineering will play an important role in CAI systems.

Acknowledgements

Australian Research Council (LP190100505). European Union's Horizon 2020 research and innovation programme under grant agreement No.857627 (CIPHR).

Author details

[1]Optical Sciences Centre and ARC Training Centre in Surface Engineering for Advanced Materials (SEAM), School of Science, Swinburne University of Technology, Hawthorn, VIC 3122, Australia. [2]Institute of Physics, University of Tartu, 50411 Tartu, Estonia. [3]School of Electrical and Computer Engineering, Ben-Gurion University of the Negev, P.O. Box 653, Beer-Sheva 8410501, Israel. [4]Tokyo Tech World Research Hub Initiative (WRHI), School of Materials and Chemical Technology, Tokyo Institute of Technology, 2-12-1, Ookayama, Meguro-ku, Tokyo 152-8550, Japan

Conflict of interest

The authors declare that they have no conflict of interest.

Received: 05 September 2021 Revised: 22 March 2022 Accepted: 24 March 2022

Accepted article preview online: 25 March 2022

Published online: 01 May 2022

References

1. Rosen, J. et al. Recent advances in self-interference incoherent digital holography. *Advances in Optics and Photonics* **11**, 1-66 (2019).
2. Dainty, J. C. Laser Speckle and Related Phenomena. (New York: Springer-Verlag, 1984).
3. Singh, A. K. et al. Scatter-plate microscope for lensless microscopy with diffraction limited resolution. *Scientific Reports* **7**, 10687 (2017).
4. Antipa, N. et al. DiffuserCam: lensless single-exposure 3D imaging. *Optica* **5**, 1-9 (2018).
5. Lee, K. & Park, Y. Exploiting the speckle-correlation scattering matrix for a compact reference-free holographic image sensor. *Nature Communications* **7**, 13359 (2016).
6. Rosen, J. et al. Review of 3D imaging by coded aperture correlation holography (COACH). *Applied Sciences* **9**, 605 (2019).
7. Horner, J. L. & Gianino, P. D. Phase-only matched filtering. *Applied Optics* **23**, 812-816 (1984).
8. Rai, M. R., Vijayakumar, A. & Rosen, J. Non-linear adaptive three-dimensional imaging with interferenceless coded aperture correlation holography (I-COACH). *Optics Express* **26**, 18143-18154 (2018).
9. Gerchberg, R. W. & Saxton, W. O. A practical algorithm for the determination of phase from image and diffraction plane pictures. *Optik* **35**, 237-250 (1972).
10. Ogura, Y., Aino, M. & Tanida, J. Design and demonstration of fan-out elements generating an array of subdiffraction spots. *Optics Express* **22**, 25196-25207 (2014).
11. Rai, M. R. et al. Resolution enhancement in nonlinear interferenceless COACH with point response of subdiffraction limit patterns. *Optics Express* **27**, 391-403 (2019).
12. Choi, Y. et al. Overcoming the diffraction limit using multiple light scattering in a highly disordered medium. *Physical Review Letters* **107**, 023902 (2011).
13. Kashter, Y., Vijayakumar, A. & Rosen, J. Resolving images by blurring: superresolution method with a scattering mask between the observed objects and the hologram recorder. *Optica* **4**, 932-939 (2017).
14. Rosen, J. & Brooker, G. Digital spatially incoherent Fresnel holography. *Optics Letters* **32**, 912-914 (2007).
15. Rai, M. R., Vijayakumar, A. & Rosen, J. Superresolution beyond the diffraction limit using phase spatial light modulator between incoherently illuminated objects and the entrance of an imaging system. *Optics Letters* **44**, 1572-1575 (2019).
16. Pullman, J. M. et al. Visualization of podocyte substructure with structured illumination microscopy (SIM): a new approach to nephrotic disease. *Biomedical Optics Express* **7**, 302-311 (2016).
17. Rey, J. J. et al. A deployable, annular, 30m telescope, space-based observatory. Proceedings of SPIE 9143, Space Telescopes and Instrumentation 2014: Optical, Infrared, and Millimeter Wave. Montréal: SPIE, 2014.
18. Bulbul, A., Vijayakumar, A. & Rosen, J. Partial aperture imaging by systems with annular phase coded masks. *Optics Express* **25**, 33315-33329 (2017).
19. Bulbul, A., Vijayakumar, A. & Rosen, J. Superresolution far-field imaging by coded phase reflectors distributed only along the boundary of synthetic apertures. *Optica* **5**, 1607-1616 (2018).
20. Tian, N., Fu, L. & Gu, M. Resolution and contrast enhancement of subtractive second harmonic generation microscopy with a circularly polarized vortex beam. *Scientific Reports* **5**, 13580 (2015).
21. Bouchal, P. & Bouchal, Z. Selective edge enhancement in three-dimensional vortex imaging with incoherent light. *Optics Letters* **37**, 2949-2951 (2012).
22. Anand, V. et al. Edge and contrast enhancement using spatially incoherent correlation holography techniques. *Photonics* **8**, 224 (2021).
23. Fenimore, E. E. & Cannon, T. M. Coded aperture imaging with uniformly redundant arrays. *Applied Optics* **17**, 337-347 (1978).
24. Wagadarikar, A. et al. Single disperser design for coded aperture snapshot spectral imaging. *Applied Optics* **47**, B44-B51 (2008).
25. Galvis, L. et al. Coded aperture design in compressive spectral imaging based on side information. *Applied Optics* **56**, 6332-6340 (2017).
26. Ghosh, G. Sellmeier coefficients and dispersion of thermo-optic coefficients for some optical glasses. *Applied Optics* **36**, 1540-1546 (1997).
27. Anand, V. et al. White light three-dimensional imaging using a quasi-random lens. *Optics Express* **29**, 15551-15563 (2021).
28. Vijayakumar, A. & Rosen, J. Spectrum and space resolved 4D imaging by coded aperture correlation holography (COACH) with diffractive objective lens. *Optics Letters* **42**, 947-950 (2017).
29. Rai, M. R., Vijayakumar, A. & Rosen, J. Extending the field of view by a scattering window in an I-COACH system. *Optics Letters* **43**, 1043-1046 (2018).
30. Hai, N. & Rosen, J. Coded aperture correlation holographic microscope for single-shot quantitative phase and amplitude imaging with extended field of view. *Optics Express* **28**, 27372-27386 (2020).
31. Vijayakumar, A. et al. Implementation of a speckle-correlation-based optical lever with extended dynamic range. *Applied Optics* **58**, 5982-5988 (2019).
32. Vijayakumar, A., Bhattacharya, S. & Rosen, J. Spatial multiplexing technique for improving dynamic range of speckle correlation based optical lever. *Scientific Reports* **9**, 16035 (2019).
33. Anand, V. et al. Single shot multispectral multidimensional imaging using chaotic waves. *Scientific Reports* **10**, 13902 (2020).
34. Rai, M. R. & Rosen, J. Noise suppression by controlling the sparsity of the point spread function in interferenceless coded aperture correlation holography (I-COACH). *Optics Express* **27**, 24311-24323 (2019).
35. Bulbul, A. & Rosen, J. Partial aperture imaging system based on sparse point spread holograms and nonlinear cross-correlations. *Scientific Reports* **10**, 21983 (2020).
36. Hai, N. & Rosen, J. Interferenceless and motionless method for recording digital holograms of coherently illuminated 3D objects by coded aperture correlation holography system. *Optics Express* **27**, 24324-24339 (2019).
37. Correa, C. V., Arguello, H. & Arce, G. R. Spatiotemporal blue noise coded aperture design for multi-shot compressive spectral imaging. *Journal of the Optical Society of America A* **33**, 2312-2322 (2016).
38. Lau, D. L., Ulichney, R. & Arce, G. R. Blue and green noise halftoning models. *IEEE Signal Processing Magazine* **20**, 28-38 (2003).
39. Vijayakumar, A. & Bhattacharya, S. Design and Fabrication of Diffractive Optical Elements with MATLAB. (Bellingham: SPIE Press, 2017), 250.
40. Anand, V. et al. Spatio-spectral-temporal imaging of fast transient phenomena using a random array of pinholes. *Advanced Photonics Research* **2**, 2000032 (2021).
41. Gottesman, S. R. & Fenimore, E. E. New family of binary arrays for coded aperture imaging. *Applied Optics* **28**, 4344-4352 (1989).
42. Anand, V. et al. Single-shot mid-infrared incoherent holography using Lucy-Richardson-Rosen algorithm. *Opto-Electronic Science* **1**, 210006 (2022).
43. Arce, G. R. et al. Compressive coded aperture spectral imaging: an introduction. *IEEE Signal Processing Magazine* **31**, 105-115 (2014).
44. Bertolotti, J. et al. Non-invasive imaging through opaque scattering layers. *Nature* **491**, 232-234 (2012).
45. Katz, O. et al. Non-invasive single-shot imaging through scattering layers and around corners via speckle correlations. *Nature Photonics* **8**, 784-790 (2014).
46. Rueda-Chacon, H. et al. Snapshot compressive ToF+spectral imaging via optimized color-coded apertures. *IEEE Transactions on Pattern Analysis and Machine Intelligence* **42**, 2346-2360 (2020).
47. Mukherjee, S., Vijayakumar, A. & Rosen, J. Spatial light modulator aided noninvasive imaging through scattering layers. *Scientific Reports* **9**, 17670 (2019).
48. Ng, S. H. et al. Invasive and non-invasive observation of occluded fast transient events: computational tools. *Photonics* **8**, 253 (2021).

Review **Open Access**

Deep holography

Guohai Situ[1,2,*]

Abstract

With the explosive growth of mathematical optimization and computing hardware, deep neural networks (DNN) have become tremendously powerful tools to solve many challenging problems in various fields, ranging from decision making to computational imaging and holography. In this manuscript, I focus on the prosperous interactions between DNN and holography. On the one hand, DNN has been demonstrated to be in particular proficient for holographic reconstruction and computer-generated holography almost in every aspect. On the other hand, holography is an enabling tool for the optical implementation of DNN the other way around owing to the capability of interconnection and light speed processing in parallel. The purpose of this article is to give a comprehensive literature review on the recent progress of *deep holography*, an emerging interdisciplinary research field that is mutually inspired by holography and DNN. I first give a brief overview of the basic theory and architectures of DNN, and then discuss some of the most important progresses of deep holography. I hope that the present unified exposition will stimulate further development in this promising and exciting field of research.

Keywords: Deep learning, deep neural networks, digital holography, computer-generated hologram, optical neural networks

Introduction

Since the pioneering works by Gabor[1], Leith and Upatnieks[2,3] and Denisyuk[4], holography has become an important and widespread technique that has found applications in various fields of optical engineering, ranging from optical imaging and microscopy[5,6], metrology[5,7] to three-dimensional (3D) display[8,9].

Physically, holography is a two-stage process: the recording, and the reconstruction, of a wavefront. Nowadays, both these processes can be performed either optically or digitally. We refer the kind of holography that the recording is performed optically by a digital camera while the reconstruction is digitally as digital holography (DH)[5-7]. In contrast, the kind of holography that the recording (synthesis) is performed digitally while the reconstruction is optically is called computer-generated holography (CGH)[8,9].

For the optical recording of a wavefront, one would prefer a light source with a certain level of coherence[10,11], in particular, for off-axis holography[3], because the lack of coherent light sources only allows interference patterns to be formed in the vicinity of the optical axis. As a result, only the hologram of a small object can be recorded by using an in-line setup. Furthermore, the reconstructed image is blurred owing to the superposition of a fuzzy defocused twin image, which was then difficult to effectively eliminate[12], although many efforts have been elaborately made[13] in the history of holography. Thanks to the invention of DH[14,15], coherence is not a fundamental limit for contemporary holographic imaging techniques any more. Light sources with short-coherence[16,17] and even incoherence[18] can be used for holographic recording.

One of the great advantages of DH is the capability of numerical reconstruction of a digitally recorded hologram.

Correspondence: Guohai Situ (ghsitu@siom.ac.cn)
[1]Shanghai Institute of Optics and Fine Mechanics, Chinese Academy of Sciences, Shanghai 201800, China
[2]University of Chinese Academy of Sciences, Beijing 100049, China

In this way, the fuzzy defocused twin image superposing with the in-focus image can be removed numerically. Conventional this can be done by physics-based approaches[19–26], phase-retrieval approaches[27–31], or more generalized inverse problem approaches[32–40].

With the recent prosperous development of a new class of optimization tools called deep neural networks (DNN)[41,42], we have witnessed the emergence of a new paradigm of solving inverse problems in various fields of optics and photonics by using DNN[43–48]. This shift of paradigm also has significant influence to the field of DH[49,50] in many aspects. Indeed, in additional to holographic reconstruction[51–57], DNN has also been proposed for phase aberration compensation[58], focus prediction[59–64], extension of depth-of-field[65], speckle reduction[66–69], resolution enhancement[70], and phase unwrapping[71–73], just to name a few.

DNN has also been used for the design of CGH[74–81], a technique whose invention was mainly attributed to Lohmann's pioneering works[82,83]. As suggested by the name, the objective of CGH is to artificially encode a target object within a space volume into a hologram called computer-generated hologram so that it can reconstruct the desired wavefront within that space volume under the illumination of a proper coherent light. The optically reconstructed wavefront can be a perfect reference of an optical surface for holographic testing[84–86] or a 3D object/scene for holographic display[8,9]. Conventional approaches for the encoding of a computer-generated hologram are either to take it as an optimization problem, which can be solved by iterative phase-retrieval algorithms[27,28,87], or non-iterative interference-specific or diffraction-specific algorithms[88,89]. Although it can be sped up by using the trick of look-up table[90], the use of DNN still promises the most dramatic increment in terms of calculation efficiency[74–81].

Holography has been used the other way around, i.e., as a way to implement optical neural networks (ONN), in particular, the Hopfield model[91,92] and fully-connected neural networks[93–97]. With the development of optical material manufacturing technologies such as 3D printing and metamaterials, multi-layers fully-connected neural networks can be implemented in a modern fashion[98,99].

These recent progresses suggest that the distinct fields of holography and deep learning have incorporated into each other, forming a new interdisciplinary field, the name of which can be coined as *deep holography*. In this article, I will give a comprehensive literature review of this emerging but exciting field. The structure of this article is organized as follows: In section 1 I will first give a concise introduction to deep neural networks. Then I will discuss in detail how DNN is used to solve various problems in holography, and *vice versa*, in section 2 and section 3, respectively. Finally, the perspective of further development will be discussed in Sec. 4.

Deep Neural Networks: A Concise Introduction

DNN can be regarded as a category of machine learning algorithms that are designed to extract information from raw data and represent it in some sort of model[42]. Specifically, a neural network (NN) is built on a collection of connected units called artificial neurons, which are typically organized in layers, an idea somehow inspired by the biological neuron in the mammalian brain. As schematically shown in Fig. 1a, a modern NN consists of three kinds of layers: the input layer, the output layer, and the hidden layers. The input layer usually represents the signal to be processed and the output layer represents the expected result that one wishes the network to produce. So the widths (P), i.e., the number of neurons, of these two layers are task-specific. Data processing is mainly performed by the hidden layers that lay between the input and the output layers. Each successive hidden layer uses the outputs from its upstream layers as its input, processes it, and then passes the result to a downstream layer. In this manuscript, we use the digit $l = 1, \ldots, L$ to enumerate the layers, where L is called the depth of the NN. A neural network is *deep* if it has many layers. The depth of modern *deep* neural networks ranges from 8 layers in AlexNet[100] to 152 layers in ResNet[101], which has the potential to increase to more than 1000 layers[102]. The requirement of computation resource dramatically increases along with the up-scaling of the DNN, i.e., the number of hidden layers and hidden neurons. For example, a neural network used for DH have a depth up to 20 layers in a typical proof-of-principle demonstration[52]. It usually takes tens of hours to train on a training set consisting of about thousands of holograms with a modern graphic workstation. Unfortunately, given a problem to be solved by DNN, it is not trivial at all to determine how deep it should be[103,104]. Hornik has proved that, for any continuous function $\boldsymbol{y} = f(\boldsymbol{x})$, where $\boldsymbol{x}$ and $\boldsymbol{y}$ are data (vectors) in the Euclidean or non-Euclidean space, there is always an NN, no matter how shallow it is, that can approximate the function f with an infinitesimal error, i.e., $\mathrm{NN}\{\boldsymbol{x}\} \rightarrow \boldsymbol{y}$, provided that it is sufficiently wide[105]. Practically, however, one still needs a good rule of thumb to configure the number of layers (L) and the numbers of neurons in each layer ($P^{(l)}$). It is commonly believed that the performance of DNN is heavily dependent on the network architecture, which is defined in part by L, $P^{(l)}$, and the types of connections between layers, the quality of the raw data, and the

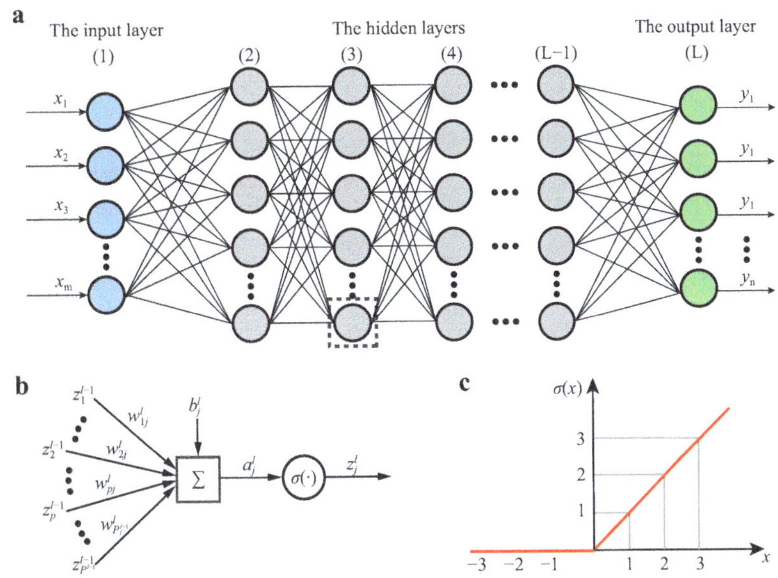

Fig. 1 A conceptual architecture of DNN.

technique to train the network on them.

Perhaps the most well-known and easiest to understand DNN is the so-called feedforward neural networks. The architectures of all the other DNNs that are widely used in holography[51–56,58–62,64–81] are developed on the base of it. Thus it is worthy of discussing it in detail.

Feedforward neural networks

As shown in Fig. 1a, a feedforward neural network, or multilayer perceptron (MLP) has one input layer, one output layer, and one or many hidden layers. Each layer may have a different number of neurons called the perceptron. The connections between the neurons in the layers form an acyclic graph[106]. The objective of a feedforward neural network is to optimize an NN model f_{NN} that approximates a continuous function f, which maps $\boldsymbol{x}$ in the input space to $\boldsymbol{y}$ in the output space through a set of parameters Θ that are learned from the raw data.

The artificial neuron

The basic unit in a DNN is the artificial neuron. As shown in Fig. 1b, an artificial neuron simply calculates the weighted sum of all the quantities outputted from the neurons in its immediately upstream layer, and passes the resulting quantities to the neurons in the next layer. Let us take the jth neuron at the l layer for example, the input to this neuron can be written as[41,42]

$$a_j^{(l)} = \sum_{p=1}^{r} w_{pj}^{(l)} z_p^{(l-1)} + b_j^{(l)} \tag{1}$$

where $z_p^{(l-1)}$ is the output from the pth neuron at the $(l-1)^{th}$ layer, $w_{pj}^{(l)}$ is the weighting factor that connects these two neurons, and $b_j^{(l)}$ is a bias. The values of the network parameters $w_{pj}^{(l)}$ and $b_j^{(l)}$ are to be learned from a set of raw data called the training set. One can think of their values as the connection strengths between the two neurons. The jth neuron at the l layer then can be activated if the quantity $a_j^{(l)}$ is significant (for example, >0), and this value is passed on to the next layer. Otherwise, this neuron is dead, and should have no contribution to the neurons in the downstream layer. Analogously, one can think of the input signal being an electric current that flows through the network from the input layer to the output layer. Each neuron in the hidden layers acts like a gate that controls the amount of incoming current that is allowed to pass through to the downstream neurons. The "gate" function in a NN is not just a simple "0" and "1" binary function as in the digital electric circuit, but has a form of an activation function. Fig. 1c plots the Rectified Linear Unit (ReLU), which is one of the most important activation function nowadays used in DNN. It is defined as[42]

$$z = \sigma(a) = \max(0, a) \tag{2}$$

ReLU is widely employed in most of the modern neural network architectures, as it has a number of benefits over other old-fashioned activation functions such as Sigmoid and Tanh[42]: (a) It can be applied to minimize the interaction effects; (b) It is simple and easy to compute, and thus leads to an increment of efficiency in the network training; (c) It helps avoid the vanishing gradient problem; and (d) It is sparsely activated because the output is zero for all negative inputs. However, ReLU sometimes dies, referring to the situation that an neuron has a zero activation value. This dying ReLU issue causes slow-learning because the optimization algorithm is gradient-based and does not adjust the unit weights if the gradient is zero in an inactive neuron. Thus, extensions and alternatives such as Leaky ReLU (or LReLU for short), exponential linear unit (ELU), and parametric ReLU (PReLU) are highly desirable when it happens[107].

The width of the input layer is the number of pixels of the image one wishes the network to process. The width of the output layer is usually task-dependent. For example, in the applications of holographic reconstruction[51–54,56], the width of the output layer is the same as the input layer. Whereas in the application of holographic autofocusing[59–62], the width of the output layer is simply 1, which gives the focusing distance. The width of each hidden layer is dependent on task in hand and the choice of the network architecture. Indeed, the width of the lth layer and that of the $(l-1)^{\text{th}}$ layer may not be the same in most of the cases. Thus, what Eq. 1 implies is that it transforms a P-dimensional signal to a J-dimensional space. This can be more clearly seen by writing Eq. 1 in the form of

$$\boldsymbol{a}^{(l)} = \boldsymbol{W}^{(l)}\boldsymbol{z}^{(l-1)} + \boldsymbol{b}^{(l)} \tag{3}$$

The substitution of Eq. 1 into Eq. 2 yields the output from the l layer

$$z^{(l)} = f^{(l)}\left(z^{(l-1)}; \boldsymbol{w}^{(l)}, \boldsymbol{b}^{(l)}\right) = \sigma\left(\boldsymbol{a}^{(l)}\right) \tag{4}$$

where $f^{(l)}$ is defined as the transform from the $(l-1)^{\text{th}}$ layer to the lth layer. From a more theoretical point of view, deep learning relies on this kind of mapping between spaces of different dimensions[108].

The feedforward neural network model

Now we can mathematically express the feedforward neural network model as

$$\boldsymbol{y} = \delta\Big(\boldsymbol{W}^{(L)}\sigma\Big(\ldots\sigma\Big(\boldsymbol{W}^{(2)}\sigma\Big(\boldsymbol{W}^{(1)}\boldsymbol{z}^{(0)} + \boldsymbol{b}^{(1)}\Big) + \boldsymbol{b}^{(2)}\Big) + \ldots\Big) + \boldsymbol{b}^{(L)}\Big) \tag{5}$$

where $\boldsymbol{z}^{(0)} \triangleq \boldsymbol{x}$ is the input signal, and $\delta(\cdot)$, the activation function at the output layer. It is not necessary to be the ReLU function as in the hidden layer. For example, it takes the form of a softmax function

$$\delta(z_j) = \frac{\exp[z_j]}{\sum_{k=1}^{K}\exp[z_k]} \tag{6}$$

for autofocusing in holography[59–61].

The set of network parameters Θ then can be defined as $\Theta \triangleq \{\boldsymbol{W}^{(1)}, \boldsymbol{b}^{(1)}, \ldots, \boldsymbol{W}^{(L)}, \boldsymbol{b}^{(L)}\}$. Then one can write the feedforward NN model in Eq. 5 in a more compact form

$$\boldsymbol{y} = f^{(L)} \circ f^{(L-1)} \circ \ldots \circ f^{(1)} = f_{\text{NN}}(\boldsymbol{x}; \Theta) \tag{7}$$

This simply tells the fact that a feedforward NN model f_{NN} is to approximate a function f and map the input $\boldsymbol{x}$ to the output $\boldsymbol{y}$ through a neural network specified by the set of parameters Θ.

The training of the network

Although the universal approximation theorem[105] guarantees that a feasible NN model f_{NN} exists for an arbitrary given training set, Eq. 7 does not provide any clue to its architecture and weight configuration. In terms of DNN, the network architecture is defined on a set of hyperparameters such as the depth L and the width $P^{(l)}$ of each layer that one needs to set up, mostly by a rule of thumb. Many efforts have been made to clarify this point, but it is still an open question[103,104]. The weighing factors $\boldsymbol{W}$ and $\boldsymbol{b}$ are to be determined by a learning process, which consists of repeated steps of optimal adjustment of the parameters in Θ.

For the supervised learning methods that are mainly used in the community of holography, the parameters in Θ are learned from a large set of labeled data $S = \{(\boldsymbol{x}_k, \boldsymbol{y}_k)\}_{k=1}^{K}$. It consists of many pairs of $(\boldsymbol{x}_k, \boldsymbol{y}_k)$ with $\boldsymbol{x}_i$ being the signal (such as a hologram) one wishes the network to process, and $\boldsymbol{y}_k$, the associated correct result (the reconstructed object, the focusing distance, etc.) that are already known. Thus it is possible to compare the calculated output, denoted by $\hat{\boldsymbol{y}}_k$, with the correct answer $\boldsymbol{y}_k$, and evaluate their difference for each neuron at the output layer. This leads one to define the loss function $\mathcal{L}[f_{\text{NN}}(\boldsymbol{x}; \Theta), \boldsymbol{y}]$. Thus one can then formulate the NN learning as the optimization of the parameters in Θ so as to minimize the loss function

$$\arg\min_{\Theta} \mathcal{L}[f_{\text{NN}}(\boldsymbol{x}; \Theta), \boldsymbol{y}] \tag{8}$$

An instinct philosophy to train a neural network is to adjust the values of $\boldsymbol{W}^{(l)}$ and $\boldsymbol{b}^{(l)}$ and see if the loss function decreases or not. An efficient and straightforward way to do this is to evaluate the gradient of the loss function with respect to Θ. Note that DNN has a layered architecture, one needs to calculate the gradient of the loss function with respect to the weights and bias one by one from the output layer back to the input layer. This can be done by the algorithm of back propagation[109].

To develop the error back propagation model, let us first define the loss function of layer l

$$\mathcal{L}^{(l)} = \mathcal{L} \circ f^{(L)} \circ f^{(L-1)} \circ \ldots f^{(l)} \tag{9}$$

Then the back gradient of the loss function with respect to the parameters $\boldsymbol{W}^{(l)}$ and $\boldsymbol{b}^{(l)}$ at layer l can be formulated by using the recurrence relation[110]

$$\frac{\partial \mathcal{L}^{(l)}}{\partial \boldsymbol{W}^{(l)}} = \frac{\partial \mathcal{L}^{(l+1)}}{\partial f^{(l)}} \frac{\partial f^{(l)}}{\partial \boldsymbol{W}^{(l)}} \tag{10}$$

$$\frac{\partial \mathcal{L}^{(l)}}{\partial \boldsymbol{b}^{(l)}} = \frac{\partial \mathcal{L}^{(l+1)}}{\partial f^{(l)}} \frac{\partial f^{(l)}}{\partial \boldsymbol{b}^{(l)}} \tag{11}$$

$$\frac{\partial \mathcal{L}^{(l)}}{\partial \boldsymbol{z}^{(l-1)}} = \frac{\partial \mathcal{L}^{(l+1)}}{\partial f^{(l)}} \frac{\partial f^{(l)}}{\partial \boldsymbol{z}^{(l-1)}} \tag{12}$$

From the recurrence relations Eq. 10 – Eq. 12, one can derive $\partial\mathcal{L}/\boldsymbol{W}^{(l)}$ and $\partial\mathcal{L}/\boldsymbol{b}^{(l)}$ using the chain rule[111]. Then the architectural parameters at layer l can be updated using the strategy of gradient descent[110]

$$\boldsymbol{W}^{(l)} \leftarrow \boldsymbol{W}^{(l)} - \eta \frac{\partial \mathcal{L}}{\partial \boldsymbol{W}^{(l)}} \tag{13}$$

$$\boldsymbol{b}^{(l)} \leftarrow \boldsymbol{b}^{(l)} - \eta \frac{\partial \mathcal{L}}{\partial \boldsymbol{b}^{(l)}} \tag{14}$$

where η is the learning rate, or step size, in the gradient descent method. It determines how many the parameters should be adjusted each time. The convergence will not get to the right place if the learning rate is either too large or too small. Thus an ideal η value is desirable in training a neural network. However, the determination of its value is yet a comprehensive theoretical study[112]. Empirically, setting the learning rate $\eta = 10^{-4}$ should work pretty well for many applications in holography[52,65]. But is can be adjusted during the iteration process as well[53,64].

The number of labeled data pairs (K) in the training set should be sufficiently large in order for the network to learn the statistics of the data. Indeed, it can easily go up to tens of thousands in a typical DNN for holographic reconstruction[52]. The calculation of the error back propagation is then extremely time-consuming. Thus a practical and intuitive way to evaluate the error is to randomly select a small batch of labeled data in each epoch (which means a period of time) and calculate the gradients of the loss function, and use them to update Θ. This is a trick called stochastic gradient descent (SGD). Furthermore, it is also possible to employ the method of adaptive moment estimation[113], or *Adam* for short, that adds a momentum term to speed up the learning process, and adaptively shrinks the learning rate along with the progress of the learning process to achieve faster convergence.

As apparently suggested by Eq. 10 – Eq. 12, the specific form of the loss function plays a crucial role in the optimization process and the explicit setting of Θ that it converges to[114]. Thus one should choose the most appropriate loss function depending on the problem in hand[62]. Some of the widely adopted loss functions in DNN include the averaged mean squared error (MSE) loss, the L_1, or mean absolute error (MAE), loss, and the cross entropy loss.The MSE loss is one of the defined as

$$\begin{aligned}\mathcal{L}[f_{\mathrm{NN}}(\boldsymbol{x};\Theta),\boldsymbol{y}] &= \mathbb{E}_S \|\boldsymbol{y}_k - f_{\mathrm{NN}}(\boldsymbol{x}_k;\Theta)\|^2 \\ &= \frac{1}{K}\sum_{k=1}^{K} \|\boldsymbol{y}_k - f_{\mathrm{NN}}(\boldsymbol{x}_k;\Theta)\|^2 \end{aligned} \tag{15}$$

where the subscript k denotes the kth pair of data in the training set, and $\|\cdot\|^2$ denotes the Euclidean distance between the correct output $\boldsymbol{y}_k$ and the calculated output $\hat{\boldsymbol{y}}_k = f_{\mathrm{NN}}(\boldsymbol{x}_k;\Theta)$ with respect to a setting of Θ. The MSE loss, and the root of it, i.e., the root mean square error (RMSE) loss, are widely used to train DNN for holographic reconstruction[52,64]. Alternatively, the MAE loss is defined as[115]

$$\mathcal{L}[f_{\mathrm{NN}}(\boldsymbol{x};\Theta),\boldsymbol{y}] = \frac{1}{K}\sum_{k=1}^{K} \|\boldsymbol{y}_k - f_{\mathrm{NN}}(\boldsymbol{x}_k;\Theta)\| \tag{16}$$

where $\|\cdot\|$ is the L_1 norm. And the cross entropy loss is defined as the inner product of $\boldsymbol{y}_k$ and $\hat{\boldsymbol{y}}_k$

$$\mathcal{L}[f_{\mathrm{NN}}(\boldsymbol{x};\Theta),\boldsymbol{y}] = -\frac{1}{K}\sum_{k=1}^{K} \boldsymbol{y}_k \cdot \log f_{\mathrm{NN}}(\boldsymbol{x}_k;\Theta) \tag{17}$$

When more than one criterion are concerned, one can defined a combined loss function that is a weighted sum of several parts[64,116]. This is in particular useful for holography because of the complex nature of an optical wavefront. For example, if one wishes to measure both the amplitude and phase of the reconstructed wavefront, he/she can define a loss function as $\mathcal{L} = \mathcal{L}_{\text{amp}} + \alpha\mathcal{L}_{\text{phase}}$, a linear combination of the errors in both the amplitude and phase[54]. Alternatively, one can also define a complex loss function[78].

I will show later on that a loss function does not have to define on the training set Θ, but on a physical model. That is, $\mathcal{L}\{H[f_{\mathrm{NN}}(\boldsymbol{y})],\boldsymbol{y}\}$, where H is a forward physical model that maps the object space to the measured image space[55].

When the training process is completed, the performance of the neural network should be validated by using a set of data that have not been used for training in any way. The performance is usually evaluated by using the test error $\mathcal{L}_{\text{test}} = \mathbb{E}_S \|\boldsymbol{y}_m, f(\boldsymbol{x}_m;\Theta)\|^2$. This metric also quantifies the ability of *generalization* of the trained network[42].

The Convolutional Neural Networks

In the feedforward NN model described by Eq. 7, the neurons in neighboring layers are typically fully connected with the weight and bias parameters independent of each other. Although DNN has been employed to solve many problems in computational imaging[117-119], ranging from ghost imaging to imaging through scatterers, there are several issues with it. First, as there are too many parameters to train, it often has the issue of overfitting. Second, it requires a large memory footprint to temporally store the parameter set Θ and thus the training usually takes a lot of time. Third, it ignores the intrinsic structure that the data to be processed may have. This is in particular important for the tasks of speech and image processing. Images, in particular, have significant intrinsic structures. For example, neighboring pixels may have similar values; the image may be shift-invariant, etc. It is therefore highly demanded to have units in a neural network to learn these features.

Inspired by the physiological mechanism of visual cortexes[120], a convolutional neural network (CNN) also has a layered structure. Indeed, it consists of an input layer, an output layer and multiple hidden layers. But the hidden layers in CNN do not have to be fully connected. Instead, each convolutional layer in CNN has a filter called kernel function, denoted by w, to convolve with the incoming data z from an upstream layer, and extract a feature map of it at a certain level of abstraction. Instead of Eq. 1, the calculated feature map $a(i, j)$ can be mathematically written as[42]

$$a(i,j)=(z*w)(i,j)=\sum_m\sum_n z(m,n)w(i-m,j-n) \quad (18)$$

where (i, j) and (m,n) stand for the neurons at two neighboring layers. Equation (18) means that the elements of the kernel function, $w(m,n)$, will apply to many neurons in the layer. In other words, all those neurons share the weighting parameters in contrast to the case of DNN that each neuron is tied to a unique weight. This parameter sharing mechanism guarantees that the network just needs to optimize a much smaller set of parameters for each layer. It is because of this reason that the requirement for memory footprint and computation efficiency can be significantly reduced in comparison to DNN[42]. Indeed, the size of the kernel function is typically from 3×3 to 5×5 for many applications in holography[52,64], which is very small in comparison to that of a layer.

Note that a natural image has various features in one level of abstraction. For example, an image of a human face may contain edges with different orientations. Thus it is preferable to use multiple filters in one layer to extract all these edge orientation features, generating multiple feature maps. Usually these feature maps are arranged in a three-dimensional volume as they are to pass to a downstream layer. Denoting the width M and height N as the transverse size of each feature map, and the depth U as the number of feature maps, the value of the $(i, j)^{\text{th}}$ pixel in the tth feature map in the lth convolutional layer ($l \geq 2$) can be written as

$$\begin{aligned}a^{(l)}(i,j;v) &= f^{(l)}\left(z^{(l-1)};w^{(l)},b^{(l)(t)}\right)\\ &=\sigma\Bigg(\sum_{u=1}^{U}\sum_{m=0}^{M-1}\sum_{n=0}^{N-1}w^{(l)}(m,n;u,v)\\ &\quad\times z^{(l-1)}(i+m,j+n;u)+b^{(l)}(t)\Bigg)\end{aligned} \quad (19)$$

where $b^{(l)}(t)$ is a bias term for the vth feature map in the l layer, u denotes the uth feature map in the $(l-1)^{\text{th}}$ layer, w is the corresponding kernel function, and $z^{(l-1)}$ is the output from the upstream layer. One can clearly see from Eq. 19 that the convolution algorithm is actually implemented by cross-correlation in CNN, in contrast to what we are familiar with in terms of Fourier optics[121]. However, this does not change the resulting feature maps except their indices. We adapt this custom and call both Eq. 18 and Eq. 19 convolution.

The numerical calculation of the convolution in Eq. 19 requires moving the filter across spatial dimensions of the input data $z^{(l-1)}$. In conventional digital image processing and numerical implementation of convolution in optics[122], the filter is moved one pixel to the right and one pixel to

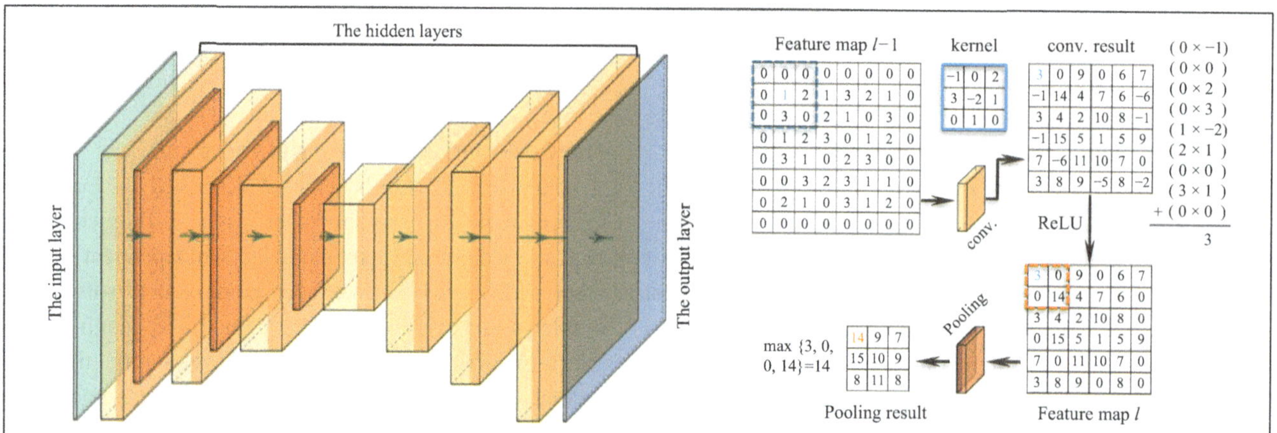

Fig. 2 A conceptual architecture of CNN.

the bottom at a time. In the language of deep learning, this means that the stride is equal to 1. But it is not necessary to be like this in CNN. Indeed, the stride of 2 is commonly used. More specifically, for an input image of size $N \times N$ and a kernel of size $M \times M$ with stride = k, the resulting output will be of size $[(N-M)/k+1] \times [(N-M)/k+1]$.

The output feature maps $a^{(l)}(i,j;t)$ are then passed through an activation function, usually the ReLU function defined by Eq. 2, to a pooling layer, which performs nonlinear down-sampling. This can be done in many ways. But the one that is most commonly used and has good performance is maxpooling[123], which partitions each incoming feature map into a set of non-overlapping rectangle regions by using a filter with the size of $\kappa \times \kappa$ and outputs the maximum value of each region. Thus, the spatial size of the resulting feature map is reduced by a factor of κ. As a consequence, the number of parameters, memory footprint and amount of computation in the network can be reduced accordingly. The reduction of network parameters will of course improve the situation of overfitting. Maxpooling also guarantees that the most significant features and their rough location relative to the other features can be passed to the downstream layer.

In a typical CNN model, the convolutional layer, the ReLU layer and the pooling layer are arranged in sequence, forming the basic building block[42]. Usually several "convolution-ReLU-pooling " blocks are arranged in cascade, each of which performs the same set of operations as described above. In the end, only the most significant features (activated features) of the input data can be retained after the data stream passes through several blocks. In the applications of image recognition[100] and focused distance determination in holographic reconstruction[62], a flatten layer is usually used to reshape the three-dimensional feature volume into a one-dimensional vector, which is then sent to fully-connected layer described by Eq. 1 for further analysis. In the applications of holographic reconstruction[51-53] and aberration compensation[58], however, one wishes to reconstruct the object function and need a path to transform the activated feature maps back to the image pixels. This can be implemented by adding a deconvolutional network[124,125], which consists of a series of unpooling (reverse maxpooling), rectifying, and transpose convolution operations that upsample the feature maps many times until it reaches the size of the input hologram. The elegant U-Net[126] operates in a similar way, except that the unpooling layers are replaced by "up-convolution " layers.

A CNN model can be trained in the way as we described in Sec.. It involves the calculation of the gradient of the loss function with respect to the weight of every kernel function, which is then used to update the weight, usually according to the Adam method[113]. The back propagation model is a little different. One can refer to Ref. 42 for more details.

When the network goes deeper, it becomes very difficult to train because of the problems such as gradient vanishing and exploding[127]. He et al. proposed the residual neural network, or ResNet for short, to address this problem[101]. The most distinguished feature of ResNet is that two distanced layers can be connected directly through the short-cut. That says, the signal goes through a series of "convolution-ReLU-convolution " blocks instead of the "convolution-ReLU-maxpooling " in CNN. The result is then added to the input of this block. Thus, the forward propagation model can be formulated as[101]

$$z^{(l)} = f^{(l)}\left(f^{(l-1)}\left(z^{(l-2)}; \boldsymbol{W}^{(l-2,l-1)}, \boldsymbol{b}^{(l-1)}\right); \ \boldsymbol{W}^{(l-1,l)}, \boldsymbol{b}^{(l)}\right) + z^{(l-2)} \tag{20}$$

where $f^{(l)}$ is the CNN forward propagation function defined in Eq. 19, $\boldsymbol{W}^{(l-1,l)}$ is the weight that connects layer $l-1$ and layer l. It is clearly seen that the input of layer $l-2$, $z^{(l-2)}$, is directly connected to the lth layer. As it does not need to undergo the nonlinear transform, the gradient will flow easily during back propagation.

Nowadays, a common and practical strategy to design a DNN to solve holographic problems is to take U-Net as a backbone, and incorporate into it the ResNet ingredient of short-cuts[53,55,58,71,128-130]. The U-Net architecture can also be extended to allow the extraction of features of different size by introducing multiple channels in the down-sampling convolutional blocks[51,52,59,131-133]. Another interesting extension of U-Net is the so-called U-Net++[134], which has a pyramid shape architecture. It essentially consists of an encoder and a decoder that are connected through a series of nested, dense skip pathways, bridging the semantic gap between the feature maps of the encoder and the decoder prior to fusion[135].

The Generative adversarial networks

Generative adversarial networks (GAN's) are NN's that learn to generate synthetic instances of data with the same statistical characteristics as the training data[136]. GAN is able to keep a parameter count significantly smaller than other methods with respect to the amount of data used to train the network. It the field of holography, GAN has been used for wavefront reconstruction[67,137,138], enhancement[139] and image classification[140]. It can be trained on paired data[139], unpaired data[67,138] or even unsupervisely[140] in some cases.

Architecturally, GAN is constituted of two neural networks, one of which is called the generator, and the other, the discriminator. As shown in Fig. 3, the two networks are pitted one against the other (and thus "adversarial") in order to generate new, synthetic instances of data that can pass for real data. Explicitly, the generator G is a deconvolutional neural network[124,125] that generates new images as real as possible from a given noise variable input z, whereas the discriminator D is a CNN-based classifier that estimates the probability of a generated image and determines if it looks like a real image from the training set or not.

To proceed, let us denote the probability distribution of the input variable z as p_z, that of the generator over data x as p_g, and that of the discriminator over real sample x as p_r. The purpose of GAN is to make sure that the discriminator's decisions over the real data are accurate by maximizing $\mathbb{E}_{x\sim p_r(x)}[\log D(x)]$, while the discriminator outputs a probability $D(G(z))$ that is close to zero by maximizing $\mathbb{E}_{z\sim p_z(z)}[\log(1-D(G(z)))]$ for a given generative data instance $G(z)$, where $z \sim p_z(z)$.

Thus, one can see that D and G are actually playing a minimax game so that the objective is to optimize the following loss function[136]

$$\begin{aligned}\min_G \max_D \mathcal{L}(D,G) =& \mathbb{E}_{x\sim p_r(x)}[\log D(x)] \\ &+ \mathbb{E}_{z\sim p_z(z)}[\log(1-D(G(z)))] \\ =& \mathbb{E}_{x\sim p_r(x)}[\log D(x)] \\ &+ \mathbb{E}_{x\sim p_g(x)}[\log(1-D(x)]\end{aligned} \quad (21)$$

where

$$\mathcal{L}(G,D) = \int_x p_r(x)\log[D(x)] + p_g(x)\log[1-D(x)]\mathrm{d}x \quad (22)$$

Note that for any $(a,b) \in \mathbb{R}_2 \backslash \{0,0\}$, the function $y \rightarrow a\log(y) + b\log(1-y)$ achieves its maximum in $[0,1]$ at $y = a/(a+b)$. It is then straightforward to obtain the best value of the discriminator[136]

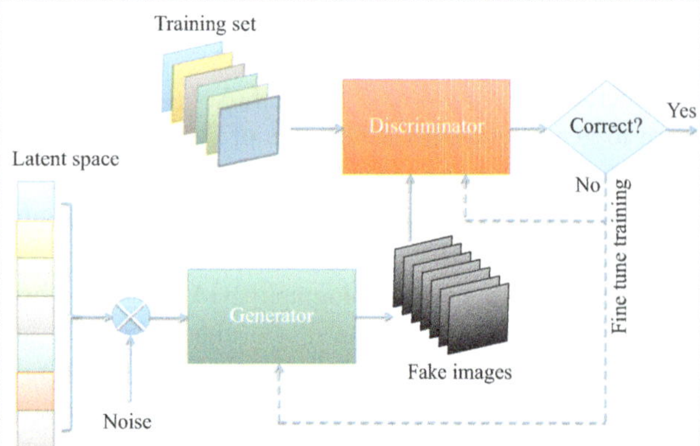

Fig. 3 A conceptual architecture of GAN.

$$D^*(x) = \tilde{x}^* = \frac{p_r(x)}{p_r(x) + p_g(x)} \in [0,1] \quad (23)$$

Once the generator is trained to its optimal, $p_g = p_r$. Thus $D^*(x) = 1/2$, and the loss function $\mathcal{L}(G,D^*) = -2\log 2$.

GAN can be trained by using SGD-like algorithm such as Adam[113] as in the case of CNN. But the discriminator and the generator should be trained against a static adversary[141]. That is, one should hold the generator values constant while training the discriminator, and *vice versa*.

There are several adaptations of GAN, among which the cycle-GAN[142] and conditional GAN (cGAN)[143] have been adopted for holography. Different from GAN, the purpose of the generator in cycle-GAN is not to generate an image from noise, but to take a hologram as its input and extract the most significant features via a series of convolutional layers, and then build a reconstructed image of the same size as the input hologram from these transformed features using a series of transpose convolutional layers. The most distinguished idea behind cycle-GAN is the introduction of a cycle consistency loss

$$\begin{aligned}\mathcal{L}_{\text{cycle}}(G,F) =& \mathbb{E}_{z\sim p_r(z)}[||F(G(z)) - z||] \\ &+ \mathbb{E}_{x\sim p_r(x)}[||F(G(x)) - x||]\end{aligned} \quad (24)$$

that imposes a constrain to the model. The generator G generates an object image x from a hologram z, and F generates the hologram z of x. Thus, the total loss function can be defined as

$$\mathcal{L}(G,F,D_x,D_z) = \mathcal{L}_{\text{GAN}_x} + \mathcal{L}_{\text{GAN}_z} + \mathcal{L}_{\text{cycle}} \quad (25)$$

where D_x and D_z are the discriminators of x and z, respectively, and $\mathcal{L}_{\text{GAN}_x}(G,D_x,x,z)$ and $\mathcal{L}_{\text{GAN}_z}(G,D_z,x,z)$, defined by Eq. 22, are the conventional GAN loss of the objects x and holograms z in the training set. It is clearly seen that x and z show up independently in each term of $\mathcal{L}(G,F,D_x,D_z)$, meaning that they do not need to pair up in the training set.

The generation of training data

The training of DNN usually requires a large set of data, the size of which is typically ranging from a few thousands to tens of thousands in a typical proof-of-concept demonstration. The amount of labeled data is far less than that is used for deep learning applications in other communities such as computer vision. For example, AlexNet[100] was trained on a set composed of 1.2 million images.

In the case of supervised training the data used for training should be labeled so that every input data $\boldsymbol{x}_k$ is paired up with a corresponding ground truth data $\boldsymbol{y}_k$. But it is not necessary to do so in some other cases[67,140] like unsupervised training. Optical acquisition of these data usually takes the most time, and requires the optical instruments in use to be stable during the long period. Otherwise, the data pairs cannot be registered well enough to match each other[51]. However, since holography is extremely sensitive to environment vibration[10], it is unavoidable to capture such vibration in the holograms during the time of acquisition (usually tens of hours depending on the number of holograms required to train the DNN), resulting in the instability of the fringe patterns. However, we have shown that DNN can be well trained on these "noisy" data[52].

An alternative and more flexible way to generate the training data is to use a numerical simulator provided that the physical system that describes the data link from the source to the detector can be accurately modeled. For example, this strategy has been applied to phase unwrapping[72] as well as speckle removal[66,68], ghost imaging[131], STORM[144,145] and diffraction tomography[146].

The raw data are mainly taken from MNIST[147], Faces-LFW[148] and CelebAMask-HQ[149], which are publicly available. In most of the proof-of-concept experiments, a spatial light modulator (SLM) is used to display these images in order to form the holograms of them by using a standard holographic system. However, in most of the practical applications of holography, it is not the 2D hand-written digits, English letters[147], or 2D human faces[148,149] that are of interest. Thus, DNN trained on these data sets is difficult to be generalized[150] to cope with most of the objects in the real world.

Recently, Ulyanov et al. have shown that the structure of a generator network can capture a great deal of low-level image statistics prior to any learning[151]. This can be generalized to a more general DNN such as the U-Net by incorporating a physical model into it, resulting in an untrained neural network that does not require any data to train[55]. It can be used to reconstruct the holograms of realistic objects. Indeed, over the past year, untrained DNN has been applied by to solve problems of holographic reconstruction[55-57], phase unwrapping[73], phase microscopy[152], diffraction tomography[153] and imaging[154], and even ghost imaging[155]. I will discuss it in more detail later on.

Physics-informed DNN

So far I have introduced several important DNN models that are widely used in optics and holography. Indeed, DNN has been shown to outperform conventional physics-based approaches. For example, DNN allows twin-image-free reconstruction from a single-shot in-line digital holography[52]. A major reason for the success is that DNN, given enough data, can learn feature hierarchies with features from higher levels of the hierarchy formed by the composition of lower level features[156] even explicit formulation of a system's exact physical nature is impossible owing to its complexity[119,132].

However, it is also well-known that DNN has a black-box issue[157]: the information stored in DNN is represented by a set of weights and connections that provides no direct clues to how the task is performed or what the relationship is between inputs and outputs[158]. When it is used to solve real-world physical problems, DNN has met with limited success due to a number of reasons: First, DNN requires a large amount of labeled data for training, which is rarely available in real application settings[159]. As discussed above, for most of the leaning-based methods for optical imaging and holography, an SLM is required to display the ground-truths. Frequently, the publicly available dataset such as the MNIST[147] database is used for demonstration. But this is hard to generalize to real-world samples owing to the issue that DNN models can only capture relationships in the available training data[150]. Second, DNN models often produce physically inconsistent results[160] when violating fundamental constraints. Third, the output is unexplainable[161].

Thus, it is highly desirable to take the benefits of both DNN models and physics models, and develop physics-informed or physics-guided DNN[162-165]. Barbastathis and coworkers[45] have concluded three different ways to incorporate a physical model into DNN, namely, recurrent physics-informed DNN, cascaded physics-informed DNN, and single-pass physics-informed DNN. In contrast, Ba and coworkers have concluded four different ways[166]: physical fusion, residual physics, physical regularization, and embedded physics. One can see that both these two ways of classification are somewhat equivalent.

According to Ba et al.[166], physical fusion is the most straightforward way. It feeds directly the solution from a physics model as (part of) the input to a DNN model. Barbastathis and coworkers[45] term this method as single-pass physics-informed DNN. This strategy has been employed in the very first work on learning-based holographic reconstruction, in which Rivenson et al.[51] used a conventional diffraction-based algorithm[5] to reconstruct a blurred wavefront from a hologram, and then used a trained DNN to improve the quality. This method has also been used for other problems such as ghost imaging[117] and phase retrieval[167].

In contrast, residual physics is to add the physical solution to the DNN output so that the DNN model only needs to learn the mismatch between the model-based solution and the ground truth[168]. Physical regularization, on the other hand, harnesses the regularization term from a set of physical constraints to penalize the network solutions. The regularization term can be appended as part of the loss function explicitly or through a reconstruction process from physics[160]. These two concepts are similar to the recurrent physics-informed DNN and cascaded physics-informed DNN discussed in[45].

More exciting is the strategy of embedded physics. As shown in Fig. 4, the central idea is to take the physical model inside the network optimization loop: the physical model takes care of the well-posed forward propagation while DNN, the ill-posed backward propagation, in each iteration[55–57,138,152–155]. The error between the forward calculated output and the measured data can be used to estimate a defined loss function, which is then used to update the weights based on an SGD-like algorithm.

Here I would also like to draw the attention of the readers to an emerging strategy, which I call network approximating physics. By the name, it is to approximate a physical model by using a DNN[78,169]. For example, Shi et al. proposed to approximate the Fresnel zone plates through successive application of a set of learned 3 × 3 convolution kernels[78] in order to build a DNN model that can approximate the Fresnel diffraction and occlusion.

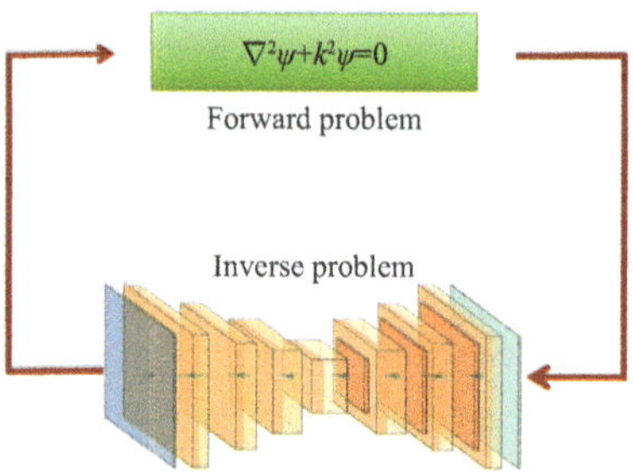

Fig. 4 A typical architecture of an embedded physics DNN.

DNN-inspired holography

After the brief introduction of deep learning neural networks in Sec. 1, now I will review some of the recent studies on the applications of deep learning in holography in this section. Before going into the detail, it is worthy of mentioning that the idea of using NN to for holography is not new. It has been proposed and demonstrated many years ago[170–174]. But the performance of neural networks was limited at that time because they were not deep enough due to the limited computation power. Indeed, one can find that some of the ideas demonstrated recently have been proposed at that time.

Digital holographic reconstruction

A hologram can be formed by the superposition of the object beam $u_o(x,y)$ that carries the information of an object of interest and a reference beam $u_r(x,y)$, where (x,y) is the spatial coordinates in the hologram plane

$$I(x,y) = |u_o(x,y) + u_r(x,y)|^2 = u_o^*(x,y)u_r(x,y) + |u_o(x,y)|^2 + |u_r(x,y)|^2 + u_o(x,y)u_r^*(x,y) \quad (26)$$

where the symbol * stands for phase conjugate.

Conventional approaches

Intuitive approaches for holographic reconstruction are based on the physical model of diffraction, i.e., the numerical calculation of the diffraction process of the wave field[6]. In the off-axis geometry with a sufficient high carrier frequency all the three terms in Eq. 26 are well separated in the Fourier space, and therefore one can simply apply a spatial filter to remove the two unwanted terms. However, spatial filtering inevitably results in the loss of high-frequency components, which greatly hinder the reconstructed image quality[32]. In addition, one can use only a small part of the spatial bandwidth product (SBP) that the camera can offer[175–177] in this case. In in-line DH the reconstructed image are overlapped with the twin-image and the zeroth-order terms. Since the removal of the zeroth-order is comparatively straightforward, most of the studies on in-line holographic reconstruction is to deal with the twin image term.

Physics-based approach relies on some physical models as suggested by the name. Back in 1951, Bragg and Rogers[13] had realized that the twin image is actually the out-of-focus copy of the reconstructed object image, and it can be eliminated by the subtraction of the defocused wavefront from the other. But this method is technically tricky, and can be implemented only after the invention of DH[178,179]. The most widely used strategy nowadays is to tune some physical parameters of the optical system and acquire the corresponding holograms so as to set up a small linear equation system that relates the recorded holograms and the tuning parameters and solve for the object wavefront. For example, one can introduce multiple phase retardations stepwise in the reference beam and acquire the phase-shifted holograms[20–22], or move the camera along the propagation direction[23–25], or slightly tune the wavelength of the illumination laser beam[26]. However, as the control of these parameters is extremely difficult for very short wavelength radiation, these methods are infeasible for electron holography[180], X-ray holography[181], or γ–ray holography[182]. In this case, one should implement the phase shift by using an amplitude element such as the Chinese Taiji lens[183] or the Greek-ladder zone plate[184].

Mathematically, the twin image artifact arises due to the missing of the phase when the hologram is recorded[185]. This suggests that the twin image artifact can be resolved if the missing phase of the hologram $|u_0(x,y)+u_r(x,y)|$ can be retrieved. This is the fundamental logic behind the phase-retrieval approach. Effectively, phase retrieval can be solved by using either a deterministic algorithm that is called now the transport-of-intensity equation (TIE)[186], or an iterative algorithm such as the Gerchberg-Saxton (GS)[27] and/or the Hybrid-Input-Output (HIO) algorithm[28]. This is in particular useful when the coherence of radiation source in used is poor (X-ray, for example). Thus the communities of X-ray holography and electron holography have made intensive studies since the late 1980s[185,187–190]. Along with the improvement of the technique and better modeling of the objective function, people now can achieve the reconstruction of the whole wavefront[29–31].

Phase retrieval is actually an inverse source problem of image reconstruction from magnitude[191,192]. It can be formulated as a more general class of inverse problems. The inverse problem approach treats the DH image reconstruction as a pure digital signal processing (DSP) problem, and solves it by using various numerical algorithms, such as statistical model[32,33], sparsity-enforcing prior[34], least squares[35], regularization[36,37], and compressive sensing[38–40]. A critical issue with it, from the computational point of view, is that the two-dimensional (2D) hologram must be rearranged as a one-dimensional (1D) vector in contrast to treating it as a 2D array in the two other aforementioned approaches. It thus requires the calculation of very large matrices, which is too heavy to do efficiently[193].

Learning-based approaches

Several strategies have been proposed to solve the problem of holographic reconstruction. The most straightforward approach is the end-to-end DNN[52,54]. For example, Wang et al.[52] took the advantages of ResNet[101] and U-Net[126], and developed an alternative approach called eHoloNet for end-to-end holographic reconstruction. eHoloNet receives the raw digital hologram as the input, and produces the artifact-free object wavefront, which is a phase profile in their study, as the output. They treat the holographic reconstruction as solving an ill-posed inverse problem for the function $\mathcal{R}$ that maps directly the hologram space to the object space

$$\mathcal{R}_{\text{learn}} = \underset{\mathcal{R}_\theta,\theta\in\Theta}{\arg\min} \sum_{n=1}^{N} \mathcal{L}(u_{o,n}, \mathcal{R}_\theta\{I_n\}) + \Psi(\theta) \qquad (27)$$

where θ is an explicit setting of the network parameters Θ, $\mathcal{L}(\cdot)$ is the loss function to measure the error between the nth phase object $u_{o,n}$, and the corresponding in-line hologram $\mathcal{R}_\theta\{I_n\}$, and $\Psi(\theta)$ is a regularizer on the parameters with the aim of avoiding overfitting[194]. They demonstrated their approach using 10,000 handcraft images from the MNIST dataset[147] and 12,651 images of the USAF resolution chart. All these images were resized to 768 × 768 pixels and displayed on a phase-only SLM (Holoeye, LETO), making them effectively phase objects. The in-line digital holograms of all the 22,651 phase objects were acquired by a Michelson interferometer. 9000 pairs of handcraft images and their holograms and 11,623 pairs of resolution charts and their holograms were used to train the eHoloNet, respectively. The lefts were used for test.

In order to reconstruct both the intensity and phase simultaneously from a single digital hologram, Wang et al. proposed a Y-shaped architecture[54]. The loss function then is defined as

$$\mathcal{L} = \lambda\mathcal{L}_I + \mathcal{L}_P \qquad (28)$$

where $\mathcal{L}_I$ and $\mathcal{L}_P$, defined according to Eq. 15, denote the loss function of the intensity and phase of the complex wavefront, and the weight $\lambda = 0.01$ in their experiments so as to enforce the significance of the phase.

The end-to-end approach can be implemented via GAN[67,137] as well. One advantage to use GAN is that the training data do not need to pair up.

The second approach is the physics fusion or single-pass physics-informed DNN[51,53]. As discussed in Sec., this is a two-step process. First, the complex wavefront was reconstructed by using the conventional numerical free space propagation back to the object plane. As aforementioned, the reconstructed wavefront is usually overlapped with the twin image, and the zeroth-order artifacts. The amplitude and phase of the reconstructed wavefront were then sent separately into a DNN, which has been trained to remove all these artifacts[51]. In their study, Rivenson et al. adopted a network architecture based on ResNet[101], as shown in Fig. 5a. The network was trained by the directly reconstructed amplitude and phase using numerical free space propagation algorithm and the corresponding ground truths (which are reconstructed by using phase retrieval algorithms from multiple holograms[195,196]). 100 image pairs were used to train the network. The results are shown in Fig. 5b. This method can be applied to off-axis DH to improve the quality of the reconstructed image as well[53].

The third one, physics-informed DNN is an exciting approach for holographic reconstruction. For example, Wang et al. have proposed a physics-enhanced DNN (PhysenNet)[55] that employs a strategy of incorporating a physical imaging model into a conventional DNN.

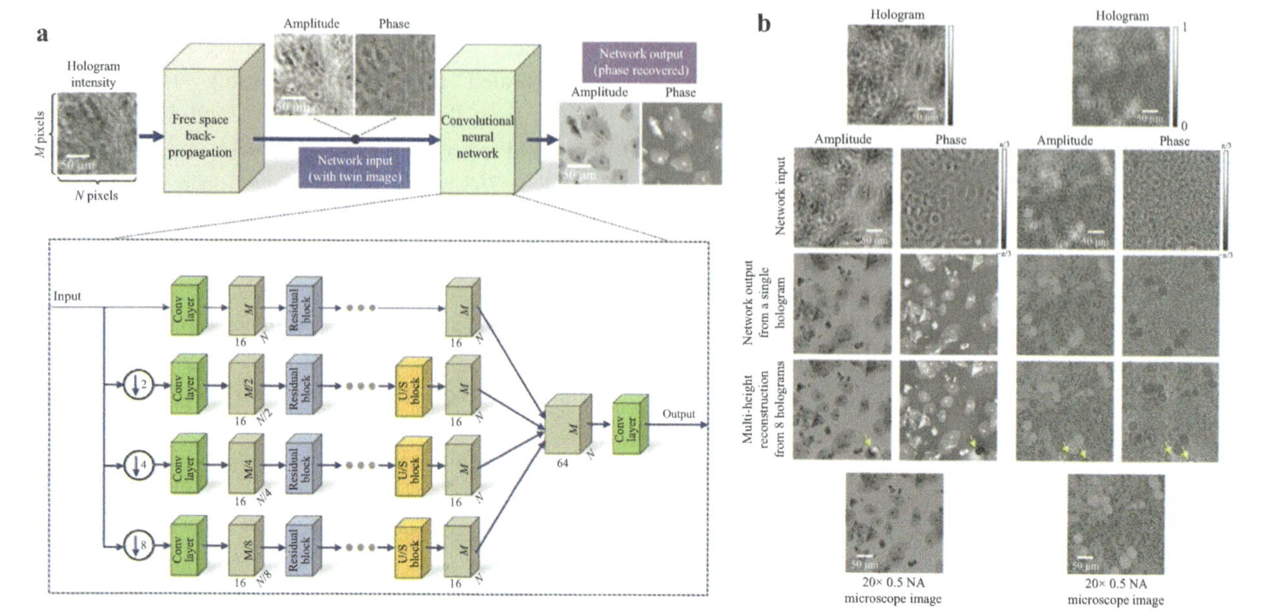

Fig. 5 **a** The DNN architecture and the pipeline for holographic reconstruction proposed by Rivenson et al.51 and **b** the experimental results. (adapted from 51).

PhysenNet has two apparent advantages. First, it does not need any data to pre-train. This can be clearly seen in the objective function

$$\mathcal{R}_{\theta^*} = \underset{\theta\in\Theta}{\arg\min}\ \mathcal{L}(H(\mathcal{R}_\theta\{I\}, I) \tag{29}$$

where I is the hologram or intensity pattern from which we wish to reconstruct the phase, and H is the physical model, which is the Fresnel transform in their explicit case. It can be any other image formation process that can be accurately modeled[55–57,73,152–155]. Eq. 29 suggests that PhysenNet just requires the data to be process (I in this case) as its input. The interplay between the physical model and the randomly initialized DNN provides a mechanism to optimize the network parameters, and produce a good reconstruction. Second, the reconstructed image satisfies the constraint imposed by the physical model so that it is interpretable[163]. The experimental results are plotted in Fig. 6.

The DNN model in PhysenNet can be replaced by other neural networks dependent on the task in hand. For example, Zhang et al. have demonstrated the incorporation of a phase imaging model into GAN[138].

Phase unwrapping

Holographically reconstructed phases are usually wrapped owing to the 2π-phase ambiguities and thus need unwrapping, which is also a typical ill-posed inverse problem. Conventional phase unwrapping techniques estimate the phase either by integrating through the confined path (referred to as path-dependent methods) or by minimizing the energy function between the wrapped phase and the approximated true phase (referred to as minimum-norm approaches)[72]. DNN provides a very feasible solution to this kind of problem because it can resolve the issues such as error accumulation, high computational time and noise sensitivity that conventional techniques frequently encounter.

Actually, the idea of using neural networks for phase unwrapping has been proposed by Takeda et al.[171] and Kreis et al.[172,173] in the 1990s. But with the developments of DNN techniques and computer power, much deeper neural networks are available now. There are ways to treat the phase unwrapping problem from the DNN point of view. A straightforward way is to take it as a regression problem, and develop a DNN to map a wrapped phase to an unwrapped phase. This can be done, for example, by using a U-Net trained on labeled data[71]. One research line is to improve the network design, aiming to enhance the phase quality. For example, Zhang et al.[198] have proposed a DNN model called DeepLabV3+, which can achieve noise suppression and strong feature representation capabilities. They demonstrated that it is out-performed the conventional path-dependent and minimum-norm algorithms. It is also possible to unwrap a phase by using an untrained DNN in a way similar to PhysenNet[55]. For example, Yang et al. have experimentally demonstrated that the proposed method faithfully recovers the phase of complex samples on both real and simulated data[73].

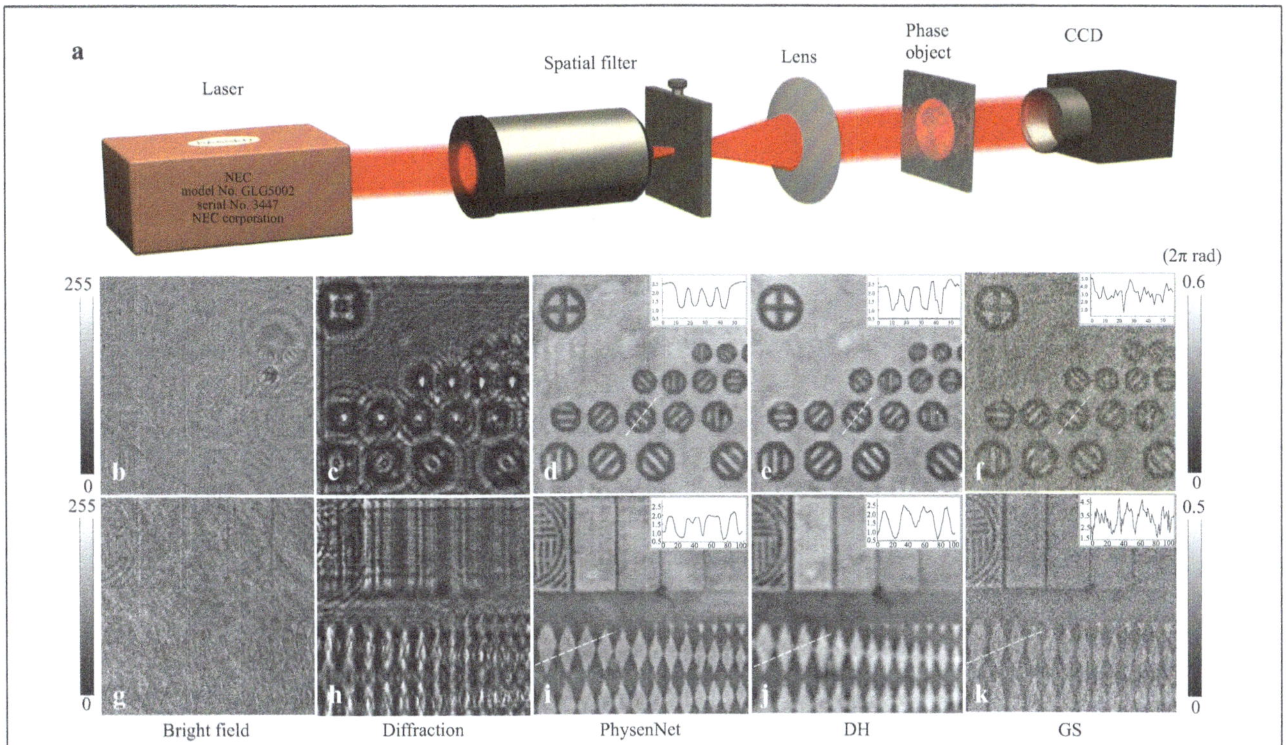

Fig. 6 Experimental results. **a** Experimental set-up. **b** and **g** show two different parts of the phase object, **c** and **h** show the diffraction patterns, **d** and **i** show the phase images reconstructed by PhysenNet, **e** and **j** show the phase images reconstructed via off-axis digital holography, and **f** and **k** show the phase images reconstructed with the GS algorithm. (after 55).

Alternatively, one can treat phase unwrapping as a classification problem. For example, Zhang et al.[197] have demonstrated it by transferring phase unwrapping into a multi-class classification problem and introduced an efficient segmentation network to identify the classes. Their experimental results are plotted in Fig. 7.

Learning-based phase unwrapping algorithms have been applied to solve the problems in many different fields of studies, such as biology[199] and Fourier domain Doppler optical coherence tomography[200].

Autofocusing

Autofocusing is about the automatic determination of the numerical calculation of the free space propagation distance of the wavefront from the hologram plane[201]. This is in particular important for the applications of DH in industrial and biological inspection[202]. Conventionally, the focused distance is determined by a criterion function with respect to the reconstruction distance. The criterion function can be defined in many ways, such as the entropy of the reconstructed image, the magnitude differential[201], and sparsity[203], and usually has a local maximum or minimum value at the focal plane.

Learning-based autofocusing algorithms employ different strategies. The prediction of the focusing distance is not made by searching a local extreme value of a criterion function, but by directly analyzing a digital hologram by using a deep neural network. One can think of autofocusing as a regression problem or a classification problem. The regression approach is to train the network by using a stack of artifact-free reconstructed images that are paired up with a hologram[59–61]. Each image in the stack is associated with a number that indicates the reconstruction distance. All these numbers are used to rectify the output layer during the training process. Taking the advantage of the U-Net[126] and ResNet[101], Wu et al[65] proposed the HIDEF (Holographic Imaging using Deep learning for Extended Focus) CNN. This allows the direct reconstruction at the correct distance when a hologram is inputted to the trained HIDEF CNN. Jaferzadeh et al. proposed a DNN model with a regression layer as the top layer to estimate the best reconstruction distance[204].

The classification approach was proposed by Ren et al[62] and Shimobaba et al[63]. For example Ren and coworkers experimentally recorded the 5000 holograms of several objects (a resolution chart, a testis slice, a ligneous dicotyledonous stems, an earthworm crosscut, etc.) at 10 different distances; and use the holograms and the associated distance values to train their neural network.

An alternative strategy is to take the focusing distance as

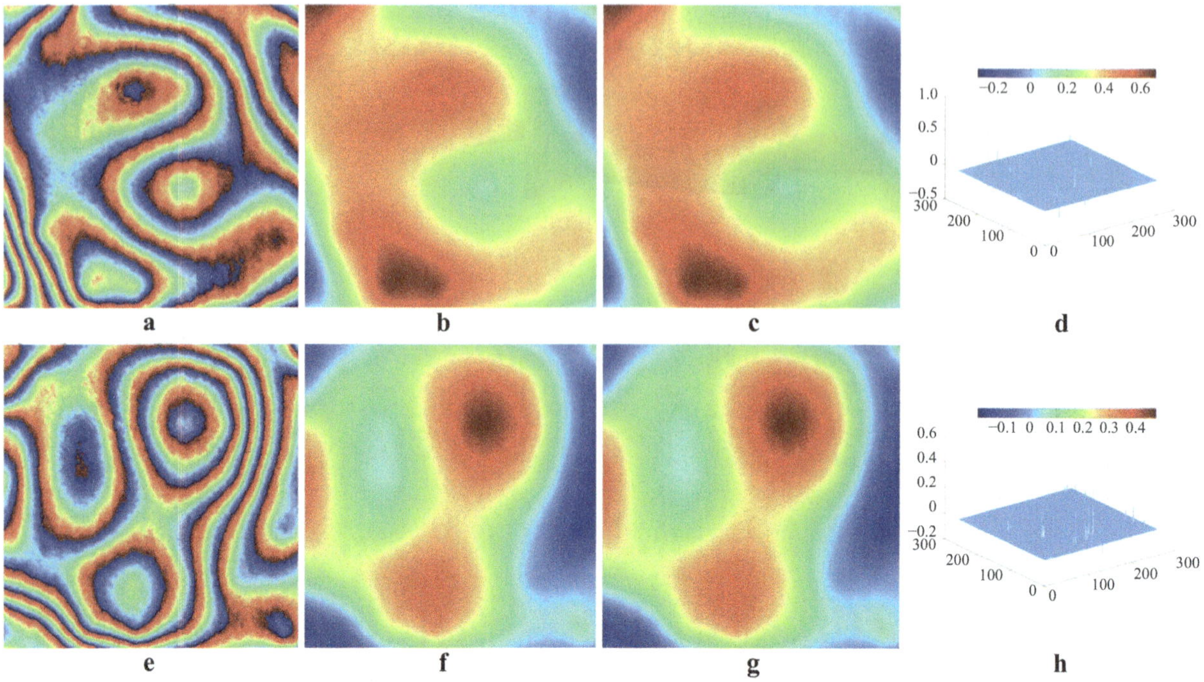

Fig. 7 Unwrapping results on real data. From left to right are: wrapped phases [input, **a**, **e**], reconstructed unwrapped phases by the DNN [**b**, **f**] and MG [**c**, **g**], and differences [**d**, **h**]. (after 197).

an uncertain parameter, and ask the neural network to optimize automatically[135]. In this case, the objective function can be written as

$$[\mathcal{R}_{\theta^*}, d] = \underset{\theta \in \Theta,\, d}{\arg\min}\, \mathcal{L}(H[\mathcal{R}_\theta(I), d], I) \qquad (30)$$

where the uncertain focusing distance d enters the the physical model H now, and will be optimized by the network.

The objective function in the form of Eq. 30 is similar to Eq. 29. This suggests that the only input required by the neural network is the hologram I, and the DNN does not need to be pre-trained on any dataset. As shown in Fig. 8, the algorithm will converge to the exact distance value along with proceeding of the iteration.

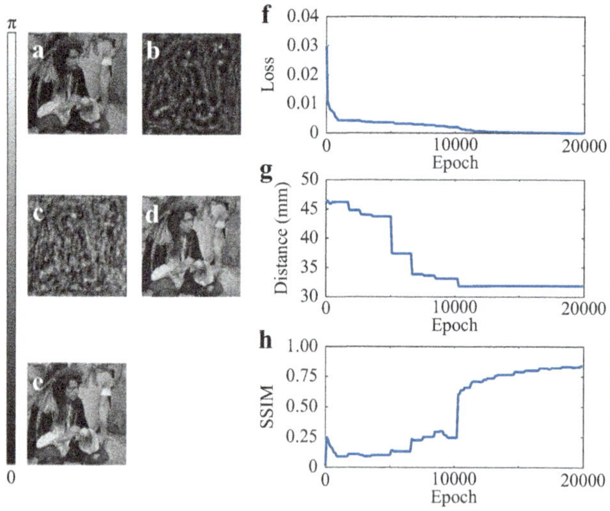

Fig. 8 The reconstruction process of a phase object. **a** the original phase object. **b** the diffraction pattern. The retrieved phase at the epoch of **c** 900, **d** 10300, and **e** 19800. The behavior of **f** the loss function, **g** estimated distance, and **h** the SSIM value as a function of the number of epoch. (after 135).

Phase Aberration Compensation

Learning-based approaches have also been used for phase aberration compensation in digital holographic microscopy[58,152,205]. Again, phase aberration compensation can be formulated as a classification[58] or a regression[152,205] problem. In the work by Nguyen et al.[58], the role DNN plays is to segment the reconstructed and unwrapped phase. The phase aberration then can be determined by Zernike polynomial fitting, and its conjugate can be numerically calculated to compensate the aberration. Nguyen et al. experimentally took the holograms of 306 breast cancer cells as the input and the corresponding manually segmented maps as the output to train their neural network, which is also a U-Net + ResNet architecture in this case. They used the softmax function defined in Eq. 6 in the last layer in their neural network to calculate the prediction probability of background/cell potential, and the cross-entropy loss defined in Eq. 17 for back propagation. The experimental results are plotted in Fig. 9.

In contrast, the regression approach proposed by Xiao et al.[205] endeavors to optimize the coefficients for constructing the phase aberration map that act as responses corresponding to the input aberrated phase image. Embedded physics DNN can be used for this problem as well. Encapsulating the image prior and the system physics, Bostan et al.[152] have proposed an untrained DNN that can simultaneously reconstruct the phase and pupil-plane aberrations by fitting the weights of the network to the captured images.

Suppression of Speckle

As a coherent imaging modality, DH reconstruction is also influenced by the coherence of the illumination laser source[11,206], which naturally results in speckle[207]. The elimination of speckle noise has been one of the main issues in DH. Conventionally, this can be done either optically or digitally. Optical methods usually require multiple measurements under different conditions. Digital methods can work on a single hologram but the reduction of speckle results in the loss of information as well. Bianco et al have given a very nice review of the most important speckle removal techniques[208].

Recently, Jeon et al.[66] have demonstrated that, by using DNN, it is possible to remove the speckle without any degradation of the image quality. The network architecture they used is again the combination of U-Net and ResNet. For supervised learning, one needs to pair up the speckled images and the corresponding speckle-free ones in order to train the network. But speckle-free images are unlikely to be obtainable from experimentally acquired holograms. So they used numerically generated speckled images from speckle-free images according to the model $\boldsymbol{y} = \mathcal{R}(\boldsymbol{x}) + \mathcal{N}(0, \varsigma^2)$, where $\mathcal{R}(\varsigma)$ is the Rayleigh distribution with scale parameter ς, and $\mathcal{N}(\mu, \varsigma^2)$ is the Gaussian distribution with the mean μ and standard deviation ς, to train their network, and test it with experimentally acquired holograms. Similar DNN can be applied to remove the speckle noise in phase image from holographic interferometry[68,69]. The strict requirement of labeled data can be released by using a more suitable network architecture such as Noise2Noise[209]. For example, Yin et al.[210] have demonstrated a speckle removal DNN without using clean data.

Computer-generated holography

CGH has been recognized as the most promising true-3D display technology since it can account for all human visual cues such as stereopsis and eye focusing[8,9,88,89] as well as a powerful tool for the test of optical elements[84–86]. In particular, for the application in holographic display, it requires the generated holograms to be reasonably large in size. But the calculation of such holograms within acceptable time has been one of the main challenges in this field[211]. Although iterative phase-retrieval algorithms[27,28,87,212] have been intensively employed for this task, modern approaches for CGH calculation are non-iterative[88,89,211,213], in combination of acceleration techniques such as look-up table[90,214] and the use of GPU[215].

The use of DNN has dramatically accelerated the calculation of CGH[74–76,78–81]. People have used U-Net-based architecture to generate phase-only holograms[74] and binary holograms[216], Y-shaped architecture to generate multi-depth

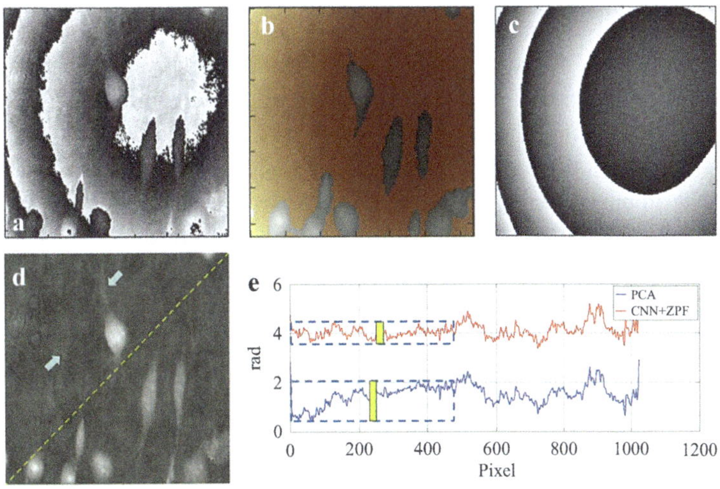

Fig. 9 a Phase aberration, **b** unwrapped phase overlaid with CNN's image segmentation mask, where background (color denoted) is fed into ZPF, **c** conjugated residual phase using CNN + ZPF, **d** fibers are visible after aberration compensation and are indicated by blue arrows, and **e** phase profile along the dash line in **d**. Yellow bars denote the flatness of region of interest. (after 58).

holograms[76], and autoencoder-based DNN for the fast generation of high-resolution holograms[80,81]. DNN has also been used to improve the quality of holographic display[217], and it allows to train in the loop[77]. Eybposh et al. have demonstrated an unsupervised learning based on GAN to achieve a fast hologram computation[75], although there is argument that this indirect training strategy may not obtain an optimal hologram[81]. The superb DNN-based algorithms allow the design of CGH to generate not just scalar but even arbitrary 3D vectorial fields in an instant and accurate manner[218].

A DNN-based CGH synthesis technique called tensor holography for true 3D holographic display has been proposed recently by Shi et al.[78]. Tensor holography is a physics-informed DNN technique. It imposes underlying physics (Fresnel diffraction) to train a CNN as an efficient proxy for both. Tensor holography was trained on MIT-CGH-4K Fresnel holograms dataset, consisting of 4000 pairs of RGB-depth (RGB-D) images and the corresponding 3D holograms that take the occlusion effect into account. Thus their DNN takes the 4-channel RGB-D image as its input, and predicts a color hologram as a 6-channel image (RGB amplitude and RGB phase), which can be used to drive three optically combined SLMs or one SLM in a time-multiplexed manner to achieve full-color holographic display.

Holography-inspired DNN

Holography has been one of the important avenues to implement optical neural networks (ONN). Early researches include the optical implementations of fully-connected neural networks[93–97] and the Hopfield model[91,92], which is the base of the recurrent neural network (RNN). Rather than using logical neurons as in the digital counterpart, holographic neural networks rely on interconnection[219] that a hologram is inherently capable of[96,97]. In a fully-connected holographic neural network, the weights are stored in the pixels (neurons) of the holograms. Each neuron in a layer (hologram) performs a simple modulation of the light that impinges onto it from an upstream layer and subsequently illuminates a downstream layer. A holographic neural networks can be created in a photorefractive crystal[93,94], which is inherently a 3D device that has a potential to store billions of weights. Thus it is in principle promising to solve large-scale inverse problems. However, interest in further developments has been on the wane owing to some unfortunate reasons[220].

Modern implementation of holographic neural networks takes the advantage of diffraction[98], and thus named diffraction deep neural networks (D^2NN). In the hardware implementation, the holograms in D^2NN are actually diffractive optical elements (DOE), which can be fabricated using 3D printing (see Fig. 10)[98], nanomanufacturing[221], or electric addressable digital micro-mirror devices (DMD)[222]. The weights are stored in the pixels as in conventional holographic neural networks[93–97]. There are good reasons to implement D^2NN on DMD and SLM: it allows the creation of parallel and reconfigurable network connections as well as the addition of nonlinearity to each layer via fast square-law detection. Such cascaded nonlinear operations strongly amplify the dimensionality of data representation so that they can leverage for challenging computations[222].

As many other optical neural networks[220], the original D^2NN engine was trained off-line[98]. But efforts have been made to implement better training strategy. For example, Zhou et al. have demonstrated an *in situ* back propagation training[223], Xiao et al. have proposed a back propagation technique that updates the unitary weights through the gradient translation from Euclidean to Riemannian space[224], and a method to implement optical dropout[225].

People have also investigated the applications of D^2NN for logic operations[226,227], optical information processing[228,229], holographic reconstruction[230], pulse shaping[231], spectrally encoded single-pixel machine vision[232].

Conclusion and perspective remarks

To conclude, I have reviewed the recent progresses on the field of *deep holography*, describing how holography and deep neural networks can benefit from each other. I would like to emphasize that this is a rapidly developing field. New and exciting results are published every few days. It is impossible to cover all the works within a single literature review article. It is also challenging to catch up with all the progresses. But I do identify a few lines of trend for further studies. For DNN-inspired holography, instead of trying different DNN architectures, an important trend is to incorporate a physical model into a DNN model[162–166]. Indeed, this idea has been received intensive attentions from the researchers in diverse fields[160–165] and will be a future direction. I have discussed five different ways of incorporating a physical model and showed their applications in solving various problems with respect to holography. Of particular interest is PhysenNet as it does not require any data to train in advance and the network prediction is satisfied with the constraint imposed by the physical model. However, the optimization is slow in comparison to conventional data-driven DNN[55,73,152] and thus lightweight network architectures and more efficient training algorithms are highly demanded. Perhaps it is also possible to take the advantages of both the physics-

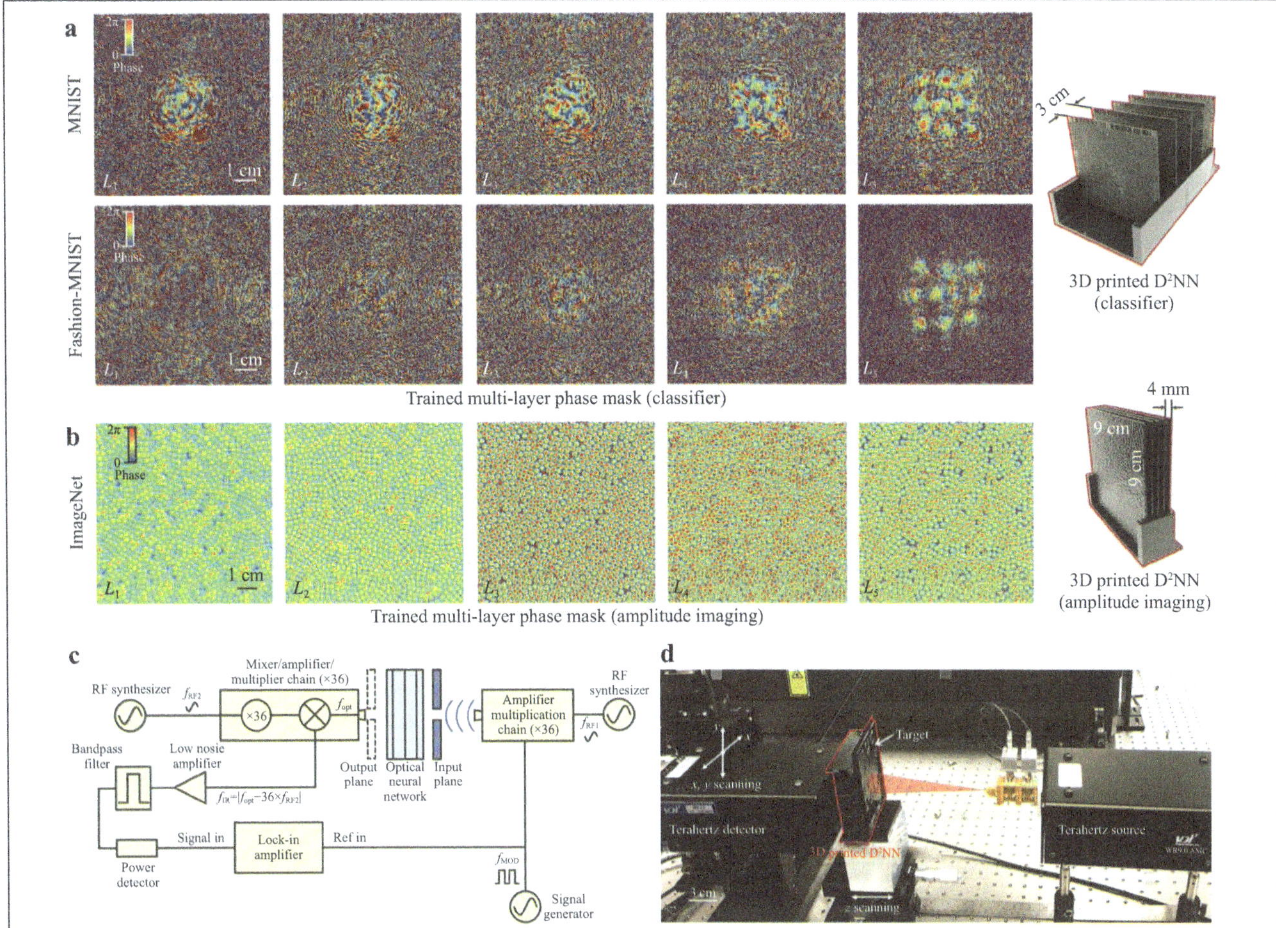

Fig. 10 Experimental testing of **a** 3D-printed D²NN. The optimized phase maps of five different layers of the MNIST classifier, fashion product classifier, and **b** the amplitude imager. Schematic **c** and photo **d** of the experimental terahertz setup. (after 98).

informed and conventional data-driven methods[233], so that the training can be more efficient and the generalization, significantly improved.

For holography-inspired DNN, most of the studies published so far focus on optical inference. The capability of light-speed processing in parallel of holographic neural networks indeed guarantees tremendous inference power, even outperforming Nvidia's top of the line Tesla V100 tensor core GPU in certain task[222]. But the advantage of optics cannot be fully utilized if on-line training of the network cannot be perform optically. Initial efforts have been making along this line[222–225]. But there are many possibilities out there to efficiently implement most of the fundamental functions in a DNN. Furthermore, the performance of D²NN is explicitly determined by the pixel number, pixel size, and frame rate of SLM as these factors are related to the scale and reconfigurability of matrix computation and the capability of interconnection. Current liquid crystal devices are too slow; DMD is fast but its pixel pitch is too large. One promising way to get around relies on the advancing of novel optical materials and devices that can modulate the wave in the sub-wavelength scale[234] in high speed.

From a higher level point of view, all the DNN methods I have discussed herein belong to a wide class of algorithms called neural computing, which itself is also fast evolving. Indeed, one can have an impression of the evolution of DNN from feedforward NN to CNN and U-Net, and so on. Neural computing provides us more powerful algorithms such as the spiking neurosynaptic networks (SNN)[235] that can model the behavior and learning potential of the brain. The applicability and potential of these new algorithms in holography are a still open question.

Nevertheless, I hope I have convinced you that the field of deep holography as a whole is rich and exciting. It is a cross-disciplinary that requires holography and neural computing and many others. The mergence of them stimulates the development of each other, and gives us a fantastic field to explore.

Acknowledgements
This work was supported by the National Natural Science Foundation of China (62061136005, 61991452), the Sino-German Center (GZ1391), and the Key Research Program of Frontier Sciences of the Chinese Academy of Sciences (QYZDB-SSW-JSC002).

Author contributions
G.S. wrote the paper.

Conflict of interest
The author declares that he has no conflict of interest.

Received: 01 September 2021 Revised: 06 February 2022 Accepted: 15 February 2022
Accepted article preview online: 16 March 2022
Published online: 30 March 2022

References

1. Gabor, D. A new microscopic principle. *Nature* **161**, 777-778 (1948).
2. Leith, E. N. & Upatnieks, J. New techniques in wavefront reconstruction. *Journal of the Optical Society of America* **51**, 1469-1473 (1961).
3. Leith, E. N. & Upatnieks, Reconstructed wavefronts and communication theory. *Journal of the Optical Society of America* **52**, 1123–1130 (1962).
4. Denisyuk, Y. N. On the reflection of optical properties of an object in a wave field of light scattered by it. *Dokl. Akad. Nauk SSSR* **144**, 1275-1278 (1962).
5. Schnars, U. & Jüptner, W. Digital Holography: Digital Hologram Recording, Numerical Reconstruction, and Related Techniques. (Berlin: Springer, 2005).
6. Kim, M. K. Digital holographic microscopy. in Digital Holographic Microscopy (ed Kim, M. K.) (New York: Springer, 2011).
7. Asundi, A. Digital Holography for MEMS and Microsystem Metrology. (Hoboken: Wiley, 2011).
8. Benton, S. A. & Bove, V. M. Jr. Holographic Imaging. (Hoboken: John Wiley & Sons, 2008).
9. Matsushima, K. Introduction to Computer Holography: Creating Computer-Generated Holograms as the Ultimate 3D Image. (Cham: Springer, 2020).
10. Caulfield, H. J. Handbook of Optical Holography. (New York: Academic Press, 1979).
11. Claus, D., Iliescu, D. & Rodenburg, J. M. Coherence requirement in digital holography. *Applied Optics* **52**, A326-A335 (2013).
12. Gabor, D. Holography, 1948-1971. *Proceedings of the IEEE* **60**, 655–668 (1972).
13. Bragg, W. L. & Rogers, G. L. Elimination of the unwanted image in diffraction microscopy. *Nature* **167**, 190-191 (1951).
14. Goodman, J. W. & Lawrence, R. W. Digital image formation from electronically detected holograms. *Applied Physics Letters* **11**, 77-79 (1967).
15. Schnars, U. & Jüptner, W. Direct recording of holograms by a CCD target and numerical reconstruction. *Applied Optics* **33**, 179-181 (1994).
16. Pedrini, G. & Tiziani, H. J. Short-coherence digital microscopy by use of a lensless holographic imaging system. *Applied Optics* **41**, 4489-4496 (2002).
17. Zhang, Y. Z. et al. Application of short-coherence lensless Fourier-transform digital holography in imaging through diffusive medium. *Optics Communications* **286**, 56-59 (2013).
18. Rosen, J. & Brooker, G. Fresnel incoherent correlation holography (FINCH): a review of research. *Advanced Optical Technologies* **1**, 151-169 (2012).
19. Cuche, E., Marquet, P. & Depeursinge, C. Spatial filtering for zero-order and twin-image elimination in digital off-axis holography. *Applied Optics* **39**, 4070-4075 (2000).
20. Yamaguchi, I. & Zhang, T. Phase-shifting digital holography. *Optics Letters* **22**, 1268-1270 (1997).
21. Meng, X. F. et al. Two-step phase-shifting interferometry and its application in image encryption. *Optics Letters* **31**, 1414-1416 (2006).
22. Liu, J. P. & Poon, T. C. Two-step-only quadrature phase-shifting digital holography. *Optics Letters* **34**, 250-252 (2009).
23. Zhang, Y. et al. Reconstruction of in-line digital holograms from two intensity measurements. *Optics Letters* **29**, 1787-1789 (2004).
24. Situ, G. et al. Generalized in-line digital holographic technique based on intensity measurements at two different planes. *Applied Optics* **47**, 711-717 (2008).
25. Das, B. & Yelleswarapu, C. S. Dual plane in-line digital holographic microscopy. *Optics Letters* **35**, 3426-3428 (2010).
26. Ryle, J. P., L i, D. & Sheridan, J. T. Dual wavelength digital holographic laplacian reconstruction. *Optics Letters* **35**, 3018-3020 (2010).
27. Gerchberg, R. W. & Saxton, W. O. A practical algorithm for the determination of phase from image and diffraction plane pictures. *Optik* **35**, 237-246 (1972).
28. Fienup, J. R. Phase retrieval algorithms: a comparison. *Applied Optics* **21**, 2758-2769 (1982).
29. Latychevskaia, T. & Fink, H. W. Solution to the twin image problem in holography. *Physical Review Letters* **98**, 233901 (2007).
30. Latychevskaia, T. & Fink, H. W. Simultaneous reconstruction of phase and amplitude contrast from a single holographic record. *Optics Express* **17**, 10697-10705 (2009).
31. Rong, L. et al. Iterative solution to twin image problem in in-line digital holography. *Optics and Lasers in Engineering* **51**, 553-339 (2013).
32. Sotthivirat, S. & Fessler, J. A. Penalized-likelihood image reconstruction for digital holography. *Journal of the Optical Society of America A* **21**, 737-750 (2004).
33. Cetin, M., Karl, W. C. & Willsky, A. S. Edge-preserving image reconstruction for coherent imaging applications. *Proceedings of the International Conference on Image Processing. Rochester*, NY, USA: IEEE, 2002.
34. Denis, L. et al. Inline hologram reconstruction with sparsity constraints. *Optics Letters* **34**, 3475-3477 (2009).
35. Kamau, E. N. et al. Least-squares based inverse reconstruction of in-line digital holograms. *Journal of Optics* **15**, 075716 (2013).
36. Bourquard, A. et al. A practical inverse-problem approach to digital holographic reconstruction. *Optics Express* **21**, 3417-3433 (2013).
37. Schretter, C. et al. Regularized non-convex image reconstruction in digital holographic microscopy. *Optics Express* **25**, 16491-16508 (2017).
38. Brady, D. J. et al. Compressive holography. *Optics Express* **17**, 13040-13049 (2009).
39. Rivenson, Y., Stern, A. & Javidi, B. Overview of compressive sensing techniques applied in holography[Invited]. *Applied Optics* **52**, A423-A432 (2013).
40. Zhang, W. H. et al. Twin-image-free holography: a compressive sensing approach. *Physical Review Letters* **121**, 093902 (2018).
41. LeCun, Y., Bengio, Y. & Hinton, G. Deep learning. *Nature* **521**, 436-444 (2015).
42. Goodfellow, I., Bengio, Y. & Courville, A. Deep Learning. (Cambridge: MIT Press, 2016).
43. Goda, K., et al. AI boosts photonics and vice versa. *APL Photonics* **5**, 070401 (2020).

44. Jin, K. H. et al. Deep convolutional neural network for inverse problems in imaging. *IEEE Transactions on Image Processing* **26**, 4509-4522 (2017).
45. Barbastathis, G., Ozcan, A. & Situ, G. On the use of deep learning for computational imaging. *Optica* **6**, 921-943 (2019).
46. Jo, Y. et al. Quantitative phase imaging and artificial intelligence: a review. *IEEE Journal of Selected Topics in Quantum Electronics* **25**, 6800914 (2019).
47. Ma, W. et al. Deep learning for the design of photonic structures. *Nature Photonics* **15**, 77-90 (2021).
48. Wiecha, P. R. et al. Deep learning in nano-photonics: inverse design and beyond. *Photonics Research* **9**, B182-B200 (2021).
49. Rivenson, Y., W u, Y. C. & Ozcan, A. Deep learning in holography and coherent imaging. *Light: Science & Applications* **8**, 85 (2019).
50. Sheridan, J. T. et al. Roadmap on holography. *Journal of Optics* **22**, 123002 (2020).
51. Rivenson, Y. et al. Phase recovery and holographic image reconstruction using deep learning in neural networks. *Light: Science & Applications* **7**, 17141 (2018).
52. Wang, H., L yu, M. & Situ, G. eHoloNet: a learning-based end-to-end approach for in-line digital holographic reconstruction. *Optics Express* **26**, 22603-22614 (2018).
53. Zhang, G. et al. Fast phase retrieval in off-axis digital holographic microscopy through deep learning. *Optics Express* **26**, 19388-19405 (2018).
54. Wang, K. Q. et al. Y-Net: a one-to-two deep learning framework for digital holographic reconstruction. *Optics Letters* **44**, 4765-4768 (2019).
55. Wang, F. et al. Phase imaging with an untrained neural network. *Light: Science & Applications* **9**, 77 (2020).
56. Lawrence, H. et al. Phase retrieval with holography and untrained priors: tackling the challenges of low-photon nanoscale imaging. arXiv: 2012.07386 (2021).
57. Niknam, F., Qazvini, H. & Latifi, H. Holographic optical field recovery using a regularized untrained deep decoder network. *Scientific Reports* **11**, 10903 (2021).
58. Nguyen, T. et al. Automatic phase aberration compensation for digital holographic microscopy based on deep learning background detection. *Optics Express* **25**, 15043-15057 (2017).
59. Pitkäaho, T., Manninen, A. & Naughton, T. J. Performance of autofocus capability of deep convolutional neural networks in digital holographic microscopy. Proceedings of the Digital Holography and Three-Dimensional Imaging 2017. JeJu Island Republic of Korea: Optical Society of America, 2017.
60. Pitkäaho, T., Manninen, A. & Naughton, T. J. Focus prediction in digital holographic microscopy using deep convolutional neural networks. *Applied Optics* 58-A208 (2019).
61. Ren, Z. B., Xu, Z. M. & Lam, E. Y. Autofocusing in digital holography using deep learning. Proceedings of SPIE 10499, Three-Dimensional and Multidimensional Microscopy: Image Acquisition and Processing XXV. San Francisco, California, United States: SPIE, 2018.
62. Ren, Z. B., X u, Z. M. & Lam, E. Y. Learning-based nonparametric autofocusing for digital holography. *Optica* **5**, 337-344 (2018).
63. Shimobaba, T., Kakue, T. & Ito, T. Convolutional neural network-based regression for depth prediction in digital holography. Proceedings of the 27th International Symposium on Industrial Electronics. Cairns, QLD, Australia: IEEE, 2018.
64. Huang, L. Z. et al. Holographic image reconstruction with phase recovery and autofocusing using recurrent neural networks. *ACS Photonics* **8**, 1763-1774 (2021).
65. Wu, Y. C. et al. Extended depth-of-field in holographic imaging using deep-learning-based autofocusing and phase recovery. *Optica* **5**, 704-710 (2018).
66. Jeon, W. et al. Speckle noise reduction for digital holographic images using multi-scale convolutional neural networks. *Optics Letters* **43**, 4240-4243 (2018).
67. Yin, D. et al. Digital holographic reconstruction based on deep learning framework with unpaired data. *IEEE Photonics Journal* **12**, 3900312 (2020).
68. Yan, K. T. et al. Fringe pattern de-noising based on deep learning. *Optics Communications* **437**, 148-152 (2019).
69. Tahon, M., Montresor, S. & Picart, P. Towards reduced CNNs for de-noising phase images corrupted with speckle noise. *Photonics* **8**, 255 (2021).
70. Ren, Z. B. et al. Fringe pattern improvement and super-resolution using deep learning in digital holography. *IEEE Transactions on Industrial Informatics* **15**, 6179-6186 (2019).
71. Wang, K. Q. et al. One-step robust deep learning phase unwrapping. *Optics Express* **27**, 15100-15115 (2019).
72. Spoorthi, G. E., Gorthi, R. K. S. S. & Gorthi, S. PhaseNet 2.0: phase unwrapping of noisy data based on deep learning approach. *IEEE Transactions on Image Processing* **29**, 4862-4872 (2020).
73. Yang, F. S. et al. Robust phase unwrapping via deep image prior for quantitative phase imaging. *IEEE Transactions on Image Processing* **30**, 7025-7037 (2021).
74. Horisaki, R., Takagi, R. & Tanida, J. Deep-learning-generated holography. *Applied Optics* **57**, 3859-3863 (2018).
75. Eybposh, M. H. et al. DeepCGH: 3D computer-generated holography using deep learning. *Optics Express* **28**, 26636-26650 (2020).
76. Lee, J. et al. Deep neural network for multi-depth hologram generation and its training strategy. *Optics Express* **28**, 27137-27154 (2020).
77. Peng, Y. F. et al. Neural holography with camera-in-the-loop training. *ACM Transactions on Graphics* **39**, 185 (2020).
78. Shi, L. et al. Towards real-time photorealistic 3D holography with deep neural networks. *Nature* **591**, 234-239 (2021).
79. Kang, J. W. et al. Deep-learning-based hologram generation using a generative model. *Applied Optics* **60**, 7391-7399 (2021).
80. Liu, S. C. & Chu, D. P. Deep learning for hologram generation. *Optics Express* **29**, 27373-27395 (2021).
81. Wu, J. C. et al. High-speed computer-generated holography using an autoencoder-based deep neural network. *Optics Letters* **46**, 2908-2911 (2021).
82. Lohmann, A. W. & Paris, D. P. Binary Fraunhofer holograms, generated by computer. *Applied Optics* **6**, 1739-1748 (1967).
83. Brown, B. R. & Lohmann, A. W. Computer-generated binary holograms. *IBM Journal of Research and Development* **13**, 160-168 (1969).
84. Birch, K. G. & Green, F. J. The application of computer-generated holograms to testing optical elements. *Journal of Physics D: Applied Physics* **5**, 1982-1992 (1972).
85. Osten, W., Baumbach, T. & Jüptner, W. Comparative digital holography. *Optics Letters* **27**, 1764-1766 (2002).
86. Pruss, C., et al. Computer-generated holograms in interferometric testing. *Optical Engineering* **43**, 2534-2540 (2004).
87. Wyrowski, F. Diffractive optical elements: iterative calculation of quantized, blazed phase structures. *Journal of the Optical Society of America A* **7**, 961-969 (1990).
88. Slinger, C., Cameron, C. & Stanley, M. Computer-generated holography as a generic display technology. *Computer* **38**, 46-53 (2005).
89. Bove, V. M. Jr. Display holography' s digital second act. *Proceedings of the IEEE* **100**, 918-928 (2012).
90. Lucente, M. Interactive computation of holograms using a look-up

table. *Journal of Electronic Imaging* **2**, 28-34 (1993).
91. Hopfield, J. J. Neural networks and physical systems with emergent collective computational abilities. *Proceedings of the National Academy of Sciences of the United States of America* **79**, 2554-2558 (1982).
92. White, H. J. & Wright, W. A. Holographic implementation of a hopfield model with discrete weightings. *Applied Optics* **27**, 331-338 (1988).
93. Psaltis, D. et al. Holography in artificial neural networks. *Nature* **343**, 325-330 (1990).
94. Hsu, K. Y., L i, H. Y. & Psaltis, D. Holographic implementation of a fully connected neural network. *Proceedings of the IEEE* **78**, 1637-1645 (1990).
95. Keller, P. E. & Gmitro, A. F. Design and analysis of fixed planar holographic interconnects for optical neural networks. *Applied Optics* **31**, 5517-5526 (1992).
96. Jutamulia, S. & Yu, F. T. S. Overview of hybrid optical neural networks. *Optics & Laser Technology* **28**, 59-72 (1996).
97. Kaikhah, K. & Loochan, F. Computer generated holograms for optical neural networks. *Applied Intelligence* **14**, 145-160 (2001).
98. Lin, X., et al. All-optical machine learning using diffractive deep neural networks. *Science* **361**, 1004-1008 (2018).
99. Yan, T., et al. Fourier-space diffractive deep neural network. *Physical Review Letters* **123**, 023901 (2019).
100. Krizhevsky, A., Sutskever, I. & Hinton, G. E. ImageNet classification with deep convolutional neural networks. Proceedings of the 25th International Conference on Neural Information Processing Systems. Lake Tahoe Nevada: Curran Associates Inc., 2012, 1097-1105.
101. He, K. M. et al. Deep residual learning for image recognition. Proceedings of 2016 IEEE Conference on Computer Vision and Pattern Recognition. Las Vegas, NV, USA: IEEE, 2016.
102. He, K. M. *et al.* Identity mappings in deep residual networks. Proceedings of the 14th European Conference on Computer Vision. Amsterdam, The Netherlands: Springer, 2016.
103. Bergstra, J. & Bengio, Y. Random search for hyper-parameter optimization. *Journal of Machine Learning Research* **13**, 281-305 (2012).
104. Lee, S. et al. Background information of deep learning for structural engineering. *Archives of Computational Methods in Engineering* **25**, 121-129 (2018).
105. Hornik, K. Approximation capabilities of multilayer feedforward networks. *Neural Networks* **4**, 251-257 (1991).
106. Patterson, J. & Gibson, A. Deep Learning: A Practitioner's Approach. (Sebastopol, CA: O'reilly, 2017).
107. Hansen, C. Activation functions explained -GELU, SELU, ELU, ReLU and more. at https://mlfromscratch.com/activation-functions-explained/.
108. Lei, N. et al. Geometric understanding of deep learning. arXiv: 1805.10451 (2018).
109. Rumelhart, D. E., Hinton, G. E. & Williams, R. J. Learning representations by back-propagating errors. *Nature* **322**, 533-536 (1986).
110. Ionescu, C., Vantzos, O. & Sminchisescu, C. Matrix backpropagation for deep networks with structured layers. Proceedings of 2015 IEEE International Conference on Computer Vision. Santiago, Chile: IEEE, 2015, 2965-2973.
111. Rojas, R. Neural Networks: a Systematic Introduction. (Berlin, Heidelberg: Springer, 1996).
112. Yuan, Y. X. Step-sizes for the gradient method. *AMS/IP Studies in Advanced Mathematics* **42**, 785-796 (2008).
113. Kingma, D. & Ba, J. Adam: a method for stochastic optimization. arXiv: 1412: 6980 (2017).
114. Zhao, H., et al. Loss functions for image restoration with neural networks. *IEEE Transactions on Computational Imaging* **3**, 47-57 (2017).
115. Menard, S. Coefficients of determination for multiple logistic regression analysis. *The American Statistician* **54**, 17-24 (2000).
116. Xue, Y. J., et al. Reliable deep-learning-based phase imaging with uncertainty quantification. *Optica* **6**, 618-629 (2019).
117. Lyu, M. et al. Deep-learning-based ghost imaging. *Scientific Reports* **7**, 17865 (2017).
118. Lyu, M. et al. Exploit imaging through opaque wall via deep learning. arXiv: 1708.07881 (2017).
119. Lyu, M. et al. Learning-based lensless imaging through optically thick scattering media. *Advanced Photonics* **1**, 036002 (2019).
120. Hubel, D. H. & Wiesel, T. N. Receptive fields and functional architecture of monkey striate cortex. *The Journal of Physiology* **195**, 215-243 (1968).
121. Goodman, J. W. Introduction to Fourier Optics. (Englewood: Roberts & Company, 2004).
122. Voelz, D. G. Computational Fourier Optics: A MATLAB Tutorial. (Bellingham: SPIE, 2011).
123. Scherer, D., Müller, A. C. & Behnke, S. Evaluation of pooling operations in convolutional architectures for object recognition. Proceedings of the 20th International Conference on Artificial Neural Networks. Thessaloniki, Greece: Springer, 2010, 92-101.
124. Zeiler, M. D. et al. Deconvolutional networks. Proceedings of 2010 IEEE Computer Society Conference on Computer Vision and Pattern Recognition. San Francisco, CA, USA: IEEE, 2010.
125. Mohan, R. Deep deconvolutional networks for scene parsing. arXiv: 1411.4101 (2014).
126. Ronneberger, O., Fischer, P. & Brox, T. U-Net: convolutional networks for biomedical image segmentation. arXiv: 1505.04597 (2015).
127. Glorot, X. & Bengio, Y. Understanding the difficulty of training deep feedforward neural networks. Proceedings of the Thirteenth International Conference on Artificial Intelligence and Statistics. Chia Laguna Resort, Sardinia, Italy: PMLR, 2010.
128. Shimobaba, T. et al. Digital holographic particle volume reconstruction using a deep neural network. *Applied Optics* **58**, 1900-1906 (2019).
129. Zhang, Z. D. et al. Holo-UNet: hologram-to-hologram neural network restoration for high fidelity low light quantitative phase imaging of live cells. *Biomedical Optics Express* **11**, 5478-5487 (2020).
130. Chang, T. et al. Calibration-free quantitative phase imaging using data-driven aberration modeling. *Optics Express* **28**, 34835-34847 (2020).
131. Wang, F. et al. Learning from simulation: an end-to-end deep-learning approach for computational ghost imaging. *Optics Express* **27**, 25560-25572 (2019).
132. Zheng, S. S. et al. Incoherent imaging through highly nonstatic and optically thick turbid media based on neural network. *Photonics Research* **9**, B220-B228 (2021).
133. Zheng, S. S. et al. Non-line-of-sight imaging under white-light illumination: a two-step deep learning approach. *Optics Express* **29**, 40091-40105 (2021).
134. Zhou, Z. W. et al. Unet++: a nested U-net architecture for medical image segmentation. Proceedings of the Deep Learning in Medical Image Analysis - and - Multimodal Learning for Clinical Decision Support - 4th International Workshop, DLMIA 2018, and 8th International Workshop, ML-CDS 2018, Held in Conjunction with MICCAI 2018. Granada, Spain: Springer, 2018.
135. Zhang, X. Y., Wa ng, F. & Situ, G. BlindNet: an untrained learning approach toward computational imaging with model uncertainty. *Journal of Physics D: Applied Physics* **55**, 034001 (2022).
136. Goodfellow, I. J. *et al.* Generative adversarial networks. Proceedings of

the 27th International Conference on Neural Information Processing Systems. Montreal, Canada, 2014.

137. Moon, I. et al. Noise-free quantitative phase imaging in Gabor holography with conditional generative adversarial network. *Optics Express* **28**, 26284-26301 (2020).

138. Zhang, Y. H. et al. PhaseGAN: a deep-learning phase-retrieval approach for unpaired datasets. *Optics Express* **29**, 19593-19604 (2021).

139. Wu, Y. C. et al. Bright-field holography: cross-modality deep learning enables snapshot 3D imaging with bright-field contrast using a single hologram. *Light: Science & Applications* **8**, 25 (2019).

140. Mangal, J. et al. Unsupervised organization of cervical cells using bright-field and single-shot digital holographic microscopy. *Journal of Biophotonics* **12**, e201800409 (2019).

141. Salimans, T. et al. Improved techniques for training GANs. Proceedings of the 30th International Conference on Neural Information Processing Systems. Barcelona, Spain: Curran Associates Inc., 2016.

142. Zhu, J. Y. et al. Unpaired image-to-image translation using cycle-consistent adversarial networks. Proceedings of 2017 IEEE International Conference on Computer Vision. Venice, Italy: IEEE, 2017.

143. Isola, P. *et al.* Image-to-image translation with conditional adversarial networks. Proceedings of 2017 IEEE Conference on Computer Vision and Pattern Recognition. Honolulu, HI, USA: IEEE, 2017.

144. Nehme, E. et al. Deep-STORM: super-resolution single-molecule microscopy by deep learning. *Optica* **5**, 458-464 (2018).

145. Weigert, M. et al. Content-aware image restoration: pushing the limits of fluorescence microscopy. *Nature Methods* **15**, 1090-1097 (2018).

146. Matlock, A. & Tian, L. Physical model simulator-trained neural network for computational 3D phase imaging of multiple-scattering samples. arXiv: 2103.15795 (2021).

147. LeCun, Y. et al. Gradient-based learning applied to document recognition. *Proceedings of the IEEE* **86**, 2278-2324 (1998).

148. Huang, G. B. et al. Labeled Faces in the Wild: A Database for Studying Face Recognition in Unconstrained Environments. (University of Massachusetts, 2007).

149. Lee, C. H. *et al.* MaskGAN: towards diverse and interactive facial image manipulation. Proceedings of 2020 IEEE/CVF Conference on Computer Vision and Pattern Recognition. Seattle, WA, USA: IEEE, 2020.

150. Neyshabur, B. et al. Exploring generalization in deep learning. Proceedings of the 31st International Conference on Neural Information Processing Systems. Long Beach, CA, USA: Curran Associates Inc., 2017.

151. Lempitsky, V., Vedaldi, A. & Ulyanov, D. Deep image prior. Proceedings of 2018 IEEE/CVF Conference on Computer Vision and Pattern Recognition. Salt Lake City, UT, USA: IEEE, 2018.

152. Bostan, E. et al. Deep phase decoder: self-calibrating phase microscopy with an untrained deep neural network. *Optica* **7**, 559-562 (2020).

153. Zhou, K. C. & Horstmeyer, R. Diffraction tomography with a deep image prior. *Optics Express* **28**, 12872-12896 (2020).

154. Yang, D. Y. et al. Dynamic coherent diffractive imaging with a physics-driven untrained learning method. *Optics Express* **29**, 31426-31442 (2021).

155. Wang, F. et al. Far-field super-resolution ghost imaging with a deep neural network constraint. *Light: Science & Applications* **11**, 1 (2022).

156. Bengio, Y. Learning deep architectures for AI. *Foundations and Trends in Machine Learning* **2**, 1-127 (2009).

157. Sjöberg, J. et al. Nonlinear black-box modeling in system identification: a unified overview. *Automatica* **31**, 1691-1724 (1995).

158. Tzeng, F. Y. & Ma, K. L. Opening the black box -data driven visualization of neural networks. Proceedings of the VIS 05. IEEE Visualization, 2005. Minneapolis, MN, USA: IEEE, 2005.

159. Karpatne, A., et al. Theory-guided data science: a new paradigm for scientific discovery from data. *IEEE Transactions on Knowledge and Data Engineering* **29**, 2318-2331 (2017).

160. Beucler, T. et al. Enforcing analytic constraints in neural networks emulating physical systems. *Physical Review Letters* **126**, 098302 (2021).

161. Buhrmester, V., Münch, D. & Arens, M. Analysis of explainers of black box deep neural networks for computer vision: a survey. *Machine Learning and Knowledge Extraction* **3**, 966-989 (2021).

162. Arrieta, A. B. et al. Explainable artificial intelligence (XAI): concepts, taxonomies, opportunities and challenges toward responsible AI. *Information Fusion* **58**, 82-115 (2020).

163. Willard, J. et al. Integrating scientific knowledge with machine learning for engineering and environmental systems. arXiv: 2003.04919v5 (2021).

164. Roscher, R. et al. Explainable machine learning for scientific insights and discoveries. *IEEE Access* **8**, 42200-42216 (2020).

165. Wijesinghe, P. & Dholakia, K. Emergent physics-informed design of deep learning for microscopy. *Journal of Physics: Photonics* **3**, 21003 (2021).

166. Ba, Y. H., Zhao, G. Y. & Kadambi, A. Blending diverse physical priors with neural networks. arXiv: 1910.00201 (2019).

167. Goy, A. et al. Low photon count phase retrieval using deep learning. *Physical Review Letters* **121**, 243902 (2018).

168. Zeng, T. J. & Lam, E. Y. Model-based network architecture for image reconstruction in lensless imaging. Proceedings of SPIE 11551, Holography, Diffractive Optics, and Applications X. SPIE, 2020.

169. Iten, R. et al. Discovering physical concepts with neural networks. *Physical Review Letters* **124**, 010508 (2020).

170. Takeda, M. & Goodman, J. W. Neural networks for computation: number representations and programming complexity. *Applied Optics* **25**, 3033-3046 (1986).

171. Takeda, M. Phase unwrapping by neural network. Proceedings of the FRINGE'93. Akademie Verlag, 1993, 136-141.

172. Kreis, T. M., Biedermann, R. & Jüptner, W. P. O. Evaluation of holographic interference patterns by artificial neural networks. Proceedings of SPIE 2544, Interferometry VII: Techniques and Analysis. San Diego, CA, United States: SPIE, 1995, 11-24.

173. Kreis, T., Jüptner, W. & Biedermann, R. Neural network approach to holographic nondestructive testing. *Applied Optics* **34**, 1407-1415 (1995).

174. Frauel, Y. & Javidi, B. Neural network for three-dimensional object recognition based on digital holography. *Optics Letters* **26**, 1478-1480 (2001).

175. Situ, G. & Sheridan, J. T. Holography: an interpretation from the phase-space point of view. *Optics Letters* **32**, 3492-3494 (2007).

176. Stern, A. & Javidi, B. Space-bandwidth conditions for efficient phase-shifting digital holographic microscopy. *Journal of the Optical Society of America A* **25**, 736-741 (2008).

177. Claus, D., Iliescu, D. & Bryanston-Cross, P. Quantitative space-bandwidth product analysis in digital holography. *Applied Optics* **50**, H116-H127 (2011).

178. Rogers, G. L. In-line soft-x-ray holography: the unwanted image. *Optics Letters* **19**, 67 (1994).

179. Xiao, T. Q. et al. Digital image decoding for in-line X-ray holography using two holograms. *Journal of Modern Optics* **45**, 343-353 (1998).

180. Tonomura, A. Applications of electron holography. *Reviews of Modern Physics* **59**, 639-669 (1987).

181. Tegze, M. & Faigel, G. X-ray holography with atomic resolution. *Nature* **380**, 49-51 (1996).
182. Korecki, P., Korecki, J. & Ślęzak, T. Atomic resolution γ-ray holography using the Mössbauer effect. *Physical Review Letters* **79**, 3518-3521 (1997).
183. Zhang, J. Y. et al. Phase-shifting lensless Fourier-transform holography with a Chinese Taiji lens. *Optics Letters* **43**, 4085-4087 (2018).
184. Zhang, S. M. et al. Phase-shifting radial-shearing digital holography with Greek-ladder zone plates. *Optics Letters* **43**, 5575-5578 (2018).
185. Liu, G. & Scott, P. D. Phase retrieval and twin-image elimination for in-line Fresnel holograms. *Journal of the Optical Society of America A* **4**, 159-165 (1987).
186. Teague, M. R. Deterministic phase retrieval: a Green's function solution. *Journal of the Optical Society of America* **73**, 1434-1441 (1983).
187. Barton, J. J. Removing multiple scattering and twin images from holographic images. *Physical Review Letters* **67**, 3106-3109 (1991).
188. Nugent, K. A. Twin-image elimination in Gabor holography. *Optics Communications* **78**, 293-299 (1990).
189. Tiller, J. B. et al. The holographic twin image problem: a deterministic phase solution. *Optics Communications* **183**, 7-14 (2000).
190. Bleloch, A. L., Howie, A. & James, E. M. Amplitude recovery in Fresnel projection microscopy. *Applied Surface Science* **111**, 180-184 (1997).
191. Levi, A. & Stark, H. Image restoration by the method of generalized projections with application to restoration from magnitude. *Journal of the Optical Society of America A* **1**, 932-943 (1984).
192. Shechtman, Y. et al. Phase retrieval with application to optical imaging: a contemporary overview. *IEEE Signal Processing Magazine* **32**, 87-109 (2015).
193. Fournier, C. et al. Inverse problem approaches for digital hologram reconstruction. Proceedings of SPIE 8043, Three-Dimensional Imaging, Visualization, and Display 2011. Orlando, Florida, United States: SPIE, 2011.
194. McCann, M. T., J in, K. H. & Unser, M. Convolutional neural networks for inverse problems in imaging: a review. *IEEE Signal Processing Magazine* **34**, 85-95 (2017).
195. Misell, D. L. An examination of an iterative method for the solution of the phase problem in optics and electron optics: I. test calculations. *Journal of Physics D: Applied Physics* **6**, 2200-2216 (1973).
196. Greenbaum, A. & Ozcan, A. Maskless imaging of dense samples using pixel super-resolution based multi-height lensfree on-chip microscopy. *Optics Express* **20**, 3129-3143 (2012).
197. Zhang, J. C. et al. Phase unwrapping in optical metrology via denoised and convolutional segmentation networks. *Optics Express* **27**, 14903-14912 (2019).
198. Zhang, T. et al. Rapid and robust two-dimensional phase unwrapping via deep learning. *Optics Express* **27**, 23173-23185 (2019).
199. Dardikman-Yoffe, G. et al. PhUn-Net: ready-to-use neural network for unwrapping quantitative phase images of biological cells. *Biomedical Optics Express* **11**, 1107-1121 (2020).
200. Wu, C. C. et al. Phase unwrapping based on a residual en-decoder network for phase images in Fourier domain Doppler optical coherence tomography. *Biomedical Optics Express* **11**, 1760-1771 (2020).
201. Lyu, M. et al. Fast autofocusing in digital holography using the magnitude differential. *Applied Optics* **56**, F152-F157 (2017).
202. Bian, Y. X. et al. Optical refractometry using lensless holography and autofocusing. *Optics Express* **26**, 29614-29628 (2018).
203. Zhang, Y. B. et al. Edge sparsity criterion for robust holographic autofocusing. *Optics Letters* **42**, 3824-3827 (2017).
204. Jaferzadeh, K. et al. No-search focus prediction at the single cell level in digital holographic imaging with deep convolutional neural network. *Biomedical Optics Express* **10**, 4276-4289 (2019).
205. Xiao, W. et al. Sensing morphogenesis of bone cells under microfluidic shear stress by holographic microscopy and automatic aberration compensation with deep learning. *Lab on A Chip* **21**, 1385-1394 (2021).
206. Gopinathan, U., Pedrini, G. & Osten, W. Coherence effects in digital in-line holographic microscopy. *Journal of the Optical Society of America A* **25**, 2459-2466 (2008).
207. Dainty, J. C. et al. Laser Speckle and Related Phenomena. (Berlin, Heidelberg: Springer 1975).
208. Bianco, V. et al. Strategies for reducing speckle noise in digital holography. *Light: Science & Applications* **7**, 48 (2018).
209. Lehtinen, J. et al. Noise2Noise: learning image restoration without clean data. Proceedings of the 35th International Conference on Machine Learning. Stockholm, Sweden: PMLR, 2018.
210. Yin, D. et al. Speckle noise reduction in coherent imaging based on deep learning without clean data. *Optics and Lasers in Engineering* **133**, 106151 (2020).
211. Blinder, D. et al. Signal processing challenges for digital holographic video display systems. *Signal Processing: Image Communication* **70**, 114-130 (2019).
212. Zhang, F. et al. An approach to increase efficiency of DOE based pupil shaping technique for off-axis illumination in optical lithography. *Optics Express* **23**, 4482-4493 (2015).
213. He, Z. H. et al. Progress in virtual reality and augmented reality based on holographic display. *Applied Optics* **58**, A74-A81 (2019).
214. Shimobaba, T. et al. Rapid calculation algorithm of Fresnel computer-generated-hologram using look-up table and wavefront-recording plane methods for three-dimensional display. *Optics Express* **18**, 19504-19509 (2010).
215. Ito, T. et al. Special-purpose computer HORN-5 for a real-time electroholography. *Optics Express* **13**, 1923-1932 (2005).
216. Goi, H., Komuro, K. & Nomura, T. Deep-learning-based binary hologram. *Applied Optics* **59**, 7103-7108 (2020).
217. Park, D. Y. & Park, J. H. Hologram conversion for speckle free reconstruction using light field extraction and deep learning. *Optics Express* **28**, 5393-5409 (2020).
218. Ren, H. R. et al. Three-dimensional vectorial holography based on machine learning inverse design. *Science Advances* **6**, eaaz4261 (2020).
219. Goodman, J. W. et al. Optical interconnections for VLSI systems. *Proceedings of the IEEE* **72**, 850-866 (1984).
220. Wetzstein, G. et al. Inference in artificial intelligence with deep optics and photonics. *Nature* **588**, 39-47 (2020).
221. Luo, X. H. et al. Metasurface-enabled on-chip multiplexed diffractive neural networks in the visible. arXiv: 2107.07873 (2021).
222. Zhou, T. K. et al. Large-scale neuromorphic optoelectronic computing with a reconfigurable diffractive processing unit. *Nature Photonics* **15**, 367-373 (2021).
223. Zhou, T. K. et al. In situ optical backpropagation training of diffractive optical neural networks. *Photonics Research* **8**, 6000940 (2020).
224. Xiao, Y. L. et al. Unitary learning for diffractive deep neural network. *Optics and Lasers in Engineering* **139**, 106499 (2021).
225. Xiao, Y. L. et al. Optical random phase dropout in a diffractive deep neural network. *Optics Letters* **46**, 5260-5263 (2021).
226. Qian, C. et al. Performing optical logic operations by a diffractive neural network. *Light: Science & Applications* **9**, 59 (2020).
227. Wang, P. P. et al. Orbital angular momentum mode logical operation using optical diffractive neural network. *Photonics Research* **9**, 2116-2124 (2021).
228. Kulce, O. et al. All-optical information-processing capacity of

diffractive surfaces. *Light: Science & Applications* **10**, 25 (2021).

229. Huang, Z. B. et al. All-optical signal processing of vortex beams with diffractive deep neural networks. *Physical Review Applied* **15**, 014037 (2021).

230. Rahman, S. S. & Ozcan, A. Computer-free, all-optical reconstruction of holograms using diffractive networks. arXiv: 2107.08177 (2021).

231. Veli, M. et al. Terahertz pulse shaping using diffractive surfaces. *Nature Communications* **12**, 37 (2021).

232. Li, J. X. et al. Spectrally encoded single-pixel machine vision using diffractive networks. *Science Advances* **7**, eabd7690 (2021).

233. Wang, F. et al. Single-pixel imaging using physics enhanced deep learning. *Photonics Research* **10**, 104-110 (2022).

234. Cai, X. D. et al. Dynamically controlling terahertz wavefronts with cascaded metasurfaces. *Advanced Photonics* **3**, 036003 (2021).

235. Lobo, J. L. et al. Spiking Neural Networks and online learning: an overview and perspectives. *Neural Networks* **121**, 88-100 (2020).

Index

A

Accuracy, 28, 32, 41, 48, 64, 74, 75, 97, 98, 105, 116–118, 120–128, 134, 155, 175, 178, 187–189, 204, 209, 223, 224, 228, 230–232, 234, 236, 267, 270, 273, 274, 277, 279, 281, 282, 285, 362, 404, 405, 407, 410, 412–414, 416–418, 427, 430, 435, 437, 444–448, 450, 451, 467, 468, 477, 519, 529, 533, 539, 543, 544, 545, 550, 552, 557, 559, 562, 566, 567, 570, 574, 586, 613, 617, 635, 644, 646, 649
Adaptive optics, 404, 437, 438, 633, 678, 679
Additive manufacturing, 222, 394, 454, 455, 458, 460, 464, 530
Aircraft, 23–25, 66, 72, 73, 92–95, 496, 497
Art conservation, 74, 466–468, 472, 484
Artificial intelligence, 152, 153, 194, 196, 466, 548, 613, 687
Artificial neuron, 705, 706
Artwork, 74, 92, 99, 466–472, 474–480, 483–485
Artwork conversation, 469
Artwork inspection, 92
Augmented Reality (AR) , 292, 334, 622, 662
Automotive, 92, 94, 95, 116, 117, 130, 196, 214, 301, 454
Autostereoscopy, 610

B

Beam shaping, 196, 331, 370, 371, 373, 374, 378–381, 383, 390–394
Bessel beams, 351–357, 370, 379
Biomedical imaging, 151, 172, 194, 197, 206, 241, 677
Biopolymer, 334, 335, 339
Blood cell, 157, 187
Butch beam, 371, 379, 381, 394, 395, 397, 398

C

Calibration, 64, 67, 96, 97, 121, 125, 131, 133, 139, 219, 220, 230, 236, 281, 319, 326, 346, 362–364, 366, 368, 406, 409, 411, 412, 417, 444, 449, 450, 506, 507, 544, 567, 569, 581, 653, 673, 699
Calibration samples, 97
Carrier frequency, 62, 63, 67–69, 106, 133, 195, 196, 243, 457, 468
Carrier frequency technique, vii
Carrier fringes, 39, 42, 43, 49, 50, 65, 67, 71, 76
Chaos engineering, 690
Classical interferometry, 39, 88, 104–106, 108–112
Closed loop, 405, 413, 602
Coded aperture imaging, 690, 691, 701
Coding standards, 197, 198
Coherence, 8, 19, 44, 51, 54, 59, 70, 76, 91, 92, 95, 105–111, 117, 118, 127, 132, 153, 172, 208, 240, 248, 310, 345, 364, 368, 435, 454, 457, 459, 460, 469, 489, 490, 514, 518, 519, 560, 566, 567, 571, 594–603, 611, 615, 638, 641, 649, 678, 682–684, 704, 714, 718
Coherence holography, 594–599, 602, 603, 682
Coherence vortex, 594, 595, 598–603
Coherent optical metrology, 100, 101
Common path DH, 111, 131
Composite materials, 73, 74, 215, 472
Compression, 71, 192–194, 196–209, 608, 609, 615, 617, 639, 641, 643, 653, 654, 655
Compressive sensing, 156, 714
Computational imaging, 577, 704, 708
Computer generated holograms, 35, 39, 46, 48, 49, 54, 197, 201, 204, 298, 300, 303, 305, 334, 340, 405, 567, 569, 572, 598, 612–614, 622–629, 632–635, 637–656, 666, 668–672, 718, 719
Conical hologram, 623
Convolutional neural networks, 160, 258–268, 270–274, 276, 278, 284, 708, 709, 715
Correlation, 62, 64, 69, 88, 89, 91, 92, 127, 137, 140, 199, 207, 208, 225, 280, 285, 338, 341, 353, 363, 364, 446, 447, 450, 451, 477, 503, 528, 536, 537, 539, 540, 541, 546, 547, 551, 561–563, 594, 596, 642, 655, 677, 680, 699, 701
Cross talk, 404, 405, 407, 410–415, 417, 513
Curvature, 19, 48, 53, 59, 62, 64, 67, 68, 71, 201, 228, 251, 259, 260, 294, 299, 301, 328, 397, 405, 480, 526, 598, 611, 698

D

3D display, 244, 334, 468, 608–613, 616, 617, 622, 627, 632, 668, 718
3D human vision, 609
3D imaging, 28, 31, 32, 34, 35, 39, 49, 52, 53, 194, 215, 230, 231, 236, 422, 428, 551, 559, 676, 677, 686, 691
3D topography, 247, 422, 424, 427–429, 437
Data representation, 205, 206, 208
Deep holography, 704, 705, 719, 720
Deep-learning, 155, 159–161, 207, 234, 235, 258–260, 262, 264, 265, 284, 485, 639, 643
Deep Neural Network (DNN) , 160, 161, 207, 208, 264, 285, 577, 643, 652, 653, 704–710, 716, 719
Deformation, 21, 24, 25, 38, 49, 50, 58, 60–71, 88, 89, 95–99, 110–112, 120, 179, 182, 188, 216, 224, 232, 324–329, 362, 364, 365, 405, 438, 444, 451, 455, 467, 470, 472, 477, 482, 491, 492, 494–497, 543, 544, 546, 547, 627, 683
Diffractive optical elements, 67, 344, 345, 348, 349, 351, 353, 354, 356, 357, 371, 407, 445, 446, 566, 719
Digital holographic microscope, 178–181, 195, 242
Digital holographic microscopy, 45, 52, 112, 150, 151, 160, 178–180, 193, 198, 230, 233, 241, 258, 259, 265, 267, 279, 285, 428, 497, 632, 717
Digital holography, 24, 25, 30, 32–34, 36, 44–47, 50–52, 94–99, 116–118, 121–123, 125, 135, 140, 142–148, 150, 166, 167, 172,

W. Osten and Y. Bai (eds.), *Holography and Speckle*,
https://doi.org/10.1007/978-3-031-36878-3

177, 192–195, 214, 216, 218–221, 223, 225, 226, 230–233, 240–244, 247, 250, 251, 254, 258, 345, 362, 363, 422, 423, 454–456, 458, 463, 464, 468, 488, 489, 492, 496, 502, 503, 537, 543, 550–552, 557, 559, 576, 633, 646, 649, 683, 684, 686, 687, 704, 712, 716
Digital processing, 43, 61
Digital shearography, 59, 63, 65, 67–69, 71, 99
Dual wavelength interferometry, 92, 414
Dynamic holographic displays, 662, 663, 668

E
Electronic speckle pattern interferometry, 58, 94, 468
Entomology, 550, 551
Erosion monitoring, 92, 98, 99
Eyebox expansion, 298, 299, 315

F
Femtosecond pulse, 358, 536, 552, 554
Field of view, 8, 36, 50, 57, 101–112, 115, 118, 135, 137, 145, 152–155, 161, 166, 178, 200, 229, 280, 285, 309, 310, 313, 314, 324, 333, 436, 475, 516, 522, 530, 531, 535–537, 539, 542, 544, 545, 565, 567, 593, 596, 608, 618, 619, 638, 647, 664, 665, 683, 684, 686
Fizeau interferometer, 27–29, 31, 34, 37
Flat optical systems, 279, 280
Flat optics, 279–281
Fourier transform, 27, 28, 30, 36, 38, 47, 52–57, 91, 117, 118, 128, 137, 142, 156, 157, 178, 180, 188, 228, 230, 244–246, 250, 253, 319, 325, 333, 340, 341, 410–412, 415, 417, 421, 434, 435, 441, 474, 489, 490, 494, 507, 546, 548, 553–555, 564, 581, 585, 597, 598, 610, 614, 621, 625, 631, 636, 667–669, 677, 678, 681, 686
Foveated holographic displays, 656
Free electron laser, 329, 330, 336, 341, 342
Fringe pattern, 5, 25, 26, 29, 31, 35, 46–48, 50, 56, 58, 60, 77, 81, 92, 95, 96, 139, 157, 408–410, 414, 451, 452, 454–457, 460–467, 469, 531, 582–584, 587, 669, 697

G
Generative adversarial networks, 145, 252, 695
Geometric phase, 355, 357, 358, 360–366, 368, 375, 503–505, 650–652

H
Harsh environment, 75, 77, 80, 83, 85, 120, 282
Helmet mounted display, 277, 306
Hermite-Gauss beams, 336, 337, 338
High speed recording, 55, 207, 208, 663
Holographic AR display, 16, 608, 610, 653, 656, 657
Holographic correloscopy, 665–670
Holographic endoscope, 309, 313, 314
Holographic image replication, 429, 430
Holographic interferometry, 3, 5–10, 26, 34–37, 43, 44, 46, 50, 73, 77, 81, 82, 89, 97, 115, 116, 123, 124, 199, 200, 217, 226, 227, 422, 440, 451–455, 457, 463, 469, 473, 480, 483, 617, 667–669, 703
Holographic multipoint method, 430
Holographic phase conjugation, 664, 665
Holographic wavefront sensor, 389–391, 394, 403
Holography, 3, 6, 8–10, 13–21, 23–32, 34–37, 39, 49, 50, 53, 58, 73, 79, 81–84, 86, 89–97, 101–103, 106, 110–112, 115, 116, 120, 125, 127–133, 135, 141, 151, 152, 157, 161, 162, 177–182, 184, 186, 187, 192–194, 199–201, 203–208, 210, 211, 213, 215–218, 225–229, 232, 235, 236, 239, 243, 283, 286, 319, 329, 330, 332, 333, 341, 342, 347–349, 352, 353, 356, 375, 376, 407–410, 422, 423, 439–441, 443, 448, 449, 452–454, 463, 464, 473–478, 481, 483, 487, 488, 518, 522, 528, 535–537, 542, 544, 561, 562, 566, 573, 579–584, 587, 588, 593, 594, 596, 598, 599, 601–603, 617–619, 623, 624, 627–629, 631, 634, 637–640, 648, 658, 661–669, 671, 672, 678–680, 689–699, 701, 703–705

I
Illumination modulation, 135, 145
Image estimation, 634
Image processing, 107, 156, 162, 164, 265, 423, 431, 432, 453, 536, 539, 546, 630, 634, 694
Image quality assesment, 177
Image restoration, 634
Imaging, 13, 16, 17, 19, 20, 23, 24, 26, 29, 30, 34, 36, 37–39, 44, 48, 50, 53, 54, 56, 59–61, 75, 94, 102, 104, 107, 116, 118–120, 124, 125, 127, 133, 135–141, 143–146, 151, 155–157, 161, 163–167, 169, 170–172, 178–180, 182, 188, 189, 191, 193, 199–201, 203, 204, 206–208, 215, 216, 219–221, 225–227, 229–234, 236–239, 244, 251–253, 257, 258, 260–262, 264–266, 270, 277, 279, 282, 283, 290, 309–316, 329, 330–333, 335, 348, 355, 364, 407, 409, 410, 412–416, 421–423, 429, 431, 433, 436, 441–444, 453, 475–479, 481–484, 487, 488, 499, 500, 503, 517, 523, 527, 528, 530, 532, 535–537, 542, 544, 545, 547, 552, 554, 555, 561–568, 570, 573, 574, 600, 601, 617, 618, 628, 639, 647, 657, 661, 662, 664–669, 671, 672, 675–684, 686, 689, 693, 697, 699–701, 703, 705
Incoherent optics, 675
Industrial application, 73, 101, 103, 106, 112, 200, 348, 429, 435, 473, 477, 481, 484, 555
Infrared illumination, 79, 94
Inkjet printing, 199, 200, 207, 220
In-lime holography, 332, 333, 561, 562, 563, 566, 573
In-process monitoring, 551–553, 556, 557, 559, 560
Inspection, 14, 44, 56–59, 73, 76–79, 84–86, 101, 103, 106–109, 113, 199, 200, 226, 258, 453, 483
Instrument transfer function, 30, 76
Integral photography, 17
Interferometry, 3, 5–10, 23–30, 34–39, 43, 44, 46, 50, 53, 56, 73, 76, 77, 79, 81, 82, 86, 89–97, 102, 105, 115, 116, 123, 124, 193, 199–206, 217, 221, 226, 227, 251, 335, 407–409, 412–414, 422, 423, 440, 451–455, 457, 463, 464, 465, 469, 473–478, 480, 481, 483, 503–505, 512, 519, 552, 580, 617, 667–703
Inverse problems, 144, 145, 341, 490, 690, 699, 704

J
Jones matrix, 375, 377–379, 385

K
Kummer beam, 382, 384, 385

L
Laser, 3, 4
Laser beam melting, 439
Laser grooving, 551, 552, 557, 559, 560
Lenticular photographs, 17
Light source, 20, 29, 30, 44, 55, 75, 77, 85, 91, 92, 95, 112, 115, 116, 117, 120, 123, 124, 142–144, 164, 204, 209, 221, 225, 234, 320, 321, 322, 326, 332, 408, 410, 412, 429, 430, 431, 434, 435, 436, 489, 515, 517, 527, 530, 536, 545, 553, 554–556, 587, 598, 609–614, 619, 620, 622, 657, 671, 689

Liquid, 50, 60, 136, 199, 200–203, 205–208, 210–213, 215, 217–221, 226, 236, 358, 448, 478, 671
Liquid crystal, 50, 52, 53, 181, 234, 286, 296, 357, 358, 360–363, 365–368, 377, 423, 503, 504, 593, 601, 632, 648, 705
Liquid micro lenses, 211, 213
Loading methods, 51

M
Machine learning, 164, 165, 172–174, 243, 264–267, 270, 310, 451, 593, 598, 628, 664, 690
Material, 5, 9, 10, 15, 18, 19, 32, 43, 51, 58, 59, 74, 75, 77, 81–83, 108, 109, 118, 123, 132, 133, 145, 151–153, 157, 174, 199–203, 206–211, 214–217, 219, 220, 225, 233, 234, 243, 277, 286, 290, 295, 303, 304, 306, 310, 312, 314, 320–322, 325, 326, 329, 333, 334, 335, 342, 355, 356, 360, 362, 363, 365–369, 378, 382, 408, 423, 439, 440, 451, 452, 454, 455–457, 460, 462–465, 468, 473, 474, 478, 481, 483, 484, 491, 499, 517, 522, 523, 528, 536, 541, 544, 551, 552, 568, 597, 598, 601, 602, 612, 622, 623, 651, 652, 665, 690, 705
Measurement system, 40, 74–77, 101, 103, 104, 112, 115, 116, 121, 123, 200, 216, 347, 348, 350, 379, 430, 433, 436, 440, 490, 521
Measurement technology, 43, 74, 76, 77, 85, 111, 112, 206, 207, 353, 490, 503
Metasurface, 277, 279, 283, 290–292, 296, 302–304, 306, 356, 357, 358, 360–368, 377, 618, 652, 659
Metrology, 34, 36, 39, 43, 73, 74, 76, 85, 86, 89, 90, 92, 97, 101, 105, 108, 115, 116, 127, 186, 193, 199, 200, 220, 243, 339, 377, 408, 409, 416, 441, 452, 453, 483, 503, 535, 537, 636, 689
Micro-Electro-Mechanical Systems (MEMS), 77, 82, 83, 200, 227, 295, 298, 301, 521, 522, 523, 527, 528, 530–532, 593, 601, 602, 655, 656
Micro-Opto-Electro-Mechanical Systems (MOEMS), 82, 200
Microscopy, 14, 17, 18, 26, 30, 32, 36, 37, 61, 89, 90, 93, 102, 103, 112, 115, 135–139, 145, 146, 151, 163, 164, 165, 178, 179, 183, 192, 193, 208, 210, 214, 215, 218, 226, 231, 235, 243, 244, 250, 252, 264–266, 270, 323, 338, 353, 413, 415, 482, 498, 551, 552, 567, 617, 618, 668, 680, 689, 697, 702
Miniaturization, 77, 82, 182, 216, 347–350, 353
Mixed Reality (MR), 277, 306
Mosquitoe, 535, 539, 541, 542, 544–548
Motion parallax, 595, 596, 598, 607, 608, 611, 627
Multicore fibre, 309, 310, 315
Multiple wavelength interferometry, 409

N
Near-eye display, 647
Neural network, 145, 192, 219, 220, 243–245, 247–251, 255–258, 260, 261, 269, 270, 562, 593, 598, 628, 638, 657–659, 664, 689–698, 700–702, 704, 705
Neurodegeneration, 498
Neurophotonics, 309, 314
Noise performance, 412, 420
Non-destructive testing, 26, 50, 57, 59, 73, 74, 80, 86, 473
Numerical aperture, 36, 105, 121, 125, 132, 136, 138, 155, 237, 280, 310, 326, 475, 503, 504, 546, 609, 652, 676, 679, 680

O
Objective quality assessment, 178, 187, 188, 193, 638
Oblique illumination, 135, 137, 138, 146, 180, 203, 226
Optical coherence tomography, 551, 552
Optical modeling, 36
Optical neural networks, 690, 704, 705
Optical tweezer, 163–165, 167, 170, 172, 174

P
Parallax, 16, 17, 19, 29, 127, 128, 131–133, 188, 464, 594–598, 600–603, 607, 608, 611, 618, 627, 653
Payload holograms, 320, 321, 324–326
Perspective, 16, 23, 25, 39, 89, 92, 127, 131, 132, 179, 185, 188, 191, 229, 258, 269, 321, 325, 355, 360, 366, 368, 375, 423, 564, 595, 597, 598, 600, 675, 686, 690, 704
Phase array device, 601
Phase imaging, 37, 140, 151, 157, 163–167, 169, 170, 179, 225, 226, 234, 244, 252, 257, 264, 407, 409, 421, 422, 562, 563, 565, 566, 573, 574, 671, 700
Phase microscopy, 135, 151, 208, 231, 234, 235, 567, 697
Phase only modulator, 676, 684
Phase retrieval, 143, 178, 180, 186, 251, 254, 257, 261–263, 333, 415, 416, 418, 422, 476, 503, 509, 561–569, 571–573, 634, 666, 678, 680, 681, 686, 690, 697, 699, 703
Phase shifting interferometry, 24, 251, 413, 454
Phase singulariies, 355, 359, 376, 579–581, 583–587
Phase space, 127–132, 621
Phase unwrapping, 26, 151–155, 157, 158, 160, 186, 190, 201–203, 221, 226, 245, 246, 250–252, 254, 257, 262–264, 269, 270, 407, 409, 411–416, 419, 422, 480, 524, 527–529, 690, 697, 700, 701
Photograph, 4, 6, 14, 16, 17, 19, 20, 26, 27, 33, 34, 55, 60, 61, 91, 107, 108, 165, 315, 445, 447, 483, 595, 609, 611, 623, 668
Photographic plate, 5, 15, 43, 46, 56, 128, 225
Photography, 16, 17, 43, 49, 55, 281, 319, 351, 352, 595, 666
Physical limits, 74, 90, 96
Physics informed DNN, 697–699, 704
Planar optical waveguide combiner, 282, 285
Point-Spread Function (PSF), 144, 132, 228, 253, 258, 260, 395, 396, 398, 401, 553, 620, 621, 625, 627, 629, 633, 676, 677, 680, 681, 683, 684
Polarization camera, 503, 504, 514, 515, 517–519
Polymer, 200, 202, 208–210, 217, 220, 286, 320–326, 334, 362–364, 367, 504
Printed micro structures, 213
PSF engineering, 680

Q
Quantitative phase imaging, 37, 157, 163–167, 169, 170, 179, 208, 225, 234, 244, 264, 421, 562, 565, 566, 573, 574, 671

R
Random media imaging, 661, 662, 672
Ray tracing, 293, 422, 596, 622–624
Regularization, 76, 143, 145, 190, 248, 255, 259, 490, 561–564, 566, 569–571, 697–699
Reproducibility, 105, 106, 110, 112, 124, 350, 351, 639
Residual stress, 59, 74, 77, 81, 82, 440
Resolution, 16, 26–31, 45, 57, 61, 73, 74, 77, 83, 85, 90, 95, 96, 104, 105, 111, 115, 117, 128, 130–132, 135–146, 152, 154, 155, 157, 159–161, 178, 180–182, 184, 185, 188, 193, 200, 201, 204, 208, 214, 216, 220, 221, 226, 230, 234, 237, 244, 245, 252, 254, 255, 257, 258, 261, 266, 278, 282, 291, 305, 306, 309–314, 330–333, 341, 347, 352, 355, 383, 389, 407, 411–413, 416–418, 423, 429–436, 443–446, 453, 469, 475, 476, 478, 479, 482–484, 487, 488, 503, 504, 508, 512, 514–516, 518, 522, 525, 527–530, 536, 537, 539, 542, 544, 547, 551–556, 559, 562, 565–568, 594, 596, 597, 599–602, 618–620, 626, 628, 631–633, 638–640, 647, 656–658, 662, 676–683, 686, 690, 699, 701
Resolution enhancement, 135, 137–142, 144–146, 305, 311, 482, 677–680, 686, 690
Resolution limit, 137, 214, 311, 432, 433, 536, 539, 542, 544, 676, 679
Resolving power, 31, 136, 330, 677, 679, 680, 682

Retina, 487, 488, 496–499, 648
Retinal organoids, 487, 488, 496–500
Root-Mean-Squared (RMS) error, 319
Rough surface, 24, 34, 39, 60, 73, 89–97, 102, 119, 423, 599

S

Scattering media, 94, 487, 499, 662, 669, 671
Sensitivity, 31, 38, 44–46, 48, 49, 52, 53, 57, 59–61, 73, 74, 80, 82, 85, 96, 97, 112, 118, 124, 179, 191, 199, 225, 233, 234, 237, 259, 312, 330–332, 363, 390, 397, 398, 409, 417, 422, 432, 434, 436, 453, 459, 482, 483, 521, 527, 529, 530, 532, 536, 553, 571, 583, 639, 650, 682, 700
Setup stability, 473, 474, 477
Shape measurement, 73, 74, 77, 84, 86, 127, 440, 521–530, 532
Shearing holography, 95, 96, 518
Shearing interferometry, 44, 73, 91, 93, 95, 96, 504, 505, 519
Shearography, 43–62, 77–80, 84, 85, 96, 97, 454, 462, 463
Signal To Noise Relation (SNR), 52, 90, 95, 136, 140, 146, 189, 190, 260, 393, 398, 399, 401–403, 413, 416–420, 496, 497, 552, 555–557, 559, 560, 570–572, 634, 684–686
Slope, 38, 44, 47–51, 53–56, 61, 89–91, 93, 95–97, 101, 102, 107, 112, 155, 161, 382, 397, 417–419, 458
Soft matter, 199–201, 219–221, 465
Space components, 479
Sparsity, 561–564, 566, 571, 572, 619, 624–627, 629, 701
Spatial Light Modulator (SLM), 18, 34, 54, 82, 85, 116, 138–140, 181, 182, 190, 233, 330, 390, 423, 484, 487, 490, 552, 593, 597, 600, 601, 607, 618, 632, 639, 653, 663, 680, 697
Speckle, 26, 28, 29, 43–47, 49, 50, 52–57, 73, 77–80, 82, 84, 85, 90–94, 96, 97, 102–105, 107, 117, 118, 120–122, 125, 132, 135, 137, 140, 146, 156, 184, 188, 189, 193, 200, 226, 233, 244, 259, 270, 347–353, 384, 420–423, 441, 444, 453, 454, 463, 473–478, 480–482, 487, 489, 491, 529–531, 538, 545, 598, 628, 631–635, 639, 657, 659, 665–671, 676–679, 681, 682, 690, 697, 703
Speckle illumination, 135, 137, 140, 146, 226
Speckle interferometry, 43, 53, 82, 105, 423, 453, 463, 474, 476, 478, 481
Speckle projection, 349, 350
Spray coating, 81
Standardization, 177, 193, 194, 640
Statistical optics, 588
Stereo photography, 17
Stereoscopy, 16, 17, 595, 602
Strain field, 44, 51, 56
Stress analysis, 73, 74, 77, 86
Stress monitoring, 217
Structured illumination, 90, 93, 116, 135, 137–140, 146, 226, 282, 347–349, 657, 680
Subjective quality assesment, 178, 187, 193, 638, 640
Super-resolution, 135, 137, 142–144, 146, 270, 355, 677, 680
Surface profilometry, 151
Surface topography, 23, 24, 28–32, 34, 36–39, 89, 91–93, 97, 101, 120, 157, 215, 411, 414, 423, 449
Synthetic aperture, 138, 140, 142, 144, 146, 226, 415, 482, 680

T

Terahertz, 59, 141, 235, 329–338, 340–342, 355, 356, 375, 473, 474, 705
Terahertz holography, 329
Thermal infrared waves, 473
Thin film measurement, 201
Topography, 23, 24, 26–32, 34, 36–39, 75, 89–94, 97, 101, 103, 108, 112, 115, 116, 118–124, 157, 215, 216, 227, 232–234, 407–409, 411–415, 422, 423, 440, 444–449, 468, 470
Transfer function, 30, 36–39, 76, 229, 278, 310, 311, 412, 441, 504, 511, 512, 518, 546, 548, 566, 570, 621, 630, 631, 678
Transmission matrix, 140, 311, 316, 487–493, 497, 499
Triangulation, 89–97, 102, 116, 151, 527–529, 532
Two-photon polymerization, 216
Twyman green interferometer, 35

U

Ultra-High speed digital holography, 521, 522
Uncertainty, 34, 75, 76, 82, 83, 89–97, 106, 115, 117, 118, 120, 121, 123, 183, 221, 270, 377, 420, 430, 435, 499, 529, 536, 626

V

Vector beam, 340, 355, 356, 359, 360, 363, 365, 368, 375, 376, 385
Vergence accommodation conflict, 181, 193, 305, 596, 648
Vertical Cavity Surface Emitting Laser (VCSEL), 115, 117–125, 138, 295, 300
Vibration, 5–10, 23, 34, 35, 44, 47, 51, 52, 56, 57, 59, 60, 73, 74, 79, 84, 104, 107, 110, 115, 123, 199, 227, 230, 235, 270, 282, 429–436, 452, 457, 461, 474, 477, 481, 503, 521, 522, 524, 526, 532, 555, 697
Vibration measurement, 6, 429, 430, 432, 433, 435, 436, 481, 522, 532
Virtual Reality (VR), 17, 59, 86, 178, 277, 279–281, 306, 599, 600, 607, 647–649, 651–654, 656–659
Visual quality assesment, 186, 187, 190, 619, 638
Volumetric displays, 597–600, 602, 639
Vortex beam, 336–342, 355–361, 364–366, 375–377, 380, 382–385

W

Wavefront correction, 313, 423, 664, 665
Wave plate, 52, 104, 161, 237, 238, 315, 316, 340, 341, 357–369, 375, 377–382, 385, 442, 482, 490, 554, 649
White light interferometry, 94, 96, 203, 204, 335, 408, 552
Wide viewing zone, 607, 611, 612

SPRINGER NATURE

GPSR Compliance

The European Union's (EU) General Product Safety Regulation (GPSR) is a set of rules that requires consumer products to be safe and our obligations to ensure this.

If you have any concerns about our products, you can contact us on ProductSafety@springernature.com

In case Publisher is established outside the EU, the EU authorized representative is:

Springer Nature Customer Service Center GmbH
Europaplatz 3
69115 Heidelberg, Germany